Fundamentals of PICOSCIENCE

Fundamentals of PICOSCIENCE

Edited by
Klaus D. Sattler

CRC Press is an imprint of the
Taylor & Francis Group, an **informa** business

CRC Press
Taylor & Francis Group
6000 Broken Sound Parkway NW, Suite 300
Boca Raton, FL 33487-2742

First issued in paperback 2020

ISBN 13 : 978-0-367-57630-1 (pbk)
ISBN 13 : 978-1-4665-0509-4 (hbk)

Library of Congress Cataloging-in-Publication Data

Fundamentals of picoscience / [edited by] Klaus D. Sattler.
pages cm
Includes bibliographical references and index.
ISBN 978-1-4665-0509-4 (hardback)
1. Nuclear structure. 2. Atomic theory. 3. Nanotechnology. 4. Nanostructured materials. I. Sattler, Klaus D., editor of compilation.

QC793.3.S8F865 2014
620'.5--dc23 2013012595

Visit the Taylor & Francis Web site at
http://www.taylorandfrancis.com

and the CRC Press Web site at
http://www.crcpress.com

Contents

PART VIII Picometer Positioning

PART IX Picoscale Devices

Preface

Nanoscale science has brought many new effects and inventions and is the basis for a worldwide surge in nanotechnology. Currently, there are more than one million scientists involved in projects with nanoscale structures and materials. From the development of new quantum mechanical methods to far-reaching applications in the electronic industry and medical diagnostics, nanoscience has inspired numerous scientists and engineers to new instrumental developments and inventions. Even the general public is now informed about many of the benefits of nanoscience. New terms such as "fullerenes," "nanotubes," and "quantum dots" are increasingly often used in public discussions. "Nano" has become the magic word for "extremely small." Every day there are reports of new effects and materials, and many surprises can be expected in the future for structures at this size.

We are entering an era of ever smaller and more efficient devices, which will rely on smaller designs and structures. One nanometer is a billionth of a meter, and it is about ten times the diameter of a hydrogen atom. Is there a size range beyond nano that is accessible to us? Do we already have instruments to probe this range? How can we develop new instruments to visualize and measure structures at the subnanometer size?

Answers to these and other questions are given in this book, *Fundamentals of Picoscience*. It describes methods and materials at the picometer-size scale, which is the next size range below nanometer, covering three orders of magnitude. One picometer is the length of a trillionth of a meter. Compared to a human cell of typically ten microns, this is about ten million times smaller. To access this extremely small size, instrumentation has been developed only in recent years, and there are efforts at many research and industry laboratories to move further into this very small range. This book corresponds to these developments and covers some of the instrumentation and experiments undertaken at the picometer-size scale.

Editor

Klaus D. Sattler performed his undergraduate and master's studies at the University of Karlsruhe and at the Nuclear Research Center in Karlsruhe, Germany. He received his PhD under the guidance of Professors G. Busch and H. C. Siegmann at the Swiss Federal Institute of Technology (ETH) in Zurich, where he was among the first to study spin-polarized photoelectron emission. In 1976, he began a group for atomic cluster research at the University of Konstanz in Germany, where he built the first source for atomic clusters and led his team to pioneering discoveries such as "magic numbers" and "Coulomb explosion." He was at UC Berkeley for three years as a Heisenberg fellow, where he initiated the first studies of atomic clusters on surfaces with a scanning tunneling microscope.

Dr. Sattler accepted a position as professor of physics at the University of Hawaii in 1988. There, he initiated a research group for nanophysics, which, using scanning probe microscopy, obtained the first atomic-scale images of carbon nanotubes, directly confirming the graphene network. In 1994, his group produced the first carbon nanocones. In collaboration with ETH Zurich, he also studied the formation of polycyclic aromatic hydrocarbons and nanoparticles in hydrocarbon flames. Other research has involved nanopatterning of nanoparticle films, charge density waves on rotated graphene sheets, band gap studies of quantum dots, and graphene foldings. His current work focuses on novel nanomaterials, nanodiamonds and graphene quantum dots, and solar photocatalysis with nanoparticles for the purification of water.

Among his many accomplishments, Dr. Sattler was awarded the prestigious Walter Schottky Prize by the German Physical Society in 1983. At the University of Hawaii, he teaches courses in general physics, solid-state physics, and quantum mechanics.

In his private time, he has worked as musical director at an avant-garde theater in Zurich, composed music for theatrical plays, and conducted several critically acclaimed musicals. He has also studied the philosophy of Vedanta. He loves to play the piano (classical, rock, and jazz) and enjoys spending time at the ocean and with his family.

Contributors

Masaaki Araidai
World Premier Institute Advanced Materials Research
Tohoku University
Sendai, Japan

and

Core Research for Evolutional Science and Technology
Japan Science and Technology Agency
Kawaguchi, Japan

and

Center for Computational Sciences
University of Tsukuba
Tsukuba, Japan

Nadjib Baadji
School of Physics
and
Center for Research on Adaptive Nanostructures and Nanodevices
Trinity College Dublin
Dublin, Ireland

Maria Mikhailovna Barysheva
Institute for Physics of Microstructures
Russian Academy of Sciences
Nizhny Novgorod, Russia

Umberto Bortolozzo
Institut Non Linéaire de Nice
Centre National de la Recherche Scientifique
University of Nice-Sophia Antipolis
Valbonne, France

Sergey I. Bozhko
Institute of Solid State Physics
Russian Academy of Sciences
Chernogolovka, Russia

Harald Brune
Institute of Condensed Matter Physics
Swiss Federal Institute of Technology
Lausanne, Switzerland

Ben C. Buchler
Department of Quantum Science
The Research School of Physics and Engineering
The Australian National University
Canberra, Australian Capital Territory, Australia

Alexander N. Chaika
Institute of Solid State Physics
Russian Academy of Sciences
Chernogolovka, Russia

Shirley Chiang
Department of Physics
University of California, Davis
Davis, California

Nikolay Ivanovich Chkhalo
Institute for Physics of Microstructures
Russian Academy of Sciences
Nizhny Novgorod, Russia

Trudo Clarysse
Interuniversity Microelectronics Centre
Leuven, Belgium

John D. Close
Department of Quantum Science
The Research School of Physics and Engineering
The Australian National University
Canberra, Australian Capital Territory, Australia

László Cser
Neutron Spectroscopy Department
Wigner Research Centre for Physics
Hungarian Academy of Sciences
Budapest, Hungary

Pierre Eyben
Interuniversity Microelectronics Centre
Leuven, Belgium

József Fortágh
Department of Physics
University of Tübingen
Tübingen, Germany

Takeshi Fukuma
Frontier Science Organization
and
Division of Electrical Engineering and Computer Science
and
Bio-AFM Frontier Research Center
Kanazawa University
Kanazawa, Japan

and

Japan Science and Technology Agency
Precursory Research for Embryonic Science and Technology
Kawaguchi, Japan

Pietro Gambardella
Catalan Institute of Nanotechnology
Barcelona, Spain

and

Department of Materials
Eidgenössische Technische Hochschule
Zürich, Switzerland

Knud Gentz
Institute of Physics and Theoretical Chemistry
University of Bonn
Bonn, Germany

Andreas Günther
Department of Physics
University of Tübingen
Tübingen, Germany

Thomas Hantschel
Interuniversity Microelectronics Centre
Leuven, Belgium

Simon J. Higgins
Department of Chemistry
School of Physical Sciences
University of Liverpool
Liverpool, England

Jean-Pierre Hiugnard
Jphopto Consultant
Palaiseau, France

Hendrik Hölscher
Institute for Microstructure Technology
Karlsruhe Institute of Technology
Eggenstein-Leopoldshafen, Germany

Jisang Hong
Department of Physics
Pukyong National University
Busan, South Korea

Aaron Hurley
School of Physics
and
Center for Research on Adaptive Nanostructures and Nanodevices
Trinity College Dublin
Dublin, Ireland

Mariko Kajima
Dimensional Standards Section
National Metrology Institute of Japan
National Institute of Advanced Industrial Science and Technology
Tsukuba, Japan

Alexei A. Kamshilin
Department of Applied Physics
University of Eastern Finland
Kuopio, Finland

Maciej Kokot
Faculty of Electronics, Telecommunications, and Informatics
Gdańsk University of Technology
Gdańsk, Poland

Gerhard Krexner
Faculty of Physics
University of Vienna
Vienna, Austria

Young Kuk
Department of Physics and Astronomy
Seoul National University
Seoul, South Korea

Heinz Langhals
Department of Chemistry
Ludwig Maximilians University of Munich
Munich, Germany

Sergio Lozano-Perez
Department of Materials
University of Oxford
Oxford, United Kingdom

Qingyou Lu
High Magnetic Field Laboratory
Chinese Academy of Sciences
and
Hefei National Lab for Physical Sciences at Microscale
University of Science and Technology of China
Hefei, Anhui, People's Republic of China

Márton Markó
Neutron Spectroscopy Department
Wigner Research Centre for Physics
Hungarian Academy of Sciences
Budapest, Hungary

Tatjana I. Mazilova
National Science Center
Kharkov Institute of Physics and Technology
Kharkov, Ukraine

Igor M. Mikhailovskij
National Science Center
Kharkov Institute of Physics and Technology
Kharkov, Ukraine

Jay Mody
IBM Systems and Technology
Hopewell Junction, New York

Janez Možina
Faculty of Mechanical Engineering
University of Ljubljana
Ljubljana, Slovenia

Aftab Nazir
Interuniversity Microelectronics Centre
and
Department of Physics and Astronomy
Katholieke Universiteit Leuven
Leuven, Belgium

Richard J. Nichols
Department of Chemistry
School of Physical Sciences
University of Liverpool
Liverpool, United Kingdom

Jouko Nieminen
Department of Physics
Tampere University of Technology
Tampere, Finland

and

Department of Physics
Northeastern University
Boston, Massachusetts

Andrew Scott Parkins
Department of Physics
University of Auckland
Auckland, New Zealand

Aleksei Evgenievich Pestov
Institute for Physics of Microstructures
Russian Academy of Sciences
Nizhny Novgorod, Russia

Tatiana Petrova
Institute of Mathematical Problems of Biology
Russian Academy of Sciences
Pushchino, Russia

Oswald Pietzsch
Institute of Applied Physics
and
Microstructure Advanced Research Center
University of Hamburg
Hamburg, Germany

Marco Pisani
Mechanics Division
National Institute for Metrological Research
Torino, Italy

Alberto Podjarny
Department of Integrative Biology
Institut de Génétique et de Biologie Moléculaire et Cellulaire
Université de Strasbourg
Illkirch, France

Rachel Poldy
Department of Quantum Science
The Research School of Physics and Engineering
The Australian National University
Canberra, Australian Capital Territory, Australia

Tomaž Požar
Faculty of Mechanical Engineering
University of Ljubljana
Ljubljana, Slovenia

Ezio Puppin
Dipartimento di Fisica
Consorzio Nazionale Interuniversitario per le Scienze Fisiche della Materia
Politecnico di Milano
Milano, Italy

Lew Rabenberg
Department of Mechanical Engineering
Texas Materials Institute
The University of Texas at Austin
Austin, Texas

Stefania Residori
Institut Non Linéaire de Nice
Centre National de la Recherche Scientifique
Institut Nonlinéaire de Nice
University of Nice-Sophia Antipolis
Valbonne, France

Nicholas P. Robins
Department of Quantum Science
The Research School of Physics and Engineering
The Australian National University
Canberra, Australian Capital Territory, Australia

Evgenij V. Sadanov
National Science Center
Kharkov Institute of Physics and Technology
Kharkov, Ukraine

Akira Saito
Department of Precision Science and Technology
Graduate School of Engineering
Osaka University
Osaka, Japan

and

RIKEN/SPring-8
Sayo, Japan

Nikolay Nikolaevich Salashchenko
Institute for Physics of Microstructures
Russian Academy of Sciences
Nizhny Novgorod, Russia

Stefano Sanvito
School of Physics
and
Center for Research on Adaptive Nanostructures and Nanodevices
Trinity College Dublin
Dublin, Ireland

Yuji C. Sasaki
Biomedical Group
SPring-8 Japan Synchrotron Radiation Research Institute
Mikazuki, Japan

and

Japan Science and Technology Agency
Core Research for Evolutional Science and Technology Sasaki Team
Tachikawa, Japan

and

Department of Advanced Materials Science
Graduate School of Frontier Sciences
The University of Tokyo
Chiba, Japan

Christian G. Schroer
Institute of Structural Physics
Technische Universität Dresden
Dresden, Germany

Andreas Schulze
Interuniversity Microelectronics Centre
and
Department of Physics and Astronomy
Katholieke Universiteit Leuven
Leuven, Belgium

Igor V. Shvets
School of Physics
and
Center for Research on Adaptive Nanostructures and Nanodevices
Trinity College Dublin
Dublin, Ireland

Alex Szakál
Neutron Spectroscopy Department
Wigner Research Centre for Physics
Hungarian Academy of Sciences
Budapest, Hungary

Akira Tamura
Graduate School of Engineering
Saitama Institute of Technology
Fukaya, Japan

Mikhail Nikolaevich Toropov
Institute for Physics of Microstructures
Russian Academy of Sciences
Nizhny Novgorod, Russia

Masaru Tsukada
World Premier Institute Advanced Materials Research
Tohoku University
Sendai, Japan

and

Core Research for Evolutional Science and Technology
Japan Science and Technology Agency
Kawaguchi, Japan

Wilfried Vandervorst
Interuniversity Microelectronics Centre
and
Department of Physics and Astronomy
Katholieke Universiteit Leuven
Leuven, Belgium

David M. Villeneuve
Joint Attosecond Science Laboratory
National Research Council of Canada
and
University of Ottawa
Ottawa, Ontario, Canada

Yusuke Wakabayashi
Division of Materials Physics
Graduate School of Engineering Science
Osaka University
Osaka, Japan

Klaus Wandelt
Institute of Physical and Theoretical Chemistry
University of Bonn
Bonn, Germany

and

Institute of Experimental Physics
University of Wroclaw
Wroclaw, Poland

Roland Wiesendanger
Institute of Applied Physics
and
Microstructure Advanced Research Center
University of Hamburg
Hamburg, Germany

Clayton C. Williams
Department of Physics and Astronomy
University of Utah
Salt Lake City, Utah

Maria Vladimirovna Zorina
Institute for Physics of Microstructures
Russian Academy of Sciences
Nizhny Novgorod, Russia

I

Picoscale Detection

1

Picometer Detection by Adaptive Holographic Interferometry

Umberto Bortolozzo
University of Nice-Sophia Antipolis

Stefania Residori
University of Nice-Sophia Antipolis

Jean-Pierre Hiugnard
Jphopto Consultant

Alexei A. Kamshilin
University of Eastern Finland

1.1 Introduction

When we want to measure small displacements, tiny changes in thickness, or, more generally, infinitesimal length disturbances, optical methods are often the most convenient ones, and, especially, interferometric techniques allow reaching unequaled sensitivities. Indeed, interferometry relies on the measurement of variations in the phase of optical beams, which constitutes a precise, well-tested, and nonperturbing protocol of detection. In order to reach maximum sensitivity, the interferometer must be optimized; so, an optoelectronic feedback or a modulation system has to be applied. In this case, the sensitivity of the interferometer is only limited by the granular noise of the photons on the detector, also known as the photon shot noise, and measurements as small as less than picometers can be, in principle, effectuated. However, in practice, this limit is reached with difficulty because of several reasons, for instance, the wave fronts of the interfering beams do not match perfectly, the beams propagate in a distorting or fluctuating medium, the surfaces of the reflecting objects are not perfectly smooth, etc. For these reasons, it is necessary to resort to adaptive systems, that is, to systems that are able to automatically compensate the low-frequency distortions and diminish the effects of the environmental noise.

In this framework, adaptive holography has been proposed as an interferometric method based on dynamical holographic recording in nonlinear media. The adaptive holographic interferometry (AHI) is based on the use of a medium that is able to react dynamically to low-frequency changes. This property allows to overcome the problems related to the active stabilization of the interferometer optical paths, this stabilization being necessary when dealing with complex interfering wave fronts, as in the case of speckles due to the light passing through scattering media, or when working in disturbed environments. In these cases, the adaptive hologram reacts by self-adjusting its phase, which is an elegant solution to get rid of the several difficulties that arise when trying to achieve optimal wave front superposition and path difference on the different speckle grains. The principle was first demonstrated by using photoconductive crystals driven by vibrating interference light patterns [1]. In this context, highly precise measurements of small displacements were reported. Successively, a similar approach was adopted by performing two-wave mixing (TWM) experiments in photorefractive materials [2]. There, adaptive holography had been proved as a useful tool for measuring small vibration amplitudes of reflecting objects (see Refs. [3–7] and the articles cited therein). More recently, a similar technique has been employed to achieve picometer detection by using liquid crystal light valves (LCLV) [8] and to realize acousto-optic imaging through highly scattering biological media [9]. Moreover, self-adaptive holography allows to reach high sensitivities by employing a low complexity in the experimental system. To have an idea of the extremely high sensitivity given by AHI, we can cite, for example, the paper from Petrov et al. [10], where the Casimir force, a physical force arising from the quantization of the electromagnetic field, is easily measured.

In this chapter, we present a review of the various AHI techniques realized with different adaptive media, discuss the performances and capabilities of the analyzed systems to detect displacements in the picometer range, and give some of the most significant examples of applications of these methods. We will start by briefly recalling the principles of the classical interferometry and the associated shot noise detection limit. Then, the basic principles of AHI will be presented, together with the main parameters defining the performances of an adaptive interferometer. Examples of different types of adaptive holographic media will be provided, namely, photorefractive crystals, optical fibers, and LCLVs. Then, applications will be briefly described, such as the detection of Casimir force, the pressure radiation measurement, and photoacoustic detection. Finally, a discussion on and a comparison between the different systems will be provided.

1.2 Theoretical Background of Adaptive Holographic Interferometry

Optical interferometry [11] has always been considered one of the most flexible and sensitive techniques for measuring mechanical vibrations or small length changes [12,13]. The number of applications of this method increased dramatically after the discovery of lasers in 1960 and the development of low-loss optical fibers in 1966. A laser as a source of coherent radiation assures that an interferometer has high sensitivity, while an optical fiber guarantees that the measuring system has compactness, light weight, immunity to electromagnetic influence, and capability to operate in hazardous conditions, such as high temperature, radiation, etc. The phase of a light field cannot be directly detected, and an interferometer detects the phase difference by combining the object wave that bears information about the measurand with the coherent reference wave and measures the changes in the resultant light flux via a photodetector. The sensitivity of an interferometer to phase difference is, in principle, limited only by the shot noise of the photoelectrons generated, for instance, in the photodiode (PD), and can be extremely high, for example, the theoretical minimum detectable displacement is of the order of 10^{-16} m/$\sqrt{\text{Hz}}$ for 10 mW of detected laser power at a wavelength of 500 nm and quantum efficiency of 0.10 [14,15]. However, two main problems should be overcome to achieve such high sensitivity in practical conditions. The first is the necessity of the precise adjustment of the object and the reference wave fronts interfering at the photodetector. This requirement makes it difficult to realize the interferometer when wave front distortions are present, for instance, when air fluctuations or thermal gradients modify locally the refractive index of the region before the sample, or, as another example, when strong phase distortions are introduced by atmospheric turbulence or when speckle fields are generated by strongly scattering media, as typically happens in biological samples. The second requirement is the need to keep constant the average phase shift between the interfering wave fronts. This problem is more fundamental than the first one because the induced phase shift, which is assumed to be proportional to the influence of the measurand on the optical paths, is nonlinearly transferred into the light flux change at the detector via the familiar cosine interference function.

A simple and elegant solution to both these problems was achieved when a conventional beam splitter, which combines the reference and the object waves, was replaced by a dynamic hologram continuously recorded in a photorefractive crystal. Since the configuration involves only two waves that interfere inside the crystal, it was termed two-wave mixing [2]. Moreover, the photorefractive dynamic hologram not only adapts the object wave front to the reference one, but it is also self-adapted to slow temporal variations in the phase difference. Therefore, this type of optical system is also referred to as adaptive interferometry [16].

Since its early developments, phase volume holography in photorefractive crystals has demonstrated great potentialities for applications in image storage and processing [17], vibration analysis [18], and continuous reconstruction of holographic interferograms [19]. After the seminal paper of Petrov et al. on AHI [20], the possibility of using the dynamic holographic recording technique to achieve highly precise measurements of small displacements has been demonstrated in several different systems [7], and AHI has been proved to be a useful tool for measuring small vibration amplitudes of reflecting objects [3–6]. A similar technique has been recently employed to achieve picometer detection by using LCLVs [8] as well as to realize acousto-optic imaging through highly scattering biological media [9].

As a general rule, the physical mechanisms that form the basis of the AHI schemes rely on the narrow frequency bandwidth of the TWM process and on the self-adapted reconfiguration of the refractive index distribution in the nonlinear medium. Applications of adaptive interferometers include numerous practical situations, such as nondestructive testing of materials and elements of technical constructions, stabilization of fiber-optical interferometric sensors, elimination of unwanted influences of the environment, real-time analysis of deformations and vibrations in industrial processes, analysis of surfaces or artworks, molecular recognition, acousto-optical imaging of biological tissues, refractive index measurement in noisy environments, optical coherence-domain reflectometry, and many other cases.

The principle of optical phase measurement as a sensitive technique for the detection of small displacements is depicted in Figure 1.1. A surface is displacing in time with an amplitude $S(t)$. If a laser beam is incident at an angle θ on the surface and k_0 is the optical wave number, then the phase shift acquired by the reflected light beam is given by

$$\Delta\varphi = 2k_0\cos(\theta)S(t), \tag{1.1}$$

that is, it carries the information on the surface displacement. In a classical interferometer, the output intensity I is proportional

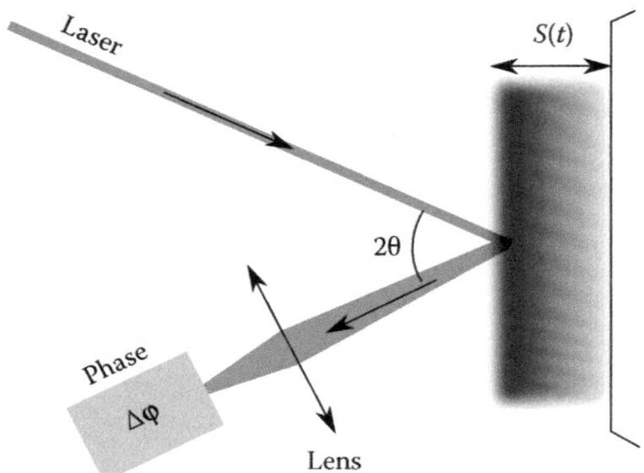

FIGURE 1.1 The principle of optical phase measurement as a sensitive technique for the detection of small displacements. The displacement $S(t)$ of the surface is transferred to a phase modulation $\Delta\varphi$ on the reflected beam.

to $\cos(\Delta\varphi + \varphi_0)$, where φ_0 is an additional phase retardation given by the optical path difference between the two arms of the interferometer. Hence, interferometry is the natural candidate to perform the detection. Moreover, if the displacement amplitude is small, $S(t) \ll 1$, and the optical path difference in the interferometer is such that $\varphi_0 = \pi/2$, then $\cos(\Delta\varphi + \varphi_0) = \sin(\Delta\varphi)$, and by developing the sinus function at the first order, we have $P_{opt} \propto S(t)$, that is, the measured optical power is directly proportional to the unknown displacement. This configuration is called the quadrature condition, and it is the condition of maximum sensitivity of the interferometer. It constitutes one of the fundamental parameters that must be taken into account when designing an interferometric detection system.

1.2.1 Noise-Limited Sensitivity of the Classical Interferometers—Shot Noise

In any interferometer, a physical parameter that has to be measured is encoded in the phase modulation of the light wave, so its sensitivity to small phase excursions is a parameter serving as a primary criterion for the comparison of different systems. In the analysis of a classical interferometer, it was shown that its sensitivity can be extremely high if the available light power is not limited and measurements of the phase modulation are carried out within a very narrow frequency band [21]. If any other source of noise is avoided, the only limitation to sensitivity is imposed by the granular structure of the measured photocurrent. The minimal detectable phase difference is then defined by the noise power of the measuring system.

There are several sources of noise in an optical interferometer: laser noise, thermal and shot noise of the photodetector, noise of amplifying electronics. When the light power reaching the photodetector is high enough, the shot noise (which is proportional to the square root of the received light power) of the photoexcited charge carriers prevails over all the other noise levels. As schematically represented in Figure 1.2, the shot noise basically arises from the interaction between light and atoms in the light source (photon shot noise $n(t)$) and the quantum efficiency in the generation of photoelectrons $n_e(t)$. The instantaneous photocurrent intensity is given by

$$i_{PD}(t) = 2\Delta f\, en_e(t), \tag{1.2}$$

where

- e is the electron charge
- $\Delta f = 1/(2T)$, the frequency bandwidth fixed by the integration time T of the detection electronics

Mathematically speaking, the shot noise is a Poisson random process, and the mean and standard deviation of the instantaneous photocurrent are given by [22]

$$I_{PD} = E[i_{PD}] = 2\Delta f\, eE[n_e], \tag{1.3}$$

$$\sigma_{PD}^2 = E[(i_{PD} - I_{PD})^2] = 4\Delta f^2 e^2 E[n_e], \tag{1.4}$$

where $E[\cdot]$ denotes the expectation value. By using these two equations, the standard deviation of the photocurrent is related to its mean value:

$$\sigma_{PD} = \sqrt{2e\Delta f\, I_{PD}}. \tag{1.5}$$

Finally, the intensity current is related to the optical power by the quantum efficiency η of the PD:

$$I_{PD} = \eta e \frac{P_{opt}}{\hbar\omega}, \tag{1.6}$$

and then the power of the shot noise is

$$\sigma_{PD} = \sqrt{\frac{2e^2\eta\Delta f}{\hbar\omega} P_{opt}}. \tag{1.7}$$

Let us now calculate the signal-to-noise ratio (SNR = i_{PD}/σ_{PD}) in a classical interferometer, as, for instance, the one shown in Figure 1.3. For simplicity, we consider a sinusoidal displacement $S(t) = \varepsilon(t)\sin(\Omega t)$ at angular frequency Ω and small amplitude $\varepsilon(t)$.

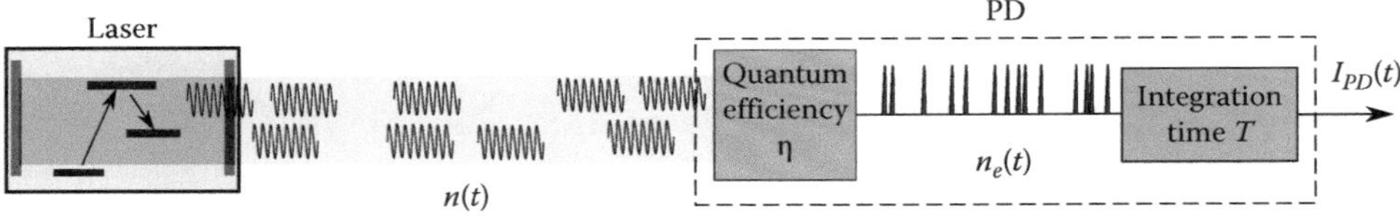

FIGURE 1.2 Illustration of the photon shot noise $n(t)$ in the laser source and correspondingly generated photoelectrons $n_e(t)$ in the detector; $I_{PD}(t)$ is the instantaneously measured photocurrent on the PD.

The highest sensitivity of the interferometer is achieved when the average phase difference between the interfering beams is equal to $\pi/2$, the so-called quadrature condition. In this case, the measured optical power is proportional to the displacement $P_{opt}(\Omega) \propto \varepsilon(t)$, and the SNR is

$$SNR = \sqrt{\frac{\eta}{2\Delta f \hbar\omega}} \frac{P_{opt}(\Omega)}{\sqrt{P_{opt}}}. \tag{1.8}$$

As depicted from the previous expression, in order to calculate the SNR of the interferometer, we need to proceed by calculating the optical mean power and the optical signal power at the considered frequency Ω. The minimal detectable displacement, called the classical homodyne detection limit ε_{lim}, is usually defined as the displacement that has the same amplitude of the shot noise, which leads to $SNR = 1$. In the following section, we will calculate the minimum detectable signal in three different cases of classical interferometers.

1.2.2 Classical Interferometric Detection

As examples of classical interferometric detection, let us consider three types of interferometers, the path-stabilized Mach–Zehnder interferometer, the active heterodyne detection, and the Fabry–Perot (FP) interferometer. We will see how each of them is most adapted for a specific type of measurement, and we will compare their performances and limitations.

1.2.2.1 Path-Stabilized Mach–Zehnder Interferometer

The setup for a Mach–Zehnder interferometer is sketched in Figure 1.3. The laser is split into two beams following two different paths before being recombined onto the PD detector. On one arm of the interferometer (reference beam) is inserted a piezoelectrically driven mirror (PZT) that allows to adjust the length difference between the two optical paths, followed by the reference and signal fields, respectively. On the other arm (signal beam), a mirror is sinusoidally vibrated at a small amplitude $\varepsilon(t)$ and with a modulation frequency Ω. The aim is to measure ε.

When arriving at the PD, the amplitudes of the reference and signal beams, E_R and E_S, respectively, can be expressed as follows:

$$\begin{aligned} E_R &= A_R e^{i(k_0 L_R - \omega t)}, \\ E_S &= A_S e^{i(k_0 L_S + 2k_0\varepsilon\sin(\Omega t) - \omega t)}, \end{aligned} \tag{1.9}$$

where

ω is the laser frequency

k_0 is the optical wave number

L_R and L_S are the optical path lengths for the reference and the signal beams, respectively

The total power at the PD is

$$P = |E_S + E_R|^2 = P_R\left[1 + K^2 + 2K\cos\left(k_0(L_R - L_S) + 2k_0\varepsilon\sin(\Omega t)\right)\right], \tag{1.10}$$

where

$$K = \sqrt{\frac{P_S}{P_R}}.$$

In the linear regime, that is, for small modulation amplitude ε, by fixing the quadrature condition

$$k_0(L_R - L_S) = m\frac{\pi}{2}, \tag{1.11}$$

we obtain the maximum sensitivity of the interferometer, which is given by

$$P = P_R\left[1 + K^2 - 4Kk_0\varepsilon\sin(\Omega t)\right], \tag{1.12}$$

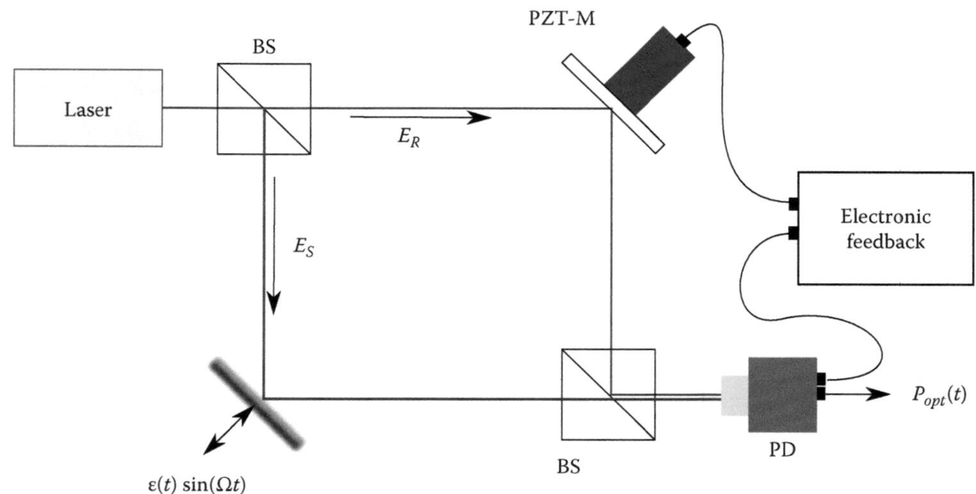

FIGURE 1.3 Setup for a path-stabilized Mach–Zehnder interferometer; the electronic feedback and the PZT-M are used to maintain the constant optical path difference between the two arms of the interferometer.

that is, the intensity measured at the PD is directly proportional to the phase change, and therefore the small displacement ε is directly measured. From the expression (1.12), we can directly obtain the optical mean power $P_{opt} = P_R(1 + K^2)$ and the optical signal power $P_{opt}(\Omega) = 4KP_Rk_0\varepsilon\sin(\Omega t)$. The SNR, then, is given by Equation 1.8:

$$SNR = \sqrt{\frac{\eta P_R}{2\Delta f\hbar\omega}}\frac{2K}{\sqrt{1+K^2}}2k_0\varepsilon. \tag{1.13}$$

Correspondingly, the detection limit is obtained when $K = 1$, and is given by

$$\varepsilon_{lim} = \frac{1}{2k_0}\sqrt{\frac{\Delta f\hbar\omega}{\eta P_R}}. \tag{1.14}$$

As an example, if the laser power is $P_R = 1$ mW, the detector quantum efficiency $\eta = 10\%$, and the laser wavelength 0.6328 nm, then, the minimum detectable displacement is of the order of $\varepsilon_{lim} \sim \times 10^{-14}\ \mathrm{m}/\sqrt{\mathrm{Hz}}$. Note that this displacement is given in terms of meters per square root hertz. Using such units implies that a bandwidth for the detection electronics has not been selected and that, in order to obtain the minimum detectable displacement by using a detection bandwidth of Δf, one must multiply ε_{lim} given earlier by $\Delta f^{1/2}$.

From the previous derivation, we see that the theoretical sensitivity of the interferometer is very high and, therefore, extremely small displacements can be measured with this method. However, in practice, the achieved sensitivity depends on numerous parameters. In particular, when low-frequency noise is present in the system or when the beams are distorted, it becomes difficult to fulfill the quadrature condition, because the two interfering wave fronts do not superpose perfectly and also because the optical path difference $(L_R - L_S)$ changes during the time. In order to get rid of the latter problem, the PZT mirror can be actively driven in such a way as to correct unwanted changes of $(L_R - L_S)$. This is, nevertheless, a complex task that requires a closed-loop control system and which fails when dealing with complex light fields, as speckles, because for this type of nonuniform wave fronts, the optical path difference varies from grain to grain of the speckle pattern.

1.2.2.2 Active Heterodyne Detection

The heterodyne detection is an interferometric technique where the frequency of one of the two interfering beams is changed by a fixed amount. This is typically achieved by using an acousto-optical modulator or Bragg cell that introduces a frequency shift ω_B on one of the two optical beams. The interference of the two beams produces an optical beating of the resultant field magnitude at the difference frequency. In active heterodyne systems, the difference in frequency between the interfering beams (the signal and reference) is achieved by actively frequency shifting the light in one of the optical paths. The stability and constancy of the optical beating at the detector depend on the character of the shifting device used [15].

As schematically represented in Figure 1.4, the signal driving the Bragg acoustic cell is also coupled to a demodulation system (a lock-in amplifier, for instance) placed after the PD, so that only the signal at the shifted frequency is detected. Since ω_B is typically of the order of tens of MHz, this method automatically filters out low-frequency noise.

When arriving at the PD, the amplitudes of the reference and signal beams can be written as

$$\begin{aligned} E_R &= A_R e^{i(k_0 L_R - (\omega+\omega_B)t)}, \\ E_S &= A_S e^{i(k_0 L_S + 2k_0\varepsilon\sin(\Omega t) - \omega t)}, \end{aligned} \tag{1.15}$$

and the total power is given by

$$P = P_R\left[1 + K^2 + 2K\cos\left(k_0(L_R - L_S) + 2k_0\varepsilon\sin(\Omega t) - \omega_B t\right)\right]. \tag{1.16}$$

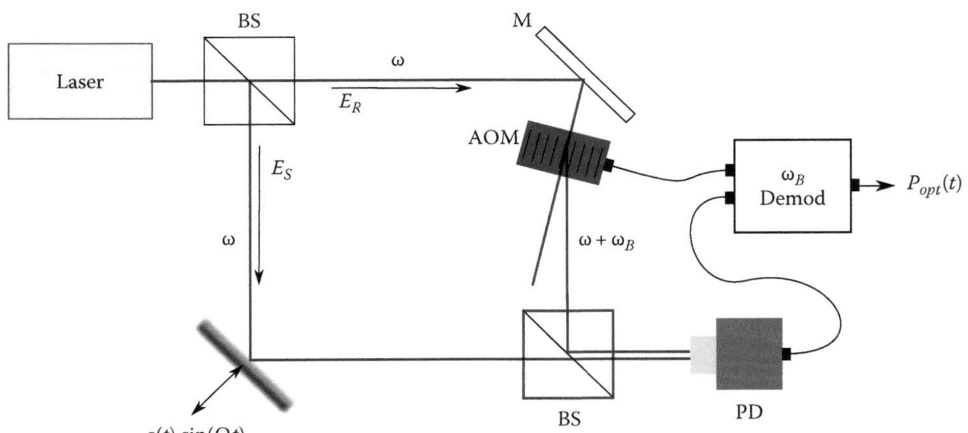

FIGURE 1.4 Setup for active heterodyne detection; AOM is an acousto-optic modulator introducing a frequency shift ω_B on the reference beam. The optical power $P_{opt}(t)$ at the shifted frequency is measured with a demodulation electronic system.

For small displacement amplitudes $\varepsilon \ll 1$, we get

$$P = 2KP_R k_0 \varepsilon\left[\cos\left(k_0\Delta - (\Omega + \omega_B)t\right) - \cos\left(k_0\Delta - (\Omega - \omega_B)t\right)\right] + P_R\left[1 + K^2 + 2K\cos(k_0\Delta - \omega_B t)\right], \quad (1.17)$$

where $\Delta \equiv L_R - L_S$. The total power has a high frequency modulation at ω_B with two side bands at $\omega_B - \Omega$ and $\Omega_B + \Omega$. The amplitude of these two side bands is proportional to the displacement. By demodulating the output signal, we have a direct measure of the displacement ε. By following the same procedure shown earlier for the case of the Mach–Zehnder interferometer, the detection limit can be derived [15] and expressed as

$$\varepsilon_{lim} = \frac{1}{2k_0}\sqrt{\frac{\Delta f \hbar\omega}{\eta P_R}}. \quad (1.18)$$

By taking the laser power $P_R = 1$ mW, the detector quantum efficiency $\eta = 10\%$, and the laser wavelength 0.6328 nm, the minimum detectable displacement is $\varepsilon_{lim} \sim \times 10^{-14}$ m/$\sqrt{\text{Hz}}$. Again, this displacement is given as an implicit function of the detection system bandwidth.

In practice, in these types of interferometers, the degree to which the quantum-noise-limited sensitivity is obtained depends on the quality of the detection electronics and on the stability of the modulation–demodulation system. However, unlike the path-stabilized interferometer, the heterodyne system does not require the quadrature condition to be fixed in order to optimize the detection. Therefore, the active heterodyne can operate without closed-loop control and can reject low-frequency drift or noise electronically, while maintaining a near-maximum sensitivity.

1.2.2.3 Fabry–Perot Interferometer

Fabry–Perot interferometers use the wavelength-transmission selectivity of a tuned FP cavity to detect the presence of dynamic (low-frequency) displacements. Optimum signal sensitivity of the FP interferometers is achieved by fixing the optical length of the cavity with respect to a reference wavelength on the cavity intensity transmission curve, so that small changes in the wavelength of light entering the cavity produce corresponding changes in the intensity of the transmitted light. In practice, the cavity is tuned to light of a given wavelength, while light of the same wavelength is reflected from the surface of interest and is frequency/wavelength Doppler shifted by the surface motion [15].

As depicted in Figure 1.5, the length L of the FP cavity is controlled by using a PZT. When the surface of interest moves, for instance, it vibrates at a modulation frequency Ω and with a small amplitude ε, then, the wavelength λ in the FP cavity is Doppler shifted and the cavity transmission changes. If λ_0 is the unshifted wavelength of the reference beam, we have

$$\lambda = \frac{\lambda_0}{\left(1 - \dfrac{2v}{c}\right)} = \frac{\lambda_0}{\left(1 - \dfrac{2\varepsilon\Omega}{c}\cos(\Omega t)\right)}. \quad (1.19)$$

Therefore, the transmitted signal can be calculated as a function of the small modulation amplitude ε [15].

Consequently, the detection limit can be derived, resulting in

$$\varepsilon_{lim} = \sqrt{\frac{\Delta f \hbar\omega}{\eta P_R}\frac{\left(1 + R^2 - 2R\cos(4\pi L/\lambda_0)\right)^{3/2}}{(1-R)(8\pi RL/\lambda_0 \sin(4\pi L/\lambda_0))}\frac{c}{\Omega}}, \quad (1.20)$$

where R is the total reflectivity of the mirror in the FP cavity. Assuming the incident power 1 mW, the detector quantum efficiency 0.10, the frequency of the light arriving at the detector 4.74 × 10^{14} Hz, the total mirror reflectivity 0.90, the optimal effective cavity length $L/\lambda_0 = 10^6 + 0.006$, we obtain $\varepsilon_{lim} \sim \times 10^{-15}$ m/$\sqrt{\text{Hz}}$.

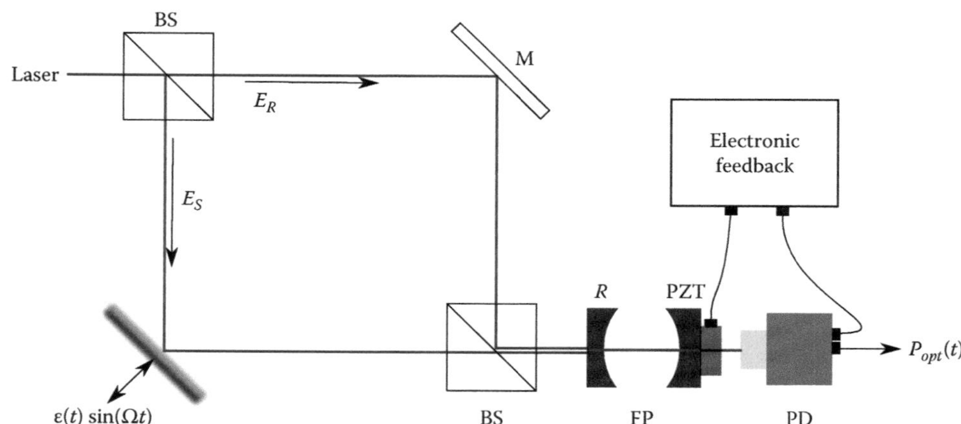

FIGURE 1.5 Setup for a FP interferometer; R indicates the reflectivity of the FP cavity. The PZT and the electronic feedback allow the adjustment of the FP cavity length.

Note that in this system the minimum detectable displacement depends heavily on the choices made for the cavity length and signal frequency. Since the choice of the cavity-length operating point depends solely on the cavity mirror reflectivity, the stabilization point set for the interferometer is independent of sample reflectivity. However, the incident power of the signal upon the cavity is directly proportional to the sample reflectivity, and no method for separating effects of the local sample reflectivity from those of the surface interaction is available in which the signal beam is used after it passes through the cavity.

1.2.2.4 Minimum Detectable Displacement

As seen earlier, the minimum detectable displacement in classical interferometric systems can be extremely small. When working under similar conditions and low power (~1 mW), the three examples of interferometers considered give each a very low limit $\varepsilon_{lim} \sim 10^{-3}\text{pm}/\sqrt{\text{Hz}}$. This corresponds to the limitation imposed by the shot noise of the light source and of the detection system. However, in practice, this limit is reached with difficulty in practical conditions, because, in real situations, all interferometers suffer from the environmental low-frequency noise and spatiotemporal distortions of the interfering optical wave fronts. We will see in the following sections how these main drawbacks can be overcome by adopting the AHI approach.

1.2.3 Adaptive Interferometry Based on Wave Mixing

The difference between the classical homodyne interferometer and the adaptive interferometer is that the latter, instead of the conventional beam combiner (beam splitter), contains a medium in which a dynamic hologram is continuously being recorded. The basic setup is based on TWM in a nonlinear holographic medium, as schematically depicted in Figure 1.6. The reference and the object beams produce interference at the nonlinear medium and create the hologram, and both are simultaneously diffracted from it. In this way, the diffracted part of the object beam propagates in the direction of the reference beam and vice versa. Owing to the basic principle of holography, the wave front of the diffracted part of the reference is the exact copy of the wave front of the nondiffracted part of the object beam. The same is true for another pair of beams in the medium. Therefore, after the nonlinear medium and in the direction of each beam, we have the coherent summation of two interfering beams with exactly the same wave fronts. In a TWM interferometer, the problem of the adjustment of the wave fronts is, thus, solved automatically. This allows the use of any complicated light wave in both arms of an adaptive interferometer.

1.2.4 Parameters Defining the Performances of an Adaptive Interferometer

The insertion of a nonlinear medium in an interferometer increases the ability of the system in rejecting noise and automatically fixes the quadrature condition also for complex wave fronts, for example, speckle beams. However, at the same time, the minimum detectable signal can be degraded. There are three main parameters that define the performance of any adaptive interferometer: the sensitivity to phase-difference excursions, the adaptability, which is determined by the response time of the nonlinear medium, and the output power of the laser required to reach these parameters.

1.2.4.1 Sensitivity

The performance of any adaptive interferometer can conveniently be compared with that of a classical interferometer by introducing the parameter

$$\delta_{rel} = \frac{\varepsilon_{lim}^{A}}{\varepsilon_{lim}^{C}} = \frac{SNR_C}{SNR_A}, \tag{1.21}$$

which is termed as the relative detection limit, where ε_{lim}^{A} is the minimal detectable phase difference in the adaptive interferometer, SNR_A is the SNR achieved in the adaptive holographic interferometer, and ε_{lim}^{C} and SNR_C are, respectively, the minimal detectable phase difference and the SNR in the case of a classical interferometer.

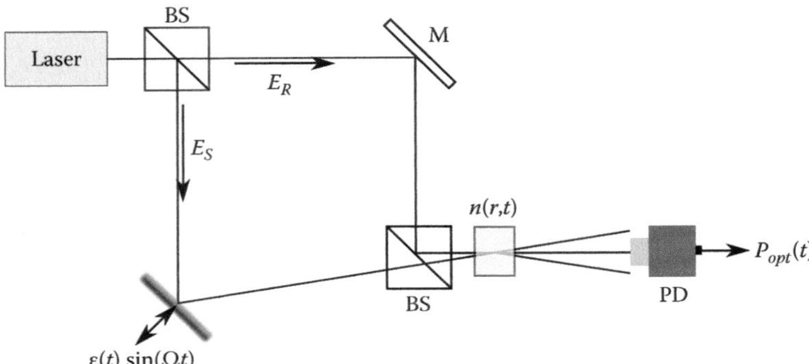

FIGURE 1.6 Setup for adaptive interferometry based on TWM in a nonlinear holographic medium; $n(r,t)$ is the refractive index grating induced in the nonlinear medium.

If the performance of the adaptive interferometer can be analytically described, the relative detection limit can be calculated theoretically. Experimentally, a direct comparison of the adaptive and classical interferometers can be carried out by using the same power of the object beam, by using the same photodetector and electronic circuit, and by maintaining the same phase difference between the interfering beams.

The relative detection limit is always greater than 1 and increases for higher optical losses. The relative detection limit δ_{lim} is a parameter that is very convenient for comparing different adaptive interferometers. Moreover, it also allows us to estimate the minimal detectable displacement ε_{lim}^{A}. On one hand, the relative detection limit δ_{rel} is the ratio of the detection limits of both interferometers; on the other hand, it shows how the response of the classical interferometer is larger than that of the adaptive interferometer.

1.2.4.2 Adaptability

Consider the case in which the phase difference between two interfering waves is occasionally changed, for instance, due to air flow or mechanical impact. Such a change leads to a displacement of the interference pattern inside the nonlinear medium and, consequently, to the recording of a new hologram. After this hologram is rerecorded, the phase difference between the matched wave fronts in the nonlinear medium output returns to its previous value. There are two different scenarios. If the time during which this phase shift takes place is shorter than the time τ_H needed to record the hologram, then this phase shift is immediately transferred into the change in the output power of the output beam. In the opposite case, when the phase difference is slowly varying compared with τ_H, the output light power remains constant. In other words, the dynamic hologram follows slow displacements of the recorded pattern, thus keeping the phase shift between the grating and the pattern nearly constant. Thanks to these properties of adaptive holography, low-frequency noise caused by external sources, such as air fluctuations or thermal deformations, is automatically compensated.

The adaptability of the adaptive interferometric system depends, therefore, on the response time of the nonlinear medium, which, on its turn determines the frequency bandwidth under which the TWM process is efficient. To quantify the adaptability of the adaptive interferometer, we can define a transfer function. For example, in the case of a Kerr-like optical nonlinearity, the dynamical equation for the refractive index reads as

$$\tau_H \frac{\partial \Delta n}{\partial t} = -\Delta n + n_2 I_{opt}, \tag{1.22}$$

where

n_2 is the Kerr-like coefficient
I_{opt} is the total incident light intensity

It is easy to show that the transfer function is similar to that of a conventional differentiating RC electronic circuit with time constant τ_H. An equivalent RC circuit at the output of the photodetector can also suppress slowly varying signals with frequencies smaller than $(2\pi\tau_H)^{-1}$. However, the RC circuit suppresses slow variations only when the phase difference is small. In contrast, the dynamic hologram accomplishes high-pass filtering of the signal in the stage of phase-to-intensity transformation. Therefore, high-amplitude, low-frequency modulations do not affect the measurement of small high-frequency excursions of the phase difference.

A convenient parameter for the quantitative estimation of interferometer adaptability is the cutoff frequency f_{cut}. For example, in the case when a photorefractive crystal acts as a dynamical hologram, f_{cut} can be expressed as [7]

$$f_{cut} = \frac{1}{2\pi\tau_H} = \frac{e}{2\pi\varepsilon\varepsilon_0} \frac{\alpha\xi\mu\tau_{PH}}{h\nu} F_K I_{opt}, \tag{1.23}$$

where

α is the optical absorption of the medium
ξ is the quantum efficiency of photoconductivity
μ and τ_{PH} are the mobility and average lifetime of photocarriers
e is the electron charge
$\varepsilon\varepsilon_0$ is the dielectric constant of the medium
$h\nu$ is the photon energy
F_K is a factor accounting for variations depending on the geometry of the TWM and on the photoexcited charge carriers in the nonlinear medium
I_{opt} is the total input intensity

1.2.4.3 Laser Power

The third parameter that has to be taken into account to optimize adaptive interferometers is the laser power. This parameter is important for three main reasons. First, the optical power determines the interferometer sensitivity. Indeed, the higher the light power reaching the photodetector, the lower is the minimal detectable phase difference achieved. Second, the cutoff frequency also depends on the optical power density. Then, the two main parameters of adaptive interferometers, sensitivity and adaptability, are directly related to the power of the laser, and, in general, they can be simultaneously improved by increasing the laser power. However, this increase usually leads to a nonproportional increase in the system cost and, additionally, makes the measuring system more energy consuming, which can be a drawback for certain applications where, for instance, an autonomous operating mode is required. This is the third reason why, when developing an adaptive interferometer, it is important to take into account the laser power and to search for a way to achieve high sensitivity and cutoff frequency using lasers with lower power.

1.3 Different Types of Adaptive Holographic Systems

1.3.1 Drift-Dominated Holograms in Photorefractive Crystals

The main mechanism of holographic recording in photorefractive crystals is well understood [23]. A schematic setup for dynamic holography in photorefractive crystals is displayed in Figure 1.7. A reference and a signal beam are simultaneously sent to interfere in the crystal, where they create an intensity fringe pattern. The photorefractive effect consists in the spatially nonuniform photoexcitation of mobile charge carriers by the recording interference pattern, their spatial redistribution via thermal diffusion or drift in an external field, and consequent trapping by the deep trapping centers. The spatially modulated electric field of these trapped charges, a space-charge field, leads to the spatial modulation of the dielectric tensor in the crystal volume via the linear electro-optic effect.

The time τ_H required to record the hologram depends on the efficiency of the charge generation and transport processes. The response time is determined by the dielectric relaxation time, τ_M, multiplied by a factor that takes into account the variations in the conductivity due to the spatially varying electron number density. Under some conditions, this factor approaches unity, and the response time can be taken as τ_M, which is given by

$$\tau_M = \frac{\varepsilon\varepsilon_0}{\sigma_0} = \frac{\varepsilon_0 h\nu}{e}\left(\frac{\varepsilon}{\alpha\xi\mu\tau_{PH}}\right)\frac{1}{I_{opt}}, \tag{1.24}$$

where

$\varepsilon\varepsilon_0$ is the static dielectric constant of the crystal
σ_0 is its average photoconductivity for the input light intensity I_{opt}
ξ is the quantum efficiency of photoconductivity
α is the optical absorption of the crystal
μ and τ_{PH} are the mobility and average lifetime of photocarriers
e is the electron charge
$h\nu$ is the photon energy

Since τ_M is inversely proportional to the intensity of the recording beams, there exists a trade-off between the hologram recording time and the sensitivity of the interferometer.

Let us briefly consider the main features of an adaptive interferometer based on the drift-dominated hologram recorded in a photorefractive crystal under a dc-electric field. Due to a half-of-the-period spatial shift or a zero shift, this hologram allows direct fulfillment of the quadrature condition depending on the direction of the external field between the input interference pattern and the refractive-index grating, which means either π or a zero phase shift between them. This phase shift is achieved when the external field E_0 is larger than the diffusion field E_D but smaller than the saturation field E_S of the crystal [18]:

$$K_g\frac{k_B T}{e} = E_D \ll E_0 \ll E_S = \frac{eN_A}{K_g\varepsilon\varepsilon_0}, \tag{1.25}$$

where

$K_g = 2\pi/\Lambda$ is the spatial frequency of the grating (Λ is its spatial period)
T is the temperature of the crystal
e is the electron charge
N_A is the concentration of the photorefractive centers

The saturation field E_S is the maximum possible amplitude of the space-charge electric field for the given K_g and N_A.

Due to automatic fulfillment of the quadrature condition, the geometry of the adaptive interferometer is quite simple. It is enough to mix two equally polarized coherent beams in the volume of the crystal and to measure the power of any beam at the output of the crystal. No polarization elements are needed. Therefore, the optical losses in the whole system are minimized and reduced to the inevitable light absorption of the crystal, providing very high sensitivity of the interferometer. The large amplitude of the space-charge field, which is proportional to the externally applied electric field E_0, is capable of providing efficient beam coupling in a thinner sample, thus diminishing the

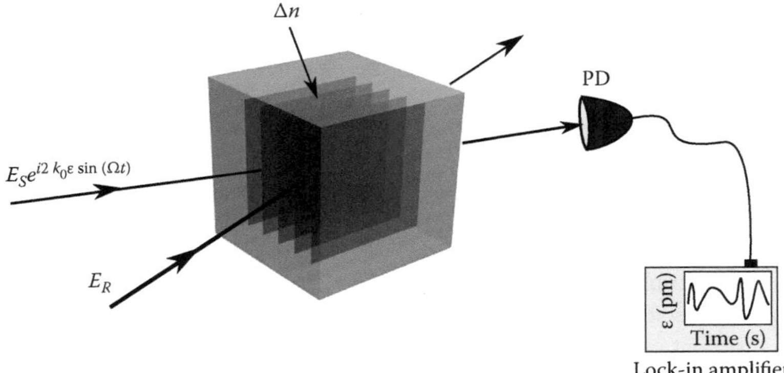

FIGURE 1.7 Schematic representation of the holographic recording in a photorefractive crystal. The reference E_R and signal E_S beams create an interference fringe pattern; correspondingly, a refractive index grating of amplitude Δn is induced in the crystal. The phase modulation of amplitude ε is directly transformed in intensity variation at the output of the crystal and measured with a PD and a lock-in amplifier.

losses associated with light absorption. All these features make two-wave coupling via a drift-dominated hologram one of the most popular configurations of the adaptive interferometer.

The main disadvantage of this approach is the screening of the external electric field. When the illumination of the crystal is spatially nonuniform, the electric field is redistributed according to the resistance of different parts of the crystal. The strongest field is applied to the darkest part, while the illuminated part is under the influence of the smallest electric field due to its high photoconductivity. Consequently, coupling of the interfering beams in this illuminated part is seriously suppressed. The screening effect can be avoided either by expanding the light beams to cover the whole interelectrode area or by using additional background illumination of the crystal. However, the first approach leads to a reduction in the light intensity inside the crystal and consequently to a decrease in the cutoff frequency. The second approach causes a diminishing visibility of the interference pattern and consequently a less efficient wave coupling. Moreover, using the background illumination, which should be of rather high intensity compared with the intensity of the interacting beams, makes the measuring system more complicated and energy consuming.

Adaptive interferometers based on drift-dominated holograms have been intensively developed during recent decades, mainly for ultrasound (US) detection under industrial conditions, which requires a measuring system possessing both high sensitivity and high cutoff frequency. Therefore, most of the proposed interferometers of this type exploit fast photoconductors such as GaAs, InP, and CdTe [24–27].

1.3.2 Diffusion-Dominated Holograms in Photorefractive Crystals

Even though the hologram recorded under an external dc-electric field usually shows a higher diffraction efficiency, which increases the sensitivity of the adaptive interferometer, the exploitation of a photorefractive crystal working without any electric field is desirable for many practical applications. The most popular configurations adopted to perform AHI with photorefractive crystals without applied dc electric field are described in the following sections. Note that, in this case, we speak of diffusion-dominated holograms, since diffusion is the main mechanism for inducing the spatial charge separation in the crystal.

1.3.2.1 External Phase Bias

The diffusion holograms do not support straightforward linear phase-to-intensity transformation due to their nonlocal response. Indeed, in the diffusion-dominated regime, the space-charge field created by the photoexcitation in the crystal is a quarter of period ($\pi/2$ phase shift) out of phase with respect to the light-intensity grating resulting from the interference of the two interacting beams. A simple solution to the problem was already proposed in the first paper devoted to adaptive interferometers with photorefractive TWM [28]. An additional $\pi/2$ phase bias can be introduced into one of the interfering waves by using an electro-optic modulator placed in the reference beam. Driving the modulator by a square-wave voltage with a quarter-wave amplitude makes it possible to instantly shift interference fringes by a quarter of their spatial period, which corresponds to the desired $\pi/2$ phase shift. The modulation frequency of the voltage applied to the modulator should be higher than the reciprocal response time τ_H in order to avoid hologram self-adaptation to the bias phase modulation. At the same time, in order to remove any switching transients related to the square-wave modulation, its frequency should be higher than the highest frequency of the signal to be detected. To this end, a low-pass filter at the photodetector output can be used. The main drawback of this technique is a noticeable reduction in the hologram diffraction efficiency due to the integration of moving fringes.

A modified technique of external phase bias has been described in Ref. [29]. In this technique, a low-frequency phase modulation of the harmonic type was imposed on one of the interfering beams. The modulation was of relatively large amplitude and was used for both the active stabilization of the interferometer via the negative feedback loop and for the linearization of the crystal response on a small phase difference. The signal of the photodetector appears in this technique in an upconverted form. This interferometer is also characterized by diminished diffraction efficiency of the dynamic hologram caused by the large-amplitude external phase modulation and consequently by reduced sensitivity. Nevertheless, the measurement of high-frequency vibrations of as low equivalent displacement of the object surface as 1 Å in the frequency band of 10 Hz was demonstrated using the $Bi_{12}TiO_{20}$ crystal and a He–Ne laser at a wavelength of 632.8 nm [29].

Recently, a similar technique of external phase modulation of the harmonic type was used for highly sensitive detection of low-frequency phase modulation [30]. Optimization of the high-frequency (5 kHz) phase modulation amplitude (1.86 rad) and the reference-to-object power ratio with properly considered amplification of the object beam led to a displacement sensitivity of 180 fm in the frequency band of 1 Hz for a signal modulated at 5 Hz. The experiment was carried out with a photorefractive $BaTiO_3$ crystal and a continuous wave (cw) laser operating at a wavelength of 532 nm with a few milliwatts of optical power.

One of the common drawbacks of the linearization technique employing the external phase bias is the necessity of using external electronic circuits for inducing both the phase modulation and output signal processing. This makes the measuring system more complicated and increases the noise level.

1.3.2.2 Anisotropic Diffraction

In the first experimental demonstration of the linear phase-to-intensity transformation using photorefractive TWM via a dynamic hologram recorded in the diffusion mode but without external phase bias [31], the reference and object waves were mixed in a photorefractive $Bi_{12}TiO_{20}$ crystal in the configuration of anisotropic diffraction [32]. This way, any diffracted beam has the polarization state orthogonal to that of the readout beam. In this experiment, the mechanism that leads to the fulfillment of the quadrature condition is probably due to the existence of internal stresses inside the crystal, which transform the polarization state of the light beams from linear to elliptical, thus ensuring the fulfillment of the quadrature condition after a properly oriented polarizer.

In 1991, Rossomakhin and Stepanov [33] described an adaptive interferometer that also exploits anisotropic diffraction from a diffusion hologram. They proposed a technique that exploits the vectorial nature of light and the anisotropic properties of dynamic holograms recorded in photorefractive crystals of cubic symmetry. Indeed, for these crystals it is possible to choose a recording geometry in which the readout and the diffracted waves are polarized differently and sometimes even orthogonally. An additional phase shift between the transmitted and the diffracted waves can be introduced by means of a quarter-wave plate installed after the crystal. Additionally, a polarizer in front of the photodetector is needed to project the orthogonal polarizations of the waves onto one common axis to let them interfere on the photodetector. A similar configuration of TWM via anisotropic diffraction was also implemented with a GaAs crystal [34].

1.3.2.3 Isotropic Diffraction

Another approach to fulfill the quadrature condition in TWM from the diffusion-dominated hologram has been proposed by Ing and Monchalin [35] by exploiting isotropic diffraction. They used a birefringent photorefractive crystal of $BaTiO_3$ in a geometry in which light diffraction occurs without a change in the polarization state. The input polarization state of the object beam was linear but tilted to 45° with respect to the linear polarization state of the reference beam. Consequently, only the component of the object beam having the polarization that coincides with that of the reference beam was enhanced via self-diffraction from the diffusion hologram. The object beam after the crystal consists of two orthogonally polarized components: one bears the phase of the reference beam, while the other bears that of the object beam. Installation of the phase retarder and a properly oriented output polarizer leads to the linear mode of phase demodulation.

A similar configuration of the adaptive interferometer was presented in 1994 by Blouin and Monchalin [36]. Instead of $BaTiO_3$ with an argon-ion laser, the authors used a faster photoconductive crystal of GaAs with a more powerful Nd:YAG (yttrium aluminum garnet) laser at a wavelength of 1064 nm that allowed them to significantly speed up the response time of the dynamic hologram. However, there are additional optical losses in TWM via isotropic diffraction associated with the diminished visibility of the interference pattern due to noncoinciding input polarization states.

1.3.2.4 Diffraction Enhancement by AC-Electric Field

In photoconductive crystals with a large mobility-lifetime product $\mu\tau_{PH}$ of the photoexcited charge carriers, the application of an external alternating electric field leads to the enhancement of the created space-charge field and correspondingly to a more efficient coupling of the interfering waves. For the efficient enhancement of wave coupling, the period of the ac field must be much shorter than the recording time of the hologram τ_H [37,38].

The important feature of wave coupling enhanced by an ac field is that the spatial shift between the interference pattern and the recorded hologram is the same as in the diffusion mode of hologram recording: a quarter of the grating period. Therefore, an additional technique for accomplishing the quadrature condition has to be used in this case. Unfortunately, the installation of either a polarizer or a polarization beam splitter is not possible here, because the external ac field modulates the polarization state of both transmitted and diffracted beams, which is transferred into intensity modulation by any polarizing element, thus seriously increasing the noise level.

The solution was found after the publication of a rigorous theory [39] of vectorial wave coupling in crystals of cubic symmetry. Analysis shows that in the geometry of anisotropic diffraction, the mixing of waves with linear and elliptical polarization allows for linear phase-to-intensity transformation. It occurs because the phase difference $\Delta\psi$ between the orthogonal components of the elliptically polarized light beam is transferred into a nonzero phase difference in the interference term between the transmitted part of the object beam and the diffracted part of the reference beam with the optimal $\Delta\psi$ being $\pi/2$. To underline the important role of vectorial wave interaction, the proposed technique was referred to as vectorial wave mixing [5]. In contrast with the previously reported techniques, no polarizing elements are installed after the crystal. Nevertheless, a small deterioration of the sensitivity is associated with the diminished visibility of the interference pattern due to the different polarization states of the input beams.

The technique was experimentally implemented with $Bi_{12}TiO_{20}$ [5], GaP [40], and CdTe [41]. Theoretical analysis of adaptive interferometers based on vectorial wave mixing was carried out in Ref. [42]. The best sensitivity among adaptive interferometers (relative detection limit of $\delta_{rel} = 1.5$) was achieved with a 30 mW He–Ne laser and $Bi_{12}TiO_{20}$ crystal under an ac field of the square-wave form at a repetition rate of 80 Hz with an amplitude of 8 kV/cm [5]. The advantages of the ac-enhancement technique are the larger efficiency of the dynamic hologram and the absence of electric field screening. The latter gives rise to the possibility of mixing tightly focused light beams, which increases the cutoff frequency at the same level of the laser power. Consequently, higher adaptability can be achieved before overheating the crystal. The main disadvantage of this method is the necessity of applying to the crystal a high voltage at a high frequency: for an interferometer with a cutoff frequency of a few kilohertz, the frequency of the external ac field must be tens of kilohertz [41].

1.3.2.5 Vectorial Wave Mixing via a Reflection Hologram

Since the phase shift between the input interference pattern and the hologram recorded under an ac electric field is the same as for a hologram created without any field, the technique of vectorial wave mixing is also applicable to a pure diffusion hologram. However, the diffraction efficiency of a diffusion hologram recorded in the conventional geometry of copropagating beams is much smaller than that under an ac field. Since the diffusion field is proportional to the spatial frequency of the interference pattern, the natural way of increasing the beam coupling in this case is by using the geometry of counterpropagating beams, which forms a reflection hologram.

An adaptive interferometer based on a vectorial wave coupling via a diffusion reflection hologram was demonstrated with a $Bi_{12}TiO_{20}$ crystal and a He–Ne laser: $\lambda = 632.8$ nm [43] and, then, with a CdTe:V crystal and a Nd:YAG laser: $\lambda = 1064$ nm [44]. It was shown that the reduced spatial period of the reflection hologram together with better beam overlapping provides sensitivity at least four times higher than that achieved with the transmission geometry in the same crystal [44].

1.3.3 Adaptive Holograms in Liquid Crystal Light Valves

While a lot of research during the last 30 years has been devoted to photorefractive materials and their applications [23], in the last decade, spatial light modulators (SLM) have attracted a great deal of attention and have emerged as important components for optical processing [45]. Indeed, SLMs are able to affect the phase modulation and the intensity of a readout beam, thus permitting the manipulation of information in the optical domain. In optically addressed SLM, the control signal is provided by an optical input beam, so that optical parallelism can be fully exploited [46]. The general structure of an optically addressed SLM comprises two components: the photoreceptor and the electro-optic material, often separated by a dielectric mirror [47]. The input beam activates the photoreceptor, which produces a corresponding charge field on the electro-optic material. The read light is modulated in its double pass through the electro-optic element in the retroreflective scheme. Since the readout can provide optical gain, this type of optically addressed SLM has also been called a light valve. Historically, the photoreceptor has been a photoconductor, such as selenium or cadmium sulfide [48], amorphous silicon [49], or GaAs [50], and the electro-optic material has been a nematic liquid crystal (LC) layer, either in parallel or twisted configuration [45–57], so that optically addressable SLMs are also more widely known as LCLVs.

An interesting type of LCLV has been realized by associating nematic LCs with photorefractive crystals. The first photorefractive LCLV has been prepared by using as a photoconductor a thin monocrystalline $Bi_{12}SiO_{20}$ (BSO) crystal [52]. The BSO, well known for its photorefractive properties, is chosen for its large photoconductivity and dark resistance, while its transparency in the visible range allows the LCLV working in transmissive configurations, where the input and readout beams may in general coincide. These features make the photorefractive LCLV optical elements attractive with capabilities for numerous applications, such as laser beam manipulation [53], coherent image amplification through dynamic holography [54], slow-light [55,56], and nonlinear optics implementations [57]. Here, we will focus on two-beam coupling experiments and on the creation of dynamical holograms. By this way, AHI can be performed with LCLV, in close analogy to that previously described with photorefractive crystals. The main advantage of the LCLV is that photoconductive and electro-optic properties are separately optimized, so that an excellent photosensitivity comes from the large photoconductivity of the BSO and a large nonlinear response comes from the high birefringence of the nematic LCs. This is particularly important in view of the adaptive holography applications, since it allows decreasing the required laser power considerably. Moreover, the possibility to obtain large (lateral size of a few centimeters) BSO crystals with a high degree of spatial homogeneity and uniform dark resistance makes LCLV very attractive for operating over large areas, thereby allowing interferometry with spatially extended optical beams.

1.3.3.1 Liquid Crystal Light Valve

The LCLV is schematically depicted in Figure 1.8a. It is made by associating a LC layer with a photorefractive BSO crystal, cut in the form of a thin plate (1 mm thickness, 20×30 mm^2 lateral size) [52,54]. The BSO is one of the confining walls, and the other wall

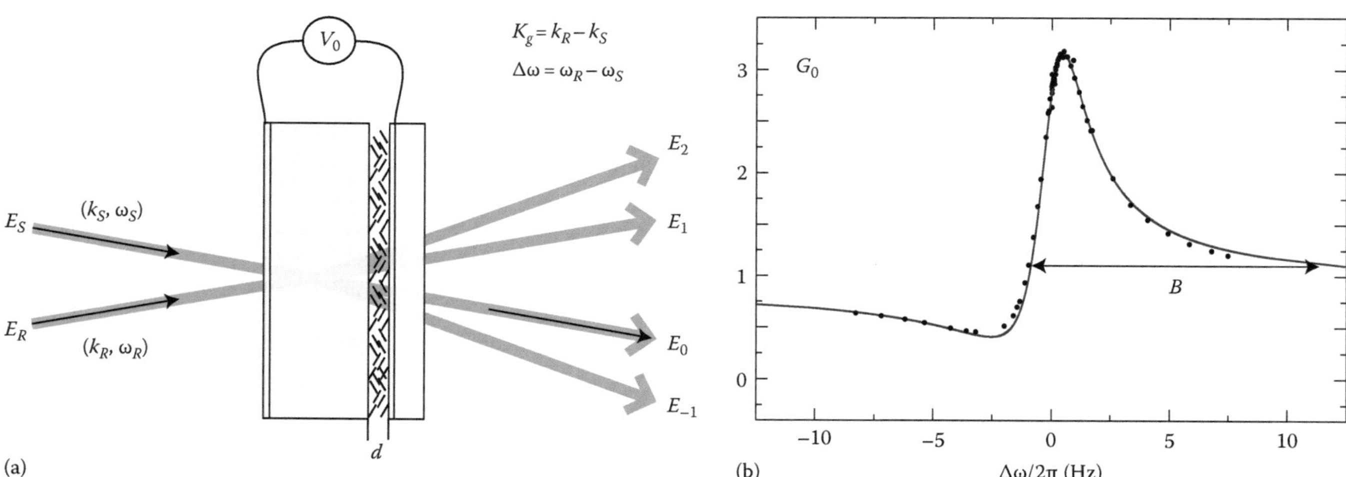

FIGURE 1.8 (a) Two-wave mixing in the LCLV. (b) Measured (points) and theoretical (line) gain $G_0 \equiv I_0/I_S$ as a function of the frequency detuning $\Delta\omega$ between the reference and signal beam. B is the frequency bandwidth characterizing the two-beam coupling process.

is a glass window. The thickness of the LC layer is typically of the order of 10 μm. While LCs are used for their large birefringence, the BSO is used for its large photoconductivity and transparency in the visible range [23]. Transparent electrodes (indium tin oxide, ITO, layers) deposited over the BSO and the glass wall allow the application of an external voltage V_0 across the LC layer. The voltage applied is ac, with an rms value from 2 to 20 V and a frequency from 50 Hz to 20 kHz. The LCs are nematics E48 (from Merck).

The nematic phase is characterized by a long-range orientational order for which all the molecules are aligned, in average, along a preferential direction, the so-called nematic director $\hat{n}_{LC}$ [58]. Because the molecules have different polarizabilities along their long and short axes, $\varepsilon_{\parallel}$ and $\varepsilon_{\perp}$ being the dielectric susceptibility parallel and orthogonal, respectively, to the long axis of the molecules, when an electric field, or a voltage V_0, is applied across the nematic layer, an induced dipole moment arises and all the molecules reorient toward the direction of the applied field.

Because of the LC birefringence, the nematic layer as a whole behaves like a strongly birefringent material, characterized by a different refractive index for a beam polarized along the long or short molecular axis, called, respectively, the extraordinary n_e and the ordinary n_o index. Typical values for nematics are $n_e = 1.7$ and $n_o = 1.5$, which give birefringence $\Delta n = n_e - n_o$ as large as $\Delta n = 0.2$. Therefore, when the LC molecules reorient under the action of an applied field, their collective motion implies a change of the principal axis of the nematic layer; hence, an incoming light field experiences a corresponding refractive index change [59]. When a light beam impinges onto the LCLV, photogeneration of charges occurs at the BSO surface because of its photoconductive properties. Therefore, the local voltage across the LC layer increases, inducing a further molecular reorientation and, thus, an additional refractive index change. As a result, at the exit of the LCLV, the light beam acquires a phase shift that is a function of the applied voltage V_0 and of the total intensity I_{in} of the incident beam.

In the linear region of its response, the LCLV behaves as a Kerr-like nonlinear medium, providing a refractive index change proportional to the input intensity $n = n_0 + n_2 I$, where n_0 is the value fixed by the applied voltage and n_2 the nonlinear coefficient. Saturation occurs when all the LC molecules are aligned along the direction of the applied electric field. Thanks to the large LC birefringence, the nonlinear coefficient, which is the slope of the linear part of the response curve, is as large as $n_2 = -6\ \text{cm}^2/\text{W}$, the minus sign accounting for the defocusing character of the nonlinearity (the refractive index changes from n_e to n_o, with $n_e > n_o$ when LC molecules reorient under the action of the electric field). The response time is dictated by the time τ_{LC} required by the collective motion of the LC molecules to establish over the whole thickness d of the nematic layer. This is given by

$$\tau_{LC} = \frac{\gamma}{K} d^2, \tag{1.26}$$

where

γ is the LC rotational viscosity
K is the splay elastic constant [58]

For $d = 14$ μm and typical values of the LC constants, τ_{LC} is of the order of 100 ms. The spatial resolution, which is the minimal size of an independently addressed area, is given by the electric coherence length of the LC,

$$l_{LC} = \sqrt{\frac{\Delta\varepsilon}{K} \frac{d}{V_0}}, \tag{1.27}$$

where $\Delta\varepsilon = \varepsilon_{\parallel} - \varepsilon_{\perp}$ is the dielectric anisotropy of the LC. For the usual values of V_0, l_{LC} is typically of the order of 10 μm.

1.3.3.2 Two-Wave Mixing in the LCLV

When TWM is performed in thin media, the beam coupling occurs in the Raman–Nath regime of optical diffraction. As against the Bragg regime, for which the phase matching condition is satisfied only in one direction, the Raman–Nath diffraction produces several output order beams [60]. The interaction scheme comprises a reference beam E_R, which is sent onto the LCLV together with a signal beam E_S (Figure 1.8a). The total electric field at the input of the LCLV can be written as

$$E_{in}(\vec{r},t) = E_S e^{i[\vec{k}_S\cdot\vec{r} - \omega_S t]} + E_R e^{i[\vec{k}_R\cdot\vec{r} - \omega_R t]} + c.c., \tag{1.28}$$

where

E_R and E_S are the amplitudes of the reference and signal waves, respectively
$\vec{k}_R$ and $\vec{k}_S$ are their respective propagation vectors
ω_R and ω_S are their respective frequencies

The two beams produce an intensity fringe pattern

$$|E_{in}(\vec{r},t)|^2 = I_T\left[1 + 2\frac{E_R E_S}{I_T}\cos(\vec{K}_g\cdot\vec{r} - \Delta\omega\cdot t)\right], \tag{1.29}$$

where

$I_T \equiv |E_S|^2 + |E_R|^2 = I_S + I_R$ is the total input intensity
$\vec{K}_g = \vec{k}_R - \vec{k}_S$ is the grating wave vector
$\Delta\omega = \omega_R - \omega_S$ is the frequency detuning between the reference and the signal

The ratio between the reference and signal intensities $\beta \equiv I_R/I_S$ is usually kept much larger than 1.

The fringe pattern induces, on its turn, a photoinduced space-charge distribution, thereby a molecular reorientation pattern in the LC layer, which creates a refractive index grating with the same wave vector $\vec{K}_g$. The spatial period of the grating $\Lambda \equiv 2\pi/K_g$ is usually larger than the thickness of the LC layer; therefore, the LC grating acts as a thin hologram, and several diffracted beams, distinguished by the numbers 0, ±1, ±2,..., ±m, are observed at the output of the LCLV. Due to self-diffraction, photons from the pump are transferred to the different output orders. The $m = 0$, +1, ±2,... orders are amplified, that is, they receive from the reference, also called pump beam, more photons than they are losing due to the scattering on the other orders. The $m = -1$ order is the reference beam that, even though depleted, remains of much higher intensity than the other beams [61].

To derive the full expression for the output field, we have to consider the evolution of the amplitude $n(\vec{r}, t)$ of the refractive index grating inside the LC layer. This is governed by a relaxation equation following the molecular orientation dynamics of the LC [58]:

$$\tau_{LC}\frac{\partial n}{\partial t} = -(1 - l_{LC}^2\nabla^2)n + n_0 + n_2 \,|\, E_{in} \,|^2, \quad (1.30)$$

where

l_{LC} = 10 μm is the transverse diffusion length

n_0 = 1.6 is the constant value of the refractive index given by the average LC orientation under the application of the voltage V_0

$n_2 \simeq -6$ cm²/W is the equivalent Kerr-like coefficient of the LCLV

By coupling Equation 1.30 with the wave propagation equation for the input electric field, one can easily show that the m output order field can be written as [61–63]

$$\tilde{E}_m = E_m e^{i(\vec{k}_m \cdot \vec{r} - \omega_m t)} + c.c., \quad (1.31)$$

where

$\omega_m = \omega_S - m\Delta\omega$ is the frequency

$\vec{k}_m = \vec{k}_S - m\vec{K}_g$ is the wave vector

The amplitude is given by

$$E_m = \left[E_S J_m(\rho) + iE_R J_{m+1}(\rho)e^{-i\Psi}\right] \cdot e^{i\left[k(n_0 + n_2 I_T)z + m\left(\frac{\pi}{2} - \Psi\right)\right]}, \quad (1.32)$$

where J_m is the Bessel function of the first kind and of order m,

$$\rho = \frac{2kn_2 E_R E_S}{\sqrt{\left(1 + l_{LC}^2 K_g^2\right)^2 + \left(\Delta\omega \cdot \tau_{LC}\right)^2}} d \quad (1.33)$$

is the grating amplitude, and

$$\tan\Psi = \frac{\Delta\omega \cdot \tau_{LC}}{1 + l_{LC}^2 K_g^2}. \quad (1.34)$$

From Equation 1.32, we see that each order m receives two contributions: one is the scattering of the signal and the other is the scattering of the reference beam onto the refractive index grating.

It is useful to write each output order field in the form

$$\tilde{E}_m = \sqrt{G_m}\, E_S e^{i\Phi_m} e^{i(\vec{k}_m \cdot \vec{r} - \omega_m t)} + c.c., \quad (1.35)$$

where we define $G_m = |E_m|^2/|E_S|^2$ as the gain factor and Φ_m as the associated nonlinear phase shift. Both G_m and Φ_m can be calculated from Equation 1.32.

Let us consider the $m = 0$ order, which coincides with the original propagation direction of the signal,

$$\tilde{E}_0 = \sqrt{G_0}\, E_S e^{i\Phi_0} e^{i(\vec{k}_S \cdot \vec{r} - \omega_S t)} + c.c. \quad (1.36)$$

The envelope amplitude can also be written as

$$E_0 = \left[E_S J_0(\rho) + iE_R J_1(\rho)e^{-i\Psi}\right] e^{i[k(n_0 + n_2 I_T)z]}, \quad (1.37)$$

where $I_T = I_R + I_S$ is the total input intensity. The two contributions, namely the scattering of the signal and the scattering of the reference beam, sum up with their relative phases. The final effect is to produce a gain with a narrow frequency bandwidth. In Figure 1.8b, the gain curve G_0 is plotted as a function of the frequency detuning $\Delta\omega$ for parameter values close to typical experimental conditions, τ_{LC} = 120 ms, l_{LC} = 15 μm, Λ = 150 μm, n_0 = 1.63, n_2 = −6 cm²/W, and β = 30, and compared with the experimental data (filled circles). The maximum gain (and, correspondingly, maximum phase shift) is obtained for $\Delta\omega \sim 0$. In correspondence, a temporally modulated signal will experience a large dispersion, due to the strong selectivity of the gain whose frequency bandwidth is approximately B = 10 Hz.

AHI is realized in the LCLV by performing TWM experiments as described earlier. The narrow frequency bandwidth of the gain makes the LCLV an ideal medium for adaptive holography. Moreover, the high nonlinear response of the LCLV allows operating the adaptive interferometer with very low incident power, another fundamental advantage for the realization of AHI systems. We will see in the next section how it is possible to easily implement picometer detection with LCLV-based interferometers [8].

1.3.4 Dynamic Holograms in Rare-Earth-Doped Optical Fibers

Dynamic Bragg gratings have been recorded by saturation of absorption in rare-earth-doped optical fibers and demonstrated to possess attractive capabilities for the realization of adaptive interferometers [64–68]. In these experiments, the nonlinear medium is the optical fiber itself, and dynamic Bragg reflectance gratings are formed as a result of saturation of fiber-optic absorption, a mechanism also called as spatial hole burning. The main part of the research, as well as related applications, was performed with erbium-doped fibers in the wavelength range 1490–1570 nm. The dynamic gratings are usually formed by two mutually coherent cw laser waves of the same polarization, counterpropagating through the doped single-mode fiber. The typical experimental setup is schematically sketched in Figure 1.9.

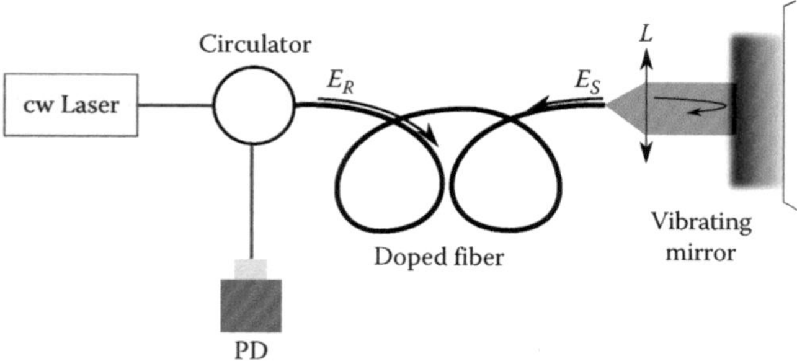

FIGURE 1.9 Setup for adaptive interferometry with erbium-doped single-mode optical fiber.

The reference E_R and signal E_S beams counterpropagate in the fiber. In the steady state, the two-beam coupling can be described by the coupled equations

$$\frac{\partial E_R}{\partial z} = -\frac{\alpha(z)}{2} E_R(z) - \frac{\Delta\alpha}{4} S(z),$$
$$\frac{\partial E_S}{\partial z} = -\frac{\alpha(z)}{2} E_S(z) - \frac{\Delta\alpha}{4} R(z), \quad (1.38)$$

where $\alpha(z)$ and $\Delta\alpha(z)$ are the average fiber absorption and the recorded grating amplitude, respectively. Analytical expression for $\alpha(z)$ and $\Delta\alpha(z)$ can be obtained from the equations describing the absorption/gain saturation in the fiber and read as [68]

$$\alpha(z) = -\frac{\alpha_0}{1+(|E_R(z)|^2 + |E_S(z)|^2)/I_{sat}},$$
$$\Delta\alpha(z) = -\frac{2\alpha_0 E_R(z) E_S(z)}{1+(|E_R(z)|^2 + |E_S(z)|^2)/I_{sat}}, \quad (1.39)$$

where I_{sat} is the characteristic saturation intensity. In the case of amplitude population grating recorded in a medium with saturable absorption, formation of the grating leads to the growth of both transmitted wave intensities. Due to self-diffraction, the transmitted recording wave and the result of diffraction of the other wave from the grating interfere constructively; this is because the interference pattern "burns" some spatial holes in the profile of the optical absorption.

In order to realize an optical fiber sensor, transient TWM has been performed with different phase modulation techniques [68]. The modulation frequencies are set to be $\Omega \gg \tau^{-1}$, where τ is the response time of the wave mixing process, and are typically in the range of 10 kHz. The adaptive interferometric optical fiber sensor can be used for the measurement of environmentally induced signals of various nature: acoustic, hydro-acoustic, seismic, electric, magnetic, thermic, etc. The external parameter to be measured has to influence the optical path length inside the fiber. This has to occur in a similar way as performed to realize the transient TWM. The dynamic population grating acts in this case as an adaptive beam splitter, able to compensate slow environmentally induced phase drifts. In this way, it also supports an optical operation point of the interferometer for the detection of fast phase modulation. Utilization of dynamic fiber gratings in AHI is especially attractive for industrial or field applications, because this method combines together (in the fiber) the sensor and the nonlinearity, thus providing a compact, closed, all-fiber, and robust setup, which is based on easily available elements commercialized for the telecom area.

1.4 Examples of Picometer Detection in Noisy Environments

We report here a few examples of picometer detection based on adaptive interferometry, such as an optical fiber strain sensor able to detect fiber elongations as small as of the order of 0.1 pm, the detection of subpicometer displacements performed with an LCLV, the measurement of the Casimir force, the detection of deformations induced by radiation pressure, and, finally, acousto-optic imaging in biological tissues.

1.4.1 Multimode-Fiber Strain Sensor with AHI in Photorefractive Crystals

Optical fibers are widely used for sensing and measuring different physical parameters such as temperature, pressure, electric current, and vibration. An optical fiber serves as a sensitive element to transform induced strain into a change of the phase $\Delta\varphi$ of the light wave propagating through the fiber. To measure small phase excursions ($\Delta\varphi \ll 1$), an interferometer is used. Most of the sensors use a single-mode optical fiber to ensure highly visible interference with the reference wave, although a multimode fiber is a better solution because of its efficient and stable coupling with the laser. However, multimode fibers are not often used in interferometers due to problems caused by random distribution of the phase and amplitude of the output light field, referred to as a dynamic speckle pattern.

In Refs. [6,44], a strain sensor in which the sensitive element is a multimode fiber and in which a reflection dynamic hologram recorded in a CdTe:V crystal compensates for instabilities and fluctuations of the speckle pattern emerging from the multimode fiber has been demonstrated. The configuration of the strain sensor is shown in Figure 1.10. A light beam generated by a cw Nd:YAG laser λ = 1064 nm, output power 500 mW is divided into a reference and an object beam. The reference beam is directly sent to a photorefractive CdTe crystal. Its polarization state is controlled by adjustable wave plates. The object beam is launched into the multimode fiber. After passing through the fiber, the light radiation is linearly polarized by polarizer P and collected into the crystal from the other side. The object beam transmitted through the crystal is entirely collected into the active area of the PD. Part of the fiber is reeled

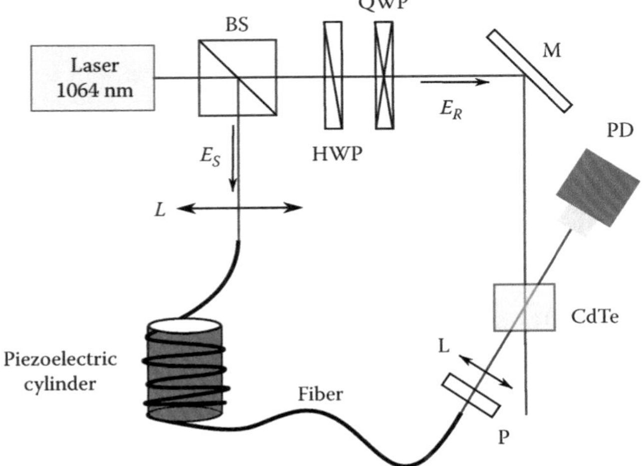

FIGURE 1.10 Optical scheme of the multimode-fiber strain sensor stabilized via wave mixing in photorefractive CdTe crystal operating in the reflection mode. QWP and HWP are quarter-wave and half-wave plates, respectively. (From Di Girolamo, S. et al., *Opt. Lett.*, 32, 1821, 2007.)

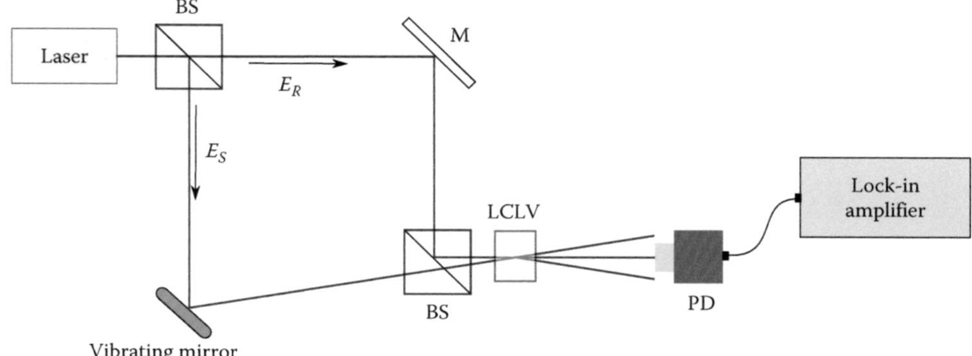

FIGURE 1.11 Experimental setup of the AHI interferometer with the LCLV as adaptive hologram. The input laser is split into a reference beam and a signal beam that are incident on the LCLV; before arriving at the LCLV, the signal is sent to a vibrating object, whose amplitude of oscillation we want to detect. The output intensity on the −1 order is measured with a PD and converted to a voltage.

onto a piezoelectric cylinder. By applying voltage to the cylinder, strains are induced on the fiber, which leads to phase modulations of the output speckled beam.

Object and reference waves are mixed in the photorefractive crystal. By using the AHI principle of beam combining, it is possible to measure phase variations under unstable environmental conditions because of the perfect and dynamic matching of different wave fronts. To achieve linear and the most sensitive phase-to-intensity transformation, light beams propagate under a small angle to the crystallographic axis of the CdTe crystal, while the polarization state of the reference beam is elliptical and the object beam is linearly polarized [12]. No external electric field was applied to the crystal. The intensity modulation of the transmitted object beam was measured by a conventional PD.

It was shown that the hologram compensates for both temporal and spatial instabilities in the intensity distribution of the speckles emerging from the fiber, but it cannot compensate for instabilities in the polarization state of speckles, the latter producing the largest source of noise affecting strain measurements. However, this noise was significantly diminished by using a fiber with a larger diameter of the core. It was shown that a 550 μm core fiber operates excellently as a strain sensor in spite of the more than 50,000 spatial modes excited in it. The modulation amplitude of 1.1 rad has been achieved when the amplitude of the dynamic elongations was 120 nm.

1.4.2 Picometer Detection with AHI in Liquid Crystal Light Valves

As we have seen in Section 1.3.3, TWM in the LCLV is characterized by a narrow frequency bandwidth B of the gain (see Figure 1.8). In a similar way as it occurs for AHI in photorefractive crystals [7], the gain resonance curve acts as an optical filter able to adapt the dynamic hologram by following low-frequency variations and noise disturbances (inside the TWM gain bandwidth B). TWM in the LCLV provides both good sensitivity and narrow frequency bandwidth, while allowing a simplified configuration to efficiently realize AHI detection systems [8].

The experimental setup for AHI in the LCLV is shown in Figure 1.11. The input beam is from a cw-doubled diode-pumped solid-state laser, of wavelength 532 nm, with the total input intensity typically less than 5 mW/cm². The laser beam is divided into a reference and a signal wave. By means of a PZT, the signal beam E_S is phase modulated with a sinusoidal oscillation at high frequency $\Omega >> B$ and small amplitudes ε. The signal beam is sent onto the LCLV together with the reference beam E_R, thus producing a thin diffraction grating. Several output beams are obtained at the exit of the LCLV. The optical power of the output beams is measured with a PD and a lock-in amplifier. The frequency of the modulation is fixed at $\Omega/2\pi = 1$ kHz, which is much greater than the bandwidth $B \simeq 10$ Hz of the TWM in the LCLV. A PD ($\eta \simeq 0.63$) is placed on one of the diffracted orders, and the output optical power is measured by using a lock-in amplifier (1 Hz bandwidth).

The total intensity distribution on the LCLV is

$$I = | E_R e^{i(\vec{\mathbf{k}}_\mathbf{R}\cdot\vec{\mathbf{r}}+k_0\Delta-\omega_0 t)} + E_S e^{i(\vec{\mathbf{k}}_\mathbf{S}\cdot\vec{\mathbf{r}}+\theta-\omega_0 t)} |^2, \tag{1.40}$$

where

- ω_0 is the laser frequency
- $\theta = 2k_0\varepsilon\sin(\Omega t)$ is the phase shift due to the vibrating object
- ε is the small displacement that we aim to detect
- Δ is the optical path difference acquired by the reference and signal before arriving at the LCLV

Given the narrow frequency bandwidth of the gain, the grating formation automatically filters out the high frequency, and the amplitude of the phase grating reads as [2]

$$\rho = 2k_0 d n_2 J_0(2k_0\varepsilon) E_R E_S. \tag{1.41}$$

Because of the large value of the LCLV nonlinear coefficient n_2, the process is very efficient and the output diffracted beams are easily detected. By solving the wave propagation equation in the Raman–Nath regime [55], we obtain for the optical power of the m output order

$$P_m = P_R e^{-\alpha D}\left(K^2 J_m^2 + J_{m+1}^2 + 2KJ_m J_{m+1}\sin(\theta)\right), \tag{1.42}$$

where

$\alpha \approx 0.3$ cm^{-1} is the total absorption coefficient of the LCLV
$D = 1$ mm is the thickness of the photoconductor
$K^2 = P_S/P_R$ is the ratio between the signal and reference power
$J_m \equiv J_m(\rho)$ is the Bessel function of the first kind and of order m

By substituting in Equation 1.42 the expression for θ, we find the component at the modulation frequency Ω:

$$\hat{P}_m(\Omega) = 4P_R e^{-\alpha D} KJ_m J_{m+1} J_1(2k_0\varepsilon)\sin(\Omega t), \tag{1.43}$$

where we have made use of the component parts of the Fourier–Bessel expansion [15]. Either the zero-order beam, which coincides with the direction of the signal, or the −1 beam, which coincides with the direction of the reference, is detected with a PD and a lock-in amplifier. These orders are also used to calibrate the system. Indeed, if $\rho \ll 1$, and for $m = -1$, we have

$$\hat{P}_m \propto J_0(2k_0\varepsilon)J_1(2k_0\varepsilon), \tag{1.44}$$

which has a maximum at 1.1 rad. As already done in other AHI setups [3,72], this property has been used in the LCLV system to find the relation between the displacement ε and the measured lock-in voltage $V_{lock\text{-}in}$.

1.4.2.1 Relative Detection Limit

For small displacements, we can approximate $J_1(2k_0\varepsilon) \approx k_0\varepsilon$, and the detection becomes linear with ε, which automatically gives the highest sensitivity of the AHI interferometer. In classical interferometers, to achieve this condition, the average phase difference between the interfering beams has to be set to $\pi/2$ (quadrature condition). Moreover, the AHI system does not require the stabilization with respect to variations of the optical path difference Δ, since the beam coupling is self-adapted. The sensitivity of the AHI system is obtained by considering the limit given by the photon shot noise. The SNR in this case can be expressed as [4,8]

$$SNR = \sqrt{\frac{2\eta P_R}{\hbar\omega\Delta f}}\, e^{-\alpha D/2} \frac{KJ_m(\rho)J_{m+1}(\rho)}{\sqrt{K^2 J_m^2(\rho) J_{m+1}^2(\rho)}} 2k\Delta, \tag{1.45}$$

where

η is the quantum efficiency of the PD
Δf is the bandwidth of the electronic detection system

The minimum detectable displacement ε_{lim} is calculated by setting $SNR = 1$. In order to compare the performances of the AHI with classical homodyne detection, the relative detection limit $\delta_{lim}^{(rel)}$ has to be considered. In the case of a classical interferometer in quadrature configuration and lossless, the ideal detection limit is [4]

$$\delta_{ideal} = \frac{1}{(2k_0)}\left[\frac{(\hbar\omega_0\Delta\nu)}{(2\eta P_R)}\right]^{1/2}, \tag{1.46}$$

and thus, we obtain

$$\delta_{lim}^{(rel)} = \frac{\sqrt{K^2 J_m^2 + J_{m+1}^2}}{KJ_m J_{m+1}} e^{\alpha D/2}. \tag{1.47}$$

The minimum $\delta_{lim}^{(rel)} \approx 1.1$ is obtained for the −1 order. Correspondingly, the maximum SNR is obtained, and the minimum detectable phase is on the order of 7 nrad/$\sqrt{\text{Hz}}$.

The signal $V_{lock\text{-}in}$ detected at the $m = -1$ order, for which the theoretical curves predict the maximum sensitivity, is plotted in Figure 1.12a as a function of the mirror displacement ε. The intensity of the signal beam was 3 mW/cm^2 and $K = 5$. In Figure 1.12b,

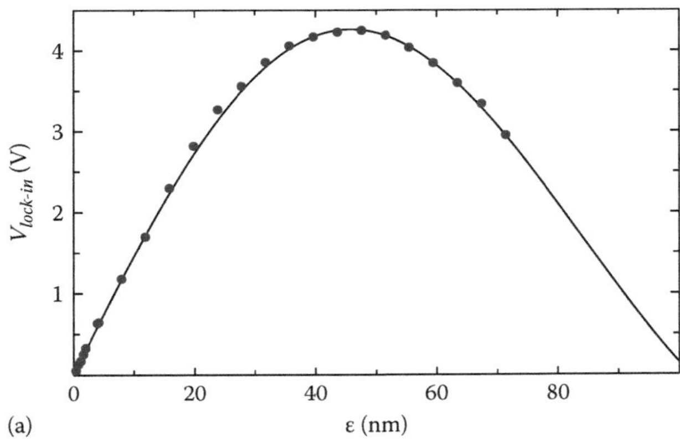

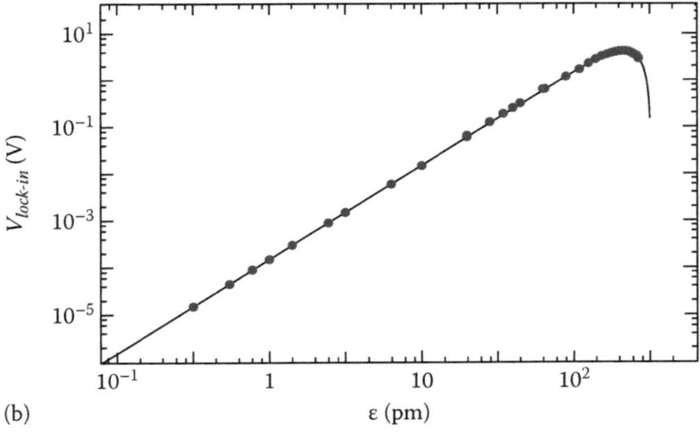

FIGURE 1.12 (a) Linear and (b) log-scale plot of the signal $V_{lock\text{-}in}$ detected at the output of the LCLV (−1 order) versus the mirror displacement ε; frequency of modulation $\Omega/2\pi = 1$ kHz, $n_2 = 4.5$ cm^2/W, $K = 5$, $P_S = 3.2$ mW; the solid curves are the fits with the theoretical curve $J_0(2k_0\varepsilon)J_1(2k_0\varepsilon)$. (From Bortolozzo, U. et al., *Opt. Lett.*, 34, 2006, 2009.)

the same data are plotted in logarithmic scale. We see that for small displacements, the detection is linear, and mirror displacements as small as 0.1 pm are efficiently detected. Similar results can be obtained on the zero order. The theoretical limit is not reached, mainly because the smallest displacement that could be achieved was limited by the sensitivity of the vibrating system. The ultimate relative detection limit, theoretically predicted to be 1.1 times that of an ideal interferometer, could be reached in future by using more sophisticated electronics and better isolated working conditions.

1.4.3 Measurement of the Casimir Force

The Casimir force is one of the most intriguing phenomena of quantum physics [69,70], and in the past years it has attracted a great deal of attention not only because of its fundamental character but also because of the rapid developments of optomechanical and optoelectronical engineering in the microsize and nanosize scale [71]. For the simplest case of two parallel metallic plates placed in vacuum and separated by a distance Z, the Casimir force, arising because of the spatial redistribution of the zero-point electromagnetic mode density with respect to the free space, gives rise to an attraction between the plates. For two infinite perfectly conducting parallel plates, the Casimir pressure (Casimir force per cm^2) at zero temperature can be expressed as

$$P_c = \frac{\pi^2 \hbar c}{240 Z^4}, \tag{1.48}$$

where

c is the speed of light

$\hbar$ is the Planck constant

The absolute value of P_c at Z = 300 nm is approximately 0.16 N/m^2. Usually, this force is measured with a torsion pendulum or by using atomic force measurement techniques. In Ref. [10], it has been demonstrated that dynamic holography can be efficiently employed to measure the Casimir force between macroscopic objects.

In the experiment, two highly conductive objects were used. One was a thin pellicle with a diameter of 7.62 cm and a thickness of 5 μm covered by a thin (120 nm) aluminum film, and the second was a spherical glass lens also coated by a thin aluminum film. These two bodies were placed in a vacuum chamber at different distances Z in different experiments. The lens was mounted on a vibrating piezodriver, so that the position of the lens could oscillate near the midpoint with an amplitude ε. The oscillations of the lens position resulted in a periodic modulation of the Casimir force and, hence, in corresponding deformations of the pellicle. The piezodriver was precisely calibrated to produce a perfect sinusoidal displacement of the lens. However, perfectly sinusoidal mechanical displacements of the lens induced not only first harmonics but also the high-order harmonics of the Casimir force due to the nonlinear dependence of P_c on Z. Therefore, one can expect a corresponding mechanical response of the pellicle at various temporal harmonics. In the experiment, the first and second harmonics of the output signal were investigated.

The mechanical deformations of the pellicle were detected by an adaptive holographic interferometer based on TWM in a photorefractive $BaTiO_3$:Co crystal. The crystal was illuminated by two coherent laser beams, one of which was phase modulated. In the experiment, the phase modulation of one of the recording beams arises because of the reflection of the beam from the vibrating pellicle. It is known that $BaTiO_3$:Co exhibits a strong diffusion holographic recording mechanism; hence, one expects a quadratic transfer function of the interferometer. As a consequence, if Ω is the frequency of the modulation of the lens position, the output signal will be at frequency 2Ω (second harmonic). This dramatically reduces the sensitivity of the setup. To provide a linear transfer function, a special technique of an artificial linearization of the photorefractive response has been realized, allowing to obtain the signal only at the first harmonic and to improve the interferometer sensitivity.

The light intensity in the interferometer was about 3–5 mW/mm^2, and the experiments were performed with $Z \sim$ 300, 400, and 600 nm. The modulation frequency was $\Omega/2\pi$ = 3.0 Hz, and the frequency band was 0.01 Hz. With these parameters, experimental data in quite a good agreement with the theoretical predictions were obtained [10].

1.4.4 Pressure Radiation Measurements

The ability of light waves to produce pressure is well known. The pressure P of normally incident cw light experienced by a macroscopic object with a plane surface has a linear dependence on the light intensity I:

$$P = \frac{I}{c}(1+R), \tag{1.49}$$

where

c is the speed of light

R is the reflectivity of the illuminated surface of the object

In Ref. [72], it was demonstrated that AHI is an efficient tool to measure light pressure radiation. In the experiment, a thin, highly reflective pellicle was used as a vibrating object. It was illuminated from one side by a periodically amplitude-modulated incident laser beam, which caused periodic mechanical displacements of the pellicle. A sensitive adaptive holographic interferometer based on the principle of dynamic holography was used to measure the amplitude of these displacements. The adaptive medium was a photorefractive BSO crystal. The data yielded a mechanical displacement of the pellicle of approximately 70 pm at the light intensity I = 9.2 mW/mm^2. Moreover, it has been verified, within an experimental error of 5% or less, that the measured light pressure did not depend on wavelength in a quite wide range (405–1560 nm).

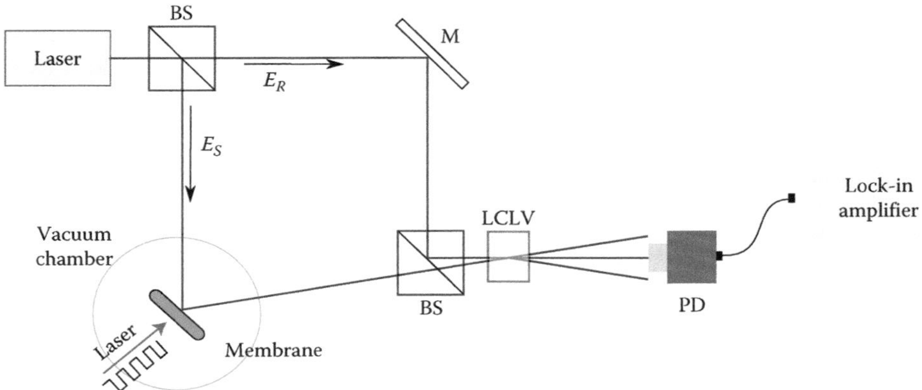

FIGURE 1.13 Setup for AHI measurement of the light radiation pressure with the LCLV acting as a dynamic hologram.

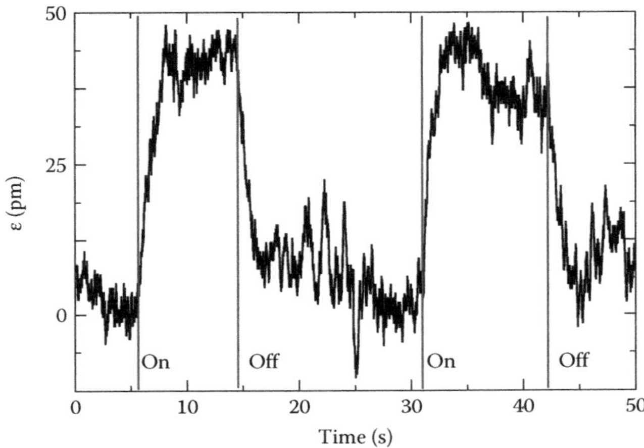

FIGURE 1.14 Laser-induced membrane displacement; on/off indicates the state of the laser.

Detection of light radiation pressure was successively realized with AHI in the LCLV. The setup is sketched in Figure 1.13. A 5 μm thick nitrocellulose membrane with a reflective coating was used as a vibrating object. The membrane was placed inside a vacuum box in order to reduce air pressure perturbations. A laser beam with power 45 mW, $\lambda = 632$ nm, and waist 3 mm impinged on the rear side of the membrane. The laser was square modulated at a frequency 10 kHz and could be switched on/off by external control.

The detected signal reflected by the membrane is shown in Figure 1.14. The laser-induced displacement can be clearly detected. The transition rate of 1 Hz is due to the bandwidth of the lock-in, whereas the noise is related to fluctuations of the membrane under the laser-induced deformation and subsequent relaxation.

1.4.5 Photoacoustic Detection and Imaging through Scattering Media

Acousto-optic imaging is based on US modulation of multiply scattered light in thick and highly scattering media, for example, in biological samples or in human breast tissues [73]. While light is highly scattered within biological tissues, US waves are ballistic. By analyzing the US-modulated photons (tagged photons), it is possible to get both the optical contrast information given by light and the spatial localization from the US longitudinal waves [1,2]. The aim is to image objects embedded within very thick biological media (few centimeters) where multiple scattering events strongly increase the average distance traveled by photons. It is thus important to use a wavelength where light exhibits its maximum penetrating depth in biological tissues. This can be achieved in the wavelength range of the so-called optical therapeutic window, which mainly corresponds to the near-IR wavelengths (700–1100 nm), where the absorption, due to either blood at shorter wavelengths or water at longer wavelengths, is minimized.

Several techniques have been used to detect US-tagged photons outcoming from a scattering media, and self-adaptive wave front holography using a photorefractive crystal is one of them. The technique has been first demonstrated with green light or with light at 1.06 μm because of the difficulty to find a photorefractive crystal matching the required specifications around 800 nm, that is, a high photorefractive gain at zero applied field for a good optical contrast, large entrance surfaces in order to optimize flux collection after a scattering media, and a short response time (approximately milliseconds) to prevent speckle decorrelation within biological tissues. Recently, a lot of progress has been achieved in the growth technology and study of tellurium-doped tin thiohypodiphosphate (SPS:Te) ferroelectric crystals. In particular, SPS:Te 1% exhibits a photorefractive gain up to 7 cm^{-1} at 780 nm with a response time of less than 10 ms, making it suitable for applications in the therapeutic window.

The principles of photorefractive detection of tagged photons rely on AHI. When two beams interfere in the crystal, the interference pattern generates a refractive index grating. The dynamic hologram engraved in the crystal contains the phase and amplitude of the object wave front, which corresponds to the multiple scattered light exiting the sample. The reference beam is diffracted by this grating, and thus it is perfectly wave front matched to the transmitted object beam. These two beams

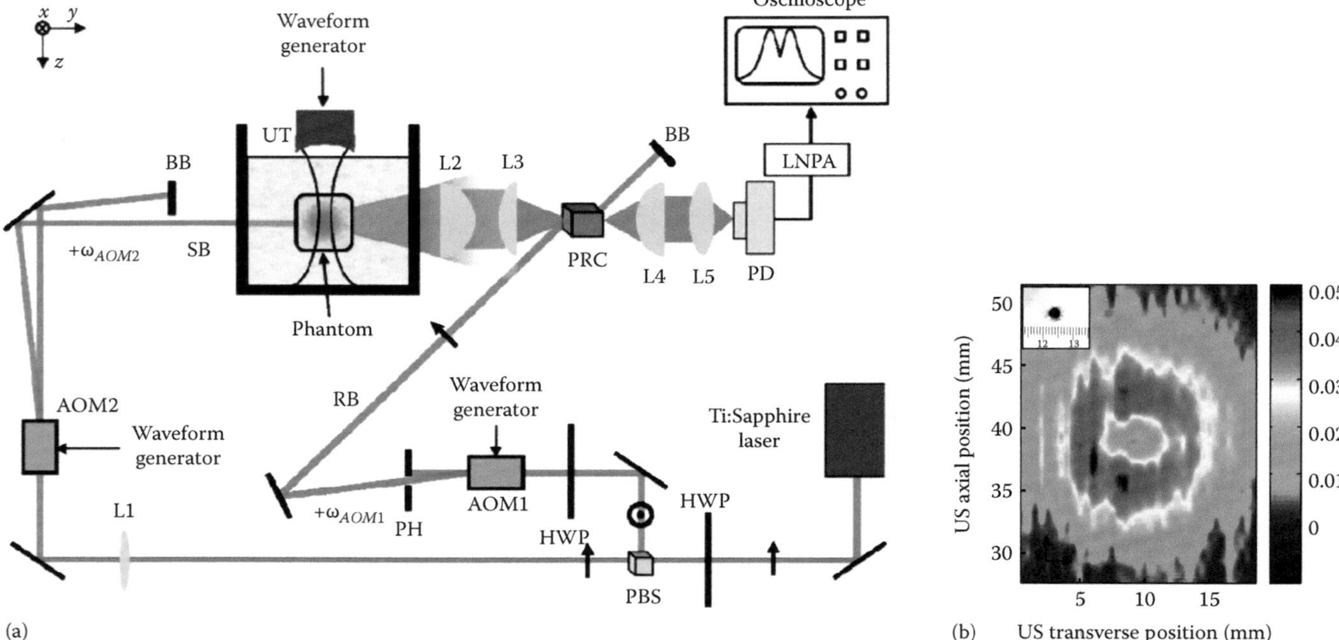

FIGURE 1.15 (a) Setup for AHI photoacoustic detection. HWP, half-wave plate; PBS, polarizing beam splitter; AOM1, AOM2, AO, modulators; ω_{AOM1}, ω_{AOM2}, shifting frequencies; RB, reference beam; SB, signal or object beam; BB, beam-blocker; L1, lens; L2, L3, L4, L5, wide-aperture aspherical lenses; PD, photodiode; LNPA, low-noise preamplifier. (b) Acousto-optic image realized with four bursts of 2.3 MHz US, that is, 2.6 mm in axial resolution. Inset: transverse section of the 10% intralipid and agar scattering phantom. (From Farahi, S. et al., *Opt. Lett.*, 35, 1798, 2010.)

are now spatially coherent with each other and can interfere constructively on a large-area array or a single detector. As a result, the phase modulation of the object beam, through its interaction with US, is converted into an intensity modulation. It is possible to perform the AHI to detect either untagged photons or tagged photons by shifting the reference beam from the US frequency.

The typical setup is presented in Figure 1.15a. The reference beam is tuned to interfere with the untagged photons of the object beam, and a short burst of US is applied to get resolved information along the US beam-propagation direction [73]. Since the US burst duration is much shorter than the crystal response time, the hologram previously recorded by the untagged photons and the reference beam remains static. When the US burst interacts with scattered light, it creates tagged photons, resulting in a decrease in the untagged photon quantity. Therefore, if the burst crosses an absorbing object, the quantity of untagged photons remains constant. By analyzing the time-domain signal issued from the detector, it is possible to localize any change of the tagged photons quantity along the US path. To record a two dimensional acousto-optic image, the US transducer is set on a translation stage, and the temporal traces of the intensity are recorded for each transverse position in the range of 2 cm. Figure 1.15b shows an example of the obtained acousto-optic image. An absorber of 3 mm × 3 mm can be clearly distinguished in the center of the image.

1.5 Comparison between the Different Systems and Conclusions

Among the different nonlinear optical phenomena, the photorefractive effect has been one of the firsts to manifest at low optical power, thus permitting the realization of AHI systems without the requirement of high-intensity lasers and without the presence of any threshold. The optimization of the AHI interferometers is based on a deep knowledge of the physical mechanisms occurring in each type of photorefractive crystal considered, as drift-dominated or diffusion-dominated holograms, requiring or not the application of an external electric field and operating in anisotropic or isotropic wave-mixing conditions. More recently, LCLVs have been added to the list of nonlinear optical media suitable for the efficient realization of AHI. The LCLV has the advantage of being of large transverse size, while being very thin along the longitudinal direction, hence reducing losses and distortions due to the passage of light inside the bulk of the medium. Moreover, LCs offer a wide tunability of their nonlinear response when driven by the electro-optic effect provided by the photorefractive substrate of the LCLV. The required power for operating LCLV systems is quite low, often lower than the typical power levels used for photorefractive crystals.

The different AHI configurations discussed in the previous sections are characterized by different combinations of

TABLE 1.1 Parameters of Adaptive Interferometers

	Configuration Parameters						Achieved		
Device	λ (nm)	*E* (kV/cm)	Coupling	P_S (mW)	P_R (mW)	I_{opt} (W/cm²)	δ_{rel}	f_{cut} (Hz)	Ref.
GaAs	1064	—	T	5	500	40	41	10,000	[36]
GaAs	1064	—	T	3.8	400	50	8	3,500	[34]
$Bi_{12}TiO_{20}$	633	—	T	1	15	0.5	NA	0.03	[33]
$Bi_{12}TiO_{20}$	633	—	T	0.2	25	3	21	NA	[5]
$BaTiO_3$	514	—	T	1	155	NA	NA	1	[35]
CdTe:V	1064	—	R	3	500	67	5.7	300	[6]
CdTe:V	1064	—	R	3	500	67	10	1,250	[6]
GaP	633	—	T	0.3	25	17.2	130	4,500	[40]
GaP	633	10—ac	T	0.3	25	17.2	5.4	1,200	[40]
$Bi_{12}TiO_{20}$	633	8—ac	T	0.2	25	3	1.5	5	[5]
CdTe:Ti	1064	9—ac	T	1	90	10	2.3	3,000	[41]
CdZnTe:V	1060	9—dc	T	NA	50	0.022	2.2	40	[4]
CdTe:V	1550	8—dc	T	0.025	NA	0.066	20	15	[26]
CdTe:Ge	1550	8—dc	T	NA	60	0.14	8.5	1,000	[74]
InP:Fe	1064	7—dc	T	NA	500	100	6.5	1,000	[24]
$Bi_{12}SiO_{20}$	532	5—dc	T	0.17	120	0.84	4.3	10	[75]
LCLV	532	21—ac	T	3.2	0.13	0.006	1.4	8	[8]

sensitivity and cutoff frequency. Besides the type of nonlinear medium employed, the input laser power is also fundamental in determining these two main parameters. The diversity of adaptive interferometers reported in the literature is shown in Table 1.1, where the following notations are used: λ is the wavelength, *E* is the externally applied dc electric field, T and R are, respectively, the transmission and reflection geometries of light wave interaction, P_S is the power of the signal beam at the photodetector, P_R is the power of the reference beam, I_{opt} is the total light intensity incident on the nonlinear medium, and NA means that data are not available.

From Table 1.1, it can be seen that both the sensitivity and cutoff frequency vary in a wide range. This makes it possible to choose the interferometer configuration that best fits the required application. Table 1.1 also shows that there is a trade-off between the sensitivity and the adaptability of the AHI interferometer. This trade-off depends both on the nonlinear medium and the configuration geometry of the TWM. Finally, another important parameter that has to be taken into account is the operating wavelength, which has to be chosen as a function of the application aimed to, and for which a suitable choice of the nonlinear medium has to be made. The development of a new system is in progress and actually under study for the extension of the AHI techniques in less-explored spectral regions. As an example, we have reported in the last section the photo-acoustic imaging implemented in the therapeutic window around 800 nm. Other promising techniques are those based on nonlinear optical fibers, since these methods combine in the same element (the fiber) the sensor itself and the nonlinearity that provides the wave-mixing process. New developments are also expected in view of increasing the lateral resolution of the adaptive interferometers, so that picometer detection could be simultaneously realized in the plane transverse to the longitudinal perturbation and a full high-resolution three-dimensional reconstruction of the object displacement could be achieved.

References

1. S. I. Stepanov, I. A. Sokolov, G. S. Trofimov, V. I. Vlad, D. Popa, and I. Apostol, *Opt. Lett.* **15**, 1239 (1990).
2. J. P. Huignard and A. Marrakchi, *Opt. Lett.* **6**, 622 (1981).
3. V. Petrov, C. Denz, J. Petter, and T. Tschudi, *Opt. Lett.* **22**, 1902 (1997).
4. L. A. de Montmorillon, P. Delaye, J. C. Launay, and G. Roosen, *J. Appl. Phys.* **82**, 5913 (1997).
5. A. A. Kamshilin and A. I. Grachev, *Appl. Phys. Lett.* **81**, 2923 (2002).
6. S. Di Girolamo, A. A. Kamshilin, R. V. Romashko, Y. N. Kulchin, and J. C. Launay, *Opt. Express* **15**, 545 (2006).
7. A. A. Kamshilin, R. V. Romashko, and Y. N. Kulchin, *J. Appl. Phys.* **105**, 031101 (2009).
8. U. Bortolozzo, S. Residori, and J. P. Huignard, *Opt. Lett.* **34**, 2006 (2009).
9. M. Lesaffre, F. Jean, F. Ramaz, A. C. Boccara, M. Gross, P. Delaye, and G. Roosen, *Opt. Express* **15**, 1030 (2007).
10. V. Petrov, M. Petrov, V. Bryksin, J. Petter, and T. Tschudi, *Opt. Lett.* **31**, 3167 (2006).
11. M. Born and E. Wolf, *Principles of Optics* (Pergamon Press, New York, 1980).
12. H. Osterberg, *J. Opt. Soc. Am. B* **22**, 19 (1932).
13. P. Hariharan, *Rep. Prog. Phys.* **54**, 339 (1991).
14. R. L. Forward, *Phys. Rev. D* **17**, 379 (1978).
15. J. W. Wagner and J. Spicer, *J. Opt. Soc. Am. B* **4**, 1316 (1987).

16. S. I. Stepanov, in *International Trends in Optics*, ed. J. W. Goodman (Academic, New York, 1991), Chapter 9, pp. 125–140.
17. M. P. Petrov, *Introduction to Optical Signal Processing with Photorefractive Materials*, ed. P. Gunter (Springer-Verlag, New York, 1987), pp. 284–290.
18. M. P. Petrov, S. I. Stepanov, and A. V. Khomenko, *Photorefractive Crystals in Coherent Optical Systems* (Springer, Berlin, Germany, 1991).
19. A. A. Kamshilin, and M. P. Petrov, *Opt. Commun.* **53**, 23 (1985).
20. M. P. Petrov, V. M. Petrov, I. S. Zouboulis, and L. P. Xu, *Opt. Commun.* **134**, 569 (1997).
21. G. S. Gorelik, Application of the modulation technique in optical interferometry, *Dokl. Acad. Nauk SSSR* **83**, 549 (1952).
22. A. Papoulis, *Probability, Random Variables, and Stochastic Processes* (McGraw-Hill, New York, 1999).
23. P. Gunter and J. P. Huignard, *Photorefractive Materials and Their Applications 1* (Springer Science, New York, 2006).
24. P. Delaye, A. Blouin, D. Drolet, L.-A. de Montmorillon, G. Roosen, and J.-P. Monchalin, *J. Opt. Soc. Am. B* **14**, 1723 (1997).
25. P. Delaye, L.-A. de Montmorillon, and G. Roosen, *Opt. Commun.* **118**, 154 (1995).
26. A. de Montmorillon, I. Biaggio, P. Delaye, J.-C. Launay, and G. Roosen, *Opt. Commun.* **129**, 293 (1996).
27. S. De Rossi, P. Delaye, J.-C. Launay, and G. Roosen, *Opt. Mater.* (Amsterdam, the Netherlands) **18**, 45 (2001).
28. T. J. Hall, M. A. Fiddy, and M. S. Ner, *Opt. Lett.* **5**, 485 (1980).
29. J. Frejlich, A. A. Kamshilin, V. V. Kulikov, and E. V. Mokrushina, *Opt. Commun.* **70**, 82 (1989).
30. S. M. Hughes and D. Z. Anderson, *Appl. Opt.* **46**, 7868 (2007).
31. A. A. Kamshilin and E. V. Mokrushina, *Sov. Tech. Phys. Lett.* **12**, 149 (1986).
32. M. P. Petrov, S. V. Miridonov, S. I. Stepanov, and V. V. Kulikov, *Opt. Commun.* **31**, 301 (1979).
33. M. Rossomakhin and S. I. Stepanov, *Opt. Commun.* **86**, 199 (1991).
34. B. Campagne, A. Blouin, L. Pujol, and J.-P. Monchalin, *Rev. Sci. Instrum.* **72**, 2478 (2001).
35. R. K. Ing and J.-P. Monchalin, *Appl. Phys. Lett.* **59**, 3233 (1991).
36. A. Blouin and J.-P. Monchalin, *Appl. Phys. Lett.* **65**, 932 (1994).
37. S. I. Stepanov and M. P. Petrov, *Opt. Commun.* **53**, 292 (1985).
38. B. I. Sturman, M. Mann, J. Otten, and K. H. Ringhofer, *J. Opt. Soc. Am. B* **10**, 1919 (1993).
39. B. I. Sturman, E. V. Podivilov, K. H. Ringhofer, E. Shamonina, V. P. Kamenov, E. Nippolainen, V. V. Prokofiev, and A. A. Kamshilin, *Phys. Rev. E* **60**, 3332 (1999).
40. A. A. Kamshilin and V. V. Prokofiev, *Opt. Lett.* **27**, 1711 (2002).
41. K. Paivasaari, H. Tuovinen, A. A. Kamshilin, and E. Raita, in *OSA Trends in Optics and Photonics (TOPS), Photorefractive Effects, Materials, and Devices*, eds. G. Zhang, D. Kip, D. D. Nolte, and J. Xu, OSA, Washington, DC, 2005, Vol. 99, pp. 681–686.
42. O. S. Filippov and B. I. Sturman, *Appl. Phys. B: Lasers Opt.* **83**, 97 (2006).
43. R. V. Romashko, Y. N. Kulchin, and A. A. Kamshilin, in *OSA Trends in Optics and Photonics (TOPS), Photorefractive Effects, Materials, and Devices*, eds. G. Zhang, D. Kip, D. D. Nolte, and J. Xu, OSA, Washington, DC, 2005, Vol. 99, pp. 675–680.
44. S. Di Girolamo, A. A. Kamshilin, R. V. Romashko, Y. N. Kulchin, and J. C. Launay, *Opt. Lett.* **32**, 1821 (2007).
45. U. Efron and G. Liverscu, *Spatial Light Modulator Technology: Materials, Devices and Applications* (Dekker, New York, 1995).
46. N. Collings, *Optical Pattern Recognition Using Holographic Techniques* (Reading, MA: Addison-Wesley, Wokingham, England 1988).
47. D. Armitage, J. I. Thackara, and W. D. Eades, *Appl. Opt.* **28**, 4763 (1989).
48. J. Grinberg, A. Jacobson, W. P. Bleha, and L. Miller, *Opt. Eng.* **14**, 217 (1975).
49. P. R. Ashley and J. H. Davis, *Appl. Opt.* **26**, 241 (1978).
50. S. A. Akhmanov, M. A. Vorontsov, and V. Yu. Ivanov, *JETP Lett.* **47**, 707 (1988).
51. U. Efron, S. T. Wu, and T. D. Bates, *J. Opt. Soc. Am. B* **3**, 247 (1986).
52. P. Aubourg, J. P. Huignard, M. Hareng, and R. A. Mullen, *Appl. Opt.* **21**, 3706 (1982).
53. N. Sanner, N. Huot, E. Audouard, C. Larat, J. P. Huignard, and B. Loiseaux, *Opt. Lett.* **30**, 1479 (2005).
54. A. Brignon, I. Bongrand, B. Loiseaux, and J. P. Huignard, *Opt. Lett.* **22**, 1855 (1997).
55. S. Residori, U. Bortolozzo, and J. P. Huignard, *Phys. Rev. Lett.* **100**, 203603 (2008).
56. U. Bortolozzo, S. Residori, and J. P. Huignard, *Opt. Lett.* **34**, 2006 (2009).
57. S. Residori, U. Bortolozzo, and J. P. Huignard, *Appl. Phys. B* **95**, 551 (2009).
58. P. G. De Gennes and J. Prost, *The Physics of Liquid Crystals*, 2nd edn. (Oxford Science Publications, Oxford, U.K., 1993).
59. I. C. Khoo, *Liquid Crystals*, Wiley Series in Pure and Applied Optics (Wiley, Hoboken, NJ, 2007).
60. A. Yariv, *Optical Waves in Crystals* (John Wiley & Sons, Hoboken, NJ, 2003).
61. U. Bortolozzo, S. Residori, and J. P. Huignard, *Phys. Rev. A* **79**, 053835 (2009).
62. U. Bortolozzo, S. Residori, and J. P. Huignard, *Laser Photon. Rev.* **1**, 1 (2009).
63. U. Bortolozzo, S. Residori, and J. P. Huignard, *J. Hologr. Speckles* **5**, 1 (2009).
64. S. Stepanov, E. Hernández, and M. Plata, *Opt. Lett.* **29**, 1327 (2004).

65. S. Stepanov, E. Hernández, and M. Plata, *J. Opt. Soc. Am. B* **22**, 1161 (2005).
66. S. Stepanov, A. Fotiadi, and P. Mégret, *Opt. Express* **15**, 8832 (2007).
67. S. Stepanov and P. Cota, *Opt. Lett.* **32**, 2532 (2007).
68. S. Stepanov, *J. Phys. D: Appl. Phys.* **41**, 224002 (2008).
69. H. B. G. Casimir, *Proc. K. Ned. Akad. Wet.* **51**, 793 (1948).
70. S. K. Lamoreaux, *Rep. Prog. Phys.* **68**, 201 (2005).
71. H. J. De Los Santos, *Principles and Applications of NanoMEMS Physics* (Springer-Verlag, Dordrecht, the Netherlands, 2005).
72. V. Petrov, J. Hahn, J. Petter, M. Petrov, and T. Tschudi, *Opt. Lett.* **30**, 3138 (2005).
73. S. Farahi, G. Montemezzani, A. A. Grabar, J.-P. Huignard, and F. Ramaz, *Opt. Lett.* **35**, 1798 (2010).
74. M. B. Klein, K. V. Shcherbin, and V. Danylyuk, in *OSA Trends in Optics and Photonics (TOPS), Photorefractive Effects, Materials, and Devices*, eds. P. Delaye, C. Denz, L. Mager, and G. Montemezzani, OSA, Washington, DC, 2003, Vol. 87, pp. 483–489.
75. T. Honda, T. Yamashita, and H. Matsumoto, *Jpn. J. Appl. Phys., Part 1* **34**, 3737 (1995).

2

Single Atom in an Optical Cavity: An Open Quantum System

John D. Close
The Australian National University

Rachel Poldy
The Australian National University

Ben C. Buchler
The Australian National University

Nicholas P. Robins
The Australian National University

2.1 Introduction

A single atom strongly coupled to a cavity mode is a fascinating example of a controllable, open quantum system. For cavities with sufficiently small volume and mirrors of sufficiently high reflectivity, the Rabi flopping frequency of a single atom interacting with a single photon can exceed both the cavity decay rate and the spontaneous emission rate into the continuum of non-cavity modes. This is a coupled system of two resonators that can, under appropriate conditions (strong coupling), display nearly coherent dynamics in the extreme quantum limit of one photon interacting with one atom. Information on the quantum state of the system (we define the system in this chapter as a cavity mode interacting with a two-level atom) is typically measured by counting photons that leak out through the cavity mirrors. In addition, photons can leave the system via spontaneous emission. The dynamics of the system exhibits coherent oscillations punctuated by stochastic jumps caused by loss of photons to the environment. The research field is referred to as cavity quantum electrodynamics or cavity QED and is the subject of this chapter.

Although difficult to realize experimentally, the strongly coupled atom–cavity system has been studied intensively in theory and experiment. This system has not only elucidated a great deal of fundamental physics but is also applicable to quantum information, metrology, and single atom counting. For a physicist, this is a veritable playground to study, understand, and exploit properties such as entanglement, quantum coherence, Hamiltonian and non-Hamiltonian evolution, and the subtleties of quantum measurement [1].

The literature on this subject is vast, and our aim here is not to review the field but rather to supplement more advanced treatments with a discussion on the subject readily accessible to graduate students and advanced undergraduates. After reading this chapter, we strongly recommend that the reader consult the excellent articles by Kimble [2], Doherty and Mabuchi [3], and Mabuchi and Doherty [4]. These articles are written at a level reasonably compatible with this chapter and complement much of the material that we present here.

Our goal in the early part of the chapter is to understand the two major building blocks, atoms and cavities, in isolation. We briefly recap on the spectroscopy of real atoms in the absence of a cavity, but quickly confine our attention to two-level atoms that can be approximately realized in the lab under appropriate conditions. We consider in some detail the interaction of a two-level atom with a single mode classical electromagnetic field. We then turn our attention to cavities and discuss their classical properties in the absence of an atom. We quantize the single mode field, introduce the Jaynes–Cummings Hamiltonian and briefly discuss photon blockade as a good example of cavity QED physics.

Up to this point in the chapter, our discussion will not have included interaction of the atom–cavity system with the environment. We will have considered the coherent evolution of an isolated two-level atom interacting with an isolated cavity mode. If we include the environment, we must describe our problem as an open quantum system through a reduced system density matrix that evolves according to the master equation. Open quantum systems and their descriptions are the focus of the final sections.

2.2 Semiclassical Atom–Cavity System

2.2.1 Quantized Atom

Hydrogen and hydrogen-like atoms and ions (alkali atoms and alkali-earth ions) have been used extensively in this field, and we embark on our discussion by considering atoms of this kind. A discussion on the spectroscopy of more complex atoms is beyond the scope of this chapter and is not our focus. Our aim in this early discussion is to motivate the use of a two-level description and to indicate how the complexity of a real atom is included in just two parameters in the model. The two-level systems that are used in cavity QED and in the closely related fields, quantum optics and quantum-atom optics, are diverse. They include neutral atoms [5], highly excited Rydberg atoms [6] and ions [7], as well as "artificial atoms" such as quantum dots [8–10] and confined Cooper pairs in superconducting circuits [11].

In our discussion, we will ignore atomic motion and only account for the atom's internal energy. If our atomic Hamiltonian includes only the kinetic energy of the electron and the $1/r^2$ Coulomb interaction between the valence electron and the nucleus, the resulting energy spectrum is characterized only by the principal quantum number n and is given by the well-known relation

$$E_n = -hcR_\infty \frac{Z_{eff}^2}{n^2}, \tag{2.1}$$

where

R_∞ is the Rydberg constant
h is Planck's constant
c is the speed of light
Z_{eff} is the effective nuclear charge that accounts for the shielding effect of inner orbital electrons on the bare nuclear charge

Although the magnitude of the nuclear charge is shielded, at this level of description, we are still considering a $1/r$ potential. The consequence of this assumption is that the energy spectrum is independent of angular momentum. This is termed an accidental degeneracy. It is not a consequence of symmetry, but rather is a consequence of the specific form of the potential. The classical analog is the conservation of the Laplace–Runge–Lenz vector in astrophysical systems [12].

If we include the mean field of the core electrons, the valence electron no longer experiences a $1/r$ potential, and the accidental degeneracy is lifted. This effect is substantial. In the alkali atoms, the first excited state is split by near infrared frequencies from the ground state by this effect, and it is the transitions between states with the same principle quantum number but different orbital angular momentum that are often exploited in these systems.

More subtle structure exists (termed fine structure) due to the interaction between the electron's spin magnetic moment and its orbital angular momentum. This interaction again is significant compared with the gross structure described by Equation 2.1 and is responsible for the well-known D1 and D2 doublet observed in all alkali atoms. The interaction of the nuclear spin and the electron angular momentum (hyperfine structure) is smaller and gives rise to the ground-state splitting in the alkali atoms. The splitting is on the order of 1–10 GHz and is exploited in the Cs atomic clock that defines the second. Applied external magnetic and electric fields change the symmetry of the Hamiltonian and can be used to lift the remaining degeneracy in the system.

An electron spin in an applied magnetic field has an interaction energy on the order of 1 MHz per gauss. Given that the spontaneous emission rate (and therefore natural line width) of a strong dipole-allowed transition in the optical regime is on the order of tens of MHz, all the interactions listed earlier, including the Zeeman effect in fields as small as 10 gauss or less, lead to resolvable structure, and all are important. In the case where any applied electric and magnetic fields lead to unresolved splittings, the underlying structure can still be important in understanding the dynamics of an atom in a driving field due to effects like optical pumping.

Even for the simplest of atoms, the energy level structure is complicated. Fortunately, if we understand the spectroscopy of a particular atom or ion well, a judicious choice of polarization of the driving electromagnetic field and the application of external fields of appropriate symmetry allows us to couple only two or, if desired, only three levels, and we can include the detailed and complex structure of real atoms in just two parameters, the splitting of the two levels of the bare (undriven) two-level atom and the transition electric dipole moment coupling of the atom to the driving electromagnetic field. Of course, if we wish theorists and experimentalists to converse successfully, it is useful if both groups are well versed in real atomic structure.

Three-level atoms coupled by two fields yield beautiful physics in the form of electromagnetically induced transparency, slow light, Raman transitions, and beam splitters for atom interferometers. An entire book could be written on this subject. Unfortunately, we do not have space to address this physics here, and we concentrate on the simpler but still very rich model of a two-level atom coupled to a driving electromagnetic field. For the two-level atom, we label the lower of the two energy eigenstates,

$$|g\rangle = \begin{pmatrix} 0 \\ 1 \end{pmatrix},$$

and the upper, or excited state,

$$|e\rangle = \begin{pmatrix} 1 \\ 0 \end{pmatrix}.$$

The energies of these states are E_g and E_e giving the atomic transition frequency $\omega_a = (E_e - E_g)/\hbar$, where $\hbar$ is Planck's constant divided by 2π. The energy difference or the atomic transition frequency $\omega_a = (E_e - E_g)/\hbar$ is the first parameter that is taken from the spectroscopy of the atom that we are describing with this two-level model. A general state vector in the Schrödinger picture is

$$|\Psi(t)\rangle_S = C_g(t)e^{-iE_g t/\hbar}\,|g\rangle + C_e(t)e^{-iE_e t/\hbar}\,|e\rangle, \tag{2.2}$$

where $|C_g|^2$ and $|C_e|^2$ represent the occupation probability of the ground and excited states, respectively. In the Schrödinger picture, the Hamiltonian for the bare atom can be written using projection operators as

$$\hat{H}_a = \frac{1}{2}\hbar\omega_a\{|e\rangle\langle e| - |g\rangle\langle g|\}. \tag{2.3}$$

We have chosen the zero of energy to lie midway between the two atomic states. This choice is arbitrary, but the symmetry is appealing. The reader can check that this is indeed the appropriate two-level Hamiltonian by operating on the energy eigenstates $|g\rangle$ and $|e\rangle$. Later in this discussion, we will change from the Schrödinger picture to the interaction picture. It is important, therefore, to know how we interpret our basis states, $|g\rangle$ and $|e\rangle$. We consider them to be the Schrödinger picture states $|g\rangle$ and $|e\rangle$ at $t = 0$. This set is time independent and forms a complete orthogonal basis.

We now consider the interaction between the two-level atom described by Equation 2.3 and a classical oscillating electric field. The Hamiltonian for this system is

$$\hat{H} = \hat{H}_a + \hat{V}_0 \cos\omega t. \tag{2.4}$$

$\hat{V}_0 = -\hat{\boldsymbol{\mu}}\cdot \boldsymbol{E}_c(\hat{\boldsymbol{r}})$ defines the electric dipole interaction, where $\hat{\boldsymbol{\mu}}$ is the dipole moment operator. For a hydrogen-like atom, $\hat{\boldsymbol{\mu}} = q_e\,\hat{\boldsymbol{r}}$, where q_e is the charge on an electron and $\hat{\boldsymbol{r}}$ is the position of the electron in a coordinate system where the nucleus is at the origin. As we are treating the field classically, there is no field term in the Hamiltonian. For a strong electric dipole-allowed transition, μ is on the order of $q_e a_0$, where a_0 is the Bohr radius.

We can express the interaction Hamiltonian in terms of the atomic basis vectors by appropriately using the property $|e\rangle\langle e| + |g\rangle\langle g| \equiv 1$ and inserting a complete set of states on the left and right sides of the Hamiltonian:

$$\begin{aligned}\hat{V} &= -\boldsymbol{\mu}\cdot \boldsymbol{E}_c(\hat{\boldsymbol{r}})\\ &= -\{|e\rangle\langle e| + |g\rangle\langle g|\}\boldsymbol{\mu}\cdot\boldsymbol{E}_c(\hat{\boldsymbol{r}})\{|e\rangle\langle e| + |g\rangle\langle g|\}\\ &= \boldsymbol{d}\cdot\boldsymbol{E}_c(\hat{\boldsymbol{r}})\{|e\rangle\langle g| + |g\rangle\langle e|\},\end{aligned}$$

where $\boldsymbol{d} = -\langle e|\hat{\boldsymbol{\mu}}|g\rangle$ is the transition dipole moment matrix element that is assumed in this discussion as a real quantity. The generalization to a complex dipole moment is straightforward and does not change any of the conclusions we draw in this chapter.

The Hamiltonian (Equation 2.4) is now expressed as

$$\hat{H} = \frac{1}{2}\hbar\omega_a\{|e\rangle\langle e| - |g\rangle\langle g|\} + \boldsymbol{d}\cdot\boldsymbol{E}_c(\hat{\boldsymbol{r}})\{|e\rangle\langle g| + |g\rangle\langle e|\}\cos\omega t. \tag{2.5a}$$

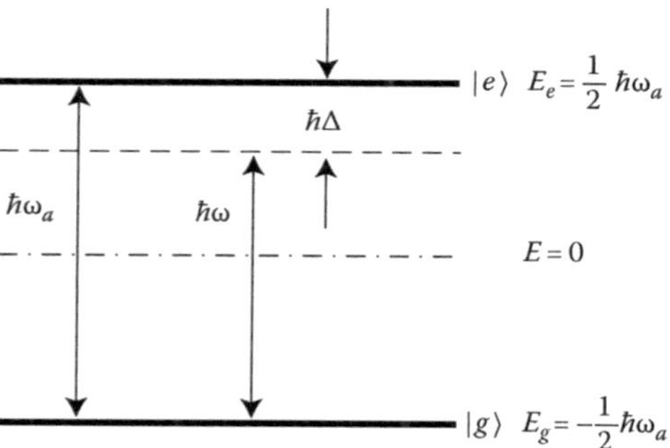

FIGURE 2.1 Schematic diagram for two atomic energy levels driven with a near-resonant laser at frequency ω. The atomic resonant frequency is ω_a, and the detuning is $\Delta_a = (\omega_a - \omega)$.

The calculation of the transition dipole moment matrix element is the second parameter that requires input from real atomic spectroscopy. In addition, the frequency of the driving electromagnetic field, polarization of the driving field, and any applied additional magnetic and/or electric fields must be chosen so that only two levels are coupled in the real atomic system that is being described by this simple but very useful two-level model. The reduction of a real system to a two-level description requires a detailed understanding and knowledge of real atomic spectroscopy. It is very easy to be tricked when comparing theoretical results based on two-level atoms with experimental results on real atoms by effects such as optical pumping to a third level, and care must be taken in the lab if a reasonable approximation to a two-level atom is to be realized. That being said, this is routinely achieved in many experiments.

With reference to Figure 2.1, we recast the Hamiltonian in the following form:

$$\hat{H} = \underbrace{\frac{1}{2}\hbar\omega\{|e\rangle\langle e| - |g\rangle\langle g|\}}_{\hat{H}_0} + \underbrace{\frac{1}{2}\hbar\Delta_a\{|e\rangle\langle e| - |g\rangle\langle g|\} + \boldsymbol{d}\cdot\boldsymbol{E}_c(\hat{\boldsymbol{r}})\{|e\rangle\langle g| + |g\rangle\langle e|\}\cos\omega t}_{\hat{V}},$$

and move to an interaction picture defined by the following transformation:

$$\left.\begin{aligned}\hat{H}_{sys} &= U_0^\dagger \hat{V} U_0\\ U_0 &= e^{-i(\hat{H}_0)t/\hbar}\end{aligned}\right\}. \tag{2.5b}$$

The reader can check that the interaction picture Hamiltonian is now

$$\begin{aligned}\hat{H}_{sys} &= \frac{1}{2}\hbar\Delta_a\{|e\rangle\langle e| - |g\rangle\langle g|\}\\ &\quad + \hbar\Omega\{|e\rangle\langle g|e^{i\omega t} + |g\rangle\langle e|e^{-i\omega t}\}\cos\omega t,\end{aligned}$$

where

$\Omega = (\boldsymbol{d}\cdot\boldsymbol{E}_c)/\hbar$

Δ_a is the detuning of the driving field from resonance

Writing the time-dependent cosine function as a sum of exponentials and dropping the rapidly oscillating terms (the rotating wave approximation [13]), we arrive at the following time-independent Hamiltonian:

$$\hat{H}_{sys} = \frac{1}{2}\hbar\Delta_a\{|e\rangle\langle e| - |g\rangle\langle g|\} + \frac{1}{2}\hbar\Omega\{|e\rangle\langle g| + |g\rangle\langle e|\} \tag{2.6}$$

or its matrix representation:

$$\hat{H}_{sys} = \frac{\hbar}{2}\begin{bmatrix} \Delta_a & \Omega \\ \Omega & -\Delta_a \end{bmatrix}. \tag{2.7}$$

This Hamiltonian describes a quantum two-level atom interacting with a classical driving field in a regime where the detuning of the driving field from resonance is small compared with the energy (frequency) difference of the bare atomic transition and in the absence of damping (spontaneous emission). In the interaction picture, the general state vector (Equation 2.2) becomes

$$\begin{aligned} |\Psi\rangle_I &= e^{i\hat{H}_0 t/\hbar}|\Psi\rangle_S \\ &= C_g(t)|g\rangle + C_e(t)|e\rangle. \end{aligned} \tag{2.8}$$

The dressed eigenstates and eigenstate energies of the atom interacting with a classical driving field are obtained by diagonalizing Equation 2.7. Both are functions of the strength of the driving field and the detuning of the field from atomic resonance. Consider the limiting case where the atom–light coupling is set to zero; the interactions between the atom and the light are turned off. At zero detuning, the dressed states will be degenerate. This rather counterintuitive observation is a result of treating the driving field classically and then transforming to the interaction picture that is rotating at the frequency of that driving field. Although the equation correctly describes the dynamics of the two-level atom in the semiclassical approximation, we violate conservation of energy. On absorption, no energy is removed from the field, and on stimulated emission, no energy is added to the field.

In the full quantum picture, which will be discussed in Section 2.3, where we treat atoms and light quantum mechanically, we arrive at the identical result but with an obvious explanation. An applied field with $n + 1$ photons resonant with an atom in the ground state is clearly degenerate with an atom in the excited state and one less photon in the field. At nonzero detuning but still zero coupling, the dressed states are separated by the detuning. Again, the explanation in the fully quantized model is obvious by similar reasoning to the resonant case discussed earlier.

At zero detuning and nonzero coupling, the dressed states are no longer degenerate and are now split in energy by the Rabi frequency. This can be seen by inspection of Equation 2.7 with $\Delta_a = 0$. This is reminiscent of two initially degenerate uncoupled classical oscillators. Coupling the oscillators leads to two normal modes with frequencies that are split precisely by the coupling term. The normal modes are the analogs of the atomic dressed states. Some readers may find the coupled oscillator problem worth revisiting in the light of this analogy. It is treated in many undergraduate mechanics books [12].

This Hamiltonian, although simple, is tremendously rich and is an important ingredient in straightforward but accurate explanations of a vast array of modern atomic physics from the ac Stark shift and optical trapping, to Sisyphus cooling, Landau–Zener dynamics at an avoided crossing, rf evaporative cooling of cold atoms in a magnetic trap, atom laser outcoupling from a Bose–Einstein condensate (BEC), and Bloch oscillations in an optical lattice, to name just a few.

The standard interaction-picture Schrödinger equation can be combined with Equations 2.7 and 2.8 to determine the dynamics of the system. Again, this is a rich problem, and the dynamics of this system is the basis of microwave atomic clocks, inertial sensors based on cold atoms, atomic magnetometers, and many other topical and important problems in modern physics. Consideration of this Hamiltonian leads naturally to a discussion of motion on the Bloch sphere. This is a problem well worth studying, and although the subjects listed earlier are outside the scope of this chapter, we encourage the reader to read widely about the applications and implications of this simple Hamiltonian in atomic physics and beyond. Details of dressed states, dressed-state eigen-energies, dynamics, and applications are discussed in most standard books on atomic physics. We give a brief discussion of these effects later in the chapter.

2.2.2 Classical Field in an Optical Cavity

In the same way, there exists a diverse range of "atoms"; there are many types of electromagnetic resonators (cavities) used in experimental quantum and atom optics to isolate and trap photons in different frequency regimes. Useful cavity fields range from the whispering gallery modes of toroidal structures and microresonators [14–16] to those of resonators using Bragg mirrors in the optical realm [17,18], polished mirror microwave cavities [19,20], as well as semiconductor heterostructures [8,9], microwave circuits [21], and defects in nanofabricated photonic crystals [10]. A review of much of the varied work that has been carried out using optical cavities is given in Ref. [3].

The boundary conditions imposed on the electromagnetic field in spherical-mirror cavities result in resonant beams that are described by the infinite set of Hermite–Gaussian modes [22]. These modes are defined by three independent integers (l,m,n) describing the intensity distribution of the field. The integers l and m characterize the beam profile in the transverse plane, while n defines the order of the longitudinal modes. The transverse modes are commonly labeled as TEM_{lm} (transverse electromagnetic). In specific experimental circumstances, higher order modes are used [23–26].

Our interest is in the lowest mode or Gaussian mode: TEM_{00}. The coupling strength of a photon in a cavity mode to an atomic transition is proportional to the transition dipole moment of the atomic transition and the electric field per photon. The electric field per photon is inversely proportional to the square root of

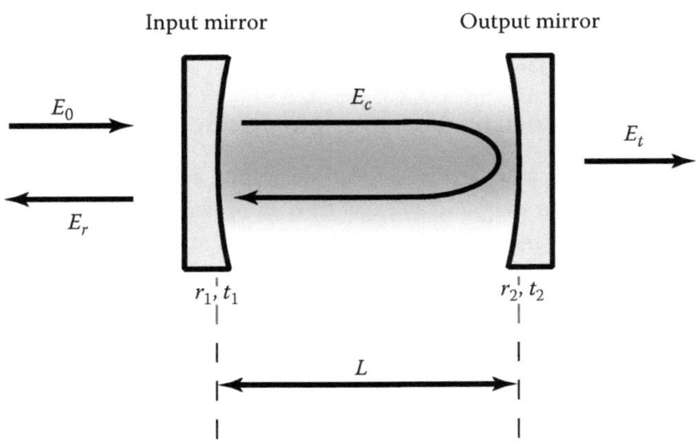

FIGURE 2.2 Schematic diagram of a Fabry–Pérot cavity showing input, E_0, reflected, E_r, circulating, E_c, and transmitted, E_t, fields.

the mode volume through the classical formula that relates the square of the electric field to the energy density. Small mode volumes produce high electric fields per photon and strong coupling. For a given cavity, the TEM_{00} mode has the smallest volume, hence our focus on this particular mode.

The simplest cavity to analyze is the Fabry–Pérot resonator, shown schematically in Figure 2.2. The intensity profile of this mode varies as a Gaussian distribution proportional to $\exp[-2(x^2 + y^2)/w(z)^2]$. The origin of our coordinate system is on the axis defined by the centers of the two cavity mirrors and is placed at the waist in the center of the cavity. The z-axis measures displacement in the axial direction, and x and y are bases that span the radial degrees of freedom. $w(z)$ is the $1/e^2$ radius of the beam intensity, which varies along the cavity axis. A standing wave multiplies this mode envelope, modulating the intensity from zero to maximal, every half-wavelength.

In a geometrically symmetric resonator, where both mirrors have equal radius of curvature, R_m, the beam is focused in the center of the cavity, and the waist size is given by

$$w_0^2 = \frac{\lambda L}{2\pi}\left(2\frac{|R_m|}{L} - 1\right)^{1/2}, \tag{2.9}$$

where L is the distance between the mirror centers. For cavity mirrors that are separated by a distance much smaller than their radius of curvature ($L \ll R_m$), the transverse radius of the beam does not vary significantly, so $w(z) \approx w(0) \equiv w_0$, and the position-dependent intracavity electric field is given by

$$\boldsymbol{E}(\boldsymbol{r}) = E_c \epsilon \cos\left(\frac{2\pi z}{\lambda}\right) \exp\left[\frac{x^2 + y^2}{w_0^2}\right] \cos \omega t \tag{2.10}$$

where

E_c is the maximum amplitude of the intracavity electric field (on axis and at an antinode of the standing wave)

ε is the polarization vector

Since the coupling of an electric-dipole-allowed atomic transition to a cavity mode is proportional to the local strength of the electric field, the coupling is maximized for an atom on the TEM_{00} mode axis at an antinode of the standing wave. If the driving field is on resonance with a cavity mode, the circulating field is enhanced by a factor proportional to the cavity finesse in comparison with the input driving field as we discuss in the following paragraphs.

As illustrated in Figure 2.2, we can write down the following equations for the amplitude of the reflected field E_r, transmitted field E_t, and circulating field E_c in terms of the amplitude of the input driving field amplitude E_0:

$$E_r = r_1 E_0 + i r_2 t_1 E_c e^{i\delta}, \tag{2.11}$$

$$E_t = i t_2 E_c, \tag{2.12}$$

$$E_c = i t_1 E_0 + r_1 r_2 E_c e^{i\delta}. \tag{2.13}$$

Here, $\delta = 4\pi L/\lambda$ is the phase accumulated in a round-trip of the cavity, for a resonator of length L, formed by an input mirror with amplitude reflectivity r_1 and transmissivity t_1, and output mirror of r_2 and t_2. The factor $i = \sqrt{-1}$ is included to take account of the π phase shift on transmission relative to reflection at a beam splitter. This is a very convenient and straightforward way to write down the equations that govern the steady-state response of a cavity to a driving field. The reader can use these equations to solve for the transmitted field and power, the reflected field and power, and the circulating field and power in terms of the input field.

For a symmetric cavity where $r_1 = r_2 = r$, the aforementioned expressions can be solved to give the transmitted intensity (proportional to the square of the transmitted field) as

$$\frac{I_t}{I_0} = \frac{(1-R)^2}{1 - 2\cos\delta + R^2}, \quad \text{where } R = r^2. \tag{2.14a}$$

It is usual to rewrite this expression as

$$\frac{I_t}{I_0} = \frac{1}{1 + (2\mathscr{F}/\pi)^2 \sin^2(\delta/2)} \tag{2.14b}$$

where $\mathscr{F}$ is a parameter known as the *finesse*. In the absence of intracavity loss, the finesse is determined completely by the quality of the cavity mirrors:

$$\mathscr{F} = \frac{\pi R^{1/2}}{1-R}.$$

The transmitted field intensity is a periodic function of the round-trip phase (Equation 2.14), and consequently depends on the wavelength of light used and its relation to the cavity length. Each time the length is changed by one-half wavelength, a new peak in transmission is found. Operating *on resonance* restricts the intracavity field to standing waves, whose wavelengths are defined by the cavity length,

$$2L = n\lambda, \tag{2.15}$$

where n is the number of nodes in the standing wave: each n defines a different *longitudinal* mode of the field. Alternatively, the resonant frequencies of the cavity take discrete values given by

$$\nu = \frac{nc}{2L}. \tag{2.16}$$

There is a constant frequency difference between adjacent resonator modes, n and $n - 1$, known as the free spectral range

$$\nu_{FSR} = \frac{c}{2L}. \tag{2.17}$$

From Equation 2.14, the full width at half the maximum intensity (FWHM), κ, of each resonance can be found. The finesse measures the ratio of the free spectral range to this width:

$$\mathcal{F} = \frac{\nu_{FSR}}{\kappa}. \tag{2.18}$$

Figure 2.3 shows the transmitted field intensity as a function of the accumulated round-trip phase, for three different values of finesse. Such a spectrum could be measured either by scanning the cavity length or the wavelength of the laser used to probe the cavity and monitoring the transmitted power with a photodetector. The spectrum shows clear resonances, provided the laser line width is small compared to the frequency width of the cavity resonance. For a narrow resonance, the transmitted power increases rapidly as the laser comes into resonance with the cavity, the reflected power dips sharply, and the circulating power builds to a maximum. There are two frequency detunings that will be important in our description of an atom in a cavity, the detuning of the laser from the nearest (in frequency) bare (as in no atom) cavity resonance and the detuning of the laser from the bare (no cavity) atom resonance. We assume that resonant frequency of only one cavity mode is sufficiently close to the bare atomic resonance to play a significant role, and we disregard all other cavity modes. This is routinely realized in the lab.

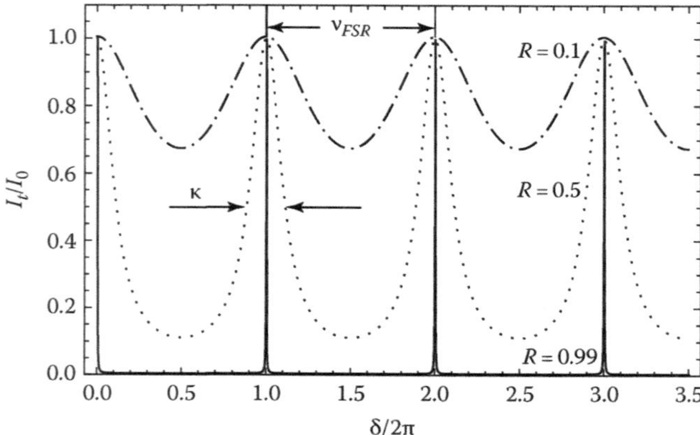

FIGURE 2.3 Fabry–Pérot transmission intensity Equation 2.14 over three free spectral ranges, for a cavity formed with mirrors of poor to good reflectance $R = r^2 = 0.1, 0.5$, and 0.99.

2.3 Fully Quantized Atom–Cavity System

2.3.1 Quantum-Mechanical Mode

Thus far, the cavity field has been described as the isolation of a classical electromagnetic wave within an optical resonator. When dealing with small numbers of photons and atoms in a cavity, as we often do, the quantized description of light is necessary [13].

The quantum-mechanical description of the harmonic energy levels of such a field can be written as*

$$\hat{H}_c = \hbar\omega_c \hat{a}^\dagger \hat{a} \tag{2.19}$$

where

$\hbar\omega_c$ is the energy of a photon of the cavity mode

$\hat{a}^\dagger$ and $\hat{a}$ are the creation and annihilation operators that add or remove photons:

$$\begin{aligned} \hat{a}\,|n\rangle &= \sqrt{n}\,|n-1\rangle \\ \hat{a}^\dagger\,|n\rangle &= \sqrt{n+1}\,|n+1\rangle. \end{aligned} \tag{2.20}$$

They obey the commutation relation $[\hat{a}, \hat{a}^\dagger] = 1$, and together give the photon number-operator $\hat{n} = \hat{a}^\dagger\hat{a}$. Such a description can be applied to any single cavity mode or more than one cavity mode. In the latter case, we would need to label the raising and lowering operators with a unique label for each mode. In this chapter, we will assume that the atom interacts with TEM_{00} and that $\hat{a}^\dagger$ and $\hat{a}$ are the raising and lowering operators for the TEM_{00}, respectively.

The intracavity energy density scales with the number of intracavity photons, and the single-photon energy density is simply the photon energy, $\hbar\omega_c$, divided by the cavity-mode volume, V_m. The mode volume is found by integrating the cavity mode (Equation 2.10) over all three spatial dimensions:

$$V_m = \frac{1}{4}\pi w_0^2 L,$$

and the local electric field can be expressed as

$$\mathbf{E}(\mathbf{r}) = \epsilon\left[\frac{\hbar\omega_c}{\epsilon_0 V_m}\right]^{1/2}(\hat{a} + \hat{a}^\dagger)U(\mathbf{r}), \tag{2.21}$$

where ϵ is the electric field polarization unit vector.

2.3.2 Jaynes–Cummings Model

At this point, it is convenient to introduce the *atomic transition operators*. These operators act to raise and lower the atomic excitation

$$\begin{aligned} \hat{\sigma}_+ &= |e\rangle\langle g| \\ \hat{\sigma}_- &= |g\rangle\langle e|, \end{aligned} \tag{2.22}$$

* We ignore the zero point energy conventionally included in the Hamiltonian of the harmonic field operator since it will not be important for the coupled system to be discussed shortly.

and the atomic inversion operator

$$\hat{\sigma}_z = |e\rangle\langle e| - |g\rangle\langle g| \\ = (\hat{\sigma}_+\hat{\sigma}_- - \hat{\sigma}_-\hat{\sigma}_+) \tag{2.23}$$

measures the atomic excitation. Although both atoms and cavities are resonant systems, they are described by subtly but importantly different physics. The commutation relation between the cavity-mode raising and lowering operators is identical to that of a quantum harmonic oscillator. Like a simple harmonic quantum oscillator, the ground state of a cavity mode (no photons present) can be raised an arbitrary number of times, and the Hilbert space describing the mode is infinite. The ground state of a two-level atom can be raised only once, and the Hilbert space is two-dimensional, one ground state and one excited state. The raising and lowering operators for a two-level atom display the commutation relation given in Equation 2.23. A key physical consequence of the difference in these commutation relations is that the cavity mode does not saturate with increasing driving field amplitude, whereas the atomic system does.

The interaction between a two-level atom and the quantized cavity field is illustrated by the Jaynes–Cummings model [27]:

$$\hat{H} = \frac{1}{2}\hbar\omega_a\hat{\sigma}_z + \hbar\omega\hat{a}^\dagger\hat{a} + \hbar g(\mathbf{r})(\hat{\sigma}_+ + \hat{\sigma}_-)(\hat{a} + \hat{a}^\dagger). \tag{2.24}$$

The first term describes the excitation of the two-level atom with atomic energy spacing, $\hbar\omega_a$, while the second relates the occupation of a mode with equally spaced energy levels separated by $\hbar\omega$. The third term once again accounts for the coupling between these "bare" states with the electric field now represented quantum-mechanically. The coupling term is proportional to the vacuum Rabi frequency $g(\mathbf{r})$:

$$g(\mathbf{r}) = -\boldsymbol{\mu}\cdot\boldsymbol{\epsilon}\left(\frac{\hbar\omega}{\epsilon_0 V_m}\right)U(\mathbf{r}). \tag{2.25}$$

A better name for this frequency may have been the "single photon Rabi frequency," but some problems in quantum electrodynamics are too hard to solve, and the name has stuck.

The Hamiltonian looks, not surprisingly, similar to the Hamiltonian we developed for a two-level atom interacting with a classical field. Of course, the difference is that there is a term describing the quantized light field and that the interaction term is written now in terms of field raising and lowering operators, whereas in the semiclassical description, the coupling term was proportional to the amplitude of the classical electric field. If the cavity field is in a coherent state with a large expectation value, n, for the number of photons, the expectation value of both the raising and lowering operators becomes the square root of n (for large n). The raising and lowering operators are multiplied by the vacuum Rabi frequency and by the transition dipole moment. In this case, we recover a coupling term very similar to our semiclassical coupling. In both cases, the coupling is the product of the transition dipole moment and the electric field. It is maximal at an antinode of the cavity field and can be increased by decreasing the cavity-mode volume (bringing the mirrors closer together and decreasing the waist size).

To proceed, we once again go to the interaction picture using the transformation in Equations 2.5, giving

$$\hat{H} = \hbar\omega\hat{a}^\dagger\hat{a} + \frac{1}{2}\hbar\omega\hat{\sigma}_z + \hbar\Delta_c\hat{a}^\dagger\hat{a} + \frac{1}{2}\hbar\Delta_a\hat{\sigma}_z \\ + \hbar g(\mathbf{r})(\hat{\sigma}_+ + \hat{\sigma}_-)(\hat{a} + \hat{a}^\dagger). \tag{2.26a}$$

In this picture, we are rotating with the driving frequency of the system (ω). This adds terms to the Hamiltonian that account for the detuning of the atom (Δ_a) and the cavity (Δ_c) from the driving frequency. These are given by

$$\Delta_a = \omega_a - \omega, \\ \Delta_c = \omega_c - \omega.$$

With the appropriate transformation, the Hamiltonian for the Jaynes–Cummings system in the interaction picture becomes

$$\hat{H}_{sys} = \hbar\Delta_c\hat{a}^\dagger\hat{a} + \frac{1}{2}\hbar\Delta_a\hat{\sigma}_z \\ + \hbar g(\mathbf{r})(\hat{a}\hat{\sigma}_+ + \hat{a}^\dagger\hat{\sigma}_- + e^{2i\hbar\omega}\hat{\sigma}_+\hat{a}^\dagger + e^{-2i\hbar\omega}\hat{\sigma}_-\hat{a}).$$

Employing the same rotating wave approximation that we exploited in the interaction of an atom with a classical field, we drop the terms that vary rapidly:

$$\hat{H}_{sys} = \hbar\Delta_c\hat{a}^\dagger\hat{a} + \frac{1}{2}\hbar\Delta_a\hat{\sigma}_z + \hbar g(\mathbf{r})(\hat{a}\hat{\sigma}_+ + \hat{a}^\dagger\hat{\sigma}_-). \tag{2.26b}$$

The coupling term is now energy conserving; the annihilation of a photon in the cavity mode via the operator $\hat{a}$ is always accompanied by the raising of the atomic energy with $\hat{\sigma}_+$ and vice versa, so the only allowed transitions are

$$|e\rangle|n-1\rangle \leftrightarrow |g\rangle|n\rangle.$$

For a given n, the dynamics of the Hamiltonian are restricted to a 2 × 2 subspace where the photon number changes by ±1:

$$\hat{H}_{sys} = \hbar\begin{bmatrix} \Delta_c(n-1) + \frac{1}{2}\Delta_a & g(\mathbf{r})\sqrt{n} \\ g(\mathbf{r})\sqrt{n} & \Delta_c n - \frac{1}{2}\Delta_a \end{bmatrix}.$$

2.3.2.1 Dressed States

As in the semiclassical case, the dressed states (atoms dressed by the photon field) are determined by diagonalizing the Hamiltonian. If we again consider the limiting case where the atoms and light are decoupled (i.e., $g(r) = 0$), and assume that the injected light is resonant with both the atom and the cavity

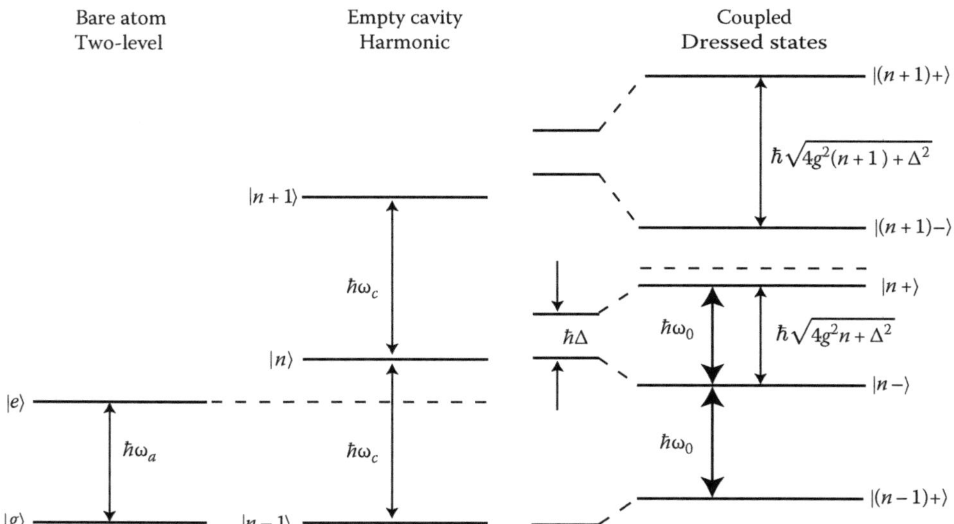

FIGURE 2.4 Energy-level splitting due to Jaynes–Cummings interaction between the two-level atom and the quantized electromagnetic field of the cavity mode. The bold arrows represent photon energies of $\hbar\omega_0$ that drive the coupled system resonantly for a unique photon occupation.

($\Delta_a = \Delta_c = 0$), then we find that the dressed states are degenerate. In other words, with all detunings set to zero, an atom in the ground state with $n + 1$ photons in the cavity mode has the same energy as an atom in the excited state with n photons in the cavity mode. If we now detune the laser from the bare atom resonance ($\Delta_a \neq 0$) but keep the laser resonant with the cavity ($\Delta_c = 0$) and again assume zero coupling, the energy splitting of the dressed states is the detuning from atomic resonance exactly as it was in the semiclassical discussion we presented earlier in this chapter. This now has a clear interpretation. An atom in the ground state with $n + 1$ photons in the cavity mode differs in energy from an atom in the excited state and n photons in the cavity mode because the light is detuned, and it clearly differs by precisely the detuning. If we now turn on the atom–light coupling, with $\Delta_a = \Delta_c = 0$, we find nondegenerate dressed state energies given by

$$E_{\pm}(n) = \pm\hbar g\sqrt{n}. \tag{2.27}$$

In general, the dressed states are linear combinations of the bare states:

$$\begin{aligned} |n+\rangle &= \sin\alpha\,|n-1\rangle\,|e\rangle + \cos\alpha\,|n\rangle\,|g\rangle \\ |n-\rangle &= \cos\alpha\,|n-1\rangle\,|e\rangle - \sin\alpha\,|n\rangle\,|g\rangle \end{aligned}. \tag{2.28}$$

The associated eigen-energies are

$$E_{\pm}(n) = \hbar\Delta_c\left(n - \frac{1}{2}\right) \pm \frac{1}{2}\hbar\sqrt{4g^2n + \Delta^2}, \tag{2.29}$$

where $\Delta = (\Delta_a - \Delta_a)$. Figure 2.4 illustrates the energy splitting arising from the interaction.

In the absence of coupling, the bare states in each 2 × 2 manifold are split by the difference of the atom and cavity detuning, and this splitting is identical in every manifold. This is no longer true in the presence of coupling, that is, when the off-diagonal terms in the Hamiltonian are nonzero. The splitting increases as n increases. The set of energy levels of the dressed states is known as the *Jaynes–Cummings ladder*, and the energy of each "rung" depends on the photon occupation (Equations 2.27 and 2.29). A photon with energy $\hbar\omega$ is resonant with the dressed-state transition $|(n-1)+\rangle \rightarrow |n-\rangle$, but the addition of another photon with the same energy does not bring the coupled system to the next rung. The photon is "blocked" from the cavity, and the phenomenon is known as the *photon blockade*. The anharmonic Jaynes–Cummings ladder has been observed experimentally with a single atom coupled to optical [28] and microwave [29] cavities as well as in analogous circuit QED systems, composed of a microwave field and a superconducting qubit as an artificial two-level atom [11].

If one knows the precise number of photons n in the system, measuring the energy splitting of the dressed states is a direct way to ascertain the strength of the atom–cavity coupling. In a situation on resonance, with the atom initially excited and the cavity mode in the ground (or "vacuum") state—that is to say that there is precisely one quantum of excitation in the system—the dressed-state doublet energies are known as *vacuum Rabi resonances*. There has been a wealth of experimental interest in measuring vacuum Rabi splitting in a wide range of real-world systems. In the time domain, the oscillations have been observed with Rydberg atoms in microwave cavities [6], and spectroscopically Rabi splitting has been measured in a range of strongly coupled optical (and infrared) systems [5,9–11,21,30,31].

2.4 Open Quantum Systems

To date in our discussion, we have assumed that the atom couples to one and only one mode of the electromagnetic field, the TEM_{00} cavity mode. Of course, in reality, the atom couples to all cavity modes, but we have assumed and will continue to assume that the TEM_{00} mode is the only cavity mode sufficiently close to the atomic resonance to be significant in the problem.

Most cavities that are employed in experiments subtend a small fraction of the full 4π solid angle as viewed by the atom. Under these conditions, the atom can and does couple to the continuum of electromagnetic vacuum modes resulting in spontaneously emitted photons that leave the system and do not return. The continuum of electromagnetic modes is a reservoir, and coupling to the reservoir causes loss of coherence in the system dynamics. In addition to spontaneous emission, photons can be lost from the cavity field by leaking through the finite reflectivity cavity mirrors. This provides a second loss or decoherence mechanism. These two loss rates are characterized by the cavity line width κ and the spontaneous emission rate γ.

Cavity QED is generally separated into two regimes, those of strong and weak coupling. These define the relative strengths of coupling within the system and to external reservoir modes. Strong coupling is the regime where the vacuum Rabi frequency is higher than both the spontaneous emission rate and the cavity decay rate. A specific coupling regime can be characterized by two dimensionless parameters, the critical photon number given in Equation 2.30 and the critical atom number given in Equation 2.31:

$$m_0 = \frac{(\gamma/2)^2}{2g^2}, \tag{2.30}$$

$$N_0 = \frac{\gamma\kappa}{g^2} = C^{-1}. \tag{2.31}$$

These two dimensionless parameters indicate the number of quanta necessary to significantly influence the system. Strong coupling occurs when N_0 and m_0 are both less than 1. The critical photon number is the number of cavity photons needed to saturate the atom in a resonant transition. The critical atom number refers to the number of atoms required to significantly affect the cavity field. N_0 is often inverted and referred to as the cooperativity parameter, C. In order to describe an open quantum system of this kind, we will need to introduce the reduced system density matrix and the equation that governs its dynamics, the master equation. This is the focus of the remainder of this chapter. This approach has broad applicability, and cavity QED is a beautiful system to gain some familiarity with these concepts and methods.

2.4.1 Composite Systems and the Reduced Density Matrix

We assume the reader is familiar with the density operator, the equation describing its Hamiltonian evolution and the concept of pure and mixed states. For those readers unfamiliar with this material, excellent discussions can be found in most graduate books on quantum mechanics and quantum statistical mechanics [32–34]. We refer the reader to those books and do not reproduce that material here.

We begin our discussion by considering a composite system comprised of two subsystems, A and B, that are coupled by some Hamiltonian. The respective Hilbert spaces are $H_A = \{|1\rangle_A, |2\rangle_A, |3\rangle_A, \ldots, |i\rangle_A\}$ and $H_B = \{.|1\rangle_B, |2\rangle_B, |3\rangle_B, \ldots, |j\rangle_B\}$. In our case, for example, the two Hilbert spaces we will consider will be the atom/cavity-mode system Hilbert space and the Hilbert space comprised of all the other electromagnetic modes. Although the details of the reservoir Hamiltonian might not be known, the coupling rates that characterize the system–reservoir interaction are assumed to be known, and in our case the cavity decay rate κ and the spontaneous emission rate γ. A quantum state in our composite Hilbert space can always be written as

$$|\Psi_k\rangle = \sum_i \sum_j c_{ij} \,|i\rangle_A\,|j\rangle_B. \tag{2.32}$$

The expectation value of an operator $\hat{O}$, that in general acts on both subspaces of our composite Hilbert space, could be calculated directly using Equation 2.32, or alternatively, Equation 2.32 could be used to construct the full composite system density operator $\hat{\rho}$, which could in turn be used to calculate the expectation value of $\hat{O}$ in the usual way, that is,

$$\langle\hat{O}\rangle = \mathrm{Tr}_{AB}\{\hat{\rho}\hat{O}\}. \tag{2.33}$$

If we wish to calculate how expectation values evolve in time, we could solve for the density operator as a function of time and use Equation 2.33. If, as is the case in this chapter, one of the Hilbert spaces is a reservoir with a very large or infinite set of modes, this approach is intractable.

Consider however an operator, $\hat{A}$, that acts only on subsystem A. In our case, this might be the cavity-mode number operator or the atomic inversion or the sum or product of these operators. The operator acts only on the system Hilbert space and not on the reservoir. We emphasize here that the system comprised of the cavity mode and the two-level atom is one subsystem in our composite Hilbert space, say H_A. The reservoir comprised of the continuum of electromagnetic modes is H_B. We calculate the expectation value of $\hat{A}$ in the usual way:

$$\begin{aligned}\langle\hat{A}\rangle &= \mathrm{Tr}_{AB}\{\hat{\rho}\hat{A}\}\\ &= \sum_i\sum_j \langle j|_B\,\langle i|_A\,\hat{\rho}\hat{A}\,|i\rangle_A\,|j\rangle_B\\ &= \sum_i \langle i|_A\left[\sum_j \langle j|_B\,\hat{\rho}\,|j\rangle_B\right]\hat{A}\,|i\rangle_A\\ &= \mathrm{Tr}_A\{\hat{\rho}_R\hat{A}\},\end{aligned} \tag{2.34}$$

where we have defined the reduced density operator $\hat{\rho}_R$ as

$$\hat{\rho}_R \equiv \mathrm{Tr}_B\{\hat{\rho}\}. \tag{2.35}$$

The expectation value of $\hat{A}$ has been expressed as the trace of the reduced density operator formed by tracing the full density operator over the reservoir. If we can solve for the dynamics of the reduced density operator, we have gained an enormous advantage in terms of reducing the size of the problem.

As a simple example, consider a system that is composed of two subsystems, labeled A and B, each of which is a two-level atom, with the familiar ground and excited states $|g\rangle$ and $|e\rangle$. In contrast to our previous discussion, there is no reservoir in this example, and our composite system is defined by two (2 × 2) two-level atom Hilbert spaces, a problem that is not too difficult. We assume that the system is initially in the entangled state:

$$\frac{1}{\sqrt{2}}\{|g\rangle_A|g\rangle_B + |e\rangle_A|e\rangle_B\}. \tag{2.36}$$

To simplify notation, we define

$$\left.\begin{aligned} |11\rangle &\equiv |g\rangle_A|g\rangle_B \\ |12\rangle &\equiv |g\rangle_A|e\rangle_B \\ |12\rangle &\equiv |g\rangle_A|e\rangle_B \\ |22\rangle &\equiv |e\rangle_A|e\rangle_B \end{aligned}\right\}. \tag{2.37}$$

The initial entangled state can be written as

$$\frac{1}{\sqrt{2}}\{|11\rangle + |22\rangle\}.$$

The density operator in the full composite Hilbert space is

$$\hat{\rho} = \frac{1}{2}\{|11\rangle\langle 11| + |11\rangle\langle 22| + |22\rangle\langle 11| + |22\rangle\langle 22|\},$$

and the matrix representation of density operator describing the composite system is

$$\hat{\rho} = \frac{1}{2}\begin{bmatrix} 1 & 0 & 0 & 1 \\ 0 & 0 & 0 & 0 \\ 0 & 0 & 0 & 0 \\ 1 & 0 & 0 & 1 \end{bmatrix}. \tag{2.38}$$

As this is the full density operator describing both subsystems and this is all there is in the problem, the density operator must describe a pure state. The reader can check this by verifying that $\mathrm{Tr}\{\hat{\rho}^2\} = 1$. Consider an operator $\hat{A}$ that acts only on the subsystem A. The expectation value of any operator can be calculated by employing the full density operator in the usual way:

$$\begin{aligned} \langle \hat{A} \rangle &= \mathrm{Tr}\{\hat{\rho}\hat{A}\} \\ &= \langle 11|\hat{\rho}\hat{A}|11\rangle + \langle 12|\hat{\rho}\hat{A}|12\rangle + \langle 21|\hat{\rho}\hat{A}|21\rangle + \langle 22|\hat{\rho}\hat{A}|22\rangle \\ &= \langle g|_A \langle g|_B (\hat{\rho}\hat{A})|g\rangle_A|g\rangle_B + \langle g|_A \langle e|_B (\hat{\rho}\hat{A})|g\rangle_A|e\rangle_B \\ &\quad + \langle e|_A \langle g|_B (\hat{\rho}\hat{A})|e\rangle_A|g\rangle_B + \langle e|_A \langle e|_B (\hat{\rho}\hat{A})|e\rangle_A|e\rangle_B \\ &= \langle g|_A \{\langle g|_B \hat{\rho}|g\rangle_B + \langle e|_B \hat{\rho}|e\rangle_B\}\hat{A}|g\rangle_A \\ &\quad + \langle e|_A \{\langle g|_B \hat{\rho}|g\rangle_B + \langle e|_B \hat{\rho}|e\rangle_B\}\hat{A}|e\rangle_A \\ &\equiv \mathrm{Tr}_A\{\hat{\rho}_R\hat{A}\}, \end{aligned} \tag{2.39}$$

where $\hat{\rho}_R$ is the reduced density matrix found by *tracing out* the subsystem B:

$$\begin{aligned} \hat{\rho}_R &= \mathrm{Tr}_B\{\hat{\rho}\} \\ &\equiv \langle g|_B \hat{\rho}|g\rangle_B + \langle e|_B \hat{\rho}|e\rangle_B \\ &= \frac{1}{2}\{|g\rangle_A\langle g|_A + |e\rangle_A\langle e|_A\} \\ &= \frac{1}{2}\begin{bmatrix} 1 & 0 \\ 0 & 1 \end{bmatrix}. \end{aligned} \tag{2.40}$$

$\mathrm{Tr}\{\hat{\rho}_R^2\} \neq 1$, and the reduced density matrix in Equation 2.40 describes a mixed state. Information has been lost by tracing over the second subsystem. Notice that the reduced density matrix $\hat{\rho}_R$ defined in Equation 2.40 is a (2 × 2) matrix, whereas the size of $\hat{\rho}$ was (4 × 4). If subsystem B is an n-dimensional space—rather than a *two-level atom*—the coupled system is represented by a $(2n \times 2n)$ density matrix. The size of the reduced density matrix, however, is still only (2 × 2). This is the benefit of tracing out a subsystem or reservoir.

Coupling of a system to the environment leads to a loss of information from the system and leads to a loss of a great deal of observed *quantum* behavior. The objective of putting an atom in a high finesse cavity in the strong coupling regime is to isolate the system (atom/cavity mode) as much as possible from the environment so that we can observe and exploit highly quantum behavior such as coherent single-photon/single-atom dynamics. In practice, all physical quantum systems are coupled to the environment, and all systems strictly are open quantum systems. Of course, we can always expand our definition of the system to include the modes of the environment, but this is not usually very useful. We may know little or nothing about the evolution of the reservoir modes. There are simply too many to keep track of, and in many cases we only have detailed information on the system Hamiltonian and some information (decay rates) on the coupling of the system to the environment.

We start by deciding how much of the universe needs to be included in the system for an adequate treatment. In choosing how much of the universe constitutes our system, we must, for example, satisfy the Markov approximation if we are to use the master equation to describe the system dynamics through the reduced density matrix. The evolution of the reduced density matrix is not as straightforward as the expression for the total system + environment. A first-order differential equation for the reduced density operator—describing the time evolution of an *open* quantum system—is therefore needed. This is the *master equation*, introduced in the following section.

2.4.2 Entanglement, Decoherence, and the Master Equation

Consider a system and reservoir that are initially unentangled; the full quantum state describing the system and reservoir can be written as a direct product of the system state vector and the reservoir state vector. If the reservoir and system are unentangled, no information about the system is lost by tracing over the reservoir modes, and the reduced density matrix describing the initial *system* quantum state is a pure state density matrix. The reader can check this statement by replacing the entangled state, Equation 2.36, in our earlier discussion with an unentangled state.

We bring the system and reservoir into contact via some Hamiltonian. We could imagine, for example, that our system is a two-level atom in its excited state, that the reservoir is the continuum of vacuum modes of the electromagnetic field, and that we "turn on" that piece of the Hamiltonian that gives rise to spontaneous emission. Alternatively, we might have a cavity mode initially populated with n photons that can leak photons through the mirrors to the vacuum modes of the electromagnetic field, or we might have a system comprised of a two-level atom coupled to a cavity mode that can leak photons by spontaneous emission and through the cavity mirrors.

In general, bringing a system into contact with a reservoir would mean that we allow the system and reservoir to exchange energy, particles, momentum, angular momentum, or any combination of these quantities. In general, the system becomes entangled with the reservoir through this process. As we saw in the example discussed earlier, tracing over the reservoir modes in this situation leads to a loss of information about the system, and the reduced density matrix now describes a mixed state. The loss of information is reflected in the loss of coherence expressed through the decay of the off-diagonal terms in the density matrix.

So far, in the Jaynes–Cummings model (Equation 2.26), we have discussed the *intra*-system coupling between the atom and cavity mode, but not the coupling to external modes via system–environment interactions. In the absence of dissipation (a closed system), the model can be solved exactly, however such a simplification brings limitations, perhaps most importantly those pertaining to system measurements; it is only via external observations that information about the internal state of the system can be gained. Almost always these measurements make use of photons transmitted through the cavity mirrors.*

A more accurate model must include the decoherence that occurs both as a result of spontaneous emission from the excited state of the two-level atom into modes other than the cavity mode and the loss of photons from the cavity mode via transmission, scattering, and absorption in the mirrors. When these dissipative processes are included in the Jaynes–Cummings model, the system dynamics are found using the reduced density matrix and solutions to the master equation.

Two standard derivations of the master equation are presented in the quantum optics literature. The first is a lengthy procedure originally introduced by Senitzky [35] and covered well in many quantum optics textbooks [3]. This method considers directly the coupling of a quantum-mechanical system and a large reservoir of modes—so large that one has no chance of keeping track of the evolution of all its degrees of freedom—and is mathematically somewhat involved.

The second is a more aphoristic approach, making use of a Monte Carlo-*esque* method. It ignores the quantized nature of the reservoir modes and avoids some of the technical difficulties of the previous approach. Instead, it considers the statistical nature of the decay of any given state of the system. We will outline the latter approach.

To motivate our discussion of the master equation, we consider the rather simple case of a cavity mode that loses photons by leaking through the cavity mirrors. We will include the two-level atom in the description at the end of our discussion. We assume that the probability of loss of a photon from the cavity in a time δt is

$$\delta P = \kappa\langle\Psi\,|\,\hat{a}^{\dagger}\hat{a}\,|\,\Psi\rangle\delta t. \tag{2.41}$$

The probability that the cavity loses a photon is assumed to be proportional to the expectation value of the number of photons in the cavity and is proportional to a decay rate κ. The decay rate characterizes the coupling of the system to the reservoir. We do not know the details of the reservoir Hamiltonian, and to a large extent, they are unimportant, which is the reason the master equation approach works in the first place. It is important, however, that photons leaked from the system through the mirror do not return to the system. This requirement is the basis of the Markov approximation, which requires correlations in the reservoir to decay quickly on the time scale of the coupling between the system and the reservoir. We do not dwell on this point but refer the reader to the literature [33,34]. This is an important requirement, one that is met in many experiments and a requirement that is not obvious in the discussion we present here. In more sophisticated derivations, the Markov approximation arises quite naturally.

The probability that there is *no* decay is given by $(1 - \delta P)$. Thus, after a time δt, there are two possible outcomes for an initial state $|\Psi(0)\rangle$. We have either lost a photon and $|\Psi(0)\rangle \rightarrow |\Psi_{loss}\rangle$ or we have not and $|\Psi(0)\rangle \rightarrow |\Psi_{no\ loss}\rangle$. To generate $|\Psi_{loss}\rangle$, we lower the photon number in the initial state and normalize:

$$\begin{aligned}|\,\Psi_{loss}\rangle &= \frac{\hat{a}\,|\,\Psi\rangle}{\langle\Psi\,|\,\hat{a}^{\dagger}\hat{a}\,|\,\Psi\rangle^{1/2}} \\ &= \frac{(\kappa\delta t)^{1/2}\hat{a}\,|\,\Psi\rangle}{P^{1/2}}.\end{aligned} \tag{2.42}$$

The evolution in the case of *no-loss* requires somewhat more careful discussion. It is tempting but incorrect to simply evolve the initial state under the system Hamiltonian. If, for example, we measure no photons emitted in a time interval δt, we gain some information on the mean photon number in the cavity. If, in successive time intervals, we continue to measure no emitted photons, our estimate of the mean number of photons in the cavity continues to decrease. We gain information about the mean number of photons by measuring no emitted photons, and we must evolve the state in the *no-loss* case to account for this gain in information. We achieve this by evolving the initial state using an effective Hamiltonian:

$$\hat{H}_{eff} = \hat{H} - i\left(\frac{\hbar\kappa}{2}\right)\hat{a}^{\dagger}\hat{a}. \tag{2.43}$$

* Sometimes the inverse may be true, and independent measurement of an atom's excitation-state is used to deduce information about the presence of intracavity photons [20].

The effective Hamiltonian is by necessity non-Hermitian to allow for loss to the reservoir modes. The first term is the system Hamiltonian. The second term in the effective Hamiltonian clearly leads to the decay of the cavity mode. We can now write the evolution in the *no-loss* case as

$$
\begin{aligned}
|\Psi_{no\,loss}\rangle &= \frac{e^{-i\hat{H}_{eff}\delta t/\hbar}|\Psi\rangle}{[\langle\Psi|e^{i\hat{H}^{\dagger}_{eff}\delta t/\hbar}e^{-i\hat{H}_{eff}\delta t/\hbar}|\Psi\rangle]^{1/2}} \\
&= \frac{e^{-i\hat{H}_{eff}\delta t/\hbar}|\Psi\rangle}{[\langle\Psi|e^{-\kappa\hat{a}^{\dagger}\hat{a}\delta t}|\Psi\rangle]^{1/2}} \\
&\approx \frac{\{1-(i/\hbar)\hat{H}\ \delta t-(\kappa/2)\hat{a}^{\dagger}\hat{a}\ \delta t\}|\Psi\rangle}{(1-P)^{1/2}}. \qquad (2.44)
\end{aligned}
$$

The evolution of the reduced density operator becomes

$$
\begin{aligned}
&|\Psi(0)\rangle\langle\Psi(0)| \rightarrow |\Psi(\delta t)\rangle\langle\Psi(\delta t)| \\
&= P|\Psi_{loss}\rangle\langle\Psi_{loss}| + (1-P)|\Psi_{no\text{-}loss}\rangle\langle\Psi_{no\text{-}loss}|. \qquad (2.45)
\end{aligned}
$$

Notice that there are no coherence terms in the equation for the evolution of the density operator. This is an important point. Under this description, the system either jumps to a state where a photon has been lost or does not. Substituting Equations 2.42 and 2.44 into Equation 2.45 leads to the master equation describing the evolution of the system (reduced) density operator:

$$
\frac{d}{dt}\hat{\rho}(t) = -\frac{i}{\hbar}[\hat{H},\hat{\rho}] + \frac{\kappa}{2}(2\hat{a}\hat{\rho}\hat{a}^{\dagger} - \hat{a}^{\dagger}\hat{a}\hat{\rho} - \hat{\rho}\hat{a}^{\dagger}\hat{a}). \qquad (2.46)
$$

The first term in the master equation describes the evolution of the reduced density matrix describing the cavity mode under the system Hamiltonian. The second term describes the interaction with the reservoir via loss of photons through the cavity mirrors. In the coupled atom–cavity system defined by the Jaynes–Cummings model (Equation 2.26), photons can be lost from the cavity mode either by transmission through the mirrors, with rate κ, or by spontaneous emission of the two-level atom into modes outside the cavity at a rate γ. Just as the probability of emission of a photon through a cavity mirror is proportional to the mean number of photons through the photon number operator $\hat{a}^{\dagger}\hat{a}$, the probability of spontaneous emission of a photon from the atom is proportional to the excited state probability through the inversion operator $\hat{\sigma}_{+}\hat{\sigma}_{-}$. The reader is encouraged to repeat the derivation of the master equation including the spontaneous emission term. The master equation describing a two-level atom interacting with a cavity mode including loss of photons from the cavity mode through the mirror and loss of photons via spontaneous emission is

$$
\begin{aligned}
\frac{d}{dt}\hat{\rho}(t) &= -\frac{i}{\hbar}[\hat{H},\hat{\rho}] + \frac{\kappa}{2}(2\hat{a}\hat{\rho}\hat{a}^{\dagger} - \hat{a}^{\dagger}\hat{a}\hat{\rho} - \hat{\rho}\hat{a}^{\dagger}\hat{a}) \\
&+ \frac{\gamma}{2}(2\hat{\sigma}_{-}\hat{\rho}\hat{\sigma}_{+} - \hat{\sigma}_{+}\hat{\sigma}_{-}\hat{\rho} - \hat{\rho}\hat{\sigma}_{+}\hat{\sigma}_{-}). \qquad (2.47)
\end{aligned}
$$

The Jaynes–Cummings Hamiltonian *with* dissipation necessarily leads to a depletion of photons from the cavity mode; the photon number decays via loss through the cavity mirrors and atomic spontaneous emission. In practice, most experiments drive the cavity mode with a laser field providing constant power at the input mirror that replenishes photons lost from the system.

The pump rate is related to the incident power and the transmissivity of the input mirror. For a resonant cavity with intracavity photon number n_{res},

$$
\varepsilon^2 = n_{res}\kappa^2.
$$

If, without changing the pumping rate ε, the cavity is detuned, then the intracavity photon number n_0 will reduce to

$$
\begin{aligned}
n_0 &= \frac{n_{res}}{1+(\Delta/\kappa)^2} \\
&= \kappa^2\frac{n_{res}}{\kappa^2+\Delta^2} \\
&= \varepsilon^2\frac{1}{\kappa^2+\Delta^2} \\
\therefore \varepsilon &= \sqrt{n_0(\kappa^2+\Delta_c^2)}. \qquad (2.48)
\end{aligned}
$$

This allows us to find a steady state for the density matrix (Equation 2.47) by including a constant pump term in the Hamiltonian, and we arrive at the *driven* Jaynes–Cummings Hamiltonian:

$$
\hat{H} = \hbar\Delta_c\hat{a}^{\dagger}\hat{a} + \hbar\Delta_a\hat{\sigma}_{+}\hat{\sigma}_{-} + \hbar g(\vec{r})(\hat{a}\hat{\sigma}_{+} + \hat{a}^{\dagger}\hat{\sigma}_{-}) + \hbar\varepsilon(\hat{a}+\hat{a}^{\dagger}). \qquad (2.49)
$$

An instructive example of the application of the driven Jaynes–Cummings Hamiltonian is provided by single atom detection using a high finesse cavity. Consider a beam of atoms that propagates through a high finesse cavity, as shown in Figure 2.5. A probe laser pumps the cavity, and the transmitted beam is detected with a photodetector. The photodetector noise is assumed to be negligible compared with photon shot noise. The presence of an atom reduces the transmitted flux. If the photon flux drops below an appropriately placed discriminator, we assume an atom was present as shown in Figure 2.5b. In order to produce a high-fidelity detector, we need the signal-to-noise ratio (SNR) for a transit to be as large as possible.

Poldy et al. have studied and optimized the SNR for single atom detection in just such a detector [36]. Expectation values for the system operators were determined from the steady-state solution to the driven Jaynes–Cummings master equation that results from substituting the Hamiltonian given in Equation 2.49 into the master equation given in Equation 2.47. The cavity mode was modeled with a truncated Fock-state basis. It was assumed that the atom flux was sufficiently low that only one atom was present in the cavity, and in the selection of data presented here, it was further assumed that the atom was located at the cavity waist and at an antinode.

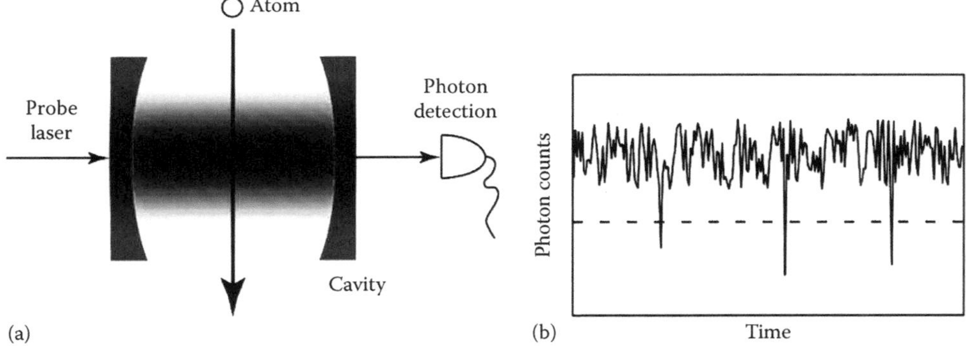

FIGURE 2.5 Single atom detection with an optical cavity showing (a) schematic diagram of the cavity setup; (b) typical photon counts for a detection event. The dashed line indicates the threshold or discriminator for an atom detection event.

Figure 2.6 shows how the SNR varies with the incident probe power and the flux. The dashed line in the figure traces the line of maximum SNR. At low finesse, the atom behaves as a saturable absorber, and the probe power decreases along the line of maximum SNR as the finesse is increased. This behavior is quite intuitive. It is reasonable to expect that in this regime, the maximum SNR will be achieved when the intracavity field is enough to just saturate the atom. As the finesse increases beyond 3500, the behavior changes, and the probe power increases along the line of maximum SNR with increasing finesse. This behavior is not obvious. In this regime, the atom can no longer be considered a saturable absorber. The atom signal is now due to the shift in resonance with an atom present in the cavity. This is the same physics that describes photon blockade.

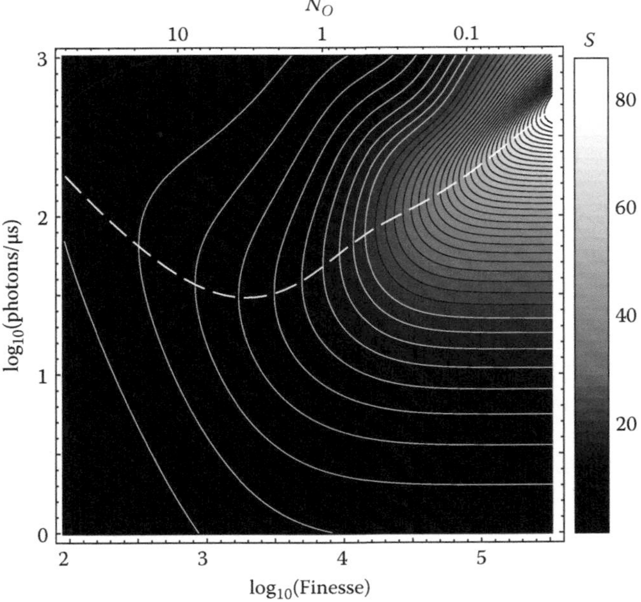

FIGURE 2.6 SNR, S, for single atom counting for resonant atom detection with ideal photon counting. A symmetric cavity was used in this model with a length of 100 μm and a waist of 20 μm. The spontaneous emission rate was $\gamma = 2\pi \times 6$ MHz. Detector integration time was set at 100 μs.

2.5 Conclusions

In this chapter, we have focused on the basic physical principles behind the coupling of atoms and photons in the setting of cavity QED. Experiments in this research field demonstrate behaviors that are inherently quantum-mechanical, and a straightforward model of an open quantum system elegantly describes experimental results with charming fidelity.

More advanced but very readable and accessible treatments of cavity QED can be found in *Optical Microcavities*, a collection of papers edited by Vahala [3], and in the books *Exploring the Quantum* by Haroche and Raimond [37]. The *Quantum Mechanics Solver* by Basdevant and Dalibard [38] and *Atomic Physics* by Budker et al. [39] have some excellent problems with solutions in atomic spectroscopy, two-level atoms, and their applications. *Introductory Quantum Optics* by Gerry and Knight gives a clear and readily accessible introduction to the quantized radiation field [13]. A more advanced treatment can be found in the classic textbook by Walls and Milburn [34]. *Photonics* by Saleh and Teich presents a clear exposition on cavity modes [22]. The first two chapters in *Laser Cooling and Trapping* by Metcalf and van der Straten give a clear discussion of the driven two-level atom [40]. These are excellent resources for students or researchers new to the field. They expand on much of the material presented here, and all are at a level similar to the level of preparation of the reader assumed in this chapter.

We note that in the quantum optics literature there exist several comprehensive reviews of cavity QED that focus on the current status of experiments within this discipline. In particular, the reader may refer to Refs. [3,4,41].

References

1. H. M. Wiseman and G. J. Milburn. *Quantum Measurement and Control*. Cambridge University Press, Cambridge, U.K., 2010.
2. H. J. Kimble. The quantum internet. *Nature*, 453(7198): 1023–1030, June 2008.

3. A. C. Doherty and H. Mabuchi. Atoms in microcavities: Quantum electrodynamics, quantum statistical mechanics, and quantum information science. *Optical Microcavities*. K. Vahala (ed.). World Scientific Publishing Co. Pte. Ltd., Singapore, Chapter 8, 2007.
4. H. Mabuchi and A. C. Doherty. Cavity quantum electrodynamics: Coherence in context. *Science*, 298:1372–1377, November 2002.
5. R. J. Thompson, G. Rempe, and H. J. Kimble. Observation of normal-mode-splitting for an atom in an optical cavity. *Phys. Rev. Lett.*, 68(8):1132–1135, May 1992.
6. J. M. Raimond, M. Brune, and S. Haroche. Colloquium: Manipulating quantum entanglement with atoms and photons in a cavity. *Rev. Mod. Phys.*, 73(3):565–582, January 2001.
7. J. I. Cirac and P. Zoller. Quantum computation with cold trapped ions. *Phys. Rev. Lett.*, 74(20):4091, May 1995.
8. G. Khitrova, H. M. Gibbs, M. Kira, S. W. Koch, and A. Scherer. Vacuum Rabi splitting in semiconductors. *Nat. Phys.*, 2:81–90, January 2006.
9. K. Hennessy, A. Badolato, M. Winger, D. Gerace, M. Atature, S. Gulde, S. Falt, E. L. Hu, and A. Imamoglu. Quantum nature of a strongly coupled single quantum dot–cavity system. *Nature*, 445:896–899, February 2007.
10. T. Yoshie, A. Scherer, J. Hendrickson, G. Khitrova, H. M. Gibbs, G. Rupper, C. Ell, O. B. Shchekin, and D. G. Deppe. Vacuum Rabi splitting with a single quantum dot in a photonic crystal nanocavity. *Nature*, 432(7014):197–200, November 2004.
11. L. S. Bishop, J. M. Chow, J. Koch, A. A. Houck, M. H. Devoret, E. Thuneberg, S. M. Girvin, and R. J. Schoelkopf. Nonlinear response of the vacuum Rabi resonance. *Nat. Phys.*, 5:105–109, January 2009.
12. H. Goldstein. *Classical Mechanics*, 2nd edn. Addison Wesley, Boston, MA, 1980.
13. C. C. Gerry and P. L. Knight. *Introductory Quantum Optics*. Cambridge University Press, Cambridge, U.K., 2005.
14. M. Rosenblit, Y. Japha, P. Horak, and R. Folman. Simultaneous optical trapping and detection of atoms by microdisk resonators. *Phys. Rev. A*, 73(063805):063805, January 2006.
15. A. Schliesser, N. Nooshi, P. Del'Haye, K. Vahala, and T. J. Kippenberg. Cooling of a micro-mechanical oscillator using radiation-pressure induced dynamical backaction. In *Conference on Lasers and Electro-Optics/Quantum Electronics and Laser Science Conference and Photonic Applications Systems Technologies*, OSA Technical Digest (CD), Optical Society of America, 2007.
16. A. A. Savchenkov, A. B. Matsko, V. S. Ilchenko, and L. Maleki. Optical resonators with ten million finesse. *Opt. Express*, 15(11):6768–6773, January 2007.
17. A. Öttl, S. Ritter, M. Köhl, and T. Esslinger. Hybrid apparatus for Bose-Einstein condensation and cavity quantum electrodynamics: Single atom detection in quantum degenerate gases. *Rev. Sci. Instrum.*, 77(6):063118, January 2006.
18. C. J. Hood, H. J. Kimble, and J. Ye. Characterization of high-finesse mirrors: Loss, phase shifts, and mode structure in an optical cavity. *Phys. Rev. A*, 64(3):033804, January 2001.
19. D. Meschede, H. Walther, and G. Müller. One-atom maser. *Phys. Rev. Lett.*, 54(6):551–552, January 1985.
20. P. Maioli, T. Meunier, S. Gleyzes, A. Auffeves, G. Nogues, M. Brune, J. M. Raimond, and S. Haroche. Nondestructive Rydberg atom counting with mesoscopic fields in a cavity. *Phys. Rev. Lett.*, 94(11):113601, January 2005.
21. A. Wallraff, D. I. Schuster, A. Blais, L. Frunzio, R.-S. Huang, J. Majer, S. Kumar, S. M. Girvin, and R. Schoelkopf. Strong coupling of a single photon to a superconducting qubit using circuit quantum electrodynamics. *Nature*, 431(7005):159–162, September 2004.
22. B. E. A. Saleh and M. C. Teich. *Fundamentals of Photonics*. John Wiley & Sons, Inc., New York, 1991.
23. J. Janousek, K. Wagner, J.-F. Morizur, N. Treps, P. K. Lam, C. C. Harb, and H.-A. Bachor. Optical entanglement of co-propagating modes. *Nat. Photonics*, 3(7):399–402, July 2009.
24. P. Maunz, T. Puppe, T. Fischer, P. Pinkse, and G. Rempe. Emission pattern of an atomic dipole in a high-finesse optical cavity. *Opt. Lett.*, 28(1):46–48, January 2003.
25. D. A. Shaddock, B. C. Buchler, W. P. Bowen, M. B. Gray, and P. K. Lam. Modulation-free control of a continuous-wave second-harmonic generator. *J. Opt. A*, 2:400–404, June 2000.
26. P. Zhang, Y. Guo, Z. Li, Y. Zhang, Y. Zhang, J. Du, G. Li, J. Wang, and T. Zhang. Elimination of the degenerate trajectory of a single atom strongly coupled to a tilted TEM_{10} cavity mode. *Phys. Rev. A*, 83(3):031804, March 2011.
27. E. T. Jaynes and F. W. Cummings. Comparison of quantum and semiclassical radiation theories with application to beam maser. *Proc. IEEE*, 51(1):89, January 1963.
28. K. M. Birnbaum, A. Boca, R. Miller, A. D. Boozer, T. E. Northup, and H. J. Kimble. Photon blockade in an optical cavity with one trapped atom. *Nature*, 436:87–90, July 2005.
29. M. Brune, F. Schmidt-Kaler, A. Maali, J. Dreyer, E. Hagley, J. M. Raimond, and S. Haroche. Quantum Rabi oscillation—A direct test of field quantization in a cavity. *Phys. Rev. Lett.*, 76(11):1800, March 1996.
30. D. Englund, A. Majumdar, A. Faraon, M. Toishi, N. Stoltz, P. Petroff, and J. Vuckovic. Coherent excitation of a strongly coupled quantum dot–cavity system. arXiv:0902.2428v1, p. 17, February 2009.
31. A. B. Mundt, A. Kreuter, C. Russo, C. Becher, D. Leibfried, J. Eschner, F. Schmidt-Kaler, and R. Blatt. Coherent coupling of a single $^{40}Ca^{+}$ ion to a high-finesse optical cavity. *Appl. Phys. B*, 76:117–124, June 2003.
32. B. Schumacher and M. Westmoreland. *Quantum Processes Systems and Information*. Cambridge University Press, Cambridge, U.K., 2010.
33. G. Auletta, M. Fortunato, and G. Parisi. *Quantum Mechanics*. Cambridge University Press, Cambridge, U.K., 2009.

34. D. F. Walls and G. J. Milburn. *Quantum Optics*, 2nd edn. Springer, New York, 2008.
35. I. R. Senitzky. Dissipation in quantum mechanics: The harmonic oscillator. *Phys. Rev.*, 119(2):670–679, March 1960.
36. R. Poldy, B. C. Buchler, and J. D. Close. Single-atom detection with optical cavities. *Phys. Rev. A*, 78(1):013640, July 2008.
37. S. Haroche and J.-M. Raimond. *Exploring the Quantum*. Oxford University Press, Oxford, U.K., 2006.
38. J.-L. Basdevant and J. Dalibard. *The Quantum Mechanics Solver*, 2nd edn. Springer Verlag, New York, 2006.
39. D. Budker, D. F. Kimball, and D. P. DeMille. *Atomic Physics: Exploration through Problems and Solutions*, 2nd edn. Oxford University Press, Oxford, U.K., 2008.
40. H. J. Metcalf and P. van der Straten. *Laser Cooling and Trapping*. Springer, New York, 1999.
41. H. Kimble. Strong interactions of single atoms and photons in cavity QED. *Phys. Scripta*, T76:127–137, January 1998.

3

Measurements of Subnanometer Molecular Layers

Maciej Kokot
Gdańsk University of Technology

3.1 Introduction

The nanometer and subnanometer molecular layers have attracted a lot of attention in the last decades. Their abilities to change totally the surface properties by a layer of only one molecule thickness provide great interest in application, production, and measurement of such layers.

One can say that the observation of such thin, and often undetectable by simple chemical analysis, layers is impossible without the use of very sophisticated methods falling within high science. In particular, this is not true. Everyone remembers the soap bubbles from one's childhood. It is impossible to produce bubbles in pure water, but the addition of very little soap results in a beautiful rainbow effect.

Soap molecules are composed of long, hydrophobic chains of carbon (14–18) and hydrogen atoms and a hydrophilic carboxyl group ($-COO^-$) at the other end of the chain. The hydrophobic long chains are expelled outside the water-to-air surface by the water molecules, and two monomolecular layers are formed on the inner and outer surfaces of a bubble (Figure 3.1). Such a molecular layer plays two roles. First, it lowers the surface tension of water, and second, the hydrophobic (greasy) ends of the soap molecules protect the bubble from rapid evaporation.

Another example of a drastic change in surface properties by one-atom-thick layers can be a silicon monocrystal surface. When the surface of a silicon plate is etched in hydrofluoric acid (in order to dissolve the thin native oxide layer), the atoms of silicon on the surface are terminated with hydrogen atoms. This makes such a plate hydrophobic—drops of water do not wet the surface. But when we wait for some time or put the plate into deionized water for a few tens of minutes, the SiH groups on the surface slowly change to SiOH, and the water starts to wet it completely. This effect of only one atom change on the surface is visible by the naked eye (Lehmann 2002).

3.2 Formation Methods of Self-Assembled Monomolecular Layers

3.2.1 Langmuir–Blodgett Films

In the first example mentioned earlier, the spontaneous formation of a monomolecular layer on the phase boundary between water and air can be observed. The important requirement to self-assembly is that the molecules must have rather long shapes and amphiphilic properties (hydrophobic tail–hydrophilic head). Usually, fatty acids are used here. The formation of monomolecular layer of amphiphilic compounds is the basis of the Langmuir–Blodgett (LB) film deposition (Langmuir 1917; Blodgett 1934, 1935; Blodgett and Langmuir 1937).

This technique has been known for almost a 100 years. On the surface of ultraclean water, a very small amount of an amphiphilic compound (usually in an organic solvent immiscible with water) is applied, and after the evaporation of the solvent, the layer is compressed by a small vertical force applied to the movable bar, which reduces the water surface occupied by molecules. The molecules are then compacted and organized into a two-dimensional crystal-like surface with all hydrophobic tails standing vertically up and in near proximity to each other.

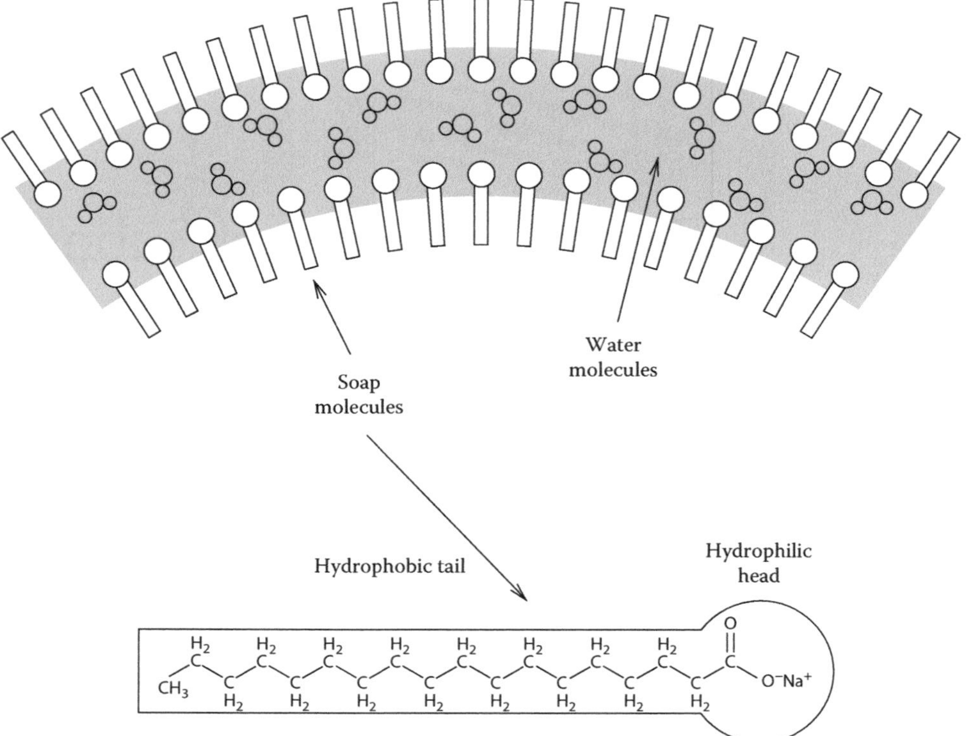

FIGURE 3.1 Example of a soap molecule—sodium stearate (below). The soap bubble cross section (above).

In the next step, a solid plate is pulled up from water, and the monomolecular layer is transferred onto the surface of the plate. In this setup, the surface of the plate should be hydrophilic in order to adsorb the hydrophilic heads of the layer. It is also possible to use a plate with a hydrophobic surface. In this opposite case, the plate is pushed down into the water through the monomolecular layer.

These steps can be repeated many times, which allows for the formation of a film consisting of a specified number of monomolecular layers (Blodgett 1935). This technique is, however, very sensitive to trace amounts of contamination, vibration, and uneven pulling of the substrate. Moreover, the molecular layers are only physisorbed to the substrate.

3.2.2 Thiols on Gold

Another, but a little younger, technique of monolayer formation utilizes the affinity of sulfur to gold (Nuzzo and Allara 1983; Bain et al. 1989; Porter et al. 1987). The precursors of molecular layers are sulfo-organic compounds—alkanethiols (Figure 3.2) or dithiols. Alkanethiols are composed of a hydrocarbon tail bound to an SH group. They are the sulfur analogs of alcohols (instead of the oxygen atom, we have an atom of sulfur). Dithiols are the analogs of ethers.

This method is much simpler than the LB film formation. The most common technique of preparing a thiol monolayer on surfaces of gold, silver, palladium, etc. involves the immersion of freshly prepared or cleaned substrate into a dilute (1–10 mM) ethanoic solution of thiol at room temperature. After a few minutes, a self-assembled monolayer is formed (Figure 3.3). In the additional time of a few hours, the reorganization of the layer provides the maximal density of the molecules and minimizes the density of defects.

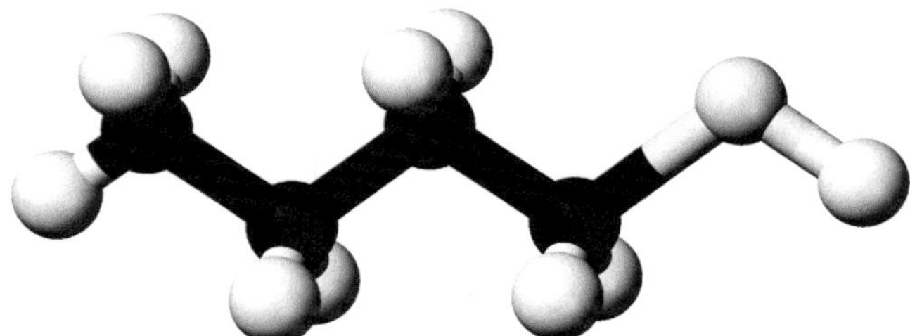

FIGURE 3.2 Ball and stick model of a butanethiol molecule.

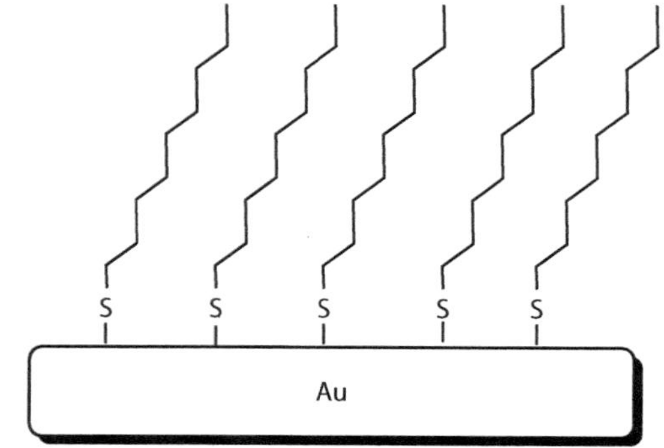

FIGURE 3.3 Schematic view of thiol monolayer on gold.

The most common substrate for the thiol layers is gold. Gold is a relatively inert metal, it does not form stable oxides in air, and its surface resists atmospheric contamination. Its strong specific interaction with sulfur enables anchoring of the thiol molecules on its surface, and the long hydrocarbon chains support the formation of densely packed crystalline monomolecular layer (Porter et al. 1987).

The thiol molecules reflect the arrangement of gold atoms on the surface. Hence, usually, an evaporated thin layer of gold on a silicon monocrystal or a freshly cleaved mica substrate is used. A very interesting property of thiol layers on gold is their *self-healing* ability. The sulfur heads are only chemisorbed to the gold atoms, and this bond is weaker than the strong covalent one. At elevated temperatures (on the order of 340–350 K), the anchored molecules can move to fill the vacancies (Bucher et al. 1994).

A very broad overview of self-assembled layers of thiolates on metals, their properties, and preparation methods can be found in Love et al. (2005).

3.2.3 Silane Monolayers on Oxides

In 1980, Sagiv presented the first self-organizing monomolecular layer using organosilicon compounds (Sagiv 1980). As the layer precursor, *n*-octadecyltrichlorosilane (OTS) was used. The most used group of silane compounds are trichlorosilanes and trialkoxysilanes. These molecules are composed of a long hydrocarbon chain and a head group that is in the form of $-SiCl_3$ or $-Si(OX)_3$, where X is an alkyl group, mainly methyl ($-CH_3$) or ethyl ($-C_2H_5$).

The substrate used here must be hydrophilic with the polar –OH groups on its surface. The mostly used surfaces are glass, silicon with native oxide, silicon nitride, or others. The most popular ones are silicon wafers with a thin, very flat layer of thermal oxide. The silicon wafers (polished thin plates made of a semiconductor-grade silicon monocrystal) are the most often used substrates thanks to their almost atomically flat surface.

The procedure of preparing such monolayers is analogous to the thiols. The clean substrate is immersed into a very dilute solution of silane in an organic solvent (methanol, ethanol, hexane, chloroform, or toluene are the most common). The duration of reaction is within the range from some minutes to a few hours. The process of formation of monolayers is slightly more complicated than that of thiols.

In the first stage (Figure 3.4a), the head groups are adsorbed onto the polar substrate surface. Next (Figure 3.4b), with the aid of traces of water (adsorbed on the surface of the substrate and in the bulk of the solvent), the head groups are hydrolyzed into the reactive high polar trihydroxysilanes ($-Si(OH)_3$). These hydroxysilane groups can form strong covalent bonds with the surface hydroxyl groups with the release of water molecules (Figure 3.4c). The last stage (Figure 3.4d) of the process is condensation between neighboring SiOH groups of the silane molecules. It is supported by van der Waals forces, causing straightening and increasing the organization of hydrocarbon chains (Aswal et al. 2006).

The monomolecular layers of silanes are not as good in quality as thiol layers, which can form two-dimensional crystalline layers on gold substrates. The cause of poor repeatability of the process is its sensitivity to the difficult to control content of trace amount of water which plays a fundamental role in the formation of layers.

3.3 Some Methods of Molecular Layer Detections and Measurements

3.3.1 Water Drop Contact Angle

This is one of the simplest methods of detecting the existence of molecular layers. We all know that very clean glass is completely wetted by pure water. But, after a few hours or days of its exposure to air, a very thin, invisible layer of hydrocarbons or fatty acids is adsorbed due to the air contamination on the glass surface. This makes the surface hydrophobic, and water does not form a layer, but many drops.

The contact angle of such pure water drop can be measured (Figure 3.5), and it provides information about the degree of hydrophobicity of the surface (de Gennes 1985). Various surface groups of atoms produce different contact angles. Some examples of wetting angles are shown in Table 3.1, cited after Aswal et al. (2006).

Sometimes, there are other wetting angles measured. The advancing and receding ones are measured when the solid liquid area increases or decreases (Figure 3.6). The significant (10 or more degrees) difference between them—the hysteresis—provides an information about the surface roughness or its inhomogeneities (de Gennes 1985).

By the change of contact angle, the existence of additional layer can be confirmed, but almost nothing can be said about the layer's thickness. Only the surface atoms of the layer have influence on its wetting properties.

This simple method is used very often, but in some cases does not provide clear results. For example, when we have a clean glass or silica surface, the hydroxyl (–OH) groups are exposed, and the water contact angle is very small (<15°). The application of monomolecular layer with hydroxy or carboxy (–COOH) tail groups produces no change in contact angle (Table 3.1).

3.3.2 Scanning Probe Microscopy

The scanning probe microscopy (SPM) is a broad class of methods of imaging the surfaces with a very high resolution. The image of the investigated surface is formed by moving the probe over the surface line-by-line (scanning) and recording the probe–surface interactions.

The basis of this method is the use of piezoelectric actuators that can execute the movements in the *x*-, *y*-, and *z*-directions with atomic precision (or even more) as a result of the electrical excitations. The second condition needed to obtain atomic

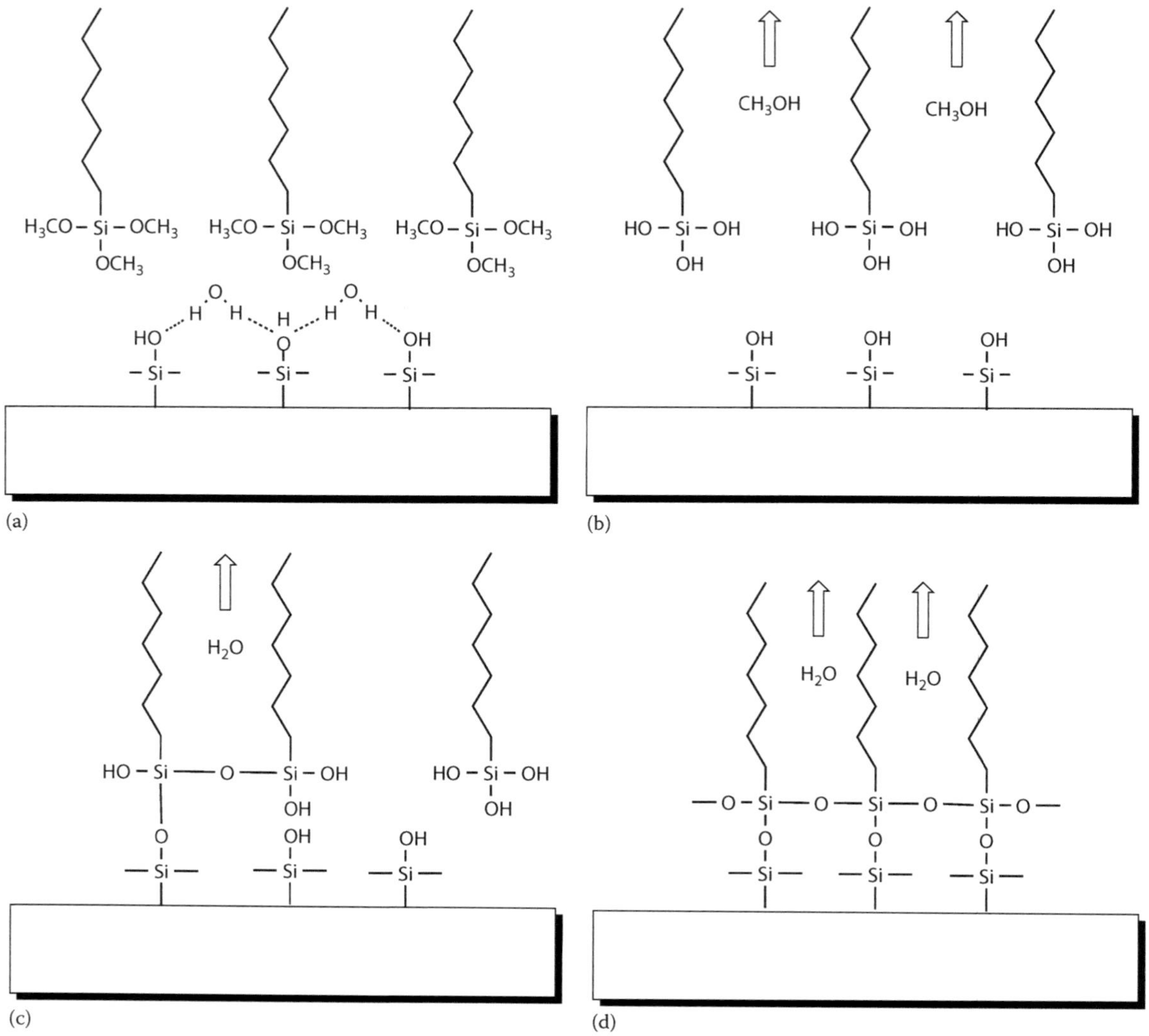

FIGURE 3.4 Schematic process of silane monomolecular layer formation. (a) Adsorbtion of head groups on the surface, (b) hydrolysis of head groups, (c) covalent bonds forming, and (d) cross-bonding.

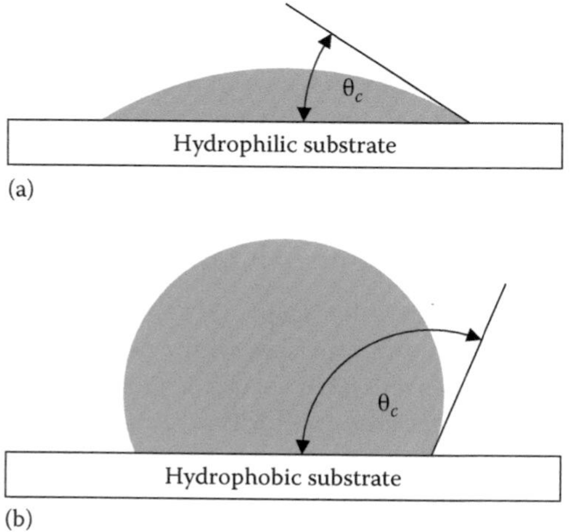

FIGURE 3.5 Water drop contact angle on the hydrophilic (a) and hydrophobic (b) surfaces.

resolution is the choice of short-range interaction (the physical value, e.g., force or electric current, decreases very rapidly with increasing distance) between the probe and the sample.

The first example of SPM is the scanning tunneling microscope (STM) invented in 1981 (Binnig et al. 1982) by the 1986 Nobel Prize laureates Gerd Binnig and Heinrich Rohre. The recorded probe–surface interaction is the tunneling electric current. The very sharp conducting metal tip is scanned over the surface in the proximity of a few tenths of a nanometer. The bias voltage is applied between the tip and the surface. The voltage is too small to pull electrons out of the atom shells, but when the distance is very short the electrons can tunnel through the barrier of potential. This tunneling current increases exponentially (very fast in short range) with decreasing distance between the tip and the surface. The only 0.1 nm distance decrease (which is in the range of the diameter of a single atom) gives about 10-fold current increase.

TABLE 3.1 Water Drop Contact Angles for Various Surface Groups

Surface Group	Contact Angle [°]
$-(CH_2)_n-CH_3$; $n > 10$	110–117
$-(CH_2)_n-CH_3$; $n < 10$	97–108
$-(CH_2)_n-CH=CH_2$; $10 < n > 20$	95–105
$-X$; X = Cl, Br, J	80–89
$-CO_2CH_3$, $-CO_2CH_2CH_3$	73–75
-SCN	73–75
-CN	68–74
-SH, -S, -S-S	65–71
$-SO_2$	50
$-SO_3H$	30
$-NH_2$	36
$-NH_3^+$	42
$-OH$, $-CO_2H$	<15

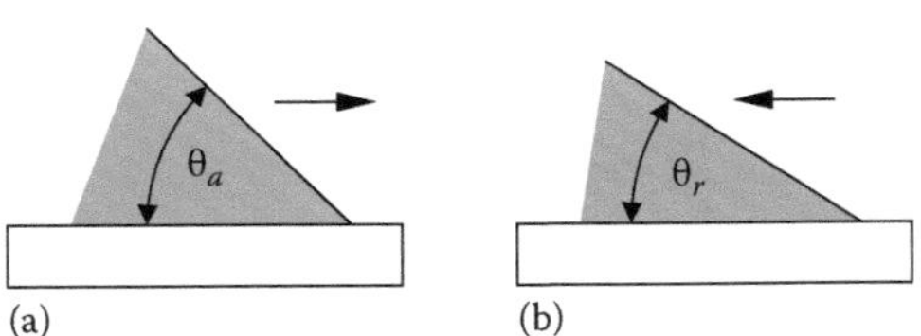

FIGURE 3.6 Contact angles with advancing (a) and receding (b) water drop volume.

There are two main modes of STM operations:

1. The constant height mode: when the tip is scanned with constant distance over the sample and the tunneling current variations are recorded.
2. The constant current mode (most popular): when the tunneling current is kept constant and the changes of the distance needed are the output information.

The change of the current or the distance in the last case reflects the topography and composition of the surface.

The resolution of STM can be very high: 0.01 nm in vertical distance and 0.1 nm in horizontal plane. The main disadvantage of this method is the need of the investigated sample to be electrically conducting (Bai 2000).

The next example of SPM is the atomic force microscopy (AFM). Its construction is similar to that of STM. The probe is a small cantilever with a sharp tip at the end. The cantilever is typically silicon or silicon nitride with the tip radius of curvature on the order of nanometers. When the tip is brought into proximity of a sample surface, forces between the tip and the sample surface lead to a deflection of the cantilever, which in turn is recorded by the laser beam reflected from the top surface of the cantilever into an array of photodiodes. In most cases, the constant force mode is used. The tip-to-sample distance is adjusted in order to keep the constant deflection of the cantilever.

The resolution of AFM is not as high as that of STM but is still in an atomic range. In contrast to the STM, the AFM method can be used also with nonconducting surfaces. Figure 3.7 shows example scans of a silicon nitride layer on the silicon plate and an octadecyltrimethoxysilane monomolecular layer prepared on the previous substrate.

3.4 ISFET Transconductance Method of Subnanometer Layer Detection and Measurements

In this section, a new method of detection and measurement with subnanometer resolution of layers adsorbed or bonded to the gate dielectric of the ion selective field effect transistor (ISFET) will be presented. This method was briefly introduced for the first time in a paper by Kokot (2011).

3.4.1 Ion Selective Field Effect Transistor—Introduction

The ISFET sensor has been known for almost 40 years (Bergveld 2003). The main field of its application was the measurement of pH (hydrogen ion concentration) in solutions. Also, in recent times, it has been used in the detection or measurement of the concentration of various species: other ions, organic substances, DNA, or proteins, usually with the aid of specially modified thin membranes on the top of the sensor. All sensors of this kind are potentiometric—the change in the surface potential (which translates to the change in ISFET's threshold voltage) is measured.

The ISFET sensor is similar to the metal oxide semiconductor field effect transistor (MOSFET) but without gate metalization. Figure 3.8 shows a very simplified explanation of the MOSFET and ISFET operation basis.

Transistor MOS is composed of two n-type doped regions, forming source and drain on the p-type silicon substrate. Over the region between the source and drain, a conducting metallic gate electrode is formed. This gate electrode, which plays the control role, is isolated from the silicon substrate (and the rest of the device) by the thin dielectric (usually SiO_2) layer.

Figure 3.8a shows the MOSFET in the off state. Drain–source voltage is positive, but there can be no electric current floating from drain to source. In the n-doped regions of drain and source, the current carriers are electrons; in the p-doped substrate between drain and source, the current carriers are holes (the electron vacancies in the silicon crystal, which can be considered just like electrons, but with positive charge). So, the electrons from the source can flow into the p layer, but they recombine with the holes there, and there will be no electrons to flow into the drain region.

Figure 3.8b shows a different state. The gate electrode is positively polarized, and the electric field attracts the minority electrons from the bulk of the substrate to the surface and repels the holes. A conducting channel is formed between the source and drain, and the drain current can flow now. When the gate–source voltage is more positive, the conducting channel is formed deeper,

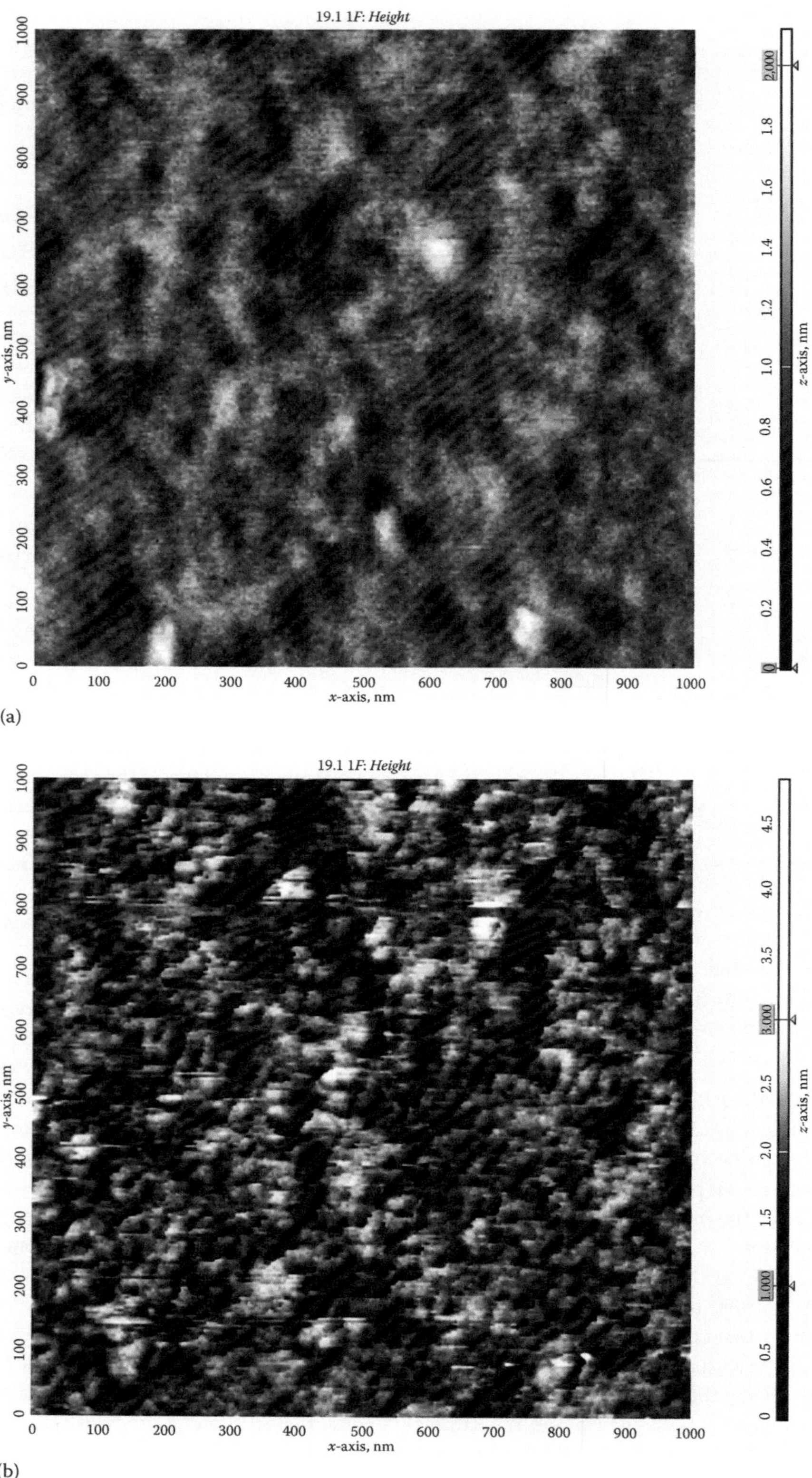

FIGURE 3.7 AFM picture of a bare silicon nitride layer on silicon plate (a), and the same silicon nitride layer with the monomolecular layer of octadecylsilane on it (b).

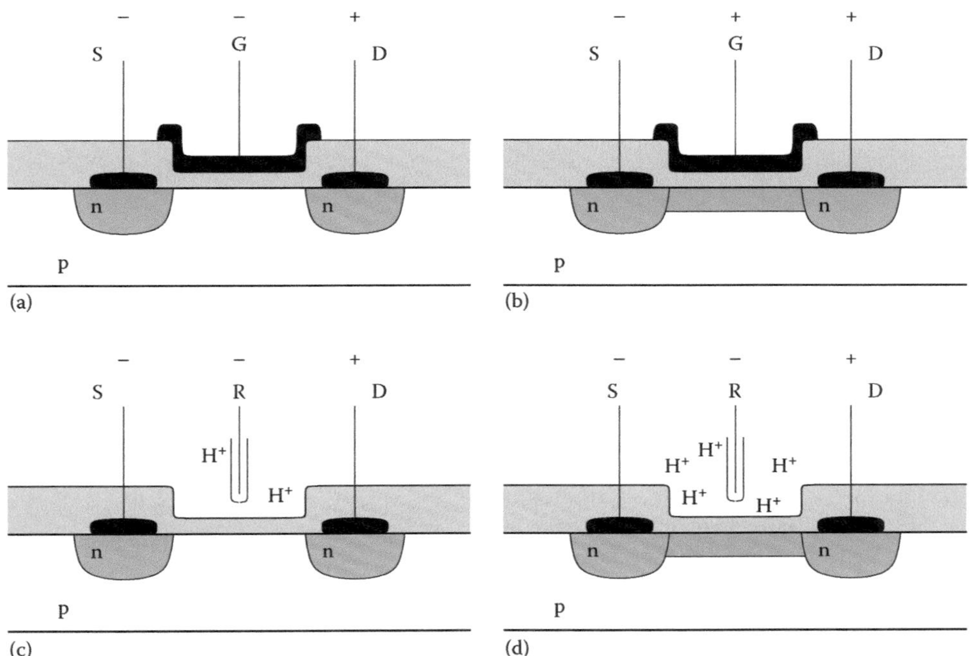

FIGURE 3.8 Schematic comparison of the MOSFET (above) and ISFET (below) devices: (a) MOSFET without the conducting channel due to low gate voltage. (b) MOSFET with the conducting channel formed by the higher gate voltage. (c), (d) Formation of conducting channel of ISFET due to rising hydrogen ion concentrations with the reference electrode (which plays the role of the gate) voltage not changed.

and the drain current will be higher with the same drain–source voltage. The gate–source voltage over which the conducting channel is formed is known as the threshold voltage V_t.

In Figure 3.8c and d, the ISFET sensor is shown in the *off* and *on* states, respectively. The gate of this device is separated from the sensor in the form of an electrochemical reference electrode inserted in a solution that is in contact with the thin gate dielectric layer. The reference electrode is a special kind of electrochemical electrode that provides constant potential drop between its contact and the solution, independent of the solution composition.

There is no change in the gate–source voltages between the figures, but the hydrogen ion concentration (pH) of the solution is different. In low pH (acid solution), the hydrogen ions interact with the dielectric layer and form a positively charged layer modifying the surface potential. The electrons are now attracted to the surface and the conducting channel is formed, which is shown in Figure 3.8d. The hydrogen ion concentration has a similar effect as the gate electrode potential change.

The general, simplified expression for the drain current of the ISFET in nonsaturated mode is qualitatively the same as that for a MOSFET:

$$I_d = \beta\left(\left(V_{rs} - E_{ref} - V_t\right)V_{ds} - \frac{1}{2}V_{ds}^2\right), \tag{3.1}$$

with

$$\beta = C_{ox}\mu\frac{W}{L}, \tag{3.2}$$

where

- V_{rs} is the voltage between the reference electrode (which plays a gate role) and source
- E_{ref} is the reference electrode constant electrochemical potential
- V_{ds} is the source–drain voltage
- C_{ox} is the oxide (gate insulation) capacity per unit area
- W and L are the width and length of the channel, respectively
- μ is the electron mobility in the channel

The threshold voltage V_t is not constant in contrast to the MOSFET. V_t is modified by the electric charge, which mainly depends on the ion concentration in the solution, formed on the surface of the gate dielectric:

$$V_t = V_{to} + K\text{pH}, \tag{3.3}$$

where

- V_{to} is the pH-independent threshold voltage
- K is the sensitivity parameter

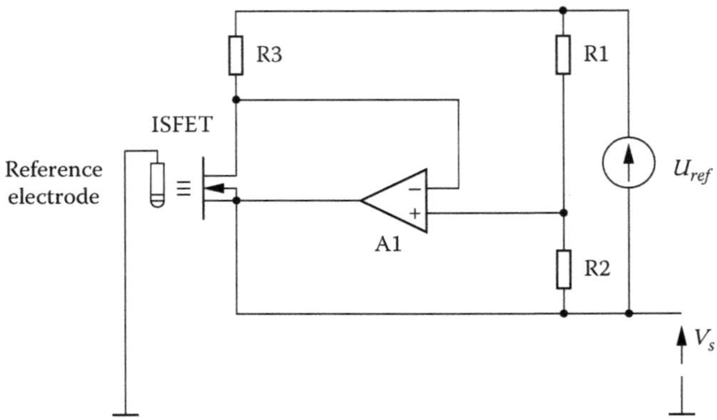

FIGURE 3.9 Typical ISFET application electrical circuit.

Typically, in most of the ISFET applications, the drain–source voltage is kept constant, and the drain current is also kept constant by the reference electrode–source voltage adjustment, which reflects the changes in the threshold voltage. The output signal V_{rs} is proportional to the electric charge on the surface of the gate dielectric (Figure 3.9).

When, instead of the reference electrode, a simple metal conductor in the solution would be used as the ISFET's gate electrode, Equation 3.1 takes the form

$$I_d = \beta\left(\left(V_{gs} - E_{gate} - V_t\right)V_{ds} - \frac{1}{2}V_{ds}^2\right), \tag{3.4}$$

where E_{gate} is the electrochemical potential of the metal gate electrode, which in contrast to E_{ref} is not constant, and depends on the solution composition, ion or oxygen concentrations, etc. The output signal V_{gs} now reflects not only the electric charge on the dielectric surface but also the E_{gate} variations. So, the reference electrode with constant electrochemical potential is a must in the typical ISFET applications (Kokot 2011).

3.4.2 ISFET Transconductance Method Basis

The gate capacitance per unit area C_{ox} has always been treated as constant. The gate dielectric thickness and material are defined once in production time. But, when the gate dielectric is open to the ambient, some species can attach to it and change the overall capacitance. With classical bias circuits, the detection of such species is unreliable—there can be some charges associated with the adsorbed species and, as can be seen from Equations 3.2 and 3.3, there is no difference between V_t (charge) and C_{ox} (thickness of gate dielectric) changes in the V_{rs} output signal.

Differentiating Equation 3.1 with respect to V_{rs} gives the transconductance

$$g_m = \frac{dI_d}{dV_{rs}} = C_{ox}\mu\frac{W}{L}V_{ds}, \tag{3.5}$$

and differentiating Equation 3.4 with respect to V_{gs} gives exactly the same value of transconductance:

$$g_m = \frac{dI_d}{dV_{gs}} = C_{ox}\mu\frac{W}{L}V_{ds}, \tag{3.6}$$

which, when the working point of ISFET is constant (I_d and V_{ds} are constant), depends only on the gate dielectric capacitance C_{ox}. Moreover, the V_t and E_{gate} variations (pH-dependent, or when the gate electrode potential in the solution is not constant) have no influence.

Because the transconductance is the first derivative of the drain current with respect to the driving voltage, there is no need for the reference electrode with stable electrochemical potential in order to detect by the transconductance changes observations any species or any additional layer attached to the gate dielectric.

In Figure 3.10, a simplified diagram of an ISFET sensor (with constant drain current and drain–source voltage provided by the additional electronic circuit, not shown here) in the solution and with an additional layer on the gate dielectric is shown. The potential levels are shown with solid lines for the two pH values. Inside the dielectric, the potential levels (common to all pH values) are shown with a single solid line. Additionally, the dotted line shows the potential values as they would be without the additional layer on the gate dielectric. The electrochemical double layer is then shifted to the right by the now nonexistent additional layer thickness (not shown here for the clarity of the figure).

The dashed lines in proximity of the potential solid lines show small signal potential variations that cause small signal drain current variations dI_d (5). The channel depth (its resistance) depends on the electric field in the gate dielectric at the channel side; the small variations of this electric field dE (3) are the causes of small channel resistance changes dR (4), which are reflected in the dI_d current part. It is clear that, in order to achieve the said constant electric field variations dE (which implies constant dI_d small signal), the potential changes dV_0 (2) on the top of the original gate dielectric and little greater dV (1) on the top of the additional dielectric layer are needed.

Summarizing, as the double layer, solution, and gate (or electrochemical reference) electrode with its own double layer are electrically conducting, there are no further changes of dV on its way (to the left side of the figure) to the gate electrode. The measured transconductance dI_d/dV depends only on the total dielectric thickness (the original gate dielectric and the additional layer) and is independent of the pH-dependent voltage drops on the dielectric–solution interface and of the voltage drop on the reference (constant E_{ref} from Equation 3.1) or gate (nonconstant E_{gate} from Equation 3.4) electrode–solution interface (Kokot 2011).

Equation 3.6 shows that with the constant operation point (I_d and V_{ds} constant) of the ISFET, its transconductance is proportional to the gate dielectric capacitance per surface area C_{ox}:

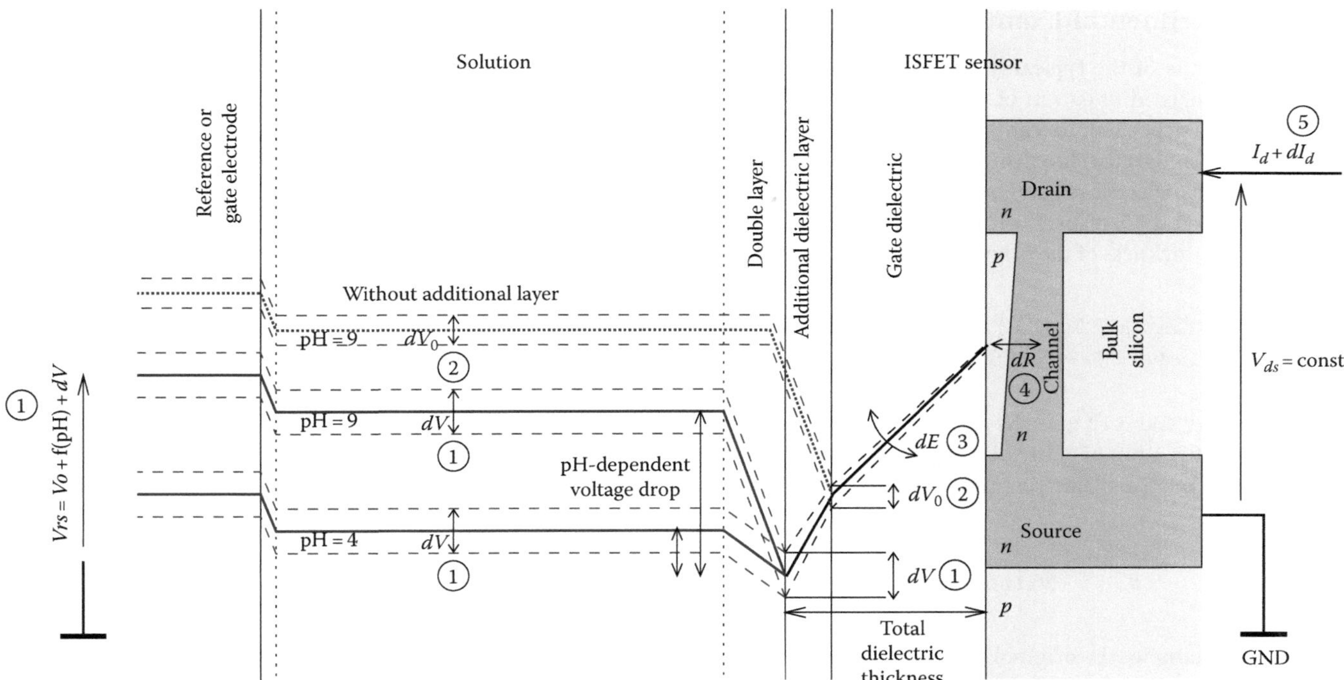

FIGURE 3.10 Simplified diagram of the ISFET operations in solutions. The solid lines show potential levels for various pH. The single solid line inside the dielectric (common to all pH values) are the potentials in the gate dielectric. Dashed lines indicate small signal part of the potentials. The dotted line indicates potentials without the additional dielectric layer. (From *Sens. Actuat. B: Chem.*, 157(2), Kokot, M., Measurement of subnanometer molecular layers with ISFET without a reference electrode dependency, 424–429, Copyright 2011 with permission from Elsevier.)

$$C_{ox} = \frac{\varepsilon_{ox}\varepsilon_0}{d_{ox}}, \tag{3.7}$$

where

- d_{ox} is the dielectric thickness
- ε_{ox} is the relative dielectric constant
- ε_0 is the dielectric constant of vacuum

When the gate dielectric is evenly coated with an additional layer, the total capacitance of the dielectric C_{oxl} is equal to the serial connection of C_{ox} and C_l, the additional layer capacitance:

$$C_{oxl} = \frac{\varepsilon_0}{d_{ox}/\varepsilon_{ox} + d_l/\varepsilon_l}, \tag{3.8}$$

where d_l and ε_l are thickness and dielectric constant of the additional layer, respectively.

The additional layer is much thinner than the original gate dielectric, so the changes of transconductance are very small. It is convenient to consider the relative change

$$\frac{\Delta g_m}{g_{ml}} = \frac{g_m - g_{ml}}{g_{ml}} = \frac{C_{ox} - C_{oxl}}{C_{oxl}}, \tag{3.9}$$

where Δg_m is the transconductance drop due to the additional layer. With Equations 3.7 and 3.8, the thickness of the additional layer is

$$d_l = d_{ox}\frac{\varepsilon_l}{\varepsilon_{ox}}\frac{\Delta g_m}{g_{ml}} \approx d_{ox}\frac{\varepsilon_l}{\varepsilon_{ox}}\frac{\Delta g_m}{g_m}. \tag{3.10}$$

Assuming that g_m and g_{ml} are very close, the approximation error will be small. For example, with 1 nm of layer over 100 nm (typical value) of gate dielectric, the error will be in the range of 1%.

Writing Equation 3.10 in another form, it can be said that the *de facto* measured value is the electrical thickness (the physical thickness divided by the dielectric constant—the thickness of the equivalent vacuum layer) of the additional layer d_{el} in relation to the original gate dielectric electrical thickness d_{eox}:

$$\frac{d_l/\varepsilon_l}{d_{ox}/\varepsilon_{ox}} = \frac{d_{el}}{d_{eox}} = \frac{\Delta g_m}{g_m}. \tag{3.11}$$

Equation 3.11 is the base of the ISFET transconductance method of measurement or detection of subnanometer layers.

3.4.3 Experimental Examples

Considering one of the typical ISFET constructions with gate dielectric composed of 67 nm of silicon dioxide and 67 nm of silicon nitride deposited, we can determine its sensitivity to the deposited (or subtracted) layer thickness.

The relative dielectric constants of silicon dioxide and silicon nitride are 3.9 and 7.5, respectively (Sze and Ng 2006). Therefore, the electrical thickness of the original gate dielectric is

$$d_{eox} = \frac{d_{ox}}{\varepsilon_{ox}} = \frac{d_{SiO_2}}{\varepsilon_{SiO_2}} + \frac{d_{Si_3N_4}}{\varepsilon_{Si_3N_4}} = \frac{67\,\text{nm}}{3.9} + \frac{67\,\text{nm}}{7.5} = 26.1\,\text{nm}. \quad (3.12)$$

Equations 3.11 and 3.12 give the relative changes of transconductance for the 1 nm of electrical thickness (the real thickness divided by its relative dielectric constant) of the detected layer:

$$\frac{\Delta g_m / g_m}{d_{el}} = \frac{1}{26.1\,\text{nm}} = 3.8\%/\text{nm}. \quad (3.13)$$

In the first example, the original upper layer of silicon nitride gate dielectric was etched sequentially for the time of 100 s in 1% solution of hydrofluoric acid. With each etching, the gate nitride dielectric layer was 0.6 nm thinner (Lehmann 2002). Figure 3.11 shows the measured transconductance values as a function of the decreasing thickness of the silicon nitride. The measurements were repeated five times for three various pH solutions, and then the next etching and the next set of measurements were done.

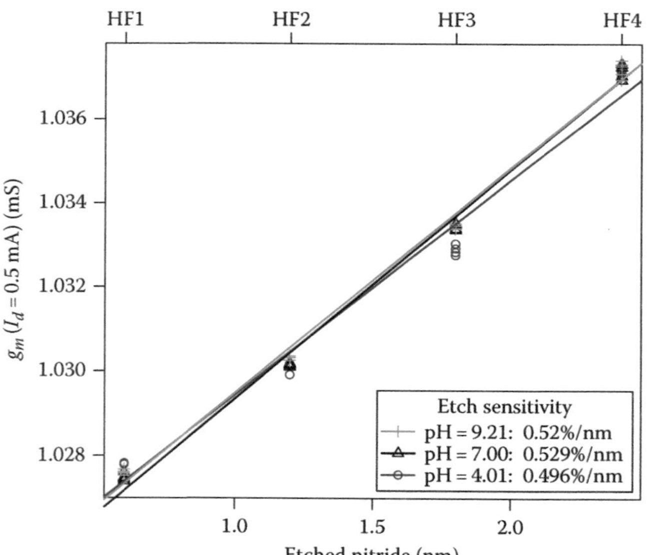

FIGURE 3.11 Transconductance values—g_m vs. etched layer thickness for various pH values. I_d = 0.5 mA, V_{ds} = 0.5 V. (Modified from *Sens. Actuat. B: Chem.*, 157(2), Kokot, M., Measurement of sub-nanometer molecular layers with ISFET without a reference electrode dependency, 424–429, Copyright 2011 with permission from Elsevier.)

It can be seen that g_m is (within the measurement accuracy) insensitive to pH variations, but the dielectric thickness changes with the resolution of a tenth of a nanometer can be clearly recognized. The small differences between points within each etch can be explained by small temperature variations and finite measured data accuracy.

The 1 nm of silicon nitride layer etched is an equivalent of 7.5 nm of electrical thickness, so the measured sensitivity of about 0.52%/nm (from Figure 3.11) gives the relative changes of transconductance for the 1 nm of electrical thickness equal to 3.9%/nm. A very good correlation with the theoretical sensitivity shown by Equation 3.13 can be observed here.

In the second example, the octadecyltrimethoxysilane monomolecular layer was produced on the top of the gate dielectric stack of ISFET. The length of the molecule bonded to the surface is 2.62 nm, and the reported film thickness of such a layer is slightly lower, near 2.52 nm, which is the effect of the hydrocarbon tails not being fully perpendicular to the dielectric surface. The dielectric constant of such silanes is about 2.6 (Aswal et al. 2006).

The electrical thickness of the layer is

$$d_{el} = \frac{d_l}{\varepsilon_l} = \frac{2.52\,\text{nm}}{2.6} = 0.97\,\text{nm}, \quad (3.14)$$

and the relative decline of transconductance, which is an effect of the additional layer, should be

$$\frac{-\Delta g_m}{g_m} = \frac{\Delta g_m / g_m}{d_{el}} \cdot d_{el} = 3.8\%/\text{nm} \cdot 0.97\,\text{nm} = 3.7\%. \quad (3.15)$$

The observed drop in transconductance due to additional silane layer, shown in Figure 3.12, is about 10 μS, which gives 0.96%. This value is only 26% of that expected (Equation 3.15). There is a need to explain such a difference.

The observed decrease in transconductance is dependent on the electrical thickness of the layer and not directly on its real thickness. Only the ratio of the thickness and dielectric constant can be measured, as indicated by Equations 3.11 and 3.14. In order to know one component, the value of another must be determined. The main suspect here is the dielectric constant of the silane layer, which appears to be over three times higher than the value known from the literature. The AFM picture of such silane layer in Figure 3.7b shows that the layer is neither ideally flat nor tightly packed and forms some kind of small aggregates.

Because measurements are made in aqueous medium, the spaces between the aggregates of hydrocarbon tails can be filled with water (with a high dielectric constant: 80) molecules. Simple calculations show that as small as 8.5% water content in the layer can be responsible for the increase of effective dielectric constant of the silane layer over three (1/0.26) times.

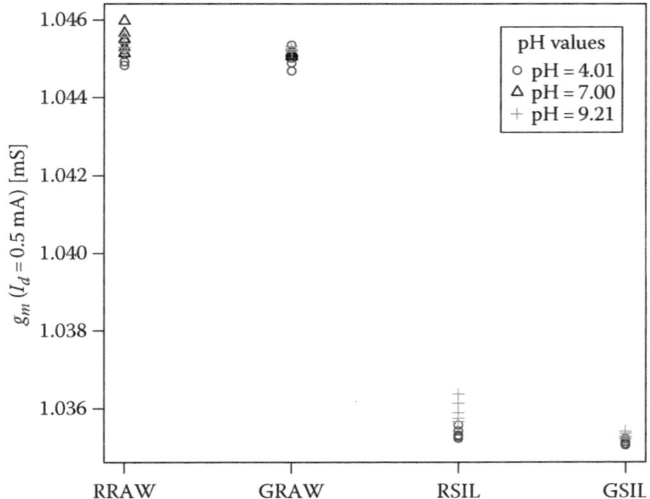

FIGURE 3.12 Transconductance change between the silane layer deposition. RRAW and GRAW without the octadecylsilane layer deposited; RSIL and GSIL with the octadecylsilane layer deposited. RRAW and RSIL with the reference electrode and GRAW and GSIL without. I_d = 0.5 mA, V_{ds} = 0.5 V. (From *Sens. Actuat. B: Chem.*, 157(2), Kokot, M., Measurement of sub-nanometer molecular layers with ISFET without a reference electrode dependency, 424–429, Copyright 2011 with permission from Elsevier.)

3.5 Final Remarks

We have discussed here only some arbitrarily chosen methods of measuring molecular layer properties. The simplest, semiquantity one is the water drop contact angle, demanding practically no specialist equipment. The methods that are most widely used are the SPM methods, but they demand very complicated measuring equipment; another popular method (not discussed here) is that of ellipsometry, which, contrary to the aforementioned methods, allows measuring of the thickness of subnanometer layers. Much space has been devoted to the promising new ISFET transconductance method (worked out by the author of this chapter), which allows to measure the thickness (mainly electrical thickness) of the layer, produced on the gate dielectric of an ISFET sensor, without the necessity of using too complicated equipment. The sensitivity of this method is high enough to detect the mean thickness variations below 0.1 nm, which is in the range of single atom diameter. Since the measurements are performed in electrolytic liquid medium, the transconductance method has the potential to detect in situ layer formation or attachments of various molecular species.

References

Aswal, D. K., S. Lenfant, D. Guerin, J. V. Yakhmi, and D. Vuillaume. 2006. Self assembled monolayers on silicon for molecular electronics. *Anal. Chim. Acta* 568(1–2) (May 24): 84–108. doi: 10.1016/j.aca.2005.10.027.

Bai, C. 2000. *Scanning Tunneling Microscopy and Its Applications*. Berlin, Germany: Springer.

Bain, C. D., E. Barry Troughton, Y. T. Tao, J. Evall, G. M. Whitesides, and R. G. Nuzzo. 1989. Formation of monolayer films by the spontaneous assembly of organic thiols from solution onto gold. *J. Am. Chem. Soc.* 111(1): 321–335. doi:10.1021/ja00183a049.

Bergveld, P. 2003. Thirty years of Isfetology: What happened in the past 30 years and what may happen in the next 30 years. *Sens. Actuat. B: Chem.* 88(1) (January 1): 1–20. doi:10.1016/S0925-4005(02)00301-5.

Binnig, G., H. Rohrer, Ch. Gerber, and E. Weibel. 1982. Tunneling through a controllable vacuum gap. *Appl. Phys. Lett.* 40(2) (January 15): 178–180. doi:10.1063/1.92999.

Blodgett, K. B. 1934. Monomolecular films of fatty acids on glass. *J. Am. Chem. Soc.* 56(2): 495. doi:10.1021/ja01317a513.

Blodgett, K. B. 1935. Films built by depositing successive monomolecular layers on a solid surface. *J. Am. Chem. Soc.* 57(6): 1007–1022. doi:10.1021/ja01309a011.

Blodgett, K. B. and I. Langmuir. 1937. Built-up films of barium stearate and their optical properties. *Phys. Rev.* 51(11): 964–982. doi:10.1103/PhysRev.51.964.

Bucher, J.-P., L. Santesson, and K. Kern. 1994. Thermal healing of self-assembled organic monolayers: Hexane- and octadecanethiol on Au(111) and Ag(111). *Langmuir* 10(4): 979–983. doi:10.1021/la00016a001.

de Gennes, P. G. 1985. Wetting: Statics and dynamics. *Rev. Mod. Phys.* 57(3): 827–863. doi:10.1103/RevModPhys.57.827.

Kokot, M. 2011. Measurement of sub-nanometer molecular layers with ISFET without a reference electrode dependency. *Sens. Actuat. B: Chem.* 157(2): 424–429. doi:16/j.snb.2011.04.079.

Langmuir, I. 1917. The constitution and fundamental properties of solids and liquids. II. Liquids.1. *J. Am. Chem. Soc.* 39(9): 1848–1906. doi:10.1021/ja02254a006.

Lehmann, V. 2002. *The Electrochemistry of Silicon: Instrumentation, Science, Materials and Applications*. Weinheim, Germany: Wiley-VCH.

Love, J. C., L. A. Estroff, J. K. Kriebel, R. G. Nuzzo, and G. M. Whitesides. 2005. Self-assembled monolayers of thiolates on metals as a form of nanotechnology. *Chem. Rev.* 105(4): 1103–1170. doi:10.1021/cr0300789.

Nuzzo, R. G. and D. L. Allara. 1983. Adsorption of bifunctional organic disulfides on gold surfaces. *J. Am. Chem. Soc.* 105(13): 4481–4483. doi:10.1021/ja00351a063.

Porter, M. D., T. B. Bright, D. L. Allara, and C. E. D. Chidsey. 1987. Spontaneously organized molecular assemblies. 4. Structural characterization of N-alkyl thiol monolayers on gold by optical ellipsometry, infrared spectroscopy, and electrochemistry. *J. Am. Chem. Soc.* 109(12): 3559–3568. doi:10.1021/ja00246a011.

Sagiv, J. 1980. Organized monolayers by adsorption. 1. Formation and structure of oleophobic mixed monolayers on solid surfaces. *J. Am. Chem. Soc.* 102(1): 92–98. doi:10.1021/ja00521a016.

Sze, S. M. and K. K. Ng. 2006. *Physics of Semiconductor Devices*. Hoboken, NJ: John Wiley & Sons, Inc.

4

Electrostatic Potential Mapping in Electron Holography

Lew Rabenberg
The University of Texas at Austin

4.1 Introduction

4.1.1 Overview

The phase shift of an electron wave is directly proportional to the electrostatic potential it encounters, integrated over its trajectory. Electron holographic mapping techniques extract these phase shifts as distributed over the specimen. A holographic map is an image of the electrostatic potential as projected through the specimen and its surroundings. This is the essence of this chapter.

Holography is implemented within a transmission electron microscope where the specimen can form high magnification images and electron diffraction patterns. The flexibility of an electron-optical column is essential in order to manipulate coherent electron waves. Forming a coherent beam, illuminating the specimen plane, creating a carrier wave, and magnifying the resulting hologram are all accomplished within slightly modified commercially available (scanning) transmission electron microscopes. This is true, even though holography relies on interference phenomena that are essentially lens-free. In practice, lenses must be involved, and lens aberrations are still important for lattice-resolution holography (Allard and Völkl, 1999). However, most electrostatic potential maps are limited by signal-to-noise issues, and wave front distortions are routinely subtracted away. In most cases, it is not necessary to address the aberrations explicitly.

The strong interactions of electrons with matter, electrostatic potentials, or magnetic potentials enable useful holography signals to be obtained from specimens that are as small as several hundred picometers. Electron holography (EH) is inherently quantitative because the relationship between phase shift and electrostatic potential is linear. Imaging is intrinsic to the holography process. Thus, holography is eminently suited to mapping potentials distributed over scales corresponding to tens to thousands of atomic diameters.

4.1.2 Coherent Interference

Electron holography in its broadest sense is any technique that uses the interference patterns from coherent electron waves in order to extract the phase and amplitude information from the waves that have been scattered by a specimen. In traditional amplitude contrast transmission electron microscopy (TEM), two electron waves, $A_1e^{i\phi 1}$ and $A_2e^{i\phi 2}$, are detected incoherently, that is,

$$I_{inc} \approx \left|A_1e^{i\phi 1}\right|^2 + \left|A_2e^{i\phi 2}\right|^2 = A_1^2 + A_2^2, \tag{4.1}$$

and all phase information is lost. In holography, they are added coherently:

$$I_{coh} \approx \left|A_1e^{i\phi 1} + A_2e^{i\phi 2}\right| = A_1^2 + A_2^2 + A_1A_2\cos\{\phi'\}, \tag{4.2}$$

so that a cosine term carrying phase information is included. In these expressions, A's are wave amplitudes, ϕ's are phases, I's are detected intensities, and $\phi' = \phi_1 - \phi_2$. In the experimental hologram, displacements of the cosine fringes correspond to changes in the phase of the electron waves caused by the interaction of the electron beam with the specimen. The primary causes of the phase changes in the electron wave are the averaged potential within the specimen and the atomic potentials when projected along the electron beam direction. Phase shifts associated with any electrostatic or magnetic potentials are superimposed on these strong interactions. In a complete reconstruction of the hologram, an amplitude image and a phase image are formed. The amplitude image is a simple product of the amplitude of a wave coherently scattered by the specimen with a reference wave that bypassed the specimen. The phase image is a map of the projected electromagnetic potential. In actual practice, this brief introduction is modified by a host of theoretical and instrumental parameters.

4.1.3 Brief History

Electron holography is not new (Tonomura, 1987; Tonomura, 1992; Möllenstedt, 1999; Völkl, 1999; Lichte and Lehmann, 2007; McCartney and Smith, 2007), but it is only in the last two decades, with the advent of coherent electron sources (Allard and Völkl, 1999), linear position-sensitive electron detectors (Krivanek and Mooney, 1993; de Ruijter and Weiss 1993), high-speed computers, flexible software (Holoworks©, Digital Micrograph© (Gatan, Inc., the United States) and self-authored Digital Micrograph© scripts), and a book of comprehensive review articles (Völkl, 1999) that it has become a useful technique outside the specialist labs.

Gabor proposed EH (Gabor, 1948, 1949) in order to extend the resolving power of electron beam instruments by totally eliminating the lenses along with their aberrations. Since his invention, electron lenses and microscopes have improved in parallel with EH so that the best current spatial resolution limits for both electron microscopy and EH are about 100 pm. It is safe to say that the magnitude of work devoted to the development of TEM, high-resolution electron microscopy (HREM), scanning transmission electron microscopy (STEM), and analytical electron microscopy (AEM) (Reimer, 1993; de Graef, 2003; Fultz and Howe, 2008; Williams and Carter, 2009; Erni, 2010) has been immense compared to that of EH. The lack of attention to EH has had the effect that many of the modes of EH that can potentially be used in electron-optical columns (Cowley, 1992; Herring and Pozzi, 1999) have never been thoroughly investigated, much less exploited. Some recent resolution-enhancement techniques (phase shifting: Ru et al., 1991; Yamamoto et al., 2010; Double biprism: Harada et al., 2004; Ikeda et al., 2011; multiple beam EH: Kawasaki et al., 1993; Hirayama et al., 1995) were anticipated, but went into obscurity because of limitations imposed by instrument instability, lack of linear area-sensitive detectors, and convenient computational facilities. From this perspective, one might expect that improvements in EH will exceed those of the broader electron microscopy (EM) techniques in the near future.

In fact, there are reasons to believe that EH may advance more rapidly than EM in the next few years. Both techniques are currently limited by microscope structural instabilities that manifest themselves in vibrations, specimen shifts, and the like. HREM has the added burden of chromatic aberration; recent advances in this area have come at a great cost in instrumental complexity (Haider et al., 2008, 2009) and severe constraints on power supplies, electron-optical columns, and the entire microscope environment. New developments in holography-related techniques can push well into the picometer range for certain specimens (Humphry et al., 2012). In any case, the fact that EH has finally escaped from the specialists' labs increases the probability that new, creative minds will push the EH limits.

4.1.4 Plan

This chapter focuses on EH and related techniques that can be used to map potential variations across a thin specimen. As a first order of business, it will try to establish principles and limitations of EH that are essential for TEM-trained individuals who are new to EH, as well as for curious scientific nonpractitioners. Second, this chapter will review the progress specific to potential mapping in EH, highlighted by a special emphasis on mapping in inhomogeneous nanowires. Finally, a few new directions are discussed in the contexts of their historical antecedents. As noted earlier, there are several excellent review articles and books on EH; it will not be necessary, nor appropriate, to do a comprehensive review of the literature here.

4.2 Essentials

4.2.1 Overview

Deriving phase information from short-wavelength radiation requires a coherent reference wave in order to form an interference pattern with the wave that passes through the specimen. Off-axis EH (Figure 4.1) is the most common way to do this. In off-axis EH, the electron beam is caused to pass near the edge of the sample such that one portion of the incident electron wave intercepts the sample and one portion does not. The specimen wave and the reference wave propagate through the objective lens (Lorentz or minilens for low magnifications) and form a crossover. A positively charged electron biprism, inserted near the usual back image plane of the objective lens of the TEM, causes the two portions of the beam to overlap and form the holographic fringe pattern (Figure 4.2). The fringes are magnified by subsequent lenses and recorded by a suitable, linear, digital camera.

After the interference pattern—the hologram—is recorded, a series of digital manipulations of the hologram fringes yields a map of the projected phase modulations imposed by the specimen on the electron wave. A digital fast Fourier transform (FT) (displayed as a power spectrum, i.e., $|FT|^2$) of the hologram consists of a central region that corresponds to the nominal spatial frequencies in the fringe pattern,

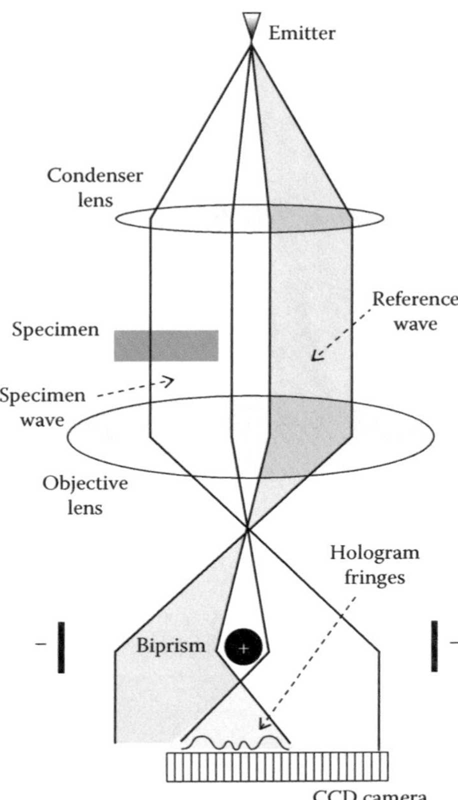

FIGURE 4.1 Schematic diagram showing the setup for off-axis EH. The projector lenses following the biprism are omitted for clarity. The emitter and condenser system create a parallel beam that illuminates the specimen (the specimen wave) and an equal area of vacuum (the reference wave). An image of the specimen plane is formed near, but not at, the biprism. The two beams are overlapped in the plane of the camera where they form a pattern of interference fringes—the hologram.

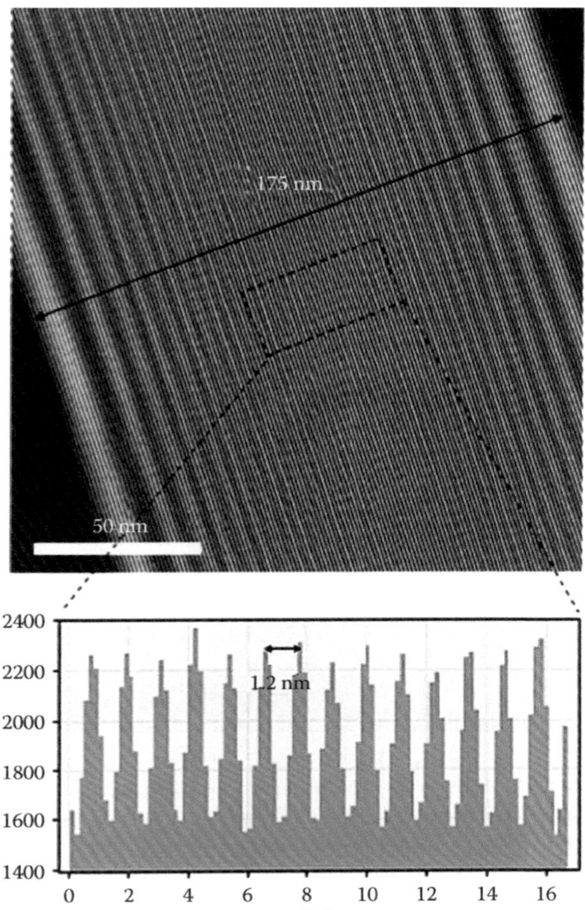

FIGURE 4.2 A typical hologram with no specimen and a line scan average across the box. In this case, the fringes are ~1.2 nm apart giving a resolution approaching 3.6 nm. The wide fringes at the perimeters are Fresnel edge fringes arising from the shadow image of the biprism. The useful area of the hologram excludes the Fresnel fringes. (From Chung, J. and Rabenberg, L., *J. Mater. Res.*, 21, 1215, 2006b.)

with two side-bands resulting from the interference of the specimen wave with the oblique carrier wave. A digital "aperture" that surrounds one of the side-bands and excludes the central region and the other side-band is inserted into the Fourier pattern. The remaining side-band is moved, digitally, to the center of the Fourier space and subjected to an inverse FT. The real part of this inverse FT is the amplitude image; the imaginary part is the phase image (Figures 4.3 and 4.4).

Described in two paragraphs, the two major stages of EH—recording and reconstruction—seem to be relatively straightforward. However, these procedures developed over more than a quarter century; many details and requirements must be addressed to do successful phase mapping, understand its limitations, examine current trends, and predict future developments. This section starts at the top, with the coherent electron emitter, passes through and around the specimen, examines the fringes, reconstructs the phase and amplitude, and constructs maps of the desired potentials and their applications. A final few paragraphs discuss other modes of EH that are likely to impact potential mapping in the foreseeable future.

4.2.2 Electron Source

4.2.2.1 Electron Emitters

Coherent illumination is essential for holography. While beams from classic W or LaB_6 thermionic emitters can be made to be spatially coherent by extreme defocusing of the condenser system, this gives inadequate illumination for most purposes, and the spread in electron energies from these emitters limits the temporal coherence. Instead, thermally assisted field emitters (Schottky emitters) and, more recently, cold field emission guns (FEGs) are universally used for EH. (Crewe and Wall, 1970; Tonomura et al., 1979; Allard and Völkl, 1999). The emitter in a Schottky gun consists of an oriented W crystal supported on a classic W hairpin. The W work function barrier to emission is reduced in magnitude and spatial extent by a 5×10^8 V/m electrostatic field applied to the tip. Heating W to a relatively modest temperature of 1800 K enables the electrons at the high-energy

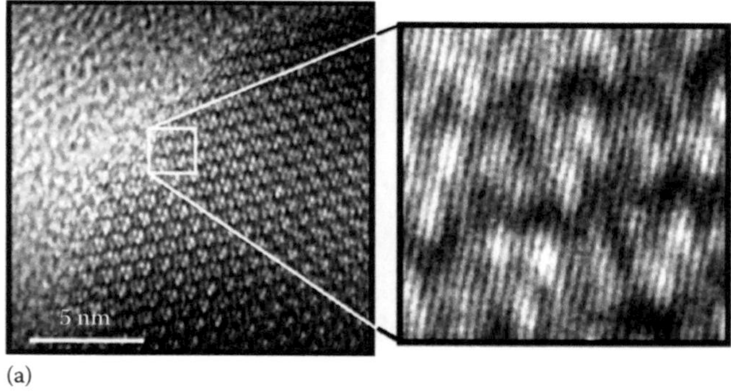

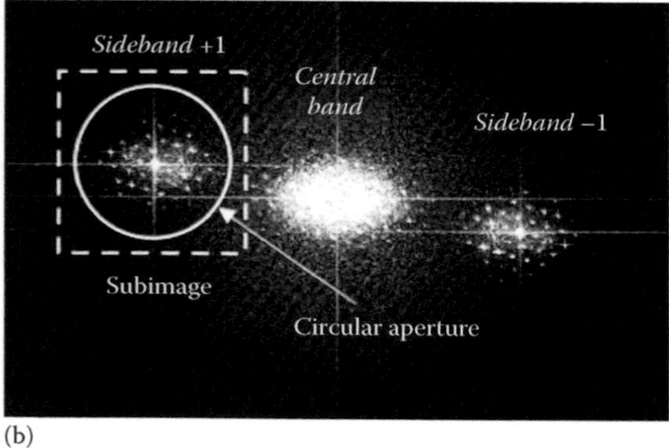

FIGURE 4.3 Overview of reconstruction process. (a) A lattice-resolution hologram showing the hologram fringes as well as the underlying periodic crystal potential. (b) The power spectrum of (a) showing the two side-bands. The periodic crystal potential is responsible for the pattern of spots in the center as well as the two side-bands. In potential maps, the periodic potential is avoided by placing the crystal at an orientation that gives greatly reduced dynamic scattering. As a result, the periodic patterns are not seen. In the reconstruction process, an aperture is placed around one side-band, excluding the central pattern and the opposite side-band. The chosen side-band is moved to the center and inverse-transformed. The real part of the inverse transform is the amplitude image; the imaginary part is the phase image. (From Lichte, H. and Lehmann, M., *Reports on Progress in Physics*, 71, 016102, 2007.)

tail of the energy distribution to tunnel through the reduced barrier and to be accelerated down the electron-optical column. This provides a small effective source size for high brightness and spatial coherence, and a relatively narrow thermal energy distribution (0.6–1.0 eV full width half maximum [FWHM]) for increased temporal coherence. In the other option, cold FEGs increase the field by an order of magnitude so that electrons tunnel through the narrowed potential barrier with no heating except the Joule heating of the beam current itself. The effective source size and the energy spread (0.2–0.3 eV FWHM) are reduced relative to those of the Schottky emitters, significantly improving the temporal and spatial coherence. Vacuum quality and the cleanliness of the emitter are critical in order to maintain a stable electric field of this magnitude. Most cold FEGs are equipped with heaters that are used to "flash" deposits and oxides from the electrodes. When operated correctly, both types

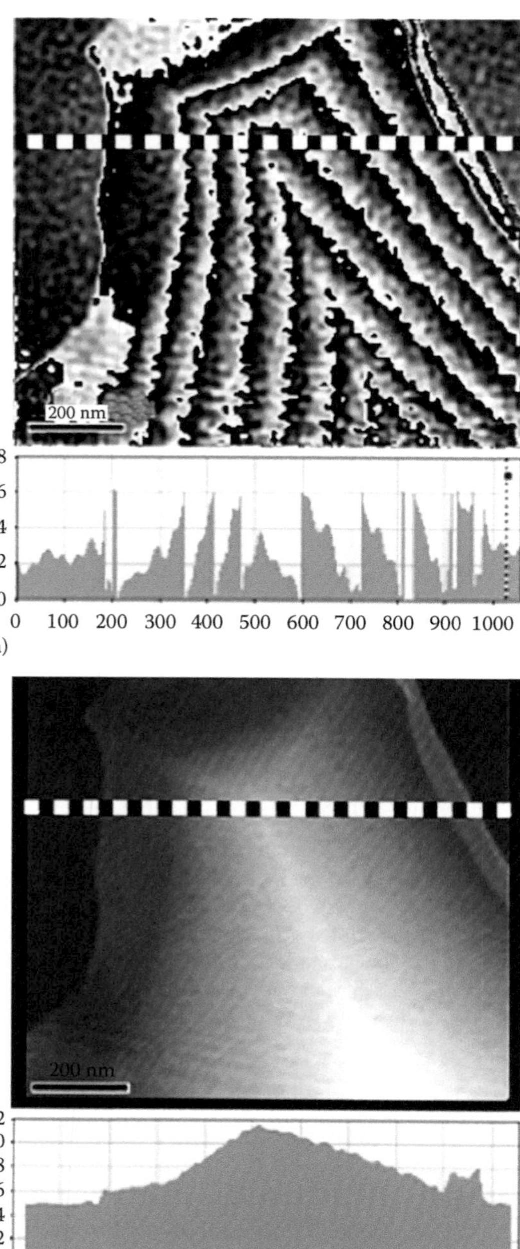

FIGURE 4.4 Phase images from a sample of varying thickness. Assuming no electrostatic potential variations exist, the phase is proportional to $V_{MIP} \times t(x, y)$ where MIP is the mean inner potential and $t(x, y)$ is the thickness as a function of spatial coordinates. Since phase is calculated between 0 and 2π, the contours in (a) represent jumps of 2π rad. In this image, the calculated value of the phase was multiplied by 4 in order to accentuate the topography. In (b), the phase was unwrapped. 2π was added to each successive contour and smoothed in order to produce a continuous image with brightness being proportional to thickness. The numerical values corresponding to the horizontal black/lines are displayed in the line scan. (From Lichte, H. and Lehmann, M., *Reports on Progress in Physics*, 71, 016102, 2007.)

of field emission sources can function well for years. Currently, carbon-based field emission sources (de Jonge et al., 2002; de Jonge and Bonard, 2004; Wang et al., 2010) are being actively investigated; they have not yet proved their potential to improve brightness significantly relative to the W systems. Generally, all types of FEGs can meet the usual brightness and coherency requirements for off-axis EH.

4.2.2.2 Brightness

Brightness, β, the usual figure-of-merit of an electron gun, is defined as the current density per solid angle at the source:

$$\beta = i_e \left(\frac{4}{\pi d_0^2} \right) \left(\frac{1}{\pi \alpha_0^2} \right), \tag{4.3}$$

where i_e represents the emission current that passes through the area, $\pi d_0^2/4$, of the gun crossover within a solid angle bounded by the planar angle α_0. For a TEM, this is approximately equal to the effective brightness, β_{eff}:

$$\beta_{eff} = i_p \left(\frac{4}{\pi d_p^2} \right) \left(\frac{1}{\pi \alpha_p^2} \right). \tag{4.4}$$

Here, the subscript, p, indicates that this is the current in a focused probe through an area, $\pi d_p^2/4$, within the solid angle, $\pi\alpha_p^2$, subtended by the condenser aperture at the specimen. Since the brightness of all electron sources is proportional to the accelerating voltage (with relativistic corrections), high-voltage guns are desirable for EH.

4.2.3 Condenser System

4.2.3.1 Coherence

For off-axis EH, the condenser lens system must deliver a parallel beam of electrons to the desired area of the specimen and an equal area of vacuum. In most columns, the first condenser lens forms a demagnified image of the source; the second condenser converts this small image at its front focal plane into a parallel beam directed though the condenser aperture. In order to provide a suitable reference wave, the coherence width,

$$X_c = \frac{\lambda}{2\pi\theta_c} \tag{4.5}$$

of the beam transferred through the aperture must be at least twice as large as the area to be mapped. λ is the electron wavelength, and θ_c is an exceedingly small half-angle (~10^{-7}–10^{-9} rad) subtended by a small, coherently illuminated, condenser aperture at the specimen, neglecting any pre-specimen lens fields. Even with FEGs, efficient use of available electrons usually requires intentional exaggeration of the astigmatism in the illuminating lenses in order to form an elliptical beam whose long axis is perpendicular to the desired hologram fringes (Smith and McCartney, 1999).

4.2.3.2 Beam Current

The coherent beam current delivered to the specimen plane can limit the spatial resolution of the hologram. A simple calculation estimates the working magnification, M as

$$M = \frac{npW}{d_{res}}, \tag{4.6}$$

where
- n is the number of fringes to resolve a distance, d_{res}, $n = 3$
- p is the number of pixels per fringe
- W is the width of a pixel in the camera

If the dark current in the camera is D, and if the signal-to-background is S/N, α is the number of counts per electron in the camera (~1000), the beam current density is

$$i_b = \frac{D(S/N)M^2}{\alpha W^2} = \frac{n^2 p^2 D(S/N)}{\alpha (d_{res})^2}. \tag{4.7}$$

If three fringes are required for a resolution, d_{res}, of 1.0 nm, if five pixels are required for fringe visibility, and if the pixels in the camera are 10 μm in size, then the working magnification must be greater than 150,000×. If the dark current is 10 counts/(pixel width)^2/s, the S/N ratio is 2, counts per electron registered in the camera are 1000, and all the other parameters remain the same, i_b must be greater than 5×10^{18} $e^-/m^2/s$. This would be a reasonable number for an inorganic solid as compared with biological material (Rez, 2003). Regardless of the specific numbers, the effective resolution is inversely related to the square root of the coherent beam current density. This is a powerful motivation for the development of higher brightness sources or high-voltage microscopes.

4.2.4 Specimen

4.2.4.1 Images

Electron holographic reconstructions produce images corresponding to (i) the amplitude of the electron wave at the detector and (ii) the shift in the phase caused by the specimen relative to that of a free electron wave.

4.2.4.2 Amplitude Image

The amplitude image is similar, but not identical, to the classic bright field TEM image. The TEM bright field (BF) image is simply detected as intensity variations when the electrons are absorbed in the detector; the reconstructed EH amplitude image is the real part of the electron wave interference pattern. Another useful distinction between EH amplitude images and classical BF images is that EH amplitude images are acquired along with a reference amplitude image. McCartney and Gajdardziska-Josifovska (1994) has shown that the EH amplitude image, A_0, normalized by the reference amplitude image, A_R, can be used to determine the total inelastic scattering parameter, t/λ_{inel}, through the following relation:

$$\frac{t}{\lambda_{inel}} = -2\ln\left\{\frac{A_0}{A_R}\right\}. \tag{4.8}$$

If the thickness, t, is known a priori, as for a uniformly wedged, cylindrical, or spherical specimen, the total inelastic mean free path, λ_{inel}, can be determined. This can be valuable for quantitative electron energy loss spectroscopy.

4.2.4.3 Phase Image

The difference in phase between electron waves that pass through the specimen and waves that propagate through the vacuum, $\Delta\phi(x, y, z)$, is given by

$$\Delta\phi(x,y,z) = C_E \int V(x,y,z)dz - \frac{e}{\hbar}\int \vec{B}(x,y,z,)dz. \tag{4.9}$$

Integrating along the electron trajectory, taken parallel to z, gives a map of the phase change over the specimen plane (x,y):

$$\Delta\phi(x,y) = C_E V(x,y) - \frac{e}{\hbar}\vec{B}(x,y), \tag{4.10}$$

where C_E is the relativistically corrected interaction constant, totally determined by the electron kinetic energy, E, and the electron rest mass energy, E_0:

$$C_E = \frac{2\pi}{\lambda E}\left(\frac{E+E_0}{E+2E_0}\right). \tag{4.11}$$

C_E is 0.00728/V/nm for a TEM operating at 200 keV; e and $\hbar$ have their usual meanings. $V(x, y, z)$ is the spatial variation in the electrostatic potentials; with some approximations, it can be divided into the mean inner potential (MIP) $V_{MIP}(x, y, z)$ and any superimposed electrostatic potentials $V_E(x, y, z)$:

$$V(x,y,z) = V_{MIP}(x,y,z) + V_E(x,y,z). \tag{4.12}$$

The MIP is the zeroth component of the Fourier series representation of the potential energy of an electron in the specimen; it is the average of the deep potential wells at nuclear positions with the shallower potential distributions over all the intervening space (Gajdardziska-Josifovska, 1999). MIP values are typically in the range 10–30 eV. Thus, in the absence of any magnetic induction, $\vec{B}(x, y, z)$, subtracting the MIP from the total measured potential yields a map of the projected electrostatic phase shift. This information is encoded in the hologram; it will form the holographic phase image after the reconstruction process. If magnetic fields, $\vec{B}(x, y, z)$, are present, only the component perpendicular to the biprism and lying in the plane of the specimen is relevant. If the biprism is along y, this component is

$$B = \vec{B}\sin\theta_x \sin\theta_z, \tag{4.13}$$

where θ_x and θ_z are the angles between the overall magnetic field direction and the x- and z- axes, respectively. For the present discussion, the absence of any magnetic phenomena will be assumed.

4.2.4.4 Mapping

For electrostatic potential mapping, the specimen (i) must not be too thick, but not too thin; (ii) it must be clean and free of surface oxides, amorphous films, and the like, without being susceptible to electron beam induced charging; (iii) it must be oriented so that no strong dynamical diffraction effects occur; (iv) it must have a simple geometry so that the MIP can be subtracted away; and (v) the area of interest must be located adjacent to the edge of the specimen.

All these requirements can be discussed with reference to the usual starting equation for experimental EH (Smith and McCartney, 1999):

$$I_{holo}(\vec{r}) = I_{inel}(\vec{r}) + 1 + A^2(\vec{r}) + 2\mu A(\vec{r})\cos\left\{2\pi\vec{q}_c\cdot\vec{r} + \phi(\vec{r}) + \theta\right\}. \tag{4.14}$$

The total intensity of the hologram, $I_{holo}(\vec{r})$, distributed over the plane, $\vec{r} = \sqrt{x^2+y^2}$, of the detected signal consists of intensity terms, $I_{inel}(\vec{r}) + 1 + A^2(\vec{r})$, plus a set of cosinusoidal fringes of amplitude $2\mu A(\vec{r})$ superimposed on a constant background. Here, $I_{inel}(\vec{r})$ is the inevitable inelastic scattering intensity that contributes to the background. The cosine term is produced when the reference wave ("carrier wave," $\vec{q}_c$) interferes coherently with the wave emanating from the bottom of the specimen; the fringes in the resulting interference pattern are parallel to the biprism. Phase shifts of the electron wave caused by the projected potential appear as displacements of the cosine fringes; $\phi(\vec{r})$ represents that shift. θ is an additional shift of the fringes usually taken as constant everywhere. μ is the signal-to-noise ratio—the contrast—of the fringes relative to the constant background ($0 < \mu < 1$) (Wang et al., 2004).

4.2.4.5 Specimen

A thick specimen depletes the electron wave amplitude by scattering electrons incoherently or at high angles. On the other hand, a sample that is too thin cannot impart a phase difference above the detection limit imposed by signal-to-noise issues and instrumental parameters. In their seminal paper on mapping dopant distributions in MOSFET transistors, Rau et al. (1999) measured an optimum specimen thickness of 200–400 nm for Si samples, excluding the ion-beam-damaged surface layers. In that study, they were able to measure changes in electrostatic potential voltages as little as 0.1 V with a spatial resolution of 10 nm. When specimens were thinned to 50 nm, the voltage discrimination was 0.3 V. From a theoretical perspective, Lichte (2008) shows that this apparent trade-off between voltage discrimination and specimen thickness can be dramatically improved by changes in instrumentation using technologies that are available today. With suitable equipment and specimens, it should be possible to move the technique into the picometer scale.

Efforts to quantify potentials within biased semiconductor junctions (Twitchett et al., 2002) or to combine electron tomographic holography (ETH) (Cooper et al., 2009) have clearly demonstrated the effects of amorphized or dead surface layers on potential mapping across Si pn junctions. The presence of these films has been recognized for many years, and specimen preparation by dimple polishing followed by wide-beam low-angle ion beam milling is preferable to focused ion beam milling (McCartney et al., 1994; Gribelyuk et al., 2002; Ikarashi et al., 2010). Even so, Si specimens are never perfectly clean, and some adjustments must be made for accurate analysis. Conversely, very clean semiconductors and insulators are susceptible to charging. The transmitted electrons excite secondary electrons from the sample; the resulting positive charges distort the potential maps. At the risk of increasing the background, it may be desirable to deposit a light C film on the specimen in order to bleed off the excess charge (McCartney, 2005). Mapping of potentials within "bottom up" nanostructures usually does not have specimen thinning problems, but charging may be observed (Chung and Rabenberg, 2006b).

Equation 4.11 assumes that the beam of electrons do not interact strongly with any periodic structures within the specimen. A crystalline sample should be tilted to avoid zone axes, two-beam conditions, and any other orientations for which strong dynamical diffraction occurs. If properly oriented, the specimen can be treated as a continuous medium such that the desired electrostatic potential is only superimposed on the smoothly varying MIPs and not the much larger periodic potentials. Formanek and Bugiel (2006) and Lubk et al. (2010) have investigated this effect as applied to mapping in MOS transistors. A graphical indication of the magnitude of this effect appears in Twitchett-Harrison et al. (2008), where the consecutive tilts in ETH inevitably pass through dynamical diffraction orientations giving outlying data points. In that study, thickness measurements using normalized amplitude images (McCartney and Gajdardziska-Josifovska, 1994) were in error by 5%–20% when crossing dynamical diffraction orientations. For a static hologram, such errors would introduce corresponding errors in compensating for the MIP.

Electrostatic potentials are pure phase objects that are only revealed clearly when the stronger electron–specimen interactions are taken away. Incoherent electron waves and electrons that are scattered inelastically cannot form interference patterns. Electrons that undergo high-angle scattering do not propagate to the detector plane. Dynamical diffraction can be minimized by controlling the specimen orientation. All that remains is the MIP. Equation 4.11 suggests that knowledge of the MIP and the thickness—the $V_{MIP} \times t(x,y)$ product—of the specimen allows one to subtract away the effects of the solid matter from the total projected potential. Experimental (Gajdardziska-Josifovska et al., 1993; Li et al., 1999; Kruse et al., 2003, 2006; Chung et al., 2007) and theoretical (Rez et al., 1994; Schowalter et al., 2005, 2006) determinations of MIP for many solids are known to two-figure accuracy, but close comparisons of the literature reveal discrepancies. It is reasonable to suspect that inadequate satisfaction of specimen requirements *vis-à-vis* exacting theories are responsible. In practice, potential mapping is usually done with prior knowledge of the specimen shape, whether it be spherical (Wang et al., 1997), cylindrical (Chung and Rabenberg, 2006a; den Hertog et al., 2009, 2010; Li et al., 2011; Wolf et al., 2011), planar (Frabboni et al., 1985; McCartney and Gajdardziska-Josifovska, 1994; Rau et al., 1999), or cuboidal (Twitchett-Harrison et al., 2008).

Off-axis holography demands a reference wave that can be used to form an interference pattern with the post-specimen electron wave. Displacements of the fringes within the interference pattern correspond to phase differences ($\phi(\vec{r})$ in Equation 4.14) between the two waves. The reference wave is simply formed by allowing part of the electron beam to bypass the specimen. The biprism overlaps the specimen wave with the reference wave, forming the interference pattern. In order to compensate for distortions of the wave fronts by the microscope lenses, it is customary to acquire a hologram without the specimen and subtract it from the experimental hologram.

The need to form an interference pattern between the post-specimen and reference waves sets a variety of practical limitations for potential mapping. Most simply, the specimen must be prepared such that the region of interest lies adjacent to the edge of the specimen. Second, the coherence width of the beam must exceed twice the size of the field of interest without serious interference with the Fresnel edge fringes. Third, the biprism should be able to operate at voltages such that narrow interference fringes are formed. This is necessary because the spatial resolution of off-axis holography can only be as small as three fringes. Fourth, in practical instruments, the biprism can limit the field of view. If the width of a typical camera is 1000 pixels; if 5 pixels are required per fringe; if 20% are lost to the edge fringes; and if 50% of the remaining fringes are in the reference wave, then only 30 fringes will remain on the sample. If the desired resolution is 1 nm, the field of view is reduced to 10 nm. Wang et al. (2004) and Lichte (2008) have discussed this and other limitations much more thoroughly.

4.2.5 Post-Specimen Electron-Optical Elements

4.2.5.1 Elements

The post-specimen optical elements typically consist of the usual TEM objective lens, the biprism (Figure 4.1), and the projector lens and camera system. An additional long focal length lens within the usual TEM objective lens gives added flexibility.

4.2.5.2 Biprism

In the conventional setup for off-axis EH, the first real image formed by the objective lens occurs at a plane below the electron biprism. That is, the biprism is located above the intermediate aperture plane, or the lens is weakened such that the focal length increases. In any case, the image of the biprism is an out-of-focus shadow image with white Fresnel edge fringes.

The biprism is the optical element that is not normally provided in the purchase of the electron microscope. Manufacturers (i) provide a separate port in order to insert the biprism or (ii) use the port intended for the intermediate lens aperture. The biprism consists of a very thin wire of platinum or gold-plated silica glass stretching across the beam. It should be rotatable about the electron beam in order to position it normal to any potential gradients where it will be most sensitive.

It is the function of the positively charged biprism (Figure 4.1) to deflect the electrons such that they overlap near the center of the electron column. If the coherence is adequate, this will cause a series of interference fringes. The information provided by the hologram is encoded within the relative displacements—the phase shifts—of the fringes. In contrast to light-optical holograms that provide three-dimensional (3D) images by the interference of lasers at large angles, the electron hologram is formed from coherent electron beams interfering at very small angles. The electronic hologram appears to be a pattern of two-dimensional (2D) interference fringes bordered by the broad Fresnel edge fringes caused by the biprism itself (Figure 4.2).

The voltage applied to the biprism wire is usually a trade-off between (i) the maximum coherence width at the specimen plane, (ii) the need to create fine fringes at high voltage, (iii) the need to increase the field of view at low voltage, and (iv) the possibility of electric discharges. All these must be achieved at adequate signal-to-noise conditions. Since three fringes define the maximum spatial resolution of EH, high voltages must be applied to the biprism in order to produce large angular deflections ($\vec{q}_c$ large) and very narrow fringes. For adequate *S/N*, the field of view will then be limited by the need to concentrate available electrons into a small area at the specimen plane. Regardless, everything will be limited by the brightness of the gun, the stability of the biprism power supply, and the design of the biprism environment against arc formation or dielectric breakdowns.

4.2.5.3 Lorentz or Minilens

If a much weaker lens—"Lorentz lens" or "objective minilens"—is installed within the coils of the objective, the flexibility in the choice of fringe spacing, fringe width, and field of view expands dramatically. Lorentz lenses were developed in order to remove the lens fields from the specimen plane for the sake of imaging the magnetic fields in ferromagnetic specimens. In that arena, the strong objective lens is shut off, and the Lorentz lens forms the primary image, albeit at lower magnification and higher aberrations. Lorentz lenses have been used as the primary imaging lenses for EH of ferromagnetic materials (McCartney et al., 1997). Objective minilenses were developed by one manufacturer as a necessary complement to their short focal length condenser/objective lens design. In either case, the ability to shorten the focal length of the conventional objective and compensate with the long-focal-lens improves the flexibility immensely (Wang et al., 2004).

4.2.5.4 Projector System

Finally, the projector lenses and CCD camera must be capable of recording narrow fringes with high *S/N* ratios. The projector lenses magnify the interference pattern that forms near the back image plane of the microscope to a scale that can be recorded with a digital camera whose pixels are on the order of 10 μm. For 10 nm spatial resolution at the specimen, this requires a bare minimum magnification within the projector system of ~20,000×.

4.2.5.5 Cameras

The cameras in use are very efficient; they have scintillators that produce many photons per fast electron impact; they record many of these photons; and they have dark currents on the order of 10 counts per second per pixel. Each electron that strikes the scintillator produces many (~1000) photons, and a fraction of the photons is channeled to a pixel in the CCD device and its immediate surroundings (de Ruijter and Weiss, 1992, 1993; Krivanek and Mooney, 1993). The image formed by low-intensity beams consists of very bright pixels; with longer exposure times, the hologram fringes emerge from the background. These cameras have dominated the TEM market for two decades; they have been constantly improved. At this time, the cameras are not limiting, but gun brightness and total instrumental stability are.

4.2.6 Hologram Reconstruction

To a novice, an electron hologram appears to be just a pattern of fringes. An expert can usually extract qualitative information from deviations from the regular pattern of fringes. But, it takes numerical reconstruction techniques in order to derive truly quantitative mapping data.

The reconstruction process (Völkl and Lehmann, 1999) begins with a Fourier transformation of the uniform region (i.e., well between the Fresnel edge fringes) of the "raw" hologram (Figure 4.3a). The transformation creates a pattern on the monitor consisting of a central region and two side-bands (Figure 4.3b). The central region is a straightforward map of the Fourier coefficients in a hologram, usually displayed as intensities, rather than as squared real parts. The two side-bands are complex conjugates of each other and contain identical information. They are displaced from the center by the carrier wave vector, $\pm\vec{q}_c$, as they represent the interference of the forward beam relative to the reference beam. In this particular example, the crystal lattice is resolved, and the central pattern and side-bands consist of intensity maxima corresponding to the Fourier components in the image. For electrostatic potential mapping, the crystal should be reoriented such that the strong lattice periodicities do not obscure the weaker signal from the electrostatic potentials.

More mathematically, the FT of the intensity distribution in the hologram (Equation 4.14) is given by (Völkl, 1999)

$$\begin{aligned} \mathrm{FT}\{I_{holo}(\vec{r})\} &= \delta(\vec{q}) + \mathrm{FT}\{I_{inel}(\vec{r}) + A^2(\vec{r})\} \\ &+ \delta(\vec{q}+\vec{q}_c)\otimes \mathrm{FT}\{\mu A(\vec{r})e^{i\phi(\vec{r})}e^{i\theta}\} \\ &+ \delta(\vec{q}-\vec{q}_c)\otimes \mathrm{FT}\{\mu A(\vec{r})e^{-i\phi(\vec{r})}e^{-i\theta}\} \end{aligned} \quad (4.15)$$

Line 1 of the right-hand side of Equation 4.15 represents the FT of the normalized intensity distribution in the hologram; it makes up the central region ($q \approx 0$) of the pattern. Lines 2 and 3 represent the fact that the finite angle between the carrier wave and the specimen wave creates a finite displacement ($q \pm q_c \neq 0$) of the pattern. The phase-sensitive interference pattern is convoluted ($\otimes$) with the finite carrier wave vectors ($\pm q_c$). The cosine function of Equation 4.14 is rewritten in terms of complex exponentials to give two equivalent side-bands.

Next, one of the side-bands is shifted to the origin of the numerical Fourier space. This sets $\vec{q}_c = 0$, such that the argument of the cosine term in Equation 4.13 is dominated by $\phi(\vec{r})$.

A numerical "aperture function"

$$\begin{aligned} a(q) &= 1, \quad \text{if } |q| \leq q_{ap} \\ a(q) &= 0, \quad \text{if } |q| > q_{ap} \end{aligned} \tag{4.16}$$

is inserted into the pattern, centered about the origin. Its radius, q_{ap}, is chosen to exclude the original central region and the other side-band. Conceptually, the numerical aperture is a function whose value is unity inside the aperture and zero everywhere else. In practice, it is usually a circular function whose magnitude is unity at the center and falls gracefully to zero at q_{ap}. In practice, a Gaussian-like function is chosen in order to minimize oscillations in subsequent processing. All information beyond q_{ap} is excluded from further analysis. It can be shown that the relative magnitudes of q_{ap} and q_c impose the spatial resolution limit for off-axis EH. The width of three fringes defines the smallest distance (taken normal to the fringes) that can be resolved.

Inverse transformation of this centered side-band results in a complex-valued image. The real part is the amplitude image described earlier; the imaginary part is the phase image. The phase image is a map of the phase shifts, modulo 2π, of the electron beam as a function of position across the specimen plane. This phase image is the desired map of the potentials within the specimen and surroundings.

In most cases, the phase image will be dominated by a pattern of phase ramps and 2π rad jumps (Figure 4.4). Phase unwrapping programs (Ghiglia and Pritt, 1998) remove the 2π jumps so that phase becomes a linear function of the projected potential. If there is a way to subtract away the phase variations caused by changes in the projected mean interatomic potential, a map of the electrostatic potentials can be produced.

Most commonly, all these manipulations are accomplished through the use of scripts written within DigitalMicrograph©, a flexible software package created for image acquisition and manipulation that is bundled with the CCD cameras sold by Gatan, Inc. Gatan sells, at additional cost, their own package of EH scripts entitled Holoworks©, but individually authored scripts, such as that by McCartney, are common. Gatan, Digital Micrograph, and Holoworks are registered trademarks of Gatan, Inc. It is necessary to mention these trademarks, because Gatan Inc. dominates the market for TEM CCD cameras and acquisition software. Other manufacturers have tried to enter the market for CCD cameras, but Gatan's complete package has enhanced TEM productivity everywhere; it has enabled EH to escape from the specialists' labs. [This is not intended as a commercial message for Gatan, Inc; it is a realistic fact.]

The phase image is a map of the electrostatic potentials encountered by the electron wave as it propagated through the specimen volume. For practical purposes, it is usually necessary to convert these potentials into dopant distributions, surface charges, and the like. Numerical solutions of Poisson's equation are usually necessary.

4.3 Mapping Electrostatic Potentials in Review

4.3.1 Prehistory

The study of electrostatic potentials by electronic holography goes back at least as far as Frabboni et al. (1985) who detected potential differences between the ends of a reverse-biased pn junction. During this time period, Tonomura and collaborators were successfully investigating magnetic (Tonomura, 1993) and superconducting phenomena (Matsuda et al., 1989). They achieved the first real confirmation of the Aharonov–Bohm effect (Aharonov and Bohm, 1959; Tonomura et al., 1986). Before 1990 and the advent of the digital CCD camera, interference micrographs were recorded on photographic film, and reconstruction involved lasers and optical lenses. Tonomura's book (Tonomura, 1993) describes the use of in line holograms with optical reconstructions.

de Ruijter and Weiss (1993) explored the possible use of CCD cameras for off-axis EH, and McCartney et al. (1994) mapped the depletion region potential of an unbiased pn junction using the new technology.

4.3.2 Historical Overview of the Development of 2D Mapping

4.3.2.1 Modern Era

Two field-changing events occurred in 1999. The publication of the "pink book" (Völkl, 1999) collected all the fundamentals of EH into a single comprehensive volume. This book made EH approachable by any TEM-trained scientist, and this review draws heavily from it; the truly interested reader is urged to consult this book early and often. The second event was the publication of a paper (Rau and Lichte, 1999) describing the successful mapping of electrostatic potentials in 0.3 μm n-MOS and p-MOS transistors. In addition, the figures in this paper show clearly defined contrast changes between p and n+ regions, and between n and p+ regions. The promise of dopant distribution profiling caused the semiconductor industry to devote its considerable resources to the development of EH as a routine device characterization technique.

4.3.2.2 Electronic Device Mapping

In the following years, devices were mapped with ever-greater sophistication. Gribelyuk et al. (2002) mapped out electrostatic potentials within deep submicron CMOS devices. Twitchett et al.

(2002) used a novel focused ion beam (FIB) technique to prepare pn junction specimens that were suitable for in situ biasing. Han et al. (2007) worked with 90 nm MOSFETs. Gribelyuk et al. (2008) systematically addressed the specimen-related difficulties that had been overcome in the previous years, such as specimen thickness variations, charging, and dead layers due to FIB damage or surface depletion effects. McCartney and Smith (2007) and Lichte et al. (2007) provide comprehensive reviews of most of this material.

In 2009, Li et al. (2009) reported the measurement of approximately 30 holes confined within a strained-Ge-on-Si quantum dot (Figure 4.5). This is a noteworthy contribution for at least two reasons. First, it demonstrates the extreme sensitivity of a coherent electron wave to very small changes in projected potential. It belies the received wisdom that EH specimens must not be too thin lest the phase shift integrated over the electron's trajectory be undetectable. Second, it shows how mature this field has become. All the instrumental considerations, the specimen issues, the reconstruction, and the modeling were dealt with systematically. The experimental techniques are well established; the results are not to be questioned.

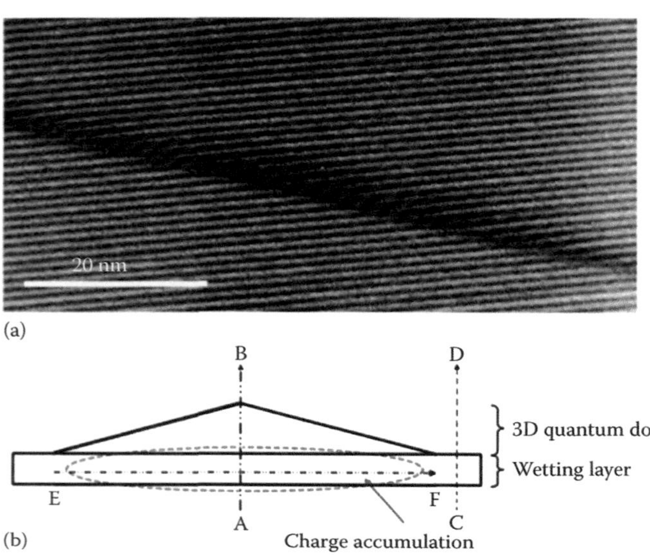

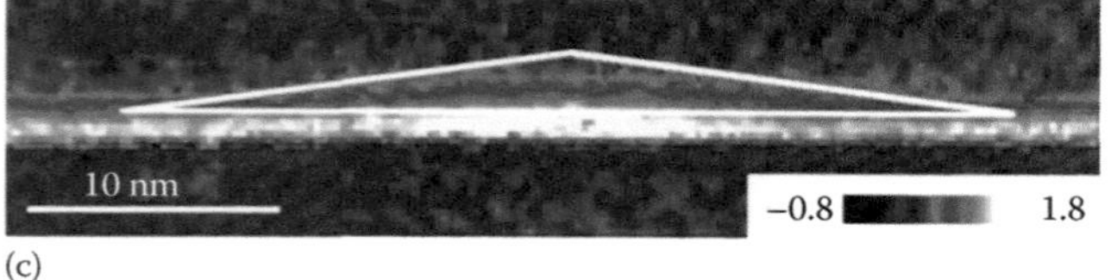

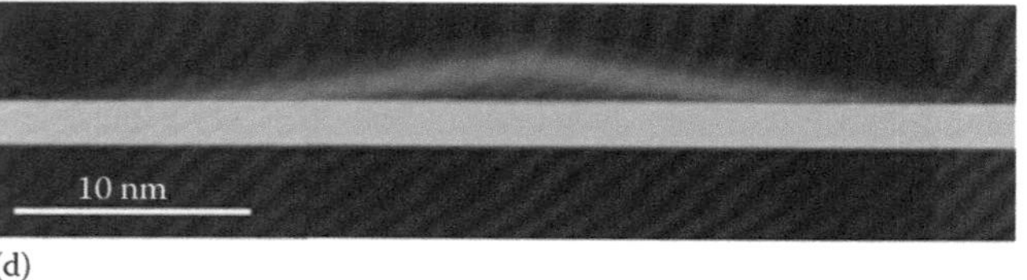

FIGURE 4.5 Detection of ~30 excess positive charges in a Ge-on-Si quantum dot that is totally enclosed within the surrounding Si. (a) The experimental hologram is rotated ~25° with respect to (b), (c), and (d). The presence of the quantum dot is visible as a thicker gray region in the center of the hologram. The offset of the fringes is greater within this area. (b) A schematic of the geometry of the quantum dot and the Ge wetting layer; the Si substrate capping layers are not shown. Note that the excess charge accumulates within the wetting layer under the quantum dot. (c) Experimental phase image and (d) simulated phase image. A small excess phase shift—the slightly darker arch within the triangle—is visible in the experimental phase image. Analysis of these images indicates that 30 excess charges are trapped within the quantum dot. (From Li, L. et al., *Appl. Phys. Lett.*, 94, 232108, 2009.)

4.3.2.3 Fields in the Vacuum

EH can also detect the concentrated electric fields associated with asperities on samples biased relative to other objects in their environment. Matteucci et al. (1992) mapped potential gradients with interference patterns recorded on photographic film. Cumings et al. (2002) studied field patterns around carbon nanotubes and determined that field emission would occur only at the end of the nanotube if it were to be incorporated into a field emission device. den Hertog et al. (2010) detected the presence of electrical potential gradients (fields) surrounding a thin Si nanowire; they attributed these fields to charges retained within the native oxide.

4.3.3 Inhomogeneous Semiconductor Nanowires

4.3.3.1 Nanowires

Geometrically inhomogeneous semiconductor nanowires emerged during the early 2000s. Bamboo (superlattice) structures consisting of alternating segments of lattice-matched, but distinct materials (Björk et al., 2002a,b; Gudiksen et al., 2002; Wu et al., 2002; Thelander et al., 2003); core-shell nanowires consisting of a cylindrical semiconductor encased within another semiconductor (Lauhon et al., 2002; Lin et al., 2003; Tateno et al., 2004; Sköld et al., 2005); and quantum dots within nanowires (Panev et al., 2003) were all synthesized within a short time span. Further refinements continue to date (Xiang et al., 2006; Chen et al., 2011; Thierry et al., 2012). Optical and transport measurements are usually conducted on these elegant nanostructures, but there is very little effort to map the internal electrostatic potentials within these inhomogeneous semiconductors.

4.3.3.2 Holography of Nanowires

Chung and Rabenberg (2006a,b) was the first to explore the possible use of EH to map potentials within core–shell nanowires (Figures 4.6 and 4.7) The specimens used consisted of a crystalline Ge core prepared using supercritical synthesis (Hanrath and Korgel, 2002, 2003) surrounded by a shell of amorphous Ge produced by remote plasma chemical vapor deposition. The nanowires in this study were very large (~200 nm in diameter), and the shells were heavily doped in order to assure detectable signals. Because of the simple cylindrical geometry, it was possible to estimate the MIPs, measure the coaxial interface potentials, and solve Poisson's equation for effective doping levels. EH of cross sections through the core–shell structure demonstrated

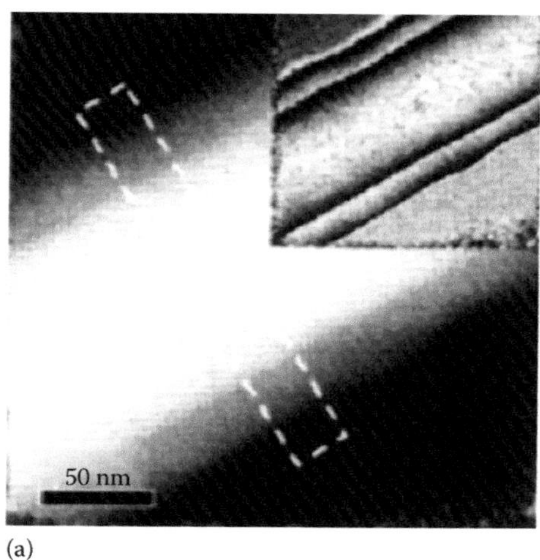

(a)

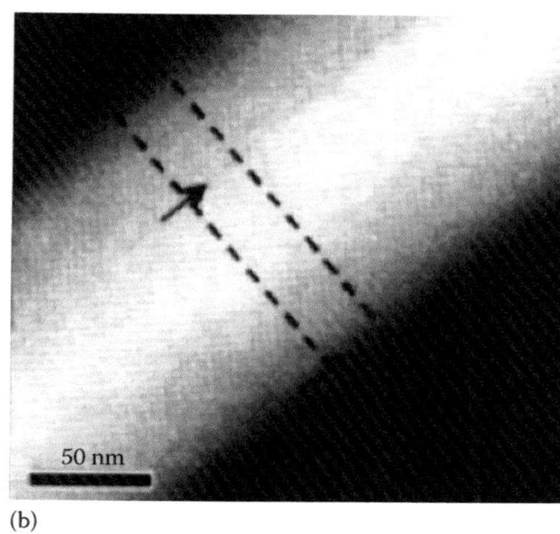

(b)

FIGURE 4.6 (a) Phase image (inset) and unwrapped phase image of an oxidized Ge nanowire. Assuming concentric cylindrical symmetry, it is possible to work out the MIP of the two substances. (b) Phase image of the Ge core–Native Ge oxide–Ge shell nanowire. The core is an intrinsic single crystal, whereas the shell is degenerately doped and poorly crystallized. The core–shell geometry is not exactly concentric. The phase contrast between the core and the shell allowed the doping level in the shell to be calculated in very good agreement with that expected from synthesis conditions. The dashed rectangles indicate the segments of the nanowires analyzed in the chapter. (From Chung, J. and Rabenberg, L., *Appl. Phys. Lett.*, 88, 013106, 2006a.)

the feasibility of measuring the density of interfacial states and depletion lengths at a spatial resolution of approximately 10 nm.

den Hertog et al. (2009, 2010) used EH to characterize thin (d = 60 nm) Si nanowires with alternating 150 nm long doped and intrinsic segments. With a stable instrument, they were able to distinguish intrinsic regions from those doped with 10^{18}, 10^{19}, and 10^{20} phosphorus atoms per cubic centimeter. Their results also indicate that the native oxide sheath in their wires contained 10^{12} charges per square centimeter. Taken at face value, these numbers are dramatic. With $\rho_{bulk} = 10^{18}/\text{cm}^3$, $\pi r^2 h\rho_{bulk} = 425\text{e}^-/$segment, and with $\rho_{surface} = 10^{12}/\text{cm}^2$, $4\pi r^2\rho_{surface} = 570\text{e}^-/$segment, the surface charges would be dominant. With a resolution of 10 nm along the wire, the 1000 excess electron charges per segment equates to a detection of roughly 70 electrons. Even for the highly doped segments (40,000 e$^-$/segment), the surface charges would be significant. Furthermore, the phase difference between intrinsic and doped ($10^{18}/\text{cm}^3$) segments along the nanowire was measured to be 0.4 rad across a junction 50 nm thick. Assuming that all the electrical fields are along the wire, this would correspond to a field magnitude of 2×10^7 V/m—approximately two-thirds of the breakdown field of Si. Adding (i) that the nanowire was supported at one end with a nanoparticle of gold at the other and (ii) that charging effects due to secondary electron production were certainly significant, this becomes a rather difficult problem. Nevertheless, it was a significant step.

Wolf et al. (2011) have presented beautiful 3D images of a GaAs-AlGaAs core-shell nanowire acquired with ETH (Figure 4.8). The MIPs of core and shell were determined, albeit with some variations from the published literature. The ETH revealed the faceting on {110} planes about the—[111] growth direction of the GaAs core, both in 3D projection and in tomographic slices perpendicular to the growth direction. The authors reported no attempts to distinguish between charging effects, internal junction effects, and MIPs.

The work of Li et al. (2011) firmly establishes the strength of EH for the determination of electrical charge distributions within inhomogeneous semiconductor nanowires (Figure 4.9). Working meticulously, they made holographic phase maps across carefully prepared Ge core–Si shell (60 nm core Ge diameter–70 nm Si shell) nanowires, and then characterized them using a suite of TEM-related techniques. The results not only corroborate transport measurements (Lu et al., 2005) but enable the reader to ask answerable questions regarding the magnitude of the compressive strains in the Si shell (People and Bean, 1985; Thompson et al., 2006; Chung et al., 2008) and their effects on electron transport in strained layers and growth instabilities in core–shell nanowires (Schmidt et al., 2008). This chapter will stimulate much more EH of inhomogeneous semiconductor nanowires.

4.4 Future Prospects and Converging Phase Fields

4.4.1 New Techniques

4.4.1.1 Revived Techniques

One can extrapolate from these trends into the near future. Some techniques are currently being developed, and some older techniques are being revived in light of newer technology. A thrust for the last several years is the combination of electron tomography with EH. Twitchett-Harrison et al. (2007, 2008), Wolf et al. (2010, 2011), and Tanigaki et al. (2012) have been developing this technique and applying it to semiconductor devices and nanowires. ETH will prove especially useful for lower dimensional specimens like complex nanowires,

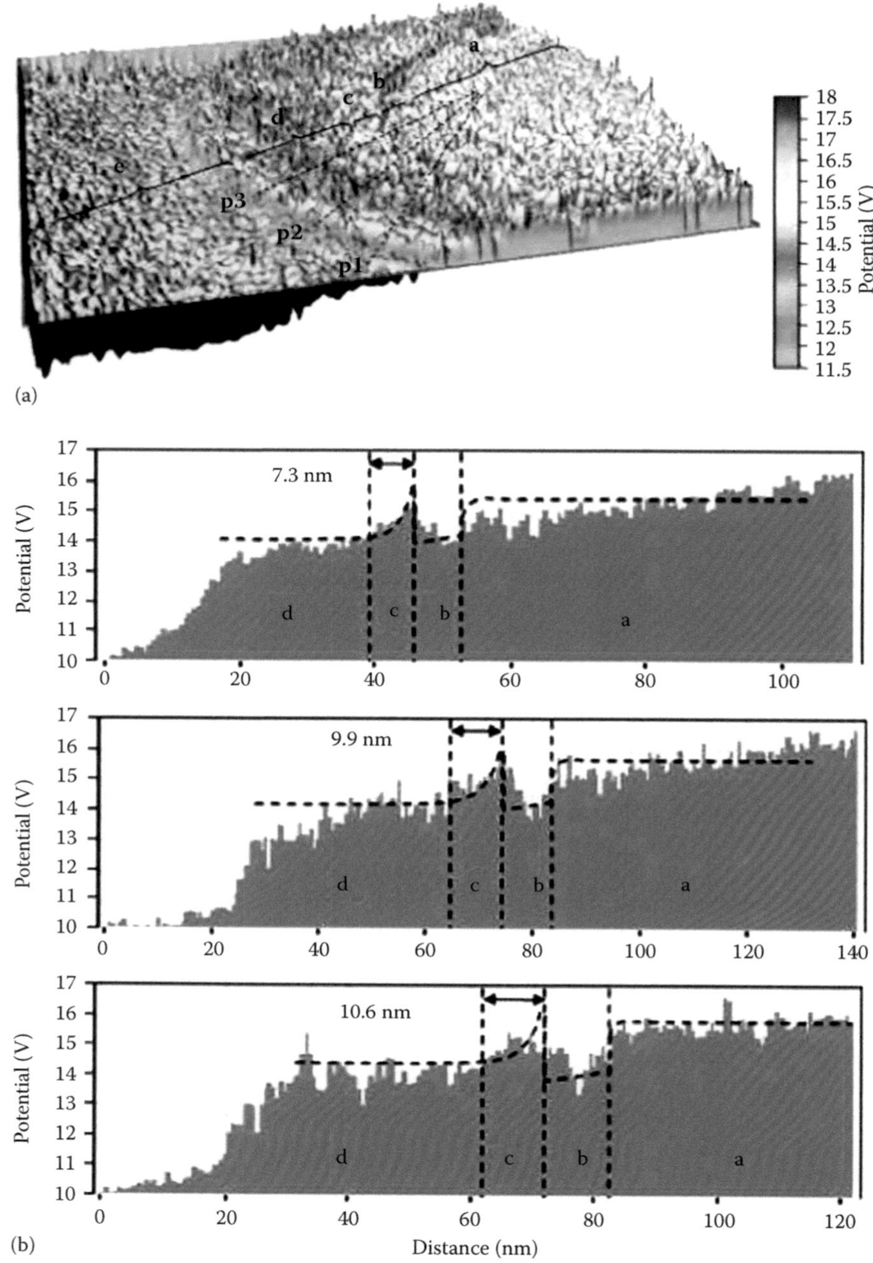

FIGURE 4.7 (a) Potential map across a cross section of the Ge core–Native Ge oxide–Ge shell nanowire of Figure 4.6. (b–d) Three radial line-profiles indicated as p1, p2, and p3 in (a). On the line-profiles, a, b, c, d represent the Ge core, the Ge oxide, the screening length in the amorphous Ge shell, and the remainder of the Ge shell, respectively. The profiles can be used to determine the number of defect states within the Ge oxide and the screening length within the heavily doped Ge shell. The white dome in the center of the wire is an effect of charging for this specimen that was embedded in insulating SiO_2. (From Chung, J. and Rabenberg, L., *J. Mater. Res.*, 21, 1215, 2006b.)

nanoparticles, and quantum dots. Phase shifting techniques (Ru et al., 1991) that have lain dormant for two decades because of experimental difficulties are being revived (Yamamoto et al., 2010). They promise higher resolution by ameliorating the problem of overlapping side-bands in the Fourier reconstruction. There may be some advantages from the use of multiple biprism arrangements (Kawasaki et al., 1993; Hirayama et al., 1995; Harada et al., 2004). In situ biasing is another technique with a long history (Frabboni et al., 1985; Cumings et al., 2002; Twitchett et al., 2002). Increased control of specimen geometry will enable in situ work to become more quantitative. Unfortunately, most of these developments will be confined to the specialists' labs that have facilities and resources to make modifications to their electron microscope columns.

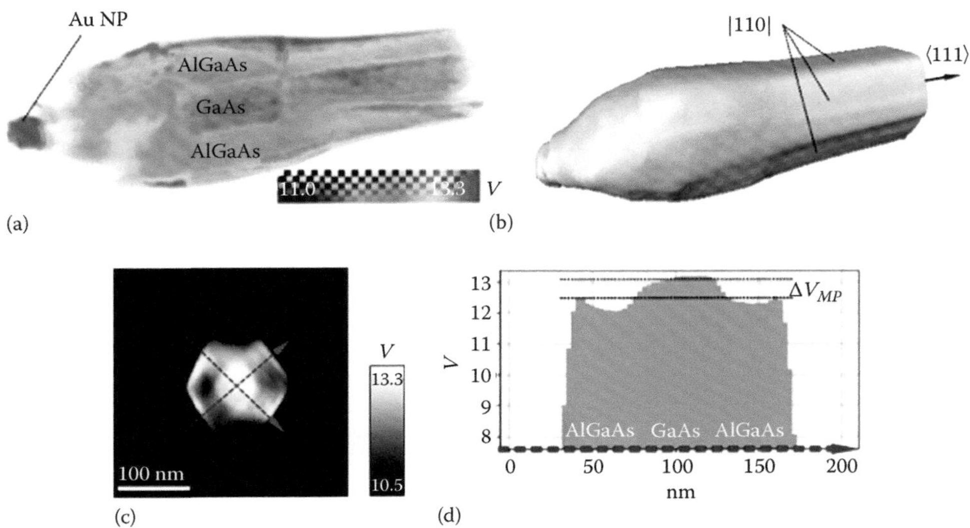

FIGURE 4.8 Electron tomography holography of a GaAs core–AlGaAs shell nanowire. (a) A volume rendering of the GaAs core, visible by its higher MIP, and the surrounding AlGaAs. The Au catalysis nanoparticle is visible because of its much larger MIP. (b) A surface within which the solid has a MIP in excess of 9 V. This is essentially the free surface of the nanowire. (c) A computed cross section through the wire. The <110> facets are clearly seen in (b) and (c). (d) A line profile through the wire with the GaAs core at a higher MIP than the AlGaAs shell. (From Wolf, D. et al., *Appl. Phys. Lett.*, 98, 264103, 2011.)

4.4.1.2 New Developments

Dark-field EH for strain mapping appeared in 2008 (Hÿtch et al., 2008), and is being continuously developed (Cooper et al., 2009, 2011; Béché et al., 2011). In this technique, a single crystal with at least two areas that are elastically strained relative to each other is set in an orientation for strong diffraction. A specific diffracted beam is passed through the TEM objective aperture and around a biprism located near the usual image plane. If the biprism is oriented parallel to the diffracting planes, the set of interference fringes can be reconstructed to build a map of the component of strain perpendicular to the diffracting planes. Using two different diffracting beams sequentially measures the 2D in-plane strain tensor. This technique is sensitive to relative displacements (strains) of 10^{-3} and absolute displacements of a few picometers. This technique has been applied to strained silicon devices (Thompson et al., 2006), where large built-in strains may have implications for electrostatic potential distributions.

4.4.1.3 Aberration Correction

The potential for EH in the new aberration-corrected instruments has not been explored. As mentioned earlier, high-resolution holography does require an analysis of the electron microscope column in terms of its aberrations (Allard and Völkl, 1999). Certainly, spherical aberration correction will be valuable, by itself, or in combination with other adaptations of the microscope, such as the placement of the biprism or the use of additional minilenses. Chromatic aberration correction may not be so valuable because of the intrinsic temporal coherency of the holographic technique. Electrons with a wide range of energies, from the gun or from inelastic collisions in the sample, are simply unable to produce interference fringes. On the other hand, a monochromator within the electron source could be valuable, but only if it conserves brightness. In any case, instruments that have been designed with improved power supplies and greater resistance to vibrations and other environmental disturbances in order to take advantage of aberration correctors will also be better for conventional EH.

Aberration correction should also have advantages in STEM holography modes, because the finer probes can lead to greater spatial coherency. Also, coupled with suitable detectors (Gribelyuk and Sum, 1999) and high-speed computation facilities, it should be possible to make more effective use of the available electrons. For example, a detector/computational facility that is capable of picking up holograms pixel by pixel could be augmented by a high angle annular dark field (HAADF) detector and an energy dispersive spectroscopy (EDS) detector for simultaneous high-resolution imaging, microanalysis, and hologram acquisition. In this case, the hologram would use the coherently scattered electrons, the STEM detector would make use of some of the incoherently scattered electrons, and the EDS detector would use the excitations from inelastic scattering events. With some advances in the reconstruction of STEM holograms, this complete materials analysis system could be automated.

4.4.2 New Materials

4.4.2.1 Zero-Dimensional Particles

The rapid development of nanometer and picometer materials science over the last decade will open up very many new applications for holographic electrical potential mapping.

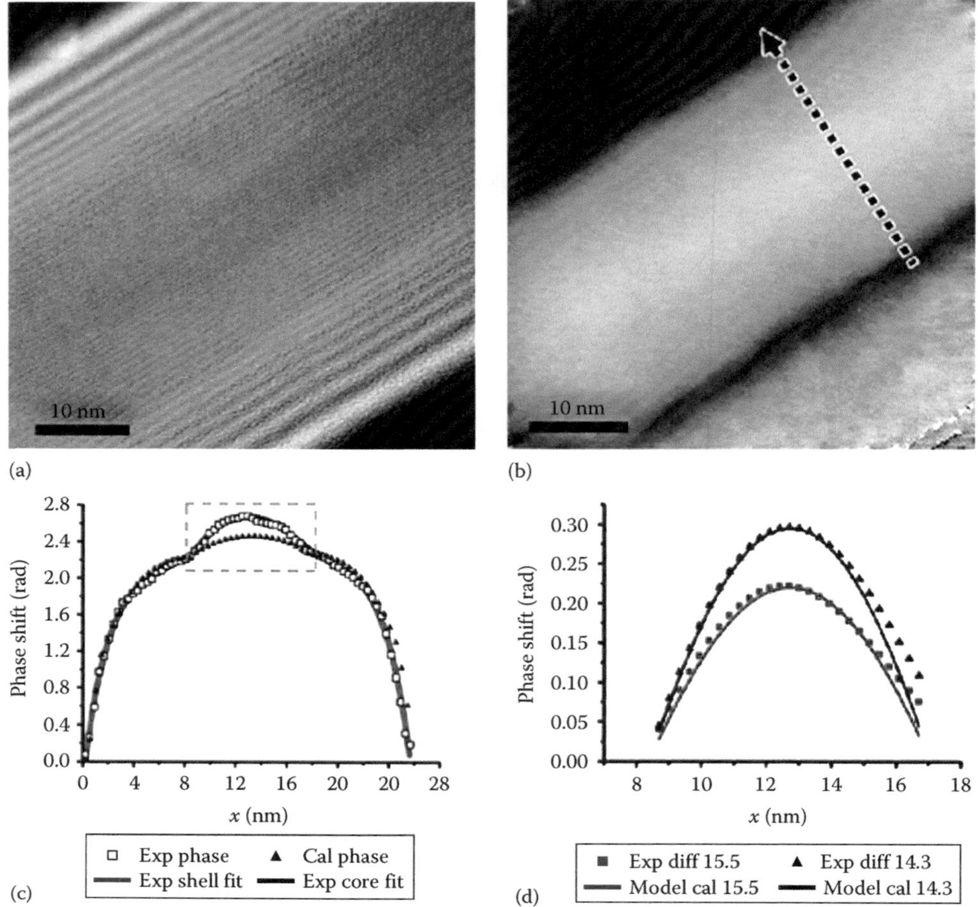

FIGURE 4.9 (a) Electron hologram of a Ge core–Si shell nanowire. (b) Unwrapped phase image with marker for line scans. (c) The experimentally measured phase profile (upper line) compared with a phase profile (lower line) calculated by extrapolation of the core and shell independently. The excess charge within the core is seen by the separation between these two curves. (d) Because of the uncertainty in the Ge MIP, the excess charge can only be known between an upper bound (Ge MIP = 14.3 eV, lower line) and a lower bound (Ge MIP = 14.3 eV, upper line). (From Li, L. et al., *Nano Lett.*, 11, 493, 2011.)

In zero dimensions, the interactions between solid catalysts and their supports, the construction of assemblies of photoactive particulates for solar energy applications, the charge transfer between quantum dots, and properties of magnetic nanoparticles will benefit from potential mapping. ETH should be well adapted for such systems where particulates are suspended in three dimensions, but there are no reports that this technique has been extended to this level.

4.4.2.2 Quasi-One-Dimensional Systems

EH of quasi-1D nanowires composed of inhomogeneous semiconductors is in its infancy. The charging and oxidation issues that den Hertog et al. (2009, 2010) encountered are certainly more challenging for small-diameter wires, but they are surmountable, based on the lessons learned from the study of 2D electronic devices. However, the truly interesting cases involve nanowires with internal structures (Thierry et al., 2012), or assemblies of such nanowires (Xu et al., 2011). The development of devices based on these structures will profit from any technique that characterizes internal potentials or potentials caused by particle/wire or wire/wire junctions.

4.4.2.3 Ferroelectrics and Related Materials

While mapping of the 2D potential distributions in Si, Ge, and III–V semiconductor devices by EH is very mature, there is surprisingly little work on other types of materials, either in bulk or in thin films. For example, $SrTiO_3$ and other perovskites have become the darlings of the high-resolution (S)TEM community; many are ferroelectric, and others display a huge panoply of complex electrical phenomena, but the literature on EH of perovskites is very limited. Zhang et al. (1993) did some preliminary studies, Lichte et al. (2002) mapped out what would be expected, and Matsumoto et al. (2008) reported a study of 90° domains in $BaTiO_3$. This should become a fruitful area of research.

4.4.3 Phase Convergence

4.4.3.1 Phase Problem

As seen earlier, EH is a technique that seeks to extract the phase of a wave scattered by a specimen. As such, it is a member of a long line of approaches aimed at solving the "phase problem" in structural analysis. That is, if all the phases of all the waves elastically scattered by a specimen were known, it would be straightforward to work back from the scattering/diffraction pattern to work out the structural aspects of any specimen. Because radiation with wavelengths comparable to interatomic spacings is detected incoherently (Equation 4.1), the phase information is normally lost. In this broad picture, EH can be thought of as one of several partial solutions of the "phase problem." They seem to be converging.

4.4.3.2 X-Rays

Attempts to work around the phase problem in the x-ray crystallography community include (i) heavy-element substitution, (ii) direct methods, and (iii) anomalous scattering (Ladd and Palmer, 2003). Under very favorable conditions, iterative phase retrieval techniques can extract wavelength-limited images from nonperiodic x-ray (Miao et al., 1999; Abbey et al., 2008) and electron (Zuo et al., 2003) diffraction patterns. In these "coherent diffractive imaging" (CDI) techniques, a diffraction pattern is acquired from a specimen surrounded by an area of fixed phase—the "support." Repetitive Fourier transformations of diffraction data that include scattering from the support converge on a function that represents the amplitude and phase of the specimen and surroundings.

4.4.3.3 Lorentz Electron Microscopy

Cohen (1967) seems to be the one who first recognized the relationship between Lorentz electron microscopy (LM) and EH. LM constitutes an ensemble of techniques that derive image contrast from the deflection of electrons by the internal fields of a ferromagnetic specimen. Features in Lorentz images, especially repeated fringes at cross-tie walls, can only be interpreted as the result of phase objects. In effect, a 180° ferromagnetic domain wall is a biprism within the specimen. Following up on work that identified inhomogeneous magnetic domain distributions as phase objects (Boersch et al., 1961; Wohlleben, 1967), Cohen worked out their imaging characteristics and compared them with Gabor's EH. Unfortunately, he did not recognize the value of holography for imaging of the magnetic domain distributions. One of his footnotes says, "Holography may not be applicable to domain walls, however, because of the large phase shifts associated with them." In an obscure paper (Rabenberg et al., 1980), it was shown that LM with extremely defocused (i.e., coherent) illumination could detect the slight phase variations caused by subtle pre-crystallization changes in an amorphous ferromagnetic alloy. Cowley and Spence (1999) has briefly described LM as a form of holography. Tonomura (1987) characterized ferromagnetic specimens and their fields with electron interference patterns. McCartney and Smith (2007) has reviewed work on mapping of magnetic fields within and outside of ferromagnetic materials. Going forward, EH might contribute to the study of the solid-state "magnetic monopoles" that have been observed in tetrahedrally coordinated spin ice (Castelnovo et al., 2008). Solid-state magnetic monopoles consist of topological defects in the spin lattice that can separate and take uncorrelated trajectories through the crystal. Of course, EH would require a cryogenic stage, because spin ice only exists at very low temperatures. In any case, LM and EH have converged such that beautiful, quantitative, phase maps around magnetic nanoparticles and within ferromagnetic thin films are available in the literature (Dunin-Borkowski et al., 2004).

4.4.3.4 High-Resolution TEM Imaging

High-resolution TEM imaging is another case of phase mapping, as applied at the level of the interatomic spacings in the specimen. In HRTEM, phase differences between electron waves propagating along atomic columns in the specimen are converted to fringes in the lattice image by the very aberrations that Gabor was trying to avoid. In fact, if the specimen were a pure phase object, and the microscope lenses were perfect, the image contrast would disappear. The aberrations serve as a subtle and complex "multiprism" that maps atomic potential distributions into intensity distributions in the image plane. While the major thrust of TEM development has always been the resolution of finer periodicities, TEM is not very sensitive to projected phase variations, such as electrical potentials and magnetic fields, extending over lengths greater than the dimensions of the unit cells.

4.4.3.5 Inline Holography

Inline holography, as it is now practiced, is a technique that combines concepts from Gabor's point projection holography (Gabor, 1948) with computationally intensive focal series reconstructions (Koch and Lubk, 2010; Laytichevskaia et al., 2010).

Gabor's original concept was (i) to use a point source of electrons to illuminate the object, (ii) to record a greatly magnified shadow pattern on photographic film, and (iii) to subsequently illuminate the film with a coherent beam of visible light. However, if the point source of electrons, object, film, and beam of visible light are collinear, two holographic images are produced, greatly obscuring one another. Off-axis holography using a biprism has been the most successful of all proposed solutions to this "twin image" problem. In HRTEM imaging, it has long been recognized that a series of images taken with systematically varying defocus can be used to extract the wave function exiting the specimen. In effect, the defocus variations fill in the gaps in the TEM contrast transfer function. Inline holography is essentially this. Each point on the exit surface is taken as a source of electrons. By varying defocus slightly, these specimen waves become controllable reference waves for each other. However, these are interfering waves subject to nonlinear diffraction, and iterative algorithms must be used to extract the amplitude and phase. In order to make these calculations manageable, this technique benefits from a coherent electron emitter, stable instrument construction, spherical aberration corrector, a thin specimen, and an electron energy filter. Inline holography does not need a biprism, and it is not necessary to position the edge of the specimen near the area of interest. Comparisons suggest the spatial resolution of this form of holography is better than off-axis holography, but approximations in the theory invoked in the focal series reconstruction make quantification less reliable. Improvements in all aspects of electron scattering theory and computational routines are required before this form of inline holography can become routine.

4.4.3.6 Ptychography

Ptychography (Maiden and Rodenburg, 2009; Hüe et al., 2010; Humphry et al., 2012) unites CDI with EH. As a phase-retrieval technique, ptychography has its antecedents within the CDI community, although it has long been recognized that it is related to Gabor's point projection holography with a scanning beam. In ptychography, the redundant information produced by overlapping beam positions serves as the "support" required by CDI techniques. That is, at each point along the beam's raster, a coherent electron far-field diffraction pattern is produced at the plane of the detector. The illuminate areas in the scan are overlapped by 70%–85% so that much of the information is redundant. Iterative algorithms exploit the redundant information to calculate amplitude and phase maps over the exit plane of the specimen (Figure 4.10).

In principle, ptychography is only limited by electron wavelength and the degree of coherence of the electron source. With typical TEM electron wavelengths of 2.5 pm, it has the potential for resolutions far into the picometer range. It is well adapted to map electrostatic and magnetic potentials around very thin specimens, such as thin bamboo nanowires. But, at present, ptychography is only applicable to very thin phase objects. Ptychography places severe demands on raster patterns, and instrumental and environmental noise within the scan generator, and the entire instrument will probably place the lower bound on the achievable resolution. Gabor's idea that the highest resolution imaging would not involve magnetic lenses at all is still very much alive.

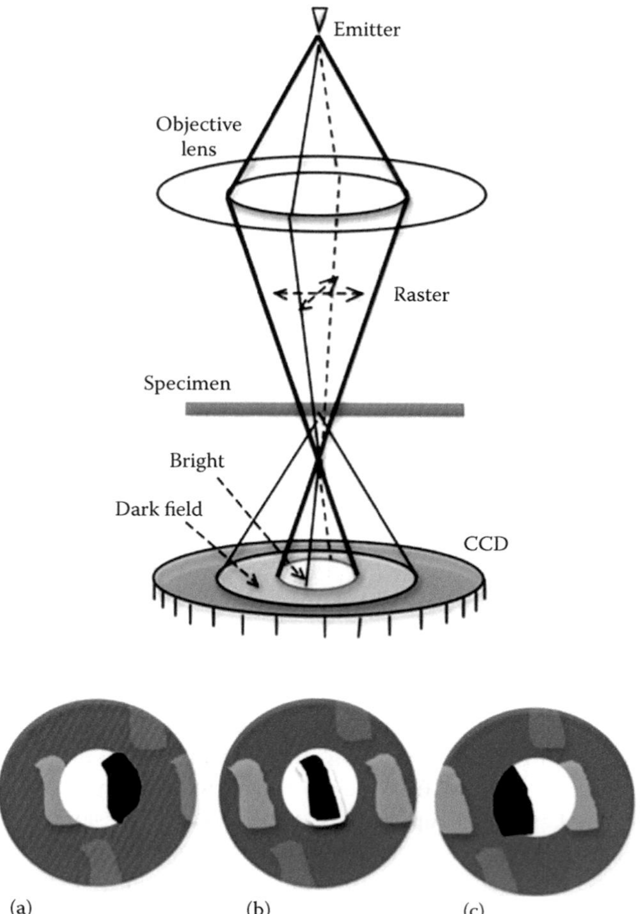

FIGURE 4.10 Fundamental setup for ptychography. An underfocused STEM probe projects a scattering pattern on the plane of detection. Magnifying lenses (not shown) are necessary in order to increase the size of the pattern to the size of the CCD camera. At each point of the scan, the intensity of the diffraction pattern is recorded. The overlap of the probe discs generates redundant information (a, b, c) in consecutive patterns. An iterative algorithm uses only the measured intensities to converge on a consistent map of amplitudes and phases emitted from the specimen. Scattering outside the forward cone can also be included to extend the technique to spacings well into the picometer range. (From Hüe, F. et al., *Phys. Rev. B*, 82, 121415, 2010; Humphry, M.J. et al., *Nat. Commun.*, 3, 730, 2012.)

4.5 Summary

The purposes of this chapter were (i) to introduce the concept of potential mapping using EH to a broader audience, (ii) to review the basic principles with a list of organized citations to the established literature so that a truly interested reader could pursue this topic efficiently, (iii) to survey the literature of EH potential mapping with a special emphasis on mapping within inhomogeneous nanowires, and (iv) to point out current directions and possibilities for future work.

Acknowledgments

Helpful discussions with J. Chung and D. J. Smith are gratefully acknowledged.

References

Abbey, B., Nugent, K.A., Williams, G.A., Clark, J.N., Peele, A.G., Pfeifer, M.A., De Jonge, M., and McNulty, I. 2008. Keyhole coherent diffractive imaging. *Nature Physics* 4: 394–398.

Aharonov, Y. and Bohm, D. 1959. Significance of electromagnetic potentials in the quantum theory. *Physical Review* 115: 485–491.

Allard, L.F. and Völkl, E. 1999. Optical characteristics of a holography electron microscope. In *Introduction to Electron Holography*, eds. E. Völkl, L. F. Allard, and D. C. Joy. New York: Kluwer Academic/Plenum Publishers, pp. 57–86.

Béché, A., Rouviere, J.-L., Barnes, J.-P., and Cooper, D. 2011. Dark field electron holography for strain measurement. *Ultramicroscopy* 111: 227–238.

Björk, M.T., Ohlsson, B.J., Sass, T., Persson, A.I., Thelander, C., Magnusson, M.H., Deppert, K., Wallenberg, L.R., and Samuelson, L. 2002a. One-dimensional heterostructures in semiconductor nanowhiskers. *Applied Physics Letters* 80: 1058–1060.

Björk, M.T., Ohlsson, B.J., Thelander, C., Persson, A.I., Deppert, K., Wallenberg, L.R., and Samuelson, L. 2002b. Nanowire resonant tunneling diodes. *Applied Physics Letters* 81: 4458–4460.

Boersch, H., Hamisch, H., Grohmann, K., and Wohlleben, D. 1961. Experimenteller Nachweiss der Phasenscheibung von Elektronenwellen durch das magnetische Vektorpotential. *Zeitschrift für Physik* 164: 79–93.

Castelnovo, C., Moessner, R., and Sondhi, S.L. 2008. Magnetic monopoles in spin ice. *Nature* 451: 42–45.

Chen, G., Gallo, E.M., Leaffer, O.D., McGucken, T., Prete, P., Lovergine, N., and Spanier, J.E. 2011. Tunable hot-electron transfer within a single core-shell nanowire. *Physical Review Letters* 107: 156802.

Chung, J., Lian, G., and Rabenberg, L. 2008. Local strain measurement in a strain engineered complementary metal-oxide-semiconductor device by geometrical phase analysis in the transmission electron microscope. *Applied Physics Letters* 93: 081909.

Chung, J. and Rabenberg, L. 2006a. Mapping of electrostatic potentials within core-shell nanowires by electron holography. *Applied Physics Letters* 88: 013106.

Chung, J. and Rabenberg, L. 2006b. Two-dimensional imaging of the potential distribution within a core/shell nanowire by electron holography. *Journal of Materials Research* 21: 1215–1220.

Chung, S., Smith, D.J., and McCartney, M.R. 2007. Determination of the inelastic mean free-path and mean inner potential for AlAs and GaAs using off-axis electron holography and convergent beam electron diffraction. *Microscopy and Microanalysis* 13: 329–335.

Cohen, M.S. 1967. Wave-optical aspects of Lorentz microscopy. *Journal of Applied Physics* 38: 4966–4976.

Cooper, D., Barnes, J.-P., Hartmann, J.-M., Béché, A., and Rouviere, J.-L. 2009a. Dark field electron holography for quantitative strain measurements with nanometer-scale spatial resolution. *Applied Physics Letters* 95: 053501.

Cooper, D., Rouviere, J.-L., Béché, A., Kadkhodazadeh, S., Semenova, E.S., Yvind, K., and Dunin-Borkowski, R.E. 2011. Quantitative strain mapping of InAs/InP quantum dots with 1 nm spatial resolution using dark field electron holography. *Applied Physics Letters* 99: 261911.

Cooper, D., Truche, R. et al. 2009b. Quantitative off-axis holography of GaAs p–n junctions prepared by focused ion beam milling. *Journal of Microscopy* 233(1): 102–113.

Cowley, J.M. 1992. Twenty forms of electron holography. *Ultramicroscopy* 41: 335–348.

Cowley, J.M. and Spence, J.C.H. 1999. Principles and theory of electron holography. In *Introduction to Electron Holography*, eds. E. Völkl, L. F. Allard, and D. C. Joy. New York: Academic/ Plenum Publishers, pp. 17–56.

Crewe, A.V. and Wall, J. 1970. A scanning electron microscope with 5 Å resolution. *Journal of Molecular Biology* 48: 375–393.

Cumings, J., Zettl, A., McCartney, M.R., and Spence, J.C.H. 2002. Electron holography of field-emitting carbon nanotubes. *Physical Review Letters* 88: 056804.

Dunin-Borkowski, R.E., Kasama, T., Wei, A., Tripp, S.L., Hÿtch, M., Snoeck, E., Harrison, R., and Putnis, A. 2004. Off-axis electron holography of magnetic nanowires and chains, rings, and planar arrays of magnetic nanoparticles. *Microscopy Research and Technique* 64: 390–402.

Erni, R. 2010. *Aberration-Corrected Imaging in Transmission Electron Microscopy: An Introduction*. London, U.K.: Imperial College Press.

Formanek, P. and Bugiel, E. 2006. On specimen tilt for electron holography of semiconductor devices. *Ultramicroscopy* 106: 292–300.

Frabboni, S., Matteucci, G., Pozzi, G., and Vanzi, M. 1985. Electron holographic observations of the electrostatic field associated with thin reverse-biased p–n junctions. *Physical Review Letters* 55: 2196–2199.

Fultz, B. and Howe, J.M. 2008. *Transmission Electron Microscopy and Diffractometry of Materials*. Berlin, Germany: Springer-Verlag.

Gabor, D. 1948. A new microscopic principle. *Nature* 161: 777–778.

Gabor, D. 1949. Microscopy by reconstructed wave-fronts. *Proceedings of the Royal Society (London) A* 197: 454–487.

Gajdardziska-Josifovska, M. and Carim, A.H. 1999. Applications of electron holography microscope. In *Introduction to Electron Holography*, eds. E. Völkl, L. F. Allard, and D. C. Joy. New York: Kluwer Academic/Plenum Publishers, pp. 267–293.

Gajdardziska-Josifovska, M., McCartney, M.R., de Ruijter, W.H., and Smith, D.J. 1993. Accurate measurements of mean inner potential of crystal wedges using digital electron holograms. *Ultramicroscopy* 50: 285–299.

Gatan Digital Micrograph. Pleasanton, C.A, Gatan Inc.: Software for acquisition and analysis of most TEM functions.

Gatan Holoworks, Gatan Inc.: Reconstructs off-axis holograms using Digital Micrograph scripts.

Ghiglia, D.C. and Pritt, M.D. 1998. *Two-Dimensional Phase Unwrapping: Theory, Algorithms, and Software*. New York: Wiley.

de Graef, M. 2003. *Introduction to Conventional Transmission Electron Microscopy*. Cambridge, U.K.: Cambridge University Press.

Gribelyuk, M.A., Domenicucci, A.G., Ronsheim, P.A., McMurray, J.S., and Gluschenkov, O. 2008. Application of electron holography to analysis of submicron structures. *Journal of Vacuum Science and Technology B* 26: 408–414.

Gribelyuk, M.A., McCartney, M.R., Li, J., Murthy, C.S., Ronsheim, P.A., Doris, B., McMurray, J.S., Hegde, S., and Smith, D.J. 2002. Mapping of electrostatic potential in deep submicron CMOS devices by electron holography. *Physical Review Letters* 89: 025502.

Gribelyuk, M.A. and Sum, J. 1999. Off-axis STEM holography. In *Introduction to Electron Holography*, eds. E. Völkl, L. F. Allard, and D. C. Joy. New York: Kluwer Academic/Plenum Publishers, pp. 231–248.

Gudiksen, M.S., Lauhon, L.J., Wang, J., Smith, D.C., and Lieber, C.M. 2002. Growth of nanowire superlattice structures for nanoscale photonics and electronics. *Nature* 415: 617–620.

Haider, M., Hartel, P., Müller, H., Uhlemann, S., and Zach, J. 2009. Current and future aberration correctors for the improvement of resolution in electron microscopy. *Philosophical Transactions of the Royal Society A* 367: 3665–3682.

Haider, M., Müller, H., Uhlemann, S., Zach, J., Loebau, U., and Hoeschen, R. 2008. Prerequisites for a Cc/Cs-corrected ultrahigh-resolution TEM. *Ultramicroscopy* 108: 167.

Han M.-G., Fejes, P., Xie, Q., Bagchi, S., Taylor, B., Conner, J., and McCartney, M.R. 2007. Quantitative analysis of 2-D electrostatic potential distributions in 90-nm Si pMOSFETS using off-axis electron holography. *IEEE Transactions on Electron Devices* 54: 3336–3341.

Hanrath, T. and Korgel, B.A. 2002. Nucleation and growth of germanium nanowires seeded by organic monolayer-coated gold nanocrystals. *Journal of the American Chemical Society* 124: 1424–1429.

Hanrath, T. and Korgel, B.A. 2003. Supercritical fluid-liquid-solid (SFLS) synthesis of Si and Ge nanowires seeded by colloidal metal nanocrystals. *Advanced Materials* 15: 4237–4440.

Harada, K., Tonomura, A., Togawa, Y., Akasho, T., and Matsuda, T. 2004. Double-biprism electron interferometry. *Applied Physics Letters* 84: 3229–3231.

Herring, R.A. and Pozzi, G. 1999. Electron holography using diffracted electron beams. In *Introduction to Electron Holography*, eds. E. Völkl, L. F. Allard, and D. C. Joy. New York: Kluwer Academic/Plenum Publishers, pp. 295–310.

den Hertog, Rouvière, J.-L. et al. 2010. Off axis holography of doped and intrinsic silicon nanowires: Interpretation and influence of fields in the vacuum. *Journal of Physics: Conference Series* 209: 012027.

den Hertog, M.I., Schmid, H., Cooper, D., Rouvière J.-L., Björk, M.T., Riel, H., Rivallin, P., Karg, S., and Reiss, W. 2009. Mapping active dopants in single silicon nanowires using off-axis electron holography. *Nano Letters* 9: 3837–3843.

Hirayama, T., Tanji, T., and Tonamura, A. 1995. Direct visualization of electromagnetic microfields by interference of three electron waves. *Applied Physics Letters* 67: 1185–1187.

Hüe, F., Rodenburg, J.M., Maiden, A.M., Sweeney, F., and Midgley, P.A. 2010. Wave-front phase retrieval in transmission electron microscopy via ptychography. *Physical Review B* 82: 121415.

Humphry, M.J., Kraus, B., Hurst, A.C., Maiden, A.M., and Rodenburg, J.M. 2012. Ptychographic electron microscopy using high-angle dark-field scattering for sub- nanometre resolution imaging. *Nature Communications* 3: 730.

Hÿtch, M., Houdellier, F., Hüe, F., and Snoeck, E. 2008. Nanoscale holographic interferometry for strain measurements in electronic devices. *Nature* 453: 1086–1089.

Ikarashi, N., Toda, A., Uejima, K., Yako, K., Yamamoto, T., Hane, M., and Sato, H. 2010. Electron holography for analysis of deep submicron devices: Present status and challenges. *Journal of Vacuum Science and Technology* B28: C1D5–C1D10.

Ikeda, M., Sugawara, A., and Harada, K. 2011. Twin-electron biprism. *Journal of Electron Microscopy* 60: 353–358.

de Jonge, N. and Bonard, J.-M. 2004. Carbon nanotube electron sources and applications. *Philosophical Transactions of the Royal Society A* 362: 2239–2266.

de Jonge, N., Lamy, Y., Schoots, K., and Oosterkamp, T.H. 2002. High brightness electron beam from a multi-walled carbon nanotube. *Nature* 420: 393–395.

Kawasaki, T., Missiroli, G.F., Pozzi, G., and Tonomura, A. 1993. Multiple-beam interference experiments with a holographic electron microscope. *Optik* 92: 168.

Koch, C.T. and Lubk, A. 2010. Off-axis and inline electron holography: A quantitative comparison. *Ultramicroscopy* 110: 460–471.

Krivanek, O.L. and Mooney, P.E. 1993. Applications of slow-scan CCD cameras in transmission electron microscopy. *Ultramicroscopy* 49: 95–108.

Kruse, P., Rosenauer, A., and Gerthsen, D. 2003. Determination of the mean inner potential in III–V semiconductors by electron holography. *Ultramicroscopy* 96: 11–16.

Kruse, P., Schowalter, M., Lamoen, D., Rosenauer, A., and Gerthsen, D. 2006. Determination of the mean inner potential in III–V semiconductors, Si and Ge by density functional theory and electron holography. *Ultramicroscopy* 106: 105–113.

Ladd, M.F.C. and Palmer, R.A. 2003. *Structure Determination by X-Ray Crystallography*. New York: Kluwer Academic/Plenum Publishers.

Latychevskaia, T., Formanek, P., Koch, C.T., and Lubk, A. 2010. Off-axis and inline electron holography: Experimental comparison. *Ultramicroscopy* 110: 472–482.

Lauhon, L.J., Gudiksen, M.S., Wang, D., and Lieber, C.M. 2002. Epitaxial core-shell and core-multishell nanowire heterostructures. *Nature* 420: 57–61.

Li, L., Ketharanathan, S., Drucker, J., and McCartney, M.R. 2009. Study of hole accumulation in individual germanium quantum dots in p-type silicon by off-axis electron holography. *Applied Physics Letters* 94: 232108.

Li, J., McCartney, M.R., Dunin-Borkowski, R.E., and Smith, D.J. 1999. Determination of mean inner potential of germanium using off-axis electron holography. *Acta Crystallographica A* A55: 652–658.

Li, L., Smith, D.J., Dailey, E., Madras, P., Drucker, J., and McCartney, M.R. 2011. Observation of hole accumulation in Ge/Si core/shell nanowires using off-axis electron holography. *Nano Letters* 11: 493–497.

Lichte, H. 2008. Performance limits of electron holography. *Ultramicroscopy* 108: 256–262.

Lichte, H., Formanek, P., Lenk, A., Linck, M., Matzeck, C., Lehmann, M., and Simon, P. 2007. Electron holography: Applications to materials questions. *Annual Review of Materials Research* 37: 539–588.

Lichte, H. and Lehmann, M. 2007. Electron holography—Basics and applications. *Reports on Progress in Physics* 71: 016102.

Lichte, H., Reibold, M., Brand, K., and Lehmann, M. 2002. Ferroelectric electron holography. *Ultramicroscopy* 93: 199–212.

Lin, H.-M., Chen, Y.-L., Yang, J., Liu, Y.-C., Yin, K.-M., Kai, J.-J., Chen, F.-R., Chen, L.-C., Chen, Y.-F., and Chen, C.-C. 2003. Synthesis and characterization of core–shell GaP@GaN and GaN@GaP nanowires. *Nano Letters* 3: 537–541.

Lu, W., Xiang, J., Timko, B., Wu, Y., and Lieber, C.M. 2005. One-dimensional hole gas in germanium/silicon nanowire heterostructures. *Proceedings of the National Academy of Science of the USA* 102: 10046–10051.

Lubk, A., Wolf, D., and Lichte, H. 2010. The effect of dynamical scattering in off-axis holographic mean inner potential and inelastic mean free path measurements. *Ultramicroscopy* 110: 438–446.

Maiden, A.M. and Rodenburg, J.M. 2009. An improved ptychographical phase retrieval algorithm for diffractive imaging. *Ultramicroscopy* 109: 1256–1262.

Matsuda, T., Hasegawa, S., Igarashi, M., Kobayashi, T., Naito, M., Kajiyama, H., Endo, J., Osakabe, N., and Todokuro, H. 1989. Magnetic field observation of a single flux quantum by electron-holographic interferometry. *Physical Review Letters* 62: 2519–2522.

Matsumoto, T., Koguchi, M., Suzuki, K., Nishimura, H., Motoyoshi, Y., and Wada, N. 2008. Ferroelectric 90° domain structure in a thin film of BaTiO3 fine ceramics observed by 300 kV electron holography. *Applied Physics Letters* 92: 072902.

Matteucci, G., Missiroli, G.F., Muccini, M., and Pozzi, G. 1992. Electron holography in the study of the electrostatic fields: The case of charged microtips. *Ultramicroscopy* 45: 77–83.

McCartney, M.R. 2005. Characterization of charging in semiconductor device materials by electron holography. *Journal of Electron Microscopy* 54: 239–242.

McCartney, M.R. and Gajdardziska-Josifovska, M. 1994. Absolute measurement of normalized thickness, t/lambda, from off-axis electron holography. *Ultramicroscopy* 53: 283–289.

McCartney, M.R. and Smith, D.J. 2007. Electron holography: Phase imaging with nanometer resolution. *Annual Review of Materials Research* 37: 729–767.

McCartney, M.R., Smith, D.J., Farrow, R.F.C., and Marks, R.F. 1997. Off-axis electron holography of epitaxial FePt films. *Journal of Applied Physics* 82: 2461–2465.

McCartney, M.R., Smith, D.J., Hll, R., Bean, J.C., Völkl, E., and Frost, B. 1994. Direct observation of potential distribution across Si/Si p–n junctions using off-axis electron holography. *Applied Physics Letters* 65: 2603–2605.

Miao, J., Charalambous, P., Kirz, J., and Sayre, D. 1999. Extending the methodology of X- ray crystallography to allow imaging of micrometre-sized non-crystalline specimens. *Nature* 400: 342–344.

Möllenstedt, G. 1999. The history of the electron biprism. In *Introduction to Electron Holography*, eds. E. Völkl, L. F. Allard, and D. C. Joy. New York: Kluwer Academic/Plenum Publishers, pp. 1–15.

Panev, N., Persson, A.I., Sköld, N., and Samuelson, L. 2003. Sharp exciton emission from single InAs quantum dots in GaAs nanowires. *Applied Physics Letters* 83: 2238–2240.

People, R. and Bean, J.C. 1985. Calculation of critical layer thickness versus lattice mismatch for GexSi1-x/Si strained-layer heterostructures. *Applied Physics Letters* 47: 322–324.

Rabenberg, L., Mishra, R.K., Thomas, G., Kohmoto, O., and Ojima, T. 1980. Electron microscopy of Co/Fe/Si/B amorphous alloys. *IEEE Transactions on Magnetics* MAG-16: 1135–1137.

Rau, W.D. and Lichte, H. 1999. High-resolution off-axis electron holography. In *Introduction to Electron Holography*, eds. E. Völkl, L. F. Allard, and D. C. Joy. New York: Kluwer Academic/Plenum Publishers, pp. 201–229.

Rau, W.D., Schwander, P., Baumann, F.H., Höppner, W., and Ourmazd, A. 1999. Two dimensional mapping of the electrostatic potential in transistors by electron holography. *Physical Review Letters* 82: 2614–2617.

Reimer, L. 1993. *Transmission Electron Microscopy: Physics of Image Formation and Microanalysis*. Berlin, Germany: Springer-Verlag.

Rez, P. 2003. Comparison of phase contrast transmission electron microscopy with optimized scanning transmission annular dark field imaging for protein imaging. *Ultramicroscopy* 96: 117–124.

Rez, D., Rez, P., and Grant, I. 1994. Dirac-Fock calculations of x-ray scattering factors and contributions to the mean inner potential for electron scattering. *Acta Crystallographica A* A50: 481–497.

Ru, Q., Endo, J., Tanji, T., and Tonomura, A. 1991. Phase-shifting electron holography by beam tilting. *Applied Physics Letters* 59: 2372–2375.

de Ruijter, W.J. and Weiss, J.K. 1992. Methods to measure properties of slowscan CCD cameras for electron detection. *Review of Scientific Instruments* 63: 4314–4321.

de Ruijter, W.H. and Weiss, J.K. 1993 Detection limits in quantitative off-axis holography. *Ultramicroscopy* 50(3): 269–283.

Schmidt, V., McIntyre, P.C., and Gösele, U. 2008. Morphological instability of misfit-strained core-shell nanowires. *Physical Reviews B* 77: 235302.

Schowalter, M., Rosenauer, A., Lamoen, D., Kruse, P., and Gerthsen, D. 2006. Ab initio computation of the mean inner Coulomb potential of wurtzite-type semiconductors and gold. *Applied Physics Letters* 88: 232108.

Schowalter, M., Titantah, J.T., Lamoen, D., and Kruse, P. 2005. Ab initio computation of the mean inner Coulomb potential of amorphous carbon structures. *Applied Physics Letters* 86: 112102.

Sköld, N., Karlsson, L.S., Larsson, M.W., Pistol M.-E., Seifert, W., Trägård, J., and Samuelson, L. 2005. Growth and optical properties of strained GaAs–GaxIn1-xP core–shell nanowires. *Nano Letters* 5: 1943–1947.

Smith, D.J. and McCartney, M.R. 1999. Practical electron holography. In *Introduction to Electron Holography*, eds. E. Völkl, L. F. Allard, and D. C. Joy. New York: Kluwer Academic/Plenum Publishers, pp. 87–106.

Tanigaki, T., Aizawa, S., Suzuki, T., and Tonomura, A. 2012. Three-dimensional reconstructions of electrostatic potential distributions with 1.5-nm resolution using off-axis electron holography. *Journal of Electron Microscopy* 61: 77–84.

Tateno, K., Gotoh, H., and Watanabe, Y. 2004. GaAs/AlGaAs nanowires capped with AlGaAs layers on GaAs(311)B substrates. *Applied Physics Letters* 85: 1808–1810.

Thelander, C., Mårtensson, T., Björk, M.T., Ohlsson, B.J., Larsson, M.W., Wallenberg, L.R., and Samuelson, L. 2003. Single-electron transistors in heterostructure nanowires. *Applied Physics Letters* 83: 2052–2054.

Thierry, R., Perillat-Merceroz, G., Jouneau, P.H., Ferret, P., and Feuillet, G. 2012. Core–shell multi-quantum wells in ZnO/ZnMgO nanowires with high optical efficiency at room temperature. *Nanotechnology* 23: 085705.

Thompson, S.E., Sun, G., Choi, Y.S., and Nishida, T. 2006. Uniaxial-process-induced strained-Si: Extending the CMOS roadmap. *IEEE Transactions on Electron Devices* 53: 1010.

Tonomura, A. 1987. Applications of electron holography. *Reviews of Modern Physics* 59: 639–669.

Tonomura, A. 1992. Electron-holographic interference microscopy. *Advances in Physics* 41: 59–103.

Tonomura, A. 1993. *Electron Holography*. New York: Springer.

Tonomura, A., Matsuda, T., Endo, J., Todokuro, H., and Komoda, T. 1979. Development of a field emission electron microscope. *Journal of Electron Microscopy* 28: 1–11.

Tonomura, A., Osakabe, N., Matsuda, T., Kawasaki, T., Endo, J., Yano, S., and Yamada, H. 1986. Evidence for Aharonov-Bohm effect with magnetic field completely shielded from electron wave. *Physical Review Letters* 58: 792–795.

Twitchett, A.C., Dunin-Borkowski, R.E., and Midgley, P.A. 2002. Quantitative electron holography of biased semiconductor devices. *Physical Review Letters* 88: 238302.

Twitchett-Harrison, A.C., Yates, T.J.V., Dunin-Borkowski, R.E., and Midgley, P.A. 2008. Quantitative electron holographic tomography for the 3D characterisation of semiconductor device structures. *Ultramicroscopy* 108: 1401–1407.

Twitchett-Harrison, A.C., Yates, T.J.V., Newcomb, S.B., Dunin-Borkowski, R.E., and Midgley, P.A. 2007. High-resolution three-dimensional mapping of semiconductor dopant potentials. *Nano Letters* 7: 2020–2023.

Völkl, E., Allard, L.F., and Joy, D.C. 1999. *Introduction to Electron Holography*. New York: Kluwer Academic/Plenum Publishers.

Völkl, E. and Lehmann, M. 1999. The reconstruction of off-axis electron holograms. In *Introduction to Holography*, eds. E. Völkl, L. F. Allard, and D. C. Joy. New York: Kluwer Academic/Plenum Publishers, pp. 125–151.

Wang, Y.C., Chou, T.M., Libera, M., and Kelly, T.F. 1997. Transmission electron holography of silicon nanospheres with surface oxide layers. *Applied Physics Letters* 70: 1296–1298.

Wang, M.-S., Golberg, D., and Bando, Y. 2010. Carbon "onions" as point electron sources. *ACS Nano* 4: 4396–4402.

Wang, Y.Y., Kawasaki, M., Bruley, J., Gribelyuk, M.A., Domenicucci, A.G., and Gaudiello, J. 2004. Off-axis electron holography with a dual-lens imaging system and its usefulness in 2-D potential mapping of semiconductor devices. *Ultramicroscopy* 101: 63–72.

Williams, D.B. and Carter, C.B. 2009. *Transmission Electron Microscopy*. New York: Springer.

Wohlleben, D. 1967. Diffraction effects in Lorentz microscopy. *Journal of Applied Physics* 38: 3341–3352

Wolf, D., Lichte, H., Pozzi, G., Prete, P., and Lovergine, N. 2011. Electron holographic tomography for mapping the three-dimensional distribution of electrostatic potential in III–V semiconductor nanowires. *Applied Physics Letters* 98: 264103.

Wolf, D., Lubk, A., Lichte, H., and Friedrich, H. 2010. Towards automated electron holographic tomography for 3D mapping of electrostatic potentials. *Ultramicroscopy* 110: 390–399.

Wu, Y., Fan, R., and Yang, P. 2002. Block-by-block growth of single-crystalline Si/SiGe superlattice nanowires. *Nano Letters* 2: 83–86.

Xiang, J., Lu, W., Hu, Y., Wu, Y., Hao, and Y., Lieber, C.M. 2006. Ge/Si nanowire heterostructures as high-performance field-effect transistors. *Nature* 441: 489–493.

Xu, F., Ma, X., Gerlein, L.F., and Cloutier, S.G. 2011. Designing and building nanowires: Directed nanocrystal self-assembly into radically branched and zigzag PbS nanowires. *Nanotechnology* 22: 265604.

Yamamoto, K., Sugawara, Y., McCartney, M.R., and Smith, D.J. 2010. Phase-shifting electron holography for atomic image reconstruction. *Journal of Electron Microscopy* 59(Suppl.): S81–S88.

Zhang, X., Joy, D.C., Zhang, Y., Hashimoto, T., Allard, L.F., and Nolan, T.A. 1993. Electron holography techniques for study of ferroelectric domain walls. *Ultramicroscopy* 51: 21–30.

Zuo, J.M., Vartanyants, I., Gao, M., Zhang, R., and Nagahara, L.A. 2003. Atomic resolution imaging of a carbon nanotube from diffraction intensities. *Science* 300: 1419–1421.

II

Picoscale Characterization

5

Interferometric Measurements at the Picometer Scale

Marco Pisani
National Institute for Metrological Research

5.1 Interferometer: An "Ideal" Measurement Tool

Whenever a high-precision dimensional measurement must be carried out, the best instrument we can use is an *interferometer.* The interferometer is an "ideal" instrument which exploits the undulatory nature of light by comparing the length we have to measure with the *wavelength* of a light source. This allows us to perform measurements for which accuracy (or uncertainty) is only limited by the knowledge of the wavelength used. With modern *laser* sources, the knowledge of the wavelength can be as accurate as one part over 10^{12} or even better, meaning that we could measure a distance of 1 km with an error of 1 nm, which is a millionth of a millimeter. Later we will see that this is really difficult, yet it is possible. Another important property of the interferometry is that it allows to perform measurements that are directly *traceable* to the *meter* (the length unit of the International System of Measurement Units, SI), that is, the metrological link between the measurement result and the length standard is very short. Indeed, the meter is defined through the physical constant (speed of light) c and the time unit s as the length of the path traveled by light in vacuum during a specific fraction of a second (CGPM 1983). So, if we measure the frequency of a laser wave, we can have an *absolute* knowledge of the laser wavelength and consequently the interferometric measurements made with the same laser are directly linked (or traceable) to the SI. No further calibration is needed. This unique property cannot be found in any other measurement instrument like capacitive transducers, LVDT, line-scales, etc., all of them needing a periodical calibration procedure.

Because of these unique properties (traceability to the SI and extremely high accuracy), interferometry is used in many fields of industrial and scientific worlds. Interferometers are used both *directly* to measure the displacement of moving parts of some machines, and *indirectly* to calibrate displacement measurement instruments, or finally to calibrate objects which in turn are used as *reference standards* to calibrate some measurement tools. Just to mention only two, here follows very classical examples of interferometer application.

If we want to calibrate a caliper having a resolution of 20 μm, we make use of some steel *gauge blocks* having given nominal values. These blocks are periodically calibrated by comparison (using a mechanical comparator having 0.1 μm uncertainty) with *reference* gauge blocks, which in turn are periodically calibrated with an interferometer with an uncertainty of 50 nm (0.05 μm).

If we want to measure the diameter of some nanospheres or other nanostructured artifacts, we make use of a *scanning probe microscope* (SPM) having sub-picometer resolution. The scanning stage used to move the sample in a deterministic way is moved by means of piezoelectric actuators under the control of capacitive sensors. These *piezocapacitive* actuators are eventually periodically calibrated (compared) with an interferometer. Alternatively, the scanning stage of the microscope can move directly under the control of two interferometers (for x and y axes). That adds complexity to the system but allows to avoid calibration procedures.

In this chapter, we will describe the working principle of the most popular interferometers, the properties and the limits of the same, and the solution adopted to realize interferometers in the field of measurements at the nanometer and picometer scale.

5.2 Interferometric Techniques

The *undulatory* nature of light is known since the beginning of the nineteenth century when Thomas Young demonstrated unmistakably with his *double slit experiment* that the light is a wave and, as a wave, experiences interference (Young 1804); later, James Clerk Maxwell formalized the electromagnetic theory once for all. As waves of any nature, two light waves can interfere in a constructive or destructive manner according to the relative phase. In its simplest form, an interferometer can be seen as a device that divides a continuous light wave (i.e., a light beam) into two equal parts; the two waves experience different paths and eventually are summed and sent to a target. The intensity of the light on the target is a function of the phase difference between the two beams. In the period from the end of the nineteenth century to the first decades of the twentieth century, many interferometric arrangements have been invented mostly with the main purpose to demonstrate physical theories. The most famous ones (and the first) are the Michelson interferometer invented by Albert Michelson and Edward Morley in 1887 to investigate the existence of *ether* and the Fabry–Perot interferometer (FPI) invented in 1899 by Charles Fabry and Alfred Perot to perform accurate measurement of wavelength of light sources (Michelson and Morley 1887; Fabry and Perot 1899). Other famous interferometers taking their names from the respective inventors that are worth to be mentioned are Mach–Zehnder (1891–1892), Jamin (1856), Fizeau (1851), and Sagnac (1913).

From here, we will only deal with the Michelson and FPIs, the first being the most popular and versatile one, the second allowing exceptional results at the picometer scale.

5.2.1 Michelson Interferometer

In Figure 5.1, the schematic of the Michelson interferometer is shown. All the existing versions of this interferometer have in common the optical elements depicted in the scheme.

First we need a monochromatic light source, that is, a source producing photons having the same wavelength. During the period of Michelson, scientists used lamps with band-pass filters to select only a narrow portion of the electromagnetic spectrum or gas discharge lamps emitting light at a specific wavelength, like *sodium* lamp and *krypton* lamp; since the invention of the *laser* (1960), we can rely on a quite ideal light source which not only generates a perfectly monochromatic light but also a light that can be collimated in a bundle of perfectly parallel rays allowing easy construction of the very long optical setup.

Our laser beam is sent to a *beam-splitter* (BS), which is an optical device behaving as a partially reflecting mirror that reflects half of the energy and allows the remaining to pass through. Now, we have two identical laser beams that are, for the sake of clarity but not necessarily, mutually orthogonal. The two beams are reflected back by two mirrors and sent again onto the BS which now acts as a *beam-combiner*. Each beam is again divided into two equal parts; let us consider the two halves going toward the *detector*. The detector sees the sum of two light beams having

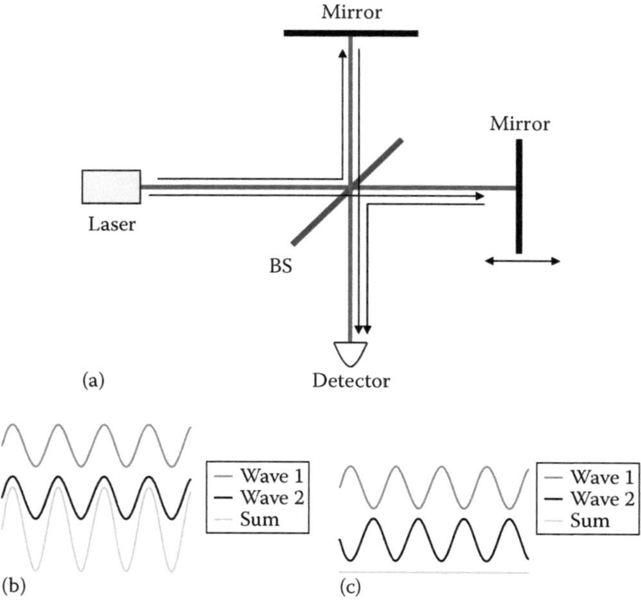

FIGURE 5.1 (a) Schematic of the Michelson interferometer. (b and c) Constructive and destructive interference of electromagnetic waves.

experienced two different optical paths. Now, D being the length difference between the two *arms* of the interferometer, the *optical path difference* (OPD) experienced by the two beams is $2D$.

The two beams can be described as oscillating electric fields having same intensity (normalized to 1), angular frequency ω, wavelength $\lambda = 2\pi \cdot c/\omega$, and the *phase difference* between them being $kx = k \cdot 2D = 2\pi/\lambda \cdot 2D$:

$$E_1 = \sin(\omega t)$$

$$E_2 = \sin(\omega t + kx) = \sin(\omega t + 2\pi/\lambda \cdot 2D).$$

Since the detector (be it your eye or a photodetector) generates a signal proportional to the *power* transported by the light and the power is proportional to the *square* of the electric field E, the interference signal S is therefore proportional to $(E_1 + E_2)^2$. Developing the product and neglecting the term oscillating with angular frequency 2ω, the signal measured by the detector (or your eye) is

$$S = 2\left(1 + \cos\left(4\pi\frac{D}{\lambda}\right)\right),$$

that is, a signal which goes between a zero (dark) and a maximum (light) when one of the mirrors experiences a $\lambda/4$ displacement, and then again to zero for the next $\lambda/4$ displacement, and so on. In other words, we have built a simple instrument acting as a ruler having infinite regular marks at a distance of $\lambda/2$ from each other. Using, as an example, a green laser having $\lambda = 500$ nm, our instrument will be able to easily resolve a displacement equal to $\lambda/4 = 125$ nm or 1/8000 mm.

With this instrument, Michelson and Benoit in 1893 compared the international standard of meter (preserved at BIPM in

Sevres, Paris) with different wavelengths emitted by a cadmium lamp. This important measurement led to the definition of meter in 1960 (CGPM 1960) based on the orange line of the krypton lamp, which in turn opened the road to the *dissemination* of the meter unit through the use of interferometry.

Our very first Michelson interferometer made with a laser, a BS, two mirrors, and one detector is capable of resolving very small displacements but has a defect which limits its use in practical displacement measurements: it is not able to distinguish the *direction of movement*. Indeed, a periodical oscillation of the detector signal could represent uniform movement of one mirror in one direction, or in the opposite one, or even could be generated by an oscillation of a mirror back and forth. This, in practice, is due by the fact that the real displacement information is in the phase of the interference signal S, but the detector only sees the projection of the said phase on the cosine axis. In order to retrieve the complete phase information, we need to extract a sin φ signal. There are many possible techniques to obtain a pair of signals *in quadrature* from a Michelson interferometer; later, one of these methods will be described in detail. Here, we will describe a simple method, although rarely used in practical implementations, but useful for the following discussion.

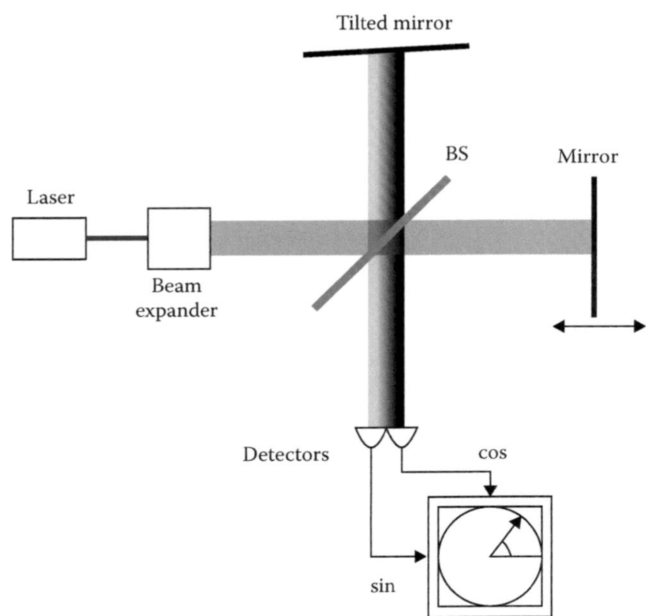

FIGURE 5.3 Tilted mirror homodyne interferometer. The two halves of the beam experience a $\lambda/4$ phase shift.

5.2.2 Homodyne Michelson Interferometer

As in Figure 5.2, the light beam has a width W. On the detector side, two detectors D1 and D2 intercept half beam each. Along one of the two arms, we place some piece of transparent material which, because of its optical properties, introduces an optical delay to half of the beam with respect to the other half not passing through it. If this optical delay is equal to $\lambda/8$, or more easily $(2N + 1)\ \lambda/8$, after passing through it back and forth, the beam experiences an odd number of $\lambda/4$ longer optical path with respect to the other half of the beam (Downs and Raine 1979). As a result, the two interference signals on detectors D1 and D2 have a $\pi/2$ phase difference (i.e., they are in quadrature) independent of the movement of the mirrors. We can compose the two signals on a Cartesian plane and reconstruct the *phase vector*. Now, it is straightforward to measure the phase angle of this vector, thus having complete information on the movement of the mirrors.

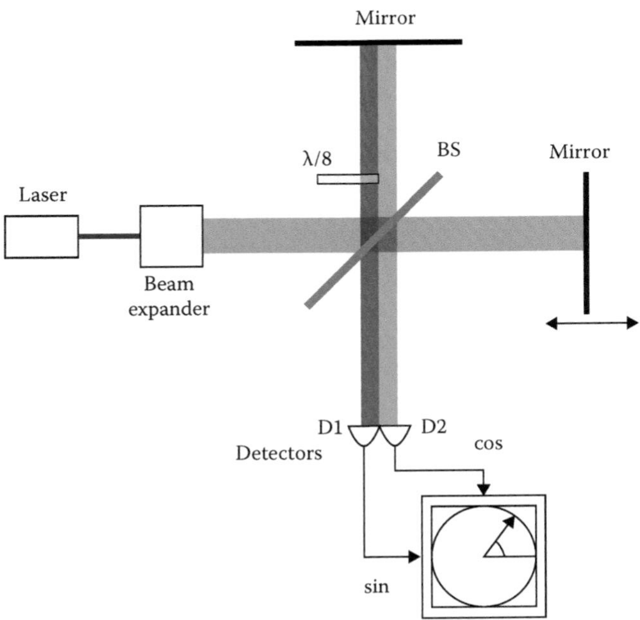

FIGURE 5.2 Practical homodyne interferometer: A retarder introduces a phase delay on one portion of the beam such that the two interference signals on detectors D1 and D2 are in quadrature.

Another easy way to obtain two quadrature signals is to tilt one of the mirrors (preferably the one not moving) in such a way that the average physical distance experienced by the two halves of the beam is $\lambda/4$, thus obtaining a similar effect on the two detectors, as depicted in Figure 5.3.

This kind of practical interferometer is called *homodyne* Michelson interferometer. The name is due to the fact that a single laser frequency is used in contrast to what happens in the heterodyne interferometer.

5.2.3 Heterodyne Michelson Interferometer

This interferometer is the most widely used because of its extreme versatility. Since the first commercial product from Hewlett–Packard in 1970 (HP 1970), several products are available today off-the-shelf at reasonable prices. The working principle exploits the phenomena of the *polarization* of light and the *beat* between two frequencies. The polarization status of a laser beam describes how the electric field of the electromagnetic radiation oscillates. Without going into details, we only need to

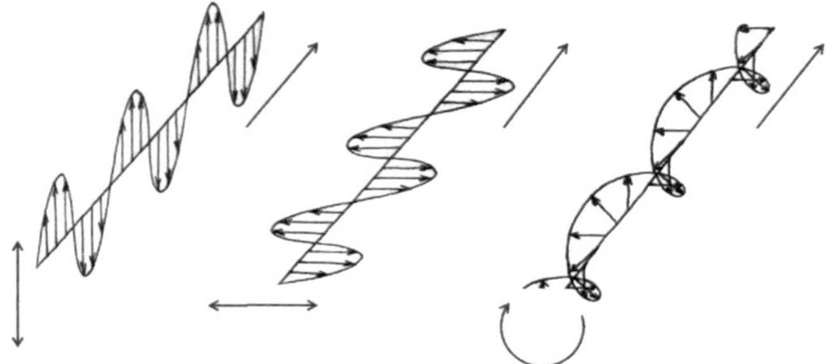

FIGURE 5.4 Vertical and horizontal linear polarized light, and quadrature combination of the two generating circular polarization.

know whether the polarization is linear or circular. In a linear polarized beam, the electric field oscillates on a plane (e.g., vertical or horizontal polarization); in a circularly polarized beam, we have an electric field which rotates along the propagation direction with a periodicity λ. A simple and useful way to imagine a circularly polarized beam is to break down the electric field into two orthogonal components (e.g., vertical and horizontal) and to introduce a $\pi/2$ (or $\lambda/4$) delay between the two components (Figure 5.4).

The simplest heterodyne setup is illustrated in Figure 5.5. A particular laser source generates two superimposed laser beams having mutually orthogonal linear polarization states and two slightly different frequencies ω_1 and ω_2. The frequency difference $\Delta\omega$ is typically of the order of several megahertz (small compared with the oscillation frequency of the electric field of the order of hundreds of terahertz). A small portion of the laser beam is extracted by means of a BS and sent toward detector D1 passing through a *polarizer*. The polarizer is an optical device that selects the components of light parallel to its axis. The axis of the polarizer is set at 45° with respect to the polarization of the two beams, so that two parallel equal-intensity portions go to the detector (only parallel polarization beams can interfere). It is easy to see, with the same electric field equations used before,

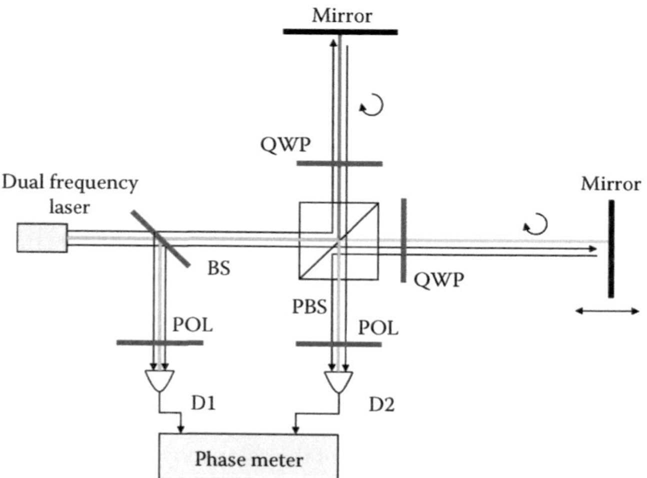

FIGURE 5.5 Heterodyne interferometer: Mirror displacement is transformed in phase delay between two beat signals.

that the interference between two laser beams having different frequencies generates a sinusoidal signal S1, called *beat signal*, having frequency equal to $\Delta\omega$. The beams now proceed and cross a *polarizing beam-splitter* (PBS) having the property of transmitting the light with horizontal linear polarization and reflecting the light having vertical polarization. This is a critical element of the heterodyne interferometer as we will see later. The two beams having frequencies ω_1 and ω_2 now have been separated and sent to the two arms; they pass through two *quarter-wave plates* (QWPs), are reflected back by the two mirrors, and pass again through the QWP before reentering the PBS. The QWPs are made of *birefringent* materials (such as quartz or mica) and have the property of transforming a linear polarized beam into a circularly polarized one by retarding one of the two orthogonal components with respect to the other by $\lambda/4$. When crossed twice, the retardation becomes $\lambda/2$ and it can be seen that the effect is to transform the beam again in linearly polarized, but rotated by 90° with respect to the initial orientation. In other words, by passing through a QWP twice, the two beams invert their polarization orientation from vertical to horizontal, and vice versa. Now, the PBS readdresses the two beams toward the detector D2 (again through a polarizer as for D1). Here, a new beat signal (S2) with frequency $\Delta\omega$ is generated. When one of the two mirrors is moved generating an OPD change, the effect is to change the phase of S2 with respect to S1. Measurement of the mirror displacement is thus performed by measuring the phase changes between signal S1 called *reference signal* and signal S2 called *measurement signal*.

We can see the heterodyne interferometer as an evolution of the homodyne interferometer. Indeed, in the homodyne interferometer, we measure the phase change induced by the moving mirror by comparing it (with the help of interference) with a portion of the *same* beam having fixed arm length and hence fixed phase (from *homo* = *same*). The same happens in the heterodyne interferometer, but to measure the phase change induced by the moving mirror, we compare it with another beam with different frequency (hence the name from *hetero* = *different, other*). This means that the phase difference is induced on an AC signal rather than on a DC signal. Obviously, phase changes of an AC signal can be measured only with respect to another AC signal where the changes are not induced, that is, taken before the OPD change.

Finally, we can also see the phase changes between S2 and S1 as induced by the *Doppler frequency shift* caused by the movement of the mirror in the measurement arm. The calculations made using different formulas lead to the same result. For this reason, such kind of interferometer is sometimes referred to as *Doppler interferometer*.

Although more complicated with respect to homodyne interferometer, the heterodyne interferometer allows simpler implementation in the measurement electronics. This is due to the fact that measuring a phase between two alternate signals can be done in a very accurate way, and, mostly, it is immune from changes in signal amplitude. On the other hand, the considerations we have given on phase measurement of the homodyne signals work well

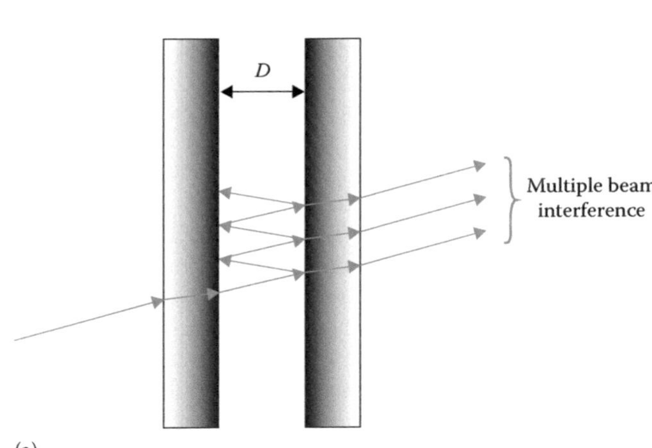

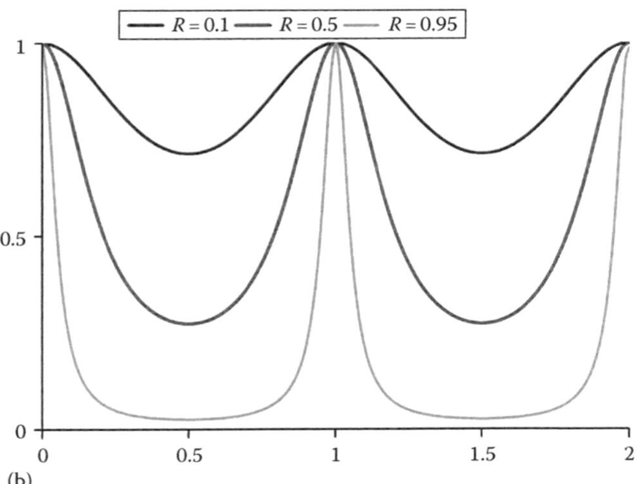

FIGURE 5.6 Fabry–Perot interferometer made from two-faced partially reflecting mirrors facing each other. (a) Multiple beam interference. The transmitted power (b) is a function of the reflectivity R of the mirrors and the ratio $2D/\lambda$.

only when the two signals are perfect: any changes in amplitude or offset (caused either by electronic drifts or by air turbulence) of one of the two would induce relatively large errors, or even the complete loss of the interferometric signal.

5.2.4 Fabry–Perot Interferometer

The FPI in its simplest version is made from a pair of partially reflecting mirrors faced and parallel. A comprehensive analysis of the F–P physics can be found in optics textbooks; for our purposes, we can consider the following simplified scheme.

Referring to Figure 5.6, the light beam coming from one side hits the first mirror (M1) and is partially reflected and partially transmitted according to a ratio given by its *reflectivity factor R*. Considering only the transmitted beam, this hits the second mirror (M2) and again is partially transmitted and partially reflected back toward M1. This process continues indefinitely. As a result, to the right side, we have a superimposition of infinite light beams having decreasing intensity and having a respective OPD equal to $2D$, D being the distance between the reflecting surfaces. This kind of interferometer is thus called *multiple beam* interferometer in contrast to the *dual beam* interferometer as the Michelson one. It is easy to see that when the distance between the mirrors is $D = N\lambda/2$, all the exiting beams have the same phase and interfere constructively, and the transmitted intensity is maximum. The result is summarized in the following formula:

$$I_T = \frac{I_0}{1 + F\sin^2(2\pi D/\lambda)}.$$

In practice, the F–P *transfer function* (called *Airy function*) is a periodical series of sharp peaks, the position of which depends on the relationship between the distance of the mirrors and to wavelength of the laser light. The sharpness of the peaks is given by the parameter F, which in turn is proportional to the reflectivity of the mirrors according to the formula:

$$F = \frac{4R}{(1-R)^2}.$$

Another parameter often used to define the quality of the interferometer (also called F–P *cavity*, or *etalon*) is the *finesse* defined as the ratio between the distance between the peaks and the half-width (in frequency) of the same peaks. The finesse is defined as

$$f = \frac{\pi\sqrt{F}}{2}.$$

A very interesting property of a high-finesse F–P cavity is that the wavelength (or the frequency) at which the maximum power is transmitted (also called *resonant frequency*) is a direct function of the mirror distance D. As we will see, this will allow very sensitive displacement measurements.

5.3 Limits of Interferometers at the Picometer Scale

As said, laser interferometers allow the realization of ideal measuring instruments having exceptional resolution and accuracy. In the real world, interferometric measurements encounter some factors that limit the ultimate performances of the instrument. In order to classify the possible error sources, it is useful to recall the nature of the interferometric measurement which in a word is a *rotating phase*. The cyclic nature of interferometric signal makes the interferometer behave as an ideal ruler with regular spaced marks having submicrometer pitch and infinite length.

This allows extremely accurate measurements at medium to long distances, that is, meters to kilometers. Indeed, in the case where micrometer resolution is sufficient, the measurement is merely a matter of counting the number of fringes (i.e., the phase revolutions). In this range of measurement, the accuracy is directly proportional to the accuracy of the wavelength, that is, factors influencing the value of λ will affect the measurement result with an error proportional to the length to be measured. Briefly, we will mention the two main causes of errors, which are the *refractive index of air* and the *stability of the laser*. The refractive index of air r is the ratio between the speed of light in vacuum c and the speed of light in the presence of atmospheric gases. This ratio is slightly larger than 1 ($r \approx 1.00025$), meaning that the wavelength in air is *shorter* than that in vacuum. r is dependent on the environmental parameters such as temperature of air (around 1 ppm per °C), relative humidity (around 0.1 ppm per %RH), air pressure, and CO_2 content. The knowledge of r is the real practical accuracy limit in interferometric measurements performed in the atmosphere. Accuracy of some parts in 10^{-8} is an ultimate limit. Measurements performed in vacuum do not suffer from this limitation because r is exactly 1. In this case, the accuracy is only limited by the knowledge of λ. The latter is limited by the stability (and accuracy) of the laser source. It is possible today to build lasers, the wavelength of which is referred (or *locked*) to some molecular or atomic energetic level transitions with accuracies of 10^{-14} and better. Just to mention an extreme example demonstrating the potentiality of interferometers in vacuum, in the *ESA-NASA space mission LISA* (a mission for the study of gravitational waves, at present under development), three satellites placed at 5 million kilometers distance from each other are under interferometric control with sensitivity in the distance variations of tens of picometers (LISA). In this case, the lasers will be locked to ultrahigh finesse F–P cavities.

But, when we deal with much smaller distance measurements, we want to measure nanometers or fractions of nanometers, and it is not sufficient anymore to *count* fringes. We need to measure very small fractions of a fringe, meaning that we have to be able to divide the 2π phase revolution into small identical parts. In principle this could seem simple, but in practice it is not. A comprehensive analysis of causes of errors associated with the subdivision of the optical fringe can be found in the study of Bobroff (1993). Here, we will briefly summarize the nature of the most important causes of errors, both in the homodyne and in the heterodyne interferometers.

Let us start first from the homodyne interferometer: we have assumed two ideal interference signals generating two electrical sine and cosine signals realizing what we will call the *phase circle*. Once converted into numbers by some digital system, we only have to calculate the *arctangent* to retrieve the phase value. In the real world, we will deal with some possible effects: (1) the amplitude and/or the offset of the signal is not constant with time and (2) the phase difference is not exactly 90°, and/or changes with time. Referring to Figure 5.7, the first two effects will cause a distortion of the phase circle which will become an ellipse not centered around zero. It is easy to see that if we try to calculate the phase from the theoretical formula (assuming the phase circle undistorted), we will make errors in the phase and hence in the displacement estimation. These errors will have cyclic nature and a periodicity that can be 2π in the case of offset errors and π in the case of amplitude unbalance. So, once again, these errors will not affect *large* measurements because the average error for each complete phase revolution is zero but are critical for *subfringe* displacements.

The amplitude and offset errors could in principle be kept under control by means of careful design of the detectors and related electronics. The second effect—not-perfect *quadrature* between sine and cosine signals—has the effect of changing the phase circle into an ellipse having 45° axis. This effect also causes errors with π periodicity. Phase delay errors are more difficult to be kept under control, because they are dependent on the optical properties of the material (e.g., thickness, temperature, incidence angle, etc.); so, even the best homodyne interferometer can have some small phase errors. As a result, in any real homodyne interferometer, we have cyclic errors also called *nonlinearity*, due to the nonideal behavior of the optical and electronic setup. Typical nonlinearity magnitude can be of several percentage corresponding to an error of several degrees in the phase estimation, in turn corresponding to nanometer error in the length measurement. In practical realization, it is very difficult to keep the error less than 1 nm without some error correction procedure.

In fact, in principle, it is possible to correct these errors to a very low level: it would be enough to know the shape of the ellipse (or whichever the distortion is) and operate some kind of transformation which corrects the phase circle, or, even easier, to implement some *lookup table* that associates the correct phase value to any pair of values coming from the detectors. This kind of approach is usually implemented with remarkable results (Heidemann 1981). Independently, in the adopted method, the general concept is that we have to move the interferometer (possibly in a controlled way), observe the shape of the phase circle, and adopt the needed corrections. The weak point is that it is not possible to undertake this procedure once and use it forever; indeed, these errors can change both with time and with the mirror position, so the calibration must be done as often as possible. In particular, when we deal with displacements much smaller than λ and we are not allowed to move the system during the measurement for the calibration procedure, uncontrolled errors can occur. As an example, if during the measurement the offset or the gain of one of the signal changes, this would be interpreted as a displacement of the mirror causing an error in the measurement. This is the main reason why heterodyne interferometer is less prone to errors caused by the instability of optical and electronic components.

Heterodyne interferometer relies on the phase difference between two AC signals in the megahertz range. This measurement is less sensitive to the quality of the signals (e.g., if one of the two signals changes with amplitude or offset, the phase measurement is not affected). Nevertheless, they suffer from a cause of error called *polarization mixing*. We have assumed that two laser beams having different frequencies are perfectly separated into

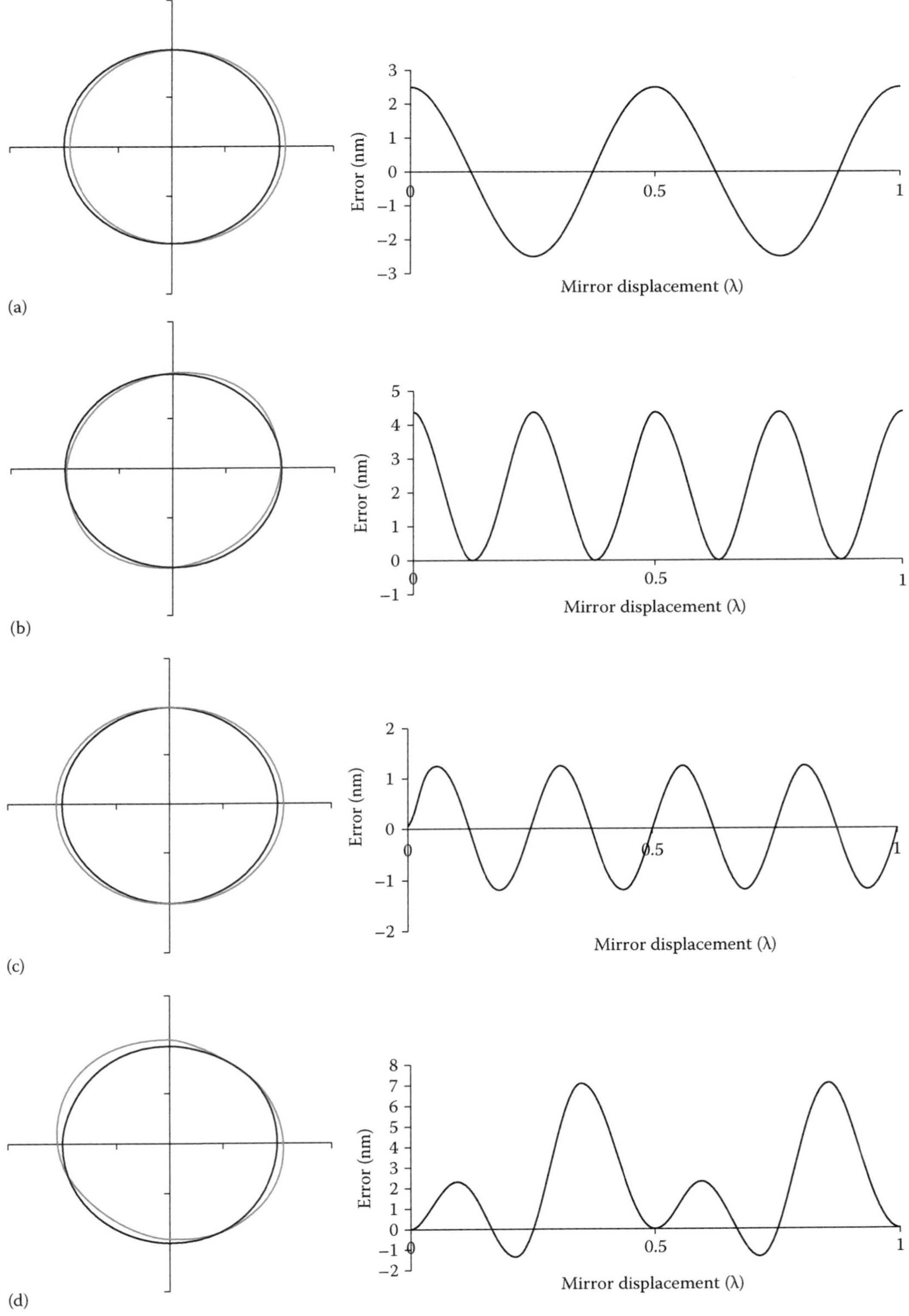

FIGURE 5.7 Effect of offset, amplitude, phase distortion on displacement evaluation errors. At the left, the distorted circle (gray); at the right, the errors in nanometers. From top left: (a) effect of 5% offset unbalance, (b) effect of 5% gain unbalance, (c) effect of 5° error with respect to perfect quadrature, (d) combination of the three.

the two interferometer arms. In fact, that would be possible only if the PBS is perfectly selective, while typical selectivity, or *extinction ratio*, or transmission/reflection ratio, is close to 99%–1% and vice versa. This means that a small portion of the beam with a given frequency runs the wrong arm. And even if the PBS were perfect, it is very difficult to have perfectly linearly polarized and orthogonal laser beams (a small ellipticity is always present); so, the polarization mixing would unavoidably occur. The consequence of this effect is the combination of a signal carrying the phase information related to the mirror displacement with a small spurious signal having fixed phase difference with respect to the reference one. As a result, again a periodical error having periodicity π appears in the measurement result. Typical polarization mixing of the order of 1% causes displacement errors of the order of a couple of nanometers. Also, in this case, correction procedures based on the preliminary evaluation of the error are carried out (Tanaka et al. 1989; Wu et al. 1999; Bong 2002).

5.4 Solutions for Error Reduction

So far, we have seen that classical interferometric techniques have typical nonlinearity errors of the order of 1 nm or more, which is far from the needs of the most demanding scientific and industrial fields where a goal of picometer resolution and 10 pm accuracy is expected. We have also seen that in principle it is possible to reduce the cyclic error by implementing correction procedures that have the drawback that the error must be continuously monitored (furthermore the error monitoring should be carried out by means of independent measurements, while in most cases the systems make use of a self-calibration procedure). In this section, we will describe special interferometric techniques and instruments designed to perform accurate measurements at the picometer scale where the errors are kept under control by means of careful optical design or error-monitoring method.

An approach to minimize the heterodyne cyclic errors has been adopted by Seppa et al. (2011). The idea is to combine a heterodyne interferometer with a *capacitive sensor*. The latter is a sensor based on the measurement of the electric capacitance of a capacitor made from two plane-parallel armatures to infer the distance between the same. The sensor has exceptional sensitivity although it is affected by rather high nonlinearities for large displacements; nevertheless, we can assume that for a displacement as low as $\lambda/2$, the nonlinearities of the capacitive sensor can be neglected. So, in principle, by combining the $\lambda/2$ interval measurements made with the interferometer and the subwavelength measurement made with the capacitive sensor, we can obtain a high-resolution error-free measurement. In practical realization (see Figure 5.8), the moving mirror of

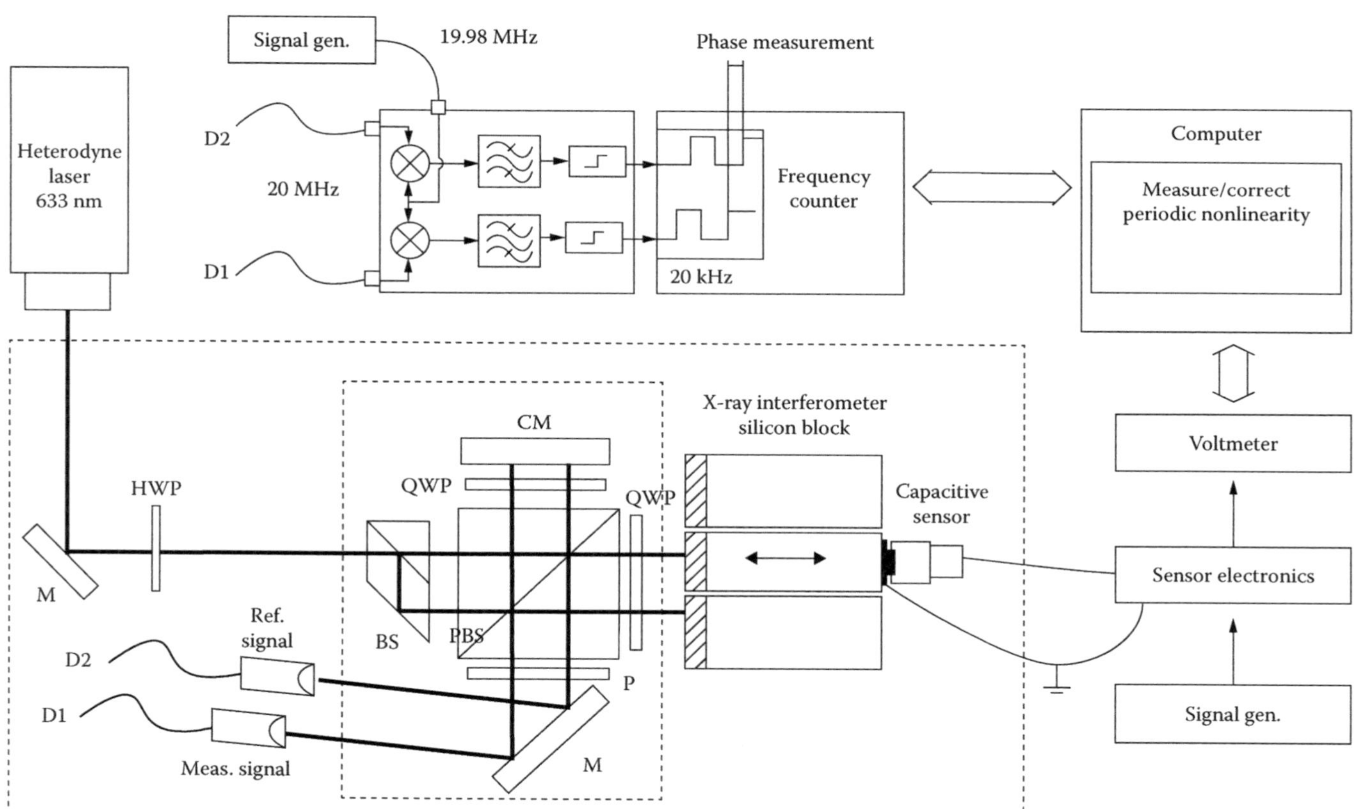

FIGURE 5.8 Combined capacitive heterodyne interferometer. In the lower block, the optical arrangement of the heterodyne interferometer having two parallel arms is visible. On the right side, one armature of the capacitive sensor is attached to the opposite side of the moving mirror. At the top, a detail of the heterodyne phase measurement is shown.

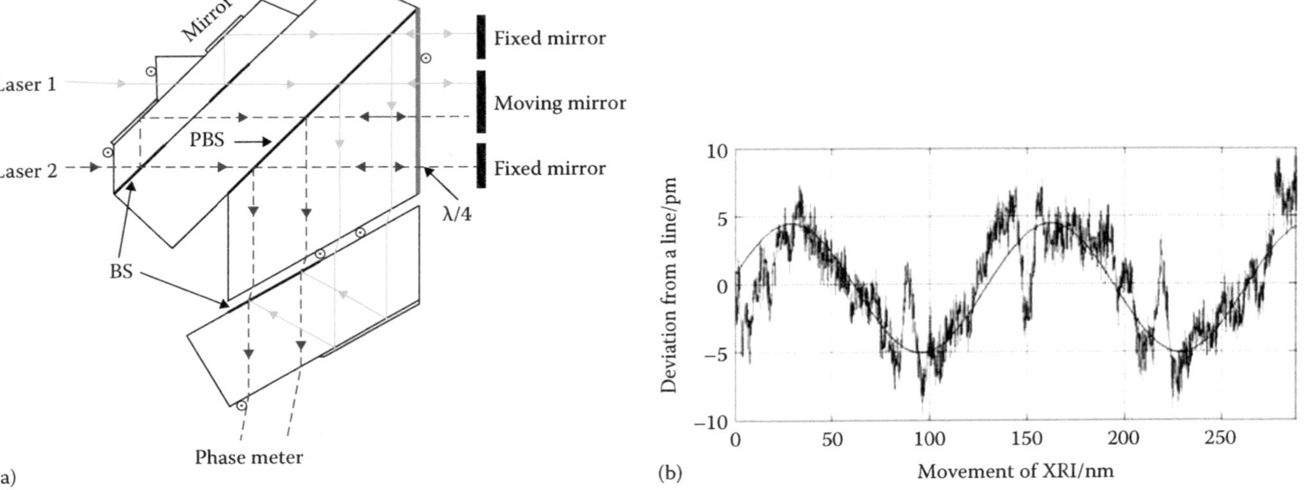

FIGURE 5.9 Separated beams heterodyne interferometer. (a) The optical BS assembly; (b) the residual errors of the interferometer showing a nonlinearity as low as ±5 pm with λ/4 periodicity.

the interferometer hosts the moving armature of the capacitive sensor in such a way that the two instruments "see" the same displacement. By comparison with an x-ray interferometer, the combined optical–capacitive interferometer has demonstrated ±30 pm error (Pisani et al. 2012).

Another approach to avoid the problem of polarization mixing occurring in heterodyne interferometers is to realize an interferometric setup with spatially separated measurement and reference beams. This approach has been adopted by Jens Fluegge at PTB and described by Pisani et al. (2012). The basic idea is to build an almost ideal heterodyne interferometer where the two orthogonally polarized laser sources are generated starting from a duplicated Nd:YAG laser emitting light at 532 nm and using two acousto-optical modulators. The two lasers are sent to the interferometer through optical fibers in such a way that the two beams are perfectly separated. The interferometer is rather complex and is illustrated in Figure 5.9. A series of prisms having surfaces coated with mirrors, with 50% BS, with PBS, and QWP (λ/4), separate and combine the laser beams. The beams are sent to three mirrors (two fixed for reference and one moving for measurement), reflected back, mixed, and eventually sent to the two detectors, which perform the phase measurement. The high level of complexity of the optical setup is justified by an almost absolute *absence of polarization mixing*. The characterization of the separated beams' interferometer against an x-ray interferometer shows an exceptionally low nonlinearity (less than 10 pm p.p.).

A method to perform nonlinearity-free optical measurement is based on the use of high-finesse FPI. We have seen that the resonance frequency of a FPI is linearly dependent on the distance between the mirrors. If we "lock" the frequency of a tunable laser to the resonance peak of the FPI, we have realized an oscillator having frequency proportional to D. Now, if we build a FPI where one of the two mirrors is fixed to the moving part that we have to control, we simply have to measure the laser frequency to know the displacement of the mirror. Thanks to the exceptionally high resolution of frequency measurements, displacements to the level of the femtometers can be easily done. Examples of applications of this measuring principle can be found in the studies of Lawall (2005) and Bisi et al. (1999). The realization of a practical measuring instrument is described by Pisani et al. (2012). A differential FPI is realized (see Figure 5.10a) with two identical *extended cavity tunable lasers* respectively locked to a pair of nominally identical FPIs. The latter are built facing two concave mirrors in front of two flat mirrors that are the reference and the measurement mirrors materializing the displacement to be measured. By mixing the beams of the two lasers, we obtain a beat signal having frequency equal to the difference between the laser frequencies. This frequency is proportional to the relative displacement of the two mirrors. A comparison with an x-ray interferometer (Pisani et al. 2012) demonstrates sensitivity and linearity to the picometer level (see Figure 5.10b).

A further method to reduce the nonlinearities of homodyne interferometer is described by Kren (2009). The idea is to build a homodyne interferometer which makes use of two different lasers with rather different wavelengths. The schematic is shown in Figure 5.11. Two lasers (a green and a red one) are superimposed by means of a *dichroic mirror*, becoming a single light beam which runs all the interferometer, and separated into two orthogonally polarized parallel arms by means of two calcite-polarizing beam displacers (PBDs) and sent toward the two mirrors (one fixed and one moving), recombined by the same PBDs and sent to the phase detector assemblies (RD and GD). Here, thanks to the dichroic mirror and two interference filters, each phase detector "sees" only one wavelength. In practice, we have realized two independent interferometers sharing the same path, hence measuring the same OPD. Since the wavelengths are different, the periodicities of the residual nonlinearities (kept low by careful optical design) are also different. This allows us to use one interferometer to characterize the error of the other and vice versa.

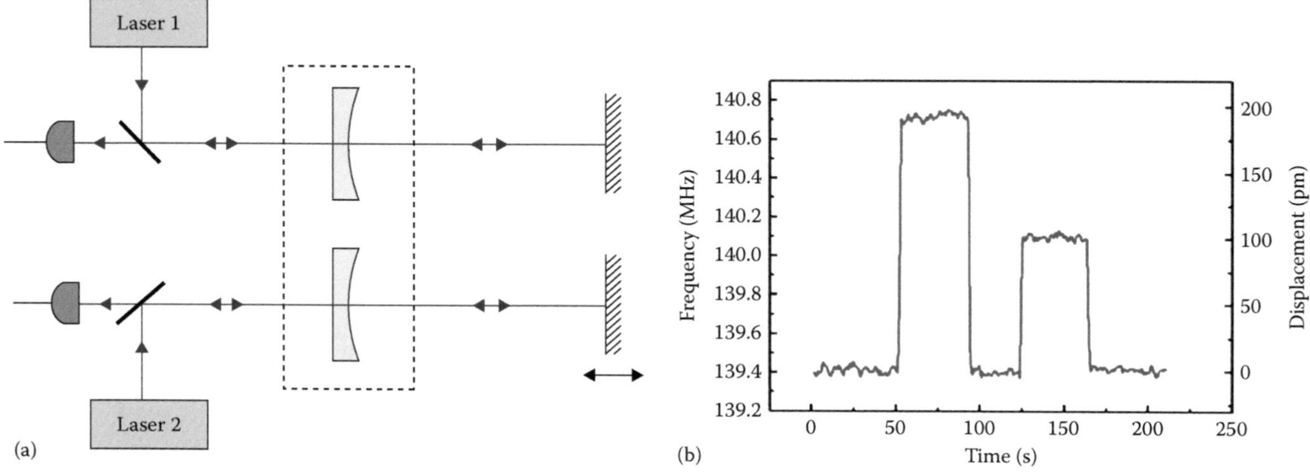

FIGURE 5.10 Differential FPI. (a) Two tunable lasers are locked to a pair of FP cavities built facing two concave mirrors in front of two flat mirrors. Measuring the frequency difference between the two lasers, the relative displacement of the two flat mirrors is obtained. (b) The movement of the mirror of an x-ray interferometer is measured with picometer resolution.

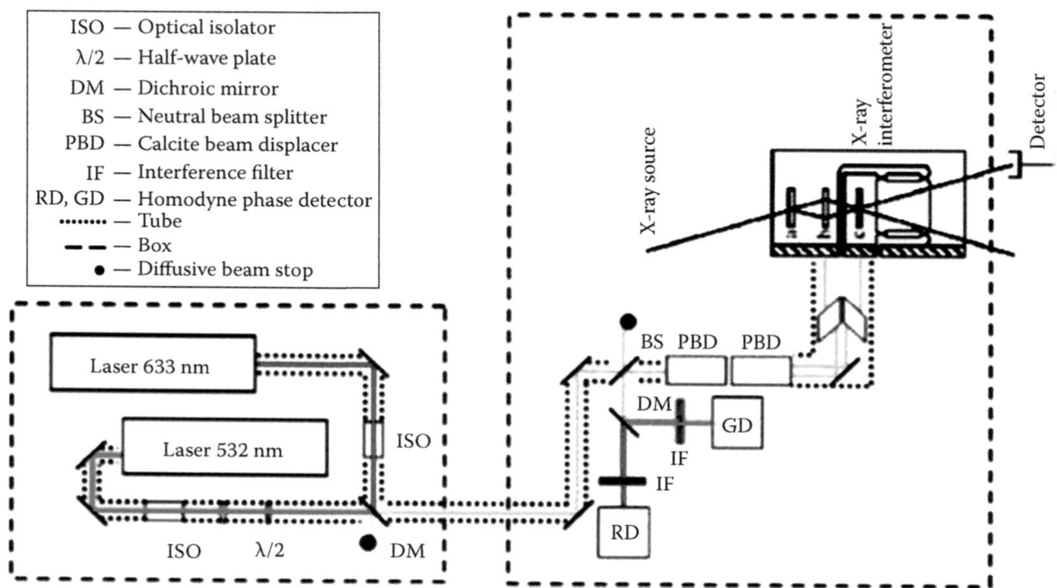

FIGURE 5.11 Schematic of the dual wavelength homodyne interferometer. Left: the two stabilized lasers having different wavelengths are combined to behave as a single beam. Right: the homodyne interferometer aiming at the two mirrors of the x-ray interferometer.

Another trick to get rid of the polarization mixing effect in heterodyne interferometer is to make the phase-meter work always at a fixed phase. In the interferometer proposed by G.B. Picotto at INRIM (Pisani et al. 2012), an electro-optical modulator (EOM) is introduced in one arm of the interferometer. The EOM is a crystal, which changes its optical properties when immersed in an electric field causing a phase delay in the light passing through it, proportional to the voltage applied across it. If during the changes in the OPD induced by the mirror displacement, we compensate the phase unbalance by applying the right voltage to the EOM, we can make the phase detector operate always at the same phase (e.g., zero phase). In this way, the phase detector operates at zero error, while the displacement is proportional to the voltage applied to the EOM. Provided the voltage-to-phase function of the EOM is linear (or else is well characterized), this method allows very good results.

An alternative approach to reduce the effect of nonlinearities, whichever is its origin, is to multiply the optical path. As a general scheme, if the measurement arm of an interferometer is reflected n times forth and back between two mirrors, when we change the distance between the mirrors by an amount d,

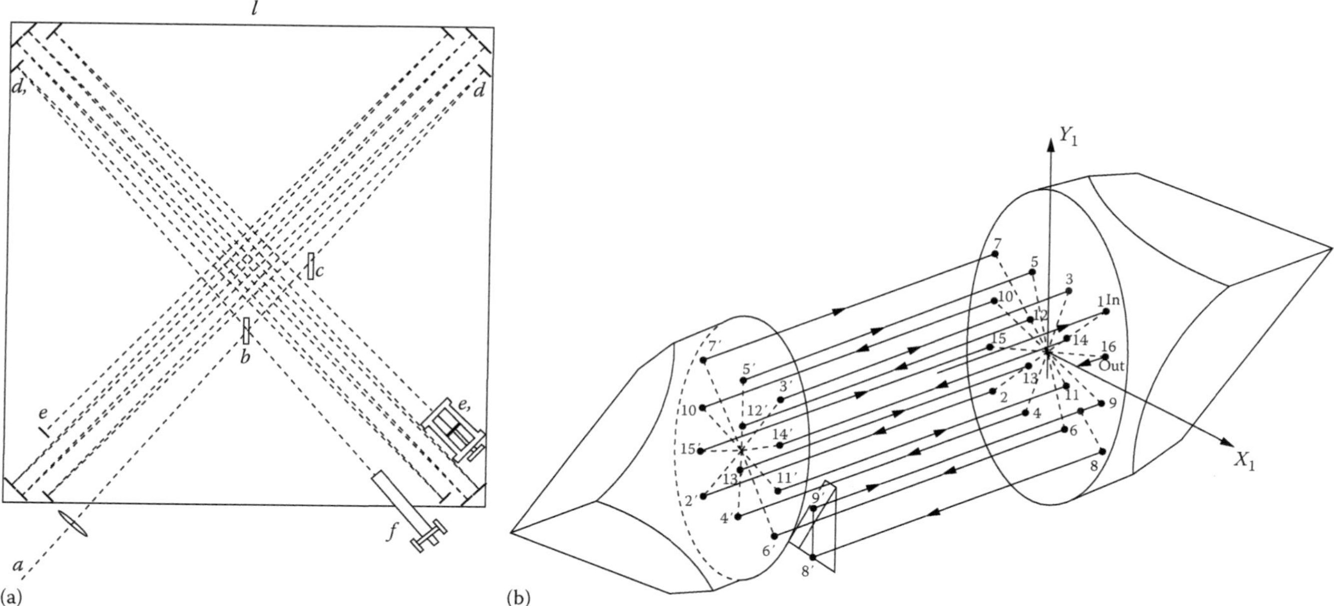

FIGURE 5.12 Adopting multiple reflection to increment interferometer resolution (and/or to reduce errors). (a) Optical setup adopted by Michelson (1887). (b) Multiple reflection between corner cubes described in the work of Vitushkin (1998).

the optical path experienced by the laser beam is $2n$ times d. Multiplication of the optical path by means of multiple reflection with the purpose of increasing the resolution has been exploited since the very first realization of the Michelson interferometer (Figure 5.12a) (Michelson and Morley 1887).

A classical multiplication setup based on multiple reflection between two mirrors is described by Vitushkin and Vitushkin (1998), where two *corner cube retroreflectors* are used to multiply the optical path by 12 (Figure 5.12b). As a result to obtain the real displacement, we have to divide the interferometer reading by the multiplication number. The consequence is that whichever is the nonlinearity of the interferometer, it is divided by n. As we will see later, n can be as large as 100; so, it is possible to achieve picometer-level nonlinearities by using a normal interferometer. Although not an optical interferometer, we have to mention the x-ray interferometer. Because of its unsurpassed resolution and accuracy, this instrument can be used as an ideal displacement actuator used to characterize any displacement measurement device, in particular, high-resolution interferometers. X-ray interferometry can be regarded as a ruler or translation stage, where the graduations or steps are based on the lattice spacing of silicon. Unlike most optical interferometers, the fringe spacing in an x-ray interferometer is independent of the wavelength of the incident radiation; it is determined by the spacing of diffraction planes in the crystal from which x-rays are diffracted: in this case, the (220) planes of silicon with a spacing of 0.192 015 497 ± 1.2 × 10^{-8} nm, at 22.5°C, and a pressure of 1 bar. Silicon is the material normally used for an x-ray interferometer since it is readily available in a pure, defect-free form with a known crystallographic orientation and lattice parameter. The fringe spacing in an x-ray interferometer is therefore not only several orders of magnitude smaller than that in an optical interferometer thereby obviating the need for fringe division as with optical interferometry, but it is also traceable.

Figure 5.13 shows a schematic diagram and photograph of the monolithic XRI developed for the combined optical and x-ray interferometer project (COXI) (Basile et al. 2000) and operative at the National Physical Laboratory (NPL) in Teddinghton (United Kingdom). It is made from a single crystal of silicon in which three thin, vertical, equally spaced lamellae were machined. A flexure stage that has a range of 10 μm and is driven by a piezoelectric transducer has been machined around the third lamella. X-rays are incident on the first lamella (B) and are Bragg-diffracted from the (220) planes. Two diffracted beams are produced; the first lamella can be thought of being analogous to the BS in an optical interferometer. The two beams diffracted from the first lamella are incident on the second lamella (M). Two more pairs of diffracted beams are produced, and one beam from each pair converges on the third lamella (A). These two beams give rise to a fringe pattern whose period is equal to that of the lattice planes from which the x-rays were diffracted. The pattern would be too small to resolve individual fringes; however, when the third lamella is translated (by the piezo) parallel to the other two lamellae, a Moiré fringe pattern between the coincident beams and the third lamella is produced with a period of 0.192 nm. Consequently, the intensity of the beams transmitted through the third lamella varies sinusoidally as the third lamella is translated. By measuring the intensity of one of these transmitted beams, it is possible to measure the displacement of the third lamella. Since the displacements are traceable, the x-ray interferometer can be regarded as a ruler with subnanometer divisions. At the sides of the interferometer,

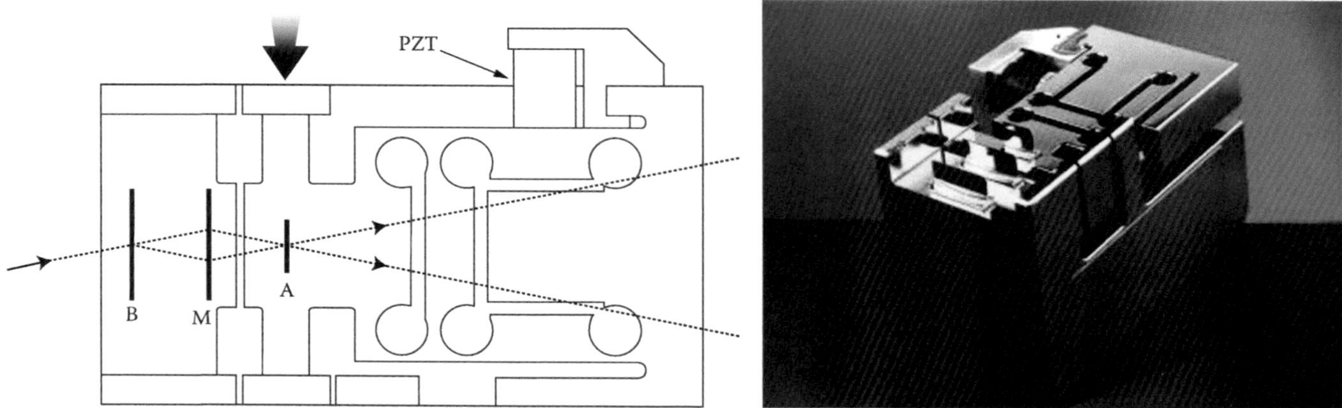

FIGURE 5.13 Schematic diagram and photograph of the monolithic x-ray interferometer.

there are three optical mirrors allowing the XRI to be interfaced to an optical interferometer. These mirror surfaces were coated with aluminum to increase their reflectivity. The x-ray interferometer is by far the most accurate traceable displacement sensor; indeed, it is used regularly for the characterization of top-level measuring instruments (Yacoot and Cross 2003), nevertheless, its use is very complex since a special laboratory equipped with x-ray sources and detector is needed.

In the previous paragraphs, we have listed a selection of the best interferometers recently developed; obviously, the list cannot be complete and the progress in the field is continuously ongoing.

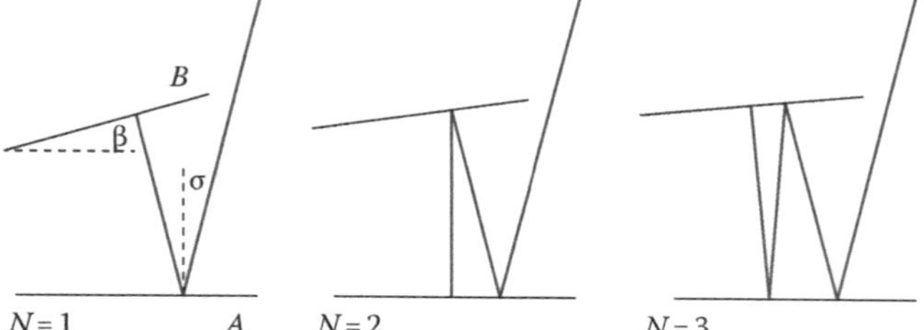

FIGURE 5.14 Ray tracing simulation of the auto-collimating optical path multiplication scheme with (from left) $N = \alpha/\beta = 1, 2,$ and 3 and $M = 2, 3,$ and 4.

5.5 Multiple Reflection Homodyne Interferometer

In this section, an interferometer based on an optical path multiplication principle designed to perform measurements at the picometer scale is described. The basic idea is the implementation of a Michelson interferometer where one of the two arms is sent to an arrangement of two quasi-parallel mirrors where it experiences multiple reflection before being sent back toward the interferometer.

The multiplication setup presented here is based on the two mirror arrangements described in Pisani and Astrua (2006) where it was used to amplify the mirror tilt in sub-nano-radian metrology applications.

A similar arrangement has been exploited in the work of Pisani (2008), where a heterodyne interferometer is described. Referring to Figure 5.14, with β the angle between the two mirrors and α the incidence angle on mirror A, once the condition $\alpha/\beta = N$ with N integer is satisfied, it is easy to see that the beam is reflected N times between the two mirrors before it impinges normally on one of the two. Hence, it is reflected back, retraces the optical path, and after further N reflection exits towards the source. As a result, when we move one of the two mirrors along its normal, we cause a change of the optical path proportional to the mirror displacement multiplied by the number of reflections on the same mirror. This can be seen as a multiplication factor M with respect to the classical Michelson interferometer. It is easy to see that $M = N + 1$ if we move mirror A and $M = N$ if we move mirror B. In this simplified description, the effect of the laser incidence angle on the moving mirror has been neglected (we will discuss this effect later). The number of reflections, and hence M, is limited by geometrical factors. Indeed, the incidence angle and the distance between the mirrors must be arranged so that the beam arrives at mirror A without being cut by the edge of mirror B, then impinges on mirror B being completely contained inside the reflecting surface. This condition, in combination with the beam diameter, limits the minimum distance between successive reflections. The latter, in combination with the mirror diameter, limits the maximum reflections that can be contained by the mirror. Furthermore, the energy loss experienced by the laser beam at each reflection must be taken into account. In Figure 5.15, the power loss versus number of reflections has been plotted for different reflectivity values. It is evident that a classic aluminum mirror having typical reflectivity between 95% and 97% would cause an unacceptable power loss degrading the interferometric signal. Special mirrors with dielectric coating exceeding 99.5% reflectivity should be used.

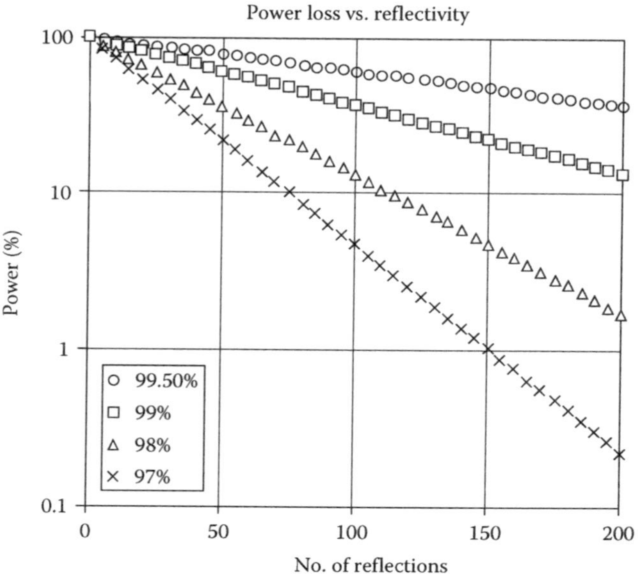

FIGURE 5.15 Energy loss versus number of reflections for different reflectivity values.

On these bases, we will build our homodyne Michelson interferometer as depicted in Figure 5.16 and as described in detail by Pisani (2009). The laser source is a He–Ne frequency-stabilized laser linearly polarized ($\lambda \approx 632.8$ nm). A Faraday isolator (FI) prevents feedback-induced instabilities and rotates the polarization by 45°. A 50% BS divides the beam into two equal parts. One (the reference beam) is sent to a fixed mirror M through a $\lambda/8$ retarder; the latter has the effect of giving a circular polarization to the beam returning toward the BS. The second beam (the measurement beam) is sent to the double mirror multiplication setup. The latter can be considered as a single flat mirror as in a classical Michelson interferometer. The beam reflected back is summed with that coming from mirror M on the BS and is sent to the detector assembly. Here, a PBS is used to separate the vertical and the horizontal components of the two beams and to address them toward photodetectors D1 and D2. The vertical and the horizontal components of the linearly polarized measurement beam are summed respectively with the sine and cosine components of the circularly polarized reference beam, generating two quadrature interference signals, which are eventually amplified and sent to the acquisition system.

In the present setup, by using two 50 mm long mirrors, a multiplication factor exceeding 100 has been obtained. In order to determine the multiplication factor M, two ways are possible: one is to find N starting from the classical Michelson condition (mirror A orthogonal to the beam) and counting the number of autocollimation conditions while gently rotating mirror A, that is, the appearing and disappearing of the signal on the detectors; a second is to make use of a calibrated translator which displaces mirror A by a known value—the integer that best approximates the ratio between the interferometer software reading and the true displacement is the gain ratio M.

Because of the nonzero incidence angle, each reflection on the moving mirror is not orthogonal to the displacement, which implies that the optical path increment is a bit larger than the mirror displacement (cosine effect); on the other hand, the mirror displacement causes a lateral shift of the whole reflection pattern toward the direction where the mirrors are closer, thus reducing the optical path (see Figure 5.17). The combination of these two effects causes a deviation from the gain calculated as if α was zero of the order of few parts per thousand. In demanding metrological applications, it is necessary to know this deviation. We will call f the ratio between the real multiplication factor and theoretical gain M calculated as shown earlier. In the work of Pisani (2009), an exact calculation for $f(N, \alpha)$ is developed. Here, we will only report the simplified

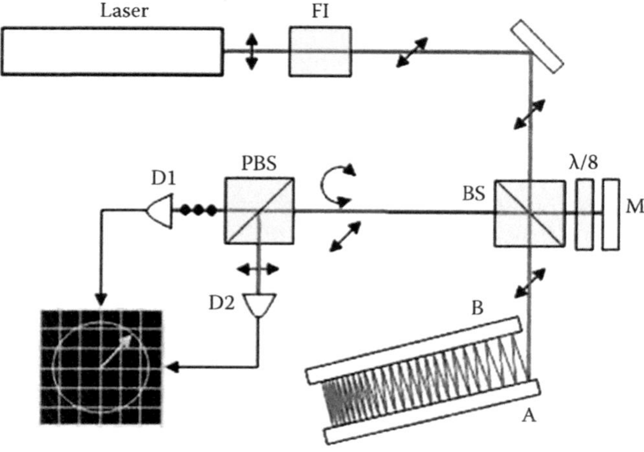

FIGURE 5.16 Simplified schematic of the multipass interferometer.

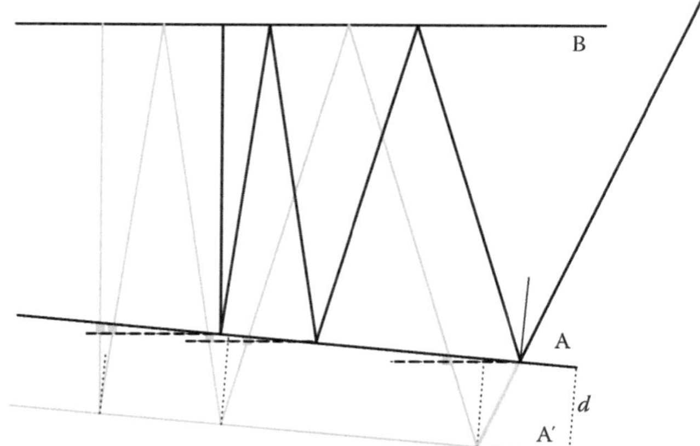

FIGURE 5.17 Evaluation of the errors due to the incidence angle. In the example, $N = \alpha/\beta = 5$. Mirror A is moved in the position A′ with a displacement d. The optical path changes from the black to the gray pattern. The optical path increment can be seen as the added parts below initial mirror position, which is larger than $d \times (N + 1)$, because a 1/cos correction (e.g., the length of the blue segment on the right is $d/\cos\alpha$). To this, because of a shift of the pattern to the left, the gray square segments must be subtracted. The net effect is a gain G slightly less than $N + 1$ (see text).

formulas giving a good approximation when considering relatively small incidence angles ($\alpha \leq 10°$) and large N ($N \geq 100$). The two values, for the cases when respectively mirror A and mirror B are moved, are

$$f_A(N,\alpha) \approx 1 - \left(\frac{\alpha 2}{6}\right)\left(1 + \frac{2}{N}\right);$$

$$f_B(N,\alpha) \approx 1 - \left(\frac{\alpha 2}{6}\right)\left(1 - \frac{1}{N2}\right).$$

The importance of the correcting factor and the accuracy of the approximate formulae can be appreciated by considering the values $N = 100$ and $\alpha = 9° = 0.15708$ rad for case A, which correspond to the experimental conditions discussed later; the exact value for f_A (100, 0.15708) = 0.995811. By using the simplified formula $f_A = 0.995805$, an estimation error of few parts in 10^6 can be tolerated at the micrometer scale.

In order to test the sensitivity of the multipass interferometer, the following tests have been carried out. The results shown herewith have been recorded with $N = 100$. One of the two mirrors is mounted on a piezocapacitive actuator (Physik Instrumente P-753) capable of generating calibrated displacement up to 25 μm. In Figure 5.18, the noise spectral density of the interferometer output is shown. The vertical scale is in nm/√Hz: the noise is input-referred, that is, the interferometer output is divided by the gain. The dark (lower) curve represents the noise spectral density of the interferometer at rest (piezocapacitive actuator switched off): the 100–1000 Hz range is dominated by residual mechanical vibration with a resonance peak at 300 Hz; from 1 Hz to 2 kHz, decreasing behavior is typical of air turbulence and drift effects; and beyond 2 kHz, a plateau due to the residual electronic noise (namely, photodiode and preamplifier noise) dominates the spectrum. The noise limit level is below 20 fm/√Hz. In order to demonstrate the potentialities of the system, a second spectrum (lighter curve) is recorded with the piezocapacitive device switched on, driven with a fixed reference voltage. If we exclude the 100–1000 Hz area where the external noise still dominates, the residual mechanical noise introduced by the active position control system is clearly visible. In particular, from 1 to 100 Hz, an $f^{-1/2}$ noise, which is typical of electronic active elements, can be observed. This characterization is useful to determine the resolution/accuracy limit of precision actuators when used in demanding applications.

Finally, a modulation with 400 pm p.p. amplitude and 0.5 Hz frequency is applied to one of the two mirrors by means of an open loop piezoelectric actuator. Figure 5.18 shows the interferometer output. The signal is sampled at 1 kHz and observed on a 10 Hz bandwidth. Although the residual mirror vibration effect is clearly visible, the sensitivity of the setup can be appreciated.

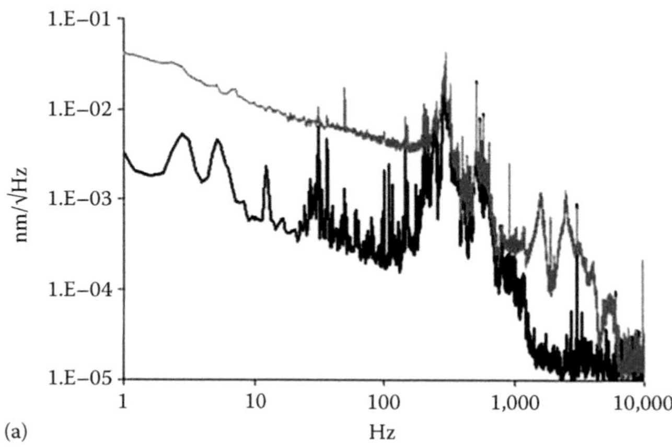

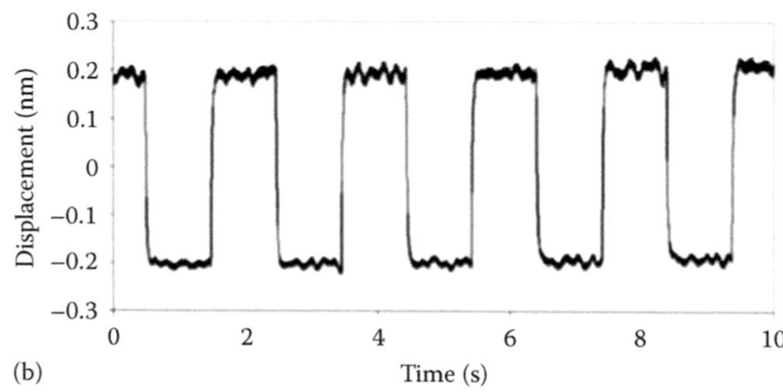

FIGURE 5.18 (a) Noise spectral density of the interferometer with the piezocapacitive actuator on (top curve) and off (bottom). When the actuator is switched on, the mechanical noise density introduced by the active electronics elements, although very low, is clearly visible. When the actuator is off, the result of the measurement is limited by mechanical vibrations and drifts up to 2 kHz and by the detector amplifier noise beyond this value. The electronics noise plateau is less than 20 fm/√Hz. (b) Response of the interferometer to a square-wave displacement having 0.5 Hz frequency and 0.4 nm p.p. amplitude, low-pass filtered at 10 Hz. The residual ripple is due to the mechanical vibration of the mirror assembly.

5.6 Realization of a Precision Displacement Actuator

Based on the interferometer described earlier, a displacement actuator, whose main purpose is the characterization of high-precision interferometers, has been realized. As we have seen, most of the high-resolution interferometers are designed to work in "differential mode"; that is, they measure the difference in displacement of two mirrors placed side by side. This allows canceling the movement of the interferometer block relative to the object to be measured. On the measurement side, one of the two mirrors is fixed to a part of the machine (e.g., the basement) supposed to be the mechanical reference and the other to the moving part to be monitored. As an example in a metrological SPM, one mirror will be placed close to the probe of the microscope (usually fixed) and the other is placed on the stage which moves the sample, thus measuring only the relative

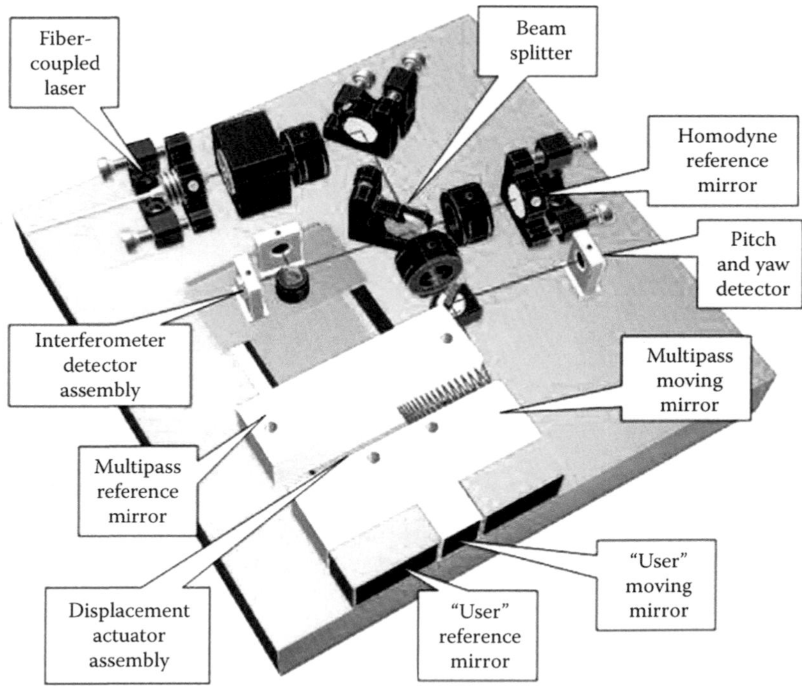

FIGURE 5.19 CAD representation of the overall structure of the actuator.

probe–sample displacement. This is obtained by folding one of the two arms of the interferometer in order to have the two arms parallel. The actuator described here has been designed to characterize this kind of interferometer that explains its overall structure visible in Figure 5.19. The main structure is a 200 × 200 × 30 mm block made from Clearceram-Z, an ultra-low expansion glass-ceramics, manufactured by Ohara (Japan): thermal expansion of $\pm 0.2 \times 10^{-7}/°C$. This allows long-term stability even in not-so-thermally stabilized environments. The various optical components are mounted on compact commercial precision tip–tilt mounts fixed to the main structure by means of epoxy glue. The laser light generated by a He–Ne-stabilized source (SIOS SL-03) having 1 mW power is brought to the bench by means of an optical fiber.

The main Clearceram blocks visible in Figure 5.20 are as follows: the main base on which the interferometers and actuators are fixed, which has a front side mirrored with aluminum coating; two parallelepiped blocks are glued to the base and mirrored in their front surface; the three previously mentioned mirrored surfaces are flat and parallel; the T-shaped block in the middle is the main element which is moved by the piezoactuators assembly described later; its front side is coated with aluminum and is nominally parallel to its backside; the latter is mirrored with high-reflectivity coating ($R > 99.5\%$); the last block is fixed to the base and is coated with the same high-reflectivity coating on its front surface. The three front mirrors are the "user" mirrors for the calibration of the interferometers. In the simplest differential interferometer, two beams are sent parallel toward the target where one mirror is fixed and one is moving, and the measurand

FIGURE 5.20 The "user side" of the actuator made of two side mirrors fixed to the base used as reference for the interferometers under test and a small mirror between them moving back and forth by means of a piezoelectric actuator placed beneath. The backside of the moving mirror is also the moving mirror of the multipass interferometer integrated in the structure.

is the relative displacement of the two mirrors. In more refined interferometers, four beams are used (two as reference and two for the measurement); the adopted fixed and moving mirrors' arrangement is versatile enough to host different interferometric configurations. In particular, the three mirrors' arrangement and dimension have been chosen to mimic the NPL x-ray interferometer described before.

The actuator used to move the mirror is based on a commercial piezoactuated linear nano-positioning stage driven in closed loop by its own driver (Mad City Lab, Inc., mod. Nano-OP100;

FIGURE 5.21 Picture of the actuator. On the foreground are visible the optical fiber coming from the laser (center), the detectors group (right) and the tilters holding the optical elements of the interferometer (left). On the background, the multiple reflection mirrors are facing an interferometer under test.

Nano-Drive™ controller), capable of 100 μm displacement with a step resolution of 0.2 nm (Figure 5.21). The closed loop control has not enough resolution to reach the 10 pm goal (10 pm resolution over 100 μm range would require an extremely high dynamic control: 10^7–23–24 bit); so, we adopted a two-step coarse/fine control introducing an auxiliary piezo for interfringe positioning, which is controlled in closed loop via the developed fringe counter software.

The interferometer used or the displacement control is basically the same as described in the previous paragraph. With respect to that setup making use of a $\lambda/8$ retarder in the reference arm, two auxiliary wave plates ($\lambda/2$ and $\lambda/4$) have been added to the optical path in order to compensate for unwanted polarization rotations occurring at the BS and at the multiple reflection level. By trimming the three wave retarders, it is possible to balance the relative intensity and phase of the two signals so that they are as close as possible to the ideal sine and cosine. A further normalization is implemented numerically.

The multipass interferometer is extremely sensitive to the tilt of the two mirrors since any rotation is amplified by the same multiple reflection mechanism. For this reason, an extremely straight movement is required. The straightness of the Nano-OP100 is not sufficient for the purpose; therefore, a system to detect and correct tip-tilt errors has been built. The detection system is based on the same laser beam used for the interferometer which undergoes multiple reflections between the two mirrors. If the interferometer is aligned and the mirror movement is perfectly parallel, the reflected beam is superimposed to the outgoing beam. As soon as the mirror rotates, the

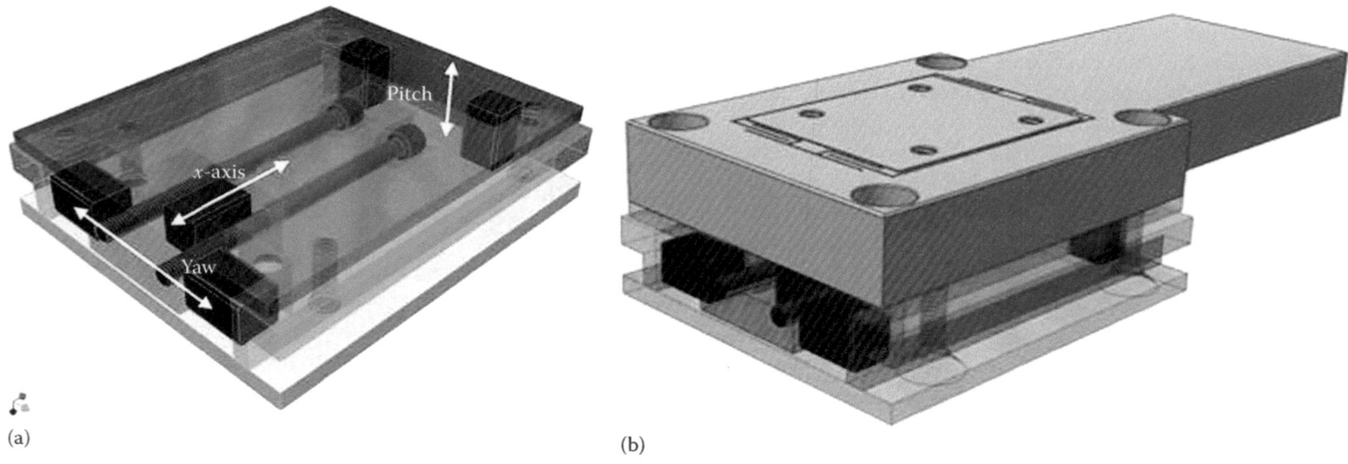

FIGURE 5.22 (a) CAD representation of the "pentapode." The lower part is fixed to the main TS structure; the upper part is adjusted by means of the five piezos (in black). (b) The OP-100 actuator mounted on the pentapode.

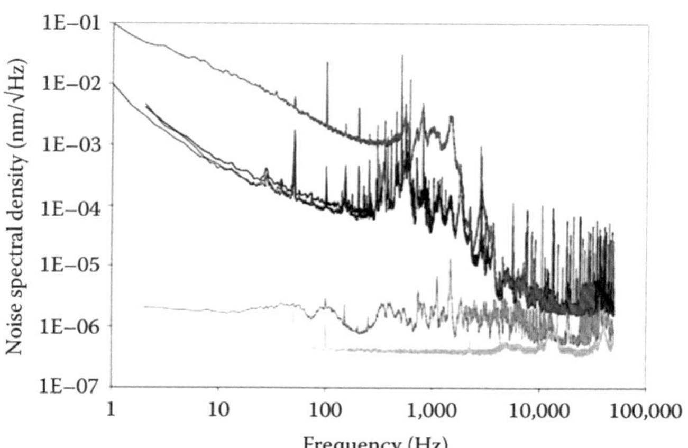

FIGURE 5.23 Noise spectral density of the interferometer. (See text for detail.)

reflected beam undergoes a rotation equal to the mirror one multiplied by the number of reflections on the same mirror. This rotation, combined with the laser path length, causes a lateral displacement of the returning beam. A small portion of the returning beam is extracted by the BS and sent to a position-sensitive detector (e.g., a quadrant detector), which generates two electrical signals proportional to the vertical and horizontal rotations of the mirror (see Figure 5.19). The two signals are used as error signals by the angle control unit.

For the correction of the errors, the actuator shown in Figure 5.22 has been built. The device is based on five piezoactuators (connected to a lower base, in turn fixed to the main base) that move an upper plate to which the OP-100 main actuator is fixed. Two out of the five actuators work in parallel to lift the back of the actuator implementing pitch angle; other two work horizontally in push–pull mode to implement yaw angle. The last piezo works horizontally to push back and forth the stage parallel to the main actuator axis (x-axis). Load springs are used to counterbalance the piezo action. The five piezoactuators of the "pentapode" have about 5 μm range with 100 V leading to a tip-tilt range of about 12 arcsec (60 μrad), enough to compensate the OP-100 errors. Furthermore, the x-axis piezo can be used to move the main mirror by small displacements to fill the gap between the OP-100 steps. The closed-loop tilt control is able to keep the rotation of the moving mirror as low as 0.1 arcsec (0.5 μrad) along the complete 100 μm scan.

The resolution of the actuator has been verified by measuring the spectral noise density of the interferometer output. In Figure 5.23, the spectra recorded in different conditions are shown. The darkest curves in the middle represent the interferometer noise with the angle control on and the pentapode on. As for the multipass interferometer described earlier, noise and vibrations are visible in the 300 Hz–3 kHz range, a noise that is well below the 1 pm/√Hz level. The noise decreases with frequency down to a few fm/√Hz plateau dominated by electronic noise. The lower curve is the detector noise with the laser off, and the lowest curve is the noise of the electronic board alone. Finally, the highest curve is the noise of the actuator when the OP-100 is switched on (the noise due to the active control, although very low, is evident).

As an application example, in Figure 5.24 we report the result of the characterization of a high-precision heterodyne interferometer. The difference between the two interferometers is recorded during a 5 μm scan. The gray curve shows a cyclic error with periodicity of $\lambda/2$ of about 0.3 nm p.p. (see figure caption for details).

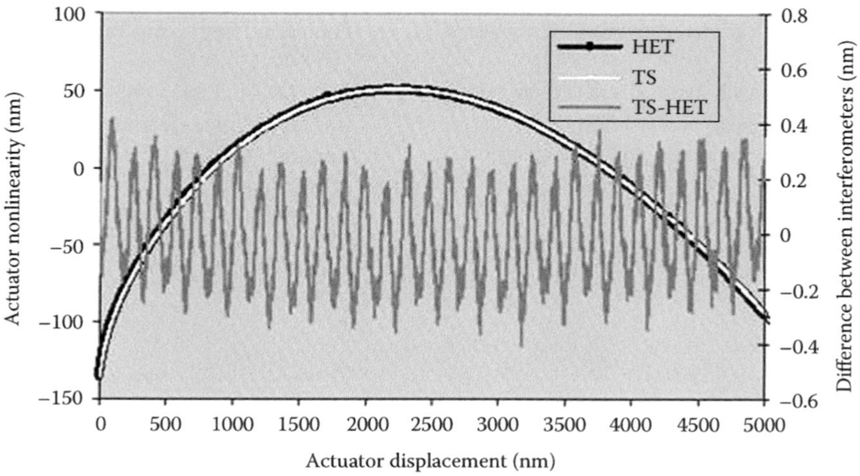

FIGURE 5.24 Calibration of a precision interferometer using the picometer actuator. The piezocapacitive actuator is moved by 5 μm, and the displacement is measured by the reference multipass interferometer and by the interferometer under test. The black and white curves (HET and TS curves, left scale) represent the difference between the readings with respect to a theoretical linear displacement (i.e., the typical second order nonlinearity of the piezoactuator) as measured by the two interferometers. The gray curve (TS-HET curve, right scale) is the difference between the two readings, that is, the errors of the interferometer under test. A cyclic error with periodicity of $\lambda/2$ of about 0.3 nm p.p. is clearly visible.

Acknowledgments

The research leading to most of the results presented here has received funding from the European Community's Seventh Framework Program, ERA-NET Plus, under grant agreement no. 217257 (EMRP). (EURAMET project NANOTRACE "New traceability routes for nanometrology".)

References

Basile G et al. (2000). Combined optical and x-ray interferometer for high precision dimensional metrology. *Proc. R. Soc. A* 456, 701–729.

Bisi M et al. (1999). Investigation of enabling interferometry technologies for the GAIA astrometric mission. In *SPIE Proc. Interferometry '99*. Pultusk Castle, Poland, Vol. 3744, pp. 31–42.

Bobroff N (1993). Recent advances in displacement measuring interferometry. *Meas. Sci. Technol.* 4, 907–926.

Downs MJ and Raine K (1979). An unmodulated bidirectional fringe-counting interferometer for measuring displacement. *Precision Eng.*, 1, 25–88.

Eom TB et al. (2002). A simple method for the compensation of the non-linearity in the heterodyne interferometer. *Meas. Sci. Technol.* 13, 222–225.

Fabry C and Perot A (1899). On the application of interference phenomena to the solution of various problems of spectroscopy and metrology. *Astrophys. J.* 9, 87–115.

Heydemann PLM (1981). Determination and correction of quadrature fringe measurement errors in interferometers. *Appl. Opt.* 20, 3382–3384.

Hewlett-Packard Journal, August 1970, 21(12), 1–16.

Jenkins FA, White HE. (1976). *Fundamental of Optics*. 4th edn. McGraw Hill.

Křen P and Balling P (2009). Common path two-wavelength homodyne counting interferometer development. *Meas. Sci. Technol.* 20, 084009.

Lawall JR (2005). Fabry–Perot metrology for displacements up to 50 mm. *J. Opt. Soc. Am. A* 22(12), 2786–2798.

LISA: Laser Interferometer Space Antenna, http://lisa.nasa.gov/

Michelson AA and Morley EW (1887). On the relative motion of the Earth and the luminiferous ether. *Am. J. Sci.* 34, 333–335.

Pisani M (2008). Multiple reflection Michelson interferometer with picometer resolution. *Opt. Express* 16, 21558–21563.

Pisani M (2009). A homodyne Michelson interferometer with sub-picometer resolution. *Meas. Sci. Technol.* 20, 084008.

Pisani M and Astrua M (2006). Angle amplification for nanoradian measurements. *Appl. Opt.* 45, 1725–1729.

Pisani M et al. (2012). Comparison of the performance of the next generation of optical interferometers. *Metrologia* 49, 455–467.

Resolution 1 of the 17th meeting of CGPM (1983). *Comptes Rendus de la 17ᵉ CGPM*, 1984, 97. http://www.bipm.org/en/CGPM/db/11/6/

Resolution 6 of the 11th meeting of the CGPM (1960). *Comptes Rendus de la 11ᵉ CGPM*, 1961, 85. http://www.bipm.org/en/CGPM/db/17/1/

Seppa J, Korpelainen V, Merimaa M, Picotto GB, and Lassila A (2011). A method for linearization of laser interferometer down to picometre level with a capacitive sensor. *Meas. Sci. Technol.* 22, 094027.

Tanaka M, Yamagami T, and Nakayama K (1989). Linear interpolation of periodic error in a heterodyne laser interferometer at subnanometer levels. *IEEE Trans. Instrum. Meas.* 38, 552.

Vitushkin AL and Vitushkin LF (1998). Design of a multipass optical cell based on the use of shifted corner cubes and right-angle prisms. *Appl. Opt.* 37, 162–165.

Wu B, Lawall J, and Deslattes RD (1999). Heterodyne interferometry with subatomic periodic nonlinearity. *Appl. Opt.* 38, 4089–4094.

Yacoot A and Cross NR (2003). Measurement of picometer non-linearity in an optical interferometer grating encoder using x-ray interferometry. *Meas. Sci. Technol.* 14, 148–152.

Young T (1804). Experimental demonstration of the general law of the interference of light. *Phil. Trans. Royal Soc. London* 94, 1–16.

6

Protein Crystallography at Subatomic Resolution

Tatiana Petrova
Institute of Mathematical Problems of Biology, RAS

Alberto Podjarny
CNRS, INSERM, Université de Strasbourg

6.1 Introduction

In the last decades, macromolecular crystallography (MX) has been the main source of structural information for macromolecular systems of biological interest. The unique power of MX is its capacity to obtain from x-ray diffraction data a direct image of electron density for nucleic acids and proteins in an aqueous environment close to that present in biological systems. The quality of an electron density map and, correspondingly, the quantity of information about a model of the protein structure are determined by the quality of the crystal and diffraction data. The major indicator of how good are the crystal and the data is a maximal resolution limit (often referred to as the resolution or the resolution of the structure). Resolution is measured in Å and is directly related to the minimal distance between different points that can be distinguished in the electron density map. The higher the resolution, the better is the quality of the electron density map, and the more information about the structure can be obtained.

Due to the flexibility inherent in macromolecules and to the presence of a significant quantity of water molecules in the crystal, the crystalline order is perturbed; therefore, the resulting diffraction is often limited to resolutions between 2 and 3 Å. Since the corresponding electron density maps in this resolution range do not show independent atomic positions, stereochemical limitations derived from studies of small molecule have to be imposed to build and refine atomic models. As a result, the stereochemical details of final models, such as bond distances and angles are strongly biased to the initially imposed ones. A typical length of a hydrogen bond between the oxygen atoms of water molecules is about 2.7 Å. Hence, at a resolution better than 2.7 Å, individual water molecules start to appear in electron density maps. Another important drawback of the limited resolution in x-ray crystallography is the impossibility to see hydrogen atoms and hence to determine the protonation state of residues in the active sites of enzymes.

In the last two decades, the technical and methodological advances changed this situation. Among these are, first of all, the availability of brilliant and focused synchrotron x-ray sources, the use of cryogenic temperatures for data collection, and the improvement of crystallographic software at all steps of structure solution, from data treatment to refinement. An increasingly great number of protein structures were determined at high (better than 2 Å) and atomic resolutions (better than 1.2 Å). The typical length of the covalent bond between C–C atoms and C–O atoms is about 1.5 and 1.2 Å, correspondingly.

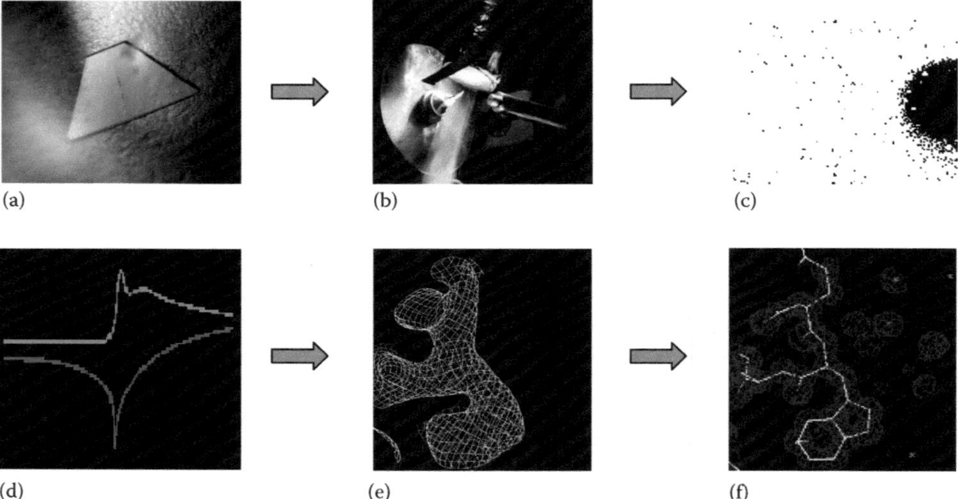

FIGURE 6.1 **(See color insert.)** The overview of solving a structure in MX. (a) Crystallization. (b) Cryocooling. View of helium blower. (c) Data collection and processing (view of diffraction intensities frame). (d) Solving phase problem (view of anomalous dispersion curves used in MAD method). (e) Calculation of electron density map using structure factor magnitudes and phases. (f) Building macromolecular structures into the electron density and refinement using crystallographic restraints.

At this resolution, atoms are clearly resolved, and geometrical restraints can be applied with lower weights. At the atomic resolution, atomic displacements can be refined anisotropically. A typical length of the covalent bond between hydrogen and more heavy atoms is about 0.9–1.0 Å. At a resolution higher than 1 Å, hydrogen atoms begin to appear in electron density maps; therefore, the protonation state of polar residues in the active site of an enzyme can be determined. Furthermore, protonation states observed can differ from those in solution for isolated amino acids. These differences may have strong implications for biological function, and their experimental determinations are therefore crucial.

The first protein structure at the subatomic resolution, the small protein crambin (resolution 0.83 Å), appeared in 1993 (Teeter et al. 1993). The next one, serine protease (resolution 0.78 Å) was solved in 1998 (Kuhn et al. 1998). Currently, Protein Data Bank (PDB) contains 67 structures of proteins at a resolution better than 0.85 Å. At this resolution level, it is possible to obtain information beyond the scope of the model of spherical atoms with anisotropic vibrations. In particular, it becomes possible to observe the deformation of electron densities that arise from chemical interactions such as covalent bonding, electrostatic interactions, and hydrogen bonds. This additional information about the electron density can be interpreted and refined using a multipolar model.

The consequences of the direct observation of the electron density from diffraction amplitudes are far-reaching. It should be possible in theory to measure the occupation of quantum orbitals and therefore to determine the reactivity of a given atom directly from the diffraction data. This goal has not yet been attained for macromolecules of biological interest, but substantial efforts have been made in this direction.

During the last 10 years, the development of MX has been excellently reported in several reviews (e.g., Esposito et al. 2002; Schmidt and Lamzin 2002; Vrielink and Sampson 2003; Dauter 2006; Wlodawer et al. 2008; Dauter et al. 2010). The purpose of this chapter is to concentrate only on the structures determined at the subatomic resolution level. The chapter includes the main methodological and technical advances made at each step of protein structure solution (see Figure 6.1), which led to the gain in resolution, the results of subatomic x-ray studies, and the perspectives they open.

6.2 Experimental Techniques

In a usual x-ray data collection experiment, the crystal is placed in a collimated x-ray beam, and, while the crystal is rotated, a series of diffraction images are recorded. The essential components of the experiment are a crystal, an x-ray source, and an x-ray detector.

6.2.1 Crystallization

The main limiting factor for MX is the production of large, well-ordered crystals. Despite the experience accumulated at many laboratories, it is still a challenge to obtain protein crystals that contain no impurities and are large enough to produce diffraction images suitable for structure solution. The main reason why growing macromolecular crystals is more complicated than growing crystals of small molecules is that the former have a high content of solvent (30%–80%), few intermolecular contacts, and a large density of defects (Malkin et al. 1996).

The experimental settings for macromolecular crystallization constitute a many-variable system in which a broad spectrum of conditions and methods has to be determined. These include chemical components (precipitant, buffer, and additives), physical parameters (pH, temperature, and pressure), different methods (sitting drop, hanging drop, batch, etc.), and protein-dependent parameters (concentration and cofactors). Due to the very complex nature of proteins, the crystallization conditions for each particular protein cannot be determined a priori, even when major efforts have been made in this direction (Chernov and DeLucas 2002). A search for rational strategies of producing crystals diffracting to higher resolution demands an understanding of the chemical and physical processes of the crystallogenesis, which has yet to be attained.

In practice, the approach is to test systematically ("screen") a great variety of combinations of different conditions for getting microcrystals. Since most of these screens have been predetermined in the previous experiments, robotic crystallization systems are widely used (Stevens 2000; Müller et al. 2001). Once the conditions for getting small crystals are found, they are optimized in order to obtain large crystals.

During crystallization, physicochemical processes occur, which include nucleation, crystal growth, and mass transport. Because of gravity, any nonhomogeneity in the sample leads to convection, which offers problems. To provide minimum convection, crystals are sometimes grown in reduced-gravity environments, for example, in space shuttles (Littke and John 1984; Kundrot et al. 2001). A good overview of the field can be found in the study of Lorber et al. (2002) and Vergara et al. (2003). Intensive efforts to stabilize the environment while growing crystals are also made on the earth, for example, investigations of crystallization conditions in gels (Robert et al. 1999), in a mixture of oils (Moreno et al. 2002), and in magnetic fields (Ataka and Wakayama 2002; Kinoshita et al. 2003).

Most proteins are flexible and adaptive molecules, which can exist in different conformations. The presence of multiple conformations and disordered parts in the molecule can also cause structural heterogeneity and affects crystallization. Therefore, another approach to improve crystal quality is to modify the protein itself. For example, in some particular cases, truncations and mutations of flexible and charged residues on the surface of the protein can be effective tools in overcoming problems during crystallization and can drastically improve the crystal quality (Mateja et al. 2002; Minasov et al. 2002; Czepas et al. 2004).

To obviate problems associated with the very close packing and large amount of solvent in protein crystals, a very promising study aimed at improving the crystalline order by crystal dehydration is under development (Schick and Jurnak 1994; Heras et al. 2003).

6.2.2 Radiation Damage and Protective Cryocooling

One of the methodological developments that allows the collection of very high-resolution x-ray diffraction data is the cryocooling of protein crystals during the experiment (Hope 1990; Garman and Schneider 1997). The main purpose of cryocooling is to reduce the destructive effects exerted by irradiation on biological samples. X-rays destroy proteins by two mechanisms. The ionization of x-ray photons through the photoelectric effect and Compton scattering ("primary" damage) leads to the ejection of highly energetic secondary photoelectrons from atoms. As these electrons travel through a biological sample, they interact with the protein and solvent and create highly reactive radiolytic species, which, in turn, cause a cascade of further ionization events ("secondary" damage) (Burmeister 2000; Ravelli and McSweeney 2000).

Irradiation deteriorates the crystal and causes the fading of the diffraction data; the high-resolution data, which provide the information about the fine details of the structure disappear first. Radiation damage can strongly influence the results of multi-wavelength anomalous dispersion (MAD) phasing methods (see Section 6.3.2). During a MAD experiment, several data sets on the same crystal have to be collected. Radiation damage causes nonisomorphism between sequential batches of diffraction data collected from the same crystal, which leads to a serious hindrance to the structure solution.

The primary damage cannot be avoided; it mostly depends on the x-ray dose. However, the diffusion of active radicals drastically decreases at cryogenic temperatures. Freezing slows down diffusion, increases the lifetime of the crystal, and enables the collection of more complete data sets from single crystals. At 100 K, as compared with room temperature, the crystal being irradiated retains its diffracting properties for a much longer time. Thus, it was reported that cryocooling from room temperature to 100 K provides a 26–113-fold increase in the lifetime of the crystal (Southworth-Davies et al. 2007).

It should be emphasized that freezing does not completely eliminate the problem of radiation damage. At cryogenic temperatures, the radiation damage to protein crystals manifests itself at different levels:

1. As local-specific chemical and structural changes in the protein molecule, for example, the reduction of bound metal ions (Adam et al. 2004; Yano et al. 2005), the disruption of disulfide bonds (Burmeister 2000; Ravelli and McSweeney 2000; Weik et al. 2000), the rupture of covalent bonds between C and some heavy atoms, such as Se, Br, I, Hg (Evans et al. 2003; Ramagopal et al. 2005; Oliéric et al. 2007), the decarboxylation of glutamate and aspartate residues (Burmeister 2000; Ravelli and McSweeney 2000; Fioravanti et al. 2007).
2. As overall structural changes of the protein molecule, for example, the movement of secondary structure elements and rotation of the protein molecule in the unit cell.
3. As an overall crystal disorder, which shows up as a degradation of crystal diffraction properties, changes in unit cell dimensions, an increase in the mosaicity and atomic displacement parameters (ADP, also referred to as atomic temperature factors).

The sensitivity of the active site residues to x-rays was also reported (Ravelli and McSweeney 2000; Rypniewski et al. 2001). Based on the data on x-ray-induced changes in bond lengths in peptide crystals and radiolysis experiments, Meents and colleagues proposed that irradiation causes the abstraction of hydrogen atoms from organic molecules with subsequent formation of gaseous hydrogen bubbles (Meents et al. 2009, 2010). For subatomic resolution structures, possible radiation damage effects should be taken into account for accurate interpretation of electron density maps, in particular, in the active site of the enzyme. At the same time, to elucidate the mechanism of radiation-induced deterioration of protein crystals, it is interesting to conduct diffraction experiments at subatomic resolution. Takeda et al. (2009) performed radiation-damage studies using crystals of tepidum, which diffracted to 0.7 and 0.72 Å. They concluded that thermal effects in the crystal occur prior to the aggravation of the data quality. They also investigated the dose-dependent disappearance of electron density that corresponded to hydrogen atoms.

The usual temperature of data collection at modern synchrotrons is 100 K. As a cooling medium in an open flow cryostat, a cold nitrogen stream is used, which is explained by its low price. A possible alternative would be to use cold helium, which might have two advantages. First, using helium instead of nitrogen at the same temperature would reduce the background scattering and, therefore, provide data of better quality (Polentarutti et al. 2004). Second, the use of helium would allow one to attain lower temperatures. At atmospheric pressure, the minimal temperature that could be reached using liquid nitrogen is 77 K, while the use of liquid helium would allow one to attain 4 K. It is believed that cooling crystals below 100 K may further decrease the radiation damage to the protein crystal. It was reported that x-ray-induced damage to the active site of metalloproteins at liquid helium temperature of 10 K decreases compared with the damage at 100 K (Yano et al. 2005), and the rate of photoreduction at 40 K is 40 times lower than that at 110 K (Corbett et al. 2007). However, it was shown in several studies that the benefits of using helium to decrease x-ray damage are, first, notable only after the crystal has absorbed a handsome dose (Moffat and Teng 2002; Petrova et al. 2010), and, second, for other kind of specific damage (not for the active site of metalloproteins) and for the global damage, they are not significant (Chinte et al. 2007; Meents et al. 2007; Petrova et al. 2010).

Last, but not least, an important advantage of using cryogenic temperatures is the facility of storage, transport, and mounting of frozen crystals (Garman and Schneider 1997).

6.2.3 Synchrotrons–Detectors

X-rays are produced by acceleration (or deceleration) of electrons. The first sources of x-rays were vacuum tubes, in which x-rays were emitted when a flow of electrons struck the anode. Rotating anodes, which allow a better dissipation of the heat released during this process, were a first improvement, which increased the brightness by two orders of magnitude. But the greatest progress came from synchrotron sources, in which electrons or positrons move at relativistic velocities in a storage ring, and photons are emitted with energies in the range of those from infrared to x-rays. Compared with x-rays emitted in tubes, synchrotron radiation is much more intense. The first synchrotrons offered a gain of brightness of 6 orders of magnitude, and the third-generation sources, such as the ESRF (Grenoble), the APS (Argonne), the Swiss Light Source (Villigen), Spring 8 (Tsukuba), and Petra (Hamburg) provide a gain of at least 14 orders of magnitude. The highest brilliance is obtained in third-generation synchrotrons using undulator beamlines. In these beamlines, periodic series of magnets are positioned in the straight sections of the ring. The undulator device forces the electrons to follow a highly curved path. This results in a concentrated emission of photons with a higher value of flux and brightness of the beam.

Since the average intensity of diffracted x-rays strongly decreases at high resolution, high source brilliance is necessary to collect these data. For this reason, atomic and subatomic resolution data from medium and large size proteins can be collected only at synchrotrons. However, the high intensity of synchrotron radiation is not the only factor that makes synchrotron an essential tool for data collection in MX. High positional stability, collimation of the beam, and the tunability of synchrotron radiation are also important, making the corresponding and specific developments in the optics of the beamline necessary. Furthermore, detectors allowing fast, accurate, and low-noise measurements of large amounts of diffraction data, efficient algorithms of data reduction, and good storage capacity are also necessary.

One of the most crucial technological needs for subatomic resolution data collection is the development of fast and large x-ray detectors. The function of a detector is to convert a diffracted x-ray beam into a signal, which is proportional to the number of x-ray photons. Today, image plate (IP) detectors and detectors based on charged coupled devices (CCD) are widely used. Both types are integrating detectors in which, a signal consisting of many photons is accumulated, and then, the total signal is read out. Image plate detectors have a large surface, but are too slow for modern synchrotrons. CCD detectors are very fast, but their sensitive surface is usually smaller. To improve the situation, efforts are currently concentrated on constructing a large and fast detector; thus, a flat panel detector was developed by MAR research group and a pixel detector (PILATUS) was developed at the Swiss Light Source. Furthermore, these detectors offer the possibility of continuous data collection which can increase the signal-to-noise ratio for weak reflections (see Appendix 6.A), and hence the resolution. For these reasons, pixel detectors are becoming more widely used.

6.3 Calculations

6.3.1 Basic Mathematical Expressions

Single crystal x-ray diffraction data leads to structure factor amplitudes. The structure factors $F(\mathbf{h})$ are the Fourier transform of the electron density $\rho(\mathbf{r})$ of the unit cell of volume V and cell parameters $\mathbf{a}_i$, $i = 1, 3$:

$$F(\mathbf{h}) = \int_{Unit\ cell} \rho(\mathbf{r}) \exp(2\pi i\,\mathbf{hr})\, d^3\mathbf{r}, \tag{6.1}$$

where

$\mathbf{r} = x\,\mathbf{a}_1 + y\,\mathbf{a}_2 + z\,\mathbf{a}_3$ is a point inside the unit cell (the "fractional coordinates," x, y, and z have values between 0 and 1)

$\mathbf{h} = h\,\mathbf{a}_1^* + k\,\mathbf{a}_2^* + l\,\mathbf{a}_3^*$ is a point of the "reciprocal lattice," defined by the unit vectors

$$\mathbf{a}_\mathrm{i}^* = (\mathbf{a}_\mathrm{j} \wedge \mathbf{a}_\mathrm{k})V^{-1}.$$

Since the crystal is triply periodic, the Fourier transform has nonzero values only for integer values of (h, k, l), known as the "Miller indices." The range of these values is limited by the resolution of the data.

The result of the diffraction experiment is the amplitude of the complex quantity $\boldsymbol{F}(\mathbf{h}) = F(\mathbf{h})\exp(i\phi(\mathbf{h}))$, and we must know both their amplitude $F(\mathbf{h})$ and phase $\phi(\mathbf{h})$ for calculating $\rho(\mathbf{r})$ by inverse Fourier transform. Once the phases are known (see Section 6.3.2), the electron density can be calculated by an inverse Fourier transform. An atomic model can then be superimposed to this density, leading to the map "interpretation." The interpretation is facilitated, since the electron density is mainly concentrated around atomic positions, and can therefore be expressed as a sum of contributions of individual atoms.

$$\rho(\mathbf{r}) = \sum_j \rho_j(\mathbf{r} - \mathbf{r}_j), \tag{6.2}$$

where

the variable j runs over all atoms

$\mathbf{r}_j$ is the position of each one of them

For an ideal crystal, without thermal vibrations and with exactly the same molecules in all unit cells, the electron density ρ_j corresponds strictly to the atomic electronic cloud, which is mostly a sharp Gaussian around the nuclei. However, in practice, there are two important contributions that widen this distribution. One of them is the thermal vibration of atoms, and the second one is the variation in atomic positions in different unit cells. For experiments conducted at a single temperature, these two effects cannot be separated and are therefore grouped in a single Gaussian distribution of the atomic centers. The width of this Gaussian distribution is defined by the value of the mean square displacement of the atomic centers from their mean position $\langle u^2 \rangle$. In real space, the observed electron density is obtained by convoluting this Gaussian distribution with the shape of the atom at rest. By taking the Fourier transform, this convolution is transformed into a product, leading to the following formula for the structure factor:

$$\boldsymbol{F}(\mathbf{h}) = \sum_j f_j(|\mathbf{h}|)\exp(2\pi i\mathbf{h}\mathbf{r}_j)\exp\left(-B_j\frac{\mathbf{h}^2}{4}\right), \tag{6.3}$$

where

$\mathbf{r}_j$ are the atomic positions

f_j is the radial part of the Fourier transform of the atomic shape of the electron density of the free neutral atom (independent atom model, IAM atomic scattering factors)

B_j is the isotropic Debye–Waller factor for each atom

The Debye–Waller factor B is related to the mean square displacement:

$$B = 8\pi^2\langle u^2 \rangle. \tag{6.4}$$

For macromolecules, the following factors contribute to the displacement parameter values:

- Dynamic disorder (vibrations of atoms)
- Static disorder (different conformations in the different cells)
- Errors in the model
- Defects and vibrations of the lattice

Isotropic displacement parameters are a crude approximation. More correct is a model where the mean square displacement is anisotropic. In this case, the Debye–Waller factor can be expressed as a symmetric [3 × 3] matrix B_j, and the structure factors are calculated by the formula:

$$\boldsymbol{F}(\mathbf{h}) = \sum_j f_j(|\mathbf{h}|)\exp(2\pi i\mathbf{h}\mathbf{r}_j)\exp\left(-\frac{1}{4}\mathbf{h}^T\mathbf{B}_j\mathbf{h}\right). \tag{6.5}$$

6.3.2 The "Phase Problem"

Since the result of the diffraction experiment is only the structure factor's amplitude, the calculation of the phases has been a major problem (the "phase problem") in crystallography during several decades, leading to developing methods such as somorphous replacement (Green et al. 1954), molecular replacement (MR; Rossmann and Blow 1962), density modification (DM; Read 2001), direct methods (Miller et al. 1993; Sheldrick and Schneider 2001), and multiple (or single) anomalous dispersion (MAD or SAD) (Karle 1980; Hendickson and Ogata 1997). Currently, the combined use of anomalous scattering methods and density modification has changed the situation to the point that the phases can be considered as an experimental result in most cases.

6.3.3 Model Building

The magnitudes and phases of structure factors obtained from the diffraction data and the solution of the phase problem are used to calculate an electron density map. After this map is calculated, a model of the protein structure has to be built into the density. If the resolution is higher than 2.3 Å, this interpretation can be done automatically, using programs like ARP/wARP (Perrakis et al. 1999), RESOLVE (Terwilliger 2001), and MAID (Levitt 2001).

In many cases, the model is already available, either for a similar or the same molecule, crystallized in different conditions. Then, a starting model can be obtained using the MR method, such as implemented, for example, in AMoRe (Navaza 1994). For high-resolution studies, the situation when the same molecule has already been solved at a lower resolution is quite common. Small parts of the model that were not originally included (e.g., an inhibitor as a part of the complex) can be built manually into the density where the difference between the observed and calculated densities is seen. A number of programs are available for the manual rebuilding of the model, like O (Jones et al. 1991), XtalView/Xfit (McRee 1999), MAIN (Turk 2000, 2004), and Coot (Emsley and Cowtan 2004). These graphic systems allow the users to manipulate the atomic coordinates of the model while maintaining the standard geometry of the polypeptide chain.

6.3.4 Refinement

Equation 6.3 is the basis for most MX refinements, which fit the observed amplitude values $F(\mathbf{h})$ using a model with four parameters per atom, that is, atomic coordinates and isotropic Debye–Waller factor, as well as adjusting other parameters, like scaling, bulk solvent corrections, etc. This fit is usually done by minimizing (usually by least squares) a "target function" $E(x\text{-}ray)$, related to the difference between the observed and calculated structure factor magnitudes.

At the usual resolution for MX (2 Å < d < 3 Å), the number of observations $F(\mathbf{h})$ is not enough for determining the total number of atomic parameters. There are two options to solve the problem. The first one is to diminish the number of parameters using constraints, for example, fixing torsion angles or bond lengths, as it is done in constrained refinement in software CORELS (Sussman et al. 1977) and in CNS torsion angle refinement (Adams et al. 1997). The second is to increase the number of observations by including the standard stereochemistry values (bond lengths, bond angles, torsion angles, planarity of aromatic rings, chiral centers, etc.). The additional restraint terms are included in the target function, which becomes of the type

$$E = wE(x\text{-}ray) + E(chem), \tag{6.6}$$

where

$E(chem)$ is related to the difference between ideal exact values of the bonds and angles and the model ones

w is the weight between these two terms

Precise ideal values (commonly termed "restraints") used in protein crystallography were derived by Engh and Huber (1991) from small-molecule crystallographic data stored in the Cambridge Structural Database. In a more recent contribution by Engh and Huber (2001) to the *International Tables for Crystallography Vol. F*, individual mean values for the 20 different amino acids are listed. They were calculated from a more recent version of the same database. In a macromolecular structure refinement at low resolution, restraints may generate a huge bias in the minimal search. Knowledge of the accurate restraints values is thus essential for refinement at low and medium resolutions.

At higher resolution, refinement can be done with only weak stereochemical restraints, and one can observe deviations to the standard geometry. If the data are at a resolution better than 1.2 Å, the isotropic temperature factor may be replaced by an anisotropic factor, leading to a higher accuracy of atomic positions which is necessary to validate shifts from the standard geometry. Note that introducing anisotropic B-factors instead of isotropic ones increases the number of parameters for each atom from 5 to 10 (three coordinates, the value of occupancy [see Appendix 6.A], and six values for B-factor instead of 1).

6.3.5 Refinement of a Model at Subatomic Resolution

At subatomic resolution, the ratio of the number of observations to the number of parameters is very high. For example, for the final model of the antifreeze protein (d = 0.62 Å), the number of refined parameters was 6,818 against 118,101 observed data used in refinement (Ko et al. 2003). This leads to the possibility of an unrestrained refinement. However, because of the presence of disordered parts and alternative conformations in the protein structure, most restraints need usually to be kept even at subatomic resolution. Several structures were reported for which the stereochemical restraints were excluded only for the well-ordered part of the structure during the last steps of refinement (Koepke et al. 2003; Kang et al. 2004; Wang et al. 2007). The refinement of the model of the bacterial photoreceptor photoactive yellow protein (PYP) (d = 0.82 Å) at the last step was performed in the absence of any geometrical restraints (Getzoff et al. 2003).

The refinement of a model at subatomic resolution requires the appropriate software, which allows one to refine ADP anisotropically, refine atoms in alternative conformations, refine occupancy values for a group of atoms and individual atoms, and include hydrogen atoms in the model. The programs SHELX (Sheldrick and Schneider 1997), Phenix.refinement (Afonine et al. 2005), and REFMAC (Murshudov et al. 1999) are most commonly used. SHELX came from crystallography of small molecules. It has a great variety of options necessary for subatomic resolution refinement, but it demands experience in mastering it. REFMAC and Phenix.refine were originally implemented for macromolecules and are in progress continuously; the number of options for subatomic-resolution refinement steadily increases. For example, Phenix.refinement has specific tools to model the redistribution of electron density due to the formation of covalent bonds (Afonine et al. 2007).

6.3.6 Deviations from Standard Stereochemistry

Until now, restraints used in crystallographic refinement were derived from small molecule structures, which do not necessarily reflect the conditions inside the protein. Statistical analysis of the geometry of the final high-resolution structures (bond lengths, bond angles, torsion angles, etc.) is of great importance for creating more accurate restraints. These restraints can then be used for the refinement of macromolecular structures at low and medium resolutions. The first step toward this aim was a detailed analysis of geometrical data for the final model of DFPase, refined at 0.85 Å (Koepke et al. 2003). The analysis revealed significant differences between the observed bond lengths and bond angles of single amino acids and the values most commonly used (Engh and Huber 1991). Some examples of deviations of the peptide bond from the planarity can be found in a very detailed review (Esposito et al. 2002). Among them are a much broader distribution and sometimes extreme values of torsion angles (Genick et al. 1998), the effect of pyramidalization of the C and N main chain atoms (Esposito et al. 2000), and the resonant states of the peptide bond (Howard et al. 2004). The latter can be revealed through the anticorrelated variations of the CN and CO bonds (see Section 6.5.3 and Figure 6.10).

The standard geometries of the chemical moiety, as observed in small molecules, are determined by the balance of internal forces. Therefore, statistically significant deviations are important since they may indicate the presence of external forces, which are possibly relevant to functional properties. For example, in the 0.82 Å resolution structure of the complex of photoreceptor PYP with a bound chromophore (Getzoff et al. 2003), it was observed that the final model of the chromophore is arched out of the plane of its aromatic ring. Generally, stereochemical restraints should predict a perfectly planar chromophore in this case. However, the observed deviations strongly exceeded those that might be caused by experimental errors. In the previously solved structure of PYP mutant (in which the tyrosine residue was replaced by phenylalanine), the chromophore molecule was arched in the opposite direction (Brudler et al. 2000). For the mutant, a remarkable increase in the fluorescence quantum yield was observed. Although there are yet no clear explanations of the mechanism, this result strongly suggests a relation between chromophore bowing and fluorescence.

6.3.7 Solvent

Usually, it is very difficult for MX to reliably locate and accurately refine water molecules, except those being in very close contact with the protein. The reason is a high level of disorder of the solvent structure, which leads to low occupancies and a large number of "multiple" locations. At the same time, water occupies 30%–70% in the unit cell and plays an important role in protein function and stabilization.

Subatomic resolution data allow one not only to refine water molecules with anisotropic B-factor values but also to identify water molecules in dual locations and refine them with fractional occupancies (Ko et al. 2003). The presence of "multiple" locations of water molecules is due to the fact that protein residues are in multiple conformations.

6.3.8 Multiple Conformations and Hybrid Configurations

At lower resolutions, disordered residues are usually modeled as single residues with high temperature factors in average positions. At subatomic resolution, it is possible to decipher the disorder and reveal multiple conformations. These are normally detected during refinement, normally by reinterpretations of the electron density. Revealing multiple conformations and accurately determining their occupancies are especially important for the residues of the active site region, because some of these conformations may be associated with the function of protein (Rypniewski et al. 2001; Getzoff et al. 2003; Kursula and Wierenga 2003).

In the case of the complex of PYP with a chromophore molecule (0.82 Å), the quality of the final model was high enough to allow an analysis of interatomic distances, which suggested that the chromophore was present in two electronic configurations (Getzoff et al. 2003). This hybrid configuration is essential for stabilizing the negative charge of the chromophore molecule.

6.3.9 Bond Densities

At subatomic resolution, the deviations of the "real" electron density from the spherical atom model can be directly observed. These density deformations can be seen by computing a crystallographic difference map (see Appendix 6.A):

$$\rho_{res}(\mathbf{r}) = \frac{1}{V} \sum_{|\mathbf{h}|<1/d_{\min}} (F^{obs}(\mathbf{h}) - F^{calc}(\mathbf{h}))\exp[i\varphi^{calc}(\mathbf{h})]\exp[-2\pi i(\mathbf{h},\mathbf{r})], \tag{6.7}$$

where the F^{calc} and φ^{calc} values are calculated from the spherical atom model.

The deviations appear at the difference map (Equation 6.7) as peaks at the middle of the individual bonds and densities surrounding an oxygen atom.

Several structures were reported for which in well-ordered parts of the model, containing atoms with Debye–Waller factor lower than 4 Å^2, peaks at the middle of the bonds between Cα–C, C–N, N–Cα and density corresponding to lone pairs on oxygen were clearly seen (Jelsch et al. 2000; Ko et al. 2003; Howard et al. 2004). In Figure 6.2, a fragment of the difference

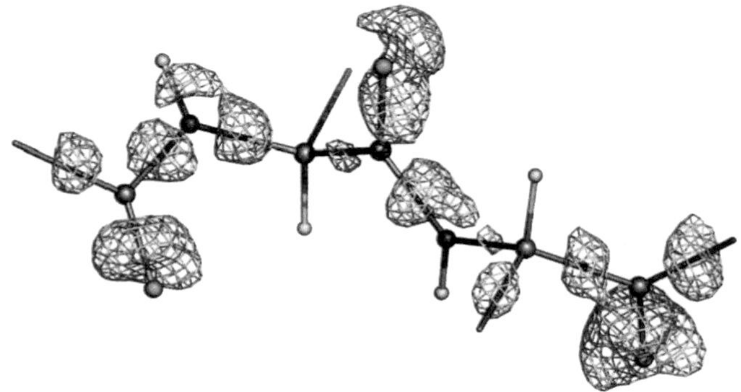

FIGURE 6.2 Fragments of the difference map (F^{obs}–F^{calc}, ϕ^{calc}) (see Appendix 6.A) for antifreeze protein RD1 calculated at 0.62 Å resolution (the main chain atoms for the residues Val5–Val6 are shown). The density is shown only in the vicinity of the presented atoms. Peaks for the bond electron density are clearly seen. The cutoff level ρ_{crit} = 0.39 e/Å^3 (3.0σ). (Reproduced with permission from Afonine, P. et al., *Acta Crystallogr. D Biol. Crystallogr. D*, 60, 260–274. Copyright 2004 International Union of Crystallography.)

map for the antifreeze protein RD1 (Ko et al. 2003), calculated by Afonine et al. is shown (Afonine et al. 2004). The observation of this bond density in the difference map can only be possible if the phases were calculated from the model that was refined at ultrahigh resolution about 0.6 Å. Refinement at a lower resolution leads to overestimation of ADPs that, in turn, causes disappearance of the signal (Afonine et al. 2004).

6.3.10 Hydrogen Atoms—Protonation States

Hydrogen atoms constitute roughly half of the atoms in a protein molecule. They are essential for stabilizing the structure of proteins. With subatomic resolution data for TEM-1 beta-lactamase, 70% of the hydrogen atoms are clearly seen at the difference map.

A very promising application of the H-atom observations in ultrahigh resolution crystallography is the elucidation of enzymatic reactions. Enzymes are proteins which catalyze a given chemical reaction by creating a favorable environment around the chemical compounds (substrate). The residues of the protein that are in the vicinity of the substrate constitute the active site of the enzyme.

Usually, an active site contains polar residues involving (fully or partially) charged atoms. Water molecules and cofactors, which are often found in the protein active site, can also participate in the reaction.

While the active site of an enzyme is usually identified from either biochemical data or the structure itself, the detailed mechanism of the enzymatic reaction is not simple to elucidate. Understanding the reaction mechanism at atomic level requires, first, that we identify the interactions between the protein active site and the substrate. These interactions are mostly of an electrostatic nature (short-range interactions: salt bridges, hydrogen bonds; long-range electrostatics), although they are also fine-tuned by apolar properties such as the van der Waals contacts and the hydrophobic effect. In particular, these electrostatic interactions play an important role in stabilizing the intermediates of the chemical reaction.

Therefore, it is important to know the charge of the polar groups of the residues in the active site. The charge of a residue depends in turn on its protonation state, that is, whether its polar groups are carrying a proton or not. The protonation state of a particular residue at a given pH is sometimes available from experiments, such as infrared methods (Dioumaev 2001), Raman spectroscopy (Ames and Mathies 1990), and NMR (McDermott et al. 1991). However, these experiments can be difficult to interpret since the protonation states of the numerous polar residues in a protein can strongly depend on each other (Balashov 2000; Harris and Turner 2002). Sometimes, the protonation states of polar residues has been attempted at lower resolutions (around 1.2–1.3 Å) by analysis of bond lengths (Ahmed et al. 2007). Even when this is possible in favorable cases, these determinations need to be confirmed (Fisher et al. 2012).

On the other hand, with subatomic resolution crystallography, the hydrogen atoms can become directly visible in the electron density maps. It is thus possible to infer the role of specific protein residues in the enzymatic reaction by analyzing their respective protonation states as well as their interactions with the substrate, both in bound and unbound forms (Minasov et al. 2002; Nukaga et al. 2003; Howard et al. 2004).

We should not forget, however, that the conditions inside the crystals are not the same as in vivo, and that even at subatomic resolution and in very ordered parts of structure the observation of hydrogen atoms might be misleading. Therefore, before making a conclusion about the reaction pathway, the crystallographic results need to be combined with other types of experimental information.

6.4 New Developments in High-Resolution Studies

6.4.1 Charge Density Refinement: The Multipolar Model

The presence of significant densities in difference maps is an indication that spherical free atom models, even with anisotropic thermal parameters, give a too crude approximation to the real electron density at ultrahigh resolution. The so-called multipolar model (Hansen and Coppens 1978; Coppens 1997; Spackman 1998), which is widely used in crystallography of small molecules, gives a better representation of electron charge densities. In this model, the total atomic electron density ρ is composed of three parts: (1) ρ_{core}, which corresponds to the spherically symmetric core electrons; (2) ρ_{val}, which corresponds to the spherical part of the valence electrons; and (3) the nonspherical part of valence electron densities caused by chemical bonding (Hansen and Coppens 1978). This third part is modeled by the sum of multipolar pseudo-atoms lying at atomic positions. The valence electron density of such a pseudo-atom is projected on the basis of real spherical harmonics functions $Y_{lm}(\theta, \phi)$ centered on each pseudo-atom. The result is given by the formula:

$$\rho(\mathbf{r}) = \rho_{core}(\mathbf{r}) + \kappa^3 P_v \rho_{val}(\kappa, \mathbf{r}) + \sum_{l=0, l\,max} \kappa'^3 R_l(\kappa', \mathbf{r}) \sum_{m=\pm 1} P_{lm} Y_{lm}(\theta, \phi). \quad (6.8)$$

The radial functions $R_l(r)$ used here are of Slater type: $R_l(r) = r^{nl} \exp(-\kappa'\xi r)$. The number of multipolar parameters P_v, P_{lm}, κ, κ' for each atom increases with the number of included spherical harmonics. Usually, nonhydrogen multipolar atoms have a minimum of 28 parameters (18 multipolar, 3 positional, 6 temperature parameters, 1 occupancy). In cases that the observation/parameter ratio is high enough, parameter values can be obtained directly from the least square refinement against the structure factor amplitudes (Guillot et al. 2001). The representation (6.8) is used to calculate the properties of the crystal and molecules, such as electrostatic potential (Ghermani et al. 1993), electric field, net charges, higher moments (Spackman 1992), and the topology of the electron density (Souhassou and Blessing 1999).

For the case of protein residues, the first approximation to the atomic multipolar atomic charge density parameters can be directly transferred from the precalculated database (Pichon-Pesme et al. 1995; Jelsch et al. 1998). This transfer results in a considerable saving of calculation time, compared with the refinement procedure. It was shown that the electrostatic potential calculated with multipolar parameters from this database leads to results being in qualitative agreement with quantum chemistry calculations and the results of multipolar refinement (Muzet et al. 2003).

As mentioned earlier, the observation/parameters ratio for the subatomic resolution data is high enough to perform the refinement of multipolar parameters against structure factor amplitudes. In this case, the parameter database can be used to obtain the starting values of multipolar parameters. The refinement of multipolar parameters is implemented in the software MoPro (Guillot et al. 2001). Starting positions and thermal parameters are taken from the output of SHELX (Sheldrick and Schneider 1997) refinement. First, a full anisotropic spherical atom refinement is performed. The next refinement steps are the following: (1) the positional and thermal motion parameters of the core electrons of non-H atoms are refined; (2) initial multipolar parameters are transferred from the database; and (3) the full multipolar charge density parameters are refined.

The multipolar refinement has been successfully applied to the 0.54 Å data of crambin (Jelsch et al. 2000), 0.96 Å data of the scorpion toxin (Housset et al. 2000), and 0.66 Å data for the complex aldose reductase (AR) with the inhibitor IDD 594 and cofactor $NADP^+$ (Muzet et al. 2003).

6.4.2 Charge Density Refinement: The Dummy Bond Electrons Model

Even at subatomic resolution, the number of experimental observations is often not great enough to assure a good observations/parameters ratio for the applicability of full multipolar refinement. One way to overcome this obstacle is a direct transfer of library parameters without their refinement. An alternative way is to use the "dummy bond electrons" model, which is more complex than the conventional model of spherical atoms with anisotropic displacement parameters but simpler than the multipolar model (Afonine et al. 2004, 2007). The main idea is to introduce additional Gaussian scatterers into the anisotropically refined model (in the middle of interatomic bonds and peaks corresponding to lone pair electrons) and refine this "mixed" model with geometrical restraints. As in the case for the multipolar model, the "dummy bond electrons" model came from the crystallography of small molecules (Brill 1960). The application of this approach to ultrahigh resolution study of macromolecules is implemented in the program phenix.refine (Afonine et al. 2005, 2007), which is a part of the crystallographic program suite PHENIX (Adams et al. 2002).

6.4.3 Quantum Modeling

By using the quantum mechanics (QM) methods, the researcher can determine the electronic density distribution from the atomic positions by solving the Schrödinger equations and applying the fact that the electron density is the product of the wave function by its conjugate. This implies that, given a perfect atomic structure including hydrogens (fixed set of the nuclei positions) and a method for solving the equations, it would be theoretically possible to determine the electrostatic potentials, and, in combination with molecular dynamics methods, to obtain the structural, dynamic, and chemical properties of proteins, including interactions in the active site of enzymes.

There are two main limitations to this idea. First, the exact positions of all nuclei are not known even at the subatomic resolution, and therefore, protonation and polarization states cannot be determined unambiguously from the model without some additional restraints or assumptions. Second, because of the limits of computer power, QM methods cannot be applied to a system as large as a protein.

Even with imperfect models, quantum modeling has been used to discriminate between different wave functions, and, hence, between alternative protonation or oxidation states. A model obtained after crystallographic refinement at the highest possible resolution, and usually restricted to the active site residues, is used as a starting geometrical template for QM calculations. Alternative assumptions are made about the protonation and oxidation states of the residues of interest. For each of them, the corresponding QM model is built. To discriminate between the alternatives, each particular QM model is optimized energetically, and the final models are compared. The most realistic model (based on different criteria) is chosen, and the corresponding protonation or oxidation state is accepted to be valid.

This approach was applied to a series of atomic and subatomic x-ray structures. In the case of the complex of horse liver alcohol dehydrogenase with NADH, researchers were able to identify a hydroxyl ion (instead of a water molecule) as the active moiety in the catalytic site and proposed a new mechanism of enzymatic activation of the cofactor (Meijers et al. 2001). Using similar ideas, the protonation states of metal-bound ligands were investigated for four enzymes (Nilsson and Ryde 2003). The application of this approach to the investigation of substrate–trypsin complexes (Schmidt et al. 2003) will be considered in detail later.

When the protonation states can be determined by complementary methods, QM modeling can be used to gain precise information about local properties of the molecule, for instance, bond lengths and angles, bond order, energy values, charge distribution, and electrostatic potential in active sites (Gogonea et al. 2001). This is particularly important if the resolution is limited.

Quantum mechanics modeling can also be applied to the modeling of interaction of substrate with catalytically important residues. Given the large dimensionality of the conformational space, small errors in the atomic coordinates may lead to significant changes in the resulting reaction mechanisms; therefore, the choice of proper starting coordinates is essential. It is in this sense that the combination of subatomic resolution crystallography and QM achieves its full power.

An example of the direct application of subatomic resolution data to QM studies of the reaction path is the investigation of the catalytic mechanism of human aldose reductase (hAR) (Cachau et al. 2000). The model of enzyme–inhibitor complex refined at subatomic resolution has been used as a starting point for molecular dynamics and quantum chemistry calculations. The available information on the presence of labile hydrogens in the ultrahigh resolution model enabled the determination of the protonation state of titratable residues during the modeling study. The key information from the ultrahigh resolution model was the orientation of the ring and the single protonation state of the active site residue histidine 110. A neutron diffraction study showed the presence of a mobile proton in the H-bond between Lys 77 and Asp 43. The combination with quantum chemistry calculations led to the proton rearrangement mechanism involving multiple residues, mainly Asp 43, Lys 77, and Tyr 48 (Blakeley et al. 2008). Another example of the combination of QM calculations and atomic resolution structures is the case of hydroxynitrile lyase (Schmidt et al. 2008). These calculations allowed the determination of the protonation states in the enzyme active site. The work showed that His 235 of the catalytic triad must be protonated in order for the catalysis to proceed, and the authors could reproduce the cyanohydrin synthesis in *ab initio* calculations. We also found evidence for the considerable *pKa* shifts, which had been hypothesized earlier.

The latest advances in obtaining ultrahigh resolution x-ray data for proteins open perspectives for the future use of QC to determine the reactivity of macromolecules. This possibility is based on the Hohenberg–Kohn theorem (Hohenberg and Kohn 1965) and Kohn and Sham equations (Kohn and Sham 1965), which established a relation between the electron density of a molecule and its wave function. This means that the detailed knowledge of the electron density could give additional information, sufficient to accurately determine the wave function. The knowledge of the wave function, in turn, is sufficient to obtain the chemical indices required to understand the reactivity of a molecule. Thus, theoretically, the knowledge of the electron density at ultrahigh resolution may provide the entire information required to fully characterize the local subtleties of its chemical nature. The main difficulties with this trend are related to the complexity of mathematical expressions and the limits in computing power. Therefore, these approaches were applied only to small molecules (Jayatilaka 1998; Grimwood and Jayatilaka 2001; Bytheway et al. 2002).

6.4.4 Electrostatic Potential Calculation

Since most drugs and inhibitors are not covalently bound to the protein, electrostatic interactions play an important role in binding mechanisms. Given a very accurate high-resolution structure, electrostatic properties can be calculated by different ways, for example, using QM methods, multipolar charge electron density (either using multipolar parameters from the database or refining these parameters), or using point charge models like those implemented in the software packages AMBER (Bayly et al. 1993) and CHARMM (Brooks et al. 1983). Although point charge models for calculations of Coulombic interactions are effective, they do not explicitly take into account the effect of dipoles and higher atomic electrostatic moments.

In cases where information about protonation state is known experimentally, for example, from kinetic measurements, or from the visual analysis of the electron density map, or when the assumptions are already made, QM calculations (which are supposed to be the most precise) can be used for computing electrostatic properties of proteins. A comparison of the results of computing of electrostatic potential using the DFT method and multipolar density refinement showed their qualitative similarity in the case of complex AR with inhibitor and cofactor $NADP^+$ (Muzet et al. 2003).

The electrostatic potential calculated for the active site for the complex AR with inhibitor and cofactor $NADP^+$, described in the results section, showed the electrostatic complementarities between the enzyme and the cofactor. These complementarities suggest a "lock and key" mechanism of binding (Muzet et al. 2003).

6.4.5 Highest Resolution Data Collection

Currently, the highest resolution diffraction data sets from protein crystals were collected for crambin at 0.48 Å (Schmidt et al. 2011) and 0.54 Å (Jelsch et al. 2000), antifreeze protein at 0.62 Å (Ko et al. 2003), lysozyme at 0.65 Å (Wang et al. 2007), and the complex of AR with inhibitor IDD594 and $NADP^+$ at 0.66 Å (Howard et al. 2004; Table 6.1).

The reasons for the high crystalline order in these particular cases are not easily explained. It has been suggested that the high quality of crystals of crambin is mainly explained by the small dimensions of the unit cell, low mobility associated to the presence of S–S bridges, strong packing contacts, and low content of the solvent (30% of the unit cell) which is almost perfectly packed. However, the crystals of AR do not satisfy these requirements. The unit cell dimensions are much larger, and the structure of AR does not contain S–S bridges; solvent occupies 43.3% of the unit cell. Packing analysis does not reveal any particularly tight intermolecular interaction. Therefore, for the time being, it does not seem to be possible to obtain general rules explaining the reasons for such high crystalline order.

6.4.6 Joint Use of High-Resolution X-Ray and Neutron Diffraction Data

Protein neutron crystallography (NC) is a powerful tool for the determination of protonation states, and it has been used to validate the results of high-resolution x-ray crystallography. This has been done in the enzymes, hAR (see Chapter 5) (Blakeley et al. 2008), D-xylose isomerase (XI) from *Streptomyces rubiginosus* (Katz et al. 2006), diisopropyl fluorophosphatase (DFPase) from *Loligo vulgaris* (Blum et al. 2007), and endothiapepsin from *Endothia parasitica* (Tuan et al. 2007). It should be noted that the information about protonation states was present in the x-ray ultrahigh resolution maps, but that it was difficult to validate due to the low scattering power of hydrogen atoms. The nuclear scattering maps obtained from

TABLE 6.1 List of 31 Protein Structures from the Protein Data Bank That Have Been Determined to Better than 0.85 Å

Structure (PDB Identification Code)	Experimental and Structure Solution Details	Information from This Subatomic Resolution Structure
Crambin (3NIR) (Schmidt et al. 2011) (1EJG) (Jelsch et al. 2000) (1CBN) (Teeter et al. 1993)	• Resolution: 3NIR 0.48 Å, 1EJG 0.54 Å, 1CBN 0.83 Å • Number of residues: 3NIR 48, 1EJG 46, 1CBN 48 • Solvent content: 3NIR and 1EJG 30%, 1CBN 32% • Refinement was done by SHELXL and MoPro • R-factor (final): 3NIR 0.127, 1EJG 0.09, 1CBN 0.106	For the 1EJG model, a deformation map averaged over 34 nondisordered peptide groups, was calculated. The map displayed a significant residual density in the bonds between nonhydrogen atoms and lone-pair peaks on the oxygen atoms. After multipolar refinement, the static deformation electron density of the average peptide residue was in quantitative agreement with the theoretical maps obtained from quantum chemistry calculations on the monopeptide. Extending the resolution limit from 0.54 (1EJG) to 0.48 Å (3NIR) increased the amount of data by a factor of 1.5. However, no additional structural features were revealed in the refined 3NIR model.
Type III Antifreeze protein RD1 (1UCS) (Ko et al. 2003)	• Resolution: 0.62 Å • Number of residues: 64 • Solvent content: 41% • R-factor/R-free: 0.137/0.155	The final difference map revealed the valence electron density in the middle of most bonds and densities corresponding to possible lone pairs. The final model contains water molecules with fractional occupancies and in "dual" locations. In the protein model, an unusual internal cavity without density and water molecules was observed. An extensive hydrogen bond network allowed one to perform an analysis of interactions between the IBS of the protein and water molecules.
Lysozyme (2VB1) (Wang et al. 2007)	• Resolution: 0.65 Å • Number of residues: 129 • Solvent content: 27% • *Ab initio* phasing with ACORN (Foadi et al. 2000) • R-factor/R-free: 0.084/0.095	At the last step of refinement, stereochemical restraints were removed for the well-ordered part of the structure, which lead to the lowering of R and R-free factors by 0.17% and 0.1%, correspondingly. The final values of bond lengths were more consistent with the values recommended by Jaskolski et al. (2007) than with those recommended by Engh and Huber (1991). The occupancy values for all water molecules were refined. Approximately 35% of the protein structure was identified as disordered and was modeled in multiple conformations.
hAR + inhibitor IDD594 + cofactor $NADP^+$ 1US0 (Howard et al. 2004) 2QXW (Blakeley et al. 2008) 2I16, 2I17 (Petrova et al. 2006)	• Resolution: 1US0 0.66 Å, 2QXW 0.80 Å, 2I16 and 2I17 0.81 Å • Number of residues: 316 • Solvent content: 43.3% • R-factor/R-free: 1US0 0.094/0.103, 2QXW 0.104/0.112, 2I16 0.079/0.089, 2I17 0.077/0.091	Densities corresponding to valence electrons, possible lone pairs, and 54% of hydrogen atoms were clearly seen in the difference maps. The final model revealed significant deviations from the standard stereochemistry. Analysis of the final difference map and the geometry of the model allowed the protonation state of the catalytic histidine and the ligand to be determined. An unusual short distance (2.97 Å) between the Br atom of the inhibitor and the hydroxyl oxygen of Thr 113 was observed. Electrostatic potential calculations for the residues of the active site were done in two different ways: (1) from multipolar parameters (either transferred from a database, or after refinement against experimental data) and (2) from quantum chemistry calculations (DFT). The results show electrostatic apoenzyme–cofactor and holoenzyme–inhibitor complementarity, as well as the polarization of the Br atom of the inhibitor. They allow further analysis of enzyme–inhibitor interaction. The reaction path between enzyme and substrate was studied by quantum modeling. Two data sets for the complex (2I16, 2I17) were collected from different parts of the same crystal at 15 and 60 K. A contribution of temperature to atomic B values (isotropic equivalents of ADPs) was estimated. It was found that, as the temperature was increased from 15 to 60 K, the differences in B values were approximately constant for well-ordered atoms (about 1.7 $Å^2$), although being slightly different for different kinds of atoms. The mean value of this difference varied according to the number of non-H atoms covalently bound to the parent atom.
Rubredoxin from *Desulfovibrio gigas* (2DSX) (Chen et al. 2006)	• Resolution: 0.68 Å • Number of residues: 52 • Solvent content: 27% • R-factor/R-free: 0.099/0.111	The bond lengths in iron–sulfur cluster differ from those in the structure solved at a lower resolution. The difference electron density map shows an extra electron density around Fe and four SG atoms. The authors proposed that the redistribution of the electron density is related to the electron transfer.
Pyrococcus abyssi rubredoxin mutant W4L/R5S (1YK4) (Bonish et al. 2005)	• Resolution: 0.69 Å • Number of residues: 52 • Solvent content: 55% • R-factor/R-free: 0.100/0.108	Unusually large deviations from planarity were revealed for eight peptide bonds. Sixty nine percent of all possible H atoms were observed. (The electron density difference map has peaks in the corresponding positions.) Six C–H···O hydrogen bonds were revealed. Six N–H···S hydrogen bonds previously found in 0.95 Å structure were confirmed. The deformation electron density was visible for approximately 60 covalent bonds. The bond electrons along Cα–C bonds were seen most clearly.

(continued)

TABLE 6.1 (continued) List of 31 Protein Structures from the Protein Data Bank That Have Been Determined to Better than 0.85 Å

Structure (PDB Identification Code)	Experimental and Structure Solution Details	Information from This Subatomic Resolution Structure
High-potential iron–sulfur protein from Thermochromatium Tepidum (3A38) (3A39) (1IUA) (Takeda et al. 2009; Liu et al. 2002)	• Resolution: 3A38 0.70 Å, 3A39 0.72 Å, 1IUA 0.80 Å • Number of residues: 83 • Solvent content: 34% • R-factor/Rfree: 3A38 0.066/0.074, 3A39 0.069/0.076, 1IUA 0.092/0.114	Most of the hydrogen atoms are seen. Ninety three water molecules were revealed for the 1IUA model, while the previous model determined at 1.5 Å contained only 43 water molecules. The final model revealed a partially hydrophobic cavity. A possible role of this cavity in electron transfer is discussed. Ultrahigh resolution data sets (3A38 and 3A39) were used to investigate x-ray-induced perturbations at the early step of radiation damage. An increase in relative B-factor for the first 60 frames was found, while the indicators of data quality such as Rsym and $I/\sigma(I)$ remained unchanged, suggesting that the heating of the crystal occurs prior to the aggravation of the data quality. As the dose increased, the electron densities of the hydrogen atoms of water molecules gradually disappeared.
The PDZ2 domain of systenin (1R6J) (Kang et al. 2004)	• Resolution: 0.73 Å • Number of residues: 82 • Solvent content:% • R-factor/R-free: 0.075/0.087	At the last step of refinement, stereochemical restraints were removed from the greatest part of the structure, excluding the minor alternative conformations. Several weak (CH···O) hydrogen bonds, in particular, between antiparallel β-strands were clearly identified. In eight places, deviations from the planarity of peptide bonds were found.
Bacillus lentus Subtilisin (1GCI) (Kuhn et al. 1998)	• Resolution: 0.78 Å • Number of residues: 269 • Solvent content: 45% • R-factor/R-free: 0.099/0.103	The ultrahigh resolution structure confirmed the existence of unusual short hydrogen bonds previously detected by NMR. The density corresponded to hydrogen atom was observed between the catalytic histidine Nδ1 and aspartate Oδ2, at 1.2 Å from Nδ*1* and at 1.5 Å from Oδ2. This kind of hydrogen bond is termed a "catalytic hydrogen bond."
Trypsin (1PQ7 and 1PQ5) (Schmidt et al. 2003)	• Resolution: 1PQ7 0.80 Å, 1PQ5 0.85 Å • Number of residues: 224 (1PQ7), 227 (1PQ5) • Solvent content: 34% • R-factor: 0.109 (1PQ7), 0.098 (1PQ5)	A series of trypsin structures containing either cleaved peptide fragments or covalently bound inhibitors was subjected to *ab initio* quantum chemical calculations and multipolar refinement. Quantum chemical calculations enabled one to determine the protonation state of active site residues. The multipolar refinement revealed the charge distribution in the active site and confirmed the results of quantum modeling. Knowledge about the protonation state of catalytic residues allowed the determination of the chemical state and the role of water molecules involved in catalysis. Possible mechanisms of the reaction were proposed.
Lysine-49 Phospholipase A2 from *Agkistrodon Acutus Venom* (1MX2) (Liu et al. 2003)	• Resolution: 0.8 Å • Number of residues: 122 • Solvent content: 15.3% • Initial phases were obtained using direct methods program SnB (Weeks and Miller 1999) • The model was built with ARP/wARP • Charge density refinement were performed with MOPRO and MOLLY (Jelsch et al. 1998) • R-factor/R-free: 0.095/0.121	Alternative conformations for disordered residues were modeled. The relation of conformational flexibility with functional properties is discussed. The values of standard deviations of the bond angles and main chain dihedral angles were found to be different from the standard stereochemical values. Peaks corresponding to about 38% of hydrogen atoms are seen in the difference electron density map. The hydrogen atom network in the vicinity of the active site is analyzed. The protonation of Nε2 atom of the catalytic histidine is determined from the difference map. Bond densities in a few places are seen. Multipolar charge density is refined for atoms with low B values.
Trypsin (1FY5,1GDN, 1FN8, 1FY4) (Rypniewski et al. 2001)	• Resolution: 0.81 Å • Number of residues: 227 • Solvent content: 34% • R-factor (final): 0.124 (1FY5), 0.108 (1GDN), 0.108 (1FN8), 0.108 (1FY4)	Electron densities for different trypsin structures in the active site were compared, which allowed the identification of components of peptides and water molecules with partial occupancies. In particular, S1, the specificity pocket of trypsin, is occupied by the arginine residue of the substrate peptide. A comparison of the electron densities of structures whose exposure times were different allowed one to observe directly the effect of radiation damage. For example, two of three S–S bridges were partially broken for the structure that was exposed longer. The models refined with subatomic resolution data were compared with the models refined at lower resolution and room temperature. The effects of freezing and difference in resolution were revealed itself in the ordering or disordering of a few particular side-chain residues.

TABLE 6.1 (continued) List of 31 Protein Structures from the Protein Data Bank That Have Been Determined to Better than 0.85 Å

Structure (PDB Identification Code)	Experimental and Structure Solution Details	Information from This Subatomic Resolution Structure
The bacterial light receptor PYP complexed with chromophore (1NWZ) (Getzoff et al. 2003)	• Resolution: 0.82 Å • Number of residues: 125 • Solvent content: 9.9% • The last steps of refinement were done without stereochemical restraints • R-factor/R-free: 0.123/0.144	Analysis of interatomic distances in the active site of the complex led to a conclusion about the deprotonated state of the cofactor (chromophore). The negative charge of the cofactor is essential for the shift of the chromophore absorption spectrum to the blue region of the visible spectrum. Analysis of the geometry revealed a hybrid electronic configuration of the cofactor. This configuration stabilizes the negative charge of the cofactor and is supposed to facilitate the double-bond isomerization required for the putative signaling function of the PYP. For the cofactor, a deviation of the geometry from the standard planar is observed (arching). A possible interrelation between this nonplanarity and the suppression of fluorescence is discussed. Based on the analysis of the anisotropy of atomic displacement parameters and multiple conformations of active site residues, the conclusions concerning the motion of the atoms were made. These motions favor the light-driven double-bond isomerization.
Triosephosphate isomerase complexed with PGN (2VXN) (Alahuhta and Wierenga 2010)	• Resolution: 0.82 Å • Number of residues: 251 • Solvent content: 29.8% • R-factor/R-free: 0.092/0.103	Two unusually short hydrogen bonds (2.69 and 2.60 Å) between PGN and catalytic Glu were found in the active site. Very precise details of the geometry of the active site of the enzyme allowed one to discuss the mechanism of the reaction of the enzyme, in particular, to address the protonation state of the first reaction intermediate and the features of the hydrogen-bond network in the active site.
Triosephosphate isomerase complexed with 2-phospho glycolate (1N55) (Kursula and Wierenga 2003)	• Resolution: 0.83 Å • Number of residues: 251 • Glu residue has been substituted to Gln residue • Solvent content: 29.8% • R-factor/R-free: 0.095/0.108	The structure is a complex of the enzyme with the inhibitor, which represents a transition state analog of the substrate. The inhibitor and the catalytic Glu residue occur in two conformations. In both of them, a very short hydrogen bond between the carboxylate group of the ligand and the Glu residue is observed. These bonds seem to be low-barrier hydrogen bonds. They indicate that the geometry of the active site favors the stabilization of the transition state required for the proton transfer. The model was refined with anisotropic displacement parameters, which allowed the analysis of the anisotropies in the thermal motion of the structure, in particular, of catalytic residues. The anisotropy of the key catalytic residues is consistent with the proton transfer mechanisms proposed earlier.
Photoactive yellow protein (3PYP) (Genick et al. 1998)	• Resolution: 0.85 Å • Number of residues: 125 • Solvent content: 32% • R-factor/R-free: 0.132/0.155	The rapid isomerization of the chromophore molecule in PYP is studied by comparing the ultrahigh resolution structures of the dark-adapted protein and an early photocycle intermediate trapped at low temperature. The difference electron density maps show that, upon light absorption, the incomplete trans-to-cis isomerization of the chromophore results in a highly distorted, transition-state-like geometry of its isomerizable double bond. Additionally, the chromophore is isomerized by flipping its thioester linkage rather than its aromatic ring, resulting in minimal perturbation of the tightly packed protein interior. The rotation of the thioester oxygen atom disrupts its hydrogen bond with the protein and places it in an electrostatically unfavorable hydrophobic environment. It is suggested that much of the initial photon energy is stored in this unstable, short-lived structure of the chromophore, which is subsequently relaxed in the protein pocket to form the next photocycle intermediates responsible for the signaling function of the protein.
Tem-1 beta-lactamase (1M40) (Minasov et al. 2002)	• Resolution: 0.85 Å • Number of residues: 263 • Solvent content: 27.3% • R-factor/R-free: 0.091/0.112	A long controversy existed concerning the question as to which residues are involved in the activation of the catalytic serine. The final difference map indicates that the catalytic glutamine is protonated and allows one to determine the protonation states of adjacent water molecules. These observations lead to a proposition of the activation pathway in which the glutamine residue activates a water molecule, which, in turn, activates the serine residue.
Squid ganglion DFPase (1PJX) (Koepke et al. 2003)	• Resolution: 0.85 Å • Number of residues: 314 • Solvent content: 43.8% • R-factor/R-free: 0.121/0.128	Forty five of 314 residues were modeled in alternative conformations. At the last step of refinement, 208 most ordered residues were refined without restraints. It is evident from the electron density map that Nε2 of His 274, which is in the Ca^{2+}-binding site, is protonated. This information, together with the precise bond lengths, can explain the stability of the Ca^{2+}-binding site. Hydrogen atoms in the hydrogen bonds between two antiparallel β-strands are clearly visible. Average values of the bond lengths and bond angles for the residues that were refined without restraints were compared with the standard values (Engh and Huber 1991), and significant differences were found.

(continued)

TABLE 6.1 (continued) List of 31 Protein Structures from the Protein Data Bank That Have Been Determined to Better than 0.85 Å

Structure (PDB Identification Code)	Experimental and Structure Solution Details	Information from This Subatomic Resolution Structure
Complex of AR with NADP+ and simultaneously bound competitive inhibitors Fidarestat and IDD594 (2PFH, 2PF8) (Cousido-Siah et al. 2012)	• Resolution: 0.85 Å • Number of residues: 316 • Solvent content: 34.6% • R-factor/R-free: 2PFH 0.082/0.095, 2PF8 0.085/0.095	The superposition of two different ligands competitively bound to the enzyme is seen in the electron density map of the active site. The residues of the active site and water molecules are also in multiple conformations. The subatomic resolution data allowed one to reveal all superimposed multiple conformations for the residues and water molecules in the vicinity of the active site and determine occupancy values for all residues in multiple conformations independently of each other. The relative occupancy values for the competitive inhibitors were further compared with the occupation ration in solution obtained by mass spectroscopy.

Note: As of May 2012, there were 69 protein structures in PDB with a maximal resolution limit better than 0.85 Å. The table contains all protein structures solved at a resolution better than 0.73 Å. The structures are sorted according to the resolution value.

neutron diffraction were essential to confirm this information. Another interesting case in this context is that of type III antifreeze protein (AFP), which binds ice nuclei through an ice-binding surface (IBS) to prevent their growth. Joint neutron (1.85 Å resolution) and x-ray (1.05 Å resolution) diffraction studies (Howard et al. 2011) have allowed to identify a tetrahedral water cluster bound to the IBS of AFP. The difficulty of this case was that one of the waters of the tetrahedral cluster has a double conformation, and the conformation reported in previous studies was not the tetrahedral one. Figure 6.3 shows the maps and models, as follows: (1) the σ_A-weighted F_o–F_c omit nuclear scattering density map (orange contour level = 2.6σ) for the tetrahedral water cluster, showing the strong signal for three of the four water molecules (1001, 1002, 1003) and a relatively weaker signal averaging the two conformations of water 1004; (2) the σ_A-weighted F_o–F_c omit electron density map, with peaks for the water 1004A (magenta contour level = 2.6σ) and 1004B (green contour level = 2.6σ). The position of 1004A puts the water cluster in an ideal tetrahedral geometry, while the second position 1004B (not tetrahedral) is the one reported previously in the PDB. In this case, the neutron structure was essential to detect the double conformation of water 1004. This tetrahedral cluster was then used to build a model of the AFP–ice interface. This model explains the preference of AFP to bind ice instead of water, as hydrophobic residues in the IBS fit inside the cavity in the middle of six water rings in the ice structure.

6.5 Biologically Important Results

From the ensemble of structures solved at a resolution better than 0.85 Å (see Table 6.1), we have chosen three examples which have high biological impact and also clearly illustrate how the subatomic resolution details can be crucial for studying the protein function and catalytic mechanism. These examples are described in detail as follows.

6.5.1 TEM-1 β-Lactamase

Class A β-lactamases are bacterial enzymes that provide resistance to the antibiotics of the β-lactam family, like penicillin and its derivatives. A detailed understanding of the precise catalytic mechanism at the atomic level is important for designing new and more specific antibiotics.

TEM-1 is one of the best studied enzymes of the class A β-lactamases. It catalyzes the hydrolysis of β-lactams. The general scheme for the reaction is the following: the active site serine attacks the carbonyl group of the lactam ring to form a transiently stable acyl–enzyme intermediate. Subsequently, this intermediate is attacked by hydrolytic water to form a deacylation intermediate, and, then, to form a hydrolyzed product.

While the second part of the mechanism is clear, the first one remained controversial. One important point was unclear: which residues are involved in the activation of the catalytic serine? Two alternative hypotheses were proposed. In one mechanism,

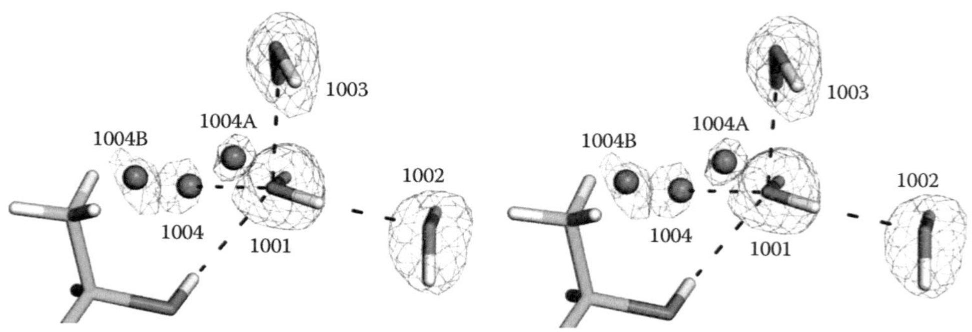

FIGURE 6.3 Superposition of the tetrahedral water cluster model and density maps for antifreeze protein.

the catalytic lysine residue is proposed to be deprotonated and to act as the base for activating the serine (Herzberg and Moult 1987; Strynadka et al. 1992; Leung et al. 1994; Strynadka et al. 1996). In the second mechanism, a glutamine acts as the catalytic base, activating the serine through bound water (Lamotte-Brasseur et al. 1991; Damblon et al. 1996; Lietz et al. 2000; Mustafi et al. 2001; see Figure 6.4a and b for both schemes).

The knowledge of the position of hydrogen atoms and the protonation states of these residues is crucial to resolve this controversy and, hence, to determine the mechanism of the reaction.

To resolve the uncertainty about the protonation states of the active site residues, intensive *pKa* computation studies (Lamotte-Brasseur et al. 1999), *pKa* measurements by NMR (Damblon et al. 1996), modeling (Ishiguro and Imajo 1996), and crystallographic structure determinations have been performed (Chen et al. 1996). Most of the crystal structures of TEM-1 were solved at around 1.7 Å, a resolution not high enough to determine the protonation states of the lysine and glutamine residues. A data set at 0.85 Å resolution was collected for the complex of TEM-1 β-lactamase with boronic acid, an acylation transition-state analog (Minasov et al. 2002). One of the main factors of the essential improvement in resolution was the use of the TEM-1 mutant for crystallization. The substitution of a methionine residue by a threonine had a stabilizing effect on the structure and, at the same time, kept the catalytic activity (Huang and Palzkill 1997).

The good quality of the data provided a very accurate, refined model, with a final *R*-factor (see Appendix 6.A) value of 0.091. In the final difference map, peaks for all main chain hydrogen atoms were found, except for residues in double conformations. It was even possible to observe hydrogen atoms for a few water molecules.

The strong positive peak clearly showed that glutamine residue was protonated (Figure 6.5).

Furthermore, the difference maps unambiguously showed hydrogen atoms of the water molecule with which the glutamine residue forms a hydrogen bond. These observations led to the following interpretation. The glutamine residue activates the water molecule which, in turn, activates the serine residue. Therefore, the activation pathway consists in a proton shuttle from the serine to the water molecule, and, then, to the glutamine residue. In this case, ultrahigh resolution solved the long-standing controversy concerning the reaction mechanism.

In an independent study of the native structure of SHV-2 class A β-lactamase at atomic resolution, a hydrogen atom from

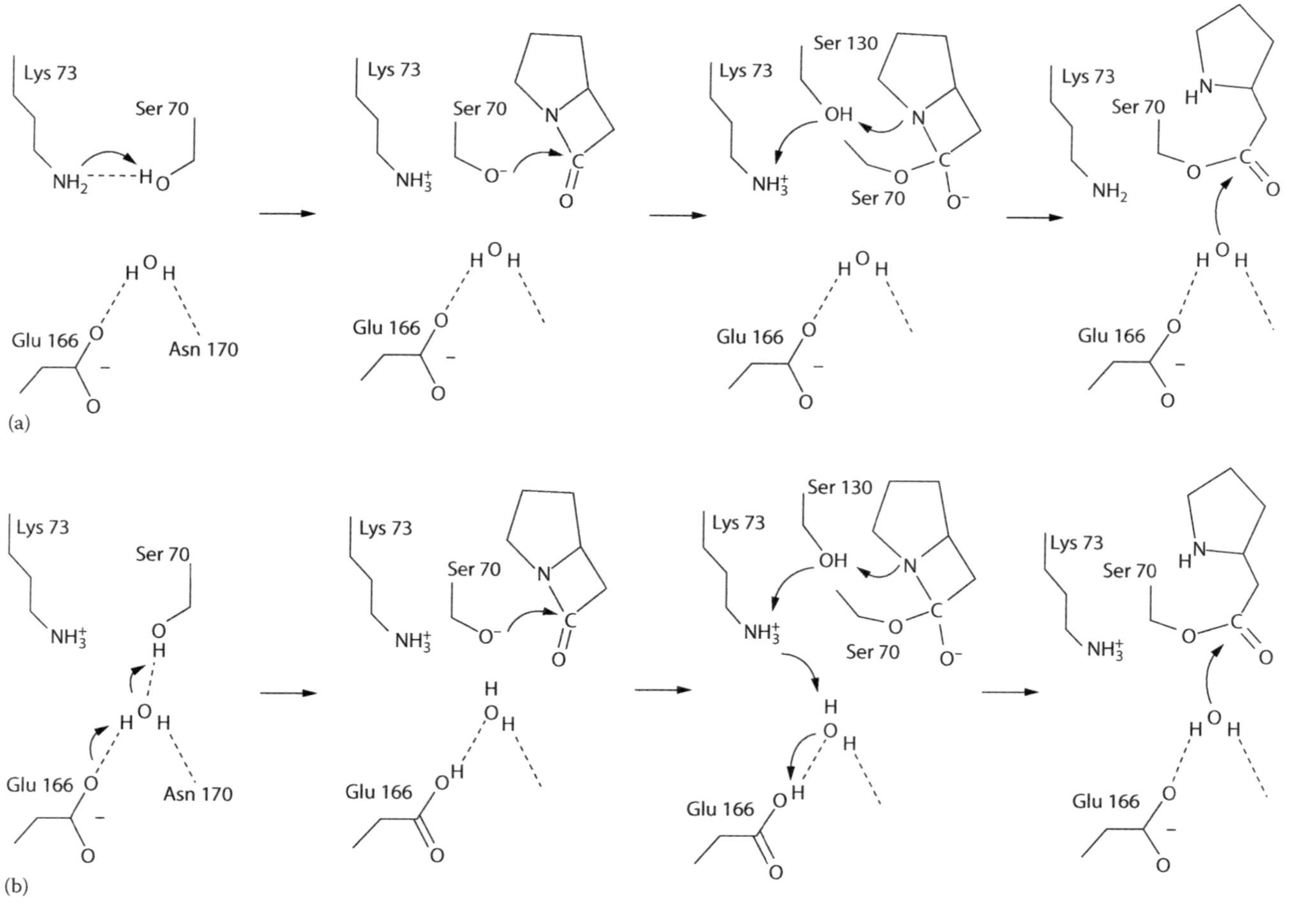

FIGURE 6.4 Two proposed alternative mechanisms of serine acylation involving either Lys 73 (scheme [a]) or Glu166 (scheme [b]) as the general base. (Reproduced from Nukaga, M. et al., *J. Mol. Biol.*, 328, 289–301. Copyright 2003.)

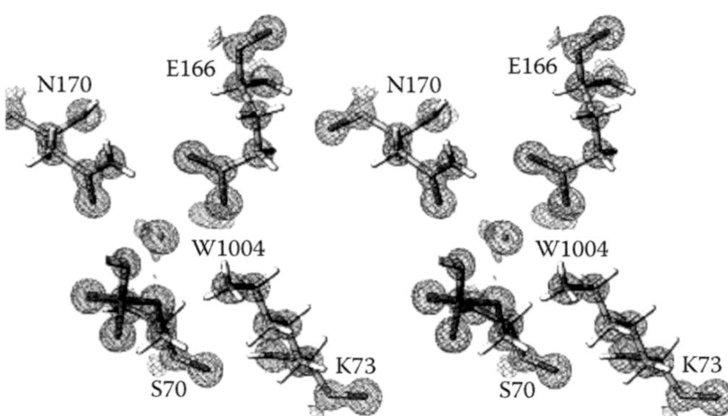

FIGURE 6.5 Stereoview of $2F^{obs}-F^{calc}$ (dark gray, cutoff = 2σ) and $F^{obs}-F^{calc}$ (light gray, cutoff = 1.4σ) electron density maps (see Appendix 6.A) in the active site of TEM M182T. The read peak at Oε2 of Glu 166 indicates that the carboxylate group is protonated. The two peaks on the water 1004 show the possible positions of hydrogen atoms. (Reproduced with permission from Minasov, F., Wang, X., and Shoichet, B.K. 2002. An ultrahigh resolution structure of TEM-1 beta-lactamase suggests a role for Glu 166 as the general base in acylation, *J. Am. Chem. Soc.*, 124, 5333–5340. Copyright 2002 American Chemical Society.)

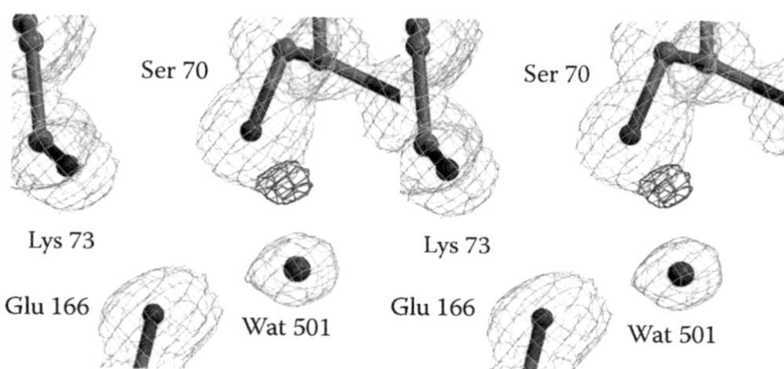

FIGURE 6.6 Stereoview of $2F^{obs}-F^{calc}$ (gray, cutoff = 2σ) and $F^{obs}-F^{calc}$ (dark gray, cutoff = 3σ) electron density maps (see Appendix 6.A) in the region in between catalytic serine and glutamine. The peak at serine 70 indicates that this residue is clearly protonated. (Reproduced from Nukaga, M. et al., *J. Mol. Biol.*, 328, 289–301. Copyright 2003.)

the reactive serine residue was clearly seen in the difference map (Nukaga et al. 2003). This hydrogen atom was found to be exposed to the water molecule activated by the glutamine and asparagine residues (Figure 6.6). This observation confirms the preceding analysis and provides further support for the second mechanism.

6.5.2 Trypsin

Trypsin is an enzyme secreted by the pancreas and works in the intestine at an optimal pH of 8.5–9.0. It cleaves peptides at the C-terminal part of arginine or lysine residues. Afterward, other enzymes cut these pieces into amino acids, which will be used throughout the body. The active site of trypsin contains three catalytic residues: serine, histidine, and asparagine, and two water molecules. The catalytic mechanism of trypsin has been studied extensively for several decades (Blow 1976; Kossiakoff and Spencer 1981; Singer et al. 1993; Katona et al. 2002; Wilmouth et al. 2003), but the exact relationship between the catalytic residues and the sequence of events during the catalysis is still not completely elucidated. In particular, the chemical state and the role in catalysis of the two water molecules in the active site are under discussion.

Recently, a series of five structures of trypsin were solved at atomic and subatomic resolutions (Schmidt et al. 2003). The crystallographic data for three native proteins crystallized at different pH were collected at resolutions of 1.0, 0.83, and 0.80 Å. Data for two other structures with covalently bound inhibitors were also collected at 1.22 and 1.23 Å. The protein was shown to be active in the crystal. A peptide fragment is always found in the active site. The C-terminal residue of the peptide is an arginine and is located at the vicinity of the catalytic serine. These structures gave a good opportunity to investigate the catalytic mechanism of trypsin. The inhibited structures mimicked the reaction intermediates.

Three models were subjected to QM calculations, which were performed with the software GAMESS (Schmidt et al. 1993). The first model comprised the natural substrate (arginine residue) and the protein active site, in which catalytic residues and water molecules were assumed to be in their standard protonation state (native state). The view of the active site for this model is shown in Figure 6.7. The second model differed from the first by the protonation state of the catalytic serine, which was treated as deprotonated (negatively charged Ser$^-$). The third model had the inhibitor PMSF in place of the natural arginine, and an attacking water molecule was modeled as a hydroxyl ion (PMSF/OH$^-$). In all these cases, the atomic coordinates of the native protein refined at 1.0 Å were used for the protein atoms. These theoretical

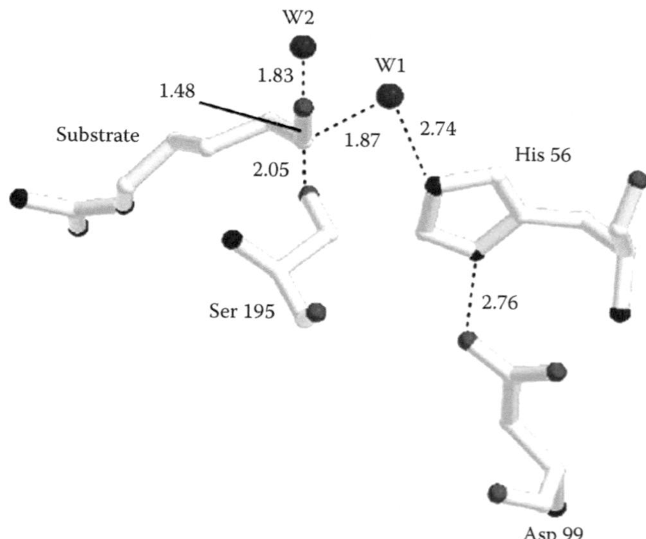

FIGURE 6.7 The active site of trypsin (native structure). The geometry around the substrate shows unusual interatomic distances. (Reproduced from Schmidt, A. et al., *J. Biol. Chem.*, 278, 43357–43362. Copyright 2003 American Society for Biochemistry and Molecular Biology.)

models were subjected to QM optimization, and their final optimized structures were compared to the crystallographically refined one. The best match was obtained for the model with the negatively charged serine, while two other models deviated significantly from the crystallographic structure. This implies that the crystal structure represents a state close to the tetrahedral transition state, where the catalytic serine is deprotonated, and the water W1 is attacking the neutral arginine carbonyl rather than a hydroxyl ion. This last result also leads to the conclusion that the catalytic histidine is deprotonated on Nε2 and acts as a Lewis base to activate the catalytic water.

Finally, these conclusions about the protonation state of catalytic residues and the chemical state of the W1 water molecule, in combination with a detailed analysis of the geometry of the active site of the native structures, as well as the structures complexed to inhibitors, were used to propose a detailed mechanism for the reaction. It was also found that the second water molecule W2 forms a very short, and therefore strong, H-bond with the substrate. This last observation was crucial to explain the role of this water molecule in catalysis. In the proposed mechanism, the water molecule W2 stabilizes the "intermediate state" and prepares the reaction, while the first one attacks the substrate.

In addition, a multipolar model was built using a native ultrahigh resolution data set of 0.8 Å. The multipolar parameters were transferred from the database. In Figure 6.8, a deformation map for the catalytic histidine suggested that Nε2 is deprotonated while Nδ1 is protonated (neutral histidine). The small peak near Nε2 is explained by the authors as the density corresponding to an electron lone pair on the nitrogen atom or a partial protonation; on the other hand, it could probably belong to Nδ1 of the second conformation of the ring. Hence, the visual analysis of the electron density map confirmed that the catalytic histidine was deprotonated on Nε2, which, in turn, confirmed the results of the QM calculations.

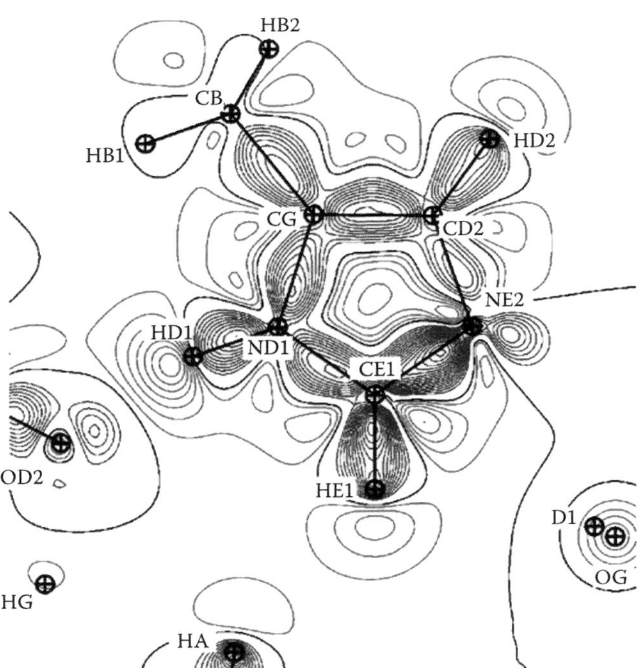

FIGURE 6.8 Deformation density for the catalytic histidine of native structure of trypsin shows its electronic state. (Reproduced from Schmidt, A. et al., *J. Biol. Chem.*, 278, 43357–43362. Copyright 2003 American Society for Biochemistry and Molecular Biology.)

The combination of ultrahigh resolution crystallography, *ab initio* QM calculations, and multipolar refinement revealed the role of one of the water molecules in catalysis and helped to propose a mechanism for the reaction.

6.5.3 Aldose Reductase

Aldose reductase (EC 1.1.1.21) is a NADPH-dependent enzyme that reduces a wide range of substrates, such as aldehydes, aldoses, and corticosteroids. As it reduces D-glucose into D-sorbitol, it is believed to cause severe degenerative complications of diabetes (Dvornik 1987; Yabe-Nishimura 1998). The catalytic reaction involves a hydride transfer from NADPH, which becomes $NADP^+$, and a proton donation from the enzyme (Wermuth 1985). Aldose reductase folds in a $(\beta/\alpha)_8$ barrel with the nicotinamide ring of the $NADP^+$ buried at the bottom of a deep cleft. The ternary complex AR–$NADP^+$–inhibitor IDD 594 has been solved to the highest resolution so far for a medium-size protein (MW: 36 kDa) (Howard et al. 2004). We will therefore describe this case in much more detail.

6.5.3.1 Crystallographic Studies

Crystals of the ternary *human* complex AR–$NADP^+$–IDD594 were prepared at pH 5.0 as described (Howard et al. 1999; Lamour et al. 1999). The diffraction quality could be significantly improved by co-crystallization with IDD594, followed by microseeding and cryotechniques (Litt et al. 1998). X-ray crystallographic data extending up to 0.62 Å were measured on the beamline ID19 at SBC-CAT, APS, from which a subset (limited to 0.66 Å) was kept for the structure refinement.

The structure was initially refined with CNS (Brünger et al. 1998), starting with the coordinates of the complex of hAR with another inhibitor, IDD 384, solved at 1.7 Å (Calderone et al. 2000). At the next step, the refinement with SHELXL, first isotropic, followed by an anisotropic refinement, then an addition of H-atoms, led to a final model comprising 313 out of the 316 residues, $NADP^+$, IDD 594, two citrate molecules, and 613 water molecules. Multiple conformations were observed for 99 residues; 198 water molecules have less than unit occupancy. The statistics for data collection and refinement are given in Table 6.2.

Fifty-four percent of all the possible hydrogen atoms in the whole protein were observed; this includes most of those riding on atoms with the lowest values of the temperature factors. The percentage of observed hydrogens is linearly correlated to the temperature factors of the covalently linked non-H atoms (Figure 6.9).

TABLE 6.2 Data Collection, Structure Refinement, and Model Statistics for the Complex AR with Inhibitor IDD594 and Cofactor NADP+

Cell dimensions, space group	a = 49.28 Å, b = 66.59 Å, c = 47.26 Å, β = 92.4°, P21
Data Collection: Resolution (Å) (last shell)	20.0–0.66 (0.68–0.66)
No. of unique reflections (last shell)	511,265 (43,107)
Completeness (%) (last shell)	89.1 (75.4)
$I/\sigma(I)$ (last shell)	14.5 (2.4)
$R(I)_{merge}$ (%) (last shell)	2.9 (27.9)
Refinement: Reflections with $F > 4\sigma_F$ (for R_{free})	405,150 (21,261)
Isotropic: R_{cryst}/R_{free} without H (%)	16.30/16.99
Anisotropic: R_{cryst}/R_{free} without H (%)	9.63/10.01
R_{cryst}/R_{free} with riding H (%)	8.42/9.34
Anisotropic using all reflections (for R_{free})	485,662 (25,585)
R_{cryst}/R_{free} without H (%)	10.56/10.94
R_{cryst}/R_{free} with riding H (%)	9.36/10.32
Model	
Estimated standard deviations of fully occupied atomic positions	
Most ordered regions (B < 3 Å^2)	<0.005 Å
Rest of the molecule	0.007 Å
R. m. s. deviations from ideal values	
Bonds (Å)	0.016
Angles (°Å) (comment: angles are restrained as 1–3 distances in SHELXL)	0.034
Mean B-factor (Å^2) (fully occupied, partially occupied)	
Protein main chain 1036/426	5.4/6.3
Protein side chain 909/597	6.9/9.9
NADP+ 48/0	3.9/n/a
Inhibitor 24/0	4.2/n/a
Citrate 0/26	n/a/6.9
Solvent 415/198	19.7/12.5

Note: See glossary of crystallographic terms in Appendix 6.A.

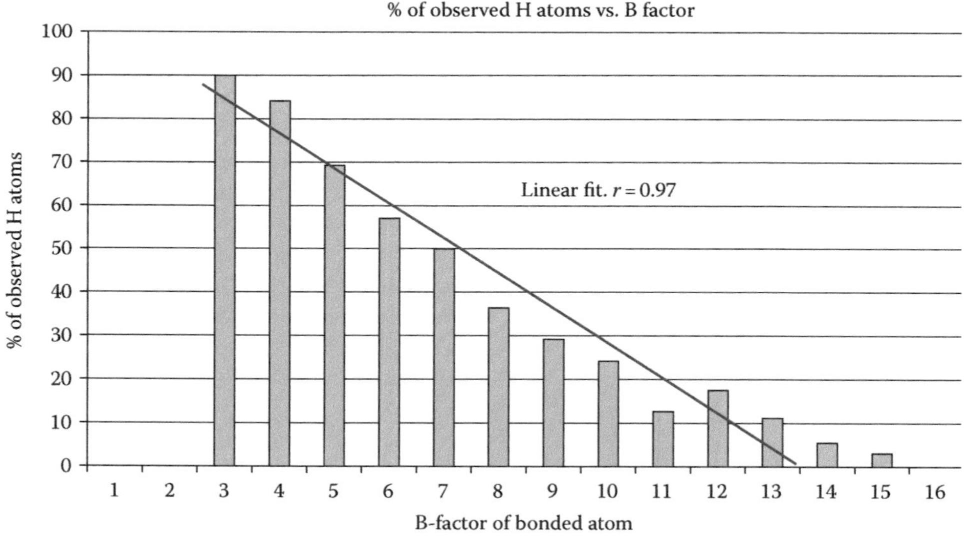

FIGURE 6.9 Histogram of percentage of observed hydrogen atoms as a function of the B-factor of the bonded heavy atom for the model of AR. (Reproduced from Howard, E.I. et al., *Proteins: Struct. Funct. Genet.*, 55, 792, 2004.)

6.5.3.2 Unusual Stereochemistry

As was mentioned earlier, the stereochemistry of the active site shows clear departures from standard values (Rick and Cachau 2000). Because of the low error values (estimated standard deviations of atomic positions around 0.005 Å), these observations are significant. For example, deviations of the peptide bond angle ω, which measures the peptide bond planarity, were observed. The usual value for a planar peptide bond is 180°. However, 27 cases were found with a deviation from this value larger than 10°. These deviations are stabilized in most parts of cases by strong H-bonds involving the peptide N atom. One of these cases is observed in the well-ordered region involving the contacts between the residues 75–77 and 44–45. The peptide bond Ser 76–Lys 77 has an ω angle of −167.3°, stabilized by a strong (2.69 Å) H-bond of the N atom of Lys 77 with the OG atom of Ser 76 (Figure 6.10). Since Lys 77 is in the active site region, this could be important for the catalytic mechanism.

The accuracy of the experimental data and of the anisotropically refined structure is sufficient to show deviations from the spherical atom model in an electron density difference map; in particular, bond densities are clearly seen in the highly ordered residues (Howard et al. 2004). The observed density corresponds to the expected σ-bond shape, as discussed in the multipolar studies of crambin (Jelsch et al. 2000).

Note that the bond densities in the peptide link 44–45 appear both in the C–N and C–O bonds; this corresponds to a similar partial bond order for the C–N and C–O bonds. On the other hand, in the link 45–46, the bond density appears strongly only

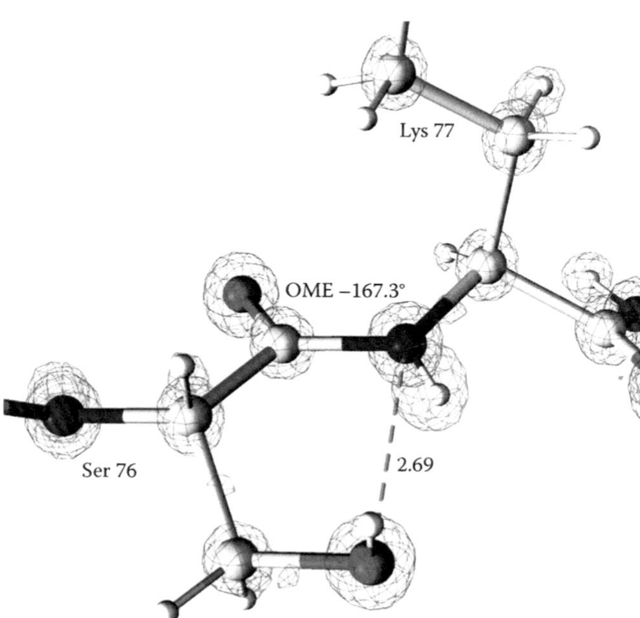

FIGURE 6.10 F^{obs}–F^{calc} (Omit-H, Sigma-weighted, red contours at 0.35 e/Å^3 (3.2σ)) and 2F^{obs}–F^{calc} (Sigma-weighted, blue contours at 4.70 e/Å^3 (4.2σ)) maps around the peptide bond Ser 76–Lys 77 of the model of AR. The strong H-bond linking Lys 77 N and Ser 76 OG stabilizes the large deviation from the standard value of the ω angle.

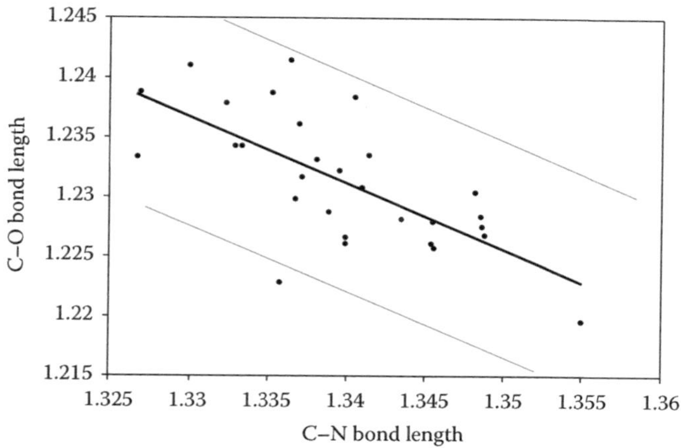

FIGURE 6.11 Graph of the C–O versus C–N bond lengths for the 29 residues with the lowest values of temperature factors for the model of AR. The thick line represents the regression curve (length[C–O] = 1.98 A − 0.55 × length[CN], correlation coefficient = 0.69); the two thin lines represent the 2σ error margins, assuming a RMSD value of 0.003 Å in both bond lengths. (Reproduced from Howard, E.I. et al., *Proteins: Struct. Funct. Genet.*, 55, 792, 2004.)

in the C–O bond and not in the C–N bond. This can indicate that the C–O bond is double and the C–N bond is single. These bond orders are confirmed by the C–O and C–N bond lengths, which are respectively (1.243[3], 1.334[3] Å) for the 44–45 peptide link and (1.219[3],1.359[3] Å) for the 45–46 peptide link. Since the resonance of the peptide bond involves a transfer of electron density from the C–O bond to the C–N one, when the double bond character of the C–N bond increases, one of the C–O bond decreases. As a double bond is shorter than a single one, this implies that the bond lengths are anticorrelated (Figure 6.11). This negative correlation is observed in all the residues where the temperature factor values of the main chain atoms are less than 3 Å^2, showing a range of different resonance states of the peptide bond.

6.5.3.3 Analysis of the Solvent Structure

Water molecules form highly organized, 3D networks in which the water oxygens maintain as much as possible the tetrahedral structure. Sometimes, these networks resemble the ice structure (e.g., in the vicinity of His 46 and of Gln 59) and at other times are organized in five-member rings, joined as portions of a dome. The networks are particularly evident on the protein surface, for example, in the environment of the side chains of Ser 2 and Arg 3 (Figure 6.12).

The organization of water in rings at the protein surface is compatible with the tetrahedral structure which emerges toward the solvent, resulting in the 3D arrangement previously described. The temperature factor of water molecules (mean value 21 Å^2) approximately increases both as a function of the distance to the protein surface and of the diminishing number of observed neighbors. The lowest B values are observed for those water molecules making several contacts with the protein (e.g., buried ones). The structure confirms the previous observations

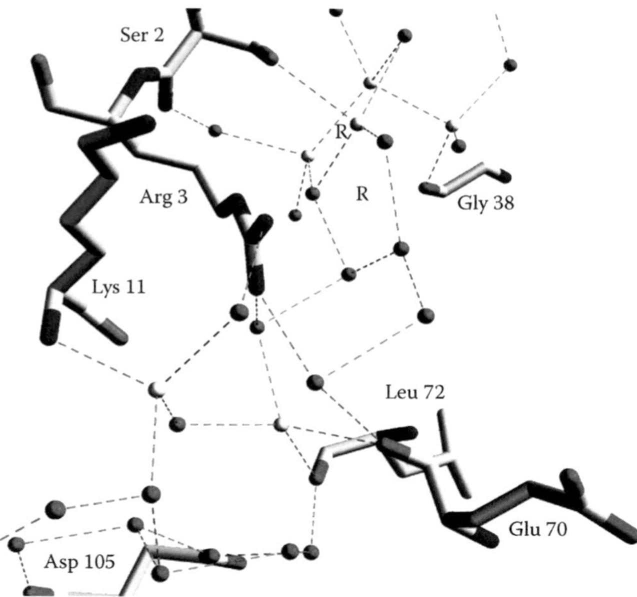

FIGURE 6.12 Water networks in the vicinity of Ser 2 and Arg 3 of the model of AR. Five-member rings formed by associated water molecules are marked with R; the tetrahedral water molecules are marked in light gray.

(Podjarny et al. 1997) that the water molecules follow closely the crevices in the protein surface. Analysis of the anisotropic temperature factors shows also that, in the crevices, the water molecules move preferentially along the crevices, in a direction parallel to the protein surface.

6.5.3.4 Analysis of the Reaction Center

The electron density map is of very high quality, due to the experimental resolution. The high accuracy of this map enables the researcher to unambiguously identify the number of electrons, and hence the chemical nature (C,N,O,F,S,Br) of all non-hydrogen atoms in the active site (Figure 6.13a).

The stereochemistry of this region shows the hydrophilic head of IDD 594 strongly bound to His 110, Tyr 48, and Trp 111 (Figure 6.13a). At this resolution, the charge state of IDD 594 can be identified by the lack of riding H-atoms in the carboxyl head of the inhibitor and to the similar distances for the two C–O bonds (1.25 and 1.26 Å), thus suggesting a charged state for the IDD 594 carboxylic head in the complex. This stereochemistry is consistent with previously observed interactions of AR with charged inhibitors, and in particular to that observed at lower resolution in the complex of the pig enzyme with the inhibitor Tolrestat at a much higher pH (=6.2) (Urzhumtsev et al. 1997). Since the interactions are similar, the charge state of the inhibitor head is likely to be the same in both complexes. Since His 110 is neutral (see the following paragraph), the charge of the inhibitor compensates the positive charge of $NADP^+$, resulting in a charge-neutral reaction center.

Histidine 110 is a key residue in the active site, and therefore, its protonation state is crucial for determining the catalytical mechanism. If it were completely surrounded by solvent, His 110 should be doubly protonated at the crystallization conditions (pH = 5). However, in the complex structure, His 110 was found to be singly protonated at the NE2 position (Figure 6.13b), which points toward the inhibitor carboxylate group making a

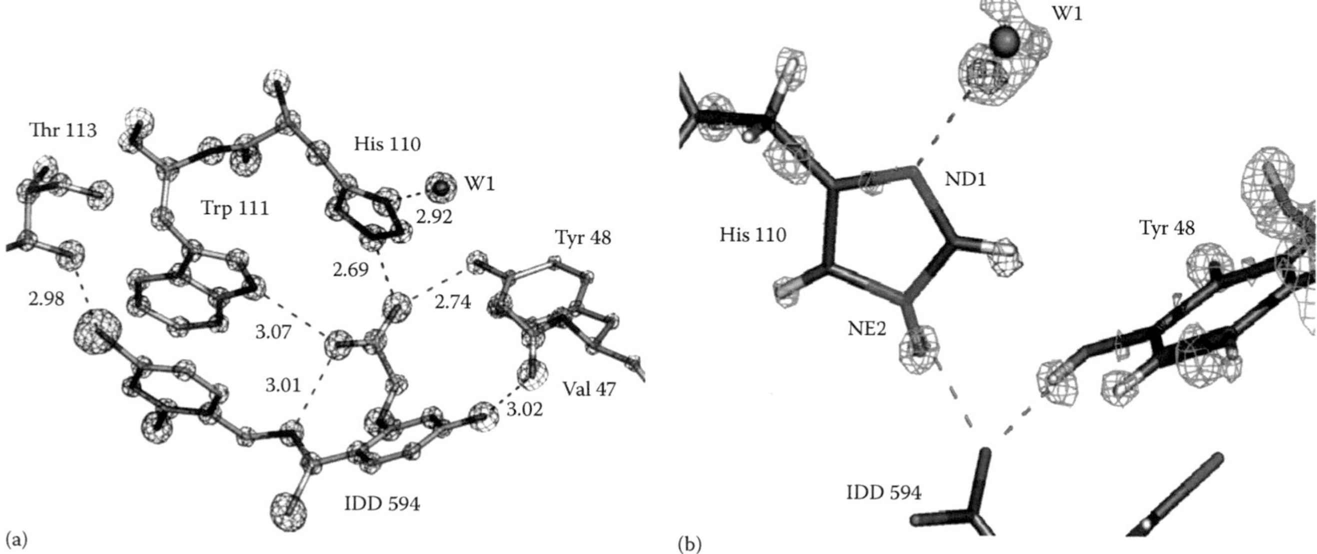

FIGURE 6.13 Inhibitor contact details for the complex of AR with inhibitor IDD 594 and cofactor $NADP^+$. (a) σA-weighted 2 F^{obs}–F^{calc} maps, contoured at 4.15 e/Å^3 (4σ). The respective peak volumes correlate well with the atomic number. (b) σA-weighted F^{obs}–F^{calc} map of the active site region with omitted hydrogen atoms, contoured at 0.44 e/Å^3 (4.0σ), 0.31 e/Å^3 (2.8σ), and 0.11 e/Å^3 (1.0σ). The single protonation of His 110 at NE2 is clearly visible, while the strong peak near the W1 water molecule indicates that W1 is the proton donor in the W1–His 110 ND1 hydrogen bond. The hydrogen atoms are observed in the maps, as interpreted by SHELXL. The 1.0σ contour is restricted to the neighborhood of the H atom. (Reproduced from Howard, E.I. et al., *Proteins: Struct. Funct. Genet.*, 55, 792, 2004.)

very short H-bond with a favorable geometry. This H-bond stabilizes the protonation at the NE2 side, while on the ND1 side the proton making the ND1–W1 H-bond belongs clearly to the water molecule. The difference between the electron density peaks at C (peak = 8.3 e/Å^3) and N (peak = 10.4 e/Å^3) positions unambiguously determines the orientation of the His 110. The Tyr 48 OH group is likely to be very strongly polarized by the close proximity of the charged Lys 77 NZ, leading to a lowering of the *pKa* value (Wilson et al. 1992). This polarization is critically dependent on the positioning of the Lys residue and suggests that the unusual value of the ω angle in the peptide bond 76–77 (described in the following) is linked to the catalytic reaction. The H-bond network between the buried water molecule W2 and Lys 77 may help to stabilize the variations in the electron density involved in this polarization.

The observations about the orientation and protonation state of His 110 do not agree with the previous proposed mechanisms, in particular, with the push–pull one (Schlegel et al. 1998), in which the orientation of the histidine residue is different. To further study the protonation state of the residues in the catalytic site, a neutron diffraction study was undertaken, using fully deuterated AR in complex with IDD 594. The structure was determined both with x-rays at helium temperature at 0.9 Å resolution and with combined neutrons/x-ray at room temperature (x-ray data resolution 1.75 Å, neutron data resolution 2.2 Å). This structure showed the existence of a mobile proton in the hydrogen bond between Asp 43 and Lys 77. This implies a neutral form of Lys 77, which becomes mobile. This information was used to propose a catalytic model based on QM calculations, which explains the donation of the catalytic proton by Tyr 48 (Blakeley et al. 2008).

6.5.3.5 Interactions of AR with the IDD 594 Specificity Pocket

The binding of IDD 594 induces a conformational change, opening a specificity pocket between Trp 111 and Ala 299, to which it binds in an orientation close to that observed in the AR–Zopolrestat complex (Wilson et al. 1993). This binding is stabilized by several interactions, including stacking between one of the inhibitor rings and Trp 111, and by an unusually short contact, Br–Thr 113 OG (2.98 Å) distance (Figure 6.13a). This distance is significantly shorter than the minimum reported in the Cambridge Data Bank (3.17 Å) for an interaction between a Br (attached to an aromatic ring) and a OH group, and is much closer to the distance observed between a Br and a carbonyl oxygen (Cambridge 2001). Such close approach of Br and OH group in the structure is facilitated by the displacement of the H-atom linked to OG–Thr 113. Instead of being on a Br–OG axis, this H atom is involved in a H-bond (2.81 Å) with its own main chain O atom. This hydrogen bond enhances the electronegative character of the OG–Thr 113 atom, while the electropositivity of the Br in the axial direction implies a strong polarization which has been confirmed by quantum calculations (Muzet et al. 2003). All these observations suggest a strong bromine–oxygen electrostatic interaction, which is important to maintain the specificity against aldehyde reductase, where Thr 113 is substituted with much bulkier tyrosine.

6.5.3.6 Multipolar Modeling and Electrostatic Potential Calculations

Based on the model described earlier, the electron density and the electrostatic potential of the active site of the complex were studied by different methods (Muzet et al. 2003):

1. Use of the multipolar charge density parameters, transferred from multipolar atoms database. As a preliminary step to this study, multipolar parameters have been modeled for NADP$^+$ based on studying the cofactor NAD$^+$, which is an analog of NADP$^+$. Then, these parameters were added to the database.
2. Multipolar refinement with MoPro for the protein main chain atoms. Only a subset of the structure including atoms with B-factor values less than 8 Å was subjected to the calculations, because the high thermal motion of atoms does not allow multipolar analysis.
3. Computing DFT potential using software SIESTA. For these calculations, the atomic positions were fixed at the crystallographic geometry.

Sixty-four residues that surrounded completely the active site region were included into the calculations.

In particular, the electrostatic potential calculations using all these methods were performed for the NADP$^+$ cofactor alone (Figure 6.14) and for the active site without NADP$^+$ (Figure 6.15a through c). First, a good qualitative agreement between results obtained by the different approaches was demonstrated (Figure 6.15). The importance of this result is that the ultrahigh resolution data electrostatic potential provided could be computed, with a quality comparable to the quality of quantum chemistry calculations but for a much less computing cost. Second, a comparison between Figures 6.14 and 6.15a through c shows that the electrostatic potentials generated by the cofactor alone and by the active site without cofactor are clearly complementary. The regions with the negative potential on Figure 6.15a through c correspond to the regions of the positive potential on Figure 6.13, and vice versa. The result of this study was the charge density demonstration of an electrostatic complementarity between the protein and cofactor NADP$^+$.

6.6 Perspectives

The very impressive success of subatomic MX during the last decade opens new wide perspectives. The most important result is the possibility of having new insights of protein function, for example, of catalytic reaction mechanisms. The very detailed information about the protonation state of catalytically important residues and the observed deviations from standard stereochemistry makes these insights possible. The shortcoming of crystallographic images, which are "photos" but not "films," can be superseded by using MX in combination with complementary experimental techniques and theoretical modeling.

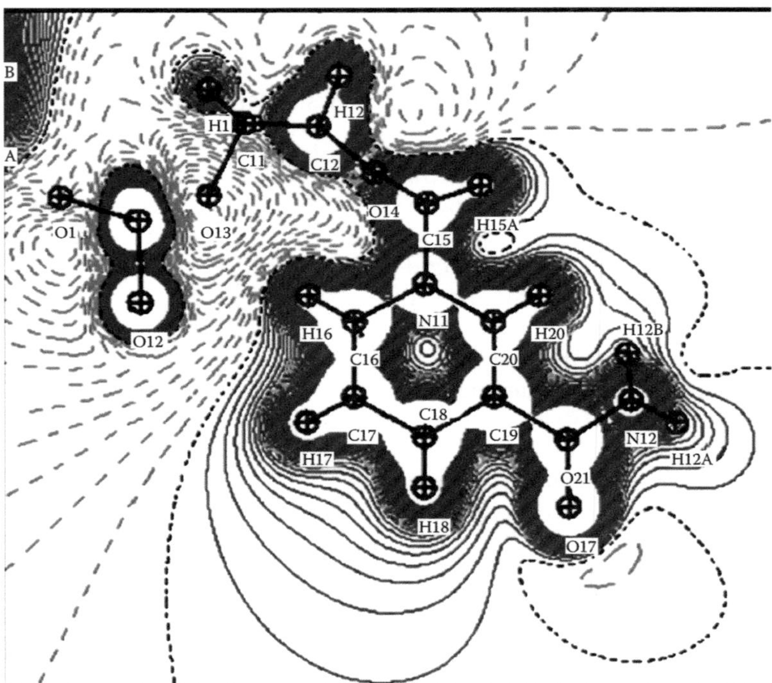

FIGURE 6.14 Electrostatic potential generated by the NADP⁺ cofactor alone in the plane of the nicotinamide ring. Full lines correspond to the region of positive value, dashed lines correspond to the region of negative values, black dotted lines correspond to zero level. (Reproduced from Muzet, N. et al., *PNAS* 100, 8742–8747. Copyright 2003 National Academy of Sciences.)

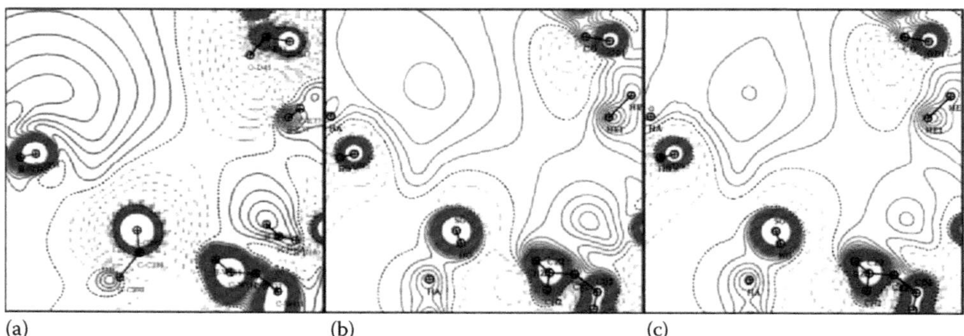

FIGURE 6.15 Electrostatic potential generated for the active site of AR in the absence of the cofactor. Calculations were performed (a) by DFT (b) using charge density parameters which were transferred directly from the multipolar database (c) using multipolar parameters obtained from the preliminary multipolar refinement. The same situation as with Figure 6.14. (Reproduced from Muzet, N. et al., *PNAS* 100, 8742–8747. Copyright 2003 National Academy of Sciences.)

In turn, the high impact of the results of subatomic MX can initiate further developments of experimental and theoretical methods. The main stumbling block is still the availability of good crystals. Significant improvement of crystal quality can be expected in the near future for two reasons: (1) the increasing power of automated crystallization systems makes it possible to test a very large number of conditions, and (2) the intensive accumulation of experience of crystallogenesis studies.

For determining the protonation states of critical residues, the use of neutron scattering in MX studies is very promising. Its main advantages are the absence of radiation damage and the visualization of hydrogen (or deuterium) nuclei at resolutions lower than 2 Å (Schoenborn 1969). Another advantage is the identification of disordered water molecules, in combination with ultrahigh resolution x-ray crystallography. An example is the identification of the tetrahedral cluster of water molecules binding antifreeze protein, mentioned earlier, which is important in the biological function (Howard et al. 2011).

The main limitations are the low flux of neutron sources, which generates a need for large crystals, and the strong background due to the diffuse scattering of hydrogen atoms (much lower for deuterium atoms). With the possibility of expression and purification of fully deuterated proteins, together with the availability of new (spallation) sources of neutrons, this form of structure determination will clearly have a much wider field of application and can complement high-resolution x-ray studies (Habash et al. 1997).

New results in quantum mechanics theory led to algorithms that allow the researcher to obtain information about the wave function directly from experimental observations. These ideas have not yet been applied to macromolecules but seem very promising. The growing power of computers has resulted in recent years in large in-roads in this area, where three distinct lines of work are being developed: (1) QM/x-ray *embedding* efforts (Parker et al. 2003); (2) restrained wave function fitting (Jayatilaka 1998; Jayatilaka and Grimwood 2001; Grimwood and Jayatilaka 2001); and (3) use of QM techniques for the direct phasing of x-ray data (Bethanis et al. 2002).

An important consequence of the improvements in multipolar refinement and the determination of protonation states by neutron refinement is the possibility of getting an accurate picture of the electrostatic energy of binding is under way. For small molecules, one example of calculation of the binding energy from the multipolar charge density can be found in the work of Li et al. (2002). In macromolecules, an interesting example, for the case of the ternary complex AR–NADP$^+$–IDD594, is the calculation of the electrostatic potentials from a charge density database, preliminary experimental electron density analysis, and DFT computations, which showed the complementarity of the ligand and the protein, as described in Section 6.4.4.

In summary, ultrahigh resolution crystallography gives a very precise description of structural details and allows the detailed modeling of interatomic interactions. This is particularly important for the study of ligand binding, where the accuracy of observations is essential to calculate association/dissociation energies. The improvement in data collection facilities has greatly facilitated the task of measuring the diffraction data. The stumbling block remains the availability of crystals diffracting to high resolution. The large improvement in the quality and quantity of information obtained at high versus medium resolution should motivate efforts for the improvement of crystallization conditions.

Acknowledgments

The authors thank Raul Cachau for thoughtful discussion about quantum chemistry methods, Bernard Lorber for thoughtful discussion about problems with protein crystallization, and Nicolas Calimet for help and useful discussion about protein structures. The authors thank Alexandra Cousido, Isabelle Hazemann, Andre Mitschler, and Federico Ruiz for sharing their insights in all aspects of the subatomic resolution work, and the staff of the IGBMC and its Integrative Biology Department for their continuing support. The Antifreeze Protein work described in this chapter was financed by the Human Frontiers Science Program. The aldose reductase work described in this chapter was supported by the Centre National de la Recherché Scientifique (CNRS), by the Institut National de la Santé et de la Recherché Médicale, and the Hôpital Universitaire de Strasbourg (H.U.S), by collaborative projects CNRS-CONICET, CNRS-CERC, CNRS-RAS, and CNRS-NSF (INT-9815595), by Ecos Sud, by Russian Foundation for Basic Research (RFBR research projects No. 10-04-00254-a and No. 13-04-00118-a), and by the Institute for Diabetes Discovery, Inc., through a contract with the CNRS, and included data collected at the SBC, APS, Argonne, Illinois, at the ESRF, Grenoble, France, and at the SLS, PSI, Villigen, Switzerland.

Appendix 6.A Glossary of Crystallographic Terms

Reflections. The diffraction picture is recorded as a number of images. Each image consists of a number of separate "spots," each corresponding to the reflection of the incident beam in one lattice plane. The number of photons recorded in a given reflection is known as the intensity (*I*). The observed structure factor magnitude is proportional to the root square of the intensity. Each spot is identified by a set of Miller indices (*h*,*k*,*l*), associated to the lattice plane and to a "reciprocal vector" perpendicular to this plane. The length of this reciprocal vector gives the resolution of the spot.

R-factor. It is a measure of agreement between experimentally measured crystallographic data (observed structure factor magnitudes, F^{obs}) and that calculated from the model (calculated structure factor magnitudes, F^{calc}). It is calculated by the formula

$$R = \frac{\sum_{hkl}\left|F_{hkl}{}^{obs}\right| - k\left|F_{hkl}{}^{calc}\right|}{\sum_{hkl}\left|F_{hkl}{}^{obs}\right|}, \tag{6.A.1}$$

and is the most used indicator of the quality of the model.

Free R-factor. It is computed by the same way as usual R-factor (formula 6.A.1), but using the randomly chosen part of data, which were excluded from the refinement. It is used to get unbiased measure of model quality.

R(merge). This is a measurement of the agreement of different experimental measures of the same Bragg diffraction spot. It is computed with a formula similar to (6.A.1) but using measured intensities (*I*) instead of structure factors (F).

Resolution shell. For statistical purposes, the different measures on structure factors are given as a function of resolution. This is done by grouping the reflections in different intervals, known as resolution shells. The important values are the overall one and that corresponding to the last interval (last shell).

Completeness. It is a measure of the ratio of the number of observed reflections versus the possible total number.

Difference map. $F^{obs}-F^{calc}$ is calculated as a Fourier transform with magnitudes $F^{obs}-F^{calc}$ and phases calculated from the Model 6.7. If the model is correct, the map should be flat. Otherwise, this map is inspected in order to locate a missing part (strong positive density), or a wrong part of the model (strong negative density) at each step of refinement.

2 F^{obs}–F^{calc} map is calculated as a Fourier transform with magnitudes $2F^{obs}$–F^{calc} and phases calculated from the model. For a correct model, the atoms follow the peaks in the map. Otherwise, the map gives the path the model should follow.

Occupancy. Occupancy is a measure of the fraction of molecules in the crystal for which an atom occupies this particular position. For example, if the atom has the same position for all the unit cells in the crystal, it has occupancy equal to 1.0.

References

Adam, V., Royant, A., Niviere, V., Molina-Heredia, F. P., and Bourgeois, D. 2004. Structure of superoxide reductase bound to ferrocyanide and active site expansion upon x-ray-induced photo-reduction. *Structure* 12(9):1729–1740.

Adams, P. D., Grosse-Kunstleve, R. W., Hung, L.-W., Ioerger, T. R., McCoy, A. J., Moriarty, N. W., Read, R. J., Sacchettini, J. C., Sauter, N. K., and Terwilliger, T. C. 2002. PHENIX: Building new software for automated crystallographic structure determination. *Acta Crystallogr. D* 58:1948–1954.

Adams, P. D., Pannu, N. S., Read, R. J., and Brunger, A. T. 1997. Cross-validated maximum likelihood enhances crystallographic simulated annealing refinement. *Proc. Natl. Acad. Sci. USA* 94:5018–5023.

Afonine, P. V., Grosse-Kunstleve, R. W., and Adams, P. D. 2005. The Phenix refinement framework. *CCP4 Newslett.* 42:contribution 8.

Afonine, P. V., Grosse-Kunstleve, R. W., Adams, P. D., Lunin, V. Y., and Urzhumtsev, A. 2007. On macromolecular refinement at subatomic resolution with interatomic scatterers. *Acta Crystallogr. D* 63:1194–1197.

Afonine, P., Lunin, V. Y., Muzet, N., and Urzhumtsev, A. 2004. On the possibility of observation of valence electron density for individual bonds in proteins in conventional difference maps. *Acta Crystallogr. D Biol. Crystallogr. D* 60(2):260–274.

Ahmed, H. U., Blakeley, M. P., Cianci, M., Cruickshank, D. W., Hubbard, J. A., and Helliwell, J. R. 2007. The determination of protonation states in proteins. *Acta Crystallogr. D Biol. Crystallogr.* 63(Pt 8):906–922.

Alahuhta, M. and Wierenga, R. K. 2010. Atomic resolution crystallography of a complex of triosephosphate isomerase with a reaction-intermediate analog: New insight in the proton transfer reaction mechanism. *Proteins* 78:1878–1888.

Ames, J. B. and Mathies, R. A. 1990. The role of back-reactions and proton uptake during the N–O transition in bacteriorhodopsin's photocycle: A kinetic resonance Raman study. *Biochemistry* 29(31):7181–7190.

Ataka, M. and Wakayama, N. I. 2002. Effects of a magnetic field and magnetization force on protein crystal growth. Why does a magnet improve the quality of some crystals? *Acta Crystallogr. D* 58:1708–1710.

Balashov, S. P. 2000. Protonation reactions and their coupling in bacteriorhodopsin. *Biochim. Biophys. Acta* 1460(1):75–94.

Bayly, C. I., Cieplak, P., Cornell, W. D., and Kollman, P. A. 1993. A well-behaved electrostatic potential based method using charge restraints for determining atom-centered charges: The RESP model. *J. Phys. Chem.* 97:10269–10280.

Bethanis, K., Tzamalis, P., Hountas, A., and Tsoucaris, G. 2002. Ab initio determination of a crystal structure by means of the Schrödinger equation. *Acta Crystallogr. A* 58:265–269.

Blakeley, M. P., Ruiz, F., Cachau, R., Hazemann, I., Meilleur, F., Mitschler, A., Ginell, S. et al. 2008. Quantum model of catalysis based on a mobile proton revealed by subatomic x-ray and neutron diffraction studies of h-aldose reductase. *Proc. Natl. Acad. Sci. USA* 105:1844–1848.

Blow, D. M. 1976. Structure and mechanism of chymotrypsin. *Acc. Chem. Res.* 9:145–152.

Blum, M. M., Koglin, A., Ruterjans, H., Schoenborn, B., Langan, P., and Chen, J. C. 2007. Preliminary time-of-flight neutron diffraction study on diisopropyl fluorophosphatase (DFPase) from Loligo vulgaris. *Acta Crystallogr. Sect. F Struct. Biol. Crystallogr. Commun.* 63(Pt 1):42–45.

Bonish, H., Schmidt, C. L., Bianco, P., and Ladenstein, R. 2005. Ultrahigh-resolution study on Pyrococcus abyssi rubredoxin. I. 0.69 A x-ray structure of mutant W4L/R5S. *Acta Crystallogr. D* 61:990–1004.

Brill, R. 1960. On the influence of binding electrons on x-ray intensities. *Acta Crystallogr.* 13:275–276.

Brooks, B. R., Bruccoleri, R. E., Olafson, B. D., States, D. J., Swaminathan, S., and Karplus, M. 1983. CHARMM: A program for macromolecular energy, minimization and dynamics calculations. *J. Comput. Chem.* 4:187–217.

Brudler, R., Meyer, T. E., Genick, U. K., Devanathan, S., Woo, T. T., Millar, D. P., Gerwert, K., Cusanovich, M. A., Tollin, G., and Getzoff, E. D. 2000. Coupling of hydrogen bonding to chromophore conformation and function in photoactive yellow protein. *Biochemistry* 39:13478–13486.

Brünger, A. T., Adams, P. D., Clore, G. M., DeLano, W. L., Gros, P., Grosse-Kunstleve, R. W., Jiang, J.-S. et al. 1998. Crystallography & NMR system: A new software suite for macromolecular structure determination. *Acta Crystallogr. D* 54:905–921.

Burmeister, W. P. 2000. Structural changes in a cryo-cooled protein crystal owing to radiation damage. *Acta Crystallogr. D Biol. Crystallogr. D* 56:328–341.

Bytheway, I., Grimwood, D. J., and Jayatilaka, D. 2002. Wavefunctions derived from experiment. III. Topological analysis of crystal fragments. *Acta Crystallogr. A* 58:232–243.

Cachau, R., Howard, E., Barth, P., Mitschler, A., Chevrier, B., Lamour, V., Joachimiak, A. et al. 2000. Model of the catalytic mechanizm of human aldose reductase based on quantum chemical calculations. *Journal de Physique* 10:3–13.

Calderone, V., Chevrier, B., Van Zandt, M., Lamour, V., Howard, E., Poterszman, A., Barth, P. et al. 2000. The structure of human aldose reductase bound to the inhibitor IDD384. *Acta Crystallogr. D* 56:536–540.

Cambridge. 2001. Cambridge Crystallographic Data Base. 2 Union Road, Cambridge CB21EZ, U.K. Cambridge Crystallographic Data Center.

Chen, C. J., Lin, Y. H., Huang, Y. C., and Liu, M. Y. 2006. Crystal structure of rubredoxin from *Desulfovibrio gigas* to ultra-high 0.68 Å resolution. *Biochem. Biophys. Res. Commun. B* 349:79–90.

Chen, C. C., Smith, T. J., Kapadia, G., Wasch, S., Zawadzke, L. E., Coulson, A., and Herzberg, O. 1996. Structure and kinetics of the beta-lactamase mutants S70A and K73H from *Staphylococcus aureus* PC1. *Biochemistry* 35:12251–12258.

Chernov, A. A. and DeLucas, L. 2002. View on biocrystallization from Jena, 2002. *Acta Crystallogr. D* 58:1511–1513.

Chinte, U., Shah, B., Chen, Y.-S., Pinkerton, A. A., Schall, C. A., and Hanson, B.L. 2007. Cryogenic (<20 K) helium cooling mitigates radiation damage to protein crystals. *Acta Crystallogr. D* 63:486–492.

Coppens, P. 1997. *X-Ray Charge Densities and Chemical Bonding*. Oxford University Press, New York.

Corbett, M. C., Latimer, M. J., Poulos, T. L., Sevrioukova, I. F., Hodgson, K. O., and Hedman, B. 2007. Photoreduction of the active site of the metalloprotein putidaredoxin by synchrotron radiation. *Acta Crystallogr. D* 63:951–960.

Cousido-Siah, A., Petrova, A., Hazemann, I., Mitschler, A., Ruiz, F. X., Howard, E., Ginell, S. et al. 2012. Crystal packing modifies ligand binding affinity: The case of Aldose Reductase. *Proteins* 80(11):2552–2561.

Czepas, J., Devedjiev, Y., Krowarsch, D., Derewenda, U., Otlewski, J., and Derewenda, Z. S. 2004. The impact of Lys→Arg surface mutations on the crystallization of the globular domain of RhoGDI. *Acta Crystallogr. D* 60:275–280.

Damblon, C., Raquet, X., Lian, L.-Y., Lamotte-Brasseur, J., Fonze, E., Charlier, P., Roberts, G. C. K., and Frere, J.-M. 1996. The catalytic mechanizm of beta-lactamases: NMR titration of an active-site lysine residue of the TEM-1 enzyme. *Proc. Natl. Acad. Sci. USA* 93:1747–1752.

Dauter, Z. 2006. Current state and prospects of macromolecular crystallography. *Acta Crystallogr. D* 62:1–11.

Dauter, Z., Jaskolski, M., and Wlodawer, A. 2010. Impact of synchrotron radiation on macromolecular crystallography: A personal view. *J. Synchrotron Radiat.* 17:433–444.

Dioumaev, A. K. 2001. Infrared methods for monitoring the protonation state of carboxylic amino acids in the photocycle of bacteriorhodopsin. *Biochemistry (Mosc)* 66(11):1269–1276.

Dvornik, D. 1987. *Aldose Reductase Inhibition: An Approach to the Prevention of Diabetic Complications*. McGraw-Hill, New York.

Emsley, P. and Cowtan, K. 2004. Coot: Model-building tools for molecular graphics. *Acta Crystallogr. D* 60:2126–2132.

Engh, R. A. and Huber, R. 1991. Accurate bond and angle parameters for x-ray protein structure refinement. *Acta Crystallogr. A* 47:392–400.

Engh, R. A. and Huber, R. 2001. Structure quality and target parameters. In *International Tables for Crystallography*, Vol. F, Rossmann, M. G. and Arnold, E., eds. Kluwer Academic Publishers, Dordrecht, the Netherlands, pp. 382–416.

Esposito, L., Vitagliano, L., and Mazzarella, L. 2002. Recent advances in atomic resolution protein crystallography. *Protein Pept. Lett.* 9:95–106.

Esposito, L., Vitagliano, L., Zagari, A., and Mazzarella, L. 2000. Experimental evidence for the correlation of bond distances in peptide groups detected in ultrahigh-resolution protein structures. *Protein Eng.* 13:825–828.

Evans, G., Polentarutti, M., Carugu, K. D., and Bricogne, G. 2003 SAD phasing with triiodide, softer x-rays and some help from radiation damage. *Acta Crystallogr. D* 59:1429–1443.

Fioravanti, E., Vellieux, F. M. D., Amara, P., Madern, D., and Weik, M. 2007. Specific radiation damage to acidic residues and its relation to their chemical and structural environment. *J. Synchrotron Radiat.* 14:84–97.

Fisher, S. J., Blakeley, M. P., Cianci, M., McSweeney, S., and Helliwell, J. R. 2012. Protonation-state determination in proteins using high-resolution x-ray crystallography: Effects of resolution and completeness. *Acta Crystallogr. D Biol. Crystallogr.* 68(Pt 7):800–809.

Foadi, J., Woolfson, M. M., Dodson, E. J., Wilson, K. S., Jia-xing, Y., and Chao-de, Z. 2000. A flexible and efficient procedure for the solution and phase refinement of protein structures *Acta Crystallogr. D* 56:1137–1147.

Garman, E. F. and Schneider, T. R. 1997. Macromolecular cryocrystallography. *J. Appl. Crystallogr.* 30:211–237.

Genick, U. K., Soltis, S. M., Kuhn, P., Canestrelli, I. L., and Getzoff, E. D. 1998. Structure at 0.85 A resolution of an early protein photocycle intermediate. *Nature* 392:206–209.

Getzoff, E. D., Gutwin, K. N., and Genick, U. K. 2003. Anticipatory active-site motions and chromophore distortion prime photoreceptor PYP for light activation. *Nat. Struct. Biol.* 10:658.

Ghermani, N. E., Lecomte, C., and Bouhmaida, N. 1993. Electrostatic potential from high resolution x-ray diffraction data: Application to a pseudopeptide molecule. *Z. Naturforsch. Teil A*. 48:91–98.

Gogonea, V., Suarez, D., Van der Vaart, A., and Merz, K. M. Jr. 2001. New developments in applying quantum mechanics to proteins. *Curr. Opin. Struct. Biol.* 11(2):217–223.

Green, D. W., Ingram, V. M., and Perutz, M. F. 1954. The structure of haemoglobin. IV. Sign determination by the isomorphous replacement method. *Proc. R. Soc. Lond. Sec. A* 225:287–307.

Grimwood, D. J. and Jayatilaka, D. 2001. Wavefunctions derived from experiment. II. A wavefunction for oxalic acid dihydrate. *Acta Crystallogr. A* 57:87–100.

Guillot, B., Viry, L., Guillot, R., Lecompte, C., and Jelsch, C. 2001. Refinement of proteins at subatomic resolution with MoPro. *J. Appl. Crystallogr.* 34:214–223.

Habash, J., Raftery, J., Weisgerber, S., Cassetta, A., Lehmann, M., Hoghoj, P., Wilkinson, C., Campbell, J., and Helliwell, J. 1997. Neutron Laue diffraction study of concavalin A. The proton of Asp28. *J. Chem. Soc., Faraday Trans.* 93(24):4313–4317.

Hansen, N. K. and Coppens, P. 1978. Testing aspherical refinement on small molecules data sets. *Acta Crystallogr.* A34:909–921.

Harris, T. K. and Turner, G. J. 2002. Structural basis of perturbed pKa values of catalytic groups in enzyme active sites. *IUBMB Life* 53(2):85–98.

Hendickson, W. A. and Ogata, C. M. 1997. Phase determination from multiwavelength anomalous diffraction measurements. *Methods Enzymol.* 276:494–523.

Heras, B., Edeling, M. A., Byriel, K. A., Alun Jones, A., Raina, S., and Martin, J. L. 2003. Dehydration converts DsbG crystal diffraction from low to high resolution. *Structure* 11(2):139–145.

Herzberg, O. and Moult, J. 1987. Molecular basis for bacterial resistance to beta-lactam antibiotics: Crystal structure of beta-lactamase from *Staphylococcus aureus* PC1 at 2.5Å resolution. *Science* 236:694–701.

Hohenberg, P. and Kohn, W. 1965. Inhomogeneous electron gas. *Phys. Rev.* 136:B864–B871.

Hope, H. 1990. Crystallography of biological macromolecules at ultra-low temperature. *Ann. Rev. Biophys. Biophys. Chem.* 19:107–126.

Housset, D., Benabicha, F., Pichon-Pesme, V., Jelsch, C., Maierhofer, A., David, S., Fontecilla-Camps, J. C., and Lecomte, C. 2000. Towards the charge-density study of proteins: A room-temperature scorpion-toxin structure at 0.96 Å resolution as a first test case. *Acta Crystallogr. D* 56:151–160.

Howard, E. I., Blakeley, M. P., Haertlein, M., Petit-Haertlein, I., Mitschler, A., Fisher, S. J., Cousido-Siah, A. et al. 2011. Neutron structure of type-III antifreeze protein allows the reconstruction of AFP-ice interface. *J. Mol. Recogn.* 24 (4):724–732.

Howard, E., Lamour, V., Mitschler, A., Barth, P., Moras, D., and Podjarny, A. 1999. Resolution improvement to 0.9A in crystals of human aldose reductase. *Acta Crystallogr. A* 55 Suplement:P09.OB.001.

Howard, E. I., Sanishvili, R., Cachau, R. E., Mitschler, A., Chevrier, B., Barth, P., Lamour, V. et al. 2004. Ultra-high resolution drug design I: Details of interactions in human aldose reductase–inhibitor complex at 0.66 Å. *Proteins: Struct. Funct. Genet.* 55:792–804.

Huang, W. and Palzkill, T. 1997. A natural polymorphism in lactamase is a global suppressor. *Proc. Natl. Acad. Sci. USA* 94:8801–8806.

Ishiguro, M. and Imajo, S. 1996. Modelling study on a hydrolitic mechanizm of calss A beta-lactamases. *J. Med. Chem.* 39:2207–2218.

Jaskolski, M., Gilski, M., Dauter, Z., and Wlodawer, A. 2007. Stereochemical restraints revised: How accurate are refinement targets and how much should protein structures be allowed to deviated from them? *Acta Crystallogr. D* 63:611–620.

Jayatilaka, D. 1998. Wave function for Beryllium from x-ray diffraction data. *Phys. Rev. Lett.* 80:798–801.

Jayatilaka, D. and Grimwood, D. J. 2001. Wavefunctions derived from experiment. I. Motivation and theory. *Acta Crystallogr. A* 57:76–86.

Jelsch, C., Pichon-Pesme, V., Lecomte, C., and Aubry A. 1998. Transferability of multipole charge-density parameters: Application to very high resolution oligopeptide and protein structures. *Acta Crystallogr. D* 54:1306–1318.

Jelsch, C., Teeter, M. M., Lamzin, V., Pichon-Pesme, V., Blessing, R. H., and Lecomte, C. 2000. Accurate protein crystallography at ultra-high resolution: Valence electron distribution in crambin. *Proc. Natl. Acad. Sci. USA* 97(7):3171–3176.

Jones, T. A., Zou, J. Y., Cowan, S. W., and Kjeldgaard, M. 1991. Improved methods for the building of protein models in electron density maps and the location of errors in these models. *Acta Crystallogr. A* 47:110–119.

Kang, B. S., Devedjiev, Y., Derewenda, U., and Derewenda, Z. S. 2004. The PDZ2 domain of syntenin at ultra-high resolution: Bridging the gap between macromolecular and small molecule crystallography. *J. Mol. Biol.* 338:483–493.

Karle, J. 1980. Some developments in anomalous dispersion for the structural investigation of macromolecular systems in biology. *Int. J. Quant. Chem. Quant. Biol. Symp.* 7:357–367.

Katona, G., Wilmouth, R. C., Wright, P. A., Berglund, G. I., Hajdu, J., Neutze, R., and Schofield, C. J. 2002. X-ray structure of a serine protease acyl–enzyme complex at 0.95-Å resolution. *J. Biol. Chem.* 277(24):21962–21970.

Katz, A. K., Li, X., Carrell, H. L., Hanson, B. L., Langan, P., Coates, L., Schoenborn, B. P., Glusker, J. P., and Bunick, G. J. 2006. Locating active-site hydrogen atoms in D-xylose isomerase: Time-of-flight neutron diffraction. *Proc. Natl. Acad. Sci. USA* 103(22):8342–8347.

Kinoshita, T., Ataka, M., Warizaya, M., Neya, M., and Fujii, T. 2003. Improving quality and harvest period of protein crystals for structure-based drug design: Effects of a gel and a magnetic field on bovine adenosine deaminase crystals. *Acta Crystallogr. D* 59:1333–1335.

Ko, T.-P., Robinson, H., Gao, Y.-G., Cheng, C.-H. C., DeVries, A. L., and Wang, A. H.-J. 2003. The refined crystal structure of an Eel pout Tipe III Antifreeze protein RD1 at 0.62-Å resolution reveals structural microheterogeneity of protein and solvation. *Biophys. J.* 84:1228–1237.

Koepke, J., Scharff, E. I., Lücke, C., Rüterjans, H., and Fritzsch, G. 2003. Statistical analysis of crystallographic data obtained from squid ganglion DFPase at 0.85 Å resolution. *Acta Crystallogr. D Biol. Crystallogr.* 59:1744–1754.

Kohn, W. and Sham, L. J. 1965. Self-consistent equations including exchange and correlation effects. *Phys. Rev.* 140:A1133–A1138.

Kossiakoff, A. A. and Spencer, S. A. 1981. Direct determination of aspartic acid-102 and histidine-57 in the tetrahedral intermediate of the serine proteases: Neutron structure of trypsin. *Biochemistry* 20:6462–6474.

Kuhn, P., Knapp, M., Soltis, S. M., Ganshaw, G., Thoene, M., and Bott, R. 1998. The 0.78 A structure of a serine protease: *Bacillus lentus* subtilisin. *Biochemistry* 37:13446–13452.

Kundrot, C. E., Judge, R. A., Pusey, M. L., and Snell, E. H. 2001. Microgravity and macromolecular crystallography. *Crystallogr. Growth Design* 1:87–99.

Kursula, I. and Wierenga, R. K. 2003. Crystal structure of triosephosphate isomerase complexed with 2-phosphoglycolate at 0.83-Å resolution. *J. Biol. Chem.* 278:9544–9551.

Lamotte-Brasseur, J., Dive, G., Dideberg, O., Charlier, P., Frere, J. M., and Ghuysen, J. M. 1991. Mechanism of acyl transfer by the class A serine beta-lactamase of Streptomyces albus G. *Biochem. J.* 279:213–221.

Lamotte-Brasseur, J., Lounnas, V., Raquet, X., and Wade, R. C. 1999. pKa calculations for class A beta-lactamases: Influence of substrate binding. *Protein Sci.* 8:404–409.

Lamour, V., Barth, P., Rogniaux, H., Poterszman, A., Howard, E., Mitschler, A., Van Dorsselaer, A., Podjarny, A., and Moras, D. 1999. Production of crystals of human aldose reductase with very high resolution diffraction. *Acta Crystallogr. D* 55:721–723.

Leung, Y. C., Robinson, C. V., Aplin, R. T., and Waley, S. G. 1994. Site-directed mutagenesis of beta-lactamase I: Role of Glu-166. *Biochem. J.* 299:671–678.

Levitt, D. G. 2001. A new software routine that automates the fitting of protein x-ray crystallographic electron-density maps. *Acta Crystallogr. D* 57:1013–1019.

Li, X., Wu, G., Abramov, Y. A., Volkov, A. V., and Coppens, P. 2002. Application of charge density methods to a protein model compound: Calculation of Coulombic intermolecular interaction energies from the experimental charge density. *PNAS* 99(19):12132–12137.

Lietz, E. J., Truher, H., Kahn, D., Hokenson, M. J., and Fink, A. L. 2000. Lysine-73 is involved in the acylation and deacylation of beta-lactamase. *Biochemistry* 39:4971–4981.

Litt, A., Arnez, J. G., Klaholz, B. P., Mitschler, A., and Moras, D. 1998. A eucentric goniometer head sliding on an extended removable arc modified for use in cryocrystallography. *J. Appl. Crystallogr.* 31:638–640.

Littke, W. and John, C. 1984. Protein single crystal growth under microgravity. *Science* 225:203–204.

Liu, Q., Huang, Q., Teng, M., Weeks, C. M., Jelsch, C., Zhang, R., and Niu, L. 2003. The crystal structure of a novel, inactive, lysine 49 PLA2 from *Agkistrodon acutus* venom: An ultrahigh resolution, ab initio structure determination. *J. Biol. Chem.* 278(42):41400–41408.

Liu, L., Nogi, T., Kobayashi, M., Nozawa, T., and Miki, K. 2002. Ultrahigh-resolution structure of high-potential iron-sulfur protein from Thermochromatium tepidum. *Acta Crystallogr. D Biol. Crystallogr.* 58(Pt 7):1085–1091.

Lorber, B., Théobald-Dietrich, A., Charron, C., Sauter, C., Ng, J. D., Zhu, D.-W., and Giegé, R. 2002. From conventional crystallization to better crystals from space: A review on pilot crystallogenesis studies with aspartyl-tRNA synthetases. *Acta Crystallogr. D* 58:1674–1680.

Malkin, A. J., Kuznetsov, Yu. G., and McPherson, A. 1996. Defect structure of macromolecular crystals. *J. Struct. Biol.* 117(2):124–137.

Mateja, A., Devedjiev, Y., Krowarsch, D., Longenecker, K., Dauter, Z., Otlewski, J., and Derewenda, Z. S. 2002. The impact of Glu→Ala and Glu→Asp mutations on the crystallization properties of RhoGDI: The structure of RhoGDI at 1.3 A resolution. *Acta Crystallogr. D* 58:1983–1991.

McDermott, A. E., Thompson, L. K., Winkel, C., Farrar, M. R., Pelletier, S., Lugtenburg, J., Herzfeld, J., and Griffin, R. G. 1991. Mechanism of proton pumping in bacteriorhodopsin by solid-state NMR: The protonation state of tyrosine in the light-adapted and M states. *Biochemistry* 30(34):8366–8371.

McRee, D. E. 1999. XtalView/Xfit—A versatile program for manipulating atomic coordinates and electron density. *J. Struct. Biol.* 125:156–165.

Meents, A., Dittrich, B., and Gutmann, S. 2009. A new aspect of specific radiation damage: Hydrogen abstraction from organic molecules. *J. Synchrotron Radiat.* 16:183–190.

Meents, A., Gutmann, S., Wagner, A., and Schulze-Briese, C. 2010. Origin and temperature dependence of radiation damage in biological samples at cryogenic temperatures. *Proc. Natl. Acad. Sci. USA* 107:1094–1099.

Meents, A., Wagner, A., Schneider, R., Pradervand, C., Pohl, E., and Schulze-Briese, C. 2007. Reduction of x-ray-induced radiation damage of macromolecular crystals by data collection at 15 K: A systematic study. *Acta Crystallogr. D* 63:302–309.

Meijers, R., Morris, R. J., Adolph, H. W., Merli, A., Lamzin, V. S., and Cedergren-Zeppezauer, E. S. 2001. On the enzymatic activation of NADH. *J. Biol. Chem.* 276:9316–9321.

Miller, R., DeTitta, G. T., Jones, R., Langs, D. A., Weeks, C. M., and Hauptman, H. M. 1993. On the application of the minimal principle to solve unknown structures. *Science* 259:1430–1433.

Minasov, F., Wang, X., and Shoichet, B. K. 2002. An ultrahigh resolution structure of TEM-1 beta-lactamase suggests a role for Glu166 as the general base in acylation. *J. Am. Chem. Soc.* 124:5333–5340.

Moffat, T.-Y. and Teng, K. 2002. Radiation damage of protein crystals at cryogenic temperatures between 40 K and 150 K. *J. Synchrotron Radiat.* 9:198–201.

Moreno, A., Saridakis, E., and Chayen, N. E. 2002. Combination of oils and gels for enhancing the growth of protein crystals. *J. Appl. Crystallogr.* 35:140–142.

Müller, U., Nyarsik, L., Horn, M., Rauth, H., Przewieslik, T., Saenger, W., Lehrach, H., and Eickhoff, H. 2001. Development of a technology for automation and miniaturisation of protein crystallisation. *J. Biotechnol.* 85:7–14.

Murshudov, G. N., Lebedev, A., Vagin, A. A., Wilson, K. S., and Dodson, E. J. 1999. Efficient anisotropic refinement of macromolecular structures using FFT. *Acta Crystallogr. D* 55:247–255.

Mustafi, D., Sosa-Peinado, A., and Makinen, M. W. 2001. ENDOR structural characterization of the acylenzyme reaction intermediate of TEM-1 ß-lactamase confirms glutamate-166 as the base catalyst. *Biochemistry* 40:2379–2409.

Muzet, N., Guillot, B., Jelsch, C., Howard, E., and Lecompte, C. 2003. Electrostatic complementarity in an aldose reductase complex from ultra-high-resolution crystallography and first-principles calculations. *PNAS* 100(15):8742–8747.

Navaza, J. 1994. AMoRe: An automated package for molecular replacement. *Acta Crystallogr. A* 50:157–163.

Nilsson, K. and Ryde, U. 2003. Quantum refinement: A method to determine protonation and oxidation states of metal sites in protein crystal structures. In *21st European Crystallographic Meeting*. Book of Abstracts. Durban, South Africa, August 24–29, 2003, p. 13.

Nukaga, M., Mayama, K., Hujer, A. M., Bonomo, R. A., and Knox, J. R. 2003. Ultrahigh resolution structure of a class A beta-lactamase: On the mechanism and specificity of the extended-spectrum SHV-2 enzyme. *J. Mol. Biol.* 328:289–301.

Oliéric, V., Ennifar, E., Meents, A., Fleurant, M., Besnard, C., Pattison, P., Schiltz, M., Schulze-Briese, C., and Dumas, P. 2007. Using x-ray absorption spectra to monitor specific radiation damage to anomalously scattering atoms in macromolecular crystallography. *Acta Crystallogr. D* 63:759–768.

Parker, C. L., Ventura, O. N., Burt, S. K., and Cachau, R. E. 2003. DYNGA: A general purpose QM-MM-MD program. I. Application to water. *Mol. Phys.* 101(17):2659–2668.

Perrakis, A., Morris, R. M., and Lamzin, V. S. 1999. Automated protein model building combined with iterative structure refinement. *Nat. Struct. Biol.* 6:458–463.

Petrova, T., Ginell, S., Mitschler, A., Hazemann, I., Schneider, T., Cousido, A., Lunin, V. Y., Joachimiak, A., and Podjarny, A. 2006. Ultra-high resolution study of protein atomic displacement parameters at cryotemperatures obtained with helium cryostat. *Acta Crystallogr. D* 62:1535–1544.

Petrova, T., Ginell, S., Mitschler, A., Kim, Y., Lunin, V. Y., Joachimiak, G., Cousido-Siah, A., Hazemann, I., Podjarny, A., Lazarki, K., and Joachimiak, A. 2010. X-ray-induced deterioration of disulfide bridges at atomic resolution. *Acta Crystallogr. D* 66:1075–1091.

Pichon-Pesme, V., Lecompte, C., and Lachekar, H. 1995. On building a databank of transferable experimental electron density parameters: Application to polypeptides. *J. Phys. Chem.* 99:6242–6250.

Podjarny, A. D., Howard, E. I., Urzhumtsev, A., and Grigera, J. R. 1997. A multicopy modeling of the water distribution in macromolecular crystals. *Proteins: Struct. Funct.* 28:303–312.

Polentarutti, M., Glazer, R., and Djinović Carugo, K. 2004. A helium-purged beam path to improve soft and softer x-ray data quality. *J. Appl. Crystallogr.* 37:319–324.

Ramagopal, U. A., Dauter, Z., Thirumuruhan, R., Fedorov, E., and Almo, S. C. 2005. Radiation-induced site-specific damage of mercury derivatives: Phasing and implications. *Acta Crystallogr. D Biol. Crystallogr.* 61(Pt 9):1289–1298.

Ravelli, R. B. and McSweeney, S. M. 2000. The 'fingerprint' that x-rays can leave on structures. *Struct. Fold Design* 8:315–328.

Read, R. J. 2001. Density modification: Theory and practice. In *Methods in Macromolecular Crystallography*. Turk, D. and Johnson, L., eds. IOS Press, Amsterdam, the Netherlands, pp. 123–135.

Rick, S. W. and Cachau, R. E. 2000. The nonplanarity of the peptide group: Molecular dynamics simulations with a polarizable two-state model for the peptide bond. *J. Chem. Phys.* 112(11):5230–5241.

Robert, M.-C., Vidal, O., Garcia-Ruiz, J.-M., and Otálora, F. 1999. Crystallization in gels and related methods. In *Crystallization of Nucleic Acids and Proteins*, Ducruix, A. and Giegé, R., eds. Oxford University Press, Oxford, U.K., pp. 149–175.

Rossmann, M. G. and Blow, D. M. 1962. The detection of subunits within the crystallographic asymmetric unit. *Acta Crystallogr.* 15:24–31.

Rypniewski, W. R., Oestergaard, P., Noerregaard-Madsen, M., Dauter, M., and Wilson, K. S. 2001. Fusarium Oxysporum trypsin at atomic resolution at 100 and 283 K: A study of ligand binding. *Acta Crystallogr. D Biol. Crystallogr. D* 57:8–19.

Schick, B. and Jurnak, F. 1994. Crystal growth and crystal improvement strategies. *Acta Crystallogr. D* 50:563–568.

Schlegel, B. P., Jez, J. M., and Penning, T. M. 1998. Mutagenesis of 3 alpha-hydroxysteroid dehydrogenase reveals a "push-pull" mechanism for proton transfer in aldo-keto reductases. *Biochemistry* 37:3538–3548.

Schmidt, M. W., Baldridge, K. K., Boatz, J. A., Elbert, S. T., Gordon, M. S., Jensen, J. H., Koseki, S. et al. 1993. The general atomic and molecular electronic structure system. *J. Comput. Chem.* 14:1347–1363.

Schmidt, A., Gruber, K., Kratky, C., and Lamzin, V. S. 2008. Atomic resolution crystal structures and quantum chemistry meet to reveal subtleties of hydroxynitrile lyase catalysis. *J. Biol. Chem.* 283(31):21827–21836.

Schmidt, A., Jelsch, C., Østergaard, P., Rypniewski, W., and Lamzin, V. S. 2003. Trypsin revisited crystallography at (sub) atomic resolution and quantum chemistry revealing details of catalysis. *J. Biol. Chem.* 278(44):43357–43362.

Schmidt, A. and Lamzin, V. S. 2002. Veni, vidi, vici—Atomic resolution unravelling the mysteries of protein function. *Curr. Opin. Struct. Biol.* 12:698–703.

Schmidt, A., Teeter, M., Weckert, E., and Lamzin, V. S. 2011. Crystal structure of small protein crambin at 0.48 A resolution. *Acta Crystallogr. Sect. F Struct. Biol. Crystallogr. Commun.* 67(Pt 4):424–428.

Schoenborn, B. P. 1969. Neutron diffraction analysis of myoglobin. *Nature* 224:143–146.

Sheldrick, G. M. and Schneider, T. R. 1997. SHELXL: High resolution refinement. In *Methods in Enzymology*. Sweet, R. M. and Carter, C. W. Jr., eds. Academic Press, Orlando, FL, Vol. 277, pp. 319–343.

Sheldrick, G. M. and Schneider, T. R. 2001. Direct methods for macromolecules. In *Methods in Macromolecular Crystallography*. Turk, D. and Johnson, L., eds. IOS Press, Amsterdam, the Netherlands, pp. 72–81.

Singer, P. T., Smalas, A., Carty, R. P., Mangel, W. F., and Sweet, R. M. 1993. The hydrolytic water molecule in trypsin, revealed by time-resolved Laue crystallography. *Science* 261:620–622.

Souhassou, M. and Blessing, R. H. 1999. Topological analysis of experimental electron densities. *J. Appl. Crystallogr.* 32:210–217.

Southworth-Davies, R. J., Medina, M. A., Carmichael, I., and Garman, E. F. 2007. Observation of decreased radiation damage at higher dose rates in room temperature protein crystallography. *Structure* 15(12):1531–1541.

Spackman, M. A. 1992. Molecular electric moments from x-ray diffractions data. *Chem. Rev.* 92:1769–1797.

Spackman, M. A. 1998. Charge densities from x-ray diffraction data. *Ann. Rep. Prog. Chem. C: Phys. Chem.* 94:177–207.

Stevens, R. C. 2000. High-throughput protein crystallization. *Curr. Opin. Struct. Biol.* 10:558–563.

Strynadka, N. C. J., Adachi, J. S. E., Johns, K., Sielecki, A., Betzel, C., Sutoh, K. H., and James, M. N. G. 1992. Molecular structure of the acyl–enzyme intermediate in lactam hydrolysis at 1.7 Å resolution. *Nature* 359:700–705.

Strynadka, N. C., Eisenstein, M., Katchalski-Katzir, E., Shoichet, B. K., Kuntz, I. D., Abagyan, R., Totrov, M. et al. 1996. Molecular docking programs successfully predict the binding of a beta-lactamase inhibitory protein to TEM-1 beta-lactamase. *Nat. Struct. Biol.* 3:688–695.

Sussman, J. L., Holbrook, S. R., Church, G. M., and Kim, S. H. 1977. A structure-factor least-squares refinement procedure for macromolecular structures using constrained and restrained parameters. *Acta Crystallogr. A* 33:800–804.

Takeda, K., Kusumoto, K., Hirano, Y., and Miki, K. 2009. Detailed assessment of x-ray induced structural perturbation in a crystalline state protein. *J. Struct. Biol.* 169(2):135–144.

Teeter, M. M., Roe, S. M., and Heo, N. H. 1993. Atomic resolution (0·83 Å) crystal structure of the hydrophobic protein crambin at 130 K. *J. Mol. Biol.* 230(1):292–311.

Terwilliger, T. C. 2001. Maximum-likelihood density modification using pattern recognition of structural motifs. *Acta Crystallogr. D* 57:1755–1762.

Tuan, H. F., Erskine, P., Langan, P., Cooper, J., and Coates, L. 2007. Preliminary neutron and ultrahigh-resolution x-ray diffraction studies of the aspartic proteinase endothiapepsin cocrystallized with a gem-diol inhibitor. *Acta Crystallogr. Sect. F Struct. Biol. Crystallogr. Commun.* 63(Pt 12):1080–1083.

Turk, D. 2000. From modeling and refinement to refinement with modeling. *Acta Crystallogr. A* 56:s27.

Turk, D. 2004. MAIN in 2004: Model building at 100 residues per minute. *Acta Crystallogr. A* 60:s16.

Urzhumtsev, A., Tete-Favier, F., Mitschler, A., Barbanton, J., Barth, P., Urzhumtseva, L., Biellmann, J. F., Podjarny, A., and Moras, D. 1997. A "specificity" pocket inferred from the crystal structures of the complexes of aldose reductase with the pharmaceutically important inhibitors tolrestat and sorbinil. *Structure* 5:601–612.

Vergara, A., Lorber, B., Zagari, A., and Giegé, R. 2003. Physical aspects of protein crystal growth investigated with the advanced protein crystallization facility in reduced-gravity environments. *Acta Crystallogr. D* 59:2–15.

Vrielink, A. and Sampson, N. 2003. Sub-angstrom resolution enzyme x-ray structures: Is seeing believing? *Curr. Opin. Struct. Biol.* 13:709–715.

Wang, J., Dauter, M., Alkire, R., Joachimiak, A., and Dauter, Z. 2007. Triclinic lysozyme at 0.65 A resolution. *Acta Crystallogr. D Biol. Crystallogr.* 63(Pt 12):1254–1268.

Weeks, C. M. and Miller, R. 1999. The design and implementation of SnB version 2.0. *J. Appl. Crystallogr.* 32:120–124.

Weik, M., Ravelli, R. B., Kryger, G., McSweeney, S., Raves, M. L., Harel, M., Gros, P., Silman, I., Kroon, J., and Sussman, J. L. 2000. Specific chemical and structural damage to proteins produced by synchrotron radiation. *Proc. Natl. Acad. Sci. USA* 97(2):623–628.

Wermuth, B. 1985. *Enzymology of Carbonyl Metabolism 2: Aldehyde Dehydrogenase, Aldo-Keto Reductase, and Alcohol Dehydrogenase.* Alan R. Liss Inc., New York.

Wilmouth, R. C., Edman, K., Neutze, R., Wright, P. A., Clifton, I. J., Schneider, T. R., Schofield, C. J., and Hajdu, J. 2003. X-ray snapshots of serine protease catalysis reveal a tetrahedral intermediate. *Nat. Struct. Biol.* 278(44):43357–43362.

Wilson, D. K., Bohren, K. M., Gabbay, K. H., and Quiocho, F. A. 1992. An unlikely sugar substrate site in the 1.65 A structure of the human aldose reductase holoenzyme implicated in diabetic complications. *Science* 257:81–84.

Wilson, D. K., Tarle, I., Petrash, J. M., and Quiocho, F. A. 1993. Refined 1.8 A structure of human aldose reductase complexed with the potent inhibitor zopolrestat. *Proc. Natl. Acad. Sci. USA* 90(21):9847–9851.

Wlodawer, A., Minor, W., Dauter, Z., and Jaskolski, M. 2008. Protein crystallography for non-crystallographers, or how to get the best (but not more) from published macromolecular structures. *FEBS J.* 275(1):1–21.

Yabe-Nishimura, C. 1998. Aldose reductase in glucose toxicity: A potential target for the prevention of diabetic complications. *Pharmacol. Rev.* 50:21–33.

Yano, J., Kern, J., Irrgang, K. D., Latimer, M. J., Bergmann, U., Glatzel, P., Pushkar, Y. et al. 2005. X-ray damage to the Mn4Ca complex in single crystals of photosystem II: A case study for metalloprotein crystallography. *Proc. Natl. Acad. Sci. USA* 102(34):12047–12052.

7

X-Ray Optics: Toward Subnanometer Focusing

Christian G. Schroer
Technische Universität Dresden

7.1 Introduction

The key strength of x-rays as a probe is their large penetration depth in matter, allowing one to investigate the interior of an object without destructive sample preparation. This is, for example, particularly interesting for in situ studies of all kinds, as the specimen can be investigated in its natural surrounding or inside a special sample environment, such as a chemical reactor or a pressure cell. In addition, x-ray imaging offers a wealth of different contrasts through the use of different x-ray analytical techniques, such as x-ray fluorescence, absorption, and scattering. In this way, x-ray microscopy is sensitive to the concentration of chemical elements, the chemical state of a given element, or the local mesoscopic and microscopic structure, respectively.

The field of x-ray microscopy has gained momentum with the advent of brilliant x-ray sources, in particular, dedicated synchrotron radiation sources of the so-called third generation, such as the European Synchrotron Radiation Facility (ESRF) in Grenoble, France, the Advanced Photon Source (APS) at Argonne National Laboratory near Chicago, and SPring-8 in Japan. In the last decade, many national synchrotron radiation sources became operational, such as PETRA III at DESY in Hamburg that is currently the storage ring with the lowest emittance. Since recently, x-ray free-electron lasers have extended the possibilities in x-ray imaging toward ultra-short timescales, allowing one to image dynamical processes in matter down to the atomic scale.

The development of x-ray optics for high-resolution imaging and spectroscopy has seen a significant boost since the advent of all these sources. With increasing brilliance, x-ray microscopy—in particular scanning microscopy—has become more and more efficient, pushing the spatial resolution from a few micrometers in the early 1990s to the nanometer regime, today. As we will show in this chapter, manipulating an x-ray beam with optics is quite challenging, in particular due to the weak elastic scattering of x-rays in matter and their relatively strong attenuation. Therefore, over the years, a large variety of x-ray optics has been designed, based on refraction, reflection, and diffraction. Most of the optics encountered today are technology-limited, that is, the theoretical performance limits based on the underlying physics are not reached, yet. The smallest one-dimensional focus, that is, 7 nm full width at half maximum (FWHM), obtained so far has been created by a hybrid optic made of a curved multilayer mirror and an aberration correcting total-reflection mirror [1].

As the three effects—refraction, reflection, and diffraction—are all based on the same physics of elastic scattering, their physical limitations are quite similar. Therefore, as technology advances, the performances of the different x-ray optics approach each other. Depending on the application, one or the other optic may be advantageous, however, it seems today that diffractive optics have the best chance of breaking the nanometer barrier.

7.2 X-Ray Nanobeams

In this section, the general principles of nanofocusing and the relevant optical parameters are discussed. In addition, a method for characterizing the nanobeams is given.

7.2.1 Optical Parameters and Beam Size

In order to generate a nanobeam, the x-rays from a source are focused to the sample position by an x-ray optic. Figure 7.1a depicts the general focusing scheme. If the optic has a focal length f and is located at a distance L_1 from the source, the image of the source, that is, the focus, is formed at a distance

$$L_2 = \frac{L_1 f}{L_1 - f}$$

behind the optic. From geometric optics, it follows that the image size b is given by

$$b = \frac{L_2}{L_1} \cdot g = M \cdot g,$$

where

- g is the source size
- M is the magnification

For nanofocusing geometries, the magnification is very small ($\ll 1$), that is, the image of the source is reduced as much as possible.

Besides the geometric image of the source, the optic itself determines the lateral size of the nanobeam by diffraction at its aperture, and potentially also by aberrations. It is obvious that the latter should be made as small as possible, imposing quite stringent requirements on the fabrication of optics (cf. Section 7.4). The diffraction at the aperture of the optic is an intrinsic limitation that is described by Abbe's equation for the resolution of an optical system [2]:

$$d_t = a \cdot \frac{\lambda}{2NA}, \tag{7.1}$$

where

- d_t is the lateral size of the diffraction-limited Airy disc
- λ is the wavelength of the x-rays
- $NA = n_a \sin\alpha$ is the so-called numerical aperture with n_a being the ambient refractive index and α being half the opening angle of the aperture as seen from the focus (Figure 7.1b)
- a is a constant close to 1 that depends on the shape of the aperture of the optic (cf. Section 7.4)

Equation 7.3 also implies that focusing to ultimately small dimensions will require the largest possible numerical apertures.

For x-rays, the ambient refractive index n_a, that is, that of air or vacuum surrounding the optic, is unity to very good approximation (cf. Section 7.3). In addition, the angle α rarely exceeds a few milliradians, such that NA can be simplified to

$$NA \approx \alpha \approx \frac{D_{\text{eff}}}{2L_2}. \tag{7.2}$$

Here, D_{eff} is the effective aperture of the optic (Figure 7.1b). It can be different from the geometric aperture, for example, if the transmission through the optic is not homogeneous.

In the absence of aberrations of the optic, the beam size in the focus is thus given by the convolution of the geometric image of the source with the Airy disc. The lateral width of this convolution depends in detail on the shape of the two contributions. In any case, however, the beam can never be smaller than the largest of these two contributions, that is, the Airy disc d_t or the size of the geometric image b. Therefore, for ultimate focusing, both these contributions need to be minimized.

Currently, it is the x-ray optic and its diffraction limit that determine the size of the smallest beams. Therefore, the smallest possible beams imply diffraction-limited focusing, that is, the geometric image of the source is smaller than the Airy disc:

$$d_t = a \cdot \frac{\lambda}{2NA} \geq \frac{L_2}{L_1} g = b. \tag{7.3}$$

Inserting Equation 7.2 into Equation 7.3 and solving for D_{eff} yields

$$D_{\text{eff}} \leq a \frac{\lambda L_1}{g} = a l_t, \tag{7.4}$$

where $l_t = \lambda L_1 / g$ is the transverse coherence length of the x-rays. With $a \approx 1$, this implies that in order to have diffraction-limited focusing, the lateral coherence length must be larger than the effective aperture D_{eff} of the optic.

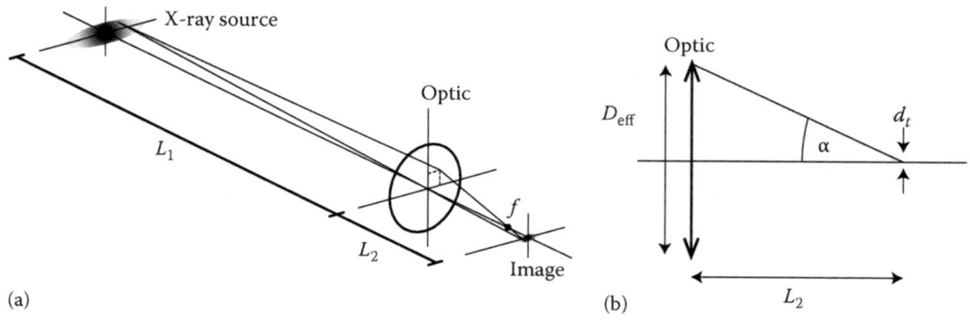

FIGURE 7.1 (a) Imaging the x-ray source onto the sample position. (b) Geometry of diffraction-limited focusing.

7.2.2 Flux and Flux Density in Diffraction-Limited Beams

In this section, we address the question of flux and flux density in a diffraction-limited nanobeam. The figure of merit of the source that describes the photon flux available for nanofocusing is the so-called brightness or brilliance Br defined as

$$Br = \frac{F}{A \cdot \Omega \cdot (\Delta E/E)},$$

where

- F is the photon flux (measured in ph/s)
- A is the area of the source (in mm^2)
- Ω is the solid angle, into which the photons are emitted (measured in $mrad^2$)
- $(\Delta E/E)$ is the energy bandwidth

The flux that is transmitted through the aperture of the focusing optic in diffraction-limited focusing geometry scales at best with

$$F_c = Br \cdot A \cdot \Omega_{opt} \cdot \frac{\Delta E}{E} \cdot T \propto Br \cdot g^2 \cdot \left(\frac{D_{eff}}{L_1}\right)^2 \cdot \frac{\Delta E}{E} \cdot T, \quad (7.5)$$

where

- A is the area of the source
- Ω_{opt} is the solid angle spanned by the optic
- T is its transmission

Inserting Equation 7.4 into Equation 7.5 yields a diffraction-limited flux of

$$F_c \propto Br \cdot \lambda^2 \cdot \frac{\Delta E}{E} \cdot T. \quad (7.6)$$

The brilliance is thus the right figure of merit of the source that determines the flux in a diffraction-limited focus.

Figure 7.2 shows the brilliance of various x-ray sources. It shows that modern synchrotron radiation sources and x-ray free-electron lasers are many orders of magnitude more brilliant than conventional x-ray tubes. Thus, the ultimate x-ray nanobeams will be available at synchrotron radiation sources and XFELs only.

For many experiments, the flux density is an important parameter. For diffraction-limited nanobeams, it scales like

$$I_c = \frac{F_c}{d_t^2} \propto Br \cdot NA^2 \cdot T \cdot \frac{\Delta E}{E}. \quad (7.7)$$

Here, we used Equations 7.3 and 7.6. Equation 7.7 shows that the flux density I_c in a diffraction-limited beam depends on the brilliance as the only source parameter. The optics influence the flux density only by their numerical aperture NA and transmission T. $\Delta E/E$ is given by the experimental parameters and typically lies in the range of 10^{-4} for monochromatic radiation.

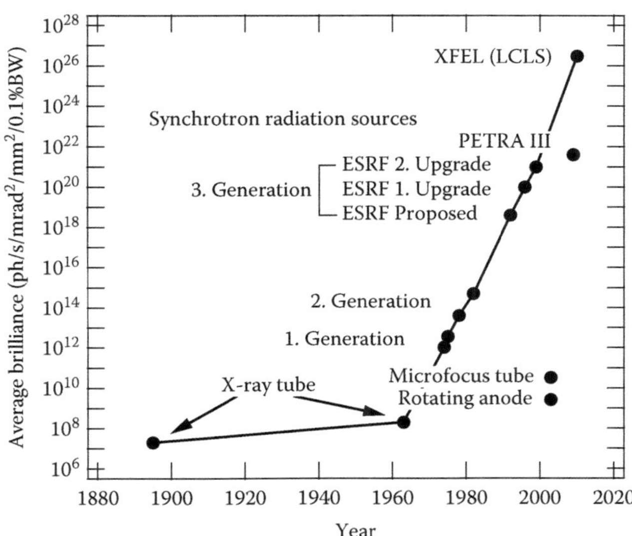

FIGURE 7.2 Average brilliance as a function of time.

7.2.3 Characterizing Hard X-Ray Nanobeams

The most common method to characterize the lateral extension of x-ray beams is the so-called knife-edge technique. In this method, an object with a sharp edge is scanned through the beam. At each position of the scan, the intensity behind the object is measured. In this way, the integrated intensity profile along the scanning direction can be determined. With decreasing x-ray beam size, the requirements on the edge in terms of roughness become more and more stringent, requiring the use of nanofabricated knife edges. Due to the reduced dimensions of the latter, the contrast in transmission is usually reduced as well. Therefore, other x-ray analytical techniques, such as x-ray fluorescence or scattering are typically used. For highest spatial resolutions, phase-shifting edges [3] and analyzers based on thin-film technology [4] are scanned through the beam. The knife-edge technique becomes more and more difficult to apply as the beam size is reduced.

Recently, a scalable alternative for beam characterization has emerged from x-ray microscopy, that is, the so-called ptychographic scanning coherent diffraction microscopy [5,6]. In this technique, an arbitrary object with small features, for example, a test pattern, is scanned through the diffraction-limited beam and thus coherent nanobeam in the two dimensions perpendicular to the optical axis, recording at each position of the scan a far-field diffraction pattern (cf. Figure 7.3). The distance between scan points is chosen such as to allow for a significant overlap between neighboring scan points [7]. From these data, the complex transmission function of the object can be reconstructed as well as the complex illuminating wave field [8–10]. The latter fully characterizes the nanobeam, giving access to its full caustic.

This is illustrated in Figure 7.4, showing the reconstructed complex wave field of a nanobeam in the plane of the object (Figure 7.4a and b). In this case, the hard x-ray beam (E = 15.25 keV) was focused by nanofocusing parabolic refractive

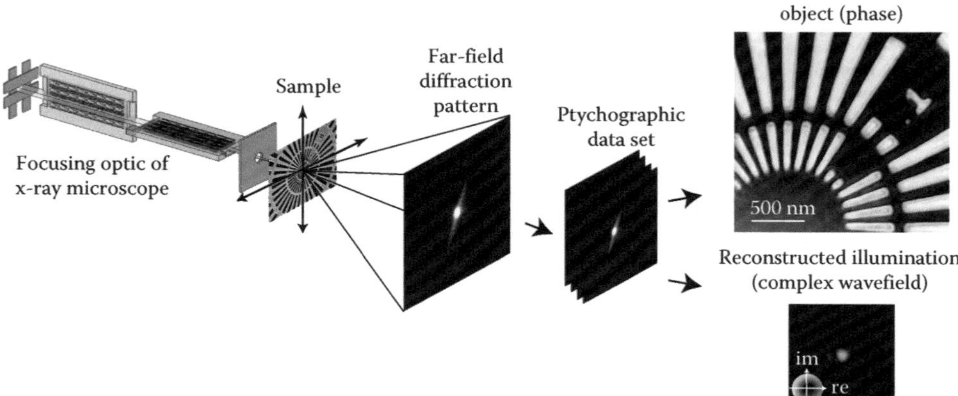

FIGURE 7.3 Characterization of the nanobeam by ptychographic imaging. A test sample is raster-scanned while recording far-field diffraction patterns at each point of the scan. From the series of diffraction patterns, the complex transmission function of the object and the illuminating wave field can be reconstructed.

x-ray lenses (cf. Section 7.4.2) and was probed with a resolution test chart by NTT-AT (model: ATN/XRESO-50HC; reconstructed object in Figure 7.3) made of tantalum (material thickness 500 nm). Details of this particular experiment can be found in [11].

The nanobeam is reconstructed in great detail, also revealing its low-intensity side lobes (Figure 7.4b). The method is sensitive to relative intensities that can vary over four to five orders of magnitude (Figure 7.4d). It has been shown to yield reliable reconstructions of the beam [12] and was used to characterize different optics, for example, ellipsoidal multilayer mirrors [13], a nanofocusing Kirkpatrick–Baez (KB) mirror system [14], nanofocusing refractive x-ray lenses [11], and Fresnel zone plates (FZPs) [15,16]. In these examples, the beam size ranges from several micrometers down to about 20 nm. Indeed, ptychography works independently of the

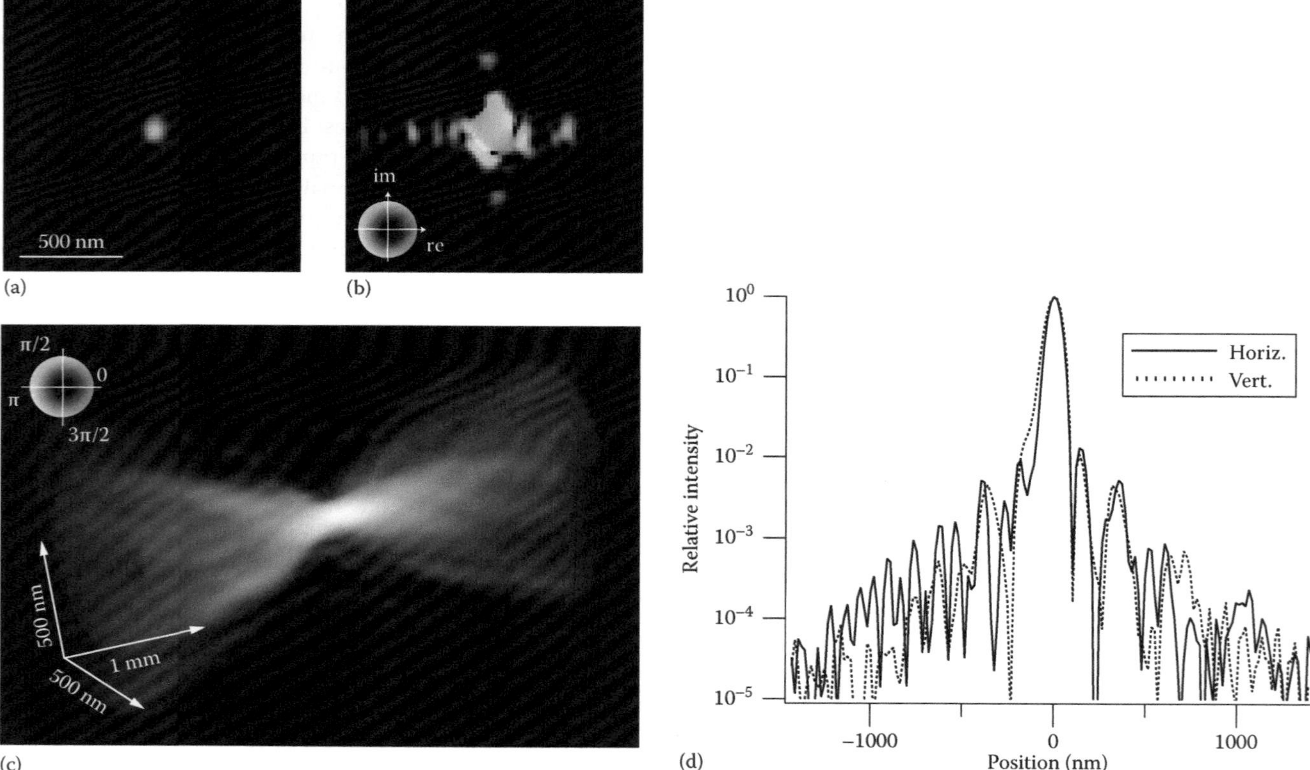

FIGURE 7.4 **(See color insert.)** (a) Complex wave field reconstructed in the plane of the test object. (b) Wave field rescaled by a factor 10 to enhance the visibility of the side maxima. The color codes the complex amplitude according to the color wheel. (c) Wave field propagated numerically along the optical axis. (d) Horizontal and vertical line scan through the focus (intensity) depicted on logarithmic scale.

beam size and will also be applicable to subnanometer beams. As it requires no a priori knowledge of the sample, the requirements on the test object are not very stringent. Indeed, the only requirement is a certain structural diversity on the length scale of the beam.

7.3 Interaction of X-Rays with Matter

In the x-ray range of the electromagnetic spectrum, the interaction is dominated by photoabsorption and by elastic and Compton scattering. While photoabsorption and Compton scattering both contribute to the attenuation in an x-ray optic, the elastic scattering is responsible for its optical, for example, focusing, effect. All x-ray optics are based on elastic scattering, relying either on refraction, reflection, or diffraction.

Refraction of x-rays in matter is very weak. The refractive index of x-rays is typically written as

$$n = 1 - \delta + i\beta, \tag{7.8}$$

where

- $\delta > 0$ is the *index of refraction decrement*
- β describes the attenuation of x-rays in matter and comprises the effects of photoabsorption and Compton scattering

Equation 7.8 and $\delta > 0$ imply a refractive index in matter that is smaller than 1. As opposed to refraction of visible light

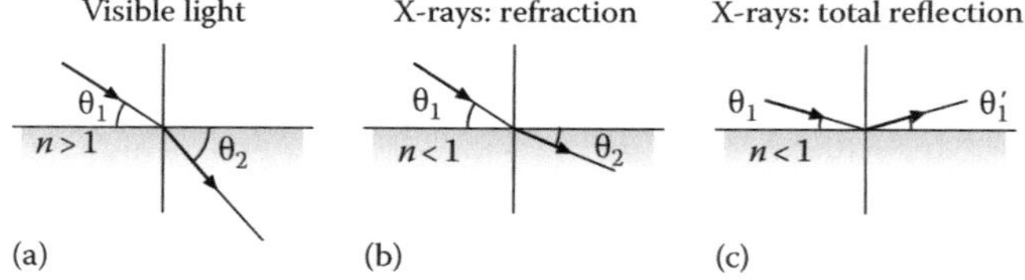

FIGURE 7.5 Refraction at the interface between vacuum and matter for visible light (a) and for x-rays (b). For x-rays, the refractive index in matter is smaller than 1, leading to a refraction away from the surface normal upon entering the medium from vacuum. (c) For sufficiently small angles θ_1, the x-rays do not propagate into the material but are reflected at the surface (external total reflection).

(Figure 7.5a), this leads to a refraction away from the surface normal upon entering a material Figure 7.5b). As δ is very small, for example, $\delta \approx 10^{-6}$ in the hard x-ray range around 10 keV, the deflection angle $\Delta\theta = \theta_2 - \theta_1$ (cf. Figure 7.5b) is also small, for example, $\Delta\theta \approx 10^{-6}$ for x-rays incident under 45° onto the surface.

The index of refraction decrement δ can be written as

$$\delta = \frac{N_a}{2\pi} r_0 \lambda^2 \rho \frac{Z + f'(E)}{A},$$

where

- N_a is Avogadro's number
- r_0 is the classical electron radius
- λ is the wavelength of the x-rays
- ρ is the density of the material
- $f(K = 0, E) = Z + f'(E)$ is the real part of the atomic form factor of the material in forward direction
- A is its atomic weight

Figure 7.6a shows δ as a function of x-ray energy. Away from absorption edges, δ/ρ is very similar for all chemical elements and follows a E^{-2} power law to good approximation.

The attenuation of x-rays in matter is typically described by Lambert and Beer's law:

$$I(x) = I_0 \cdot \exp(-\mu \cdot x) \tag{7.9}$$

where

- $I(x)$ is the intensity after a piece of material of thickness x, when I_0 is the incident intensity
- μ is the attenuation coefficient and is related to β in Equation 7.8 by

$$\mu = \frac{4\pi\beta}{\lambda}.$$

Figure 7.6b shows the specific attenuation coefficient μ/ρ as a function of energy for different materials. Photoabsorption

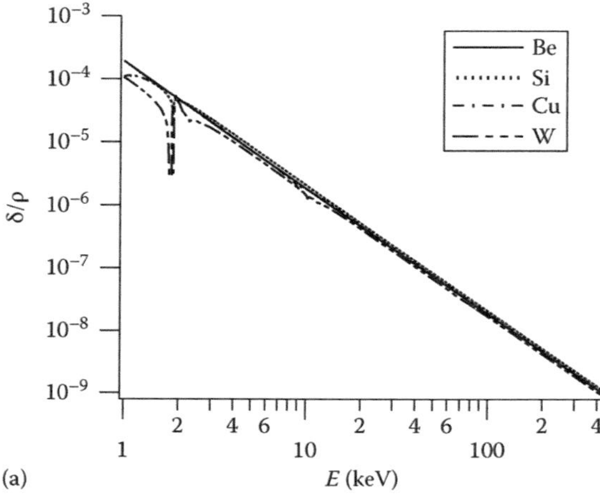

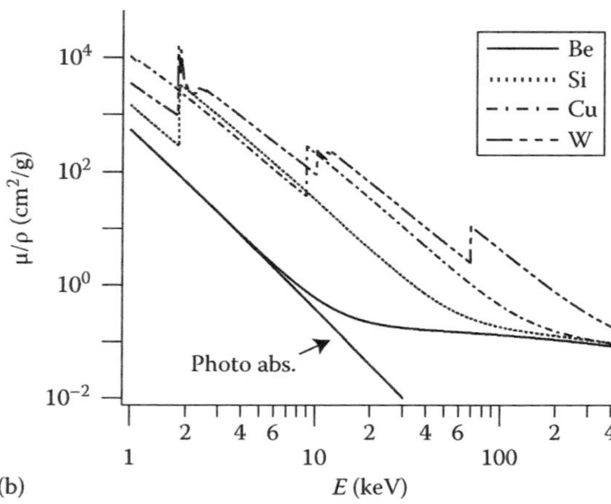

FIGURE 7.6 (a) Specific refractive index decrement δ/ρ and (b) specific attenuation coefficient μ/ρ as a function of x-ray energy for beryllium, silicon, copper, and tungsten.

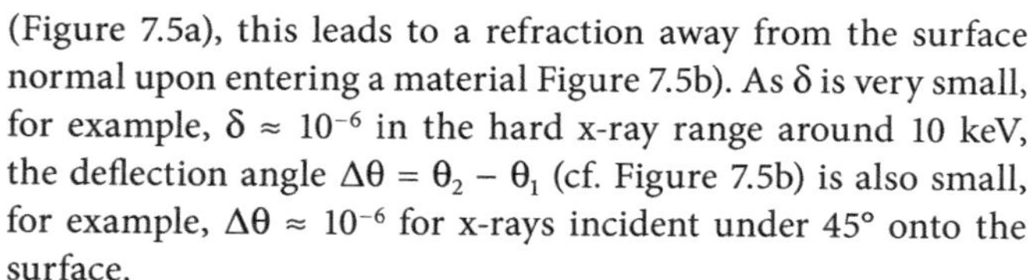

is strongly dependent on the atomic species and grows with Z^{α} ($\alpha \approx 3$). At the same time, it decreases with an approximate E^{-3} power law away from absorption edges. At higher x-ray energies, Compton scattering dominates the attenuation and limits the specific attenuation μ/ρ from below to approximately 0.1 cm^2/g. As a result, there is no material for x-rays that is as transparent as glass is for visible light. This is the main limitation to making refractive optics with large numerical aperture (cf. Section 7.4.2).

A weak refraction is accompanied by a small reflectivity. Therefore, the reflection of x-rays is negligible at high reflection angles. Significant reflectivity is only observed for incidence angles θ_1 below a critical angle $\theta_c = \sqrt{2\delta}$ (Figure 7.5c). In this case, the x-rays cannot propagate inside the material and are reflected by external total reflection. With $\delta \approx 10^{-6}$ around 10 keV, the critical angle θ_c typically lies in the range of milliradians. Therefore, reflective optics for x-rays always involve grazing incidence (cf. Section 7.4.1).

Refraction and reflection at a single surface both result in small deflection angles. This ultimately limits the numerical aperture and thus the spatial resolution of optics based on these two effects. This limitation can be overcome by diffraction at sufficiently fine structures. FZP or multilayer mirrors are examples of optics based on diffraction (cf. Section 7.4.3).

7.4 X-Ray Optics

In this section, the different types of x-ray optics are discussed in view of nanofocusing.

7.4.1 Reflective Optics: Mirrors and Waveguides

X-ray mirrors and waveguides are both based on external total reflection (cf. Section 7.3). These optics will be discussed here in view of their physical limitations.

In order to focus x-rays in one dimension from point to point with a mirror, its shape needs to be an ellipse as shown in Figure 7.7a. Source and focus are located at the two foci of the ellipse. As the reflection angle has to be below the critical angle of external total reflection θ_c, the ellipse is typically highly eccentric. In order to focus not only in one dimension, a single bounce mirror optic would have to be an ellipsoid of rotation. Due to the high eccentricity, the curvatures in azimuthal and meridional directions would have to be quite different, making such an optic very difficult to fabricate with an accuracy high enough for nanofocusing. Therefore, the typical scheme is to combine two one-dimensionally focusing mirrors in the so-called KB geometry shown in Figure 7.7b [17]. Here, the mirrors need to be curved to a segment of an ellipse in one dimension only. The segments must be chosen and aligned in such a way as to have a common focus at the sample position. The imaging geometry is nearly independent of the x-ray energy, provided the largest reflection angles in the imaging geometry are smaller than the critical angle for the whole x-ray spectrum. Therefore, these optics are intrinsically achromatic making them very attractive for x-ray microscopy with absorption spectroscopic contrast.

The short wavelength of x-rays lays high demands on the roughness and shape accuracy of mirrors, and their fabrication is quite expensive. In order to suppress aberrations due to shape errors and roughness, the figure accuracy needs to be below 3 nm from peak to valley on large length scales on the mirror (>10 mm), and the high-frequency figure error and the surface roughness need to be less than 0.2 nm (rms). Such values are technically feasible today, yielding diffraction-limited focusing by mirror optics [18] down to 25 nm. As the critical angle θ_c and the length of the mirror limit the numerical aperture, no significantly smaller foci are expected to be possible with single bounce mirror systems in KB geometry. Future improvements could result from rotational ellipsoids and potentially from multi-bounce mirror systems.

An x-ray waveguide consists of a sandwich structure made of an optically denser channel surrounded by an optically less dense cladding material (cf. Figure 7.8) [19,20]. Ideally, the channel is filled with air; however, it often consists of a low-Z low-density material. The cladding is made of a material of high mass density ρ. In this way, x-rays can travel along the channel, being confined in the transverse direction by external total reflection. Depending on the way the x-rays are coupled into the waveguide, different transverse modes can be excited.

The transverse confinement of the x-ray beam can be modeled by a square potential well, whose width is given by the transverse dimension of the channel and whose potential depth is determined by the difference in index of refraction between the channel and the cladding. For large transverse dimensions, the potential well holds several modes. As the transverse dimension is reduced, the number of transverse modes is reduced, as well, until only a single mode is confined in the waveguide. When the dimension of the waveguide is reduced further, the single confined transverse mode persists. However, it was shown by Bergemann et al. [21] that lateral size of the mode cannot

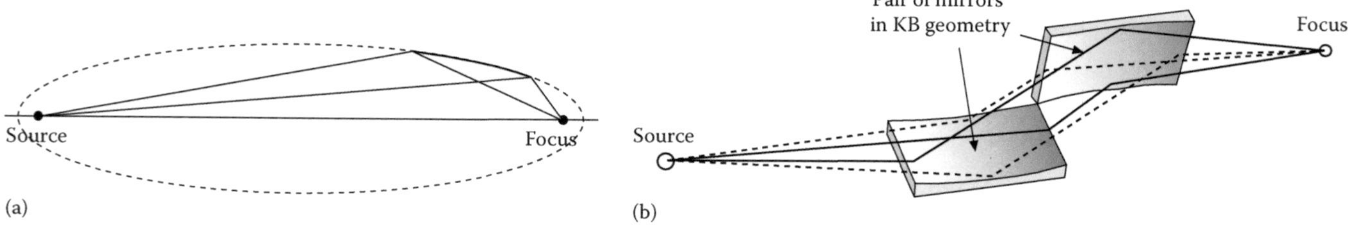

FIGURE 7.7 (a) Geometry to generate a focused x-ray beam using a curved mirror. (b) Two-dimensional focusing using two crossed curved mirrors in the so-called KB geometry. (From Kirkpatrick, P. and Baez, A., *J. Opt. Soc. Am.*, 38, 766, 1948.)

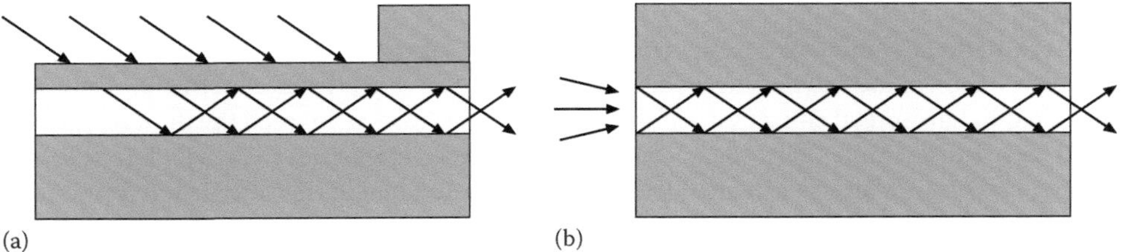

FIGURE 7.8 X-ray waveguide: (a) A parallel x-ray beam is coupled into the waveguide through the cap layer. (b) Focused x-rays are directly coupled into the waveguide from the side.

be squeezed below $\lambda/2\theta_c$. In analogy to Equation 7.1, θ_c is the numerical aperture of the waveguide, limiting the size of the transverse mode to approximately 10 nm.

While the first waveguides were confining the beam only in one dimension, several schemes have been developed to fabricate two-dimensional waveguides or to produce a quasi-point source [22–24]. Experimentally, beams down to slightly below 10 nm were confined in optimized waveguides [24,25], approaching the limit for numerical aperture described theoretically by Bergemann et al. [21].

In conclusion, reflective optics are technologically well advanced and have reached the physical limits of their performance in view of nanofocusing. With the current optical designs, subnanometer focusing will not be possible with reflective optics.

7.4.2 Refractive Optics: Lenses

Although abundant in the visible light regime, refractive optics have long been considered not feasible in the x-ray range. W. C. Röntgen made first experiments trying to refract x-rays on prisms and lenses [26], but could not resolve the weak refraction effect. Later, Kirkpatrick and Baez [17] were discouraged from using refractive optics by the weak refraction and relatively strong attenuation of x-rays in matter. Discussions about the feasibility of refractive x-ray optics persisted [27,28] until the first refractive x-ray lenses were realized in the mid-1990s [29].

Since then, refractive x-ray lenses have developed quickly with a great variety of different realizations. The main strategy of all these schemes is to compensate the weak refraction of x-rays in the lens material by stacking a large number of strongly curved lenses behind each other. Note that due to the refractive index $n = 1 - \delta$ (cf. Section 7.3) being smaller than 1, focusing lenses have to be concave. In order to keep the attenuation in the lens material minimal, the lenses need to be made of the most transparent lens materials, that is, materials composed of elements with low atomic number Z, such as Be, B, C, Al, or Si [30]. Other aspects affecting the choice of lens material are the stability in the x-ray beam, homogeneity of the material, and its machinability.

As the refraction is weak, the curvature of individual lens surfaces is often made as strong as possible. This implies that the spherical approximation that is fulfilled when the lens aperture is small against the radius of curvature is not applicable for refractive x-ray lenses. Highest spatial resolution in x-ray focusing is reached only by the use of aspherical optics, that is, with parabolic shape [30–32] for compound optics. Figure 7.9 shows

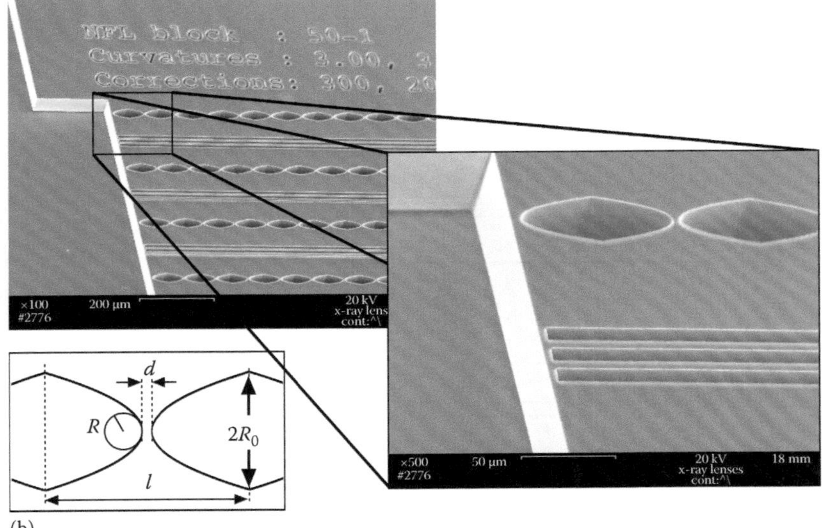

FIGURE 7.9 (a) Rotationally parabolic refractive x-ray lenses made of beryllium. Individual lenses are stacked behind each other inside a V groove to form a compound lens. (b) Nanofocusing refractive x-ray lenses made of silicon. They are made by lithography and deep reactive ion etching. Individual lenses are parabolic cylinders, focusing the beam in one dimension only. Therefore, in order to focus in two dimensions, two such lenses need to be aligned behind each other in crossed geometry (cf. Figure 7.3).

two different realizations of parabolic refractive x-ray lenses. In Figure 7.9a, a stack of rotationally parabolic refractive x-ray lenses made of beryllium are shown. Individual lenses embossed into beryllium are centered in hard-metal coins that are aligned on a common optical axis by stacking them in a V groove (Figure 7.9a). Depending on the number of individual lenses in a stack and on their curvature, these optics typically have focal lengths in the range between about 10 cm and several meters, and are routinely used at many synchrotron radiation sources for microfocusing, high-resolution imaging, and beam conditioning in the hard x-ray range [33,34].

For refractive x-ray lenses, the numerical aperture scales as $NA \propto 1/\sqrt{f}$ for focal lengths f in the range above, thus growing with decreasing focal length. This led to the development of nanofocusing refractive x-ray lenses with focal lengths in the range of 10 mm. These short focal lengths require both extreme curvatures of the lens surfaces and short lens stacks that can only be made by nanofabrication techniques, such as lithography and deep reactive ion etching. Figure 7.9b shows a series of nanofocusing lenses made of silicon [35,36]. As the focal length of these optics is reduced to the centimeter range, the numerical aperture NA grows, approaching the limiting value of $\sqrt{2\delta}$ from below. Thus, the numerical aperture of compound refractive x-ray lenses is limited by the same upper bound as that of reflective optics (cf. Section 7.4.1), and the focal spot sizes cannot fall below 10 nm [35].

At first glance, it seems that refractive x-ray optics are subject to the same limitations as reflective optics (cf. Section 7.4.1 and [21]). However, there are refractive optics, for which the numerical aperture can exceed the limit of $\sqrt{2\delta}$, the so-called adiabatically focusing lenses (AFLs) [37]. In these optics, the aperture of the individual lenses is matched to the converging beam inside the optic. Thus, toward the end of the lens, the refractive power per unit length can be significantly increased, also increasing the numerical aperture. It has been shown in numerical simulations that ideal AFLs can generate an x-ray focus of about 5 nm [37]. Figure 7.10a illustrates the design of an AFL. If combined with the concept of kinoform lenses, AFLs can, in principle, generate hard x-ray nanobeams with 2 nm lateral extension [37]. So far, the concept of AFLs has not been demonstrated experimentally, mainly due to the strict requirements imposed on their fabrication. The fabrication of the tiny lens structures at the exit of the AFL will ultimately limit the spatial resolution that can be obtained with these optics. A focus of a few nanometers is conceivable, but subnanometer focusing seems to be out of reach.

Currently, the performance of refractive optics is still technology-limited. The smallest foci reached so far lie slightly below 50 nm [36,38]. As nanofabrication techniques are developed further, these optics still have significant room for improvement.

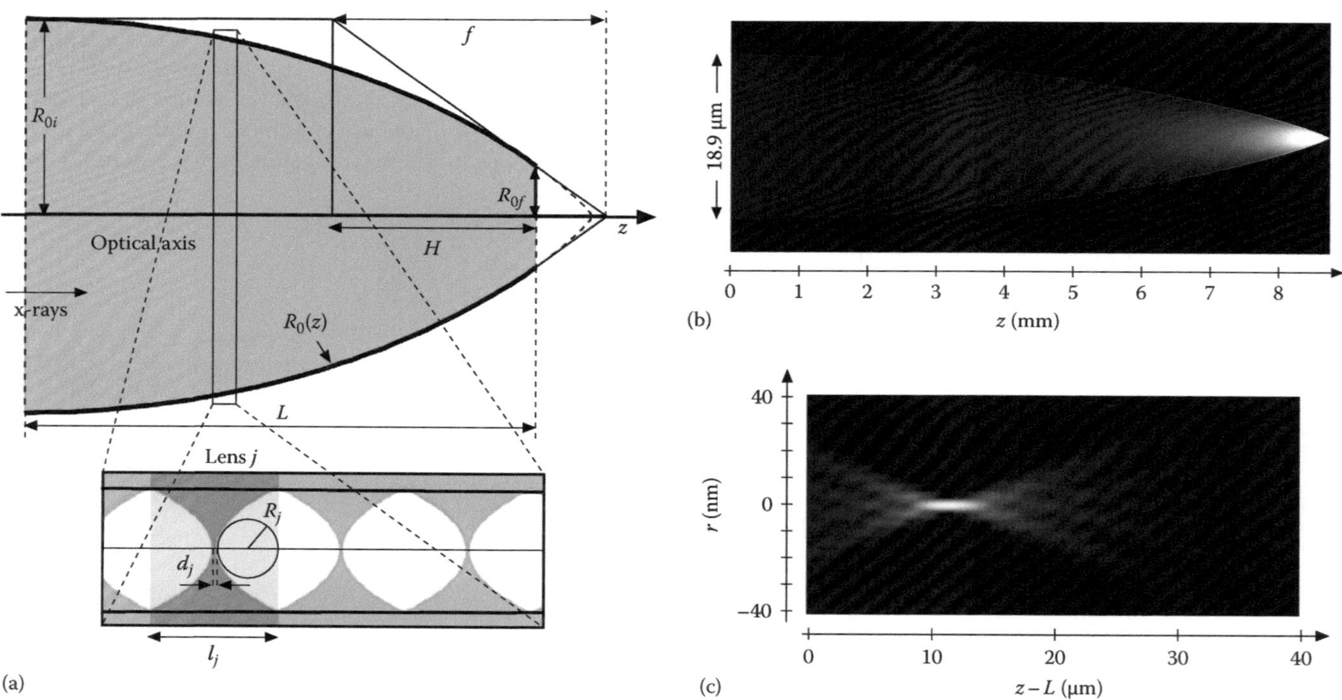

FIGURE 7.10 (a) Schematic diagram of an AFL: the optic is made of a large number of individual lenses, whose aperture $R_0(z)$ is smoothly adapted to the converging beam, ranging from $2R_{0i}$ at the entrance down to $2R_{0f}$ at the exit of the lens. The focus is formed at a short distance behind the optic. The focal length f as well as the distance H of the principal plane from the exit of the lens are shown. (b) Intensity of the converging x-ray beam inside an ideal AFL. (c) Simulated intensity around the focus of the AFL that has a lateral extension of 4.7 nm.

7.4.3 Diffractive Optics: Fresnel Zone Plates, Multilayer Mirrors, and Multilayer Laue Lenses

Diffractive optics have the highest potential for subnanometer focusing, mainly due to the fact that x-rays can be diffracted to large angles by small periodic structures. We can distinguish two types of diffractive optics, that is,

- Multilayer mirrors that diffract x-rays in reflection (Bragg) geometry
- FZPs or multilayer Laue lenses (MLLs) that diffract the x-rays in transmission (Laue) geometry

Typically, a multilayer is a planar structure, in which alternating layers of two (or more) different materials are deposited on a flat substrate. Figure 7.11a depicts a periodic multilayer structure made of 40 double layers of molybdenum and silicon.

The reflectivity of x-rays at a surface and under angles larger than the critical angle θ_c is very small due to the weak refraction of x-rays in matter (cf. Section 7.3). The reflectivity of a multilayer structure, however, can be significantly enhanced, when the Bragg condition for diffraction at the multilayer is fulfilled, that is, the diffraction amplitudes between all double layers in the stack are in phase (Figure 7.11b):

$$n\lambda = 2d\sin\theta \qquad (7.10)$$

where

$n = 1, 2, 3, \ldots$ is an integer
d is the thickness of a double layer in the stack (Figure 7.11b)
θ is the angle of incidence of the x-rays

To have a good reflectivity, the refractive power of the multilayer materials should be very different. As the specific refractive index is nearly the same for all materials, this is realized by depositing two materials of significantly different mass density ρ (cf. Section 7.3). As most high-density materials are strongly absorbing for x-rays, a compromise must be found between attenuation and contrast in refractive index. A detailed account on the x-ray optics of multilayers is given in [40].

In order to use multilayers for nanofocusing, similar schemes as for mirror optics are pursued. Figure 7.11c shows an elliptically curved multilayer for point-to-point imaging. As the reflection angle is different for different positions on the surface of the elliptical multilayer, the d spacing needs to vary accordingly to fulfill the Bragg condition (Equation 7.10) at each point. Such laterally graded multilayer optics (also termed Göbel mirrors [41]) have been perfected to generate nanobeams [1,42,43]. As the numerical aperture exceeds that of total reflection mirrors, the requirements on figure errors and roughness become more and more stringent. Therefore, multilayer optics are currently technology-limited. As larger numerical apertures are achieved with smaller d spacings, the higher are the demands on figure error and roughness.

Using a combination of an adaptive total reflection mirror and a multilayer optic, a 7 nm focus was recently achieved in one dimension [1]. In this case, the adaptive mirror was used to compensate wave front distortions of the multilayer.

Alternatively, x-rays can be focused using transmission optics, that is, FZPs. These optics consist of a series of concentric rings (zones) with radius r_n ($n = 0, 1, \ldots, N$), aligned according to the Fresnel construction

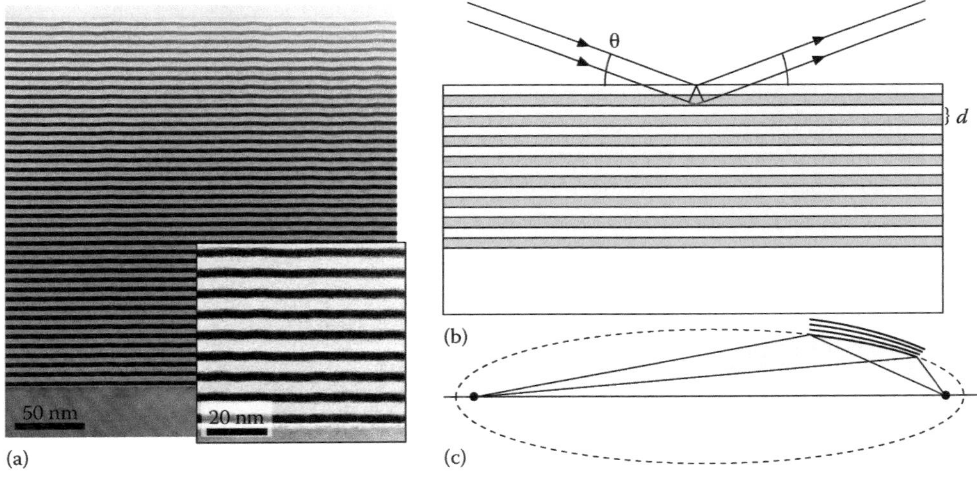

FIGURE 7.11 (a) High-resolution transmission electron micrograph of a multilayer optic consisting of $N = 40$ double layers of Mo and Si. The lattice spacing is $d = 7$ nm. The inset shows a detail of the multilayer including the interface with the monocrystalline substrate. (b) Schematic drawing of a multilayer optic. The x-rays are reflected from the multilayer if the Bragg condition (Equation 7.10) is fulfilled, that is, the optical path length difference between the two paths shown is an integer multiple of the x-ray wavelength λ. (c) Elliptical graded multilayer. The period d changes as a function of position on the optic, locally adjusting the Bragg angle to the angle of reflection. (a: Courtesy of S. Braun, Fraunhofer IWS, Dresden, Germany; Braun, S. and Mai, H., in *Metal-Based Thin Films for Electronics*, eds. K. Wetzig and C.M. Schneider, Wiley-VCH, Weinheim, Germany, 2006.)

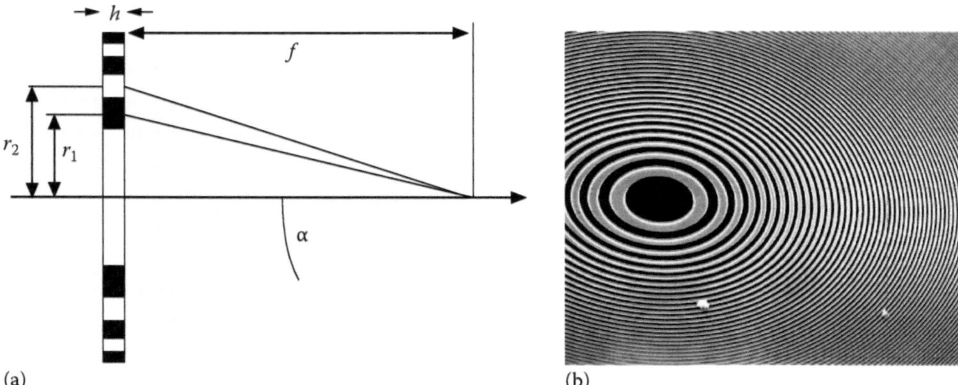

FIGURE 7.12 (a) Scheme of a thin FZP with focal length f. Odd zones r_i either block the incident x-rays by absorption or shift their phase by $\pi/2$ by refraction. (b) Scanning electron micrograph of a FZP.

$$r_n = \sqrt{n\lambda f + \frac{n^2\lambda^2}{4}} \quad (7.11)$$

where
f is the focal length
λ is the wavelength of the x-rays

Figure 7.12a shows a scheme of focusing with FZPs. The radii of the zones are chosen such that the optical path of a ray passing through their center is shifted by $n\lambda/2$ with respect to the ray traveling along the optical axis.

In an absorption zone plate, each second zone is made of an absorbing material, blocking the x-rays there. The zones in between are not blocked, letting the x-rays pass. Thus, only the rays that interfere more or less constructively at the focal position are transmitted. Those that would contribute destructively to the interference are blocked. For absorption zone plates, about 10% of the incident radiation is ideally concentrated on the focus at distance f. Due to their binary structure, x-rays are also diffracted into other diffraction orders. To single out the first diffraction order at the focal distance f, a combination of central beam stop and an order sorting aperture close to the focal position are needed. A detailed account on focusing with FZPs is given in [44]. The absorption zone plate is mostly used in the soft x-ray regime, as the zone plate materials become more and more transparent in the hard x-ray range. There, another scheme is pursued: rather than absorbing the rays that would destructively contribute to the interference in the focus, they are shifted in phase by $\pi/2$ in every second zone. In this way, they too, contribute constructively to the focus. Therefore, these phase-shifting zone plates are more efficient, focusing up to 40% of the incident radiation into the first diffraction order.

Independent of the type of zone plate, its numerical aperture $NA = \sin\alpha$ (Figure 7.12a) is given by

$$NA = \frac{\lambda}{2\Delta r_N},$$

where Δr_N is the width of the outermost zone. For a circular FZP ($\alpha = 1.22$ in Equation 7.1), this leads to a diffraction-limited focus of

$$d_t = 1.22\frac{\lambda}{2NA} = 1.22 \cdot \Delta r_N.$$

Thus, the width of the outermost zone determines the diffraction-limited focal size. In order to make the diffraction-limited spot size as small as possible, the zone plate needs to be designed to have the smallest possible outermost zone. For example, for fixed focal length f, this can be achieved by increasing the number N of zones (cf. Equation 7.11).

FZPs are usually made by electron beam lithography and electroplating. The zones need to have a certain height h (Figure 7.12a) along the optical axis in order to block or phase-shift the x-rays efficiently. Thus, with decreasing width of the outermost zones, the aspect ratio between height and width increases, making the fabrication of these optics evermore challenging. So far, zone plates for soft x-rays with an outermost zone width of 15 nm have been made [45], and—using a scheme of zone doubling—outermost zones with a width of about 12 nm have been reached [46]. In the hard x-ray range, where the zones need to be higher, slightly larger outmost zones are realized in this way [16].

The optical scheme of the FZP discussed earlier implies that the thickness h is small enough to be able to neglect the propagation effects of the x-rays inside the optic. As the outermost zones become smaller and smaller, propagation effects inside the zone plate start to play a more and more important role, reducing the diffraction efficiency in the small zones, effectively limiting the numerical aperture of these optics to the 10 nm range.

To overcome this limitation, the zone plate structures have to be tilted to fulfill a local Bragg condition (Equation 7.10) as shown in Figure 7.13a [47]. Several numerical studies have simulated these optics, all coming to the conclusion that subnanometer focusing should be feasible [48–50]. Indeed, focusing

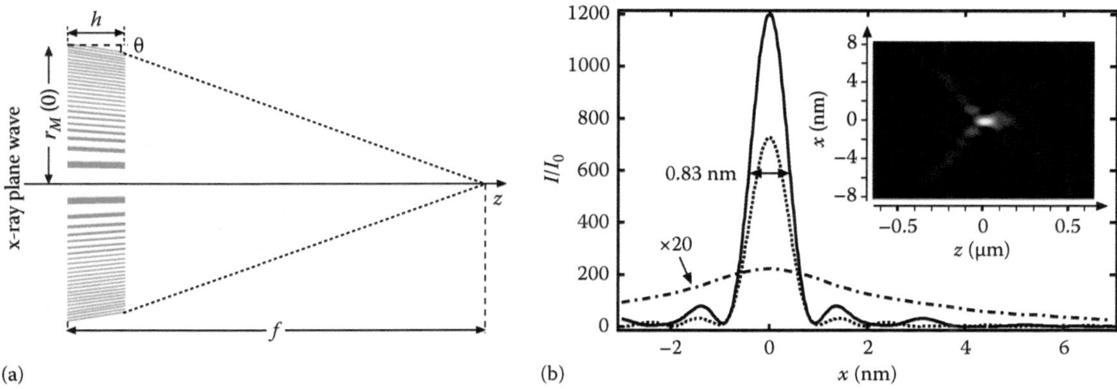

FIGURE 7.13 (a) Thick zone plate with zones tilted to fulfill a Bragg condition locally. (b) Simulation of a subnanometer focus generated by an ideal thick zone plate. (From Schroer, C.G., *Phys. Rev. B*, 74, 033405, 2006.)

x-rays to the atomic level is conceivable if the zones of the FZP are elliptically bent [50]. In addition, the ideally tilted optic has high diffraction efficiency, as other diffraction orders are suppressed.

So far, it is not possible to fabricate such ideally tilted zone plates, but several approaches to approximate them have been pursued. One is the fabrication of the so-called multilayer Laue lenses [51–53]. These optics are made out of the thin slice of a multilayer that is tilted to match the Bragg condition for the smallest zones. Theoretically, beam sizes down to 5 nm are expected with these optics [52], and a focal size of 16 nm has been reached for one-dimensional focusing, so far [4]. Two-dimensional focusing is possible by crossing two MLLs [54].

Ideal thick zone plates have the potential of focusing x-rays to the atomic level. So far, the fabrication of these optics is technology-limited and will surely improve in the coming years.

7.5 Conclusion and Outlook: X-Ray Imaging with Atomic Resolution?

As the refraction of x-rays in matter is weak and the attenuation is strong compared to visible light in glass, many different x-ray optical designs have been made based on reflection, refraction, and diffraction, trying to cope with these difficulties. The short wavelength of x-rays imposes in general stringent requirements on the quality of all these optics, making them quite expensive. Diffractive optics have the potential to outperform the other optics and reach subnanometer diffraction-limited beam sizes. Technologically, however, there is still significant need for improvement to reach this goal.

X-ray imaging with atomic resolution may, however, become possible before the focal spot size reaches atomic dimensions. It could be achieved by ptychography (cf. Section 7.2.3), that is, scanning coherent diffraction microscopy with a strongly focused beam. However, imaging single atoms requires quite high radiation doses [55]. This either limits the application to particularly radiation hard samples or will require fast data acquisition, for example, using an x-ray free-electron laser.

References

1. H. Mimura, S. Handa, T. Kimura, H. Yumoto, D. Yamakawa, H. Yokoyama, S. Matsuyama et al., *Nat. Phys.* **6**, 122 (2010).
2. E. Hecht, *Optics* (Addison-Wesley, Reading, MA, 1987).
3. H. Mimura, H. Yumoto, S. Matsuyama, S. Handa, T. Kimura, Y. Sano, M. Yabashi et al., *Phys. Rev. A* **77**, 015812 (2008).
4. H. C. Kang, H. Yan, R. P. Winarski, M. V. Holt, J. Maser, C. Liu, R. Conley, S. Vogt, A. T. Macrander, and G. B. Stephenson, *Appl. Phys. Lett.* **92**, 221114 (2008).
5. J. M. Rodenburg and H. M. L. Faulkner, *Appl. Phys. Lett.* **85**, 4795 (2004).
6. J. M. Rodenburg, A. C. Hurst, A. G. Cullis, B. R. Dobsen, F. Pfeiffer, O. Bunk, C. David, K. Jefimovs, and I. Johnson, *Phys. Rev. Lett.* **98**, 034801 (2007).
7. O. Bunk, M. Dierolf, S. Kynde, I. Johnson, O. Marti, and F. Pfeiffer, *Ultramicroscopy* **108**, 481 (2008).
8. P. Thibault, M. Dierolf, A. Menzel, O. Bunk, C. David, and F. Pfeiffer, *Science* **321**, 379 (2008).
9. P. Thibault, M. Dierolf, O. Bunk, A. Menzel, and F. Pfeiffer, *Ultramicroscopy* **109**, 338 (2009).
10. A. M. Maiden and J. M. Rodenburg, *Ultramicroscopy* **109**, 1256 (2009).
11. A. Schropp, P. Boye, J. M. Feldkamp, R. Hoppe, J. Patommel, D. Samberg, S. Stephan et al., *Appl. Phys. Lett.* **96**, 091102 (2010).
12. S. Hönig, R. Hoppe, J. Patommel, A. Schropp, S. Stephan, S. Schöder, M. Burghammer, and C. G. Schroer, *Opt. Express* **19**, 16325 (2011).
13. C. M. Kewish, P. Thibault, M. Dierolf, O. Bunk, A. Menzel, J. Vila-Comamala, K. Jefimovs, and F. Pfeiffer, *Ultramicroscopy* **110**, 325 (2010).
14. C. M. Kewish, M. Guizar-Sicairos, C. Liu, J. Qian, B. Shi, C. Benson, A. M. Khounsary et al., *Opt. Express* **18**, 23420 (2010).
15. J. Vila-Comamala, A. Diaz, M. Guizar-Sicairos, S. Gorelick, V. A. Guzenko, P. Karvinen, C. M. Kewish et al., Characterization of a 20-nm hard x-ray focus by

ptychographic coherent diffractive imaging, in *Advances in X-Ray/EUV Optics and Components VI*, Vol. 8139 of *Proceedings of the SPIE*, C. Morawe, A. M. Khounsary, and S. Goto (eds.) (SPIE Optical Engineering Press, Bellingham, WA, 2011), p. 81390E.
16. J. Vila-Comamala, A. Diaz, M. Guizar-Sicairos, A. Mantion, C. M. Kewish, A. Menzel, O. Bunk, and C. David, *Opt. Express* **19**, 21333 (2011).
17. P. Kirkpatrick and A. Baez, *J. Opt. Soc. Am.* **38**, 766 (1948).
18. H. Mimura, H. Yumoto, S. Matsuyama, Y. Sano, K. Yamamura, Y. Mori, M. Yabashi et al., *Appl. Phys. Lett.* **90**, 051903 (2007).
19. S. Lagomarsino, A. Cedola, P. Cloetens, S. Di Fonzo, W. Jark, G. Soullié, and C. Riekel, *Appl. Phys. Lett.* **71**, 2557 (1997).
20. W. Jark, A. Cedola, S. Di Fonzo, M. Fiordelisi, S. Lagomarsino, N. V. Kovalenko, and V. A. Chernov, *Appl. Phys. Lett.* **78**, 1192 (2001).
21. C. Bergemann, H. Keymeulen, and J. F. van der Veen, *Phys. Rev. Lett.* **91**, 204801 (2003).
22. F. Pfeiffer, C. David, M. Burghammer, C. Riekel, and T. Salditt, *Science* **297**, 230 (2002).
23. A. Jarre, C. Fuhse, C. Ollinger, J. Seeger, R. Tucoulou, and T. Salditt, *Phys. Rev. Lett.* **94**, 074801 (2005).
24. S. P. Krueger, K. Giewekemeyer, S. Kalbfleisch, M. Bartels, H. Neubauer, and T. Salditt, *Opt. Express* **18**, 2010 (2010).
25. S. P. Krueger, H. Neubauer, M. Bartels, S. Kalbfleisch, K. Giewekemeyer, P. J. Wilbrandt, M. Sprung, and T. Salditt, *J. Synchrotron Rad.* **19**, 227 (2012).
26. W. C. Röntgen, Ueber eine neue Art von Strahlen, Sitzungsberichte der physikal.-medizin. Gesellschaft, 132, 1895.
27. S. Suehiro, H. Miyaji, and H. Hayashi, *Nature* (London) **352**, 385 (1991).
28. A. Michette, *Nature* (London) **353**, 510 (1991).
29. A. Snigirev, V. Kohn, I. Snigireva, and B. Lengeler, *Nature* (London) **384**, 49 (1996).
30. B. Lengeler, J. Tümmler, A. Snigirev, I. Snigireva, and C. Raven, *J. Appl. Phys.* **84**, 5855 (1998).
31. B. Lengeler, C. G. Schroer, M. Richwin, J. Tümmler, M. Drakopoulos, A. Snigirev, and I. Snigireva, *Appl. Phys. Lett.* **74**, 3924 (1999).
32. B. Lengeler, C. Schroer, J. Tümmler, B. Benner, M. Richwin, A. Snigirev, I. Snigireva, and M. Drakopoulos, *J. Synchrotron Rad.* **6**, 1153 (1999).
33. C. G. Schroer, M. Kuhlmann, B. Lengeler, T. F. Günzler, O. Kurapova, B. Benner, C. Rau, A. S. Simionovici, A. Snigirev, and I. Snigireva, Beryllium parabolic refractive x-ray lenses, in *Design and Microfabrication of Novel X-Ray Optics*, Vol. 4783 of *Proceedings of the SPIE*, D. C. Mancini (ed.) (SPIE, Bellingham, WA, 2002), pp. 10–18.
34. B. Lengeler, C. G. Schroer, M. Kuhlmann, B. Benner, T. F. Günzler, O. Kurapova, F. Zontone, A. Snigirev, and I. Snigireva, *J. Phys. D: Appl. Phys.* **38**, A218 (2005).
35. C. G. Schroer, M. Kuhlmann, U. T. Hunger, T. F. Günzler, O. Kurapova, S. Feste, F. Frehse et al., *Appl. Phys. Lett.* **82**, 1485 (2003).
36. C. G. Schroer, O. Kurapova, J. Patommel, P. Boye, J. Feldkamp, B. Lengeler, M. Burghammer et al., *Appl. Phys. Lett.* **87**, 124103 (2005).
37. C. G. Schroer and B. Lengeler, *Phys. Rev. Lett.* **94**, 054802 (2005).
38. C. G. Schroer, A. Schropp, P. Boye, R. Hoppe, J. Patommel, S. Hönig, D. Samberg et al., Hard x-ray scanning microscopy with coherent diffraction contrast, in *The 10th International Conference on X-Ray Microscopy*, Vol. 1365 of *AIP Conference Proceedings*, I. McNulty, C. Eyberger, and B. Lai (eds.) (AIP, Melville, New York, 2011), pp. 227–230.
39. S. Braun and H. Mai, Multilayer and single-surface reflectors for x-ray optics, in *Metal-Based Thin Films for Electronics*, K. Wetzig and C. M. Schneider (eds.) (Wiley-VCH, Weinheim, Germany, 2006).
40. J. Underwood and T. Barbee Jr., *Appl. Opt.* **20**, 3027 (1981).
41. M. Schuster and H. Göbel, *J. Phys. D: Appl. Phys.* **28**, A270 (1995).
42. O. Hignette, P. Cloetens, G. Rostaing, P. Bernard, and C. Morawe, *Rev. Sci. Instrum.* **76**, 063709 (2005).
43. C. Morawe, O. Hignette, P. Cloetens, W. Ludwig, C. Borel, P. Bernard, and A. Rommeveaux, *Proc. SPIE* **6317**, 63170F (2006).
44. D. Attwood, *Soft X-Rays and Extreme Ultraviolet Radiation* (Cambridge University Press, Cambridge, U.K., 1999).
45. W. Chao, B. D. Harteneck, J. A. Liddle, E. H. Anderson, and D. T. Attwood, *Nature* **435**, 1210 (2005).
46. J. Vila-Comamala, K. Jefimovs, J. Raabe, T. Pilvi, R. H. Fink, M. Senoner, A. Maassdorf, M. Ritala, and C. David, *Ultramicroscopy* **109**, 1360 (2009).
47. J. Maser, Theoretical description of the diffraction properties of zone plates with small outermost zone width, in *X-Ray Microscopy IV*, V. V. Aristov and A. I. Erko (eds.) (Institute of Microelectronics Technology, Chernogolovka, Russia, 1994), pp. 523–530.
48. C. G. Schroer, *Phys. Rev. B* **74**, 033405 (2006).
49. F. Pfeiffer, C. David, J. F. van der Veen, and C. Bergemann, *Phys. Rev. B* **73**, 245331 (2006).
50. H. Yan, J. Maser, A. Macrander, Q. Shen, S. Vogt, G. B. Stephenson, and H. C. Kang, *Phys. Rev. B* **76**, 115438 (2007).
51. C. Liu, R. Conley, A. T. Macrander, J. Maser, H. C. Kang, M. A. Zurbuchen, and G. B. Stephenson, *J. Appl. Phys.* **98**, 113519 (2005).
52. H. C. Kang, J. Maser, G. B. Stephenson, C. Liu, R. Conley, A. T. Macrander, and S. Vogt, *Phys. Rev. Lett.* **96**, 127401 (2006).
53. T. Koyama, S. Ichimaru, T. Tsuji, H. Takano, Y. Kagoshima, T. Ohchi, and H. Takenaka, *Appl. Phys. Express* **1**, 117003 (2008).
54. H. Yan, V. Rose, D. Shu, E. Lima, H. C. Kang, R. Conley, C. Liu et al., *Opt. Express* **19**, 15069 (2011).
55. A. Schropp and C. G. Schroer, *New J. Phys.* **12**, 035016 (2010).

III

Picoscale Imaging

8

Imaging Small Molecules by Scanning Probe Microscopy

Shirley Chiang
University of California, Davis

8.1 Introduction

The invention of the scanning tunneling microscope (STM) by Gerd Binnig and Heinrich Rohrer in 1981 at the IBM Zurich Research Laboratory led to a renaissance in the study of surfaces, because it allowed the measurement of real-space images of flat surfaces on an atomic scale.[1–4] Binnig and Rohrer won half of the Nobel Prize in Physics in 1986 for this marvelous invention. The microscope is based on the principle of quantum mechanical tunneling of current between a sharp tip and a conducting or semiconducting surface, and the exponential decrease of current with the increase in the separation distance is responsible for the high resolution, as described in further detail later. The tunneling tip is scanned over the surface in an x–y raster pattern using a piezoelectric scanner. Since the tunneling process in the STM always involves tunneling between electronic states of the tip and the sample, high-resolution imaging of molecules tends to show the images of hybridized molecular orbitals and not the structure of individual atoms within a molecule. Binnig and his collaborators used the STM to produce pictures of biological material quite early, publishing some images of a bacteriophage and a virus.[5,6]

The author and her coworkers successfully made high-resolution images of benzene adsorbed on the Rh(111) surface, which showed internal structure related to the molecular orbitals of the molecule.[7] Since the separation of benzene molecules in the (3×3) unit cell on the Rh(111) surface is 0.8 nm, the internal molecular features are clearly resolved on the pm scale. That work began the field of high-resolution images of molecules using STM. Since that time, STM and the related technique of atomic force microscopy (AFM), both of which were classified as scanning probe microscopes (SPM), have been used to image many different types of small molecules on surfaces. The STM has become a tool in every surface science laboratory, and many uses of the instrument relate to molecular imaging, with applications to studies of molecular adsorption and chemical reactions. The following section will give a brief tutorial about these SPM techniques, and then the chapter will show many applications of SPM to high-resolution imaging of molecules on surfaces. Since this field has become very large, with thousands of references, this chapter cannot be a comprehensive review but will show some particularly interesting images and applications.

8.2 Introduction to Techniques and Instrumentation

SPM refers to all types of microscopes which use a probe tip to investigate the surface. After the invention of the STM, Binnig, Quate, and Gerber invented the atomic force microscope in 1986,[8] which is discussed in more detail later.

8.2.1 Scanning Tunneling Microscopy

Figure 8.1 shows a schematic diagram of an STM. The separation distance between the tip and sample must be in the range of 0.5 nm in order to have a quantum mechanical tunneling current of ~1 nA between the tip and sample for a voltage bias of ~1 V. The tip is mounted onto a single tube piezoelectric scanner,[9] which allows the motion of the tip in three dimensions. The tube has four electrodes on the outside, which control the motion of the tip in the lateral plane by bending the tube sideways, and the z-voltage which controls the tip–sample separation is applied to the inner electrode of the tube. Because the maximum motion of the tip using the piezoelectric scanner is on the order of 1 μm, another positioning device is needed to move the sample until the z-range of the scanner can put the tip into tunneling range. Thus, the sample is mounted onto a coarse positioning device, which can be a piezoelectric or mechanical translator. This device is used for the approach process, which moves the sample toward the tip with a step size less than the z-range of the scanner. Then the sample can be moved one step toward the tip, and the tip scanner can be extended to try to find a tunneling current. If there is no current, the electronics direct the sample to walk an additional step, and the process repeats until a tunneling current is sensed.

The tunneling current depends exponentially on the tip–sample distance. For two infinite flat parallel electrodes, the tunneling current would have the following dependence:

$$I = V \exp(-A\sqrt{\Phi}\, s),$$

where

s is the tip–sample distance
V is the voltage
A is a constant
Φ is the average work function of the tip and sample

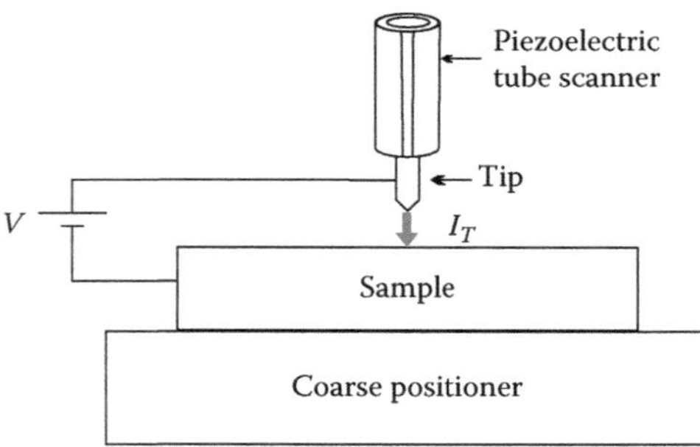

FIGURE 8.1 Schematic diagram showing operation of an STM. The tip is mounted onto a piezoelectric tube scanner, with four electrodes around the outside so that the tube bends when voltages are applied to the outside, producing a lateral scan. The z-voltage is applied to the inside of the tube, causing a lengthwise extension or contraction. A bias voltage V is applied between the tip and the sample, causing a quantum mechanical tunneling current I_T (gray arrow) to flow from the tip to the sample, when the tip–sample distance is sufficiently small. The sample is mounted onto a coarse positioner, which acts as both a z-approach mechanism and a large lateral translation mechanism to move the tip to a different area of the sample.

An electronic feedback circuit is used to detect the tunneling current. As the tip is scanned laterally over the surface, the feedback electronics maintain the current as a constant by changing the z-piezoelectric voltage, thus keeping the tip–sample distance constant. A computer can then be used to plot the z-voltage as a function of the x and y scanning voltages, making a three-dimensional (3D) topographic map of the surface. This method of operating the STM is called the constant current mode.

For sufficiently flat samples, it is also possible to operate the STM in the constant height mode, in which the x–y scan is faster than the feedback loop can follow, and the tunneling current is then roughly proportional to the height of the surface corrugation.

In the previous equation, all electronic effects are incorporated into the work function term. Tersoff and Hamann[10,11] applied the simple perturbation theory to the operation of the STM and found that the tunneling current is proportional to the local density of states (LDOS) of the sample at the position of the tip. Thus, the STM often measures the atomic positions for metals,[12] but it measures dangling bonds and filled or empty states on semiconductors. For GaAs, the STM measures the position of Ga atoms when positive sample bias is used to image unoccupied sample states, but the images show the position of As atoms when the opposite bias is used.[13] For molecules on metals, the STM image will show molecular orbitals hybridized with electronic states of the metal substrate.[14]

The typical lateral resolution of an STM is 200 pm, on the order of the atomic spacing in solids, and the vertical resolution is <10 pm. In order to maintain high resolution in an STM, the tip-to-sample distance must be very stable during the measurement. Thus, an STM will usually have some method of vibration isolation, so that environmental vibrations do not disturb the measurement. The simplest method of vibration isolation, developed by Binnig and Rohrer in the first STMs, is to mount the STM stage onto springs and to damp the vibrations of the springs using eddy current damping.

8.2.2 Atomic Force Microscopy

The AFM uses a probe tip mounted onto a cantilever. As a result of the force interaction between the tip and surface, the end of the cantilever deflects. The deflection is measured and used to determine the amount of the interaction force. Various methods can be used to measure the deflection of the cantilever. The first AFM had a diamond tip mounted onto a cantilever, with an STM behind the cantilever to sense the deflection,[8] but this method could be prone to difficulties because there are two junctions in the instrument which involve tips. Other methods for sensing the deflection include capacitance[15] and optical interferometry.[16]

Figure 8.2 shows a schematic diagram of an AFM which uses the most common method of sensing the lever deflection, the detection of a reflected laser beam from the back of the cantilever. This reflected beam falls on a position-sensitive detector, which is a photodiode with either two or four active sections. A differential amplifier can then be used to measure the difference in signal between sections when the beam on the detector

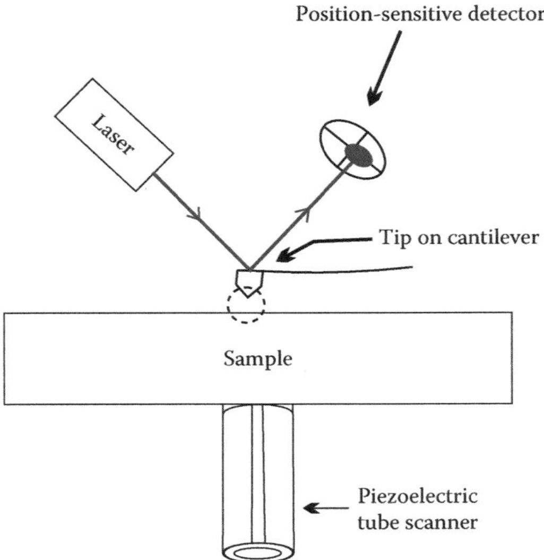

FIGURE 8.2 Schematic diagram showing operation of an AFM. The tip is mounted onto a cantilever, which is positioned near the sample. Interaction forces between the tip and the sample that occur within the dashed circle are sensed by measuring the deflection of the reflected laser beam using the position-sensitive detector. The sample is mounted onto a piezoelectric tube scanner to allow its lateral position under the tip to change.

shifts due to the deflection of the cantilever. Usually, the sample is mounted onto a piezoelectric scanner so that it can be moved under the tip. An electronic feedback loop is used to maintain one parameter, such as force or force derivative, as a constant while the scanning is performed.

The tip can be pushed against the surface, so that it is effectively dragged across the surface while in contact. This is the repulsive mode of operation, which can cause damage to the sample from the dragging of the tip. In a noncontact mode of operation, the weak van der Waals force can be kept constant, resulting in a larger tip–sample distance during the scanning of the sample under the tip. If the cantilever is vibrated with an amplitude less than 10 nm near its resonance frequency, the small change in frequency due to the force interaction between the tip and sample can be sensed, and the force derivative can be kept constant during scanning. Finally, a tapping mode was developed to circumvent the problem of the tip sticking to the surface due to a liquid meniscus layer in ambient conditions. For this mode of operation, the cantilever is oscillated with large amplitude, typically 100–200 nm, near its resonance frequency. As the tip approaches the surface, the interaction force can be detected through the change in the resonance frequency, and the image consists of measurements of the force during intermittent contacts with the sample.

The interaction between the tip and sample can be sensitive to various types of forces. Unlike STM, the sample in an AFM does not need to be electrically conducting. For example, with a magnetic tip and a magnetic sample, the instrument becomes a magnetic force microscope. If it is used to sense electric forces, it is an electrostatic force microscope.

8.3 Molecules Adsorbed on Metals

The author has written three earlier reviews on STM imaging of molecules[17–19] and a recent historical perspective.[20]

8.3.1 Small Aromatic Molecules Adsorbed on Metals

8.3.1.1 Benzene Adsorbed on Metal Surfaces

Mate and Somorjai had found that CO induced the ordering of adsorbed benzene molecules on Rh(111), resulting in a strongly chemisorbed layer,[21] with two different ordered overlayers, a (3×3) hexagonal structure and a $c(2\sqrt{3} \times 4)$ rect one. Therefore, we used STM to measure benzene and CO coadsorbed together on Rh(111).[7] Our STM images of the 3×3 overlayer of benzene and CO on Rh(111) clearly show individual molecules of benzene as ringlike structures (Figure 8.3). As the first images

(a)

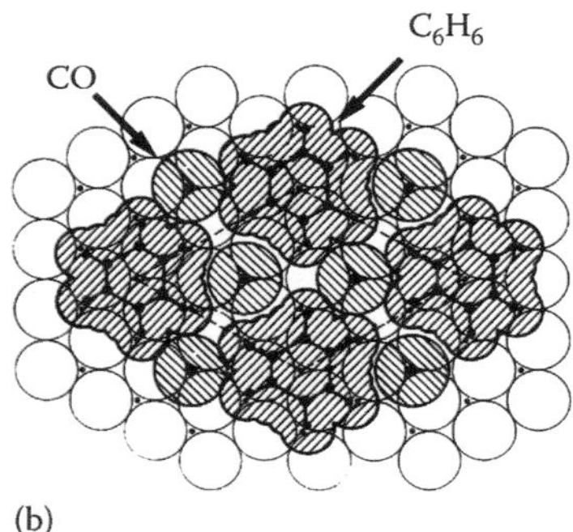

(b)

FIGURE 8.3 Benzene coadsorbed with CO on Rh(111). (a) 3D view of the STM image of the 3×3 overlayer, with ringlike benzene molecules. (b) 3×3 structure determined by LEED.[23] Large circles and small dots represent first- and second-layer metal atoms, respectively. (Reprinted with permission from Ohtani, H., Wilson, R.J., Chiang, S., and Mate, C.M., *Physical Review Letters*, 60, 2398–2401, Copyright 1988 by the American Physical Society.)

which showed the structure of a molecule, these attracted much attention, including from the popular press, even though the chemists had known from Kekulé's work over a century earlier that benzene had a ring structure,[22] after he had reported having a dream of a snake chasing its tail. Not surprisingly, the rings in the STM image are not completely round, as a result of the chemical bonding of the molecule to the metal surface resulting in hybridized molecular orbitals with a somewhat triangular shape. The structural model from low-energy electron diffraction (LEED) (Figure 8.3b)[23] shows one benzene molecule and two CO molecules per unit cell, with the COs bonded vertically in threefold hollow sites with the carbon atom nearer to the surface. The STM image only shows a bump at the location of one of the CO sites, because the other site is occluded by a part of the benzene ring. Although the LEED model suggested very small atomic distortions in the benzene ring, STM data are sensitive to electronic structure and not atomic position, and thus could not confirm these.

The LEED model for the c($2\sqrt{3}\times 4$) rect overlayer of benzene and CO on Rh(111) has one benzene and one CO molecule per unit cell (Figure 8.4c).[24] For this overlayer, the high-resolution STM images again show the benzene molecules as triangular rings, and, in addition, one protrusion in the unit cell appears at the site expected for CO (Figure 8.4a and b).[25] Because this molecular arrangement is rectangular on a hexagonal lattice, three different rotational domains can occur on the surface. All three were observed in the images, together with domain boundaries and the arrangement of molecules near the step edges.[26]

Several years later, we measured STM images of benzene and CO coadsorbed on Pd(111) at low temperature[27] and found many similarities to the earlier work on adsorption of these molecules on Rh(111).[7,26] If CO is deposited first onto the Pd(111) surface, it acts as an ordering agent, and two different ordered structures can form varying benzene coverage when CO and benzene are subsequently dosed onto the surface simultaneously. The ($2\sqrt{3}\times 2\sqrt{3}$)R30° LEED pattern is observed for low benzene coverage, followed by a 3×3 pattern for higher benzene coverage. Individual benzene molecules are clearly resolved in the STM image of the ($2\sqrt{3}\times 2\sqrt{3}$)R30° overlayer in Figure 8.5a. On the other hand, a rectangular ordered structure is formed if the Pd(111) surface is dosed first with benzene, followed by dosing with Co and benzene together. Three different types of rectangular domains are clearly observed in Figure 8.5b.[27]

Weiss and Eigler used a low-temperature STM at 4 K to measure images of benzene on Pt(111).[28] They observed three different shapes for the molecules: a three-lobed structure, a volcano with a dip in the center, and a simple bump (Figure 8.6). By comparing with the calculated images of Sautet and Bocquet,[29] these different images were ascribed to the different binding sites of the molecules to the surface, that is, hcp threefold hollow site, on-top site,

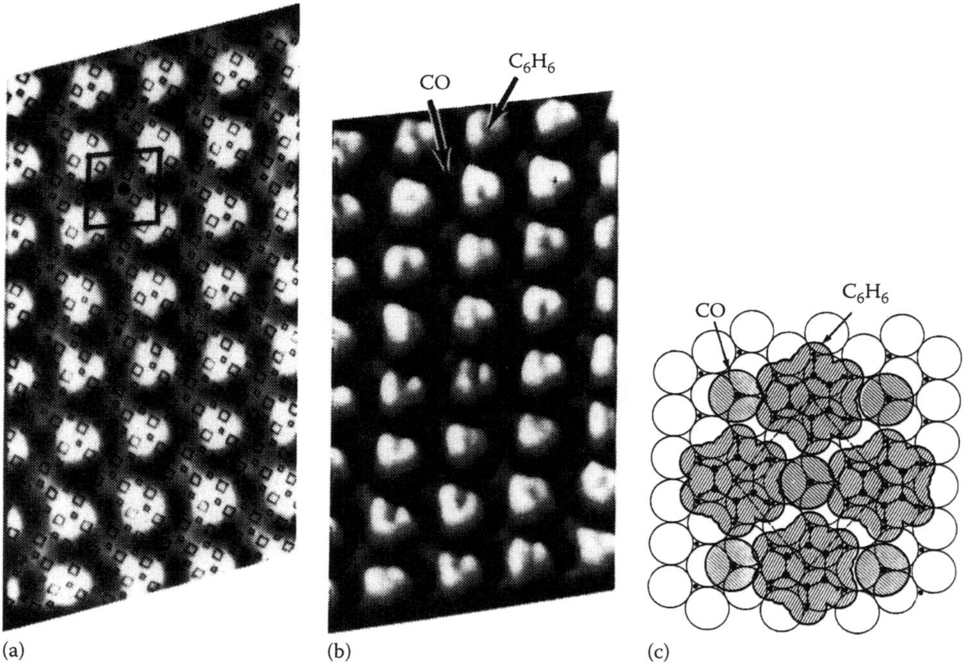

FIGURE 8.4 (a) Top view of STM image of the c($2\sqrt{3}\times 4$) rect overlayer on benzene and CO coadsorbed on Rh(111), with brightness proportional to the height of the tip above the surface. The mesh, with large (small) diamonds indicating top (second) layer Rh atoms, was overlaid on the data according to the LEED model.[24] Solid lines show the primitive unit cell, with benzenes at corners and CO molecule in the center. (b) 3D view of the data shown in (a), with threefold benzene features ~0.6 Å high and smaller CO protrusions ~0.2 Å high. (c) c($2\sqrt{3}\times 4$) LEED model, with the same symbols as in Figure 8.3b. (From Chiang, S., Wilson, R.J., Mate, C.M., and Ohtani, H.: Real space imaging of co-adsorbed CO and benzene molecules on Rh(111). *Journal of Microscopy-Oxford*. 1988. 152. 567–571. Copyright Wiley-VCH Verlag GmbH & Co. KGaA. Reproduced with permission.)

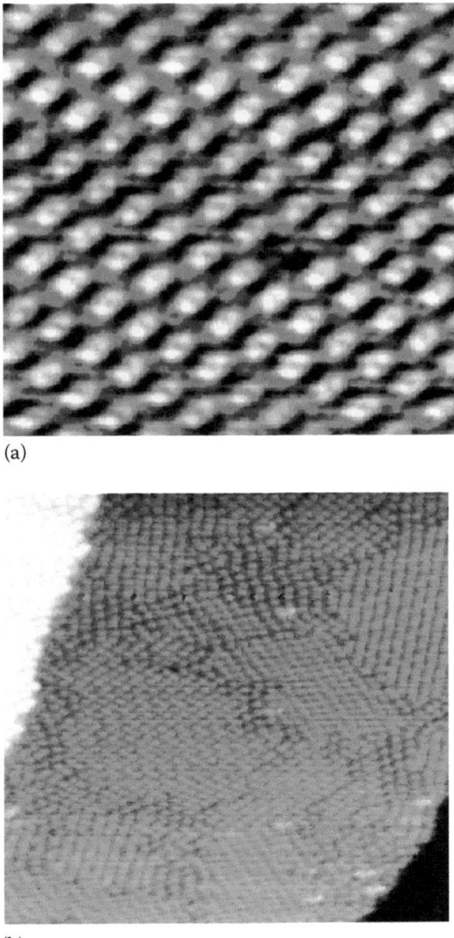

FIGURE 8.5 (a) STM image of the $(2\sqrt{3}\times 2\sqrt{3})R30°$ structure of CO and benzene coadsorbed on Pd(111), measured at 95 K. The ordered arrangement and internal ringlike structure of the benzene molecules are clearly evident in the image. 75 Å × 75 Å, $V_S = 0.1$ V, $I_T = 0.7$ nA. (b) STM image of the three-domain rectangular structure of CO and benzene coadsorbed on Pd (111), measured at 100 K. The benzene molecules are tightly arranged within each domain, while the domain boundaries display disorder. 250 Å × 250 Å, $V_S = 0.1$ V, $I_T = 0.5$ nA. (a: Reprinted with permission from Chiang, S., Imaging atoms and molecules on surfaces by scanning tunnelling microscopy, *Journal of Physics D-Applied Physics*, 44, 464001, Copyright 2011 American Institute of Physics; b: Reprinted from *Materials Science and Engineering B*, 96, Pearson, C., Anderson, G.W., Chiang, S., Hallmark, V.M., and Melior, B.J., A low temperature scanning tunneling microscope designed for imaging molecular adsorbates, 209–214, Copyright 2002 with permission from Elsevier.)

and bridge site, respectively. Sautet and Joachim developed their electron scattering quantum chemistry theory to simulate STM images of molecules and applied it first to benzene on Rh(111),[30] obtaining very good agreement with our experimental data.[7,25]

Other studies have shown STM images of adsorbed benzene at low temperature on Cu(111)[31,32] and on Ag(110).[33] We also used simple Hückel molecular orbital theory to calculate predicted STM images for benzene on Pt(111) and on Pd(111).[34]

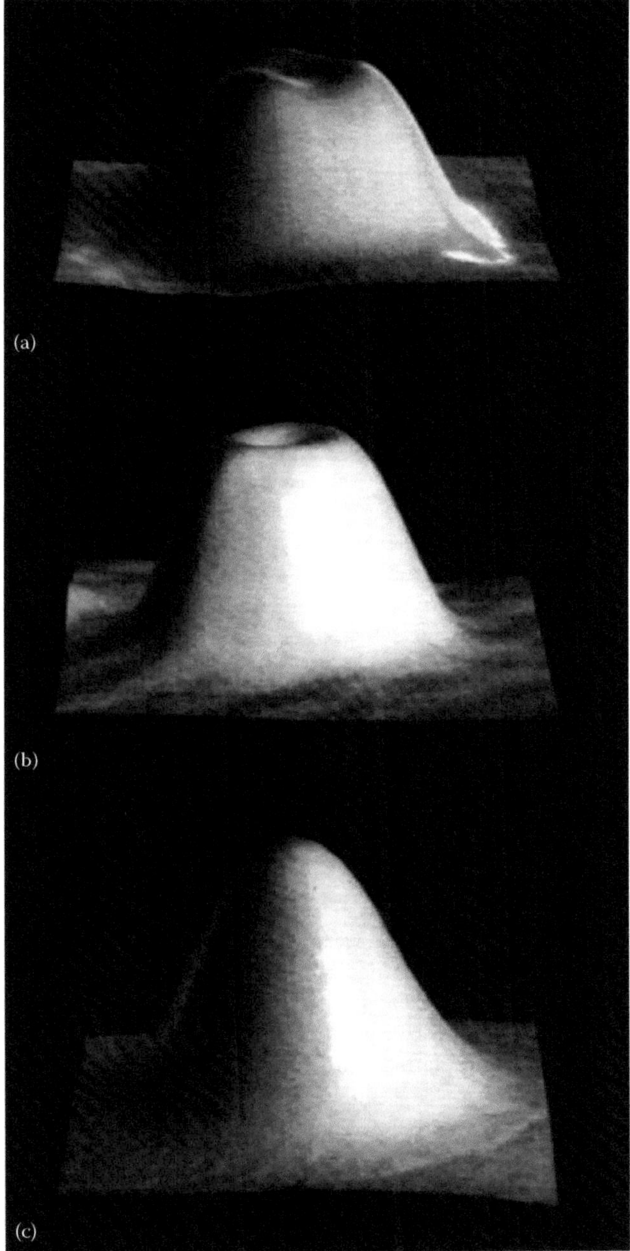

FIGURE 8.6 STM image of three different 15 Å × 15 Å regions, each showing a single adsorbed benzene molecule on Pt(111). These images have been assigned to (a) hcp threefold hollow site; (b) on-top site; and (c) bridge site; respectively. The images were recorded with (a) $V_{bias} = -0.050$ V, $I_T = 100$ pA; (b) $V_{bias} = -0.010$ V, $I_T = 1$ nA; (c) $V_{bias} = -0.010$ V, $I_T = 100$ pA. The minimum to maximum height differences in the images are (a) 0.58 Å, (b) 0.72 Å, and (c) 0.91 Å, respectively. The observed images of the individual molecules did not change qualitatively for a wide range of tunneling parameters. (Reprinted with permission from Weiss, P.S. and Eigler, D.M., Site dependence of the apparent shape of a molecule in scanning tunneling microscope images—Benzene on Pt(111), *Physical Review Letters*, 71, 3139–3142, Copyright 1993 American Physical Society.)

8.3.1.2 Naphthalene, Azulene, and Methylazulenes on Pt(111)

The study of a series of related aromatic molecules on Pt(111) elucidated many details about molecular imaging with STM. Naphthalene consists of two adjoining benzene rings. LEED studies had previously found that annealing produced an ordered molecular layer characterized by a (6×3) pattern with a glide plane, leading to a proposed herringbone arrangement of the molecules.[35,36] Examination of the real-space STM images showed that the molecules were organized in quasi-3×3 arrays, with about 40% of the molecules satisfying the required glide plane symmetry.[37] The naphthalene molecules typically appeared as double-lobed or oblong structures in three rotational orientations (Figure 8.7d). In addition, we examined regions of both ordered and disordered naphthalene on Pt(111).[38]

By simultaneously imaging two different molecules observed onto the Pt(111) surface with the same tunneling tip, we could isolate the effects of the tip from other causes of differences in resolution between molecular images. We imaged naphthalene and its isomer azulene, which consists of a seven-membered ring connected to a five-membered ring.[39] In STM images, azulene usually appeared as a round disk without other distinguishing features (Figure 8.7h) and sticks far less well to the surface than naphthalene, with the sticking coefficient down by a factor of 4. With a particularly good tip, the naphthalene molecules would occasionally appear as double rings (Figure 8.7e), while azulene molecules would appear as single rings (Figure 8.7i).[40] We also saw evidence for molecular motion at room temperature.

We then did a more systematic study of STM imaging of related molecules on Pt(111). In addition to naphthalene and azulene, we also examined a series of azulene isomers with substituted methyl groups, which we obtained from collaborators Meinhardt and Hafner at Darmstadt. In addition to three isomers of monomethyl azulene, we used the STM to image dimethyl and trimethyl azulene. The shapes of the monomethyl azulene isomers were noticeably different, with 1-methyl azulene having a kidney-bean shape and 2-methyl azulene having the shape of a pear (Figure 8.7p).[40] The highest resolution images were obtained for 6-methyl azulene, for which some internal molecular features were resolved (Figure 8.7q).[41,42] Trimethyl azulene usually had the shape of a clover leaf with one bright lobe (Figure 8.7v), while dimethyl azulene appeared as a distorted oblong with a bright spot (Figure 8.7w). Imaging empty states of the sample usually yielded higher resolution images. The STM data were also used to study molecular resolution effects and to determine molecular adsorption parameters.[41]

8.3.1.3 Calculations of Predicted STM Images of Molecules on Metal Surfaces

We developed our simple Hückel calculations to simulate STM images for molecules on metal surfaces. Our method assumed that the Tersoff and Hamann model for theory of STM[10,11] was sufficient and calculated only the LDOS of the surface, using a molecule plus a cluster of metal atoms. These calculations therefore had no tip and no tunneling process. The calculations of the lowest unoccupied molecular orbital (LUMO) or highest occupied molecular orbital (HOMO) for isolated molecules did not fit the observed STM images as well as simulations which included a cluster of metal atoms. This behavior was particularly conspicuous for simulated images of the monomethyl azulenes, for which the calculated images for the three isomers looked very similar for isolated molecules.[42] Once the metal cluster was added, the calculated images for the three isomers appeared different. To compare with low-resolution data, we computed the LDOS at a height of 2 Å above the molecule, while we computed the LDOS at a height of 0.5 Å above the molecule to compare with higher resolution data. This method worked quite well, allowing the comparison of the calculated low-resolution image (Figure 8.7b) with double-lobed experimental images, while the high-computed image (Figure 8.7c) could be compared with the experimental image of double rings.

A more systematic study of the parameters for the Hückel calculations considered the effects of different adsorption geometries and different sizes, shapes, and number of layers for the metal clusters.[14] In addition, to simulate the effects of the bias voltage, we considered the effect of different numbers of either empty or filled molecular orbitals.

8.3.1.4 Xylene on Pd(111) and Rh(111)

Cernota et al.[43] observed the molecular ordering and orientations of para-xylene (benzene with two substituted methyl groups) and meta-xylene on Rh(111). Our calculated images from Hückel's theory[44] agreed well with their experimental observations of molecular shapes.[43]

Our group studied the adsorption of xylene on Pd(111), both experimentally and theoretically. We found that the images of meta-xylene and para-xylene on Pd(111) have very different molecular shapes. Meta-xylene appears somewhat triangular and prefers to adsorb along step edges,[45] while para-xylene forms more ordered arrays on the surface. On the other hand, the shape of adsorbed para-xylene is asymmetric (Figure 8.8b). We compared the observed molecular shape with the calculated images from Hückel's theory for a molecule on a Pd cluster. The experimental shape appears similar to the simulated image which has an angle of 15° between the molecular axis and the overlayer close-packed direction (Figure 8.8a).[46] Experimentally, the angle between the close-packed direction of the overlayer structure and the orientation of the para-xylene molecules was measured to be $15° \pm 5.2°$, which is consistent with the energy calculations; this suggests that the molecules adsorb on the hollow sites at angles between 15° and 30° with respect to the substrate close-packed direction.

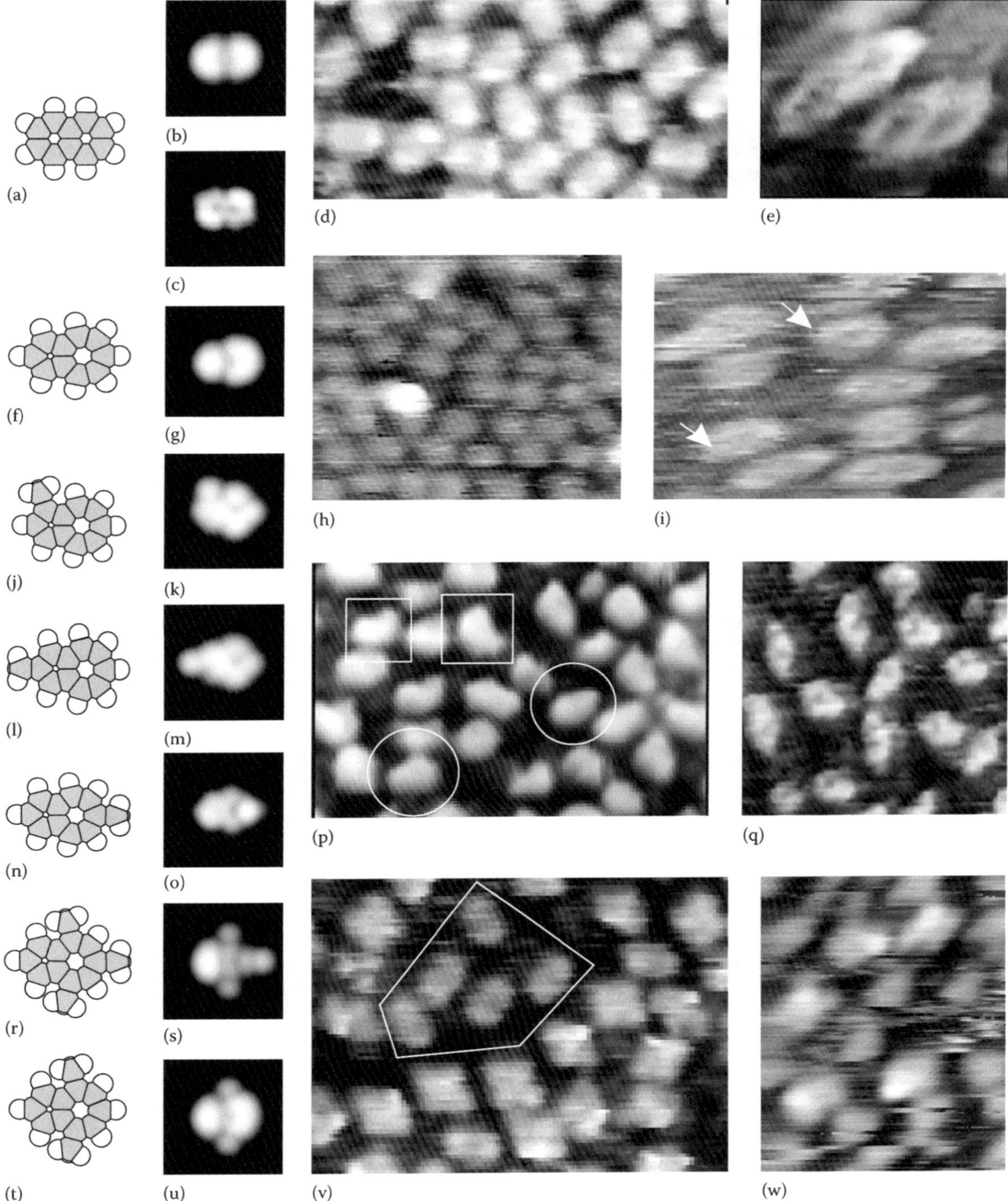

FIGURE 8.7 Predicted and measured STM images for a series of related molecules on Pt(111). (a) Naphthalene diagram. (b) Naphthalene ρ_H (LDOS of unoccupied sample states) at 2 Å above the molecule. (c) Naphthalene ρ_H at 0.5 Å. (d) Low-resolution image of array of naphthalene molecules. Molecular van der Waals length is 8.1 Å. (e) High-resolution STM image of naphthalene. (f) Azulene model. (g) Azulene ρ_H at 2 Å. (h) Low-resolution azulene data near 1 monolayer (ML) coverage. Molecular spacing within a (3×3) domain is 8.3 Å. (i) High-resolution STM image of mixed molecular layer, with two azulenes (marked by arrows) among naphthalene neighbors. (j) 1-methylazulene (1-MA) model. (k) 1-MA ρ_H at 2 Å (l) 2-MA model. (m) 2-MA ρ_H at 2 Å. (n) 6-MA model. (o) 6-MA ρ_H at 0.5 Å. (p) Low-resolution image of mixed 1-MA, ~20% coverage (in squares) and 2-MA (in circles) near saturation coverage. (q) High-resolution image of 6-MA. (r) 4,6,8-Trimethylazulene (TMA) model. (s) TMA ρ_H at 2 Å. (t) 4,8-Dimethylazulene (DMA). (u) DMA ρ_H at 2 Å. (v) STM image for mixed TMA and naphthalene (inside marked area) layer. (w) STM image of DMA molecules at less than full coverage; noisy areas indicate molecules which moved during the scan duration. (Reprinted with permission from Hallmark, V.M., Chiang, S., Meinhardt, K.P., and Hafner, K., *Physical Review Letters*, 70, 3740–3743, Copyright 1993 by the American Physical Society.)

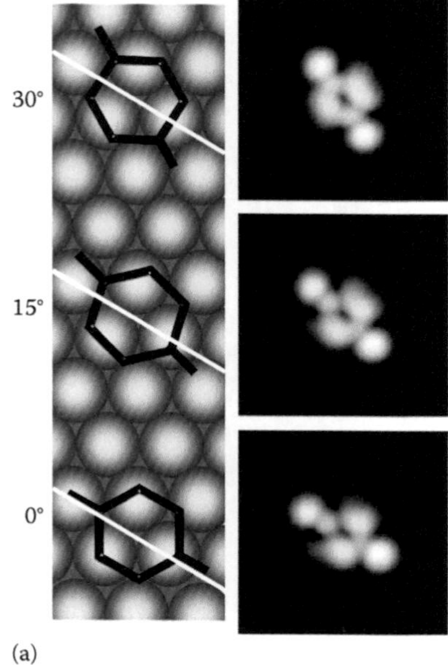

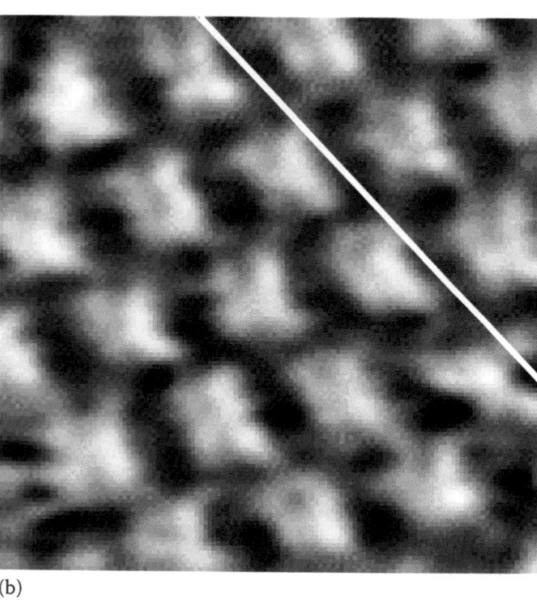

FIGURE 8.8 (a) Schematic drawings illustrating the relative molecular orientation for para-xylene on the surface and associated calculated images, for decreasing angular difference between the molecular axis and the overlayer close-packed direction denoted in white. As the angular difference decreases, the image displays an increasingly symmetric profile. (b) STM image of para-xylene molecules, with the overlayer close-packed direction denoted in white. (Reprinted with permission from Futaba, D.N., Landry, J.P., Loui, A., and Chiang, S., *Physical Review B*, 65, 045106, Copyright 2002 by the American Physical Society.)

8.3.1.5 Other Aromatic Molecules on Various Substrates

Numerous studies have used STM to measure the properties of other aromatic molecules on various substrates. These include thiophene, 2,5-diemthylthiophene, and 2,2′-bithiophene on Ag(111) at 120 K[47,48]; perylene-3,4,9,10-tetra-carboxylic-dianhydride (PTCDA) on graphite[49,50]; PTCDA and diimide (PTCDI) on graphite and MoS_2[50]; and naphthalene-1,4,5,8-tetracarboxylic-dianhydride (NTCDA) on graphite and MoS_2.[51] In addition, our group calculated predictions of STM images for thiophene on Pd(111)[34] and for furan and pyrrole on Pd(111).[52]

8.3.2 Porphyrins on Metals

The first STM images showing the internal structure of individual molecules adsorbed on a surface were obtained for copper phthalocyanine (Cu-phth) on Cu(100).[53] Figure 8.9a shows a model of the molecule on the rotated copper lattice. The Cu-phth molecules adsorb onto the surface in a flat orientation with two different rotational orientations. The internal

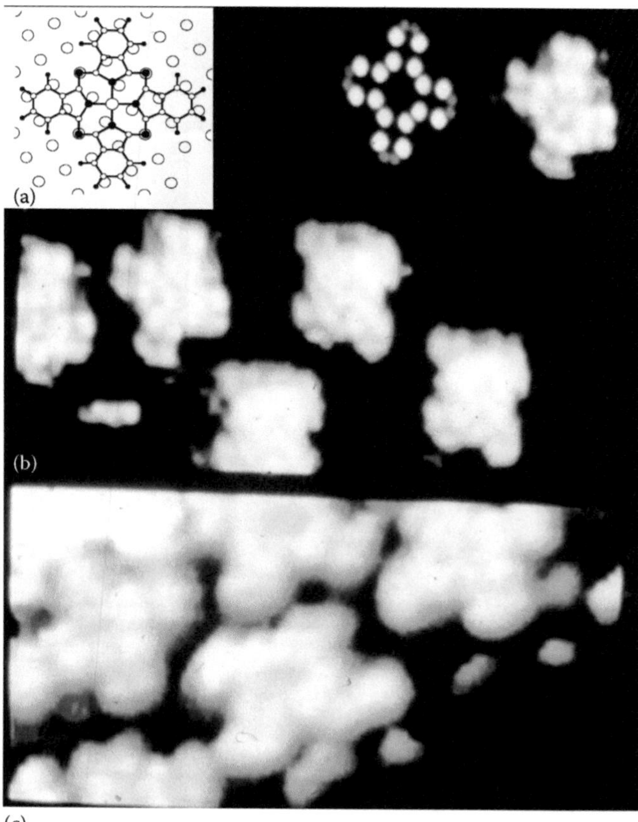

FIGURE 8.9 (a) Model of the Cu-phthalocyanine molecule above a Cu(100) surface. Small (large) open circles are C (Cu) atoms and small (large) filled circles are H (N) atoms. The Cu(100) lattice is shown rotated by 26.5°. (b) High-resolution STM image of Cu-phthalocyanine molecules on Cu(100) at submonolayer coverage, $V_T = -0.15$ V, $I_T = 2$ nA. Fine structure has been emphasized by baseline subtraction, and a grayscale representation of the HOMO, evaluated 2 Å above the molecular plane, has been embedded in the image. (c) High-resolution STM image near 1 ML coverage with $V_T = -0.07$ V, $I_T = 6$ nA. (Reprinted with permission from Lippel, P.H., Wilson, R.J., Miller, M.D., Wöll, C., and Chiang, S., *Physical Review Letters*, 62, 171–174, Copyright 1989 by the American Physical Society.)

structure observed in the STM images shows strong resemblance to charge density contours at 2 Å above the molecular plane for the HOMO and LUMO, which were both calculated using simple Hückel molecular orbital theory for an isolated molecule. Figure 8.9b shows the HOMO embedded into a high-resolution STM image, which shows the internal structure of several individual molecules at submonolayer coverage. In this image, the two rotational orientations of the molecule have different internal structure, perhaps due to strong tip asymmetry. Figure 8.9c shows an additional high-resolution image near 1 monolayer (ML) coverage. For this coverage, other images show packed molecular arrays, usually with two rotational domains, consistent with the previous LEED data.[54] Occasionally, the tip induces motion of the molecules. Isolated molecules could sometimes be observed above the atomically resolved metal surface.

High-resolution STM images were obtained by Jung et al.[55] for the related molecule of Cu-tetra(3,5, di-tertiary-butyl-phenyl) porphyrin (Cu-TBP-prophyrin) on Cu(100) (Figure 8.10). This molecule has four di-tertiary butyl phenyl (DTP) substituents which act as legs for the molecule. The tunneling tip could then be used to "push" the molecule across the surface at room temperature. The legs of the molecule had strong enough surface interactions to prevent thermally activated diffusional motion but could flex and permit the controllable translation of the molecule when repulsion between the tip and sample was used to "push" the molecule. Calculations of the internal molecular mechanics demonstrated the role of flexure of the molecular legs during the translation process.

FIGURE 8.10 STM topograph of Cu-TBP-porphyrin on Cu(100), showing islands with a (√58 × √58) superstructure. Each individual molecule is imaged as four bright lobes corresponding to the four DTP substituents. $V_T \sim 2200$ mV and $I_T \sim 80$ pA. Image area is 21 nm × 21 nm. (From Jung, T.A., Schlittler, R.R., Gimzewski, J.K., Tang, H., and Joachim, C., Controlled room-temperature positioning of individual molecules: Molecular flexure and motion, *Science*, 271, 181–184, Copyright 1996. Reprinted with permission of American Association for the Advancement of Science.)

Other STM studies of Cu-phth have measured images of the molecule on numerous substrates, including GaAs(110),[56] Si(100) 2 × 1,[57] Si(100) and Si(111),[58] $SrTiO_3$ and Cu(111),[59,60] and highly oriented pyrolytic graphite (HOPG) and MoS.[61] Pb-phth has also been studied on MoS_2,[62] as has diphenylporphyrin on Ag(111).[63]

8.4 Molecules on Semiconductor Substrates

Many STM studies of molecular systems on semiconductors have involved the Si(100) surface. Hamers and Avouis studied the dissociative reaction of ammonia with Si(100) and investigated the electronic effects of the formation of Si–H bonds.[64–66] Maki Kawai's group studied the adsorption of ketones and the formation of molecular lines on Si(100).[67–69]

Figure 8.11 shows STM images from the adsorption of 1,5-cyclooctadiene (COD), a molecule for which the C=C groups are not coplanar, on Si(001).[70] Figure 8.11b shows the STM image of the arrangement of the molecules on the surface, and Figure 8.11c is a high-resolution image in which one can see the internal structure of the molecules. High-resolution images were also obtained for the adsorption of 1,3,5,7-cyclooctatetraene (COT), a conjugated, antiaromatic molecule with two sets of coplanar C=C groups on the Si(001) surface.[71]

8.5 Self-Assembled Molecular Monolayers

As the literature has over 700 papers on self-assembled monolayers measured by STM, the author highlights only one interesting example here. Cai and Bernasek studied the adsorption of octadecanol on HOPG.[72] Although this molecule is achiral, they found that the alkane chain undergoes distortion as it adsorbs onto the graphite surface, leading to the observed asymmetry in the resultant arrangement of the molecules on the surface (Figure 8.12). The system involves supramolecular assembly by hydrogen bonding. The adsorbate conformation on the surface is different from that of the molecule in solution. The asymmetric bending of the molecule due to the interaction with the substrate breaks the achiral symmetry of the substrate.

Mark describes an interesting STM study visualizing 2D chirality of adsorbed molecular layers directly,[73] with the chirality induced by the adsorption of numerous molecules onto Cu(110).

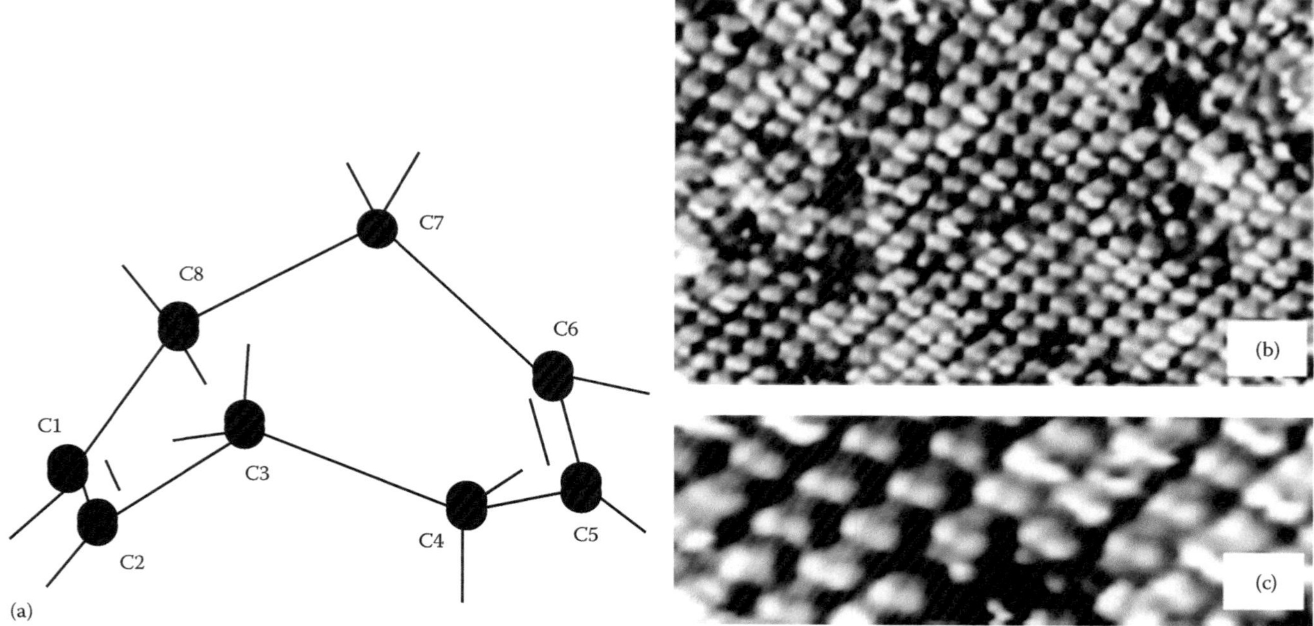

FIGURE 8.11 Adsorption of 1,5-cyclooctadiene on Si(001). (a) Schematic illustration of 1,5-cyclooctadiene (COD) molecule. (b) STM image of Si(001) after saturation exposure to COD, showing ordered molecular overlayer. (c) High-resolution STM image of Si(001) exposed to COD, showing internal structure of individual molecules. (Reprinted from *Surface Science*, 402, Hovis, J.S., Liu, H., and Hamers, R.J., Cycloaddition chemistry and formation of ordered organic monolayers on silicon (001) surfaces, 1–7, Copyright 1998 with permission from Elsevier.)

FIGURE 8.12 STM image of octadecanol SAM. Scan area 86.0 Å × 85.0 Å, V_{bias} = 1.17 V, I_T = 0.63 nA. Upper right shows model of octadecanol chevron structure overlaid on the STM image. The inset illustrates the asymmetric distortion in the chevron pair. (Reprinted with permission from Cai, Y.G. and Bernasek, S.L., Adsorption-induced asymmetric assembly from an achiral adsorbate, *Journal of the American Chemical Society*, 126, 14234–14238, Copyright 2004 American Chemical Society.)

8.6 Imaging Molecules at Low Temperature

8.6.1 Adsorption of CO on Metals at Low Temperature

Numerous studies of the adsorption of CO on metal surfaces at low temperature have been performed with the STM. The first images of CO on Pt(111) at 4 K were shown by Eigler and Coworkers.[74] They observed some molecules as a "bump" and others as a "sombrero." By comparing with theoretical calculations of STM images,[75] they subsequently assigned the bump state to CO bonded to an on-top site and the sombrero to a bridge site.

Meyer, Neu, and Rieder observed ordered structures of CO on Cu(221) at temperatures between 30 and 80 K.[76] The CO molecules usually appeared as bumps in high-coverage images and as dips for low coverage. They therefore inferred that the chemical character of the tip affected the observed images, with a metallic tip producing images of molecules as dips, and a tip with an adsorbed molecule producing images which appeared as bumps. They also observed several ordered structures for CO on Cu(221), including a (2 × 1) phase with ½ ML coverage, a (3 × 1) phase with 2/3 ML, and a (4 × 1) phase with ¾ ML. They were also able to move molecules laterally with the tunneling tip to form rows of molecules, in a process described in more detail in the next section.[77]

Salmeron and coworkers used STM to study adsorbed CO on both Pd(111)[78,79] and Rh(111)[80] at low temperature. For CO on Pd(111), they observed three different ordered structures,

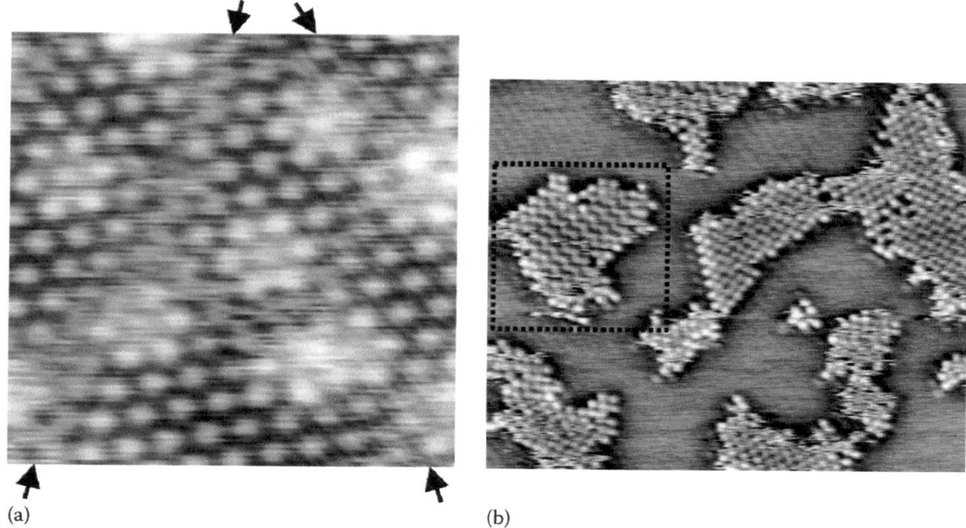

FIGURE 8.13 (a) STM image of $(\sqrt{3}\times\sqrt{3})R30°$ CO structure on the surface of a Pd(111) crystal with high subsurface defect density. The CO coverage is slightly below 1/3 ML. Hexagonal modulations of the CO corrugation are due to subsurface impurities located beneath surface fcc hollow sites. The phase of each CO domain is fixed by the local distribution of impurities. Several antiphase domain boundaries cut through the adlayer. The intersections of these domain boundaries with the edges of the image are indicated by arrows. 4.5 × 4.5 nm, 120 K, 0.20 nA, −110 mV sample bias. (b) STM image of c(4 × 2)-2CO islands at 120 K. Two types of c(4 × 2) ordering are observed. In one type, CO molecules occupy fcc and hcp threefold hollow sites, while in the other both CO molecules in the unit cell bind to the bridge sites. An island with the fcc–hcp structure is indicated by the dashed box. (Reprinted from *Surface Science*, 512, Rose, M.K., Mitsui, T., Dunphy, J., Borg, A., Ogletree, D.F., Salmeron, M., and Sautet, P., Ordered structures of CO on Pd(111) studied by STM, 48–60, Copyright 2002 with permission from Elsevier.)

including $(\sqrt{3}\times\sqrt{3})R30°$, c(2 × 4)-2CO, (2 × 2)-3CO. They also inferred that CO occupied several different types of binding sites depending on coverage, and they used density functional theory (DFT) calculations to support their assignments. Subsurface impurities were found to act as nucleation and pinning sites for ordered $(\sqrt{3}\times\sqrt{3})R30°$ structures of CO (Figure 8.13a). Two types of c(4 × 2) islands of CO occurred at 120 K (Figure 8.13b), with hcp and fcc binding sites in one type of domain, and only bridge sites in the other. Transitions were observed between the two types of domains on a 1 min timescale, and the regions between the CO domains may contain a two-dimensional (2D) gas of H atoms.

8.6.2 Motion of Molecules at Low Temperature

At low temperature (4 K), Eigler et al. not only imaged individual atoms of Xe on a Ni(110) surface,[81] but they found that they could move atoms on the surface with the tunneling tip.[82] They were able to position individual Xe atoms on the Ni(110) lattice with atomic precision and made the first atomic scale advertisement by spelling “IBM” in xenon atoms.[82] Later, they were able to construct a quantum corral out of Fe atoms arranged in a circle on a Cu(111) surface and imaged the LDOS associated with the states of an electron trapped in a round, 2D box.[83] Their work demonstrated the principles for laterally moving atoms and molecules by using the voltage applied to the tunneling tip to pick up and drop atoms, allowing them to be repositioned into atomic-scale structures.[74,82,84] The typical procedure involves reducing the tunneling resistance to ~1 MΩ in order to bring the tip close to the atom to be moved, then moving the tip parallel to the surface so that it drags or pushes the atom with it, and then withdrawing the tip slightly to the typical scanning distance. A new STM image is then measured to verify the results of the atomic manipulation.

One example of molecular motion induced by a tunneling tip was demonstrated by Ohara et al.[85] who deposited methyl thiolate (CH_3S) molecules onto Cu(111) at <50 K. They then observed the molecules at 4.7 K and found that the electric field of the tip could be used to control the molecular motion. An individual CH_3S molecule in a vibrationally excited state will hop away from a negatively charged tip and hop toward a positively charged one. Figure 8.14 shows the resulting images when the CH_3S molecules are arranged to form the letters “STM.” In a related study, the same authors used the tunneling tip to dissociate $(CH_3S)_2$ into two CH_3S molecules and then used the electric field of the tip to induce hopping of the molecules so that they formed into a one-dimensional chain.[86]

Stipe et al. used the tunneling tip in a low-temperature STM to induce and view the rotational motion of a single molecule of O_2 on Pt(111).[87] They observed three different orientations of these molecules on the surface and used tunneling electrons to induce motion from one rotational orientation to another. Because the rotation rate depended linearly on current for sufficiently high sample bias,

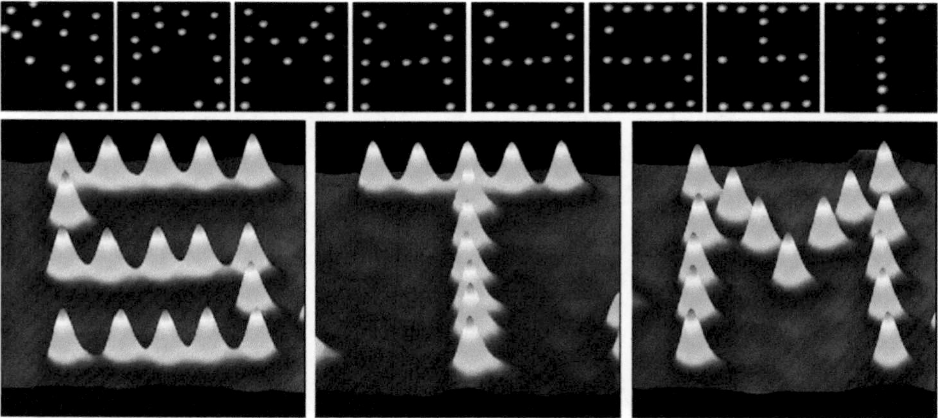

FIGURE 8.14 Sequential STM images showing the gradual construction of the letters and 3D images of the letters S, T, and M with CH_3S molecules on Cu(111). Scan area for each image is 6×6 nm^2. (Reprinted with permission from Ohara, M., Kim, Y., and Kawai, M., *Physical Review B*, 78, 201405(R), Copyright 2008 by the American Physical Society.)

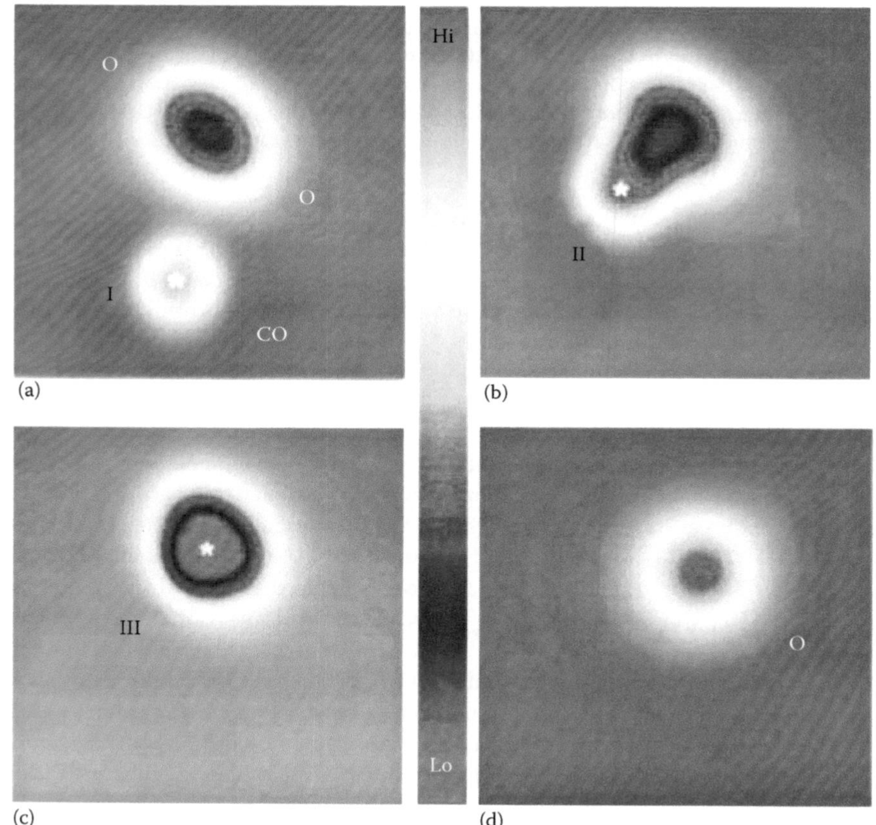

FIGURE 8.15 STM topographical images obtained with a bare tip, 70 mV sample bias, and 1 nA tunneling current, showing the manipulation of a CO molecule toward two O atoms coadsorbed on Ag(110) at 13 K. (a) A single CO molecule and two O atoms. (b) The CO was moved toward O atoms by applying sample bias pulses (1240 mV) after positioning the tip over it. This movement prevented the measurement of C–O stretch (267 meV). (c) The CO was moved to the closest distance from the two O atoms to form the O–CO–O complex. (d) An additional voltage pulse applied to the CO side of the complex led to an image of the remaining O atom on the surface. Scan area of (a)–(d) is 29 Å×29 Å. (Reprinted with permission from Hahn, J.R. and Ho, W., *Physical Review Letters*, 87, 166102, Copyright 2001 by the American Physical Society.)

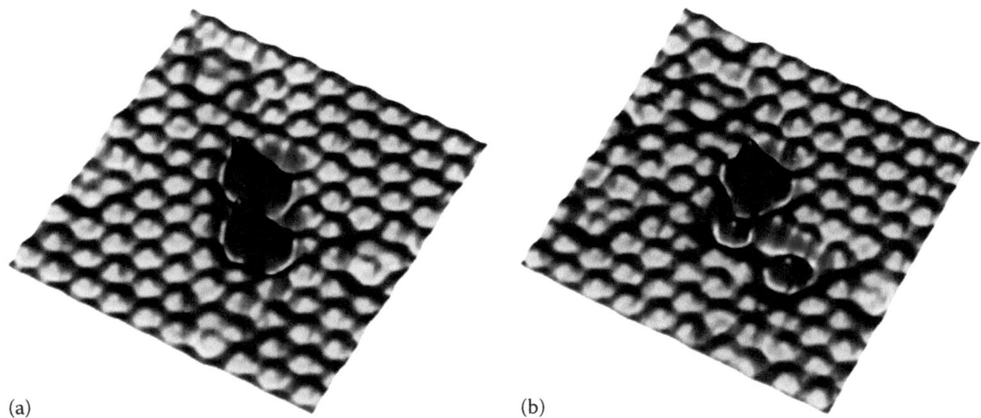

FIGURE 8.16 Breaking a single bond with inelastic tunneling electrons. (a) The STM image of two oxygen molecules on the Pt(111) surface. The molecules are adsorbed on the fcc threefold hollow sites of the surface. By careful positioning of the STM tip, tunneling current of the proper energy was passed through the lower molecule until it dissociated. (b) The resulting two oxygen atoms (sombrero-shaped features), while the top molecule was left undisturbed. (Reprinted from *Current Opinion in Solid State and Materials Science*, 4, Stipe, B.C., Tuning in to a single molecule: Vibrational spectroscopy with atomic resolution, 421–428, Copyright 1999 with permission from Elsevier.)

they proposed a single-step mechanism involving inelastic electron scattering by means of an adsorbate-induced resonance.

8.6.3 Formation and Dissociation of Molecules by Atom Manipulation

Not only can atoms be pushed into specific nearby positions on the surface, but the electrons from the tunneling tip can be used to induce either the formation or the dissociation of molecules. The tip-induced oxidation of a single CO molecule to form CO_2 on an Ag(110) surface at 13 K is shown in Figure 8.15.[88]

Tip-induced molecular dissociation is shown in Figure 8.16. Here, the tip was used to dissociate a single molecule of O_2 on Pt(111) into two O atoms.[89,90] This process has been demonstrated at temperatures from 40 to 150 K. Another example of tip-induced dissociation of molecules was demonstrated by C–H bond breaking in HCCH on Cu(001) at 9 K, resulting in a CCH molecule; in another tip-induced process, that molecule could then be further dehydrogenated to form CC (dicarbon).[91]

8.6.4 Vibrational Spectroscopy of Molecules on Surfaces

From the time of the invention of the STM, the measurement of vibrational spectroscopy through inelastic tunneling was discussed as a possible way to identify molecules on the surface. Inelastic electron tunneling spectra had previously been demonstrated to be sensitive to vibrational modes for molecules in a large sandwich tunnel junction with areas on the order of 0.1 mm^2.[92]

Stipe et al. first demonstrated inelastic tunneling vibrational spectroscopy in a low-temperature STM from isolated molecules of acetylene C_2H_2 and deuterated acetylene C_2D_2 on Cu(100).[93] With the tip positioned directly over the molecule, the plot of second derivative d^2I/dV^2 with respect to voltage had a peak at 358 mV corresponding to the C–H stretching vibration, with the mode shifting to 266 mV for the C–D stretch in C_2D_2. These values correspond well to vibrational spectra obtained using high-resolution electron energy loss spectroscopy (HREELS). Figure 8.17 shows spectroscopic spatial imaging of the inelastic channels for one molecule each of C_2H_2 and C_2D_2. The C_2H_2 molecule is clearly evident in the d^2I/dV^2 image recorded at the C–H vibrational energy, and the C_2D_2 molecule is evident in a similar image recorded at the C–D vibrational energy.

A related study by the same authors found that the STM could be used to determine the orientation of individual C_2HD molecules on Cu(100) at 8 K.[94] In addition, the spatial distribution of the C–D stretching vibration within the molecule could be determined relative to the normal constant-current STM image.

8.7 Molecular Chemical Reactions at Surfaces

8.7.1 Ethylene on Metals

One of the first STM studies which followed a chemical reaction as a function of temperature used the adsorption system of ethylene on Pt(111).[95–97] STM imaging was used to follow all of the well-known reaction steps[98–100]: (1) $T<230$ K: adsorption with the C=C bond parallel to the surface; (2) 230 K $<T<$ 450 K: conversion to ethylidyne (CCH_3), with the C–C bond perpendicular to the surface; (3) 450 K $<T<$ 770 K: dehydrogenation of ethylidyne to form "carbidic" carbon on the surface; (4) above 800 K: carbon converts to graphitic structure. At 230 K, ethylene appears as well-ordered sharper structures in STM images measured during the reaction, while ethylidyne appears fuzzier and more disordered.

A more recent study of ethylene adsorbed at 50 K on Pt(111) and Pd(110) found evidence for two forms of molecules, which were attributed to π-bonded and di-σ-bonded ethylene, with switching between the two forms occurring under the influence of the tunneling tip.[101] The hydrogenation and dehydrogenation of cyclohexene on Pt(111) has also been studied.[102]

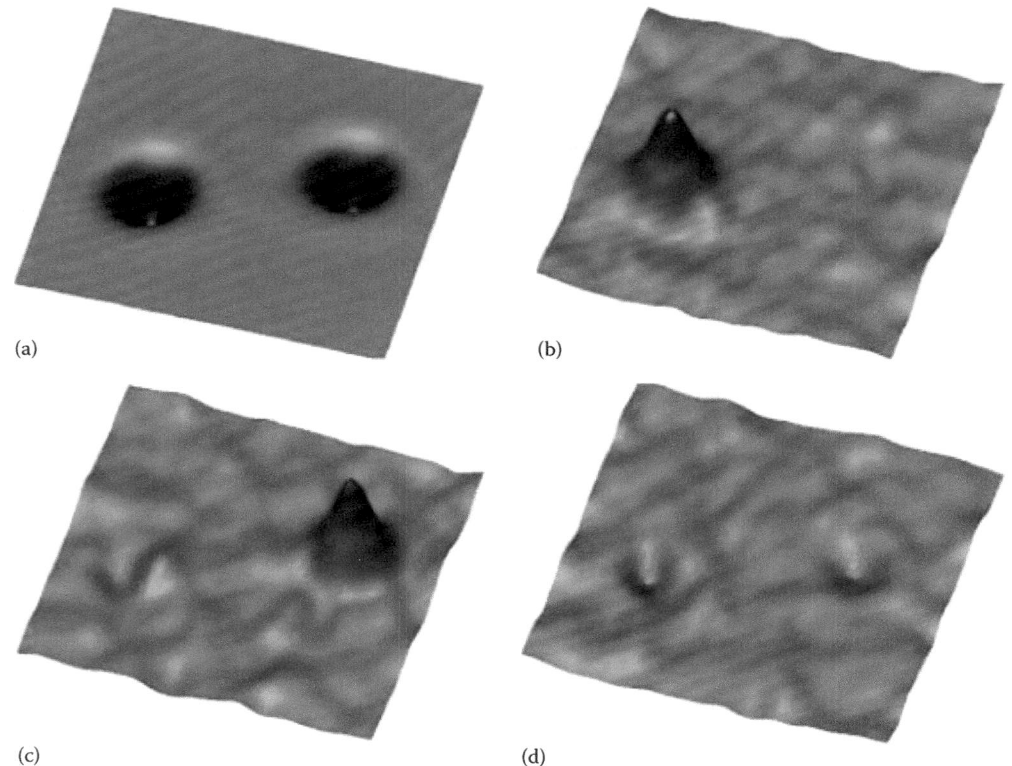

FIGURE 8.17 Spectroscopic spatial imaging of the inelastic channels for C_2H_2 and C_2D_2. (a) Regular (constant current) STM image of a C_2H_2 molecule (left) and a C_2D_2 molecule (right). Data are the average of the STM images recorded simultaneously with the vibrational images. The imaged area is 48 Å × 48 Å. d^2I/dV^2 images of the same area recorded at (b) 358 mV, (c) 266 mV, and (d) 311 mV are the average of four scans of 25 min each with a bias modulation of 10 mV. All images were scanned at 1 nA dc tunneling current. The symmetric, round appearance of the images is attributable to the rotation of the molecule between two equivalent orientations during the experiment. (From Stipe, B.C., Rezaei, M.A., and Ho, W., Single-molecule vibrational spectroscopy and microscopy, *Science*, 280, 1732–1735, Copyright 1998. Reprinted with permission of American Association for the Advancement of Science.)

8.7.2 Furan Decomposition on Pd(111)

We have used STM to study the adsorption and decomposition of furan, C_4H_4O, on Pd(111).[103] In the STM images, furan appears as a heart-shaped molecule. Comparison with the predicted image calculated from the Hückel molecular orbital theory[52] indicates that the oxygen atom is located at the dip in the shape of a heart (Figure 8.18b). The molecule prefers to adsorb along the step edges with the O atom oriented either toward or away from the step. Figure 8.18c shows an STM image of furan molecules near a Pd step. Land and Coworkers had previously studied the decomposition reaction of furan on Pd(111) and found evidence that the molecule decomposes just above room temperature into H, CO, and a C_3H_3 fragment on the surface.[104] They also found that the two C_3H_3 fragments dimerize in a low-yield process to form benzene around 350 K. Using a low-temperature STM, we observed adsorption and diffusion of furan molecules at 225 K and found some molecular species above room temperature which were consistent with C_3H_3 fragments.[105,106] To determine the preferred binding sites and molecular orientations, we also performed calculations using DFT for furan and C_3H_3 on Pd(111).[105,107] Woodruff and coworkers have recently studied this system using photoelectron diffraction, near-edge x-ray adsorption fine structure, and DFT.[108–111]

8.7.3 Ammonia and Oxygen on Cu(110)

Madix et al. were pioneers in building and using a variable temperature STM to follow chemical reactions. Some of the systems they studied were as follows: CO adsorption and oxidation on oxygen-precovered Cu(110) at 150 K[112]; CO_2 + O and carbonate formation on Ag(110)[113,114]; and reaction of ammonia with Ag(110)-p(2 × 1)O.[115]

They published several papers on the reaction of ammonia with oxygen on Cu(110) at 300 and 400 K.[116–118] Atomic oxygen on Cu(110) can form two different structures, added –Cu–O– rows at low coverage and p(2 × 1) islands at high coverage. Figure 8.19 shows a series of images during ammonia oxydehydrogenation at 300 K, with ammonia ambient

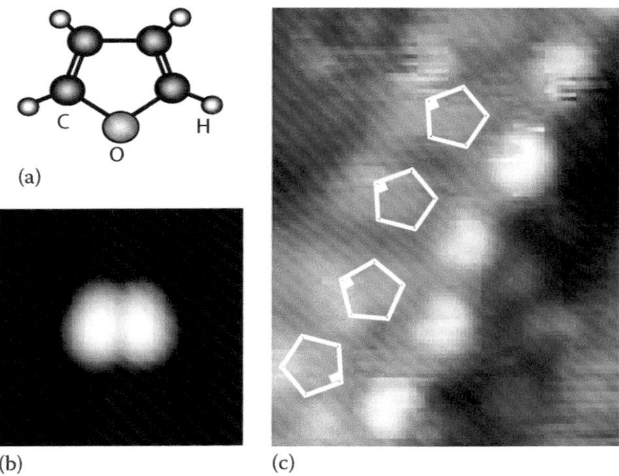

FIGURE 8.18 (a) Schematic representation of furan molecule, with atomic species indicated. (b) Base 10 logarithmic plot of integrated Fermi-level electron density at 2 Å above molecular plane.[52] (c) STM image of furan on Pd(111) at 225 K. The features populating the upper step edge are chiefly oriented with the heteroatom facing toward the terrace. Pentagons representing the furan molecule, with the oxygen atom marked, are added to guide the eye. 30Å × 50 Å, $V_S = -1.0$ V, $I_T = 0.13$ nA. (Reprinted from *Applied Surface Science*, 237, Loui, A. and Chiang, S., Investigation of furan on vicinal Pd(111) by scanning tunneling microscopy, 559–564. Copyright 2004 with permission from Elsevier.)

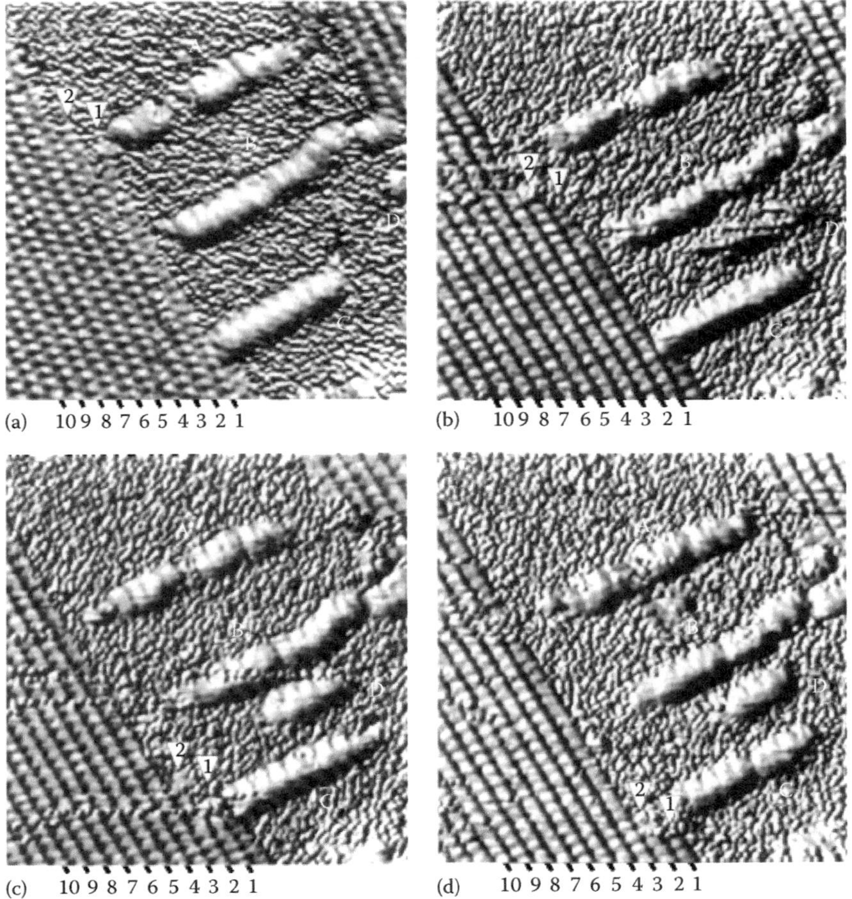

FIGURE 8.19 A sequence of STM images (205 s per image) during ammonia oxydehydrogenation at 300 K under an ammonia ambient pressure of 2×10^{-9} Torr at 300 K. From (a) to (d) the –Cu–0– rows indicated by arrows 1 and 2 are being consumed. The STM images were taken at 300 K with a tunneling current of 1.24 nA at a sample bias of –84.9 mV. (From Guo, X.C. and Madix, R.J., Atom-resolved investigation of surface reactions: Ammonia and oxygen on Cu(110) at 300 and 400 K, *Faraday Discussions*, 105, 139–149, Copyright 1996. Reproduced by permission from The Royal Society of Chemistry.)

pressure of 2×10^{-9} Torr, on a surface which had high oxygen precoverage. The -Cu-O- rows are labeled 1–10, and the thick structures A-D are imide NH(a) rows. Rows 1 and 2 of the -Cu-O- rows clearly get shorter as the reaction proceeds, consuming the oxygen. Since these rows are at the edge of p(2×1)-O islands, the images demonstrate that the reaction starts from the end of -Cu-O- rows attached to the island and then proceeds along the row.

8.8 High-Resolution Atomic Force Microscopy of Molecules

AFM has also recently been used to make very high-resolution images of molecules. The comparison of STM and AFM images of pentacene on Cu(111) clearly shows that the AFM image has higher resolution (Figure 8.20).[119] In this study, the short-range chemical forces were probed with noncontact force microscopy. In order to obtain the highest resolution, the apex of the AFM tip had to be functionalized with a suitable termination, such as a CO molecule. The AFM images in Figure 8.20c and d were measured in constant height mode, with the frequency shift Δf recorded to make the image. This mode of operation allowed stable imaging for the case here, in which Δf is a nonmonotonic function of tip height z. Unlike the STM image in Figure 8.20b, the AFM images in Figure 8.20c and d clearly resolve all five hexagonal rings of the pentacene molecule. Comparisons of images of pentacene on 2 ML of NaCl on Cu(111) with different tip modifications showed that a CO molecule at the end of the tip gave higher resolution than the tips with Ag or Cl atoms, or a pentacene molecule.

The continuation of this work in a very recent paper demonstrated that a CO-functionalized tip used in a noncontact AFM can distinguish different bond orders of individual C–C bonds in polycyclic aromatic hydrocarbons and fullerenes.[120] Figure 8.21a shows the model of hexabenzocoronene, with the i-bonds within the central ring having greater bond order than the j-bonds which connect the central ring to the outside rings. The AFM images for two different z-heights are shown in Figure 8.21b and c, and the i-bonds are imaged as brighter and shorter than the j-bonds. The calculated electron density from DFT in Figure 8.21d corroborates this assignment.

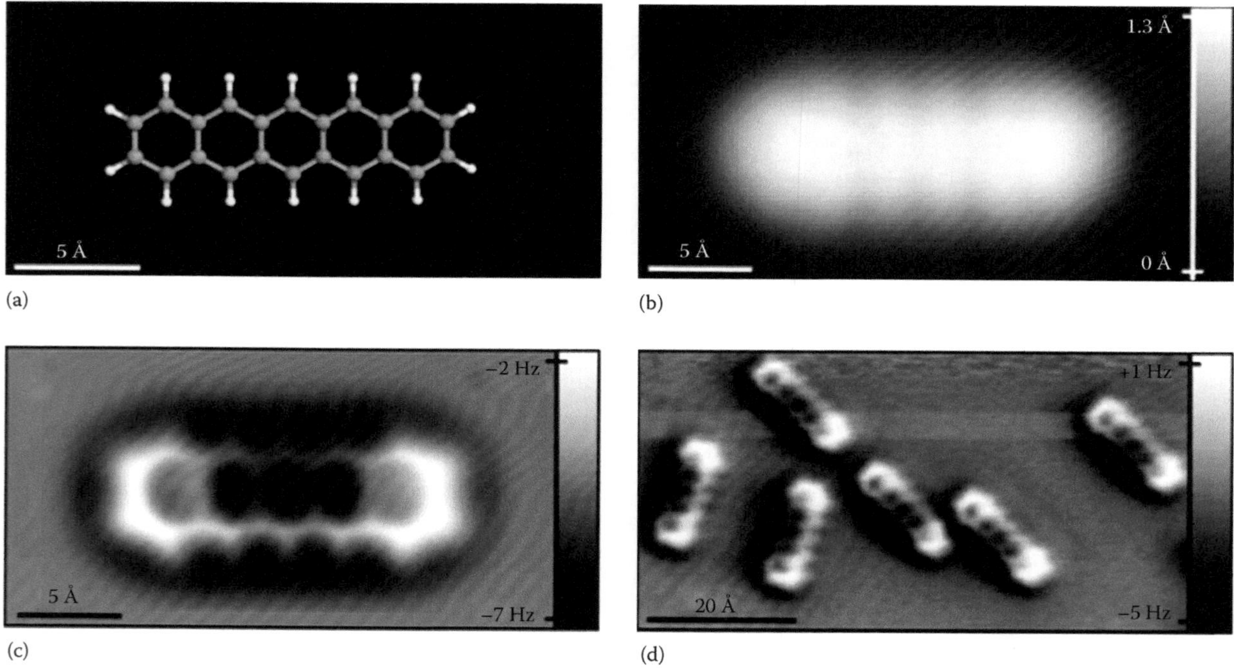

FIGURE 8.20 STM and AFM imaging of pentacene on Cu(111). (a) Ball-and-stick model of the pentacene molecule. (b) Constant-current STM and (c and d) constant-height AFM images of pentacene acquired with a CO-modified tip. Imaging parameters are as follows: (b) set point $I = 110$ pA, $V = 170$ mV; (c) tip height $z = -0.1$ Å [with respect to the STM set point above Cu(111)], oscillation amplitude $A = 0.2$ Å; and (d) $z = 0.0$Å, $A = 0.8$ Å. The asymmetry in the molecular imaging in (d) (showing a "shadow" only on the left side of the molecules) is probably caused by asymmetric adsorption geometry of the CO molecule at the tip apex. (From Gross, L., Mohn, F., Moll, N., Liljeroth, P., and Meyer, G., The chemical structure of a molecule resolved by atomic force microscopy, *Science*, 325, 1110–1114, Copyright 2009. Reprinted with permission of American Association for the Advancement of Science.)

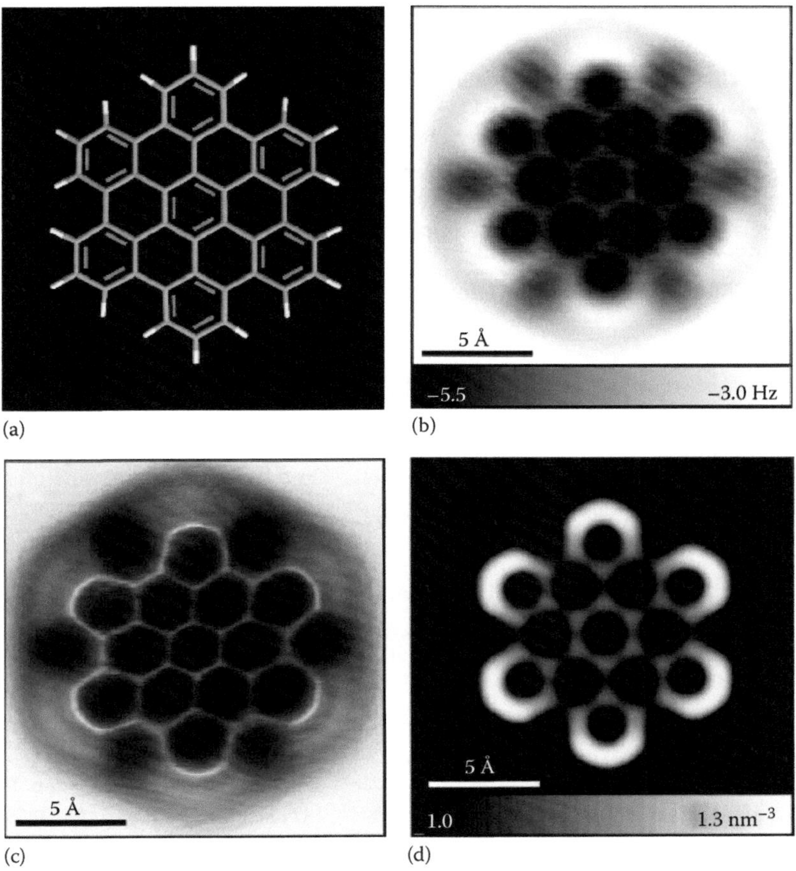

FIGURE 8.21 Hexabenzocoronene (HBC) model (a) and constant height AFM measurements ($A = 0.35$ Å) on HBC on Cu(111) at $z = 3.7$ Å (b) and 3.5 Å (c). In (c), a pseudo-3D representation is shown to highlight the local maxima. (d) Calculated electron density at a distance of 2.5 Å above the molecular plane. Note that *i*-bonds are imaged brighter (b) and shorter (c) compared with *j*-bonds. (From Gross, L., Mohn, F., Moll, N., Schuler, B., Criado, A., Guitián, E., Peña, D., Gourdon, A., and Meyer, G., Bond-order discrimination by atomic force microscopy, *Science*, 337, 1326–1329, Copyright 2012. Reprinted with permission of American Association for the Advancement of Science.)

8.9 Conclusion

Because the STM and AFM have become such widespread tools for examining the organization and structure of molecules on surfaces, this chapter has only given a brief survey of some interesting examples of high-resolution imaging at the picometer scale on a wide variety of molecular systems.

References

1. G. Binnig and H. Rohrer, Scanning tunneling microscopy, *Helvetica Physica Acta*, 55, 726–735 (1982).
2. G. Binnig, H. Rohrer, C. Gerber, and E. Weibel, Surface studies by scanning tunneling microscopy, *Physical Review Letters*, 49, 57 (1982).
3. G. Binnig and H. Rohrer, Scanning tunneling microscopy, *Surface Science*, 126, 236–244 (1983).
4. G. Binnig and H. Rohrer, Surface imaging by scanning tunneling microscopy, *Ultramicroscopy*, 11, 157–160 (1983).
5. A.M. Baro, R. Miranda, J. Alaman, N. Garcia, G. Binnig, H. Rohrer, C. Gerber, and J.L. Carrascosa, Determination of surface-topography of biological specimens at high-resolution by scanning tunnelling microscopy, *Nature*, 315, 253–254 (1985).
6. G. Binnig and H. Rohrer, The scanning tunneling microscope, *Scientific American*, 253, 50–56 (1985).
7. H. Ohtani, R.J. Wilson, S. Chiang, and C.M. Mate, Scanning tunneling microscopy observations of benzene molecules on the Rh(111)-(3 × 3)(C_6H_6 + 2CO) surface, *Physical Review Letters*, 60, 2398–2401 (1988).
8. G. Binnig, C.F. Quate, and C. Gerber, Atomic force microscope, *Physical Review Letters*, 56, 930–933 (1986).
9. G. Binnig and D.P.E. Smith, Single-tube 3-dimensional scanner for scanning tunneling microscopy, *Review of Scientific Instruments*, 57, 1688–1689 (1986).
10. J. Tersoff and D.R. Hamann, Theory and application for the scanning tunneling microscope, *Physical Review Letters*, 50, 1998 (1983).

11. J. Tersoff and D.R. Hamann, Theory of the scanning tunneling microscope, *Physical Review B*, 31, 805 (1985).
12. V.M. Hallmark, S. Chiang, J.F. Rabolt, J.D. Swalen, and R.J. Wilson, Observation of atomic corrugation on Au(111) by scanning tunneling microscopy, *Physical Review Letters*, 59, 2879–2882 (1987).
13. R.M. Feenstra, J.A. Stroscio, J. Tersoff, and A.P. Fein, Atom-selective imaging of the GaAs(110) surface, *Physical Review Letters*, 58, 1192–1195 (1987).
14. V.M. Hallmark and S. Chiang, Predicting STM images of molecular adsorbates, *Surface Science*, 329, 255–268 (1995).
15. D. Sarid, *Scanning Force Microscopy, with Applications to Magnetic, Electric, and Atomic Forces*. 1994, Oxford University Press: New York, p. 263.
16. R. Erlandsson, G.M. McClelland, C.M. Mate, and S. Chiang, Atomic force microscopy using optical interferometry, *Journal of Vacuum Science & Technology A*, 6, 266–270 (1988).
17. S. Chiang, Scanning tunneling microscopy imaging of small adsorbed molecules on metal surfaces in an ultrahigh vacuum environment, *Chemical Reviews*, 97, 1083–1096 (1997).
18. S. Chiang, Molecular imaging by STM, in *Scanning Tunneling Microscopy I*, H.-J. Güntherodt and R. Wiesendanger (Eds.), 1992, Springer-Verlag: Berlin, Germany, pp. 181–205.
19. S. Chiang, Recent developments in molecular imaging by STM, in *Scanning Tunneling Microscopy I*, H.-J. Güntherodt and R. Wiesendanger (Eds). 1994, Springer-Verlag: New York, pp. 258–267.
20. S. Chiang, Imaging atoms and molecules on surfaces by scanning tunnelling microscopy, *Journal of Physics D-Applied Physics*, 44, article no. 464001, 1–19 (2011).
21. C.M. Mate and G.A. Somorjai, Carbon-monoxide induced ordering of benzene on Pt (111) and Rh (111) crystal-surfaces, *Surface Science*, 160, 542–560 (1985).
22. A. Kekulé, Sur la constitution des substances aromatiques, *Bulletin de la Societe Chimique de Paris*, 3, 98–110 (1865).
23. R.F. Lin, G.S. Blackman, M.A. Vanhove, and G.A. Somorjai, LEED intensity analysis of the structure of coadsorbed benzene and CO on Rh(111), *Acta Crystallographica B*, 43, 368–376 (1987).
24. M.A. Van Hove, R.F. Lin, and G.A. Somorjai, Surface-structure of coadsorbed benzene and carbon-monoxide on the rhodium(111) single-crystal analyzed with low-energy electron-diffraction intensities, *Journal of the American Chemical Society*, 108, 2532–2537 (1986).
25. S. Chiang, R.J. Wilson, C.M. Mate, and H. Ohtani, Real space imaging of co-adsorbed CO and benzene molecules on Rh(111), *Journal of Microscopy-Oxford*, 152, 567–571 (1988).
26. S. Chiang, R.J. Wilson, C.M. Mate, and H. Ohtani, Observations of domain boundaries and steps on the Rh(111)-C($2\sqrt{3}\times 4$)rect (C_6H_6+CO) surface by scanning tunneling microscopy, *Vacuum*, 41, 118–120 (1990).
27. C. Pearson, G.W. Anderson, S. Chiang, V.M. Hallmark, and B.J. Melior, A low temperature scanning tunneling microscope designed for imaging molecular adsorbates, *Materials Science and Engineering B*, 96, 209–214 (2002).
28. P.S. Weiss and D.M. Eigler, Site dependence of the apparent shape of a molecule in scanning tunneling microscope images—Benzene on Pt(111), *Physical Review Letters*, 71, 3139–3142 (1993).
29. P. Sautet and M.L. Bocquet, A theoretical-analysis of the site dependence of the shape of a molecule in STM images, *Surface Science*, 304, L445–L450 (1994).
30. P. Sautet and C. Joachim, Calculation of the benzene on rhodium STM images, *Chemical Physics Letters*, 185, 23–30 (1991).
31. S.J. Stranick, M.M. Kamna, and P.S. Weiss, Interactions and dynamics of benzene on Cu(111) at low temperature, *Surface Science*, 338, 41 (1995).
32. S.J. Stranick, M.M. Kamna, and P.S. Weiss, Nucleation, formation, and stability of benzene islands on Cu{111}, *Nanotechnology*, 7, 443–446 (1996).
33. J.I. Pascual, J.J. Jackiw, Z. Song, P.S. Weiss, H. Conrad, and H.P. Rust, Adsorption and growth of benzene on Ag(110), *Surface Science*, 502–503, 1–6 (2002).
34. D.N. Futaba and S. Chiang, Calculations of scanning tunneling microscopic images of benzene on Pt(111) and Pd(111), and thiophene on Pd(111), *Japanese Journal of Applied Physics Part 1*, 38, 3809–3812 (1999).
35. D. Dahlgren and J.C. Hemminger, Symmetry extinction of LEED beams for naphthalene adsorbed on Pt(111), *Surface Science*, 109, L513–L518 (1981).
36. D. Dahlgren and J.C. Hemminger, Chemisorption and thermal chemistry of azulene and naphthalene adsorbed on Pt(111), *Surface Science*, 114, 459–470 (1982).
37. V.M. Hallmark, S. Chiang, J.K. Brown, and C. Wöll, Real-space imaging of the molecular-organization of naphthalene on Pt(111), *Physical Review Letters*, 66, 48–51 (1991).
38. V.M. Hallmark, S. Chiang, and C. Wöll, Molecular imaging of ordered and disordered naphthalene on Pt(111), *Journal of Vacuum Science & Technology B*, 9, 1111–1114 (1991).
39. V.M. Hallmark, S. Chiang, J.K. Brown, and C. Wöll, How well can the scanning tunneling microscope distinguish between two very similar molecules?, in *Synthetic Microstructures in Biological Research*, J.M. Schnur and M. Peckerar (Eds.), 1992, Plenum Press: New York, pp. 79–90.
40. V.M. Hallmark and S. Chiang, Imaging structural details in closely related molecular adsorbate systems, *Surface Science*, 286, 190–200 (1993).
41. S. Chiang, V.M. Hallmark, K.P. Meinhardt, and K. Hafner, Imaging molecular adsorbates—Resolution effects and determination of adsorption parameters, *Journal of Vacuum Science & Technology B*, 12, 1957–1962 (1994).

42. V.M. Hallmark, S. Chiang, K.P. Meinhardt, and K. Hafner, Observation and calculation of internal structure in scanning tunneling microscopy images of related molecules, *Physical Review Letters*, 70, 3740–3743 (1993).
43. P.D. Cernota, H.A. Yoon, M. Salmeron, and G.A. Somorjai, The structure of para-xylene and meta-xylene adsorbed on Rh(111) studied by scanning tunneling microscopy, *Surface Science*, 415, 351–362 (1998).
44. D.N. Futaba and S. Chiang, Calculations of scanning tunneling microscope images of xylene on Rh(111), *Surface Science*, 448, L175–L178 (2000).
45. D.N. Futaba, J.P. Landry, A. Loui, and S. Chiang, Scanning tunneling microscopy study of the molecular arrangement of meta- and para-xylene on Pd(111), *Journal of Vacuum Science & Technology A*, 19, 1993–1995 (2001).
46. D.N. Futaba, J.P. Landry, A. Loui, and S. Chiang, Experimental and theoretical STM imaging of xylene isomers on Pd(111), *Physical Review B*, 65, 045106 (2002).
47. X. Chen, E.R. Frank, and R.J. Hamers, Spatially and rotationally oriented adsorption of molecular adsorbates on Ag(111) investigated using cryogenic scanning tunneling microscopy, *Journal of Vacuum Science & Technology B*, 14, 1136–1140 (1996).
48. E.R. Frank, X.X. Chen, and R.J. Hamers, Direct observation of oriented molecular adsorption at step edges—A cryogenic scanning tunneling microscopy study, *Surface Science*, 334, L709–L714 (1995).
49. C. Kendrick, A. Kahn, and S.R. Forrest, STM study of the organic semiconductor PTCDA on highly-oriented pyrolytic graphite, *Applied Surface Science*, 104, 586–594 (1996).
50. C. Ludwig, B. Gompf, J. Petersen, R. Strohmaier, and W. Eisenmenger, STM investigations of PTCDA and PTCDI on graphite and MoS_2—A systematic study of epitaxy and STM image-contrast, *Zeitschrift für Physik B—Condensed Matter*, 93, 365–373 (1994).
51. R. Strohmaier, C. Ludwig, J. Petersen, B. Gompf, and W. Eisenmenger, STM investigations of NTCDA on weakly interacting substrates, *Surface Science*, 351, 292–302 (1996).
52. D.N. Futaba and S. Chiang, Predictions of scanning tunneling microscope images of furan and pyrrole on Pd(111), *Journal of Vacuum Science & Technology A*, 15, 1295–1298 (1997).
53. P.H. Lippel, R.J. Wilson, M.D. Miller, C. Woll, and S. Chiang, High-resolution imaging of copper-phthalocyanine by scanning-tunneling microscopy, *Physical Review Letters*, 62, 171–174 (1989).
54. J.C. Buchholz and G.A. Somorjai, Surface-structures of phthalocyanine monolayers and vapor-grown films—Low-energy electron-diffraction study, *Journal of Chemical Physics*, 66, 573–580 (1977).
55. T.A. Jung, R.R. Schlittler, J.K. Gimzewski, H. Tang, and C. Joachim, Controlled room-temperature positioning of individual molecules: Molecular flexure and motion, *Science*, 271, 181–184 (1996).
56. R. Möller, R. Coenen, A. Esslinger, and B. Koslowski, The topography of isolated molecules of copper-phthalocyanine adsorbed on GaAas(110), *Journal of Vacuum Science & Technology A* 8, 659–660 (1990).
57. Y. Maeda, T. Matsumoto, M. Kasaya, and T. Kawai, Adsorption structure of copper-phthalocyanine molecules on a Si(100)2×1 surface observed by scanning tunneling microscopy, *Japanese Journal of Applied Physics Part 2*, 35, L405–L407 (1996).
58. M. Kanai, T. Kawai, K. Motai, X.D. Wang, T. Hashizume, and T. Sakura, Scanning tunneling microscopy observation of copper-phthalocyanine molecules on Si(100) and Si(111) surfaces, *Surface Science*, 329, L619–L623 (1995).
59. H. Tanaka and T. Kawai, Scanning tunneling microscopy observation of copper-phthalocyanine and nucleic acid base molecules on reduced $SrTiO_3$(100) and Cu(111) surfaces, *Japanese Journal of Applied Physics Part 1*, 35, 3759–3763 (1996).
60. M.C. Cottin, J. Schaffert, A. Sonntag, H. Karacuban, R. Möller, and C.A. Bobisch, Supramolecular architecture of organic molecules: PTCDA and CuPc on a Cu(111) substrate, *Applied Surface Science*, 258, 2196–2200 (2012).
61. C. Ludwig, R. Strohmaier, J. Petersen, B. Gompf, and W. Eisenmenger, Epitaxy and scanning tunneling microscopy image-contrast of copper phthalocyanine on graphite and MoS_2, *Journal of Vacuum Science & Technology B*, 12, 1963–1966 (1994).
62. R. Strohmaier, C. Ludwig, J. Petersen, B. Gompf, and W. Eisenmenger, Scanning tunneling microscope investigations of lead-phthalocyanine on MoS_2, *Journal of Vacuum Science & Technology B*, 14, 1079–1082 (1996).
63. B.E. Murphy, S.A. Krasnikov, A.A. Cafolla, N.N. Sergeeva, N.A. Vinogradov, J.P. Beggan, O. Lubben, M.O. Senge, and I.V. Shvets, Growth and ordering of Ni(II) diphenylporphyrin monolayers on Ag(111) and Ag/Si(111) studied by STM and LEED, *Journal of Physics—Condensed Matter*, 24, article no. 045005, 1–7 (2012).
64. R.J. Hamers, P. Avouris, and F. Bozso, Imaging of chemical-bond formation with the scanning tunneling microscope NH_3 dissociation on Si(001), *Physical Review Letters*, 59, 2071–2074 (1987).
65. P. Avouris, F. Bozso, and R.J. Hamers, The reaction of Si(100) 2x1 with NO and NH_3—The role of surface dangling bonds, *Journal of Vacuum Science & Technology B*, 5, 1387–1392 (1987).
66. R.J. Hamers, P. Avouris, and F. Bozso, A scanning tunneling microscopy study of the reaction of Si(001)-(2x1) with NH_3, *Journal of Vacuum Science & Technology A*, 6, 508–511 (1988).
67. M.Z. Hossain, H.S. Kato, and M. Kawai, Selective chain reaction of acetone leading to the successive growth of mutually perpendicular molecular lines on the Si(100)-(2x1)-H surface, *Journal of the American Chemical Society*, 129, 12304–12309 (2007).

68. M.Z. Hossain, H.S. Kato, and M. Kawai, Self-directed chain reaction by small ketones with the dangling bond site on the Si(100)-(2×1)-H surface: Acetophenone, a unique example, *Journal of the American Chemical Society*, 130, 11518–11523 (2008).
69. M.Z. Hossain, H.S. Kato, and M. Kawai, Valence states of one-dimensional molecular assembly formed by ketone molecules on the Si(100)-(2×1)-H surface, *Journal of Physical Chemistry C*, 113, 10751–10754 (2009).
70. J.S. Hovis, H. Liu, and R.J. Hamers, Cycloaddition chemistry and formation of ordered organic monolayers on silicon (001) surfaces, *Surface Science*, 402, 1–7 (1998).
71. J.S. Hovis and R.J. Hamers, Structure and bonding of ordered organic monolayers of 1,3,5,7-cyclooctatetraene on the Si(001) surface: Surface cycloaddition chemistry of an antiaromatic molecule, *Journal of Physical Chemistry B*, 102, 687–692 (1998).
72. Y.G. Cai and S.L. Bernasek, Adsorption-induced asymmetric assembly from an achiral adsorbate, *Journal of the American Chemical Society*, 126, 14234–14238 (2004).
73. A.G. Mark, M. Forster, and R. Raval, Direct visualization of chirality in two dimensions, *Tetrahedron: Asymmetry*, 21, 1125–1134 (2010).
74. J.A. Stroscio and D.M. Eigler, Atomic and molecular manipulation with the scanning tunneling microscope, *Science*, 254, 1319–1326 (1991).
75. M.L. Bocquet and P. Sautet, STM and chemistry: A qualitative molecular orbital understanding of the image of CO on a Pt surface, *Surface Science*, 360, 128–136 (1996).
76. G. Meyer, B. Neu, and K.H. Rieder, Identification of ordered CO structures on Cu(211) using low temperature scanning tunneling microscopy, *Chemical Physics Letters*, 240, 379–384 (1995).
77. L. Bartels, G. Meyer, and K.H. Rieder, Basic steps involved in the lateral manipulation of single CO molecules and rows of CO molecules, *Chemical Physics Letters*, 273, 371–375 (1997).
78. M.K. Rose, T. Mitsui, J. Dunphy, A. Borg, D.F. Ogletree, M. Salmeron, and P. Sautet, Ordered structures of CO on Pd(111) studied by STM, *Surface Science*, 512, 48–60 (2002).
79. P. Sautet, M.K. Rose, J.C. Dunphy, S. Behler, and M. Salmeron, Adsorption and energetics of isolated CO molecules on Pd(111), *Surface Science*, 453, 25–31 (2000).
80. P. Cernota, K. Rider, H.A. Yoon, M. Salmeron, and G. Somorjai, Dense structures formed by CO on Rh(111) studied by scanning tunneling microscopy, *Surface Science*, 445, 249–255 (2000).
81. D.M. Eigler, P.S. Weiss, E.K. Schweizer, and N.D. Lang, Imaging Xe with a low-temperature scanning tunneling microscope, *Physical Review Letters*, 66, 1189–1192 (1991).
82. D.M. Eigler and E.K. Schweizer, Positioning single atoms with a scanning tunneling microscope, *Nature*, 344, 524–526 (1990).
83. M.F. Crommie, C.P. Lutz, and D.M. Eigler, Confinement of electrons to quantum corrals on a metal surface, *Science*, 262, 218 (1993).
84. P. Zeppenfeld, C.P. Lutz, and D.M. Eigler, Manipulating atoms and molecules with a scanning tunneling microscope, *Ultramicroscopy*, 42, 128–133 (1992).
85. M. Ohara, Y. Kim, and M. Kawai, Electric field response of a vibrationally excited molecule in an STM junction, *Physical Review B*, 78, 201405(R) (2008).
86. M. Ohara, Y. Kim, and M. Kawai, Controlling the reaction and motion of a single molecule by vibrational excitation, *Chemical Physics Letters*, 426, 357–360 (2006).
87. B.C. Stipe, M.A. Rezaei, and W. Ho, Inducing and viewing the rotational motion of a single molecule, *Science*, 279, 1907–1909 (1998).
88. J.R. Hahn and W. Ho, Oxidation of a single carbon monoxide molecule manipulated and induced with a scanning tunneling microscope, *Physical Review Letters*, 87, article no. 166102, 1–4 (2001).
89. B.C. Stipe, M.A. Rezaei, W. Ho, S. Gao, M. Persson, and B.I. Lundqvist, Single-molecule dissociation by tunneling electrons, *Physical Review Letters*, 78, 4410–4413 (1997).
90. B.C. Stipe, Tuning in to a single molecule: Vibrational spectroscopy with atomic resolution, *Current Opinion in Solid State and Materials Science*, 4, 421–428 (1999).
91. L.J. Lauhon and W. Ho, Control and characterization of a multistep unimolecular reaction, *Physical Review Letters*, 84, 1527–1530 (2000).
92. P.K. Hansma, ed., *Tunneling Spectroscopy: Capabilities, Applications, and New Techniques*. 1982, Plenum Press: New York.
93. B.C. Stipe, M.A. Rezaei, and W. Ho, Single-molecule vibrational spectroscopy and microscopy, *Science*, 280, 1732–1735 (1998).
94. B.C. Stipe, M.A. Rezaei, and W. Ho, Localization of inelastic tunneling and the determination of atomic-scale structure with chemical specificity, *Physical Review Letters*, 82, 1724–1727 (1999).
95. T.A. Land, T. Michely, R.J. Behm, J.C. Hemminger, and G. Comsa, STM investigation of the adsorption and temperature-dependent reactions of ethylene on Pt(111), *Applied Physics A—Materials Science & Processing*, 53, 414–417 (1991).
96. T.A. Land, T. Michely, R.J. Behm, J.C. Hemminger, and G. Comsa, Direct observation of surface reactions by scanning tunneling microscopy: Ethylene → ethylidyne → carbon particles → graphite on Pt(111), *Journal of Chemical Physics*, 97, 6774–6783 (1992).
97. T.A. Land, T. Michely, R.J. Behm, J.C. Hemminger, and G. Comsa, STM investigation of single layer graphite structures produced on Pt(111) by hydrocarbon decomposition, *Surface Science*, 264, 261–270 (1992).
98. R.J. Koestner, J. Stohr, J.L. Gland, and J.A. Horsley, Orientation and bonding of ethylene and ethylidyne on Pt(111) by means of near-edge x-ray absorption fine structure spectroscopy, *Chemical Physics Letters*, 105, 332–335 (1984).

99. H. Steininger, H. Ibach, and S. Lehwald, Surface reactions of ethylene and oxygen on Pt(111), *Surface Science*, 117, 685–698 (1982).
100. C.L. Pettiette-Hall, D.P. Land, R.T. McIver, and J.C. Hemminger, Kinetics of the ethylene to ethylidyne conversion reaction on Pt(111) studied by laser-induced thermal-desorption fourier-transform mass-spectrometry, *Journal of Physical Chemistry*, 94, 1948–1953 (1990).
101. T. Okada, Y. Kim, Y. Sainoo, T. Komeda, M. Trenary, and M. Kawai, Coexistence and interconversion of di-sigma and pi-bonded ethylene on the Pt(111) and Pd(110) surfaces, *Journal of Physical Chemistry Letters*, 2, 2263–2266 (2011).
102. M. Montano, M. Salmeron, and G.A. Somorjai, STM studies of cyclohexene hydrogenation/dehydrogenation and its poisoning by carbon monoxide on Pt(111), *Surface Science*, 600, 1809–1816 (2006).
103. A. Loui and S. Chiang, Investigation of furan on vicinal Pd(111) by scanning tunneling microscopy, *Applied Surface Science*, 237, 559–564 (2004).
104. T.E. Caldwell, I.M. Abdelrehim, and D.P. Land, Furan decomposes on Pd(111) at 300 K to form H and CO plus C_3H_3, which can dimerize to benzene at 350 K, *Journal of the American Chemical Society*, 118, 907–908 (1996).
105. A. Loui, An experimental and theoretical study of furan decomposition on Pd(111) using scanning tunneling microscopy and density functional theory, PhD dissertation in Physics, University of California, Davis, CA (2013).
106. A. Loui, D.N. Futaba, and S. Chiang, Direct imaging of the decomposition reaction of furan on Pd(111) scanning tunneling microscopy, to be submitted (2013).
107. A. Loui, C.Y. Fong, and S. Chiang, Density functional calculations of furan adsorbed on Pd(111), to be submitted (2013).
108. M.J. Knight, F. Allegretti, E.A. Kroger, M. Polcik, C.L.A. Lamont, and D.P. Woodruff, The adsorption structure of furan on Pd(111), *Surface Science*, 602, 2524–2531 (2008).
109. M.J. Knight, F. Allegretti, E.A. Kroger, M. Polcik, C.L.A. Lamont, and D.P. Woodruff, A structural study of a C_3H_3 species coadsorbed with CO on Pd(111), *Surface Science*, 602, 2743–2751 (2008).
110. M.K. Bradley, J. Robinson, and D.P. Woodruff, The structure and bonding of furan on Pd(111), *Surface Science*, 604, 920–925 (2010).
111. M.K. Bradley, D.A. Duncan, J. Robinson, and D.P. Woodruff, The structure of furan reaction products on Pd(111), *Physical Chemistry Chemical Physics*, 13, 7975–7984 (2011).
112. W.W. Crew and R.J. Madix, CO adsorption and oxidation on oxygen precovered Cu(110) at 150 K: Reactivity of two types of adsorbed atomic oxygen determined by scanning tunneling microscopy, *Surface Science*, 356, 1–18 (1996).
113. X.C. Guo and R.J. Madix, Carbonate on Ag(110): A complex system clarified by STM, *Surface Science*, 489, 37–44 (2001).
114. X.C. Guo and R.J. Madix, CO_2 + O on Ag(110): Stoichiometry of carbonate formation, reactivity of carbonate with CO, and reconstruction-stabilized chemisorption of CO_2, *Journal of Physical Chemistry B*, 105, 3878–3885 (2001).
115. X.C. Guo and R.J. Madix, Structural and morphological changes accompanying the reaction of ammonia with Ag(110)-p(2 × 1)-O: An STM study, *Surface Science*, 501, 37–48 (2002).
116. X.C. Guo and R.J. Madix, Atom-resolved investigation of surface reactions: Ammonia and oxygen on Cu(110) at 300 and 400 K, *Faraday Discussions*, 105, 139–149 (1996).
117. X.C. Guo and R.J. Madix, Site-specific reactivity of oxygen at Cu(110) step defects: An STM study of ammonia dehydrogenation, *Surface Science*, 367, L95–L101 (1996).
118. X.C. Guo and R.J. Madix, In situ STM imaging of ammonia oxydehydrogenation on Cu(110): The reactivity of preadsorbed and transient oxygen species, *Surface Science*, 387, 1–10 (1997).
119. L. Gross, F. Mohn, N. Moll, P. Liljeroth, and G. Meyer, The chemical structure of a molecule resolved by atomic force microscopy, *Science*, 325, 1110–1114 (2009).
120. L. Gross, F. Mohn, N. Moll, B. Schuler, A. Criado, E. Guitián, D. Peña, A. Gourdon, and G. Meyer, Bond-order discrimination by atomic force microscopy, *Science*, 337, 1326–1329 (2012).

9

Neutron Holographic Imaging with Picometer Accuracy

László Cser
Wigner Research Centre for Physics

Gerhard Krexner
University of Vienna

Márton Markó
Wigner Research Centre for Physics

Alex Szakál
Wigner Research Centre for Physics

9.1 Introduction

A long-standing problem of materials science consists in obtaining detailed information on crystal lattice distortions in the vicinity of impurity atoms. To this day, such information is still derived mostly from indirect techniques (e.g., diffuse scattering). Only about a decade ago, at the turn of the century, an attempt was made to tackle this problem by proposing a new method (Cser et al. 2001) called atomic resolution neutron holography (ARNH). In this chapter, the results of the efforts applied in developing this approach are outlined.

In view of the relatively low intensity of available neutron sources, one can ask the question why one cannot be satisfied with the well-known holographic methods based on x-rays or electrons which, at first sight, seem to be much more readily available? The answer is, in the first instance, that both radiations mentioned interact with the electron cloud of the atoms extending over distances of the order of Angstroms, that is, $\approx 10^{-10}$ m. On the other hand, neutrons are scattered by nuclei having a size of $\approx 10^{-15}$ m. From this it becomes clear why, in aiming at an improvement in the determination of interatomic—or, for neutrons, rather internuclear—distances the chances to succeed are more promising for neutron holography.*

In addition, the electromagnetic interaction is weak for atoms possessing only a small number of electrons in the atomic shell, and thus the very first elements of the periodic table are out of scope for x-rays and electrons while neutrons are equally well scattered by light elements, including even hydrogen and its isotopes, and heavy nuclei. There is no systematic dependence on atomic number, however, neutron scattering amplitudes are sensitive to isotopic composition even for nuclei having the same atomic number. This feature allows the use of isotopic replacement, which does not change the chemical properties of the sample but provides a method to emphasize or shadow some parts of the object under investigation.

The concept of matter waves as postulated by de Broglie induced a series of experiments using electrons in order to prove its validity. It is obvious that neutrons should behave in a similar way, and this expectation was later on proven by the observation of neutron optical phenomena, such as refraction, total external reflection, Fraunhofer and Fresnel diffraction, interference effects applied, for example, in interferometry, etc. Toward the end of the twentieth century, holography represented the only notable exception among these analogs to classical light optics which had not been reproduced in neutron experiments.

Holographic imaging techniques were conceived by Dénes Gábor in an attempt to attain atomic resolution and are based on

* The term "holography" was derived by Dénes Gábor from the two Greek words όλος (whole) and γράφειν (write).

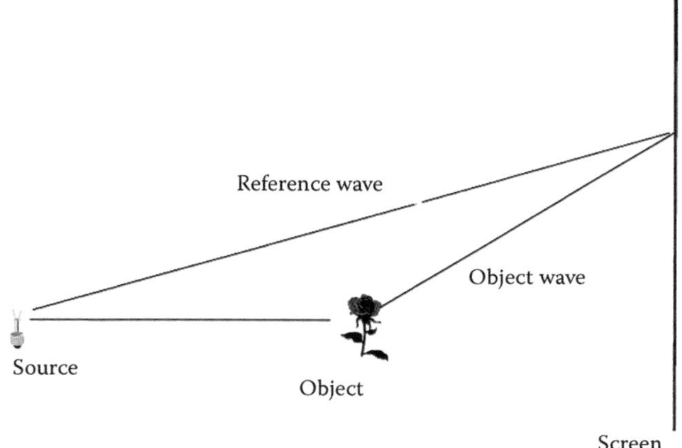

FIGURE 9.1 Schematic of a holographic imaging setup.

the recording of the interference pattern of two coherent waves emitted by the same point source. The first wave that reaches the detector directly serves as the reference wave; the second one is scattered by the object of interest and subsequently interferes with the reference wave (Figure 9.1). However, this concept which, in its original form, makes use of a point-like, monochromatic and coherent radiation source positioned outside of the sample was never implemented for atomic resolution holography due to the limitations of presently available experimental techniques.

In search of a suitable source, atomic nuclei are appealing since they are small enough to be represented by a delta-function and, geometrically, can be considered as point-like objects. Yet, the kinetic energies of neutrons emerging from nuclear reactions lie in the range of several MeV corresponding to very short wavelengths (~20 fm) thereby creating obstacles for the recording of holograms which are insurmountable by the present experimental techniques. The de Broglie wavelengths of neutrons slowed down to thermal energies, however, are suitable for holographic studies because they are commensurate with typical interatomic distances in condensed matter. Also, coherent and monochromatic thermal neutron beams have been used for decades in neutron scattering experiments.

It turns out that, in solving the source problem, incoherent elastic scattering processes play a key role. The redistribution of incoming neutrons into spherical waves originating from a scattering nucleus effectively imitates a point-like source. In addition, due to the optical reciprocity law, source and detector can be interchanged if, in doing so, a detector is used whose size is comparable with the source. These considerations suggest two experimental approaches to neutron holography, applying the so-called internal source and internal detector techniques, respectively.

Since the applied concepts deviate in various respects from the approach typically chosen in neutron scattering experiments and in neutron optics, we devote the following three sections to working out the basic principles of ARNH in order to provide the necessary background for practical applications.

9.2 Ways and Means of Implementing ARNH

9.2.1 Internal Source Holography

Let us assume that a way has been found to realize a point-like source of monochromatic spherical waves of slow neutrons which is located at the origin. A wave emitted by this source will be partly scattered by the nuclei in the sample and partly propagated without being scattered. Both the scattered and the unscattered parts of the wave will be registered by a detector positioned at some distance from the origin in the direction of the wave vector k of the outgoing beam defined with reference to the coordinate system of the sample.

The question arises on how to implement a point-like source of thermal neutrons. The essential point is that a monochromatic incident neutron wave undergoing a scattering process at a point-like incoherent scatterer is isotropically redistributed into the solid angle 4π as is illustrated in Figure 9.2. In view of their large incoherent scattering cross sections ($\sigma_{inc} \sim 80$ barns), hydrogen nuclei, that is, protons, suggest themselves in this context. Other incoherent scatterers can be used as well, but with less efficacy.

In addition, it is important to keep in mind that in order to record a holographic interference pattern there is no need of a coherent incoming neutron beam. The interference pattern appears, because every single neutron interferes with itself. This effect which continues to be somewhat mysterious to this day is one of the peculiar properties of all particles obeying the rules of quantum mechanics. (The phenomenon was experimentally demonstrated for single photons, for elementary particles such as electrons, and even for large molecules such as C_{60}. The successful observation of holograms provides one more independent—though, of course, expected—evidence proving this property for neutrons, too.)

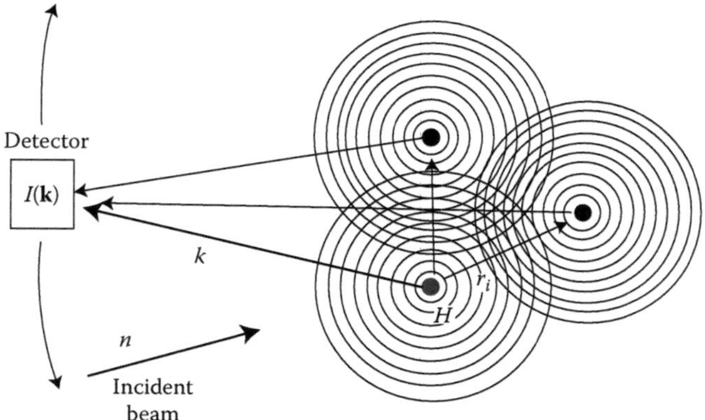

FIGURE 9.2 A monochromatic incident neutron beam illuminates the sample. Neutrons incoherently scattered by hydrogen nuclei form spherical waves. The detector which is placed at a larger distance can be reached by these waves either directly or after having undergone a coherent scattering process by the neighboring nuclei. By moving the detector along the surface of a sphere around the sample, the interference pattern of these waves can be recorded.

Thus, each neutron scattered incoherently by a hydrogen nucleus finally can reach a remote detector in two ways: one part of the spherical wave arrives at the detector directly (the reference wave) and the remaining part reaches the detector after having been scattered coherently by the nuclei of the sample surrounding the hydrogen nucleus (the object wave). A detector scanning a spherical surface around the sample, therefore, will register the interference pattern of these two waves and record a hologram of the nuclei surrounding the hydrogen nucleus. Further, by applying a proper mathematical reconstruction procedure, the 3D holographic image of the atoms can be subsequently restored.

9.2.2 Internal Detector Holography

An alternative way of carrying out ARNH measurements is by the so-called internal detector holography (IDH) approach. This method inverts the inside source technique, based on the optical reciprocity law. According to this law, the positions of the source and the detector are interchangeable without modification of the equations describing the situation. Thus, a plane wave from a far-field source will reach a particular nucleus serving as a microscopic neutron detector within the sample either directly or after having been scattered from the neighboring nuclei forming the crystal lattice.

The initial neutron wave coming from a distant source outside the sample serves as the reference beam. The object beams are those parts of the initial beam that are scattered coherently by atoms in the neighborhood of the detector atom. The detector atom detects the wave field resulting from interference of the reference beam and the object beams at the position of the detector atom. The hologram is defined by the positions of the atoms to be imaged and the wave vector of the incoming beam $\mathbf{k}_i$ defined with reference to the coordinate system of the sample (Figure 9.3).

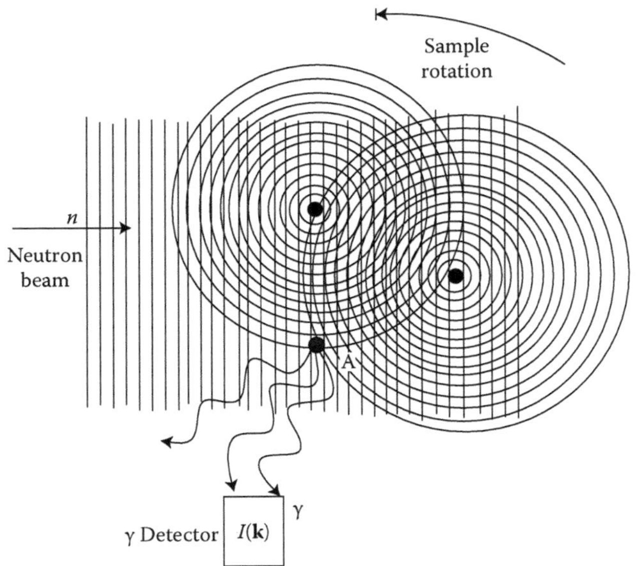

FIGURE 9.3 Schematic of the inside detector concept.

TABLE 9.1 Neutron Capture Data for Some Strongly Absorbing Isotopes (Absorption Cross Sections at $v_{neutron}$ = 2200 m/s)

Isotope (σ_a in barn)	Type of Nuclear Reaction
^{3}He (5333)	^{3}He + $n \rightarrow p$ (573 keV) + ^{3}H (191 keV)
^{6}Li (940)	^{6}Li + $n \rightarrow \alpha$ (2.05 MeV) + ^{3}H (2.727 MeV)
^{10}B (3835)	^{10}B + $n \rightarrow \alpha$ (1.47 MeV) + ^{7}Li (0.83 MeV)
^{113}Cd (20 600)	^{113}Cd + $n \rightarrow$ ^{114}Cd + γ (mainly 0.576 MeV)
^{157}Gd (259 000)	^{157}Gd + $n \rightarrow$ ^{157}Gd (0.83 MeV) + γ $\rightarrow$ conversion electrons (29–182 keV)

The "detector nucleus" should possess a high cross section for neutron capture, and the absorption process should trigger a prompt nuclear reaction generating either a γ-ray cascade or other nuclear products (e.g., α- or β-radiation). The measured quantity is the total nuclear reaction yield. Some typical highly neutron-absorbing isotopes are listed in Table 9.1 (Cser et al. 2001).

It is evident that isotopes producing α-particles, conversion electrons or protons, due to their very small path lengths in condensed matter, can be used only for surface investigations. For the study of bulk matter, one has to utilize samples containing isotopes emitting γ-rays following neutron absorption. If such isotopes are not present in the sample, suitable specimens can be prepared, for example, in the form of alloys or convenient compounds.

9.2.3 The Case of Mixed Holograms

In neutron holography, both IDH and internal source holography (ISH) are based on the use of a probe nucleus converting part of the primary radiation into secondary radiation that cannot interfere with the primary beam. (We apply the term "probe" nuclei to subsume both the source nuclei used in the ISH and the detector nuclei required for the IDH technique.) For example, neither prompt γ-rays emitted subsequently to a neutron absorption process nor a spherical neutron wave resulting from an incoherent scattering process will interfere with the primary neutron beam.

However, there is an important complication to be taken into account. Let us consider an ISH experiment: the probability of an incoherent scattering process to take place at a probe nucleus is proportional to the intensity of the incoming neutron beam. Yet, as discussed in the previous section on IDH, the wave field arriving at the probe nucleus is modified by the interference of the undisturbed part of the incoming wave, with the parts scattered coherently by the nuclei in the sample. The strength of this modulation at the position of the probe nucleus varies according to the orientation of the crystal with respect to the primary beam. So, effectively, the source nucleus will be exposed to a modulated wave field which is equivalent to the one probed by the detector nuclei in an IDH setup. The internal detector hologram implicitly created in this way will be superimposed on the primary beam so that the intensity of the spherical waves created by incoherent scattering at the probe nuclei in the IS setup will not be constant but depend on the crystal orientation. An ISH recorded in this way, therefore, will be inherently "contaminated" by an IDH.

A similar situation, however, with inverted roles arises when recording holograms using the IDH setup.

As a matter of fact, in actual experiments, detector nuclei act at the same time as source nuclei and, conversely, source nuclei act as detector nuclei as well. Thus, usually the recorded holographic image is a superposition of both ISH and IDH. On applying the reconstruction algorithm for just one type of hologram, the simultaneous presence of the other hologram will unavoidably cause distortions of the reconstructed image.

To overcome these difficulties, a profoundly new approach, the so-called double reconstruction (DR), has been developed. In a later section, it will be shown how the DR approach suppresses the mutual distortions caused by the coexistence of the two kinds of holograms and extracts the information stored in the measured hologram in a more complete way. In addition, this novel approach reduces the level of distortion arising from elastic diffuse scattering and statistical noise by a particular choice of experimental conditions. Thus, our method is able to largely separate holographic and elastic diffuse scattering, thereby enabling neutron holographic measurements free of contaminations belonging to other scattering processes. (Note that the efficiency of the DR method can be significantly enhanced by employing 2D area detectors.)

9.2.4 Mathematical Background of ISH and IDH

In order to describe ARNH in a quantitative way, we follow the considerations presented in the following. We point out that the real space and the **k**-space coordinate systems are chosen in such a way that the directions of the respective axes coincide. This simplifies the situation greatly due to the fact that position vectors of atoms in real space can be used at the same time to define the directions of vectors in k-space. Both $\mathbf{k}_i$ and $\mathbf{k}_f$ are defined with respect to the same coordinate system. Nevertheless, $\mathbf{k}_i$-space and $\mathbf{k}_f$-space have to be carefully distinguished, and both of them are not to be confounded with **Q**-space which is frequently used in discussing neutron scattering experiments ($\mathbf{Q} = \mathbf{k}_i - \mathbf{k}_f$).

Let us assume that a way has been found to realize a point-like source of monochromatic spherical waves of slow neutrons, which is located at the origin. A wave emitted by this source will be partly scattered by nuclei j located at $\mathbf{r}_j$ and partly propagated without being scattered. Both the scattered and the unscattered parts of the wave will be registered by a detector positioned at a distance R from the origin in the direction of the wave vector **k** of the outgoing beam defined with reference to the coordinate system of the sample. For both ISH and IDH, the intensity of the wave field $I(\mathbf{k})$ registered at the detector is described by similar expressions originally given by Barton (1988, 1991), and later adapted for neutrons (Cser et al. 2001):

$$I(\mathbf{k}) = \frac{I_o}{R^2}\left|1+\sum_{j}^{n}\frac{-b_j}{r_j}\exp i(kr_j - \mathbf{kr}_j)\right|^2, \tag{9.1}$$

where

b_j is the neutron scattering length (which for nuclear scattering of slow neutrons has no angular dependence)

$\mathbf{r}_j$ is the position vector of the scatterer with respect to the source

The factor $1/R^2$ describes the dependence on the distance R between the sample and the detector.

Expanding the brackets in Equation 9.1, one gets

$$I(\mathbf{k}) = \frac{I_o}{R^2}\left(1+\sum_{j=1}^{n}\frac{-2b_j}{r_j}\cos(kr_j - \mathbf{kr}_j) + \left|\sum_{j=1}^{n}\frac{-b_j}{r_j}\exp i(kr_j - \mathbf{kr}_j)\right|^2\right)$$

$$= \frac{I_o}{R^2}\left(1+\chi(\mathbf{k}) + |O(\mathbf{k})|^2\right). \tag{9.2}$$

The first term inside the brackets represents the reference beam, and the third one represents the object beam, while the second term is composed of both the reference and the object amplitudes, and is called the holographic term accordingly. The reference beam for ISH is represented by the part of the neutron wave scattered by the source atom, while for IDH it is given by the part of the incoming beam reaching the detector nucleus.

Since the typical value for b/r is of the order 10^{-4}–10^{-5}, the third term is negligible. The second term, that is, the holographic one, can be written as

$$\chi(\mathbf{k}) = \sum_{j=1}^{n}\frac{-2b_j}{r_j}\cos(kr_j - \mathbf{kr}_j). \tag{9.3}$$

9.2.5 Reconstruction of the ISH and IDH Holograms

According to Equation 9.3, the holographic information is distributed over the reciprocal 3D k-space. The holographic image of a nucleus placed at the distance r with respect to the origin of the coordinate system in k-space is represented by surface paraboloids defined by the relation $\phi = kr - \mathbf{kr}$. The focus point of these paraboloids coincides with the origin, and the directrix planes are perpendicular to the position vector of the given nucleus in agreement with the earlier definition of coordinate systems in real space and in k-space.

For the reconstruction of the object, a modified Fourier integral is applied. Although the definition of χ in Equation 9.3 differs from that in Barton's work (Barton 1988), the generally valid reconstruction procedure for a polychromatic hologram leads to an expression which is essentially equivalent to the one of Barton:

$$U(\mathbf{R}) = \frac{1}{V_{Mn}}\int_{M_n}\chi(\mathbf{k})\exp(i(\mathbf{kR} - kR))d^n\mathbf{k}. \tag{9.4}$$

where

$U(\mathbf{R})$ is the complex amplitude of the reconstructed holographic image at point R in real space

M_n is the n-dimensional domain of the measurement in k-space, where $n = 3$ in real 3D measurements, that is, polychromatic (time-of-flight) neutron holography, and $n = 2$ in other cases

V_M is the n-dimensional volume of M_n

If $n=2$, M_n is part of the surface of a sphere with radius $2\pi/\lambda$, and $d^n\mathbf{k}$ means the differential surface element on the sphere. Obviously, the corresponding intensity of the holographic signal is equal to the squared absolute value of the amplitude $U(\mathbf{R})$:

$$I(\mathbf{R}) = |U(\mathbf{R})|^2. \tag{9.5}$$

For two nuclei n and m occupying symmetrical positions ($\mathbf{r}_n=-\mathbf{r}_m$), this relation allows rewriting Equation 9.3 as a sum over separate symmetrical pairs in the form

$$\chi(\mathbf{k}) = 2\sum_n \frac{b_n}{r_n}[\cos(kr_n - \mathbf{kr}_n) + \cos(kr_n + \mathbf{kr}_n)], \tag{9.6}$$

where the summation now is made over symmetrical atomic pairs. Using a simple trigonometric transformation, one obtains

$$\chi(\mathbf{k}) = 4\sum_i \frac{b_n}{r_n}\cos(kr_n)\cos(\mathbf{kr}_n). \tag{9.7}$$

Now, applying the Helmholtz–Kirchhoff transformation as in Equation 9.4,

$$\begin{aligned} U(\mathbf{R}) &= \int_{\sigma_{\bar{k}}} \chi(\mathbf{k})\exp(i\mathbf{kr}_n)d\sigma_{\bar{k}} \\ &= 4\sum_n \frac{b_n}{r_n}\cos(kr_n)\int_{\sigma_{\bar{k}}} \cos(\mathbf{kr}_n)\exp(i\mathbf{kr})d\sigma_{\bar{k}}. \end{aligned} \tag{9.8}$$

The observed intensity of the reconstructed hologram is obtained from the squared absolute value of $U(\mathbf{R})$. Then, Equation 9.8 gets separated into the product of two terms. The first one reflects the modulation of the maximum value of the intensity proportional to

$$I(\mathbf{R}) = |U(\mathbf{R})|^2 \sim \left[\frac{\cos(kr_n)}{r_n}\right]^2. \tag{9.9}$$

Relation 9.9 will be referred to as the "cosine rule."

9.2.6 Description of Mixed IDH and ISH

In this section, we present a brief treatment of the general case of mixed ISH and IDH already introduced in Section 9.2.3. A rigorous derivation can be found in the work of Markó et al. (2010a). As mentioned earlier, both IDH and ISH are based on the use of a probe nucleus converting part of the primary radiation into secondary radiation being unable to interfere with the primary beam. In both methods, the detected signal is the intensity of this secondary radiation.

In IDH, the detector nucleus can be either a neutron absorber or a strong incoherent scatterer. In the first case, the total emitted prompt γ-radiation is the detector signal. The detector nuclei are point sources of γ-radiation, so that theoretically one could also record an internal source γ-hologram which, however, is unobservable due to the extremely short wavelength of the prompt γ-rays and the insufficient angular resolution of the γ-detector. Thus, in this type of measurement, the inside source γ-hologram is smeared and, consequently, does not disturb the internal detector neutron hologram.

The situation is quite different in the second case when a strong incoherent scatterer such as the proton acts as a detector. Then, the total intensity of the incoherently scattered (secondary) beam contains the internal detector hologram. According to Equations 9.2 and 9.3, the IDH is described as

$$I^{tot}(\mathbf{k}^P) = \frac{I_0^P}{R^2}\left(1 + \chi^d(\mathbf{k}^P) + |O(\mathbf{k}^P)|^2\right), \tag{9.10}$$

where

- k is the wave number of the primary beam (in the following designated $\mathbf{k}^P$)
- I_0 is proportional to the intensity of the primary beam
- R is the distance between the macroscopic detector and the sample
- I^{tot} is the total intensity appearing around the sample
- I_0^P is the part of the primary beam scattered only by the proton
- χ^d is the normalized IDH

In ISH, on the other hand, the proton serves as the source of the secondary radiation (i.e., the incoherently scattered neutrons). Thus, the intensity of the secondary radiation observed in different directions will contain the ISH. Internal source hologram is described by Equations 9.2 and 9.3 as well, but here I_0 is proportional to the neutron intensity at the position of the proton and $\mathbf{k}^s$ is the wave number of the secondary beam. The resulting intensity measured in the direction of $\mathbf{k}^s$ is

$$I(\mathbf{k}^s) = \frac{I_0^s}{R^2}\left(1 + \chi^s(\mathbf{k}^s) + |O(\mathbf{k}^s)|^2\right). \tag{9.11}$$

where

- I_0^s is the total incoherently scattered intensity
- χ^s is the normalized ISH

Note that $I_0^s = I^{tot}(\mathbf{k}^P)$; thus, the detected intensity is

$$\begin{aligned} I(\mathbf{k}^P, \mathbf{k}^s) &= \frac{I^{tot}(k^P)}{R^2}\left(1 + \chi^s(\mathbf{k}^s) + |O(\mathbf{k}^s)|^2\right) \\ &= \frac{I_0^P}{R^2}\left(1 + \chi^P(\mathbf{k}^P) + |O(\mathbf{k}^P)|^2\right)\left(1 + \chi^s(\mathbf{k}^s) + |O(\mathbf{k}^s)|^2\right). \end{aligned} \tag{9.12}$$

The second or higher order terms are negligible. Thus, Equation 9.12 can be rewritten as

$$I(\bar{\mathbf{k}}^p, \mathbf{k}^s) \approx \frac{I_0^p}{R^2}\left(1 + \chi^p(\mathbf{k}^p) + \chi^s(\mathbf{k}^s)\right). \tag{9.13}$$

In other words, the observed holographic intensity is represented as the sum of IDH and ISH. It is important to note that the vectors $\mathbf{k}^p$ and $\mathbf{k}^s$ are defined in different vector spaces. The first one is the wave vector of the incoming beam, whereas the second one is the wave vector of the measured beam relative to the sample orientation. Thus, changing one of the rotation angles of the spectrometer entails different changes of the two vectors. The space of $\mathbf{k}^p$ corresponds to the IDH space and the space of $\mathbf{k}^s$ to the ISH space.

In principle, by properly choosing the experimental setup, IDH and ISH can be observed independently so that Barton's reconstruction (Barton 1988) (Equation 9.4) can be applied over the two spaces separately. However, it turns out that in this case the presence of diffraction effects (Bragg peaks, diffuse scattering) is unavoidable (Hayashi et al. 2008; Sur et al. 2001). On the other hand, minimizing such effects leads to typical experimental conditions where the IDH contribution appears as a distortion in the image reconstructed from ISH space and vice versa.

We call a holographic image "native" if it is reconstructed from its own space (IDH from IDH space, and ISH from ISH space) and "parasitic" if it is reconstructed from the other space (IDH from ISH space, or ISH from IDH space). In the course of reconstruction, the appearance of parasitic holograms may cause spurious peaks in the image.

Adding the two reconstructed holograms, one gets the so-called doubly reconstructed holographic image. Using a 2D position-sensitive detector (PSD) makes the DR method even more efficient. So, the combined evaluation of ISH and IDH synergetically improves the quality of the resulting hologram, while at the same time the parasitic effects get suppressed in the sense that the distortion of the DR remains comparable with the distortions of the two separate reconstructions (IDR and ISR) while the native holographic signal gets doubled (i.e., its intensity becomes four times larger). Experimental demonstration of this statement can be found in Section 9.4.3.

9.2.7 Theoretical Considerations on Instrument Performance

In this section, we discuss several issues which may strongly influence the quality of the holographic images so that their proper understanding is highly important in planning experiments. In particular, we consider effects due to statistical noise and their consequences for an optimum choice of the signal-to-noise-ratio (SNR), further examine the influence of the instrumental resolution, provide some qualitative remarks on the effects resulting from the application of Fourier filtering, and finally give an exact derivation of the results due to the use of symmetry operations applied to the measured hologram. Altogether, this forms a powerful toolbox in aiming at optimum performance of the instrument for recording holographic images.

9.2.7.1 Calculation of the Noise Level in the Reconstructed Image

The statistical noise superimposed on a hologram during the recording procedure will, in turn, induce a corresponding amount of noise in the reconstructed image in real space. Until now, this effect was either neglected or considered in a misleading and unsatisfactory way. Some authors (Busetto et al. 2000) used SNR in the measured hologram (in k-space) as a criterion for the successful reconstruction, but did not take into account error propagation.

First, it is crucial to use correct statistics in assessing experimental limitations for the observation of ARNH. A full mathematical treatment of the problem including the effects of error propagation is given in the study of Markó et al. (2009). It was proven that the level of the statistical noise in the reconstructed hologram is proportional to the square root of the total number of counts collected, implying that a successful holographic reconstruction is possible even if the statistical noise of the measured hologram exceeds the holographic signal. As a result, the variance of amplitude of the image ($U(\mathbf{R})$) caused by statistical noise is given by

$$\sigma^2_{|U(\mathbf{R})|} \approx \frac{I_{tot}}{N^2}. \tag{9.14}$$

Here

N is the number of measured data points forming the hologram
I_{tot} is the total number of counts collected

Then, SNR of the reconstructed image at the position of an atom can be expressed as

$$\mathrm{SNR} \approx \sqrt{2I_{tot}}\,\frac{b}{R}X(\mathbf{R}), \tag{9.15}$$

where $X(\mathbf{R})$ is a factor taking into account the twin effect and ranging from 0 to 2.

The reconstruction of the measured hologram can be successfully carried out even at a noise level in k-space which largely exceeds the holographic oscillations, that is, when $\mathrm{SNR} \geq \sqrt{1/(2N)}$ (N denotes the number of measurement points). In order to illustrate the previous statement, results of model calculations are shown in Figure 9.4. From these calculations, it is seen that the holographic reconstruction still provides quite reasonable results even if the SNR is as unfavorable as to equal 1/30 in k-space.

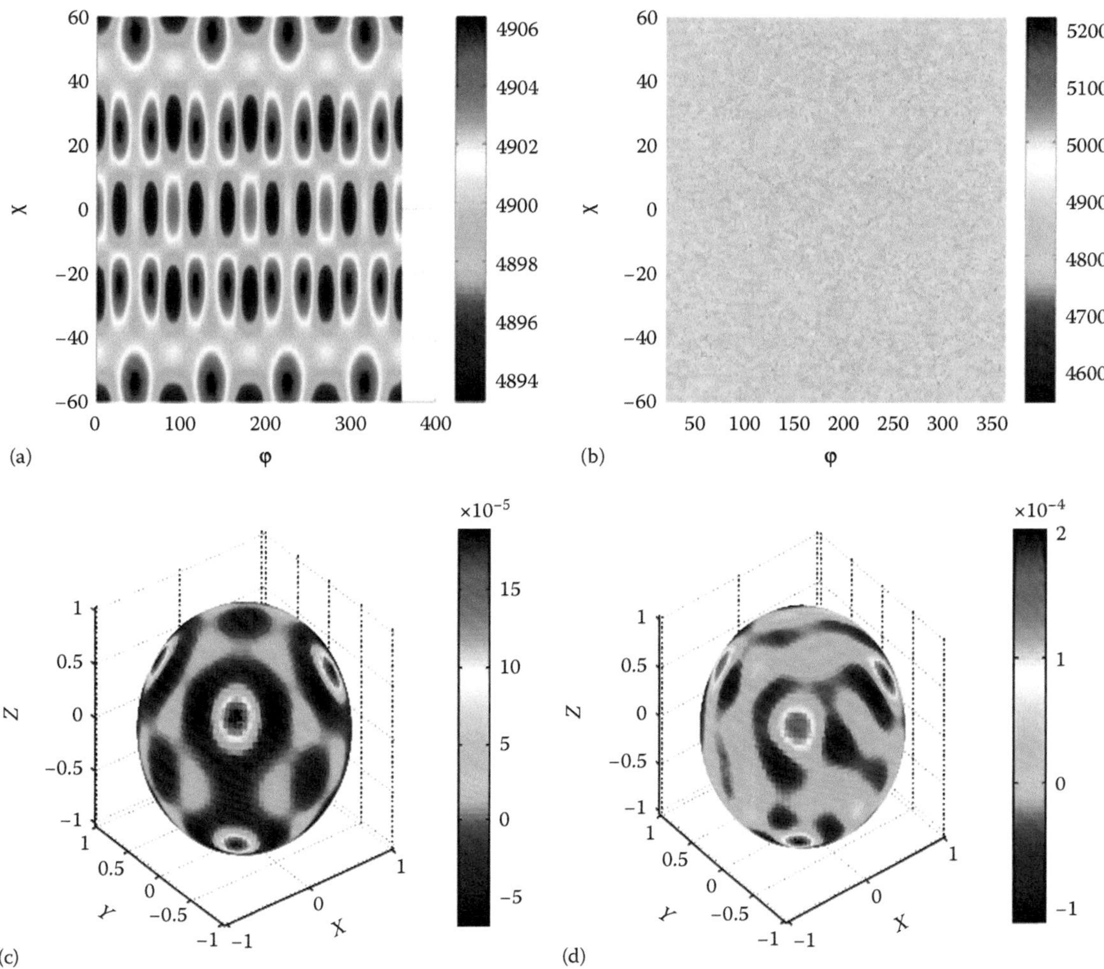

FIGURE 9.4 Model calculation of the inside source hologram of the first atomic shell of a BCC lattice and its reconstruction. The wavelength is 1 Å, and the first neighbor distance is 2 Å. (a) The hologram in k-space in angular coordinates without noise; (b) the same hologram, however, with a noise level exceeding the holographic modulations by a factor 30; (c) and (d) show the images reconstructed from (a) and (b), respectively. The color version of figure can be seen in Markó et al. (2009).

9.2.7.2 Optimization of SNR

On the basis of the earlier considerations and taking into account the imperfections of the experimental setup, we provide a way of optimization in carrying out holographic measurements. Let us suppose that the instrumental resolution in k-space has spherical symmetry and can be described by a Gaussian with standard deviation σ_k. Then, if the sample does not possess central symmetry, the X factor in Equation 9.15 is (Markó et al. 2006)

$$X(\mathbf{R}) = \exp\left(-\frac{\sigma_k^2 R^2}{2}\right). \tag{9.16}$$

If the sample has spherical symmetry, then in the monochromatic case,

$$X(\mathbf{R}) = 2\left|\cos(\mathbf{kR})\right|\exp\left(-\frac{\sigma_k^2 R^2}{2}\right), \tag{9.16a}$$

where σ_k means the total distortion of the measured hologram in k-space and consists of three contributions: $\sigma_k^2 = \sigma_s^2 + \sigma_W^2 + \sigma_{ang}^2$. The first term, σ_s, is the distortion caused by the mosaicity of the sample but not affecting the detected intensity; σ_w takes its origin from the wavelength distribution; and σ_{ang} reflects the angular resolution of the measurement, which in IDH is defined by the divergence of the direct beam and in ISH by the aperture of the detector. Generally, the beam intensity will change proportional to the wavelength resolution and proportional to the square of the angular resolution. Consequently, the detected intensity is

$$I_{det} = n_0 + \sigma_w \sigma_{ang}^2. \tag{9.17}$$

Here, n_0 is the density of the detected counts in 3D k-space. Now, on assuming that the measurement time is constant, the resolution dependence of SNR at the position of the lth atom in the reconstructed holographic image is calculated from Equation 9.13:

$$\mathrm{SNR}_{t=const}(R_j,\sigma_w,\sigma_{ang}) = \mathrm{SNR}_0 \exp\left(-\frac{(\sigma_w^2+\sigma_{ang}^2)R_j^2}{2}\right)\sqrt{\sigma_w}\,\sigma_{ang}. \tag{9.18}$$

Here, $\mathrm{SNR}_0 = \sqrt{2n_0}\,b/R$ is independent of the instrumental distortion. After derivation with respect to σ_w and σ_{ang}, one can easily get the optimum wavelength and angular resolution:

$$\sigma_w = \frac{1}{R_j\sqrt{2}}, \tag{9.19}$$

$$\sigma_{ang} = \frac{1}{R_j}. \tag{9.20}$$

If one wants to observe only the first neighbors of the probe nuclei and the wavelength used is equal to the first neighbor distance, then the optimum relative resolutions are $\sigma_w/k = 1/(2\pi\sqrt{2})$ and $\sigma_{ang}/k = 1/(2\pi)$, which means that for the optimized experiment, the wavelength spread $\Delta\lambda/\lambda = 11\%$ and the angular aperture equals 9%.

9.2.7.3 Filtering Effects

It is a widespread idea that Fourier filtering can cure many imperfections of atomic resolution holograms. There are different kinds of Fourier filtering such as 2D Fourier filtering in (χ, φ)-space (χ and φ refer to the angles used in a spherical coordinate system) or 1D general Fourier filtering on every main circle of the k-sphere (Tegze 2006). The influence of Fourier filtering of the measured data on the reconstructed holographic pattern is discussed in several papers (e.g., Fanchenko et al. 2004; Tegze 2006). The effect of filtering on SNR was investigated only by Tegze (2006) where it was found that Fourier filtering does not increase the noise level. In spite of this, some authors, in order to get rid of the fluctuations caused by the statistical error of the experimental hologram, follow the idea of smoothing methods like low-pass Gaussian filtering (e.g., Korecki et al. 1997).

Later on, it was shown that Fourier-like filtering techniques have no significant effect on the SNR of the reconstructed hologram (Markó et al. 2009). In the opinion of the present authors, Fourier filtering helps mainly to get rid of the baseline modulation (Markó et al. 2010a), where vanishing of the first two Fourier components is usually sufficient to assure satisfactory results. Note that by applying too strict Fourier filtering resulting in the removal of almost all frequencies with the exception of those expected to show up in the measured hologram, one can create a false holographic image as an artifact of the noise contained in a measured "hologram."

9.2.7.4 Symmetries

If the sample possesses any lattice symmetries, one can apply symmetry operations on the measured hologram. In this case, the normalized holographic signal remains unchanged provided that the symmetry operations do not change the measurement domain. If the system under investigation possesses n-fold symmetry, then a symmetry operation transforms the position vector of the jth pixel, belonging to the measurement domain $|\mathbf{k}_j^1|$ in k-space, into $(n-1)$ replicas $|\mathbf{k}_j^L|$ $(L = 2\ldots n)$, where $|\mathbf{k}_j^1| = |\mathbf{k}_j^L| = |\mathbf{k}_j|$.

After applying n independent symmetry operations, the reconstructed hologram can be expressed as

$$U^n(\mathbf{R}) = \frac{1}{N}\sum_{j=1}^{N}\left(\chi(\mathbf{k}_j)\sum_{l=1}^{n}\frac{1}{n}\exp(i(\mathbf{k}_j\mathbf{R}-k_jR))\right). \tag{9.21}$$

The variance of the real part of the reconstructed image $\sigma^2_{\Re U^n(\mathbf{R})}$ is (Markó 2009)

$$\sigma^2_{\Re U^n(\mathbf{R})} = \frac{1}{N}\sum_{j=1}^{N} I_0\left|\sum\frac{1}{n}\cos(\mathbf{k}_j\mathbf{R}-k_jR)\right|^2, \tag{9.22}$$

and the relative change of the variance is (Markó 2009)

$$\frac{\sigma^2_{\Re U^n(\mathbf{R})}}{\sigma^2_{\Re U(\mathbf{R})}} = \frac{\sum_{j=1}^{N}\sum_{l=1}^{n}\left|\frac{1}{n}\cos(\mathbf{k}_j\mathbf{R}-k_jR)\right|^2}{\sum_{j=1}^{N}\left|\cos(\mathbf{k}_j\mathbf{R}-k_jR)\right|^2}, \tag{9.23}$$

$$\frac{\sigma^2_{\Re U^n(\mathbf{R})}}{\sigma^2_{\Re U(\mathbf{R})}} < 1. \tag{9.24}$$

Application of this method to the imaginary part of the hologram gives the same result. From Equation 9.24, it follows that symmetry operations increase SNR. The value of this increase is determined by the size and position of the measurement domain, however, the variance after n symmetry operations will be larger than $1/n$ times the original variance. If the symmetry operation increases the domain used for reconstruction in k-space, then Equation 9.23 still remains valid, but the variance can increase even more. Nevertheless, the reconstructed holographic signal will contain more information (Markó et al. 2006) due to the additional information from the symmetry.

Note that symmetries appearing in the reconstructed image do not necessarily belong to the symmetry group of the crystal lattice. For example, there are situations where the source/detector atoms may occupy several positions in the unit cell and the hologram reflects their average neighborhood. We use the term "holographic symmetry" for the set of all symmetries present in the hologram. It is important to note that the DR can be considered as a "special symmetry operation" connecting IDH and ISH, and thus, as a result of this operation, the SNR in the reconstructed image gets improved.

9.3 Experimental Methods

9.3.1 Basic Requirements for the Experimental Setup

As detailed earlier, in neutron holographic experiments, the recorded image is generally a superposition of both ISH and IDH. Methods for measuring only one type of hologram (Hayashi et al. 2008; Sur et al. 2001) require moving the detector, which means that the measurement has to cover a large part of the reciprocal space causing an additional modulation due to diffraction effects. In order to minimize the measured volume in reciprocal space, the detector should be held in a fixed position and the DR method applied (Markó et al. 2010a).

In this case, one has to rotate the sample around two perpendicular axes. These rotations can be performed by using a Eulerian cradle. For adjusting the plane of the cradle, an additional ω-table is needed to rotate the cradle about a vertical axis. The most important angles and vectors used in neutron holography are shown in Figure 9.5. The definitions of the angles and the calculation of the different vectors can be found in the work of Busing and Levy (1967).

In γ-conversion neutron holography (GNH), γ-ray detectors are used instead of neutron detectors, since the signal emitted by the detector nucleus is a prompt γ-ray following neutron capture. So, in order to decrease the γ-background, it is recommended to keep the volume, defined by the sample position and the solid angle subtended by the detector, free of any components of the shielding emitting γ-rays created by neutron capture (e.g., boron).

In the case of IDH, the optimum value of ω is 90°, that is, the plane of the Eulerian cradle should be parallel to the incoming beam which is not achievable on some diffractometers. For reaching the best SNR of the reconstructed hologram within a given measurement time, the resolution of the measurement has to be optimized. Assuming Gaussian resolution with dispersion σ, the optimal angular (σ_{ang}) and wavelength (σ_{wl}) resolution of the instrument during a holographic measurement is defined by the maximum distance from the probe nucleus (R_{max}), up to which atoms in the reconstructed image are to be observed. The corresponding values are $\sigma_{ang} \approx \lambda/(2\pi\ R_{max})$ and $\sigma_{wl} \approx \lambda^2/(2\pi\ R_{max})$, where λ is the wavelength used.

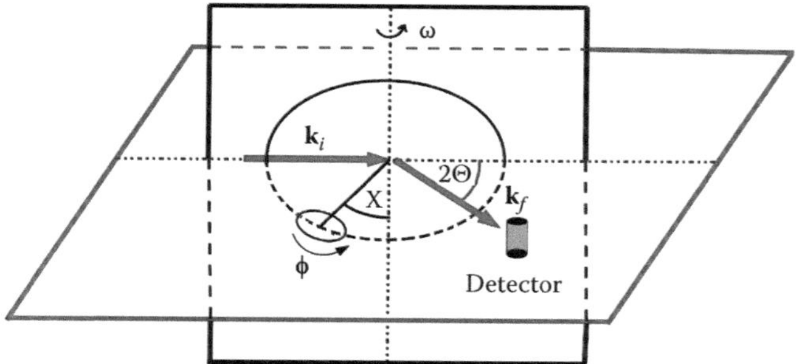

FIGURE 9.5 Schematic of the setup in a neutron holography experiment indicating the sample rotation angles adjusted by using ω-table and Eulerian cradle along with the directions of the wave vectors used. The ω-angle in the figure is equal to 90°.

In general, the optimal resolution in k-space (in the space of the vectors $\mathbf{k}_i$ and $\mathbf{k}_f$ in IDH and ISH, respectively) is realized for $\sigma_k R_{max} = 1$. For example, at a wavelength of 1 Å and an intended viewing distance of $R_{max} = 5$ Å, the optimal resolution parameters are $\sigma_{ang} \approx 1.8°$ and $\sigma_{wl} \approx 0.03$ Å, that is, the corresponding full widths at half maximum are 4.3° and 0.075 Å, respectively. Following the considerations in the last section, the optimization is based on the assumption that the detected intensity I depends on the two components contributing to the total resolution as $I \sim \sigma_{wl}$ and $I \sim \sigma_{ang}^2$. Note that in the IDH case, the angular resolution is defined by the divergence of the direct beam, while in the ISH case it depends on the solid angle subtended by the detector.

9.3.2 Description of Instruments Used for Neutron Holography

In principle, holographic experiments can be carried out at any diffractometer or three-axis spectrometer available at major neutron labs provided that measurement time is allocated for that purpose. However, this is the case only from time to time, since neutron holography experiments require a special setup, usually being not quite compatible with standard applications and, therefore, lie largely outside the scope of such instruments. So far, successful experiments were performed, for example, at the diffractometer D9 of the ILL (Grenoble, France) (Cser et al. 2002), at an instrument of the Chalk River Laboratory (Canada) (Sur et al. 2001), and at the 6T2 four-circle diffractometer at the Orphée research reactor (Laboratoire Léon Brillouin, Saclay) (Cser et al. 2004).

These instruments are constructed for satisfying various requirements but in no way designed for an optimum performance in holographic experiments. In order to approach optimal conditions for ARNH experiments, we, therefore, have built up a dedicated holographic instrument being installed at the eighth radial channel of the Budapest Research Reactor (Markó et al. 2010b). There are three different setups meeting the requirements of ISH, IDH, and mixed holography (DR), respectively, as illustrated in the schematic views in Figures 9.6 through 9.8.

An arrangement optimized for an ISH experiment is shown in Figure 9.6.

The IDH arrangement is somewhat different (see Figure 9.7).

A setup for mixed holography using a PSD is displayed in Figure 9.8.

The schematic view of the arrangement of our instrument is shown in Figure 9.9. It consists of a double focusing copper (200) monochromator with a useful surface area of 15 × 15 cm² and a mosaicity of 30′ providing a monochromatic beam with a relative wavelength resolution $\Delta\lambda/\lambda = 2.7\%–6.2\%$ at 1 Å depending

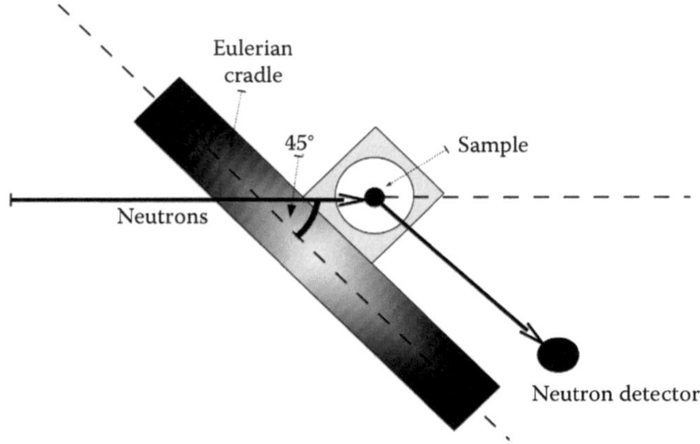

FIGURE 9.6 Schematic of the experimental setup for ISH.

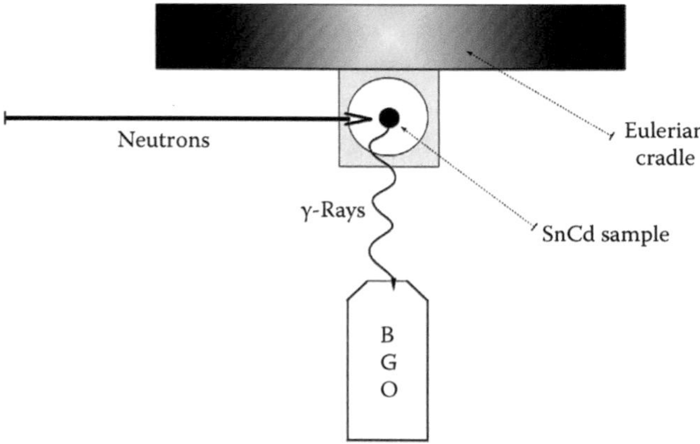

FIGURE 9.7 Schematic of the experimental setup for IDH.

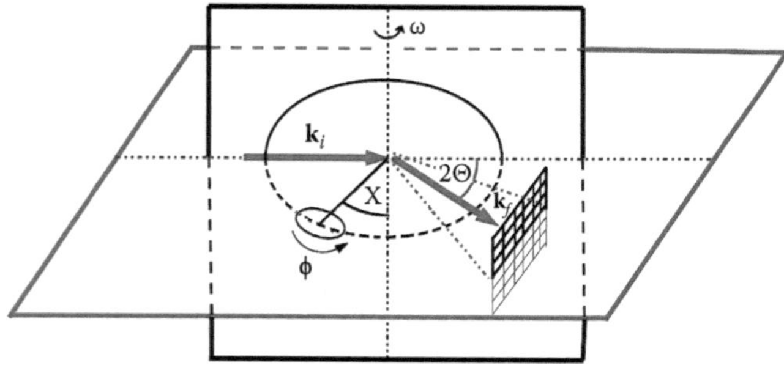

FIGURE 9.8 Schematic of a measurement setup using a 2D PSD. The angle 2Θ is fixed (the detector is not moving).

on the focusing curvature. The monochromator is surrounded by a 27.5 metric ton concrete shielding with a 6 × 6 cm² outlet window for the monochromatized neutron beam. A fission chamber monitor counter is mounted on this window. The monochromator–sample distance is about 130 cm corresponding to a divergence half angle of 1.3°. Since the detector size is 5 cm at a sample detector distance of about 50 cm, one obtains 5.6° for the acceptance angle of the detector. This value increases

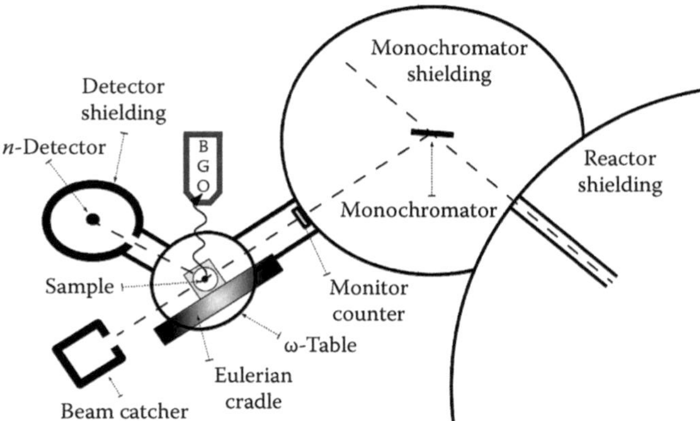

FIGURE 9.9 Schematic view of the instrument dedicated to neutron holography.

the data collection rate by more than a factor 30 compared to an acceptance angle of about 1°, which is typical for high-resolution diffractometers (we recall that the holographic modulations generally are relatively smooth as exemplified in the earlier numerical example for optimizing the resolution and, therefore, do not require high-resolution detectors).

The sample holder is mounted on a Eulerian cradle installed on an ω-table. Depending on the task planned, three different detectors can be used. For registration of scattered neutrons, a ^{3}He–gas filled single tube detector with 98% efficiency for 1 Å neutrons is in use. In order to improve the efficiency of the instrument, a 2D PSD with an active area of 200 × 200 mm² is available with a spatial resolution of 1.8 mm. The efficiency of the area detector is 57% for the wavelength of 1 Å. The area detector is also highly efficient in adjusting the focusing monochromator and for checking the orientation of the sample. For detecting secondary γ-radiation, a 3 in. BGO (Bi_4GeO_{12}) scintillator detector is installed on a separate support whose position can be adjusted manually, independent of the other parts of the instrument.

Due to the massive shielding, the minimum distance between the sample and the γ-detector is 15 cm. Both single tube and BGO detectors are surrounded by proper screening for suppressing the background. The single neutron counting tube has a 30 cm thick boronated plastic shielding. The collimator between the sample and the detector is a Cd tube surrounded by boronated plastic against the fast neutrons escaping from the monochromator shielding. The BGO detector has a sandwich-type shielding to eliminate both fast and thermal neutrons and γ-background as well. The outer part of the shielding is composed of paraffin mixed with boric acid. Under the paraffin shielding, Cd plates are used against the thermalized fast neutrons.

For shielding against γ-background a 20 cm thick Pb block is applied. The scintillator and the photomultiplier housing are coated by ^{6}Li-containing plastic sheets preventing the detector housing from getting activated. The nuclei of the isotope ^{6}Li absorb neutrons with high efficiency and decay to ^{3}H and α without γ-photon emission. In order to get rid of the background originating from the neighboring neutron beam channels, the whole equipment is carefully surrounded by a 2 cm thick boric acid wall. This careful shielding is required due to the rather small amplitude (about 10^{-4}) of the holographic signal. For suppressing the neutron and γ-background entering the viewing angle of the detector behind the sample, a sandwich wall was constructed consisting of 10 cm thick boric acid containing paraffin and 5 cm thick Pb blocks. The direct beam is trapped by a massive beam catcher hole.

The remote control of the instrument is performed via dedicated electronics made of commercially available hardware elements. Further, a control and data acquisition software package have been developed and tested by the authors.

9.4 Test Measurements

In order to test the three types of holographic techniques, three samples were chosen, namely single crystals of ammonium chloride (NH_4Cl) salt, a SnCd alloy with low Cd content, and a PdH metal–hydrogen system.

9.4.1 ISH Measurement on a NH_4Cl Sample

This crystal is a challenging object, because it has a rather complicated structure containing a large number of atoms in the elementary cell. Despite these complications, NH_4Cl was considered as a suitable sample for demonstrating the INH approach due to its high hydrogen content.

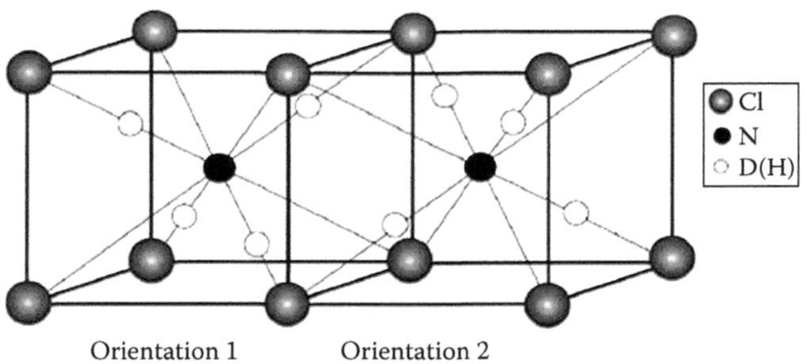

FIGURE 9.10 The crystal structure of the NH_4Cl illustrating the two possible orientations of the hydrogen tetrahedron surrounding the nitrogen atom.

The structure of NH_4Cl was studied by Goldschmidt and Hurst (1952). It was proven that the nitrogen atom is surrounded by eight atoms of chlorine adopting a CsCl structure. At a closer distance, four hydrogen atoms are forming a tetrahedron around the nitrogen atom (see Figure 9.10). The tetrahedron can adopt two different configurations. At room temperature, these two positions are equally occupied by ammonium ions (Frenkel model, Goldschmidt and Hurst 1952). The corresponding space group is $P\bar{4}3m$. The first neighbors of the hydrogen atoms are nitrogen atoms at a distance of 1.03 Å. The second neighbor shell is occupied by hydrogen atoms at a distance of 1.63 Å, while chlorine atoms form the third neighbor shell at 2.32 Å.

The hydrogen atom serves as the origin of the reconstructed hologram, thus, in contrast to the crystal space group, the system possesses simple cubic holographic symmetry. The sample was a spherically shaped NH_4Cl single crystal of 7 mm diameter mounted onto the tip of a thin quartz needle.

The wavelength of the neutrons used was 1 Å. The angular resolution is close to the optimal resolution required to see the first three neighboring shells of the hydrogen atom in an ISH (Markó et al. 2009). The scheme of the experimental setup is shown in Figure 9.11. Altogether, $37(\chi) \times 72(\varphi) = 2664$ steps were carried out, and a total of 2×10^8 counts were collected.

Four unavoidable Bragg peaks were cut out. The background modulation was determined applying a low-pass Fourier filter using the first two components in the χ-direction and the first three components along the direction for determination of the background level. Then the measured data were divided pixel by pixel by the calculated values of the background. The real parts of the measured and modeled holographic images (after applying the fourfold symmetry around the *z*-axis) are shown in Figure 9.11. The signs and positions of the atomic peaks are in agreement with the results obtained from the model calculations. As shown in Figure 9.11, the coincidence of the measured and modeled holograms convincingly demonstrates the validity of the results.

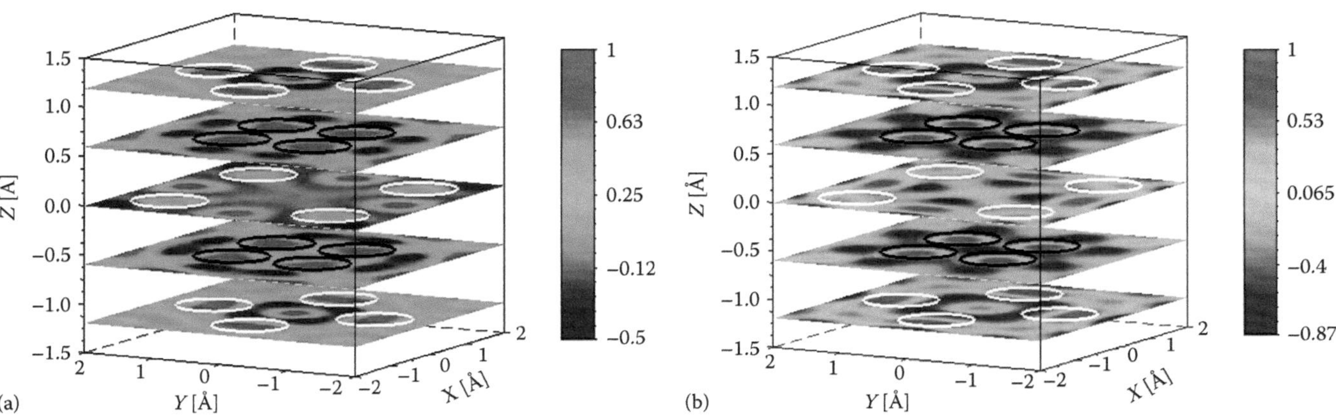

FIGURE 9.11 Holographic reconstruction of NH_4Cl. (a) Occupation of the lattice sites as derived from the measured data; (b) model calculation. The positions of nitrogen and hydrogen atoms are marked by black and white rings, respectively. The color version of figure can be seen in Markó et al. (2010).

9.4.2 IDH Study of a $SnCd_{0.0026}$ Alloy

This measurement (Markó et al. 2010b) serves to demonstrate a typical GNH-type holographic experiment. The Cd atoms embedded in the tin crystal lattice are ideal detectors. The concentration of the Cd atoms is so low that each Cd atom will be entirely surrounded by tin atoms, and no interaction between Cd atoms has to be considered.

The tin crystal has a tetragonal structure (space group I41/amd) with lattice parameters $a = b = 5.18$ Å and $c = 3.819$ Å. Figure 9.12 shows the elementary cell of the tin crystal. The primitive unit cell contains only two atoms. If the Cd atoms randomly occupy both sites with equal probability, then this gives rise to a central and fourfold symmetry about the *c*-axis which is not present in the tetragonal symmetry of the crystal (see Section 9.2.7.4).

The experimental setup is shown in Figure 9.7. The characteristic γ-rays of the ^{113}Cd nucleus were registered by a BGO γ-detector. The spherically shaped sample of 7 mm diameter was rotated about the χ-axis over an angular range of −60° to +60°, and at each χ-position a full rotation through 360° about the φ-axis was performed. Altogether, a total of more than 10^9 counts were collected.

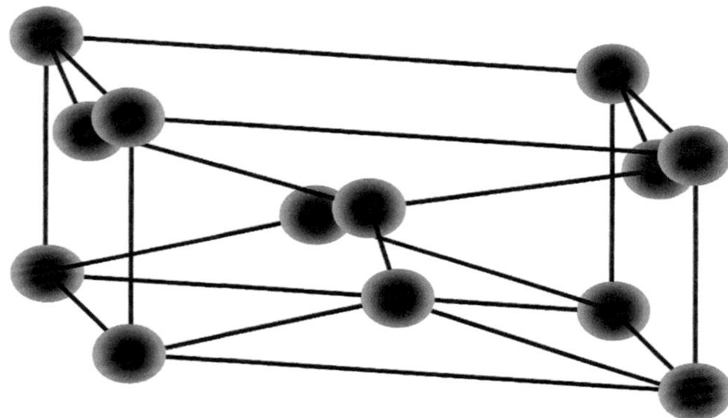

FIGURE 9.12 Structure of the elementary cell of the tin crystal.

The background modulation was determined using a low-pass Fourier filter applied to the data line-by-line at each χ-position. The recorded hologram was reconstructed using the Helmholtz–Kirchhoff integral as proposed by Barton (1988).

Here, instead of the intensity, we display the amplitudes of the reconstructed hologram holding the information about the sign of the atomic peaks, as seen in Figure 9.13. In the same figure, the model calculation of the expected reconstructed holographic image is shown. The comparison of the measured and calculated holograms demonstrates that the arrangement of the Sn atoms surrounding the Cd atom coincides with the regular Sn crystalline structure. A model calculation imitating the holographic measurement (see Figure 9.13b) supports our observation. No indication of the presence of Cd atoms on interstitial positions was observed. Consequently, it follows that Cd atoms prefer the substitutional positions instead of interstitial ones.

9.4.3 Mixed Holography Using PSD and DR

This novel method was proposed by Markó et al. (2010a) (see Section 9.2.6). The setup is outlined in Figure 9.8. In this case, one obtains the ISH by collecting 2D data as a function of the detector position in $\mathbf{k}_f$ space and the IDH by registering the sum of all counts in the detector as a function of the sample orientation. Because each sample orientation defines one point in IDH space, the IDH (which is the parasitic hologram in this case) appearing in $\mathbf{k}_f$ space will be convoluted with a box function whose width equals the viewing angle of the detector. The same happens if one plots the measured data in IDH space (here, the ISH image will be smeared). The proposed method will suppress the reconstructed parasitic image (Markó et al. 2009).

In order to prove the efficiency of the DR method, we carried out a holographic measurement on a PdH system. Earlier, two holographic experiments had been performed on this system but both of them were not fully satisfactory. In the first one (Cser et al. 2005), only the first neighbors of the H-atom

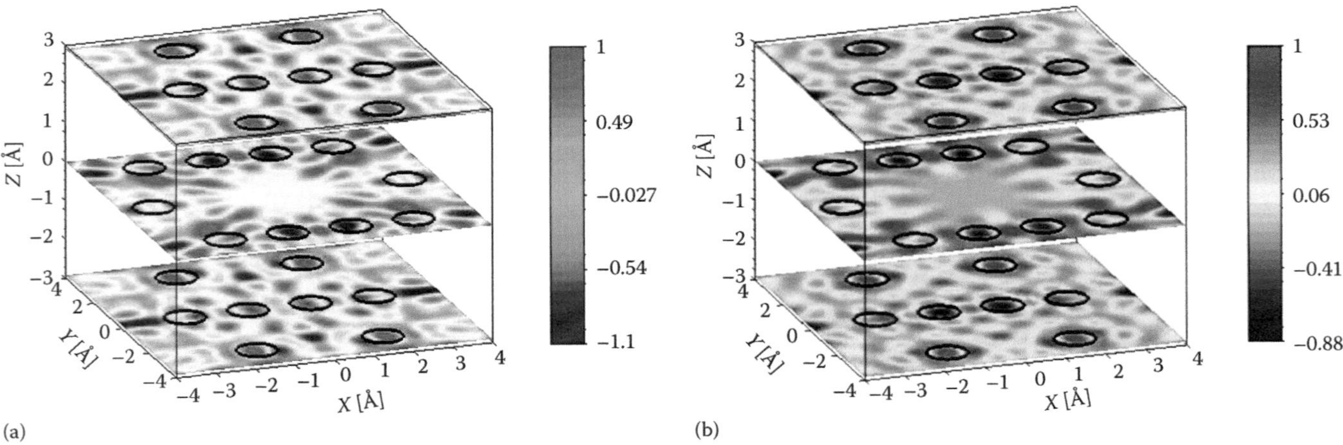

FIGURE 9.13 The reconstructed hologram of the SnCd alloy. The (a) measured holographic image and (b) result of model calculation. The amplitude distribution of the hologram is shown in the planes being perpendicular to the (100) direction. The expected atomic positions are marked by black circles. The color version of figure can be seen in Markó et al. (2010).

were observed. In this case, due to the small atomic distance–wavelength ratio, the distortion of the mixed hologram was only faintly present. In another experiment (Hayashi et al. 2008), a detector array was used allowing to measure only the ISH. In this measurement, the diffuse scattering caused serious distortions.

The neutron holographic experiment was carried out at the diffractometer TriCS (Schefer et al. 2000) at the Swiss spallation source SINQ (PSI) which provides a neutron wavelength of 1.18 Å. The sample was a spherically shaped $PdH_{0.78}$ single crystal with a diameter of about 7 mm. To avoid hydrogen loss, the sphere was electrolytically coated with a thin copper film (thickness ~ 10 μm).

Applying the procedure introduced in Section 9.2.6 to the experimental data leads to the holographic image in a doubly reconstructed form. Figure 9.13 allows comparison of the reconstruction techniques ISR, IDR, and DR. White circles mark the hydrogen octahedral interstitial sites and black circles mark the palladium lattice sites.

Comparing the results of the three reconstruction techniques, the most attractive advantage of the DR method consists in the nearly complete elimination of spurious peaks. Figures 9.14a

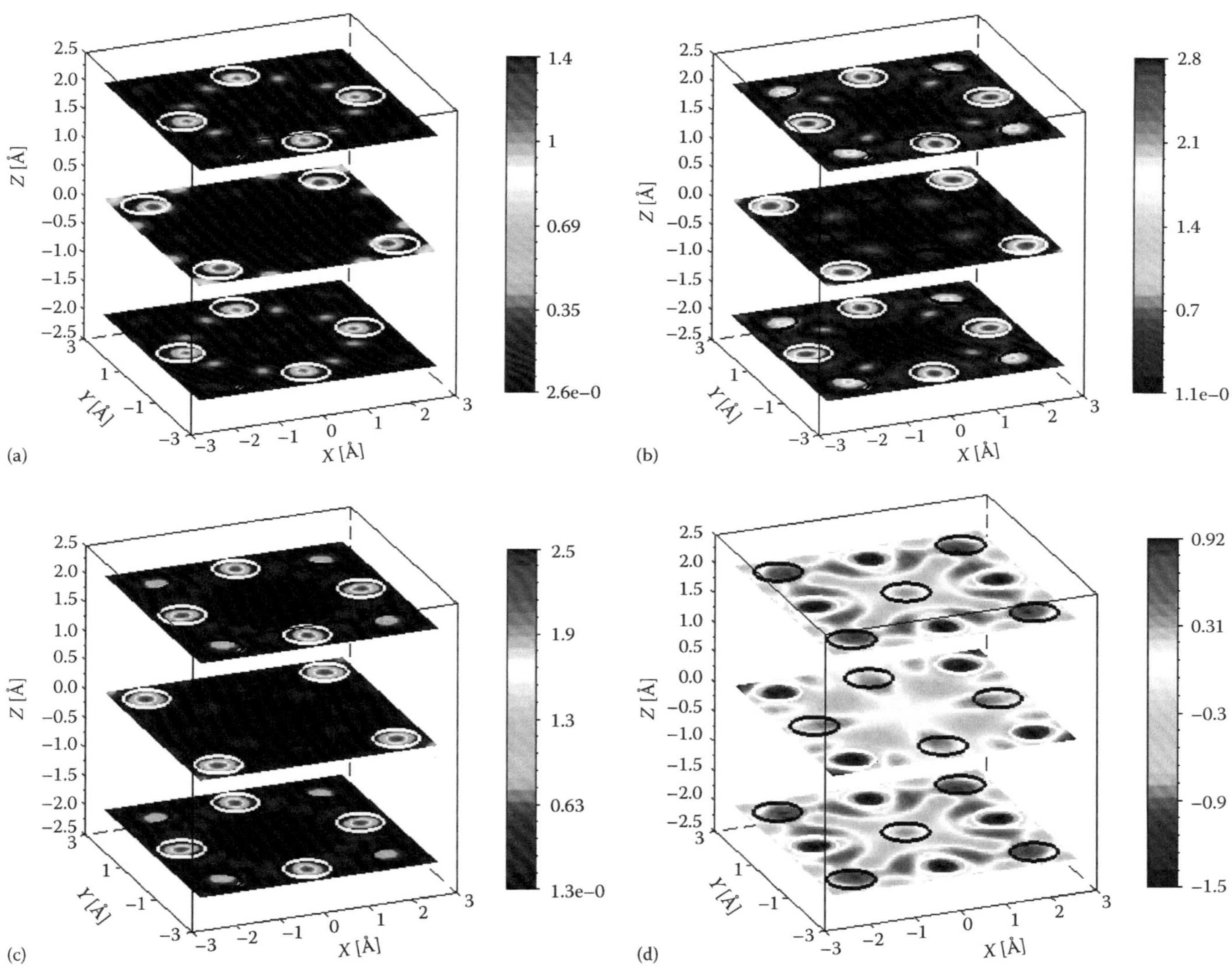

FIGURE 9.14 The intensities of holographic images of the system $PdH_{0.78}$ reconstructed by ISR (a), IDR (b), and DR (c), and the amplitudes of the holographic image reconstructed by the DR method (d). The layers of the reconstructed images are parallel to the (001) lattice planes at $Z = -2$, 0, and 2 Å, respectively. The positions of the 6 first and 8 second neighbor Pd atoms (marked by black circles) and the 12 first neighbor H atoms (marked by white circles) arranged around the hydrogen atom at the center of the coordinate system (not shown) are indicated. In (c), the first H and second Pd neighbors can be clearly identified, whereas the spurious intensities present in (a) and (b) are largely suppressed due to the application of the DR method. In (d), the spots surrounded by black circles represent the Pd atom positions and the spots surrounded by white circles give the positions of the H atoms. The centers of the circles coincide with the atomic positions of an ideal, undistorted crystal lattice. Two weak spots in the middle of the upper and lower planes in (d) hint at the presence of the first neighbor Pd atoms. The color version of figure can be seen in Markó et al. (2010).

through c show intensity distributions, Figure 9.14d displays the amplitudes of the reconstructed wave functions: hydrogen peaks have negative amplitudes, while palladium peaks have positive ones corresponding to the tabulated values. The observed positions of the hydrogen atoms prove that they occupy octahedral interstitial sites.

In all, the holography measurement carried out on the PdH system proves the advantages of the simultaneous application of both the DR method and the 2D detector. The 6 first Pd neighbors around the octahedral interstitial hydrogen sites were clearly visible in the reconstructed image. The reconstruction gave the positions of first Pd neighbors and, in addition, second hydrogen neighbors around the octahedral interstitial sites. By comparison, in the present experiment, the 8 second Pd neighbors and the 12 first hydrogen neighbors could be observed for the first time.

The data presented in Figure 9.14 prove that according to our expectation the use of the DR method effectively suppresses the spurious spots observed on applying exclusively either ISR or IDR. The reconstructed amplitudes displayed in Figure 9.14d demonstrate convincingly that holography provides not only the intensity but also the phase information. It is worth noting that the signs of the positive and negative amplitudes are well pronounced, thereby allowing one to distinguish nuclei with positive and negative scattering lengths. It should be noted that, in the present investigation, the successful use of 2D multidetectors for neutron holography was demonstrated for the first time.

9.5 Direct Observation of Local Distortion of a Crystal Lattice with Picometer Accuracy Using ARNH

The means and ways of elaboration of ARNH have been considered in detail in Section 9.2. It was proven that ARNH undoubtedly is a valuable tool for the observation and investigation of the local neighborhood of a probe nucleus serving either as a source or a detector. Now, we are in a position to demonstrate the ability of this tool as follows.

Impurity atoms embedded in a regular host lattice generally give rise to local lattice distortions. If the induced atomic displacements occur on surfaces, they can, at least in principle, be viewed directly. Yet, in the bulk usually one has to draw an indirect evidence, and there is no practical way to unambiguous determination of lattice distortions due to single atoms. One of the commonly accepted approaches to their study is diffuse x-ray or neutron scattering around Bragg peaks (Dederichs 1973). However, this method is based on continuum approximation, and, therefore, provides information only about average distortions on the nanoscale (Krexner et al. 2003) while it breaks down in the immediate neighborhood of the defect where the atomistic structure of the lattice has to be taken into account. The interpretation is intimately related to the semi-macroscopic notion of the displacement field (Krivoglaz 1996) and gives reliable quantitative results only at a certain distance from the impurity. The determination of the displacements of individual atoms belonging to the closest atomic shells surrounding the impurity requires an experimental approach with higher resolution. Theoretically, a reconstruction of the atomic positions would be feasible by scanning the elastic diffuse scattering intensity in large regions of reciprocal space and calculating its Fourier transform. *In praxi*, however, the accessible momentum range is strongly restricted, and, consequently, experimental data can be compared only with calculations based on preconceived models.

The novel approach, ARNH, provides detailed information about the positions of individual atoms surrounding the probe nuclei in the bulk. In the work of Cser (2002), the successful reconstruction of the 3D holographic image of the first neighboring shell of Cd impurity atoms occupying regular lattice sites in a single crystal of Pb was described. At that time, the main goal of the experiment was to demonstrate the feasibility of the concept of ARNH. Later on, these data were used for the precise determination of the positions of about 50 individual Pb atoms making up the first 4 neighboring shells surrounding the Cd atom. The composition of the sample was $Pb_{99.74}Cd_{0.26}$ (i.e., nearly 400 Pb atoms for 1 Cd atom) so that the Cd atoms can be considered as noninteracting impurity atoms on an FCC lattice of Pb atoms (Cser 2006).

Applying the Helmholtz–Kirchhoff transformation (Barton 1988) to the experimental data, the positions of the Pb atoms in the first, second, third, and fourth neighboring shells were recognized and identified. In order to illustrate this result, the first neighboring shell of Pb atoms around the Cd site is shown in Figure 9.15. In contrast to x-ray holography which provides images of the electronic shells, the nuclei become visible in a neutron experiment. The successful holographic imaging of atomic positions beyond the nearest neighbors was achieved in this experiment for the first time. For the sake of simplicity, only

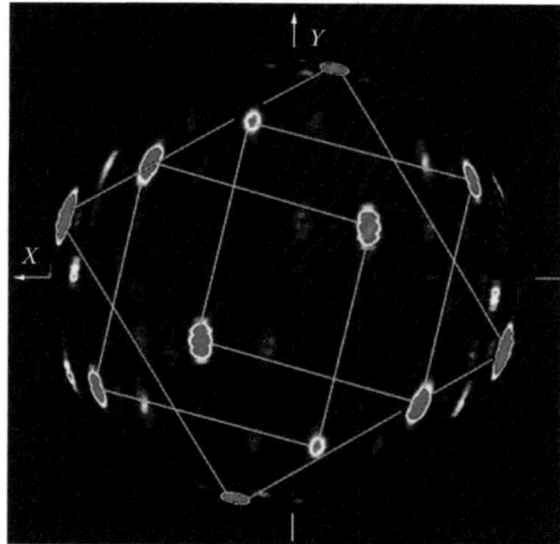

FIGURE 9.15 Stereoscopic view of the first Pb neighbors of a Cd nucleus.

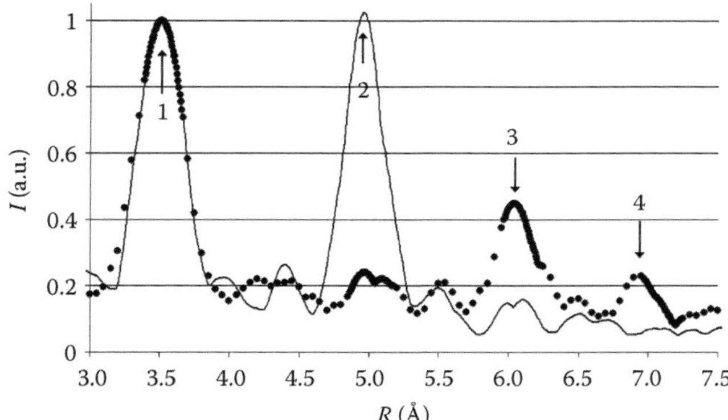

FIGURE 9.16 The maximum intensity distribution of the restored holographic image as a function of the distance R from the Cd nucleus. Experimental values (•) and model calculation for a nondistorted crystal involving 100 surrounding shells (continuous line). The intensity is normalized to the first peak. Vertical arrows mark the positions of atoms in a nondistorted Pb crystal. Oscillations between the peaks are inherent to the holographic restoring procedure.

the maximum value of the restored holographic intensity as a function of the distance from the Cd nucleus was considered (see Figure 9.16).

In Figure 9.16, the intensities of the first, third, and fourth neighbors are clearly visible on the oscillating background. According to model calculations which were carried out on the supposition that all Pb atoms in the PbCd sample occupy the same positions as in the pure Pb crystal, the expected intensity of the second neighbors should be even higher than that of the first one. This fact is a serious argument in favor of the hypothesis that the local arrangement of the crystal is deformed due to the presence of alloyed Cd. In order to explain the deviation of the observed intensities from the calculated ones, let us consider the following arguments: The holographic term of the scattering intensity distribution is given by Equation 9.3.

The integral (Equation 9.8) forming the second term describes the shape of the given peak. The intensity variation according to Equation 9.9 is illustrated in Figure 9.16. Relation 9.9, referred to as the "cosine rule," allows us to obtain the positions of the holographic spots, that is, the positions of the Pb nuclei from the variation of the peak intensity maximum. The distance dependence of this value shown in Figure 9.17 is not normalized.

This means that, from the experimental data, we are able to determine only the ratio of the maximum values. In order to overcome this obstacle, we carried out a numerical fit of 3D Gaussians $G(\chi, \varphi, R)$ to the independent peaks belonging to the first neighbors—both for the experimental results and the model calculation—and accepted the positions of the maxima of the Gaussians as the positions of the holographic spots. Since the points to be fitted are statistically not independent, the quality of the fit was characterized by the coefficient of determination amounting in our case to 0.99962. (This quantity—usually marked as R^2—shows how close the points are to the Gaussian line.)

Using this procedure, all independent positions of first neighbors were obtained. Their average value is equal to 3.5318 ± 0.0024 Å. The error was calculated as the standard deviation from this average value. The oscillations appearing in the model calculation due to the rather small number of shells taken into account do not allow applying a similar approach to more distant neighbors. Yet, starting from the result for the first neighbors, we can determine the relation between their position and intensity, and from the intensity ratio between the first and the other three neighbors, one can calculate their positions using Equation 9.9.

In our case, $\lambda = 0.84$ Å; thus, $|\mathbf{k}| = 2\pi/\lambda = 7.48$ Å^{-1}. The results are given in Table 9.2. From the data given in Table 9.2, it is clear that the displacement varies nonmonotonically with increasing shell number. For the explanation of this observation, we propose the following reasoning: There are two effects, the balance of which determines the perturbation of interatomic distances actually found. One is the ratio of the atomic radii and the second is an electrostatic force derived from the valence difference of the host and the impurity atoms. The atomic radius calculated from the bond length of Pb is equal to 175 pm, while that of Cd is 149 pm (WebElements 2004). The ratio of these two values would favor a contraction of the next neighbor distance; however, the experimental data for the first neighbors actually show an increase. From this, we conclude that the electrostatic force is dominating over the size effect.

Indeed, the outer shell of a Cd atom consists of $5s^2$ electrons while the Pb atom has the electron configuration $6s^2\ 6p^2$. This results in charge depletion at the position of the Cd atom which is equal to −2. In other words, by using a simple qualitative explanation we can tell that Cd atoms embedded in the Pb lattice carry a negative extra charge, causing a repulsive force acting on the p electrons of the first neighboring Pb atoms. As a result, the whole Pb atom suffers a repulsion leading to an increase of the Pb–Cd distance. This fact immediately suggests applying Friedel's theory (Friedel 1952; Kittel 1963). In this approach, the conduction electrons of Pb scatter at the extra charge of the impurity (in the present case, the Cd atom) and form an oscillating charge distribution around it (Kittel 1963). In Figure 9.18, the oscillating term of the p-electrons of the Pb atoms is displayed as calculated using Friedel's description.

At the position of the first neighbor shell formed by the Pb atoms surrounding the Cd impurity, the electron density modulation $\Delta\rho(R)$ has a positive sign. This also means that a repulsive force appears between Cd and Pb atoms, in accordance with the expectation based on the foregoing qualitative consideration. Concerning the more distant neighbors, the variation of the Cd–Pb distance also follows the charge density curve (see the black squares in Figure 9.18). As a consequence, the correlation between the charge density oscillation and the Cd–Pb distance variation becomes obvious. Evidently, the forces caused by the electron density variation around the Cd atom are predominating over the simple internal stress effect due to the size difference. Nevertheless, Figure 9.18 shows that the average local deformation of the first four shells around the Cd atom is primarily contractive.

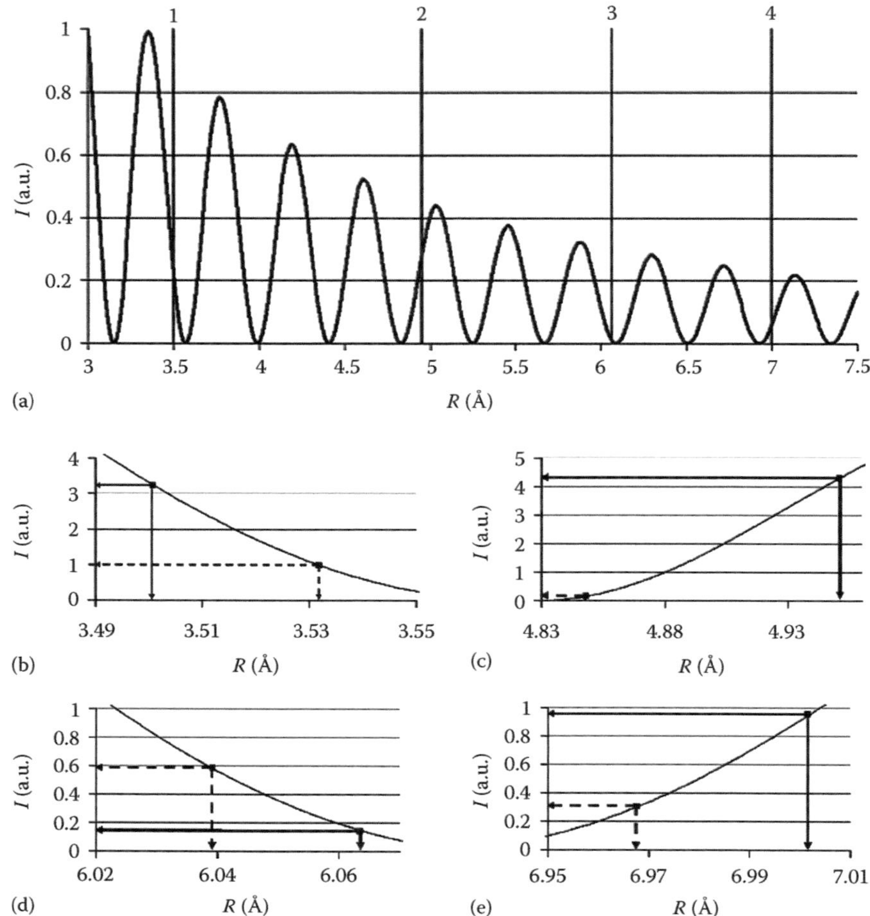

FIGURE 9.17 (a) The variation of the holographic intensity as a function of the distance R from the probe atom calculated using Equation 9.9. The vertical solid lines show the expected positions of the Pb atoms for the perfect crystal. Figures (b), (c), (d), and (e) show a close-up view of the vicinity of the first, second, third, and fourth neighbors, respectively. The vertical dashed lines show the real positions of the Pb atoms obtained from the experimental data.

In conclusion, we underline that, using neutron holography, we have entered a range of interatomic distance measurements which is not accessible to conventional diffuse scattering, neither with x-rays nor with neutrons. This method provides information about the local lattice distortion on an atomic scale. The very fact that there is a direct connection between the intensity of the holographic peaks and their position provides a new approach to the measurement of atomic distances with accuracy in the picometer range. Moreover, from the relationship (Equation 9.9), it can be immediately seen that, by proper choice of the wavelength, the intensity for a particular neighboring shell can be adjusted in such a way as to coincide with the zone of steepest slope with respect to its dependence on the product (**kr**). Thus, the variation of the holographic intensity associated with particular atoms may serve as a very sensitive probe for the determination of strains and, more generally, for detecting changes of interatomic distances as a function of variables such as

TABLE 9.2 Intensity of the Holographic Signal of the First Four Neighbor Shells around a Cd Probe Atom and the Corresponding Displacements from Their Positions in a Perfect Pb Crystal

Neighbor Shell	Relative Intensity	Observed Position (Å)	Undistorted Position (Å)	Shift (Å)
1	1	0.35318	0.35007	0.00311 ± 0.0002
2	0.155	0.48481	0.49508	−0.01027 ± 0.0011
3	0.583	0.60392	0.60635	−0.00243 ± 0.0011
4	0.301	0.69676	0.70015	−0.00339 ± 0.0013

Note: All distances are given in Å.

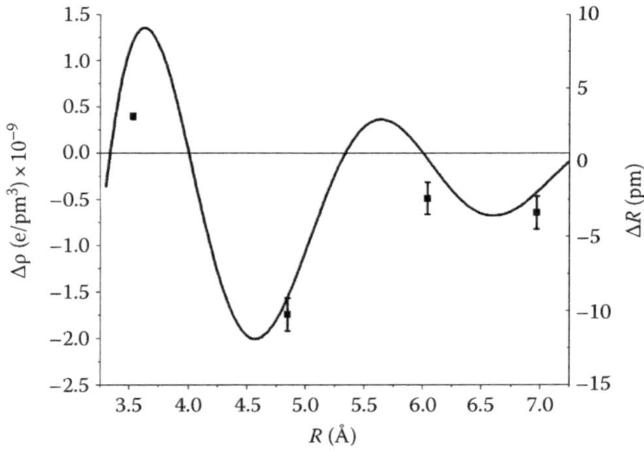

FIGURE 9.18 Continuous line: calculated electron density modulation around a Cd atom (refer to the left scale). Black squares: experimentally observed local distortion as a function of the distance from the Cd nucleus (refer to right scale). The error bars are estimated from the average value of the oscillations modulating the background line.

pressure, temperature, magnetic field, etc., or, for example, due to an incipient phase transition.

9.6 Conclusions and Future Prospects

The authors hope to have been able to convince the reader that ARNH is a promising new tool for studying the 3D local structure of crystalline materials, with extremely high accuracy. Existing spallation neutron sources (ISIS, SNS, JSNS) are powerful pulsed sources of neutrons. It is expected that in the not-too-distant future, the European Spallation Source (ESS) will also come into operation. As it is well-known, the so-called time-of-flight techniques allow to make use of the total quasi-Maxwellian spectrum of neutrons produced by the target of pulsed sources. By exploiting this opportunity, multiwavelength ARNH using the DR method together with position-sensitive area detectors will become feasible. Experiments of this type will be performed efficiently and with high quality, thus allowing us to measure and evaluate both ISH and IDH simultaneously.

Amongst others, by relying on the cosine relationship, it can be anticipated that specific holographic peaks will be measured within those ranges of the wavelength spectrum where they will emerge most distinctly above the background. At the same time, the side oscillations can be suppressed which is very important both for observing weak spots and peaks associated with scatterers occupying positions at a short distance from the probe nuclei. In all, it appears justified to expect a further improvement of the accuracy in measuring interatomic distances even beyond the picometer range. Finally, it should be noted that the high neutron beam intensity of the mentioned new sources will allow us to extend the application of ARNH to samples consisting of nuclei for which the cross sections, both for incoherent scattering and absorption are rather small. This offers a wide range of applications in materials science—including the investigation of magnetic systems—in the not-too-distant future.

Conversion note: 1 Å = 0.1 nm

References

Barton, J.J. 1988. Photoelectron holography, *Phys. Rev. Lett.* 61: 1356–1359.

Barton, J.J. 1991. Removing multiple scattering and twin images from holographic images, *Phys. Rev. Lett.* 67: 3106–3109.

Busetto, E., Kopecky, M., Lausi, A., Menk, R.H., Miculin, M., and Savoia, A. 2000. X-ray holography: A different approach to data collection, *Phys. Rev. B* 62: 5273.

Busing, W.R. and Levy, H.A. 1967. Angle calculations for 3- and 4-circle x-ray and neutron diffractometers, *Acta Crystallogr.* 22: 457–464.

Cser, L., Krexner, G., Prem, M., Sharkov, I., and Török, Gy. 2005. Neutron holography of metal–hydrogen systems, *J. Alloys Comp.* 404–406: 122–125.

Cser, L., Krexner, G., and Török, Gy. 2001. Atomic-resolution neutron holography, *Europhys. Lett.* 54: 747–752.

Cser, L., Török, Gy., Krexner, G., Prem, M., and Sharkov, I. 2004. Neutron holographic study of palladium hydride, *Appl. Phys. Lett.* 85: 1149–1151.

Cser, L., Török, Gy., Krexner, G., Sharkov, I., and Faragó, B. 2002. Holographic imaging of atoms using thermal neutrons, *Phys. Rev. Lett.* 89: 175504.

Cser, L., Török, Gy., Krexner, G., Sharkov, I., and Faragó, B. 2006. Direct observation of local distortion of a crystal lattice with picometer accuracy using atomic resolution neutron holography, *Phys. Rev. Lett.* 97: 255501–255504.

Dederichs, P.H. 1973. The theory of diffuse x-ray scattering and its application to the study of point defects and their clusters, *J. Phys. F* 3: 471–496.

Fanchenko, S.S., Tolkiehn, M., Novikov, D.V., Schley, A., and Materlik, G. 2004. Reply to comment on "Invalidity of low-pass filtering in atom-resolving x-ray holography," *Phys. Rev. B* 70: 106102–106105.

Friedel, J. 1952. The distribution of electrons round impurities in monovalent metals, *Philos. Mag.* 43: 153–189.

Goldschmidt, G.H. and Hurst, D.G. 1952. The structure of ammonium chloride by neutron diffraction, *Phys. Rev.* 83: 88–97.

Hayashi, K., Ohoyama, K., Orimo, S., Nakamori, Y., Takahashi, H., and Shibata, K. 2008. Neutron holography measurement using multi array detector, *Jpn. J. Appl. Phys.* 47: 2291–2293.

Kittel, C. 1963. *Quantum Theory of Solids.* John Wiley & Sons, New York.

Korecki, P., Korecki, J., and Slezak, T. 1997. Atomic resolution γ-ray holography using the Mössbauer effect, *Phys. Rev Lett.* 79: 3518–3521.

Krexner, G., Prem, M., Beuneu, F., and Vajda, P. 2003. Nanocluster formation in electron-irradiated Li_2O crystals observed by elastic diffuse neutron scattering, *Phys. Rev. Lett.* 91: 135502–135506.

Krivoglaz, M.A. 1996. *X-Ray and Neutron Diffraction in Nonideal Crystals*. Springer-Verlag, Berlin, Germany.

Markó, M., Cser, L., Krexner, G., and Sharkov, I. 2006. Instrumental distortion effects in atomic resolution neutron holography, *Physica B* 385–386: 1200–1202.

Markó, M., Cser, L., Krexner, G., and Török, Gy. 2009. Theoretical consideration of the optimal performance of atomic resolution holography, *Meas. Sci. Technol.* 20: 015502.

Markó, M., Krexner, G., Schefer, J., Szakál, A., and Cser, L. 2010a. Atomic resolution holography using advanced reconstruction techniques for 2d detectors, *New J. Phys.* 12: 063036.

Markó, M., Szakál, A., Török, Gy., and Cser, L. 2010b. Construction and testing of the instrument for neutron holographic study at the Budapest Research Reactor, *Rev. Sci. Instrum.* 81: 105110.

Schefer, J., Könnecke, M., Murasik, A., Czopnik, A., Strässle, Th., Keller, P., and Schlumpf, N. 2000. Single-crystal diffraction instrument TriCS at SINQ, *Physica B* 276–278: 168– 169.

Sur, B., Rogge, R.B., Hammond, R.P., Anghel, V.N.P., and Katsaras, J. 2001. Atomic structure holography using thermal neutrons, *Nature* 414: 525–527.

Tegze, M. 2006. Effect of low-pass filtering on atomic-resolution x-ray holography, *Phys. Rev. B* 73: 214104–214110.

WebElements. 2004. *TM Periodical Table (Professional Edition)*, www.webelements.com, accessed on September 23, 2005.

10

Subnanometer-Scale Electron Microscopy Analysis

Sergio Lozano-Perez
University of Oxford

10.1 Brief Introduction to Electron Microscopy

Electron microscopy has been an invaluable technique since it was discovered in the 1930s (Knoll and Ruska 1932). With wavelengths in the range of just a few picometers, it is an ideal tool to discover and characterize the nanoworld and has enabled a true atomic description of a great variety of materials. There is still a long way to go until we can fully exploit such small wavelengths. Optical microscopes have been operating with resolutions very close to their theoretical limit for almost a century. In contrast, electron microscopes still offer a resolution which is, at best, one order of magnitude greater than their wavelength. This is mostly due to the imperfect nature of magnetic lenses. We have, however, seen great improvements recently, especially with the arrival of aberration correctors and monochromators. This chapter will concentrate on the techniques that are capable of extracting sub-nm information, covering imaging and microanalysis. For a comprehensive overview of the technique, many readings are recommended (Buseck et al. 1988; Williams and Carter 1996; Hawkes and Spence 2007).

10.1.1 Types of Electron Microscopes

There are two basic types of electron microscopes: scanning electron microscopes (SEMs) and transmission electron microscopes (TEMs). In a SEM, a small electron probe is scanned over the area of interest, while a detector reads the intensity of emitted or scattered electrons at regular intervals. Each of these measurements will be used to populate the different pixels that will form the final image. Although field emission (FE) SEMs can form probes as small as 0.5 nm, in practice, images never achieve that spatial resolution due to the greater size of the interaction volume from which the detected signal is generated (Gauvin et al. 2006; Schatten 2011). Although the following example might not strictly count as SEM, the incorporation of high-efficiency secondary electron (SE) detectors to scanning TEMs (STEMs) has enabled the acquisition of atomic resolution images with SEs (Konno et al. 2011). This is basically achieved by using a TEM operating in scanning mode with a SE detector above the sample, mimicking SEM operation. This achievement has only been possible due to the use of samples of reduced thicknesses (<50 nm, typical of TEMs), which together with electron channeling, drastically reduces the previously mentioned issue of the interaction volume in conventional SEMs. In Figure 10.1, an atomic resolution image of $SrTiO_3$ acquired in STEM SE mode can be observed.

In the rest of this chapter, only TEM-related techniques will be considered, since SEM is still not capable of providing reliable information in the sub-nm scale. Next, the two most common TEM operation modes will be reviewed.

10.1.2 TEM Imaging Modes

A TEM can be operated in several ways. The most common modes are known as TEM (or image) mode and STEM mode. TEM mode involves using a broad beam to illuminate the sample and acquire images through image plates or charge-coupled devices (CCDs). Scanning TEM mode relies on using a focused beam which is scanned across the sample in a pixel-to-pixel basis, where we would say that it is operated in "scanning mode." The transmitted signal intensity is sequentially recorded with a photodetector(s) for each pixel.

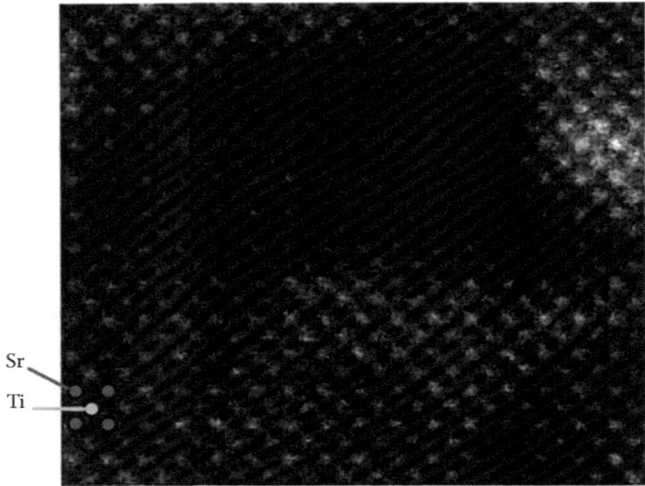

FIGURE 10.1 High-resolution STEM SE image taken at 200 kV along the $SrTiO_3$ [001] zone axis. (Reproduced from Konno, M. et al., *Microsc. Microanal.*, 17, 930, 2011. With permission.)

10.1.2.1 TEM Mode

Sub-nm resolution (and therefore atomic) has been a usual feature for TEMs since the 1970s. Interpretation, however, was not trivial, and full understanding was only achieved when combined with image simulation. As Scherzer stated many years ago, it is not possible to design a static and round lens without axial aberrations. Therefore, spherical (Cs) and chromatic (Cc) aberrations in electromagnetic round lenses will always impose a maximum achievable resolution in a TEM (Scherzer 1936). A "direct" interpretation of the atomic positions and element identification became easier when aberrations were calculated and/or removed. Two approaches can be used and will be described next.

10.1.2.1.1 Software-Based Aberration Removal

This approach is based on an indirect method that allows the recovery of structural information at resolutions beyond those offered by conventional imaging. It relies on the acquisition of a series of images with different defoci and some sort of computer postprocessing to calculate all the experimental parameters and produce a restored

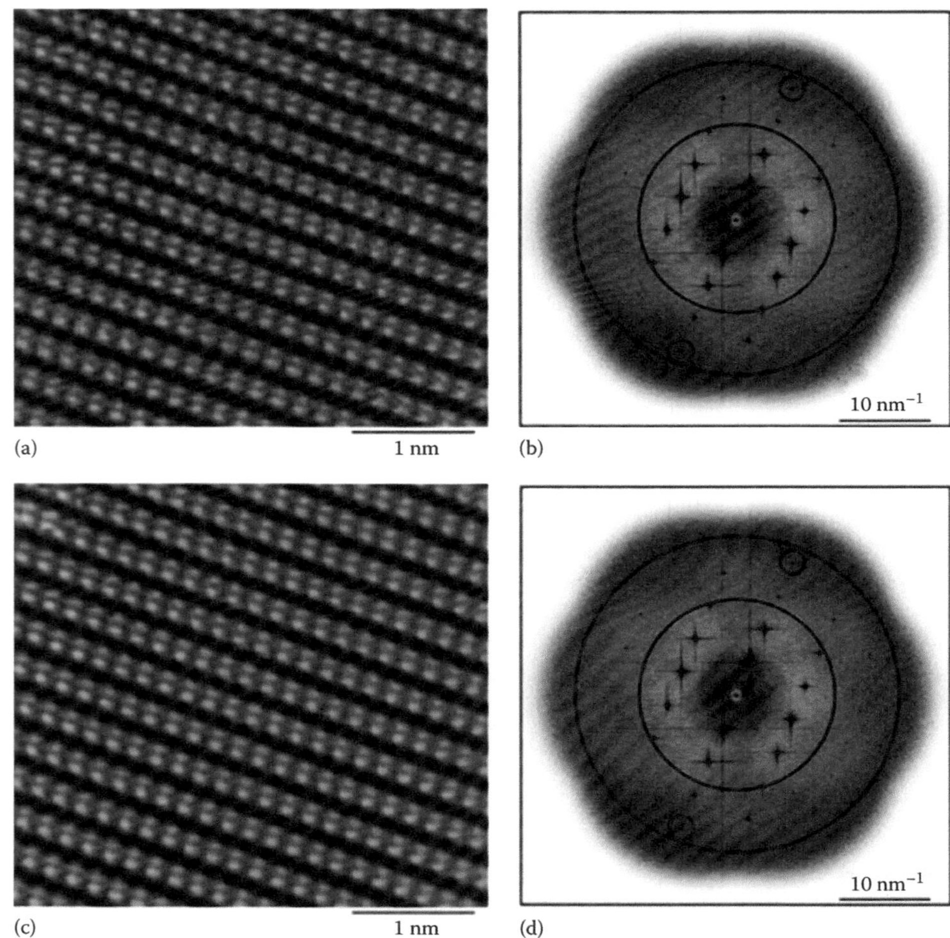

FIGURE 10.2 Identical regions from the phase of the ⟨112⟩ silicon exit wave restored from a tilt–defocus series of 27 images with a maximum tilt magnitude of 16.0 mrad demonstrating the effect of higher order axial compensation. (a) Before higher order aberration correction and (c) after compensating for the calculated higher order axial aberrations. The moduli of the Fourier transforms for panels (a) and (c) are shown in panels (b) and (d), respectively. (Reproduced from Haigh, S. J. et al., *Microsc. Microanal.*, 16, 409, 2010. With permission.)

image of the exit wave function. The number of images acquired can range from just 3 up to a 100, and the restoration of the object plane wave function (modulus and phase, since it is a complex function) will be possible because the different images contain enough information for the calculations (actually, more than enough in most cases). Phase images are very useful to image the true atomic location and, in some cases, allow for the separation of atomic columns of different atomic numbers. When the coherence of the electron source is maintained, as it is the case with field emission gun (FEG) microscopes, and the inelastically scattered electrons are eliminated by using energy filters, the attainable resolution with this technique can be greatly improved (Meyer et al. 2002).

An even more powerful method involves the calculation of the exit-wave function restoration using a combined tilt–defocus series. This way, the restored exit wave can be extended beyond the axial imaging limit. By using spherical aberration corrected microscopes, larger tilts can be used to obtain an improved resolution. In the example shown in Figure 10.2, exit waves have been restored for a range of experimental tilt–defocus datasets acquired using tilts up to 18 mrad. The improvement is demonstrated using a silicon foil orientated along a ⟨112⟩ direction. The method is capable of resolving 0.078 nm separations ({444} reflections) in the phase image. Compensation of the higher order aberrations, by using a Cs-corrector, improves the visibility of the bright peaks in the phase and corrects the symmetry of the atomic columns (Figure 10.2a and c; Haigh et al. 2010).

10.1.2.1.2 Hardware-Based Aberration Removal

This method involves the development of new hardware components, in the form of sets of extra lenses that are added to the microscope. By achieving a resolution down to ~0.1 nm, one can directly characterize the atomic structure of defects, interfaces, and grain boundaries. At "medium" accelerating voltages, this can only be achieved by incorporating aberration correctors to the TEM column. The first Cs-correctors were designed and demonstrated in Scherzer's group in Darmstadt (Haider et al. 2010). Cs-correction results in a more interpretable image of the structure, by reducing contrast delocalization. This has traditionally been the limiting factor in understanding the images of defects, interfaces, and nanoparticles. The influence of Cs on contrast delocalization can be seen in Figure 10.3, where a Si/$CoSi_2$ interface is imaged with/without Cs-correction (Kabius et al. 2002).

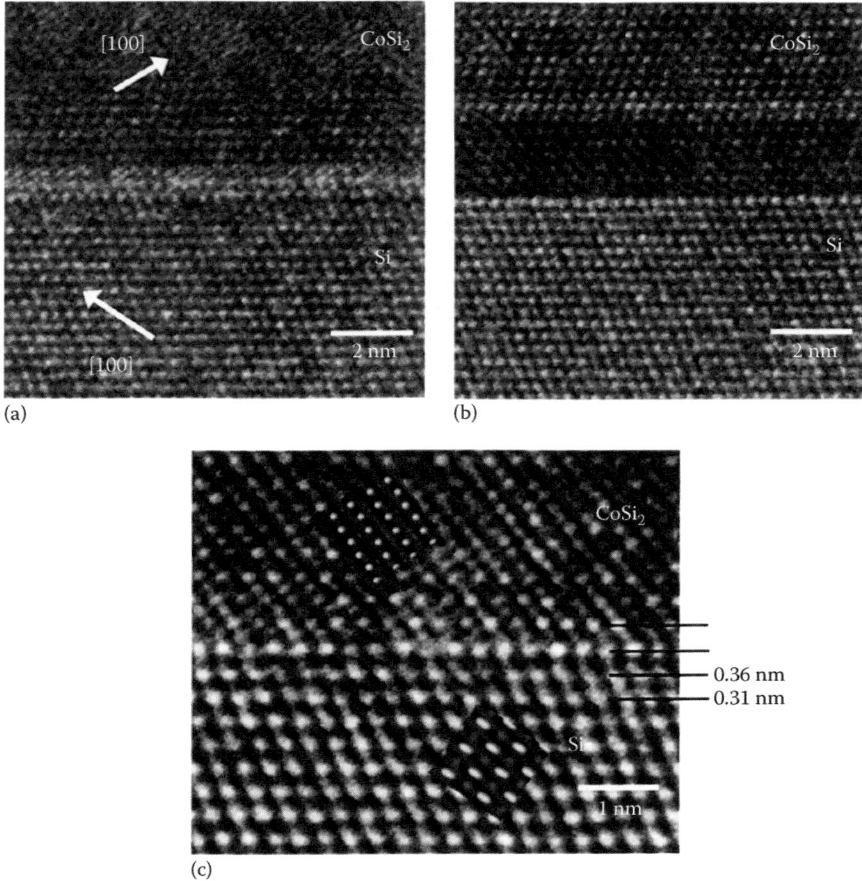

FIGURE 10.3 High-resolution images of an epitaxial Si (111)/$CoSi_2$ interface demonstrating the influence of the spherical aberration on contrast delocalization. Images (a) and (b) were taken with a Cs of 1.2 mm at −67 and −257 nm, respectively. Image (c) was recorded in the aberration-corrected state at a defocus of −12 nm and a Cs value of 50 μm. (From Kabius, B., Haider, M., Uhlemann, S., Schwan, E., Urban, K., and Rose, H., *Journal of Electron Microscopy*, 2002, by permission of Oxford University Press.)

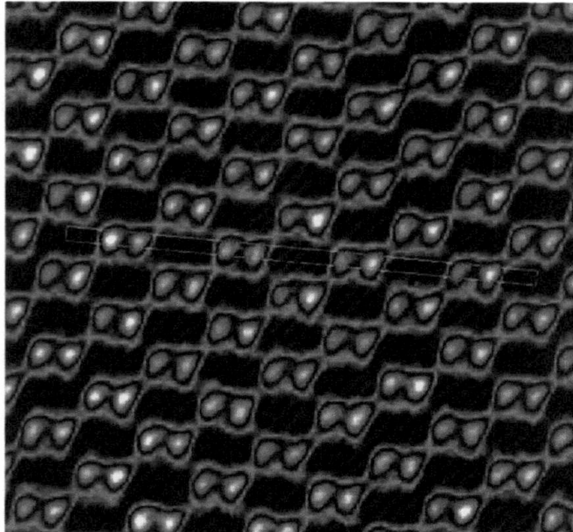

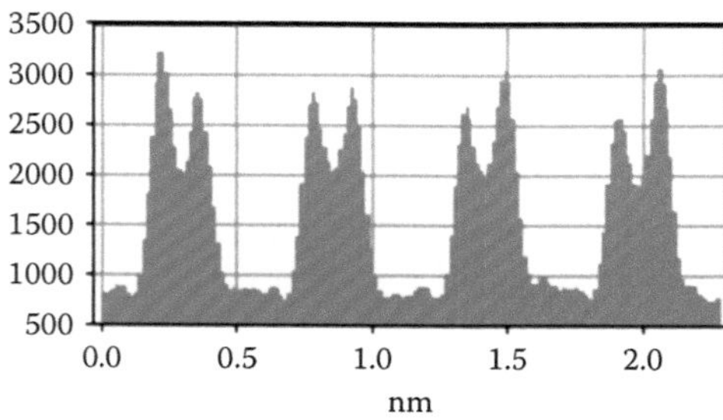

FIGURE 10.4 TEM image of Ge ⟨110⟩ (dumbbell spacing 1.41 Å), using a monochromated beam with $dE = 0.13$ eV. (Reprinted from *Ultramicroscopy*, 114, Tiemeijer, P. C., Bischoff, M., Freitag, B., and Kisielowski, C., Using a monochromator to improve the resolution in TEM to below 0.5Å. Part I: Creating highly coherent monochromated illumination, 72–81, Copyright 2012 with permission from Elsevier.)

Nowadays, it is possible to design (S)TEMs with incorporated Cc- and Cs-correctors (see Section 10.3). Monochromators have also become a reality and offer an alternative way of reducing the effects of chromatic aberrations. As demonstrated by Tiemeijer et al. (2012), when the benefits of spherical aberration correction (reaching resolutions up to 0.7 Å) are combined with the effects of a monochromator, attainable resolution can be dropped to 0.5 Å. This resolution enabled the structure of Ge ⟨110⟩ dumbbells to be resolved with great accuracy, as shown in Figure 10.4.

An increasing demand to characterize carbon-based and 2D materials means that sub-nm resolution should also be possible at low accelerating voltages. This is required if knock-off damage is to be avoided during imaging. Recent work at Ulm (Kaiser et al. 2011) has demonstrated this to be possible when combining low-keV TEM with Cs-correction and monochromation (Figure 10.5).

10.1.2.2 STEM Mode

Sub-nm resolution in STEM mode usually relies on electron channeling. Therefore, it is only available in crystalline samples along well-aligned crystallographic axes or in 1D or 2D materials, where only one layer of atoms is imaged in projection. By using high-angle annular dark field (HAADF) detectors, Z-contrast (with Z being the atomic number) can be used to obtain chemical information with atomic resolution. This was demonstrated during the late 1980s (Pennycook 1989), and since then, the technique has become very popular, with new microscopes pushing the resolution limits appearing almost every year. A detailed review of the technique can be found in the work of Liu (2005). A major development was achieved when the first probe spherical aberration correctors were available over a decade ago. By using a sub-nm Cs-corrected probe, Z-contrast can, under certain circumstances, be used to fully resolve the composition of a sample in projection. Recent examples showed that by calibrating the signal collected by the HAADF detector, single atoms can be identified (E et al. 2010). Another great example of application by Krivanek et al. (2010) illustrates how HAADF signal calibration can be used to identify individual atoms in 2D materials. In Figure 10.6, it can be seen how a 2D layer of boron nitride (BN), doped with N and O, can be imaged and the intensities of single atoms measured, providing unique signatures that enable their identification.

10.2 Chemical Analysis

As with imaging, high-resolution chemical analysis can be attained in TEM or STEM mode. The two approaches are, however, very different. In TEM mode, only energy-filtered TEM (EFFTEM) is available, while spectrum imaging (SI) in STEM mode can offer energy-dispersive x-ray spectroscopy (EDS) or electron energy loss spectroscopy (EELS).

10.2.1 Energy-Filtered TEM

Energy-filtered TEM involves the collection of elastically and inelastically scattered electrons through an image filter which will separate them according to energies using a magnetic prism. A slit is then used to select only the electrons with the desired energy range. Two sorts of magnetic prisms can be used. One is integrated in the column and performs all the aberration corrections in a geometrical way. There are several designs available, but the Ω-filter design (Tsuno 1999; 2004) is the most popular. The other involves the use of a postcolumn filter (located at the end of the electron column) (Gubbens and Krivanek 1993; Krivanek et al. 1993). The main difference with the in-column design is that all aberration corrections are done actively through the use of complex lenses (quadrupoles, hexapoles, etc.). The two approaches have advantages and disadvantages. The Ω-filter is compact, practically alignment-free (since most corrections are geometrical and predesigned), and allows to acquire EFTEM images at the same magnification as on the viewing screen. This makes the acquisition straightforward and fast. On the other hand, geometric corrections only cover aberrations up to the second order, and performance is not as impressive as with postcolumn image filters. In particular, EELS acquisition is penalized and energy resolution is not

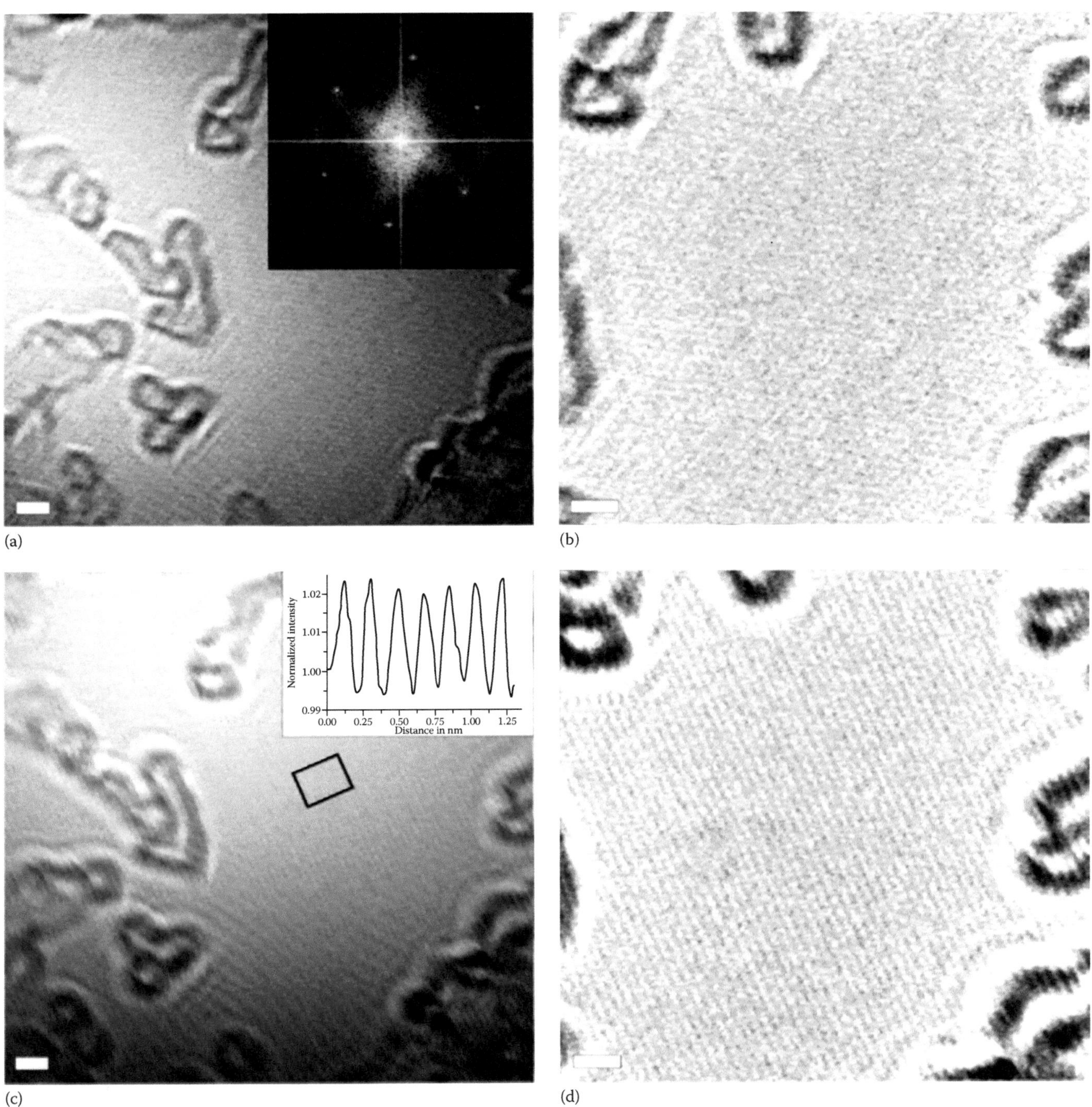

FIGURE 10.5 HRTEM images of single-layer graphene, acquired at 20 kV. (a) Unfiltered exposure, with beam tilt and defocus adjusted to transfer all six primary lattice reflections of graphene (corresponding to a 213 pm spacing) with similar intensity. Inset shows Fourier transform of this image. (b) Background-subtracted section of the image (filtered to remove the uneven illumination). (c) Unfiltered single exposure, with conditions adjusted to maximize transfer of one of the lattice reflections. Inset shows that a modulation of 2.5% (normalized to the mean intensity) was achieved. (d) Flat-filtered section of image (c). All scale bars are 1 nm. (Reprinted from *Ultramicroscopy*, 111, Kaiser, U., Biskupek, J., Meyer, J. C., Leschner, J., Lechner, L., Rose, H., Stöger-Pollach, M. et al., Transmission electron microscopy at 20 kV for imaging and spectroscopy, 1239–1246, Copyright 2011 with permission from Elsevier.)

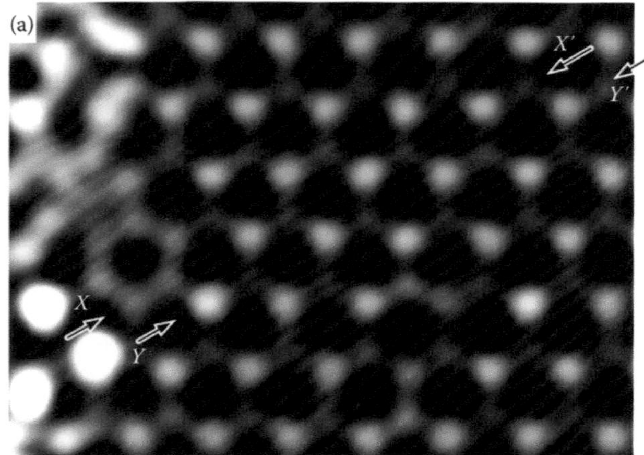

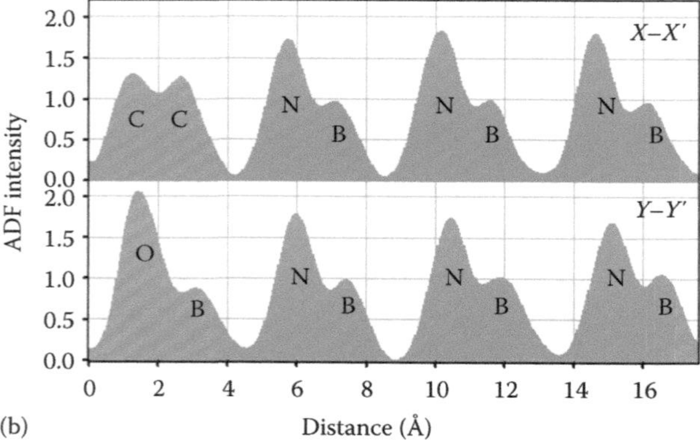

FIGURE 10.6 (a) ADF STEM image of monolayer BN corrected for distortion, smoothed, and deconvolved to remove probe tail contributions to nearest neighbors. (b) Line profiles showing the image intensity (normalized to equal one for a single boron atom) as a function of position in image (a) along *X–X′* and *Y–Y′*. The elements giving rise to the peaks seen in the profiles are identified by their chemical symbols. (Reprinted by permission from Macmillan Publishers Ltd. Krivanek, O. L., Chisholm, M. F., Nicolosi, V., Pennycook, T. J., Corbin, G. J., Dellby, N., Murfitt, M. F. et al., Atom-by-atom structural and chemical analysis by annular dark-field electron microscopy, *Nature*, 464, 571–574, 2010. Copyright 2010.)

optimum. Postcolumn filters require frequent lens alignments that are, fortunately, fully automated. Modern models can correct up to fourth order aberrations, optimizing EFTEM and EELS acquisition. They have their dedicated CCD. Their main disadvantage is that they acquire images at a different magnification from the viewing screen, and this can make locating small features a bit challenging. A good introduction to the technique can be found in the works of Reimer (1998) and De Bruijn et al. (1993).

In the "early" days, EFTEM elemental maps were mostly obtained by acquiring one (for jump-ratio images) or two pre-edge images and one postedge. This approach could greatly simplify the computational requirements to make the background subtraction possible and was widely used. As computers became more powerful, the benefits of acquiring a series of EF images which contained several pre-edge and post-edge images became obvious (Thomas and Midgley 2001). Background fitting could then be performed with higher confidence, and the optimum postedge integration region could be chosen after a careful examination of the dataset, thus improving the final result.

Energy-filtered TEM images suffer from "extra" sources of degradation compared to their nonfiltered counterparts. Krivanek and Egerton have characterized them carefully (Krivanek et al. 1995; Egerton and Crozier 1997), identifying the main sources as a contribution from chromatic and spherical aberrations, diffraction by the objective (or equivalent) aperture, and delocalization of the electrons as they interact with the nuclei. This way, expected image degradations depending on the experimental conditions used can be predicted. The choice of experimental parameters include: primary electron beam energy, objective aperture diameter, slit width, energy loss of the acquired images, or aberration coefficients. The effects of these parameters can be predicted and simulated, so that the right choice is made beforehand (Lozano-Perez and Titchmarsh 2007).

In a typical EELS or EFTEM acquisition, where a sample would have a thickness well below its inelastic mean free path, most of the electrons involved would just pass through it without suffering any inelastic interaction. This is easily verified by checking how tall the zero-loss peak is compared to the rest of the spectrum. A direct consequence of this behavior would be the lack of electrons in EF images, particularly those acquired from core-loss regions (typically above 200 eV loss). In order to overcome this problem, there are two approaches:

1. Use a brighter beam, which is achieved by either using a brighter source (e.g., LaB_6) or by increasing the aperture or the slit size. The first approach might increase beam damage if the sample is sensitive, and the second will decrease the spatial resolution of the images.
2. Use longer exposure times, which will only be successful if the drift is kept under control.

In general, in addition to using a microscope with good imaging performance, it is often found that a bright source is what ultimately makes a difference. A LaB_6 filament for medium magnifications (up to 300 k×) or Cs-corrected FEG-TEMs are good options. In addition, it was recently demonstrated that, although the information of sub-nm features might be present on the data, the use of multivariate statistical analysis (MSA) tools might be necessary to reveal it, since it will probably be masked by the statistical noise (Lozano-Perez et al. 2009). This will be illustrated in Section 10.3.

Atomic high-resolution information, formed by wave interference, can also be present in EFTEM images, as proven by Lugg et al. (2010) and shown in Figure 10.7. This can be useful in some cases, but it must not be considered to be truly chemical information.

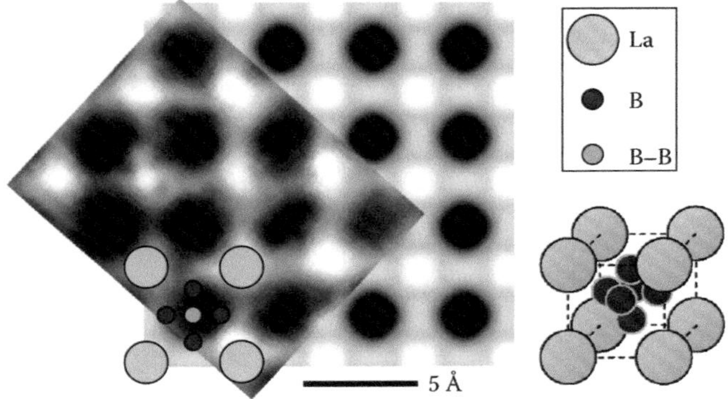

FIGURE 10.7 Experimental EFTEM image and theoretical simulation (underlay) for lanthanum $N_{4,5}$ edge (~100 eV loss) in LaB_6. The sample thickness was assumed to be 60 Å. The projected structure is indicated, and we distinguish between columns containing one (B) and two (B–B) boron atoms per unit cell. (Reprinted from *Ultramicroscopy*, 110, Lugg, N. R., Freitag, B., Findlay, S. D., and Allen, L. J., Energy-filtered transmission electron microscopy based on inner-shell ionization, 981–990, Copyright 2010 with permission from Elsevier.)

10.2.2 Spectrum Imaging

An alternative approach to microanalysis can be achieved in STEM mode, where a small probe is scanned across the sample and the attainable spatial resolution is controlled by the interaction volume of the electrons and the sample. A review of how interaction volumes are calculated can be found in the study of Williams and Carter (1996). When working in STEM mode, two different methods are available to extract chemical information: EELS and EDX. They both present the same experimental challenges. The electron probe has to be small enough and positioned with enough precision over an atomic column (or single atom) for sufficient time so that a meaningful spectrum can be collected. Potential problems will be as follows:

1. Specimen drift that can be corrected in situ with the right software/hardware.
2. Beam damage on the sample if the accelerating voltage is too high (unless hydrolysis is to be avoided) or the beam is too bright or scanned too slowly.
3. Slow spectrum read-out that slows down the scanning and imposes a minimum time between consecutive acquisitions. This has, however, been greatly improved in more recent spectrometers/detectors.

The acquired EDX/EELS spectra will typically contain very few counts, and signal extraction can be improved by MSA processing, as will be discussed in Section 10.3. A recent review of the capabilities of high-resolution analytical STEM can be found in the work of Stroppa et al. (2012).

As discussed in Section 10.1.2.2, electron channeling is the key to obtaining sub-nm or atomic resolution in STEM mode. The concept, although simple, could lead to the wrong interpretation of the results in 3D samples if it is not properly understood. Though, a small enough probe can be located on a single atomic column that will "channel" the electrons. What recent experiments and simulation have revealed is that, as the electrons travel through the sample, they can transfer to the neighboring columns (Allen et al. 2006; Oxley et al. 2007; Lugg et al. 2011). This dechanneling effect will have consequences for understanding the real origin of the observed signal (HAADF, EELS, or EDX), and it should not be ignored.

10.2.2.1 EELS SI

EELS acquisitions can be performed through a Ω-filter, a dedicated parallel spectrometer or a postcolumn image filter operated in spectroscopy mode. In all cases, the electrons will be detected with a fixed collection angle after interacting with the sample. An excellent review of the technique can be found in the work of Pennycook et al. (2009).

Sub-nm resolution can only be achieved for 2D or 1D materials (single atoms) where only one atom is excited in projection, or in 3D materials oriented in a particular zone-axis (with limited thickness), where electron channeling keeps most of the electrons traveling through a single atomic column (Spence 2006).

As an example of atomic resolution EELS in 3D materials, an EELS SI from a region of a layered perovskite manganite, $La_{1.2}Sr_{1.8}Mn_2O_7$, is shown in Figure 10.8. Extracted maps of La (edges M and N), Mn, and O reveal the location of the atoms. It should be reminded that these types of experiments require a crystal with special atomic arrangements. Thus, when the right zone-axis is used, along certain atomic columns, only one type of atom is found. This is the case for the perovskite shown along the [010] direction (Kimoto et al. 2007).

When analyzing 2D materials such as graphene, beam damage becomes the main issue, and low-voltage operation is crucial for success. As can be seen in Figure 10.9, an accelerating voltage of 60 keV combined with the use of a specially designed spectrometer is capable of acquiring EEL spectra from individual carbon atoms. The results show how the different bonding arrangements influence the fine structure of the carbon K edge (Suenaga and Koshino 2010).

10.2.2.2 EDX SI

When the beam of electrons interacts with the atoms in the sample, x-rays are generated in all directions (covering a 4π steradian solid angle). Traditional side-entry EDX detectors will only collect x-rays emitted from the sample over a solid angle of ~0.2 sr. This, unlike in EELS acquisitions where the collection angles are big enough to detect most inelastic electrons, will mean that only a small fraction of the produced x-rays will ever reach the detector. Another problem which is not present in EELS is fluorescence. X-rays generated inside the interaction volume (directly by the electron beam) will travel through the sample toward the detector (located at a certain take-off angle, usually ~20°) and might "lend" their energies to excite the electrons from atoms in their way which will then de-excite by emitting a fluorescent x-ray. In Figure 10.10,

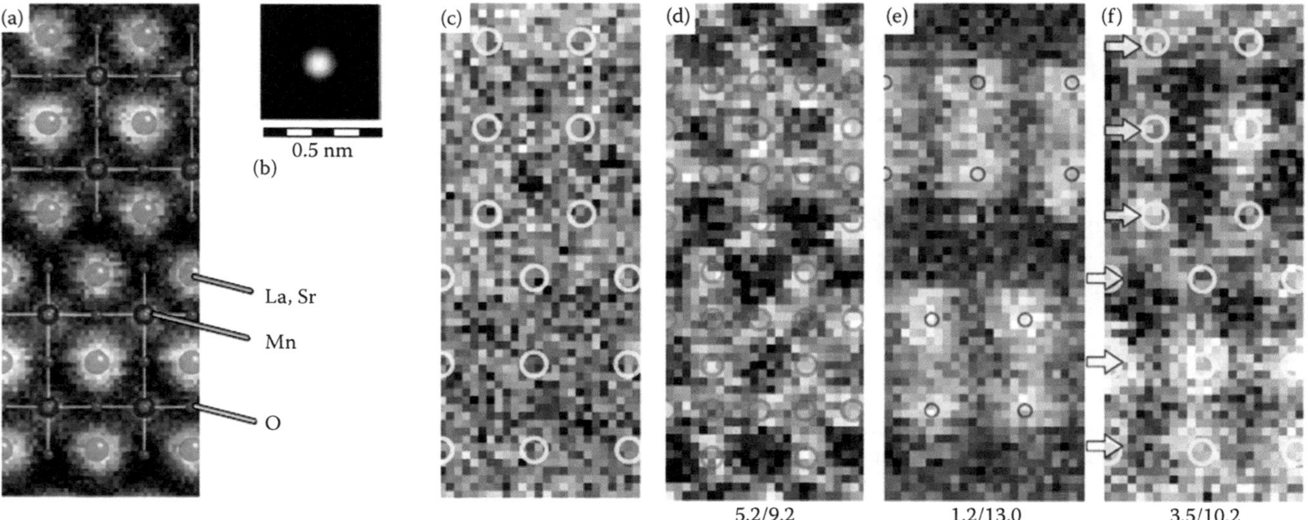

FIGURE 10.8 Atomic-column imaging using STEM and EELS. (a) Enlarged ADF image of analyzed area overlapped with crystal structure. (b) Calculated incident probe for EELS and ADF observations. (c–f) Core-loss images of La N_{45}, O K, Mn $L_{3,2}$, and La $M_{5,4}$. Numbers below panels (d–f) indicate the minimum/maximum of the display range. Circles in panels (c–f) indicate the corresponding positions of atomic columns. (Reprinted by permission from Macmillan Publishers Ltd. Kimoto, K., Asaka, T., Nagai, T., Saito, M., Matsui, Y., and Ishizuka, K., Element-selective imaging of atomic columns in a crystal using STEM and EELS, *Nature*, 450, 702–704, 2007. Copyright 2007.)

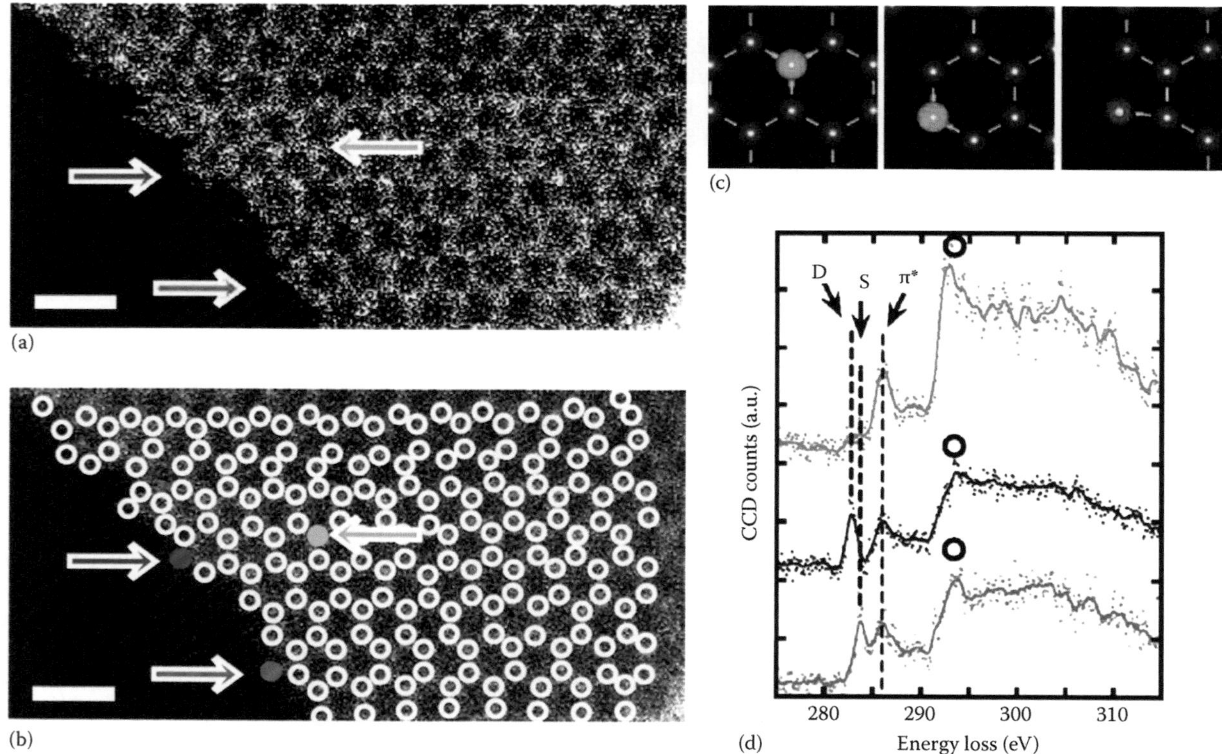

FIGURE 10.9 **(See color insert.)** Graphene edge spectroscopy. (a) ADF image of single graphene layer at the edge region. No image processing has been done. Atomic positions are marked by circles in a smoothed image (b). Scale bars, 0.5 nm. (d) ELNES of carbon K (1s) spectra taken at the color-coded atoms indicated in (c). Green, blue, and red spectra correspond to the normal sp2 carbon atom, a double-coordinated atom, and a single-coordinated atom, respectively. These different states of atomic coordination are marked by colored arrows in (a) and (b) and illustrated in (c). (Reprinted by permission from Macmillan Publishers Ltd. Suenaga, K. and Koshino, M., Atom-by-atom spectroscopy at graphene edge, *Nature*, 468, 1088–1090, 2010. Copyright 2010.)

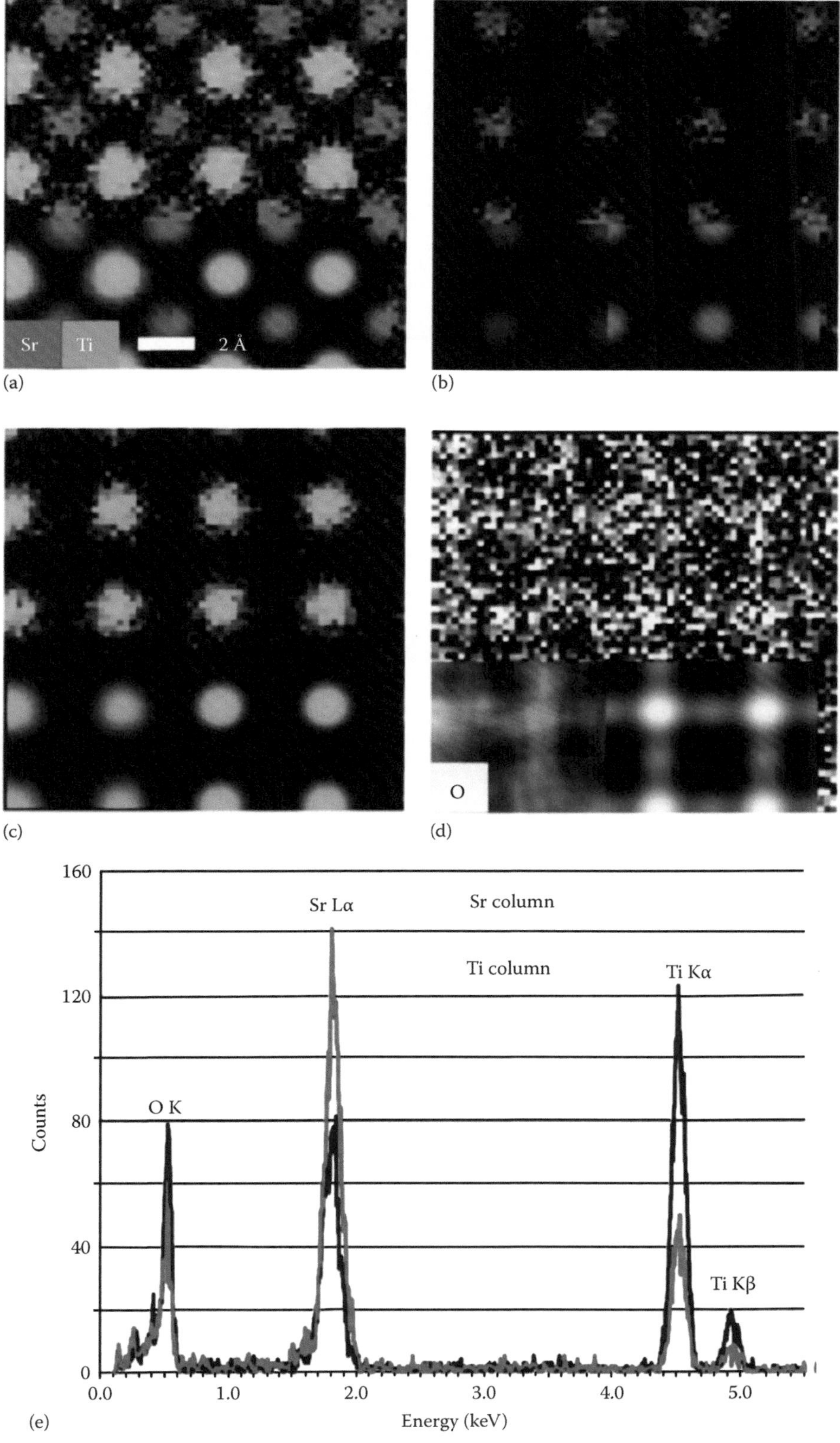

FIGURE 10.10 (a) Composite Sr/Ti chemical map with an averaged map obtained from several such maps, inset at bottom left, and a simulation to the right of that; the top half of the map is the raw data from a single scan. (b) The Sr (K- plus L-shell) contribution to (a). (c) The Ti K-shell contribution to (a). (d) O K-shell map with average and simulations inset as before. (e) Typical EDS with the probe placed above the Sr (light gray) and Ti-O (dark gray) columns. (Reproduced from Allen, L. J. et al., *MRS Bull.*, 37, 47, 2012; Klenov, D. and Lazar, S., *Microsc. Anal.*, 82, 3, 2011. With permission.)

it can be seen how by using an improved collection solid angle, individual atomic columns in $SrTiO_3$ can be identified by EDX SI (Allen et al. 2012). This new development will be covered in the next section. A good introduction to this technique can be found in the work of Williams et al. (1995).

10.3 New Developments

10.3.1 High Solid-Angle EDX Detectors

In the last few years, a new generation of x-ray detectors has become available in TEMs. They are based on silicon drift detector systems, so they do not need to be cooled by liquid nitrogen and offer improved collection efficiencies by using solid angles of collection of up to 1 sr. This has been achieved by using multiple detectors integrated around the upper objective pole-piece (FEI™ ChemiSTEM and SuperX) (Von Harrach et al. 2010) or by using a single detector that can be located much closer to the sample than before (Jeol™ Centurio or Oxford Instruments™ X-Max) (Rowlands and Burgess 2009).

The application of this new technology enables the acquisition of EDX SI in a much shorter time (or in the same time but with more pixels or longer pixel times) (Von Harrach et al. 2010) and the detection of impurities in smaller amounts (Pantel 2011, as can be seen in Figure 10.11, where *P* segregation below 0.1 wt.% is detected in a flash memory component. A recent review, where excellent results obtained by 1 sr EDX detectors are shown can be found in the work of Allen et al. (2012) (see the example in Figure 10.10).

10.3.2 Chromatic Aberration Correctors

Although chromatic aberration correction was the first to be implemented (in an SEM), it has not been until very recently when it has finally become a reality in a TEM (Zach 2009; Haider et al. 2010). Chromatic correction might change the way we do electron microscopy, since most microscopes were designed to minimize its effects through high electron energies, highly excited objective lenses, and the use of FEGs. The TEAM project in Argonne National Lab (the United States) was one of the first to implement a Cc-correction in a TEM back in 2007.

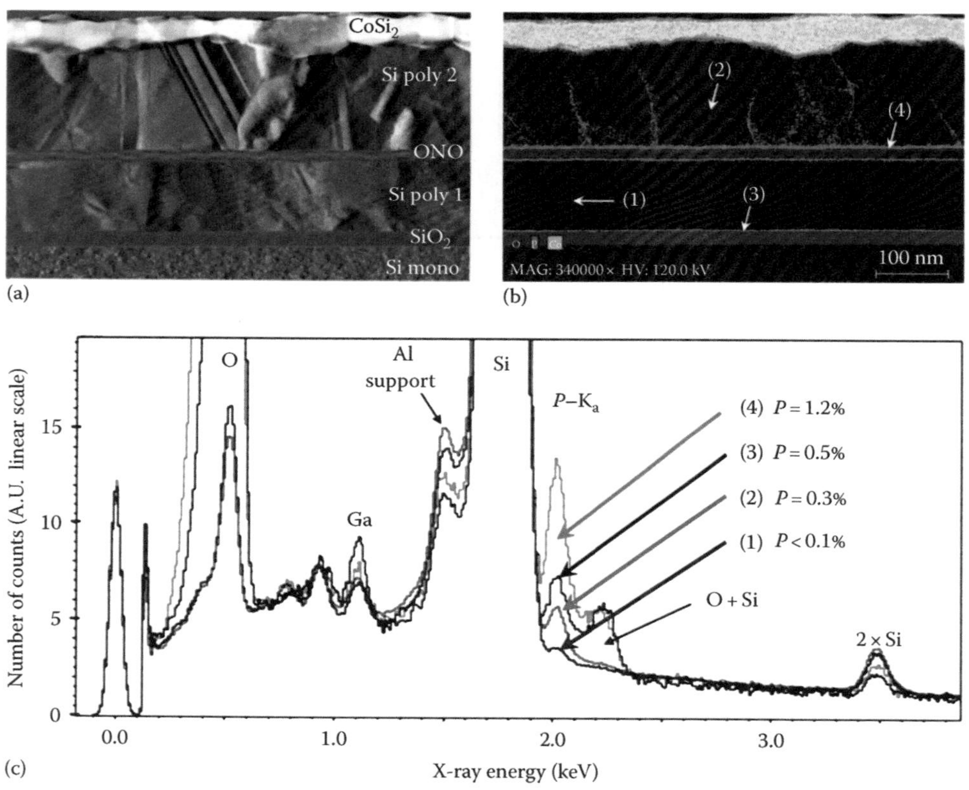

FIGURE 10.11 (a) STEM dark field image of a gate of flash memory cross section. (b) STEM EDX elemental maps showing, O, P, and Co in the same cross section. (c) Zoom detailing EDX spectra around the phosphorus P–K$_a$ line. The spectra are acquired at different points indicated in (b) (labeled 1, 2, 3, 4). Minor fluorescence and artifacts peaks are also indicated. (Reprinted from *Ultramicroscopy*, 111, Pantel, R., Coherent Bremsstrahlung effect observed during STEM analysis of dopant distribution in silicon devices using large area silicon drift EDX detectors and high brightness electron source, 1607–1618, Copyright 2011 with permission from Elsevier.)

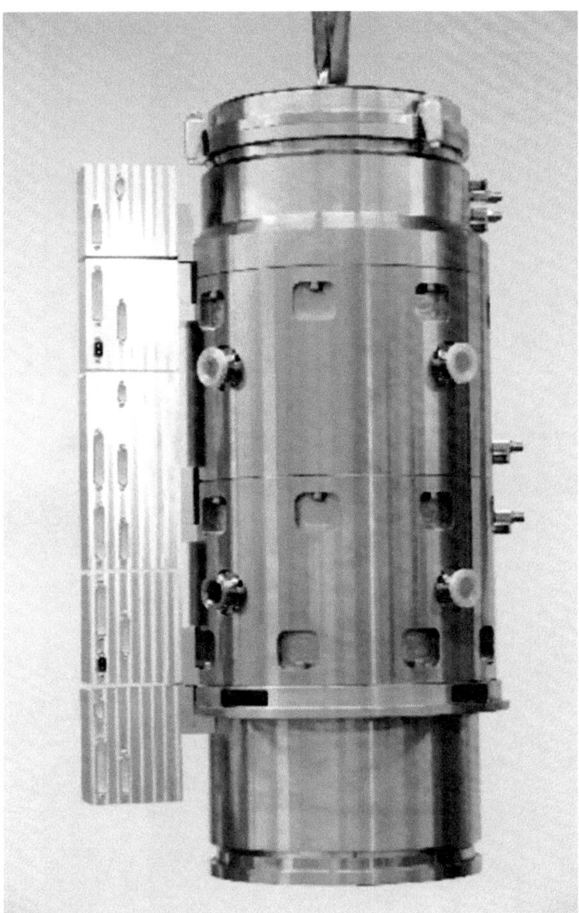

FIGURE 10.12 Picture of the CEOS Cc-/Cs-corrector C-COR hanging at a crane just before it was incorporated into a FEI™ Titan column. At the left side, the connector boxes for the current and voltage supplies and, almost at the center, four vacuum ports can be seen. The total height of the corrector is 823 mm. (Reproduced from Haider, M. et al., *Microsc. Microanal.*, 16, 393, 2010. With permission.)

They have now demonstrated resolutions of 0.5 Å (Shenkenberg 2007). Currently, Cs-/Cc-correctors can be fitted to electron columns, such as the one offered by CEOS™ (Figure 10.12; Haider et al. 2010).

10.3.3 Fast-Spectrum Image Acquisitions

Traditionally, chemical mapping by spectrum imaging has been limited by the relatively long acquisition times required in order to obtain elemental maps with enough pixel density. Early EEL spectrometers had long read-out times (~1 s) preventing any acquisition with a high number of pixels and EDX spectrometers, with their small collection angles, also required a long acquisition time per pixel to accumulate signal with an acceptable signal-to-noise ratio (SNR). For this reason, line profiles were often used instead. This has now changed, and modern hardware is fully prepared for the challenge of atomic resolution analytical mapping. When this was first demonstrated, "bulk" specimens such as $SrTiO_3$ with simple atomic arrangements down certain columns were used as a proof of principle. Nowadays, we expect chemically resolved atomic resolution from nanoparticles, and this could only be achieved by making sure that the atoms in the particle were not affected by the interaction with the electron beam. This interaction is not easy to avoid and there might be thresholds for almost all experimental parameters involved: electron beam energy, electron current, or pixel time. Fast scanning to minimize pixel time does not generate enough signal, and multiple passes/frames are required. This means that the hardware and software used have to be fast enough to collect a full spectrum in the short time when the beam is on a single pixel. Drift also needs to be measured and compensated for. As can be seen from Figure 10.13, choosing the right instrument and experimental conditions can provide with truly atomically resolved EDX maps (Herzing et al. 2008).

A similar approach enables simultaneous acquisition of low- and core-loss EELS at the required speed for atomic mapping (Scott et al. 2008).

10.3.4 Multivariate Statistical Analysis

Although we should not consider MSA to be a "recent" development, it is only in the last couple of years that it has started becoming a mainstream technique. Since Trebbia demonstrated its power in his classic papers (Trebbia 1988; Trebbia and Bonnet 1990; Trebbia and Mory 1990), MSA has been used in various forms to improve the interpretation or the noise levels of the data, including principal component analysis (PCA) (Bentley et al. 2006; Bosman et al. 2006; Burke et al. 2006; Kotula et al. 2006; Titchmarsh 1999) or independent component analysis (ICA) (Colliex et al. 2010; Cooper et al. 2011; de la Peña et al. 2011).

MSA post-acquisition processing offers several advantages. Datasets that usually contain millions of data-points and that are not trivial to analyze can be fully decomposed into simpler components that might offer a direct and easier interpretation after ICA. Since this procedure is purely mathematical and therefore "unbiased," it can pick up trends in the data that might be unnoticed even for a specialist. In addition, once the data is decomposed, each component can be sorted in terms of its relevance and the components that contain information separated from those which only contain statistical noise. If only the relevant components are used to reconstruct the original dataset, the SNR of any feature in the dataset will be increased. The results can be astonishing and features which were originally "masked" by the statistical noise fully revealed. In the example provided in Figure 10.14, the author was able to visualize, via chemical mapping, sub-nm Yttrium-rich nanoclusters using EFTEM + MSA processing (Lozano-Perez et al. 2009).

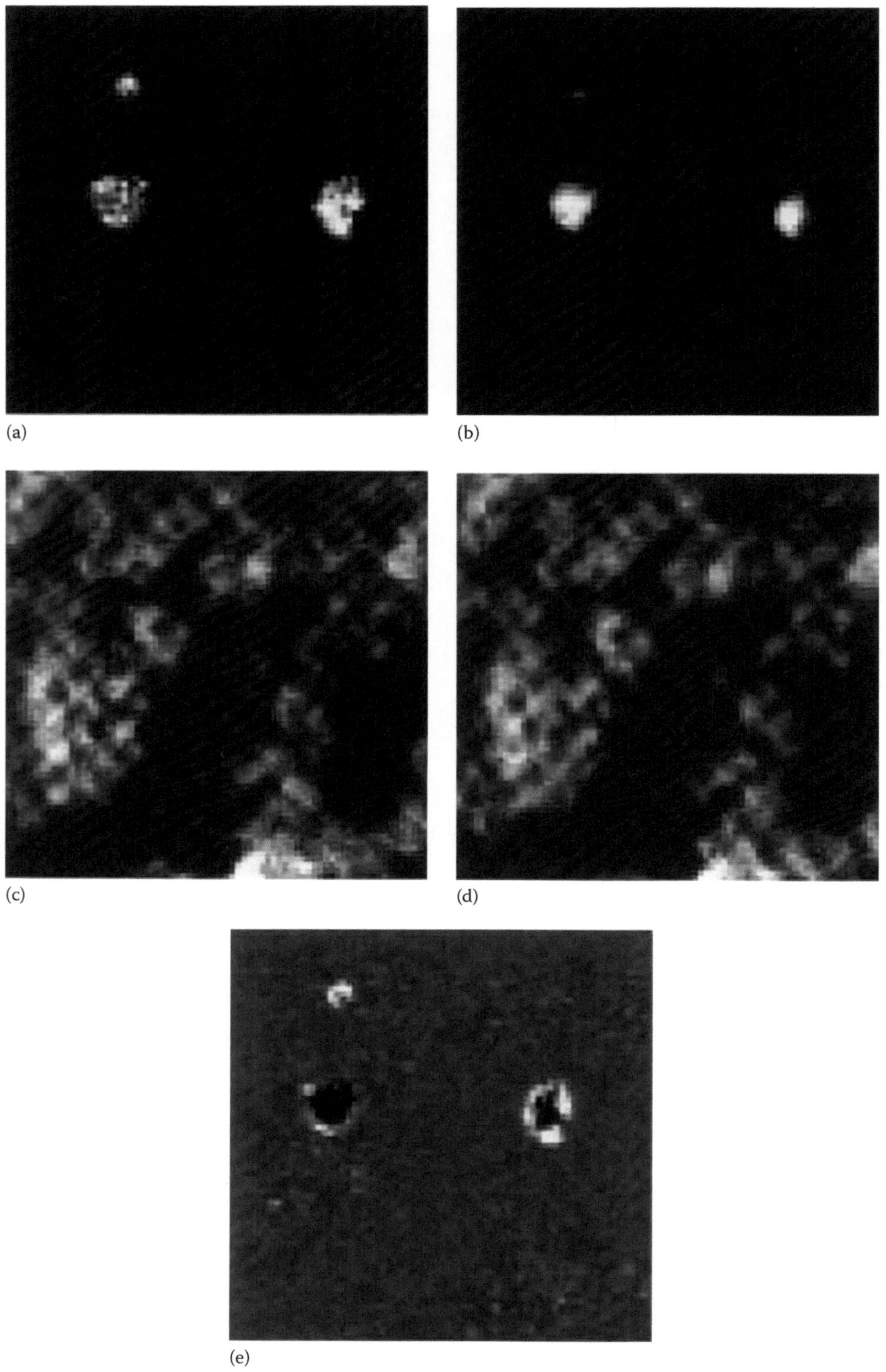

FIGURE 10.13 HAADF image (top) and MSA-reconstructed EDX maps of (a) Pd, (b) Au, (c) Fe, (d) O, and (e) fully isolated Pd signal. (From Herzing, A. A., Watanabe, M., Edwards, J. K., Conte, M., Tang, Z. -R., Hutchings, G. J., and Kiely, C. J., Energy dispersive x-ray spectroscopy of bimetallic nanoparticles in an aberration corrected scanning transmission electron microscope, *Faraday Discussions*, 138, 337–351, 2008. Reproduced by permission of The Royal Society of Chemistry.)

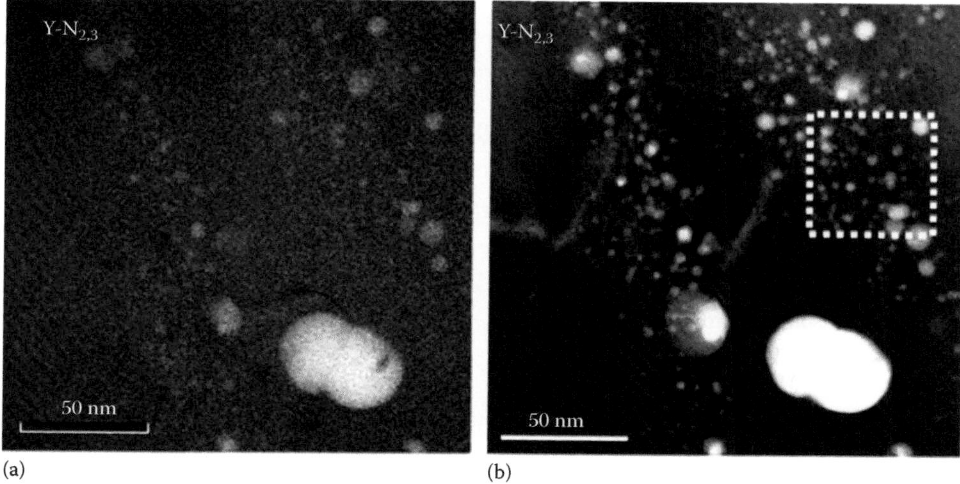

(a) (b)

FIGURE 10.14 EFTEM elemental maps from raw data (a) and after MSA "noise-free" reconstruction (b). (Reprinted from *Ultramicroscopy*, 109, Lozano-Perez, S., de Castro Bernal, V., and Nicholls, R. J., Achieving sub-nanometre particle mapping with energy-filtered TEM, 1217–1228, Copyright 2009 with permission from Elsevier.)

10.4 Conclusions

In this chapter, an overview of the capabilities of electron microscopy for unveiling the sub-nm world has been presented, demonstrating that the technique is perfectly suited for the challenge. More importantly, this is a fast-developing field, and every year new developments are presented. The twenty-first century has seen many exciting and important advances, but there are surely many others to come. Through its many different modes of operation, an electron microscope can provide information on the microstructure and atomic arrangement with a resolution in tens of picometers. More importantly, the analytical side is quickly catching up and sub-nm resolution is finally a reality. Through this chapter, many of the common problems and artifacts that affect both the acquisition and the interpretation of the results have been discussed. This has highlighted the importance of fully understanding the beam–sample interaction and how to test it by modeling. When the limits of resolution and detectability are pushed, no data will be easy enough for a quick and direct interpretation.

References

Allen, L. J., A. J. D'Alfonso, B. Freitag, and D. O. Klenov. 2012. Chemical mapping at atomic resolution using energy-dispersive X-ray spectroscopy. *MRS Bulletin* 37 (1): 47–52.

Allen, L. J., S. D. Findlay, M. P. Oxley, C. Witte, and N. J. Zaluzec. 2006. Modelling high-resolution electron microscopy based on core-loss spectroscopy. *Ultramicroscopy* 106 (11–12 SPEC. ISS.): 1001–1011.

Bentley, J., S. R. Gilliss, C. B. Carter, J. F. Al Sharab, F. Cosandey, I. M. Anderson, and P. J. Kotula. 2006. Nanoscale EELS analysis of oxides: Composition mapping, valence determination and beam damage. *Journal of Physics: Conference Series* 26 (1): 69–72.

Bosman, M., M. Watanabe, D. T. L. Alexander, and V. J. Keast. 2006. Mapping chemical and bonding information using multivariate analysis of electron energy-loss spectrum images. *Ultramicroscopy* 106 (11–12): 1024–1032.

Burke, M. G., M. Watanabe, D. B. Williams, and J. M. Hyde. 2006. Quantitative characterization of nanoprecipitates in irradiated low-alloy steels: Advances in the application of FEG-STEM quantitative microanalysis to real materials. *Journal of Materials Science* 41 (14): 4512–4522.

Buseck, P., J. M. Cowley, and L. Eyring. 1988. *High-Resolution Transmission Electron Microscopy and Associated Techniques*. New York: Oxford University Press.

Colliex, C., L. Bocher, F. De La Peña, A. Gloter, K. March, and M. Walls. 2010. Atomic-scale STEM-EELS mapping across functional interfaces. *JOM* 62 (12): 53–57.

Cooper, D., F. De La Peña, A. Béché, J.-L. Rouvière, G. Servanton, R. Pantel, and P. Morin. 2011. Field mapping with nanometer-scale resolution for the next generation of electronic devices. *Nano Letters* 11 (11): 4585–4590.

De Bruijn, W. C., C. W. J. Sorber, E. S. Gelsema, A. L. D. Beckers, and J. F. Jongkind. 1993. Energy-filtering transmission electron microscopy of biological specimens. *Scanning Microscopy* 7 (2): 693–709.

E, H., P. D. Nellist, S. Lozano-Perez, and D. Ozkaya. 2010. Towards quantitative analysis of core-shell catalyst nano-particles by aberration corrected high angle annular dark field STEM and EDX. *Journal of Physics: Conference Series* 241.

Egerton, R. F. and P. A. Crozier. 1997. The effect of lens aberrations on the spatial resolution of an energy-filtered TEM image. *Micron* 28 (2): 117–124.

Gauvin, R., K. Robertson, P. Horny, A. M. Elwazri, and S. Yue. 2006. Materials characterization using high-resolution scanning-electron microscopy and x-ray microanalysis. *JOM* 58 (3): 20–26.

Goldstein, J. 1992. *Scanning Electron Microscopy and X-Ray Microanalysis: A Text for Biologists, Materials Scientists, and Geologists*. New York: Plenum Press.

Gubbens, A. J. and O. L. Krivanek. 1993. Applications of a post-column imaging filter in biology and materials science. *Ultramicroscopy* 51 (1–4): 146–159.

Haider, M., P. Hartel, H. Mller, S. Uhlemann, and J. Zach. 2010. Information transfer in a TEM corrected for spherical and chromatic aberration. *Microscopy and Microanalysis* 16 (4): 393–408.

Haigh, S. J., H. Sawada, K. Takayanagi, and A. I. Kirkland. 2010. Exceeding conventional resolution limits in high-resolution transmission electron microscopy using tilted illumination and exit-wave restoration. *Microscopy and Microanalysis* 16 (4): 409–415.

Hawkes, P. W. and J. C. H. Spence. 2007. *Science of Microscopy*. New York: Springer.

Herzing, A. A., M. Watanabe, J. K. Edwards, M. Conte, Z.-R. Tang, G. J. Hutchings, and C. J. Kiely. 2008. Energy dispersive x-ray spectroscopy of bimetallic nanoparticles in an aberration corrected scanning transmission electron microscope. *Faraday Discussions* 138: 337–351.

Kabius, B., M. Haider, S. Uhlemann, E. Schwan, K. Urban, and H. Rose. 2002. First application of a spherical-aberration corrected transmission electron microscope in materials science. *Journal of Electron Microscopy* 51 (suppl.): S51–S58.

Kaiser, U., J. Biskupek, J. C. Meyer, J. Leschner, L. Lechner, H. Rose et al. 2011. Transmission electron microscopy at 20 kV for imaging and spectroscopy. *Ultramicroscopy* 111 (8): 1239–1246.

Kimoto, K., T. Asaka, T. Nagai, M. Saito, Y. Matsui, and K. Ishizuka. 2007. Element-selective imaging of atomic columns in a crystal using STEM and EELS. *Nature* 450 (7170): 702–704.

Klenov, D. and S. Lazar. 2011. *Microscopy and Analysis (Asia Pacific Issue)* 82: 3.

Knoll, M. and E. Ruska. 1932. Das elektronenmikroskop. *Zeitschrift Für Physik* 78 (5–6): 318–339.

Konno, M., Y. Suzuki, H. Inada, and K. Nakamura. 2011. High-resolution SEM imaging with aberration correction for high precise measurement of semiconductors. *Microscopy and Microanalysis* 17 (suppl. 2): 930.

Kotula, P. G., M. R. Keenan, and J. R. Michael. 2006. Tomographic spectral imaging with multivariate statistical analysis: Comprehensive 3D microanalysis. *Microscopy and Microanalysis* 12 (1): 36–48.

Krivanek, O. L., M. F. Chisholm, V. Nicolosi, T. J. Pennycook, G. J. Corbin, N. Dellby et al. 2010. Atom-by-atom structural and chemical analysis by annular dark-field electron microscopy. *Nature* 464 (7288): 571–574.

Krivanek, O. L., A. J. Gubbens, M. K. Kundmann, and G. C. Carpenter. 1993. Elemental mapping with an energy-selecting imaging filter. Paper presented at the *Proceedings—Annual Meeting*, Microscopy Society of America.

Krivanek, O. L., M. K. Kundmann, and K. Kimoto. 1995. Spatial resolution in EFTEM elemental maps. *Journal of Microscopy* 180: 277–287.

Liu, J. 2005. Scanning transmission electron microscopy and its application to the study of nanoparticles and nanoparticle systems. *Journal of Electron Microscopy* 54 (3): 251–278.

Lozano-Perez, S., V. de Castro Bernal, and R. J. Nicholls. 2009. Achieving sub-nanometre particle mapping with energy-filtered TEM. *Ultramicroscopy* 109 (10): 1217–1228.

Lozano-Perez, S. and J. M. Titchmarsh. 2007. EFTEM assistant: A tool to understand the limitations of EFTEM. *Ultramicroscopy* 107: 313–321.

Lugg, N. R., S. D. Findlay, N. Shibata, T. Mizoguchi, A. J. D'Alfonso, L. J. Allen, and Y. Ikuhara. 2011. Scanning transmission electron microscopy imaging dynamics at low accelerating voltages. *Ultramicroscopy* 111 (8): 999–1013.

Lugg, N. R., B. Freitag, S. D. Findlay, and L. J. Allen. 2010. Energy-filtered transmission electron microscopy based on inner-shell ionization. *Ultramicroscopy* 110 (8): 981–990.

Meyer, R. R., A. I. Kirkland, and W. O. Saxton. 2002. A new method for the determination of the wave aberration function for high resolution TEM: 1. Measurement of the symmetric aberrations. *Ultramicroscopy* 92 (2): 89–109.

Oxley, M. P., M. Varela, T. J. Pennycook, K. Van Benthem, S. D. Findlay, A. J. D'Alfonso, L. J. Allen, and S. J. Pennycook. 2007. Interpreting atomic-resolution spectroscopic images. *Physical Review B—Condensed Matter and Materials Physics* 76 (6).

Pantel, R. 2011. Coherent Bremsstrahlung effect observed during STEM analysis of dopant distribution in silicon devices using large area silicon drift EDX detectors and high brightness electron source. *Ultramicroscopy* 111 (11): 1607–1618.

de la Peña, F., M.-H. Berger, J.-F. Hochepied, F. Dynys, O. Stephan, and M. Walls. 2011. Mapping titanium and tin oxide phases using EELS: An application of independent component analysis. *Ultramicroscopy* 111 (2): 169–176.

Pennycook, S. J. 1989. Z-contrast stem for materials science. *Ultramicroscopy* 30 (1–2): 58–69.

Pennycook, S. J., M. Varela, A. R. Lupini, M. P. Oxley, and M. F. Chisholm. 2009. Atomic-resolution spectroscopic imaging: Past, present and future. *Journal of Electron Microscopy* 58 (3): 87–97.

Reimer, L. 1998. Energy-filtering imaging and diffraction. *Materials Transactions, JIM* 39 (9): 873–882.

Rowlands, N. and S. Burgess. 2009. Energy dispersive analysis in the TEM. *Materials Today* 12 (suppl.): 46–48.

Schatten, H. 2011. Low voltage high-resolution SEM (LVHRSEM) for biological structural and molecular analysis. *Micron* 42 (2): 175–185.

Scherzer, O. 1936. Über einige fehler von elektronenlinsen. *Z Physik* 101: 593.

Scott, J., P. J. Thomas, M. MacKenzie, S. McFadzean, J. Wilbrink, A. J. Craven, and W. A. P. Nicholson. 2008. Near-simultaneous dual energy range EELS spectrum imaging. *Ultramicroscopy* 108 (12): 1586–1594.

Shenkenberg, D. L. 2007. Team develops electron microscope with 0.5-Å resolution. *Photonics Spectra* 41 (11): 108.

Spence, J. C. H. 2006. Absorption spectroscopy with sub-angstrom beams: ELS in STEM. *Reports on Progress in Physics* 69 (3): 725–758.

Stroppa, D. G., L. F. Zagonel, L. A. Montoro, E. R. Leite, and A. J. Ramirez. 2012. High-resolution scanning transmission electron microscopy (HRSTEM) techniques: High-resolution imaging and spectroscopy side by side. *ChemPhysChem* 13 (2): 437–443.

Suenaga, K. and M. Koshino. 2010. Atom-by-atom spectroscopy at graphene edge. *Nature* 468 (7327): 1088–1090.

Thomas, P. J. and P. A. Midgley. 2001. Image-spectroscopy—I. The advantages of increased spectral information for compositional EFTEM analysis. *Ultramicroscopy* 88 (3): 179–186.

Tiemeijer, P. C., M. Bischoff, B. Freitag, and C. Kisielowski. 2012. Using a monochromator to improve the resolution in TEM to below 0.5Å. Part I: Creating highly coherent monochromated illumination. *Ultramicroscopy* 114: 72–81.

Titchmarsh, J. M. 1999. Detection of electron energy-loss edge shifts and fine structure variations at grain boundaries and interfaces. *Ultramicroscopy* 78 (1–4): 241–250.

Trebbia, P. 1988. Unbiased method for signal estimation in electron energy loss spectroscopy, concentration measurements and detection limits in quantitative microanalysis: Methods and programs. *Ultramicroscopy* 24 (4): 399–408.

Trebbia, P. and N. Bonnet. 1990. EELS elemental mapping with unconventional methods. I. Theoretical basis: Image analysis with multivariate statistics and entropy concepts. *Ultramicroscopy* 34 (3): 165–178.

Trebbia, P. and C. Mory. 1990. EELS elemental mapping with unconventional methods. II. Applications to biological specimens. *Ultramicroscopy* 34 (3): 179–203.

Tsuno, K. 1999. Optical design of electron microscope lenses and energy filters. *Journal of Electron Microscopy* 48 (6): 801–820.

Tsuno, K. 2004. Evaluation of in-column energy filters for analytical electron microscopes. *Nuclear Instruments and Methods in Physics Research, Section A Accelerators, Spectrometers, Detectors and Associated Equipment* 519 (1–2): 286–296.

Von Harrach, H. S., P. Dona, B. Freitag, H. Soltau, A. Niculae, and M. Rohde. 2010. An integrated multiple silicon drift detector system for transmission electron microscopes. *Journal of Physics: Conference Series* 241.

Williams, D. B. and C. B. Carter. 1996. *Transmission Electron Microscopy: A Textbook for Materials Science*. New York: Plenum.

Williams, D. B., J. Goldstein, and D. Newbury. 1995. *X-Ray Spectrometry in Electron Beam Instruments*. New York: Plenum.

Zach, J. 2009. Chromatic correction: A revolution in electron microscopy? *Philosophical Transactions of the Royal Society A: Mathematical, Physical and Engineering Sciences* 367 (1903): 3699–3707.

11

Atomic-Scale Imaging of Dielectric Point Defects

Clayton C. Williams
University of Utah

11.1 Introduction

Dielectric materials play a central role in many important nanotechnologies today. For example, silicon-based devices used in computer memory and logic chips are critically dependent on thin dielectric films. In silicon field effect transistors, gate dielectrics provide the physical separation between the gate and the semiconducting channel region, providing a structure by which the gate voltage can modulate the conductance of the channel by the "field effect," without injecting charge into the channel. These ultra-thin gate dielectric films also provide essential passivation of unterminated or "dangling" bonds at the silicon-oxide interface above the channel region. Furthermore, dielectric films in computer chips are also needed to separate information-carrying metallic lines above the devices (interlayer dielectric films). In FLASH memory devices, they provide the nonconducting barrier which inhibits leakage of the charge representing the "ones" and "zeros" associated with the memory function.

Dielectrics films in silicon devices are typically composed of amorphous oxides and nitrides. These materials contain point defects which can act as electronic trap states. Many studies of these point defect states have been performed over many decades (Pacchioni et al. 2000). One of the most common electronic defect states observed in silicon dioxide is the E′ center. The E′ center is believed to be composed of an oxygen vacancy and a captured hole (Lenahan and Conley 1998). Oxygen vacancies are generally produced during film growth (Conley et al. 1997). Many variants of this defect exist (Pantelides et al. 2008). When the density of these defects is high enough, electrons can tunnel or hop through these states, creating a leakage path for charge.

11.2 Standard Characterization Methodologies

Many characterization methods have been developed over the years to study electronic defects in dielectric films. When a measurable leakage current is detectable (density of defect states is high enough), the electronic trap states can be characterized by current–voltage (Ribes et al. 2006; Bersuker et al. 2007; Young et al. 2009) and charge pumping methods (Kerber and Cartier 2009). They can also be studied using capacitance–voltage (Placidi et al. 2010) and optical absorption measurements (Nguyen et al. 2005; Price et al. 2007; Hoppe and Aita 2008; Park et al. 2008). These standard methods do not provide atomic scale spatial information and do not have the sensitivity to detect individual trap states.

The scanning tunneling microscope (STM) (Binnig et al. 1982), on the other hand, does provide atomic spatial resolution. However, it typically requires a minimum tunneling current greater than 0.1 pA to operate. This current corresponds to an electron tunneling rate greater than 10^6 electrons/s. Therefore, the STM cannot directly image trap states for which the dwell time of an injected electron in the state is greater than approximately 1 μs. Any state that has an electron dwell time longer than this will not be able to sustain a steady current above the STM's detectible limit. Thus, the STM is not useful for characterizing trap states in dielectric films which cannot provide a detectible current.

Multiple attempts have been made to produce electron tunneling images of dielectric surfaces over the years. Kochanski (1989) and Stranick et al. (1993) both attempted to do a form of ac tunneling to dielectric surfaces. These efforts did not produce convincing atomic scale imaging results, and the work was not continued. Additionally, STM and Ballistic Electron Emission Microscopy (BEEM) have been applied to the imaging of ultra-thin dielectric films (Kaczer et al. 1986; Welland and Koch 1986). In STM and BEEM, however, electrons do not tunnel directly to the trap states, but rather through the trap states or directly to the substrate. Both require a measurable current.

The atomic force microscope (AFM) can provide atomic scale spatial resolution on dielectric surfaces (Binnig et al. 1986). While this method is sensitive to topographic and chemical information, it is not sensitive to the electronic properties which can be addressed by tunneling. Kelvin probe force microscopy (KPFM), electrostatic force microscopy (EFM), and scanning capacitance microscopy (SCM) have also been used to investigate electronic defects and charge in insulating films (Martin et al. 1988; Terris et al. 1989; Schonenberger and Alvarado 1990; Zhu et al. 2005; Barth and Henry 2008; Naitou et al. 2008). In charge imaging by KPFM and EFM, sensitivity exists only to the charge present at the surface, not to the electronic states themselves; so uncharged states cannot be observed directly by these methods. The SCM can also sense the presence of charged trap states, but has not to date achieved atomic spatial resolution (Bussmann and Williams 2004). For many years, KPFM and EFM were limited in spatial resolution to a value on the order of the tip radius, but recent KPFM imaging has shown beautiful atomic scale results (Gross et al. 2009; Nony et al. 2009; Sadewasser et al. 2009; Barth et al. 2011; Mohn et al. 2012).The conducting atomic force microscope (c-AFM) does provide high spatial resolution (Kremmer et al. 2005), but does not typically achieve atomic scale spatial resolution or single defect detection. By combining EFM and optical pumping, changes in the location of charge caused by the absorption of light have been mapped (Ludeke and Cartier 2001). This method, however, only shows where the change in charge occurs, and does not identify the trap states in which the charge is unchanged.

In the following, a recently developed scanning probe microscopy (SPM) method will be described which overcomes many of the limitations of the approaches described earlier. It is based upon direct tunneling of single electrons between a metallic SPM probe tip and individual trap states in the dielectric surface. Each tunneling event is detected by the change in the Coulomb force or force gradient that the SPM probe tip experiences, as the net charge at the surface of the dielectric film changes due to the electron tunneling event.

11.3 Force-Detected Tunneling

The electrostatic force microscope (EFM) (Martin et al. 1988; Stern et al. 1988; Terris et al. 1989; Schonenberger and Alvarado, 1990) was derived from the *noncontact* AFM (Martin et al. 1987). It measures the electrostatic force acting on an AFM tip due to the Coulomb interaction with nearby charge distributed on a sample surface. Imaging of charge deposited on dielectric (insulating) surfaces by corona discharge (Stern et al. 1988; Schonenberger and Alvarado 1990) and contact charging (Terris et al. 1989) have been demonstrated. Nanoclusters have also been electrostatically characterized by EFM (Schaadt et al. 1999). In one of the corona discharge experiments (Schonenberger and Alvarado 1990), small discrete steps in post-discharge charge decay traces were ascribed to be due to single electron recombination events. In these early studies, the total charge transferred to the sample by discharge or contact was both difficult to control and challenging to quantify due to the complex nature of the charging process and uncertainty about the electrostatic properties of the tip and the sample.

Some years later, the author began to think about alternative ways to detect tiny amounts of current well below the detection noise of current amplifiers in scanning probe measurements. The interpretation of the small steps observed in the corona discharge experiment mentioned earlier provided inspiration to consider the possibility of using an electrostatic force to measure not just the charge at the surface, but the *transfer* of ultra-small amounts of charge *between the tip and the sample by quantum tunneling*.

The first idea was to build a small, electrically isolated metallic island or dot at the end of an AFM probe tip (Figure 11.1). The benefit of the metallic island is that if an electron were to leave the island (by tunneling), the charge and electrostatic potential of the island would change, causing a modification of the electrostatic force on the AFM probe tip. This potential change would not occur if the AFM probe was composed of just a standard metal tip. By measuring the electrostatic force change, ultra-small amounts of charge leaving the island could be detected. This force-detection approach has now been developed into a means to controllably tunnel single electrons to and from electronic trap states. When the electrons are transferred to or from the surface by quantum tunneling, atomic scale spatial resolution can be achieved, as in STM.

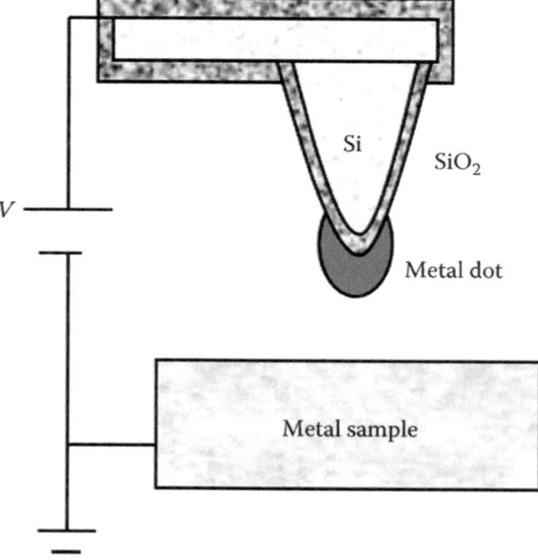

FIGURE 11.1 Conceptual diagram of an AFM probe tip with a metallic dot at the tip apex, isolated from the silicon by a thick silicon dioxide film.

The specially fabricated probes with electrically isolated metallic islands were built, and the concept of force-detected tunneling was first demonstrated in 2000 (Klein et al. 2000). The first force-detected single electron tunneling events to metallic surfaces were reported in 2001 (Klein and William 2001). In this early work with the specially fabricated probes, it was noted that ultrasmall currents on the order of 1 electron/s (0.1 attoampere) could be detected when tunneling to (from) a metallic (conducting) surface (Klein et al. 2000). After the demonstration of the force-detected tunneling concept using the specially fabricated probes, it was realized that *standard metallic AFM probes* could be used when tunneling to states in *nonconducting* dielectric surfaces. In this case, the trap state at the dielectric surface acts as a metallic island or dot on the previously developed AFM probe tips. When charge enters or leaves the trap state, the change causes a modification of the electrostatic force between the tip and the sample. The use of standard metallic probe tips made the technique much simpler and more accessible. Single electron tunneling to a dielectric surface using standard metallic coated AFM probes was demonstrated in 2002 (Klein and William 2002). All the force-detected tunneling work done by Klein was performed using an amplitude detection mode.

It was later realized that there are significant advantages to using a frequency modulation (FM) detection mode (Giessibl et al. 2000) rather than the previous amplitude detection mode. In 2004, an FM tunneling force detection mode was demonstrated which showed single electron tunneling detection sensitivity (Bussmann et al. 2004). Figure 11.2 shows a probe tip over the surface of a dielectric film. The prime advantages of the FM mode include the fact that the oscillation amplitude of the AFM cantilever probe tip is maintained constant by active control (feedback loop). This allows the tunneling measurements to be performed without having to deal with the frequency-shift-induced amplitude changes observed in the prior measurements (Klein and William 2002). It also eliminates the dynamical instability observed in the single electron tunneling amplitude detection method (Klein and William 2004). The FM detection method has now become the standard approach.

The single electron tunneling measurements by FM detection are typically performed with an Omicron Multiprobe S AFM under a vacuum below 10^{-9} Torr at room temperature. A platinum-coated AFM probe with 10–40 nm oscillation amplitude, 300 kHz natural resonance frequency, and 50 N/m stiffness is brought within a couple of nanometers (closest approach) to a dielectric surface. A dc voltage between the tip and back contact to the dielectric sample is applied. As noted earlier, the oscillation amplitude is kept constant by a feedback loop during the single electron tunneling force measurements.

To measure single electron tunneling events, the probe is positioned within a few nanometers of the dielectric surface and is scanned toward the surface. The shift of the natural resonance frequency due to the electrostatic interaction between the tip and the sample is recorded versus the displacement of the probe. When the minimum gap between the probe and the sample is within the tunneling range (typically less than 2 nm), abrupt steps occasionally appear in the frequency shift versus distance curves, as shown in Figure 11.3. The frequency steps are caused by single electron tunneling events to or from individual trap states in the surface. A charge tunneling in the direction of the applied field causes a positive frequency step as seen in the trace (a) of Figure 11.3, while a charge tunneling against the applied field causes a negative frequency step. Trace (b) shows two forward events and one reverse event. The forward events typically outnumber the reverse events by more than an order of magnitude.

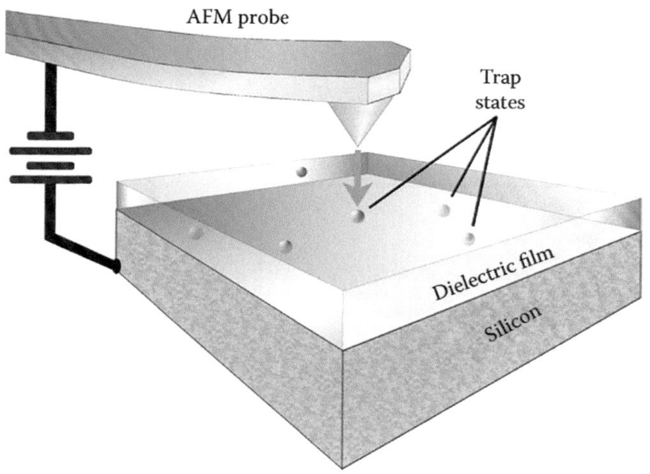

FIGURE 11.2 Dynamic tunneling force microscope cartoon, illustrating the geometry of the oscillating AFM cantilever over the dielectric film containing the electronic trap states. (Reprinted with permission from Bussmann, E. and Williams, C.C., *Appl. Phys. Lett.*, 88, 263108, Copyright 2006 American Institute of Physics.)

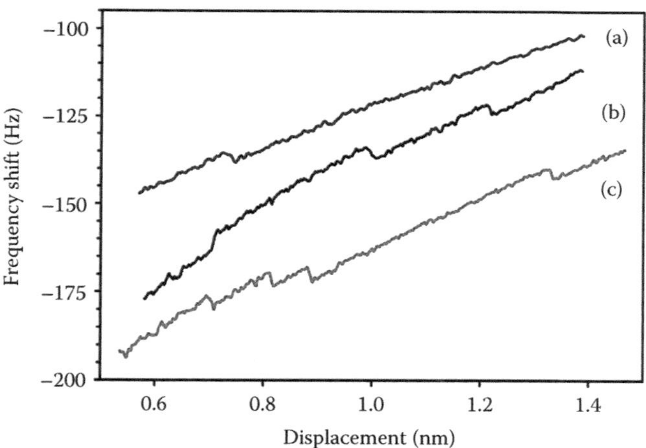

FIGURE 11.3 Frequency shift curves showing steps caused by single electron tunneling between probe tip and the surface of two different silicon dioxide films. Curve (a) was acquired near a 4.6 nm film, and (b) and (c) were acquired near a 10 nm film. (Reprinted with permission from Bussmann, E., Kim, D.-J., and Williams, C.C., *Appl. Phys. Lett.*, 85, 2538, Copyright 2004 American Institute of Physics.)

The steps in Figure 11.3 do not appear absolutely abrupt because of the finite detection bandwidth (10 Hz) of the measurements. Each data point in Figure 11.3 has been numerically averaged with its two neighbors to reduce noise. Frequency steps have been observed on 4.6, 10, and 20 nm thick SiO_2 layers and 3.9 nm thick HfO_2 layers on Si substrates. In many locations, no tunneling events (steps) are detected. However, in some locations, a single step is seen. Rarely, a multiple step sequence is observed as shown in traces (b) and (c). How often the events are observed depends upon the spatial distribution and density of trap states in the dielectric surface, the applied voltage, and the tip radius.

In Figure 11.3, the multiple steps are most likely due to several electrons separately tunneling to several electronic states in the vicinity of the tip apex. There are several reasons to believe that the steps in the data shown in Figure 11.3 correspond to single electrons. First, these events never occur with a tip–sample gap which is greater than 2 nm, even at higher electric field strength (Bussmann 2004). Second, the step size (frequency shift associated with an event) for a given tip and dielectric film are typically all of one size, with an occasional step size of twice the magnitude of the other events. Third, calculations of the expected frequency shift due to a single electron tunneling event match well with the observed frequency step size (Bussmann 2004).

The ability to inject (extract) single electrons to (from) the surface by quantum tunneling provides a means to control the occupancy of individual trap states. This was demonstrated in 2005 (Bussmann et al. 2005). In this work, it was shown that single electrons could be injected into a single trap state and then by reversing the polarity of the voltage, the electrons could then be removed, leaving the surface in its original state. This reversible manipulation of single electrons to and from the single trap states in a dielectric surface by a scanning probe tip represents a new level of atomic scale electronic control.

Figure 11.4 shows two EFM images (150 × 150 nm) of a silicon dioxide surface after a single electron has been injected (a) and then extracted (b). The images are obtained by first imaging the surface by EFM (which measures the local surface potential or charge), then injecting a single electron into the center region of that area, and then reimaging the area by EFM. A difference image (Figure 11.4a) is produced by subtracting the

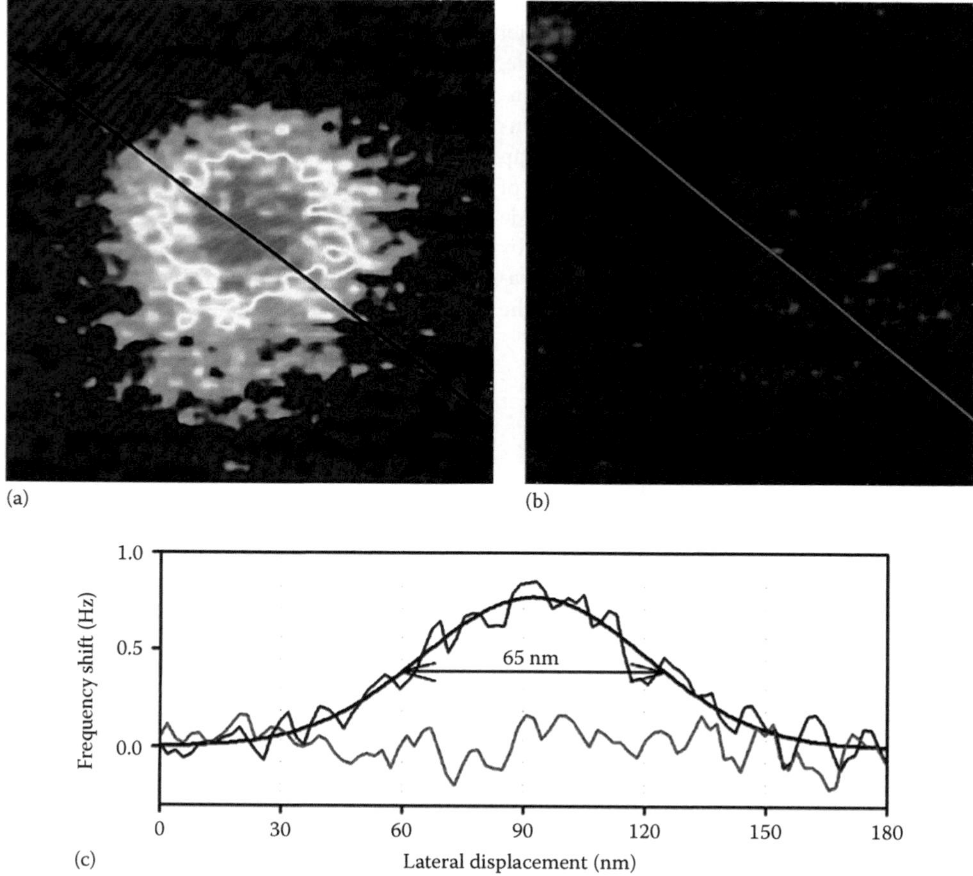

FIGURE 11.4 Two EFM images (150 × 150 nm) obtained in the same location on a silicon dioxide surface, after an electron is injected (a) and extracted (b). (a) The surface potential change due to a single electron injected into the surface near the center of the image. (b) The surface potential after the single injected electron is removed. A line cut through each image shows the frequency shift with and without the injected electron. The two line cuts are superimposed in (c).

second EFM image from the first. The difference image shows only the contribution of the injected single electron charge. The probe tip is then repositioned to the center region of the image area, and a reverse bias voltage is applied. After an electron is extracted by tunneling, the surface is then reimaged by EFM and a difference image is produced by subtracting the first and last EFM images (Figure 11.4b). Since from the magnitude of the individual tunneling events (steps heights) it is known that a single electron is injected and extracted, the two images are known to represent a map of the surface potential with and without one electron. Line cuts from both images are shown in Figure 11.4c. They show that the presence of the single electron causes a frequency shift of approximately 1 Hz in these measurements.

Note that the spatial resolution displayed in the EFM difference images is rather poor (full width at half maximum is approximately 65 nm). In these images, the spatial resolution is limited by the radius of the tip and the tip–sample gap used. The EFM images are simply maps of the local charge or surface potential, convolved with the effective area of the metallic probe tip. The image contrast is not related to quantum tunneling, and therefore atomic spatial resolution is not expected. Atomic scale imaging of trap states by tunneling will be described in a later section.

11.4 Spectroscopy

The STM is powerful not only for its imaging capability, but also for its ability to perform atomically resolved tunneling spectroscopy measurements, that is, to measure the local density of states of the sample surface (Weisendanger 1994). Fortunately, this capability is also available in force-detected tunneling measurements.

Trap states have energies that fall between the valence and conduction bands of the dielectric material (Figure 11.5). When a metallic probe tip is brought well within tunneling range of an unfilled trap state in the dielectric surface, and the Fermi level of the tip (controlled by the applied voltage) is greater than the energy of the unfilled trap state, an electron will tunnel from the tip to the trap state in the sample surface. After the state is filled, the Fermi level of the probe tip can be lowered by the applied voltage so that it is below the filled trap state. The electron in the filled trap state will then tunnel back to the empty states in the tip. The voltage applied to the tip can thus control the occupancy of the trap.

By monitoring tunneling events as a function of tip–sample bias voltage at fixed tip–sample gap, the energy of a particular trap state in the dielectric film can be determined (Bussmann and Williams 2006). In these spectroscopic measurements,

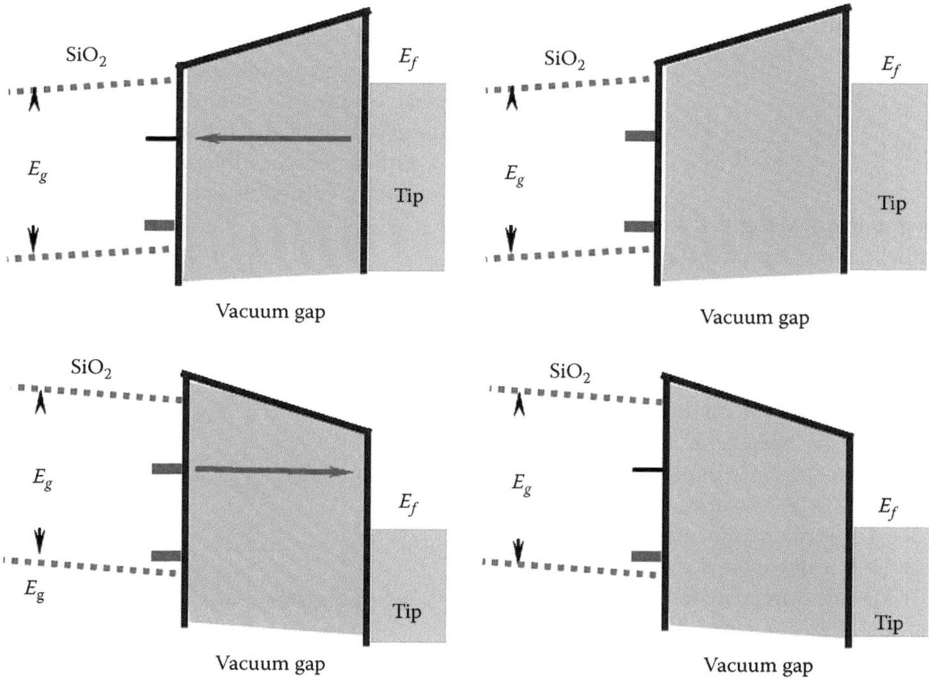

FIGURE 11.5 An energy diagram showing the energy levels of the conduction and valence bands of the dielectric surface, trap state energy levels in the dielectric, vacuum level, and Fermi-level of the probe tip under conditions of electron injection from tip to dielectric (upper) and electron extraction (lower). The figures on the right show the occupation of the states after the tunneling event on the left occurs. The empty states are shown as thin dashes, and the filled states are shown as thick dashes. In the top left, the Fermi-level of the probe tip is raised and a tunneling event occurs filling the upper trap state (top right). In the lower left, the Fermi-level of the probe is lowered, and an electron tunnels out of the upper state, leaving it empty (lower right). (Adapted with permission from Bussmann, E., Zheng, N., and Williams, C.C., *Appl. Phys. Lett.*, 86, 163109, Copyright 2005 American Institute of Physics.)

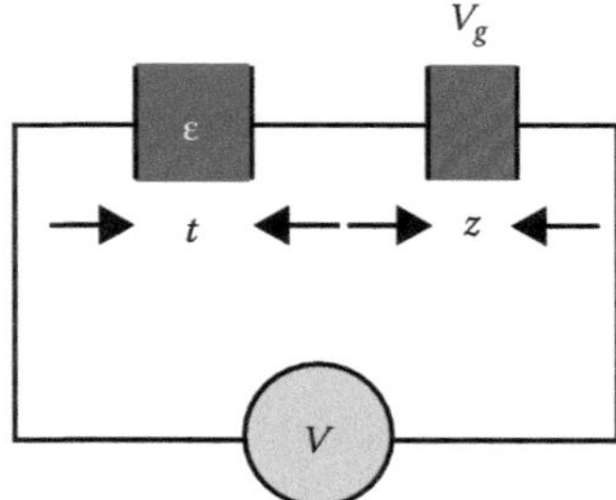

FIGURE 11.6 One-dimensional equivalent circuit model to calculate the voltage dropped in the tip–sample gap V_g. Part of the applied voltage V is dropped in the oxide film of thickness t and dielectric constant ε, and part is dropped in the vacuum gap z.

the applied voltage is ramped, with the probe tip at fixed height, while the tunneling events are monitored. Since part of the applied voltage is dropped in the oxide film and part in the tip–sample gap, a simple 1D electrostatic model is used to determine the potential difference between the tip and states near the surface of the dielectric sample (Figure 11.6). It is this voltage drop which determines the shift of the tip Fermi-level relative to trap states near the dielectric surface (eV_g). The electrostatic model provides a means to calibrate the energy scale in tunneling measurements for a given tip–sample gap z, dielectric constant ε, and thickness t of the dielectric film. The relation between the applied voltage V and the tip Fermi-level shift relative to the trap states at the dielectric surface eV_g is given by the following equation:

$$eV_g = eV\left(\frac{z}{z+(t/\varepsilon)}\right).$$

The measurement methodology described earlier is referred to as single electron tunneling force spectroscopy (SETFS). SETFS is performed by fixing the probe height z above the surface (typically between 0.4 and 1.0 nm) and ramping the applied probe–sample dc voltage V while recording the cantilever frequency shift. Figure 11.7 shows a spectroscopic measurement performed at one location above a silicon dioxide film of 20 nm thickness. The voltage scale has been calibrated using the electrostatic model for the particular tip–sample gap (0.7 nm) and oxide thickness (20 nm), with an assumed dielectric constant (silicon dioxide) of 3.9. For these parameters, electrostatic calculations indicate that 12% of the applied voltage is dropped in the tip–sample gap. In the measurement shown in Figure 11.6, the voltage applied to the sample is ramped from −2.5 to −4.16 V (and back up to −2.5 V), which corresponds to a Fermi-level movement of −0.4 to −0.5 eV (and back to −0.4 eV) relative to the trap states at the dielectric surface. As the Fermi-level moves, the occupancy of a single trap state goes from occupied (−0.4 eV), to intermittently occupied, to fully occupied (−0.5 eV). As the voltage scan direction is reversed, the state occupation reverses. The intermittent region shows a random telegraph signal (RTS). In this voltage range, the electron tunnels back and forth randomly between the probe tip and the trap state. Similar RTSs, due to random occupation of charge traps in oxide layers, have been observed in the drain current of field-effect transistors (Ralls et al. 1984).

If it is assumed that the probe Fermi-level at zero applied voltage is located near the middle of the silicon dioxide band gap, the 1D electrostatic model can be used to relate the experimental spectroscopic data to a calibrated energy scale, so that the energy of a particular trap state be determined. For the particular data shown in Figure 11.7, the calibration places the energy level of the state at ~0.5 eV below the middle of the gap. As can be seen in the figure, the energy width of the RTS transition region can be seen to be comparable to the thermal energy (25 meV) at room temperature. The energy width of the Fermi distribution of the probe, the phonon broadening of the state, and the tunneling rate may all contribute to this transition width.

In the SETFS measurements shown in Figure 11.7, the characteristic RTS charge shuttling time is on the order of 1 s. This time appears to be determined by the tunneling rate, that is, the tip–sample gap. If the probe–sample gap is reduced, the characteristic RTS shuttling time decreases considerably, until it is not measurable due to the finite measurement bandwidth. By fixing the applied voltage so that the probe Fermi-level has the same energy as the trap state, the RTS noise can be observed for up to 1 min or more, before the lateral tip–sample drift moves the probe off the state.

The spectroscopic measurement results vary from place to place on the SiO_2 sample. At some locations, no tunneling is observed, indicating that there is no state at that position that is within the accessible energy range of the probe tip. At other locations, reversible tunneling events are observed, meaning

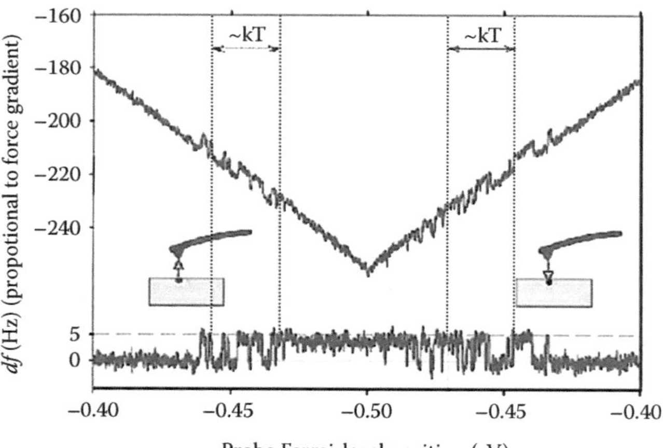

FIGURE 11.7 SETFS measurement of the energy of an individual trap state. The frequency shift of the cantilever is plotted against the probe tip Fermi-level position, as calibrated using a simple 1D electrostatic model. A RTS is observed as the Fermi-level of the probe is ramped through the energy level of the particular trap state.

that as the applied voltage is ramped up and down, trap states at the surface are reversibly filled and emptied in that energy range. However, at other locations, it is observed that when a probe tip first approaches a surface, a tunneling event is sometimes observed which is not reversible with a small voltage change. For example, if a probe tip with a small applied voltage (tip Fermi-level assumed to be near the middle of the dielectric band gap) is initially brought close to a surface, an electron trapped in a state several eV above the middle of the gap may tunnel out of the surface. To reoccupy this state, a large voltage must be applied to move the tip Fermi-level up to that state. Another possible scenario occurs when an electron is injected into a state by tunneling and it hops to nearby trap states which are outside of tunneling range. These electrons will not return under any bias voltage.

The spectroscopic method described here has been improved, modified, and applied to other systems. In 2010, a variant of the method was applied to the characterization of single monolayer protected gold clusters (Zheng et al. 2010). In that work, electronic spectra of 25 atom gold clusters surrounded by an alkane–thiol ligand shell were obtained by SETFS. The results were compared with electrochemical measurements with reasonable agreement. In 2011, systematic studies of hafnium dioxide (Winslow et al. 2011a), and silicon dioxide and silicon nitride films (Winslow and Williams 2011b) were performed using the SETFS approach. The results were compared with previous theoretical and experimental results produced by the more standard (spatially averaging) methods with good agreement.

We note here that other researchers have also employed force detection to perform single electron spectroscopic measurements (Dana and Yamamoto 2005; Stomp et al. 2005). In their research, however, the probe tip is not used to inject (extract) charge to (from) the surface, but rather to detect the charge which tunnels from the substrate into the quantum states in the dielectric film. In those measurements, atomic spatial resolution is not achieved, because the injection and extraction occur from a nonlocalized source (the substrate). Other researchers have employed force detection to observe single-electron tunneling between states within a conducting sample (Woodsides and McEuen 2002; Zhu et al. 2005). Single-electron charging of a localized state through which an average tunneling current flows from a scanning probe tip to a conducting substrate has also been demonstrated by electrostatic force detection (Suganuma et al. 2002; Azuma et al. 2006).

In SETFS measurements, there is a coupling between the apparent energy level and physical depth of a trap state. This is due to the fact that the applied voltage is partly dropped between the tip and the sample state and partly dropped between the sample state and the underlying substrate. When a bias voltage is applied, the movement of the tip Fermi-level is smaller for states nearer the surface than those which are deeper. To address this, an additional spectroscopic capability was developed which provides a means to quantitatively determine both the energy and the physical depth of a trap state by performing spectroscopic measurements as a function of probe height (Johnson et al. 2011). When combined with imaging (as described in the following), this quantitative energy/depth analysis method will enable the simultaneous determination of the energy and three-dimensional spatial distribution of the trap states in a dielectric surface.

In summary, the SETFS method provides a powerful approach which is capable of determining the energy level of individual trap states in nonconducting dielectric surfaces. The spectra are obtained by tunneling directly to and from the state. With appropriate calibration, the absolute energy of these defect states relative to the band edges of the dielectric can be determined. The method has also been applied to metal clusters on dielectric surfaces, showing that any quantum state which can donate or accept an electron can be characterized by the SETFS method.

11.5 Imaging

One of the primary motivations for developing force-detected tunneling is to *image* the spatial distribution of quantum states in or on *nonconducting surfaces* with atomic spatial resolution. Since STM cannot image surfaces which do not provide a detectable current, thick insulating films or bulk insulating substrates are not accessible for characterization by that method. Also, molecules and nanostructures that sit on completely nonconducting substrates and mid-gap states in semiconductors are also out of STM's measurement domain. Force-detected tunneling provides a new avenue to image many interesting materials and nanoscale systems which are out of the STM's reach.

The beautiful atomic scale imaging capabilities of the STM are due to the strong exponential dependence of the tunneling current on tunneling gap. Under typical STM conditions (UHV), and with typical work functions of the order of 5 eV, the tunneling current changes by an order of magnitude per Angstrom change in tip/sample vacuum gap. This means that if the probe tip has an atom which sits 1 Å closer to the sample surface than all of the other tip atoms, the tunneling current which flows between the tip and the sample will be dominated by the current flowing through that one atom. Thus, as the tip is scanned over the sample, the surface is imaged with a "point spread function" of Angstrom size. Since the force-detected tunneling methods described here are also based upon quantum tunneling to or from a metallic probe tip, the same spatial resolution achieved in STM imaging can be achieved using the force-based tunneling methods. Theoretical calculations of the expected tunneling rate between a metallic probe tip and a localized, electronically isolated quantum trap state have been performed. The results show the same strong exponential dependence of the tunneling rate upon the tip–trap state gap and provide quantitative calculations of the absolute tunneling rate as a function of tip–sample gap, energy and depth of the trap state, and other parameters (Zheng et al. 2007).

11.5.1 Single Electron Tunneling Force Microscopy

The demonstrated ability to manipulate electrons to and from individual trap states found in a dielectric surface was eventually applied to the problem of producing a tunneling image of a surface. The initial idea was to attempt injection and extraction of electrons at a 2D array of points to form an image. An earlier version of the surface potential measurement method referenced earlier (Zheng et al. 2010) was used to detect the amount of charge injected and extracted at each location in the 2D array. The surface charge difference between the injection and extraction attempts at each point in the image is measured and acquired by a computer and displayed as a pseudo-color image. In locations where there are no available states at the surface, the charge difference signal will be zero (no tunneling has occurred). In locations where a state does exist, a charge difference will be nonzero because an electron has tunneled to (from) the surface with the injection (extraction) attempts indicating the presence of a trap state. This method is called single electron tunneling force microscopy (SETFM).

By this method, 2D images of the trap state distribution at a silicon dioxide surface were produced (Bussmann et al. 2006). Figure 11.8a contains two repeated 5 × 5 nm images of a silicon dioxide film. Each image is obtained by performing tunneling injection/extraction attempts at 144 locations on a 12 × 12 array of pixels. As can be seen in the image, there are regions in which no trap states exist (dark regions). In other locations, there is a clear tunneling charge difference signal, indicating the presence of trap states.

Figure 11.8b shows a histogram of the image values, that is, the charge difference signal, found in the two images in Figure 11.8a. The quantization of the amount of charge injected (extracted) to (from) the surface by tunneling is apparent, indicating that almost all of the contrast in the image is truly due to single electron tunneling events. Note that there are some tunneling events which correspond to two electrons being transferred between the tip and the sample.

While much of the tunneling signal repeats well in the two images, there are excess variations between the two images which make the repeatability less than desirable. The origin of this noise requires further study. The modest repeatability in the imaging results motivated a search for a better approach.

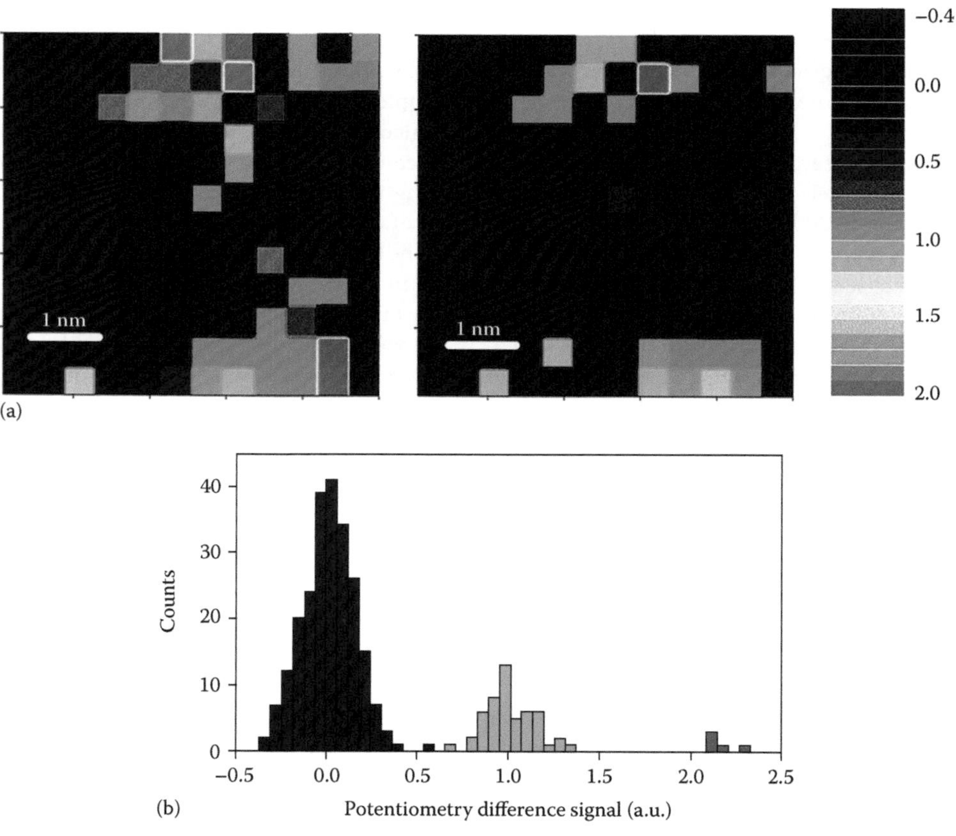

FIGURE 11.8 (a) Repeated (5 × 5 nm^2) SETFM images of a region of a 20 nm thick SiO_2 film. The contrast in the image is proportional to the charge or potential difference in the surface produced by a tunneling injection and extraction attempt. (b) A histogram of the charge or potential difference signal in the SETFM images shown in (a). As can be seen, there is clear quantization in the amount of charge injected and extracted from the surface. Note that there are also some events which correspond to two electrons tunneling to and from the surface. (Reprinted with permission from Bussmann, E., Zheng, N., and Williams, C.C., *Nano Lett.*, 6, 2577, Copyright 2006 American Chemical Society.)

A method called Dynamic Tunneling Force Microscopy (DTFM) was invented which improves the imaging performance significantly. This approach is described in the following section.

11.5.2 Dynamic Tunneling Force Microscopy

In DTFM, single electrons are *dynamically* shuttled to and from electron trap states by a bipolar applied voltage, and the periodic tunneling events are detected by a lock-in amplifier at the frequency of the applied bipolar voltage. This method provides a means to spectrally isolate the dynamic tunneling signal from low-frequency charge fluctuations. This dynamic tunneling approach provides beautiful, repeatable, and direct images of individual trap states in completely nonconducting surfaces with subnanometer spatial resolution (Johnson et al. 2009).

In DTFM measurements, electrons are dynamically shuttled to and from trap states. Detecting this electron shuttling involves monitoring small but periodic changes in the resonance frequency of the oscillating AFM probe. The same cantilevers and UHV system are used in these measurements as described earlier. A periodic shuttling voltage waveform at 300 Hz is applied to the sample (with the tip grounded), consisting of a positive voltage (typically 3–5 V) for 85% of its duty cycle and a negative pulse voltage (3–5 V) for the remaining 15% (Figure 11.9). The positive and negative voltage levels (such as +3 and −3 V) are chosen to be equal in magnitude and opposite in sign with respect to the flat band voltage. The flat band voltage is experimentally defined as the dc voltage applied between the tip and the sample at which the electrostatically induced frequency shift of the oscillating AFM probe is minimized. The measurement of the flat band voltage is performed at a fixed height with the tip just outside of the tunneling range. Note that the cantilever frequency shifts are proportional to the square of the applied voltage with respect to the flat band voltage (Bussmann 2004). Therefore, when the bipolar applied voltage levels are symmetric with respect to the flat band voltage, the background frequency shift of the cantilever is constant during the periodic applied shuttling waveform. Under this condition, the only frequency shift detected in these measurements at 300 Hz is due to the shuttled charge.

The tip–sample gap is simultaneously modulated with a sinusoidal height modulation of 3 nm at exactly twice the electron shuttling voltage frequency (600 Hz). The 3 nm height modulation value is chosen to bring the tip in and out of tunneling range. Under this condition, the voltage applied to the sample has an opposite polarity each time the tip moves into tunneling range (Figure 11.9). This provides the conditions for an electron to be shuttled between the tip and state each time the gap modulation brings the tip within the tunneling range. Note that during these measurements, the cantilever itself is always oscillating at its resonance frequency near 300 kHz, with an amplitude of ~10 nm. So, the 3 nm gap modulation simply brings the closest approach of the tip into and out of tunneling range.

When an electron is shuttled to and from a trap state at the sample surface, it periodically alters the local electrostatic surface

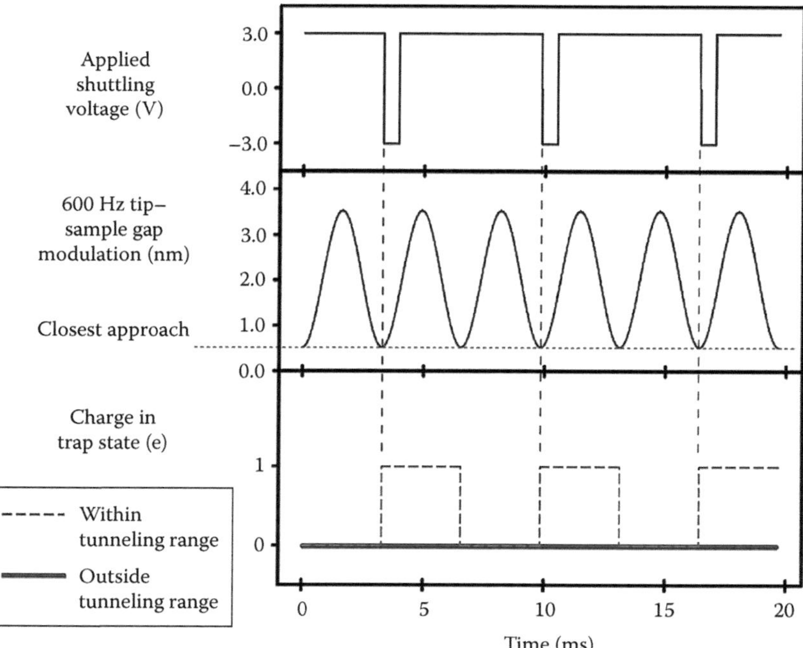

FIGURE 11.9 Electron shuttling voltage and height modulation waveforms and tunneling range. Note the time relationship between the applied 300 Hz shuttling voltage, the 600 Hz tip–sample gap modulation, and charge which may shuttle between the tip and a trap state in DTFM measurements. (Reprinted with permission from Johnson, J.P., Zheng, N., and Williams, C.C., *Nanotechnology*, 20, 055701, Copyright 2009 IOP Publishing Ltd.)

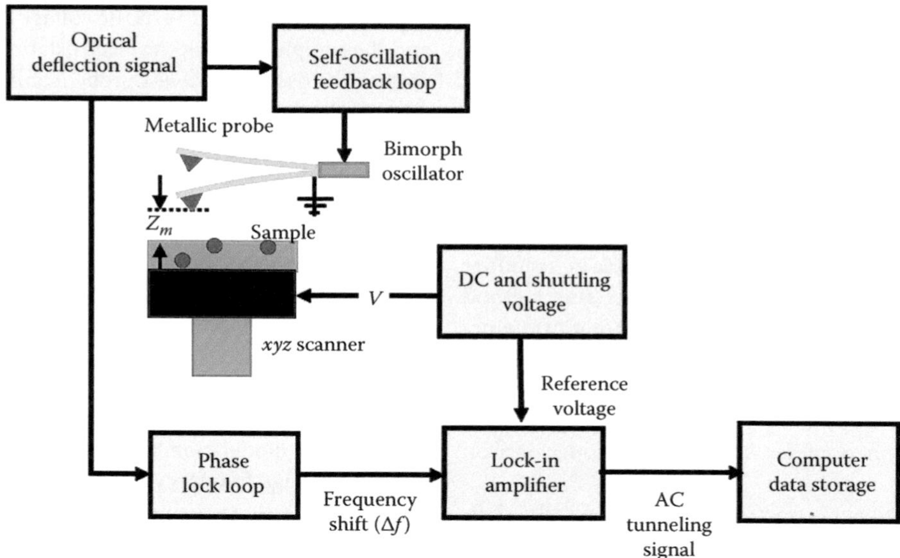

FIGURE 11.10 Block diagram of the experimental system for performing DTFM measurements.

potential of the sample. The oscillating AFM probe experiences a periodic frequency shift in its resonance frequency, associated with the shuttled charge. This periodic frequency shift is detected by a phase lock loop (PLL). The PLL output signal is sent to a lock-in amplifier appropriately phase-referenced at 300 Hz, producing a signal proportional to the amount of charge being shuttled between the tip and the surface. The block diagram for the experimental detection system is shown in Figure 11.10.

The phase of the shuttling voltage waveform is adjusted so that the initial edge of the negative voltage pulse appears just as the tip height modulation brings the tip to its closest approach, as shown in Figure 11.9. This causes the DTFM signal due to the shuttled electron to be approximately 90° out of phase with a background signal (at 300 Hz) that is proportional to the local surface potential. As the tip is scanned over the surface, variations in local surface potential cause this background signal to change. The phase-sensitive detection provides a way to separate the electron shuttling signal from the background surface potential variations because they are in quadrature (Johnson et al. 2009).

As a demonstration of the dynamic tunneling method, an oscillating tip is situated over a trap state in a dielectric film at a single location, and the tip is slowly scanned directly toward the surface. The lock-in amplifier (DTFM) signal is plotted versus the closest approach probe height (Figure 11.11). Initially, when the tip is out of dynamic tunneling range, the lock-in amplifier signal is near zero. When the tip is close enough to allow consistent shuttling of an electron to and from the trap state, the lock-in amplifier signal rises. After an initial rise, the dynamic tunneling signal saturates, because the amount of shuttled charge saturates, indicating that there are no additional states which are in tunneling range.

To form a DTFM image, the tip is raster-scanned at a constant height (no height feedback) over the surface, while the

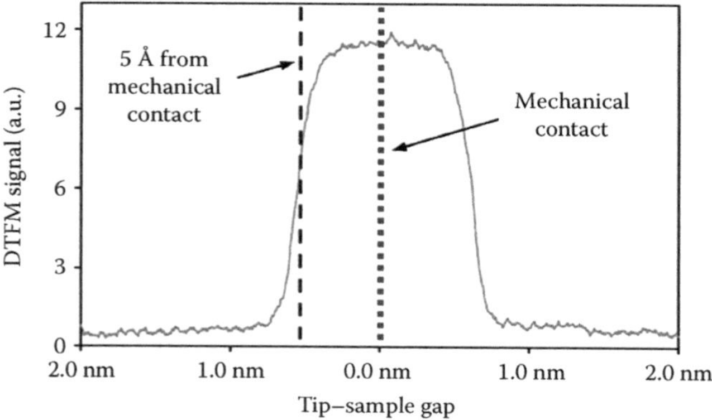

FIGURE 11.11 Measured lock-in amplifier electron shuttling signal (DTFM signal) obtained by scanning the tip toward (and away from) a hafnium silicate surface. The electron shuttling turns on approximately 0.5 nm from contact. (Reprinted with permission from Johnson, J.P., Zheng, N., and Williams, C.C., *Nanotechnology*, 20, 055701, Copyright 2009 IOP Publishing Ltd.)

lock-in amplifier (DTFM) signal is recorded at each point, producing an image with contrast proportional to the DTFM signal (electron shuttling signal). The typical image acquisition time is 3 min.

Images of trap states in a silicon dioxide film are shown in Figure 11.12. The figure contains two repeated 10 × 10 nm DTFM images of a 20 nm thick silicon dioxide film. As can be seen in the image, there is excellent repeatability, in contrast to the previous imaging attempts described earlier using the SETFM approach. This type of repeatability is characteristic of all DTFM images. Figure 11.13 contains two 20 × 20 nm DTFM images of hafnium silicate. These two images were acquired in the same location, but not with the same tip–sample gap. Nominally, the gap is 0.1 nm larger for the right image as compared to the left image.

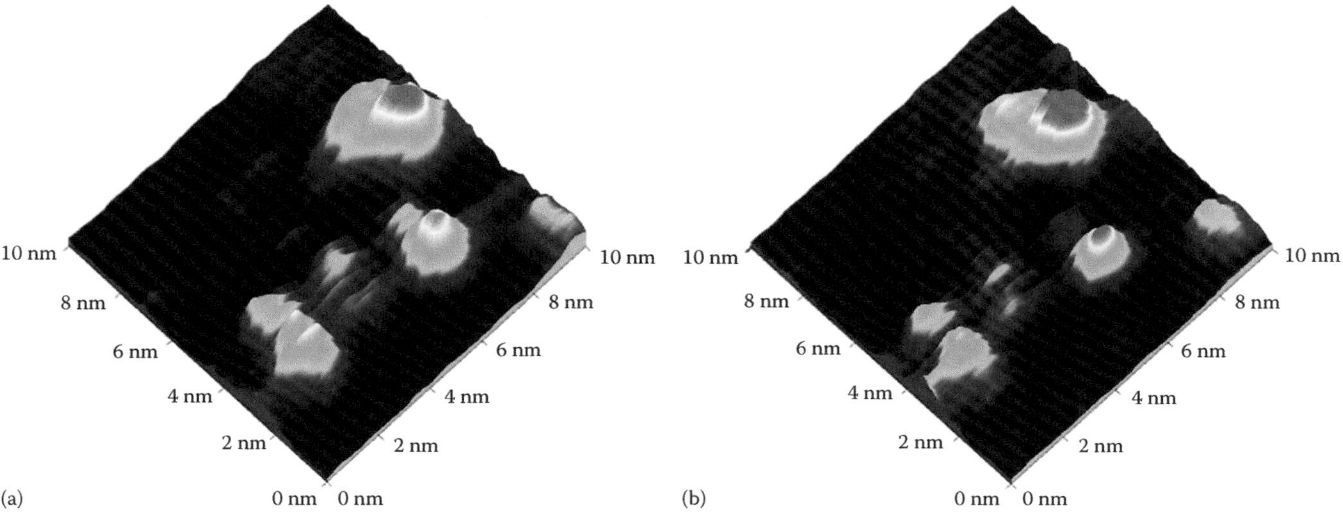

FIGURE 11.12 DTFM images of electronic trap states in SiO_2. (a) 10 nm × 10 nm DTFM scan of a 20 nm thick SiO_2 film on silicon. Resolution is about 0.5 nm. (b) Repeat scan of the same area demonstrating the repeatability of this technique. (Adapted with permission from Johnson, J.P., Zheng, N., and Williams, C.C., *Nanotechnology*, 20, 055701, Copyright 2009 IOP Publishing Ltd.)

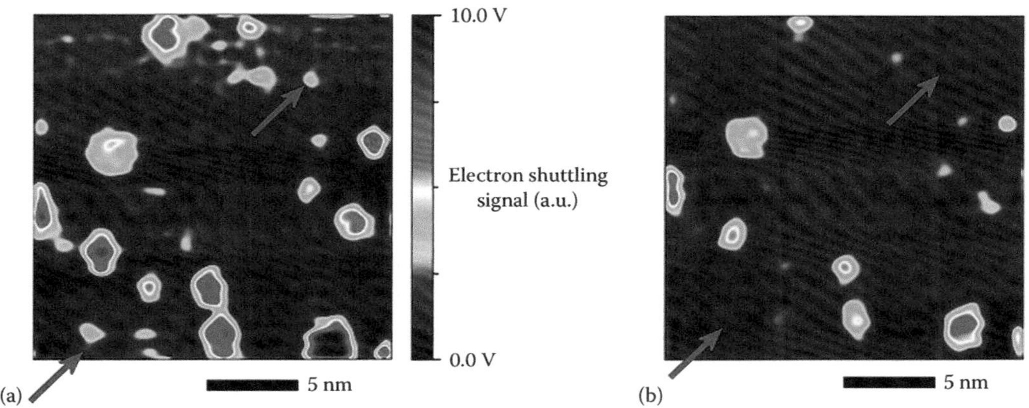

FIGURE 11.13 Imaging of electronic trap states in hafnium silicate. Two consecutive 20 nm × 20 nm scans of the same area of a 4.7 nm thick HfSiO*x* film on silicon, but at different tip-sample gaps. The tip–sample gap for the image on the right was nominally 0.1 nm greater than for the image shown on the left. Note that in the right image, the states appear smaller than in the left image, and some of the states have completely disappeared. The arrows point to two states which appear in the image with the smaller gap (a) and do not appear in the image with the larger tunneling gap (b). This effect is associated with the physical depth of the trap states in the dielectric film. (Reprinted with permission from Johnson, J.P., Zheng, N., and Williams, C.C., *Nanotechnology*, 20, 055701, Copyright 2009 IOP Publishing Ltd.)

Upon careful inspection, it is observed that while most of the states seen in the left image do appear in the right image, the states in the right image appear generally smaller in lateral size and some disappear completely (see arrows). This effect is due to the physical depth of the trap states.

As with STM, the DTFM images are the result of a convolution between the effective tip tunneling area and the states being imaged. With a fixed probe scan height, a trap state that is physically deep may only be accessed when tunneling from the very apex of the probe tip. A state that is closer to the surface may be accessed by tunneling from both the tip apex and tip regions near the apex but somewhat further up the tip (Figure 11.14). This effect causes the apparent lateral size of a trap state in the image to be related to the physical depth of the state. States which appear larger in a DTFM image are closer to the surface than states which appear smaller. Note that the states in the right image that have completely disappeared due to the larger tip height also have the smallest lateral extent in the left image, meaning that those states have the greatest depth (see arrows). Of course, there is always the possibility that some of the larger states are actually larger defects, but this can be determined by acquiring several images at different heights to see how their lateral extent changes with height. By imaging at different tip heights and differentially subtracting pairs of images, the 3D spatial distribution of the trap states in the surface can be measured. The method for quantitative depth determination was discussed briefly in the previous section on spectroscopy (Johnson et al. 2011).

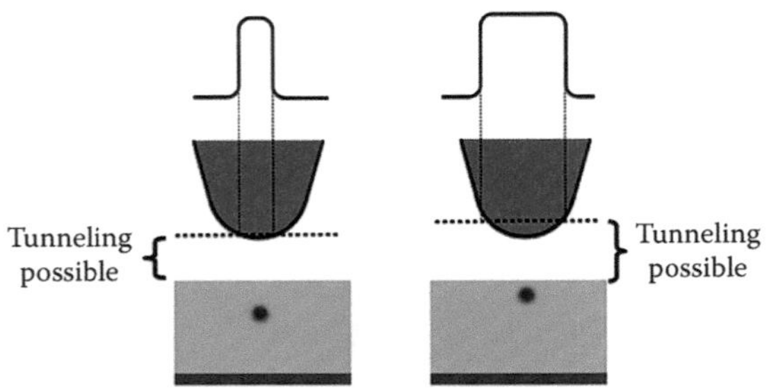

FIGURE 11.14 Illustration of the physical origin of the lateral size effect in DTFM images. Note that states with smaller physical depth will have a larger apparent size in DTFM images. This is due to the fact that states closer to the probe tip can be reached by tunneling from a larger region of the probe tip than states deeper in the surface.

11.6 Summary

It is the hope of the author that this chapter has provided a useful introduction to the various capabilities of force-detected single electron tunneling methods. The capabilities include detection of single electron tunneling events between a metallic scanning probe tip and a nonconducting dielectric surface, controlling the occupancy of single trap states, imaging of the spatial extent, depth and distribution of trap states with atomic scale spatial resolution, and direct measurement of the energy level of individual trap states in dielectric surfaces. The methods are also viable for exploring the electronic properties of other quantum states which exist on the surface of nonconducting substrates. These methods open a new window through which the electronic properties of many semiconductor and dielectric materials, currently inaccessible to STM, can be explored with atomic scale spatial resolution.

Acknowledgments

The author would like to acknowledge the significant contributions of Dr. L. Klein, Dr. E. Bussmann, Dr. N. Zheng, Dr. J. Johnson, and Dr. D. Winslow. As graduate students, they performed the challenging work of experimentally demonstrating for the first time the force-detected tunneling capabilities described in this chapter. Appreciation is also expressed to P. Rahe for his careful reading of the manuscript and G. Wang for his contributions to the monolayer protected cluster work.

References

Azuma, Y., M. Kanchara, T. Teranishi, and Y. Majima, *Phys. Rev. Lett.* 96, 016108, 2006.

Barth, C. and C. Henry, *Phys. Rev. Lett.* 100, 096101, 2008.

Barth, C., A.S. Foster, C.R. Henry, and A.L. Shluger, *Adv. Mater.* 23, 477, 2011.

Bersuker, G., J.H. Sim, C.S. Park et al., *IEEE Trans. Dev. Mat. Rel.* 7, 138, 2007.

Binnig, G., H. Rohrer, Ch. Gerber, and E. Weibel, *Phys. Rev. Lett.* 49, 57, 1982.

Binnig, G., C.F. Quate, and Ch. Gerber, *Phys. Rev. Lett.* 56, 930, 1986.

Bussmann E. and C.C. Williams, *Rev. Sci. Instrum.* 75, 422, 2004.

Bussmann, E., D.-J. Kim, and C.C. Williams, *Appl. Phys. Lett.* 85, 2538, 2004.

Bussmann, E., N. Zheng, and C.C. Williams, *Appl. Phys. Lett.* 86, 163109, 2005.

Bussmann, E. and C.C. Williams, *Appl. Phys. Lett.* 88, 263108, 2006.

Bussmann, E., N. Zheng, and C.C. Williams, *Nano Lett.* 6, 2577, 2006.

Conley, J.F. Jr., P.M. Lenahan, B.D. Wallace, and P. Cole, *IEEE Trans. Nucl. Sci.* 44, 1804, 1997.

Dana, A. and Y. Yamamoto, *Nanotechnology* 16, S125, 2005.

Giessibl, F., S. Hembacher, H. Bielefeldt, and J. Mannhart, *Science* 289, 422, 2000.

Gross, L., F. Mohn, P. Liljeroth, J. Repp, F.J. Giessibl, and G. Meyer, *Science*, 324, 1428, 2009.

Hoppe, E.E. and C.R. Aita, *Appl. Phys. Lett.* 92, 141912, 2008.

Johnson, J.P., N. Zheng, and C.C. Williams, *Nanotechnology* 20, 055701, 2009.

Johnson, J.P., D.W. Winslow, and C.C. Williams, *Appl. Phys. Lett.* 98, 052902, 2011.

Kaczer, B., Z. Meng, and J.P. Pelz, *Phys. Rev. Lett.* 48, 724, 1986.

Kerber, A. and E.A. Cartier, *IEEE Trans. Dev. Mat. Rel.* 9, 147, 2009.

Klein, L.J., C.C. Williams, and J. Kim, *Appl. Phys. Lett.* 77, 3615, 2000.

Klein, L.J. and C.C. Williams, *Appl. Phys. Lett.* 79, 1828, 2001.

Klein, L.J. and C.C. Williams, *Appl. Phys. Lett.* 81, 4589, 2002.

Klein, L.J. and C.C. Williams, *J. Appl. Phys.* 95, 2547, 2004.

Kochanski, G.P. *Phys. Rev. Lett.* 10, 2285, 1989.

Kremmer, S., H. Wurmbauer, C. Teichert et al., *J. Appl. Phys.* 97, 074315, 2005.

Lenahan, P.M. and J.F. Conley, Jr., *J. Vac. Sci. Technol. B* 16, 2134, 1998.

Ludeke, R. and E. Cartier, *Appl. Phys. Lett.* 78, 3998, 2001.

Martin, Y., C.C. Williams, and H.K. Wickramasinghe, *J. Appl. Phys.* 61, 4723, 1987.

Martin, Y., D.W. Abraham, and W.K. Wickramasinghe, *Appl. Phys. Lett.* 52, 1103, 1988.

Mohn, F., L. Gross, N. Moll, and G. Meyer, *Nat. Nanotechnol.* 7, 227, 2012.

Naitou, Y., H. Arimura, N. Kitano et al., *Appl. Phys. Lett.* 92, 012112, 2008.

Nguyen, N.V., A.V. Davydov, D. Chandler-Horowitz, and M.M. Frank, *Appl. Phys. Lett.* 87, 192903, 2005.

Nony, L., A.S. Foster, F. Bocquet, and C. Loppacher, *Phys. Rev. Lett.* 103, 036802, 2009.

Pacchioni, G., L. Skuja, and D.L. Griscom, eds., *Defects in SiO_2 and Related Dielectrics: Science and Technology*, Kluwer Academic Publishers, Dordrecht, the Netherlands, 2000.

Pantelides, S.T., Z.-Y. Lu, C. Nicklaw, T. Bakos et al., *J. Non-Cryst. Sol.* 354, 217, 2008.

Park, J.W., D.K. Lee, D. Lim, H. Lee, and S.H. Choi, *J. Appl. Phys.* 104, 033521, 2008.

Placidi, M., A. Constant, A. Fontserè et al., *J. Electrochem. Soc.* 157, 1008, 2010.

Price, J., P.S. Lysaght, S.C. Song, H.J. Li, and A.C. Diebold, *Appl. Phys. Lett.* 91, 061925, 2007.

Ralls, K.S., W.J. Skocpol, L.D. Jackel et al., *Phys. Rev. Lett.* 52, 228, 1984.

Ribes, G., S. Bruyère, D. Roy et al., *IEEE Trans. Dev. Mat. Rel.* 6, 132, 2006.

Sadewasser, S., P. Jelinek, C.-K. Fang et al., *Phys. Rev. Lett.* 103, 266103, 2009

Schonenberger, C. and S.F. Alvarado, *Phys. Rev. Lett.* 65, 3162, 1990.

Schaadt, D.M., E.T. Yu, S. Sankar, and A.E. Berkowitz, *Appl. Phys. Lett.* 74, 472, 1999.

Stern, J.E., B.D. Terris, H.J. Mamin, and D. Rugar, *Appl. Phys. Lett.* 53, 2717, 1988.

Stranick, S.J., P.S. Weiss, A.N. Parikh, and D.L. Allara, *J. Vac. Sci. Technol. A* 11, 739, 1993.

Stomp, R., Y. Miyahara, S. Schaer et al., *Phys. Rev. Lett.* 94, 056802, 2005.

Suganuma, Y., P.-E. Trudeau, and A.-A. Dhirani, *Phys. Rev. B* 66, 241405, 2002.

Terris, B.D., J.E. Stern, D. Ruger, and H.J. Mamin, *Phys. Rev. Lett.* 63, 2669, 1989.

Welland, M.E., and R.H. Koch, *Appl. Phys. Lett.* 48, 724, 1986.

Wiesendanger, R., *Scanning Probe Microscopy and Spectroscopy*, Cambridge University Press, Cambridge, U.K., 1994.

Winslow, D.W., J.P. Johnson, and C.C. Williams, *Appl. Phys. Lett.* 98, 172903, 2011a.

Winslow, D.W. and C.C. Williams, *J. Appl. Phys.* 110, 114102, 2011b.

Woodsides, M. T. and P. L. McEuen, *Science* 296, 1098, 2002.

Young, C.D., Y. Zhao, D. Heh, R. Choi, B.H. Lee, and G. Bersuker, *IEEE Trans. Electron. Dev.* 56, 1322, 2009.

Zheng, N., C.C. Williams, E.G. Mishchenko, and E. Bussmann, *J. Appl. Phys.* 101, 093702, 2007.

Zheng, N., J.P. Johnson, C.C. Williams, and G. Wang, *Nanotechnology* 21, 295708, 2010.

Zhu, J., M. Brink, and P.L. McEuen, *Appl. Phys. Lett.* 87, 242102, 2005.

12

Picometer-Scale Dynamical Single-Molecule Imaging by High-Energy Probe

Yuji C. Sasaki
SPring-8 Japan Synchrotron Radiation Research Institute
and
Japan Science and Technology Agency
and
The University of Tokyo

12.1 Picometer-Scale Imaging Concept

Recently, the measuring methods exceeding diffraction limits have been appearing from the field of biophysics. In those methodologies, many scientists pay their attention to individual single-molecule measurement technology [1–3]. Figure 12.1 shows the difference between a diffraction limit and center position's accuracy of individual single spot. I have distinguished these two different concepts from the imaging concept and the tracking concept.

The imaging concept has been adapted by the methodology called the usual microscope. In the newest single-molecule measurements, highly precise individual single-molecule pursuit is enabled by the latter tracking concept. Thus, single-molecule measurement was attained in the measurement accuracy of the nanometer scale using visible light. The single-molecule measurement using the x-rays introduced later is very simply adapted for an x-ray wavelength region in the tracking concept [4–6]. This simple idea has enabled realizations of picometer movement observations for individual single molecules.

What is the essential difference between single-molecule measurements and the information probably acquired by multimolecules (bulk) measurements? Figure 12.2 should explain this difference using crystal conditions. Structural biology has succeeded in explosive progress because protein molecules crystallize also by chance. However, the state where the lattice is constructed should add various restriction conditions in the evaluation of function of protein molecules. For example, a ligand cannot be combined with a protein molecule and the interactions with other protein molecules cannot be measured at all. It is ideal when single-molecule measurement observes interactions of the other molecules with molecular movements.

The final purpose of single-molecule measurement is in vivo measurements. Many scientists are interested in what a functional protein molecule carries out an interaction to other molecules within a living cell. Additionally, it has been expected that the translational motions of molecule and rotational movements would be measurable. It is not difficult to measure simple translation motion. Although it is difficult, measuring intramolecular rotation movements is directly linked with the functions, and it is very important information.

Figure 12.3 shows the difficulty that measures the rotation motions inside single molecule to each individual single molecule.

Since the size of a general protein molecule is about 10 nm, the internal rotational movement as shown in Figure 12.3 is usually approximately 1%–5% (0.1–0.5 nm) at the maximum values. The motion measurement of picometer accuracy is needed. Additionally, this important movement is not standing it still stably. In order to acquire unstable structural information stably, the time-resolved measurement technique is needed, as shown in Figure 12.4.

Therefore, in order to monitor molecule internal motion correctly, measurement of the individual single-molecule measurement technique equipped with the following three characteristics must be attained within a cell: high speed, high sensitivity, and high accuracy. I found out the possibility to the methodology using high-energy probe, for example, x-ray, electron, and neutron.

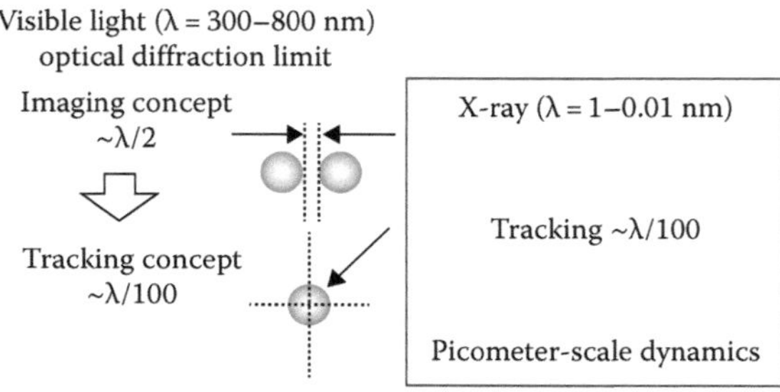

FIGURE 12.1 Difference between the imaging concept and the tracking one. The single-molecule technology is usually applied to the tracking concept. Both DXT and DET methods are the first demonstration of single-molecular detection systems to be adapted in the tracking concept in x-ray wavelength region.

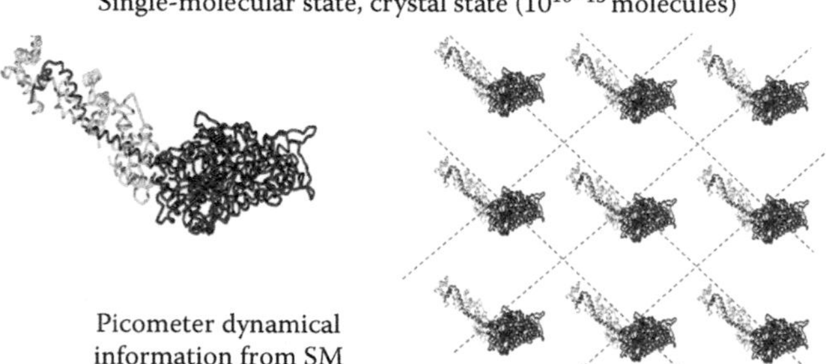

FIGURE 12.2 The single-molecule measurement can observe the molecular movement information which probably is not acquired by multimolecule measurements. A crystal condition is a wonderful sample system which can acquire highly precise structural information. However, it is impossible to extract the random molecular movements which cannot synchronize from the crystal conditions.

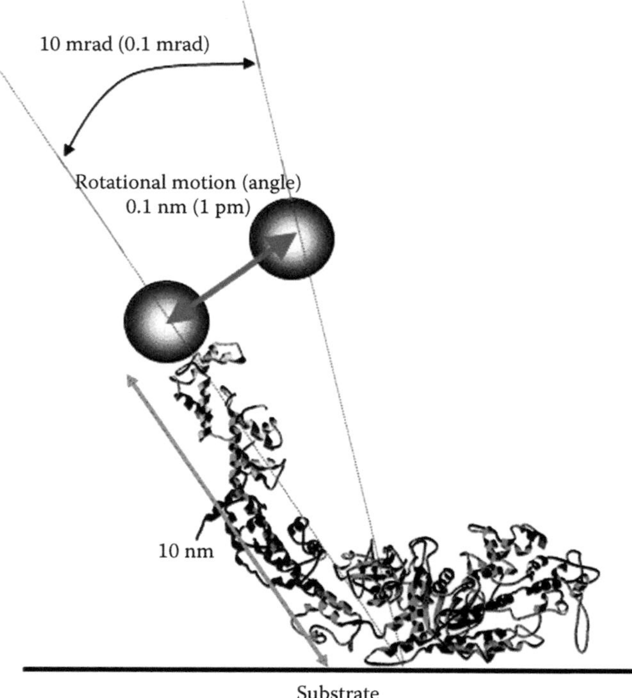

FIGURE 12.3 A protein molecule 10 nm in diameter carries out functional revelation when the end of the molecule moves about 1%–5%. For example, the width of dynamical motions which a ligand reacts to protein or penetrates ions guide only movements below 0.1 nm. In order to understand some protein functions from a molecular movement, the measurement capability of a picometer level must be needed.

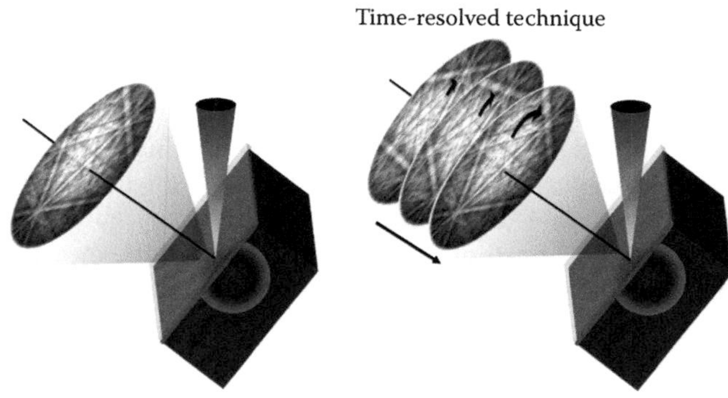

FIGURE 12.4 In order to measure intramolecular motion correctly, it is indispensable that time-resolved measurement and single-molecule measurement can be performed. The function of a protein molecule determines the time zone. Thus, all time scales are also needed, for example, millisecond, microsecond, and nanosecond.

12.2 Diffracted X-Ray Tracking

Dynamical single-molecule observation of protein in the living cell has progressed with the development of several techniques. Single-molecule observation techniques using fluorescent probes have proved valuable in solving many basic but important questions in biology and biophysics, as these methods provide positional information on single molecules with ultrahigh accuracy, far beyond the optical diffraction limit of half of the wavelength, λ/2. In particular, single-molecule fluorescence resonance energy transfer (FRET), which relies on the distance-dependent transfer of energy from donor to acceptor fluorophores, is one of the few tools available for measuring nanometer-scale distances and changes in distance, or the intermolecular orientation between the two fluorophores, both in vitro and in vivo. However, measuring intramolecular structural changes of single protein molecules by single-molecule FRET is extremely difficult due to the lack of monitoring precision and the instability of signal intensity under physiological conditions. In addition, FRET is insufficient for detecting the conformational change of a single protein molecule because this method provides only distance information. For that purpose, information about both the orientation and distance, collected simultaneously, is needed at the angstrom or picometer scale and with milli- or microradian precision. One approach to improving the positional accuracy is to shorten the wavelength by using, for example,

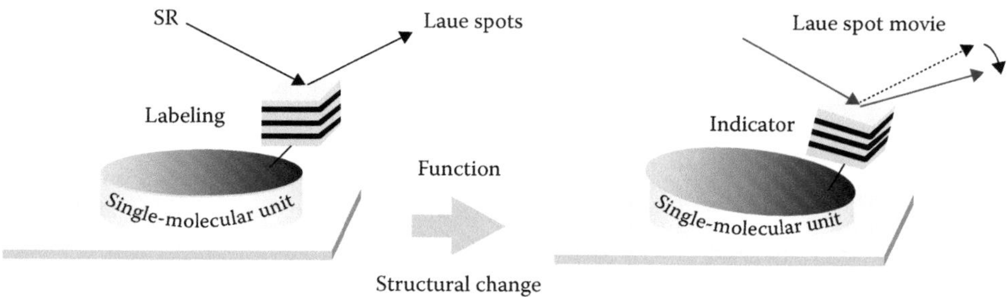

FIGURE 12.5 Schematic drawing of the single-molecular detection system using x-rays (not to scale). DXT monitors the behavior of a single special domain with the guidance of diffraction spots from the nanocrystal which is tightly coupled to the special domain. But we assumed that motions of a specific site in individual proteins are equal to the observed orientations of nanocrystals. From many DXT data, it is judged that this assumption is almost right. There is certainly a labeling position at least which forms this assumption.

x-rays, electrons, neutrons, and other accelerated ion probes. We believe that the probe that should be tried first is x-ray, because both the damage to samples and the flux of the utilized probe should be considered in regard to the probe of new single-molecule detection system.

Most existing experimental techniques using x-rays are based on the averages of observations of many molecules, and thus the behavior of a single molecule cannot be determined. Several years ago, we proposed the direct observation of the rotating motion of an individual nanocrystal in aqueous solution by time-resolved Laue diffraction. In 2000, we successfully conducted in vitro time-resolved x-ray observations of picometer-scale slow Brownian motions of individual protein molecules in aqueous solutions [4]. In this single-molecule detection system, which we call diffracted x-ray tracking (DXT), we observed the rotating motions of an individual nanocrystal labeled to a specific site in individual protein molecules by using the time-resolved Laue diffraction technique. DXT can monitor the dynamics of individual molecules or specific sites in individual protein molecules (Figure 12.5). Here, the labeled gold nanocrystal is a sign that carries out the index of the direction of movement in the adsorbed protein molecules.

In DXT, measurements of individual protein molecules with ultrahigh accuracy are theoretically possible. In vivo observations in visible light region have greatly progressed due to the remarkable development of single-molecule detection fluorescence techniques. These single-molecule techniques have provided positional information with an accuracy of about $\lambda/100$, far beyond the optical diffraction limit ($\sim\lambda/2$). These methods utilize tracking as opposed to imaging. By using this tracking technique, we were able to achieve time-resolved x-ray ($\lambda \approx 0.1$ nm) observations of picometer-scale ($\lambda/100$) slow Brownian motions in individual protein molecules.

DXT is the method of pursuing one diffraction spot from one gold nanocrystal. It is explained why the diffraction spot was utilized in DXT. Usually, single-molecule measurements using visible light detect the fluorescence from the labeled quantum dots or the labeled fluorescence molecule. It is not high sensitivity although there is a physical phenomenon (absorption) of fluorescence x-rays also in an x-ray region as shown in Figure 12.6. Since a cross section of elastic scattering in x-ray region will increase rapidly, high-sensitivity detection is attained. In addition, since an x-ray diffraction phenomenon is very sensitive to the diffraction angles, it becomes an advantage extremely when orientation accuracy is important.

There is a big advantage when DXT technology can be utilized for the internal motion measurement of a protein molecule. It may be said that it is difficult to determine the structure of a giant protein molecule as one of the difficult problems of structural biology field. Many of the giant molecules are formed after the subunit with the same structure domain has symmetrically condensed as shown in Figure 12.7. In DXT, since the size of the labeled gold nanocrystal is 20–40 nm of the diameter when protein is large, the influence of the labeling process should be smaller and also possible to evaluate a motility characteristic from the symmetry of labeling positions. Additionally, this symmetry in the big protein molecule must be paving the way for real-time observations of the allosteric effect, which is the basic concept of protein dynamics.

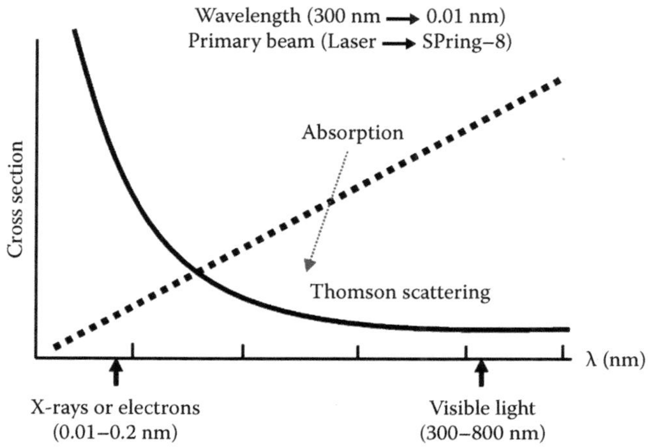

FIGURE 12.6 A visible region differs in the cross-sectional area of some physical phenomena from an x-ray wavelength region. Using an x-ray diffraction phenomenon is the best selection that realizes high sensitivity measurement in x-ray region. Theoretically, x-ray single-molecule measurement can also perform by using x-ray light source of a laboratory level.

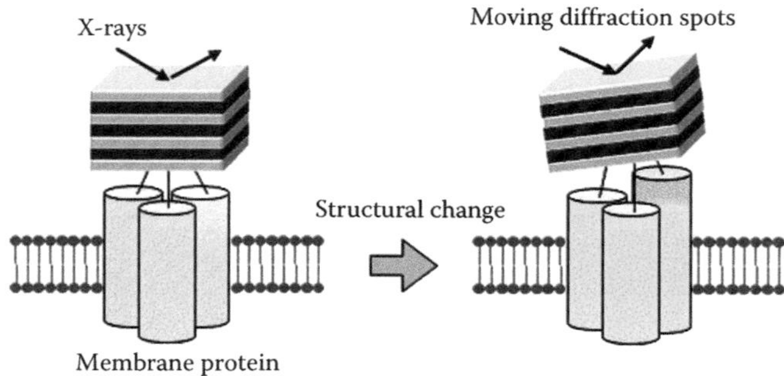

FIGURE 12.7 Most of the proteins or enzymes are composed of multi meric proteins, such as dimer, trimer, or higher-order structures. DXT realizes collaboration motion measurements using this structural symmetry.

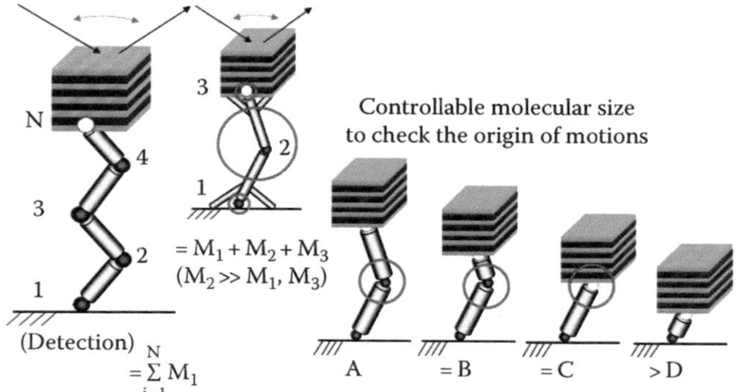

FIGURE 12.8 DXT can perform only overall movement measurement of a molecule. Thus, DXT can be determined as a local structure factor by utilizing the imperfect structure, or controlling protein motions. If the part C is a main part of movement, DXT can detect the big difference between the motility characteristic of A and B, and the motility characteristic of D.

We can explain what is measured with DXT as shown in Figure 12.8. A protein molecule is explained as a candidate for DXT observations. The protein molecule consists of primary polymer of amino acids. The protein structure can be assigned as both inflexible rods and flexible nodes from secondary structural information. DXT can monitor the total motions with these rods and nodes. DXT monitors all the movements from these rods and nodes as a total information as shown in Figure 12.8A. For example, a protein molecule consists of two rods and single node as shown in Figure 12.8B. Nodes 1 and 3 can control movements by chemical processing. Thus, DXT can monitor the only motions of node 2. Additionally, it is mentioned to one of the advantages of protein study that the technology which produces a mutant is established. For example, it can perform simply producing the mutant that made a part of protein molecule suffer a loss as shown in Figure 12.8C. From the DXT measurement of all these mutants, we can determine whether the measurement result from Figure 12.8B has been original movements from node 2. Therefore, in DXT measurement, it may be not able to judge of which structure origin it is movement by one-sample system measurement. Known structural information is certainly required of both the determination of a label position and DXT data analysis.

The labeling process of this gold nanocrystal has not only a fault but a possibility of becoming a new interface to the functional protein molecules. In 2006, we reported the first observations of x-ray radiation pressure force on individual single nanocrystals using DXT [7]. In fields dealing with microscopic phenomena, the optical radiation pressure force from visible light lasers has recently become an important research tool in the fields of biology, physical chemistry, and soft condensed matter physics. On an astronomical scale, the presence of x-ray radiation pressure forces is fundamental to explaining the physical properties of black holes or neutron stars. However, on a nanoscopic scale, x-ray radiation pressure forces have not yet been observed because of the limitation on the number of x-ray photons and the absence of an ultrasensitive detection device. DXT can not only monitor slight internal motions of protein molecules but can also detect weak x-ray force fields as shown in Figure 12.9.

In single-molecule science and technology, it is crucial to control dynamical Brownian motions of individual functional protein molecules. Optical traps and tweezers using visible light gave the first demonstration of controlling micrometer-sized particles or molecules. In order to improve both the monitoring and control of internal molecular motions of single-molecular units with improved ultrahigh precision under in vitro or in vivo physiological conditions, we proposed new single-molecule control techniques using x-rays, electrons, and neutrons. We found that x-ray radiation pressure forces are dependent upon x-ray intensity and the size of the gold nanocrystal under observation. To our surprise, we estimated the observed force using DXT to be on the atto-Newton level.

In DXT, we used white x-rays or quasi-white x-rays (e.g., energy peak width of 15% using undulator radiation, BL40XU, SPring-8, Japan) to record time-resolved Laue diffraction spots from labeled gold nanocrystals. Additionally, to detect the individual labeled gold nanocrystals, high photon flux at the sample

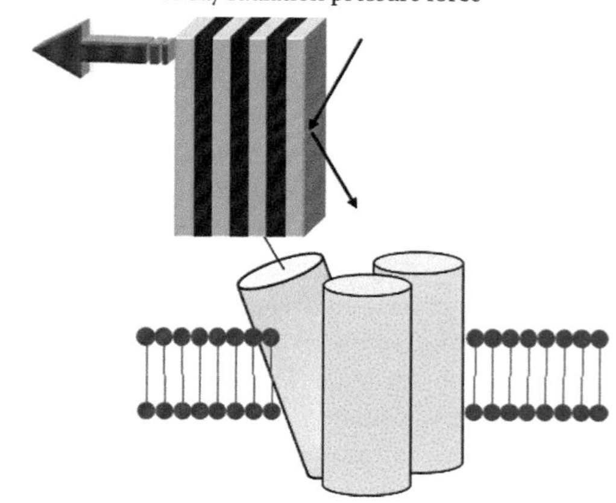

FIGURE 12.9 Schematic drawing of the observed x-ray radiation pressure force on individual single gold nanocrystal which is linked to the functional membrane protein molecules. This technology, which introduces modification into a part of protein structure, is certainly required for the future analysis of protein molecules.

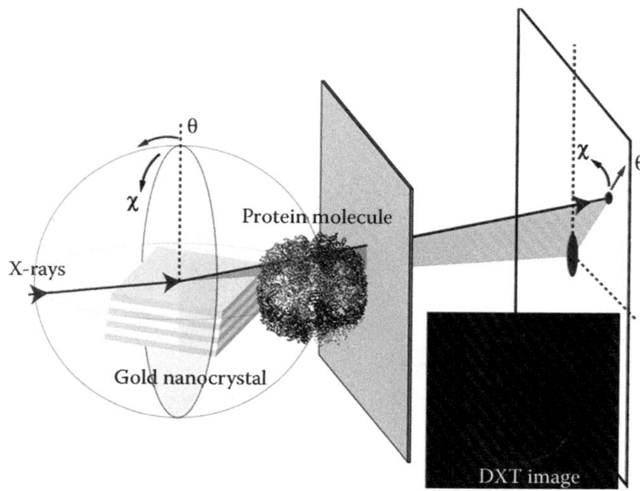

FIGURE 12.10 Schematics instrumentation of DXT. The nanocrystal on the protein is used as a tracer for dynamic motion of protein. The movement of crystal can be monitored by irradiating x-ray on it and trajectory of diffracted spot from the crystal was investigated. In this study, a mutant group II chaperonin from Thermococcus strain KS-1 with a cysteine residue at the tip of the helical protrusion was immobilized on the gold-coated substrate surface and was labeled with a gold nanocrystal through gold–thiol bond. If the ring twists, the diffracted spot should be moved to concentric circle direction, as χ direction on this figure.

position is required to be of the order of 10^{16-17} photons/s/mm^2. A diffraction spot in DXT was monitored with an x-ray image intensifier (V5445P, Hamamatsu Photonics, Japan) and a CCD camera (C4880-82, Hamamatsu Photonics, Japan) with a resolution of 656 × 494 pixels or a CMOS camera (FASTCAM SA 1.1, Photron) with 128 × 112 pixels. We also utilized high-speed x-ray shutters. The average exposure time in a single DXT shot was between 1 μs and 30 ms.

In another beam line, white x-rays from the beamline Photon Factory (PF)-AR NW14A (KEK, Japan) were used to record time-resolved Laue diffraction spots from the individual gold nanocrystals on the single functional protein molecules (Figure 12.7). The x-ray energy range was 20–30 keV ($\Delta E/E = 15\%$), and the spot size of the beam at the sample was 70 μm (vertical) and 450 μm (horizontal) (full width at half-maximum). The diffracted spots from each nanocrystal were monitored with an x-ray image intensifier (V7739P, Hamamatsu Photonics, Japan) and a CCD camera (C4880-80, Hamamatsu Photonics, Japan). The specimen-to-sample distance was around 70 mm and was calibrated by the diffraction from gold thin film. The dynamics of an individual single protein were monitored through the trajectory of the Laue spots from the nanocrystal, which labeled the target protein in DXT.

Figure 12.10 shows a typical DXT arrangement. Fundamentally, DXT is equal to technique for time-resolved Laue method without crystallization of protein molecules. Thus, important factors are the θ and χ direction of the moving diffraction spots from the labeled gold nanocrystal as shown in Figure 12.10. Figure 12.11 shows the photograph of DXT setup in the beam line KEK_AR. In the sample system, there is the layer of a very thin aqueous solution. Thus, the greatest attention is paid to an x-ray stopper's position and size in order to reduce dispersion of the elastic or inelastic x-ray scattering from the aqueous solution layer as much as possible.

DXT enables to measure milliradian resolved rotating and tilting motion of single protein molecules using high-brilliant and wide-band energy x-ray source. Tracking the diffraction spot from gold nanocrystals attached functional group with protein was developed for revealing the detailed fluctuation and functionalized motions in protein reaction processes at the beamline BL28B2, and BL40XU at the SPring-8 and NW14A at the PF-AR, KEK. Recently, we installed the x-ray toroidal mirror at the beamline BL28B2. The wide-angle diffraction spot tracking

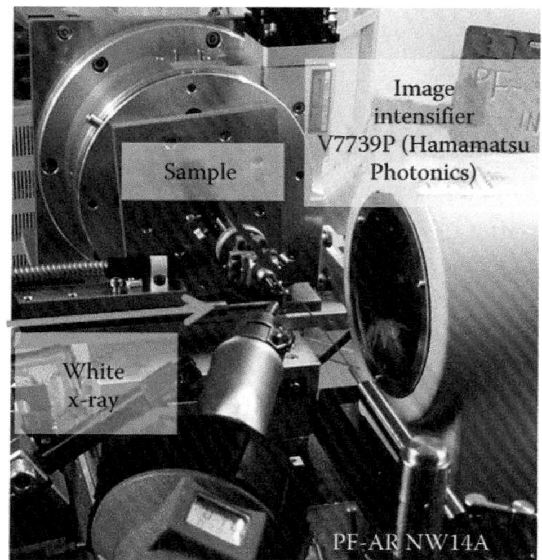

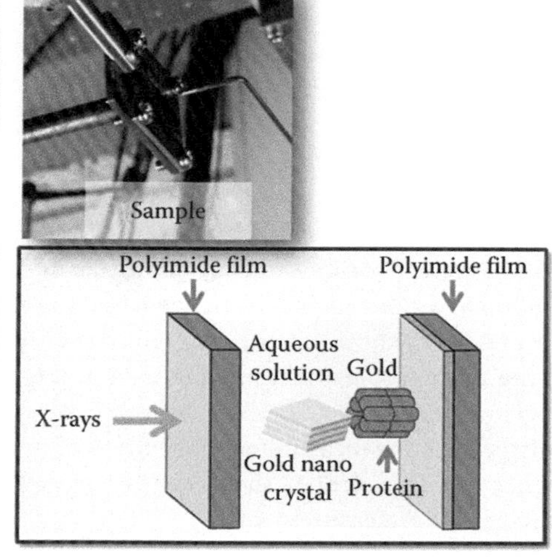

FIGURE 12.11 A photograph of the instrumental arrangements for DXT at the beamline Photon Factory-AR NW14A (KEK, Japan). The sectional view of the circumference of a sample is shown. Although x-rays have penetrability, there is a limit in the thickness of an aqueous layer (5–10 μm).

SPring-8 BL40XU

KEK-PF-AR NW-14A

SPring-8 BL28B2

FIGURE 12.12 Now, in Japan, DXT can be measured by three beamlines. The directions that suited the feature themselves of three beamlines are practiced in three beamlines.

of gold nanocrystal markers from 10 to 20 keV can be observed in this beamline. The undulator beamline BL40XU allows high-speed DXT in microsecond time region. In addition, new DXT measurement system, which was synchronized with pulse laser system, has succeeded to observe the open–close motion of membrane protein in reaction process at the beamline NW14A. Thus, the biological science applications of DXT can make high profits from observation of the in vivo functionalized motions of single protein molecules. In the beamline BL28B2, we, designed and specialized in the normal speed measurement about large movement in a single protein molecule. Three beamline organizations of DXT distinguish the measurement purpose and are going to be used effectively. Figure 12.12 shows the photograph of the three beamline for DXT measurements.

The methodology of this DXT is reclassified according to a big and important concept. Our final goal is to understand in vivo biological interesting phenomena at the single molecule levels. Therefore, we need high-speed dynamical observations of functional individual single molecules in living cell. I think that these are two strategies of measurement techniques to achieve our final goals, for example, top-down optics using mirror or zone plate, and another one is bottom-up optics using nanoprobe or nanoparticles as shown in Figure 12.13. For example, there is x-ray microscope as a typical method of top-down optics. Since a microscope has various researches also historically, I think that development of the bottom-up optics will become important from now on. Especially, we call, single-molecular detection techniques. As mentioned earlier, x-ray single-molecular detection system, we called DXT, can obtain effective dynamical information for individual single biomolecules and localized soft materials. An important thing is not that this method determines the structure of individual single protein, but is that it is the purpose to detect dynamical movements of structure. Molecular motions are directly connected functional revelation

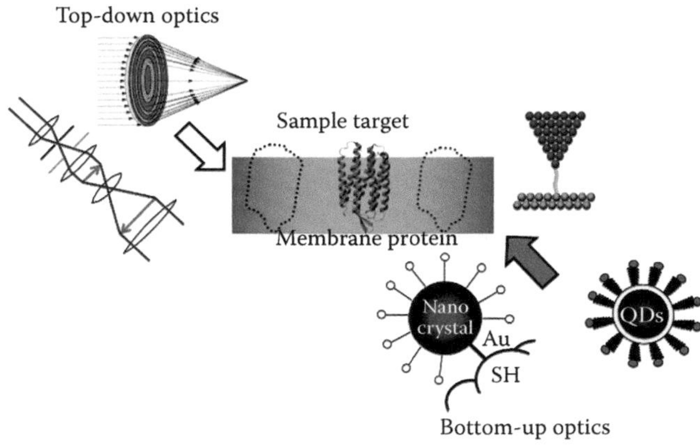

FIGURE 12.13 In visible light technologies, there are two methods used in microscopes. The top-down optical microscope is a typical optical microscope using optical concentrating systems. The bottom-up optical microscope is a fluorescence microscope using many fluorophores. DXT can be considered as a new first method for the bottom-up optical microscope.

of protein molecules. As the next stage, we need both high speed measurements and the minimizations of the labeled gold nanocrystals. There are a lot of interesting biological phenomena at the level of some ns–μs. Additionally, it is very important to fabricate a smaller crystal, because we are aiming at in vivo nonleveling measurements.

12.3 Manipulation of Gold Nanocrystal

All biological and chemical reactions are directed intramolecular structural changes. Macromolecular detection methods and crystallography cannot measure time trajectories and reaction pathways of individual molecules in nonequilibrated systems. A recent advancement in single-molecular detection systems operating at room temperature by laser-induced fluorescence offers new tools for the study of individual molecules under physiological conditions. However, it is difficult to observe the behavior of individual conformational changes using the wavelength region of visible light (300–700 nm) because of the limitations of wavelength properties. We were able to make observations of intramolecular structural changes of single molecules using x-rays (~0.1 nm), which are shorter than visible lights by two or more orders of magnitude. We utilized the x-ray diffraction phenomena from nanocrystals because the signal-to-noise rate is the highest of many x-ray physical phenomena.

Since x-ray diffraction phenomena are the interferences of the reflected x-rays by periodic structures, for example, crystals, a single-molecular detection system using x-rays requires that a single nanocrystal be tightly labeled with individual biomolecules. When the dynamics of individual single molecules were linked to each observed diffraction spot from the nanocrystal, the behavior of each molecule could be determined. Therefore, in order to achieve DXT, the fabrication of dispersive nanocrystals is a very important technology.

We expect that the sizes (~nm level) of these crystals are equal to those of the objective protein molecules because individual single nanocrystals are labeled with each single molecule. Since DXT utilizes time-resolved Laue diffraction, nanocrystals must have complete crystallinity to obtain the high intensity of diffracted x-rays from nanocrystals.

The synthesis of monodispersive gold nanoparticles, which we call colloidal golds, has been investigated intensively for a long time, and colloidal golds of ~nm level size are commercially available. However, we confirmed that Laue diffraction spots with single shot (ms) were unable to be obtained from individual colloidal golds. I devised the two producing methods as shown in Figure 12.14.

In this first work, we fabricated Mo/Si one-dimensional nanocrystals by magnetron sputtering and reactive ion etching (RIE). The most striking characteristic of one-dimensional nanocrystals is the number of observed nanocrystals, because individual one-dimensional nanocrystals reflect only a single or one pair of x-ray diffraction spot. The fabrication of one-dimensional nanocrystals requires three important phenomena: (a) fabrications of high x-ray reflectivity from the Mo/Si multilayer; (b) selectivity of the resist materials for RIE ions; and (c) detachment of nanocrystals from the substrate. In order to lead high reflectivities from one-dimensional nanocrystals, we introduced the evaporated Si layer between a polyvinylalcohol (PVA) layer and Mo/Si multilayer. If Mo/Si multilayers were directly deposited on the PVA using the sputtering process, the reflectivity from the Mo/Si multilayer decreased remarkably, because the bombardment of sputter particles (Mo, Si) destroys the flat surface of the PVA film and leads to decreased reflectivity from the Mo/Si multilayer. Recently, various lithography processes containing the RIE system have received considerable attention for nanometer-fabrication because these techniques are able to faithfully replicate nanometer scale features [8–10]. However, it is difficult to

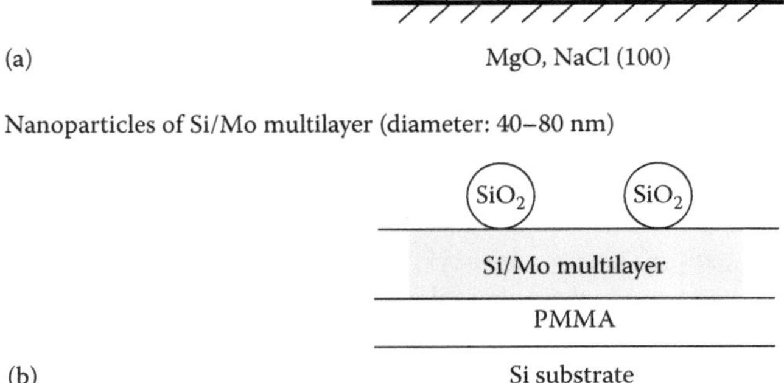

FIGURE 12.14 In order to perform the DXT as super-high-sensitive methodology, the most important and most difficult technology is the production of a nanoprobe, which is labeled to the functional protein molecules. We developed two processes for completing perfect nanocrystal. (a) Three-dimensional nanocrystal; (b) one-dimensional nanoparticle.

fabricate nanometer-scale structures over macroscopic areas (~cm^2) at a time. We investigated lithograph techniques as follows: we considered colloidal beads (e.g., gold, polystyrene, and silicon dioxide) as resist materials for protecting the Mo/Si multilayer from RIE ions. Since gold is characteristically more ductile than the other metals, colloidal gold cannot be used as the resist material of the Mo/Si multilayer under RIE, because the form of the colloidal gold changes after ion bombardment. In addition, we were not able to use polystyrene beads, because the etching rate of these beads is very fast. As a final result, we utilized silicon dioxide (SiO_2) beads, which are durable to RIE ions.

In order to detach nanocrystals, the Mo/Si multilayer is deposited on the dissoluble material (e.g., NaCl, polymethylmethacrylate [PMMA], and PVA). We used water-soluble PVA, because coated PMMA, which is a dissoluble organic solvent, may denature the biomolecule. In addition, we confirmed that deposited NaCl could not be used, because there is large roughness on the NaCl surface. Consequently, we succeeded in monitoring well-defined diffracted x-ray spots from the nanocrystals of fabricated Mo/Si three-layers (d-space = 5.0 nm) on the water-soluble PVA film, which was coated on the Si wafer in order to detach the nanocrystals from the substrate. One-dimensional (Mo/Si) nanocrystals as artificial crystals were fabricated by a sequential process using polished Si (100) wafers. At first, a water-soluble PVA film was coated on a Si wafer to separate the fabricated nanocrystals from the Si substrate. Next, in order to protect the flatness of the PVA films from the bombardment of sputter particles, the Si layer was evaporated on the PVA film under the vacuum condition (8.0×10^{-4} Pa). Mo/Si multilayers were deposited on the evaporated Si layer using radio-frequency (RF) and direct-current (DC) magnetron sputtering systems (L-332S-FH, ANELVA) in argon of 99.999% purity. The vacuum systems reached a base pressure of 4.0×10^{-4} Pa prior to Mo/Si deposition, and the argon pressure was fixed at 2.0×10^{-1} Pa during Mo/Si deposition. Each target was set with 150 mm below the sample substrate. The RF and DC power were held constant at 50 and 100 W, respectively. The thickness of each film was controlled using the shutters, which were covered with both the Si and Mo targets. The substrate holder was rotated above the Si and Mo targets at about 1000 rpm. The deposition rates for Si (99.999%) and Mo (99.98%) were approximately 0.023 and 0.079 nm/s, respectively. These values were determined by x-ray reflectivity.

One of the final goals in the fabrication of nanocrystals is to label them with biomolecules. In order to react with the active site (e.g., amino group, carboxyl group, and sulfhydryl group [SH]) of the individual single molecules, the gold layer on the Mo/Si multilayers was deposited by vacuum evaporation because gold is well known to directly and strongly attach to the SH groups of cysteine in the biomolecules by the Au–S bonds [11–13].

A method of RIE, which offers the potential of controlled anisotropy etching and adequate selectivity of materials, was used to complete the fabrication of the microstructure on the Mo/Si multilayer. SiO_2 beads, used as the resistant material of the RIE, were coated on this substrate using a spin-coater (1H-D7, MIKASA). The average size of the SiO_2 beads was about 40 nm. Samples were etched by RIE (power densities 100 mW/cm^2) with a mixture of 150 sccm carbon fluoride and 30 sccm oxygen gasses at 2.0×10^{-2} Pa.

A scanning electron microscope (SEM) was employed to observe the etching profile, the etching rate, and the selectivity for obtaining the parameters of optimal fabrication. The x-ray reflectivity of the fabricated Mo/Si multilayer and one-dimensional nanocrystals was measured at a wavelength of 0.154 nm using an x-ray diffractometer (RINT2000, RIGAKU). To dissolve the PVA film and detach the one-dimensional Mo/Si nanocrystal from the substrate (Si wafer), the substrates were soaked in super-pure water. In order to fabricate nanosize and highly reflective one-dimensional nanocrystals, we determined the optimal d-space and stacking periods of the Mo/Si multilayer and etching conditions. Evaluations of the reflectivity of the multilayer utilized the first-order reflection, since there are clear peaks of Bragg reflection resulting from the structure of the multilayer in the low angle area. The diffraction pattern in Figure 12.15 was obtained from a 10-bilayer Mo/Si (Mo:Si = 2:5, d-space = 4.0 nm) using CuKα (0.154 nm) radiation. In this case, the angle of the first-order reflectivity is 2.20° (2θ).

We confirmed the relationship between the Mo/Si bilayer thickness (d-space) on the Si substrate and the first-order reflectivity of the Mo/Si multilayer, whose total thickness was fixed at approximately 30 nm. An increase in the d-space means a decrease in the stacking periods because the total thickness of the Mo/Si multilayer was fixed at about 30 nm. The Mo layer thickness was 0.4 times the periodic length of the Si thickness. An increase in the stacking periods is attended by an increase in the reflectivity of the multilayer. In addition, a decrease in the d-space is attended by a decrease in the reflectivity of the multilayer because of the augmentation of the interface roughness. In the case of d-space = 5.0 nm or less, the reflectivity of the Mo/Si multilayer increased strongly, in spite of a decrease in stacking periods. This result shows that the reflectivity of the

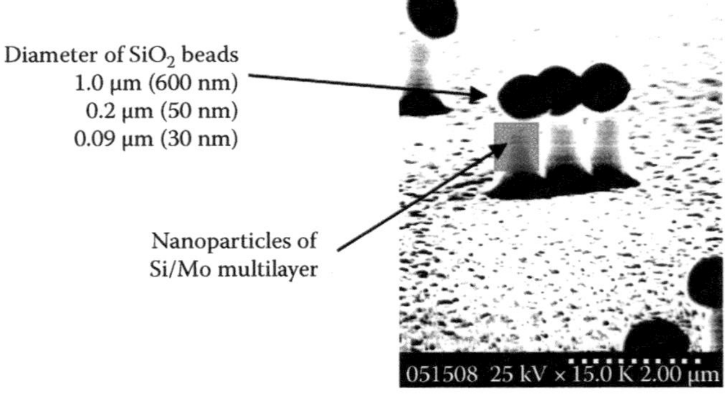

FIGURE 12.15 SEM image shows the etched Mo/Si multilayer structure after etching time 40 s. The under darkest part: PVA layer, the second darkest circles: SiO_2 beads, and white part: Mo/Si one-dimensional nanoparticle. The size of a nanoparticle is controllable by the size of a silica bead.

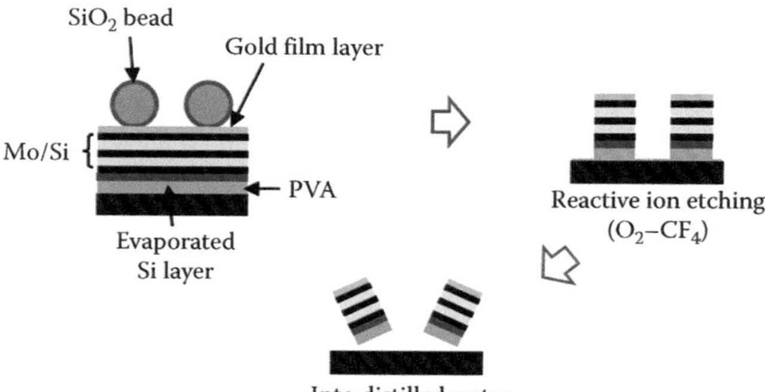

FIGURE 12.16 Models for the fabrication method of one-dimensional nanocrystals. Cross-sectional view of the multilayer structure to fabricate Mo/Si one-dimensional nanocrystals. In order to detach nanocrystals from the substrate, water-soluble polyvinyl alcohol (PVA) layer was introduced in the multilayer structure. After a RIE, the local area of the Mo/Si multilayer under the presence of SiO_2 beads is protected from etching ion. Fabricated nanocrystals are detached from the substrate in distilled water because PVA layer was dissolved in water.

Mo/Si multilayer is influenced by the interface roughness. In the case of d-space = 5.0 nm or more, the reflectivity was stable at approximately 6.3% ± 0.6%, in spite of a decrease in the stacking periods of the Mo/Si multilayer. Therefore, we chose d-space = 5.0 nm of the Mo/Si multilayer to fabricate the smallest one-dimensional nanocrystals.

We measured the relationship between the first-order reflectivity of the Mo/Si multilayer at d = 5 nm and the Mo/Si stacking periods. The symbols show the experimental results and the solid curve shows the theoretical results. The calculated reflectivities, which were performed using Fresnel equations and Henke's optical data [14], are in fair agreement with the experimental ones, because the calculations assumed that the multilayers have smooth and sharp boundaries. The inset in Figure 12.16 shows the x-ray diffraction profile of the Mo/Si 2.5 bilayers (Mo/Si/Mo/Si/Mo) and 7.5 bilayers. The reflectivity of the Mo/Si three layers showed 3.1% ± 0.2%. We determined a 2.5 bilayer by considering the smaller nanocrystals.

We evaluated the reflectivity of the multilayer grown on the Si substrate. The PVA, which is water-soluble, was coated on the Si substrate using a spin-coater because one-dimensional nanocrystals must be detached from the Si substrate. Although the Mo/Si six bilayers with a 5.0 nm period on the Si substrate showed reflectivity of 6.3% ± 0.5%, those on the PVA film were 0.07%. Thus, sputtering particles break the flat surface of the PVA film and the Mo/Si multilayer has large roughness. In general, it is known that the average energy of the sputtering particle is about 100 times that of the evaporated atom by vacuum deposition [15]. Therefore, the Si layer was evaporated on the coated PVA film using vacuum deposition, because the deposited Si layer on the PVA film has a flat surface and is physically resistant to bombardment from sputtering particles (Mo and Si).

We checked the relationship between an evaporated Si-layer thickness on the PVA films and the reflectivity of the Mo/Si multilayer of six stacking periods with d = 5.0 nm on an evaporated Si layer. A significant increase in reflectivity was observed for the evaporated Si thickness over 2.5 nm on the PVA films. For a Si thickness of 5.0 nm or above, these reflectivities were stable at approximately 2.5%. Therefore, we chose an evaporated Si layer thickness of 5.0 nm. The reflectivity of the multilayer (six stacking periods, d-space = 5.0 nm) grown on the Si substrate is 6.3% ± 0.5%. The difference of these reflectivities, which is caused by the existence of the PVA, means that the roughness of the surface of the evaporated Si is not equal to that on the Si substrate. As experimental results, we determined that the Mo/Si multilayer condition is d-space = 5.0 nm, 2.5 bilayers (Mo/Si/Mo/Si/Mo: total thickness ≈ 11.4 nm) and evaporated Si layer = 5.0 nm on the PVA film.

The gold layer (10 nm) on the Mo/Si multilayers was deposited by vacuum evaporation because it must react with the active site of the objective individual molecules. In RIE, SiO_2 beads (approximately 40 nm in diameter), used as resistant materials, which protect the local multilayer structure, were coated on the substrate using the spin-coater.

We confirmed the relationship between the etching times of the Mo/Si multilayer structure and the reflectivity of those by RIE with or without the presence of SiO_2 beads. The Mo/Si multilayer structure consisted of a PVA layer (50 nm) with a Si wafer, an evaporated Si layer (5.0 nm), Mo/Si three layers (d-space = 5.0 nm), an evaporated Au (10 nm) and SiO_2 beads. The total thickness of the multilayer structure was approximately 30 nm, except for the PVA film. In spite of the presence of the SiO_2 beads, both reflectivities of the multilayer structure at the nonetching condition were 1.6% ± 0.2%. This result indicates that x-ray absorption of SiO_2 beads minimally influences the reflectivity of the Mo/Si multilayer structure. The calculations of x-ray absorption indicated a similar result. These reflectivities of both Mo/Si multilayer structures at an etching time below 25 s were decreased slowly at the same slopes and were independent of the existence of the SiO_2. The slow decrease of reflectivity at an etching time below 25 s means that the gold layer protects the Mo/Si multilayer structure from the etching ions and the form of the gold is discontinuous. This etching time (25 s) of a 10 nm gold layer on the Mo/Si multilayer is in good agreement with the etching time that had been obtained by the 10 nm gold layer on the Si substrate. When SiO_2 beads were not present, the reflectivities of the multilayer structure at an etching time of over 25 s began to decrease strongly and were hardly exhibited at over 40 s (0.003%). This result indicates that the Mo/Si multilayer at an etching time of over 25 s began to be etched and, at an etching time of 40 s, the entire Mo/Si multilayer structure was etched completely. SEM has confirmed that the multilayer structure and SiO_2 beads are completely etched at 40 s, respectively. In other words, at an etching time of 40 s, nanocrystals are fabricated perfectly because the multilayer of the area of the absent SiO_2 beads at 40 s is completely etched and the multilayer under the SiO_2 beads is protected.

The SEM image in Figure 12.15 shows the top view of the etched multilayer structure at an etching time of 40 s. The darkest part indicates the PVA layer, the second darkest circles, which are approximately 30 nm in diameter, indicate the Mo/Si multilayer (one-dimensional nanocrystals), and the white circles, with about a 15 nm diameter, indicate the gold layer. Figure 12.16 shows the cross-sectional view model of the fabricated one-dimensional nanocrystal. We confirmed that the diameter of the top (gold layer) of the one-dimensional nanocrystals is approximately 15 nm, the bottom is approximately 30 nm, and the total thickness is about 25 nm. We estimated the effective area of the Mo/Si multilayer by comparing the initial reflectivity (1.61%) of the Mo/Si multilayer structure with the reflectivity (0.077%) at an etching time of 40 s. The result indicates that the effective area is approximately 4.8% after etching at 40 s. This value is in fair agreement with the obtained effective area (5.5%) by SEM photograph. Therefore, we succeeded in creating one-dimensional nanocrystals with approximately 25 nm diameters.

After etching at 40 s, the nanofabricated one-dimensional nanocrystals were detached from the substrate in distilled water to dissolve the PVA layer. This solution, including these nanocrystals, was dialyzed overnight with the distilled water to exclude the dissolved PVA. This solution, including the dispersive one-dimensional Mo/Si nanocrystals, was evaporated to dryness on the quartz substrate.

Figure 12.17 shows an image of four diffracted x-ray spots from individual, fabricated one-dimensional nanocrystals. According to the Bragg conditions ($2d\sin\theta = n\lambda$), the order of the recorded diffraction spots is $n = 6, 7, 8$, and 9 because the value of 2d for the one-dimensional nanocrystals is 10 nm, the region of the x-ray wavelength is $\lambda = 0.04–0.18$, nm and the Bragg angles of the four diffraction spots are $2\theta a = 297$ mrad, $2\theta b = 334$ mrad, $2\theta c = 270$ mrad, and $2\theta d = 281$ mrad. These spot intensities approximately agreed with the analytical data from fabricated nanocrystals (25 nm diameter). This means that DXT can quantitatively measure x-ray intensities from single nanocrystals. All the protocols for one-dimensional nanoparticle production are shown in Figure 12.16.

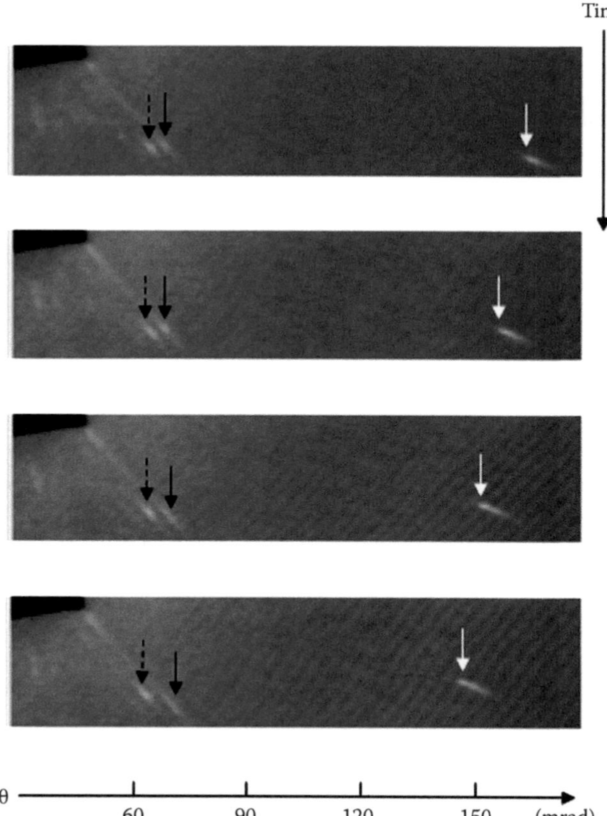

FIGURE 12.17 DXT monitors the motion of a single particle with the guidance of a diffraction spot from an individual particle (Si/Mo multilayer nanocrystal). The detectable displacement is dependent on the distance between the particle and the rotating center. Comparison between DXT and other systems for single-molecular detection, for example, SPT by measuring the viscosity of the supercooled liquid water.

By contrast with an earlier paper, we have shown that fabricated one-dimensional nanocrystals with a diameter of 25 nm can be monitored as well-defined diffracted spots by introducing the evaporated Si layer between a PVA and Mo/Si multilayer and directly link one nanocrystal to active site of individual single biomolecules without treating the surface of nanocrystals because of introducing the deposited gold layer on the Mo/Si multilayer. Furthermore, this fabricated one-dimensional nanocrystal enabled the observation of Brownian motions of single DNA molecules in aqueous solutions at a picometer level. The improvements of this technique, which created a nanometer-scale structure over a macroscopic area, not only lead to the improvement of DXT performance but will also make important contributions to nanotechnology, for example, using several lithography technologies, nanocompact disks using the integrated technology of several lithography technologies and a surface emitting laser using a microcavity structure in an optical system.

Nanoparticles and nanocrystals consisting of inorganic materials are of great interest in the field of materials science. Examples of their potential uses include semiconductor materials, ferromagnetic materials, and ferroelectric materials because nanoparticles and nanocrystals have unique size- and crystallinity-dependent characteristics that are significantly different from those of bulk materials. Furthermore, these particles have also been employed as probes for solving many basic problems of biomolecules in bioscience.

We have fabricated two different nanocrystals: Mo/Si multilayer nanocrystals fabricated by magnetron sputtering and RIE, and gold nanocrystals fabricated by vacuum evaporation. We succeeded in fabricating Mo/Si multilayer nanocrystals with diameters of approximately 25 nm. However, these multilayer nanoparticles exhibit low reactivity with cysteine in protein molecules because the reaction site on the nanocrystals is extremely small. Moreover, it is difficult to fabricate these nanocrystals with a diameter of less than 25 nm because diameter depends on the size of the SiO_2 beads used as resist materials for RIE. To overcome the disadvantages of multilayer nanoparticles, gold nanocrystals were fabricated.

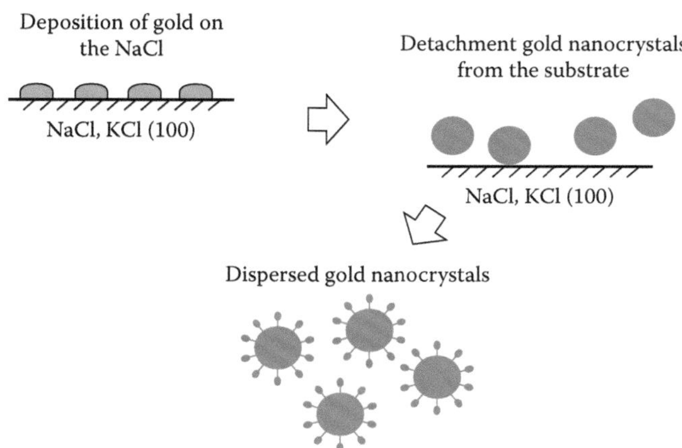

FIGURE 12.18 The preparation scheme of labeled gold nanocrystals.

Gold is a suitable material for use as a probe in biophysical studies because it can be directly and strongly linked to the thiol groups of cysteine in protein molecules. Colloidal gold, one of the most extensively studied nanoparticles, is commercially available in all sizes. However, these nanoparticles are not suitable for DXT experiments because Laue spots cannot be obtained from individual colloidal gold even when using a third-generation synchrotron radiation facility ($\sim 10^{15}$ photons/s/mm^2). The aim of this work is to prepare stable dispersive gold nanocrystals with higher crystallinity in aqueous solutions.

Gold nanocrystals were fabricated by the sequential processes as shown in Figure 12.18 [16]. First, vacuum evaporation was used to deposit a super thin layer of gold on a cleaved NaCl (100). The deposited gold formed islands on the substrate and grew epitaxially. Then, to disperse the deposited gold (gold nanocrystals) in aqueous solutions without aggregation, the substrate was dissolved by adding the detergent (3-[(3-cholamidopropyl) dimethyl-ammonio] propane sulfonic acid [CHAPS]) solutions (50 mM, pH 7.0).

Samples were prepared by thermal vacuum evaporation (ANELVA L-300EK) of 99.95% gold from a tungsten basket in vacuum of approximately 2.0×10^{-4} Pa. The cleaved NaCl (100) (10 × 10 mm) was used as a substrate, which was kept at approximately 450°C during deposition. The distance between the filament and the substrate was 150 mm. Gold was deposited at 0.3 nm/min onto the substrate, and a constant filament temperature of 1486.7 ± 4.3 K, determined using a radiation thermometer (CHINO Co. Ltd., IR-AH), was used in these experiments. Atomic force microscopy (AFM) (Nanoscope IIIa, Digital Instruments) in the tapping mode was employed to observe the form of the gold thin film deposited on the substrate. Cantilevers (125 μm long) with a spring constant of c = 42 N/m (SSS-NCH) purchased from Digital Instruments were used. To detach the gold nanocrystals from the NaCl (100) substrates, the substrates were dissolved in CHAPS solution (50 mM, pH 7.0).

We used the dynamic light scattering (DLS) (DLS-8000DS, Otsuka Electronics) device with a 488 nm Ar-ion laser to measure the particle size distributions and stability of gold nanocrystals in aqueous solutions. The scattering data were analyzed by the cumulant method. The scattering angle was 90°, and the water circulator provided temperature control with a stability of ±0.1°C.

We used both the monochromatic x-ray (12 keV) and white (Laue) x-ray (7–30 keV) modes of beamline BL44B2 (RIKEN Structural Biology, SPring-8, Japan) to confirm the epitaxial growth of gold nanocrystals on the cleaved NaCl (100) substrates and record Laue diffraction spots from individual single gold nanocrystals, respectively. The photon flux at the sample position was estimated to be approximately 10^{12} photon/s/mm^2 (20 keV, monochromatic mode), and 1015 photons/s/mm2 (7–30 keV, white mode), while the x-ray beam's focal size was 0.1 mm (horizontal) × 0.1 mm (vertical) in both modes. In the monochromatic x-ray mode, we observed a diffraction pattern from gold nanocrystals under the conditions of nearly x-ray total diffraction on the NaCl (100) substrate to suppress x-ray scattering from the NaCl substrate. To monitor Laue diffraction spots from gold nanocrystals using the white x-ray mode, dispersed gold nanocrystals were dried on a polymer film after dialysis to remove the salt and detergent from the dispersion containing the nanocrystals.

Figure 12.19 shows an AFM image of the 2.0 nm thick gold deposited at 450°C on the NaCl (100) substrate. The deposited gold formed islands on the NaCl (100) substrate, and these particles (gold nanocrystals) were not perfect spheres but hemispherical structures. The average diameter of the gold nanocrystals observed on the NaCl substrate was estimated to be 15.6 nm (averaged length: 19.8 nm, averaged height: 8.5 nm). The root mean square value, which is the standard deviation of the height value, was about 3.2 nm and the maximum height was about 13.7 nm.

We confirmed that the increase in gold deposition thickness over about 1.0 nm lead to increase in not the number of gold

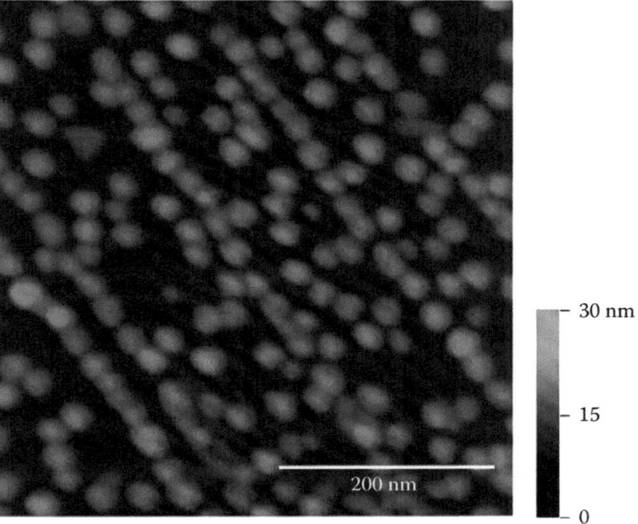

FIGURE 12.19 AFM image of the gold deposited 2.0 nm thick at substrate temperature 400°C on the cleaved NaCl (100) substrate.

nanocrystals but the size of ones. This result indicates that an evaporated gold cluster is electrically neutral because gold particles already deposited on a nonconducting substrate can serve as seeds for further gold cluster deposition. The hemispherical island growth of gold nanocrystals on the substrate might be the result of this phenomenon. Furthermore, it is expected that deposited gold clusters nucleate in the gas phase. In this experiment, we confirmed using AFM that under a lower filament temperature (under ~1500 K), the gold nanocrystals fabricated on the substrate were smaller and had a narrower distribution than those obtained at a higher temperature, and hence the adjustment of evaporation flux is another means of controlling the size of gold nanocrystals on a substrate. This indicates that the nucleation and growth rates of clusters in the gas phase affect the size of gold nanocrystals on a substrate. The x-ray diffraction pattern obtained in the monochromatic x-ray mode showed a strong peak at θ = 242 mrad, which was due to the (200) reflection of gold nanocrystals, and a weak peak at θ = 190 mrad, which was due to the (200) reflection of the NaCl substrate. There were no diffraction patterns corresponding to gold nanocrystals except in the (200) direction. This result indicates that gold nanocrystals fabricated on a substrate are expected to be not three-dimensional but one-dimensional. The explanation for this is as follows: small amorphous clusters, which evaporate at low filament temperatures, nucleate in the gas phase and are deposited on the substrate. Once the clusters have been deposited, they adopt the orientation optimized to the conditions in this system owing to their sufficiently small size. Other small clusters can then be deposited and adopt the structure of the clusters that have already been deposited. The 450°C substrate temperature used in this experiment assists in this process. In this way, the orientation of the original cluster is retained. Larger clusters would have their own orientation due to the larger number of atoms which would be retained upon deposition. The orientation of the films or nanocrystals resulting from the deposition of these larger clusters was more random. Therefore, the island growth with a preferred orientation reported here may be the result of epitaxial sticking of small gold clusters.

The stability and ensemble size of the gold nanocrystals dispersed in the CHAPS were confirmed by DLS observations. Figure 12.20 shows six DLS histograms of the gold nanocrystals deposited with different thicknesses (1.0–6.0 nm), and we measured the average diameter of each histogram as functions of the total thickness of the deposited gold. The increase in deposited gold thickness led to an increase in the average diameter and diameter distribution of the fabricated gold nanocrystals. The average diameter observed of the gold nanocrystals due to the variations in deposited gold thickness (1.0, 2.0, 3.0, 4.0, 5.0, and 6.0 nm) are 13.7, 16.2, 19.0, 26.1, 30.2, and 39.2 nm, respectively. For a deposited gold thickness of 2.0 nm, the average diameter (16.2 nm) agreed with that obtained from AFM images. The average diameters of the gold deposited at thicknesses of more than 3.0 nm began to increase markedly, whereas those of the gold deposited at thickness of less than 3.0 nm increased

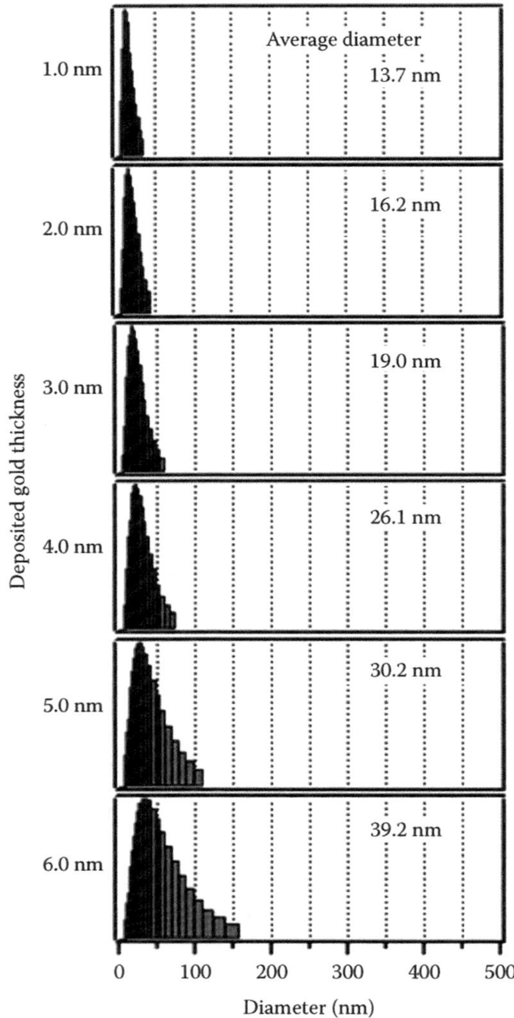

FIGURE 12.20 Three dynamical light scattering (DLS) histograms of gold nanocrystals dispersed in detergent solutions. The average diameters of the nanocrystals are determined to be 13.7, 16.2, and 19.0 nm for the deposited gold thickness of 1.0, 2.0, and 3.0 nm, respectively.

slightly. This indicates that the gold nanocrystals on the substrate coalesced closely throughout the 3 nm thick deposit because the spaces between the nanocrystals were decreased by increasing the thickness of the deposited gold.

Furthermore, we confirmed that the DLS distributions and average diameters of gold nanocrystals can remain constant for a sufficient time (more than 24 h) to enable the labeling of gold nanocrystals with individual biomolecules at room temperature. These results indicate that gold nanocrystals fabricated in vacuum conditions can be stably dispersed in an appropriate aqueous solution.

Figure 12.21 shows an image of the Laue spots (open circles) diffracted from approximately 16.2 nm gold nanocrystals (deposited gold thickness = 2.0 nm) using the white x-ray mode. The exposure time was about 10 ms. In general, the advantage of the measurement using x-rays is that it can quantitatively measure x-ray intensity. Therefore, as the intensities of these spots

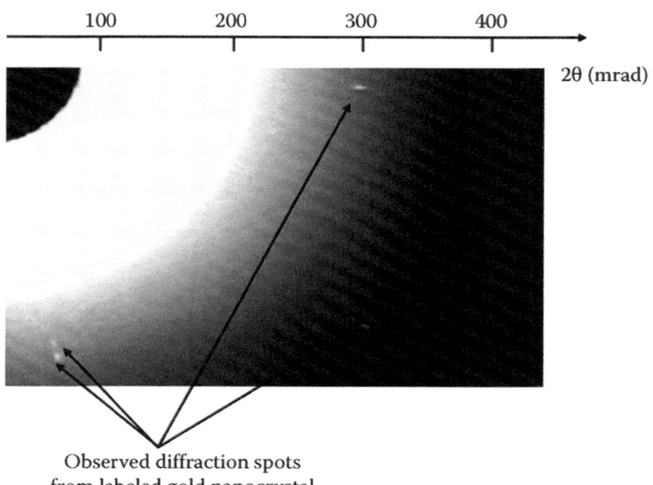

FIGURE 12.21 Diffracted x-ray spots from fabricated gold nanocrystals using the white x-ray mode.

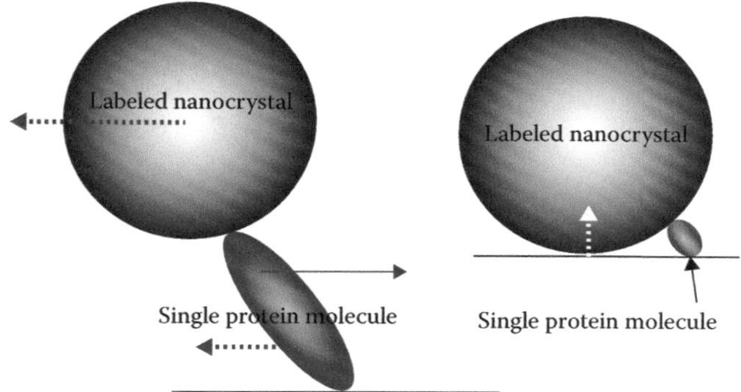

FIGURE 12.22 According to Navier–Stokes equations, the labeled gold nanocrystal does not have big influences on protein motions in a remarkable size. However, by an interaction between the labeled nanocrystal and other parts (e.g., the substrate surface), the size of the labeled nanocrystal may control the movement of protein molecules.

are all the same, the obtained diffracted spots can be regarded as originating from nanocrystals of similar sizes. Furthermore, considering the preferred orientation of gold nanocrystals, individual single nanocrystals produce only one or two (one pair) diffraction spots. Since two spots existing in pairs reflected from single nanocrystals show exactly the same motion in the real-time tracking of Laue spots, the pairs of spots from single nanocrystals are easily distinguishable from the other spots. We have succeeded in the fabrication of one-dimensional gold nanocrystals with a diameter of approximately 16 nm on cleaved NaCl (100) substrates, and the stable dispersion of such nanocrystals in aqueous solutions. Furthermore, we could monitor the diffracted x-ray spots from individual fabricated gold nanocrystals with an exposure time of less than 10 ms in the SPring-8 BL44B2 beamline. In DXT experiments, knowledge of the number of diffracted spots from a one-dimensional nanocrystal linked to a single biomolecule enables us to determine the number of biomolecules because individual one-dimensional nanocrystals reflect only one or two (one pair) x-ray diffraction spots. This proves there has been significant increase in DXT experimental efficiency.

We considered the size of the labeled nanocrystals. If the size of the labeled nanocrystal can be enlarged, high-sensitive and high-speed observations will become easier. The relationship between the size of the detected motions in the protein molecules and the size of the labeled nanocrystal is very important for DXT observations. Since we can control the size of the labeled nanocrystal, we can evaluate this relation experimentally. In the observation of the adsorbed protein molecule on the substrate surface, it should be careful that movement may be blocked as shown in Figure 12.22.

The labeling process itself must also consider affecting observed motions of protein molecules. Fundamentally, DXT can monitor time-tracking measurements of the Brownian motions of the labeled nanocrystal. Thus, a mean square displacement (MSD) curve can estimate the detailed movement form of the labeled nanocrystal. For example, on the labeling conditions which are restraining movements of the observed protein molecules, an MSD curve becomes the saturated form as shown in Figure 12.23. Therefore, only when linear relationship has a MSD curve, the labeling process of the nanocrystal can disregard DXT measurement for the first time completely. Thus, it is important for DXT analysis that the experimental data which can be checked are acquirable.

12.4 DNA Dynamics

DNA was chosen as the first demonstration of DXT. In order to detect intra-molecular Brownian motions in individual single biological molecules on the picometer scale, we utilized individual x-ray diffraction spots from a nanocrystal, which was tightly linked to the DNA molecules under observation as shown in Figure 12.24. In this experiment, we observed single DNA (18 mer) molecules. DXT, a single-molecule experiment with x-rays, was used to monitor the rotating motions, rather than the translational motions, of the labeled nanocrystal. Random rotational Brownian motions are regarded as Brownian motions on a two-dimensional flat surface. Therefore, the rotational angle (= $\Delta(2\theta)$) of the labeled nanocrystal can be converted to the displacement (= Δz) at the edge of the observed molecules by $L\Delta(2\theta)/2 = \Delta z$, where L is the length of the observed molecules. To distinguish individual nanocrystals in the image, the spots from the nanocrystals must not overlap, and the number of nanocrystals must coincide with the number of diffracted spots. In consideration of this, we used one-dimensional (Si/Mo) nanocrystals, and the number of diffracted spots from the one-dimensional nanocrystals had to be very small. In single-molecular detection, it is not important to observe a two- or three-dimensional "image," but rather to continuously determine the center position of the single spot under observations. Therefore, the detection limit

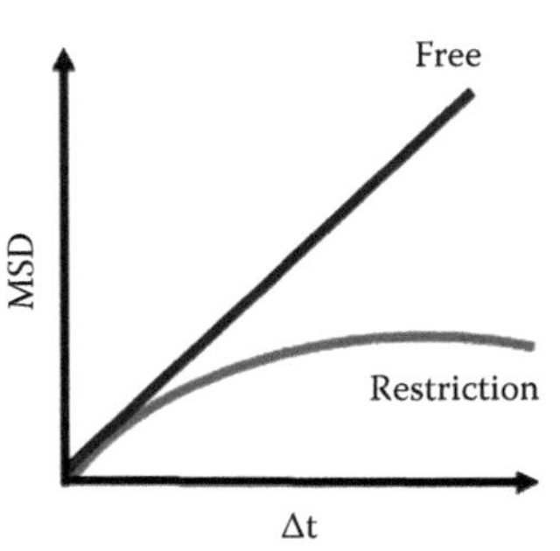

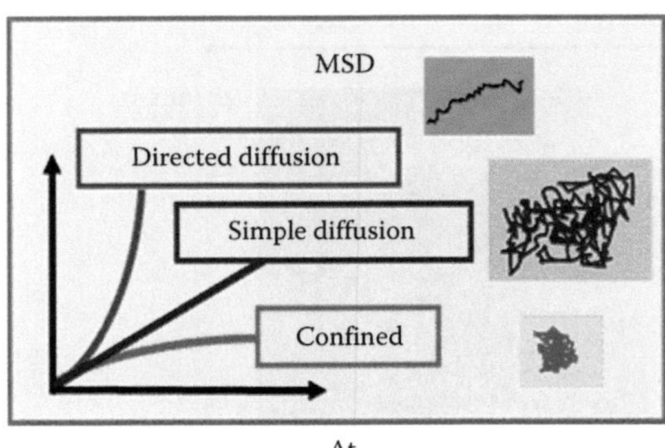

FIGURE 12.23 In a simple Brownian diffusion, the $(\Delta\theta^2)$–Δt plots are linear with a slope of 4D, where $(\Delta\theta^2)$ is the MSD of the observed spots, Δt is time interval, and D is the two-dimensional diffusion coefficient. The simple diffusion mode can be expressed as $(\Delta\theta^2)L^2 = 4D(\Delta t)$. However, from the relationship between $(\Delta\theta^2)$ and Δt, the observed displacement of θ is assigned as directed by the Brownian movement. This relationship is known as the directed diffusion mode: $(\Delta\theta^2)L^2 = 4D(\Delta t) + v^2(\Delta t)^2$. This shows that nanocrystals move in a direction at a constant drift velocity v with a diffusion coefficient of D. Since the values of D and v can be used to analyze Brownian motions, these parameters (D and v) are physically important ones. From these values, Brownian motions are classified as follows: Brownian motions and non-Brownian motions. In addition, the physical dominant factor on single-molecular scales under solution systems is not the mass (weight) but the viscosity. The relative viscosity can be estimated from the value of D through the Einstein–Stokes law. If we observe local values of D in all single molecules by using data from the different positions of the labeled nanocrystals, we can discuss not the absolute values of viscosities but the relative values. Moreover, we may be able to obtain information about the kinetics.

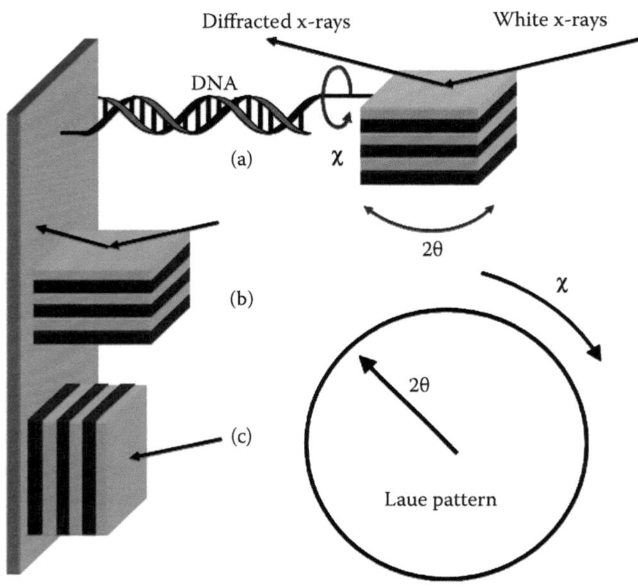

FIGURE 12.24 (a) Schematic drawing of the x-ray single-molecular detection system for individual DNA molecules in aqueous solutions. DXT can trace the motions of the single diffracted x-ray spots from the one-dimensional Si/Mo nanocrystal, which is linked to the individual DNA molecules. The artificial one-dimensional Si/Mo nanocrystal was fabricated by a sequential dry-etching process. The evaporated Au thin layer at the surface of the Si/Mo nanocrystal was coupled to the thiol groups (SH) at the end of the DNA. (b) Schematic drawing of the sample of physical adsorbed Si/Mo nanocrystals on the quartz. This is normal Laue Diffraction Sample's arrangement. (c) We cannot detect when the direction of x-rays and the direction of the labeled nanocrystal are in agreement.

of the observed displacements in DXT is only dependent on the accuracy and stability of the observed diffraction angles. In order to measure the total stability of the DXT system, we can observe the stopped diffraction spots from physical adsorbed nanocrystals (Figure 12.25) on Au/quartz under dry conditions. This is equivalent to the measurement of the detection limit ($\Delta 2\theta_{limit}$) in DXT. Figure 12.25b and c shows a diffraction spot from a physical adsorbed nanocrystal. We confirmed that the measurements of the diffraction angles (2θ) from the physical adsorbed nanocrystal on the substrate were stabilized within ($\Delta 2\theta_{limit}$ =) 1.5 mrad during 1 s. DXT could monitor Δz_{limit} = 4.5 pm when the nanocrystal was tightly coupled to the DNA molecule under observation (L_{DNA} = 6 nm), since the limited displacement Δz_{limit} in this DNA system was expressed

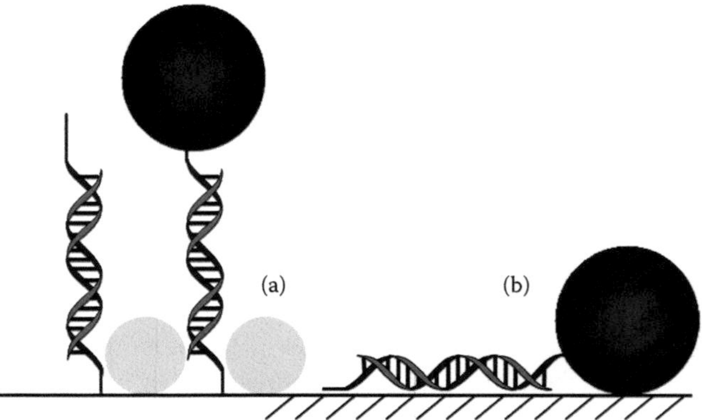

FIGURE 12.25 The arrangement of the adsorbed DNA molecules. (a) Ideal orientation conditions. (b) Movements of a DNA molecule cannot be measured when the whole DNA molecule is adsorbed to the substrate surface.

by $\Delta z_{limit} = 0.75 \times 10^{-3}$ L. We observed a histogram of intensities of individual diffracted spots. Each single diffracted x-ray spot was quantized at about 500 counts per 3 ms. This shows that DXT can quantitatively measure x-ray intensities from single nanocrystals [6], since the signals from DXT are independent of chemical conditions.

Here, we used multi-layer-(Si/Mo)-nanocrystals, because many nanocrystals are not perfectly crystallized. The number of diffracted spots from the multilayer nanocrystals enabled us to determine the number of observed nanocrystals, since a diffracted x-ray spot was only reflected from a single multilayer nanocrystal. We chose both the lattice constants and stacking period (d =5 nm, 4 pairs) for Si/Mo. In order to achieve reactions with thiol groups at the end of the observed DNA, we deposited an Au thin film (5 nm) on the surface of the Si/Mo multilayer by vacuum evaporation. Therefore, the total Si/Mo thickness (the vertical size) was 25 nm. We confirmed this total thickness value by electron scanning microscopy (SEM, S-5000, Hitachi Ltd.) and x-ray reflectometry. The Si/Mo nanocrystals were fabricated by a sequential process using a silicon substrate, silicon dioxide beads, and microprocessing techniques including RIE [4]. We confirmed that the averaged diameter (the horizontal size) of the Si/Mo one-dimensional nanocrystal was about 30 nm by SEM. We also confirmed that there were no diffracted spots from the deposited Au film on both the quartz substrate and the nanocrystals, because the Au film was amorphous and not crystalline. Unfortunately, we were not able to monitor a Si/Mo one-dimensional nanocrystal of the same size was fabricated by the microprocessing techniques of our previous work. In this work, however, we did succeed in fabricating high reflective Si/Mo nanocrystals by decreasing the roughness in the interface of the Si/Mo multilayer.

In order to label a single DNA molecule to a single nanocrystal, we restricted the size of the reactive area in the Si/Mo nanocrystal. Strictly speaking, the form of such a nanocrystal depends not on the column but on the cone because of the over-etching process. We confirmed that the upper and lower diameters of the cone were about 10 and 40 nm, respectively, by SEM. Therefore, a decrease of the reactive area, which is the upper Au surface of the conical Si/Mo nanocrystal, enables the reaction between a single nanocrystal and a single DNA molecule.

The DNA sequences were 5′-SH-CAGTCAGGCAGTCAGTCA-3′ and 5′-NH2-TGACTGACTGCCT GACTG-3′ [17]. We first used the chemical coupling made between the amino groups (NH_2) at the end of the DNA (100 μL, 0.1 nM/mL) and the surface of the evaporated Au thin film on the quartz (Au/quartz, 10 × 10 mm) by a cross-linking reagent (SPDP, Dojindo Lab.) in 50 mM Tris/HCl (pH = 8.0) for 12 h at 4°C. The chemical reaction between the amino groups in the DNA molecule and the cross-linking reagent started after the surface of the evaporated Au thin film on the quartz became saturated with the cross-linking reagent (SPDP (1 mM)) in N,N-dimethylformamide (DMF) for 12 h at 4°C. Because of this, the thiol groups (SH) at the other end of the DNA molecules were unable to react with the surface of the evaporated Au on the quartz substrate. Next, the evaporated Au thin film on the surface of the Si/Mo nanocrystal was coupled to the thiol groups at the end of the DNA in 50 mM MOPS (pH = 7.0) for 6 h at 4°C. Then, 7 μL of solution (50 mM Tris/HCl, pH = 8.0, 4°C) was mounted on the adsorbed DNA molecules.

When the adsorption efficiency of the DNA molecules on the substrate was about 1%, the estimated density of the adsorbed DNA molecules was expected to be about 900 (about 30 × 30 nm) nm^2/molecule. Since this occupied area of the adsorbed DNA molecules was in approximate agreement with the geometric size of the Si/Mo one-dimensional nanocrystal, the single nanocrystal was expected to react with a single DNA molecule. In general, the adsorption efficiency of molecules on such a surface is less than 1% [18,19]. We confirmed that the average density of Si/Mo one-dimensional nanocrystals was about 4000 (about 63 nm × 63 nm) nm^2/nanocrystal by SEM.

Here, we discuss the effects of labeled nanocrystals. The nanometer-level analysis (called single particle tracking [SPT] [20,21]) of movements of membranes with gold nanoparticles was demonstrated for the first time in 1985–1988. Ever since then, there have been a lot of discussions about the effects of labeled small particles [22]. As a result, many researchers have concluded that analyses by using labeled nanoparticles can provide significant information about biological events [23,24]. In addition, we confirmed in our previous work [4] that a viscosity below 273 K is able to be determined from DXT data. Because the polymer local chains in beaded agarose gel are labeled with Si/Mo nanocrystals, the dynamics of these local molecules is not affected by labeling nanocrystals. Accordingly, we estimated that the behavior of the adsorbed DNA molecules in this work would not be dominated by the effect of labeled nanocrystals.

We used the white x-ray mode (Laue mode) of beam line BL44B2 (RIKEN Structural Biology II, SPring-8, Japan) to record Laue diffraction spots from one-dimensional Si/Mo nanocrystals on a Au/quartz (70 μm) substrate as shown in Figure 12.26. The photon flux at the sample position was estimated to be about 1015 photon/s/mm^2 in the energy range of 7–30 kV. The x-ray beam's focal size was 0.2 mm (horizontal) × 0.2 mm (vertical). A diffraction spot was monitored with an x-ray image intensifier (Hamamatsu Photonics, V5445P) and a CCD camera (Hamamatsu Photonics, C4880-82) with 656 × 494 pixels. The average exposure time was 1 s. The detector's effective size was 150 mm in diameter with a 150 mm sample-to-detector distance. We calibrated the movements of the observed diffracted spots from the fixed Si/Mo nanocrystals on the sample holder when the goniometer was moved during (2θ =) 0–500 mrad. Figure 12.27 shows the cross-sectional view of the sample cell system. In this experiment, we used flat type to reduce a background as much as possible.

Figure 12.28 shows the movements of diffracted angles θ from a single nanocrystal coupled to a single DNA molecule (18 mer) at 4°C. The motions of diffracted spots can be clearly distinguished between the direction of diffraction angles θ and circumference χ in Figure 12.24. In this figure, the motions of θ and χ are in accord with those of adsorbed DNA molecules in real space, which is assigned to tilting mode and rotation mode, respectively. The observed spots randomly move along

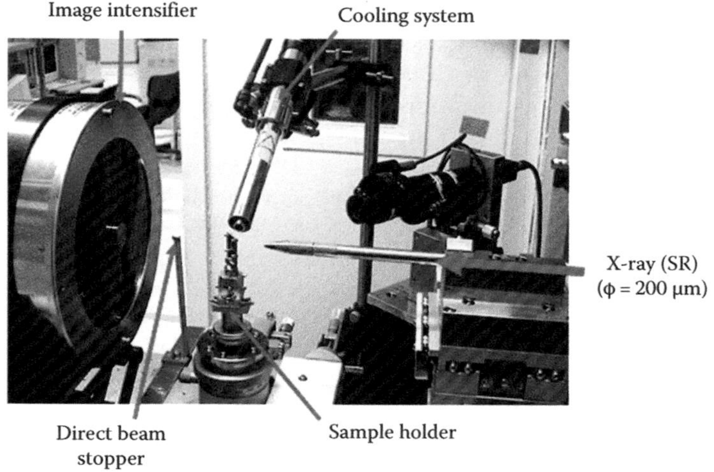

BL44B2 (white x-ray mode) at SPring-8 (photon flux = -10^{15} photon/s/mm^2)
= 1.7 Å (7 keV) – 0.4 Å(30 keV)
Detector: x-ray image intensifier (Hamamastu, V5445P) with CCD
specimen-detector distance: 10 cm, exposure time: within 1 s
(1–20 ms × 33)

FIGURE 12.26 A photograph of the instrumental arrangements for DXT at the beamline SPring-8 BL44b2.

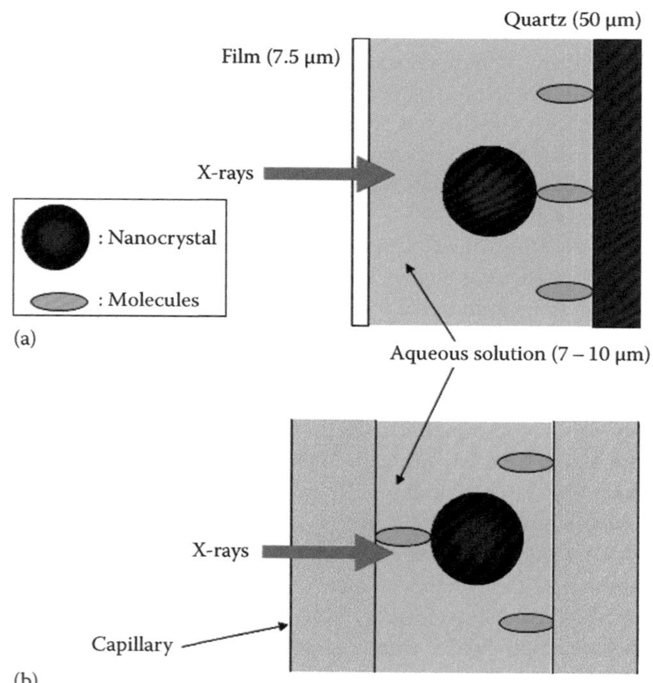

FIGURE 12.27 The sectional view of the circumference of a sample is shown. Although x-rays have penetrability, there is a limit in the thickness of an aqueous layer (5–10 μm). We have two types of sample holders: (a) flat type and (b) capillary type.

the direction of θ, and there is no displacement of the diffracted spots along the direction of χ. This indicates that the nanocrystal is tightly coupled to the observed DNA molecule and that the main motions of the nanocrystal are assigned to tilting displacement as shown in Figure 12.29.

In a simple Brownian diffusion, the $(\Delta\theta_2)$–Δt plots are linear with a slope of 4D, where $(\Delta\theta_2)$ is the MSD of the observed spots, Δt is time interval, and D is the two-dimensional diffusion coefficient. The simple diffusion mode can be expressed as $(\Delta\theta_2)L^2 = 4D(\Delta t)$. However, from the relationship between $(\Delta\theta_2)$ and Δt (Figure 12.30), the observed displacement of θ is assigned as directed by the Brownian movement [22]. This relationship is known as the directed diffusion mode: $(\Delta\theta_2)L^2 = 4D(\Delta t) + v^2(\Delta t)^2$. This shows that nanocrystals move in a direction at a constant drift velocity v with a diffusion coefficient of D. Since the values of D and v can be used to analyze Brownian motions, these parameters (D and v) are physically important ones. From these values, Brownian motions are classified as follows: Brownian motions and non-Brownian motions [22]. In addition, the physical dominant factor on single-molecular scales under solution systems is not the mass (weight) but the viscosity. The relative viscosity can be estimated from the value of D through the Einstein–Stokes law [22]. If we observe local values of D in all single molecules by using data from the different positions of the labeled nanocrystals, we can discuss not the absolute values of viscosities but the relative values. Moreover, we may be able to obtain information about the kinetics and structures of the individual single molecules by using our DXT.

From the curve fitting, we obtained the values of the diffusion constant D = 95 ± 12 mrad^2/s and the constant drift velocity v = 26 ± 8 mrad/s. The observed parabolic MSD curve showed that there is no steric inhibitor to the motions of the adsorbed DNA molecules in this system. If there were steric restricted motions of the adsorbed DNA molecules due to the presence of the substrate surface or the adsorbed Si/Mo nanocrystal, the relationship between $\Delta\theta^2$ of the diffracted spots and Δt would have shown an asymptotic curve [22].

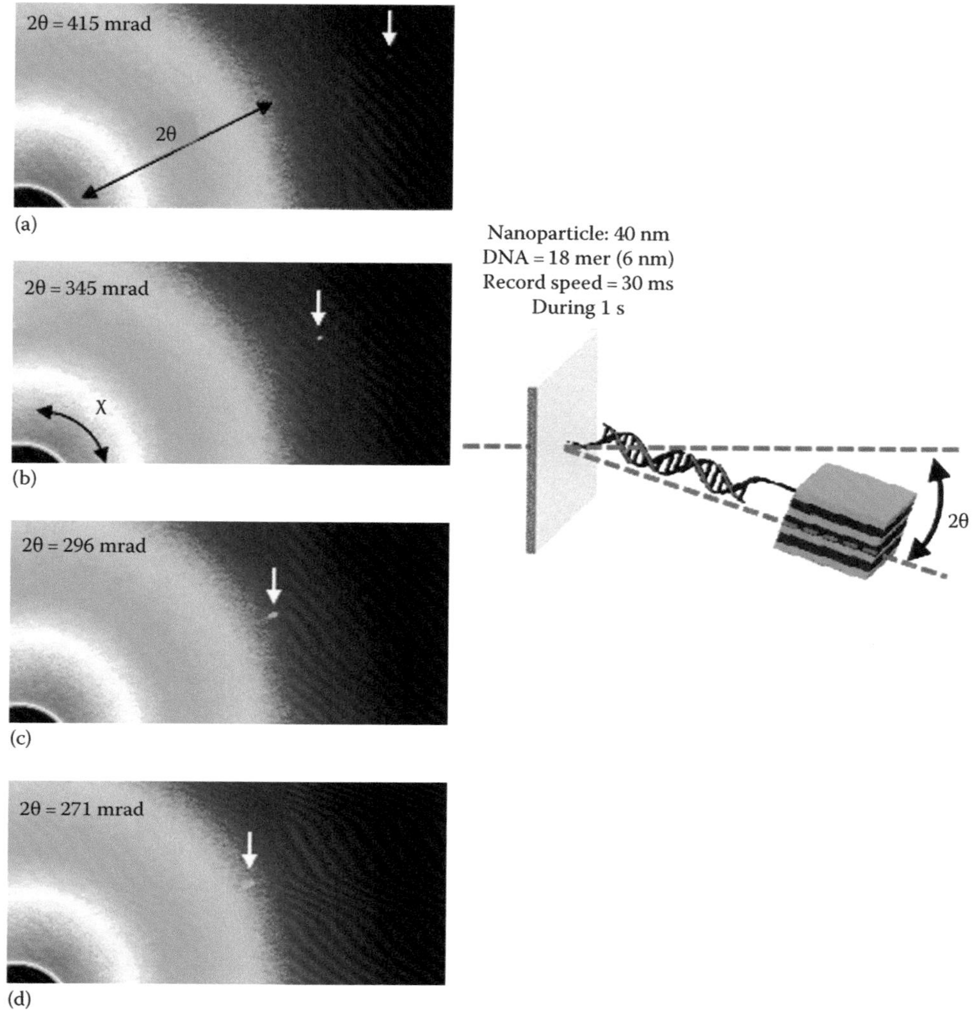

FIGURE 12.28 Examples of the diffracted spots from the single Si/Mo nanocrystal in aqueous solutions appeared as brightly shining dots (white). These images were recorded at (a) 30 ms, (b) 210 ms, (c) 420 ms, and (d) 630 ms after DXT observation time of onset. Frames are spaced at 180 ms intervals. The exposure time was 30 ms. Detected 2θ is convertible as translation motion with picometer-scale accuracy.

I consider a physical meaning at this drift velocity v. The constant drift velocity v can be regarded as the torque at the center of the labeled gold nanocrystal. In order to rotate the labeled gold nanocrystal, which is rigidly attached to the adsorbed DNA molecules in an aqueous solution, movement with this drift velocity v has emitted forces. We can call the x-ray radiation pressure forces [25]. This is first observations with x-rays. The observed gold nanocrystals are linked to the adsorbed protein molecules. We observed the directed Brownian motion of individual linked nanocrystals as shown in Figure 12.31. The observed force is estimated at about 0.13–0.63 aN. In the future, we can control and measure dynamics of micro- or nanocrystalline materials using x-ray radiation pressure force.

On an astronomical scale, the presence of x-ray radiation pressure forces is fundamental to explanations of physical properties of black holes or neutron stars. In microscopic fields, optical radiation pressure from visible light lasers has recently become an important research tool in the fields of biology, physical chemistry, and soft condensed matter physics. However, on a nanoscopic scale, x-ray radiation pressure forces have not yet been observed because of the limitation on the number of x-ray photons and the absence of an ultra-sensitive detection device. This x-ray radiation pressure force was hidden into Brownian motions as shown in Figures 12.31 and 12.32. However, we observed x-ray radiation pressure force on individual single nanocrystals using an x-ray single-molecular methodology. In addition, these observed forces cannot be assigned to a type of van der Waals interaction [26] and the Casimir force [27]. In the future, we can control and measure dynamics of micro- or nanocrystalline materials using x-ray radiation pressure force.

This observed tiny force (aN) should be compared to the gravitational force on the gold nanocrystal. However, the gravitational force is inconsequential for this x-ray radiation pressure force. Because the gravitational force has one direction and the

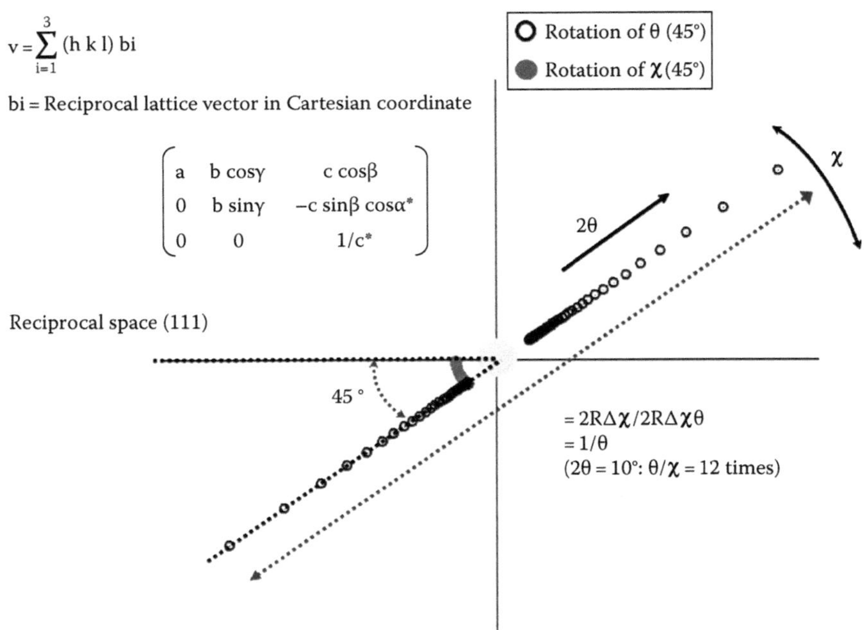

FIGURE 12.29 Typical observed Laue diffraction pattern from DXT technique. The important factors are 2θ and χ. These parameters are not the same scales. The parameter 2θ is comparatively good in sensitivity.

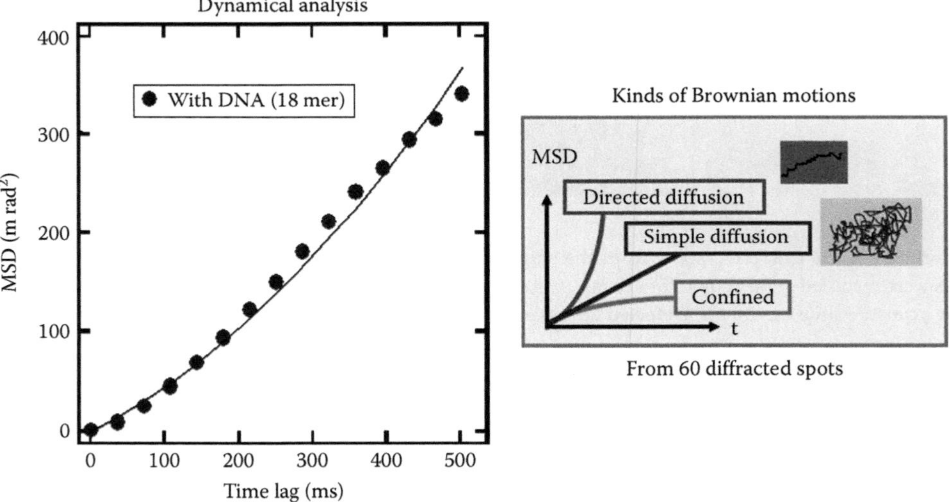

FIGURE 12.30 MSD curve $\langle \Delta\theta^2 \rangle$ as function of time interval Δt. These data were obtained from about 60 diffracted spots. This MSD curve is an expression very fundamental to analyzing Brownian motions.

movement direction of the radial angles (2θ) is random. The directional relationship between the movement direction of the radial angles (2θ) and the direction of the radiation pressure force is constant in all perpendicular angles (χ). Therefore, the gravitational force undergoes extinction.

Over the next decade, x-ray Free-electron Lasers (FELs) will offer over 10 orders of magnitude more photons per pulse than the most intense synchrotrons currently available [28]. In such ultra short pulses of high-intensity x-rays, it is obvious that x-ray radiation pressure forces can control dynamic behaviors of micro- or nanocrystalline materials under super-high accuracy.

Radiation damage was caused by x-ray photons depositing energy into the sample system. We confirmed the reproducibility of the data up to an exposure time of 3 s. At exposure times above 4 s, the motions of the nanocrystals stopped or

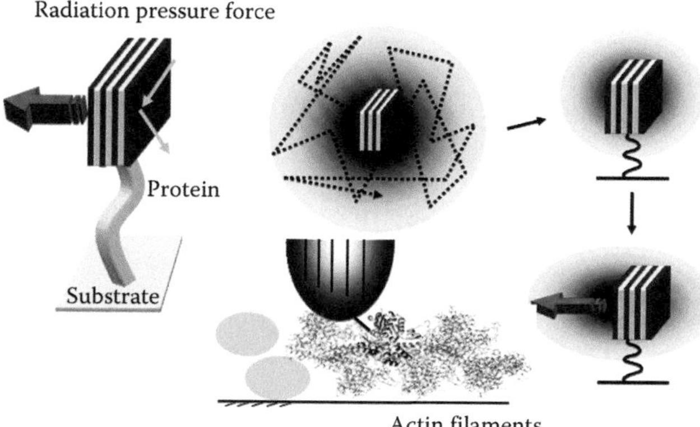

FIGURE 12.31 Schematic drawing of the observed x-ray radiation pressure force on individual single gold nanocrystal, which is linked to the adsorbed biomolecules. Schematic drawing of cross-sectional view of adsorbed Actin filaments (F-actin) to detect x-ray radiation pressure forces. The gold nanocrystals are reacted with the C-terminal of G-actin through the SH–gold covalent bonds. In order to adsorb F-actin onto an amorphous gold substrate, we utilized the interaction between the reactive amino residue in F-actin and the amorphous gold surface with a cross-linking reagent. The observed x-ray radiation pressure force is usually hidden by normal Brownian motions in the aqueous solutions.

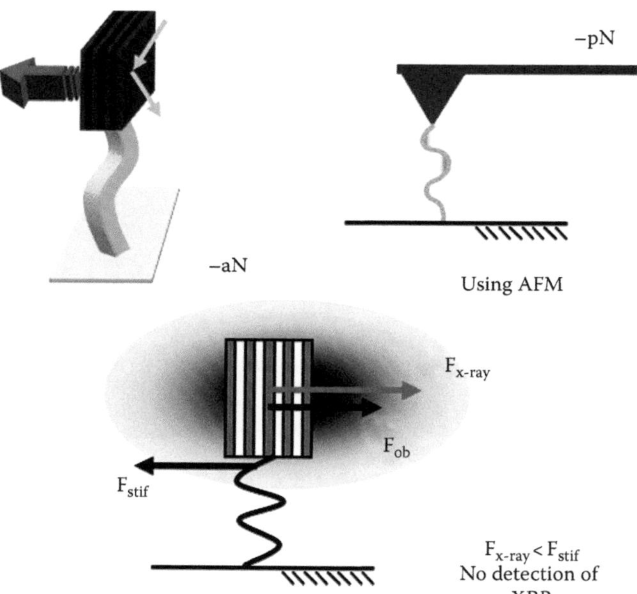

FIGURE 12.32 The measured x-ray radiation pressure force is influenced by the hardness of the protein molecule. Conversely, the hardness or stiffness of the adsorbed protein molecules can be evaluated from MSD curves of DXT measurements.

slowed down. Although it was difficult to determine the origin of the radiation damage, we can say that the modification of the observed motions may have been influenced not by the radiation effect of the DNA molecules, but by the ionization of the nanocrystals, which consisted of heavier elements than those of the DNA molecules. In addition, the nanocrystals were larger than the DNA molecules. On the other hand, we confirmed the presence of a constant drift in the solution from the obtained MSD curve. This may have caused the observed local rise in the temperature of the solution due to radiation.

In fields dealing with microscopic phenomena, the optical radiation pressure force from visible light lasers has recently become an important research tool in the fields of biology, physical chemistry, and soft condensed matter physics. On an astronomical scale, the presence of x-ray radiation pressure forces is fundamental to explaining the physical properties of black holes or neutron stars. However, on a nanoscopic scale, x-ray radiation pressure forces have not yet been observed because of the limitation on the number of x-ray photons and the absence of an ultrasensitive detection device. In 2006, we reported the first observations of x-ray radiation pressure force on individual single nanocrystals using DXT. DXT can not only monitor slight internal motions of protein molecules but can also detect weak x-ray force fields.

In single-molecule science and technology, it is crucial to control dynamical Brownian motions of individual functional protein molecules. Optical traps and tweezers using visible light gave the first demonstration of controlling micrometer-sized particles or molecules. In order to improve both the monitoring and control of internal molecular motions of single-molecular units with improved ultrahigh precision under in vitro or in vivo physiological conditions, we proposed new single-molecule control techniques using x-rays, electrons, and neutrons. We found that x-ray radiation pressure forces are dependent upon x-ray intensity and the size of the gold nanocrystal under observation. To our surprise, we estimated the observed force using DXT to be on the atto-Newton level.

Energy recovery linac (ERL) or x-ray FELs will offer over 10 orders of magnitude more photons per pulse than the most intense synchrotrons currently available. In high-intensity x-ray sources, it is obvious that x-ray radiation pressure forces can control the dynamic behavior of micro- or nanocrystalline materials with ultrahigh accuracy.

Until now, we observe many single-molecule systems as shown in Figure 12.33. In the future, the biggest challenge will be to observe individual and rare biological processes in living cells. DXT can be used to monitor not translational motions but orientational ones on picometer scales. DXT can be expected to observe the structural changes accompanying the activation of ion channels in living cells. Such changes are known as tilting or small orientational motions of the helix in channel pores [29,30]. DXT can also be expected to monitor the dynamics of the ion channels through ionic flux measurements by the patch-clamp technique [31]. At present, some groups are trying to observe the dynamics of ion channels by using single-molecular fluorescence techniques. Unfortunately, it has been very difficult to observe structural changes because of the lack of a sufficient monitoring precision.

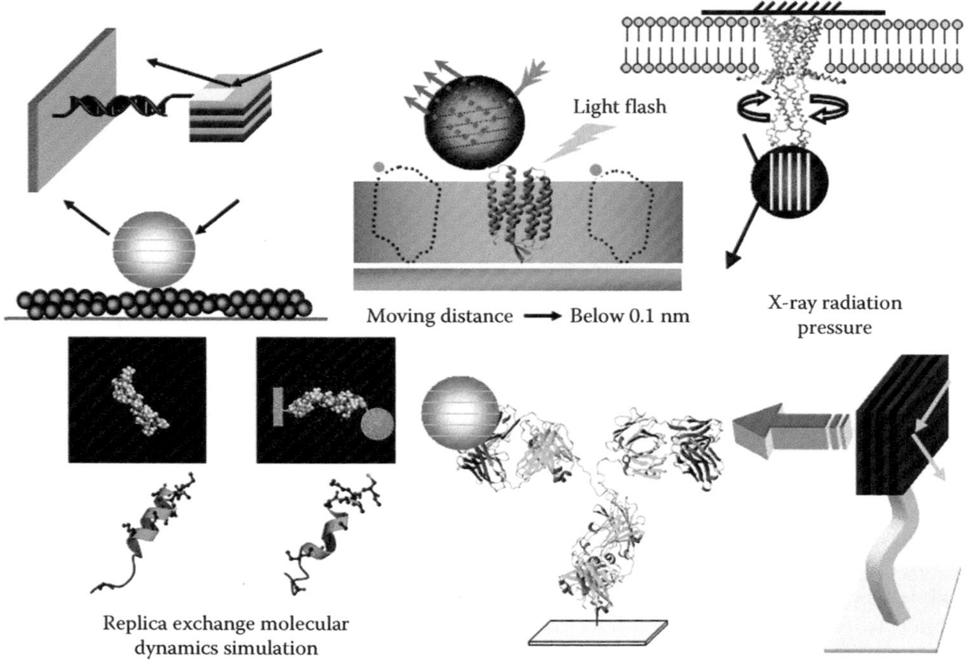

FIGURE 12.33 Schematic drawing of DXT samples. Until now, we observed individual DNA molecules, denatured proteins, antigen–antibody [31] interactions protein membranes (Bacteriorhodopsin [32], KcsA [33], KvAP, AChBP, nAChR, etc.), and actin filament. It is possible to observe in vivo observations of individual single protein molecules, because there are no diffraction spots form living cell.

12.5 Diffracted Electron Tracking

DXT was developed for obtaining information about the dynamics of single molecules. This method can observe the rotating motions of an individual nanocrystal that is linked to a specific site in units of single protein molecules by using a time-resolved Laue diffraction technique. However, this method requires a very strong x-ray source, such as SPring-8. The facilities for synchrotron orbital radiation as a strong x-ray primary beam must be utilized. Therefore, for this method to be more available to researchers, the device configuration must be limited to a small scale. To that end, electron beam technologies are advantageous, and diffracted electron tracking (DET) is clearly possible as shown in Figure 12.34. However, because DET might damage protein molecules or soft nanomaterials, we will need to investigate the processes of damages to protein molecules and soft nanomaterials.

We began developing a compact instrument for monitoring the rotation of single protein molecules by using an electron beam instead of x-rays because electron beams have cross-sectional values that are 1000 times larger than those of x-rays. Instead of Laue diffraction using white x-rays, the electron backscattered diffraction pattern (EBSP) was applied to monitor the crystal orientation of nanocrystals linked to single protein molecules. Additionally, this EBSP using electron beams contains three-dimensional information on the orientation motions. Since x-ray diffraction spots usually have two-dimensional data, obtaining three-dimensional information from DXT observations is very difficult. Table 12.1 shows the comparison between DXT and DET method.

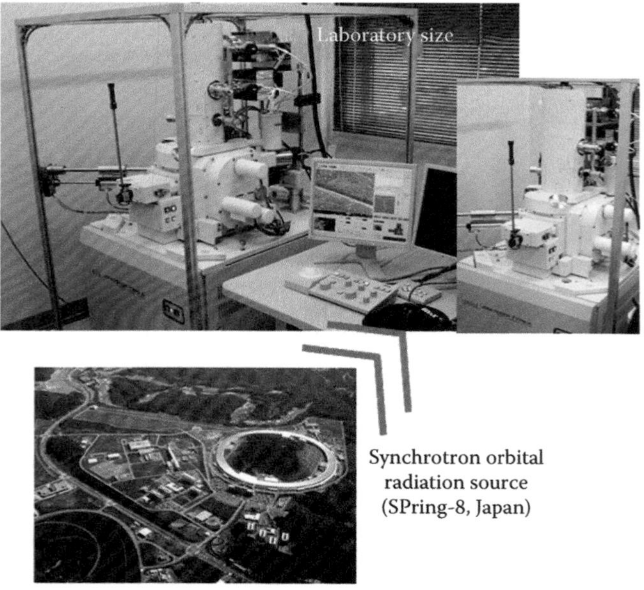

FIGURE 12.34 Although the light source of x-rays is very large like a mountain, an electron beam goes into a laboratory enough.

DET using electron diffraction patterns from labeled nanocrystals can be applied for the first electron bottom-up microscope. For this purpose, it is necessary to achieve (1) a wet cell

TABLE 12.1 Comparison of DXT and DET

	X-ray (DXT)	Electron (DET)
Accuracy	0.01 deg	0.1 deg
Dimension	2 axis	3 axis
Speed	1 μs	60 ms
Label size	20–60 nm	10–40 nm
Quality of crystal	Very good	Not so good
Damage	Small	Very large
Instrument	Large institute	Lab. level
Machine time	1 day/month	Everyday

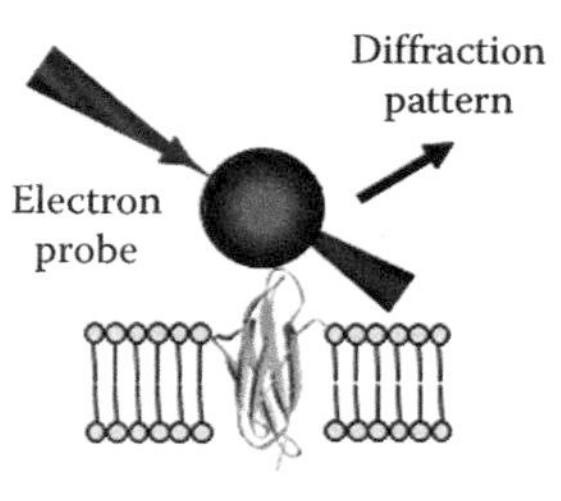

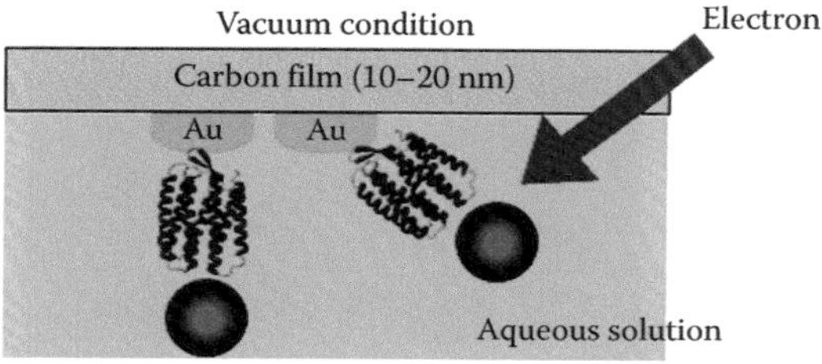

FIGURE 12.35 The arrangement of DET sample in the vacuum conditions. The typical diffraction patterns of the labeled gold nanocrystal, we called electron backscattering diffraction (EBSD or EBSP) from the labeled gold nanocrystal. An electron beam can also progress to some extent in aqueous solutions.

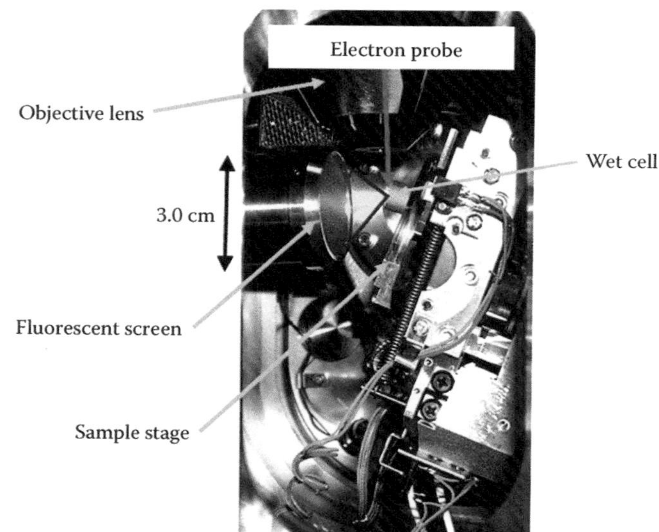

FIGURE 12.36 A photograph of the instrumental arrangements for DET. Except for using the wet cell, it is unchanging with the usual SEM.

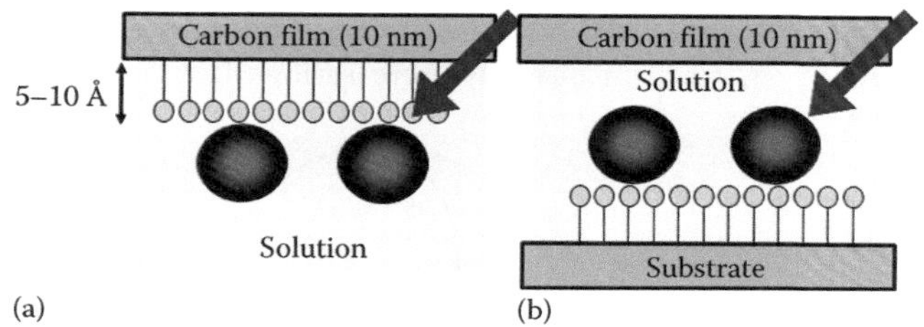

FIGURE 12.37 The sectional view of the circumference of a sample is shown. (a) is very easy for a sample set. But, the chemical fixation of sample's protein molecules on the carbon film is very difficult. Although electrons have penetrability, there is a limit in the thickness of an aqueous layer (0.1–1 μm). We have two types of sample holders. Control of the thickness of an aqueous layer is difficult for (b).

with very thin sealing film as shown in Figure 12.35, (2) an EBSP system with high sensitivity, (3) a damage less electron irradiation technique, and (4) perfect gold nanocrystals. Developing these technologies may be expected to be quite difficult.

Figure 12.36 is a photograph of the equipment arrangement within the vacuum condition in DET. Since all pieces of equipment must be put in a vacuum, the distance of a sample and a detector is short. Thus, it becomes difficult to raise angular accuracy in DET.

Figure 12.37 shows the cross-sectional view of the sample cell system. In this experiment, we used Figure 12.37a type to reduce a background as much as possible. The Figure 12.37b type has some advantages to reduce the damage process from the primary electron prove, since an electron beam does not penetrate the sample itself once. The disadvantage is that control of distance between the carbon film and the sample substrate is difficult.

Figure 12.38 shows that the crystal direction is exercising from the polar-coordinates display. The used samples are the

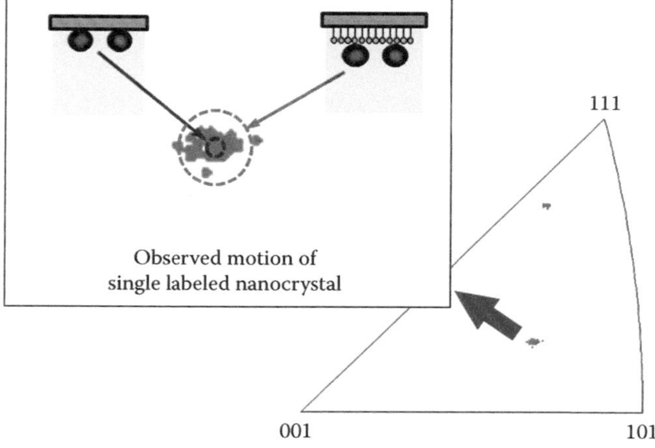

FIGURE 12.38 The typical EBSD analysis utilizes the polar-coordinates display. It is the first adaptation that evaluates a motility characteristic of the adsorbed molecules from this display.

physical adsorption gold nanocrystal and the gold nanocrystal on the self-assemble film. The movement of the gold nanocrystal on a self-organization film is clearly larger. We can obtain the MSD curves in the three axes as shown in Figure 12.39. The three-dimensional DET will work very effective in future intramolecular motional investigations in biophysics and nanomaterial science.

Figure 12.40 shows the typical experiment protocol in DET. The completely different analysis techniques from the usual electron microscope are utilized in DET method. The most important information in the DET method is the direction of the labeled crystal and not images from a high-resolution electron microscope. It is wonderful that the comparison examinations of the movements of the molecule in the inside of a vacuum and an aqueous solution can be carried out in detail for the first time in this DET experiment.

Figure 12.41 shows the cross-sectional view of the sample cell system, which is containing the protein molecules. In this experiment, we used Figure 12.41a type to reduce a background as much as possible. The Figure 12.41b type has some advantages to reduce the damage process from the primary electron prove, since an electron beam does not penetrate the protein molecule itself once. The damage of the protein molecule by an electron beam is more serious than an organic molecule system. As for the methodology with a labeling process like DET or DXT, the presence of the labeled nanocrystal tends to be considered to be a fault factor. However, the damage phenomenon which was a serious problem may be able to be avoided by irradiating only a labeling object with an electron probe or x-ray ones.

The penetrability of an electron beam in aqueous solutions is long beyond anticipation as shown in Figure 12.42. However, since the penetrability in the inside of the labeled gold nanocrystal is very short, high acceleration in DET may be needed.

12.6 High-Speed Picometer Scale Dynamics

In typical DXT experiments, we utilized adsorbed protein molecules which reduce the gold nanocrystals in aqueous solutions. We observed the free Brownian motions of individual gold nanocrystals in aqueous solutions by using normal-speed DXT at less than the millisecond level. In a new experiment of ultrahigh-speed DXT, a diffraction spot can be monitored with both an x-ray image intensifier (Hamamatsu Photonics, V5445P, Japan) and a CMOS camera (FASTCAM SA1.1, Photron, Japan). We also utilized high-speed x-ray shutters before x-ray irradiation of the sample to reduce the damage

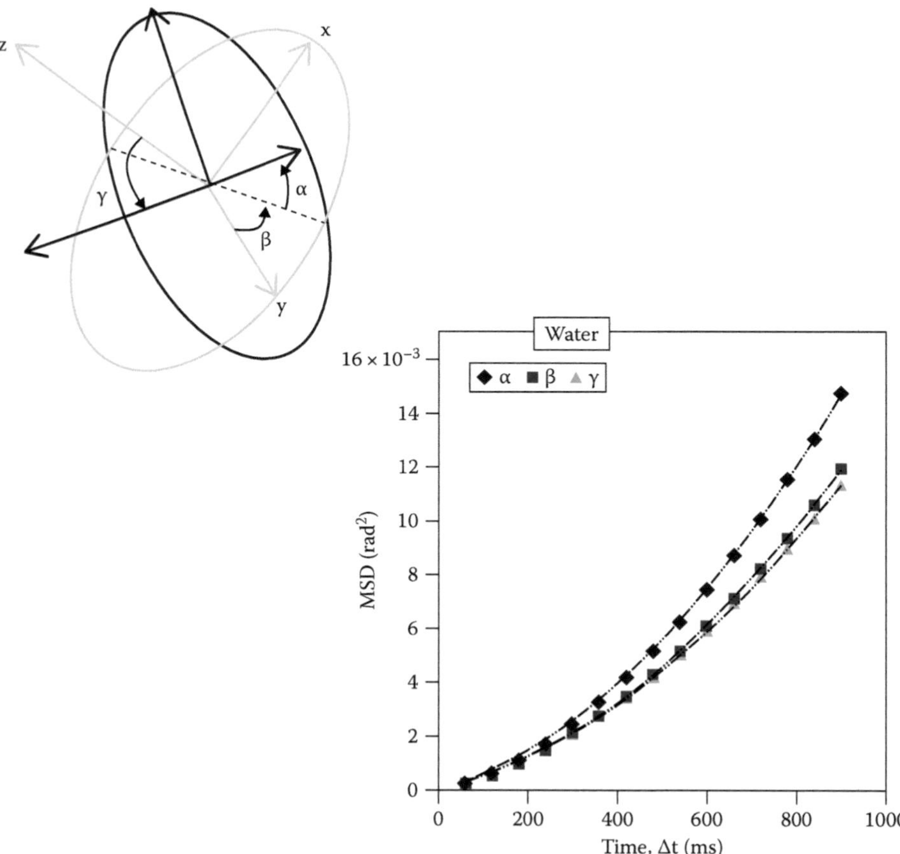

FIGURE 12.39 Three-dimensional MSD curves as function of time interval Δt. These data were obtained from about 60 diffracted patterns. This MSD curve is an expression very fundamental to analyzing Brownian motions.

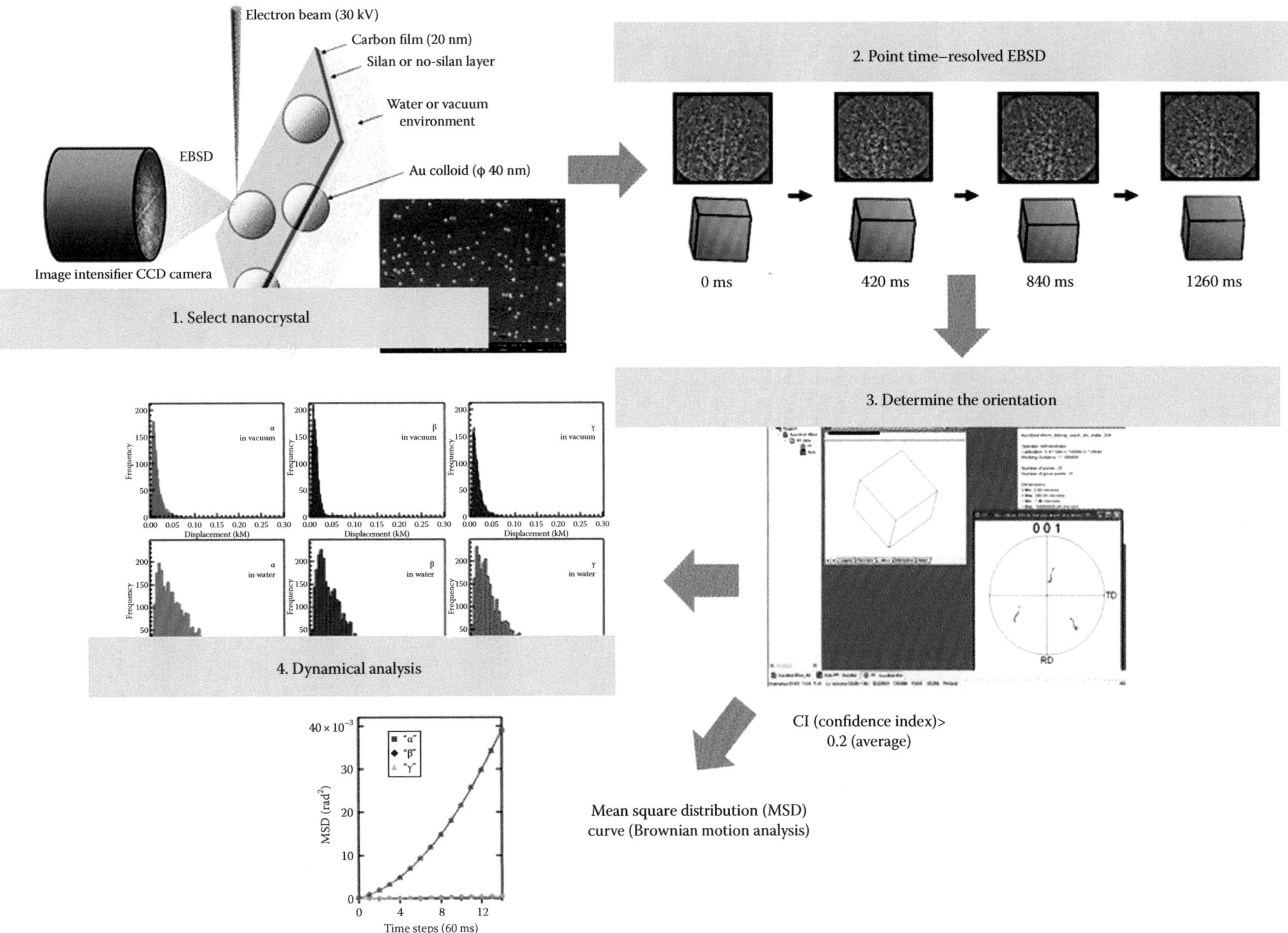

FIGURE 12.40 This is the typical experiment protocol in DET. The completely different analysis techniques from the usual electron microscope are utilized in DET method.

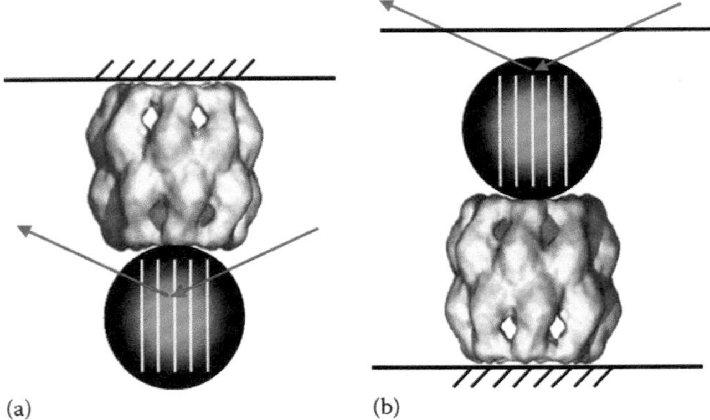

FIGURE 12.41 (a) This is the first goal to observe the three-dimensional motions of functional protein molecules. The (b) type has the ideal arrangement to reduce the damage process from the primary electron prove, since an electron beam does not penetrate the protein molecule itself once.

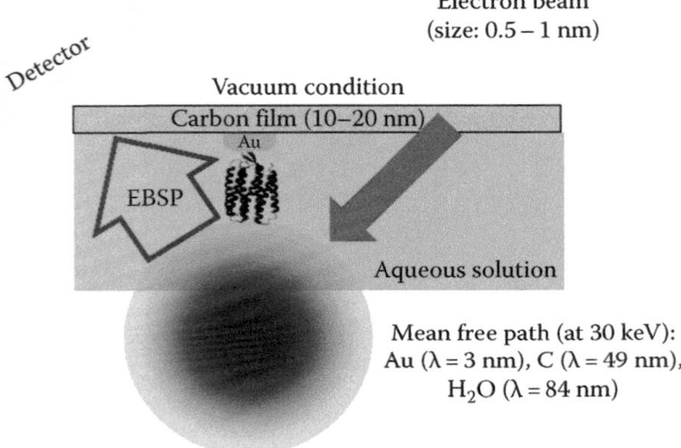

FIGURE 12.42 The penetrability of an electron beam in aqueous solutions is long beyond anticipation.

caused by x-rays to the greatest extent possible. The average exposure time of the ultrahigh-speed DXT in a single shot was between 1 and 100 μs. In these experiments, we controlled the size of the gold nanocrystals and the energy peak-width of the quasi-white x-rays.

For example, nicotinic acetylcholine receptor (nAChR) is a pentameric ligand-gated ion channel in the central and peripheral nervous systems. The nAChR channel family represents a group of important membrane protein molecules, because the channels play a central role in membrane physiology. The dynamic gating mechanism of nAChR is still unclear since the structure of nAChR in the presence of acetylcholine has not yet been determined. In addition, the single-molecule dynamics of nAChR in the presence of acetylcholine (ACh) has not been observed with single-molecule detection systems. Thus, now, ultrahigh-speed DXT was applied to the nAChR system. When collected by the patch clamp technique, data on the nAChR system show that movements of the nAChR molecular system are extremely fast, on a sub-microsecond level. However, it is possible to track the movements of nAChR molecular system by using DXT. The method enables tracking of the x-ray diffraction spot from the gold nanocrystal labeled on an individual nAChR and observation of the intermolecular dynamics of the nAChR in real time and space. In the first step of our experiments, acetylcholine-binding protein (AChBP) was used as a model of nAChR. AChBP is a structural and functional homologue of the extracellular ligand-binding domain of nAChR. We succeeded in measuring the high-speed internal motions of AChBP in the absence and presence of ACh by DXT and found that ACh played a significant role in activating the motions of AChBP. DXT results showed that the internal motions related to ligand binding may initiate vigorous molecular fluctuations in AChBP. Thus, we clarified that ultrahigh-speed DXT methodology is required for observing the dynamic gating mechanisms of nAChR and AChBP, giving positional accuracy beyond the angstrom level.

Our next step, for application to the dynamic intramolecular motions of nAChR in the membrane, is to make improvements to DXT such as greater efficiency in the preparation of gold nanocrystals and higher temporal resolution of DXT (100 ns–1 μs). Additionally, we will obtain data on three-dimensional dynamic motions of adsorbed protein molecules when we can control the two-dimensional orientation of adsorbed nAChR and AChBP.

We demonstrated real-time intramolecular dynamic observations of specific sites of single-molecular protein units. DXT was proved to be adaptable to not only an isolated single molecule on a substrate but also to membrane proteins in a membrane. DET using electron probes will be a very effective new single-molecular technique.

The intensity of x-ray light sources will strengthen rapidly in the near future. Research on increasing the luminance of electron beams will advance. In addition, this work can be said to have

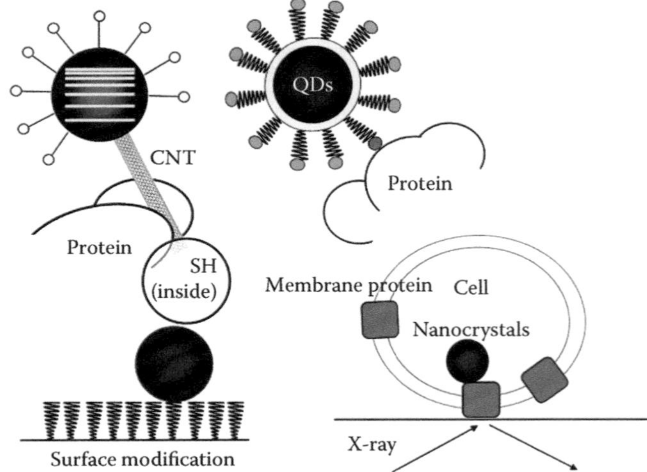

FIGURE 12.43 DXT and DET can be observed in vivo observations of individual single protein molecules, because there are no diffraction spots from living cell. If a phosphor is utilized, simultaneous observations can be carried out with fluorescence single-molecule techniques.

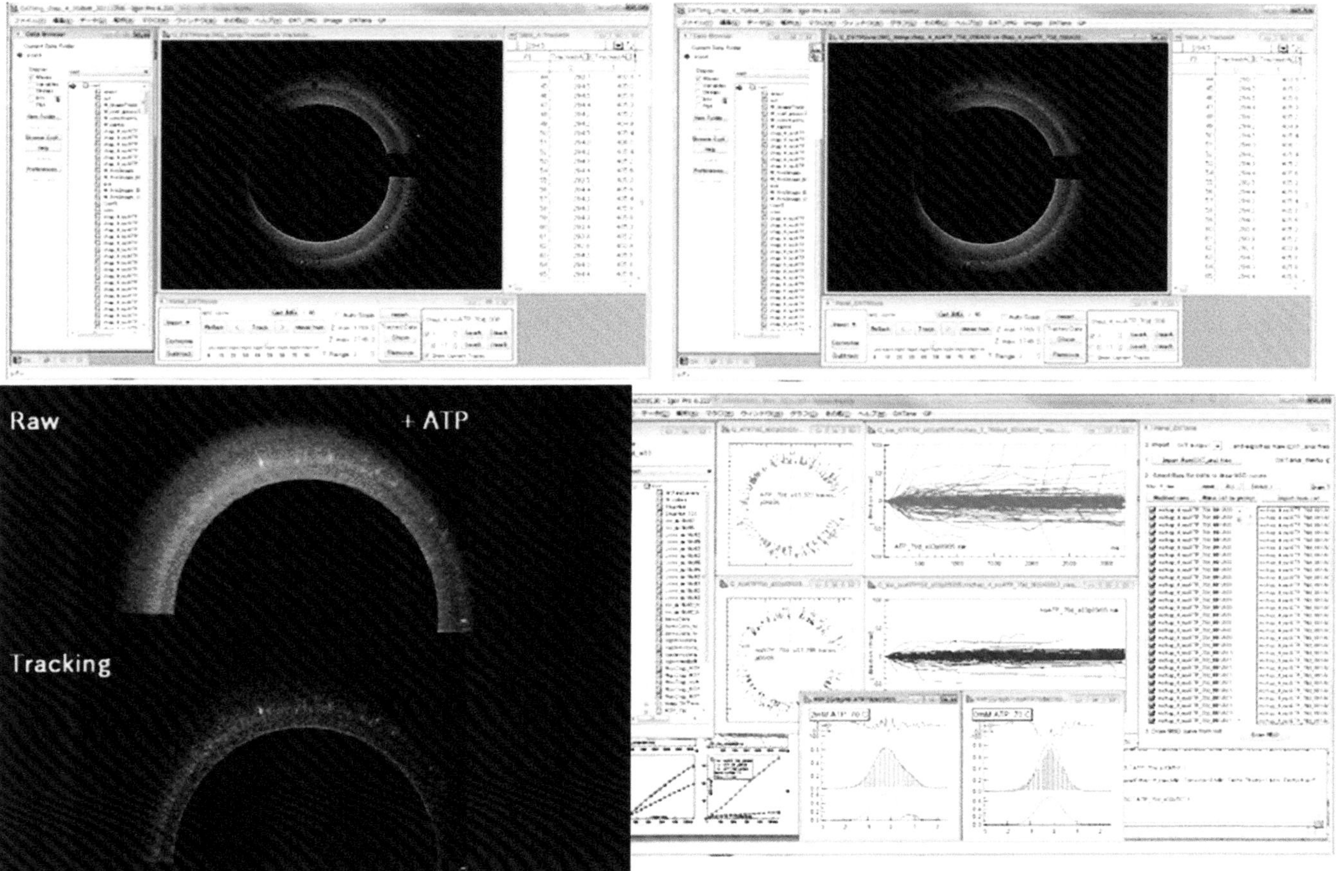

FIGURE 12.44 In order to spread a measuring method (DXT and DET), development of easier high-level analysis software is indispensable.

opened the door for utilization of neutron probes. Therefore, the future possibilities of high-energy probes are high. In particular, methods for single-molecule measurements using high-energy probes may be developed to have high precision and high speed. The goals that may be achieved in the near future are as follows:

1. Stable structural determination of single protein molecules.
2. Partial dynamic structural determination of single protein molecules under in vivo conditions.
3. High-speed motions (ps–ns) of specific sites in single protein molecules.
4. Dynamic molecular interactions between two protein molecules.

Imaging methodologies in cell biology involve both ultrafast time resolution and in vivo observations of the structural changes in single biomolecules under atomic-scale precision. For ultra fast time resolution, we need to generate ultra short-pulsed sources. With x-rays, FEL-type hard x-ray sources are used to estimate the potential for imaging single protein molecules and small assemblies. DXT and DET are labeling methodologies. Thus, they have several faults. However, in the future, development of no-labeling techniques should advance based on new technologies cultivated by the aforementioned labeling measurement methods. In this work, the first bottom-up method utilizing high-energy probes was designed. A new single-molecular method may result from new fusion of the bottom-up concept with conventional methods. For example, a method combining Q-dots or GFP by DXT or DET might be designed as shown in Figure 12.43. AFM fused with DXT may be suitable for observing single molecules with ultra super-high accuracy.

In order to spread these methodologies, analysis software also needs to be developed as shown in Figure 12.44. We also have to make a gold nanocrystal easy to produce. The diffused methodology will produce many progresses again. In such a direction, individual single-molecule measurement with picometer accuracy will become a methodology more important in future.

References

1. S. Weiss, *Science* 283, 1676 (1999).
2. W. E. Moerner, M. Orrit, *Science* 283, 1670 (1999).
3. P. R. Selvin, *Nat. Struct. Biol.* 7, 730 (2000).
4. Y. C. Sasaki, Y. Suzuki, N. Yagi, S. Adachi, M. Ishibashi, H. Suda, K. Toyota, M. Yanagihara, *Phys. Rev. E* 62, 3843 (2000).

5. Y. C. Sasaki, Y. Okumura, S. Adachi, Y. Suzuki, N. Yagi, *Nucl. Instrum. Methods Phys. Res. A* 467, 1049 (2001).
6. Y. C. Sasaki, Y. Okumura, S. Adachi, H. Suda, Y. Taniguchi, N. Yagi, *Phys. Rev. Lett.* 87, 248102-1 (2001).
7. Y. C. Sasaki, T. Higurashi, T. Miyazaki, Y. Okumura, N. Oishi, *Appl. Phys. Lett.* 89, 053121 (2006).
8. Y. Okumura, Y. Taniguchi, Y. C. Sasaki, *J. Appl. Phys.* 92, 7469 (2002).
9. H. Zhou, B. K. Chong, P. Stopford, G. Mills, A. Midha, J. M. R. Weaver, *J. Vac. Sci. Technol. B* 18, 3594 (2000).
10. P. R. Krauss, S. Y. Chou, *Appl. Phys. Lett.* 71, 3174 (1997).
11. Y. C. Sasaki, K. Yasuda, Y. Suzuki, T. Ishibashi, I. Satoh, Y. Fujiki, S. Ishiwata, *Biophys. J.* 72, 1842 (1997).
12. K. Tokashiki, K. Sato, N. Aoto, E. Ikawa, *J. Vac. Sci. Technol. B* 11, 2284 (1993).
13. E. B. Troughton, C. D. Bain, G. M. Whitesides, R. G. Nuzzo, D. L. Allara, M. D. Porter, *Langmuir* 4, 365 (1988).
14. J. H. Underwood, T. W. Barbee, Jr., *AIP Conf. Proc.* 75, 131 (1981).
15. L. Henke, J. C. Davis, E. M. Gullikson, R. C. C. Perera, A preliminary report on x-ray photoabsorption coefficients and atomic scattering factors for 92 elements in the 10–10000 eV region, LBL-26259, UC-411 (1988).
16. Y. Okumura, T. Miyazaki, Y. Taniguchi, Y. C. Sasaki, *Thin Solid Film* 471, 91–95 (2005).
17. A. P. Alivisatos, K. P. Johnsson, X. Peng, T. E. Wilson, C. J. Loweth, M. P. Bruchez, P. G. Schultz, *Nature* 382, 609 (1996).
18. H. Suda, Y. C. Sasaki, N. Oishi, N. Hiraoka, K. Sutoh, *Biochem. Biophys. Res. Commun.* 261, 276 (1999).
19. Y. C. Sasaki, K. Yasuda, Y. Suzuki, T. Ishibashi, I. Satoh, Y. Fujiki, S. Ishiwata, *Biophys. J.* 72, 1842 (1997).
20. M. De Brabander, G. Geuens, R. Nuydens, M. Moeremans, J. De May, *Cytobios* 43, 273 (1985).
21. J. Gelles, B. J. Schnapp, M. P. Sheetz, *Nature* 331, 450 (1988).
22. M. J. Saxton, K. Jacobson, *Annu. Rev. Biophys. Biomol. Struct.* 26, 373 (1997).
23. Y. Harada, O. Ohara, A. Takatsuki, H. Itoh, N. Shimamoto, K. Kinoshita, *Nature* 409, 113 (2001).
24. R. Yasuda, H. Noji, M. Yoshida, K. Kinoshita, H. Itoh, *Nature* 410, 898 (2001).
25. Y. C. Sasaki, T. Higurashi, T. Miyazaki, Y. Okumura, N. Oishi, *Appl. Phys. Lett.* 89, 053121 (2006).
26. P. Johansson, P. Apell, *Phys. Rev. B* 56, 4159 (1997).
27. S. K. Lamoreaux, *Phys. Rev. Lett.* 78, 5 (1997).
28. N. Patel, *Nature* 415, 110 (2002).
29. R. Brown, *Philos. Mag.* 4, 161 (1828).
30. A. Einstein, *Investigations on the Theory of the Brownian Movement*, Dover, New York, 1956.
31. T. Sagawa, T. Azuma, Y. C. Sasaki, *Biochem. Biophy. Res. Commun.* 335, 770 (2007).
32. Y. Okumura, T. Oka, M. Kataoka, Y. Taniguchi, Y. C. Sasaki, *Phys. Rev. E* 70, 021917 (2004).
33. H. Shimizu, M. Iwamoto, F. Inoue, T. Konno, Y. C. Sasaki, S. Oiki, *Cell* 132, 67 (2008).

IV

Scanning Probe Microscopy

13

Atomic-Resolution Frequency Modulation

Takeshi Fukuma
Kanazawa University
and
Japan Science and Technology Agency

13.1 Introduction

Scanning probe microscopy (SPM) is a family of surface analysis techniques that uses a sharp tip for probing the tip–sample interactions and measuring the surface corrugations and properties. The history of SPM started with the invention of scanning tunneling microscopy (STM) in 1981 by Binnig and Rohrer [1]. In STM, tunneling current flowing through the gap between a sharp tip and a sample surface is detected. The tip–sample distance is regulated such that the detected tunneling current is kept constant. Due to this tip–sample distance regulation, the vertical position of the tip follows the surface corrugations when laterally scanned over a sample surface. From the tip trajectory during the scan, a surface topographic image is obtained. In spite of the simple operation principle and experimental setup, STM is capable of imaging atomic-scale structures of a sample surface.

Although the invention of STM has opened up various applications, its applicability has been limited to the imaging of conductive materials. To overcome this limitation, atomic force microscopy (AFM) was invented in 1986 by Binnig et al. [2]. In AFM, a microfabricated cantilever with a sharp tip at its end is used as a force sensor. Instead of tunneling current, the tip–sample interaction force (F_t) is detected and used for the tip–sample distance regulation. Therefore, AFM can be used for imaging insulators as well as conductive materials. In addition, AFM has enabled quantitative measurements of an interaction force with piconewton-order force sensitivity and nanoscale spatial resolution. Due to this unique capability, AFM has been used not only for imaging surface structures but also for investigating various interactions between atoms, molecules, and surfaces.

To date, various operation modes of AFM have been developed. Among them, the one used for the first AFM is referred to as contact-mode AFM (CM-AFM). In CM-AFM, a tip is placed in contact with a surface and laterally scanned. During the scan, the cantilever deflection Δz induced by F_t is detected and used for the tip–sample distance regulation. The advantages of CM-AFM include its simple experimental setup and straightforward interpretation of the measured force. Namely, the quantitative F_t values can be calculated from the measured Δz values using a simple equation:

$$F_t = k\Delta z. \tag{13.1}$$

Although CM-AFM has enabled nanoscale investigations of insulating materials, its applicability has been limited to the studies on relatively hard samples. In CM-AFM, the tip is scanned in contact with a surface so that a relatively large friction force is applied to the sample. Such a lateral force often prevents nondestructive imaging of soft materials such as polymers and biological molecules.

To overcome this limitation, dynamic-mode AFM was invented in 1987 by Martin et al. [3]. In this method, a cantilever is mechanically oscillated at a frequency near the cantilever resonance (f_0). The tip–sample interaction force is detected as the reduction in the cantilever oscillation amplitude (A). This operation mode is now referred to as amplitude modulation AFM (AM-AFM). In AM-AFM, the tip intermittently contacts the surface during the tip scan. Thus, the lateral friction force is much smaller than that in CM-AFM and hence can be used for imaging soft insulating materials such as polymers [4] and biological systems [5].

One of the major drawbacks of AM-AFM is the difficulty in the operation in vacuum. In vacuum, the Q factor of the cantilever resonance becomes very high (1,000–100,000) due to the small energy dissipation caused by the friction between the cantilever and surrounding medium. In general, a higher Q factor gives a higher force sensitivity in AM-AFM. Therefore, the operation in vacuum is desirable for high-resolution imaging. However, the time response (τ_{AM}) of AM-AFM increases in proportion to the Q factor as given by the following equation [6]:

$$\tau_{AM} \simeq \frac{2Q}{\omega_0}, \tag{13.2}$$

where $\omega_0 = 2\pi f_0$.

The resonance characteristics of the cantilever are immediately changed by the tip–sample interaction. However, the reduction of A takes time to dissipate the vibration energy. In high-Q environments, the energy dissipation path is so narrow that the time required for damping A is much longer than that in low-Q environments. For a typical cantilever used in AM-AFM, τ_{AM} is more than 30 ms in vacuum. Such a slow time response is not acceptable in many of the practical applications.

To overcome this difficulty, frequency modulation AFM (FM-AFM) was invented in 1991 by Albrecht et al. [6]. In FM-AFM, the tip–sample interaction force is detected by measuring the shift of cantilever resonance (Δf). As the time response of FM-AFM (τ_{FM}) is given by $\tau_{FM} \simeq 1/f_0$, it is independent of the Q factor. Due to the high Q factor and clean surface conditions in vacuum, FM-AFM operated in vacuum has enabled reproducible true atomic resolution imaging of various surfaces [7,8].

Although FM-AFM can be used for investigating a wide range of materials, its operating environment has traditionally been limited to an ultrahigh vacuum. This has prevented its applications in biology and chemistry, where direct imaging of molecules at a solid/liquid interface is desirable. Recently, however, Fukuma et al. presented a way to overcome this limitation. Using a low-noise cantilever deflection sensor [9] and optimizing cantilever stiffness (k) and A, they have enabled operation of FM-AFM in liquid with true atomic resolution [10].

Now, FM-AFM has become a powerful tool for surface analysis. It can be operated in air [11], vacuum [6], and liquid [12]. It is capable of imaging insulators [13] as well as conductive materials [7,8]. The loading force during the imaging is small enough to image soft biological systems without destruction [14–16]. Furthermore, various surface property measurement techniques have been developed based on FM-AFM [17]. In this chapter, the author introduces instrumentation and applications of FM-AFM.

13.2 Basic Principle and Experimental Setup

Figure 13.1a shows a typical experimental setup of FM-AFM. In FM-AFM, a cantilever is mechanically oscillated at f_0. The cantilever oscillation is typically excited by applying an excitation signal $A_{ex}\cos(\omega t)$ to a piezoactuator placed near the cantilever. The induced cantilever oscillation is detected by a deflection sensor. The most commonly used method for the cantilever deflection measurement is the optical beam deflection (OBD) method. In this method, a focused laser beam is irradiated to the backside of a cantilever. The reflected laser beam is detected with a position-sensitive photodetector (PSPD). The PSPD typically consists of a four-element photodiode array. The laser spot on the PSPD is aligned at the center of the four elements so that the photo-induced current from the upper two elements (i_A) equals that from the lower two elements (i_B). The cantilever deflection induces displacement of the laser spot on the PSPD surface. Consequently, the difference (i_{A-B}) between i_A and i_B changes in proportion to the cantilever deflection. The current difference i_{A-B} is converted to a voltage signal with a pre-amplifier to obtain a deflection signal.

The deflection signal is used for two different purposes. On one hand, it is used for preparing the cantilever excitation signal. On the other hand, it is used for the detection of cantilever vibration frequency. For cantilever excitation, the phase

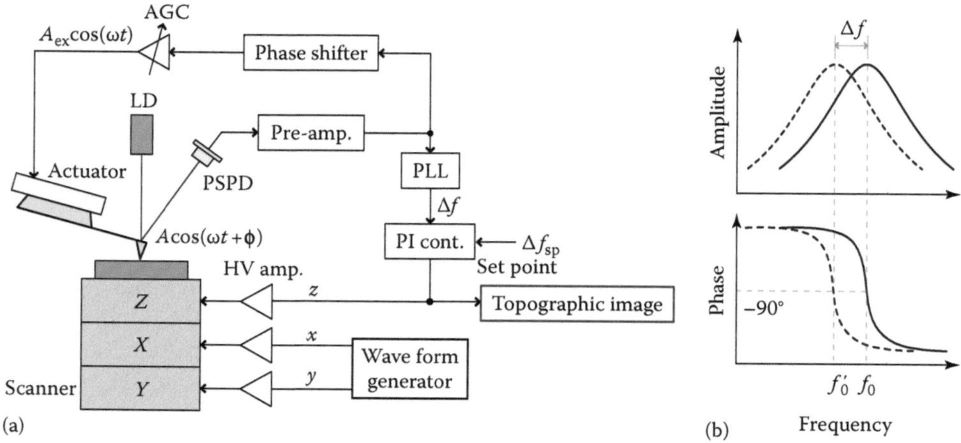

FIGURE 13.1 (a) Typical experimental setup of FM-AFM. (b) Amplitude and phase response of a cantilever.

and amplitude of the deflection signal is adjusted by a phase shifter and an automatic gain control (AGC) circuit, respectively. The obtained excitation signal is used for driving a piezoactuator. The loop consisting of a cantilever, pre-amp., phase shifter, AGC, and piezoactuator is referred to as a self-excitation loop. In the loop, a cantilever serves as a mechanical resonator which determines the oscillation frequency. At the cantilever resonance, the phase difference between the cantilever excitation signal and cantilever oscillation (ϕ) equals −90° as shown in Figure 13.1b. Thus, by adjusting the phase delay at the phase shifter such that ϕ equals −90°, the cantilever is always driven at its resonance frequency.

As the tip approaches the surface, F_t induces a shift in the cantilever resonance. Owing to the resonance tracking capability of the self-excitation circuit, the cantilever vibration frequency is also shifted by the same amount (Δf). Δf is detected from the deflection signal typically by a phase-locked loop (PLL) circuit. The detected Δf signal is fed into a proportional-integral (PI) controller, which adjusts the output signal such that Δf equals a given set point value (Δf_{sp}). The output from the PI controller is amplified by a high-voltage (HV) amplifier and applied to a Z piezoactuator. The loop consisting of a cantilever, pre-amp., PLL, PI controller, HV-amp, and Z piezoactuator is referred to as a distance feedback loop. Owing to the function of this feedback loop, the tip–sample distance is kept almost constant.

With the distance feedback regulation turned on, the tip is laterally scanned by applying triangular and ramp signals to X and Y piezoactuators, respectively. During the tip scan, the height of the Z scanner changes to follow the surface corrugations. Thus, by recording the Z driving voltage with respect to the tip XY position, a 2D image of the surface topography is obtained.

13.3 FM-AFM in Vacuum

13.3.1 Imaging of Surface Structures

13.3.1.1 Inorganic Systems

The first true atomic-resolution FM-AFM images were obtained by imaging Si(7×7) reconstructed surfaces in 1995 [7,8]. Although atomic-scale imaging of the same surface had been already achieved by STM, it was the first successful AFM imaging of a reactive surface with true atomic resolution. Subsequently, several research groups started to investigate the imaging mechanism and applicability of FM-AFM. These works revealed that FM-AFM can be used for atomic-resolution imaging of various samples such as semiconductors [18–23], metals [24,25], and metal oxides [26].

One of the distinctive features of FM-AFM is its capability of true atomic-resolution imaging of insulating samples. This capability was first demonstrated by imaging NaCl (001) surfaces by Bammerlin et al. in 1997 [13]. In contrast to the other applications described earlier, alkali halides cannot be imaged by STM. Therefore, the result highlighted distinctive features of FM-AFM for the first time. This capability has led to the idea of imaging organic molecules with high spatial resolution.

13.3.1.2 Organic Molecules

Organic molecules are mostly nonconductive and easily destructed by an applied force. Therefore, molecular-resolution imaging of organic molecules without destruction requires capability of imaging insulators with high spatial resolution and low loading force. In FM-AFM, the vertical tip position is regulated in the noncontact or atomic-scale contact regime with respect to the surface. Thus, the loading force during the imaging is extremely small. Due to this feature, FM-AFM has been considered to be suitable for molecular-scale imaging of organic molecules.

In 1997, Fukui et al. presented the first molecular-resolution FM-AFM images by imaging formate ions adsorbed on a TiO_2 (110) surface. Subsequently, many molecular-resolution FM-AFM images of different organic systems were reported [27–32]. Similar to the studies on inorganic systems, the early studies on FM-AFM imaging of organic systems were mostly performed on conductive samples whose structures and properties have been studied by STM.

In 1999, Kitamura et al. obtained the first molecular-resolution images of an insulating organic system by imaging a thick polypropylene film [35]. Subsequently, several other examples have been reported on the imaging of organic systems that cannot be imaged by STM such as self-assembled monolayers (SAMs) consisting of long-chain alkanethiols (Figure 13.2a) [33], organic ferroelectric thin films formed on an alkali halide substrate (Figure 13.2c) [34], and thick crystals of Cu-phthalocyanins formed on a conductive substrate [36]. These examples demonstrated the wide applicability of FM-AFM to the investigations on insulating organic systems.

13.3.2 Measurements of Surface Properties

13.3.2.1 Surface Potential Measurements

F_t contains different force components such as van der Waals force (F_{vdW}), electrostatic force (F_{es}), and chemical interaction force (F_{chem}). Among them, F_{es} can be easily modulated by applying an ac bias voltage so that it can be independently measured by a lock-in amplifier. This idea has led to the development of various surface potential measurement techniques based on AFM. Among them, Kelvin-probe force microscopy (KFM) has been one of the most widely used methods.

In KFM, ac and dc bias voltages are applied between the tip and sample (Figure 13.3a). The tip–sample potential difference (V_{ts}) is given by

$$V_{ts} = V_{dc} - V_s + V_{ac}\cos(\omega_m t), \tag{13.3}$$

where

V_{dc} and V_{ac} are the magnitudes of the dc and ac bias voltages
ω_m is the ac bias modulation frequency
V_s is the surface potential

F_{es} induced by the bias application is described by

$$F_{es} = \frac{1}{2}\frac{\partial C_{ts}}{\partial z}V_{ts}^2, \tag{13.4}$$

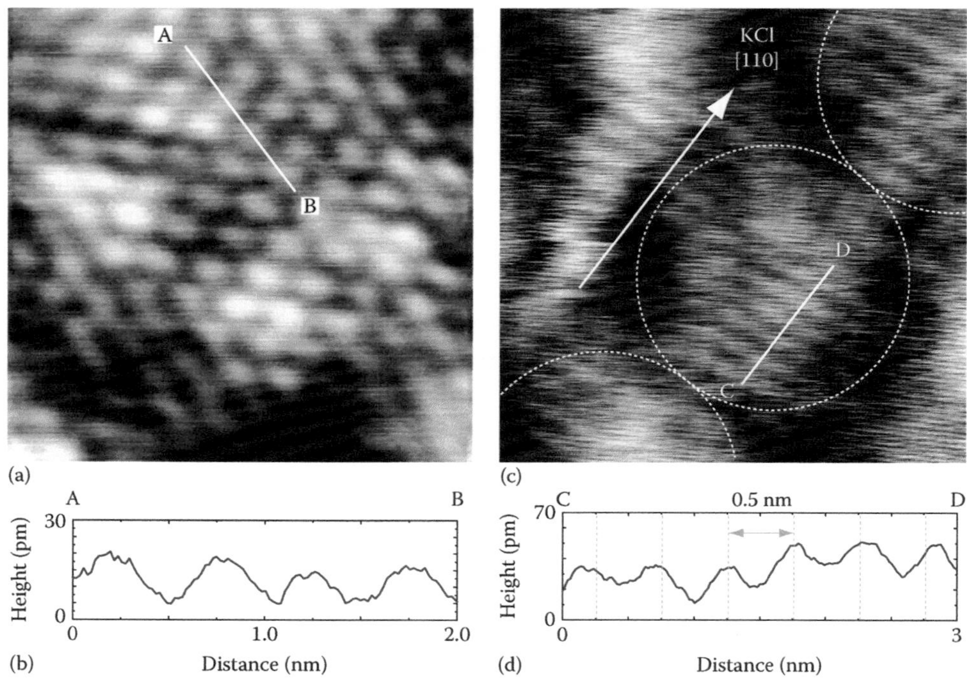

FIGURE 13.2 (a) FM-AFM image of a hexadecanethiol self-assembled monolayer (SAM) on a Au(111) surface (5 nm × 5 nm). (b) Cross-sectional profile measured along line A–B in (a). (c) FM-AFM image of a vinylidene fluoride oligomer thin film formed on a KCl substrate (10 nm × 10 nm) [34]. (d) Cross-sectional profile measured along line C–D in (d). (Reprinted with kind permission from Springer Science + Business Media: *Appl. Phys. A*, 72, 2001, S109, Fukuma, T., Kobayashi, K., Horiuchi, T., Yamada, H., and Matsushige, K., Copyright 2001; Reprinted from *Surf. Sci.*, 516, Fukuma, T., Kobayashi, K., Noda, K., Ishida, K., Horiuchi, T., Yamada, H., and Matsushige, K., 103, Copyright 2002 with permission from Elsevier.)

where C_{ts} is the tip–sample capacitance. This quadratic dependence is also schematically shown in Figure 13.3b.

From Equations 13.3 and 13.4, F_{es} is given by

$$F_{es} = \frac{1}{2}\frac{\partial C_{ts}}{\partial z}\left[(V_{dc} - V_s)^2 + 2(V_{dc} - V_s)V_{ac}\cos(\omega_m t) + V_{ac}^2\cos^2(\omega_m t)\right]. \tag{13.5}$$

This equation shows that F_{es} contains dc, ω_m, and $2\omega_m$ components. The ω_m component $F_{es,m}$ is proportional to $(V_{dc} - V_s)$. If we control V_{dc} so that $F_{es,m}$ is minimized, V_{dc} always equals V_s. Therefore, a surface potential image is obtained by recording V_{dc} during the tip scan over the surface.

To obtain high spatial resolution, precise regulation of the tip–sample distance is essential. When KFM was first introduced by Nonnenmacher et al. in 1991 [37], the tip–sample distance regulation was made by AM mode. In this implementation, the spatial resolution was limited to a nanometer-scale. In 1998, Kitamura and Iwatsuki reported a KFM setup combined with FM-AFM for the first time [17]. Owing to the high spatial resolution provided by FM-AFM, this setup has made it possible to obtain true atomic-scale resolution even in potential measurements.

Figure 13.3c shows a typical experimental setup for KFM combined with FM-AFM. In this setup, $F_{es,m}$ is detected from the Δf signal using a lock-in amplifier. The output of the lock-in amplifier is fed into a PI controller which controls V_{dc} such that the ω_m component of the Δf signal is minimized. The tip–sample distance regulation is made by the FM detection mode. Note that some components used for the tip–sample distance regulation are omitted in Figure 13.3c for clarity. By recording V_{dc} during the scan, a surface potential image is obtained.

13.3.2.2 Energy Dissipation Measurements

The tip–sample interaction force (F_t) detected by FM-AFM consists of conservative and dissipative components. In vacuum, owing to the high Q factor, the cantilever motion can be regarded as a simple harmonic oscillation given by $A\sin(\omega t)$. As F_t depends on the tip–sample distance, variation of F_t is synchronized with the sinusoidal tip motion with the same frequency ω. The in-phase component of the force $[F_{tc}\sin(\omega t)]$ is referred to as conservative force, which does not dissipate cantilever vibration energy when it is averaged over an oscillation cycle. The quadrature-phase $[F_{td}\cos(\omega t)]$ component is referred to as dissipative force as it causes energy dissipation.

The energy dissipation rate (P) caused by F_{td} is given by [38]

$$P = P_0\left(\frac{A_{ex}}{A_{ex0}} - 1\right), \tag{13.6}$$

where A_{ex} and A_{ex0} are the excitation amplitudes near and far from the surface. P_0 is given by

$$P_0 = \frac{\pi k A^2 f_0}{Q}. \tag{13.7}$$

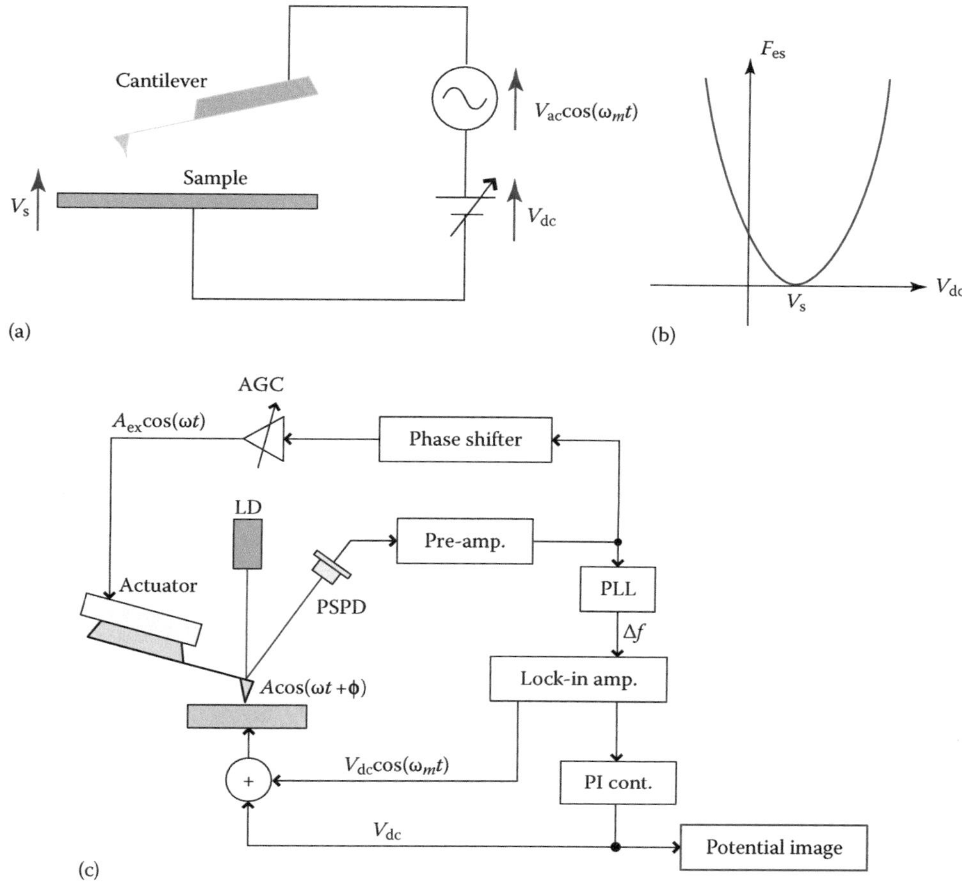

FIGURE 13.3 (a) Basic principle of KFM. (b) Schematic illustration of V_s dependence of F_{es}. (c) Typical experimental setup for KFM combined with FM-AFM.

The averaged energy dissipation in one oscillation cycle can be calculated by dividing P by f_0. Equation 13.6 shows that a dissipation image can be obtained by recording A_{ex} during the tip scan.

In one cycle of the cantilever oscillation, a tip approaches and retracts from the surface. If the force profiles during the approach and retraction processes are exactly the same, there should not be any energy dissipation. Therefore, the dissipative force should originate from the hysteresis of the tip–sample interaction force, which is typically caused by instabilities or bistabilities of the sample or tip.

In 1997, Lüthi et al. reported the first true-atomic resolution dissipation image by imaging Si(7×7) reconstructed surface [39]. The atomic-scale dissipation contrasts have been mainly attributed to the instabilities or bistabilities in the atomic-scale arrangements of the sample surface [40–43]. The energy dissipation is often caused by the instabilities or bistabilities of the tip apex atoms. However, the atomic-scale contrasts observed in a dissipation image should reflect the variation in the surface properties. Although quantitative interpretation of the energy dissipation value is often difficult, qualitative interpretation of dissipation contrast is possible in some cases. In particular, when it comes to the atomic or molecular-scale contrasts, they should originate from short-range interaction force, which helps to narrow down the possible origins for the energy dissipation contrast.

The energy dissipation signal is very sensitive to the mechanical properties of the surface and can be measured with true atomic resolution. These experimental facts have lead to an idea that the signal may be used for mapping the distribution of mechanical properties. This possibility has been intensively explored in the applications of molecular imaging [36,44–47] as molecular systems tend to show a large variation in the mechanical properties depending on the molecular conformations and packing arrangements.

Figure 13.4 shows one of the examples of such applications [44]. The figure shows topographic and dissipation images of the c(4×2) structure of a dodecanethiol SAM. The dissipation image of the zigzag-phase domain (Figure 13.4b) shows molecular-scale contrasts, while the rectangular-phase domain (Figure 13.4d) shows almost no contrasts in the dissipation image. This difference suggests that the zigzag-phase domain should have a larger variation in the molecular stability compared with the rectangular-phase domain. In fact, the authors suggest that the high-resolution imaging of the zigzag-phase domains is more difficult than that of the rectangular-phase domains due to the instability of the tip–sample interaction.

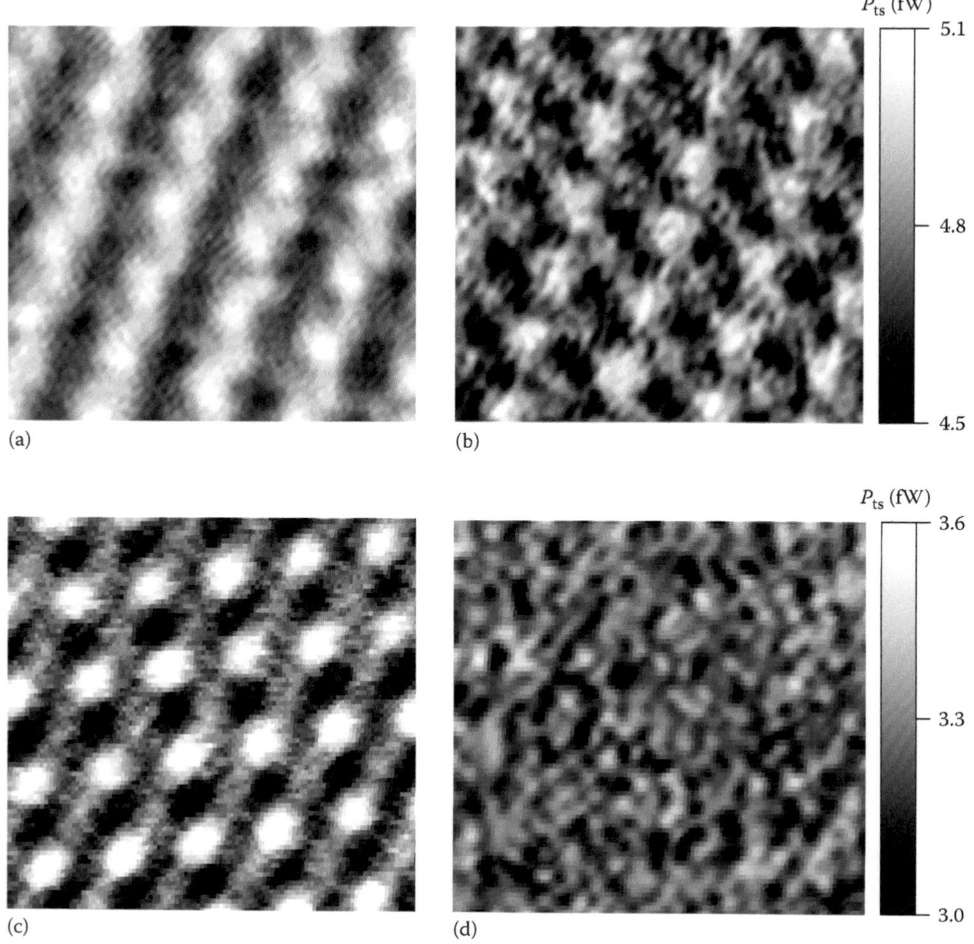

FIGURE 13.4 FM-AFM images of a dodecanethiol SAM on a Au (111) surface. (a) Topographic and (b) dissipation images of the δ-phase domain (4.5 nm × 4.5 nm). (c) Topographic and (d) dissipation images of the ε-phase domain (4.5 nm × 4.5 nm). (Reprinted with permission from Fukuma, T., Ichii, T., Kobayashi, K., Yamada, H., and Matsushige, K., *J. Appl. Phys.*, 95, 1222, Copyright 2004 American Institute of Physics.)

13.3.3 Recent Progress

13.3.3.1 Small Amplitude Operation and Quartz Sensor

In the early stages of the FM-AFM development, it was common to use A of 5–10 nm. However, the short-range force (F_{SR}), which mainly contributes to the formation of atomic-scale contrasts, has an interaction length of 0.2–0.3 nm, which is much shorter than the typical A. Thus, F_{SR} is detected only in a fraction of time in an oscillation cycle. This implies that there is a large room to improve the sensitivity to F_{SR} by reducing A.

While the reduction of A increases the sensitivity to F_{SR}, it also increases the noise in the Δf signal. As a result, there is an optimal A which gives the best sensitivity to F_{SR}. Giessibl et al. first pointed out this idea in 1999 [48]. According to their calculation, the optimal A approximately equals the length scale of the force component to be detected. In the case of F_{SR}, this corresponds to 0.2–0.3 nm. However, stable operation of FM-AFM with such a small A requires the use of a relatively high k to avoid the tip–sample adhesion.

To avoid the tip–sample adhesion known as jump-to-contact, one of the following equations should be met:

$$k > \left|\frac{\partial F_t}{\partial z}\right|, \tag{13.8}$$

$$kA > F_{ad}, \tag{13.9}$$

where F_{ad} is the adhesion force. Equation 13.8 shows that k should be higher than the effective stiffness of the tip–sample interaction (k_{ts}). Equation 13.9 shows that the restoring force of the cantilever spring should be higher than the adhesion force. These equations imply that we should use a cantilever either with high k or large A to avoid jump-to-contact.

In the case of large amplitude FM-AFM, k of 20–40 N/m is typically used. In this case, the restoring force is $kA = 100$–400 nN, which is sufficiently larger than F_{ad}. In the case of small amplitude FM-AFM, if we use the same k, the restoring force should become $kA = 2$–20 nN, which can be smaller than F_{ad}. Therefore, we should use a relatively large k for satisfying Equation 13.8.

In vacuum, k_{ts} can be as high as 30 N/m at the vicinity of the sample surface due to the attractive interactions such as electrostatic and van der Waals' interactions. For stable imaging, k should be sufficiently larger than this value. In practice, the safety margin required for ensuring the stability depends on various factors such as surface roughness, scan speed, and bandwidth of the tip–sample distance regulation. Although it is difficult to quantitatively predict the required k, our experience suggests that a safety factor of ~10 is typically enough for a stable operation, which corresponds to ~300 N/m in this case.

Based on this idea, Giessibl et al. proposed to use a quartz sensor referred to as "qPlus sensor." The sensor is made from a quartz tuning fork [49]. They fixed one of the two prongs to a plate so that the other prong can behave as a cantilever. The sensor has a relatively high k (typically 1800 N/m), which allows small amplitude operation of FM-AFM in vacuum. With the enhanced sensitivity to F_{SR}, they were able to visualize subatomic-scale features in FM-AFM images that reflect spatial distribution of electron orbitals [50].

Quartz is one of the piezoelectric materials. Thus, the deflection of a qPlus sensor can be electrically measured by amplifying the charge generated by the cantilever bending. In addition, the sensor can be used for simultaneous imaging of Δf and tunneling current images. Therefore, the invention of qPlus sensor has significantly lowered the barrier to convert an STM to an FM-AFM and expanded the application field of this technique.

13.3.3.2 Atom Manipulation and Identification

The capability of SPM techniques is not limited to the imaging of surface structures and properties. They can also be used for fabricating nanoscale and atomic-scale structures. The first demonstration of atom manipulation was reported by Eigler and Schweizer in 1991 using STM at low temperature [51]. They manipulated individual Xe atoms to form atomic-scale alphabets on a Ni surface.

For the atom manipulation, precise control of the tip position is essential. A major problem in this regard is the tip drift. For a typical AFM operated at room temperature, the XY drift rate is on the order of nanometers per minute. For atom manipulation, it is desirable to have more than 10 times smaller drift rate. This usually requires the use of low-temperature AFM or STM.

In 2005, Sugimoto et al. presented another way to solve this problem [52]. They used the atom-tracking technique, which was first reported by Phol and Möller [53], for compensating linear drift. In this technique, the XY position of a surface protrusion (or depression) is tracked by the tip for a while. During the tracking process, the tip moves in accordance with the drift. By fitting a linear function to the tip trajectory, we can precisely estimate the drift rate. Once the drift rate is known, the linear component of the drift can be easily compensated by applying a constant ramp signal to the X and Y drive signals of the scanner. In this way, they managed to suppress the drift and made it possible to manipulate individual Sn atoms on a Ge surface even at room temperature [52].

Such a precise control of the tip position has also enabled chemical species of individual atoms to be identified [54]. For atom identification, inelastic tunneling spectroscopy (IETS) has been used. However, this requires the operation at low temperature in principle. In 2007, Sugimoto et al. presented another way to identify atom species by taking force curves on individual atoms and analyzing them [54]. This was also enabled by the low-drift operation by the atom-tracking technique.

13.4 FM-AFM in Liquid

13.4.1 Special Requirements

Compared with the operation of FM-AFM in vacuum, its operation in liquid imposes additional requirements on its instrumentation. Here, the author summarizes these requirements and their solutions.

13.4.1.1 Thermal Vibration

For true atomic-resolution imaging, the tip front atom must predominantly interact with a surface atom. To satisfy this condition, the position of the tip apex atom should be stably controlled with a precision of ~10 pm. The thermal vibration amplitude of a cantilever (δz_{th}) is given by [6]

$$\delta z_{\text{th}} = \sqrt{\frac{k_B T}{k}}, \tag{13.10}$$

where k_B and T denote Boltzmann's constant and absolute temperature, respectively. While a relatively soft cantilever with a k of 0.4 N/m gives a δz_{th} of 100 pm, a relatively stiff cantilever with a k of 40 N/m gives a δz_{th} of 10 pm. The result of these calculations and our experience suggest that k should be higher than ~10 N/m for atomic-scale imaging in liquid.

13.4.1.2 Small Amplitude Operation

As discussed earlier, small amplitude operation is effective to enhance the sensitivity to F_{SR} [48]. In vacuum, it is necessary to use a relatively stiff cantilever (k > 300 N/m) for operating small-amplitude FM-AFM without jump-to-contact. In liquid, k_{ts} of the attractive force is typically much smaller than that in vacuum due to the screening of the electrostatic and van der Waals forces by the solvent. Although k_{ts} depends on the nature of the tip apex and surface, it is typically less than 2–3 N/m. This result suggests that we can operate small-amplitude FM-AFM even with k as small as ~10 N/m. This advantage partially compensates the disadvantage of having a low Q factor in liquid and provides the sufficient force sensitivity to F_{SR} for true atomic-resolution imaging.

13.4.1.3 Force Sensitivity

For true atomic-resolution imaging by FM-AFM, surface corrugations of ~10 pm should be measured (see Figure 13.2b and d). To achieve this vertical resolution (δz), the precision of the tip–sample distance regulation should be better than this value. Assuming that k_{ts} is ~1 N/m, the required force sensitivity ($\delta F = k_{ts}\delta z$) is ~10 pN. These values and our experience suggest that true atomic-resolution typically requires a force sensitivity of ~10 pN.

The minimum detectable force F_{min} obtained by FM-AFM is determined by the thermal vibration of the cantilever. Assuming that A is small compared to the length scale of the interaction force, F_{min} is given by [6]

$$F_{min} = \sqrt{\frac{4kk_BTB}{\pi f_0 Q}}, \tag{13.11}$$

where B is the measurement bandwidth. For a typical cantilever used in an atomic-resolution FM-AFM imaging, $k = 30$ N/m, $f_0 = 130$ kHz, and $Q = 8$ in liquid. From these parameters and Equation 13.11, F_{min} is ~4 pN at B of 100 Hz.

The result suggests that F_{min} obtained by the present FM-AFM is just as much as required for atomic-resolution imaging (~10 pN). This implies that there is only little margin in the force sensitivity in such applications. Therefore, such applications require the optimal F_{min} limited only by the thermal noise. To achieve this goal, the deflection noise arising from the cantilever deflection sensor should be negligible compared with the thermal noise.

The deflection noise density arising from the thermal cantilever vibration (n_{zB}) is given by [6]

$$n_{zB} = \sqrt{\frac{2k_BT}{\pi f_0 kQ} \frac{1}{[1-(f/f_0)^2]^2 + [f/(f_0 Q)]^2}}. \tag{13.12}$$

From this equation, the noise density around the thermal noise peak is given by

$$n_{zB}(f_0) = \sqrt{\frac{2k_BTQ}{\pi f_0 k}}. \tag{13.13}$$

For a typical cantilever, this value is ~74 fm/$\sqrt{\text{Hz}}$.

In FM-AFM, deflection noise at the frequency range between $f_0 - B$ and $f_0 + B$ predominantly contributes to the measurements. Thus, the deflection noise density arising from the deflection sensor (n_{zs}) should be sufficiently smaller than $n_{zB}(f_0)$. According to the calculated value and our experience, n_{zs} should be smaller than ~20 fm/$\sqrt{\text{Hz}}$ for obtaining the optimal F_{min} with a typical cantilever.

13.4.1.4 Cantilever Deflection Sensor

To date, various methods for the cantilever deflection measurement have been proposed [2,3,55–60]. Among them, the optical beam deflection (OBD) method [56] is the most widely used because of the simplicity and high sensitivity.

Major noise sources in the OBD sensor include noise from laser beam instability, shot noise at the PSPD, and Johnson noise at the I–V converter in the preamplifier. Theoretically, the PSPD shot noise should determine the optimal performance. However, in practice, the performance of the OBD sensor is often limited by the laser beam instability.

The noise arising from the laser beam instability consists of laser interference noise and optical feedback noise. These noise components become evident especially in liquid environment applications. In the case of the liquid environment AFM, a laser beam propagates through glass/air and glass/water interfaces several times. At each interface, a laser beam is partially reflected and scattered. On one hand, some of the reflected beam goes back to the laser diode to induce optical feedback noise. On the other hand, some of the scattered beam interferes with a laser beam bounced at the cantilever backside at the surface of the PSPD, leading to an increase in the optical interference noise.

To suppress these noise components associated with a laser beam, it is effective to reduce the coherence of the laser beam. The coherence of a laser beam can be reduced by using a self-pulsating laser diode, a multi-mode laser diode or a super-luminescent diode. In 2005, Fukuma et al. presented another way to reduce the coherence by modulating the power of a laser beam with a radio frequency (RF) signal [9]. This method is found to be more effective than the other methods.

Using the RF modulation technique, Fukuma et al. succeeded in reducing n_{zs} to less than ~20 fm/$\sqrt{\text{Hz}}$ and achieved true atomic resolution in liquid in 2005 [9]. Figure 13.5 shows the reported images of mica obtained in water [10]. The honeycomb-like structure of the mica surface is clearly visualized with atomic-scale irregular protrusions (Figure 13.5c). The result demonstrates that FM-AFM has true atomic resolution in liquid.

Subsequently, several groups also reported atomic-resolution FM-AFM images of mica obtained by low-noise cantilever deflection sensors with different designs. Hoogenboom et al. developed a low-noise Fabry–Perot interferometer [61] and presented atomic-resolution images of mica [14]. Kawai et al. developed a low-noise Doppler interferometer [62]. With this deflection sensor, Nishida et al. obtained true atomic-resolution images of mica [63]. These previous works have proven that a low-noise deflection sensor is essential for atomic-scale FM-AFM experiments in liquid.

13.4.2 Biological Applications

13.4.2.1 Nanoscale Imaging of Biological Systems

One of the major motivations to operate FM-AFM in liquid is its applications to the high-resolution imaging of biological systems. However, as a relatively stiff cantilever ($k > 10$ N/m) is used in FM-AFM, it was a concern if the method can be used for imaging soft biological systems. At that time, it was common to use CM-AFM or AM-AFM with a soft cantilever ($k < 0.1$ N/m) for imaging biological systems. Thus, possible damaging of soft biological systems caused by a stiff cantilever was a concern.

The applicability of FM-AFM to the imaging of biological systems was confirmed by experiments soon after the first demonstration of true atomic-resolution imaging by FM-AFM in liquid. In 2005, Fukuma et al. presented molecular-resolution images of a purple membrane consisting of trimers of bacteriorhodopsins (bR), which has proven the applicability of FM-AFM to the imaging of biological systems [16,64] (Figure 13.6). They also demonstrated the molecular-resolution images of GroELs adsorbed on mica. In this experiment,

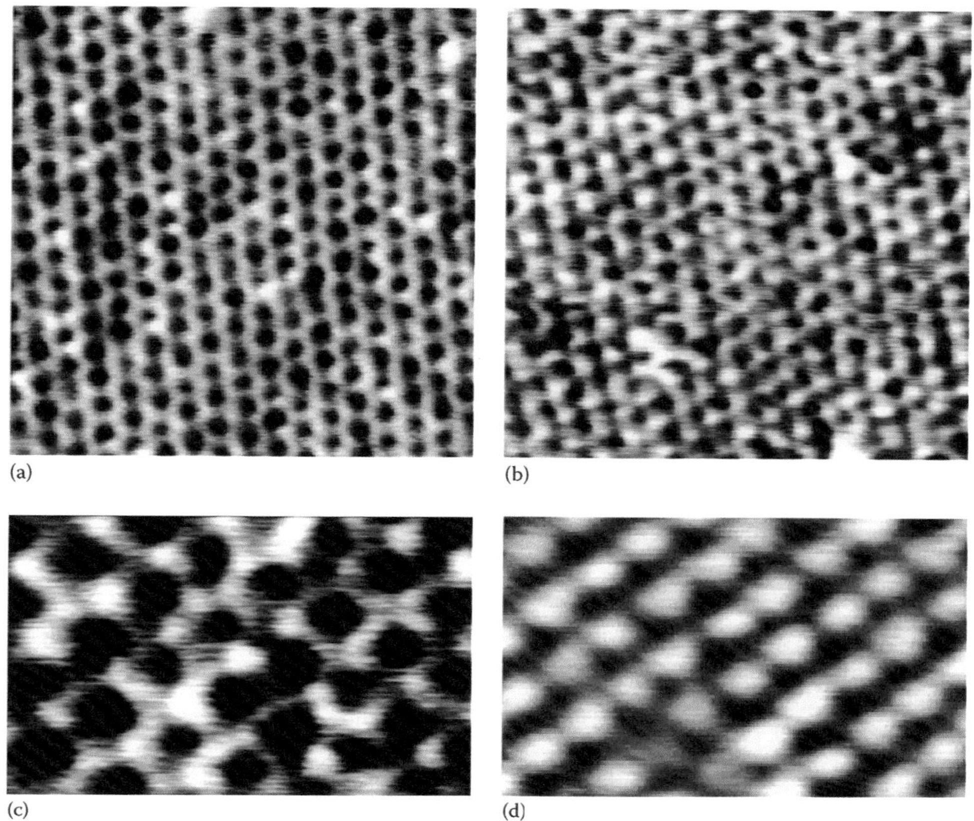

FIGURE 13.5 FM-AFM images of mica obtained in water. (a), (b) 8 nm × 8 nm. (c), (d) 4 nm × 2.5 nm. (Reprinted with permission from Fukuma, T., Kobayashi, K., Matsushige, K., and Yamada, H., *Appl. Phys. Lett.*, 87, 034101, Copyright 2005 American Institute of Physics.)

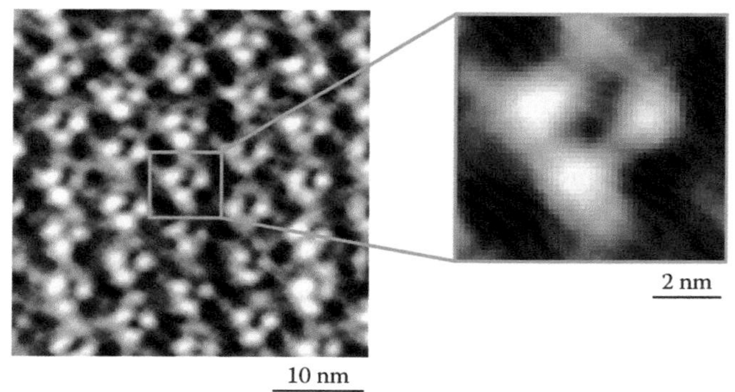

FIGURE 13.6 FM-AFM images of a purple membrane obtained in phosphate-buffered saline (PBS) solution. The images show molecular-scale contrasts corresponding to the trimers of bRs. (Reprinted from Yamada, H., et al., *Appl. Phys. Express*, 2, 095007, 2009. With permission. Copyright 2009 The Japan Society of Applied Physics.)

isolated molecular systems weakly attached to the surface were imaged without destruction. Although (2D) crystals of proteins such as a purple membrane can be imaged by CM-AFM with molecular-resolution, CM-AFM cannot be used for imaging isolated biomolecules. Thus, the imaging of GroELs highlighted the advantage of FM-AFM over the existing CM-AFM technique in this respect. In 2006, Hoogenboom et al. also presented molecular-resolution images of bRs [14], which further confirmed the applicability of FM-AFM to biological systems.

13.4.2.2 Subnanometer-Scale Imaging of Biological Systems

Although the early works described earlier have shown the applicability of FM-AFM to biological systems, the resolutions of the images were on the order of nanometers. As molecular-scale imaging of bRs was possible even with AM-AFM [65], these results were not sufficient to demonstrate the distinctive feature of FM-AFM, namely subnanometer-scale resolution.

In 2006, Higgins et al. showed the first subnanometer-scale image of biological systems obtained in liquid by imaging lipid bilayers [15]. In the reported images, individual lipid molecules separated by approximately 0.5 nm are clearly visualized (Figure 13.7). Furthermore, in 2007, Fukuma et al. showed FM-AFM images of lipid–ion complexes formed at the lipid/water interface with a resolution as small as 90 pm [66]. These results highlighted the enhanced spatial resolution obtained by FM-AFM.

13.4.2.3 Subnanometer-Scale Imaging of Proteins

Lipids are one of the simplest biomolecules. In addition, a lipid bilayer in gel phase is uniform, flat, and stable. Compared to this

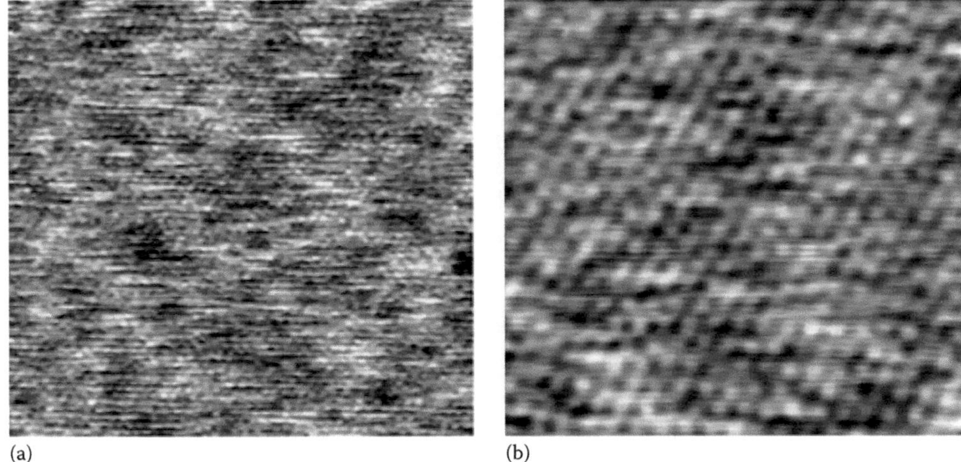

FIGURE 13.7 FM-AFM images of a lipid bilayer obtained in water. (a) 25 nm × 25 nm. (b) 10 nm × 10 nm. The images show molecular-scale contrasts corresponding to the individual lipid headgroups separated by ~0.5 nm. (Reprinted from Higgins, M. et al., *Biophys. J.*, 91, 2532, 2006. With permission. Copyright 2006 The Biophysical Society.)

special case, most of the biological systems generally have larger corrugations and inhomogeneity. Therefore, it was questioned if more complicated biomolecules such as proteins can be imaged by FM-AFM.

In 2008, Fukuma et al. first confirmed this capability by imaging the surface of amyloid fibrils [67]. The reported FM-AFM images show periodic stripes with a spacing of 0.5 nm and an orientation perpendicular to the fibrillar axis. The stripes correspond to the individual β-strands constituting a β-sheet. The images directly confirmed the arrangement of β-strands known as cross-β structure.

In 2011, Asakawa et al. imaged the surface of tubulin protofibrils to show subnanometer-scale features of individual tubulin molecules [68]. The image showed periodic structures with a spacing of 0.53 nm (Figure 13.8). This structure corresponds to the α-helices at the surface of tubulins. By comparing this image with the known structural model of a tubulin heterodimer, it is possible to determine the orientation of the molecule. Once the molecular orientation is known, we can identify a subnanometer-scale contrast as a specific portion of the molecule. In this example, some of the subnanometer-scale protrusions were identified as C-terminus of the molecules. Furthermore, the determination of the molecular orientation also makes it possible to identify the orientation of the molecular system. In this case, the positive and negative ends of the protofilaments are determined from the image.

These results show that FM-AFM can be used for visualizing subnanometer-scale features of proteins such as α-helices and β-sheets. With nanometer-scale resolution, it is possible to determine the molecular arrangement in a molecular complex. With subnanometer-scale resolution, it becomes possible to determine the arrangement of subnanometer-scale features such as α-helices and β-sheets in a molecule. This in turn enables us to determine the orientation of molecules and their assembly in real space.

13.4.3 Imaging of Hydration Structures

In an aqueous environment, an AFM tip interacts not only with the solid surface but also with the surrounding water molecules. Thus, the tip–water interaction should exert influence on the force detected by AFM.

If the water distribution around the tip is perfectly uniform, a tip scan in any direction should give no change in the force. However, at a solid/liquid interface, water molecules interact with a solid surface to present nonuniform distribution, namely, hydration structure. Therefore, if a tip is scanned at a solid/water interface, the tip–water interaction force should be changed. Furthermore, the distribution of the measured force should be correlated with the hydration structure.

13.4.3.1 One-Dimensional Spectroscopy of Hydration Forces

The simplest way to probe the force associated with the hydration structure is to take a force curve at a solid/water interface (Figure 13.9a). In the method, a tip is scanned in the Z direction near the surface. During the tip approach and retraction processes, a Δf signal is recorded with respect to the tip–sample distance. Here we refer to this method as 1D scanning force microscopy (1D-SFM).

Figure 13.9 shows an example of such an experiment [69]. In this case, a force curve was obtained at a DPPC/water interface (Figure 13.9b). The obtained Δf curve (Figure 13.9c) shows an oscillatory profile with two peaks. The Δf curve can be mathematically converted into a quantitative force versus distance curve (Figure 13.9d) using the formula reported by Sader and Jarvis [70]. The converted force curve reveals that the detected force variation is on the order of piconewtons. The result shows that FM-AFM has a piconewton order force sensitivity. The distance between the two peaks is approximately 0.25 nm, which agrees with the diameter of a single water molecule. This indicates that

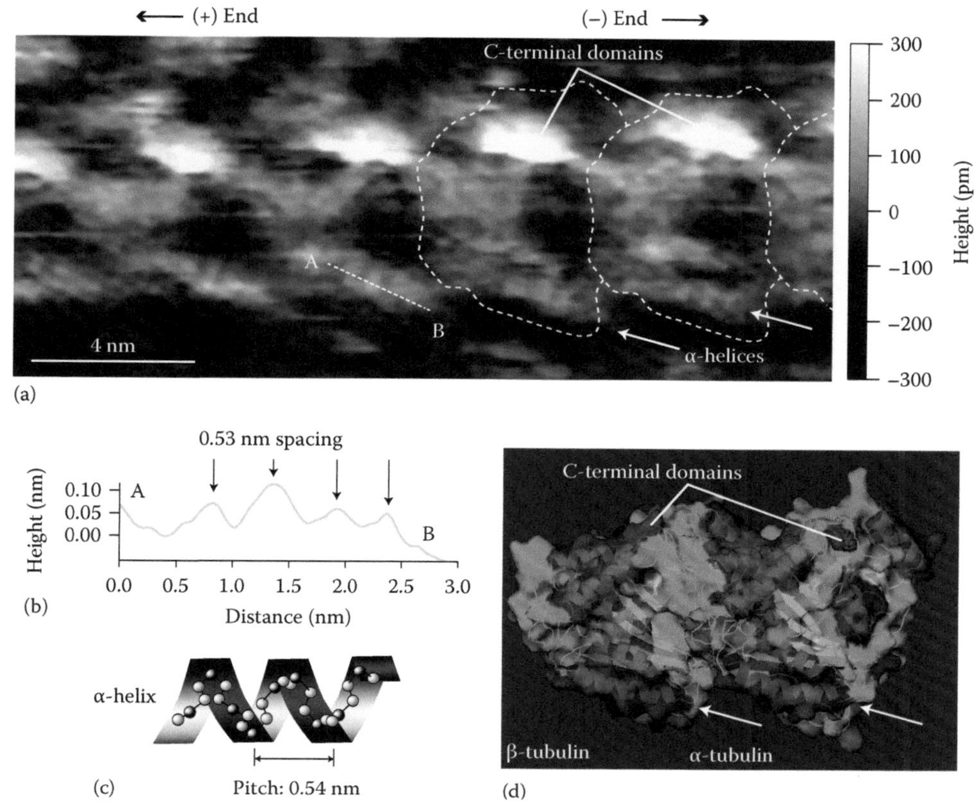

FIGURE 13.8 (a) FM-AFM image of a tubulin protofilament in a sheet-like structure ($\Delta f = 3.0$ Hz, $A = 0.30$ nm). (b) Height profile measured along line A–B in (a) (average spacing: 0.53 nm, standard deviation: 0.056 nm, $n = 12$). (c) Schematic illustration of the α-helix backbone. (d) Structural model of a tubulin heterodimer (PDB ID: 1JFF). (Reprinted from Asakawa, H. et al., *Biophys. J.*, 101, 1270, 2011. With permission. Copyright 2011 The Biophysical Society.)

the oscillatory force profile reflects the nonuniform distribution of water molecules at the DPPC/water interface.

Such an oscillatory profile associated with the hydration force has been measured by surface force apparatus (SFA) between mica surfaces [71]. However, in SFA, two opposing flat surfaces with a micrometer-scale dimension are brought close to each other. Therefore, the measured force shows a global force profile averaged over the micrometer-scale area. In addition, the water molecules confined between such a large area are forced to form a layered structure. Such geometrical confinement effect may prevent investigating the intrinsic hydration structure.

In the case of AFM, the interaction area is on the order of nanometers. Thus, the geometrical confinement effect should be much smaller than that in SFA. The measured force is more likely to reflect the local and intrinsic hydration structure at a solid/liquid interface. This is particularly true under the conditions where a true atomic-resolution image can be obtained by the tip. For obtaining a true atomic-resolution image, the interaction between the tip front atom and the surface atom should predominantly contribute to the variation of Δf signal. In such a situation, an atomic-scale protrusion should exist at the apex of the mesoscopic-scale AFM tip. With such an atomically sharp tip, the geometrical confinement is unlikely to be caused by the tip approach.

Another possible origin of the oscillatory force profile is the hydration structure at the tip apex. In most of the high-resolution AFM experiments, a Si tip is used. Thus, the tip apex should be covered with an oxidized Si layer, which is likely to be hydrophilic. Therefore, it is expected that there are hydration layers formed at the surface of the AFM tip. The measured force may reflect the hydration structure at the tip apex. Complete understanding of this contribution may require detailed comparison of the results obtained by experiments and simulation. This is one of the current research topics in this field.

13.4.3.2 Two-Dimensional Imaging of Hydration Structures

Although 1D-SFM makes it possible to visualize vertical distribution of water, the lateral distribution of water cannot be investigated by this method. One possible way to visualize the lateral distribution of water was presented by Fukuma et al. in 2007 [69]. Here, the author refers to the method as 2D-SFM and explains its principle by using the result shown in Figure 13.9c. When a Δf curve has an oscillatory profile, the tip–sample distance regulation can be operated at multiple feedback positions for a given Δf set point. For example, if we set the Δf set point at the value indicated by the dotted line in Figure 13.9c, there

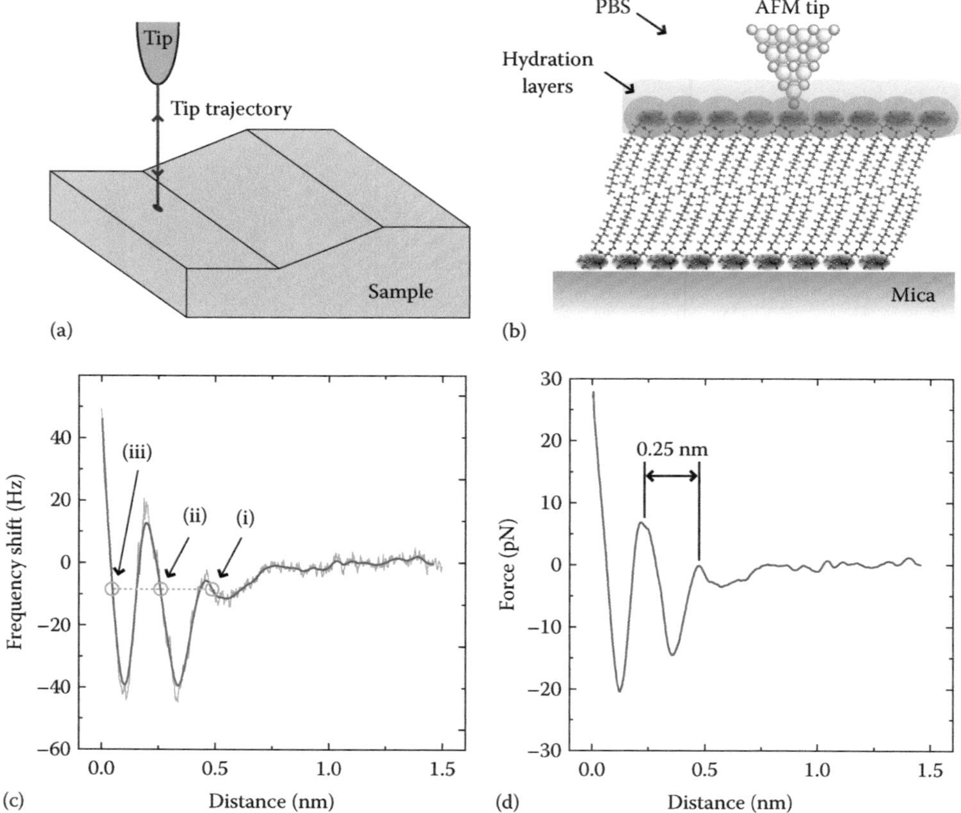

FIGURE 13.9 (a) Basic principle of 1D-SFM. (b) Schematic illustration of DPPC/water interface with an AFM tip. (c) Δf versus distance curve obtained at a DPPC/water interface in PBS solution. (d) Force versus distance curve calculated from the Δf curve shown in (c). (Reprinted from Fukuma, T. et al., *Biophys. J.*, 92, 3603, 2007. With permission. Copyright 2007 The Biophysical Society.)

are three feedback positions as indicated by the arrows (i)–(iii). When a tip is laterally scanned under this condition, the tip position may spontaneously jump from one feedback position to the other.

Such an example is shown in Figure 13.10c. The FM-AFM image was taken on a DPPC/water interface. In the image, a tip is scanned from the lowest terrace (Terrace 1), which corresponds to the lipid headgroups. During the tip scan, the tip jumps to Terrace 2 corresponding to the primary hydration layer. Furthermore, the tip shows another jump from Terrace 2 to 3, which corresponds to the secondary hydration layer. The heights of these jumps are in the range of 0.2–0.3 nm, which again agree with the diameter of a single water molecule. Figure 13.10b shows a schematic illustration of the tip trajectory during this imaging.

From the technical point of view, the result shows that the loading force during FM-AFM imaging is extremely small. Thus, the tip can be stably scanned even on a hydration layer. From biological aspect, the result shows that a stable hydration layer having a nanometer-scale lateral extent can exist on a soft biological membrane. This is not a trivial fact and has been debated for a long time. The surface of a lipid bilayer consists of an array of lipid headgroups, which are expected to be thermally fluctuating. Thus, it has been questioned if a stable hydration layer can be formed on such an unstable interface. However, the result shown here clearly demonstrates that such a hydration layer can exist even on a lipid membrane.

13.4.3.3 Three-Dimensional Imaging of Hydration Structures

2D-SFM has made it possible to visualize 2D distribution of water. However, it does not allow imaging both vertical and lateral distribution at the same time. For understanding the hydration structure having vertical and lateral extent, it is essential to develop a method to visualize 3D distribution of water.

The simplest way to achieve this goal is to use 3D force spectroscopy. In the method, many force curves are taken at arrayed XY positions. In this way, the force distribution within the whole 3D interfacial space can be recorded. However, the method takes relatively long time, which is typically on the order of hours. Therefore, such 3D spectroscopy has been used mainly in vacuum at low temperature where the influence of the tip drift is almost negligible. In vacuum and air, the drift rate can be greatly reduced by using the atom-tracking technique. However, in liquid, nonlinear drift component is often significant and hence it cannot be easily compensated even with the atom-tracking technique. Thus, it is essential to develop a high-speed 3D imaging technique for visualizing 3D hydration structure.

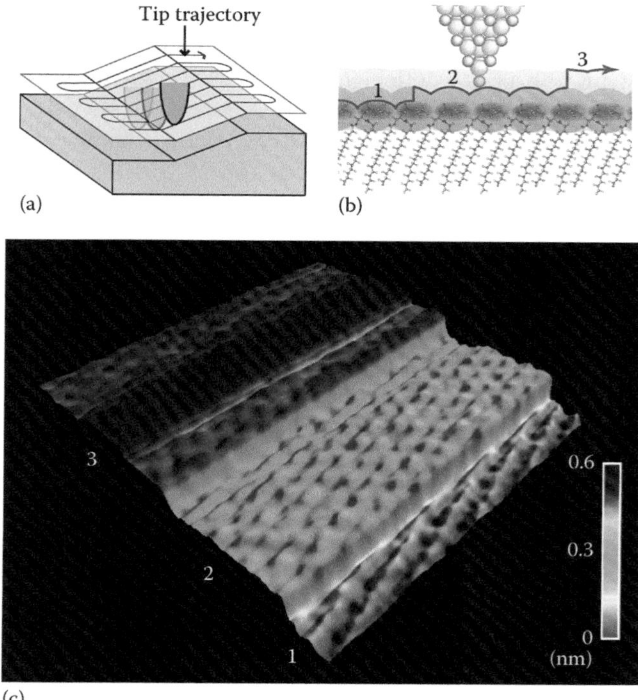

FIGURE 13.10 (a) Basic principle of 2D-SFM. (b) Schematic illustration of the DPPC/water interface during 2D-SFM imaging. (c) FM-AFM image of the DPPC/water interface obtained in PBS solution (8 nm × 8 nm). (Reprinted from Fukuma, T. et al., *Biophys. J.*, 92, 3603, 2007. With permission. Copyright 2007 The Biophysical Society.)

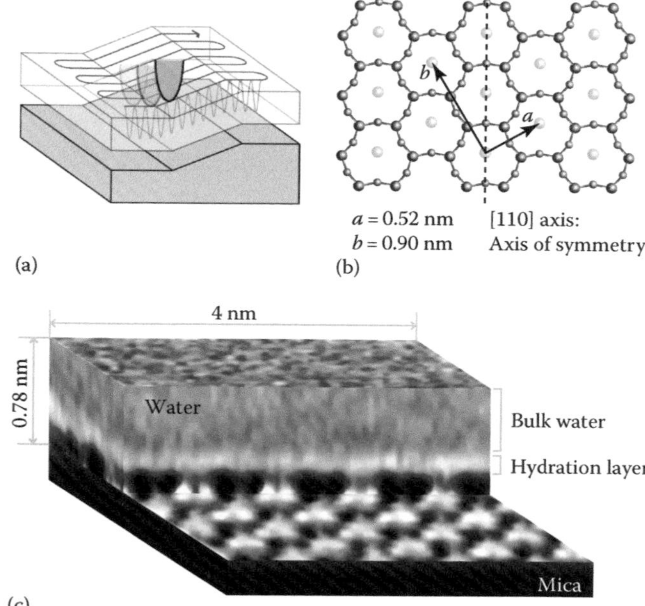

FIGURE 13.11 (a) Basic principle of 3D-SFM. (b) Atomic-scale model of cleaved mica surface. (c) 3D-SFM image of mica/water interface (4 nm × 4 nm × 0.78 nm). (Reprinted from Fukuma, T., *Sci. Technol. Adv. Mater.*, 11, 033003, 2010. With permission. Copyright 2010 National Institute for Materials Science.)

In 2010, Fukuma et al. presented a way to overcome this difficulty, which is referred to as 3D-SFM [72]. In the method, a tip is scanned laterally just as in the case of normal FM-AFM imaging. However, at the same time, the vertical tip position is modulated with a sinusoidal signal at a frequency much faster than the bandwidth of the tip–sample distance regulation (Figure 13.11a). By recording the Δf variation induced by the Z modulation, we can construct a 3D Δf image. In the meantime, the tip–sample distance is regulated such that the averaged Δf value is kept constant. Thus, even if there is any tip drift, surface tilt, or corrugations, the tip does not crash into the surface.

The conventional 3D spectroscopy is developed based on 1D force spectroscopy. Thus, the tip trajectory during the measurement is complicated, and the measurement takes relatively long time. In contrast, 3D-SFM is developed based on the 2D imaging technique. Therefore, the tip trajectory during the imaging is smooth and continuous, making it possible to scan fast without giving a large impulsive force to the AFM system. Owing to the fast imaging capability of 3D-SFM, it allows imaging of 3D force distribution in less than a minute. This is essential for visualizing 3D hydration structure without significant distortion.

Figure 13.11c shows a 3D-SFM image obtained at a mica/water interface [72]. Once a 3D force image is obtained, any XY and Z cross-sectional images can be derived from it. In fact, the model shown in Figure 13.11c consists of some of the XY and Z cross sections obtained from the 3D-SFM image. In the XY cross section, individual Si atoms are clearly visualized as bright spots. In the Z cross sections, layer-like contrast uniformly distributed in the lateral direction is observed. This layer-like contrast corresponds to the first hydration layer formed at the mica/water interface. In addition, the Z cross section shows periodic bright spots. The pairs of brighter spots should reflect the repulsive interaction between the tip apex atom and the surface topmost atoms. The darker spots sandwiched between the brighter spots should correspond to the adsorbed water molecules at the center of the honeycomb ring of the cleaved mica surface.

In the previous studies using x-ray reflectometry [74] and Monte Carlo simulation [75], the existence of adsorbed water molecules and formation of a hydration layer have been reported. The result obtained by 3D-SFM is consistent with these previous results. This result suggests that 3D force distribution measured by 3D-SFM is closely related to the 3D hydration structure.

13.5 Future Prospects

FM-AFM has become a powerful tool for surface analyses and manipulation. However, there still remains a large room for further development. Here, the author summarizes future prospects of this unique technique.

13.5.1 Atomic-Scale Tip Preparation

There are two requirements for achieving true atomic resolution by FM-AFM. First, the tip–sample interaction should be localized to an atomic-scale cross section. Second, the localized interaction should be detected with a sufficient sensitivity. In vacuum, the force sensitivity of FM-AFM is much higher than required for atomic-resolution imaging due to the high Q factor. However, the reproducible control of the atomic-scale structure of the tip apex is not established yet.

In 2004, Hembacher et al. used a carbon atom on the sample surface as a light-atom probe to minimize the cross section of the tip. They succeeded in visualizing electron orbitals within a single tungsten atom at the tip apex [76]. In 2009, Gross et al. used a CO molecule as a probe. In this experiment, they picked up a single CO molecule from the surface in a well-controlled manner and used it as an atomic-scale probe. With the CO tip, the atomic-scale structure of a pentacene molecule is visualized with an unprecedented resolution [77]. Although this method may not be applicable to all the experiments, the result suggests that we should develop a reproducible way to modify the tip apex with an appropriate molecule. This should also help to improve the reproducibility of the experiments. Furthermore, this should help to compare the experimental results and theoretical simulations.

13.5.2 Atomic-Scale Simulation of Liquid-Environment FM-AFM

So far, 1D force profiles and 3D force images obtained at a solid/liquid interface [69,72] have been compared with water distribution obtained by molecular dynamics (MD) or MC simulation [75,78]. Such comparisons have suggested that the force contrast obtained by FM-AFM is closely related to the water distribution. However, the models used in these simulations do not include an AFM tip so that the influence of the tip on the hydration structure is not taken into account. To understand the influence of the tip, detailed comparison between atomic-scale experiment and theory is required.

Atomic-scale simulation of FM-AFM has been possible but only for the operation in vacuum. Research on the simulations of liquid-environment FM-AFM has just started recently. Watkins and Shluger recently reported atomic-scale simulation of a CaF_2 surface by FM-AFM in liquid [79]. Such a theoretical study and its comparison with an experimental result should be very important to understand the influence of the tip and to establish the imaging technique of the hydration structure.

13.5.3 Improvement of Operation Speed

Due to the historical background of the FM-AFM development, it has not been seriously pursued to improve the operation speed of FM-AFM. Thus, there is a large room for the improvement of its operation speed. This is especially important for the operation in liquid. For biological applications, a sample surface often has large fluctuations, corrugations, and inhomogeneities compared to the samples used in vacuum. Therefore, improvement of the operation speed is essential.

For improving the FM-AFM operation speed, the bandwidth or resonance frequency of all the components involved in the tip–sample distance regulation feedback should be improved. Among the most critical components is the PLL circuit used for Δf detection. Mitani et al. recently developed a high-speed PLL circuit with a bandwidth higher than 100 kHz [80]. By combining such a fast PLL with a high-speed AFM system [81], it should become possible to operate FM-AFM with a scanning speed much faster than the present system.

13.5.4 Improvement of Force Sensitivity

In vacuum, due to the high Q factor of the cantilever resonance, the force sensitivity is much higher than required for true atomic-resolution imaging. In liquid, however, the force sensitivity is just as much as required for true atomic-resolution imaging. In addition, 3D force imaging generally requires much faster time response so that the force sensitivity is not necessarily sufficient for detecting the detailed hydration structure.

To improve the force sensitivity, there have been pursued two possible ways. One is the use of a bulky resonator for enhancing the Q factor in liquid. However, in this setup, k tends to become very high so that the improvement in F_{min} is not significant. Another way is to use a small cantilever to enhance f_0 without changing Q and k. Recently, Fukuma et al. reported sevenfold improvement in F_{min} using a small cantilever [82]. Although the current availability of such a small cantilever is limited, it should be expanded in the near future.

References

1. G. Binnig, H. Rohrer, Ch. Gerber, and E. Weibel. *Phys. Rev. Lett.*, 49:57, 1982.
2. G. Binnig, C. F. Quate, and Ch. Gerber. *Phys. Rev. Lett.*, 56:930, 1986.
3. Y. Martin, C. C. Williams, and H. K. Wickramasinghe. *J. Appl. Phys.*, 61:4723, 1987.
4. Q. Zhong, D. Inniss, K. Kjoller, and V. B. Elings. *Surf. Sci.*, 290:L688, 1993.
5. P. K. Hansma, J. P. Cleveland, M. Radmacher, D. A. Walters, P. E. Hilner, M. Bezanilla, M. Fritz, D. Vie, and H. G. Hansma. *Appl. Phys. Lett.*, 64:1738, 1994.
6. T. R. Albrecht, P. Grütter, D. Horne, and D. Ruger. *J. Appl. Phys.*, 69:668, 1991.
7. F. J. Giessibl. *Science*, 267:68, 1995.
8. S. Kitamura and M. Iwatsuki. *Jpn. J. Appl. Phys. Part II*, 34:L145, 1995.
9. T. Fukuma, M. Kimura, K. Kobayashi, K. Matsushige, and H. Yamada. *Rev. Sci. Instrum.*, 76:053704, 2005.
10. T. Fukuma, K. Kobayashi, K. Matsushige, and H. Yamada. *Appl. Phys. Lett.*, 87:034101, 2005.
11. T. Fukuma, T. Ichii, K. Kobayashi, H. Yamada, and K. Matsushige. *Appl. Phys. Lett.*, 86:034103, 2005.
12. T. Fukuma, K. Kobayashi, K. Matsushige, and H. Yamada. *Appl. Phys. Lett.*, 86:193108, 2005.
13. M. Bammerlin, R. Lüthi, E. Meyer, A. Baratoff, J. Lü, M. Guggisberg, C. Gerber, L. Howald, and H.-J. Güntherodt. *Probe Microsc.*, 1:3, 1997.

14. B. W. Hoogenboom, H. J. Hug, Y. Pellmont, S. Martin, P. L. T. M. Frederix, D. Fotiadis, and A. Engel. *Appl. Phys. Lett.*, 88:193109, 2006.
15. M. Higgins, M. Polcik, T. Fukuma, J. Sader, Y. Nakayama, and S. P. Jarvis. *Biophys. J.*, 91:2532, 2006.
16. H. Yamada, K. Kobayashi, T. Fukuma, Y. Hirata, T. Kajita, and K. Matsushige. *Appl. Phys. Express*, 2:095007, 2009.
17. S. Kitamura and M. Iwatsuki. *Appl. Phys. Lett.*, 72:3154, 1998.
18. S. Kitamura and M. Iwatsuki. *Jpn. J. Appl. Phys. Part II*, 35:L668, 1996.
19. H. Ueyama, M. Ohta, Y. Sugawara, and S. Morita. *Jpn. J. Appl. Phys. Part II*, 34:L1086, 1995.
20. Y. Sugawara, M. Ohta, H. Ueyama, and S. Morita. *Science*, 270:1646, 1995.
21. P. Güthner. *J. Vac. Sci. Technol. B*, 14:2428, 1996.
22. R. Lüthi, E. Meyer, M. Bammerlin, A. Baratoff, T. Lehmann, L. Howald, Ch. Gerber, and H.-J. Güntherodt. *Z. Phys. B*, 100:165, 1996.
23. N. Nakagiri, M. Suzuki, K. Okiguchi, and H. Sugimura. *Surf. Sci.*, 373:329, 1997.
24. S. Orisaka, T. Minobe, T. Uchihashi, Y. Sugawara, and S. Morita. *Appl. Surf. Sci.*, 140:243, 1999.
25. Ch. Loppacher, M. Bammerlin, M. Guggisberg, F. Battiston, R. Bennewitz, S. Rast, A. Baratoff, E. Meyer, and H.-J. Güntherodt. *Appl. Surf. Sci.*, 140:287, 1999.
26. K. Fukui, H. Onishi, and Y. Iwasawa. *Phys. Rev. Lett.*, 79:4202, 1997.
27. K. Fukui, H. Onishi, and Y. Iwasawa. *Appl. Surf. Sci.*, 140:259, 1999.
28. B. Gotsmann, C. Schmidt, C. Seidel, and H. Fuchs. *Eur. Phys. J. B*, 4:267, 1998.
29. T. Uchihashi, T. Okada, Y. Sugawara, K. Yokoyama, and S. Morita. *Phys. Rev. B*, 60:8309, 1999.
30. T. Uchihashi, T. Ishida, M. Komiyama, M. Ashino, Y. Sugawara, W. Mizutani, K. Yokoyama, S. Morita, H. Tokumoto, and M. Ishikawa. *Appl. Surf. Sci.*, 157:244, 2000.
31. K. Kobayashi, H. Yamada, T. Horiuchi, and K. Matsushige. *Appl. Surf. Sci.*, 140:281, 1999.
32. K. Kobayashi, H. Yamada, T. Horiuchi, and K. Matsushige. *Jpn. J. Appl. Phys. Part II*, 38:L1550, 1999.
33. T. Fukuma, K. Kobayashi, T. Horiuchi, H. Yamada, and K. Matsushige. *Appl. Phys. A*, 72:S109, 2001.
34. T. Fukuma, K. Kobayashi, K. Noda, K. Ishida, T. Horiuchi, H. Yamada, and K. Matsushige. *Surf. Sci.*, 516:103, 2002.
35. S. Kitamura, K. Suzuki, and M. Iwatsuki. *Appl. Surf. Sci.*, 140:265, 1999.
36. T. Fukuma, K. Kobayashi, H. Yamada, and K. Matsushige. *J. Appl. Phys.*, 95:4742, 2004.
37. M. Nonnenmacher, M. P. O'Boyle, and H. K. Wickramasinghe. *Appl. Phys. Lett.*, 58:2921, 1991.
38. B. Gotsmann, C. Seidel, B. Anczykowski, and H. Fuchs. *Phys. Rev. B*, 60:11051, 1999.
39. R. Lüthi, E. Meyer, M. Bammerlin, A. Baratoff, L. Howald, Ch. Gerber, and H.-J. Güntherodt. *Surf. Rev. Lett.*, 4:1025, 1997.
40. M. Gauthier and M. Tsukada. *Phys. Rev. B*, 60:11716, 1999.
41. N. Sasaki and M. Tsukada. *Jpn. J. Appl. Phys. Part II*, 39:L1334, 2000.
42. C. Loppacher, R. Bennewitz, O. Pfeiffer, M. Guggisberg, M. Bammerlin, S. Schär, V. Barwich, A. Baratoff, and E. Meyer. *Phys. Rev. B*, 62:13674, 2000.
43. R. Bennewitz, A. S. Foster, L. N. Kantrovich, M. Bammerlin, Ch. Loppacher, S. Schär, M. Guggisberg, and E. Meyer. *Phys. Rev. B*, 62:2074, 2000.
44. T. Fukuma, T. Ichii, K. Kobayashi, H. Yamada, and K. Matsushige. *J. Appl. Phys.*, 95:1222, 2004.
45. T. Yoda, T. Ichii, T. Fukuma, K. Kobayashi, H. Yamada, and K. Matsushige. *Jpn. J. Appl. Phys.*, 43:4691, 2004.
46. T. Ichii, H. Kawabata, T. Fukuma, K. Kobayashi, H. Yamada, and K. Matsushige. *Nanotechnology*, 16:S22, 2005.
47. T. Ichii, T. Fukuma, K. Kobayashi, H. Yamada, and K. Matsushige. *Appl. Surf. Sci.*, 210:99, 2003.
48. F. J. Giessibl, H. Bielefeldt, S. Hembacher, and J. Mannhart. *Appl. Surf. Sci.*, 140:352, 1999.
49. F. J. Giessibl. *Appl. Phys. Lett.*, 76:1470, 2000.
50. F. J. Giessibl, S. Hembacher, H. Bielefeldt, and J. Mannhart. *Science*, 289:422, 2000.
51. D. M. Eigler and E. K. Schweizer. *Nature*, 344:524, 1990.
52. Y. Sugimoto, M. Abe, S. Hirayama, N. Oyabu, O. Custance, and S. Morita. *Nat. Mater.*, 4:159, 2005.
53. D. Pohl and R. Möller. *Rev. Sci. Instrum.*, 59:840, 1988.
54. Y. Sugimoto, P. Jelinek, R. Pérez, S. Morita, P. Pou, M. Abe, and Ó. Custance. *Nature*, 446:64, 2007.
55. G. M. McClelland, R. Erlandsson, and S. Chiang. *Review of Progress in Quantitative Non-Destructive Evaluation*, vol. 6B, pp. 1307–1314. Plenum, New York, 1988.
56. G. Meyer and N. M. Amer. *Appl. Phys. Lett.*, 53:1045, 1988.
57. C. Schönenberger and S. F. Alvarado. *Rev. Sci. Instrum.*, 60:3131, 1989.
58. G. Neubauer, S. R. Cohen, G. M. McClelland, D. Horne, and C. M. Mate. *Rev. Sci. Instrum.*, 61:2296, 1990.
59. M. Tortonese, H. Yamada, R. C. Barrett, and C. F. Quate. *The Proceedings of Transducers '91*, pp. 448–451. IEEE, Pennington, NJ, 1991. Publication No. 91 CH2817-5.
60. T. Itoh and T. Suga. *Nanotechnology*, 4:218, 1993.
61. B. W. Hoogenboom, P. L. T. M. Frederix, J. L. Yang, S. Martin, Y. Pellmont, M. Steinacher, S. Zäch, E. Langenbach, and H.-J. Heimbeck. *Appl. Phys. Lett.*, 86:074101, 2005.
62. S. Kawai, D. Kobayashi, S.-I. Kitamura, S. Meguro, and H. Kawakatsu. *Rev. Sci. Instrum.*, 76:083703, 2005.
63. S. Nishida, D. Kobayashi, T. Sakurada, T. Nakazawa, Y. Hoshi, and H. Kawakatsu. *Rev. Sci. Instrum.*, 79:123703, 2008.
64. T. Fukuma, Y. Hirata, T. Ichii, K. Kobayashi, T. Kajita, and H. Yamada. In *Abstracts of the 13th International Conference on STM/STS and Related Techniques*, 2005.
65. C. Möller, M. Allen, V. Elings, A. Engel, and D. J. Müller. *Biophys. J.*, 77:1150, 1999.
66. T. Fukuma, M. J. Higgins, and S. P. Jarvis. *Phys. Rev. Lett.*, 98:106101, 2007.
67. T. Fukuma, A. S. Mostaert, and S. P. Jarvis. *Nanotechnology*, 19:384010, 2008.

68. H. Asakawa, K. Ikegami, M. Setou, N. Watanabe, M. Tsukada, and T. Fukuma. *Biophys. J.*, 101:1270, 2011.
69. T. Fukuma, M. J. Higgins, and S. P. Jarvis. *Biophys. J.*, 92:3603, 2007.
70. J. E. Sader and S. P. Jarvis. *Appl. Phys. Lett.*, 84:1801, 2004.
71. J. N. Israelachvili and G. E. Adams. *Nature*, 262:774, 1976.
72. T. Fukuma, Y. Ueda, S. Yoshioka, and H. Asakawa. *Phys. Rev. Lett.*, 104:016101, 2010.
73. T. Fukuma. *Sci. Technol. Adv. Mater.*, 11:033003, 2010.
74. L. Cheng, P. Fenter, K. L. Nagy, M. L. Schlegel, and N. C. Sturchio. *Phys. Rev. Lett.*, 87:156103, 2001.
75. S.-H. Park and G. Sposito. *Phys. Rev. Lett.*, 89:085501, 2002.
76. S. Hembacher, F. J. Giessibl, and J. Mannhart. *Science*, 305:380, 2004.
77. L. Gross, F. Mohn, P. Liljeroth, J. Repp, F. J. Giessibl, and G. Meyer. *Science*, 324:1428, 2009.
78. M. L. Berkowitz, D. L. Bostick, and S. Pandit. *Chem. Rev.*, 106:1527, 2006.
79. M. Watkins and A. L. Shluger. *Phys. Rev. Lett.*, 105, 2010.
80. Y. Mitani, M. Kubo, K. Muramoto, and T. Fukuma. *Rev. Sci. Instrum.*, 80:083705, 2009.
81. T. Ando, T. Uchihashi, and T. Fukuma. *Progress Surf. Sci.*, 83:337, 2008.
82. T. Fukuma, K. Onishi, N. Kobayashi, A. Matsuki, and H. Asakawa. *Nanotechnology*, 23:135706, 2012.

14

Theory for Picoscale Scanning Tunneling Microscopy

Jouko Nieminen
Tampere University of Technology
and
Northeastern University

14.1 Introduction

The purpose of this chapter is to help the reader to get familiar with useful theoretical tools for modeling scanning tunneling microscopy (STM). Especially, the following set of questions is good to keep in mind:

- What is the relation between the images in tunneling microscopy and the local density of electronic states (LDOS) of the sample?
- What is the relation between the density of states and electron propagation?
- What do these relations reveal about the origin of the tunneling signal observed in an experiment?

14.2 Do We Need a Theory of STM?

Topographic imaging using STM provides the most accurate information available on the local geometrical structure of a solid surface. The resolution may reach subnanometer scale, with a lateral resolution of $\propto$100 pm, and the vertical resolution may be one decade better [1,2]. This means a resolution at the atomic scale. The first atomic scale images were published very soon after the advent of STM [3–5]. In the very early days of STM, it was generalized to scanning tunneling spectroscopy (STS), where the LDOS of the sample is probed by measuring its local differential conductance, $\sigma = dI/dV$, as a function of bias voltage [6].

Topographic imaging of solid samples by STM is generally taken to provide the real geometry of the real object. In spite of the atomic scale resolution, all atoms are not what they seem in STM imaging. Sometimes, on one hand, structural information of the surface seems to be lost; on the other hand, features that originated at the subsurface structures may be seen. Understanding STM experiments at the atomic scale requires understanding the theory of electronic structure and physics of tunneling in the miniature world of nanometer and subnanometer scale structures.

The purpose of this chapter is to discuss the theoretical framework of electron tunneling in STM, starting from the standard LDOS picture, and deepen the theory by introducing Green's function methods. Especially, the concept of tunneling channels is introduced in connection with the latter methods. The theoretical subtleties are alleviated by considering simple and tractable examples.

Although the theoretical methods discussed here are rather standard, the author attempts to take a somewhat exceptional viewpoint to the atomic scale imaging in STM by rationalizing experimental observations in terms of tunneling paths and channels. Instead of considering the resolution due to properties of the tip and its distance to the sample, the concept of Green's function is utilized to analyze the tunneling route of an electron between the sample and the tip. Green's function is a rather involved mathematical tool, but nevertheless it can be used in a very illustrative way.

In the work of Hofer et al. [2], the computational methods have been classified according to their complexity, and four categories are listed: (i) the standard Tersoff–Hamann theory [7], based on the LDOS of the surface at Fermi-energy; (ii) Bardeen approximation, which explicitly includes the tip wave functions [8]; (iii) Landauer–Büttiker method, which provides the basis for looking at STM from the point of view of tunneling channels and paths [9]; (iv) nonequilibrium (Keldysh) Green's function (NEGF) methods [10]. In the same category as (iii) can be included electronic scattering quantum chemistry (ESQC) [11–13], where tunneling is treated as a scattering phenomenon. The Lippmann–Schwinger equation for quantum dynamic scattering [14] is the common factor within this category. In practice, increasing the level of sophistication increases the explanatory power of tunneling calculations. Tersoff–Hamann gives a useful description in terms of the LDOS; in the more involved methods, the transmission channels and many-body effects can be attached.

In order to build a step-by-step introduction to Green's function theory of tunneling, the author has chosen the elegant derivation by Pendry et al. [15] as the principal reference. Very similar derivations are given by Todorov et al. [16] as well as Ness and Fisher [17]. This approach apparently falls into category (ii), but as will be shown, it is possible to extend it to category (iii) or Landauer–Büttiker-type methods. A closer look at these methods leads to formalisms resembling NEGF as discussed in, for example, Ref. [10]. It will also be shown that these theories are compatible with category (i).

Although utilizing Green's functions requires a further step toward theoretical subtlety, they provide a handy machinery to define tunneling paths and channels for rationalizing experimental results, especially in the case of molecular adsorbates. Although the concept of eigen channels is derived from more general transport theories (see, e.g., Refs. [18,19]), the idea of tunneling paths in STM can be attributed to the work by Joachim, Sautet, and many others [12,13,20]. This has been further elaborated by Nieminen and Niemi [21–25].

14.3 Why LDOS and Tunneling Paths?

The reputation of STM is based on its ability to show the real space geometry of the sample and, at best, reach the atomic scale resolution. The birth of STM as an atomic scale probe in topographic imaging can be attributed to studies of 7 × 7 reconstruction of Si (111) surface [4], where the atomic structure of the top layer of the sample was observed by STM. Followed by the early success of this atomic scale imaging, a huge variety of systems with or without adsorbate molecules have been investigated. For clean systems, it has become apparent at very early stages that the atomic scale resolution does not guarantee the visibility of all the surface atoms. This was the case already in the very early study by Binnig et al. [4], but it has been a recurrent feature of STM images.

Depositing adsorbate molecules onto the surface complicates the view further. There are good examples of molecules, where the STM image follows the shape of the molecules, just to mention the well-known case of O_2 on Pt(111) in Ref. [26]. However, adsorbates as simple as CO molecules provide an anomalous image. As is shown by Bartels et al. [27], the molecules on Cu (111) surfaces are seen as dark protrusions, if a clean metal tip is used in imaging. However, if a CO molecule is picked up from the surface to the apex of the tip, the microscope tip is functionalized, and this makes the remaining CO molecules visible. It is interesting that for other adsorbate molecules, such as oxygen, their contrast is not changed by functionalization. This means that while functionalization retains the subnanometer resolutions, it improves the chemical resolution of the microscopy. The same phenomenon has been observed in a more complicated form in the case of clusters of CO molecules [28].

In another category, with somewhat unexpected results, fall the rather recent studies of high-temperature superconductors (HTS), where the superconducting CuO_2 layers are covered by poorly conducting metal oxide layers [29–32]. In those cases, the metal atoms, such as bismuth, appear as bright spots, whereas the surface oxygen atoms are completely invisible in the topographic images. This can be attributed to the LDOS near Fermi-energy at the oxygen sites. However, the tunneling channel picture is the alternative paradigm [25]: the metal atoms at the surface contribute to the eigenstates spatially found mainly within the cuprate layers, which open a vertical tunneling channel between the superconducting layers and the surface. In contrast to this, the orbitals of the oxygen atoms couple strongly to horizontal eigenstates, and there is very weak coupling to the wave functions in cuprate layers. For pristine materials (or even those with oxygen doping), the channel approach seems to have marginal relevance, only. However, if copper atoms of the subsurface cuprate layers are substituted by magnetic or nonmagnetic impurities, the fourfold symmetry of tunneling channels through the insulating layers can be seen.

14.4 Simple Ideas about STM: Electron Tunneling through a Potential Barrier

In the standard textbook models (see, e.g., Ref. [33]), tunneling current through a potential barrier is described in terms of transmission coefficient, $T \propto e^{-\kappa z}$. Hence, the current across the vacuum decreases exponentially, $j \propto |T|^2 \propto e^{-2\kappa z}$, where the decay coefficient κ is proportional to the square root of the effective work function ϕ:

$$\kappa = \frac{\sqrt{2m\phi}}{\hbar}. \tag{14.1}$$

While Bardeen [8] derived the tunneling matrix element, M, from the general expression for current density operator, Tersoff and Hamann explicitly recovered this exponential decay for tunneling current between a microscope tip and a solid metallic surface [7]. This simple result makes the Tersoff–Hamann theory so useful in practical calculations.

In fact, this exponential dependence sets some limits to the resolution of tunneling measurements. Since ϕ is of the order of a few electron volts, κ is of the order of Angstrom^{-1}, that is, the decay length is of the order of 100 pm. This value does not, as itself, tell the lateral or vertical resolution, but it sets the scale for them. The exponential decay of the tunneling current is also recovered in more material-specific calculations, for example, in the best-known and widely applied Tersoff–Hamann method [7]. Since the work function ϕ is, in general, roughly 5 eV, a rule of thumb is that for every increased Angstrom in the tip–sample distance, the current decreases roughly one decade.

The implications of this to the horizontal resolution have been widely discussed (see Refs. [1,2]), and it has been suggested that for microscope tips with spherically symmetric orbitals, atomic scale resolution is hard to obtain. This resolution seems to be saved, if the tip orbitals are "elongated," making the tip effectively sharper than its geometric shape. Chen [34–36] discusses the role of different tip orbitals in the microscopy resolution, and gives derivative rules relating the tip symmetry and the lateral resolution of the probe. In tight-binding (TB) basis utilizing, for example, Slater–Koster rules [37], the derivative rules can equivalently be given in the directionality of the hopping integrals between the tip and surface orbitals [38].

14.5 Tunneling Channels and Tunneling Paths

In addition to matching with the experimental resolution, a good theory answers the questions: "where does the signal come from, and how is the original state of the electron it modified?" An attempt to deal with this question is by decomposing the theoretically calculated signal into tunneling channels or paths.

The idea of tunneling channels has been presented in various forms. A significant contribution for intuitive understanding of tunneling through nanostructures has been provided by TB-based nonequilibrium Green's function theories [18,19]. As the Green's function can be understood as an electron propagator, its presentations can directly be used to visualize the tunneling channels either in terms of real space paths or atomic orbitals. The idea of eigenchannel is based on diagonalization of this product; this brings us back to the matrix element $|T|^2$, that is, transmission probability in simplistic description of tunneling through a potential barrier.

Sautet et al. [12,13] reformulated the tunneling channel approach in the form of a scattering problem in the framework of ESQC. In these studies, a decomposition into *through-vacuum* and *through-adsorbate* paths is explicitly given. In the studies published in Refs. [25,39,40], Green's functions in TB basis are used to obtain spatially confined tunneling channels which, on the other hand, follow the symmetries of ordinary atomic orbitals.

The studies discussed earlier are mainly based on *equilibrium* Green's functions, but more generally tunneling could be understood as a nonequilibrium process. Calculating STM genuinely in an NEGF framework is quite uncommon, due to high computational costs. Nevertheless, there are attempts to incorporate these studies into *ab initio* electronic structure calculations in a localized basis; to mention an example, the GPAW package has been used [41–43].

In the tunneling channel approach, it is profitable to express the electronic structure in a local basis, such as linear combination of atomic orbitals. First, that makes constructing transmission matrix elements more feasible than using a continuum basis. Second, local basis—either in atomic or molecular orbitals—allows conceiving tunneling channels in real space. Hence, a sufficient localization of basis orbitals makes the tunneling channels spatially confined, which makes theoretical analysis of picoscale features possible.

14.6 Theories for STM: From LDOS to NEGF

The standard method to calculate STM topographic maps is Tersoff–Hamann theory of tunneling [7]. It straightforwardly models topographic imaging by assuming the differential conductance at a given bias voltage V_b to be proportional to the LDOS of electrons, ρ, of the sample at the position of the microscope tip:

$$\left.\frac{dI}{dV}\right|_{V_b} \propto \rho(r_t, E_F + eV_b). \tag{14.2}$$

For small bias voltages, it is assumed that it is sufficient to consider LDOS at Fermi-energy, E_F. In their classic article, Tersoff and Hamann calculate the distance-dependent overlap between a spherical tip orbital and a k-dependent surface wave function and apply the result to Bardeen formalism. The result is quite elegant:

$$I \propto V_b R^2 e^{2\kappa R} \rho(r_t, E_F), \tag{14.3}$$

simultaneously bringing in the radius of curvature, R, of the microscope tip. In detail, the method utilizes Bardeen theory to the tunneling matrix elements between the tip orbital (which is assumed to be spherically symmetric) and the surface wave functions. In principle, the tip can be more realistically modeled by calculating the tunneling matrix elements for a tip wave function expanded in spherical harmonics. This is how Chen [34–36] arrives at the derivative rules for the tip symmetry.

Taking a closer look at LDOS opens another track to follow, since it can be expressed in terms of *Green's functions*. In practical usage, the imaginary part of the retarded (causal) Green's function, $I[G(E)]$, gives directly the LDOS [44]:

$$\rho(\mathbf{r},E) = -\frac{1}{\pi}\Im[G^{+}(\mathbf{r},\mathbf{r},E)], \tag{14.4}$$

where

$$G^{\pm}(\mathbf{r},\mathbf{r}',E) = \sum_k \frac{\psi_k(\mathbf{r})\psi_k^{*}(\mathbf{r}')}{E - E_k \pm i\eta}. \tag{14.5}$$

Here, the summation is taken over all the eigenstates of the Hamiltonian of the system, H, and $\psi_k(\mathbf{r})$ are the corresponding eigenfunctions, and E_k the eigenenergies. Formally, we can consider Green's function $G^{\pm}(E) = (E - H \pm i\eta)^{-1}$ as an operator giving the eigenvalues of the Hamiltonian as sharp peaks. Hence, the imaginary part of Green's function gives the electronic spectrum of the sample. In practical calculations, all the operators, such as Hamiltonian and Green's functions are utilized in the matrix form and, in mathematical terms, quantum mechanical calculations turn into matrix algebra.

There is more to a Green's function than meets the eye—the very essence of a Green's function is its action as a *propagator* describing the possible paths taking a particle from an initial position and time a to another position and time b. This opens up a powerful technique to interpret tunneling microscopy and spectroscopy. And if one is not afraid of its theoretical subtleties, an unexpectedly illustrative view to tunneling can be derived.

In order to introduce the concept of tunneling channels, we first have to discuss Green's function methods, and how they are related to the dynamics of tunneling electrons. But that requires understanding of the matrix representation of quantum mechanics, that is, operators as matrices and state functions as vectors. In addition, hopping integrals form the central set of input parameters for calculations in TB basis or in utilizing linear combination of atomic orbitals. Good general references to electronic structure in solids described in linear combinations of atomic orbitals (LCAO) are, for example, Refs. [45] and [46].

14.7 Theoretical Prerequisites

14.7.1 Operators as Matrices and Wave Functions as Vectors

While tunneling calculations beyond the Tersoff–Hamann approach can, in principle, be performed by directly solving scattering equations, it is necessary in practical calculations to use matrix algebra. Generally, a set of local basis functions, $\{\varphi_i\}$, is required. Often, the wave functions are expressed as LCAO, $\psi = \sum_i c_i\varphi_i$, and the set of expansion coefficients can be treated as a set of components of a vector, $\mathbf{c}$. Correspondingly, any operator such as Hamiltonian, H, a perturbation potential, V, or Green's function G can be expressed on this basis as a matrix with elements, such as $H_{ij} = \int \varphi_i^{*} H\varphi_j d\tau$. Then, expressions such as $GV\psi$ should be considered as multiplication of a vector by a product of two matrices and a vector: $\sum_{kj} G_{ik}V_{kj}c_j$.

Regularly, the matrix elements of Hamiltonian or perturbation potentials are given as the input parameters to the calculations; as an output arise Green's functions and possible self-energies. Green's function is then a matrix which possesses the available information about the electronic structure and transport properties of the modeled system. The matrix elements of the Hamiltonian and the potentials may be calculated using real-space integrals, but there are parameterized methods, such as Slater–Koster formulas [37]. They automatically give the correct angular dependence and symmetry properties, but still require a suitable scaling for distance dependence and its amplitude, such as that given in Ref. [45]. While very handy in real calculations, the atomic orbitals (or TB orbitals) tend to have a too short spatial extension [2], which is a drawback in striving for a quantitative accuracy to tip–sample overlap.

14.7.2 Sign and Symmetry of Hopping Integrals

In many spectroscopies, for example, in photoemission, one attempts to apply symmetry-based selection rules, which explain invisible and specially pronounced features. Such selection rules appear rather naturally in nonlocal methods which operate in the reciprocal space of electron momentum. As STM is a local real-space method, definitive selection rules are more difficult to establish. Nevertheless, the symmetry of the wave functions of the microscope tip may be very important in the formation of an STM topographic image or appearing features in STS spectra.

Symmetry properties are inherited from hopping integrals, which formally are defined as matrix elements of the Hamiltonian:

$$V_{\alpha\beta} = \int \varphi_{\alpha}^{*}(\mathbf{r})H\varphi_{\beta}(\mathbf{r})d\tau, \tag{14.6}$$

but, in practice, the matrix elements are parameters fitted to semi-empirical data or *ab initio* calculations. Preferably, the symmetry of the basis orbitals is, nevertheless, retained in fitting, and an example of symmetry preserving parameterization is shown in the Slater–Koster table [37,45], where the angle dependence is rigorously derived from Equation 14.6 assuming a spherically symmetric Hamiltonian in the region, where the overlap of the orbitals is significant. Derivative rules by Chen [34–36] also take into account the symmetry of the tip wave functions, and they are applicable generally to STM calculations, irrespectively to the chosen computational basis. Either of the approaches is essential in understanding the selective role of the tip symmetry.

Since the TB orbitals are bound state solutions of a local Hamiltonian, the Hamiltonian is effectively negative in Equation 14.6, and hence the hopping integral between two s-orbitals is negative. More generally, if the hopping integral is taken between orbitals with the same sign for the overlap region, the matrix element is negative; if they have the opposite sign, the matrix element is positive. Hence, if two p-orbitals form a σ-bond, the overlapping lobes of the orbital have an opposite sign, and the outcome of the hopping integral is positive. In the case of a π-bond, orbital lobes with the same sign overlap, and hence the hopping integral is negative. In addition, the symmetry of orbitals may also make certain hopping integrals zero. In the case of, for example, a diatomic molecule and two p-orbitals of different atoms, the hopping integral is zero between one of the orbitals parallel to the interatomic bond and the other perpendicular.

14.8 Green's Functions, Propagators and All That

Utilizing Green's function might, at the first sight, look like a mathematical subtlety, which would make the tunneling problem rather abstract. However, there are very strong advantages in using Green's functions. If the reader feels intimidated by the assumed mathematical difficulties, the author recommends a nice introduction, *Non-Equilibrium Green's Functions for Dummies: Introduction to One Particle NEGF Equations*, by Paulsson [47]. To imply that introduction to the real world, it is shown in Ref. [43], how NEGF can be applied to STM using the *ab initio* package GPAW. More advanced treatises are, for example, Refs. [44,48], and much of the following discussion about Green's functions, Dyson's equation, and self-energies is based on them.

In simple terms, Green's function, $G(a,b,E)$ is an entity, which quantifies the probability amplitude for transition of an electron from state a (position, atomic orbitals, etc.) to state b. There are two features to be vigilant about: the absolute value and the complex phase of $G(a,b,E)$. Total Green's function includes all the possible paths between the original and final positions, but it may be possible to decompose Green's function into those which are attributed to chosen paths of itinerary. This gives tools of interpretation, since the absolute value measures the *probability amplitude* of the chosen path. Furthermore, since electrons have a wave nature, *the phase* of the matrix element of a chosen path is important due to *interference effects* between different paths.

First, formulating a tunneling problem in terms of Green's functions is a surprisingly illustrative way of describing transport of electrons. The basic interpretation for Green's function gives the probability amplitude for an electron to propagate from point a to point b, more generally between times t_a and t_b.

Second, Green's function techniques provide suitable methods to include "external" degrees of freedom or correlation terms as effective self-energies to the propagator. The standard example of self-energy techniques is seen in nonequilibrium Green's function theories of transport through nanostructures. In that case, the nanostructure is attached to two (or more) conducting electrodes, the details of which are not essential. These electrodes can, in principle, be modeled using simplified self-energy terms. On the other hand, interactions within the nanostructure can be included in Green's function, and hence also inelastic propagation of the electron can be modeled in this framework. (See Meir–Wingreen [10]) Of course, these self-energy techniques are applicable to STM calculations as well.

14.8.1 Basics of Green's Function

It is useful to consider Green's function as a *propagator* (see Figure 14.1) of an electron from the position x_a at time t_a to a position x_b at time t_b. Then, assume $\psi^*(x_a,t_a)$ acts as a deposition (creation) of the electron at a given position and $\psi(x_b,t_b)$ denotes removal (annihilation) of the electron at a later moment. Then the probability amplitude of the electron for traveling between these two positions is

$$P(b,a) = \psi(x_b,t_b)\psi^*(x_a,t_a)\Theta(t_b - t_a). \tag{14.7}$$

The step function, $\Theta(t_b - t_a)$, is there to ensure that the process is causal, that is, the creation of the particle takes place before its annihilation. Figure 14.1 describes this interpretation.

Let us further assume that both the wave functions correspond to an energy eigenstate E_k, that is, we can write $\psi(x,t) = \sum_k \psi_k(x)\exp(-(i/\hbar)E_k t)$. If the propagator is

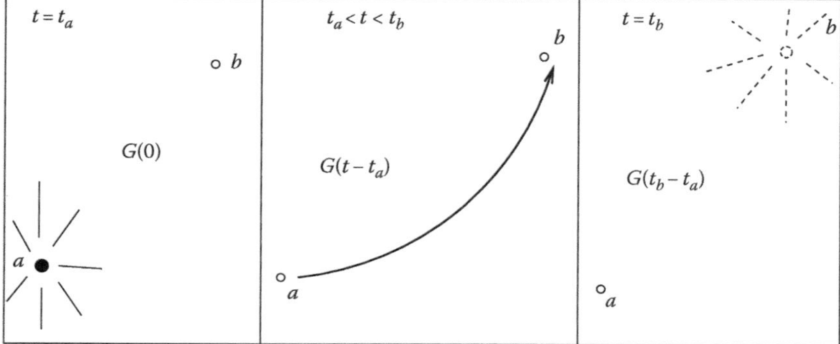

FIGURE 14.1 Green's function as a propagator. At time t_a, an electron appears at position a. It propagates in space and time until at time t_b it disappears with some probability at position b. Green's function contains information about the probability amplitude of this event.

Fourier-transformed with respect to the time difference $t_b - t_a$ of Equation 14.7, it gives us the *retarded* (or *causal*) Green's function of

$$G^+(x_b,x_a,E)=\sum_k \frac{\psi_k(x_b)\psi_k^*(x_a)}{E-E_k+i\eta}, \quad (14.8)$$

where η is a small positive number to ensure the convergence of the Fourier transformation. In the later discussion of self-energies, we will show that the convergence parameter, η, can be attributed to coupling to external degrees of freedom. Note that Equation 14.8 is essentially the same as Equation 14.5.

More generally, Green's function can be expressed in any basis: between states a and b:

$$G^+(a,b,E)=\sum_k \frac{\langle a|k\rangle\langle k|b\rangle}{E-E_k+i\eta}=\sum_k \frac{c_k(a)c_k^*(b)}{E-E_k+i\eta}, \quad (14.9)$$

where k is a quantum number attributed to the eigenvalue E_k of the system. Hence, Green's function is peaked at energies close to the eigenstates of the Hamiltonian. On the other hand, the linear combination coefficients, $c_k(a/b)$, tell how strongly the initial and final states a and b contribute to the eigenstates.

It is noteworthy that despite the mathematical equivalence of Equations 14.7 and 14.8, they show physically two separate things. As the former equation describes the probability amplitude of a path from position + time a to position + time b, the latter emphasizes the eigenstate of the Hamiltonian involved in the route. Hence, each eigenstate presents an easy transfer channel, in terms of energy, between initial and final states.

14.8.2 Lippmann–Schwinger and Dyson Equations

Mathematically, Green's function is a response function which tells what would be the solution of a differential equation in the case of a perturbation with an impulse nature. Let us look at the time-dependent Schrödinger equation

$$\frac{\hbar}{i}\frac{\partial\phi}{\partial t}+H_0\phi=0, \quad (14.10)$$

and let us assume that one can solve $\phi(t)$. Then we switch on a perturbation potential V, and our equation has a new solution ψ:

$$\frac{\partial\psi}{\partial t}+\frac{i}{\hbar}H_0\psi=-\frac{i}{\hbar}V\psi. \quad (14.11)$$

This equation can be solved by the Lippmann–Schwinger equation [14]

$$\psi(t)=\phi(t)+\int G^0(t-t')V\psi(t')dt', \quad (14.12)$$

where $G^0(t-t')$ is the solution of Green's function equation:

$$\frac{\partial}{\partial t}G^0(t-t')+\frac{i}{\hbar}H_0G^0(t-t')=-\frac{i}{\hbar}\delta(t-t'). \quad (14.13)$$

The superscript "0" refers to Green's function of the nonperturbed Hamiltonian H_0.

The interpretation is very simple: if the system is perturbed by an impulse at the moment t', the wave function of the electron is changed by $G^0(t-t')$. If we bring in a perturbation potential V, we can think that Green's function propagates the cumulative effect of the perturbation to the electron wave function.

Green's functions are easier to deal with in energy representation, thence. Let us take the Fourier transformation with respect to the time difference from Equation 14.13:

$$(E-H_0)G^0(E)=I, \quad (14.14)$$

where $E=\hbar\omega$. This should be compared to Equation 14.4, which connects LDOS to Green's function. Hence, a propagator in time gives the spectral function in energy. Also, Lippmann–Schwinger equation, Equation 14.12, can be Fourier-transformed to the energy plane:

$$\psi=\phi+G^0V\psi, \quad (14.15)$$

which allows us to get rid of the convolution integral. Note that we drop off the argument E if the variable in operations is energy instead of time. Even then, the time-dependent wave functions are expressed in the energy plane. Another advantage here is the possibility to include also retarded perturbations which describe external systems, such as vibrational modes or the effect of electrodes to which the structure under scrutiny is coupled.

The great advantage of Equation 14.15 is that it is very straightforward to solve the perturbed wave function ψ from the equation

$$\psi=(1-G^0V)^{-1}\phi.$$

However, normally we wish to distinguish between the original wave function ϕ and the variation due to the perturbation potential. Hence, we insert the solution back in Equation 14.15:

$$\psi=\phi+G^0T\phi, \quad (14.16)$$

where

$$T=V(1-G^0V)^{-1} \quad (14.17)$$

is the transition matrix.

We can, on the other hand, write Green's function for the interacting system: $(E-H_0-V)G=G(E-H_0-V)=I$, which can be rewritten as $(G^0)^{-1}G-VG=I$. This leads to Dyson's equation:

$$G=G^0+G^0VG=G^0+GVG^0=G^0+G^0TG^0, \quad (14.18)$$

where T is the transition matrix defined earlier. Let us multiply Equation 14.17 by $(1-G^0V)$, which leads to $T+TG^0V=V$. Since $TG^0=VG$, we obtain a useful equation

$$T=V+VGV. \quad (14.19)$$

This gives an explicit expression how multiple scattering effectively alters the direct perturbation potential causing the scattering. This is the key equation in tunneling path formulations in the remaining sections.

14.8.3 Self-Energies—What Are They?

There is a small convergence parameter η in Equation 14.8 which can be attributed to the lifetime of an eigenstate. Mathematically, it is responsible for the broadening of the energy levels of a quantum mechanical system. In more physical terms, it can be taken as coupling to an external system, such as lattice vibrations or electrodes coupling a nanosystem to a current circuit. More specifically, it is the dissipation part of this coupling which, again mathematically, can be modeled as a self-energy term in Green's function theory. These interactions can be gathered to external potentials or the so-called self-energy terms Σ, which in general are energy-dependent since they are retarded in time, and complex since they are dissipative.

In terms of time-dependent Schrodinger equation, the general origin of self-energy terms comes from time-dependent retarded perturbation of the quantum system:

$$\frac{\partial \psi}{\partial t} + \frac{i}{\hbar} H_0 \psi = -\frac{i}{\hbar} V \psi - \frac{i}{\hbar} \int_{-\infty}^{t} \Sigma(t-t')\psi(t')dt', \quad (14.20)$$

which is a generalization of Equation 14.11. An example of this is the coupling of electrons to vibrations of lattice or molecule: the electron state affects the vibrations, which affect back the electron state. Hence, there is a causal connection between the electron state at time t' and at a later time t.

Due to the properties of Fourier transformation, it is easier to deal with Equation 14.20 in energy plane: $(E - H_0 - V - \Sigma)G = G(E - H_0 - V - \Sigma) = I$, where Σ is a complex, energy-dependent function. Using the shorthand notation $H = H_0 + V$, one can write the Fourier-transformed counterpart of Equation 14.20:

$$(E - H - \Sigma)\psi = 0. \quad (14.21)$$

In contrast to Equation 14.14, we consider H as the unperturbed Hamiltonian, and now defining $G^0 = (E - H)^{-1}$, and applying Dyson's equation, we obtain

$$G = G^0 + G^0 \Sigma G. \quad (14.22)$$

In order to relate Green's functions to the density of states, we need the following equations:

$$G^R(E - H - \Sigma) = I$$

for the *causal* or *retarded* Green's function, G^R. As an adjunct counterpart to this, one can define the *advanced* Green's function, G^A,

$$(E - H - \Sigma^*)G^A = I.$$

Let us multiply the first equation from the right by G^A and the latter from the left by G^R and subtract the equations from each other. Hence, we obtain a very useful relation:

$$G^A - G^R = G^R(\Sigma - \Sigma^*)G^A = 2iG^R \Gamma G^A, \quad (14.23)$$

where $\Gamma = \Im[\Sigma]$. On the other hand, when changing to a diagonal basis, it is easy to see that

$$\Im[G^R] = -G^R \Gamma G^A, \quad (14.24)$$

which means that the density of states function can be written as

$$\rho = -\frac{1}{\pi}\Im[G^R] = \frac{1}{\pi} G^R \Gamma G^A = \frac{1}{2\pi i}(G^R - G^A). \quad (14.25)$$

This is a very useful relation for coupling the density of states to Green's function formalism used in transport calculations.

14.9 From Scattering to Tunneling: Derivation by Pendry

After these somewhat technical derivations, we finally end up in a method to calculate tunneling currents in Green's function formalism.

Here, we shall review the STM theory by Pendry et al. [15]. The idea is to divide space into two regions: the tip (τ) and sample (σ). In the beginning, the regions are totally isolated from each other, and both of them have a separate set of eigenstates, E_I and corresponding eigenfunctions, ϕ_I, where $I = \tau$ or σ. The eigenstates are nonzero only within their own region.

Suppose that the two regions are brought into contact with each other so that tunneling is possible between the two of them. Formally, this is accompanied by a change in potential for electrons of both the regions, and a nonzero matrix element of the potential $v_{\sigma\tau}$ between orbitals in τ and σ regions appears.

Tunneling from the sample to the tip can be treated as a scattering due to the perturbation potential, where the scattered wave function can be written as a sum of the original wave function and its modification: $\psi = \phi + \delta\phi$. Dividing the solution between the two regions, and comparing this to the Lippmann–Schwinger equation yields a contribution $\delta\phi_{\tau\sigma}$ of a σ orbital to the modified wave functions in the τ region:

$$\delta\phi_{\tau\sigma} = G^0_{\tau\tau} T_{\tau\sigma} \phi_\sigma, \quad (14.26)$$

where G^0_{II} is Green's function of the isolated region I:

$$G^0_{II} = \frac{1}{E - E_I - \Sigma_I}, \quad (14.27)$$

and the T-matrix can be derived:

$$T_{IJ} = v_{IJ}(1 - G^0_{II} v_{IJ} G^0_{JJ} v_{JI})^{-1}. \quad (14.28)$$

Hence, the potential $v_{\tau\sigma}$ is a scattering potential for electrons in the sample region, which takes the electrons to the tip side; $T_{\tau\sigma}$ is that potential modified to take into account multiple scattering; and $G^0_{\tau\tau}$ models propagation of the electrons within the tip region.

The interpretation for $\delta\phi_{\tau\sigma}$ is the number of electrons of the sample region that "leak" into the tip region. This "leakage" is the origin of the current flow between the regions.

To derive the equations for the tunneling currents, one needs to realize that $\delta\phi_{\tau\sigma}$ is a solution to the Fourier-transformed time-dependent Schrödinger equation within region τ:

$$E\delta\phi_{\tau\sigma} = -\frac{\hbar^2}{2m}\nabla^2\delta\phi_{\tau\sigma} + V\delta\phi_{\tau\sigma} + \Sigma_\tau\delta\phi_{\tau\sigma}, \tag{14.29}$$

where a self-energy term Σ_τ is included. Let us multiply Equation 14.29 by $\delta\phi^*_{\tau\sigma}$, take the complex conjugate of the resulting equation, and subtract the latter from the former. Following further Ref. [15], integrate both sides of the equation over whole of the tip region, r_τ, and the result is

$$2i\int_\tau \delta\phi^*_{\tau\sigma}\Gamma_\tau\delta\phi_{\tau\sigma}dr_\tau = \frac{\hbar^2}{2m}\int_\tau\left(\delta\phi^*_{\tau\sigma}\nabla^2\delta\phi_{\tau\sigma} - \delta\phi_{\tau\sigma}\nabla^2\delta\phi^*_{\tau\sigma}\right)dr_\tau, \tag{14.30}$$

where

$$\Gamma_\tau = \frac{1}{2i}(\Sigma_\tau - \Sigma^*_\tau) = \Im[\Sigma(E)]. \tag{14.31}$$

In Ref. [15] Green's theorem is applied to convert the right-hand side to a surface integral, the definition of current density, and combining to Equation 14.26, a simple formula for current between state σ and τ can be obtained:

$$I_\sigma = \frac{2e}{\hbar}\int_\tau \phi^*_\sigma T_{\sigma\tau}G^{0-}_{\tau\tau}\Gamma_\tau G^{0+}_{\tau\tau}T_{\tau\sigma}\phi_\sigma dr_\tau. \tag{14.32}$$

If we consider current from all the eigenstates of the sample region, we can arrange the corresponding orbital in the form:

$$\sum_\sigma \phi_\sigma\delta(E-E_\sigma)\phi^*_\sigma = -\frac{1}{\pi}\sum_\sigma \Im\left(\frac{\phi_\sigma\phi^*_\sigma}{E-E_\sigma+i\eta}\right) = \rho^0_\sigma(E). \tag{14.33}$$

So, if we calculate the current within a certain energy range $E \to E + dE$, we have to take into account all the possible sample states:

$$dI = \sum_\sigma I_\sigma\delta(E-E_\sigma)dE,$$

where $\sum_\sigma \delta(E-E_\sigma)dE$ gathers together all the states of region σ which contribute to the current within the energy range. Coupling the two previous equations, the differential current is written as

$$dI = \frac{2e}{\hbar}\sum_{\sigma\tau} T_{\sigma\tau}G^{0-}_{\tau\tau}\Gamma_\tau G^{0+}_{\tau\tau}T_{\tau\sigma}\rho^0_\sigma(E)dE. \tag{14.34}$$

Note that from here on, we replace the integration over r_τ by summation over τ. This means using a discretized basis of the tip-side eigenfunctions.

Following Equation 14.25, this can be transformed to

$$dI = \frac{4e}{h}\sum_{\sigma\tau} T_{\sigma\tau}G^{0-}_{\tau\tau}\Gamma_\tau G^{0+}_{\tau\tau}T_{\tau\sigma}G^{0+}_{\sigma\sigma}\Gamma_\sigma G^{0-}_{\sigma\sigma}dE. \tag{14.35}$$

From Dyson's equation, $G^{0+}_{\tau\tau}T_{\tau\sigma}G^{0+}_{\sigma\sigma} = G^+_{\tau\sigma}$, and

$$dI = \frac{4e}{h}\sum_{\sigma\tau} G^-_{\sigma\tau}\Gamma_\tau G^+_{\tau\sigma}\Gamma_\sigma dE. \tag{14.36}$$

If we take into account the voltage difference between the tip (or its background) and the sample (or, again, the electrode it has been coupled to), the total current has the following form:

$$I = \frac{4e}{h}\int dE\,\mathrm{Tr}[(f_\sigma(E) - f_\tau(E))G^-_{\sigma\tau}\Gamma_\tau G^+_{\tau\sigma}\Gamma_\sigma], \tag{14.37}$$

which is the equilibrium limit of the Landauer formula derived by Meir and Wingreen [10]. The notation Tr means summation over all the states of regions σ and τ where summation is replaced by integration in the case of continuous variables. This is the equation useful in transport calculations in nanostructures, where semi-infinite electrodes are attached to a scattering region which consists of the tip and the sample.

This is, however, somewhat heavy machinery, and not readily available in *ab initio* packages. Another, and here, more appropriate track to follow is to convert Equation 14.34 into the form

$$I_\sigma = \frac{2\pi e}{\hbar}\sum_\tau T_{\sigma\tau}\rho^0_\tau T_{\tau\sigma}\rho^0_\sigma \tag{14.38}$$

or

$$I = \frac{2\pi e}{\hbar}\int dE\,\mathrm{Tr}[(f_\sigma(E) - f_\tau(E))T_{\sigma\tau}\rho^0_\tau T_{\tau\sigma}\rho^0_\sigma]. \tag{14.39}$$

In transport calculations, the arguments of the occupation function, $f(E)$, and the density function, $\rho(E)$, would be written as $f_i(E) = f(E-\mu_i)$ and $\rho^0_i = \rho^0(E-\mu_i)$, where the chemical potential due to the voltage difference is $\mu_i = \mu^0_i - eV_b$. Hence, the current shall be written as

$$I = \frac{2\pi e}{\hbar}\int dE\,\mathrm{Tr}[(f_\sigma(E-E_F+eV_b) - f_\tau(E-E_F))T_{\sigma\tau}\rho^0_\tau(E)T_{\tau\sigma}\rho^0_\sigma(E)], \tag{14.40}$$

if we assume that the sample is at the voltage V_b.

Hence, this means that at positive voltages, STM probes sample states above Fermi-energy E_F, that is, the unoccupied states, and with the reversed polarity, the occupied states are probed.

It is notable that the differential conductance dI/dV is rather simple to approximate, if we take the occupation function as a step function with respect to E_F. We should end up in the form

$$\frac{dI}{dV} = \frac{2\pi e}{\hbar}\mathrm{Tr}[T_{\sigma\tau}\rho^0_\tau(E_F)T_{\tau\sigma}\rho^0_\sigma(E_F + eV_b)]. \tag{14.41}$$

On the other hand, if the coupling between the tip and the sample is weak, we recover the Tersoff–Hamann limit:

$$\frac{dI}{dV} \propto |V_{\sigma\tau}|^2 \rho^0_\sigma(E_F + eV_b). \tag{14.42}$$

Hence, the derivation by Pendry can be, on one hand, generalized toward Landauer–Büttiker-type equations or, on the other hand, simplified toward the Tersoff–Hamann equation.

14.9.1 Transition Matrix and Tunneling Channels

The tunneling channel approach by Sautet et al. can be taken a step further by noting that tunneling current is proportional to the transition matrix, T. More explicitly, in the Todorov–Pendry formalism, the tunneling current (at a chosen voltage) is proportional to

$$I(E) \propto \rho^0_{\tau\tau'}(E - eV)T_{\tau'\sigma}\rho^0_{\sigma\sigma'}(E)T^\dagger_{\sigma'\tau}. \tag{14.43}$$

Hence, while the current is a convolution of the density of states of the tip and the sample, its important determinant is the product $T_{\tau'\sigma}T^\dagger_{\sigma'\tau}$.

On the other hand, using Dyson's equation, this can be decomposed into separate terms which can be attributed to tunneling channels. Let us assume here that there is a direct hopping, $V_{\sigma\tau}$, between the tip and a metal substrate, which simultaneously acts as an electrode. In addition, an adsorbate molecule is deposited onto the surface, and it causes multiple scattering of tunneling electrons. Hence, the direct hopping term must be replaced by a transition matrix term $T_{\sigma\tau}$.

Remembering that the transmission matrix between the substrate and the tip can be written in the form $T = V + VGV$ (Equation 14.19), let us construct a simplistic model to study the phase difference between the vacuum channel, $V_{\tau\sigma}$, and a through-molecule channel $V_{\tau\mu}G_{\mu\nu}V_{\nu\sigma}$. The aim is to get an idea of in which conditions the channels interfere constructively and in which conditions they interfere destructively. It is obvious that in the case of resonant tunneling, opening of an easy channel through an adsorbate dominates the transmission so strongly that phase difference between through-molecule and through-vacuum channels is not significant in the vicinity of the molecule. On the other hand, the amplitudes of the channels depend on the distance between the tip and the orbitals related to the channels, and hence different channels may compete at a larger distance from the adsorbate.

The innocent-looking term, VGV, seems mainly to bring an energy (bias voltage) dependence to the tunneling probability, since V itself is energy-independent. In addition to the coupling interaction terms V, Green's function matters, since it has very high value close to eigenstates, which may be modified by the substrate below. Hence, tunneling through the molecule may be dominant compared to the direct tunneling, if the bias voltage coincides with the resonance levels. But this molecule term may be significant also in nonresonant case, since being a complex quantity, Green's function brings in a phase difference between direct and molecule paths.

14.9.2 Filtering by a Thin Insulating Layer

The idea of transition matrices as a determinant of tunneling channels can be modified to quite another purpose. One can consider a sample where everything interesting is buried below thin insulating films. The technique of ultrathin insulating films on metal substrates has been applied to make molecular orbitals of an adsorbate molecule visible [49,50]. In the case of pentacene molecules on NaCl-covered Cu substrates, the idea is to decouple the molecular orbitals from the substrate below, which enhances the through-molecule path. However, the interesting issues here are the molecular orbitals and their modifications, as additional atoms or groups are attached to them.

As will be discussed in Section 14.10, in the case of STM experiments on high-temperature superconducting cuprate materials, the superconducting cuprate layers are hidden below one or two poorly conducting metal oxide layers [29]. Naturally, the experimentalists are interested in the electronic and geometric order of the superconducting layers than in the rest of the material.

Let us denote an arbitrary orbital of the sample with α, and the surface layers with s; symbols f and c denote orbitals of filter layers and cuprate layer, respectively, overlapping across the interface. For superconducting materials, in addition to the regular Green's function, G, also another Green's function, F, indicating creation and annihilation of superconducting Cooper pairs is necessary [51] (detailed discussion of the filtering effect in Ref. [25]):

$$G^+_{s\alpha} = G^{0+}_{sf}V_{fc}G^+_{c\alpha} \quad \text{and} \quad F^+_{s\alpha} = G^{0+}_{sf}V_{fc}F^+_{c\alpha}.$$

Hence, the terms in Equation 14.19 can be written as

$$T_{t\alpha} = -\frac{2e^2}{\hbar}\sum_{t'cc'}\rho_{tt'}(E_F)M_{t'c}(G^+_{c\alpha}\Gamma_\alpha G^-_{\alpha c'} + F^+_{c\alpha}\Gamma_\alpha F^-_{\alpha c'})M^\dagger_{c't}, \tag{14.44}$$

where

$$M_{tc} = V_{ts}G^{0+}_{sf}V_{fc}, \tag{14.45}$$

which gives the filtering amplitude between the cuprate layer and the tip, and constitutes a multiband generalization of filtering function of Ref. [52]. Similarly, the matrix element of the density of states operator $\rho_{cc'}$ within the cuprate plane can be recovered in terms of the spectral function:

$$\frac{dI}{dV} = \frac{2\pi e^2}{\hbar}\sum_{tt'cc'}\rho_{tt'}(E_F)M_{t'c}\rho_{cc'}(E_F + eV)M^\dagger_{c't}. \tag{14.46}$$

Hence, the tunneling signal can be written in terms of the DOS of the cuprate layer $\rho_{cc'}(E)$ and filtering matrix elements, $M^\dagger_{c't}M_{t'c}$.

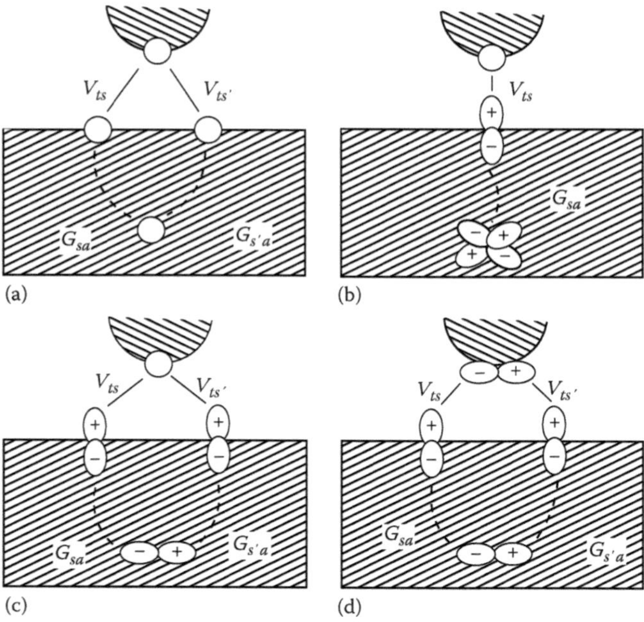

FIGURE 14.2 Filtering effects as a kind of selection rules in imaging an impurity below surface layers. (a) The signal from a buried impurity may reach the tip through different channels. In the matrix elements of the two channels, there is a constructive interference, and hence a strong signal. (b) Due to symmetry reasons, some essential orbitals of the impurity may not share the same eigenstates as the surface orbitals are overlapping strongly with the tip. This makes the orbital invisible. (c) Due to the symmetry of an orbital of the impurity, there may be tunneling channels in the opposite phase, which makes the impurity invisible. (d) However, the symmetry of the tip orbitals may reverse the phase difference, and an invisible orbital becomes visible. (From Ziman, J.M., *The Elements of Advanced Quantum Theory*, Cambridge University Press, Cambridge, U.K., 1988.)

As will be shown in Section 14.10.2, this filtering term may be very selective in terms of the symmetries of the tip orbital, on one hand, and the cuprate orbitals, on the other. Figure 14.2 depicts how the effective selection rules arise from filtering effects.

14.9.3 A Tutorial Example: Transmission through a Toy Molecule on a Simplified Surface

In this section, we demonstrate how a transmission through a toy molecule can be calculated, and how the through-vacuum and through-molecule channels may interfere with each other.

The procedure is as follows:

- Construct a Green's function, G^0, for a noninteracting two-atom molecule from eigenstates of independent atoms.
- Use a model Green's function derived from the density of states to the substrate.
- Treat the molecule and the substrate as separate systems; the substrate can be incorporated into the interacting Green's function as a self-energy term. Construct the interacting self-energy.
- Construct the Hamiltonian which connects the adsorbate molecule to the substrate (in practical calculations, for example, using Slater–Koster hopping integrals).
- Manipulate the transmission matrix $T = V + VGV$ and get the through-vacuum and through-molecule terms for tunneling.
- Check how the properties of hopping integrals between atomic orbitals affect the transmission.
- Give a physical interpretation to the mathematical results.

Let us consider a very naive picture of a molecule whose molecular orbitals can be constructed using two atomic orbitals ϕ_1 and ϕ_2, with eigenenergies E_1 and E_2. Let us assume that the two orbitals are coupled by an effective interaction potential v_{12}. This is analogous to construction of a simple model of hydrogen molecule in elementary textbooks of quantum mechanics.

For the purpose of considering tunneling channels, we are not that much interested in the eigenstates of the constructed molecules, but rather in Green's function matrix elements between different orbitals of the molecule.

Green's function for the atomic orbitals, g, is obviously

$$g_{ii} = \frac{1}{E - E_i + i\eta}. \tag{14.47}$$

According to Dyson's equation, we can construct Green's function G^0 for a noninteracting molecule:

$$G_{11}^0 = g_{11} + g_{11}v_{12}G_{21}^0$$
$$G_{21}^0 = g_{22}v_{21}G_{11}^0.$$

It is straightforward to solve this pair of equations:

$$G_{11}^0 = \frac{g_{11}}{1 - g_{11}v_{12}g_{22}v_{21}}$$
$$G_{21}^0 = \frac{g_{22}v_{21}g_{11}}{1 - g_{11}v_{12}g_{22}v_{21}}.$$

Inserting Equation 14.47, we obtain

$$G_{12}^0 = \frac{1}{(E - \bar{E} + i\eta)^2 - (\Delta^2 + |v_{12}|^2)}, \tag{14.48}$$

where

$\bar{E} = (1/2)(E_1 + E_2)$
$\Delta = (1/2)(E_1 - E_2)$

The poles of Green's function give two solutions: bonding level $E_b = \bar{E} - \sqrt{\Delta^2 + |v_{12}|^2}$ and antibonding level $E_b = \bar{E} + \sqrt{\Delta^2 + |v_{12}|^2}$.

It is easy to see that the real part of the matrix element is negative for $E_b < E < E_a$, that is, between the two molecular eigenstates, and positive otherwise. The imaginary part is negative if $E < \bar{E}$, but otherwise it is positive. At energies far from the molecular eigenstates, the imaginary part is very small, and in such cases we can assume almost completely real Green's function.

Next we apply Dyson's equation again to obtain Green's function G for the coupled system. When discussing tunneling channel, we have to compare the through-vacuum hopping V_{ts} to the through-molecule term $V_{t1}G_{12}V_{2s}$. The final phase of this channel depends not only on the phase of Green's function but also on the hopping integrals between the molecular orbitals 1 and 2 and the tip and the sample, $V_{t1}V_{2s}$.

It is easy to show that as coupled to a substrate,

$$G_{12} = \frac{v_{12}}{(E - E_1)(E - E_2 - \Sigma_{22}) - |v_{12}|^2},$$

and

$$G_{2s} = \frac{1}{V_{s2}} \frac{(E - E_1)\Sigma_{22}}{(E - E_1)(E - E_2 - \Sigma_{22}) - |v_{12}|^2},$$

where $\Sigma_{22} = V_{2s}G_{ss}^0(E)V_{s2}$.

Hence, the model for a through-molecule channel can be written as

$$T_{ts}^{mol} = V_{t1}G_{12}V_{2s} = \frac{V_{t1}v_{12}V_{2s}}{(E - E_1)(E - E_2 - \Sigma_{22}) - |v_{12}|^2},$$

and the scattering part of the through-vacuum channel is

$$T_{ts}^{s2} = V_{ts}G_{s2}V_{2s} = \frac{V_{ts}(E - E_1)\Sigma_{22}}{(E - E_1)(E - E_2 - \Sigma_{22}) - |v_{12}|^2}.$$

Coupling the through-vacuum tunneling with and without scattering, one obtains the total through-vacuum term:

$$T_{ts}^{vac} = V_{ts} + V_{ts}G_{s2}V_{2s} = \frac{(E - E_1)(E - E_2) - |v_{12}|^2}{(E - E_1)(E - E_2 - \Sigma_{22}) - |v_{12}|^2} V_{ts}.$$

Hence, this elementary case allows us to compare the phase of the terms $V_{t1}v_{12}V_{2s}$ and $V_{ts}[(E - E_1)(E - E_2) - |v_{12}|^2]$ in order to recognize the interference effects between the two channels. Note that the term in square brackets is negative if the energy is between the bonding and antibonding states, and otherwise positive.

It is important to distinguish between resonant and nonresonant tunneling. In the case of resonant tunneling, the denominator of both the previous transmission terms goes close to zero, and hence these terms dominate over the direct vacuum tunneling. In that case, the phase differences of the tunneling channels are not the important factors, but rather the amplitude of each term.

On the other hand, in the case of nonresonant tunneling (or tunneling through the gap), Green's function tends to have generally a low value. Especially, the imaginary part of T_{ts}^a is not significant, since the density of states is low at these energies [21–23]. Thence, it is the sign of the real part of each T_{ts}^a which determines the phase of the path. Hence, if V_{ts} and the real part of T_{ts}^{12} have the opposite signs, they interfere with each other destructively. Otherwise, the interference is constructive. The latter case may lead to a dark image of the molecule instead of a bright protrusion. In Figure 14.3, we apply the simple arguments given in this section. As it is seen, there are two cases with essentially the same molecular eigenstates, but the phase of the interatomic matrix element of Green's function within the molecule determines whether electrons can tunnel through the adsorbate system.

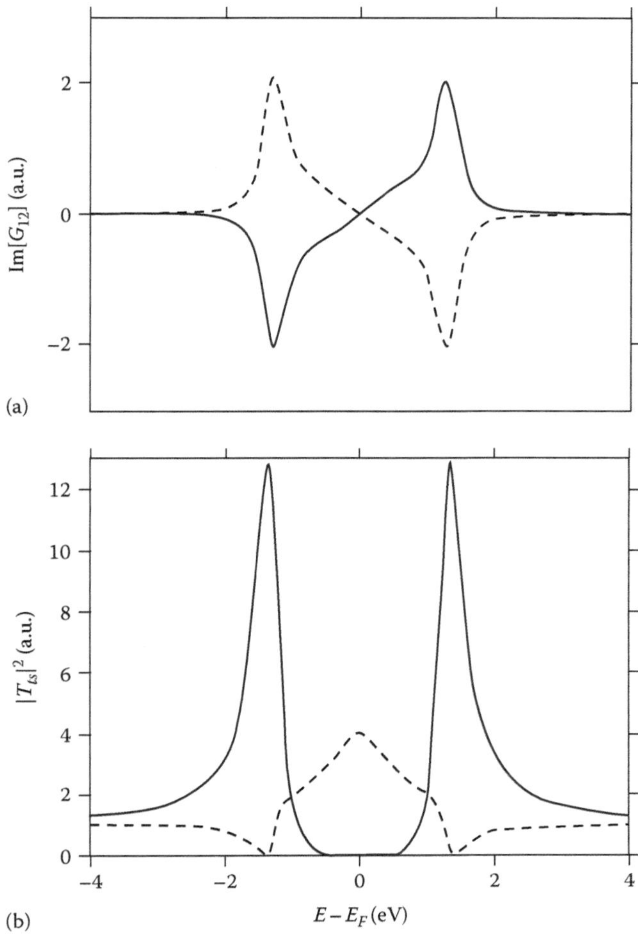

FIGURE 14.3 The toy molecule on a substrate with $v_{12} < 0$ (solid line) and $v_{12} > 0$ (dashed line). (a) The imaginary parts of the off-diagonal Green's function G_{12}, and (b) the total transmission coefficient for both cases of the intermolecular hopping integral. Note that the eigenstates are the same for both the hopping integrals, but due to the phase change, the transmission near E_F is zero for $v_{12} < 0$ while it is significant for $v_{12} > 0$.

In the following section, we give two case studies where we explain subnanometer observations in terms of tunneling channels.

14.10 Two Examples of Tunneling Channel Analysis

In this section, we consider two case studies, where the experimental topographic maps reveal atomic scale structures, but the interpretation of the results is not straightforward. In the first case, we consider CO molecules on a metal surface—in this case the Cu substrate—where interesting contrast changes are seen for different adsorbate configurations and varied condition of the microscope tip. In the second case, high-temperature superconducting Bismuth Strontium Copper Oxide material

$Bi_2Sr_2CaCu_2O_8$ (Bi2212) is studied. This compound is an example of a cuprate material with a layered structure, where most of the interesting issues are hidden below poorly conducting oxide layers. CO on Cu is a very good example of how Green's function approach is applied to study *tunneling channels*, while the same theoretical machinery reveals *filtering effects* of oxide layers when applied to Bi2212.

14.10.1 Tunneling through Adsorbate Molecule: CO on Cu (111)

Bartels et al. [27] studied manipulation of CO molecules on Cu (111) surfaces with a sharp metal tip. As the surface was intact, wide dark patches and narrow dark depressions, a couple of Angstroms wide, were observed in the STM topographic map. A further study revealed that the dark depressions were images of CO molecules, while the wider patches obviously were oxygen atoms. It was possible to pick up a CO molecule and anchor it to the apex of the tip and then rescan the surface. In the rescanned image, one of the dark depressions was missing, which indicated that the CO molecule of that site was attached to the tip. Furthermore, the contrast of all the rest of the dark depressions were inverted to bright protrusions, while the originally dark wide adsorbate images remained intact, as is seen in Figure 14.4a.

In the studies by Niemi et al. [21–23,40], the appearance of CO molecules on metal surfaces in different configurations is explained in terms of tunneling channels. Understanding the contrast variations starts from considering the nature of the hopping between the tip and sample orbitals. Let us first assume the tip is in a position, where it has a significant overlap only

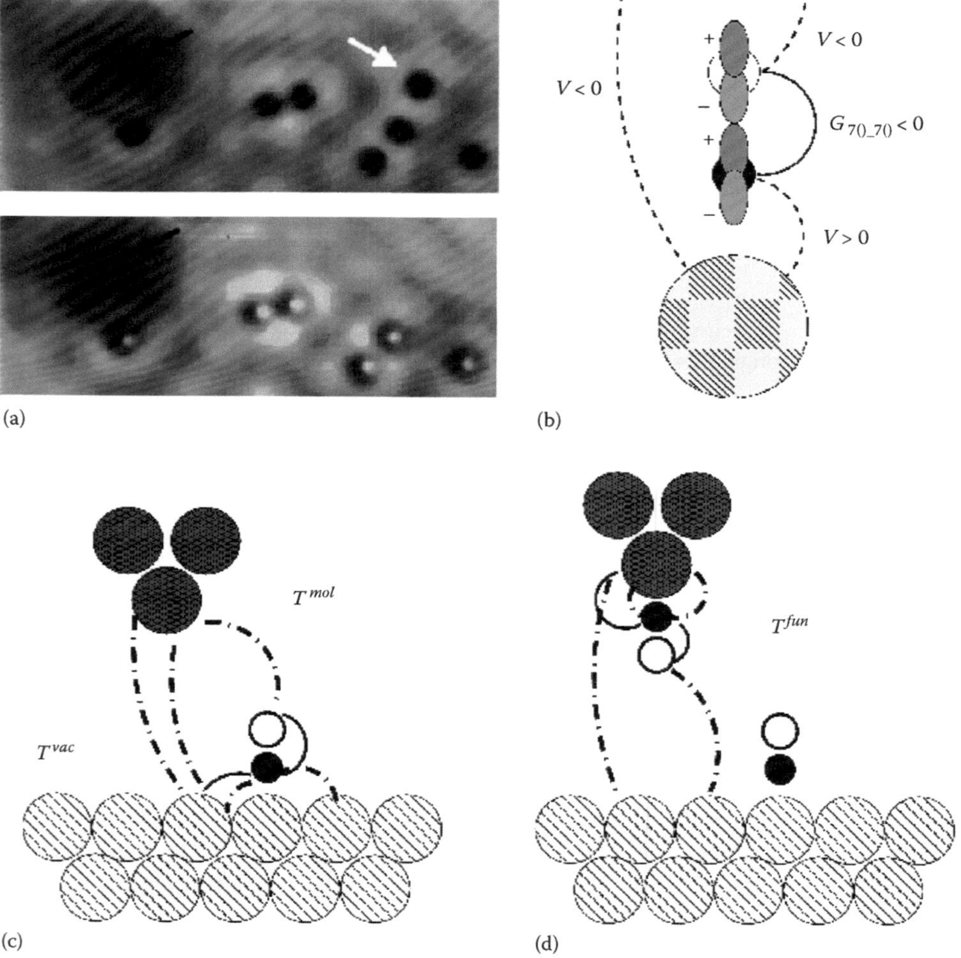

FIGURE 14.4 Tunneling paths through adsorbed CO and the functionalized tip. (a) Above: The experimental image of CO molecules (dark patches indicated by a white arrow) and O atoms (indicated by a black arrow) on Cu(111) imaged with a clean tip. Below: The same arrangement imaged with a functionalized tip having captured the CO molecule pointed by the white arrow. (b) Depicition of how the sign of through-vacuum and through-molecule channels are formed in the case of the σ-orbital of CO. (c) The through-vacuum, T^{vac}, and through-molecule, T^{mol}, channels. The dash-dotted lines indicate hopping integrals, and the solid lines denote matrix elements of Green's function. (d) The channel for the functionalizing molecule, T^{fun}, as in the previous subfigure. (a: Reprinted with permission from Bartels, L., Meyer, G., and Rieder, K.-H., *Appl. Phys. Lett.*, 71, 213, Copyright 1997 American Institute of Physics.)

with the σ-type molecular orbitals, which consist of p-type atomic orbitals of C and O atoms. If the tip is right at the position of a molecule, the hopping integral is zero with π-type molecular orbitals, since the latter is odd with respect to rotations of π rad. This construction is depicted in Figure 14.4b.

In Figure 14.4c and d is described the idea of tunneling paths through vacuum, molecule, and the functionalizing molecule. When considering the through-vacuum channel we assume that both the tip and surface orbitals are of the s-type. In that case, the direct hopping between the orbitals is $V_{ts} < 0$. For an upright molecule, the through-molecule tunneling takes place between the bonding 5σ and antibonding 6σ states. Let us also simplify the picture so much that we can assume that both the σ-orbitals arise from the overlap of two p-type atomic orbitals. Since the mutual hopping integral of two p-orbitals is positive, $v_{12} > 0$, one obtains a factor $[(E - E_1)(E - E_2) - |v_{12}|^2] < 0$, and hence the total through-vacuum channel with multiple scattering yields $V_{ts}[(E - E_1)(E - E_2) - |v_{12}|^2] > 0$.

The sign for the through-molecule channel is constructed from $V_{t1}v_{12}V_{2s}$. Since the tip overlaps with the positive lobe of the p-orbital of the upper atom and the sample with the negative lobe of the lower atom, we find that $V_{t1}V_{2s} < 0$, and hence $V_{t1}v_{12}V_{2s} < 0$. This means that T^{vac} and T^{mol} have the opposite signs. The resulting *destructive interference* should lead to a decreased intensity right above the molecule.

These simple arguments apply to the functionalized tip as well. The phase for the channel through the functionalizing molecule, T^{fun}, and a surface molecule should arise similarly, leading to a *constructive interference*, but off the molecule, the tunneling through vacuum and the functionalizing molecule weaken each other.

For tilted molecules, such as in Ref. [28], where dimers and clusters of CO molecules are observed, also π-type molecular orbitals can overlap with the tip orbital, especially when the tip is located outside the molecular dimer. Since the molecular orbital is constructed from horizontal p-orbitals, one should obtain $V_{t1}v_{12}V_{2s} > 0$, and hence the outer fringes of the molecular dimers are seen as bright protrusions, while there is a dark contrast in the middle of the molecule. This has also been shown in Refs. [21–23].

14.10.2 Tunneling through a Filtering Layer: Subsurface Impurities in $Bi_2Sr_2CaCu_2O_8$

The next example deals with a situation where the interesting part of the sample is not directly exposed to the microscope tip, but is covered by a poorly conducting oxide layer.

Bi2212 is the most common of the HTS materials studied by STM [29]. The topographic STM images are not as themselves very revealing, since the samples are invariably cleaved BiO layer on the top, while the superconducting CuO_2 layers are hidden at the third layer and deeper. In picometer scale, the topographic STM image only shows the Bi atoms of the surface layer, and long wavelength reconstructions are observed too. Especially, the *d-wave symmetry* of the superconducting pairing observed by, for example, photoemission studies, does not express itself in topographic maps.

Interesting information about the electronic structure of the superconducting layers is obtained by taking differential conductance (dI/dV) maps as the first order approximation of the LDOS of the sample (see, e.g., Ref. [30]). With a varying hole doping by electronegative interstitial oxygen atoms, a rich phase diagram arises. The dI/dV spectra and maps can be used to study not only the superconducting phase, but also the mysterious pseudogap phase at low oxygen doping.

In the superconducting state, the dI/dV spectrum shows a gap within a few tens of meVs around the Fermi-energy, reflecting the formation of Cooper pairs. dI/dV maps do not merely show the spatial evolution of the superconducting gap, but they also expose effects of irregularities hidden below the surface layers. By means of these maps, driving mechanisms of different phases have been pursued.

Filtering effects can be demonstrated by scrutinizing fingerprints of substitutional impurities within the CuO_2 layer. From the point of view of the possible mechanism of superconductivity, embedding magnetic (Ni, [31,53]) and nonmagnetic (Zn, [32]) atoms to substitute Cu atoms might shed light on superconducting of pairing.

A curious atomic scale observation is the appearance of a Ni atom as a fourfold symmetric image in topographic dI/dV maps, as shown in Figure 14.5a. These "clover-leaf" patterns appear at

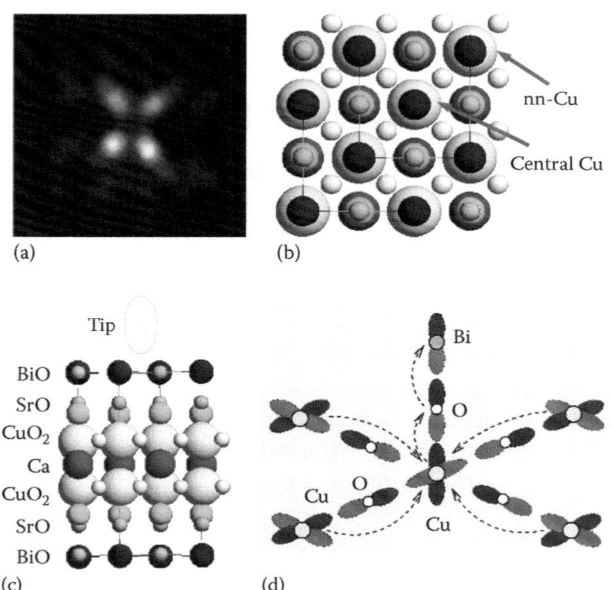

FIGURE 14.5 Tunneling paths in Bi2212. (a) An experimental STS image of a substitutional Ni impurity at a Cu site. (b) A top view of a computation slab for Bi2212. Note that the position of the two bright spots in (a) is the central Cu corresponding to the site of the substitutional impurity (dark center in (a)) and its nearest neighbors (nn-Cu) coincide with the bright wings of (a). (c) The side view of the sample shows that the cuprate layers are below two metal oxide layers. (d) The fourfold symmetry of tunneling channels from the CuO_2 layer to the surface is depicted. (a: Reprinted by permission from Macmillan Publishers Ltd. *Nature*, Hudson, E.W., Lang, K.M., Madhavan, V., Pan, S.H., Eisaki, H., Uchida, S., and Davis, J.C., 411, 920, 2001, Copyright 2001.)

certain bias voltages, and in the case shown here, the elongated "clover-leaf" seems to cover the first and the third neighbors of the central atom. It is tempting to think that *d*-wave symmetric atomic orbitals are seen, but again it is necessary to confirm this hypothesis by considering tunneling channels and filtering effects, especially as the origin of the signal is in the superconducting *third layer* of the sample.

Symmetry arguments compatible with the idea of tunneling channels are utilized in the study by Martin et al. [52,54]. The analysis is based on the known electronic structure, and an explicit form of a *d*-wave symmetric filtering function, *M*, corresponding to Equation 14.45 was constructed accordingly. These symmetry arguments give qualitatively correct *dI*/*dV* maps.

An accurate simulation of tunneling channels still awaits to be done, but the finding is in accord with the findings in Ref. [25], where channels in pristine Bi2212 have been analyzed. As is seen in Figure 14.5b through d, the signal reaches the surface through the p_z-orbitals of the Bi and O atoms of the oxide layers. Due to symmetry reasons, the dominant $d_{x^2-y^2}$-orbital of the Cu atom below does not couple with this channel. Hence, a significant part of the tunneling signal is mediated from the four neighboring Cu atoms through the central rotationally symmetric d_{z^2} orbital, and the p_z-orbitals of the O and Bi atoms on the top of the central Cu atom. Therefore, if one of the Cu atoms is substituted, the corresponding signal is probably seen at the neighboring sites, and not the impurity site itself.

14.11 Conclusions and Discussion

STM is able to reveal atomic scale structures of a solid sample, but it is not quite faithful to the "hard sphere" geometry. A further understanding of topographic imaging is obtained by electronic structure calculations. Calculating STM image from the LDOS and convoluting that with the tip wave function will give a quite satisfactory description of the topographic image in terms of the local electronic structure. However, from the experimental point of view, STM can be extended to spectroscopic applications as well, which provide information on both the geometric and electronic order. The most straightforward theoretical approach is to, for example, equate the local differential conductance with the LDOS.

Nevertheless, it is possible to obtain further understanding of the tunneling process and the origin of different features in topographic maps, tunneling spectra, and conductance maps by using a theoretical tool slightly improved from the standard methods. The aim of this chapter is to show how Green's functions not only provide the LDOS, but also give a deeper insight into the propagation of electrons in solid samples and adsorbate molecules. Furthermore, the idea of Green's functions can be understood without immersing oneself in too bad subtleties of theoretical physics.

The two examples show that it is profitable to go beyond the LDOS picture in theoretical studies. The experimental observations make more sense when rationalized with concepts such as tunneling channels and filtering effects. This, on the other hand, requires a certain level of mastering Green's function theory describing the local electronic structure.

Acknowledgments

The author wishes to express his warmest gratitude to Dr. Sami Paavilainen for valuable comments. Prof. Eric Hudson and Prof. Ludwig Bartels are gratefully acknowledged for their kind permission to reuse figures from Refs. [31] and [27].

References

1. R. Wiesendanger, *Scanning Probe Microscopy and Spectroscopy, Methods and Applications*, Cambridge University Press, Cambridge, U.K., 1998.
2. W.A. Hofer, A.S. Foster, and A.L. Shluger, *Rev. Mod. Phys.* **75**, 1287 (2003).
3. G. Binnig, H. Rohrer, Ch. Gerber, and E. Weibel, *Phys. Rev. Lett.* **49**, 57 (1983).
4. G. Binnig, H. Rohrer, Ch. Gerber, and E. Weibel, *Phys. Rev. Lett.* **50**, 120 (1983).
5. G. Binnig and H. Rohrer, *Rev. Mod. Phys.* **71**, S324 (1999).
6. J.A. Stroscio, R.M. Feenstra, and A.P. Fein, *Phys. Rev. Lett.* **57**, 2579 (1986).
7. J. Tersoff and D.R. Hamann, *Phys. Rev. B* **31**, 805 (1985).
8. J. Bardeen, *Phys. Rev. Lett.* **6**, 57(1961).
9. M. Büttiker, Y. Imry, R. Landauer, and S. Pinhas, *Phys. Rev. B* **31**, 6207 (1985).
10. Y. Meir and N.S. Wingreen, *Phys. Rev. Lett.* **68**, 2512 (1992).
11. P. Sautet and C. Joachim, *Phys. Rev. B* **38**, 12238–12247 (1988).
12. P. Sautet, *Surf. Sci.* **374**, 374 (1997).
13. J. Cerda, M. van Hove, P. Sautet, and M. Salmeron, *Phys. Rev. B* **56**, 15885 (1997).
14. B.A. Lippman and J. Schwinger, *Phys. Rev.* **79**, 469480 (1950).
15. J.B. Pendry, A.B. Prêtre, and B.C.H. Krutzen, *J. Phys.: Condens. Matter* **3**, 4313 (1991).
16. T.N. Todorov, G.A.D. Briggs, and A.P. Sutton, *J. Phys.: Condens. Matter* **5**, 2389 (1993).
17. H. Ness and A.J. Fisher, *Phys. Rev. B* **56**, 12469 (1997).
18. M. Brandbyge, N. Kobayashi, and M. Tsukada, *Phys. Rev. B* **60**, 17064–17070 (1999).
19. M. Paulsson and M. Brandbyge, *Phys. Rev. B* **76**, 115–117 (2007).
20. C. Joachim, J.K. Gimzewski, and A. Aviram, *Nature* **408**, 541 (2000).
21. E. Niemi and J. Nieminen, *Chem. Phys. Lett.* **397**, 200 (2004).
22. J. Nieminen, E. Niemi, K.-H. Rieder, *Surf. Sci.* **552**, L47(2004).
23. E. Niemi and J. Nieminen, *Chem. Phys. Lett.* 397 (2004).
24. A. Korventausta, S. Paavilainen, E. Niemi, and J. Nieminen, *Surf. Sci.* **603**, 437 (2009).
25. J. Nieminen, I. Suominen, R.S. Markiewicz, H. Lin, and A. Bansil, *Phys. Rev. B* **80**, 134509 (2009).
26. B.C. Stipe, M.A. Rezaei, W. Ho, S. Gao, M. Persson, and B.I. Lundqvist, *Phys. Rev. Lett.* **78**, 4410 (1997).

27. L. Bartels, G. Meyer, and K.-H. Rieder, *Appl. Phys. Lett.* **71**, 213 (1997).
28. A.J. Heinrich, C.P. Lutz, J.A. Gupta, and D.M. Eigler, *Science* **298**, 1381 (2002).
29. O. Fischer, M. Kugler, I. Maggio-Aprile, Chr. Berthod, and Chr. Renner, *Rev. Mod. Phys.* **79**, 353 (2007).
30. K. McElroy, J. Lee, J.A. Slezak, D.-H. Lee, H. Eisaki, S. Uchida, and J.C. Davis, *Science* **309**, 1048 (2005).
31. E.W. Hudson, K.M. Lang, V. Madhavan, S.H. Pan, H. Eisaki, S. Uchida, and J.C. Davis, *Nature* **411**, 920 (2001).
32. S.H. Pan, E.W. Hudson, K.M. Lang, H. Eisaki, S. Uchida, and J.C. Davis, *Nature* **403**, 746 (2000).
33. E. Merzbacher, *Quantum Mechanics*, John Wiley & Sons, New York, 1970.
34. C.J. Chen, *J. Vac. Sci. Technol. A* **6**, 319 (1988).
35. C.J. Chen, *Phys. Rev. Lett.* **65**, 448 (1990)
36. C.J. Chen, *Phys. Rev. B* **42**, 8841 (1990).
37. J.C. Slater and G.F. Koster, *Phys. Rev. Lett.* **94**, 1498 (1954).
38. I. Suominen, J. Nieminen, R.S. Markiewicz, and A. Bansil, *Phys. Rev. B* **84**, 014528 (2011).
39. J. Nieminen, S. Lahti, S. Paavilainen, and K. Morgenstern, *Phys. Rev. B* **66**, 165421 (2002).
40. E. Niemi and J. Nieminen, *Surf. Sci.* **600**, 2548 (2006).
41. J.J. Mortensen, L.B. Hansen, and K.W. Jacobsen, *Phys. Rev. B* **71**, 035109 (2005).
42. J. Enkovaara et al. Electronic structure calculations with GPAW: A real-space implementation of the projector augmented-wave method, Scientific Highlight of the month, *Psi-k Newsletter* 98 (2010).
43. H. Lin, J.M.C. Rauba, K.S. Thygesen, K.W. Jacobsen, M.Y. Simmons, and W.A. Hofer, *Front. Phys. China* **5**(4), 369 (2010).
44. J.M. Ziman, *The Elements of Advanced Quantum Theory*, Cambridge University Press, Cambridge, U.K., 1988.
45. W.A. Harrison, *Electronic Structure and the Properties of Solids: The Physics of the Chemical Bond*, Dover, New York, 1986.
46. A.P. Sutton, *Electronic Structure of Materials*, Oxford University Press, Oxford, U.K., 1993.
47. M. Paulsson, Non Equilibrium Green's Functions for Dummies: Introduction to the One Particle NEGF Equations, arXiv:cond-mat/0210519v2, 2006.
48. G.D. Mahan, *Many Particle Physics*, 3rd edn., Kluwer Academic/Plenum Publishers, New York, 2000.
49. J. Repp, G. Meyer, S. Stojkovi, A. Gourdon, and C. Joachim, *Phys. Rev. Lett.* **94**, 026803 (2005).
50. J. Repp, G. Meyer, S. Paavilainen, F. Olsson, and M. Persson, *Science* **312**, 1196 (2006).
51. A.A. Abrikosov, L.P. Gorkov, and I.E. Dzyaloshinski, *Methods of Quantum Field Theory in Statistical Physics*, Dover, New York, 1975.
52. I. Martin, A.V. Balatsky, and J. Zaanen, *Phys. Rev. Lett.* **88**, 097003 (2002).
53. M.H. Hamidian, I.A. Firmo, K. Fujita, S. Mukhopadhyay, J.W. Orenstein, H. Eisaki, S. Uchida, M.J. Lawler, E.-A. Kim, and J.C. Davis, *New J. Phys.* **14**, 053017 (2012).
54. A.V. Balatsky, I. Vekhter, and J.-X. Zhu, *Rev. Mod. Phys.* **75**, 373 (2006).

15

Electrochemical STM: Atomic Structure of Metal/ Electrolyte Interfaces

Knud Gentz
University of Bonn

Klaus Wandelt
University of Bonn
and
University of Wroclaw

15.1 Introduction

Surface science under ultra-high vacuum (UHV) conditions has made huge progress over the past four decades and has led to a deep understanding of properties and processes at solid surfaces, which are the basis for many modern technologies like, for instance, electronics and heterogeneous catalysis. Inasmuch as the UHV condition was a fundamental prerequisite for this progress in surface science, it also implies a serious limitation, because most of the investigations in UHV have "only" model character. In reality, most processes take place at the interfaces between more or less dense phases. This has two implications. First, most likely the results obtained in UHV are not straightforwardly transferable to these more realistic interfaces. Second, the "hiding" of the interface between two phases prevents the application of all electron-, ion-, or atom-beam based analytical techniques which require UHV.

Some of the large-scale industrial processes such as flotation, metal-recovery, refinement, galvanization, electrolysis, passivation, etc. are based on the phenomena at solid–liquid interfaces. Macroscopically, these phenomena are well described. Their characterization at the atomic and molecular levels, however, represents a new dimension of complexity and, thus, scientific challenge [1]. Likewise, electrocatalysis, for example, in fuel cells, and redox processes in batteries are increasingly important in future energy conversion and storage; electrodeposition and etching down to the nanoscale are fundamental in, for example, modern electronic chip production; and deposition and self-assembly of organic layers, increasingly important in the production of thin functional films, are best done in solution. In order to understand and control these processes, solid–liquid interface investigations are necessary that reach the standards set by UHV surface science.

In principle, there are two ways to study solid–liquid interfaces. The ex situ approach relies on the rich arsenal of UHV-based methods. This, however, requires the transfer of the sample from the liquid environment into a UHV apparatus, preferably without contact with air. Even though such transfer chambers exist, it remains to be shown in each case that the properties of the immediate interface are the same before and after transfer. More appropriate, of course, is an in situ approach, which seeks for spectroscopic and microscopic information directly from the solid–liquid interface and, hence, requires experimental probes which permeate either one (or both) of the two condensed phases, primarily the liquid one. These methods are based on photon beams and scanning probe techniques. This chapter gives an insight into the status of electrochemical scanning tunneling microscopy (ECSTM) and its capabilities to contribute to an atomic scale analysis of metal–electrolyte interfaces. Whenever necessary, the ECSTM results are complemented by in situ and ex situ spectroscopic measurements.

Binnig and Rohrer [2,3] invented scanning tunneling microscopy (STM) in the early 1980s and revolutionized the local surface analysis first under UHV conditions. However, only 4 years after the invention the first report on STM measurements in liquids, namely oil and liquid nitrogen, was published [4]. But, again only 1 year later Sonnenfeld and Hansma [5] managed to image the surface of graphite and a gold film on glass in water and aqueous electrolyte solutions.

In these early experiments, a mere "two-electrode" setup (metallic tip and electrode surface) was used, with a given bias voltage between them. This arrangement lacked the capability to control the electrode potential, and thereby also that of the tip, with respect to a known reference. Since the tip also constitutes an electrode, this arrangement involves the risk that polarization effects and even electrochemical charge-transfer processes can occur at the tip. The corresponding capacitive and exchange currents would disturb the detection of the mere tunneling current or, even worse, could lead to a destruction of the tip. In 1988, Siegenthaler and Coworkers [6] introduced the three-electrode setup by combining the tip and the sample (working electrode) with an unpolarizable reference electrode (in this case, Ag/AgCl) and controlling both potentials with a bipotentiostat. Only shortly afterward, Itaya and Tomita [7] described a four-electrode design in which a counter-electrode, commonly used in electrochemical measurements, was also added. This arrangement permits continuous scanning while the potential of the working electrode is changed, and, thus, established true in situ ECSTM.

Since then several groups have entered the field of ECSTM, following the pioneering work of Bard [8,9], namely Behm [10,11], Endres [12], Ertl [13,14], Kolb [11,15], Magnussen [11,16,17], Soriaga [18], Wandelt [19,20], Wandlowski [21], and Weaver [22], and the range of possible electrolytes was expanded even including ionic liquids [22]. Furthermore, the enormous progress in computing capacity and microprocessor technology nowadays allows very fast data acquisition. Magnussen and coworkers, for instance, built the first video ECSTM which allowed reaching frame rates of 10–30 images/s [16].

15.2 Scanning Tunneling Microscopy in UHV and in Liquid Environment

In classical physics, a particle with a finite energy cannot pass an energy barrier higher than the total energy of the particle. In quantum mechanics, on the other hand, the particle can also be described by a wave function $\Psi(x,y,z)$. This wave function can penetrate into the barrier, however therein it decreases exponentially as a function of barrier width. According to quantum mechanics, a particle will be reflected only by a barrier of infinite height or width while with a barrier of finite height and width the described wave function continues on the other side of the barrier with smaller amplitude but at the same frequency.

The probability of a particle tunneling through a noninfinite barrier is given by the Gamov equation [23], which shows that the tunneling events should decrease with increasing particle mass m, barrier height V_B, and width z. If two electrodes are brought in close proximity to each other, electrons will tunnel through the gap of width z between them from the filled states of *each* electrode to empty states of the other electrode, so that no net current can be measured. If the Fermi level of one electrode is permanently shifted upward by the application of a bias voltage U_B, electrons will dominantly tunnel from that electrode to the empty states of the other electrode, leading to a resulting tunneling current

$$I \sim U_B \cdot \mathrm{e}^{-z\sqrt{8mV_B/\hbar^2}} \rho(E_F),$$

where $\rho(E_F)$ is the local density of states at the tip position. Thus, the tunneling current changes exponentially with the width z of the tunneling gap, which allows the changes of the distance between the tip and surface to be measured very precisely and, thereby, a topographical image of the surface with atomic resolution in x- and y-directions to be obtained (if to a first approximation, an x,y-dependence of $\rho(E_F)$ is neglected). The scanning can be performed either in the "constant height mode" or in the "constant current mode" as depicted in Figure 15.1 [24]. In the latter case, a feedback loop controls the tip–sample distance such that the tunneling current I remains constant.

Scanning tunneling microscopy quickly became a standard method in the surface science arsenal and was applied with great success to various metallic and semiconducting surfaces in vacuum. In a liquid, on the other hand, several difficulties had to be overcome. First, the liquid is obviously much more denser than vacuum or even air, and so the tip experiences a stronger resistance during its movement through the electrolyte. This has to be considered in the feedback loop of the STM. Preferentially, the feedback loop should be adjustable to different liquids. Second, if the liquid is an electrolyte, the applied bias voltage U_B will cause the flow of a Faradaic current between the tip and the sample, which is usually several orders of magnitude larger than the tunneling current. Since both currents are flowing through the tunneling tip, they are superimposed and cannot be distinguished by the STM controller. Therefore, it becomes necessary to shield the tip in order to minimize the Faradaic current. Coating the tip with nail polish, apiezon wax, or hot melting glue as well as encasing the tip in glass, leaving just the very tip open, have been successfully done for this purpose [25] (see inset upper left corner in Figure 15.2).

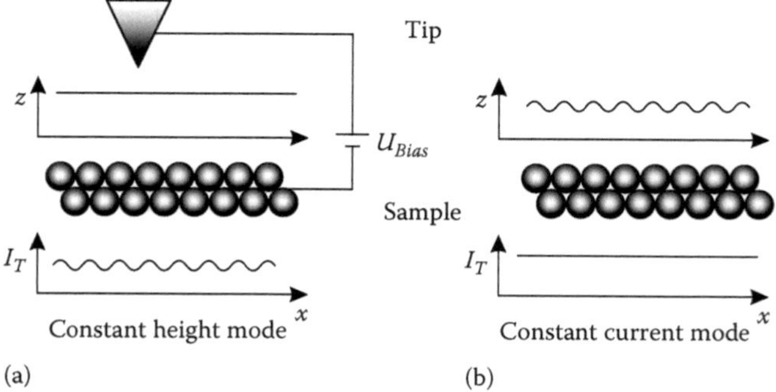

FIGURE 15.1 Constant height and constant current tunneling conditions: (a) the absolute height of the tip is held constant, and the resulting change in tunneling current is recorded; (b) the height is adjusted to follow the surface corrugation so that the resulting tunneling current is constant.

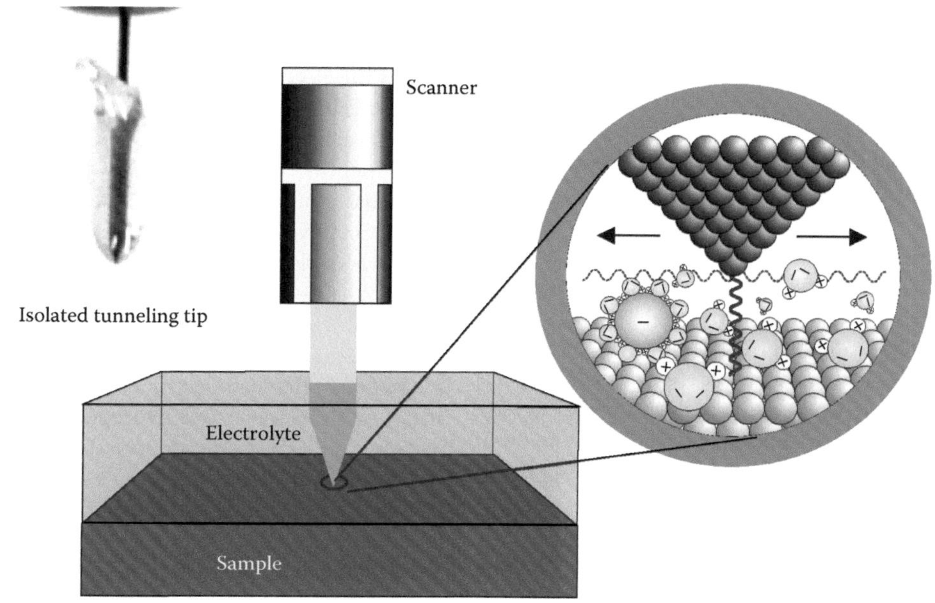

FIGURE 15.2 Tunneling through an electrolyte requires insulation of the scanner (see inset in upper left corner).

Other sources of perturbations on the tunneling current that can lead to an increase in the noise level of the tunneling signal are currents induced by electromagnetic fields in the experimental chamber, or a current flux due to the high electrical fields generated by the piezo electrodes where voltages up to 100 V are applied to move the scanner piezo. While the former can be avoided by surrounding the electrochemical cell and the scanner with aluminum [6,19], the latter can be minimized by adding a grounded shield ring at the front of the scanner piezo (4 in Figure 15.3).

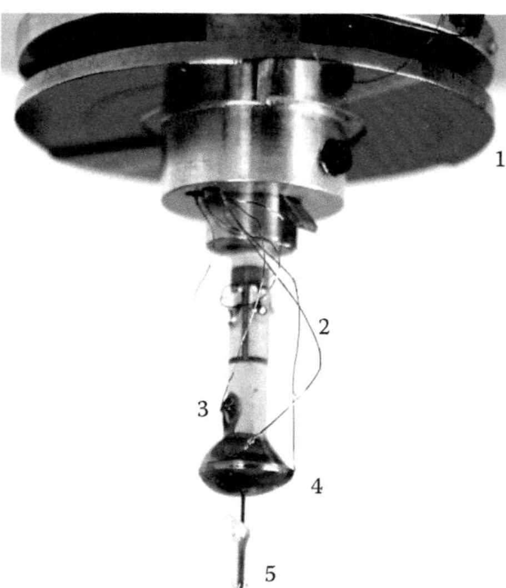

FIGURE 15.3 Single tube piezo scanner with (1) circular ramp for the coarse approach, (2) x and y electrode connections, (3) z electrode, (4) grounded shield ring to reduce interference, (5) coated tunneling tip (see text).

Despite several efforts to explain the principles of tunneling through an electrolyte, the processes involved are still not completely understood. Experimentally measured tunneling barriers are significantly lower than in UHV, and the barrier height is also modified by a 2D network of maxima and minima in the potential-energy surface. The first effect can be explained by the adsorption of the polar water molecules at the electrode surface, which commonly decreases the work function of a metal surface. The modulation of the potential-energy surface is the result of the fluctuations of hydrogen and oxygen atoms in the molecular water layer between the tip and the surface [26]. While the water molecules in the electrolyte between the tip and surface move about and thereby cause a continuous change in the potential-energy surface, for an individual tunneling event, the tunneling barrier remains static since the time an electron needs to tunnel through an assumed tunneling gap of ~10 Å is at least two orders of magnitude faster than the movement of the water molecules [27]. In order to register a tunneling current in the STM controller, on the other hand, a large number of tunneling events are necessary, and the computing capability of the controller, that is, the sampling rate, limits the speed with which the given information can be processed. Therefore, the variation in the barrier height averages over a number of tunneling events, and the recorded tunneling current is not significantly influenced by the distribution of the water molecules, but an effective (averaged) barrier height can be deduced from the registered data. One last important difference is that in UHV the tunneling current depends strictly exponentially on the distance from the tip to surface (see Equation 15.1). In an electrochemical environment, however, several groups have detected oscillations within the current–distance correlation measured by conduction–distance tunneling spectroscopy [28,29]. These oscillations are ascribed to individual layers of water molecules parallel to the surface.

15.3 Description of the Dedicated Bonn-ECSTM

15.3.1 Instrumental Details

Most of the studies selected for this overview have been conducted with a home-built ECSTM that has been developed and built in our group in Bonn [19]. The instrument is basically a modified Besocke-type [30] STM with a single-tube scanner (Figure 15.3) [31], the latter, of course, pointing downward as shown in Figure 15.2. The complete setup is shown in Figure 15.4 and consists of the STM head (2), STM base (5), and the electrochemical cell (6). The STM head rests on three spacer bolts (7) and holds the ramp-ring (1 in Figure 15.3) with the scanner (3), and houses the preamplifier and necessary circuitry. A screw at the top (1 in Figure 15.4) serves to settle the ramp-ring down on the piezo-actuators (4), which are mounted on the STM base (5) for the coarse approach, or to lift it off. Also attached to the base is the electrochemical cell (6) where various electrodes can be connected as well as the tubing system which contains the electrolyte flood- and drain-tubes. Apart from the sample at the bottom of the cell (Figure 15.5), which is electrically connected to virtual ground, an *internal* reference electrode, for example, Ag/AgCl in hydrochloric solution, can be used, and the counter-electrode is also connected via the metallic cell holder. Finally, a generator electrode can also be installed in the cell (not shown here), allowing in situ dissolution and, thereby, dosing of this electrode metal into the electrolyte can be controlled. The use of an *external* reference electrode, for example, Hg/Hg_2SO_4 ("calomel") or RHE (reversible hydrogen electrode) connected via a Luggin capillary, is also supported by this setup.

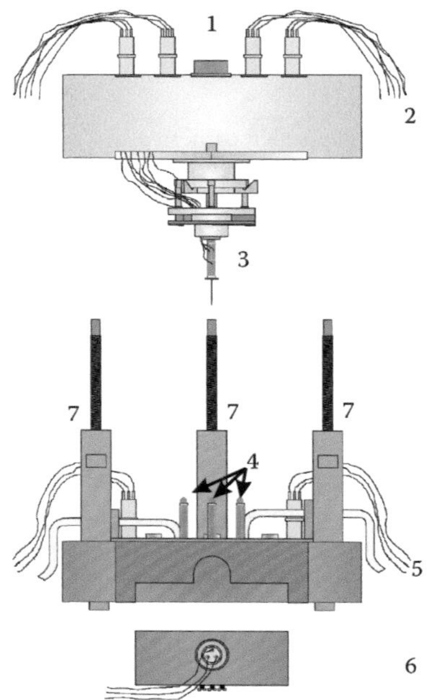

FIGURE 15.4 Schematic of the Bonn-ECSTM: (1) lift screw for the STM scanner, (2) STM head with preamplifier, (3) piezo scanner with coated STM tip, (4) coarse approach actuators, (5) STM base with electrolyte supply system, (6) electrochemical cell, and (7) bolt spacers (see text).

The whole system is placed on a stack of brass plates separated by rubber spacers and is mounted inside an aluminum cube to avoid any external electromagnetic influence. The cube, in turn, rests again on a rubber mat on a heavy granite slab which is suspended from the ceiling via four steel springs. This combination of damping elements filters out any vibrations that could otherwise affect the measurements. This design offers some key advantages over many commercially available instruments:

- The STM has a built-in first order thermal drift compensation due to identical piezos being used for the scanner and the coarse approach actuators (4 in Figure 15.4), so that all of them have identical thermal expansion coefficients. In lateral directions, the drift compensation comes from the symmetrical setup of the actuators. The coarse approach also allows sideways movement over the surface so that every spot on the sample can be reached by the tip.
- The compact setup of a Besocke-type STM reduces vibrational sensitivity and allows for a reduced size of the final instrument.
- The STM head (2 in Figure 15.4) contains a preamplifier, which is connected to the tunneling tip by short and very thin wires, and is shielded by aluminum from electromagnetic interference to minimize the noise on the tunneling signal.
- The electrochemical cell (Figure 15.5) contains a volume of about 2.5 mL of electrolyte, which is sufficient to allow

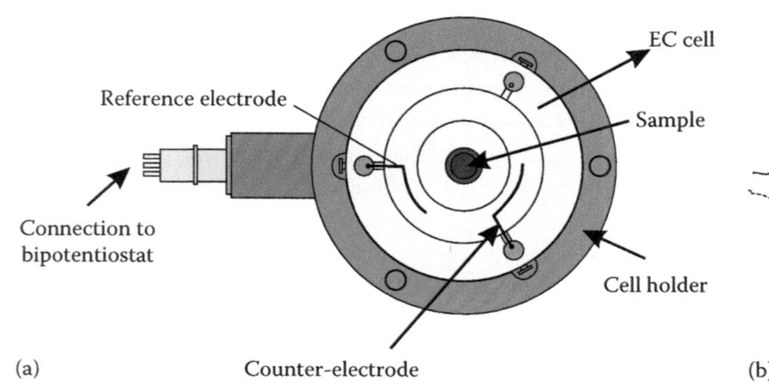

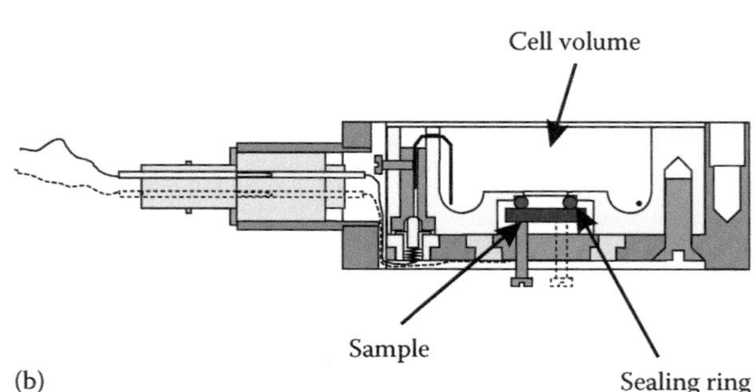

FIGURE 15.5 (a) The electrochemical setup of the Bonn-ECSTM; (b) cell volume 2.5 mL.

the formation of a full Nernst diffusion layer within the cell and reduces the concentration effects due to evaporation of the electrolyte. This improves the quality of results gained by concomitant electrochemical measurements such as cyclic voltammetry or chronoamperometry, performed in the very same cell. The sample is connected to the cell via a sealing ring in a cutout in the cell bottom so that only a defined area of the sample surface is in contact with the electrolyte (and not the sample edge).

- The electrolyte flow system allows the exchange of the electrolyte without opening the surrounding aluminum cube, which provides great experimental flexibility and reduces contamination. The aluminum housing has also a gas inlet to allow air-free operation under an inert gas atmosphere.

A more detailed description of the instrument and the controller systems can be found in Ref. [19].

15.3.2 Potentiostatic, Potentiodynamic, and Spectroscopic Detection Methods

The Bonn-ECSTM uses a four-electrode setup (working electrode, counter-electrode, reference electrode, and tip) so that the potentials of the STM tip and the sample (working electrode) versus the reference electrode are controlled independently by the bipotentiostat. On the one hand, in the so-called potentiostatic mode, both the sample potential and the bias voltage are kept constant, leading to the classical imaging. By varying the bias voltage and measuring, at constant sample potential, the resulting tunneling current, it is possible to record the corresponding dI/dV curves through which it is possible to gain insight into the local density of states under the tip [24]. During potentiodynamic measurements on the other hand, the potential of the working electrode is changed during the acquisition of an STM image. If the tip potential would be held constant with respect to the working electrode, it would thereby shift with respect to the reference electrode, leading to changing Faradaic currents through the tip and to disturbances of the image properties. Therefore, there is an option to set the tip potential as constant versus the reference electrode. In this case, the tip–sample voltage is not constant over the course of the measurement, but the effect of this is smaller than that of a change in Faradaic current through the tip. Such measurements are desirable since they allow the direct correlation of current waves in the cyclic voltammogram to changes in the surface structure visible in the *simultaneously* acquired STM image.

15.3.3 Sample Preparation

Most of the usual sample preparation methods employed in UHV such as sputtering and annealing are not feasible for preparation in an electrochemical environment, except in more complex systems which allow the transfer from UHV to an ambient pressure chamber without contact to air. To achieve a well-defined and reproducible surface for ECSTM measurements, alternative preparation strategies have been devised.

15.3.3.1 Flame Annealing

A simple preparation method is to expose the sample to a butane-air flame and heat the crystal to a slightly reddish hue. This procedure removes organic compounds by oxidation and heals surface defects. In his original procedure, Clavilier et al. [32] subsequently quenched the heated sample in water. A less drastic method is to slowly cool the crystal to room temperature in an Argon-flow under careful exclusion of oxygen, in order to avoid the formation of surface oxides. This method has been successfully applied to Au single crystals [33,34], but is unsuitable for more reactive and low-melting-point metals such as copper.

15.3.3.2 Electrochemical Etching and Annealing

Another commonly applied method for the preparation of samples in an electrochemical environment is electrochemical etching. For this procedure, the crystal surface is brought in contact with an electrolyte, and an AC voltage of several volts is applied. The top layers of the crystal are dissolved along with impurities and contaminations, if any. The crystal is subsequently protected from reoxidation by covering it with a drop of de-aerated electrolyte. The main disadvantage of this method, especially for single crystals, is the loss of material from the surface and the concomitant roughening of the surface. Therefore, this method is best used for systems where the crystal can heal itself in the electrolyte. This kind of "electrochemical annealing" has been described, for example, for Au [35], Cu [36], Ru [37], and Fe [38] as well as conducting organic films [39], and has recently been theoretically described by Schmickler et al. [40,41].

15.3.3.3 UHV-EC Transfer

If flame annealing and electrochemical etching are not successful, the only remaining possibility is the preparation in UHV by sputtering the sample with Ar ions followed by annealing at elevated temperatures and subsequent transfer into the electrolyte. The preparation in UHV has also the significant advantage that the crystallographic perfectness as well as the absence of contaminants can be checked by standard UHV analysis methods such as low-energy electron diffraction (LEED), photoelectron or Auger electron spectroscopy (UPS, XPS, AES), or ion scattering spectroscopy (ISS). If the UHV preparation chamber is directly attached to the electrochemical cell by a buffer chamber, then it is possible to transfer the prepared sample into the electrochemical cell without contact to ambient atmosphere [42,43].

15.4 Examples of In Situ ECSTM Studies

15.4.1 Atomic Structure of Electrode Surfaces: Copper and Gold Single Crystal Electrodes

Figure 15.6a shows the cyclic voltammogram of a Cu(111) electrode in aqueous 10 mM HCl solution. Between the cathodic limit (hydrogen evolution reaction, HER) and the anodic limit (here, before the onset of the copper dissolution reaction, CDR; see also

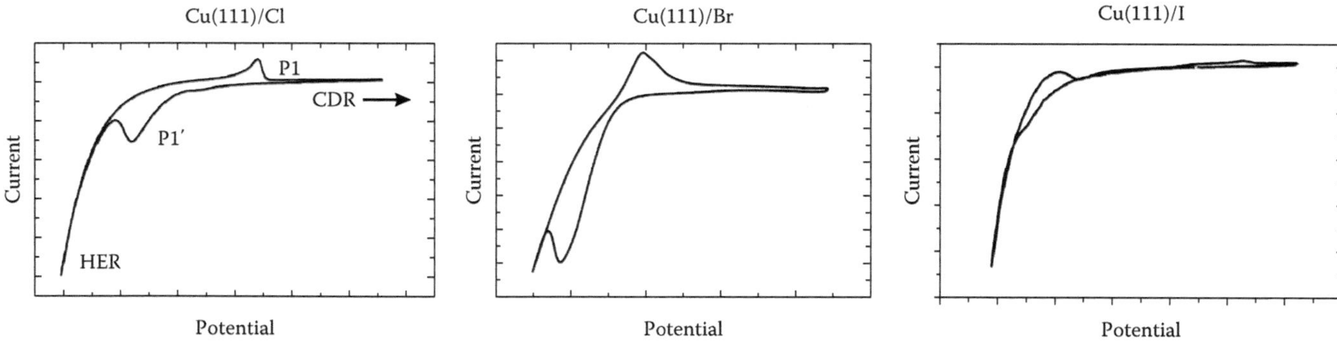

FIGURE 15.6 Cyclic voltammograms measured with a Cu(111) single crystal electrode in 10 mM HCl-, HBr-, and HI-solutions, respectively; HER, hydrogen evolution reaction; CDR, copper dissolution reaction.

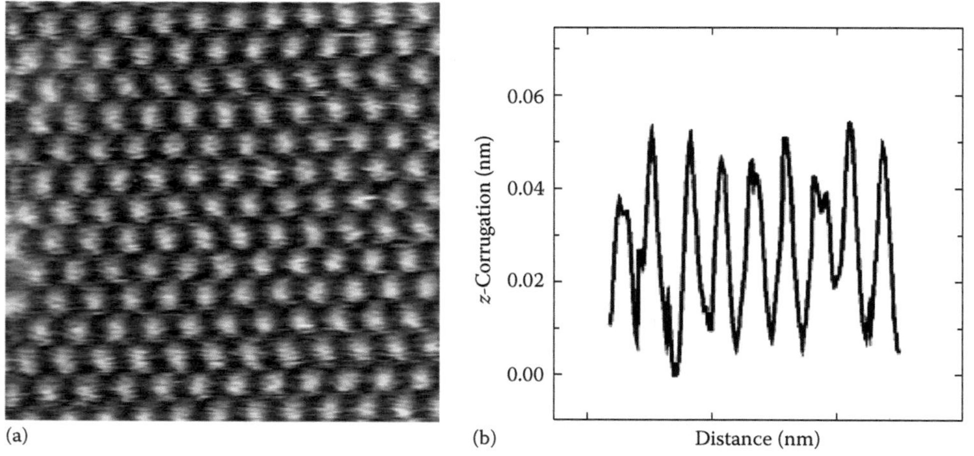

FIGURE 15.7 In situ STM image (a) of the anion-free Cu(111) electrode surface with atomic resolution; (b) line profile along a row of Cu atoms. (From Springer Science+Business Media: *Solid-Liquid Interfaces: Macroscopic Phenomena—Microscopic Understanding*, Topics in Applied Physics, Vol. 85, 2003, p. 141, Broekmann, P., Wilms, M., Arenz, M., Spänig, A., and Wandelt, K.)

Figure 15.12), the voltammogram exhibits two characteristic current peaks. The anodic peak P1 (at more positive potential) indicates the adsorption of chloride anions, while the cathodic peak P1′ (at more negative potential) represents the chloride desorption. Thus, on the positive side of both peaks, the Cu(111) electrode surface is covered with "specifically adsorbed" (see following paragraphs) chloride anions, while within the potential regime negative of both peaks, the surface is adsorbate-free, which can actually be verified by ex situ photoelectron spectroscopy and ISS in a "transfer chamber" [44].

Figure 15.7a shows an STM image of the Cu(111) electrode taken in solution in the adsorbate-free potential regime. This unfiltered image shows clearly the expected hexagonal symmetry of the surface with an interatomic distance of 256 pm and a corrugation of 7 pm [45], as indicated by the line profile in panel (b).

The Au(111) surface is known to be reconstructed under UHV conditions. Figure 15.8 reproduces results from Dretschkow and Wandlowski [46], which show the large-scale (a) as well as the atomically resolved (b) ($\sqrt{3} \times 22$) reconstruction of an Au(111) electrode surface in 0.05 mM H_2SO_4 solution at negative potentials, and the unreconstructed Au(111)(1 × 1) surface (c, d) at positive potentials. The surface atomic density of the reconstructed surface is ~4% higher than that of the normal hexagonal unreconstructed (111) surface, Thus, upon deconstruction, 4% of Au atoms are displaced onto the (1 × 1) surface, forming small islands (white dots in Figure 15.8c). This structural transition is accompanied by measurable current peaks as seen in the corresponding cyclic voltammogram shown in Figure 15.8. Unlike in UHV, however, in solution, this structure transition is fully reversible, depending on the electrode potential, a feature unique to the electrified interface [47].

15.4.2 Anion Adsorption on a Cu(111) Electrode

The first step in surface electrochemical processes is, like in UHV, the adsorption of reactants (here, the adsorption of ions from the electrolyte) on the electrode surface. Unlike in UHV, however, the ionic species are always accompanied by their respective counterions, which may influence the process under investigation.

While spectroscopic and electrochemical methods average over the properties of the whole surface, the ECSTM can add

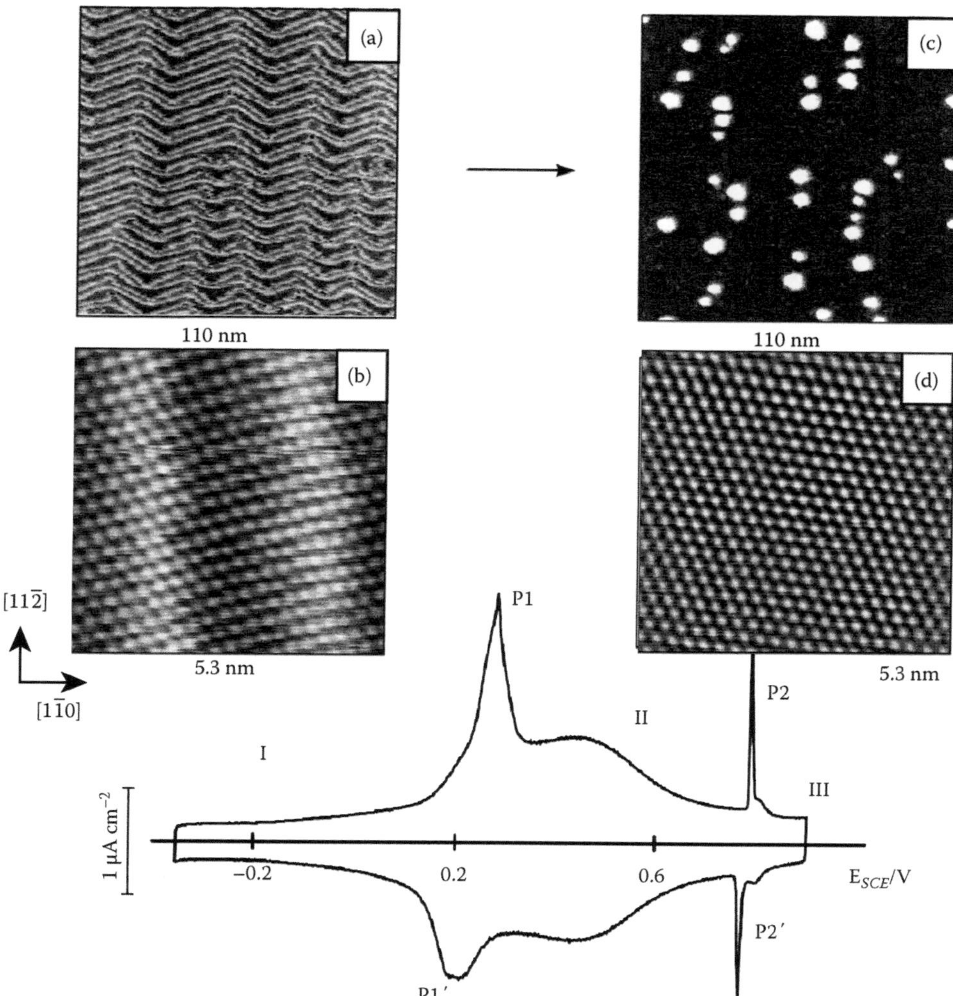

FIGURE 15.8 In situ STM images and cyclic voltammogram of an Au(111) electrode in 0.05 M H_2SO_4 solution: large-scale (a) and atomically resolved (b) images of the reconstructed surface at negative potentials; large-scale (c) and atomically resolved (d) images of the unreconstructed surface at positive potentials (see text and Ref. [47]).

high-resolution information about the local structure and the adsorption processes at an atomic or molecular level.

A basic concept of the solid–electrolyte interface is the model of the "electrochemical double layer," which is depicted in Figure 15.9 for an electrode at positive potential (with respect to the reference potential). Anions adsorb on the surface, either strongly, in direct contact with the surface losing (at least part of) their hydration sphere (panel (b), "inner Helmholtz plane"), or weakly, with intact hydration sphere (panel (a), "outer Helmholtz plane"). ECSTM provides insight into the structure of the adsorbed anion layer, for example, as a function of the electrode potential.

15.4.2.1 Halide Adsorption

The cyclic voltammogram of a Cu(111) surface in dilute hydrochloric acid already has been shown in Figure 15.6a. While in the range negative of the anodic peak P1, the bare Cu(111) surface could be imaged (Figure 15.7a), in the range positive of the anodic peak, chloride is adsorbed in a highly ordered ($\sqrt{3}\times\sqrt{3}$) Cl superstructure [44,48]. In order to gain insight into the correlation between the structure of adsorbate and substrate, one might think of taking an image of the chloride-covered surface, then change the potential to a regime negative of the desorption peak, take another image of the chloride-free surface, and then superimpose the two images. The problem with this approach, however, is that some drift between the two images prevents positional correlation between the image of the adsorbate and the substrate, respectively. This difficulty can be circumvented by exploiting the spectroscopic capabilities of the STM. If both the adsorbate and the substrate have density of states near the Fermi energy, they both will contribute to the total STM signal. Thus, by varying the bias voltage such that both the substrate and the adsorbate layers contribute to a composite image, it is possible to separate the different periodicities of both layers by

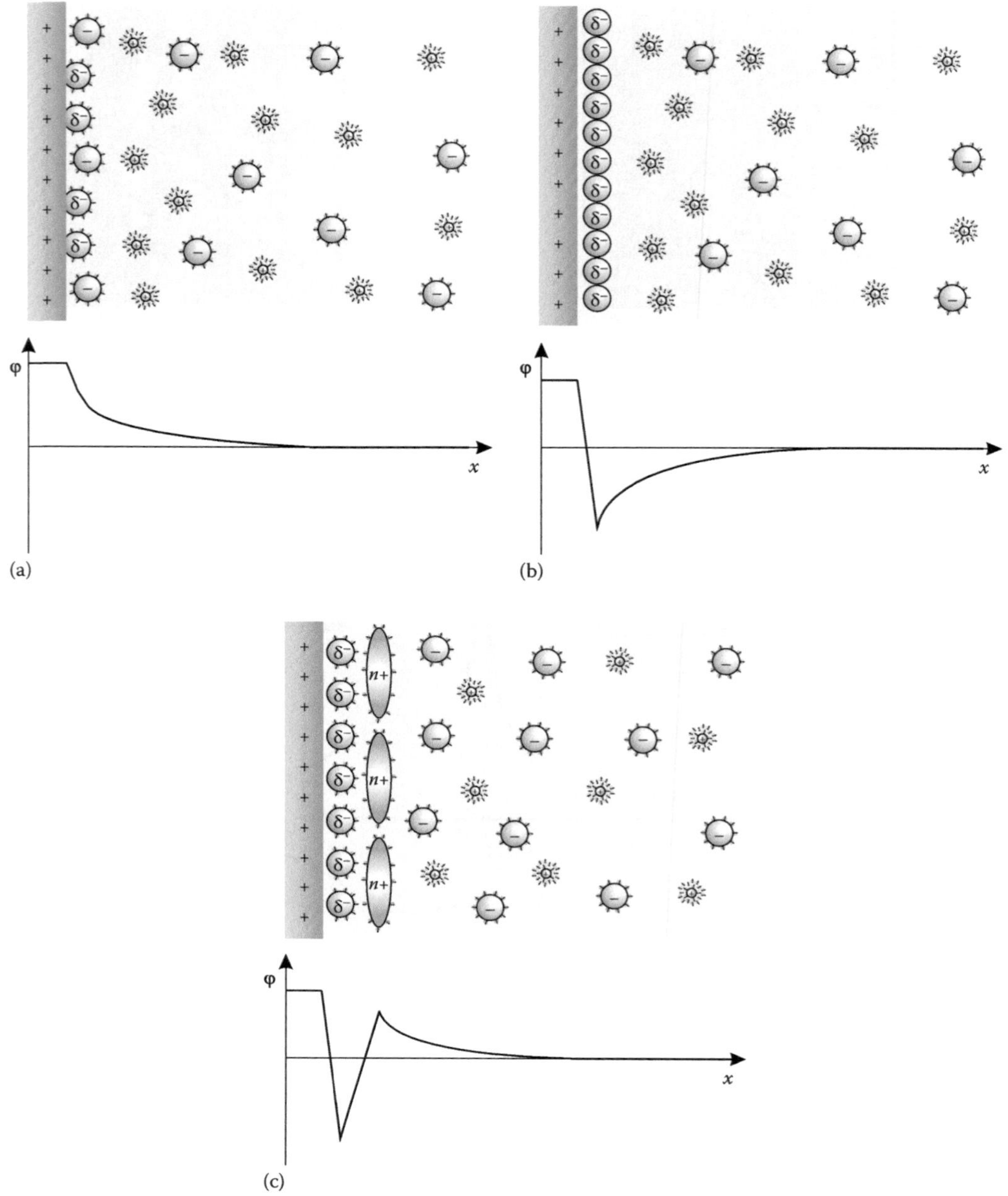

FIGURE 15.9 Model of the "electrochemical double layer": (a) without specific adsorption, (b) with specific adsorption leading to a charge reversal at the surface, and (c) adsorption of organic cations on the anion-modified surface.

Fourier transformation [44]. Since the separated images were obtained simultaneously with exactly the same tip position during the scan, their superposition unambiguously shows the relative atom positions of the adsorbate and the substrate. The results obtained by this approach for the Cl-covered Cu(111) surface confirm that the chloride ions are adsorbed in the threefold hollow sites on the Cu(111) substrate, like in UHV (Figure 15.10). It shall be mentioned that at very positive potentials, due to a further increase in coverage, this $(\sqrt{3}\times\sqrt{3})$Cl structure undergoes a structure transition to a uniaxially incommensurate structure, similar to the one described in the following for bromide.

Bromide is larger than chloride, which has several consequences. First, the charge density of bromide ions in solution is lower than that of chloride ions. Consequently, its hydration sphere is bound weaker, so that a less positive electrode potential suffices to adsorb Br ions specifically on Cu(111), as manifested by Figure 15.6b. Second, bromide anions do not fit into an unstrained $(\sqrt{3}\times\sqrt{3})$ structure on the Cu(111) surface. As a result, the STM image in Figure 15.11 shows a 1D incommensurate structure, in which bromide ions are in registry with the Cu substrate only in the $[\bar{2}11]$ direction, while in the perpendicular direction the bromide anions sit in inequivalent positions,

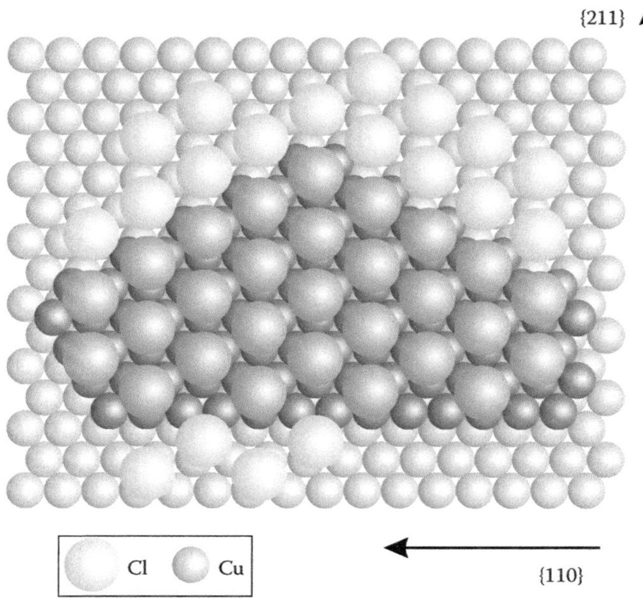

FIGURE 15.10 ($\sqrt{3}\times\sqrt{3}$) superstructure of chlorine adsorbed on a Cu(111) electrode surface as deduced from potentiostatic and potentiodynamic ECSTM measurements (see text).

which leads to the long-ranged wavy superstructure visible in Figure 15.11a and b. Due to the threefold symmetry of the substrate surface, this uniaxially incommensurate structure always occurs in three coexisting rotational domains, rotated by 120°. Note also in Figure 15.11a that the step edges are stabilized along the close-packed rows of anions. This unidirectional incommensuracy explains also the anisotropic etch- and growth-structures after some dissolution (Figure 15.11d) and redeposition (Figure 15.11c) of copper on a Cu(111) electrode surface in HBr-containing electrolyte [49]. Note also the angles of 120° between step edges in these growth- and etch-structures.

What has been shown here for bromide is particularly true for iodide, which is even larger. Iodide anions release their hydration sphere even easier than bromide and therefore specifically adsorb readily at the onset of the hydrogen evolution reaction (Figure 15.6c). Iodide also forms rotational domains of uniaxially incommensurate structures like bromide, which due to the higher polarizability of the iodide anions show the phenomenon of "electrocompressibility," that is, the wavelength of the superstructure changes reversibly with the electrode potential, because the iodide coverage increases/decreases with increasing/decreasing potential [50].

15.4.2.2 Sulfate Adsorption

The Cu(111) surface in contact with dilute sulfuric acid (5 mM H_2SO_4) represents a particularly interesting case [45,51]. Figure 15.12 shows the cyclic voltammogram for this system; note the large hysteresis between the SO_4^{2-} adsorption (at −50 mV vs. RHE) and SO_4^{2-} desorption peak (at −250 mV). This indicates a strong kinetic hindrance of one of the two, most likely the adsorption process. This is supported by in situ STM measurements shown in Figure 15.13. The images in panels (a)–(f) are registered sequentially over a period of 15 min at a constant potential near −0.58 V versus RHE. Panel (a) shows no particular structure except a rounded step edge, even though in situ IR spectroscopy proves a saturation coverage of SO_4^{2-} anions at this potential [52]. Over time, panels (b–f) show the slow development of a hexagonal structure (arrow in panel b) which, in panel (f), covers almost the whole surface. Several details are worthwhile to be mentioned. First, each panel represents a surface area of 101 nm × 101 nm; thus, the lattice spacing of the (nearly) hexagonal structure is ~3 nm [45,51], that is, by far not of atomic dimensions. In fact, the visible structure corresponds to a long-range Moiré-type superstructure as will be explained later. Second, the newly created step edges are straight and parallel to the rows of dark Moiré-dots. Third, in panel (f), an island has formed on top of the surrounding terrace, both being covered by the Moiré-superstructure.

Figure 15.14 shows a close-up of the Moiré structure. The rows of bright dots correspond to adsorbed SO_4^{2-} anions which form a "quasi" ($\sqrt{3}\times\sqrt{7}$) R30° structure, similar to the findings on other fcc(111) electrode surfaces [45]. The zigzag rows of weaker bright dots between the sulfate anions are also observed on these other sulfate-covered surfaces and have been ascribed to coadsorbed water molecules (or hydronium ions), which in the case of ($\sqrt{3}\times\sqrt{7}$) R30°/Au(111) was actually verified by Ataka and Osawa [53] using surface-enhanced infrared absorption spectroscopy. Unlike all other fcc(111) surfaces, however, Figure 15.13 shows additional long-range Moiré-type modulations (darker regions). A straightforward explanation for this peculiarity of the SO_4^{2-}-covered Cu(111) surface is given in Figure 15.15. Here again, advantage was taken of the spectroscopic mode of STM measurements with our ECSTM. Panels (a)–(c) are taken from the SO_4^{2-}-covered surface, each under different tunneling conditions, with the result that panel (a) shows enhanced the positions of the sulfate anions, panel (b) accentuates the positions of the copper atoms underneath the SO_4^{2-} layer, and panel (c) not only includes contributions of both but even resolves the broad sulfate features from panel (a) into two bright dots. Comparison of the Cu–Cu interatomic distance as measured in panel (a) with that from the SO_4^{2-}-free Cu(111) surface (Figure 15.7) indicates that the first Cu layer underneath the adsorbed sulfate layer is expanded by about 6%. Superposition of the unit cell of the SO_4^{2-} anion lattice from Figure 15.15a on the Cu mesh of Figure 15.15b proves the ($\sqrt{3}\times\sqrt{7}$) coincidence. Thus, the SO_4^{2-} overlayer lattice is in registry with the *expanded* first Cu layer, but the expanded first Cu layer is not in registry with the subsurface Cu layers, giving rise to the long-range Moiré-superstructure (see Figure 15.15a and b). Moreover, the two bright dots representing one SO_4^{2-} anion in Figure 15.15c strongly suggest a twofold adsorption symmetry of the SO_4^{2-} anions as sketched in Figure 15.15d, in contrast to the conclusions from IR measurements [53]. Finally, it is the expansion of the first Cu layer, that is, its reduction in surface atomic density, which causes the crowd out of atoms

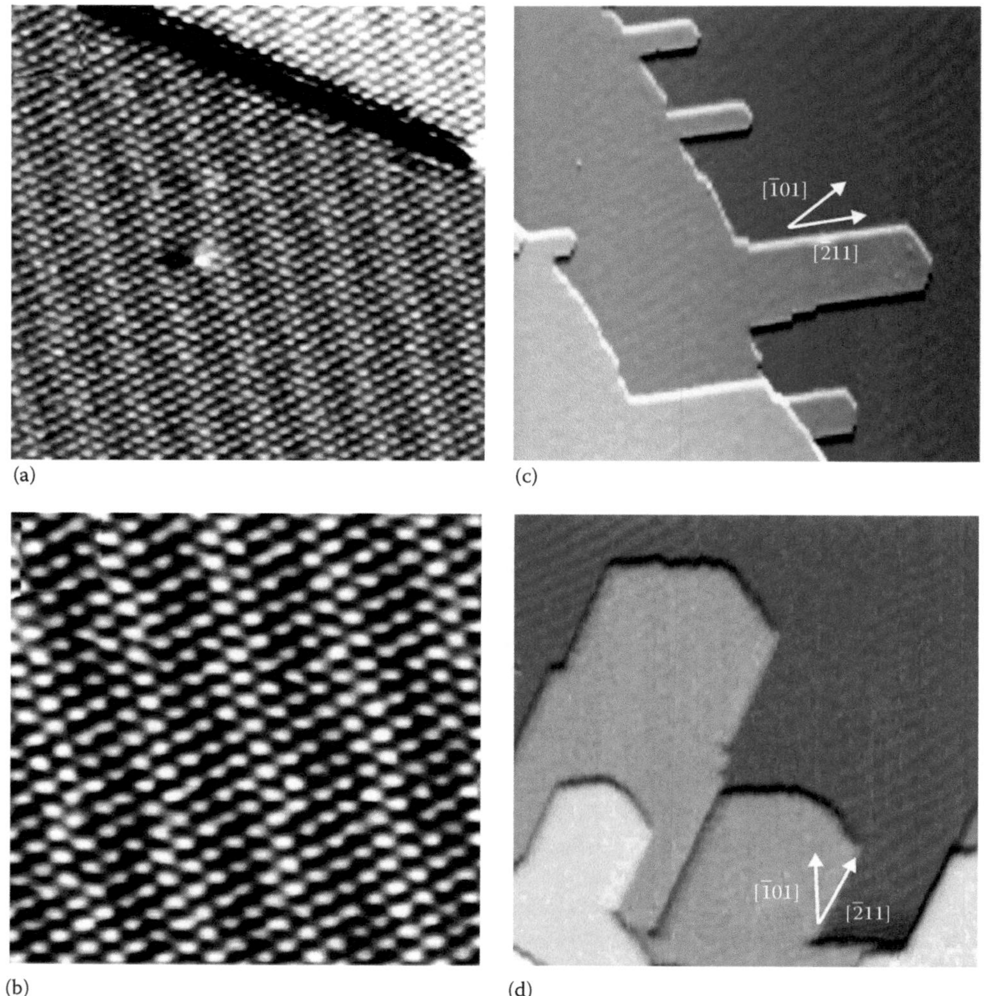

FIGURE 15.11 In situ ECSTM images of bromide specifically adsorbed on a Cu(111) electrode: (a) large-scale image of the uniaxially incommensurate (wavy) superstructure; note the straight step edge in the upper right corner; (b) close-up of panel (b); (c) monolayer thick deposition structures of Cu; (d) monolayer deep etch structures of Cu in HBr solution; arrows give crystallographic directions (see text).

and the formation of islands on top of the surface (arrow in Figure 15.13f). These islands are again Moirè-covered, and it is interesting to note that the islands are "quantized" in shape and size in units of the Moiré structure [45,51]. It must be concluded that the SO_4^{2-}–Cu interaction is very strong, and that this strong interaction enforces the massive reconstruction of the Cu(111) surface. The significant mass transport involved in this restructuring process explains the large hysteresis in the cyclic voltammogram in Figure 15.12 as well as the strong dependence of the order of the Moiré structure on the potential scan rate. The faster the anodic scan rate, the less well ordered is the resulting Moiré structure. It is worth mentioning that SO_4^{2-} adsorption on the UPD (underpotential deposition) layer of Cu on Au(111) does *not* lead to the formation of a Moiré superstructure, because the Cu layer is pseudomorphic to the gold substrate and, thus, readily expanded [54]. The phenomenon of UPD is described in the following section.

15.4.3 Metal Underpotential Deposition

The structure and growth mechanism of thin metal films on the surface of another metal have been a major research interest under vacuum conditions. Technologically, however, thin metal film deposition from an electrolyte is much easier and more economic. Another unique feature of electrochemical deposition is the capability to deposit just a single monolayer (or submonolayer) of foreign metals at potentials *more positive* than the corresponding reversible Nernst potential (UPD). In this process, UPD layers of a metal with a lower work function can be deposited on a metal substrate with a higher work function, and the shift in the UPD peak is correlated to the difference in the work functions.

15.4.3.1 Cu/Au

The cyclic voltammogram of a Au(111) surface in sulfuric acid containing $CuSO_4$ shows two distinct cathodic peaks assigned to

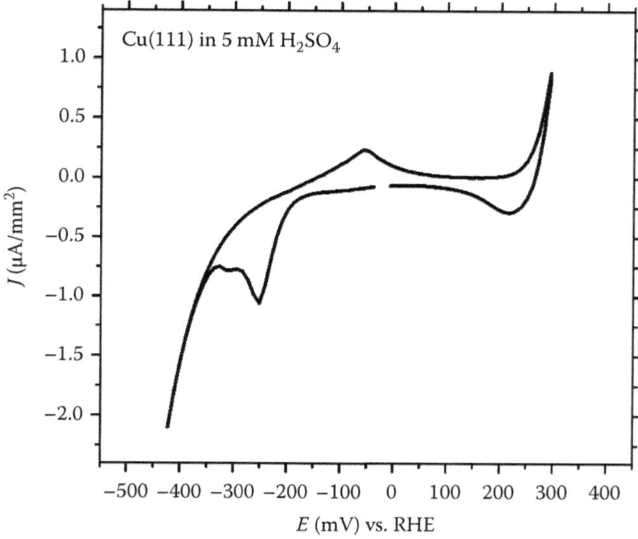

FIGURE 15.12 Cyclic voltammogram measured with a Cu(111) electrode in 5 mM H_2SO_4 solution between the hydrogen evolution reaction at negative and the Cu dissolution reaction at positive potentials.

Cu deposition, even though both peaks occur at potentials more positive than the Cu Nernst reduction peak (Figure 15.16). For several years, there has been a discussion on what exactly causes these two deposition peaks and, moreover, what structure exists on the Au(111) surface at the potentials *between* those two peaks [55]. In the early literature, a $(\sqrt{3} \times \sqrt{3})R30°$ Cu superstructure, corresponding to one third of a Cu monolayer was proposed. Behm and coworkers were able to use an in situ STM to obtain the first real-space images of the suspected $(\sqrt{3} \times \sqrt{3})R30°$ Cu superstructure [56]. Unfortunately, this case also shows one of the major limitations of STM, namely the lack of chemical sensitivity. While the $(\sqrt{3} \times \sqrt{3})R30°$ structure was clearly visible in the STM data, there was no proof that the structure was actually formed by Cu atoms. Indeed, it was proven later by in situ x-ray diffraction and chronocoulometric experiments that this structure is *not* formed by adsorbed Cu atoms but by sulfate anions coadsorbed *within* an otherwise pseudomorphic (1×1) Cu monolayer [57–59].

Also on Au(100) single crystal electrodes, Cu has been found to form a pseudomorphic (1×1) monolayer with a coverage being dependent on the applied potential, rising with a

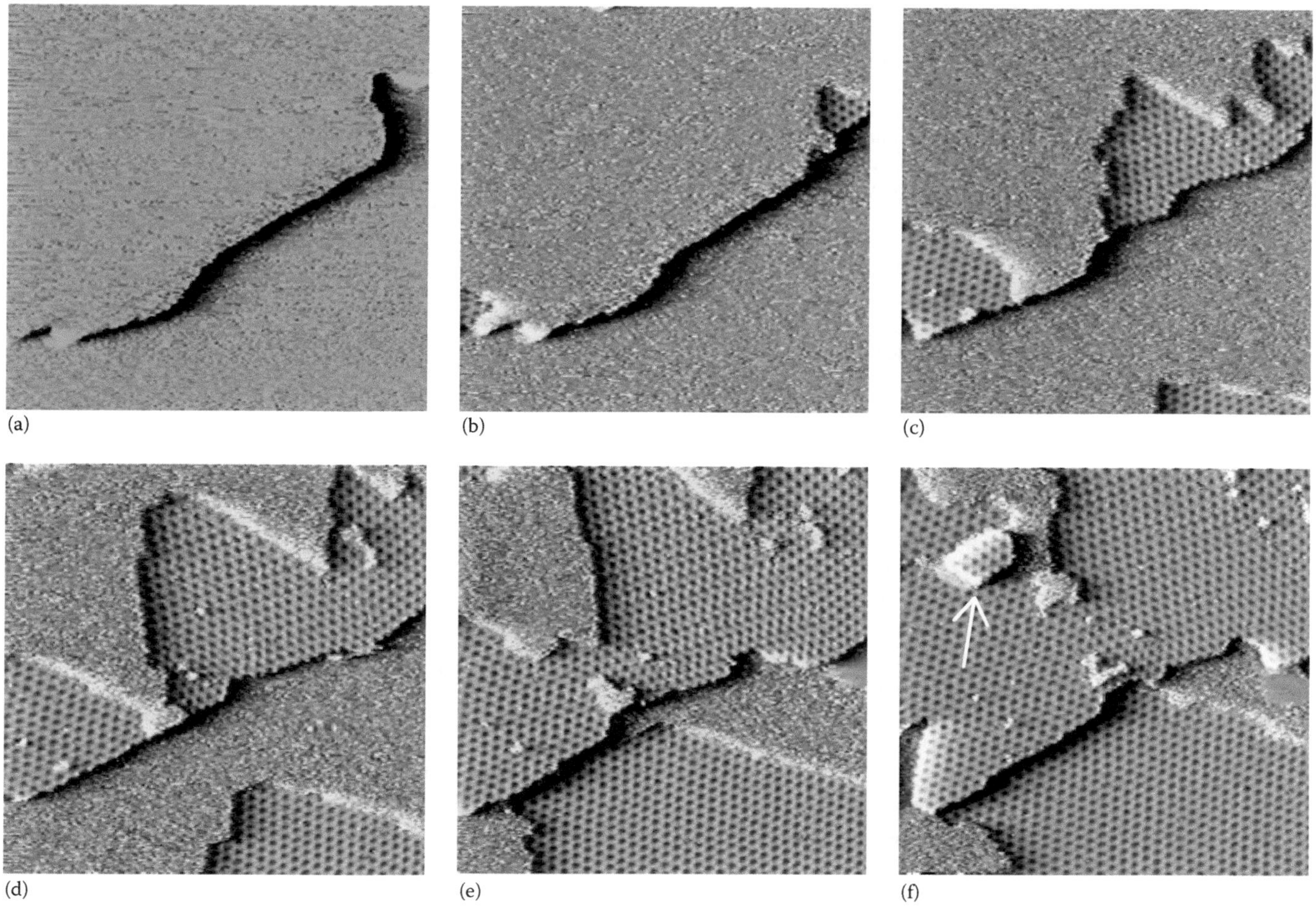

(a) (b) (c)

(d) (e) (f)

FIGURE 15.13 Series of in situ ECSTM images showing the evolution of the sulfate-induced reconstruction of the Cu(111) surface with time (images sizes 110×110 nm²), leading to a Moiré-type superstructure (see text and Refs. [45,52]).

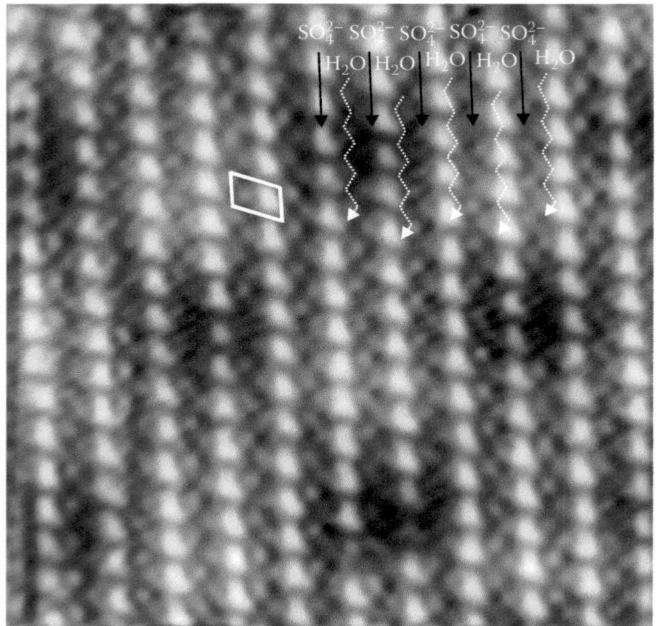

7.6 nm × 7.6 nm
I_t = 1 nA, U_b = −120 mV, E = −56 mV

FIGURE 15.14 In situ ECSTM image of the coadsorption layer of sulfate anions (large bright spots) and water molecules (zigzag chains of weak dots between the sulfate rows). (After Broekmann, P. et al., *Progr. Surf. Sci.*, 67, 59, 2001; From Springer Science+Business Media: *Solid-Liquid* Interfaces: Macroscopic *Phenomena—Microscopic Understanding*, Topics in Applied Physics, Vol. 85, 2003, p. 141, Broekmann, P., Wilms, M., Arenz, M., Spänig, A., and Wandelt, K.)

more negative potential and reaching a full monolayer at 0.1 V versus SCE [60]. Above that potential, the repulsively interacting Cu adatoms are highly mobile and form a quasi-2D gas phase. In the absence of copper, the mobility of the gold atoms is strongly reduced, so that an Au(100) surface, which has been prepared by the flame-annealing procedure described earlier, and immersed in sulfuric acid at 0 V versus SCE, shows the usual well-ordered surface reconstruction. On the other hand, a $CuSO_4$ concentration of 5×10^{-5}mol/L added to the electrolyte is sufficient to lift the reconstruction and small gold islands with a diameter of 20 nm begin to form on the surface. A quasi-hexagonal structure found in earlier experiments [56] originates from contamination with chloride ions. Again, the STM in itself offers no chemical sensitivity, but the combination with electrochemical measurements often leads to a conclusive picture of the situation at the surface. A model proposed by Tourillon et al. [61] which indicated an 8% contraction of the topmost gold layer covered with a Cu layer of similar geometry could clearly be ruled out by the STM results, because the lattice observed was that expected for a Au(100) surface, while an 8% contraction of the topmost layer should have clearly been observable as a long-range Moiré-type modulation in the STM. Recent experiments have shown that in the positive scan range just before the onset of the dissolution, the structure of the Cu UPD layer changes and forms a striped phase with a commensurate $c(3 \times 1)$ structure [62]. In this phase, out of any group of the three adjacent atomic rows, the outer two rows move slightly closer together, giving rise to the appearance of stripes.

15.4.3.2 Cu/Pt

Despite the difference in lattice constant of about 4%, the electrodeposition of Cu on Pt(111) resembles quite closely that of its neighbor in the PSE, gold. When Cu is deposited from a sulfate-containing electrolyte, a $(\sqrt{3} \times \sqrt{3})$R30° superstructure with a corresponding coverage of $\Theta = 2/3$ is observed, which is attributed to coadsorbed sulfate ions [63,64]. In the presence of halides on the other hand, different structures are again observed. In a chloride-containing electrolyte, Cu has been found to form a (4×4) superstructure, while in the presence of bromide, a $(\sqrt{3} \times \sqrt{3})$R30° has been observed [65].

While Cu UPD on Pt(111) has been the focus of a number of groups, the deposition of Cu on Pt(100) has received significantly less attention. Kolb et al., in their earlier work [66], used optical and electrochemical techniques and found a two-step mechanism of Cu deposition. The first half monolayer forms a $c(2 \times 2)$ structure with weak adatom interaction. For coverages higher than $\Theta = ½$, the vacant sites are slowly filled, resulting in a final (1×1) layer. Later, a bilayer structure was proposed, where on the first half monolayer of Cu with the $c(2 \times 2)$ structure, a second identical layer adsorbs in the fourfold hollow sites of the first one, leaving the $c(2 \times 2)$ structure of the first layer intact [67], Later, however, it was found that the $c(2 \times 2)$ structure forms only in the presence of chloride ions, while, if a very weakly adsorbing counterion like $HClO_4^-$ is present in the electrolyte, Cu adsorbs on the Pt(100) surface as 2D islands with a $p(1 \times 1)$ structure [68,69]. Interesting enough, this growth is in agreement with the structures found for the deposition of thin Cu films in UHV, where naturally no coadsorption of anions takes place [70]. In the presence of bromide ions, Cu was found to form a full $p(1 \times 1)$ monolayer, on top of which bromide adsorbs in a $c(2 \times 2)$ close-packed layer [71].

The deposition of Cu on a Pt(110) surface leads to a diffuse (1×1) structure [72], again quite similar to the results obtained for gold.

15.5 Self-Assembly of Organic Molecules

Organic materials, in particular, thin organic films, play an ever-increasing role in modern material science, because both their growth structure and their functionality can easily be "tuned" by the design of their building blocks, that is, by the synthesis of the organic molecules. One disadvantage, however, is the fact that at least more complex molecules may not be intact volatile for vapor deposition in UHV. In this case, a promising, if not the only, route is deposition from solution, preferably from aqueous solution.

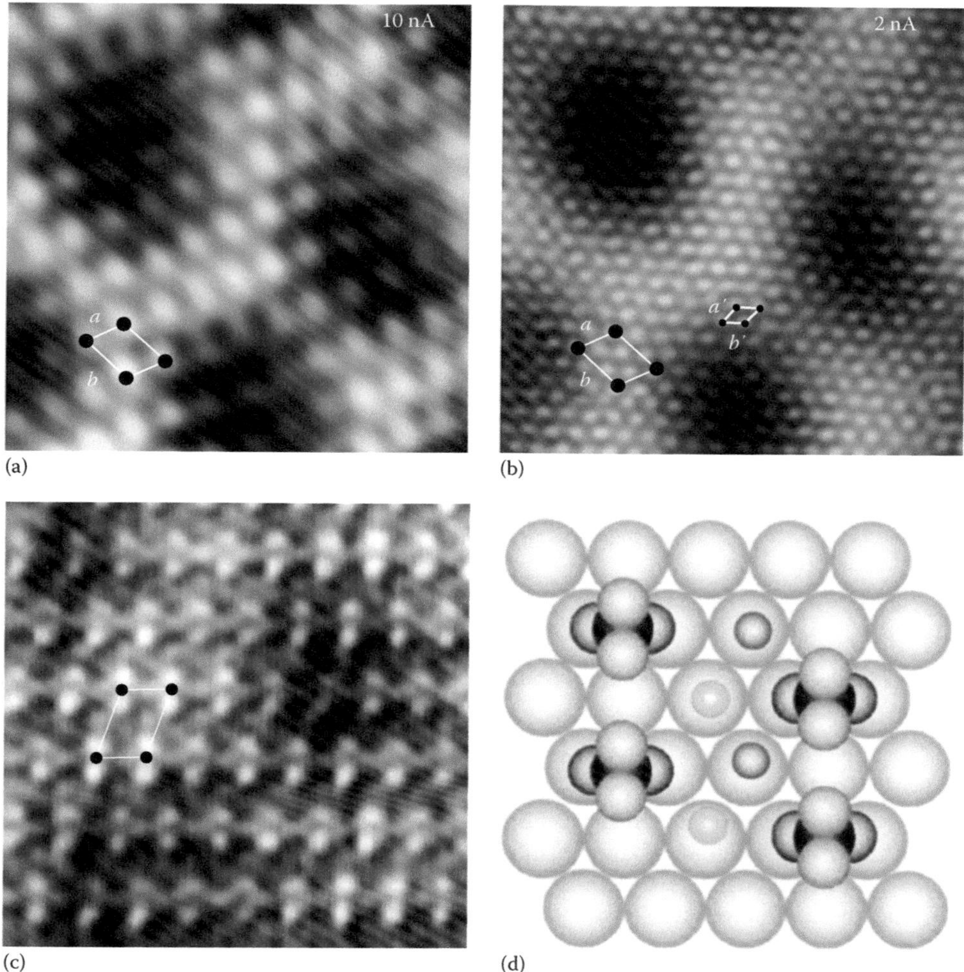

FIGURE 15.15 In situ ECSTM image of a sulfate-covered Cu(111) electrode surface measured with different tunneling conditions (i.e., tunneling currents (a) 10 nA, (b) 2 nA, (c) 40 nA) leading to an enhanced imaging of (a) sulfate states, (b) copper states, (c) higher resolution of the sulfate states, (d) model of the twofold sulfate adsorption configuration (see text).

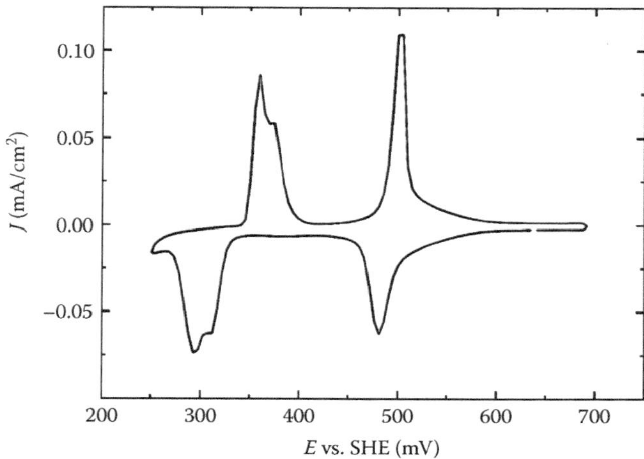

FIGURE 15.16 Cyclic voltammogram of the underpotential deposition of copper on an Au(111) surface.

For this approach, the organic molecules must be sufficiently polar or form ions in solution. In the latter case, the organic components may be used in the form of an appropriate salt which dissociates in water. Of course, it needs to be shown whether or not the counterions have influence on the ultimate properties of the organic film. In the following, as an example, the adsorption and self-assembly of cationic viologen molecules, that is, di-substituted bipyridinium ions, will be presented.

Viologen molecules are redox-active. In aqueous solution, they form dications, which depending on the potential of the electrode may undergo two one-electron reduction steps, first to the monocation radical and then to the neutral molecule. In principle, the dissolved dications would be attracted by a negatively polarized electrode, but might also spontaneously be reduced, for example, to the neutral molecules, which are no longer specifically attracted by the surface. More interesting is to precover the electrode with a strongly adsorbing anion

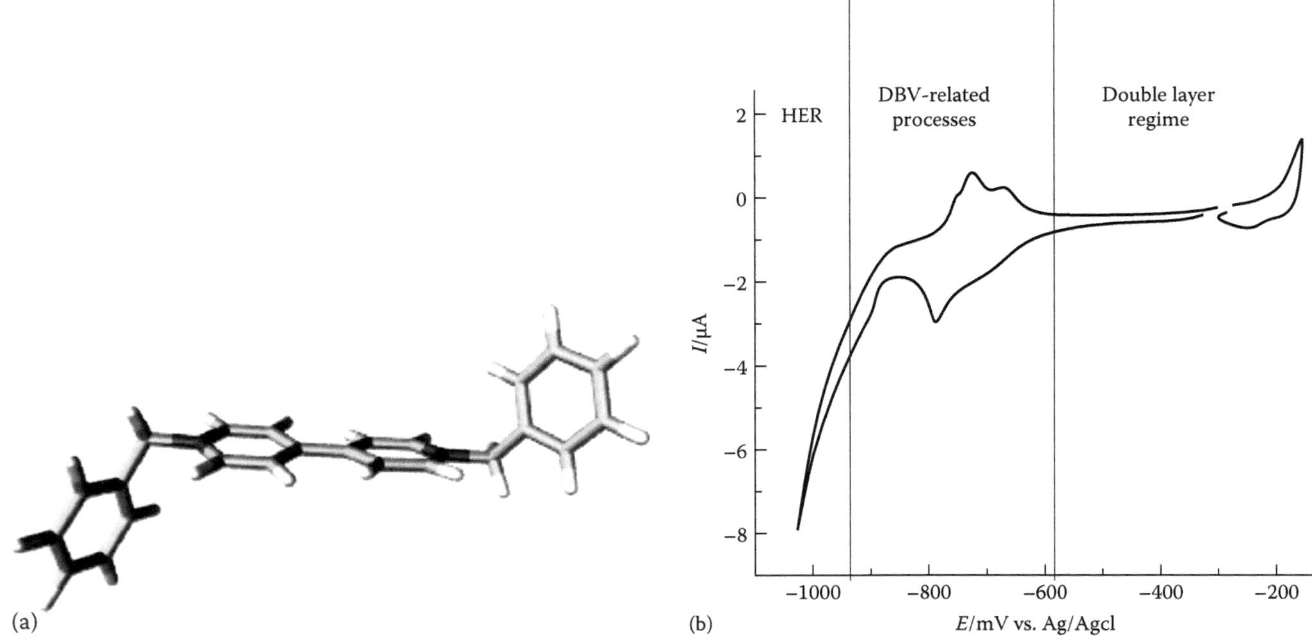

FIGURE 15.17 (a) Model of a dibenzyviologen molecule; (b) cyclic voltammogram of a Cu(111) electrode in DBV containing hydrochloric acid solution (see text).

layer which remains on the surface over a wide potential range. This has two advantages. On the one hand, the excess of negative anion charge on the electrode attracts the organic cations, and on the other hand, the preadsorbed anion layer acts as a spacer layer which prevents direct contact between the organic cations and the metal, and, thereby, immediate electron transfer reactions. This situation is sketched in Figure 15.9c. Moreover, by sufficiently changing the electrode potential, it may still be possible to induce redox processes of the organic cations, now in a controlled way. Most likely, different oxidation states of the same molecule will influence the structural and chemical properties of the organic layer. Electrochemical scanning tunneling microscopy appears most suited to investigate in situ the adsorption and self-assembly of such layers and to follow potential- and redox-dependent structural transitions.

It is well-known, for example, that chloride anions form a specifically adsorbed c(2 × 2) adlayer on a Cu(100) electrode surface [73,74]. On top of this chloride layer, redox-active (1,1′)-dibenzyl-(4,4′)-bipyridinium- (in short, dibenzylviologen or DBV^{2+}) cations (Figure 15.17a) are adsorbed in a highly regular pattern of square moieties at potentials above −600 mV versus Ag/AgCl (see Figure 15.18a). The individual

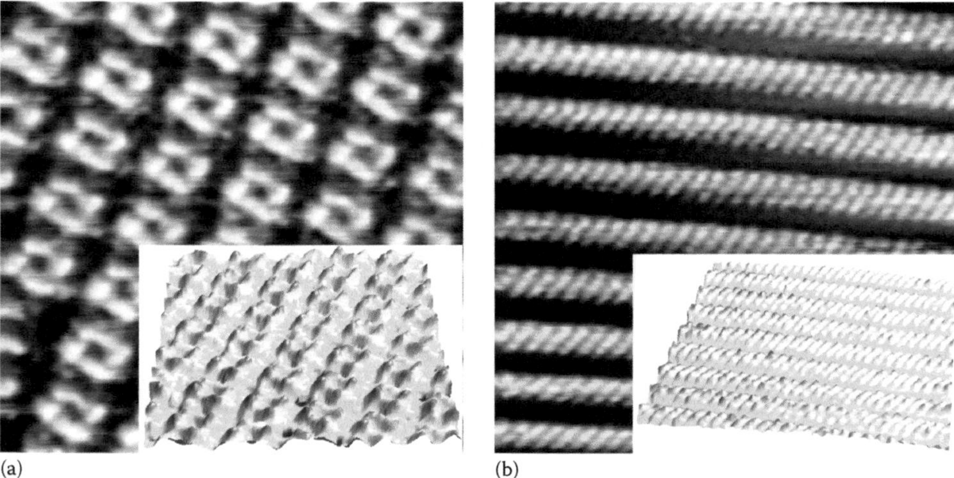

FIGURE 15.18 In situ ECSTM images of dibenzyl viologen (DBV) molecules adsorbed on a Cl-modified Cu(100) surface: (a) self-assembled “cavitand-structure” of DBV^{2+} dications, each square motif consists of four DBV^{2+} dications; inset 3D representation; (b) self-assembled “stripe structure” of π–π stacked $DBV^{\bullet+}$ monocation radicals; inset 3D representation.

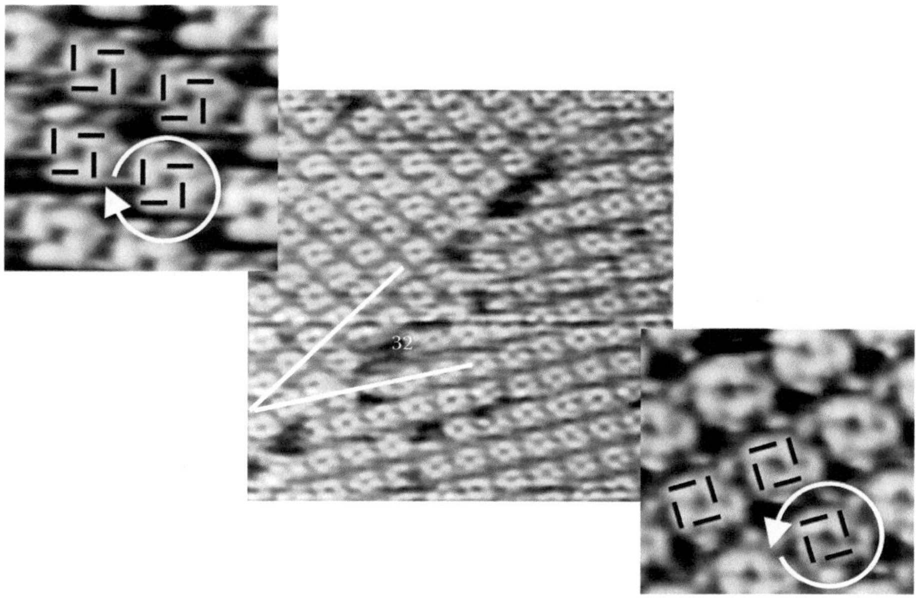

FIGURE 15.19 In situ ECSTM images of chiral mirror domains of the cavitand structure adsorbed DBV^{2+} dications on a Cl-modified Cu(100) electrode surface.

squares are formed by four single DBV^{2+} dications giving rise to a cavitand with a diameter of 1.1 nm; the inset in Figure 15.18a shows a 3D representation of this structure. The individual molecules can be arranged either in a right-handed or left-handed order (see Figure 15.19), which gives rise to a planar chirality of the cavitand structure. Since neither the molecules nor the substrate show any chirality, both enantiomers should be observable on the surface, which is indeed the case. The enantiomers do not mix on the surface but form mirror domains instead (see Figure 15.19), with an angle of 32° with respect to each other. If the electrode potential is decreased below −600 mV, the DBV^{2+}dications are reduced to the radical cation $DBV^{\bullet+}$, which is indicated by the current wave in the cyclic voltammogram at −700 mV (Figure 15.17b; the current peak at −800 mV arises from chloride desorption). This reduction is accompanied by a significant structural change as well as an increase in surface coverage which can be observed in the STM images. The resultant stripe structure in Figure 15.18b is formed by π–π stacking of the radical cations, and is characteristic of the monocation radicals of most bipyridinium derivatives [75–78].

In order to relate the structure of the organic overlayer to the c(2 × 2)Cl structure underneath, two strategies are possible in this case. First, if the bias voltage and tunneling current are tuned such that the tip approaches the surface, the tip may act like a brush, removing the organic cations, so that the underlying substrate structure can be observed. If this technique is applied during the acquisition of one STM image, so that the first half of the image displays still the organic overlayer, while the second half shows the structure of the anion underlayer, an extrapolation of the latter into the first half of the image will give insight into the structure relation between the adsorbate and the substrate. Second, as already described for chloride on Cu(111) and shown in the case of sulfate (Figure 15.15), a very careful tuning of the tunneling parameters may lead to contributions of both substrate and adsorbate states, to the STM image, which may be separated by Fourier filtering and subsequently superimposed. In this case, no drift has to be considered between the two images, making the structure correlation even more reliable. For instance, the relationship between the "cavitand" structure of the DBV^{2+} layer and the c(2 × 2)Cl underneath may be described by a $\begin{pmatrix} 2 & 7 \\ \bar{7} & 2 \end{pmatrix}$ matrix, which corresponds to a coverage of $\Theta = 0.075$ and an area of 160 Å^2 per cavitand. The coverage of the radical–monocationic stripe phase, in turn, is $\Theta = 0.2$, hence, much denser.

The adsorption of redox-active species shows another remarkable measurement option of the ECSTM. By changing the voltage during the acquisition of an STM image, it is possible to see the oxidized and reduced species in one and the same image, so that structural differences can immediately be observed without having to consider drift between images. In this way, it is possible to gain additional insight into the kinetics of the phase transition/redox process. The change from the cavitand structure of the DBV dications to the π–π stacked structure (Figure 15.18) undergoes an order–disorder–order transition. In the case of DBV, the transition $DBV^{2+} \leftrightarrow DBV^{\bullet+}$ is a slow process. After passing the reduction potential near −700 mV (see Figure 15.17b), first a disordered phase forms on the surface, from which the stripe structure emerges over a period of several minutes. On the other hand, the reduction of a related molecule (diheptylviologen) on the other hand is extremely fast. Figure 15.20a shows a "dot" structure formed by individual dicationic species. In Figure 15.20b, a potential step from $U_{work} = 0$ mV to $U_{work} = -120$ mV is made after about one third

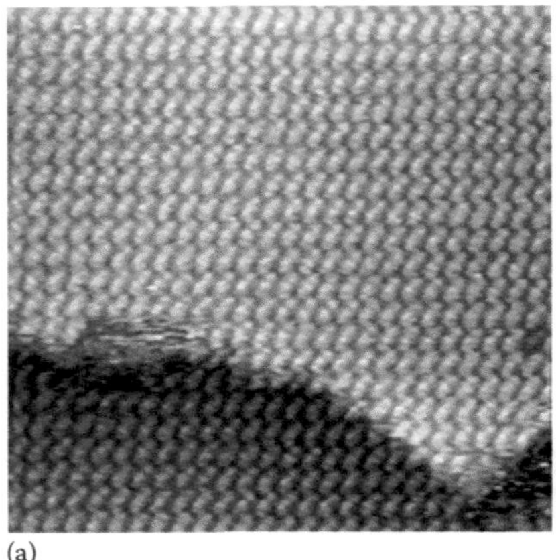
(a)

(b)

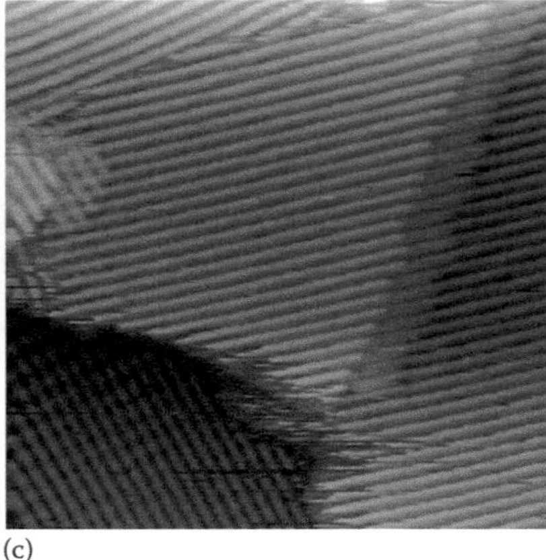
(c)

FIGURE 15.20 Transition from the dication- to the monocation-radical structure of diheptylviologen adsorbed on a Cl-modified Cu(100) electrode surface: (a) "Dot"- structure of the adsorbed dications at positive potentials; (b) transition to the "stripe" structure of π–π stacked $DBV^{\cdot+}$ monocation radicals induced by a potential step toward negative potentials; (c) fully developed "stripe structure" of monocation radicals (see text).

of the STM image. The transition from the dicationic dot structure to the stripe structure of the radical cation (Figure 15.20c) is almost instantaneous.

When interpreting the results of adsorption of organic molecules, one finds another caveat against STM. For molecules, the equivalent to the local density of states is the arrangement of molecular orbitals. Depending on the position of the highest occupied and the lowest unoccupied molecular orbital and, obviously, the tunneling direction, organic molecules can be imaged either brighter or darker than the substrate though they must obviously be topologically higher than the surface on top of which they are adsorbed. Also, the STM offers no chemical sensitivity. This means neither the type of atoms or molecules nor their charge can be deducted directly from STM experiments. If this kind of information is required, STM has to be combined with theoretical calculations and other ex situ methods such as photoelectron spectroscopy.

15.6 Summary and Conclusions

In this chapter, we have shown that ECSTM has undoubtedly become a very valuable probe to study, in real space, structures at metal–electrolyte interfaces with atomic resolution. This was demonstrated for bare and anion-covered metal surfaces, for metal monolayers electrodeposited on an unlike metal substrate, and for the adsorption and self-assembly of organic molecules, namely viologen species, on a chloride precovered Cu single crystal surface. The specific capabilities of the applied, home-designed and home-built "Bonn-ECSTM" also enabled us to follow structural phase transitions at the interfaces as a function of the electrochemical potential of the respective electrode, as well as to separate contributions from the electronic structure of the substrate and the adsorbate, respectively.

Though all this is important information, it is obvious that ECSTM alone cannot provide an exhaustive picture of the properties of metal–electrolyte interfaces. In order to achieve this, the ECSTM-method has to be (1) developed further itself, and (2) combined with complementary methods. Necessary developments of the ECSTM method are, first of all, a better theoretical description of the tunneling process through the electrolyte-filled gap as a prerequisite for a more detailed interpretation of ECSTM data. Also, faster imaging would be desirable for kinetic studies. The basic disadvantage of ECSTM of not being sensitive to the chemical nature of the imaged species requires the use of additional in situ techniques based on photons, like infrared- and x-ray-based spectroscopic methods. The definite need for a better understanding of solid–liquid interfaces in general and metal–electrolyte interfaces in particular will undoubtedly drive these developments.

References

1. K. Wandelt and S. Thurgate (Eds.), *Solid-Liquid Interfaces: Macroscopic Phenomena—Microscopic Understanding*, Topics in Applied Physics, Vol. 85 (Springer-Verlag, Heidelberg, Germany, 2003, ISBN 3-540-42583-7).
2. G. Binnig, H. Rohrer, C. Gerber, and E. Weibel, *Phys. Rev. Lett.*, 1982, 40(2), 178.
3. G. Binnig and H. Rohrer, *Helv. Phys. Acta*, 1982, 55(2), 128.
4. B. Drake, R. Sonnenfeld, J. Schneir, P.K. Hansma, G. Slough, and R.V. Coleman, *Rev. Sci. Instrum.*, 1986, 57(3), 441.
5. R. Sonnenfeld and P.K. Hansma, *Science*, 1986, 232(4747), 211.
6. P. Lustenberger, H. Rohrer, R. Christoph, and H. Siegenthaler, *J. Electroanal. Chem. Interf. Electrochem.*, 1988, 243(1), 225.
7. K. Itaya and E. Tomita, *Surf. Sci.*, 1988, 201(3), L 507.
8. D.W. Suggs and A.J. Bard, *J. Phys. Chem.*, 1995, 99(20), 8349.
9. D.W. Suggs and A.J. Bard, *J. Am. Chem. Soc.*, 1994, 116(23), 10725.

10. J. Wiechers, T. Twomey, D.M. Kolb, and R.J. Behm, *J. Electroanal. Chem.*, 1988, 248(2), 451.
11. O.M. Magnussen, J. Hotlos, G. Beitel, D.M. Kolb, and R.J. Behm, *J. Vac. Sci. Technol. B*, 1991, 9(2), 969.
12. F. Endres, *Phys. Chem. Phys.*, 2001, 3(15), 3165.
13. A.M. Bittner, J. Wintterlin, B. Beran, and G. Ertl, *Surf. Sci.*, 1995, 335(1), 291.
14. A.M. Bittner, J. Winttterlin, and G. Ertl, *J. Electroanal. Chem.*, 1995, 388(1–2), 225.
15. D.M. Kolb, R. Ullmann, and T. Witt, *Science*, 1997, 275(5303), 1097.
16. T. Tansel and O.M. Magnussen, *Phys. Rev. Lett.*, 2006, 96(2), 026101.
17. O.M. Magnussen, J. Hagebeck, J. Hotlos, and R.J. Behm, *Faraday Discuss.*, 1992, 94, 329.
18. K. Sashikata, J. Matsui, K. Itaya, and M.P. Soriaga, *J. Phys. Chem.*, 1996, 100(51), 20027.
19. M. Wilms, M. Kruft, G. Bermes, and K. Wandelt, *Rev. Sci. Instrum.*, 1999, 70(9), 3641.
20. M. Wilms, P. Broekmann, C. Stuhlmann, and K. Wandelt, *Surf. Sci.*, 1998, 416(1–2), 121.
21. C. Li, I. Pobelov, T. Wandlowski, A. Bagrets, A. Arnold, and F. Evers, *J. Am. Chem. Soc.*, 2008, 130(1), 318.
22. X.P. Gao and M.J. Weaver, *J. Am. Chem. Soc.*, 1992, 114(22), 8544.
23. G. Gamow, *Z. Phys.*, 1928, 51.
24. A. Kühnle and M. Reichling, in: *Surface and Interface Science, Vol. 1, Concepts and Methods*, Ed. K. Wandelt (Wiley-VCH, Weinheim, Germany, 2012, ISBN 978-3-327-41156-6), p. 427.
25. H. Siegenthaler, in: *Scanning Tunneling Microscopy II*, Eds. H.R. Wiesendanger and H.J. Güntherodt (Springer, Berlin, Germany, 1992), p. 7.
26. W. Schmickler, *Chem. Rev.*, 1996, 96(8), 3177.
27. K.L. Sebastian and G. Doyen, *Surf. Sci. Lett.*, 1993, 290(3), L703.
28. W. Schindler and M. Hugelmann, *Surf. Sci.*, 2003, 541(1–3), L643.
29. T. Wandlowski and G. Nagy, *Langmuir*, 2003, 19(24), 10271.
30. J. Frohn, J.F. Wolf, K. Besocke, and M. Teske, *Rev. Sci. Instrum.*, 1989, 60(6), 1200.
31. G. Binnig and D.P.E. Smith, *Rev. Sci. Instrum.*, 1986, 57(8), 1688.
32. J. Clavilier, R. Faure, G. Guinet, and R. Durand, *J. Electroanal. Chem.*, 1980, 107(1), 205.
33. T. Dretschkow, D. Lampner, and T. Wandlowski, *J. Electroanal. Chem.*, 1998, 458(1–2), 121.
34. D. Friebel, C. Schlaup, P. Broekmann, and K. Wandelt, *Surf. Sci.*, 2006, 600(13), 2800.
35. L.B. Goetting, B.M. Huang, T.E. Lister, and J.L. Stickney, *Electrochim. Acta*, 1995, 40(1), 143.
36. P. Broekmann, M. Wilms, M. Kruft, C. Stuhlmann, and K. Wandelt, *J. Electroanal. Chem.*, 1999, 467(1–2), 307.
37. M.S. Zei and G. Ertl, *Phys. Chem. Chem. Phys.*, 2000, 2(17), 3855.
38. D.S. Kong, S.H. Chen, L.J. Wan, and M.J. Han, *Langmuir*, 2003, 19(6), 1954.
39. J.A. Last and M.D. Ward, *Adv. Mater.*, 1996, 8(9), 730.
40. M. Giesen, G. Beltramo, S. Dieluweit, J. Müller, H. Ibach, and W. Schmickler, *Surf. Sci.*, 2005, 595(1–3), 127.
41. N.B. Luque, H. Ibach, K. Potting, and W. Schmickler, *Electrochim. Acta*, 2010, 55(19), 5411.
42. F. Richarz, B. Wohlmann, H. Hoffschulz, U. Vogel, and K. Wandelt, *Surf. Sci.*, 1995, 335, 361; F. Richarz, Elektrochemisch erzeugte Pt, Ru und PtRu-Elektroden: Charakterisierung und Elektrooxidation von Kohlenmonooxid, PhD thesis, University of Bonn, Bonn, Germany, 1995.
43. M.V. Lebedev, T. Mayer, and W. Jaegermann, *Surf. Sci.*, 2003, 547(1–2), 171.
44. C. Stuhlmann, W. Wohlmann, Z. Park, M. Kruft, P. Broekmann, and K. Wandelt, in: *Solid-Liquid Interfaces: Macroscopic Phenomena—Microscopic Understanding*, Topics in Applied Physics, Vol. 85, Eds. K. Wandelt and S. Thurgate (Springer-Verlag, Heidelberg, Germany, 2003, ISBN 3-540-42583-7), p. 199.
45. P. Broekmann, M. Wilms, M. Arenz, A. Spänig, and K. Wandelt, in: *Solid-Liquid Interfaces: Macroscopic Phenomena—Microscopic Understanding, Topics in Applied Physics*, Vol. 85, Eds. K. Wandelt and S. Thurgate (Springer-Verlag, Heidelberg, Germany, 2003, ISBN 3-540-42583-7), p. 141.
46. T. Dretschkow and T. Wandlowski, in: *Solid-Liquid Interfaces: Macroscopic Phenomena—Microscopic Understanding*, Topics in Applied Physics, Vol. 85, Eds. K. Wandelt and S. Thurgate (Springer-Verlag, Heidelberg, Germany, 2003, ISBN 3-540-42583-7), p. 259
47. D.M. Kolb, *Prog. Surf. Sci.*, 1996, 51(2), 109.
48. O.M. Magnussen, *Chem. Rev.*, 2002, 102, 679.
49. A. Spänig, P. Broekmann, and K. Wandelt, unpublished results.
50. B. Obliers, P. Broekmann, and K. Wandelt, *J. Electroanal. Chem.*, 2003, 554–555, 183.
51. P. Broekmann, M. Wilms, A. Spänig, and K. Wandelt, *Prog. Surf. Sci.*, 2001, 67, 59.
52. Lennartz, P. Broekmann, M. Arenz, C. Stuhlmann, and K. Wandelt, *Surf. Sci.*, 1999, 442, 215.
53. K.-I. Ataka and M. Osawa, *Langmuir*, 1998, 14, 951.
54. D. Friebel, PhD thesis, University of Bonn, Bonn, Germany, 2007.
55. Y. Nagai, M.S. Hei, D.M. Kolb, and G. Lehmpfuhl, *Ber. Bunsengesellschaft—Phys. Chem. Chem. Phys.*, 1984, 88(4), 340.
56. O.M. Magnussen, J. Hotlos, R.J. Nichols, D.M. Kolb, and R.J. Behm, *Phys. Rev. Lett.*, 199, 64(24) 2929.
57. M. Nakamura, O. Endo, T. Ohta, M. Ito, and Y. Yoda, *Surf. Sci.*, 2002, 514(1–3), 227.
58. M.F. Toney, J.N. Howard, J. Richter, G.L. Borges, J.G. Gordon, O.R. Melroy, D. Yee, and L.B. Sorensen, *Phys. Rev. Lett.*, 1995, 75(24), 4472.

59. Z.C. Shi and J. Lipkowski, *J. Electroanal. Chem.*, 1994, 365(1–2), 303.
60. O.M. Magnussen, J. Hotlos, R.J. Behm, N. Batina, and D.M. Kolb, *Surf. Sci.*, 1993, 296(3), 310.
61. G. Tourillon, D. Guoy, and A. Tadjeddine, *J. Electroanal. Chem.*, 1990, 289(1–2), 263.
62. C. Schlaup and K. Wandelt, to be published.
63. K. Sastikala, N. Furuya, and K. Itaya, *J. Electroanal. Chem.*, 1991, 316(1–2), 361.
64. Y. Shingaya, M. Matsumoto, H. Ogasawara, and M. Ito, *Surf. Sci.*, 1995, 335(1–3), 23.
65. H. Matsumoto, J. Inukai, and M. Ito, *J. Electroanal. Chem.*, 1994, 379(1–2), 223.
66. D.M. Kolb, R. Kolz, and K. Yamamoto, *Surf. Sci.*, 1979, 87(1), 20.
67. R. Durand, R. Faure, D. Aberdann, C. Salam, G. Tourillon, D. Guay, and M. Ladoucer, *Electrochim. Acta*, 1992, 37(11), 1977.
68. N.M. Markovic and P.N. Roes, *Langmuir*, 1993, 9(2), 580.
69. D. Aberdam, Y. Gauthier, R. Durand, and R. Faure, *Surf. Sci.*, 1994, 306(1–2), 114.
70. B. Schaefer, M. Nohlen, and K. Wandelt, *J. Phys. Chem. B*, 2004, 108(38), 14663.
71. N.M. Markovic, B.N.Grgur, C.A. Lucas, and P. Ross, *Electrochim. Acta*, 1998, 44(6–7), 1009.
72. G. Bertel, O.M. Magnussen, and R.J. Behm, *Surf. Sci.*, 1995, 336(1–2), 19.
73. D.W. Suggs and A.J. Bard, *J. Phys. Chem.*, 1995, 99(20), 8349.
74. M.R. Vogt, F.A. Moller, C.M. Schilz, O.M. Magnussen, and R.J. Behm, *Surf. Sci.*, 1996, 367(2), L33.
75. T. Dretschkow and T. Wandlowski, *Electrochim. Acta*, 1999, 45(4–5), 731.
76. Y.X. Diao, M.J. Han, L.J. Wan, K. Itaya, T. Uschida, H. Micake, A. Yamakata, and M. Osawa, *Langmuir*, 2006, 22(8), 3640.
77. T. Dretschkow and T. Wandlowski, *J. Electroanal. Chem.*, 1999, 467(1–2), 207.
78. D.T. Pham, S.L. Tsay, K. Gentz, C. Zoerlein, S. Kossmann, J.S. Tsay, B. Kirchner, K. Wandelt, and P. Broekmann, *J. Phys. Chem. C*, 2007, 111(44), 16428.

16

Cold-Atom Scanning Probe Microscopy: An Overview

Andreas Günther
University of Tübingen

Hendrik Hölscher
Karlsruhe Institute of Technology

József Fortágh
University of Tübingen

16.1 Introduction

During the last few decades, scanning probe microscopy (SPM) became one of the key techniques in modern nanoscience because true atomic resolution is routinely achieved with scanning tunneling microscopy (STM) and atomic force microscopy (AFM). The simultaneous imaging of electronic, magnetic, thermal, and chemical properties of nanostructures is also possible and gave valuable insight into the development of new nanodevices. In this chapter, we give an introduction to a new scanning probe technique called cold-atom scanning probe microscopy (CA-SPM). Since the basic idea resembles that of AFM, we shortly review its basic principles. In CA-SPM, gaseous probe tips confined in an electromagnetic trap are used to analyze the sample. We describe topographical imaging and ultra-sensitive force measurements with clouds of thermal atoms and Bose–Einstein condensates (BEC) and outline some prospects of this new technology.

16.2 Atomic Force Microscopy

16.2.1 Basic Principles

The basic idea of an AFM or scanning force microscope (SFM) is quite simple and resembles that of a profiler or record player (Figure 16.1). A sharp tip mounted on a cantilever scans over the sample surface and detects the tip–sample forces via a spring. In AFM, the "spring" is a bendable cantilever with a stiffness between 0.01 and 10 N/m. Since intra-atomic forces are in the range of some nN, the cantilever will be deflected by 0.01–100 nm. Consequently, precise detection of these small cantilever deflections is the key feature in AFM.

Research over the last decades has led to the development of many AFM setups in laboratories, and today AFMs are commercially available from various manufacturers. Although most of these instruments are designed for specific applications and environments, they are typically based on the following types of sensors, detection methods, and scanning principles.

16.2.1.1 Sensors

Most AFM cantilevers are produced by standard microfabrication techniques [8,79], typically from silicon or silicon nitride as rectangular or V-shaped cantilevers. Spring constants and resonance frequencies of cantilevers depend on the actual mode of operation. In a typical force microscope, cantilever deflections in the range from 0.1 Å to a few micrometers are measured. This corresponds to a force sensitivity ranging from 10^{-13} to 10^{-5} N.

Figure 16.2 shows two scanning electron microscope (SEM) images of a typical rectangular silicon cantilever. Using this imaging technique, the length (l), width (w), and thickness (t) of the cantilever can be precisely measured, and the spring constant k can then be calculated from these values [56]. Since the dimensions of cantilevers given by the manufacturer are only average values, high-accuracy calibration of the spring constant requires the measurement of length, width, and thickness for each individual cantilever. The length and width can be measured with sufficient accuracy using an optical microscope; however the thickness requires high-resolution techniques such as SEM. In order to avoid this time- and cost-consuming measurement, several researchers developed many other methods to calibrate

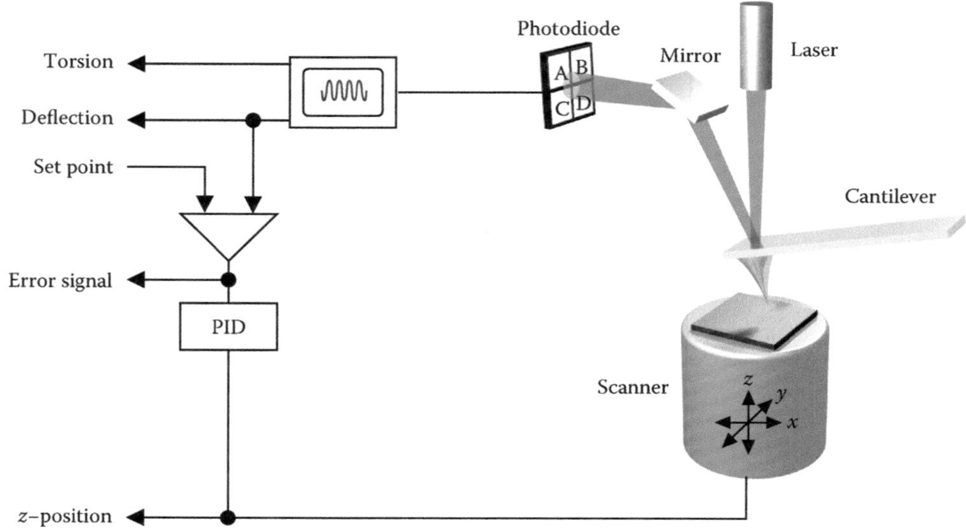

FIGURE 16.1 Operating principle of an AFM using the laser beam deflection method. Deflection (normal force) and torsion (friction) of the cantilever are measured simultaneously by measuring the lateral and vertical deflections of a laser beam while the sample is scanned in the x–y-plane. The deflection of the laser beam is determined using a four-quadrant photodiode: If A, B, C, and D are proportional to the intensity of the incident light of the corresponding quadrant, the signal (A + B) − (C + D) is a measure of the deflection and (A + C) − (B + D) a measure of the torsion of the cantilever. A schematic of the feedback system is shown by solid lines. The actual deflection signal of the photodiode is compared with the set point chosen by the experimentalist. The resulting error signal is fed into the PID controller which moves the z-position of the scanner in order to minimize the deflection signal. (From Hölscher, H., Atomic force microscopy and spectroscopy, in: *Dynamical Force Spectroscopy and Recognition*, eds. A.R. Bizzari and S. Cannistraro, Taylor & Francis Group, Boca Raton, FL, 2012, pp. 51–91.)

the forces measured with an AFM [7,14,28,72,78]. Today, most commercial AFM setups include software routines to calibrate the vertical spring constant via thermal noise analysis [11,14,40] in a convenient way.

16.2.1.2 Detection Schemes

Today, the majority of commercial AFMs use the so-called *laser beam deflection* scheme illustrated in Figure 16.1. The bending and torsion of cantilevers can be detected using a laser beam reflected from the backside of the cantilever [2,57], while the reflected laser spot is detected with a sectioned photodiode. The different photodiode sections are read out separately. Usually a four-quadrant diode is used to detect the normal as well as the torsional movements of the cantilever. With the cantilever at equilibrium, the position of the laser spot is adjusted such that the upper and the lower sections show the same intensity. If the cantilever bends up or down, the spot moves, and the difference signal between the upper and lower sections is a measure of the bending. A detailed analysis of the optimal position where to focus the laser spot on the backside of the cantilever was given by Schäffer and Fuchs [70].

The sensitivity can be even improved by interferometer systems adapted by several research groups [3,44,61,68]. This detection scheme is mainly used in ultra-high vacuum (UHV) systems. It is also possible to use cantilevers with integrated deflection sensors based on piezoresistive films [51,77,80]. Since no optical parts are needed in the experimental setup of an AFM with self-sensing cantilevers, their design can be very compact [73]. However, since it is very difficult to produce piezoresistive cantilevers with a high, consistent quality, they are rarely used.

16.2.1.3 Scanning Principle

During scanning of the sample surface, the deflection of the cantilever is kept constant by a feedback system, which controls the vertical movement of the scanner. A schematic of the feedback system is drawn in Figure 16.1. The operation principle is as follows: the current signal of the photodiode is compared with a preset value. The feedback system including a proportional, integral, and differential (PID) controller varies the z-movement of the scanner to minimize the difference. As a consequence, the tip–sample force is kept practically constant for an optimal setup of the PID parameters.

While the cantilever is moving relative to the sample in the x–y-plane of the surface by a piezoelectric scanner, the current z-position of the scanner is recorded as a function of the lateral x–y-position with (ideally) sub-Ångström precision. The obtained data represents a map of equal forces, which is analyzed and visualized by computer processing.

In principle, every type of force can be measured with an AFM. The obtained contrast and sensitivity, however, depend on the operational mode and the actual tip–sample interactions, that is, magnetic force can be only detected with magnetized tips. In general, the effective tip–sample interaction force will be a sum of different force contributions [10,42,69].

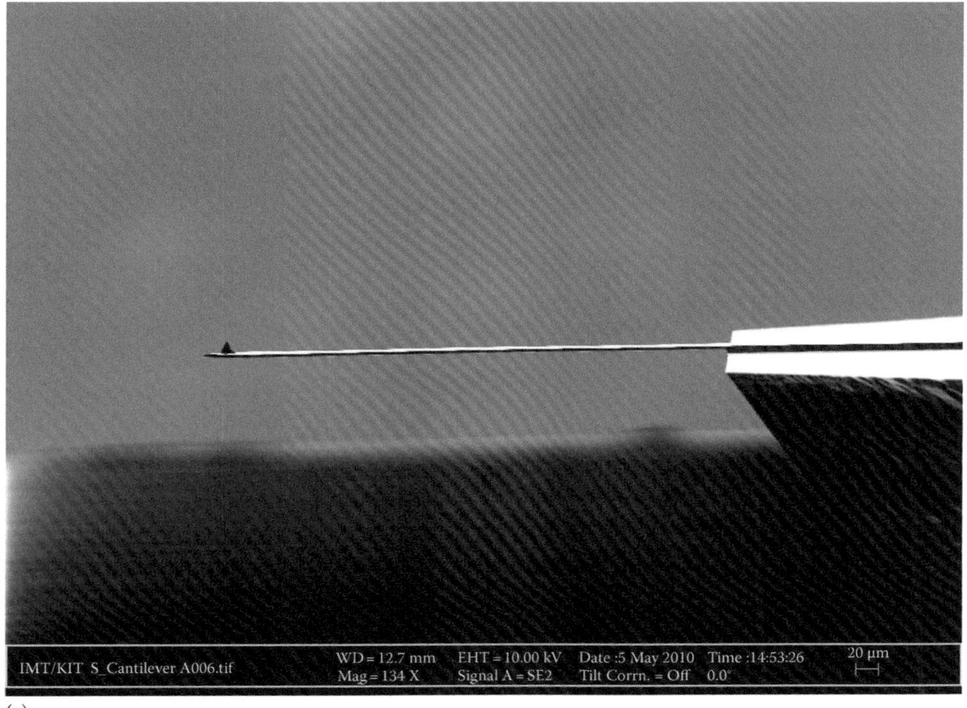

(a)

(b)

FIGURE 16.2 (a) Scanning electron micrograph of a rectangular silicon cantilever. (b) A closer view of the tip reveals its pyramidal shape obtained by the anisotropic etching of silicon. (From Hölscher, H., Atomic force microscopy and spectroscopy, in: *Dynamical Force Spectroscopy and Recognition*, eds. A.R. Bizzari and S. Cannistraro, Taylor & Francis Group, Boca Raton, FL, 2012, pp. 51–91.)

16.2.2 Modes of Operation

16.2.2.1 Contact Mode

An AFM can be driven in different modes of operation; the historically oldest is the contact mode. In order to distinguish it from the later introduced dynamic modes, the contact mode is also sometimes referred to as static mode. While static mode operation employs a very straightforward operating principle, it can still be used to easily obtain nanometer resolution images on a wide variety of surfaces. Furthermore, it has the advantage that not only the deflection, but also the torsion of the cantilever can be measured. As shown by Mate et al. [55], the lateral force can be directly correlated to the friction between tip and sample, thus extending AFM to *friction force microscopy* (FFM) [37].

16.2.2.2 Dynamic Modes

Despite the success of contact mode AFM, the resolution is limited in many cases (in particular for soft samples) by lateral forces acting between the tip and the sample. In order to avoid this effect, the cantilever can be oscillated in vertical direction near the sample surface. Imaging with vibrating cantilever is often denoted as *dynamic force microscopy* (DFM).

The most common scheme of cantilever excitation in DFM imaging is the external driving of the cantilever at a fixed excitation frequency exactly at or very close to the cantilever's first resonance [53,54,64,81]. For this driving mechanism, different detection schemes measuring either the change of the oscillation amplitude or the phase shift were proposed. Amplitude modulation (AM) or "tapping" mode, where the oscillation amplitude is used as a measure of the tip–sample distance, has developed into the most widely used technique for imaging under ambient conditions and liquids.

Under the influence of tip–sample forces, the resonant frequency (and consequently also amplitude and phase) of the cantilever will change and serve as the measurement parameters. If the tip approaches the surface, the oscillation parameters, amplitude and phase are influenced by the tip–surface interaction, and can therefore be used as feedback channels. A certain set point, for example, the amplitude, is given, and the feedback loop will adjust the tip–sample distance such that the amplitude remains constant. The control parameter is recorded as a function of the lateral position of the tip with respect to the sample, and the scanned image essentially represents the surface topography.

Figure 16.3 is a sketch of the experimental setup of a DFM utilizing the AM technique. As in the contact mode, the deflection of the cantilever is typically measured using the laser beam deflection method. During operation in conventional AM mode, the cantilever is driven at a fixed frequency by a constant amplitude signal originating from an external function generator, while the resulting oscillation amplitude and/or the phase shift are detected by a lock-in amplifier. The function generator supplies not only the signal for the dither

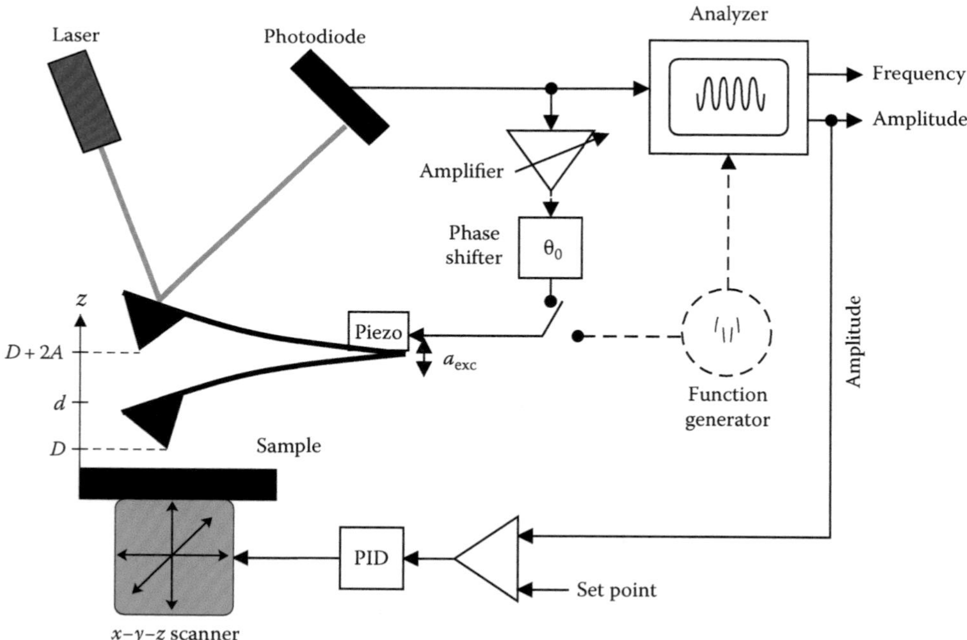

FIGURE 16.3 Schematic drawing of the experimental setup of a DFM where the driving of the cantilever can be switched between the AM mode (solid lines) and FM mode (dashed lines). While the cantilever in the AM mode is externally driven with a frequency generator, the FM mode exhibits a feedback loop consisting of a time ("phase") shifter and an amplifier. In both cases, we assume that the laser beam deflection method is used to measure the oscillation of the tip which oscillates between the nearest tip–sample position D and $D + 2A$. The equilibrium position of the tip is denoted as d. (From Hölscher, H., Atomic force microscopy and spectroscopy, in: *Dynamical Force Spectroscopy and Recognition*, eds. A.R. Bizzari and S. Cannistraro, Taylor & Francis Group, Boca Raton, FL, 2012, pp. 51–91.)

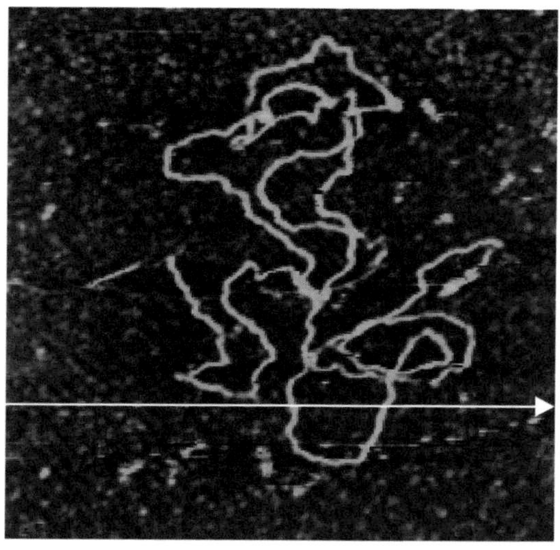

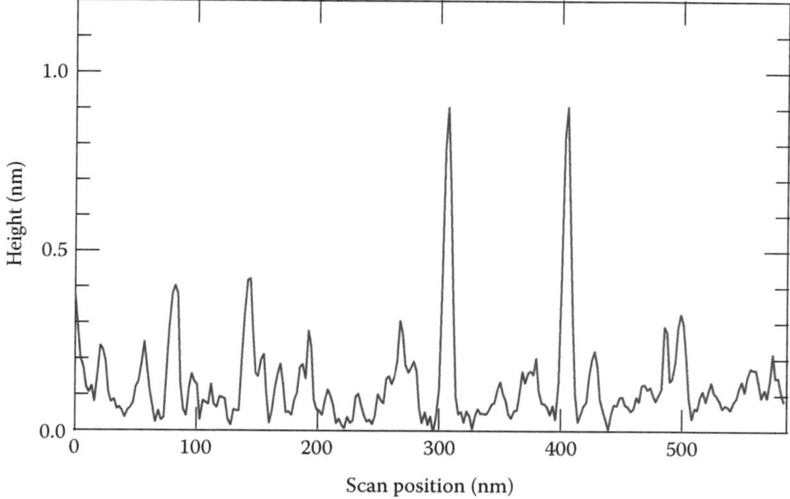

FIGURE 16.4 Topography of DNA adsorbed on mica imaged in buffer solution by tapping mode AFM. The right graph shows a single scan line obtained at the position marked by an arrow in the left image. (From Hölscher, H., Atomic force microscopy and spectroscopy, in: *Dynamical Force Spectroscopy and Recognition*, eds. A.R. Bizzari and S. Cannistraro, Taylor & Francis Group, Boca Raton, FL, 2012, pp. 51–91.)

piezo; its signal serves simultaneously as a reference for the lock-in amplifier in the analyzer electronics.

The tapping mode can be operated in air and in liquids. A typical image for high-resolution imaging of DNA in liquid solution is shown in Figure 16.4.

To obtain high-resolution images with an AFM, it is most important to prepare clean sample surfaces free from unwanted adsorbates. Therefore, these experiments are usually performed in UHV with pressures as low as 1×10^{-10} mbar. As a consequence, most DFM experiments in vacuum utilize the so-called *frequency modulation* (FM) detection scheme introduced by Albrecht et al. [1]. In this mode, the cantilever is self-oscillated [36], and a feedback circuit consisting of an amplifier and a phase shifter (dashed lines in Figure 16.3) is used to oscillate the cantilever via the dither piezo. This specific driving mechanism ensures that the cantilever instantaneously adapts to changes in the resonance frequency.

The feedback signal in FM mode is the change of the resonance frequency $\Delta f = f - f_0$ caused by the tip–sample interaction. Typically, this frequency shift is quite small (some Hz) compared to the eigenfrequency of the cantilever (some 100 kHz). Nevertheless, the effect can be precisely measured and is stable enough to obtain true atomic resolution on various surfaces.

The FM technique enables the imaging of single point defects on clean sample surfaces in vacuum and its resolution is comparable with the STM, while not restricted to conducting surfaces [24,27,56,59]. In the years after the invention of the FM technique, the term *noncontact atomic force microscopy* (NC-AFM) was established for AFM in vacuum, because it is commonly believed that a repulsive, destructive contact between tip and sample is prevented by this technique.

The excitement about the NC-AFM technique in UHV was driven by the first results of Giessibl [26] who succeeded in imaging the true atomic structure of the Si(111)-7 × 7-surface with this technique in 1995. In the same year, Sugawara et al. [75] observed the motion of single atomic defects on InP with true atomic resolution. However, imaging on conducting or semiconducting surfaces is also possible with STM. The true potential of NC-AFM lies in the imaging of nonconducting surfaces with atomic precision. A long-standing question about the surface reconstruction of the technological by relevant material aluminum oxide could be answered in this way by Barth and Reichling [6], who imaged the atomic structure of the high-temperature phase of α-Al_2O_3 (0001). The manipulation of single Sn-atoms with NC-AFM was nicely demonstrated by Sugimoto et al. [76] who manipulated single Sn-atoms on a Ge(111)-c(2 × 8) semiconductor surface. By pushing single Sn-atoms from one lattice site to the other, they finally succeeded to write the chemical symbol "Sn" with single atoms (Figure 16.5). Today, true atomic resolution is routinely

FIGURE 16.5 Final topographic NC-AFM image of the process of rearranging single Sn-atoms on a Ge(111)-c(2 × 8) semiconductor surface at room temperature. (From Sugimoto, Y. et al., *Nat. Mater.*, 4, 156, 2005.)

obtained (for a review, see, e.g., [24,27,56,59,60]). Recent efforts concentrate on the analysis of functional organic molecules, since in the field of nanoelectronics it is anticipated that, in particular, organic molecules will play an important role as the fundamental building blocks of nanoscale electronic device elements [30].

16.3 Cold-Atom Probe Tip

The heart of an AFM is the cantilever and its attached probe tip. Thereby, the cantilever serves not only as a support for the probe tip, but also as a sensitive spring for detecting the deflection and oscillation of the tip via bending of the cantilever. The performance of an AFM is not only given by the size of the probe tip, defining the spatial resolution, but also by the characteristics of the cantilever; first, the cantilever's force constant k, which defines, in combination with the smallest measurable deflection, the force sensitivity of the microscope, and second, the cantilever's fundamental resonance frequency ω, limiting the microscope's operational speed. An ideal AFM would thus combine an ultra-sharp probe tip with a cantilever of low force constant and high resonance frequency. While force constant and resonance frequency are connected via Hook's law, $k = m\omega^2$, the key for increasing the microscope's force sensitivity lies in reducing the mass m of the cantilever and its attached probe tip. Ultimately, one would therefore replace the tip and cantilever by a single atom, strongly confined in a harmonic trapping potential [48]. This would allow not only for an almost "perfect" spatial resolution, but also tremendously reduce the mass of the macroscopic cantilever to a single atom, that is, by about 12 orders of magnitude. While trapping and manipulating single atoms are still challenging, it has become possible over the last few years to precisely manipulate and trap small ensembles of few thousand ultra-cold atoms in magnetic micro- and nanopotentials (for reviews, see [20,23,35,66]). These developments lead to the realization of CA-SPM [25].

In a CA-SPM, the probe tip is a gas of atoms that is trapped in a magnetic trap (Figure 16.6). The atom cloud can be either in a thermal state (classical gas) or a quantum gas (Bose–Einstein condensate or degenerate Fermi-gas) [4,9,18,19]. A typical cold-atom cloud consists of up to 10^3–10^5 atoms at temperatures in the range of 10 nK–1 μK and densities of 10^{12}–10^{14} cm^{-3}, eight orders of magnitude below the density of conventional AFM tips.

These dilute probe tips have exceptional purity: the trapped atoms are of the same isotope; they are all in the same internal quantum state, and in the case of a Bose–Einstein condensate, all atoms are indistinguishable and share even the same motional state. The size and shape of the probe tip are determined by the harmonic trapping potential (see Section 16.4), the thermodynamic properties of the atom cloud (atom number, temperature, density), and from interatomic interactions. A thermal cloud in the harmonic trap realizes a Gaussian tip profile with a density distribution given by

$$n(\vec{r}) \sim \prod_{i=1}^{3} e^{-\frac{x_i^2}{2\sigma_i^2}}, \quad \sigma_i = \sqrt{\frac{k_B T}{m\omega_i^2}}, \quad i = 1,\ldots 3, \tag{16.1}$$

where

ω_i are the trap frequencies
k_B is the Boltzmann constant
m is the atomic mass
T is the cloud temperature

The classical probe tip is thus characterized by 1/e tip radii of $R_i = \sqrt{2}\sigma_i$.

A Bose–Einstein condensate with repulsive interaction between the atoms (e.g., in the case of ^{87}Rb) realizes a probe tip with a parabolic density distribution:

$$n \sim \max\left\{0, 1 - \sum_{i=1}^{3} \frac{x_i^2}{R_i^2}\right\}, \quad R_i = \sqrt{\frac{2\mu}{m\omega_i^2}}. \tag{16.2}$$

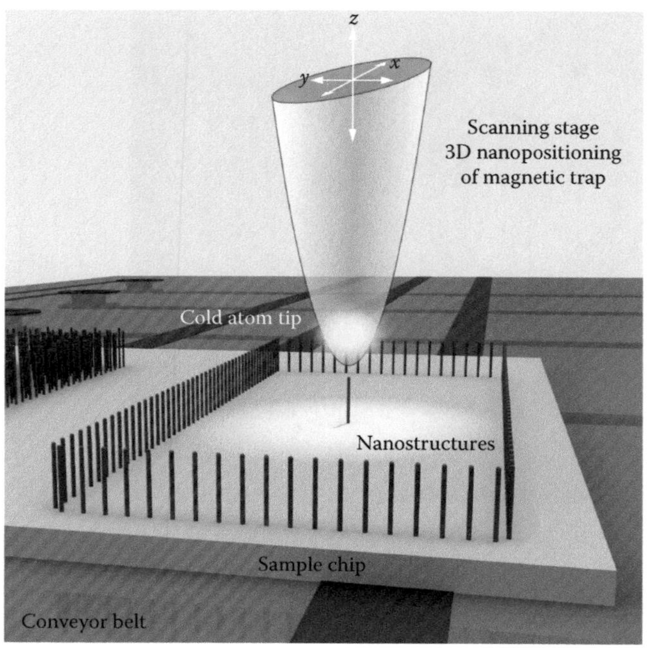

FIGURE 16.6 Working principle of a CA-SPM: The probe tip of the microscope consists of a dilute gas of ultra-cold atoms confined to a magnetic trap. The magnetic trap allows for full three-dimensional positioning of the atom cloud and serves as a microscope scanning stage. In contact mode, the surface topography is determined by scanning the cloud across the surface of interest, while monitoring atom losses from the cloud. In dynamic mode, changes in the center-of-mass oscillation of the probe tip are observed. (From Gierling, M. et al., *Nat. Nanotechnol.*, 6, 446, 2011.)

The tip radii $R_i = \sqrt{2\mu/(m\omega_i^2)}$ are given by the chemical potential

$$\mu = \frac{\hbar\omega_{\text{ho}}}{2}\left(\frac{15Na}{a_{\text{ho}}}\right)^{2/5}, \tag{16.3}$$

where

N is the atom number

a is the s-wave scattering length characterizing the interaction between atoms [17]

State-of-the-art magnetic traps have trapping frequencies in the range between tens of Hertz and hundreds of kilohertz, giving large flexibility for the tip size. In addition, interatomic interactions can be varied by means of Feshbach resonances [13,16,41]. For reaching high spatial resolution, small probe tips are of particular interest. The tip radii demonstrated so far were in the micron range [25]. The tip size, however, shrinks when increasing the trap frequency (see Equations 16.1 and 16.2). The size of the smallest possible tip can be estimated by the spatial spread of a single atomic wave function in the ground state of the harmonic oscillator potential, $a_{\text{ho}} = \sqrt{\hbar/(m\omega_{\text{ho}})}$, with the mean trap frequency, $\omega_{\text{ho}} = \prod_{i=1}^{3} \omega_i^{1/3}$. This so-called harmonic oscillator length a_{ho} gives the fundamental resolution limit of the microscope. For trapping frequencies close to 1 MHz, a tip size and spatial resolution of tens of nanometers may be reached.

The cold-atom tip is also ultra-soft. Not only in terms of its low density, but also in terms of its force constant, $k = m\omega^2$. The low density is advantageous for nondestructive topographical imaging of fragile nano-objects in the contact mode (see Section 16.5). Soft force constants (down to 10^{-19} N/m for ^{87}Rb atoms in a $\omega = 2\pi \times 80$ Hz trap), on the other hand, allow for ultra-sensitive force measurements in the dynamic mode (see Section 16.6). A force resolution on the sub-YoctoNewton-level (1 yN = 10^{-24} N) has already been demonstrated [25]. We note that force sensitivity and spatial resolution scale opposite to the trapping frequency. Thus, force mapping requires compromising these two measures. Nevertheless, even at trap frequencies of 1 MHz, the force constant of a cold-atom scanning probe tip is multiple orders of magnitude smaller than for conventional AFMs.

Besides these classical operational modes, the quantum nature of the cold-atom tip can be exploited to develop nonclassical modes. Atoms can be prepared in a coherent superposition of states. Any differential shift between the energy levels can be measured using Ramsey spectroscopy [65]. Such "atomic clocks" as ultra-soft probe tips may complement solid quantum probes. In addition, the quantum nature of the probe tip might be used for developing more accurate detection schemes for the tip position.

16.4 Microscope Setup

The CA-SPM is operated in UHV which is required for cold-atom trapping (Figure 16.7). The vacuum background limits the lifetime of the atom cloud to a few hundred seconds at 10^{-11} mbar (few seconds at 10^{-9} mbar) and limits the measurement time with this probe tip. There are standard techniques of magneto-optical trapping (MOT) and magnetic trapping for the preparation of cold-atom clouds (see, e.g., [21]) on the timescale of seconds. Clouds of up to 10^8 atoms at typical temperatures of 100 μK and densities of 10^{11} atoms/cm^3 are loaded from MOT into the conservative potential of a magnetic trap [23] or a dipole trap [29] in which the atom cloud is further cooled by forced evaporation [47]. For our demonstration experiment that follows, we use ^{87}Rb atoms in the $5S_{1/2}$, $F = 2$, $m_F = 2$ hyperfine ground state [74].

The scanning stage of the CA-SPM—optical tweezers [12,32] or a magnetic conveyor belt [31]—is then loaded with typically 10^6 atoms at 1 μK and densities of 10^{13} cm^{-3}. Depending on the requirements of the probe tip (shape, size, density), the cloud is further cooled to temperatures in the nanokelvin range. In the following, we describe nanopositioning of atom clouds by means of a magnetic conveyor belt [25].

Magnetic trapping of neutral, paramagnetic particles is a well-known technique [15,33,34,45,49,58,62]. It requires a local minimum of the magnetic field modulus $|\vec{B}|$ in free space. In vacuum, magnetic traps levitate atoms and keep them thermally isolated from the environment. In such perfect Dewar containers, atoms can be held at nanokelvin temperatures for minutes. The potential energy of a paramagnetic atom with magnetic moment $\vec{\mu}$ in the field of magnetic induction $\vec{B}$ is given by the Zeeman–Hamiltonian [23]

$$U = -\vec{\mu}\cdot\vec{B} = g_F\mu_B m_F\,|\vec{B}|. \tag{16.4}$$

Here

μ_B is the Bohr magneton

g_F and m_F are the Landé g-factor and the magnetic quantum number of the spin state, respectively

The magnetic moment is at the same time precessing around the local field vector with the Larmor-frequency:

$$\omega_L = \frac{g_F\mu_B\,|\vec{B}|}{\hbar}. \tag{16.5}$$

Even cold atoms do not rest in the trap; thus, their magnetic moment turns with the magnetic field along the trajectory of motion. If an atom is in a "low field seeking state" with the magnetic moment antiparallel to $\vec{B}$ (i.e., $g_F m_F > 0$), it remains trapped if at any position in the trap, the condition $d\omega_L/dt < \omega_L^2$ is fulfilled. This condition of "adiabatic following" is violated in magnetic quadrupole fields with $B = 0$ at the center, as $\omega_L = 0$ at this position. However, it is fulfilled in the so-called Ioffe–Pritchard trap configurations [5,63] with $B \neq 0$ at the trap center. Then m_F is a good quantum number and conserved throughout the atomic motion in the trap. Such a configuration can be achieved, for example, by adding an offset field to a linear quadrupole trap as illustrated and explained in Figure 16.8. The magnetic field of this configuration is calculated with the Biot–Savart law, and a

FIGURE 16.7 (a) Vacuum chamber (10^{-11} mbar) with viewports and optics for trapping and detection of cold atoms. The "microscope head" is attached to the top flange. The electric feed-throughs on top supply electric currents (ampere range) for the electromagnets. (b) "Microscope head" on the UHV flange. The electromagnets are used for trapping atoms (c). The magnetic conveyor belt ("scanning stage") (d) is mounted upside down between the trapping magnets.

Taylor expansion of the field magnitude around the minimum gives a harmonic trapping potential with a nonvanishing offset field in the center [23,31].

If B_{off} is homogeneous, we get a translation invariant "waveguide potential" with a harmonic radial confinement. With an inhomogeneous B_{off}, as given by the pair of parallel conductors in Figure 16.8, we get a three-dimensional harmonic trap. The oscillation frequencies ω_r and ω_a give the quantum-mechanical level spacing of the potential ($\hbar\omega$). It is now just a small step to construct a magnetic conveyor belt as shown in Figure 16.9: the field of the trapping wire (R2) is biased by the field B_{bias} of parallel wires (R1 and R3) to form a linear quadrupole. The pair of

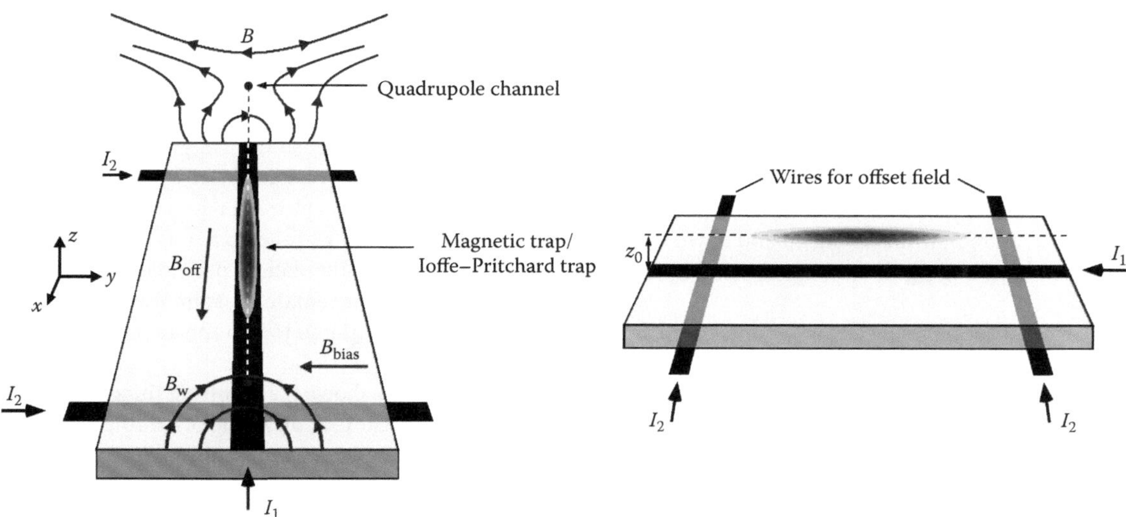

FIGURE 16.8 Magnetic trapping on a chip. The magnetic field of a current-carrying wire (I_1) is biased with the homogeneous field B_{bias} to form a line of $B = 0$ parallel to the wire at a distance $z_0 = \frac{\mu_0}{2\pi}\frac{I_1}{B_{bias}}$ above the chip. Starting from this line, the magnetic field modulus increases linearly in the transverse directions (y and z), thus forming a linear quadrupole trap for ultra-cold atoms with a radial field gradient $a_r = \frac{2\pi}{\mu_0}\frac{B_{bias}^2}{I_1}$. If an inhomogeneous magnetic offset field B_{off} is applied along x (the field of a pair of parallel conductors carrying the same current I_2) the total field near the quadrupole reads $\vec{B} = (B_{off}, a_r y, a_r z)$. Close to the quadrupole channel, that is, for $a_r^2(y^2 + z^2) \ll B_{off}$, the field modulus $|\vec{B}| = [a_r^2(y^2 + z^2) + B_{off}^2]^{1/2}$ can thus be Taylor-expanded to $|\vec{B}| \approx B_{off} + \frac{1}{2}\frac{a_r^2}{B_{off}}(y^2 + z^2)$. Hence, a "Ioffe–Pritchard" trap is formed with three-dimensional harmonic confinement and a nonvanishing offset field in the trap center.

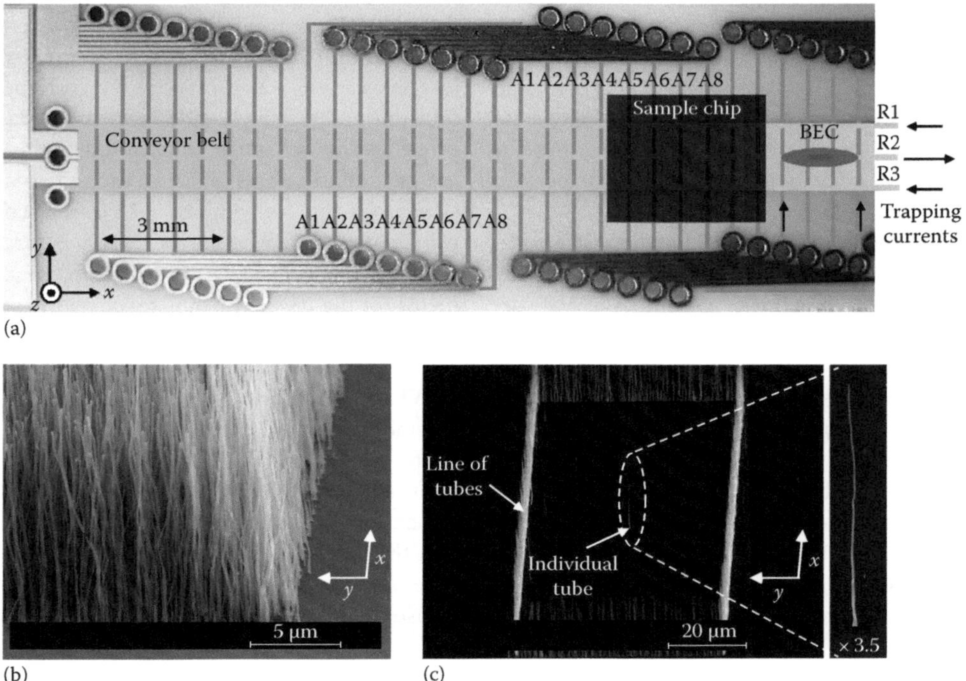

FIGURE 16.9 Scanning stage of the CA-SPM. A magnetic conveyor belt is used for nanopositioning cold-atom probe tips near the nanostructure of interest. (a) The chip contains a full set of microfabricated wires for excitation-free transport and nanopositioning of atom clouds near sample chips. (b) and (c) Structures on the sample chip that are used to demonstrate the operation of the microscope: a "forest"' of carbon nanotubes and a freestanding carbon nanotube surrounded by a nanotube fence. (From Gierling, M. et al., *Nat. Nanotechnol.*, 6, 446, 2011.)

perpendicular wires is complemented by another set of parallel wires (A1–A8), and the set is periodically repeated along the chip. The trap is formed by applying currents to R1, R2, R3, and a perpendicular pair, for example, A3 and A6 (see Figure 16.9). The trap is translated along the chip (x-direction) by changing to another pair of perpendicular wires, for example, A4 and A7. Translation in the y-direction is achieved by changing the current ratio in R1, R2, and R3, while lifting up and down the cloud along z is achieved by changing the current in R2 or in the biasing pair R1 and R2.

We have developed special algorithms for optimized smooth transport of atom clouds [31]. Using high-precision current drivers, three-dimensional nanopositioning of atom clouds is possible in a large volume (here, 20 mm^2 laterally and 500 μm in height) above the chip surface. This precise positioning system is the "scanning stage" for the CA-SPM. On top of the conveyor belt, several sample chips can be mounted. The one illustrated in Figure 16.9 contains carbon nanotube structures that are used to illustrate the performance of the microscope.

16.5 Contact Mode

The contact mode of the CA-SPM allows for three-dimensional topographical imaging and dispersion force mapping of nanostructured surfaces and fragile nano-objects. In this operation mode, the probe tip is brought into partial overlap with the surface of interest, resulting in well-defined atom losses from the cloud. Scanning the probe tip across the surface, while monitoring these losses, provides information about the position and geometry of nanoscaled objects [25]. Time-resolved measurements of the losses reveal additional information about the actual loss processes and allows insight into dispersion forces between nano-objects and probe tip atoms. These measurements can be used to determine the scattering radii of nanoscaled objects and local dispersion force measurements [71].

The contact mode of the CA-SPM has been demonstrated by topographically imaging the position of a single freestanding carbon nanotube, surrounded by a line of nanotubes [25]. Figure 16.9c shows a scanning electron micrograph of the corresponding nanotube pattern. All tubes are multiwalled and aligned perpendicular to the underlying silicon sample chip surface. The nanotubes are grown via plasma-enhanced chemical vapor deposition [52,67], which allows for precise positioning of each individual tube. As seen from the micrograph, the line of carbon nanotubes is formed in a rectangular shape with edge lengths of 50 and 150 μm. A freestanding tube is placed directly in the center of this rectangle. It has a length of 10.25 μm and a varying width from bottom to top between 275 and 40 nm.

For topographically imaging this tube pattern, we use the CA-SPM with a BEC probe tip, which we laterally scan across the sample. At each lateral position (x, y), we prepare the probe tip at a distance of 50 μm above the surface, where no interactions yet take place. We then move the BEC within 300 ms to the desired position d above the surface, where the dilute probe tip starts interacting with the surface structures. After 2 s of interaction, we speedily move the cloud away from the surface and measure the number of remaining probe tip atoms by standard absorption imaging techniques [46]. Therefore, the BEC is allowed to freely expand for a given time-of-flight, before it is illuminated by a resonant laser beam. The resulting shadow image of the BEC is captured on a CCD camera, from which the number of remaining probe tip atoms is deduced. Repeating this procedure for different lateral positions (x, y) across the surface and monitoring the remaining atom fraction in the probe tip, we gain a two-dimensional topographical image of the sample chip surface at height d.

Figure 16.10a shows the resulting image for a total scan area of 100 μm × 250 μm, at a height $d = 10$ μm above the sample chip surface. The image contrast is given by the remaining atom fraction in the probe tip. If at a given lateral position there is no nanotube, the cloud is almost fully recovered after the interaction time (see white areas in Figure 16.10a). However, if the probe tip touches a nanotube, atoms are removed due to inelastic scattering. The remaining atom fraction in the probe tip is then reduced depending on the interaction time and the size of the overlap between tip and surface (see black areas in Figure 16.10a). The image clearly shows strong losses at the position of the nanotube lines (dashed line) and unveils an isolated loss area directly in the center, corresponding to the single freestanding carbon nanotube. The image allows for clear separation of the single tube from its surrounding tube lines.

The maximum contrast is reached for a linescan in the y-direction across the position of the single tube. Figure 16.10a shows this linescan for a classical probe tip (open circles) and a quantum tip (black dots). Both measurements clearly show the loss areas at $y = \pm 25$ μm and $y = 0$ corresponding to the nanotube line and the single tube, respectively. However, the classical probe tip shows a much reduced image contrast as compared to the quantum tip. This is due to the rather large size of the classical probe tip (1/e width of ≈ 22 μm) as compared to the BEC tip (Thomas–Fermi width ≈ 8 μm). Future realizations of the CA-SPM might be capable of realizing much stronger confinements for the probe tip, thus reducing the probe tip size by several orders of magnitude. Lateral resolutions in the 10 nm regime could thus become possible (see Section 16.3).

While the CA-SPM probe tip is dilute, contact mode measurements are not limited to two dimensions. As there is no real "hard body" contact, the surface topography can be probed in arbitrary heights above the surface. This allows for full three-dimensional imaging of nanostructured objects and surfaces. It is thus possible not only to measure the position of a single tube but also its height. Figure 16.11 shows the corresponding measurement. While the lateral BEC probe tip position is fixed at the single nanotube ($x = 0$, $y = 0$), the vertical probe position d is scanned. As before, the remaining atom fraction is measured after 2 s of interaction. As the probe tip moves deeper and deeper into the single tube, the remaining atom fraction drops, until it reaches zero at about $d \approx 273$ μm. This is when the top of the tube touches the center of the probe tip. Repeating the same

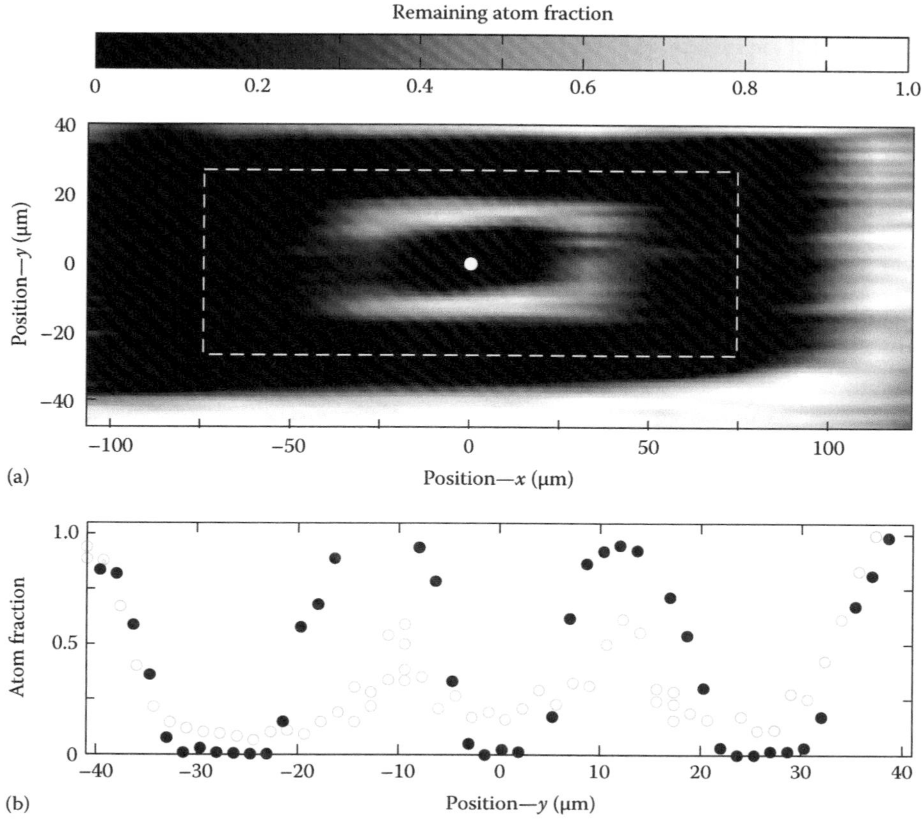

FIGURE 16.10 (a) Two-dimensional CA-SPM image of a single carbon nanotube surrounded by a nanotube fence. The position of the single tube at $x = 0$, $y = 0$ is clearly resolved. (b) One-dimensional linescan across the single tube, revealing three areas of atom losses, corresponding to the single tube at $y = 0$ and the tube lines at $y = \pm 25$ μm. Comparing the linescans for a classical probe tip (open circles) and a quantum probe tip (black dots), the latter shows largely increased image contrast due to the reduced size of a BEC as compared to a thermal gas. (From Gierling, M. et al., *Nat. Nanotechnol.*, 6, 446, 2011.)

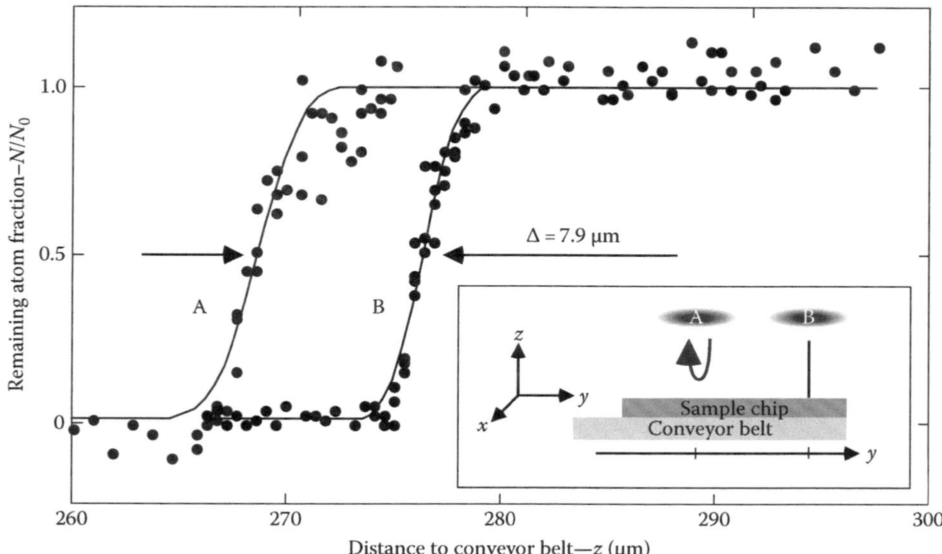

FIGURE 16.11 Scan of the CA-SPM in vertical direction above the plain surface (A) and above a single freestanding nanotube (B). The tube causes the losses to start far away from the surface. The displacement of the loss curves allows for determining the length of the carbon nanotube after taking attractive surface interactions into account. We note that d corresponds to the distance above the conveyor belt which includes the thickness of the sample chip. (From Gierling, M. et al., *Nat. Nanotechnol.*, 6, 446, 2011.)

measurement above the plain surface, we see that the loss curve is shifted by about 7.9 μm toward the surface. However, this loss does not occur directly at the sample chip surface, but slightly above due to the attractive Casimir–Polder interactions between the probe tip atoms and the silicon surface [39,50], causing an opening of the magnetic trap potential. From calculations, we find that the trap fully opens at a distance of 2.3 μm above the surface [25]. The length of the nanotube can thus be estimated to 7.9 + 2.3 = 10.2 μm, which is in perfect agreement with the SEM micrographs.

While the interaction time has certainly a strong influence on the CA-SPM image contrast, it can be used to gain additional information on the interaction between the probe tip and the nanosized objects. As the probe tip is chemically pure and very dilute, such measurements give insight into the fundamental dispersion forces between single atoms and nanoscaled objects. In the first approach, ultra-cold classical probe tips (T = 40 nK) have been used to measure the cloud losses with temporal resolution at an individual freestanding tube. Therefore, the probe tip has been prepared at a given spatial overlap with the nanotube. From the loss data, it was possible to extract the velocity-dependent scattering radius of the carbon nanotube and the strength of the underlying Casimir–Polder interaction [71]. It had been shown that the scattering radius for the ultra-cold atoms is approximately 220 nm, which was at least three times larger than the tube's geometrical radius. The corresponding dispersion force at this scattering radius is approximately 3×10^{-24} N. Such force measurements are far beyond the scope of current AFMs.

16.6 Dynamic Mode

While in "contact mode" the probe tip suffers strong atom losses due to direct contact interaction with the surface structures of interest, the "dynamic mode" of the CA-SPM allows for almost loss-free measurements [25]. In this mode, the probe tip is excited to a center-of-mass oscillation within the trap. The frequency and amplitude of this oscillation are then monitored while scanning the probe tip across the sample. If the probe tip comes close to the structures of interest, attractive dispersion forces (i.e., van der Waals or Casimir–Polder forces) influence and deform the tip's trapping potential. This results in a change in the tip's oscillation frequency and amplitude, even if the tip is not in direct contact with the nano-object. If the tip's oscillation parameters are monitored with high accuracy, this allows for loss-free force imaging of surfaces and nano-objects.

Similar to the loss measurements, the oscillation frequency and amplitude are determined from optical absorption images. The probe tip position is determined by fitting a model function for the tip's density profile to these images, yielding the tip's center-of-mass position with a resolution limited by the optical imaging system. In the given experiment, this resolution is ~6 μm. However, a simple trick allows for the determination of much smaller oscillation amplitudes of the probe tip. If the probe tip is allowed to freely expand for some time-of-flight τ before imaging, an in situ harmonic oscillation $x(t)$ of the probe tip with frequency ω, amplitude A, and phase φ

$$x(t) = A\cos(\omega_r t + \varphi) \tag{16.6}$$

is then (after time-of-flight τ) observed as harmonic oscillation

$$X(t) = x(t) + \dot{x}(t)\tau = \tilde{A}\cos(\omega_r t + \tilde{\varphi}), \tag{16.7}$$

showing the same frequency ω, but different amplitude $\tilde{A}$ and phase $\tilde{\varphi}$

$$\tilde{A} = A\sqrt{1+\omega_r^2\tau^2}, \quad \tilde{\varphi} = \arctan\frac{\sin\varphi + \omega_r\tau\cos\varphi}{\cos\varphi - \omega_r\tau\sin\varphi}. \tag{16.8}$$

Depending on the oscillation frequency ω and the free expansion time τ, the oscillation amplitude is thus increased by a factor of $\sqrt{1+\omega^2\tau^2}$ [31]. With a radial trap frequency of $\omega_r \approx 2\pi \times$ 80 Hz and typically τ = 15 ms time-of-flight, the oscillation amplitude is therefore amplified by a factor of 7.6. Thus, oscillation amplitudes as small as 6 μm/7.6 = 0.8 μm can, in principle, be detected.

The dynamic mode of the CA-SPM has been demonstrated using the same freestanding carbon nanotube as that used for the contact mode measurements. A BEC probe tip has been prepared 20 μm above the chip surface and scanned in the y-direction across the single tube. At each lateral position, the magnetic trap has been suddenly displaced toward the surface, initiating a center-of-mass oscillation of the BEC probe tip in the trap's radial direction. The oscillation parameters of this oscillation are determined by taking a time series of absorption images, from which the tip position $x(t)$ is extracted. Amplitude, frequency, and phase of this oscillation can be determined according to Equation 16.8, by fitting the model function (Equation 16.7) to the data.

Figure 16.12 shows the relative frequency shift and the oscillation amplitude as a function of the tip's lateral position y. The oscillation frequency is typically determined with a relative accuracy of about 1×10^{-3}, which could in principle be further increased by extending the oscillation measuring time. The amplitude is determined with an accuracy of about 500 nm, limited by the time-of-flight and the optical imaging system. Both the relative frequency shift and the oscillation amplitude show a clear signature of the single nanotube crossed at $y = 0$. While the Casimir–Polder potential is attractive, the tip's resonance frequency at the nanotube is decreased by about 1%. However, the oscillation amplitude is increased by more than 30%. This indicates a strong anharmonicity of the trapping potential due to the nanotube interaction. According to $F = m\omega^2 x$, such a change in oscillation amplitude is expected from a force F of about 2×10^{-25} N acting on each atom.

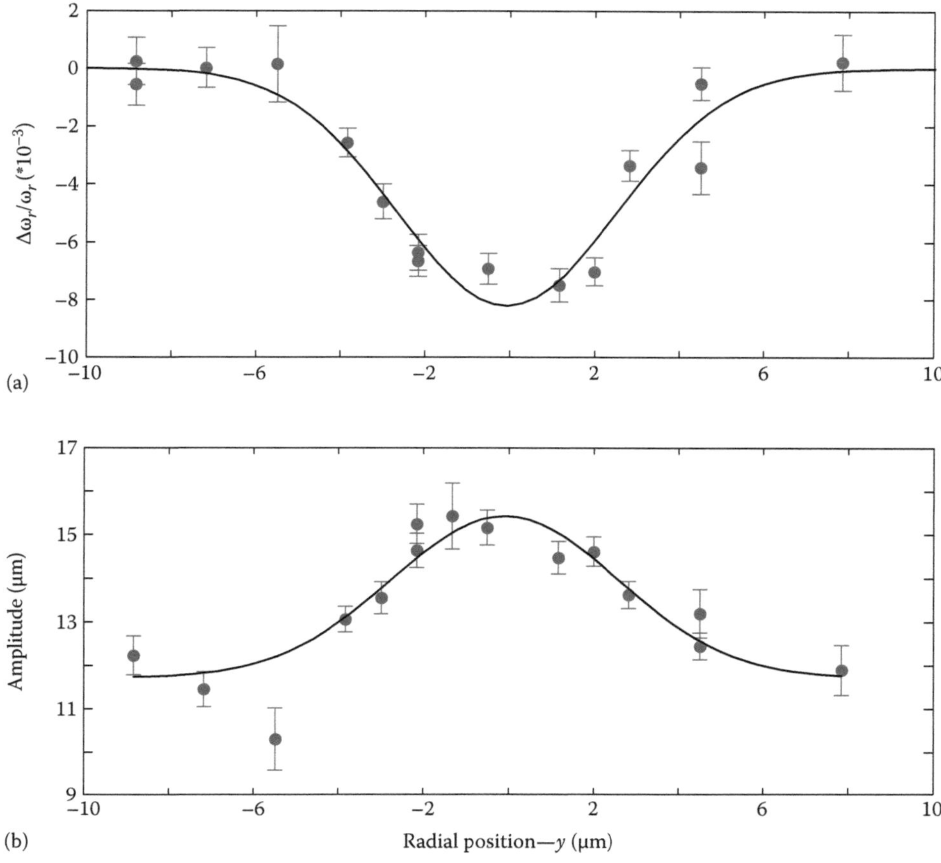

FIGURE 16.12 Dynamic mode of the CA-SPM. An oscillating BEC is scanned across a single nanotube positioned at $y = 0$. The attractive potential between the nanotube and BEC results in a clear signature of the nanotube tip's oscillation frequency (a) and amplitude (b). (From Gierling, M. et al., *Nat. Nanotechnol.*, 6, 446, 2011.)

16.7 Discussion and Outlook

The development of CA-SPMs has just started. In future, it should be possible to transfer some of the procedures and techniques already developed for the AFM to CA-SPM in order to increase performance and resolution. The extreme purity of the probe tip and the quantum control over the atomic states open up additional possibilities of SPM with nonclassical probe tips. Besides imaging techniques, the CA-SPM establishes an interface between quantum gases and solid nanodevices. Nanopositioning atoms near a nanoscaled object aids the development of hybrid quantum sensors based, for example, on atoms and nanowires [43].

References

1. T. R. Albrecht, P. Grütter, D. Horne, and D. Rugar. Frequency modulation detection using high-Q cantilevers for enhanced force microscope sensitivity. *J. Appl. Phys.* **69**, 668–673 (1991).
2. S. Alexander, L. Hellemans, O. Marti, J. Schneir, V. Elings, and P. K. Hansma. An atomic-resolution atomic-force microscope implemented using an optical lever. *J. Appl. Phys.* **65**, 164–167 (1988).
3. W. Allers, A. Schwarz, U. D. Schwarz, and R. Wiesendanger. A scanning force microscope with atomic resolution in ultra-high vacuum and at low temperatures. *Rev. Sci. Instrum.* **69**, 221–225 (1998).
4. M. H. Anderson, J. R. Ensher, M. R. Matthews, C. E. Wieman, and E. A. Cornell. Observation of Bose–Einstein condensation in a dilute atomic vapor. *Science* **269**, 198–201 (1995).
5. V. S. Bagnato, G. P. Lafyatis, A. G. Martin, E. L. Raab, R. N. Ahmad-Bitar, and D. E. Pritchard. Continuous stopping and trapping of neutral atoms. *Phys. Rev. Lett.* **58**, 2194–2197 (1987).
6. C. Barth and M. Reichling. Imaging the atomic arrangement on the high-temperature reconstructed α-Al_2O_3 (0001) surface. *Nature* **414**, 54–57 (2001).
7. P. Bilas, L. Romana, B. Kraus, Y. Bercion, and J. L. Mansot. Quantitative characterization of friction coefficient using lateral force microscope in the wearless regime. *Rev. Sci. Instrum.* **75**, 415–421 (2004).
8. G. Binnig, C. Gerber, E. Stoll, T. R. Albrecht, and C. F. Quate. Atomic resolution with atomic force microscope. *Europhys. Lett.* **3**, 1281 (1987).

9. C. C. Bradley, C. A. Sackett, J. J. Tollett, and R. G. Hulet. Evidence of Bose–Einstein condensation in an atomic gas with attractive interactions. *Phys. Rev. Lett.* **75**, 1687–1690 (1995).
10. H.-J. Butt and M. Kappl. *Surface and Interfacial Forces*. Wiley-VCH, Weinheim, Germany, 2010.
11. H.-J. Butt and M. Jaschke. Calculation of thermal noise in atomic force microscopy. *Nanotechnology* **6**, 1 (1995).
12. D. Cano, H. Hattermann, B. Kasch, C. Zimmermann, R. Kleiner, D. Koelle, and J. Fortágh. Experimental system for research on ultracold atomic gases near superconducting microstructures. *Eur. Phys. J. D* **63**, 17–23 (2011).
13. C. Chin, R. Grimm, P. Julienne, and E. Tiesinga. Feshbach resonances in ultracold gases. *Rev. Mod. Phys.* **82**, 1225–1286 (2010).
14. S. M. Cook, T. E. Schäffer, K. M. Chynoweth, M. Wigton, R. W. Simmonds, and K. M. Lang. Practical implementation of dynamic methods for measuring atomic force microscope cantilever spring constants. *Nanotechnology* **17**, 2135–2145 (2006).
15. E. A. Cornell and C. E. Wieman. Nobel lecture: Bose–Einstein condensation in a dilute gas, the first 70 years and some recent experiments. *Rev. Mod. Phys.* **74**, 875–893 (2002).
16. Ph. Courteille, R. S. Freeland, D. J. Heinzen, F. A. van Abeelen, and B. J. Verhaar. Observation of a Feshbach resonance in cold atom scattering. *Phys. Rev. Lett.* **81**, 69–72 (1998).
17. F. Dalfovo, S. Giorgini, L. P. Pitaevskii, and S. Stringari. Theory of Bose–Einstein condensation in trapped gases. *Rev. Mod. Phys.* **71**, 463–512 (1999).
18. K. B. Davis, M. O. Mewes, M. R. Andrews, N. J. van Druten, D. S. Durfee, D. M. Kurn, and W. Ketterle. Bose–Einstein condensation in a gas of sodium atoms. *Phys. Rev. Lett.* **75**, 3969–3973 (1995).
19. B. DeMarco and D. S. Jin. Onset of Fermi degeneracy in a trapped atomic gas. *Science* **285**, 1703–1706 (1999).
20. R. Folman, P. Krüger, J. Schmiedmayer, J. Denschlag, and C. Henkel. Microscopic atom optics: From wires to an atom chip. *Adv. At. Mol. Opt. Phys.* **48**, 263–356 (2002).
21. J. Fortágh, H. Ott, S. Kraft, A. Günther, and C. Zimmermann. Bose–Einstein condensates in magnetic waveguides. *Appl. Phys. B* **76**, 157–163 (2003).
22. J. Fortágh, and C. Zimmermann. Toward atom chips. *Science* **307**, 860–861 (2005).
23. J. Fortágh and C. Zimmermann. Magnetic microtraps for ultracold atoms. *Rev. Mod. Phys.* **79**, 235–289 (2007).
24. R. Garcia and R. Pérez. Dynamic atomic force microscopy methods. *Surf. Sci. Rep.* **47**, 197–301 (2002).
25. M. Gierling, P. Schneeweiss, G. Visanescu, P. Federsel, M. Häffner, D. P. Kern, T. E. Judd, A. Günther, and J. Fortágh. Cold-atom scanning probe microscopy. *Nat. Nanotechnol.* **6**, 446–451 (2011).
26. F.-J. Giessibl. Atomic resolution of the silicon (111)-(7 × 7) surface by atomic force microscopy. *Science* **267**, 68 (1995).
27. F.-J. Giessibl. Advances in atomic force microscopy. *Rev. Mod. Phys.* **75**, 949–983 (2003).
28. C. P. Green, H. Lioe, J. P. Cleveland, R. Proksch, P. Mulvaney, and J. E. Sader. Normal and torsional spring constants of atomic force microscope cantilevers. *Rev. Sci. Instrum.* **75**(6), 1988–1996 (2004).
29. R. Grimm, M. Weidemüller, and Y. B. Ovchinnikov. Optical dipole traps for neutral atoms. *Adv. At. Mol. Opt. Phys.* **42**, 95–170 (2000).
30. L. Gross, F. Mohn, N. Moll, P. Liljeroth, and G. Meyer. The chemical structure of a molecule resolved by atomic force microscopy. *Science* **325**, 1110–1114 (2009).
31. A. Günther, M. Kemmler, S. Kraft, C. J. Vale, C. Zimmermann, and J. Fortágh. Combined chips for atom optics. *Phys. Rev. A* **71**, 063619 (2005).
32. T. L. Gustavson, A. P. Chikkatur, A. E. Leanhardt, A. Görlitz, S. Gupta, D. E. Pritchard, and W. Ketterle. Transport of Bose–Einstein condensates with optical tweezers. *Phys. Rev. Lett.* **88**, 020401 (2001).
33. H. F. Hess, G. P. Kochanski, J. M. Doyle, N. Masuhara, D. Kleppner, and T. J. Greytak. Magnetic trapping of spin-polarized atomic hydrogen. *Phys. Rev. Lett.* **59**, 672–675 (1987).
34. T. W. Hijmans, O. J. Luiten, I. D. Setija, and J. T. M. Walraven. Optical cooling of atomic hydrogen in a magnetic trap. *JOSA B* **6**(11), 2235–2243 (1989).
35. E. A. Hinds and I. G. Hughes. Magnetic atom optics: Mirrors, guides, traps, and chips for atoms. *J. Phys. D: Appl. Phys.* **32**, R119–R146 (1999).
36. H. Hölscher, B. Gotsmann, W. Allers, U. D. Schwarz, H. Fuchs, and R. Wiesendanger. Comment on "damping mechanism in dynamic force microscopy." *Phys. Rev. Lett.* **88**, 019601 (2002).
37. H. Hölscher, A. Schirmeisen, and U. D. Schwarz. Principles of atomic friction: From sticking atoms to superlubric sliding. *Phil. Trans. R. Soc. A* **366**, 1869 (2008).
38. H. Hölscher. Atomic force microscopy and spectroscopy. In *Dynamical-Force Spectroscopy and Recognition*, eds. A. R. Bizzari and S. Cannistraro, pp. 51–91, Taylor & Francis Group, Boca Raton, FL (2012).
39. D. Hunger, S. Camerer, T. W. Hänsch, D. König, J. P. Kotthaus, J. Reichel, and P. Treutlein. Resonant coupling of a Bose–Einstein condensate to a micromechanical oscillator. *Phys. Rev. Lett.* **104**, 143002 (2010).
40. J. L. Hutter and J. Bechhofer. Calibration of atomic-force microscope tips. *Rev. Sci. Instrum.* **64**, 1868 (1993).
41. S. Inouye, M. R. Andrews, J. Stenger, H. J. Miesner, D. M. Stamper-Kurn, and W. Ketterle. Observation of Feshbach resonances in a Bose–Einstein condensate. *Nature* **392**, 151–154 (1998).
42. J. N. Israelachvili. *Intermolecular and Surface Forces*. Academic Press, London, U.K., 1992.
43. O. Kalman, T. Kiss, J. Fortágh, and P. Domokos. Quantum galvanometer by interfacing a vibrating nanowire and cold atoms. *Nano Lett.* **12**, 435–439 (2012).

44. H. Kawakatsu, S. Kawai, D. Saya, M. Nagashio, D. Kobayashi, H. Toshiyoshi, and H. Fujita. Towards atomic force microscopy up to 100 MHz. *Rev. Sci. Instrum.* **73**(6), 2317–2320 (2002).
45. W. Ketterle. Nobel lecture: When atoms behave as waves: Bose–Einstein condensation and the atom laser. *Rev. Mod. Phys.* **74**, 1131–1151 (2002).
46. W. Ketterle, D. S. Durfee, and D. M. Stamper-Kurn. Making, probing and understanding Bose–Einstein condensates. In *Bose–Einstein Condensation in Atomic Gases, Proceedings of the International School of Physics—Enrico Fermi, Course CXL*, eds. M. Inguscio, S. Stringari, and C. E. Wieman, pp. 67–176, IOS Press, Amsterdam, the Netherlands (1999).
47. W. Ketterle and N. J. van Druten. Evaporative cooling of trapped atoms. *Adv. At. Mol. Opt. Phys.* **37**, 181–236 (1996).
48. C. Kollath, M. Köhl, and T. Giamarchi. Scanning tunneling microscopy for ultracold atoms. *Phys. Rev. A* **76**, 063602 (2007).
49. K. J. Kügler, W. Paul, and U. Trinks. A magnetic storage ring for neutrons. *Phys. Lett. B* **72**, 422–424 (1978).
50. Y. Lin, I. Teper, C. Chin, and V. Vuletić. Impact of the Casimir-Polder potential and Johnson noise on Bose–Einstein condensate stability near surfaces. *Phys. Rev. Lett.* **92**, 050404 (2004).
51. R. Linnemann, T. Gotszalk, I. W. Rangelow, P. Dumania, and E. Oesterschulze. Atomic force microscopy and lateral force microscopy using piezoresistive cantilevers. *J. Vac. Sci. Technol. B* **14**(2), 856–860 (1996).
52. R. Löffler, M. Häffner, G. Visanescu, H. Weigand, X. Wang, D. Zhang, M. Fleischer, A. J. Meixner, J. Fortágh, and D. P. Kern. Optimization of plasma-enhanced chemical vapor deposition parameters for the growth of individual vertical carbon nanotubes as field emitters. *Carbon* **49**, 4197–4203 (2011).
53. Y. Martin and H. K. Wickramasinghe. Magnetic imaging by "force microscopy" with 1000 Å resolution. *Appl. Phys. Lett.* **50**, 1455–1457 (1987).
54. Y. Martin, C. C. Williams, and H. K. Wickramasinghe. Atomic force microscope—Force mapping and profiling on a sub 100-Å scale. *J. Appl. Phys.* **61**, 4723–4729 (1987).
55. C. M. Mate, G. M. McClelland, R. Erlandsson, and S. Chiang. Atomic-scale friction of a tungsten tip on a graphite surface. *Phys. Rev. Lett.* **59**(17), 1942–1945 (1987).
56. E. Meyer, H.-J. Hug, and R. Bennewitz. *Scanning Probe Microscopy—The Lab on a Tip*. Springer-Verlag, Berlin, 2004.
57. G. Meyer and N. M. Amer. Novel optical approach to atomic force microscopy. *Appl. Phys. Lett.* **53**, 1045–1047 (1988).
58. A. L. Migdall, J. V. Prodan, W. D. Phillips, T. H. Bergeman, and H. J. Metcalf. First observation of magnetically trapped neutral atoms. *Phys. Rev. Lett.* **54**, 2596–2599 (1985).
59. S. Morita, R. Wiesendanger, and E. Meyer, eds. *Noncontact Atomic Force Microscopy*. Springer-Verlag, Berlin, Germany, 2002.
60. S. Morita, F. J. Giessibl, and R. Wiesendanger, eds. *Noncontact Atomic Force Microscopy*, Vol. **2**. Springer, Berlin, Germany, 2009.
61. A. Moser, H.-J. Hug, T. Jung, U. D. Schwarz, and H.-J. Güntherodt. A miniature fibre optic force microscope scan head. *Meas. Sci. Technol.* **4**, 769–775 (1993).
62. W. Paul. Electromagnetic traps for charged and neutral particles. *Rev. Mod. Phys.* **62**, 531–540 (1990).
63. D. E. Pritchard. Cooling neutral atoms in a magnetic trap for precision spectroscopy. *Phys. Rev. Lett.* **51**, 1336–1339 (1983).
64. C. A. J. Putman, K. O. Vanderwerf, B. G. Degrooth, N. F. Vanhulst, and J. Greve. Tapping mode atomic force microscopy in liquid. *Appl. Phys. Lett.* **64**, 2454–2456 (1994).
65. N. F. Ramsey. Experiments with separated oscillatory fields and hydrogen masers. *Science* **248**, 1612–1619 (1990).
66. J. Reichel. Microchip traps and Bose–Einstein condensation. *Appl. Phys. B: Lasers Opt.* **74**, 469–487 (2002).
67. Z. F. Ren, Z. P. Huang, J. W. Xu, J. H. Wang, P. Bush, M. P. Siegal, and P. N. Provencio. Synthesis of large arrays of well-aligned carbon nanotubes on glass. *Science* **282**, 1105–1107 (1998).
68. D. Rugar, H. J. Mamin, and P. Guethner. Improved fiber-optic interferometer for atomic force microscopy. *Appl. Phys. Lett.* **55**(25), 2588–2590 (1989).
69. D. Sarid. *Scanning Force Microscopy—With Applications to Electric, Magnetic, and Atomic Forces*. Oxford University Press, New York, 1994.
70. T. E. Schäffer and H. Fuchs. Optimized detection of normal vibration modes of atomic force microscope cantilevers with the optical beam deflection method. *J. Appl. Phys.* **97**, 083524 (2005).
71. P. Schneeweiss, M. Gierling, G. Visanescu, D. P. Kern, T. E. Judd, A. Günther, and J. Fortágh. Dispersion forces between ultracold atoms and a carbon nanotube. *Nat. Nanotechnol.* **7**, 515–519 (2012).
72. U. D. Schwarz, P. Köster, and R. Wiesendanger. Quantitative analysis of lateral force microscopy experiments. *Rev. Sci. Instrum.* **67**, 2560–2567 (1996).
73. U. Stahl, C. W. Yuan, A. L. de Lozanne, and M. Tortonese. Atomic force microscope using piezoresistive cantilevers and combined with a scanning electron microscope. *Appl. Phys. Lett.* **65**(22), 2878–2880 (1994).
74. D. A. Steck. Rubidium 87 D Line Data. Available online at http://steck.us/alkalidata (revision 2.1.3, December 23, 2010).
75. Y. Sugawara, M. Otha, H. Ueyama, and S. Morita. Defect motion on an InP(110) surface observed with noncontact atomic force microscopy. *Science* **270**, 1646 (1995).

76. Y. Sugimoto, M. Abe, S. Hirayama, N. Oyabu, O. Custance, and S. Morita. Atom inlays performed at room temperature using atomic force microscopy. *Nat. Mater.* **4**, 156–159 (2005).
77. M. Tortonese, R. C. Barrett, and C. F. Quate. Atomic resolution with an atomic force microscope using piezoresistive detection. *Appl. Phys. Lett.* **62**(8), 834–836 (1993).
78. M. Varenberg, I. Etsion, and G. Halperin. An improved wedge calibration method for the lateral force in atomic force microscopy. *Rev. Sci. Instrum.* **74**, 3362 (2003).
79. O. Wolter, Th. Bayer, and J. Greschner. Micromachined silicon sensors for scanning force microscopy. *J. Vac. Sci. Technol. B* **9**, 1353 (1991).
80. C. W. Yuan, E. Batalla, M. Zacher, A. L. de Lozanne, M. D. Kirk, and M. Tortonese. Low temperature magnetic force microscope utilizing a piezoresistive cantilever. *Appl. Phys. Lett.* **65**(10), 1308–1310 (1994).
81. Q. D. Zhong, D. Inniss, K. Kjoller, and V. B. Elings. Fractured polymer/silica fiber surface studied by tapping mode atomic force microscopy. *Surf. Sci. Lett.* **290**, L688–L692 (1993).

17

Atomic Resolution Ultrafast Scanning Tunneling Microscope

Qingyou Lu
Chinese Academy of Sciences
and
University of Science and Technology of China

It has been 30 years since the invention of the atomically resolved scanning tunneling microscope (STM) by G. Binnig in 1981 [1], but one of its most severe drawbacks still remains: slowness. Because of this, the STM is, although considered a powerful real-space imaging tool, never regarded as a real-time microscope and the vast majority of today's STM applications are still in steady state imaging, thus missing the much more wonderful and important dynamic world. This chapter provides some thinking on how the STM can image faster while the real-space atomic resolution is still preserved.

17.1 Why Fast?

Before we discuss how to run an STM fast, let us briefly make clear why we need to. There are two reasons: less distortion and better real time, which are discussed next.

17.1.1 Less Distortion

Even if you just want to take some static images (do not care if the images can form a coherent movie), you still prefer a fast STM because drifting (due to temperature variation, creeping effect of the piezoelectric scanner [2–6], stress release of the tip–sample mechanical loop, mechanical vibration, circuit instability, etc.) will distort the images, which can be so severe that the imaged atoms form a completely different lattice pattern compared with the real arrangement in the sample. Figure 17.1a [7] shows an example in which the hexagonal lattice of a highly oriented pyrolytic graphite (HOPG) sample is heavily skewed and becomes rectangular, which is a totally different lattice type. This is a very serious mistake because you will draw a wrong conclusion if the sample is unknown. In addition to the distortion in scan plane, there will also be instability in tunneling current due to tip–sample-gap unsteadiness (1 Å drift in junction gap will cause roughly 10 times change in tunneling current) [8,9] and electronic drift, which lowers the image quality. While imaging in a temperature or magnetic field varying environment or in other harsh conditions (such as high temperature, high pressure, liquid, reactive gas, etc.), which reflects the real world (thus more important), the drifting issue will be much more severe.

17.1.2 Toward Real-Time Imaging

The importance of the STM is that it provides real-space atomic resolution, meaning that it measures every single atom in the scan area, including any individual atomic defects. This is not like the x-ray diffraction measurement of lattice structure or TEM images showing lattice structure, which are the average effects of multiple atoms. So, the STM can potentially shoot a movie to show how a single atomic defect (such as a dopant atom or an absorbed atom on the sample) behaves (diffuses or reacts with other atoms). Unfortunately, these behaviors are typically very fast—too fast for the STM to track by continuously imaging the same area. For instance, bromine diffusion on flat Cu (111) surface with a barrier of $E = 0.06$ eV has a hopping frequency of 4.8×10^{10} Hz at 150 K [10]. To capture this important rapid surface process as a movie in real time, we have to use a STM capable of imaging at 10^{10} frames/s, which corresponds to a line scan rate of $\sim 10^{12}$ lines/s (1000 GHz, in other words). Previously, the fastest STM with atomic resolution could scan at only 10.2 kHz [11]. We have boosted the scan rate to 26 KHz [12], which is still not

(a)

(b)

FIGURE 17.1 (a) The line-by-line scanned raw data STM image (3.04×2.45 nm^2, sample biased at +100 mV) of a HOPG sample, showing a big distortion in which the hexagonal lattice is deformed into almost rectangular type (wrong conclusion can be drawn). (b) We can correct the left image by deducting the impact of drifting since the correct lattice structure (hexagonal) is known (but this will not work for any unknown samples). (From Wang, J.T. et al., *Rev. Sci. Instrum.*, 81, 073705, 2010.)

even close to the aforementioned requirement. This severe disadvantage of the STM has made it very difficult to study many fast dynamic phenomena such as phase transition, absorption, nucleating, diffusion, and so on [11,13–15]. The development of fast STMs is hence attracting more and more attention.

17.2 How to Speed Up?

The simplest STM (constant height mode STM) consists of (1) a scan head, which implements the tip–sample coarse approach and drives a tip to scan on a sample surface with angstrom (Å) to nanometer (nm) tip–sample gap; (2) a preamplifier (preamp), which converts the weak tunneling current I_{TC} into an appropriate voltage signal V_{TC} to be acquired as image data; (3) a controller, which adjusts the tip–sample gap (via V_z signal) and generates the scan signals (V_x and V_y), thus controlling the scan method (linear scan or spiral scan, say), scan rate, and scan area; and (4) a data-acquisition (DAQ) system to collect image data (tunneling current data V_{TC} plus the corresponding position data V_x and V_y). STM images are formed by plotting the digital data converted from (V_x, V_y, V_{TC}). We will discuss in detail how we have successfully built the world's fastest STM with 26 KHz scan rate (without losing atomic resolution, of course) [12] by scrutinizing all these four aspects, plus some overall consideration of the whole system. Some easy but feasible future improvements that may increase the imaging speed further are also suggested based on these analyses together with our experimental results.

A more typical STM can also operate in constant current mode, which is more complicated (needs a PID feedback control unit to maintain a constant tunneling current by adjusting the tip–sample gap) and slower. Some severe tradeoff is hence needed if a constant current mode STM needs to speed up [11].

17.2.1 Scan Head

17.2.1.1 Scan Device

We surely need a mechanism that can scan (deform) as fast as possible. This is a limiting factor since it is difficult to mechanically move (vibrate) a macroscopic tip or sample very fast. However, in a conventional line-by-line scan (LLS) mode (interchangeably called XY scan mode here), only one direction needs to scan fast, which is herewith defined as X scan. The other scan (Y scan) is much slower and does not need special consideration. As a good try for fast scan, we use a commercial quartz tuning fork (QTF) with a characteristic frequency of $\omega_0 = 32.768$ kHz, which reduces to about $\omega_{0,+tip} = 24$ kHz after a STM tip is glued to the free prong of the fork (the other prong is fixed).

17.2.1.2 Fully Low-Voltage STM and Its Scan Head

A fast scan device (such as QTF) is typically small in size, which requires high voltage to scan in order to achieve a large scan area. However, high-voltage circuits are usually slow with poor precision and cannot drive the fast scan. Our solution is to scan (oscillate) a QTF at (or near) its resonant frequency where maximum amplitude (hence large scan range) is obtained at a low driving voltage. This is not easy though because we need to build an STM that can image at very high speed (about $\omega_{0,+tip} = 24$ kHz or higher).

There are other high voltages in a normal STM, which may diminish imaging speed. As a matter of fact, most, if not all, of STMs incorporate high-voltage devices such as high-voltage operational amplifiers (op amps) and transistors in their coarse approach, fine approach and feedback control circuits, where high voltage means >20 V, the absolute upper limit for all low-voltage op amps. The reasons are (i) the coarse approach using a piezoelectric motor typically relies on high voltage to provide sufficient step length so as to overcome friction and backlash; (ii) the fine approach that follows has to use high voltage to push the tip and sample close enough to create tunneling current, because of the large tip–sample distance left over by the high-voltage coarse approach; (iii) in cases where the sample surface corrugation is large, a substantial tip–sample regulation range is needed for the tip to follow the surface topography, which also requires high voltage. The aforementioned high-voltage devices are typically of low quality: slow, noisy, expensive with high leakage current and instability and low precision, etc., which greatly

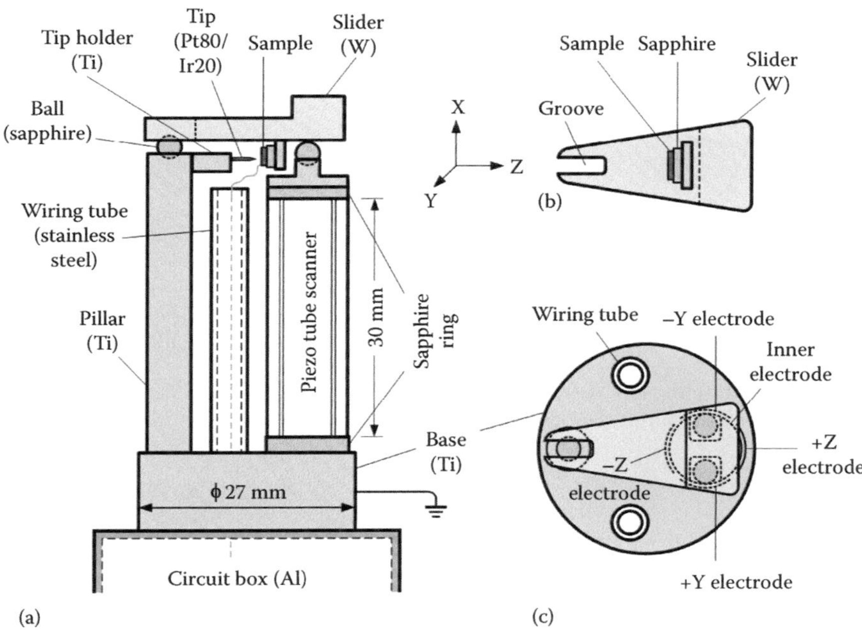

FIGURE 17.2 Schematic view of the STM head with the definitions of X, Y, and Z directions: (a) side view; (b) bottom view of the slider; (c) top view. The materials used for the important parts are given in the parentheses. (From Pang, Z., *Meas. Sci. Technol.*, 20, 065503, 2009.)

demote imaging speed (and quality) of the STM, particularly when constant current mode is used since slow and low-quality high-voltage signals will have to play the important role of maintaining sensitive tunneling current constant.

Having considered all these issues, we decided to implement the first fully low-voltage STM [16,17] from the very beginning. Its scan head in the final version [17] is shown in Figure 17.2, where titanium is the main material used for its low thermal expansion and high hardness (to make the scan head less drifting and firm enough against external vibrations). A piezoelectric scanner tube (PT130.24 from Physik Instrumente with length L = 30 mm, outer diameter O.D. = 10 mm, wall thickness = 0.5 mm) and a titanium pillar are mounted in parallel on a titanium base [12]. The scanner tube can bend toward the pillar (in $\mathbf{H}_J$ direction with **H** standing for horizontal and **J** for junction) by one pair of push–pull electrodes. One polished sapphire ball of 3 mm diameter is glued on the pillar top and two polished sapphire balls of 2.5 mm diameter are glued on the scanner top. These three balls form a narrow triangle and support a horizontal triangular slider piece. The sample stage is vertically glued to the bottom of the slider and the tip is attached to the pillar via a tip holder (see the description given next).

One prong of a QTF is glued on a tiny sapphire piece, which is attached on the tip holder with the fork prongs pointing to the sample and the fork plane (spanned by the two prongs) lying horizontal (see Figure 17.3). The STM tip is a hand-cut 0.15 mm thick

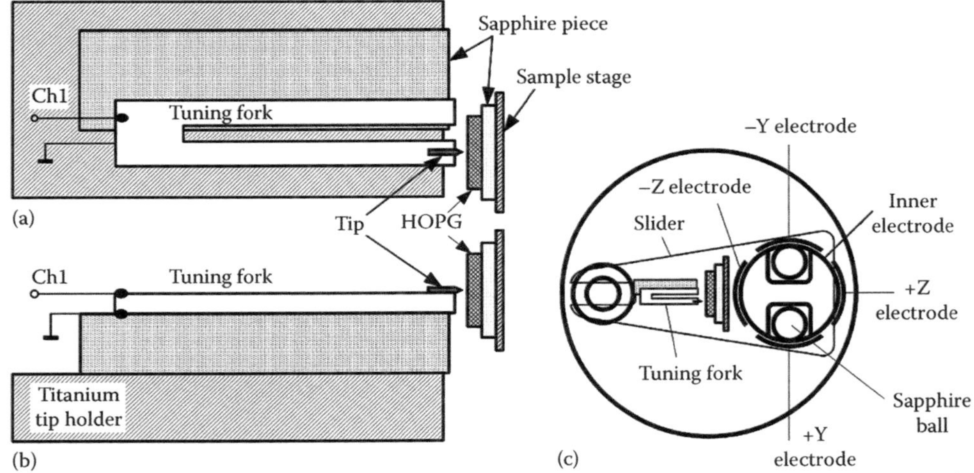

FIGURE 17.3 Top view (a) and side view (b) of the tuning fork (fast scanner) and its position relative to the sample; (c) top view of the STM's mechanical setup. (From Li, Q. and Lu, Q., *Rev. Sci. Instrum.*, 82, 053705, 2011.)

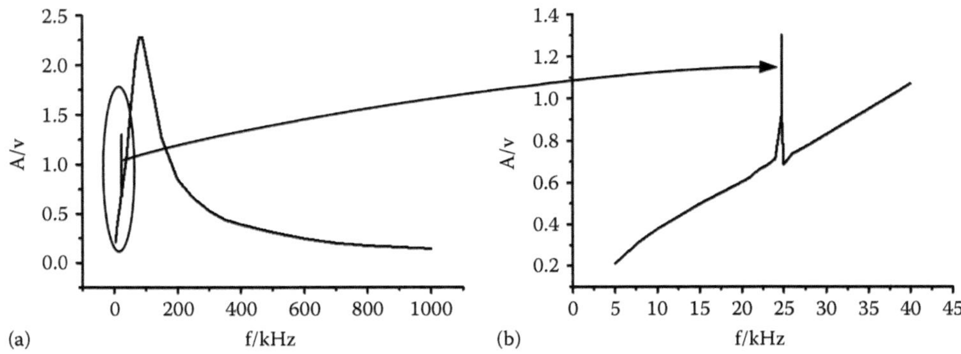

FIGURE 17.4 Resonance curves of the tuning-fork fast scanner measured with the STM preamplifier (100 kHz bandwidth): (a) covers large frequency range from 0 to 1 MHz and (b) is a zoom-in of (a). (From Li, Q. and Lu, Q., *Rev. Sci. Instrum.*, 82, 053705, 2011.)

Pt90/Ir10 wire, which is lengthwise glued on the free prong of the fork with the tip protruding from the prong toward the sample. The quality factor Q is roughly 700 in air at room temperature (ambient conditions) and the fork can oscillate well even in the range of hundreds of kilohertz (see Figure 17.4). Here, the QTF is responsible for the fast scan (along its vibration direction, $\mathbf{H}_F$, where H stands for horizontal direction and F stands for fast scan) only, whereas the slow scan (in direction $\perp_S$, where $\perp$ stands for vertical direction and S stands for slow scan) is implemented by the scanner tube's axial displacement (faster than the conventional lateral-bending scan, [18,19] which is another advantage in achieving higher imaging speed). The mass center of the slider is at the scanner side so that the slider motion well follows the scanner's control. The $\mathbf{H}_J$ direction oriented slot at the narrow end of the slider piece serves as the guide for the slider to slide toward the pillar ball by inertial stepping (coarse approach) [12]. Because the tip–sample junction gap (in $\mathbf{H}_J$ direction) is now adjusted by the scanner tube's lateral bending (large displacement per unit voltage) instead of the conventional axial deformation (small displacement per unit voltage), the coarse and fine approaches can be operated at very low voltages (<4 V), thus the whole STM is built with exclusively low-voltage devices (power supply voltages < the industrial standard ±15 V), [16,17] which is vastly favorable to fast imaging.

17.2.2 Preamplifier

The tunneling current I_{TC} that the STM measures to form an image is fairly weak, generally in the range of pA to nA. A large feedback resistor, R_F (ranging from MΩ to GΩ), is typically needed to convert weak I_{TC} into a comfortable (not weak) voltage signal V_{TC} for further processing, thus limiting the bandwidth of the preamp because any nonvanishing parasitic capacitance C_F on the large R_F will form a low-pass filter with very narrow bandwidth $B = 1/(2\pi R_F C_F)$. Reducing R_F can definitely expand the bandwidth, but may cause the STM to lose atomic resolution because the signal-to-noise ratio (SNR) in the measurement of the weak I_{TC} by R_F will reduce accordingly (since the noise is proportional to $\sqrt{R_F}$, but the signal is proportional to R_F). For an excellent electrical conductive sample such as graphite, this is not an issue because our experimental data show that we can

FIGURE 17.5 Atomically resolved HOPG STM image taken with an ultra-low feedback resistor of 10 kΩ, showing the possibility of achieving atomic resolution with an ultra-faster preamplifier. Image size: 1.72×1.23 nm^2. (From Li, Q. and Lu, Q., *Rev. Sci. Instrum.*, 82, 053705, 2011.)

reduce R_F to as low as 10 kΩ without losing atomic resolution (see Figure 17.5 [12], in which the atoms are flake-like and the lattice is skewed into square type from hexagonal, which is due to distortion and can be corrected by spiral scan [20]—see later for spiral scan discussion). This has a potential wide bandwidth of above 5 MHz, which is well in the range of radio frequency.

However, for poorly conductive samples such as semiconductors and nearly insulating materials where R_F has to be large so as to ensure atomic resolution, we need a completely different solution. We can, for instance, use the STM junction itself as the R_F to convert the weak I_{TC} into a strong V_{TC} with an enhanced bandwidth [21] since the junction in general has high resistance but very low parasitic capacitance (the tip–sample capacitance can be as low as 1 fF, see Ref. [22]). Another example to boost the preamp bandwidth to about 10 MHz is to embed the tip–sample junction (with junction resistance R_J) into a simple LC circuit to form a LCR network, where the change in R_J will cause a change in the damping of the resonant LC circuit [23]. The change in R_J is in the end converted into the change in the reflected power from the resonant LC circuit, which can be measured to form the STM image.

17.2.3 Control and Data Acquisition

17.2.3.1 Line-by-Line Scan

A common way of scanning an STM image in constant height mode is as follows: a controller outputs a pair of scan voltages (V_X^C, V_Y^C, where C stands for "controlled" to mean a given or set value) to move the tip to a sample point neighboring to the previous point ⇄ measures the tunneling current data V_{TC}^{M} at the preamp output, and records (V_X^C, V_Y^C, V_{TC}^M) as the image data for that sample point. These two steps are repeated until enough sample area is measured and the set of (V_X^C, V_Y^C, V_{TC}^M) data is used to create the image. Here, the superscripts C and M stand for controller output data (prearranged, not measured) and measured data (unknown if not measured), respectively, and the two-way arrow ⇄ denotes a repeated switch between these two data processing modes. We will show next that this is actually a big mistake in fast imaging. The problem is that the "⇄" action is time-consuming and occurs too often (once per image pixel), leading to an exceptionally slow overall imaging speed.

Our improvement [12] is (see Figure 17.6): exploit a double-channel (Ch1 and Ch2) function generator to blindly vibrate the tuning fork (by signal V_X from Ch1, whose waveform is sinusoidal) for fast scan and blindly deform the scanner tube in its axial direction (by signal V_Y from Ch2, whose waveform is sinusoidal) for slow scan (the frequencies of the fast and slow scan satisfy $f_{fast} = N \times f_{slow}$, where N is the number of rows per frame) and blindly measure and record (V_X^M, V_Y^M, V_{TC}^M) as image data (all are measured, no prearranged data). The point is that there is no switch ("⇄" action) between the "output mode" (controller outputting V_X^C and V_Y^C) and the "input mode" (measuring V_{TC}^M). This is time saving.

In the actual setup (see Figure 17.6), channel Ch1 of the function generator (Tektronix AFG3000) sends a sinusoidal wave V_X to a noninverting reducer (by a factor of 0.18) circuit whose output drives the tuning fork to perform fast scan (in $\mathbf{H}_F$ direction) through the mechanical vibration of the free prong. Channel Ch2 outputs another (slower) sinusoidal wave V_Y, which is first attenuated by an inverting reducer (by a factor of 0.1) circuit and then applied to the inner electrode of the scanner tube (see Figure 17.3) for slow scan. A real-time computer (National Instruments PXI-8106 RT) blindly measures and records (V_X^M, V_Y^M, V_{TC}^M) from Ch1, Ch2, and the preamp output, respectively. The preamp has a bandwidth of about 100 kHz with a $R_F = 10\ M\Omega$ feedback resistor. The computer only acquires data. There is no switch between acquiring and outputting; therefore the imaging speed can be greatly increased. We can see that no high-voltage devices or instruments are used in this setup. All voltages are less than ±15 V (which are the industrial standard voltage supply for low-voltage operational amplifiers).

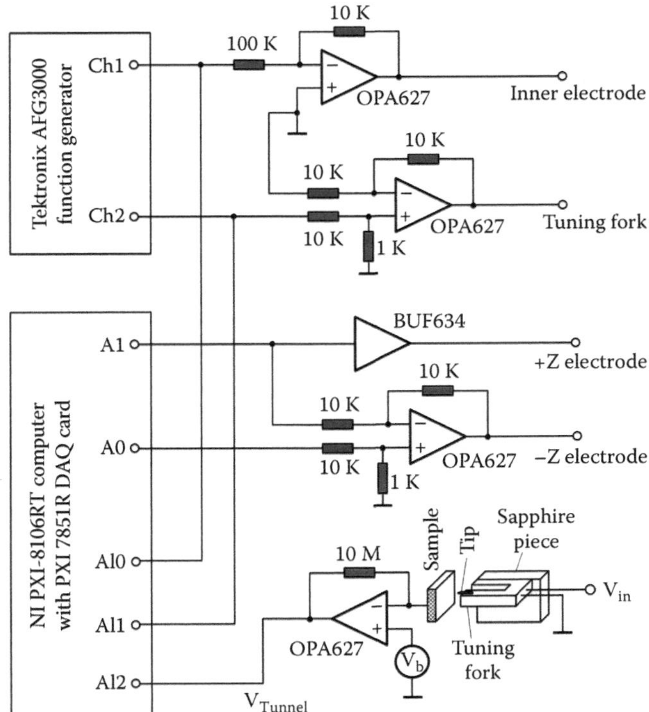

FIGURE 17.6 Schematic drawing of the electronic system that drives the electrodes of the scanner tube and tuning fork and collects the imaging data. (From Li, Q. and Lu, Q., *Rev. Sci. Instrum.*, 82, 053705, 2011.)

Our real-time computer is in fact quite slow: its maximum data acquisition speed is only 520,000 data per channel per second (520 kHz/Ch). We have indeed tried the conventional point-by-point scan-and-measure imaging mode (with a switch between each data acquisition and data output) using this computer and found the maximum data acquisition speed drops to 40 kHz/Ch. We will also demonstrate in the next section that in this mode the atoms in the image become completely invisible (losing atomic resolution) at 3 kHz scan rate due to the lack of enough pixels (only 3 pixels per pixel line are acquired). However, with our new blind-scan-blind-measure method, the atoms are still clearly seen even at a scan rate as high as 26 KHz, which is beyond the resonant frequency, $\omega_{0,+tip} = 24$ kHz, of the tuning fork scanner. This also means that using very small scan device (thus very fast) and purely low voltage (thus very fast) to achieve large scan area becomes very practical since we are now fast enough to take advantage of resonance effect.

17.2.3.2 Achieving the New Record of Fast Imaging

Our aforementioned ultra-fast STM with tuning fork scanner, fully low-voltage operability and blind-scan-blind-measure imaging method was tested by imaging a HOPG sample in ambient conditions. For comparison [12], we first scanned the sample in traditional mode: using the computer as the controller to periodically (1) output a pair of scan voltages (V_X and V_Y) to move the tip to a new position and then (2) measure and record V_{TC} data. Figure 17.7 [12] shows the so-obtained images at a series of increasing scan rates: 57, 909, 2000, and 3300 Hz. Each atomic row contains just three atoms because the driving voltage applied on the tuning fork is low and the scan rate is far below the resonant frequency of the fork, which cannot invoke a large prong vibration amplitude.

It is clearly seen from Figure 17.7 that as the scan rate goes higher and higher, the number of pixels per pixel-line reduces, causing the atoms to look more and more mosaicked. This failure mode implies that the data acquisition rate is too low if the traditional scan mode is used. Our solution is to get rid of all the switches between V_{TC} acquisition and scan action by using a function generator to blindly scan the tip and letting the computer do nothing but

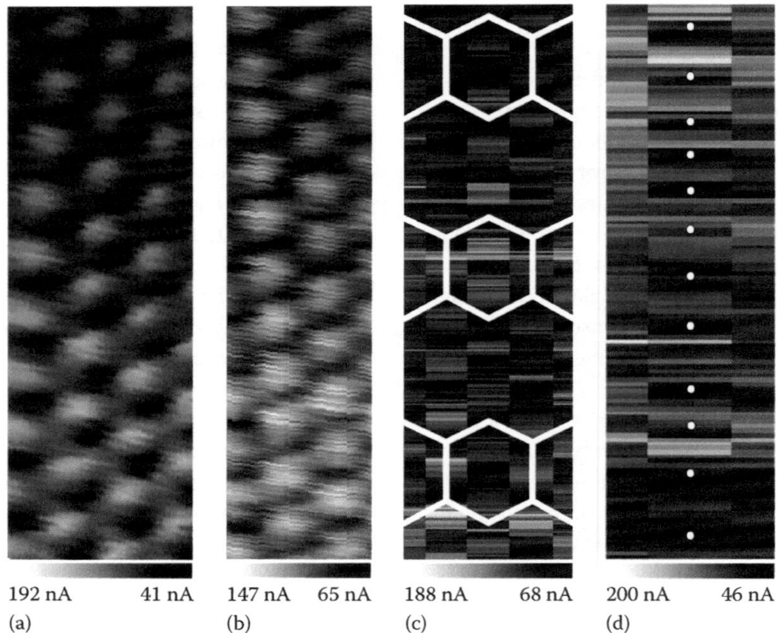

FIGURE 17.7 A series of HOPG STM images (raw data) taken in ambient conditions with increasing scan rate using the traditional point-by-point "scan ⇄ data acquisition" mode. Image conditions are (a) 57 Hz, 0.68×2.34 nm^2, 177×477 pixels; (b) 909 Hz, 0.50×2.58 nm^2, 11×600 pixels; (c) 2 kHz, 0.59×2.71 nm^2, 5×800 pixels, lattice barely seen and (d) 3.3 kHz, 0.52×3.08 nm^2, $3 \times 10{,}000$ pixels, atomic lines barely seen. (From Li, Q. and Lu, Q., *Rev. Sci. Instrum.*, 82, 053705, 2011.)

blindly acquire all the image data (scan voltages plus V_{TC}). Using this "blind-scan and blind-acquire" technology, we first attained a HOPG atomic image at 13 kHz scan rate (Figure 17.8, constant height mode with 50 mV sample positive bias voltage) [12], which is not only the highest scan rate in the literature that still holds atomic resolution but also exhibits improved image quality.

FIGURE 17.8 The atomically resolved HOPG STM image (raw data) obtained in ambient conditions using the proposed "uninterrupted scan and uninterrupted data acquisition" mode (under conditions: 13 kHz scan rate, 1.07×0.98 nm^2 scan size, 52 Hz frame rate, tuning fork driven by $V_{P-P} = 1.82$ V sine wave, 20×250 pixels, 520 kHz sampling rate). (From Li, Q. and Lu, Q., *Rev. Sci. Instrum.*, 82, 053705, 2011.)

Now, we are in a position to increase the scan rate further. Figure 17.9 is an atomically resolved HOPG image at 26 kHz [12], which already exceeds the resonant frequency ($\omega_{0,+tip} = 24$ kHz) of the tuning fork resonator (tip attached). The scan range in the fast scan direction is small because the scan frequency is too much above the resonant frequency (not at resonance) and the fork driving voltage is as low as only 1.2 V. This STM's large area searching capability [17] across the entire sample should solve the issue of small scan size by allowing us to join several small area images to form a larger one. We can also achieve larger scan area by increasing the fork driving voltage or reducing the scan rate to 24 kHz where the fork resonates.

Since the graphite atoms can still be resolved at 26 kHz scan rate and the only issue seen at this high scan speed is lower pixel number per pixel-line, we surely expect that the scan rate can go even higher by simply using a faster data acquisition card. It will be very interesting to see what failure mode at what scan rate will show up first if we use a data acquisition card much faster than 520 kHz. Note that the STM preamp is not yet a limiting factor for the HOPG sample because we have obtained atomic resolution HOPG images (see Figure 17.5) with a feedback resistor as low as $R_F = 10$ kΩ, which has a potential bandwidth (in MHz) much higher than that of the preamp used here ($R_F = 10$ MΩ, the bandwidth ~100 kHz).

17.2.3.3 Spiral Scan and Universal Distortion Correction

We have accomplished the new record of 26 kHz ultra-fast atomically resolved STM imaging employing just traditional LLS scan mode. The biggest advantage of LLS in the construction of

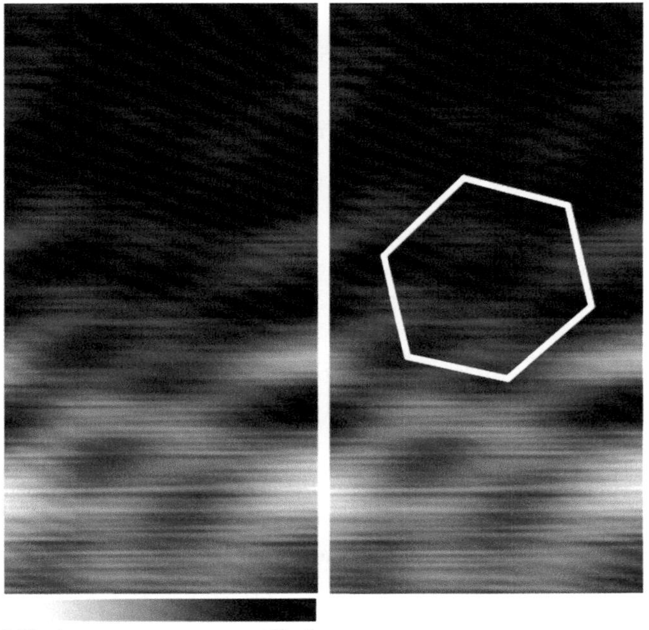

FIGURE 17.9 Atomically resolved HOPG STM image (raw data) attained in ambient conditions at 26 kHz scan rate. The right-hand side image is identical to the left one except that one unit cell is marked. Image size is 0.53×1.23 nm^2, tuning fork driving voltage is $V_{P-P} = 0.22$ V sine wave, pixel number is 10×250, sampling rate is 520 kHz. (From Li, Q. and Lu, Q., *Rev. Sci. Instrum.*, 82, 053705, 2011.)

ultra-fast STM is that only one direction needs to scan fast, which makes things a lot simpler. However, LLS has its own limitations [20], including (i) the forward and backward scanned data in each line do not match well and cannot be used together to construct an image. The change of scan direction at the turn-around point can generate a jump in the image data. Thus, half of the imaging time is wasted, meaning that the scan speed is reduced by a factor of 2. (ii) Hysteresis and creeping (H.C.) distort the images severely in LLS mode (although there are many models [2–6] and feedback-controlled compensations proposed to reduce the H.C., they usually bring more complexity [24,25] and the accuracy is often a problem). (iii) The scan speeds in the fast and slow scan directions are extremely asymmetric: when the fast scan runs close to the resonant frequency of the scanner and cannot go faster, the slow scan still has big room to increase its speed. This is also a big waste of imaging speed. (iv) LLS gives a square scan area, which is not the maximum area (should be a circle) achievable with the same scan voltage, leading to a waste of scan area. (v) LLS mode takes longer time to scan one complete lattice unit because to complete the scan of one lattice unit we have to complete one whole row of lattice units. (vi) Worse still, line-by-line scanned images may be distorted such that the atomic rows are still in straight lines (therefore look normal) but the lattice structure is deformed into a totally wrong type, which is still allowable in reality and can be mistakenly believed to be the true structure. As an example, Figure 17.1a is a line-by-line scanned STM image of a HOPG sample that shows a rectangular lattice instead of the well-known hexagonal lattice. Since the rectangular lattice is also an allowable arrangement in reality, it would be mistakenly regarded as the actual sample structure if the sample was unknown. Here, of course, we all know that there is a severe distortion in the image and the rectangular lattice should not be considered as the true HOPG structure, because HOPG is a known sample whose true lattice structure is known to be hexagonal. Indeed, Figure 17.1a can also be skewed back to the correct hexagonal lattice as shown by Figure 17.1b. But, for unknown samples, we do not know how (or even whether necessary) to correct it.

In conclusion, LLS is intrinsically not appropriate to fulfill the two most important goals of fast STM: less distortion and more real time (i.e., as fast as possible).

Spiral scan mode seems to be able to solve all the aforementioned issues associated with the LLS mode: (i) It scans all the points only once with no time wasted and no sudden U-turns to produce any jumps in the image data. (ii) The state change between neighboring points is ignorable, thus minimizing the effects of H.C. (iii) The scanner moves in both X and Y directions synchronously, therefore it can scan up to the theoretical maximum scan rate (set by the limitation of the resonant frequency). (iv) Its scanned area is a circle, which has the biggest area under the same scan voltages. (v) The lattice unit at the center is always completed first before any other lattice units are scanned (if the spiral scan starts at the center, which is normally the case), meaning that the spiral scan is in principle much faster (hence having minimum distortion) than the LLS mode in terms of finishing scanning one whole lattice unit. In other words, the central lattice unit is likely distortion free in a spiral-scanned image. (vi) And, more importantly, we will show that drifting (and other distortions) will make the atomic rows bent in a spiral scanned image [20]. Since atomic rows should not be curved in reality, the existence of distortion is consequently self-manifested if curved atomic rows are seen. The drifting strength and direction can be further deduced from the curvature information. The correction of the distorted image will become particularly simple and universal even for an unknown sample: just straighten the curved atomic rows and the true atomic lattice structure will show up automatically.

Our method to realize the spiral scan is to synchronously scan in both orthogonal directions that span the scan plane, making the tip start from the center and smoothly pass every measured point only once in the whole centrifugal scan process. The spiral trajectory in polar coordinates can be described by the following equations:

$$\frac{dR}{d\theta} > 0, \quad \frac{d}{d\theta} R(2\pi + \theta) = \frac{d}{d\theta} R(\theta) \tag{17.1}$$

$$\left(\frac{dR}{dt}\right)^2 + \left(R\frac{d\theta}{dt}\right)^2 = \text{constant} \tag{17.2}$$

where

R is the radius from the origin
θ is the argument
t is the time [20]

Equation 17.1 requires that R increase all the time and the distance between sequential circles remain constant. Equation 17.2 ensures an "approximately" constant linear velocity (spiral scan with exactly constant linear velocity is very complicated and not even analytic).

In Cartesian rectangular coordinates,

$$X = aR\cos(\theta) \tag{17.3}$$

$$Y = bR\sin(\theta) \tag{17.4}$$

where

a and b are the parameters that adjust the scan sizes in the X and Y directions, respectively

R and θ satisfy Equations 17.1 and 17.2

The X and Y values are then converted into analog voltages: ±X volts and Y volts. The ±X volts are applied to one pair of the push–pull electrodes in our homemade STM [16] to bend a piezoelectric tube scanner in the X direction and the Y + X and Y − X volts are applied to the same electrode pair to deform the tube scanner in the Y direction simultaneously. The tube scanner now scans in a spiral itinerary as a function of time.

We imaged a HOPG sample in ambient conditions under the spiral scan mode with 100 mV sample positive bias. The X and Y signals were generated by a LABVIEW program (see Figure 17.10) in National Instruments PXI-8106RT and are the approximate solutions of Equations 17.1 through 17.4:

$$X = X_r C_{oe}\sqrt{t}\cos(A_{rg}\sqrt{t}) \tag{17.5}$$

$$Y = Y_r C_{oe}\sqrt{t}\sin(A_{rg}\sqrt{t}) \tag{17.6}$$

where

X_r and Y_r ($X_r = Y_r$) represent the X and Y scan ranges and are multiplied by a factor C_{oe} for dimension match

A_{rg} is also introduced for dimension consistency in the argument

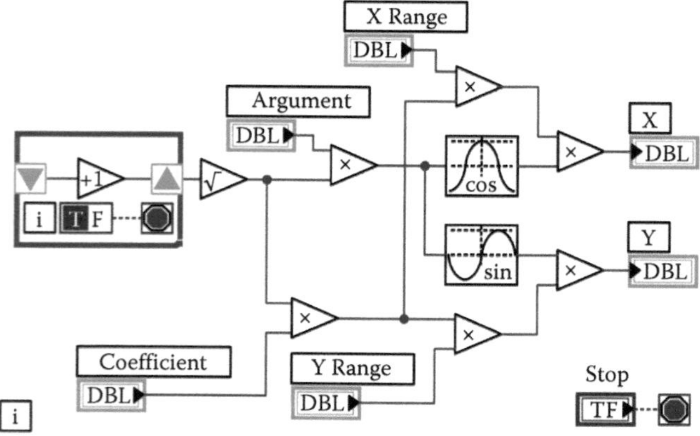

FIGURE 17.10 The LABVIEW program that generates the spiral scan signals defined by Equations 17.5 and 17.6. (From Wang, J.T. et al., *Rev. Sci. Instrum.*, 81, 073705, 2010.)

Equations 17.5 and 17.6 are the approximate solutions of Equations 17.1 and 17.2 because they can be reformed as

$$\begin{cases} R = \sqrt{X^2+Y^2} = X_r C_{oe}\sqrt{t} \\ \theta = A_{rg}\sqrt{t} \end{cases}$$

which satisfy Equation 17.1. However, they only approximately satisfy Equation 17.2, because

$$\left(\frac{dR}{dt}\right)^2 + \left(R\frac{d\theta}{dt}\right)^2 = \left(\frac{dX}{dt}\right)^2 + \left(\frac{dY}{dt}\right)^2 = \frac{(X_r C_{oe})^2}{4t} + \frac{(X_r C_{oe} A_{rg})^2}{4}$$

which is not strictly a constant, meaning that the linear velocity only gradually becomes a constant as time t increases.

The first scanned point is achieved by converting the first coordinate pair (Y + X, Y − X) to an analog voltage pair by PXI-7851R (National Instruments), which bends the tube scanner to reach the first scan point. Similarly, a series of coordinate pairs (Y + X, Y − X) are generated and the corresponding analog signals scan the tube scanner in a spiral track (see Figure 17.11). The tunneling current and coordinate data are collected. The whole scanned area is divided into 170 × 170 grids with each grid containing several sets of tunneling current data. The data average of a grid is used to represent the tunneling current of that grid.

Raw data images of different scan rates (Figure 17.12a and b) for a HOPG surface (4.07 × 4.07 nm^2) were obtained under the spiral scan in constant height mode [20]. Figure 17.12a and b are scanned at 20,000 and 6,000 points/s, respectively. In Figure 17.12b,

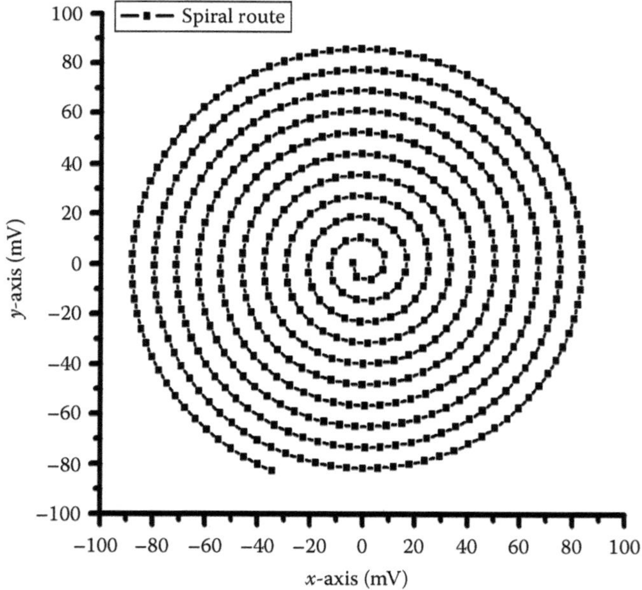

FIGURE 17.11 A segment of spiral scan route generated by the program in Figure 17.10.

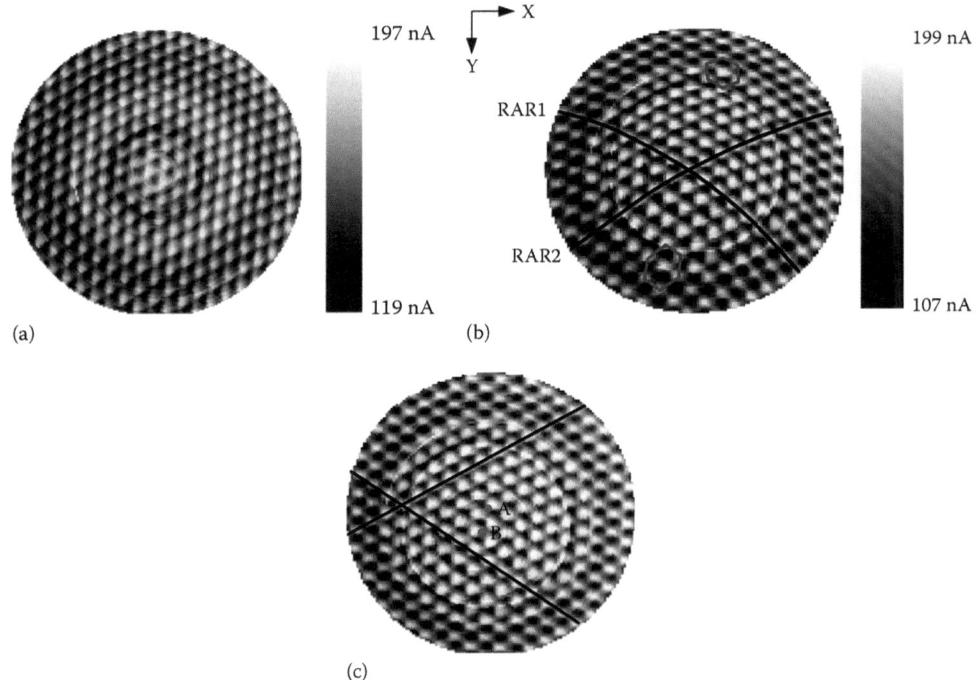

FIGURE 17.12 (a) is the spiral scanned raw data image (4.07 nm × 4.07 nm, sample biased at +100 mV) of HOPG at a higher scan speed of 20 k points/s, showing less distortion. (b) is similar to (a) except that the scan speed is lower (6 k points/s), showing curved atomic rows and distorted hexagonal lattice. (c) is the corrected image from (b) after deducting the drift impact, in which the curved atomic rows are straightened. A and B are the starting points of the spiral scans before and after the correction, respectively. (From Wang, J.T. et al., *Rev. Sci. Instrum.*, 81, 073705, 2010.)

curved atomic rows are clearly seen, which must be unnatural and contain distortion because atomic rows should be in straight lines or otherwise the periodicity of atomic arrangement would be lost (see Figure 17.12b in which the two marked unit cells are not identical, therefore losing periodicity. Note, however, that the central unit cell is almost distortion free since it is the first completed entire unit cell, implying an ultra high scan speed for that cell). Other reasons why "curved atomic rows" signify "distorted image" include the following: (i) in the faster spiral-scanned image (Figure 17.12a), the atomic rows bend more weakly, which can be explained that if the imaging time is shorter, the total accumulated distortion (drifting) is lower (thus, again, confirming that faster imaging results in less distortion); (ii) even in the slower spiral-scanned image (Figure 17.12b), the hexagonal arrangement at the center is much less distorted than those near the edge, which can be interpreted that the central hexagon is finished in a much shorter time (less drifting time) than the edge ones (each of which needs greater scan length to complete). If the bending of atomic rows is owing to the existence of drifting, then the bending direction should reveal the drifting direction and the curvature should correlate to the strength of the drifting. To confirm these, assume that the drifting strength is D nm/s and the drifting angle is θ with respect to a reference atomic row (RAR, defined as any bent atomic row that passes through the origin). The simplest RAR is the atomic row on the *x*-axis (called *x*-axis RAR) and satisfies

$$\begin{cases} X = X_r C_{oe}\sqrt{t}\cos(A_{rg}\sqrt{t}), \quad A_{rg}\sqrt{t} = m\pi \ (m = 0, 1, 2\ldots). \\ Y = 0 \end{cases} \tag{17.7}$$

A tilted RAR can be described as

$$\begin{cases} X = X_r C_{oe}\sqrt{t}\cos(A_{rg}\sqrt{t}), \quad A_{rg}\sqrt{t} = m\pi \,(m = 0, 1, 2,\ldots). \\ aX + bY = 0 \end{cases} \tag{17.7'}$$

The *x*-axis RAR is bent by the drifting as

$$\begin{cases} X = X_r C_{oe}\sqrt{t}\cos(A_{rg}\sqrt{t}) + Dt\cos\theta, \quad A_{rg}\sqrt{t} = m\pi \\ \qquad (m = 0, 1, 2,\ldots). \\ Y = Dt\sin\theta \end{cases} \tag{17.8}$$

Its curvature K can be calculated as [26]

$$K = \left| \frac{\dot{X}\ddot{Y} - \ddot{X}\dot{Y}}{(\dot{X}^2 + \dot{Y}^2)^{3/2}} \right|. \tag{17.9}$$

Combining Equations 17.8 and 17.9, the curvature becomes

$$K = \left| \frac{X_rC_{oe}D\left[\cos(A_{rg}\sqrt{t}) + A_{rg}\sqrt{t}\sin(A_{rg}\sqrt{t}) + A_{rg}^2 t\cos(A_{rg}\sqrt{t})\right]\sin\theta}{4\left\{\left[(X_rC_{oe}/2)\cos(A_{rg}\sqrt{t}) - (X_rC_{oe}A_{rg}\sqrt{t}/2) \times \sin(A_{rg}\sqrt{t}) + D\sqrt{t}\cos\theta\right]^2 + D^2 t\sin^2\theta\right\}} \right|$$

$$\overset{A_{rg}\sqrt{t}=m\pi}{=} \left| \frac{X_rC_{oe}D(1+m^2\pi^2)\sin\theta}{4\left[(X_r^2C_{oe}^2/4) + (-1)^m(X_rC_{oe}Dm\pi/A_{rg}) \times \cos\theta + (D^2m^2\pi^2/A_{rg}^2)\right]^{3/2}} \right|. \quad (17.10)$$

We can treat a tilted RAR as the *x*-axis RAR with a new drifting angle and use Equation 17.10 again to get its curvature at the origin. Two RARs together can solve for D and θ. To make things simpler, we introduce a useful theorem: the tangent S_{RAR} of a curved RAR at the origin is the RAR's corresponding undrifted atomic row. To prove it, assume that the RAR is the *x*-axis RAR described by Equation 17.8. Its undrifted atomic row is described by Equation 17.7. The S_{RAR} is written as

$$S_{RAR} = \frac{dY}{dX} = \frac{dY/dt}{dX/dt} = \frac{D\sqrt{t}\sin\theta}{D\sqrt{t}\cos\theta - (X_rC_{oe}/2)\left(\cos(A_r\sqrt{t}) + \sqrt{t}\sin(A_r\sqrt{t})\right)} \quad (17.11)$$

which is zero not only when $t=0$ (and $D\neq 0$) but also when $D=0$ (undrifted case). This is true regardless of θ, implying that this conclusion also holds for tilted RARs. Now, we can compare the theoretical Equation 17.10 with the experiment. We choose two RARs, RAR1 and RAR2, as shown in Figure 17.12b. Their curvatures (K_1, K_2) and slopes (S_1, S_2) at the origin can be calculated by polynomial fitting the two RARs. The aforementioned theorem tells that S_1 and S_2 correspond to the sample's actual atomic arrangement (without drifting distortion). Set $K_1 = K(\theta, D)$ and $K_2 = K(\theta - \beta, D)$ in Equation 17.10, where β is the angle formed by the tangent lines of RAR1 and RAR2 at the origin: $\tan\beta = (S_2 - S_1)/(1 + S_1 \times S_2)$, and solve for D and θ. The results are D = 0.02 nm/s and θ ≈ 100°. This pair of D and θ can also be used to correct the entire distorted image of Figure 17.12b, which produces the image of Figure 17.12c, where all the atomic rows are close to straight lines and the undistorted graphite unit cells of hexagonal symmetry are seen everywhere in the image.

17.2.4 Overall Consideration of the Whole System

17.2.4.1 Impedance Matching

In general, the preamp circuit should be located very close to the scan head (as close possible) so as to shorten the wires connecting the scan head and minimize the leakage current and interference signals that go into these wires and get significantly amplified by the preamp. It is therefore very often that the preamp circuit box is directly fixed with the scan head. In this case, the controller and the DAQ system should be connected to the preamp with soft cables to avoid transferring vibration to the preamp (i.e., scan head). Since we are dealing with high-frequency signals, the output impedance of the preamp (first stage), the input impedance of the second stage, and the impedance of the cables connecting first and second stages should all match, otherwise signal reflection will happen at the interface and cause noises.

17.2.4.2 Driving Capacitive Load

The cables connecting different stages typically have rather high capacitance. When they are driven by the output circuit of a certain stage, the driving operational amplifiers should be carefully selected because only very few operational amplifiers can be used to drive capacitive load at high frequency. The failure modes if wrong operational amplifiers are used include circuit oscillation, severe distortion, low signal-to-noise ratio, etc.

One example of handling both capacitive load and impedance matching is given in Figure 17.13, in which 75 Ω cable is used. Buf634 of Texas Instruments is a wideband powerful buffer to drive moderate capacitive load. Two 75 Ω resistors at the first stage output and the next stage input, respectively, are utilized to match the impedance of the 75 Ω cable.

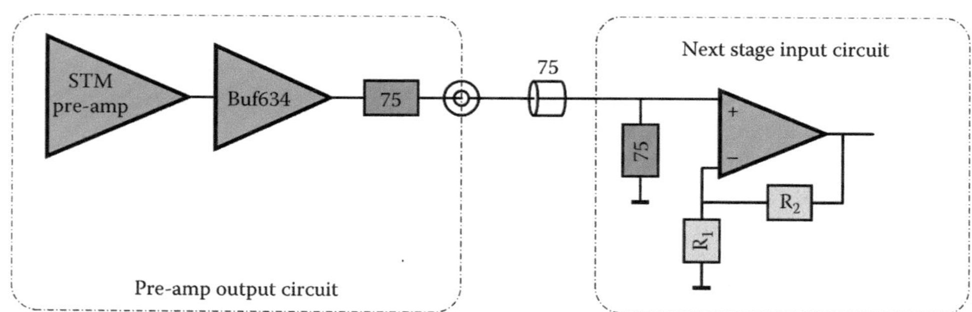

FIGURE 17.13 High-frequency interface circuit system that can deal with both capacitive load and impedance matching (using 75 Ω cable).

17.3 Constant Current Mode

A general purpose STM can also image in constant current mode, which needs a PID feedback control unit to maintain a constant tunneling current by adjusting the tip–sample gap using the amplified error signal between the tunneling-current voltage signal V_{TC} and the set voltage V_{set}. This reduces the imaging speed greatly. There is no good solution presently to boost a constant current STM. Some researchers propose a hybrid mode between the fast constant height and slow constant current modes [11]. This is apparently a tradeoff, which leads to a maximum scan speed of 10 kHz. Beyond this scan speed, the atomic resolution is lost.

17.4 Future Improvement

17.4.1 Faster Scan Device

Using smaller quartz tuning fork, the resonant frequency can be in MHz range. Actually, even with the well-known 32 kHz tuning fork, we can achieve pretty large oscillation amplitude (i.e., large scan area) at frequencies much larger than the resonant frequency as shown in Figure 17.4. So, scan device should not be an issue in achieving MHz scan rate. However, GHz scan rate with fairly large scan area will be a severe issue presently.

17.4.2 Higher Preamplifier Bandwidth

As discussed earlier, a MHz bandwidth STM preamp is accomplishable by reducing R_F of the preamp to 10 kΩ [12] (or even lower), by using the junction itself as the R_F or by embedding the tip–sample junction into a simple LC circuit, [23] etc. There are other ways of improving the preamp bandwidth such as using an AC boosting second-stage amplifier to recover the higher frequency above the 3 dB cutoff of the first stage [27] or replacing the large R_F preamp with a few low R_F preamps connected in series at the cost of reduced SNR, etc. These are well worth an immediate trial. However, boosting the preamp bandwidth to GHz range needs some further investigation such as utilizing a combination of several aforementioned methods. Anyway, a GHz wideband preamp is probably not a big barrier compared with a GHz scan device.

17.4.3 Multicard DAQ

From the aforementioned discussion, it is clear that the data-acquisition speed is critical in boosting the scan speed. The simplest future improvement is to use a much faster DAQ card since the DAQ card we used is not very fast and has a large room for improvement. In case the fastest DAQ card is used and there is no room to improve the DAQ rate, we can always use multiple DAQ cards (in one or multiple computers) to acquire data simultaneously. The final DAQ rate will be the fastest single card DAQ rate multiplied by the number of cards used. This could be unlimited.

17.4.4 Firmness and Compactness of the Scan Head

Our aforesaid 26 kHz ultra-fast STM uses piezoelectric inertial slider to implement coarse approach. Its disadvantage is that the scan head is not very firm because the inertial force produced by a piezoelectric inertial slider is small in general. This brings a severe issue for MHz–GHz scan rate because the vibration force produced by the scanner (resonator) is proportional to the product of scan range (A) and square of the scan rate (ω): $m_r A\omega^2$ (m_r is the effective mass of the resonator), which will become large at very high scan rate ($\sim$1 g × 10 nm × MHz^2 = 0.1 N, which is larger than the friction force ($\sim \mu m_s g \sim 0.06$ N) that holds the sliding piece (assume its mass $m_s \sim 20$ g and $\mu \sim 0.3$) and will cause instability of the junction. Actually, firmness and compactness are always highly desired for any STM with atomic resolution including the slow ones). Up to now, there are many other kinds of piezomotors, including Inchworm, [28] beetle type [29], and shear piezo stepper [30]. However, they all have severe drawbacks: each needs three or more piezo actuators to operate, which is too complicated in both structure and control. Their reliability and applicability in small space (such as in extreme condition environments) and weak signal measurements all become an issue.

In 2006, we filed an invention patent [31] which we name "PandaDrive" here. It discloses a firm, compact, and simple piezomotor ideal for coarse approach: two piezoelectric tubes are mounted in series and three spring pads fixed at both ends and between the two piezo tubes, respectively, hold a central shaft (Figure 17.14). The three friction forces between the three pads and shaft satisfy the condition that any one friction force is smaller than the sum of the other two (friction force conditions. The ideal case is that the three friction forces are equal). To operate it, shrink one piezo tube first (substep 1), then expand the other piezo tube (substep 2), finally both piezo tubes simultaneously change to the opposite displacement status (substep 3).

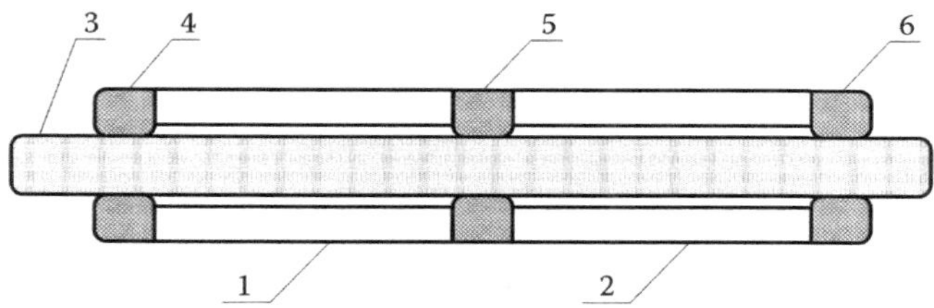

FIGURE 17.14 PandaDrive piezomotor; 1: left piezo tube, 2: right piezo tube, 3: shaft, 4–6: three spring pads. (From Lu, Q. and Hou, Y., China invention patent authorization number ZL200610161477.3, filed on December 15, 2006, disclosed on July 11, 2007 and authorized on October 7, 2009; available at http://211.157.104.87:8080/sipo/zljs/hyjs-yx-new.jsp?recid=CN200610161477.3)

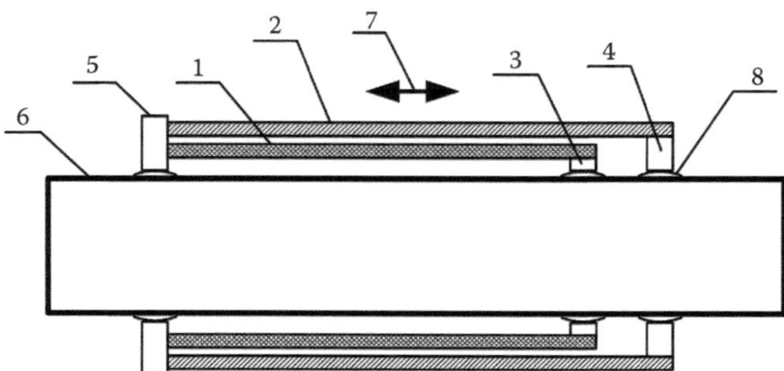

FIGURE 17.15 A variant of PandaDrive; 1: inner piezo tube, 2: outer piezo tube, 3–5: protective rings, 6: shaft, 7: piezo deformation direction, 8: spring pads. (From Lu, Q. and Wang, Q., China patent application number 201010254442.0, filed on August 17, 2010, available at http://211.157.104.87:8080/sipo/zljs/hyjs-yx-new.jsp?recid=CN201010254442.0; Lu, Q. and Wang, Q., PCT patent application number PCT/CN2011/074697, received at International Bureau, June 13, 2011, publication number WO2012/019477; available at http://www.wipo.int/patentscope/search/en/detail.jsf?docId=WO2012019477)

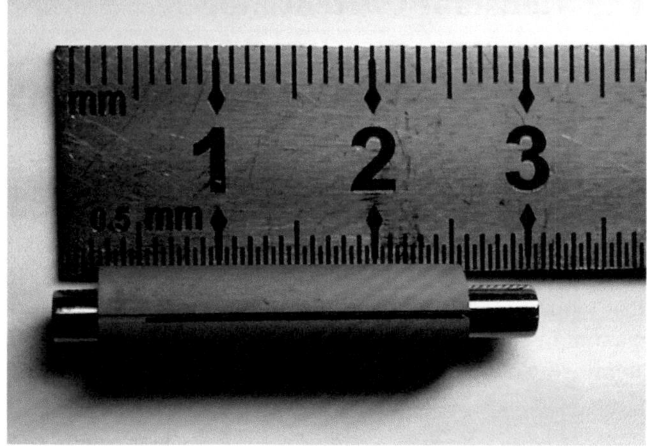

FIGURE 17.16 Photo of a home-made firm and compact piezoelectric motor of the two-piezo-three-friction type suitable for ultra-fast STM, where the two driving piezos are the two half-tubular piezos cut from one whole piezo tube. The round central shaft is pushed against the end rings by a spring strip between the shaft and the slit of the piezo tube.

These three substeps accomplish one step of motion, in which the friction force conditions guarantee that each substep moves (in the same direction) only one pad on the shaft. A similar principle can also move the shaft in the opposite direction. The maximum driving force of PandaDrive is associated with the blocking force of each driving piezo, which is much stronger than the inertial force a piezo can produce. So, PandaDrive can be made very rigid, compact, and simple (controlled by only two piezos).

Figure 17.15 shows a variant of PandaDrive: [32,33] two piezo tubes are mounted in parallel with one tube coaxial fixed in the other and three springs at the mount and the free ends, respectively, hold a central shaft by three friction forces (satisfying the same aforesaid friction force conditions). Another important design of the two-piezo-three-friction type motor (which we call "GeckoDrive") is to mount two identical piezo plates (or two stacks of piezo plates) in parallel (face to face) or axially cut one piezo tube into almost two halves as shown in Figure 17.16 [34] so that the two driving piezos form a fork which spring holds, a central shaft with the holding points being at the mount and the free ends of the two prongs, respectively. This does not require that the holding points be on the same straight line, thus reducing the requirement on the machine precision. It also allows us to use piezo stacks, which enhances firmness, compactness, and the pushing force.

In conclusion, increasing the STM imaging speed from the current record of 26 kHz to MHz scan rate without losing atomic resolution should be achievable in the near future without any major hurdles. Extremely fast atomically resolved STM in the range of GHz is not impossible if corresponding GHz high quality factor crystal forks (to ensure certain scan area) can be fabricated. The spiral scan with two-dimensional resonators is more preferable than the traditional line-by-line scan in attaining high-quality ultra-fast STM with distortion-free atomic resolution.

References

1. G. Binnig, H. Rohrer, C. Gerber, and E. Weibel, *Phys. Rev. Lett.* **49**, 57, 1982.
2. P. Ge and M. Jouaneh, *Precision. Eng.* **17**, 211, 1995.
3. M. Goldfarb and N. Celanovic, *ASME J. Dyn. Syst. Meas. Control* **119**, 478, 1997.
4. R. B. Mrad and H. Hu, *IEEE/ASME Trans. Mechatron.* **7**, 479, 2002.
5. S. H. Lee and T. J. Royston, *J. Acoust. Soc. Am.* **108**, 2843, 2000.
6. I. Mayergoyz, *Mathematical Models of Hysteresis*, Springer, New York, 1991.
7. J. T. Wang, J. H. Wang, Y. B. Hou, and Q. Y. Lu, *Rev. Sci. Instrum.* **81**, 073705, 2010.
8. J. C. Vickerman and I. S. Gilmore, *Surface Analysis—The Principal Techniques*, 2nd edn., Wiley, New York, 2009.
9. H. Neddermeyer, *Rep. Prog. Phys.* **59**, 701, 1996.
10. D. M. Rampulla, A. J. Gellman, and D. S. Sholl, *Surf. Sci.* **600**, 2171, 2006.
11. M. J. Rost et al., *Rev. Sci. Instrum.* **76**, 053710, 2005.
12. Q. Li and Q. Lu, *Rev. Sci. Instrum.* **82**, 053705, 2011.
13. H. J. Mamin, H. Birk, P. Wimmer, and D. Rugar, *J. Appl. Phys.* **75**, 161, 1994.
14. S. O. Reza Moheimani, *Rev. Sci. Instrum.* **79**, 071101, 2008.
15. T. Ando, *Roadmap of Scanning Probe Microscopy*, Springer, Berlin, Germany, 2007.
16. Y. Hou, J. Wang, and Q. Lu, *Rev. Sci. Instrum.* **79**, 113707, 2008.
17. Z. Pang, J. Wang, and Q. Lu, *Meas. Sci. Technol.* **20**, 065503, 2009.
18. C. R. Ast, M. Assig, A. Ast, and K. Kern, *Rev. Sci. Instrum.* **79**, 093704, 2008.

19. S. A. Z. Jahromi, M. Salomons, Q. Sun, and R. A. Wolkow, *Rev. Sci. Instrum.* **79**, 076104, 2008.
20. J. T. Wang, J. H. Wang, Y. B. Hou, and Q. Y. Lu, *Rev. Sci. Instrum.* **81**, 073705, 2010.
21. Q. Lu, China invention patent authorization number ZL200610097197.0.
22. S. Kurokawa and A. Sakai, *J. Appl. Phys.* **83**, 7416, 1998.
23. U. Kemiktarak, T. Ndukum, K. C. Schwab, and K. L. Ekinci, *Nature* **450**, 85, 2007.
24. A. Sebastian and S. M. Salapaka, *IEEE Trans. Control Syst. Technol.* **13**, 868, 2005.
25. B. M. Chen, T. H. Lee, H. Chang-Chieh, Y. Guo, and S. Weerasooriya, *IEEE Trans. Control Syst. Technol.* **7**, 160, 1999.
26. USTC Teaching and Research Group on Advanced Mathematics, *Introduction to Advanced Mathematics*, Vol. 238, USTC Press, Hefei, China, 2007.
27. D.-J. Kim and J.-Y. Koo, *Rev. Sci. Instrum.* **76**, 023703, 2005.
28. Burleigh Instruments, Inc., U.S. Patent No. 3,902,084, 1975; R. A. Wolkow, *Rev. Sci. Instrum.* **63**, 4049, 1992; P. E. Tenzer and R. Ben Mrad, *IEEE/ASME Trans. Mechatron.* **9**, 427, 2004; J. Frank, G. H. Koopmann, W. Chen, and G. A. Lesieutre, *Proc. SPIE* **3668**, 717, 1999; J. Ni and Z. Zhu, *IEEE/ASME Trans. Mechatron.* **5**, 441, 2000; K. Duong and E. Garcia, *Proc. SPIE* **2443**, 782, 1995; J. E. Miesner and J. P. Teter, *Proc. SPIE* **2190**, 520, 1994.
29 T. H. Chang, C. H. Yang, M. J. Yang, and J. B. Dottellis, *Rev. Sci. Instrum.* **72**, 2989, 2001; J. H. Ferris, J. G. Kushmerick, J. A. Johnson, M. G. Yoshikawa Youngquist, R. B. Kessinger, H. F. Kingsbury, and P. S. Weisse, *Rev. Sci. Instrum.* **69**, 2691, 1998; N. Pertaya, K.-F. Braun, and K.-H. Rieder, *Rev. Sci. Instrum.* **75**, 2608, 2004; L. A. Silva, *Rev. Sci. Instrum.* **68**, 1300, 1997; B. Koc, S. Cagatay, and K. Uchino, *IEEE Trans. Ultrason. Ferroelectr. Freq. Control* **49**, 495, 2002; M. Bexell and S. Johansson, *Sens. Actuat. A* **75**, 118, 1999; J. Frohn, J. F. Wolf, K. Besocke, and M. Teske, *Rev. Sci. Instrum.* **60**, 1200, 1989.
30 Chr. Wittneven, R. Dombrowski, S. H. Pan, and R. Wiesendanger, *Rev. Sci. Instrum.* **68**, 3806, 1997; T. Hanaguri, *J. Phys.: Conf. Ser.* **51**, 514, 2006; S. H. Pan, E. W. Hudson, and J. C. Davis, *Rev. Sci. Instrum.* **70**, 1459, 1999; A. K. Gupta and K.-W. Ng, *Rev. Sci. Instrum.* **72**, 3552, 2001; S. H. Pan, International Patent Publication No. WO 93/19494, 1993.
31. Q. Lu and Y. Hou, China invention patent authorization number: ZL200610161477.3 (filed on December 15, 2006, disclosed on July 11, 2007 and the authorized on October 7, 2009). This patent is still active now. The link to this patent is http://211.157.104.87:8080/sipo/zljs/hyjs-yx-new.jsp?recid=CN200610161477.3
32. Q. Lu and Q. Wang, China patent application number: 201010254442.0 (filed on August 17, 2010). The link to this patent is http://211.157.104.87:8080/sipo/zljs/hyjs-yx-new.jsp?recid=CN201010254442.0
33. Q. Lu and Q. Wang, PCT patent application number: PCT/CN2011/074697 (received at International Bureau: June 13, 2011), publication number: WO2012/019477 (publication date: February 16, 2012). The link to this PCT patent is http://www.wipo.int/patentscope/search/en/detail.jsf?docId=WO2012019477
34. Q. Wang and Q. Lu, *Rev. Sci. Instrum.* **80**, 085104, 2009

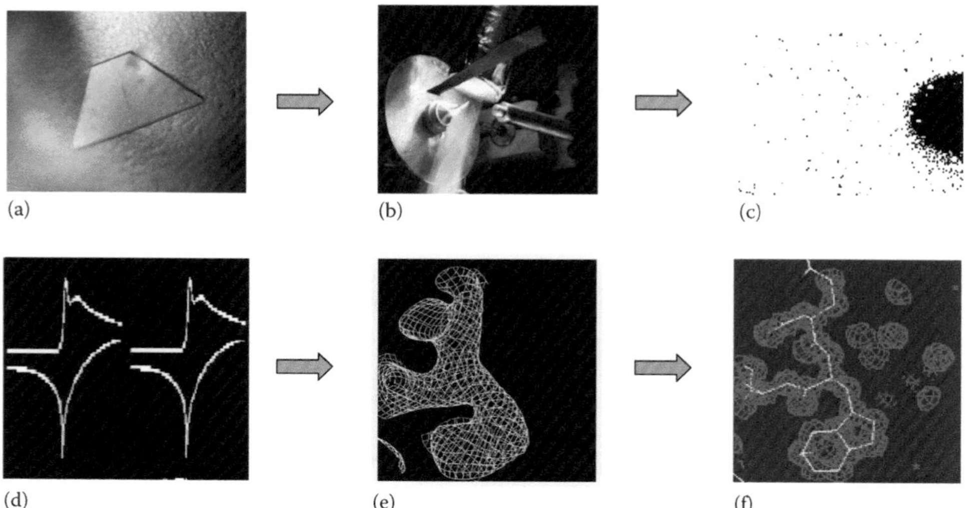

FIGURE 6.1

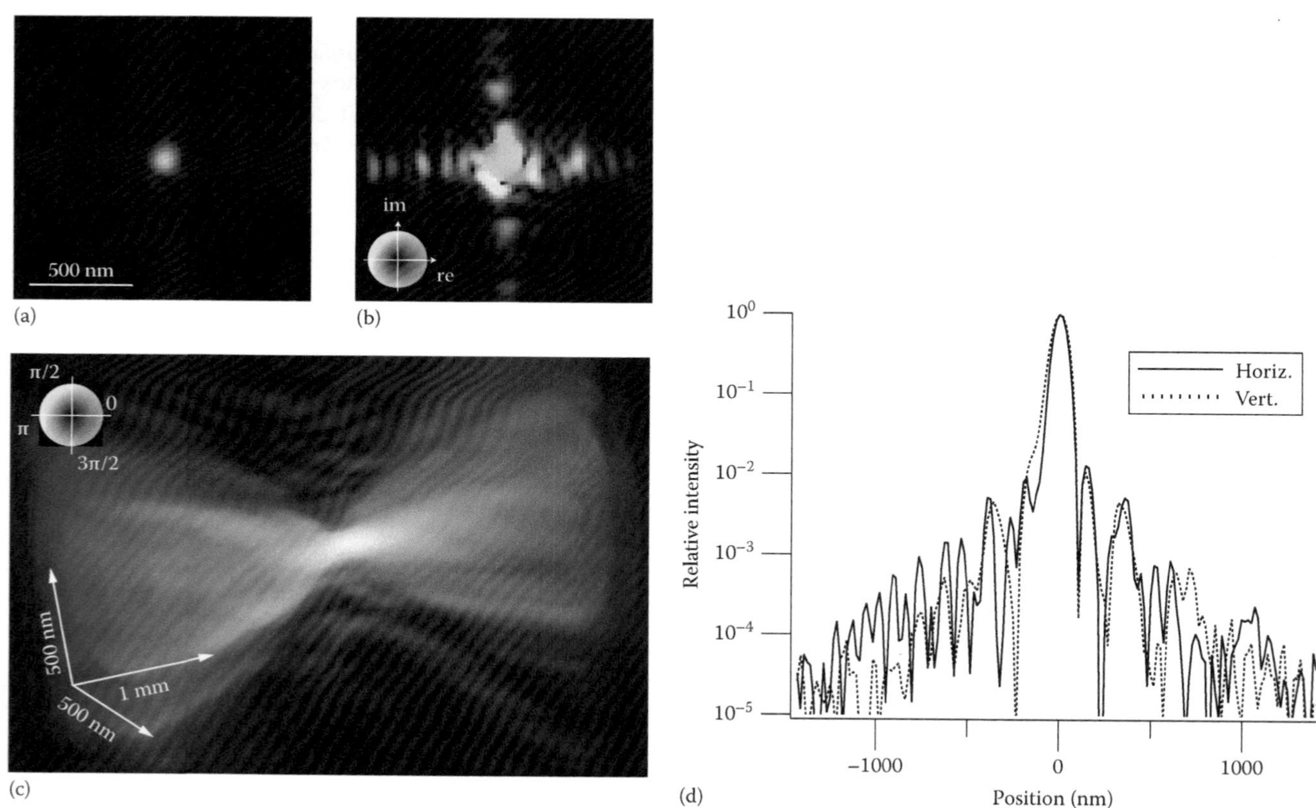

FIGURE 7.4

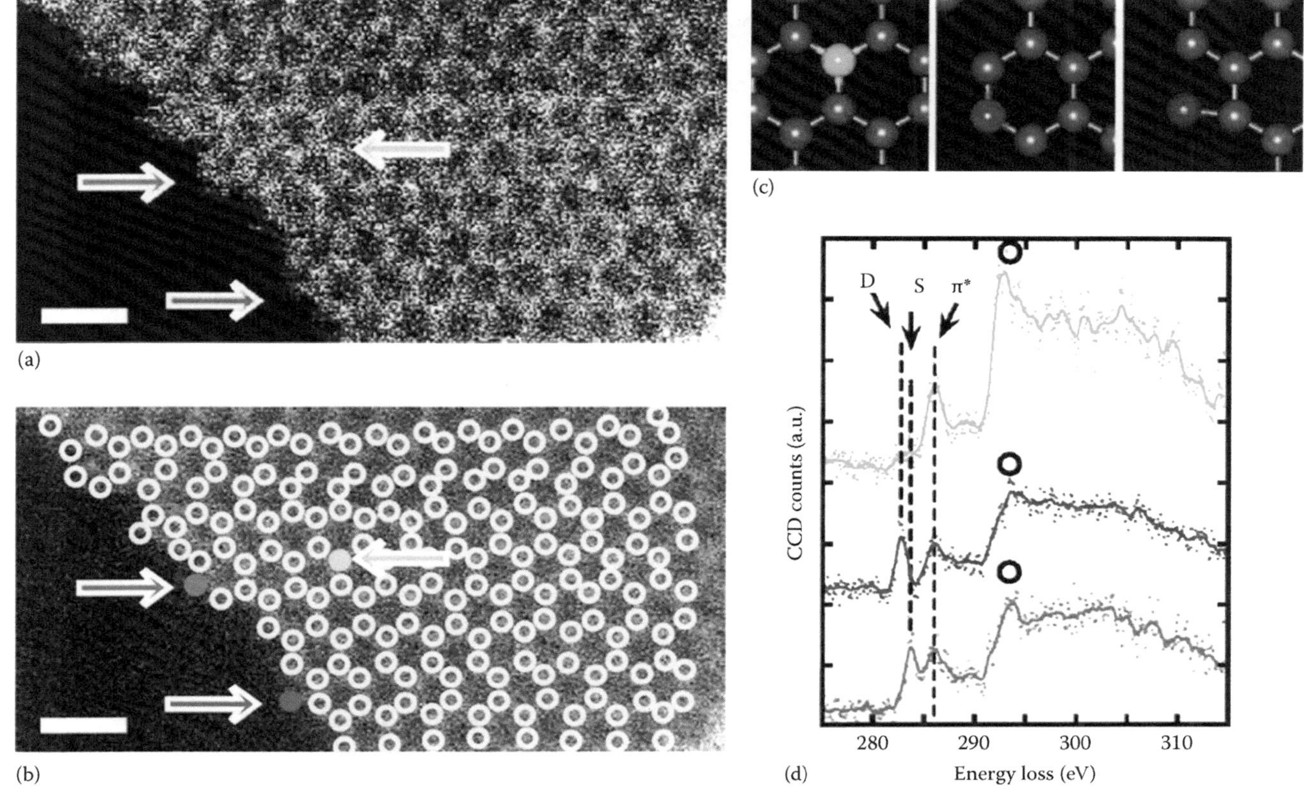

FIGURE 10.9

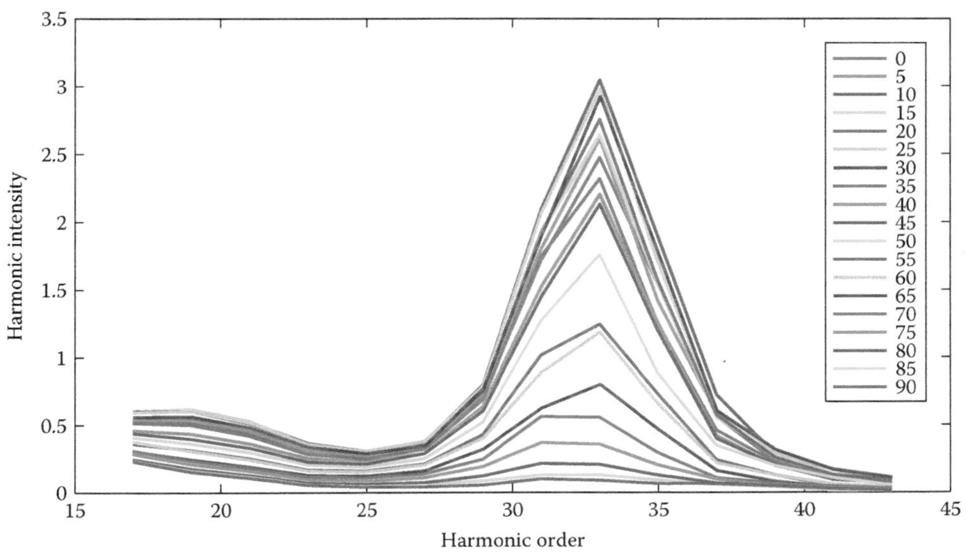

FIGURE 20.10

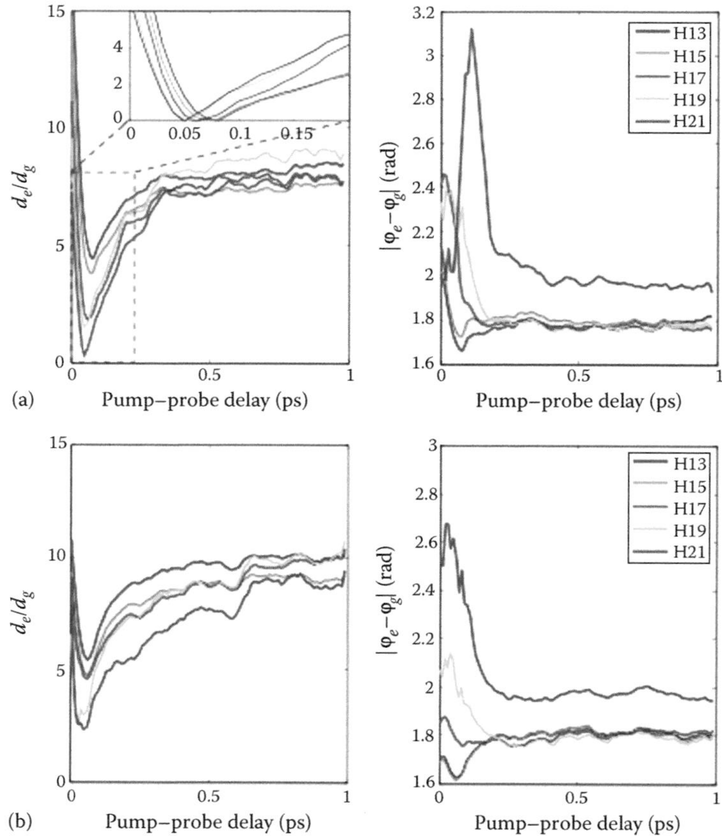

FIGURE 20.15

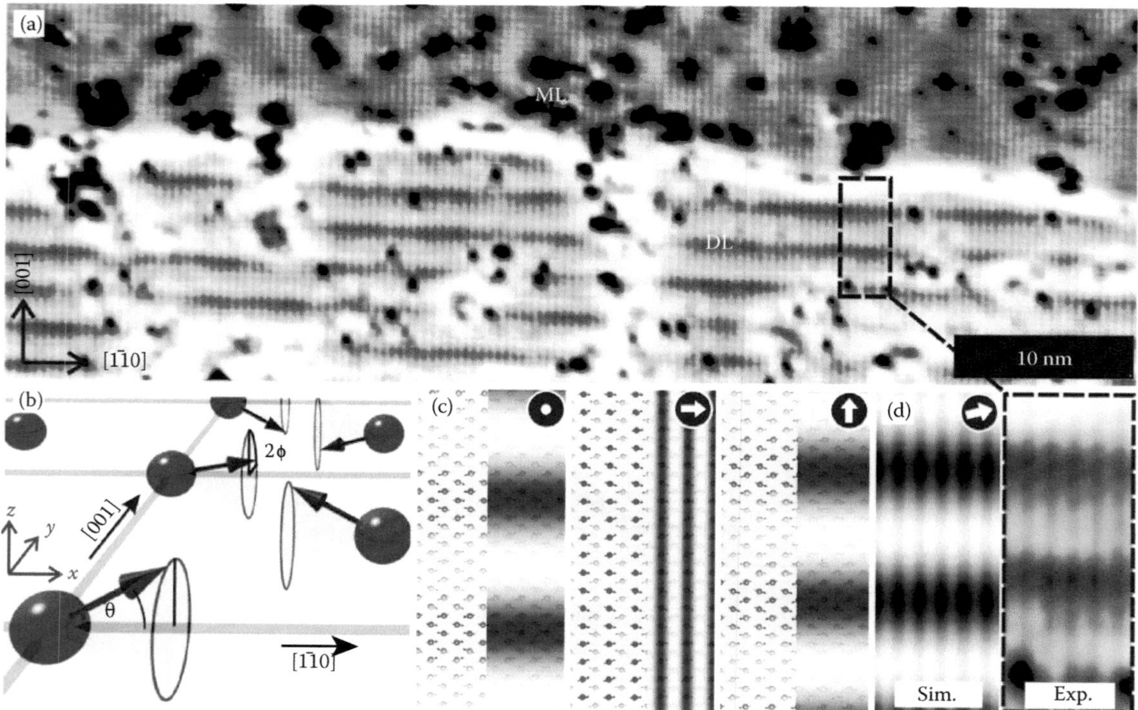

FIGURE 22.25

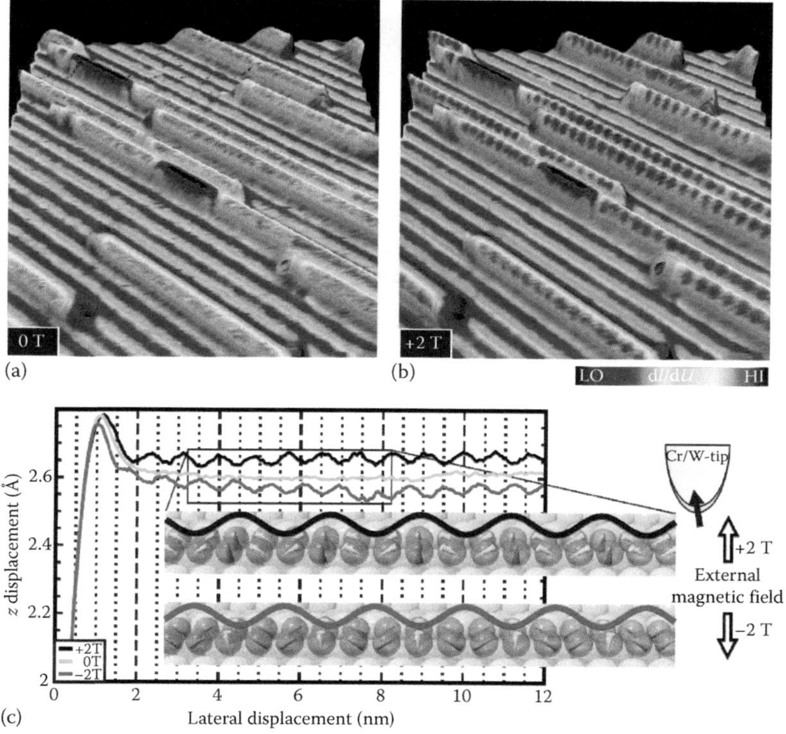

FIGURE 22.26

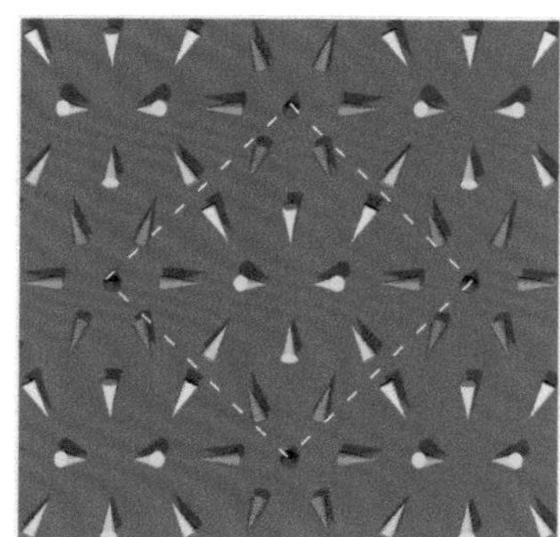

FIGURE 22.30

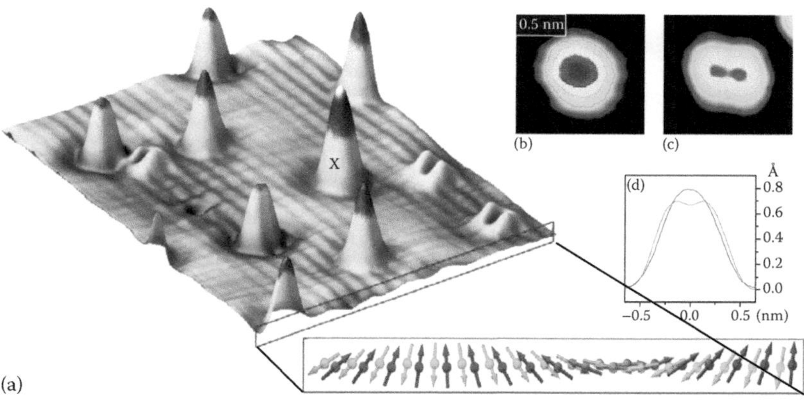

FIGURE 22.32

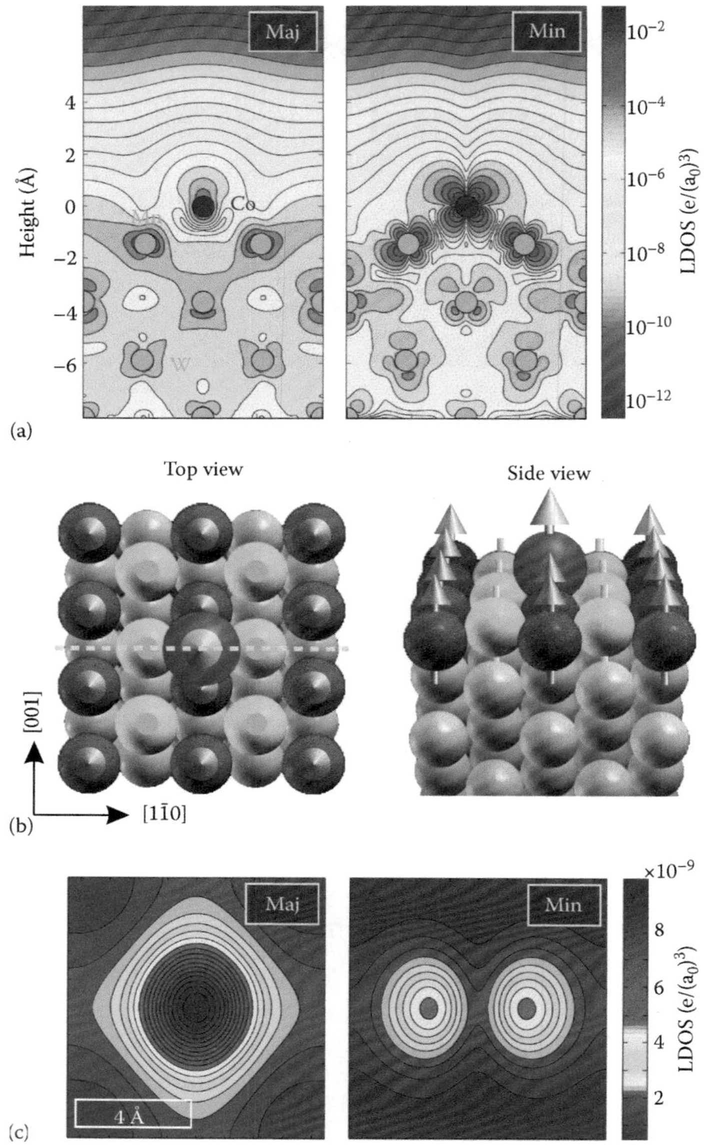

FIGURE 22.33

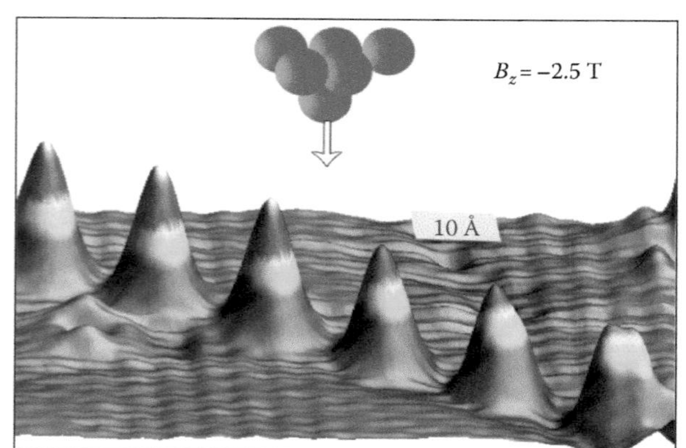

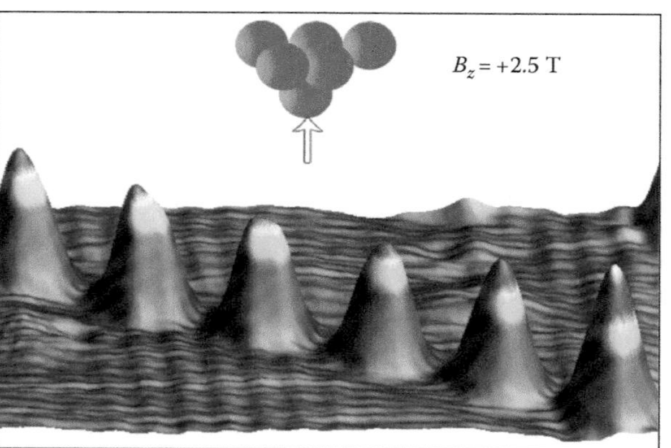

FIGURE 22.34

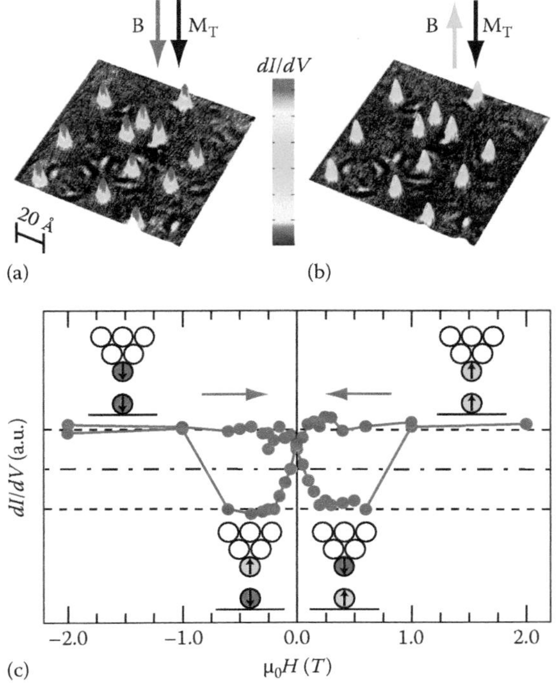

FIGURE 23.3

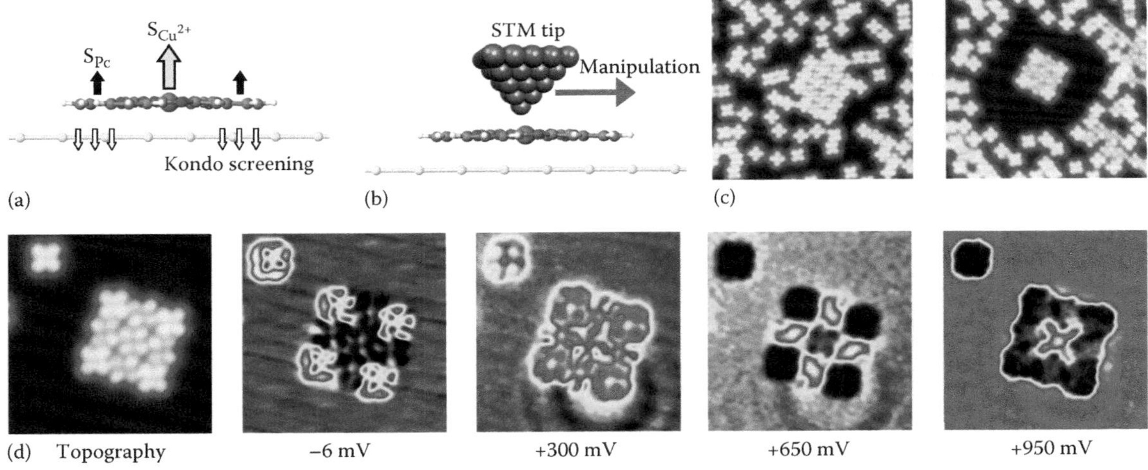

FIGURE 23.14

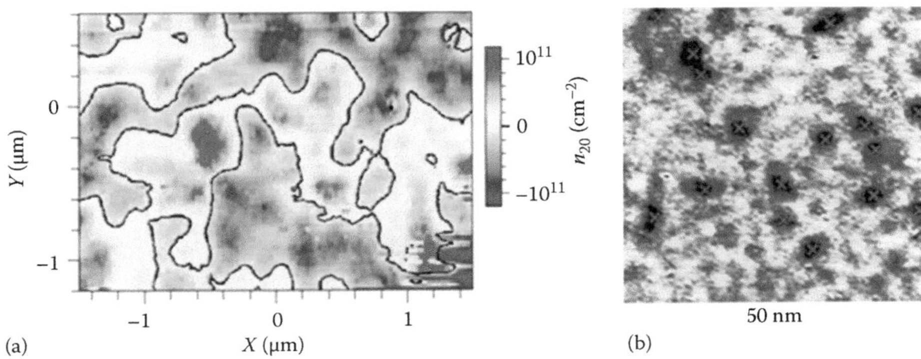

FIGURE 35.33

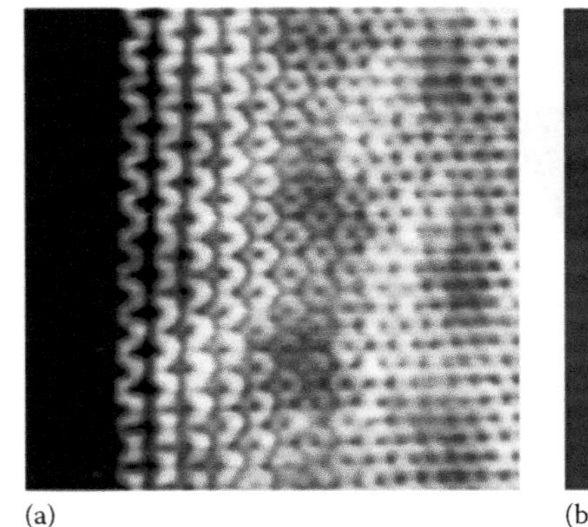

(a)

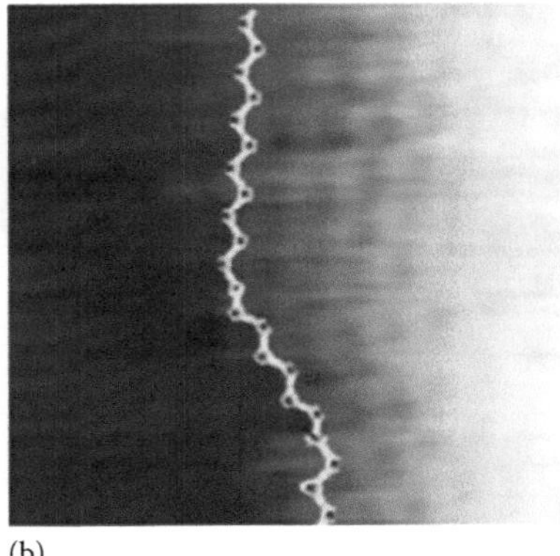

(b)

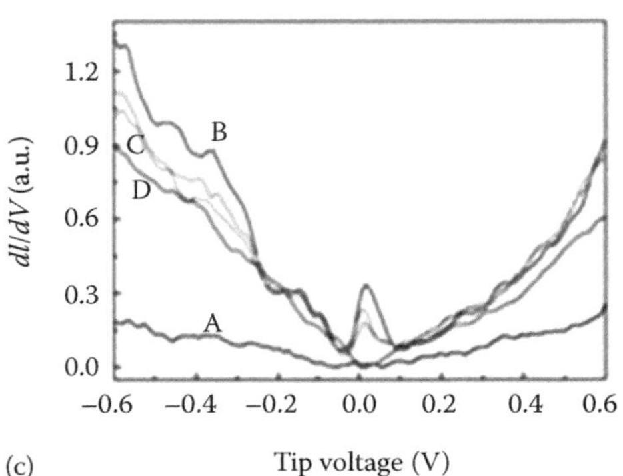

(c)

FIGURE 35.35

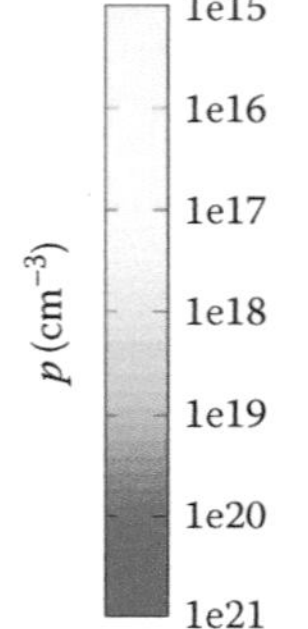

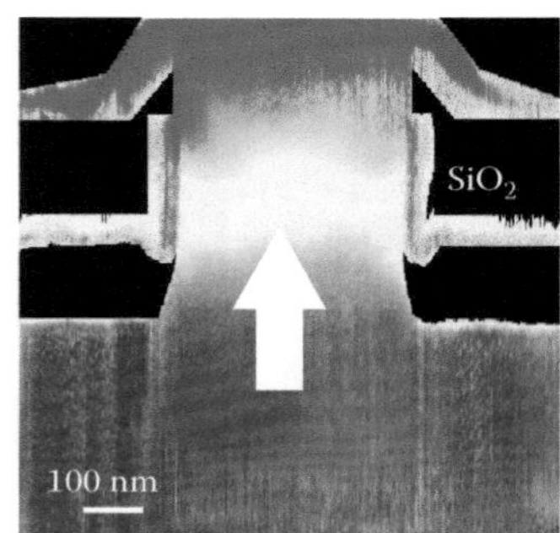

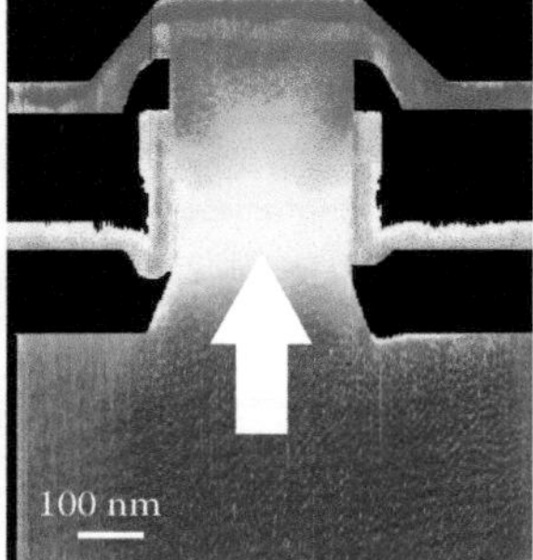

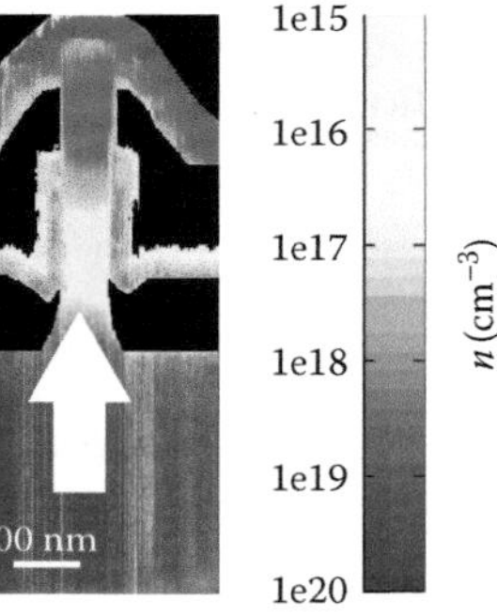

FIGURE 36.39

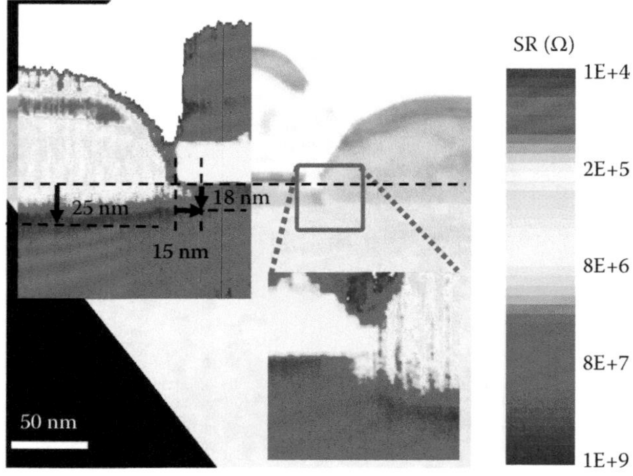

FIGURE 36.46

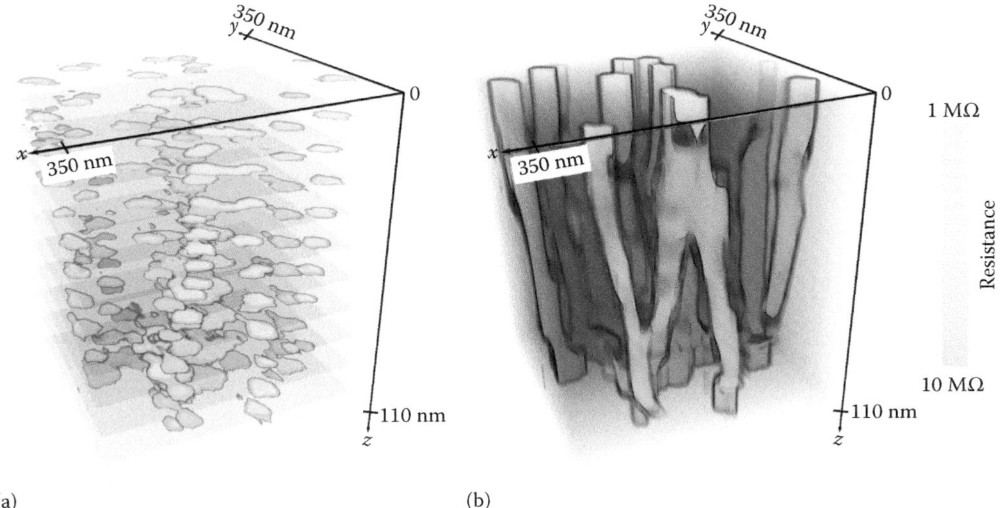

FIGURE 36.49

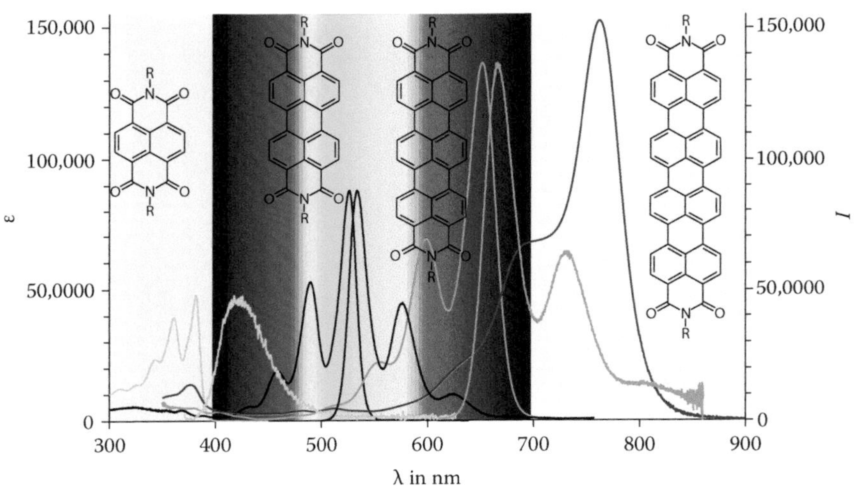

FIGURE 37.2

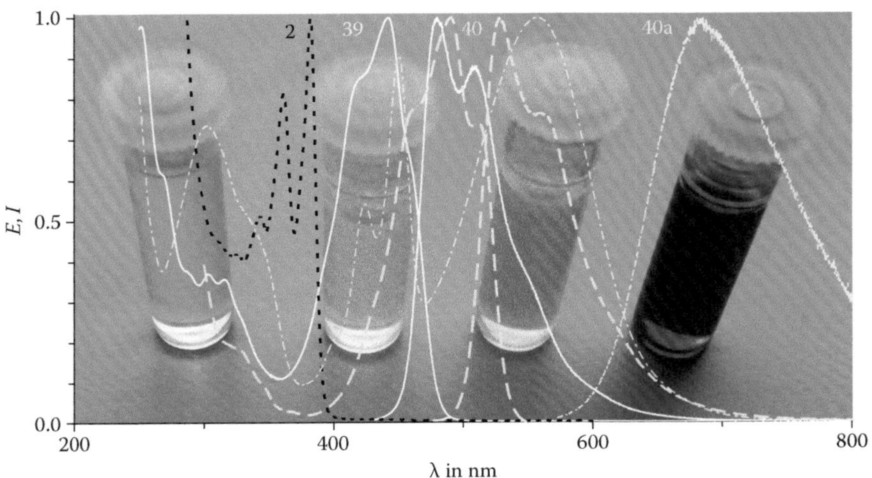

FIGURE 37.3

V

Electron Orbitals

18

Imaging Atomic Orbitals with Scanning Tunneling Microscopy

Alexander N. Chaika
Institute of Solid State Physics
Russian Academy of Sciences

Sergey I. Bozhko
Institute of Solid State Physics
Russian Academy of Sciences

Igor V. Shvets
Trinity College Dublin

18.1 Introduction

Apparently, the first concept of the smallest units (atoms), constructing all substances, was proposed in the fifth century BC. Different combinations of atoms (from Greek ἄτομος—indivisible) were believed to be responsible for the existing variety of organic and inorganic compounds. The atomistic model of matter persisted through 25 centuries despite the inability of direct visualization of atomic-scale objects. However, the progress in understanding the details of the atomic structure was achieved only in the last century. First, the divisibility of atoms was proposed by J. J. Thomson (1897) who proved the existence of small negatively charged subatomic particles (electrons) and suggested that an atom consists of a massive positively charged sphere of about 100 pm in diameter with electrons embedded like raisins in a pudding (Thomson 1904). Then, this model was replaced by the more appealing planetary model (Rutherford 1911, Bohr 1913). With the invention of quantum mechanics (Born and Jordan 1925, Heisenberg 1925, 1927, Schrödinger 1926), the concept of electrons rotating around the nucleus on elliptic orbits was replaced by the concept of atomic orbitals. A single-electron state in an atom was described by a set of wave functions $\psi_{nlm}(r)$ associated with particular electron energy, orbital momentum, spin, and momentum projections on the quantization axes. Spatial distribution of these wave functions (atomic orbitals) determines the probability for an electron with quantum numbers n, l, m_l, and m_s to be detected in a particular volume of space. Although the models of electron orbitals were developed in the beginning of the quantum mechanics era and became one of the most common illustrations of our intuitive understanding of nature, visualization of atomic orbitals, for a long time, was out of the reach of experimentalists. For that reason, the famous talk of Richard Feynman (1960) was sometimes claimed as "completely vacuous so far as the real world is concerned" (Freiser and Marcus 1969). In his lecture, Feynman anticipated the information storage on individual atoms, and working with systems "involving the quantized energy levels." This prediction of atomic-scale technologies seemed to be highly speculative 50 years ago. The situation changed with the development of nonoptical microscopes resolving molecular and atomic-scale objects. The development of transmission (TEM) and scanning (SEM) electron microscopes (Knoll and Ruska 1932, Knoll 1935) and field electron emission (FEEMs) and field ion emission (FIMs) microscopes (Muller 1936, Muller 1951, Muller and Tsong 1969) pushed the detection limit of electron microscopy down to picoscale. The atomic structures of specially prepared sharp tips and individual atoms on surfaces was successfully visualized with FIM (Tsong and Muller 1969), FEEM (Brodie 1978), and TEM (Crewe et al. 1978). However, a new vision of Feynman's lecture appeared

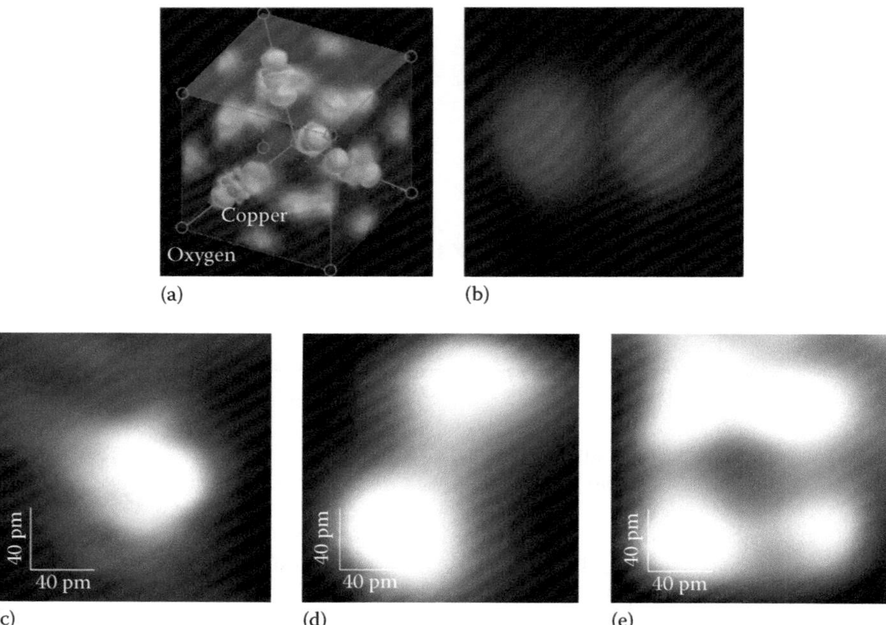

FIGURE 18.1 Electron orbital resolution achieved with different techniques. (a) Experimental and theoretical difference maps between static crystal charge density and superimposed spherical O^{2-} and Cu^{+} ions in Cu_2O. (b) FEEM image of the electron p-orbital of a carbon atom. (c–e) Electron d-orbitals of a tungsten tip atom imaged by carbon atomic orbitals using the STM technique. (a: Reprinted with permission from Macmillan Publishers Ltd., *Nature*, Zuo, J.M., Kim, M., O'Keeffe, M., and Spence, J.C.H., 401, 49–52, 1999. Copyright 1999; b: Reprinted with permission from Mikhailovskij, I.M., Sadanov, E.V., Mazilova, T.I., Ksenofontov, V.A., and Velicodnaja, O.A., *Phys. Rev. B*, 80, 165404, Copyright 2009 American Physical Society; c–e: From Chaika, A.N. et al., *EPL*, 92, 46003, 2010.)

with the invention of scanning tunneling microscope (STM) (Binnig and Rohrer 1982, Binnig et al. 1982a,b), which allowed direct imaging of atomic reconstructions on metallic (Binnig et al. 1983b) and semiconducting (Binnig et al. 1983a) surfaces. The invention of STM was a starting point for the development of a variety of scanning probe microscopy (SPM) methods. The invention of atomic force microscope (AFM) (Binnig et al. 1986) allowed direct atomically resolved imaging of conducting, semiconducting (Giessibl 1995), and insulating (Bammerlin et al. 1997, 1998, Livshits et al. 1999) surfaces. The development of SPM methods allowed spin-sensitive atomic-scale imaging (Wiesendanger et al. 1990, 1992, Wiesendanger 2009) and atomically resolved chemical-selective imaging of multicomponent surfaces (Feenstra et al. 1987, Stroscio et al. 1988, Schmid et al. 1993, Hofer et al. 1998, Abe et al. 2005, Sugimoto et al. 2005, 2007), investigations of vibrational spectra of single molecules (Stipe et al. 1998), manipulations by individual atoms on surfaces (Eigler and Schweizer 1990), and imaging of electron standing waves in artificially fabricated "quantum corral" structures (Crommie et al. 1993). Rapid success of SPM methods brought focus to an earlier fundamental question: Can one visualize the interior of an individual atom? Of course, one could not anticipate direct imaging of both amplitudes and phases associated with particular atomic orbitals in quantum mechanics, but could intra-atomic charge density maps, proportional to $|\psi_{nlm}(r)|^2$, be measured experimentally? Until recently, the answer was rather "no" than "yes."

Apparently, the first experimental maps of electron orbitals were drawn at the end of the 1990s (Zuo et al. 1999) based on x-ray diffraction (XRD) data (Figure 18.1a). These charge density distribution maps revealed electron d-holes in the copper atoms of Cu_2O crystal. Shortly afterward, SPM showed the ability to directly visualize the subatomic orbital structure for the first time. Two hybrid sp^3 orbitals of the silicon atom at a [001]-oriented Si-terminated tip were resolved in AFM experiments (Giessibl et al. 2000). Later, AFM proved the ability to resolve even the finer details of electron orbital structures with lateral resolution below 100 pm (Hembacher et al. 2004). For several years, this kind of resolution seemed to be unachievable for an STM. However, recent experimental visualizations of two lobes of the d_{xz} electron orbitals of manganese (Chaika et al. 2007, Murphy et al. 2007) and copper (Chaika et al. 2008) atoms and two (Figure 18.1d) and four (Figure 18.1e) lobes of the tungsten atom d_{xz} and d_{xy} orbitals (Chaika et al. 2010) demonstrated the possibility to experimentally measure subatomic electron orbital maps and reach a lateral resolution below 100 pm in STM experiments even at room temperature (Figure 18.1c through e). Note that during the past decade, similar spatial resolution in the picometer range has also been achieved with TEM (Erni et al. 2011, Takayanagi et al. 2011) and FEEM (Mikhailovskij et al. 2009, Sadanov et al. 2011). Some of the FEEM images (an example is shown in Figure 18.1b) spontaneously revealed patterns reproducing the shape of the carbon atom p_{xy} electron orbital (Mikhailovskij et al. 2009) discussed in the literature (Manini and Onida 2009, Mikhailovskij et al. 2010).

The possibility to reach a spatial resolution in the subatomic range has been demonstrated in a number of experimental work using different methods. However, XRD experiments (Zuo et al. 1999) cannot provide direct visualization of atomic orbitals, while electron microscopy experiments demand extremely high voltages to be applied to the studied sample, which can destroy the system or at least raise serious questions on the stability of the subatomic imaging. At present, only SPM methods provide the unique possibilities to study atomic-scale systems in direct space without destruction of the investigated samples. To date, the subatomic resolution on the level of individual electron orbitals has been claimed only in several AFM (Giessibl et al. 2000, Hembacher et al. 2004) and STM (Herz et al. 2003, Chaika et al. 2007, 2010, Murphy et al. 2007, Chaika and Myagkov 2008) experimental studies. STM and AFM methods are based on distance dependence of the tunneling current and interaction forces between the atoms of the studied surface and the front atom of a sharp probe tip. Both these dependences are strong at distances between interacting atoms below 1 nm. However, stronger exponential dependence of the tunneling current and nonmonotonous dependence of the interaction forces suggest that experimental realization of controllable selective electron orbital imaging of conducting and semiconducting surfaces can be achieved somewhat easier with STM.

In this chapter, we shall focus on the theoretical and experimental studies aimed at detailed understanding of the role of selected electron orbitals in STM experiments and development of high-resolution imaging at the level of individual electron orbitals. We shall discuss the conditions necessary for probing atomic orbitals and the potential outcomes of selective orbital imaging.

18.2 Spatial Resolution of Scanning Tunneling Microscopy

18.2.1 Basic Principle of STM

In STM, a sharp conducting tip and a conducting sample are separated by a vacuum gap, forming a barrier for electrons. If the vacuum gap is small enough (about 1 nm and below), electrons can travel through it because of the tunneling effect (Landau and Lifshitz 1977). When a bias voltage $\boldsymbol{V_b}$ is applied between the tip and the sample, a tunneling current $\boldsymbol{I_t}$ is produced. The total tunneling current in a one-dimensional case (Chen 2008) can be described as

$$I_t = G_0 \frac{4\pi^2}{e} \int_0^{eV_b} \rho_t(E_F - eV + \varepsilon)\rho_s(E_F + \varepsilon)M^2 d\varepsilon \qquad (18.1)$$

where

G_0 is the conductance quantum

$\rho_t(E)$ and $\rho_s(E)$ are the electron densities of states (DOS) of the tip and surface, respectively

M is a tunneling matrix element

In the case of piecewise constant potential presented in Figure 18.2, the tunneling matrix element is determined by exponential decay of electron wave function in the 0–Z region

$$|M|^2 \sim e^{-kz} \qquad (18.2)$$

$$k = \frac{\sqrt{2m(U - E)}}{h} \qquad (18.3)$$

where

k is the inverse decay length of the electron wave functions in vacuum

h is the Plank constant

m is the mass of an electron

$\boldsymbol{I_t}$ depends exponentially on the tip–sample distance. Simple estimation using the typical values of barrier height (work function) for metals and semiconductors shows that every 1 Å increase in tip–sample separation leads to a decrease in the tunneling current by a factor of 10. This is the origin of the unique high spatial resolution of an STM. According to existing literature data, the vertical and lateral resolution of the best contemporary instruments can be below 1 pm (Gawronski et al. 2008) and 100 pm (Chaika et al. 2008, 2010), respectively.

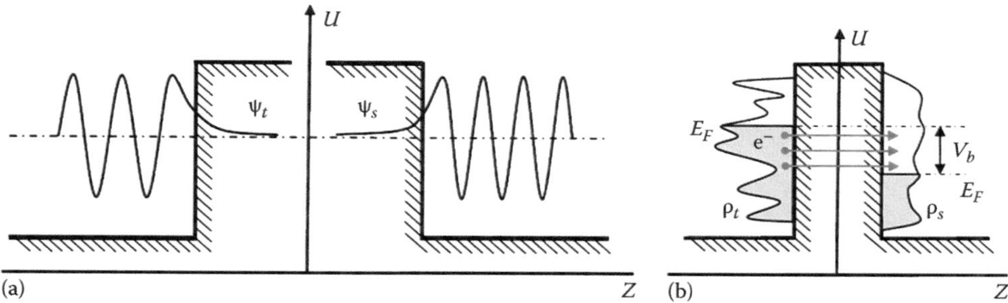

FIGURE 18.2 The Bardeen tunneling theory in the one-dimensional case. (a) Tip and surface are far away, and the wave functions decay into vacuum. (b) Tip and surface are separated by a small gap, and tunneling of electrons takes place (indicated by gray arrows). Occupied electron states are marked in gray.

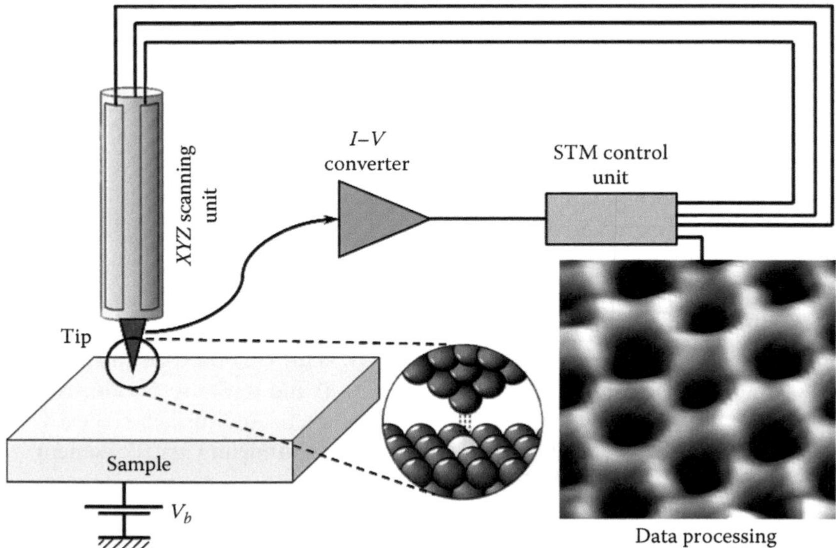

FIGURE 18.3 Schematic view of an STM unit.

Figure 18.3 demonstrates a typical technical realization of an STM (Binnig and Rohrer 1982). In order to control the barrier width and the lateral position of the tip, it is mounted on a piezoelectric ceramic *XYZ* scanning unit. A control unit usually governs *z*-piezodrive in such a way that I_t is maintained constant during the tip displacement in *XY* directions. This is the most commonly used constant current mode (CCM) of STM imaging. In the CCM, the tip height *z* is plotted versus the scan lateral position (x, y). Alternatively, in the constant height mode (CHM), a tip is scanned across a surface at nearly constant height, while I_t is monitored and plotted versus lateral tip position. It is evident from Equation 18.1 that images measured in both modes should simultaneously reflect the surface topography and convolution of a tip and a sample local density of states (LDOS).

The surface DOS distribution can be probed directly in scanning tunneling spectroscopy (STS) experiments based on accurate measurements of current–voltage dependences during scanning across the surface. From Equation 18.1, one can suggest that DOS distribution can be obtained from the derivatives of $I_t(V)$ curves if the tip DOS does not change appreciably in the energy interval of interest. Note that this can be realized experimentally only when the tip DOS as a function of energy is known a priori.

18.2.2 Reciprocity Principle of STM

The electronic structure of a tip and a surface enters into Equation 18.1 in a symmetric manner. This is a basis of the reciprocity principle of STM (Chen 1990a): if the electronic states of the tip and sample are interchanged, the image should be the same. In other words, STM images may be interpreted either as probing the sample states with a tip state or as probing the tip state with sample states. As an example, Figure 18.4 illustrates the equivalence between high-resolution imaging of the surface d_{xz}-orbitals by the tip d_{z^2}-orbital and imaging of the surface d_{z^2}

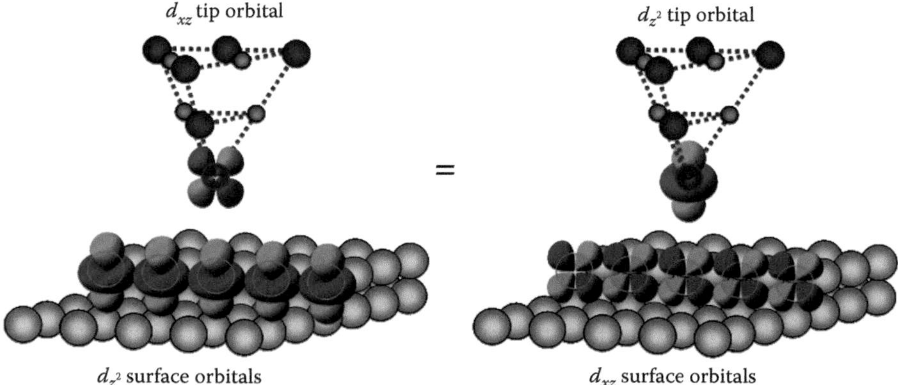

FIGURE 18.4 Schematic presentation of the reciprocity principle of STM. Imaging of d_{xz} atomic orbitals of the surface atoms by d_{z^2} tip state is equivalent to imaging of surface d_{z^2} electron states by the d_{xz} orbital of the front tip atom.

electron states by the tip d_{xz}-state. This is a simplification of more complex realistic tip and surface electronic structures, which usually include a combination of different atomic orbitals with different quantum numbers n, l, and m lying in the energy range between the Fermi level and the applied bias voltage. Accurate calculations of the tunneling current and simulations of the STM/STS images for real tip–surface systems usually demand substantial theoretical efforts and include contribution of all atomic orbitals relevant to experiments. We would refer readers to books (e.g., Foster and Hofer 2006, Chen 2008) or extensive theoretical reviews (Hofer et al. 2003, Hofer 2003) for details. In this section, we only mention more simplified theories considering the contribution of certain electron orbitals of the tip and their influence on the resulting image, which nevertheless give realistic predictions for specific systems. Note that because of the reciprocity principle, these theories can also be useful for the analysis of the partial orbital contributions of the surface atoms.

18.2.3 STM Imaging with Different Tip Orbitals

The problem of evaluating the tunneling matrix elements in an STM junction has been investigated by many authors (Tersoff and Hamann 1983, 1985, Baratoff 1984, Chug et al. 1987, Chen 1988, 1990a,b, Lawunmi and Payne 1990, Tersoff and Lang 1990, Sacks and Noguera 1991a,b, Sacks 2000). Details of different approaches can be found in original papers and STM books (e.g., Chen 2008). We shall discuss the results of only several most relevant studies considering the role of the tip electron orbitals with different l and m in high-resolution STM experiments.

18.2.3.1 Tersoff and Hamann Approach: Resolution with a Spherically Symmetric s-Wave Tip

One of the first theoretical explanations of atomic contrast in STM experiments was given by Tersoff and Hamann who considered a spherically symmetric s-wave STM tip (Tersoff and Hamann 1983, 1985). It was demonstrated that in this case, the resulting image should reflect the surface DOS distribution. This simplification of a realistic tip structure allowed to us avoid the difficulties related to thorough calculations of simultaneous contribution of all tip atomic orbitals. In most cases, this model allows to reach a good agreement between the calculated and experimental images. Therefore, it still remains a "working horse" in STM simulations.

However, there are two problems that cannot be fully explained within this model. First, the lateral resolution R according to Tersoff and Hamann (1983) was estimated as

$$R = \sqrt{d \cdot 2k^{-1}} \tag{18.4}$$

where d is the distance between the cores of the nearest tip and surface atoms.

Since the $2k^{-1} \approx 2$ Å, one can find that the ultimate lateral resolution for the tip–sample distance of 1 nm should be in the range 0.4–0.5 nm. In contrast to this, lateral resolution achieved in early STM experiments on metals (Hallmark et al. 1987, Wintterlin et al. 1988) was below this resolution limit. The spatial resolution can be enhanced at smaller tip–sample distances, but even in this case, the estimation according to Tersoff and Hamann cannot explain the latest experimental data demonstrating lateral resolution below 100 pm (Chaika et al. 2008, 2010).

Another serious problem with the model is related to giant atomic corrugations observed on close-packed metal surfaces. The estimation based on the s-wave tip model predicted corrugations about two orders of magnitude smaller in comparison with those observed in experiments by Hallmark et al. (1987) and Wintterlin et al. (1988, 1989). The enhancement of corrugations in STM experiments can be explained by the tip and surface atom relaxations (Hofer et al. 2004) and essential contribution of the electron orbitals with nonzero l and m quantum numbers. For example, the calculations of Tersoff and Lang (1990) for different tip atoms (Mo, Na, Ca, Si, and C) on a jellium surface demonstrated that corrugations can vary from 10 to 100 pm, depending on the relative contribution of the electron states with different l and m. A more thorough analysis of the corrugation problem was given in later theoretical studies of C. J. Chen and W. Sacks who considered a single atom tip having d-orbitals or a combination of s, p, and d-orbitals.

18.2.3.2 Theories of C.J. Chen and W. Sacks: STM Imaging with $m \neq 0$ Tip States

In the works of C. J. Chen (1990a,b), it was shown that the atomic resolution and large corrugations in STM images of metallic surfaces are most probably observed due to d_{z^2} and p_z tip states. According to the derivative rule (Chen 1990a), the tunneling matrix elements for these states are proportional to the z derivative of the surface atom wave function at the center of the tip apex atom. Therefore, it was predicted that the best candidates for STM tips are d-band metals possessing d_{z^2} states at the apex atom and semiconductors with p_z dangling bonds that coincided with the experimental results (Demuth et al. 1988). Later (Chen 1992) it was shown using the derivative rule that nonzero m tip states can provide enhanced but inverted corrugations with multiple protrusions instead of a single atomic maxima.

In the work of Sacks (2000), the spherical harmonic expansion was used to represent the tip wave functions in the gap region. The electron orbitals of the tip were characterized by atomic-like quantum numbers, l and m. Figure 18.5a shows the density plots for different electron orbitals in the (x, z) plane (Sacks 2000). d-Band metals (W and Pt–Ir alloy) are most frequently used in STM as tip material. Therefore, it is natural to consider the role of d-electron states in STM. Indeed, the contribution of s-states to DOS at the Fermi level does not exceed 5% for Pt and Ir (Papaconstantopoulos 1986). In this case, the tip state in the vicinity of the Fermi level should be a linear combination of different d-orbitals.

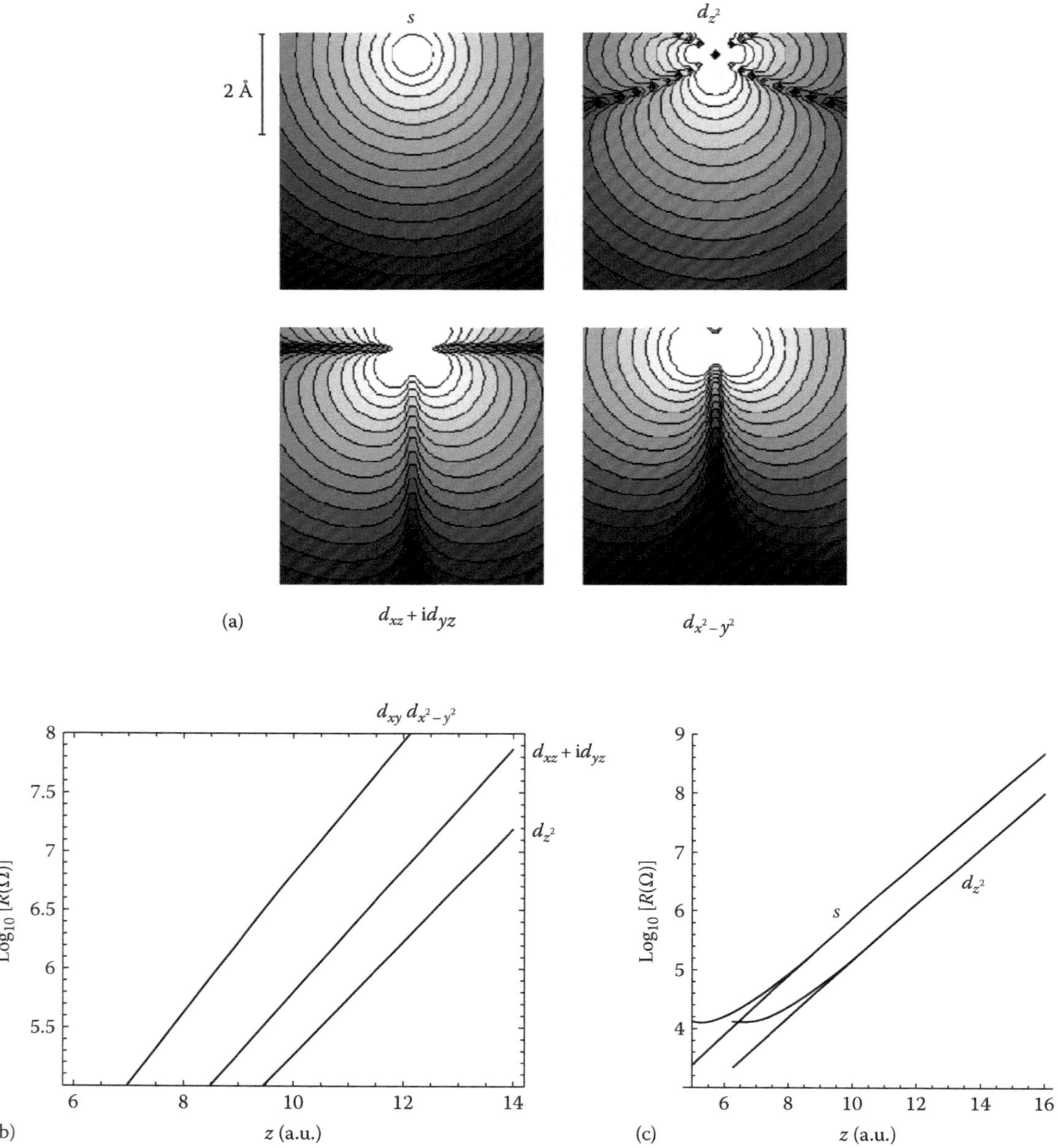

FIGURE 18.5 (a) The density plots for different tip orbitals in the (x,z) plane. Both the gray level, corresponding to a fixed logarithmic scale, and the contours reveal distinctly the nodal lines for the $m \neq 0$ (d_{xz}, d_{yz}, and $d_{x^2-y^2}$) tip states along the vertical z-axis. (b) $R(z)$ curves for each of the tip d-orbitals, having the same spectral weight, and in the nearly free electron model for the surface. These show different asymptotic slopes for the d_{z^2} tip state and $m \neq 0$ ones. Due to the near axial symmetry of the surface, $R(z)$ is the same for the d_{xy} and $d_{x^2-y^2}$ cases. (c) Tunneling gap resistance $R(z)$ curves for the two tip orbitals, s and d_{z^2}. Here, z is the distance between the surface and the tip muffin-tin center. For a fixed resistance, the d_{z^2} tip must be further away from the surface than the s tip. (Reprinted with permission from Sacks, W., Tip orbitals and the atomic corrugation of metal surfaces in scanning tunneling microscopy, *Phys. Rev. B*, 61, 7656–7668, Copyright 2000 American Physical Society.)

Figure 18.5b and c demonstrates the dominant contribution of the d_{z^2} orbital to the tunneling current. It is evident from Figure 18.5b that the asymptotes for both s- and d-tips have the same slope and thus the same orbital decay in vacuum, which is confirmed in Figure 18.5a showing the orbital density plots in the (x, z) plane. It is important that the $R(z)$ curve for d_{z^2} orbital be shifted with respect to the s curve, toward a larger z, by nearly a Bohr radius. One of the possible explanations is related to the shifting of the center of gravity of the tip d_{z^2} orbital away from the tip center (Chen 1990a,b, 1992). Conductivity of the tunneling gap in the case of a tip composed of d-orbitals with $m \neq 0$ is also depressed in comparison with the case of a pure d_{z^2} orbital (Figure 18.5b). In the orientation proposed in Figure 18.5a, the $m \neq 0$ states, having a nodal line along the z-axis, give a reduced tunneling probability and R is much larger than for the d_{z^2} tip.

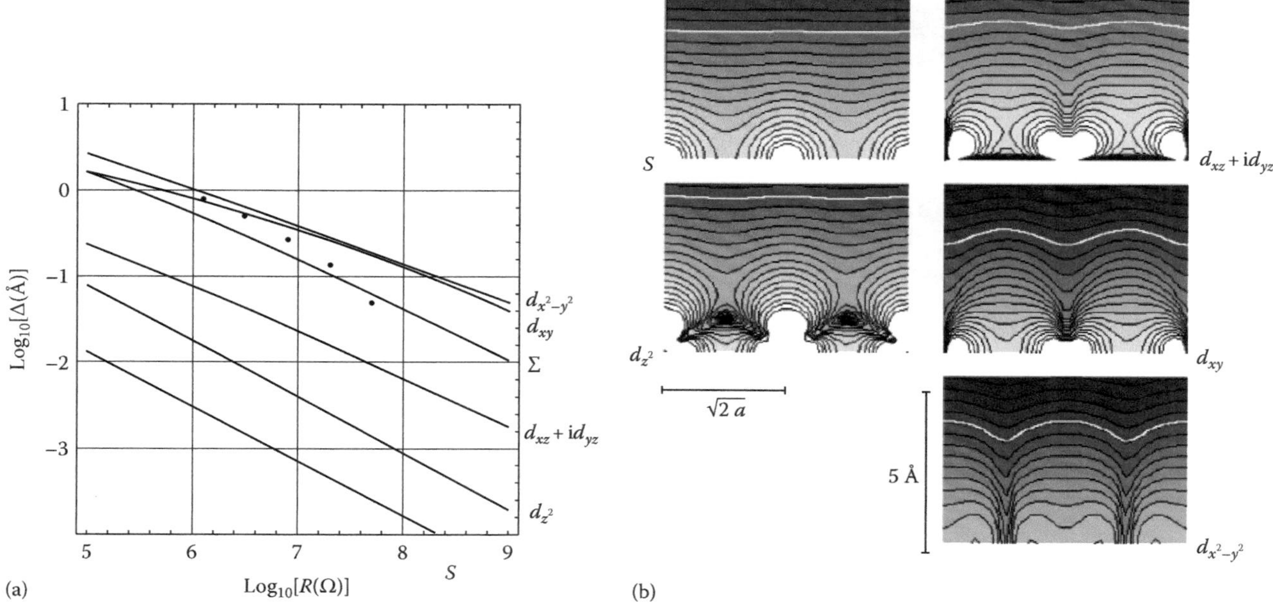

FIGURE 18.6 (a) Magnitude of the corrugation for *s* and *d*-orbital tips as a function of tunneling gap resistance *R*. (b) Gray level representation of the surface LDOS in the plane normal to the surface for the five different tip orbitals. The contours in white correspond to the same value of resistance, $R = 10^5\ \Omega$. Corrugation is slightly larger for d_{z^2} but increasingly enhanced for the orbitals of higher *m*. (Reprinted with permission from Sacks, W., Tip orbitals and the atomic corrugation of metal surfaces in scanning tunneling microscopy, *Phys. Rev. B*, 61, 7656–7668, Copyright 2000 American Physical Society.)

The dependence of corrugation $\Delta(x)$ on tunneling gap resistance as calculated for different electron orbitals (Sacks 2000) is presented in Figure 18.6a. The $m \neq 0$ tip states produce enhanced corrugations in comparison with the *s* tip state, which also increase with decreasing gap resistance. Noteworthily, for the tip orbitals of higher angular momentum, the corrugation gains more than one order of magnitude for each increase in *m*. $\Delta(R)$ shows a gain of more than one order of magnitude, for the d_{xz} and d_{yz} tips, or even two orders of magnitude, for the d_{xy} or $d_{x^2-y^2}$ tips. Curve Σ shows that the $d_{x^2-y^2}$ corrugation reduced by the superposition of only 2% d_{z^2}, which is in good agreement with experimental data presented by dots in Figure 18.6a. To obtain qualitative dependence of the corrugations on a tunneling gap, we note that $\log_{10}[R(\Omega)]$ is proportional to *z* for all orbitals, as indicated in Figure 18.5a and b, resulting in a linear dependence of $\log_{10}[\Delta(Å)]$ on *z*. Hence, corrugations exponentially decay as tip–surface distance increases. Figure 18.6b shows the calculated LDOS of the surface in proximity to the tips with different orbital structure. The contours corresponding to the same gap resistance illustrate that nonzero *m* (d_{xz}, d_{yz}, d_{xy}, and $d_{x^2-y^2}$) states can produce multiple protrusions above the atom positions at very small tunneling gaps and yield enhanced atomic corrugations in comparison with those produced by *s* and d_{z^2} tip orbitals.

18.2.4 Preparation of Tips and Samples for STM Experiments

The processes in an STM junction and the role of the tip electron orbital structure are still not completely understood. However, it was evident from the invention of STM that only sharp single-atom-terminated tips can provide reproducible high-resolution surface imaging. Even though there are some indications that multiple-atom-terminated tips can also provide atomic resolution on metal surfaces (e.g., Jurczyszyn et al. 1998), the ultimate resolution at the level of individual atomic orbitals can be anticipated only with extremely sharp tips collecting the tunneling current through the electron orbitals of one apex atom closest to the studied surface. Therefore, the material of the STM tip, quality of the apex (aspect ratio and radius of curvature), and its electronic structure are crucial for the advancement of spatial resolution and development of picoscale orbital imaging in STM experiments.

Sharp STM tips suitable for atomically resolved studies are typically fabricated from low-cost polycrystalline noble-metal (Nagahara et al. 1989, Iwami et al. 1998) and tungsten (Ibe et al. 1990) wires using electrochemical etching, which allows to obtain significantly sharper apex compared with mechanical cutting (e.g., see comparison in Figure 18.7). For specific (superconducting, spin-polarized, etc.) studies, functionalized probes of other materials can be fabricated. For example, GaAs (Nunes and Amer 1993, Prins et al. 1996), InAs (Sutter et al. 2003), Nb (Kohen et al. 2005), MnNi (Murphy et al. 1999a), MnPt (Murphy et al. 1999b), Cr (Minakov and Shvets 1990, Wiesendanger et al. 1991, Shvets et al. 1992), and magnetic-metal-covered tungsten probes (Wiesendanger 2009, and references therein) were successfully applied in certain high-resolution STM/STS experiments.

To reach stable and reproducible atomic resolution imaging, STM tips are usually cleaned and sharpened in ultra-high

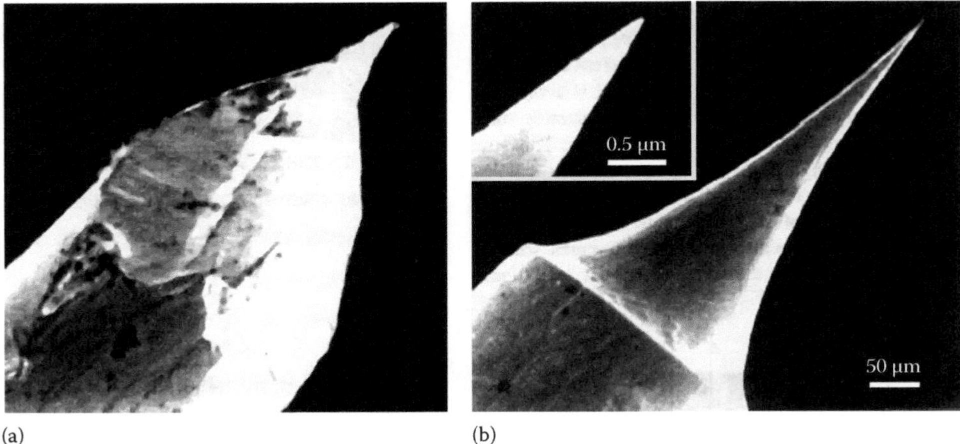

FIGURE 18.7 Comparison of the shape of STM tips obtained by mechanical cutting (a) and electrochemical etching (b). (Reprinted from *Ultramicroscopy*, 30, Stemmer, A., Hefti, A., Aebi, U., and Engel, A., Scanning tunneling and transmission electron microscopy on identical areas of biological specimens, 263–280. Copyright 1989 with permission from Elsevier.)

vacuum (UHV) using annealing, field evaporation (Fink 1986, Stroscio et al. 1987, Neddermeyer and Drechsler 1988), electric field restructuring (Wintterlin et al. 1989, Chen 1991), co-axial ion bombardment (Biegelsen et al. 1987, 1989, Morishita and Okuyama 1991, Eltsov et al. 1996), and controllable crash using either voltage pulses (Murphy et al. 1999a) or gentle contact between the tip and sample, inducing mass transport toward/from the apex (Hansma and Tersoff 1987, Demuth et al. 1988). In some studies, it was demonstrated that enhanced spatial resolution can be achieved using conducting tips terminated by carbon nanotubes (Dai et al. 1996), fullerene (Kelly et al. 1996), organic (Repp et al. 2005), hydrogen (Temirov et al. 2008, Weiss et al. 2010), and carbon monoxide (Gross et al. 2011) molecules having atoms of light elements at the apex. Unfortunately, most of the commonly used tip treatments lead to uncontrollable modifications of the apex on an atomic scale. Even though lateral resolution below 100 pm, allowing us to resolve intramolecular chemical bonds (Weiss et al. 2010, Gross et al. 2011), has been reached using molecule-terminated probes, the applied functionalization procedures are hardly capable of producing stable tips with a controllable atomic and electronic structure. Besides, foreign atoms or molecules at the apex of a metal tip can produce two spatially separated tips working at different bias voltage polarities (e.g., Tromp et al. 1988). This may complicate the interpretation of the atomically resolved STM and STS data obtained with such tips.

To date, control of apex atomic structure during STM experiments has been achieved only in rare studies combining STM and FIMs (Cross et al. 1998). In situ tip characterization using FIM or FEEM represents the best way to control the atomic structure of tips used in high-resolution studies, but this combination can hardly be common experimental practice. In addition, even well-characterized STM tips can be modified during scanning. Such modifications are totally unpredictable in experiments with polycrystalline probes. Although it was demonstrated that tungsten tips fabricated from cold-drawn polycrystalline wires are preferably oriented along the [110] axis (Biegelsen et al. 1987), the orientation of the apex can be assumed to be a priori known only in STM experiments with single crystalline tips. In the latter case, one can control the orbital structure of the front tip atom that, due to the exponential dependence of the tunneling matrix elements, collects up to 90% of the total tunneling current and therefore determines the spatial resolution in experiments. It was shown that chemically etched refractory metal single crystalline tips produce single asperity at the apex (Fink 1986, Biegelsen et al. 1989, Fink et al. 1990, Morishita and Okuyama 1991) instead of multitips frequently obtained from polycrystalline wires (Biegelsen et al. 1987). As an example, Figure 18.8a and b show electron microscopy images of a chemically etched [001]-oriented single crystalline tungsten tip sharpened in UHV using electron beam heating and co-axial ion bombardment (Chaika et al. 2009). The main advantage of oriented, single crystalline tips is illustrated in Figure 18.8c: Even for the nonideal geometry of the apex, its atomic structure can sometimes be predicted from the well-defined apex orientation and tip behavior. In this case, reversible apex atom hoppings, identified in consecutive atomically resolved STM images, allow one to suggest two equivalent positions at the apex of the STM tip (Figure 18.8c).

Despite the evident advantages of oriented single crystalline tips, they can hardly be considered as routinely used probes for high-resolution STM studies. For example, according to estimations given by Greiner and Kruse (2007), single crystalline tungsten tips are roughly 10,000 times more expensive than cold-drawn polycrystalline tips. However, this high price can be paid in particular experiments demanding picometer-scale precision (Chaika et al. 2010) and detailed knowledge of the apex electronic structure, which is highly important both for STM and STS experiments. Here, we should mention, however, that sharp STM tips yielding highest (orbital) resolution in a topography mode are commonly believed to be not suitable for STS experiments demanding relatively blunt tips.

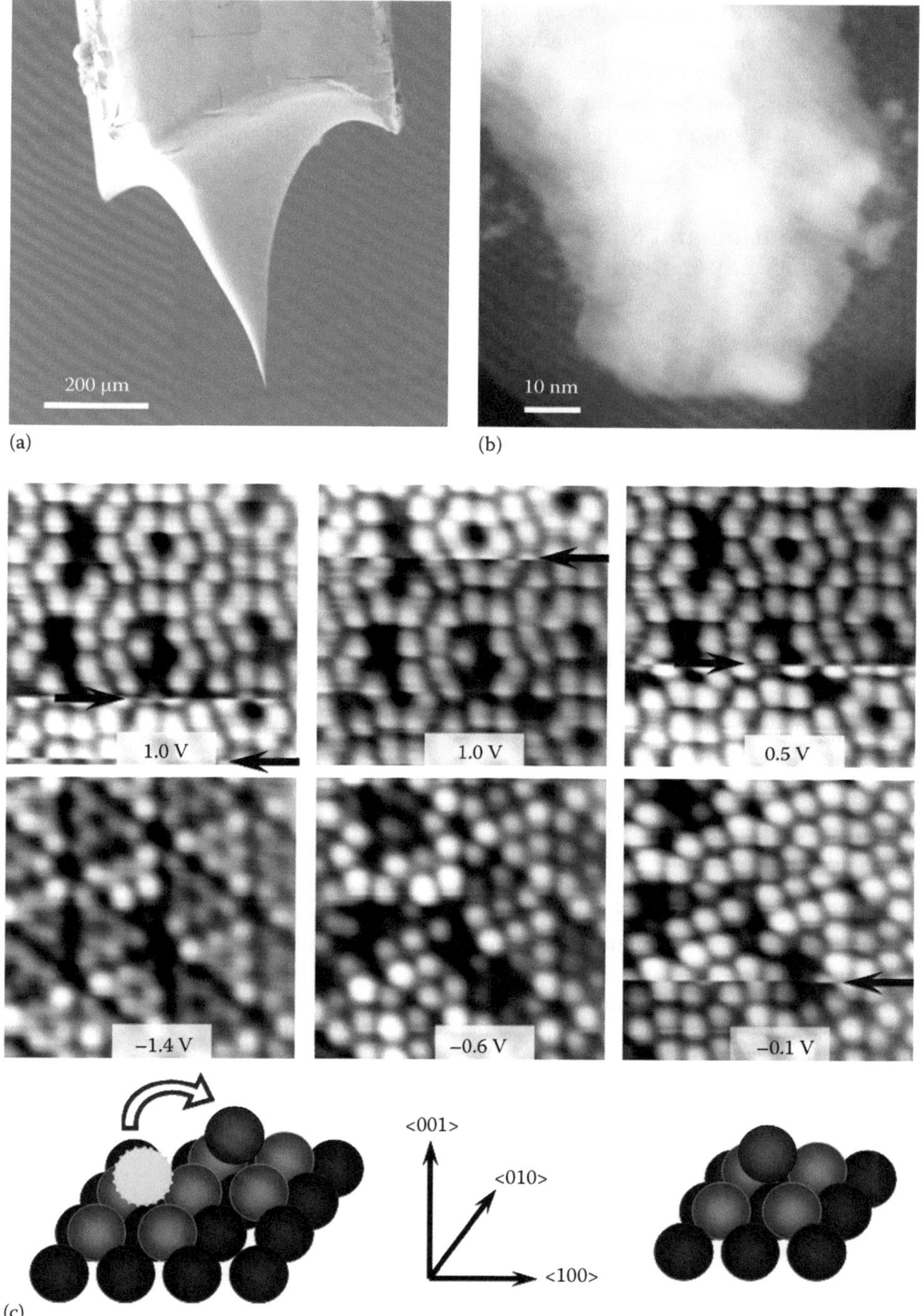

FIGURE 18.8 (a and b) SEM (a) and TEM (b) images of a [001]-oriented single crystalline W tip after cleaning by electron beam heating and ion sputtering. (c) Consecutive 7×7 nm^2 STM images of the Si(111)7×7 surface measured with a [001]-oriented W tip at $I = 40$ pA and different voltages as indicated (top) and models of different W[001] tips (bottom). The arrows indicate the switches of the tip state due to the apex atom hoppings in opposite directions. The scanning was from left to right and from bottom to top of the images. The bottom panel shows ideal (right) and realistic apex configurations, with atom hopping between two equivalent positions (left).

While appreciating the role of the STM tip treatment, one should not underestimate the importance of sample surface preparation in achieving high lateral resolution. Usually, the surface should be extremely well ordered to be imaged by STM with atomic resolution. Although single-atom defects and adsorbed atoms on surfaces are frequently used to demonstrate the sharpness of the tip, they can substantially increase the probability of modifying, and sometimes leading to loss of, the tip with a well-defined orbital structure at the apex during scanning. Clean and well-ordered surfaces can be achieved using in situ cleavage in the STM chamber. This procedure is typically used for preparation of low-index surfaces of semiconductors and layered materials, but can hardly be used for preparation of metallic and high-index (vicinal) semiconducting surfaces, which usually require thorough cleaning procedures consisting of annealing and ion bombardment cycles.

18.2.5 Tip–Sample Separation and Spatial Resolution in STM Experiments

According to the theory (Tersoff and Hamman 1983, Sacks 2000), an enhancement of lateral resolution should be anticipated with decreasing tip–sample separation. The theoretical calculations also predict an increase in atomic corrugations with decreasing tunneling gap (Figure 18.6). Therefore, small tip–sample separations are crucial for enhancement of spatial resolution in STM experiments. Indeed, it has been demonstrated that extremely sharp atomic features reproducing the atomic orbital distribution maps can be resolved with clean (Chaika et al. 2010) and molecule-terminated (Gross et al. 2011) tungsten tips at distances below 400 pm. At such small distances, interactions between the tip and surface atoms can change the tunneling barrier height (Binnig et al. 1984), induce tip and surface atom relaxations leading to atomic corrugation enhancement (Wintterlin et al. 1989, Zheng and Tsong 1990, Hofer et al. 2004), and modify the surface and tip electronic structure (Hofer 2003, Hofer et al. 2003, Jelinek et al. 2008, Chaika et al. 2010, Polok et al. 2011). The interaction can change the orbital structure of the tip and surface atoms, and, consequently, the resulting atomically resolved STM image symmetry. There is still no detailed information on possible changes of the partial density of electron states (PDOS) associated with certain atomic orbitals in STM atom–atom contact at extremely small tunneling gaps. However, even in the simplest case of a nonperturbed electronic structure of the interacting tip and surface atoms, the theoretical calculations (Sacks 2000) suggest that appearance of atomic features in STM images can be distance dependent when atomic orbitals with nonzero m dominate at the apex of the probe tip. For example, as illustrated in Figure 18.6b, d_{xz}, d_{yz}, d_{xy}, and $d_{x^2-y^2}$ electron orbitals of the tip can produce multiple subatomic features at small tip–sample separations and single protrusions at tunneling gaps exceeding 300 pm. Taking into account the reciprocity principle of STM, similar distance-dependent contributions can be predicted for the electron orbitals of the surface atoms. Therefore, one could anticipate anomalous tip–sample distance dependencies of atomically resolved STM images measured on d-metal surface reconstructions.

Apparently, the first distance dependence of atomically resolved STM images was observed in the experiments on gold-sputtered graphite surface (Bryant et al. 1986) where either regular graphite surface lattice or subsurface defects induced by the sputtering were imaged at different tip–sample distances. Then, pronounced dependence of the STM image contrast on the gap resistance was experimentally observed in a number of STM studies. For example, distance dependence of atomically resolved STM images of W(110)/C-R(15 × 3) (Bode et al. 1996, Wiesendanger et al. 1996), Al(001) (Jurczyszyn et al. 1998), Ru(0001)–O (Calleja et al. 2004), and Cu(014)–O (Murphy et al. 2007) surface reconstructions and chemical contrast in PtRh and PtNi alloys (Hofer et al. 1998) were explained by spatial distribution of surface wave functions and change of the relative contributions of different surface atomic orbitals at varying tip–sample distance. In some studies, anomalous distance dependence of STM images measured at very small tunneling gaps was attributed to elastic interaction between the tip and the surface atoms. As an example, Figure 18.9 shows sharp distance dependence of atomically resolved STM images measured at certain tip states on an InAs(110) surface (Klijn et al. 2003). Images taken at a small positive sample bias voltage of 50 mV and lower tunneling currents (larger gap resistances) demonstrate a regular array of spherically symmetric features corresponding to the positions of As atoms on the surface (Figure 18.9a). However, at higher currents (smaller gap resistances), STM images measured with the same tip demonstrate asymmetric subatomic features (Figure 18.9b through d). Their appearance is essentially dependent on the tunneling current and fast scan axis direction. Note that transformation from spherically symmetric to asymmetric subatomic features was reproducibly observed in a very narrow interval of the tunneling currents between 450 and 530 pA. These values prove the necessity to control the gap spacing with picometer precision during high-resolution STM experiments. The subatomic features in Figure 18.9, which could not be identified with the well-known shapes of p- or d-electron orbitals, were explained by elastic interactions between the tip and surface atoms. Note that tip–sample interaction forces at small distances can be directly related to the orbital structure of the interacting atoms because the electron charge distribution around the nuclei is responsible for strong Pauli repulsion at extremely small distances (Weiss et al. 2010) where the highest spatial resolution can be achieved in SPM experiments.

18.3 Atomic Orbital Resolution in STM and STS Experiments

18.3.1 Orbital Imaging on Semiconducting Surfaces

The first SPM resolution at the level of separate atomic orbitals was achieved in the early experiments on the Si(111)7 × 7 surface (Binnig et al. 1983a). This fascinating reconstruction (Takayanagi et al. 1985) became a milestone in the development of SPM techniques and orbital imaging capability. It is now accepted that the atomic structure of the Si(111)7 × 7 reconstruction corresponds to the dimer–adatom–stacking fault (DAS) model (Takayanagi et al. 1985). The topmost layer of the Si(111)7 × 7 surface contains 12 adatoms having p_z dangling bonds. Because of a large distance between the adatoms (~0.7 nm), these dangling bonds can be separately resolved in STM experiments in a wide range of bias voltages and tip–sample distances. Therefore, the most frequently observed 7 × 7 STM images (e.g., Figure 18.10a)

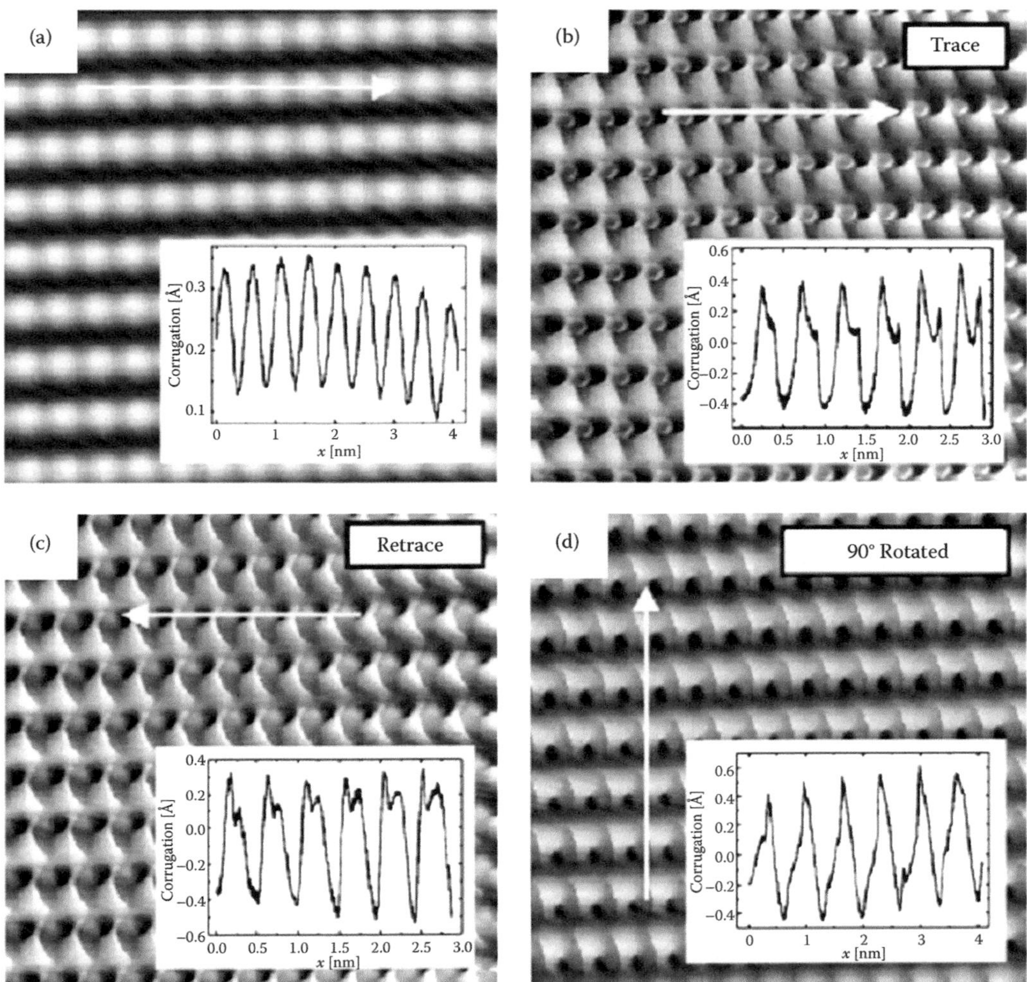

FIGURE 18.9 CCM images recorded with an unchanged tip on InAs(110): (a) Conventional atomic resolution. $V = 50$ mV, $I = 450$ pA. (b–d) Images demonstrating anomalous subatomic-scale features. (b) $V = 50$ mV, $I = 710$ pA. (c) Same as in (b), but scan direction is reversed. (d) $V = 50$ mV, $I = 530$ pA; scan direction is rotated by 90° relative to that in (b) and (c). Insets show line sections along the marked arrows. The arrows indicate the scan direction ("trace"). (Reprinted with permission from Klijn, J., Sacharow, L., Meyer, C., Blugel, S., Morgenstern, M., and Wiesendanger, R., STM measurements on the InAs(110) surface directly compared with surface electronic structure calculations, *Phys. Rev. B*, 68, 205327, Copyright 2003 American Physical Society.)

can be considered as regularly arranged images of individual p_z orbitals associated with the top layer silicon atoms. It is known that six rest atoms of the Si(111)7 × 7 unit cell (Figure 18.10c) also have dangling bonds, but they can be resolved only at particular bias voltages using extremely sharp single-atom-terminated tips (Wang et al. 2004). An example of the image revealing both the adatom and rest-atom p_z electron states is shown in Figure 18.10b. The cross section in Figure 18.10d demonstrates the unit cell asymmetry and nonequivalence of different surface layer atoms, which can be identified in high-resolution STM images (Figure 18.10a and b). Shortly after publication of the first atomically resolved STM data on the Si(111)7 × 7 surface (Binnig et al. 1983a), bias voltage–dependent STM and STS experiments (Hamers et al. 1986, 1987a, Tromp et al. 1986) resulted in selective visualization of the top layer atom orbitals, backbond electron states, rest atom dangling bonds, and even the corner hole p_z-states associated with the silicon atoms located in the third layer. As an example, Figure 18.11 shows more recent spectroscopic images of certain atomic orbitals of the 7 × 7 surface taken at different bias voltages (Paz et al. 2005).

It is known that various atomic reconstructions can be fabricated on cleaved or ion-etched and annealed Si(111) surfaces can demonstrate other than 7 × 7 atomic reconstructions. Figure 18.12 shows the experimental CCM STM images and spectroscopic LDOS maps of the Si(111)2 × 1 surface (Garleff et al. 2004). Apart from the features associated with surface atom nuclei positions, subatomic features, which could not be identified either with atom positions or with the tip LDOS, are clearly discernible in the spectroscopic maps measured at tunneling voltages of ±2.0 V. Similar to imaging different atomic orbitals of the Si(111)7 × 7 reconstruction, the appearance of the atomic and subatomic features in Si(111)2 × 1 STS maps was bias-voltage dependent.

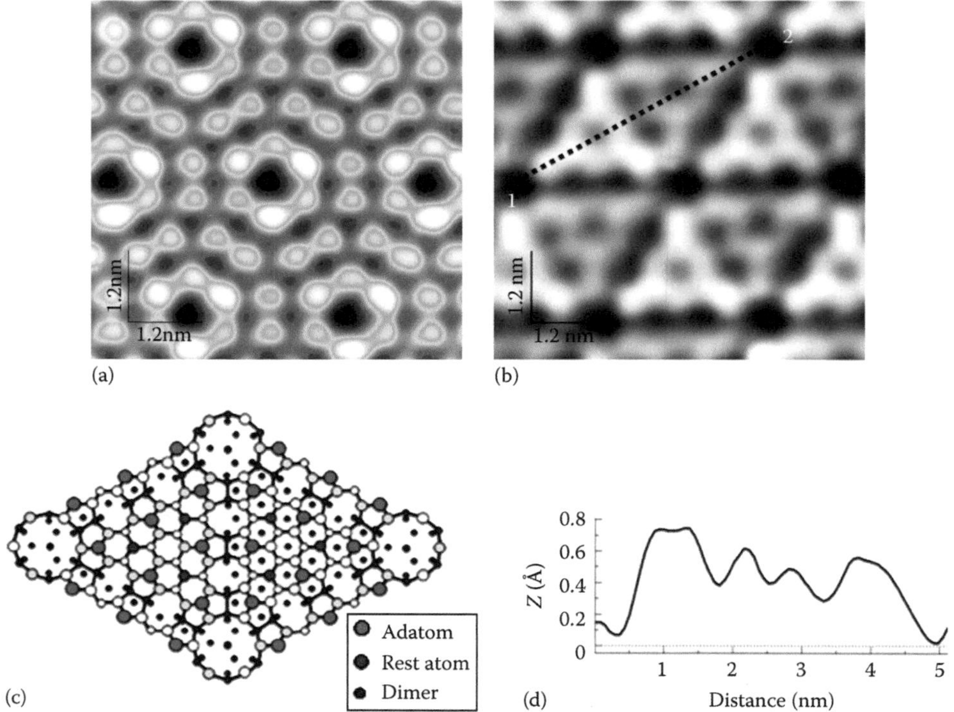

FIGURE 18.10 (a and b) Empty (a) and occupied (b) state CCM STM images of the Si(111)7 × 7 surface. The images were measured with tungsten tips at V_b = 1 V and I = 80 pA (a) and V_b = −1.4 V and I = 50 pA (b). (c) Top view of the DAS model. (d) Cross section 1–2 of the image in panel (b) taken along the dotted line. (c: From Takayanagi, K. et al., *J. Vac. Sci. Technol. A*, 3, 1502, 1985.)

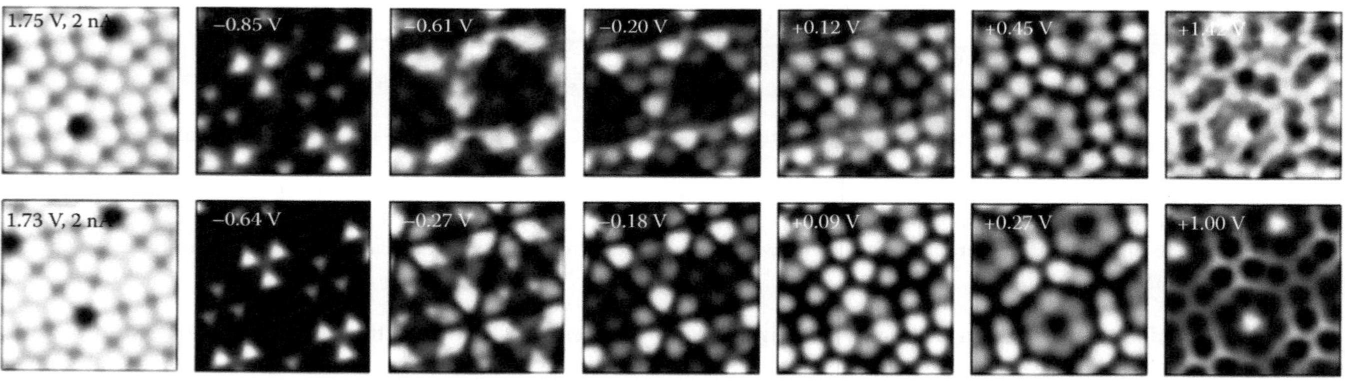

FIGURE 18.11 Experimental current-imaging-tunneling spectroscopy maps of the Si(111)7 × 7 surface (top pictures) and simulations using a silicon tip (bottom pictures). The left panels show CCM STM images at I = 2 nA and V_b = 1.75 V. The remaining images represent dI/dV *maps* as a function of V_b. (Reprinted with permission from Paz, O., Brihuega, I., Gomez-Rodriguez, J.M., and Soler, J.M., Tip and surface determination from experiments and simulations of scanning tunneling microscopy and spectroscopy, *Phys. Rev. Lett.*, 94, 056103, Copyright 2005 American Physical Society.)

Figures 18.10 through 18.12 illustrate the role of bias voltage in STM experiments on semiconducting surfaces. Here, the choice of bias voltage is crucial for discrimination of occupied/empty surface electron states associated with nonequivalent surface atoms. As a result, backbond/dangling bond states of the same surface atoms and electron orbitals of nonequivalent atoms could be resolved at different bias voltages. Figure 18.13 shows an example of bias-dependent atomic resolution studies (Dubois et al. 2005) of the Si(001) surface. The atomic and electronic structure investigations (Hamers et al. 1987b, Wolkow 1992) showed that the topmost Si(001) surface layer contains buckled dimer atoms having one dangling bond. As can be seen from Figure 18.13, either the *z*-directed dangling bonds of the topmost dimer atoms or in-plane aligned bonds can be resolved

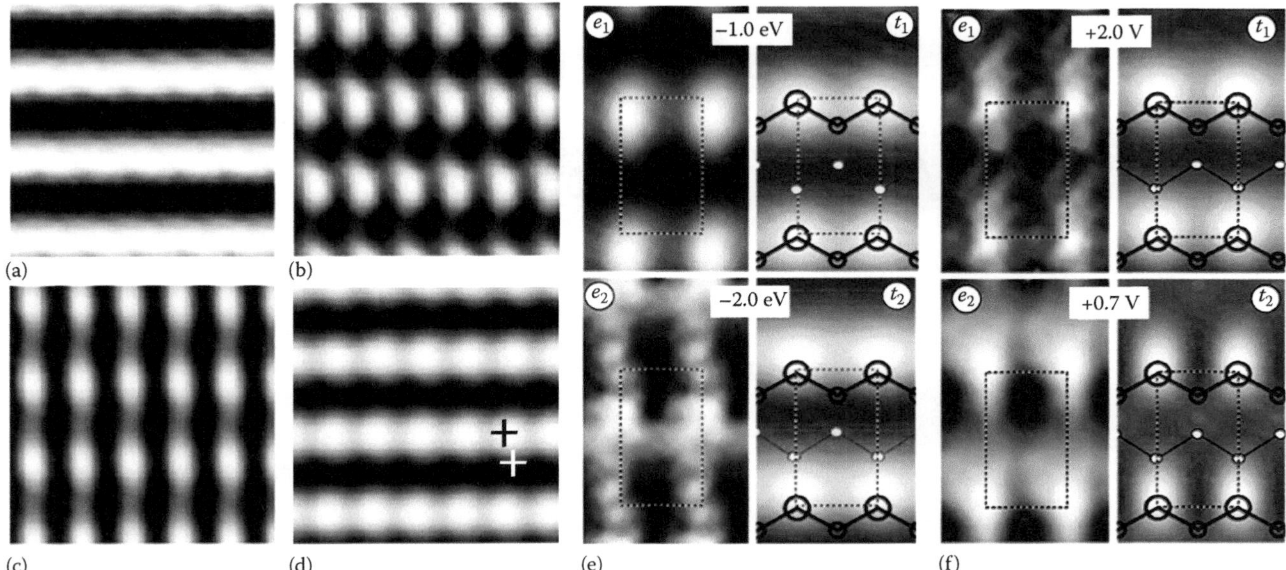

FIGURE 18.12 (a–d) 45 × 45 Å^2 CCM images of Si(111)-2 × 1 at different tunneling parameters: +1.4 V, 0.5 nA (a); +0.6 V, 0.5 nA (b); −0.3 V, 0.1 nA (c); and −1.1 V, 0.2 nA (d). (e and f) Laterally resolved LDOS maps of the empty (e) and filled (f) states from theory (t_i) and experiment (e_i) at typical energies (white, high LDOS; black, low LDOS). In the calculated images, the underlying atomic lattice is shown by large (small) open circles indicating the up (down) atoms of the *p*-bonded chains and the gray filled circles marking the zigzag rows in between. (Reprinted with permission from Garleff, J.K., Wenderoth, M., Sauthoff, K., Ulbrich, R.G., and Rohlfing, M., 2 × 1 reconstructed Si(111) surface: STM experiments versus *ab initio* calculations, *Phys. Rev. B*, 70, 245424, Copyright 2004 American Physical Society.)

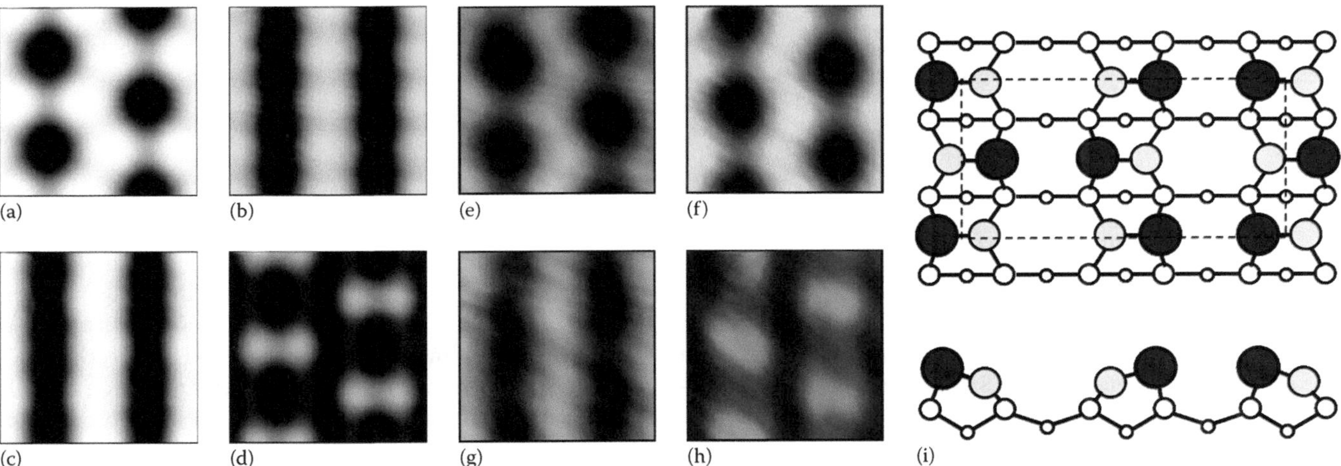

FIGURE 18.13 (a–d) Simulated STM images obtained on the Si(001)-c(4 × 2) phase for sample voltages of −1.0 V (a), +0.7 V (b), +1.6 V (c) and +2.1 V (d). (e–h) Experimental STM images measured on the c(4 × 2) phase of an *n*-type Si(001) surface for sample voltages of −1.5 V (e), +0.7 V (f), +0.7 V (g), and +1.5 V (h). (Reprinted with permission from Dubois, M., Perdigao, L., Delerue, C., Allan, G., Grandidier, B., Deresme, D., and Stievenard, D., Scanning tunneling microscopy and spectroscopy of reconstructed Si(100) surfaces, *Phys. Rev. B*, 71, 165322, Copyright 2005 American Physical Society.) (i) Schematic model of the Si(001)-c(4 × 2) reconstruction.

in STM images under specific tunneling conditions. The experimental data in Figure 18.13 demonstrate that qualitatively the same images of the Si(001)-c(4 × 2) surface can be measured at different bias voltages (Figure 18.13e and f) as well as different images can be resolved at the same bias voltage (Figure 18.13f and g). These data show that selective visualization of electron orbitals of the Si(001)-c(4 × 2) surface is controlled by both bias voltage and gap resistance (i.e., tip–sample distance).

The bias voltage dependence of atomic-scale STM images has been emphasized in numerous studies of semiconducting surfaces. For multicomponent surfaces, this can lead to selective imaging of atoms of different chemical nature (Feenstra et al. 1987). Figure 18.14 shows one of the first examples of atomically resolved chemical discrimination with STM. In this case, selective visualization of the electron orbitals associated with either Ga or As atoms on the GaAs(110) surface was achieved at

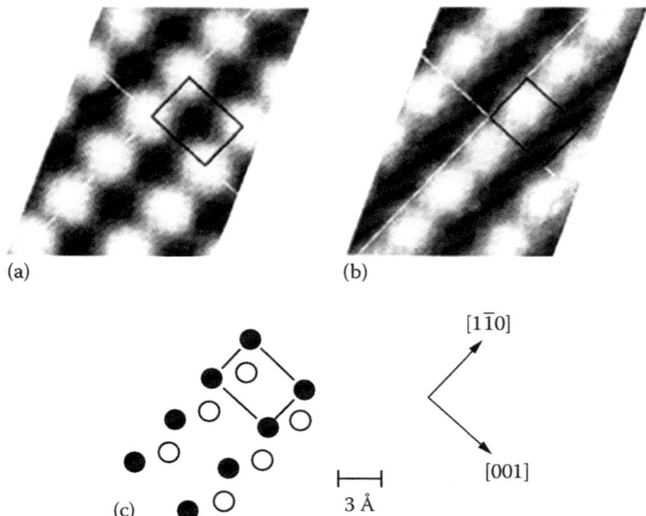

FIGURE 18.14 Chemical-selective STM imaging of the GaAs(110) surface: (a and b) CCM STM images measured at sample bias voltages of +1.9 V (a) and −1.9 V (b). (c) Top view of the surface atoms. As and Ga atoms are represented by open and closed circles, respectively. (Reprinted with permission from Feenstra, R.M., Stroscio, J.A., Tersoff, J., and Fein, A.P., Atom-selective imaging of the GaAs(110) surface, *Phys. Rev. B*, 58, 1192, 1987–1195. Copyright 1987 American Physical Society.)

different bias voltage polarities. Note that such an energy separation of the surface atom electron orbitals can hardly take place for metal oxides (Ruan et al. 1993, Chaika and Bozhko 2005) and metallic alloys (Schmid et al. 1993, Hebenstreit et al. 1999) where the effects of chemical-selective imaging can be observed at the same tunneling voltages. Understanding these images requires detailed analysis of the relative contributions of electron orbitals with different momentum projections on the quantization axes (Hofer et al. 1998, Murphy et al. 2007, Chaika et al. 2008).

18.3.2 Imaging Atomic Orbitals in Molecular Structures

Molecular ensembles on metallic and semiconducting surfaces represent another system where occupied and empty atomic orbitals that differ in shapes and energy values can be resolved in bias-dependent STM/STS experiments. The molecular structures have extensively been studied because of possible applications in organic-based nanoelectronics. However, the intramolecular electronic structure is frequently not resolved in STM experiments. It is believed (Swart et al. 2011) that usual transition metal STM tips having highly localized d-electron states at the tip apex are not suitable for submolecular resolution, which can only be achieved using functionalized probes with molecules of light elements at the apex. STM images of the pentacene molecule shown in Figure 18.15a demonstrate an enhancement of spatial resolution in STM experiments with a molecule-terminated probe having an atom of a light element at the apex. In some recent works (Weiss et al. 2010, Gross et al. 2011), it was demonstrated that spatial resolution in STM experiments on molecular structures can be enhanced at extremely small tunneling gaps when short-range forces dominate in the tip–sample interaction. Therefore, the possibility to fabricate a sharp and stable light element–terminated probe appears to be one of the crucial issues for STM visualization of molecular orbitals with ultimate spatial resolution. Figure 18.15b shows an alternative way to probe the intramolecular electronic structure. In this case, direct visualization of certain molecular orbitals could be achieved in energy-resolved STS experiments.

18.3.3 Orbital Resolution in Distance-Dependent STM Experiments on Multicomponent Surfaces

Despite a large number of STM studies of various semiconducting and molecular structures, the existing atomically resolved data rarely reveal drastic dependence of the images on gap resistance (e.g., as shown in Figure 18.9) and appearance of asymmetric subatomic features, which could be directly associated with decisive contribution of particular electron orbitals. Theoretical calculations (Figure 18.6b) suggest that pronounced dependence of STM images on the tip–sample distance can be anticipated when higher momentum electron states with nonzero m are responsible for the tunneling current. This can take place in experiments on metal surface reconstructions. Indeed, the gap resistance dependence of atomically resolved STM images of multicomponent metallic surfaces has been reported in a number of studies (Bode et al. 1996, Wiesendanger et al. 1996, Jurczyszyn et al. 1998, Calleja et al. 2004, Murphy et al. 2007). However, because of simultaneous contributions of different electron orbitals of both the tip and surface atoms, the role of electron orbitals with different momentum projections in STM imaging is still not completely understood. Nevertheless, the experimental and theoretical studies prove that spatial distribution of the tip and surface wave functions is one of the most important factors responsible for the distance

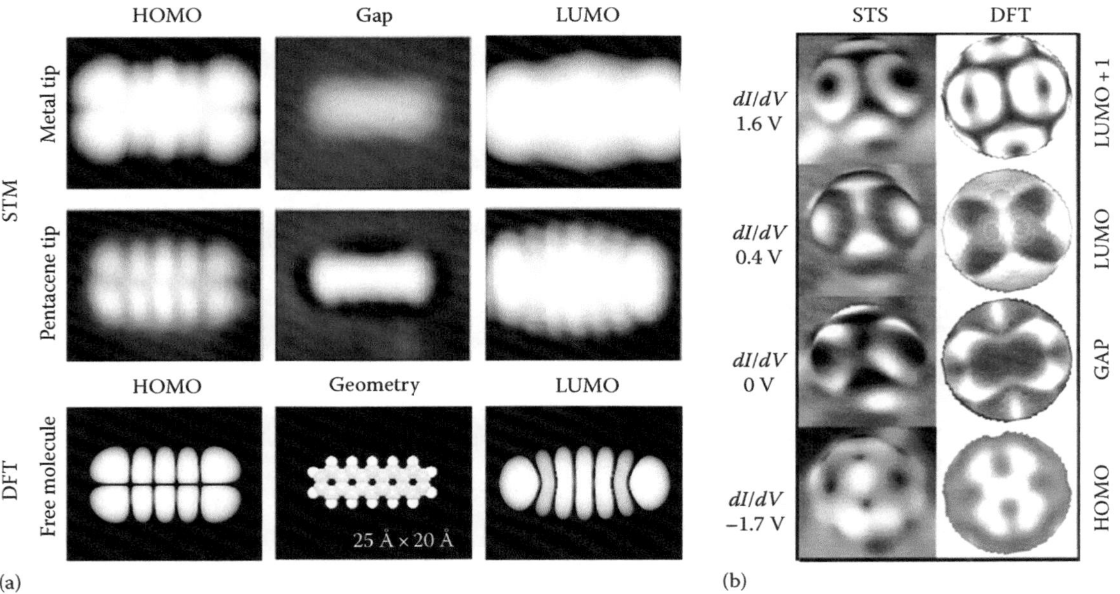

FIGURE 18.15 (a) STM images of occupied (HOMO) and empty (LUMO) orbitals of a pentacene molecule measured with a metal and a pentacene-terminated tip, and DFT calculations of the electron density for the HOMO and LUMO of the free molecule. (b) Experimental and theoretically calculated STS images of a single C_{60} molecule on Ag(100). Left column shows experimental dI/dV maps taken at different voltages. Right column shows the results of DFT calculations of energy-resolved electron density maps at E = 1.2, 0.3, 0.0, and −1.6 eV. (a: Reprinted with permission from Repp, J., Meyer, G., Stojkovic, S.M., Gourdon, A., and Joachim, C., *Phys. Rev. Lett.*, 94, 026803, Copyright 2005 American Physical Society; b: Reprinted with permission from Lu, X., Grobis, M., Khoo, K.H., Louie, S.G., and Crommie, M.F., *Phys. Rev. Lett.*, 90, 096802, Copyright 2003 American Physical Society.)

dependence of the image symmetry and sometimes for the chemical contrast in atomically resolved STM experiments on multicomponent surfaces.

Figure 18.16 shows the gap resistance dependence of the images measured on the W(110)/C-R(15 × 3) reconstruction (Bode et al. 1996, Wiesendanger et al. 1996). The observed reproducible transformation of atomic contrast was explained by spatial inhomogeneities in the decay length of the surface atom wave functions, which could lead to qualitative changes in the images.

Figures 18.17 and 18.18 show two examples of gap resistance dependence of atomically resolved STM images of oxygen-induced reconstructed metallic surfaces. For the Ru(0001)–O system (Figure 18.17), the observed gradual transformation of the STM images with increasing current (decreasing distance) was explained by a change in the relative contribution of the ruthenium s and p_z orbitals and oxygen p_{xy} orbitals. It was suggested (Calleja et al. 2004) that ruthenium orbitals dominate at larger distances (honeycomb pattern), whereas the oxygen orbitals are responsible for the hexagonal pattern observed at small distances (large tunneling currents). Note that the change in STM contrast for the W(110)/C-R(15 × 3) and Ru(0001)–O systems (Figures 18.16 and 18.17) was observed in a wide range of gap resistances. In contrast, STM studies of the stepped Cu(014)–O surface (Murphy et al. 2007) revealed sharp distance dependence of the atomically resolved images at less than one order of magnitude decrease in the gap resistance (Figure 18.18a). At certain tunneling gaps and tip states, three typical atomically resolved

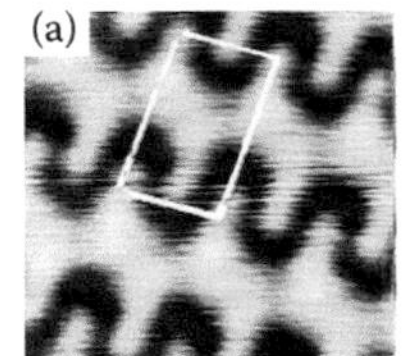

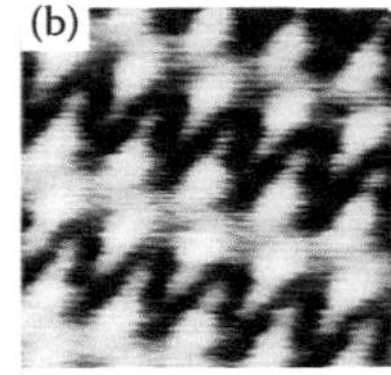

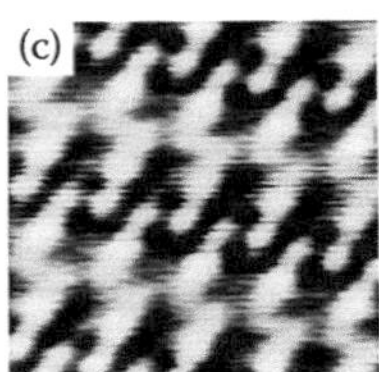

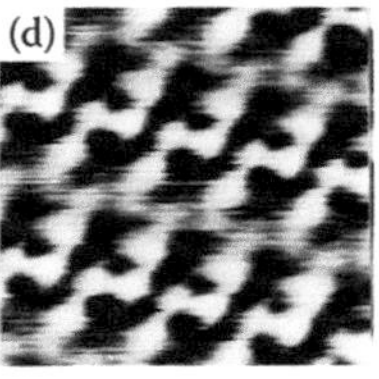

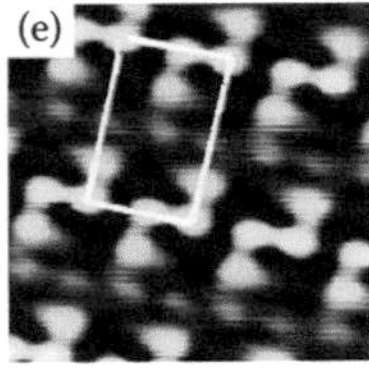

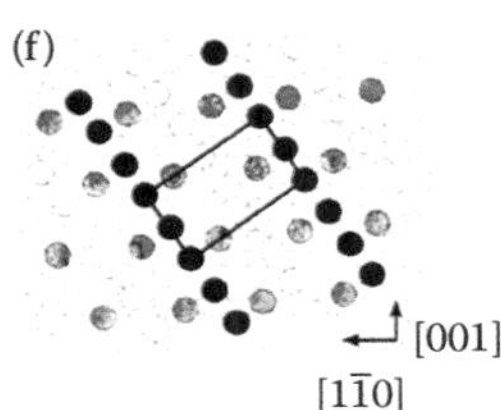

FIGURE 18.16 (a–e) Five CCM STM images of the same size 3.1 × 3.1 nm^2 of the W(110)/C-R(15 × 3) reconstruction measured at different gap resistances: 1.72 × 10^{10} Ω (a); 2.27 × 10^9 Ω (b); 1.0 × 10^8 Ω (c); 1.2 × 10^7 Ω (d); 2.11 × 10^6 Ω (e). (f) Schematic model of the W(110)/C-R(15 × 3) reconstruction. (Reprinted with permission from Wiesendanger, R., Bode, M., Rascal, R., Allers, W., and Schwarz, U.D., Issues of atomic-resolution structure and chemical analysis by scanning probe microscopy and spectroscopy, *J. Vac. Sci. Technol. A*, 14, 1161–1167, Copyright 1996 American Vacuum Society.)

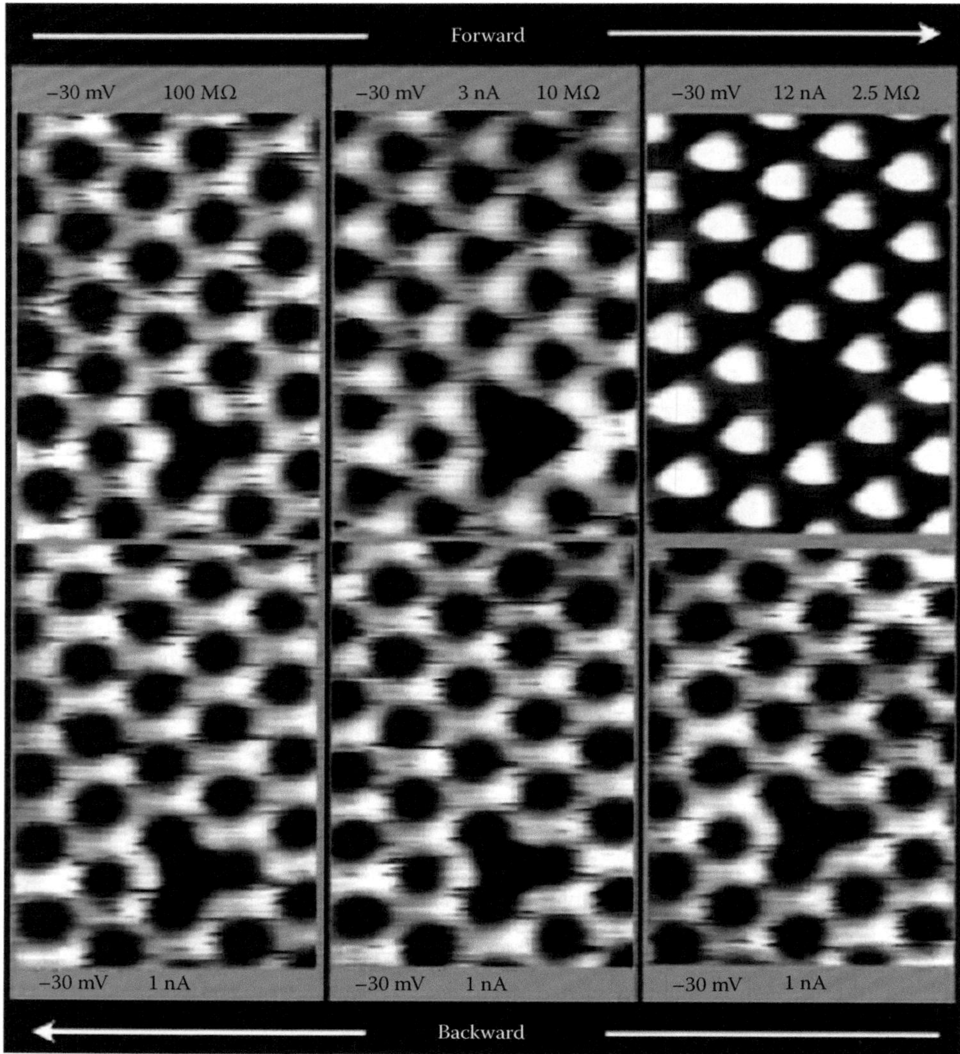

FIGURE 18.17 A series of 2.2 × 3.2 nm² dual-mode STM images of the oxygen-induced O(2 × 2)/Ru(0001) surface reconstruction. The upper panels (forward channel) show images recorded at a constant sample voltage of −30 mV and decreasing gap resistances (from left to right, 100, 10, and 2.5 MΩ). The reference images (backward channel) in the lower panels were all measured at a sample voltage of −30 mV and a constant gap resistance of 30 MΩ to exclude possible tip changes. (Reprinted with permission from Calleja, F., Arnau, A., Hinarejos, J.J., Vazquez de Parga, A.L., Hofer, W.A., Echenique, P.M., and Miranda, R., Contrast reversal and shape changes of atomic adsorbates measured with scanning tunneling microscopy, *Phys. Rev. Lett.*, 92, 206101, Copyright 2004 American Physical Society.)

STM images of the Cu(014)–O surface could be measured with W and MnNi tips (Chaika and Bozhko 2005, Murphy et al. 2007, Chaika et al. 2008), as illustrated in Figure 18.18b. The image shown in Figure 18.18c demonstrates that minor modifications of the tip state at the same tunneling parameters can switch STM contrast in high-resolution STM experiments on the Cu(014)–O surface. The gap resistance dependence in Figure 18.18a proves that all typical images of this surface reconstruction can be measured with an unchanged tip at different tunneling gaps. The observed dependence was explained by the decisive contribution of the d_{z^2} electron orbital of copper atoms at larger distances (visualization of copper atomic rows), d_{xy,x^2-y^2} orbitals at the smallest distances (imaging two oxygen atomic rows), and $d_{xz,yz}$ atomic orbitals at intermediate distances (visualization of one copper atomic row within four atomic row-wide terraces). This qualitative interpretation, based on the well-known shapes of different *d*-orbitals, was supported by the tight binding (TB) calculations of the Cu(014)–O surface electronic structure (Chaika et al. 2008) and high-resolution STM images measured with tungsten tips. For example, as shown in Figure 18.18d, all atomic features within the well-resolved down-step copper row are essentially asymmetric: The shape of these double features qualitatively reproduces the electron density distribution in the Cu d_{xz} atomic orbital that supports the interpretation of the gap resistance dependence (Figure 18.18a) based on selective orbital imaging. In this case, the change of the relative contribution of electron

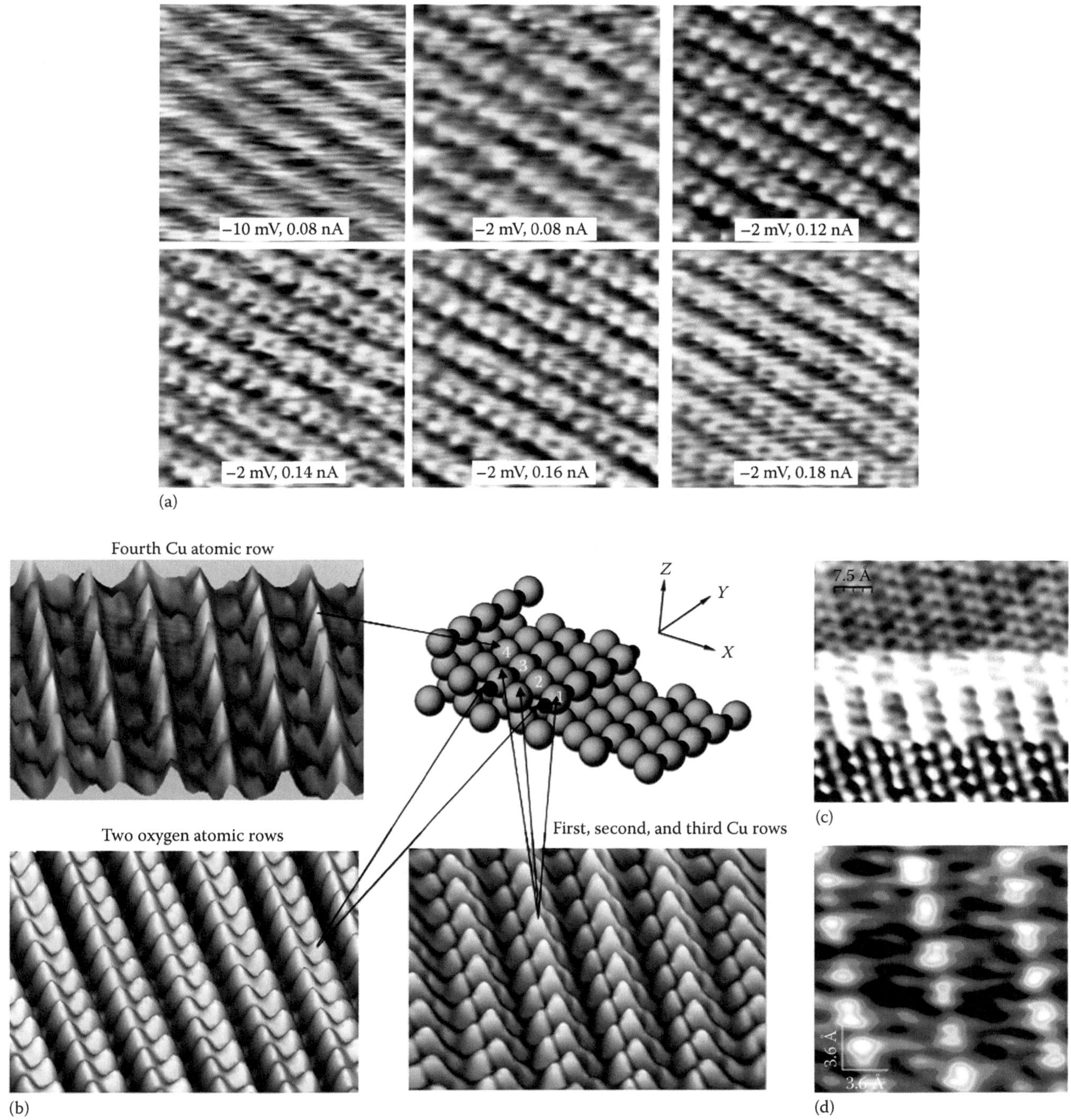

FIGURE 18.18 (a) Gap resistance dependence of the STM images of the an oxygen-reconstructed Cu(014) surface. The dependence is explained by the decisive contributions of the surface atom *d*-orbitals with different *m*. (b) Schematic model and three-dimensional presentations of three typical atomically resolved STM images of the Cu(014)–O surface measured with polycrystalline tungsten tips. (c) A 3.8 × 3.8 nm STM image of the Cu(014)–O surface taken at $V_b = -2$ mV and $I = 120$ pA. Three typical patterns are resolved within one image after switchings of the tip state during scanning. (d) A 2.1 × 2.1 nm STM image of the Cu(014)–O surface illustrating the distinct asymmetry of the atomic features associated with the fourth (down-step copper row) resolved at a certain polycrystalline W tip state and $V_b = -5$ mV and $I = 100$ pA. The double features along the bright atomic row reproduce the shape of the Cu d_{xz} electron orbital. The sample was rotated between the measurements shown in panels (a) and (b)–(d). (a: From Murphy, S. et al., *Phys. Rev. B*, 76, 245423, 2007; b: From Chaika, A.N. and Bozhko, S.I., *JETP Lett.*, 72, 416, 2005; d: From Chaika, A.N. et al., *Surf. Sci.*, 602, 2078, 2008.)

orbitals with different m could lead to chemical-selective imaging of the copper and oxygen atomic rows of the Cu(014)–O surface at the same sample bias voltages (Chaika and Bozhko 2005, Chaika et al. 2008). However, these data do not explain why the partial contribution of the d_{z^2} orbital dominating at larger distances can be substantially suppressed at such a small decrease in the gap resistance (tip–sample distance). Note that according to the theoretical calculations of Sacks (Figure 18.5), the conductivity through the d_{z^2} orbital at all distances relevant to STM experiments should exceed the conductivity through all other d-orbitals.

18.4 Subatomic Electron Orbital Resolution in STM Experiments

The image shown in Figure 18.18d represents an example of the resolution that can be achieved with STM when a selected set of electron orbitals of the surface atoms is responsible for the subatomic resolution imaging of the surface electronic structure. Although the first indication of orbital channels in the tunneling conductance of single-atom contacts was reported more than a decade ago (Scheer et al. 1998), only a few examples of SPM imaging with subatomic lateral resolution have been published up to date. In some cases (e.g., Figures 18.9 and 18.12), the origin of the subatomic features is not completely understood. At the same time, the atomic orbital features, observed in several studies, are apparently related to a direct visualization of the tip atom electron states by more localized surface atom electron states in accordance with the reciprocity principle. In this section, we shall discuss the STM data that can be directly associated with certain electron orbital contribution. Although the goal of SPM experiments is high-resolution study of the surface atomic and electronic structure, these data may shed light upon the possible reasons for anomalous distance dependencies in atomically resolved STM experiments and atomic-scale chemical contrast observed on metallic surfaces at the same sample bias voltages. These still rare results with subatomic electron orbital resolution can provide the key for future reproducible selective imaging of separate atomic orbitals in STM experiments.

18.4.1 Imaging Atomic Orbitals of the Si-Terminated Tip Atom Using Dangling Bonds of the Si(111)7 × 7 Surface

Similar to atomic resolution STM imaging, the Si(111)7 × 7 surface can be considered as a milestone for development of subatomic electron orbital resolution SPM imaging. Figure 18.19a shows an AFM image of the Si(111)7 × 7 surface demonstrating regular splitting of the surface atomic features (Giessibl et al. 2000). This image is unlikely to be caused by a double tip effect because it reveals well-resolved corner holes (lower panel in Figure 18.19a). Since the tip for the experiments was prepared by a gentle contact of a tungsten probe with a silicon surface, the observed crescents were explained (Giessibl et al. 2000) by direct visualization of two hybrid sp^3 atomic orbitals of the silicon atom at the apex of a [001]-oriented Si tip as schematically shown in Figure 18.20c. This interpretation of the subatomic features was disputed in the literature (Giessibl et al. 2001, Hug et al. 2001, Huang et al. 2003, Chen 2006, Campbellova et al. 2011). One of the arguments against the orbital origin of the observed features was related to the absence of similar results in numerous STM experiments on the Si(111)7 × 7 reconstruction, which is frequently used as a substrate for growth of metallic overlayers

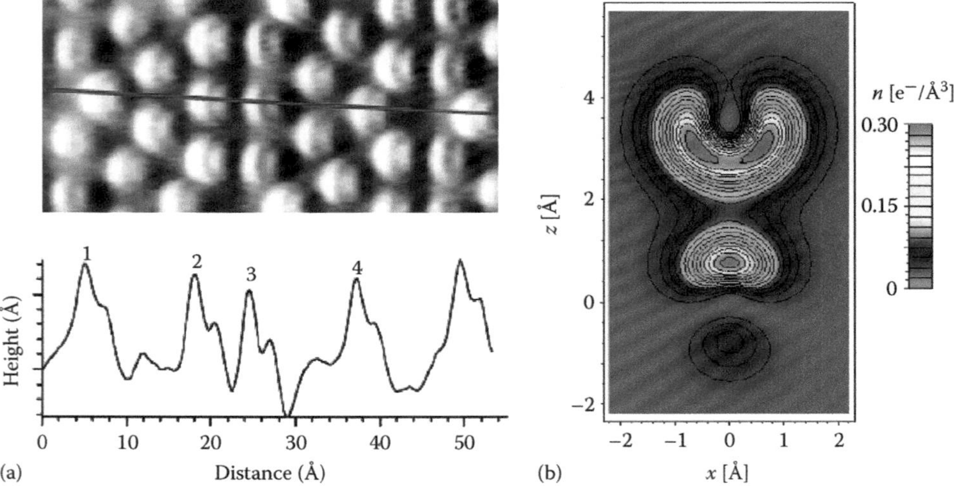

FIGURE 18.19 (a) AFM image of the Si(111)-(7 × 7) unit cell measured with a silicon-terminated tungsten tip (top panel) and cross section of the image (lower panel). (b) Charge density for one sp^3 orbital originating from an adatom located at $x = 0$ and $z = 0$ and two sp^3 orbitals originating from a tip atom located at $x = 0$ and $z = 3.5$ Å. (a: From Giessibl, F.J., Hembacher, S., Bielefeldt, H., and Mannhart, J., *Science*, 289, 422–425, 2000. Copyright 2000. Reprinted with permission of AAAS; b: From Giessibl, F.J., Bielefeldt, H., Hembacher, S., and Mannhart, J.: *Annal. Phys.* 2001. 10. 887–910. Copyright Wiley-VCH Verlag GmbH & Co. KGaA. Reprinted with permission.)

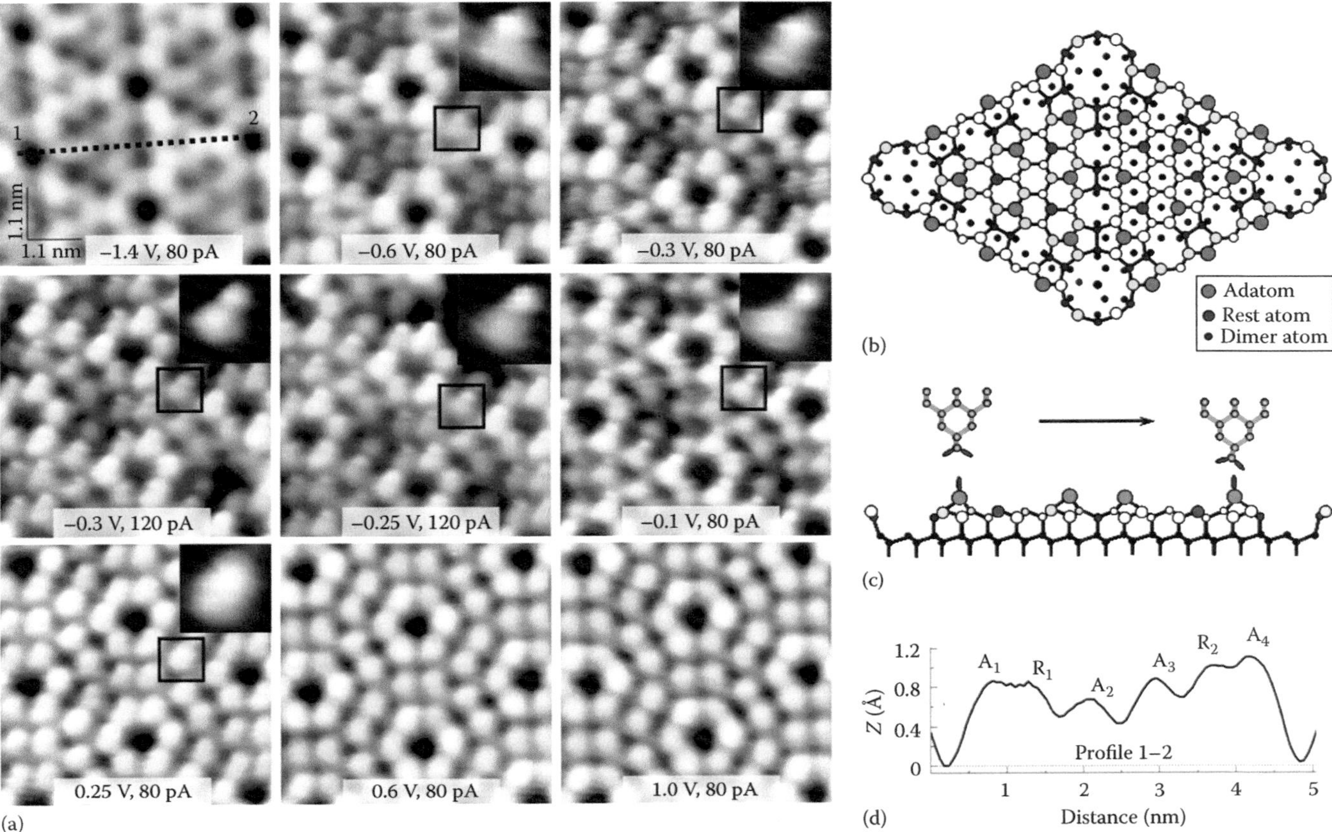

FIGURE 18.20 (a) 5.5 × 5.5 nm STM images of the Si(111)7 × 7 surface measured at different bias voltages and tunneling currents (shown at the bottom part of each frame) using an unchanged silicon-terminated tungsten tip. The images were corrected to a 7 × 7 unit cell using the WSXM software (Horcas et al. 2007) to remove drift distortions. Insets in panels show a magnified view of the individual atomic features indicated by square frames. The fast scanning direction was from left to right. (b) Top view of the Si(111)7 × 7 unit cell. (c) Side view of a [001]-oriented silicon tip over the Si(111)7 × 7 surface; the right panel depicts a tip bending opposite to the scan direction (indicated by arrow). (d) The cross section (1–2) of the STM image in panel (a).

and routinely used for standard UHV scanner tests and calibrations. Because of the proportionality between tunneling current and tip–surface interaction force (Hofer and Fisher 2003, Chen 2008), it was anticipated that similar subatomic features can be observed both in AFM and STM experiments. However, this was not demonstrated till 2008 (Chaika and Myagkov 2008).

Despite the arguments against the orbital interpretation of the subatomic features (Hug et al. 2001, Campbellova et al. 2011), independent theoretical calculations (Giessibl et al. 2001, Huang et al. 2003) showed that it can certainly be valid. The theoretical calculations showed that visualization of the asymmetric charge distribution around the apex atom of the [001]-oriented Si tip (Figure 18.19b) can be achieved only at very small (250–400 pm) tip–surface separations that are not frequent in STM experiments on silicon surfaces. Later, theoretical simulations (Zotti et al. 2006) proved that double features in STM images of the Si(111)7 × 7 surface can be experimentally measured with the [001]-oriented Si tip at high negative sample bias voltages (Figure 18.21). The double features, qualitatively reproducing the subatomic features resolved in AFM experiments (Figure 18.19a), were actually observed in STM experiments with a silicon-terminated tungsten tip (Figure 18.20a). However, the tunneling parameters necessary for imaging these unusual features were different from the theoretically predicted ones (Zotti et al. 2006).

The gap resistance dependence of the Si(111)7 × 7 STM images measured with the silicon-terminated tip (Figure 18.20a) demonstrates that at large negative bias voltages, a (7 × 7) pattern with well-resolved adatoms, rest atoms, and corner holes is observed, which is only possible with extremely sharp single-atom-terminated tips (Wang et al. 2004). At small negative bias voltages (smaller distances), the regular splitting of the adatom features becomes discernible. The effect cannot be explained by a two-atom-terminated tip because of simultaneous sharpness of the corner holes, rest atoms, and single-atom defects on the surface. The images in Figure 18.20a demonstrate that the double features became sharper with the decrease in gap resistance. Additionally, one can note that the subatomic features are almost symmetric at larger gap resistances and essentially asymmetric at smaller gap resistances. The changes in the shapes of the double features were observed both with varying tunneling current at fixed voltage and decreasing/increasing voltage at the same current. The experiments validated the

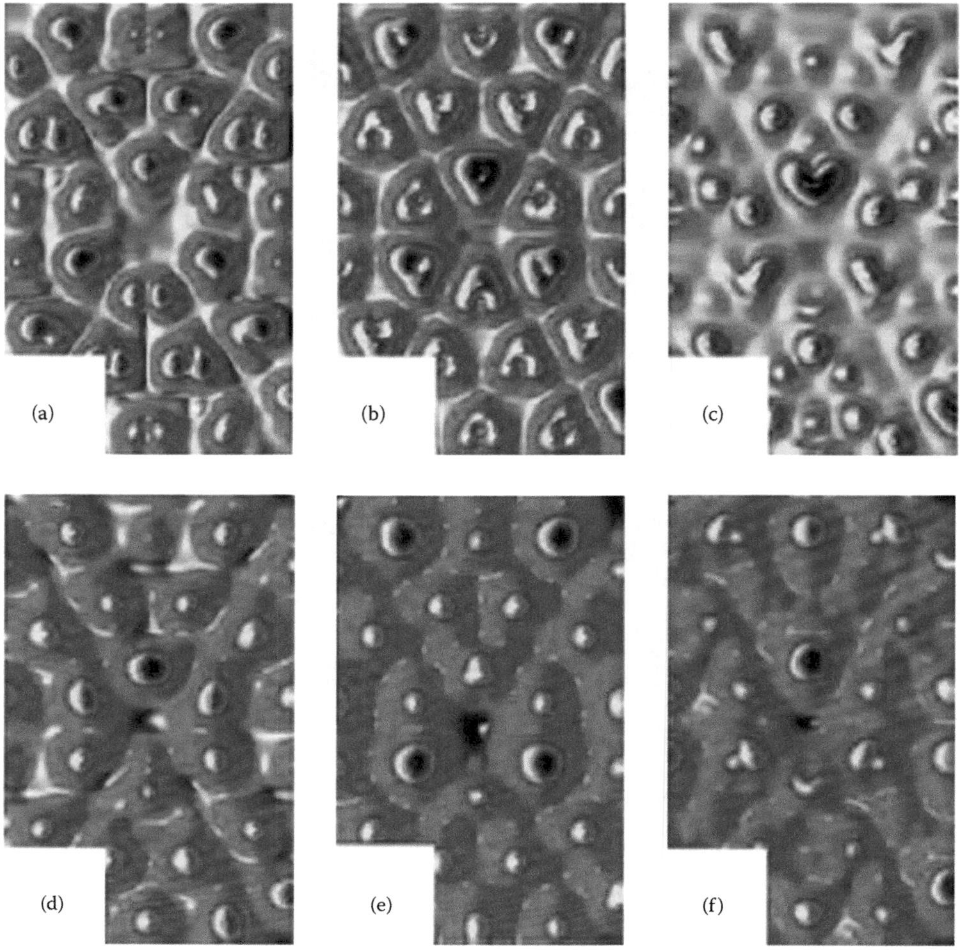

FIGURE 18.21 Conductance contours d*I*/d*V* (x, y, z_0) for negative bias voltages of −2.0, −1.5, and −1.0 V (frames a–c) and for positive bias voltages +1.0, +1.5, and +2.0 V (frames d–f). The double features in the conductance maps are only visible in the negative bias regime. (Reprinted from *Chem. Phys. Lett.*, 420, Zotti, L.A., Hofer, W.A., and Giessibl, F.J., Electron scattering in scanning probe microscopy experiments, 177–182. Copyright 2006 with permission from Elsevier.)

interpretation of the double features originating from the two dangling bonds of the silicon atom at the apex (Figure 18.20c). The asymmetry in this case can be explained by a relaxation of the apex atom, which should be more pronounced at smaller tip–surface separations. The distances between two maxima within the subatomic features lie in the range 210–250 pm (Chaika and Myagkov 2008) that is in agreement with the charge density distribution map calculated for the [001]-oriented Si tip atom interacting with the Si(111) surface atom (Figure 18.19b). It should be mentioned, however, that because of uncontrollable modification of the tip apex during contact with the silicon surface, some other interpretations cannot be totally excluded. For example, the calculations of C.J. Chen (2006) showed that splitting of the atomic features at very small distances can even be related to imaging individual tetrahedral hybrid orbitals of a silicon atom at the apex that is tilted to the surface normal. Apparently, an unambiguous interpretation of the subatomic contrast observed in STM (Figure 18.20) and AFM (Figure 18.19) experiments on the Si(111)7 × 7 surface using silicon-terminated tips can only be given on the basis of distance-dependent experimental data obtained with well-defined single crystalline tips.

18.4.2 Evidence of Sm 4*f* Electron Orbital in STM Images of the Si(111)7 × 7 Surface

Another example of imaging of tip apex atom electronic structure by spatially separated p_z orbitals of the Si(111)7 × 7 surface atoms was reported by Herz et al. (2003). The authors used a dynamic-STM mode with an oscillating probe to decrease the tip–sample distance and consequently enhance the spatial resolution in experiments. The increase in stability of the tip at extremely small gaps in this mode was explained by substantial decrease of the lateral forces in tip–sample contact because of the vertical movements of the probe. Figure 18.22 shows an example of high-resolution Si(111)7 × 7 STM images measured with a Co_6Fe_3Sm tip. At first glance, this is a conventional image of the (7 × 7) reconstruction. However, one can note that the appearance of the adatoms in Figure 18.22a is rather unusual.

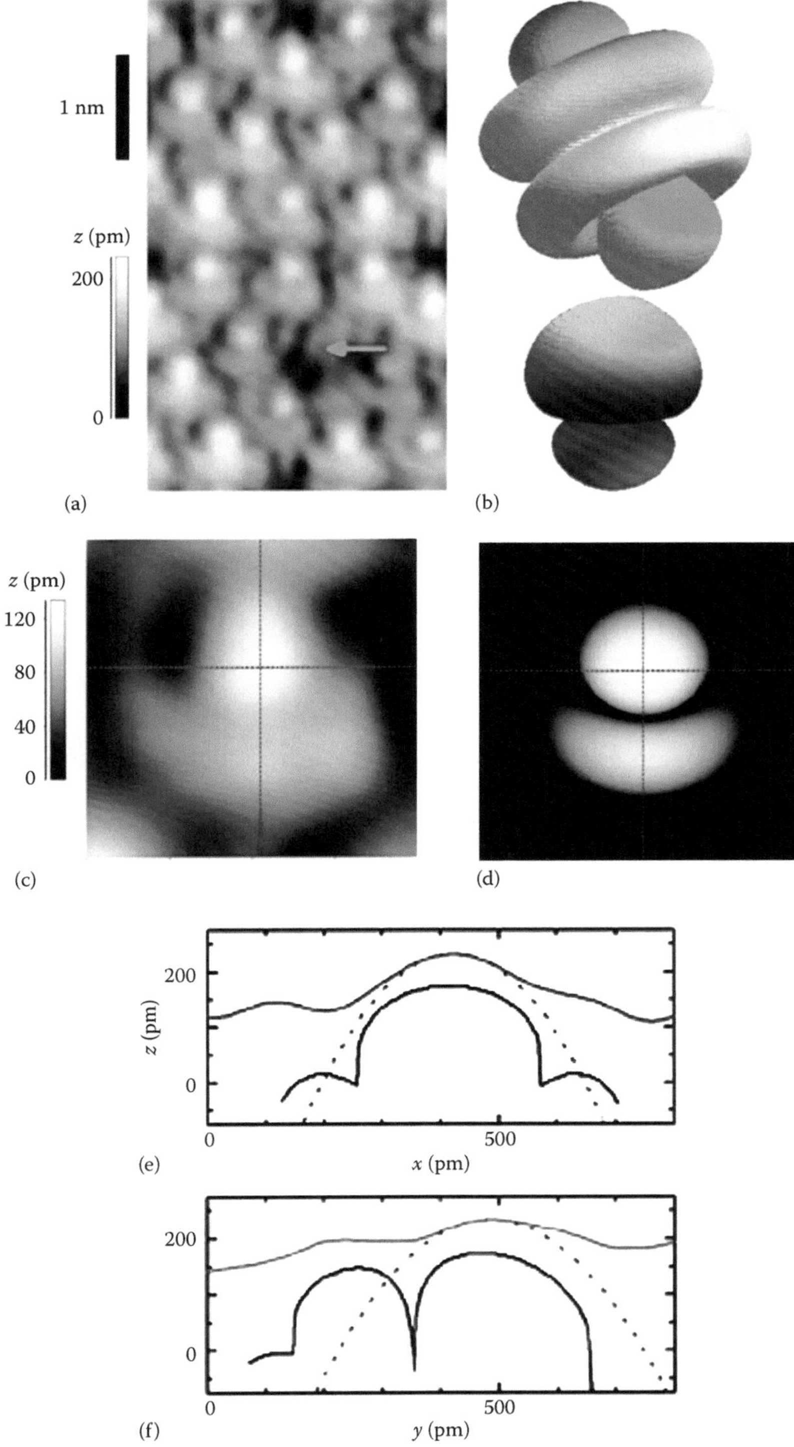

FIGURE 18.22 (a) Dynamic STM image of an Si(111)-(7 × 7) surface, acquired with a Co_6Fe_3Sm tip. The arrow indicates an atomic defect (missing adatom). (b) Schematic presentation of the current-carrying atomic states leading to the observed image shown in (a). The sp^3 silicon states are tunneling mainly into an Sm $4f_{z^3}$ tip state tilted by a fixed angle determined to be ~37°. (c) Experimental image of a single Si adatom. (d) Calculated dynamic STM topography image for a silicon $3p_z$ sample state and an Sm $4f_{z^3}$ tip state inclined 37° with respect to the z-axis. (e) Profile through the maximum of the image in the x direction (gray line). The paraboloid fitted to the profile has an apex radius $R_x = 0.12$ nm. (f) Profile through the maximum of the image in the y direction (gray line). The paraboloid fitted to the trace has an apex radius $R_y = 0.15$ nm. The black lines in (e) and (f) show the corresponding cross sections of the constant current surface at the main peak of the theoretical image. (Reprinted with permission from Herz, M., Giessibl, F.J., and Mannhart, J., Probing the shape of atoms in real space, *Phys. Rev. B*, 68, 045301, Copyright 2003 American Physical Society.)

The adatoms are imaged as extremely sharp spherically symmetric features surrounded by lower lying crescents. This asymmetry of atomic features was explained by a convolution of p_z electron states of the surface atoms and f_{z^3} states of Sm atom at the apex (Figure 18.22b). The comparison of experimental (Figure 18.22c and e) and theoretical (Figure 18.22d and f) features calculated in an assumption of the Sm f_{z^3} tip orbital tilted to the surface normal by 37° adds to the validity of this interpretation. Again, the well-resolved single-atom defect on the surface, indicated in Figure 18.22a, proves that the tip is atomically sharp and does not possess spherically symmetric DOS distribution at the apex atom at some specific tunneling conditions. As a result of this asymmetry, the STM image reveals a (7 × 7) pattern of the tilted Sm $4f_{z^3}$ electron orbitals.

18.4.3 First STM Images of Nonzero *m* Electron *d*-States

According to the theories of C.J. Chen and W. Sacks, asymmetric atomic features with distinct multiple protrusions can be observed experimentally if the tip with dominating nonzero *m* electron states at the apex were employed for high-resolution STM imaging. For a long time, this prediction could be considered as having only some theoretical interest. The first clear evidence of STM imaging using nonzero *m* electron states of the tip was reported only few years ago (Chaika et al. 2007). Figure 18.23a shows the STM image of the Cu(014)–O surface measured with a polycrystalline MnNi tip at certain tunneling parameters and tip state. Remarkably, in this case, the STM image of the Cu(014)–O surface demonstrates regular doubling of the copper atomic features along the close-packed [1–10] direction. As a result, the number of atomic maxima along this direction, resolved within the terraces of the stepped surface, exceeds the actual number of atomic rows within the terraces (see the model in the top panel in Figure 18.23a) by a factor of 2. The experimental image displays a period corresponding to the regular stepped surface terraces of 7.2 ± 0.2 Å width, with step edges directed along the [100] direction and an additional fine structure on the terraces. The image in Figure 18.23a also displays a single-atom defect (i.e., kink site), proving the sharpness

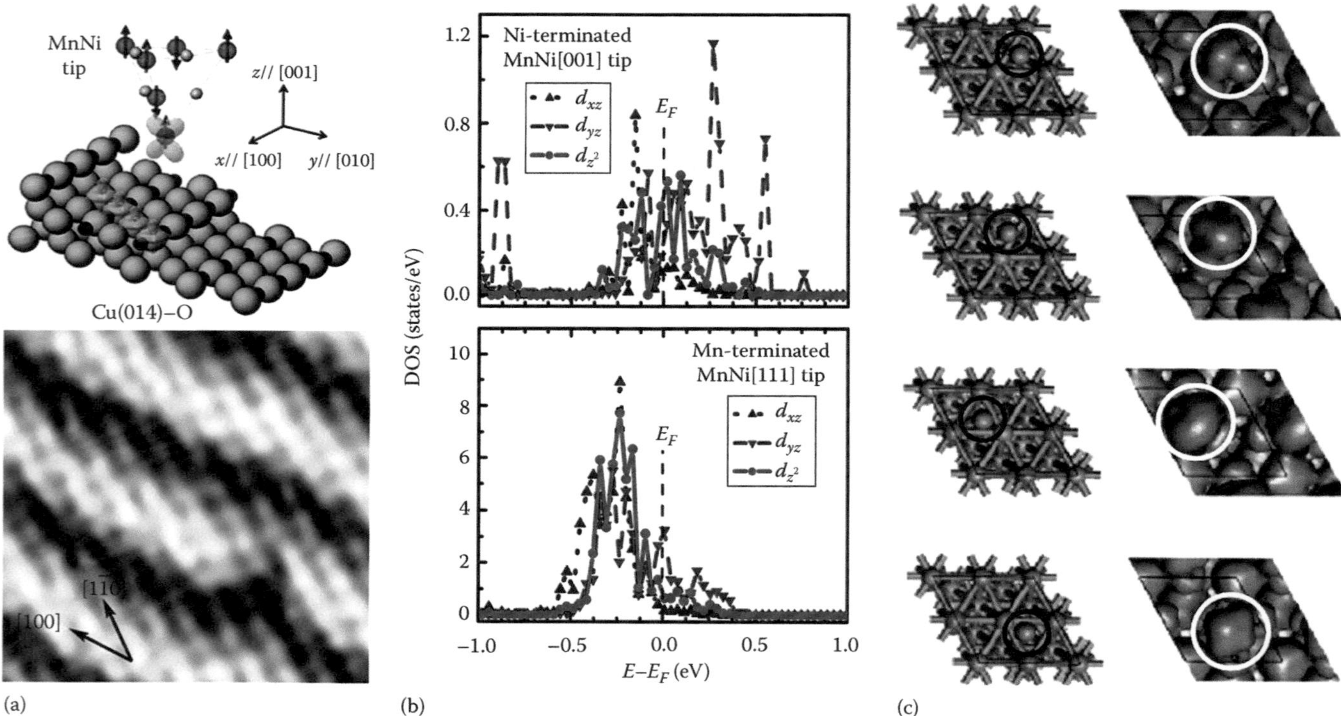

FIGURE 18.23 (a) Regular doubling of atomic features in STM images of the Cu(014)–O surface measured with a MnNi tip (lower panel) and a schematic model of the STM tip d_{yz} orbital scanning over the d_{z^2} orbitals of the surface atoms (top panel). The image was taken at $V = -30$ mV and $I = 80$ pA. (b) TB calculation of the PDOS associated with different *d*-orbitals of the apex atoms of the Ni-terminated MnNi[001] tip (top) and Mn-terminated MnNi[111] tip. (c) Schematic models and calculated electron density isosurfaces for four different [111]-oriented MnNi tip configurations (Murphy et al. 2007). The model of the tip is presented on the left and shows the relative position of the tip atom on the (111) surface. The corresponding isosurface is shown on the right and displays the calculated electron density at 2.4×10^{-3} electrons/Å^3 in the $E_F \pm 0.22$ eV energy range. The apex atom positions are indicated by black and white circles in the left and right panels, respectively; Parts (a) and (c). (Reprinted with permission from Murphy, S., Radican, K., Shvets, I.V., Chaika, A.N., Semenov, V.N., Nazin, S.S., and Bozhko, S.I., Asymmetry effects in atomically resolved STM images of Cu(014)–O and W(100)–O surfaces measured with MnNi tips, *Phys. Rev. B*, 76, 245423, Copyright 2007 American Physical Society.)

of the MnNi tip. The doubling of atomic features along one of the close-packed directions was clearly observed in only some experiments for certain tunneling tip states and scanning parameters. In particular, well-resolved images with clearly seen doubling were measured at small negative sample bias voltages between −30 and −50 mV in a very narrow range of the tunneling currents. The sharp dependence of the effect from the applied tunneling parameters can be easily understood from the distance dependence of the Cu(014)–O STM images shown in Figure 18.18a and electronic structure calculations carried out for different MnNi tip configurations (Murphy et al. 2007). For example, Figure 18.23b shows the PDOS associated with the Ni and Mn atoms at the apexes of the [001]- and [111]-oriented MnNi tips calculated using the TB method. For the [111]-oriented MnNi tip (lower panel in Figure 18.23b), PDOS demonstrates domination of the d_{yz} orbital near the Fermi level. This is not the case for the Ni-terminated tip (top panel in Figure 18.23b). Thorough density functional theory (DFT) calculations (Murphy et al. 2007) demonstrated that distinct asymmetry, which could lead to the regular doubling of atomic features, is only observed for rare MnNi(111) tip configurations. As an example, the charge density maps shown in Figure 18.23c illustrate that only one of four possible configurations of the Mn-terminated MnNi(111) tip could produce pronounced doubling with almost symmetric double features.

For the Cu(014)–O surface interacting with the MnNi tip, a regular pattern of the double features reproducing the shape of the Mn atom d_{yz} electron orbital could be resolved only in assumption of the imaging of the surface atom d_{z^2} orbitals by the tip atom d_{yz} orbital as shown in Figure 18.23a. However, because of the unknown crystallographic orientation of the polycrystalline MnNi tip, one could not control the orbital structure of the tip during the experiments. The surface complexity and drastic distance dependence of atomically resolved STM images of the Cu(014)–O surface did not allow us to study the dependence of the relative contribution of the d_{yz} orbital as a function of the applied bias voltage and gap resistance. Therefore, it was impossible to reveal the conditions necessary to maximize the the contribution of certain electron orbitals of the tip in this case.

18.4.4 Distance Dependence of the W[001] Tip Atom *d*-Orbital Contribution in STM Experiments on Graphite

The possibility to control the electron orbital structure and relative contribution of different *d*-orbitals of transition metal STM tips has recently been demonstrated in high-resolution experiments with [001]-oriented single crystalline tungsten tips (Chaika et al. 2010). In this case, the *s*- and p_z-orbitals of the graphite (0001) surface atoms were used to probe the electronic structure of the tip and study the relative contributions of different *d*-orbitals of the W[001] tip at different bias voltages and tip–sample separations. According to the PDOS calculations for an isolated W[001] tip (Chaika et al. 2008), the d_{z^2} and $d_{x^2-y^2}$ states can dominate in the occupied and empty electron state regions near E_F, respectively. Indeed, extremely sharp spherically symmetric features and fourfold split subatomic features (Figure 18.24b) were reproducibly resolved at small positive and negative bias voltages, respectively, in a series of high-resolution STM experiments. Additionally, twofold split features, reproducing the shape of $d_{xz,yz}$ electron orbitals, were also resolved at small negative bias voltage and some intermediate tip–surface distances. These three cases may correspond to imaging different W[001] tip *d*-orbitals by carbon atomic orbitals as shown in Figure 18.24a.

In these experiments, we only employed tungsten tips made out of single crystals with the tip axis along the [001] direction. To eliminate possible modifications of STM contrast by

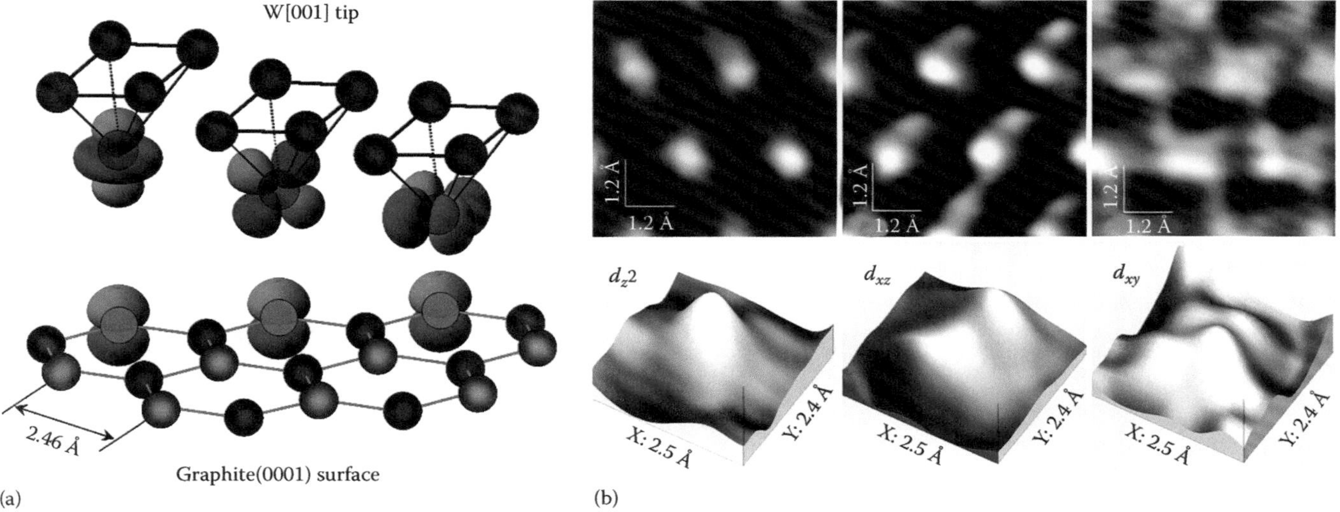

FIGURE 18.24 (a) Schematic model of a [001]-oriented W tip over a graphite(0001) surface. (b) 6 × 6 Å² (top row) and 2.5 × 2.4 Å² (bottom row) STM images measured with single crystalline W[001] tips at different tunneling parameters: V = 23 mV and I = 2.7 nA (left panels), V = −35 mV and I = 6.8 nA (middle panels), V = −100 mV and I = 1.7 nA (right panels).

undesirable contaminants at the apex, these tips were cleaned by flash heating and ion sputtering in UHV before scanning the (0001) graphite surface. The electron microscopy images of a W[001] tip shown in Figure 18.8a and b prove that this treatment produces an oriented atomic-scale pyramid at the apex. The twofold and fourfold split subatomic features in Figure 18.24b can hardly be explained by the electronic structure of the surface, which is threefold symmetrical along the tip axis. Therefore, the subatomic features could only be explained by a direct visualization of the tip atom electron orbital structure that could be modified because of tip–sample interactions. Thorough experimental studies showed that different

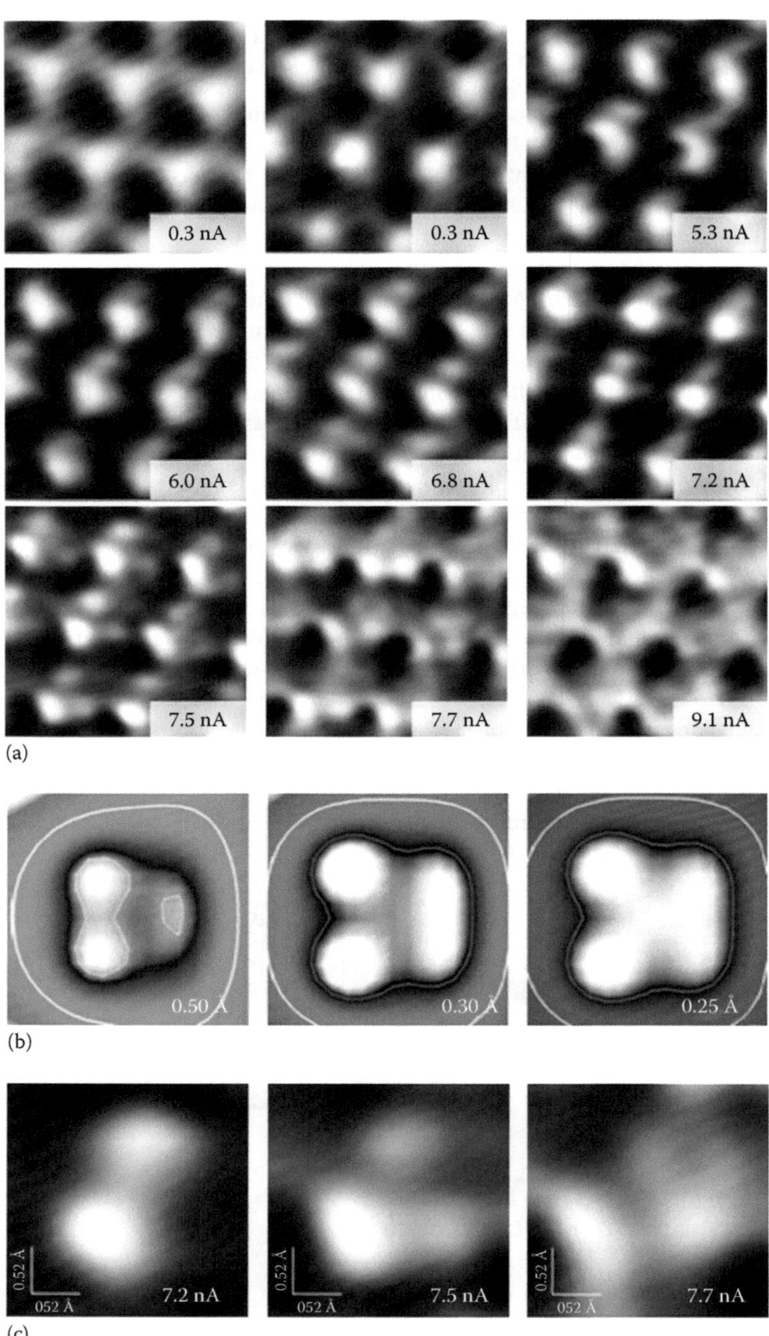

FIGURE 18.25 (a) Gap resistance dependence of 7 × 7 Å^2 HOPG(0001) STM images measured with an unchanged W[001] tip at $V = -35$ mV and different tunneling currents (indicated in each particular frame). (b) Calculated charge density maps in the "W[001] tip–graphite (0001)" interface. (001) slices are located 0.5, 0.3, and 0.25 Å above the apex atom (the W–C gap spacing is 1.75 Å). (c) 2.1 × 2.1 Å^2 regions of the images in panel (a).

subatomic features may be resolved at very small tip–sample distances using the same [001]-oriented single crystalline tungsten tip.

Figure 18.25a shows a gap resistance dependence of atomically resolved STM images of a graphite surface measured with a W[001] tip at a small negative bias voltage of −35 mV. At small currents (larger distances), a typical hexagonal pattern is observed. One can see that, with increasing current (decreasing distance), this pattern becomes sharper. The apparent diameter of the features at $I = 2.7$ nA suggests that they can be caused by imaging of carbon atomic orbitals by a tungsten atom d_{z^2} electron orbital. With further increase in the tunneling current, the spherically symmetric features are consecutively transformed into two-, three-, and fourfold split subatomic features at tunneling currents between 5.3 and 9.1 nA. In accordance with Equation 18.3, such an increase in the tunneling current could correspond to a very small difference in the gap spacing of the order of 10–20 pm. Note that similar subatomic features were also resolved in AFM experiments with polycrystalline tungsten probes (Hembacher et al. 2004). Because of the unknown tip atomic structure and orientation of the apex in those experiments, the AFM features were ascribed to the orbital structure of the tungsten tips with three different crystallographic orientations. The gap resistance dependence in Figure 18.25a, measured with an unchanged [001]-oriented tungsten tip, clearly demonstrates that these orbital images belong to the electronic structure of the same atom interacting with a graphite surface. This is also proven by DFT calculations of the electronic structure of the interacting tip–surface system. Figure 18.25b shows the charge density maps in three planes perpendicular to the W[001] tip axis (parallel to the graphite surface) located 50, 30, and 25 pm above the apex atom core. The charge density maps in Figure 18.25b qualitatively reproduce the experimental images of the orbital features magnified in Figure 18.25c. The strong dependence of the charge density maps on the distance from the apex atom core agrees with the experimentally observed tunneling current dependence of the STM images presented in Figure 18.25a.

The magnified views of the two- and fourfold split subatomic features in Figure 18.24b show that their shapes at correctly adjusted tunneling parameters reproduce the well-known electron density distribution associated with the $d_{xz,yz}$ and d_{xy,x^2-y^2} orbitals in a tungsten atom. This is likely the contribution of a

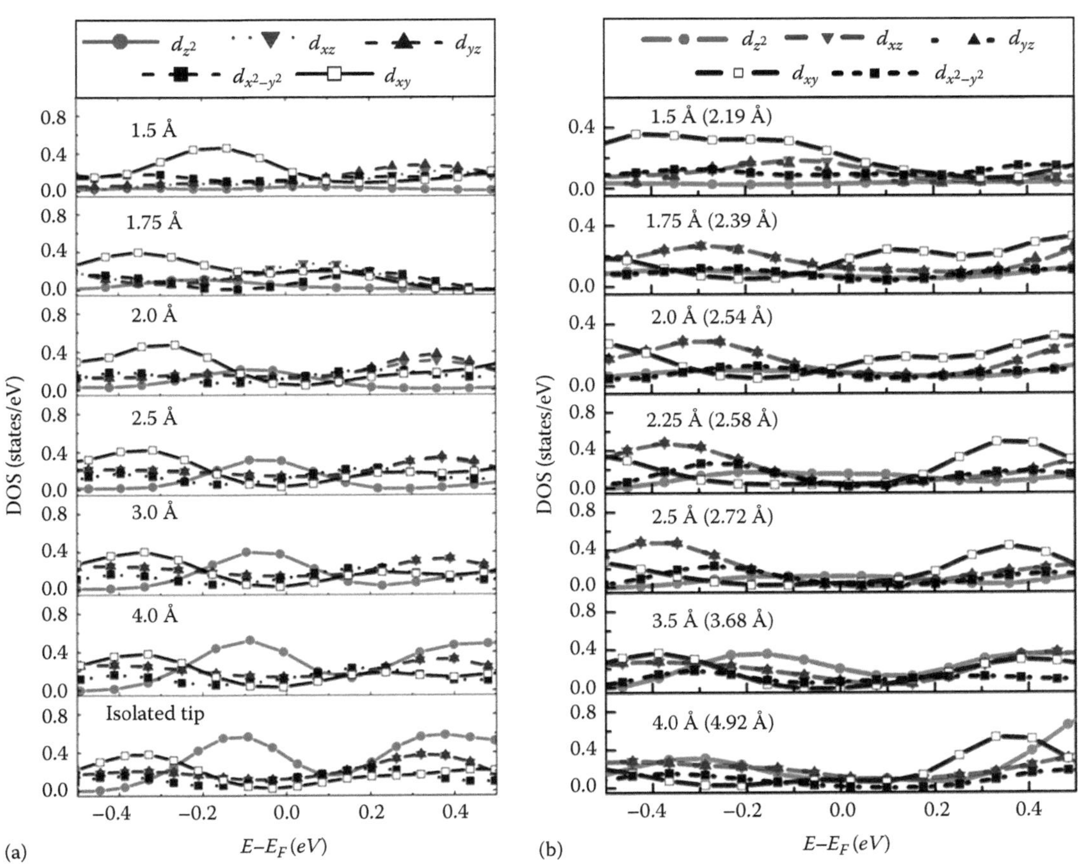

FIGURE 18.26 PDOS associated with different d-orbitals of the W[001] tip apex atom interacting with the graphite(0001) surface calculated for constrained (a) and fully relaxed (b) tip–surface slabs. The distances between interacting tip and surface atom nuclei are indicated in each panel. For the fully relaxed calculations, the initial (final) distances are indicated.

single electron orbital of the tungsten tip atom collecting most of the tunneling current. DFT calculations of PDOS associated with electron orbitals of the apex atom at different tip–surface distances (Chaika et al. 2010) reveal the origin of the essential increase of the relative contribution of $m \neq 0$ tungsten atomic orbitals with decreasing distance. DFT calculations for the constrained and fully relaxed "W[001] tip–graphite surface" system (Figure 18.26) show a suppression of d_{z^2} electron states near the Fermi level at tip–sample separations below 300 pm when the overlapping of tungsten d_{z^2} electron orbital with carbon atomic orbitals becomes substantial. The domination of the d_{xy} electron orbital near E_F is especially pronounced in a fully relaxed case (Figure 18.26b) at distances between 225 and 255 pm. However, the behavior of the "W[001] tip–graphite" system could be qualitatively predicted from substantially less time-demanding DFT calculations for a constrained tip–surface slab. The PDOS calculations presented in Figure 18.26 disclose the distances necessary for atomic orbital imaging (i.e., selection of the tungsten tip atomic orbital) in this case. The graphite surface imaging by tungsten the d_{z^2} orbital can be realized at tunneling gaps of about 300 pm, whereas a d_{xy} electron orbital can yield a maximum contribution at distances between 200 and 250 pm. Recent theoretical calculations (Wright and Solares 2011) confirmed the asymmetric electronic structure of the W[001] tip interacting with the graphite surface in this range of tip–surface distances. The calculations shown in Figure 18.26 prove that the appearance of asymmetrical features at small distances is related to modification of the PDOS associated with different electron d-orbitals of the tungsten tip atom. Note that the suppression of d_{z^2} electron orbital near E_F and the modifications of local electronic structure of the surface atoms interacting with the STM tip at small tip–surface distances below 400 pm have also been reported in a number of recent theoretical studies (Jelinek et al. 2008, Polok et al. 2011).

18.5 Conclusions and Outlook

There is an increasing number of STM studies demonstrating selective visualization of individual electron orbitals and images with subatomic resolution. These can reveal unusual patterns that can be directly compared with the known shapes of electron orbitals. To date, SPM studies carried out with subatomic resolution mostly demonstrate probing tip atomic orbitals by surface electron orbitals that have greater localization. However, the distance-dependent experiments with a single crystalline tungsten tip on a graphite surface (Chaika et al. 2010) and high-resolution SPM data presented in this chapter suggest the way to reverse the situation: one should use an STM tip with highly localized electron states symmetrical along the z-axis.

It is evident from the examples of subatomic electron orbital imaging and theoretical studies that direct STM imaging of intra-atomic orbital structures of separate atoms or selective imaging of electron orbitals of surface atoms can only be achieved at finely adjusted bias voltages and tip–sample distances. Recent experiments and DFT calculations show that stable and predictable selective orbital imaging can be achieved using well-defined single crystalline tips. For further development of atomic orbital imaging capability, the stability of the tip–sample separation, comparable with typical inter-atomic distances in solids, should be at about 1 pm or below. The issues of tip stability and accurate control of gap spacing are of particular importance for both high-resolution STM and AFM experiments, which can reveal qualitatively the same atomic orbital images at extremely small tip–sample distances. Latest experimental data show that the tip and tunneling gap stability can be enhanced by using single crystalline tips producing single asperity at the apex.

Finally, we outline three important reasons for developing selective orbital imaging capability with precise control of the tunneling gap. The first reason is developing probes with a controllable electron orbital structure for picometer-scale STM imaging. Second, the ability to perform imaging using electron orbitals with a selected quantum number m can be important for development of spin-polarized STM. Indeed, it has been shown recently (Serrate et al. 2010) that spin direction of individual atoms can be mapped via spin-polarized orbitals with different orbital momentum projections on the quantization axes (namely, d_{z^2} and $d_{xz,yz}$ orbitals). Third, selective imaging of electron orbitals is related to the development of chemical-selective imaging. There are examples of chemical-selective STM imaging achieved at fixed bias voltages (Schmid et al. 1993, Hebenstreit et al. 1999, Chaika et al. 2008, Yashina et al. 2012). This could take place if the PDOS associated with the surface atoms are modified at varying distances because of overlapping of the tip and surface atomic orbitals. Detailed information on such PDOS redistributions in a tunneling junction and controlled chemical-selective imaging on an atomic scale can only be achieved with accurate control of tip–surface distance in STM experiments using well-characterized tips.

Acknowledgments

This work was supported by the Russian Academy of Sciences, Russian Foundation for Basic Research, Science Foundation Ireland (Walton Fellowship Programme), and Marie Curie International Incoming Fellowship project within the 7th European Community Framework Programme. The authors are especially grateful to S. N. Molotkov, S. S. Nazin, V. N. Semenov, A. M. Ionov, V. Yu. Aristov, M. G. Lazarev, A. N. Myagkov, S. Murphy, O. Lübben, K. Radican, and S. A. Krasnikov for their help and fruitful discussions over a long period of collaborative work.

References

Abe, M., Sugimoto, Y., and Morita, S. 2005. Imaging the restatom of the Ge(111)-c(2 × 8) surface with noncontact atomic force microscopy at room temperature. *Nanotechnology* 16: S68–S72.

Bammerlin, M., Luthi, R., Meyer, E., Baratoff, A., Guggisberg, M., Gerber, C., Howard, L., and Guntherodt, H.-J. 1997. True atomic resolution on the surface of an insulator via ultrahigh vacuum dynamic force microscopy. *Probe Microscopy* 1: 3–9.

Bammerlin, M., Luthi, R., Meyer, E., Lü, J., Guggisberg, M., Loppacher, C., Gerber, C., and Guntherodt, H.-J. 1998. Dynamic SFM with true atomic resolution on alkali halide surfaces. *Appl. Phys. A* 66: S293–S294.

Baratoff, A. 1984. Theory of scanning tunneling microscopy—Methods and approximations. *Physica B* 127: 143–150.

Biegelsen, D. K., Ponce, F. A., and Tramontana, J. C. 1989. Simple ion milling preparation of <111> tungsten tips. *Appl. Phys. Lett.* 54: 1223–1225.

Biegelsen, D. K., Ponce, F. A., Tramontana, J. C., and Koch, S. M. 1987. Ion milled tips for scanning tunneling microscopy. *Appl. Phys. Lett.* 50: 696–698.

Binnig, G., Garcia, N., Rohrer, H., Soler, J. M., and Flores, F. 1984. Electron-metal-surface interaction potential with vacuum tunneling: Observation of the image force. *Phys. Rev. B* 30: 4816–4818.

Binnig, G. and Rohrer, H. 1982. Scanning tunnelling microscopy. *Helv. Phys. Acta* 55: 726–735.

Binnig, G., Rohrer, H., Gerber, Ch., and Weibel, E. 1982a. Tunneling through a controllable vacuum gap. *Appl. Phys. Lett.* 40: 178–180.

Binnig, G., Rohrer, H., Gerber, Ch., and Weibel, E. 1982b. Surface studies by scanning tunneling microscopy. *Phys. Rev. Lett.* 49: 57–61.

Binnig, G., Rohrer, H., Gerber, Ch., and Weibel, E. 1983a. (7 × 7) reconstruction on Si(111) resolved in real space. *Phys. Rev. Lett.* 50: 120–123.

Binnig, G., Rohrer, H., Gerber, Ch., and Weibel, E. 1983b. (111) facets as the origin of reconstructed Au(110) surfaces. *Surf. Sci.* 131: L379–L384.

Binnig, G., Quate, C. F., and Gerber, Ch. 1986. Atomic force microscope. *Phys. Rev. Lett.* 56: 930–933.

Bode, M., Pascal, R., and Wiesendanger, R. 1996. Distance-dependent STM-study of the W(110)/C-R(15 × 3) surface. *Z. Phys. B* 101: 103–107.

Bohr, N. 1913. On the constitution of atoms and molecules, Part I. *Phil. Mag.* 26: 1–24.

Born, M. and Jordan, P. 1925. On quantum mechanics. *Zeits. F. Phys.* 34: 858–888.

Brodie, I. 1978. The visibility of atomic objects in the field electron emission microscope. *Surf. Sci.* 70: 186–196.

Bryant, A., Smith, D. P. E., Binnig, G., Harrison, W. A., and Quate, C. F. 1986. Anomalous distance dependence in scanning tunneling microscopy. *Appl. Phys. Lett.* 49: 936–938.

Calleja, F., Arnau, A., Hinarejos, J. J., Vazquez de Parga, A. L., Hofer, W. A., Echenique, P. M., and Miranda, R. 2004. Contrast reversal and shape changes of atomic adsorbates measured with scanning tunneling microscopy. *Phys. Rev. Lett.* 92: 206101.

Campbellova, A., Ondracek, M., Pou, P., Perez, R., Klapetek, P., and Jelinek, P. 2011. 'Sub-atomic' resolution of non-contact atomic force microscope images induced by a heterogeneous tip structure: A density functional theory study. *Nanotechnology* 22: 295710.

Chaika, A. N. and Bozhko, S. I. 2005. Atomic structure of the Cu(410)-O surface: STM visualization of oxygen and copper atoms. *JETP Lett.* 72: 416–420.

Chaika, A. N. and Myagkov, A. N. 2008. Imaging atomic orbitals in STM experiments on a Si(111)-(7 × 7) surface. *Chem. Phys. Lett.* 453: 217–221.

Chaika, A. N., Nazin, S. S., and Bozhko, S. I. 2008. Selective STM imaging of oxygen-induced Cu(115) surface reconstructions with tungsten probes. *Surf. Sci.* 602: 2078–2088.

Chaika, A. N., Nazin, S. S., Semenov, V. N., Bozhko, S. I., Lubben, O., Krasnikov, S. A., Radican, K., and Shvets, I. V. 2010. Selecting the tip electron orbital for scanning tunneling microscopy imaging with sub-angstrom lateral resolution. *EPL* 92: 46003.

Chaika, A. N., Semenov, V. N., Glebovskiy, V. G., and Bozhko, S. I. 2009. Scanning tunneling microscopy with single crystal W[001] tips: High resolution studies of Si(557)5 × 5 surface. *Appl. Phys. Lett.* 95: 173107.

Chaika, A. N., Semenov, V. N., Nazin, S. S., Bozhko, S. I., Murphy, S., Radican, K., and Shvets, I. V. 2007. Atomic row doubling in the STM images of Cu(014)-O obtained with MnNi tips. *Phys. Rev. Lett.* 98: 206101.

Chen, C. J. 1988. Theory of scanning tunneling spectroscopy. *J. Vac. Sci. Technol. A* 6: 319–322.

Chen, C. J. 1990a. Origin of atomic resolution on metal surfaces in scanning tunneling microscopy. *Phys. Rev. Lett.* 65: 448–451.

Chen, C. J. 1990b. Tunneling matrix elements in three-dimensional space: The derivative rule and the sum rule. *Phys. Rev. B* 42: 8841–8857.

Chen, C. J. 1991. Microscopic view of scanning tunneling microscopy. *J. Vac. Sci. Technol. A* 9: 44–50.

Chen, C. J. 1992. Effects of $m \neq 0$ tip states in scanning tunneling microscopy: The explanation of corrugation reversal. *Phys. Rev. Lett.* 69: 1656–1659.

Chen, C. J. 2006. Possibility of imaging lateral profiles of individual tetrahedral hybrid orbitals in real space. *Nanotechnology* 17: S195–S200.

Chen, C. J. 2008. *Introduction to Scanning Tunneling Microscopy*. Oxford University Press, New York.

Chug, M. S., Feuchtwang, T. E., and Cutler, P. H. 1987. Spherical tip model in the theory of the scanning tunneling microscope. *Surf. Sci.* 187: 559–568.

Crewe, A. V., Wall, J., and Langmore, J. 1970. Visibility of single atoms. *Science* 168: 1338–1340.

Crommie, M. F., Lutz, C. P., and Eigler, D. M. 1993. Confinement of electrons to quantum corrals on a metal surface. *Science* 262: 218–220.

Cross, G., Schirmeisen, A., Stalder, A., Grutter, P., and Durig, U. 1998. Adhesion interaction between atomically defined tip and sample. *Phys. Rev. Lett.* 80: 4685–4688.

Dai, H., Hafner, J. H., Rinzler, A. G., Colbert, D. T., and Smalley, R. E. 1996. Nanotubes as nanoprobes in scanning probe microscopy. *Nature* 384: 147–150.

Demuth, J. E., Koehler, U., and Hamers, R. J. 1988. The STM learning curve and where it may take us. *J. Microsc.* 152: 299–316.

Dubois, M., Perdigao, L., Delerue, C., Allan, G., Grandidier, B., Deresme, D., and Stievenard, D. 2005. Scanning tunneling microscopy and spectroscopy of reconstructed Si(100) surfaces. *Phys. Rev. B* 71: 165322.

Eigler, D. M. and Schweizer, E. K. 1990. Positioning single atoms with a scanning tunneling microscope. *Nature* 344: 524–526.

Eltsov, K. N., Shevlyuga, V. M., Yurov, V. Y., Kvit, A. V., and Kogan, M. S. 1996. Sharp tungsten tips prepared for STM study of deep nanostructures in UHV. *Phys. Low-Dim. Struct.* 9–10: 7–14.

Erni, R., Rossell, M. D., Kisielowski, C., and Dahmen, U. 2009. Atomic-resolution imaging with a sub-50-pm electron probe. *Phys. Rev. Lett.* 102: 096101.

Feenstra, R. M., Stroscio, J. A., Tersoff, J., and Fein, A. P. 1987. Atom-selective imaging of the GaAs(110) surface. *Phys. Rev. Lett.* 58: 1192–1195.

Feynman, R. 1960. There's plenty of room at the bottom. *Eng. Sci.* 23: 22–36.

Fink, H.-W. 1986. Mono-atomic tips for scanning tunneling microscopy. *IBM J. Res. Develop.* 30: 460–465.

Fink, H.-W., Stocker, W., and Schmid, H. 1990. Coherent point source electron beams. *J. Vac. Sci. Technol. B* 8: 1323–1324.

Foster, A. and Hofer, W. 2006. *Scanning Probe Microscopy. Atomic Scale Engineering by Forces and Currents*. Springer Science+Business Media, LLC, Berlin, Germany.

Freiser, M. J. and Marcus, P. M. 1969. A survey of some physical limitations on computer elements. *IEEE Trans. Magn.* 5(2): 82–90.

Garleff, J. K., Wenderoth, M., Sauthoff, K., Ulbrich, R. G., and Rohlfing, M. 2004. 2 × 1 reconstructed Si(111) surface: STM experiments versus ab initio calculations. *Phys. Rev. B* 70: 245424.

Gawronski, H., Mehlhorn, M., and Morgenstern, K. 2008. Imaging phonon excitation with atomic resolution. *Science* 319: 930–933.

Giessibl, F. J. 1995. Atomic-resolution of the silicon (111)-(7 × 7) surface by atomic-force microscopy. *Science* 267: 68–71.

Giessibl, F. J., Bielefeldt, H., Hembacher, S., and Mannhart, J. 2001. Imaging of atomic orbitals with the atomic force microscope—Experiments and simulations. *Ann. Phys. (Leipzig)* 10: 887–910.

Giessibl, F. J., Hembacher, S., Bielefeldt, H., and Mannhart, J. 2000. Subatomic features on the silicon (111)-(7 × 7) surface observed by atomic force microscopy. *Science* 289: 422–425.

Greiner, M. and Kruse, P. 2007. Recrystallization of tungsten wire for fabrication of sharp and stable nanoprobe and field-emitter tips. *Rev. Sci. Instrum.* 78: 026104.

Gross, L., Moll, N., Mohn, F., Curioni, A., Meyer, G., Hanke, F., and Persson, M. 2011. High-resolution molecular orbital imaging using a *p*-wave STM tip. *Phys. Rev. Lett.* 107: 086101.

Hallmark, V., Chiang, S., Rabalt, J., Swalen, J., and Wilson, R. 1987. Observation of atomic corrugation on Au(111) by scanning tunneling microscopy. *Phys. Rev. Lett.* 59: 2879–2882.

Hamers, R. J., Tromp, R. M., and Demuth, J. E. 1986. Surface electronic structure of Si (111)-(7 × 7) resolved in real space. *Phys. Rev. Lett.* 56: 1972–1975.

Hamers, R. J., Tromp, R. M., and Demuth, J. E. 1987a. Electronic and geometric structure of Si(111)-(7 × 7) and Si(001) surfaces. *Surf. Sci.* 181: 346–355.

Hamers, R. J., Tromp, R. M., and Demuth, J. E. 1987b. Scanning tunneling microscopy of Si(001). *Phys. Rev. B* 34: 5343–5357.

Hansma, P. K. and Tersoff, J. 1987. Scanning tunneling microscopy. *J. Appl. Phys.* 61: R1–R23.

Hebenstreit, E. L. D., Hebenstreit, W., Schmid, P., and Varga, P. 1999. $Pt_{25}Rh_{75}$(111), (110), and (100) studied by scanning tunneling microscopy with chemical contrast. *Surf. Sci.* 441: 441–453.

Heisenberg, W. 1925. Quantum-theoretical re-interpretation of kinematic and mechanical relations. *Zeitschr. F. Phys.* 33: 879–893.

Heisenberg, W. 1927. On the perceptual content of quantum theoretical kinematics and mechanics. *Zeitschr. F. Phys.* 43: 172–198.

Hembacher, S., Giessibl, F. J., and Mannhart, J. 2004. Force microscopy with light-atom probes. *Science* 305: 380–383.

Herz, M., Giessibl, F. J., and Mannhart, J. 2003. Probing the shape of atoms in real space. *Phys. Rev. B* 68: 045301.

Hofer, W. A. 2003. Challenges and errors: Interpreting high resolution images in scanning tunneling microscopy. *Prog. Surf. Sci.* 71: 147–183.

Hofer, W. A. and Fisher, A. J. 2003. Signature of a chemical bond in the conductance between two metal surfaces. *Phys. Rev. Lett.* 91: 036803.

Hofer, W. A., Foster, A. S., and Shluger, A. L. 2003. Theories of scanning probe microscopes at the atomic scale. *Rev. Mod. Phys.* 75: 1287–1331.

Hofer, W. A., Garcia-Lekue, A., and Brune, H. 2004. The role of surface elasticity in giant corrugations observed by scanning tunneling microscopes. *Chem. Phys. Lett.* 397: 354–359.

Hofer, W. A., Ritz, G., Hebenstreit, W., Schmid, M., Varga, P., Redinger, J., and Podloucky, R. 1998. Scanning tunneling microscopy of binary-alloy surfaces: Is chemical contrast a consequence of alloying? *Surf. Sci.* 405: L514–L519.

Horcas, I., Fernandez, R., Gomez-Rodriguez, J. M., Colchero, J., Gomez-Herrero, J., and Baro, A. M. 2007. WSXM: A software for scanning probe microscopy and a tool for nanotechnology. *Rev. Sci. Instrum.* 78: 013705.

Huang, M., Cuma, M., and Liu, F. 2003. Seeing the atomic orbital: First-principles study of the effect of tip termination on atomic force microscopy. *Phys. Rev. Lett.* 90: 256101.

Hug, H. J., Lantz, M. A., Abdurixit, A., van Schendel, P. J. A., Hoffmann, R., Kappenberger, P., and Baratoff, A. 2001. Subatomic features in atomic force microscopy images. *Science* 291: 2509a.

Ibe, J. P., Bey, P. P., Brandow, S. L., Brizzolara, R. A., Burnham, N. A., DiLella, D. P., Lee, K. P., Marrian, C. R. K., and Colton, R. J. 1990. On the electrochemical etching of tips for scanning tunneling microscopy. *J. Vac. Sci. Technol.* 8: 3570–3575.

Iwami, M., Uehara, Y., and Ushioda, S. 1998. Preparation of silver tips for scanning tunneling microscopy imaging. *Rev. Sci. Instrum.* 69: 4010–4011.

Jelinek, P., Shvec, M., Pou, P., Perez, R., and Chab, V. 2008. Tip-induced reduction of the resonant tunneling current on semiconductor surfaces. *Phys. Rev. Lett.* 101: 176101.

Jurczyszyn, L., Mingo, N., and Flores, F. 1998. Influence of the atomic and electronic structure of the tip on STM images and STS spectra. *Surf. Sci.* 402–404: 459–463.

Kelly, K. F., Sarkar, D., Hale, G. D., Oldenburg, S. J., and Halas, N. J. 1996. Threefold electron scattering on graphite observed with C60-adsorbed STM tips. *Science* 273: 1371–1373.

Klijn, J., Sacharow, L., Meyer, C., Blugel, S., Morgenstern, M., and Wiesendanger, R. 2003. STM measurements on the InAs(110) surface directly compared with surface electronic structure calculations. *Phys. Rev. B* 68: 205327.

Knoll, M. 1935. Charging potential and secondary emission of bodies under electron irradiation. *Z. Tech. Phys.* 16: 467–475.

Knoll, M. and Ruska, E. 1932. The electron microscope. *Z. Phys.* 78: 318–339.

Kohen, A., Noat, Y., Proslier, T., Lacaze, E., Aprili, M., Sacks, W., and Roditchev, D. 2005. Fabrication and characterization of scanning tunneling microscopy superconducting Nb tips having highly enhanced critical fields. *Physica C* 419: 18–24.

Landau, L. D. and Lifshitz, L. M. 1977. *Quantum Mechanics*, 3rd edn., Pergamon Press, Oxford, U.K.

Lawunmi, D. and Payne, M. C. 1990. Theoretical investigation of the scanning tunnelling microscope image of graphite. *J. Phys.: Condens. Matter* 2: 3811–3821.

Livshits, A. I., Shluger, A. L., Rohl, A. L., and Foster, A. S. 1999. Model of noncontact scanning force microscopy on ionic surfaces. *Phys. Rev. B* 59: 2436–2448.

Lu, X., Grobis, M., Khoo, K. H., Louie, S. G., and Crommie, M. F. 2003. Spatially mapping the spectral density of a single C60 molecule. *Phys. Rev. Lett.* 90: 096802.

Manini, N. and Onida, G. 2010. Comment on imaging the atomic orbitals of carbon atomic chains with field-emission electron microscopy. *Phys. Rev. B* 81: 127401.

Mikhailovskij, I. M., Sadanov, E. V., Mazilova, T. I., Ksenofontov, V. A., and Velicodnaja, O. A. 2009. Imaging the atomic orbitals of carbon atomic chains with field-emission electron microscopy. *Phys. Rev. B* 80: 165404.

Mikhailovskij, I. M., Sadanov, E. V., Mazilova, T. I., Ksenofontov, V. A., and Velicodnaja, O. A. 2010. Reply to comment on imaging the atomic orbitals of carbon atomic chains with field-emission electron microscopy. *Phys. Rev. B* 81: 127402.

Minakov, A. A. and Shvets, I. V. 1990. On the possibility of resolving quantization axes of surface spins by means of a scanning tunneling microscope with a magnetic tip. *Surf. Sci.* 236: L377–L381.

Morishita, S. and Okuyama, F. 1991. Sharpening of monocrystalline molybdenum tips by means of inert-gas ion sputtering. *J. Vac. Sci. Technol. A* 9: 167.

Muller, E. W. 1936. Theory of electron emission under the act on strong fields. *Z. Tech. Phys.* 17: 412–416.

Muller, E. W. 1951. The field ion microscope. *Z. Phys. B* 131: 136–142.

Muller, E. W. and Tsong, T. T. 1969. *Field Ion Microscopy, Principles and Applications*, American Elsevier Publishing Co., Inc., New York.

Murphy, S., Osing, J., and Shvets, I. V. 1999a. Atomically resolved p(3 × 1) reconstruction on the W(100) surface imaged with magnetic tips. *J. Magn. Magn. Mater.* 199: 686–688.

Murphy, S., Osing, J., and Shvets, I. V. 1999b. Fabrication of submicron-scale manganese-nickel tips for spin-polarized STM studies. *Appl. Surf. Sci.* 144–145: 497–500.

Murphy, S., Radican, K., Shvets, I. V., Chaika, A. N., Semenov, V. N., Nazin, S. S., and Bozhko, S. I. 2007. Asymmetry effects in atomically resolved STM images of Cu(014)-O and W(100)-O surfaces measured with MnNi tips. *Phys. Rev. B* 76: 245423.

Nagahara, L. A., Thundat, T., and Lindsay, S. M. 1989. Preparation and characterization of STM tips for electrochemical studies. *Rev. Sci. Instrum.* 60: 3128–3130.

Neddermeyer, H. and Drechsler, M. 1988. Electric field-induced changes of W(110) and W(111) tips. *J. Microsc.* 152: 459–466.

Nunes, G. and Amer, N. M. 1993. Atomic resolution scanning tunneling microscopy with a gallium arsenide tip. *Appl. Phys. Lett.* 63: 1851–1853.

Papaconstantopoulos, D. A. 1986. *Handbook of the Structure of Elemental Solids*, Plenum, New York.

Paz, O., Brihuega, I., Gomez-Rodriguez, J. M., and Soler, J. M. 2005. Tip and surface determination from experiments and simulations of scanning tunneling microscopy and spectroscopy. *Phys. Rev. Lett.* 94: 056103.

Polok, M., Fedorov, D. V., Bagrets, A., Zahn, P., and Mertig, I. 2011. Evaluation of conduction eigenchannels of an adatom probed by an STM tip. *Phys. Rev. B* 83: 245426.

Prins, M. W. J., Jansen, R., and Van Kempen, H. 1996. Spin-polarized tunneling with GaAs tips in scanning tunneling microscopy. *Phys. Rev. B* 53: 8105–8113.

Repp, J., Meyer, G., Stojkovic, S. M., Gourdon, A., and Joachim, C. 2005. Molecules on insulating films: Scanning tunneling microscopy imaging of individual molecular orbitals. *Phys. Rev. Lett.* 94: 026803.

Ruan, L., Besenbacher, F., Stensgaard, I., and Laegsgaard, E. 1993. Atom resolved discrimination of chemically different elements on metal surfaces. *Phys. Rev. Lett.* 70: 4079–4082.

Rutherford, E. 1911. The scattering of α and β particles by matter and the structure of the atom. *Phil. Mag.* 21: 669–688.

Sacks, W. 2000. Tip orbitals and the atomic corrugation of metal surfaces in scanning tunneling microscopy. *Phys. Rev. B* 61: 7656–7668.

Sacks, W. and Noguera, C. 1991a. Generalized expression for the tunneling current in scanning tunneling microscopy. *Phys. Rev. B* 43: 11612–11622.

Sacks, W. and Noguera, C. 1991b. Beyond Tersoff and Hamann: A generalized expression for the tunneling current. *J. Vac. Sci. Technol. B* 9: 488–491.

Sadanov, E. V., Mazilova, T. I., Mikhailovskij, I. M., Ksenofontov, V. A., and Mazilov, A. A. 2011. Field-ion imaging of nano-objects at far-subångstrom resolution. *Phys. Rev. B* 84: 035429.

Scheer, E., Agrait, N., Cuevas, J. C., Yeyati, A. L., Ludoph, B., Martin-Rodero, A., Bollinger, G. R., Van Ruitenbeek, J. M., and Urbina, C. 1998. The signature of chemical valence in the electrical conduction through a single-atom contact. *Nature* 394: 154–157.

Schmid, M., Stadler, H., and Varga, P. 1993. Direct observation of surface chemical order by scanning tunneling microscopy. *Phys. Rev. Lett.* 70: 1441–1444.

Schrödinger, E. 1926. An undulatory theory of the mechanics of atoms and molecules. *Phys. Rev.* 28: 1049–1070.

Serrate, D., Ferriani, P., Yoshida, Y., Hla, S.-W., Menzel, M., von Bergmann, K., Heinze, S., Kubetzka, A., and Wiesendanger, R. 2010. Imaging and manipulating the spin direction of individual atoms. *Nat. Nanotechnol.* 5: 350–353.

Shvets, I. V., Wiesendanger, R., Bürgler, D., Tarrach, G., Günterodt, H.-J., and Coey, J. M. D. 1992. Progress towards spin-polarized scanning tunneling microscopy. *J. Appl. Phys.* 71: 5489–5499.

Stemmer, A., Hefti, A., Aebi, U., and Engel, A. 1989. Scanning tunneling and transmission electron microscopy on identical areas of biological specimens. *Ultramicroscopy* 30: 263–280.

Stipe, B. C., Rezaei, M. A., and Ho, W. 1998. Single-molecule vibrational spectroscopy and microscopy. *Science* 280: 1732–1735.

Stroscio, J. A., Feenstra, R. M., and Fein, P. A. 1987. Local density and long-range screening of adsorbed oxygen atoms on the GaAs(110) surface. *Phys. Rev. Lett.* 58: 1668–1671.

Stroscio, J. A., Feenstra, R. M., Newns, D. M., and Fein, A. P. 1988. Voltage-dependent scanning tunneling microscopy imaging of semiconductor surfaces. *J. Vac. Sci. Technol. A* 6: 499–507.

Sugimoto, Y., Abe, M., Yoshimoto, K., Custance, O., Yi, I., and Morita, S. 2005. Non-contact atomic force microscopy study of the Sn/Si(111) mosaic phase. *Appl. Surf. Sci.* 241: 23–27.

Sugimoto, Y., Pou, P., Abe, M., Jelinek, P., Perez, R., Morita, S., and Custance, O. 2007. Chemical identification of individual surface atoms by atomic force microscopy. *Nature* 446: 64–67.

Sutter, P., Zahl, P., Sutter, E., and Bernard, J. E. 2003. Energy-filtered scanning tunneling microscopy using a semiconductor tip. *Phys. Rev. Lett.* 90: 166101.

Swart, I., Gross, L., and Liljeroth, P. 2011. Single-molecule chemistry and physics explored by low-temperature scanning probe microscopy. *Chem. Commun.* 47: 9011–9023.

Takayanagi, K., Kim, S., Lee, S., Oshima, Y., Tanaka, T., Tanishiro, Y., Sawada, H. et al. 2011. Electron microscopy at a sub-50 pm resolution. *J. Electron Microsc.* 60: S239–S244.

Takayanagi, K., Tanishiro, Y., Takahashi, M., and Takahashi, S. 1985. Structural analysis of Si(111)-(7 × 7) by UHV-transmission electron diffraction and microscopy. *J. Vac. Sci. Technol. A* 3: 1502–1506.

Temirov, R., Soubatch, S., Neucheva, O., Lassise, A. C., and Tautz, F. S. 2008. A novel method achieving ultra-high geometrical resolution in scanning tunnelling microscopy. *New J. Phys.* 10: 053012.

Tersoff, J. and Hamann, D. R. 1983. Theory and application for the scanning tunneling microscope. *Phys. Rev. Lett.* 50: 1998–2001.

Tersoff, J. and Hamann, D. R. 1985. Theory of the scanning tunneling microscope. *Phys. Rev. B* 31: 805–813.

Tersoff, J. and Lang, N. D. 1990. Tip-dependent corrugation of graphite in scanning tunneling microscopy. *Phys. Rev. Lett.* 65: 1132–1135.

Thomson, J. J. 1897. Cathode rays. *The Electrician* 39: 104–109.

Thomson, J. J. 1904. On the structure of the atom: An investigation of the stability and periods of oscillation of a number of corpuscles arranged at equal intervals around the circumference of a circle; with application of the results to the theory of atomic structure. *Phil. Mag.* 7: 237–265.

Tromp, R. M., Hamers, R. J., and Demuth, J. E. 1986. Atomic and electronic contributions to Si(111)-(7 × 7) scanning tunneling-microscopy images. *Phys. Rev. B* 34: 1388–1391.

Tromp, R. M., Van Loenen, E. J., Demuth, J. E., and Lang, N. D. 1988. Tip electronic structure in scanning tunneling microscopy. *Phys. Rev. B* 37: 9042–9045.

Tsong, T. T. and Muller, E. W. 1969. Effects of static-field penetration and atomic polarization on the capacity of a capacitor, field evaporation, and field ionization processes. *Phys. Rev.* 181: 530–534.

Wang, Y. L., Gao, H.-J., Guo, H. M., Liu, H. W., Batyrev, I. G., McMahon, W. E., and Zhang, S. B. 2004. Tip size effect on the appearance of a STM image for complex surfaces: Theory versus experiment for Si(111)-(7 × 7). *Phys. Rev. B* 70: 073312.

Weiss, C., Wagner, C., Kleimann, C., Rohlfing, M., Tautz, F. S., and Temirov, R. 2010. Imaging Pauli repulsion in scanning tunneling microscopy. *Phys. Rev. Lett.* 105: 086103.

Wiesendanger, R. 2009. Spin mapping at the nanoscale and atomic scale. *Rev. Mod. Phys.* 81: 1495–1550.

Wiesendanger, R., Bode, M., Rascal, R., Allers, W., and Schwarz, U. D. 1996. Issues of atomic-resolution structure and chemical analysis by scanning probe microscopy and spectroscopy. *J. Vac. Sci. Technol. A* 14: 1161–1167.

Wiesendanger, R., Bürgler, D., Tarrach, G., Schaub, T., Hartmann, U., Güntherodt, H.-J., Shvets, I. V., and Coey, J. M. D. 1991. Recent advances in scanning tunneling microscopy involving magnetic probes and samples. *Appl. Phys. A: Solids Surf.* 53: 349–355.

Wiesendanger, R., Güntherodt, H.-J., Güntherodt, G., Gambino, R. J., and Ruf, R. 1990. Observation of vacuum tunneling of spin-polarized electrons with the scanning tunneling microscope. *Phys. Rev. Lett.* 65: 247–250.

Wiesendanger, R., Shvets, I. V., Bürgler, D., Tarrach, G., Güntherodt, H. J., Coey, J. M. D., and Gräser, S. 1992. Topographic and magnetic-sensitive scanning tunneling microscope study of magnetite. *Science* 255: 583–586.

Wintterlin, J., Brune, H., Hofer, H., and Behm, R. 1988. Atomic scale characterization of oxygen adsorbates on Al(111) by scanning tunneling microscopy. *Appl. Phys. A* 47: 99–102.

Wintterlin, J., Wiechers, J., Brune, H., Gritsch, T., Hofer, H., and Behm, R. J. 1989. Atomic-resolution imaging of close-packed metal surfaces by scanning tunneling microscopy. *Phys. Rev. Lett.* 62: 59–62.

Wolkow, R. A. 1992. Direct observation of an increase in buckled dimmers on Si(001) at low temperatures. *Phys. Rev. Lett.* 68: 2636–2639.

Wright, C. A. and Solares, S. D. 2011. On mapping subångstrom electron clouds with force microscopy. *Nano Lett.* 11: 5026–5033.

Yashina, L. V., Püttner, R., Volykhov, A. A., Stojanov, P., Riley, J., Vassiliev, S. Y., Chaika, A. N. et al. 2012. Atomic geometry and electron structure of the GaTe(10-2) surface. *Phys. Rev. B* 85: 075409.

Zheng, N. J. and Tsong, I. S. T. 1990. Resonant-tunneling theory of imaging close-packed metal surfaces by scanning tunneling microscopy. *Phys. Rev. B* 41: 2671–2677.

Zotti, L. A., Hofer, W. A., and Giessibl, F. J. 2006. Electron scattering in scanning probe microscopy experiments. *Chem. Phys. Lett.* 420: 177–182.

Zuo, J. M., Kim, M., O'Keeffe, M., and Spence, J. C. H. 1999. Direct observation of *d*-orbital holes and Cu-Cu bonding in Cu_2O. *Nature* 401: 49–52.

19

STM of Quantum Corrals

Akira Tamura
Saitama Institute of Technology

19.1 Primary Stage of STM and Afterward

In 1981, Binnig and Rohrer invented the scanning tunneling microscope (STM). After 1 year, they published a series of articles that exhibited the validity of STM (Binnig and Rohrer, 1982, Binnig et al., 1982a,b, 1983, 1987). STM is a powerful instrument that makes it feasible to observe atomic configurations of material surfaces in a high spatial resolution less than nm (n:10^{-9}) through the current flowing between the STM tip and the specimen. The central part of STM is a sharpened STM tip and the controlling system of it. To scan the tip in three-dimensional directions, they used a piezoelectric device that converts voltage to mechanical strains and vice versa. Keeping the STM current constant of the order of nA, they observed STM images of metal surfaces by controlling the tip position. It should be emphasized that they acquired a controlling technology that is feasible to move smoothly the tip at a speed of pm (p:10^{-12}) per second. To demonstrate the validity of the STM further, they observed the reconstructed 7 × 7 structure of an Si(111) surface. The 7 × 7 structure is a typical superstructure of the Si(111) surface. The atomic configuration of the 7 × 7 structure has never been determined since Schreier and Farnsworth (1959) obtained the LEED (low-energy electron diffraction) pattern. Takayanagi et al. (1982, 1985) had carried out TED (transmission electron diffraction) measurement of a quite thin Si film and proposed the DAS (dimer · adatom · stackingfault) model by using the Patterson function obtained from experimental TED patterns. With STM, Binnig et al. (1983) found 12 protruded Si atoms (adatoms) in the 7 × 7 unit and 4 holes at 4 corners out of 49 atom sites. Among several structure models proposed by that time, the observed STM image lends support to the DAS model. For many reconstructed structures of Si, Ge, compound semiconductors, and those adsorbed surfaces, STM has been providing important information of atomic configurations. Many kinds of metal surfaces have also been studied with STM and those results confirmed the validity of STM (Wiesendanger, 1994, Besenbacher, 1996, Hofer et al., 2003a,b, Chen, 2008).

Though STM images had been observed at normal temperatures at the early stage, many researchers observed STM images at low temperatures near liquid helium temperature (4.2 K) to reduce thermal drift of the STM tip and movements of adsorbed atoms. As an application to another field of research, STM has been used to specify vortex states of oxide superconducting materials (Hess et al., 1989, 1990; Maggio-Aprile et al., 1995, Fisher et al., 2007). At present, it is possible to observe STM images at temperatures up to 1300 K. As an environmental condition for observing clean surfaces, it is necessary to keep the pressure in the vacuum chamber at less than 10^{-10} Pa.

STM is a compact instrument that needs no electron lens system of large size in contrast to TEM (transmission electron microscope) and SEM (scanning electron microscope). It is indispensable, however, to equip with a set of systems against mechanical vibrations that disturb data accumulation.

When the STM was invented, the main feature was to observe atomic configurations of surface atoms in real space. A couple of years later, it was demonstrated that the STM provides scanning tunneling spectrum (STS) through bias voltage dependence of the STM current (Binnig et al., 1985). This means that the STM became a powerful instrument that gives physical quantities in energy space. Though electron energy loss spectra (EELS), with an incident energy less than 100 eV, have been measured for many kinds of material surfaces, it is difficult to obtain EELS in a quite small area. The STS provides information on the position dependence of energy states of material surfaces: local density of states (LDOS). To specify surface electron states, we should obtain both STM images and STS profile of the surface.

In the early stages, STM had been used to specify atomic and electronic structures of conductive materials. After that, Binnig et al. (1986) invented a new technology that enables us to observe nonconductive materials via the atom force exerted between the tip (cantilever) and the surface. They named it AFM (atomic force microscope). Since then many researchers have been using the AFM to observe insulative materials and genetic materials such as DNA (García and Pérez, 2003). To obtain TEM images of organisms, we usually prepare freeze-dried specimens with gold decorations to avoid charging effects or replicas of them, but we can observe living organisms with the AFM. The AFM has also been used to observe atomic configurations of conductive materials such as metals, semiconductors, and vibrational modes of surface atoms. Contacting and noncontacting AFMs have come into wide use as powerful instruments. The AFM also provides information of chemical reactions that occur on surfaces (Meyer et al., 1996, Barterls et al., 1997, 1998). To observe spin configurations on surfaces of magnetic materials, the magnetic force microscope (MFM) was developed (Wiesendanger et al., 1990, 2009; Pietzsch et al., 2001, 2006; Bode, 2003). Now the term SPM (scanning probe microscope) is widely used, including STM, AFM, MFM, and so on.

With TED, we cannot directly determine atomic configurations of material surfaces because the measured diffraction patterns are in reciprocal lattice space. By assuming many kinds of lattice structure model, we select the most probable model that reproduces the experimental TED pattern and the rocking curve. In this sense, TED is an indirect method for determining atomic configurations. By converging electron beams to a small area of nm size, a single atom can be observed. The area hit by the convergent beam of high incident energy, however, may suffer damage. As another instrument, field ion microscope (FIM) provides the image of the tip in real space: magnified image of tip atoms. The situation surrounding the atom on the tip is not normal because the tip is exposed to a very high electric field. The STM, in contrast to those instruments, is a nondestructive instrument because the STM current is in the order of nA and the applied bias voltage is less than a few electron volts.

19.2 Physical Quantities Obtained by STM

The most important physical quantity supplied by STM is the current that flows between the STM tip and the specimen surface. A weak STM current of the order of nA transmits a barrier that separates the vacuum and the bulk. The sign of the bias voltage is determined with respect to the Fermi level E_F. When the bias voltage V is positive, electrons in the STM tip make the transition to the specimen surface. When V is negative, electrons in the specimen surface make the transition to the tip. Because an electron charge is negative, the flowing direction of the STM current is reverse to the flow of electrons. For a positive V, the STM current provides information of unoccupied electron states of the specimen surface, while it provides information of occupied electron states for a negative V.

Observing modes of STM images are mainly of two kinds: topographical mode and differential conductance mode. In the topographical mode, we gain the tip-height data by keeping the STM current constant with a feedback circuit system while we scan the tip over the surface. In the differential conductance mode (dI/dV), we differentiate the STM current with respect to the bias voltage V at each scanned position. Topographical STM image provides the probability density summed over electron states occupying the energy range between E_F and $E_F + eV$. The dI/dV image has been thought to give the LDOS. For quantum corrals (QCs), the dI/dV image, however, is not merely proportional to the LDOS, which will be explained in Section 19.4.1.2.

At present, it is difficult to identify species of adsorbed atoms on surfaces with STM. This comes from the situation where we observe STM images and STS in the neighborhood of the Fermi level E_F. When we extend the energy range far from E_F, STM images and STS become more complex and it becomes difficult to identify the atom species. In ELS (energy loss spectrum), AES (Auger electron spectrum), and XPS (x-ray photoelectron spectrum), we generally obtain information an atom species from spectra of core excitations. For the known adsorbed atom, we can easily identify the species by investigating differences in the contrast of an STM image and in the STS profile. When the atom species is unknown, we need to accumulate a lot of data of the adsorbed system. If the adsorbate couples strongly with a substrate, the electron states change remarkably from those of the atom in a vacuum. In such a case we need to develop the STM that has high spatial and energy resolution. Using AFM, it is possible to identify some kinds of species by analyzing dynamical atom forces exerted between surface atoms and the cantilever (Giessibl, 2003, García and Pérez, 2003, Sugimoto et al., 2007).

19.3 STM Current

Tersoff and Herman (1983, 1985) derived the STM current on the basis of Bardeen's model Hamiltonian (Bardeen, 1961) as

$$I = \frac{4\pi e}{\hbar} \int_{-\infty}^{\infty} [f(E-eV) - f(E)] D_{sp}(E) D_{tip}(E-eV) \,|\, M_{\mu\nu} \,|^2 \, dE, \tag{19.1}$$

where

$f(E)$ is the Fermi distribution function: $f(E) = 1/\{1 + \exp[(E - E_F)/(k_B T)]\}$
V is a positive bias voltage with respect to E_F
$D(E)$ represents the density of states (DOS)

Subscripts "tip" and "sp" label the STM tip and the specimen, respectively. $M_{\mu\nu}$ represents the tunneling matrix element defined as

$$M_{\mu\nu} = \frac{\hbar^2}{2m} \int_S (\varphi_\mu^* \nabla \psi_\nu - \psi_\nu^* \nabla \varphi_\mu) \cdot dS \tag{19.2}$$

where ψ_ν and φ_μ are wave functions of the specimen and the STM tip, respectively. The integration ∫ has to be carried out over the surface lying entirely within the vacuum barrier region that separates the tip and the specimen surface. Lang (1985, 1986) also derived a similar expression for the STM current. Selloni et al. (1985) showed that the voltage-dependent STM current can be expressed as

$$I_+(\rho,z,V) \propto \int_{-\infty}^{+\infty} D_{3D}(\boldsymbol{r},E)[f_{tip}(E-eV)-f_{sp}(E)]T(E-E_F,V,z)\mathrm{d}E, \quad (19.3)$$

where we neglected a contribution from the electron states of the tip. A subscript "+" represents a positive bias voltage: electrons flow from the STM tip to the specimen. $D_{3D}(\boldsymbol{r}, E)$ represents a three-dimensional LDOS of an electron:

$$D_{3D}(\boldsymbol{r},E) = \sum_{\sigma nm} |\Psi^{3D}_{nm}(\boldsymbol{r})|^2 \frac{\Gamma_{nm}}{2\pi[(E-E^{3Dr}_{nm})^2 + (\Gamma_{nm}/2)^2]} \quad (19.4)$$

where

- $\Psi^{3D}_{nm}(\boldsymbol{r})$ is the eigenfunction
- n and m denote quantum numbers which specify the electron state
- Γ_{nm} is the energy width associated with an eigenenergy E^{3Dr}_{nm}
- $T(E,V,z)$ represents the transition probability of an electron between the tip and the specimen

Within WKB approximation, we obtain

$$T(E,V,z) = \exp\left[-z\sqrt{\left(\frac{4m}{\hbar^2}\right)(W_{sp}+W_{tip}+eV-2E)}\right]. \quad (19.5)$$

19.4 Quantum Corrals

When the STM was invented in 1981, it was a passive instrument that enable one to observe atomic configurations of many kinds of material surfaces. In 1990, Eigler's group in IBM lined up Xenon atoms by controlling an STM tip. They fabricated artificial nanostructures (Eigler and Schweitzer, 1990, Stroscio and Eigler, 1991). That was the instant when STM evolved into an active instrument, which made it feasible to construct new materials in nano scale we had never imagined. Since then the STM started to play the new role: we have acquired a new technology of controlling a single atom at will and we can make drawings in atomic scale. They made a Chinese character and geometrical figures on noble metal surfaces with atoms (STM Image Gallery in IBM). The most famous one is a circle made of 48 Fe atoms fabricated on a Cu(111) surface at 4.3 K (Crommie et al., 1993, Heller et al., 1994). They observed the STM image that shows a wavy pattern inside the corral depicted the wave property of an electron. They also depicted its making process in four STM images (Crommie et al., 1995). They named the sequence of Fe atoms QC. The STM image shows a circular profile despite the fact that the equilateral triangular lattice forms on the ideal Cu(111) surface. This indicates that nonlocalized electrons build up stationary states within the QC. Manipulation of molecules on surfaces has been extensively studied with STM (Hla, 2005).

19.4.1 STM Images and STS of a Quantum Corral

19.4.1.1 Shockley Electron

There exist Shockley electrons on noble metal surfaces such as Cu, Ag, and Au. Corral atoms of transition metals such as Fe and Mn fabricated on those surfaces confine Shockley electrons inside and those electrons exhibit wavy patterns. The Shockley electron behaves as a free electron in the direction parallel to the surface and stays near the surface in the direction normal to the surface (Goodwin, 1939, Ashcroft and Mermin, 1976, Smith, 1985, Kevan, 1986, Kevan and Gaylord, 1987, Desjonquere and Spanjaard, 1996). This means that the Shockley electron state exists in the band gap projected on the surface plane. For the (111) surface of the fcc lattice, the origin of the wave number parallel to the surface is set at L point at the zone boundary.

Using a nearly free-electron model, we outline the situation. Wave functions outside and inside the crystal can be assigned as

$$\psi_{out}(\boldsymbol{r}) = \exp(i\boldsymbol{k}_{\parallel}\cdot\rho - \kappa z), \quad (19.6)$$

and

$$\psi_{in}(\boldsymbol{r}) = \exp(i\boldsymbol{k}_{\parallel}\cdot\rho)\left[\exp[(ik_0+\kappa')z]c_k + \exp\{[i(k_0-g)+\kappa']z\}c_{k-g}\right] \quad (19.7)$$

where

- $\boldsymbol{r}$ represents a 3D position vector $\boldsymbol{r} = (\rho, z)$ in which ρ is a two-dimensional vector parallel to the surface and z is the coordinate pointing to the vacuum
- $\boldsymbol{k}_{\parallel}$ represents the wave number vector parallel to the surface
- Both κ and κ' are real numbers which yield decaying wave functions on both sides of the surface
- k_0 is the wave number in the bulk normal to the surface

Expanding the periodic potential inside the crystal as Fourier series,

$$U(\boldsymbol{r}) = U_0 + \sum_{g\neq 0} U_g \exp(i\boldsymbol{g}\cdot\boldsymbol{r}), \quad (19.8)$$

with a reciprocal lattice vector $\boldsymbol{g}$, and imposing boundary conditions that the wave function is continuous and smooth at the surface $z = z_0$, we obtain the energy band at the zone boundary $(1/2)g$ as follows:

$$E(k_{\parallel}) = U_0 + \frac{\hbar^2}{2m}(k_{\parallel}^2 - \kappa'^2 + \tfrac{1}{4}g^2) \pm \left(U_g^2 - \frac{\hbar^4}{4m^2}g^2\kappa'^2\right)^{1/2}. \quad (19.9)$$

Through a complex wave number $i\kappa'$, the upper band (+) and the lower band (−) are continuously connected at the zone boundary. We obtain a relation between κ and κ' as

$$\kappa + \kappa' = \frac{1}{2}g\tan\left(\frac{1}{2}gz_0 + \delta\right). \quad (19.10)$$

Assuming that $z_0 = 0$, we find

$$\kappa = \frac{1}{2}g\left(\frac{U_0}{U_0 + U_g}\right)\tan\delta, \quad (19.11)$$

where δ represents the phase difference between $c_{\boldsymbol{k}}$ and $c_{\boldsymbol{k}-\boldsymbol{g}}$: $c_{\boldsymbol{k}} = c_{\boldsymbol{k}-\boldsymbol{g}}\exp(i2\delta)$. The energy $E(k_{\parallel}) = [\hbar^2/(2m)](k_{\parallel}^2 - \kappa^2)$ shows a parabolic dispersion relation. In reality, the electron in the direction parallel to the surface has an effective mass smaller than the free electron mass (Smith, 1985, Kevan, 1986, Kevan and Gaylord, 1987). In this manner, the Shockley state comes from the barrier between the vacuum and the band gap in the bulk. This situation is similar to impurity levels of p-type and n-type semiconductors that exist in the band gap. An electron occupying those impurity levels cannot move freely in the crystal but only stay there long time.

In general, we observe STM images of a semiconductor surface at bias voltages of several volts above and below E_F. In contrast, we observe STM images of QCs in the narrower range near E_F. In that voltage range, a wave number $k_{\parallel}$ is much smaller than the zone boundary. These are reasons why we cannot see a sequence of a crystal lattice of the surface atoms inside a QC. In this manner the Shockley electron stays isotropically over the surface in contrast to a covalent electron that is anisotropically bonded on semiconductor surfaces. Hence, the bias voltage for observing semiconductor surfaces is much higher than that for the QC in two orders of magnitude.

19.4.1.2 Topographical and Differential Conductance STM Images

Because we observe STM images of a QC at low bias voltages, $e|V| << (E_F + V_0)$, we can approximate the transition probability to $T(E,V,z) = \exp(-\alpha z)$ with $\alpha = \sqrt{4m^*(W_{sp} + W_{tip})/\hbar^2}$. Here we concentrate on the two-dimensional electron state confined in a QC. We incorporate the term except $\hbar^2 k_{\parallel}^2/(2m)$ in Equation 19.9 into $-V_0$, and substitute E_{nm}^{2Dr} for E_{nm}^{3Dr} in Equation 19.4. Using a relation $|\Psi_{nm}^{3D}(\boldsymbol{r})|^2 = |\Psi_{nm}^{2D}(\rho)|^2\exp(-2\kappa z)$ and defining the two-dimensional LDOS of a QC as

$$D_{2D}(\rho,E) = \sum_{\sigma nm}|\Psi_{nm}^{2D}(\rho)|^2 \frac{\Gamma_{nm}^{2D}}{2\pi[(E - E_{nm}^{2Dr})^2 + (\Gamma_{nm}^{2D}/2)^2]} \quad (19.12)$$

we arrive at the STM current expressed as

$$I_+(\rho,z,V) = K_0\exp(-\alpha' z)] \sum_{\sigma nm}|\Psi_{nm}^{2D}(\rho)|^2[S_{nm}(V) - S_{nm}(0)] \quad (19.13)$$

where $\alpha' = \alpha + 2\kappa$, $S_{nm}(V) = \mathrm{Arctan}\left[\frac{2(E_F + eV - E_{nm}^{2Dr})}{\Gamma_{nm}^{2D}}\right]$ and K_0 is a constant. The weighting factor $[S_{nm}(V) - S_{nm}(0)]$ restricts the summation over n and m to the energy range between E_F and $E_F + eV$.

When the STM current $I_+(\rho, z(\rho), V)$ is constant, we obtain $z(\rho,V)$ from Equation 19.13 as

$$z(\rho,V) = \alpha'^{-1}\left\{C_0 + \log_e\sum_{\sigma nm}|\Psi_{nm}^{2D}(\rho)|^2[S_{nm}(V) - S_{nm}(0)]\right\}, \quad (19.14)$$

where C_0 is a constant. The $z(\rho, V)$ is the vertical distance between the specimen and the apex of the tip at a lateral position $\rho = (x, y)$, and $z(\rho, V)$ yields the topographical STM image as a function of ρ. The STM image strongly depends on V. It is obvious that $z(\rho, V)$ is neither proportional to the probability density $\sum_{\sigma nm}|\Psi_{nm}^{2D}(\rho)|^2$ nor to the LDOS. Because $z(\rho, V)$ includes a logarithmic function in its expression, spatial variation of $z(\rho, V)$ is small.

From the STM current $I_+(\rho, z(\rho), V)$, the differential conductance is written as

$$\begin{aligned}\frac{d}{dV}I_+(\rho,z(\rho),V) \propto \\ &\int_{-\infty}^{+\infty} D_{2D}(\rho,E)\left\{\frac{\partial}{\partial V}[f_{tip}(E - eV) - f_{sp}(E)]\right\}T(E - E_F,V,z(\rho))dE \\ &+ \int_{-\infty}^{+\infty} D_{2D}(\rho,E)[f_{tip}(E - eV) - f_{sp}(E)]\frac{\partial}{\partial V}T(E - E_F,V,z(\rho))dE.\end{aligned} \quad (19.15)$$

Because a derivative $df_{tip}(E - eV)/d(eV)$ yields a delta-function $\delta(E - (E_F + eV))$ at 0 K and the term including $\partial T(E - E_F, eV, z(\rho))/\partial V$ makes a negligible contribution to dI_+/dV, we have

$$\begin{aligned}\frac{dI_+}{dV} &\propto D_{2D}(\rho, E_F + eV)\exp[-\alpha' z(\rho)] \\ &\propto F(\rho,V)D_{2D}(\rho, E_F + eV)\end{aligned} \quad (19.16)$$

where

$$F(\rho,V)=\left\{\sum_{\sigma nm}|\Psi_{nm}^{2D}(\rho)|^2[S_{nm}(V)-S_{nm}(0)]\right\}^{-1}. \quad (19.17)$$

In this manner, we find analytical expressions for topographical and dI/dV images.

It is apparent that the dI/dV image does not correspond to the LDOS image obtained from $D_{2D}(\rho, E_F + eV)$. The $F(\rho, V)$ acts as a filtering function and greatly modifies the LDOS image. The dI/dV image at a certain bias voltage is rather different from the topographical image at the same bias voltage. Differential conductance dI/dV measured at a certain position gives the STS. From the position dependence of STS, we obtain symmetry of the wave function at a certain eigenenergy. When an STM tip is far from a specimen surface and the variation induced by $z(\rho, V)$ is negligible, the dI/dV is proportional to the LDOS.

Both STM images and the STS are indispensable for analyzing experimentally electron states of surfaces. In addition, a theoretical analysis is required to provide both STM images and the STS in a wide range of bias voltage to show the validity of the theory.

19.4.2 Several Kinds of Quantum Corral

Here we show a list of quantum corrals classified by geometrical shape; one dimensional (Yokoyama and Takayanagi, 1999, Nilius et al., 2002, 2005, Fölsch et al., 2004), triangular (Braun and Rieder, 2002, Rieder et al., 2003, Veuillen et al., 2003, Lagoute et al., 2005, Pietzsch et al., 2006, Kumagai and Tamura, 2008), rectangular (Kliewer et al., 2001, Kumagai and Tamura, 2009, Tamura, 2011), hexagonal (Li et al., 1998, Kliewer and Berndt, 2001, Jensen et al., 2005, Niebergall et al., 2006), circular (Crommie et al., 1993, Heller et al., 1994, Harbury and Porod, 1996, Kliewer et al., 2001, Donner et al., 2005), elliptical (Manoharan et al., 2000, Fiete et al., 2001, Fiete and Heller, 2003), triangular InAs QDs (Kanisawa et al., 2001a,b, Kumagai and Tamura, 2008b), stadium (Heller et al., 1994, Fiete et al., 2001), adsorbed rectangular QC (Kliewer et al., 2000, 2001, Mitsuoka and Tamura, 2012), and steps on metal surfaces (Bürgi et al., 1998, Mitsuoka and Tamura, 2011a). We discuss some of them next.

19.4.2.1 Circular Quantum Corral

On a clean surface, a Shockley electron forms a continuous dispersion curve against a parallel wave number $k_{||}$ in the projected band structure. The electron confined by a QC has discrete wave numbers and energy levels because of quantum size effect. As a simple analysis, Crommie et al. (1993) considered a circular potential wall having an infinite height and derived stationary states confined inside the wall. They explained the wavy pattern as that built up with stationary electron states expressed by the first kind of Bessel function $J_l(k\rho)$. The fixed boundary condition at the wall edge leads to a sequence of zero points: $J_l(k_{nl}R) = 0$, where R represents the radius. Parameters l and n denote an angular momentum and a quantum number in the radial direction, respectively. We arrive at the normalized eigenfunction expressed as

$$\Psi_{nl}(\rho,\theta)=\frac{J_l(k_{nl}\rho)\exp(\pm i l\theta)}{\sqrt{\pi}R\,|J_{l+1}(k_{nl}R)|}. \quad (19.18)$$

The pair $\pm l$ means the difference in rotational direction of an electron in the QC. To explain the observed STM image, Crommie et al. (1993) superposed probability densities of each electron state as

$$z(\rho)=\sum_{nl}g(l)c_{nl}\,|\Psi_{nl}(\rho)|^2 \quad (19.19)$$

and determined coefficients c_{nl} to reproduce their experimental profile. $g(l)$ denotes the degeneracy associated with angular momenta $\pm l$ and spin degrees of freedom: $g(0) = 2$ for $l = 0$ and $g(l) = 4$ for $l > 0$. They did not derive STM images from the STM current. To explain the measured STS, they only assigned peak positions from discrete energy levels $E_{nl}=\hbar^2k_{nl}^2/(2m^*)-V_0$ in which $-V_0$ is the bottom of the potential and m^* is the effective mass of the electron.

Here we derive the STM image from the STM current and the STS from LDOS. Figure 19.1 shows the distribution of eigenenergies as a function of l where we chose $-V_0 = -5.419$ eV, $E_F = -4.980$ eV, $R = 7.13$ nm and $m^* = 0.38\ m_e$. A horizontal line represents the Fermi level E_F. In Figure 19.2, we show a topographical STM image calculated at $V = 10$ mV from $z(\rho, V)$ of Equation 19.14, in which we carried out summation over l from 0 to 11 and took account of lifetime effect by introducing a phenomenological energy width $\Gamma_{nl}^{2D}=0.1(E_{nl}+V_0)+\Gamma_0$ with $\Gamma_0 = 25$ meV, for simplicity. The Γ_{nl}^{2D} reflects a situation where the confined electron temporarily stays in a QC and slips through coral atoms afterwards. A peak at the center and four concentric rings are reproduced well. In addition, the second and the fourth

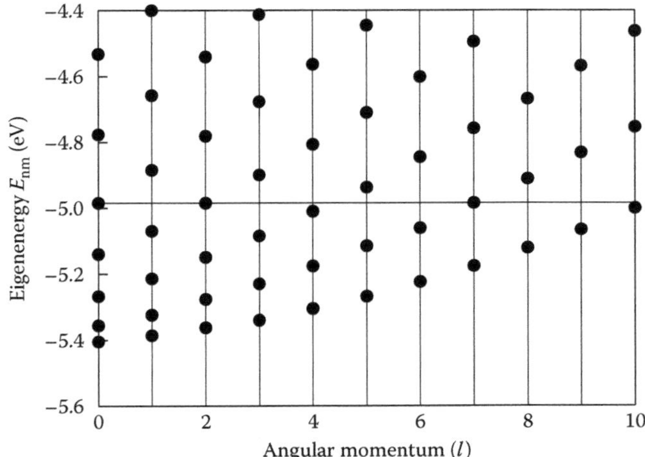

FIGURE 19.1 Distribution of eigenenergy.

FIGURE 19.2 Topographical STM image.

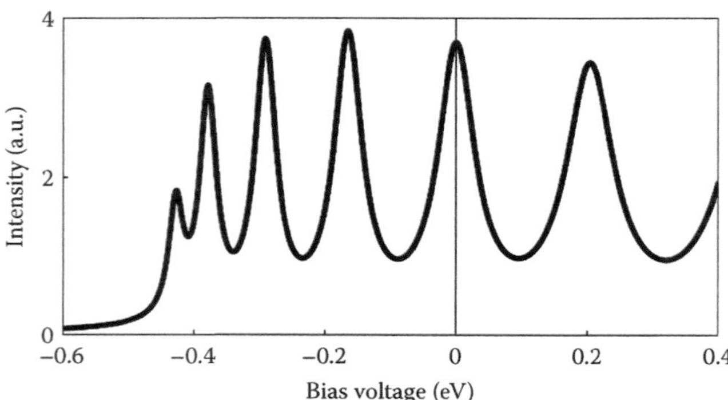

FIGURE 19.3 STS calculated at the center.

circular grooves from the center are deeper than others, which also reproduces experimental STM image. Figure 19.3 shows the STS calculated at the center and it is consistent with the measured STS. These peak positions correspond to a sequence of E_{n0}. At a position shifted from the center, electron states with $l > 0$ contribute and yield additional peaks between peaks shown in Figure 19.3. These results basically corroborate a claim made by Crommie et al. (1993) that STM images and the STS can be explained by Shockley electrons confined in the circular wall. It is important, however, that the STS can be explained by introducing energy widths of electron states. Without energy widths, the STS turns to be merely a sequence of spikes.

Heller et al. (1994) calculated STM images and the STS from LDOS, considering that the electron wave supplied by the STM tip undergoes scatterings from each corral atom and yields a lot of circular waves. They concluded that those scattered waves gather in the wavy profile inside the circular corral. They assumed that each corral atom is a perfect black dot which absorbs incoming waves perfectly. As they pointed out, it should be noted that diffraction occurs even if the scatterer is a perfect absorber for incoming waves. Though they derived only LDOS, their calculated results reproduce experimental ones well and they claimed that corral atoms reflect a quarter portion of electron waves and a half portion penetrates into the bulk through scattering with corral atoms. It is desirable to derive the topographical image from the STM current. In addition, there remains a problem that the experimental peak position of the sixth peak is lower than that of the calculated STS. On this point Harbury and Porod (1996) supposed that this reduction comes from an enlarged corral circle while scanning the STM tip. The increase in the radius, however, reduces all energy levels of the confined electron. This problem remains unsettled.

19.4.2.2 Triangular Quantum Corral

Rieder et al. (2003) observed dI/dV images of a triangular QC of Ag atoms fabricated on an Ag(111) surface. As a theoretical analysis to explain their STM images, Kumagai and Tamura (2008a) obtained eigenstates of the electron confined in a triangular QC surrounded by three walls having an infinite height. They considered a process that an electron undergoes sequential scatterings against the three walls and forms a closed trajectory. They found two kinds of eigenfunctions labeled with integers n and m for the triangle whose length of one side is L. One is

$$\begin{aligned}\Psi_{nm}(\rho) &= A\exp[\mathrm{i}G(2n-m)x]\sin(\sqrt{3}Gmy)\\ &\quad - A\exp[\mathrm{i}G(2m-n)x]\sin(\sqrt{3}Gny)\\ &\quad + A\exp[-\mathrm{i}G(n+m)x]\sin[\sqrt{3}G(n-m)y] \qquad (19.20)\end{aligned}$$

and the other is its complex conjugate $\Psi^{*}_{nm}(\rho)$ in which $G = 2\pi/(3L)$. Two indices n and m are natural numbers and A is a normalization constant; $A = \left(\sqrt{8\sqrt{3}}\right)/(3L)$. They obtained the eigenenergy expressed as

$$E_{nm} = \frac{2\hbar^2 G^2}{m^*}(n^2 - mn + m^2) - V_0 \qquad (19.21)$$

where $-V_0$ represents the bottom of the potential well. Equation 19.21 is the same as that obtained by Krishnamurthy et al. (1982). Because the eigenenergy E_{nm} is invariant under exchange of n for m, there exists a duplicate degeneracy in eigenstates; $E_{13} = E_{23}$, $E_{14} = E_{34}$, $E_{25} = E_{35}$ and so forth. When $n = m$, both $\Psi_{nm}(\rho)$ and $\Psi^{*}_{nm}(\rho)$ are zero anywhere inside the triangle because of the relation $\Psi_{nm}(\rho) = -\Psi_{mn}(\rho)$. Hence, electron states with $n = m$ have to be omitted from summation over electron states. They used parameters of the QC as $L = 24.5$ nm, $-V_0 = -4.802$ eV, $E_F = -4.740$ eV, and $m^* = 0.42\ m_e$. They took the energy width as $\Gamma^{QC}_{nm} = 0.24(E^{2Dr}_{nm} + V_0) + \Gamma_0$ with $\Gamma_0 = 8$ meV. Figure 19.4 shows calculated dI/dV images. Those images are remarkably consistent with experimental ones observed by Rieder et al. (2003)—and those images are reproduced in Chen 2008 on page lvi. It should be emphasized that those images are quite different depending on the sign of bias voltages. At positive bias voltages, dI/dV patterns show only a sequence of dark spots and their number increases with the bias voltage. In contrast, at negative bias voltages, those patterns are complex. Figure 19.5 shows the relation between the LDOS image and the filtering function F(ρ,

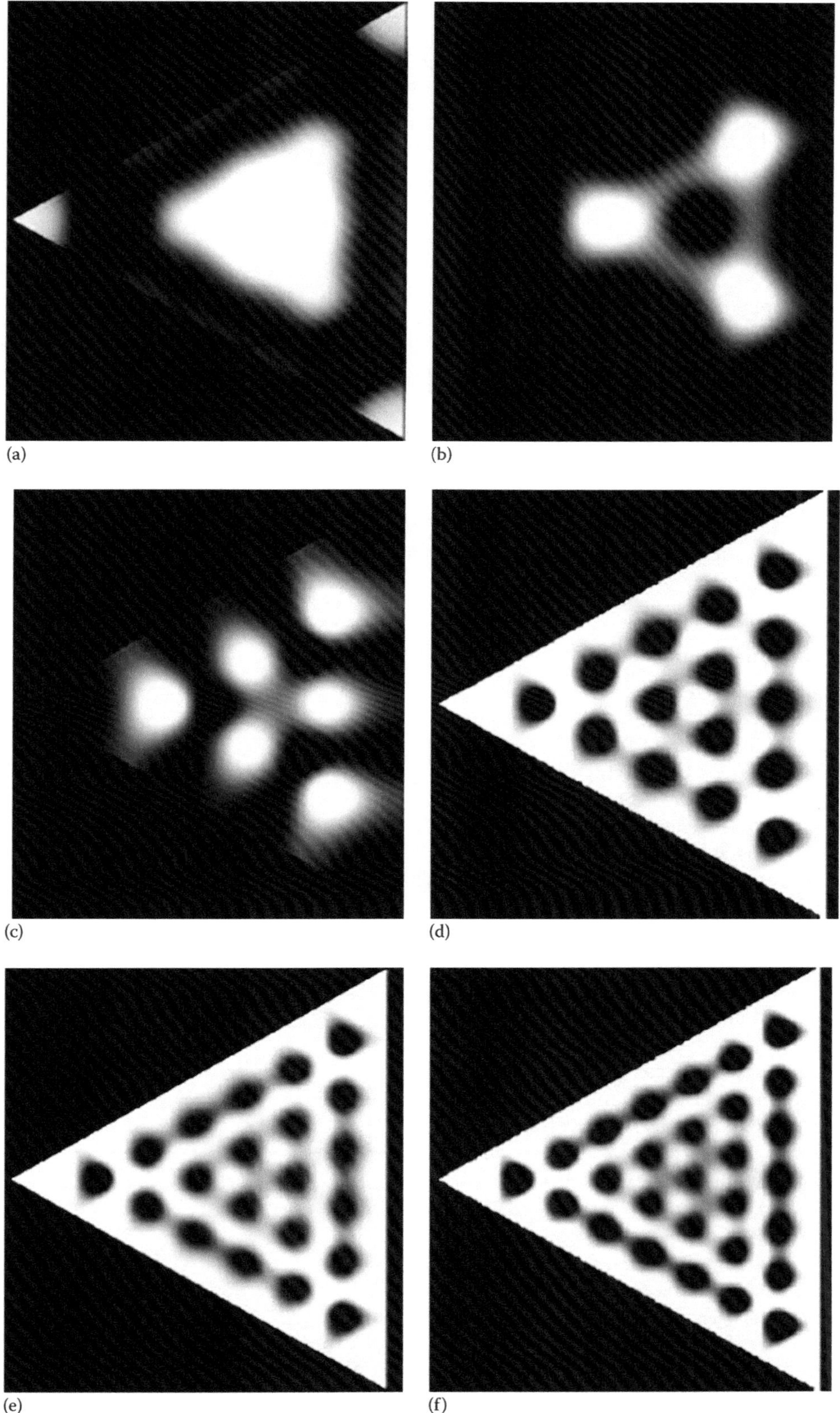

FIGURE 19.4 dI/dV images at six bias voltages. For negative bias voltages the low-intensity region of each image is cut off: (a) −55 mV, (b) −39 mV, (c) −25 mV, (d) +54 mV, (e) +89 mV, and (f) +125 mV.

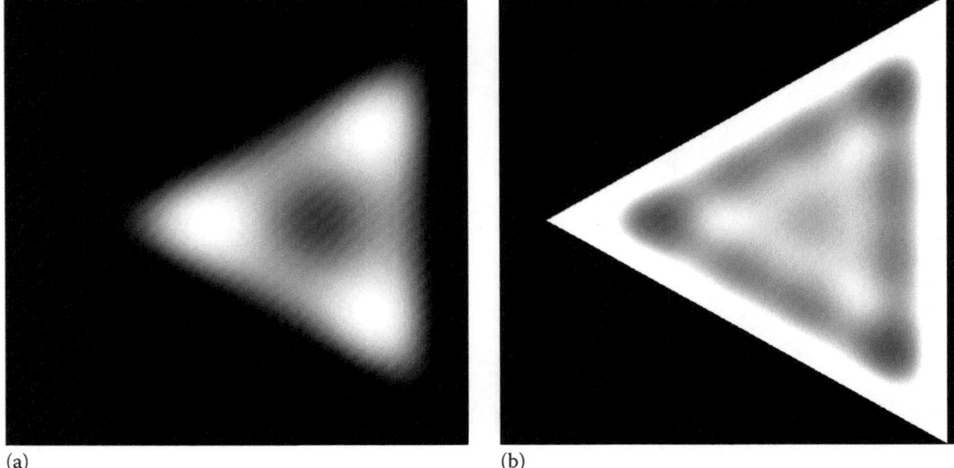

FIGURE 19.5 (a) LDOS image and (b) the image of the filtering function $F(\rho, V)$ at -39 mV.

V) calculated at −39 mV. Intensity product of these images gives rise to the dI/dV pattern. It is clear that the dI/dV image differs much from the LDOS image.

19.4.2.3 Rectangular Quantum Corral

Kliewer et al. (2001) manipulated 28 Mn atoms on an Ag(111) surface and fabricated a rectangular QC of 9 nm × 10 nm in size. They observed STM images and STS for the rectangular QC at 4.6 K and calculated them by deriving the Green function of s-waves scattered multiply from individual corral atoms. In the same manner as Heller et al. (1994) considered, they regarded corral atoms as black dots. To analyze their experimental results, Tamura (2011) considered the QC as that surrounded by four long walls with a finite width and height. It was claimed that the confined electron stays temporarily in the rectangular QC and the electron finally fades away from there. Tamura (2011) derived complex eigenenergies of the electron and called the electron state a quasi-stationary state.

As far as we treat corral atoms as components of a potential barrier, the probability density of the confined electron is low near corral sites. Hence, we cannot specify electron states at corral atoms. To do this, it is necessary to consider electron states of corral atoms and elucidate the STM current flowing at corral sites. We often discuss electron states by making corral atoms be scatterers.

Here we show a viewpoint of the quasi-stationary states (Kumagai and Tamura, 2009, Tamura, 2011). At first, we discuss electron states confined between repulsive barriers shown in Figure 19.6.

Figure 19.7 shows a cross section of the two-dimensional potential energy along the direction normal to barriers. We set the bottom of $V(x)$ at $-V_0$ and the top of two barriers at $-V_1$.

We focus on a time-dependent Schrödinger equation for a position vector $\rho = (x, y)$:

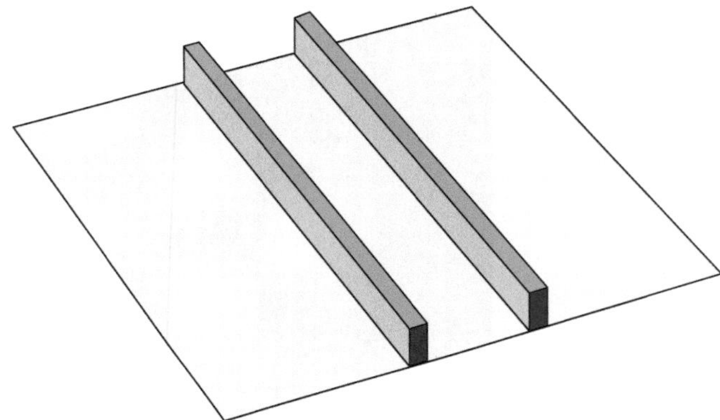

FIGURE 19.6 3D illustration of an elongated double barrier potential energy.

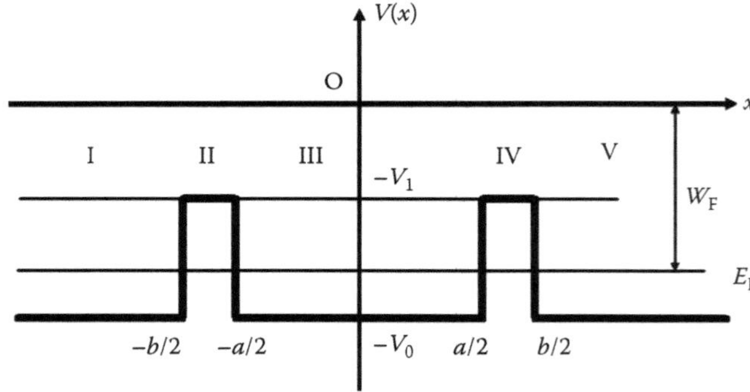

FIGURE 19.7 Cross section of the elongated potential energy.

$$i\hbar\frac{\partial}{\partial t}\Psi(\rho,t) = H\Psi(\rho,t) = \left[-\frac{\hbar^2}{2m^*}\left(\frac{\partial^2}{\partial x^2}+\frac{\partial^2}{\partial y^2}\right)+V(x)\right]\Psi(\rho,t), \tag{19.22}$$

where H represents the two-dimensional Hamiltonian of the present system. The potential energy shown in Figure 19.6

makes us use a method of separation of variables x and y. The wave function can be expressed as $\Psi(\boldsymbol{\rho}, t) = \psi(x)\varphi(y)\exp(-iEt/\hbar)$. Imposing a periodic boundary condition on $\varphi(y)$ with a period L_y, we find $\varphi_m(y) = \left(1/\sqrt{L_y}\right)\exp(ik_{ym}y)$ where $k_{ym} = 2\pi m/L_y$ with an integer m. The electron freely moves in the y-direction and has the eigenenergy $E_{ym} = 2\pi^2\hbar^2 m^2/(m^* L_y^2)$. Here we consider the case where an electron is initially supplied in the region III. Hence, we assign the wave function $\psi(x)$ in each region as $\psi_{\text{I}}(x) = A\exp(-ikx)$, $\psi_{\text{II}}(x) = B\exp(qx) + C\exp(-qx)$, $\psi_{\text{III}}(x) = D\sin kx + F\cos kx$, $\psi_{\text{IV}}(x) = G\exp(qx) + H\exp(-qx)$, and $\psi_{\text{V}}(x) = J\exp(ikx)$. In regions I and V, we assign only outgoing waves because the electron supplied in the region III tunnels through two barriers. The energy E is expressed as $E = E_{ym} + \hbar^2k^2/(2m^*) - V_0$. Hereafter we omit the term of E_{ym}, for simplicity.

To obtain eigenvalues of k and q, we impose boundary conditions that the wave function and its first derivative are continuous at four boundaries; $x = \pm a/2$ and $x = \pm b/2$. Because the number of equations coming from these boundary conditions is eight and the number of coefficients of wave functions is also eight, we obtain two kinds of eigenvalue equations. One is

$$q\cos\frac{1}{2}ka - k\sin\frac{1}{2}ka = e^{-q(b-a)}\frac{q+ik}{q-ik}\left(q\cos\frac{1}{2}ka + k\sin\frac{1}{2}ka\right) \tag{19.23}$$

for the symmetric wave function ($D = 0$) with respect to the $V(x)$ axis, and the other is

$$k\cos\frac{1}{2}ka + q\sin\frac{1}{2}ka = -e^{-q(b-a)}\frac{q+ik}{q-ik}\left(k\cos\frac{1}{2}ka - q\sin\frac{1}{2}ka\right) \tag{19.24}$$

for the anti-symmetric wave function ($F = 0$). In the limit $b \to \infty$, we reproduce eigenvalue equations for stationary states in which eigenenergies are real numbers. Because of an imaginary unit "i" appearing an the right-hand sides of Equations 19.23 and 19.24, eigenvalues of k and q become complex numbers, and so do eigenenergies. By making both k and q be $k = k^{\text{r}} + ik^{\text{i}}$ and $q = q^{\text{r}} + iq^{\text{i}}$, we can numerically solve eigenvalue Equations 19.23 and 19.24, and we obtain discrete eigenvalues k_n. An eigenfunction $\psi_n(x)$ has a complex eigenenergy expressed as $E_n = \hbar^2(k_n^{\text{r}} + ik_n^{\text{i}})^2/(2m^*) - V_0 = E_n^{\text{r}} - i\Gamma_n^{\text{QC}}/2$, where the imaginary part is $\Gamma_n^{\text{QC}} = -2\hbar^2 k_n^{\text{r}} k_n^{\text{i}}/m^*$. Because $k_n^{\text{i}} < 0$, Γ_n^{QC} is positive.

For the quasi-stationary state mentioned earlier, the Hamiltonian H is not a hermitian operator for an eigenfunction $\Psi_{nm}(\rho, t)$. This nonhermiticity comes from outgoing waves assigned in regions I and V. Using Dirac's notation, we have a nonzero integral of $\langle H\Psi_{nm}|\Psi_{nm}\rangle - \langle\Psi_{nm}|H\Psi_{nm}\rangle$:

$$\langle H\Psi_{nm}|\Psi_{nm}\rangle - \langle\Psi_{nm}|H\Psi_{nm}\rangle = \left(E_{nm}^{\text{r}} + i\frac{1}{2}\Gamma_n^{\text{QC}}\right)\langle\Psi_{nm}|\Psi_{nm}\rangle$$
$$-\left(E_{nm}^{\text{r}} - i\frac{1}{2}\Gamma_n^{\text{QC}}\right)\langle\Psi_{nm}|\Psi_{nm}\rangle = i\Gamma_n^{\text{QC}}\langle\Psi_{nm}|\Psi_{nm}\rangle \neq 0. \tag{19.25}$$

For an electron supplied in the region III, regions I and V are external systems in the x-direction. In such a case, the present system should be regarded as an open system. If H is a hermitian operator, the integral $\langle H\Psi_{nm}|\Psi_{nm}\rangle - \langle\Psi_{nm}|H\Psi_{nm}\rangle$ vanishes and $\Gamma_n^{\text{QC}} = 0$; the electron has an infinite lifetime in the well. Such a system is a closed one.

To determine the running direction of a progressive wave, we consider time and position dependence of its phase. When a bias voltage of the STM tip is positive, the tip supplies the well with an electron. When a bias voltage is negative, the STM tip extracts an electron occupying a quasi-stationary state in the well and creates a hole there. Electrons outside the corral enter the well via resonant tunneling to fill the hole formed in the well. Under such a resonant condition, the corral wall does not repel the electron coming from outside. These processes can be explained by reversing the time-evolution process applied to the case of a positive bias voltage. That is, when the applied bias voltage is negative, the hole created in the well moves away with time in the same manner as an electron moves away from the well when the bias voltage is positive. Hence, at negative bias voltages, a peak width of STS means the energy width of the hole associated with the relevant energy level. The inverse of the peak width represents lifetime of the hole created there.

Considering these situations, we normalize the eigenfunction, taking account of the time-dependence of a quasi-stationary state. By focusing only on the x-direction, we introduce a wave function defined as $\chi(x, t) = \psi(x)\exp[-i(E - E_{\text{F}})t/\hbar]$. The $\chi(x, t)$ represents the state having the energy $E - E_{\text{F}}$. In the region V, the wave function tunneling from the well to the far right can be written as

$$\chi_{\text{V}}(x,t) = J\exp\left\{i\left[\frac{k^{\text{r}}x - (E^{\text{r}} - E_{\text{F}})t}{\hbar}\right]\right\}\exp[|k^{\text{i}}|(x - \upsilon^{\text{r}}t)] \tag{19.26}$$

where $\upsilon^{\text{r}} = \hbar k^{\text{r}}/m^*$. The progressive wave $\chi_{\text{V}}(x, t)$ has the phase velocity $(E^{\text{r}} - E_{\text{F}})/(\hbar k^{\text{r}})$ inside the envelope $\exp[|k^{\text{i}}|(x - \upsilon^{\text{r}}t)]$ that moves in the $+x$ direction when $E^{\text{r}} > E_{\text{F}}$. Because $\chi_{\text{V}}(x, t)$ has a wave front at $x = \upsilon^{\text{r}}t$ and the wave outgoing from the QC cannot pull ahead of its wave front, we normalize $|\chi_{\text{V}}(x, t)|^2$ in the region from $(1/2)b$ to $\upsilon^{\text{r}}t$. If we do not impose a restriction $x < \upsilon^{\text{r}}t$, the wave function diverges at $x \to \infty$. At a finite time t we have

$$\int_{b/2}^{\upsilon^{(\text{r})}t} |\chi_{\text{V}}(x,t)|^2 dx = \frac{|J|^2}{2|k^{\text{i}}|}\left\{1 - \exp[2|k^{\text{i}}|(\tfrac{1}{2}b - \upsilon^{\text{r}}t)]\right\}. \tag{19.27}$$

Because an electron supplied in the well moves away in the limit $t \to \infty$ and the potential energy is symmetric with respect to the $V(x)$ axis, the integrated value should be 1/2, yielding $|J| = \sqrt{|k^{\text{i}}|}$. Similarly, we have $|A| = \sqrt{|k^{\text{i}}|}$ in the region I. Hence, when $t \gg b/(2\upsilon^{\text{r}})$, coefficients of the wave function are independent of time. As other normalization methods of wave functions, two kinds of δ-function, $\delta(k - k')$ and $\delta(E - E')$, have often been used. If we use these normalization methods we obtain inadequate STS whose peaks in the low energy range are considerably enhanced and

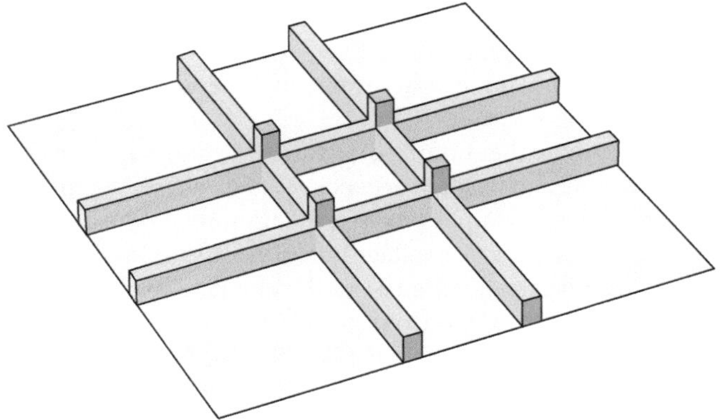

FIGURE 19.8 3D illustration of our two-dimensional potential energy for the rectangular QC.

peaks in the higher range are extremely reduced. In this manner, coefficients $|J| = \sqrt{|k^{\mathrm{i}}|}$ and $|A| = \sqrt{|k^{\mathrm{i}}|}$ are indispensable for deriving the proper STS of a QC (Kumagai and Tamura, 2009).

Defining the eigenfunction as $\Psi_{nm}(\rho) = \psi_n(x)\psi_m(y)$, we apply the method mentioned earlier to the rectangular QC shown in Figure 19.8. We consider the electron confined in the area surrounded by four walls. Using a separation of variables method, we obtain eigenenergy of the electron confined in the rectangular QC as follows:

$$E_{nm}^{\mathrm{2Dr}} = \frac{\hbar^2}{(2m^*)}(k_{nx}^{\mathrm{r}\,2} + k_{my}^{\mathrm{r}\,2} - k_{nx}^{\mathrm{i}\,2} - k_{my}^{\mathrm{i}\,2}) - V_0 \tag{19.28}$$

and

$$\Gamma_{nm}^{\mathrm{QC}} = -\left(\frac{2\hbar^2}{m^*}\right)(k_{nx}^{\mathrm{r}}k_{nx}^{\mathrm{i}} + k_{my}^{\mathrm{r}}k_{my}^{\mathrm{i}}). \tag{19.29}$$

Figure 19.9 shows $\Gamma_{nm}^{\mathrm{QC}}$ as a function of $(E_{nm}^{\mathrm{r}} + V_0)/e$, in which parameters are chosen as $-V_0 = -4.802$ eV, $-V_1 = -4.350$ eV, d = 0.274 nm (diameter of an Ag atom: Wells 1993) and $m^* = 0.42$ m. At E_{nm}^{r}, we define the energy width as the sum $(\Gamma_{nm}^{\mathrm{QC}} + \Gamma_0)$, where $\Gamma_{nm}^{\mathrm{QC}}$ is associated with leakage of the electron temporarily trapped in the well and Γ_0 is associated with the lifetime of

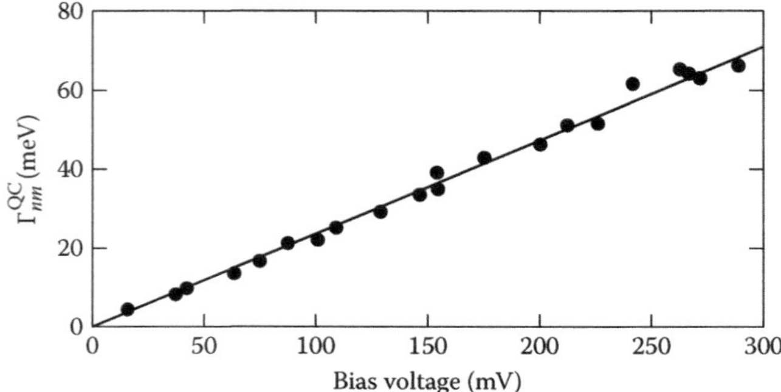

FIGURE 19.9 Energy width $\Gamma_{nm}^{\mathrm{QC}}$. A fitting curve is shown as a straight line.

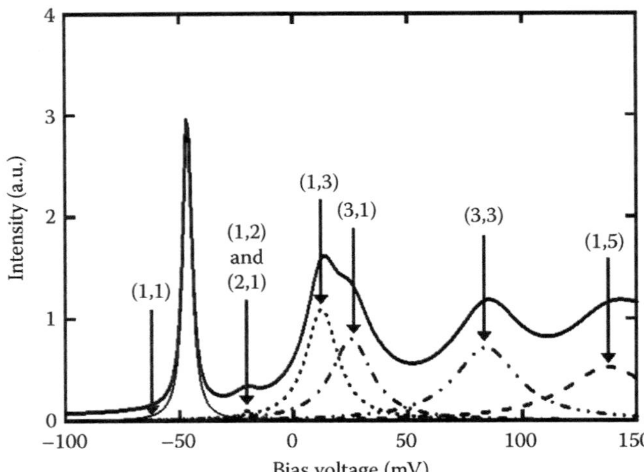

FIGURE 19.10 STS calculated at a position (0.4 nm, 0.2 nm) as a function of bias voltage.

the confined electron except tunneling through corral atoms. Kliewer et al. (2001) calculated STM images and STS using the experimental energy width measured for a circular QC. Figure 19.10 shows the STS calculated at an off-centered position. Work functions of the Ag(111) surface and the tip (tungsten) have been given as $W_{\mathrm{sp}} = 4.74$ eV for the Ag(111) surface and $W_{\mathrm{tip}} = 4.55$ eV for tungsten (Michelson, 1977). The Γ_0 is chosen at 3 mV to reproduce the highest peak at about −47 mV. The overall profile is consistent with the experimental one. Peaks labeled (n, m) show components of the STS. The smallness of a peak composed of (1, 2) and (2, 1) states comes from symmetry of the wave function because the state with an even number n or m has an anti-symmetric wave function. Other peaks with odd numbers come from symmetric wave functions resulting in large peaks.

Figure 19.11 shows topographical images calculated at eight bias voltages. Because $T(E - E_{\mathrm{F}}, V, z)$ has a weak dependence on $(E - E_{\mathrm{F}})$ and eV, we approximate it as $T(E - E_{\mathrm{F}}, V, z) = \exp(-\alpha z)$, where α amounts to 22 nm^{-1}. All images are consistent with experimental ones.

Figure 19.12 shows LDOS images calculated from quasi-stationary states mentioned earlier and they closely resemble those calculated by Kliewer et al. (2001).

Figure 19.13 shows dI/dV images obtained from Equation 19.16, and Figure 19.14 shows filtering functions $F(\rho, V)$. It should be emphasized that dI/dV images are quite different from LDOS images at relevant bias voltages because filtering functions are not uniformly distributed over the rectangle.

Because the perimeter of the rectangle is 38 nm and the number of Mn atoms is 28, the average distance between Mn atoms is about 1.4 nm. On the other hand, the wave number $k_{\parallel}$ of the confined electron is about 0.85 nm^{-1} for $n = 3$ state and its wavelength $2\pi/k_{\parallel}$ amounts to about 7.4 nm. Hence, confined electron states are insensitive to the discrete configuration of corral atoms.

Analyses made by Heller et al. (1994) and Kliewer et al. (2011) carry the advantage that electron states confined in any

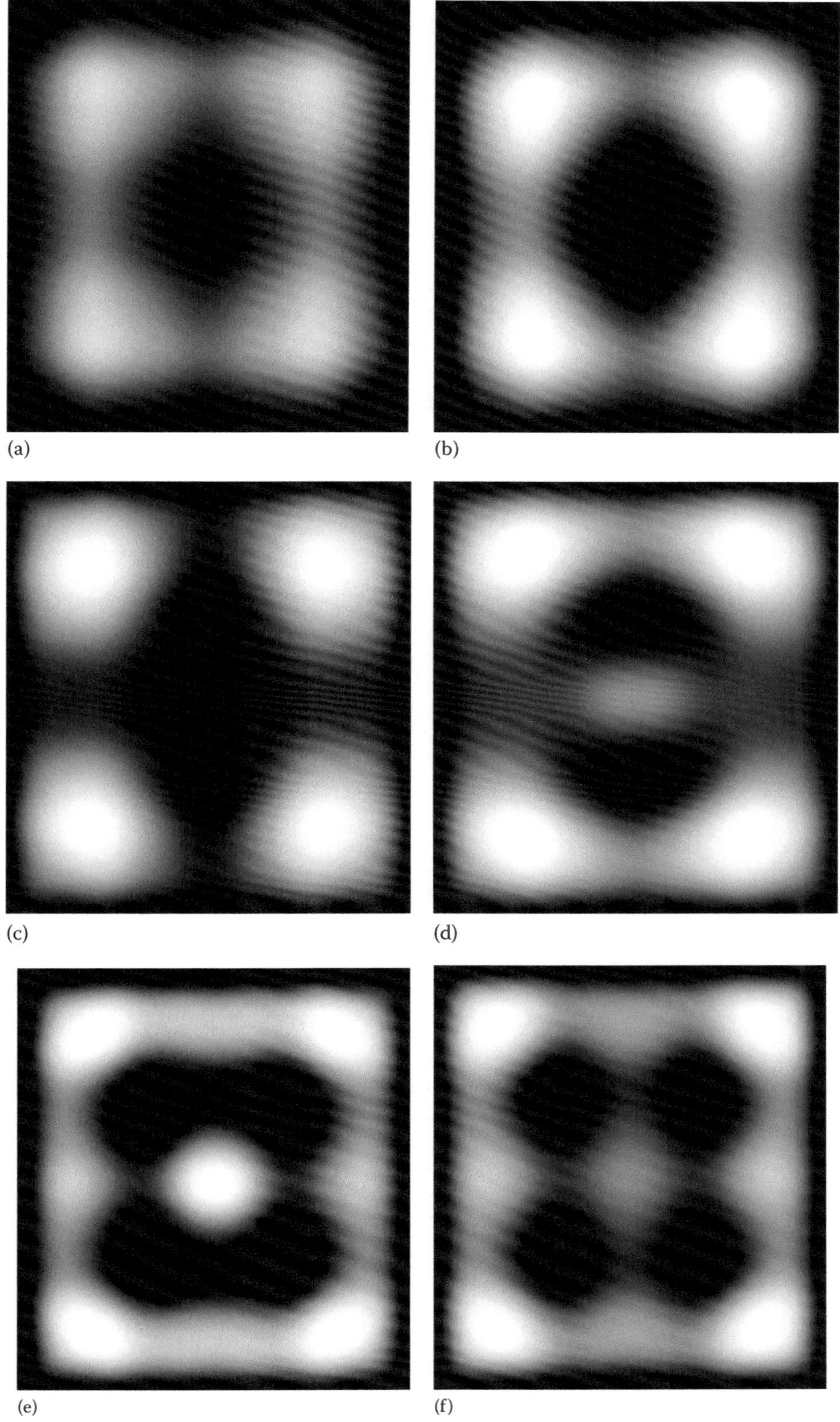

FIGURE 19.11 Topographical images obtained at eight bias voltages: (a) −50 mV, (b) −30 mV, (c) −10 mV, (d) +10 mV, (e) +30 mV, and (f) +50 mV.

(continued)

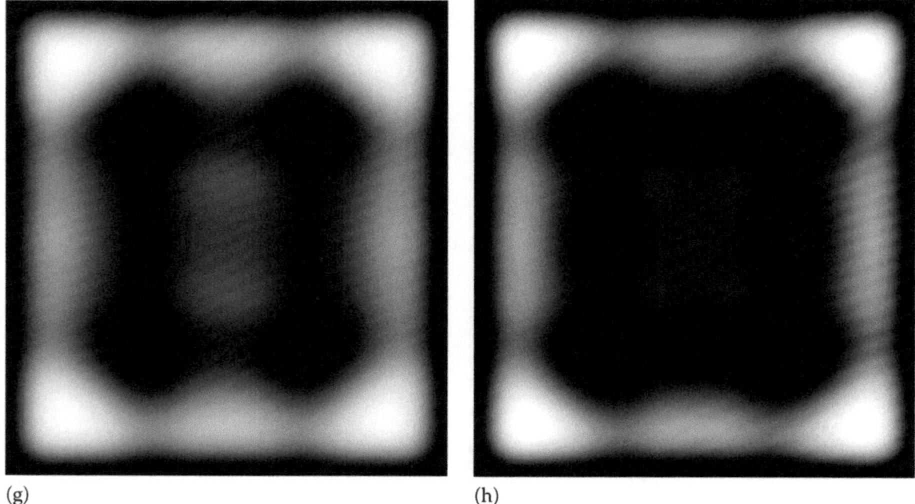

FIGURE 19.11 (continued) Topographical images obtained at eight bias voltages: (g) +80 mV and (h) +100 mV.

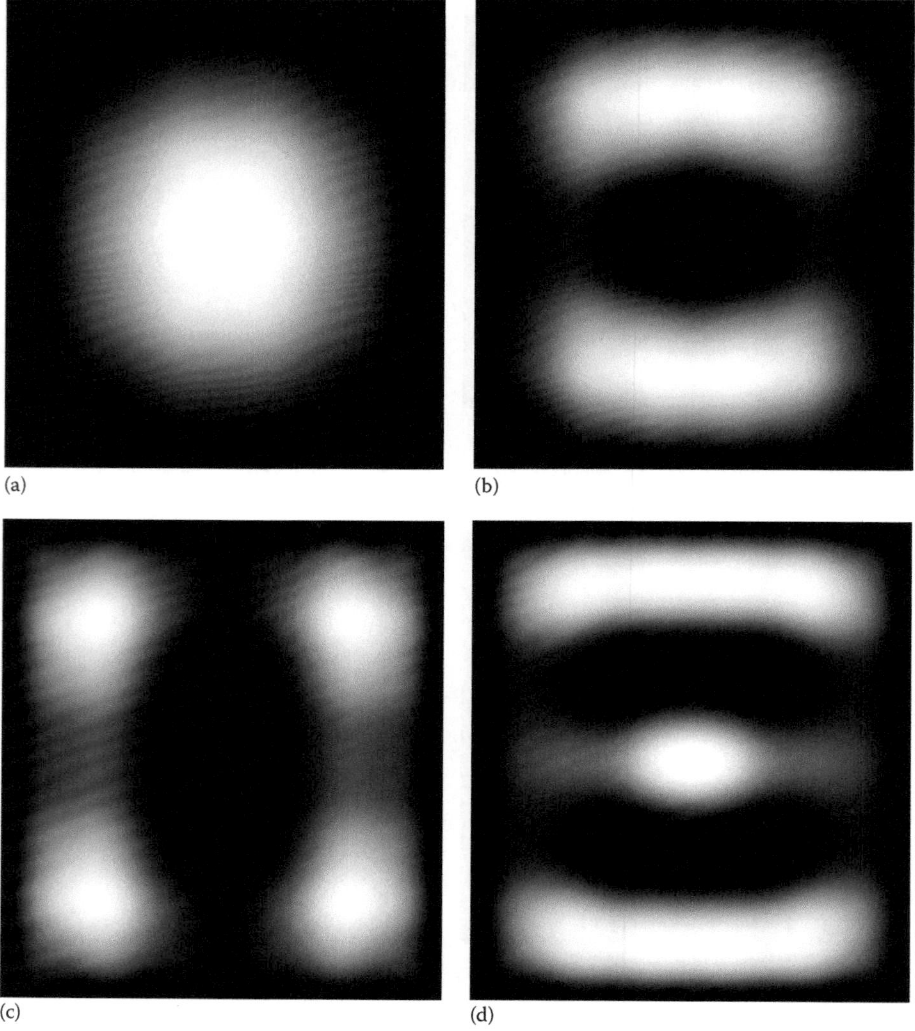

FIGURE 19.12 LDOS images obtained at eight bias voltages: (a) −50 mV, (b) −30 mV, (c) −10 mV, and (d) +10 mV.

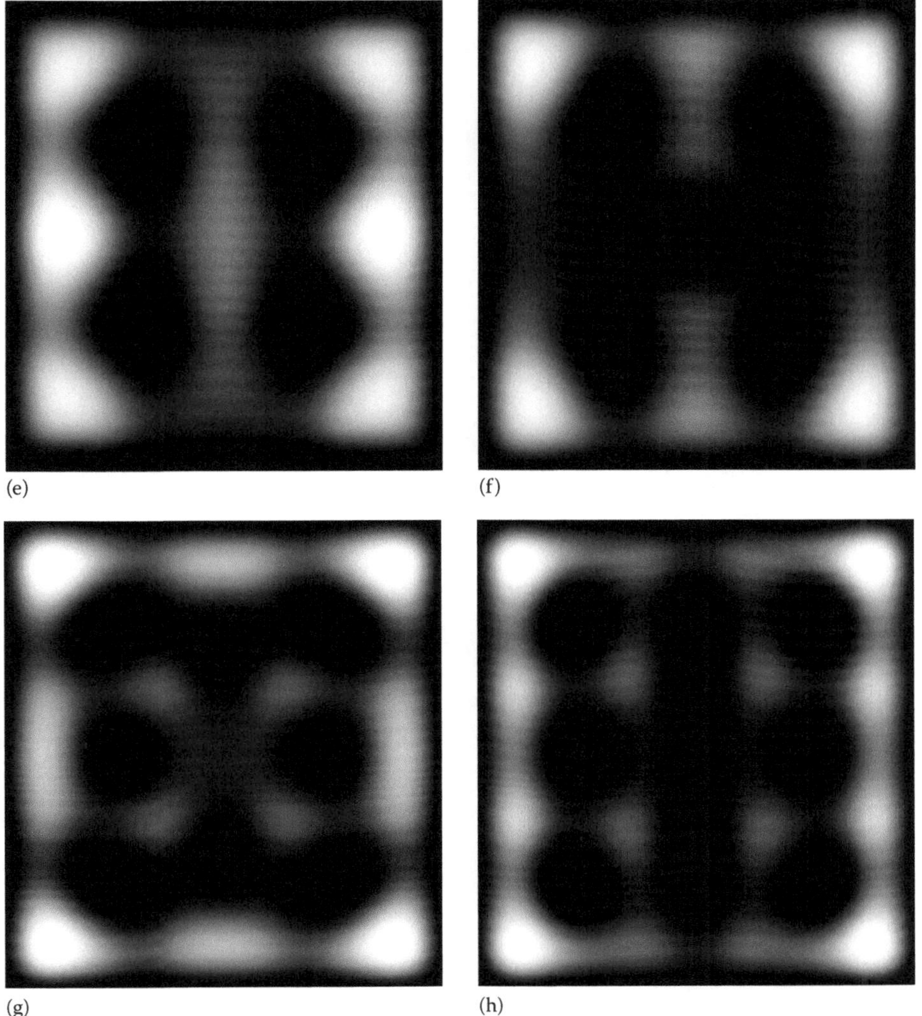
(e) (f) (g) (h)

FIGURE 19.12 (continued) LDOS images obtained at eight bias voltages: (e) +30 mV, (f) +50 mV, (g) +80 mV, and (h) +100 mV.

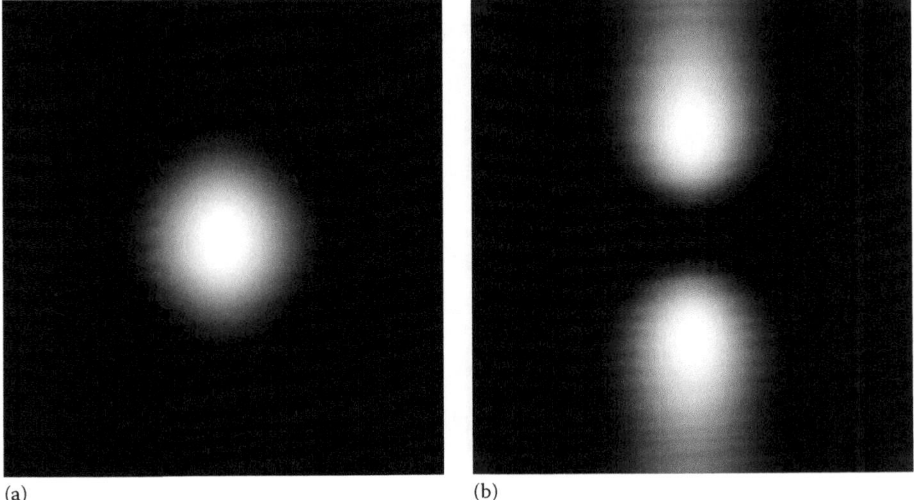
(a) (b)

FIGURE 19.13 dI/dV images obtained at eight bias voltages: (a) −50 mV and (b) −30 mV.

(continued)

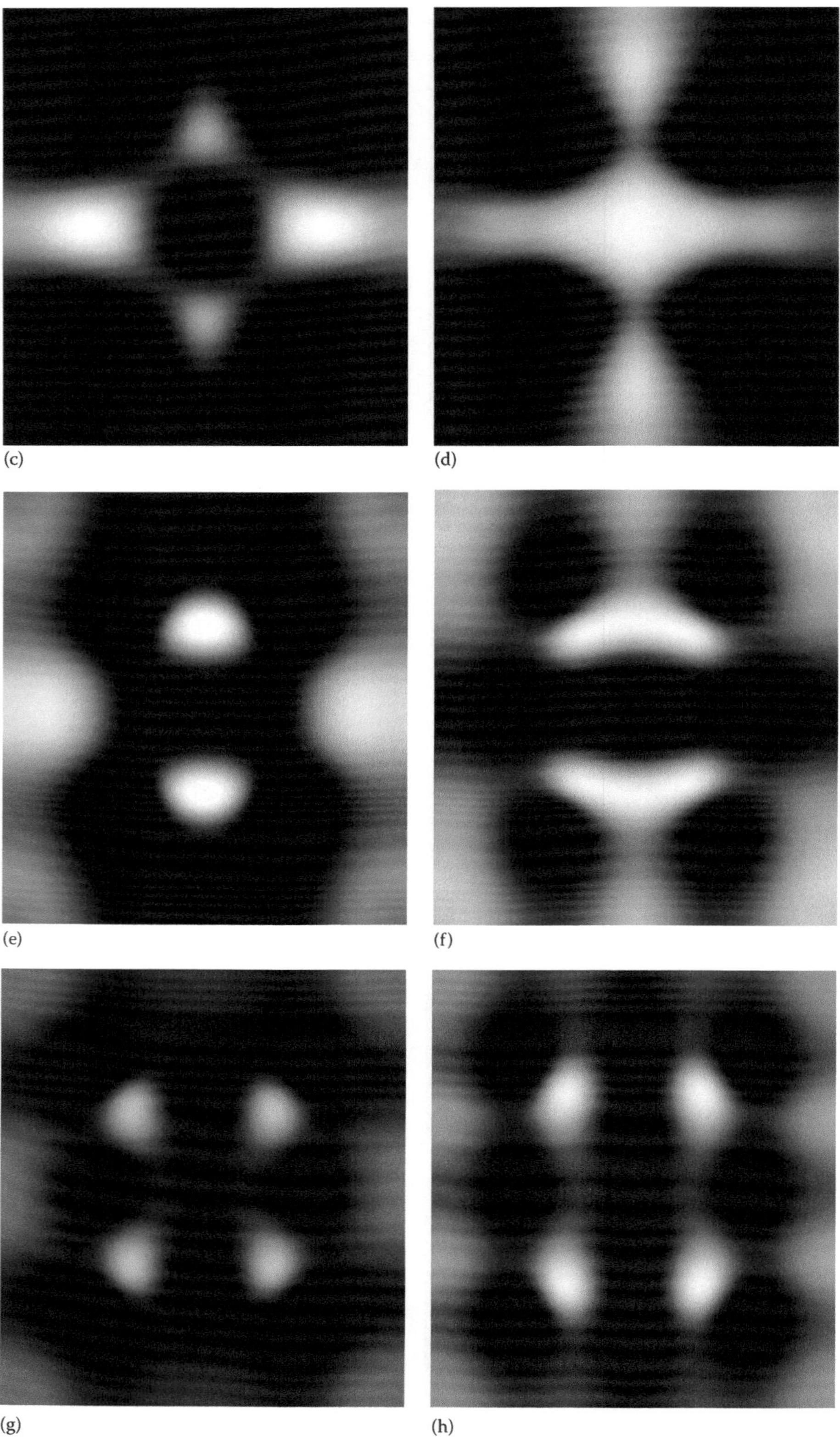

FIGURE 19.13 (continued) d*I*/d*V* images obtained at eight bias voltages: (c) −10 mV, (d) +10 mV, (e) +30 mV, (f) +50 mV, (g) +80 mV, and (h) +100 mV.

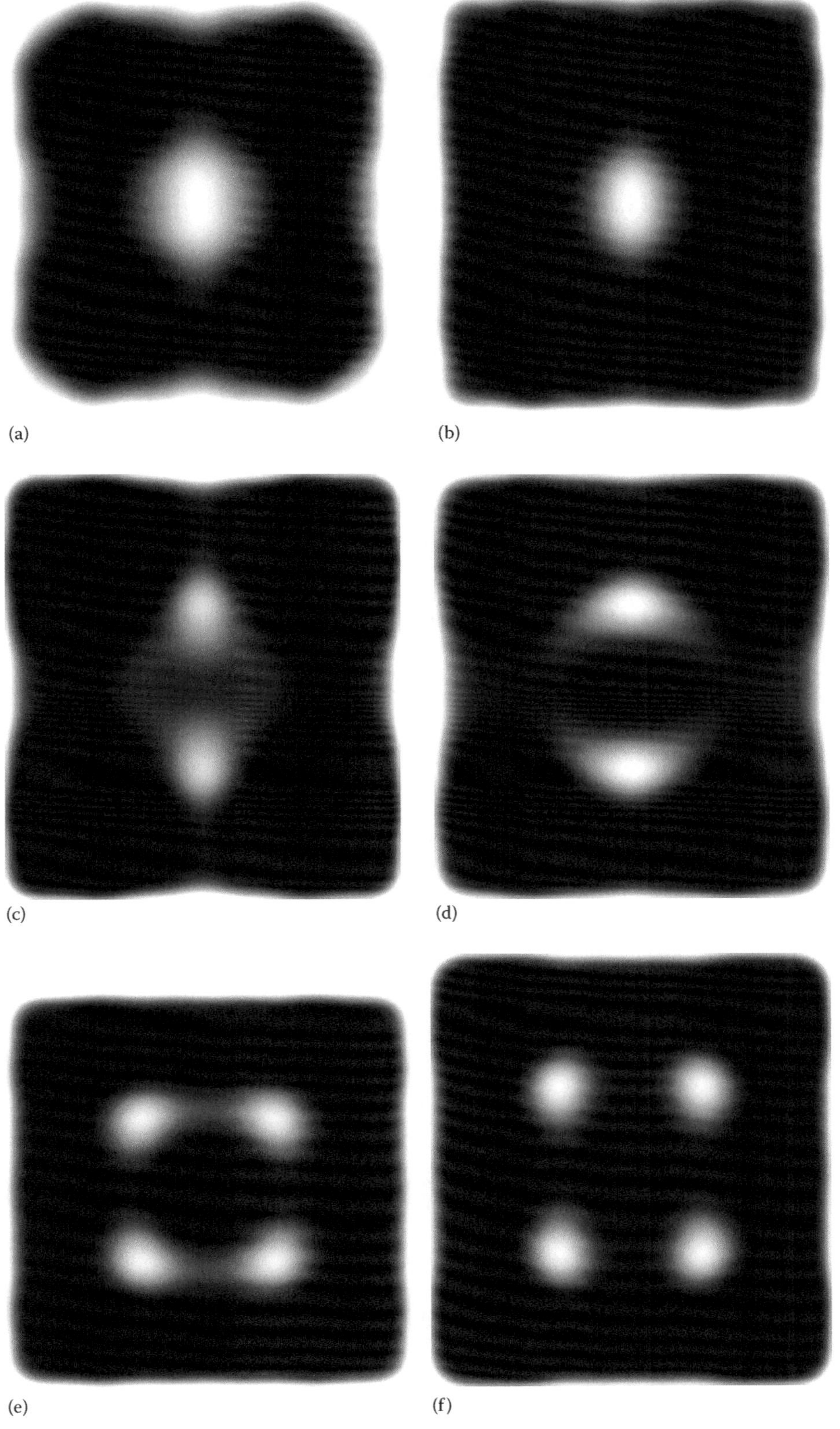

FIGURE 19.14 Images of the filtering function $F(\rho, V)$ obtained at eight bias voltages: (a) −50 mV, (b) −30 mV, (c) −10 mV, (d) +10 mV, (e) +30 mV and (f) +50 mV.

(continued)

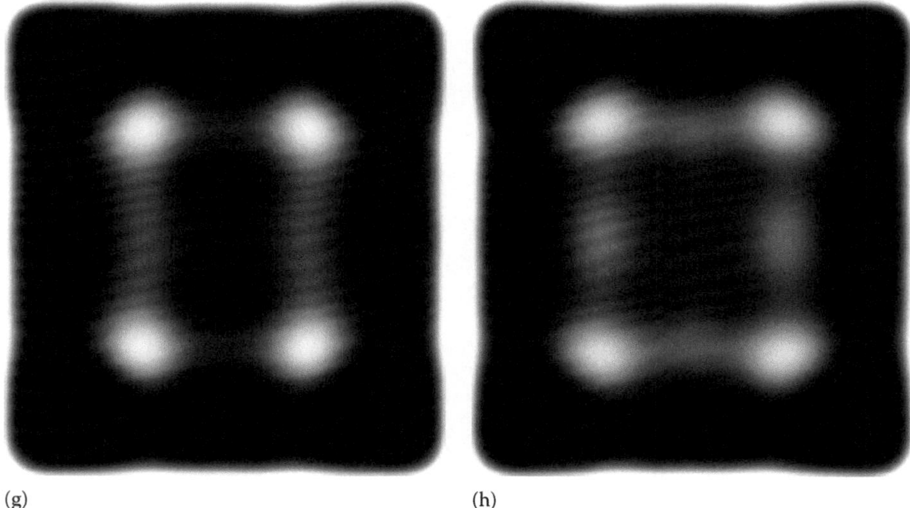

FIGURE 19.14 (continued) Images of the filtering function $F(\rho, V)$ obtained at eight bias voltages: (g) +80 mV and (h) +100 mV.

geometric shapes of QC can be obtained. For the QC having a remarkable number of corral atoms, however, such as long arrayed QCs for instance, their analyses depend on the capability of numerical calculations. In addition, symmetries of the wave function that reflects symmetry of the QC cannot be obtained, while the analysis given by Tamura (2011) makes it feasible to pick out components that construct STM images and the STS from wave functions having the symmetry of the QC.

To specify decaying features of a Shockley electron, inelastic scatterings induced by electron–electron and electron–phonon interactions have been discussed for clean surfaces of noble metals (Echenique et al., 2004). It should be emphasized that an electron confined in a QC has a set of discrete energy levels and wave numbers which restrict those inelastic scatterings inside the QC. Hence, those inelastic scatterings do not affect the energy width of the confined electron much. In addition, phonon contributions are low at a liquid helium temperature. We may consider that the energy width Γ_0 contains such contributions. For the rectangular QC, Kumagai and Tamura (2008) and Tamura (2011) used the value Γ_0 = 3 meV which is smaller than 6 meV estimated for the clean Ag(111) surface (Echenique, 2004).

19.4.2.4 Quantum Dot Similar to a QC

It is well known that the MOS (metal/oxide/semiconductor) structure possesses the two-dimensional electron gas (2DEG) in the direction parallel to the boundary between the oxide and the semiconductor (Ando et al., 1982). The electron has a wave function decaying on both sides of the boundary. This feature is similar to a Shockley electron on noble metal surfaces. Kanisawa et al. (2001a,b) fabricated a triangular QD of InAs and observed STM images for the In enriched InAs (111) surface. Differently from a Shockley electron on a metal surface, there exist multiple sub-bands induced by an attractive potential just inside the semiconductor. This situation is more complex than that for the QCs on noble metal surfaces. On the InAs (111) surface, two sub-bands form and the electron state in the lower sub-band is strongly localized near the surface and that in the upper one has a long tail toward the bulk. Considering these sub-bands, Kumagai and Tamura (2008b) calculated STM images on the basis of the STM current given by Equation 19.13 and obtained STM images consistent with experimental ones.

19.4.3 Coupled Quantum Corrals

A number of studies have been carried out on a single quantum corral. Here we focus on coupled quantum corrals: a doubly coupled QC and a triply coupled QC (Mitsuoka and Tamura, 2011b). The QC system can be regarded as a giant molecule composed of single QCs. Both coupled QCs have features similar to a molecule having bonding and anti-bonding states.

First we consider a potential problem of double barriers composed of delta-functions. Figure 19.15 shows a schematic diagram of the potential $V_\delta(x) = V_a d[\delta(x + L/2) + \delta(x - L/2)] - V_0$. To derive eigenenergies, we assign wave functions in three regions as follows: $\psi_{\mathrm{I}} = A\exp(-ikx)$, $\psi_{\mathrm{II}} = B\sin kx + C\cos kx$, and $\psi_{\mathrm{III}} = D\exp(ikx)$, where $k = [2m^*(E + V_0)/\hbar^2]^{1/2}$. As discussed in previous sections, we treat the case that the STM tip supplies the well with an electron in the region II.

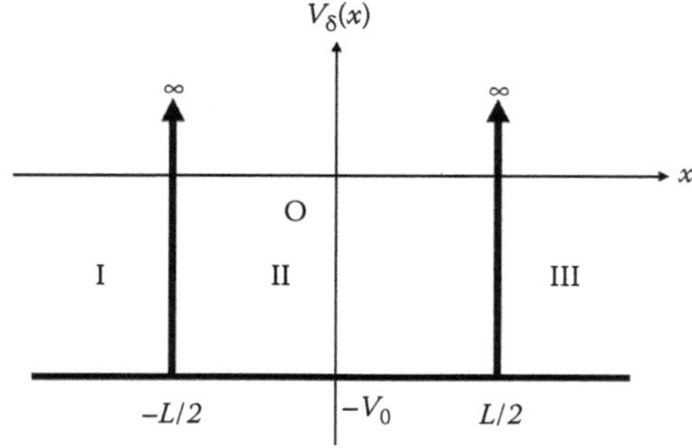

FIGURE 19.15 One-dimensional potential having two barriers of δ-function type.

A boundary condition that the wave function is continuous at $x = \pm(1/2)L$ leads to relations $\exp(-ikL/2)\,A = -\sin(kL/2)\,B + \cos(kL/2)\,C$ and $\sin(kL/2)\,B + \cos(kL/2)\,C = \exp(ikL/2)\,D$. In addition to these, we obtain other relations, $\exp(-ikL/2)A = -\sin(kL/2)B + \cos(kL/2)C$ and $\sin(kL/2)\,B + \cos(kL/2)C = \exp(-ikL/2)D$, by integrating the Schrödinger equation in two regions $-L/2 - \varepsilon < x < -L/2 + \varepsilon$ and $L/2 - \varepsilon < x < L/2 + \varepsilon$ and subsequently making a limit $\varepsilon \to 0$. Thus we find two kinds of eigenvalue equation. One is

$$k\sin\left(\frac{kL}{2}\right) = (\alpha - ik)\cos\left(\frac{kL}{2}\right) \tag{19.30}$$

resulting in symmetric wave functions, and the other is

$$k\cos\left(\frac{kL}{2}\right) = -(\alpha - ik)\sin\left(\frac{kL}{2}\right) \tag{19.31}$$

resulting in anti-symmetric wave functions, in which $\alpha = 2m^*V_d d/\hbar^2$. These eigenvalue equations are simpler than those of Equations 19.23 and 19.24. The imaginary unit 'i' included in Equations 19.30 and 19.31 makes k a complex number; $k = k^r + ik^i$. The eigenvalue of kL can be determined by one parameter αL. By numerically solving Equations 19.30 and 19.31 for k^r and k^i, we obtain a complex eigenenergy as $E_n = E_n^r - i\Gamma_n/2$, where $E_n^r = [\hbar^2/(2m^*)](k_n^{r2} - k_n^{i2}) - V_0$ and $\Gamma_n = -(2\hbar^2/m^*)\,k_n^r k_n^i$. The Γ_n is the energy width associated with E_n^r, and is positive since $k^r > 0$ and $k^i < 0$. We can determine $\psi_{II}(x)$ through $|A| = \sqrt{|k^i|}$ or $|D| = \sqrt{|k^i|}$ in the same manner as $\psi_{III}(x)$ in Section 19.4.2.3. Figure 19.16 shows Γ_n as a function of E_n^r where parameters for Ag(111) surface are chosen as $-V_0 = -4.802$ eV, $V_d = 0.452$ eV, $d = 0.274$ nm: the diameter of a Mn atom, and $m^* = 0.42m_e$. In the low-energy range, the sequence of Γ_n can be approximated as $\Gamma_n = 0.24(E_n^r + V_0)$. There exists a kind of universality in E_n^r dependence of Γ_n as far as the area $V_d d$ remains the same. Crosses show Γ_n calculated for double barriers of δ-function. Obviously the sequence of Γ_n calculated for a square barrier approaches that for a δ-function barrier at low energy levels. This feature enables us to use a δ-function potential barrier instead of a square potential barrier.

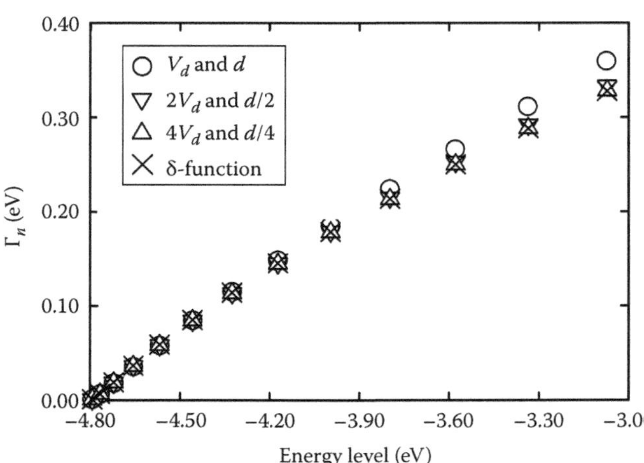

FIGURE 19.16 Γ_n for the one-dimensional potential barrier against E_n^r.

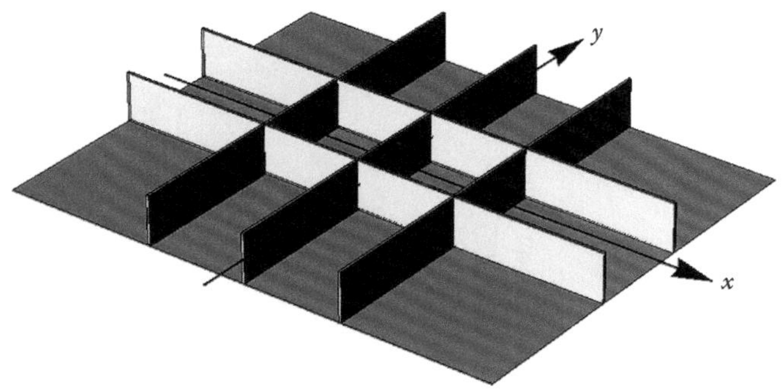

FIGURE 19.17 3D image of the potential of the coupled QCs.

Next we consider electron states of the two QCs coupled with each other. We presume that an electron is confined in the potential defined as

$$V_{\delta 2}(\rho) = V_d d[\delta(x + a_x) + \delta(x) + \delta(x - a_x)] + V_d d[\delta(y + \tfrac{1}{2}a_y) + \delta(y - \tfrac{1}{2}a_y)] - V_0. \tag{19.32}$$

Figure 19.17 shows a schematic 3D image of the potential, and Figure 19.18 shows the cross section in the x-direction. a_x denotes the width of individual QCs. Long bars do not affect electron states inside the QC because wave functions near long bars have negligible amplitudes (Tamura, 2011, Mitsuoka and Tamura, 2011b).

By the separation of variables method, we obtain eigenvalue equations in the x-direction as

$$[\alpha(\alpha - ik) - 2k^2]\sin ka_x = -k[\alpha + 2(\alpha - ik)]\cos ka_x \tag{19.33}$$

for the symmetric wave function and

$$k\cos ka_x = -(\alpha - ik)\sin ka_x \tag{19.34}$$

for the anti-symmetric wave function. Equation 19.34 is the same as Equation 19.31 except that $(1/2)L$ is replaced by a_x

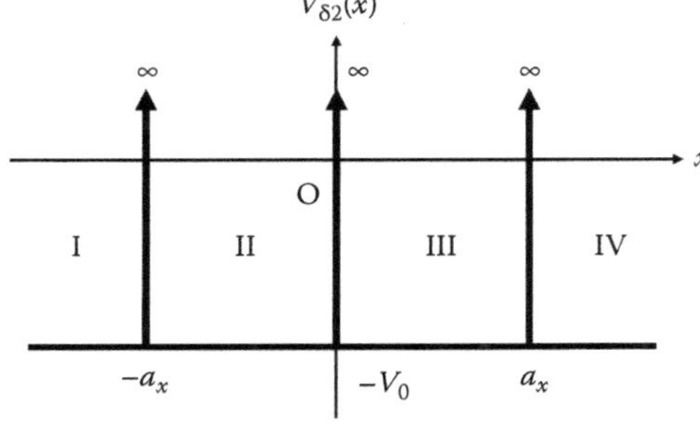

FIGURE 19.18 Cross section of the δ-function potential.

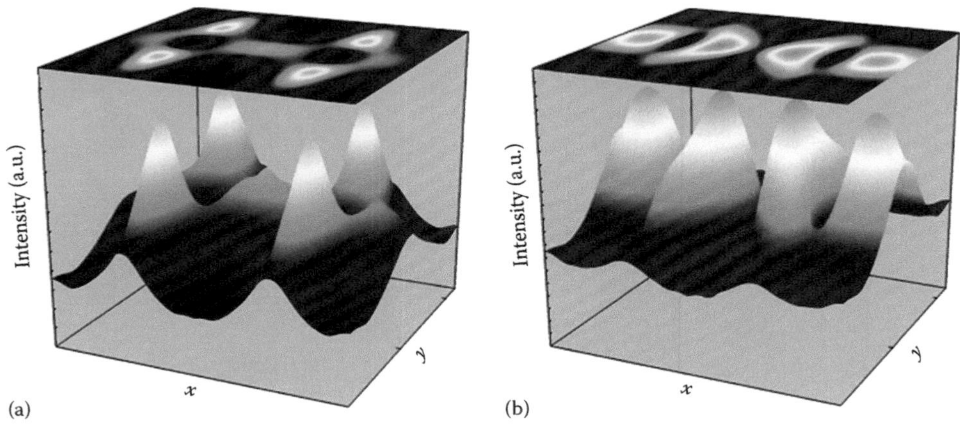

FIGURE 19.19 dI/dV images for the doubly coupled QC whose side length in the x-direction is (9 nm + 9 nm) and 10 nm in the y-direction: (a) −20 mV and (b) −15 mV.

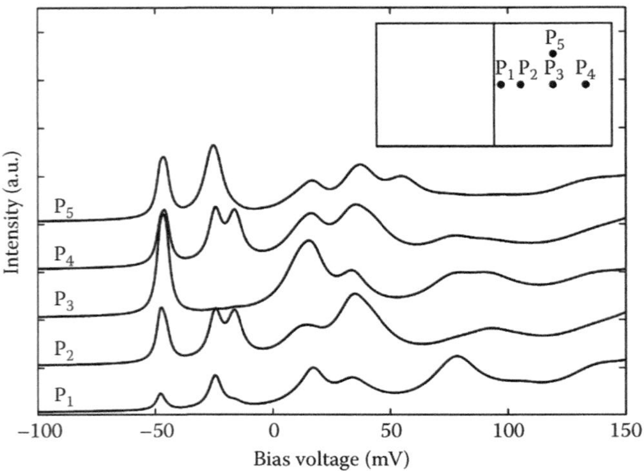

FIGURE 19.20 STS for the coupled QC (9 nm + 9 nm) × 10 nm.

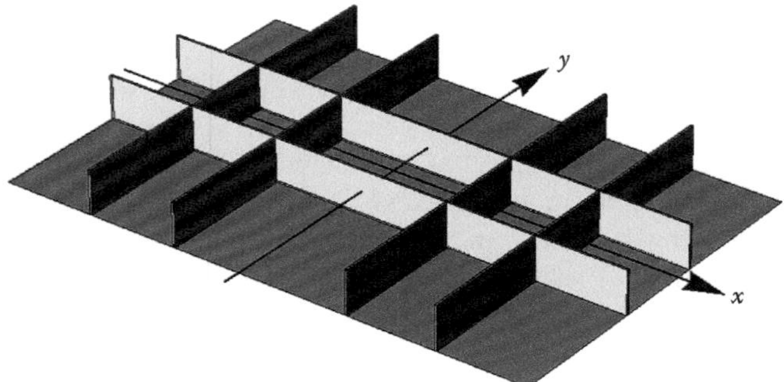

FIGURE 19.21 3D image of the potential of a triply coupled QC.

because the anti-symmetric wave function of the doubly coupled QC has a node at $x = 0$ nm.

Applying those eigenvalue equations to an electron confined in the two-dimensional potential shown in Figure 19.17, we obtained dI/dV images shown in Figure 19.19: density plots and three-dimensional profiles. Figures 19.19a and b correspond to a bonding state and an anti-bonding state, respectively. Reflecting the anti-bonding state, the dI/dV image of Figure 19.19b shows zero intensity at the boundary between two QCs.

Figure 19.20 shows STS for the doubly coupled QC. Among five positions, the profile at P_3 is the closest to that of the single QC shown in Figure 19.10. At P_2 and P_4, (2, 1)-state and (1, 2)-state contribute and yield double peaks near −25 mV. At P_5 a peak of the (1, 2) state is higher than that of the (2, 1) state because the (2, 1) state has a node there in the x-direction.

For a triply coupled QC shown in Figure 19.21, Mitsuoka and Tamura (2011b) obtained dI/dV images. Figure 19.22 shows dI/dV images calculated at −30 mV for the triply coupled QC whose width of the middle QC is 18 nm and that of QCs at both ends is 9 nm. The width in the y-direction is 10 nm. At −30 mV,

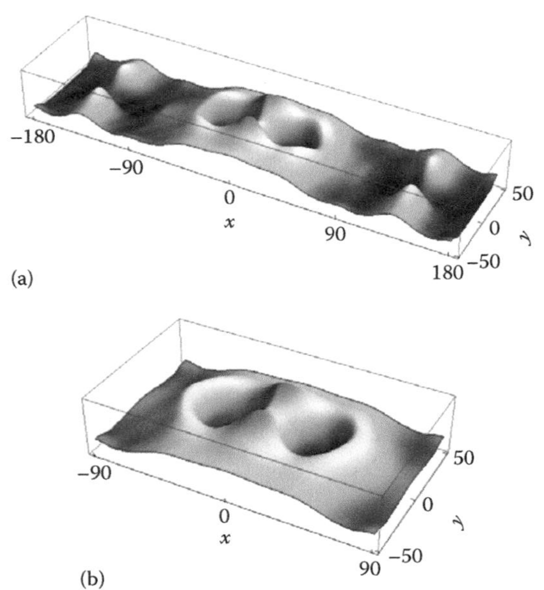

FIGURE 19.22 dI/dV image at −30 mV for (a) the triply coupled QC where the width of the middle QC is 18 nm and (b) the single QC with 18 nm in width. Numerals labeled near axes are in units of 0.1 nm.

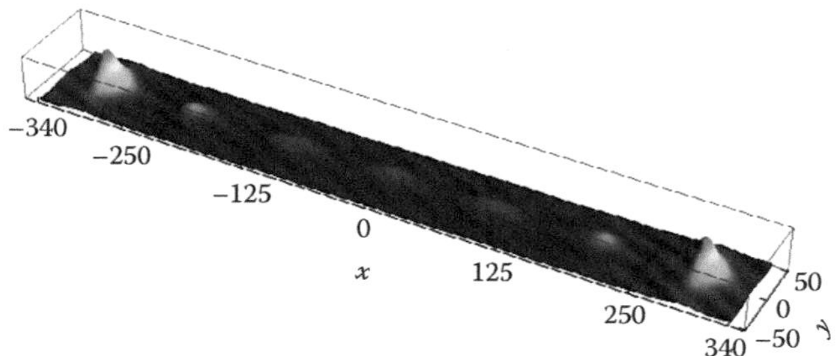

FIGURE 19.23 dI/dV image at −47 mV for the triply coupled QC where the width of the middle QC is 50 nm. Numerals labeled near axes are in units of 0.1 nm.

each image in three areas is almost the same as that of an isolated QC. At different bias voltages, dI/dV images change from those of isolated QCs because of coupling between them. Figure 19.23 shows the dI/dV image calculated at −47 mV for the triply coupled QCs the width of the middle QC being 50 nm. It is clear that resonance occurs between two QCs at both ends and peak heights in between are considerably lowered.

19.5 Future of Quantum Corrals

Fabrication methods of nanomaterials can be classified into two: the top-down method and the bottom-up method. The top-down method is based on cutting the bulk material into small pieces and/or subtracting atoms or molecules from the bulk, while the bottom-up method is based on adding them or self-assembling them. Theoretical analyses of electron states confined in a QC can be also classified into two approaches: the bottom-up approach and the top-down approach. Analyses given by Heller et al. (1994) and Kliewer et al. (2001) are of a bottom-up-type since they considered the process in which an incident electron wave is scattered by each corral atom and derived LDOS by Lippmann–Schwinger formalism for the Green's function. The tight-binding approach has been used for analyzing electron states of one-dimensional QC (Fölsch et al., 2004) and triangular QC (Lagoute et al., 2005). In contrast to a nearly free electron method, the tight-binding method is based on a bottom-up viewpoint because the whole wave function is taken as a linear combination of atomic orbitals. For the QC built up with transition metal atoms, the LDOS and spin configurations within the QC have been obtained by setting the Newns-Anderson Hamiltonian (Fiete and Heller, 2003). A more elaborate ab initio method including Coulomb interactions and spin correlations has been made assuming some kinds of density functional form (Stepanyuk et al., 2005, 2007). To specify the stability of a QC, it is indispensable to carry out such an ab initio calculation. With a top-down approach it is impossible to clarify the stability of the QC system.

Many researchers have considered top-down approaches to study electron states confined in a QC. Nilius et al. (2002, 2005) regarded the finite 1D atom-chain built up with Au and Pt atoms as that having 1D electron gas bounded by atoms at both ends. They explained dI/dV spectra by eigenenergies and eigenfunctions calculated by stationary electron states in the 1D box. Analyses given by Crommie et al. (1993), Kumagai and Tamura (2008a,b), and Tamura (2011) for example are based on the top-down approach; continuous wave numbers and energy dispersion of a Shockley electron become discrete ones because of the confinement by corral walls. In those analyses we obtain wave functions having the symmetry of the QC which cannot be obtained by summation of all electron waves scattered by individual corral atoms. In this manner, both bottom-up and top-down approaches are requisite for specifying electron states of QCs because of the respective merits and demerits.

Manoharan et al. (2000) fabricated an elliptic QC with a Co atom adsorbed at one focus. They observed its STM image and pointed out that the Co atom induces a signal at another focus and they call it a mirage. This comes from a situation where the distance between two foci via one point on the ellipse is the same irrespective of the point on the ellipse. Waves created from the Co atom adsorbed at one focus accumulate at another focus. They proposed that this system can be used as a remote control device for signal transmission in conjunction with the Kondo effect (Fiete and Heller 2003) that reflects interaction between the localized spin moment of the adsorbed atom and Shockley electrons.

In Section 19.4.3, we showed STM images for a triply combined QC. It is possible to transmit the electron confined in the QC at one end to the other end. By making an $N \times M$ lattice, the electron resonance device can be extended to a two-dimensional system. Such coupled QCs forming multiple lattices can be a candidate for the two-dimensional electron device by changing the size, the number of lattices, and band structures. Controlling technology of the magnetism of the arrayed magnetic atoms is likely to give birth to a data storage device (Loth et al., 2012). Various and new aspects of QC systems will be developed more in the future.

References

Ando, T., Fowler, A. B., and Stern, F. 1982. Electronic properties of two-dimensional systems. *Reviews of Modern Physics* **54**: 437–672.

Ashcroft, N. W. and Mermin, N. D. 1976. *Solid State Physics*, Holt Reinhart, Philadelphia, PA. p. 369.

Bardeen, J. 1961. Tunnelling from a many-particle point of view. *Physical Review Letters* **6**: 57–59.

Bartels, L., Meyer, G., and Rieder, K. H. 1997. Basic steps of lateral manipulation of single atoms and diatomic clusters with a scanning tunneling microscope tip. *Physical Review Letters* **79**: 697–700.

Bartels, L., Meyer, G., Rieder, K. H., Velic, D., Knoesel, E., Hotzel, A., Wolf, M., and Ertl, G. 1998. Dynamics of electron-induced manipulation of individual CO molecules on Cu(111). *Physical Review Letters* **80**: 2004–2007.

Besenbacher, F. 1996. Scanning tunneling microscopy studies of metal surfaces. *Report on Progress in Physics* **59**: 1737–1802.

Binnig, G., Frank, K. H., Fuchs, H., García, N., Reichl, B., Rohrer, H., Salvan, F., and William, A. R. 1985. Tunneling spectroscopy and inverse photoemission: Image and field states. *Physical Review Letters* **55**: 991–994.

Binnig, G. and Rohrer, H. 1982. Scanning tunneling microscopy. *Helvetica Physica Acta* **55**: 726–735.

Binnig, G. and Rohrer, H. 1987. Scanning tunneling microscopy—From birth to adolescence. *Reviews of Modern Physics* **59**: 615–625.

Binnig, G., Rohrer, H., Gerber, Ch., and Weibel, E. 1982a. Tunneling through a controllable vacuum gap. *Applied Physics Letters* **40**(2): 178–180.

Binnig, G., Rohrer, H., Gerber, Ch., and Weibel, E. 1982b. Surface studies by scanning tunneling microscopy. *Physical Review Letters* **49**: 57–61.

Binnig, G., Rohrer, H., Gerber, Ch., and Weibel, E. 1983. 7 × 7 reconstruction on Si(111) resolved in real space. *Physical Review Letters* **50**: 120–123.

Binnig, G., Quate, F., and Gerber, Ch. 1986. Atomic force microscope. *Physical Review Letters* **56**: 930–933.

Bode, M. 2003. Spin-polarized scanning tunneling microscopy. *Report on Progress in Physics* **66**: 523–582.

Braun, K. F. and Rieder, K. H. 2002. Engineering electronic lifetimes in artificial atomic structures. *Physical Review Letters* **88**: 096801-1–096801-4.

Bürgi, L., Jeandupeux, O., Hirstein, A., Brune, H., and Kern, K. 1998. Confinement of surface state electrons in Fabry–Pérot resonators. *Physical Review Letters* **81**: 5370–5373.

Chen, C. J. 2008. *Introduction to Scanning Tunneling Microscopy*, 2nd edn., Oxford University Press, New York.

Crommie, M. F., Lutz, C. P., and Eigler, D. M. 1993. Confinement of electrons to quantum corrals on a metal surface. *Science* **262**: 218–220.

Crommie, M. F., Lutz, C. P., Eigler, D. M., and Heller, E. J. 1995. Quantum corrals. *Physica* D **83**: 98–108.

Desjonquères, M. C. and Spanjaard, D. 1996. *Concepts in Surface Physics*, 2nd edn, Springer, Berlin, Germany. p. 197.

Donner, B., Kleber, M., Bracher, C., and Kreuzer, H. J. 2005. A simple method for simulating tunneling images. *American Journal of Physics* **73**: 690–700.

Echenique, P. M., Berndt, R., Chulkov, E. V., Fauster, T., Goldmann, A., and Höfer, U. 2004. Decay of electronic excitations at metal surfaces. *Surface Science Reports* **52**: 219–317.

Eigler, D. M. and Schweizer, E. K. 1990. Positioning single atoms with a scanning tunnelling microscope. *Nature* **344**: 524–526.

Fiete, G. A. and Heller, E. J. 2003. Theory of quantum corrals and quantum mirages. *Reviews of Modern Physics* **75**: 933–948.

Fiete, G. A., Hersch, J. S., Heller, E. J., Manoharan, H. C., Lutz, C. P., and Eigler, D. M. 2001. Scattering theory of Kondo mirages and observation of single Kondo atom phase shift. *Physical Review Letters* **86**: 2392–2395.

Fisher, Ø., Kugler, M., Maggio-Aprile, I., and Berthod, C. 2007. Scanning tunneling spectroscopy of high-temperature superconductors. *Reviews of Modern Physics* **79**: 353–419.

Fölsch, S., Hyldgaard, P., Koch, R., and Ploog, K. H. 2004. Quantum confinement in monoatomic Cu chains on Cu(111). *Physical Review Letters* **92**: 056803-1–056803-4.

García, R. and Pérez, R. 2003. Dynamic atom force microscopy methods. *Surface Science Report* **47**: 197–301.

Giessibl, F. J. 2003. Advances in atomic force microscopy. *Reviews of Modern Physics* **75**: 949–983.

Goodwin, E. T. 1939. Electronic states at the surfaces of crystals I. The approximation of nearly free electrons. *Proceeding of Cambridge Philosophical Society* **35**: 205–220.

Harbury, H. K. and Porod, W. 1996. Elastic scattering theory for electronic waves in quantum corrals. *Physical Review* B**53**: 15455–15458.

Heller, E. J., Crommie, M. F., Lutz, C. P., and Eigler, D. M. 1994. Scattering and absorption of surface electron waves in quantum corrals. *Nature* **369**: 464–466.

Hess, H. F., Robinson, R. B., Dynes, R. C., Valles Jr, J. M., and Waszczak, J. V. 1989. Scanning tunneling microscope observation of the Abrikosov flux lattice and the density of states near and inside a fluxoid. *Physical Review Letters* **62**: 214–216.

Hess, H. F., Robinson, R. B., and Waszczak, J. V. 1990. Vortex-core structure observed with a scanning tunneling microscope. *Physical Review Letters* **64**: 2711–2714.

Hla, S.-W. 2005. Scanning tunneling microscopy single atom/molecule manipulation and its application to nanoscience and technology. *Journal of Vacuum Science and Technology* B**23**: 1351–1360.

Hofer, W. A. 2003. Challenges and errors: Interpreting high resolution images in scanning tunneling microscopy. *Progress in Surface Science* **71**: 147–183.

Hofer, W. A., Foster, A. S., and Shluger, A. L. 2003. Theories of scanning probe microscopes at the atomic scale. *Reviews of Modern Physics* **75**: 1287–1331.

Jensen, H., Kröger, J., and Berndt, R. 2005. Electron dynamics in vacancy islands: Scanning tunneling spectroscopy Ag(111). *Physical Review* B**71**: 155417-1–155417-4.

Kanisawa, K., Butcher, M. J., Tokura, Y., Yamaguchi, H., and Hirayama, Y. 2001a. Local density of states in zero-dimensional semiconductor structures. *Physical Review Letters* **87**: 196804-1–196804-4.

Kanisawa, K., Butcher, M. J., Yamaguchi, H., and Hirayama, Y. 2001b. Imaging of Friedel oscillation patterns of two-dimensionally accumulated electrons at epitaxially grown InAs (111)A surfaces. *Physical Review Letters* **86**: 3384–3387.

Kevan, S. D. 1986. Effective-mass theory of simple surface states. *Physical Review* B**34**: 6713–6718.

Kevan, S. D. and Gaylord, R. H. 1987. High-resolution photoemission study of the electronic structure of the noble-metal (111) surfaces. *Physical Review* B**36**: 5809–5818.

Kliewer, L. and Berndt, R. 2001. Scanning tunneling spectroscopy of Na on Cu(111). *Physical Review* B**65**: 035412-1–035412-6.

Kliewer, J., Berndt, R., and Crampin, S. 2000. Controlled modification of individual adsorbate electronic structure. *Physical Review Letters* **85**: 4936–4939.

Kliewer, J., Berndt, R., and Crampin, S. 2001. Scanning tunnelling spectroscopy of electron resonators. *New Journal of Physics* **3**: 22.1–22.11.

Krishnamurthy, H. R., Mani, H. S., and Verma, H. C. 1982. Exact solution of the Schrödinger equation for a particle in a tetrahedral box. *Journal of Physics A: Mathematical and General* **15**: 2131–2137.

Kumagai, T. and Tamura, A. 2008a. Analysis of STM images and STS of electrons confined in equilateral-triangular quantum corrals. *Journal of the Physical Society of Japan* **77**: 014601-1–014601-7.

Kumagai, T. and Tamura, A. 2008b. Scanning tunneling microscopy images of a triangular quantum dot of InAs. *Journal of Physics: Condensed Matter* **20**: 285220-1–285220-6.

Kumagai, T. and Tamura, A. 2009. Scanning tunneling spectrum of electrons confined in a rectangular quantum corral. *Journal of Physics: Condensed Matter* **21**: 225004-1–225004-7.

Lagoute, J., Liu, X., and Fölsch, S. 2005. Link between adatom resonances and the Cu(111) Shockley surface state. *Physical Review Letters* **95**: 136801-1–136801-4.

Lang, N. D. 1985. Vacuum tunneling current from an adsorbed atom. *Physical Review Letters* **55**: 230–233.

Lang, N. D. 1986. Spectroscopy of single atoms in the scanning tunneling microscope. *Physical Review* B**34**: 5947–5950.

Li, J., Schneider, W. D., Berndt, R., and Crampin, S. 1998. Electron confinement to nanoscale Ag islands on Ag(111): A quantitative study. *Physical Review Letters* **80**: 3332–3335.

Loth, S., Baumann, S., Lutz, C. P., Eigler, D. M., and Heinrich, A. J. 2012. Bistability in atomic scale antiferromagnetism. **335**: 196–199.

Maggio-Aprile, J., Renner, Ch., Erb, A., Walker, E., and Fisher, Ø. 1995. Direct vortex lattice imaging and tunneling spectroscopy of flux lines on $YBa_2Cu_3O_{7-\delta}$. *Physical Review Letters* **75**: 2754–2757.

Manoharan, H. C., Lutz, C. P., and Eigler, D. M. 2000. Quantum mirages formed by coherent projection of electronic structure. *Nature* **403**: 512–515.

Meyer, G., Zöphel, S., and Rieder, K. H. 1996. Scanning tunneling microscopy manipulation of native substrate atoms: A new way to obtain registry information on foreign adsorbates. *Physical Review Letters* **77**: 2113–2116.

Michelson, H. B. 1977. The work function of the elements and its periodicity. *Journal of Applied Physics* **48**: 4729–4733.

Mitsuoka, S. and Tamura, A. 2011a. Electron states confined within nano-steps on metal surfaces. *Journal of Physics: Condensed Matter* **23**: 045008-1–045008-15.

Mitsuoka, S. and Tamura, A. 2011b. Scanning tunneling microscopic images and scanning tunneling spectra for coupled rectangular quantum corrals. *Journal of Physics: Condensed Matter* **23**: 275302-1–275302-11.

Mitsuoka, S. and Tamura, A. 2012. STM images and STS for a rectangular quantum corral constructed with δ-function barriers and the effect of an adsorbed atom on STM images and STS. *Physica E* **44**: 1410–1419.

Niebergall, L., Rodary, G., Ding, H. F., Sander, D., Stepanyuk, V. S., Bruno, P., and Kirschner, J. 2006. Electron confinement in hexagonal vacancy islands: Theory and experiment. *Physical Review* B**74**: 195436-1–195436-6.

Nilius, N., Wallis, T. M., and Ho, W. 2002. Development of one-dimensional band structure in artificial gold chains. *Science* **297**: 1853–1856.

Nilius, N., Wallis, T. M., and Ho, W. 2005. Realization of a particle-in-a-box: Electron in an atomic Pd chain. *The Journal of Physical Chemistry B* **109**: 20657–20660.

Pietzsch, O., Kubetzka, A., Bode, M., and Wiesendanger, R. 2001. Observation of magnetic hysteresis at the nanometer scale by spin-polarized scanning tunneling spectroscopy. *Science* **292**: 2053–2056.

Pietzsch, O., Okatov, S., Kubetzka, A., Bode, M., Heinze, S., Lichtenstein, A., and Wiesendanger, R. 2006. Spin-resolved electronic structure of nanoscale cobalt islands on Cu(111). *Physical Review Letters* **96**: 237203-1–237203-4.

Rieder, K. H., Meyer, G., Braun, K. F., Hla, S. W., Moresco, F., Morgenstern, K., Repp, J., Foelsch, S., and Bartels, L. 2003. STM as an operative tool: Physics and chemistry with single atoms and molecules. *Europhysics News* **34**(3), 95.

Schlier, R. E. and Farnsworth, H. E. 1959. Structure and adsorption characteristics of clean surfaces of germanium and silicon. *Journal of Chemical Physics* **30**: 917–926.

Selloni, A., Carnevali, P., Tosatti, P., and Chan, C. D. 1985. Voltage-dependent scanning tunneling microscopy of a crystal surface: Graphite. *Physical Review* B**31**: 2602–2605.

Smith, N. V. 1985. Phase analysis of image states and surface states associated with nearly-free-electron band gaps. *Physical Review* B**32**: 3549–3555.

STM Image Gallery in IBM, Home page: http://researcher.watson.ibm.com/researcher/view_project.php?id=4245, accessed on April 25, 2013.

Stepanyuk, V. S., Niebergall, L., Hergert, W., and Bruno, P. 2005. *Ab initio* study of mirages and magnetic interactions in quantum corrals. *Physical Review Letters* **94**: 187201-1–187201-4.

Stepanyuk, V. S., Negulyaev, N. N., Niebergall, L., and Bruno, P. 2007. Effect of quantum confinement of surface electrons on adatom–adatom interactions. *New Journal of Physics* **9**: 388-1–388-15.

Stroscio, J. A. and Eigler, D. M. 1991. Atomic and molecular manipulation with the scanning tunneling microscope. *Science* **254**: 1319–1326.

Sugimoto, Y., Pou, P., Abe, M., Jelinek, P., Pérez, R., Morita, S., and Custance, Ó. 2007. Chemical identification of individual surface atoms by atom force microscopy. *Nature* **446**: 64–67.

Takayanagi, K., Tanishiro, Y., Takahashi, M., and Takahashi, S. 1982. Structural analysis of Si(111)-7 × 7 by UHV-transmission electron diffraction and microscopy. *Journal of Vacuum Science and Technology* A**3**: 1502–1506.

Takayanagi, K., Tanishiro, Y., Takahashi, M., and Takahashi, S. 1985. Structure analysis of Si(111)-7 × 7 reconstructed surface by transmission electron diffraction. *Surface Science* **164**: 367–392.

Tamura, A. 2011. Quasi-stationary states of an electron confined in a rectangular quantum corral and STM images. *Journal of Physics: Condensed Matter*, **23**: 025303-1–025303-13.

Tersoff, J. and Hamann, D. R. 1983. Theory and application for the scanning tunneling microscope. *Physical Review Letters* **50**: 1998–2001.

Tersoff, J. and Hamann, D. R. 1985. Theory of scanning tunneling microscope. *Physical Review* B**31**: 805–813.

Veuillen, J. V., Mallet, P., Magaud, L., and Pons, S. 2003. Electron confinement effects on Ni-based nanostructures. *Journal of Physics: Condensed Matter* **15**: S2547–S2574.

Wells, A. F. 1993. *Structural Inorganic Chemistry*, 5th edn., Oxford University Press, Oxford, U.K., p. 1288.

Wiesendanger, R. 1994. *Scanning Probe Microscopy and Spectroscopy*, Cambridge University Press, Cambridge, U.K.

Wiesendanger, R. 2009. Spin mapping at the nanoscale and atomic scale. *Reviews of Modern Physics* **81**: 1495–1550.

Wiesendanger, R., Güntherrodt, H. J., Güntherrodt, G., Gambino, R. J., and Ruf, R. 1990. Observation of vacuum tunneling of spin-polarized electrons with the scanning tunneling microscope. *Physical Review Letters* **65**: 247–250.

Yokoyama, T. and Takayanagi, K. 1999. Size quantization of surface-state electrons on the Si(001) surface. *Physical Review* B**59**: 12232–12235.

20

Attosecond Imaging of Molecular Orbitals

David M. Villeneuve
National Research Council of Canada
and
University of Ottawa

20.1 Introduction and Background

Knowledge of the microscopic world is one of the great advances of the nineteenth and twentieth centuries. Understanding the structure of atoms and molecules has led to much of what we take for granted in the modern world, from electronics to medicines. Many of these advances were made possible by imaging techniques that enable us to view matter on a scale much smaller than the wavelength of light, such as spectroscopy [1], electron diffraction [2], x-ray diffraction [3], nuclear magnetic resonance (NMR) [4], and scanning tunneling microscopy (STM) [5]. X-ray crystallography has revealed the double helix of DNA and the structure of thousands of proteins. Today's challenge is turning from structure to function. Even when we know the structure of a molecule at the atomic level, we may not understand its function. Consider the retinal molecule, part of the larger rhodopsin molecule that is found in the retina of the eye. When retinal absorbs a photon, it begins a rapid conformational change that occurs on a 200 fs timescale [6,7]. This starts a chain reaction that eventually results in an electrical nerve impulse that we perceive as vision. Elucidating such dynamical events is a grand challenge of modern physics, chemistry, and biochemistry.

Yet, we require even more information to truly understand these processes beyond the positions of the atoms within the molecule. The atoms are held together by electrons that form the chemical bonds. The valence electrons are responsible for many of the properties of molecules, particularly the chemical dynamics, yet only a few techniques exist to observe the rearrangement of the electrons, which occurs on a timescale of hundreds of attoseconds [8]. We seek new measurement techniques that are sensitive to the electronic configuration of the molecule while having a very fast response time. This is where the new field of attosecond science [9–11] shows such promise. There are two parallel approaches to using attosecond techniques: (1) using the attosecond-duration, extreme ultraviolet (XUV) pulse to photoionize the sample and observe the response on the attosecond or femtosecond timescale and (2) using the process that produces the attosecond pulse as the probe. Both of these approaches are considered as hot topics, judging by the number of high-impact publications [12–20].

For centuries, scientists have sought to observe objects that are smaller than what can be seen by the naked eye. Since human

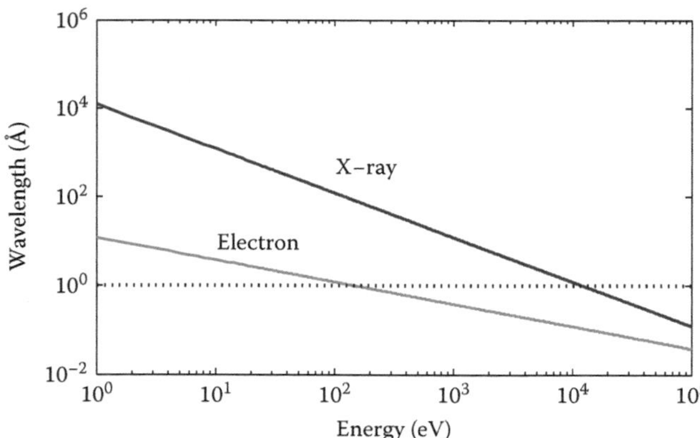

FIGURE 20.1 A comparison of the wavelength of photons and electrons versus energy. The photon has a wavelength determined by the relation $E = hc/\lambda$, whereas the electron has a de Broglie wavelength determined by the relation $\lambda = h/p$. Here, h is Planck's constant, c is the speed of light, E is the photon energy, and p is the electron momentum. To achieve a wavelength of 1 Å (100 pm), a 12.4 keV photon is required, compared with only 150 eV for an electron. Therefore, an electron might be a better probe of atomic structure than a photon.

vision is based on photons, it is natural that scientists would use photons to observe smaller and smaller objects. The smallest size that can be resolved is determined by the wavelength of light used. This has led to the use of ultraviolet and x-ray radiation to image successively smaller features. The ultimate resolution for imaging materials is at the atomic level. The electron cloud surrounding an atom has a diameter on the order of 1 Å or 100 pm. An x-ray with a wavelength of 1 Å has a photon energy of 12.4 keV. On the other hand, an electron with a de Broglie wavelength of 1 Å has a kinetic energy of 151 eV. See Figure 20.1 for a comparison of the wavelengths of electrons and photons.

Of course, there are a variety of reasons for choosing one particle versus another. A 50 keV photon can be used to determine the structure of a molecule by locating the position of the atoms. Unfortunately, such a photon will photoionize inner shell electrons and so will not provide a picture of the outer electrons. Yet, it is the outer valence electrons that are responsible for the formation of chemical bonds that allow molecules to exist. The valence electrons, not the inner shell electrons, define the chemical properties of molecules. An electron with kinetic energy of 150 eV, also with a wavelength of 1 Å, will largely interact with the valence electrons and so might be a better probe of the shape of the electron cloud in a molecule.

In this chapter, we will describe a surprising and unexpected approach to determine the shape and structure of valence electron orbitals in molecules. It makes use of an intense femtosecond laser pulse to remove a valence electron from an atom or molecule. This molecule is pulled away from the ion by the laser's electric field, and then within a single period of the oscillation of the optical field, the electron is pushed back toward the parent ion. The electron can elastically scatter from the ion and thereby provide a diffraction image of the location of the atoms within the molecule [21,22].

A much more unlikely process permits the electron to recombine with the parent ion, thereby emitting a photon. Radiative recombination in plasma will occur on a timescale of nanoseconds; however, this direct recollision of the electron mediated by the electric field of the laser results in immediate recombination with a time precision of tens of attoseconds. Radiative recombination often leaves the atom or molecule in an excited electronic state that will eventually decay to the ground state. In the process that we call high harmonic generation (HHG), only phase matched emission is detected, meaning that all recombination events that do not immediately return the parent atom or molecule to its initial state are not phase matched. By detecting only emission that is coherent and collimated, only the events of immediate recombination to the ground state are seen. All incoherent processes produce emission into 4π sr and will not compete with the collimated emission.

Another advantage of this technique is that the femtosecond laser pulse can only detach a valence electron; the probability of detaching a more deeply bound electron is exponentially dependent on the binding potential. Therefore, HHG only involves valence electrons and is a wonderfully selective probe of the electrons that are important to chemistry.

Finally, the femtosecond laser pulse has a duration of 5–30 fs. By using a pump–probe configuration, one is able to initiate a chemical process and probe it with 5–30 fs resolution. This is the timescale of rearrangements within molecules.

We will show that the technique of HHG using aligned-in-space molecules permits one to make a "photograph" of a single molecular orbital wave function. Thus, this approach has sufficient spatial resolution to resolve the electronic orbital structure of a molecule. And by utilizing the pump–probe approach, we will show that we can follow simple chemical reactions in time.

20.2 Attosecond Science and High Harmonic Generation

20.2.1 Atoms in Intense Laser Fields

At intensities lower than 10^{16} W/cm^2, a laser pulse can be described by its electric field in time, $E(t)$. At higher intensities, the magnetic field must also be considered. For pulsed lasers, the field is a product of a carrier and an envelope, $E(t) = A(t)\cos(\omega t)$. For most pump–probe type experiments, the temporal resolution is given by the width of the envelope A. Today's femtosecond lasers are usually based on titanium:sapphire (Ti:Sa) gain medium with a wavelength of 800 nm. Pulse durations less than 5 fs full width at half maximum (FWHM) are now routine. Yet, for many of the experiments to be described in this chapter, it is not the duration of the pulse envelope that is important but the duration of a single optical cycle, $\tau = 2\pi/\omega$. For Ti:Sa lasers at 800 nm, $\tau = 2.66$ fs.

Consider a hydrogen atom exposed to an intense femtosecond laser field, as shown in Figure 20.2. The electric field associated with the laser can exceed the electric field that binds the electron to the atom. Quantum mechanically, this leads to either

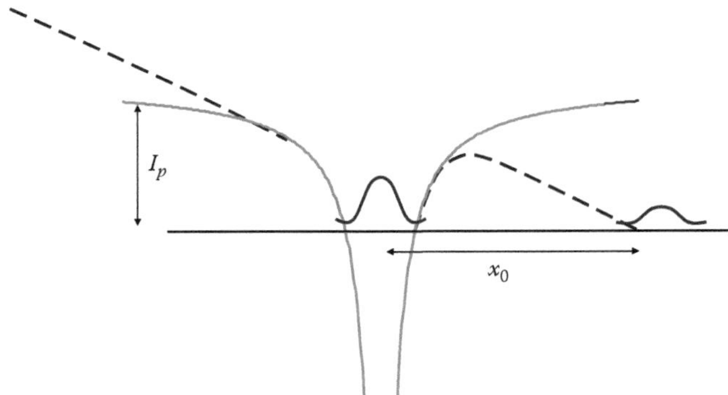

FIGURE 20.2 A hydrogen-like atom is shown, with a − 1/*r* potential. The bound state electron wave function is shown. When an intense laser field is applied, it creates an additional potential *Ex*, shown as the dotted line. If the laser field is strong enough, the barrier is suppressed, and the electron wave function can tunnel through the barrier into the continuum. This is tunnel ionization.

multiphoton ionization or tunnel ionization [23–26]. Ionization is most probable near the peak of each optical cycle but is nonzero at all times. Once the electron is detached from the atom, its motion is almost entirely determined by the laser field.

20.2.2 Three-Step Model of High Harmonic Generation

Following ionization, the trajectory of the electron is almost classical. If the electron is freed from the atom at time t', its trajectory can be written as

$$a(t) = \frac{e}{m}E(t) = \frac{eE_0}{m}\cos\omega t \tag{20.1}$$

$$v(t) = \frac{eE_0}{m\omega}(\sin\omega t - \sin\omega t') \tag{20.2}$$

$$x(t) = -\frac{eE_0}{m\omega^2}(\cos\omega t - \cos\omega t' + (t-t')\sin\omega t'). \tag{20.3}$$

Here, we have applied the initial conditions $x(t') = v(t') = 0$. There are certain trajectories that return to the parent ion at later times, i.e., $x(t) = 0$.

When the electron returns to the parent ion, three things can occur. The electron can elastically scatter from the ion, the electron can inelastically scatter from the ion, and the electron can photorecombine with the ion. The latter process is very unlikely but, nevertheless, is responsible for the field of attosecond science. During photorecombination, the kinetic energy that the electron has picked up from the laser field is converted into a single photon that is emitted by the atom.

This process is called the three-step model [27]. The laser field detaches the electron, the electron is accelerated by the field, and the electron photorecombines with the parent ion. The three-step model can be described quantum mechanically by the time-reversed S-matrix method [28], also called the strong field approximation (SFA):

$$\boldsymbol{d}(t) = -i\int d\mathbf{k}\int_0^t dt'\langle\psi\,|\,\boldsymbol{r}\,|\,\boldsymbol{k}+\boldsymbol{A}(t)\rangle e^{-iS}\boldsymbol{E}(t')\cdot\langle\boldsymbol{k}+\boldsymbol{A}(t')\,|\,\boldsymbol{r}\,|\,\psi\rangle. \tag{20.4}$$

Here

d is the dipole response of the atom
k is the canonical momentum
A is the vector potential of the laser field
ψ is the single electron wave function describing the bound electron in the atom
$S = \int_{t'}^{t} dt''(\boldsymbol{k}+\boldsymbol{A}(t''))^2/2 + I_p(t-t')$ is the classical action, with I_p being the ionization potential

Reading this expression from right to left, ionization occurs at time t', the electron accumulates a Volkov phase in the continuum, described by the classical action *S*, and recombination back to ψ occurs at time *t*.

The radiated electromagnetic spectrum is the square of the Fourier transform of the second derivative of $d(t)$, namely,

$$S(\Omega) = \Omega^4\left|\int d(t)e^{-i\Omega t}dt\right|^2.$$

This is the single-atom part of the process of HHG. Because the process begins and ends with the atom in the state described by ψ, the phase of ψ cancels out. This means that the process is coherent, independent of the phase of each atom. Recombination to excited states is also possible, but does not result in coherent emission. Thus, HHG is very sensitive to the wave function of the atom and ignores other processes that do not end at the initial state.

20.2.3 Properties of HHG: Photon Energy, Coherence, Collimation

HHG usually takes place in a gas medium, as shown in Figure 20.3. The femtosecond laser pulse is focused into the gas medium, and each atom produces emission according to the electric field that it experiences. Because the emission is locked to the femtosecond laser field's optical cycle, the emission from each atom has a fixed phase relationship. Each atom's emitted electric field is added coherently with all the other atoms, leading to macroscopic phase matching. The emission is therefore coherent and collimated in the forward direction [29–31].

The photon energy Ω is given by the kinetic energy of the recolliding electron plus the ionization potential of the atom, $\hbar\Omega = K + I_p$. The kinetic energy is given by the velocity of the returning electron (Equation 20.3). For typical laser conditions, the emitted spectrum lies in the XUV or soft x-ray portion of the electromagnetic spectrum.

In the time domain, the emission occurs in an attosecond-duration pulse during each half optical cycle of the driving laser field. This can be seen in Figure 20.4. The Fourier transform gives the spectrum of the emission, shown in Figure 20.5. Because the

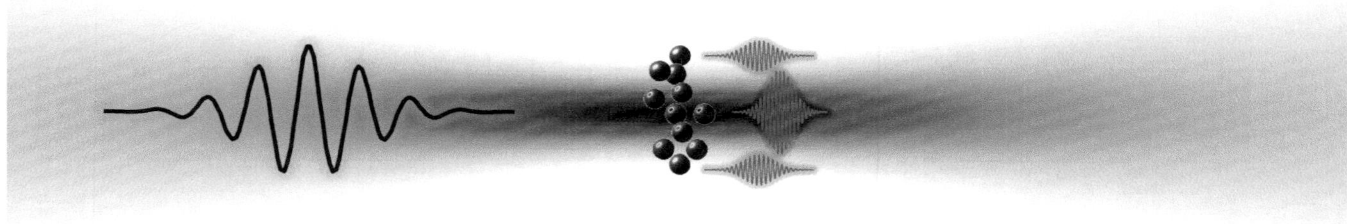

FIGURE 20.3 Sketch of the high harmonic generation process. The femtosecond laser pulse, with an intensity typically 1×10^{14} W/cm^2, comes from the left and focuses into the gas jet containing the target atoms. Each atom emits an XUV field according to the three-step model. The emission from each atom adds coherently and results in a phase matched emission going to the right. The XUV emission then enters a spectrometer (not shown) to the right, which disperses the spectrum and records it. For high harmonic spectroscopy, this is the output of the experiment.

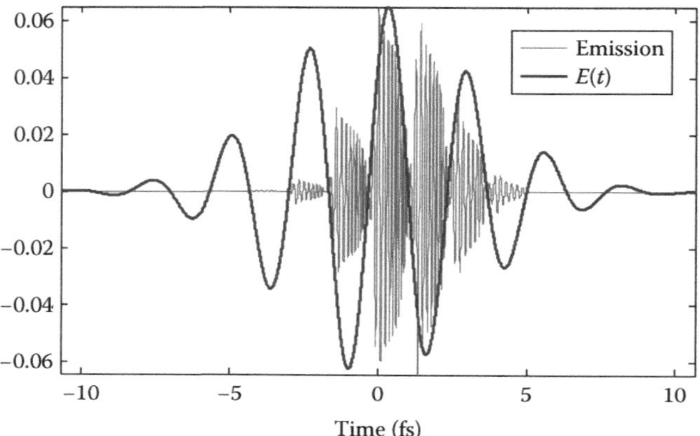

FIGURE 20.4 The thin curve is the electric field of a short laser pulse with a wavelength of 800 nm and an intensity of 1.5×10^{14} W/cm^2. The thick curve is the electric field of the high harmonic emission from an atom exposed to the laser field. The emission is a series of attosecond-duration bursts in each optical half cycle.

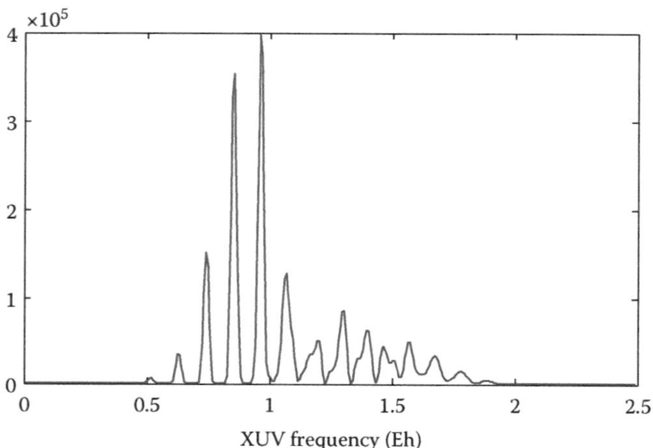

FIGURE 20.5 This is the power spectrum of the attosecond pulse train shown in Figure 20.4. The spectrum is composed of peaks at odd integer multiples of the laser frequency, giving rise to the name high harmonic generation, or HHG, for the process. 1 Eh = 27.2 eV photon energy.

attosecond burst repeats every half optical cycle (period $\tau/2$), the spectrum will be modulated with peaks separated by $\Delta\Omega = 2\pi/(\hbar\tau/2) = 2\omega$. Because the sign of the emitted field alternates for each half optical cycle, the emission peaks at odd harmonics of the driving laser field ω. This is why it is called HHG, because the emission appears as high-order harmonics of ω. If the driving laser pulse is so short that it contains only a single optical cycle, then a single attosecond pulse can be emitted; its spectrum then becomes continuous.

20.2.4 Factorizing the Three Steps to Provide Recombination Dipole Moments

The expression in Equation 20.4 does not clearly show that the HHG spectrum contains information about the target atom. Examination of Equation 20.4 shows a transition dipole moment $\langle\psi_0|\boldsymbol{r}|\boldsymbol{k}+\boldsymbol{A}(t)\rangle$ between the bound state ψ and a set of continuum wave functions labeled by momentum k. In the SFA, these continuum states are plane waves. Therefore, the HHG process in some way contains information on the transition dipole moment of the state that was ionized.

It has been shown that the HHG three-step process can indeed be factorized into three steps, not just in the time domain, but in the frequency domain [32–35]. From [35], the power spectrum $S(E_\Omega)$ can be approximately factored as

$$S(E_\Omega) = I(F,\omega)W(E)\sigma^r(E) \tag{20.5}$$

where

I represents the ionization step
W represents the acceleration of the electron in the continuum
σ^r is the photorecombination cross section
E is the kinetic energy of the recolliding electron and $E_\Omega = E + I_p$

The approach used in [32] was to determine the product of I and W using a reference atom for which σ^r is known. Then S is a measure of σ^r for the target atom or molecule.

20.2.5 Relationship to Photoelectron Spectroscopy

In photoelectron spectroscopy, a single photon detaches an electron from an atom. The direction and the kinetic energy of the photoelectron give information about the bound state, in particular, the photoionization cross section σ^i. HHG is a time-reversed version of photoionization. A valence electron is detached by the intense laser field and then returns to the parent ion. The incoming electron photorecombines and emits a photon whose energy is the sum of the kinetic energy of the electron and the binding energy of the active electron.

In femtosecond photoelectron spectroscopy, single-photon absorption creates a photoelectron, which encodes information of the molecule's electronic structure in its spectrum [36,37]. In femtosecond high harmonic spectroscopy (HHS), similar detailed information [33,38] is carried in the emitted photon. In photoelectron spectroscopy, different initial (neutral) and final (ionic) electronic states are distinguished through the photoelectron energy. In HHS, the broadband recolliding electron wave can only recombine to vacant states, which are selectively created by the tunnel ionization step. Tunnel ionization is exponentially sensitive to the binding energy of each ionization channel. The simultaneous measurement of a broad photon spectrum and the selection of the probed state by tunneling both point to greater simplicity for HHS. However, the emitted photons originating from different initial electronic states overlap spectrally, adding an undesirable complexity to the interpretation of HHS. We show later that this apparent complexity becomes an advantage as the unexcited molecules can serve as a local oscillator against which we measure the excited state dynamics. Just as in a radio receiver, the local oscillator makes a weak signal that would be otherwise difficult to observe readily visible.

The differential photoionization cross section σ^i and the differential photorecombination cross section σ^r both contain the transition dipole moment between the bound state and the continuum state and only differ by prefactors [33,39]:

$$\frac{d^2\sigma^r}{\omega^2 d\Omega_n d\Omega_k} = \frac{d^2\sigma^i}{c^2k^2 d\Omega_n d\Omega_k}. \tag{20.6}$$

Therefore, HHG has the ability to provide some of the same information as the established field of photoelectron spectroscopy. This new field is called high harmonic spectroscopy or HHS. It has certain advantages over photoelectron spectroscopy:

- HHG is highly parallel, in that a large number of photon frequencies happen simultaneously. The entire spectrum is generated in a single laser shot.
- The tunnel ionization process is very selective for the weakest bound electrons. The probability of ionization is exponential with binding energy. Photoelectron spectroscopy looks at deeper bound electrons as the photon frequency increases.
- The photons emitted by HHG are characterized by photon energy, polarization, and phase and thus contain more information than an electron.
- The HHG process occurs on the timescale of tens of femtoseconds or less. Impulsive molecular alignment techniques [40,41] can be applied to make molecular frame measurements.

20.3 High Harmonic Generation in Atoms

20.3.1 Cooper Minimum in Argon

To demonstrate the spectroscopic capabilities of HHS, we apply it to a well-known electronic structure in argon, known as the Cooper minimum. A Cooper minimum occurs in photoionization when the transition dipole matrix elements become small due to cancelation of different partial wave components of the continuum wave function [42].

The experiment [38] comprised a laser source, a gas jet target, and an XUV spectrometer. The Ti:Sa laser system (KM Labs Red Dragon) produced 2 mJ 800 nm 30 fs pulses at a repetition rate of 1 kHz. The pulses were spectrally broadened in a hollow core fiber [43] and compressed with chirped dielectric mirrors to produce pulses with a duration of 6–8 fs. These pulses were focused into a pulsed gas jet target containing argon (see Figure 20.6). The gas jet was less than 1 mm long in the laser propagation direction to minimize the effects of phase mismatch in the HHG process [44–46]. The forward propagating high harmonic emission was detected by an XUV spectrometer with a variable spacing grating (Hitachi) and a microchannel plate detector and camera. The resulting XUV spectra were recorded on a computer. Lineouts of the spectra are shown in Figure 20.7.

Such a minimum in argon has been commented upon since the early days of high harmonics [47]. It agrees closely to the location of the minimum reported in photoionization experiments [48]. This experiment verifies the conjecture stated in Section 20.2.4 that the high harmonic spectrum contains the photorecombination cross section.

This result also verified another observation. In modeling the photoionization process, the continuum is characterized as field-free scattering states [33,49–51]. Until recently, plane waves were used to characterize the continuum, as in the SFA [28] (see Equation 20.4). In the SFA, after the electron is removed from an atom, its motion is determined solely by the laser field, until it recombines. Yet, the use of field-free scattering states suggests that the laser field can be ignored during the photorecombination process in HHG. This is a remarkable conclusion and leads to the utility of HHS in studying the electronic properties of atoms and molecules.

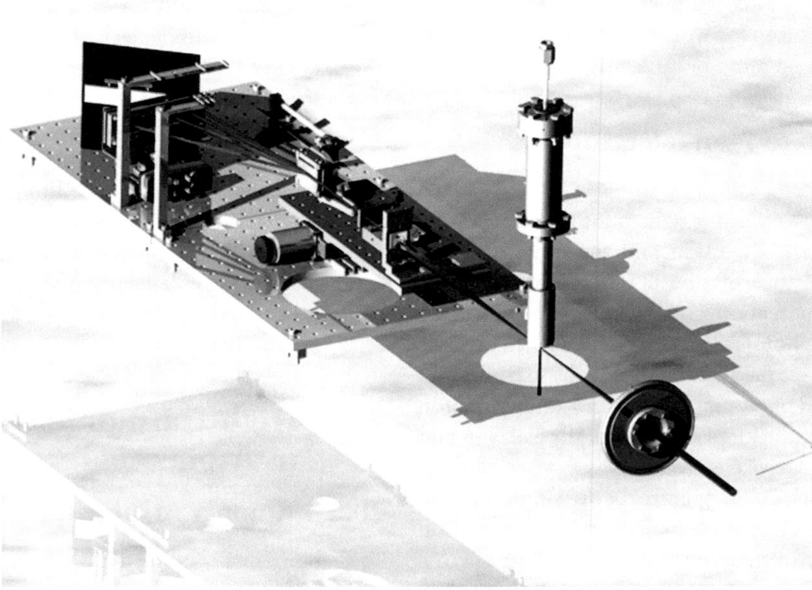

FIGURE 20.6 A cutaway diagram of a high harmonic experiment. The apparatus shown is enclosed in a vacuum system with a pressure of 10^{-6} Torr or better. The vertical cylinder is a pulsed gas jet that emits a supersonic atomic or molecular beam. The laser beam enters from the bottom right and is focused into the gas jet. The XUV radiation is generated in the gas jet in a phase matched process and propagates along with the driving laser into an XUV spectrometer. The spectrometer includes a slit, a variable groove spacing Hitachi grating, a microchannel plate detector, and a CCD camera outside the vacuum chamber. The grating disperses the spectrum onto the detector.

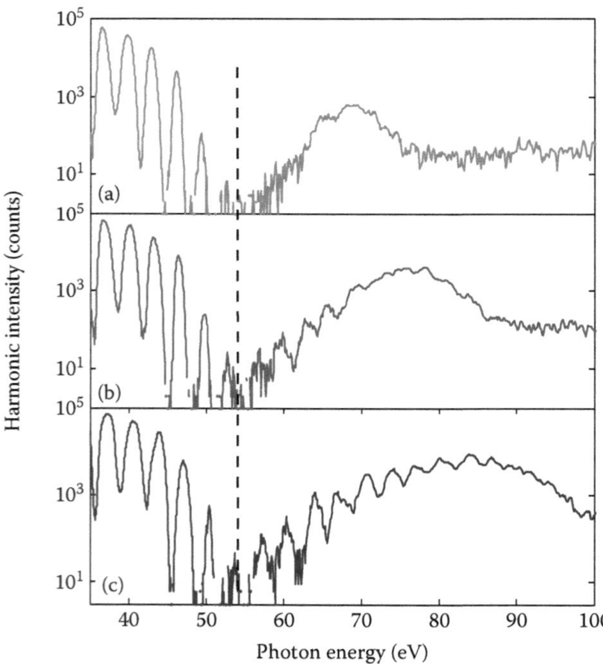

FIGURE 20.7 High harmonic spectra of argon at various laser intensities. The horizontal axis is the photon energy in eV as detected by the XUV spectrometer. The vertical axis is the signal level on a log scale. The Cooper minimum is shown at 54 eV. The three panels show an increasing laser intensity, which moves the highest photon energy higher; the fact that the Cooper minimum does not change with laser intensity verifies that this is an electronic feature of the atom and not related to the laser. (Derived from Wörner, H.J. et al., *Phys. Rev. Lett.*, 102, 103901, 2009.)

The use of scattering states can be understood through the analysis of photoionization and scattering processes [52]. In photoionization, the continuum is modeled as in incoming spherical wave and an outgoing plane wave that is detected. In scattering, the reverse process applies: an incoming plane wave represents the incoming electron, and the scattered wave is represented by an outgoing spherical wave. The case of HHG is very much like scattering in that the electron as it returns toward the ion looks like a plane wave that becomes distorted by the ion potential. See Figure 20.1 [53] and Figure 20.2 [38] for an idea of what the continuum wave function looks like.

20.3.2 Giant Resonance in Xenon and Interchannel Couplings

It was stated in Section 20.2.5 that only the most weakly bound electron is detached by the laser field. This is generally true, but in the case of molecules with closely spaced valence electron levels, more than one orbital can be ionized [20,54,55]. In this section, we show that electron–electron interactions can also occur, providing information on deeper orbitals.

The experiment [59] used a novel laser source that is almost ideal for spectroscopic studies [60,61], having a wavelength of 1.8 μm and a duration of less than 2 optical cycles. With its long wavelength, it creates a recollision electron whose energy can exceed 100 eV, even for low ionization potential systems such as small organic molecules. The photon cutoffs shown here are 150 eV.

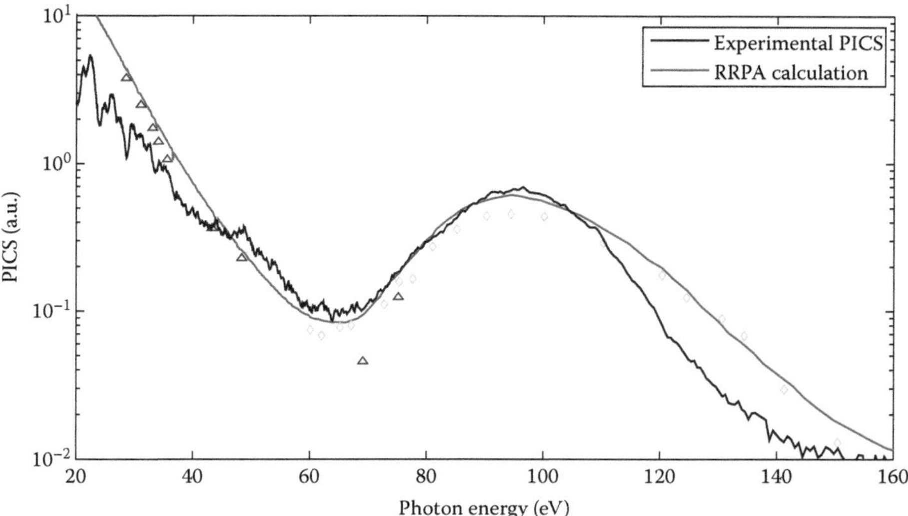

FIGURE 20.8 HHG results for xenon. Experimental HHG spectrum divided by the krypton wave packet (dark line) and the calculation of the xenon photoionization cross section (PICS) by Kutzner et al. [56] (light line). The triangles and diamonds are PICS measurements by Fahlmann et al. [57] and Becker et al. [58], respectively, each weighted using the anisotropy parameter calculated by Kutzner et al. [56]. The xenon photoionization cross section as determined by the high harmonic experiment closely matches previous synchrotron results and detailed calculations. (Derived from Shiner, A.D. et al., *Nat. Phys.*, 7, 464, 2011.)

First, the HHG spectrum from krypton $S_{Kr}(E_\Omega)$ was measured, and the photoionization cross section $\sigma^i_{Kr}(E_\Omega)$ from the literature [62] was used to accurately determine the recolliding electron wave packet spectrum (see Equation 20.5) $W(E)$ by setting $W(E) = S_{Kr}(E_\Omega)/\sigma^i_{Kr}(E_\Omega)$, using a similar approach to that used by Itatani et al. [32]. The measured HHG spectrum for xenon was then divided by this term, to extract the photoionization cross section for xenon: $\sigma^i_{Xe} = S_{Xe}(E_\Omega)/W(E)$. Because this procedure divides one spectrum by another, experimental details such as grating reflectivity and detector response cancel out. The experimentally derived σ^i_{Xe} is plotted in Figure 20.8, together with the photoionization cross section from synchrotron experiments. The excellent agreement shows that the HHG spectrum contains detailed information about the electronic structure of atoms, imprinted through the photorecombination cross section. As was the case with argon in Section 20.3.1, it is remarkable that the intense laser field can be neglected.

The large bump in the cross section around 100 eV seen in Figure 20.8 is unexpected in the HHG experiment if only the valence 5p electron is removed by the infrared laser field. The explanation for the bump in the photoionization cross section [56] is an electron–electron interaction with a 4d electron that has a strong photoionization cross section near 100 eV due to a shape resonance. The fact that this 5p–4d electron interaction occurs in photorecombination in HHG shows that HHG is very much a time-reversed photoionization process.

The probability that the infrared laser field can tunnel ionize a 4d electron, with a 70 eV binding energy, is 10^{-50}. The explanation is shown in Figure 20.9. The left half shows the three-step model (see Section 20.2.2) in which a 5p electron is removed by the laser and then photorecombines to the 5p hole. The right half shows the process that occurs in xenon. The returning 5p electron has a Coulomb interaction with all the electrons in the ion, but in particular can collisionally excite a 4d electron to fill the 5p hole. The 5p electron then recombines to the 4d hole. In both cases, a 100 eV photon is emitted.

20.4 High Harmonic Generation in Molecules

20.4.1 Impulsive Alignment and Orientation of Molecules Using Lasers

It is possible to align gas-phase molecules in space by means of non-resonant laser fields [41]. It has been demonstrated that medium-sized molecules can be held in space in three dimensions using elliptically polarized fields [63]. Field-free alignment has been demonstrated using an adiabatic pulse that abruptly turns off [64], and by short pulses that produce rotational wave packets that periodically rephase [40,41,65]. Furthermore, polar molecules can be oriented in a particular direction [66,67].

In the following sections, we use impulsive alignment to achieve molecular frame measurements. A stretched Ti:Sa laser pulse, typically 100 fs in duration, is used to give a kick to the molecules. Due to the periodicity of the rotational eigenstates, the molecules will periodically line up parallel to the aligning laser's polarization direction. Since the laser pulse has long gone, the molecules are aligned without the presence of a laser field. By rotating the polarization of the aligning laser pulse relative to the polarization of the pulse that generates HHG, the molecular response can be mapped out in the molecular frame [68–70].

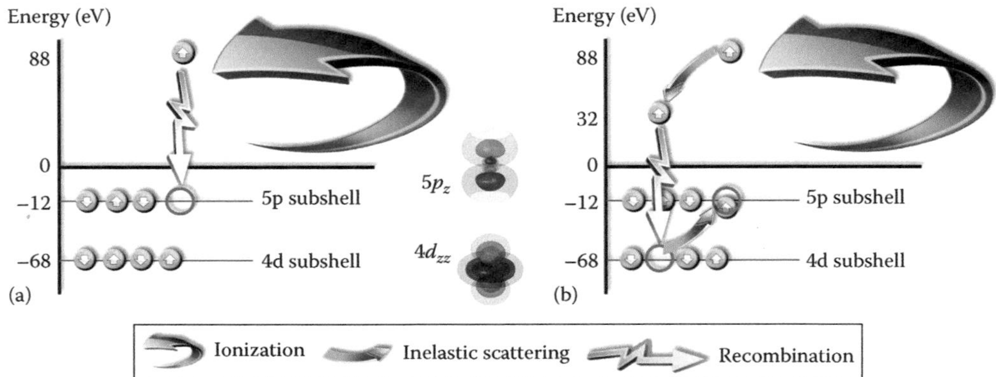

FIGURE 20.9 Steps for harmonic generation. In the usual three-step model, an electron is tunnel ionized from the valance shell, accelerates in the continuum, and then recombines to the state from which it came (a). With inelastic scattering, the returning electron can promote a lower lying electron into the valance band and then recombine to the vacancy in the lower lying state (b). In both cases, a 100 eV photon is emitted by recombination to a 5p vacancy (a) or a 4d vacancy (b).

20.5 Tomographic Reconstruction of Molecular Orbitals

20.5.1 Reality of Molecular Orbitals

We have shown up to this point that HHS is a measure of the transition dipole matrix elements between a bound state and a set of continuum wave functions. We will now proceed to show that, by recording a complete set of such matrix elements at a series of angles in the molecular frame, it is possible to reconstruct a spatial image of the orbital that was ionized. This is called molecular orbital tomography [32].

Some interpretations of quantum mechanics hold that wave functions do not exist in reality and that only the square of the wave function has any meaning. Nevertheless, scientists use the concept of wave functions to help visualize the electronic structure of atoms and molecules [71]. Whereas a single quantum system cannot be observed, an ensemble of quantum systems can be measured [72].

Single-electron molecular orbital wave functions are mathematical constructs that are used to describe the multi-electron wave function of molecules. The highest lying orbitals are of particular interest since they are responsible for the chemical properties of molecules. Yet, they are elusive to observe experimentally. Using the highly nonlinear process of tunnel ionization in an intense, femtosecond infrared laser field, we selectively remove the highest occupied molecular orbital (HOMO) electron, and then recombine this electron about 2 fs later. This results in the emission of high harmonics from the molecule that contains information about the shape of the molecular orbital. By aligning the gas-phase molecule at a set of angles, the resulting HHG spectra can be tomographically inverted to yield the two-dimensional orbital wave function. The coherent interference between the free electron wave function and the molecular orbital wave function, a form of homodyne detection, enables us to see the actual wave function, not its square [32].

Only a few methods can currently "see" the highest molecular orbitals—electron momentum spectroscopy (EMS) [73] and STM [74–76]. These experiments have provided valuable data that can be compared with various theoretical descriptions, e.g., Hartree–Fock, Kohn–Sham, and Dyson orbitals [77]. Other techniques, such as electron scattering or x-ray diffraction, measure the total electron density of the molecule, not specific orbitals. Yet it is the frontier orbitals that give the molecule its chemical properties.

It has been suggested that the HHG spectrum from molecules might contain information on the internuclear separation [78,79]. We will show that, by recording a series of HHG spectra from molecules held at fixed angles, it is possible to tomographically reconstruct the shape of the highest electronic orbital, including the relative phase of the wave function.

20.5.2 Measurement of Transition Dipoles of Dinitrogen

A proof-of-principle experiment [32] was performed on a simple molecule, N_2, whose HOMO is known to be a $2p\sigma_g$ orbital. More precisely, the $2p\sigma_g$ is the Dyson orbital that is associated with ionization of N_2 to the ground state of the cation. The next highest orbitals, about 1 eV below the HOMO, are $2p\pi_u$, and so have distinctly different symmetry. Evidence for emission from lower orbitals of N_2 has been reported [54].

In the experiment [32], the output of a Ti:Sa laser system (10 mJ, 27 fs, 800 nm, and 50 Hz) was split into two pulses with a variable delay. The first pulse served to produce a rotational wave packet in the nitrogen gas emanating from a nozzle [65]. Its intensity was low enough ($<10^{14}$ W/cm^2) that no harmonics were generated. The second laser pulse (3×10^{14} W/cm^2) produced the HHG spectrum that was detected with an XUV spectrometer.

During the rotational revivals, there were two distinct times at which the molecules had a clear spatial orientation—parallel (4.094 ps) and perpendicular (4.415 ps) to the laser polarization [65]. We used the parallel alignment time and rotated the molecular

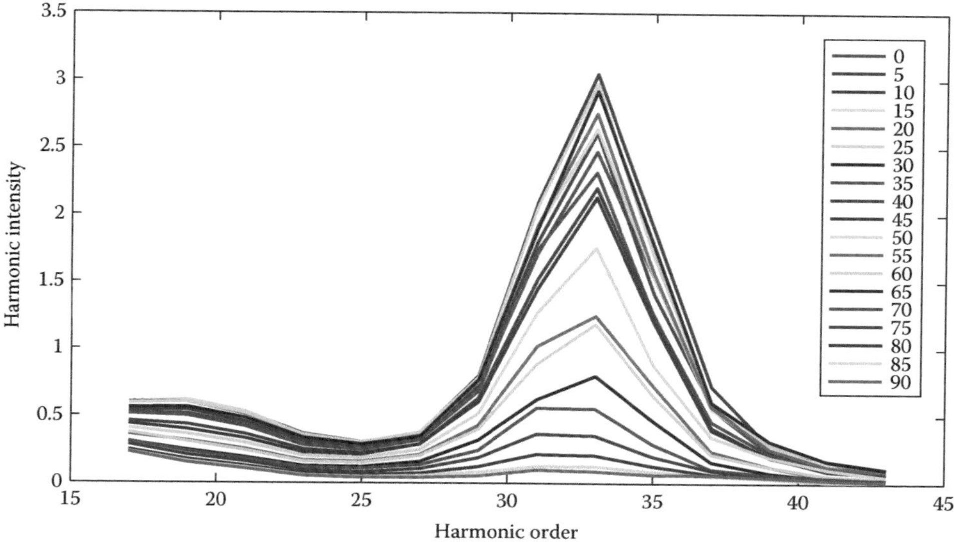

FIGURE 20.10 (See color insert.) High harmonic spectra of N_2 recorded at five degree increments for the angle between the molecular axis and the laser polarization direction. Each spectrum has been divided by the linearized argon reference spectrum to remove the amplitude of the recolliding wave packet. The horizontal axis is the harmonic order of the emission; 1 order corresponds to 1.5 eV.

axis relative to the polarization of the HHG pulse using a half-wave plate. In order to remove the sensitivity of the XUV spectrometer and other systematic effects, we have normalized the N_2 spectrum to that of argon. Argon has nearly the same ionization potential as nitrogen, and so the ionization process will be very similar. This calibration is shown to be valid for other atoms as well [80].

We recall that the high harmonic process can be factored into three steps (see Section 20.2.4) [32–35]. The harmonic signal will be proportional to the square of the dipole moment induced by the returning electron:

$$\mathbf{d} = \langle \psi_m(\mathbf{r}) \,|\, \mathbf{r} \,|\, \psi_e(\mathbf{r}) \rangle. \qquad (20.7)$$

Here, ψ_m is the molecular orbital wave function that was ionized. The outgoing electronic wave function is a Volkov wave [28], and upon return $\psi_e(\mathbf{r})$ is described by a plane wave, e^{ikx}. More recent models have described the continuum state as a field-free scattering state (see Section 20.3.1). For the sake of simplicity in analysis, we will use the plane wave model. For each harmonic number n, we know that the corresponding momentum k_n of the returning electron is $\hbar k_n = \sqrt{2m_e(nE_L - E_i)}$.

In Figure 20.10, we show the high harmonic spectra recorded for aligned N_2 molecules, divided by the argon reference signal. The molecular axis has been rotated in five degree increments between each measurement, and the fact that each spectral amplitude is different indicates the high degree of alignment achieved.

20.5.3 Tomographic Reconstruction Procedure

We now go on to show that the HHG signal can be tomographically inverted to yield a picture of the molecular orbital. We will assume that the laser polarization axis is in the x direction, and that the molecular axis is at an angle θ with respect to x. For a non-planar molecule, this angle can be replaced by the Euler angles that will completely describe its orientation. We then use a rotated version of the wave function to represent the rotated molecule, $\psi_m(\mathbf{r},\theta)$.

From Equation 20.7, the dipole magnitude for the nth harmonic can be written as an integral:

$$\mathbf{d}_n(\theta) = \int_{-\infty}^{\infty}\int_{-\infty}^{\infty} \psi_m(\mathbf{r},\theta)\mathbf{r}e^{ik_n x}dxdy \qquad (20.8a)$$

$$= FT\left\{\int_{-\infty}^{\infty} \psi_m(\mathbf{r},\theta)\mathbf{r}dy\right\} \qquad (20.8b)$$

where we have dropped the third dimension z for clarity. Note that $\mathbf{d}$ is a complex vector. It can be seen that this is a spatial Fourier transform in direction x of an integral along y of the molecular wave function.

The Fourier slice theorem [81] shows that the Fourier transform of a projection P is equal to a cut at angle θ through the two-dimensional transform F of the object. This is the essence of computed tomography based on the inverse Radon transform. Our dipole is the Fourier transform of a projection of the wave function, and so the parallel between the HHG process and computed tomography is strong.

The result of the tomographic deconvolution is shown in Figure 20.11. The upper panel shows the experimental reconstruction, while the lower panel is the calculated shape of the $3\sigma_g$ orbital of N_2. Note that the color scale includes both positive and negative values, indicating that we are measuring a wave function as opposed to the square of the wave function. The lobes

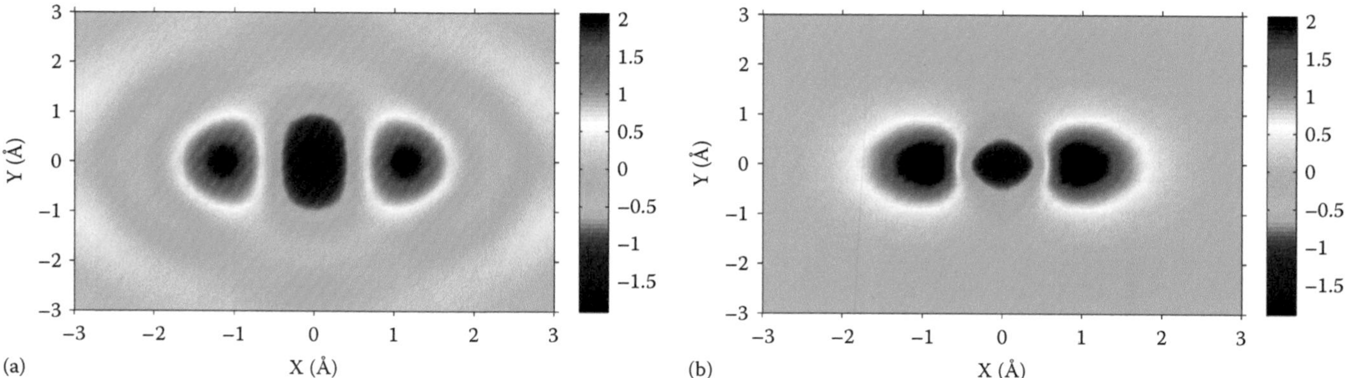

FIGURE 20.11 (a) Reconstructed highest occupied molecular orbital of N_2 as measured by high harmonic spectroscopy. The color scale contains both positive and negative values, indicating that the measurement is of a wave function and not an electron density. (b) Ab initio calculation of the $3\sigma_g$ orbital of N_2. (Derived from Itatani, J. et al., *Nature (London)*, 432, 867, 2004.)

at which the wave function passes through zero are well reproduced in the reconstruction.

As discussed earlier, it may seem that quantum mechanics forbids the observation of a wave function [71]. We effectively record the dipole matrix elements of a transition between two states of the system, Equation 20.7. Since this is an expectation value of a Hermitian operator, it is an observable in quantum mechanics. The image of the orbital is interpreted within a Hartree–Fock model of orbital wave functions. We determine the molecular orbital wave function to within an arbitrary phase, with a normalization constant and a DC term, because it is effectively a homodyne measurement; part of the bound state electron wave function is removed by the laser and then interferes with itself when it recombines.

The spatial resolution is limited to the electron wavelength corresponding to the highest attainable harmonic order. This is determined by the ionization potential E_i and the intensity at which the molecule ionizes (not necessarily the peak laser intensity). For $I = 2 \times 10^{14}$ W/cm^2 at 800 nm wavelength, $nE_L = 38$ eV and $\lambda_e = 2$ Å. By going to longer wavelengths, as described in Section 20.3.2, the cutoff energy is increased. For a laser wavelength of 1.8 μm, $\lambda_e = 0.9$ Å, and the highest spatial frequency component is $k = 7$ Å^{-1}. This is usually sufficient to resolve typical molecular orbitals [82].

The mathematical underpinnings of this reconstruction technique have been studied in more detail in [53,83]. Rather than considering just a single molecular orbital, the more detailed approaches consider all electrons in the system to be indistinguishable, leading to concepts of Dyson orbitals and electron exchange upon recombination. These factors have an effect on the interpretation of the reconstruction and lead to even better agreement between theory and experiment than shown in Figure 20.11.

More recent experiments have shown that several molecular orbitals may be imaged at the same time [84]. This approach relies on the fact that several orbitals with closely separated energy levels (HOMO and HOMO-1, with σ_g and π_u symmetries, respectively) are both ionized by the laser field. A phase shift between the emissions from each orbital permits them to be simultaneously imaged. A plane wave model is necessary for this to work.

20.6 Following a Chemical Reaction Using High Harmonic Spectroscopy

20.6.1 Advantages of High Harmonic Spectroscopy

We extend HHS from probing static molecular structure to probing photochemical dynamics. Using the impulsive photodissociation of molecular bromine, we show that the electromagnetic interference between high harmonics generated from the molecular ground state and the excited state occurs on the attosecond timescale. The coherent addition of the emitted radiation results in high visibility of the excited state dynamics despite the low excited state fraction.

Simultaneous imaging of the geometric and electronic structure of a molecule as it undergoes a chemical reaction is one of the main goals of modern ultrafast science. Techniques based on diffraction [85,86] measure the position of the atoms within the molecule with high accuracy but are much less sensitive to the electronic structure of the molecule, particularly the valence shell in which the chemical transformations originate. A new complementary approach exploits the rescattering of an electron removed from the molecule by a strong laser field to measure the structure of the molecule [21]. The associated recollision also leads to HHG that encodes the structure of the orbital to which the electron recombines [32,84,87]. So far, these methods have only been applied to the electronic ground state of molecules [88–91]. Ultrafast dynamics in molecules occur predominantly in excited electronic states.

A significant problem in femtochemistry is the difficulty in creating a significant population of excited molecules. To ensure that each molecule absorbs one excitation photon in a femtosecond pulse requires a laser intensity that is so high that many molecules experience multiphoton ionization. Therefore, it is one of the most sought techniques that can deal with a relatively low fraction of excited molecules in the presence of a majority of unexcited molecules. Fortunately, HHS is a coherent process in which emission from all molecules, both excited and unexcited, add coherently to produce the observed signal. Because of this coherence, the presence of many unexcited molecules is actually an advantage, as it permits both amplitude and phase of the emission from the small excited state population to be determined. Moreover, the coherent detection provides a high sensitivity to the phase of the radiation, which reflects the evolution of the ionization potential along the dissociation coordinate.

20.6.2 Following the Dissociation of Br_2 Molecules

The experimental setup [92] consists of a chirped-pulse amplified titanium–sapphire femtosecond laser system, a high harmonic source chamber equipped with a pulsed valve, and an XUV spectrometer. The laser system provides 800 nm pulses of 32 fs duration (FWHM). The laser beam is split into two parts of variable intensities using a half-wave plate and a polarizer. The minor part of the energy is used to generate 400 nm radiation in a type I BBO of 60 μm thickness. The major part is sent through a computer-controlled delay stage and is recombined with the 400 nm radiation using a dichroic beam splitter. The combined beams are focused into the chamber using a $f = 50$ cm spherical mirror.

High-order harmonics are generated in a supersonic expansion of Br_2 seeded in two bars of helium. The helium carrier gas is sent through liquid Br_2 kept at room temperature. Bromine molecules are excited by single-photon absorption at 400 nm to the repulsive C $^1\Pi_{1u}$ state and high harmonics are generated in the strong 800 nm field. The focus of both beams is placed ~1 mm before the pulsed molecular jet expanding through a nozzle of 250 μm diameter. This setup minimizes the effect of phase mismatch and reabsorption of the high harmonic radiation and leads to the observation of the single-molecule response [93].

Figure 20.12 shows the relevant potential energy curves of Br_2 and Br_2. Single-photon excitation at 400 nm from the X $^1\Sigma_g^+$ ground state leads almost exclusively to the repulsive C $^1\Pi_{1u}$ state that dissociates into two bromine atoms in their ground spin-orbit state ($^2P_{3/2}, m_J = 1/2$) [94]. The figure also shows the shape of the vibrational wave function in the ground state and the calculated nuclear wave packet on the excited state surface at selected delays Δt after excitation by a 40 fs pump pulse centered at 400 nm. The $^2\Pi_{3/2g}$ ground state curve of Br_2^+ is also shown to illustrate the variation of the ionization potential with the internuclear distance.

Figure 20.13 shows the observed harmonic and ion signals in a pump–probe experiment with perpendicular polarizations. The power of H19 decreases during the excitation, reaches a minimum after the peak of the 400 nm pulse, and then recovers to its initial power level. In contrast, the ion yield increases, reaching its maximum after the peak of the 400 nm pulse, and subsequently decreases to its initial level. The maximum increase in ion yield amounts to 7%, whereas the harmonic signal is depleted by up to 30%. The signal of H19 and the ion yield have been normalized to unity at negative delays. This signal level corresponds to all molecules being in the ground electronic state. Temporal overlap of

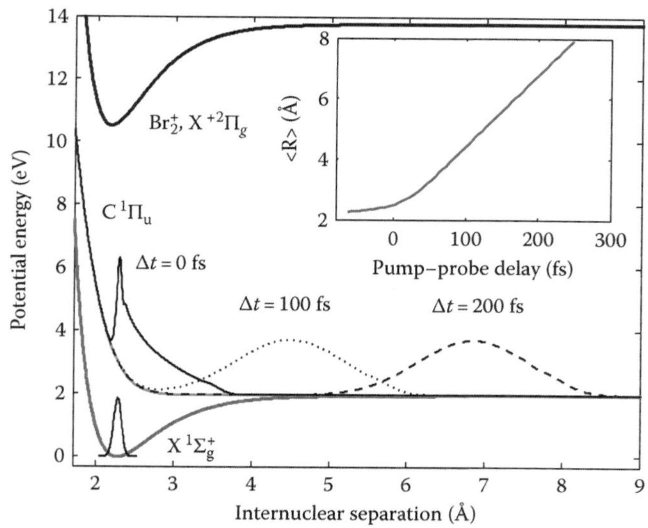

FIGURE 20.12 Potential energy curves of Br_2 (X $^1\Sigma_g^+$ ground state and C $^1\Pi_{1u}$ excited state) and Br_2^+ (X $^{+2}\Pi_g$ ground state). The shape of the nuclear wave packet in the excited state after selected delays Δt is also shown. The wave packets were obtained by numerical propagation, assuming a 40 fs excitation pulse centered at 400 nm. The inset shows the expectation value of the internuclear separation as a function of the pump–probe delay. (Reproduced from Wörner, H.J. et al., *Phys. Rev. Lett.*, 105, 103002, 2010. With permission.)

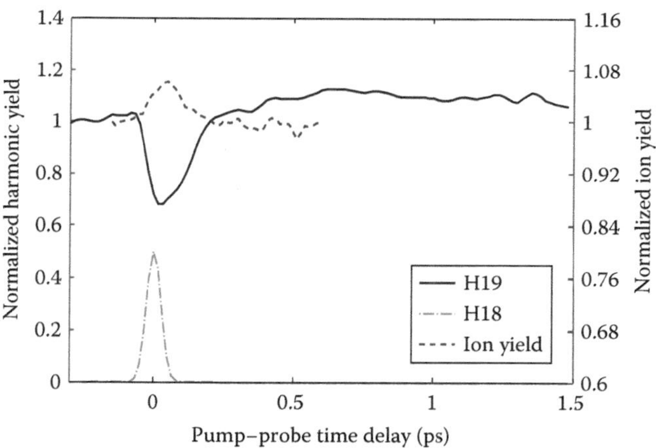

FIGURE 20.13 Intensity of harmonics 19 (full line) and 18 (dash-dotted line) from excited bromine molecules as a function of the delay between a 400 nm pump pulse and a perpendicularly polarized 800 nm pulse generating high harmonics (left-hand axis). The temporal overlap of 400 and 800 nm pulses leads to the emission of even order harmonics, like H18. The total ion yield (dotted line, right-hand axis) shown as dashed line was measured under identical conditions but with higher statistics. (Reproduced from Wörner, H.J. et al., *Phys. Rev. Lett.*, 105, 103002, 2010. With permission.)

the 800 and 400 nm pulses leads to the appearance of even-order harmonics [95], which provide the time origin and a high-order cross-correlation (~50–60 fs).

When Br_2 is excited to the C $^1\Pi_{1u}$ state, the ionization potential for the removal of the most weakly bound electron is reduced from 10.5 to 7.5 eV, explaining the observed increase in the ion yield. The rising part of the ion yield curve reflects the build-up of the excited state population during the excitation pulse. As Br_2 dissociates along the repulsive C $^1\Pi_{1u}$ state, the ionization potential increases from 7.5 to 11.8 eV, resulting in a decreasing ionization rate of the excited state. Since the ionization rate increases at early delays, one might expect that the harmonic yield would also increase. However, the opposite is observed. Moreover, the variation of the harmonic signal is much larger than that of the ion signal and exceeds the excitation fraction by a factor of 2.

These results clearly demonstrate a destructive interference between harmonics emitted by the excited molecules and those emitted by the ground state molecules. Destructive interference is the origin of the opposite behavior of ion and harmonic yield. Since the interference between the excited and unexcited molecules involves both phase and amplitude of emission, it is impossible to determine both parameters with a single measurement. In the following section, we utilize a different geometry in order to extract both amplitude and phase.

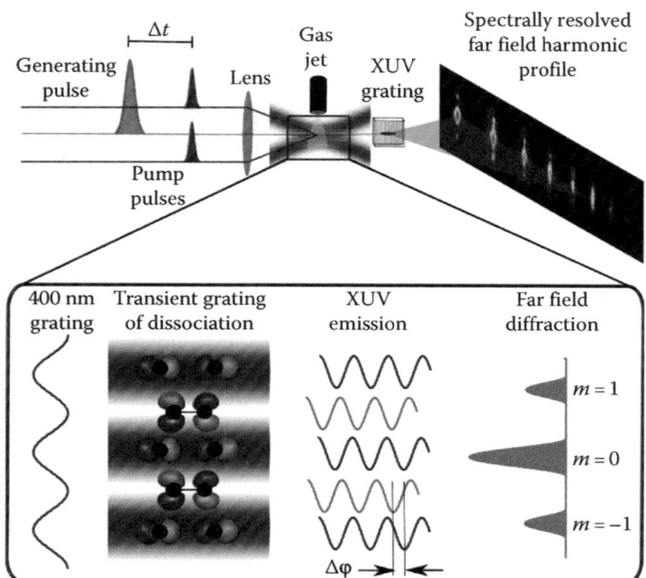

FIGURE 20.14 Schematic of the transient grating excitation geometry. Two 400 nm pump pulses set up a transient grating of excitation in the seeded molecular beam of Br_2 molecules. A delayed 800 nm pulse generates high harmonics. The periodic modulation of the high harmonic amplitude and phase in the near field results in first-order diffraction in the far field. (Derived from Wörner, H.J. et al., *Nature*, 466, 604, 2010.)

20.7 Transient Grating High Harmonic Spectroscopy

In the previous section, XUV emission from both excited and unexcited molecules added coherently to produce the detected high harmonic spectrum. Destructive interference between the species was observed, but it was difficult to determine whether the reduction in signal was due to amplitude or phase effects. In this section, we utilize a transient grating excitation geometry that enables us to extract both amplitude and phase of the excited species relative to the unexcited ones [96], using Br_2 molecules.

We form a sinusoidal grating of excited Br_2 molecules using two pump beams that cross in the medium, as shown in Figure 20.14. Horizontal planes of excited molecules alternate with planes of unexcited molecules. We generate high harmonics from this grating with a delayed 800 nm laser pulse (probe). The zero time delay and the cross-correlation time of 50 fs are evident through the appearance of even-order harmonics.

From the zeroth- and first-order diffracted signals, we can uniquely extract the harmonic amplitudes d_e/d_g and phases $|\varphi_e - \varphi_g|$ of the excited state relative to the ground state (where $d_{g,e}$ and $\varphi_{g,e}$ are the harmonic amplitudes and phases of the ground (g) and excited (e) states, respectively). We show the experimentally determined values in Figure 20.15, when the pump and probe pulses are parallel (a) or perpendicular (b).

The different time evolution for the amplitude and the phase is striking. While the phase reaches its asymptotic value after ≈ 150 fs, the amplitude takes more than 300 fs. It is also striking that the time response of the amplitude changes with the relative polarization of the excitation and harmonic generation pulse. In contrast, while the time-dependent phase is different for the different polarizations, it reaches the same asymptotic value at the same time delay. We will first concentrate on the phase, then discuss the amplitude.

The phase of high harmonic radiation has two main contributions: (1) The electron and the ion accumulate a relative phase between the moment of ionization and recombination. The phase shift between the same harmonic order q being emitted by two electronic states differing in ionization potential by ΔI_p can be expressed as $\Delta\phi_q \approx \Delta I_p \bar{\tau}_q$ [97], where $\bar{\tau}_q$ is the average transit time of the electron in the continuum. (2) When the electron recombines, the transition moment imposes an amplitude and a phase on the radiation [20,38]. The first contribution depends on the electron trajectory (determined by the laser parameters) and the ionization potential. The second contribution characterizes the electronic structure of the molecule. It depends on the emitted photon energy and the angle of recombination in the molecular frame [32].

The time evolution of the reconstructed phase in Figure 20.15 can be split into two regions: the first 150 fs, where the phase undergoes a rapid variation, and the subsequent flat region, where the phase is independent of the relative polarizations. The rapid variation of the phase reflects the fast variation of the ionization potentials with delay. The strong dependence of the phase on the relative polarizations (Figure 20.15a vs. b) also shows that the phase traces the evolution of the electronic structure of the

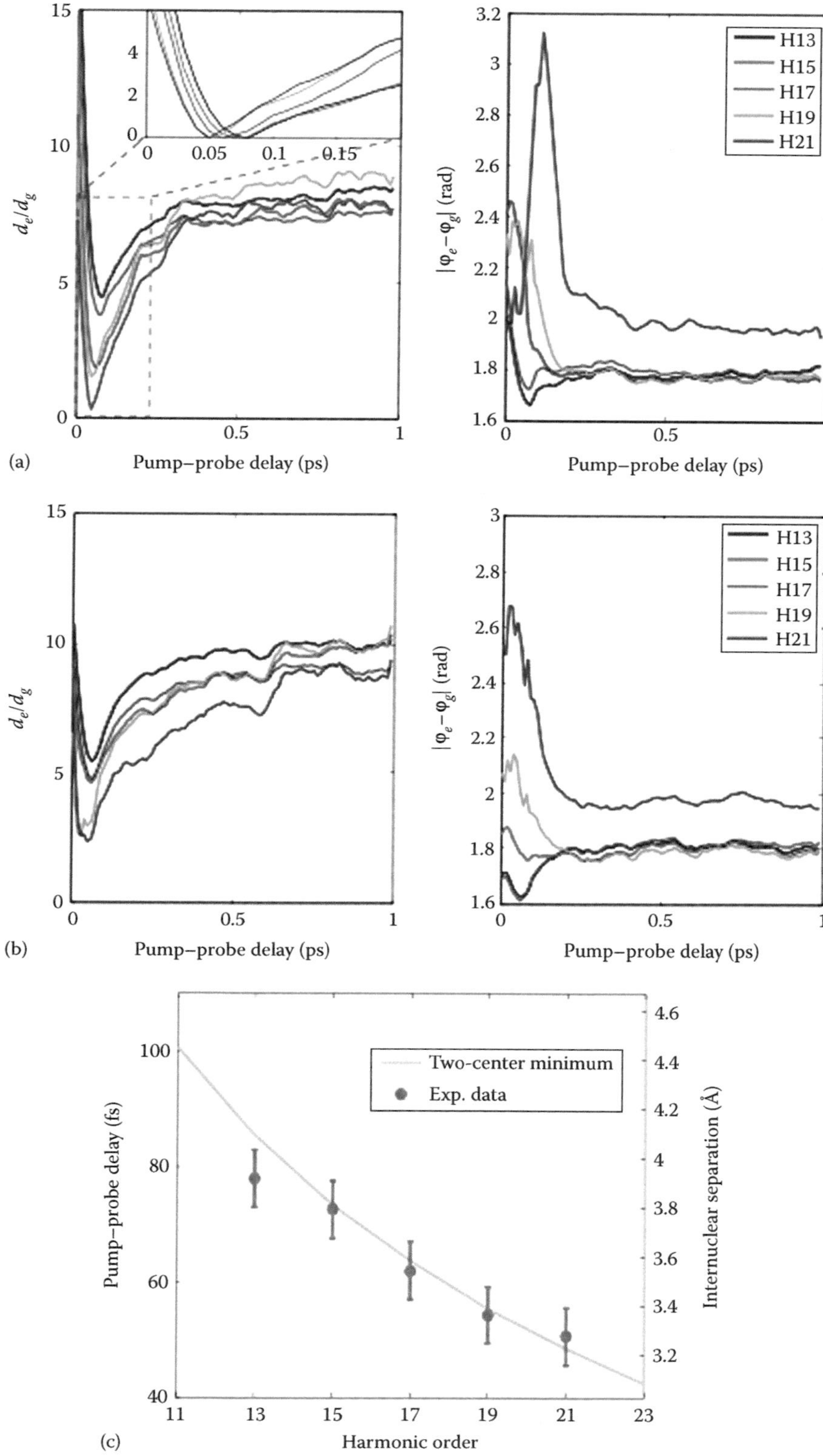

FIGURE 20.15 **(See color insert.)** Reconstruction of high harmonic amplitude and phase of excited Br_2 molecules. (a) shows the reconstructed amplitude (left) and phase (right) of the excited state emission relative to the ground state, for parallel pump–probe polarizations. (b) is the same as (a) except that the pump–probe polarizations are perpendicular. (c) shows the measured internuclear separations as determined by the two-center interference condition for each harmonic order. (Derived from Wörner, H.J. et al., *Nature*, 466, 604, 2010.)

molecule as it dissociates. This variation occurs because the electrostatic potential into which the electron recombines changes significantly. At asymptotic delays, we measure Br atoms relative to ground state molecules. The phase shift is independent of the direction of recombination, consistent with the fact that Br_2 dissociates into atoms in the $|m_j| = 1/2$ magnetic sublevel [94]. The phase shift is 1.8 rad for H13. Using the relation $\Delta\phi_q \approx \Delta I_p \bar{\tau}_q$, we obtain $\Delta I_p = 1.3$ eV, in good agreement with the known ionization potentials of Br_2 and Br.

We now turn to the temporal evolution of the amplitudes. All harmonics go through a deep minimum at an early time delay that depends on the harmonic order (see inset of Figure 20.15a). The minimum occurs between 51 ± 5 fs (H21) and 78 ± 5 fs (H13). It measures the stretching of the orbital as the molecule dissociates. When the electron recombines to the initial state, its de Broglie wavelength λ_q can destructively interfere with the initial state wave function. Destructive interference occurs when $R = (2n + 1)/2\lambda_q$ (n integer) in the case of a σ_u orbital [78]. Using $n = 1$ and the relation $\Omega = k^2/2$ between the photon energy Ω and the electron momentum k, we translate the minimum of H21 to a bond length of 3.3 Å and that in H13 to 3.9 Å—in good agreement with wave packet calculations. Thus, we trace the bond length as a function of time using quantum interference. This is illustrated in Figure 20.15c.

As the molecule dissociates, additional minima corresponding to destructive interference with $n > 1$ could be expected. Instead, we observe a slow rise of the amplitude. At delays larger than 150 fs, the four valence molecular orbitals of Br_2 formed from the 4p atomic orbitals of Br become nearly degenerate, and HHG becomes essentially atomic in character. In this regime, the gerade or ungerade states of the ion are degenerate—there should be no quantum interference in ionization or in recombination. Thus, we observe the transition from molecular to atomic character. Classically, one could say that when the atoms form a molecule, the electrons are shared and cannot be distinguished, leading to quantum interference; when the atoms become separate, one can remove an electron from, say, the right atom, and it cannot recombine to the left atom that has not been ionized.

At later times, it is only the propagation of the electron in the laser field that is affected by the second atom. For perpendicularly polarized pump and probe beams, the interaction of the ionized electron with the neighboring atom is maximized since the electron trajectory between tunneling and recollision lies in the plane of the disc of dissociating atoms. The slower recovery of the amplitude in Figure 20.15b reflects this fact. This property of HHS is analogous to extended x-ray absorption fine structure (XAFS) and may be useful to probe the chemical environment of a low-I_p species (e.g., a molecule in a helium droplet).

Time-resolved photoelectron measurements of the dissociation of Br_2 have demonstrated how the binding energies shift as the atoms move apart. In Refs. [37,98,99], the time delay for the appearance of an atomic-like photoelectron spectrum was in the range of 40–85 fs. The minima between 50 and 80 fs show that the electron recombines to a two-center molecular wave function. The absence of such minima in the range of 100–150 fs suggests that at longer delays the recombination of the electron has become essentially atomic. An analogous transition between two- and one-center signatures has recently been observed in core-shell photoionization of a static molecule [100].

20.8 Conclusions

HHG has many future applications. It is a laboratory-scale source of XUV and soft x-ray photons with femtosecond and attosecond durations that can be used to form microscopic images of materials or to time-resolve the motion of electrons within atoms and molecules. HHS can be applied to understanding the electronic structure of atoms and molecules. In the case of molecular dynamics, knowledge of the amplitude and phase of the photorecombination moment relative to a fully characterized ground state reference [32] will allow the dynamic imaging of orbitals in a chemical reaction. The unique properties of HHS will lead to other applications in femtochemistry, from simple dissociation dynamics, to proton transfer, to non-adiabatic reaction dynamics, to complex photochemical processes. For example, the change in electronic structure associated with the crossing of a conical intersection [101–103] will be mapped into the harmonic radiation. In all these cases, the sensitivity of HHS to the electronic structure will provide new insight.

Acknowledgments

The material presented in this chapter represents the work of many people over many years. I list most of them here alphabetically and apologize to those who I inadvertently missed: Bert Avery, Heidi Bandulet, Julien Bertrand, Thomas Brabec, Daniel Comtois, Paul Corkum, David Crane, Pat Dooley, Nirit Dudovich, Mathieu Giguère, Paul Hockett, Jiro Itatani, Misha Ivanov, Nathaniel Kajumba, Daniil Kartashov, Jean-Claude Kieffer, Kevin Lee, François Légaré, Jérôme Levesque, Igor Litvinyuk, Yann Mairesse, Moritz Meckel, Andrei Naumov, Hiromichi Niikura, Serguei Patchkovskii, Domagoj Pavičić, Henri Pépin, David Rayner, Bruno Schmidt, Andrew Shiner, Olga Smirnova, Mike Spanner, André Staudte, Albert Stolow, Carlos Trallero-Herrero, Jonathan Underwood, Caterina Vozzi, Marc Vrakking, Hans Jakob Wörner, Gennady Yudin, and Dirk Zeidler.

References

1. G. Herzberg. *Molecular Spectra and Molecular Structure, Volume I, Spectra of Diatomic Molecules*, 2nd edn. (Krieger Publishing Company, Malabar, FL, 1989).
2. I. Hargittai and M. Hargittai. *Stereochemical Applications of Gas-Phase Electron Diffraction* (Wiley-VCH, New York, 1988).
3. L. V. Azaroff, R. Kaplow, N. Kato, R. J. Weiss, A. J. C. Wilson, and R. A. Young. *X-Ray Diffraction* (McGraw-Hill, New York, 1974).

4. G. S. Rule and K. T. Hitchens. *Fundamentals of Protein NMR Spectroscopy* (Springer, Berlin, Germany, 2006).
5. R. Wiesendanger and H.-J. Guntherodt, eds. *Scanning Tunneling Microscopy III: Theory of STM and Related Scanning Probe Methods* (Springer, Berlin, Germany, 1996).
6. M. Ben-Nun, F. Molnar, H. Lu, J. C. Phillips, T. J. Martnez, and K. Schulten. Quantum dynamics of the femtosecond photoisomerization of retinal in bacteriorhodopsin. *Faraday Discuss.* 110, 447 (1998).
7. V. I. Prokhorenko, A. M. Nagy, S. A. Waschuk, L. S. Brown, R. R. Birge, and R. J. D. Miller. Coherent control of retinal isomerization in bacteriorhodopsin. *Science* 313, 1257 (2006).
8. J. Breidbach and L. S. Cederbaum. Universal attosecond response to the removal of an electron. *Phys. Rev. Lett.* 94, 033901 (2005).
9. P. B. Corkum and F. Krausz. Attosecond science. *Nat. Phys.* 3, 381–387 (2007).
10. A. Scrinzi, M. Y. Ivanov, R. Kienberger, and D. M. Villeneuve. Attosecond physics. *J. Phys. B* 39, R1–R37 (2006).
11. F. Krausz and M. Y. Ivanov. Attosecond physics. *Rev. Mod. Phys.* 81, 163 (2009).
12. M. Hentschel, R. Kienberger, C. Spielmann, G. A. Reider, N. Milosevic, T. Brabec, P. Corkum, U. Heinzmann, M. Drescher, and F. Krausz. Attosecond metrology. *Nature* 414, 509 (2001).
13. A. Baltuska, T. Udem, M. Uiberacker, M. Hentschel, E. Goulielmakis, C. Gohle, R. Holzwarth et al. Attosecond control of electronic processes by intense light fields. *Nature* 421, 611–615 (2003).
14. M. Drescher, M. Hentschel, R. Kienberger, M. Uiberacker, V. Yakolev, A. Scrinzi, T. Westerwalbesloh, U. Kleineberg, and F. Krausz. Time-resolved atomic inner-shell spectroscopy. *Nature* 419, 803 (2002).
15. E. Goulielmakis, Z.-H. Loh, A. Wirth, R. Santra, N. Rohringer, V. S. Yakovlev, S. Zherebtsov et al. Real-time observation of valence electron motion. *Nature* 466, 739–743 (2010).
16. R. Kienberger, E. Goulielmakis, M. Uiberacker, A. Baltuska, V. Yakovlev, F. Bammer, A. Scrinzi et al. Atomic transient recorder. *Nature* 427, 817–821 (2003).
17. M. Uiberacker, T. Uphues, M. Schultze, A. J. Verhoef, V. Yakovlev, M. F. Kling, J. Rauschenberger et al. Attosecond real-time observation of electron tunnelling in atoms. *Nature* 446, 627–632 (2007).
18. H. Niikura, F. Légaré, R. Hasbani, A. D. Bandrauk, M. Y. Ivanov, D. M. Villeneuve, and P. B. Corkum. Sub-laser-cycle electron pulses for probing molecular dynamics. *Nature* 417, 917 (2002).
19. H. Niikura, F. Légaré, R. Hasbani, M. Ivanov, D. M. Villeneuve, and P. B. Corkum. Probing molecular dynamics with attosecond resolution using correlated wave packet pairs. *Nature* 421, 826 (2003).
20. O. Smirnova, Y. Mairesse, S. Patchkovskii, N. Dudovich, D. Villeneuve, P. Corkum, and M. Y. Ivanov. High harmonic interferometry of multi-electron dynamics in molecules. *Nature (London)* 460, 972–977 (2009).
21. M. Meckel, D. Comtois, D. Zeidler, A. Staudte, D. Pavičić, H. C. Bandulet, H. Pépin et al. Laser induced electron tunnelling and diffraction. *Science* 320, 1478–1482 (2008).
22. C. I. Blaga, J. Xu, A. D. DiChiara, E. Sistrunk, K. Zhang, P. Agostini, T. A. Miller, L. F. DiMauro, and C. D. Lin. Imaging ultrafast molecular dynamics with laser-induced electron diffraction. *Nature* 483, 194 (2012).
23. A. Perelemov, V. Popov, and M. Terent'ev. Ionization of atoms in an alternating electric field. *Sov. Phys. JETP* 23, 924 (1966).
24. M. V. Ammosov, N. B. Delone, and V. P. Krainov. Tunnel ionization of complex atoms and of atomic ions in an alternating electromagnetic field. *Zh. Eksp. Teor. Fiz.* 91, 2008 (1986).
25. N. B. Delone and V. P. Krainov. Tunneling and barrier-suppression ionization of atoms and ions in a laser radiation field. *Uspekhi Fizicheskikh Nauk* 41, 469–485 (1998).
26. G. L. Yudin and M. Y. Ivanov. Nonadiabatic tunnel ionization: Looking inside a laser cycle. *Phys. Rev. A* 64, 013409 (2001).
27. P. B. Corkum. Plasma perspective on strong field multiphoton ionization. *Phys. Rev. Lett.* 71, 1994 (1993).
28. M. Lewenstein, P. Balcou, M. Y. Ivanov, A. L'Huillier, and P. B. Corkum. Theory of high-harmonic generation by low-frequency laser fields. *Phys. Rev. A* 49, 2117 (1994).
29. P. Salières, A. L'Huillier, and M. Lewenstein. Coherence control of high-order harmonics. *Phys. Rev. Lett.* 74, 3776 (1995).
30. Y. Mairesse, A. de Bohan, L. J. Frasinski, H. Merdji, L. C. Dinu, P. Monchicourt, P. Breger et al. Attosecond synchronization of high-harmonic soft x-rays. *Science* 302, 1540 (2003).
31. C.-G. Wahlström, J. Larsson, A. Persson, T. Starczewski, S. Svanberg, P. Salières, P. Balcou, and A. L'Huillier. High-order harmonic generation in rare gases with an intense short-pulse laser. *Phys. Rev. A* 48, 4709–4720 (1993).
32. J. Itatani, J. Levesque, D. Zeidler, H. Niikura, H. Pépin, J. C. Kieffer, P. B. Corkum, and D. M. Villeneuve. Tomographic imaging of molecular orbitals. *Nature (London)* 432, 867–871 (2004).
33. A.-T. Le, R. R. Lucchese, S. Tonzani, T. Morishita, and C. D. Lin. Quantitative rescattering theory for high-order harmonic generation from molecules. *Phys. Rev. A* 80, 013401–013423 (2009).
34. M. V. Frolov, N. L. Manakov, T. S. Sarantseva, M. Y. Emelin, M. Y. Ryabikin, and A. F. Starace. Analytic description of the high-energy plateau in harmonic generation by atoms: Can the harmonic power increase with increasing laser wavelengths? *Phys. Rev. Lett.* 102, 243901–243904 (2009).
35. M. V. Frolov, N. L. Manakov, T. S. Sarantseva, and A. F. Starace. Analytic formulae for high harmonic generation. *J. Phys. B* 42, 035601 (2009).

36. L. Nugent-Glandorf, M. Scheer, D. A. Samuels, A. M. Mulhisen, E. R. Grant, X. Yang, V. M. Bierbaum, and S. R. Leone. Ultrafast time-resolved soft x-ray photoelectron spectroscopy of dissociating Br_2. *Phys. Rev. Lett.* 87, 193002 (2001).
37. P. Wernet, M. Odelius, K. Godehusen, J. Gaudin, O. Schwarzkopf, and W. Eberhardt. Real-time evolution of the valence electronic structure in a dissociating molecule. *Phys. Rev. Lett.* 103, 013001 (2009).
38. H. J. Wörner, H. Niikura, J. B. Bertrand, P. B. Corkum, and D. M. Villeneuve. Observation of electronic structure minima in high-harmonic generation. *Phys. Rev. Lett.* 102, 103901 (2009).
39. L. D. Landau and E. M. Lifshitz. *Quantum Mechanics Non-Relativistic Theory*, volume 3 of *Course of Theoretical Physics*, 3 edn. (Pergamon Press Ltd., New York, 1977).
40. F. Rosca-Pruna and M. J. J. Vrakking. Experimental observation of revival structures in picosecond laser-induced alignment of i_2. *Phys. Rev. Lett.* 87, 153902 (2001).
41. H. Stapelfeldt and T. Seideman. Aligning molecules with strong laser pulses. *Rev. Mod. Phys.* 75, 543–557 (2003).
42. J. W. Cooper. Photoionization from outer atomic subshells. A model study. *Phys. Rev.* 128, 681 (1962).
43. J. S. Robinson, C. A. Haworth, H. Teng, R. A. Smith, J. P. Marangos, and J. W. G. Tisch. The generation of intense, transform-limited laser pulses with tunable duration from 6 to 30 fs in a differentially pumped hollow fibre. *Appl. Phys. B* 85, 525–529 (2006).
44. P. Balcou, P. Salières, A. L'Huillier, and M. Lewenstein. Generalized phase-matching conditions for high harmonics: The role of field-gradient forces. *Phys. Rev. A* 55, 3204–3210 (1997).
45. T. Popmintchev, M.-C. Chen, A. Bahabad, M. Gerrity, P. Sidorenko, O. Cohen, I. P. Christov, M. M. Murnane, and H. C. Kapteyn. Phase matching of high harmonic generation in the soft and hard x-ray regions of the spectrum. *Proc. Natl. Acad. Sci. USA* 106, 10516–10521 (2009).
46. V. S. Yakovlev, M. Ivanov, and F. Krausz. Enhanced phase-matching for generation of soft x-ray harmonics and attosecond pulses in atomic gases. *Opt. Express* 15, 15351–15364 (2007).
47. A. L'Huillier and P. Balcou. High-order harmonic generation using intense femtosecond pulses. *Phys. Rev. Lett.* 70, 774 (1993).
48. J. A. R. Samson and W. C. Stolte. Precision measurements of the total photoionization cross-sections of He, Ne, Ar, Kr, and Xe. *J. Electron Spectrosc. Relat. Phenom.* 123, 265 (2002).
49. Z. Chen, A.-T. Le, T. Morishita, and C. D. Lin. Quantitative rescattering theory for laser-induced high-energy plateau photoelectron spectra. *Phys. Rev. A* 79, 033409 (2009).
50. A.-T. Le, R. R. Lucchese, M. T. Lee, and C. D. Lin. Probing molecular frame photoionization via laser generated high-order harmonics from aligned molecules. *Phys. Rev. Lett.* 102, 203001 (2009).
51. M. Okunishi, T. Morishita, G. Prumper, K. Shimada, C. D. Lin, S. Watanabe, and K. Ueda. Experimental retrieval of target structure information from laser-induced rescattered photoelectron momentum distributions. *Phys. Rev. Lett.* 100, 143001–143004 (2008).
52. A. F. Starace. Theory of atomic photoionization. In: *Handbuch der Physik*. Ed. W. Mehlhorn, Vol. 31, pp. 1–121 (Springer, Berlin, Germany, 1981).
53. S. Patchkovskii, Z. Zhao, T. Brabec, and D. M. Villeneuve. High harmonic generation and molecular orbital tomography in multielectron systems. *J. Chem. Phys.* 126, 114306 (2007).
54. B. K. McFarland, J. P. Farrell, P. H. Bucksbaum, and M. Gühr. High harmonic generation from multiple orbitals in N_2. *Science* 322, 1232–1235 (2008).
55. H. J. Wörner, J. B. Bertrand, P. Hockett, P. B. Corkum, and D. M. Villeneuve. Controlling the interference of multiple molecular orbitals in high-harmonic generation. *Phys. Rev. Lett.* 104, 233904 (2010).
56. M. Kutzner, V. Radojević, and H. P. Kelly. Extended photoionization calculations for xenon. *Phys. Rev. A* 40, 5052–5057 (1989).
57. A. Fahlman, M. O. Krause, T. A. Carlson, and A. Svensson. Xe 5*s*, 5*p* correlation satellites in the region of strong interchannel interactions, 28–75 ev. *Phys. Rev. A* 30, 812–819 (1984).
58. U. Becker, D. Szostak, H. G. Kerkhoff, M. Kupsch, B. Langer, R. Wehlitz, A. Yagishita, and T. Hayaishi. Subshell photoionization of xe between 40 and 1000 ev. *Phys. Rev. A* 39, 3902–3911 (1989).
59. A. D. Shiner, B. Schmidt, C. Trallero-Herrero, H. J. Wörner, S. Patchkovskii, P. B. Corkum, J.-C. Kieffer, F. Légaré, and D. M. Villeneuve. Probing collective multi-electron dynamics in xenon with high-harmonic spectroscopy. *Nat. Phys.* 7, 464 (2011).
60. B. E. Schmidt, P. Béjot, M. Giguère, A. D. Shiner, C. Trallero-Herrero, É. Bisson, J. Kasparian et al. Compression of 1.8 μ m laser pulses to sub two optical cycles with bulk material. *Appl. Phys. Lett.* 96, 121109 (2010).
61. B. E. Schmidt, A. D. Shiner, P. Lassonde, J.-C. Kieffer, P. B. Corkum, D. M. Villeneuve, and F. Légaré. Cep stable 1.6 cycle laser pulses at 1.8 um. *Opt. Express* 19, 6858 (2011).
62. K. N. Huang, W. R. Johnson, and K. T. Cheng. Theoretical photoionization parameters for the noble gases argon, krypton, and xenon. *At. Data Nucl. Data Tables* 26, 33–45 (1981).
63. J. J. Larsen, K. Hald, N. Bjerre, H. Stapelfeldt, and T. Seideman. Three dimensional alignment of molecules using elliptically polarized laser fields. *Phys. Rev. Lett.* 85, 2470 (2000).
64. J. G. Underwood, M. Spanner, M. Y. Ivanov, J. Mottershead, B. J. Sussman, and A. Stolow. Switched wave packets: A route to nonperturbative quantum control. *Phys. Rev. Lett.* 90, 223001 (2003).

65. P. W. Dooley, I. V. Litvinyuk, K. F. Lee, D. M. Rayner, M. Spanner, D. M. Villeneuve, and P. B. Corkum. Direct imaging of rotational wave-packet dynamics of diatomic molecules. *Phys. Rev. A* 68, 23406 (2003).
66. H. Sakai, S. Minemoto, H. Nanjo, H. Tanji, and T. Suzuki. Controlling the orientation of polar molecules with combined electrostatic and pulsed, nonresonant laser fields. *Phys. Rev. Lett.* 90, 083001 (2003).
67. L. Holmegaard, J. H. Nielsen, I. Nevo, and H. Stapelfeldt. Laser-induced alignment and orientation of quantum-state-selected large molecules. *Phys. Rev. Lett.* 102, 023001 (2009).
68. K. F. Lee, D. M. Villeneuve, P. B. Corkum, A. Stolow, and J. G. Underwood. Field-free three-dimensional alignment of polyatomic molecules. *Phys. Rev. Lett.* 97, 173001 (2006).
69. I. V. Litvinyuk, K. F. Lee, P. W. Dooley, D. M. Rayner, D. M. Villeneuve, and P. B. Corkum. Alignment-dependent strong field ionization of molecules. *Phys. Rev. Lett.* 90, 233003 (2003).
70. D. Pavičić, K. F. Lee, D. M. Rayner, P. B. Corkum, and D. M. Villeneuve. Direct measurement of the angular dependence of ionization for N_2, O_2, and CO_2 in intense laser fields. *Phys. Rev. Lett.* 98, 243001 (2007).
71. W. H. E. Schwarz. Measuring orbitals: Provocation or reality? *Angew. Chem. Int. Ed.* 45, 1508 (2006).
72. A. Royer. Measurement of quantum states and the Wigner function. *Found. Phys.* 19, 3 (1989).
73. C. E. Brion, G. Cooper, Y. Zheng, I. V. Litvinyuk, and I. E. McCarthy. Imaging of orbital electron densities by electron momentum spectroscopy—A chemical interpretation of the binary (e,2e) reaction. *Chem. Phys.* 270, 13 (2001).
74. M. F. Crommie, C. P. Lutz, and D. M. Eigler. Confinement of electrons to quantum corrals on a metal surface. *Science* 262, 218 (1993).
75. L. C. Venema, J. W. G. Wildöer, J. W. Janssen, S. J. Tans, H. L. J. T. Tuinstra, L. P. Kouwenhoven, and C. Dekker. Imaging electron wave functions of quantized energy levels in carbon nanotubes. *Science* 283, 52 (1999).
76. J. I. Pascual, J. Gómez-Herrero, C. Rogero, A. M. Baró, D. Sánchez-Portal, E. Artacho, P. Ordejón, and J. M. Soler. Seeing molecular orbitals. *Chem. Phys. Lett.* 321, 78 (2000).
77. E. K. Gross and R. M. Dreizler. *Density Functional Theory* (Plenum Press, New York, 1995).
78. M. Lein, N. Hay, R. Velotta, J. Marangos, and P. Knight. Role of the intramolecular phase in high-harmonic generation. *Phys. Rev. Lett.* 88, 183903 (2002).
79. M. Lein, P. P. Corso, J. P. Marangos, and P. L. Knight. Orientation dependence of high-order harmonic generation in molecules. *Phys. Rev. A* 67, 23819 (2003).
80. J. Levesque, D. Zeidler, J. P. Marangos, P. B. Corkum, and D. M. Villeneuve. High harmonic generation and the role of atomic orbital wave functions. *Phys. Rev. Lett.* 98, 183903 (2007).
81. A. C. Kak and M. Slaney. *Principles of Computerized Tomographic Imaging* (Society for Industrial and Applied Mathematics, New York, 2001).
82. V.-H. Le, A.-T. Le, R.-H. Xie, and C. D. Lin. Theoretical analysis of dynamic chemical imaging with lasers using high-order harmonic generation. *Phys. Rev. A* 76, 013414 (2007).
83. S. Patchkovskii, Z. Zhao, T. Brabec, and D. M. Villeneuve. High harmonic generation and molecular orbital tomography in multielectron systems: Beyond the single active electron approximation. *Phys. Rev. Lett.* 97, 123003 (2006).
84. S. Haessler, J. Caillat, W. Boutu, C. Giovanetti-Teixeira, T. Ruchon, T. Auguste, Z. Diveki et al. Attosecond imaging of molecular electronic wavepackets. *Nat. Phys.* 6, 200 (2010).
85. R. Neutze, R. Wouts, D. van der Spoel, E. Weckert, and J. Hajdu. Potential for biomolecular imaging with femtosecond x-ray pulses. *Nature* 406, 752–757 (2000).
86. H. Ihee, V. A. Lobastov, U. M. Gomez, B. M. Goodson, R. Srinivasan, C.-Y. Ruan, and A. H. Zewail. Direct imaging of transient molecular structures with ultrafast diffraction. *Science* 291, 458–462 (2001).
87. S. Baker, J. S. Robinson, M. Lein, C. C. Chirila, R. Torres, H. C. Bandulet, D. Comtois et al. Dynamic two-center interference in high-order harmonic generation from molecules with attosecond nuclear motion. *Phys. Rev. Lett.* 101, 053901–053904 (2008).
88. T. Kanai, S. Minemoto, and H. Sakai. Quantum interference during high-order harmonic generation from aligned molecules. *Nature (London)* 435, 470–474 (2005).
89. C. Vozzi, F. Calegari, E. Benedetti, J.-P. Caumes, G. Sansone, S. Stagira, M. Nisoli et al. Controlling two-center interference in molecular high harmonic generation. *Phys. Rev. Lett.* 95, 153902 (2005).
90. N. L. Wagner, A. Wüest, I. P. Christov, T. Popmintchev, X. Zhou, M. M. Murnane, and H. C. Kapteyn. Monitoring molecular dynamics using coherent electrons from high harmonic generation. *Proc.Natl. Acad. Sci.* 103, 13279–13285 (2006).
91. W. Li, X. Zhou, R. Lock, S. Patchkovskii, A. Stolow, H. C. Kapteyn, and M. M. Murnane. Time-resolved dynamics in N2O4 probed using high harmonic generation. *Science* 322, 1207–1211 (2008).
92. H. J. Wörner, J. B. Bertrand, P. B. Corkum, and D. M. Villeneuve. High-harmonic homodyne detection of the ultrafast dissociation of Br_2 molecules. *Phys. Rev. Lett.* 105, 103002 (2010).
93. A. D. Shiner, C. Trallero-Herrero, N. Kajumba, H.-C. Bandulet, D. Comtois, F. Légaré, M. Giguere, J.-C. Kieffer, P. B. Corkum, and D. M. Villeneuve. Wavelength scaling of high harmonic generation efficiency. *Phys. Rev. Lett.* 103, 073902–073904 (2009).

94. T. P. Rakitzis and T. N. Kitsopoulos. Measurement of cl and br photofragment alignment using slice imaging. *J. Chem. Phys.* 116, 9228–9231 (2002).
95. H. Eichmann, A. Egbert, S. Nolte, C. Momma, B. Wellegehausen, W. Becker, S. Long, and J. K. McIver. Polarization-dependent high-order two-color mixing. *Phys. Rev. A* 51, R3414 (1995).
96. H. J. Wörner, J. B. Bertrand, D. V. Kartashov, P. B. Corkum, and D. M. Villeneuve. Following a chemical reaction using high-harmonic interferometry. *Nature* 466, 604–607 (2010).
97. T. Kanai, E. J. Takahashi, Y. Nabekawa, and K. Midorikawa. Destructive interference during high harmonic generation in mixed gases. *Phys. Rev. Lett.* 98, 153904 (2007).
98. L. Nugent-Glandorf, M. Scheer, D. A. Samuels, A. M. Mulhisen, E. R. Grant, X. Yang, V. M. Bierbaum, and S. R. Leone. Ultrafast time-resolved soft x-ray photoelectron spectroscopy of dissociating Br_2. *Phys. Rev. Lett.* 87, 193002 (2001).
99. L. Nugent-Glandorf, M. Scheer, D. A. Samuels, V. M. Bierbaum, and S. R. Leone. Ultrafast photodissociation of Br_2: Laser-generated high-harmonic soft x-ray probing of the transient photoelectron spectra and ionization cross-sections. *J. Chem. Phys.* 117, 6108 (2002).
100. B. Zimmermann, D. Rolles, B. Langer, R. Hentges, M. Braune, S. Cvejanovic, O. Geszner et al. Localization and loss of coherence in molecular double-slit experiments. *Nat. Phys.* 4, 649–655 (2008).
101. W. Domcke, D. R. Yarkony, and H. Köppel, eds. *Conical Intersections: Electronic Structure, Dynamics and Spectroscopy*. Vol. 15. *Adv. Ser. in Phys. Chem.* (World Scientific, Singapore, 2004).
102. P. H. Bucksbaum. The future of attosecond spectroscopy. *Science* 317, 766–769 (2007).
103. H. J. Wörner, J. B. Bertrand, B. Fabre, J. Higuet, H. Ruf, A. Dubrouil, S. Patchkovskii et al. Conical intersection dynamics in NO_2 probed by homodyne high-harmonic spectroscopy. *Science* 334, 208 (2011).

21

Picoscale Electron Density Analysis of Organic Crystals

Yusuke Wakabayashi
Osaka University

21.1 Introduction

The atomic arrangement within a crystal defines the overall features of the system, such as one-, two-, or three-dimensional systems. Thus, structural information provides a fundamental idea of the physical understanding of a material. In some cases, precise structural information makes it possible to perform theoretical calculations that explain the physical/electronic properties. The band structure of organic conductors is well explained by molecular orbital calculations based on the structure, and therefore x-ray structure analysis has been done for many materials [1,2]. Another characteristic phenomenon in organic systems is the periodic arrangement of valence electrons, i.e., charge density waves or charge ordering. Here, we take α-(BEDT-TTF)$_2$I$_3$ as an example. This compound shows a metal–insulator transition at 135 K, and the low-temperature insulator phase is a charge-ordered phase. Theoretical work [3] predicted that the valence arrangement of the Bis(ethylenedithio) tetrathiafulvalene (BEDT-TTF) molecule can show various structures as a function of the ratio between the on-site, U, and inter-site, V, Coulomb interactions. The valence arrangement was clarified by an x-ray structure analysis [4], and the resulting structure agreed with the theoretical predictions with a reasonable U/V ratio.

While x-ray diffraction is widely used to determine the atomic arrangement in materials, what is directly observed by this method is the Fourier transformation of the electron density. In that sense, electron density analysis is the first step in x-ray structure analysis. Another perspective is that because of the large flexibility of the structural model that can be treated, electron density analysis is the final stage of x-ray structure analysis.

In principle, analysis of electron density provides the information of the chemical bond or the valence of a molecule. For bulk crystals, electron density analysis of this quality was successfully performed on several molecular systems. Section 21.3 explains how we can obtain such electron densities. When we then turn our attention to surfaces or interfaces, another method of x-ray diffraction—crystal truncation rod (CTR) scattering—is used. Since CTR is less popular than the common x-ray structure analysis method, only the electron density analysis in the first meaning is successfully performed for the organic material surfaces. Section 21.4 is dedicated to the observation of the near-surface structure. In order to provide a concrete background, we begin with a description of the fundamentals of x-ray scattering.

21.2 Fundamentals of X-Ray Diffraction

21.2.1 X-Ray Scattering from Crystals

When an x-ray electromagnetic wave illuminates an electron, the electron is vibrated by the electric field. The electron vibration creates an oscillating electric field of the same frequency as the incident one. This dipole radiation process is the elastic scattering of x-rays by an isolated electron.

Let us think about two electrons at positions $\boldsymbol{r}_1$ and $\boldsymbol{r}_2$ and an incident wave having the wavevector $\boldsymbol{k}$. The wavevector of the scattered x-ray is $\boldsymbol{k}'$. While the scattering wave emanates in every direction, the magnitude k' is the same as that of the incident wave, $k = |\boldsymbol{k}| = 2\pi/\lambda$, where λ is the wavelength. The situation is depicted in Figure 21.1.

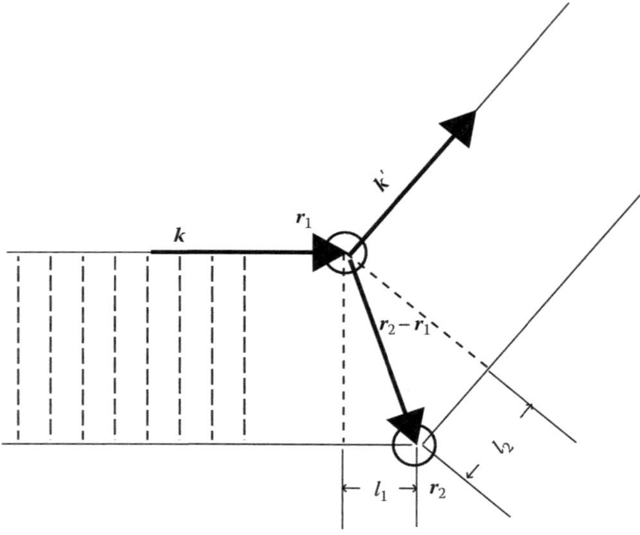

FIGURE 21.1 X-ray scattering from electrons. Optical path difference is $l_1 + l_2 = -\boldsymbol{Q} \cdot (\boldsymbol{r}_2 - \boldsymbol{r}_1)$.

Using unit vectors $\hat{\boldsymbol{k}}$ and $\hat{\boldsymbol{k}}'$, which are pointing in the directions of $\boldsymbol{k}$ and $\boldsymbol{k}'$, we can write l_1 and l_2 in Figure 21.1 as

$$\begin{aligned} l_1 &= \hat{\boldsymbol{k}} \cdot (\boldsymbol{r}_2 - \boldsymbol{r}_1), \\ l_2 &= -\hat{\boldsymbol{k}}' \cdot (\boldsymbol{r}_2 - \boldsymbol{r}_1). \end{aligned} \tag{21.1}$$

Thus, the optical path difference δ and the phase difference ϕ are

$$\delta = l_1 + l_2 = (\hat{\boldsymbol{k}} - \hat{\boldsymbol{k}}') \cdot (\boldsymbol{r}_2 - \boldsymbol{r}_1), \tag{21.2}$$

$$\phi = \frac{2\pi}{\lambda}\delta = k\delta = (\boldsymbol{k} - \boldsymbol{k}') \cdot (\boldsymbol{r}_2 - \boldsymbol{r}_1). \tag{21.3}$$

We define the scattering vector $\boldsymbol{Q}$ as $\boldsymbol{k}' - \boldsymbol{k}$ for simplicity. It denotes the momentum transfer from the sample to the x-ray photon. $\boldsymbol{Q}$ has a dimension of the inverse of the length, and thus the Q-space is called reciprocal space. $|Q| = 2k \sin\theta = 4\pi \sin\theta/\lambda = 2\pi/d$, where d is the lattice spacing.

A superposition of waves, $\exp(i\omega t - i\boldsymbol{Q}\cdot\boldsymbol{r})$, is expressed as the sum of the amplitude as the waves interfere. The scattering amplitude F is given by the following formula:

$$F(\boldsymbol{Q}) = f(\exp[i\boldsymbol{Q}\cdot\boldsymbol{0}] + \exp[i\boldsymbol{Q}\cdot(\boldsymbol{r}_2 - \boldsymbol{r}_1)]) \tag{21.4}$$

$$= f(1 + \exp[i\boldsymbol{Q}\cdot(\boldsymbol{r}_2 - \boldsymbol{r}_1)]), \tag{21.5}$$

where f denotes the scattering amplitude from an electron, and the time-dependent term $\exp(i\omega t)$ was omitted. So far, we have treated two electrons. In the case of an arbitrary number of electrons, the scattering amplitude is given by

$$F(\boldsymbol{Q}) = \sum_j f e^{i\boldsymbol{Q}\cdot\boldsymbol{r}_j}. \tag{21.6}$$

In reality, each electron distributes as a probability cloud, and the summation turns into an integral on the electron distribution $\rho(\boldsymbol{r})$, and we obtain

$$F(\boldsymbol{Q}) \propto \int_{all} \rho(\boldsymbol{r}) e^{i\boldsymbol{Q}\cdot\boldsymbol{r}} d\boldsymbol{r}. \tag{21.7}$$

As one can see, the scattering amplitude is the Fourier transform of the electron density.

Next, we derive the Bragg reflections. The electron density in a unit cell $\rho^c(\boldsymbol{r})$ is written as

$$\rho^c(\boldsymbol{r}) = \sum_{j=1}^{n} \rho_j^a(\boldsymbol{r} - \boldsymbol{r}_j), \tag{21.8}$$

where

$\rho_j^a(\boldsymbol{r})$ is the electron density around the jth atom at the origin
$\boldsymbol{r}_j$ denotes the positional vector of the jth atom
n is the number of atoms in a unit cell

Similarly, the total electron density distribution $\rho(\boldsymbol{r})$ is given by

$$\rho(\boldsymbol{r}) = \sum_{n_1,n_2,n_3} \rho^c[\boldsymbol{r} - (n_1\boldsymbol{a} + n_2\boldsymbol{b} + n_3\boldsymbol{c})], \tag{21.9}$$

where $\boldsymbol{a}$, $\boldsymbol{b}$, and $\boldsymbol{c}$ are the unit translational vectors, and the summations of n_1–n_3 are taken over the entire crystal. Using this formula, the scattering amplitude F given in Equation 21.7 is

$$F(\boldsymbol{Q}) \propto \int_{all} \sum_{n_1,n_2,n_3} \rho^c[\boldsymbol{r} - (n_1\boldsymbol{a} + n_2\boldsymbol{b} + n_3\boldsymbol{c})] e^{i\boldsymbol{Q}\cdot\boldsymbol{r}} d\boldsymbol{r}.$$

Let $\boldsymbol{r} - (n_1\boldsymbol{a} + n_2\boldsymbol{b} + n_3\boldsymbol{c})$ be $\boldsymbol{R}$, which gives

$$\begin{aligned} F(\boldsymbol{Q}) &\propto \sum_{n_1,n_2,n_3} \int_{all} \rho^c(\boldsymbol{R}) \exp[i\boldsymbol{Q}\cdot(\boldsymbol{R} + n_1\boldsymbol{a} + n_2\boldsymbol{b} + n_3\boldsymbol{c})] d\boldsymbol{R} \\ &= \sum_{n_1,n_2,n_3} \exp[i\boldsymbol{Q}\cdot(n_1\boldsymbol{a} + n_2\boldsymbol{b} + n_3\boldsymbol{c})] \int_{all} \rho^c(\boldsymbol{R}) e^{i\boldsymbol{Q}\cdot\boldsymbol{R}} d\boldsymbol{R}. \end{aligned} \tag{21.10}$$

Although it appears different from the popular expression, the integral part is called the crystal structure factor. Using Equation 21.8,

$$\begin{aligned} \int_{all} \rho^c(\boldsymbol{R}) e^{i\boldsymbol{Q}\cdot\boldsymbol{R}} d\boldsymbol{R} &= \int_{all} \left[\sum_{j=1}^{n} \rho_j^a(\boldsymbol{R} - \boldsymbol{r}_j) \right] e^{i\boldsymbol{Q}\cdot\boldsymbol{R}} d\boldsymbol{R} \\ &= \sum_{j=1}^{n} \left[\int_{all} \rho_j^a(\boldsymbol{R} - \boldsymbol{r}_j) e^{i\boldsymbol{Q}\cdot(\boldsymbol{R} - \boldsymbol{r}_j)} d(\boldsymbol{R} - \boldsymbol{r}_j) \right] e^{i\boldsymbol{Q}\cdot\boldsymbol{r}_j} \\ &\equiv \sum_{j=1}^{n} f_j(\boldsymbol{Q}) e^{i\boldsymbol{Q}\cdot\boldsymbol{r}_j}, \end{aligned} \tag{21.11}$$

we obtain the popular expression. $f_j(\mathbf{Q})$ is the atomic form factor, which is defined by the Fourier transform of the electron density of an atom.

The summation in Equation 21.10 can be separated into a product of three summations:

$$\sum_{n_1,n_2,n_3} \exp[i\mathbf{Q}\cdot(n_1\mathbf{a}+n_2\mathbf{b}+n_3\mathbf{c})] = \sum_{n_1} e^{in_1\mathbf{Q}\cdot\mathbf{a}} \sum_{n_2} e^{in_2\mathbf{Q}\cdot\mathbf{b}} \sum_{n_3} e^{in_3\mathbf{Q}\cdot\mathbf{c}}. \tag{21.12}$$

Let us calculate the $\mathbf{a}$ component, $\sum_n \exp[in\mathbf{Q}\cdot\mathbf{a}]$, with limits of n from $-N/2$ to $N/2$. This formulation is a geometrical series with the first term of $\exp[-i(N/2)\mathbf{Q}\cdot\mathbf{a}]$ and a geometric ratio of $\exp[-i\mathbf{Q}\cdot\mathbf{a}]$. Using the formula for a geometric series $a + ar + ar^2 + \cdots + l = (rl - a)/(r - 1)$, one obtains

$$\begin{aligned}\sum_{n=-N/2}^{N/2} e^{in\mathbf{Q}\cdot\mathbf{a}} &= \frac{e^{i\mathbf{Q}\cdot\mathbf{a}} e^{i\frac{N}{2}\mathbf{Q}\cdot\mathbf{a}} - e^{-i\frac{N}{2}\mathbf{Q}\cdot\mathbf{a}}}{e^{i\mathbf{Q}\cdot\mathbf{a}} - 1} \\ &= \frac{e^{i\mathbf{Q}\cdot\mathbf{a}/2} e^{i\frac{N}{2}\mathbf{Q}\cdot\mathbf{a}} - e^{-i\mathbf{Q}\cdot\mathbf{a}/2} e^{-i\frac{N}{2}\mathbf{Q}\cdot\mathbf{a}}}{e^{i\mathbf{Q}\cdot\mathbf{a}/2} - e^{-i\mathbf{Q}\cdot\mathbf{a}/2}} \\ &= \frac{\sin[(N+1)\mathbf{Q}\cdot\mathbf{a}/2]}{\sin[\mathbf{Q}\cdot\mathbf{a}/2]}.\end{aligned} \tag{21.13}$$

If we let $h = \mathbf{Q}\cdot\mathbf{a}/2\pi$, the equation is a function of h and has peaks at integer h positions. This is called the Laue function $L(h)$, and its shape is depicted in Figure 21.2.*

The value of N is huge in crystals, and thus the Laue function appears to be a periodic series of δ-functions. In three-dimensional space, the product of the three Laue functions is practically zero for most cases, returning a large value only when all three of the $h = \mathbf{Q}\cdot\mathbf{a}/2\pi$, $k = \mathbf{Q}\cdot\mathbf{b}/2\pi$, and $l = \mathbf{Q}\cdot\mathbf{c}/2\pi$ values are integer quantities. These spots are known as Bragg reflections.

The scattering amplitude $F(\mathbf{Q})$ is the Fourier transform of the electron density. As we have seen, the translational symmetry produces Bragg reflections. Therefore, the inverse Fourier transform of the Bragg amplitude (not intensity) provides the electron density having the translational symmetry.

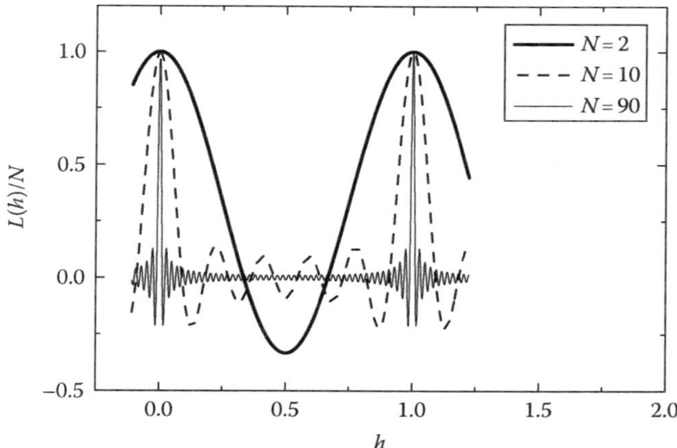

FIGURE 21.2 Laue function.

* In some textbooks, the Laue function is defined as $|L(h)|^2$.

21.2.2 X-Ray Scattering from Surfaces

Think about a crystal having a flat surface parallel to the c-plane. The vector $\mathbf{c}$ points inward to the crystal. The presence of the well-defined surface is expressed by introducing an absorption term, $e^{-\mu_c n_3}$, to Equations 21.10 and 21.11:

$$\begin{aligned}F(\mathbf{Q}) &= \sum_{n_1=-\infty}^{\infty}\sum_{n_2=-\infty}^{\infty}\sum_{n_3=0}^{\infty}\sum_{j=1}^{n} f_j \exp(i\mathbf{Q}\cdot(\mathbf{r}_j+n_1\mathbf{a}+n_2\mathbf{b}+n_3\mathbf{c}))\exp(-\mu_c n_3) \\ &= L(h)L(k)\sum_{j=1}^{n} f_j \exp(i\mathbf{Q}\cdot\mathbf{r}_j)\sum_{n_3=0}^{\infty}\exp(i\mathbf{Q}\cdot n_3\mathbf{c})\exp(-\mu_c n_3).\end{aligned}$$

The summation over n_3 is a geometrical series with a first term of 1 and a geometric ratio of $\exp(i\mathbf{Q}\cdot\mathbf{c})\exp(-\mu_c)$, and thus we can straightforwardly perform the calculation:

$$F(\mathbf{Q}) = \frac{L(h)L(k)}{1-\exp(i\mathbf{Q}\cdot\mathbf{c})\exp(-\mu_c)}\sum_{j=1}^{n} f_j \exp(i\mathbf{Q}\cdot\mathbf{r}_j). \tag{21.14}$$

μ_c denotes the absorption due to the thickness of c, and thus it usually is a tiny value. In the limit of $\mu_c \to 0$, the intensity can be written as

$$|F(\mathbf{Q})|^2 = \frac{|L(h)L(k)\sum_{j=1}^{n} f_j \exp(i\mathbf{Q}\cdot\mathbf{r}_j)|^2}{2-2\cos(\mathbf{Q}\cdot\mathbf{c})}. \tag{21.15}$$

This formula depicts a scattering intensity distribution that is as narrow as a δ-function in the h and k directions and is spread in the l-direction as $1/(1 - \cos(2\pi l))$. Such scattering due to the termination of crystals at the surface is called CTR scattering. One example of CTR scattering from a $SrTiO_3$ (001) surface is presented in Figure 21.3. Rod-shaped scattering perpendicular to the surface spread from Bragg reflections is clearly seen. The intensity in the middle of the Bragg reflections is as intense as the scattering from a single atomic layer. If an adsorbate is present on the sample, the scattering amplitude from the adsorbate interferes with the CTR scattering from the crystal. Even if the adsorbate is only one atomic layer thick, its scattering amplitude is comparable with the CTR amplitude, which results in a significant change in the intensity profile. For this reason, the CTR scattering profile is extremely sensitive to the surface structure. Similar discussion can be made for interfaces if each side of the interface has a large contrast in electron density. Therefore, for some cases, the interfacial structure can be well studied in the same manner.

The surface structure is reflected in the CTR signal. Therefore, if one obtains the amplitude of the CTR scattering, the electron

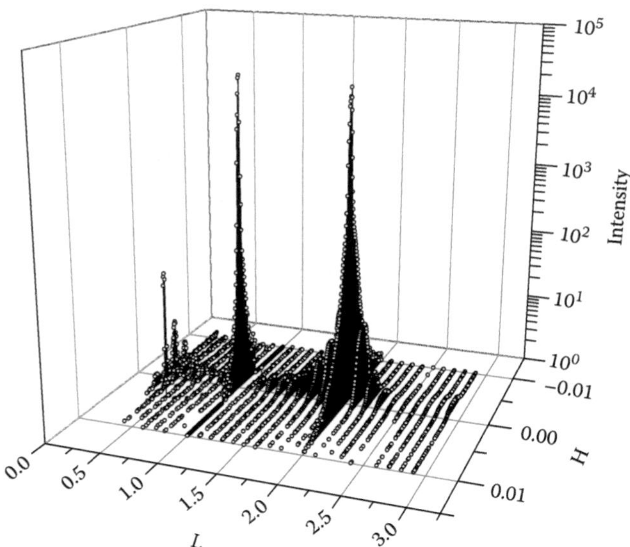

FIGURE 21.3 CTR profile of a $SrTiO_3$ (001) surface.

density distribution near the surface is straightforwardly calculated by performing the inverse Fourier transform. This technique provides unique structural information, the electron density profile along the depth direction, in a non-destructive manner. For example, low-energy electron diffraction or scanning tunneling microscopy (STM) techniques observe in-plane periodicity. Interfacial structure is sometimes studied by cross-sectional transmission electron microscopy (TEM), which requires thin slices of the sample. Pico-meter shifts in the atomic positions, which are often seen in electric polarization, are rarely observed in such microscopic techniques. In contrast, the CTR method easily observes electric polarization, and it allows us to observe deeply buried interfaces.

21.2.3 Phase Information

As we have seen earlier, x-ray diffraction can be used to observe the electron density distribution directly. The value observed by this method is clear. The only barrier for utilization of the x-ray scattering method is the lack of phase information. In real experiments, only the scattered intensity, which is the modulus square of the amplitude, is observed—phase information is missing. Therefore, electron density analysis is equivalent to the phase retrieval analysis. The phase retrieval method for bulk crystals is the main body of the x-ray crystal structure analysis, and various software such as SIR [5] have been developed for this task. Phase retrieval methods for surfaces have shown significant progress over the past decade, and this is the main topic of Section 21.4.

Let us see the importance of the phase information. The two pictures shown in Figure 21.4a and b are Fourier transformed into panels (c)–(f). Panels (c) and (e) show the modulus of the amplitude, and panels (d) and (f) show the phase. Inverse Fourier transform of (c) × (d) gives (g), and that of (e) × (f) gives (i), respectively. In these cases, everything is recovered. When we perform inverse transformation of (c) × (f) and (e) × (d), we obtain the pictures shown in panels (g) and (h), which look like original pictures for phase information.

In fact, this example exaggerates the phase importance. In this example, the complexity of the amplitude in (c) and (e) is great, and a wide area in the Q-space is covered with noticeable intensity. As a result, the intensity distributions shown in panels (c) and (e) have large similarity, which increases the importance of the phase. In order to make this effect clearer, a similar presentation for small lattices is shown in Figure 21.5. In this case, panels (c) and (e) show Bragg reflections. The mixed inverse Fourier transforms (h) and (i) show images close to the simple average of panels (a) and (b). The former situation, wide Q-range is covered with finite intensity, sometimes happens in powder x-ray diffraction data in high scattering angle region.

While it is easy for low frequency cases such as AC current, observation of scattering waves with phase information is extremely difficult for x-rays. Even if a single photon detector that can observe the x-ray phase is available, the observed phase varies significantly by the difference in position where the photon interacts with the detector owing to the time-dependent factor $\exp(i\omega t)$ that we ignored in our derivations. Therefore, we need some technique to determine the phase.

There are two ways of phase retrieval. One is a computational method that involves large numbers of iterations or trials. The other method utilizes the interference. Ordinary direct methods for x-ray structure analysis are classified into the former, and the x-ray standing wave method is the latter.

21.3 Electron Density Analysis of Bulk Organic Crystals

In ordinary x-ray structure analysis, we refine the structural parameters based on an assumption that the atoms are spherical, while the spherical symmetry is usually broken in real solids. This symmetry breaking makes a systematic deviation between the experimentally observed intensities of Bragg reflections and calculated ones, because the value x-ray scattering reflects is the electron density, as was seen in the previous section. One possible expansion of the ordinary analysis is using the multipole expansion of the atomic electron density [6]. While this method has been used for decades, it has not gained popularity because the number of fitting parameters increases greatly. Another method is the electron density analysis that refines the electron density distribution directly. The main topic of this section is the latter.

An example of electron density analyses is presented in Figure 21.6. It shows electron density maps of Si for the region exhibited in panel (a). Panel (b) shows the map obtained by the inverse Fourier transform of the scattering amplitude. Since the phase of a Bragg reflection is limited to 0 or π for systems having inversion symmetry, it is easy to obtain the scattering amplitude with the phase factor, which is sometimes called the

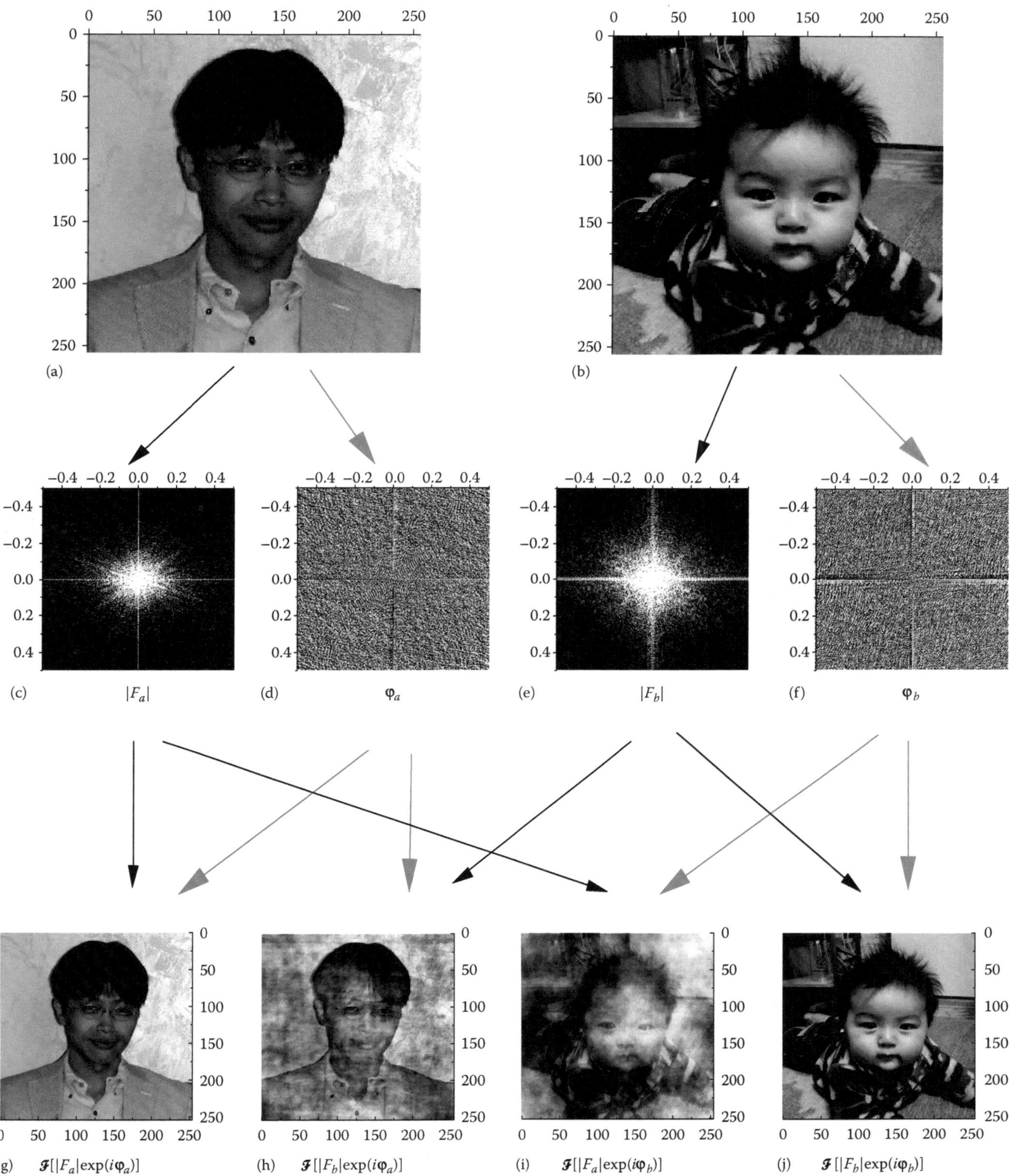

FIGURE 21.4 Fourier transform of photographs. The amplitude and the phase components of the Fourier transform of (a) are shown in (c) and (d), and those of (b) are in (e) and (f). (g–j): Inverse Fourier transform of (c and d), (e and d), (c and f), and (e and f).

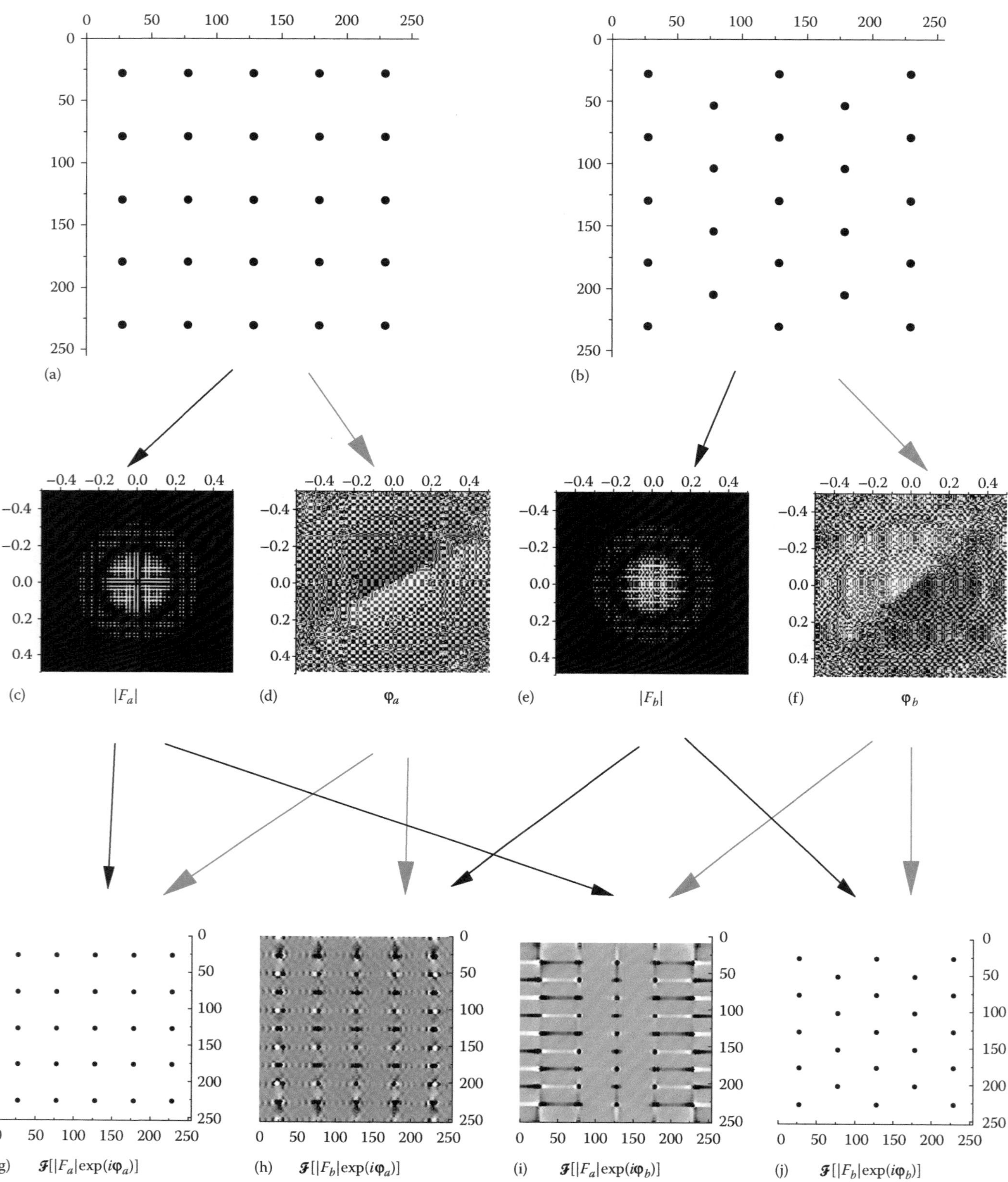

FIGURE 21.5 Fourier transform of lattices. The amplitude and the phase components of the Fourier transform of (a) are shown in (c) and (d), and those of (b) are in (e) and (f). (g–j): Inverse Fourier transform of (c and d), (e and d), (c and f), and (e and f).

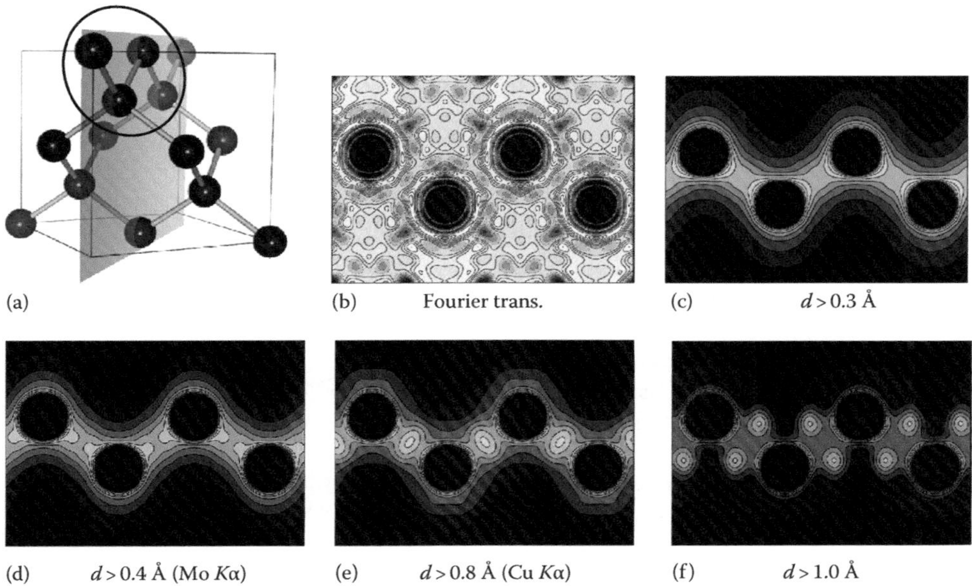

FIGURE 21.6 (a) Crystal structure of Si. The circle indicates the region depicted in the following panels. (b) Electron density map obtained by the inverse Fourier transform. (c–f) Electron density map obtained by the MEM applied to diffraction data over various *Q*-ranges. (Courtesy of Prof. E. Nishibori and Prof. H. Sawa, Nagoya University, Nagoya, Japan.)

complex amplitude, from the experimental result; the modulus of the amplitude is the square root of the intensity, and the phase is calculated based on the diamond structure. Although the atomic positions are clear in Figure 21.6b, regions with low electron densities show only noise and carry no information. This noise is caused by the termination error of the Fourier transform. Ordinary Fourier synthesis treats the complex amplitudes of the reflections that are not measured as zero, which makes a significant termination error. This termination error is not a problem for the observation of heavy elements but presents a serious barrier to observe bonding electrons. One solution for this problem is to observe all of the Bragg reflections up to exceedingly large values of *Q*. However, such an experiment is practically impossible. Instead, aided by information theory, a method that extracts the most probable electron density derived from the experimental data set has been developed: maximum entropy method (MEM). We present the advantages of this method first. Figure 21.6c shows the electron density map for the same region with panel (b) obtained by the MEM. While the atomic positions are unchanged, the information for the low electron density region significantly increases, and the bonding electron can easily be seen.

In order to obtain correct information down to such a low electron density region, high-quality experimental results as well as an appropriate analysis method are required. The popular MEM software package [7,8] allows us to study the electron density and to perform electron density analysis on materials with complex structures. On the other hand, the popularity of such software packages can increase false results because of careless use. In this section, we present the theoretical background, application, and tips for MEM electron density analysis.

21.3.1 Maximum Entropy Method

The "entropy" in the name "maximum entropy method" is not the entropy described in the thermodynamics sense, but is instead the information entropy in information theory. Using this entropy, one can deduce the most likely electron density distribution from the experimental result, which necessarily contains uncertainty, through the Bayesian inference process. The MEM determines the most flat electron density that reproduces the experimentally observed Bragg intensities within the error bars. We start from an introduction to Bayesian inference and proceed to the use of information entropy. After that, the application of MEM to the analysis of experimental diffraction data will be discussed.

The Bayesian probability used in Bayesian inference is an interpretation of the probability. It is conceptually different from the common frequency probability interpretation, whose value is determined only after an infinite number of trials and is therefore independent of the limitation of knowledge we can use at a certain moment. Bayesian probability takes into account the limitation of knowledge, and hence it can sometimes treat an object that is usually treated as non-probabilistic. For example, based on the Bayesian probability interpretation, Laplace said that the mass of Saturn is 1/3512 that of the Sun and that the odds are 1:11,000 that his estimate is off by more than 1% of the computed mass of Saturn [9]. Although the mass of Saturn is a fixed value, 1/3499 of the mass of the Sun, the uncertainty due to the limitation of his knowledge was treated as the probability distribution in the Bayesian probability interpretation. Another characteristic example is distorted dice. Think about a distorted six-sided die. What is the probability of getting six? When we use the frequency probability, we can say nothing based on only this information. When we use the Bayesian probability, the probability is one-sixth. This value is refined when we obtain additional information.

The definition of the information entropy S is

$$S = -\sum_i P_i \ln \frac{P_i}{Q_i}, \quad (21.16)$$

where P_i denotes the probability of getting the value i. Q_i is referred to as the prior probability, which reflects the probability distribution obtained a priori. Let us use the distorted dice as example. Fix Q_i to 1/6 and calculate S as a function of P_6. Figure 21.7a shows the value of S under a constraint of $P_i = (1 - P_6)/5$ $(i \neq 6)$; without this constraint, we need six dimensions to plot it. S has a maximum at $P_6 = 1/6$. This is the value we mentioned earlier. Let us add an item of information to the system. The die is so flat that only one or six can be obtained. Now, we have an additional constraint, $P_1 = 1 - P_6$ and $P_i = 0$ $(i = 2\text{–}5)$. The S value under these conditions is shown in panel (b). P_6 is maximized at 1/2. As can be seen, the maximum S gives the most probable situation with the knowledge available.

Let us rewrite the expression for the diffraction experiment analysis. There are several ways to apply MEM to the diffraction experiment. Here, we follow the method of Collins [10] and Sakata and Sato [11]. Divide the unit cell into M voxels. The electron density normalized by the total number of electrons for the ith voxel before and after the Bayesian inference are τ_i and ρ_i, respectively. Replacing Q and P in Equation 21.16 with τ_i and ρ_i, we obtain the entropy as

$$S = -\sum_{i=1}^{M} \rho_i \ln \frac{\rho_i}{\tau_i}. \quad (21.17)$$

Lack of the information maximizes the entropy with a flat electron density. The electron density is obtained by performing the inferences from an a priori electron density (in many cases, flat electron density is used as it implies no information) based on the given experimental information, i.e., the complex amplitude and the standard deviation. The experimental result is accounted for as a constraint. The constraint is

$$C = \frac{1}{N} \sum_{\mathbf{Q}} \frac{|F_o(\mathbf{Q}) - F_c(\mathbf{Q})|^2}{\sigma(\mathbf{Q})^2} = 1, \quad (21.18)$$

where

- N denotes the number of reflections
- $F_o(\mathbf{Q})$ is the experimentally observed scattering amplitude
- $F_c(\mathbf{Q})$ is the complex amplitude that is given by the Fourier transform of ρ_i
- $\sigma(\mathbf{Q})$ denotes the standard deviation of $F_o(\mathbf{Q})$

$C = 1$ implies that the MEM electron density reproduces the experimental result within the experimental error. Similarly, $C > 1$ means the MEM density gives larger error than the quality of the data, and $C < 1$ means the MEM density reproduces the experimental result too much, and experimental error affects the electron density. Note that Equation 21.18 is not a constraint that requires all the reflections to be reproduced within the error. It requires the MEM density to reproduce the whole data set to within the standard deviation of the data as a whole. The electron density distribution that gives a maximum S under the $C = 1$ condition is the most probable distribution.

The electron density can be calculated by using Lagrange's method of undetermined multipliers. Using the Lagrange multiplier λ, we can find the electron density that provides the maximum value of $S(\lambda)$:

$$S(\lambda) = -\sum_i \rho_i \ln \frac{\rho_i}{\tau_i} - \frac{\lambda}{2}(C - 1). \quad (21.19)$$

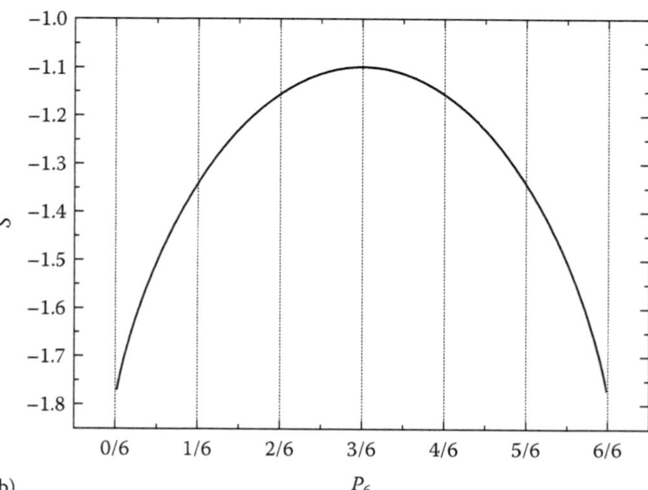

FIGURE 21.7 (a) Information entropy S for a distorted die as a function of P_6. P_i $(i \neq 6)$ are equal to each other. (b) S for the case that the die is as flat as a coin. In both cases, the most probable situation gives a maximum value of S.

The condition for ρ that provides extreme values of S is

$$\frac{\partial S(\lambda)}{\partial \rho_i} = 0. \tag{21.20}$$

Solving this, one obtains

$$\rho_i = \exp\left[\ln \tau_i + \frac{\lambda F_0}{N} \sum_{Q} \frac{1}{\sigma_i^2} [F_o(\boldsymbol{Q}) - F_c(\boldsymbol{Q})] \exp(i\boldsymbol{Q} \cdot \boldsymbol{r}_i)\right], \tag{21.21}$$

where

$\boldsymbol{r}_i$ is the vector that points the ith voxel

F_0 is the total number of electrons in a unit cell

In practice, Equation 21.21 cannot be solved directly because the F_c term on the right hand side is given by the Fourier transform of ρ, the left hand side. Instead, a numerical solution is obtained by substituting F_c by the Fourier transform of τ and performing an iterative calculation. The prior electron density τ for the next cycle is updated using ρ. The phase of F_o can be either fixed until the end of the iteration or updated by the phase of F_c in the previous cycle. The iteration is continued until C reaches unity to obtain the final MEM electron density.

The resulting MEM electron density contains Fourier components that are not included in the experimental data, since the real space structure has a tendency toward the uniform electron density to make S larger, and this tendency modifies the reciprocal space structure. The accuracy of the "inference" is presented in Figure 21.6. In panels (c through f), we show the MEM electron density of Si based on various 2θ-ranges of powder diffraction data. When we use these data down to small d-value ($d = \lambda/(2\sin\theta)$, small d means large 2θ-range), physically sound covalent bonds are observed (see panel c). Using a smaller range of reciprocal space, the shape of the covalent bond appears to be modified. Panel (e) shows the MEM electron density based on the data for $d > 0.8$ Å, which corresponds to the region accessible with Cu $K\alpha$ x-ray sources. There is a spurious peak in the middle of the atoms. As can be seen, achieving stable MEM densities and observation of the bonding electrons requires experimental data taken over a wide reciprocal space range. The "inference" by Equation 21.21 provides the flattest electron density that reproduces the experimental result within the error, with no consideration for physics. It is important to use as much information as possible to make little room for inference.

21.3.2 Electron Density Analysis of Organic Crystals

Electron density analyses using MEM have been applied on various materials including silicon, diamond, metal oxides, fullerenes and their variants (such as endohedral fullerenes), organic conductors, and ferroelectrics. In the last section, we presented the theoretical formulation of the MEM. In this section, after showing the experimental background, some examples of electron densities obtained from the MEM are presented.

There are two ways of collecting data for MEM electron density analysis: powder diffraction and single crystal diffraction. While there are merits and demerits for both methods, the latter has a significant advantage of peak discrimination. For the MEM electron density analysis, peak indices *hkl*, intensity, and the standard deviation are required as the experimental data. These values are directly observed in a single crystal diffraction experiment. In contrast, there are many processes required to obtain them from a powder diffraction result, because the peaks overlap.

Usually, MEM on powder data utilizes Rietveld analysis to resolve the overlapped peaks. Rietveld analysis involves a least-squares fitting for the powder diffraction profile by using structural parameters as well as peak shape parameters. When there are overlapped peaks, Rietveld software calculates the intensities of each peaks from the structural model and determines the *observed* intensities for each peak based on the calculated intensity ratio for each 2θ. This process works properly if the structure model is correct. However, if the structural model is incorrect, this process provides a bias toward the wrong structural model. Figure 21.8a shows three adjacent peaks. Peak positions are marked as short vertical bars, and the plots "Obs." show the result of the superposition of the three peaks labeled as "Real." Think about a model that expects the peak intensities of "Calc. (separate)." This provides a calculated profile of "Calc.(total)," which is similar to the observation, while the expected relative peak intensity is very different from the real one. Rietveld analysis splits the peaks based on the structural model in order to calculate the R-factor or to use MEM, and the *observed* intensities that the Rietveld derives are the dashed curves $I_{\text{"o"}}$. They are far from the real intensities and very close to those from the structural model. When peaks are superposed, Rietveld analysis is affected by the initial structural model significantly, and provides biased *observed* intensities. This results in biased MEM electron densities. Assuming the structural model in the real space is identical with assuming a phase set in the reciprocal space. Therefore, using an incorrect structural model with heavily overlapping peaks gives a real space structure similar to the initial, wrong structural model through a process analogous to that seen in Figure 21.4. Proper Rietveld analysis is essential for electron density analysis on powder diffraction data.

Let us see the degree of peak overlap in a real powder pattern. As we have seen in Figure 21.6, MEM requires data up to small d values. A powder pattern of a perovskite oxide $YTiO_3$ measured at the SPring-8 is presented in Figure 21.8b. The peak positions are marked with short vertical bars. The panel on the right shows a magnified view for $d \simeq 0.34$ Å. There are 264 peaks within two degrees in 2θ, and it is impossible to resolve each peak. The *observed* intensity should be affected greatly by the estimation of the background level.

A single crystal of the same material is also measured by using a curved imaging plate camera. One of the oscillation photographs is presented in panel (c). On the right hand side, a white

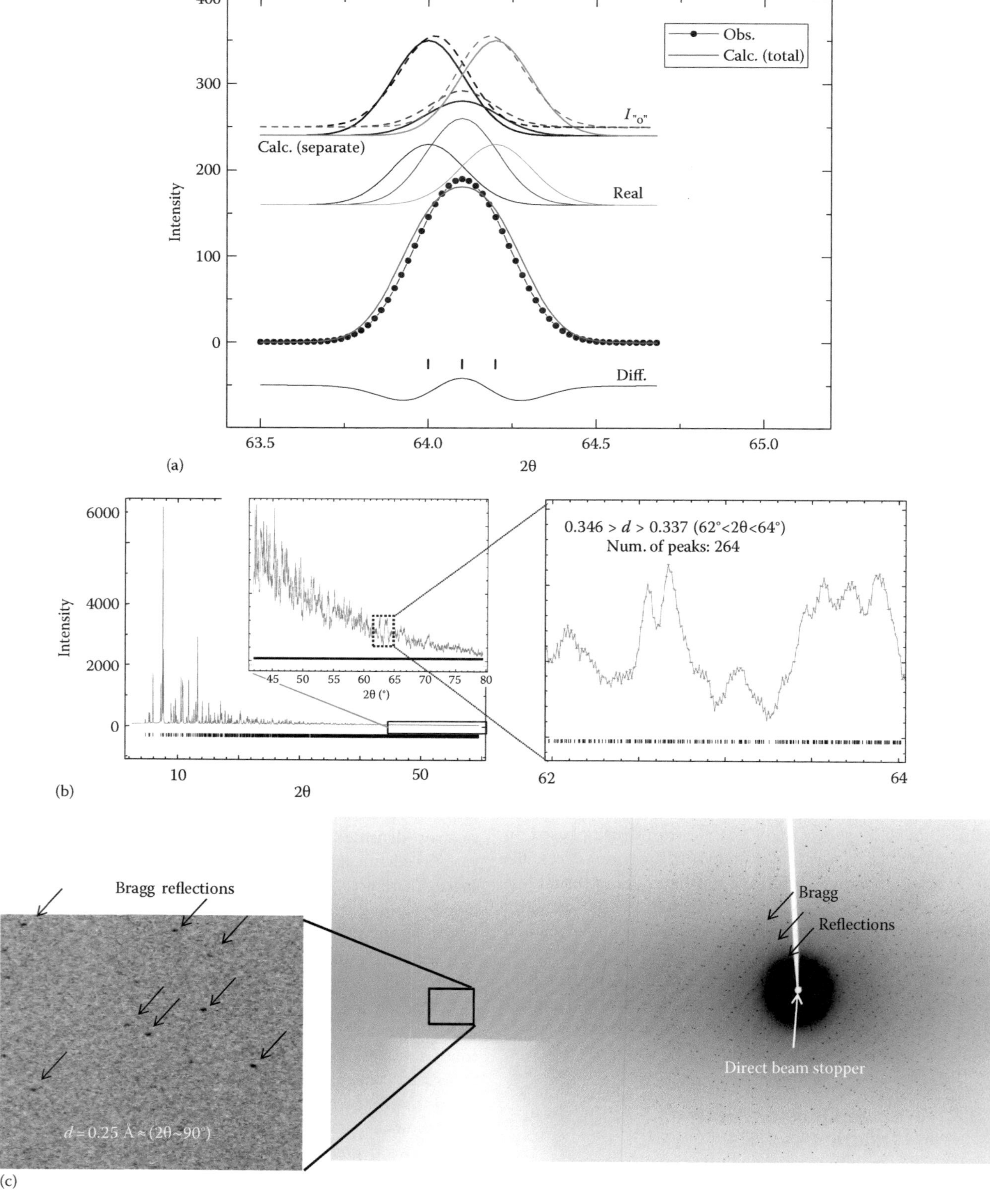

FIGURE 21.8 (a) Overlapped peaks are split using the structural model in Rietveld analysis. (b) Powder diffraction profile of perovskite $YTiO_3$. (c) Oscillation photograph of a single crystal of $YTiO_3$. (Courtesy of Prof. E. Nishibori and Prof. H. Sawa, Nagoya University, Nagoya, Japan.)

silhouette of the direct beam stopper is seen in the middle of the big black circle of forward air scattering background. Small Bragg reflections form lines in that region. The high scattering angle region corresponds to the area away from the beam stopper, and the small square on the left hand side corresponds to $d \simeq 0.25$ Å, $2\theta \simeq 90°$. The magnified view of this region clearly shows that each Bragg reflection is well separated, and the Bragg intensities are well observed in this region. In terms of peak separation, the advantage of the single crystal data is obvious. Instead, other aspects of data degradation, such as data completeness, anisotropic absorption, and extinction, are only problematic in the single crystal case, and need proper correction or special care. Currently, using single crystal data with special care to account for these factors is the best way to achieve precise electron densities.

An exemplar of MEM electron density analysis on powder diffraction data is a systematic study of endohedral fullerenes. The interior space of a fullerene is larger than the atom, which provides encapsulated atoms with the potential to adopt a wide variety of positions within the cage. The electron density of $Y@C_{82}$ ($A@C_n$ denotes a fullerrene C_n that encloses small ion or molecule A) was reported in 1995 [12]. After that, electron densities of many metallofullerenes have been reported. Since x-rays interact with electrons, lighter elements are harder to observe. Nevertheless, an endohedral hydrogen molecule in $H_2@C_{60}$ was successfully observed by this method [13] as shown in Figure 21.9. While metals generally stick to the inner walls of fullerenes, the H_2 molecule positions itself at the center of the interior space.

Using single crystal data and the MEM technique, the electron density distribution of $[Li^+@C_{60}]\ (PF_6)^-$ was reported [14]. Figure 21.10 shows the electron density distribution measured at 400 and 22 K. The Li ion is seen as a dot in the center of the 22 K panel. Significant differences in shape are seen in the fullerene and the PF_6 molecules between the two temperatures. Molecular shape is directly seen at the low temperature, while a totally different shape is found at the high temperature. This reflects the differences in thermal motion. Fullerenes rotate at high temperatures, and thus the electron density appears to be a sphere. PF_6 molecules also have a large disorder. Detailed observation of the electron density tells us that (1) the rotation of the fullerene in this material is not a free rotation but a hindered rotation, and (2) the positional distribution of Li ion has maxima near the PF_6^- molecules, showing some interaction between Li and PF_6. Electron density analysis helps to understand a structure that is hard to parameterize.

Next, we present an example of the observation of C–C bonding electrons in an organic material. MEM analysis on single crystal diffraction data of α-(BEDT-TTF)$_2$I$_3$, the charge ordering material we mentioned at the beginning of this chapter, reveals the molecular shape as shown in Figure 21.11. In the ordinary structure analysis, only the anisotropy of each atom

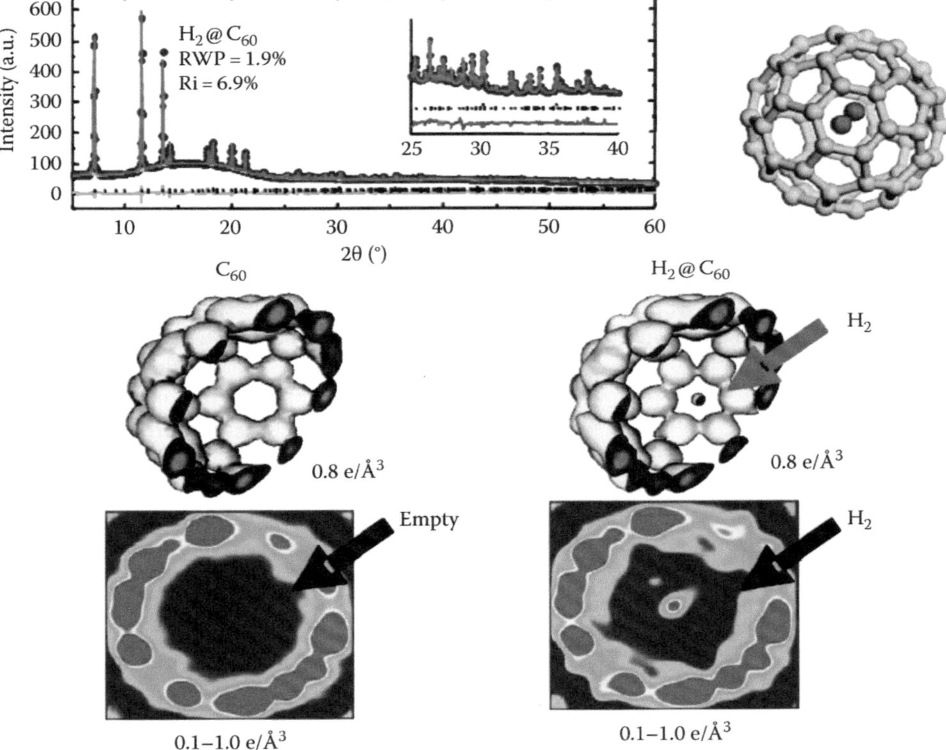

FIGURE 21.9 Powder diffraction patterns and equidensity surfaces of C_{60} and H_2C_{60}. (Taken from Kohama, Y., Rachi, T., Jing, J., Li, Z., Tang, J., Kumashiro, R., Izumisawa, S., Kawaji, H., Atake, T., Sawa, H., Murata, Y., Komatsu, K., and Tanigaki, K., *Phys. Rev. Lett.*, 103, 073001, Copyright 2009 American Physical Society.)

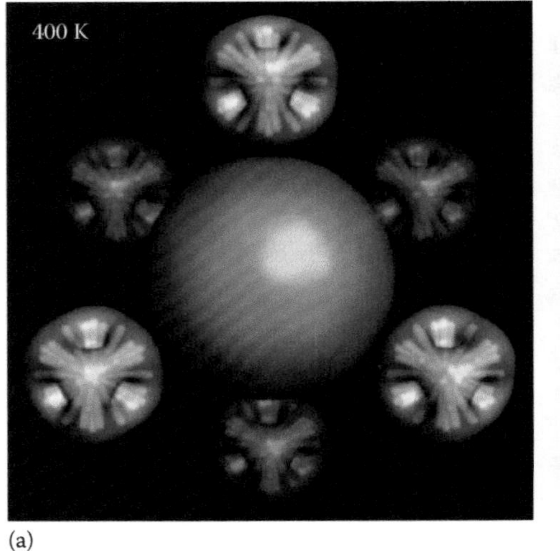

(a)

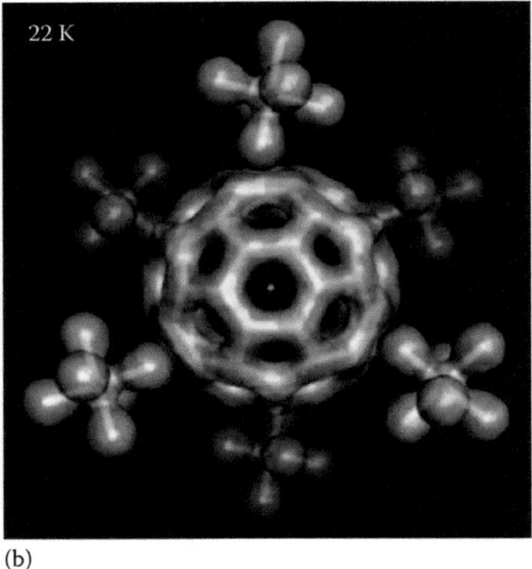

(b)

FIGURE 21.10 Equidensity surfaces of $[Li^+@C_{60}]$ $(PF_6)^-$ at 0.8 e/Å^3 viewed from [111] at (a) 400 K and (b) 22 K. (Taken from Aoyagi, S., Sado, Y., Nishibori, E., Sawa, H., Okada, H., Tobita, H., Kasama, Y., Kitaura, R., and Shinohara, H. *Angew. Chem.* 2012. 124. 3377. Copyright Wiley-VCH Verlag GmbH & Co. KGaA.)

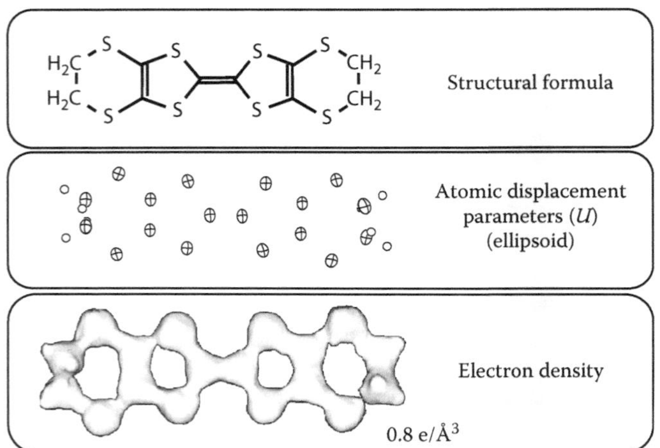

FIGURE 21.11 Structural formula, anisotropic atomic displacement parameters, and equidensity surfaces of a BEDT-TTF molecule in α-$(BEDT\text{-}TTF)_2I_3$. (Courtesy of Dr. T. Kakiuchi and Prof. H. Sawa.)

is expressed by the ellipsoid-shaped anisotropic atomic displacement parameters. By contrast, electron density depicts the difference between single bonds and double bonds. Double bonds are observed as areas of "thicker" density, and hydrogen atoms connected to carbon atoms are observed as small bumps in the electron density. Overall, the electron density gives more information.

The properties of organic systems are often dominated by the dynamics of protons. A change in proton dynamics that relates to the ferroelectric–paraelectric transition was reported in an organic ferroelectric [15]. The characteristic part of the material's crystal structure is shown in Figure 21.12a. Two protons connect the two molecules through hydrogen bonds, and the protons hop between the molecules. Proton hopping, accompanied by molecular twisting, is found only in the paraelectric phase. In the ferroelectric phase, the protons are static and the molecular twisting is also frozen. In order to visualize the positions of the hydrogen atoms, a differential MEM plot is presented in Ref. [15]. This plot utilizes a MEM density map deduced from a simulated intensity distribution. An electron density map was made from a set of Bragg intensities calculated from structural parameters with the target hydrogens omitted. Subtracting this map from the MEM density map based on the experimental result, one obtains the electron density for the target hydrogen atoms. Figure 21.12b and c show the differential MEM maps of the boxed area in panel (a) for the ferroelectric and paraelectric phases. The hydrogen atom splits in the paraelectric phase, while the electron density is disproportionate to one side in the ferroelectric phase.

21.3.3 Tips for MEM Analysis

Electron density analysis treats larger degrees of freedom than ordinary structure analysis. Since the sphericity of atoms is not assumed, deformation of the original data results in deformation of the atoms, which is physically impossible. In this section, experimental/analysis requirements and tips are listed.

1. Electron density analysis demands data collected over a wide dynamic range. Intensity lists obtained with excellent linearity are one aspect of "less biased" data. From this perspective, the classical combination of a point detector and a four-circle diffractometer is ideal. However, organic materials have long lattice parameters, which make the number of Bragg reflections enormous, and utilization of a two-dimensional detector system is favorable. The dynamic range of charge-coupled device (CCD) cameras is three to four orders of magnitude, which is insufficient for observation of the bonding electrons. Imaging plates have seven orders of magnitude of dynamic range, which is sufficient for this purpose.

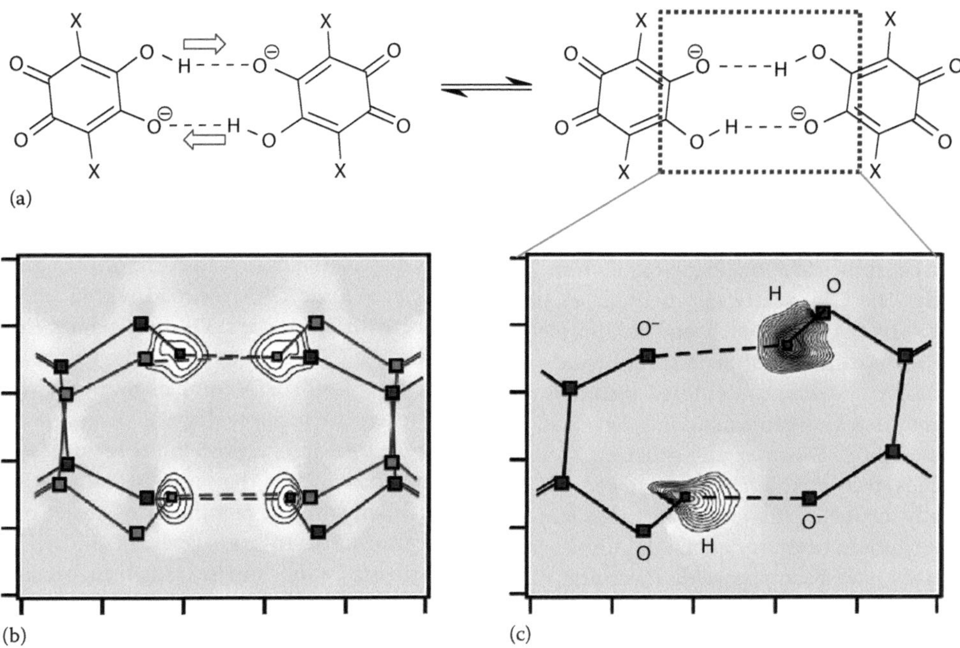

FIGURE 21.12 (a) Structural changes accompanied by proton hopping. (b) Electron density for hydrogen atoms in the paraelectric phase and (c) in the ferroelectric phase. (Taken from Horiuchi, S., Kumai, R., Tokunaga, Y., and Tokura, Y., *J. Am. Chem. Soc.* 130, 13382, Copyright 2008 American Chemical Society.)

2. Collect the data in a spherical region of the reciprocal space. This is another aspect of less biased data. MEM electron density based on a spherically measured low-Q data set is reported to be better than that based on data with several additional Bragg peaks in a high-Q-region [16]. Powder diffraction automatically bypasses this problem if the preferred orientation is avoided. In the case of single crystal diffraction with a two-dimensional detector, oscillation photographs taken with various rotating axes are needed to achieve completeness = 1. This method also increases redundancy of the data. Larger redundancy helps to enhance the quality of the data, since the effects which are hard to control or predict, such as multiple scattering, can be examined using larger datasets.
3. Another source of distortion in data is anisotropic absorption. Take a plate-shaped single crystal with thickness t and in-plane size L for example. Using the linear absorption coefficient of μ, the out-of-plane and in-plane absorptions are characterized by the product μt and μL. If $\mu L \gg 1$ and $\mu t \ll 1$, the incident/scattered x-ray along the in-plane direction suffers significant absorption, while an x-ray beam along the out-of-plane direction remains nearly intact. One experimental solution is to use a small crystal so that $\mu L < 1$. When μL is not too large, an absorption correction can be made with popular software. Similarly, extinction is often severe for single crystals, and proper corrections are needed.
4. Similar to ordinary structure analysis, the choice of the space group is important. Twin or multidomain samples should be avoided for use in the single crystal case.
5. The standard deviation of the intensity is also needed as input for the MEM electron density analysis. It is as important as the intensity itself. This requirement is different from ordinary structure refinement, in which the standard deviation is merely used for weighting of the least squares refinement. In some crystallographic analysis software, the standard deviation is not treated properly. The treatment of the standard deviation is also related to the detector properties. In the case of an imaging plate camera, the intensity of each pixel is read by either a high- or low-sensitivity photomultiplier. Pixels that receive strong signal intensities are read by the latter, and those that receive weak intensities are read by the former. Therefore, the statistical error has a discontinuity at intermediate intensity values. The image processing software has to treat the data with the photomultiplier selection pixel-by-pixel.

   Related to the standard deviation, higher background gives greater standard deviation for background intensity, which results in a decrease of information. Low-background measurement is always preferable.
6. As discussed in a previous section, the Rietveld analysis must be done properly for powder samples. The MEM electron density reflects the result of the Rietveld refinement, and thus the obtained structure must be examined to ensure that it is physically/chemically reasonable.
7. Iterative analysis generally gives better results. Starting from ordinary structure analysis, use MEM to find a better structure model, and perform least-squares refinement with the modified model. This cycle improves the phase information, and thus the final electron density.

21.4 Electron Density Analysis of the Surface of Organic Crystals

In the previous section, precise electron density analysis of bulk crystals was discussed. When we measure bulk crystalline samples, bonding electrons are easily observed. Now we focus our attention on surfaces or interfaces, which are important to modern electronic devices. The vast majority of these devices are based on Si, and surface studies of Si or silicon oxides have been performed extensively. The CTR scattering method, we introduced in Section 21.2, is often used for such studies. This method has been utilized in surface studies of Si, Au, Ag, liquids, and many other rather simple systems. By contrast, materials with complicated structures, such as organic materials, have not been examined using this method. However, the situation is changing. Organic solar cells and organic electroluminescence displays are commercially available, ink-jet printers that fabricate organic transistors are built in laboratories, and highly sensitive, flexible pressure sensors have been reported. These and future applications are likely to increase the importance of obtaining information about the surfaces of organic materials. In this section, the method and results of near-surface electron density analysis of organic semiconductor single crystals, which were first reported in 2010 [17], are introduced.

21.4.1 Phase Retrieval for Surface Scattering

There are various methods for studying surface structure, such as scanning probe microscopy, electron diffraction, and cross-sectional TEM. Most of the surface probes allow observation in the in-plane direction or are sensitive to only one atomic/molecular layer. Methods that can obtain structural information in the depth direction in non-destructive way are limited, and as explained in Section 21.2, the CTR method provides such information.

As mentioned in Section 21.2, CTR scattering provides the Fourier transform of the sharp step in the electron density at the surface or interface. Even for a continuum having no periodicity, flat surfaces provide step-function shaped electron density profiles along the surface normal direction, with scattering amplitudes expected to be proportional to $1/Q_\perp$, which is provided by the Fourier transform of the step function ($Q_\perp$ denotes the surface normal component of the scattering vector). All we have to know to obtain the electron density is the phase of the scattered x-ray.

In this decade, significant improvement has been achieved on the phasing of CTR scattering. Several groups [18,19] implement the "oversampling method [20,21]," which is an iterative method and has been used for x-ray diffraction microscopy. Another approach is in-line holography, which utilizes the fact that the CTR scattering amplitude is dominated by that from a known structure. In this section, the two methods are introduced. Since the phase retrieval is a purely mathematical process, readers who are not interested in the details can skip the rest of this section.

Oversampling is an analysis method that makes use of the fact that a fine-step measurement in the reciprocal space corresponds to a measurement for a wide volume in real space. Let us examine the amount of information for the measurement of a nanometer scale one-dimensional sample. There is an electron density profile along a line within the length of A. When we digitize the electron density by splitting the sample into N parts, we can write the density as $\rho(r_i)$ with $0 \leq i < N$. The discrete Fourier transform of $\rho(r_i)$ gives the discrete complex amplitude $G(Q_i)$ defined within $2\pi N/A$ in the reciprocal space with $2\pi/A$ steps. On first glance, the information in the real space (N real numbers in the real space) is doubled by the Fourier transform (N complex numbers in the reciprocal space). However, the reciprocal space always has inversion symmetry, and the total information in the reciprocal space is $N/2$ complex numbers, which is the same as that of N real numbers. This is consistent with the fact that the Fourier transform is an invertible transform. In practice, however, half of the information is missing from the Fourier transform, because we cannot observe the phase. In the real experiment, we can measure the reciprocal space as fine as the instrumental resolution allows. When we measure $C \cdot N$ points within the range of $2\pi N/A$ in the reciprocal space, the corresponding real space is as wide as $C \cdot A$, while the resolution remains A/N. For reference, the real space resolution is defined by the range of reciprocal space, or the value of N. In the case of $C > 1$, regions outside of the sample are observed. Using the knowledge that no electrons reside outside of the sample, one can obtain sufficient information on the real space structure only from the intensity if we choose a large enough C. Concretely speaking, this situation is sufficiently met when $C > 2$ for one dimension, or, in general, when $(C \cdot N)^n > 2$ for n dimensions with digitization of $(C \cdot N)^n$ pixels. This requirement is called the oversampling condition. The large C results in a wide zero-density region in the observable range. One can choose the phase so that the zero-density region has no electrons.

While the oversampling condition is merely a necessary condition for a phase retrieval, it is known that the procedure shown in Figure 21.13 practically provides a stable solution for two- or higher dimensional cases [20]. Starting from a random electron density, the phase is recovered during the iteration that uses the constraints in the real and reciprocal spaces. $\rho_n(\boldsymbol{r})$ and $\rho_{n+1}(\boldsymbol{r})$ denote the electron

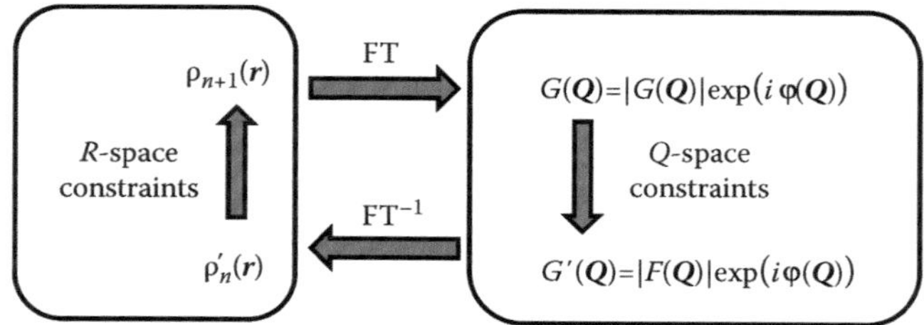

FIGURE 21.13 Phase retrieval scheme in the oversampling method. $\rho'_n(\boldsymbol{r})$ and $\rho_{n+1}(\boldsymbol{r})$ denote the electron density before and after the application of the real space constraint in the nth iteration, $G(\boldsymbol{Q})$ is the complex amplitude provided by the Fourier transform of $\rho_{n+1}(\boldsymbol{r})$, and $|F(\boldsymbol{Q})|$ is the square root of the experimental intensity.

density before and after the application of the real space constraint in nth iteration, and the real space constraint is

$$\rho_{n+1}(\boldsymbol{r}) = \begin{cases} \rho'_n(\boldsymbol{r}), & \boldsymbol{r} \text{ points inside of the sample and } \rho'_n(\boldsymbol{r}) > 0, \\ \rho_n(\boldsymbol{r}) - \beta\rho'_n(\boldsymbol{r}), & \text{otherwise}, \end{cases} \tag{21.22}$$

where β is a parameter whose value is ~ 1. The constraint in the reciprocal space is just a replacement of the magnitude of the calculated complex amplitude by the square root of the measured intensity as shown in Figure 21.13. At a certain n, inverse Fourier transform of the complex amplitude $G'(\mathbf{Q})$, which satisfies the reciprocal space constraint, gives the electron density $\rho'_n(\boldsymbol{r})$, which may have finite electron density outside of the sample. The real space constraint reduces the electron density outside of the sample at the cost of breaking the reciprocal space constraint. Iterative calculation of this procedure provides a solution.

X-ray diffraction microscopy requires a nanometer sized sample, which allows us to use the real space constraint, and a coherent x-ray whose coherence length is longer than the sample size. It is applicable to rather limited systems. When we apply the algorithm to the analysis of CTR scattering, the range of probable systems is extended. CTR scattering shows a continuous intensity distribution perpendicular to the surface, and thus one can collect the data that satisfy the oversampling condition. The real space constraint should be modified; one side is vacuum and the other side is the bulk structure.

The uniqueness of the solution is not very obvious when we use the oversampling method. It is reported that in two- or higher dimensional cases, solution multiplicity is pathologically rare [22]. In one-dimensional cases, oversampling often provides a false solution [23]. However, it is also reported [24] that phase retrieval of the CTR scattering amplitude is effectively stable and yields reproducible, well-defined interfacial structures when the real space constraints of vacuum, bulk structure, and positivity are properly used. There remains room for study of the uniqueness of the CTR phase solution.

Another phasing method, the holography method, is simpler than the oversampling method. The CTR amplitude from a real sample is the result of an interference between the CTR amplitude from a flat, perfect surface for the bulk structure $F_B(\mathbf{Q})$ and the amplitude from a thin unknown structure that involves the surface reconstruction or relaxation $F_S(\mathbf{Q})$. The intensity is given by $|F_B(\mathbf{Q}) + F_S(\mathbf{Q})|^2$, where the modulus and the phase of F_B can be calculated from the known structure. Therefore, information of the phase of $F_S(\mathbf{Q})$ can be extracted by using the interference with F_B. While there are several holographic methods, here we use the most simple method, the Takahashi method [25] as an example.

The difference between the observed CTR intensity $I(\mathbf{Q})$ and the calculated intensity based on an ideal surface (sudden termination of a bulk structure without any surface effect) $I_c(\mathbf{Q})$ is

$$I(\mathbf{Q}) - I_c(\mathbf{Q}) = F_B^*(\mathbf{Q})F_S(\mathbf{Q}) + F_B(\mathbf{Q})F_S^*(\mathbf{Q}) + |F_S(\mathbf{Q})|^2. \tag{21.23}$$

Here, we assume F_S is small. This assumption is always correct in the vicinity of the Bragg reflections. In this case, the last term can be neglected. Dividing both sides by $F_B^*(\mathbf{Q})$ gives

$$\frac{I(\mathbf{Q}) - I_c(\mathbf{Q})}{F_B^*(\mathbf{Q})} = F_S(\mathbf{Q}) + \frac{F_B(\mathbf{Q})F_S^*(\mathbf{Q})}{F_B^*(\mathbf{Q})}. \tag{21.24}$$

Fourier transform of the first term, $F_S(\mathbf{Q})$, gives the electron density corresponding to the unknown structure, which must be a real number. Although the Fourier transform of the second term gives some noise, it is not necessarily a real value. Thus, the real part of the Fourier transform provides information regarding surface adsorbents or atomic displacements.

21.4.2 Electron Density Analysis near the Surface

In this section, we present several examples of electron density analysis based on CTR scattering profiles. While there are several CTR studies on organic materials, only one electron density analysis has been reported so far. By contrast, there are many examples of studies of semiconductor and metal oxide surfaces. Here, we take one example from a metal oxide ultrathin film [26].

Figure 21.14 shows the electron density distribution of a five-unit-cell thick $LaAlO_3$ (LAO) epitaxial film fabricated on a $SrTiO_3$ (STO) substrate. Both LAO and STO have an ABO_3 perovskite structure. The top panel shows the electron density of an AO plane, and the bottom panel shows that of a BO_2 plane. The electron density is extracted from 14 inequivalent CTR profiles by one of the holographic methods—coherent Bragg rod analysis (COBRA) [27]. While the COBRA method requires a good initial guess of the surface structure, it does not provide ghost noise like the Takahashi method. The main purpose of the electron density analysis on a system having rather simple structure is to obtain an excellent starting point for the refinement. The reason why such a technique is required is the large number of structural parameters. In ordinary perovskite relatives, there are only 5–10 structural parameters. The number of parameters refined in Ref. [26] was 90, which is far greater than in the bulk case. This increase in parameter number arises from breaking translational symmetry. Atoms in the first layer feel a different environment than in the second layer, and thus the structure varies as a function of the depth. As a result, large numbers of atoms have to be treated as unique. In cases of metal oxides, however, this problem remains manageable. In fact, in some studies [26,28], least-squares fitting could be performed to find precise atomic positions.

By contrast, for studies of organic material surfaces, the electron density analysis is the final destination of the surface structure analysis. The reason is, again, the number of parameters. Figure 21.15 shows the structure of the molecular semiconductor, rubrene, as well as its CTR scattering intensity profile.

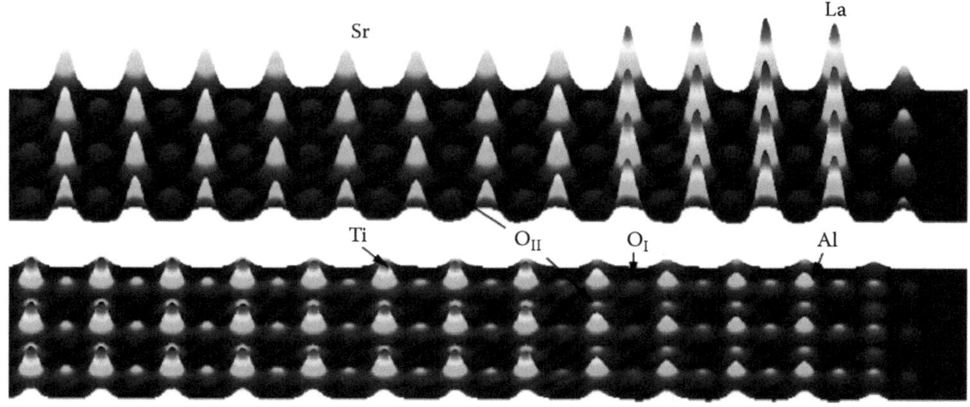

FIGURE 21.14 Result of the electron density analysis of the $LaAlO_3/SrTiO_3$ interface. The top panel shows the electron density on an *A* O plane, and the bottom panel shows that on a BO_2 plane. (Taken from Willmott, P.R., Pauli, S.A., Herger, R., Schlepütz, C.M., Martoccia, D., Patterson, B.D., Delley, B., Clarke, R., Kumah, D., Cionca, C., and Yacoby, Y., *Phys. Rev. Lett.* 99, 155502, Copyright 2007 American Physical Society.)

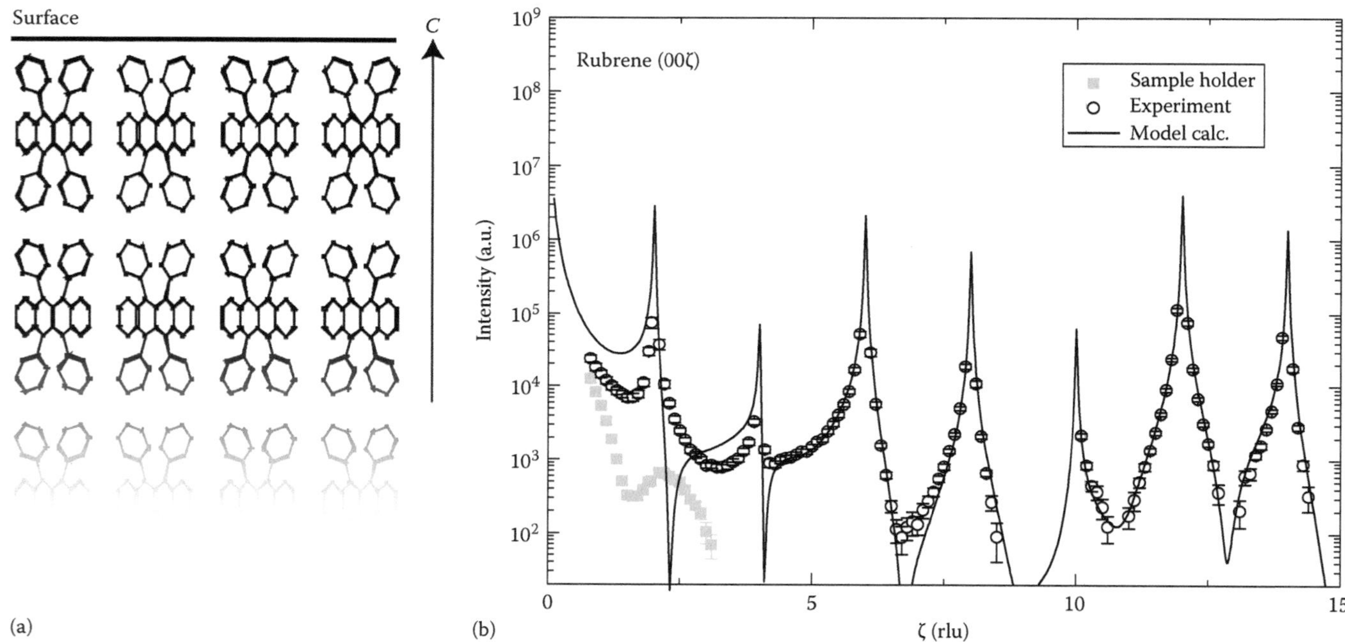

FIGURE 21.15 (a) Crystal structure of rubrene. (b) CTR scattering profile of rubrene single crystal. (From Wakabayashi, Y. et al., *Phys. Rev. Lett.*, 104, 066103, 2010.)

The rubrene molecule, $C_{42}H_{28}$, contains 70 atoms. If we refine the *z* coordinate and the isotropic atomic displacement parameter for three molecular layers, we have 420 parameters to deal with, but the available information with which to fit these parameters is only that shown in Figure 21.15b. A least-squares refinement over such a parameter set is impossible.

Model calculations were performed to find overall features of the surface structure of rubrene. Although it is impossible to treat all of the atoms uniquely, it is feasible to try several parameters for the *molecule*. The effects of molecular distortion or molecular displacement parameters on the CTR profile were systematically examined. The results of the calculation are summarized in Figure 21.16. In addition to the ideal surface, (a) molecular displacement due to thermal vibration or surface reconstruction, (b) molecular expansion/contraction (contraction may mean the tilt of the molecule), and (c) surface adsorption of some light elements were assumed for the calculations. Molecular expansion around the surface, which is called surface relaxation, causes asymmetric profiles between two Bragg reflections [29], while molecular displacement decreases the CTR intensity in a symmetrical manner. In both cases, the high-*Q* region suffers larger effects. Conversely, surface adsorbates modify the low-*Q* intensities. The structure model that best reproduces the experimental result is that in which there

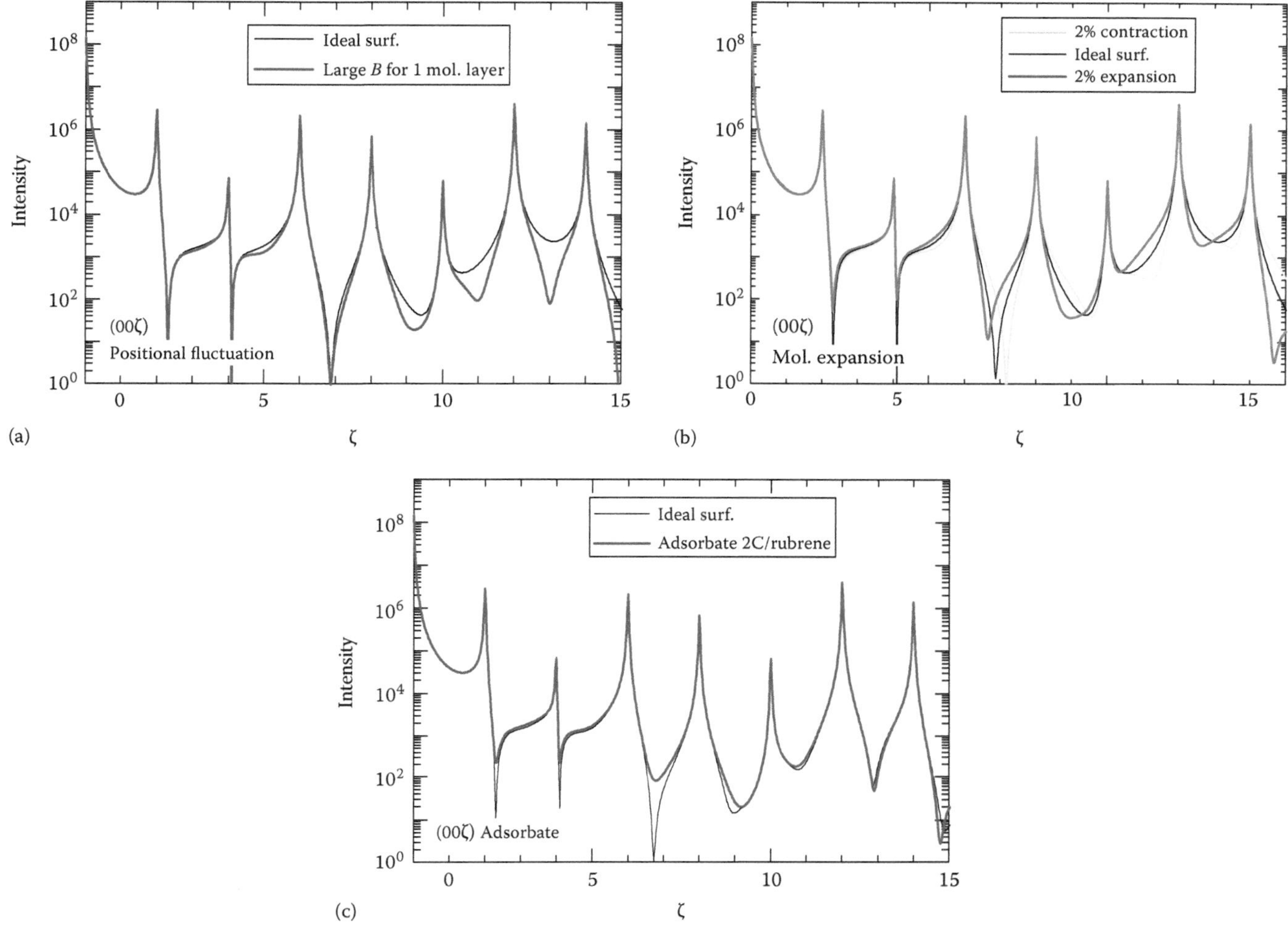

FIGURE 21.16 Model calculations of the CTR profile of rubrene along the (00ζ) axis with various assumptions of the surface behavior. (a) Larger thermal vibration in the first molecular layer, (b) expansion and contraction of the first molecular layer, and (c) adsorption of some light element are assumed for the calculation.

is no adsorbate, large molecular displacements in the outermost layer, and 1% of molecular expansion perpendicular to the surface. The calculated CTR profile is shown in Figure 21.15b by a solid curve.

The electron density profile obtained by the COBRA method applied to Figure 21.15b is presented in Figure 21.17. What we can read from the electron density are (1) each molecule is observed as six peaks, i.e., each six-membered ring is seen as two peaks, and (2) only the first molecular layer has noticeable differences from the bulk structure.

The reliability of this analysis is examined. As mentioned in the last section, one-dimensional phase retrieval sometimes provides a false solution. There are several ways to rule out the false solution: using many CTR profiles to form two- or three-dimensional data, or exclusion of unphysical results such as negative electron densities or over-deformation of chemical bonds. In this case, let us examine the number of electrons within the phenyl groups and tetracene backbone as a function of the depth. They have to be constant. Figure 21.18a shows the electron numbers of the surface-side phenyl group, the tetracene backbone, and the bulk-side phenyl group as a function of the depth. All of them show a flat dependence down to 180 Å in depth, which justifies the electron density profile.

Next, let us see the surface relaxation. In order to perform detailed comparisons, the electron densities of the first ($\rho 1(z)$) and third molecular layer ($\rho 3(z)$) are superposed in Figure 21.18b. One can see two features. The molecule in the first layer is larger than that in the third layer, and the electron density of the first layer displays lower contrast. The difference in contrast reflects the positional distribution of the molecules. As shown by the dashed curve in Figure 21.18b, a distribution of ±0.25 Å in the molecular position reproduces this contrast. However, the molecular size is not reproduced by this positional distribution. While there can be various modes of molecular distortion, we focus our attention on the tetracene backbone, where most of the highest occupied molecular orbital (HOMO) density resides, as shown in Figure 21.18c. Since only the depth dependence of the electron density is provided, only one mode of the

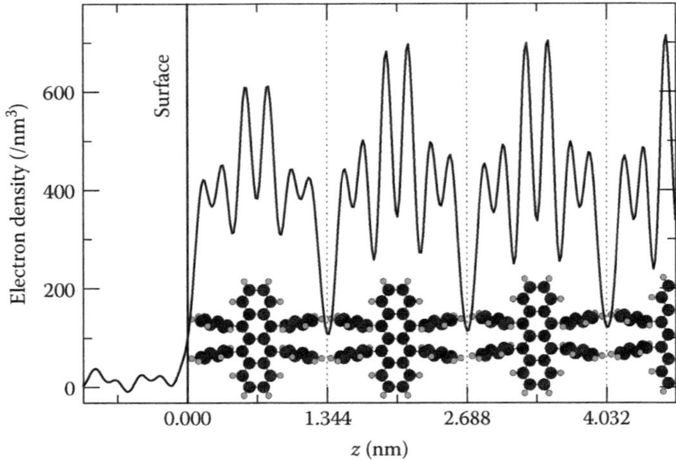

FIGURE 21.17 Electron density profile of rubrene single crystal. (From Wakabayashi, Y. et al., *Phys. Rev. Lett.*, 104, 066103, 2010.)

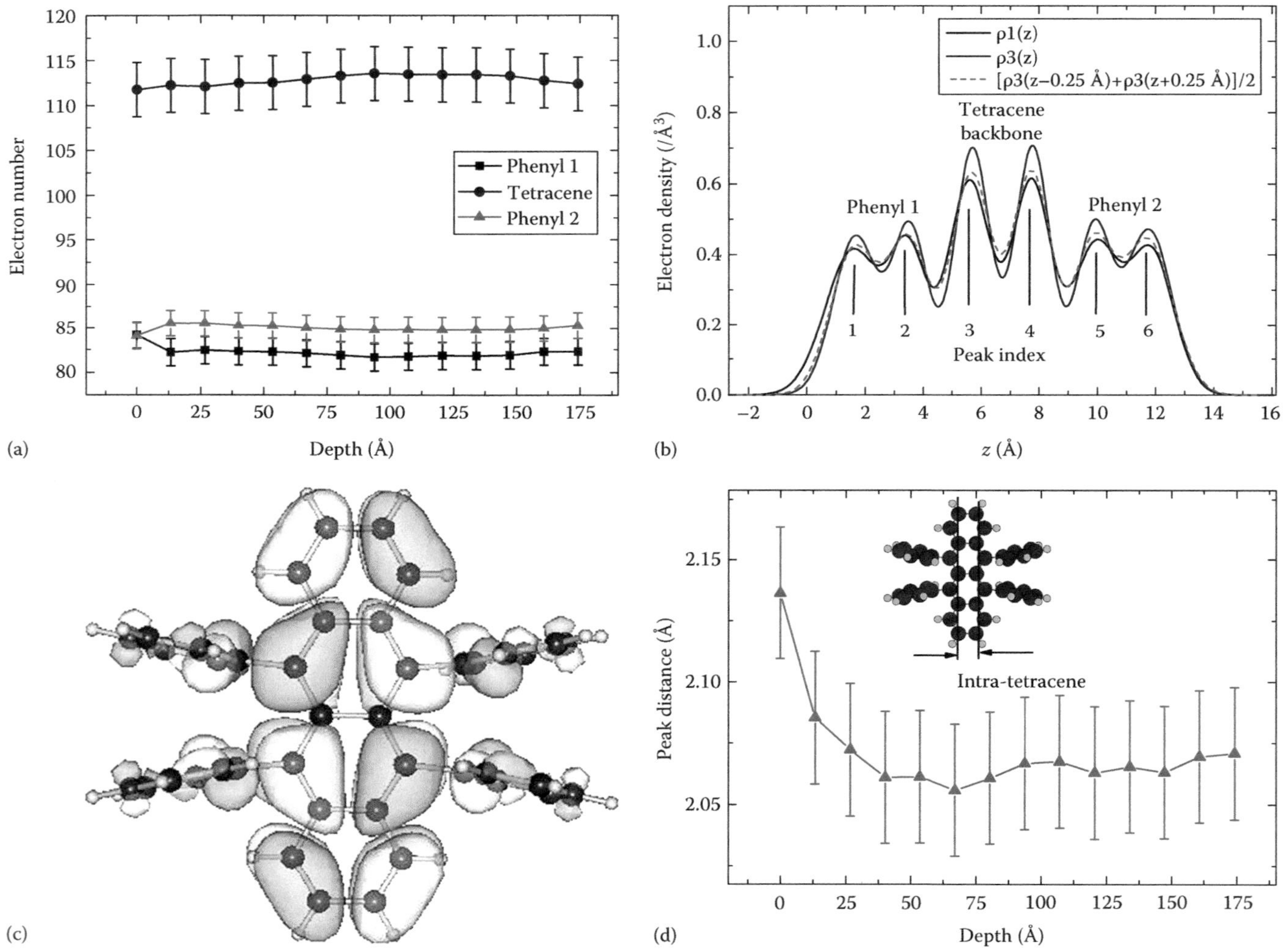

FIGURE 21.18 (a) Electron numbers of each phenyl group and tetracene backbone as a function of the depth. (b) Electron density profile of the molecules belonging to the first layer $\rho1(z)$ (thin solid curve) and third layer $\rho3(z)$ (thick curve). The dashed curve shows $[\rho3(z - 0.25 \text{ Å}) + \rho3(z + 0.25 \text{ Å})]/2$, which illustrates the electron density for molecules having a height distribution of ±0.25 Å. (c) HOMO of the rubrene molecule. (d) Depth dependence of the intra-tetracene peak distance.

distortion related to the tetracene backbone can be obtained, i.e., the distance between peak 3 and peak 4 in panel (b). The depth dependence of this peak distance is shown in panel (d). The plot demonstrates that some deformation of the tetracene backbone happens in the molecule in the first layer, and the magnitude is 0.1 Å. A molecular orbital calculation was performed under an assumption that all of the C–C bonds connecting peaks 3 and 4 were elongated by this amount. The result indicated that the HOMO energy decreases by 0.1 eV, which subsequently would decrease the number of carriers (holes) at the surface. As shown in this section, electron density analysis of the surface provides insight into the surface behavior.

21.5 Concluding Remarks

Two entirely different methodologies for electron density analysis are introduced. The first method is used to observe the electron density distribution in bulk crystalline samples, and the second method is for the near-surface region. In both cases, subatomic resolution is achieved. The former technique provides information about the bonding electrons, which is impossible to obtain from ordinary structure analysis. Note that the ordinary single crystal structure analysis provides highly accurate atomic positions, and the standard deviations of these atomic positions are often as low as 0.01 pm. This precise knowledge of atomic positions provided by ordinary structure analysis allows us to extract the bonding electron information as the discrepancy from the spherical atom approximation. The latter method is useful to clarify near-surface structural information of organic materials, which has oftentimes been left unknown because of the complex nature of the interface. Both methods allow us to interrogate aspects of the systems we could not previously observe and provide highly flexible information. These features require high-quality experimental data. When an ordinary crystallographic analysis is applied to distorted data, no atomic arrangement reproduces the distorted data well, which provides a larger R-value. In the case of electron density analysis, a distorted intensity dataset produces a distorted electron density. Collecting diffraction data in wide reciprocal space with little distortion is a technically challenging task. Nevertheless, high-quality data collection is vitally important to obtain reliable electron densities.

It is also important that we are careful when using the resulting electron density. There is no established way to estimate the error in electron density distributions. This is a severe limitation of electron density analysis. Especially in the MEM electron density analysis, statistical errors of the Bragg intensities were also used as information input. This makes for a complicated case for error estimation. In the meantime, these methods provide beautiful three-dimensional electron density distribution, details of which can be convincing but misleading, i.e., a feature in the electron density that is lower than the significance level may be treated to be significant. The reliability of the result has to be examined especially if one wants to discuss subtle features in the electron density.

Once the result is found to be reliable, a range of information can be obtained from the electron density. Electron density analysis can provide fundamental information to understand the electronic, magnetic, or thermal properties of materials.

Acknowledgments

The author is grateful to Prof. H. Sawa and Prof. E. Nishibori for helpful discussions and providing graphical materials. Financial support from the Toray Science Foundation, kakenhi (23684026), and the Global COE Program (G10) are also acknowledged.

References

1. T. Mori et al., *Bull. Chem. Soc. Jpn.* **57**, 627 (1984).
2. M. Watanabe, Y. Nogami, K. Oshima, H. Ito, T. Ishiguro, and G. Saito, *Synth. Met.*, **103**, 1909 (1999).
3. H. Seo, *J. Phys. Soc. Jpn.* **69**, 805 (2000).
4. T. Kakiuchi, Y. Wakabayashi, H. Sawa, T. Takahashi, and T. Nakamura, *J. Phys. Soc. Jpn.*, **76**, 113702-1-4 (2007).
5. M. C. Burla et al., *J. Appl. Crystallogr.* **38**, 381 (2005).
6. N. K. Hansen and P. Coppens, *Acta Crystallogr. A* **34**, 909–921 (1978).
7. F. Izumi and R. A. Dilanian, *Recent Research Developments in Physics*, Vol. 3, Part II, Transworld Research Network, Trivandrum, India (2002), pp. 699–726.
8. H. Tanaka, M. Takata, E. Nishibori, K. Kato, T. Iishi, and M. Sakata, *J. Appl. Crystallogr.* **35**, 282 (2002).
9. J. Tabak, *Probability and Statistics: The Science of Uncertainty*, Chapter 4, Facts on File, Inc., New York (2004).
10. D. M. Collins, *Nature* **298**, 49 (1982).
11. M. Sakata and M. Sato, *Acta Crystallogr. A* **46**, 263 (1990).
12. M. Takata, B. Umeda, E. Nishibori, M. Sakata,Y. Saito, M. Ohno, and H. Shinohara, *Nature* **377**, 46–49 (1995).
13. Y. Kohama, T. Rachi, J. Jing, Z. Li, J. Tang, R. Kumashiro, S. Izumisawa et al., *Phys. Rev. Lett.* **103**, 073001 (2009).
14. S. Aoyagi, Y. Sado, E. Nishibori, H. Sawa, H. Okada, H. Tobita, Y. Kasama, R. Kitaura, and H. Shinohara, *Angew. Chem.* **124**, 3377 (2012).
15. S. Horiuchi, R. Kumai, Y. Tokunaga, and Y. Tokura, *J. Am. Chem. Soc.* **130**, 13382 (2008).
16. M. Takata and M. Sakata, *Acta Crystallogr. A* **52**, 287 (1996).
17. Y. Wakabayashi, J. Takeya, and T. Kimura, *Phys. Rev. Lett.* **104**, 066103 (2010).
18. R. Fung, V. L. Shneerson, P. F. Lyman, S. S. Parihar, H. T. Johnson-Steigelman, and D. K. Saldin, *Acta Crystallogr. A* **63**, 239 (2007).
19. M. Björck, C. M. Schlepütz, S. A. Pauli, D. Martoccia, R. Herger, and P. R. Willmott, *J. Phys. Condens. Matter* **20**, 445006 (2008).
20. J. R. Fienup, *Appl. Opt.* **21**, 2758 (1982).
21. J. Miao, P. Charalambous, J. Kirz, and D. Sayer, *Nature* **400**, 342 (1999).
22. R. Barakat and G. Newsam, *J. Math. Phys.* **25**, 3190 (1984).

23. Yu M. Bruck and L. G. Sodin, *Opt. Commun.* **30**, 304 (1979).
24. P. Fenter and Z. Zhang, *Phys. Rev. B* **72**, 081401(R) (2005).
25. T. Takahashi, K. Sumitani, and S. Kusano, *Surf. Sci.* **493**, 36 (2001).
26. P. R. Willmott, S. A. Pauli, R. Herger, C. M. Schlepütz, D. Martoccia, B. D. Patterson, B. Delley et al., *Phys. Rev. Lett.* **99**, 155502 (2007).
27. M. Sowwan, Y. Yacoby, J. Pitney, R. MacHarrie, M. Hong, J. Cross, D. A. Walko, R. Clarke, R. Pindak, and E. A. Stern, *Phys. Rev. B* **66**, 205311 (2002).
28. R. Yamamoto, C. Bell, Y. Hikita, H. Y. Hwang, H. Nakamura, T. Kimura, and Y. Wakabayashi, *Phys. Rev. Lett.* **107**, 036104 (2011).
29. E. Vlieg, J. F. van der Veen, S. J. Gurman, C. Norris, and J. E. Macdonald, *Surf. Sci.* **210**, 301 (1989).

VI

Atomic-Scale Magnetism

22

Atomic-Scale Magnetism Studied by Spin-Polarized Scanning Tunneling Microscopy

Oswald Pietzsch
University of Hamburg

Roland Wiesendanger
University of Hamburg

22.1 Introduction

The understanding of magnetism at the ultimate, atomic, length scale is one of the current frontiers in solid state physics. It is a key to future applications in spin electronics and highest density data storage. Spin-polarized scanning tunneling microscopy (SP-STM) and spectroscopy (SP-STS) are powerful tools to access magnetic phenomena on a scale all the way down to the very atoms.

Scanning tunneling microscopy (STM) has revolutionized surface science ever since its invention in 1982 by Binnig and Rohrer [1] (Nobel Prize in 1986), and it was the capability of the STM to produce atomically resolved images of conducting surfaces that fascinated people most strongly. The spin-polarized version of the STM is based on the idea to detect not only the flow of electrical *charge* in the tunneling current but also to make the STM sensitive to the *spin* of the tunneling electrons, in order to combine the ultimate resolution capability of the STM with magnetic sensitivity. During the past decade, a large variety of surprising magnetic structures were discovered by SP-STM. Competing magnetic interactions effective at the atomic length scale give rise to unexpected ordered structures of great complexity in monolayers (MLs) of magnetic atoms. The main part of this chapter will present an overview of SP-STM research on two-dimensional (2D) antiferromagnetism, spin spirals, and other noncollinear magnetic structures with periodicities at the atomic level. Beforehand, the basic principles of STM and SP-STM will be introduced, together with theoretical studies that contributed strongly to the progress in this field. As a central technical ingredient, the preparation of magnetically sensitive tips is discussed in detail.

22.2 Principles of SP-STM and Theoretical Background

A convenient starting point to understanding the unsurpassed resolution capability of STM, ultimately down to the individual atoms, is the effect of *tunneling* in one dimension, like it is introduced in virtually all basic quantum mechanics textbooks. The effect to be described is a result of the wave–particle dualism, which is unknown to classical physics.

If a particle of total energy E impinges upon a potential barrier of height V_0 and finite width d it will, according to the laws of

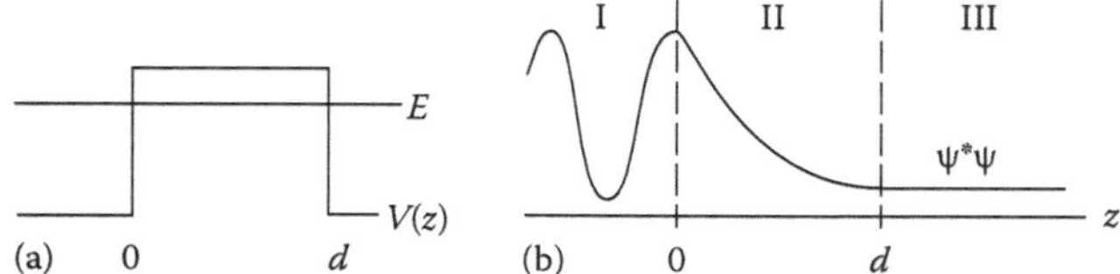

FIGURE 22.1 Schematics of the one-dimensional tunneling problem. (a) Particle energy E is lower than the barrier potential energy $V(z)$. (b) Probability density function of a particle incident from the left. Incoming and reflected amplitudes combine to a standing wave in region I. Inside the barrier region II the amplitude is exponentially damped but remains finite at d. The particle can tunnel through the barrier.

classical physics, only be able to pass the barrier if E is greater than V_0, otherwise it will be reflected. If the particle is of microscopic dimensions as, e.g., an electron, it must be described in terms of quantum physics, and the result is completely different. Even for the case $E < V_0$, there is a certain probability to find the particle behind the barrier, and this phenomenon is known as *tunneling*. The most simple situation is a single particle, let us say an electron of kinetic energy E, incident from the left upon a one-dimensional (1D) potential barrier (cf. Figure 22.1). The electron is described by its wave function $\psi(z)$, which is a solution of the time-independent Schrödinger equation. We can distinguish three regions: region I left of the barrier, $z < 0$; region II the barrier itself, $0 < z < d$; and region III right of the barrier, $z > d$:

$$\psi(z) = \begin{cases} e^{ikz} + re^{-ikz} & z < 0 \\ ae^{ik'z} + be^{-ik'z} & 0 < z < d \\ te^{ikz} & d < z \end{cases} \tag{22.1}$$

r, a, b, and t are arbitrary constants. In regions I and III, $V(z) = 0$, and the electron wave function is that of a free particle, $k = \sqrt{2mE}/\hbar$, with m the electron mass and $\hbar$ Planck's constant divided by 2π. Inside the barrier, that is $0 < z < d$, $V(z) = V_0$, and therefore we have $k' = \sqrt{2m(E - V_0)}/\hbar$. The total energy is negative in this region; thus, k' is complex, and the exponents become real:

$$\kappa^2 = -k'^2 = \frac{2m(V_0 - E)}{\hbar^2}. \tag{22.2}$$

Therefore, the exponentials are *real* functions describing waves decaying exponentially within the barrier.

$\psi(z)$ and also its first derivative $(d/dz)\psi(z)$ are required to be continuous for all z, and by matching the partial solutions found for the respective regions at the points $z = 0$ and $z = d$ (*wave matching method*), we can obtain a set of equations that allows us to determine the values of the constants. Now we can gain an exact expression for the transmission coefficient, which is the ratio of the transmitted and the incident probability flux j_T and j_0, respectively,

$$T = \frac{j_T}{j_0} = \frac{1}{1 + (k^2 + \kappa^2)^2 / 4k^2\kappa^2 \sinh^2(\kappa d)}. \tag{22.3}$$

In the limit of $\kappa d \gg 1$ (high and/or thick barrier), this formula reduces to

$$T \approx 16 \frac{k^2\kappa^2}{(k^2 + \kappa^2)^2} \exp(-2\kappa d). \tag{22.4}$$

When this last expression is a good approximation, T is extremely small. The most important result, however, is the exponential dependence of T on the width d of the potential barrier. It is this relationship that is exploited in the scanning tunneling microscope. It is the key to the extremely high spatial resolution that allows for a study of conducting sample surfaces on a scale where individual atoms can be made visible.

22.2.1 Tunneling Process in an STM

In an STM measurement, a fine metallic tip is approached as close as a few Å (1 Å = 10^{-10} m) to the surface of a conducting sample (a metal for simplicity). The tip is then scanned line by line across the surface by means of a piezoelectric scanner (Figure 22.2), and, with a small bias voltage U applied, a tunneling current $I(U)$ can be measured, which will, according to Equation 22.4, vary exponentially as a function of the distance between tip and sample. This is an example of metal–vacuum–metal tunneling. The tunneling barrier between the two electrodes in this case is the vacuum gap of width d, and the height V_0 of the barrier is given by the work function W of the electrode, i.e., the (classical) energy

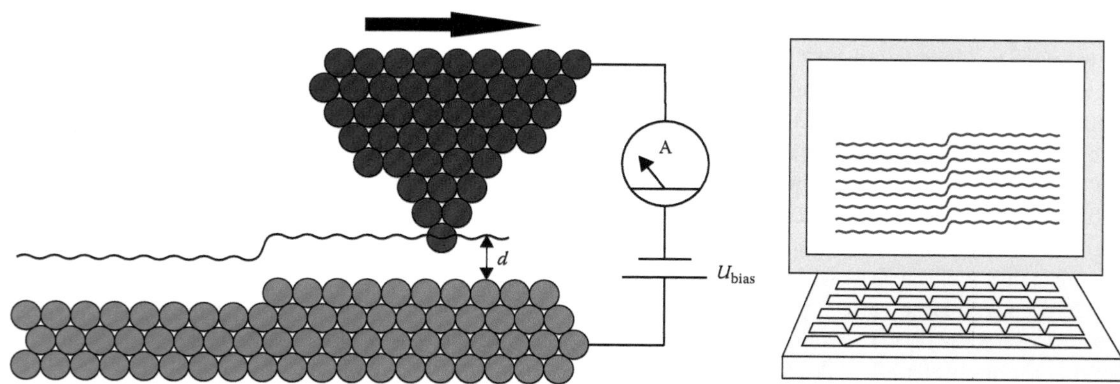

FIGURE 22.2 Scanning the tip across the surface, line by line.

required to extract an electron out of the metal surface into the vacuum, which is a material parameter.

Inside tip and sample, the electrons may be described in the free-electron-gas model. All electronic states are occupied up to the Fermi level E_F (for simplicity, we assume a temperature of 0 K resulting in a sharp edge in the Fermi function, separating occupied and unoccupied states). At tunneling distance, the Fermi levels of tip and sample will level out. If we apply a small bias voltage U_{bias} between tip and sample, the Fermi levels of the electrodes will shift accordingly with respect to each other. In the following, we will use the convention that the tip potential is always held grounded. Thus, for positive sample bias, electrons will tunnel from occupied states of the tip into unoccupied states of the sample, and for negative sample bias, the electrons come from occupied sample states and go to unoccupied tip states. Thus the direction of the current flow depends on the polarity of the applied bias voltage U_{bias}.

If we consider the limit of small bias voltage, i.e., $E = eU_{bias} \ll \phi$, the energy of the tunneling electrons is approximately equal to the Fermi energy E_F. Inside the barrier, the wave function of an electron decays:

$$\psi(z) = \psi(0)\exp(-\kappa d), \quad \kappa = \sqrt{\frac{2m \cdot W}{\hbar^2}}, \tag{22.5}$$

with κ the so-called *decay constant*. We can determine the probability density w of finding an electron on the other side of the barrier by taking the square of the wave function:

$$w = |\psi(d)|^2 = |\psi(0)|^2 \exp(-2\kappa d). \tag{22.6}$$

For a typical metal, we may assume $W \approx 4$ eV. This results in a decay constant $\kappa \approx 1$ Å^{-1}. As a consequence, a given value of a tunneling current will be reduced by about an order of magnitude if the gap between tip and sample is increased by 1 Å. These numbers illustrate the enormous vertical resolution that can be achieved by the STM. Furthermore, we can conclude from Equation 22.5 the important fact that the tunneling current will be carried almost exclusively by the outermost atom at the tip apex, while contributions from atoms of the next atomic layer within the tip can in most cases be neglected. Therefore, the tunneling process in an STM is highly localized in the sense that it occurs between one atom at the tip and the sample spot right below it. Thus, when scanned across a sample surface, the tip probes *local* properties of the sample with a lateral and vertical resolution that allows us, in general, to resolve individual atoms.

22.2.2 Tersoff–Hamann Theory of STM

Until now, we did not discuss the properties of the probing tip at all. It was introduced just as a conducting electrode being located a distance d away from the sample surface. In an STM experiment, the ideal of a nonintrusive measurement would be a point probe with an arbitrarily localized wave function [2]. A realistic tip, however, is made from a certain material, having its atoms at the apex arranged in a particular way, i.e., it has a certain geometry in space and a more or less extended wave function. In other words, the tip has an electronic structure that has to be accounted for in a three-dimensional (3D) approach. As will be discussed in Section 22.3, tunneling tips are prepared in a way that does not really allow a detailed control of the tip at the atomic scale; furthermore, spontaneous rearrangements of the apex atoms are frequently experienced during STM measurements, showing up as sudden changes in imaging quality. Therefore, in order to include the tip into theory of the STM, some reasonable approximations need to be made.

A very successful approach was introduced by Tersoff and Hamann [2,3]. Here, we will present the basic assumptions, main results, and limitations of their theory. They considered a model tip of arbitrary shape but with its lower end forming a spherical potential well with effective radius R (not necessarily restricted to the ultimate case of the radius of a single atom), the center of curvature located at position $\vec{r}_0$. The simplest possible tip wave function, a spherically symmetric s-wave function, was assumed; wave functions with angular dependence ($l \neq 0$) were neglected. The tip density of states (DOS) was considered structureless. The limits of low temperature and small bias were applied. With these approximations, the tunneling current as a function of the bias voltage can be expressed as an energy integral

$$I(\vec{r}_0, U) \propto n_t \int_{E_F}^{E_F+eU} n_s(\vec{r}_0, \varepsilon)d\varepsilon \tag{22.7}$$

of the tip DOS n_t times the *local density of states* (LDOS) of the sample n_s, evaluated at the center of curvature $\vec{r}_0$ of the effective tip. As a last assumption, the tip DOS is taken to be constant. The formula is given here in the $T = 0$ K limit; at finite temperatures, the integration limits are smeared out.

The Tersoff–Hamann model leads to some remarkable results. The most important one is that the tunneling current is determined by sample properties alone, while the role of the tip is simply reduced to that of a probe. The tunneling current is proportional to the energy integral over the sample's LDOS in the energy interval from E_F to $E_F + eU$ determined by the applied bias voltage, evaluated in the vacuum at the position $\vec{r}_0$ of the tip. The sample LDOS

$$n_s(\vec{r}_0, \varepsilon) = \sum_{\upsilon} |\psi_{\upsilon}(\vec{r}_0)|^2 \, \delta(E_{\upsilon} - \varepsilon) \tag{22.8}$$

is the central quantity accessed in any STM experiment, with all sample states ψ_{υ} within the aforementioned energy interval. The sample wave functions decay exponentially into the vacuum (the surface normal is taken as z direction). Through this, we can relate our measurements to the electronic structure of the sample surface under study.

In a typical STM experiment, the tunneling current, Equation 22.7, is fed into a feedback circuit that regulates the vertical tip position $z_0 + \Delta z$ as to keep the current equal to a setpoint value while the tip is scanned across the surface, see Figure 22.3. This is called the *constant current mode*. The voltage

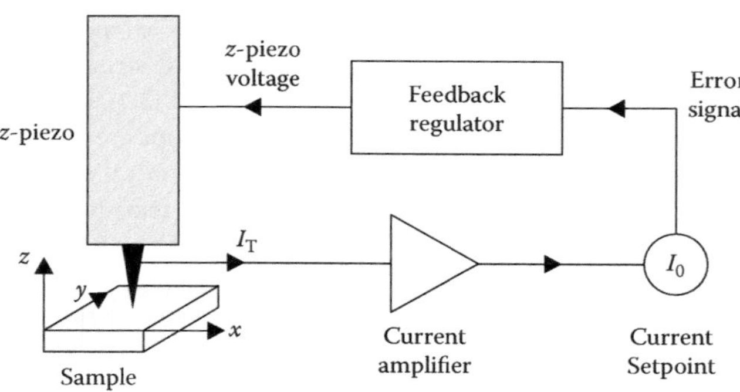

FIGURE 22.3 Schematics of the feedback loop used in the constant current mode of STM. In a topographic image, the height information $z(x,y)$ is proportional to the voltage applied to the z-piezo in order to keep the tunneling current at the setpoint value.

variations applied to the z-piezo correspond to the corrugation Δz. We thus can plot a map $\Delta z(x,y)$ as a function of lateral tip position (x,y). Constant-current STM images can now be interpreted as contour maps of constant sample LDOS, and to a first approximation these contours follow the *topography* of the sample surface. In this way, the details of the surface landscape like atomically flat terraces with monatomic step edges, islands, defects, surface reconstructions, etc. can be made visible in real space. After its invention, the STM was most widely applied in the study of structural properties of surfaces. This new microscopy technique allowed one to address questions that were previously not accessible by other surface sensitive techniques that rely on the diffraction of electromagnetic or matter waves at periodic structures and typically average over surface areas of about 1 mm^2.

The Tersoff–Hamann model is applied successfully in many cases to interpret experimental results qualitatively or even quantitatively. However, when it comes to atomic resolution imaging of densely packed metal surfaces, the model is still capable of reproducing the lattice periodicity but predicts a corrugation amplitude that is much smaller than the measured experimental values. This deficiency is well understood and can be attributed to the restriction to the s-wave-tip model. An extension of the theory was developed by Chen [4] who proposed a simple derivative rule, allowing other orbitals, in particular those with a charge density distribution more strongly localized along the tip axis (p_z,d_{z^2}) to be included. Chen introduced corrugation enhancement factors specific to each particular orbital type. For a p_z orbital, the factor is $[1 + (q^2/\kappa^2)]$, and for a d_{z^2} orbital it is $[1 + 3q^2/2\kappa^2]^2$ with $q = G_{\|}/2$, $G_{\|}$ being the length of a reciprocal surface lattice vector, $G_{\|} = 2\pi/a_{\|}$. In the case of a typical close-packed metal surface with $a_{\|} \approx$ 0.25 nm, the corrugation enhancement due to a d_{z^2} orbital is about an order of magnitude, giving a much better agreement with experimental results. Intuitively, this can be understood by considering that such orbitals exhibit lobes being much "sharper" than a s sphere. With increasing structural length scale, the enhancement factor is reduced, and the model ultimately becomes independent of the type of orbital.

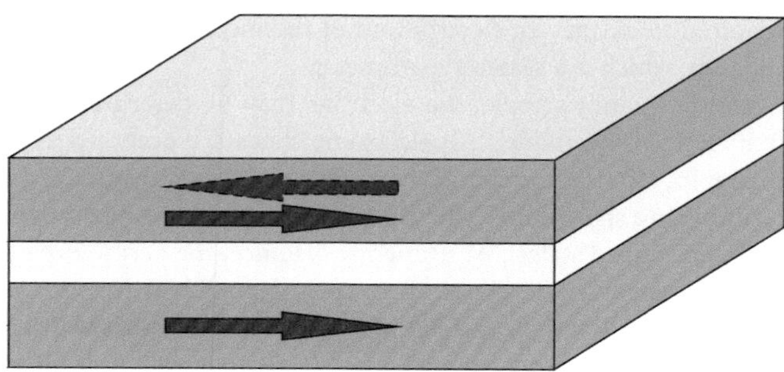

FIGURE 22.4 A planar ferromagnetic tunneling junction, consisting of a hard and a soft magnetic layer separated by an insulating barrier layer.

22.2.3 Spin-Polarized Electron Tunneling

Spin-polarized electron tunneling between two magnetic electrodes was first observed in 1975 by Jullière [5]. He investigated the tunneling conductance G of planar tunnel junctions consisting of two ferromagnetic (FM) Fe and Co films separated by a thin insulating Ge layer, Figure 22.4. Exploiting the difference in coercivities of the FM films, their magnetizations could be brought into a parallel or an antiparallel arrangement by application of an external magnetic field. The tunnel conductance of the parallel case clearly exceeded that of the antiparallel one. Assuming spin conservation during tunneling, that is, majority electrons can only tunnel into empty majority states, and minority electrons only into empty minority states, Jullière proposed a model in which the tunneling probability is proportional to the DOS at the Fermi energy E_F; but the DOS is different for spin-up and spin-down electrons in a ferromagnet. Hence, the relative conductance variation is given by

$$\frac{\Delta G}{G} = \frac{2P_1P_2}{1+P_1P_2}. \tag{22.9}$$

Here, P_1 and P_2 denote the spin polarizations of the two FM electrodes, respectively. In the Stoner model of band ferromagnetism, the electronic band structure is split into two subbands for majority and minority spin electrons, respectively, as schematically depicted in Figure 22.5. These subbands are shifted relative to each other by the exchange energy E_{ex}. The occurrence of magnetism in band magnets is thus based on an imbalance in occupation between spin-up and spin-down electronic states* near the Fermi level, giving rise to the spin polarization, which is defined as

$$P = \frac{n_\uparrow - n_\downarrow}{n_\uparrow + n_\downarrow} \tag{22.10}$$

* By convention, the meaning of the labels spin-up (↑) and spin-down (↓) is equivalent to that of majority and minority spin, respectively. In any case, these labels refer to a quantization axis in *spin space*, not necessarily to up and down directions in *real space*; if the spin quantization axis happens to lie in the surface plane, one can still speak of spin-up and spin-down electrons.

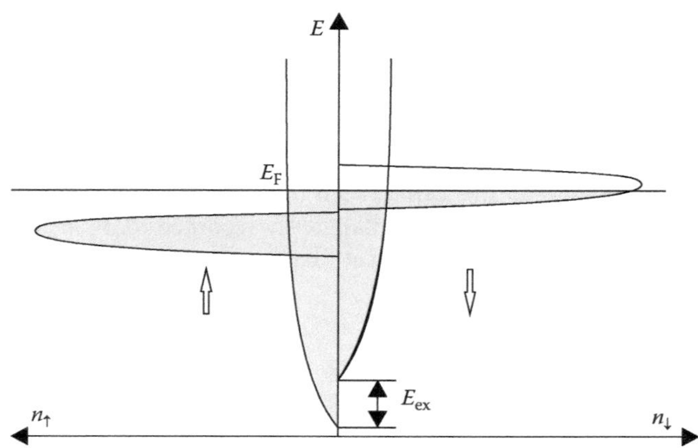

FIGURE 22.5 Schematic exchange-split density of states. Flat *d*-bands near the Fermi energy E_F contribute strongly to the spin polarization. Gray areas indicate occupied states, unoccupied states are found above E_F.

with $n\uparrow$ and $n\downarrow$ being the densities of spin-up and spin-down states, respectively. This definition thus describes the fraction of excess spins over the total number of states.

Jullière's planar magnetic tunnel junction (MTJ) was the first example of a so-called spin valve based on the *tunneling magneto-resistance* (TMR) effect. Mainly because the effect initially was observed at 4.2 K but not at room temperature, little attention was paid to Jullière's results for a decade. In the late 1980s yet another MR effect, coined the *giant magneto-resistance* (GMR) effect [6,7], was discovered in ferromagnet–metal–ferromagnet multilayer structures with typical 1 nm layer thicknesses (Nobel Prize for P. Grünberg and A. Fert in 2007). Again, the resistance depends on the relative alignment of the magnetization in adjacent FM layers. Soon it was recognized that GMR has a high potential for applications. GMR elements can be used as highly susceptible magnetic field sensors. The resistance variations achievable with GMR are much larger than those of the traditionally known anisotropic magnetoresistance. In 1997, less than 10 years after the discovery of GMR, IBM presented a first magnetic hard disk drive (HDD) exploiting the new effect in a GMR-based read head, introducing GMR into a billion dollar mass market.

In the wake of the renewed strong interest in magnetism at the nanoscale, also MTJs were revisited. Miyazaki and Tezuka [8] and Moodera et al. [9] found TMR ratios of 18% and 12%, respectively, at room temperature, using amorphous Al_2O_3 tunneling barriers. Continuing progress arose from improvements in barrier materials and interface quality. Since 2005, TMR read heads are being used in the next generation of high-density HDDs that employ the perpendicular magnetic recording scheme. Nowadays, a very vivid field of device-oriented research on planar magnetic tunneling junctions has developed, in particular aiming at non-volatile magnetic random access memory (MRAM) for future data storage and processing applications. For a review see [10].

The experiments by Miyazaki and Tezuka [8] brought about another result, which is important in the context of our discussion. The tunneling conductance was experimentally shown to vary with the cosine of the angle θ between the magnetizations of the FM layers. The magnetic signal is thus proportional to the projection of the magnetization vectors of the two electrodes.

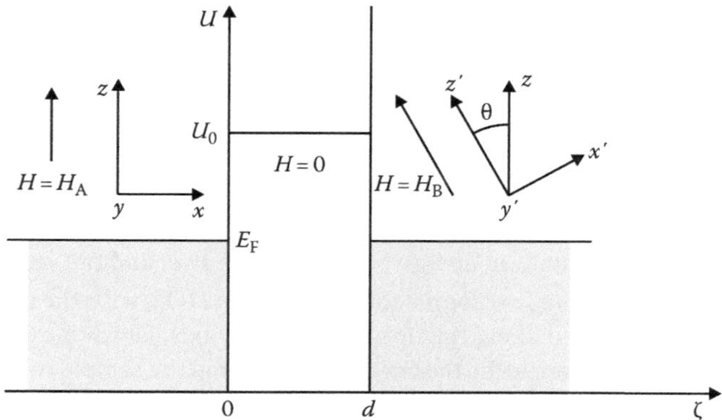

FIGURE 22.6 Generalized magnetic tunneling junction with arbitrary relative orientation of the quantization axes determined by internal molecular fields $H_{A,B}$, as analyzed by Slonczewski [11].

This angular dependence had already been predicted in an earlier theoretical paper by Slonczewski [11]. He considered the general case of a tunneling junction with the two ferromagnets possessing quantization axes being arbitrarily oriented to each other, determined in each of the ferromagnets by their respective internal molecular fields H, see Figure 22.6. In the limits of small bias voltage and $T = 0$ K, he gave an expression for the spin-dependent tunneling conductance $G = I/V$,

$$G = G_{\mathrm{fbf'}}(1 + P_{\mathrm{fb}}P_{\mathrm{f'b}} \cdot \cos\theta) \qquad (22.11)$$

with fb and f′b denominating the first and second ferromagnet/barrier interfaces, respectively, and fbf′ the complete tunnel junction. Again, spin conservation during tunneling is implied.* According to this formula, the tunneling junction conductance can be subdivided into a contribution that is not spin-polarized and a second part that is proportional to the product of the polarizations of the two electrodes times the cosine of the angle between the quantization axes inside the ferromagnets.

22.2.4 Spin-Polarized STM

Slonczewski's expression is valid also in SP-STM where one of the magnetic electrodes is replaced by the magnetic tip, and the insulating barrier layer by the vacuum gap. In a typical SP-STM experiment, the tip is scanned across a magnetic surface. Usually, the tip magnetization direction will remain fixed during the scan, as expressed by its unit vector $\mathbf{m}_t$ aligned along its quantization axis. But the tip may scan areas of varying magnetization $\mathbf{m}_s$ on the sample surface if there is, e.g., a magnetic domain structure. The spin-polarized contribution to the conductance is thus dependent on the projection of the local sample

* Interestingly, Slonczewski expressly mentioned vacuum as a possible barrier.

magnetization onto the tip magnetization, hence the angle is a function of the lateral tip position, $\theta = \theta\ (x,y)$. The extremal contrasts will be achieved in the collinear configurations, $\theta = 0$ and $\theta = \pi$, and spin contrast will vanish if $\mathbf{m}_t$ and $\mathbf{m}_s$ are orthogonal.

A spin-polarized version of the Tersoff–Hamann model was proposed by Wortmann et al. [12]. In analogy to Tersoff and Hamann, the spin-up, $n_t^{\uparrow}(\varepsilon)$, and spin-down, $n_t^{\downarrow}(\varepsilon)$, tip DOS was assumed constant in energy but different in size, and the vector $\mathbf{m}_t = \left(n_t^{\uparrow} - n_t^{\downarrow}\right)\mathbf{e}_{M,t}$ accounts for the tip's *spin* DOS, with the unit vector oriented along the tip's quantization axis, which is taken to be the reference for the tip–sample system. The sample wavefunctions, describing the varying local magnetization of the surface, will then be spin-mixed:

$$\Psi_{\mu}^{s} = \begin{pmatrix} \Psi_{\mu\uparrow}^{s} \\ \Psi_{\mu\downarrow}^{s} \end{pmatrix}. \qquad (22.12)$$

The tunneling current can now be expressed as

$$\begin{aligned} I(\mathbf{R}_t,U,\theta) &= I_0(\mathbf{R}_t,U) + I_P(\mathbf{R}_t,U,\theta) \\ &\propto n_t\tilde{n}_s(\mathbf{R}_t,U) + \mathbf{m}_t\tilde{\mathbf{m}}_s(\mathbf{R}_t,U). \end{aligned} \qquad (22.13)$$

The quantities $\tilde{n}_s(\mathbf{R}_t,U)$ and $\tilde{\mathbf{m}}_s(\mathbf{R}_t,U)$ marked by a tilde are the integrated LDOS (ILDOS) and the vector of the integrated local magnetization DOS $\tilde{\mathbf{m}}(\mathbf{R}_t,U)$, respectively, taken as the energy integral over all sample states n_s and $\mathbf{m}_s$ in the energy interval from E_F to $E_F + eU$,

$$\tilde{\mathbf{m}}_s(\mathbf{R}_t,U) = \int d\varepsilon\ \mathbf{m}_s(\mathbf{R}_t,\varepsilon), \qquad (22.14)$$

and similar for $\tilde{n}_s(\mathbf{R}_t,U)$. Just as in Slonczewski's expression, Equation 22.11, the current consists of a non-polarized and a polarized part, the latter being dependent on the projection of the sample magnetization onto that of the tip.

As pointed out by Wortmann et al., it is always desirable to maximize the spin-polarized part over the non-polarized one. This can easily be achieved by measuring the differential conductance dI/dU:

$$\frac{dI}{dU}(\mathbf{R}_t,U) \propto n_t n_s(\mathbf{R}_t,E_F + eU) + \mathbf{m}_t\mathbf{m}_s(\mathbf{R}_t,E_F + eU). \qquad (22.15)$$

Here, the quantities n_s and $\mathbf{m}_s$ of Equation 22.15 are *not* energy integrated; hence, dI/dU is directly proportional to n_s and $\mathbf{m}_s$ at an energy $E_F + eU$. The crucial difference between Equations 22.13 and 22.15 is demonstrated in Figure 22.7. At finite bias, the energy interval taken for integration may contain states of varying polarization, $P = P(\varepsilon)$. In particular, the polarization can even change sign as a function of energy, possibly leading to a degradation or even cancellation of the spin-polarized contribution to the overall integral signal (Figure 22.7c). Also, the spin-averaged part of the tunneling current will increase with increasing bias, whereas the polarized part may stay constant, thereby reducing the polarization-dependent corrugation variation. On the other hand, in a dI/dU measurement, one can, by choosing a proper bias voltage, directly address certain highly polarized states of the sample's spin-split band structure (Figure 22.7d). In this way, strong magnetic contrasts can be achieved. This can be seen by comparison of the topographic image and the simultaneously recorded dI/dU map in Figure 22.7a and b: Virtually all magnetic information is contained in the dI/dU map.

22.2.5 Magnetic Imaging at the Atomic Length Scale

The technique of dI/dU mapping has been used successfully to investigate the magnetic structure of a variety of surfaces [13], thin films [14–16], nanowires [17–21], and islands [22–28]. However, when it comes to imaging periodic magnetic superstructures at the ultimate, atomic, length scale such as antiferromagnetic (AFM) surfaces or noncollinear patterns, it is, surprisingly, the simple constant current mode that proves to be the superior method. One might expect that the spin-polarized contribution just adds a slight modification to an ordinary atomically resolved STM image. But the effect is indeed quite dramatic. This was explained theoretically by Heinze et al. [29] and in the aforementioned work of Wortmann et al. [12].

The constant current image $\Delta z(\mathbf{r}_{\|},U,\theta)$ of a surface is determined by the variation of the tunneling current ΔI. If the surface has 2D translational symmetry, ΔI can be expanded into a 2D Fourier series:

$$\Delta I(\mathbf{r}_{\|},z,U,\theta) = \sum_{n\neq 0} \Delta I_{\mathbf{G}_{\|}^{n}}(z,U,\theta)\cdot e^{i\mathbf{G}_{\|}^{n}\mathbf{r}_{\|}}. \qquad (22.16)$$

$\mathbf{G}_{\|}^{n}$ denotes the reciprocal lattice vectors parallel to the surface, and $\Delta I_{\mathbf{G}_{\|}^{n}}(z,U,\theta)$ is the nth expansion coefficient, which depends on the tip–sample distance z, the bias U, and the magnetization angle θ. These expansion coefficients decay exponentially with increasing length of $G_{\|}^{n}$. The constant current image will thus be dominated by the contribution from the coefficient corresponding to the shortest available reciprocal lattice vector, and all other expansion terms are exponentially more attenuated.

If we now consider the surface of an elementary 2D antiferromagnet, see Figure 22.8, it will consist of a lattice of chemically equivalent atoms. But magnetically, the atoms are inequivalent, possessing magnetic moments pointing in opposite directions. Using a non-magnetic tip, i.e., $P_t = 0$, the tunneling current is the ordinary spin-averaged current I_0, and the image will represent the chemical surface unit cell, with maxima at the positions of the atoms. In a spin-polarized measurement, we have to take into account the two inequivalent magnetic orientations. The magnetic surface unit cell is larger than the chemical one. Hence, shorter reciprocal lattice vectors become available. The spin-polarized current I_P is exponentially larger than I_0, and the image will be dominated by the

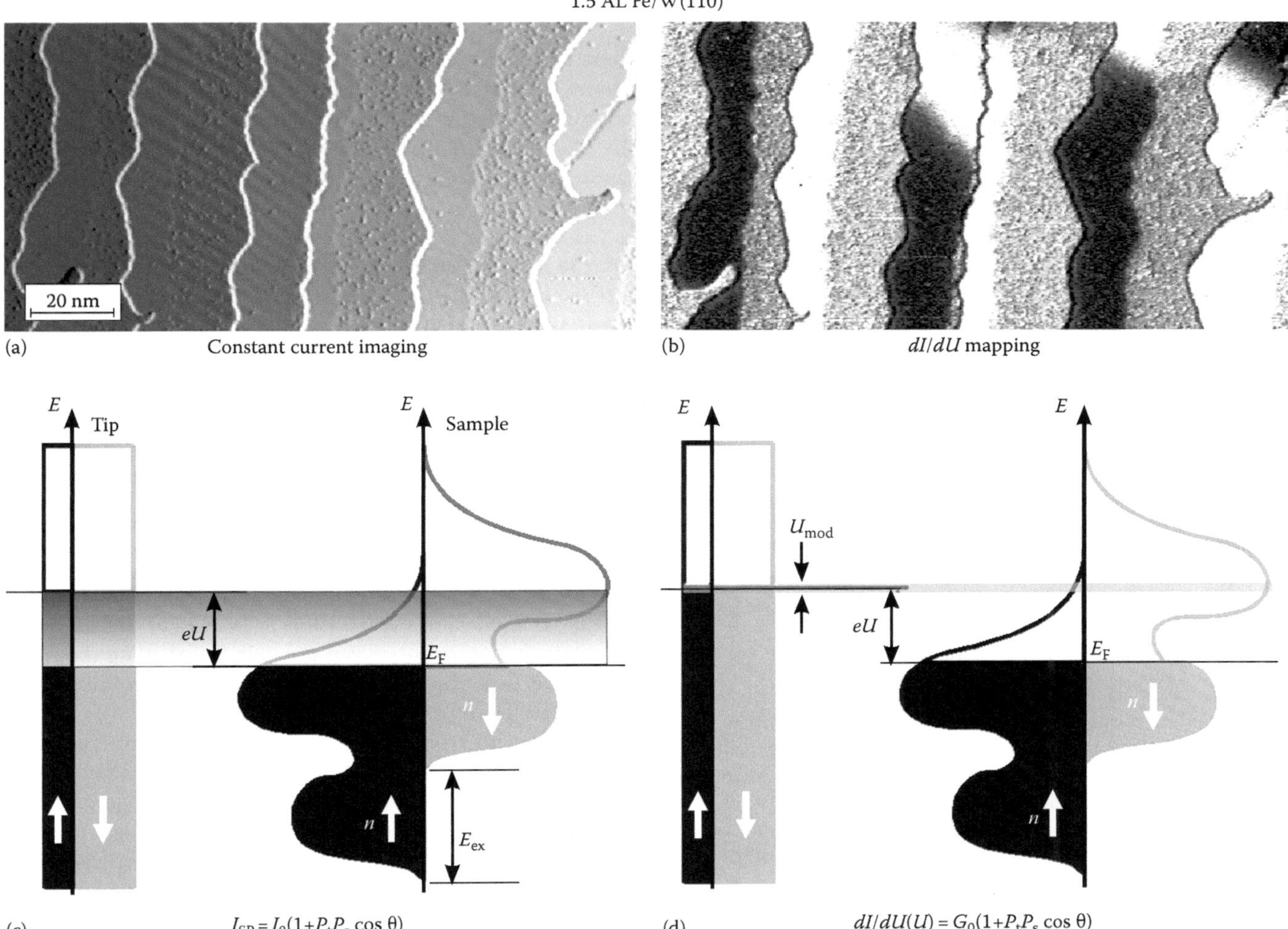

FIGURE 22.7 (a) Constant current image of 1.5 atomic layers Fe on W(110), (b) map of the differential conductance $dI/dU(U)$, measured simultaneously. The spin-polarized current according to Equation 22.13 is fed into the feedback loop to record the sample topography. In (b) strong black-and-white contrasts reveal the perpendicular magnetic domain structure in stripes of local double-layer coverage, which is not seen in the topography. (c and d) Schematics of a spin-split electronic band structure. (c) The signal responsible for the topographic image is the energy integral from E_F to $E_F + eU$ (shaded rectangle). Contributions of positive and negative spin-polarization may be present such that the polarized part of the tunneling current may be too small to be visible as a significant height variation between the magnetic domains. (d) The differential conductance $dI/dU(U)$ signal is *not* energy integrated but is measured directly at an energy $E_F + eU$. One can adjust U_{bias} as to address highly polarized states in the sample's electronic structure. This is the reason why dI/dU-maps can show quite strong magnetic contrasts. The signal does not contain any height information.

spin-polarized contribution. The effect of the different decay lengths is so strong that even in the case of small effective spin polarization (θ close to 90°) the image will still be dominated by the magnetic superstructure rather than the chemical lattice, with a strongly enhanced corrugation amplitude.

As will be discussed in detail in Section 22.4.1, this theoretical approach was verified experimentally by SP-STM measurements.

22.2.6 Simulating SP-STM Images: Independent-Orbital Approximation

Panels (c) and (f) of Figure 22.8 allow an easy and intuitive understanding of the difference between the conventional atomic contrast and that arising from the magnetic superstructure. Such simulated real space images revealing typical patterns depending on the relative alignment of tip magnetization and local sample magnetization are very useful in facilitating a comparison of theoretical and experimental STM images, and the agreement of theory and experiment can be judged. The images shown were calculated based on the spin-polarized extension of the Tersoff–Hamann model. The problem is to accurately know the spin-dependent electronic structure of the sample and the decay of the wavefunctions into the vacuum. This knowledge can be obtained from density functional theory (DFT) calculations, the state-of-the-art method to predict theoretically the electronic and magnetic properties of low-dimensional systems.

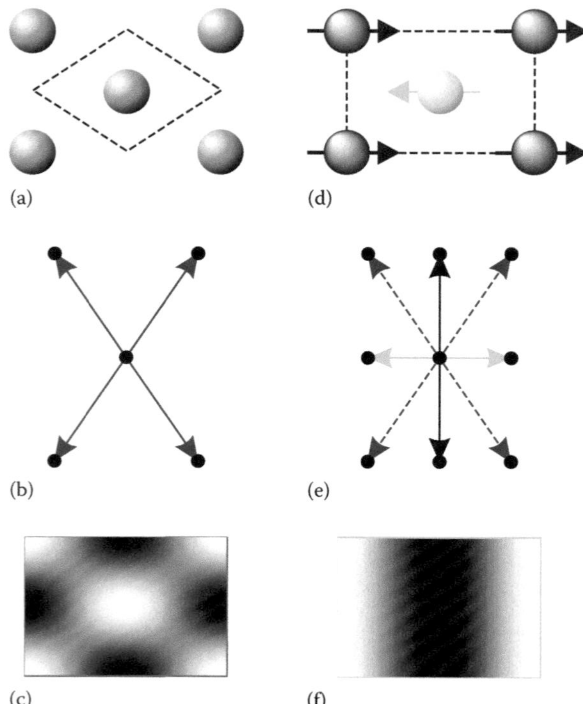

FIGURE 22.8 (a) Chemical unit cell of a bcc(110) surface. (b) Corresponding reciprocal lattice vectors. (c) Calculated STM image, showing maxima at the atom positions. (d) Magnetic unit cell of the $c(2 \times 2)$ antiferromagnetic superstructure. (e) Two inequivalent pairs of reciprocal vectors (black and light gray) corresponding to the magnetic structure become available; the vectors of the chemical unit cell are shown dashed. The calculated image (f) is dominated by a stripe pattern attributed to the shortest, light gray, vectors in (e).

On the other hand, DFT is computationally very demanding. This is particularly true for noncollinear spin structures where the magnetic unit cell can become arbitrarily large. Such systems easily pose a challenge even for current supercomputers.

In 2006, a simplified scheme to simulate SP-STM images, coined *independent-orbital approximation*, was introduced by Heinze [30], based on the constant current mode as discussed in the previous section. Despite its simplicity, the scheme has proven very successful in capturing the key features of atomic scale SP-STM images of complex noncollinear magnetic structures at atomic length scales, but also at nanometer scale structures, and even of non-periodic structures such as domain walls of a 2D antiferromagnet. Examples will be presented along with experimental results in the remaining sections. Here, we intend to just elucidate the basic ideas of the scheme.

In order to achieve a minimal computational load, the images are calculated without taking the electronic structure fully into account. There are two simplifying assumptions: (1) The spin-dependent local DOS of every surface atom α can be given with respect to a local quantization axis that is rotated by an angle θ_α with respect to the tip magnetization $\mathbf{e}_t$. (2) The contribution of atom α to the local DOS in the vacuum is approximated by the spherical tail of an atomic wavefunction.

The latter assumption neglects *d*-orbitals with their distinct directional symmetries. These types of orbitals are quite common for magnetic transition metal atoms. However, since often several *d* orbitals with a variety of symmetries are present, on average, a spherical decay is a reasonable assumption. The gain in computing time is dramatic: Using the independent orbital approximation, simulations of STM images of sophisticated noncollinear magnetic structures become a matter of minutes on an office PC, as opposed to several days on a supercomputer for the full DFT calculation. The images shown in Figure 22.9 are in excellent agreement with those from the full first-principles calculations [30]. Still, in special cases, the applicability of the scheme may be limited, and the accurate electronic structure has to be considered explicitly.

With these assumptions, the vacuum decay can be described in the low bias limit as $h(\mathbf{r}) = \exp(-2\kappa|\mathbf{r}|)$, and $\kappa = (2m\phi/\hbar^2)^{1/2}$, which is the result of Tersoff and Hamann [2,3], and the tunneling current is

$$I(\mathbf{R}_t) \propto h(\mathbf{R}_t - \mathbf{R}_\alpha)[1 + P_t P_s \cos\theta_\alpha]. \quad (22.17)$$

If the effective spin polarization $P_{eff} = P_t P_s$ of the tunneling junction is known, it is even possible to calculate the maximum vertical tip displacement (corrugation amplitude) quantitatively.

As a benefit from the strongly reduced use of computing resources, it is easily possible to simulate more than one image for a given problem, assuming different tip magnetizations. An example is presented in Figure 22.9. Such images can then be easily used to check the consistency with the experimental findings, or to patterns obtained from Monte Carlo (MC) simulations. More examples will be shown in the experimental sections.

22.3 Preparation of Magnetic Tips

In SP-STM experiments, a key ingredient is a tip that not only provides a good spatial resolution—ultimately down to the atomic level—but is, at the same time, sensitive to the spins of the tunneling electrons. After two decades of successful SP-STM, there is, however, still no systematic study available that would allow us to give one general recipe of how to obtain a proper tip. Groups working in the field usually develop their own recipes, largely based on experience and accounting for their available preparation conditions. The rich variety of magnetic structures found at the nanoscale pose different requirements for magnetic tips. Here, we will present a few basic concepts.

In the early days of magnetic STM imaging, two different approaches were proposed. Following the idea of Pierce [31], several groups used GaAs tips, inducing a spin polarization by irradiating them with circularly polarized light. Despite years of efforts in the 1990s, no major breakthrough was achieved.

Wiesendanger and co-workers took a different route. They used FM tips and employed the tunneling magnetoresistance

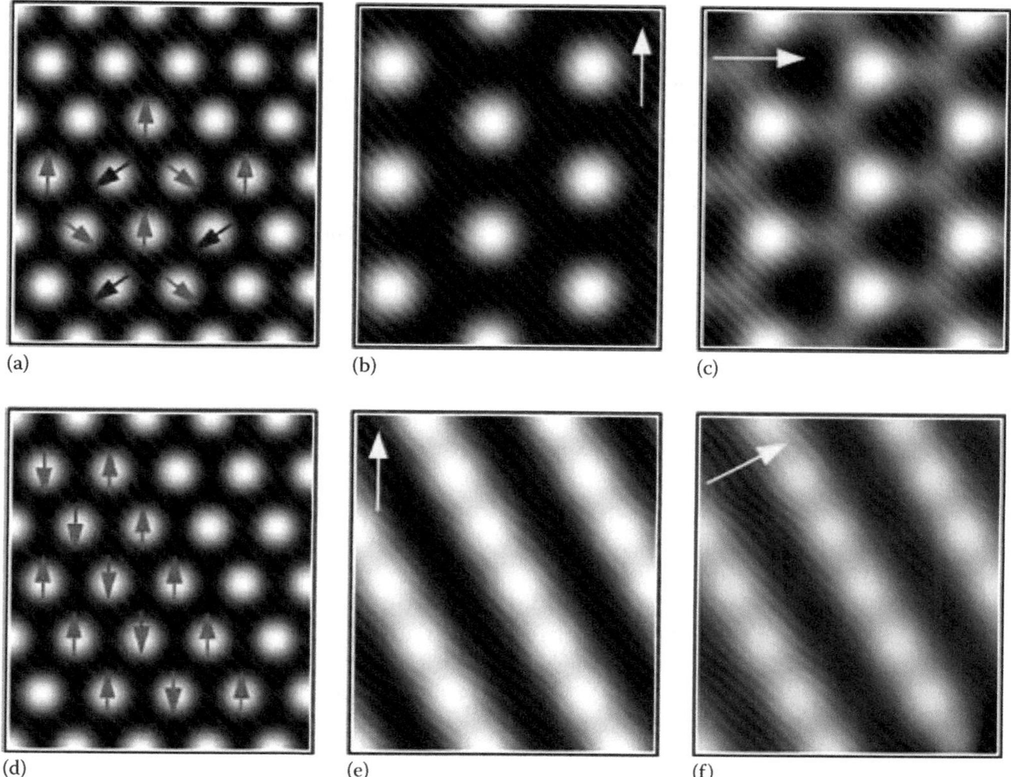

FIGURE 22.9 Calculated SP-STM images for two possible ground states of a 2D antiferromagnet on a hexagonal lattice of chemically equivalent atoms. Top: Frustrated Néel structure with 120° angles between magnetic moments of adjacent atoms. Bottom: Row-wise AFM state. (a) and (d) show images for a non-magnetic tip, arrows indicate the magnetic structure. In panels (b and c) and (e and f), the orientation of the tip magnetization is indicated by white arrows, resulting in distinct patterns. (After Springer Science + Business Media: *Applied Physics A*, Simulation of spin-polarized scanning tunneling microscopy images of nanoscale non-collinear magnetic structures. 85(4), September 2006, 407–414, S. Heinze.)

effect in order to achieve spin contrast combined with a high lateral resolution down to the atomic level. The first spin-polarized STM measurements were achieved with tips consisting of a CrO_2 thin film deposited on cleaved Si(111), offering a very high degree of spin polarization [32]. Later, bulk FM tips were prepared in a two-step procedure as described in Ref. [33]. First, a polycrystalline FM Fe wire was electrochemically etched until a narrow constriction was obtained. This procedure was carried out at ambient conditions, leaving the wire contaminated with an oxide film. The wire was then on to the STM. Under vacuum conditions, it was then torn apart by flipping back the scan head [34]. Immediately before breaking, the wire material starts to viscously flow, eventually providing a clean and sharp tip capable of atomic resolution [33].

A major disadvantage of a bulk FM tip is that it inevitably exerts significant stray fields. Due to shape anisotropy, the tip magnetic moments will align parallel to the tip axis, and at a tunneling distance of about 5 Å, the stray fields from such a bar magnet will interact strongly with the sample underneath. Therefore, the use of this type of tip is limited to samples being insensitive to external fields, which is the case for AFM materials.

In order to minimize the stray field problem, Bode et al. [22] introduced non-magnetic tungsten tips coated with ultrathin films of magnetic material. Typical film thicknesses are 5–10 atomic layers (≈ 2 nm). Thus, the amount of magnetic material is greatly reduced, and so is the stray field emanating from the tip. On the other hand, coating of tips requires a reliable tip exchange mechanism allowing the tip red out of the scanner and transfer to be taken it to the preparation facilities without breaking the vacuum. In Ref. [35], such a tip exchange mechanism and transport device are described. A typical tip preparation procedure is as follows: the tungsten tip is electrochemically etched ex situ from a polycrystalline tungsten wire. After transferring into ultrahigh vacuum (UHV), it needs to be purged from oxide layers. Short flashes to temperatures as high as 2000°C effectively remove any surface contamination and enhance the sticking properties for a magnetic thin film. Caused by the heat treatment, it is not unusual that a tip gets quite blunt. Tip diameters can be as large as 1 μm. This is not a problem if atomically flat surfaces are to be scanned; there will always be a cluster of a few atoms eventually carrying the tunneling current. In the next step, thin film deposition of magnetic material can be carried out, followed by annealing to improve the crystallinity of the coating. After insertion into the STM, the tip is ready for use.

The anisotropy of the thin film defines the tip's quantization axis and hence its sensitivity to the components of the sample magnetization. Contrary to intuition, it is not the pointed shape of the tip that determines the magnetic orientation of the tip. A film with a thickness of about 2 nm on a tip with a radius of curvature of 500 nm will "feel" as if it were deposited on a flat surface. One has already a good guess if one knows the easy axis of the same film on the (110) surface of a W single crystal—parallel to the sample plane or perpendicular, that is, along the tip axis. This is a material property. The desirable magnetic orientation of the tip of course depends on the magnetic structure of the sample to be studied and can be chosen by selecting the proper material. Bode et al. [22] used an Fe-coated tip to image the in-plane domain structure of a network of thin Gd islands and were able to switch the tip magnetization by pulses of an external magnetic field. In order to image the perpendicularly magnetized domains of double-layer (DL) Fe nanostripes on stepped W(110), a film of seven atomic layers Gd at the tip provided the required out-of-plane sensitivity [17]. On the same system, an in-plane sensitive Fe coated tip revealed strong contrasts at the positions of the domain walls between oppositely magnetized domains where the magnetic moments locally rotate through an in-plane orientation [18].

While most materials exhibit a clear preference for either in-plane or out-of-plane magnetization, it is not unusual that the tip magnetization is somewhat canted. This can occasionally be exploited to observe both components of a sample magnetic structure at the same time. As illustrated by Figure 22.10, the tilt angle has to be taken into account if magnetization profiles are to be concluded from experimental data [36].

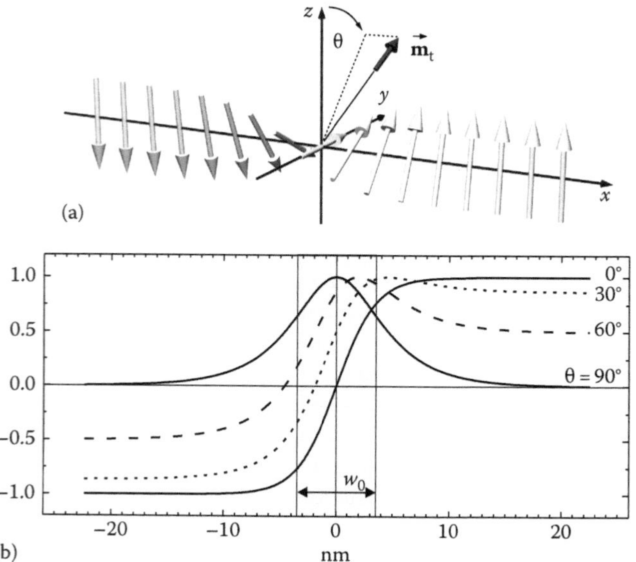

FIGURE 22.10 (a) Schematic representation of a Bloch wall. (b) Measured wall profiles strongly depend on the tip's magnetization $\mathbf{m}_t$. The relevant angle θ is the one between the *z* direction and the projection of $\mathbf{m}_t$ onto the *y*–*z* plane.

A more subtle effect has been observed in the *dI/dU* mode. Depending on the applied bias voltage, the tip sensitivity is found to change reversibly between perpendicular and in-plane [37]. This was interpreted in terms of an *intra-atomic noncollinear magnetism*: while the global magnetization of the effective tip atom is unchanged, different atomic orbital contributions with differently oriented moments are addressed by the variable bias.

As already mentioned, FM tips will of course respond to an externally applied magnetic field. With both tip and sample being FM, they can be forced into a parallel configuration. While, in general, a parallel alignment will result in a higher tunneling signal, this cannot be taken for granted because the spin polarization P is a function of energy, $P = P(\varepsilon)$, and may even change its sign with energy. In *dI/dU* maps, this is frequently observed as a contrast reversal depending on the bias voltage. The only way to unambiguously determine whether a high or low signal at a given bias represents a parallel alignment is to apply an external magnetic field strong enough to saturate both magnetic electrodes of the tunneling junction.

Thin film FM tips were introduced as a solution to the stray field problem of bulk FM tips, as discussed earlier. However, even the strongly reduced amount of magnetic material gives rise to stray fields, which can be significant on a local scale. In the study of the remagnetization of DL Fe stripes on W(110) by a perpendicular field, it has been found that sample areas that had been scanned by the Gd-coated tip were remagnetized much more efficiently than areas that had not been scanned by the tip [20]. This observation gave the initial motivation to coat a tip with chromium, which is AFM. An antiferromagnet has two sublattices of oppositely oriented magnetic moments, ideally all moments canceling each other such that there is no net magnetization and hence no stray field. Still, the outermost atom at the very end of the tip that carries the tunneling current should be spin-polarized. And that is all what is needed. In contrast to magnetic force microscopy (MFM), another high resolution scanning probe technique based on the detection of magnetic stray fields, SP-STM does not rely on a net magnetization but solely on the spin polarization of electronic states. By comparison of measurements taken with the Gd- versus the Cr-coated tip, the effective contribution of the Gd tip's stray field to the external field inducing the remagnetization could be estimated to roughly 300 mT [20].

With Cr as an alternative tip coating material, even extremely delicate magnetic structures like magnetic vortex cores can be imaged successfully. Fe islands of a thickness $d \geq 7.5$ nm exhibit an in-plane magnetization circulating around a vortex core with a perpendicular moment right at the center [24]. Such a vortex core has no fixed position on the island but can be easily moved around by smallest fields acting on it. In the presence of a stray field from an FM tip, it is virtually impossible to image the vortex core because it will evade the approaching tip. It took a tip coated with a thin (≈ 35 atomic layers) Cr to image it at its equilibrium position. In-plane sensitivity was achieved with a thick (≈ 200 ML) Cr coating, allowing the domains of

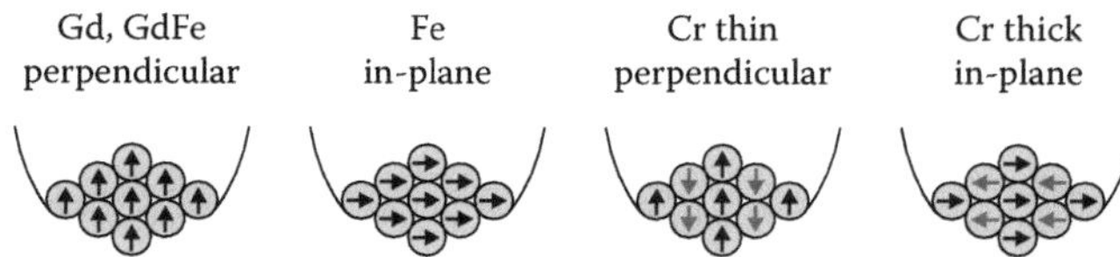

FIGURE 22.11 Variety of magnetic tip coating materials. The desired magnetic properties and sensitivity to a specific sample magnetization component can be chosen by selecting the proper material and film thickness.

the same sample to be imaged. Figure 22.11 summarizes the tip characteristics discussed so far.

Having both AFM and FM tips available provides many new opportunities for measurements in external fields. One may remagnetize the FM sample without remagnetizing the AFM tip as was done, e.g., for Co islands on Cu(111) [25] or Co islands on Pt(111) [28]. Or, an FM tip on an AFM sample can be remagnetized by an external field without affecting the sample. In Section 22.5, examples of this kind will be presented which are particularly useful in the study of noncollinear structures because the tip magnetization can be continuously rotated from in-plane to out-of-plane by virtue of a perpendicular field.

One disadvantage of thin film magnetic tips is owing to the limited amount of magnetic material. Such tips need to be prepared repeatedly, and a tip exchange mechanism as discussed earlier is indispensable. This requirement is obsolete with a bulk AFM tip in use. In contrast to FM bulk tips, no significant stray fields are expected. As described by Schlenhoff et al. [38], Cr tips made from polycrystalline bulk Cr were applied successfully to image the various magnetization components of 1.5 ML Fe on bcc W(110). Their tip exhibited a canted magnetization and gave simultaneously contrasts for all three spatial directions, that is, on single Fe layer regions with in-plane $[1\bar{1}0]$ easy axis, DL areas with perpendicular (110) magnetization, and at DL domain walls where the spins locally point along the (001) axis. The actual orientation of the tip magnetization in this case depends on the accidental orientation of the crystallite that happens to form the tip apex.

The great advantage of bulk AFM tips is their virtually infinite resource of magnetic material. As every STM practitioner knows, if a tip happens to be too blunt one can improve the imaging quality by applying voltage pulses up to 10 V in order to modify it. This will blast away some tip material and result in a changed effective microtip. With a magnetic thin film coating, there is a chance to end up with a tip that has lost its spin polarization. This cannot happen to a bulk AFM tip.

Another way to obtain a spin-polarized tip is to cautiously dip a non-magnetic tip into the magnetic material on the sample surface. This will most likely result in a few atoms or clusters of that material sticking to the tip apex, providing spin contrast. Loth et al. [39] used vertical atom manipulation to pick up a single Mn atom with a non-magnetic tip in a controlled fashion and excited the spin of another individual Mn atom at the sample surface by spin-polarized currents, thereby exploring the quantum spin states of the latter.

22.4 Atomic Resolution Imaging with Spin Sensitivity

22.4.1 First Results: Fe_3O_4

Magnetic STM imaging at the atomic level was first demonstrated in 1992 by Wiesendanger et al. [40], using a bulk FM Fe tip. As already mentioned, because of the inevitable stray field problem associated with these tips, only robust magnetic structures are suited for this purpose. They used a natural single crystal of magnetite (Fe_3O_4), which is a ferrimagnet containing Fe^{3+} as well as Fe^{2+} ions. It has a cubic inverse spinel structure (see Figure 22.12) and a Curie temperature of 860 K, allowing for magnetic STM measurements at room temperature.

Two completely different surface topographies were observed. The first one is characterized by terraces predominantly separated by steps 4.2 Å high, corresponding to one half of the height of the conventional unit cell, Figure 22.12. While atomic resolution of the square surface lattice of A-site Fe^{3+} ions was achieved, no difference was observable between images taken with nonmagnetic W or magnetic Fe tips.

The second type of topography, much less frequently found in merely small areas, exhibited step heights of only 2 Å. The terraces showed straight atomic rows being 6 Å apart from each other, changing their orientation by 90° from one terrace to the next. This observation is compatible with the Fe B-sites occupied by Fe^{2+} and Fe^{3+} within the Fe-O planes, while the O atoms remained invisible in the images, see Figure 22.13a. In this topography, a clear distinction was found between images taken with either tungsten or iron tips. While W tips did not resolve any periodic structure along the atomic rows, enhanced corrugation with a period of 12 Å was observed with Fe tips, which is four times the atomic period in that direction, see Figure 22.13b. From a simple topographic image, a period of just 3 Å was to be expected, but in a spin-polarized measurement, the different spin configurations need to be taken into account, $3d^5{\uparrow}3d{\downarrow}$ for Fe^{2+} and $3d^5{\uparrow}$ for Fe^{3+}, giving rise to different magnetic moments. Fe_3O_4 has a low temperature ordered phase where the repeat period is indeed 12 Å, as sketched at the bottom of Figure 22.13b. The measurement temperature, however, was well above the

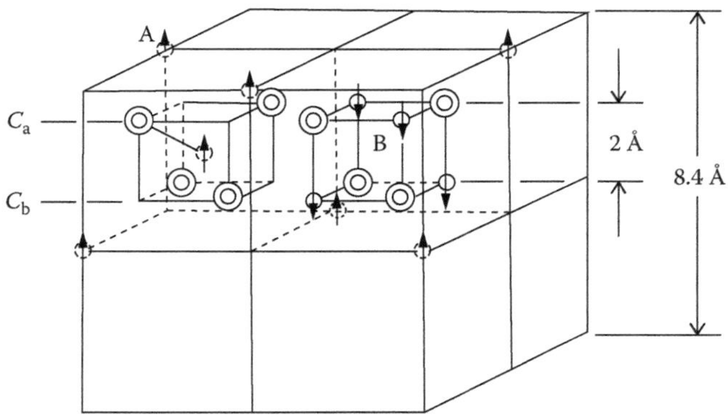

FIGURE 22.12 Conventional unit cell of magnetite (↓ octahedrally coordinated Fe^{2+} and Fe^{3+} sites, ↑ tetrahedrally coordinated Fe^{3+} sites).

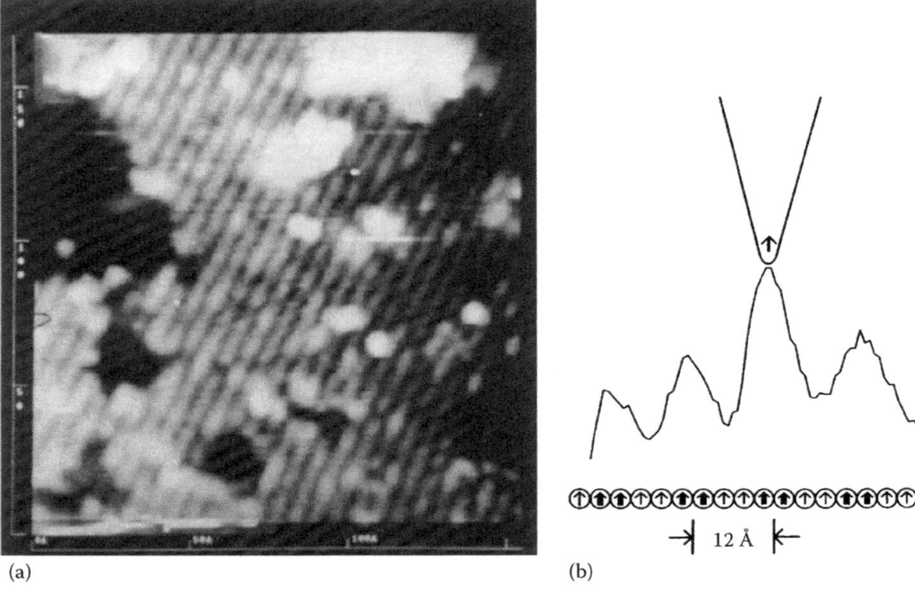

FIGURE 22.13 (a) Atomic rows of magnetite. (b) Line section taken along an atomic row of octahedrally coordinated Fe^{2+} and Fe^{3+} sites.

temperature of the order-disorder (Verwey) transition (~120 K in the bulk), providing evidence for an enhanced Verwey transition temperature at the Fe_3O_4 (001) surface.

It is noteworthy that this early experiment was performed at a time when no theory of magnetic STM imaging was available. As seen *ex post*, the interpretation of the measurement is in full agreement with the theory developed much later by Heinze, Wortmann, and coworkers as discussed in Section 22.2 along with Equation 22.16 and Figure 22.8, predicting a strong corrugation enhancement in topographic measurements arising from the larger magnetic repeat period.

22.4.2 Antiferromagnetism in Two Dimensions: Monolayer Mn on W(110)

A new chapter of SP-STM research started in 1998 with the introduction of the spectroscopic mode by Bode et al. [22]. Based on 2D *dI/dU* maps with magnetic contrasts, the FM domain structure of a Gd(0001) film on W(110) was revealed. One important prerequisite was to carry out the measurement with the sample held at a temperature below the Curie temperature T_C, which is strongly reduced for thin films as compared to bulk materials. Further improvements of imaging quality resulted from a new dedicated instrument where the whole STM including the tip was cooled to 14 K [35]. All measurements shown throughout this article were taken at this or even lower temperatures. The STM has been equipped with a versatile tip exchange mechanism and mounted inside the bore of a superconducting magnet, providing additional degrees of freedom to perform magnetic measurements in fields up to 2.5 T, perpendicular to the sample plane [18].

To demonstrate simultaneously atomic resolution and spin sensitivity, an ML of chemically equivalent atoms with magnetic moments arranged antiferromagnetically, that is, adjacent moments pointing in opposite directions, was used as model-type system. The existence of such 2D AFM systems, such as V, Cr, and Mn on noble metal substrates, had been predicted more than 10 years earlier by Blügel et al. on the basis of first-principles calculations [41]. An experimental verification is quite challenging because the antiferromagnetism is at the atomic scale, the total magnetization is zero, and the Néel temperature is unknown. Some experimental evidence was available from photoemission experiments [42], but an unambiguous proof was lacking.

In a joint theoretical and experimental study, the system of 1 ML Mn/W(110) was investigated [29]. Three possible magnetic configurations were scrutinized by first-principles calculations, the FM and two different AFM solutions, $p(2 \times 1)$ and $c(2 \times 2)$, see Figure 22.14 upper row panels. The $c(2 \times 2)$ structure was found clearly lowest in energy. The spin–orbit interaction couples the direction of the magnetic moments to the crystal lattice and is responsible for the magnetic anisotropy. Including this interaction, the calculations revealed a preference for an in-plane orientation along the $[1\bar{1}0]$ direction. On this basis SP-STM constant-current images were simulated as shown in Figure 22.14 lower panels, making use of the spin-polarized version of the Tersoff–Hamann model presented in Section 22.2. In particular, the two AFM solutions, both giving stripe patterns, are easily discerned by their variance in stripe directions, running along (001) or $(1\bar{1}1)$, respectively.

After deposition, Mn forms a pseudomorphic overlayer without any alloying, i.e., the Mn atoms acquire the lattice constant of the W substrate. Using a non-magnetic W tip, atomic resolution images were obtained, resolving the diamond-shaped chemical unit cell as shown in Figure 22.15a. In a second experiment, a magnetic tip was applied. To account for the expected in-plane magnetization, an Fe-coated tip was used. This time, a

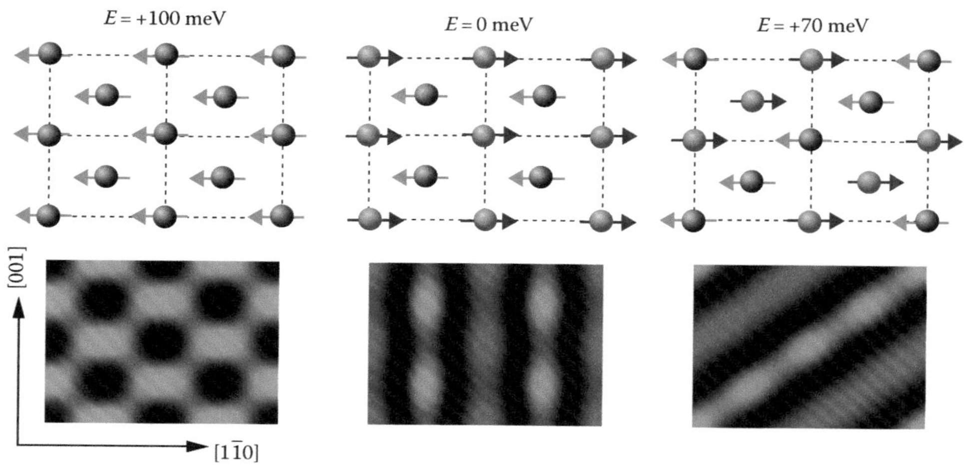

FIGURE 22.14 Upper panels: three magnetic configurations considered theoretically (left to right): FM, AFM $c(2 \times 2)$, and AFM $p(2 \times 1)$. $c(2 \times 2)$ is lowest in energy. Lower panels: simulated STM images, assuming a tip with a magnetization pointing to the right.

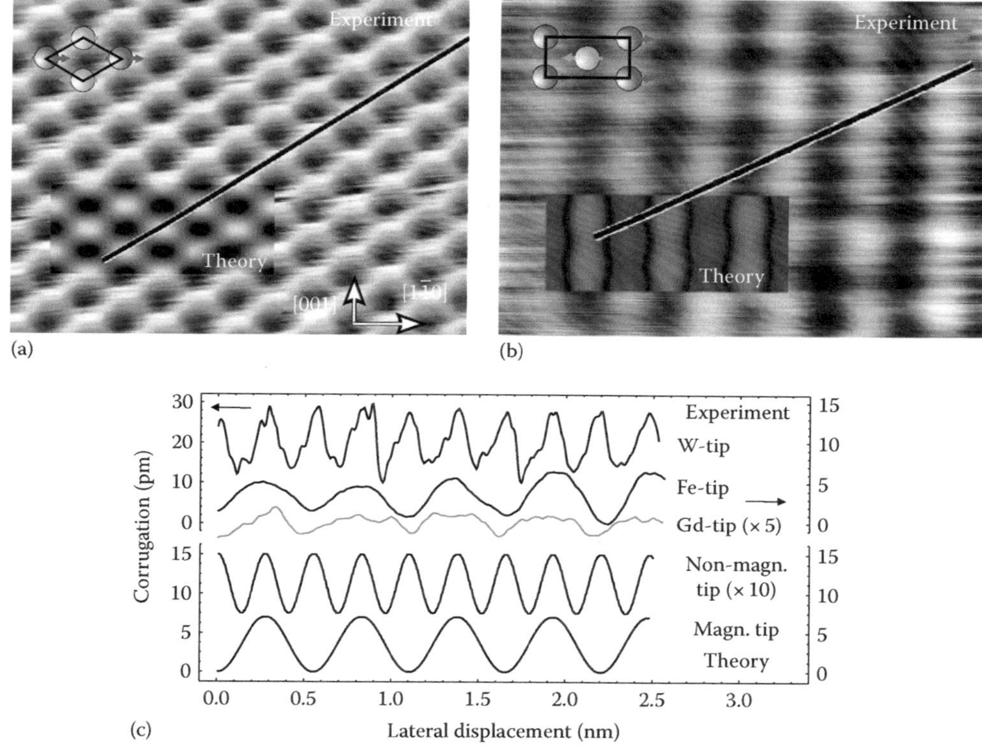

FIGURE 22.15 Atomically resolved Mn monolayer on W(110) as imaged with (a) non-magnetic W tip, (b) magnetic Fe tip. Insets allow a comparison to images calculated from theory. (c) Line profiles taken along the densely packed [1$\bar{1}$1] atom rows, showing the period doubling for the magnetic measurement.

stripe pattern was observed, reflecting the larger magnetic unit cell. It is instructive to compare line sections taken along the [1$\bar{1}$1] nearest neighbor direction, Figure 22.15c. The magnetic measurement doubles the period as expected. This is true even if a Gd-coated tip is in use, which is almost perpendicularly magnetized, i.e., nearly orthogonal to the sample magnetization, rendering the spin-polarized contribution to the tunneling current very small. Still, the magnetic signal is dominating, albeit with a largely reduced corrugation. This observation strongly corroborates the ideas developed in Section 22.2. Also, the direct comparison of the experimental with the simulated images (see insets) is very favorable. The only disagreement is found in the much too small corrugation as derived from the non-magnetic simulation, which is owing to the well-known deficiency of the Tersoff–Hamann model at the atomic length scale, as already discussed in Section 22.2.

22.4.3 Monolayer Fe on W(001)

One might not be too much surprised to find Mn AFM as an ML—after all, Cr and Mn are well known to be AFM in the bulk, whereas the other magnetic elements of the $3d$ transition metals, body-centered cubic (bcc) Fe, hexagonal close-packed (hcp) Co, and face-centered cubic (fcc) Ni, are the classical ferromagnets. Both the chemical symbol Fe as well as the term "ferromagnetic" stem from iron's Latin name *ferrum*, rendering this element the prototypical ferromagnet. Therefore, much excitement was raised when Fe was found AFM in a metastable fcc phase [43,44], which is difficult to obtain.

For the development of SP-STM, Fe thin films grown on W(110) played an important role as a reference system, providing a stunning variety of magnetic properties (see Ref. [45] and references therein), such as the coexistence of in-plane magnetization in the ML and out-of plane magnetization in the DL, eventually being recognized as an inhomogeneous spin spiral [46,47]. The magnetism of Fe films was also studied on the (001) face of tungsten [14,48]. Fe film growth starts by forming a pseudomorphic wetting layer, followed by growth of islands of two and more ML thickness. The second and third layer islands have in-plane easy axes along the <110> directions (diagonal of the square surface unit cell). A spin reorientation by 45° is observed to the <100> bulk orientation (edge of the unit cell) starting with the fourth layer, all coverages exhibiting FM order.

The first layer Fe on W(001) remained a puzzle for a long time. Numerous experiments confirmed an absence of remanent magnetization but were unable to clarify whether the film is AFM, FM with a very low Curie temperature, or non-magnetic at all. Also, DFT calculations gave inconsistent results, depending on the choice of the exchange correlation approximations.

The magnetic structure of the ML film was investigated by SP-STM and DFT calculations [49]. Figure 22.16 shows a 100 nm × 100 nm 3D view of 1.3 ML Fe on W(001), with the height information from the topographic image, colorized with the simultaneously recorded dI/dU signal, the latter revealing the magnetic contrasts. Islands of DL thickness display four stages of contrast representing the four equivalent easy in-plane magnetization directions as measured with an in-plane sensitive Fe-coated tip. Figure 22.16b and c present a zoom into a 2.5 nm × 2.5 nm region on the ML as indicated in Figure 22.16a, constant-current and dI/dU signal, respectively. This measurement was taken with a strong external magnetic field of 2.5 T applied perpendicular to the sample plane. Any AFM structure will remain unaffected by the field. In contrast, the magnetization of the FM tip is forced into the field direction, making it sensitive to the out-of-plane magnetization. The contrasts now appearing in Figure 22.16b and c are thus caused by the sample moments pointing up and down with a period twice that of the substrate atoms, exposing a $c(2 \times 2)$ AFM supercell as sketched in Figure 22.16b. The two images highlight the fact that both imaging techniques can be used equally well to resolve atomic-scale magnetic structures. Notably, there is a phase shift of 180° observed between line sections extracted from topographic (Figure 22.16b) and differential conductance dI/dU data (Figure 22.16c) as plotted in Figure 22.16d. This phase shift can be easily understood: in the constant current topograph, a maximum in $z(x,y)$ means the tip being retracted from the surface because of maximum local DOS; at the same time, due to the enlarged tip–sample distance, the differential conductance is reduced, and *vice versa*.

What is the cause of the AFM ordering? This question was addressed by DFT calculations. This method allows us to scrutinize the electronic band structure of an Fe ML on W(001) as a function of the interlayer distance d between film and substrate. Figure 22.17 shows a plot of the total energy and magnetic moment per atom versus distance for the non-magnetic, FM, and AFM solutions. At the largest distance considered (6 a.u.), the Fe film hardly "feels" the presence of the W atoms, and the DOS is very similar to that of a free-standing, unsupported ML, exhibiting a very large non-magnetic DOS right at the Fermi energy. But this matches the Stoner criterion for ferromagnetism, and indeed at this distance, the $p(1 \times 1)$ FM solution has the lowest energy. Reducing d increases the Fe-W hybridization, and the energies of FM and $c(2 \times 2)$ AFM solutions become degenerate. Relaxing the film further to its equilibrium d results in a huge energy difference in favor of the AFM solution. Due to Fe-W hybridization, the DOS distribution near the Fermi energy is strongly modified and resembles that of prototypical AFM MLs of Cr and Mn on noble metal (001) surfaces [50]. One can now explain why the Fe ML is FM on W(110) and AFM on W(001): the effect of hybridization is stronger on the (001) than on the (110) surface because the number of nearest-neighbor substrate atoms is four for each Fe atom on (001) but only two in the (110) case. It is well known that an induced spin polarization is present in the W atoms when Fe is deposited on the (110) surface, contributing to the in-plane anisotropy of the Fe ML on that surface. This effect is absent on the (001) oriented surface: for symmetry reasons the W atoms below the AFM layer cannot be polarized.

Summarizing, an Fe ML on a bcc W crystal can be prepared FM and in-plane, or AFM and out-of-plane, depending on the choice of the crystal face. Motivated by this finding, a systematic theoretical survey of $3d$ transition metal MLs on the W(001) surface was carried out by first-principles calculations [51]. The surprising result of this study is that V, Cr, and Mn are predicted to exhibit an FM ground state, whereas the prototypical ferromagnets Fe and Co favor a $c(2 \times 2)$ AFM ordering, this trend again explained by strong $3d - 5d$ hybridization due to the presence of the interface to the W substrate.

22.4.4 Antiferromagnetic Domain Walls

In FM thin films of perpendicular anisotropy, it is common to find domains of up and down magnetization with transition regions between them where the atomic spins rotate from one orientation through an in-plane direction to the other. These transition regions are called domain walls (DW). Domain walls cost energy. But this is overcompensated by the gain in dipolar stray field energy that can be saved by subdividing the

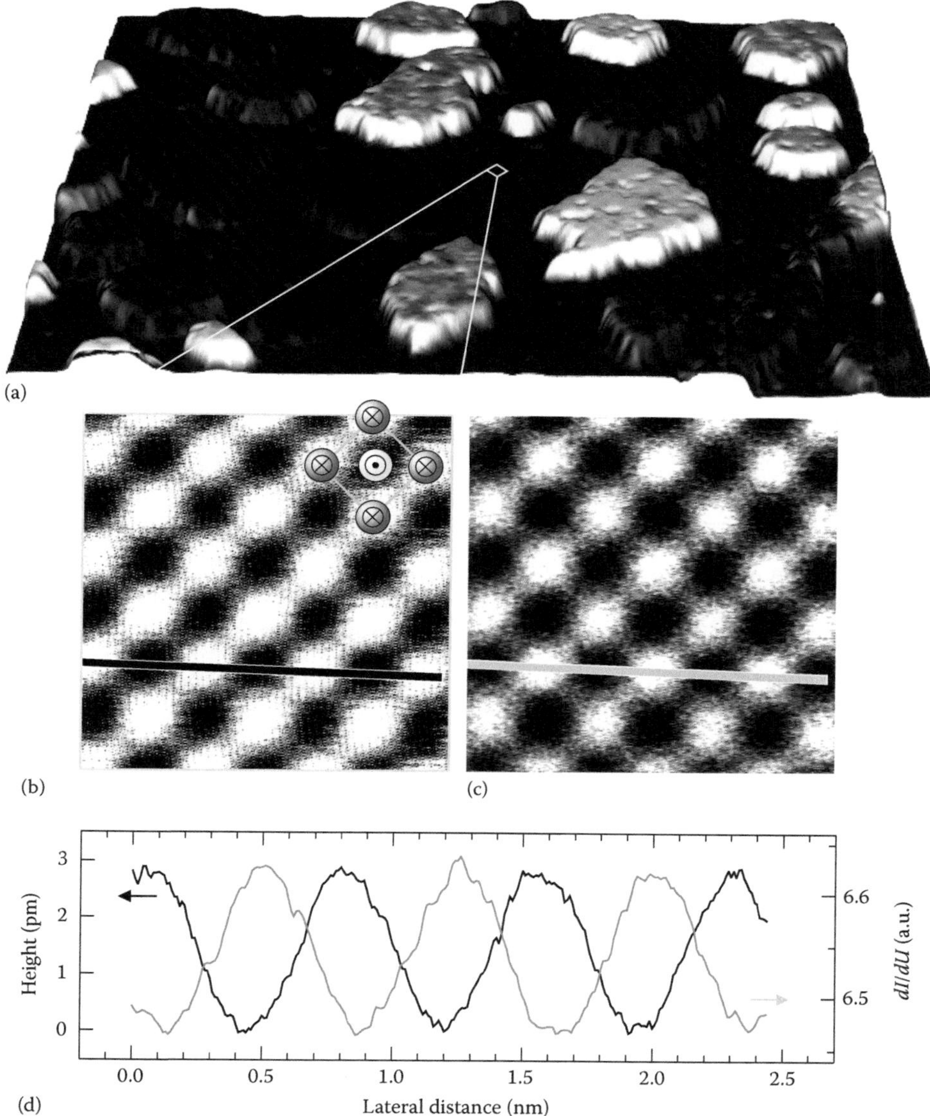

FIGURE 22.16 (a) Perspective view of 1.3 ML Fe on W(001), 100 $nm^2 \times 100$ nm^2, composed with height information from topography and colorization from the simultaneously recorded dI/dU map, using an Fe tip ($I = 2$ nA, $U = 500$ mV). Double layer islands show a four stage magnetic contrast. (b and c) At the atomic length scale, a $c(2 \times 2)$ superstructure is resolved on the ML when a perpendicular magnetic field (+2.5 T) forces the tip magnetization to the vertical direction ($I = 3$ nA, $U = -100$ mV). (d) Constant-current and dI/dU signals are phase shifted by ≈ 180°.

homogeneous FM film into ever smaller domains. Domain walls are introduced until at some point a balance in energy is achieved.

For an AFM film, the situation is very different. Ideally, all spins are compensated, and there is no dipolar energy that can be saved. Hence, a domain wall is energetically not favorable at all. In an AFM film like the Fe/W(001) ML, a domain wall would delimit two areas where the spin order is shifted by just one lattice constant, thereby introducing uncompensated spins along the wall. It is believed that uncompensated spins play an important role in the exchange bias effect, which is of great technological relevance for magnetic data storage and MRAM applications. A magnetically hard FM layer is needed as a reference against a soft layer that can be switched, and the hardening is achieved by coupling the FM layer to an AFM layer, which then strongly pins the FM layer.

AFM domain walls in the Fe/W(001) ML were investigated experimentally by SP-STM and theoretically by MC simulations [52]. For energetic reasons, AFM domain walls are very rare. A large area of 2 μm × 1 μm of virtually perfect AFM order was observed, despite the presence of numerous point defects, step edges, and second layer islands; a small portion is presented in Figure 22.18a. Only after intentionally increasing the defect density, the short DW shown in Figure 22.18b and c was found, running in a <010> direction, being pinned between two second layer the islands in Figure 22.18b was taken with a magnetic field of 2 T applied perpendicular to the sample plane, aligning the tip magnetization vertically, revealing

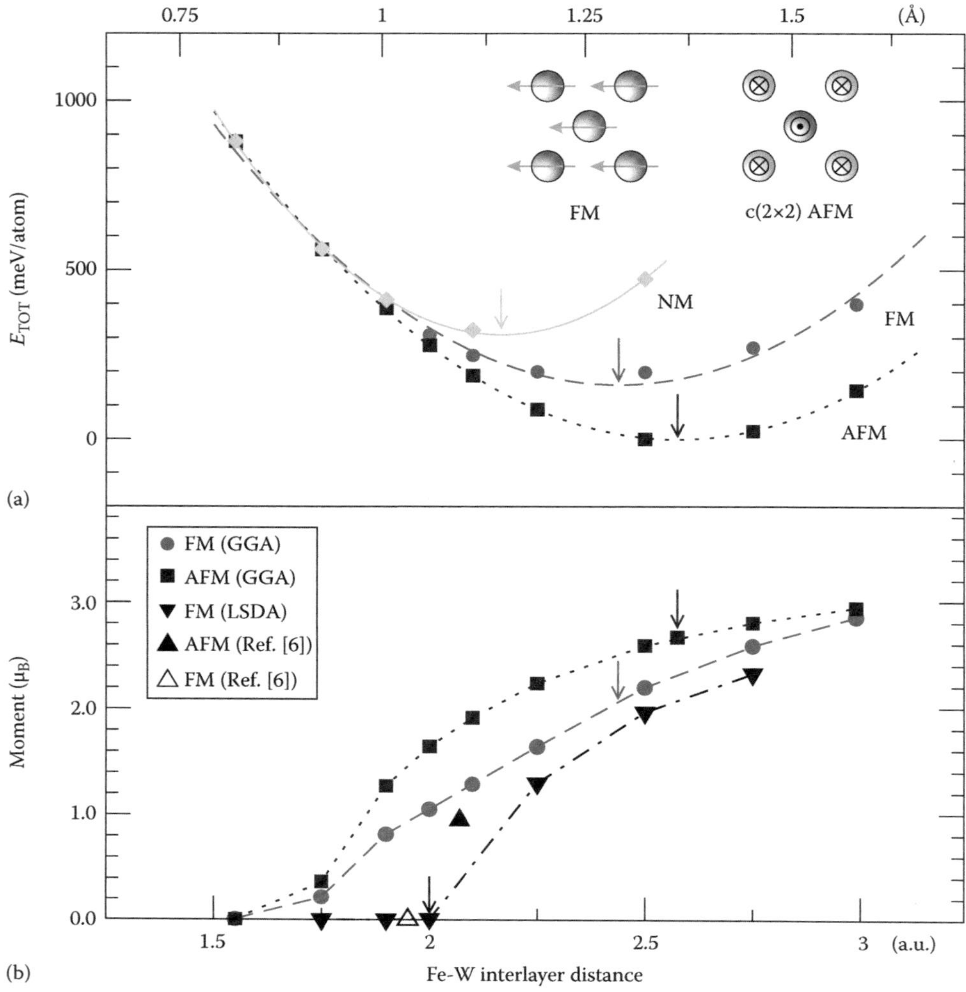

FIGURE 22.17 (a) Total energy and (b) magnetic moment per Fe atom as a function of interlayer distance, calculated for the non-magnetic, FM, and AFM solution.

the checkerboard pattern we already know. When the sample spins go in-plane at the wall, the magnetic contrast gets blurred because tip and sample magnetizations are orthogonal. The phase shift across the wall can be clearly seen by following the two straight dashed lines in Figure 22.18b, which pass through the minima of the pattern in the upper part of the image and match the maxima in the lower part. Field removal lets the tip magnetization relax into its preferred in-plane orientation, see Figure 22.18c. Now the domain checkerboard vanishes; instead a pattern emerges right at the DW, this time arising from the sample's in-plane spins.

Figure 22.19 allows a comparison of results for a DW oriented along a <110> direction as obtained by (a) MC simulations, (b) a simulated SP-STM image of the same spin configuration, (c) experimental SP-STM image, and (d) line profiles extracted from (c) as indicated. In MC simulations, a rapid thermal quench down to the measurement temperature of 13 K had to be applied in order to prevent the DWs to be driven out of the investigated area, underlining the strong tendency to avoid DWs at all. The configuration shown here has the center of the DW not at the position of an atomic row but halfway between two rows. The spin rows closest to the wall center have a predominant in-plane orientation (polar angle $\Theta \geq 65°$) but quickly tilt into a vertical orientation with increasing distance from the wall center. The in-plane component integrated over the wall width cancels but, interestingly, the integrated out-of-plane component does not, leaving a net non-vanishing magnetic moment. This theoretical result is in good qualitative agreement with the experimental line profiles shown in Figure 22.19d. The black line, indicating the sum of the blue and green profile and representing the averaged perpendicular magnetic signal, appears mirror-symmetric, just like the red profile taken along the nearest-neighbor direction (bottom panel). As in the MC simulations, the total wall width is about 6–8 atomic rows (≈ 1.8 nm).

22.5 Noncollinear Spin Structures

The 2D AFM systems studied in the previous section are examples of collinear spin order—the magnetic moments, pointing in opposite directions, align with a common axis. There are

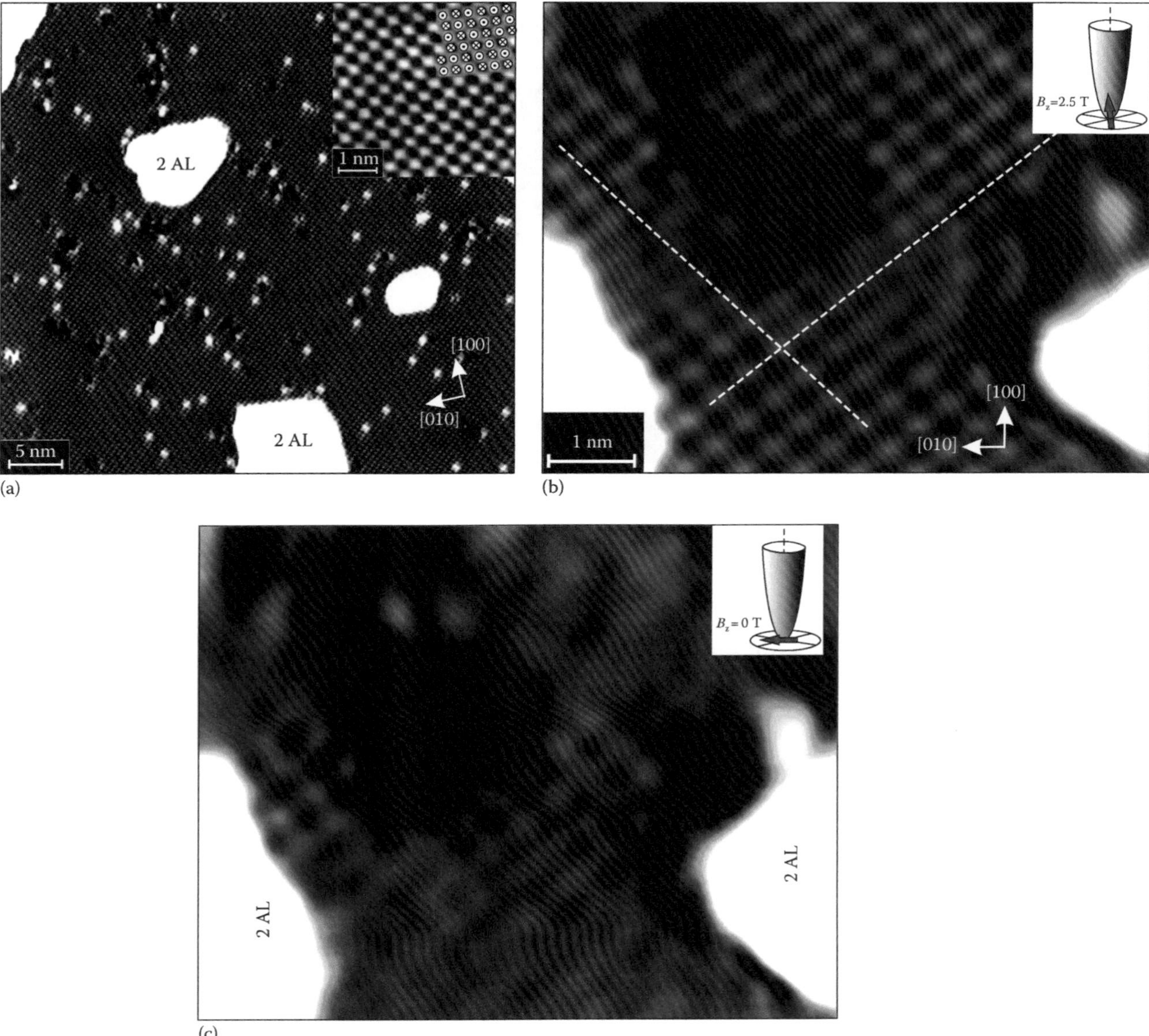

FIGURE 22.18 (a) Despite numerous point defects, the antiferromagnetic order of an ML Fe on W(001) is almost perfect. (b) A rare phase domain wall, pinned between two second-layer islands. The contrast from the perpendicularly magnetized tip gets blurred. (c) After field removal, the tip magnetization goes back in-plane, giving contrast at the domain wall where the spins locally are in-plane.

many configurations, however, where a pairwise antiparallel alignment cannot be accommodated, causing spin frustration. One particularly simple situation is given by three spins to be arranged on a triangular lattice (e.g., the (111) face of an fcc crystal) as sketched in Figure 22.20a. A collinear solution would be a row-wise AFM structure with spins coupled FM within one row and AFM coupling between neighboring rows. Another—noncollinear—solution is the so-called Néel structure where the magnetic moments attain an angle of 120°, giving rise to a $(\sqrt{3}\times\sqrt{3})$ magnetic superstructure, Figure 22.20b. Wortmann et al. predicted such a structure for one layer of Cr on Ag(111) and calculated SP-STM images by assuming certain orientations of the tip magnetization [12], see also Figure 22.9. The Néel structure was indeed observed with SP-STM by Gao et al. in one ML Mn on Ag(111) [53]. Similar results were obtained for a Cr ML on Pd(111) by Waśniowska et al. [54].

While this type of noncollinear magnetic structure caused by geometrical frustration may appear quite intuitive, there are other causes for noncollinearity that are more subtle and less obvious, arising from a complex interplay of competing magnetic interactions, and this is what will be illustrated in the following section. It is only recently that the importance of noncollinear structures in thin films and nanostructures has been getting recognized. At the atomic length scale, more and more systems with long range order of spin spirals, vortex structures, and even skyrmion-like spin configurations emerge. In the end,

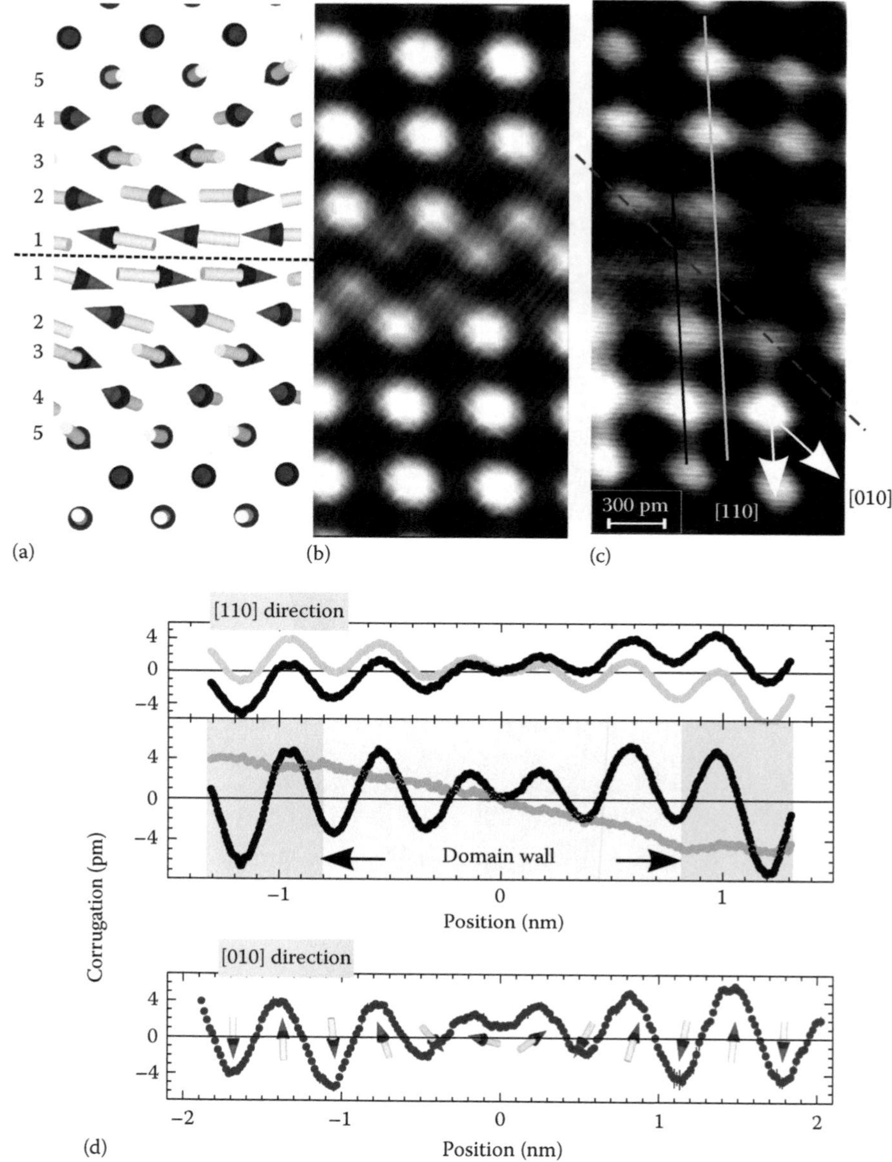

FIGURE 22.19 (a) Spin configuration from Monte Carlo simulation. (b) Simulated SP-STM image, assuming a perpendicular tip magnetization. (c) Experimental SP-STM image, revealing the phase shift across the domain wall. (d) Top: line section taken along [110] direction, see (c). Center: sum (black line) and difference (gray line) of the two profiles. Bottom: profile along the nearest-neighbor [010] direction, arrows indicating the spin orientations.

it may turn out that noncollinear magnetism at surfaces and thin films could be the rule rather than the exception.

22.5.1 Energetics of Magnetic Interactions

Magnetism in metals is, ultimately, a consequence of electrostatic interaction and the Pauli exclusion principle.* Itinerant spin-polarized electrons hop across the crystal lattice and exert the Heisenberg exchange interaction between spin moments **S** located at sites i and j. The energy related to exchange is expressed as

$$E_{\mathrm{H}} = -\sum_{i,j} J_{ij}\mathbf{S}_i \cdot \mathbf{S}_j, \tag{22.18}$$

with J_{ij} the exchange coupling constant between pairs of spin moments. The sign of J determines the type of favored magnetic order, FM (AFM) for positive (negative) sign. In any case, the spins will tend to align collinear—parallel or antiparallel, and any deviation from collinearity is associated with an energy penalty. The Heisenberg term has no directional preference. Also, there is no

* The quantum mechanical origin of magnetism in condensed matter is the subject of numerous textbooks, e.g., Ref. [55], and cannot be elaborated in detail here.

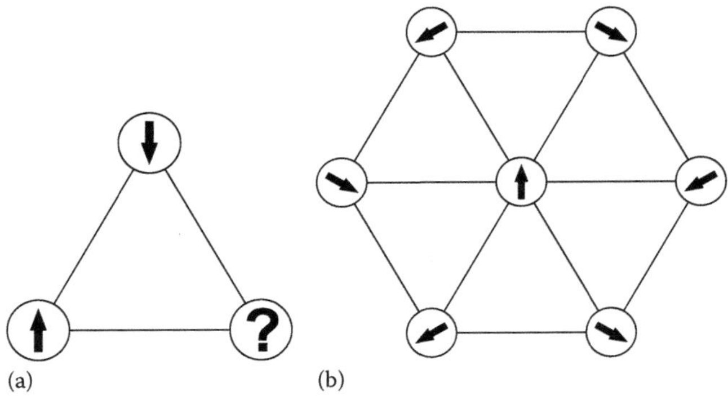

FIGURE 22.20 (a) Antiferromagnetism on a triangular lattice causes geometrical spin frustration. (b) One possible solution: noncollinear 120° Néel structure.

difference in energy whether a deviation be right-handed (↖↗) or left-handed (↗↖)—the interaction is symmetric.

Another energy term to be considered is the anisotropy. It is a relativistic effect of spin–orbit interaction and couples the magnetization to certain crystallographic directions. It is responsible for the occurrence of easy and hard magnetic axes. The anisotropy term is of the form

$$E_{\text{ani}} = \sum_i K_i \cdot \sin^2 \varphi_i \tag{22.19}$$

where

K_i is the anisotropy constant

φ_i is the angle between the easy axis and the local magnetization at site *i*

In an FM system, long-range magnetostatic interactions have to be included, which also contribute to the *effective* anisotropy via the shape anisotropy.*

None of the aforementioned symmetric contributions to the total energy lend themselves to generating a magnetic ground state that is a spin spiral with a unique sense of spin rotation. Such a term was introduced by Dzyaloshinsky [57] and Moriya [58], hence Dzyaloshinsky–Moriya interaction (DMI):

$$E_{\text{DMI}} = \sum_{i,j} \mathbf{D}_{ij}(\mathbf{S}_i \times \mathbf{S}_j). \tag{22.20}$$

This term describes an antisymmetric interaction. It is known to arise from spin–orbit coupling (SOC), again relating the crystal lattice with the spin symmetry. If the lattice exhibits inversion symmetry, that is $(x, y, z) \mapsto (-x, -y, -z)$, DMI immediately vanishes. This is the case in most bulk systems, and therefore DMI did not receive much attention in the past. However, at the *surface* of a crystal inversion symmetry is always broken, and DMI can be relevant. Comparing the Heisenberg and DMI terms, Equations 22.18 and 22.20, respectively, we note that pairs of spin vectors at sites *i* and *j* are concatenated by a scalar product in the first case and a vector product in the second. We already saw that Equation 22.18 favors a collinear configuration. In contrast, Equation 22.20 tells us that energy can be saved by introducing an angle between the spins. Hence, the two terms directly compete. Furthermore, DMI will save energy only if the spin rotation is in the proper direction (depending on the sign of **D**), while rotation in the opposite direction increases energy; DMI thus provides a mechanism to select one sense of spin rotation over the other. Generally, DMI is a weak effect compared to Heisenberg exchange, thus only a small deviation from the collinear spin structure can be expected in most cases. However, we will explore examples where nearest neighbor Heisenberg exchange is softened due to competing interactions, such that DMI can be of similar order.

22.5.2 Revisiting Mn/W(110): Spiraling a 2D Antiferromagnetic Structure

In Section 22.4.2, the Mn ML on W(110) was introduced as an example of a 2D AFM system, cf. Figure 22.15. This sample was recently revisited on a larger length scale by experiments in applied magnetic fields, accompanied by spin density theory calculations [59]. Figure 22.21a shows an Mn island at 0.77 ML total Mn coverage, imaged in constant-current mode by a Cr-coated tip. In addition to the pattern of bright-and-dark lines with atomic-scale period that we interpreted in Section 22.4.2 as a $c(2 \times 2)$ AFM structure, an additional modulation of the corrugation is observed, with a much larger distance of about 6 nm between the nodes. Areas of strong corrugation change periodically with areas of weak corrugation, see the zoom in panel (b) and the line profile (black symbols) in panel (c). A comparison of the line profile to a sine function (red curve in (c)) reveals a phase shift of 180° across a node, disclosing a true period of about 12 nm. In order to investigate the origin of the long wavelength modulation measurements were performed with an Fe-coated tip in variable magnetic fields up to $\mu_0 H = 2$ T, applied perpendicular to the sample plane (μ_0 the magnetic field constant). The AFM Mn film is virtually not affected by the field because all magnetic moments are compensated. However, the magnetization of the FM Fe tip, being in-plane at zero field, follows the increasing field successively from a horizontal into a vertical orientation, altering its sensitivity from in-plane to canted to out-of plane, as sketched in Figure 22.22. Taking any of the adsorbed defects as a position reference, one can see in the images that this translates into a lateral shift to the left of the long-wavelength feature. This observation can be explained by a spin spiral in the Mn film with a propagation direction along $[1\bar{1}0]$. While, in zero field, high contrast is achieved at locations where sample moments lie in the sample plane, these contrasts get weaker there with the tip magnetization forced successively into the vertical direction. But now contrasts are amplified at areas where sample moments point up and down.

* While these dipolar interactions do not play a role in the compensated antiferromagnetic systems discussed here, they may tip the scales between collinear or noncollinear order in ferromagnetic structures. An example is the inhomogeneous spin spiral observed in double layer Fe on W(110) [47,56].

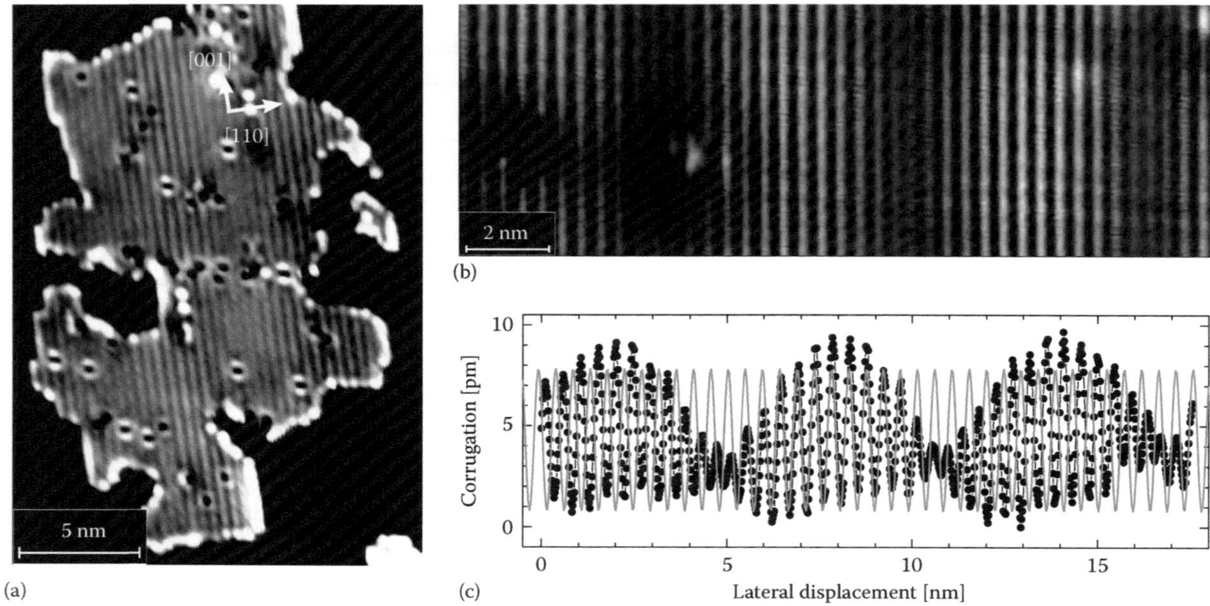

FIGURE 22.21 (a) Monolayer Mn island on W(110), imaged with a Cr-coated tip (parameters: $I = 15$ nA, $U = 3$ mV). The stripe pattern with sub-nm period arises from a $c(2 \times 2)$ AFM structure, cf. Section 22.4.2. (b) The corrugation is attenuated periodically. (c) Comparison of the line profile taken along $[1\bar{1}0]$ and a sine function reveals a 180° phase shift across a node.

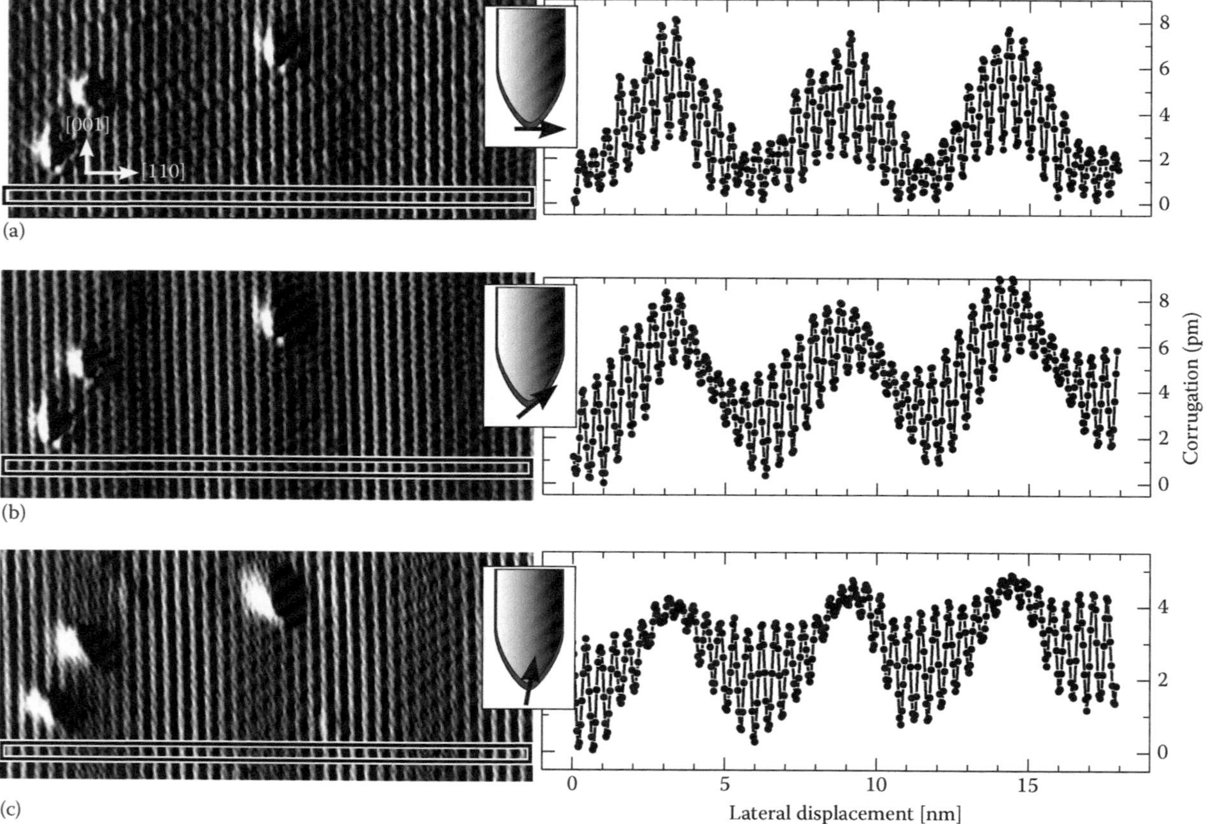

FIGURE 22.22 Measurements in perpendicular magnetic fields with an Fe-coated tip. (a) 0 T, (b) 1 T, (c) 2 T. Line profiles extracted from the indicated boxes. The applied field rotates the tip magnetization successively from in-plane to out-of-plane. Areas of large corrugation appear to shift to the left. Tunneling parameters: $I = 2$ nA, $U = 30$ mV.

Two types of spin spirals are consistent with the experimental observation: a helical or a cycloidal spiral. If z is the surface normal and the spiral propagation direction **q** is along the x coordinate, then a helical spiral is characterized by a spin rotation entirely in the y,z plane, with the chirality vector $\mathbf{C}_i = \mathbf{S}_i \times \mathbf{S}_{i+1}$ pointing along **q**. For a cycloid, in contrast, the spins rotate in the x,z plane, and $\mathbf{C}_i$ is perpendicular to **q**. The instrumental setup used in the experiment does not provide control over the azimuthal orientation of the tip magnetization; thus, the experiment cannot discriminate the two types of spirals.* There are, however, certain symmetry rules introduced by Moriya [60] that can be applied if the spiral propagates along a high symmetry line of the surface. According to these conditions, **D** lies (1) in the surface plane and (2) perpendicular to **q**. From Equation 22.20, it is clear that the energy gain by E_{DMI} is maximized for **D** being antiparallel to $\mathbf{C}_i$ and vanishes if $\mathbf{D} \perp \mathbf{C}_i$. This rules out a helical spiral.

Figure 22.23 shows results from total energy calculations based on spin DFT. In a first step, the energy per Mn atom was calculated for various spiral periods λ without SOC (x symbols). The lowest energy was found for $\lambda = \infty$, i.e., the collinear AFM solution. Including SOC has two effects. Energy is increased by an averaged anisotropy contribution $\bar{K}$ that cannot be avoided for any finite spiral period. DMI, on the other hand, causes a new energy minimum to appear at a finite λ_0, shifted to the left from the origin; negative values, $\lambda < 0$, indicate a left-handed spin rotation. The theoretical pitch of $|\lambda| = 8$ nm is found in reasonable agreement with the experimental value of 12 nm, translating into a 7° deviation from the collinear (↑↓) arrangement.

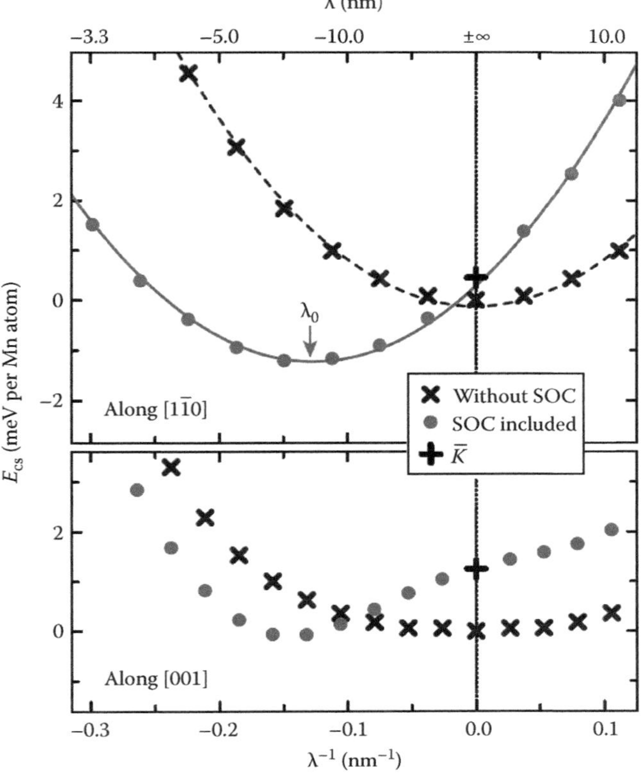

FIGURE 22.23 Total energy calculations without (x symbols) and with spin–orbit coupling (SOC) (dots) as a function of (inverse) spiral period. Without SOC, the minimum of the energy parabola occurs at infinite period, i.e., the collinear solution. For any finite spiral period, SOC adds a contribution from the average magnetic anisotropy $\bar{K}$, shifting the parabola upward (+ symbol). Due to DMI, a new absolute energy minimum at λ_0 appears with negative sign (left-handed spiral).

22.5.3 Spiraling a Ferromagnetic Structure

The spin spiral system Mn/W(110) just discussed is an example for DMI acting on a 2D magnet with AFM Heisenberg exchange. However, DMI can also act on a FM system; as mentioned earlier, this is equivalent to a sign inversion of the Heisenberg coupling constant J_{ij} in Equation 22.18. Such a system was studied recently by Ferriani et al. [61], and again it was an ML Mn on W—this time on the (001) rather than the (110) surface. A representative constant current STM image is shown in Figure 22.24a, recorded with an Fe-coated tip in a perpendicular magnetic field of +2 T; the tip is, hence, sensitive to out-of-plane spins. While the $p(1 \times 1)$ atomic lattice is clearly resolved, the dominating contrasts stem from a wave-like pattern of bright and dark lines running in <110> directions, i.e., diagonal of the square surface unit cell. The lateral period is about 2.2 nm, this value being close to five times the diagonal. There are two rotational domains visible, in accordance with the structural symmetry of the surface, and the domain boundaries are found to move very easily. Upon field removal, a very similar picture is observed (not shown), this time with a tip being sensitive to the in-plane spins. This observation is a strong indication that the structure is indeed a spin spiral. This conclusion is corroborated by total energy calculations based on DFT. Theory revealed some remarkable features of this system: Heisenberg exchange is found particularly softened because not only nearest-neighbor interactions J_1 contribute but also interactions with next-nearest neighbors, J_2, and more distant ones, up to J_5, play a role. Interestingly, the leading term is FM, $J_1 > 0$, contributions beyond nearest neighbors favor AFM, $J_{2...5} < 0$. Given this scenario, DMI interaction is able to induce a large deviation from the collinear state with about 36° between adjacent moments, giving rise to the quite short spiral period.

The labyrinth structure arising from the coexistence of the two rotational domains was scrutinized by MC simulations, see Figure 22.24b and c, using parameters obtained from the DFT calculations. The agreement with experimental observations is almost perfect, reproducing the spiral period and the overall domain structure. In FM films, domain formation is typically driven by a gain in long-range dipolar interaction energy. This is not the case for the spin spiral here where magnetic moments cancel over one spiral period. It is concluded, instead, that domain formation in this system is entropy driven.

* An *experimental* identification of a spin spiral type and also its unique rotational sense was recently achieved for the first time by SP-STM installed inside a 3D vector field magnet. The magnet was used to controllably vary the tip's in-plane magnetization [46].

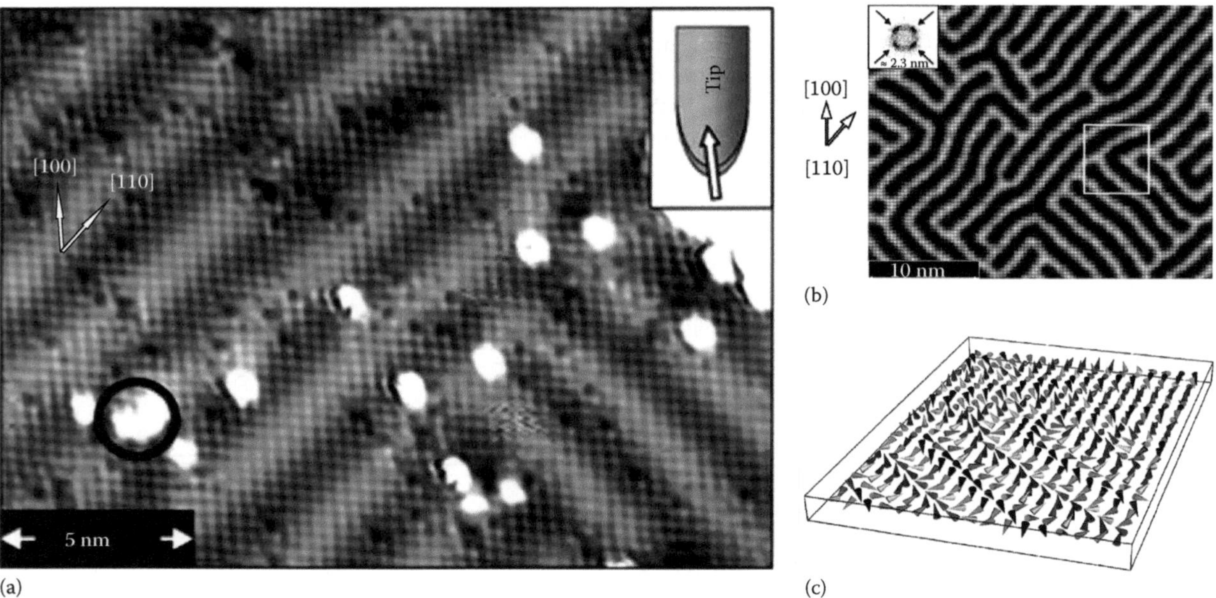

FIGURE 22.24 Labyrinth-like magnetic structure of an ML Mn on W(001) arising from spin spirals propagating along two equivalent directions, forming rotational domains. (a) Spin-resolved constant-current measurement, revealing the spiral as well as the atomic lattice. Parameters: $U = -0.1$ V, $I = 1$ nA, Fe tip, $B_z = +2$ T. (b and c) Monte Carlo simulations based on interaction parameters obtained by *ab initio* calculations.

22.5.4 Spiraling Spins on a Cone

Yet another step further toward increased complexity of magnetic order has been described very recently for DL areas of Mn on W(110) by Yoshida et al. [62]. We recall that the first layer spin spiral with AFM character on this surface unveiled itself by a periodic modulation of the atomic-scale magnetic contrast, cf. Figure 22.22. In Figure 22.25a, this pattern can again be seen in the top part as fine vertical lines with twice the atomic lattice period, vanishing every 6 nm. The lower part of this constant current image shows an area covered by two layers of Mn. Here, the fine vertical lines are also visible but without the aforementioned periodic modulation. Instead, a very different additional contrast shows up as broad (several rows of atoms) lines extending along $[1\bar{1}0]$, with a repeat period of 2.4 nm in (001) direction. This system has been identified as a conical spin spiral as sketched in Figure 22.25b. As demonstrated by simulated SP-STM images, Figure 22.25c and d, using the independent-orbital approximation (cf. Section 22.2.6), the experimentally observed contrasts can be reproduced by assuming a superposition of contributions from two sources: (1) row-wise alternating spin orientations at the atomic length scale along $[1\bar{1}0]$ and (2) spins advancing by a certain angle on a cone as one progresses in (001) direction. A magnetic tip that has a magnetization in the plane but oriented a few degrees off the principal axes samples both contributions, thus explaining the observed contrasts (see Figure 22.25d). By sophisticated DFT calculations it was shown that, even without SOC, a noncollinear magnetic structure is favored due to higher order interaction terms beyond Heisenberg exchange, inducing an optimum canting angle $\theta = 30°$ (opening angle of the transverse cone) and a rotation angle $\phi = 32°$ corresponding to a period of $\lambda_{(001)} = 1.8$ nm, in reasonable agreement with the experimental value of 2.4 nm. DMI contributes an additional energy gain and warrants a unique rotational sense.

22.5.5 Transmitting Information by a Spiral in a Biatomic Spin Chain

The spin spirals discussed so far were found in more or less extended 2D films of ML or DL thickness, supported by a nonmagnetic substrate. A spiral can exist, however, also in a 1D magnetic structure such as an atomic chain. The (5 × 1) reconstructed surface of Ir(001) forms a grove-like template pattern on which biatomic chains of Fe atoms of variable lengths can be grown by self-assembly. The spiral magnetic order in these Fe chains has been studied in a very recent publication by means of SP-STM, DFT calculations, and MC simulations by Menzel et al. [63]. One particular difficulty in observing this spin spiral, even at $T = 8$ K, is that the whole spiral is not stable in time but fluctuates due to thermal excitations. Contrasting to its unsurpassed *spatial* resolution capability, the *temporal* resolution of STM usually is rather limited. To experimentally access the spiral, it had to be stabilized. It turns out, surprisingly, that the fluctuating spiral teaches us a new paradigm of how information may be transferred to a remote place without the need of a flow of a charge current—just by a dissipationless spin current; but this is exactly one of the great promises associated with the emergent field of spintronics.

Figure 22.26a and b presents SP-STM measurements of a 30 nm^2 × 30 nm^2 area with several biatomic Fe chains on Ir(001). The tip in use was coated with an Fe thin film, the

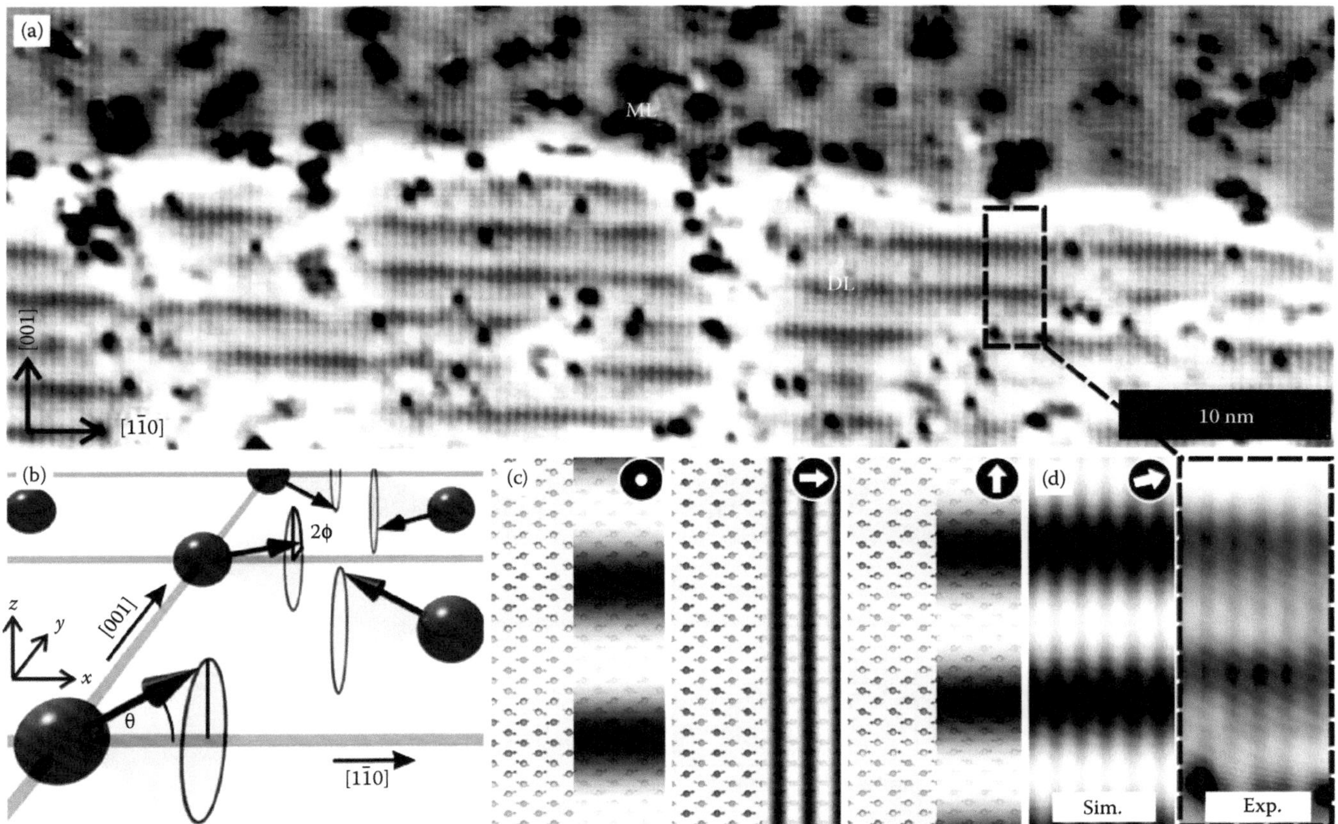

FIGURE 22.25 **(See color insert.)** (a) Top half: ML Mn on W(110) with the spin spiral as discussed along Figure 22.21. Bottom half: bilayer Mn. The fine line contrasts at the atomic scale are preserved, but now additional contrasts occur as lines perpendicular to the fine ones, induced by a conical spin spiral propagating in (001) direction. Constant current image, $U = -10$ mV, $I = 2$ nA, in-plane sensitive tip. (b) Sketch of the conical spiral (only top layer is shown). (c) Calculated SP-STM images assuming tip magnetizations oriented purely perpendicular, in-plane along [1 $\bar{1}$0], in-plane along (001), see arrows. (d) The experimental observation is best reproduced by assuming an intermediate in-plane tip magnetization angle.

magnetic sensitivity depending on the presence of the external magnetic field. Image (a) was recorded at zero Tesla, while for (b) a $B_z = +2$ T field was applied. The main difference between images (a) and (b) is found in contrasts with a period of three lattice constants along the chains in (b), these contrasts being absent in (a). This behavior can also be seen in the line profiles shown in (c), recorded with a Cr tip. The featureless blue profile was measured without a field applied, whereas the sinusoidal black profile was taken at a field of +2 T. Upon field reversal, the amplitude of the profile is inverted (red profile). We note that the magnetization of the Cr tip is not affected by the field due to its AFM order [25], but the orientation of the chain spins is reversed, cf. the suggested underlying spin structure in the inset. It is obvious that the field stabilizes the spiral in time, whereas the featureless blue curve arises from a time-averaged fluctuating signal, obscuring the magnetic structure of the chain.

The magnetic ground state of the biatomic Fe chain was investigated by *ab initio* spin-density calculations. At zero temperature the freestanding (unsupported) chain was clearly found to be FM. Upon relaxation on the Ir surface, however, the strong Heisenberg exchange (≈75 meV per Fe atom) is almost completely quenched due to Fe-Ir hybridization, rendering the AFM state almost degenerate (difference only ≈ 1 meV/Fe). As we saw earlier, a weak Heisenberg exchange is a condition favorable for the competing antisymmetric DMI to induce particularly large spin rotations, resulting here in a clockwise cycloidal spiral with an angle between adjacent spin moments of 120°, in agreement with the experimentally observed three atoms period. Remarkably, this is the minimum periodicity to form a noncollinear state.

To theoretically access the thermal fluctuations, a heat bath MC scheme was employed on a model chain of 30 Fe atoms in length and 2 atoms in width. At low temperature, the magnetic order is found to be that of an almost ideal clockwise 120° spin spiral at any time. As expected, thermal fluctuations increase with increasing temperature until, at about 100 K, the order of neighboring moments, averaged over time and chain length as expressed by the scalar parameter $\langle \mathbf{S}_i \cdot \mathbf{S}_{i+1} \rangle$, gets lost. Surprisingly, the spiral character of the chain, related to the chiral parameter $\langle \mathbf{S}_i \times \mathbf{S}_{i+1} \rangle$, is preserved even for $T \gg 100$ K, decaying much more slowly with temperature, see Figure 22.27.

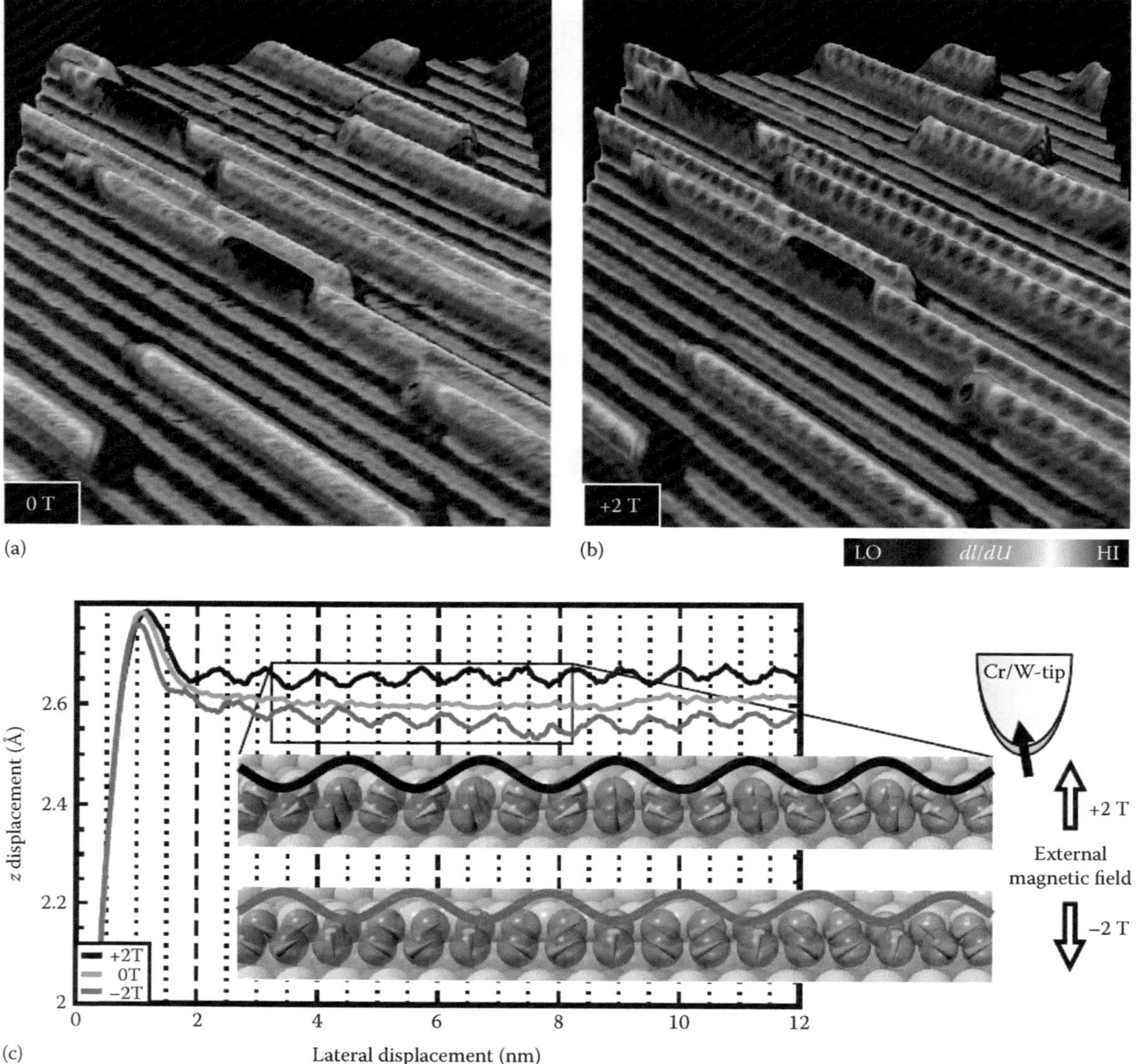

FIGURE 22.26 **(See color insert.)** Biatomic Fe chains on the (5 × 1) reconstructed Ir(001) surface. (a) Without applied magnetic field, (b) B_z = +2 T applied. U = +500 mV, I = 5 nA, and T = 8 K, Fe tip. (c) Line profiles taken with a Cr tip along a chain at fields as indicated. Insets illustrate the proposed spin structure, which fluctuates without field stabilization.

The spirals in Figure 22.26b were stabilized by an external magnetic field. Another way to achieve a stable spiral has been demonstrated experimentally by attaching a FM particle, e.g., a short Co chain, to one end of the Fe chain. By virtue of the strong direct exchange interaction, the end Fe atom pair is prevented from fluctuating, and the intrachain coupling is strong enough to immediately suppress fluctuations in the whole chain, fixing its spiral state in space and time. In other words, in order to know the magnetic state of the Co nanoparticle and monitor its changes, it is sufficient to "read out" the other end of the chain, which may be tens of nanometers away from the ferromagnetic particle. Hence, the fluctuating spiral provides a new mechanism of information transfer that may be of great value in future spintronics circuits.

22.5.6 Two-Dimensional Spin Texture: An Atomic-Scale Magnetic Skyrmion Lattice

All DMI-induced spin spirals discussed so far are 1D objects, in the sense that they have one distinct propagation direction (or, by symmetry, another equivalent one, as in the case Fe/W(001)) and can be characterized by one single vector **Q** in reciprocal space. The skyrmion lattice found recently in an ML Fe on Ir(111) can be viewed as a 2D analogon. The concept of skyrmions was first described by Skyrme [64] in the context of theoretical particle physics as topologically protected field configurations with particle-like properties. It has since become a general concept in various branches of physics, including magnetism [65,66]. A skyrmion is classified by

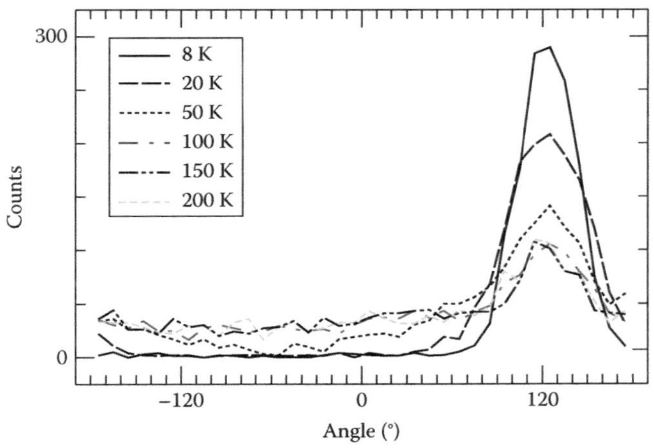

FIGURE 22.27 Distribution of spin rotation angles at variable temperatures, averaged over time and chain length, from heat bath MC calculations of a chain of 30 atom pairs. Even at $T \gg 100$ K, the characteristic vector spin chirality is preserved, with a positive 120° angle clearly favored.

an integer skyrmion number S obtained from integrating the spin rotation angle over the 2D unit cell. A trivial FM or AFM configuration has $S = 0$, whereas it is $S = +1$ for a skyrmion and $S = -1$ for an antiskyrmion (Figure 22.28).

When first investigated in 2006 with SP-STM by von Bergmann et al. [67], the Fe layer on Ir(111) was observed to display a nanometer-sized periodic square lattice superimposed on the hexagonal atomic lattice, with a magnetic unit cell containing 15 Fe atoms. This peculiar structure exists in three equivalent rotational domains, similar to Figure 22.29. Because in-plane magnetic contrast was not observed in this early work, a collinear magnetic ground state was proposed, with eight magnetic moments pointing up, and seven pointing down, coined the (7:8) mosaic state. However, the authors already speculated that the true magnetic ground state might be noncollinear, made up from a so-called multiple-Q state, a superposition of more than one spin spiral.

This conjecture has recently been successfully verified in a complementary theoretical and experimental study by Heinze et al. [68]. Figure 22.29a presents an image recorded with an Fe tip

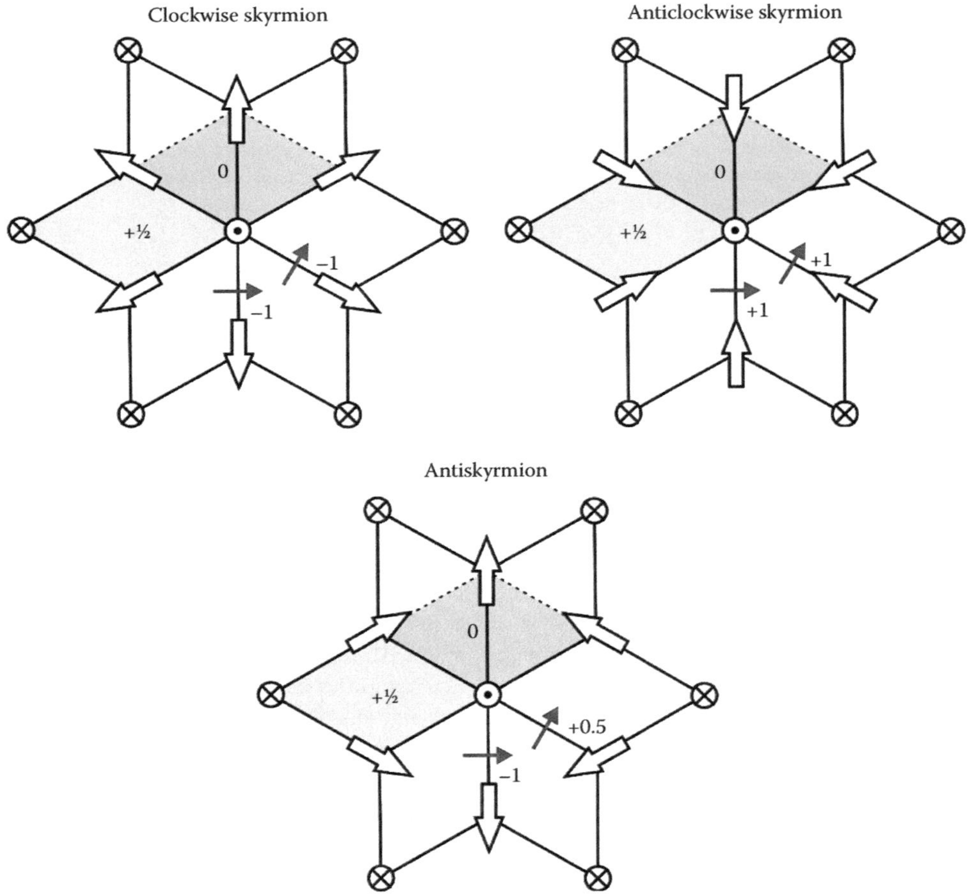

FIGURE 22.28 Schematics of possible skyrmion configurations on a 2D hexagonal lattice. The in-plane projection of the DM vector **D** for pairs of nearest neighbors is indicated by small arrows. Numbers inside the shaded areas represent the four-spin term with four adjacent spins in a diamond.

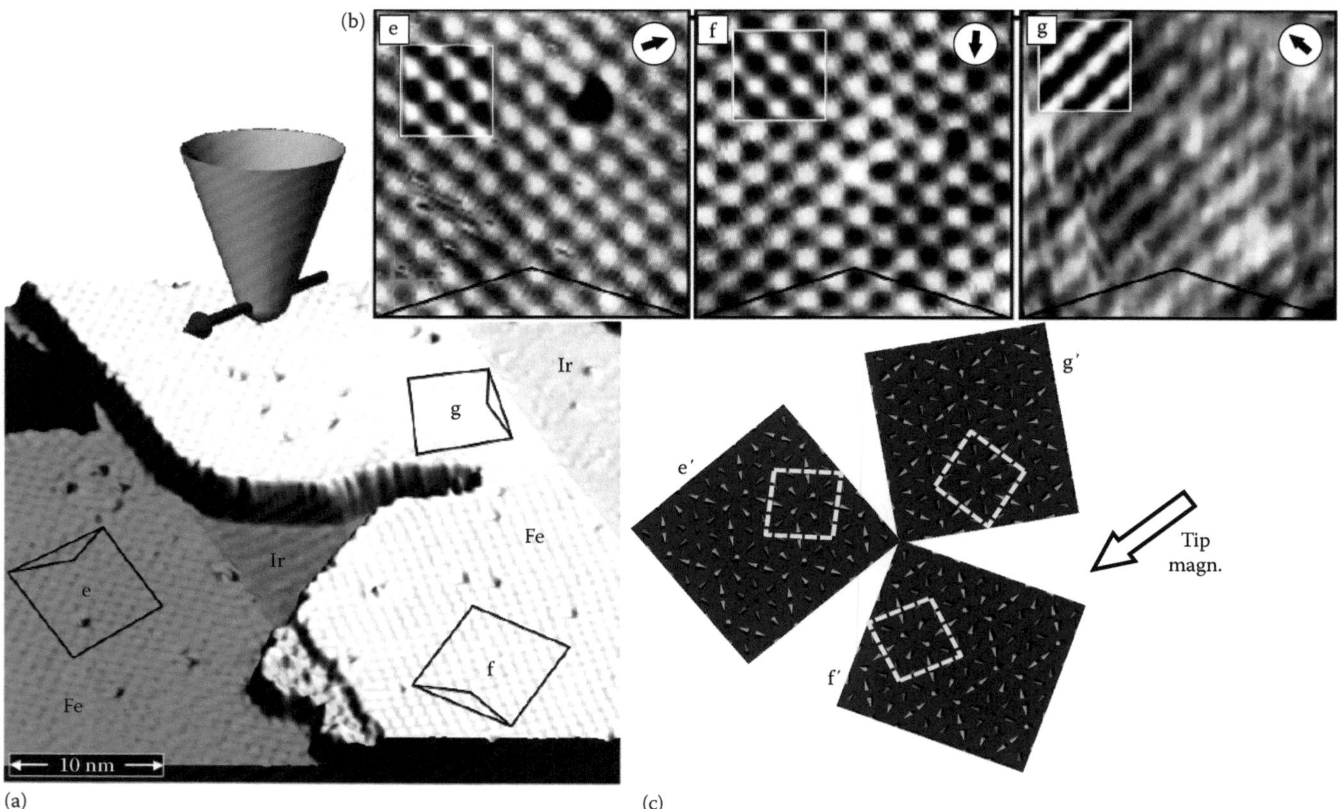

FIGURE 22.29 (a) The Fe ML on the close-packed Ir(111) surface exhibits a square lattice magnetic superstructure with a unit cell of 1 nm² × 1 nm² containing 15 Fe atoms. Three rotational domains marked e–g coexist. (U = +5 mV and I = 0.2 nA). (b) Zoom into the rotational domains. Note the triangular orientation marks as given in (a) and the corresponding tip magnetization direction, as also indicated in (c). Insets in (b): Simulated SP-STM images. (c) Sketch of the three rotational domains with the magnetic unit cells marked by dashed squares.

magnetized in-plane as sketched. All three rotational domains can be found in this image, respective areas of equal size are marked e–g, and the magnetic contrasts can be compared in magnifications shown in (b). During the scan, the magnetization of the tip of course did not change, leading to three different relative alignments of tip and local sample magnetizations in the three domains, giving rise to three distinct patterns. The insets in (b) show simulated images using the independent orbital approximation (Section 22.2.6), derived from a nanoskyrmion structure on a discrete atomic lattice as schematically shown in (c) and in Figure 22.30, with cones representing the in-plane directions of atomic magnetic moments, colors red and green indicating opposite out-of-plane components (see Figure 22.30). The simulations are a quite tough test of the assumed spin structure because the simulated patterns have to match the experimental results in all three domains simultaneously. Several spin textures of equal complexity were tested—single-Q spirals, antiskyrmions, multi-Q states, and vortex states [69]. Interestingly, for an out-of-plane sensitive tip some of these structures were found indistinguishable in these simulations, and in-plane contrasts had to be analyzed for consistency.

A key to an understanding of this structure is revealed in the 2D Fourier transform of the experimental SP-STM images, Figure 22.31. Inside the hexagonal 2D Brillouin zone of the close-packed atomic lattice, four additional spots are observed due to the square superstructure. These latter two pairs of spots can be associated with reciprocal vectors $\mathbf{Q}_1$ and $\mathbf{Q}_2$ of length $Q = |\mathbf{Q}_1| = |\mathbf{Q}_2|$, enclosing an angle θ of nearly 90°, belonging to two spin spirals. Extending the Heisenberg model of direct exchange interaction and DMI as discussed in Section 22.5.1 by an additional four-spin interaction term describing electron hopping between four adjacent sites (see Reference [68] for details), it can be shown that this four-spin interaction plays a crucial role in coupling the spin spirals in such a way that the resulting superposition state is lower in energy than a single spin spiral. The contributions of exchange, four-spin, and DM interactions to the total energy were investigated by first-principles calculations. Direct exchange is particularly weak in this system, once more due to strong hybridization of Fe 3d with Ir 5d states. While four-spin interaction usually is weak enough to be ignored in typical magnets, here its strength is of the same order as the two-spin Heisenberg exchange. The four-spin interaction is responsible for the occurrence of the 2D magnetic structure manifested in the nanoscale square superstructure, the periodicity and symmetry depending on Q and θ. However, the

FIGURE 22.30 **(See color insert.)** Skyrmion spin structure with square magnetic unit cell as used to calculate the simulated images shown as insets in Figure 22.29. Cones represent in-plane components of discrete atomic magnetic moments, their red and green colors denoting opposite out-of-plane components.

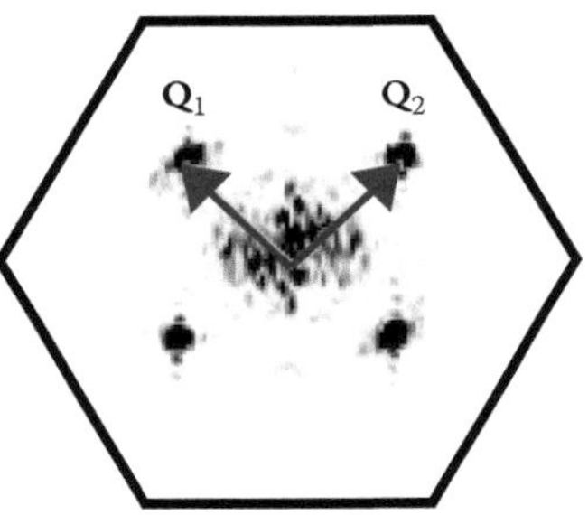

FIGURE 22.31 Fourier transform of an SP-STM image, out-of plane tip sensitivity. Inside the 2D surface Brillouin zone of the hexagonal atomic lattice four extra spots are found. Reciprocal vectors $\mathbf{Q}_1$ and $\mathbf{Q}_2$ are related to the nanometer-scale square superstructure (see text).

four-spin interaction does not distinguish between skyrmions and antiskyrmions. This degeneracy is lifted by DMI. Because SOC and hence DMI scales with the substrate's nuclear number, DMI is enhanced for the late 5*d* element Ir. Thus, considerable energy is gained by favoring one unique sense of spin rotation, ultimately establishing the skyrmion lattice as the magnetic ground state.

A careful comparison of experimental Fourier transforms with those obtained from transforming simulated images revealed a subtle yet significant discrepancy. From a slight offset of certain spots, it is concluded that the skyrmion lattice is not strictly commensurate with the atomic lattice but has a unit cell compressed by about 10% relative to the commensurate one.

22.6 Controlling Individual Adsorbate Spins on a Spin Spiral

When a single magnetic atom is adsorbed on the surface of a non-magnetic substrate, its magnetic moment will rapidly fluctuate due to thermal excitations at frequencies much too high to be resolved in a typical SP-STM experiment. Co atoms on a Pt(111) surface were found to switch their magnetization even at temperatures as low as 300 mK [70]. Without stabilizing the atom's magnetic moment, the measured SP-STM signal will be time-averaged.

In order to access the magnetic properties of individual adsorbate atoms by SP-STM, different strategies can be employed. Meier et al. [70] used a variable external magnetic field to create an imbalance in residency time of the spin in its up and down states, which translates into a net spin-polarized signal, the degree of polarization depending on the field strength; in this way, magnetization curves of individual Co atoms on Pt(111) were obtained [71].

Another stabilization strategy is to directly exchange couple the adsorbates' magnetic moments to a stationary magnetic structure. Yayon et al. [72] studied individual Fe and Cr atoms adsorbed on perpendicularly magnetized Co islands grown on Cu(111), using SP-STM at low temperatures. From their tunneling spectra, they conclude that Fe atoms couple FM to the supporting Co island, while the coupling is AFM for Cr atoms.

The use of a spin spiral as a template for individual magnetic atoms provides new opportunities to control and manipulate the orientation of the adsorbates' magnetic moments. This has been demonstrated by Serrate et al. [73]. As a crucial new element in this work, the technique of atom manipulation has been employed for the first time in a spin-polarized experiment. Atom manipulation is a well-established mode of STM operation first demonstrated in 1990 by Eigler and Schweizer [74]. The tunneling tip is used to move individual atoms in a controlled fashion across the surface, allowing to construct artificial structures atom-by-atom [75–78]. The challenge here was to carry out the manipulation without losing the spin sensitivity of the tip.

Figure 22.32 once again shows a constant current image of an ML of Mn on W(110) as imaged with a tip sensitive to the out-of-plane magnetization. As discussed in detail in Section 22.5.2, the spiral magnetic order as sketched in the lower panel of the figure can be recognized by the atomic scale stripe pattern arising from alternating rows of oppositely magnetized atoms, the stripe corrugation vanishing roughly every 6 nm when the spiral spins go orthogonal to the tip magnetization. The protrusions visible on top of the spiral are individual Co atoms. These atoms were deposited at low temperatures ($T < 20$ K), in order to prevent them from diffusing across the surface and forming

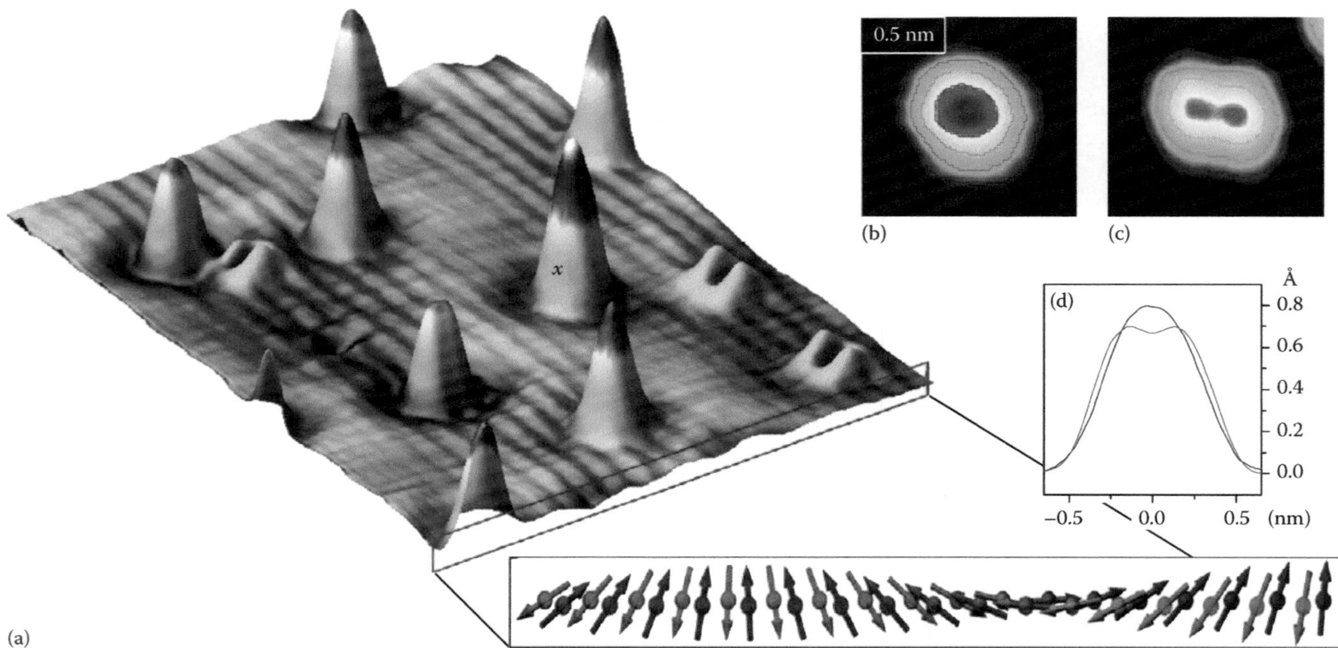

FIGURE 22.32 **(See color insert.)** (a) Co atoms on a Mn spin spiral on W(110) (U = −10 mV and I = 2 nA). The monomers differ in height and shape (a rare dimer is x-marked). (b) Co atom exhibiting rotational symmetry, sitting on a bright Mn row. (c) Co atom on adjacent dark Mn row, appearing with twofold symmetric shape. (d) Line profiles taken across the atoms in (b and c).

larger clusters. All of these protrusions are monomers (one rare dimer is x-marked). Surprisingly, these monomers exhibit variations in height and shape, depending on their position on the surface. The difference in appearance is particularly obvious in panels (b) and (c), showing top views of two Co atoms; the one in (b) exhibits rotational symmetry while the one in (c) clearly has a twofold symmetry with a dip at the center, also visible in the line profiles shown in (d). The key to understanding the variation in shape is the adsorption site: the atom in (b) was found on top of a bright Mn stripe and that in (c) on top of the dark stripe right next to it. In other words, the spins of the neighboring Mn rows hosting the Co monomers change by nearly 180°, suggesting a magnetic origin of the peculiar phenomenon.

This conjecture was indeed confirmed by spin-resolved *ab initio* calculations. For these calculations, a perfect AFM arrangement in the Mn layer was assumed, neglecting the 7° deviation from the collinear state. In a first step, the preferred adsorption site was determined, which is the bridge site of the carrier Mn row along the (001) direction, see Figure 22.33b. Second, the Co moments are found to couple FM to that row, so that the adsorbate spin directions follow that of the Mn row they sit upon. The DFT method allows to study separately the majority and minority LDOS in the vacuum, eventually revealing the cause of the observed distinct symmetries. As shown in Figure 22.33a, the states available for tunneling in the majority channel are hybridized from s, p_z, and d_{z^2} orbitals, all of them exhibiting rotational symmetry about an axis normal to the surface. In contrast, the minority LDOS is dominated by states of d_{xz} symmetry, with two lobes pointing off the normal axis, and a node at the center. Tunneling into these states gives rise to an elongated shape in the image. Figure 22.33c presents a plot of the calculated vacuum LDOS 5.2 Å above the surface, nicely reproducing the experimentally observed LDOS distribution depicted in Figure 22.33 b and c.

The configuration shown in Figure 22.33 corresponds to the idealized limiting case of the Mn magnetic moments pointing up and down, collinear to the surface normal. Taking other sites along the spin spiral into account, the FM coupled adsorbate atoms are expected to acquire spin orientations with varying in-plane components. In order to study the resulting position-dependent variations of height and shape in a systematic manner, the tunneling tip was used to move the Co atoms to well-defined sites on the Mn lattice, assembling a chain of six atoms, equidistantly positioned at a pitch of six Mn rows in $[1\bar{1}0]$ direction. This corresponds to an angular progression of about 42° from one Co atom to the next, or ≈210° along the full chain. Figure 22.34 presents the chain as measured with an Fe-coated tip. The tip magnetization is controlled by an applied magnetic field of 2.5 T as indicated, which neither affects the AFM structure nor the strongly exchange coupled moments of the Co atoms. An analysis of the changes in apparent heights gives access to the absolute value of the spin rotation angle $|\theta|$ relative to the tip magnetization, and the sign of θ can be deduced from the asymmetric shapes occurring off the normal spin orientation. Thus, the spin orientation of a Co atom in a plane spanned by the propagation direction of the Mn spin spiral and the surface normal can be chosen deliberately by moving the atom to the appropriate lattice position. In this way, a new level of control of individual atomic spins has been demonstrated.

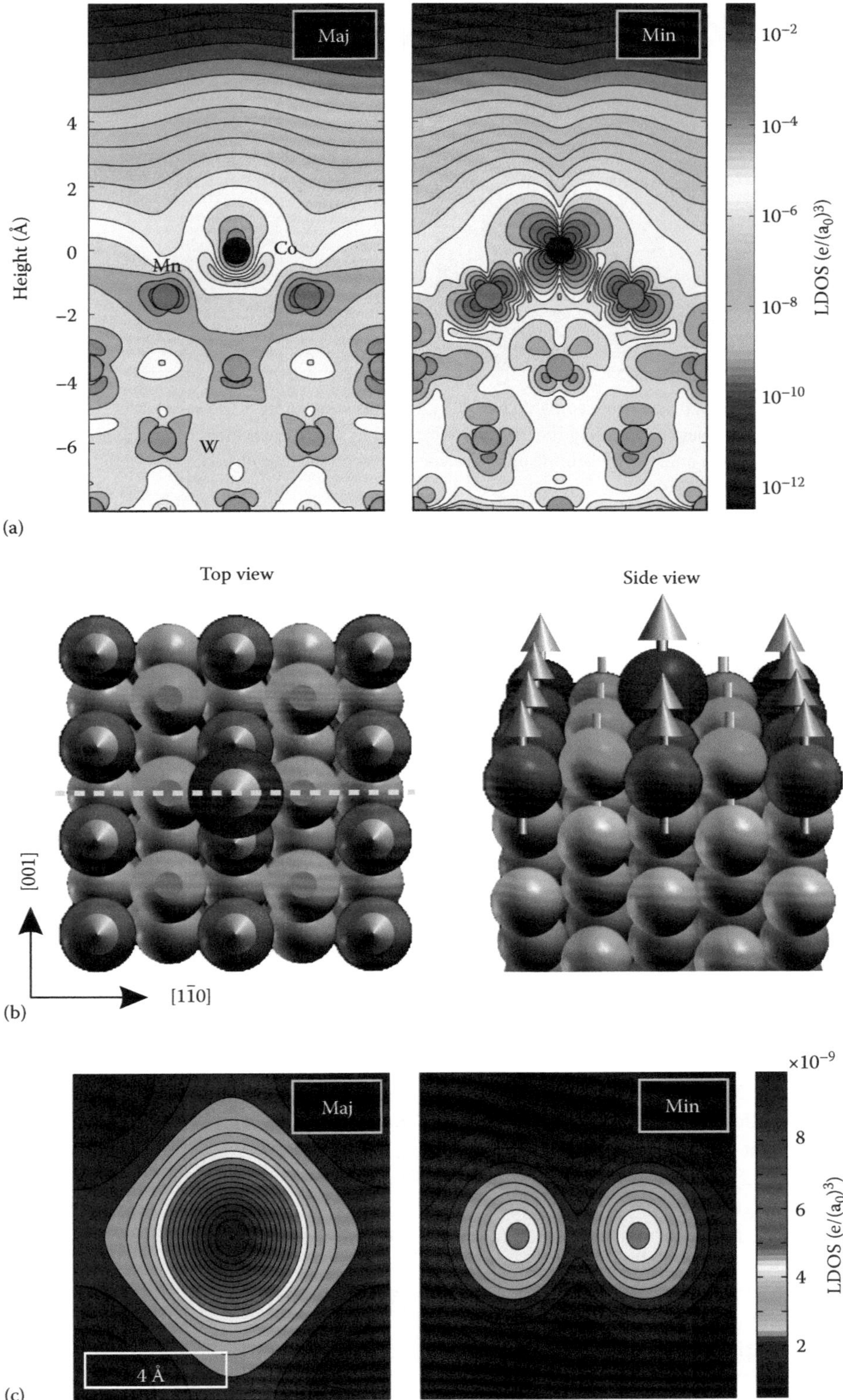

FIGURE 22.33 (See color insert.) (a) Calculated LDOS for majority and minority states in a vertical cut containing the Co atom (see the dashed line in (b)). While a bump is found in the majority channel, it is a dip in the minority channel. (b) Co atoms occupy bridge sites and couple FM to their respective carrier row. (c) Spin-resolved vacuum LDOS distribution calculated 5.2 Å above the surface according to (a).

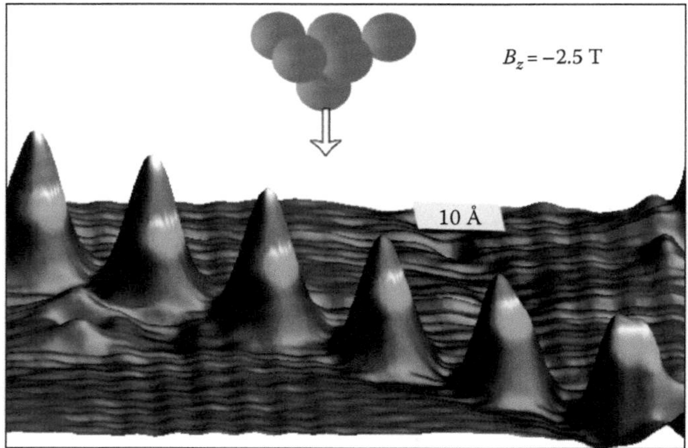

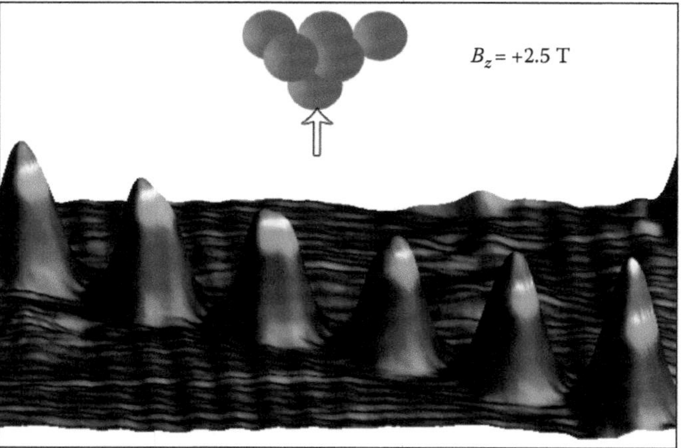

FIGURE 22.34 **(See color insert.)** Artificially assembled chain of Co atoms, spaced by 6 Mn rows in [1$\bar{1}$0] direction, offset by four rows in (001) ($U = -10$ mV and $I = 2$ nA). The magnetization of the Fe tip is controlled by an external magnetic field of 2.5 T as indicated. From the variation of apparent heights and shapes, the angle of the atomic magnetic moments can be determined.

22.7 Conclusions and Outlook

Magnetic imaging at the atomic length scale has developed into a mature technique, and spin contrasts can nowadays be achieved on a routine basis. Samples of magnetic MLs, small clusters, or even single atoms on a substrate can be prepared with a high degree of control. Magnetic tips with a large variety of properties are available and provide a bouquet of experimental conditions to choose from. Next to these improvements on the experimental side, one particularly important factor of progress has been the close collaboration of experiment and theory. Magnetism in condensed matter is ultimately rooted in its spin-resolved electronic structure, which is the subject matter of DFT. Competing magnetic interactions can now be calculated at an ever refined level, and tools have been developed that address specific problems related to STM measurements, being of great benefit for the interpretation of magnetic contrasts. It is the local DOS in the vacuum, calculated from theory and measured in an STM experiment, which virtually acts as the point of overlap between experiment and theory. Another important theoretical contribution comes from advanced MC simulations based on a lattice of discrete atomic magnetic moments, thus perfectly being adapted to the structures under investigation.

In spite of the achievements of the past decade, the study of magnetism at the atomic length scale is certainly still at the beginning. So far, the structures observed have become ever more complex, and the wealth of noncollinear magnetic spin textures found up to date is far from being exhausted. On the other hand, the ultimate vision is to use the spin of individual atoms to store a bit of information, or to apply a magnetic atom as a switch in some future spintronic circuits. SP-STM has proven its capability to *read* magnetic information at the ultimate, atomic scale. First successful attempts have been made to *control* individual magnetic moments (Section 22.6) and to *transmit* spin information by means of a spin chain (Section 22.5.5). While these examples rely on direct exchange, another promising direction of current research employs the more subtle indirect Ruderman–Kittel–Kasuya–Yosida (RKKY) interaction between individual magnetic atoms, which is mediated by the conduction electrons of the non-magnetic substrate. Strength and sign of the interaction can be tuned by adjusting the interatomic distances. Recently, Khajetoorians et al. [79] constructed a logic OR gate, made from 10 individual Fe atoms on a Cu(111) surface, assembled by atom manipulation into two spin leads, which share one final atom acting as the read-out. This concept inherently offers a large variety of possible configurations. It does not require any charge transport for information processing.

There are a number of prospective routes in SP-STM-based research. One is to overcome the limitations in temporal resolution of the STM. Loth et al. of the IBM Almaden Research Center have recently presented an all-electronic pump–probe measurement scheme allowing to study the spin relaxation time of individual atoms with nanosecond time resolution [80]. Another intriguing field will be that of spin transport, i.e., the injection, detection, and manipulation of spin-polarized currents passing through nanostructures. By application of two or even four magnetic probes STM can provide contacts to the nanoworld. The interaction of magnetic atoms, clusters, and nanostructures with novel Dirac materials like graphene [81] or topological insulators [82] is a new field of great current interest. SP-STM can be expected to deliver important contributions.

Acknowledgments

The authors gratefully acknowledge major contributions to the research, figures, and artwork presented in this chapter by these collaborators: S. Blügel, G. Bihlmayer, and M. Heide (Institute of Solid State Research, Forschungszentrum Jülich, Germany); S. Heinze, P. Ferriani, and S. Schröder (Spintronics Theory Group, Christian-Albrechts-Universität zu Kiel, Germany); K. von Bergmann, M. Bode, S.-W. Hla, A. Kubetzka, M. Menzel,

D. Serrate, E. Vedmedenko, R. Wieser, and Y. Yoshida (Scanning Probe Methods Group, Institute of Applied Physics, University of Hamburg, Germany). Financial support from the Center of Excellence SFB 668 of the Deutsche Forschungsgemeinschaft, from the ERC Advanced Grant FURORE, from the Cluster of Excellence NANOSPINTRONICS funded by the Hamburgische Stiftung für Wissenschaft und Forschung, and the Partnerships for International Research and Education (PIRE) program of the National Science Foundation, Arlington, VA, is gratefully acknowledged.

References

1. G. Binnig, H. Rohrer, Ch. Gerber, and E. Weibel. Surface studies by scanning tunneling microscopy. *Physical Review Letters*, 49(1):57–61, July 1982.
2. J. Tersoff and D. R. Hamann. Theory of the scanning tunneling microscope. *Physical Review B*, 31(2):805–813, January 1985.
3. J. Tersoff and D. R. Hamann. Theory and application for the scanning tunneling microscope. *Physical Review Letters*, 50(25):1998–2001, June 1983.
4. C. J. Chen. *Introduction to Scanning Tunneling Microscopy*, 2nd edn. Oxford University Press, Oxford, U.K., 2008.
5. M. Jullière. Tunneling between ferromagnetic films. *Physics Letters*, 54A(3):225–226, 1975.
6. M. N. Baibich, J. M. Broto, A. Fert, F. Nguyen Van Dau, and F. Petroff. Giant magnetoresistance of (001)Fe/(001)Cr magnetic superlattices. *Physical Review Letters*, 61(21):2472–2475, November 1988.
7. G. Binasch, P. Grünberg, F. Saurenbach, and W. Zinn. Enhanced magnetoresistance in layered magnetic structures with antiferromagnetic interlayer exchange. *Physical Review B*, 39(7):4828–4830, March 1989.
8. T. Miyazaki and N. Tezuka. Giant magnetic tunneling effect in Fe/Al2O3/Fe junction. *Journal of Magnetism and Magnetic Materials*, 139(3):L231–L234, January 1995.
9. J. S. Moodera, L. R. Kinder, T. M. Wong, and R. Meservey. Large magnetoresistance at room temperature in ferromagnetic thin film tunnel junctions. *Physical Review Letters*, 74(16):3273–3276, April 1995.
10. S. Yuasa and D. D. Djayaprawira. Giant tunnel magnetoresistance in magnetic tunnel junctions with a crystalline MgO(0 0 1) barrier. *Journal of Physics D: Applied Physics*, 40(21):R337–R354, November 2007.
11. J. Slonczewski. Conductance and exchange coupling of two ferromagnets separated by a tunneling barrier. *Physical Review B*, 39(10):6995–7002, April 1989.
12. D. Wortmann, S. Heinze, Ph. Kurz, G. Bihlmayer, and S. Blügel. Resolving Complex Atomic-Scale Spin Structures by Spin-Polarized Scanning Tunneling Microscopy. *Physical Review Letters*, 86(18):4132–4135, April 2001.
13. M. Kleiber, M. Bode, R. Ravlić, and R. Wiesendanger. Topology-induced spin frustrations at the Cr(001) surface studied by spin-polarized scanning tunneling spectroscopy. *Physical Review Letters*, 85(21):4606–4609, November 2000.
14. K. von Bergmann, M. Bode, and R. Wiesendanger. Magnetism of iron on tungsten (001) studied by spin-resolved scanning tunneling microscopy and spectroscopy. *Physical Review B*, 70(17), 174455, 2004.
15. S. Krause, L. Berbil-Bautista, T. Hänke, F. Vonau, M. Bode, and R. Wiesendanger. Consequences of line defects on the magnetic structure of high anisotropy films: Pinning centers on Dy/W (110). *Europhysics Letters (EPL)*, 76:637, 2006.
16. L. Berbil-Bautista, S. Krause, M. Bode, and R. Wiesendanger. Spin-polarized scanning tunneling microscopy and spectroscopy of ferromagnetic Dy(0001)/W(110) films. *Physical Review B*, 76(6):1–10, August 2007.
17. O. Pietzsch, A. Kubetzka, M. Bode, and R. Wiesendanger. Real-space observation of dipolar antiferromagnetism in magnetic nanowires by spin-polarized scanning tunneling spectroscopy. *Physical Review Letters*, 84(22):5212–5215, May 2000.
18. O. Pietzsch, A. Kubetzka, M. Bode, and R. Wiesendanger. Observation of magnetic hysteresis at the nanometer scale by spin-polarized scanning tunneling spectroscopy. *Science*, 292(5524):2053–2056, June 2001.
19. M. Pratzer, H. J. Elmers, M. Bode, O. Pietzsch, A. Kubetzka, and R. Wiesendanger. Atomic-scale magnetic domain walls in quasi-one-dimensional Fe nanostripes. *Physical Review Letters*, 87(12):127201, 2001.
20. A. Kubetzka, M. Bode, O. Pietzsch, and R. Wiesendanger. Spin-polarized scanning tunneling microscopy with antiferromagnetic probe tips. *Physical Review Letters*, 88(5):057201, January 2002.
21. K. von Bergmann, M. Bode, A. Kubetzka, M. Heide, S. Blügel, and R. Wiesendanger. Spin-polarized electron scattering at single oxygen adsorbates on a magnetic surface. *Physical Review Letters*, 92(4), 046801, January 2004.
22. M. Bode, M. Getzlaff, and R. Wiesendanger. Spin-polarized vacuum tunneling into the exchange-split surface state of Gd(0001). *Physical Review Letters*, 81(19):4256–4259, November 1998.
23. A. Kubetzka, O. Pietzsch, M. Bode, and R. Wiesendanger. Magnetism of nanoscale Fe islands studied by spin-polarized scanning tunneling spectroscopy. *Physical Review B*, 63(14):140407, March 2001.
24. A. Wachowiak, J. Wiebe, M. Bode, O. Pietzsch, M. Morgenstern, and R. Wiesendanger. Direct observation of internal spin structure of magnetic vortex cores. *Science*, 298(5593):577–580, October 2002.
25. O. Pietzsch, A. Kubetzka, M. Bode, and R. Wiesendanger. Spin-polarized scanning tunneling spectroscopy of nanoscale cobalt islands on Cu(111). *Physical Review Letters*, 92(5):057202, February 2004.
26. O. Pietzsch, S. Okatov, A. Kubetzka, M. Bode, S. Heinze, A Lichtenstein, and R. Wiesendanger. Spin-resolved electronic structure of nanoscale cobalt islands on Cu(111). *Physical Review Letters*, 96(23):237203, June 2006.

27. L. Berbil-Bautista, S. Krause, T. Hänke, M. Bode, and R. Wiesendanger. Spin-polarized scanning tunneling microscopy through an adsorbate layer: Sulfur-covered Fe/W(110). *Surface Science*, 600(3):L20–L24, February 2006.
28. F. Meier, K. von Bergmann, P. Ferriani, J. Wiebe, M. Bode, K. Hashimoto, S. Heinze, and R. Wiesendanger. Spin-dependent electronic and magnetic properties of Co nanostructures on Pt(111) studied by spin-resolved scanning tunneling spectroscopy. *Physical Review B*, 74(19), 195411, 2006.
29. S. Heinze, M. Bode, A. Kubetzka, O. Pietzsch, X. Nie, S. Blügel, and R. Wiesendanger. Real-space imaging of two-dimensional antiferromagnetism on the atomic scale. *Science*, 288(5472):1805–1808, 2000.
30. S. Heinze. Simulation of spin-polarized scanning tunneling microscopy images of nanoscale non-collinear magnetic structures. *Applied Physics A*, 85(4):407–414, September 2006.
31. D. T. Pierce. Spin-polarized electron microscopy. *Physica Scripta*, 38:291–296, 1988.
32. R. Wiesendanger, H.-J. Güntherodt, G. Güntherodt, R. Gambino, and R. Ruf. Observation of vacuum tunneling of spin-polarized electrons with the scanning tunneling microscope. *Physical Review Letters*, 65(2):247–250, July 1990.
33. R. Wiesendanger, D. Bürgler, G. Tarrach, T. Schaub, U. Hartmann, H. J. Güntherodt, I. V. Shvets, and J. M. D. Coey. Recent advances in scanning tunneling microscopy involving magnetic probes and samples. *Applied Physics A Solids and Surfaces*, 53(5):349–355, November 1991.
34. R. Wiesendanger, G. Tarrach, D. Bürgler, T. Jung, L. Eng, and H.-J. Güntherodt. An ultrahigh vacuum scanning tunneling microscope for surface science studies. *Vacuum*, 41:386–388, 1990.
35. O. Pietzsch, A. Kubetzka, D. Haude, M. Bode, and R. Wiesendanger. A low-temperature ultrahigh vacuum scanning tunneling microscope with a split-coil magnet and a rotary motion stepper motor for high spatial resolution studies of surface magnetism. *Review of Scientific Instruments*, 71(2):424–430, 2000.
36. A. Kubetzka, O. Pietzsch, M. Bode, and R. Wiesendanger. Spin-polarized scanning tunneling microscopy study of 360° walls in an external magnetic field. *Physical Review B*, 67(2):020401, January 2003.
37. M. Bode, O. Pietzsch, A. Kubetzka, S. Heinze, and R. Wiesendanger. Experimental evidence for intra-atomic noncollinear magnetism at thin film probe tips. *Physical Review Letters*, 86(10):2142–2145, March 2001.
38. A. Schlenhoff, S. Krause, G. Herzog, and R. Wiesendanger. Bulk Cr tips with full spatial magnetic sensitivity for spin-polarized scanning tunneling microscopy. *Applied Physics Letters*, 97(8):083104, 2010.
39. S. Loth, K. von Bergmann, M. Ternes, A. F. Otte, C. P. Lutz, and A. J. Heinrich. Controlling the state of quantum spins with electric currents. *Nature Physics*, 6(5):340–344, March 2010.
40. R. Wiesendanger, I. V. Shvets, D. Bürgler, G. Tarrach, H. J. Güntherodt, J. M. D. Coey, and S. Gräser. Topographic and magnetic-sensitive scanning tunneling microscope study of magnetite. *Science*, 255(5044): 583–586, January 1992.
41. S. Blügel, M. Weinert, and P. Dederichs. Ferromagnetism and antiferromagnetism of 3d-metal overlayers on metals. *Physical Review Letters*, 60(11):1077–1080, March 1988.
42. C. Krembel, M. C. Hanf, J. C. Peruchetti, D. Bolmont, and G. Gewinner. Growth of an ordered Cr monolayer on Ag(100): Evidence of two-dimensional antiferromagnetism. *Journal of Magnetism and Magnetic Materials*, 93:529–533, February 1991.
43. S. Abrahams, L. Guttman, and J. Kasper. Neutron diffraction determination of antiferromagnetism in face-centered cubic (γ) iron. *Physical Review*, 127(6):2052–2055, September 1962.
44. U. Gonser, C. J. Meechan, A. H. Muir, and H. Wiedersich. Determination of Neel temperatures in fcc iron. *Journal of Applied Physics*, 34(8):2373–2378, 1963.
45. M. Bode. Spin-polarized scanning tunnelling microscopy. *Reports on Progress in Physics*, 66(4):523–582, April 2003.
46. S. Meckler, M. Gyamfi, O. Pietzsch, and R. Wiesendanger. A low-temperature spin-polarized scanning tunneling microscope operating in a fully rotatable magnetic field. *Review of Scientific Instruments*, 80(2):023708, February 2009.
47. S. Meckler, O. Pietzsch, N. Mikuszeit, and R. Wiesendanger. Micromagnetic description of the spin spiral in Fe double-layer stripes on W (110). *Physical Review B*, 85:024420, 2012.
48. W. Wulfhekel, F. Zavaliche, R. Hertel, S. Bodea, G. Steierl, G. Liu, J. Kirschner, and H. Oepen. Growth and magnetism of Fe nanostructures on W(001). *Physical Review B*, 68(14):144416, October 2003.
49. A. Kubetzka, P. Ferriani, M. Bode, S. Heinze, G. Bihlmayer, K. von Bergmann, O. Pietzsch, S. Blügel, and R. Wiesendanger. Revealing antiferromagnetic order of the Fe monolayer on W(001): Spin-polarized scanning tunneling microscopy and first-principles calculations. *Physical Review Letters*, 94(8):87204, March 2005.
50. S. Blügel, D. Pescia, and P. H. Dederichs. Ferromagnetism versus antiferromagnetism of the Cr(001) surface. *Physical Review B*, 39(2):1392–1394, January 1989.
51. P. Ferriani, S. Heinze, G. Bihlmayer, and S. Blügel. Unexpected trend of magnetic order of 3d transition-metal monolayers on W(001). *Physical Review B*, 72(2), 024452, July 2005.
52. M. Bode, E. Y. Vedmedenko, K. von Bergmann, A. Kubetzka, P. Ferriani, S. Heinze, and R. Wiesendanger. Atomic spin structure of antiferromagnetic domain walls. *Nature Materials*, 5(6):477–481, June 2006.
53. C. L. Gao, W. Wulfhekel, and J. Kirschner. Revealing the 120° antiferromagnetic Néel structure in real space: One monolayer Mn on Ag(111). *Physical Review Letters*, 101(26), 267205, December 2008.

54. M. Waśniowska, S. Schröder, P. Ferriani, and S. Heinze. Real space observation of spin frustration in Cr on a triangular lattice. *Physical Review B*, 82(1):12402, July 2010.
55. S. J. Blundell. *Magnetism in Condensed Matter*. Oxford University Press, Oxford, NY, 2001.
56. S. Meckler, N. Mikuszeit, A. Preßler, E. Y. Vedmedenko, O. Pietzsch, and R. Wiesendanger. Real-space observation of a right-rotating inhomogeneous cycloidal spin spiral by spin-polarized scanning tunneling microscopy in a triple axes vector magnet. *Physical Review Letters*, 103(15):157201, October 2009.
57. I. Dzyaloshinsky. A thermodynamic theory of weak ferromagnetism of antiferromagnetics. *Journal of Physics and Chemistry of Solids*, 4(4):241–255, 1958.
58. T. Moriya. New mechanism of anisotropic superexchange interaction. *Physical Review Letters*, 4(5):228–230, 1960.
59. M. Bode, M. Heide, K. von Bergmann, P. Ferriani, S. Heinze, G. Bihlmayer, A. Kubetzka, O. Pietzsch, S. Blügel, and R. Wiesendanger. Chiral magnetic order at surfaces driven by inversion asymmetry. *Nature*, 447(7141):190–193, 2007.
60. T. Moriya. Anisotropic superexchange interaction and weak ferromagnetism. *Physical Review*, 120(1):91–98, 1960.
61. P. Ferriani, K. von Bergmann, E. Y. Vedmedenko, S. Heinze, M. Bode, M. Heide, G. Bihlmayer, S. Blügel, and R. Wiesendanger. Atomic-scale spin spiral with a unique rotational sense: Mn monolayer on W(001). *Physical Review Letters*, 101(2), 027201, July 2008.
62. Y. Yoshida, S. Schröder, P. Ferriani, D. Serrate, A. Kubetzka, K. von Bergmann, S. Heinze, and R. Wiesendanger. Conical spin-spiral state in an ultrathin film driven by higher-order spin interactions. *Physical Review Letters*, 108(8):087205, February 2012.
63. M. Menzel, Y. Mokrousov, R. Wieser, J. E. Bickel, E. Y. Vedmedenko, S. Blügel, S. Heinze, K. von Bergmann, A. Kubetzka, and R. Wiesendanger. Information transfer by vector spin chirality in finite magnetic chains. *Physical Review Letters*, 108(19), 197204, May 2012.
64. T. H. R. Skyrme. A non-linear field theory. *Proceedings of the Royal Society of London. Series A, Mathematical and Physical Sciences*, 260(1300):127–138, 1961.
65. U. K. Rößler, A. N. Bogdanov, and C. Pfleiderer. Spontaneous skyrmion ground states in magnetic metals. *Nature*, 442(7104):797–801, August 2006.
66. S. Mühlbauer, B. Binz, F. Jonietz, C. Pfleiderer, A. Rosch, A. Neubauer, R Georgii, and P Böni. Skyrmion lattice in a chiral magnet. *Science*, 323(5916):915–919, 2009.
67. K. von Bergmann, S. Heinze, M. Bode, E. Y. Vedmedenko, G. Bihlmayer, S. Blügel, and R. Wiesendanger. Observation of a complex nanoscale magnetic structure in a hexagonal Fe monolayer. *Physical Review Letters*, 96(16), 167203, April 2006.
68. S. Heinze, K. von Bergmann, M. Menzel, J. Brede, A. Kubetzka, R. Wiesendanger, G. Bihlmayer, and S. Blügel. Spontaneous atomic-scale magnetic skyrmion lattice in two dimensions. *Nature Physics*, 7(9):713–718, July 2011.
69. S. Heinze, M. Menzel, J. Brede, R. Wiesendanger, and G. Bihlmayer. Supplementary information to: Spontaneous atomic-scale magnetic skyrmion lattice in two dimensions. *Nature Physics*, 7:1–18, 2011.
70. F. Meier, L. Zhou, J. Wiebe, and R. Wiesendanger. Revealing magnetic interactions from single-atom magnetization curves. *Science*, 320(5872):82–86, April 2008.
71. J. Wiebe, L. Zhou, and R. Wiesendanger. Atomic magnetism revealed by spin-resolved scanning tunnelling spectroscopy. *Journal of Physics D: Applied Physics*, 44(46):464009, November 2011.
72. Y. Yayon, V. W. Brar, L. Senapati, S. C. Erwin, and M. F. Crommie. Observing spin polarization of individual magnetic adatoms. *Physical Review Letters*, 99(6):067202, August 2007.
73. D. Serrate, P. Ferriani, Y. Yoshida, S.-W. Hla, M. Menzel, K. von Bergmann, S. Heinze, A. Kubetzka, and R. Wiesendanger. Imaging and manipulating the spin direction of individual atoms. *Nature Nanotechnology*, 5(5):350–353, May 2010.
74. D. M. Eigler and E. K. Schweizer. Positioning single atoms with a scanning tunnelling microscope. *Nature*, 344(6266):524–526, April 1990.
75. M. F. Crommie, C. P. Lutz, and D. M. Eigler. Confinement of electrons to quantum corrals on a metal surface. *Science*, 262(5131):218–220, October 1993.
76. H. C. Manoharan, C. P. Lutz, and D. M. Eigler. Quantum mirages formed by coherent projection of electronic structure. *Nature*, 403(6769):512–515, February 2000.
77. S.-W. Hla, K.-F. Braun, and K.-H. Rieder. Single-atom manipulation mechanisms during a quantum corral construction. *Physical Review B*, 67(20):201402, May 2003.
78. S.-W. Hla. Scanning tunneling microscopy single atom/molecule manipulation and its application to nanoscience and technology. *Journal of Vacuum Science & Technology B: Microelectronics and Nanometer Structures*, 23(4):1351–1360, 2005.
79. A. A. Khajetoorians, J. Wiebe, B. Chilian, and R. Wiesendanger. Realizing all-spin-based logic operations atom by atom. *Science*, 332:1062–1064, May 2011.
80. S. Loth, M. Etzkorn, C. P. Lutz, D. M. Eigler, and A. J. Heinrich. Measurement of fast electron spin relaxation times with atomic resolution. *Science*, 329(5999):1628–1630, September 2010.
81. M. Gyamfi, T. Eelbo, M. Waśniowska, T. Wehling, S. Forti, U. Starke, A. Lichtenstein, M. Katsnelson, and R. Wiesendanger. Orbital selective coupling between Ni adatoms and graphene Dirac electrons. *Physical Review B*, 85(16):161406, April 2012.
82. J. Honolka, A. A. Khajetoorians, V. Sessi, T. O. Wehling, A. I. Lichtenstein, Ph. Hofmann, K. Kern, and R. Wiesendanger. In-plane magnetic anisotropy of Fe atoms on Bi_2Se_3(111). *Physical Review Letters*, 108(25):256811, June 2012.

23

Atomic and Molecular Magnets on Surfaces

Harald Brune
Institute of Condensed Matter Physics
Swiss Federal Institute of Technology

Pietro Gambardella
Catalan Institute of Nanotechnology
and
ETH Zürich

23.1 Introduction

This chapter provides a description of atomic-scale magnets fabricated by the deposition and assembly of metal atoms and metal–organic molecules on different substrates. We describe systems composed of individual magnetic atoms, small clusters, and molecular networks as well as methods to investigate and control their magnetization, anisotropy, and temperature-dependent magnetic behavior. The experimental techniques reviewed in these pages, x-ray magnetic circular dichroism (XMCD) and scanning tunneling microscopy (STM), represent state-of-the-art probes that have a very large degree of complementarity and potential for future improvements.

Magnetic atoms on surfaces represent the ultimate limit of monodisperse magnets. Although individual atoms generally present paramagnetic behavior, their investigation provides clues to fundamental and practical issues in magnetism, such as the dependence of the magnetization on system size, atomic coordination, and composition [1]. The spin and orbital magnetic moments, exchange coupling, and magnetic ordering of such systems can be tuned by tiny changes in dimensions and coupling to the environment [2–8]. The magnetic anisotropy energy per atom can be increased by up to three orders of magnitude with respect to bulk materials, leading to metastable (blocked) magnetic states at low temperature [2,3,8]. Depending on the system size and interaction with host media, the magnetization can behave as a classical vector or be quantized, leading to effects such as magnetic hysteresis, the typical macroscale property of a magnet, as well as quantum tunneling and phase interference effects, which are characteristic of microscopic systems [9–11]. Moreover, magnets made of one or a few atoms or molecules organized into regular patterns allow for the investigation of the ultimate limits of magnetic storage and quantum computation in novel materials [12].

23.2 Magnetic Interactions on the Atomic Scale

Regardless of the size of each system, the basic interactions that lead to magnetism are short-range, with typical lengthscales of a few Angstroms (Figure 23.1). The magnitude of the magnetization depends on the spin (m_S) and orbital (m_L) magnetic moment of each atom, as well as on the coupling between the atomic moments due to the exchange interaction, which is responsible for magnetic order. The stability of the magnetization and its preferential orientation depend on the magnetic anisotropy energy, which has two contributions: the first, called magnetocrystalline anisotropy (MCA), depends on the local environment of the magnetic ions, the second, called dipolar or shape anisotropy, depends on long-range magnetostatic interactions. Magnetic interactions in atomic-scale structures, however, can be considerably different compared to macroscopic samples. Even more importantly in some cases, interactions forbidden by symmetry in extended three-dimensional (3D) systems can suddenly become relevant [13]. Enhanced electron correlation

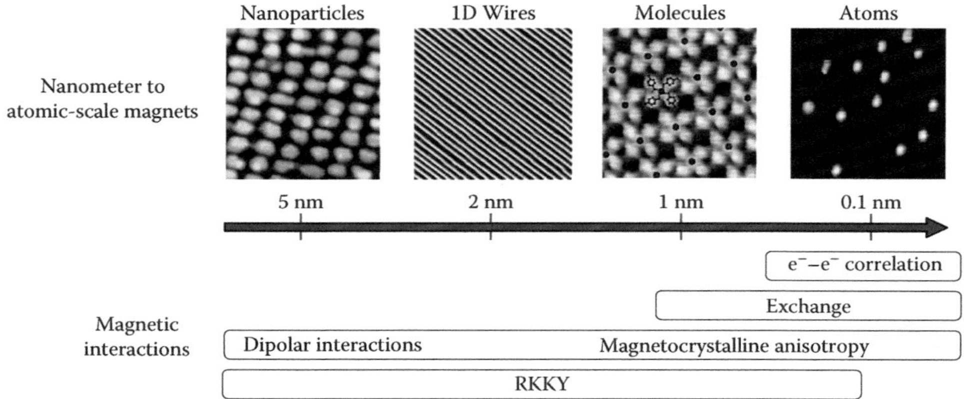

FIGURE 23.1 Schematics of magnetic structures between the nanometer and atomic scales. The numbers indicate the average dimensions of the magnetic unit cell. The range of relevant magnetic interactions is also shown.

effects in low dimensions generally lead to an increase of m_S and m_L [1], the magnetic anisotropy energy becomes a main factor determining the stability of the magnetization [3,4,14], and spin–orbit coupling (SOC) can compete with the exchange interaction and induce exotic noncollinear spin structures [15]. On nonmagnetic substrates, weak, Ruderman–Kittel–Kasuya–Yosida (RKKY) interactions mediated by conduction electrons [16–18] can induce relatively long-range coupling between small magnetic structures and significantly modify their behavior in external fields [8].

Free atoms have magnetic moments determined by the vector sum of the spin and orbital moments of electrons belonging to unfilled shells. The electron spin and orbital moment in each shell couple according to the first and second Hund's rules, respectively, which reflect the antisymmetry of the many-electron wavefunction (Pauli principle) and Coulomb repulsion effects. According to the first rule, configurations corresponding to maximum total spin (i.e., to parallel alignment of the individual spins in each shell) are favored because each electron can then occupy a different orbital state. According to the second rule, the total orbital moment is also maximized to minimize electron repulsion, compatibly with first rule. These rules allow one to predict m_S and m_L for free atoms with reasonable accuracy. The third Hund's rule determines the parallel or antiparallel alignment of m_L to m_S, depending on the sign of the SOC. In molecules and crystals, however, the atomic magnetic moments are usually strongly reduced owing to (i) the delocalization of electrons due to the overlap of the wavefunctions of neighbor atoms, which attenuates the correlation effects responsible for the first and second Hund's rule and (ii) the crystal-field potential, i.e., the electrostatic potential produced by the charges surrounding each atom, which imposes symmetry restrictions on the orbital character of the wavefunctions that influence both m_L and m_S. This is why nearly all of the transition-metal elements possess a nonzero magnetic moment as free atoms, whereas only five (Cr, Mn, Fe, Co, and Ni) retain a local moment in their bulk crystalline phases.

The appearance of magnetism in structures made by more than one magnetic atom requires the existence of local moments as a necessary condition, but also that some kind of coupling exists between them. This coupling is induced by the *interatomic* exchange interaction that arises as the wavefunction of neighboring atoms overlap and mix, leading to either ferromagnetic (FM) or antiferromagnetic (AFM) correlations between the atomic magnetic moments. Such an exchange interaction between neighboring moments is often represented, in a simplified way, by the Heisenberg model, whose Hamiltonian is

$$\mathcal{H}_{exc} = -\sum_{i,j=1}^{N} J_{i,j}\mathbf{S}_i \cdot \mathbf{S}_j. \quad (23.1)$$

In the aforementioned expression, the indices *i* and *j* run over the nearest neighbor atoms of the system ($i \neq j$), $\mathbf{S}_i$ is the spin moment of the *i*th atom, and $J_{i,j}$ the exchange coupling constant. As the coupling depends on the type of bond and atomic distance, $J_{i,j}$ may change from site to site. The type of coupling is determined by the sign of $J_{i,j}$, positive for FM and negative for AFM. The magnitude of J determines the robustness of magnetic order in a given system: the larger J is the stronger the tendency of the atomic moments to stay aligned to each other. Using a mean field approximation, it can be shown that the Curie temperature of an FM is given by

$$T_c = \frac{nJ(S+1)}{3k_B S}, \quad (23.2)$$

where n is the number of neighbors of each spin and k_B the Boltzmann constant. In reality, however, T_c depends not only on n but also on the dimensionality and size of the system. From a thermodynamic point of view, for example, the tendency to form magnetic ordered structures gradually decreases from 3D to two-dimensional (2D) and one-dimensional (1D) structures. Consider, for example, a 1D atomic chain consisting of N moments described by the Ising Hamiltonian $\mathcal{H}_{Ising} = -J\sum_{i=1}^{N-1} S_{zi}S_{zi+1}$ with $J > 0$, which is the uniaxial anisotropic limit of the Heisenberg model (Equation 23.1). The ground state energy of this system is $E_0 = -J(N-1)$ and corresponds to the situation where all the moments are aligned. The lowest lying excitations are those in which a single break occurs at any one of the N sites, as shown in the following:

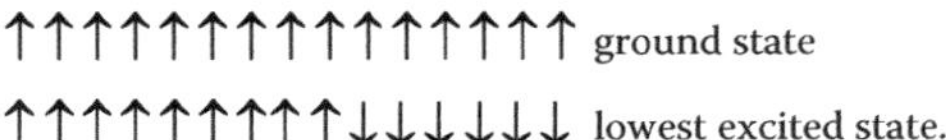

There are $N - 1$ such excited states, all with the same energy $E = E_0 + 2J$. At finite temperature T, the change in free energy due to these excitations is $\Delta G = 2J - k_B T \ln(N-1)$, where the last term represents the magnetic entropy of the chain. For $N \to \infty$, we have $\Delta G < 0$ at any $T > 0$, and the FM state becomes unstable against thermal fluctuations. Only for small systems satisfying $(N-1) < e^{2J/k_BT}$, FM order is thermodynamically stable in 1D [19].

In addition to interatomic exchange, the so-called spin Hamiltonian of a magnetic system includes single-ion terms:

$$\mathcal{H} = -\sum_{i,j=1}^{N} J_{i,j}\mathbf{S}_i \cdot \mathbf{S}_j + \sum_{i=1}^{N} h_i, \tag{23.3}$$

where

$$h_i = -KS_z^2 - E\left(S_x^2 - S_y^2\right) - g\mu_B \mathbf{S}\cdot\mathbf{B} \tag{23.4}$$

Here, the terms proportional to K and E represent the uniaxial and transverse magnetic anisotropy energy of the system (up to second order) and the last term the Zeeman interaction with an external magnetic field $\mathbf{B}$. The g-factor is such that $gS = m_L + m_S$. Note that, depending on the model, $\mathbf{S}$ represents either the true spin of the atom or an effective spin vector that yields the minimum multiplicity $(2S + 1)$ of the lowest lying energy states necessary to model its properties.

The MCA energy per atom is typically orders of magnitude smaller compared to the energy gained by the formation of local magnetic moments (~ 1 eV), interatomic exchange ($J_{i,j} \sim 20$ meV, $nJ \sim 200$ meV), and even minor lattice relaxations (~ 50 meV). For bulk FM, the MCA energy ranges from about 0.3 μeV/atom in fcc Ni to 2.5 μeV/atom in bcc Fe, and 45 μeV/atom in hcp-Co. Despite being so small, however, the MCA plays a fundamental role in stabilizing the magnetization with respect to external perturbations, for example, caused by temperature or external magnetic fields. This is true for macroscopic as well as for nanosized magnets. In extended systems, the large number of atoms ensures that the total MCA energy (proportional to NK) is always relatively large. In atomic-scale structures, on the other hand, NK is often comparable or even smaller than k_BT, meaning that thermal excitations can easily induce fluctuations of the magnetization. According to the Néel–Brown model of magnetization reversal [9], the relaxation time of the magnetization of a single-domain particle is given by an Arrhenius law of the form

$$\tau = \tau_0 \exp\left(\frac{NK}{k_B T}\right), \tag{23.5}$$

where τ_0 is a prefactor of the order of 10^{-9} s. The energy NK can thus be considered as a barrier that hinders magnetization reversal. This barrier is extremely important for magnetic storage systems, which typically require relaxation times in excess of 10 years, i.e., $NK \gtrsim 35k_BT$. At room temperature ($k_B T \approx 25$ meV), this turns out to be a very stringent requirement for nanosized magnets, even if K can increase up to a few meV/atom in structures with low atomic coordination. Systems that behave according to Equation 23.5 are called *superparamagnetic*, because they behave similarly to a paramagnet but have a total magnetic moment that is N times that of a single atom. Such systems present magnetic hysteresis only on timescales shorter than τ. Alternatively, for a given τ, one can say that their magnetization is stable below a *blocking* temperature defined as $T_B = NK/[k_B \ln(\tau/\tau_0)]$. The miniaturization of magnetic structures thus brings new challenges for the stability of the magnetization, which must be addressed either by design or by reducing their operational temperature.

23.3 Experimental Probes

We introduce two techniques apt to the study of atomic-scale magnetic structures. Spatially averaging techniques, such as x-ray circular magnetic dichroism and x-ray linear dichroism, reveal in an element specific way the orbital and spin moments, as well as magnetic anisotropy and coupling effects. Local probe techniques, such as STM, can now access effective spin moments, magnetization curves, spin relaxation times, and exchange coupling energies atom by atom. Extended reviews of these methods can be found in References [20] and [21].

23.3.1 X-Ray Magnetic Circular Dichroism

X-ray dichroism is based on polarization-dependent x-ray absorption spectroscopy (XAS), a synchrotron radiation technique, which exploits the intense x-ray beam emitted by relativistic electrons forced to follow a curved trajectory by a bending magnet or a so-called undulator insertion device, see Figure 23.2a). In XAS, the energy of the x-rays is tuneable and the direction of the electric field vector can change from linear to right (R) or left (L) circularly polarized. The x-ray beam is monochromatized and focused before being directed onto the sample. The sample is kept in ultra-high-vacuum to avoid the absorption of x-rays from ambient gas. In a typical experiment, the x-ray energy is scanned across one or more of the absorption edges of the elements contained in the sample. As photons are absorbed, the intensity of the x-ray absorption spectra is recorded by measuring the electric current passing from ground to the sample to replace the photoemitted electrons. This method, schematized in Figure 23.2a, is called total electron yield (TEY). Other methods to measure the x-ray absorption intensity are based on the fluorescence yield or transmission yield. In the soft x-ray range used in most XMCD experiments, TEY is the most sensitive technique and, due to the limited escape depth of the electrons (~1 nm), also inherently surface sensitive.

In the simplest picture of XAS, an x-ray photon is absorbed by transferring its energy to a core electron, which is excited into an unoccupied state just above the Fermi level of the sample. TEY detects the Auger and secondary electrons that escape the sample for a core hole that decays via an Auger process. The dipole-allowed transitions of interest for magnetic studies are the core $2p \rightarrow$ valence $3d$ excitations of $3d$-transition-metal elements and the core $3d \rightarrow$ valence $4f$ transitions of the rare-earths. Figure 23.2b shows the x-ray absorption process for the spin–orbit split $2p$-core level of Co: as the x-ray energy matches the binding energy of the $2p_{3/2}$ level, an intense peak (the L_3 edge) is observed in the absorption spectrum, followed by a second one (the L_2 edge) as the x-ray energy increases up to that of the $2p_{1/2}$ level. The larger intensity of the L_3 edge compared to the L_2 edge is due to the double degeneracy of the $2p_{3/2}$ level as well as to final state effects related to SOC. A correct description of XAS beyond one-electron model requires to consider excitations from a many-electron ground state, such as $2p^6 3d^n$, to all the possible final state configurations, $2p^5 3d^{n+1}$, allowed by the dipole selection rules. In systems with localized d- or f-states, the interaction of the core hole with the valence electrons gives rise to a series of multiplets that appear as sharp features in the XAS lineshape (see Figure 23.6 for example). Such features can be used to identify the ground state of the element under investigation using ligand field multiplet theory [22,23]. We also note that the XAS intensity depends on the symmetry and orientation of the final states relative to the x-ray polarization direction, which gives rise to *linear* dichroism effects. Here, however, we will concentrate on *circular* dichroism effects of interest for the investigation of magnetic systems.

The absorption of polarized light by a magnetized sample depends on the orientation of the magnetization **M** relative to the light polarization direction. XMCD is defined as the difference in the x-ray absorption coefficients for parallel and antiparallel orientation of the magnetization direction of the sample with respect to the helicity of circularly polarized x-rays. A qualitative understanding of XMCD can be given using a two-step model [20]. In the *first step*, R or L circularly polarized photons are absorbed and transfer their angular momentum ($\Delta m = \pm 1$, respectively) to the excited photoelectron. If the photoelectron originates from a spin–orbit split level, for example, the $p_{3/2}$ level, the angular momentum of the photon can be transferred, in part, to the spin through SOC (note: the $\Delta S = 0$ selection rule of dipole transitions holds only for ls coupling). R-polarized photons transfer opposite angular momentum to the electron than L-polarized photons, and hence one obtains large transition

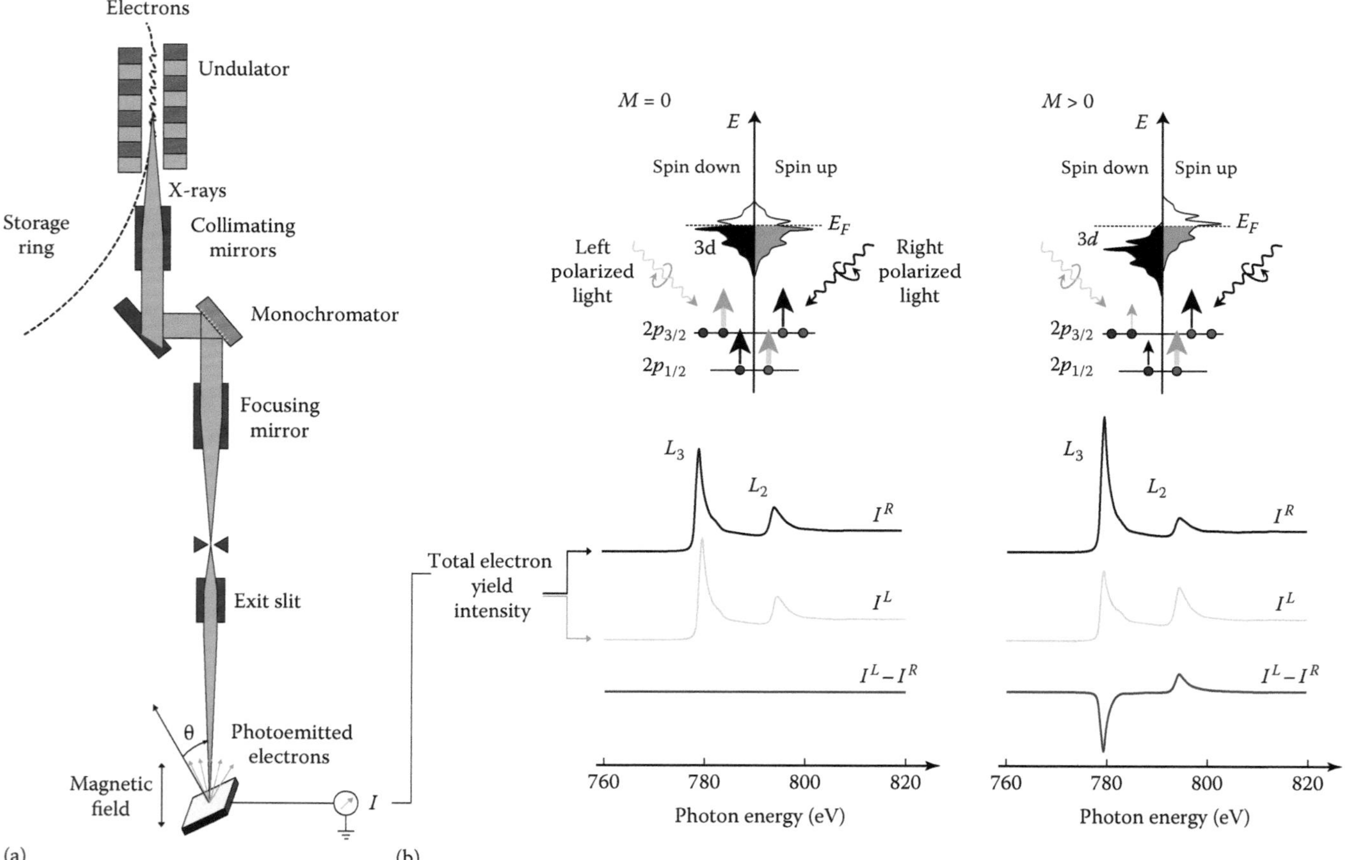

FIGURE 23.2 (a) Layout of a soft x-ray beamline. (b) Diagrams showing the excitation of $2p$ core electrons following the absorption of circularly polarized x-rays for a nonmagnetic (left) and magnetic (right) $3d$ metal. Examples of x-ray absorption spectra and XMCD are shown for the specific case of Co.

matrix elements between final states of opposite spin polarization in the two cases (see arrows in Figure 23.2b). In other words, for a given initial state, R- and L-polarized photons excite photoelectrons having opposite spin polarization. Since the $p_{3/2}$ and $p_{1/2}$ levels have opposite SOC ($l + s$ and $l - s$, respectively), the spin polarization will be opposite at the L_3 and L_2 edge.

The magnetic properties enter in the *second step*, i.e., in the promotion of the excited photoelectron to an unfilled 3*d*-state of the valence band. In the absence of a net magnetization, the number of excited electrons for R or L polarization is the same for each of the L_3, L_2 edges because the total spin polarization of both the $p_{3/2}$ and $p_{1/2}$ manifolds is equal to zero. Suppose now that the 3*d* band is exchange split; the empty 3*d* states have predominantly minority character (spin up with the convention used in Figure 23.2b). The favored transitions are thus those that involve initial states with predominant minority spin. As we will show later, these are the L_3 transitions for antiparallel and the L_2 transitions for parallel direction of the magnetization with respect to the photon helicity. The exchange split final states therefore act as a filter for the spin of the excited photoelectrons. The quantization axis of the filter is that of the sample magnetization, which should be aligned with the x-ray beam direction in order to obtain the maximum XMCD intensity. We recall that the photon helicity is positive (negative) when it is parallel (antiparallel) to the light propagation direction, corresponding to R (L) polarization. The XMCD spectrum is finally obtained by taking the difference $I^L - I^R$ of two consecutive absorption spectra taken with opposite polarization. From that mentioned earlier, it follows that the XMCD intensity, normalized by the total absorption intensity $I^L + I^R$, is proportional to the inbalance between the minority and majority spins of the unoccupied states of the element under consideration, which in turn is proportional to the magnetic moment per atom.

Due to the fact that x-ray absorption is a localized process that obeys dipole selection rules, powerful sum rules exist that relate the shell-specific ground-state expectation value of m_L and m_S projected on the direction of the incident photon beam [24–26]. In the case of $2p \rightarrow 3d$ transitions, one has

$$m_L = -\frac{4}{3}\frac{\int_{L_3+L_2} dE(I^R - I^L)}{\int_{L_3+L_2} dE(I^R + I^L)} n_h, \tag{23.6}$$

and

$$m_S + m_T = -\frac{6\int_{L_3} dE(I^R - I^L) - 4\int_{L_2} dE(I^R - I^L)}{3\int_{L_3+L_2} dE(I^R + I^L)} n_h, \tag{23.7}$$

where

E is the x-ray energy
n_h is the number of holes in the d shell
m_T is the intraatomic magnetic dipole moment

m_T arises from the multipole expansion of the spin density over the atomic volume and reflects the anisotropy of the spin distribution in the atomic cell. It is usually negligible in bulk samples [26], but large for single atoms [3] and molecules [27].

XMCD is thus a very powerful quantitative magnetometry tool. The XMCD intensity is element-specific and proportional to the magnetization projected on the x-ray beam direction. It is extremely sensitive [3] and, unique among magnetooptical techniques, allows for the separate measurement of m_L and m_S. Also, although not shown here, the short wavelength of x-rays offers spatial resolution down to a few tens of nm in x-ray microscopy experiments, whereas the sub-ns time structure of synchrotron x-ray beams can be exploited for ultrafast measurements [20].

23.3.2 Spin-Sensitive Scanning Tunneling Microscopy

STM has now become a mature technique for the quantitative study of the magnetism of individual atoms, molecules, and nanoscale islands adsorbed on surfaces. The first signatures of magnetism that have been detected by means of STM for individual adatoms have been midgap states for magnetic adatoms placed on a superconductor [28,29], or when superconducting tips were used for the detection of magnetism in atoms adsorbed on normal metals [1,30]. This has been closely followed by Kondo resonances in the differential conductance *dI/dV* [31–33] and by the suppression of one spin conductance channel for reversible STM quantum point contacts across magnetic atoms [1,30,34]. These effects reveal that the atoms possess a magnetic moment but do not determine the size of this moment and its anisotropy landscape.

Parallel to these efforts on single adatoms, spin-polarized (SP) STM has delivered valuable magnetic information on thin films and surface adsorbed nanostructures. This technique has been demonstrated for the first time on Cr(100), where an SP tip revealed the spin-contrast from terrace to terrace of this layered antiferromagnet [35]. SP-STM is now a well-established technique [21,36,37]. It has unraveled noncollinear spin ground states in thin films, which would have been difficult to guess without atomic-scale spatial magnetic resolution [15]. Furthermore, our understanding of current-induced magnetization reversal has been improved by varying the position where the SP current is injected into a nanoscale island [38], and a giant magnetoresistance of 800% has been reported [39] and explained by adsorbates in the tunnel junction [40].

Recent experiments pushed the magnetic information gained by STM significantly beyond this. As we will describe in this section, magnetization curves can now be recorded by SP-STM on single adatoms, their effective spin moments, exchange, and anisotropy energies can be deduced by spin-excitation spectroscopy (SES), and finally their spin coherence times can be determined with spin-pumping and SP-STM pump–probe experiments. Here, we describe the principles of these spin-sensitive STM measurements, while we elaborate on examples for single atoms and dimers in Section 23.4.

The magnetic contrast of SP-STM relies on the spin-valve effect, in particular, on the tunnel magnetoresistance of an STM junction formed by a magnetic tip and a magnetic adsorbate. We explain how entire magnetization curves can be measured on isolated adatoms with this contrast. We use the example of Co/Pt(111) since for this system the first such single-atom magnetization curves have been reported [41]. In order not to be hampered by small differences in the apparent height of the adatoms, the authors used the so-called SP-STS mode, where maps of the differential conductance dI/dV, such as the ones shown in Figure 23.3a and b are recorded for given external magnetic fields. These figures show a clear spin-contrast between the two field orientations. The out-of-plane spin contrast has been achieved with Cr-coated W-tips, which were dipped into a Co ML-film on Pt(111).

dI/dV is composed of a non-polarized (np) and a polarized (pol) part, the latter being maximized for parallel and minimized for antiparallel alignment of tip and sample magnetization, thus one can write $dI/dV = dI/dV_{np} + \mathbf{M}_{tip} \cdot \mathbf{M}_{sample}\, dI/dV_{pol}$. In analogy with the polarization of the current itself, $P = (I_p - I_{ap})/(I_p - I_{ap})$, where p denotes parallel and ap denotes antiparallel tip and sample magnetization, one defines a differential polarization $p = ((dI/dV)_p - (dI/dV)_{ap})/((dI/dV)_p - (dI/dV)_{ap})$.

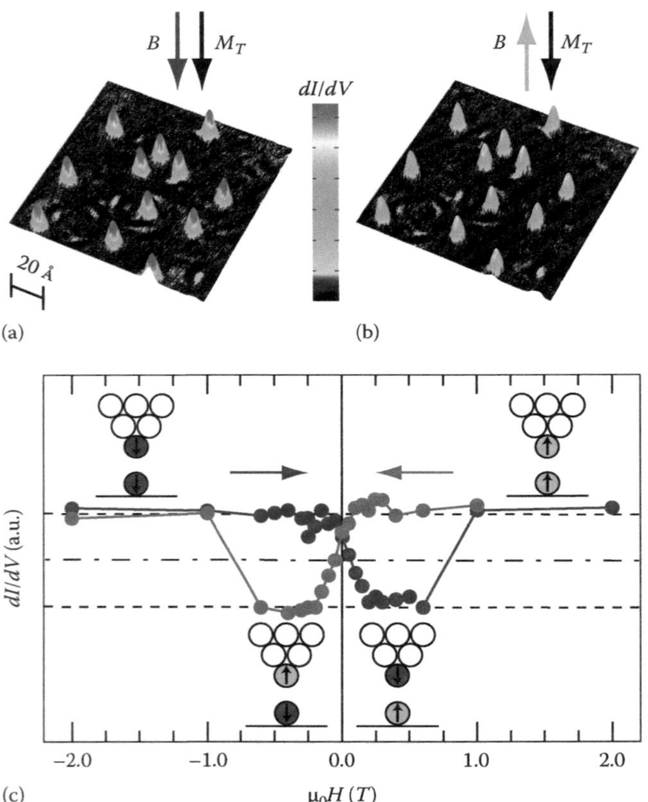

FIGURE 23.3 **(See color insert.)** (a) and (b) 3D view of constant current STM images of Co atoms on Pt(111). Color code shows magnetic contrast from dI/dV for parallel (a) and antiparallel (b) field direction and tip magnetization ($V_t = 0.3$ V, $B_z = \pm 0.5$ T, and $T = 0.3$ K). (c) Magnetic-field-dependent differential conductance of soft magnetic tip over a Co atom on Pt(111). Red curves show up and blue ones down sweep of magnetic field ($I_t = 0.8$ nA, $V_t = 0.3$ V, $V_{mod} = 20$ mV, and $T = 0.3$ K). The insets indicate the relative orientation of adatom and tip **M**. The tip reverses its magnetization at around 0.8 T. The curves are shifted horizontally due to the stray field of the tip. (Adapted from Meier, F. et al., *Science*, 320, 82–86, 2008.)

The $M(H)$-curves shown in Figure 23.3c have been obtained by recording dI/dV images for several out-of-plane fields and laterally averaging the signal over 5 Å × 5 Å squares centered at the Co atoms. As expected for magnetic tunnel junctions with soft magnetic tips, the curves have the shape of a butterfly, caused by magnetization reversal of both electrodes at different fields [41,42]. The positive field sweep (red) starts at −2 T with the magnetization of both electrodes pointing down, until at + 0.1 T one turns up reducing dI/dV, and the second electrode turns up between + 0.7 and + 1.0 T, recovering the initial dI/dV value, which stays constant up to + 2 T. The down sweep (blue) is symmetric with respect to the zero field line. From reference measurements with the same tip on a Co monolayer stripe adsorbed on the same surface, the electrode switching at ± 0.8 T is the tip. For symmetry reasons, switching of the atom takes place at exactly half way between dI/dV_p and dI/dV_{ap}, see the horizontal dash-dotted line in Figure 23.3c. The atom switches at an external field of + 0.1 T for the forward and at −0.1 T for the backward sweep. These shifts are caused by the stray field of the tip changing sign with the field sweep direction. Stray field and tip reversal have been corrected for in the curves shown in Section 23.4. SP-STM magnetization curves over individual atoms have been reported in Refs. [8,12,41,43] and over nanoscale islands in Refs. [42,44,45].

SES with the STM has recently emerged as a tool to quantify in a complementary way to $M(H)$-curves the magnetic properties of individual atoms and of very small clusters [5,46–51]. The technique relies on inelastic scanning tunneling spectroscopy measuring excitation energies with high lateral resolution and inspired by inelastic electron tunneling spectroscopy (IETS) in planar tunnel junctions [52,53]. The excitations can be vibrations [54] or changes in the magnetic state [5] of atoms or molecules in the tunnel junction, as well as surface magnons [55]. With increasing tunnel voltage, one observes a stepwise conductance increase, with a step occurring each time the tunneling electrons reach the threshold energy needed to excite a vibrational or spin degree of freedom of the system since this opens a new inelastic conductance channel. A conductance profile and the inelastic conductance channels (red arrows) are schematically represented in Figure 23.4 for the case of a $S = 5/2$ magnetic adatom prepared in the $m = +5/2$ state by a high external magnetic field pointing up. For magnetic atoms, the energies and amplitudes of the conductance steps deliver valuable information on the magnetic ground state and anisotropy of the adatoms [5,47–49], and in the case of clusters also on the Heisenberg exchange coupling between the atoms in the cluster [46,56].

Figure 23.4b shows that the symmetry of the curves about E_F bears information on the spin polarization of the tip. The tunnel current is composed of three parts, two elastic and one inelastic. The two elastic conductance channels are tunneling between tip and substrate, bypassing the magnetic atom (b_0), and tunneling into the magnetic adatom without changing its magnetic quantum state (E^- and E^+ for negative and positive tunnel voltage, V_t, respectively). The spin-polarization of the tip is schematically

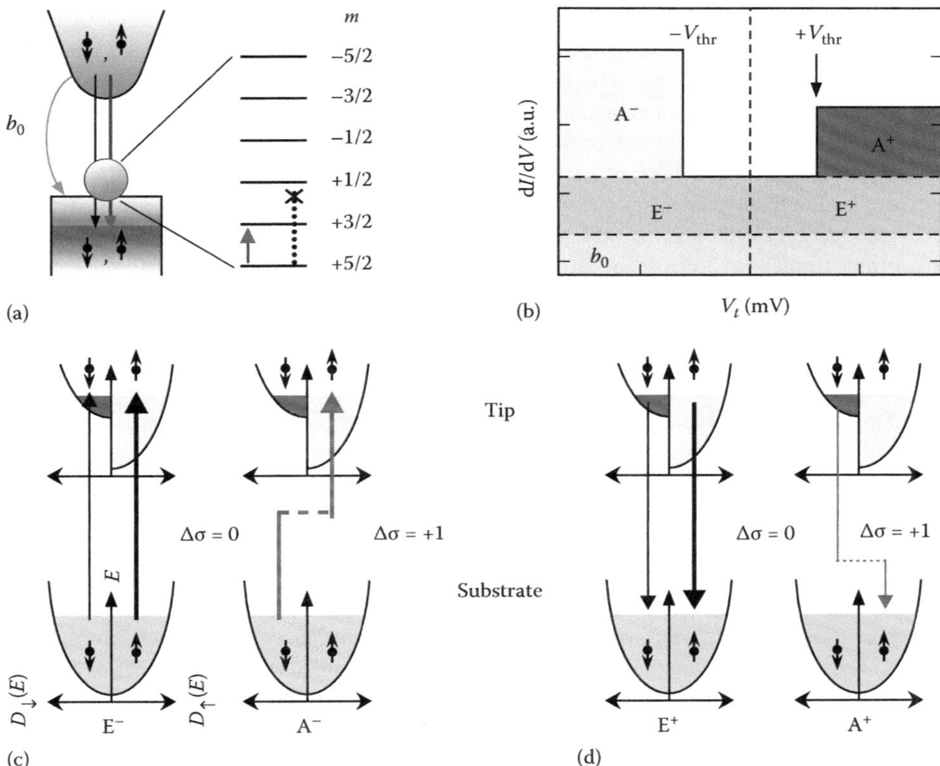

FIGURE 23.4 Spin excitation spectroscopy with a spin-polarized tip for an $S = 5/2$ magnetic impurity, such as Mn atoms adsorbed on $C_2N/Cu(100)$ $-c(2 \times 2)$, prepared by a large out-of-plane magnetic field in the $m = +5/2$ state. (a) Schematic diagram of the possible tunnel paths when the STM tip is positioned over a Mn atom: inelastic spin-dependent tunneling (gray), elastic spin-dependent tunneling (black) and spin-independent background tunneling, (b_0). Only the inelastic tunneling excites from the $m = +5/2$ to the $m = +3/2$ state. Direct excitations into higher states are forbidden (dotted line). (b) Idealized conductance spectrum with contributions for each tunnel path: inelastic spin-dependent tunneling, A^+ and A^-; elastic spin-dependent tunneling, E^+ and E^-; spin-independent background tunneling, b_0. (c) Sketch of the tunnel paths in the spin-dependent densities of states of the STM tip (upper row) and Cu substrate (lower row) at negative voltage. The number of available states with correct spin in the tip and the substrate determines the relative strength of each tunnel path (indicated by the arrow thickness). (d) The same plot as in (c), but for positive voltage. (Adapted from Loth, S. et al., *N. J. Phys.*, 12, 125021, 2010.)

shown in the density of states plots with spin-down electrons and spin-up electrons. If the magnetic state is unchanged (black arrows), tunneling takes place between identical spin states of both electrodes ($\Delta\sigma = 0$), and even though there is a higher current for the majority spin, the total current does not depend on the sign of the bias voltage (V_t). This is different for the third conductance channel where the tunnel electrons undergo a spin-flip, $\Delta\sigma = \pm 1$. For a Mn atom prepared in the $m = +5/2$ state, the only change can be $\Delta m = -1$, therefore the electrons can make a spin-flip only from spin down to spin up ($\Delta\sigma = 1$). For negative polarity, the final state of this inelastic channel is a majority state, while for positive polarity the initial state is a minority state leading to the shown asymmetry for energies above and below E_F [57,58].

As will be shown later, this spin-polarization can be used to progressively excite the magnetic system in the tunnel junction, which is referred to as spin-pumping. The decay of the *dI/dV* signal for energies beyond the threshold for SES and as function of tunnel current gives access to the life-time of the excited states [58]. As we will show in the following section, this spin-relaxation time can also be accessed by a SP-STM pump–probe experiment where a pump voltage pulse excites the system and a probe pulse measures its magnetic state a delay time after the probe [59].

These examples illustrate how the initial satisfaction to detect some signature of magnetism with the STM has now been replaced by the capacity of pinning down magnitudes of magnetic moments, anisotropy energies, magnetic ground and excited state configurations, entire magnetization curves, and spin relaxation times. This evolution of STM as a sensor of atomic magnetism is moving on rapidly and is expected to have strong impact on our understanding of quantum magnetism, magnetic impurities in solids, and magnetism of entities composed of a few atoms or molecules only.

23.4 Magnetism of Single Atoms and Clusters on Surfaces

Although most transition-metal atoms possess a magnetic moment in the gas-phase, the survival of this moment when an atom is placed on a nonmagnetic substrate is not granted. Owing to their reduced atomic coordination, surface adatoms can be viewed as a bridge between the atomic and solid state, with many of their electronic and magnetic properties determined by the competition between the Coulomb energy and the kinetic energy associated with electrons hopping from site to site in the lattice [60]. For moderate hybridization between adatom and host electron states, the Anderson model [61] describes well the formation of a magnetic moment. Many-body spin-flip processes may also lead to the Kondo effect, the screening of the local moment by conduction electron spins, which can be directly visualized by scanning tunneling spectroscopy [31,33].

From a different perspective, controlling the bonding, diffusion, and nucleation processes of adatoms at surfaces offers countless opportunities to tune the adatom–substrate interaction as well as to construct multiatom magnetic clusters of tailored shape and dimensions [62]. Adatoms and clusters may further be considered as the precursors of thin films, as the growth of magnetic mono- and multilayers is typically initiated by the deposition of transition-metal atoms from the vapor phase onto a nonmagnetic substrate. Investigating substrate–impurity hybridization and coordination effects thus provides basic understanding and useful guidelines to tailor the magnetization and magnetic anisotropy of nanomagnets and optimize sensitive interface properties that govern the performances of magnetic storage media and electron transport in spintronic devices. In the perspective of this book, adatoms and clusters on nonmagnetic surfaces represent the ultimate paradigm of atomic-scale magnets, whose behavior can approach that of classical or quantum magnets depending on the adatom–substrate and adatom–adatom interactions.

23.4.1 Hybridization with the Substrate

The first effect noticed when a transition-metal atom is deposited on a metallic substrate is a reduction of m_S and m_L compared to the free atom case. Figure 23.5 shows a set of density functional calculations of the atomic moment of 3*d*, 4*d*, and 5*d* adatoms adsorbed on the (100) face of an Ag substrate [63–65]. The adatoms still follow the Hund's first rule with maximal moments at the center of each series; however, their magnetic moment assumes smaller, noninteger values relative to the gas phase. For example, a Co free atom with seven electrons in the *d*-shell has $m_S = 3$ and $m_L = 3\ \mu_B$, which reduce to about $m_S = 2$ and $m_L = 1\ \mu_B$ on Pt(111) [2] and Ag(100) [65] owing to charge transfer from the substrate into the Co 3*d*-states and hybridization. The larger relative decrease of m_L compared to m_S is related to the hierarchy of the Hund's rules, as electron delocalization reduces Coulomb repulsion effects within the *d*-states. Note that the local moment tends to decrease further with increasing atomic coordination, as shown by the calculations performed for monolayer films. Interestingly, elements that are nonmagnetic in the bulk display sizeable magnetic moments as adatoms, including 4*d* and 5*d* elements. This allows for a much broader choice of elements in the construction of atomic-scale magnets. The reduction of the magnetic moments due to the increase of hybridization with coordination number is greater for 5*d* and 4*d* elements because of the larger extension of their wavefunctions compared to 3*d* metals. As a rule of thumb, it can be shown that the atomic magnetic moment decreases with increasing atomic coordination proportionally to the inverse of the width of the *d*-band [66].

XMCD [1,3] and photoemission [60] studies show that the degree of hybridization of the 3*d*-states of single adatoms depends very significantly on the electron density of the substrate, even for simple *sp* metals like the alkalis. However, even for very small mixing between the 3*d*-states and substrate conduction electron bands, charge transfer can occur and affect the magnitude of m_S and m_L. This is nicely seen by analyzing the multiplet features of the x-ray absorption spectra of Co adatoms

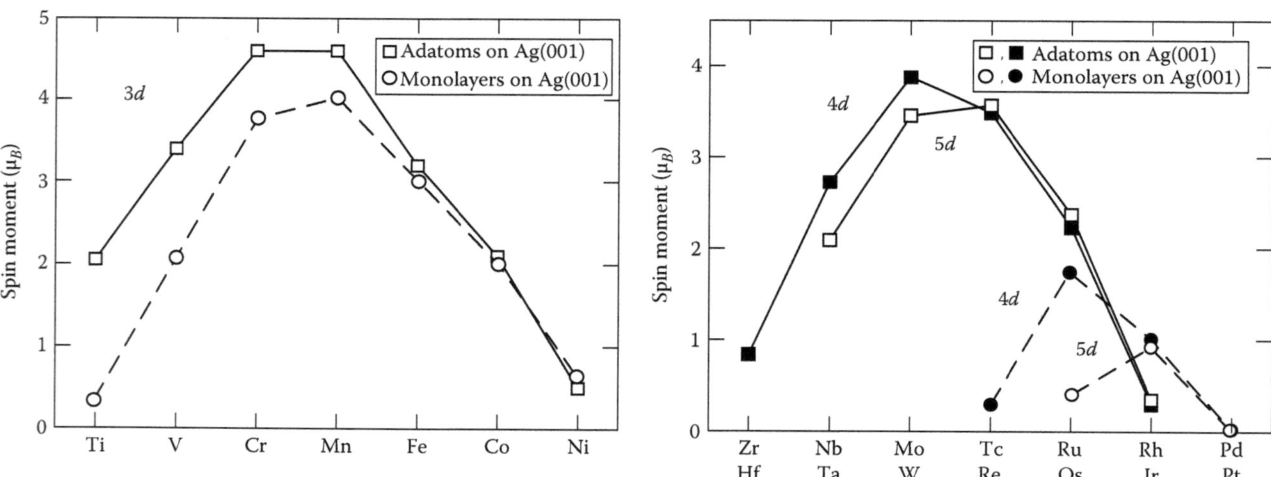

FIGURE 23.5 Spin magnetic moments calculated for (a) 3*d* and (b) 4*d* and 5*d* transition metal adatoms at the hollowsite of Ag(100) (squares connected by full lines) and monolayers on Ag(100) (circles connected by dashed lines). (Adapted from Blügel, S., *Phys. Rev. Lett.*, 68, 851, 1992; Lang, P. et al., *Solid State Commun.*, 92, 755, 1994; Nonas, B. et al., *Phys. Rev. Lett.*, 86, 2146, 2001; Blügel, S., Magnetism goes nano: 36th IFF Spring School, In: S. Blügel, T. Brückel, C. M. Schneider eds., Ch. *Reduced Dimensions I: Magnetic Moment and Magnetic Structure*, Forschungszentrums Jülich, Jülich, 2005. With permission from the authors.)

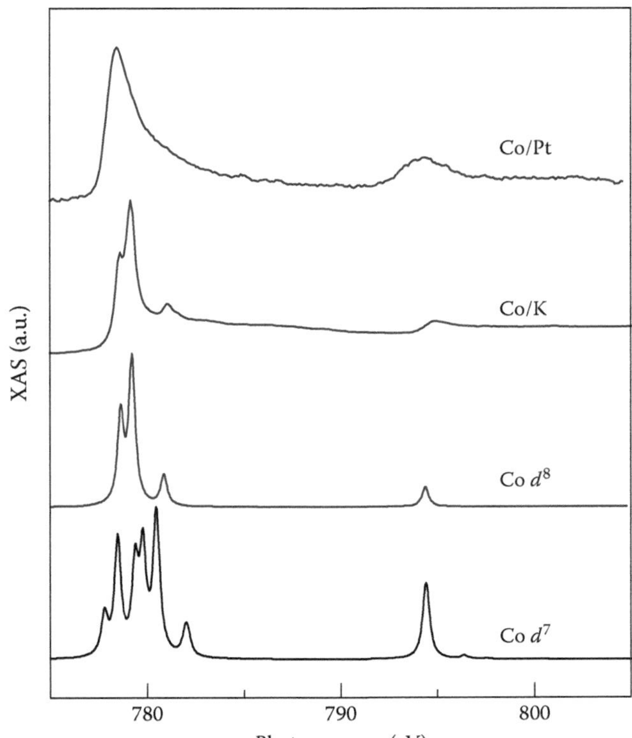

FIGURE 23.6 Experimental and theoretical x-ray absorption spectra of Co impurities in the $L_{3,2}$ edge region. The experimental spectra were recorded as the sum of the absorption intensity in total electron yield for parallel (I^R) and antiparallel (I^L) alignment of the photon helicity with respect to the magnetic field $B = 7$ T at an angle $\theta = 0°$ relative to the surface normal. The temperature of the Pt(111) and K substrates was 6 K and 10 K, respectively. The theoretical spectra were calculated using an atomic multiplet model [23] for d^7 and d^8 configurations. (Adapted from Gambardella, P. et al., *Phys. Rev. Lett.*, 88, 047202, 2002.)

on a K surface, as shown in Figure 23.6. The narrow XAS multiplet structure observed for Co_1/K indicates that the Co ground state has a strongly localized character, specifically a $3d^8$ configuration with nearly atomic-like $m_S \approx 2$ and $m_L \approx 3$ μ_B [3]. Note that the spectra expected for the d^7 and d^8 ground states of free Co atoms are very different, indicating that the occupation of the Co $3d$-states increases by one electron upon deposition on K. The hybridization of the Co d-states changes drastically from being very weak for Cs to much stronger for Li along the alkali group, as the electronic density of alkali metals increases for the lighter species. Transition-metal substrates differ from free-electron like metal surfaces not only due to their larger conduction electron density but also for the presence of unfilled d-states crossing the Fermi level, which heavily affect most of their magnetic properties (e.g., the susceptibility and magnetoresistance) as well as the nonmagnetic ones (e.g., cohesion, diffusion barriers, catalytic activity, etc.). As an example of a strongly interacting substrate, we present data for isolated Co adatoms on the (111) surface of Pt [2]. The x-ray absorption spectra of Co_1/Pt(111) are much broader and present no clear multiplet feature compared to those of Co_1/K (Figure 23.6) as the adatom $3d$-states hybridize strongly with both the $5d$- and $6s$-states of the substrate. This is accompanied by a substantial decrease of m_L to about 1.1 μ_B, whereas m_S remains close to 2 μ_B.

23.4.2 Magnetization Curves for Ensembles vs. Individual Atoms

Given the pronounced anisotropic spatial extension of the d-orbitals, the Co–Pt admixture of $3d$- and $5d$-states may lead to unequal filling of electronic states with different symmetry, and hence to a strong anisotropy of the orbital magnetization. Due to the strong SOC between m_S and m_L, a strong magnetic anisotropy of the overall magnetization is to be expected. Figure 23.7a shows the intensity of the XMCD Co L_3 minimum measured for **B** and x-ray beam oriented out-of-plane (black) and at 70° with respect to the surface normal (red). In order to account for the dependence of the cross section with angle, the signal has been normalized by the XAS intensity. It is seen that the magnitude of the saturation signal differs by more than 60% for the two field directions, revealing the presence of extraordinary magnetic anisotropy.

The fixed-energy L_3 intensity depends on the linear combination $m_S + 3m_L + m_T$, Equations 23.6 and 23.7, but it can be demonstrated that the three moments m_S, m_L, and m_T follow the same field dependence, i.e., that the XMCD L_3 signal is proportional to the total Co magnetic moment projected onto the beam, respectively, field direction. Therefore, the curves shown in Figure 23.7a are XMCD magnetization curves that can be fitted in the framework of a classical model, where the time-averaged projection of the total impurity moment $\langle m \rangle$ on the magnetic field direction is given by

$$\langle m \rangle = m_0 \frac{\int \widehat{\mathbf{m}}\cdot\widehat{\mathbf{B}} \exp[(\mathbf{m}\cdot\mathbf{B} + K(\widehat{\mathbf{m}}\cdot\hat{\mathbf{e}})^2)/k_B T]\mathrm{d}\Omega}{\int \exp[(\mathbf{m}\cdot\mathbf{B} + K(\widehat{\mathbf{m}}\cdot\hat{\mathbf{e}})^2)/k_B T]\mathrm{d}\Omega}. \tag{23.8}$$

Here, m_0 stands for the saturation value of the Co plus induced Pt moment, $\hat{\mathbf{e}}$, $\widehat{\mathbf{m}}$, and $\widehat{\mathbf{B}}$ represent the unit vectors of the easy axis, the magnetic moment, and the field direction, respectively; K is a uniaxial magnetic anisotropy barrier, and the integration is carried out over the solid angle Ω of the magnetic moment in spherical coordinates. The solid lines represent fits of the data by means of numerical integration of Equation 23.8 with m and K as free parameters fitted simultaneously for the two curves.

Note that, in such a model, all directions are in principle allowed for the Co magnetic moment, showing that a classical description is well-suited to describe a strongly hybridized impurity system. Moreover, owing to the strong Stoner enhancement factor of Pt, the substrate atoms are highly polarized by $3d$ transition-metal species, developing a significant intrinsic magnetization that decays exponentially away from the impurity site. In dilute bulk CoPt alloys with 1% at Co concentration, the total moment per Co atom is of the order of 10 μ_B [67]. In our case of surface dilute impurities, the fit of Equation 23.8 yields $m = 5.0 \pm 0.6$ μ_B and $K = 9.3 \pm 1.6$ meV.

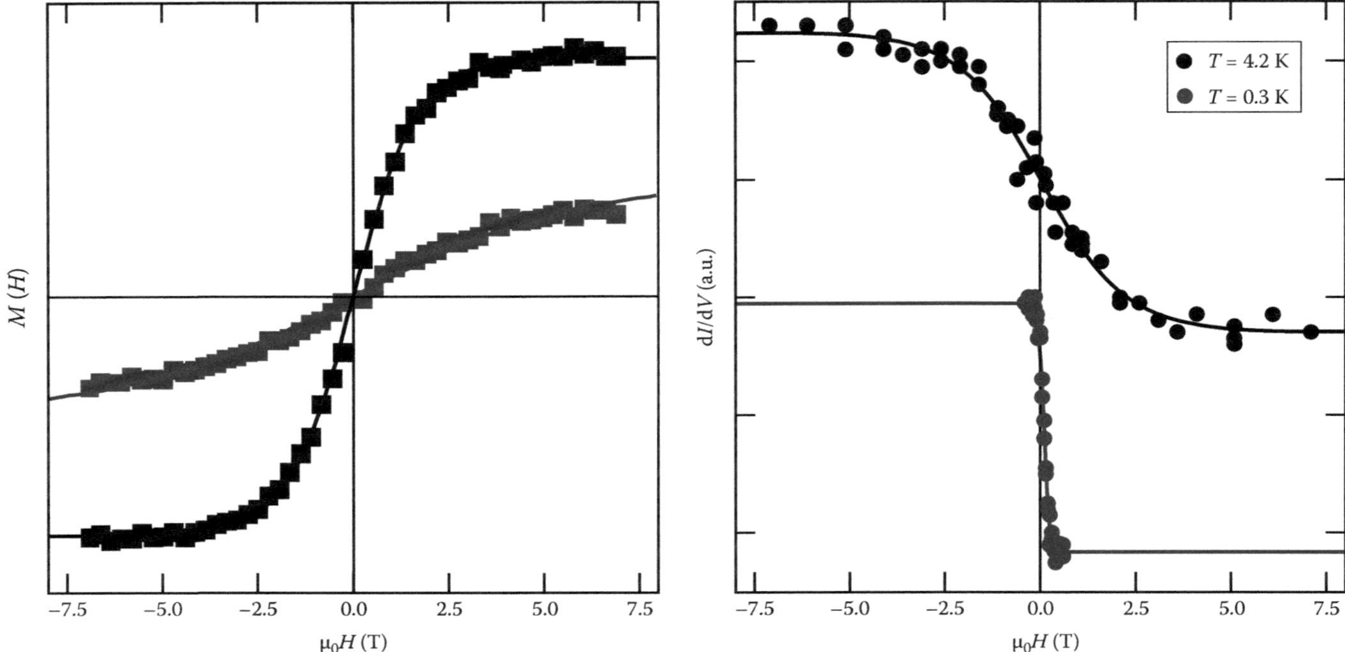

FIGURE 23.7 Comparison of magnetization curves of ensembles of adatoms measured with XMCD and of an individual adatom measured with SP-STM, both for Co/Pt(111). (a) Field dependence of the Co L_3-XMCD signal normalized to the total XAS intensity. The out-of-plane projection of the magnetization is shown in black, the one where B and x-ray beam are at 70° with respect to the surface normal in gray ($\Theta = 0.03$ ML, $T_{dep} = 5.5$ K, and $T_{XMCD} = 5.5$ K). (Adapted from Gambardella, P. et al., *Science*, 300, 1130, 2003.) (b) $M(H)$-curves on an individual atom at $T = 4.2$ K (black) and 0.3 K (gray) ($I_t = 0.8$ nA, $V_t = 0.3$ V, and $V_{mod} = 20$ mV). (Adapted from Meier, F. et al., *Science*, 320, 82–86, 2008.) Symbols in (a) and (b) represent experimental data, solid lines are fits according to the functions described in the text.

These values remain unchanged on samples with Co coverage comprised between 0.007 ML and 0.03 ML, showing that magnetic or electronic interactions between the adatoms are negligible. We note that weak inelastic features at energies close to the expected 9 meV have been reported in dI/dV spectra recorded above Co adatoms on Pt(111) at high tunnel current [49]. These features have been assigned to spin excitations; however, an unequivocal proof would require to study their magnetic field dependence, which is hampered by their large energetic width.

$M(H)$-curves measured by SP-STM for the same system using the method described in Section 23.3.2 are shown in Figure 23.7b. STM can measure such curves over individual atoms and therefore report on variations associated by the atomic environment of the respective atom. It cannot measure the angular dependence of the magnetization, but $M(H)$ curves at different temperatures over the very same Co atom, as shown for 4.2 (black) and 0.3 K (red). The absence of hysteresis down to 0.3 K is not expected for an uniaxial MCA energy barrier of the order of 10 meV. $M(H,T)$ fits were performed using a simple Hamiltonian $\mathcal{H} = -m(B - B_{tip})\cos(\theta) - K\cos^2(\theta)$, where B_{tip} is the stray field from the tip and θ the angle between the magnetic moment m and the sample normal. The fit parameters are m, the saturation value of the spin-dependent dI/dV contrast M_{sat}, and B_{tip}; $K = 9.3$ meV was taken according to the XMCD measurements presented earlier. SP-STM on about 80 different Co atoms evidenced narrow m-distributions at 4.2 K with a mean value of 3.7 μ_B. This is lower than the total moment detected in XMCD, but consistent with the sum of the Co spin and orbital moment, meaning that SP-STM is sensitive to this sum but not to the moments induced in the Pt substrate. However, the moment distribution obtained for the different atoms at 0.3 K is rather broad ($2\ \mu_B \leq m \leq 6\ \mu_B$). This has been attributed to substrate-mediated RKKY interactions with a coupling energy ranging from about 10 to 180 μeV [41,43]. At 4 K, these interactions are smaller than thermal excitations and at 0.3 K they become apparent.

These results show that single Co atoms on Pt(111) behave as paramagnetic impurities when placed in an STM junction biased at 0.3 V. This voltage has been chosen in order to obtain comparable magnetic contrast for Co atoms adsorbed on the two non-equivalent threefold Pt(111) hollow sites [41,68]. The absence of remanence might therefore well be due to spin excitations induced by the tunnel electrons. The atoms are subject to substrate-mediated RKKY interactions, which can be detected down to 10 μeV resolution by recording the exchange field stemming from a FM monolayer high stripe attached to a substrate step. Once single magnetic atoms are subject to such interactions, remanence becomes apparent in SP-STM also with electron energies beyond the threshold for spin-excitations [12].

A second source for the absence of remanence can be quantum tunneling of the magnetization. The likelihood of this process depends on the amplitude of the anisotropy energy variation in a plane perpendicular to the designated magnetization axis.

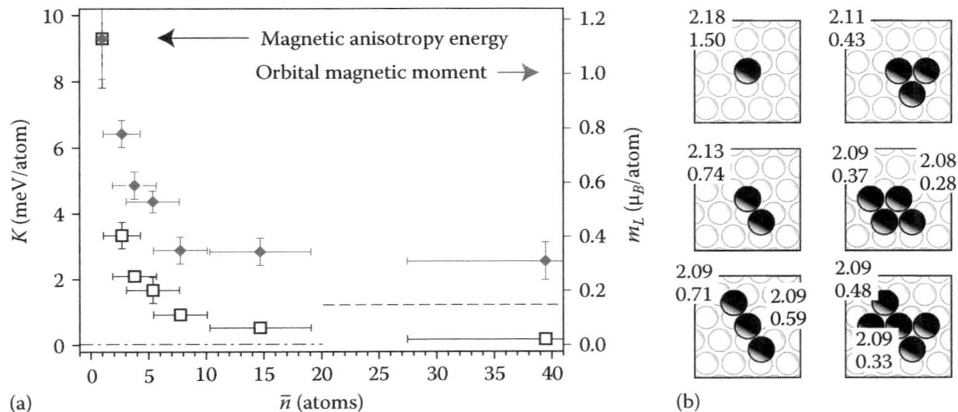

FIGURE 23.8 (a) Orbital moment m_L (diamonds) and anisotropy energy K (open squares) as a function of the average cluster size $\bar{n}$ for Co_n/Pt(111). For comparison, the dashed and dashed-dotted lines show the MCA energy per Co atom of the $L1_0$ CoPt alloy and hcp-Co, respectively. The error bars on the horizontal scale represent the standard deviation of the size distribution determined by STM. (b) m_S (top) and m_L (bottom) calculated using density functional theory for the differently coordinated atoms in the Co_n/Pt(111) clusters. (Adapted from Gambardella, P. et al., *Science*, 300, 1130, 2003.)

The present case of Co/Pt(111) has sixfold transverse symmetry implying higher order terms in the transverse anisotropy and thus the matrix element for magnetization reversal by tunneling is expected to be small. Therefore, XMCD measurements at $T < 2$ K, or SP-STM $M(H)$ curves recorded at $V_t < 10$ meV, might well reveal remanence for a single atom for this system.

XMCD experiments and ab initio calculations also show that m_L and the MCA of small Co_N clusters are dramatically sensitive to unit changes of the atomic coordination [2,14], as shown in Figure 23.8. This tendency, governed by lateral adatom–adatom interactions as well as substrate–adatom interactions, continues from adatoms to clusters and thin films until eventually the bulk values of $m_L = 0.15\ \mu_B$ and $m_S = 1.6\ \mu_B$ are reached.

23.4.3 Magnetic Anisotropy from STM Spin Excitation Spectroscopy

The first STM recorded differential conductance steps stemming from spin excitations were reported for Mn atoms adsorbed on an oxide monolayer on a metal surface [5]. This layer has been the self-limiting $Al_{10}O_{13}$ layer formed by exposing a NiAl(110) surface at high temperature to oxygen [69]. The conductance steps appeared only in the presence of an external magnetic field. Due to the small excitation energies involved, the measurements had to be performed at 0.6 K. The step energies were proportional to the magnetic field, $E_{\text{step}} = g\mu_B B$, yielding the Landé g-value for Mn atoms as a function of their adsorption site. The results were with $g = 1.88 \pm 0.02$ for Mn on the oxide and 2.01 ± 0.03 for Mn at the edge of the oxide close to the free electron value of 2.0023. The amplitudes of the conductance steps were with 20%–60% significant and very sensitive to the atomic environment. Mn atoms adsorbed directly onto NiAl(110) did not show the steps. Therefore, the insulating layer enhances the cross section for spin scattering. However, STM-SES also works for magnetic adatoms directly adsorbed onto a metal surface. Inelastic conductance steps with a height of 5% and the magnetic field shift identifying them as magnetic excitations have been reported for Fe/Cu(111) [51].

We outline how quantitative information on the anisotropy landscape and effective spin moments can be gained from SES taking different transition metal atoms on copper-nitride monolayer patches on a copper (100) surface as example [47,48]. Figure 23.9a shows an STM image of square Cu_2N-islands with a $c(2 \times 2)$ structure separated by clean Cu bands appearing as narrow vertical and wider horizontal ridges. The island pattern is typical for chemisorbed N on Cu(100) [70]. The island size is determined by an optimum between strain and edge energy, and their equidistant arrangement is established by long-range elastic interactions mediated by the substrate [71]. Note that the islands are not exactly forming a $c(2 \times 2)$ structure, STM measurements reveal that the Cu_2N lattice constant is slightly larger than $\sqrt{2}\, d_{\text{nn, Cu}}$ [72]. Density functional theory (DFT) calculations show that the N atoms are almost at the same height as the first atomic Cu plane [47]. One therefore speaks of a Cu_2 N-layer rather than of a N/Cu(100) adlayer. The adsorbates visible as protrusions are adsorbed Mn, Fe, and Co atoms.

Spectra taken on Mn show clear (10%–20%) conductance steps at very low energies. As determined from assembling atomic chains of these atoms with AFM order [46], Mn has a spin of $S = 5/2$, as in the gas phase. The fact that there is a conductance step at zero field, together with the fact that it shifts in the presence of a magnetic field, signifies that the atoms have a magnetocrystalline anisotropy energy. SES for different field directions can be analyzed in terms of the anisotropy energy landscape by using the spin-Hamiltonian, which we give here in the nomenclature introduced for molecular magnets:

$$H = g\mu_B \mathbf{B}\cdot\mathbf{S} + D\hat{S}_z^{\,2} + E\left(\hat{S}_x^{\,2} - \hat{S}_y^{\,2}\right). \tag{23.9}$$

The first term is the Zeeman energy, and the following two represent the axial and transverse anisotropies D and E, respectively. The assignment of the axes is such as to maximize $|D|$

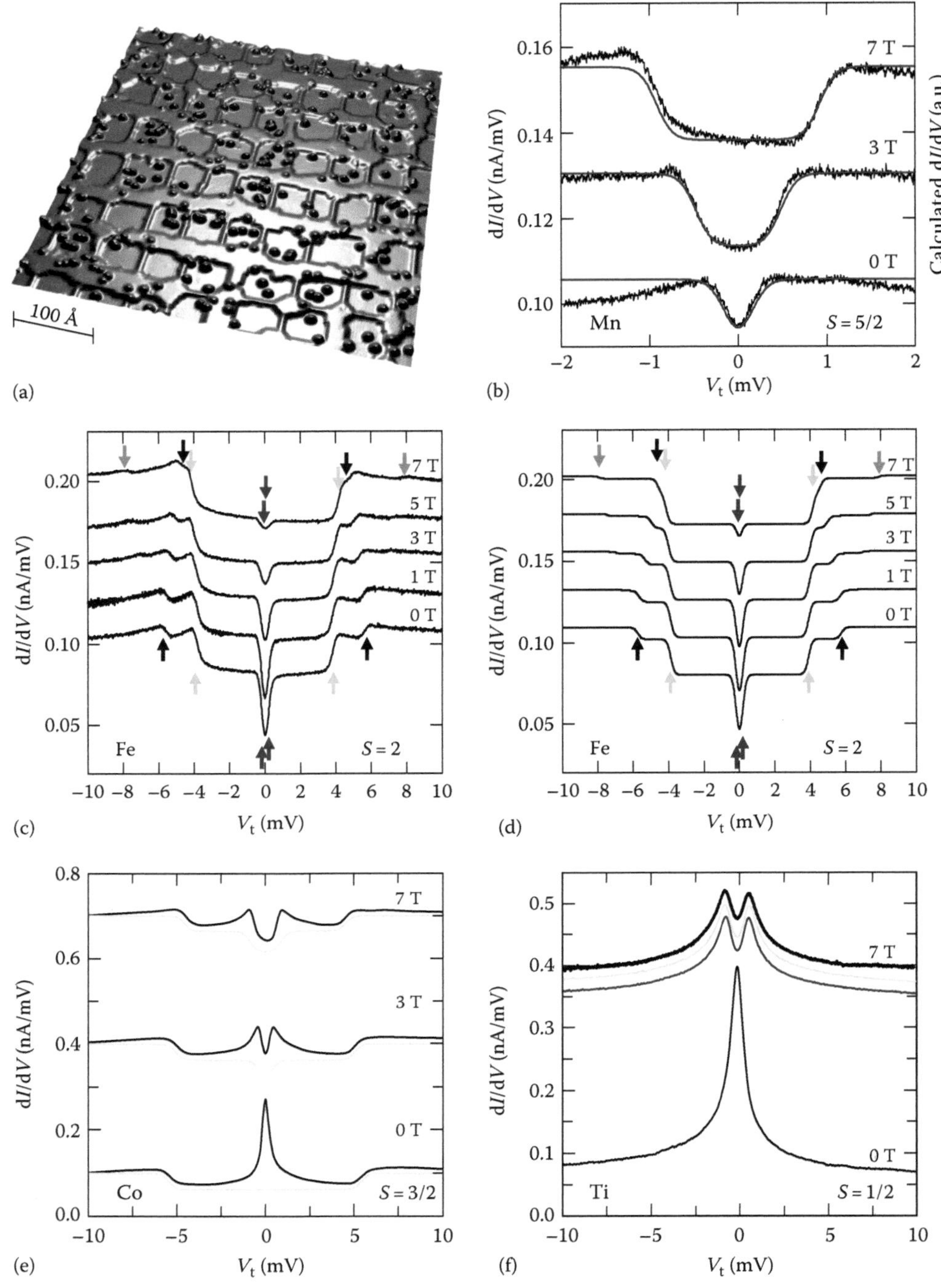

FIGURE 23.9 (a) Constant current STM image showing square Cu_2N-patches on Cu(100) appearing lower than the clean Cu ridges. The adatoms are Mn, Fe, and Co (500 Å × 500 Å, V_t = 10 mV, I_t = 0.5 nA, and T = 5 K). (b) Field-dependent dI/dV over Mn together with calculated spin-excitation spectrum in gray. (c) Spectra over Fe. (d) Calculated SES. (e) Spectra over Co together with calculated SES in light gray (for (b), (c), and (e) tip stabilized at V_t = 10 mV, I_t = 1.0 nA, $V_{mod, rms}$ = 20 μ V, f_{mod} = 800 Hz, and T = 0.5 K). (f) dI/dV for Ti showing the Kondo resonance and its isotropic splitting for magnetic field along three orthogonal directions (light gray in-plane, dark gray out-of-plane, $V_{mod, rms}$ = 50 μ V, f_{mod} = 745 Hz, and T = 0.5 K). The lower curves give absolute dI/dV, the upper ones have been offset for clarity. (Adapted from Otte, A.F. et al., *Nat. Phys.*, 4, 847, 2008; Hirjibehedin, C.F. et al., *Science*, 317, 1199–1203, 2007.)

and to yield $E > 0$. By diagonalization of Equation 23.9, one finds the eigenvectors Ψ_i and calculates the spin excitation spectrum considering the selection rule that initial and final state are connected by $\Delta m = \pm 1$ given by the spin-flip of the tunnel electrons. The resulting red curves in Figure 23.9b fit very well the energies and step heights for the shown out-of-plane field direction. Fitting these curves for all three orthogonal field directions and using $S = 5/2$ reveals that Mn adsorbed onto a Cu site on $Cu_2N/Cu(100) - c(2 \times 2)$ has out-of-plane easy axis anisotropy with $D = -0.039 \pm 0.001$ meV, a very small transverse term of $E = 0.007 \pm 0.001$ meV, and a Landé factor of $g = 1.90 \pm 0.01$.

One can reproduce the essential features neglecting E. This simplifies the picture as then the eigenstates are pure $m = -5/2$, $-3/2$, $-1/2$, $+ 1/2$, $+ 3/2$, and $+5/2$, with m being the z-projection of the magnetization. For $B = 0$, there are three sets of doubly degenerate energy levels. From low to high energy these are $m = \pm 5/2$, $\pm 3/2$, and $\pm 1/2$. At $T = 0.5$ K, the thermal population of the higher energy levels can be ignored. The step seen at 0 T therefore corresponds to the two transitions $m = \pm 5/2 \rightarrow \pm 3/2$. Their energy difference is $D((5/2)^2 - (3/2)^2) = 0.16$ meV in agreement with the observed excitation energy. For large B, all of the levels are non-degenerate and the levels are separated by much more than the thermal energy. Therefore, only the $-5/2$ ground state is populated and the step seen in the upper two curves marks the excitation from this state to $m = -3/2$. Transitions to higher lying levels are forbidden by conservation of the total angular momentum. The outward shift of the conductance steps with out-of-plane field is due to the field stabilizing the ground state. The expected shift is the Zeeman energy for $\Delta S = 1$ being 0.77 meV for a field of 7 T, thus positioning the step at 0.93 meV, again in agreement with observation.

Figure 23.9c shows the spin excitation spectrum for an Fe atom on an identical adsorption site as Mn. The zero field spectrum shows three with respect to E_F symmetric pairs of conductance steps located at 0.2, 3.8, and 5.7 meV. Applying a magnetic field along the in-plane direction where the Fe atom has no nearest N neighbors, the so-called hollow direction, some steps move and some out. Under the assumption that $S = 2$, all step positions and step heights for all field values and directions could perfectly be fitted by $g = 2.11 \pm 0.05$, $D = -1.55 \pm 0.01$ meV, $E = 0.32 \pm 0.01$ meV, and with z along the in-plane direction where the Fe atom has N atoms as nearest neighbors. The agreement with the calculated curves shown in Figure 23.9d is striking. Note that there is a fourth peak expected from the calculations (purple arrow), which can also be guessed from the 7 T spectrum. In order to compare D with the uniaxial anisotropy K, used in bulk, thin films and nanostructures and referred to in Section 23.3.1, one has to multiply D by S^2 and thus obtain anisotropies almost as large as the ones of Co/Pt(111) [2]. Easy axis anisotropy implies that the zero field ground state has large and identical weights in the $m = \pm 2$ eigenstates with very little weight also in $m = 0$ since the finite transverse E-term mixes states of different m. This gives rise to more than one excitation step. DFT calculations confirm the choice of $S = 2$ but also show that Fe pushes the underlying Cu atom deep below the first atomic plane and therefore forms an adsorption complex with the neighboring N atoms and the underlying Cu, having similarities with the configuration in molecular magnets [47].

The zero field spectrum of Co reproduced in Figure 23.9e exhibits conductance steps around ±6 meV, together with a prominent Kondo peak centered at E_F [48]. Analysis of the field-dependent step energies, identical to the one described earlier for Mn and Fe, leads to $S = 3/2$, $D = 2.75 \pm 0.05$ meV, and $g = 2.19 \pm 0.09$. The positive D-term signifies easy plane anisotropy. Therefore, the lowest energy states have small projection onto the designated z-axis, and the magnetic ground state is a twofold degenerate $m = \pm 1/2$ doublet. This is the necessary condition for the Kondo effect of a high-spin impurity as these two states are linked by $\Delta m = \pm 1$, enabling Kondo scattering of the conduction electrons. Mn and Fe are high-spin impurities with easy axis anisotropy and therefore the lowest energy states are separated by $\Delta m > 1$, excluding first-order Kondo scattering and naturally explaining the absence of a Kondo feature for these adsorbates. These low-T STM observations highlight the role of magnetic anisotropy in the Kondo effect for high-spin impurities.

The last example is a Ti atom having $S = 1/2$. The spectrum of Figure 23.9f shows a large Kondo peak with isotropic field splitting, but no indication of spin-excitation steps. Ti is adsorbed on the same site than the other transition metal atoms; thus, it is subject to the same crystal field that obviously induces no magnetocrystalline anisotropy in a low-spin impurity. We close this part on SES of single adatoms with the remark that the anisotropy deduced from the conductance steps for Fe/Cu(111) is in perfect agreement with the one inferred from single atom SP-STM magnetization curves [51].

We now turn to ensembles containing a few magnetic atoms and illustrate how textbook examples of low-dimensional magnetism, such as Heisenberg chains, can be assembled and studied by means of STM. Vertical atom transfer [73,74] was used to assemble straight Mn_n chains with the STM, again on Cu_2N/Cu(100). The chains displayed a striking parity dependence on their SES. Chains with even number of atoms n had no conductance step close to 0 eV, but large steps at several meV energy, while chains with odd n displayed zero-energy steps together with less pronounced ones at higher energy. The absence of low energy spin excitation in even chains implies a ground state with $S = 0$ and their presence in odd chains an $S \neq 0$ ground state. Therefore, the chains were AFM and hence a realization of Heisenberg chains with finite length [75]. The dimer conductance steps split up into three in an external magnetic field. Accordingly, they were attributed to transitions from the $S_{tot} = 0$ singlet to the $S_{tot} = 1$ triplet state with magnetic quantum numbers $m = 0, \pm 1$. The step positions were used to derive the exchange energy $J = 6.2$ meV from a Heisenberg Hamiltonian. The spin per Mn atom could be inferred to $S = 5/2$ from the position of the IETS step of Mn_3. Note that $S = 5/2$ is also the spin of a free Mn atom. From the known J and S values, the spin transitions for all chain lengths can be derived from a Heisenberg open-chain model, and the values were in excellent agreement with experiment for chain length up to six atoms. The most recent example of small clusters with AFM order are chains of Fe atoms on the same substrate that represent the smallest reported AFM with stable magnetization state [56]. Note that SES-STM has also been used to explore the magnetic anisotropy in surface adsorbed metal–organic molecules [76,77] and to probe super-exchange interaction in molecular magnets [78].

23.4.4 STM Spin Pumping

We illustrate in this section how STM can be used to manipulate the magnetic quantum state and access the spin relaxation times of individual atoms or molecules adsorbed onto surfaces. Successive tunnel electrons injected from a SP tip at time intervals smaller than the spin-relaxation time can pump the magnetic system through the momentum transfer during the course of inelastic spin excitations into higher excited states. This can best be observed if the magnetic system is prepared by an external magnetic field in a ground state from which excitation can only be done with one sign of Δm such that a suitable SP tip leads to a net excitation only for one sign of the tunnel voltage. The average time between successive electrons determines how many excitations are possible before relaxation and therefore the non-equilibrium spin-population. For high tunnel currents, this can lead to the inversion of the spin population compared with the one favored by the external field. The SP conductance through the magnetic impurity depends on its magnetic quantum state and therefore allows this state to be read out. Note that this dependence is not the regular SP-STM contrast described by the SP-local density of states of tip and sample, but it is due to SP tunnel electrons that interact with m but do not change it. The excitation and the simultaneous read-out of the magnetic state lead to a strong current dependence of the dI/dV signal that can be used to infer the spin-relaxation time τ_1 of excited states [58].

As schematically outlined in Figure 23.4 and explained in the corresponding section earlier, SP-STM tips give rise to asymmetric spin-excitation step heights in dI/dV. Figure 23.10a shows such asymmetries for SES recorded with a SP tip over Mn atoms adsorbed on Cu_2N, whereas these asymmetries are absent for a non-SP tip as seen in Figure 23.10b. The SP tip has been created by picking up a Mn atom from the surface to the tip apex, and the non-SP tip by dropping it off and picking up a Cu atom. Since Mn has a very weak out-of-plane anisotropy and $S = 5/2$,

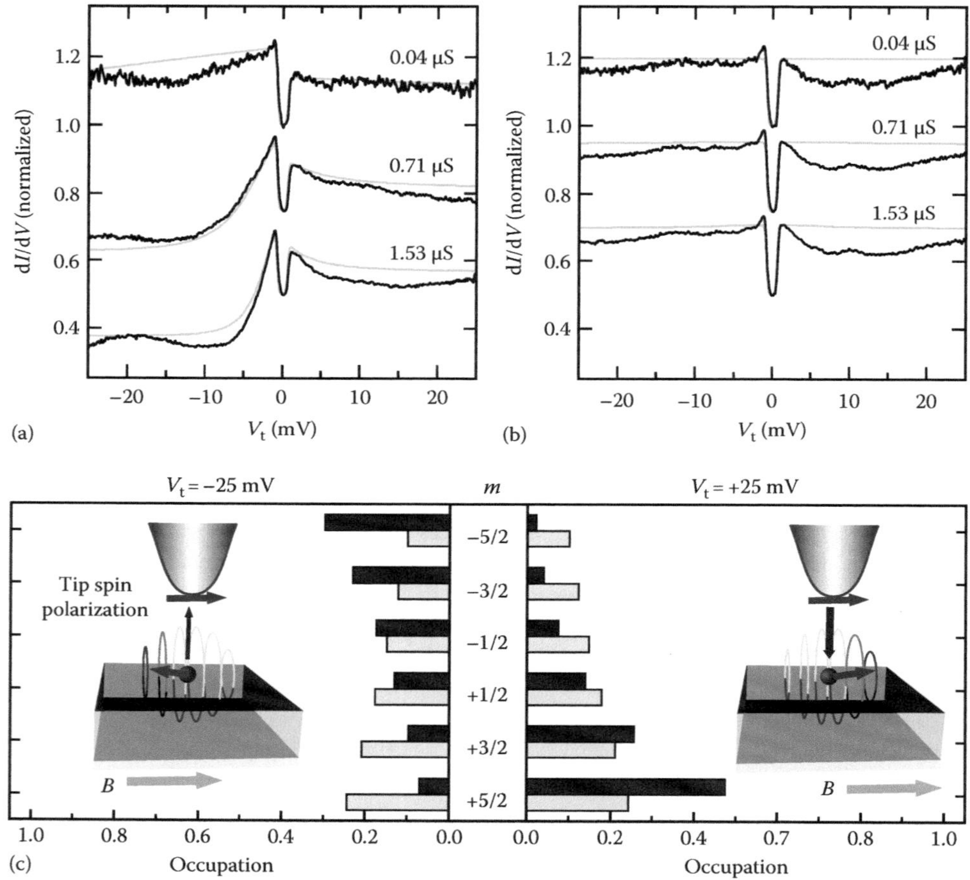

FIGURE 23.10 Spin-pumping for Mn/Cu_2N/Cu(100)-$c(2 \times 2)$. Tunnel spectra over the same Mn atom, (a) recorded with a spin-polarized and (b) with a non-spin-polarized tip (black measured, gray calculated spectra, $B = 7$ T in-plane, $T = 0.5$ K, dI/dV normalized to 1 at $V_t = 0$ V, spectra offset by −0.3 for clarity). The spectra are labeled by σ_0, the value of the conductance at $V_t = 0$. (c) Populations of magnetic quantum states for positive and negative tunnel bias and for SP (dark gray) and non-SP (light gray) STM tip. (Adapted from Loth, S. et al., *Nat. Phys.*, 6, 340, 2010.)

application of a small magnetic field in any direction prepares it in the $m = +5/2$ state. The tip atom is a paramagnetic impurity with low anisotropy that saturates at 1 T at 0.5 K, thus fields beyond this value fully magnetize tip and Mn atom alike. For negative polarity, the majority electrons can excite the Mn atom to the $m = +3/2$ state, while for positive polarity of the tunnel junction, this excitation can only be done by minority electrons. This difference results in the different inelastic conduction step heights in Figure 23.10a, which can directly be read out in terms of the spin-polarization of the tip, yielding $\eta = 0.24 \pm 0.04$ in (a) and evidently $\eta = 0$ in (b).

The negative polarity side of the spectra recorded with the SP tip levels off very strongly for high tunnel conductance, ($\sigma_0 \geq 0.71$ μS), reaching an asymptotic value already at $V_t = -10$ mV. This signifies that the Mn atom has a spin-population with large weight in the $m = -5/2$ state, where the magnetization points opposite to the field. For that state, the elastic conductance is 0.99 spin sensitive, explaining the large drop of the overall conductance. Note that also the conductance on the positive polarity side changes slightly with increasing tunnel current, being due to spin-excitations by tip minority electrons. Figure 23.10c shows the spin population for the SP (dark gray bars) and the non-SP tip (light gray bars) for both junction polarities. Note that, in the present case, all excitations have the same energy, therefore the different m states form a spin ladder with equidistant states and all excitations appear at a single energy. This is different for a Mn dimer that has been used to distinguish the different excitations [58].

Fits of the I_t decay at $V_t < -V_{exc}$ give lifetimes τ_1 of the first excited state, i.e., for $m = 3/2 \rightarrow m = 5/2$, of $\tau_1 = 0.25 \pm 0.04$ ns at 7 T and $\tau_1 = 0.73 \pm 0.10$ ns at 3 T. The spin relaxation takes place by interaction with substrate conduction electrons. Their number is proportional to the energy of the inelastic excitation, which is by Zeeman energy proportional to the magnetic field. To first order, the spin relaxation time τ_1 is therefore expected to be inversely proportional to B, which is indeed what is found. Spin-phonon coupling plays no significant role as decay mechanism of the magnetic state.

Another example of the manipulation of the magnetic quantum state of an adatom with a magnetic STM tip has been through exchange interaction between adatom and tip [79]. Note also a theory paper discussing the spin torque induced change of magnetization of adatom [80], which has experimentally been realized and discussed earlier for Mn/Cu_2N.

23.4.5 Magnetic Relaxation Times from STM Pump–Probe Experiments

The time resolution of STM is limited by the bandwidth of current-to-voltage amplifier used to measure the tunnel current. This bandwidth is sufficient to resolve the flicker noise caused by thermal magnetization reversal of magnetic islands in SP-STM [38,81,82]. However, it is orders of magnitude slower than the relaxation times of magnetic quantum states of individual atoms and molecules. As sketched in Figure 23.11a, it suffices to use a high bandpass electronics for the tunnel voltage [59]. As discussed earlier, a magnetic system saturated by an external field can be excited to higher lying states by injection of SP electrons with energy beyond the threshold for spin excitations, eV_{thr}.

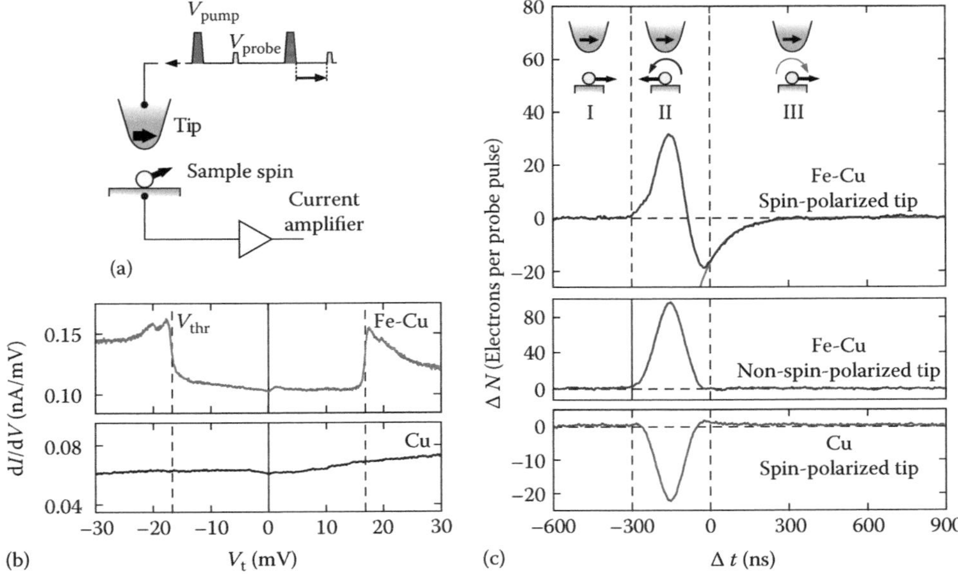

FIGURE 23.11 STM pump–probe experiments on an FeCu dimer on Cu_2 N/Cu(100)–$c(2 \times 2)$. (a) Sketch of the sequence of tunnel voltage pulses, the delay time Δt is from the end of the pump to the beginning of the probe pulse, as indicated by the arrow. (b) SES steps at ±16.7 mV in dI/dV over the dimer, whereas the reference spectrum over Cu adatom shows no inelastic features, both spectra have been recorded with a non-polarized tip. (c) $\Delta N(\Delta t)$: = $N(\Delta t) - N(-600$ ns) for spin-polarized tip above FeCu dimer (upper panel), with a non-polarized tip above the FeCu dimer (middle), and with a spin-polarized tip above the Cu atom (lower panel) ($V_{pulse} = 36.5$ mV, FWHM = 100 ns, rise and fall times 50 ns, $V_{probe} = -4.0$ mV). (b) and (c) $B = 7$ T out-of-plane and $T = 0.6$ K. (Adapted from Loth, S. et al., *Science*, 329, 1628, 2010.)

These electrons may stem from a paramagnetic tip that is saturated by the same external field. A pump pulse with $V_t > V_{thr}$ excites the system and a probe pulse with $V_t < V_{thr}$ follows after a delay time to read out the magnetic state by its m-dependent conductance. Inside the tunnel, voltage is zero such that the magnetic state outside the pulses does not contribute to the tunnel current. The DC tunnel current averaged over many pump–probe sequences depends on the delay time since the mean conductance depends on whether the magnetic system is still in its excited state when the probe pulse comes, or whether it has had time to decay into the ground state.

This method has been demonstrated on an FeCu dimer on $Cu_2N/Cu(100)$-$c(2 \times 2)$ [59]. This dimer has $S = 2$ and easy axis magnetic anisotropy as single Fe atoms on this surface; however, the dimer anisotropy is close to out-of-plane instead of in-plane, and D is of much larger magnitude. The latter causes the first spin excitation energy to be at much higher tunnel voltage, as seen from the inelastic conductance steps in Figure 23.11b V_{thr} = 16.7 mV. Figure 23.11c shows the difference $\Delta N(\Delta t)$ between the number of electrons detected per probe pulse for a given delay time $N(\Delta t)$ and this number for a delay time of $\Delta t = -600$ ns, i.e., for a probe pulse preceding the pump pulse. The upper panel shows this quantity for an SP tip above the dimer. In region I $\Delta N(\Delta t) = 0$ since the dimer is in its ground state, the pump pulse influences the system long after the probe pulse. In region II, there is a large positive peak for -300 ns $\leq \Delta t \geq -100$ ns, followed by a strong decrease in the conductance. The positive peak is caused by the overlap of pump and probe pulse, their sum is applied to the tunnel junction. The conductance decrease is caused by the magnetization of the dimer pointing opposite to the one of the tip during the probe pulse and therefore the tunnel-magnetoresistance of the junction is high. For positive delay times, the conductance recovers exponentially with a time constant of $\tau_1 = 87 \pm 1$ ns, as deduced from the fit of ΔN in region III of Figure 23.11c. This is the dimers spin-relaxation time at $T = 0.6$ K and $B = 7$ T in-plane field.

The reference measurements shown in the lower two panels of Figure 23.11c demonstrate that the exponential decay for $\Delta t \geq 0$ is only obtained for a SP tip above a magnetic impurity. An identical pump–probe sequence applied with a non-polarized tip above the dimer gives only the positive Gaussian peak but no sign change and exponential decay since the tunnel junction is insensitive to the orientation of the dimer magnetization. The $\Delta N(\Delta t)$-curve obtained with an SP-tip above a Cu atom shows a dip as a sign of the cross-correlation between pump and probe pulse. When both pulses overlap, $|V_t|$ is higher and the associated I_t change depends on d^2I/dV^2, which seems to have opposite signs for FeCu and Cu at the relevant energy. The authors ensured that the STM observation does not influence the dimer's relaxation time. τ_1 is independent of V_{pump}, as long as $|V_{pump}| \geq |V_{thr}|$; it is further independent of the pump pulse length and tip-sample distance. Both polarities of V_{pump} work; however, negative polarity works better due to spin momentum transfer from the tip, increasing the spin excitation cross section.

The state from which the dimer decays within τ_1 is the $m = -2$ state. This can be explained as follows. The highest energy excitation is from the ground state $m = +2$ to $m = +1$ requiring 16.7 meV. Since all other excitations require less energy, all states, including $m = 0$, are accessible once $|V_{pump}| \geq |V_{thr}|$. The rate-limiting step in the de-excitation is the one with the highest barrier, i.e., $m = -2 \rightarrow +2$. The relaxation mechanism is likely magnetic tunneling and not thermally assisted due to the absence of a T-dependence of τ_1 for $1\text{ K} \leq T < 10$ K at $B = 5.5$ T. Only from 10 K onward, one sees evidence for the decrease of τ_1, which can be due to onset of thermal reversal. The B-dependence of τ_1 is an increase up to 6 T and a decrease beyond this value. The first is due to the reduced tunnel matrix element by increased Zeeman splitting of the $m = \pm 2$ states and the second by transverse field component mixing of states.

Altogether, these measurements of single adatoms and of very small clusters, being them taken out by means of XMCD for mono-disperse ensembles or on individual objects by means of STM, show that an increase by one to two orders of magnitude in magnetic anisotropy energy with respect to bulk or 2D films can be obtained by reducing the size of magnetic particles to a few tens of atoms or less on suitable substrates. This is very beneficial to increase the stability of the magnetic moment, according to Equation 23.5. However, while this holds for a single atom, it is obvious that the overall stability of the magnetization of an N-atom cluster is governed by the sum over the MCA energy contributions from each atom. As more atoms are assembled together to fabricate clusters with a large total magnetic moment and a total MCA strong enough to stabilize FM behavior against thermal fluctuations, this gain is countered by the decrease of the MCA per atom with increasing size (Figure 23.8a). The problem, however, can be circumvented by noting that the atomic coordination rather than the absolute particle size is the key parameter that governs the magnitude of the MCA, m_L, and m_S, see, e.g., Figure 23.8b). Surface supported atomic-scale structures where the shape and composition are tuned so as to control the coordination of the magnetic atoms and maximize useful interface effects, such as in core-shell 2D particles [83], 1D atomic wires [4,84], 2D metal–organic networks [85], and single molecules [6,86,87], offer very interesting opportunities to exploit such effects. Examples for the latter two are discussed in the next section.

23.5 Magnetism of Molecular Networks and Single Molecules on Surfaces

Individual atoms are difficult to arrange in thermally robust regular patterns on surfaces. Moreover, their magnetic properties are dominated by the type of substrate because of electronic hybridization. These problems can be overcome by embedding the magnetic atoms into a planar molecular framework [85]. However, whereas the chemistry of metal–organic complexes is well established, little is known about the electronic and magnetic properties of atomically thin metal–organic grids interfaced with a metallic substrate or electrode. We analyze here two approaches to this problem, one relying on the supramolecular synthesis of 2D metal–organic networks on appropriate surfaces and the other on the deposition of integral molecules such as metal phthalocyanines [7,27,88,89] and single molecule magnets (SMMs) [6].

23.5.1 Self-Assembled Supramolecular Spin Networks

The codeposition of transition metal ions and organic ligands on crystalline surfaces offers the potential to design supramolecular grids with programmable structural and chemical features, where the interaction with the substrate is used to stabilize a planar geometry [90,91]. For example, Fe atoms coadsorbed with terephthalic acid (TPA) molecules on Cu(100) in ultra-high-vacuum constitute a prototypical 2D hetero-assembled system forming a variety of mono- and bi-nuclear network structures, whose morphology is determined by the Fe:TPA stoichiometry, substrate symmetry, and annealing temperature [90]. Figure 23.12a shows a hard sphere model of a square planar $Fe(TPA)_4$ network obtained by sequential deposition of TPA and Fe on Cu(100), where each Fe atom is coordinated to four TPA molecules through Fe-carboxylate bonds, with the supramolecular $Fe(TPA)_4$ units organized in a (6 × 6) unit cell with respect to the underlying Cu lattice [85]. Weak hydrogen bonding interactions between the complexes favor long-range order extending over entire terraces of the substrate. The resulting superlattice of individual Fe atoms has perfect 15 × 15 Å periodicity (Figure 23.12b). XAS shows that the $3d$-states of the Fe ions in the molecular network are highly localized compared to single Fe adatoms adsorbed on Cu(100), and have an almost pure d^6 character (Fe^{2+}), with maximum 14% d^7 weight [85]. These results indicate that coordination bonds have formed between the Fe centers and carboxylate ligands, with partial decoupling of Fe from the metal substrate. Such bonds, which involve Fe $3d$ and O $2p$ states, explain the fourfold coordination geometry as well as the thermal stability of these complexes. XAS also proves that the $Fe(TPA)_4$ network is formed by high-spin Fe^{2+} ions, which is interesting because Fe^{2+} is expected to favor large MCA (i.e., zero field splitting) and anisotropic g-factors. Indeed, contrary to Fe atoms on Cu(100) that have very small MCA, XMCD measurements of $Fe(TPA)_4$ show a very strong anisotropic magnetization with in-plane easy axis (Figure 23.12c).

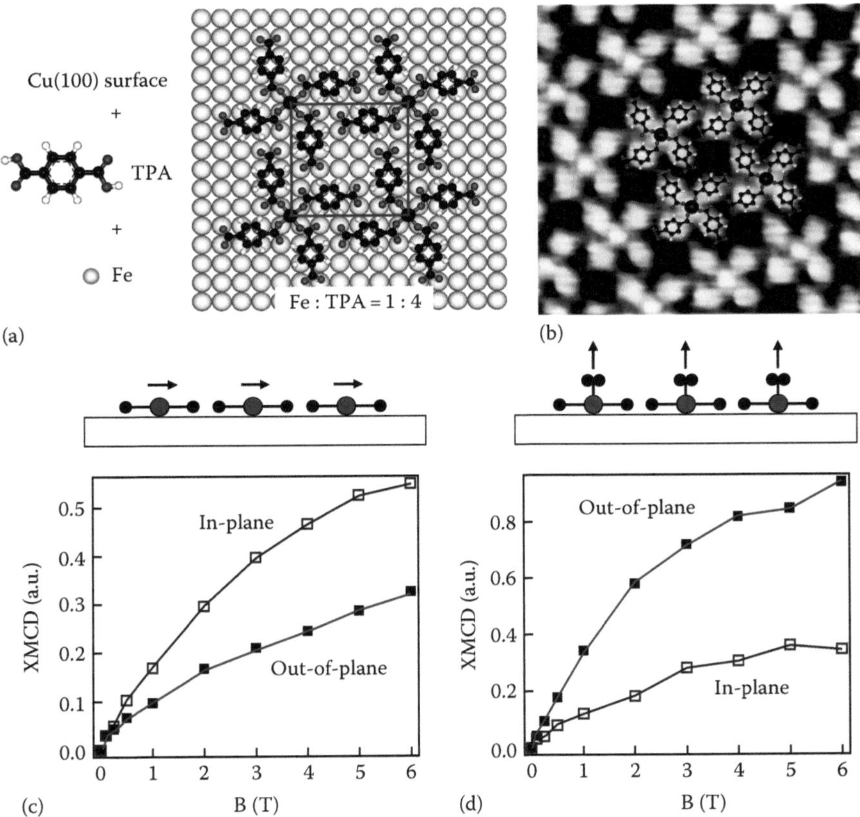

FIGURE 23.12 Planar supramolecular networks of $Fe(TPA)_4$ complexes self-assembled on Cu(100). (a) Ball and stick model and (b) STM image of the $Fe(TPA)_4$ array. Blue dots indicate the position of Fe atoms, red O, black C, white H. (c) Element-selective magnetization curves of the Fe centers of $Fe(TPA)_4$ and (d) O_2-$Fe(TPA)_4$ measured by XMCD at $T = 8$ K with magnetic field applied out-of-plane (θ = 0°, full symbols) and close to the in-plane direction (θ = 70°, open symbols). The data points represent the Fe XMCD intensity integrated over the L_3 edge and normalized by the corresponding L_3 integrated XAS intensity. The diagrams show the easy axis magnetization direction. (Adapted from Gambardella, P. et al., *Nat. Mater.*, 8, 189, 2009.)

Such networks are of interest not only as a way to organize individual spins in a regular pattern but also because of the possibility to tune their magnetic susceptibility by chemical means. The coordination of the metal ions with lateral molecular ligands yields stable but unsaturated coordination bonds, which allow for the chemical modification of the electronic and magnetic properties of the magnetic atoms independently from the substrate. This was shown by exposing an $Fe(TPA)_4$ network to O_2, which binds as an axial ligand on top of the active Fe sites [85]. A prolonged exposure to O_2 results in a saturated O_2-$Fe(TPA)_4$ network that has distinct properties with respect to $Fe(TPA)_4$. Although the formal Fe oxidation state remains 2^+, the symmetry of the ligand field acting on Fe changes from square-planar in $Fe(TPA)_4$ to pyramidal in O_2-$Fe(TPA)_4$ and electron localization effects increase gradually with the number of ligands, favoring electronic decoupling of the Fe atoms from the substrate. Density functional calculations show that the formation of the supramolecular complexes is accompanied by a substantial increase of the Fe-substrate distance, calculated as 2.32 Å for individual Fe atoms on Cu(100), 2.71 Å for $Fe(TPA)_4$, and 3.32 Å for O_2-$Fe(TPA)_4$ [85]. XMCD reveals the presence of large local magnetic moments on Fe, and in particular of increasing orbital moments going from Fe_1/Cu(100) (0.18 ± 0.03 μ_B), to $Fe(TPA)_4$ (0.42 ± 0.06 μ_B), and to O_2-$Fe(TPA)_4$ (0.55 ± 0.07 μ_B), measured at 8 K parallel to the easy magnetization direction. Most importantly, angle-dependent XMCD measurements show that O_2 adsorption at the Fe sites drives an abrupt spin reorientation transition, rotating the Fe easy axis out-of-plane (Figure 23.12d). This easy axis switch can be explained as the axial ligand induces a change of the Fe ground state from A_{1g} to E_g in O_2-$Fe(TPA)_4$, as expected based on symmetry arguments. The E_g term is an orbitally degenerate doublet with nonzero m_L pointing along the principal symmetry direction, which explains the tendency of O_2-$Fe(TPA)_4$ to magnetize out-of-plane, together with its enhanced orbital moment compared to $Fe(TPA)_4$.

Controlling the magnetic anisotropy of spin networks independently from the substrate is a key issue in the development of molecule–metal interfaces for spintronic applications, both at the level of single molecules and extended layers. With respect to bulk molecular crystals, the planar and open coordination structure of 2D supramolecular array make such systems extremely sensitive to ligand modifications, providing a handle to the spin orientation and enhanced chemical sensitivity of the magnetization. The capability to fabricate 2D arrays of monodisperse spin centers with nanometer spacing, and to understand and control their magnetic properties at the interface with a metal substrate, constitutes a basic step toward the exploitation of single spin phenomena in small scale devices.

23.5.2 Arrays of Metal Phthalocyanines

A different strategy to isolate and arrange single spins on surfaces is to directly sublimate magnetic molecules by means of thermal evaporation in vacuum. This works particularly well for small and compact molecules, which are usually chemically stable and can withstand temperatures of the order of a few hundreds of degrees C. Metal phthalocyanines (MPc) constitute a model system for investigating the properties of magnetic molecules on surfaces using this approach [7,27,88,89,92,93]. MPc are well-known metal–organic complexes with applications in organic electronics and photovoltaics. They are formed by a central metal ion (M) coordinated by an aromatic macrocyclic ligand (Pc), as shown in Figure 23.13a. More than 70 metal ions have been found to be able to coordinate to the Pc ligand, giving rise to a broad variety of chemical, electronic, and magnetic properties. When deposited on metal surfaces, MPc adopt a flat adsorption geometry placing both the M and Pc species in contact with the substrate. Moreover, van der Waals molecular interactions and spontaneous registry with the substrate lead to the self-assembly of ordered monolayers [77], as shown in Figure 23.13b for CuPc on Ag(100). The lattice spacing of a monolayer of MPc is about 15 Å, similar to that of $Fe(TPA)_4$.

As for single metal atoms, the first question to address in this case is if and how much of the pristine MPc magnetic moment survives adsorption on a substrate. There is no unique answer to this point, depending on the type of substrate as well as on the symmetry of the unoccupied d-states of the M ions [27,88,89,93]. It is generally believed that charge transfer from the substrate to metal d-states extending perpendicularly to the Pc plane leads to total or partial quench of the molecular magnetic moment. However, this is but a partial picture of the interaction between surface and molecules. We present here a case study of CuPc deposited on Ag(100) to highlight some additional aspects. According to XMCD measurements, CuPc complexes forming a compact molecular layer on Ag(100) conserve their gas-phase magnetic moment, $m_S = 1$ and $m_L = 0.1\ \mu_B$, corresponding to a Cu^{2+} ion with a $b_{1g}(d_{x^2-y^2})$ ground state, $S = 1/2$, and small SOC-induced orbital moment [27]. This agrees with the weak Cu-Ag hybridization expected for the $d_{x^2-y^2}$ orbital of Cu due to its planar symmetry. However, STM measurements reveal a delocalized Kondo resonance on isolated CuPc molecules (Figure 23.14), evidencing the presence of an additional spin on CuPc induced by charge transfer from Ag to the $2e_g$ orbital of the Pc ligand (the gas-phase lowest unoccupied molecular orbital of CuPc) [7]. Thus, the magnetic moment of single CuPc molecules increases rather than decrease after adsorption, leading to a triplet ground state with $S = S_M + S_{Pc} = 1$ (Figure 23.14a). A similar phenomenon occurs on NiPc, which is diamagnetic in the gas phase, leading to $S = S_{Pc} = 1/2$ [7].

STM also offers the possibility to manipulate individual molecules with atomic-scale precision (Figure 23.14b). Molecular structures of arbitrary shape and size can be assembled in this way, such as the 3 × 3 "molecular sudoku" (shown in Figure 23.14c). Besides playing LEGO with molecules, such an approach is useful to study the effect of lateral interactions between molecules on their electronic structure. The spectroscopy maps reported in Figure 23.14d reveal that the Kondo resonance (−6 meV) due to the unpaired spin found in the $2e_g$ Pc orbital disappears in molecules with more than two nearest neighbors due to the gradual upshift of the $2e_g$ state from about −0.3 eV in single CuPc (not shown)

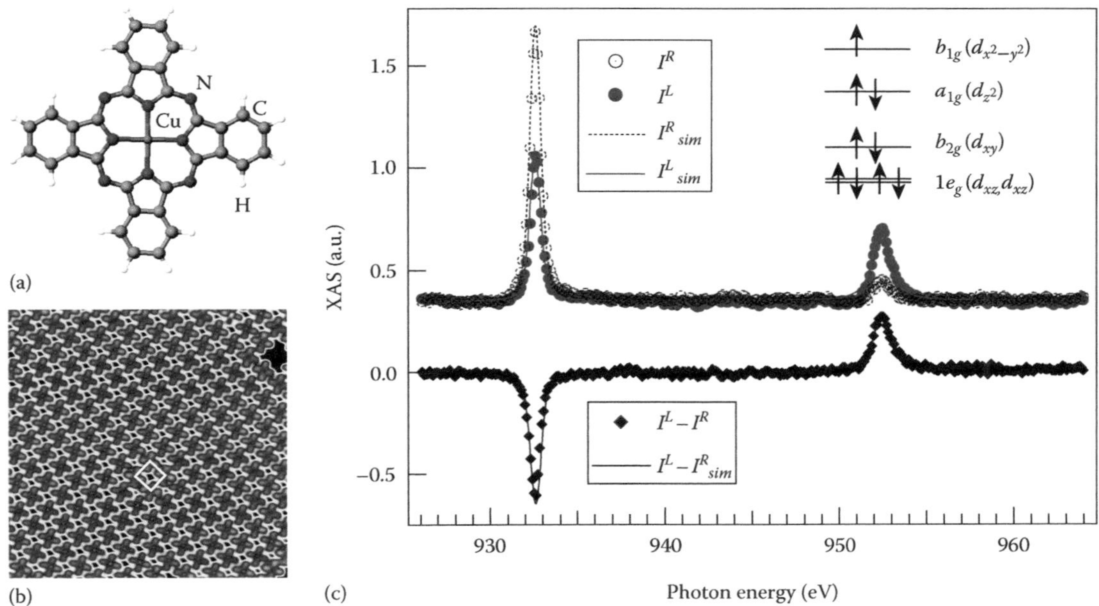

FIGURE 23.13 (a) Ball and stick model of CuPc. (b) STM image of one monolayer CuPc on Ag(100), image size 170 Å^2 × 170 Å^2. The square indicates the CuPc unit cell. (c) Circularly polarized $L_{2,3}$ XAS and XMCD of one monolayer CuPc/Ag(100) recorded at normal incidence at $B = 5$ T and $T = 6$ K. Solid and dashed lines are simulated spectra using ligand field multiplet theory, which gives the energy diagram of the Cu 3d-states shown in the inset. (Adapted from Stepanow, S. et al., *Phys. Rev. B*, 82, 014405, 2010.)

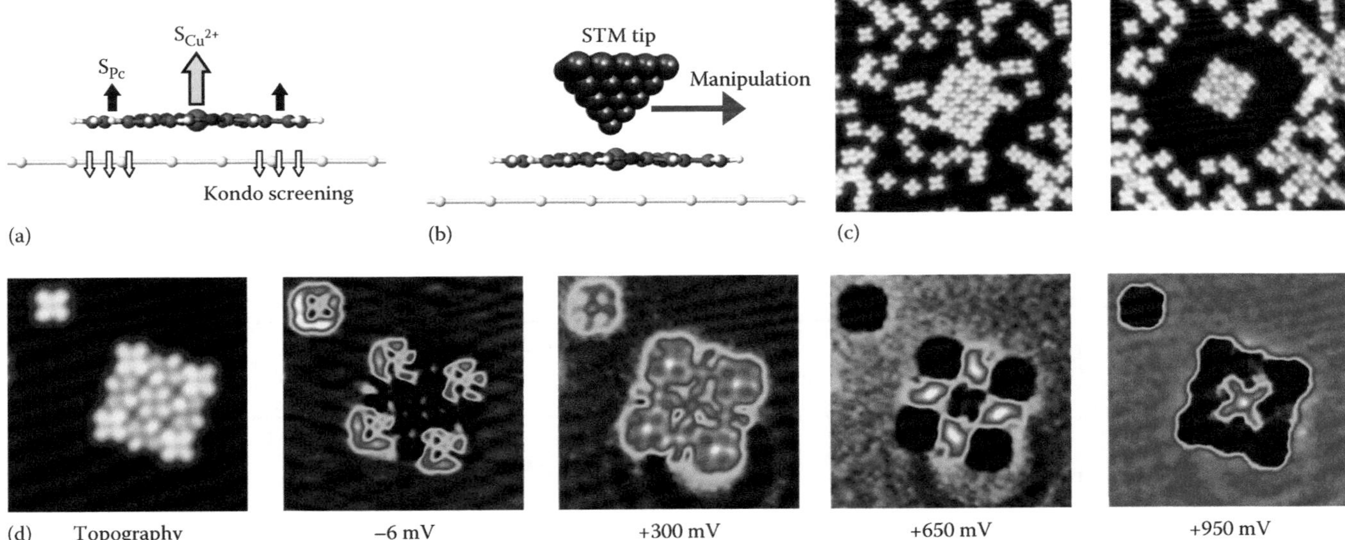

FIGURE 23.14 (See color insert.) (a) Diagram showing the spin distribution and couplings of CuPc induced by adsorption on Ag. (b) Schematic of molecule manipulation using the tip of an STM. (c) STM images showing the fabrication of a 3 × 3 CuPc cluster. (d) Topography of the cluster and spectroscopy maps (d^2I/dV^2 at −6 meV and dI/dV for the remaining maps) recorded at different energy at $T = 4.8$ K. Note that the Kondo resonance at −6 meV can be observed only for molecules with the number of lateral bonds smaller than three. The maps also show the correlation between molecular coordination, energy shift of the lowest unoccupied molecular orbital, which is responsible for the disappearance of the ligand spin. (Adapted from Mugarza, A. et al., *Nat. Commun.*, 2, 490, 2011.)

to +0.95 eV in the fourfold coordinated CuPc at the center of the cluster. The CuPc spin changes accordingly, from $S = 1$ to 1/2 [7].

Thus, the interaction between molecules and metal substrates brings about substantial changes to the molecular electronic structure and, with it, to the magnetic moment and electrical conductance. As demonstrated here, such changes are not only due to the symmetry-allowed matching of substrate and molecular orbitals but also to charge transfer effects. Lateral interactions between molecules may play a role analogous to that of an electrostatic gate, inducing strong shifts of the molecular orbitals with respect to the Fermi level of the substrate. The magnetic moment as well as the spatial extension of the spin density in a small molecular cluster can depend on the molecule position.

23.5.3 Single Molecule Magnets on Surfaces

The molecular systems discussed earlier are essentially paramagnets. Although some molecules present large MCA, this is not sufficient to induce the blocking of the magnetic moment at reasonable temperatures. Some molecular species, however, usually including multiple transition-metal sites or rare-earth ions, present MCA energy barriers so large that their relaxation time (Equation 23.5) increases so much at low temperature that they behave similar to a ferromagnet [94]. For this reason, they are called SMMs and represent ideal candidates for both magnetic storage and quantum computing applications [95,96]. Examples are the archetypal Mn_{12} compound [94] and mononuclear Tb double-decker complexes ($TbPc_2$) [97], whose relaxation time becomes slow, compared to the time scale of observations, below a few degrees K. Measurements of $TbPc_2$, for example, show that magnetic hysteresis is measured at $T \leq 2$ K, but not above [87,98,99]. Increasing the magnetic stability independently of temperature is thus one of the greatest challenges faced by SMMs.

A useful approach to stabilize the magnetic moment of paramagnetic molecules against thermal fluctuations is that of depositing them on FM substrates. This has been demonstrated in the case of metal porphyrins and phthalocyanines using both XMCD and SP-STM [86,100–102]. However, because of the close proximity of the metal ions to the substrate, the magnetic moment of these molecules couples rigidly to the substrate magnetization, making it impossible to control their magnetic state independently from the substrate. SMMs such as $TbPc_2$, however, appear to behave in a different way [6].

Figure 23.15 shows a schematic of $TbPc_2$ deposited on FM Ni films with either out-of-plane (a) or in-plane (b) MCA. XMCD measurements allow one to probe the magnetization of Tb separately from Ni. If the easy magnetization axis of $TbPc_2$, which is out-of-plane, coincides with that of the Ni film, the magnetic moment of the molecule is effectively stabilized by the interaction with the substrate, resulting in a square magnetization hysteresis curve with nearly saturated magnetic remanence at zero applied field (Figure 23.15a). Finite magnetic remanence persists up to 100 K, a temperature that is two orders of magnitude higher compared to isolated $TbPc_2$ [6]. Depending on the strength of the applied magnetic field, we observe that both antiparallel and parallel magnetic configurations can be reached, as the Zeeman interaction compensates and eventually overcomes the exchange coupling between Tb and Ni. Moreover, if the easy

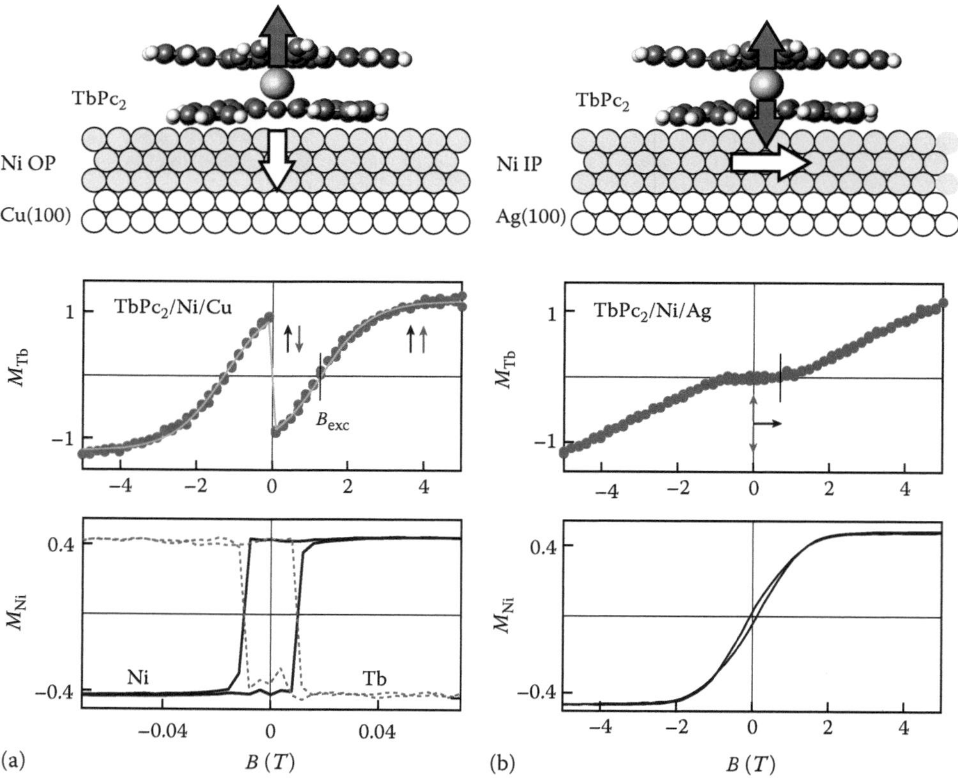

FIGURE 23.15 $TbPc_2$ single molecule magnets coupled to ferromagnetic Ni films with (a) out-of-plane and (b) in-plane easy magnetization axis. The plots show element-resolved out-of-plane magnetization loops of Ni (blue) and Tb (red) obtained by XMCD. The arrows indicate the relative orientation of the molecule and substrate magnetic moments. (Adapted from Lodi-Rizzini, A. et al., *Phys. Rev. Lett.*, 107, 177205, 2011.)

magnetization axis of the Ni film is orthogonal to that of $TbPc_2$, pronounced frustration effects are observed as the molecule magnetization cannot align with the substrate at equilibrium, and exhibits zero remanence at zero field (Figure 23.15b). Given that Tb and Ni are physically separated by a Pc ligand, an indirect exchange mechanism mediated by electrons hopping back and forth between the Pc macrocycle and Tb on one side and Ni on the other (Tb-Pc-Ni superexchange) may be responsible for the magnetic coupling. The strength of this coupling, which is in the meV range, can be tuned by electron or hole doping of the molecule–substrate interface, which is expected to change the occupation of the Pc electron orbitals [6]. This behavior also shows how the interface chemistry and magnetic response are intimately related in such systems.

In principle, SMMs can be used to store one bit of information in an extremely small volume or act as molecular-scale spin filters and injectors. The results reported earlier demonstrate enhanced thermal stability of the $TbPc_2$ magnetic moment and the possibility to orient it parallel or antiparallel to a macroscopic FM layer. Thus SMMs behave as coupled but separate magnetic units from an underlying FM substrate, which is a prerequisite for exploiting their spin as a state variable in future devices.

Acknowledgments

We gratefully acknowledge S. Blügel, C. F. Hirjibehedin, S. Loth, A. J. Heinrich, J. Wiebe, and R. Wiesendanger for allowing us to reproduce figures from their work. We acknowledge C. Carbone, A. Mugarza, and S. Stepanow for interesting discussions and C.-L. Bandelier for preparation of the figures. P.G. received financial support through the Ministerio de Ciencia e Innovación (MAT2010-15659), Agència de Gestió d'Ajuts Universitaris i de Recerca (2009 SGR 695), and European Research Council (StG 203239 NOMAD).

References

1. H. Brune and P. Gambardella, Magnetism of individual atoms adsorbed on surfaces, *Surf. Sci.* 603 (2009) 1812.
2. P. Gambardella, S. Rusponi, M. Veronese, S. S. Dhesi, C. Grazioli, A. Dallmeyer, I. Cabria et al., Giant magnetic anisotropy of single Co atoms and nanoparticles on Pt, *Science* 300 (2003) 1130.
3. P. Gambardella, S. S. Dhesi, S. Gardonio, C. Grazioli, P. Ohresser, and C. Carbone, Localized magnetic states of Fe, Co, and Ni impurities on alkali metal films, *Phys. Rev. Lett.* 88 (2002) 047202.
4. P. Gambardella, A. Dallmeyer, K. Maiti, M. C. Malagoli, S. Rusponi, P. Ohresser, W. Eberhardt, C. Carbone, and K. Kern, Oscillatory magnetic anisotropy in one-dimensional atomic wires, *Phys. Rev. Lett.* 93 (2004) 077203.
5. A. J. Heinrich, J. A. Gupta, C. P. Lutz, and D. M. Eigler, Single-atom spin-flip spectroscopy, *Science* 306 (2004) 466–469.
6. A. Lodi-Rizzini, C. Krull, T. Balashov, J. J. Kavich, A. Mugarza, P. S. Miedema, P. K. Thakur, V. Sessi, S. Klyatskaya, M. Ruben, S. Stepanow, and P. Gambardella, Coupling single molecule magnets to ferromagnetic substrates, *Phys. Rev. Lett.* 107 (2011) 177205.
7. A. Mugarza, C. Krull, R. Robles, S. Stepanow, G. Ceballos, and P. Gambardella, Spin coupling and relaxation inside molecule? metal contacts, *Nat. Commun.* 2 (2011) 490.
8. L. Zhou, J. Wiebe, S. Lounis, E. Vedmedenko, F. Meier, S. Blügel, P. H. Dederichs, and R. Wiesendanger, Strength and directionality of surface RKKY-interaction mapped on the atomic scale, *Nat. Phys.* 6 (2010) 187.
9. W. Wernsdorfer, Classical and quantum magnetization reversal studied in nanometer-sized particles and clusters, *Adv. Chem. Phys.* 118 (2001) 99.
10. W. Wernsdorfer, Quantum dynamics in molecular nanomagnets, *C. R. Chimie* 11 (2008) 1086.
11. J. S. Miller and D. Gatteschi, Molecule-based magnets, *Chem. Soc. Rev.* 40 (2011) 3065.
12. A. A. Khajetoorians, J. Wiebe, B. Chilian, and R. Wiesendanger, Realizing all-spin-based logic operations atom by atom, *Science* 332 (2011) 1062.
13. C. Carbone, S. Gardonio, P. Moras, S. Lounis, M. Heide, G. Bihlmayer, N. Atodiresei et al., Self-assembled nanometer-scale magnetic networks on surfaces: Fundamental interactions and functional properties, *Adv. Funct. Mater.* 21 (2011) 1212.
14. P. Gambardella, S. Rusponi, T. Cren, N. Weiss, and H. Brune, Magnetic anisotropy from single atoms to large monodomain islands of Co/Pt(111), *C. R. Phys. 6* (2005) 75–87.
15. M. Bode, M. Heide, K. v. Bergmann, P. Ferriani, S. Heinze, G. Bihlmayer, A. Kubetzka, O. Pietzsch, S. Blügel, and R.Wiesendanger, Chiral magnetic order at surfaces driven by inversion asymmetry, *Nature* 447 (2007) 190.
16. M. A. Ruderman and C. Kittel, Indirect exchange coupling of nuclear magnetic moments by conduction electrons, *Phys. Rev.* 96 (1954) 99.
17. T. Kasuya, A theory of metallic ferro-and antiferromagnetism on Zener's model, *Prog. Theor. Phys.* 16 (1956) 45.
18. K. Yosida, Magnetic properties of Cu-Mn alloys, *Phys. Rev.* 106 (1957) 893.
19. A. Vindigni, A. Rettori, M. G. Pini, C. Carbone, and P. Gambardella, Finite-sized Heisenberg chains and magnetism of one-dimensional metal systems, *Appl. Phys. A* 82 (2006) 385.
20. J. Stöhr and H. C. Siegmann, *Magnetism—From Fundamentals to Nanoscale Dynamics*, Vol. 152, Springer, Berlin, Germany, 2006.
21. R. Wiesendanger, Spin mapping at the nanoscale and atomic scale, *Rev. Mod. Phys.* 81 (2009) 1495.
22. B. T. Thole, G. v. d. Laan, J. C. Fuggle, G. A. Sawatzky, R. C. Karnatak, and J. M. Esteva, 3d x-ray-absorption lines and the 3d94fn+1 multiplets of the lanthanides, *Phys. Rev. B* 32 (1985) 5107.

23. G. v. d. Laan and B. T. Thole, Strong magnetic x-ray dichroism in 2p absorption spectra of 3d transition-metal ions, *Phys. Rev. B* 43 (1991) 13401–13411.
24. B. T. Thole, P. Carra, F. Sette, and G. v. d. Laan, X-ray circular dichroism as a probe of orbital magnetization, *Phys. Rev. Lett.* 68 (1992) 1943–1946.
25. P. Carra, B. T. Thole, M. Altarelli, and X. Wang, X-ray circular dichroism and local magnetic fields, *Phys. Rev. Lett.* 70 (1993) 694–697.
26. C. T. Chen, Y. U. Idzerda, H. J. Lin, N. V. Smith, G. Meigs, E. Chaban, G. H. Ho, E. Pellegrin and F. Sette, Experimental confirmation of the X-ray magnetic circular dichroism sum rules for iron and cobalt, *Phys. Rev. Lett.* 75 (1995) 152–155.
27. S. Stepanow, A. Mugarza, G. Ceballos, P. Moras, J. Cezar, C. Carbone, and P. Gambardella, Giant spin and orbital moment anisotropies of a Cu-phthalocyanine monolayer, *Phys. Rev. B* 82 (2010) 014405.
28. A. Yazdani, B. A. Jones, C. P. Lutz, M. F. Crommie, and D. M. Eigler, Probing the local effects of magnetic impurities on superconductivity, *Science* 275 (1997) 1767–1770.
29. E. W. Hudson, K. M. Lang, V. Madhavan, S. H. Pan, H. Eisaki, S. Uchida, and J. C. Davis, Interplay of magnetism and high-Tc superconductivity at individual Ni impurity atoms in $Bi_2Sr_2CaCu_2O_8$+delta, *Nature* 411 (2001) 920–924.
30. L. Bürgi, *Scanning Tunneling Microscopy as Local Probe of Electron Density, Dynamics, and Transport at Metal Surfaces*, Ph.D. thesis, Swiss Federal Institute of Technology, Lausanne, Switzerland, (1999).
31. J. Li, W. D. Schneider, R. Berndt, and B. Delley, Kondo scattering observed at a single magnetic impurity, *Phys. Rev. Lett.* 80 (1998) 2893.
32. M. Ternes, A. J. Heinrich, and W. D. Schneider, Spectroscopic manifestations of the Kondo effect on single adatoms, *J. Phys.: Condens. Matter* 21 (2009) 053001.
33. V. Madhavan, W. Chen, T. Jamneala, M. F. Crommie, and N. S. Wingreen, Tunneling into a single magnetic atom: Spectroscopic evidence of the kondo resonance, *Science* 280 (1998) 567–569.
34. J. Kröger, N. Néel, and L. Limot, Contact to single atoms and molecules with the tip of a scanning tunnelling microscope, *J. Phys.: Condens. Matter* 20 (2008) 223001.
35. R. Wiesendanger, H. J. Güntherodt, G. Güntherodt, R. J. Gambino, and R. Ruf, Observation of vaccuum tunneling of spin-polarized electrons with the scanning tunneling microscope, *Phys. Rev. Lett.* 65 (1990) 247.
36. M. Bode, Spin-polarized scanning tunnelling microscopy, *Rep. Prog. Phys.* 66 (2003) 523.
37. W. Wulfhekel and J. Kirschner, Spin-polarized scanning tunneling microscopy of magnetic structures and antiferromagnetic thin films, *Ann. Rev. Mater. Res.* 37 (2007) 69.
38. S. Krause, L. Berbil-Bautista, G. Herzog, M. Bode, and R. Wiesendanger, Current-induced magnetization switching with a spin-polarized scanning tunneling microscope, *Science* 317 (2007) 1537.
39. S. Rusponi, N. Weiss, T. Cren, M. Epple, and H. Brune, High tunnel magnetoresistance in spin-polarized scanning tunneling microscopy of Co nanoparticles on Pt(111), *Appl. Phys. Lett.* 87 (2005) 162514.
40. W. A. Hofer, K. Palotás, S. Rusponi, T. Cren, and H. Brune, Role of hydrogen in giant spin polarization observed on magnetic nanostructures, *Phys. Rev. Lett.* 100 (2008) 026806.
41. F. Meier, L. Zhou, J. Wiebe, and R. Wiesendanger, Revealing magnetic interactions from single-atom magnetization curves, *Science* 320 (2008) 82–86.
42. G. Rodary, S. Wedekimd, D. Sander, and J. Kirschner, Magnetic hysteresis loop of single co nano-islands, *Jpn. J. Appl. Phys.* 47 (2008) 9013.
43. A. A. Khajetoorians, J. Wiebe, B. Chilian, S. Lounis, S. Blügel, and R. Wiesendanger, Atom-by-atom engineering and magnetometry of tailored nanomagnets, *Nat. Phys.* 8 (2012) 497.
44. H. Oka, K. Tao, S. Wedekind, G. Rodary, V. Stepanyuk, D. Sander, and J. Kirschner, Spatially modulated tunnel magnetoresistance on the nanoscale, *Phys. Rev. Lett.* 107 (2011) 187201.
45. S. Ouazi, S. Wedekind, G. Rodary, H. Oka, D. Sander, and J. Kirschner, Magnetization reversal of individual co nanoislands, *Phys. Rev. Lett.* 108 (2012) 107206.
46. C. F. Hirjibehedin, C. P. Lutz, and A. J. Heinrich, Spin-coupling in engineered atomic structures, *Science* 312 (2006) 1021.
47. C. F. Hirjibehedin, C. Y. Lin, A. F. Otte, M. Ternes, C. P. Lutz, B. A. Jones, and A. J. Heinrich, Large magnetic anisotropy of a single atomic spin embedded in a surface molecular network, *Science* 317 (2007) 1199–1203.
48. A. F. Otte, M. Ternes, K. v. Bergmann, S. Loth, H. Brune, C. P. Lutz, C. F. Hirjibehedin, and A. J. Heinrich, The role of magnetic anisotropy in the Kondo effect, *Nat. Phys.* 4 (2008) 847.
49. T. Balashov, T. Schuh, A. F. Takács, A. Ernst, S. Ostanin, J. Henk, I. Mertig, P. Bruno, T. Miyamachi, S. Suga, and W. Wulfhekel, Magnetic anisotropy and magnetization dynamics of individual atoms and clusters of Fe and Co on Pt(111), *Phys. Rev. Lett.* 102 (2009) 257203.
50. A. A. Khajetoorians, B. Chilian, J. Wiebe, S. Schuwalow, F. Lechermann, and R. Wiesendanger, Detecting excitation and magnetization of individual dopants in a semiconductor, *Nature* 467 (2010) 1084.
51. A. A. Khajetoorians, S. Lounis, B. Chilian, A. T. Costa, L. Zhou, D. L. Mills, J. Wiebe, and R. Wiesendanger, Itinerant nature of atom-magnetization excitation by tunneling electrons, *Phys. Rev. Lett.* 106 (2011) 037205.
52. J. Lambe, R. C. Jaklevic, Molecular vibration spectra by inelastic electron tunneling, *Phys. Rev.* 165 (1968) 821.
53. E. L. Wolf, *Principles of Electron Tunneling Spectroscopy*, Oxford University Press, New York, 1989.
54. B. C. Stipe, M. A. Rezaei, and W. Ho, Single molecule vibrational spectroscopy and microscopy, *Science* 280 (1998) 1732.

55. T. Balashov, A. F. Takacs, W. Wulfhekel, and J. Kirschner, Magnon excitation with spin-polarized Scanning Tunneling Microscopy, *Phys. Rev. Lett.* 97 (2006) 187201.
56. S. Loth, S. Baumann, C. P. Lutz, D. M. Eigler, and A. J. Heinrich, Bistability in atomic-scale antiferromagnets, *Science* 335 (2012) 196.
57. S. Loth, K. v. Bergmann, C. P. Lutz, and A. J. Heinrich, Spin-polarized spin excitation spectroscopy, *N. J. Phys.* 12 (2010) 125021.
58. S. Loth, K. v. Bergmann, M. Ternes, A. F. Otte, C. P. Lutz, and A. J. Heinrich, Controlling the state of quantum spins with electric currents, *Nat. Phys.* 6 (2010) 340.
59. S. Loth, M. Etzkorn, C. P. Lutz, D. M. Eigler, and A. J. Heinrich, Measurement of fast electron spin relaxation times with atomic resolution, *Science* 329 (2010) 1628.
60. C. Carbone, M. Veronese, P. Moras, S. Gardonio, C. Grazioli, P. H. Zhou, O. Rader et al., Correlated electrons step by step: Itinerant-to-localized transition of Fe impurities in free-electron metal hosts, *Phys. Rev. Lett.* 104 (2010) 117601.
61. P. W. Anderson, Localized magnetic states in metals, *Phys. Rev.* 124 (1961) 41.
62. H. Brune, Microscopic view of epitaxial metal growth: nucleation and aggregation, *Surf. Sci. Rep.* 31 (1998) 121.
63. S. Blügel, Two-dimensional ferromagnetism of 3d, 4d, and 5d transition metal monolayers on noble metal (001) substrates, *Phys. Rev. Lett.* 68 (1992) 851.
64. P. Lang, V. S. Stepanyuk, K. Wildberger, R. Zeller, and P. H. Dederichs, Local moments of 3d, 4d, and 5d atoms at Cu and Ag(001) surfaces, *Solid State Commun.* 92 (1994) 755.
65. B. Nonas, I. Cabria, R. Zeller, P. H. Dederichs, T. Huhne, and H. Ebert, Strongly enhanced orbital moments and anisotropies of adatoms on the Ag(100) surface, *Phys. Rev. Lett.* 86 (2001) 2146.
66. S. Blügel, Magnetism goes nano: 36th IFF Spring School, In: S. Blügel, T. Brückel, C. M. Schneider eds., Ch. *Reduced Dimensions I: Magnetic Moment and Magnetic Structure*, Forschungszentrums Jülich, Jülich, 2005.
67. G. J. Nieuwenhuys, Magnetic behaviour of cobalt, iron and manganese dissolved in palladium, *Adv. Phys.* 24 (1975) 515.
68. Y. Yayon, X. H. Lu, and M. F. Crommie, Bimodal electronic structure of isolated Co atoms on Pt(111), *Phys. Rev. B* 73 (2006) 155401.
69. G. Kresse, M. Schmid, E. Napetschnig, M. Shishkin, L. Köhler, and P. Varga, Structure of the ultrathin aluminum oxide film on NiAl (110), *Science* 308 (2005) 1440.
70. H. Ellmer, V. Repain, M. Sotto, and S. Rousset, Prestructured metallic template for the growth of ordered, square-based nanodots, *Surf. Sci.* 511 (2002) 183.
71. B. Croset, Y. Girard, G. Prévot, M. Sotto, Y. Garreau, R. Pinchaux, and M. Sauvage-Simkin, Measuring surface stress discontinuities in self-organized systems with X rays, *Phys. Rev. Lett.* 88 (2002) 56103.
72. T. Choi, C. D. Ruggiero, and J. A. Gupta, Incommensurability and atomic structure of $c(2\times2)$N/Cu(100): A scanning tunneling microscopy study, *Phys. Rev. B* 78 (2008) 035430.
73. D. M. Eigler, C. P. Lutz, and W. E. Rudge, An atomic switch realized with the scanning tunneling microscope, *Nature* 352 (1991) 600–603.
74. L. Bartels, G. Meyer, and K. H. Rieder, Basic steps of lateral manipulation of single atoms and diatomic clusters with a scanning tunneling microscope, *Phys. Rev. Lett.* 79 (1997) 697–700.
75. H. Brune, Assembly and probing of spin chains of finite size, *Science* 312 (2006) 1005.
76. N. Tsukahara, K. I. Noto, M. Ohara, S. Shiraki, N. Takagi, Y. Takata, J. Miyawaki, M. Taguchi, A. Chainani, S. Shin, and M. Kawai, Adsorption-induced switching of magnetic anisotropy in a single iron(II) phthalocyanine molecule on an oxidized Cu(110) surface, *Phys. Rev. Lett.* 102 (2009) 167203.
77. A. Mugarza and P. Gambardella, Orbital specific chirality and homochiral self-assembly of achiral molecules induced by charge transfer and spontaneous symmetry breaking, *Phys. Rev. Lett.* 105 (2010) 115702.
78. X. Chen, Y. S. Fu, S. H. Ji, T. Zhang, P. Cheng, X. C. Ma, X. L. Zou, W. H. Duan, J. F. Jia, and Q. K. Xue, Probing superexchange interaction in molecular magnets by spin-flip spectroscopy and microscopy, *Phys. Rev. Lett.* 101 (2008) 197208.
79. K. Tao, V. S. Stepanyuk, W. Hergert, I. Rungger, S. Sanvito, and P. Bruno, Switching a single spin on metal surfaces by a STM tip: ab initio studies, *Phys. Rev. Lett.* 103 (2009) 057202.
80. F. Delgado, J. J. Palacios, and J. Fernandez-Rossier, Spin-transfer torque on a single magnetic adatom, *Phys. Rev. Lett.* 104 (2010) 026601.
81. M. Bode, O. Pietzsch, A. Kubetzka, and R. Wiesendanger, Shape-dependent thermal switching behavior of superparamagnetic nanoislands, *Phys. Rev. Lett.* 92 (2004) 067201.
82. S. Krause, G. Herzog, T. Stapelfeldt, L. Berbil-Bautista, M. Bode, E. Y. Vedmedenko, and R. Wiesendanger, Magnetization reversal of nanoscale islands: How size and shape affect the arrhenius prefactor, *Phys. Rev. Lett.* 103 (2009) 127202.
83. S. Rusponi, T. Cren, N. Weiss, M. Epple, P. Buluschek, L. Claude, and H. Brune, The remarkable difference between surface and step atoms in the magnetic anisotropy of 2D nanostructures, *Nat. Mater.* 2 (2003) 546.
84. P. Gambardella, A. Dallmeyer, K. Maiti, M. C. Malagoli, W. Eberhardt, K. Kern, and C. Carbone, Ferromagnetism in one-dimensional monatomic metal chains, *Nature* 416 (2002) 301–304.
85. P. Gambardella, S. Stepanow, A. Dmitriev, J. Honolka, F. M. F. d. Groot, M. Lingenfelder, S. S. Gupta et al., Supramolecular control of the magnetic anisotropy in two-dimensional high-spin Fe arrays at a metal interface, *Nat. Mater.* 8 (2009) 189.
86. M. Bernien, J. Miguel, C. Weis, M. E. Ali, J. Kurde, B. Krumme, P. M. Panchmatia et al., Tailoring the nature of magnetic coupling of Fe-porphyrin molecules to ferromagnetic substrates, *Phys. Rev. Lett.* 102 (2009) 047202.

87. S. Stepanow, J. Honolka, P. Gambardella, L. Vitali, N. Abdurakhmanova, T. C. Tseng, S. Rauschenbach et al., Spin and orbital magnetic moment anisotropies of monodispersed bis(phthalocyaninato) terbium on a copper surface, *J. Am. Chem. Soc.* 132 (2010) 11900.
88. A. Mugarza, R. Robles, C. Krull, R. Korytr, N. Lorente, and P. Gambardella, Electronic and magnetic properties of molecule-metal interfaces: Transition-metal phthalocyanines adsorbed on Ag(100), *Phys. Rev. B* 85 (2012) 155437.
89. S. Stepanow, P. S. Miedema, A. Mugarza, G. Ceballos, P. Moras, J. C. Cezar, C. Carbone, F. M. F. d. Groot, and P. Gambardella, Mixed-valence behavior and strong correlation effects of metal phthalocyanines adsorbed on metals, *Phys. Rev. B* 83 (2011) 220401.
90. M. A. Lingenfelder, H. Spillmann, A. Dmitriev, S. Stepanow, N. Lin, J. V. Barth, and K. Kern, Towards surface-supported supramolecular architectures: Tailored coordination assembly of 1,4-Benzenedicarboxylate and Fe on Cu(100), *Chem. Eur. J* 10 (2004) 1913.
91. S. Stepanow, N. Lin, and J. V. Barth, Modular assembly of low-dimensional coordination architectures on metal surfaces, *J. Phys.: Condens. Matter* 20 (2008) 184002.
92. L. Gao, W. Ji, Y. B. Hu, Z. H. Cheng, Z. T. Deng, Q. Liu, N. Jiang et al., Site-specific kondo effect at ambient temperatures in iron-based molecules, *Phys. Rev. Lett.* 99 (2007) 106402.
93. J. Brede, N. Atodiresei, S. Kuck, P. Lazic, V. Caciuc, Y. Morikawa, G. Hoffmann, S. Blügel, and R. Wiesendanger, Spin- and energy-dependent tunneling through a single molecule with intramolecular spatial resolution, *Phys. Rev. Lett.* 105 (2010) 047204.
94. R. Sessoli, D. Gatteschi, A. Caneschi, and M. A. Novak, Magnetic bistability in a metal-ion-cluster, *Nature* 365 (1993) 141.
95. L. Bogani and W. Wernsdorfer, Molecular spintronics using single-molecule magnets, *Nat. Mater.* 7 (2008) 179.
96. M. N. Leuenberger and D. Loss, Quantum computing in molecular magnets, *Nature* 410 (2001) 789.
97. N. Ishikawa, M. Sugita, T. Ishikawa, S. Koshihara, and Y. Kaizu, Lanthanide double-decker complexes functioning as magnets at the single-molecular level, *J. Am. Chem. Soc.* 125 (2003) 8694.
98. L. Margheriti, D. Chiappe, M. Mannini, P. Car, P. Sainctavit, M. A. Arrio, F. B. d. Mongeot et al., X-ray detected magnetic hysteresis of thermally evaporated terbium double-decker oriented films, *Adv. Mater.* 22 (2010) 5488.
99. R. Biagi, J. Fernandez-Rodriguez, M. Gonidec, A. Mirone, V. Corradini, F. Moro, V. De Renzi, U. del Pennino, J. C. Cezar, D. B. Amabilino, and J. Veciana, X-ray absorption and magnetic circular dichroism investigation of bis(phthalocyaninato) terbium single-molecule magnets deposited on graphite, *Phys. Rev. B* 82 (2010) 224406.
100. A. Scheybal, T. Ramsvik, R. Bertschinger, M. Putero, F. Nolting, and T. A. Jung, Induced magnetic ordering in a molecular monolayer, *Chem. Phys. Lett.* 411 (2005) 214.
101. C. Iacovita, M. V. Rastei, B. W. Heinrich, T. Brumme, J. Kortus, L. Limot, and J. P. Bucher, Visualizing the spin of individual cobalt-phthalocyanine molecules, *Phys. Rev. Lett.* 101 (2008) 116602.
102. S. Javaid, M. Bowen, S. Boukari, L. Joly, J. B. Beaufrand, X. Chen, Y. J. Dappe et al., Impact on interface spin polarization of molecular bonding to metallic surfaces, *Phys. Rev. Lett.* 105 (2010) 077201.

24

Spin Inelastic Electron Spectroscopy for Single Magnetic Atoms

Aaron Hurley
Trinity College Dublin

Nadjib Baadji
Trinity College Dublin

Stefano Sanvito
Trinity College Dublin

24.1 Introduction

The interaction between conduction electrons and localized spins in transition metals with partially filled d-shells is central to many low-temperature spin effects, which may underpin the development of spintronics and quantum information technology. The continuous advances in low-temperature scanning tunneling microscopy (STM) have enabled us to detect excitations of spin origin, a spectroscopy, which is usually named spin-flip inelastic electron tunneling spectroscopy (SF-IETS). Crucially, this allows us to characterize the elementary spin excitations of magnetic nanostructures at the atomic level. When adsorbed on the surface of a metallic host, magnetic transition metal atoms exhibit various distinctive features in the conductance spectrum of SF-IETS experiments, which are indicative of many-body scattering between the conduction electrons and localized spins. These manifest themselves as conductance steps at voltages corresponding to the quasi-particle energies of specific magnetic excitations and as zero-bias conductance peaks, known as Kondo resonances. Many-body scattering events have been detected for Mn [1], Fe [2], and Co [3,4] adatoms adsorbed on a CuN insulating substrate, with the latter exhibiting a Kondo peak at zero bias.

The recent rapid growth in the experimental activity has been matched by an equally fast explosion of theoretical works. A general and now standard approach to calculate the conductance spectra of the various possible magnetic nanostructures is that of combining a master equation solver for the quantum transport problem with a model Hamiltonian describing the magnetic interaction [5]. This method, however, is not very amenable to implementation with first principle electron transport approaches, and it remains dependent on the parameters of the model Hamiltonian. As such, although very useful for developing a qualitative understanding and for testing new ideas, it is not a fully predictive scheme. In an attempt to make a theory for SF-IETS compatible with accurate electronic structure methods, we propose to describe quantum transport by the non-equilibrium Green's function (NEGF) formalism [6], whose mean-field version can be combined with density functional theory (DFT) to produce efficient and predictive algorithms [7]. Importantly, for this discussion, inelastic contributions to the elastic current can also be included within the NEGF formalism. In the case of scattering to phonons, the problem is usually treated perturbatively by constructing an appropriate self-energy at the level of either the first or the self-consistent Born approximation [8]. In this chapter, we will describe a similar approach to the case of spin excitations.

Our theoretical analysis is based on a tight-binding Hamiltonian for electrons, locally magnetically exchange-coupled to quantum spins. As such, our formulation works by assuming an adiabatic separation between the transport electrons, *s*, and the electrons occupying the atomic shell responsible for the formation of the magnetic moment, *d*; this is the *s-d* model for magnetism [9,10]. We then proceed by constructing an appropriate self-energy for the electronic degrees of freedom, first up to second order in the perturbation expansion and then extended to the third. It will be shown that the conductance steps in the SF-IETS spectra are already described at the second-order level, while the third-order contribution gives access to Kondo physics. This self-energy is implemented in the standard NEGF scheme for transport and applied to describe SF-IETS in atoms and atomic chains, demonstrating a good agreement with experiments. An important aspect of our perturbation theory is that it also describes the non-equilibrium state of the local spins, namely, we are able to predict how a current changes the spin direction of an atom or a small magnetic nanostructure.

The chapter is organized as follows. We begin our discussion by recalling the general many-body Green's functional formalism at zero temperature and how this can be applied to the electron transport problem. We then discuss the case of the *s-d* model and construct a perturbative expansion over the electron-spin coupling Hamiltonian. Finally, we show several examples of the theory at work and draw some conclusions.

24.2 Zero-Temperature Green's Function Formalism

In general, one uses a perturbative Green's function (GF) approach for problems that cannot be solved exactly. Let us assume that the Hamiltonian of interest, H, comprises a part H_0 for which an exact solution can be found and an unspecified perturbation V:

$$H = H_0 + V, \tag{24.1}$$

where we assume the non-commutativity of each part $[H_0, V] \neq 0$. The system is initially described by H_0 and then the perturbation V is switched on at a subsequent time.

24.2.1 Interaction Representation

In this section, we describe how a quantum mechanical system evolves in time after it is perturbed at an initial time t_0. It is convenient to tackle this problem by using the *interaction representation*, in which a generic operator, O, and wavefunctions, φ, written in the Schrödinger representation assume an explicit time dependence, given by

$$\hat{O}(t) = e^{iH_0 t} O e^{-iH_0 t}, \tag{24.2}$$

$$\hat{\psi}(t) = U(t)\phi, \tag{24.3}$$

where we have defined the unitarity operator $U(t) = e^{iH_0 t}$. Note that all operators in the interaction picture will be denoted by '^'.

This choice ensures that the time dependence of the wave functions is governed by the interaction $\hat{V}(t)$ through the equation of motion

$$\frac{\partial}{\partial t}\hat{\psi}(t) = -i\hat{V}(t)\hat{\psi}(t). \tag{24.4}$$

The unitarity operator also obeys the differential equation

$$\frac{\partial}{\partial t}U(t) = -i\hat{V}(t)U(t), \tag{24.5}$$

and by imposing the initial condition $U(t_0 = 0) = 1$, one has the solution

$$U(t) = \sum_{n=0}^{\infty}(-i)^n \int_0^t dt_1 \int_0^{t_1} dt_2 \ldots \int_0^{t_{n-1}} dt_n \hat{V}(t_1)\hat{V}(t_2)\ldots\hat{V}(t_n). \tag{24.6}$$

After introducing the time ordering operator, T, which arranges operators in order of decreasing time from left to right, and after having redefined the time integration variables, we arrive at the following expression for the unitarity operator:

$$U(t) = 1 + \sum_{n=1}^{\infty}\frac{(-i)^n}{n!}\int_0^t dt_1 \int_0^t dt_2 \ldots \int_0^t dt_n T[\hat{V}(t_1)\hat{V}(t_2)\ldots\hat{V}(t_n)], \tag{24.7}$$

$$U(t) = T\exp\left[-i\int_0^t dt_1\hat{V}(t_1)\right]. \tag{24.8}$$

In addition to the unitarity operator, we can also define the S-matrix as the operator, which propagates the wave function $\hat{\psi}(t)$ from the time t' to the time t:

$$\hat{\psi}(t) = S(t,t')\hat{\psi}(t'). \tag{24.9}$$

The S-matrix also satisfies the differential equation

$$\frac{\partial}{\partial t}S(t,t') = -i\hat{V}(t)S(t,t'), \tag{24.10}$$

with solution

$$S(t,t') = T\exp\left[-i\int_{t'}^t dt_1\hat{V}(t_1)\right]. \tag{24.11}$$

The S-matrix is important when dealing with the ground state of the full interacting Hamiltonian, for which we have no information. The ground-state wave function of the unperturbed system (denoted by $\hat{\phi}_0$), however, can be found exactly. It can be shown

[11] that the relation between the ground states of the interacting and non-interacting systems can be defined as

$$\hat{\psi}(0) = S(0,-\infty)\hat{\phi}_0, \tag{24.12}$$

so that $\hat{\psi}(-\infty) = \hat{\phi}_0$. This tells us that at $t = -\infty$, the system is in the non-interacting ground state, $\hat{\phi}_0$. The operator $S(0, -\infty)$ then propagates adiabatically the wave function up to the present ($t = 0$), following the evolution dictated by the interaction potential $\hat{V}$.

24.2.2 Green's Function Formalism

We are now in the position to introduce the GF formalism to treat the system perturbation. At zero temperature, the electron GF, G, is defined as

$$G_{\lambda,\lambda'}(t,t') = -i\left\langle | T\left\{ \hat{c}_\lambda(t)\hat{c}^\dagger_{\lambda'}(t')\right\} | \right\rangle, \tag{24.13}$$

where $|\rangle$ is the ground state of the full Hamiltonian, H. The c's are the electron creation ($c^\dagger$) and annihilation (c) operators with the indexes λ labeling the eigenstates of the Hamiltonian $H = \sum_\lambda \varepsilon_\lambda c^\dagger_\lambda c_\lambda$ with eigenvalues ε_λ, that is, they are the operators that fully diagonalize the interacting problem. In order to convert this expression to the interaction representation, we use the S-matrix result from the previous section, Equation 24.12, which gives us

$$G_{\lambda,\lambda'}(t,t') = -i\frac{{}_0\left\langle | T\left\{ \hat{c}_\lambda(t)\,\hat{c}^\dagger_{\lambda'}(t')\hat{S}(\infty,-\infty)\right\} | \right\rangle_0}{{}_0\left\langle | \hat{S}(\infty,-\infty) | \right\rangle_0}. \tag{24.14}$$

In Equation 24.14, the GF is the expectation value of an operator calculated over the non-interacting ground state $|\rangle_0$, where the operators $\hat{c}$ and $\hat{c}^\dagger$ are now written in the interaction representation. By using the S-matrix expansion provided in Equation 24.11, we arrive at the final expression for the fully interacting GF, which reads

$$G_{\lambda,\lambda'}(t,t') = \sum_{n=0}\frac{(-i)^{n+1}}{n!}\int_{-\infty}^{\infty} dt_1 \ldots \int_{-\infty}^{\infty} dt_n \frac{{}_0\langle | T\{\hat{V}(t_1)\ldots\hat{V}(t_n)\hat{c}_\lambda(t)\hat{c}^\dagger_{\lambda'}(t')\} | \rangle_0}{{}_0\langle | \hat{S}(\infty,-\infty) | \rangle_0}. \tag{24.15}$$

At this point, the formalism encounters a problem. In fact, although the ground state of the system at $t = -\infty$ is well defined [from Equation 24.12], we have no means of describing the ground state at the time $t = +\infty$. This problem was overcome by Schwinger [12], who proposed to write the time integral in the S-matrix in terms of two contributions, namely, $(-\infty, \tau)$ and $(\tau, -\infty)$. This process ensures that eventually τ will reach $+\infty$. The advantage of this idea is that the integration begins and ends with a known ground state $\psi(-\infty) = \varphi_0$. This is called the *time loop method*, and it is the primary way to deal with non-equilibrium systems. The Schwinger contour describes systems at zero temperature and it will be used throughout. This is easier to deal with than the Keldysh contour [13], which accounts for finite temperature. We will include finite temperature only in the definition of the electronic functions describing the electrodes. With this at hand, the S-matrix becomes

$$\hat{S}(-\infty,-\infty) = T_C \exp\left[-i\int_C d\tau_1 \hat{V}(\tau_1)\right], \tag{24.16}$$

where we have defined the time-loop contour C as the one that runs over the time interval $(-\infty, \tau_1)$ and then $(\tau_1, -\infty)$, and the operator T_C as the time ordering operator along the entire loop. The full GF expansion finally becomes

$$G_{\lambda,\lambda'}(\tau,\tau') = \sum_{n=0}\frac{(-i)^{n+1}}{n!}\int_C d\tau_1 \ldots \int_C d\tau_n \frac{{}_0\left\langle | T_C\left\{ \hat{V}(\tau_1)\ldots\hat{V}(\tau_n)\hat{c}_\lambda(\tau)\hat{c}^\dagger_{\lambda'}(\tau')\right\} | \right\rangle_0}{{}_0\langle | \hat{S}(-\infty,-\infty) | \rangle_0}. \tag{24.17}$$

Although the introduction of the C contour solves the issue of the $t = +\infty$ limit, it also brings the drawback of introducing six new GFs, depending on how the times t and t' are positioned over C. These are schematically represented in Figure 24.1 and in the Heisenberg representation for operators:

$$G^<_{\lambda,\lambda'}(t,t') = i\left\langle | \left\{ c^\dagger_{\lambda'}(t')c_\lambda(t)\right\} | \right\rangle, \tag{24.18}$$

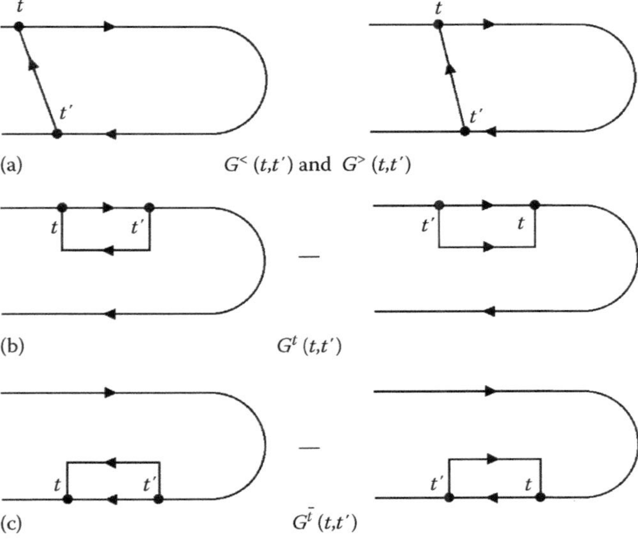

FIGURE 24.1 Feynman rules for non-equilibrium Green's functions calculated over the contour running from $-\infty$ to τ and from τ to $-\infty$. The four time loops represent the lesser and greater Green's functions (a) and the time ordered (b) and time anti-ordered (c) Green's functions.

$$G^{>}_{\lambda,\lambda'}(t,t') = -i\left\langle | \left\{ c_\lambda(t) c^\dagger_{\lambda'}(t') \right\} | \right\rangle, \quad (24.19)$$

$$G^{t}_{\lambda,\lambda'}(t,t') = \theta(t-t')G^{>}_{\lambda,\lambda'}(t,t') + \theta(t'-t)G^{<}_{\lambda,\lambda'}(t,t'), \quad (24.20)$$

$$G^{\bar{t}}_{\lambda,\lambda'}(t,t') = \theta(t'-t)G^{>}_{\lambda,\lambda'}(t,t') + \theta(t-t')G^{<}_{\lambda,\lambda'}(t,t'), \quad (24.21)$$

$$G^{r}_{\lambda,\lambda'}(t,t') = G^{t}_{\lambda,\lambda'}(t,t') - G^{<}_{\lambda,\lambda'}(t,t'), \quad (24.22)$$

$$G^{a}_{\lambda,\lambda'}(t,t') = G^{t}_{\lambda,\lambda'}(t,t') - G^{>}_{\lambda,\lambda'}(t,t'). \quad (24.23)$$

Here, $G^<$ and $G^>$ are, respectively, the lesser and greater GFs, G^t and $G^{\bar{t}}$ are the time-ordered and anti-time-ordered GFs, and G^r and G^a are retarded and advanced ones. More information about the properties of these six GFs can be obtained by considering the unperturbed Hamiltonian, that is, by looking at the non-interacting case, which can be solved exactly. After diagonalization, the unperturbed Hamiltonian takes the form $H_0 = \sum_\lambda \varepsilon_\lambda c^\dagger_\lambda c_\lambda$, where ε_λ are the eigenvalues. The GFs are then evaluated over the known ground state $|\rangle_{\lambda=0}$, whose energy is ε_0. In this case, the explicit time dependence of the creation and annihilation operators can be found from the Heisenberg equation. For instance, for the annihilation operator, we have

$$\frac{\partial}{\partial t} c_\lambda(t) = -i[H_0, c_\lambda(t)], \quad (24.24)$$

which has the solution (the initial condition is $c_\lambda(0) = c_\lambda$)

$$c_\lambda(t) = c_\lambda \exp(-i\varepsilon_\lambda t). \quad (24.25)$$

Likewise, the solution for the creation operator $c^\dagger_\lambda(t)$ is

$$c^\dagger_\lambda(t) = c^\dagger_\lambda \exp(+i\varepsilon_\lambda t). \quad (24.26)$$

If now one inserts such solutions in the expressions for the lesser and greater GFs provided by Equations 24.18 and 24.19, these can be written as

$$G^{<}_{\lambda,\lambda'}(t,t') = -i\delta_{\lambda,\lambda'}\langle n_\lambda \rangle \exp[-i\varepsilon_\lambda(t-t')], \quad (24.27)$$

$$G^{>}_{\lambda,\lambda'}(t,t') = i\delta_{\lambda,\lambda'}(1-\langle n_\lambda \rangle)\exp[-i\varepsilon_\lambda(t-t')], \quad (24.28)$$

where we have introduced the expectation value over the contour of electron number operator, $n_\lambda = c^\dagger_\lambda c_\lambda$, and have used the fermions anticommutation relations. We note that at thermal equilibrium, the number operator average is given by the Fermi–Dirac distribution $\langle n_\lambda \rangle = f(\varepsilon_\lambda) = 1/[\exp(\beta\varepsilon_\lambda) + 1]$, which has an occupation of 1 up to the Fermi energy ε_F, and $\beta = k_B T$ with k_B being the Boltzmann constant. By using these relations, we can construct the full energy resolved dependence of all six GFs, which is obtained by Fourier transform of their time-dependent counterparts:

$$G^<(\lambda,\omega) = 2\pi i f(\varepsilon_\lambda)\delta(\omega-\varepsilon_\lambda), \quad (24.29)$$

$$G^>(\lambda,\omega) = -2\pi i[1-f(\varepsilon_\lambda)]\delta(\omega-\varepsilon_\lambda), \quad (24.30)$$

$$G^t(\lambda,\omega) = \frac{1}{\omega-\varepsilon_\lambda+i\delta_\lambda}, \quad (24.31)$$

$$G^{\bar{t}}(\lambda,\omega) = \frac{-1}{\omega-\varepsilon_\lambda-i\delta_\lambda}, \quad (24.32)$$

$$G^r(\lambda,\omega) = \frac{1}{\omega-\varepsilon_\lambda+i\delta}, \quad (24.33)$$

$$G^a(\lambda,\omega) = \frac{1}{\omega-\varepsilon_\lambda-i\delta}, \quad (24.34)$$

where δ is positive and satisfies $\delta \to 0^+$, while δ_λ is positive for $\varepsilon_\lambda > \varepsilon_F$ and negative for $\varepsilon_\lambda < \varepsilon_F$.

24.2.3 Wick's Theorem

In order to tackle the interacting problem, we need to return to Equation 24.17 and ascertain the rules, which determine the contour time ordering of the product of the $\hat{V}$'s. These are different depending on whether the $\hat{V}$'s describe electrons, phonons, or, as in this case, spins. Assuming that the perturbation is electronic in nature (for instance, a density–density operator representing Coulomb repulsion), then one needs to deal with time ordering products of the form

$${}_0\left\langle | T_C \left\{ \hat{c}_1(\tau_1)\hat{c}^\dagger_{1'}(\tau_{1'})\hat{c}_2(\tau_2)\hat{c}^\dagger_{2'}(\tau_{2'})\ldots\hat{c}_\lambda(\tau)\hat{c}^\dagger_{\lambda'}(\tau') \right\} | \right\rangle_0, \quad (24.35)$$

where there is always an equal number of creation and annihilation operators and an operator that creates a state is always followed by an operator that destroys one. In this case, the task is that of re-expressing Equation 24.35 in terms of products of time ordered pairs ${}_0\left\langle | T_C \left\{ \hat{c}_\lambda(\tau)\hat{c}^\dagger_{\lambda'}(\tau') \right\} | \right\rangle_0$. Clearly, as we have λ pairs of operators, there are $\lambda!$ different possible orderings for such pairs. There are then three simple rules for ordering the operators (Wick's theorem):

1. There is a sign change whenever two neighboring Fermi operators are interchanged. Therefore, the number of interchanges must be noted.
2. If there are a mixture of particles in the time ordering bracket, that is, if there exists operators that do not commute, these can be separated into different time ordering brackets.

3. We recognize the time ordering of a product of operators with different time arguments as the unperturbed GF, namely,

$$ {}_0\langle|\,T_C\{\hat{c}_\lambda(\tau)\hat{c}^\dagger_{\lambda'}(\tau')\}\,|\rangle_0 = iG_{0,\lambda,\lambda'}(\tau,\tau'). \quad (24.36)$$

These rules will be implemented when performing the full time ordering of the interaction Hamiltonian, which will be introduced later in the chapter. This will show Wick's theorem in action.

24.2.4 Langreth's Theorem and Dyson's Equation

Wick's theorem essentially expands the fully interacting GF in terms of the non-interacting ones up to a given order, n. This can be generalized at any order, $n = \infty$, by *Dyson's equation*. Dyson's equation is a reformulation of the general perturbation expansion of the full interacting GF, Equation 24.17, in terms of a quantity called the *self-energy*, that itself can be evaluated perturbatively and provides a compact way to describe the effect of a perturbation on a non-interacting system. Dyson's equation takes the form

$$G(\tau,\tau') = G_0(\tau,\tau') + \int_C d\tau_1 \int_C d\tau_2 G_0(\tau,\tau_1)\Sigma(\tau_1,\tau_2)G(\tau_2,\tau'), \quad (24.37)$$

where
G_0 is the unperturbed GF
Σ is the self-energy

In order to find the real time components of this relation, we refer to Haug and Jauho [14], who describe Langreth's theorem for the evaluation of "contour-convolutions." It can be shown that the retarded/advanced and the lesser/greater components of the GF are, respectively,

$$G^{r(a)}(\tau,\tau') = G_0^{r(a)}(\tau,\tau') + \int_C d\tau_1 \int_C d\tau_2 G_0^{r(a)}(\tau,\tau_1)\Sigma^{r(a)}(\tau_1,\tau_2)G^{r(a)}(\tau_2,\tau'), \quad (24.38)$$

$$G^{\lessgtr}(\tau,\tau') = G^{\lessgtr}{}_0(\tau,\tau') + \int_C d\tau_1 \int_C d\tau_2 G^r(\tau,\tau_1)\Sigma^{\lessgtr}(\tau_1,\tau_2)G^a(\tau_2,\tau'), \quad (24.39)$$

whose energy resolved components can be calculated by Fourier transform

$$G^{r(a)}(\omega) = G_0^{r(a)}(\omega) + G_0^{r(a)}(\omega)\Sigma^{r(a)}(\omega)G^{r(a)}(\omega), \quad (24.40)$$

$$G^{\lessgtr}(\omega) = G^{\lessgtr}{}_0(\omega) + G^r(\omega)\Sigma^{\lessgtr}(\omega)G^a(\omega). \quad (24.41)$$

24.2.5 General Expression for the Current Flowing through an Interacting Region

We will demonstrate the power of the GF formalism by describing the perturbation to a close system originating from the coupling to two semi-infinite leads. This is a problem that can be solved exactly within the approximations that we are about to introduce. Such an approach coincides with the widely used NEGF method extensively employed in the electron transport problem [6,7,13]. Here, we will discuss how to construct a transport theory starting from the GFs that we have defined in the previous sections. At this point, let us introduce the specific Hamiltonian that describes an interacting region sandwiched between M semi-infinite non-interacting leads:

$$H = H_{\text{int}}\left(\{c^\dagger_\lambda; c_\lambda\}\right) + \sum_i^M H^i_{\text{lead}} + \sum_i^M H^i_{\text{coupling}}, \quad (24.42)$$

$$H^i_{\text{lead}} = \sum_\alpha \varepsilon_{\alpha i} b^\dagger_{\alpha i} b_{\alpha i}, \quad (24.43)$$

$$H^i_{\text{coupling}} = \sum_{\alpha,\lambda}\left(V_{\alpha i,\lambda} b^\dagger_{\alpha i} c_\lambda + \text{h.c.}\right). \quad (24.44)$$

Here, H^i_{lead} is the Hamiltonian of the ith lead with $b^\dagger_{\alpha i}(b_{\alpha i})$ creating (annihilating) electrons at the energy $\varepsilon_{\alpha i}$, while $H_{\text{int}}\left(\{c^\dagger_\lambda; c_\lambda\}\right)$ describes the interacting region, and it is constructed with the creation and annihilation operators $c^\dagger_\lambda$ and c_λ (orthonormal). Finally, H^i_{coupling} couples the ith leads with the interacting region, through the hopping matrix elements $V_{\alpha i,\lambda}$. Let us assume that, initially, the leads are completely decoupled from the scattering region, that is, $V_{\alpha i,\lambda} = 0$ for every i. The rate of change of the expectation value of the number operator in the ith lead, $\hat{n}^i = \sum_\alpha b^\dagger_{\alpha i} b_{\alpha i}$, gives the current, I^i, flowing from/to the interacting region to/from the ith lead. This is simply

$$I^i = -\frac{\partial}{\partial t}\langle n^i\rangle = -i\langle[H, n^i]\rangle = i\sum_{\alpha,\lambda}\left(V_{\alpha i,\lambda}\langle b^\dagger_{\alpha i} c_\lambda\rangle(t) - V^*_{\alpha i,\lambda}\langle c^\dagger_\lambda b_{\alpha i}\rangle(t)\right). \quad (24.45)$$

Clearly, one now can recast Equation 24.45 in terms of the GFs introduced before. In this particular case, we can define the *hybrid* real-time GFs describing the overlap between the ith lead and interacting region as

$$G^<_{\lambda,\alpha i}(t,t') = i\left\langle b^\dagger_{\alpha i}(t) c_\lambda(t')\right\rangle, \quad (24.46)$$

$$G^<_{\alpha i,\lambda}(t,t') = i\left\langle c^\dagger_\lambda(t) b_{\alpha i}(t')\right\rangle, \quad (24.47)$$

so that the current can be written as

$$I^i = \sum_{\alpha,\lambda}\left(V_{\alpha i,\lambda} G^<_{\lambda,\alpha i}(t,t) - V^*_{\alpha i,\lambda} G^<_{\alpha i,\lambda}(t,t)\right), \quad (24.48)$$

$$= \int_{-\infty}^{\infty} \frac{d\omega}{2\pi} \sum_{\alpha,\lambda}\left(V_{\alpha i,\lambda} G^<_{\lambda,\alpha i}(\omega) - V^*_{\alpha i,\lambda} G^<_{\alpha i,\lambda}(\omega)\right). \quad (24.49)$$

At this point, we still have no expression for the lesser *hybrid* GFs. Therefore, in order to proceed, we must define the following contour ordered GFs

$$G_{\lambda,\alpha i}(\tau,\tau')=-i\left\langle T_C\left\{c_\lambda(\tau)b_{\alpha i}^\dagger(\tau')\right\}\right\rangle, \quad (24.50)$$

$$G_{\alpha i,\lambda}(\tau,\tau')=-i\left\langle T_C\left\{b_{\alpha i}(\tau)c_\lambda^\dagger(\tau')\right\}\right\rangle, \quad (24.51)$$

$$G_{\lambda,\lambda'}(\tau,\tau')=-i\left\langle T_C\left\{c_\lambda(\tau)c_{\lambda'}^\dagger(\tau')\right\}\right\rangle, \quad (24.52)$$

$$G_{\alpha i,\alpha i}(\tau,\tau')=-i\left\langle T_C\left\{b_{\alpha i}(\tau)b_{\alpha i}^\dagger(\tau')\right\}\right\rangle. \quad (24.53)$$

Finally, by using Dyson's equation, we obtain two expressions for $G_{\lambda,\alpha i}$ and $G_{\alpha i,\lambda}$, in which the coupling between the interacting region and the leads appears explicitly:

$$G_{\lambda,\alpha i}(\tau,\tau')=\sum_{\lambda'}\int_C d\tau_1 G_{\lambda,\lambda'}(\tau,\tau_1)V_{\alpha i,\lambda'}^\star G_{\alpha i,\alpha i}(\tau_1,\tau'), \quad (24.54)$$

$$G_{\lambda,\alpha i}(\tau,\tau')=\sum_{\lambda'}\int_C d\tau_1 G_{\alpha i,\alpha i}(\tau,\tau_1)V_{\alpha i,\lambda'}G_{\lambda',\lambda}(\tau_1,\tau'). \quad (24.55)$$

Now, the contour integration rules and the relations defined in Equations 24.18 through 24.23 allow us to re-write the real time expression for the hybrid GFs as

$$G_{\lambda,\alpha i}^<(t,t')$$
$$=\sum_{\lambda'}V_{\alpha i,\lambda'}^\star\int_{-\infty}^{\infty}dt_1\left[G_{\lambda,\lambda'}^{\mathrm{r}}(t,t_1)G_{\alpha i,\alpha i}^<(t_1,t')+G_{\lambda,\lambda'}^<(t,t_1)G_{\alpha i,\alpha i}^{\mathrm{a}}(t_1,t')\right], \quad (24.56)$$

$$G_{\alpha i,\lambda}^<(t,t')$$
$$=\sum_{\lambda'}V_{\alpha i,\lambda'}\int_{-\infty}^{\infty}dt_1\left[G_{\alpha i,\alpha i}^{\mathrm{r}}(t,t_1)G_{\lambda,\lambda'}^<(t_1,t')+G_{\alpha i,\alpha i}^<(t,t_1)G_{\lambda,\lambda'}^{\mathrm{a}}(t_1,t')\right], \quad (24.57)$$

which after Fourier transform becomes

$$G_{\lambda,\alpha i}^<(\omega)=\sum_{\lambda'}V_{\alpha i,\lambda'}^\star\left[G_{\lambda,\lambda'}^{\mathrm{r}}(\omega)G_{\alpha i,\alpha i}^<(\omega)+G_{\lambda,\lambda'}^<(\omega)G_{\alpha i,\alpha i}^{\mathrm{a}}(\omega)\right], \quad (24.58)$$

$$G_{\alpha i,\lambda}^<(\omega)=\sum_{\lambda'}V_{\alpha i,\lambda'}\left[G_{\alpha i,\alpha i}^{\mathrm{r}}(\omega)G_{\lambda,\lambda'}^<(\omega)+G_{\alpha i,\alpha i}^<(\omega)G_{\lambda,\lambda'}^{\mathrm{a}}(\omega)\right]. \quad (24.59)$$

We can now return to the expression for the current, which can now be written as

$$I^i=\int_{-\infty}^{\infty}\frac{d\omega}{2\pi}\sum_{\alpha,\lambda,\lambda'}V_{\alpha i,\lambda}^\star V_{\alpha i,\lambda'}\left[G_{\alpha i,\alpha i}^<(\omega)G_{\lambda,\lambda'}^>(\omega)-G_{\alpha i,\alpha i}^>(\omega)G_{\lambda,\lambda'}^<(\omega)\right]$$
$$=\int_{-\infty}^{\infty}\frac{d\omega}{2\pi}\sum_{\lambda\lambda'}\left[\Sigma_{\lambda,\lambda'}^{i,<}(\omega)G_{\lambda,\lambda'}^>(\omega)-\Sigma_{\lambda,\lambda'}^{i,>}(\omega)G_{\lambda,\lambda'}^<(\omega)\right]$$
$$=\int_{-\infty}^{\infty}\frac{d\omega}{2\pi}\mathrm{Tr}\{[\Sigma^{i,<}(\omega)][G^>(\omega)]-[\Sigma^{i,>}(\omega)][G^<(\omega)]\}, \quad (24.60)$$

where the square brackets denote the matrix form of the operator describing the interacting region. In the same equation, we have defined the lesser and greater parts of the self-energy in the same region as

$$\Sigma_{\lambda,\lambda'}^{i,\lessgtr}(\omega)=\sum_\alpha V_{\alpha i,\lambda}^\star G_{\alpha i,\alpha i}^\lessgtr(\omega)V_{\alpha i,\lambda'}. \quad (24.61)$$

These represent the rates at which a particle with energy ω leaves (<) or enters (>) the ith lead. Equation 24.60 provides a rather transparent interpretation for the current, which is nothing but the difference between the in-scattering rate and the out-scattering one to and from the scattering region.

24.2.6 Current through a Single Non-Interacting Level

As an illustration of the method developed so far, we consider the simple case of a non-interacting region consisting of a single energy level ε_0 (note that the index λ is no longer necessary and it has been dropped), whose Hamiltonian is

$$H_{\mathrm{int}}=\varepsilon_0 c^\dagger c. \quad (24.62)$$

We further assume that the level is coupled to only two semi-infinite, non-interacting leads, one to the left, $i=L$, and one to the right, $i=R$. These are described by a single-orbital tight-binding model with hopping parameter γ_0 and on-site energy $\varepsilon_i=0$. The self energy associated to each lead is then

$$\Sigma_i^\lessgtr(\omega)=\sum_\alpha V_{\alpha i}^\star G_{\alpha i,\alpha i}^\lessgtr(\omega)V_{\alpha i}. \quad (24.63)$$

As the leads are semi-infinite and under the condition that the density of states (DOS) of each of the lead, $\rho_i(\varepsilon_\alpha)$, is known, we can replace the sum over α with an integration over energy:

$$\Sigma_i^\lessgtr(\omega)=\int_{-\infty}^{\infty}d\varepsilon_\alpha\rho_i(\varepsilon_\alpha)V_i^\star(\varepsilon_\alpha)G_{\alpha i,\alpha i}^\lessgtr(\omega)V_i(\varepsilon_\alpha) \quad (24.64)$$

$$=\begin{cases} if_i(\omega)\Gamma_i(\omega) \\ -i[1-f_i(\omega)]\Gamma_i(\omega)\end{cases}, \quad (24.65)$$

where we have used the GF derived in Equation 24.18 for non-interacting fermions. A bias, V, is applied between the left-hand side and the right-hand side lead so that the resulting chemical potential in each lead is shifted by $\mu = eV/2$ (e is the electron charge), namely, $\mu_L = \varepsilon_F + \mu$ and $\mu_R = \varepsilon_F - \mu$ (note that each lead in isolation is in equilibrium). Here, the Fermi function is defined at the relative chemical potential, namely, $f_i(\omega) = 1/\left[\exp\left(\frac{\omega-\mu_i}{k_B T}\right)+1\right]$. In the expression for the self-energy, Equation 24.64, we have also introduced the broadening function $\Gamma_i(\varepsilon_\alpha) = 2\pi\rho_i(\varepsilon_\alpha)|V_i(\varepsilon_\alpha)|^2$, which gives a finite lifetime to the energy level. Clearly, such a broadening and so the lifetime depends on the strength of coupling to the leads. It can be shown [6] that the broadening function for the single-site, one-dimensional tight-binding model is given by

$$\Gamma_i(\omega) = \frac{2\gamma_i^2}{\gamma_0}\sqrt{1-\left(\frac{\omega-\varepsilon_0}{2\gamma_0}\right)^2}, \tag{24.66}$$

where γ_i is the hopping from the closest atom in the semi-infinite chain to the interacting region, that is, there is only a single non-vanishing $V_{i\alpha} = \gamma_i$. In the wide-band limit, $\gamma_0 \to \infty$, which is assumed throughout this work, the broadening function becomes a constant, $\Gamma_i = 2\gamma_i^2/\gamma_0$. The resulting retarded and advanced self-energies in the leads can be found from the relation

$$\Sigma_i^{r(a)}(\omega) = \sum_\alpha V_{\alpha i}^{*} G_{\alpha i,\alpha i}^{r(a)}(\omega) V_{\alpha i}. \tag{24.67}$$

After substituting Equations 24.33 and 24.34, the energy integral can be evaluated to obtain

$$\Sigma_i^{r(a)}(\omega) = \pm i\frac{\Gamma_i}{2}. \tag{24.68}$$

We now have the full artillery to calculate the energy resolved GFs for the interacting region, which in turn gives us access to the current. From Equations 24.40 and 24.41, we have the following energy-resolved relations in terms of the unperturbed (or unconnected) GFs of the interacting region (denoted with a subscript 0):

$$G^{r(a)}(\omega) = G_0^{r(a)}(\omega) + G_0^{r(a)}\Sigma^{r(a)}(\omega)G^{r(a)}(\omega) = \frac{1}{\omega-\varepsilon_0-\Sigma^{r(a)}}, \tag{24.69}$$

$$G^{\lessgtr}(\omega) = G^{r}(\omega)\Sigma^{\lessgtr}(\omega)G^{a}(\omega), \tag{24.70}$$

where the total self-energy contains contributions from both leads $\Sigma(\omega) = \sum_i^{L,R}\Sigma_i(\omega)$. The final expression for the current through the energy level takes the form

$$I_i = \int_{-\infty}^{\infty}\frac{d\omega}{2\pi}\left\{\Sigma_i^{<}(\omega)G^{>}(\omega) - \Sigma_i^{>}(\omega)G^{<}(\omega)\right\} \tag{24.71}$$

$$= \int_{-\infty}^{\infty} d\omega \frac{\Gamma_L\Gamma_R}{\Gamma}\rho(\omega)[f_L(\omega) - f_R(\omega)], \tag{24.72}$$

where the total broadening is $\Gamma = \sum_i^{L,R}\Gamma_i$ and the DOS in the interacting region is $\rho(\omega) = (\Gamma/2\pi)/[(\omega-\varepsilon_0)^2 + \Gamma^2]$. We have then recovered a standard result from scattering theory, namely, the Landauer–Büttiker formula [15]. In fact, the current is simply the product of the transmission coefficient $T(E) = \frac{\Gamma_L\Gamma_R}{2\pi[(\omega-\varepsilon_0)^2+\Gamma^2]}$ and the difference between the Fermi functions of the two leads. In this particular case, the transmission coefficient has the form of a Breit–Wigner resonance at the energy level, ε_0. This relation tells us that current will flow only when the energy level is within the bias window. The corresponding current–voltage plot (see Figure 24.2) presents a characteristic current step when the energy level first enter the bias window. With our convention for the chemical potential in the two leads, this happens when $V = 2\varepsilon_0/e$.

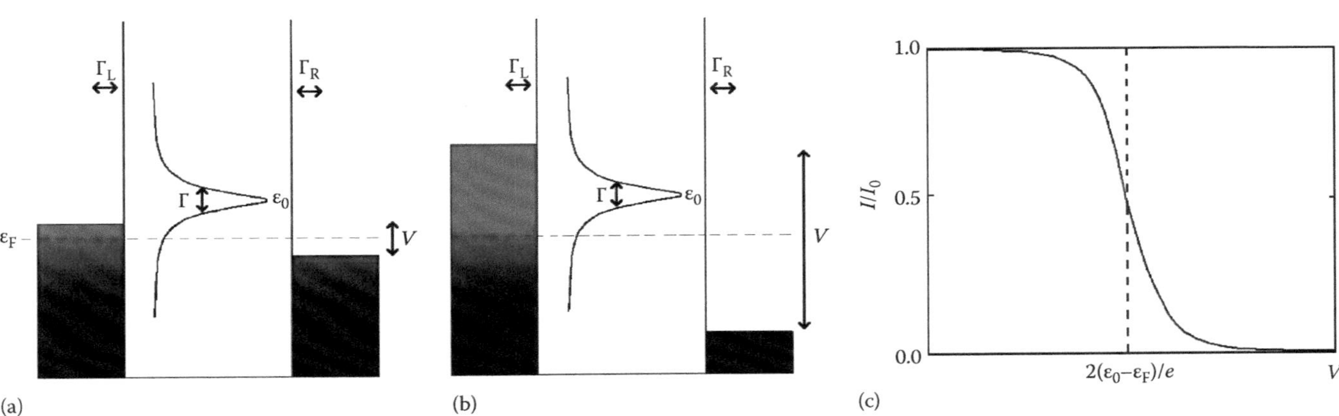

FIGURE 24.2 Schematic representation of resonant tunneling through a single onsite level when: (a) the bias window is small and does not include the level and (b) the bias window includes the level. The current–voltage I–V profile is shown in (c) where the current is calculated in units of the saturation current $I_0 = 2\Gamma_L\Gamma_R/\Gamma$.

24.3 Calculation of the Interacting Spin Self-Energy

In this section, we introduce the specific interaction between electrons and spins and formulate a perturbation theory for SF-IETS. Our theoretical analysis is based on a tight-binding Hamiltonian for the conducting electrons (as in the example of Section 24.2.6), locally exchange coupled to quantum spins. As such, we use the assumption of an adiabatic separation between the transport electrons and those contributing to form local spins, that is, the *s*-*d* model [9,10]. We then proceed with constructing an appropriate self-energy for the electronic degrees of freedom up to the third order in the perturbation expansion and use this in the standard NEGF scheme for transport. Our methodology is then applied to describing SF-IETS in atoms and atomic chains, and the results are compared with experiments.

24.3.1 Hamiltonian of Scattering Region

The typical setup considered here is that of an STM experiment, that is, it comprises an STM tip positioned above one of the atoms of a magnetic nanostructure, which in turn is weakly coupled to a metallic substrate across an insulating barrier. We model this system by a pair of non-interacting semi-infinite leads sandwiching a scattering region, as outlined in Figure 24.3. The left-hand side lead, the scattering region, and the right-hand side lead represent, respectively, the STM tip, the magnetic nanostructure, and the substrate, and they are described by the Hamiltonian H_{tip}, H_S, and H_{sub}. For simplicity, we assume an identical electronic structure for both the leads (i.e., they are made of the same material), which we describe by a one orbital per site tight-binding model with nearest neighbor interaction (as in Section 24.2.6).

To fix the ideas, let us assume that the scattering region consists of N atoms arranged in a chain structure (note that for our discussion the spatial arrangement of the atoms does not necessarily need to be a chain form). The ith atom carries a quantum mechanical spin $\mathbf{S}_i$ and it is characterized by an on-site energy ε_0.

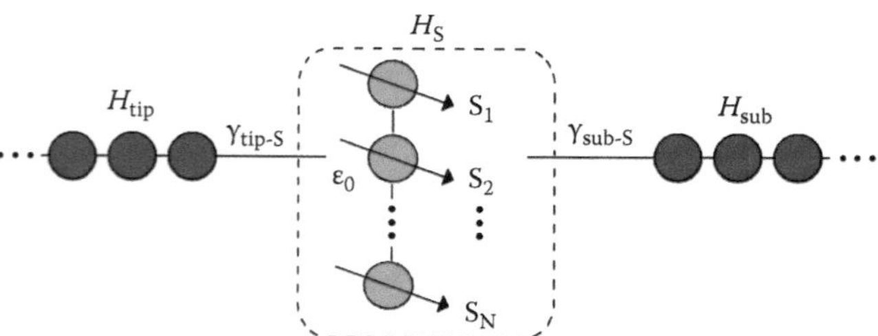

FIGURE 24.3 Schematic representation of the device investigated in this work. A scattering region, comprising N spin-carrying atoms (light gray circles) and described by the Hamiltonian H_S, is sandwiched in between two semi-infinite electrodes (dark-gray circles). These mimic the substrate and the tip in a typical STM experiment. The electrodes are non-spin-polarized, and they are described by the Hamiltonian H_{sub} and H_{tip}. In the scattering region, the transport electrons are exchange-coupled to local quantum spins $\mathbf{S}_i$.

We assume that the tip and substrate can only couple to one atom at the time in the scattering region, that is, that only one atom of the magnetic nanostructure is in electronic contact with the electrodes. Such a coupling is given by the two hopping integrals $\gamma_{\text{tip-S}}$ and $\gamma_{\text{sub-S}}$. This means that the electronic states of the scattering region are broadened by the interaction with the electrodes by $\Gamma_{\text{tip-S}} = 2\gamma^2_{\text{tip-S}}/\gamma_0$ and $\Gamma_{\text{sub-S}} = 2\gamma^2_{\text{sub-S}}/\gamma_0$, where γ_0 is again the hopping parameter within the leads. We assume that $\varepsilon_0 \gg \Gamma_{\text{tip-S}}$ and also $\varepsilon_0 \gg \Gamma_{\text{sub-S}}$, leading to a constant DOS of the scattering region at the Fermi energy. This assumption has two main consequences. On the one hand, we can neglect the electronic interaction among the atoms in the scattering region, as this will generate states far enough from the Fermi energy to ensure a constant DOS. On the other hand, we can simplify the interaction between the atoms in the chain and the substrate to an electronic coupling to a single atom, as additional coupling (as far as it remains weak) will not introduce additional DOS around ε_F.

The Hamiltonian of the scattering region then contains three parts $H_S = H_e + H_{sp} + H_{e\text{-}sp}$, where H_e is the tight-binding electronic part, H_{sp} is the spin part, and $H_{e\text{-}sp}$ describes the electron–spin interaction; more explicitly these are

$$H_e = \varepsilon_0 \sum_{\lambda;\alpha} c^\dagger_{\lambda\alpha} c_{\lambda\alpha}, \tag{24.73}$$

$$H_{sp} = 2J_{dd} \sum_{\lambda}^{N-1} \mathbf{S}_\lambda \cdot \mathbf{S}_{\lambda+1} + \sum_{\lambda}^{N} \left\{ g\mu_B \mathbf{B} \cdot \mathbf{S}_\lambda + D\left(S^z_\lambda\right)^2 + \omega\left[\left(S^x_\lambda\right)^2 - \left(S^y_\lambda\right)^2\right]\right\}, \tag{24.74}$$

$$H_{e\text{-}sp} = J_{sd} \sum_{\lambda\alpha,\alpha'} \left(c^\dagger_{\lambda\alpha} [\sigma_\lambda]_{\alpha\alpha'} c_{\lambda\alpha'}\right) \cdot \mathbf{S}_\lambda. \tag{24.75}$$

The electronic part comprises only an on-site potential, consistent with the idea of neglecting the electron hopping between the sites. The electron ladder operators $c^\dagger_{\lambda\alpha}/c_{\lambda\alpha}$ create/annihilate an electron at site λ with spin α (= ↑, ↓). We model the spin–spin interaction between the localized spins $\{\mathbf{S}_\lambda\}$ by a nearest neighbor Heisenberg Hamiltonian with coupling strength J_{dd}. Furthermore, we include interaction with an external magnetic field $\mathbf{B}$ (μ_B is the Bohr magneton and g the gyromagnetic ratio) and both uni-axial and transverse anisotropy of magnitude D and E, respectively [2,9]. This form for H_{sp} has been proposed before to describe some of the SF-IETS experiments appeared in literature [1]. Note, however, that our formalism does not depend on the particular choice of H_{sp}, and additional terms, as, for instance, Dzyaloshinskii–Moriya interaction, can be included. The particular choice of H_{sp} determines the spectrum of the system and then the shape to the IETS spectrum, but does not require any modifications to the formalism.

The electron–spin interaction Hamiltonian, $H_{e\text{-}sp}$, couples the conducting electrons with the quantum spins, $\{\mathbf{S}_\lambda\}$, through a local Heisenberg exchange term. The electronic spins are described by the operator $c^\dagger_{\lambda\alpha}[\sigma_\lambda]_{\alpha\alpha'}c_{\lambda\alpha'}$, with $\boldsymbol{\sigma}$ being the vector of Pauli matrices. The interaction strength is determined by a single interaction parameter, J_{sd}, which can be used to develop a perturbation theory (note, however, that J_{sd} is not the perturbation expansion parameter, as it will be explained later). This is possible because of our assumption of adiabatic separation between the electrons carrying the current and those responsible for the spins. Such an approximation is valid in the limit of weak electronic coupling between the electron reservoirs and the scattering region, as in the case of the STM measurements that we aim to describe here. Note that going beyond such an approximation will require formulating an entirely electronic theory for inelastic spin transport. The Anderson-like impurity model is usually a starting point for such a task [16], but this will require abandoning the perturbative approach, that is, it is outside the scope of the this work.

24.3.2 Many-Body Green's Function: Second-Order Expansion

The Hamiltonian for the scattering region contains two terms, H_e and H_{sp}, which independently can be diagonalized exactly, so that the problem is easily solvable for $J_{sd} = 0$. However, the electron–spin interaction transforms the system in an intrinsic many-body one, for which we will now proceed to deriving a perturbation theory. Our strategy is that of first constructing the electronic many-body GF at the second order [17–20] in $H_{e\text{-}sp}$ and then, by Dyson's equation, to evaluate the interacting self-energy [21]. In particular, we follow closely the procedure laid out in reference [17]. Our starting point is again the contour-ordered spin-dependent single-body GF in the many-body ground state (note that, for simplicity, we have dropped the "^" symbol denoting the interaction representation):

$$[G(\tau,\tau')]_{\sigma\sigma'} = -i\left\langle \left| T_C\left\{c_\sigma(\tau)c^\dagger_{\sigma'}(\tau')\right\}\right|\right\rangle, \tag{24.76}$$

where the time-average is performed over the full interacting ground state $|\rangle$ (note that we have also dropped the site index λ, which will be explicitly included only when necessary, and maintained the spin index σ). By following the procedure highlighted in the previous sections, Equation 24.76 can be expanded up to the nth order in $H_{e\text{-}sp}$ (see Equation 24.17) as

$$[G(\tau,\tau')]_{\sigma\sigma'} = \sum_n \frac{(-i)^{n+1}}{n!}\int_C d\tau_1 \ldots \int_C d\tau_n \times \frac{{}_0\langle| T_C\{H_{e\text{-}sp}(\tau_1)\ldots H_{e\text{-}sp}(\tau_n)c_\sigma(\tau)c^\dagger_{\sigma'}(\tau')\}|\rangle_0}{{}_0\langle| S(-\infty,-\infty)|\rangle_0}, \tag{24.77}$$

where the S-matrix is defined in Equation 24.16 and the time averages are now over the known non-interacting ($J_{sd} = 0$) ground state, $|\rangle_0$. The time integration over τ is ordered on the contour C going from $-\infty$ to $+\infty$ and then returning from $+\infty$ to $-\infty$, since the ground state of the non-equilibrium system can only be defined at $-\infty$ [14]. If the expansion is truncated to the first order, one obtains a Zeeman-like term, which can be neglected as long as $\varepsilon_0 \gg \gamma_{\text{tip-S(sub-S)}}$. The first contribution of interest then appears at the second order. This can be obtained by inserting the explicit expression for $H_{e\text{-}sp}$ (t) [Equation 24.75] into Equation 24.76:

$$\begin{aligned}[G(\tau,\tau')]^{(2)}_{\sigma\sigma'} &= \frac{(-i)^3}{2!}J^2_{sd}\sum_{i,\alpha,\alpha',j,\beta,\beta'}\int_C d\tau_1\int_C d\tau_2 \\ &\times {}_0\left\langle\left|T_C\left\{c_\sigma(\tau)c^\dagger_\alpha(\tau_1)c_{\alpha'}(\tau_1)c^\dagger_\beta(\tau_2)c_{\beta'}(\tau_2)c^\dagger_{\sigma'}(\tau')\right\}\right|\right\rangle_0 \\ &\times {}_0\left\langle\left|T_C\{S^i(\tau_1)S^j(\tau_2)\}\right|\right\rangle_0[\sigma^i]_{\alpha\alpha'}[\sigma^j]_{\beta\beta'},\end{aligned} \tag{24.78}$$

where the indices i and j of the Pauli matrices run over the Cartesian coordinates x, y, and z.

A full contour-ordered expansion must now be performed on both the electron bracket and the spin bracket. The electron bracket has six different time ordering combinations, which are explicitly listed as follows:

$$\begin{aligned}&{}_0\left\langle\left|T_C\left\{c_\sigma(\tau)c^\dagger_\alpha(\tau_1)c_{\alpha'}(\tau_1)c^\dagger_\beta(\tau_2)c_{\beta'}(\tau_2)c^\dagger_{\sigma'}(\tau')\right\}\right|\right\rangle_0 \\ &= {}_0\left\langle\left|T_C\left\{c_\sigma(\tau)c^\dagger_\alpha(\tau_1)\right\}\right|\right\rangle_0 \times {}_0\left\langle\left|T_C\left\{c_{\alpha'}(\tau_1)c^\dagger_\beta(\tau_2)\right\}\right|\right\rangle_0 \\ &\times {}_0\left\langle\left|T_C\left\{c_{\beta'}(\tau_2)c^\dagger_{\sigma'}(\tau')\right\}\right|\right\rangle_0 + {}_0\left\langle\left|T_C\left\{c_\sigma(\tau)c^\dagger_\beta(\tau_2)\right\}\right|\right\rangle_0 \\ &\times {}_0\left\langle\left|T_C\left\{c_{\alpha'}(\tau_1)c^\dagger_{\sigma'}(\tau')\right\}\right|\right\rangle_0 \times {}_0\left\langle\left|T_C\left\{c_\alpha(\tau_2)c^\dagger_{\beta'}(\tau_1)\right\}\right|\right\rangle_0 \\ &+ {}_0\left\langle\left|T_C\left\{c_\sigma(\tau)c^\dagger_\alpha(\tau_1)\right\}\right|\right\rangle_0 \times {}_0\left\langle\left|T_C\left\{c_{\alpha'}(\tau_1)c^\dagger_{\sigma'}(\tau')\right\}\right|\right\rangle_0 \\ &\times {}_0\left\langle\left|T_C\left\{c^\dagger_\beta(\tau_2)c_{\beta'}(\tau_2)\right\}\right|\right\rangle_0 + {}_0\left\langle\left|T_C\left\{c_\sigma(\tau)c^\dagger_\beta(\tau_2)\right\}\right|\right\rangle_0 \\ &\times {}_0\left\langle\left|T_C\left\{c_{\beta'}(\tau_2)c^\dagger_{\sigma'}(\tau')\right\}\right|\right\rangle_0 \times {}_0\left\langle\left|T_C\left\{c^\dagger_\alpha(\tau_1)c_{\alpha'}(\tau_1)\right\}\right|\right\rangle_0 \\ &+ {}_0\left\langle\left|T_C\left\{c_\sigma(\tau)c^\dagger_{\sigma'}(\tau')\right\}\right|\right\rangle_0 \times {}_0\left\langle\left|T_C\left\{c^\dagger_\alpha(\tau_1)c_{\alpha'}(\tau_1)\right\}\right|\right\rangle_0 \\ &\times {}_0\left\langle\left|T_C\left\{c^\dagger_\beta(\tau_2)c_{\beta'}(\tau_2)\right\}\right|\right\rangle_0 - {}_0\left\langle\left|T_C\left\{c_\sigma(\tau)c^\dagger_{\sigma'}(\tau')\right\}\right|\right\rangle_0 \\ &\times {}_0\left\langle\left|T_C\left\{c_{\alpha'}(\tau_1)c^\dagger_\beta(\tau_2)\right\}\right|\right\rangle_0 \times {}_0\left\langle\left|T_C\left\{c_{\beta'}(\tau_2)c^\dagger_\alpha(\tau_1)\right\}\right|\right\rangle_0.\end{aligned} \tag{24.79}$$

The first and the second terms represent Fock-like Feynman diagrams, while the third and the fourth correspond to Hartree-like ones (note that Hartree-like diagrams vanish because of the spin selection rules as discussed later in this section). Both these pairs are equal under the exchange of the indexes. Finally, the last two combinations can be eliminated since they represent unconnected Feynman diagrams, which vanish in the averaging process [17]. This leaves us with a simplified expression, which, when compared to Equation 24.77, gives us

$$
{}_0\left\langle \left| T_C\left\{c_\sigma(\tau)c^\dagger_\alpha(\tau_1)c_{\alpha'}(\tau_1)c^\dagger_\beta(\tau_2)c_{\beta'}(\tau_2)c^\dagger_{\sigma'}(\tau')\right\}\right|\right\rangle_0
$$

$$
= 2i^3\delta_{\sigma\alpha}\delta_{\alpha'\beta}\delta_{\beta'\sigma'}[G_0(\tau,\tau_1)]_{\sigma\sigma}[G_0(\tau_1,\tau_2)]_{\alpha'\alpha'}[G_0(\tau_2,\tau')]_{\sigma'\sigma'}
$$

$$
+ 2i^3\delta_{\sigma\alpha}\delta_{\alpha'\sigma'}\delta_{\beta\beta'}[G_0(\tau,\tau_1)]_{\sigma\sigma}[G_0(\tau_1,\tau')]_{\sigma'\sigma'}[G_0(\tau_2,\tau_2)]_{\beta\beta}. \tag{24.80}
$$

In this case, since the averaging bracket is over the non-interacting ground state, G_0 represents the non-interacting electronic GF and can be calculated exactly (see Section 24.2.6).

We then return to Equation 24.78 and evaluate the spin bracket. The ground state of the spin system alone ($J_{sd} = 0$) can be found by diagonalizing exactly H_{sp}. This is achieved by constructing the full spin basis $\{|n\rangle\}$ where $n = -S, -S + 1, \ldots, +S$. Note that this step does not require any particular form for H_{sp}, although the details of the spin Hamiltonian determine the nature of the spin states and how these interact with the conducting electrons. Note also that, in the discussion of the results, we will keep labeling the eigenvalues of H_{sp} with the z-component of the total spin S, which in general is not a good quantum number because of the transverse anisotropy. However, such an anisotropy is small so that our notation remains approximately valid. The resulting eigenvectors, $|m\rangle$, and eigenvalues, ε_m, satisfy the Schrödinger equation $H_{sp}|m\rangle = \varepsilon_m|m\rangle$, and they can be used to re-write the operators $S^i(\tau)$, for $i = \{x, y, z\}$, as

$$
S^i(\tau) = \sum_{m,n}\langle m|S^i|n\rangle d^\dagger_m(\tau)d_n(\tau). \tag{24.81}
$$

Here, d_n $\left(d^\dagger_n\right)$ is an annihilation (creation) operator for a spin quasi-particle. These are then assumed to be fermionic in nature [22,23] so that they obey the anticommutation rules $\{d^\dagger_m, d_n\} = \delta_{mn}$ and $\{d^\dagger_m, d^\dagger_n\} = \{d_m, d_n\} = 0$. Such an assumption is valid as long as the excitations considered are always around the ground state, that is, under the condition that the spin system can always efficiently relax back to the ground state between two spin-flip events. Note that, in this situation, only a single spin state, $|m\rangle$, can be excited at a time, so that the particular particle statistics becomes immaterial. We can then define a contour-ordered spin GF (spin propagator) as follows:

$$
[D(\tau,\tau')]_{n,m} = -i\left\langle\left|T_C\{d_n(\tau)d^\dagger_m(\tau')\}\right|\right\rangle. \tag{24.82}
$$

By inserting the expressions in the Equations 24.81 and 24.82 into the spin bracket and by computing the time-ordered contraction, we finally obtain

$$
{}_0\left\langle\left|T_C\{S^i(\tau_1)S^j(\tau_2)\}\right|\right\rangle_0 = -\sum_{m,n}\left\langle m|S^i|n\right\rangle\left\langle n|S^j|m\right\rangle[D_0(\tau_1,\tau_2)]_{n,n}
$$

$$
\times[D_0(\tau_2,\tau_1)]_{m,m}, \tag{24.83}
$$

where D_0 is the non-interacting spin GF.

The set of Equations 24.80 and 24.83 can now be incorporated into the expression for the second-order contribution to the many-body electronic GF, Equation 24.78. Then, by using Dyson's equation [17], one can finally write the second-order contribution to the interacting self-energy as

$$
[\Sigma_{int}(\tau_1,\tau_2)]^{(2)}_{\sigma\sigma'} = -J^2_{sd}\sum_{i,j,\beta}\{[\sigma^i]_{\sigma\beta}[\sigma^j]_{\beta\sigma'} + [\sigma^i]_{\sigma\sigma'}[\sigma^j]_{\beta\beta}\}\times[G_0(\tau_1,\tau_2)]_{\beta\beta}
$$

$$
\times\sum_{m,n}\left\langle m|S^i|n\right\rangle\left\langle n|S^j|m\right\rangle[D_0(\tau_1,\tau_2)]_{n,n}[D_0(\tau_2,\tau_1)]_{m,m}. \tag{24.84}
$$

If we now assume that the non-interacting electronic ground state is spin degenerate, that is, $[G_0]_{\uparrow\uparrow} = [G_0]_{\downarrow\downarrow}$, then the only quantity of interest is the trace of the self-energy over the spin indices. By performing such a trace, the spin-independent self-energy finally reads

$$
\Sigma_{int}(\tau_1,\tau_2)^{(2)} = -2J^2_{sd}\sum_{i,m,n}\left|\left\langle m|S^i|n\right\rangle\right|^2 G_0(\tau_1,\tau_2)[D_0(\tau_1,\tau_2)]_{n,n}
$$

$$
\times[D_0(\tau_2,\tau_1)]_{m,m}, \tag{24.85}
$$

where we have used the results $\mathrm{Tr}[\sigma^i\sigma^j] = \delta_{ij}$ and $\mathrm{Tr}[\sigma^i] = 0$. Note that the relation $\mathrm{Tr}[\sigma^i] = 0$ guarantees that the Hartree-like diagrams do not contribute to the self-energy. There is no analogous relation when the interaction of the conducting electrons is with atomic vibrations (phonons); hence, in that situation, the Hartree-like diagrams cannot be neglected a priori. Interestingly, for the phonon case, the Hartree-like diagrams drive polaronic distortions [24], which are expected not to have an equivalent in the case of spin scattering.

At this point, we can calculate the real-time quantities, such as the < (>) self-energies, by using Langreth's theorem for the time ordering over the contour, $\tau_1 \in C_1(C_2)$ and $\tau_2 \in C_2(C_1)$ [14]. C_1 is the time ordering contour from $-\infty$ to $+\infty$, and C_2 is the time anti-ordering contour from $+\infty$ to $-\infty$. We find

$$
\Sigma^{\lessgtr}_{int}(t_1,t_2)^{(2)}
$$

$$
= -2J^2_{sd}\sum_{i,m,n}\left|\left\langle m|S^i|n\right\rangle\right|^2 G^{\lessgtr}_0(t_1,t_2)\left[D^{\lessgtr}_0(t_1,t_2)\right]_{n,n}\left[D^{\gtrless}_0(t_2,t_1)\right]_{m,m}
$$

$$
= -2J^2_{sd}\sum_{i,m,n}\left|\left\langle m|S^i|n\right\rangle\right|^2 G^{\lessgtr}_0(t_1,t_2)P_n(1-P_m)e^{\pm i(\varepsilon_m-\varepsilon_n)(t_1-t_2)}, \tag{24.86}
$$

where, in the second step, we have written explicitly $D_0^{\lessgtr}(t_1,t_2)$ in terms of the spin-level occupations, $P_n = \left\langle d^\dagger_n d_n\right\rangle$. The dependence of $\Sigma^{\lessgtr}_{int}$ over the energy, E, can be found by simple Fourier transform:

$$
\Sigma^{\lessgtr}_{int}(\omega)^{(2)} = -2J^2_{sd}\sum_{i,m,n}\left|\left\langle m|S^i|n\right\rangle\right|^2 P_n(1-P_m)G^{\lessgtr}_0(\omega\pm\Omega_{mn}), \tag{24.87}
$$

where $\Omega_{mn} = \varepsilon_m - \varepsilon_n$ and + (−) symbol corresponds to $\Sigma^<$ ($\Sigma^>$).

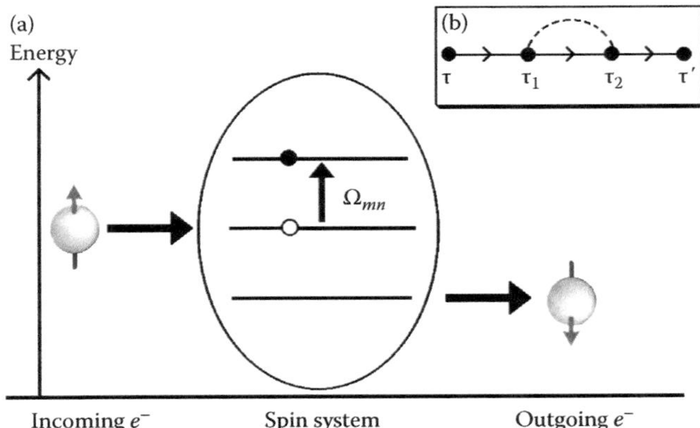

FIGURE 24.4 Cartoon of the inelastic process described by the greater self-energy, $\Sigma_{\text{int}}^{>}$. In (a), an incoming electron scatters against a localized spin and decreases its energy by Ω_{mn}. This is transferred to the local spin system, which undergoes a spin transition $|n\rangle \rightarrow |m\rangle$. In the inset (b), we display a Feynman Fock-like diagram describing the interaction in the time domain. (Figure adapted from Hurley, A. et al., *Phys. Rev. B*, 84, 035427, 2011.)

Electron-spin scattering events are now fully incorporated into $\left[\Sigma_{\text{int}}^{\lessgtr}(\omega)\right]$. In particular Equation 24.87 describes the process where an incoming electron with energy E experiences a spin-flip process, which changes its energy by $\pm\Omega_{mn}$. This is the result of the electron–spin interaction with the local spins. Such a process is schematically represented in Figure 24.4. Note that the probability for an excitation to occur is determined by the prefactors $\langle m|S^i|n\rangle|^2 P_n(1 - P_m)$, that is, by the state of the spin system.

We conclude this section by discussing the limits of validity of our perturbative expansion. At a first glance, Equation 24.87 suggests that the strength of the second order Σ_{int} depends solely on J_{sd}^2. This will indeed result in a large perturbation parameter and thus in a possibly divergent perturbative expansion (see Table 24.1 for an estimate of the various parameters). However, both the < and > GFs are proportional to the DOS of the scattering region, ρ. In the weak coupling limit, one has $\rho \sim \Gamma/\varepsilon_0^2$, where ε_0 is the onsite energy of the atom under consideration (the one under the tip) and Γ is the total broadening due to the electrodes, $\Gamma = \Gamma_{\text{tip-S}} + \Gamma_{\text{sub-S}}$. As a consequence, we have $\Sigma_{\text{int}}^{\lessgtr}(\omega)^{(2)} \sim \alpha J_{\text{sd}}$, where $\alpha = \rho J_{\text{sd}}$ is a dimensionless quantity. By continuing the expansion to the third order (see next section), one will find an additional contribution to $\Sigma_{\text{int}}^{\lessgtr}$ proportional to $\alpha^2 J_{\text{sd}}$, that is, it will be discovered that the perturbation expansion parameter is the dimensionless quantity α. Interestingly, α is the product between the Fermi level DOS at the spin site and the exchange parameter J_{sd}, that is, it is essentially the Stoner parameter [25]. This means that the perturbative expansion is justified when the system is away from developing a magnetic instability. The convergence of the perturbation series is then guaranteed by our weak coupling condition, which establishes that ρ is small and then $\alpha \ll 1$.v

TABLE 24.1 Empirical Parameters Used in the Numerical Simulations Presented in This Work and Their Assigned Numerical Values

Quantity	Symbol	Value	Origin [References]
Atomic spin	S	$\frac{5}{2}$	Exp. [30]
Temperature	T	0.6 K	Exp. [30]
d-d exchange	J_{dd}	+6.2 meV	Exp. [30]
s-d exchange	J_{sd}	+500 meV	DFT [31,32]
Fermi energy	ε_{F}	0 meV	DFT [31]
Lead hopping integral	γ_0	10,000 meV	DFT [31]
Spin on-site energy	ε_0	1000 meV	DFT [31]
Tip-spin hopping	$\gamma_{\text{tip-S}}$	50 meV	DFT [31]
Substrate-spin hopping	$\gamma_{\text{sub-S}}$	500 meV	DFT [31]
Axial anisotropy	D	−0.037 meV	Exp. [1]
Transverse anisotropy	E	0.007 meV	Exp. [1]

24.3.3 Third-Order Expansion

Having described the second-order expansion, it is now interesting to continue further to the third order and to examine what effects this has on the conductance spectra. Our starting point is again the interacting contour-ordered GF expanded up to order n, Equation 24.77:

$$\left[G(\tau,\tau')\right]_{\sigma,\sigma'}^{(3)} = \frac{(-i)^4}{3!} J_{\text{sd}}^3 \sum_{\alpha\alpha',\beta\beta',\gamma\gamma'} \int_C d\tau_1 \int_C d\tau_2 \int_C d\tau_3$$
$$\times \langle T_C\{c_\sigma(\tau)c_\alpha^\dagger(\tau_1)c_{\alpha'}(\tau_1)c_\beta^\dagger(\tau_2)c_{\beta'}(\tau_2)c_\gamma^\dagger(\tau_3)c_{\gamma'}(\tau_3)c_{\sigma'}^\dagger(\tau')\}\rangle$$
$$\times \sum_{i,j,k} \langle T_C\{S^i(\tau_1)S^j(\tau_2)S^k(\tau_3)\}\rangle [\sigma^i]_{\alpha\alpha'}[\sigma^j]_{\beta\beta'}[\sigma^k]_{\gamma\gamma'}. \tag{24.88}$$

We first consider the spin operators. Following the procedure presented in the previous section, these are decomposed into products of quasi-particle operators obeying the Fermi–Dirac statistics, thus that the ith component ($i = x, y$, and z) of the spin operator **S** is written according to Equation 24.81 and the corresponding GF is that of Equation 24.82. By substituting these expressions into Equation 24.88, we obtain

$$\left[G(\tau,\tau')\right]_{\sigma,\sigma'}^{(3)} = \frac{(-i)^4}{3!} J_{\text{sd}}^3 \sum_{\alpha\alpha',\beta\beta',\gamma\gamma'} \int_C d\tau_1 \int_C d\tau_2 \int_C d\tau_3$$
$$\times \left\langle T_C\left\{c_\sigma(\tau)c_\alpha^\dagger(\tau_1)c_{\alpha'}(\tau_1)c_\beta^\dagger(\tau_2)c_{\beta'}(\tau_2)c_\gamma^\dagger(\tau_3)c_{\gamma'}(\tau_3)c_{\sigma'}^\dagger(\tau')\right\}\right\rangle$$
$$\times \sum_{mm',nn',ll'} \langle T_C\{d_m^\dagger(\tau_1)d_{m'}(\tau_1)d_n^\dagger(\tau_2)d_{n'}(\tau_2)d_l^\dagger(\tau_3)d_{l'}(\tau_3)\}\rangle$$
$$\times \sum_{i,j,k} \langle m|S^i|m'\rangle\langle n|S^j|n'\rangle\langle l|S^k|l'\rangle [\sigma^i]_{\alpha\alpha'}[\sigma^j]_{\beta\beta'}[\sigma^k]_{\gamma\gamma'}. \tag{24.89}$$

We now perform the chronological contractions by using Wick's theorem for both the electron and spin brackets. We have previously stated that the electron brackets, the spin selection rules, and the electronic spin-degeneracy yield a vanishing contribution to any "fermion loop" contraction of the form $\langle c^\dagger(\tau)c(\tau)\rangle$. Therefore, one needs only to consider the three Fock-like contributions to the electron bracket, which are all equal under exchange of contour indexes. Furthermore, the spin bracket brings two Fock-like terms. Here, we retain only one of them in order to simplify the discussion, but both have been included in the numerical simulations that we will show later. Then Equation 24.89 can be written in terms of the known non-interacting electron and spin GFs:

$$\begin{aligned}
\left[G(\tau,\tau')\right]^{(3)}_{\sigma,\sigma'} &= i\frac{J^3_{\mathrm{sd}}}{2}\sum_{\beta,\gamma}\int_C d\tau_1\int_C d\tau_2\int_C d\tau_3\\
&\times\left[G_0(\tau,\tau_1)\right]_{\sigma\sigma}\left[G_0(\tau_1,\tau_2)\right]_{\beta\beta}\left[G_0(\tau_2,\tau_3)\right]_{\gamma\gamma}\left[G_0(\tau_3,\tau')\right]_{\sigma'\sigma'}\\
&\times\sum_{m,n,l}\left[D_0(\tau_1,\tau_2)\right]_{n,n}\left[D_0(\tau_2,\tau_3)\right]_{l,l}\left[D_0(\tau_3,\tau_1)\right]_{m,m}\\
&\times\sum_{i,j,k}\langle m|S^i|n\rangle\langle n|S^j|l\rangle\langle l|S^k|m\rangle[\sigma^i]_{\sigma\beta}[\sigma^j]_{\beta\gamma}[\sigma^k]_{\gamma\sigma'}.
\end{aligned} \tag{24.90}$$

We can now use Dyson's equation to extract from Equation 24.90 the third-order contribution to the interacting self energy, which takes the form

$$\begin{aligned}
\Sigma_{\mathrm{int}}(\tau_1,\tau_3)^{(3)} &= 2iJ^3_{\mathrm{sd}}\int_C d\tau_2 G_0(\tau_1,\tau_2)G_0(\tau_2,\tau_3)\\
&\times\sum_{m,n,l}D_n(\tau_1,\tau_2)D_l(\tau_2,\tau_3)D_m(\tau_3,\tau_1)\\
&\times\sum_{i,j,k}(2i\varepsilon_{ijk})\langle m|S^i|n\rangle\langle n|S^j|l\rangle\langle l|S^k|m\rangle.
\end{aligned} \tag{24.91}$$

Note that we have simplified the notation by writing the diagonal elements of the non-interacting spin GF as $D_m(\tau,\tau')$. We have also taken into account the electron spin degeneracy ($[G_0]_{\uparrow\uparrow} = [G_0]_{\downarrow\downarrow}$) and traced over the spin indices, $\mathrm{Tr}[\sigma^i\sigma^j\sigma^k] = 2i\varepsilon_{ijk}$.

Equation 24.91 now needs to be written in terms of the real times (t, t'), so that a close expression for the energy resolved $\Sigma^{\lessgtr}_{\mathrm{int}}$ can be explicitly written. Such a derivation is based on the Keldysh formalism for the evaluation of time-contour integrals [13,14] and it has been illustrated in the appendix of reference [26]. Finally, the complete expression for the interacting self-energies is obtained by adding the second-order term derived before, see Equation 24.87, to give

$$\begin{aligned}
\Sigma^{\lessgtr}_{\mathrm{int}}(\omega) &= -2J^2_{\mathrm{sd}}\sum_{m,n,l}P_l(1-P_m)G^{\lessgtr}_0(E\pm\Omega_{ml})\Bigg\{\delta_{nl}\sum_i|\langle m|S^i|n\rangle|^2\\
&+2i(\rho J_{\mathrm{sd}})\sum_{ijk}\varepsilon_{ijk}\langle m|S^i|n\rangle\langle n|S^j|l\rangle\langle l|S^k|m\rangle\\
&\times\left[\ln\left|\frac{W}{\sqrt{(E+V\pm\Omega_{mn})^2+(k_BT)^2}}\right|\right.\\
&\left.+\ln\left|\frac{W}{\sqrt{(E+V\pm\Omega_{nl})^2+(k_BT)^2}}\right|\right]\Bigg\},
\end{aligned} \tag{24.92}$$

where the plus (minus) sign corresponds to < (>). If we now assume that the scattering region is much more strongly coupled to the substrate than to the STM tip ($\gamma^{\mathrm{sub\text{-}S}} \gg \gamma^{\mathrm{tip\text{-}S}}$) and $\varepsilon_0 \gg \varepsilon_F$, we can approximate the tip DOS around ε_F with a constant, $\rho = (\Gamma/2\pi)/\left[\varepsilon_0^2+\Gamma^2\right]$. The weak coupling to the STM tip also ensures that the spin system remains always close to equilibrium [27], that is, in its ground state, so that $P_0 \sim 1$. A crucial feature of the third-order contribution to the self-energies is the appearance of a zero-temperature logarithmic divergence at the excitation energies Ω_{mn}. This is the fingerprint of the Kondo effect and will be the key ingredient to describe zero-bias anomalies in the conductance spectrum as well as the details of the spectrum lineshape. Note that our formalism and derivation leading to Equation 24.92 has been proposed first for describing the Kondo effect in quantum dots [22,23].

24.3.4 Additional Lineshape Features

In the following section, we will extend our formalism for spin scattering to include some additional features necessary to describe transport with spin-polarized electrodes and/or when the current is intense. In particular, we will generalize the electronic self-energy to the spin-polarized case, and we will construct a second-order perturbation expansion for the spin propagator, so that the non-equilibrium occupation of the various spin states can be evaluated.

24.3.4.1 Spin-Polarized Electron Self-Energy

The starting point for generalizing the theory to spin-polarized electrodes is once again the perturbation expansion of the spin-resolved contour-ordered interacting GF, namely, Equation 24.77. Here, we consider only the tip to be spin-polarized, with the spin-polarization being described by a single parameter, η. This means that the tip-induced electronic broadening is different for the two spin species, namely, $[\Gamma_{\mathrm{tip\text{-}S}}]_{\uparrow\uparrow} = (1+\eta)\Gamma_{\mathrm{tip\text{-}S}}/2$ and $[\Gamma_{\mathrm{tip\text{-}S}}]_{\downarrow\downarrow} = (1-\eta)\Gamma_{\mathrm{tip\text{-}S}}/2$, where

$\Gamma_{\text{tip-S}}$ is the non-polarized broadening. As a result, we have $[G_0^{\lessgtr}(E)]_{\uparrow\uparrow} \neq [G_0^{\lessgtr}(E)]_{\downarrow\downarrow}$. One then needs to carry out the same steps as in the derivation of the second-order expansion, but without taking the trace over the spin indices. We then obtain the spin-resolved components of the interacting self-energy:

$$\begin{aligned}[\Sigma_{\text{int}}^{\lessgtr}(E)]_{\uparrow\uparrow}^{(2)} &= -J_{\text{sd}}^2 \sum_{m,n} [G_0^{\lessgtr}(E \pm \Omega_{mn})]_{\uparrow\uparrow} \\ &\quad \times \left(\delta_{nm}\chi P_n S_{mn}^z + P_n(1-P_m)|S_{mn}^z|^2\right) \\ &\quad - J_{\text{sd}}^2 \sum_{m,n} [G_0^{\lessgtr}(E \pm \Omega_{mn})]_{\downarrow\downarrow} P_n(1-P_m)|S_{mn}^+|^2, \end{aligned} \quad (24.93)$$

$$\begin{aligned}[\Sigma_{\text{int}}^{\lessgtr}(E)]_{\downarrow\downarrow}^{(2)} &= -J_{\text{sd}}^2 \sum_{m,n} [G_0^{\lessgtr}(E \pm \Omega_{mn})]_{\downarrow\downarrow} \\ &\quad \times \left(-\delta_{nm}\chi P_n S_{mn}^z + P_n(1-P_m)|S_{mn}^z|^2\right) \\ &\quad - J_{\text{sd}}^2 \sum_{m,n} [G_0^{\lessgtr}(E \pm \Omega_{mn})]_{\uparrow\uparrow} P_n(1-P_m)|S_{mn}^-|^2, \end{aligned} \quad (24.94)$$

where we have introduced the inelastic ratio, $\chi = \varepsilon_0/J_{\text{sd}}$, which is typically in the range of 1–2. This is related to the so-called magnetoresistive elastic term of the Hamiltonian, Equation 24.73, which becomes relevant for spin-polarized transport. The < (>) self-energy describes the process where an incoming (outgoing) electron can excite (relax) the spin system by Ω_{mn} with probability dependent on the occupation of the spin levels P_m, P_n, and on the spin selection rules $S_{mn}^{z,+,-}$ (note: $S^{\pm} = S^x \pm iS^y$). The first term in both Equations 24.93 and 24.94 preserves the electron spins in the scattering event and it is associated to the magnetoresistive elastic term of Equation 24.73. The other contributions are inelastic in nature and depend on the spin orientation of the incoming/outgoing electron from/to the tip.

24.3.4.2 The Spin Propagator

So far, we have assumed that the spin system is always in its ground state before any scattering event, that is, that there is no build up of spin population. This is justified by the usual small current in STM experiments, by the low temperature and by the fact that the electrodes are not spin-polarized. The situation, however, changes for spin-polarized electrodes and large current density, since a spin-flip scattering event can be followed by a second one without the spin system having enough time to relax to its ground state. In this case, a realistic transport description should include the calculation of the non-equilibrium spin population.

Let us first consider the non-interacting case, $J_{\text{sd}} = 0$, at finite temperature. We then assume that the spin system is adiabatically coupled to a heat-bath kept at temperature T, which generates a weak broadening of the single spin states ε_m of magnitude k_BT. In this non-interacting case, the spin GF, Equation 24.83, simply becomes

$$[D_0^{\lessgtr}(E)]_{m,n} = \frac{[\Pi_0^{\lessgtr}(E)]_{m,n}}{(E-\varepsilon_m)^2 + (k_BT)^2}, \quad (24.95)$$

where $[\Pi_0^{>}(E)]_{m,n} = \delta_{m,n}(1-P_m^0)k_BT$ and $[\Pi_0^{<}(E)]_{m,n} = \delta_{m,n}P_m^0 k_BT$, with P_m^0 being the ground state population at $T = 0$.

If we now switch on the electron–spin interaction, the spin population will become bias-dependent as spins can be pumped from the magnetic electrodes. The task is then that of calculating the non-equilibrium spin population, P_n, as these are the relevant quantities entering the total electronic GF. By combining the first- and second-order contributions to the spin self-energy, we can derive (see reference [28] for details) a master equation for P_n, in terms of the total spin self-energy $\Pi^{\lessgtr}(E)$,

$$\frac{dP_n}{dt} = \frac{1}{\hbar} \sum_m \int_{-\infty}^{+\infty} dE \left\{ [\Pi^{>}(E)]_{nm} [D_0^{<}(E)]_{mn} - [\Pi^{<}(E)]_{nm} [D_0^{>}(E)]_{mn} \right\}. \quad (24.96)$$

After some rearrangement, this can be written in more compact form

$$\frac{dP_n}{dt} = \sum_l [P_n(1-P_l)W_{ln} - P_l(1-P_n)W_{nl}] + (P_n^0 - P_n)/\beta, \quad (24.97)$$

where the bias-dependent transition rate from an initial state l to a final state n is given by (see the appendix of [28] for a full description of the calculation)

$$\begin{aligned} W_{nl} &= -4\frac{(\rho J_{\text{sd}})^2}{\Gamma} \sum_{\eta,\eta'} \zeta(\mu_\eta - \mu_{\eta'} + \Omega_{ln}) \\ &\quad \times \{\chi S_{nn}^z([\Gamma_\eta]_{\uparrow\uparrow}[\Gamma_{\eta'}]_{\uparrow\uparrow} - [\Gamma_\eta]_{\downarrow\downarrow}[\Gamma_{\eta'}]_{\downarrow\downarrow}) \\ &\quad + |S_{nl}^z|^2 ([\Gamma_\eta]_{\uparrow\uparrow}[\Gamma_{\eta'}]_{\uparrow\uparrow} + [\Gamma_\eta]_{\downarrow\downarrow}[\Gamma_{\eta'}]_{\downarrow\downarrow}) \\ &\quad + |S_{nl}^+|^2 [\Gamma_\eta]_{\downarrow\downarrow}[\Gamma_{\eta'}]_{\uparrow\uparrow} + |S_{nl}^-|^2 [\Gamma_\eta]_{\uparrow\uparrow}[\Gamma_{\eta'}]_{\downarrow\downarrow}\}, \end{aligned} \quad (24.98)$$

where $\zeta(x) = x/[1-e^{-x/k_BT}]$ and μ_η is the chemical potential in lead η = {tip, sub}. The form of $\zeta(x)$ is such that for $\eta = \eta'$ the resulting transition rates W_{nl} are bias independent and do not contribute to the current. They do, however, contribute to the spin relaxation time, that is, to the time taken for the spin to relax back to its equilibrium state. This relaxation time gets shorter if the coupling between the sample and the leads increases. Also, the smaller the inelastic energy transition Ω_{mn}, the longer the spin will remain in its excited state before relaxing back to equilibrium. Again, we assume that the onsite energy is large enough so that the DOS of the sample remains constant in the small energy window of interest, that is, $\rho = \Gamma/(\varepsilon_0^2 + \Gamma^2)$.

Returning to Equation 24.97, it is worth noting that we are interested only in the steady state, so that the relevant quantities are the non-equilibrium steady-state spin populations at a given bias. We can then set $dP_n(t)/dt = 0$ and solve Equation 24.97 for the steady state, by simply iterating from an initial trial population $\left(P_l = P_l^0\right)$. Finally, the converged populations are used to evaluate the electronic spin-scattering self-energy.

24.3.5 NEGF Method for Electron Transport

Finally, before showing a sample of results obtained by applying the theory discussed so far, we need to generalize the electron transport scheme introduced in Sections 24.2.5 and 24.2.6 to include the effect of the inelastic electron–spin interaction. We consider a two-probe device, divided into three distinct regions: two semi-infinite leads representing, respectively, the STM tip and the substrate and a scattering region (see Figure 24.1). As mentioned before, the leads act as charge reservoirs and they are characterized by their chemical potentials, respectively, μ_{tip} and μ_{sub}. An external bias is introduced in the form of a relative shift (symmetric) of the two chemical potentials. The underlying assumption of our method is that under the external bias there is no rearrangement of the electronic structure of the leads, that is, that the electron screening and the spin relaxation in the leads are efficient. This simplifies the problem to that of calculating the retarded (advanced) GF of the scattering region [6,7] only:

$$G^{\text{r}}(\omega) = \lim_{\delta\to 0}[(\omega - i\delta)I - H_{\text{e}} - \Sigma^{\text{r}}(\omega)]^{-1}. \tag{24.99}$$

Here, $\Sigma^{\text{r}}(\omega)$ is the total retarded self-energy, which now incorporates the effects of the leads and of the inelastic interaction:

$$\Sigma^{\text{r}}(\omega) = \Sigma^{\text{r}}_{\text{tip}}(\omega) + \Sigma^{\text{r}}_{\text{sub}}(\omega) + \Sigma^{\text{r}}_{\text{int}}(\omega), \tag{24.100}$$

where $\Sigma^{\text{r}}_{\text{tip}}(\omega)$ and $\Sigma^{\text{r}}_{\text{sub}}(\omega)$ are, respectively, the STM tip and substrate retarded self-energies, while $\Sigma^{\text{r}}_{\text{int}}$ is the scattering self-energy describing the electron–spin interaction. Formally, the action of the electron–spin interaction is similar to that of a current–voltage electrode, so that $\Sigma^{\text{r}}_{\text{int}}$ can be interpreted as describing a fictitious lead, conserving the total current but breaking the electron and the spin phase coherence [29]. The leads' self-energies are defined in Equation 24.64. Finally, the retarded electron-spin scattering self-energy is found from the Hilbert transform of the < and > counterparts [20]:

$$\Sigma^{\text{r}}_{\text{int}}(\omega) = \mathcal{PV}\int_{-\infty}^{\infty}\frac{d\omega'}{2\pi}\frac{\Sigma^{>}_{\text{int}}(\omega') + \Sigma^{<}_{\text{int}}(\omega')}{\omega' - \omega} - \frac{i}{2}\left[\Sigma^{>}_{\text{int}}(\omega) + \Sigma^{<}_{\text{int}}(\omega)\right], \tag{24.101}$$

where $\mathcal{PV}$ denotes the principal value and where $\Sigma^{\lessgtr}_{\text{int}}(\omega)$ have been given before.

In addition to the retarded self-energy, also the < and > ones and the GF can be expressed as a sum over all three contributions:

$$\Sigma^{\lessgtr}(\omega) = \Sigma^{\lessgtr}_{\text{tip}}(\omega) + \Sigma^{\lessgtr}_{\text{sub}}(\omega) + \Sigma^{\lessgtr}_{\text{int}}(\omega), \tag{24.102}$$

$$G^{\lessgtr}(\omega) = G^{\text{r}}(\omega)[\Sigma^{\lessgtr}(\omega)]G^{\text{a}}(\omega). \tag{24.103}$$

The external bias, introduced as a shift of the leads chemical potentials $\mu_{\text{tip}} = \varepsilon_{\text{F}} + eV/2$ and $\mu_{\text{sub}} = \varepsilon_{\text{F}} + eV/2$, enters in the leads self-energies via the replacement $E \to \omega \pm eV/2$. Finally, the current can be calculated at finite V at any of the leads i:

$$I_i = \int_{-\infty}^{\infty}\frac{d\omega}{2\pi}\left\{\Sigma^{<}_{i}(\omega)G^{>}(\omega) - \Sigma^{>}_{i}(\omega)G^{<}(\omega)\right\}, \tag{24.104}$$

while the V-dependent conductance, $G = \mathrm{d}I_i/\mathrm{d}V$, is found by numerical differentiation.

In concluding this section, we would like to discuss the expected magnitude of the inelastic contribution to $\mathrm{d}I_i/\mathrm{d}V$ with respect to the elastic one. The ratio between two such contributions essentially corresponds to the ratio between the interacting and the non-interacting ($J_{\text{sd}} = 0$) GFs. A simple calculation shows that the unperturbed GF differs by a factor of $\Sigma_{\text{int}}/\Gamma$ from the fully interacting one. Previously, we have shown that $\Sigma^{(2)}_{\text{int}} \sim \alpha J_{\text{sd}}$; therefore, to the second order, the ratio between the elastic and inelastic contributions to $\mathrm{d}I_i/\mathrm{d}V$ turns out to be proportional to the dimensionless factor $\alpha(J_{\text{sd}}/\Gamma) \sim (J_{\text{sd}}/\varepsilon_0)^2$. Analogously, the third-order contribution will account for a factor $\alpha^2(J_{\text{sd}}/\Gamma)$. With this in hands and by using the experimental parameters (see Table 24.1), we can conclude that the contribution originating from the second-order expansion will be significant, while that from the third order will be small.

24.4 Examples of SF-IETS Experiments

All the theoretical machinery developed so far will be now put to work to describe a recent range of experiments, where an STM tip probes a magnetic nanostructure deposited on an insulating surface. We will present results by following the same strategy used for the theoretical part, namely, by looking at increasingly complex situations, where increasingly higher levels of theory are necessary.

24.4.1 Mn Mono-Atomic Chains on CuN: Second-Order Theory

We start our discussion by looking at the SP-IETS spectrum of Mn mono-atomic chains of different lengths deposited on a CuN surface and probed, in a low current mode, by a non magnetic STM tip [30]. The main features of the experiments can be captured by our second-order perturbation theory for non magnetic leads and equilibrium spin populations. In Table 24.1, we list all the parameters needed for our simulations and their assigned values.

These have been either inferred from experiments [30] or have been estimated by DFT calculations [31]. The local Mn spin is set to be 5/2, as proposed in the original experimental works [1,30], confirmed by DFT [31] and expected from the nominal Mn valence. The spin–spin coupling parameter J_{dd} corresponds to an antiferromagnetic order between the neighboring Mn, a feature verified in the experimental conductance spectra. The lead on-site energy is suitably set to zero and simply defines the reference potential, while J_{sd} is determined from theory to be of the order of 500 meV [31,32]. We evaluate the Fermi functions of the leads at the small temperature of 0.6 K. This allows us to include minor thermal smearing of the electron gas in the leads and consequently of the conductance profile. Finally, we notice that the scattering region is expected to be significantly more strongly coupled to the substrate than to the tip.

The left panel of Figure 24.5 shows the calculated dI/dV normalized against the elastic contribution, G_{el}, [calculated for $\Sigma_{int}(E) = 0$] for N-atom long Mn chains ($N \leq 4$) in no external magnetic field. The most relevant feature is the presence of a number of conductance steps characteristic of a specific chain, which appear at a well-defined V. These correspond to critical voltages where a magnetic excitation becomes possible. Such an excitation opens an additional conducting channel (inelastic) and the conductance increases. Note that, in general, one does not necessarily expect the conductance to increase at the excitation bias threshold. In fact, in the case of scattering to phonons, both conductance enhancements and suppressions have been observed, with the latter originating from the suppression of the elastic channel following the opening of the inelastic one. In general, there are no strict rules determining whether the conductance should become larger or smaller at the excitation threshold. However, there exist *propensity rules*, which essentially suggest that when the elastic transmission coefficient is small (large) the conductance should increase (decrease) due to inelastic scattering [33]. As in SF-IETS experiments, the elastic channel is tunneling in nature, one expects always a low current situation and therefore an increase of the conductance at the inelastic step.

From the figure, we observe that the relative conductance increase due to the inelastic contribution is of the order of 1/4 (for N = 2, where the amplitudes of the spin matrix elements of the self-energy are approximately unity). For our choice of parameters, the scaling factor $(J_{sd}/\varepsilon_0)^2$ is one-fourth, so that precisely a relative conductance step of one-fourth is expected. Such good agreement constitutes a strict validity test for the theory and demonstrates that our perturbative expansion is the right tool to tackle these problems. For N = 3,4 the spectrum is calculated with the STM tip placed above the second atom in the chain, but in Figure 24.5, we also show results for the tip above the first one (for N = 3). Notably, the two spectra are relatively different as the size of the conductance step at ~17 meV depends on the specific atom probed by the tip. A similar occurrence is seen for N = 4.

When the calculated conductance profiles are compared with the experimental ones (Figure 24.5 right panel), a good qualitative agreement emerges. In particular, we notice the intriguing dependence of the conductance profile over the parity of the chains, with chains comprising an odd number of atoms exhibiting a conductance dip at around V = 0, which is absent for even chains. It is worth noting, however, that the spectra for N = 3 and N = 4 contrast slightly with the experimental ones, which are asymmetric with respect to the bias

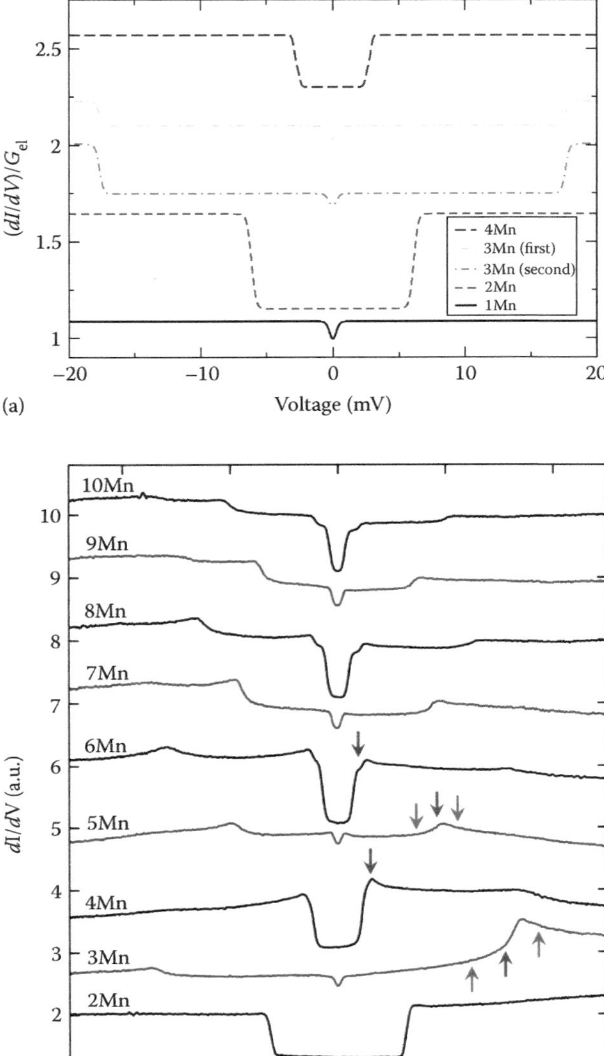

FIGURE 24.5 SF-IETS conductance traces of Mn mono-atomic chains deposited on a CuN surface. In the panel (a), we show the calculated conductance spectra for Mn chains of different lengths, N. The various spectra, except for N = 1, are offset for clarity. The tip is placed above the second atom of the chain for chains with N > 2 and also over the first atom in the case of N = 3 (light-gray dashed line). In the panel (b), we present the measured conductance traces for the same system. (Figure (a) reproduced from Hurley, A. et al., *Phys. Rev. B*, 84, 035427, 2011; figure (b) reproduced from Hirjibehedin, C.F. et al., *Science*, 312, 1021, 2006.)

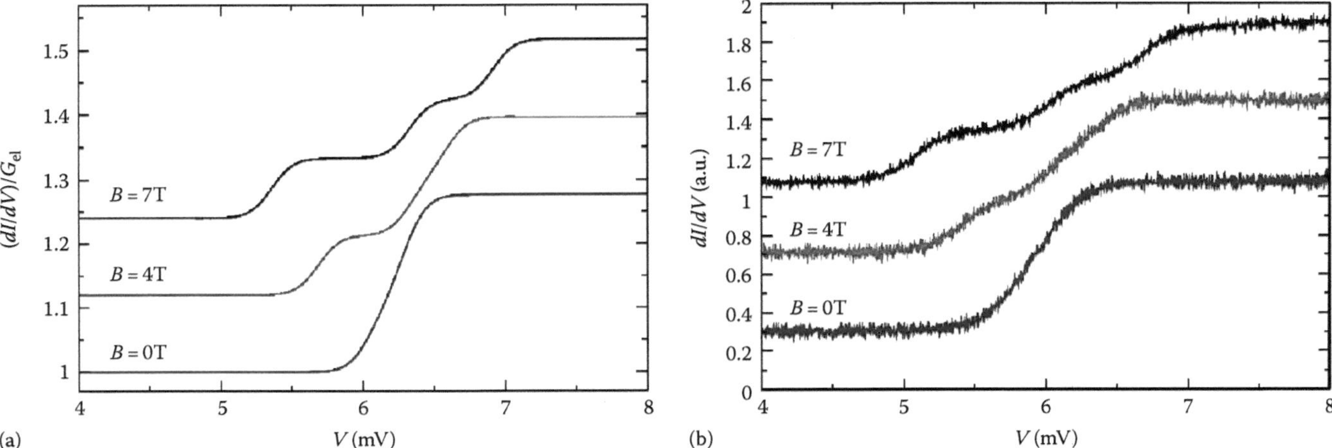

FIGURE 24.6 SF-IETS conductance traces of a Mn dimer deposited on a CuN surface in an external longitudinal magnetic field: (a) theory and (b) experiments. Note the magnetic field induced splitting of the conductance step, originating from the Zeeman splitting of the first $S_{tot} = 1$ excited state. (Panel (a) reproduced from Hurley, A. et al., *Phys. Rev. B*, 84, 035427, 2011; panel (b) reproduced from Hirjibehedin, C.F. et al., *Science* 312, 1021, 2006.)

and also exhibit some slope at the conductance steps. We argue that such minor deviations are simply due to fine features in the DOS that will not be addressed here [28]. We would like to remark here that in order to recreate the conductance profile of the trimer (3Mn), an additional ferromagnetic second-nearest-neighbor interaction of strength $\sim J_{dd}/2$ between the local spins at the edge of the chain must be included in the model [34]. The inclusion of such term changes the position of the conductance step from a second excited state at 27 mV to a first excited state at 16 mV. This correction is also included in the calculations for $N = 4$, again giving good agreement with experiments.

Going into more details of the conductance spectra, let us first discuss the case of odd-numbered chains and their zero-bias conductance anomaly. The ground state of each odd chain has a net total spin $S_{tot} = 5/2$. The transverse and axial anisotropy lift the ground state degeneracy and allow a transition between the ground state and the excited state. The excitation of such a transition results in a conductance step at a voltage corresponding to the transition energy and height corresponding to the transition rates $|\langle m|S^i|n\rangle|^2$. Since the anisotropy is small (~0.01 meV, Table 24.1), the excitation energy is small as well and features in the conductance profile near $V = 0$. More precisely, such an excitation corresponds to scattering events that produce the transition $|m = 5/2\rangle \rightarrow |m = 3/2\rangle$ (m is the magnetic quantum number) but that also preserves the total spin $S_{tot} = 5/2$. In the case of a single Mn ion ($N = 1$), this is the only transition available from the ground state so that no further conductance steps appear.

Let us take the $N = 2$ case. Now the zero-bias anomaly is not present, in agreement with the fact that the $N = 2$ ground state is a singlet ($S_{tot} = 0$), but there is a second conductance step at $V = J_{dd}/2e = 6.2$ mV. This is now due to a transition, which does not conserve the total spin, namely, a transition from the $S_{tot} = 0$ ground state to the first $S_{tot} = 1$ excited state. Also in this case, there is only one transition available from the ground state, so that no further conductance features can be found in the dI/dV plots. Finally, a glance at chains with $N > 2$ reveal that multiple conductance steps are present, indicating that several final states are accessible from the ground state.

From this discussion, one can appreciate the unique power of SF-IETS in investigating the elementary excitations of an atomic-scaled magnet. Further information may be gathered by examining how the conductance spectrum gets modified by the application of an external magnetic field. Figure 24.6 shows a comparison between the theoretical results and the experimental data for a $N = 2$ chain in a magnetic field (applied in the longitudinal direction), for different field strengths. The field lifts the degeneracy of the Mn dimes first excited state ($S_{tot} = 1$), which separates into three Zeeman-split levels. As such, there are now three equally spaced transitions from the ground state and consequentially the conductance step at 6.2 mV splits into three steps. Again note the almost perfect agreement between theory and experiments.

24.4.2 Kondo Effect and Improved Lineshape: Third-Order Calculations

In order to illustrate the effects of continuing the perturbation expansion to third order, we now discuss the conductance spectra of Co and Fe deposited on the CuN surface. This section therefore aims at rationalizing the experiments described in reference [4], where both Co and Fe ions were investigated by low-temperature STM spectroscopy. Intriguingly in the experiments it is possible to place the two ions at different distances from each other. Thus, one can investigate how the spectrum of the Kondo active Co is affected by the presence of Fe and vice versa.

This time, the parameters of the simulations are set as follows. The level broadening due to the coupling to the substrate (we neglect the one to tip) is set at $\Gamma_{\text{sub-S}} = 0.1$ eV for both Co and Fe. In order to ensure consistency in our approximations [$\Gamma_{\text{sub-S}} = (\gamma_{\text{sub-S}})^2/W$], we also set $\gamma_{\text{sub-S}} = 1.5$ eV with the substrate bandwidth being $W = 20$ eV. We then choose $\varepsilon_0 = 1$ eV to fulfill the criterion $\varepsilon_0 \gg \varepsilon_F = 0$. The magnitude of J_{sd} for both Fe and Co is held constant at 0.5 eV [32] (the same value used for the Mn chains). Finally, the axial and transverse anisotropy parameters are taken from the experimental fits of References [2,3] and are $D_{\text{Co}} = 2.75$ meV, $E_{\text{Co}} = 0$ meV, $D_{\text{Fe}} = -1.53$ meV, and $E_{\text{Fe}} = 0.31$ meV, while the adsorbed atoms spins are $S_{\text{Co}} = 3/2$ and $S_{\text{Fe}} = 2$.

Let us start the discussion by looking at the conductance traces calculated for isolated Co and Fe (i.e., for $J_{dd} = 0$), which are displayed, respectively, in Figure 24.7a and b together with their experimental counterparts from Reference [4]. In the case of Co full diagonalization of H_{sp} gives us a set of four ($2S_{\text{Co}} + 1$) eigenstates. In particular, the presence of a hard-axis anisotropy results in the following energy manifold $\varepsilon_m^{\text{Co}} = \{0.69, 0.69, 6.19, 6.19\}$ meV, that is, in a doubly degenerate ground state. A transition between the two degenerate ground states is forbidden by the second-order perturbation expansion, but it becomes allowed on inclusion of the third-order term of Equation 24.92. This is because of the selection rules imposed by the theory through the matrix elements $\langle m|S^i|n\rangle$. Such a transition appears in the spectrum of Figure 24.7a in the form of a zero-bias Kondo peak, whose intensity can be shown to increase as the value of the perturbation expansion parameter, α, gets larger (not displayed here, see reference [26]). As such, the enhancement of the Kondo peak intensity is directly associated to the relative growth of the logarithmic divergence contained in the self-energy of Equation 24.92. The same logarithmic divergence produces a second distinctive feature in the dI/dV traces, namely, the rise of the conductance following an inelastic excitation. This can be, for instance, seen in the conduction step at 6 meV. Such a step originates from the transition from the ground state to the first excited state. One may then note that at first the conductance rises sharply at the voltage corresponding to the excitation energy and then slowly decays. Such a feature is absent when the perturbation expansion is truncated to the second order (see previous section) for which the conductance step is essentially square.

Moving to the case of an isolated Fe atom, we note that H_{sp} has the five eigenvalues $\varepsilon_m^{\text{Fe}} = \{-6.30, -6.12, -2.46, -0.60, 0.18\}$ meV, so that the ground state is nondegenerate. At zero magnetic field, all the transitions allowed by the third-order expansion are resolved in the conductance traces (see Figure 24.7b). In this case, there is no Kondo peak at $V = 0$, as the ground state is nondegenerate, but instead, we now find a conductance dip, corresponding to an inelastic transition between the first two lower lying spin states. Furthermore, as in the case of Co, also here we observe the presence of a logarithmic conductance increase at the inelastic steps (note the well pronounced one at ~4 mV). The same feature is observed experimentally in all non-Kondo active adatoms (e.g., Mn and CuN), demonstrating the good level of description provided by the third-order expansion.

We now simulate the case in which the two atoms, Co and Fe, are brought into proximity so that to be coupled via exchange reaction. For small, but finite, values of J_{dd}, we notice a change in the dI/dV traces of both Co and Fe, which becomes more pronounced as J_{dd} increases. For Fe, both the conductance steps around $V = 0$ and that at 0.18 mV decrease in intensity with increasing J_{dd}. In contrast, Fe itself acts as an effective magnetic field that splits the zero-bias Kondo resonance present in the spectrum of Co. Both these effects are observed in the experiments of Reference [4]. Notably, Fe does not simply act as a source

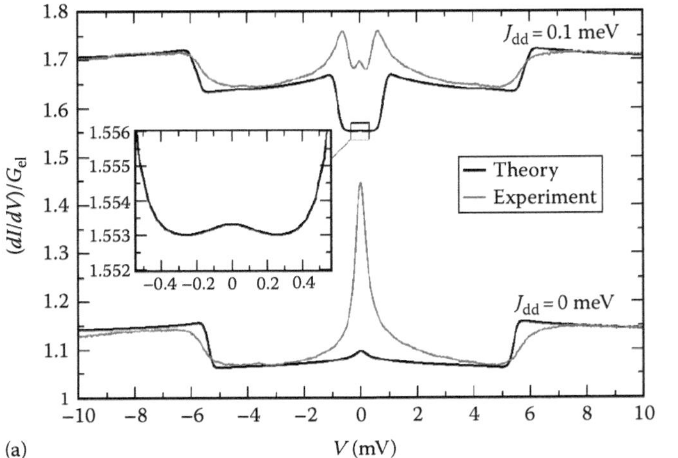

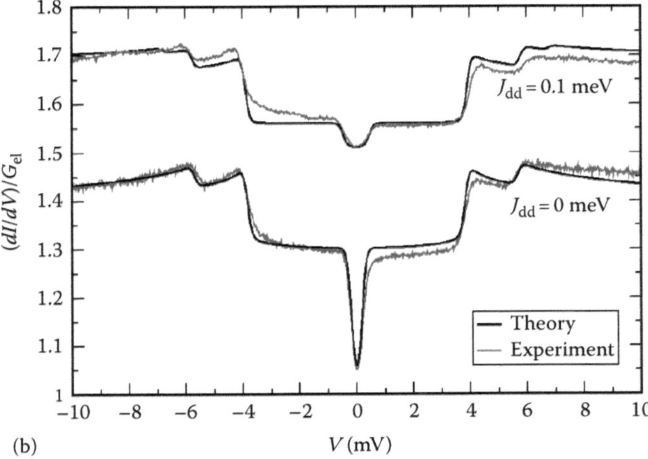

FIGURE 24.7 SF-IETS conductance traces (normalized) for both Co and Fe deposited on a CuN surface: (a) Co, (b) Fe. Here, $J_{dd} = 0$ means that there is no magnetic coupling between Co and Fe, so that the traces correspond to those of the atoms in isolation. In contrast, at finite J_{dd}, there is a magnetic coupling, and the two spectra get modified. In both the panels, the black curves are for the calculations, while the gray ones correspond to the experimental data of Reference [4]. (Figure reproduced from Hurley, A. et al., *Phys. Rev. B*, 84, 115435, 2011.)

of magnetic field on Co, as seen in the inset of Figure 24.7a for $J_{dd} = 0.1$ meV. In the figure, one can clearly observe an additional Kondo peak emerging at zero-bias in between the two principally split peaks. This is a unique feature of the exchange coupling between Co and Fe. In fact, the exchange coupled Fe-Co dimer possesses $(2S_{Co} + 1) \times (2S_{Fe} + 1) = 20$ eigenvalues and additional allowed transitions appear at each of the atomic sites. For instance, for large J_{dd}, the zero-bias region of the Co spectrum becomes completely dominated by a conductance dip. This originates from the opening of a spin transition between the ground state at −5.686 meV and the first excited state at −5.379 meV. Such a transition, absent for the isolated Co adatom, has a spectral intensity much larger than that of the Kondo resonance, which therefore disappears from the spectrum.

We now wish to compare our data to the corresponding experimental spectra of Reference [4]. This is done again in Figure 24.7. Notably, whereas the calculated spectrum of Fe is in excellent quantitative agreement with the experimental one, the same cannot be said for that of Co, which only reproduces the experimental features at a qualitative level. In particular, the experimental Kondo resonance is much more pronounced than the calculated one (note that the parameter α has been set in order to reproduce the experimental step in the spectrum at ~4 mV). At this point, we can only speculate on the reasons for such a disagreement. We note that the *s-d* model is valid only in the limit where the tunneling matrix element, *t*, is small with respect to the adatom charging energy, *U*. This is the case in which a Hubbard-like model can be mapped onto the *s-d* one [35]. Such a limit might not be satisfied for Co on CuN. In the event of a large *t*/*U* ratio, a more rigorous two-body (Hubbard-like) approach needs to be employed to describe electrons in the localized *d*-states [36], that is, the adiabatic decoupling between the conducting electrons and those generating the local spins breaks down. This suggests that the disagreement that we find between theory and experiments may be mainly due to a failure of the underlining *s-d* model and not to that of the perturbative approach. Still, it remains the fact that the theory developed here is capable to qualitatively explain all the known results.

24.4.3 Spin-Pumping

In this closing section, we discuss two cases in which the spin-population is driven out of equilibrium by the current [37]. In the first one, this is achieved by varying the tip-to-sample distance, while in the second, the tip is spin-polarized. In both situations, we now need to evaluate also the spin-propagator, that is, the non-equilibrium state of the spin system.

24.4.3.1 Intense Current Density

We return now to the case of an exchange-coupled Mn dimer deposited on a CuN surface and probed by a non magnetic STM tip and explore how the conductance spectrum varies with the intensity of the tunneling current. This simulates the experiments of Reference [37], where the current was modulated by tuning the height of the STP tip over the dimer. Here, we use again the parameters of Table 24.1, which gives us an antiferromagnetic, $S_{tot} = 0$, ground state. The first excited state is a triplet with total spin $S_{tot} = 1$ and its splitting from the ground state is exactly J_{dd}. The next excited state is the quintuplet with total spin $S_{tot} = 2$ and it is split from the first excited state by $2J_{dd}$. This pattern continues throughout the spin manifold.

We now look at the conductance traces calculated for a range of values of $\Gamma_{tip\text{-}S}$ going from 0.125 meV to 200 meV, displayed in Figure 24.8a. The changes in the conductance lineshape are a direct result of driving the system out of equilibrium as the tip is drawn closer to the sample ($\Gamma_{tip\text{-}S}$ gets larger). When the tip is far from the Mn dimer ($\Gamma_{tip\text{-}S} = 0.125$ meV), the spin system remains always in thermal equilibrium. Therefore, any incoming electron from the tip will encounter the Mn dimer in its ground state and the single transition from the ground to the first excited state will occur at the energy J_{dd}. This is precisely the same situation presented in Figure 24.5, which yield the conductance step at $V = J_{dd}/e = 6.2$ mV.

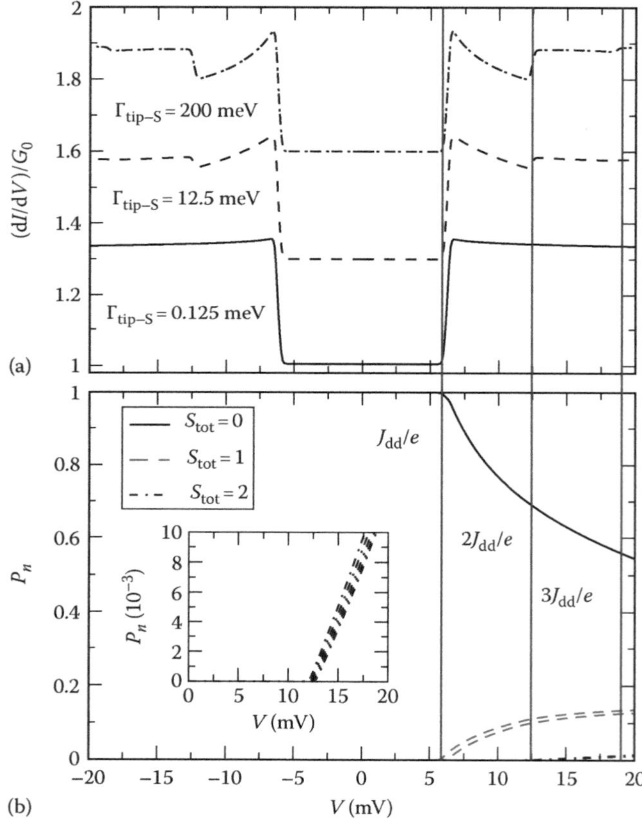

FIGURE 24.8 Normalized SF-IETS conductance traces (a) for a Mn dimer deposited on a CuN surface and probed by a current of growing intensity. The current intensity is regulated by the magnitude of the tip-to-sample coupling parameter, $\Gamma_{tip\text{-}S}$. We notice that the stronger is the coupling, the more the system is driven out of equilibrium so that additional spin transitions appear in the conductance spectrum. In panel (b), we show the non-equilibrium population of the Mn dimer singlet ($S_{tot} = 0$), triplet ($S_{tot} = 1$), and quintuplet ($S_{tot} = 2$) states. The inset shows a magnified view of the quintuplet population.

As the tip is brought closer to the sample (Γ_{tip-S} = 12.5 meV), the first excited state starts to populate. This opens up the possibility for second transition of energy $2J_{dd}$ from the S_{tot} = 1 to the S_{tot} = 2 state, which appears in the conductance spectrum as a step at $V = 2J_{dd}/e$. Note that a direct spin transition from the ground state to the quintuplet S_{tot} = 2 is highly unlikely with a single electron tunneling process, and in fact it becomes completely forbidden if there is no magnetic anisotropy. Importantly, the (S_{tot} = 1) → (S_{tot} = 2) transition becomes more probable as the initial state populates, that is, as there is little time for the system to relax back to the ground state. As such, the characteristic conductance step at $V = 2J_{dd}/e$ becomes more pronounced as the current becomes more in intense, i.e., as the spin is driven far from equilibrium.

Finally, when the current density is increased further (Γ_{tip-S} = 200 meV), a third conductance step appears at $3J_{dd}/e$. This is associated to the (S_{tot} = 2) → (S_{tot} = 3) transition and it becomes possible only if the occupation of the quintuplet does not vanish. In order to understand how the steady-state spin configuration of the system evolves with bias, in Figure 24.8b, we plot the population of the various spin states (up to S_{tot} = 2) as a function of V for strong tip-to-sample electronic coupling Γ_{tip-S} = 200 meV. Clearly one can note that, as the bias gets larger, the population of the S_{tot} = 0 state is reduced in favor of those of the closest excited states. These in turns start to become finite as soon as the bias is large enough for the particular excitation to be activated in the transport process. Overall, the effect is that of changing the spin configuration of the local spins, i.e., of driving the system away from its ground state. This is known as spin-pumping.

24.4.3.2 Spin-Polarized Current

A second way to induce spin-pumping, which does not necessarily require the use of intense current densities, is that of using spin-polarized electrodes. In this case, the current injected into the spin-system carries a spin-imbalance, i.e., there are more electrons of a particular spin-specie. This means that spin-flip transitions are likely to change the local spin always in the same direction. If only one of the two electrodes is spin-polarized (for instance the tip), an asymmetry in the dI/dV (V) curve is expected, since the spin-polarization of the current, and consequently the non-equilibrium spin-population, depends on the bias polarity.

Here, we aim at reproducing a set of experiments from Loth et al. [37], in which a STM is made magnetic by placing a single Mn atom at its apex. Such a tip is then used to probe a single Mn atom deposited on a CuN substrate in the presence of a strong magnetic field (3 T and 7 T). Due to the weak anisotropy of Mn on CuN (we use here the parameters of Table 24.1), the magnetic field effectively produces a Zeeman split of the six levels in the $S = 5/2$ manifold. The direction of the magnetic field in these experiments is chosen so that the ground state of Mn corresponds to the spin quantum number $m = +5/2$, with the tip and the atom to probe remaining spin-collinear.

In Figure 24.9a, we show the SP-IETS spectra of the system described earlier, for two magnetic fields of strengths, 3 T and

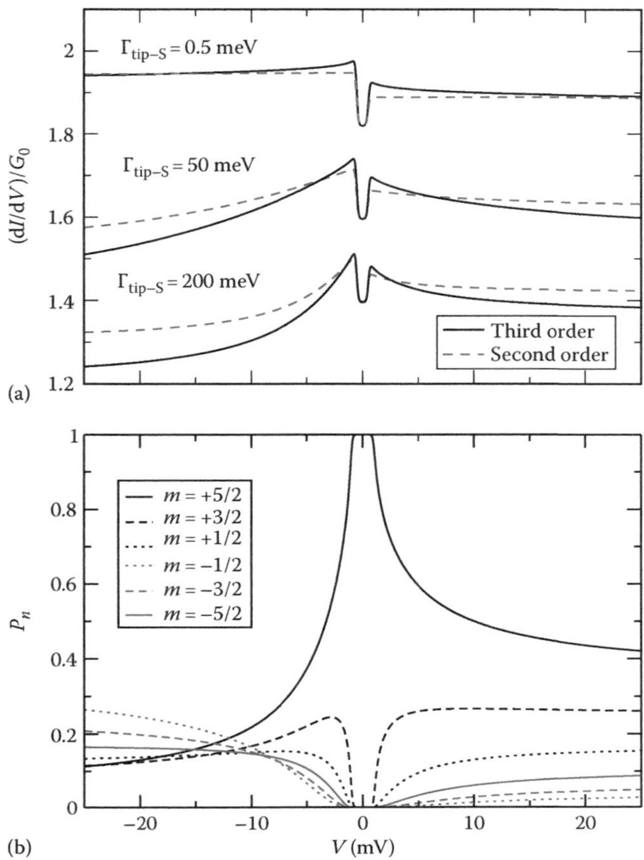

FIGURE 24.9 Normalized SF-IETS conductance traces (a) for a Mn atom deposited on a CuN surface and probed with a spin-polarized STM tip in a magnetic field (3 T and 7 T). The conductance traces are calculated for different tip-to-sample electronic couplings, Γ_{tip-S}. In panel (b), we show the evolution of the non-equilibrium population of the various spin states of the Mn atom for the strong tip-to-sample sample coupling of Γ_{tip-S} = 200 meV. Note the development of an asymmetric bias dependence of the populations with the bias polarity.

7 T, and either weak (Γ_{tip-S} = 0.5 meV), intermediate (Γ_{tip-S} = 50 meV), or strong (Γ_{tip-S} = 200 meV) tip-to-sample couplings. The tip spin-polarization constant and the inelastic ratio that best fit the experimental data are, respectively, $\eta = -0.3$ and $\chi = 1.5$. In the weak coupling limit (when the local spin is close to equilibrium), there is a little, but non-negligible, asymmetry in the conductance trace. This is due to the fact that in a magnetic field the local Mn spin resides almost entirely in its $m = +5/2$ ground state. Therefore, due to the spin-exchange selection rules, only minority (spin-down) carriers can excite the local spin out of its ground state. Since the tip has a spin-polarization of $\eta = -0.3$, there are more spin-down electrons that come from the tip than those that come from the substrate. As a result, the intensity of the inelastic interaction will change depending on the direction of the current, thus creating the expected bias-asymmetry. Note that the conductance spectrum shows the characteristic lineshape observed before, with a decrease in intensity following a

conductance step (here we have only the zero-bias step discussed before, which is now broadened by the magnetic field). Such a feature, in good agreement with experiments, is provided by the third-order self-energy, which produces a logarithmic decay at the conductance steps.

When the local spin is driven further away from equilibrium, in particular in the strong coupling limit, the bias asymmetry becomes more pronounced. This is again due to spin-pumping, which can be appreciated better by looking at Figure 24.9b, where we plot the populations of the six m levels as a function of V for $\Gamma_{\text{tip-S}}$ = 200 meV. The figure clearly shows that increasing the bias causes an enhancement of the population of the excited states at the expenses of the $m = +5/2$ ground state. Note that already at around $V = -10$ mV the population of the $m = -5/2$ level is larger than that of the ground state. Importantly, it turns out that the average magnetization direction of the local spin, which can be calculated as $\langle S^z \rangle = \sum_m P_m S^z_{mm}$, is about $\langle S^z \rangle = -1$ at −25 mV, meaning that one needs a voltage of that magnitude to effectively flip the spin against the magnetic field. Finally, note that the non-equilibrium spin populations are different depending on the bias polarity, an effect that produces the conductance asymmetry. In particular, for negative bias when the spin is essentially reversed, there is an increase in back-scattering for the minority spins, which produces a reduction in the conductance.

24.5 Conclusion

In this chapter, we have constructed a perturbative approach to SF-IETS and applied it to the description of low-temperature STM experiments for magnetic atoms deposited on insulating surfaces. This has proved to be a very valuable tool for understanding the elementary magnetic excitations of an atomic scaled magnetic nanostructure. Importantly, in magnetism, there is hardly another experimental strategy to address the single spin in a controllable way, so that SF-IETS remains a unique platform to study magnetic excitations at the most fundamental level.

The theory as it stands still depends of model Hamiltonian and parameters that need to be inferred from experiments or calculated from electronic structure methods. However, the formalism in itself may be made amenable [38] to a direct implementation within DFT and attempts in this direction are currently under way. This would be a very welcome development, which will allow us to finally have a fully predictive theory for magneto-transport. Such a theory could be then applied to more complex systems, such as magnetic molecules and molecule complexes [39–41] or spin crossover compounds [42].

Another intriguing prospective is represented by the possibility of expanding the theory to a continuum of excitations, that is, to constructing a self-energy for the interaction between conducting electrons and spin-waves. This may be then applied to problems such as spin-relaxation in magnetic tunnel junctions, leading to a fully *ab initio* transport theory for magnetic nanostructure at finite temperature.

Acknowledgment

This work is sponsored by the Irish Research Council for Science, Engineering & Technology (IRCSET). N.B. and S.S. thank Science Foundation of Ireland (grant No. 07/IN.1/I945) and CRANN for financial support. Computational resources have been provided by the Trinity Centre for High Performance Computing (TCHPC). We wish to thank Cyrus Hirjibehedin for making the experimental data shown in Figures 24.7 available to us.

References

1. C.F. Hirjibehedin, C. Lin, A.F. Otte, M. Ternes, C.P. Lutz, B.A. Jones, and A.J. Heinrich, *Science* **317**, 1199 (2007).
2. A.F. Otte, M. Ternes, S. Loth, C.P. Lutz, C.F. Hirjibehedin, and A.J. Heinrich, *Phys. Rev. Lett.* **103**, 107203 (2009).
3. A.F. Otte, M. Ternes, K. Bergmann, S. Loth, H. Brune, C.P. Lutz, C.F. Hirjibehedin, and A.J. Heinrich, et al., *Nature Phys.*, **4**, 847 (2008).
4. A.F. Otte, M. Ternes, S. Loth, C.P. Lutz, C.F. Hirjibehedin, and A.J. Heinrich, *Phys. Rev. Lett.* **103**, 107203 (2009).
5. C. Romeike, M.R. Wegewijs, and H. Schoeller, *Phys. Rev. Lett.* **96**, 196805 (2006).
6. S. Datta, *Electronic Transport in Mesoscopic Systems*, Cambridge University Press, Cambridge, U.K., 1995.
7. A.R. Rocha, V.M. Garcia-Suarez, S. Bailey, C. Lambert, J. Ferrer, and S. Sanvito, *Phys. Rev. B* **73**, 085414 (2006).
8. M. Galperin, M.A. Ratner, and A. Nitzan, *J. Chem. Phys.* **121**, 11965 (2004).
9. K. Yosida, *Theory of Magnetism*, Springer-Verlag, Berlin Germany, 1998.
10. M. Stamenova, T.N. Todorov, and S. Sanvito, *Phys. Rev. B* **77**, 054439 (2008).
11. M. Gell-Mann and F. Low, *Phys. Rev.* **84**, 350 (1951).
12. J. Schwinger, *J. Math. Phys.* **2**, 407 (1961).
13. L.V. Keldysh, *Sov. Phys. JETP* **20**, 1018 (1965).
14. H. Haug and A.P. Jauho, *Quantum Kinetics in Transport and Optics of Semiconductors*, Springer, Berlin, Germany, 1996.
15. M. Büttiker, Y. Imry, R. Landauer, and S. Pinhas, *Phys. Rev. B* **31**, 6207 (1985).
16. X. Wang, C.D. Spataru, M.S. Hybertsen, and A.J. Millis, *Phys. Rev. B* **77**, 045119 (2008).
17. G. D. Mahan, *Many-Particle Physics*, 2nd edn., Plenum Press, New York, 1990).
18. T. Frederiksen, M. Brandbyge, N. Lorente, and A.P. Jauho, *J. Comput. Electron.* **3**, 423 (2004).
19. P. Hyldgaard, S. Hershfield, J.H.Davies, and J.W. Wilkins, *Ann. Phys. (N.Y.)* **236**, 1 (1994).
20. A. Yanik, G. Klimeck, and S. Datta, *Phys. Rev. B* **76**, 045213 (2007).
21. A. Hurley, N. Baadji, and S. Sanvito, *Phys. Rev. B* **84**, 035427 (2011).
22. J. Paaske, A. Rosch, and P. Wölfle, *Phys. Rev. B* **69**, 155330 (2004).
23. J. Paaske, A. Rosch, J. Kroha, and P. Wölfle, *Phys. Rev. B* **70**, 155301 (2004).

24. W. Lee, N. Jean, and S. Sanvito, *Phys. Rev. B* **79**, 085120 (2009).
25. J.M.D. Coey, *Magnetism and Magnetic Materials*, Cambridge University Press, Cambridge, U.K., 2009.
26. A. Hurley, N. Baadji, and S. Sanvito, *Phys. Rev. B* **84**, 115435 (2011).
27. B. Sothmann and J. König, *New J. Phys.* **12**, 083028 (2010).
28. A. Hurley, N. Baadji, and S. Sanvito, *Phys. Rev. B* **86**, 125411 (2012).
29. M. Büttiker, *Phys. Rev. B* **33**, 3020 (1986).
30. C.F. Hirjibehedin, C P. Lutz, and A.J. Heinrich, *Science* **312**, 1021 (2006).
31. R. Zitko and T. Pruschke, *New J. Phys.* **12**, 063040 (2010).
32. P. Lucignano, R. Mazzarello, A. Smogunov, M. Fabrizio, and E. Tosatti, *Nature Mater.*, **8**, 563 (2009).
33. M. Paulsson, T. Frederiksen, H. Ueba, N. Lorente, and M. Brandbyge, *Phys. Rev. Lett.* **100**, 226604 (2008).
34. J. Fernandez-Rossier, *Phys. Rev. Lett.* **102**, 256802 (2009).
35. J.R. Schrieffer and P.A. *Wolff, Phys. Rev.* **149**, 491 (1966).
36. O. Újsághy, J. Kroha, L. Szunyogh, and A. Zawadowski, *Phys. Rev. Lett.* **85**, 2557 (2000).
37. S. Loth, C.P. Lutz, and A.J. Heinrich, *Nature Phys.* **6**, 340 (2010).
38. M. Persson, *Phys. Rev. Lett.* **103**, 050801 (2009).
39. A. Mugarza, C. Krull, R. Robles, S. Stepanow, G. Ceballos, and P. Gambardella, *Nature Commun.* **2**, 490 (2011).
40. A. Mugarza, R. Robles, C. Krull, R. Korytár, N. Lorente, and P. Gambardella, *Phys. Rev. B* **85**, 155437 (2012).
41. S.-H. Chang, N. Baadji, K. Clark, J.-P. Klöckner, M.-H. Prosenc, S. Sanvito, R. Wiesendanger, G. Hoffmann, and S.-W. Hla, *Nano Lett.* **12**, 3174 (2012).
42. V. Meded, A. Bagrets, K. Fink, R. Chandrasekar, M. Ruben, F. Evers, A. Bernand-Mantel, J.S. Seldenthuis, A. Beukman, and H.S.J. van der Zant, *Phys. Rev. B* **83**, 245415 (2011).

VII

Picowires

25

Ferromagnetism in One-Dimensional Atomic Chains

Jisang Hong
Pukyong National University

25.1 Introduction

Magnetism in low dimension has been receiving great amount of research efforts because the magnetic properties found in low-dimensional nanostructured materials can be utilized for magnetic nanodevice applications such as magnetic field sensors, spin filtering sensors, magnetic random access memories, and high-density magnetic recording media. Thus, thin film magnetism has long been flourished in the last few decades. All these novel properties have indeed originated from peculiar electronic structure, not observed in bulk materials and most of the previous studies have been performed with two-dimensional geometry.

If an electronic structure of material is changed, physical properties change as well. This implies that the electronic structure in one-dimensional (1D) geometry will be dramatically altered compared with that of two-dimensional material and this feature may bring many interesting quantum phenomena. For instance, 1D nanowires can be used for nanoelectronics because a transport channel in 1D geometry will be very narrow and this will result in very intriguing quantum transport property. Also, the 1D magnetic nanochains can have potential application for smallest magnetic information storage unit. In this respect various physical properties of nanowires, nanopillars, or nanodots are explored. But they are quasi 1D materials because the materials have certain diameters. It is obvious that more sophisticated experimental techniques are required when a true 1D atomic chain is considered.

Advanced atomic manipulation techniques have been remarkably improved and these make it possible to grow 1D nanostructures. For instance, a 1D Au atomic chain on NiAl(110) surface was grown through scanning tunneling microscopy (STM) tip manipulation and the electronic structure was investigated [1]. In addition, it was reported that a ferromagnetic (FM) ordering could be observed in a single Co magnetic atom or finite size of 1D atomic chain [2,3]. The magnetism in 1D chain is of particular interest because there is a chance to find a magnetic state even if a material is formed with nonmagnetic elements in bulk state because the electronic state will be substantially modified in 1D configuration. Thus, we can anticipate more rich physics including the spin degree of freedom. Besides, the magnetism in 1D chain brings a fundamental question due to the argument that there is no FM ordering at finite temperatures [4], but this prediction is based on an isotropic Heisenberg model calculation. As mentioned earlier, however, it is really shown that the FM ground state is achievable in true 1D structure. Here, a magnetic anisotropy plays an essential role for the magnetism in 1D because the magnetic anisotropy can induce a ferromagnetism even in pure 1D system.

Currently, most of the studies in magnetism have been focusing on 2D nanostructured materials, but 1D physics is now becoming a more and more important field because of novel physical properties. Nonetheless, it is still rare to find either experimental or theoretical studies for true 1D systems [5–10]. Despite this situation, experimental tools are being rapidly improved and one will be able to easily control the growing of artificial 1D structures in the near future. Thus, it is time to explore the physical properties of 1D structures. From a theoretical point of view, it is well known that the physical properties are strongly sensitive to the change of electronic band structure and one can obtain desirable properties by tailoring the band structure. To understand the 1D physics, it is essential to describe the exact electronic structures of materials. In particular, the magnetic anisotropy is a very important quantity to understand the ferromagnetism in 1D geometry because a long-range spin ordering cannot be achievable without it. The magnetic anisotropy stems from two different mechanisms such as magnetic dipolar interaction and spin–orbit coupling (SOC). Between these two factors, the SOC is of particular interest because many intriguing magnetic properties such as magnetic anisotropy, magnetic circular dichroism, and magneto-optical kerr effect are originated from this SOC. Thus, it is necessary to describe the SOC interaction very accurately if one considers the magnetic properties of materials. In this work, we will discuss the magnetic properties of various 1D atomic chains using the state-of-the-art first principles method.

25.2 Theory

25.2.1 Electronic Band Structure

One can approach in two different ways to explore the physical properties of materials such as (1) model calculation and (2) first principles calculation. Each method has its own advantage and disadvantage. Model calculation takes advantage of its simplicity because a couple of parameters are employed in this approach. Thus, one can easily handle these factors and extract the role of each parameter. However, the validity of parameters considered in model should be checked whether they are suitable in real materials. Moreover, it is hardly successful to predict unknown physical properties of artificially grown nanostructured materials. On the other hand, the first principles method does not employ any arbitrary parameters and performs fully self-consistent way. Due to this feature, it is rather difficult to understand the physical properties of materials in a straightforward way using few simple physical quantities like in model calculation. Despite this disadvantage, the first principles method is capable of dealing with new structures even if they are not yet experimentally verified. To date, the first principles method to understand the physics of nanostructured materials has become an essential tool in material sciences.

All the first principles methods are based on density functional theory (DFT). The basic principle of DFT is proposed by Hohenberg–Kohn [11]. The essential idea is that "the ground state of total energy is a functional of the ground state density only and all ground state properties can be expressed as a function of this density." The total energy of physical system can be written as

$$E = \int V_{Ext}(\vec{r})\rho(\vec{r})d\vec{r} + F[\rho],$$

where the $F[\rho]$ describes the kinetic energy of electrons and the Coulomb interaction among electrons and V_{Ext} is a unique potential producing the density ρ as its ground state. Unfortunately, with the given form, the exact $F[\rho]$ cannot be obtained due to the intrinsic many body problem. To resolve this obstacle, Kohn and Sham have replaced the many-body interaction by a fictitious one-electron system with the same density as the real system has [12]. With this scheme,

$$F[\rho] = T[\rho] + \frac{1}{2}\int \frac{\rho(\vec{r})\rho(\vec{r}')}{|\vec{r}-\vec{r}'|}d\vec{r}d\vec{r}' + E_{XC}[\rho].$$

The first term describes a kinetic energy and the second term is for conventional Coulomb interaction. The last one denotes an exchange-correlation and the effect of complicated many-body interaction is taken into account by this term. Finally, one should solve the Schrodinger equation given by

$$\left[\frac{-1}{2\nabla^2} + V_{Ext} + V_C + V_{XC}\right]\Psi = E\Psi.$$

By solving this equation, the energy eigenvalues and eigenfunctions can be obtained and the total energy will be calculated. All the procedures are performed self-consistently.

25.2.2 FLAPW and Spin–Orbit Coupling Hamiltonian

Various methods are currently available to calculate the electronic structure of material. If one is only interested in the magnetic ground state and the magnetic moment of a specific atom, any first principles method can provide fairly accurate results. But the following quantities such as magnetic anisotropy, magnetostriction, magnetic circular dichroism, and magnetocrystalline anisotropy are more intriguing properties because these quantities are very important for both fundamental interest and innovative device applications. Such magnetic properties can be seen in the presence of spatially broken symmetry of spin alignment and it results from the SOC. Thus, in most of the cases for studying of magnetic materials, the key issue is how to accurately describe the spin–orbit interaction. Among different versions of DFT-based first principles methods, the full potential linearized augmented plane wave (FLAPW) method is the most reliable and accurate method and our calculations have been performed with this method [13–16]. The most essential part in FLAPW method compared with other first principles schemes is in dealing with the potential energy as its name implies. In FLAPW method, there is no shape approximation in potential, charge, wavefunction expansions. Due to this feature, the FLAPW has become the most trustable calculation for magnetic materials. We first describe the basic treatment for electronic structure calculations and show the way of SOC coupling treatment.

As shown in Figure 25.1, the real space is divided into three different regions, namely, muffin-tin (MT) spheres around the

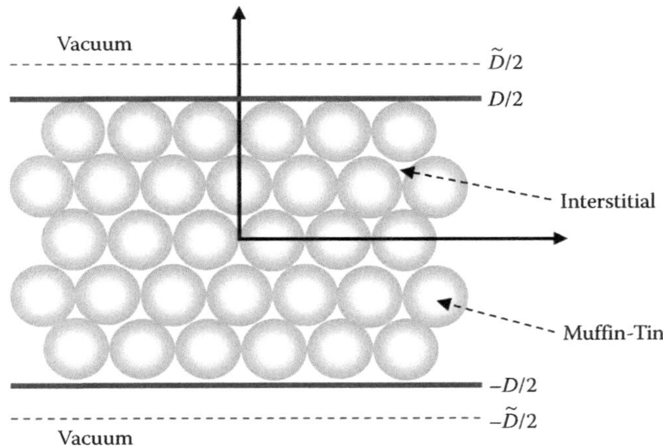

FIGURE 25.1 Schematic illustration of thin film structure considered using FLAPW method.

nuclei, a vacuum region on each side of the slab and the remaining interstitial region. In each region, we expand the augments plane wave basis function as following

$$\Psi_{k=G} = \begin{cases} e^{i(\vec{k}+\vec{G})\cdot\vec{r}/\sqrt{\Omega}} \\ e^{i(\vec{k}+\vec{G})\cdot\vec{R}_i}\sum_{l,m} i^l[a^i_{lm}u_l(r)+b^i_{lm}\dot{u}_l(r)]Y_{lm}(\hat{r}) \\ e^{i(\vec{k}+\vec{G})_{\|}\cdot\vec{r}_{\|}}\left[a^v_{(\vec{k}+\vec{G})}u_{(\vec{k}+\vec{G})_{\|}}(z)+b^v_{(\vec{k}+\vec{G})_{\|}}\dot{u}_{(\vec{k}+\vec{G})}(z)\right]. \end{cases}$$

Here, $\vec{k}$ and $\vec{G}$ stand for wave vectors in reciprocal space, while the volume of unit cell is represented by Ω and the first line is used for interstitial region. In the second line, R_i represents the position ith atom, $u_l(r)$ is the solution of radial semi-relativistic Kohn–Sham equation in MT region with angular moment l, and $\dot{u}_l(r)$ is the energy derivative of radial function. In the third line, $u_{(\vec{k}+\vec{G})_{\|}}(z)$ and $\dot{u}_{(\vec{k}+\vec{G})_{\|}}(z)$ are the solution and its derivative of the one-dimensional Kohn–Sham equation in vacuum region. The a and b are expansion coefficients and these can be obtained by using conventional boundary conditions at each boundary. Here, the SOC has not been considered. But the magnetic anisotropy, XMCD, magneto-optical kerr effect are indeed occurring because the rotational symmetry is broken and the broken symmetry in space is due to SOC because this SOC is only the source that breaks the rotational symmetry in Hamiltonian. It is necessary to include the effect of SOC when one wants to explore the magnetic properties described earlier. The SOC Hamiltonian can be written as

$$H^{SOC} = \xi(r)\hat{S}\cdot\hat{L} \quad \text{where } \xi(r) = \frac{1}{4m^2c^2r}\frac{dV(r)}{dr}$$

and the SOC operator has

$$\begin{aligned}\hat{S}\cdot\hat{L} = \hat{S}_n(\hat{l}_z\cos\vartheta + 1/2\hat{l}_+e^{-i\varphi}\sin\theta + 1/2\hat{l}_-e^{i\varphi}\sin\theta) \\ + 1/2\hat{s}_+(-\hat{l}_z\sin\vartheta - \hat{l}_+e^{-i\varphi}\sin^2\theta/2 + \hat{l}_-e^{i\varphi}\cos^2\theta/2) \\ + 1/2\hat{s}_-(-\hat{l}_z\sin\vartheta + \hat{l}_+e^{-i\varphi}\cos^2\theta/2 - \hat{l}_-e^{i\varphi}\sin^2\theta/2)\end{aligned}$$

where ϑ and the φ stand for the polar and azimuthal angles of the magnetic moment. In matrix form, the SOC part can be rewritten as

$$\hat{S}\cdot\hat{L} = \begin{pmatrix} \uparrow\uparrow & \uparrow\downarrow \\ \downarrow\uparrow & \downarrow\downarrow \end{pmatrix}$$

$$= \begin{pmatrix} \frac{A_+ + A_-}{2}\sin\vartheta + \hat{l}_z\cos & \{A\_\cos^2\vartheta/2 - A_+\sin^2\vartheta/2 \\ & -\hat{l}_z\sin\vartheta\}e^{-i\varphi} \\ \{A\_\cos^2\vartheta/2 - A_+\sin^2\vartheta/2 & -\frac{A_+ + A_-}{2}\sin\vartheta - \hat{l}_z\cos\vartheta \\ -\hat{l}_z\sin\vartheta\}e^{-i\varphi} & \end{pmatrix}.$$

Here, $A_+ = e^{-i\phi}(\hat{l}_x + i\hat{l}_y)$ and $A_- = e^{i\phi}(\hat{l}_x - i\hat{l}_y)$. By solving the equation, the wavefunction, charge density, potential, energy band, and total energy will be updated and angel dependent. We should diagonalize the $N\times N$ eigenvalue problem given by [17]

$$\begin{pmatrix} \varepsilon(\uparrow) & 0 \\ 0 & \varepsilon(\downarrow) \end{pmatrix} + \xi(r)\hat{S}\cdot\hat{L} = EI.$$

We treat the SOC Hamiltonian in a second variational way, based on the ground state properties achieved from semi-relativistic calculations [18]. Note that the effect of SOC coupling for charge density, spin density, and spin magnetic moment is rather weak for 3D transition metal systems, therefore it is not important to treat the SOC fully self-consistently. However, one should be careful if we consider rather heavy elements such as 5D atoms. In this case, one may need to take into account the SOC self-consistently from the beginning. In the earlier equation, the boundary D-tilta is employed for numerical calculation purposes by adding some artificial vacuum.

25.2.3 Magnetic Anisotropy

In any magnetic material, the spin of electron prefers alignment along a specific direction, the so-called easy axis, and the energy cost is necessary to rotate the alignment direction to the hard axis. The energy difference between these two hard and easy directions is defined as a magnetic anisotropy energy. In the field of magnetism, the magnetic anisotropy, which determines the direction of magnetization, is one of the most important physical quantities because this factor is closely related to novel device application. For instance, the thermal stability of magnetization, coercivity field for permanent magnet, and high-density magnetic recording media require the understanding of magnetic anisotropy of materials. The magnetic anisotropy is originated from two different physical origins: (1) magnetic dipole–dipole interaction and (2) SOC. The dipolar interaction has a very simple form and one can find it in any elementary textbook of electromagnetism and the spin alignment due to this dipolar interaction always prefers in-plane direction in 2D geometry. In real materials, the direction of magnetization varies according to the environment. For instance, the magnetization direction changes from in-plane to perpendicular direction and changes to in-plane direction once again when the film thickness caries. This is the so-called spin reorientation transition (SRT) and the SRT is a famous phenomenon found in various magnetic materials. In order to calculate the magnetic anisotropy, three different schemes are usually adopted, namely, force theorem, total energy method, and torque method [19]. As mentioned earlier, the strength of SOC is rather weak and consequently the energy difference depending on the direction of spin alignment along specific direction is about sub-meV order. Thus, it is required to treat very carefully. Among these three methods, the torque approach gives the most stable results with the help of Hellman-Feynman theorem. We employ this torque method.

25.3 Some Examples

We now present various magnetic properties found in purely 1D systems. Here, the following issues such as (1) magnetic ground state and (2) magnetic anisotropy will be discussed. A magnetic ground state can be found through the total energy calculations and this is a straightforward procedure. However, one should be cautious for magnetic anisotropy calculation because the spin–orbit interaction participates in this phenomenon and this is explained earlier. We now study the magnetic properties including magnetic anisotropy of several 1D systems. To find a magnetic ground state, we calculate total energies of FM and antiferromagnetic (AFM) spin configurations and find the energy difference of these two magnetic states. This procedure is quite simple and one can easily obtain the result. Therefore, we focus on the magnetic anisotropy of materials. First, both finite and infinite chains are discussed. Of course, the influence of substrate materials will also be taken into account. We briefly describe how to obtain the magnetic anisotropy energy in 1D system. If a finite 1D chain is placed along the z-direction in free space, one can write the total energy by $E=E_0+E_1\sin^2\theta+E_2\sin^4\theta$ in the lowest two orders. Here, the polar angle is measured from the z-axis. With this expression, the magnetic anisotropy energy defined by $E_x-E_z=E_1+E_2$ will be found via torque method. A positive magnetic anisotropy energy means that the direction of magnetization is along the chain axis, whereas a negative energy stands for the magnetization perpendicular to the chains axis. For infinite 1D chain, it is more convenient to place the chain along the x-axis because it is necessary to explore the influence of substrate materials for realistic calculations. In this case, we can write the total energy $E=E_0+\sin^2\theta[E_1+E_2\sin^2\phi]$ where the ϑ is the polar angle measured from the chain axis and the ϕ denotes the azimuthal angle measured from the z-axis (surface normal). Through this expression, one can easily obtain the energy differences between the three different magnetization directions. For instance, $E_x-E_z=-E_1$, $E_x-E_y=-(E_1+E_2)$, and $E_y-E_z=E_2$. Note that the E_2 becomes zero for freestanding case because of rotational symmetry.

25.3.1 Finite and Infinite Co Atomic Chains

We first discuss the magnetic properties of finite size of 1D Co atomic chains. The main issues are to explore the size dependence of magnetic moments and magnetic anisotropy. The Co clusters consisting of two to seven atoms are assumed to be placed along the z-direction [20]. The schematic illustration of finite size of Co chain is displayed in Figure 25.2. In each circle, the spin and orbital magnetic moments are presented. Also, the average spin, orbital magnetic moments, and the total magnetic anisotropy energy of each chain are shown in Figure 25.3. As is shown, both spin and orbital magnetic moments have relatively weak size dependence. However, we find an oscillatory behavior in magnetic anisotropy energy. For instance, a magnetization perpendicular to the chain axis is found when the Co chain is made of 5 and 7 atoms, while we observe a magnetization along the chain axis in other cases. The interpretation of magnetic

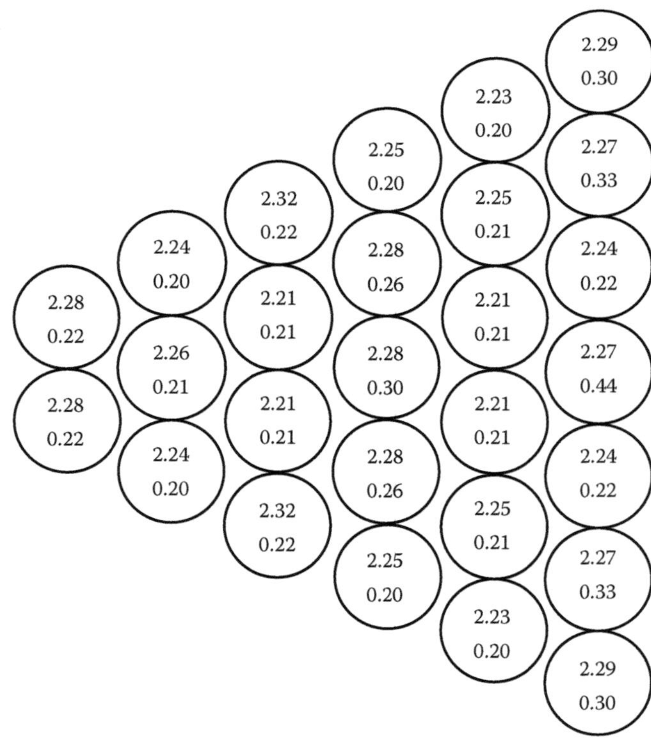

FIGURE 25.2 Calculated spin and orbital magnetic moments of each Co atom.

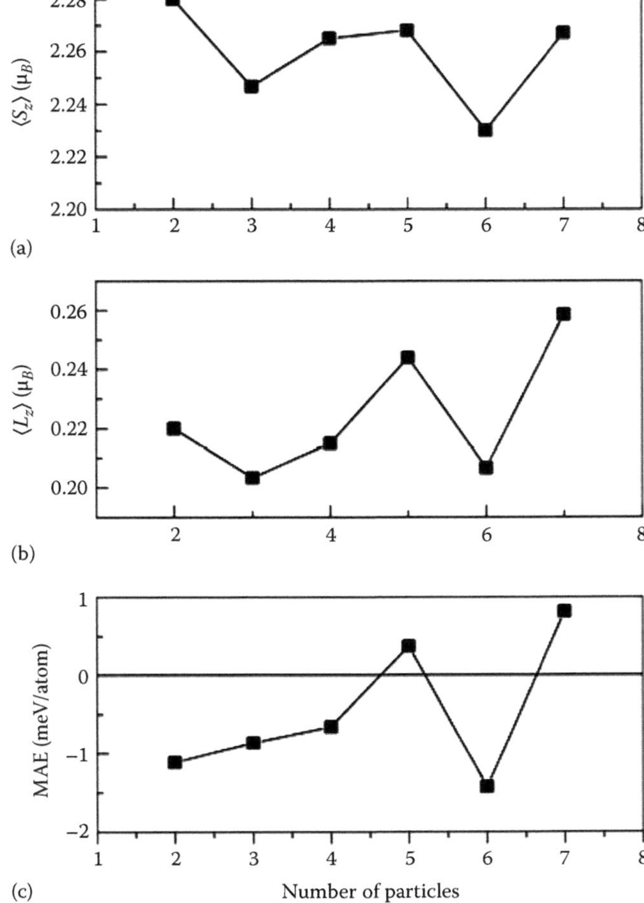

FIGURE 25.3 Size dependence of averaged (a) $\langle S_z \rangle$, (b) $\langle L_z \rangle$, and (c) MAE. The averaged quantities are obtained by simply dividing the total value quantity with the number of particles.

anisotropy can be done in terms of orbital anisotropy if the SOC coupling through minority spin channels dominates [21]. In Table 25.1, we thus present the calculated total magnetic anisotropy energy, orbital anisotropy, and the spin channel resolved magnetic anisotropy energies.

One can see that the anisotropy energy arising from spin flip contribution (↓↑) is comparable to that from minority spin channel (↓↓). This implies that there is no clear correlation between magnetic anisotropy and orbital anisotropy. The size dependence of magnetic properties will be weaken with increasing the chain length and our calculations indicate that the finite chain consisting of 7 atoms is still short to manifest stabilized magnetic state although the magnetic moment seems stable.

TABLE 25.1 Calculated MAE (meV) and $\langle L_z \rangle - \langle L_x \rangle$ in μ_B

System	E_{MAE} (meV)	$\langle L_z \rangle - \langle L_x \rangle$	E_{MAE} (↓↓)	E_{MAE} (↓↓)
Co_2	−2.23	0.05	−0.919	−1.217
Co_3	−2.590	0.06	−1.314	−1.155
Co_4	−2.62	−0.02	0.336	−2.67
Co_5	1.86	−0.18	4.38	−2.09
Co_6	−8.59	0.18	−5.59	−2.47
Co_7	5.70	−0.43	10.18	−3.62

TABLE 25.2 Spin, Orbital Magnetic Moment (in μ_B), and Magnetic Anisotropy Energies (in meV/atom)

System	$\langle S_z \rangle$	$\langle L_z \rangle$	E_1	E_2	E_{shape}
Co (1 × 3)	2.22	0.26	−1.15		0.114
Co (1 × 5)	2.23	0.28	−1.29		0.115
Co/Cu(001) (1 × 3)	1.91	0.15	+0.35	−0.22	0.084
Co/Cu(001) (1 × 5)	1.92	0.15	+0.39	−0.37	0.085
Co/Pt(001) (1 × 3)	2.15	0.10	+1.02	−2.58	0.075

We now discuss the infinite 1D Co chain on Cu(001) and Pt(001) substrate [22]. In this case, we have explored both free-standing and supported systems to understand the effect of interaction with substrate material. It is required to put periodic boundary condition in DFT calculation, thus it is of interest to understand the influence of an artificial interaction with neighboring unit cell on the magnetic properties of material. The 1D atomic chain is placed along the x-axis and we have considered two different unit cells such a 1 × 3 and 1 × 5. In Table 25.2, the first two rows show calculated results for freestanding case and the last three rows are for supported systems.

The spin and orbital magnetic moments of freestanding Co atom are almost the same in two different cell sizes. On Cu(001) substrate, both spin and orbital magnetic moments in Co/Cu(001) are greatly reduced and this is obviously due to hybridization with substrate. In contrast, a reduction of spin moment in Co/Pt(001) is less appreciable. One can also find that the direction of magnetization is substantially altered. For the freestanding Co chain, the magnetic moment aligns perpendicular to the chain axis. But the supported Co chain on Cu(001) shows the magnetization along the chain axis (x-axis). For Co/Pt(001), we find y-axis magnetization. To understand the origin of this behavior, the band structure and the distribution of magnetic anisotropy energy at each k-point for freestanding 1D Co chain is shown in Figure 25.4.

A strong positive contribution to the magnetic anisotropy occurs at $k_x = 0.17(2\pi/a)$. At this k-point, both occupied and unoccupied wavefunctions are depicted in Figure 25.5. It is clearly shown that the occupied state has the $d_{x^2-y^2}$ feature, while the unoccupied one has the d_{xy}. They have the same quantum number of m = ±2, and couple through the L_z operation. This results in strong positive contribution to the magnetic anisotropy. For the supported case, the band structures are very complicated and it is much more difficult to attribute the magnetic anisotropy energy to a single pair of states.

25.3.2 1D W Chain

As mentioned earlier, the conventional 3d transition-metal elements such as Mn, Fe, Co, and Ni have been focused because it is believed that the itinerant magnetism is originated from the 3d electrons. However, the possibility of magnetism due to 4d or 5d elements is also an interesting issue. In this sense, we have

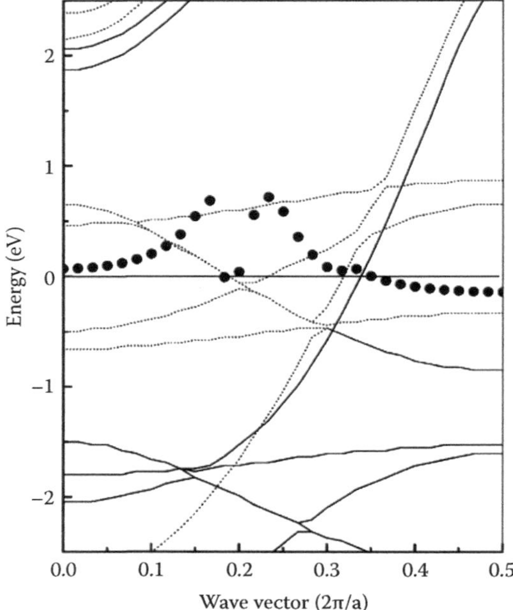

FIGURE 25.4 Electronic band structure and the distribution of E_{MCA} along the x-axis. The solid lines represent majority spin bands and the dashed lines stand for minority spin bands. The circles are the distribution of E_{MCA}.

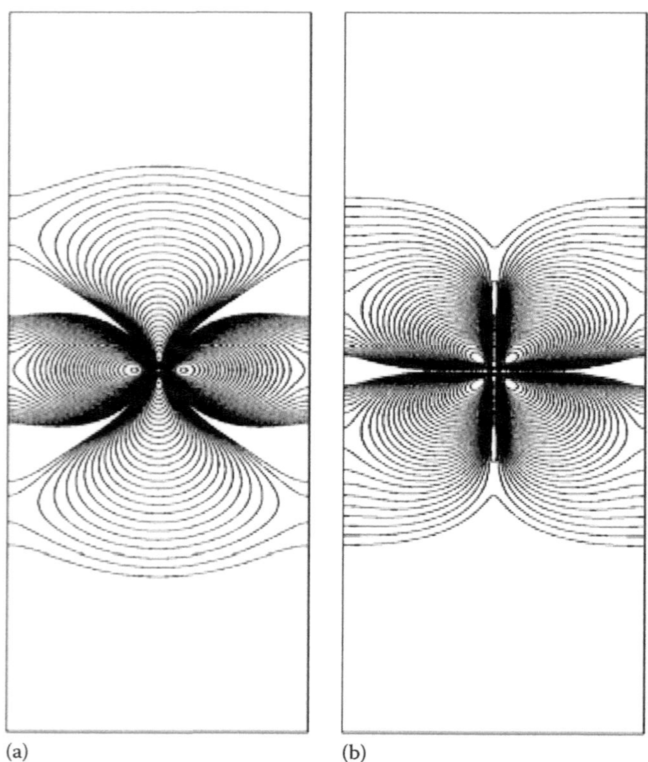

FIGURE 25.5 Wave functions of two key states at $k_x = 0.17$ in the x–y plane: (a) the occupied state and (b) the unoccupied state.

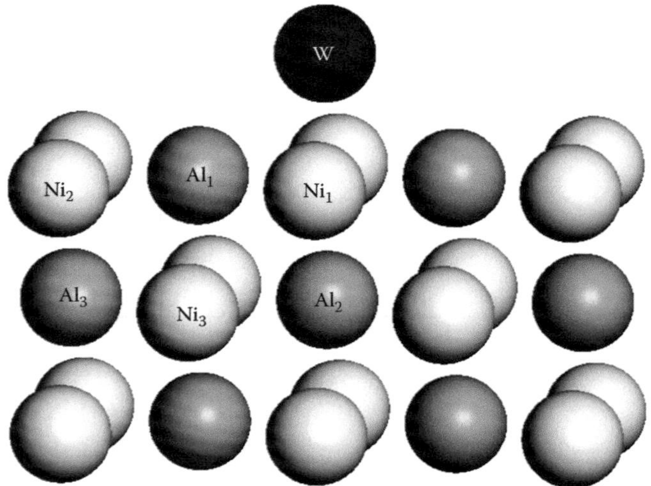

FIGURE 25.6 Schematic illustration of unit cell of W/NiAl(110).

calculated the magnetic properties of 1D W chain on NiAl(110) surface [22]. Note that the NiAl(110) film is nonmagnetic. The schematic illustration of unit cell structure considered in this calculation is shown in Figure 25.6. For freestanding case, the 1D W chain has an AFM state with a magnetic moment of 2.76 μ_B. Interestingly, the 1D W chain on nonmagnetic NiAl(110) surface displays a FM ground state with a magnetic moment of 0.81 μ_B. In addition, the induced magnetic moment of 0.19 μ_B in surface Ni atom is observed. The coupling of W atoms is substantially altered in the presence. In particular, the FM coupling is mediated by surface Ni atom. One can anticipate large magnetic anisotropy in late transition metal element because the strong SOC is expected. Thus, the magnetic anisotropy is also an interesting issue in this system. Through the total energy expression given by $E = E_0 - \sin^2\theta[E_1 + E_2\cos^2\phi]$, where the polar angle is measured from the chain axis and azimuthal angle is counted from the y-axis. It has been observed that the magnetization direction of W/NiAl(110) system is parallel to the film surface, but perpendicular to the chain axis (y-axis). The energy barrier for spin rotating in the in-plane direction is 3.509 meV/atom, which is huge in supported 1D system. In Figure 25.7, the distribution of the magnetic anisotropy energies E_1 and E_2 along the 1D band direction is displayed.

Both E_1 and E_2 have oscillatory behaviors along the 1D Brillouin zone, but there is no correlation since they are distributed independently. Besides, there is no single dominant contribution to magnetic anisotropy arising from a specific k-point. Instead, the net magnetic anisotropy energy is determined through the summation of each contribution. This indicates that the analysis of anisotropy of 1D W chain cannot be achieved in a simple manner although the geometry is rather straightforward.

25.3.3 Ti Chain

Among 3d transition metal elements, it is believed that the magnetic state can be found from V to Ni and the Sc and the Ti are believed to have nonmagnetic states. However, the band

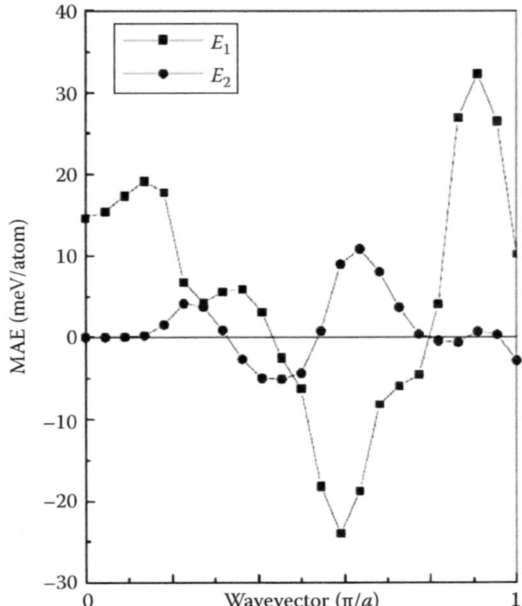

FIGURE 25.7 Distribution of magnetic anisotropy energy along the 1D Brillouin zone.

width in one dimension will be very narrow and this can cause spin polarized state in Ti chain. Indeed, for freestanding state, the magnetism has been predicted, but no explicit calculations for supported system have been performed in early 3d transition metal wires. In this respect, the magnetic property of 1D Ti chain on NiAl(110) surface has been studied [23]. The schematic illustration of Ti/NiAl(110) structure considered in this calculation is the same as in W/NiAl(110). The FM state in freestanding Ti chain is found and the magnetic moment of 1.88 μ_B. On NiAl(110) surface, we still find a FM state and the calculated magnetic moment in Ti atom is about 1.24 μ_B. Unlike in previous W/NiAl(110) case, there is no induced magnetic moment in surface Ni. To reveal the origin of magnetism in Ti atom, the m-resolved density of states is shown in Figure 25.8. In freestanding state, the major contribution to magnetic moment stems from $|m| = 1$ and $|m| = 2$ states. But, on NiAl(110) surface the charge transfer due to the hybridization effect mainly occurs in d_{xz} and d_{yz} states and the $d_{3z^2-r^2}$ orbital is almost intact. Consequently, the $d_{x^2-y^2}$ and d_{xy} orbitals are the most important in ferromagnetism in 1D Ti atomic chain on NiAl(110). We have also calculated the magnetic anisotropy of Ti chain. We find the surface normal magnetization and the magnetic anisotropy energy is about 20 µeV. Indeed, the shape anisotropy due to magnetic dipolar interaction is also calculated and it becomes 33 µeV. Since the shape anisotropy energy is larger than that due to the SOC coupling, the net magnetization is along the chain axis. In most of the cases, we have obtained that the magnetic anisotropy arising from the SOC is larger than that arising from the dipolar interaction. But, in this case, we have observed the reversed behavior [24].

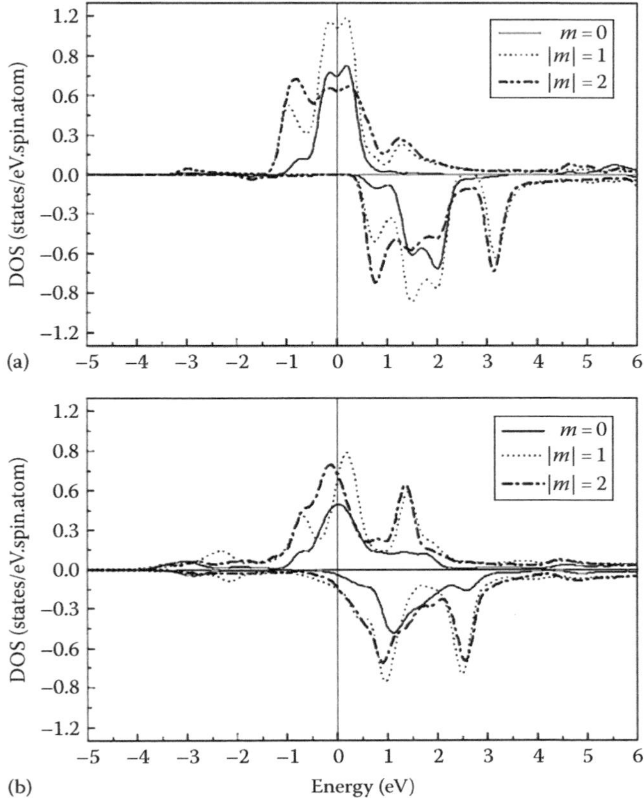

FIGURE 25.8 m-Resolved DOS of Ti. (a) Freestanding Ti and (b) Ti/NiAl(110).

25.3.4 CuO Chains

It has long been believed that the metallic magnetism stems from the 3d transition metal elements. Therefore, the 3d element has always been existed in conventional magnetism, regardless of sample size, sample geometry, etc. However, the electronic structure in one-dimensional geometry can be substantially altered and this may bring an exotic magnetic behavior. Very interestingly, the spin polarized signal is observed in 1D Cu nanowire, which is mechanically stretched. The schematic illustration of experimental system is shown in Figure 25.9.

To understand the potential magnetism stretched 1D Cu nanowires, we have calculated the magnetic moment in Cu atom elongating the inter-atomic distances and have found no sign of spin polarization in pure 1D Cu nanowires. Since the experiment has been performed in oxygen environment, we have investigated the influence of oxygen atom on the magnetism. To this aim, the total energy of 1D CuO diatomic chain varying the Cu–O inter-atomic distance is calculated. In Figure 25.10, the calculated total energy and yield stress are displayed. The equilibrium distance between Cu and O atoms is 3.23 atomic unit (a.u.) and the stress is linearly increased up to a distance of 3.8 a.u., and reaches its maximum at around 4.1 a.u., where we expect that the wire breaks down.

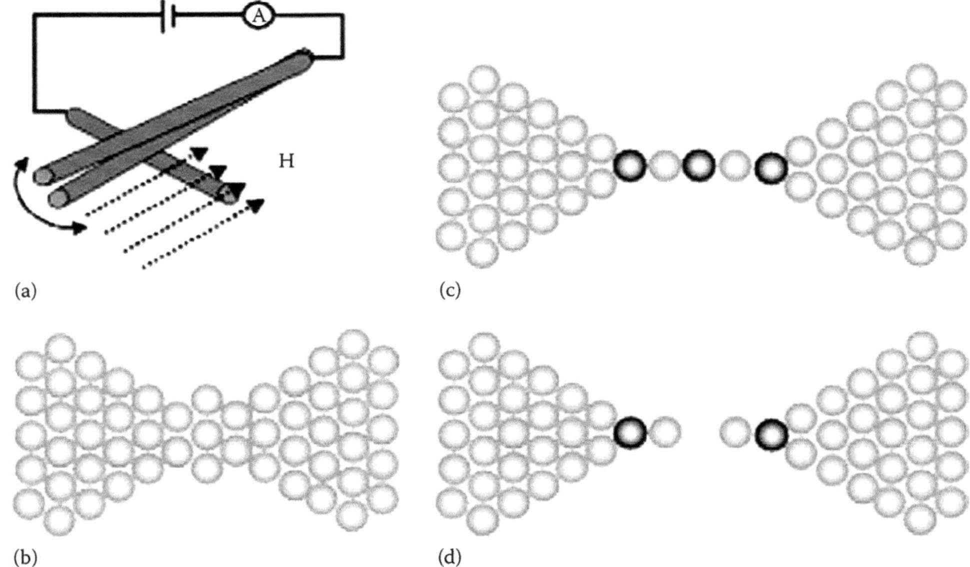

FIGURE 25.9 An outline of the experimental system used. (a) A schematic diagram of the equipment; the nanowire is made where the wire come into contact. The magnetic field is applied in the direction of the dashed arrows. (b–d) An illustration of how the nanowires are created. When the macroscopic wires are in contact, atoms bind to both wires (b); as the wires separate the metal forms a neck which stretches out until it is atomic size (c), then it snaps (d). The quantum conduction is seen in region (c) close to (d) where the wires snap.

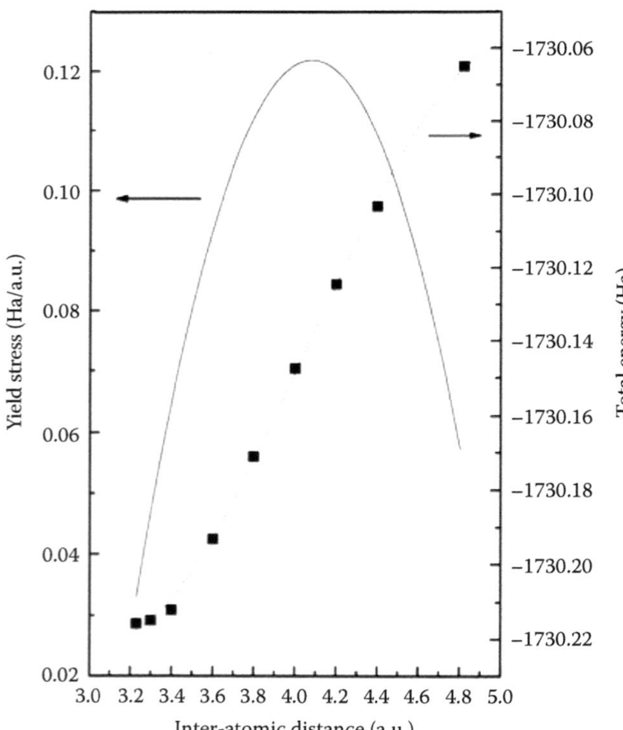

FIGURE 25.10 The calculated total energy and one-dimensional stress (the first derivative of the total energy curve with respect to the interatomic distance) for a CuO monatomic wire.

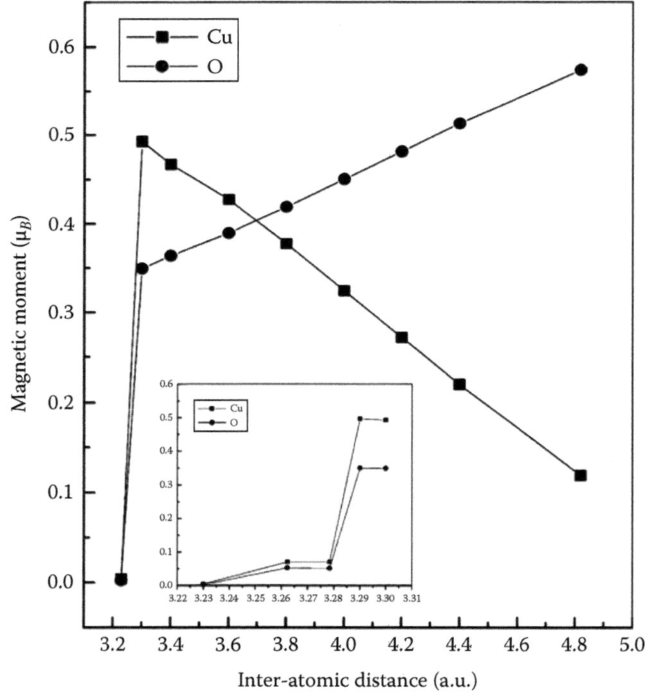

FIGURE 25.11 The calculated magnetic moment of Cu and O in the CuO diatomic wire as a function of interatomic distance. The inset shows the detailed variation of the moments near the equilibrium separation.

In Figure 25.11, the calculated magnetic moment of Cu and O atoms are shown. Surprisingly, we have found the spin polarized state in both atom, except at equilibrium distance. This result may imply that the CuO chain is formed with an elongated Cu–O distance when the Cu chain is mechanically stretched in oxygen environment [25].

25.4 Summary

In conclusion, we have discussed the magnetism of various 1D atomic chain structures. It has been found that the 1D system displays peculiar magnetic properties. For instance, the magnetic moment seems rather stable in finite Co chains, but the magnetic anisotropy is oscillating until the chain size is made of 7 Co atoms. This indicates that the finite size of chain should be longer than at least 7 atomic size to manifest bulk properties. It is also shown that the magnetic state can be even achieved in late transition metal element such as W chain, whereas the bulk W material becomes nonmagnetic. In this system, we find very large magnetic anisotropy energy and this is due to the strong SOC in 5d element. But the importance of magnetic dipolar interaction is also seen Ti atomic chain.

Acknowledgments

This research was supported by the Convergin Research Center Program through the Ministry of Education, Science and Technology (no. 2012K001312), and by the Korean Center for Artificial Photosynthesis (KCAP) located in Sogang University funded by the Ministry of Education, Science and Technology (MEST) through the National Research Foundation of Korea (NRF-2011-C1AAA001-2011-003028).

References

1. N. Nilius, T. M. Walli, and W. Ho, *Science*, V297, 1853 (2002).
2. P. Gambardella, A. Dallmyer, K. Maiti, M. C. Malagoli, W. Eberhardt, K. Kern, and C. Carbone, *Nature*, V416, 201 (2002).
3. P. Gambardella, S. Rusponi, M. Veronese, S. S. Dhesi, C. Grazioli, A. Dallmyer, I. Cabria et al., *Science*, V300, 1130 (2003).
4. N. D. Mermin and H. Wagner, *Phys. Rev. Lett.*, V17, 1133 (1966).
5. P. Gambardella, A. Dallmyer, K. Maiti, M. C. Malagoli, W. Eberhardt, K. Kern, and C. Carbone, *Phys. Rev. Lett.*, V93, 077203 (2004).
6. D. Spisak and J. Hafner, *Phys. Rev. B*, V67, 214416 (2003).
7. A. Delin, E. Tosatti, and R. Weht, *Phys. Rev. Lett.*, V92, 057201 (2004).
8. J. Guo, Y. Mo, E. Kaxiras, Z. Zhang, and H. H. Weitering, *Phys. Rev. B*, V73, 193405 (2006).
9. J. Prokop, A. Kukunin, and H. J. Elmers, *Phys. Rev. Lett.*, V95, 187202 (2005).
10. V. Rodrigues, J. Bettini, P. C. Silva, and D. Ugarte, *Phys. Rev. B*, V91, 096801 (2006).
11. P. Hohenberg and W. Kohn, *Phys. Rev.*, V136, B864 (1964).
12. W. Kohn and L. J. Sham, *Phys. Rev.*, V140, A1133 (1965).
13. E. Wimmer, H. Krakauer, M. Weinert, and A. J. Freeman, *Phys. Rev. B*, V24, 864 (1981).
14. M. Weinert, W. Wimmer, and A. J. Freeman, *Phys. Rev. B*, V26, 4571 (1982).
15. J. P. Perdew, K. Burke, and M. Ernzerhof, *Phys. Rev. Lett.*, V77, 3865 (1996).
16. J. P. Perdew, K. Burke, and Y. Wang, *Phys. Rev. B*, V54, 16533 (1996).
17. R. Q. Wu, Z. Yang, and J. Hong, *J. Phys.: Condens. Matter*, V15, S587 (2003).
18. D. D. Koelling and B. N. Marmon, *J. Phys. C: Solid State Phys.*, V10, 3107 (1977).
19. X. Wang, R. Q. Wu, D. S. Wang, and A. J. Freeman, *Phys. Rev. B*, V54, 61 (1996).
20. J. Hong and R. Q. Wu, *Phys. Rev. B*, V70, 060406(R) (2004).
21. P. Bruno, *Phys. Rev. B*, V39, 865 (1989).
22. J. Hong and R. Q. Wu, *Phys. Rev. B*, V67, 020406(R) (2003).
23. J. Hong, *Phys. Rev. B*, V76, 092403 (2007).
24. J. Hong, *J. Appl. Phys. Phys.*, V101, 09G505 (2007).
25. D. M. Gillingham, C. Muller, J. Hong, R. Q. Wu, and J. A. C. Bland, *J. Phys. Condens. Matter*, V18, 9135 (2006).

26

Carbon Atomic Chains

Igor M. Mikhailovskij
Kharkov Institute of Physics and Technology

Evgenij V. Sadanov
Kharkov Institute of Physics and Technology

Tatjana I. Mazilova
Kharkov Institute of Physics and Technology

26.1 Introduction

Carbon atomic chains (CACs) have attracted remarkable interest in recent years because of their fascinating properties from a basic scientific viewpoint as well as their great potential in forthcoming technological applications. With their one-dimensional (1D) structures, CACs have an extremely large surface area–volume ratio as compared to bulk materials, making the structure energetically expensive with physical properties quite different from those of bulk materials. CACs appear to have been first observed by accident in 1944 by Otto Hahn during nuclear experiments. In the 1970s, they have been rediscovered by radio astronomers in carbon stars. It became clear that the CACs are closely related to the diffuse interstellar bands and carbon dust blown out from the shells of carbon-rich giant red stars (Kroto et al. 1987, Kroto 1994, Allamandola et al. 1999). This carbon allotrope aroused great interest in theoreticians. First principle calculations have predicted a wide range of remarkable properties, which can make CACs a potential material for nanotechnological applications (Rinzler et al. 1995, Lou and Nordlander 1996, Breda et al. 1998, Abdurahman et al. 2002, Tongay et al. 2005, Wang et al. 2009, Onida et al. 2010). Free-standing CACs were successfully prepared by an electromigration method of shrinking carbon nanotubes (CNTs) to monatomic diameter. These chains connecting the two sections of nanotubes revealed negative differential resistance in perfect agreement with theoretical predictions for carbon atomic wires (Yuzvinsky et al. 2006). Short CACs in bridge-like configurations were also prepared in break-junction experiments, which are the ultimate junctions just before breaking. The authors have treated the reason of CACs formation upon stretching of CNTs or graphene nanoribbons. Computer simulations with realistic many-body interatomic potentials indicated a very large intrinsic breaking strain of CNTs and graphene with the late stages of fracture involving the formation of a bundle of monatomic carbon chains. Stable CACs have been recently formed by removing rows of carbon atoms from graphene via energetic electron irradiation techniques. Firm junctions between a CAC and CNTs were also produced in situ and directly imaged by electron microscopy. Quantum-mechanical calculations support the revealed stability of such molecular links and also show that CACs anchored to other carbon structures are cumulenes (Chuvilin et al. 2009, Jin et al. 2009). The CAC–CNT junctions can be regarded as archetypical building blocks for all-carbon molecular electronics.

Linear carbon chains attached to the sharpened carbon fibers consisting of more than 10 atoms can be produced in situ in a field-ion microscope (FIM) during low-temperature field evaporation. The key findings of these works are the mechanical stability of the chains emanating from graphite by unraveling in ultrahigh electric fields (Mikhailovskij et al. 2007). The experimental procedures used in FIM carbon chain studies are reviewed and the results in relation to the atomistics of unraveling processes are discussed. Molecular dynamics simulations and high-resolution FIM experiments are performed to assess the evaporation of CACs under high-field conditions. Carbon exhibits a very rich dynamics of bond breaking that allows transformation from graphene to atomic chains. High-field experiments, theories leading to carbon chain formation, and methods to extract quantitative information on a variety of chain–surface interactions are described in detail. This very appealing high-field scenario was confirmed by independent experiments. Isolated atomic carbon chains can be obtained at different temperatures, pulling speeds, and forces.

A recently developed high-field technique of a self-standing CAC preparation has made it possible to attain the ultrahigh resolution of a field-emission microscope (FEM), which can be used for direct imaging of the intra-atomic electronic structure. By applying cryogenic FEM, we are able to resolve the spatial configuration of atomic orbitals, which correspond to quantized states of the end atom in free-standing CACs. Knowledge of the intra-atomic structure will make it possible to visualize generic aspects of quantum mechanics and also lead to approaches for a wide range of nanotechnological applications. Our experimental results have indicated that the resolution of FIM below 40 pm can be realized for sub-nano-objects (Sadanov et al. 2011). The unprecedented spatial resolution propels microscopy into a picometer realm of structural characterization and opens a way to direct real-space imaging of linear monatomic chains, subnanoribbons, and primary tubular structures.

This chapter mainly discusses free-standing monatomic chains hanging between two electrodes of the same material and self-standing anchored CACs. In this case, the absence of interaction with a substrate, or between the chains themselves, simplifies the atomic and electron structure of the CACs. Descriptions of ground-state electronic structure and basic properties of CACs are given in Section 26.2. In order to give the reader an introduction in the field of CACs, it will limit the discussion to several key items, where we focus on the exceptional physics of these 1D systems. An overview of preparation of CACs by various research groups are summarized in Section 26.3. Section 26.4 reviews experiments on field-ion and field-electron emission from CACs, including imaging the atomic orbitals and recent progress in the development of high-field techniques to visualize and measure structures at the picometer scale. Recent experimental and theoretical findings on unique mechanical properties of CACs are surveyed in Section 26.5 followed by summary and conclusions (Section 26.6).

26.2 Brief Background on Atomic and Electronic Structures

Progress over the past decade in the physics of 1D quantum confined systems together with technical advances in this field helped set the stage for efficiently accessing CAC properties. The physics of monatomic carbon chains was proved to be highly influenced by the 1D character of these structures. The thinnest possible electrical conductors, the monatomic carbon chains, are considered as an ideal component in molecular electronics (Lang and Avouris 1998, 2000, 2003). First principle calculations predicted a wide range of remarkable electronic properties, which can make CACs a potential material for nanotechnological applications. However, these early results have been somewhat overshadowed lately by the success of physics, chemistry, and technology of fullerenes and CNTs. The fabrication of the first free-standing CACs in ground-breaking experiments by Yuzvinsky et al. (2006) and Jin et al. (2009) has triggered a great interest in further theoretical study of this 1D structure.

26.2.1 Electronic Properties

The electronic band structure of CACs is characterized by the presence of the σ band composed of $2s$ and $2p_z$ orbitals that is located below the Fermi level. The bonding combinations of $2p_x$ and $2p_y$ atomic orbitals contribute to the doubly degenerate π-bands crossing the Fermi level. Due to π-bonds with nodes at the sites of 1D lattice, CACs behave as 1D nearly free electron system with the electron effective mass equal to the free electron mass. Cumulene is usually considered as metallic with a quantum ballistic conductance of $4e^2/h$ due to two degenerate, half-filled bands crossing the Fermi level (Tongay et al. 2004a,b, 2005). A self-consistent ab initio calculation of the conductance and current–voltage characteristics of capped polyynes has revealed that the conductance of polyynes is an order of magnitude larger compared with other oligomers (Crljen and Baranović 2007, Chen et al. 2009). The conductance is weakly bias dependent and almost independent of the length of the chain; hence, polyynes should be nearly perfect molecular wires, irrespective of the presence of bond alternation. A major issue in the use of them is the quality of electronic transmission, as obtaining the maximum conductance is critical to accessing their inherent electric properties.

In recent years, a great interest has focused on investigating the nonlinear transport phenomena such as negative differential resistance of CACs. The current characteristics of short carbon chains exhibit even–odd behavior, and negative differential resistance is predicted for both even- and odd-numbered chains. Odd CACs carry currents orders of magnitude larger than that in even CACs, as a consequence of the difference in conduction mechanisms (Khoo et al. 2008). Amongst others, ab initio calculations of spin-dependent transport in CACs bridging graphene nanoribbons have showed that CACs coupled to graphene electrodes are perfect spin filters with the orbital matching to graphene leads and great bias-dependent magnetoresistance independent of the length of carbon chains. An atomic carbon chain joining two graphene sheets demonstrates spin polarization of the transmission in large energy ranges, which can be controlled chemically or by applying an external electric field. The spin filter and spin valve conserved in a single device open a new approach to the application of all-carbon spintronics (Durgun et al. 2006a,b, 2008, Fürst et al. 2010, Rivelino et al. 2010).

Several other interesting phenomena have been investigated for CACs. Giant thermopower was found for CACs connected with two metallic electrodes, which changes sign for even–odd number of carbon atoms. The thermal conductance is positive but reveals a similar oscillation between even and odd number of carbon atoms. The comparison of the heat conduction obtained from the microscopic calculation with that estimated by considering the chain as a cylinder characterized by the macroscopic heat conduction and cross-sectional area of 3.5 Å^2 reveals that the classical model overestimates the heat conduction of single CACs by about an order of magnitude (Segal and Nitzana 2003). The discovery of molecular forms of carbon has led to great interest in various physical and chemical properties of this element. The intrinsic magnetism of carbon nanostructures and potential influence of magnetic contaminants has been a matter of debate for some years (Makarova et al. 2001). CACs are presently considered as promising building elements for the implementation of the smallest and highly integrated nanoelectronic circuits (Wang et al. 2009), novel nonvolatile memories (Akdim and Pachter 2011), graphene-based atomic scale switches (Standley et al. 2008), and high-efficient tunable infrared laser (Lin et al. 2012).

Castelli et al. (2012) have studied such carbon-only objects, whose electronic structure they address by a standard ab initio method based on the density-functional theory (DFT). The detailed information on the geometry, cohesive energy, dynamical stability, and electronic and magnetic properties of CACs attached to graphene fragments was obtained. CACs bound to graphene edges are characterized by polyyne character attenuated to a value transitional between those usual to cumulenes and polyynes. The chemical connections of CACs to the graphene edges are exceptionally stable with the binding energy near 6 eV per bond between every chain and the sp^2-layers. Thermal-induced failure of the CAC typically has probability similar to those for the anchor point at the graphene edge, which indicates a very strong attachment. An odd CAC inside a nanohole in a graphene sheet displays a metallic behavior, with at least one band pinned to the Fermi level, whereas an even CAC has overlap with the states at the Fermi level. Even CACs are insulating and nonmagnetic. Odd CACs are instead associated with nonzero magnetization related to a spin state of the π-bonds. A consistent picture of CACs stabilized by sp^2-terminations provided by ab initio DFT calculations (Ravagnan et al. 2009) shows the sensibility to torsional deformation. This effect allows to alter the conductive states near the Fermi level and to switch the on-chain π-electron magnetism.

26.2.2 Energetics and Stability

An obvious advantage for technological applications is an exceptional degree of reversible deformation of CACs. Linear atomic structures fall into the class of ultra-strength materials. Covalent bonding of sp-hybrid orbitals along the chain and π-bonding of perpendicular $2p_x$ and $2p_y$ orbitals are responsible for the high tensile strength and overall structural design of linear chains. Analysis of several 1D chain structures of carbon including dumbbell and zigzag configurations (with atoms displaced in the direction perpendicular to the axis of the chain) has shown that the cumulene structure has the lowest total energy, and is therefore the most stable conformation (Tongay et al. 2005). All other considered conformations of CACs were transformed during relaxation into the linear one with a bond length of $c = 0.127$ nm and the cohesive energy of 8.6 eV/atom. Comparison of this value for CACs having twofold coordination and graphene and diamond having three- and fourfold coordinations, respectively, indicates the striking distinctions in bonding (Alkorta and Elguero 2006, Hu 2009). The cohesive energy of CACs is larger than that of graphene and only 8% less than the calculated cohesive energy of diamond (9.4 eV/atom).

The smaller coordination number in CACs is compensated by stronger atomic bonds. High binding energy of CACs, their superior radiation (Jin et al. 2009), and mechanical (Mazilova et al. 2010) resistance can also be explained by the double bonding and bond shortening in CACs. Comprehensive ab initio study of the effect of electron and hole doping on the equilibrium geometry and electronic structure of CACs (Okano and Tománek 2007) showed that, independent of doping CACs, they are metallic. The atomic and electronic structure of CACs changes severely upon electron and hole doping. In the absence of doping, the most stable equilibrium configuration of CACs was found to be linear, resulting from covalent bonding of sp-hybrid orbitals along the chain axis together with π-bonding of p_x and p_y orbitals. The π-bands of CACs behave as a 1D free electron system. The variation in energy as a function of dimerization shows that a cumulene structure is metastable, which easily transforms to polyyne structure (Cahangirov et al. 2010). Distributions of bond lengths exhibit even–odd disparity depending on the number of carbon atoms in the chain and on the type of saturation of carbon atoms at both ends. A local displacement of atoms at the middle of a long CAC induces oscillations of atomic forces and charge density, which are identified as long-ranged Friedel oscillations in 1D systems.

Using a developed method based on the ab initio calculation of the static potential, Lin et al. (2011) have predicted that monatomic carbon chains are very stable at room temperature. Nanostructured carbon thin films produced at room temperature by supersonic cluster beam deposition have been proven to include an extensive and durable sp-component (Casari et al. 2007). Nevertheless, the CACs are metastable and have the tendency to undergo cross-linking reactions to form sp^2-phases. Moreover, the linear carbon allotrope is the potentially explosive member of the nanocarbon family. Effective stabilization of the CACs can be achieved by their isolation in inert matrices or the capping of carbon chain ends by the neutral atomic and molecular species. Based on DFT calculations, Lin et al. (2012) showed that short B-doped CACs derived from single-layer graphene can serve as working medium for tunable infrared lasers. Such doped CACs, like pure carbon monatomic chains, appeared to be stable at room temperature in vacuum conditions. High electro-optical conversion efficiency can be achieved by applying voltage on the CAC ends. Their band gap is rather stretching controllable with laser wavelength covering broad infrared range.

An exhaustive survey of the theory of carbon chains is far beyond the scope of this chapter. The reader finding a background overview on atomic and electronic structure of CACs may read for basic information the seminal works of Bianchetti et al. (2002), Tongay et al. (2005), and Castelli et al. (2012).

26.3 Preparation of Atomic Chains

Over the past several years, numerous chemical approaches have been developed to synthesize CACs (Roth and Fischer 1996, Tsuji et al. 2003, Zhao et al. 2003, Nishide et al. 2006, Inoue et al. 2010). Recently, a linear structure consisting of 44 carbon atoms has been produced (Chalifoux and Tykwinski 2010). But in most cases, the resulting material is ill-defined, and the specific basic properties of CACs thus remain unrealized. Here, we present the routes for the fabrication and characterization of free-standing monatomic chains hanging between two electrodes and self-standing anchored CACs.

26.3.1 Unraveling of Nanotubes

Richard Smalley and his group succeeded in obtaining the first controlled field-electron emission from CNTs promoted by laser irradiation (Rinzler et al. 1995). The field emission of CNTs has been observed to depend significantly not only on the electric field strength but also on a laser beam that would extensively assist in heating the tube. Based on these data, it was assumed that large electron emission, in order of microamperes, was associated with linear monatomic carbon chains pulled out from the open tube edge, being extremely stretched by the external electric field. Heating the closed, dome-shaped ends of the nanotubes with a laser caused them to open and made the edges jagged and sharp. The authors described the carbon chains pulling away from the nanotubes "in a process that resembles unraveling the sleeve of a sweater." The large field-emission currents from the chain attached to the tube end could then be explained by a local enhancement of the electric field near sharp structures with a high aspect ratio. Although the authors cannot observe such structures directly, the carbon chains were clamed as the basis for the ultimate atomic scale field emitters.

Transport of electrons from the CNT to the tip of the negatively charged CAC is rather facile. The delocalized, cylindrically symmetrical π-bonding is responsible for a nearly metallic screening of the external electrostatic field (Lou and Nordlander 1996, Kim et al. 1997, Lee et al. 1997). The local DFT calculations of CACs in a uniform electric field have shown these chains to screen the applied field as well as a metal bar of the same configuration. As a result, the phenomenon of local field enhancement at the tip of the chain takes place. The C–C bonds were found to be extremely tough, undergoing very little extension until the peak electric field near the end atom has increased to more than 100 V/nm. We note that this critical field strength substantially exceeds that of all other materials (Miller et al. 1996, Poncharal et al. 2010).

In molecular dynamics (MD) studies with the Tersoff–Brenner interatomic potential, Yakobson et al. (1997) have simulated the mechanical response of CNTs at large deformations. At the late stages of failure, the nanotube fragments were connected by long and robust linear sequences of *sp*-hybridized atoms. The authors have treated the reason of CACs formation upon stretching of CNTs. Monatomic carbon chains were suggested to be produced in the manner similar to the unraveling of chains in field-emission experiments of Rinzler et al. (1995). These results were supported by high-resolution transmission electron microscopy (TEM) investigations and MD simulations of the tensile deformation and fracture behavior of CNTs (Marques et al. 2004, Asaka and Kizuka 2005), which have revealed the appearance of medium-sized monatomic carbon chains as the latest stage of plastic deformation. At critical deformation, the nanotube breaks and begins contracting quickly. This process is accompanied by an abrupt decrease of potential energy and a sudden rise of the temperature to 8,000–14,000 K. At this stage, nanotubes display the formation of CACs during the complex cooperative bond-breaking and bond-rearrangement processes. The nature of the C–C bonds in these CACs was found to be of cumulene type, that is, without the bond alternations. This mechanism is analogous to that previously proposed to the monatomic chain unraveling (Rinzler et al. 1995). These findings provide a reliable insight into the remarkable mechanical behavior of CNTs and their transformation into CACs, as well as an effective approach toward bond rearrangements that are beyond the scope of direct experimental observations. Investigation of disintegration mechanism of CNTs in high electric fields using the DFT formalism (Lee at al. 1997) has specified the previously developed scenario of chain unraveling. The unraveling of chains terminates at sites where adjacent graphene walls were bridged by covalent bonds. They demonstrated that at external field of about 30 V/nm, one would expect continuing unraveling of CNTs without the field evaporation and rupture of chains, resulting in linear atomic chains attached to the CNTs.

26.3.2 High-Energy Irradiation

Troiani et al. (2003) and Caudillo et al. (2005) have demonstrated the possibility of forming a linear chain of carbon atoms from a thin amorphous carbon film by electron irradiation. The irradiation was carried out with the beam of a field-emission TEM operated at 200 keV. The carbon film was first irradiated with current densities greater than 10^3 A/cm^2 reached in highly focused areas of about 1 nm^2, to produce 30–40 nm diameter holes by radiation damage such as sputtering and atomic displacement. The amorphous carbon in the vicinity of the holes has been graphitized, at least in the bridge separating the adjacent holes. This phase transformation is possible owing to radiation-induced self-organization processes. The remaining carbon film in the bridge region can be effectively considered as a carbon nanofiber. Electron microscopy patterns revealed that the nanofibers underwent a radiation-induced necking process at lower current densities between 0.1 and 100 A/cm^2. The bridge-thinning process under the soft irradiation conditions was usually completed by the formation of a CAC. This original method to prepare free-standing CACs gives a possible route toward the all-carbon molecular electronics.

Yuzvinsky et al. (2006) have reported a method to controllably shrink CNTs under transport current and exposure to the TEM electron beam. Point defect formation during electron bombardment and simultaneous resistive heating causes the nanotube to a gradual decrease in diameter. Electron irradiation effectively sputters carbon atoms from the surface of the CNT by knock-on displacement. For sp^2-bonded carbon, knock-on of lattice atoms takes place at electron energies above 120 keV. Under focused electron irradiation at high fluxes, the atom extraction occurs rapidly and heterogeneously. Nonuniform atom removal results in inhomogeneous tube forming, and local necking. If the thinning process is allowed to continue at the thickness less than 1 nm, the CNT breaks, leaving a very thin link connecting the two sections of nanotube. These links were not stable under the electron beam and their TEM images cannot be resolved. As the diameter of the CNT was reduced to near zero, negative differential resistance was observed, as expected for a carbon chain–like structure. Yuzvinsky et al. (2006) have conjectured that CACs were likely produced at the last stage of the nanotube shrinkage.

With the help of high-resolution TEM, Jin et al. (2009) and Chuvilin et al. (2009) demonstrated that free-standing CACs can be experimentally realized by removing carbon atoms from graphene membranes through a well-controlled radiation treatment by energetic electrons (Figure 26.1). These authors have chosen the graphite nanoflakes imaged by the electron optics as the starting material. At the first stage of their in situ treatment, an intensive electron irradiation–induced thinning was performed to obtain a single-layer graphene. Further electron irradiation at high intensity produced two neighboring holes in the graphene layer, separated by a graphene nanoribbon. The energy of the edge atoms of the ribbon is so higher than those at the center that the edge atoms are preferentially removed during the radiation treatment. A surface atom sputtering mechanism should dominate such a thinning process. As a result, the graphene bridge narrows as the holes grow during irradiation. The transition from a graphene nanoribbon to a single carbon chain bridge has been directly observed at atomic resolution in real time. These free-hanging atomic chains are surprisingly stable under the extreme conditions of electron irradiation, and have been observed to survive for more than 100 s. This corresponds to a dose of the order of 10^9 nm^{-2}, which is of several orders of magnitude higher than radiation dose limits for organic molecules. It has been conjectured (Chuvilin et al. 2009) that the high conductivity of the graphene and CACs reduced the effect of ionization damage. The high radiation resistance of CACs creates the necessary prerequisites for an effective electron beam synthesis of quasi-1D devices at the sub-nanometer scale. Chuvilin et al. (2009) have also revealed some structural reconstructions preceding the chain formation, involving transformations of hexagons into unexpectedly stable pentagon–heptagon network.

These experiments have demonstrated successive radiation-induced transformations from graphite to graphene, from graphene to a graphene nanoribbon, and at last from the nanoribbon to atomic chains. The observed spontaneous decomposition of two atomic ribbons and the formation of a double chain have been explained by the authors in terms of the total energy for

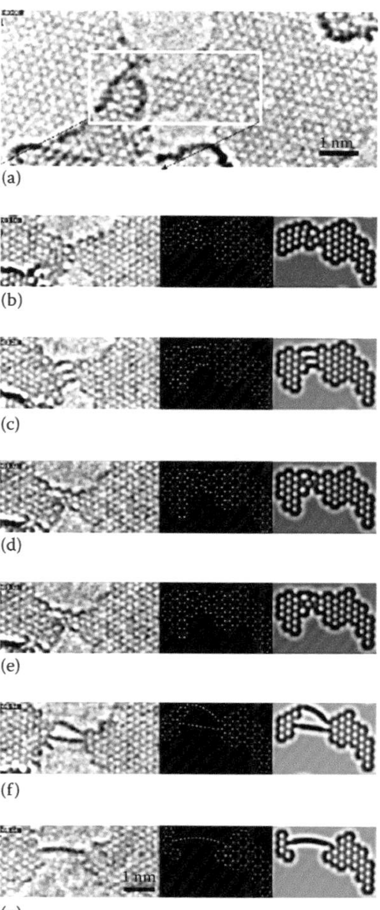

FIGURE 26.1 Formation of free-standing carbon atomic chains through continuous electron beam irradiation. (a) HRTEM image of the initial graphene ribbon configuration. (b–g) Time evolution of the bridge, in the experimental image (left), atomistic model (center), and corresponding image simulation (right). Carbon chains are present in panels (c, f, g). (Reprinted from Chuvilin, A. et al., *New J. Phys.*, 11, 083019-1, 2009. With permission.)

the chains, which is lower than that for the ribbon. This conclusion was clearly supported by the density-functional theory calculations. Jin et al. (2009) also emphasized that the CACs never break in the middle. It was also found that the chains were usually detached from the graphene edge. Under typical experimental conditions, contact between the chains and the graphene edges was not stable, and occasionally the chain ends migrated discretely along the graphene edge with atomic steps (Figure 26.2). This phenomenon is analogous to the bond-drifting process revealed in computer experiments with CACs obtained by pulling the edge atoms of a graphene nanoribbon (Wang et al. 2007). The experimental data on formation, migration, and breakage of CACs were clearly supported by atomistic modeling of such structures. The DFT calculations have explained the relatively low tensile strength of the end part of carbon chains. This was explained by the difference in C–C bond length with *sp*- and sp^2-hybridizations. The C–C bonds at the connection of

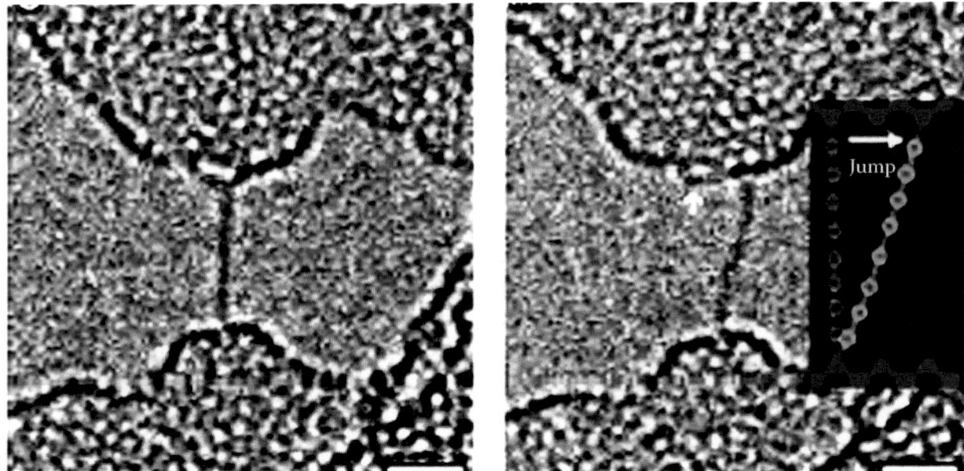

FIGURE 26.2 Consecutive HR-TEM images showing the jump of the CAC along the graphene edge with a changing of edge bonding during electron beam irradiation. The inset is a representative scheme. (Adapted from Jin, C. et al., *Phys. Rev. Lett.*, 102, 205501-1, 2009. With permission).

the carbon chain and the graphene are appreciably longer than the others in the middle of the chain. The bond breakage at the anchor point needs a lower energy than that in the middle of the chain by about 1.0 eV. Both these factors explain that the covalent bonds with graphene are less stable and relatively easy to break. Briefly reviewed radiation methods for preparing free-standing CACs from different "precursor" material ranges from amorphous to crystalline carbon have demonstrated that the chains are produced efficiently by a rather universal self-organization process.

Recently, a new approach to produce single carbon-chain/carbon-nanotube molecular junctions has been elaborated by Börrnert et al. (2010a,b). A few atom self-standing carbon chains were produced at the nanotube wall in situ during irradiation with electrons via aberration-corrected low-voltage TEM (Figure 26.3). It has been shown that free-standing CACs can be produced inside of CNTs filled by fullerenes and reliably imaged in situ using TEM. DFT-based calculations of carbon chains bridging the two CNTs have supported this finding and demonstrated that, in contrast to previous studies, such chains are semiconducting cumulenes. First principles quantum chemical considerations indicate that the CAC/CNT junctions are stable up to 1000 K. Revealed junctions between CACs and other carbon structures can be considered as representative building blocks for all-carbon nanoelectronic devices. The cohesive energy of short carbon chains is comparable with that of small diamond cluster of having sp^3-bonding. The radiation-induced growth of such chains attached perpendicularly to the CNT surface is in agreement with a mechanism of the evolution of self-interstitial ensembles on graphene proposed by Tsetseris and Pantelides (2009). While CACs have been previously produced by various chemical methods, the technique based on the controlled energetic electron irradiation is more in line with the original approaches for making other 1D nanostructures and may provide an avenue for integrating the CACs into carbon-based electronic nanodevices. Discussed recent reports on successful preparation of single CACs bridging graphene sheets as well as CNTs potentially provide a protocol for new methods of structurally assembling all-carbon systems.

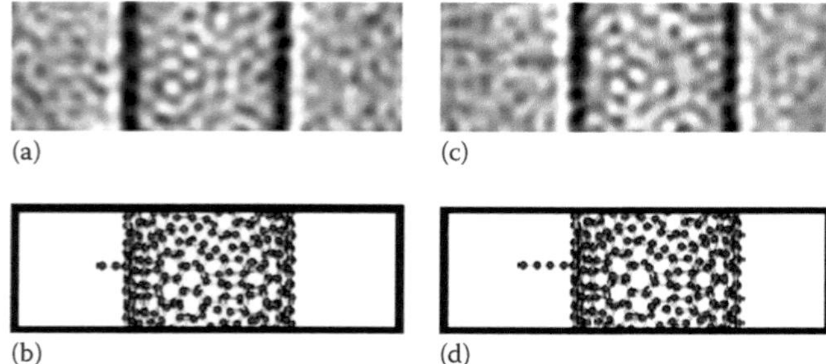

FIGURE 26.3 A few atom self-standing linear clusters at the nanotube wall. (a) and (b) Simulation and model of a two-atom protrusion, and (c) and (d) a four-atom protrusion. (Adapted from Börrnert, F. et al., *Phys. Rev. B*, 81, 201401-1, 2010. With permission.)

26.3.3 Unraveling of Graphite

Preparation of long atomic carbon wires is of immense importance for both technological and scientific research. The results of mathematical simulations with the Brenner potential showed that free atomic carbon wires with macrolength could be obtained by pulling the edge atoms of a graphene sheet with speeds lower than 30 m/s at about 300 K (Wang et al. 2007, 2009). A monatomic wire was formed with the carbon atoms that burst forth out of the graphene sheet in a process that resembles unraveling a knitted scarf (Figure 26.4). In this process, the graphene bonds near the chain anchor point may break (bond breaking) with release of one or more atoms to join the chain (Figure 26.4b and c), or the end of the chain may drift from one graphene atom

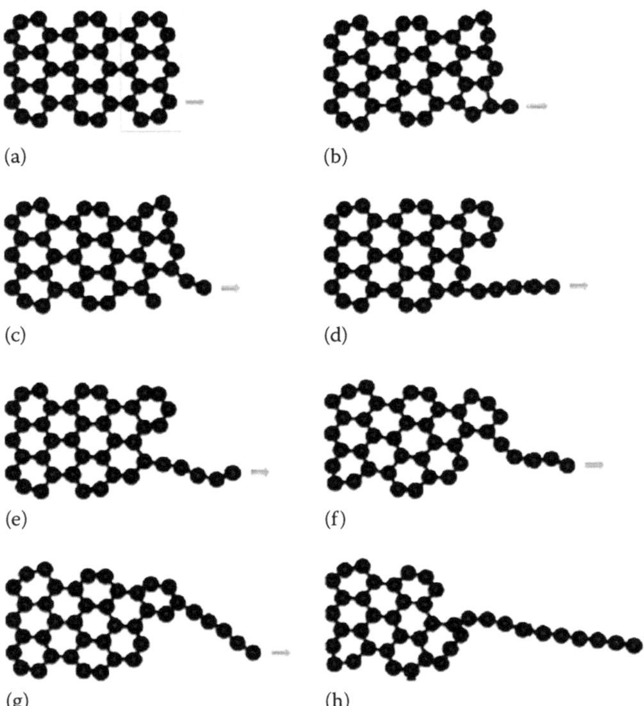

FIGURE 26.4 Snapshots for pulling chain from a graphene flake: (a) initial configuration, (b) and (c) bond-breaking process, (d–g) bond-drifting process, and (h) final structure in MD simulation. (Reprinted from Wang, Y. et al., *Phys. Rev. B*, 76, 165423-1, 2007. With permission.)

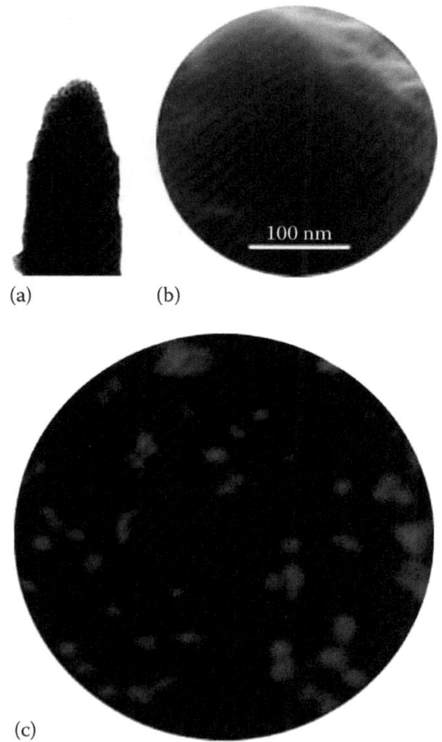

FIGURE 26.5 Electron microscope shadow graph (a), top-view SEM (b), and FIM (c) images of a carbon tip after low-temperature surface treatment in electric field of 100 V/nm. (Reprinted from Mikhailovskij, I.M. et al., *Nanotechnology*, 18, 475705-1, 2007. With permission.)

to another nearby (bond drifting) without lengthening the chain (Figure 26.4d–g). The feasibility of the bond-drifting process has been clearly demonstrated by direct experiments performed with the help of high-resolution TEM by Jin et al. (2009) and Chuvilin et al. (2009).

The computer experiments indicate that the entire graphite atomic layer is unraveled at about 1000 K to form macroscopically long monatomic wires. More sophisticated tight-binding MD calculations have confirmed these classical simulations based on the bond-order potential. In the bond-breaking process at constant temperature, failure of the chain itself is hardly probable because the energy of the *sp*-bond in a monatomic wire is larger than that of the sp^2 in graphene and the joining bond that connect the carbon chain and the graphene sheet. In adiabatic MD calculations, transverse vibrations of a CAC grow to be more and more intensive during unraveling. This process is accompanied by rising temperature and by subsequent failure of the wire. The rupture in the constant pulling speed regime is observed in the middle part of the wire but not at the junction of the wire and the graphene sheet, in contrast to that in the constant force regime (Mazilova et al. 2010).

Using field-ion and field-electron microscopy and mass spectrometry, it has been found that the presence of CACs at the surface of carbon nanotips treated by electric fields is of the order of 10^{11} V/m (Ksenofontov et al. 2007, Mikhailovskij et al. 2007). Self-standing CACs consisting of more than 10 atoms have been produced in situ in a FIM using low-temperature pulsed-voltage field evaporation. The process of field evaporation occurs with an anomalously high rate of about 10^{11} atomic layers per second. The TEM and scanning electron microscope (SEM) observations showed that carbon nanotips with initial radii of 20–50 nm were blunted up to 200–1000 nm during field evaporation (Figure 26.5). It was shown that the bright FIM image spots (Figure 26.5c) corresponded to the end atoms of invisible in the TEM and SEM carbon chains produced during a high-field treatment. After such a treatment, the tip surface is covered by self-standing CACs with density in the range 10^{15}–10^{16} m^{-2}.

By using MD simulation, it was shown that C-chains can be formed during the high-field unraveling of graphite tips. Under the simulation conditions, the unraveling of graphite layers is adiabatic and does not occur at a constant temperature. At the beginning of the tensile loading until the bond breaking starts, the graphene sheets are at about 0 K. During high-field loading, the potential energy rises gradually with the bond stretching until the first bond at the CAC–graphene junction breaks up. The consequent contraction of the nearest atomic bonds causes an abrupt decrease of potential energy and anharmonic atomic oscillations with the period of about 30 fs. The average amplitude of oscillations corresponds to temperature of about 10,000 K. The graphene-based carbon structure is incapable of fast evacuating the heat. As a result, the unraveling of the graphene layers proceeds in an ultrahigh-temperature surface region.

The possibility of explosive local overheating above the critical point of carbon was also demonstrated in MD simulations of the CNT breaking (Marques et al. 2004).

The mechanical unraveling of graphite at a constant pulling force (Wang et al. 2007) is accompanied by the relatively smaller temperature rise (up to 3000 K). Note that the mechanical loading regimes of constant stretching rate and constant puling force regime are mutually reciprocal. A constant pulling rate regime (Wang et al. 2007, Qi et al. 2010, Erdogan et al. 2011) could be experimentally realized with ideally stiff testing machines which have two heads, one of which is driven to change the distance between them and thus to impose a definite rate of elongation. A stiff system requires minute deflection for adjustment of the load and can consequently follow the drop in load required for elongation in the unstable region of unraveling responsible for overheating the chains. The high-field loading of linear atomic chains (Mikhailovskij et al. 2007) corresponds to the soft method based on "deadweight loading" through a lever. In this case, the load cannot fall and remains above the load required for stretching or unraveling of specimens. Moreover, mechanical testing in the FIM alone is an ideally soft testing method ensured the absolutely constant pulling force regime. In this regime, the unraveling is accompanied by a more intensive heating of atomic chains.

The rupture of CACs and subsequent significant shortening of interatomic bonds leads to an abrupt decrease of potential energy, and an additional temperature rising far above 10,000 K. Figure 26.6 illustrates the breakage of CAC in a high electric field and the subsequent instantaneous dissociation of the chain induced by an explosive increase of its temperature (Mazilova et al. 2010). It can be conjectured that magic carbon clusters usually observed in the low-temperature field evaporation spectra of graphite (Ksenofontov et al. 1983, 2007, Tsong 1990, Nishikawa et al. 2000, Nishikawa and Taniguchi 2005) are the result of the decomposition of long CACs into smaller atomic clusters because of the ultrahigh-temperature excitation during unraveling and breakage of CACs.

26.4 Visualization and Measurement at Picometer-Scale Resolution

26.4.1 Field-Emission Microscopy of Atomic Chains

As was shown in the previous section, an in situ high-field treatment of carbon tips resulted in the formation of a "hairy" surface made up of normally aligned CACs. Figure 26.7 shows the field-electron (a) and field-ion (b–d) images of the carbon surface covered by self-standing CACs produced during high-field forming at 5 K. The operating voltages were (a) −0.520, (b) 4.24, (c) 4.91, and (d) 6.00 kV. The positions of the field-electron image spots (a) are convincingly correlated with that in the initial field-ion pattern (b). The voltage increase is accompanied by the growth of density of FIM image spots (Figure 26.8) corresponding to the apexes of anchored atomic chains of various lengths produced during high-field unraveling.

It was evidently demonstrated (Mikhailovskij et al. 2007) that hairy carbon structures can be used as multipoint field emitters for which the surface field is much greater than that for perfectly smooth tips. The typical current–voltage dependence for the multipoint field cathode with about 20 emitting CACs at 21 and 77 K is shown in Figure 26.8. The observed suppression of current instabilities at 21 K, as seen in Figure 26.8, can be due to a decrease in the rate of surface diffusion of adatoms. The greatest field-emission current per one CAC in these experiments was about 10 μA. The electron current density in CACs can reach up to 10^{10} A/cm^2.

The FEM images are not exactly sharp, because emitted electrons have a transverse velocity, which results in a scattering spot on the screen. The resolution δ is defined as the minimal diameter of the image spot, divided by the image magnification M. The resolution of the FEM is limited by the velocity of an electron near the Fermi level and the momentum uncertainty (Gomer 1993). The last dominates the resolution of FEM images of nano-objects characterized by $M > 10^6$. In this case,

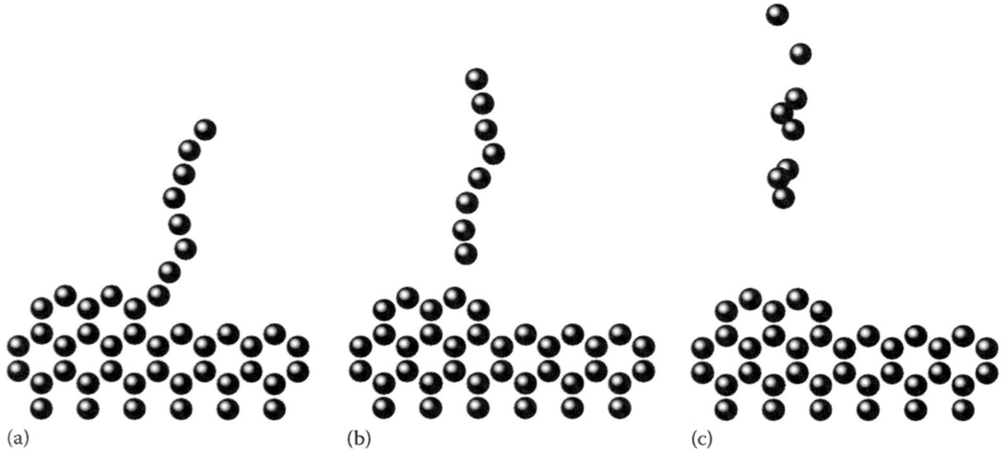

FIGURE 26.6 Fragmentation of an unraveled CAC induced by an explosive increase of its temperature. Snapshots (a) at the beginning, (b) after 6.4×10^{-14} s, and (c) 7.1×10^{-14} s. (Adapted from Mazilova, T.I. et al., *Int. J. Nanosci.*, 9, 151, 2010. With permission.)

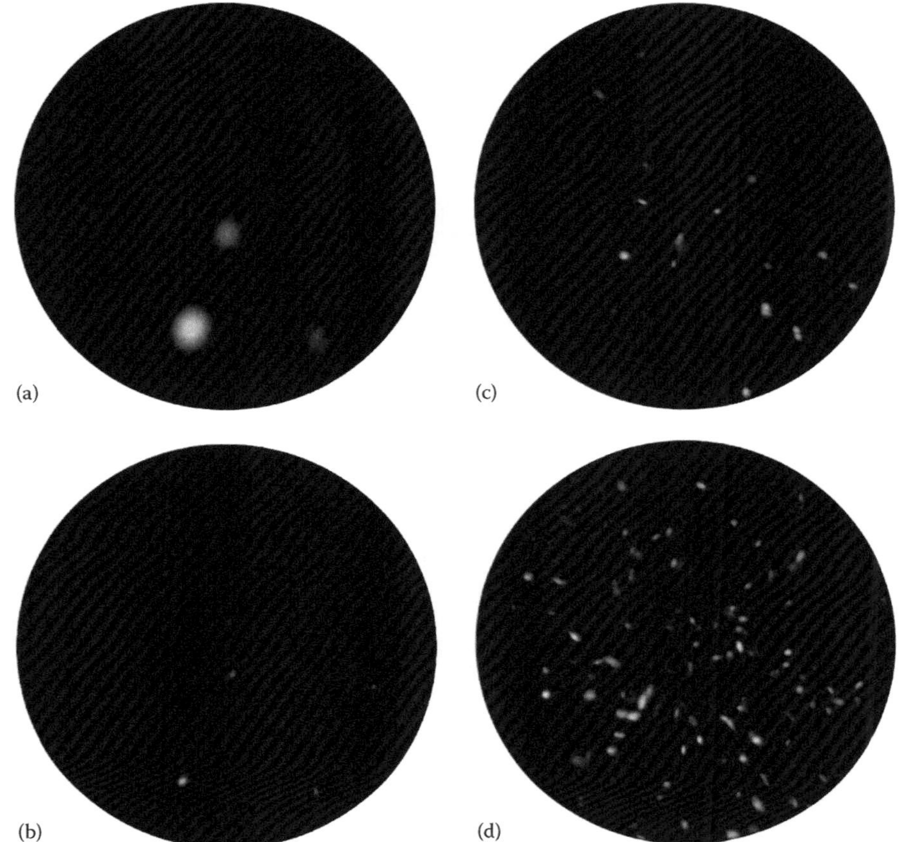

FIGURE 26.7 Field-electron (a) and field-ion (b–d) patterns of the carbon tip after a high-field treatment at 5 K. Micrographs (b–d) correspond to the various stages of field evaporation. (Reprinted from Sadanov, E.V. et al., *Phys. Rev. B*, 84, 035429-1, 2011. With permission.)

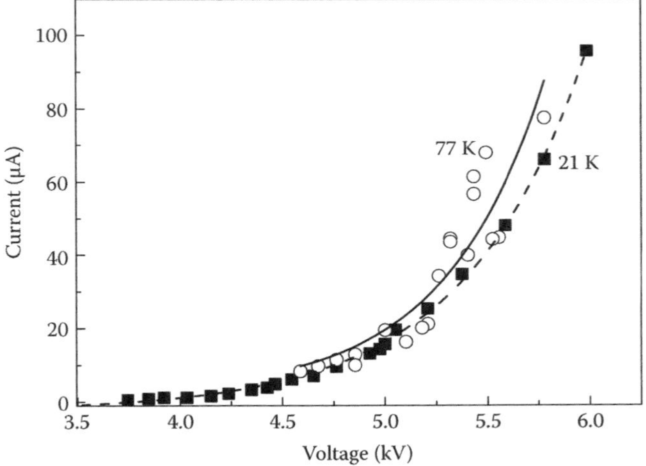

FIGURE 26.8 The field-emitted current of the multipoint cathode versus the applied voltage. (Reprinted from Mikhailovskij, I.M. et al., *Nanotechnology*, 18, 475705-1, 2007. With permission.)

the resolution is given by $\delta = (2\hbar\tau/m_e M)^{1/2}$, where m_e is the mass of the electron and τ is the time of flight from tip to screen. The time τ is almost exactly equal to the flight time of electrons at full energy eV, where e is the charge of the electron and V is the applied potential. To calculate the resolution of FEM images of the CAC anchored on the carbon tip, the "post on a paraboloid" model (Figure 26.9a) can be used. In this model, the chain having a cylindrical shape of height l and closed with a hemispherical cap with radius $\rho_0 = 0.12$ nm (Lorenzoni et al. 1997) stands normally on the tip with the radius r_0 (Figure 26.9a). The magnification of FEM is equal to $M = R/\beta\rho_0$, where R is the specimen-to-screen distance R and β is the image compression factor given by (Mazilova et al. 2009)

$$\beta = \xi\left(\frac{r_0}{L}\right)^{1/2} \tag{26.1}$$

where

L is the total distance of the apex of the hemisphere from the paraboloid surface ($L = l + \rho_0$)

$\xi = 1.145$

The apex field-enhancement factor for the chain is given by

$$\gamma = 1.05\left(2 + \frac{L}{\rho_0}\right)^{0.99}. \tag{26.2}$$

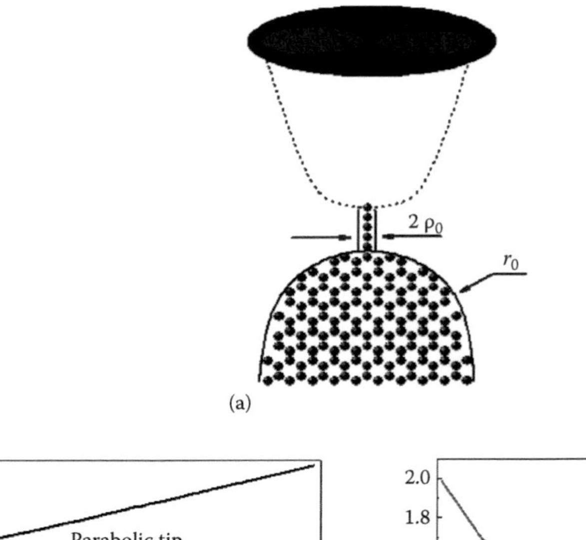

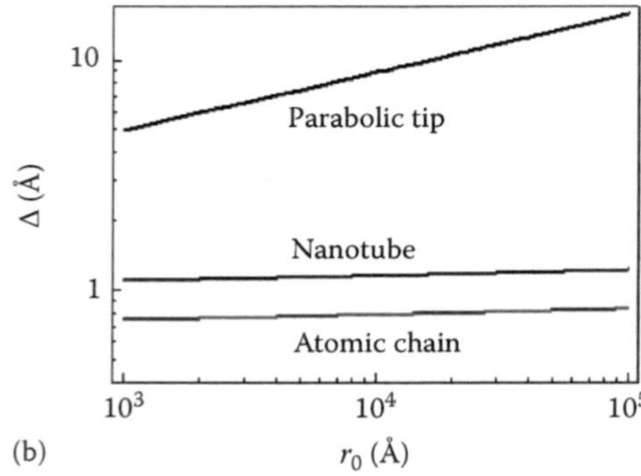

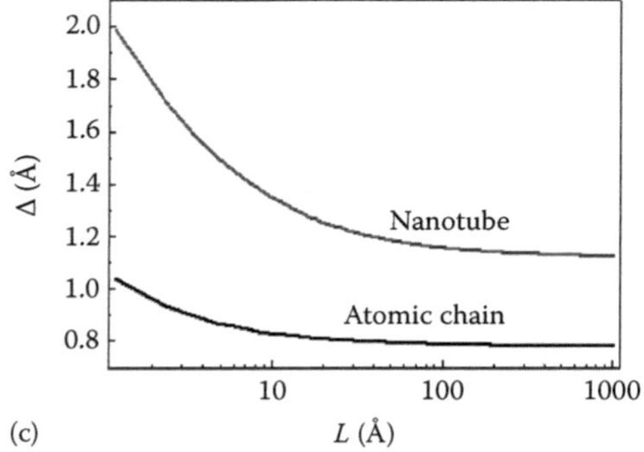

FIGURE 26.9 High-resolution field-emission electron microscopy. (a) A schematic drawing of electron emission from a self-standing atomic chain anchored at the carbon tip. (b) The dependence of FEM resolution on the radius of the supporting parabolic. (c) The FEM resolution as a function of the specimen length. (Reprinted from Mikhailovskij, I.M. et al., *Phys. Rev. B*, 80, 165404-1, 2009. With permission.)

The enhanced field strength F at the end of the chain was shown to be

$$F = \frac{2\gamma V}{r_0} \ln\left(\frac{2R}{r_0}\right). \qquad (26.3)$$

Using these expressions, one obtains the equation for the spatial resolution of FEM images of self-standing linear nano-objects:

$$\delta = (2\hbar\xi\rho_0)^{1/2} \left[\frac{em_e}{\gamma} LF \ln\left(\frac{2R}{r_0}\right)\right]^{-1/4}. \qquad (26.4)$$

The resolution given as a function of the radius of the supporting electrode for atomic chains, nanotubes, and parabolic specimens is shown in Figure 26.9b. Here, the lateral resolution Δ determined by the minimal distance between resolved image spots is equal to 0.46 δ. The FEM resolution calculated for typical conditions: $r_0 = 1 \times 10^{-6}$ m, $R = 5 \times 10^{-2}$ m, and $F = 5 \times 10^9$ V/m is determined largely by the specimen radius ρ_0 and length L (Figure 26.9c). The application of self-standing 1D specimens in FEM improves the resolution to the subangstrom level; thus, not only detecting a single atom is possible, but obtaining its spatial image can also be expected. The field-emission current is usually calculated by multiplying the impingement rate of electrons at the surface by the Gamow penetration factor (Gomer 1993, Roman et al. 1998). As only the electronic states lying near the Fermi level contribute to the field-emission current, the supply of tunneling electrons is proportional to the density of electronic state. Therefore, a two-dimensional FEM imaging of the local density of states corresponds to a spatial mapping of wave-function probability densities. The vast majority of FEM images of the end atoms of CACs look like axisymmetric bright spots (Figure 26.7a). In the general case, the FEM patterns have symmetries corresponding to singlets and doublets of bright spots (Figure 26.10). The singlet images are found to be the most stable patterns. Erasure or displacement of only a portion of a FEM pattern was never observed: such an image should be considered as a single whole entity. At the current greater than 10 pA, singlets sporadically change to doublets and vice versa (Figure 26.10c and d). These transformations correspond to $s \rightarrow p$ conversion of the electron orbital of the end atoms.

The observation of a stable doublet pattern in the FEM of CACs requires a stable mechanism breaking the axial symmetry which was not identified by Mikhailovskij et al. (2009). Using the DFT in the local-density approximation, Manini and Onida (2010) attributed the observed FEM image transformations to the symmetry breaking produced by the ligand and carbon π-bonding alternation. The type of the FEM pattern of carbon atoms was found be dependent on the hybridization of the

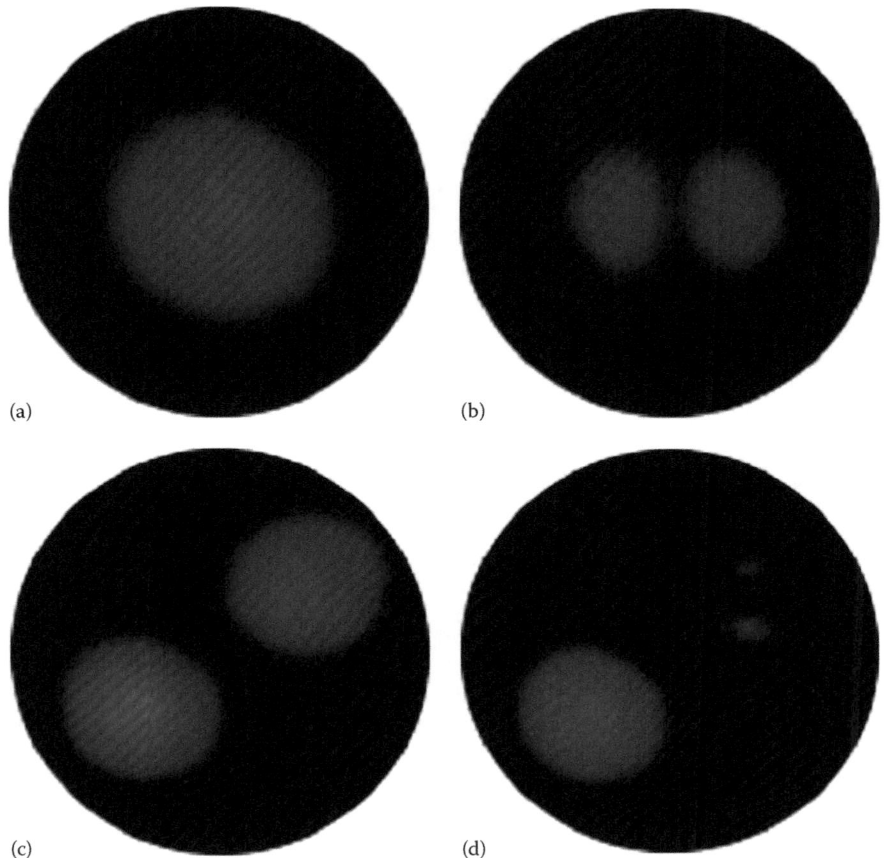

FIGURE 26.10 FEM images of the end atoms of CACs. (a) Singlet and (b) doublet of bright spots were acquired with a voltage 425 V. (c) *s*-like images of two atoms at the end of chains. (d) Spontaneous *s*→*p* transformation of the FEM image at constant voltage of one of atoms. (Reprinted from Mikhailovskij, I.M. et al., *Phys. Rev. B*, 80, 165404-1, 2009. With permission.)

anchored end of the chain. Whenever the anchored end of the chain binds a sp^2-hybridized atom of the carbon tip, the CAC acquires a partial cumulene character. In this case, the memory of the orientation of the sp^2-termination transmits along the chain axis through an alternating orientation of π-bonds. Figure 26.11a shows the highest-occupied molecular orbital, made of the four π-bonds oriented normally to the sp^2-plane. Field-electron emission from the end atom is dominated by this asymmetric π-orbital and displays an apparent nodal plane. The anchored end with sp^3-hybridization (Figure 26.11b) would induce a polyyne-type electronic structure of the CAC with alternating single/triple bonds and an essentially unbroken axial symmetry. At high emission currents, switching from one type of FEM pattern to another can be induced by exciting a jump of the chain attaching point, like those revealed by Wang et al. (2007) and Jin et al. (2009). Such jumps of anchored CACs accompanied by the mutual transformation of singlets and doublets have been observed in the FEM experiments (Mikhailovskij et al. 2010). The FEM images shown in Figure 26.12 illustrate a spontaneous transformation of the doublet to singlet patterns and a jump of image spots (about 10 nm) pointed by the arrow. Hence, the proposed mechanism by Manini and Onida (2010) of the transformation of quantum states of the end atoms in CACs is consistent with FEM observations.

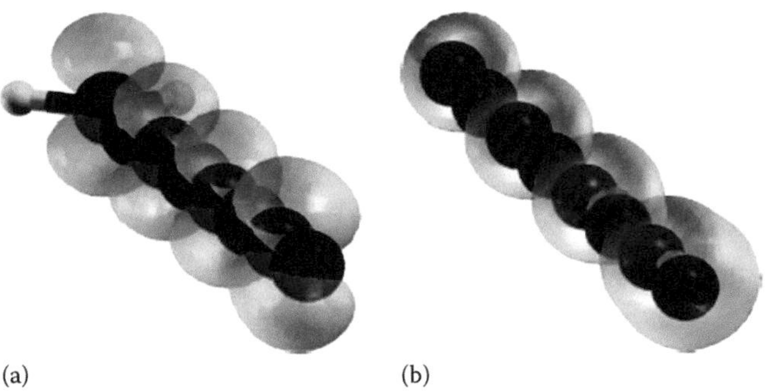

FIGURE 26.11 The HOMO of a CAC (dark balls) attached to (a) two ligands or (b) one ligand (hydrogen, clear balls). (Reprinted from Manini, N. and Onida, G., *Phys. Rev. B*, 81, 127401-1, 2010. With permission.)

It ought to be noted that the subatomic spatial resolution of 3D objects was clearly demonstrated in pioneering experiments with an atomic force microscope (Giessibl et al. 2000, Hembacher et al. 2004). Chaika and colleagues (2007, 2008a,b) have shown that picometer-scale features in STM images were superior resolved at low bias voltages and small tip–specimen separations. Obtained results prove that the electronic structure

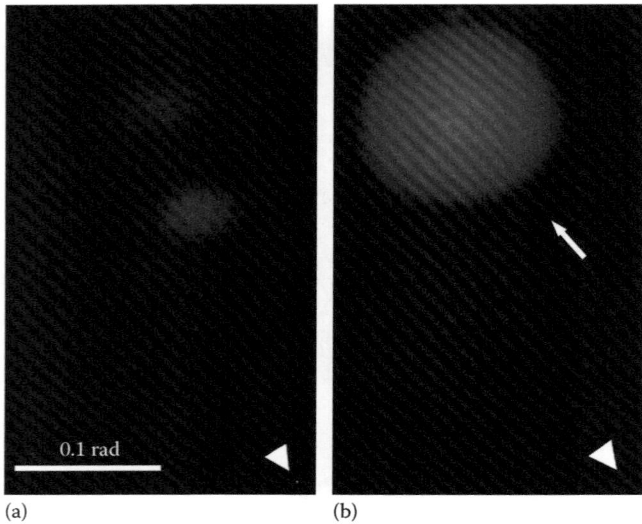

FIGURE 26.12 FEM images of the CAC (a) before and (b) after transformation of the FEM pattern at constant voltage (340 V). (Reprinted from Mikhailovskij, I.M. et al., *Phys. Rev. B*, 81, 127402-1, 2010. With permission.)

of the outer shell orbitals of the tip apex atoms plays a crucial role in the formation of high-resolution STM images. This is confirmed by reproducing the configurations of different tip electron orbitals in real space (Chaika et al. 2010).

26.4.2 Pauli Exclusion Principle and the Far-Subangstrom Resolution

The FIM invented by Ervin Müller (1956, 1965) enabled the atomic structure of solids to be seen directly for the first time. While the spatial resolution of its forerunner, the FEM, was limited to only about 20 Å because of electron diffraction, the resolution of the FIM was in the range of 2.5–3.0 Å. Recently, FIM and FEM have presented unique means to characterize 1D structures. Owing to the well-known trend toward improvement of the spatial resolutions with miniaturization of specimens, these microscopes have been found to be especially useful in the visualization of CNTs (Saito et al. 2002, Martin and Schwoebel 2007) and CACs (Mikhailovskij et al. 2007, Mazilova et al. 2009). The spatial resolution of the FEM can further be improved to 3 Å and below, so that the FEM images reveal the atomic scale structure of nanospecimens (Bakai et al. 2002, Saito et al. 2010). A FEM has recently demonstrated the possibility of resolving atomic orbitals at a lateral resolution greater than 1 Å (Section 26.4.1). Our experiments have indicated that for CACs spatial the resolution of FIM in a far-subangstrom region can be realized. The experiments have been carried out in an ultrahigh vacuum FIM at 5 K. Low temperatures are required to suppress the thermal motion of carbon atoms. Experimental procedures included in situ formation of CACs anchored at a parabolic carbon microtip described in previous section. The lateral magnification was determined by comparison of the separation between adjacent image spots with the theoretically determined interatomic spacing. The FIM is a lensless microscope and it does not require special suppression of mechanical vibration, temperature variations, or electromagnetic field.

Figure 26.13a illustrates schematically the ionization process, responsible for image formation in a FIM. Rare gas atoms are polarized in the high electric field and attracted to the CAC. Thus, the gas flux reaching the CAC has the kinetic energy equal to the sum of the thermal energy and the polarization energy given by

$$W = \frac{\alpha}{8\pi\varepsilon_0} F^2 \qquad (26.5)$$

where

α is the Gaussian polarizability
ε_0 is the electric constant
F is the field strength

The interactions of image gas atoms with the chain induce high-energy oscillations (Figure 26.13b). In the accommodation period (about 500 fs), the gas atoms execute a series of hops of decreasing height until field ionization occurs (Figure 26.13a). The FIM image spots are formed by ions originating above the chain-end atom. Field ionization involves the transfer of the gas-atom electron into a vacant electronic state in the chain (the state at or above the Fermi level). The ionization boundary ρ_{cr} (Figure 26.13a), the so-called critical surface, is the locus of gas-atom positions such that the electron transfers directly to the chain Fermi level. At closer distances, tunneling is prohibited at low temperatures by the Pauli exclusion principle, because there are no vacant electronic states of appropriate energy within the imaged specimen to receive electrons. The forbidden zone ($\rho < \rho_{cr}$) is usually defined by the expression $eF\Delta\rho_{cr} \approx I - \varphi$, where $\Delta\rho_{cr} = \rho_{cr} - \rho$, I is the ionization energy, φ is the electronic work function of the emitter surface, and ρ is the apex radius of a 1D specimen (Miller et al. 1996). For carbon chains, the work function is $\varphi \approx |E_{HOMO}|$, where E_{HOMO} is the energy of the highest-occupied molecular orbital. If the field strength F as a function of the distance r from the chain-end nucleus is taken into account, the forbidden gap is given by

$$e\int_{\rho}^{\rho_{cr}} F(r)dr = I - |E_{HOMO}|. \qquad (26.6)$$

Figure 26.13c shows the calculated critical distance $\Delta\rho_{cr}$ and relative curvature radius of the critical surface $\Delta\rho_{cr}/\rho$ versus the apex radius ρ of sharply pointed objects for helium atoms ($F = 22$ B/nm).

Figure 26.14 illustrates the possibility to observe with a FIM the atomic configuration of dimer CACs. The images are from the same dimer chain (pointed by the arrowhead) taken at a threshold image voltage of 13.4 kV (b) and at higher voltages: 13.7 (c) and 13.90 kV (d). The solid line in Figure 26.14a represents the electrical surface of the dimer, and the broken line denotes the

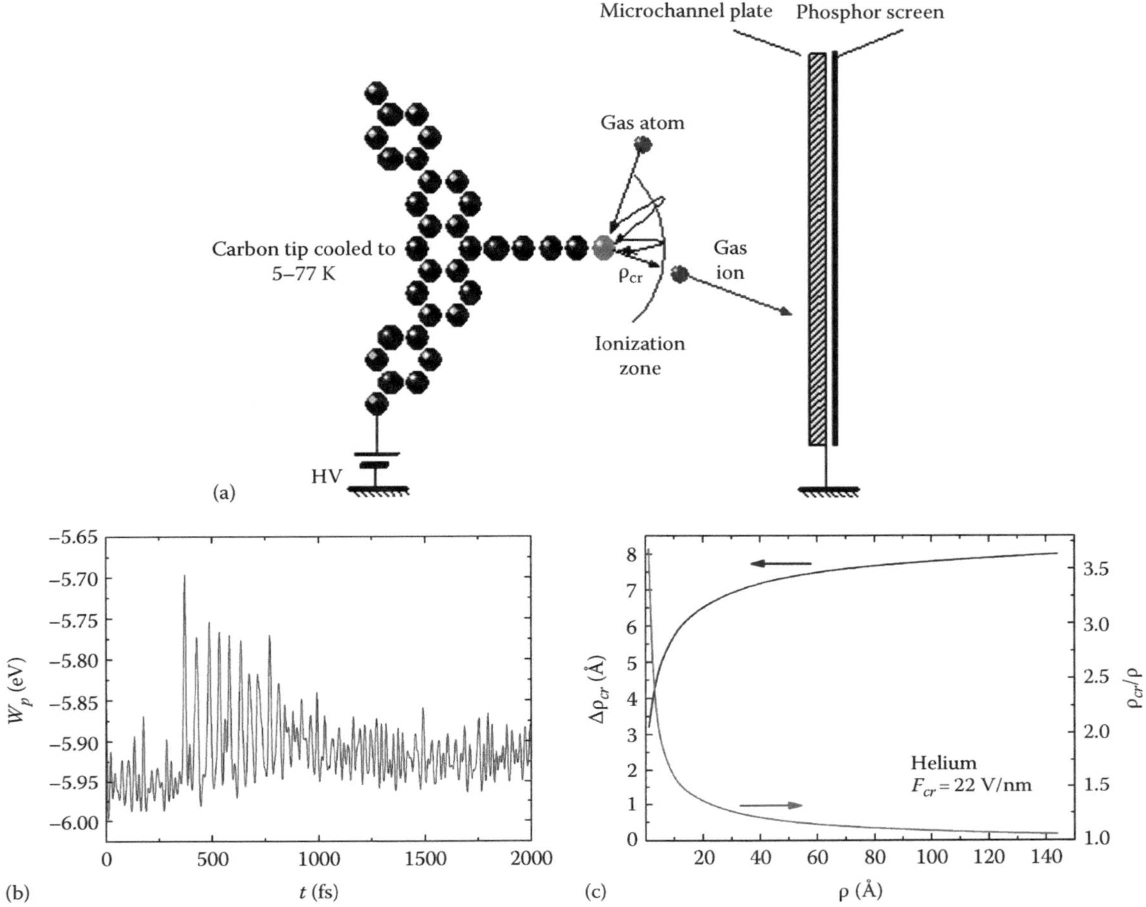

FIGURE 26.13 Field-ion microscopy of a self-standing CAC anchored at the graphite tip. (a) Field ionization above a free-standing CAC. The semicircular line shows the critical surface. (b) Oscillations of the potential energy of the CAC end atom. (c) Critical distance and radius of the critical surface versus the apex radius of nanowire. (Reprinted from Sadanov, E.V. et al., *Phys. Rev. B*, 84, 035429-1, 2011. With permission.)

critical surface. The electrical reference surface is defined as a "surface, where the electrical field appears to start" (Lang and Kohn 1973, Forbes 1999, 2003). For monatomic carbon chains, the location of the electrical surface relative to the nucleus of the end atom following to Lorenzoni et al. (1997) can be described by a hemisphere of radius ρ, equal to 1.2 Å. The FIM patterns corresponding to dimer chains were stable at a constant voltage, without obvious changes in intensity and configuration within 10–30 min.

For an experimental resolution criterion, consider points *M* and *N* that correspond to the field strength maximum at the critical surface, and point *O* on the axis of the chain, directly above the position midway between the end atoms of the dimer chain. The adjacent atomic images will be resolved if the ionization–density ratio *I*(*N*,*O*) is greater than a certain minimum value I_{min}. For field-ion images, the Raleigh criterion corresponds to $I_{min}=1.5$ (Forbes 1985). For all the FIM images shown in Figure 26.14, the ratio *I*(*N*,*O*) is much greater than I_{min}. The full width at half maximum (FWHM) of the intensity profile as a function of distance from the center of each image spot is 0.37 ± 0.05 Å at the threshold image voltage (Figure 26.14b). At higher voltages, one can observe a blurring effect and a reduction in contrast, but *I*(*N*,*O*) remains larger than I_{min}. The lateral resolution (FWHM) in this case is 0.46 ± 0.03 Å.

This lateral resolution is less than half its theoretical limit for conventional helium FIM (Miller et al. 1996). The resolution is determined by the transverse thermal velocity of an ion, the Heisenberg uncertainty principle, and the cross section of the ionization zone δ_0 above a given atom. A determining factor for a cryogenic FIM in the sub-nanoscale region is δ_0, which generally can be taken to be of about the radius of the image gas atom (1.1 Å for helium). A great disparity between the experimental and theoretical resolutions may be attributed to the nature of the local electric field distribution at ρ_{cr} above the low-dimensional nanospecimen that has not been accounted for quantitatively in the existing theory of resolution.

Whereas for conventional FIM specimens (ρ > 50 Å) represented by smooth paraboloids, $\Delta\rho_{cr}$ is much less than the specimen radius ρ, for sub-nano-objects we have $\Delta\rho_{cr} > \rho$ and the relative curvature radius of the critical surface increases from 1 to 3.75 (Figure 26.13c). This difference plays a significant role in determination of the resolution and magnification of FIM

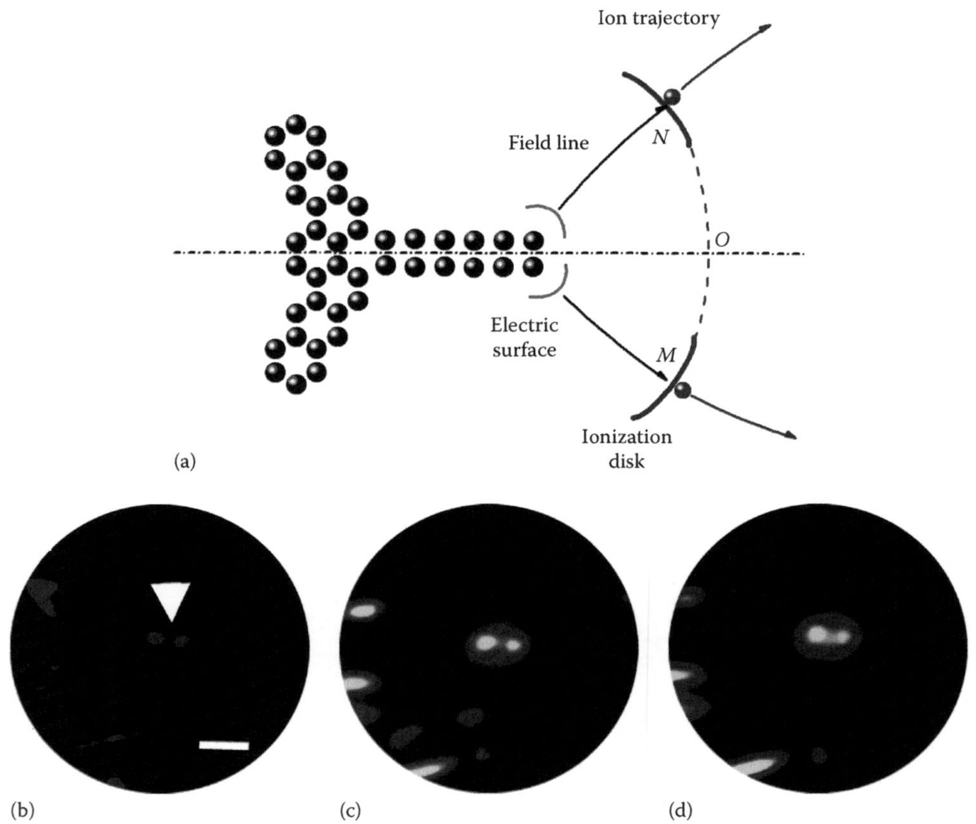

FIGURE 26.14 Field-ion microscopy of dimer carbon chains. (a) A dimer in a high electric field. The curved arrows denote the trajectories of image ions. (b–d) FIM images of the dimer (indicated by the arrowhead) acquired at voltages of (b) 13.4, (c) 13.7, and (d) 13.9 kV. Scale bars in (b–d), 2 Å. (Reprinted from Sadanov, E.V. et al., *Phys. Rev. B*, 84, 035429-1, 2011. With permission.)

images of sub-nano-objects. It leads to an anomalous extra magnification for sub-nano-objects. The high-field segment of the critical surface in the vicinity of points *M* and *N* runs roughly parallel to the reference electric surface. These segments of equipotentials can be considered as radially related. One could say that the end atoms of the nanospecimen are "projected" onto the critical surface with the extra magnification characterized by the scale factor *S*. The extra magnification of the radial projection image, which is essentially a mapping of the local electric field right at the critical surface, is determined by the ratio $S(\rho,\rho_{cr}) = \rho_{cr}/\rho$. Thus, the resolution of the FIM for nanoscale specimens can be expressed as

$$\delta = \frac{\delta_c(\rho_{cr})}{S(\rho,\rho_{cr})}. \tag{26.7}$$

Here, $\delta_c(\rho_{cr})$ is the resolution of the images of 1D nanospecimen protruding from the mesoscopic tips determined on the basis of calculations done by Mazilova et al. (2009), where the apex radius of the specimen is replaced by ρ_{cr}. Taking into account the phenomenon of extra magnification, the resolution for 5 K helium FIM images of a carbon atomic wire is calculated to be 0.3 Å. This value is in satisfactory agreement with the experimental resolutions for the carbon dimer chains shown in Figure 26.14b through d.

26.4.3 Novel Picoscale 1D Structures

The subangstrom lateral resolution opens the way to direct structural characterization of primary (elementary) tubular structures. Figure 26.15 shows sequential carbon sub-nanotube FIM images acquired at various stages of the atom-by-atom field evaporation at 5 K, alongside with corresponding atomic models discussed later. It can be seen that the patterns for the pentagonal tube are noticeably asymmetric. The azimuthal distortion of the FIM images can be due to the asymmetry of the supporting tip and/or to the deviation of the tube axis from the surface normal. All the atoms are resolved in the FIM images. In this case, it appears possible to perform a point-by-point reconstruction through plotting the position of each atom in a succession of layers, using the mapping techniques, coupled with controlled atom-by-atom field evaporation of the nanospecimen. The analysis shows that this nano-object has a quasi-1D morphology based on a pentagonal close packing of linear carbon chains. This structure is composed of pentagonal layers of carbon stacked in a top-to-top configuration (without staggering) with a rectangular structure of the lateral walls. The pentagonal tubular structures, suggested in the previous theoretical work (Tongay et al. 2005), have never been observed experimentally until now. The C–C bond distances in the

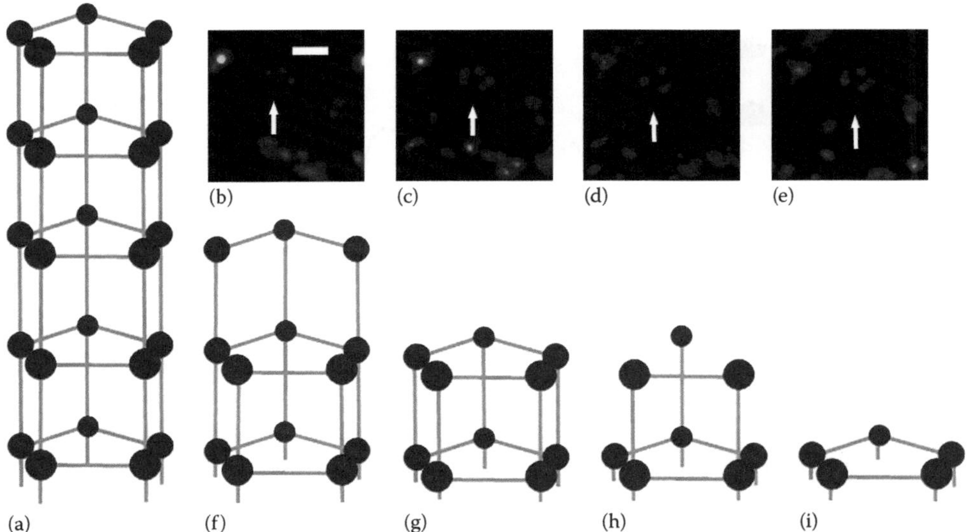

FIGURE 26.15 FIM insight into the elementary carbon sub-nanotube. (a) Atomic configuration of the pentagonal tubular structure. (b–e) Snapshots for the sub-nanotube (indicated by an arrow) corresponding to various stages of atom-by-atom field evaporation. The FIM images with atomic resolution of the end cap of the primary carbon tube were acquired at voltages of (b) 14.3, (c) 14.60, (d) 14.60, and (e) 14.8 kV. (f–i) Structural models related to these images. A scale bar in (b), 3 Å, is the same for all four images. (Reprinted from Sadanov, E.V. et al., *Phys. Rev. B*, 84, 035429-1, 2011. With permission.)

pentagonal tube are estimated to be $d_{C-C} = 1.5$ Å. Although the cases where the carbon sub-nanotubes show, by chance, a normal to surface orientation are few in number, this is indeed the first structural analysis of primary carbon atomic tubes. The lateral FIM resolution for this primary, ultimately thin carbon tube is 0.34 ± 0.05 Å.

A high spatial resolution enables a direct visualization of conformational transitions in a kinked monatomic chain (Figure 26.16a). FIM images of carbon-based kinked chains in different stereo-conformational states are shown in Figure 26.16b through d. In the experiments, hydrogen was additionally supplied through a gas inlet system up to 2×10^{-5} Pa. The simulated images shown in Figure 26.16d were formed by a molecular mechanics building-up process starting from a linear carbon chain, which is then kinked during the hydrogen adsorption by the second carbon atom (Figure 26.16a); the first (end) atom remaining bare. The field strength above the end atoms (about 100 V/nm) is significantly greater than the field required to ionize hydrogen and other gas contaminations (Miller et al. 1996). In fact, it serves to protect a clean specimen apex surface against contamination with chemically reactive residual gases in the FIM. Really, the FIM images reveal no sign of chemisorption at the end atoms, and hence, they correspond to pristine carbon surfaces. In contrast to the dumbbell pattern of dimer chains, the FIM images of kinked atomic chains are unstable. The two-spot field-ion images of the chain-end atom shown in Figure 26.16b have spontaneously turned to a single-spot pattern (Figure 26.16c). The minimal image diameter in this case is 0.24 ± 0.08 Å. Figure 26.16b can be interpreted as a two-spot image corresponding to the kinked chain, which resides in two states flipping back and forth between two well-defined stereo-conformations. The temporal resolution of the FIM is insufficient to study the picosecond-scale dynamic characteristic of the conformational transition. The mean time required to record a single image is around 1 s. Each second, about 10^3–10^4 ions originate from the ionization disk above the end atom of the chain. Thus, the image spot is formed on the phosphor screen which represents the time-averaged image of the end atom. A local brightness of the FIM pattern represents an averaged contribution of given conformation, which is proportional to the fraction of the time necessary for the chain to adopt that conformation. These patterns can be regarded as the probability densities of stereo-conformations of the bistable kinked carbon chain.

The field-ion image of the chain-end atom shown in Figure 26.16d appears as a blurry ellipse. This elliptic pattern of the chain arises from the time-averaged image of two configurations owing to a high switching frequency at high field strength. A fast rotational motion of the kink atom between opposite stereo-conformations is responsible for the blurry appearance. Despite a purely 1D nature of sp^2-terminated cumulene chains, the last ones display a nonvanishing torsional stiffness. For graphene termination of chains, the torsional barrier is of about 0.1 eV (Ravagnan et al. 2009). The polarized image gas atoms are scattered by the carbon chain, losing in this case about 0.25 eV (Figure 26.13a and b). This can cause various intramolecular changes such as atomic vibration, bond rotation, and reversible conformational changes. Comparison between experimental and calculated FIM patterns (Figure 26.16) shows that the intramolecular conformation of an individual carbon-based chain can be switched through the field-induced rotation of C–C bonds. This comparison, demonstrating a clear rotational motion, is regarded as an

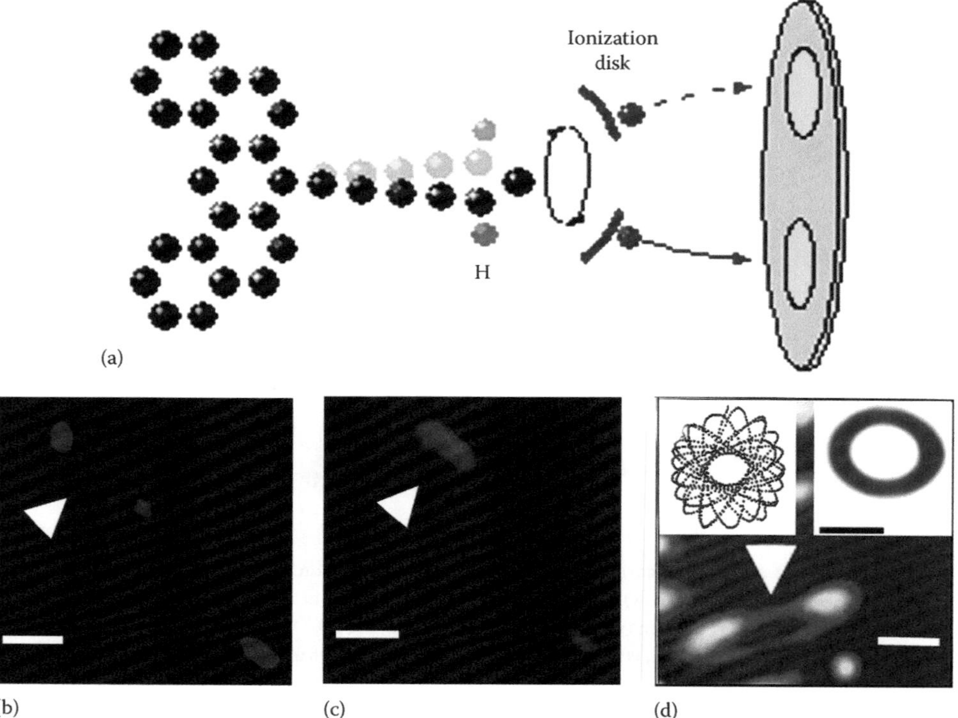

FIGURE 26.16 Field-ion patterns from carbon-based kinked chains in different stereo-conformational states. (a) The atomic model and the schematic diagram showing how a two-spot field-ion image of the chain-end carbon atom is formed. Black dots correspond to C atoms, red and pale red dots to hydrogen atoms, and blue dots to helium ions. The kinked chain can reversibly be switched into two possible conformational states by rotation at picosecond timescale. The image spots correspond to the time-averaged location of the kink atom related to different stereo-conformational states. (b) The two-spot field-ion image of the chain-end atom (shown by the arrowhead). The minimal image diameter is 0.24 Å. (c) The same chain after the structural transformation. (d) The FIM image of rotation motions of the kink atom between opposite stereo-conformations. Left inset: Calculated trajectories of the chain-end atom rotating between two conformational states. Right inset: A computer-simulated image of time-averaged rotational motion of the chain end. Scale bars in (b–d), 1 Å.

indirect evidence for the kinked chain model, accounting for the observed FIM patterns. Conversion of polarization energy of the image atom to mechanical motion of the atomic chain offers new possibilities for direct real-space imaging and controlling conformational states of the kinked carbon chain.

The achievement of a far-subangstrom spatial resolution owes much to several aspects inherent to cryogenic FIM in the subnanometer range. These are the general trend toward improvement of FIM resolutions with miniaturization of nanospecimens and the drastic increase in the relative curvature radius of the critical surface enveloping the forbidden zone of ionization. It has been shown that the Pauli exclusion principle is a direct cause of the revealed phenomenon of the anomalous extra magnification, crucial for the resolution limit of the FIM in the sub-nanometer range. The utmost spatial resolution opens the way to direct real-space imaging of primary tubular structures and other quasi-1D nanostructures, to a direct visualization with subangstrom resolution of stereo-conformational transitions in monatomic chains, and to determination of the dynamic mechanisms of elementary conformational switches in carbon-based chains.

26.5 Mechanical Properties

26.5.1 Theoretical Modeling of Mechanical Response

Designing and controlling of the mechanical characteristics of advanced structural and functional materials has long been the key subject of materials science. The mechanical characteristics of solid materials—elasticity, tensile properties, breaking strength, elastic moduli, etc.—are ultimately dependent on the strength of interatomic bonds. But this relationship is complicated by lattice defects playing a critical role in the material response to a large force at plastic deformation or fracture (Ruoff et al. 2003). The strength of atomic bonding itself is indirectly involved in the strength of bulk materials. The strength of structural materials substantially depends on its microstructure, which in turn is sufficiently size dependent. The tensile strength of perfect solids is about $E/10$, where E is Young's elastic modulus. All real 3D structural materials have much lower tensile strength, in the range 10^{-3}–$10^{-2}\,E$, due to available grain boundaries, dislocations, point defects, etc.

The picometer-scale transverse size of CACs and their elongated form, with the aspect ratio over 10, makes the mechanical properties of CACs particularly fascinating due to potential applications as structural materials in nano- and picotechnology. While the jury is still out regarding the convenient implementation of these applications, an additional stimulus comes from the basic materials science (Zhu and Li 2010). CACs possess almost translation-invariant symmetry along their nanometer-scale lengths to perform as 1D crystals with defined lattice parameters. In many aspects, these dual structures behave like well-defined structure materials with properties described in terms of modules, strength, stiffness, flexibility, and other conventional mechanical characteristics. CACs, due to their exceptional simplicity and atomically precise morphology, offer us the opportunity to precise theoretical determination of their mechanical response. Contrary to macroscopic crystalline materials where the evolution of ubiquitous dislocations and grain boundaries under mechanical loading determine the plasticity and strength, CACs can possess ideal structure showing the zero level of plasticity and the ultimate tensile strength. Due to potential use of CACs as interconnects in future quantum devices, there has been great interest in their electronic properties, motivating earlier theoretical studies. However, the mechanical characteristics of CACs, crucial to their successful incorporation in such nanodevices, were not determined until recently. It raises fundamental issues on the physics of failure and reversible deformation of CACs.

The tensile force is defined as $F_T = \partial E_T / \partial a$, where E_T is the total energy per unit cell and a is the interatomic distance. At small strain values, the $F_T(a)$ dependence is linear, corresponding to the elastic regime. In the highly anharmonic region near the lattice instability point where the tension reaches its maximum, the sound velocity decreases to zero. The reversible deformation of the CACs enhances dimerization both energetically and spatially. Cahangirov et al. (2010) showed that the tensile deformation of 8% corresponds to the energy difference between polyyne and cumulene CACs increased to 40 meV as compared to 2 meV difference in the equilibrium states. The spatial dimerization is three times higher in 8% deformed chains than that in the unstrained CACs. The increase in dimerization is accompanied by an order of magnitude increase in the band gap up to 2.87 eV at a critical strain $\varepsilon_c = 18\%$. The dependence of the band gap on strain can make CACs a potential candidate for use in a variety of strain-gauge nanodevice applications.

Atomistic simulations, using the second-generation reactive empirical bond order Brenner–Tersoff potential, were performed to investigate the uniaxial deformation of CACs and the unraveling of CNTs at different temperatures (Ragab and Basaran 2011). The dynamical effects in the tensile behavior of CACs were reported to play a key role in the unraveling processes. It was found that the primary step of the atomic chain separation from the nanotube is mainly due to the impulsive forces in the chain due to the addition of a new atom and rarely due to the steady forces in the chain. The temperature in the range of 300–1200 K can only affect the rate of the unraveling process, but does not affect the unraveling forces. The obtained force–strain relation shows that the thermal fluctuation in the axial force at 1200 K increases substantially compared to that at 300 K and the maximum average force that can be sustained in the chain decreases from 16.7 eV/Å at 300 K to 14 eV/Å at 1200 K. But the absolute maximum force is the same at both temperatures (18.6 eV/Å) and the maximum strain in the CAC is 32%. These values are rather larger than previous estimates by Mazilova et al. (2010), Cahangirov et al. (2010), and Hobi et al. (2010). This discrepancy can be partly explained by the dependence of the estimated mechanical response on the cutoff distance. The conventional cutoff function for the reactive empirical bond-order potential typically generates spurious bond forces near the cutoff distances due to discontinuity in its second derivative of the cutoff function (Lu et al. 2011).

CACs have potential application as beams of picoscale cross section that connect electronic or mechanical elements, so it is also essential to determine exactly their bending stiffness and vibration spectrum (Nair et al. 2011). The fundamental frequencies of vibration of the fixed–fixed CACs obtained from the displacement–time map are 6 and 0.625 THz for the shortest (5 Å) and longest (64 Å) CACs, respectively. Therefore, the control of CAC length can attain a wide range of resonating frequencies in the terahertz spectral region. Such CACs can be potentially used as an active element of nanomechanical resonators and extremely susceptible sensors. Assuming the chain behaves as an elastic beam, the frequency is determined by the equation $\omega = 2\pi f = \alpha^2 \sqrt{EI/mL^3}$, where m is the mass, L is the length, EI is the bending stiffness, and $\alpha = 4.43$ assuming a fixed–fixed boundary conditions for the chain. The length–frequency analysis yields bending stiffnesses of 1.36×10^{-28} Nm^2 and 2.02×10^{-28} Nm^2 for CAC of 8 and 10 Å lengths, respectively. These values are of two orders of magnitude lower than that of a CNT ($EI = 6.65 \times 10^{-26}$ Nm^2).

Quantum motions with the frequency proportional to the electric field are detected and analyzed with subangstrom lateral resolution using a FIM (Mazilova et al. 2009). Electric fields above 10 V/nm can be used for control of a transverse vibration mode of CACs in the terahertz spectral range. In MD modeling, the electric field–induced force was in the range of 3.5–7.0 nN. The main natural frequency for characteristic at field strength of 58.5 V/nm ($F = 6.4$ nN) was found to be of 1.72 THz. A crucial factor for reaching the quantum limit is the thermal occupation number n, set by the temperature and the mode angular frequency ω, $n = kT/\hbar\omega$. The quantum regime in nanomechanical oscillators is correspondent to $n \rightarrow 1$. A CAC with the natural resonant frequency equal to 1.72 THz enters the quantum regime at $T \approx 83$ K. Therefore, experimental investigation of the ground state at temperatures of liquid nitrogen and below (Mazilova et al. 2009) corresponds to a substantially quantum approach. These results clearly demonstrate that the atomic chains in high electric fields can be considered as light and short beams possessing zero-point motion at low temperatures. Electric fields above 10 V/nm can be used for force control of transverse vibrations of monatomic chains at terahertz frequencies.

26.5.2 High-Field Mechanical Testing of Chains

There is rapidly growing theoretical information indicating that CACs have certainly ultimate mechanical properties. However, the technical difficulties involved in the manipulation of these picometer-scale objects make the experimental determination of their strength a rather hard task. A singular carbyne is an exceptionally inconvenient object for mechanical tests due to its extremely small atomic dimensions. A direct and reliable measurement of their strength remains an important challenge for nanotechnology and materials physics.

There are only a few experimental data about the mechanical properties of individual nano-objects. The earlier observations of ultra-stress phenomena in nanocrystals were made using the FIM (Müller and Tsong 1969, Mikhailovskij et al. 1981). The FIM has made it possible to directly observe the atomic structure of nano-object under well-controlled crystallographic conditions combined with in situ mechanical loading. In these experiments, the nanocrystals subjected to high electric fields were fractured under tensile stresses close to the values of theoretical strength of solids. A unique advantage of this method is the possibility to avoid the inherent problem in nanoscale mechanical measurements caused by the need for assigning the effective cross-sectional area on which the force is applied. The tensile stress and strength of nano-objects at high-field mechanical testing could be determined by using only the field strength.

In such in situ mechanical testing, field strength is a crucial parameter and the calibration of the surface field is one of the most needed factors. The surface field $F_{es} = \beta V$ is determined by the applied voltage V measured with any desired accuracy. Direct calculation of the voltage-to-field conversion factor β is only possible for the rather unrealistic approximations of electrode geometry by confocal semi-infinite hyperboloids and paraboloids (Miller et al. 1996). Experimental values of the factor β and field strength F are usually based on the comparisons with the best image field F_{BI}. The best image conditions corresponded to the sharpest ion patterns with the best spatial resolution and contrast. This basic parameter for FIM, atom-probe tomography, and high-field nanotechnology has been repeatedly specified by experimental and theoretical methods for several decades. Assuming the validity of the Fowler–Nordheim equation, the F_{BI} for tungsten tips in helium was found to be of 45 V/nm. Considering some theoretical modifications and emendations of used methods, the field strength F_{BI} should be taken equal to 44 V/nm for W, Mo, Ir, and other refractory materials (see, e.g., Tsong 1990). The accuracy of F_0 and hence β determinations is within the range of ±(2–5)%. Taking into account that the mechanical stress at the surface is proportional to the square of the field strength, the systematic measurement error of the high-field mechanical testing of nanotips is within the interval of ±(4–10)%. This value is substantially less than a typical spread of mechanical properties of ultra-strength nanocrystals (Minor et al. 2006, Shan et al. 2008) and CNTs (Yu et al. 2000a,b).

Fabrication and high-field tensile testing of CACs were carried out in an ultrahigh vacuum FIM at 77 K (Kotrechko et al. 2012). The residual gas pressure in the working chamber of the microscope ranged from 1×10^{-6} to 5×10^{-7} Pa, and the imaging gas (helium) pressure was equal to 10^{-3} Pa. Needle-shaped specimens with an initial radius of curvature of 15–60 nm were prepared from commercial carbon polyacrylonitrile-based fibers by electrochemical etching with an ac voltage of 4–10 V in a 1 N KOH aqueous solution. After placement in the microscope, the carbon tips were subjected to field evaporation until a mesoscopically smooth tip was formed. During the evaporation process, self-standing CACs were produced at the tip surface.

Ionization onset voltage V_0 has been defined as the voltage corresponding to the threshold field-ion current strongly enhanced beyond the V_0. The experimental value of the ionization threshold field F_0 for helium was determined by comparison with the best image field F_{BI} using the same nanotip. Our experiments with tungsten, molybdenum, nickel, and iridium tips at 77 K showed that appreciable field-ion emission (>0.1 pA) is observed at fields greater than (0.50 ± 0.02) F_{BI}, consistent with earlier theoretical and experimental values F_0 of about 22 V/nm (Miller et al. 1996) nearly independent of the specimen size and material. The field strength at the electrical surface of nano- and sub-nanoscale object at the ionization threshold conditions $F_{es}^{\star}$ is not equal to F_0.

Mechanical testing of a carbon monatomic chain fixed at one end and loaded at its free end by field-induced mechanical forces directed parallel to the chain axis (Moy et al. 2011). The mechanical tensile stress at arbitrary voltage V was determined by using the known value at the threshold voltage V_0 from the following equation:

$$\sigma = \varepsilon_0 \frac{F_{es}^{\star 2}}{2} \left(\frac{V}{V_0} \right)^2 \tag{26.8}$$

where

the permittivity of free space ε_0 may be taken as 8.8542×10^{-12} C/Vm

F is in units of V/m

Using this equation for the determination of stress allows to avoid usual ambiguity in determining the effective diameter of CACs and which finally relates to expressions involving only experimentally measurable quantities like the CAC energy, critical tensile stress, and strain.

According to the theory of local contrast in field-ion images (Tsong 1990), the field strength at the electrical surface of nanotips at the ionization threshold conditions $F_{es}^{\star}$ is much larger than F_0 corresponding to the field strength at the critical surface. In conventional FIM of specimens with radii in the nanometer range, the critical surface runs roughly concentrically with the geometrical and electric surface, and is in the range of 0.5–1.0 nm away from the end-atom nucleus under normal FIM imaging conditions. Its exact position depending on the applied voltage

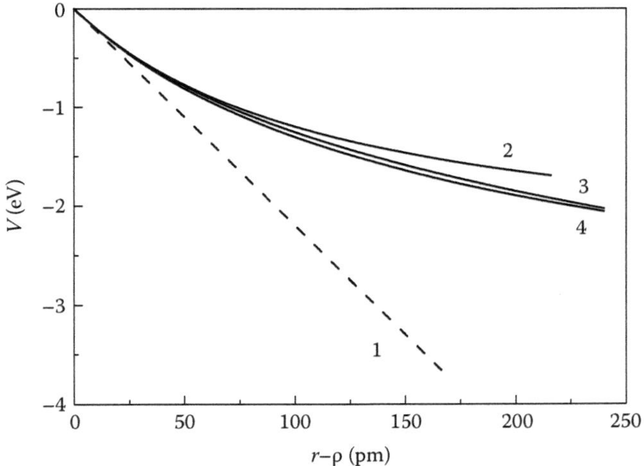

FIGURE 26.17 Calculated variations of the field potential with distance from the electrical surface of a carbon chain.

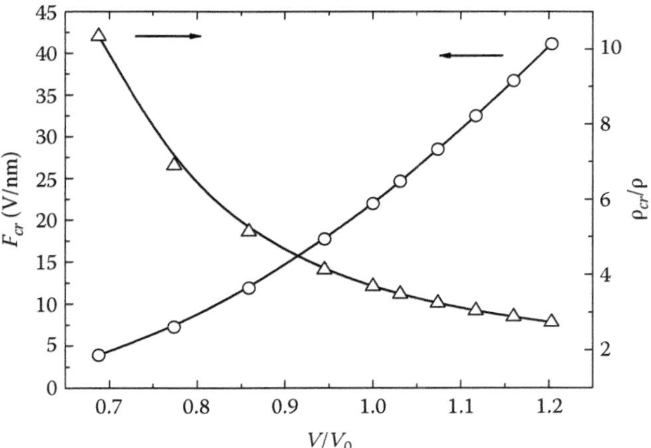

FIGURE 26.18 Electric field strength F_{cr} at the critical surface and the reduced radius ρ_{cr}/ρ of the critical surface versus the ratio of operation voltage V to the imaging gas ionization threshold voltage V_0. (Reprinted from Kotrechko, S.A. et al., *Tech. Phys. Lett.*, 38, 132, 2012. With permission.)

can be calculated using Equation 26.6. The solution to Laplace's equation near the apex of the tips of different shapes shows that the field strength falls off as the inverse square of the distance r, as well as for the ideal metal sphere. Taking for a spherical cap where the classical field potential is found to be zero at $r=\rho$, one obtains the field potential of a charge e in the general form:

$$V_F = -eF\rho\left(1-\frac{\rho}{r}\right). \tag{26.9}$$

Figure 26.17 shows the variations in the field potential near the apex of the tip approximated by the flat surface (1) and ideal metal sphere (2) calculated for $\rho=120$ pm and $F=22$ V/m. These potential distributions become similar near the apex of the tip where r approaches ρ. In planar geometry, V_F varies more rapidly than its spherical equivalent. At larger values of z, the spherical approximation (2) is characterized by a distinct lowering of the local field strength. In Figure 26.17, the curve 3 corresponds to more reliable analytical expression for potential distribution that has been obtained by Edgcombe (2005) for the near-surface regions of emitters approximated by a hemisphere supported on a long cylindrical shank:

$$V_F = -\frac{eF\rho}{2}\left(\left(\frac{r}{\rho}\right)^{1/2} - \left(\frac{r}{\rho}\right)^{-3/2}\right). \tag{26.10}$$

The curve 4 shows the exact numerical solution for the chain of cylindrical shape closed with a hemispherical cap approximated by the equipotential surface for a point charge situated at the end of a uniformly charged filament (Velikodnaya et al. 2007). The difference between Edgcombe's approximation (3) and the numerical solution (4) is indiscernible in the vicinity of the specimen surface. So we use analytical Equation 26.4 in the near-surface region for determination of the field potential that governs the field-ionization process.

Figure 26.18 shows the reduced critical distance ρ_{cr}/ρ and field strength F_{cr} at the critical surface versus the normalized voltage V/V_0 calculated from Equation 26.6 using Equation 26.10 for the work function ϕ taken as 5 eV equal to ϕ reported for several forms of carbon (diamond, graphite, fullerenes, and nanotubes) and the ionization energy I of helium equal to 24.6 eV (Miller et al. 1996). Using the data shown in Figure 26.18, the field strength at the electrical surface of the CAC F_{es}^* corresponding to the ionization threshold conditions was calculated from Equation 26.6. For typical reduced operation voltages, the field-induced stresses are an order of magnitude higher than those at high-field testing of micro- and nanotips (Zhu and Li 2010).

Figure 26.19 shows the typical FIM images of CACs formed by high-field unraveling at various operation voltages V. Initially, at $V=3.78$ kV, the imaged part of the tip shows a single CAC (1) in Figure 26.19a. As the applied voltage is increased, the initial image spot disappears and the images of other chains (2 and 3) appear (Figure 26.19b). These patterns also subsequently disappear when the voltage is increased. These transformations of FIM images are indicative of the fracture of monatomic chains loaded by the applied voltage. Calculations performed using Equation 26.8 and the numerical data shown in Figure 26.18 give the value of the fracture stress equal to 179 GPa (chain 2) and 181 GPa (chain 3). It should be noted that the obtained maximum strength of CACs at 77 K amounts to 70% of the value calculated for an ideal chain (Cahangirov et al. 2010, Hobi et al. 2010, Mazilova et al. 2010).

Figure 26.20 summarizes experimental measurements of the tensile strength of carbon structures as a function of specimen dimension. It shows that nanometer-scale carbon solids achieve a significant fraction of ideal strength, exhibiting the commonly known phenomenon "smaller is stronger" (Ruoff et al. 2003, Topsakal and Ciraci 2010, Zhu and Li 2010). The strength of carbon structures shows a strong sample size effect: the tensile

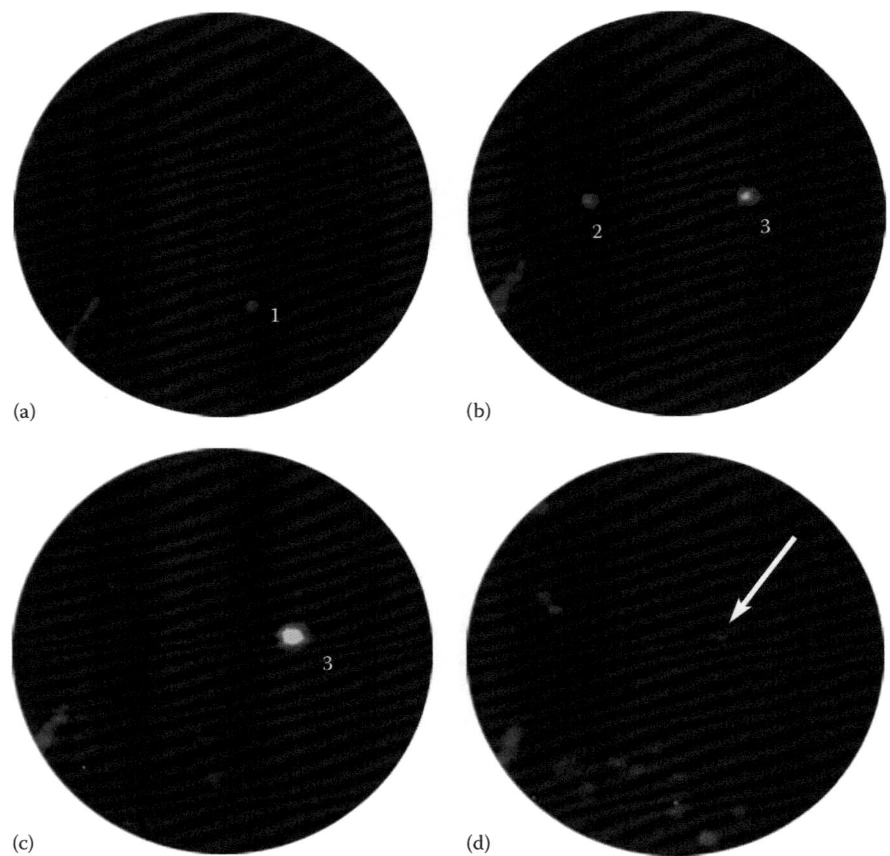

FIGURE 26.19 FIM images of CACs 1–3 on the surface of a carbon tip observed at voltages $V = 3.78$ (a), 3.83 (b), 4.40 (c), and 4.43 kV (d). The arrow shows the image of chain 3 after its fracture at electric field strength of 202 V/nm. (Reprinted from Kotrechko, S.A. et al., *Techn. Phys. Lett.*, 38, 132, 2012. With permission.)

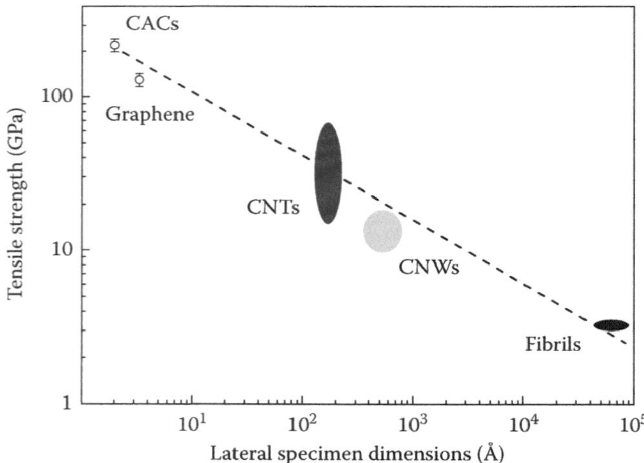

FIGURE 26.20 Experimental measurement of the tensile strength of carbon structures as a function of lateral specimen dimension.

strengths of individual CACs and graphenes with picoscale lateral dimensions (diameters and thicknesses) are much higher than the typical strength of micrometer-sized carbon fibrils (Mordkovich 2003). In experiments with CNTs (Yu et al. 2000a,b, Ruoff et al. 2003), graphenes (Lee et al. 2008), and nanowires (Gurin et al. 2007, Li et al. 2008), the strikingly large observed strength is dominated by atomic bonding, because of the absence of lattice defects and sources of them. The strength of smaller samples should converge to limit of the ideal periodic structures without lattice defects, while surface effects should dominate in so small objects such as monatomic chains. The upper strength revealed by CACs should be caused not only by their perfect structure but also by different bonding nature due to low dimensionality.

26.6 Summary and Conclusions

CACs have attracted remarkable interest in recent years because of their fascinating properties from a basic scientific viewpoint, as well as from their great potential in forthcoming technological applications in a diverse range of fields. This chapter gave a brief introduction to the fundamental electronic structure and basic properties of carbon chains. CACs have an extremely large surface area/volume ratio as compared to bulk materials, making the structure energetically expensive with physical properties quite different from those of bulk materials. The physics in monatomic carbon chains was proved to be highly influenced by the 1D character of these structures. A major issue in the use of monatomic carbon chains is the quality of electronic transmission, as obtaining the maximum quasi-ballistic conductance is critical to accessing their inherent electric characteristics. Amongst others, ab initio calculations of spin-dependent transport in carbon chains bridging graphene nanoribbons have showed that the chains are perfect spin filters with the orbital matching to graphene leads and great bias-dependent magnetoresistance independent on the length of carbon chains. The spin filter and spin valve conserved in a single device opens a new approach to the application of all-carbon spintronics. The controlled way of producing such rigid linear chains can dictate the design of novel building-block components for electronic devices at the molecular scale.

The special problems, related to mechanical response and field-emission properties of CACs, are described in detail in this chapter. Aside from the theoretical modeling, the basic experimental findings in this field have been summarized. In the last few years, the developed theory was successfully applied to interpret some recent experimental observations on these topics. It was found that the field-electron current density in CACs can reach up to 10^{10} A/cm^2. This result lends strong support to the conjecture of Richard Smalley that linear carbon chains may provide the ultimate atomic scale field emitters. We have briefly reviewed recent experimental findings and the current state of fundamental research of mechanical properties of carbon chains and have demonstrated a systematic trend of the increase in experimental yield strength of carbon nanostructures as the specimen sizes are reduced toward the picometer scale. The mechanical response of juvenile carbon chains to a large applied tensile force, especially their breaking strength, is directly dependent on the strength of interatomic C–C bonds. The revealed occurrence of strong mechanical response in the terahertz regime is of paramount technological application. It should be emphasized that the experimentally established value of the breaking stress for an atomic carbon chain notably exceeds the tensile strength reported for all ultra-strong materials, including graphene.

Because of their unique properties, it is believed that atomic chains will be utilized as the next-generation structural materials, interconnects, data storage, and quantum computation. The challenge of developing such devices is inseparably linked with the ability to prepare and characterize CACs in order to exploit their unique electronic and mechanical properties. Using a cryogenic FIM, we demonstrated a possibility to observe the carbon chains at the spatial resolution advanced far into the subangstrom scale. The Pauli exclusion principle is a direct cause of the revealed phenomenon of the anomalous extra magnification crucial for the resolution gain in the picometer range, as discussed in this chapter. The exceptional spatial resolution opens a way to direct imaging of linear monatomic chains, subnanoribbons, and primary CNTs and to real-space visualization of stereo-conformational transitions in a single monatomic carbon chain at the picometer scale. These revealed tubular structures are the most elementary form of CNT. The unique resolution attained also reveals a novel route for investigating the conformational changes of carbon-based chains using high-field schemes. The ability to reversibly switch between conformational states and determine the underlying mechanisms is important for our understanding of primary physical, chemical, and biological processes in a subangstrom realm.

Acknowledgments

The authors are grateful to R. Forbes and N. Wanderka for fruitful discussion on many of the topics of this chapter. This work is supported partly by the program “Basic Problems of Nanostructured Systems, Nanomaterials, and Nanotechnology” of the National Academy of Sciences of Ukraine (Project No. 15/11-N) and by the State program “Nanotechnology and Nanomaterials” (Project No. 1.1.1.48). IMM acknowledges the hospitality of the Hahn-Meitner-Institut Berlin and Deutsche Forschungsgemeinschaft.

References

Abdurahman, A., A. Shukla, and M. Dolg. 2002. Ab initio many-body calculations on infinite carbon and boron-nitrogen chains. *Phys. Rev. B* 65: 115106-1–115106-7.

Akdim, B. and R. Pachter. 2011. Switching behavior of carbon chains bridging graphene nanoribbons: Effects of uniaxial strain. *ACS Nano* 5: 1769–1774.

Alkorta, I. and J. Elguero. 2006. The carbon–carbon bond dissociation energy as a function of the chain length. *Chem. Phys. Lett.* 425: 221–224.

Allamandola, L.J., D.M. Hudgins, C.W. Bauschlicher, Jr., and S.R. Langhoff. 1999. Carbon chain abundance in the diffuse interstellar medium. *Astron. Astrophys.* 352: 659–664.

Asaka, K. and T. Kizuka. 2005. Atomistic dynamics of deformation, fracture, and joining of individual single-walled carbon nanotubes. *Phys. Rev. B* 72: 115431-1–115431-5.

Bakai, A.S., S.A. Bakai, I.M. Mikhailovskii, I.M. Neklyudov, P.I. Stoev, and M.-P. Macht. 2002. On the nature of the Kaiser effect in metallic glasses. *JETP Lett.* 76: 218–221.

Bianchetti, M., P.F. Buonsante, F. Ginelli, H.E. Roman, R.A. Broglia, and F. Alasia. 2002. Ab-initio study of the electromagnetic response and polarizability properties of carbon chains. *Phys. Rep.* 357: 459–513.

Börrnert, F., C. Börrnert, S. Gorantla et al. 2010b. Single-wall-carbon-nanotube/single-carbon-chain molecular junctions. *Phys. Rev. B* 81: 085439-1–085439-5.

Börrnert, F., S. Gorantla, A. Bachmatiuk et al. 2010a. In situ observations of self-repairing single-walled carbon nanotubes. *Phys. Rev. B* 81: 201401-1–201401-4.

Breda, N., G. Onida, G. Benedek, G. Colo, and R.A. Broglia. 1998. Bond-charge-model calculation of vibrational properties in small carbon aggregates: From spherical clusters to linear chains. *Phys. Rev. B* 58: 11000–11008.

Cahangirov, S., M. Topsakal, and S. Ciraci. 2010. Long-range interactions in carbon atomic chains. *Phys. Rev. B* 82: 195444-1–195444-5.

Casari, C.S., V. Russo, A. Li Bassi et al. 2007. Stabilization of linear carbon structures in a solid Ag nanoparticle assembly. *Appl. Phys. Lett.* 90: 013111-1–013111-3.

Castelli, I.E., N. Ferri, G. Onida, and N. Manini. 2012. Carbon *sp* chains in graphene nanoholes. *J. Phys.: Condens. Matter.* 24: 104019-1–104019-18.

Caudillo, R., H.E. Troiani, M. Miki-Yoshida, M.A.L. Marques, A. Rubio, and M.J. Yacaman. 2005. A viable way to tailor carbon nanomaterials by irradiation-induced transformations. *Radiat. Phys. Chem.* 73: 334–339.

Chaika, A.N. and A.N. Myagkov. 2008a. Seeing the atomic orbitals in STM images of a Si(111)-(7×7) surface. *J. Phys.: Conf. Series* 100: 012020-1–012020-4.

Chaika, A.N. and A.N. Myagkov. 2008b. Imaging atomic orbitals in STM experiments on a Si(111)-(7×7) surface. *Chem. Phys. Lett.* 453: 217–221.

Chaika, A.N., S.S. Nazin, and V.N. Semenov et al. 2010. Selecting the tip electron orbital for scanning tunneling microscopy imaging with sub-Ångström lateral resolution. *EPL* 92: 46003-1–46003-6.

Chaika, A.N., V.N. Semenov, S.S. Nazin, and S.I. Bozhko. 2007. Atomic row doubling in the STM images of Cu(014)-O obtained with MnNi tips. *Phys. Rev. Lett.* 98: 206101-1–206101-4.

Chalifoux, W.A. and R.R. Tykwinski. 2010. Synthesis of polyynes to model the *sp*-carbon allotrope carbyne. *Nat. Chem.* 2: 967–971.

Chen, W., A.V. Andreev, and G.F. Bertsch. 2009. Conductance of a single-atom carbon chain with graphene leads. *Phys. Rev. B* 80: 085410-1–085410-9.

Chuvilin, A., J.C. Meyer, G. Algara-Siller, and U. Kaiser. 2009. From graphene constrictions to single carbon chains. *New J. Phys.* 11: 083019-1–083019-10.

Crljen, Ž. and G. Baranović. 2007. Unusual conductance of polyyne-based molecular wires. *Phys. Rev. Lett.* 98: 116801-1–116801-4.

Durgun, E., S. Ciraci, and T. Yildirim. 2008. Functionalization of carbon-based nanostructures with light transition-metal atoms for hydrogen storage. *Phys. Rev. B* 77: 085405-1–085405-9.

Durgun, E., R.T. Senger, H. Mehrez, S. Dag, and S. Ciraci. 2006a. Nanospintronic properties of carbon–cobalt atomic chains. *Europhys. Lett.* 73: 642–648.

Durgun, E., R.T. Senger, H. Mehrez, H. Sevincli, and S. Ciraci. 2006b. Size-dependent alternation of magnetoresistive properties in atomic chains. *J. Chem. Phys.* 125: 121102-1–121102-4.

Edgcombe, C.J. 2005. Development of Fowler–Nordheim theory for a spherical field emitter. *Phys. Rev. B* 72: 045420-1–045420-7.

Erdogan, E., I. Popov, C.G. Rocha, G. Cuniberti, S. Roche, and G. Seifert. 2011. Engineering carbon chains from mechanically stretched graphene-based materials. *Phys. Rev. B* 83: 041401-1–041401-4.

Forbes, R.G. 1985. The origins of local contrast in field-ion images. *J. Phys. D: Appl. Phys.* 18: 973–1018.

Forbes, R.G. 1999. The electrical surface as centroid of the surface induced charge. *Ultramicroscopy* 79: 25–34.

Forbes, R.G. 2003. Field electron and ion emission from charged surfaces: A strategic historical review of theoretical concepts. *Ultramicroscopy* 95: 1–18.

Fürst, J.A., M. Brandbyge, and A.-P. Jauho. 2010. Atomic carbon chains as spin-transmitters: An ab initio transport study. *EPL* 91: 37002-1–37002-5.

Giessibl, F.J., S. Hembacher, H. Bielefeldt, and J. Mannhart. 2000. Subatomic features on the silicon (111)-(7×7) surface observed by atomic force microscopy. *Science* 289: 422–425.

Gomer, R. 1993. *Field Emission and Field Ionization*. New York: American Institute of Physics.

Gurin, V.A., I.V. Gurin, V.V. Kolosenko et al. 2007. Mechanical strength of carbon nanofibers obtained by catalytic chemical vapor deposition. *Techn. Phys. Lett.* 33: 534–536.

Hembacher, S., F.J. Giessibl, and J. Mannhart. 2004. Force microscopy with light-atom probes. *Science* 305: 380–383.

Hobi, E., Jr., R.B. Pontes, A. Fazzio, and A.J.R. da Silva. 2010. Formation of atomic carbon chains from graphene nanoribbons. *Phys. Rev. B* 81: 201406-1–201406-4.

Hu, Y.H. 2009. Stability of *sp* carbon (carbyne) chains. *Phys. Lett. A* 373: 3554–3557.

Inoue, K., R. Matsutani, T. Sanada, and K. Kojima. 2010. Preparation of long-chain polyynes of $C_{24}H_2$ and $C_{26}H_2$ by liquid-phase laser ablation in decalin. *Carbon* 48: 4209–4211.

Jin, C., H. Lan, L. Peng, K. Suenaga, and S. Iijima. 2009. Deriving carbon atomic chains from graphene. *Phys. Rev. Lett.* 102: 205501-1–205501-4.

Khoo, K.H., J.B. Neaton, Y.W. Son, M.L. Cohen, and S.G. Louie. 2008. Negative differential resistance in carbon atomic wire–carbon nanotube junctions. *Nano Lett.* 8(9): 2900–2905.

Kim, S.G., Y.H. Lee, P. Nordlander, and D. Tomanek. 1997. Disintegration of finite carbon chains in electric fields. *Chem. Phys. Lett.* 264: 345–350.

Kotrechko, S.A., A.A. Mazilov, T.I. Mazilova, E.V. Sadanov, and I.M. Mikhailovskij. 2012. Experimental determination of the mechanical strength of monatomic carbon chains. *Techn. Phys. Lett.* 38: 132–134.

Kroto, H.W. 1994. Smaller carbon species in the laboratory and space. *Int. J. Mass Spectr. Ion Process.* 138: 1–15.

Kroto, H.W., J.R. Heath, S.C. O'Brien, R.F. Curl, and R.E. Smalley. 1987. Long carbon chain molecules in circumstellar shells. *Astrophys. J.* 314: 352–355.

Ksenofontov, V.A., V.B. Kulko, and I.M. Mikhailovskij. 1983. Field emission microscopy and mass-spectrometry of carbon fiber. *Sov. Phys. Tech. Phys.* 28: 973–977.

Ksenofontov, V.A., T.I. Mazilova, I.M. Mikhailovskij, E.V. Sadanov, O.A. Velicodnaja, and A.A. Mazilov. 2007. High-field formation and field ion microscopy of monatomic carbon chains. *J. Phys.: Condens. Matter.* 19: 466204-1–466204-10.

Lang, N.D. and Ph. Avouris. 1998. Oscillatory conductance of carbon-atom wires. *Phys. Rev. Lett.* 81: 3515–3518.

Lang, N.D. and Ph. Avouris. 2000. Electrical conductance of parallel atomic wires. *Phys. Rev. B* 62: 7325–7329.

Lang, N.D. and Ph. Avouris. 2003. Understanding the variation of the electrostatic potential along a biased molecular wire. *Nano Lett.* 3: 737–740.

Lang, N.D. and W. Kohn. 1973. Theory of metal surfaces: Induced surface charge and image potential. *Phys. Rev. B* 7: 3541–3550.

Lee, Y.H., S.G. Kim, and D. Tomanek. 1997. Field-induced unraveling of carbon nanotubes. *Chem. Phys. Lett.* 265: 667–672.

Lee, C., X. Wei, J.W. Kysar, and J. Hone. 2008. Measurement of the elastic properties and intrinsic strength of monolayer graphene. *Science* 321: 385–388.

Li, H., F.W. Sun, Y.F. Li, X.F. Liub, and K.M. Liew. 2008. Theoretical studies of the stretching behaviour of carbon nanowires and their superplasticity. *Scripta Mater.* 59: 479–482.

Lin, Z.Z., W.F. Yu, Y. Wang, and X.J. Ning. 2011. Predicting the stability of nanodevices. *EPL* 94: 40002-1–40002-4.

Lin, Z.Z., J. Zhuang, and X.J. Ning. 2012. High-efficient tunable infrared laser from monatomic carbon chains. *EPL* 97: 27006-1–27006-4.

Lorenzoni, A., H.E. Roman, F. Alasia, and R.A. Broglia. 1997. High-current field emission from an atomic quantum wire. *Chem. Phys. Lett.* 276: 237–241.

Lou, L. and P. Nordlander. 1996. Carbon atomic chains in strong electric fields. *Phys. Rev. B* 54: 16659–16662.

Lu, Q., W. Gao, and R. Huang. 2011. Atomistic simulation and continuum modeling of graphene nanoribbons under uniaxial tension. *Model. Simul. Mater. Sci. Eng.* 19: 054006-1–054006-16.

Makarova, T.L., B. Sundqvist, R. Höhne et al. 2001. Magnetic carbon. *Nature* 413: 716–718.

Manini, N. and G. Onida. 2010. Comment on imaging the atomic orbitals of carbon atomic chains with field-emission electron microscopy. *Phys. Rev. B* 81: 127401-1–127401-2.

Marques, M.A.L., H.E. Troiani, M. Miki-Yoshida, M. Jose-Yacaman, and A. Rubio. 2004. On the breaking of carbon nanotubes under tension. *Nano Lett.* 4: 811–815.

Martin, G.L. and P.R. Schwoebel. 2007. Field electron emission images of multi-walled carbon nanotubes. *Surf. Sci.* 601: 1521–1528.

Mazilova, T.I., S. Kotrechko, E.V. Sadanov, V.A. Ksenofontov, and I.M. Mikhailovskij. 2010. High-field formation of linear carbon chains and atomic clusters. *Int. J. Nanosci.* 9: 151–157.

Mazilova, T.I., I.M. Mikhailovskij, V.A. Ksenofontov, and E.V. Sadanov. 2009. Field-ion microscopy of quantum oscillations of linear carbon atomic chains. *Nano Lett.* 9: 774–778.

Mikhailovskij, I.M., P.Y. Poltinin, and L.I. Fedorova. 1981. The tensile fracture of microcrystalline tungsten. *Sov. Phys. Solid State* 23: 757–759.

Mikhailovskij, I.M., E.V. Sadanov, T.I. Mazilova, V.A. Ksenofontov, and O.A. Velicodnaja. 2009. Imaging the atomic orbitals of carbon atomic chains with field-emission electron microscopy. *Phys. Rev. B* 80, 165404: 1–7.

Mikhailovskij, I.M., E.V. Sadanov, T.I. Mazilova, V.A. Ksenofontov, and O.A. Velicodnaja. 2010. Reply to comment the atomic orbitals of carbon atomic chains with field-emission electron microscopy. *Phys. Rev. B* 81: 127402-1–127402-2.

Mikhailovskij, I.M., N. Wanderka, V.A. Ksenofontov, T.I. Mazilova, E.V. Sadanov, and O.A. Velicodnaja. 2007. Preparation and characterization of monoatomic C-chains: Unravelling and field emission. *Nanotechnology* 18: 475705-1–475705-6.

Miller, M.K., A. Cerezo, M.G. Hetherington, and G.D.W. Smith. 1996. *Atom-Probe Field Ion Microscopy*. Oxford, U.K.: Oxford University Press.

Minor, A.M., S.A.S. Asif, Z. Shan et al. 2006. A new view of the onset of plasticity during the nanoindentation of aluminium. *Nat. Mater.* 5: 697–702.

Mordkovich, V.Z. 2003. Carbon nanofibers: A new ultrahigh-strength material for chemical technology. *Theor. Found. Chem. Engineer.* 37: 429–438.

Moy, C., G. Ranzi, T.C. Peterson, and S. Ringer. 2011. Macroscopic electrical field distribution and field-induced surface stresses of needle-shaped field emitters. *Ultramicroscopy* 111: 397–404.

Müller, E.W. 1956. Resolution of the atomic structure of a metal surface by the field ion microscope. *J. Appl. Phys.* 27: 474–476.

Müller, E.W. 1965. Field ion microscopy. *Science* 149: 591–601.

Müller, E.W. and T.T. Tsong. 1969. *Field Ion Microscopy. Principles and Application*. New York: Elsevier.

Nair, K., S.W. Cranford, and M.J. Buehler. 2011. The minimal nanowire: Mechanical properties of carbine. *EPL* 95: 16002-1–16002-5.

Nishide, D., H. Dohi, T. Wakabayashi et al. 2006. Single-wall carbon nanotubes encaging linear chain $C_{10}H_2$ polyyne molecules inside. *Chem. Phys. Lett.* 428: 356–360.

Nishikawa, O., Y. Ohtani, K. Maeda, M. Watanabe, and K. Tanaka. 2000. Atom-by-atom analysis of diamond, graphite, and vitreous carbon by the scanning atom probe. *J. Vac. Sci. Technol. B* 18: 653–660.

Nishikawa, O. and M. Taniguchi. 2005. Atom-by-atom analysis of non-metallic materials by the scanning atom probe. *Chin. J. Phys.* 43: 111–123.

Okano, S. and D. Tománek. 2007. Effect of electron and hole doping on the structure of C, Si, and S nanowires. *Phys. Rev. B* 75: 195409-1–195409-5.

Onida, G., N. Manini, L. Ravagnan, E. Cinquanta, D. Sangalli, and P. Milani. 2010. Vibrational properties of *sp* carbon atomic wires in cluster-assembled carbon films. *Phys. Status. Solidi.* 4: 1–5.

Poncharal, P., P. Vincent, J.-M. Benoit et al. 2010. Field evaporation tailoring of nanotubes and nanowires. *Nanotechnology* 21: 215303-1–215303-4.

Qi, Z., F. Zhao, X. Zhou, Z. Sun, H.S. Park, and H. Wu. 2010. A molecular simulation analysis of producing monatomic carbon chains by stretching ultranarrow graphene nanoribbons. *Nanotechnology* 21: 265702-1–265702-7.

Ragab, T. and C. Basaran. 2011. The unravelling of open-ended single walled carbon nanotubes using molecular dynamics simulations. *J. Electron. Packag.* 133: 020903-1–020903-7.

Ravagnan, L., N. Manini, E. Cinquanta et al. 2009. Effect of axial torsion on *sp* carbon atomic wires. *Phys. Rev. Lett.* 102: 245502-1–245502-4.

Rinzler, A.G., J.H. Hafner, P. Nikolaev et al. 1995. Unraveling nanotubes: Field emission from an atomic wire. *Science* 269: 1550–1553.

Rivelino, R., R.B. dos Santos, F. de Brito Mota, and G.K. Gueorguiev. 2010. Conformational effects on structure, electron states, and Raman scattering properties of linear carbon chains terminated by graphene-like pieces. *J. Phys. Chem. C* 114: 16367–16372.

Roman, H.E., A. Lorenzoni, and R.A. Broglia. 1998. Field emission properties of linear carbon clusters. *Czech. J. Phys.* 48: 817–820.

Roth, G. and H. Fischer. 1996. On the way to heptahexaenylidene complexes: Trapping of an intermediate with the novel M=C=C=C=C=C=C=CR_2 moiety. *Organometallics* 15: 5766–5768.

Ruoff, R.S., D. Qian, and W.K. Liu. 2003. Mechanical properties of carbon nanotubes: Theoretical predictions and experimental measurements. *C. R. Phys.* 4: 993–1008.

Sadanov, E.V., T.I. Mazilova, I.M. Mikhailovskij, V.A. Ksenofontov, and A.A. Mazilov. 2011. Field-ion imaging of nano-objects at far-subangstrom resolution. *Phys. Rev. B* 84: 035429-1–035429-7.

Saito, Y., T. Matsukawa, K. Asaka, and H. Nakahara. 2010. Field emission microscopy of Al-deposited carbon nanotubes: Emission stability improvement and image of an Al atom cluster. *J. Vac. Sci. Technol. B* 28(C2A5): 1–4.

Saito, Y., R. Mizushima, and K. Hata. 2002. Field ion microscopy of multiwall carbon nanotubes: Observation of pentagons and cap breakage under high electric field. *Surf. Sci.* 499: L119–L123.

Segal, D. and A. Nitzana. 2003. Thermal conductance through molecular wires. *J. Chem. Phys.* 119: 6840–6855.

Shan, Z.W., R. Mishra, S.A.S. Asif, O.L. Warren, and A.M. Minor. 2008. Mechanical annealing and source-limited deformation in submicrometre-diameter Ni crystals. *Nat. Mater.* 7: 115–119.

Standley, B., W. Bao, H. Zhang, J. Bruck, C.N. Lau, and M. Bockrath. 2008. Graphene-based atomic-scale switches. *Nano Lett.* 8: 3345–3349.

Tongay, S., S. Dag, E. Durgun, R.T. Senger, and S. Ciraci. 2005. Atomic and electronic structure of carbon strings. *J. Phys.: Condens. Mater.* 17: 3823–3836.

Tongay, S., E. Durgun, and S. Ciraci. 2004a. Atomic strings of group IV, III–V, and II–VI elements. *Appl. Phys. Lett.* 85: 6179–6181.

Tongay, S., R.T. Senger, S. Dag, and S. Ciraci. 2004b. Ab-initio electron transport calculations of carbon based string structures. *Phys. Rev. Lett.* 93: 136404-1–136404-4.

Topsakal, M. and S. Ciraci. 2010. Elastic and plastic deformation of graphene, silicene, and boron nitride honeycomb nanoribbons under uniaxial tension: A first-principles density-functional theory study. *Phys. Rev. B* 81: 024107-1–024107-6.

Troiani, H.E., M. Miki-Yoshida, G.A. Camacho-Bragado et al. 2003. Direct observation of the mechanical properties of single-walled carbon nanotubes and their junctions at the atomic level. *Nano Lett.* 3: 751–755.

Tsetseris, L. and S.T. Pantelides. 2009. Adatom complexes and self-healing mechanisms on graphene and single-wall carbon nanotubes. *Carbon* 47: 901–908.

Tsong, T.T. 1990. *Atom-Probe Field Ion Microscopy*. Cambridge, U.K.: Cambridge University Press.

Tsuji, M., S. Kuboyama, T. Matsuzaki, and T. Tsuji. 2003. Formation of hydrogen-capped polyynes by laser ablation of C-60 particles suspended in solution. *Carbon* 41: 2141–2148.

Velikodnaya, O.A., V.A. Gurin, I.V. Gurin et al. 2007. Multiemitter field ion source based on a nanostructural carbon material. *Tech. Phys. Lett.* 33: 583–585.

Wang, Y., Z.-Z. Lin, W. Zhang, J. Zhuang, and X.-J. Ning. 2009. Pulling long linear atomic chains from graphene: Molecular dynamics simulations. *Phys. Rev. B* 80: 233403-1–233403-4.

Wang, Y., X.-J. Ning, Z.-Z. Lin, and P. Li. 2007. Preparation of long monatomic carbon chains: Molecular dynamics studies. *Phys. Rev. B* 76: 165423-1–165423-4.

Yakobson, B.I., M.P. Campbell, C.J. Brabec, and J. Bernholc. 1997. High stain rate fracture and C-chain unraveling in carbon nanotubes. *Comput. Mater. Sci.* 8: 341–348.

Yu, M.-F., B.S. Files, S. Arepalli, and R.S. Ruoff. 2000b. Tensile loading of ropes of single wall carbon nanotubes and their mechanical properties. *Phys. Rev. Lett.* 84: 5552–5555.

Yu, M.-F., O. Lourie, M.J. Dyer, K. Moloni, T.F. Kelly, and R.S. Ruoff. 2000a. Strength and breaking mechanism of multiwalled carbon nanotubes under tensile load. *Science* 287: 637–640.

Yuzvinsky, T.D., W. Mickelson, S. Aloni, G.E. Begtrup, A. Kis, and A. Zettl. 2006. Shrinking a carbon nanotube. *Nano Lett.* 6: 2718–2722.

Zhao, X., Y. Ando, Y. Liu, M. Jinno, and T. Suzuki. 2003. Carbon nanowire made of a long linear carbon chain inserted inside a multiwalled carbon nanotube. *Phys. Rev. Lett.* 90: 187401-1–187401-4.

Zhu, T. and J. Li. 2010. Ultra-strength materials. *Prog. Mater. Sci.* 55: 710–757.

27

Single-Atom Electromigration in Atomic-Scale Conductors

Masaaki Araidai
Tohoku University
and
Japan Science and Technology Agency
and
University of Tsukuba

Masaru Tsukada
Tohoku University
and
Japan Science and Technology Agency

27.1 Introduction

27.1.1 Electromigration in Atomic Scale

Electromigration is the directed migration of atoms caused by electric currents, and it is well known as a major failure mechanism in the operation of electronic devices. So far, many researches have been done both experimentally and theoretically [1,2]. Traditionally, the origin of electromigration has been discussed in terms of electrostatic forces and momentum transfers by electrons to atoms in current flow. The latter is often called as "electron wind force," which is a characteristic force in electromigration. Joule heating around the migrating atoms is also one of the most important factors because it facilitates the atomic diffusions.

Electron flow in macroscopic conductors is essentially diffusive and it can be regarded as a particle flow. Many electrons in the conductors collide against the larger and heavier atoms, leading to the momentum transfers. This is the conventional picture of the electron wind force, and many researches have so far been devoted to the elucidation of the microscopic origin [1]. On the other hand, electron flow becomes *ballistic* in atomic-scale conductors whose dimension is smaller than the mean free path of electrons [3], which means that electrons behave as a wave in the conductors. Accordingly, electromigration in atomic-scale conductors should be essentially different from the conventional one. We focus exclusively on the quantum-mechanical nature of electromigration in this chapter.

Many theoretical researches have so far been devoted to elucidate the microscopic origin of the driving force of electromigration, especially the electrostatic and wind forces [1]. However, such a division depends on the analytical approaches and leaves certain issues somewhat ambiguous. On the other hand, in recent computational studies under nonequilibrium conditions, current-induced forces on atoms in nanoscale conductors have been calculated without making such a division. Di Ventra and coworkers discussed the origin of the forces [4–6] and investigated the stability of carbon-based molecular junctions against electric currents [7]. They calculated the so-called Hellmann–Feynman force that includes both the electrostatic and wind forces using the adiabatic approximation [4,8]. Brandbyge et al. [9] described how a small gold chain would be distorted by an applied current, and Mingo et al. [10] investigated the forces experienced by ions in the vicinity of conducting carbon nanotubes. While these studies focused on how the current modifies the bonding strength between the atoms, we pay attention to atomic diffusion processes in electromigration.

27.1.2 Single-Atom Electromigration

Electromigration is a dominant cause of failures of integrated electronic devices, but it provides us a useful method of fabricating metallic electrodes with a nanosized gap between them [11–18]. Such nanogap electrodes can be used in various areas of nanotechnology, such as a fabrication of single molecular devices in which an organic molecule is sandwiched between the electrodes [19–22]. More recently, in the fabrication of nanogap electrodes by the electromigrated break junction technique, Umeno and Hirakawa have observed an intriguing stepwise decrease over time of the electrical conductance of gold junctions [23]. This behavior resulted from the disappearance of a conducting channel (6*s* orbital) of a gold atom at the junction and implied that the electromigration of *single* gold atoms takes place steadily at a threshold voltage (0.3–0.4 V).

Recently, two groups reported the elemental process within ballistic regime [16,23]. Wu et al. performed four-terminal measurements for electromigration at gold nanojunctions and found that the junction resistance depends only slightly on temperature [16]. Umeno and Hirakawa performed a feedback-controlled break junction process at gold nanojunctions to fabricate nanogap electrodes and observed the decrease of a critical power dissipation to initiate electromigration at the junctions [23]. The remarkable point from both experiments is that the origin of electromigration at gold junctions is nonthermal within ballistic regime. This implies that an additional electromigration mechanism should be involved, in particular for nanoscale ballistic transport systems, beyond the conventional adiabatic processes, in which electromigration takes place as the atomic motion overcomes the barrier of the adiabatic potential. The additional migration mechanism may be of nonadiabatic origin, and would be caused by finite quantum jumps of ballistic electrons, which kick up the motion of migrating atoms. Such kind of nonadiabatic mechanism of electromigration would be dominant at lower temperatures and for ballistic electron transport systems. Therefore, for researches on the formation of nanogap electrodes or atomic chains, the theoretical elucidation of such nonconventional electromigration mechanisms is essential.

Accordingly, in this chapter, we introduce the nonadiabatic electron migration mechanism due to the ballistic electron flow in atomic-scale systems, and calculate the electromigration rate for a Au atomic chain model by this mechanism. Since the nonadiabatic electromigration process is taking place in parallel with the conventional adiabatic process, we also calculate the adiabatic-over-barrier electromigration rate with the same model, compare the characteristics of two processes, and discuss in what situations the nonadiabatic contributions are dominant compared with the adiabatic ones.

27.1.3 Contents

Under the circumstance, we expound the mechanism of single-atom electromigration in atomic-scale conductors, with case studies of a Au atom migration along a single Au atom chain, based on our recent theoretical researches [24,25].

In Section 27.2, theoretical bases to treat nonequilibrium electronic states are discussed within both adiabatic and nonadiabatic pictures. Adiabatic potential-energy surfaces (PES) under bias voltage are calculated to form a starting point of our analyses. The time-dependent perturbation approach is also employed for analyses of nonadiabatic electromigration.

In Section 27.3, electromigration of a single atom along a Au atom chain is investigated within the adiabatic picture. The potential barrier along the migration pathway is found to decrease with the increase of the bias voltage, resulting in the same migration direction as electron flow. The electron flow around the migrating atom is shown to be responsible for single-atom electromigration.

In Section 27.4, nonadiabatic processes of electromigration are investigated, taking the same model as in Section 27.3, that is, a Au atom moving along a chain of Au atoms. Here the mechanism of electromigration is an electron-stimulated kicking out of the adatom from the potential well by a finite quantum jump of a ballistic electron. We observe that the calculated electromigration rate shows almost linear behavior beyond a threshold and the electronic scattering takes place via the *d* orbitals of a migrating Au atom. Through a comparison of the electromigration rates between thermal and nonadiabatic migration, we clarify that the nonadiabatic contribution is dominant at low temperature and low bias beyond the threshold voltage.

Section 27.5 is devoted to summary and outlook.

27.2 Theoretical Bases

27.2.1 Adiabatic Treatments

27.2.1.1 Nonequilibrium Electronic States

Electromigration is a kind of atomic diffusion process under electric currents. Thus, we need to calculate the nonequilibrium electronic states under finite bias voltages. Here, we briefly explain the nonequilibrium Green function (NEGF) method which has been successfully applied to study the electron transport through molecular junctions and point contacts. For details, please refer to [3,26–28].

The premise of the NEGF method is to divide the system into three regions: the left and right electrodes in thermal equilibrium and the central region in which electrons originating from an electrode travel toward the opposite electrode. Further, we assume no direct interactions between the left and right electrodes to specify from which electrodes the electron scattering states originated. According to the earlier partitioning and assumption, the Hamiltonian of the full system can be written as a 3×3 block matrix:

$$\mathbf{H}=\begin{bmatrix}\mathbf{H}_{\mathrm{L}} & \mathbf{H}_{\mathrm{LC}} & 0\\ \mathbf{H}_{\mathrm{CL}} & \mathbf{H}_{\mathrm{C}} & \mathbf{H}_{\mathrm{CR}}\\ 0 & \mathbf{H}_{\mathrm{RC}} & \mathbf{H}_{\mathrm{R}}\end{bmatrix}. \tag{27.1}$$

The retarded Green function with the same block structure as the Hamiltonian is determined from

$$[(E+i\eta)\mathbf{1}-\mathbf{H}]\mathbf{G}(E)=\mathbf{1}, \tag{27.2}$$

where

$\mathbf{1}$ is a unit matrix
η is a positive infinitesimal

The earlier equation can be solved for the block of the Green function referring only to the central region:

$$\mathbf{G}_{\mathrm{C}}(E)=[E\mathbf{1}_{\mathrm{C}}-(\mathbf{H}_{\mathrm{C}}+\Sigma_{\mathrm{L}}(E)+\Sigma_{\mathrm{R}}(E))]^{-1}, \tag{27.3}$$

where

$$\Sigma_{\alpha}(E)=\mathbf{H}_{\mathrm{C}\alpha}\,\mathbf{g}_{\alpha}(E)\mathbf{H}_{\alpha\mathrm{C}},\quad \alpha=\mathrm{L}\text{ or }\mathrm{R} \tag{27.4}$$

with the retarded Green function of electrode α,

$$\mathbf{g}_{\alpha}(E)=[(E+i\eta)\mathbf{1}_{\alpha}-\mathbf{H}_{\alpha}]^{-1}. \tag{27.5}$$

There are some techniques to calculate Σ_{α} or $\mathbf{g}_{\alpha}$ efficiently [29–32]. From $\mathbf{G}_{\mathrm{C}}$, we can evaluate the nonequilibrium electron density through the density matrix in the central region defined as

$$\mathbf{D}_{\mathrm{C}}=-\frac{1}{\pi}\int_{-\infty}^{+\infty}[\mathbf{G}_{\mathrm{C}}(E)\mathrm{Im}[\Sigma_{\mathrm{L}}(E)]\mathbf{G}_{\mathrm{C}}^{\dagger}(E)f(E-\mu_{\mathrm{L}})$$
$$+\mathbf{G}_{\mathrm{C}}(E)\mathrm{Im}[\Sigma_{\mathrm{R}}(E)]\mathbf{G}_{\mathrm{C}}^{\dagger}(E)f(E-\mu_{\mathrm{R}})]dE, \tag{27.6}$$

where
- f is the Fermi–Dirac distribution function
- μ_{α} is the Fermi energy of electrode α

In addition, $\mathbf{G}_{\mathrm{C}}$ enters as a central ingredient into the Fisher–Lee relationship [33] for the calculation of the transmission function. The electron charge distribution in the central region is calculated based on the density matrix $\mathbf{D}_{\mathrm{C}}$, and then the Hamiltonian $\mathbf{H}_{\mathrm{C}}$ is revised to be consistent with the obtained charge distribution. This procedure is the same as that conventionally used for the first-principles density-functional theory (DFT).

It should be noted that the NEGF method is an outer frame to calculate nonequilibrium electronic states in open system. We can arbitrarily select the calculation method of the Hamiltonian depending on the desired accuracy.

27.2.1.2 Atomic Forces under Electric Currents

Forces acting on nuclei have been exclusively calculated by the Hellmann–Feynman theorem [34,35]. In such calculations, the electronic system is kept in its instantaneous ground state, which means that atoms move on the adiabatic PES. However, it is not trivial whether the Hellmann–Feynman force directly applies to steady-state transport problems in which the electronic system is not in the ground state. Di Ventra and Pantelides proved that the Hellmann–Feynman force is applicable to such a problem if the contributions of the continuum states originating from the semi-infinite electrodes are taken into account [4]. It has been the starting point of computational analyses of current-induced forces [7,9,10,36].

As the Hamiltonian in the central region $\hat{H}$ and the associated density matrix $\hat{D}$ are determined by the NEGF technique, the force acting on the ith atom $\boldsymbol{F}_i$ is readily calculated as

$$\boldsymbol{F}_i=\mathrm{Tr}[\hat{F}_i\hat{D}]-\frac{\partial\hat{H}_{ii}}{\partial\boldsymbol{R}_i}\quad\text{with}\quad\hat{F}_i=-\frac{\partial\hat{H}}{\partial\boldsymbol{R}_i}, \tag{27.7}$$

where
- $\hat{F}_i$ is the force operator with respect to the position of ith atom $\boldsymbol{R}_i$
- $\hat{H}_{ii}$ is the interaction between nuclei

27.2.1.3 Prescription for Calculations

Along with the prescription given in the previous section, we will numerically estimate the electromigration rate of an atom on Au atomic chain due to thermal adiabatic process. In the present study, we used the commercially available Atomistix ToolKit (ATK) [37] which is based on the NEGF method combined with the DFT [38] with the pseudo-potential approximation [39]. We employed double-zeta orbitals with polarization as numerical atomic orbitals and the local density approximation for the exchange-correlation potential [40].

The calculation model is shown in Figure 27.1. The left and right electrodes are semi-infinite gold chains and the scattering region is a portion of the chain with 14 gold atoms and a single adatom (gold or sulfur) that migrates along the chain. Although the stable structure of a chain of gold atoms has been found from DFT calculations to have a zigzag shape [41], we adopted a straight chain for simplicity. The bond length between the gold atoms in the chain was fixed at 2.5875 Å, which was evaluated from DFT calculations under periodic boundary conditions using the ATK.

The procedure of the theoretical analysis is as follows. First, we calculate the forces acting on the migrating atom under bias voltages. Next, PES for the migrating atom are evaluated from the relation between force and potential energy $U(\boldsymbol{r})$:

$$U(\boldsymbol{r})=-\int_{\mathrm{C}}\boldsymbol{F}(\boldsymbol{s})\cdot d\boldsymbol{s}, \tag{27.8}$$

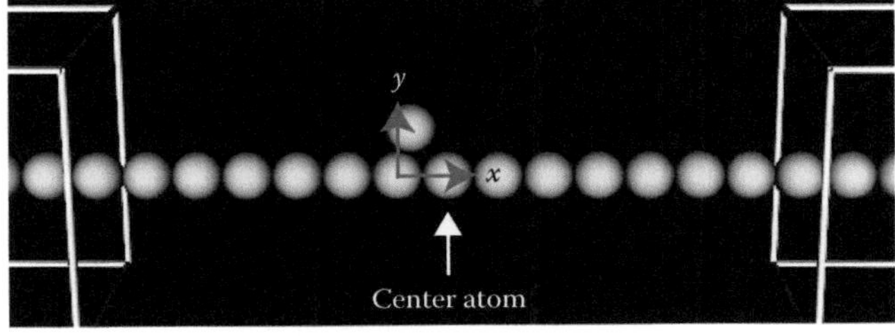

FIGURE 27.1 Calculation model for single-atom electromigration along a gold atomic chain. White boxes denote the electrode regions. A single gold or sulfur atom migrates along the chain.

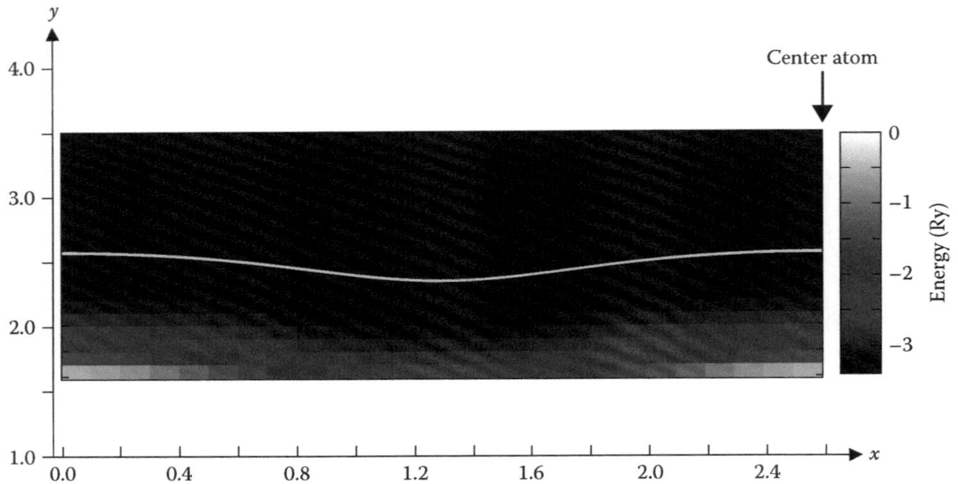

FIGURE 27.2 Potential-energy surface (PES) for the migrating atom. (The PES shown here is the case for a gold atom at zero bias. The x- and y-axes are defined in Figure 27.1 and expressed in units of Å. The bright line indicates the migration pathway that is determined as a valley of the PES.

where C is an integration route. Although the total energy is not precisely defined for an open system, the potential energy can be calculated by the earlier equation. If the current-induced force is not conservative, then the potential energy cannot be safely associated to them. The fundamental question of whether current-induced forces are conservative is one of the controversial issues among some problems in electromigration [6,42,43]. Nevertheless, our analysis by the PES is reasonable because in our situation the above integral was *not* path-dependent within the calculation accuracy, namely, the calculated force is conservative. The PES for the gold atom at a bias of 0 V is given in Figure 27.2. From these PES, we can estimate the migration pathway indicated by a bright solid line in Figure 27.2, and also obtain some valuable information concerning the electromigration mechanism.

27.2.2 Nonadiabatic Processes

27.2.2.1 Time-Dependent Perturbation Approach

The nonadiabatic process we assumed is as follows: a gold atom is occasionally kicked out from the local potential well by ballistic electrons, and migrates toward the descent direction of the overall potential. The Au atom is kicked out by a stimulation due to a quantum transition of an electron from an initial higher to a final lower ballistic state accompanied by an excitation of Au atom vibration from the ground state $\Phi_{\varepsilon_0}(\boldsymbol{R})$ to a continuum state $\Phi_{\varepsilon}(\boldsymbol{R})$ above the barrier. Figure 27.3 shows the schematic diagram of the single-atom electromigration of a gold atom along one-dimensional gold chain within the nonadiabatic process. The excitation rate per unit of time is given by the following equation:

$$\Gamma(V) = \frac{2\pi}{\hbar}\sum_{\boldsymbol{k},i}\sum_{\boldsymbol{k}',i'}\int d\varepsilon\,|\gamma^{\varepsilon\leftarrow\varepsilon_0}_{\boldsymbol{k}',i'\leftarrow\boldsymbol{k},i}|^2\,[f(E_{\boldsymbol{k},i}) - f(E_{\boldsymbol{k}',i'} + eV)]$$
$$\times\,\delta(E_{\boldsymbol{k}',i'} + \varepsilon - (E_{\boldsymbol{k},i} + \varepsilon_0)). \quad (27.9)$$

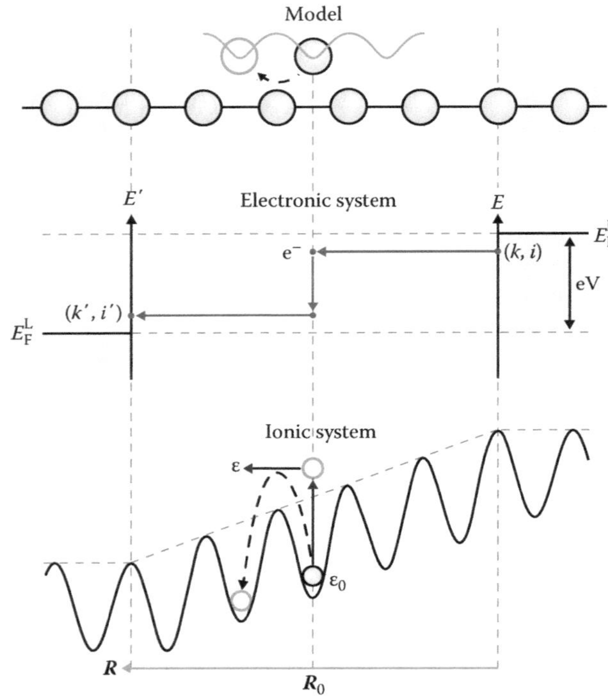

FIGURE 27.3 Schematic diagram of nonadiabatic electromigration. The top panel is the calculation model. Note that the actually calculated number of gold atoms in the straight chain is the same as in Figure 27.1. The corrugated curve is the migration pathway. The middle and bottom panels are energy diagrams of electronic and ionic systems. An electronic state $\Psi_{E_{k,i}}$ comes into the scattering region from the right electrode, is scattered by the migrating atom with (without) the ionic excitation from the ground Φ_{ε_0} to continuum Φ_{ε} states, and goes out to a state $\Psi_{E_{k',i'}}$ in the left electrode with (without) the energy loss. The corrugated curve in the bottom panel is the adiabatic potential-energy curve of the migrating atom on the pathway subject to an applied bias voltage, which will be discussed in Section 27.3. In the adiabatic case, the atom is confined in the potential and cannot migrate without sufficient thermal energy.

In the earlier equation, $f(E)$ is the Fermi–Dirac distribution function of electrons, $E_{k,i}$ and $E_{k',i'}$ are electron energies of the initial and final states, and $\gamma^{\varepsilon\leftarrow\varepsilon_0}_{k',i'\leftarrow k,i}$ is the matrix element between the initial and final state of the transition:

$$[\Psi_{E_{k',i'}}(\boldsymbol{r}),\Phi_{\varepsilon}(\boldsymbol{R})\leftarrow\Psi_{E_{k,i}}(\boldsymbol{r}),\Phi_{\varepsilon_0}(\boldsymbol{R})]$$

$$\gamma^{\varepsilon\leftarrow\varepsilon_0}_{k',i'\leftarrow k,i}=\int d\boldsymbol{r}\Psi^{*}_{E_{k',i'}}(\boldsymbol{r})\Psi_{E_{k,i}}(\boldsymbol{r})W_{\varepsilon\leftarrow\varepsilon_0}(\boldsymbol{r}). \quad (27.10)$$

In the earlier equation, $\Psi_E(\boldsymbol{r})$ is an electron wave function with an energy E and $W_{\varepsilon\leftarrow\varepsilon_0}(\boldsymbol{r})$ is the scattering potential to cause electronic transitions:

$$W_{\varepsilon\leftarrow\varepsilon_0}(\boldsymbol{r})=\int d\boldsymbol{R}\Phi^{*}_{\varepsilon}(\boldsymbol{R})\Phi_{\varepsilon_0}(\boldsymbol{R})[V(\boldsymbol{r}-\boldsymbol{R})-V(\boldsymbol{r}-\boldsymbol{R}_0)], \quad (27.11)$$

where

$\boldsymbol{R}_0$ is the most stable position of the migrating atom

$V(\boldsymbol{r})$ is an individual atomic potential

After some mathematical manipulations with the following equations

$$\delta(E_{k',i'}-E_{k,i}+\varepsilon-\varepsilon_0)$$
$$=\iint dEdE'\delta(E-E_{k,i})\delta(E'-E_{k',i'})\delta(E'-E+\varepsilon-\varepsilon_0), \quad (27.12)$$

$$\sum_{k,i}\Psi_{E_{k,i}}(\boldsymbol{r})[\Psi_{E_{k,i}}(\boldsymbol{r}')]^{*}\delta(E-E_{k,i})=-\frac{1}{\pi}\mathrm{Im}[G(\boldsymbol{r},\boldsymbol{r}';E)], \quad (27.13)$$

we can rewrite Equation 27.9 using the Green function of electron as

$$\Gamma(V)=\frac{4}{h}\iiiint dEd\varepsilon\, d\boldsymbol{r}d\boldsymbol{r}'\mathrm{Im}[G^{*}(\boldsymbol{r},\boldsymbol{r}';E)]W^{*}_{\varepsilon\leftarrow\varepsilon_0}(\boldsymbol{r})$$
$$\times\mathrm{Im}[G(\boldsymbol{r},\boldsymbol{r}';E+\varepsilon_0-\varepsilon)]W_{\varepsilon\leftarrow\varepsilon_0}(\boldsymbol{r}')$$
$$\times[f(E)-f(E+\varepsilon_0-\varepsilon+eV)]. \quad (27.14)$$

If the Green function is expanded by atomic basis functions $\varphi_\mu(\boldsymbol{r})$, then Equation 27.14 reduces to the matrix representation as

$$\Gamma(V)=\frac{4}{h}\iint dEd\varepsilon\mathrm{Tr}[\mathrm{Im}[\mathbf{G}^{\dagger}(E)]\mathbf{W}^{\dagger}(\varepsilon)\mathrm{Im}[\mathbf{G}(E+\varepsilon_0-\varepsilon)]\mathbf{W}(\varepsilon)]$$
$$\times[f(E)-f(E+\varepsilon_0-\varepsilon+eV)], \quad (27.15)$$

where

$\mathbf{G}(E)$ is the matrix of $G(\boldsymbol{r},\boldsymbol{r}';E)$ expressed by the atomic bases

$\mathbf{W}(\varepsilon)$ is the matrix of $W_{\varepsilon\leftarrow\varepsilon_0}(\boldsymbol{r})$ whose elements are defined as

$$W_{\mu\nu}(\varepsilon)=\int d\boldsymbol{r}\varphi_\mu(\boldsymbol{r})\varphi_\nu(\boldsymbol{r})W_{\varepsilon\leftarrow\varepsilon_0}(\boldsymbol{r}). \quad (27.16)$$

We considered no electronic excitation or no heat absorption in our calculations. The microscopic mechanisms for our nonadiabatic electromigration are quantum jumps. Then, the electromigration should take place at a finite voltage beyond the threshold, even if the magnitude of the local electric field around the migrating atom is negligibly small or the scattering region is infinitely long. At the same time, the potential profile around the atom is also the key factor controlling the magnitude of $\Gamma(V)$ because the details of the potential variation have a crucial influence on $\Gamma(V)$ via $\Phi(\boldsymbol{R})$, which can be estimated from the potential.

27.2.2.2 Prescription for Calculations

We use Equation 27.15 to evaluate excitation rate $\Gamma(V)$ as a function of applied voltage. For the purpose, we need to obtain the matrices $\mathbf{G}$ and $\mathbf{W}$. It is relatively easy to calculate the matrix $\mathbf{G}$ because the calculation technique of the Green function has been currently established. We employ the NEGF scheme based on the self-consistent tight-binding method [44] to obtain the matrix $\mathbf{G}$. On the other hand, it is difficult to calculate the matrix $\mathbf{W}$ owing to the quantum mechanical treatment of ionic states. Although the ionic states $\Phi(\boldsymbol{R})$ can be evaluated from the calculated PES of the migrating atom by solving the Schrödinger equation numerically, it is computationally demanding. Accordingly, we tackle it with somewhat ad hoc but physically reasonable treatments as a first step.

Following the adiabatic case, we consider that an atom migrates only along the migration pathway C on the x-z plane, that is, the plane including the Au atom chain and the adatom. Therefore, the coordinate $\boldsymbol{R}$ of the migrating Au atom is taken as $\boldsymbol{R}=(R_x,0,R_z)$ with R_x and R_z located only on C. In such a case, Equation 27.11 becomes

$$W_{\varepsilon\leftarrow\varepsilon_0}(\boldsymbol{r})=\int_C ds\Phi^{*}_{\varepsilon}(s)\Phi_{\varepsilon_0}(s)[V(\boldsymbol{r}-\boldsymbol{R}(s))-V(\boldsymbol{r}-\boldsymbol{R}_0)], \quad (27.17)$$

where s is the coordinate on the pathway C and the origin is set to be the most stable position of the migrating atom $\boldsymbol{R}_0[=\boldsymbol{R}(s=0)]$ on the pathway C, as shown in Figure 27.4. The ionic wave function of the ground state Φ_{ε_0} is simply assumed to be a Gaussian function because the potential-energy curve around the stable position of the migrating atom was well approximated as a harmonic potential. The continuum states Φ_ε on the migration pathway can be determined from the analogy with the solutions of Schrödinger equation for a particle confined within a triangular potential, which are the Airy functions Ai. Our case corresponds to the triangular potential case that the infinitely high barrier is positioned at $s=-\infty$, and the continuum states on the migration pathway were represented by the Airy functions orthogonalized by the ground-state wave function. Accordingly they can be represented as

$$\Phi_{\varepsilon_0}(s)=\left(\frac{M\omega_0}{\pi\hbar}\right)^{1/4}\exp\left[-\frac{M\omega_0 s^2}{2\hbar}\right], \quad (27.18)$$

$$\Phi_{\varepsilon}(s)=N\left[\mathrm{Ai}\left(\frac{eFs-\varepsilon}{\tilde{\varepsilon}}\right)-\langle\Phi_{\varepsilon_0}\,|\,\mathrm{Ai}\rangle\Phi_{\varepsilon_0}(s)\right], \quad (27.19)$$

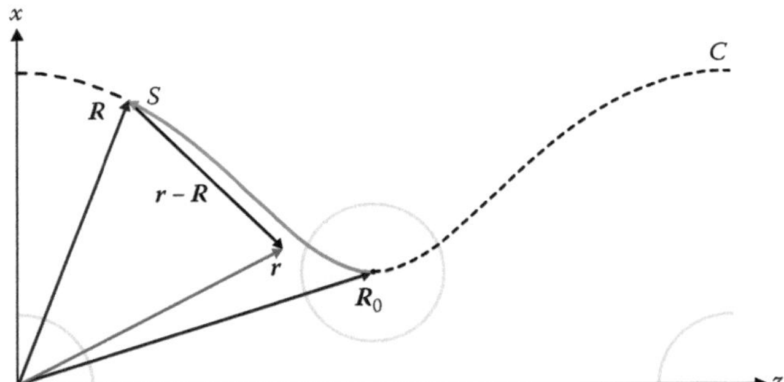

FIGURE 27.4 Definition of the coordinates in our calculations. The dashed curve is the electromigration pathway *C*. The circles are gold atoms. $\boldsymbol{R}_0$ is the stable position of the migrating atom. *s* is the length along the pathway *C* from $\boldsymbol{R}_0$. The chain axis is parallel to the *z*-direction.

where

$$\varepsilon_0 = \frac{1}{2}\hbar\omega_0, \tilde{\varepsilon} = \left[\frac{(eF\hbar)^2}{2m}\right]^{1/3}, \quad (27.20)$$

with the atomic mass of gold *M*. ε_0 and *F* are the ionic ground-state energy and the global gradient of the potential felt by the migrating atom as a function of applied bias, which is obtained from the adiabatic calculations as shown later. The atomic potential $V(\boldsymbol{r})$ in Equation 27.17 is also evaluated from the ab initio calculations. *N* is the normalization constant which can be determined such that the local density of states in the zero-field limit corresponds to that of free ions, as seen in the analysis of two-dimensional electron gas at hetero interface [45]. Once $W_{\varepsilon\leftarrow\varepsilon_0}(\boldsymbol{r})$ is obtained, we can evaluate the matrix **W** from Equation 27.16.

27.2.2.3 Explanatory Remarks

The essential feature of the time-dependent perturbation approach is the fully quantum treatment of ionic states and thus it is possible to capture the quantum transitions certainly. Some dynamical approaches, such as [46,47], also capture a part of the transitions. However, our interest in the present study is the *steady* processes of atomic diffusions in electromigration, which may be detected by experiments. Then, our approach would be more beneficial than the dynamical ones from the viewpoints of the computational effort. Although the difficulty of our approach is attributed to be many building blocks to be extracted from reliable calculations, the numerical treatment is relatively straightforward.

In this section, we considered only one ionic state confined in the potential well. However, the number of the ionic states in the potential well is not necessarily one. If we take into account the multiple states, then there should be some climbing up and down processes on the states. The climbing up process, however, is expected to be dominant because the energy is steadily supplied from the electron wind. Accordingly, the results with the multiple states would be qualitatively the same as the present results. And also, we consider the electromigration only on the migration pathway. This approximation becomes better, if the fluctuations around the migration pathway are smaller when the atom migrates along the pathway. Actually, the fluctuations can be expected to be small because the migration pathway was defined as the bottom of the valley of the PES.

27.3 Electromigration within Adiabatic Picture

27.3.1 Bias Dependence of Electromigration Barrier and the Migration Direction

Figure 27.5a shows potential-energy curves for a gold atom that migrates around the center of the chain. The central part of the model is formed by 14 gold atoms and a migrating adatom. The horizontal axis corresponds to the length along the migration pathway determined as a valley in the PES. (See Figure 27.2 for the pathway at zero bias.) The dotted, dashed, and solid curves denote the results for biases of 0.0, 0.1, and 0.2 V, respectively. The gray triangles indicate the positions of the chain atoms which are projected onto the pathway. The local minima of the potential-energy curves, depicted by A and B in Figure 27.5a, correspond to the stable positions

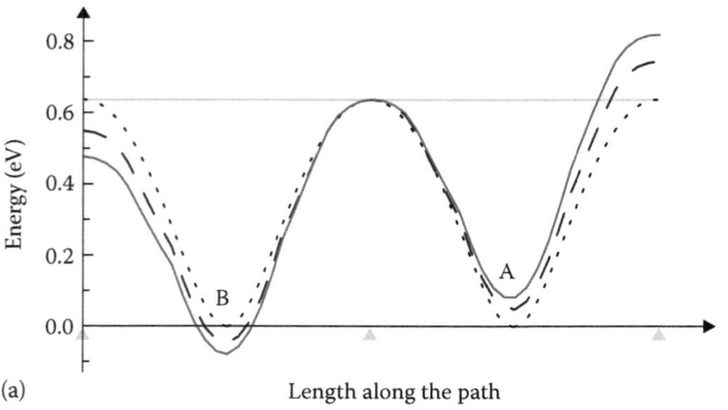

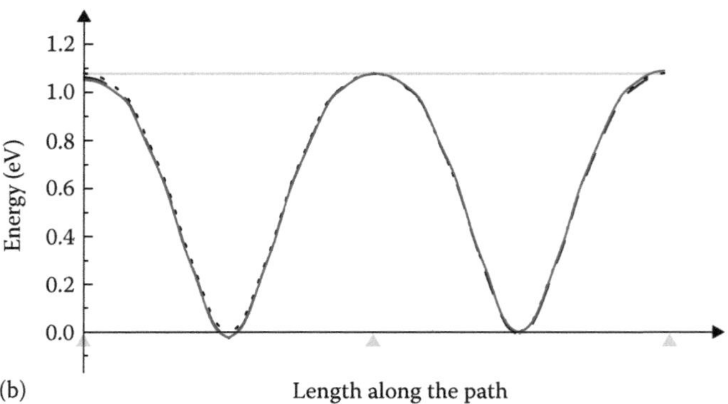

FIGURE 27.5 Potential-energy curves along the migration pathways for (a) gold and (b) sulfur atoms at 0.0 (dotted curve), 0.1 (dashed curve) and 0.2 (solid curve) V. (The migration pathway at zero bias is depicted as the bright line in Figure 27.2.) Gray triangles indicate atomic positions of the gold atoms in the chain projected onto the pathway. The central triangle is the center of the chain.

at which the migrating atom forms a triangle with two gold atoms in the chain. This result is reasonable considering that the stable surface structure of gold is the (111) surface with a triangular lattice.

We immediately observe in Figure 27.5a that the potential barrier for electromigration decreases steadily with an increase of applied bias voltage. The values of the migration barrier heights are listed in Table 27.1. It is found from Figure 27.5a and Table 27.1 that the bias dependence of the barrier reduction (A → B) deviates slightly from a linear relation. A similar nonlinear relation between the bias and current-induced force has been reported by Yang and Di Ventra [5]. The most important observation from Figure 27.5a is that the potential curves show a downward trend toward the left side of the graph. Note that the bias voltage V_{bias} is related to the difference between the Fermi energies of the left and right electrodes. In the present case, the Fermi energy of the right electrode is eV_{bias} higher than that of the left electrode, meaning that electrons flow from the right to left electrodes. This means that the gold atom migrates in the same direction as the electron flow. This result has been confirmed by recent experimental observations [48].

TABLE 27.1 Barrier Height (eV) for Electromigration for a Gold Atom at Each Bias Voltage (V) Applied to the Central Part Composed of 14 Gold Atoms

Bias	0.0	0.1	0.2
A → B	0.64	0.59	0.56
B → A	0.64	0.68	0.71

Note: A and B are defined as local minima of the potential-energy curves close to the center of the chain, as shown in Figure 27.5a.

If the contribution of the external electric field to the electromigration is distinguished from the contribution of electric current as in the traditional theory of electromigration, then in the present situation the electric-field contribution is opposite to the direction of electron flow because the gold atom is always positively charged on the migration pathway (+0.19*e* at the local minimum A in Figure 27.5a). Thus, considering the overall appearance of the potential curves in Figure 27.5a, we find that the contribution of the electric current is dominant in the electromigration of the gold atom.

In order to verify that the electron flow is responsible for the electromigration, we also performed calculations for the electromigration of a single sulfur atom along the same chain. The 3*p* orbitals of the sulfur atom hybridized with 5*d* orbitals of the gold atom have a much lower energy than the Fermi level. Therefore, it is expected that for electromigration of the sulfur atom the electric-current contribution would be small due to the absence of electrons around the bias window. Figure 27.5b shows the potential-energy curves along the migration pathway of the sulfur atom. Obviously, the bias dependence is negligibly small. To clarify the contribution of the electron flow, the effective electron potentials $V_{\text{eff}}(\boldsymbol{r})$ around the migrating atom at a bias of 0.2 V are given in Figure 27.6. $V_{\text{eff}}(\boldsymbol{r})$ within the bias window indicates the path through which electrons flow. In the case of sulfur (Figure 27.6b), the current path is very narrow around the sulfur atom, leading to a small contribution from the electric current, as predicted earlier. On the other hand, the current path around the gold atom is much wider, as seen in Figure 27.6a. It seems reasonable from the distribution of $V_{\text{eff}}(\boldsymbol{r})$ that the electron flow around the migrating atom is responsible for single-atom electromigration.

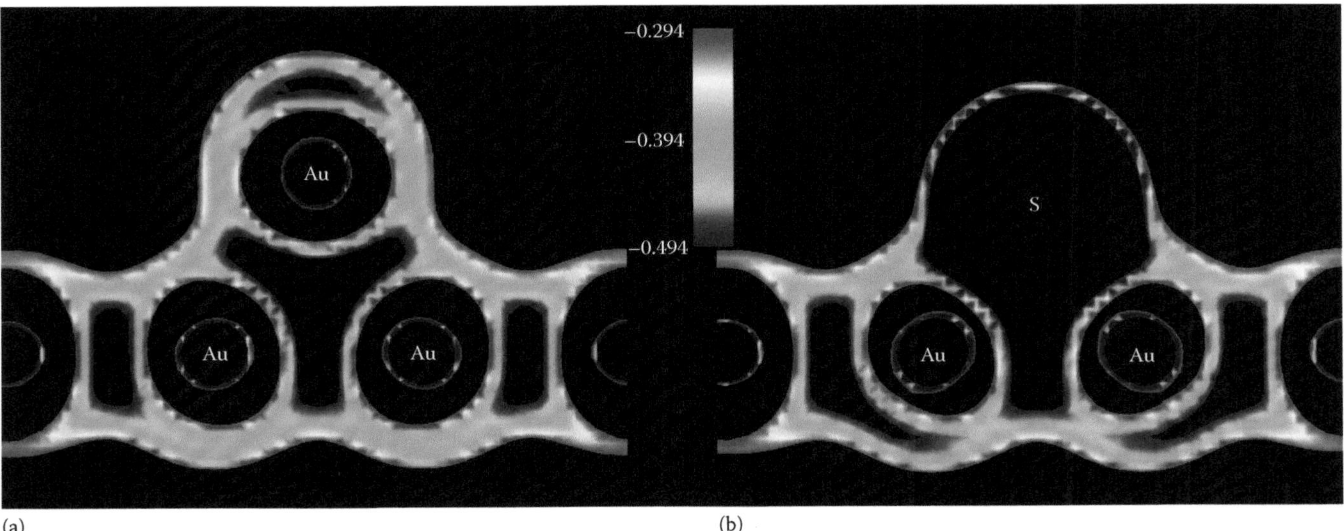

FIGURE 27.6 The effective potentials around the migrating (a) gold and (b) sulfur atom at a bias of 0.2 V. In both cases, the migrating atom is placed at the position of the local minimum which appears on the right side in Figure 27.5a and b. The range of the contour corresponds to the bias window. The value −0.394 eV in the color bar is the averaged Fermi energy of the left and right electrodes.

27.3.2 Thermal Electromigration Rate

In the previous section, we showed that the electromigration barrier for a gold atom steadily decreases as the applied voltage increases. However, the barrier remains high enough to prevent the gold atom from migrating along the chain. It is expected that this barrier is overcome by the thermal energy associated with Joule heating. Accordingly, we next estimate the electromigration rate as a function of the temperature.

The migration rate Γ is obtained from the equation as follows:

$$\Gamma = \nu \exp\left(\frac{-E}{k_B T}\right), \tag{27.21}$$

where ν, E, k_B, and T are the attempt frequency, potential-barrier height, Boltzmann constant, and temperature, respectively. In this case, we have two events 1 and 2 as defined in the schematic diagram shown in Figure 27.7. The barrier heights for the events, E_1 and E_2, are readily evaluated from the PES given in Figure 27.5a as $E_1 = 0.56$ eV and $E_2 = 0.74$ eV. The attempt frequencies, ν_1 and ν_2, obtained by fitting a quadratic function to the local minimum of the potential-energy curve as shown by the dashed curve in Figure 27.7 are $\nu_1 = \nu_2 = 2.13$ THz.

The values of Γ as a function of T are listed in Table 27.2. An exponential increase in both Γ and migration velocity with increasing temperature can be clearly seen. Migration rates of more than one per second have already been experimentally observed [23,48]. Taking these observations into account, we can predict from the theoretical calculations that the single-atom electromigration takes place at temperatures higher than room temperature.

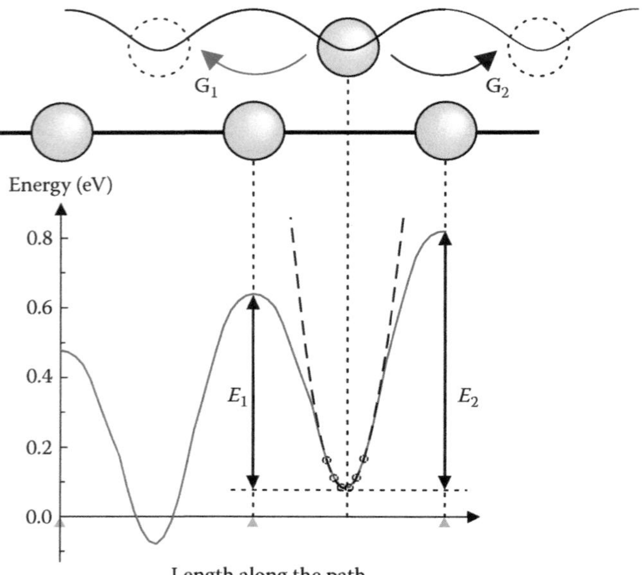

FIGURE 27.7 Potential-energy curve along the migration pathway for a gold atom at a bias of 0.2 V, together with the atomic configurations during the migration events. The parabola indicated by the dashed curve denotes the fitting curve used to estimate the spring constant.

TABLE 27.2 Electromigration Rates (1/s) for Each Temperature (K)

T	Γ_1	Γ_2	V
100	1.30×10^{-16}	1.11×10^{-25}	3.36×10^{-16} Å/s
300	8.34×10^{2}	7.96×10^{-1}	215.80 nm/s
500	4.84×10^{6}	7.45×10^{4}	125.24 μm/s

Notes: The events 1 and 2 are defined in Figure 27.7. The migration velocities V for event 1 are also listed.

27.4 Nonadiabatic Electromigration

27.4.1 Electromigration Rate

The major issue of electromigration in atomic-scale systems is how and to what extent the nonadiabatic processes caused by the ballistic electrons are involved in the migration behavior. The dependence on various parameters such as temperature, bias voltage, and size of the central part should also be clarified and compared with the adiabatic (thermal) process in the previous section. We studied these problems taking the same model as in Section 27.3 by the theoretical method described in Section 27.2.

Figure 27.8a shows the electromigration rate of a Au atom on a Au atomic chain by the nonadiabatic contribution as a function of applied bias voltage. We immediately find from Figure 27.8a that the calculated electromigration rate exhibits almost linear behavior beyond a specific threshold voltage ($V_{th} \sim 0.42$ V). The monotonically increasing behavior can be expected from the form of Equations 27.9, 27.14, or 27.15 because the electromigration rate depends explicitly on the width of the bias window. The magnified figure around the V_{th} is given in the inset of Figure 27.8a. At the bias voltage $V = V_{th}$, the energy difference between the ionic ground state and the height of the lower barrier adjacent to $\boldsymbol{R}_0$ in the inset of Figure 27.3 is about 0.42 eV. It means that the nonadiabatic transition by ballistic electrons starts when the applied bias voltage becomes larger than the activation barrier of thermal migration at the bias.

The correlation between the V_{th} and the activation barrier for thermal migration can be verified from the mathematical formulation. We consider no electronic excitation or no heat adsorption in our calculations, namely, the electromigration rate in Equation 27.15 has a finite value when $f(E) - f(E + \varepsilon_0 - \varepsilon + eV) > 0$, where E, ε, and ε_0 are energies of electrons, ionic continuum, and ground states, respectively. When the energy origin of the ionic states at each bias voltage is set to the top of the lower (left) barrier adjacent to $\boldsymbol{R}_0$ in the inset of Figure 27.3, then $-\varepsilon_0$ can be basically considered as the barrier height of the thermal migration at each voltage, $U_b(V)$, and ε must be $0 < \varepsilon < \varepsilon_0 + eV$ in order to meet the condition $f(E) - f(E + \varepsilon_0 - \varepsilon + eV) > 0$. This means that $V_{th} = -\varepsilon_0(V_{th})/e \sim U_b(V_{th})/e$. Therefore, the electron-mediated excitations from the ionic ground state toward the continuum states are triggered over the V_{th}.

To clarify what is the trigger of the nonadiabatic transition, the magnitudes of the matrix elements of $\mathbf{W}$ represented in terms of electron atomic orbitals located over three atoms

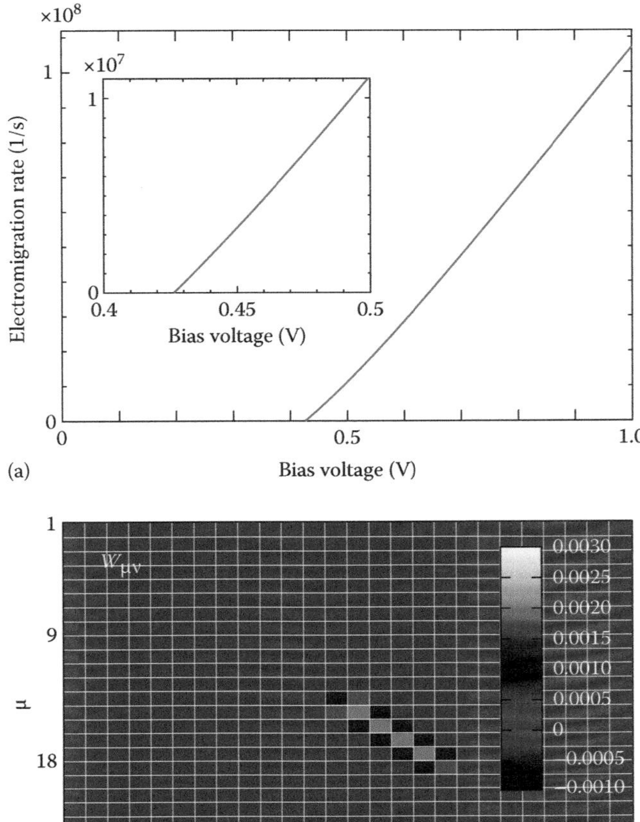

FIGURE 27.8 (a) Electromigration rate as a function of applied bias voltage. The inset shows the magnification around the threshold voltage. (b) Visual representation of the matrix **W** at the lowest energy in the continuum states. In our calculations, a gold atom has nine orbitals, namely $6s$, $6p_y$, $6p_z$, $6p_x$, $5d_{xy}$, $5d_{yz}$, $5d_{3z^2-r^2}$, $5d_{xz}$, and $5d_{x^2-y^2}$ in order. The matrix elements within 10 to 18 correspond to those of the migrating atom.

in Figure 27.4 are shown in Figure 27.8b. The orbital indices μ or ν with the numbers within 10 to 18 indicate the atomic orbitals of the migrating atom. We observe from Figure 27.8b that the matrix elements between orbitals on the migrating atom itself have larger values. The scattering potential to cause electronic transitions $W_{\varepsilon\leftarrow\varepsilon_0}(\boldsymbol{r})$ in Equation 27.11 is a short-ranged function because the atomic potential $V(r)$ in Equation 27.11 has a finite value within a few angstroms at most. As a result, the localized $5d$ orbitals of the migrating atom are considerably overlapped with the $W_{\varepsilon\leftarrow\varepsilon_0}(\boldsymbol{r})$ compared with the valence orbitals and the orbitals of the neighboring atoms, and the matrix elements of the $5d$ orbitals becomes larger than the others. Although the presence of the density of states originating from the $5d$ orbitals around the Fermi energy is a requisite condition, the trigger is the electronic scattering by the $5d$ orbitals.

27.4.2 Crossover between Nonadiabatic and Thermal Migrations

In the previous section, we investigated the electromigration rate by the thermal activation associated with Joule heating around the migrating atom. It was found that the gold atom steadily migrates if the local temperature around the migrating atom is higher than room temperature. The previous results are summarized in Figure 27.9. In the present section, we will focus on the voltage or temperature dependence of the nonadiabatic electromigration rate to discuss whether the nonadiabatic process is dominant or not compared with the thermal activation process.

We can clearly see in Figure 27.9 that the electromigration rate by the thermal activation at each voltage increases almost exponentially with temperature for narrower temperature range. The increase of the voltage also strongly enhances the migration rate. Comparing with the magnitude of the electromigration rate by the nonadiabatic contribution in Figure 27.8 and local heating in Figure 27.9, we can find the crossover between the nonadiabatic and thermal migration regimes.

To clearly see the crossover, the nonadiabatic electromigration rates are given in Figure 27.10 together with the electromigration rate by the thermal activation. Figure 27.10a shows the electromigration rate at 0.5 V, which is the voltage slightly over V_{th} and experimentally intriguing, especially in nanogap formation by electromigration [23]. We observe the crossover at about 360 K in Figure 27.10a. Note that the nonadiabatic electromigration rate at a voltage is independent of temperature in the present formulation. An interesting finding is that the nonadiabatic contribution predominates even in the range somewhat over room temperature. Figure 27.10b represents the electromigration rate at 200 K. We can see that the crossover appears at 0.89 V and the nonadiabatic contribution is dominant in this case. Taking the earlier observations

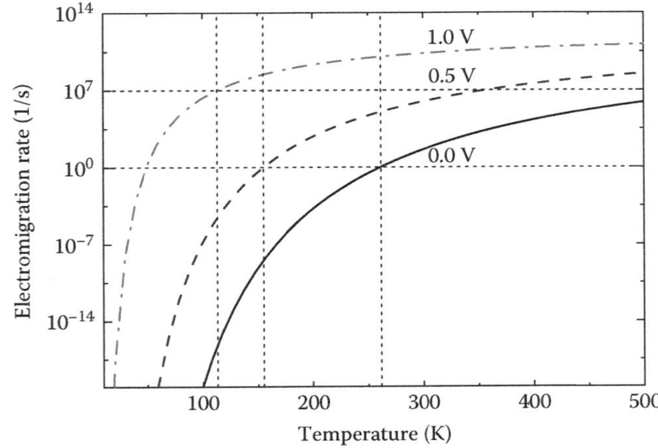

FIGURE 27.9 Electromigration rate by thermal activation associated with Joule heating around the migrating atom. The electromigration rates at 0.0, 0.5, and 1.0 V are depicted by the solid, dashed, and dotted–dashed curves, respectively.

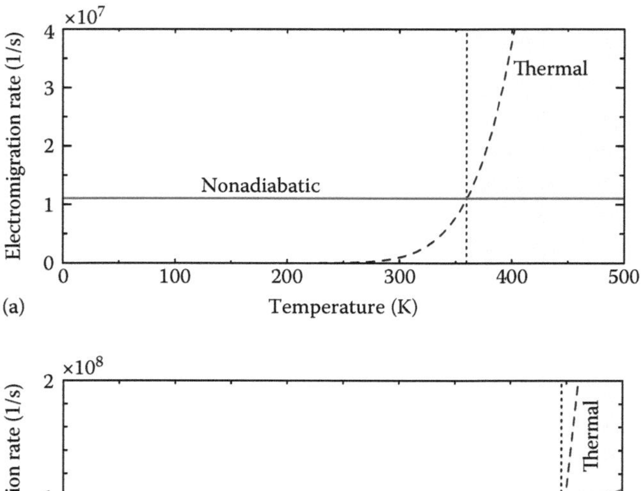

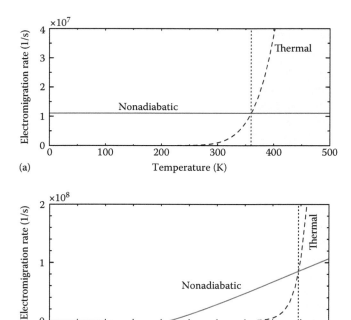

FIGURE 27.10 Nonadiabatic electromigration rates together with the electromigration rate by thermal activation at (a) 0.5 V and (b) 200 K. The solid lines and dashed curves are the electromigration rates by the nonadiabatic and thermal contributions. The vertical dotted lines represent the thresholds.

into account, the nonadiabatic contribution has an advantage in electromigration at low temperature and low bias beyond V_{th}. Electromigration with nonthermal origin in gold junctions has also been observed in recent experiments [16,23].

27.5 Summary and Outlook

Electromigration is the directed migration of atoms caused by a large electric-current density, and it is one of the key issues for nanoscale conductors since it is a major failure mechanism in the operation of electronic devices with a larger current density than their macroscopic counterparts. It has been commonly believed that electromigration is driven by electrostatic forces, momentum transfers by electrons to atoms (electron wind force), and local Joule heating around migrating atoms to facilitate the atomic diffusion, within the diffusive regime of electron transport. Recently, electron transport within the ballistic regime has been experimentally achieved due to the progress of the microfabrication technology, and some issues remain unclear with respect to electromigration within the ballistic regime. Accordingly, we studied the single-atom electromigration within the ballistic regime. The computational methods to treat electromigration of thermal (adiabatic) and nonthermal (nonadiabatic) origins were explained in Section 27.2.

In Section 27.3, single-atom electromigration along a chain of gold atoms was investigated by first-principles calculations based on NEGF+DFT. We found from the potential energy along the migration pathway that the applied voltage steadily reduces the potential barrier for electromigration. The gradient of the potential curves indicates that the gold atom migrates in the same direction as the electron flow. This has recently been confirmed experimentally in gold junction systems. By comparing electromigration of gold and sulfur atoms, we verified that electron flow around the migrating atom is responsible for single-atom electromigration. We also calculated the electromigration rate to theoretically estimate the temperature around the migrating gold atom. The results showed that single-atom electromigration takes place at temperatures higher than room temperature.

In Section 27.4, nonadiabatic electromigration of a gold atom along one-dimensional gold chain was theoretically investigated. We observed that the calculated electromigration rate exhibits almost linear behavior and the nonadiabatic transition by electron wind starts when the applied bias voltage becomes larger than the activation barrier of thermal migration at the bias. It was found that an inelastic electronic scattering to a lower ballistic state exciting vibration states of the migrating atom works to kick the atom and induces the migration. In this process, the *d* orbitals with larger lobes in the migration direction played a dominant role. Furthermore, we compared the electromigration by nonadiabatic transitions with thermal activations and found the crossover between them. From the analysis of the crossover, we clarified that the nonadiabatic contribution becomes dominant at low temperature and low bias beyond V_{th}. In this section, we considered nonadiabatic excitations only from an ionic ground state. In next stage, we will treat multiple excitation and de-excitation processes.

In this chapter, we treated electromigration of thermal and nonthermal origins *independently*. We are developing the analytical method to calculate the transition rate by phonon–phonon interactions. Combining it with the nonadiabatic transition rate in Section 27.4, both thermal and nonthermal electromigrations could be treated on equal footing. Our ultimate goal is to simulate the dynamics and/or kinetics of atomic diffusions in electromigration using the calculated electromigration rates.

References

1. R. S. Sorbello, Theory of electromigration, *Solid State Phys.*, 51, 159–231, 1997, and references therein.
2. K. N. Tu, Recent advances on electromigration in very-large-scale-integration of interconnects, *J. Appl. Phys.*, 94, 5451, 2003, and references therein.
3. S. Datta, *Electronic Transport in Mesoscopic Systems*, Cambridge University Press, Cambridge, U.K., 1995.
4. M. Di Ventra and S. T. Pantelides, Hellmann–Feynman theorem and the definition of forces in quantum time-dependent and transport problems, *Phys. Rev. B*, 61, 16207–16212, 2000.
5. Z. Yang and M. Di Ventra, Nonlinear current-induced forces in Si atomic wires, *Phys. Rev. B*, 67, 161311(R), 2003.
6. M. Di Ventra, Y.-C. Chen, and T. N. Todorov, Are current-induced forces conservative? *Phys. Rev. Lett.*, 92, 176803, 2004.

7. M. Di Ventra, S. T. Pantelides, and N. D. Lang, Current-induced forces in molecular wires, *Phys. Rev. Lett.*, 88, 046801, 2002.
8. T. N. Todorov, J. Hoekstra, and A. P. Sutton, Current-induced forces in atomic-scale conductors, *Philos. Mag. B*, 80, 421–455, 2000.
9. M. Brandbyge, K. Stokbro, J. Taylor, J. L. Mozos, and P. Ordejón, Origin of current-induced forces in an atomic gold wire: A first-principles study, *Phys. Rev. B*, 67, 193104, 2003.
10. N. Mingo, L. Yang, and J. Han, Current-induced forces upon atoms adsorbed on conducting carbon nanotubes, *J. Phys. Chem. B*, 105, 11142–11147, 2001.
11. H. Park, A. K. L. Lim, A. P. Alivisatos, J. Park, and P. L. McEuen, Fabrication of metallic electrodes with nanometer separation by electromigration, *Appl. Phys. Lett.*, 75, 301, 1999.
12. D. R. Strachan, D. E. Smith, D. E. Johnston, T.-H. Park, M. J. Therien, D. A. Bonnell, and A. T. Johnson, Controlled fabrication of nanogaps in ambient environment for molecular electronics, *Appl. Phys. Lett.*, 86, 043109, 2005.
13. G. Esen and M. S. Fuhrer, Temperature control of electromigration to form gold nanogap junctions, *Appl. Phys. Lett.*, 87, 263101, 2005.
14. A. A. Houck, J. Labaziewicz, E. K. Chan, J. A. Folk, and I. L. Chuang, Kondo effect in electromigrated gold break junctions, *Nano Lett.*, 5, 1685–1688, 2005.
15. R. Sordan, K. Balasubramanian, M. Burghard, and K. Kern, Coulomb blockade phenomena in electromigration break junctions, *Appl. Phys. Lett.*, 87, 013106, 2005.
16. Z. M. Wu, M. Steinacher, R. Huber, M. Calame, S. J. van der Molen, and C. Schönenberger, Feedback controlled electromigration in four-terminal nanojunctions, *Appl. Phys. Lett.*, 91, 053118, 2007.
17. K. O'Neill, E. A. Osorio, and H. S. J. van der Zant, Self-breaking in planar few-atom Au constrictions for nanometer-spaced electrodes, *Appl. Phys. Lett.*, 90, 133109, 2007.
18. M. Tsutsui, M. Taniguchi, and T. Kawai, Fabrication of 0.5 nm electrode gaps using self-breaking technique, *Appl. Phys. Lett.*, 93, 163115, 2008.
19. H. Park, J. Park, A. K. L. Lim, E. H. Anderson, A. P. Alivisatos, and P. L. McEuen, Nanomechanical oscillations in a single-C_{60} transistor, *Nature (London)*, 407, 57–60, 2000.
20. J. Park, A. N. Pasupathy, J. I. Goldsmith, C. Chang, Y. Yaish, J. R. Petta, M. Rinkoski, J. P. Sethna, H. D. Abruña, P. L. McEuen, and D. C. Ralph, Coulomb blockade and the Kondo effect in single-atom transistors, *Nature (London)*, 417, 722–725, 2002.
21. E. A. Osorio, K. O'Neill, M. Wegewijs, N. Stuhr-Hansen, J. Paaske, T. Bjørnholm, and H. S. J. van der Zant, Electronic excitations of a single molecule contacted in a three-terminal configuration, *Nano Lett.*, 7, 3336–3342, 2007.
22. Y. Noguchi, R. Ueda, T. Kubota, T. Kamikado, S. Yokoyama, and T. Nagase, Observation of negative differential resistance and single-electron tunneling in electromigrated break junctions, *Thin Solid Films*, 516, 2762–2766, 2008.
23. A. Umeno and K. Hirakawa, Nonthermal origin of electromigration at gold nanojunctions in the ballistic regime, *Appl. Phys. Lett.*, 94, 162103, 2009.
24. M. Araidai and M. Tsukada, Diffusion processes in single-atom electromigration along a gold chain: First-principles calculations, *Phys. Rev. B*, 80, 045417, 2009.
25. M. Araidai and M. Tsukada, Nonadiabatic electromigration along a one-dimensional gold chain, *Phys. Rev. B*, 84, 195461, 2011.
26. M. Brandbyge, J. L. Mozos, P. Ordejón, J. Taylor, and K. Stokbro, Density-functional method for nonequilibrium electron transport, *Phys. Rev. B*, 65, 165401, 2002.
27. S.-H. Ke, H. U. Baranger, and W. Yang, Electron transport through molecules: Self-consistent and non-self-consistent approaches, *Phys. Rev. B*, 70, 085410, 2004.
28. T. Ozaki, K. Nishio, and H. Kino, Efficient implementation of the nonequilibrium Green function method for electronic transport calculations, *Phys. Rev. B*, 81, 035116, 2010.
29. F. Guinea, C. Tejedor, F. Flores, and E. Louis, Effective two-dimensional Hamiltonian at surfaces, *Phys. Rev. B*, 28, 4397–4402, 1983.
30. M. P. López Sancho, J. M. López Sancho, and J. Rubio, Quick iterative scheme for the calculation of transfer matrices: Application to Mo(100), *J. Phys. F: Met. Phys.*, 14, 1205–1215, 1984.
31. T. Ando, Quantum point contacts in magnetic fields, *Phys. Rev. B*, 44, 8017–8027, 1991.
32. P. A. Khomyakov, G. Brocks, V. Karpan, M. Zwierzycki, and P. J. Kelly, Conductance calculations for quantum wires and interfaces: Mode matching and Green's functions, *Phys. Rev. B*, 72, 035450, 2005.
33. D. S. Fisher and P. A. Lee, Relation between conductivity and transmission matrix, *Phys. Rev. B*, 23, 6851–6854, 1981.
34. H. Hellmann, Einführung in die Quantenchemie, *Franz Deuticke*, Sec. 54, 1937.
35. R. P. Feynmann, Force in molecules, *Phys. Rev.*, 56, 340–343, 1939.
36. R. Zhang, I. Rungger, S. Sanvito, and S. Hou, Current-induced energy barrier suppression for electromigration from first principles, *Phys. Rev. B*, 84, 085445, 2011.
37. Atomistix ToolKit, QuantumWise A/S, www.quantumwise.com
38. W. Kohn, Nobel lecture: Electronic structure of matter-wave functions and density functionals, *Rev. Mod. Phys.*, 71, 1253–1266, 1999.
39. N. Troullier and J. L. Martins, Efficient pseudopotentials for plane-wave calculations, *Phys. Rev. B*, 43, 1993–2006, 1991.
40. J. P. Perdew and A. Zunger, Self-interaction correction to density-functional approximations for many-electron systems, *Phys. Rev. B*, 23, 5048–5079, 1981.
41. J. Nakamura, N. Kobayashi, S. Watanabe, and M. Aono, Structural stability and electronic states of gold nanowires, *Surf. Sci.*, 482–485, 1266–1271, 2001.

42. D. Dundas, E. J. McEniry, and T. N. Todorov, Current-driven atomic waterwheels, *Nat. Nanotechnol.*, 4, 99–102, 2009.
43. T. N. Todorov, D. Dundas, and E. J. McEniry, Nonconservative generalized current-induced forces, *Phys. Rev. B*, 81, 075416, 2010.
44. M. Elstner, D. Porezag, G. Jungnickel, J. Elsner, M. Haugk, Th. Frauenheim, S. Suhai, and G. Seifert, Self-consistent-charge density-functional tight-binding method for simulations of complex materials properties, *Phys. Rev. B*, 58, 7260–7268, 1998.
45. J. H. Davies, *The Physics of Low-Dimensional Semiconductors*, Cambridge University Press, Cambridge, U.K., 1998.
46. C. Verdozzi, G. Stefanucci, and C.-O. Almbladh, Classical nuclear motion in quantum transport, *Phys. Rev. Lett.*, 97, 046603, 2006.
47. E. J. McEniry, D. R. Bowler, D. Dundas, A. P. Horsfield, C. G. Sánchez, and T. N. Todorov, Dynamical simulation of inelastic quantum transport, *J. Phys. Condens. Matter*, 19, 196201, 2007.
48. Y. Oshima, private communication.

VIII

Picometer Positioning

28

Picometer Positioning Using a Femtosecond Optical Comb

Mariko Kajima
National Institute of Advanced Industrial Science and Technology (AIST)

28.1 Introduction

In the high-technology industries, such as semi-conductor industry, liquid-crystal panel industry, and high-precision processing industry, high-resolution positioning is required. For example, the minimum line width of the semi-conductors reached to the order of 10 nm. For manufacturing 10 nm level structure on the semiconductor substrate, precise positioning with the resolution of nanometer to subnanometer is necessary. The size of the structure is getting small, and it will reach to several nanometers level in the near future. Subnanometer to picometer resolution positioning is required for nanometer-level manufacturing.

For the precise positioning, accurate measurement of the displacement or position of the positioning stage is necessary. Although there are many commercial positioning stages that just improve their positioning resolution, there are few positioning stages that have high resolution along with high accuracy and SI traceability. Especially the demands for the SI traceability, that is, determine the measurement result and accuracy compared to International System of Units (SI), with high accuracy is increasing because of the globalization of manufacturing.

Laser interferometers are widely used for the accurate measurement of the displacement and position of the positioning stage. Since the standard of "meter" is defined using the speed of light, the laser interferometer, which uses the wavelength of the light as a ruler of it, is one of the instruments that achieves SI-traceable measurement. Although AFM and STM are also the instruments that have the ability of high resolution [1,2], they are at disadvantage because of their small measurement range. X-ray interferometer is another option of picometer measurement, but from the viewpoint of safety and handling, laser interferometer is superior to x-ray interferometer. AFM, STM, and x-ray interferometer have another drawback of indirect traceability to SI.

In the laser interferometer, the change of the interference phase caused by the displacement of the positioning stage is counted. The interference phase is sinusoidal against the displacement, and the period of the phase variation is the half of the wavelength of the optical source of the laser interferometer. It means that the scale of the laser interferometer is the wavelength of the optical source. Since the wavelength of the optical source is the scale of the displacement measurement, the wavelength should be accurately determined and be stable.

Femtosecond optical comb has a great feature of the accurate determination of the wavelength and high stability of the wavelength [3]. Femtosecond optical comb is a new and great technology for the optical source of the laser interferometer. In this chapter, the precise positioning methods that used the femtosecond optical comb are introduced.

28.1.1 Displacement Measurement Using Laser Interferometer

Understanding the concept of displacement measurement using laser interferometer is important [4].

Figure 28.1 shows the basic concept of a two-beam homodyne interferometer. The Michelson interferometer is a typical two-beam laser interferometer that is used for displacement measurement. Simple Michelson interferometer consists of a laser, a beam splitter, two reflectors, and a photo detector. Laser beam from the

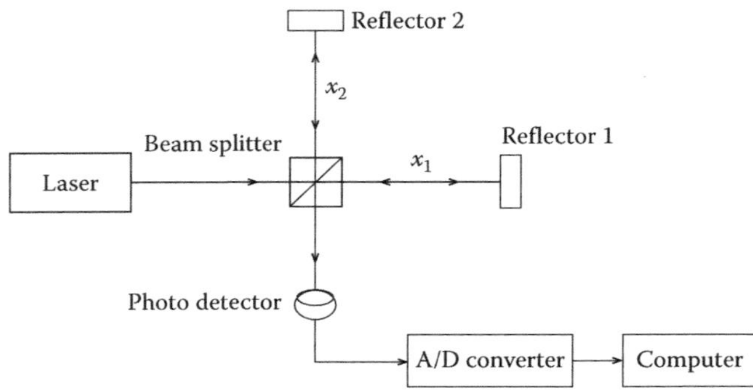

FIGURE 28.1 The basic concept of two-beam homodyne laser interferometer. The two-beam laser interferometer consists of a laser, a beam splitter, two reflectors and a photo detector.

laser optical source incidents on the beam splitter and is divided into two beams. One of the beams is reflected by Reflector 1, and the other beam is reflected by Reflector 2. Two beams are incident on the beam splitter again and detected by the photo detector.

The incident laser beam on the beam splitter is defined as

$$A_{in} = A_0 e^{i\phi_{in}}. \tag{28.1}$$

The returned beams that were traveled to each arm of interferometer and reflected by each reflector at the beam splitter are explained as

$$A_1 = A_0 e^{i\phi_1}, \tag{28.2}$$

$$A_2 = A_0 e^{i\phi_2}. \tag{28.3}$$

The light, which incidents on the photo detector, is superposition of the light that propagated the optical path with Reflector 1 and the light that propagated the optical path with Reflector 2. Then the light on the photo detector is explained as

$$A_{detect} = A_1 e^{i\phi_1} + A_2 e^{i\phi_2}. \tag{28.4}$$

The photo detector generates a current in proportion to the incident light on it. The photo current generated in the photo detector is explained as

$$I \propto \left|A_{detect}\right|^2 = I_1 + I_2 \cos(\phi_1 - \phi_2). \tag{28.5}$$

In case the optical source is a laser, the light is assumed as the monochromatic plane wave. Since the optical phase of the monochromatic plane wave can be explained as $\phi = 2\pi(ft + 2nx/\lambda + \phi_0)$, ϕ_1 and ϕ_2 are represented as

$$\phi_1 = 2\pi\left(ft + \frac{2nx_1}{\lambda} + \phi_1\right), \tag{28.6}$$

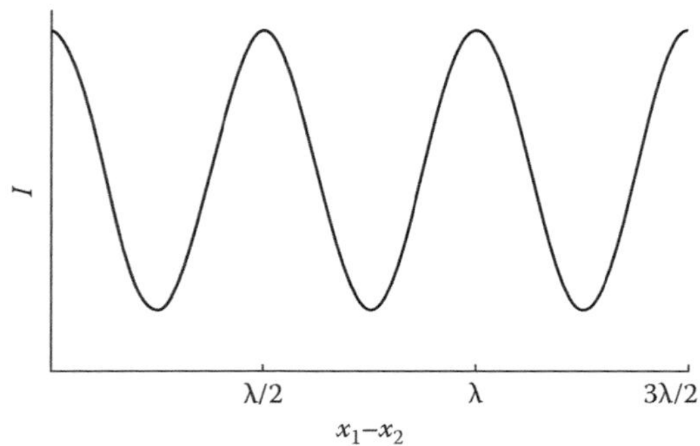

FIGURE 28.2 The photo current generated by the interferometer. The signal shows the sinusoidal behavior with the cycle of $x_1 - x_2 = \lambda/2$.

$$\phi_2 = 2\pi\left(ft + \frac{2nx_2}{\lambda} + \phi_2\right) \tag{28.7}$$

where

- n is the refractive index of the air around the optical path
- λ is the wavelength in vacuum of the laser
- x is the length of the optical path from the beam splitter to the reflector

Consequently, Equation 28.5 is rewritten as

$$I = I_1 + I_2 \cos\left(\frac{4\pi n(x_1 - x_2)}{\lambda} + \phi_1 - \phi_2\right). \tag{28.8}$$

Equation 28.8 shows that the interference signal is the cyclic signal with the cycle of the half of the length difference between two arms of the interferometer. Figure 28.2 shows the characteristic of the interfered light. By analyzing the phase shift of the photo current of the photo detector, the displacements of reflectors are able to be measured.

The voltage, that is converted from the photo current of the photo detector, is converted to the digital signal by the A/D convertor and analyzed by the digital counter. The resolution of displacement measurement using laser interferometer depends on the performance of A/D converter. When the 8-bit A/D converter is used for signal detection, and the laser source is He–Ne laser with the wavelength of 633 nm, the displacement resolution is $(\lambda/2)/F = 633$ nm/512 = 1.2 nm, and the nanometer-level displacement measurement is realized.

The major error sources of the displacement measurement that worsen the accuracy are the determination error of the wavelength of the laser, λ, the error of compensation of air refractive index around the optical path, n, and the imperfection of the cosine signal.

For the purpose of utilizing stable and accurate wavelength, λ, stabilized He–Ne laser with the wavelength of 633 nm is widely used. Six hundred and thirty-three nanometer oscillation wavelength of He–Ne laser has the stability of 10^{-6} even without stabilization mechanism. For more accurate measurement, stabilized He–Ne laser is used. There are a lot of stabilization methods for He–Ne laser, such as Zeeman stabilization method and thermal stabilization method. By applying the Zeeman stabilization method, the wavelength of He–Ne laser is stabilized to 10^{-8} in 1 hour. The absolute wavelength is able to be defined by calibrating the wavelength using the standard He–Ne laser, which is traceable to the national length standard. The accuracy of the absolute wavelength reached to 10^{-8}. For example, in the case that we measure 1 mm length by using the laser interferometer, the minimum measurement error caused by the wavelength determination is the square sum of the measurement errors as $\sqrt{(1\,\text{mm}\times 10^{-8})^2+(1\,\text{mm}\times 10^{-8})^2}=14\,\text{pm}$.

The error of compensation of the refractive index of air around the optical path is one of the sources of the measurement error of the length. When the optical paths of the interferometer are in the air, the refractive index of air, n, must be compensated. The causes of the variation of refractive index of air are air temperature, air pressure, gas distribution, and so on. The refractive index is able to be estimated and is able to be compensated by measured air temperature, air pressure, humidity, and gas composition. The refractive index of air for 633 nm wavelength in the temperature of 20 degree, the air pressure of 1013 hPa, humidity of 50 % and CO_2 content of 400 ppm, is 1.00027. Ciddor's equation [5] is known as the equation that accurately estimates the refractive index of air from the measured air temperature, air pressure, humidity, and CO_2 content. The accuracy of the Ciddor's equation is 10^{-8}. Including the measurement accuracy of environmental factors, such as air temperature and so on, the accuracy of determination of refractive index of air is about 5×10^{-8}. The measurement accuracy in the case of measuring 1 mm is 50 pm.

The difficult error source of displacement measurement using phase-measuring-type laser interferometer is the phase error caused by imperfection of cosine signal [7–10]. This imperfection is caused by the nonuniformity of the optical component in the laser interferometer and nonorthogonality of the laser it used. The mixing of the signal light and the noise causes a phase error with the cyclic behavior. This error is called "cyclic error" or nonlinear error. Since this error is caused by usage of optical components such as beam splitters and reflectors, and those components are necessary for constructing phase-measuring-type laser interferometer, cyclic error is difficult to reduce. Even in the case of using high-quality optical components, the size of the cyclic error is in the order of one to several nanometers. This error is too large to measure picometer length. The great efforts to avoid the cyclic error have been made. Length measurement using optical resonator instead of phase-measuring-type laser interferometer is one of the distinguishable methods that is able to prevent cyclic error. Femtosecond optical comb also acts as the good technology for cyclic error reduction.

28.2 Femtosecond Optical Frequency Comb

28.2.1 Femtosecond Optical Comb Generation

Femtosecond optical frequency comb is generated by a mode-locked laser [11]. The output signal of the mode-locked laser is a sequence of ultra-short pulses where the pulse intervals are equal and stable [12]. In the frequency domain, the frequency spectrum of this optical signal shows a number of lines of longitudinal modes of laser cavity. The mode lines form a row at regular intervals, and the modes are phase-coherent. Since this spectrum looks like a comb, this optical signal is called "optical frequency comb (fs-comb)."

The optical frequency of mth mode in the fs-comb is able to be explained (Figure 28.3) as

$$f_m = m\cdot f_{\text{rep}} + f_{\text{CEO}}, \tag{28.9}$$

where

f_{rep} is the mode-beat frequency (repetition rate frequency)
f_{CEO} is the carrier-envelope-offset frequency
m is an integer

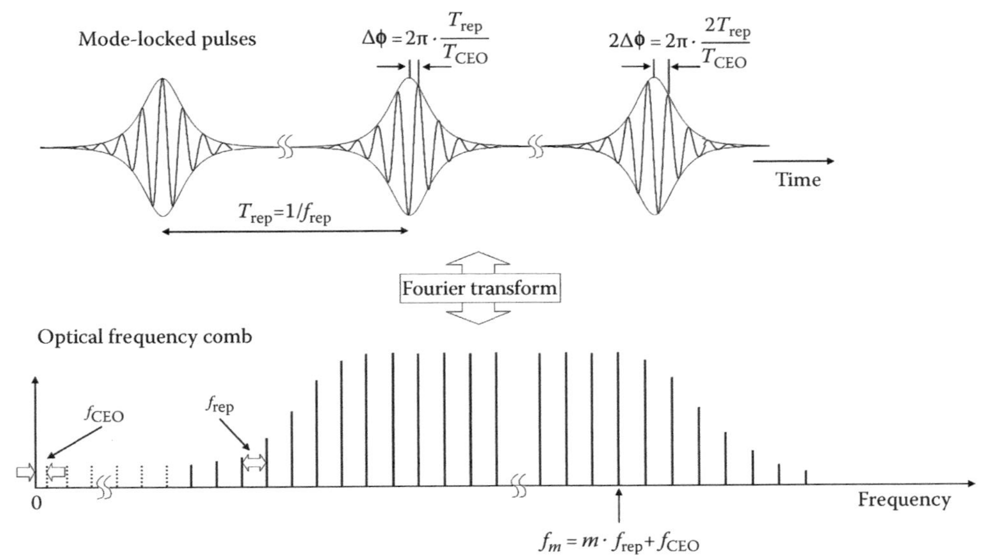

FIGURE 28.3 The scheme of the optical frequency comb.

f_{CEO} is the remainder of the optical frequency of mth mode divided by the repetition frequency, f_{rep}. Equation 28.9 shows that the optical frequency of mth mode in the fs-comb is able to be decided by knowing the repetition frequency, f_{rep}, and the carrier-envelope-offset frequency, f_{CEO}.

The optical pulses in the time domain and the comb-like spectrum in the frequency domain are in the relation of Fourier transform. The time interval of the pulses, T_{rep}, is decided by the round trip time of the laser cavity as $T_{rep} = v_g/2L$, where L is the cavity length and v_g is the group velocity in the cavity. The spacing of the modes in the fs-comb, f_{rep}, is the inverse of the time interval of the pulses. The mode-beat frequency, f_{rep}, is called repetition frequency.

The envelope of the pulse propagates by the group velocity. On the other hand, the carrier wave propagates at the phase velocity. The difference between the group velocity and the phase velocity induces the phase shift of the carrier compared to the envelope by a phase of $\Delta\phi$. This phase shift, $\Delta\phi$, is called carrier-envelope-offset phase. Since the time domain and the frequency domain are in the relation of Fourier transform, the carrier-envelope-offset phase is related to the carrier-envelope-offset frequency as

$$f_{CEO} = \left(\frac{\Delta\phi}{2\pi}\right) f_{rep}. \tag{28.10}$$

The repetition frequency (mode-beat frequency), f_{rep}, is able to be measured easily by measuring the self-beat frequency of the fs-comb with a fast photodiode. Since the mode beat frequency is from tens of megahertz to several gigahertz, it is able to be detected by a fast photodiode. Measuring the carrier-envelope-offset frequency, f_{CEO}, is rather difficult. For the measurement of the carrier-envelope-offset phase, interferometric measurement is necessary, and the precise measurement by the interferometer is difficult. However, $f_{CEO} < f_{rep}$ holds true. The integer, m, is able to be determined by dividing approximate optical frequency of the mth mode in fs-comb by the repetition frequency, f_{rep}. The integer, m, is about 10^5–10^6.

28.2.2 Control of Optical Frequency Comb

According to Equation 28.9, to determine the absolute frequency of the fs-comb, the repetition frequency, f_{rep}, and the carrier-envelope-offset frequency, f_{CEO}, are required. Since f_{rep} is the beat frequency between adjacent modes and that frequency is from several tens of megahertz to several gigahertz, it is able to be determined by simply detecting the beam of the fs-comb with fast photodiode. On the other hand, determining the carrier-envelope-offset frequency, f_{CEO}, is not easy because interferometric measurement is necessary for measurement of carrier-envelope-phase relation between pulses.

However, f_{CEO} comes to be able to be measured by the self-reference method taking advantage of widely spanning optical comb that has the spectrum spans an octave in frequency [13–15]. By using the frequency doubler such as a second harmonic crystal, the frequency of the fs-comb is doubled and an octave frequency spectrum, that is, the frequency of the highest frequency of the comb is larger than twice the frequency of the lowest frequency in one fs-comb, is obtained.

The frequency of the second harmonic of mth mode in the fs-comb is explained as

$$2f_m = 2mf_{rep} + 2f_{CEO}. \tag{28.11}$$

On the other hand, the frequency of the $2m$th mode is expressed as

$$f_{2m} = 2mf_{rep} + f_{CEO}. \tag{28.12}$$

The difference between the frequencies of the second harmonic of mth mode and the $2m$th mode is f_{CEO}. The beat frequency between the second harmonics of mth mode and the $2m$th mode in one fs-comb represents f_{CEO}. That beat frequency is able to be detected by the high-speed photodiode. As a result, the carrier-envelope-offset frequency, f_{CEO}, is able to be determined, and the absolute frequency of a mode in the fs-comb is able to be determined.

The accurate absolute frequencies of arbitrary modes in the fs-comb are obtained by controlling the frequencies of f_{rep} and f_{CEO} so that these frequencies are equal to reference frequencies such as the frequencies of the synthesizer, which is phase locked to the atomic clock. The absolute frequencies of the modes in the fs-comb are able to be utilized as an optical-frequency ruler.

The absolute measurement of optical frequency, which is approximately several hundreds of terahertz, using the fs-comb is able to be realized. The heterodyne frequency between the frequency of the laser to be measured, f_a, and the fs-comb is the beat frequency between f_a and the frequency of the nearest mode,

$$f_{beat} = f_a \pm (mf_{rep} + f_{CEO}). \tag{28.13}$$

The sign ± depends on which frequency is higher f_a or the nearest mode. The absolute optical frequency is obtained as

$$f_a = f_{beat} \pm f_m. \tag{28.14}$$

In this way, using the optical frequency comb as a frequency ruler, the accurate measurement using a laser is realized. Since modes in the fs-comb are coherent, each mode works as a cw laser. However, since the total laser power of fs-comb is divided by a numbers of modes, the illumination power of each mode is very weak. It is hard to use one mode of the fs-comb directly as an optical source for length measurement. By using fs-comb as a reference laser of optical source or by using it with a resonator, the benefit of fs-comb is able to be utilized.

28.3 Picometer Positioning Using a Zooming Interferometer

For the purpose of improving the resolution and accuracy of general laser interferometer, the optical zooming laser interferometer was proposed [16–18]. The optical zooming interferometer uses two laser sources with slightly different wavelengths and two positioning stages. One of the positioning stages (fine stage) is controlled by the zooming interferometer so that displaces proportionally reduced displacement of the other stage (coarse stage). The positioning resolution and accuracy are improved by this reduction principle. The reduction rate of the displacement from the coarse stage to the fine stage is determined by the wavelengths of two laser sources. A fs-comb was used for the stabilization of the wavelength of the laser sources. Since the arbitral wavelengths are stabilized by using the fs-comb, the reduction rate of the displacement is variable.

By using the optical zooming interferometer, a picometer positioning stage was developed. The positioning system is used as a part of a length calibrator with the positioning resolution of smaller than 50 pm and the calibration uncertainty of 0.06 nm.

28.3.1 Principle of the Optical Zooming Interferometer

As mentioned in Section 28.1, resolution of the general laser interferometer is typically 1 nm and the accuracy of it is larger than 1 nm. To obtain the positioning stage with the resolution and the accuracy of picometer-level, special mechanism is necessary. Especially the mechanism that reduces the cyclic error in the laser interferometer is absolutely necessary. The optical zooming interferometer was originally proposed as a special laser interferometer for avoiding the cyclic error [16]. Figure 28.4 shows the schematic of the zooming interferometer. The optical zooming interferometer consists of two movable stages and two laser interferometers with two optical sources which have different wavelengths, λ_1 and λ_2. One of the movable stages, Stage 1, acts as the coarse displacement stage, and the other movable stage, Stage 2, acts as the fine displacement stage. This interferometer has three reflectors. The first mirror, M1, is set on Stage 1, the second mirror, M2, is set on Stage 2, and the third mirror, M3, is fixed on the interferometer's base so that it does not move. The laser beam with the wavelength of λ_1 constructs one of the laser interferometers with the reflectors of M1 and M3

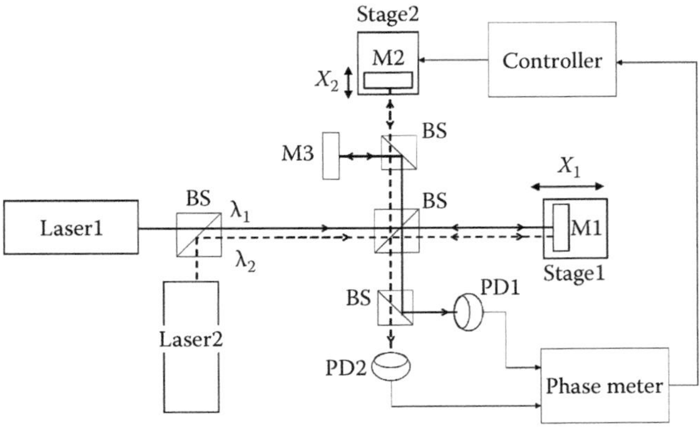

FIGURE 28.4 The scheme of the optical zooming laser interferometer. M, mirror; BS, beam splitter; PD, photo diode.

and a photo detector, PD1. On the other hand, the laser beam with the wavelength of λ_2 constructs the other partial laser interferometer with the reflectors of M1 and M2 and a photo detector, PD2. The interference signals of each partial interferometer are detected by each photo detectors, and the phase difference between the two interference signals is observed by the phase meter. The interference signal of the first partial interferometer is explained as

$$I_1(t) = A_1 + B_1 \cos\left(\frac{4\pi n_1 (X_1 - X_3)}{\lambda_1}\right), \tag{28.15}$$

where n_1, X_1, and X_3 are the refractive index of air for the wavelength of λ_1, the displacement of M1, and the displacement of M3, respectively. Since M3 is fixed, X_3 is zero.

$$I_1(t) = A_1 + B_1 \cos\left(\frac{4\pi n_1 \cdot X_1}{\lambda_1}\right). \tag{28.16}$$

The interference signal of second partial interferometer is explained as

$$I_2(t) = A_2 + B_2 \cos\left(\frac{4\pi n_2 (X_1 - X_2)}{\lambda_2}\right). \tag{28.17}$$

where n_2, X_1, and X_2 are the refractive index of air for the wavelength of λ_2, the displacement of M1, and the displacement of M2, respectively. The phase difference between the two interference signals, $\Delta\phi$, which is observed by the phase meter, is described as

$$\Delta\phi(t) = 4\pi\left(\frac{n_1 \cdot X_1}{\lambda_1} - \frac{n_2 (X_1 - X_2)}{\lambda_2}\right). \tag{28.18}$$

When M1 moves by the displacement of X_1, the value of $\Delta\phi$ changes. Then the displacement of M2 is controlled so that it compensates the phase shift, $\Delta\phi$, to be zero.

$$\Delta\phi(t) = 4\pi\left(\frac{n_1 \cdot X_1}{\lambda_1} - \frac{n_2 (X_1 - X_2)}{\lambda_2}\right) = 0. \tag{28.19}$$

As a result, the displacement of M2, X_2, is defined as

$$X_2 = \frac{\lambda_1 - (n_1/n_2)\lambda_2}{\lambda_1} \cdot X_1. \tag{28.20}$$

Equation 28.20 represents that X_2 is proportionally smaller than X_1, and the ratio is determined by the wavelengths of the optical sources that are used in the zooming interferometer. The ratio of X_2 compared to X_1 is called the zooming ratio. The refractive indices, n_1 and n_2, are spontaneously decided by the wavelengths. In the original zooming interferometer, a Nd:YAG laser with the wavelength of 1064 nm and the second harmonic of the fundamental wave were used as optical sources [14,16]. In that case, the interference signals were made in the same wavelength of 1064 nm, but the phase difference was made by the difference of refractive indices of fundamental wave and the second harmonic. Since the difference of the refractive indices was very small, the zooming ratio was 4×10^{-6}. It means that when M1 displaces 1 mm, M2 only displaces 4 nm in the zooming interferometer. By this principle, high resolution was realized. However, from the view of positioning accuracy, it becomes a problem that the position is not traceable to SI unit because of the absence of national standard of air refractive index.

The zooming interferometers that are simply based on the wavelength difference were proposed [17,18]. In these interferometers, optical sources with slightly different wavelengths, such as two wavelengths of a heat-stabilized dual-wavelength He–Ne laser or two diode lasers with slightly-distant wavelengths, were used. Since the wavelengths were almost the same in the case of those wavelength-base zooming interferometer, the difference of refractive indices was negligibly small, $n_1/n_2 \approx 1$. Then Equation 28.20 was rewritten as

$$X_2 = \frac{\lambda_1 - \lambda_2}{\lambda_1} \cdot X_1. \tag{28.21}$$

In the case of using the dual-wavelength He–Ne laser, since the frequency difference was only 1 GHz, the zooming ratio was about 1/440,000. This interferometer realized the resolution of 0.2 nm for the measurement range of 350 nm. The problem of that system was that too small zooming ratio made the measurement range small. It means that the small zooming ratio realizes high-resolution positioning but limits the positioning range.

Using two diode lasers with slightly different wavelengths, appropriate zooming ratio is able to be generated. Considering that not only the high positioning resolution but also the wide positioning range are demanded for the picometer positioning stage, appropriate zooming ratio exists. The appropriate zooming ratio is around 1/1000. For example, selecting the diode lasers with the wavelengths of 780 and 781 nm, the zooming ratio is 1/780. When M1 moves the displacement of $X_1 = 780$ nm/2 = 390 nm, the interference phase of the first partial interferometer shifts by 2π. By the displacement of M1, the interference phase of the second partial interferometer also shifts, but the phase shift is $2\pi(780/781)$. To compensate the phase difference between the first partial interferometer and the second one, M2 is displaced 0.5 nm. In this example, the zooming ratio is 1/781. This condition is appropriate for picometer positioning with wide positioning range. SI traceability is also satisfied by using diode lasers by calibrating the wavelengths of them.

However, usually diode lasers do not have sufficient stability nor the accuracy of wavelengths. For the purpose of stable and accurate zooming ratio, $\lambda_1 - \lambda_2$ and λ_1 should be stabilized.

28.3.2 Stabilization of the Diode Lasers Using fs-Comb

For the purpose of stabilizing the wavelengths of diode lasers, fs-comb is used. In a fs-comb, a lot of oscillation modes exist. All the frequency spacing between arbitral two modes next to each other are the same and stable. Taking this advantage, two diode lasers are simultaneously stabilized by one fs-comb [18]. Figure 28.5 shows the stabilizing diagram of stabilizing two

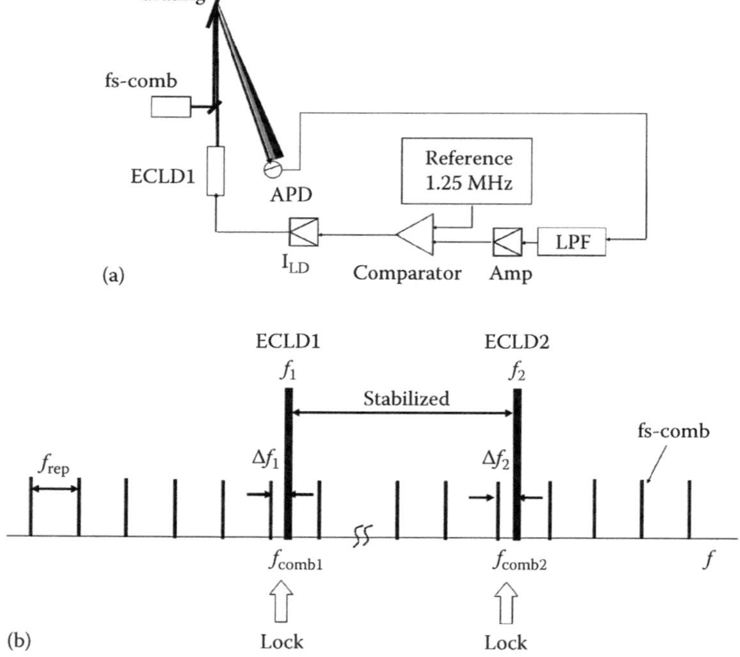

FIGURE 28.5 The stabilizing diagram of two wavelengths of two diode lasers. (a) The optical system for stabilizing ECLD1. ECLD2 was stabilized as the same way with ECLD1. (b) The scheme of stabilizing two ECLDs using a fs-comb.

wavelengths of two diode lasers. For the diode lasers, external cavity diode lasers are appropriate. The external cavity diode laser (ECLD) is a diode laser with wavelength-tuning mechanism. By changing the driving current and driving voltage of the Piezo-electric actuator, which adjusts the cavity length of the ECLD, the emitting wavelength of the laser is able to be tuned.

The stabilization of the laser will be discussed in the frequency domain instead of the wavelength domain. The center frequency of the ECLD with the wavelength, λ_1, is explained as $f_1 = c/\lambda_1$, where c is the speed of light. The center frequency of the ECLD with the wavelength, λ_2, is expressed as $f_2 = c/\lambda_2$.

To stabilize an ECLD, the laser beam of ECLD and the laser beam of fs-comb are aligned in one beam and incident on a high-speed photo diode (Figure 28.5a). The high-speed photo diode detects the beat frequency between the center frequency of the ECLD, $f_{1 \text{ or } 2}$, and a mode in the fs-comb which have a frequency nearest to the center frequency of ECLD, $f_{\text{comb1 or comb2}}$, $\Delta f_{1 \text{ or } 2} = f_{1 \text{ or } 2} - f_{\text{comb1 or comb2}}$, where $f_{\text{comb1 or comb2}}$ is expressed as $f_{\text{comb1 or comb2}} = m_{1 \text{ or } 2} \cdot f_{\text{rep}} + f_{\text{CEO}}$. Since the frequency of the fs-comb is stable, the center frequency of the ECLD is able to be stabilized by stabilizing the beat frequency between a mode of fs-comb and the center frequency of ECLD, $\Delta f_{1 \text{ or } 2}$ (Figure 28.5b). By applying the frequency-lock method for stabilizing the beat frequency between a mode of fs-comb and the center frequency of ECLD, the center frequency of the ECLD is stabilized to the stability of 10^{-8}. By applying this stabilizing method to both ECLDs, they are stabilized simultaneously using one fs-comb.

The wavelength difference of two ECLDs, $\lambda_1 - \lambda_2$, is stabilized simultaneously by stabilizing the center frequencies of ECLDs. The stability of $\lambda_1 - \lambda_2$ is determined by the stability of repetition frequency of the fs-comb. The frequency difference between center frequencies of two ECLDs is explained as $f_{\text{difference}} = f_1 - f_2 = m \cdot f_{\text{rep}} + \Delta f_1 + \Delta f_2$. The wavelength difference between two ECLDs is able to be calculated from the number of comb modes, m, which exist between the modes that are used for the stabilization of ECLDs, $m = m_1 - m_2$. The number of comb modes that exist between the frequencies of ECLDs is calculated as follows; the center frequencies of the ECLDs are roughly measured by using the wavemeter. By dividing the frequency difference between two ECLDs by repetition frequency of the fs-comb, the number of comb modes that exist between two ECLDs is able to be determined.

By stabilizing two ECLDs to two modes in one fs-comb as the method mentioned earlier, $\lambda_1 - \lambda_2$ and λ_1 are stabilized, and consequently, the zooming ratio is to be stabilized. When the stabilized fs-comb, in which repetition rate and carrier-envelope-offset frequency are stabilized, is used, $\lambda_1 - \lambda_2$ and λ_1 is able to be determined accurately.

The frequency-lock method was used in the stabilization method mentioned earlier. By applying the phase-lock method, the stability will be improved. Using an optical resonator for stabilizing an ECLD to one mode in the fs-comb is the other good way for stabilizing an ECLD. By using optical resonators, good stabilities of λ_1 and λ_2 are obtained, but only one wavelength is able to be selected with one resonator. In this case, the zooming ratio is fixed to one value. However, the zooming ratio is variable by changing the wavelength of ECLDs when the frequency-lock or phase-lock method is used. By using the frequency-lock or phase-lock method, the zooming ratio is adjustable to appropriate zooming ratio for picometer resolution and suitable measurement range.

28.3.3 Picometer Positioning Stage Using Optical Zooming Interferometer

The precision measurement with the picometer resolution and picometer accuracy, picometer measurement, is required, recently. Picometer length measurement is used for determining the length or position of nanometer-scale structures in semiconductors exposure, liquid crystal manufacturing, nanometer fabrication, and such as those high-technology with nanometer level manufacturing. The important thing in those high-tech industries is SI (International System of Units) traceability. As the globalization of production, universal standards of measuring units for industrial products are necessary. In the 2000s, those measurement standards for picometer scale were not so much important, because very few picometer measurements were realized, but in the 2010s, the importance of picometer measurement has rapidly increased. For confirming the SI traceability, the length measuring tools that are used for the picometer measurement should be calibrated. The picometer length calibrator that has the picometer resolution and picometer accuracy is required. The zooming laser interferometer is an appreciable technique for realizing the picometer length measuring system. By using the fs-comb, SI traceability is able to be satisfied.

A picometer positioning stage was applied to a precision length calibrator [19]. Figure 28.6 shows the schematic diagram of the picometer positioning stage that was constructed for precision length calibrator. The coarse stage was a high-resolution stepping motor stage, and the fine stage was a Piezo-electric actuator stage. The length measuring tool that is to be calibrated was set on the fine stage. By comparing the displacement of the fine stage and the reading of the length measuring tool, the length measuring tool is able to be calibrated. The optical sources of this system were ECLDs with the wavelengths of approximately $\lambda_1 = 781$ nm and $\lambda_2 = 780$ nm. By substituting these wavelengths to Equation 28.21, the zooming ratio was calculated as 0.0017. The theoretical positioning resolution was 20 pm. However, the practical positioning resolution was limited by the electric noise of the servo system, mechanical noise in the positioning system, and fluctuation of environment. The practical positioning resolution was about 50 pm.

The interferometer was constructed on a low-thermal expansion base plate. The base plate was set on vibration-absorption elastomer and covered by a sound-absorption box. It was set on the vibration-absorption optical table in a plastic booth for environment stabilization. Air temperature in the calibration room was controlled to 20°C ± 0.5°C. By those vibration absorption and environment control, repeatability of the measurement

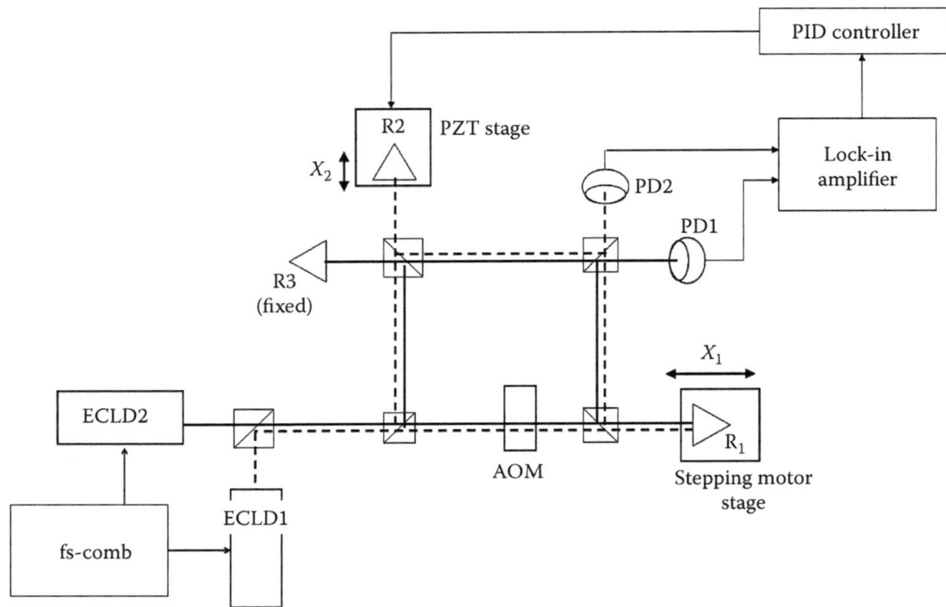

FIGURE 28.6 The scheme of the picometer length calibrator based on the optical zooming interferometer using fs-comb.

was improved. The measurement uncertainty of this calibrator reached to 0.6 nm.

For the traceability to SI, the displacement of the coarse stage was measured by a SI traceable He–Ne laser. The displacement of the fine stage was determined as the product of the displacement of the coarse stage and the zooming ratio. The wavelengths of optical sources were determined by a wavemeter that was also calibrated. By utilizing the traceable He–Ne laser and traceable wavemeter, the traceability to SI of this length calibrator was ensured.

A more simple way to ensure traceability is using the traceable fs-comb for the reference of stabilization of fs-comb. As mentioned in Section 28.2, the oscillation frequency of a mode in the fs-comb is able to be determined accurately and traceable to SI when the fs-comb, in which repetition frequency and carrier envelope offset frequency are stabilized and determined, is used. The frequency accuracy of a modes in the frequency defined fs-comb is better than 10^{-10}. However, even when the theoretical uncertainty of length calibration is improved by accurate optical sources, the most effective factor for the length calibration is environment fluctuation and mechanical error in the calibrator.

28.4 Picometer Measurement Using a Fabry–Perot Resonator

The Fabry–Perot (F–P) interferometer is a major method for precise displacement measurement along with the two-beam interferometer. The Fabry–Perot interferometer is the frequency-based length measurement method. The F–P interferometer consists of a F–P resonator. When the wavelength of the laser, which is incident on the F–P resonator, is matched with the divisor of the resonator length, the F–P interferometer is in the resonance condition. The intensity of light from the F–P resonator shows sharp peak at the resonance condition. By detecting the wavelength shift of emitted light from the resonator, the change of resonator length is able to be recognized. Although the two-beam interferometers have the ability of measuring the large displacement with subnanometer resolution, unavoidable error, cyclic error, disturbs the displacement measurement with subnanometer accuracy. Since the cyclic error is caused in the phase measurement in the two-beam interferometer, hard efforts are necessary to avoid this error. The F–P interferometer is able to prevent the cyclic error because it does not measure the sinusoidal phase changes. In the F–P interferometer, the shift of the resonance frequency indicates the displacement. For the accurate measurement of the resonant frequency, fs-comb is useful because of its feature as an optical frequency ruler and its stability.

28.4.1 Principle of Displacement Measurement Using Fabry–Perot Interferometer

Figure 28.7 shows the schematic of basic displacement measuring system using a F–P interferometer [20]. F–P interferometer for measuring precise displacement consists of a F–P cavity and

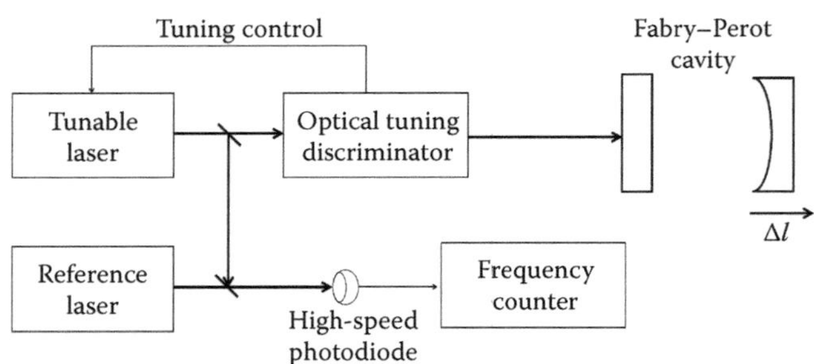

FIGURE 28.7 The schematic of basic displacement measuring system using a Fabry–Perot interferometer.

tunable laser. The tunable laser illuminates the F–P cavity. When one of the end mirrors of the F–P cavity moves, the resonant frequency of the F–P cavity shifts. The tunable laser tracks the shift of the resonant frequency of the F–P cavity. By observing the frequency shift of the tunable laser, the displacement of the mirror in the F–P cavity is able to be determined. For the purpose of detecting the frequency shift of the tunable laser, a reference laser was used. Since the frequency of the reference laser was fixed, the frequency shift of the tunable laser is determined from the shift of beat frequency between the tunable laser and the reference laser. For example, the tunable laser in Ref. [18] was an external-cavity diode laser, and the reference laser was a stabilized He–Ne laser.

The resonance condition of the F–P interferometer is satisfied in the condition that

$$N \cdot \frac{\lambda}{2} = nl, \tag{28.22}$$

where

N is an integer
λ is the wavelength which incident on the F–P cavity
n is the refractive index of air at the wavelength of λ
l is the cavity length

When the cavity length displaces Δl, the wavelength of the tunable laser changes by $\Delta\lambda$. Since the frequency of the tunable laser, f, is $f = c/\lambda$, the frequency change that is caused by the change of the cavity length, Δf, is

$$\frac{\Delta f}{f} = \frac{\Delta l}{l} + \frac{\Delta n}{n}. \tag{28.23}$$

Δf is equal to the change of the beat frequency between the tunable laser and the reference laser. By measuring the beat frequency using frequency counter, the displacement of one of the mirror in the F–P cavity is able to be determined. Since the resolution of measuring frequency is high, usually better than 10^{-10}, the resolution of displacement measurement using F–P interferometer is high. The theoretical resolution of measuring the displacement reaches to several attometer levels. However, for the long-range measurement of displacement, tunable laser with wide tuning range is necessary. Furthermore, more than one reference lasers are necessary for wide range tuning of tunable laser. Hard effort is required to obtain the wide range tunable laser and some reference lasers.

28.4.2 Application of fs-Comb for Displacement Measurement with Fabry–Perot Interferometer

An accurate and high-resolution displacement measurement was performed by using a F–P interferometer with a fs-comb [21,22]. The fs-comb was used as a reference laser of tuning the tunable laser. As the same way with the basic displacement measurement using F–P interferometer, that system consisted of a F–P cavity (FP1), a tunable laser, and a reference laser (a fs-comb). In addition, another F–P cavity, FP2, was used for environment compensation. A Nd:YAG laser (MISER) illuminated on FP2. The cavity length of FP2 was fixed, and the optical length was changed by the change of the refractive index of air in the F–P cavity. Since FP2 was set near FP1, and these cavities are in the open air, the air condition in FP2 was similar to the air condition in FP1. The resonant frequency of FP2 was changed by the change of optical length of FP2. The MISER was locked to FP2. The fs-comb was locked to the MISER in such a way that one of the modes in the fs-comb was locked to the MISER. The frequency of the mode in the fs-comb was locked to the frequency of the MISER by adjusting the repetition frequency. The beat frequency between the tunable laser and a mode in the fs-comb was measured along with the continuous change of tunable laser. The frequency change of the tunable laser reflected the displacement of the mirror of FP1 and the change of the refractive index of air in FP1. The frequency change of a mode in the fs-comb reflected the change of the refractive index of air in FP2. Since FP2 was set beside FP1, the refractive indices of air in FP1 and FP2 are almost the same value. By subtracting the frequency change of FP2 from the frequency change of FP1, the displacement of the mirror that was compensated by the refractive index of air was determined (Figure 28.8).

It is difficult to measure the frequency difference between the MISER and the tunable laser directly. By using the fs-comb that was locked to the MISER and the tunable laser, the comparison of the frequencies of the MISER and the tunable laser was realized.

By using FP2 as the compensation part for the refractive index of air, and the fs-comb acted as "the wavelength comb," excellent stability was obtained. The stability of the displacement measurement setup was sub-pm for the integration time of up to 1 minute. The resolution of the displacement measurement reached sub-pm.

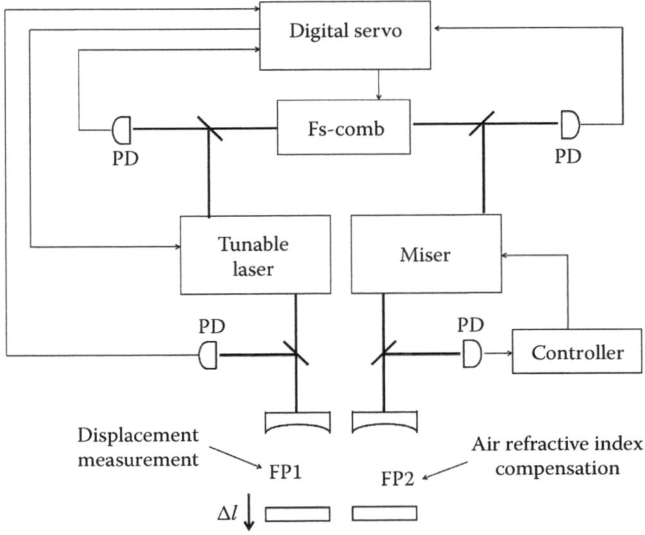

FIGURE 28.8 The picometer measuring system using Fabry–Perot cavities.

References

1. I. Misumi, S. Gonda, Q. Huang, T. Keem, T. Kurosawa, A. Fujii, N. Hisata et al., Sub-hundred nanometer pitch measurements using an AFM with differential laser interferometers for designing usable lateral scales, *Meas. Sci. Technol.* 16, 2080–2090 (2005).
2. M. Aketagawa, H. Honda, M. Ishige, and C. Patamaporn, Two-dimensional encoder with picometer resolution using lattice spacing on regular crystalline surface as standard, *Meas. Sci. Technol.* 18, 342–349 (2007).
3. Th. Udem, R. Holzwarth, and T. W. Hänsch, Optical frequency metrology, *Nature* 416, 233–237 (2002).
4. T. Yoshizawa, *Handbook of Optical Metrology*, CRC Press, Boca Raton, FL (2009).
5. P. E. Ciddor, Refractive index of air: New equations for the visible and near infrared, *Appl. Opt.* 35, 1566–1573 (1996).
6. P. E. Ciddor and R. J. Hill, Refractive index of air. 2. Group index, *Appl. Opt.* 38, 1663–1667 (1999).
7. C. M. Wu and R. D. Deslattes, Analytical modeling of the periodic nonlinearity in heterodyne interferometry, *Appl. Opt.* 37, 6696–6700 (1998).
8. N. Bobroff, Recent advances in displacement measuring interferometry, *Meas. Sci. Technol.* 4, 907–926 (1993).
9. L. Chassagne, S. Topcu, Y. Alayli, and P. Juncar, Highly accurate positioning control method for piezoelectric actuators based on phase-shifting optoelectronics, *Meas. Sci. Technol.* 16, 1771–1777 (2005)
10. C. M. Wu and C. S. Su, Nonlinearity in measurements of length by optical interferometry, *Meas. Sci. Technol.* 7, 62–68 (1996).
11. J. Ye and S. T. Cundiff, *Femtosecond Optical Frequency Comb: Principle, Operation, and Application*, Springer, New York (2005).
12. A. Yariv, *Optical Electronics in Modern Communications*, 5th edn., Oxford University Press, New York (1997).
13. Th. Udem, J. Reichert, R. Holzwarth, and T. W. Hänsch, Absolute optical frequency measurement of the cesium D_1 line with a mode-locked laser, *Phys. Rev. Lett.* 82, 3568–3571 (1999).
14. S. A. Diddams, D. J. Jones, J. Ye, S. T. Cundiff, J. L. Hall, J. K. Ranka, R. S. Windeler, R. H. Holzwarth, Th. Udem, and T. W. Hänsch, Direct link between microwave and optical frequencies with a 300 THz femtosecond laser comb, *Phys. Rev. Lett.* 84, 5102–5105 (2000).
15. D. J. Jones, S. A. Diddams, J. K. Ranka, A. Stentz, R. S. Windeler, J. L. Hall, and S. T. Cundiff, Carrier-envelope phase control of femtosecond mode-locked lasers and direct optical frequency synthesis, *Science* 288, 635–639 (2000).
16. H. Matsumoto and K. Minoshima, High-accuracy ultrastable moving stage using a novel self-zooming optical scale, *Opt. Commun.* 132, 417–420 (1996).
17. Y. Zhao, X. H. Cheng, and D. C. Li, Dual-wavelength parallel interferometer with superhigh resolution, *Opt. Lett.* 27, 503–505 (2002).
18. M. Kajima and H. Matsumoto, Picometer positioning system based on a zooming interferometer using a femtosecond optical comb, *Opt. Express* 16, 1497–1506 (2008).
19. M. Kajima and K. Minoshima, Optical zooming interferometer for subnanometer positioning using an optical frequency comb, *Appl. Opt.* 49, 5844–5850 (2010).
20. L. Howard, J. Stone, and J. Fu, Real-time displacement measurements with a Fabry–Perot cavity and a diode laser, *Precis. Eng.* 25, 321–335 (2001).
21. T. R. Schibli, K. Minoshima, Y. Bitou, F. L. Hong, H. Inaba, A. Onae, and H. Matsumoto, Displacement metrology with sub-pm resolution in air based on a fs-comb wavelength synthesizer, *Opt. Express* 14, 5984–5993 (2006).
22. Y. Bitou, T. R. Schibli, and K. Minoshima, Accurate wide-range displacement measurement using tunable diode laser and optical frequency comb generator, *Opt. Express* 14, 644–654 (2006).

29

Detection of Subnanometer Ultrasonic Displacements

Tomaž Požar
University of Ljubljana

Janez Možina
University of Ljubljana

29.1 Introduction

Ultrasound is understood as mechanical motion with frequency content above 20 kHz, the upper limit of human hearing [1], and below 10 THz, the beginning of hypersound [2]. Although the definition of the lower bound originates from sound, it can also be used for mechanical wave propagation in liquids and solids. Specifically for solids, ultrasound is considered as different types of high-frequency mechanical waves that propagate with various velocities, reflect, refract, interfere, disperse, mode convert, and attenuate within the sample [3].

Ultrasound in solids can be generated by many diverse sources. With respect to location, sources can induce ultrasound on the surface [4–6] or in the interior [7,8] of samples. They can be localized to a very small volume. Such sources are called point sources [9]. On the other hand, they can also generate ultrasound over an extended volume. For this reason, these sources are dubbed extended sources [10]. With respect to the duration of the emission of ultrasound and consecutively to its frequency content, ultrasonic sources can be classified as impulsive (wide bandwidth) or harmonic (narrow bandwidth) or as sources with an arbitrary temporal distribution lying between the two aforementioned extremes. The shortest ultrasonic sources are due to either a mere absorption or ablation of the surface of solid samples with Q-switched laser pulses [11,12]. The frequency content of thus-induced waves can reach up to several 100 MHz [13,14]. Higher surface acoustic wave (SAW) frequencies, up to 1 GHz, can be generated with picosecond laser pulses [2] and even higher, up to 90 GHz, with femtosecond lasers [15]. On the contrary, harmonically vibrating piezoelectric transducers in contact with the solid sample enable generation of very narrow-band ultrasonic waves [16]. Such waves, especially when they are standing waves, are often referred to as vibrations. Displacement-measuring sensors that detect vibrations are called vibrometers.

Ultrasonic sources may be of large scale, such as earthquakes, where a sudden release of energy in the Earth's crust, due to the motion of tectonic plates, creates seismic waves [17,18]. Their miniature counterpart, acoustic emission [18–22], is a phenomenon that arises from a rapid release of stress energy in the form of ultrasound within or on the surface of a material. Acoustic emission sources may be point defects, slips, or dislocations in crystals, twinning or grain boundary movement of polycrystallines, corrosion, fatigue cracks, plastic deformations, phase transformations, creation and collapsing of voids, crushing of inclusions, initiation and growth of cracks in materials, friction, cavitation, leaks, and realignment or growth of magnetic domains. Often, as is the case of the previous examples, the onset time on ultrasound is unknown. Ultrasound can also be generated by mechanical impacts with solids [23,24] through a linear momentum transfer from the impacting body to the mechanical waves [25].

Ultrasound can be generated deliberately for various applications with ultrasonic actuators [26–29] or other types of ultrasonic sources [28]. In general, ultrasonic actuators are slightly modified versions of the same devices that are also used to sense ultrasonic motion. Their principle of operation will be described in the following section. Examples of artificial wide-bandwidth, impulsive ultrasonic sources are small elastic ball impacts [30–32], electric sparks [33–36], expanding plasmas [37,38], and laser-pulse ablation of uncoated [11,39–44] or constrained surfaces [11,45,46]. Artificial ultrasonic sources with wide-bandwidth, step-like temporal dependence include radially loaded glass capillary fractures [30,47–49], pencil lead breaks (Hsu–Nielsen source) [30,31,50–52], fractures of small grains [53], and thermoelastic generation with laser pulses [39–43,54]. Helium gas jet impact may be employed as a continuous white noise generator of ultrasound [55–57]. Particle impacts, such as electrons from a scanning electron microscope, are used in electron-acoustic microscopy to generate ultrasound in solids [58–60]. Charged particles [61] and even ultrahigh-energy cosmic-ray neutrinos [62] may also generate ultrasound of detectable amplitude. Ultrasound generation may be achieved by electromagnetic radiation at wavelengths other than those emitted by lasers, for example, by pulsed x-ray radiation from synchrotrons [63], microwaves [64,65], and radio frequencies [66].

This chapter deals with the detection of minute displacements whose amplitudes are smaller than 1 nm and are caused by ultrasonic motion. The main attention will be given to present various means of sensing the out-of-plane (normal or vertical, denoted by u) component of the surface displacement vector in solids. These displacements are the result of reflections of ultrasonic waves from the boundaries of a solid body. Even though most detectors are predominantly responsive to a single component, either the in-plane (tangential or horizontal, denoted by w) or out-of-plane component, of the displacement vector, it is often the case that the measured displacement is deteriorated by the other components of displacement. On the other hand, some detectors are capable of measuring both the in- and out-of-plane displacement simultaneously [67–71], but the majority of detectors respond only to the out-of-plane component. Ultrasonic time-dependent displacements can also be obtained by integration of time derivatives of displacement, especially velocity and acceleration. Piezoelectric detectors, for instance, are known to respond to a frequency-dependent mixture of displacement, velocity, and acceleration [72]. Such detectors which are not directly linked to a single physical quantity are difficult to calibrate. Detector that measures a linear combination of displacement and its time derivatives is therefore not suitable for absolute measurements, but can still be used for qualitative measurements: to determine the arrival times of ultrasonic waves or to perform frequency count of ultrasound-emitting events. Ideal displacement-measuring detectors should thus be linearly sensitive to a single component of a single physical quantity, preferably to displacement itself. Additionally, its frequency response should be flat in the frequency band of interest. Special design is often required to approach these requirements [72,73]. Ideal linear detectors with flat frequency response are fully characterized by a frequency-independent figure of merit called sensitivity S. The units of sensitivity are V/m, because most commonly one reads the output in volts for a given displacement in meters. Determination of the value of sensitivity is called absolute calibration. Practically, sensitivity does not have a constant value and has to be expressed as a function of frequency. To achieve sufficiently large sensitivity, the measured signal often needs to be amplified. Amplification, however, has its own transfer function and adds additional noise to the system.

Ultrasonic displacement-measuring detectors can be characterized by the following features: *sensitivity*, *minimal detectable displacement* also called *noise-equivalent displacement*, *resolution*, *dynamic range*, and *frequency characteristics* (ultrasonic signal bandwidth, compensation bandwidth, and resonant behavior). Ultrasound-measuring detectors demand a special design when they are used in hostile and harsh environments, such as when measuring ultrasound at elevated temperatures [74], in toxic, in acid/basic, or in radioactive environments. Optical detectors are often preferred in such cases. The contact nature of detectors also varies [75]. Some need direct contact with the measuring surface (piezoelectric sensors). Their sensitivity is enhanced with a thin layer of liquid couplant that provides a better acoustic impedance match between the sample and the sensor. Capacitive and electromagnetic detectors operate in close proximity to the sample surface. Capacitive ones have a gap of a few micrometers [76], while electromagnetic detectors work up to a liftoff distance of 2 mm [77]. True contactless, standoff detectors are optical devices and can be separated from the measuring surface for up to 2 m. These detectors are also nonperturbing, which means that they have negligible influence on the ultrasound propagation, while others, especially those of contact nature, alter surface motion by their presence. Small fraction of ultrasonic energy is also transmitted to the air above the surface of the sample. If piezoelectric and capacitive sensors are lifted from the surface while they are still capable of detecting ultrasound-induced air pressure changes, they are *air coupled*. The advantage of air coupling is that the measurement becomes noncontact through the detection of airborne leaky waves leaked from material surfaces. However, the sensitivity is reduced due to a large acoustic impedance mismatch between air and solid. Moreover, the absorption of ultrasound in the air becomes severe for high frequencies and goes as the square of the frequency. Due to this reason, air-coupled sensors are used below 1 MHz. To increase the sensitivity and improve impedance mismatch, samples are rather *water coupled*, that is, immersed in water.

Ultrasonic surface displacements cannot be measured on all solid samples with all types of detectors. Material properties of the solid determine the possible detection principles. For instance, transparent samples cannot be inspected with optical techniques when insufficient light is returned toward the detector. Samples with rough surfaces cause problem to all detectors. The performance of optical detectors is severely degraded by high surface absorptivity of light (laser beam), highly scattering surfaces, and surface tilts. On the other hand, electromagnetic sensors require the samples to be either conductive or ferromagnetic. Ultrasound in conductive materials can be detected with Lorentz force mechanism. When the material is ferromagnetic, magnetostriction is the underlying detection principle. Generally, capacitive sensors of ultrasound demand conductive samples, but they can also be adapted to measure ultrasound on nonconductive materials. The active area of the detector size is also an important parameter. Its detecting area (aperture) is directly connected with the maximum detectable frequency and thus sets the limit to the detectable signal bandwidth. From geometrical reasoning, a detector with a diameter d cannot discern SAWs propagating tangentially to the sensor with velocity c with frequencies larger than $f = c/d$. When at high frequencies multiple wavelengths are averaged over the area of contact, the amplitude of the recorded wave decreases. This is called an aperture effect [78,79] and depends on the direction of the incoming ultrasonic wave, being most significant for waves propagating tangentially to the sensor face, and has no effect on plane waves with wave fronts parallel with the measuring surface. Frequency response upper limit of a displacement detector is also determined by the detector and auxiliary electronics. Further, when detectors are used in arrays [80,81], single detector units often need to be miniaturized. If needed, classical sensors are nowadays produced as very small devices called *microelectromechanical system (MEMS) sensors* either as *capacitive micromachined ultrasonic transducers (CMUTs)* [82–86] or their piezoelectric counterparts *PMUTs* [87–91]. Optical devices can be made more compact and robust with optical fibers [92,93]. Both small detector size and delivery of light through optical fibers in optical detectors enable access to remote and difficultly accessible places that other, larger detectors cannot reach. Finally, not all detectors have reached industrial and commercial stages. Some are used mainly in laboratories. Perhaps, the most widely used are piezoelectric detectors since they have the largest sensitivity and are not expensive. There is also a distinction between absolute sensors and calibrated sensors. Interferometers are self-calibrated, absolute sensors, because the displacement history of objects with adequate reflectance can be accurately and precisely measured since the measured relative length is based on counting fractions of the well-defined wavelength of the stabilized laser source. Other sensors need to be precalibrated, say, with interferometers or other calibration methods [79].

Although ultrasonic displacements can be much larger than 1 nm [12], it is possible to measure displacements even below 10 fm using a sensitive lock-in detection scheme for bandwidth reduction [94,95]. For comparison, the diameter of an atomic nucleus is 2.4 fm $A^{1/3}$, where A is the nucleon number. Thus, the diameter of ^{56}Fe is ~10 fm. The measured displacement is not the displacement of a single atom making up a solid. It is rather the average value of displacements of a multitude of atoms in contact with the sensor or having remote effect on the measurement. It is worth keeping in mind that the measured ultrasonic displacement can be in the order of some nucleic radii, but this value is averaged over an area consisting of usually several orders of magnitude >10^9 atoms. For example, in laser-based detection of ultrasound, a laser spot focused to a 10 μm diameter covers about 10^9 atoms on the surface of body-centered cubic crystal structure of α-iron, where the closest separation between atoms is 0.25 nm.

The minimal detectable displacement of any displacement-measuring sensor is set by the fundamental limit called the *thermal rattle*, also called the *phonon shot noise*. The surface atoms experience a thermally induced random motion which is macroscopically manifested as a temperature-dependent mechanical noise. This is a mechanical equivalent of the well-known Johnson–Nyquist noise used to calculate the voltage and current fluctuations in a resistor. The root-mean-square (RMS) out-of-plane displacement fluctuation of a surface area on a solid, isotropic, half-space is approximately

$$u_{\min} \doteq \frac{1}{\pi f}\sqrt{\frac{k_B T \Delta f}{\mathrm{Re}(Z(f))}}. \tag{29.1}$$

This expression [96–98] is valid for $k_B T >> hf$, where f is the frequency, k_B is the Boltzmann constant, T is the absolute temperature, Δf is the frequency bandwidth, h is the Planck constant, and $Z = F/v$ is the complex mechanical impedance defined as the ratio of an applied point force F and the resultant surface velocity v. Using the expression for Z given in [99] and a circular surface area of 1 mm diameter on Al, the value of $u_{\min}$ is between 10^{-16} and 10^{-17} m/√Hz in the bandwidth between 100 kHz and 1 MHz [96]. As an example, the thermal rattle at 1 MHz and bandwidth of 100 Hz gives a 0.1 fm RMS amplitude of the out-of-plane displacement.

Applications in science and industry where detection of sub-nanometer ultrasonic displacements plays a major role will not be discussed in this chapter. It is however worth to mention some fields where minute ultrasonic displacements often need to be measured: optodynamics [5,12,25,100–103], monitoring of laser material processing [104–106], laser-based ultrasonics [43,107–111], acoustic emission [18–22], nondestructive testing of materials [20,112,113], microseismology [81], and medicine [114].

29.2 Detection Principles and Sensors

Various physical phenomena can be exploited as transduction mechanisms to detect ultrasonic displacements of solid surfaces. They are primarily gathered in four distinct detection groups depending on the underlying physics: piezoelectric, electrostatic, electromagnetic, and optical. Within each detection group, the transduction principles are first concisely introduced, followed by a description of the most common sensor types that make use of the presented detection method. The performance characteristics, advantages, and deficiencies of each sensor of ultrasound are given. Detectors are compared within each group and between groups. In-depth explanation was intentionally omitted to keep the description simple.

29.2.1 Piezoelectric Detection

Piezoelectric ultrasonic sensors use the *piezoelectric effect* or *piezoelectricity* as a transduction mechanism to measure a frequency-dependent mixture of displacement, velocity, and acceleration of surfaces by converting these physical quantities to an electrical charge. A charge-to-voltage converter called charge amplifier is then used to proportionally convert electrical charge to voltage. Piezoelectric effect, discovered by Jacques and Pierre Curie in 1880, is a phenomenon where electrical charge accumulates on the surface of certain ceramics (e.g., lead-zirconate-titanate) and crystals (e.g., quartz) in response to an

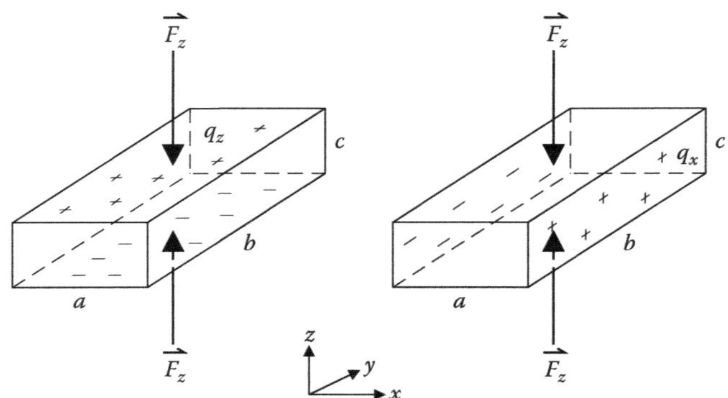

FIGURE 29.1 Longitudinal (left) and transverse (right) mode of PZT operation.

applied mechanical stress and consequent mechanical deformation. The accumulated charge is usually highly proportional to the applied force.

There are three distinguishable modes of operation depending on how a piezoelectric material (PZT) is cut: longitudinal, transverse, and shear. Imagine a rectangular cuboid of PZT with edge lengths a, b, and c in x, y, and z axes, respectively. In the *longitudinal* mode of operation (Figure 29.1, left sketch), the charge is generated on the same surfaces where the force is applied and is independent of the geometrical dimensions and shape of the PZT. A charge

$$q_z = d_{zz}F_z \tag{29.2}$$

released in z-direction by a force F_z which is also applied along z-direction is characterized by a piezoelectric coefficient d_{zz} with units of C/N. In the *transverse* mode (Figure 29.1, right sketch), the charge is generated on lateral faces in respect to the applied force. Here, the amount of charge

$$q_x = -d_{xz}F_z\frac{a}{c} \tag{29.3}$$

depends on the dimensions a and c of the piezoelectric cuboid and is determined by a different piezoelectric coefficient d_{xz}. We assumed that the charge was generated only in x-direction by a force F_z applied perpendicularly in z-direction. In the *shear* mode, the charge is again highly proportional to the applied forces and is independent of the size and shape of the PZT.

In addition to the piezoelectric effect, piezoceramic materials commonly show the ability to generate an electrical signal also when the temperature of the sensing element changes. This effect is called *pyroelectricity.*

Basic construction of a classical piezoelectric ultrasonic sensor [79,115,116] mounted on the surface of a solid is shown in Figure 29.2. Thin slab of PZT, cut in such a way that it operates in a longitudinal mode, is placed in contact with the measuring surface. The contacting surfaces are lubricated by a liquid couplant with acoustic impedance that matches the impedance of the PZT. Piezoelectric slab is backed by a large damping mass and sealed in a suitable housing. Thus, built sensor is mainly sensitive to ultrasonic out-of-plane displacements u. The whole sensor is pressed against the measuring surface so that PZT will be statically preloaded. Normal displacements of the measuring surface will either compress or extend the PZT depending on the polarity of the incoming ultrasonic wave (compression or rarefaction). Time-varying stress in the PZT will cause a proportional time-varying accumulation of electrical charge on opposite sides of the PZT slab. Two electrodes are attached to the charged PZT surfaces. The outside surface of the outer electrode has to be electrically shielded so that both conductive and nonconductive solids can be measured. Often, the signal from these electrodes is preamplified already within the casing to reduce signal noise due to parasitic capacitance. The output voltage is in general not proportional to normal ultrasonic displacement u, because classical ultrasonic sensors have complex transfer functions and do not have a flat frequency response (see Figure 29.3). They can be designed to sense only a portion of the whole frequency range of interest,

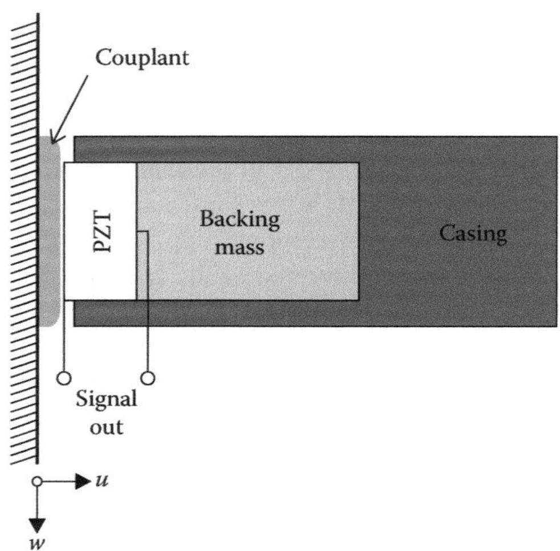

FIGURE 29.2 Cutaway schematic drawing of a classical piezoelectric sensor.

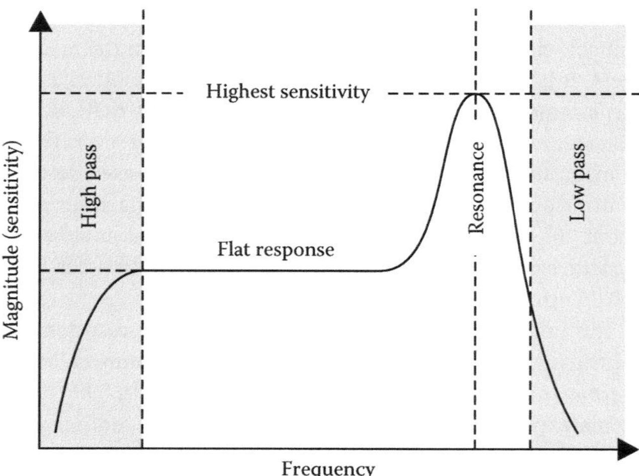

FIGURE 29.3 Typical frequency response of a classical piezoelectric sensor.

usually between 10 kHz and 1 MHz, by choosing the appropriate dimensions (usually thickness) of the piezoelectric element.

Piezoelectric ultrasonic sensors can measure ultrasound with sensitivity of about 1 V/nm as displacement sensors and few V/(mm/s) as velocity sensors [72,117]. The noise-equivalent displacement is often below 1 pm. The dynamic amplitude range of piezoelectric ultrasonic sensors is around 120 dB or 10^6. They can be used in environments with temperatures up to 600°C if aluminum nitride is used as a PZT [118]. They have another useful property when measuring ultrasound, that is, to measure only dynamic events and so automatically compensate for low-frequency motion of measuring objects that are caused by environmental vibrations. This intrinsic low-frequency cutoff is a consequence of the leakage of the accumulated charge. As seen in Figure 29.3, this acts as a high-pass filter which determines the low-frequency cutoff (compensational bandwidth) through a time constant given by the capacitance and resistance of the device. Additionally, piezoelectric ultrasonic sensors are also insensitive to electromagnetic fields and radiation, enabling measurements under harsh conditions.

The most common commercially available piezoelectric ultrasonic sensors are of a resonant type, because they provide greater sensitivity near the resonance frequency (see Figure 29.3). Unfortunately, resonant devices lack the bandwidth needed to analyze incoming waveforms. They are not suitable for absolute measurements, but can still be used for qualitative measurements: to give a reasonably accurate estimate of the arrival times of ultrasonic waves or to perform frequency count of ultrasound-emitting events. However, beyond the direction of surface motion caused by the earliest ultrasonic wave, the received signal is more a function of the sensor than of the true displacement history. When the ultrasonic wave excites the measuring surface of the solid it often sets the sensor into vibration, thus masking the desired signal. This effect is called ringing. The frequency content of such signal for the most part reflects merely the normal modes of vibration of the solid and the sensor.

To avoid frequency-dependent effects of classical PZT sensors, such as their resonant behavior at high frequencies and reflections of ultrasonic waves within the PZT slab and backing mass [119], specially designed, wideband *conical* piezoelectric ultrasonic sensors were developed [73,78,81,88,89,99,117, 119–127]. Conical piezoelectric ultrasonic sensors were introduced by the National Institute of Standards and Technology (NIST) for use in the field of wideband quantitative acoustic emission [78]. They are used where the actual displacement is to be measured with precision and accuracy. A generalized cross-sectional scheme of the conical piezoelectric ultrasonic sensors is displayed in Figure 29.4. The conical design of the sensor PZT element, usually made from a lead–zirconate–titanate composition *PZT-5a*, eliminates the aperture effect by keeping the contact area small. The contacting face of the truncated PZT cone is around 0.5–2 mm in diameter [72,117] in contrast to a few 10 mm diameter of the sensing element of the classical piezoelectric sensors. This sets the upper limit of frequency response for, for example, detection of SAWs at $f = (3000\ \text{m/s})/(1\ \text{mm}) = 3$ MHz. The wideband frequency response (both phase and magnitude) of conical piezoelectric

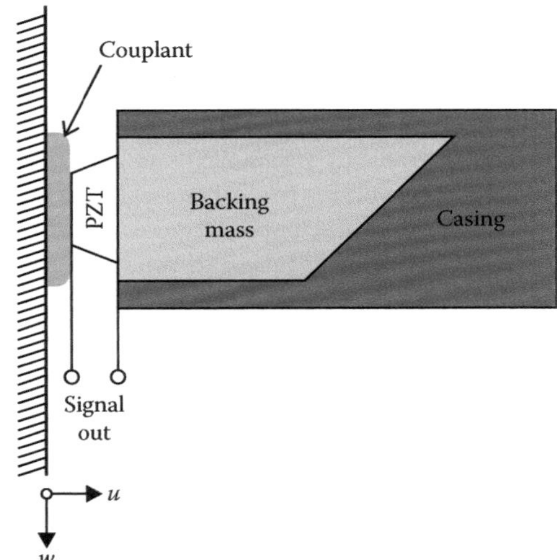

FIGURE 29.4 Cutaway schematic drawing of a conical piezoelectric sensor.

sensors is practically flat within ±3 dB from 10 kHz to 1 MHz. Moreover, such detectors are sensitive only to the normal component of displacement vector and their output is directly proportional to displacement in the frequency range from 10 kHz to 1 MHz [73,122]. Theoretical analysis of the proportionality between the output voltage and the out-of-plane displacement of the measured surface is beyond the scope of this section, but can be found in [99]. The asymmetric design of the PZT cone also reduces the degeneracies of the normal modes of the usual piezoelectric disk element [99,120]. The signal is amplified already in the vicinity of the PZT so that it is not corrupted by electromagnetic noise and capacitive loading from the cable between the sensor and the A/D converter. The material of the backing mass (brass or lead alloy) needs to be large to reduce resonances at lower frequencies. Its acoustic impedance has to match the PZT material and has to prevent back-reflections of the passing ultrasonic wave to return to the PZT cone. The former is important, because the ultrasonic wave has to pass the PZT element unaffected. The latter is achieved by choosing a backing material with high internal acoustic attenuation and by cutting the backside of the backing mass at an angle to prevent direct reflections. In general, all mechanical parts of the wideband piezoelectric sensor that are affected by the ultrasound call for an asymmetric design to reduce possible mechanical resonances [119]. For instance, the NIST conical reference transducer's backing block has no parallel faces and no right angles to ensure that only high-order multiple reflected elastic waves can reenter the PZT element [127].

The minimal detectable displacement measured with the optimized conical piezoelectric transducers outmatches other ultrasonic displacement sensors by 10–100 dB and is <10 dB above the thermal rattle limit. The measured value of the minimum displacement of such a superior conical piezoelectric detector can be as small as $3 \times 10^{-17}\ \text{m}/\sqrt{\text{Hz}}$ in the bandwidth from 0.5 to 1 MHz [96,100]. The noise model for piezoelectric sensors can be found in [128,129].

Laser-based micromachining allows the construction of even smaller conical PTZ elements with 0.2 mm aperture diameter which extends the frequency bandwidth up to few MHz [88,89]. Further miniaturization of the PZT cone is possible, but it would make it more vulnerable to mechanical damage. PMUTs fabricated as laminated structures in the bending mode using beam and plate structures have the upper-frequency cutoff as high as 25 MHz [90,91,128,130,131]. Microfabrication nowadays allows the production of resonant piezoelectric ultrasonic sensors with highest sensitivity near 100 MHz [132].

Tangential (in-plane) wideband piezoelectric sensors were also developed [133,134]. Although almost all piezoelectric sensors demand surface placement, specially designed piezoelectric sensors can also be buried inside the solid body that needs to be monitored [117]. In place of the couplant, an acoustic waveguide, usually a metal rod, may be inserted to provide a thermal and mechanical distance between the sensor and the measuring surface [135]. Liquid couplants automatically preclude sensing of tangential (in-plane) motion, because liquids do not transmit shear motion.

Piezoelectric transducers in their convectional form (see Figures 29.2 and 29.4) need direct contact with the measuring surface of a solid and do not perform well in air, because of an acoustic impedance mismatch between the air and the PZT. Special air-coupled piezoelectric ultrasonic transducers that operate in air at high frequencies were developed [136–140]. They can be detached from the sample by as much as 80 mm [139] and may also be used to measure surface displacements of nonconductive materials such as polymers and composites with maximum sensitivity in the range of 1 MHz. The acoustic mismatch was overcome by an introduction of multiple matching layers or radiating membranes. Such detection is desirable where contact or near-contact between the sensor and the measuring surface is either undesirable or impossible, such as when the measuring solid is moving or is very hot.

29.2.2 Electrostatic Detection

Transduction mechanism of *capacitance ultrasonic sensors*, also called *capacitive sensors* or *electrostatic acoustic transducers (ESATs)*, is based on the change of electrical charge δq on the terminals of a capacitor due to the motion of one of its terminals. The static terminal of the capacitor is a part of the sensor while the other, subjected to ultrasonic motion, is often the measured surface of the conducting solid itself [76,141,142]. When the measuring surface is nonconducting, a thin conducting layer has to be attached on the top of the solid.

Consider the simplest capacitor, illustrated in Figure 29.5, made of a parallel conductive plate (terminal, electrode, or probe) with surface area A separated by a distance x from the polished surface whose normal displacement $u=\delta x$, which is much smaller than x, one wants to measure. The space between the electrode and the measured surface consists of a dielectric, often air, with absolute permittivity ε. A constant DC voltage U is applied to the plate, while the measured conducting surface is grounded.

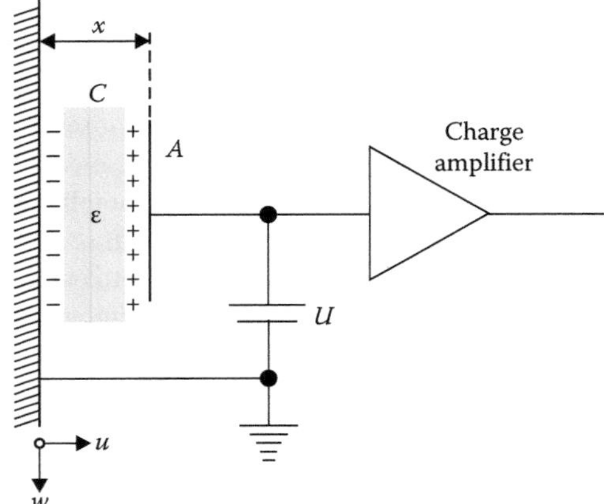

FIGURE 29.5 Schematic drawing of a capacitance sensor.

A reflection of the ultrasonic wave from the surface under the plate causes the gap to change by δx. This leads to a change in the capacitance

$$\delta C=\delta\left(\varepsilon\frac{A}{x}\right)=-\varepsilon\frac{A}{x^2}\delta x=-\frac{C}{x}u \tag{29.4}$$

between the surface and the plate. Consequently, the change in the capacitance results in the creation of an additional electrical charge δq on the surface of the plate:

$$\delta q=\delta(CU)=U\delta C=-\frac{\varepsilon UA}{x^2}u=-\frac{UC}{x}u. \tag{29.5}$$

As for the piezoelectric ultrasonic sensor, the change in charge δq is proportionally converted to voltage via a low-noise charge amplifier. Thus, the output voltage V is directly proportional to the temporal dependence of the out-of-plane motion u of a flat surface caused by ultrasound.

Typically, the capacitance ultrasonic sensor is operated in air ($\varepsilon=\varepsilon_0=8.85\times10^{-12}$ F/m) with a circular plate with 6 mm diameter, $x=5$ μm gap, $U=100$ V constant voltage bias, and the sensitivity of charge amplifier $S_{CA}=\delta U/\delta q=0.35$ V/pC. This yields the following sensitivity of the sensor:

$$S=\frac{\delta U}{\delta x}=\frac{\delta U}{\delta q}\frac{\delta q}{u}=-\frac{UC}{x}S_{CA}=-\frac{\varepsilon_0 UA}{x^2}S_{CA}=0.35\text{ V/nm}. \tag{29.6}$$

The calibration of the detector can be done theoretically using Equation 29.6 or experimentally by measuring the capacitance as a function of the gap width.

Note that the smaller the gap x and the larger the values of ε_0, U, A, and S_{CA}, the larger the sensitivity S. The sensitivity can be increased by choosing plate electrodes with larger surface areas, but this goes at the expense of frequency bandwidth due to the aperture effect. Again, the upper frequency cutoff highly depends on the angle of the incoming wave, being the largest for normal incidence and the smallest for surface waves (5 MHz for ball-type probes [143]). Maintaining the air gap as small as possible improves the bandwidth of the sensor as well as maximizes the sensitivity, because the lowest resonance frequency $f = c/2x$ of the air gap, where sound propagates with velocity of about $c = 340$ m/s, is inversely proportional to the plate separation x. For example, for a 5 μm gap, the lowest resonance frequency is about 34 MHz, and this is much higher than the limits placed by other factors on the bandwidth [141]. Due to the electrical discharge between the electrodes at normal conditions in air, the ratio of the applied DC voltage U and the spacing x has to be smaller than about 50 V/μm [141,143]. The sensitivity is also proportional to the absolute permittivity, which cannot be significantly increased when the dielectric is gas. Inserting a nonconducting liquid or solid in the gap increases the absolute permittivity, but also leads to undesirable low-frequency resonances and reflections of ultrasonic waves that deteriorate the transfer function of the capacitance ultrasonic sensor [47,144]. Insertion of a solid dielectric in between the surface and the probe may also serve as a spacer with a well-known thickness, thus mitigating demanding placements of the probe in the close proximity of the surface [144]. Moreover, the value of the breakdown threshold between the electrodes can also be enlarged by the insertion of a solid dielectric [82,145].

The main advantages of the capacitance ultrasonic sensor over the piezoelectric one are that the former is inherently a normal displacement-sensitive device over a wide bandwidth (from 10 kHz up to about 10 MHz [47,141,142,146]), free of ringing, so that its output can be calibrated, and that it does not require a couplant which otherwise makes measurements irreproducible. A typical sensitivity of capacitance ultrasonic sensors is of the same order of magnitude, or an order of magnitude less, compared to the sensitivity of the piezoelectric ultrasonic sensors. Its noise-equivalent displacement is in the order of several picometers [141,142] and its dynamic range is at least 80 dB or 10^4. A critical part of the measurement with capacitance ultrasonic sensor is a proper alignment of the probing electrode. Several improvements were made in order to simplify their placement near the surface [146].

The probing electrode of the capacitance ultrasonic sensor may be of different shapes such as the previously described plate electrode positioned parallel with the measured surface [47,76,82,141,145,146], cylindrical electrode with the cylinder axis parallel with the surface [142], thin wire electrode with a hemispherical ending [147], and spherical ball bearing electrodes [143,144].

Particularly for air-coupled capacitive transducers (condenser microphones), the detectable displacement is limited by the friction of the air against the membrane. The minimal detectable displacement is given by

$$u_{\min} = \sqrt{\frac{2k_B T \Delta f d}{\pi f \gamma P_0 A}} \tag{29.7}$$

where

k_B is the Boltzmann constant
T is the absolute temperature
Δf is the frequency bandwidth
d is the thickness of the membrane
f is the frequency
γ is the specific heat ratio
P_0 is the ambient pressure
A is the membrane area [96,148]

For a typical air-coupled capacitive transducers, the value of $u_{\min}$ is around 10^{-16} m/√Hz in the bandwidth between 100 kHz and 1 MHz.

29.2.3 Electromagnetic Detection

The underlying physical mechanism of *electromagnetic ultrasonic sensors*, also known as *electromagnetic acoustic transducers* (*EMATs*), is of a magnetic nature employing in most cases the Lorentz force and the magnetostriction effect [27,149]. Even though these two mechanisms are often additive and may have comparable coupling magnitude, EMATs are designed so that they exploit a single transduction mechanism. Those based on the Lorentz force are lifted from the measuring surface, while magnetostriction-based sensors may detect either ultrasonic motion directly as standoff devices or in a contact manner with a sandwiched layer of highly magnetostrictive material between the measuring surface and the sensor.

Electromagnetic ultrasonic sensors [150,151], which are also used to generate ultrasonic waves [152,153], are contactless devices capable of measuring surface displacements of electrically conducting and nonmagnetic solids. They are built of two main components: a permanent magnet which produces a static bias magnetic flux density **B** and a pickup coil which detects eddy currents caused by the motion of an ultrasonic wave through induction. There is an air gap, 100 μm to 2 mm wide [77], between the magnet and the surface of a specimen. The coil is placed near the surface in the gap between the magnet and the surface. In general, the directional sensitivity is determined by the design of both, the magnet and the coil [154–156]. When one wants to measure the out-of-plane motion u of the surface, field lines of the bias magnets have to be parallel with the measuring surface, as depicted in Figure 29.6. On the contrary, Figure 29.7 shows that the field lines are required to be normal to the measuring surface when measuring in-plane displacements w.

The inverse Lorentz force mechanism which occurs in any electrically conducting medium is as follows. The arrival of ultrasonic waves to the surface of a conducting solid causes this initially motionless surface to move in a stationary magnetic field. This motion induces eddy currents in the skin-depth-thick layer under the surface of the specimen. Eddy currents give rise to a time-varying magnetic field, much smaller in magnitude than the static bias magnetic field. The time-varying magnetic flux density is then sensed by the induced voltage in a pickup coil placed closely above the surface via Faraday's law of induction.

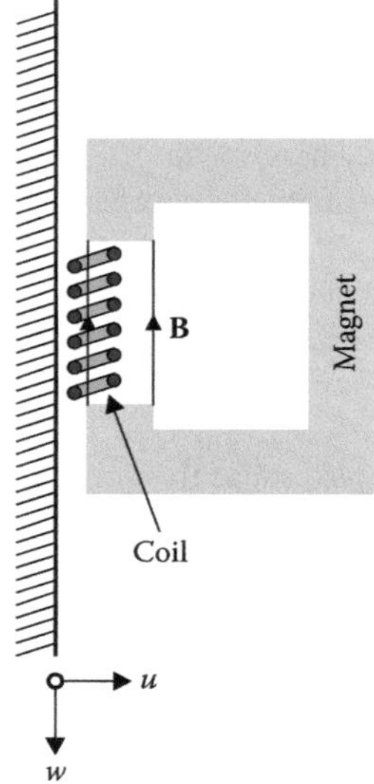

FIGURE 29.6 Schematic drawing of an electromagnetic sensor for measuring *out-of-plane* component of surface displacement.

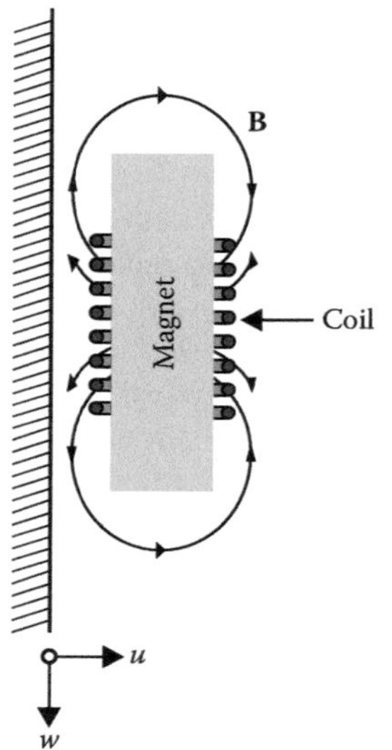

FIGURE 29.7 Schematic drawing of an electromagnetic sensor for measuring *in-plane* component of surface displacement.

Thus, the eddy currents that are inductively coupled to the coil produce a received signal that is proportional to the displacement caused by an ultrasonic wave.

Magnetostriction, on the other hand, only occurs in ferromagnetic materials in which a magnetic field produces strain. This effect is analogous to the piezoelectric effect, in which an electric field produces strain. The reverse is also true. In piezoelectrics, a strain gives rise to an electric field which causes electrical charge to accumulate on the surface of the PZT. Similarly, in the inverse magnetostriction, or the Villari effect, a stress, due to the arrival of an ultrasonic wave, produces a change in the magnetization. As a consequence, this results in a time-varying magnetic flux density that induces voltage in the coil because of Faraday's law. Magnetostriction of a polycrystalline metal occurs due to the motion of magnetic domain walls and the rotation of domains to align along the applied bias magnetic field lines. These microscopic realignments result in a mechanical deformation. In contrast with both piezoelectricity and the Lorentz force, magnetostriction is highly nonlinear [149].

Apart from the Lorentz force mechanism and magnetostriction, a direct interaction between the magnetization of the material and the applied field may also be used as a sensing mechanism in magnetic materials [151]. A detailed collective description of electromagnetic receiving mechanisms can be found in [157,158].

Assuming that the main loss mechanism in EMATs is due to Joule heating by eddy currents, the minimal detectable displacement can be estimated based on the fluctuation dissipation theorem [159]. For a lossless coil and noiseless preamplifier, the minimal detectable displacement is

$$u_{\min} = \frac{1}{\pi f}\sqrt{\frac{k_B T \Delta f \rho}{B^2 \delta A}} \tag{29.8}$$

where
- k_B is the Boltzmann constant
- T is the absolute temperature
- Δf is the frequency bandwidth
- ρ is the resistivity
- f is the frequency
- B is the magnitude of the magnetic flux density
- δ is the skin depth
- A is the area [96]

Even though the minimal detectable displacement of an ideal EMAT experiencing the noise due to Joule heating is only by about 20 dB above the noise floor of the contacting piezoelectric sensor, its typical practical realization is for another 60 dB worse. This is because the open-circuit voltage fluctuation of an ideal EMAT is considerably smaller than the corresponding fluctuations of a typical amplifier and that of the coil resistance [96].

29.2.4 Optical Detection

Optical detection of ultrasound is a truly remote, non-perturbing, and broadband frequency response technique for measuring ultrasonic surface displacements. Its contactless nature enables measurements in hostile and harsh environments where contact or near-contact detectors cannot be used. Optical detectors can be used without any damage to the detector at elevated temperatures, vacuum, high ambient pressures, plasmas, in hard to reach places, on rapidly moving samples, in toxic and in radioactive environments, all from distances up to 2 m away from the surface under inspection. They can even be used to monitor surface displacements of specimens that are closed within a chamber if a laser beam is allowed to pass through a transparent window in the chamber wall. Despite their noncontact transduction and proneness to rapid scanning, optical detectors of ultrasound are more expensive than piezoelectric sensors, are more difficult to handle due to optical alignment issues, and are less sensitive than their piezoelectric counterparts. Moreover, their performance is often severely degraded by high surface absorptivity of the laser beam, highly scattering surfaces (rough surfaces), and surface tilts (deflection issues).

Optical detection of ultrasound from solid surfaces has been reviewed several times by many authors as the field progressed [43,160–177]. The first review paper emerged in 1969 [160], while a more recent extensive one, covering the interferometric detection of ultrasound, appeared in 1999 [173]. Since then, the main breakthrough has been made with the advent of adaptive interferometers, that is, laser ultrasonic receivers based on two-wave mixing and on photo-EMF detection [108,177]. Laser interferometric sensors of ultrasound that have reached industrial applications were reviewed by Monchalin in 2007 [176].

The light source of all optical ultrasonic detectors is always a laser operating in the visible or near-infrared portion of the electromagnetic spectrum. For this reason, optical detection of ultrasound could as well be called laser detection of ultrasound. The photosensitive detector for the visible light is almost always a Si photodiode, except for adaptive interferometers based on the non-steady-state photoelectromotive force (photo-EMF) where other, photorefractive semiconductor materials are used. In general, to detect surface displacement, the surface needs to be illuminated by a laser beam. The laser light could be either continuous or pulsed. In the latter case, the pulse duration has to be sufficiently long (several 10 μs) to capture the whole duration of displacement history of interest. The reflected or diffusely scattered light is then collected by a photosensitive receiver. To improve their sensitivity, optical sensors demand a significant portion of this scattered light to be collected. The light-collection efficiency of such an optical sensor is characterized by the figure of merit called étendue or light-gathering power or optical throughput [173].

Various transduction mechanisms employing laser light can be used to detect surface displacements. When a laser beam is focused on the surface so that its diameter is much smaller than the wavelength of ultrasonic waves which cause displacements, transduction mechanisms based on the *temporal modulation* of laser beam can be used. These conditions can be met for ultrasonic waves up to frequencies as high as 1 GHz if the diameter of the laser spot is some diffraction-limited diameters wide. In such cases, the illuminated flat surface moves in phase and stays flat. It can move to and fro in the direction along its normal or it may tilt around its tangent. In the former case, the time-dependent motion of the surface can be encoded in the phase of a single frequency, coherent light beam. This phase cannot be directly detected, because the electric field in the visible laser beam oscillates with a frequency of a few 10^{14} Hz and no detector is capable to respond to such high frequencies. However, the phase can be extracted by the use of different types of laser interferometers. In general, interferometric detection consists of three successive steps. First, an ultrasonic displacement is converted into an optical phase. Second, the optical phase modulation is transformed into an optical intensity modulation. And third, the optical intensity is converted into an electrical signal. In some laser interferometric schemes, the last two steps are interchanged.

Tilting of the surface occurs when, for example, a SAW with a wavelength much longer than the laser spot diameter passes the illuminated area, thus changing its slope with respect to the probing beam. A tilted polished surface deflects the beam at a different angle. When a physical barrier, such as a knife edge, is placed in front of the photodetector so that it partially obstructs the reflected beam, tilting of the surface will cause a temporal variation of the laser power reaching the active part of the photodetector.

When the wavelength of ultrasonic waves is only a few times larger than the cross section of the probing beam, the illuminated surface becomes curved due to the surface bulge or concavity that is caused by the wave. This warping of the surface acts on the reflected beam as a curved mirror by focusing or defocusing which is related to the curvature of the surface displacement. Even though this optical transduction mechanism can be employed to measure surface displacements, it has not been developed into practical detection schemes.

So far, only temporal modulation of the laser beam was considered. Pure *spatial modulation* takes place when a standing periodic pattern of SAWs is illuminated by a probing laser beam with a diameter much larger than the acoustic wavelength. The effect of the periodic displacement across the illuminated area is equivalent to that of a phase grating, causing multiple diffraction orders to be formed in the reflected beam. The intensity of the diffracted beams as well as the angles between the diffracted beams and the zero-order beam depend on the amplitude and on the wavelength of the standing ultrasonic waves.

A *temporal and spatial modulation* of the probing laser beam occurs when short-wavelength ultrasonic waves move under a larger illuminated area.

With the advent of femtosecond laser sources and various pump-probe techniques [15], another transduction mechanism can now be applied to measure surface displacements indirectly. As mentioned so far, at frequencies below 1 GHz the mechanism for detection is based on the motion of the illuminated surface while ultrasonic waves of frequencies above 1 GHz produce strain that considerably changes the surface reflectivity of the solid. Relative changes of reflectivity can reach values between 10^{-6} and 10^{-4}. The probing ultrashort laser pulse thus reflects from the surface of which time-varying reflectivity depends on the strain induced by the passing ultrasonic wave.

As is the case with nonoptical sensors of ultrasonic displacements, laser detection may also be air coupled [178,179]. When an ultrasonic wave is reflected at the surface within the solid sample, a small fraction of ultrasonic energy is transmitted to the air above the surface as sound waves. Pressure variations accompanying the propagating sound wave change the index of refraction of the air. The probing laser beam is passed parallel with the surface through the region affected by the emitted sound. When the laser beam encounters transversal gradients of the index of refraction, or equivalently, gradients of air density, it is slightly deflected and displaced to a new course. This change is then detected by a position-sensitive photodetector such as a quadrant photodiode and converted to an electrical signal. This enables detection of ultrasound on rough and light-absorbing surfaces for frequencies up to 10 MHz [178].

Under ideal circumstances, all optical detection techniques, regardless whether they are interferometric or noninterferometric, have a comparable minimal detectable displacement. This displacement is limited by a photon shot noise. However, as explained in Section 29.1, the ultimate limit is determined by the thermal rattle (phonon shot noise, Equation 29.1), which is about two orders of magnitude smaller than the photon noise, because the phonon frequency is many orders of magnitude lower than the photon frequency. The limiting resolution of the detected displacement due to quantum laser amplitude noise (photon shot noise) is independent of the frequency. Photon shot noise dominates the detector noise if the laser power is about 1 mW. The minimal detectable displacement is thus given by

$$u_{\min} = \frac{\lambda}{4\pi}\sqrt{\frac{2hc\Delta f}{\lambda\eta P}} \tag{29.9}$$

where

- h is the Planck's constant
- c is the speed of light in a vacuum
- Δf is the frequency bandwidth
- η is the detector quantum efficiency
- P is the output power of the laser beam of wavelength λ [180–182]

Even though Equation 29.9 was developed for optical detection of ultrasound with a Michelson interferometer (MI), it holds for any optical detector to within an order of magnitude. As seen from Equation 29.9, the minimal detectable displacement can be improved by using more optical power. However, using too much power may have unwanted side effects, such as heating of the illuminated area or other more destructive effects: melting, ablation. For these reasons, optical sensors of ultrasound have about 100 times smaller signal-to-noise ratio (SNR), or 100 times larger minimal detectable displacement, than the ultimate achievable thermal rattle limit [96–98].

Novel techniques are nowadays being researched that can overcome the photon shot noise limit of optical sensors by passively amplifying the amplitude of ultrasound with mechanical resonators [183] and cheap optical transducers (CHOTs) [184,185].

The aforementioned transduction mechanisms can be used in various noninterferometric and interferometric schemes. The most important one will be henceforth shortly described.

29.2.4.1 Laser Beam Deflection

Laser beam deflection technique is a noninterferometric optical method that enables measuring ultrasonic surface tilts or angular deflections on relatively smooth and highly reflective opaque solids [186–193]. This technique has been reviewed in several papers [43,167–169]. It is also employed in atomic force microscopy to measure cantilever deflections [194].

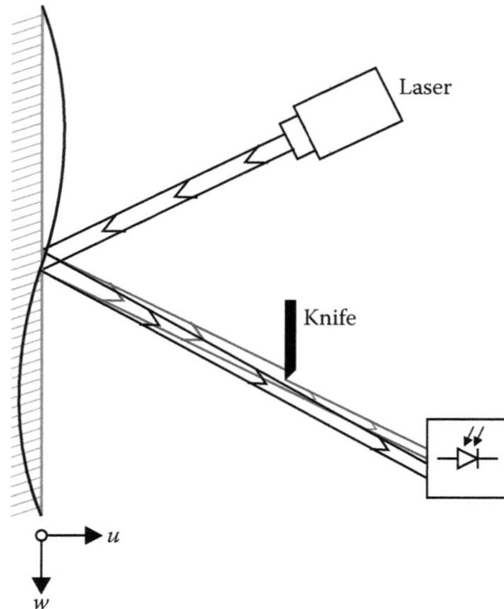

FIGURE 29.8 Scheme of a laser beam deflection probe in the plane of angular deflection.

The simplest realization depicted in Figure 29.8 uses a few milliwatt He–Ne laser source with a Gaussian laser profile. The beam is made to reflect from the surface of a polished solid by an angle. It is required that the diameter of the beam on the surface is less than the wavelength of the ultrasonic wave. The power of the reflected beam is collected by a photodiode placed behind an obstacle. Usually, a razor blade (knife edge) is used to obstruct half of the beam's cross section [186–189]. When the knife edge obstructs half the reflected beam, the sensitivity of the laser beam deflection probe is at its maximum. Any ultrasonic wave, especially SAW, that changes the slope of the surface under the illuminated area, causes the reflected beam to deflect from its path when the surface is still. The deflection is then recorded by the photosensitive element as variations of the light power passing beyond the knife edge.

Instead of using a sharp obstacle such as a knife edge, amplitude modulation can be achieved as well with a receiving bundle of optical fibers [190,193]. Additional improvement can be achieved with a balanced scheme which is also called differential detection. Such a scheme reduces the noise from intensity fluctuations of the laser beam and can be implemented with a half-split bundle of optical fibers [190,193] or with a quadrant photodiode [195]. Monitoring of the reflected beam with a quadrant photodiode enables distinguishing a deflection of the beam in two perpendicular directions with frequency bandwidth of 200 MHz [195]. This can be used to infer the direction of propagation of SAWs.

A time history of the change of surface slope can be converted to out-of-plane displacement [189]. The primary difficulty of this method is that flat, specularly reflecting surfaces are required. The method also assumes that the reflectivity of the surface does not change due to the ultrasonic strain. Since it is noninterferometric, because it is based on amplitude modulation of the transmitted beam, the laser beam does not need to be coherent.

As for all optical techniques, the minimal detectable displacement u_{min} is determined by the shot noise. Its practically achievable, frequency-independent value in laser deflection probes is around 10^{-14} m/√Hz, about an order of magnitude larger than the ideal shot noise limit. As an example, for a 1 MHz detection bandwidth and 10 mW He–Ne laser this value is around 1 pm. Practical minimal detectable displacement with 1 MHz bandwidth obtained with a laser beam deflection method is then in the order of 10 pm [166,186]. The highest detectable frequency of the SAW is limited by the focal spot size and by the frequency bandwidth of the photodiode. Laser beam deflection probes therefore sense angular deflections in the frequency range between DC and few 100 MHz.

29.2.4.2 Laser Interferometry

29.2.4.2.1 Homodyne Interferometry

The principles of homodyne interferometry will be described by the help of an MI, which is the most common type of two-beam homodyne laser interferometers.

MI is schematically illustrated in Figure 29.9. An amplitude-stabilized coherent laser beam of the interferometric laser is divided into two arms by a beam splitter (BS). The beam in the measuring arm is directed perpendicularly to the measuring surface where it is phase-modulated by the presence of the out-of-plane ultrasonic motion of the surface. The other one, in the reference arm, is reflected from the mirror. Due to coherence and wave-front superposition issues, both arms have to be of about the same length L. When the beams in both arms are reflected back toward the BS, they are recombined. Two combined beams emerge from the BS. One returns to the laser, and the other one travels toward the photodiode, which detects the power of the interfering light.

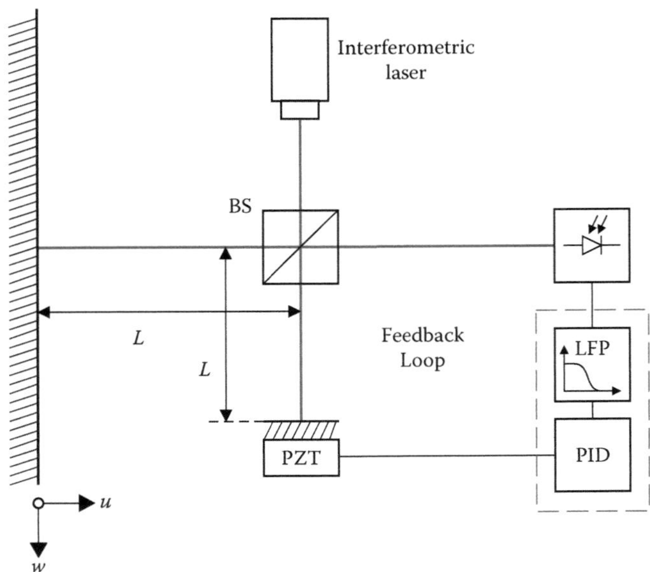

FIGURE 29.9 Arm-compensated Michelson interferometer.

Regarding the three-step process of laser interferometers that was described earlier, the first step occurs when the beam in the measuring arm is reflected from the perturbed surface, the second happens when the beams from both arms are recombined and interfere, and the third is executed by the photodiode and its accompanying amplification electronics.

The optical phase difference between the beams from separate arms, the phase p in short, is related to the out-of-plane surface displacement u as

$$p(t) = \frac{4\pi n u(t)}{\lambda} \tag{29.10}$$

Here, n is the index of refraction of the medium above the measuring surface and λ is the wavelength of the interferometric laser light in a vacuum. When measuring sub-nanometer ultrasonic surface motion of solid objects in air, n can be set to unity. The time-dependent output voltage signal $V(t)$ taken from the photodiode varies as a harmonic function if the measuring surface experiences a uniform motion along the path of the laser beam [43,173]. Assuming that the photodiode has a linear response, the interference signal on the photodiode has the following general form:

$$V(t) = V_{off} + V_{amp} \sin p(t) \tag{29.11}$$

Here, V_{off} stands for the DC offset and V_{amp} for the AC amplitude of the signal. Ideally, $V_{off} = V_{amp} = V_0/2$, but for several reasons, such as unequal beam powers and unequal wave-front curvatures in the returning beams from the reference and the measurement arm, the visibility is $V_{amp}/V_{off} < 1$. V_0 is the output photodiode voltage if the entire laser output was collected by the photodiode.

The sensitivity of MI

$$S = \frac{dV}{du} = \frac{4\pi V_{amp}}{\lambda} \cos p = S_{max} \cos p \tag{29.12}$$

changes with the optical phase. This is an undesired property of the MI. Its largest value S_{max} is reached when $\cos p = 1$ (see Equation 29.12). This corresponds to the steepest slope in the interference curve when $\sin p = 0$ (see Equation 29.11). This occurs midway between the maximum V_{max} and the minimum V_{min} of the detected signal. With a typical AC amplitude of about 1 V, and the wavelength of a He–Ne laser $\lambda = 632$ nm, the highest sensitivity is $S_{max} = 20$ mV/nm.

The phase is decoded from Equation 29.11 as follows:

$$p(t) = \arcsin \frac{V(t) - V_{off}}{V_{amp}}. \tag{29.13}$$

For small ultrasonic amplitudes $u \ll \lambda$ or $\delta V \ll V_{amp}$, Equation 29.13 can be linearized in the vicinity of its quadrature, that is, the point of maximum sensitivity:

$$u(t) = \frac{\lambda}{4\pi}\arcsin\frac{(\delta V(t) + V_{off}) - V_{off}}{V_{amp}} = \frac{\lambda}{4\pi V_{amp}}\delta V(t) = \frac{\delta V(t)}{S_{max}}. \tag{29.14}$$

It appears that $V_{amp} = (V_{max} - V_{min})/2$ and $V_{off} = (V_{max} + V_{min})/2$ have to be determined before the measurements of minute ultrasonic motion are made, because the displacement is not large enough that the extrema V_{max} and V_{min} of the interference are reached. The value of V_{amp} is needed for the purpose of absolute calibration (see Equation 29.14), while the knowledge of V_{off} is required to lock the interferometer to this point. Thus, a synthetic displacement surpassing one fringe must be accomplished prior to the measurements of ultrasound to acquire the normalization parameter V_{amp} and the midpoint V_{off}. To achieve such a displacement, the mirror in the reference arm is translated by a piezoelectric actuator (PZT) as shown in Figure 29.9. The same actuator can be used to lock the interferometer. Such locking is also referred to as arm compensation or stabilization of the interferometer.

Due to long-term mechanical vibrations and drift, the starting point of a MI may be anywhere on its interference curve. Thus, the MI has to be locked to the point of the highest sensitivity by a feedback loop that compensates for low-frequency ambient displacements. There are many ways to realize this compensation. An example is shown in Figure 29.9. Here, the low-frequency part (<1 kHz) of the signal is obtained by a low-pass filter (LPF). Then, it is fed to the proportional integral-derivative (PID) controller. The controller's output is used to drive the PZT. It should be noted that the low-frequency components of the measured displacement are not visible in the detected signal, but can, nevertheless, be reproduced by monitoring the feedback signal delivered to the PZT. Low-frequency compensation can also be performed with electro-optic compensation [43].

The frequency response of an arm-compensated MI is flat in between the lower and the upper cutoff. The lower bound is determined by the upper frequency of compensation loop and is usually in the order of few kHz. The upper limit, which can go as high as 1 GHz, is set primarily by the photodiode response characteristics and its subsequent amplification electronics.

Various multiple-pass homodyne interferometers were proposed in order to make the interferometer more sensitive to the out-of-plane motion [196–201]. A drawback of these realizations is that such interferometers cannot perform measurement on a single localized spot. Moreover, due to multiple reflections, the upper limit of frequency response is significantly reduced.

Simple optical adaptations of the previously presented MI can be made so that common-mode noise arising from the amplitude instability of the interferometric laser can be subtracted using two balanced photodiodes [43,202]. Similar interferometers can be made with optical fibers. Instead of using electro-mechanic or electro-optic low-frequency stabilization, it is possible to realize a homodyne laser interferometer in such a way that its sensitivity becomes independent of optical phase (see Equation 29.12). This type of interferometers are called homodyne quadrature laser interferometers, which demand at least two photodetectors and give two output signals that are in quadrature, that is, 90° out of phase [182,202–207].

The optical arrangement of homodyne interferometers can also be altered in such a way so that they are sensitive to the in-plane displacement w.

An example of a homodyne interferometer capable of measuring tangential motion is shown in Figure 29.10. The exiting light from the interferometric laser is linearly polarized. The polarization plane is set 45° with respect to the plane of the paper. The polarizing BS (PBS) splits the initial beam into two perpendicularly polarized beams of equal powers. The transmitted one is polarized in the plane of the paper, while the reflected one is polarized perpendicularly in the plane of the optical layout. Both beams pass through the focusing optics (FO) and are made to interrogate the same area on the surface of the measuring sample, each by the same angle, but from a different side. The scattered light is then collected by the collecting optics (CO). Usually, the reflection of the beams from the surface does not change their polarization states. Therefore, a polarizer with the direction of polarization inclined by 45° with respect to the polarization plane of either of the beams has to be placed in front of the photodiode in order to make the beams interfere. Two light beams of mutually perpendicular polarizations do not interfere. The interference signal is finally detected by the photodiode.

The normal component of the surface motion u is encoded in both beams, but is canceled out and does not change the interference pattern. This is analogous to the MI when both arms experience the same normal motion toward the BS [208]. On the contrary, the effect of the in-plane motion w on the phase is additive.

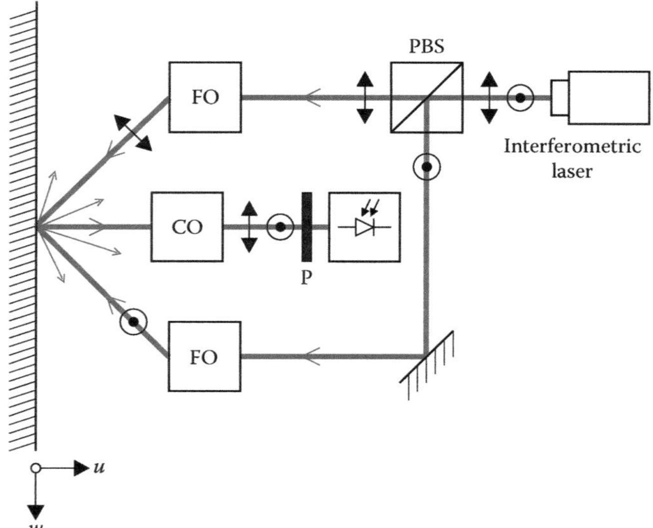

FIGURE 29.10 Polarizing homodyne interferometer sensitive to in-plane motion.

Apart from the much larger minimal detectable displacement limit of optical detectors in comparison with the contact ultrasonic sensors, there are two main drawbacks of the presented homodyne interferometers. Displacement measurements are only possible on mirror-polished surfaces and compensation systems have problems following large environmental vibrations.

29.2.4.2.2 Heterodyne Interferometry

In contrast to the homodyne interferometry which uses a single-frequency laser source, heterodyne interferometry uses a laser source with two close frequencies separated by Δf. This can be achieved with Zeeman lasers that give two output beams separated in frequency by few MHz [209,210], by two longitudinal mode He–Ne lasers which produce two mutually perpendicular linearly polarized output beams with mode spacing of several 100 MHz [211], or by an acousto-optical modulator (AOM), such as Bragg cell, which shifts the frequency of the laser beam by several 10 MHz, either the one in the measuring or the other in the reference arm [173,210].

Consider, for instance, a two-beam heterodyne ultrasonic displacement-measuring laser interferometer [67,212] schematically shown in Figure 29.11. Collinear output beams from the interferometric laser possess a frequency shift Δf and have mutually perpendicular linear polarization states. The vertically polarized beam is deflected by the PBS toward the mirror. When passing twice through the quarter-wave plate ($\lambda/4$), its polarization plane is rotated by 90° becoming a horizontally polarized beam. This beam is now transmitted through the PBS. The other beam is first transmitted by the PBS, then reflected from the surface perturbed by the ultrasonic motion, and finally deflected by the PBS. Its frequency is Doppler shifted by $\delta f(u)$, due to the normal motion u of the measuring surface. This statement is equivalent to the description that the optical phase of the beam in the measuring arm is modulated by the normal surface motion [176]. When the beams are recombined at the PBS, a single compound beam emerges from the PSB. In this case, no light is reentering the laser. Since the beams have perpendicular polarization prior to being collected by the photodiode, a polarizer (P) is inserted to make the beams interfere.

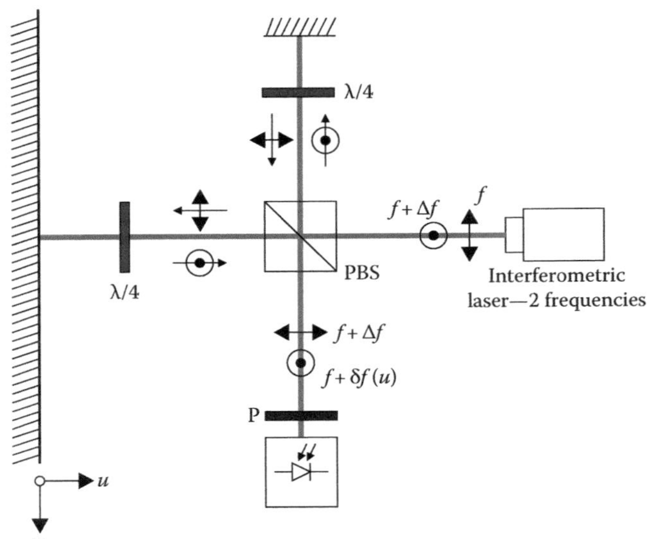

FIGURE 29.11 Heterodyne interferometer.

The output voltage signal taken from the photodiode is

$$V(t) = V_{\text{off}} + V_{\text{amp}} \sin\left(2\pi\Delta f t + p(t)\right). \tag{29.15}$$

The optical phase p and normal displacement u are again linearly dependent as given in Equation 29.10. Comparing Equation 29.15 with Equation 29.11, one notes that a term $2\pi\Delta ft$ is here added to the optical phase p [173]. Evidently, if there was no frequency shift between the beams, this interferometer becomes a classical MI with no light returning toward the laser. The “missing beam” is now absorbed by the polarizer.

Expanding Equation 29.15, removing its DC component by a high-pass filter that also removes environmental vibrations, and approximating for $u \ll \lambda$, one gets

$$V(t) = V_{\text{amp}}\left[\sin\left(2\pi\Delta f t\right) + \frac{4\pi}{\lambda}u(t)\cos\left(2\pi\Delta f t\right)\right]. \tag{29.16}$$

This signal can be decoded in two distinct ways: by a band-pass filter or by a phase-locked loop [173]. Using a band-pass filter, the interferometer becomes a velocity sensor. On the other hand, using the phase-locked loop, the interferometer measures out-of-plane displacements directly. In the latter case, the output voltage is

$$V(t) = \frac{2\pi V_{\text{amp}}}{\lambda}u(t). \tag{29.17}$$

Here, absolute calibration demands the value of V_{amp} to be known. As compared to the homodyne technique where a fringe-long synthetic displacement has to be performed for the sake of absolute calibration, in heterodyne technique the maximum and minimum values of the AC term in Equation 29.15 are always reached, because of the $2\pi\Delta ft$ term, even when there is no ultrasonic surface displacement.

The upper limit of the frequency bandwidth of heterodyne interferometer is set by the frequency shift Δf between the two laser beams. This permits ultrasonic surface motion with frequencies up to several 100 MHz to be detected.

The main advantage of heterodyne laser interferometer over their homodyne counterparts is that even when the amplitude of the AC changes due to variations of the signal-beam light power, the auto-gain-controlled amplifier keeps the value of V_{amp} constant. In homodyne interferometers, a change of the AC amplitude demands a repeatable precalibration. Heterodyne interferometers do not need any active stabilization. They are however slightly less sensitive, because some of the signal is discarded during signal processing, and less accurate due to polarization-mixing cross talk [213]. Heterodyne interferometers also necessitate a more sophisticated laser source producing two-frequency output or an external AOM. They can also detect in-plane displacements [67] and make advantage of the differential configuration to cancel the common-mode noise [43].

29.2.4.2.3 Confocal Fabry–Pérot Interferometry

Unlike homodyne and heterodyne laser interferometers which belong to the class of two-beam interferometer, a confocal Fabry–Pérot interferometer (CFPI) is a single-beam interferometer. CFPI is also a time-delay interferometer. First, the concept of time-delay or long-path-difference interferometry will be shortly introduced.

Two-beam interferometers operate reasonably well as long as the wave front of the beam is not significantly distorted during its reflection from the surface under inspection. The reflected beam from a rough surface forms a speckle pattern and when it is combined with a nearly planar waveform of the beam coming from the reference arm, the visibility of the interference is significantly reduced. In other words, the amplitude of modulation V_{amp} of a poor interference is much smaller than the DC offset V_{off} (see Equations 29.11 and 29.15). This can be remedied by interfering the laser-measuring beam with the distorted wave front with itself.

A long-path-difference interferometer is shown schematically in Figure 29.12. An interferometric laser source with the power of few 100 mW illuminates a rough surface that is perturbed by the ultrasound. The out-of-plane motion u is encoded in the optical phase. A portion of the reflected light with a distorted wave front enters the interferometer and is collimated before it is divided into equal parts by the first BS. The deflected beam travels a shorter path before it reaches the second BS, while the transmitted one takes a much longer journey. As seen from Figure 29.12, the total difference of the paths L between the two beams equals twice the distance between the BS and the mirror, because the distances between the BS and between the mirrors are subtracted. Then, the second BS recombines the direct and the delayed beam. Two beams emerge from it. They are in phase opposition, because of the energy conservation requirement. Each compound beam is then collected by the corresponding photodiodes PD_1 and PD_2. The output signal from the interferometer equals the difference between the signal outputs from each photodiode. The configuration is known as the balanced scheme.

When there are no surface ultrasonic displacements present, the output signal from the interferometer depends on the total difference in the paths L. As in the arm-compensated MI, this distance must be locked to the quadrature point midway between the maximum and the minimum of the signal so that the sensitivity of the interferometer is optimized. When the surface experiences a very rapid transient ultrasonic pulse but is for all other times still, the transient motion is detected by the long-path-difference interferometer two times. The time delay between the two detected pulses is $t_d = L/c$, where c is the speed of light in a vacuum. The measurement duration capability of such interferometers is practically limited by the distance L. Many practical deficiencies of the long-path-difference interferometers were eliminated by CFPIs.

CFPIs were initially developed by Monchalin and his group [68,214–217], but soon others [218–226] realized its practicability for measuring ultrasound on industrial, rough solid surfaces.

CFPI is based on the rotationally symmetrical Fabry–Pérot optical cavity (also called resonator or etalon) formed by two identical semitransparent concave spherical mirrors, facing each other with their concave parts, which are separated by a distance equal to their radius of curvature. As seen in Figure 29.13, the focal points of the mirrors coincide on the axis of symmetry exactly in the middle of the resonator.

Within the confocal Fabry–Pérot cavity, the entering beam, which is nearly parallel with the axis of symmetry and enters

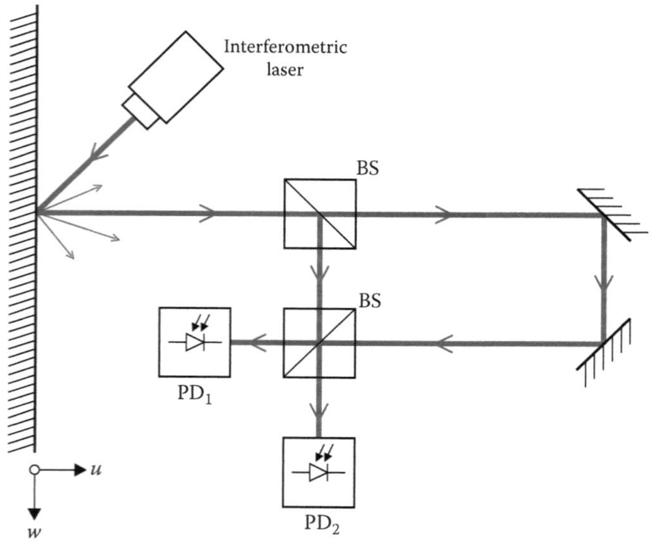

FIGURE 29.12 Long-path-difference interferometer.

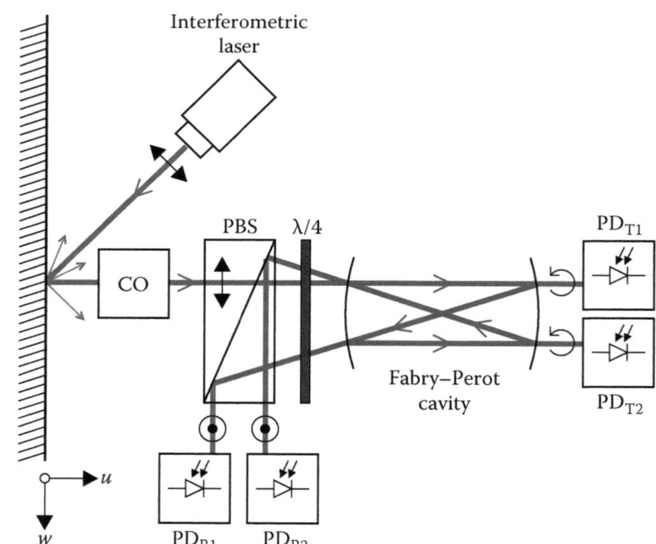

FIGURE 29.13 Confocal Fabry–Pérot laser interferometer.

the cavity off the axis, experiences multiple reflections. Its path is repeated after every four reflections. At each reflection, the beam partially exits the cavity. This occurs at four separate locations. Two are located on the transmission side (the right mirror in Figure 29.13) and two at the reflection side (the left mirror in Figure 29.13). Multiple beam interference takes place at each beam, exiting the cavity. CFPI works as an optical frequency filter which transmits a very narrow band of frequencies (band-pass filter) and reflects all the other frequencies (band-stop filter). The transmission frequency band pass is very sensitive to the length of the cavity L and can be selected by its tuning. Constructive interference occurs if the transmitted beams are in phase. This corresponds to a high-transmission peak of the etalon. The periodic peaks in the transmission are separated by the free spectral range given by $\mathrm{FSR}=c/(4L)$. The full-width half-maximum (FWHM) of the peak is the smallest, the largest the reflectivity of the mirrors. It is the half-maximum of the transmission peak where the CFPI is locked [219]. At this point, the slope of the transmission versus the frequency is the largest. Any change of the frequency due to the Doppler shift induced by the ultrasonic surface motion will result in the change of the transmitted as well as the reflected light. This is the simplest explanation of the principle of operation of the CFPI.

Consider again the scheme of one of the possible realizations of the CFPI depicted in Figure 29.13. A linearly polarized coherent laser beam illuminates the rough surface perturbed by ultrasonic motion. The reflected scattered light with a distorted wave front carries the information of the surface motion encoded in the phase modulation or equivalently in the frequency modulation. This scattered light is collected and collimated by a collimating optics (CO) making a parallel beam before it enters the cavity. In the depicted realization of the interferometer, a PBS is placed in front of the cavity so that it transmits the incoming beam. A quarter-wave retarder ($\lambda/4$) changes the polarization state of the beam from a linear to circular polarization. Two photodiodes PD_{T1} and PD_{T2} on the transmission side of the cavity monitor the transmitted beams. The other two beams exiting the cavity on its reflection side undergo another passage through the quarter-wave plate, now becoming linearly polarized but perpendicularly to the initial beam. The beams on the reflection side of the cavity are then deflected by the PBS toward the photodiodes PD_{R1} and PD_{R2}.

The frequency response of the CFPI is a complex function of the parameters describing the system [215,223]. In short, the response of the CFPI drops to zero at low frequencies. This makes the CFPI inherently insensitive to low-frequency environmental vibrations. Up to the frequency that corresponds to the FWHM of the transmission peak, CFPI operates as a velocity sensor. Above this value, CFPI has a nearly flat frequency response and works approximately as a displacement sensor, especially when the signals are obtained from the reflection side of the cavity [216]. At every multiple of FSR, the flat response experiences drops in its sensitivity and repeats its near-zero frequency behavior. Unlike homodyne and heterodyne interferometers, CFPIs have a much higher low-frequency cutoff, usually in the order of 1 MHz and are preferentially used to measure high-frequency ultrasound.

Many different optical arrangements of CFPIs can be realized. For instance, a CFPI with a totally reflective mirror in its "transmission side" [223] operates with an enhanced sensitivity. Differential schemes can also be implemented [217].

Other cavity types, such as the solid Fabry–Pérot planar etalon, can be used instead of the confocal one to form the optical filter used as a demodulator in a similar interferometer [227]. Such cavities can also be embedded within the optical fiber to form an all-fiber Fabry–Pérot sensor for ultrasound detection [228].

29.2.4.2.4 Feedback Interferometry

It is well known that light return (optical feedback) into the active medium in the laser resonator may give rise to severe amplitude and frequency fluctuations of the output beam. This unwanted effect is present in optical schemes, such as the one given in Figure 29.9, but can be well avoided with polarization schemes, for example, Figure 29.11. To prevent any optical feedback, optical Faraday isolators are placed in front of the laser.

The same undesirable effect is exploited in laser feedback interferometry (also called self-mixing or induced-modulation interferometry) to measure ultrasonic displacements [229–233]. A simplified drawing of a self-mixing feedback interferometer is shown in Figure 29.14. A fraction of the output beam from a laser diode is reflected from the measuring surface, back into the laser. The reflected light is phase-modulated by the ultrasonic out-of-plane motion. Once it reenters the cavity, it interferes with the light generated inside the laser. This generates changes in the optical and electrical properties of the laser. Optical changes, such as modulation of the amplitude and frequency, are monitored by a photodetector which measures interference of the light leaking through the backside of the cavity. This photodetector

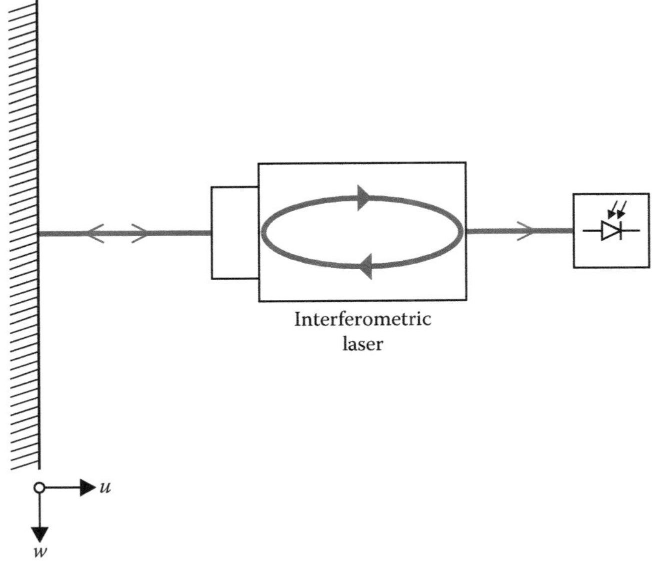

FIGURE 29.14 Simple scheme of a feedback interferometer.

is already integrated within the laser diode package as a monitor photodiode, but it is, for presentation purposes, depicted in Figure 29.14 outside the laser on its backside.

As it is clear from Figure 29.14, feedback interferometry does not need any external interferometer. The interferometer which demodulates phase-encoded ultrasonic displacement is the source laser itself. This reduces the number of components for ultrasound detection to merely a single laser diode source. The laser used in feedback interferometry is most commonly a single-mode Fabry–Pérot laser diode. This implies that only homodyne detection is possible [232]. However, the amplitude modulation term is not a harmonic function as in classical homodyne interferometry (see Equation 29.11), but a much more complicated one that depends on the so-called feedback parameter C. In general, there are four feedback regimes that determine the shape of the amplitude modulation function: the very weak, the weak, the moderate, and the strong feedback regime [230]. The feedback parameter C depends on many factors, but perhaps the most important ones when dealing with the detection of ultrasonic displacement are the relative power of the reflected beam $A = P_R/P_0$, where P_0 is the emitted power and the distance from the laser to the measuring surface s. It can be shown that C is proportional to $s/\sqrt{A}$ which implies that the shape of the modulation term depends on the standoff distance s and on the attenuation coefficient A [219]. In the moderated and the strong feedback regimes, the shape of the modulation function is asymmetric which allows for a clear discrimination of the direction of ultrasonic displacements.

For ultrasonic displacement-measuring purposes, the laser diode is operated in the moderate feedback regime. The corresponding modulation of the self-mixing signal is sawtooth-like, providing a linear dependence between the measured signal and the displacement u. The interferometer is then locked in the midst of the triangular ramp. By means of a feedback loop acting on the wavelength of the laser diode, the unwanted environmental vibration can be canceled out.

The most restrictive property of feedback interferometers is the fact that their minimal detectable displacement is about 100 worse than what could be obtained with conventional interferometry (see Equation 29.9) [230]. Experimentally, a sensitivity of 10^{-11} m/$\sqrt{\text{Hz}}$ has already been obtained [233]. This implies that sub-nanometer displacements can as well be measured with feedback interferometers when frequency bandwidth does not exceed 10 kHz. Even though laser-diode-based feedback interferometers are the most simple and the cheapest optical detectors of ultrasound, other optical methods surely perform better in terms of their minimal detectable displacement.

29.2.4.2.5 *Adaptive Interferometry*

Two-beam homodyne and heterodyne interferometers do not perform well on rough surfaces, because they cannot efficiently solve the difficulty related to wave-front distortions due to dynamic speckle changes. CFPIs remedy this problem, but their cavity needs accurate active length stabilization. They also have poor sensitivity for frequencies below 1 MHz. And finally, feedback interferometers are the least sensitive of the presented optical detectors or ultrasound.

The most recent optical solutions are adaptive interferometers exploiting photorefractive effects. They got this name because they are able to adapt the reference beam to the reflected probing beam having a speckle-distorted wave front. The low-frequency compensation for environmental vibrations and atmospheric turbulences is innately built in the adaptive interferometers and they can collect the reflected light over a large solid angle.

Two types of adaptive interferometer reached the level where they can be used not only in research laboratories but also for industrial applications. They are based either on *two-wave mixing* [71,108,234–239] or on the detection of *non-steady-state photo-EMF* [108,240–244]. Each interferometer will be briefly described. The third type of adaptive interferometers is based on phase conjugation [173]. Here, an optical element called phase conjugator corrects a distorted wave front into a planar one. The two-wave mixing and photo-EMF detector–based interferometers proved more robust than those exploiting phase conjugation.

A drawing of the self-diffraction, two-wave mixing adaptive interferometer is shown in Figure 29.15. A coherent source laser of up to few W of output power in continuous mode or several 100 W in the pulsed mode of operation illuminates the rough surface which is disturbed by ultrasonic waves. A small portion of the beam which exits the interferometric laser is used as a reference beam and has a smooth, nearly planar wave front. The probing beam is scattered by the surface, collected and collimated by the CO, and then directed toward the photorefractive semiconductor crystal which acts as an adaptive BS. The probing beam with a deteriorated wave front and the reference beam with a planar wave front intersect each other by an angle within the photorefractive crystal forming a dynamic hologram (a photorefractive grating). The reference beam exits the crystal collinearly with the probing beam having a changed wave front

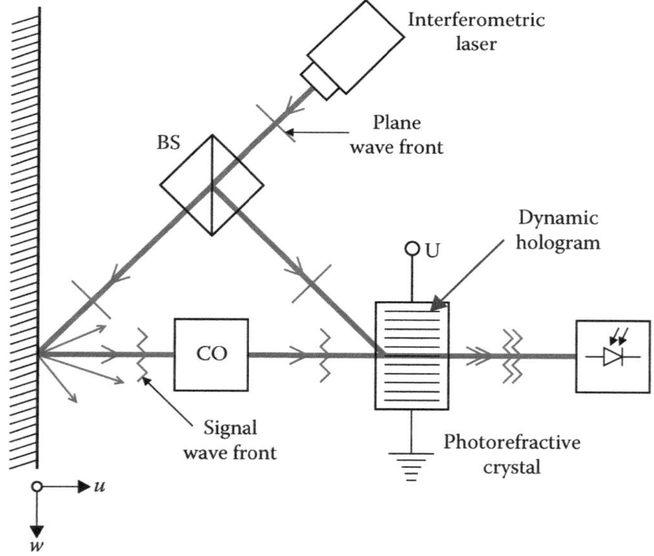

FIGURE 29.15 Two-wave mixing adaptive interferometer.

that matches the distorted wave front. Deflection and wave front adaptation occurs due to the interaction with the hologram. Finally, both beams interfere on the photodetector. The DC electric field applied to the crystal provides for the optimal sensitivity by setting the interference in quadrature.

Slow changes (up to few kHz) of the phase and wave front of the reflected probing beam also change the hologram in the crystal, while faster variations result in an intensity modulation that is detected by the photodiode. The response time of the photorefractive grating, the grating-writing time, is highly material dependent and linearly dependent on the intensity of the incident beams. It controls the wave-front distortion compensation bandwidth. On the other hand, the upper frequency limit of the detection is not limited by the crystal but solely by the response characteristics on the photodiode. This frequency cutoff can be as high as 1 GHz [71].

As far as optical arrangement is concerned, photo-EMF adaptive interferometers are no different than the two-wave mixing adaptive interferometers. Comparing Figures 29.15 and 29.16, one concludes that the two main parts of the two-wave mixing system, the photorefractive crystal and the photodiode, are here replaced by a single element—the photo-EMF detector. This is the main advantage on the photo-EMF approach. In this case, the photo-EMF element performs the dual function of the ultrasonic surface displacement detection as well as optical distortion compensation within the semiconductor crystal.

The interference of the probing beam and the reference beam in the photorefractive material causes a spatially modulated conductivity pattern. This produces a spatially periodic space charge field through the normal carrier migration and trapping process. Ultrasonic modulation of the phase of the probing beam causes a lateral vibration of the periodic free carrier grating, which induces an AC current that is proportional to the modulation amplitude and the total power. The grating has its own characteristic relaxation time that determines the low-frequency cutoff. No AC current is produced when the frequency of the phase-encoded ultrasonic motion is lower than the grating-relaxation cutoff frequency. In other words, the grating self-adapts to the low-frequency modulation of the optical phase.

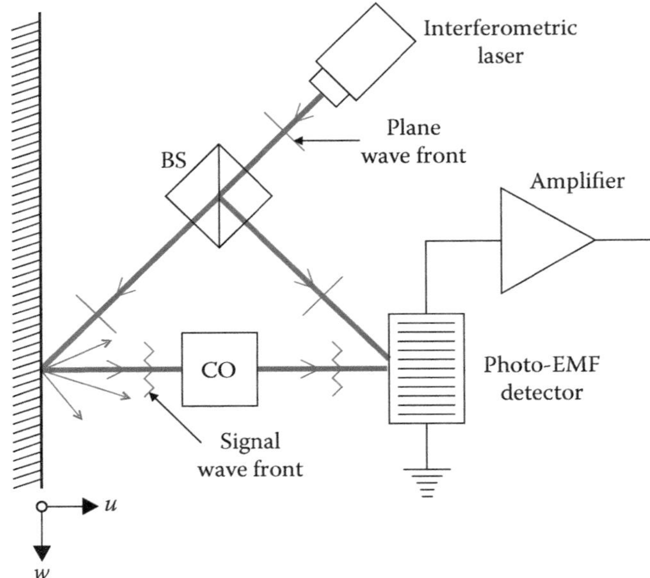

FIGURE 29.16 Photo-EMF adaptive interferometer.

In contrast to the two-mixing interferometers, the upper limit of the frequency response of the photo-EMF detectors is now determined by the material of the photo-EMF semiconductor and is given by the recombination rate, which is typically below 100 MHz. The high frequency limit is reduced, but the low-frequency compensation cutoff can be significantly increased up to 1 MHz.

29.3 Conclusions

This chapter presented an overview of the current detection principles and the most commonly used sensors for the measurements of sub-nanometer ultrasonic displacements of solid surfaces. Based on the underlying physics, detection methods were divided into four main groups: piezoelectric, electrostatic, electromagnetic, and optical. Within each group, the operation of several, most commonly encountered sensors was described.

An accompanying discussion about the pros and cons of each detector was given in a concise manner. Their comparison was provided. Although detailed comparative studies were omitted, an interested reader can find more information in the literature [29,96,115,193,245–248]. This chapter is also equipped with an extensive list of references that can serve as a springboard for those interested to know more about a certain topic covered.

The specially emphasized property of each detector was its minimal detectable displacement. A rough conclusion can be made that the most sensitive are contact piezoelectric sensors, followed by capacitive ones, then electromagnetic and, lastly, optical. As the thermal rattle sets the ultimate limit of the minimal detectable ultrasonic displacement, future researchers are given a clear goal to bring detectors' performance as close to it as it is experimentally possible.

Acknowledgment

We would like to thank Mr. Andraž Vene for his assistance with the preparation of figures.

References

1. Elert, G. Frequency range of human hearing, in *The Physics Factbook*, http://hypertextbook.com/facts/2003/ChrisDAmbrose.shtml
2. Hess, P. (2002). Surface acoustic waves in materials science. *Physics Today* 55(3): 42–47.
3. Rose, J. L. (2004). *Ultrasonic Waves in Solid Media*. Cambridge, U.K.: Cambridge University Press.

4. Pekeris, C. L. (1955). The seismic surface pulse. *Proceedings of the National Academy of Sciences of the United States of America* 41(7): 469–480.
5. Možina, J. and J. Diaci (2012). Optodynamics: Dynamic aspects of laser beam–surface interaction. *Physica Scripta* T149, 014077, 1–5.
6. Wang, J. J., Y. F. Shi et al. (2012). Analysis of laser-generated ultrasonic force source at specimen surface and display of bulk wave in transversely isotropic plate by numerical method. *Applied Surface Science* 258(6): 1919–1923.
7. Pekeris, C. L. (1955). The seismic buried pulse. *Proceedings of the National Academy of Sciences of the United States of America* 41(9): 629–639.
8. Smith, A. V. and B. T. Do (2008). Bulk and surface laser damage of silica by picosecond and nanosecond pulses at 1064 nm. *Applied Optics* 47(26): 4812–4832.
9. Pan, Y., M. Perton et al. (2006). The transient response of a transversely isotropic cylinder under a laser point source impact. *Ultrasonics* 44: E823–E827.
10. Chang, C. and W. Sachse (1985). Analysis of elastic wave signals from an extended source in a plate. *Journal of the Acoustical Society of America* 77(4): 1335–1341.
11. Dewhurst, R. J., D. A. Hutchins et al. (1982). Quantitative measurements of laser-generated acoustic waveforms. *Journal of Applied Physics* 53(6): 4064–4071.
12. Požar, T., P. Gregorčič et al. (2009). Optical measurements of the laser-induced ultrasonic waves on moving objects. *Optics Express* 17(25): 22906–22911.
13. Cosenza, C., S. Kenderian et al. (2007). Generation of narrowband antisymmetric lamb waves using a formed laser source in the ablative regime. *IEEE Transactions on Ultrasonics Ferroelectrics and Frequency Control* 54(1): 147–156.
14. Kozhushko, V. V. and H. Krenn (2012). Detection of laser-induced nanosecond ultrasonic pulses in metals using a pancake coil and a piezoelectric sensor. *IEEE Transactions on Ultrasonics Ferroelectrics and Frequency Control* 59(6): 1233–1238.
15. Schubert, M., M. Grossmann et al. (2012). Spatial-temporally resolved high-frequency surface acoustic waves on silicon investigated by femtosecond spectroscopy. *Applied Physics Letters* 101(1), 013108, 1–4.
16. Leschek, W. C. (1975). Acoustic-emission transducer calibrator. *Materials Evaluation* 33(2): 41–48.
17. Lee, W. H. K., P. Jennings et al. (2002). *International Handbook of Earthquake and Engineering Seismology.* New York: Academic Press.
18. Hardy, H. R. (2003). *Acoustic Emission, Microseismic Activity: Principles, Techniques, and Geotechnical Applications.* Balkema Publishers, Lisse, Abingdon, Exton, Tokyo.
19. Sachse, W., K. Yamaguchi et al. (1991). *Acoustic Emission: Current Practice and Future Directions*, ASTM, Charlotte, NC.
20. Miller, R. K., E. v. K. Hill et al. (2005). *Nondestructive Testing Handbook—Acoustic Emission Testing.* Columbus, OH: American Society for Nondestructive Testing.
21. Grosse, C. U. and M. Ohtsu (2008). *Acoustic Emission Testing.* Springer, Berlin Heidelberg.
22. Sikorski, W. (2012). *Acoustic Emission.* InTech, http://www.intechopen.com/books/acoustic-emission.
23. Buttle, D. J. and C. B. Scruby (1990). Characterization of particle impact by quantitative acoustic-emission. *Wear* 137(1): 63–90.
24. Boukria, Z., P. Perrotin et al. (2011). Experimental impact force location and identification using inverse problems: Application for a circular plate. *International Journal of Mechanics* 5(1): 48–55.
25. Požar, T. and J. Možina (2008). Optodynamic description of a linear momentum transfer from a laser induced ultrasonic wave to a rod. *Applied Physics A—Materials Science & Processing* 91(2): 315–318.
26. Moran, T. J. (1979). Electromagnetic ultrasonic transducers. *Nondestructive Evaluation of Materials.* J. J. Burke and V. Weiss, eds. New York: Plenum Press, pp. 283–298.
27. Sachse, W. and N. N. Hsu (1979). Ultrasonic transducers for materials testing and their characterization. *Physical Acoustics: Principles and Methods*, Vol. XIV. W. P. Mason and R. N. Thurston, eds. New York: Academic Press, pp. 277–406.
28. Hsu, N. N. and F. R. Breckenridge (1981). Characterization and calibration of acoustic-emission sensors. *Materials Evaluation* 39(1): 60–68.
29. Dewhurst, R. J., C. E. Edwards et al. (1987). Comparative-study of wide-band ultrasonic transducers. *Ultrasonics* 25(6): 315–321.
30. Michaels, J. E., T. E. Michaels et al. (1981). Applications of deconvolution to acoustic-emission signal analysis. *Materials Evaluation* 39(11): 1032–1036.
31. Naber, R. R. and H. Bahai (2007). Analytical and experimental validations of a numerical band-limited Green's function approach for modelling acoustic emission waves. *Advances in Engineering Software* 38(11–12): 876–885.
32. McLaskey, G. C. and S. D. Glaser (2010). Hertzian impact: Experimental study of the force pulse and resulting stress waves. *Journal of the Acoustical Society of America* 128(3): 1087–1096.
33. Feng, C. C. (1974). Acoustic emission transducer calibration-spark impulse calibration method. Eng. Report 74-7-C, Dunegam/Endevco.
34. Cooper, J. A., R. J. Dewhurst et al. (1984). High-voltage spark discharge source as an ultrasonic generator. *IEEE Proceedings A—Science Measurement and Technology* 131(4): 275–281.
35. Korolev, S. V., V. A. Krasilnikov et al. (1987). Mechanism of sound generation in a solid by a spark discharge near the surface. *Soviet Physics Acoustics—USSR* 33(4): 451–452.
36. Omkar, S. N. and K. U. Raghavendra (2008). Rule extraction for classification of acoustic emission signals using Ant Colony Optimisation. *Engineering Applications of Artificial Intelligence* 21(8): 1381–1388.
37. Dixon, S., C. Edwards et al. (1996). Generation of ultrasound by an expanding plasma. *Journal of Physics D—Applied Physics* 29(12): 3039–3044.

38. Dixon, S., C. Edwards et al. (2001). Ultrasonic generation using a plasma igniter. *Journal of Physics D—Applied Physics* 34(7): 1075–1082.
39. Hutchins, D. A., R. J. Dewhurst et al. (1981). Laser generation as a standard acoustic source in metals. *Applied Physics Letters* 38(9): 677–679.
40. Hutchins, D. A. (1988). Ultrasonic generation by pulsed lasers. *Physical Acoustics* 18: 21–123.
41. Aussel, J. D., A. Lebrun et al. (1988). Generating acoustic waves by laser: Theoretical and experimental study of the emission source. *Ultrasonics* 26(5): 245–255.
42. Scruby, C. B., R. J. Dewhurst et al. (1982). Laser generation of ultrasound in metals. *Research Techniques in Nondestructive Testing,* Vol. V. R. S. Sharpe, ed. New York: Academic Press, pp. 281–327.
43. Scruby, C. B. and L. E. Drain (1990). *Laser Ultrasonics: Techniques and Applications*. Adam Hilger, Bristol, U.K.
44. Murray, T. W. and J. W. Wagner (1999). Laser generation of acoustic waves in the ablative regime. *Journal of Applied Physics* 85(4): 2031–2040.
45. Hutchins, D. A., R. J. Dewhurst et al. (1981). Laser generated ultrasound at modified metal-surfaces. *Ultrasonics* 19(3): 103–108.
46. Berthe, L., R. Fabbro et al. (1997). Shock waves from a water-confined laser-generated plasma. *Journal of Applied Physics* 82(6): 2826–2832.
47. Breckenridge, F. R., C. E. Tschiegg et al. (1975). Acoustic emission: Some applications of Lamb's problem. *Journal of the Acoustical Society of America* 57(3): 626–631.
48. Kim, Y. H. and H. C. Kim (1993). Source function determination of glass-capillary breaks. *Journal of Physics D—Applied Physics* 26(2): 253–258.
49. Cho, S. I., J. K. Lee et al. (2008). PZT transducer response to an epicentral acoustic emission signal during glass capillary breakage. *Journal of the Korean Physical Society* 53(6): 3213–3219.
50. Hsu, N. N. (1977). Acoustic emission simulator. US Patent 4,018,084: 1–5.
51. Hsu, N. N. and S. C. Hardy (1978). Experiments in acoustic emission waveform analysis for characterization of AE sources, sensors and structures. *Elastic Waves and Non-Destructive Testing of Materials*. H.-Y. Pao, ed. San Francisco, CA: The American Society of Mechanical Engineers, 29: 85–106.
52. Hutchins, D. A., D. R. Billson et al. (2011). Structural health monitoring using polymer-based capacitive micromachined ultrasonic transducers (CMUTs). *Ultrasonics* 51(8): 870–877.
53. Pardee, W. J. and L. J. Graham (1978). Frequency analysis of two types of simulated acoustic emissions. *Journal of the Acoustical Society of America* 63(3): 793–799.
54. Scruby, C. B., R. J. Dewhurst et al. (1980). Quantitative studies of thermally generated elastic-waves in laser-irradiated metals. *Journal of Applied Physics* 51(12): 6210–6216.
55. Mcbride, S. L. and T. S. Hutchison (1976). Helium gas-jet spectral calibration of acoustic-emission transducers and systems. *Canadian Journal of Physics* 54(17): 1824–1830.
56. Hutchison, T. S. and S. L. McBride (1977). Excitation and spectral calibration of acoustic emission systems. US Patent 4,064,735: 1–12.
57. Mcbride, S. L. and T. S. Hutchison (1978). Absolute calibration of helium gas-jet noise source. *Canadian Journal of Physics* 56(5): 504–507.
58. White, R. M. (1963). Elastic wave generation by electron bombardment or electromagnetic wave absorption. *Journal of Applied Physics* 34(7): 2123–2124.
59. Brandis, E. and A. Rosencwaig (1980). Thermal-wave microscopy with electron-beams. *Applied Physics Letters* 37(1): 98–100.
60. Cargill, G. S. (1981). Electron-acoustic microscopy. *Physics Today* 34(10): 27–32.
61. Learned, J. G. (1979). Acoustic radiation by charged atomic particles in liquids—Analysis. *Physical Review D* 19(11): 3293–3307.
62. Rothenberg, M. S. (1979). Progress on deep-sea neutrino detection. *Physics Today* 32: 18–19.
63. Sachse, W. and K. Y. Kim (1983). Observation of x-ray generated ultrasound. *1983 Ultrasonics Symposium*, Atlanta, GA.
64. Nunes, O. A. C., A. M. M. Monteiro et al. (1979). Detection of ferromagnetic-resonance by photoacoustic effect. *Applied Physics Letters* 35(9): 656–658.
65. Nasoni, R. L., J. Evanoff, G. A. et al. (1984). Thermoacoustic emission by deeply penetrating microwave radiation. *IEEE 1984 Ultrasonics Symposium*.
66. Melcher, R. L. (1980). Thermoacoustic detection of electron-paramagnetic resonance. *Applied Physics Letters* 37(10): 895–897.
67. Monchalin, J. P., J. D. Aussel et al. (1989). Measurement of in-plane and out-of-plane ultrasonic displacements by optical heterodyne interferometry. *Journal of Nondestructive Evaluation* 8(2): 121–133.
68. Cand, A., J. P. Monchalin et al. (1994). Detection of in-plane and out-of-plane ultrasonic displacements by a two-channel confocal Fabry–Perot interferometer. *Applied Physics Letters* 64(4): 414–416.
69. Jian, X., S. Dixon et al. (2008). Electromagnetic acoustic transducers for in- and out-of plane ultrasonic wave detection. *Sensors and Actuators A—Physical* 148(1): 51–56.
70. Rosli, M. H., R. S. Edwards et al. (2012). In-plane and out-of-plane measurements of Rayleigh waves using EMATs for characterising surface cracks. *NDT&E International* 49: 1–9.
71. Bossa Nova Technologies. Tempo—laser ultrasound, http://www.bossanovatech.com/tempo.htm
72. McLaskey, G. C. and S. D. Glaser (2012). Acoustic emission sensor calibration for absolute source measurements. *Journal of Nondestructive Evaluation* 31(2): 157–168.
73. Proctor, T. M., F. R. Breckenridge et al. (1983). Transient waves in an elastic plate—Theory and experiment compared. *Journal of the Acoustical Society of America* 74(6): 1905–1907.

74. Dewhurst, R. J., C. Edwards et al. (1988). A remote laser system for ultrasonic velocity-measurement at high-temperatures. *Journal of Applied Physics* 63(4): 1225–1227.
75. Thompson, R. B. (1977). Noncontact transducers. *Ultrasonics Symposium.*
76. Hutchins, D. A. and J. D. Macphail (1985). A new design of capacitance transducer for ultrasonic displacement detection. *Journal of Physics E—Scientific Instruments* 18: 69–73.
77. Dutton, B., S. Boonsang et al. (2006). A new magnetic configuration for a small in-plane electromagnetic acoustic transducer applied to laser-ultrasound measurements: Modelling and validation. *Sensors and Actuators A—Physical* 125(2): 249–259.
78. Proctor, T. M. (1982). An improved piezoelectric acoustic-emission transducer. *Journal of the Acoustical Society of America* 71(5): 1163–1168.
79. Eitzen, D. G. and F. R. Breckenridge (2005). Acoustic emission transducers and their calibration (Chapter 2, Part 4). *Nondestructive Testing Handbook—Acoustic Emission Testing.* P. O. Moore, ed. Columbus, OH: American Society for Nondestructive Testing, 6: 51–60.
80. Michaels, J. E. (2008). Detection, localization and characterization of damage in plates with an in situ array of spatially distributed ultrasonic sensors. *Smart Materials & Structures* 17(3), 035035, 1–15.
81. McLaskey, G. C. and S. D. Glaser (2010). Mechanisms of sliding friction studied with an array of industrial conical piezoelectric sensors. *Proceedings of the SPIE, Sensors and Smart Structures Technologies for Civil, Mechanical, and Aerospace Systems 2010*. San Diego, CA: SPIE.
82. Kim, K. Y., L. Niu et al. (1989). Miniaturized capacitive transducer for detection of broad-band ultrasonic displacement signals. *Review of Scientific Instruments* 60(8): 2785–2788.
83. Oralkan, O., A. S. Ergun et al. (2002). Capacitive micromachined ultrasonic transducers: Next-generation arrays for acoustic imaging? *IEEE Transactions on Ultrasonics Ferroelectrics and Frequency Control* 49(11): 1596–1610.
84. Ozevin, D., D. W. Greve et al. (2006). Resonant capacitive MEMS acoustic emission transducers. *Smart Materials & Structures* 15(6): 1863–1871.
85. Hutchins, D. A., D. R. Billson et al. (2011). Structural health monitoring using polymer-based capacitive micromachined ultrasonic transducers (CMUTs). *Ultrasonics* 51(8): 870–877.
86. Harris, A. W., I. J. Oppenheim et al. (2011). MEMS-based high-frequency vibration sensors. *Smart Materials & Structures* 20(7), 075018, 1–9.
87. Lee, Y. C. and S. H. Kuo (2001). A new point-source/point-receiver acoustic transducer for surface wave measurement. *Sensors and Actuators A—Physical* 94(3): 129–135.
88. Lee, Y. C. and S. H. Kuo (2001). Miniature conical transducer realized by excimer laser micro-machining technique. *Sensors and Actuators A—Physical* 93(1): 57–62.
89. Lee, Y. C. and Z. Lin (2006). Miniature piezoelectric conical transducer: Fabrication, evaluation and application. *Ultrasonics* 44: E693–E697.
90. Muralt, P. (2008). Recent progress in materials issues for piezoelectric MEMS. *Journal of the American Ceramic Society* 91(5): 1385–1396.
91. Tadigadapa, S. and K. Mateti (2009). Piezoelectric MEMS sensors: State-of-the-art and perspectives. *Measurement Science & Technology* 20(9), 092001, 1–30.
92. Dorighi, J. F., S. Krishnaswamy et al. (1995). Stabilization of an embedded fiber optic Fabry–Perot sensor for ultrasound detection. *IEEE Transactions on Ultrasonics Ferroelectrics and Frequency Control* 42(5): 820–824.
93. Udd, E. (1995). An overview of fiberoptic sensors. *Review of Scientific Instruments* 66(8): 4015–4030.
94. Murray, T. W. and O. Balogun (2004). High-sensitivity laser-based acoustic microscopy using a modulated excitation source. *Applied Physics Letters* 85(14): 2974–2976.
95. Bramhavar, S., B. Pouet et al. (2009). Superheterodyne detection of laser generated acoustic waves. *Applied Physics Letters* 94(11): 3.
96. Fortunko, C. M. and E. S. Boltz (1996). Comparison of absolute sensitivity limits of various ultrasonic and vibration transducers. *Nondestructive Characterization of Materials* VII (Pts 1 and 2) 210–212: 471–478.
97. Boltz, E. S., C. M. Fortunko et al. (1995). Absolute sensitivity of air, light and direct-coupled wideband acoustic emission transducers. *Review of Progress in Quantitative Nondestructive Evaluation*. D. O. Thompson and D. E. Chimenti, eds. New York: Plenum Press, 14: 967–974.
98. Boltz, E. S. and C. M. Fortunko (1995). Absolute sensitivity limits of various ultrasonic transducers. *Ultrasonics Symposium, 1995. Proceedings.* IEEE.
99. Greenspan, M. (1987). The NBS conical transducer: Analysis. *Journal of the Acoustical Society of America* 81(1): 173–183.
100. Možina, J. and R. Hrovatin (1996). Detection of excimer laser induced sub-picometer ultrasonic displacement amplitudes. *Ultrasonics* 34(2–5): 131–133.
101. Možina, J. and R. Hrovatin (1996). Optodynamics—A synthesis of optoacoustics and laser processing. *Progress in Natural Science* 6: S709–S714.
102. Požar, T., R. Petkovšek et al. (2008). Dispersion of an optodynamic wave during its multiple transitions in a rod. *Applied Physics Letters* 92(23): 234101.
103. Možina, J. and J. Diaci (2011). Recent advances in optodynamics. *Applied Physics B—Lasers and Optics* 105(3): 557–563.
104. Grad, L., J. Diaci et al. (1993). Optoacoustic monitoring of laser beam focusing. *Lasers in Engineering* 1: 275–282.
105. Petkovšek, R., I. Panjan et al. (2006). Optodynamic study of multiple pulses micro drilling. *Ultrasonics* 44: E1191–E1194.
106. Možina, J. and J. Diaci (2010). On-line optodynamic monitoring of laser materials processing. *Advanced Knowledge Application in Practice*. I. Fuerstner, ed. InTech, pp. 37–60.

107. Hopko, S. N. and I. C. Ume (1999). Laser ultrasonics: Simultaneous generation by means of thermoelastic expansion and material ablation. *Journal of Nondestructive Evaluation* 18(3): 91–98.
108. Klein, M., B. Pouet et al. (2001). Semiconductor-based receivers aid industrial inspection and process control. *Intech* 48(10): 38–40.
109. Murray, T. W., S. Bramhavar et al. (2008). Theory and applications of frequency domain laser ultrasonics. *1st International Symposium on Laser Ultrasonics: Science, Technology and Applications.* Montreal, Canada.
110. Jain, N. (2011). Laser ultrasonics—The next big nondestructive inspection technology? Frost & Sullivan Market Insight.
111. Pierce, S. G., A. Cleary et al. (2011). Low peak-power laser ultrasonics. *Nondestructive Testing and Evaluation* 26(3–4): 281–301.
112. Hrovatin, R. and J. Možina (2000). Non-destructive and non-contact materials evaluation by means of the optodynamic method. *Insight* 42(12): 801–804.
113. Thompson, R. B. and D. O. Thompson (1985). Ultrasonics in nondestructive evaluation. *Proceedings of the IEEE* 73(12): 1716–1755.
114. Dalhoff, E., D. Turcanu et al. (2007). Distortion product otoacoustic emissions measured as vibration on the eardrum of human subjects. *Proceedings of the National Academy of Sciences of the United States of America* 104(5): 1546–1551.
115. Kline, R. A., R. E. Green et al. (1978). Comparison of optically and piezoelectrically sensed acoustic-emission signals. *Journal of the Acoustical Society of America* 64(6): 1633–1639.
116. Sachse, W. and A. Ceranoglu (1980). Absolute ultrasonic measurements with piezoelectric transducers. *IEEE Transactions on Sonics and Ultrasonics* 27(3): 153–153.
117. Glaser, S. D., G. G. Weiss et al. (1998). Body waves recorded inside an elastic half-space by an embedded, wideband velocity sensor. *Journal of the Acoustical Society of America* 104(3): 1404–1412.
118. Noma, H., E. Ushijima et al. (2006). Development of high-temperature acoustic emission sensor using aluminium nitride thin film. *Advanced Materials Research* 13–14: 111–116.
119. Cho, S. I., J. K. Lee et al. (2008). PZT transducer response to an epicentral acoustic emission signal during glass capillary breakage. *Journal of the Korean Physical Society* 53(6): 3213–3219.
120. Proctor, T. M. (1982). Some details on the NBS conical transducer. *Journal of Acoustical Emission* 1(3): 173–178.
121. Hutchins, D. A., S. B. Palmer et al. (1987). Thick conical piezoelectric transducers for NDE. *IEEE 1987 Ultrasonics Symposium.*
122. Chang, C. and C. T. Sun (1988). A new sensor for quantitative acoustic emission measurement. *Journal of Acoustical Emission* 7(1): 21–29.
123. Hutchins, D. A., L. F. Bresse et al. (1989). Measurements with a thick conical piezoelectric transducer. *Journal of the Acoustical Society of America* 85(6): 2417–2422.
124. Yan, T., P. Theobald et al. (2002). A self-calibrating piezoelectric transducer with integral sensor for in situ energy calibration of acoustic emission. *NDT&E International* 35(7): 459–464.
125. Yan, T., P. Theobald et al. (2004). A conical piezoelectric transducer with integral sensor as a self-calibrating acoustic emission energy source. *Ultrasonics* 42(1–9): 431–438.
126. Cervena, O. and P. Hora (2008). Analysis of the conical piezoelectric acoustic emission transducer. *Applied and Computational Mechanics* 2(1): 13–24.
127. Fick, S. E. and T. M. Proctor (2011). Long-term stability of the NIST conical reference transducer. *Journal of Research of the National Institute of Standards and Technology* 116(6): 821–826.
128. Fortunko, C. M., M. A. Hamstad et al. (1992). High-fidelity acoustic-emission sensor/preamplifier subsystems: Modeling and experiments. *Proceedings of the IEEE Ultrasonics Symposium,* 1992.
129. Boltz, E. S. and C. M. Fortunko (1996). Determination of the absolute sensitivity limit of a piezoelectric displacement transducer. *Review of Progress in Quantitative Nondestructive Evaluation.* D. O. Thompson and D. E. Chimenti, eds. New York: Plenum Press, 15: 939–945.
130. Belgacem, B., F. Calame et al. (2007). Piezoelectric micromachined ultrasonic transducers with thick PZT sol gel films. *Journal of Electroceramics* 19(4): 369–373.
131. Chen, Y., X. P. Jiang et al. (2010). High-frequency ultrasonic transducer fabricated with lead-free piezoelectric single crystal. *IEEE Transactions on Ultrasonics Ferroelectrics and Frequency Control* 57(11): 2601–2604.
132. Zipparo, M. J., K. K. Shung et al. (1997). Piezoceramics for high-frequency (20 to 100 MHz) single-element imaging transducers. *IEEE Transactions on Ultrasonics Ferroelectrics and Frequency Control* 44(5): 1038–1048.
133. Proctor, T. M. (1988). A high fidelity piezoelectric tangential displacement transducer for acoustic emission. *Journal of Acoustical Emission* 7(1): 41–47.
134. Wu, T. T. and J. S. Fang (1997). A new method for measuring in situ concrete elastic constants using horizontally polarized conical transducers. *Journal of the Acoustical Society of America* 101(1): 330–336.
135. Thapa, I., D. Burhan et al. (2005). In situ monitoring of solid-liquid interface of aluminum alloy using high-temperature ultrasonic sensor. *Japanese Journal of Applied Physics Part 1—Regular Papers Brief Communications & Review Papers* 44(6B): 4370–4373.
136. Manthey, W., N. Kroemer et al. (1992). Ultrasonic transducers and transducer arrays for applications in air. *Measurement Science & Technology* 3(3): 249–261.
137. Hutchins, D. A., W. M. D. Wright et al. (1994). Air-coupled piezoelectric detection of laser-generated ultrasound. *IEEE Transactions on Ultrasonics Ferroelectrics and Frequency Control* 41(6): 796–805.

138. Wright, W. M. D. (1996). *Air-Coupled Ultrasonic Testing of Materials*. Department of Engineering, University of Warwick, Coventry, UK.
139. Wright, W. M. D., D. A. Hutchins et al. (1996). Ultrasonic imaging using laser generation and piezoelectric air-coupled detection. *Ultrasonics* 34(2–5): 405–409.
140. Schindel, D. W., D. A. Hutchins et al. (1996). Capacitive and piezoelectric air-coupled transducers for resonant ultrasonic inspection. *Ultrasonics* 34(6): 621–627.
141. Scruby, C. B. and H. N. G. Wadley (1978). Calibrated capacitance transducer for detection of acoustic-emission. *Journal of Physics D—Applied Physics* 11(11): 1487–1494.
142. Breckenridge, F. R. and M. Greenspan (1981). Surface-wave displacement—Absolute measurements using a capacitive transducer. *Journal of the Acoustical Society of America* 69(4): 1177–1185.
143. Aindow, A. M., J. A. Cooper et al. (1987). A spherical capacitance transducer for ultrasonic displacement measurements in Nde. *Journal of Physics E—Scientific Instruments* 20(2): 204–209.
144. Wright, W. M. D., D. W. Schindel et al. (1994). Studies of laser-generated ultrasound using a micromachined silicon electrostatic transducer in air. *Journal of the Acoustical Society of America* 95(5): 2567–2575.
145. Cantrell, J. H. and M. A. Breazeale (1977). Elimination of transducer bond corrections in accurate ultrasonic-wave velocity-measurements by use of capacitive transducers. *Journal of the Acoustical Society of America* 61(2): 403–406.
146. Kim, K. Y. and W. Sachse (1986). Self-aligning capacitive transducer for the detection of broad-band ultrasonic displacement signals. *Review of Scientific Instruments* 57(2): 264–267.
147. Boler, F. M., H. A. Spetzler et al. (1984). Capacitance transducer with a point-like probe for receiving acoustic emissions. *Review of Scientific Instruments* 55(8): 1293–1297.
148. Tarnow, V. (1987). The lower limit of detectable sound pressures. *Journal of the Acoustical Society of America* 82(1): 379–381.
149. Ribichini, R., F. Cegla et al. (2011). Study and comparison of different EMAT configurations for SH wave inspection. *IEEE Transactions on Ultrasonics Ferroelectrics and Frequency Control* 58(12): 2571–2581.
150. Frost, H. M. (1979). Electromagnetic-ultrasound transducers: Principles, practice, and applications. *Physical Acoustics: Principles and Methods*, Vol. XIV. W. P. Mason and R. N. Thurston, eds. New York: Academic Press, pp. 179–275.
151. Thompson, R. B. (1990). Physical principles of measurements with EMAT transducers. *Physical Acoustics: Ultrasonic Measurement Methods*. R. N. Thurston and A. D. Pierce, eds. New York: Academic Press, pp. 157–200.
152. Gaerttner, M. R., W. D. Wallace et al. (1969). Experiments relating to the theory of magnetic direct generation of ultrasound in metals. *Physical Review* 184(3): 702–704.
153. Dobbs, E. R. (1973). Electromagnetic generation of ultrasonic waves. *Physical Acoustics: Principles and Methods*. W. P. Mason and R. N. Thurston, eds. New York: Academic Press, pp. 127–191.
154. Maxfield, B. W. and C. M. Fortunko (1983). The design and use of electromagnetic acoustic-wave transducers (EMATs). *Materials Evaluation* 41(12): 1399–1408.
155. Maxfield, B. W., A. Kuramoto et al. (1987). Evaluating EMAT designs for selected applications. *Materials Evaluation* 45(10): 1166–1183.
156. Alers, G. A. and L. R. Burns (1987). EMAT designs for special applications. *Materials Evaluation* 45(10): 1184–1189.
157. Ogi, H. (1997). Field dependence of coupling efficiency between electromagnetic field and ultrasonic bulk waves. *Journal of Applied Physics* 82(8): 3940–3949.
158. Hirao, M. and H. Ogi (2003). *EMATs for Science and Industry: Noncontacting Ultrasonic Measurements*. Springer.
159. Callen, H. B. and T. A. Welton (1951). Irreversibility and generalized noise. *Physical Review* 83(1): 34–40.
160. Whitman, R. L. and A. Korpel (1969). Probing of acoustic surface perturbations by coherent light. *Applied Optics* 8(8): 1567–1576.
161. White, R. M. (1970). Surface elastic waves. *Proceedings of the Institute of Electrical and Electronics Engineers* 58(8): 1238–1276.
162. Stegeman, G. I. (1976). Optical probing of surface-waves and surface-wave devices. *IEEE Transactions on Sonics and Ultrasonics* 23(1): 33–63.
163. Palmer, C. H. and R. E. Green (1979). Optical probing of acoustic emission waves. *Nondestructive Evaluation of Materials*. J. J. Burke and V. Weiss, eds. New York: Plenum Press, pp. 347–378.
164. Palmer, C. H. (1980). The measurement and generation of ultrasound by lasers. Ultrasonic materials characterization. *Proceedings of the First International Symposium on Ultrasonic Materials Characterization*. H. Berger and M. Linzer, eds. Gaithersburg, MD: U.S. Department of Commerce, National Bureau of Standards. NBS Special Publication, 596: 627–630.
165. Birnbaum, G. and G. S. White (1984). Laser techniques in NDE. *Research Techniques in Nondestructive Testing*. R. S. Sharpe, ed. New York: Academic Press, pp. 259–365.
166. Sontag, H. and A. C. Tam (1986). Optical-detection of nanosecond acoustic pulses. *IEEE Transactions on Ultrasonics Ferroelectrics and Frequency Control* 33(5): 500–506.
167. Monchalin, J. P. (1986). Optical-detection of ultrasound. *IEEE Transactions on Ultrasonics Ferroelectrics and Frequency Control* 33(5): 485–499.
168. Wagner, J. W. (1990). Optical detection of ultrasound. *Physical Acoustics: Ultrasonic Measurement Methods*. R. N. Thurston and A. D. Pierce, eds. New York: Academic Press, pp. 201–266.
169. Dewhurst, R. J. (1990). Optical sensing of ultrasound. *Nondestructive Testing and Evaluation* 5(2–3): 157–169.

170. Monchalin, J. P. (1993). Progress towards the application of laser-ultrasonic in industry. *Review of Progress in Quantitative Nondestructive Evaluation*. D. O. Thompson and D. E. Chimenti, eds. New York: Plenum Press, 12: 495–506.
171. Royer, D., M. H. Noroy et al. (1994). Optical-generation and detection of elastic-waves in solids. *Journal de Physique IV* 4: 673–684.
172. Dewhurst, R. J. (1994). Optical ultrasonic sensors for monitoring from industrial surfaces. *Proceedings SPIE*. San Diego, CA: SPIE.
173. Dewhurst, R. J. and Q. Shan (1999). Optical remote measurement of ultrasound. *Measurement Science & Technology* 10(11): R139–R168.
174. Sorazu, B., G. Thursby et al. (2003). Optical generation and detection of ultrasound. *Strain* 39: 111–114.
175. Monchalin, J. P. (2004). Laser-ultrasonics: From the laboratory to industry. *AIP Conference Proceedings*, Green Bay, WI, http://proceedings.aip.org/resource/2/apcpcs/700/1/3_1?isAuthorized=no.
176. Monchalin, J. P. (2007). Laser-ultrasonics: Principles and industrial applications. *Ultrasonic and Advanced Methods for Nondestructive Testing and Material Characterization*. C. H. Chen, ed. Singapore: World Scientific, 79–115.
177. Kamshilin, A. A., R. V. Romashko et al. (2009). Adaptive interferometry with photorefractive crystals. *Journal of Applied Physics* 105(3), 031101, 1–11.
178. Caron, J. N., Y. Q. Yang et al. (1998). Gas-coupled laser acoustic detection at ultrasonic and audio frequencies. *Review of Scientific Instruments* 69(8): 2912–2917.
179. Caron, J. N. (2008). Displacement and deflection of an optical beam by airborne ultrasound. *AIP Conference Proceedings*, Golden, CO, http://proceedings.aip.org/resource/2/apcpcs/975/1/247_1?isAuthorized=no
180. Kwaaitaal, T., B. J. Luymes et al. (1980). Noise limitations of Michelson laser interferometers. *Journal of Physics D—Applied Physics* 13(6): 1005–1015.
181. Wagner, J. W. and J. B. Spicer (1987). Theoretical noise-limited sensitivity of classical interferometry. *Journal of the Optical Society of America B—Optical Physics* 4(8): 1316–1326.
182. Požar, T., P. Gregorčič et al. (2011). A precise and wide-dynamic-range displacement-measuring homodyne quadrature laser interferometer. *Applied Physics B—Lasers and Optics* 105(3): 575–582.
183. Ashkenazi, S., C. Y. Chao et al. (2004). Ultrasound detection using polymer microring optical resonator. *Applied Physics Letters* 85(22): 5418–5420.
184. Stratoudaki, T., J. A. Hernandez et al. (2007). Cheap optical transducers (CHOTs) for narrowband ultrasonic applications. *Measurement Science & Technology* 18(3): 843–851.
185. Arca, A., T. Stratoudaki et al. (2011). Evanescent CHOTs for the optical generation and detection of ultrahigh frequency SAWs. *Journal of Physics: Conference Series* 269(1): 012012.
186. Adler, R., A. Korpel et al. (1968). An instrument for making surface waves visible. *IEEE Transactions on Sonics and Ultrasonics* 15(3): 157–160.
187. Korpel, A. and P. Desmares (1969). Rapid sampling of acoustic holograms by laser-scanning techniques. *The Journal of the Acoustical Society of America* 45(4): 881–884.
188. Noui, L. and R. J. Dewhurst (1990). Two quantitative optical-detection techniques for photoacoustic lamb waves. *Applied Physics Letters* 57(6): 551–553.
189. Noui, L. and R. J. Dewhurst (1993). A laser-beam deflection technique for the quantitative detection of ultrasonic lamb waves. *Ultrasonics* 31(6): 425–432.
190. Williams, B. A. and R. J. Dewhurst (1995). Differential fiberoptic sensing of laser-generated ultrasound. *Electronics Letters* 31(5): 391–392.
191. Dewhurst, R. J. and B. A. Williams (1996). A study of Lamb wave interaction with defects in sheet materials using a differential fibre-optic beam deflection technique. *Materials Science Forum (Nondestructive Characterization of Materials)* VII (Pts 1 and 2) 210–213: 597–604.
192. Dewhurst, R. J. and B. A. Williams (1997). Fibre optic system for the monitoring of asymmetric Lamb wave modulation in thin films. *Electronics Letters* 33(21): 1813–1815.
193. Murfin, A. S., R. A. J. Soden et al. (2000). Laser-ultrasound detection systems: A comparative study with Rayleigh waves. *Measurement Science & Technology* 11(8): 1208–1219.
194. Hoummady, M., E. Farnault et al. (1997). Simultaneous optical detection techniques, interferometry, and optical beam deflection for dynamic mode control of scanning force microscopy. *Journal of Vacuum Science & Technology B* 15(4): 1539–1542.
195. Petkovšek, R., P. Gregorčič et al. (2007). A beam-deflection probe as a method for optodynamic measurements of cavitation bubble oscillations. *Measurement Science & Technology* 18(9): 2972–2978.
196. Holloway, A. J. and D. C. Emmony (1988). Multiple-pass Michelson interferometry. *Journal of Physics E—Scientific Instruments* 21(4): 384–388.
197. Pisani, M. (2008). Multiple reflection Michelson interferometer with picometer resolution. *Optics Express* 16(26): 21558–21563.
198. Pisani, M. (2009). A homodyne Michelson interferometer with sub-picometer resolution. *Measurement Science & Technology* 20(8): 084008.
199. Ahn, J., J. A. Kim et al. (2009). High resolution interferometer with multiple-pass optical configuration. *Optics Express* 17(23): 21042–21049.
200. Lee, J., H. Yoon et al. (2011). High-resolution parallel multipass laser interferometer with an interference fringe spacing of 15 nm. *Optics Communications* 284(5): 1118–1122.
201. Wei, R. Y., X. M. Zhang et al. (2011). Designs of multipass optical configurations based on the use of a cube corner retroreflector in the interferometer. *Applied Optics* 50(12): 1673–1681.
202. Greco, V., G. Molesini et al. (1995). Accurate polarization interferometer. *Review of Scientific Instruments* 66(7): 3729–3734.

203. Mitra, B., A. Shelamoff et al. (1998). An optical fibre interferometer for remote detection of laser generated ultrasonics. *Measurement Science & Technology* 9(9): 1432–1436.
204. Vilkomerson, D. (1976). Measuring pulsed picometer-displacement vibrations by optical interferometry. *Applied Physics Letters* 29(3): 183–185.
205. Reibold, R. and W. Molkenstruck (1981). Laser interferometric measurement and computerized evaluation of ultrasonic displacements. *Acustica* 49(3): 205–211.
206. Gregorčič, P., T. Požar et al. (2009). Quadrature phase-shift error analysis using a homodyne laser interferometer. *Optics Express* 17(18): 16322–16331.
207. Požar, T. and J. Možina (2011). Enhanced ellipse fitting in a two-detector homodyne quadrature laser interferometer. *Measurement Science & Technology* 22(8), 085301, 1–8.
208. Požar, T. and J. Možina (2012). Dual-probe homodyne quadrature laser interferometer. *Applied Optics* 51(18): 4021–4027.
209. Bobroff, N. (1993). Recent advances in displacement measuring interferometry. *Measurement Science & Technology* 4(9): 907–926.
210. Zygo. Laser heads, www.zygo.com/?/met/markets/stageposition/zmi/laserheads/SL%2002_e_2013.pdf
211. SIOS Meßtechnik. Stabilized HeNe Laser SL 02-Series, www.sios.de/ENGLISCH/PRODUKTE/SL%2002_e_2013.pdf
212. Bartakova, Z. and R. Balek (2006). Use of the application of heterodyne laser interferometer in power ultrasonics. *Ultrasonics* 44: E1567–E1570.
213. Wu, C. M. and C. S. Su (1996). Nonlinearity in measurements of length by optical interferometry. *Measurement Science & Technology* 7(1): 62–68.
214. Monchalin, J. P. (1985). Optical-detection of ultrasound at a distance using a confocal Fabry–Perot-interferometer. *Applied Physics Letters* 47(1): 14–16.
215. Monchalin, J. P. and R. Heon (1986). Laser ultrasonic generation and optical-detection with a confocal Fabry–Perot-Interferometer. *Materials Evaluation* 44(10): 1231–1237.
216. Monchalin, J. P., R. Heon et al. (1989). Broad-band optical-detection of ultrasound by optical sideband stripping with a confocal Fabry-Perot. *Applied Physics Letters* 55(16): 1612–1614.
217. Blouin, A., C. Padioleau et al. (2007). Differential confocal Fabry–Perot for the optical detection of ultrasound. *Review of Quantitative Nondestructive Evaluation* 26, 193–200.
218. Hoyes, J. B., Q. Shan et al. (1991). A noncontact scanning system for laser ultrasonic defect imaging. *Measurement Science & Technology* 2(7): 628–634.
219. Shan, Q., S. M. Jawad et al. (1993). An automatic stabilization system for a confocal Fabry–Perot-interferometer used in the detection of laser-generated ultrasound. *Ultrasonics* 31(2): 105–115.
220. Dewhurst, R. J. and Q. Shan (1994). Modeling of confocal Fabry–Perot interferometers for the measurement of ultrasound. *Measurement Science & Technology* 5(6): 655–662.
221. Shan, Q., C. M. Chen et al. (1995). A conjugate optical confocal Fabry–Perot-interferometer for enhanced ultrasound detection. *Measurement Science & Technology* 6(7): 921–928.
222. Shan, Q., C. M. Chen et al. (1996). Detection of laser-generated ultrasound with a conjugate interferometer scheme. *Ultrasonics* 34(2–5): 173–175.
223. Shan, Q., A. S. Bradford et al. (1998). New field formulas for the Fabry–Perot interferometer and their application to ultrasound detection. *Measurement Science & Technology* 9(1): 24–37.
224. di Scalea, F. L. and R. E. Green (1999). High-sensitivity laser-based ultrasonic C-scan system for materials inspection. *Experimental Mechanics* 39(4): 329–334.
225. Park, S. K., S. H. Baik et al. (2011). Depth detection of a thin aluminum plate in laser ultrasonic testing using a confocal Fabry–Perot laser interferometer. *Journal of the Korean Physical Society* 59(5): 3262–3266.
226. Reitinger, B., J. Roither et al. (2011). Remote ultrasound detection with a quasi-balanced confocal Fabry–Perot interferometer. *Nondestructive Testing and Evaluation* 26 (3–4): 229–236.
227. Arrigoni, M., J. P. Monchalin et al. (2009). Laser Doppler interferometer based on a solid Fabry–Perot etalon for measurement of surface velocity in shock experiments. *Measurement Science & Technology* 20(1): 015302.
228. Dorighi, J. F., S. Krishnaswamy et al. (1995). Stabilization of an embedded fiber optic Fabry–Perot sensor for ultrasound detection. *IEEE Transactions on Ultrasonics Ferroelectrics and Frequency Control* 42(5): 820–824.
229. Donati, S., G. Giuliani et al. (1995). Laser-diode feedback interferometer for measurement of displacements without ambiguity. *IEEE Journal of Quantum Electronics* 31(1): 113–119.
230. Giuliani, G., M. Norgia et al. (2002). Laser diode self-mixing technique for sensing applications. *Journal of Optics A—Pure and Applied Optics* 4(6): S283–S294.
231. Giuliani, G., S. Bozzi-Pietra et al. (2003). Self-mixing laser diode vibrometer. *Measurement Science & Technology* 14(1): 24–32.
232. Giuliani, G. and S. Donati (2005). Laser interferometry. *Unlocking Dynamical Diversity: Optical Feedback Effects on Semiconductor Lasers*. D. M. Kane and K. A. Shore, eds. Wiley, Chichester, England, pp. 217–255.
233. Giuliani, G., M. Norgia et al. (2008). Self-mixing laser diode vibrometer for the measurement of differential displacements. *Proceedings of SPIE—Eighth International Conference on Vibration Measurements by Laser Techniques: Advances and Applications*. San Diego, CA: SPIE.

234. Ing, R. K. and J. P. Monchalin (1991). Broad-band optical-detection of ultrasound by two-wave mixing in a photorefractive crystal. *Applied Physics Letters* 59(25): 3233–3235.
235. Blouin, A. and J. P. Monchalin (1994). Detection of ultrasonic motion of a scattering surface by two-wave mixing in a photorefractive gaAs crystal. *Applied Physics Letters* 65(8): 932–934.
236. de Montmorillon, L. A., P. Delaye et al. (1997). Novel theoretical aspects on photorefractive ultrasonic detection and implementation of a sensor with an optimum sensitivity. *Journal of Applied Physics* 82(12): 5913–5922.
237. Stepanov, S., V. Petrov et al. (2001). Directional detection of laser-generated ultrasound with an adaptive two-wave mixing photorefractive configuration. *Optics Communications* 187(1–3): 249–255.
238. Wartelle, A., B. Pouet et al. (2011). Non-destructive testing using two-component/two-wave mixing interferometer. *AIP Conference Proceedings* 1335(1): 265–272.
239. Intelligent Optical Systems (IOS). Laser Ultrasonics, www.intopsys.com/products/laser/laserultrasound.html
240. Korneev, N. A. and S. I. Stepanov (1994). Nonsteady-state photo-EMF in thin photoconductive layers. *IEEE Journal of Quantum Electronics* 30(11): 2721–2725.
241. Klein, M., B. Pouet et al. (2000). Photo-EMF detector enables laser ultrasonic receiver. *Laser Focus World* 36(8): S25–S27.
242. Sokolov, I. A. (2000). Adaptive photodetectors: Novel approach for vibration measurements. *Measurement* 27(1): 13–19.
243. Murfin, A. S. and R. J. Dewhurst (2002). Estimation of wall thinning in mild steel using laser ultrasound Lamb waves and a non-steady-state photo-EMF detector. *Ultrasonics* 40(1–8): 777–781.
244. Sokolov, I. A. (2003). Adaptive photodetectors for vibration monitoring. *Nuclear Instruments & Methods in Physics Research Section A—Accelerators Spectrometers Detectors and Associated Equipment* 504(1–3): 196–198.
245. Palmer, C. H. and R. E. Green (1977). Materials evaluation by optical detection of acoustic-emission signals. *Materials Evaluation* 35(10): 107–112.
246. Scruby, C. B., H. N. G. Wadley et al. (1981). A laser-generated standard acoustic-emission source. *Materials Evaluation* 39(13): 1250–1254.
247. Hutchins, D., J. Hu et al. (1986). A comparison of laser and EMAT techniques for noncontact ultrasonics. *Materials Evaluation* 44(10): 1244–1253.
248. Thursby, G., C. McKee et al. (2010). A comparison of three optical systems for the detection of broadband ultrasound. *Smart Sensor Phenomena, Technology, Networks, and Systems 2010. Proceedings of the SPIE*, eds. K. J. Peters, W. Ecke, T. E. Matikas. Vol. 7648, article id. 76480T, pp. 1–10.

30

Picometer-Scale Optical Noninterferometric Displacement Measurements

Ezio Puppin
Politecnico di Milano

From a variety of different points of view, ranging from everyday life to fundamental science and economics, one of the most important units of measure is length. The international standard of length is the meter (m), one of the seven base units of the modern International System of Units (SI). The evolution of length metrology went hand in hand with the progress of science and technology. The definition of the standard length unit changed three times over less than one century, and in its present form, dating back to 1983, 1 m is defined as the distance traveled by a ray of electromagnetic (EM) energy through a vacuum in 1/299,792,458 ($3.33564095 \times 10^{-9}$) of a second. The presently available accuracy in defining meter is in the order of 10^{-10}, which roughly corresponds to the size of a single atom.

Along with a better definition of the standard unit, many achievements have been obtained as far as precision and accuracy in length measurements are concerned and the present frontier in this field is represented by the advent of the so-called nanotechnologies that motivated the search of new methods for the measurement of small displacements in the picometer range [1]. For instance, the state of the art in microelectronic device production is represented by field effect transistors with a channel length of a few tens of nanometers and this length scale imposes the necessity of precise and accurate methodologies in the picometer range. Also in the liquid crystal display manufacturing process, the need for precise and accurate methods for the measurement of small displacements has a relevant role. In other fields of basic research such as astronomy, this need is also present ad poses new challenges to length metrology such as, for instance, in the Space Interferometry Mission or the GAIA project.

Most of the methods presently available for dealing with picometer length measurements are based on interferometry due to the many advantages of this approach: the wide dynamic range, the possibility of noncontact measurements, and a clearly defined traceability route to the meter definition. Interferometric methods are universally used in a broad spectrum of different applications and recently one of their major drawbacks, cost, has been overcome by the appearance on the market of a compact, low-cost instrument for displacement measurement with an accuracy in the tens of picometers range. This instrument consists of a solid-state laser whose light is sent through a single-mode optical fiber to a sensing head. The photons coming out from this head are reflected by the object whose displacements are under measurement and collected by the same head. The interference pattern produced inside the fiber can be measured, and in this way it is possible to obtain a sensitivity of 25 pm over a dynamic range of 400 mm. These performances are even more impressive if the small size and the relatively low cost of the apparatus are considered. In order to further improve the resolution of interferometric methods, the possibility to use lower-wavelength x-rays in place of the visible light is presently considered [2].

Even though interferometry probably offers the most promising features in measuring small displacement, nevertheless it might be convenient to have other methods for special applications. An optical, non-interferometric method has been recently proposed [3]. A scheme of the principle is illustrated in Figure 30.1. In the upper part, a beam of light is shown, having a circular profile with a diameter D. The beam impinges on a photodetector such as, for instance, a photodiode, which generates a signal proportional to the incoming radiation intensity. The beam is partially intercepted by a blade and in this way only part of the light reaches the photodetector. The blade is mounted on a linear motion stage and can be displaced in a direction z perpendicular to the beam. The intensity V of the photodetector signal depends on the blade position and each displacement dz of the blade causes a corresponding variation dV of the signal. This technique is commonly used for laser beam characterization and is better known as "knife-edge method" [4].

The relevant orders of magnitude of the method can be understood with this simple argument. Let us define the experimental sensitivity S as the ratio between the minimum variation in the

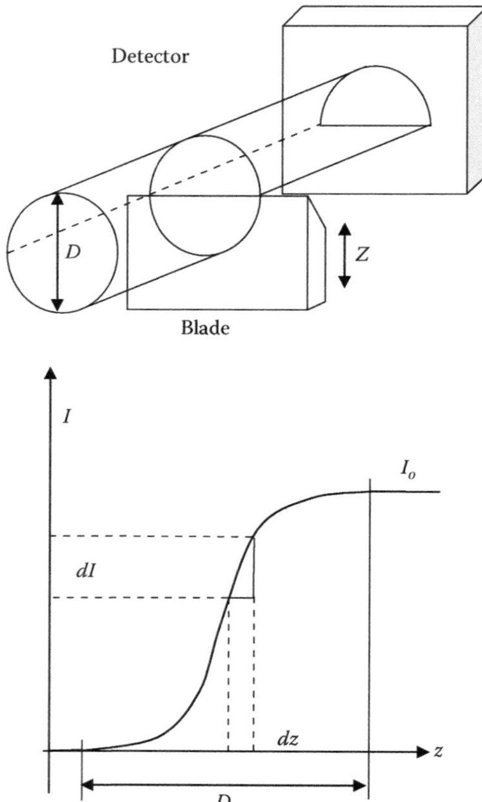

FIGURE 30.1 Upper part: the laser beam having size D is intercepted by a blade mounted on a linear stage acting along the z direction. The z displacement of the blade induces variations in the photodetector signal I. The I versus z plot is shown in the lower part.

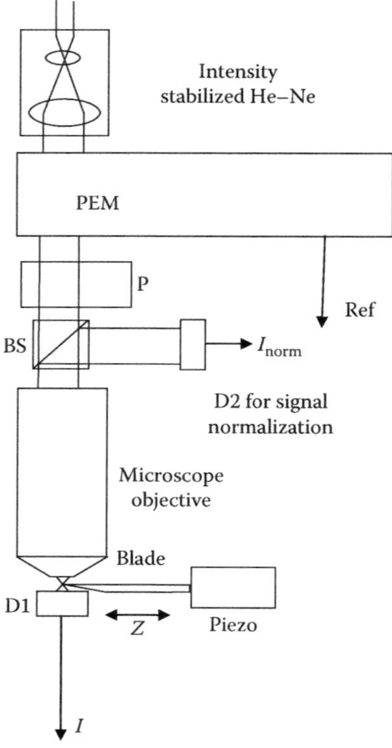

FIGURE 30.2 Basic layout of the experimental apparatus.

signal intensity (dV_{min}) appreciated by the instrument and V_{max}, the end of scale of the electronic chain: $S = dV_{min}/V_{max}$. In the following, we will assume that dV_{min} is equal to the RMS noise in the output signal observed over the measurement time. The corresponding sensitivity in the measurement of z can be roughly estimated by considering the beam size (its diameter D) and the sensitivity S previously defined. The minimum displacement detectable, in this way, is simply the product of S and D: $dz_{min} = S \times D$. In the experiment described here, the beam diameter is in the order of 3×10^{-6} m whereas the resolution of the acquisition system is better than 10^{-5}. With these values, the ultimate resolution attainable (dz_{min}) is therefore in the 10 pm range.

A more detailed description of the experimental apparatus is shown in Figure 30.2. The source is an intensity-stabilized He–Ne laser (Melles Griot 05STP903) with a continuous power output of 1.5 mW. The beam generated by this source has a Gaussian profile whose $1/e^2$ size is 0.5 mm. The size of the beam profile is first increased with a beam expender in order to obtain, in the final focusing stage, a smaller spot. After the expansion, the $1/e^2$ size is increased by a factor of 10, reaching therefore a value of 5 mm. In order to increase the signal-to-noise ratio, we modulate the beam and the modulation stage consists in a photoelastic modulator (PEM) followed by a Glan–Thomson polarizer (P). The modulation frequency is fixed and equal to 50 kHz. The combination of the PEM (which modulates the degree of circular polarization of the light) followed by a polarizer generates a square sinusoidal modulation in the light intensity at a frequency of 100 kHz (the intensity modulation frequency is doubled since the photodetector signal is proportional to $\sin^2(2\omega t)$). The modulated light is then separated into two portions with a beam splitter cube.

The straight beam is sent into a microscope objective (40×, NA = 0.85) that focuses the beam down to a size of ~3 μm (FWHM). The blade schematically shown in Figure 30.1 is placed in correspondence with the focus of the objective. The blade is first moved along the beam direction in order to place it in the focal plane. The linear stage used for this focusing operation is mechanical and is not shown in Figure 30.2 for clarity. In order to identify the focal plane, two different procedures can be used. The first is based on the fact that the spot size has been measured in several positions along the beam axis. In this way, it is possible to set the blade position along the beam by choosing the focal plane position previously identified. Another procedure, which has been verified to be consistent with the measured spot size, is to place the blade in the position corresponding to the larger fringes in the diffraction pattern produced by the blade.

In order to obtain reproducible results, an important factor to consider is the quality of the knife-edge blade that has been checked by observing the diffraction pattern produced by the focused laser beam with the blade placed in the center of the spot. By moving the blade in a direction perpendicular to the beam, the shape of the diffraction pattern remained always the same, that is, well defined and symmetrical, aside from negligible variations. This behavior demonstrates that the quality of the blade was excellent with a roughness well below the laser wavelength. The stage used for the horizontal motion of the blade (along the beam direction) was made with a combination of a mechanical stage coupled to a piezoelectric actuator for generating the small displacements. The piezoelectric motion is a low voltage ring actuator made by Piezomechanik (HPSt 150/20-15/12) with a maximum stroke of 16 μm with an applied voltage between −30 and 180 V.

The photodetector (D1) is a large size (1 cm) silicon photodiode placed at a very short distance from the focal plane in order to collect all the light of the strongly diverging beam. All the optical components have been mounted on an optical table suspended on air springs. In order to ensure the maximum stiffness to the assembly, all the components have been mounted as close as possible. The photocurrent generated in the photodetector is transformed in a voltage signal with a simple resistor (3.3 kΩ). This modulated signal is sent to a lock-in amplifier where it is demodulated at 100 kHz.

The first step in the operation of the experimental apparatus previously described has been the measurement of the calibration curve that relates the light intensity onto the photodetector and the blade position. By moving the blade with the piezoelectric actuator, we first measured the intensity of the beam onto the photodetector and the resulting curve is shown in Figure 30.3. The signal is expressed in μV and represents the output of the lock-in amplifier. The offset in the experimental data due to the dark current of the diode has been removed in order to have a value of zero when the blade completely intercepts the beam. The maximum value of the signal is reached when the blade is completely removed. The zero of the horizontal axis (the blade position) has been chosen corresponding to a value of 50% of the maximum intensity of the photosignal. In this position, the blade cuts half of the light intensity. By considering small

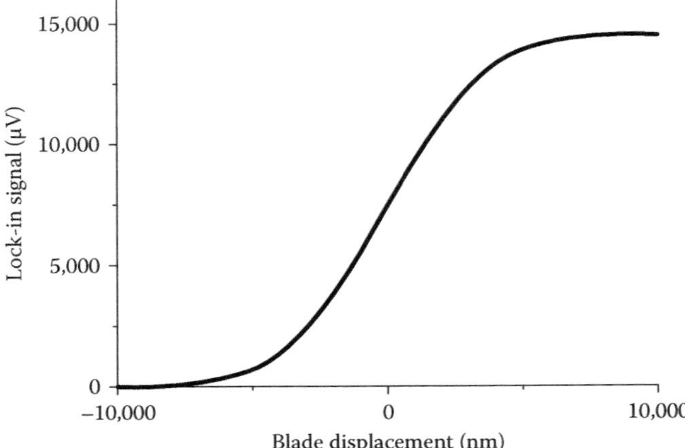

FIGURE 30.3 Actual measurement of the intensity I of the signal detected versus the z displacement of the blade.

displacements around the $z=0$ position, the V versus z curve becomes linear with a slope of 1.925 μV/nm.

The ultimate resolution of the apparatus depends, in principle, on several possible sources of noise and systematic error. Several tests have been conducted in order to identify the possible sources of noise. A few of them have been discarded since their contribution to the observed noise is negligible. Beam pointing stability, for instance, has a value of 30 μRad and therefore does not produce a significant displacement of the spot in the focal plane due to the compact assembly of the optical components. The piezoelectric stability is only determined by the noise of the driving voltage. In order to minimize this problem, the voltage applied to the piezo has been generated with a battery followed by a resistive voltage divider.

Without entering in the details of the multiple tests conducted on all the components of the system, the most serious source of inaccuracy came out to be the noise present in the output signal of the lock-in amplifier. A detailed investigation of the voltage fluctuations present in the output signal has been carried out. It came out that this noise is random with a spectral distribution proportional to f^{-2}.

In turn, in order to explain the origin of this noise in the output generated by the lock-in amplifier, all the components of the electronic chain were tested by reaching the conclusion that the only physical source of noise capable of explaining the voltage fluctuations in the output signal is the laser intensity fluctuations. These fluctuations have been extensively characterized and both their spectral distribution and absolute intensity indicate that they are indeed the origin of output voltage noise.

From a quantitative point of view, we observed that the signal-to-noise ratio in the output signal is equal to 10^5 if considered over a period of time of 1 s. By increasing the duration of the experiment up to a few hours, the same ratio decreases down to a value of 10^3. This is due to the fact that the noise has a $1/f^2$ spectrum and that low frequency drifts represent the most relevant issue to address. The presence of these fluctuations corresponds to an error in the displacement measurement of 15 pm over 1 s and 1.5 nm over several hours. The modulation of the laser intensity described earlier is obviously useless since we are talking about the noise present in the source itself and therefore it is necessary to take different actions. To this end, we performed two different kinds of measurements. The first, in the domain of time, is based on a real-time normalization of the laser intensity. The second technique is more complex and involves a second modulation stage. Both techniques allow a signal-to-noise value of 10^5, the one corresponding to a sampling interval of 1 s.

In order to directly correct the laser intensity fluctuations, we placed a beam splitter immediately before the position where the interesting physical effect, that is, the intensity variation due to the blade displacement, is taking place. By considering our experimental setup, the best position of the beam splitter is between the PEM and the microscope objective. The secondary beam generated by the beam splitter is sent to a second photodetector (D2) identical to the other (D1) and followed by the same resistor in series that converts the photocurrent in a voltage signal. A 6½ digit electrometer (Keithley model 2000) is then used for sampling the laser intensity signal. The sampled values are successively used, by the data acquisition program, for normalizing the output signal of the lock-in amplifier. The sampling frequency of the laser beam intensity was set at 1 Hz and also the time constant of the lock-in amplifier was set at 1 s. The average value of the beam intensity between each sampling of the blade position has been used for normalizing the output signal. By keeping the position of the blade fixed, we observed that the residual fluctuations in the output voltage noise were in the order of 10^{-5}. This blade position inaccuracy due to this fluctuation of the output voltage noise is around 15 pm.

In an alternative setup, in order to reduce the output noise, we modulated the blade position at a relatively low frequency, that is, 1 Hz. This procedure is, in a sense, equivalent to the other based on the normalization of the beam intensity. In fact, in both cases a narrow band-pass filter is used centered around the frequency of 1 Hz. This frequency is low but nevertheless much higher compared to the characteristic frequencies of the physical mechanisms responsible for the long-term drifts observed in the output signal. In order to describe these two techniques, however, it is more convenient to adopt different points of view. Time domain is the most convenient choice for the normalization technique whereas frequency domain is better suited for the blade modulation method.

The experimental layout of the electronic chain used for the dual modulation method is shown in Figure 30.4. The

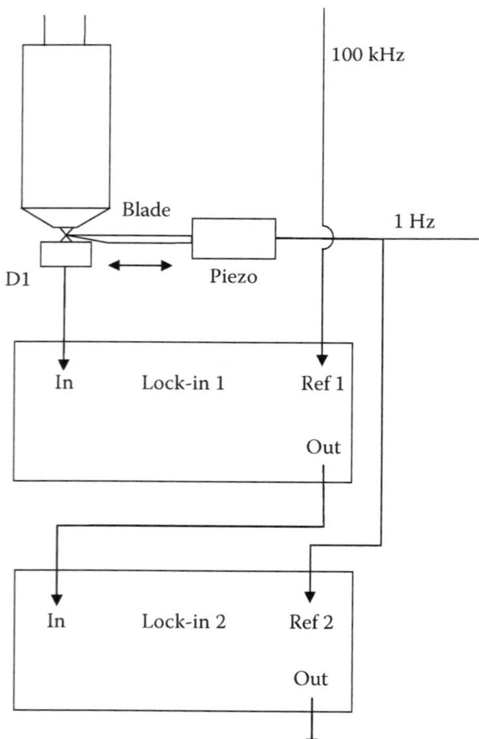

FIGURE 30.4 Experimental apparatus with a double modulation stage for increasing signal-to-noise ratio.

modulation of the blade position is performed by applying a sinusoidal wave at the piezoelectric actuator whose sensitivity is 76 nm/V. If the sinusoidal signal applied to the piezo is $V_0 \sin\omega t$, the maximum value of the induced displacement is equal to $(76 \times V_0)$ nm. This modulation of the light intensity is obviously observed in the output signal of lock-in 1. In order to follow the 1 Hz oscillation frequency, the time constant of lock-in 1 must be set to a value much lower than 1 s, used in the setup previously described with a single modulation stage. In the double modulation configuration, the time constant of lock-in 1 is equal to 1 mS.

The modulated signal generated at the output of lock-in 1 is then sent to lock-in 2 where it is demodulated at the modulation frequency of the blade position (1 Hz). The time constant of lock-in 2 must be higher with respect to the blade oscillation period, and the value we choose is 10 s.

By using this double modulation layout, we observed a strong reduction of the output noise. More precisely, in order to check the stability of the modulation stage, we monitored the output of lock-in 2 in static conditions (i.e., with the blade modulation kept at a constant value) and we noticed that noise fluctuations remained below 30 nV over a period of 10^4 s. This corresponds to the limit of 15 pm in the measurement of dz, the value imposed at this sampling frequency by the fluctuations in the laser intensity.

The double modulation layout heavily depends on the stability of the blade modulation stage whose quality can be degraded by two possible physical effects. Probably, the most serious of them is the possible presence of hysteresis in the response of the piezoelectric actuator to a variable applied voltage. In order to verify if this effect has a role in our experiment, we measured the response of the piezoelectric actuator by using the previous method based on signal normalization. In this way, we recorded the variation of the blade position, rigidly connected to the piezo, as a function of the voltage applied at the piezo itself. This test did not allow to detect any hysteresis within the experimental sensitivity.

Another effect that is necessary to take into account is the stability of the voltage used for modulating the piezo. In our experiment, we used a waveform generator followed by a passive voltage divider in order to generate a sinusoidal wave of 200 mVpp. In such conditions, the broadband noise measured with a digital voltmeter in the driving signal is in the order of 1 mV. This value would correspond to a displacement of 76 pm, three times larger compared to the resolution we are claiming. This figure is a factor of five times higher compared to the resolution of 15 pm claimed earlier. However, this apparent contradiction can be easily explained by the fact that the measurement error is related only to the spectral fraction of the noise centered around the modulation frequency (1 Hz) and within a band width determined by the lock-in time constant (10 s). By considering only this region of the frequency spectrum, the corresponding modulation noise is much smaller than 1 mV and therefore the corresponding error in the displacement measurement is much lower, consistent with the previous claim of 15 nm.

To conclude, the measurement of displacements in the pm region is relevant from the metrological and technological point of view. In this field, the most powerful and versatile approach is probably represented by devices based on interferometry. However, the possibility to use a different method could result more useful in particular situations. One such alternative possibility is represented by direct measurement of the intensity of a laser beam partially cut by a blade in a refined version of the well-known knife-edge method used for beam profiling.

In order to make of this idea a reliable measurement system capable to maintain the picometric resolution in different experimental situation, much work still has to be done. Among the relevant issues to be addressed, the long-term stability of the apparatus, which in our work has been tested only in a limited timescale, in the order of 1 h, is worth mentioning. Over a larger timescale, the role of other effects, such as thermal stability, might reduce resolution and therefore should be carefully investigated.

References

1. D. A. Swyt, *J. Res. Natl. Inst. Stand. Technol.* 106, 1 (2001).
2. G. N. Peggs and A. Yacoot, *Phil. Trans. R. Soc. Lond. A* 360, 953 (2002).
3. E. Puppin, *Rev. Sci. Instr.* 76, 105107-1 (2005).
4. A. H. Firester, M. E. Heller, and P. Sheng, *Appl. Opt.* 16, 1971 (1976); J. M. Khosrofian and B. A. Garetz, *Appl. Opt.* 21, 3406 (1983); M. Cywiak, M. Servin, and F. M. Santoyo, *Appl. Opt.* 40, 4947 (2001).

31

Direct Observation of X-Ray-Induced Atomic Motion

Akira Saito
Osaka University and RIKEN/SPring-8

31.1 Introduction

The interaction of x-ray photons with materials plays an essential role in various fields such as imaging, diffraction, and spectroscopy. As higher photon density of the x-rays by synchrotron radiation (SR) is required for higher throughput of measurements, stronger irradiation effects can be caused in either positive or negative meaning. The former example is a potential in material fabrication such as crystal growth or processing [1], because their high penetrating power enables us to deal with thick films or bulk materials. Processing with ultrahigh resolution can be expected using the interference or standing wave fields in crystals. The x-ray irradiation effect, which is most outstandingly different from conventional photo-induced processes by visible laser, is that photons can excite specific inner-shell electrons of atoms [2]. Thus, element-selective treatment is possible using the energy of x-rays tunable to an absorption coefficient dependent on the material. On the other hand, a negative example is the damage on a sample that limits the resolution of structural studies. Simulation analyses revealed that an extreme light source from an x-ray free electron laser (XFEL) with peak intensity of 3.8×10^6 photons/Å^2 at 12 keV can break a protein molecule in a timescale of 50 fs after exposure to an x-ray pulse with a FWHM of 2 fs [3].

Then, the irradiation effect of brilliant x-rays on materials has been investigated over the past few decades, and x-ray-induced chemical reactions or damages were observed in liquid, air, or low-vacuum conditions. However, direct observation of the atomic motion caused purely by hard x-rays in an ultrahigh vacuum (UHV) has not yet been reported at an atomic scale, although the STM observation of electron irradiation effect under UHV condition has been reported [4]. Using scanning tunneling microscope (STM) dedicated to in situ observation at an SR facility, we have recently directly observed the atomic motion caused by x-ray photons in UHV.

In this chapter, we will show the basic behavior of the x-ray-induced atomic motion with its dependence on beam parameters such as incident energy and photon density. We will discuss about principles of the phenomena and potential merits from the viewpoint of applications. Also we report a method, apparatus, and system to visualize a track of the atomic motion in STM images.

31.2 Experimental

To observe precisely x-ray-induced atomic motion, it is essential to compare the atomic arrangements on the surface within the same area before and after x-ray irradiation. Thus, the experiments require an in situ STM system with x-ray irradiation, because long-distance transfer of the sample between x-ray irradiation and STM observation should be avoided in order to compare the common area at an atomic scale. The in situ SR-based STM (SR-STM) system had already been installed for our past researches dedicated to chemical analysis with nanometer resolution assisted by element-selective core-hole excitation by the energy-tunable SR (Figure 31.1). An attempt to combine STM with x-rays appeared attractive because it contains various possibilities for original and important applications. Inner-shell excitation of a specific level under STM observation provides a possibility not only to analyze elements but also to control local reactions with the spatial resolution of STM. The details of the apparatus have been presented in our past reports [5–7].

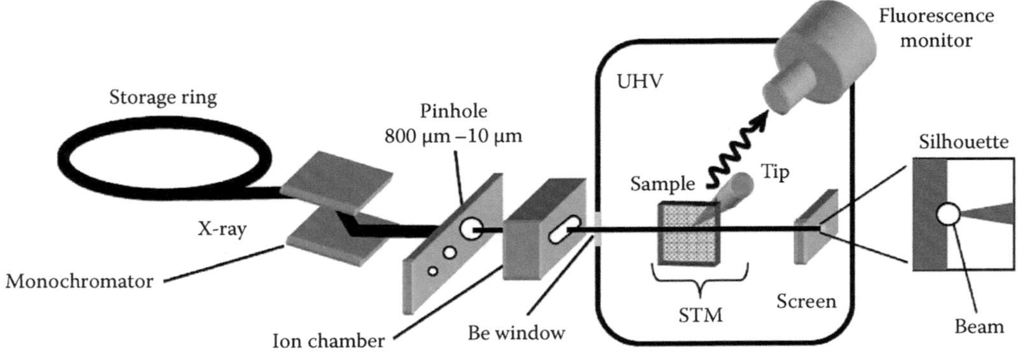

FIGURE 31.1 Schematic view of the SR-STM system.

The STM controller and scanner were based on a conventional STM system with a UHV chamber having a base pressure of 1.0×10^{-8} Pa (JEOL Co., Ltd, Tokyo, Japan). The STM tip was made from tungsten wire (0.3 mm in diameter) by electrochemical etching with NaOH solution. To overcome a small efficiency of core excitation by x-rays, we used SPring-8, which is an SR facility providing highly brilliant x-rays. Furthermore, we installed the STM system at the beamline BL19LXU [8] that can provide the highest brilliance of hard x-rays from a 27 m long undulator and focused the beam two-dimensionally to increase the photon density. To avoid excessive heat load by the brilliant x-ray irradiation, as small as possible (10 μm in diameter) incident x-rays were used under the condition of total reflection (i.e., grazing incident angle of the x-rays ~0.15°) that can effectively reduce the penetration depth of the x-rays into the sample up to ~2 nm. The x-ray beam was controlled with an accuracy of ~1 μm in UHV, which was enabled by a specially fabricated alignment system.

The Ge(111)c-(2 × 8) clean surface [9] was used from two viewpoints: first, it is a standard well-defined stable clean surface, and second, Ge atoms can be effectively excited by hard x-rays of the beamline.

In all experiments, during x-ray irradiation, the STM tip was out of the tunneling condition (~800 nm apart from the sample surface) and the sample bias (*Vs*) was kept at 0 V to avoid the atomic motion caused by STM manipulation that is produced by the electric field between the sample and the tip. Also, the STM observations were performed only under beam-off condition before and after x-ray irradiation. Thus, it was ensured that the atomic motions were not produced by the electric field around the STM tip under x-ray irradiation, but solely by x-ray irradiation.

31.3 Results and Discussion

31.3.1 Evidence of X-Ray-Induced Atomic Motion

Figure 31.2 shows the STM images in the same area on a Ge(111) c-(2 × 8) clean surface before (upper) and after (lower) x-ray irradiation for 3 min (11.119 keV > Ge K-absorption edge, photon density = 2×10^{15} photons/s/mm^2). Any difference in the surface structure is hardly observed in low-magnification images (Figure 31.2a and b), even by observing the atomic arrangement in images. This means that the rate of x-ray-induced atomic motion is so low that structural changes are hardly detectable by other surface analysis techniques such as diffraction giving averaged information, even under such high photon density. However, highly magnified STM atomic images revealed a clear change of the atomic structures (Figure 31.2c and d), where the open circles indicate the area in which the shift of Ge atoms is observed. The direct observation with atomic scale shows that the atomic motions occur mainly around defects on the surface, whereas the area composed of a closely packed 2 × 8 structure does not show any change in the atomic arrangement. The origin of the atomic motion is suggested to be an effect of surface diffusion, which is different from desorption caused by core excitation [2]. This is because the number of atoms does not change (i.e., does not decrease) after x-ray irradiation, as interpreted by comparison between Figure 31.2c and d.

31.3.2 Track Observation of Atomic Motion

After several observations as shown in Figure 31.2, we developed a technique to recognize atomic motions more clearly, which allows us to effectively comprehend their behavior. By merging an STM image of a surface area before x-ray irradiation (Figure 31.2a) into that after irradiation (Figure 31.2b), tracks of the atomic motion could be newly presented as several continuous lines (Figure 31.3), whereas other stable atoms are presented as spheres. By comparing Figure 31.3 with Figure 31.2a and b, an advantage of the merged image is shown clearly, because we can easily recognize new tracks, which were very difficult to find by comparing Figure 31.2a and b. Normally in Figure 31.3, the tracks are found as proof of atomic motion. The atomic tracks that appeared are considered as direct evidence and visualized information of the atomic diffusion at an atomic scale.

31.3.3 Characteristics of Atomic Motion

Using the visualized atomic track, we compared dependencies of the atomic motion with some physical parameters in order to know how to approach the behavior of the phenomena.

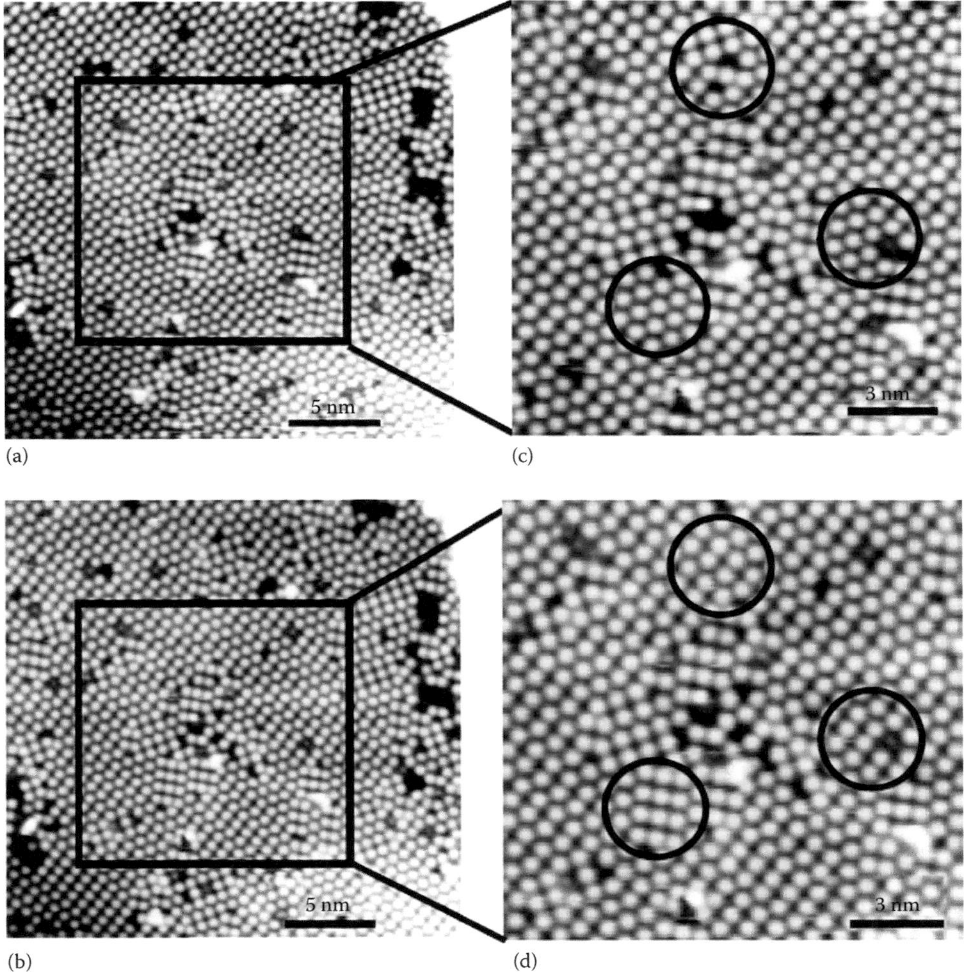

FIGURE 31.2 STM images with different scales within the same area on a Ge(111) clean surface before (a, c) and after (b, d) x-ray irradiation (3 min). Open circles indicate the area in which the Ge atomic shift is observed. $V_s = 2.0$ V, tunneling current $I_t = 0.3$ nA.

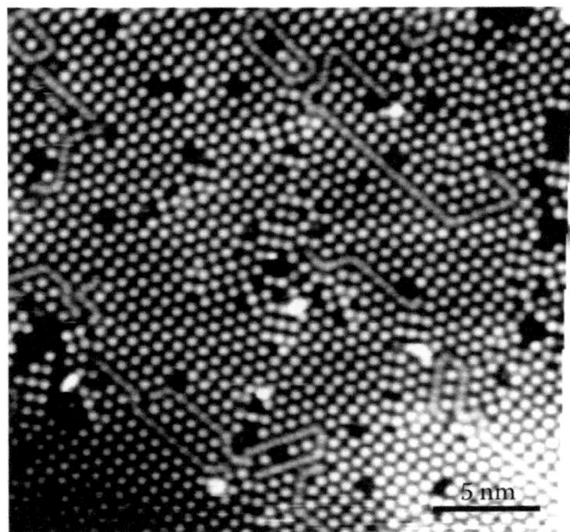

FIGURE 31.3 Merging the STM images before (Figure 31.2a) and after (Figure 31.2b) x-ray irradiation, the atomic motion tracks can be newly presented.

31.3.3.1 Confirmation: Atomic Diffusion without X-Rays at Room Temperature

The atomic motion on the Ge(111)c-(2×8) surface has long been investigated mainly from an interest in the phase transition c- $(2 \times 8) \rightarrow 1 \times 1$ around 300°C and in the origin of anisotropy in the structure [10]. Subsequently, studies have been developed around room temperature (RT) with interest in the behavior of vacancies contributing to make the specific surface structure [11–13]. Actually, we can observe the atomic motion with a minute rate in STM image at RT without x-rays on a surface containing defects or domain boundaries that give a low potential barrier of diffusion to the atoms. Here, it is worth noting that a possibility to drive atomic manipulation by an electric field around the STM tip during a scan [14] is negligibly small under the tunneling conditions used in our experiments.

Thus, for checking the premise of our discussion, we compared the atomic tracks under the x-ray off and on conditions.

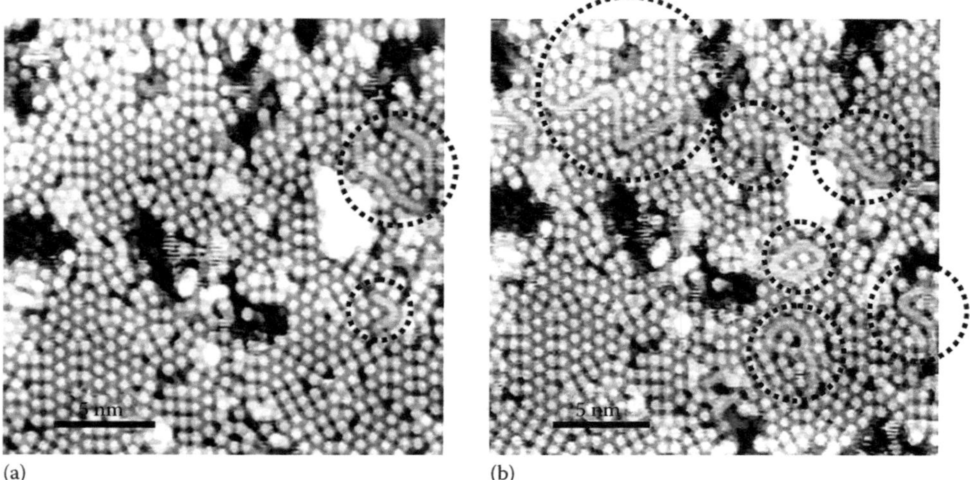

FIGURE 31.4 Comparison of the atomic tracks between the surfaces without (a) and with (b) x-ray irradiation (5 min, photon energy = 11.119 keV, photon density = 1.2×10^{15} photons/s/mm^2). The surfaces have the same defect density. Open circles indicate the area in which the shift of Ge atoms was observed. $V_s = 2.0$ V, $I_t = 0.2$ nA for both.

Figure 31.4 shows a comparison of the atomic tracks between common surfaces without (a) and with (b) the x-ray irradiation (5 min, 11.119 keV, photon density = 1.2×10^{15} photons/s/mm^2). AQ3 Figure 31.4a is a simple merged image of two STM images taken in the same area with an interval of 5 min. Open circles indicate the area in which the shift of Ge atoms is observed.

The atomic tracks in Figure 31.4a and b show a clear difference in their total length and number of circles. Although a general quantitative estimation on the increase of the track is still difficult because the atomic motion depends on the initial defect density on the surface, an increase of the atomic motion due to an irradiation effect was obviously confirmed using the surface having a common defect density. In this case, the ratio of increase in total track length is 19:64 (more than three times) for Figure 31.4a and b.

31.3.3.2 Dependence on the Incident Photon Energy

First, as an essential parameter, the dependence of the atomic motion on the incident photon energy was investigated. According to the obtained STM images, the surface atoms look to slightly change their position, but do not change the number of atoms on the surface. Thus, the atomic motion cannot be attributed to atomic desorption, but to diffusion. However, as an origin of the atomic diffusion, we can consider either the thermal effect or the core-excitation effect followed by the cascade Auger relaxation processes [2]. Thus, the track images for x-ray irradiation of the energies across the Ge K-absorption edge (11.103 keV) were compared. Figure 31.5 shows the track images on a common Ge(111) surface taken for the different incident photon energies ((a) 11.089 keV, (b) 11.119 keV, photon density = 2.4×10^{15} photons/s/mm^2, irradiation time = 5 min for both). Open circles

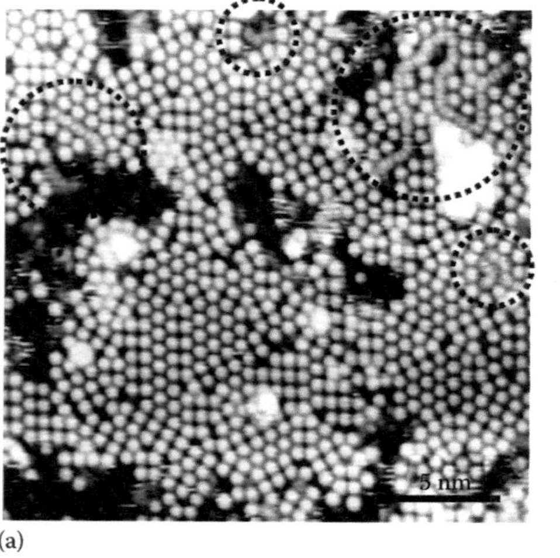

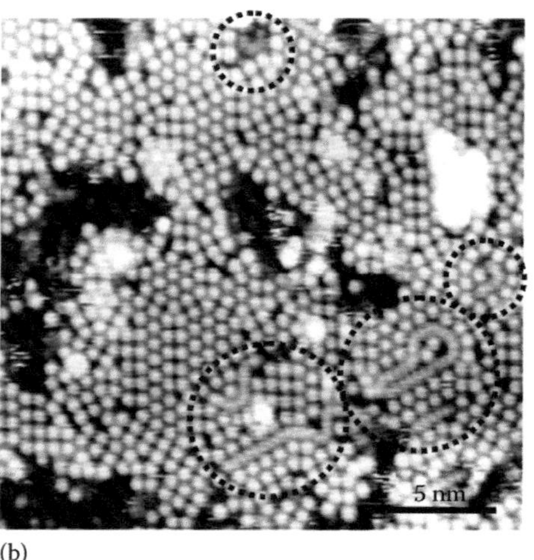

FIGURE 31.5 Dependence of the atomic tracks on the incident photon energy ((a) 11.089 keV and (b) 11.119 keV). Irradiation time = 5 min, photon density = 2.4×10^{15} photons/s/mm^2 for both). Open circles indicate the area in which the shift of Ge atoms was observed. $V_s = 2.0$ V, $I_t = 0.2$ nA for both.

indicate the area in which the shift of Ge atoms is observed. The result does not show a meaningful difference between them. As long as we repeated the measurements for the photon energy dependence, we did not obtain the results showing the different rates of atomic motion across the Ge K-absorption edge.

Thus, initially, the origin of the atomic motion was suggested to be a classical thermal effect (e.g., that concerns outer-shell excitations) [15], which is not the diffusion caused by the quantum effect such as repulsion of the highly ionized ions produced by the cascade Auger relaxation processes after core excitation. As a result, we assigned the main reason of the atomic motion to the thermal effect in our past report [15]. However, after thermal calculation, the result of the independence of the atomic motion on the incident photon energy was found to be possibly attributed to the lack of the incident brilliance of x-rays, as shown in Section 31.3.5. Therefore, further trials are necessary to check the dependence on the incident photon energy, which are not yet achieved due to our present experimental limitations.

31.3.3.3 Dependence on the Incident Photon Density

Next, to understand the origin of the atomic motion, we obtained the dependence of the atomic motion on the incident photon density. Figure 31.6 shows the track images on a common Ge(111) surface area for the different incident photon densities ((a) 1.2×10^{15}, (b) 4.8×10^{15} photons/s/mm^2, irradiation time = 5 min, photon energy = 11.119 keV > Ge K-absorption edge, for both). Open circles indicate the area in which the shift of Ge atoms is observed.

A set of Figure 31.6 shows a clear dependence on the incident photon density. Here, the ratio of increase in total track length is ~70:30 for Figure 31.6b and a, which does not absolutely correspond to the ratio of the photon density. The results of the repeated measurements were qualitatively reproducible, although a general quantitative estimation is difficult due to the dependence of the phenomena on the initial defect density. However, the clear dependence of the atomic motion on the photon density gives evidence of the irradiation effect. Furthermore, we found that a photon density of $\sim1\times10^{14}$ photons/s/mm^2 (lower by one order than that in Figure 31.6a) does not provide an increase in the atomic motion under any condition. It is then suggested that the brilliant x-rays from the long undulator [8] with two-dimensional focusing is marginal to observe such atomic motion under RT and UHV conditions. This might be a reason why there has scarcely been a report on such atomic motion by other SR experiments in UHV. By using a surface with a well-defined defect density, we will be able to obtain a precise functional form of the atomic motion on the photon density.

31.3.4 Estimation of the Event Rate

Further examinations with quantitative estimation were attempted to consider the origin of the atomic motion. The atomic motion rate was evaluated mainly from the viewpoints of core-hole excitation (including the effect of electronic transitions following the core-hole event) or the thermal effect.

For the core excitation, the event rate can be compared with the obtained atomic motion rate. The photon densities in our experiments were $1.2–4.8\times10^{15}$ photons/s/mm^2 (Figures 31.2 through 31.6), which were reduced by ~1/400 because of the grazing incidence angle that was used to suppress the penetration depth of x-rays into the sample, but increased by ~4 times due to the enhancement of the total reflection field near the surface. Multiplying the photon density at the sample surface and quantum efficiency of the core excitation at the Ge K-edge ($\sim6\times10^{-5}$), the core-excitation event rate is estimated to be $1–4\times10^{-4}$ event/s/atom. Since the number of atoms shown in each of Figures 31.4 through 31.6 is ~2000, the total event rate is to be 0.2–0.8 event/s for each image area. Thus, the total number of core-hole events by irradiation for 5 min (in each

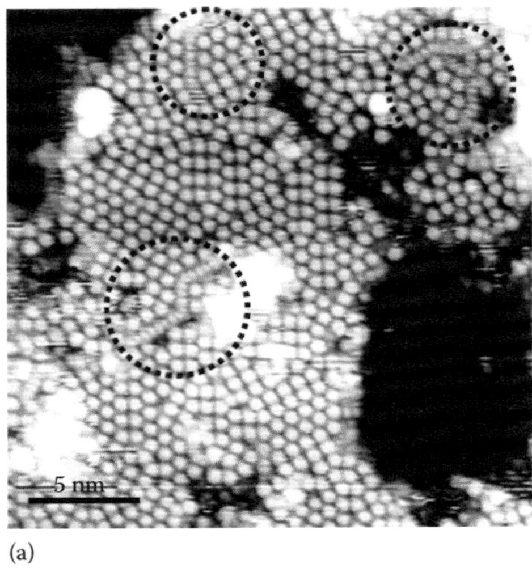

(a)

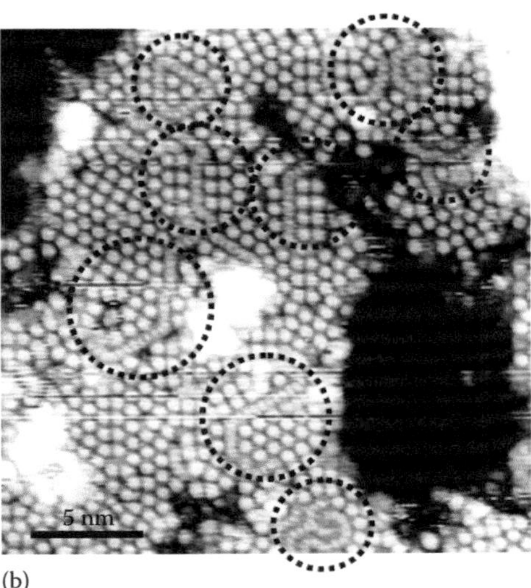

(b)

FIGURE 31.6 Dependence of the atomic tracks on the incident photon density ((a) 1.2×10^{15} photons/s/mm^2 and (b) 4.8×10^{15} photons/s/mm^2). Irradiation time = 5 min, photon energy = 11.119 keV for both). Open circles indicate the area in which the shift of Ge atoms was observed. $V_s = 2.0$ V, $I_t = 0.2$ nA for both.

of Figures 31.4 through 31.6) is 60–240. In comparison with the estimated event rate, our observed atomic motion rate (30–70 atoms in total track length) looks very higher than that expected, because one core-hole event looks to correspond to one atomic motion.

The next possibility should be considered about the atomic motion induced by electronic transitions stimulated by impacts of photo-, Auger, and secondary electrons generated under x-ray irradiation. This is because the x-ray-induced electron-stimulated event number is much larger than the direct core-excitation event number because one core excitation induces a number of photo-, Auger, and secondary electrons [16]. Actually, x-ray-induced electron-stimulated desorption is the most popular process in x-ray region (mainly in the soft x-ray region) [17], of which the methodology has also developed for these decades [18]. Moreover, the control of site-selective reaction on a specific molecular bonding has been suggested [19] and investigated [20].

Although the x-ray-induced electron-stimulated desorption versus photon-stimulated desorption has long been discussed, most of the cases were about desorption of the "adsorbed molecules" on solid surfaces [21] such as NH_3 on Ni(110) for soft x-rays [16]. On the other hand, our target is the atomic motion on a single Ge(111) surface, which looks the atomic "pileup" event in a chain form rather than the random desorption, and caused by hard x-rays. Thus, our observation may be considered different from the conventionally studied photodesorption or photodissociation process about adsorbed molecules on solid surfaces or photostimulated ion desorption from solid surfaces by soft x-rays [22]. The probability of the atomic motion versus core excitation by hard x-rays on a single crystal surface system is then still ambiguous and scarcely reported, but a comparison of such estimation with the experimental data will give us a reference for physical consideration on the phenomena.

31.3.5 Thermal Estimation

Finally, the thermal effect, in which the quantum events of core excitation are included as averaged absorption effect for mass volume, was estimated from incident heat flux, specific heat of the sample irradiated, and thermal diffusion. Concretely, from the incident photon density ($1.2–4.8 \times 10^{15}$ (photons/s/mm^2)) with the energy of 11.089 or 11.119 keV, the heat flux q was estimated to be $2–8 \times 10^{-2}$ (J/s/mm^2) = (W/mm^2). This result is derived by considering the photon density (e.g., 1.2×10^{15}) at $\sim 11.1 \times 10^3$ (eV) under the total reflection condition with the grazing incident x-ray angle that decreases the photon density on the surface by ~1/400 and increases the wave field density by the factor of ~4 (e.g., $1.2 \times 10^{15} \times 11.1 \times 10^3 \times 1.6 \times 10^{-19} \times 1/400 \times 4 = \sim 2 \times 10^{-2}$ (J/s/mm^2)).

Next, considering the following relationship:

$$Q = q \times \alpha \times S \times \tau$$

$$= m \times c \times \Delta T$$

where

Q is the heat absorbed to the sample surface
α is the absorption ratio [23]
S is the irradiated area on the surface
τ is the effective absorption time taking into consideration diffusion time ($\sim 10^{-5}$ s)
m is the mass of irradiated volume = (density 5.32×10^{-3} g/mm^3) $\times S \times$ (penetration depth ~2 nm)
c is the specific heat (0.32 J/g·K)
ΔT is the temperature increase from RT at sample surface

The local increase of temperature (ΔT) at surface was estimated in each case as shown in Figures 31.5 and 31.6. In the case showing the incident energy dependence (Figure 31.5), considering the incident photon density (2.4×10^{15} (photons/s/mm^2)) and the absorption ratio (α = 4.4% for 11.089 keV and 40% for 11.119 keV), ΔT was 5 K for 11.089 keV, and 46 K for 11.119 keV. Our results showed that this level of the temperature increase did not affect the atomic motion rate. On the other hand, in the case showing the incident photon density dependence (Figure 31.6), considering the incident energy of 11.119 keV (α = 40%), ΔT was estimated to be 23 and 92 K for the incident photon density of 1.2×10^{15} and 4.8×10^{15} (photons/s/mm^2), respectively. This difference in the temperature affected the atomic motion rate differently from the case shown in Figure 31.5. This difference between the effects shown in Figures 31.5 (ΔT = 5 K vs. 46 K) and 31.6 (ΔT = 23 K vs. 92 K) looks reasonable, because the thermal phenomena obey an exponential form depending on the temperature.

31.3.6 Further Discussion

It is worth comparing our results with a past STM observation on the Ge(111)c-(2 × 8) clean surface at the temperatures of 150°C–220°C and around 300°C [24]. It was reported that the atomic motion rate was quite limited at the temperatures below 220°C, where the surface adatoms were clearly seen with modification of the domain boundaries, whereas the entire surface was broken above 300°C. These observed results can be understood partly and approximately in consistency with our results, in which the change in the atomic motion rate was not observed at RT + 46 K (i.e., ~70°C) (Figure 31.5) and confirmed at RT + 92 K (~120°C) (Figure 31.6) with clear individual atomic images. However, the structural change in our observation occurs in a temperature still far from the brake at ~300°C [24].

Other important aspects in our results are the locality and the anisotropy of the atomic motion. In the past report on the surface atomic motion produced by the classical thermal effect [24], the atomic motion was found to occur in the form of domain (Figure 31.7) en masse together and begin at ~220°C. However, our results show the atomic track having a local chain distribution, which is not randomly distributed on the surface. This anisotropy in diffusion can concern the anisotropic surface structure, and considering the locality in the atomic track, the origin of the atomic motion can be attributed to core excitation. In fact, in consideration of the increase in the temperatures ΔT

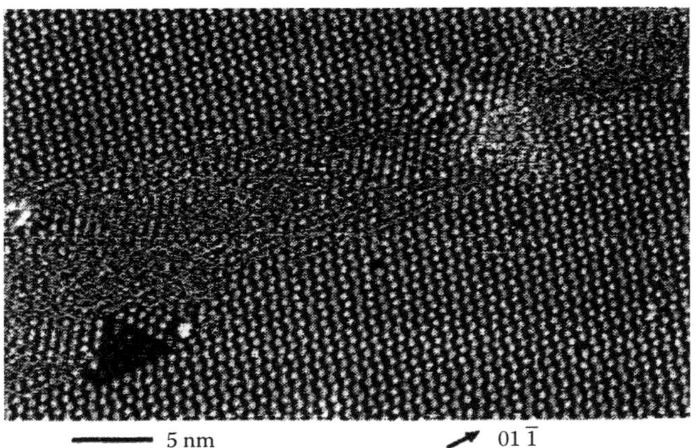

FIGURE 31.7 STM image of the Ge(111) clean surface, obtained at 235°C. A band of disordered surface area, located at a domain boundary, extends through the center of the image. (Reproduced with permission from Feenstra, R.M., Slavin, A.J., Held, G.A., and Luzt, M.A., *Phys. Rev. Lett.*, 66, 3257, 1991. Copyright 1991 by the American Physical Society.)

(92 K from RT) for Figure 31.6b, our atomic motion rate looks too high in comparison with the past report in which the domain motion begins at ~220°C. On the other hand, if we consider the atomic motion phenomena from another viewpoint, that is, the thermal diffusion of vacancy by referring to other report [13], the atomic motion rate for $\Delta T = 92$ K looks too small. This is because the diffusion coefficient of vacancies on this surface has been estimated to be different by one order between 300 and 324 K (0.308 Å²/s at 300 K and 4.0 Å²/s at 324 K [13]). Since there are several features that cannot be explained clearly at the level of the thermal order estimation, the total aspects on the atomic motion still need to be discussed.

The photon-stimulated structural change by x-rays attracts much interest in advanced structural analysis by XFEL. In XFEL, an x-ray pulse with an FWHM of 2 fs is anticipated to break a protein molecule by Coulomb explosion with an integrated x-ray intensity of 3×10^{12} (12 keV) photons per 100 nm diameter spot (3.8×10^{6} photons/Å²) [3]. Also, conventionally, the damage barrier has already been derived from the dynamics of damage formation analyses (~200 photons/Å² with x-rays of 12 keV) [25].

In our experiments, even at a much lower photon density (12–48 photons/Å² with x-rays of 11.119 keV) than the conventional barrier, the surface structure was found to possibly start to break. As far as we know, such atomic scale structural changes have never been observed directly for hard x-rays in UHV at RT. The observation of new features on the atomic motion was probably marginal for the brilliance of the x-rays from the long undulator with two-dimensional focusing. Actually, the photon density of lower than 1×10^{14} photons/s/mm² scarcely caused an increase in the atomic motion. Our observation of the damage barrier may contain potential importance. Since many examples of studies with long-time x-ray exposure (>10 min) can be found easily in recent x-ray studies, especially for small or low-dimensional specimens such as quantum dots or nanowires, our results can provide a warning of damage threshold in the near future. An increase of the photon density up to an order of ~10^{15} photons/s/mm² will make the observation itself have a strong effect onto the sample even in UHV condition. On the other hand, the highly dense x-rays are also suggested to have new applications too, for example, nanofabrication by nanometer-size brilliant x-ray beam and direct x-ray lithography under electron beam–free condition.

31.4 Conclusion

X-ray-induced atomic motion was directly observed using an SR-STM system with atomic resolution under UHV condition at RT. The atomic motion was not concerned with the STM tip bias, but caused solely by the x-ray photons. The atomic motion was visualized as a track image by newly attempted STM imaging. This technique to visualize the atomic track will serve in the diagnoses of various optical devices under the highly brilliant x-rays such as XFEL. Using the tracking image, the atomic mobility was found to be strongly affected by defects on the surface, but not dependent on the incident x-ray energy and clearly dependent on the photon density. However, the independence of the atomic motion on the incident x-ray energy was found to be probably attributed to the lack of the photon density. Also, some clearly different aspects of the atomic motion were found between our case and the past report on the classical thermal annealing. It is clear that the origin of the atomic motion is not desorption but diffusion, but the diffusion may be a synergic effect of the thermal effect by x-ray irradiation and the core excitation that can be accompanied with highly ionized ion production after the cascade Auger processes. It was possible to break the crystal surface structure even in UHV at lower photon density than the conventional barrier. This result can provide a warning to the x-ray studies with high photon density, especially for small or low-dimensional samples that need high photon density or long-time exposure of x-rays. On the other hand, the highly dense x-rays are also suggested to have new applications too. Also, the obtained results show a new availability of the in situ SR-STM system.

Acknowledgments

A.S. thanks all SPring-8 staff for experimental support, Dr. Y. Yoda (SPring-8) for providing software to control the apparatus, and Prof. M. Tsukada (Tohoku University) for useful discussion. We thank Dr. H. Takada and Dr. T. Ogawa (JEOL Co., Ltd.) for technical support. This study was performed with RIKEN Beamline Project (SPring-8) Proposal Number 20070115, 20080048, 20090060, 20100064, and 20110041 also supported mainly by PRESTO from the Japanese Science and Technology Agency, in part by a Grant-in-Aid for Scientific Research (No. 14702021, 23656033) from the Ministry of Education, Culture, Sports, Science and Technology. This work has been achieved by collaborating with Prof. Kuwahara's group of the Osaka University, Dr. Ishikawa's group of RIKEN/SPring-8 center, and Dr. Aono of the National Institute for Materials Science.

References

1. K. M. Yua, W. Walukiewicz, S. Muto, H.-C. Jin, and J. R. Abelson, *Appl. Phys. Lett.* 75, 2032 (1999).
2. F. Sato, N. Saito, J. Kusano, K. Takizawa, S. Kawado, T. Kato, H. Sugiyama, Y. Kagoshima, and M. Ando, *J. Electrochem. Soc.* 145, 3063 (1998).
3. R. Neutze, R. Wouts, D. van der Spoel, E. Weckert, and J. Hajdu, *Nature* 406, 752 (2000).
4. K. Nakayama and J. H. Weaver, *Phys. Rev. Lett.* 82, 980 (1999).
5. A. Saito, J. Maruyama, K. Manabe, K. Kitamoto, K. Takahashi, K. Takami, Y. Tanaka, et al., *J. Synchrotron Rad.* 13, 216 (2006).
6. A. Saito, J. Maruyama, K. Manabe, K. Kitamoto, K. Takahashi, K. Takami, Y. Tanaka et al., *Jpn. J. Appl. Phys.* 45, 1913 (2006).
7. A. Saito, Y. Takagi, K. Takahashi, H. Hosokawa, K. Hanai, T. Tanaka, M. Akai-kasaya et al., *Surf. Interface Anal.* 40, 1033 (2008).
8. M. Yabashi, T. Mochizuki, H. Yamazaki, S. Goto, H. Ohashi, K. Takeshita, T. Ohata, T. Matsushita, K. Tamasaku, Y. Tanaka, and T. Ishikawa, *Nucl. Instrum. Methods A* 467–468, 678 (2001).
9. R. S. Becker, B. S. Swartzentruber, J. S. Vickers, and T. Klitsner, *Phys. Rev. B* 39, 1633 (1989).
10. N. Takeuchi, A. Selloni, and E. Tosatti, *Phys. Rev. B* 49, 10757 (1994).
11. N. Takeuchi, A. Selloni, and E. Tosatti, *Phys. Rev. B* 51, 10844 (1995).
12. P. Molinas, A. Mayne, and G. Dujardin, *Phys. Rev. Lett.* 80, 3101 (1998).
13. I. Brihuega, O. Custance, and J. M. Gomez Rodriguz, *Phys. Rev. B* 70, 165410 (2004).
14. A. Kobayashi, F. Grey, R. S. Williams, and M. Aono, *Science* 259, 1724 (1993).
15. A. Saito, T. Tanaka, Y. Takagi, H. Hosokawa, H. Notsu, G. Ohzeki, Y. Tanaka et al., *J. Nanosci. Nanotechnol.* 11, 2873 (2011).
16. R. Jaeger, J. Stöhr, and J. T. Kendelewicz, *Surf. Sci.* 134, 547 (1983).
17. G. Betz and P. Varga, eds., *Desorption Induced by Electronic Transitions, DIET I~VI,* Springer-Verlag, Berlin, Germany, 1983–1995.
18. K. Mase, M. Nagasono, S. Tanaka, M. Kamada, T. Urisu, and Y. Murata, *Rev. Sci. Instrum.* 68, 1703 (1997).
19. W. Eberhardt, T. K. Sham, R. Carr, S. Krummacher, M. Strongin, S. L. Weng, and D. Wesner, *Phys. Rev. Lett.* 50, 1038 (1983).
20. S. Wada, R. Sumii, K. Isari, S. Waki, E. O. Sako, T. Sekiguchi, T. Sekitani, and K. Tanaka, *Surf. Sci.* 528, 242 (2003).
21. R. Franchy, *Rep. Prog. Phys.* 61, 691 (1998).
22. S. Tanaka, K. Mase, and S. Nagaoka, *Surf. Sci.* 572, 43 (2004).
23. B. L. Henke, E. M. Gullikson, and J. C. Davis, *Atomic Data and Nuclear Data Tables* 54(2), 181 (1993); Web site of the Center of X-ray Optics, Lawrence Berkeley National Laboratory, USA (http://henke.lbl.gov/optical_constants/).
24. R. M. Feenstra, A. J. Slavin, G. A. Held, and M. A. Luzt, *Phys. Rev. Lett.* 66, 3257 (1991).
25. R. Henderson, *Q. Rev. Biophys.* 28, 171 (1995).

Picoscale Devices

32

Mirrors with a Subnanometer Surface Shape Accuracy

Maria Mikhailovna Barysheva
Russian Academy of Sciences

Nikolay Ivanovich Chkhalo
Russian Academy of Sciences

Aleksei Evgenievich Pestov
Russian Academy of Sciences

Nikolay Nikolaevich Salashchenko
Russian Academy of Sciences

Mikhail Nikolaevich Toropov
Russian Academy of Sciences

Maria Vladimirovna Zorina
Russian Academy of Sciences

32.1 Introduction

Current interest in supersmooth and super-high-precision optical elements and systems is related to the development of a number of fundamental and applied fields, such as nanophysics and nanotechnology, x-ray microscopy in the spectral ranges of water and carbon windows, the projection nanolithography in the extreme ultraviolet (EUV), and soft x-ray spectral ranges [1–7]. To achieve a nanometer-scale space resolution (diffraction quality imaging), normal-incidence multilayer mirrors are used in this spectral range. The mirror quality is completely defined by substrate distortions and mostly by medium and high frequencies roughness. The surface deviations of spatial frequencies 10^{-6} to 10^{-3} μm^{-1} (low spatial frequency roughness [LSFR]) distort the image form; the frequencies from 10^{-3} to 1 μm^{-1} (medium spatial frequency roughness [MFSR]) blur the image; the high spatial frequencies roughness (HFSR) of 1–10^3 μm^{-1} reduce the mirrors reflectivity [8].

As it was shown in a number of papers, the quality requirements for substrates for the short-wavelength range are about 0.2–0.3 nm root-mean-square (r.m.s.) roughness in all discussed spectrum range [1,8], which is a challenge not only for manufacturers but also for measurers. So this chapter is devoted to the fundamental problems of manufacturing and surface roughness and figures errors measurements of the atomic level smooth and accurate substrates. Only the most adequate (from the authors' point of view) measurement techniques, like atomic force microscopy (AFM), white light interferometry (WLI), x-ray diffuse scattering (XRDS), and point diffraction interferometry (PDI), are considered. The latest results of scientific and technological research in this area, carried out in IPM RAS, are represented.

32.2 Surface Roughness Measurements in MSFR and HSFR Ranges

The key parameter, determining the resolution and efficiency of optical systems, providing the diffraction quality image, is the elements, (lenses or mirrors) surface roughness. For example, modern laser gyros and nanolithography systems with working wavelength of 193 nm require roughness level (r.m.s.) from a fraction to a few nanometers in a spatial frequency range from 10^{-3} to

10 μm^{-1}. For the next-generation lithography with a working wavelength of 13.5 nm and x-ray microscopy, the range of roughness spatial frequencies, influencing drastically on the properties of optical systems, expands to several hundred reversed micrometers.

Traditionally, WLI method is used for measurements in MSFR range ($\nu = 10^{-3}$–1 μm^{-1}) and AFM—for HSFR ($\nu = 2\times10^{-2}$–10^{2} μm^{-1}) [8–10]. An intermediate position takes XRDS with operating range $\nu = 10^{-2}$–10^{1} μm^{-1} covers AFM's short-wave region and WLI's long-wave one [11]. This intersection allows a reliable estimation of each method's possibilities, as well as studying the correlation between measured substrate surface roughness [12] and reflection coefficients of mirrors, deposited on these substrates, because, due to interference from a large number of borders, the roughness impact is amplified [13].

Analysis of published data, as well as the authors' own work, including cross tests, provided in other laboratories, revealed limitations of applying AFM and some obvious contradictions when using WLI for the attestation of ultrasmooth (atomically smooth) surfaces [14,15]. In this section, based on authors' own work, the possibility and the main limitations of these methods for measuring surface roughness are studied. Specific character of their application for various tasks is discussed and recommendations for application of these methods to solve specific problems are provided.

32.2.1 Surface Roughness General Description

Speaking of surface examination in an extended spectrum range, it is convenient to describe it by density function $PSD(\nu)$, characterizing the spatial frequency ν contribution to surface roughness. If $\vec{\rho}$ a radius vector located in a plane and $z(\vec{\rho})$ random surface height deviation from zero level ($\langle z\vec{\rho}\rangle = 0$), the correlation function for homogeneous surface is $C(\vec{\rho}) = \langle z(\vec{\rho})z(0)\rangle$, where angular brackets denote ensemble averaging. Two-dimensional spectral function PSD_{2D} is a Fourier transform of $C(\vec{\rho})$:

$$PSD_{2D}(\vec{\nu}) = \int_0^{\infty} C(\vec{\rho})\exp(2\pi i\vec{\nu}\vec{\rho})d\vec{\rho},$$

$$C(\vec{\rho}) = \frac{1}{2\pi}\int_0^{\infty} PSD_{2D}(\vec{\nu})\exp(-2\pi i\vec{\nu}\vec{\rho})d\vec{\nu}. \qquad (32.1)$$

For isotropic surface $C(\vec{\rho}) \equiv C(\rho)$ and 32.1 modification is

$$PSD_{2D}(\nu) = 2\pi\int_0^{\infty} C(\rho)J_0(2\pi i\nu\rho)\rho d\rho,$$

$$C(\rho) = 2\pi\int_0^{\infty} PSD_{2D}(\nu)J_0(-2\pi i\nu\rho)\nu d\nu. \qquad (32.2)$$

Dealing with isotropic surface, it is also helpful to use one-dimensional $PSD_{1D}(\nu)$ function as cos transformation:

$$PSD_{1D}(\nu) = \int_0^{\infty} C(\rho)\cos(2\pi\nu\rho)d\rho, \quad C(\rho) = \int_0^{\infty} PSD_{1D}(\nu)\cos(2\pi\nu\rho)d\nu. \qquad (32.3)$$

In [11,16] one can find exact solutions and approximate equations, allowing getting $PSD_{1D}(\nu)$ from $PSD_{2D}(\nu)$ and vice versa, so through this chapter we will generally not emphasize the difference.

It is also useful to discuss shortly the connection between *PSD* function and r.m.s. value, widely used to describe surface properties in some experiments. The standard deviation σ can be mathematically correctly defined from $\sigma^2 = C(0)$, which also, corresponding to 32.2 and 32.3, means

$$\sigma^2 = 2\pi\int_0^{\infty} PSD_{2D}(\nu)\nu d\nu = \int_0^{\infty} PSD_{1D}(\nu)d\nu. \qquad (32.4)$$

In reality, $PSD(\nu)$ can be experimentally determined only in a finite range of spatial frequencies, which means, instead of σ someone can only get the effective deviation [10,15]

$$\sigma_{eff}^2 = 2\pi\int_{\nu_{min}}^{\nu_{max}} PSD_{2D}(\nu)\nu d\nu = \int_{\nu_{min}}^{\nu_{max}} PSD_{1D}(\nu)d\nu, \qquad (32.5)$$

which depends on ν_{min}, ν_{max}. All surface investigation techniques are limited by their own spatial frequency ranges, which means comparing just the "deviation values" without taking into account method's frequency limitations can be quite confusing. That is why in this chapter, we will generally describe the substrates just by their *PSD* functions.

In the case of using AFM or WLI, the *PSD* function can be determined directly from a measured surface topography. In the case of XRDS, the specificity of this approach will be discussed later.

32.2.2 Problems of AFM Measurement of a Surface Roughness

Traditionally, AFM with proper choosing the probe is considered to be the most direct and reliable method for measuring surface roughness. In this case, the typical scan size varies from tens of nanometers to 100 μm, thereby providing roughness measurement in the spatial frequencies range from 10^{-2} to 10^{3} μm^{-1}. However, the practical application of AFM to study atomically smooth curved substrates revealed a number of problems.

The first problem is that, despite the wide range of AFM existing on the market today, allowing one to study samples with sizes from a few millimeters to tens of centimeters, strictly speaking, a special device designed to study the optical elements is missing. The existing microscopes allow one to study in fact only flat surfaces, while the imaging optic elements are curved ones. The problem is that in order to save subatomic resolution of the roughness height, a range of probe vertical movement during the scanning in plane must not exceed Dz = 1 μm. In the case of inclination between the surface and the axis of the probe (Figure 32.1), the range of possible lateral scanning is reduced, which restricts the range of spatial frequency roughness detected by this method. For example, the angle of only 3° leads to scan size no more than 19 μm. The real optics can have local slopes of a surface to an axis an order of magnitude greater. That is

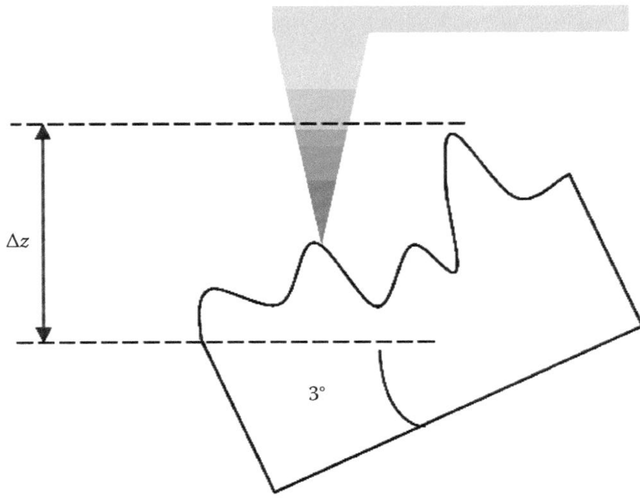

FIGURE 32.1 The problem of application AFM to study curved surfaces.

why for adequate use AFM in this case, the sample table must be designed as goniometer with *XY* scanning in the plane and tilts of ϕ and θ around these axes. Once again, as far as we know, devices on the market do not currently have this capability. So, one has to use nonstandard measuring procedures such as the direct installation of the AFM head onto the sample, which leads to large influence of vibration on the measurement results and the risk of damage of the unique substrates, or designing some specific goniometers, not fully integrated into a microscope.

The fundamental limitation of AFM abilities in supersmooth surfaces attestation is scanner motion nonlinearity. Figure 32.2 shows the AFM image with scan size of 30 × 30 μm^2 and its cross section (right) for super-polished quartz substrate, made in "Kompozit" Company (Moscow). A wave with amplitude of ~0.3 nm and a period of 20 μm can be clearly seen in the figure. The roughness, calculated for 2 × 2 μm^2 scan, turned out to be <0.1 nm. The effective roughness integrated over all scans was about 0.2 nm. If we formally put this roughness value into all scattering and attenuating factors for intensity, reflected from the surface, we can assume that the substrate will satisfy all the basic requirements for the diffraction quality optical substrates. But, if one divides the wave magnitude of 0.35 nm by its length of 20 μm, he will get angle error about 1.8×10^{-5} rad. For example, in a case of nanolithograph projection scheme with a distance between the last mirror and the wafer with photoresist of 338 mm [17] this error leads to the beam deflection from the set point of 6 μm, which is almost three orders of magnitude greater than the acceptable value.

In fact, this wave is not a property of the surface, but the property of the AFM scanner. The same wave can be seen for other substrates, like Cr/Sc multilayer structure deposited onto super-polished silicon substrate (Figure 32.3). It can be seen that the wave becomes less noticeable due to greater surface roughness (the effective roughness is about 0.3 nm). Rotating the sample by 90° does not change the wave, which confirmed its artifact nature.

It is quite clear that such nonlinearity cannot be easily subtracted from the experimental data, because according to the estimation made earlier critical waves have extremely small, sub-angstrom amplitudes. The fact that such waves are not hypothetical has been shown in several studies [18,19] (e.g., see Figure 32.4). Their registration was made possible only by exceeding their amplitude of the nonlinearity of the scanner. For practical application of the AFM method, one should consider these restrictions when analyzing the measurement results.

32.2.3 Problems of WLI Measurement of a Surface Roughness

Traditionally, surface roughness in the MFSR range is measured by WLI technique. Nowadays, there are a lot of WLI, for instance [9]. They advocate that these devices provide angstrom

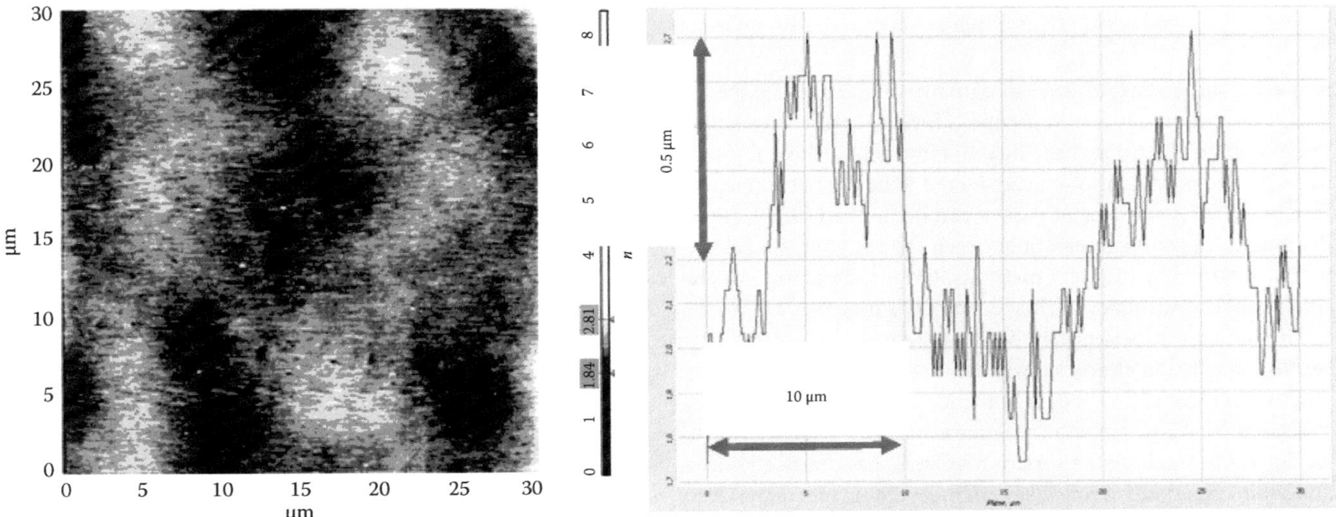

FIGURE 32.2 AFM image (30 × 30 μm^2) of quartz substrate surface and its cross section.

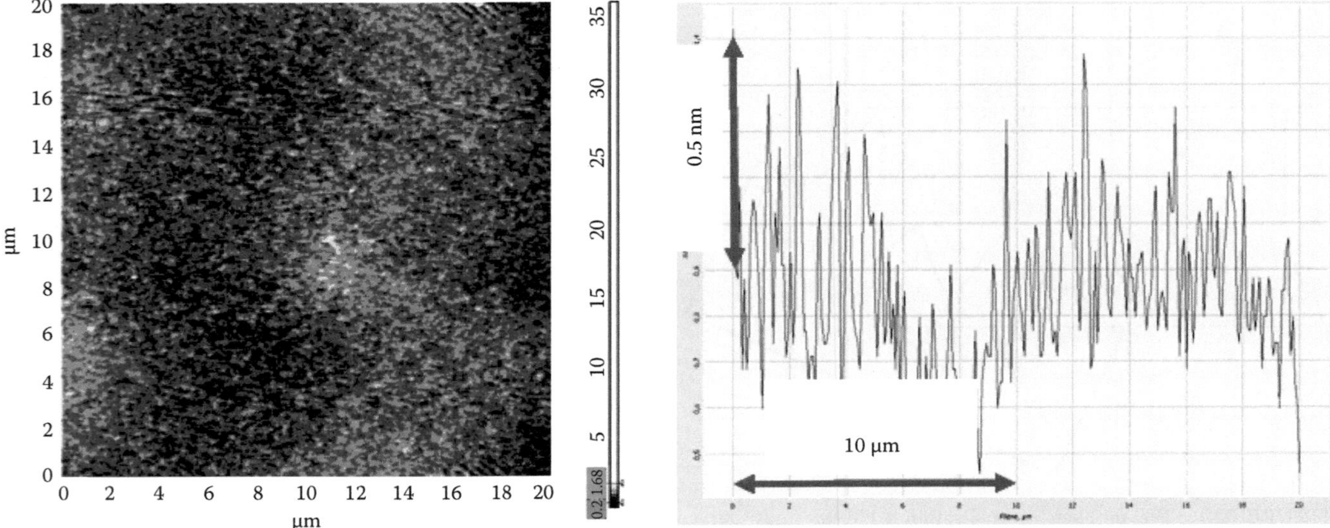

FIGURE 32.3 AFM image (20 × 20 μm²) of Cr/Sc multilayer structure deposited onto super-polished silicon substrate and its cross section. The effective roughness is 0.3 nm.

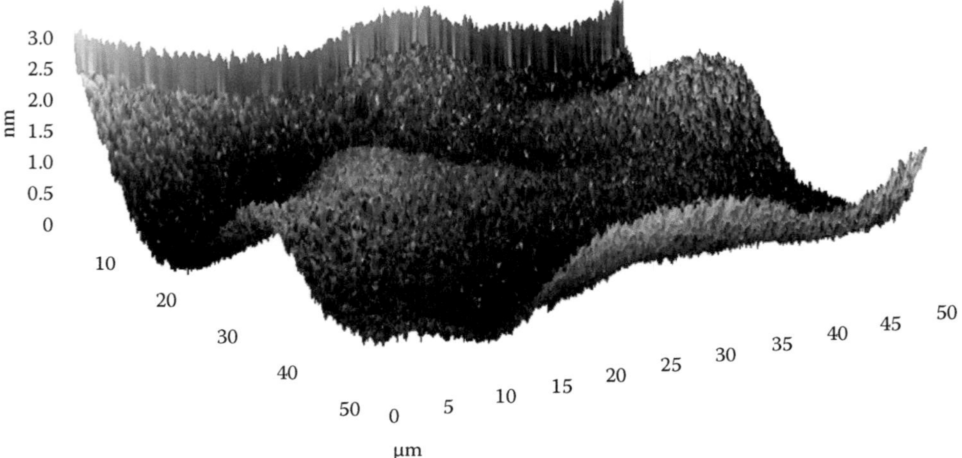

FIGURE 32.4 AFM image (50 × 50 μm²) of Si surface after ion-beam etching. The wave magnitude is about 1 nm.

and even sub-angstrom level measurement accuracy. Since the lateral resolution of the microscope is limited by the wavelength of light (about 1 μm) and the field of view extends to a few millimeters, the method is often considered to be the natural extension of AFM. A number of papers can be defined [8,20], which demonstrates good agreement between the roughness measurements obtained by different methods. On this basis, the conclusion about the adequacy of this technique is made.

The authors' experience in WLI application to supersmooth surfaces attestation clearly shows that there is a problem with the method. In Figure 32.5a, the *PSD* functions of three silicon samples, measured by AFM, are shown. Sample number 4 was etched by Ar^+ ions. After etching the surface roughness has drastically risen, and submicron pores clearly visible in the AFM image (center) were formed, but for WLI all samples are the same (Figure 32.5b). In this experiment, Talysurf CCI 2000 device was used.

The result was also confirmed in a joint experiment of IPM RAS and Rigaku Innovative Technologies with fused silicon samples (Figure 32.6). The curve 1 corresponds to measurements made in IPM RAS using AFM and XRDS. The curves 2–8 are *PSD* functions, measured by WLI ZYGO (3–8) and AFM (2) in Rigaku Innovative Technologies. Note that curves (3–8) were obtained using different microscope objectives. It can be easily seen that the results of measurements, performed by AFM and XRDS in two different laboratories, agree very well with each other, while the WLI measurements made by one instrument, with the same reference, but using various objectives are very different. The reliable agreement between XRDS, AFM, and WLI data is observed only for frequencies below 10^{-2} μm^{-1}.

We assume two main reasons for possible errors when WLI method is used for roughness measurements. First of all, it is the presence of the reference in WLI optical scheme, which requires

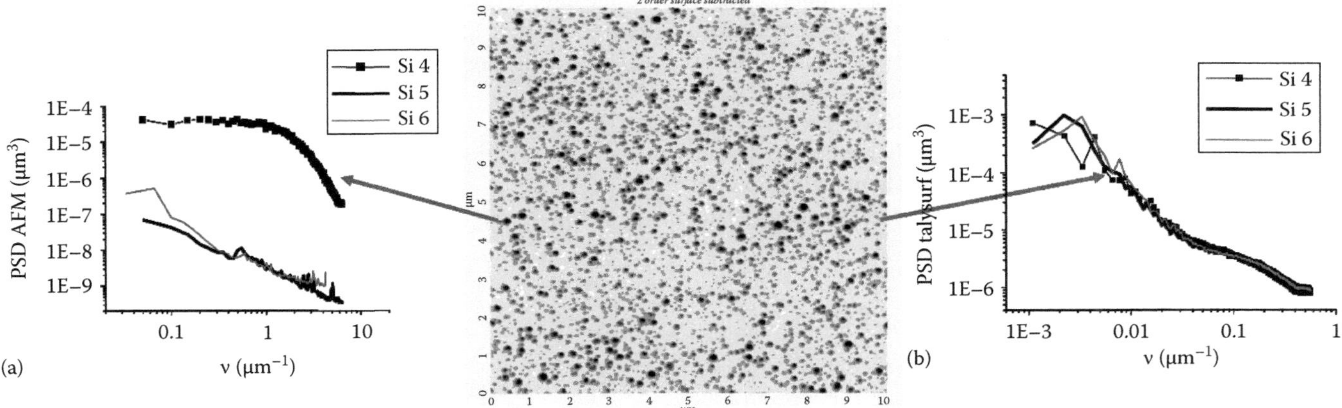

FIGURE 32.5 *PSD* function for three Si samples, measured by AFM (a) and WLI Talysurf CCI 2000 (b).

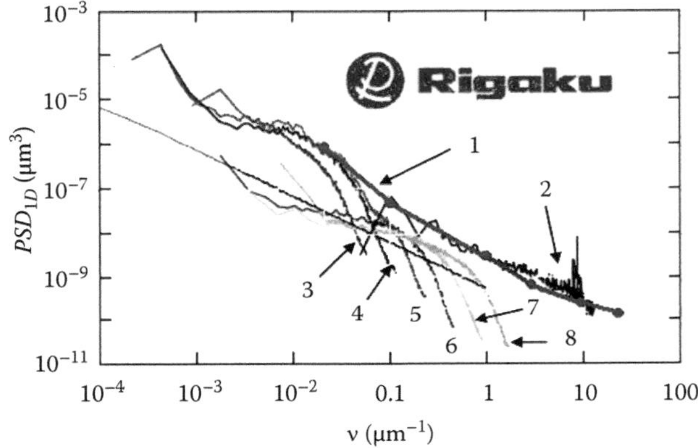

FIGURE 32.6 PSD_{1D} functions of SiO_2 substrates measured in IPM RAS (1) and in Rigaku (2–8): (1) XRDS and AFM; (2) AFM; (3) WLI Zygo 1.25×; (4) 2.5×; (5) 5×; (6) 10×; (7) 20×; and (8) 40×.

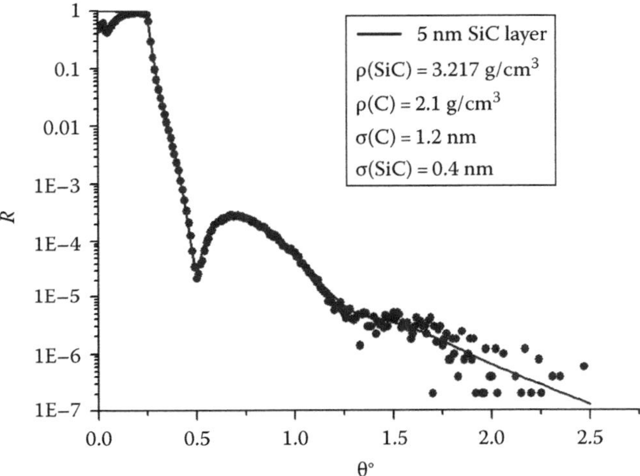

FIGURE 32.7 The angular dependence of the reflection coefficient of the x-ray radiation with a $\lambda = 0.154$ nm for WLI Talysurf reference. The symbols represent the experiment and the solid line is for calculation.

an independent attestation for subtracting errors introduced by them from experimental data. This procedure by itself implies the existence of a certain reference method. For example in Figures 32.7 and 32.8, the results of studying the surface of the WLI Talysurf reference are represented. According to the device specification, the reference is a silicon carbide substrate with a silicon carbide layer deposited by CVD method. Figure 32.7, where an angular dependence of the reflection coefficient for the reference (x-ray radiation wavelength was 0.154 nm) is given, shows that there is a layer on the surface, with a depth of about 5 nm and reduced density of 2 g/cm³. The best fit of calculated and experimental data was obtained for roughness of the boundaries 0.4 and 1.2 nm, respectively. Experimental data for the reference attestation by AFM, XRDS, and WLI can be seen in Figure 32.8. The effective roughness in the spatial frequency ranging from 10^{-2} to 60 μm^{-1} is about 0.9 nm. This is very far from the device's claimed characteristics.

The second major cause of measurement error for WLI method is the presence of a large number of optical elements, working and reference wave fronts have to pass through. Although the "wedge" between the fronts is small and they go only by slightly

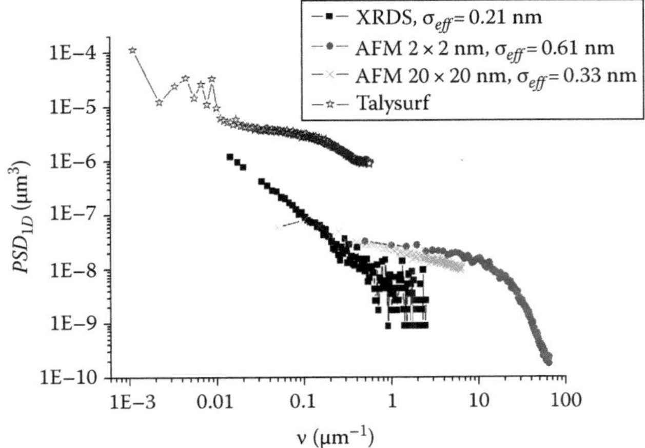

FIGURE 32.8 *PSD* functions for the WLI Talysurf reference, measured by AFM, XRDS, and WLI. The effective surface roughness in spatial frequency range from 10^{-2} to 6×10^{1} μm^{-1} is about 0.9 nm.

different ways, however, as it was shown in a number of papers devoted to PDI (e.g., [21]), it is enough for them to make uncontrollable nanometer (sub-nanometer)-scale phase shifts. Now we see intensive attempts to modify the techniques by using different averaging, compensating procedures [20,27]. However, since it does not take into account the previously mentioned systematic errors, we take the possibilities of WLI application for attestation of supersmooth optical elements in the frequency range above 10^{-2} μm^{-1} quite skeptically.

32.2.4 Surface Attestation Using XRDS

XRDS is an alternative popular and widely used method for surface studying in MSFR (and partly HSFR) range. It is given a special place because it belongs to the "ab initio" methods, because it is based on well-studied fundamental laws of interaction of electromagnetic radiation with matter, and in some sense it can be considered as a referent method. Scheme of XRDS experiment is shown in Figure 32.9. It is not a direct method, so several approaches were made to establish conformance between the scattered signal intensity and *PSD* or correlation functions. These restrictions and advantages of methods were discussed in details in [16]. In this chapter, we use the equations, based on a perturbation theory, which in the case of large correlation length correct for smooth surfaces and small angles of incidence θ_0:

$$a \gg \lambda, \quad \frac{2\pi}{\lambda}\sigma\sin\theta_0 \ll 1, \quad \frac{2\pi}{\lambda}\sigma\sqrt{\varepsilon-\cos^2\theta_0} \ll 1, \tag{32.6}$$

where

λ is the x-ray wavelength
a is the correlation length
$\varepsilon = 1 - \delta + i\beta$ is the material dielectric permittivity

Inequalities (Equation 32.6) can be easily fulfilled in hard x-ray range, where θ_0 is usually less than angle of total external reflection θ_c. For example, for $\lambda = 0.154$ nm (Cu Kα) and SiO_2 substrate $\theta_c = 0.22°$, $\theta_0 = 0.18°$, so 32.6 means $\sigma \ll 8$ nm. The limitations (Equation 32.6) seem to be very natural for supersmooth surfaces. The important advantage of this approach is a simple proportional conformance between scattering indicatrix $\Phi(\theta,\phi) = (1/W_{inc})(dW/d\Omega)$ (W_{inc} is an incident power of radiation and dW is the one scattered into a solid angle $d\Omega$) and $PSD_{2D}(\nu)$:

$$\Phi(\theta,\varphi) = \frac{\pi^2|1-\varepsilon|^2|t(\theta_0)t(\theta)|^2}{\lambda^4\sin\theta_0}PSD_{2D}(\nu), \tag{32.7}$$

where $t(\theta)$ is a Fresnel propagation coefficient for flat surface:

$$t(\theta) = \frac{2\sin\theta}{\sin\theta+\sqrt{\varepsilon-\cos^2\theta}}. \tag{32.8}$$

So the proportional coefficient depends on material optical properties only, which allows direct extraction $PSD_{2D}(\nu)$ from experimental data without any additional assumption about its shape.

For "hard" x-ray experiments, the angle of incidence is usually quite small ($\theta_0 < \theta_c \ll 1$), that is why the scattering indicatrix has a narrow form: if a is a correlation length, the indicatrix's width can be estimated as $\delta\theta \sim \lambda/\pi a\sin\theta_0$, $\delta\phi \sim \lambda/\pi a$, so $\delta\theta \gg \delta\phi$. In practice, a detector angular size in ϕ direction is $\delta\phi_{det} \gg \delta\phi$, so instead of $\Phi(\theta,\phi)$, one-dimensional indicatrix $\Pi(\theta) = \int_0^{2\pi}\Phi(\theta,\varphi)d\varphi$ is usually measured. According to [11],

$$\Pi(\theta) = \frac{|\pi(1-\varepsilon)t(\theta_0)t(\theta)|^2}{2\lambda^3\sin\theta_0\sqrt{\cos\theta\cos\theta_0}}PSD_{1D}(\nu), \qquad \nu = \frac{|\cos\theta-\cos\theta_0|}{\lambda}. \tag{32.9}$$

To use Equation 32.9 for extracting *PSD* functions from x-ray scattering data, note that measured intensity is a result of integrating the indicatrix by finite detector angular sizes $\delta\theta_{det}$, $\delta\phi_{det}$:

$$W(\theta) = \int_{\theta-\delta\theta_{det}/2}^{\theta+\delta\theta_{det}/2}\Pi(\theta)\cos(\theta)d\theta, \tag{32.10a}$$

$$W(\theta,\varphi) = \int_{\theta-\delta\theta_{det}/2}^{\theta+\delta\theta_{det}/2}\cos(\theta)d\theta\int_{-\delta\varphi_{det}/2}^{\delta\varphi_{det}/2}\Phi(\theta,\varphi)d\varphi. \tag{32.10b}$$

In hard x-ray experiments for θ_0, $\theta \ll 1$ (Equation 32.10a) can be simplified to $W(\theta) \cong \Pi(\theta)\cdot\delta\theta_{det}$, $\delta\theta_{det} \cong D_\theta/D$, where D_θ is the detector size in θ direction and D is a sample–detector length (Figure 32.9).

Spatial frequencies, available for XRDS method, are limited by detector and beam sizes from low-frequency edge, because it is impossible in this area to separate scattered signal and the specular reflected one. In our experiments, we usually have $\nu_{min} \sim 10^{-2}$ μm^{-1}. From high frequencies, the limitation is defined by signal/noise ratio, that is, the apparatus dynamic range, which usually in laboratory is about 10^5–10^6. Obviously, the quality of a surface also affects ν_{max} value, for supersmooth substrate we examine ν_{max} is about 2–3 μm^{-1}.

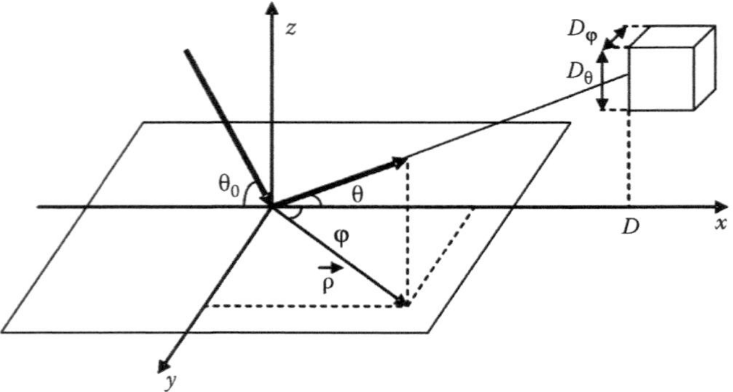

FIGURE 32.9 Some geometrical parameters and x-ray diffusion scattering experiment scheme.

In many works, XRDS is reported to be very adequate and convenient method for surface investigations; also, very good agreement between XRDS's and AFM's *PSD* functions was observed (e.g., see [11]). We have got the same result for Si and SiO_2 substrates, specially prepared for diffraction quality imaging optics purposes: polished or additionally etched by ions for small depths (Figure 32.10a and b). Note that ion-beam etching is now considered to be a very perspective method for final refinement of substrate surface shape. Etching for 0.1–0.5 μm depth is successfully used for correcting form in LSFR range [19]; 1–10 μm etching can be applied for substrate aspherezation [22]; mechanical methods are less accurate and also produce undesirable MSFR roughness. In Figure 32.10c, *PSD* functions obtained by AFM and XRDS techniques for deeply etched substrate demonstrate notable difference: surface seems to be smooth from AFM point of view, but in x-ray experiment intensive scattering is observed. X-ray specular reflection data also indicates smooth surface ($\sigma \approx 0.45$ nm), and multilayer structure, deposited onto the substrate, demonstrate quite low interfacial roughness. This means we face XRDS method's limitation, not allowing a correct examination of the actual surface.

In [15] we have shown that the anomalous large x-ray scattering can be caused by bulk inhomogeneities in undersurface layer, produced by high-energy ion bombardment and the argon atoms intrusion into the atomic lattice. To detect the presence of the layer we suggest simple experiment, based on the analysis of diffuse scattering for various incidence angles (Figure 32.11). Different θ_0 mean different penetration depths L; according to the wave equation,

$$L \cong \left(\frac{4\pi}{\lambda} \mathrm{Im} \sqrt{\varepsilon - \cos^2 \theta_0} \right)^{-1}, \quad (32.11)$$

for SiO_2 at $\lambda = 0.154$ nm the dielectric permittivity $\varepsilon = 1 - 1.42 \times 10^{-5} + i \times 1.84 \times 10^{-7}$ [13], so the minimum depth $L(\theta_0 \to 0) \cong$ AQ1 3 nm. Thus, when θ_0 increases, the distorted layer effective influence weakens, so *PSD* function goes down and approaches to the non-etched surface one (Figure 32.11). For comparison, we also present the similar *PSD* curves for as-prepared (non-etched) substrate and there is no notable difference observed for various θ_0.

Thus, when XRDS method is used for studying the substrate, it is necessary to control the presence of bulk inhomogeneities (e.g., damaged layers, caused by ion-beam etching); otherwise, the result may be incorrect.

The main disadvantage of hard x-ray scattering method is caused by small angles of total external reflection (0.2°–0.4°), which means it can only be used for flat surface attestation, whereas the imaging systems are composed by nonplanar (convex or concave) optical elements with radii from 10 mm to 1 m, deflection from a plane up to 30 mm, and entrance angles up to 10°–20°. Therefore, there is a need to develop an alternative

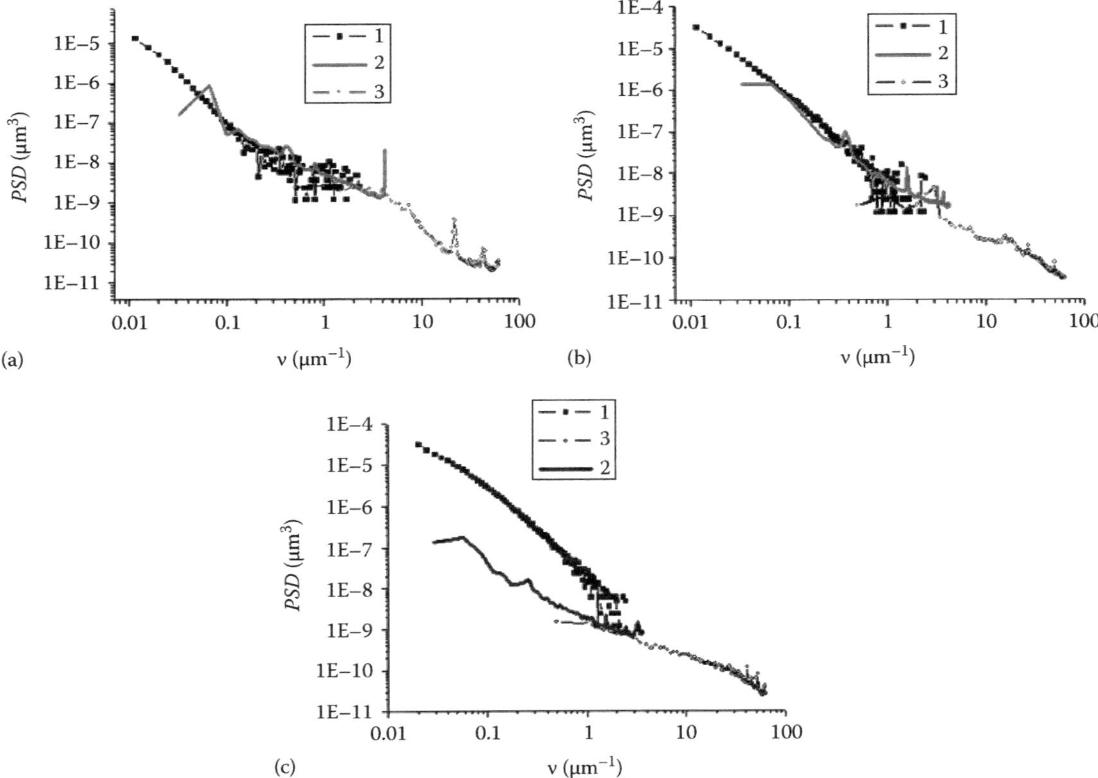

FIGURE 32.10 PSD_{1D} for SiO_2 substrate after polishing (a), ion-beam etching for 0.5 μm (b), and ion-beam etching for 11 μm (c): (1) XRDS; (2) and (3) AFM with scan sizes of 20 × 20 and 2 × 2 μm².

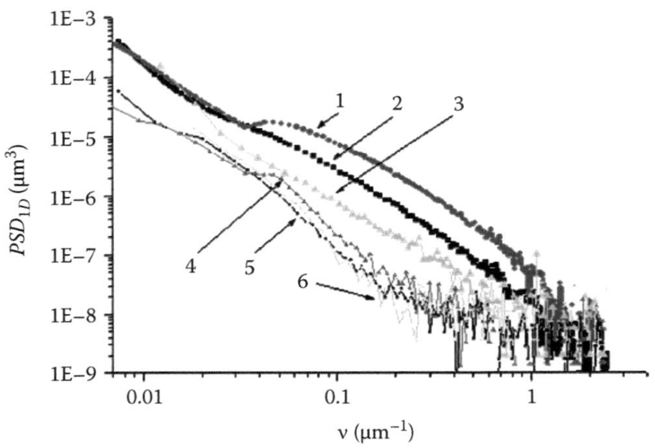

FIGURE 32.11 PSD_{1D} functions obtained by XRDS with different angles of incidence for SiO_2 substrates etched for 11 μm (1–3) and as-prepared (4–6); (1, 4) $\theta_0 = 0.1°$ ($L \sim 4$ nm); (2, 5) $\theta_0 = 0.18°$ ($L \sim 6$ nm); (3, 6) $\theta_0 = 0.3°$ ($L \sim 0.5$ μm).

approach to study supersmooth nonplanar substrates. We propose the diffuse scattering technique of soft x-rays (1–20 nm). Choosing the optimal wavelength, generally speaking, is a compromise between the need to obtain large angles of incidence (10°–20°), which requires an increase in wavelength, and the desire to explore small-scale surface roughness, which naturally requires its reduction. Let us illustrate the possibilities of a reflectometer, described later, by Figure 32.12, showing some calculated scattering curves for soft x-ray radiation with wavelengths of 13.5 nm (Si Lα), 6.7 nm (B Kα), and 4.47 nm (C Kα), scattered by quartz supersmooth surface with surface *PSD* function described by the so-called ABC—a model [11]:

$$PSD_{1D}(\nu) = \frac{2}{\sqrt{\pi}} \frac{\Gamma(h+1/2)}{\Gamma(h)} \frac{\sigma^2 a}{\left(1+a^2\nu^2\right)^{h+1/2}}, \qquad (32.12)$$

where

$a = 100$ μm is a correlation length
$h = 0.2$ is fractal Hirst parameter
$\sigma = 0.5$ nm is r.m.s. roughness

The intensity of the scattering signal recorded by detector is calculated as

$$I_{scatt}(\theta) = \int_{-\Delta\varphi/2}^{\Delta\varphi/2} d\varphi \int_{\theta-\Delta\theta/2}^{\theta+\Delta\theta/2} \cos(\theta)\Phi(\theta,\varphi)d\theta, \qquad (32.13)$$

for $\Delta\theta = 3$ mm/136 mm and $\Delta\varphi = 10$ mm/136 mm are the detector's angular sizes in θ and φ directions, respectively, and the incident angle of radiation $\theta_0 = 10°$. The minimum spatial frequency available for research in this scheme is approximately $\nu_{min} \cong 0.4$ μm^{-1}, the maximum one is directly determined by a dynamic range of the device and a surface roughness, as $I_{scatt} \sim \sigma^2$.

It can be concluded from Figure 32.12 that expanding the range of recorded spatial frequencies up to 20–50 μm^{-1} (typical for AFM) requires the dynamic range of the detected radiation intensities up to 10^8–10^{10}. Such signals could not be produced under laboratory conditions using traditional monochromators, even using laser-plasma sources of soft x-rays (10^6–10^7 in [23]).

To solve this problem, we suggest using Mo/Si multilayer x-ray mirrors with resonant wavelength of $\lambda = 13.5$ nm, instead of diffraction grating monochromators. As it was shown in [24], the intensity of the probe beam can be increased more than three orders of magnitude, all factors being the same. The device's photo and optical scheme can be seen in Figure 32.13. As a monochromator and a probe beam shaper, we apply spherical Schwarzschild objective with the input angular aperture $NA = 0.1$ and 10× zoom. The mirrors' reflection coefficients at

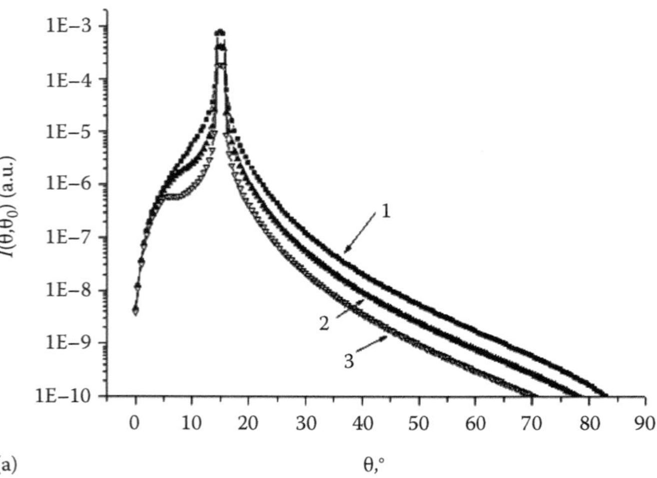

FIGURE 32.12 Calculated scattering curves for soft x-rays with wavelengths of 13.5 nm (Si Kα), 6.7 nm (B Kα), and 4.47 nm (C Kα), scattered by quartz surface: (a) angular dependence and (b) frequency dependence.

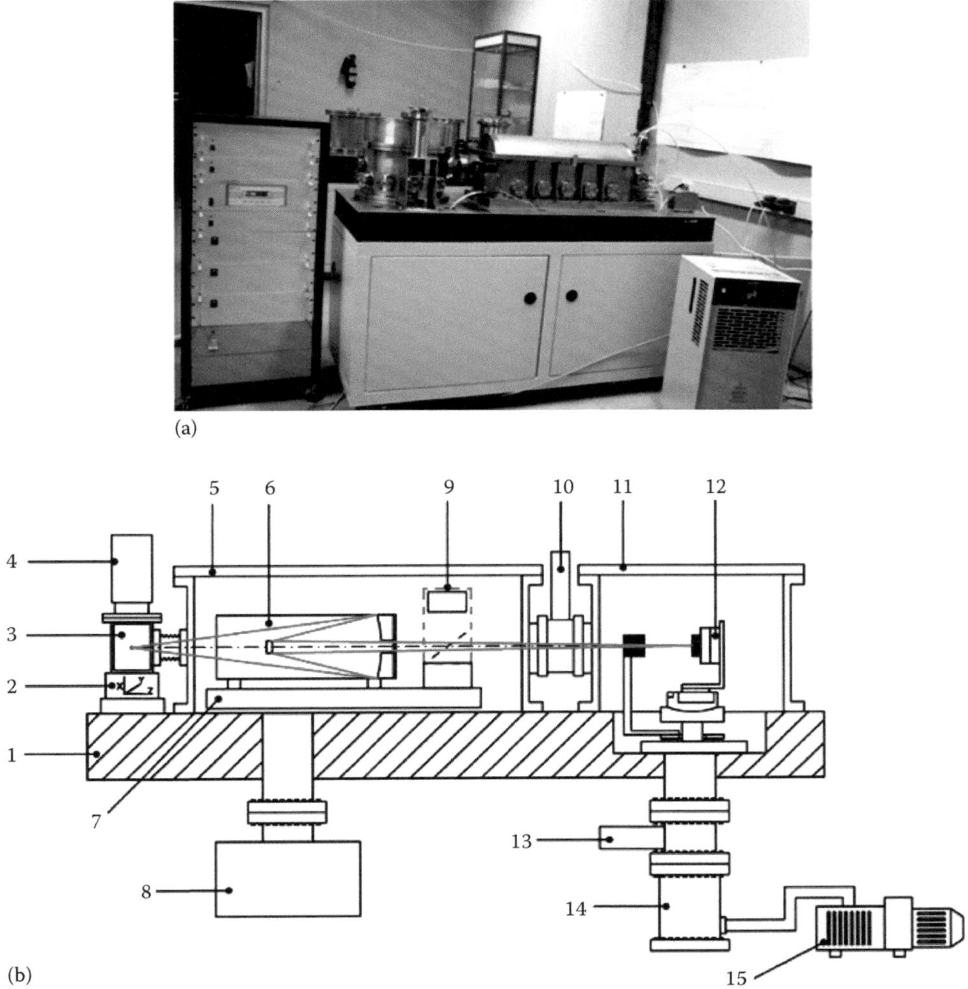

FIGURE 32.13 Photo (a) and scheme (b) of the soft x-ray reflectometer. (1) Base plate; (2) 3D table for x-ray source; (3, 4) x-ray source; (5, 11) monochromator and sample chambers; (6) Schwarzschild lens; (7) mount; (8, 14, 15) magnetic discharge, turbomolecular, and fore pumps; (9) intensity monitor; (10, 13) valves; (12) five-axis goniometer.

$\lambda = 13.5$ nm are 65%. Currently, a dismountable x-ray tube with a Si target and a power in the electron beam of 10 W is used as the radiation source. The intensity of the x-ray emission is measured by an electron channeltron SEM-6 (Russia) with CsI photocathode. The first experimental results are shown in Figure 32.14 for fused silica substrates. The specular reflection curve (Figure 32.14a) demonstrates a significant difference between tabulated $\varepsilon_{tab} = 1 - 0.044 + 0.022i$ and measured $\varepsilon_{tab} = 1 - 0.054 + 0.029i$ values of optical constants of SiO_2 near Si anomalous dispersion area ($\lambda = 13.5$ nm). The measured value was obtained by experimental data fitting for $\sigma = 0$ nm (the roughness has no effect near the critical angle) and it describes the experimental results more reliably. Note that tabular optical constants, incorrectness near Si anomalous dispersion area was observed in our experiments not only for SiO_2, but also for Si samples.

In the scattering signal measuring experiment (Figure 32.14b), the dynamic range of the reflectometer was about 10^8, which is no less than the best world analogs with synchrotron radiation sources demonstrated. Obtained *PSD* function (Figure 32.14c) is slightly different from hard x-ray scattering and AFM data, which can be possibly explained by using the approach (Equation 32.1) of the "infinite narrow" beam in φ direction. However, the effective roughness values about 0.3 nm got by AFM and XRDS are in good agreement with each other. The range of spatial frequencies, available for studying by this method, lasts from 0.4 to 70 μm^{-1} that is close to the capabilities of AFM.

Note that the diffuse scattering is also a method for examining statistical characteristics of multilayer x-ray mirror's interfacial boundaries. Thereby the reflectometer could be applied to establish a correlation between the properties of a substrate and a deposited multilayer coating in MSFR and HSFR range. Despite its obvious importance, this problem has not been given sufficient attention to date.

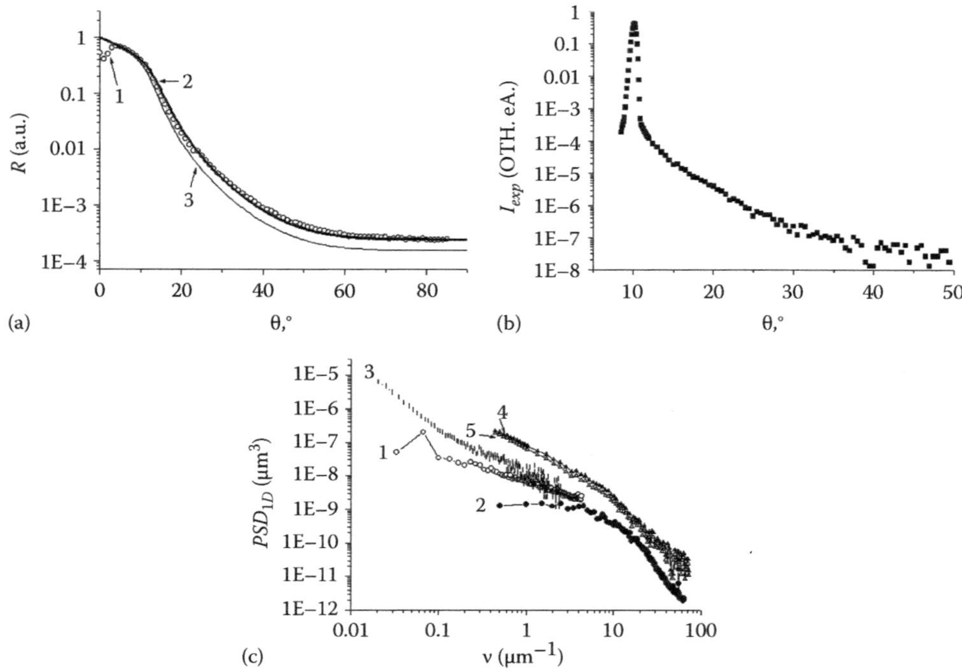

FIGURE 32.14 Experimental data for quartz: reflection (a): (1) experimental and (2, 3) theoretical curves, for $\varepsilon_{tab} = 1 - 0.054 + 0.029i$ and $\varepsilon = 1 - 0.044 + 0.022i$, and scattering (b) of x-rays with $\lambda = 13.5$ nm, PSD_{1D} functions (c) obtained by methods: (1, 2) AFM scans 30 × 30 and 2 × 2 μm^2; (3) XRDS, $\lambda = 0.154$ nm, $\theta_0 = 0.18°$; (4, 5) XRDS, $\lambda = 13.5$ nm, $\theta_0 = 10°$, tabular and measured dielectric permittivity, correspondingly.

32.2.5 Conclusions on Studying Supersmooth Surfaces' Roughness

This study has shown the real possibility of existing methods and their limitations when they are used for the attestation of supersmooth surface roughness. In case when ultrahigh spatial resolution is not required (no need to consider angular errors of less than 10^{-4}–10^{-5}) and the samples have a flat shape, by choosing the proper probe and using capacitive sensors to control the movement of piezo-scanner, one can use the standard AFM devices and scan size up to 100 × 100 μm^2 is available. A scanner's nonlinear contribution to measured roughness (r.m.s. parameter) would not be critical in terms of the surface's reflection coefficients.

In the case of imaging optics with moderate spatial resolution, all previous conclusions are true, but the microscope must be equipped with a four-axis goniometer to set sample surface's local normal coaxially with the AFM probe.

If ultrahigh spatial resolution is needed, AFM can measure accurately the r.m.s. roughness with lateral sizes up to tens of micrometers, but from surface's imaging properties prediction point of view, the measurement's result can be trusted for very small scan sizes, less than 1 × 1 μm^2, only because this roughness does not effect on image quality, only reduces its illumination. The last statement is connected to the fact that in x-ray range multilayer interference mirrors (Bragg reflectors) are used, so x-ray waves, scattered at roughness with lateral dimensions <1 μm (approximately, the value depends on the actual conditions of an experiment), leave the Bragg peak area and that is why there is no interference amplification for such waves.

This is a fundamental problem and interferometric methods for the probe position control could not help, because, since we are talking about sub-angstrom wave amplitudes, when intrinsic noise of a scanner and an interferometer is significantly higher (first of all, thermal noise associated with fluctuations of the interferometer arms' length). One possible partial solution is to create a plane reference, whose angular errors can be certified using synchrotron radiation with a high angular resolution, like 10^{-6}–10^{-7}. Using this reference, one can periodically calibrate scanner nonlinearity. Now such an experiment is planned to be conducted.

We are skeptical about the real possibilities of WLI, and we believe that it can be trusted at frequencies <10^{-2} μm^{-1}.

The XRDS can be treated as "ab initio" method, since it is based on well-studied fundamental laws of electromagnetic radiation interaction with a matter, and under certain reservations it can be considered as a referent method. Therefore, in a diagnostic laboratory it is desirable to have both AFM and XRDS devices, and periodically do the cross tests, because AFM measurement results can be greatly affected by difficult controlled and time-varying parameters, such as the nonlinearity of a piezo-scanner, the quality of a probe, the surface pollution, vibration, etc. The result of supersmooth surface attestation is reliable only if these two methods' measured data match. For studding nonplanar surfaces, soft x-rays diffuse scattering can be used. For practical use of the XRDS method, one should verify the absence of disturbed undersurface layer, which can significantly affect the measurement results.

32.3 PDI for Studying Surface Shape and Aberrations

To the last time there was a serious restriction, limiting the use of short wavelength of EUV and soft x-rays in order to achieve nanometer spatial resolution: the lack of high-aperture precision optics, allowing forming images with resolution limited by diffraction. According to the Mareshal criterion to achieve the diffraction-limited resolution of an optical system, a r.m.s. aberration of the system wave front RMS_{obj} must satisfy the ratio [25]:

$$RMS_{obj} \leq \frac{\lambda}{14}, \tag{32.14}$$

where λ is a wavelength of light. Since the errors (distortions) of elements of a complex optical system are statistically independent, the required accuracy RMS_1 of manufacturing an individual optical component is

$$RMS_1 \leq \frac{\lambda}{\left(14 \cdot \sqrt{N}\right)}, \tag{32.15}$$

where N is the number of components in the optical system. For instance, in a case of a six-mirror objective, typical for EUV lithography at the wavelength of $\lambda = 13.5$ nm, the reasonable error of individual mirror RMS_1 should not exceed 0.4 nm. For water-window microscopy (spectral region of $\lambda \approx 3$ nm), these requirements are even tougher and reach 0.1–0.2 nm. Therefore, a necessary condition for such optics creation is an existence of surfaces' form and aberrations metrology of angstrom and in some cases sub-angstrom accuracy. The problem is further complicated by the optic's aspheric form.

Traditional interferometric methods to control optical surfaces' form use the wave front, reflected from an etalon surface [26], as a reference; thus they have got quite high (λ/1000 and higher) sensitivity to wave fronts' shape changes. However, the absolute precision of the actual front's shape measurement does not exceed $\lambda/20$–$\lambda/50$ [27] (λ is a working wavelength of the interferometer). This happens due to both the etalon surface quality and specific design of the interferometers, where the light beams pass through a number of optical elements, gathering uncontrolled additional phase shifts [21].

To realize the possibility of high-precision measurements of a surface shape, it was necessary to change traditional methods of calibration using reference surfaces to development of the methods, based on fundamental physical principles, when the main characteristics of the device (method) can be measured in a physically clear experiment with the possibility of a reliable estimation of the measurement errors. For optical surfaces' form attestation, the PDI, proposed in 1933 [28] and actually demanded in the early 1990s in relation with the EUV lithography program, has become such a method. It is based on forming the reference spherical wave as a result of light diffraction on a pinhole. The classical problem of light diffraction on a pinhole in a screen of infinite conductivity and zero thickness indicates that in the diffraction peak area (maximum angular width of $\pm\lambda/d$, where d is the pinhole diameter) the wave phase surface is an "ideal" sphere. This section focuses on the challenges and opportunities of modern interferometers.

32.3.1 PDI Based on Pinhole

For the first time, a measurement accuracy of a lens aberration about 0.5 nm was achieved by PDI [29] with a single-mode fiber core of ≈5 μm diameter for aperture. So, the angular width of the diffraction peak was about ±0.1, which is insufficient for today's applications. Moreover, the drastic change in the intensity of the diffracted wave in the diffraction peak area affects negatively on the accuracy of the diffraction pattern extremum's coordinates definition [30]. Therefore, the most widespread interferometers are the ones with a pinhole in a metallic screen used as a reference spherical wave source. Depending on operating wavelength, diameters of the pinholes vary from tens of nanometers to fractions of a micrometer. Two versions of the device are the most widely used. An interferometer installed on ALS (Berkeley, USA) [31] operates in the following way. The radiation falls onto a small pinhole gap, installed in the object plane of the lens under study. A diffracted "ideal" spherical wave passing through the lens is distorted according to its aberrations and is collected in the image plane. Between the lens and the image plane a diffraction grating is installed, splitting the wave front into two ones, corresponding to zero and to first diffraction orders. The zero-order beam passes through the hole whose diameter exceeds a focusing spot thus preventing the distortion of the wave front. The first-order beam is focused onto a small-sized pinhole. Due to the diffraction behind the screen, "an ideal" spherical wave is generated. The reference wave expands toward a CCD camera where it interferes with a wave passed through the lens under test. Movement the grating leads to phase modulation of the wave. To estimate the accuracy of the measurements, the screen with a "large" hole and the pinhole is replaced with analogous screen with two pinholes, and then the interference patterns made by two "reference" spherical waves can be analyzed in an experiment similar to the Young's one. In [32], it is reported on the achievement of the sub-angstrom accuracy of measurement of the wave-front aberrations in the numerical aperture of $NA \approx 0.1$.

The main advantage of the interferometer is using central, the least aberrated parts of the wave fronts. The disadvantage is that it can be applied for the final certification of the lenses but not for direct by studying the shape of individual surfaces. Therefore, it is ineffective at the manufacturing stage of optical elements.

To study the reflecting surfaces, PDI, described in [33], can be applied. The interferometer's disadvantage is that a sample and the registering system are irradiated by side parts of the spherical front, that is, the most aberrated ones.

Both interferometers' disadvantage is a quite big aberration of the referent spherical wave, growing strongly with increasing the angle of observation, due to interaction between secondary waves

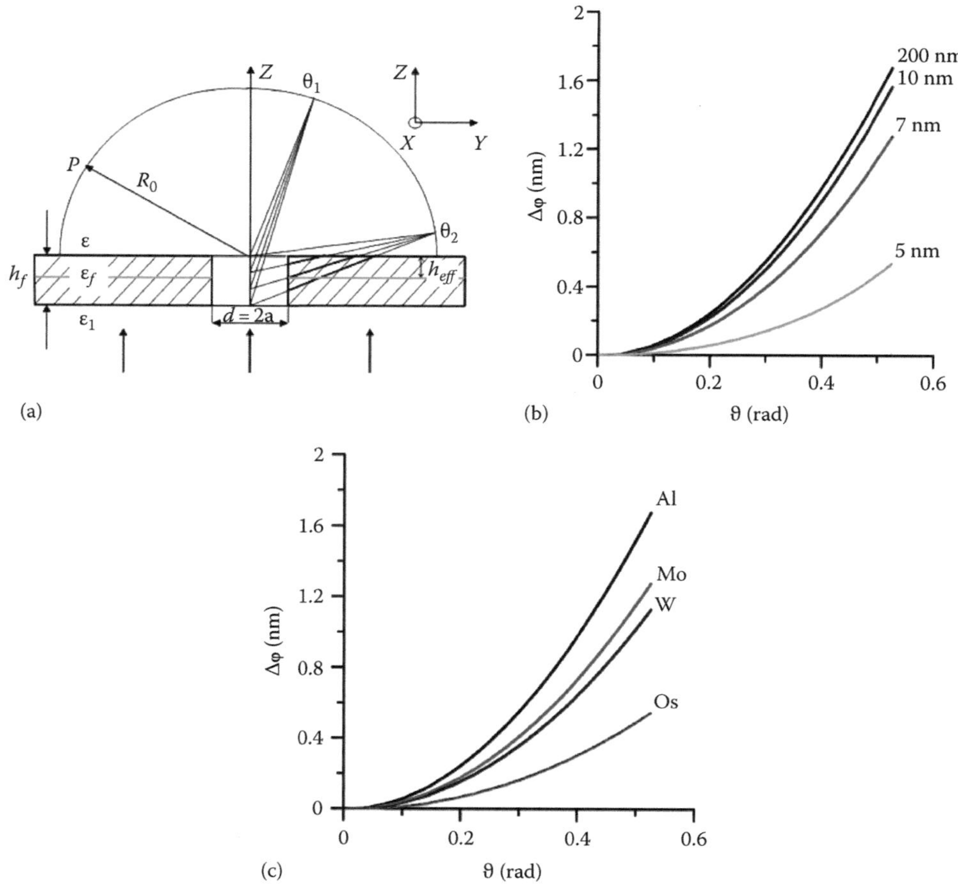

FIGURE 32.15 Geometry of the experiment. h_f, film thickness; h_{eff}, effective film thickness (characterized by skin layer thickness for the material); d, pinhole diameter; ε, ε_1, and ε_f, dielectric permittivity for media and the film (a). Phase shift $\Delta\varphi(\theta) = (\varphi(0) - \varphi(\theta))\cdot\lambda/2\pi$ on the polar angle for different thicknesses of Al film ($\varepsilon_{Al} \approx -54{,}2-i21{,}8$), a = 150 nm (b). Phase shift on the polar angle for different materials (c).

and material of the screen [30,34]. In Figure 32.15a, the geometry of the experiment can be seen. There are also plots illustrating the influence of the screen's finite width (*Al*, Figure 32.15b) and material (Figure 32.15c) on diffracted wave aberration as a function of the angle of observation. The entire phase shift is accumulated in the skin layer with a thickness of $\delta \approx c/\Omega_p \approx 10 \div 20$ nm for good conductors, characterized by plasma frequencies of $\Omega_p \sim 10^{16}$ rad/c. According to the geometry (Figure 32.15a), the greater the polar angle, the stronger the role of the film material, because the radiation is partially propagated through an optically denser material, which leads to a larger phase shift. Other effects, leading to a phase front distortion, are the excitation of plasmon–polariton and waveguide modes in the neighborhood of the hole.

Smaller phase shifts are observed at the working length of the interferometer λ = 13.5 nm ($\varepsilon \approx 1$), see, for example [32]; however, due to the short wavelength and also due to a narrow diffraction peak, the working aperture cannot be increased. Because of these effects the declared aberration of reference wave of 0.1 nm can be provided only in case of relatively small, about $NA \approx 0.1$, numerical apertures. Therefore, for high-aperture optical elements certification one has to "stitch" the results of measurements by the zones, which complicates the measurement procedure and leads to additional errors.

There is also a strong sensitivity of the diffracted wave's aberration to the primary optics aberration and the accuracy of adjustment of the pinhole and the focused beam axis [33]. These problems prevent the wide application of this type of interferometer.

32.3.2 PDI Based on Tipped Fiber with a Sub-Wavelength Output Aperture

The mentioned problems can be partly solved by using a tipped fiber with an output aperture of reduced to sub-wavelength size as the source of the reference spherical wave (TFSWS), developed in IPM RAS in 2008 [35]. The electron-microscopic image of the source can be seen in Figure 32.16c. The experimental study of the aberrations of generated wave was conducted according to the Young's method, when the interference of waves from two similar sources was studied. The experimental procedure is described in detail in [30,35].

Typical interferogram and the map of aberrations corresponding to a couple of TFSWS can be seen in Figure 32.17.

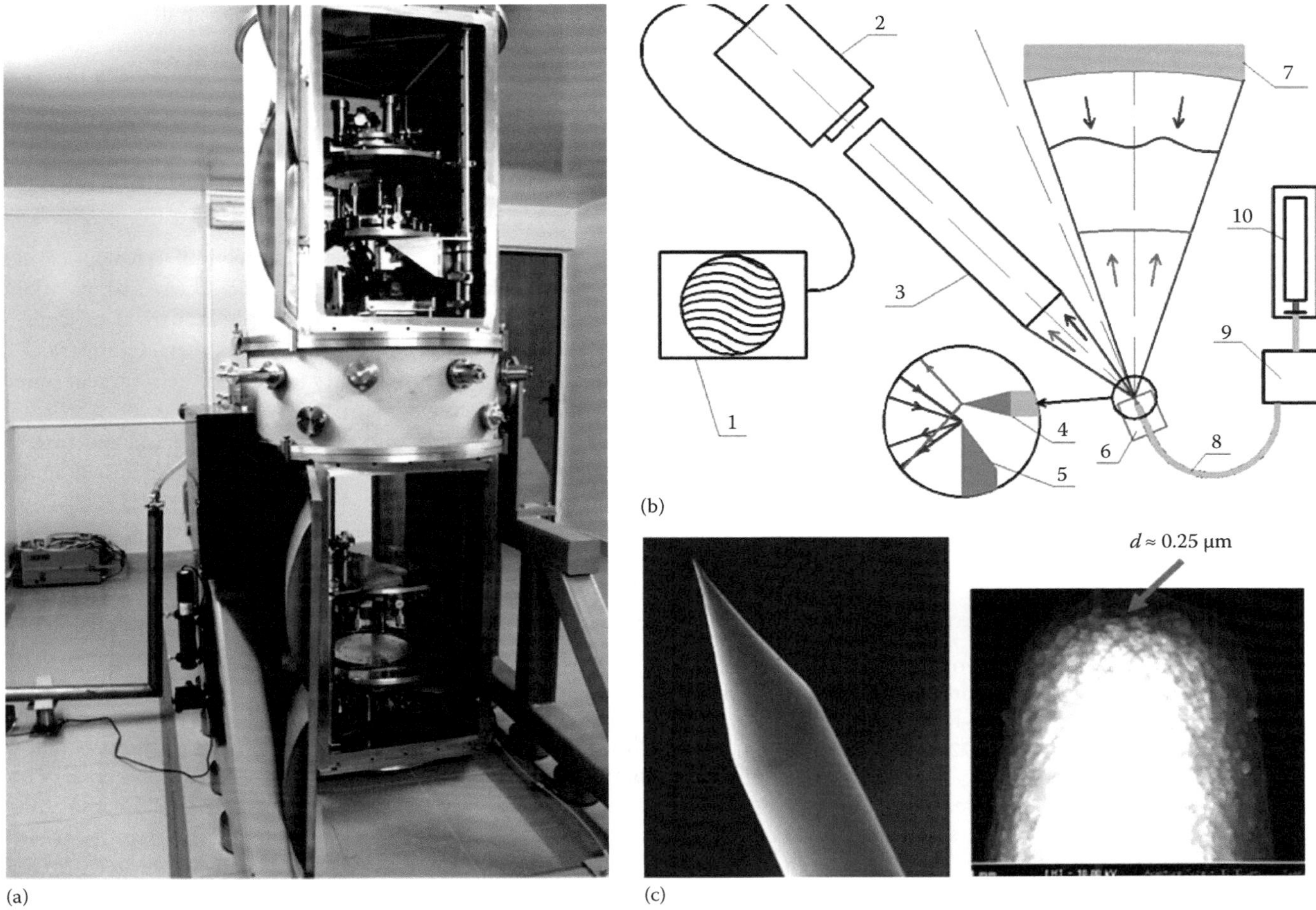

FIGURE 32.16 Photo (a) and optical scheme (b) of vacuum PDI and the electron-microscopic image of the TFSWS with an output aperture reduced to sub-wavelength size (c). (1) PC, (2) CCD, (3) observation system, (4) TFSWS, (5) flat mirror, (6) 3D table, (7) surface under investigation, (8) single-mode fiber, (9) polarization controller, and (10) laser.

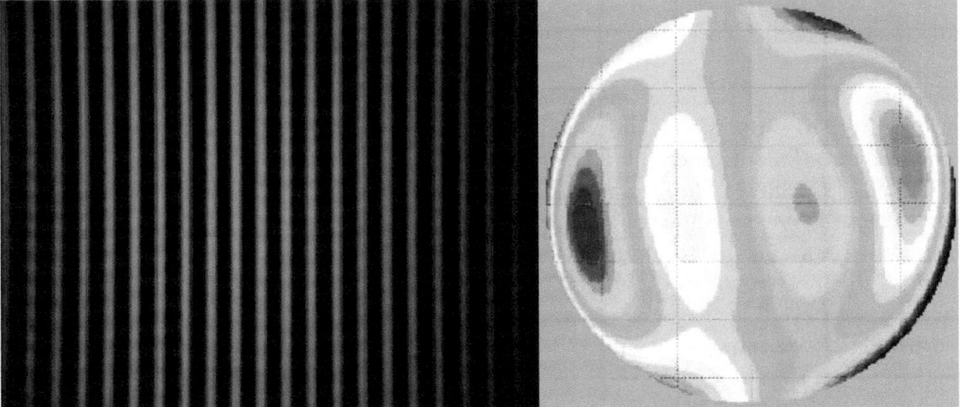

FIGURE 32.17 A typical interferogram and a wave aberration map observed in the experiments at a wavelength $\lambda = 530$ nm.

The main aberration is coma, determined not by the quality of the interfering fronts, but by the geometry of the experiment only: two sources are located off-axis of the optical system and a flat CCD detector is used for the registration. The actual mean-square wave-front aberration dependence on the numerical aperture is shown in Figure 32.18 (right curve). The measurements were made at a wavelength of 532 nm, using the second harmonic of the Nd:YAG laser. The star corresponds to a record value of a wave aberration, obtained with the traditional source of a spherical wave pinhole at the synchrotron ALS (Berkeley,

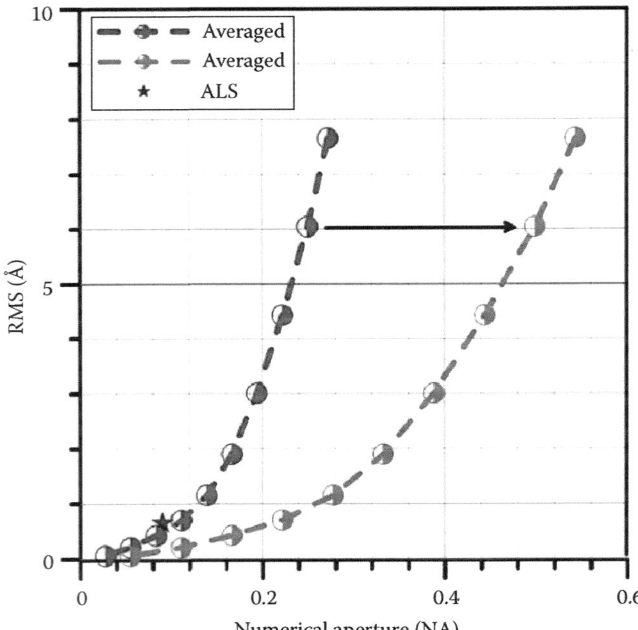

FIGURE 32.18 The mean-square aberration of TFSWS wave front—right curve; the aberration for the source in the interferometer—left curve; ALS's data for pinhole source—the star (★). (Data for ★ from Naulleau, P.P. et al., *Appl. Opt.*, 38(35), 7252, 1999.)

USA) at a wavelength of 13.5 nm [32]. It can be seen from the figure that the TFSWS has twice smaller aberration for the same numerical aperture. Moreover, due to the small operating wavelength the aperture of the ALS interferometer is limited to $NA \approx 0.1$, while a fiber-based source has sub-nanometer aberration for $NA > 0.5$. Advantages of this method of forming the reference spherical wave are defined by a convex shape of the source, the absence of the screen behind it, and a high intensity of the diffracted wave, because the well-established technique of inputting the radiation into the fiber core (diameter of 4 μm) is applied.

A vacuum interferometer, based on such type of a spherical wave source, was produced and has been successfully working in IPMRAS for >4 years now [30,36]. Photo and optical scheme of the IPM RAS's interferometer can be seen in Figure 32.16a and b. The interferometer's scheme is similar to the one described in [33] (the sample under investigation and the recording system are irradiated by side parts of the wave front), so the aberrations of the reference spherical wave behavior due to numerical aperture of the sample under investigation is shown in Figure 32.18 (left curve).

This disadvantage can be eliminated by using an optical scheme with two sources and a low-coherence light, proposed in [37]. Currently, a prototype of such interferometer has already been created and it confirmed the efficiency of the principles on the basis of the device.

In addition to the record technical characteristics, this interferometer has a number of excellent operational properties: high intensity of the reference wave, a well-developed fiber-optic infrastructure, the interferometric part's small size, tolerance to aberrations of the primary optics powering the fiber, low requirements for alignment, etc. All these allow using this type of device in production, that is, at a factory.

32.3.3 Applying PDIs to Ultraprecision Optics Characterization

Let us discuss the main schemes of applying PDIs for optical elements and systems attestation. The basic scheme, which allows studying the concave spherical or slightly aspherical surfaces, is shown in Figure 32.16. To study the convex surfaces, spherical or slightly aspherical, it is necessary to convert the divergent spherical front into a convergent one (see Figure 32.19). According to the required accuracy and the optics' numerical aperture for these purposes, single-lens (spherical or aspherical) or multilens systems can be used.

Investigation of the aberrations for transparent optical elements (single lens or objective) can be done in two ways: by using the working wave-front reflection from a mirror (Figure 32.20a) or without any mirror, due to TFSWS's small overall dimensions, a second source can be placed into the focusing area of the investigated optical element (Figure 32.20b).

Despite the impressive success of interferometry with a diffraction reference wave, strictly speaking, the problem of optics attestation with sub-nanometer precision can be considered solved only for spherical surfaces and lenses. Direct testing of aspherical surfaces by using an interferometer with the spherical reference wave is often difficult or even impossible. For example, Figure 32.21a shows the interferogram for a concave aspherical substrate with about 7 μm deviation of the surface from the nearest sphere. The measurements were carried out by using the scheme shown in Figure 32.16b. It is evident that a large number of patterns make the position determination of the minima impossible. Moreover, the upper part of the interferogram cannot be seen, because the diameter

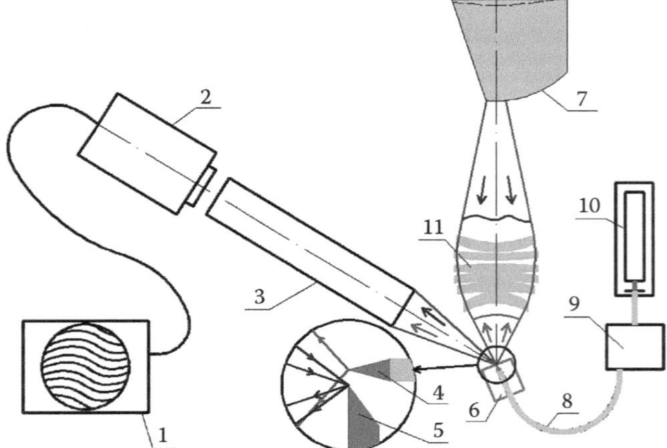

FIGURE 32.19 Optical scheme for certification of the convex surfaces. Designations on the scheme correspond to Figure 32.18b except for item 11 which is a lens.

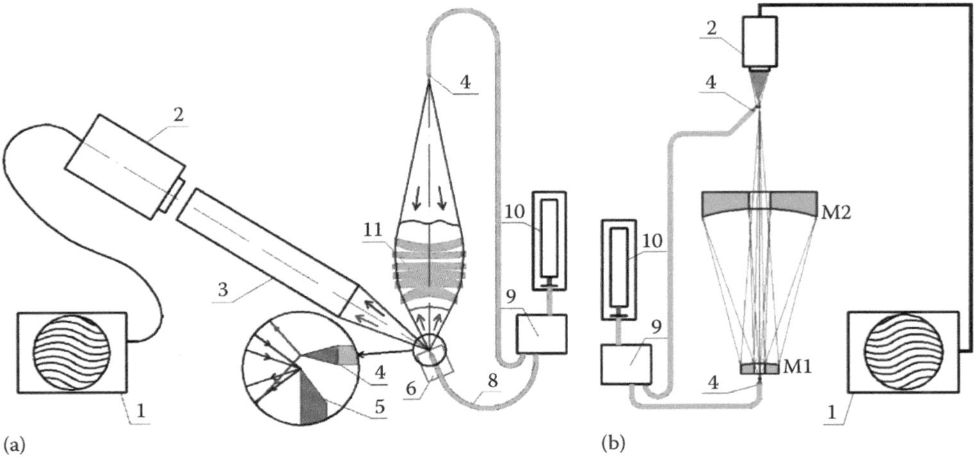

FIGURE 32.20 Measurement schemes of transparent optical elements with using the working wave-front reflection from a flat mirror (a) and without any mirror (b). Designations on the scheme correspond to Figure 32.18b except for items M1 and M2 which form an objective.

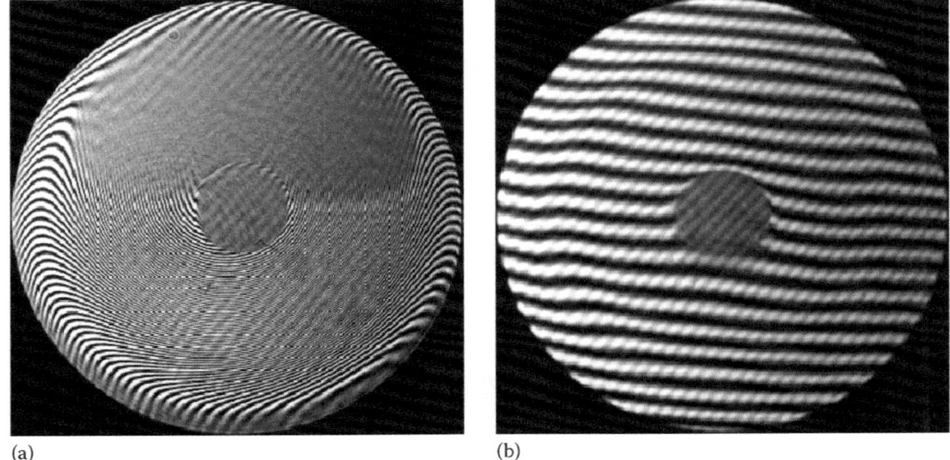

FIGURE 32.21 Interferograms from the aspherical concave mirror, obtained without wave-front corrector (a) and by using corrector (b).

of the focusing spot is about 100 μm, so the beam partially propagates above the edge of mirror (item 5, Figure 32.16b) and misses the measurement system.

One of the most effective solutions to this problem is the use of wave-front correctors, that is, special optical elements that transform a reference spherical wave front into an aspherical one with a shape, identical to the shape of the surface under investigation. To take into account the wave-front distortions, provided by corrector's own errors, its surfaces should be only spherical. In this case, it is possible to test these surfaces with sub-nanometer accuracy by using PDIs.

The scheme for testing the concave aspherical surface by using a corrector is presented in Figure 32.22. The parameters of the corrector, suitable for testing the concave aspherical surface, that is, radii of the surfaces, thickness, and the distance from the reference wave source to the top of the first surface, were calculated by minimizing the difference of longitudinal aberrations of the tested aspherical surface and the corrector (the procedure described in [26,38] in detail). It can be clearly seen from

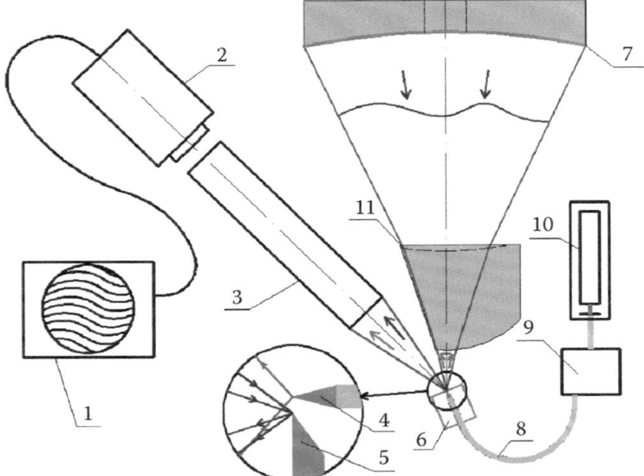

FIGURE 32.22 Optical scheme for attestation of concave aspherical surfaces using a corrector. Designations on the scheme correspond to Figure 32.18b except for item 11 which is a corrector.

Figure 32.21b that the interferogram from the aspherical surface, obtained by using such corrector, can be reliably processed. Accounting of wave-front deformation produced by the correctors is quite another challenging problem. The only exceptions are ellipsoids, which have two focuses and then can be tested by PDI developed by the authors without any corrector, just by placing two TFSWSs in the ellipse's focuses.

Convex aspheric surfaces certification is complicated by the necessity of converting diverging spherical front to a converging aspherical one with a shape that coincides with an aspherical surface under investigation. It was shown that the problem cannot be solved by a single lens with spherical surfaces. In our works, we study such surfaces using the scheme shown in Figure 32.20b; a specially designed aspherical concave mirror is applied as the wave-front corrector.

Certification of telescopes, having one of their focuses "at infinity," can be performed by using the autocollimator scheme with a planar reference (see Figure 32.23a). And the reference itself may be studied in a way as shown in Figure 32.23b.

Summarizing this paragraph we can state that with all mentioned reservations about working numerical apertures and errors, produced by wave-front correctors, PDIs can be widely used to study the aberrations practically of all types of optical components and systems with a high accuracy up to subangstrom level. In practice, a special attention has to be paid to the mounting samples under study into the interferometer; otherwise, the surface may be significantly deformed just by fixing or gravity. A great contribution to measurement error makes the observation optical system. Moreover, even a thin glass window on the CCD matrix provides notable distortion to the registered wave fronts [21,39]. Therefore for ultraprecision optics attestation, the interfering front's registration must be done directly on the matrix, without any optical elements on the front's way.

32.4 Shape Correction Methods for Optical Surfaces

Ultraprecision optics manufacturing consists of two stages. On the first stage, standard methods of polishing and surface's form and roughness attestation applied to manufacturing a supersmooth surface. On the second stage, the form shape is corrected to sub-nanometer accuracy using the methods of local ion-beam etching and/or thin film vacuum deposition; the attestation is made by PDI. There are two problems to be solved by this method of correction. The first one is fixing local errors, that is, elimination of the "hills" in the case of ion-beam etching or "valleys" in the case of thin film deposition. The second problem is aspherization of originally spherical samples. This aspherization method has obvious benefits over traditional one based on mechanical cutting and polishing by low-sized tool. When polishing spherical and flat surfaces of the polishing tool and the polished detail are in contact with each other over wide area, the surfaces lap each other. In this case, the best quality of the surface can be achieved. But an aspherical surface has a variable radius, which makes this method inapplicable. So, in practice an aspherical shape is formed by a low-sized tool, which leads to surface roughness development in the MSFR range. Thus, the aspherization of spherical substrates by methods like ion-beam etching creates the preconditions for producing aspheric surfaces of higher quality.

32.4.1 Correction Schemes and Equipment

The aspherization scheme for axial symmetric surfaces, using ion-beam etching, is shown in Figure 32.24. On the ion beam's way there is a diaphragm with a profile, specially designed for the desired aspherical shape and the radial dependence of the ion current in a beam (neutralized in the case of etching dielectrics). The detail rotates continuously around its axis.

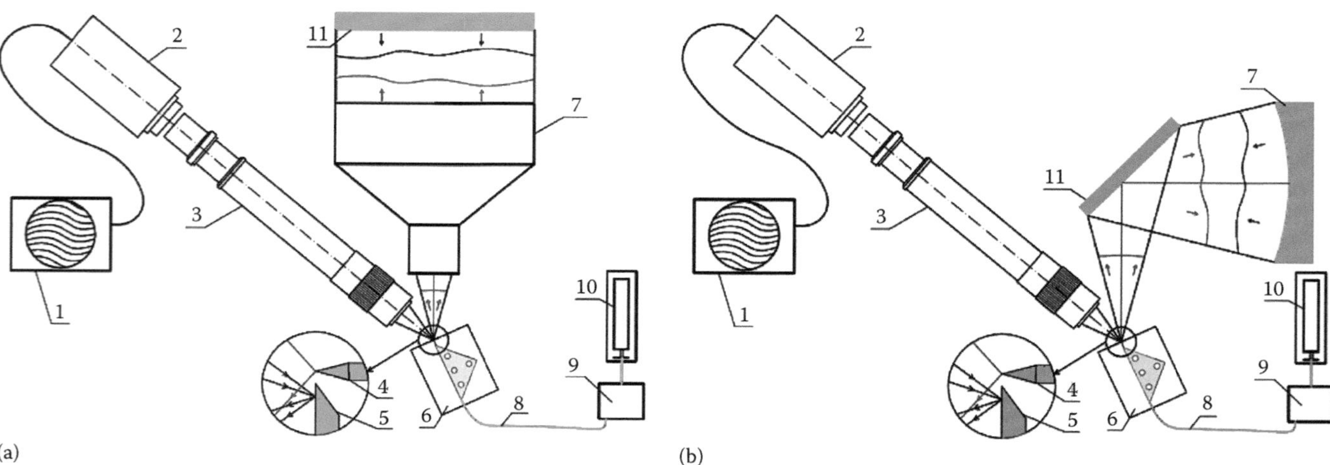

FIGURE 32.23 Schemes for studying a telescope's aberrations (a) and a plane reference mirrors' distortions (b). Designations on the schemes correspond to Figure 32.18b except for items 7 which is a telescope in part (a) and a reference spherical mirror in part (b) and 11—a plane reference mirror.

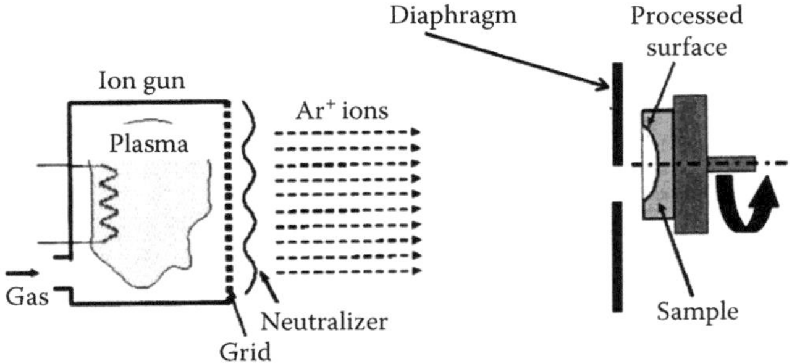

FIGURE 32.24 The aspherization scheme by using ion-beam etching.

In the case of correcting some local errors, a flat metal mask with the holes in front of the surface's "hills" is set. To smooth the sharp edges on the surface, we use retouching by focused ion beam. The photo and the scheme of the ion-beam correcting apparatus, developed in IPM RAS, can be seen in Figure 32.25. The previous generation unit [30,40] was upgraded by equipping with an ion source with cold cathode, which allows working with chemically active gases. The goniometer's size and movements allow us to correct samples with any surface's shape and diameter up to 300 mm.

For samples correction by thin films deposition, the apparatus with magnetron sputtering can be used. Some information about its construction and operating modes can be found in [30,41].

32.4.2 Studying Etching Effects on Surface Roughness

A key requirement to the process of correction is keeping roughness on the initial atomically smooth level. Thus, there is a wide front of studying the effect of technological process' parameters on surface roughness. The analysis of a large number of works shows that the experimental results are quite controversial. In particular, in some papers the recommended parameters for ion-beam etching are small grazing angles of incidence and energies of ions (neutral atoms) above 1 keV. A strong dependence of the parameters of the etching process on the substrate material is also noticed [42–44].

Our studies of the ion-beam parameters and etching modes influence on surface roughness have shown that the surface roughness of supersmooth Cr/Sc multilayer structures and fused silica remains at the initial stage (see Figure 32.26) if the ion's angle of incidence is close to the normal. At the same time, the energy of the ions (or neutral atoms) should not exceed 200 eV for Cr/Sc and 800 eV for fused silica. On the contrary, grazing incidence of the ion beam leads to the development of roughness in the whole spectrum of spatial frequencies (see Figure 32.26b).

In our experiments with fused silica, the mentioned regularities were confirmed up to 11 μm etching depths (the dynamics of *PSD* functions vs. etching depth shown in Figure 32.27), which indicates the possibility of a sufficiently deep substrates aspherization for this material.

This behavior of roughness can be explained qualitatively by the fact that for normal angles of incidence the penetration depth of ions increases, so decreases the fraction of energy transmitted to surface atoms. In this case, only the least bounded atoms escape the surface (etching process), thus the peaks flatten or at least, the surface roughness does not develop. The similar effect can be observed when ions (atoms) with low energies are used in etching process.

The described influence of ion beams' parameters on surface roughness is confirmed by the analysis of the specular reflection coefficients of hard x-rays from both the substrate and the short-period multilayer mirror deposited on them. For mirrors, the effect of roughness is magnified due to the wave interference from a large number of boundaries. In [13], an obvious clear correlation between roughness data of etched substrates (Figure 32.26) and reflection coefficients of multilayer mirrors deposited on the substrates was shown.

Etching of some materials with extremely low coefficients of thermal expansion, like Zerodur and Sitall (Russian analog of Zerodur), was also studied. *PSD* function's evolution during

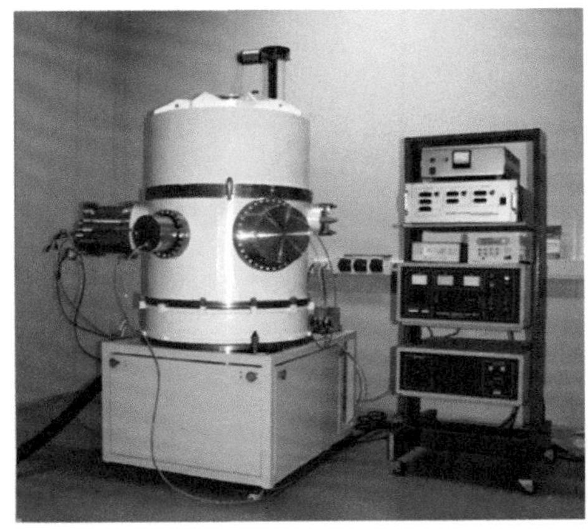

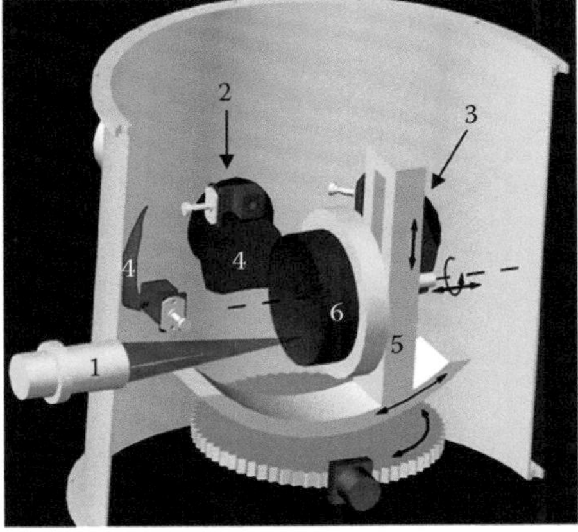

FIGURE 32.25 The photo and the scheme of the ion-beam correcting apparatus, developed in IPM RAS. (1) Focusing ion-beam gun, (2) source with cold cathode, (3) source with the filament cathode, (4) shutter, measuring the ion-beam current, (5) five-axis goniometer, and (6) processed sample.

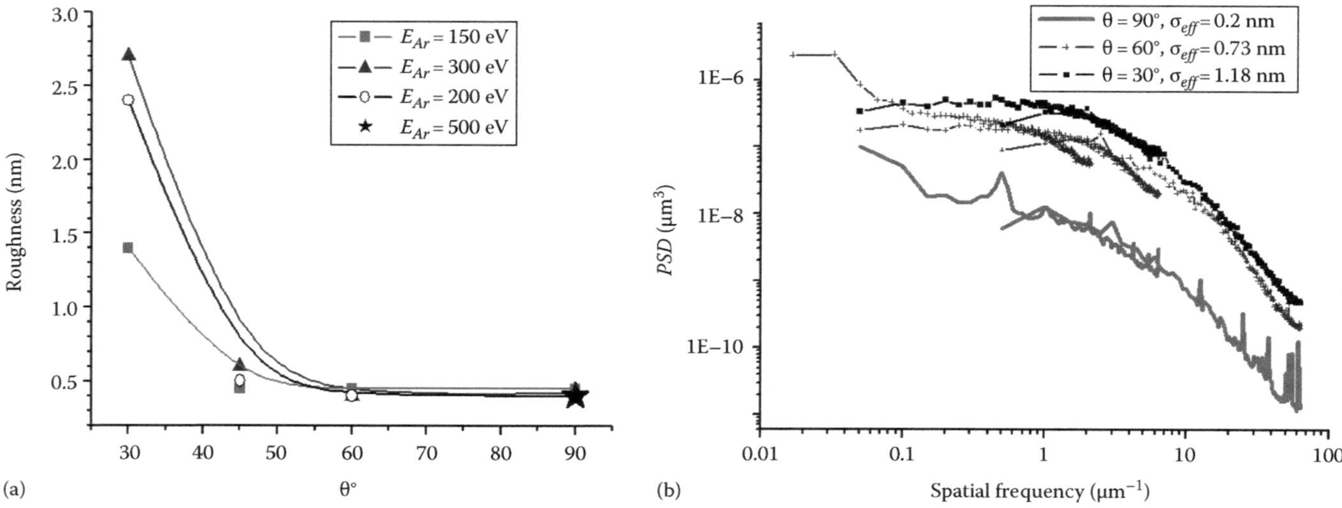

FIGURE 32.26 The dependence of the fused silica's surface roughness on accelerated Ar atoms' angle of incidence for energies E_{Ar} 150, 200, 300, and 500 eV (a) and the dynamics of *PSD* functions for Cr/Sc multilayer structure etched by Ar with E_{Ar} = 200 eV (b).

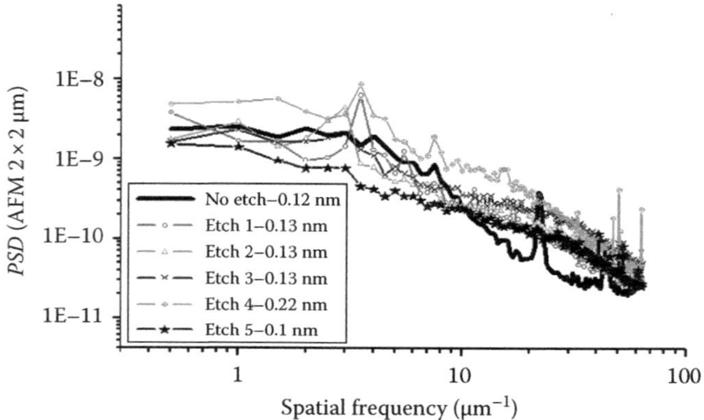

FIGURE 32.27 The dynamics of *PSD* functions for fused silica versus etching depth. Ar atoms' energy of 300 eV, normal incidence. The material removals are etch 1, 1.8 μm; etch 2, 3.3 μm; etch 3, 5.1 μm; etch 4, 8.2 μm; and etch 5, 11 μm.

etching is shown in Figure 32.28. It can be seen from Figure 32.28a that after etching Sitall to 1 μm the effective roughness σ_{eff} = 0.44 nm slightly improved in comparison with the initial one. After etching to a depth of 3.8 μm, the effective roughness increased to σ_{eff} = 0.51 nm mainly due to the HSFR part. The MSFR are practically the same. Figure 32.28b shows the *PSD* functions for Zerodur substrate. It can be seen that in contrast to Sitall, development of surface roughness even for "small" material removal is observed, and the roughness growth can be seen in both HSFR and MSFR ranges. The effective roughness changed from σ_{eff} = 0.57 nm to σ_{eff} = 1.08 nm for etching depth of 2 μm. Thus, it can be seen that despite the similar functions, these are two materials with differing physical and chemical properties.

A situation is more complicated for monocrystalline Si. As for the fused silica and Cr/Sc, its surface roughness remains on the same level or even can be improved if the angle of etching

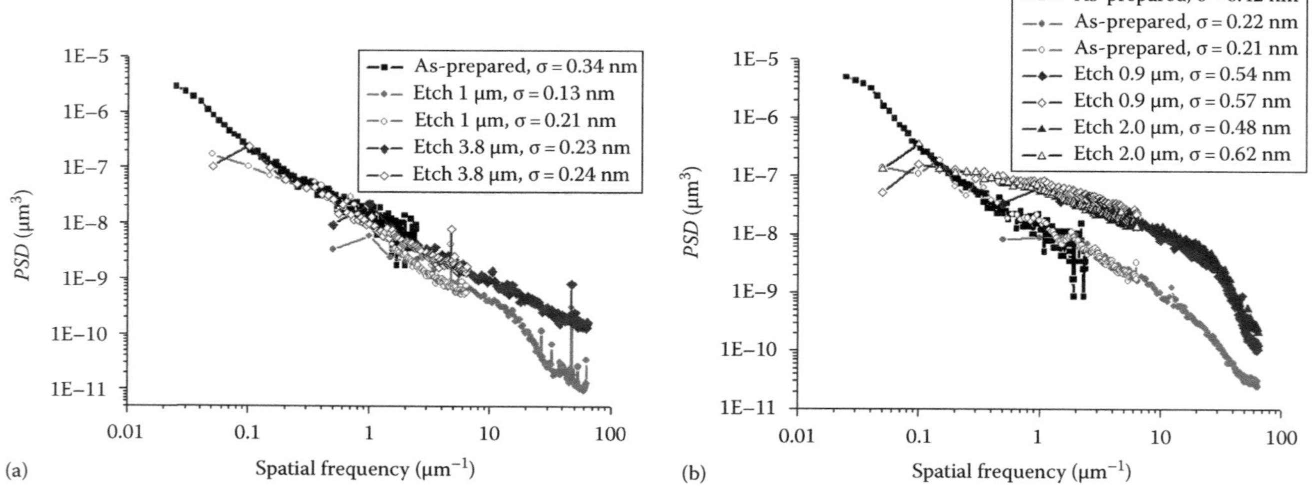

FIGURE 32.28 *PSD* functions of Sitall (a) and Zerodur (b) versus etching depths.

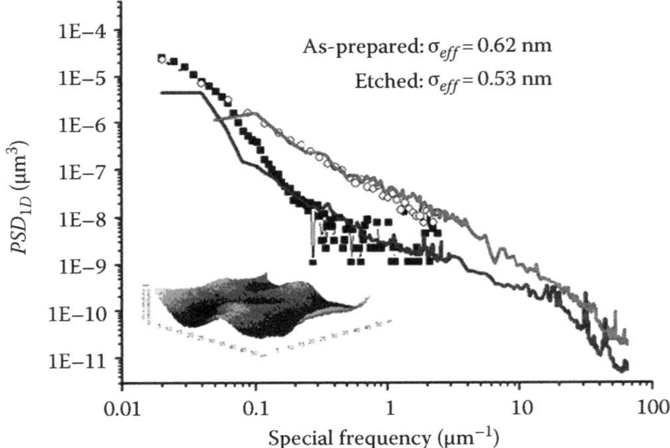

FIGURE 32.29 *PSD* functions of monocrystalline Si before (upper curves) and after (lower curves) etching by Ar^+ ions, and the AFM image of the surface after etching.

ion's incidence is close to a normal [44]. However, the range of acceptable angles for this effect is quite narrow, 50°–60° [44,45]. Also in this case, an additional relief is formed on the surface: the waves with a period $L \sim 20$ µm (see Figure 32.29). Although the amplitude (h) of the wave is very small, usually a fraction of a nanometer, this can lead to a rather large angular error ($h/L \sim 10^{-4}$–10^{-5}) for x-ray radiation wave front reflected from the surface. Thus, currently developed technology of the ion-beam correction is not applicable for high-precision silicon-based optics production, and requires some significant improvement.

In general, according to the literature data and authors' experiments, it can be concluded that at present there are well-developed technology for manufacturing ultraprecision supersmooth optics based on fused silica and optical ceramics like Zerodur, ULE, and Sitall. Figure 32.30 illustrates the capabilities of ultraprecision optics production end and meteorology techniques, developed in IPM RAS. In Figure 32.30a, there is photo of a spherical mirror in a frame with NA = 0.25, and the map of surface deformation (Figure 32.30b). The surface form deviations from an ideal sphere are PV = 7.3 nm and RMS = 0.6 nm.

32.5 Conclusion

Currently, we are only at the beginning of the practical application of optics, manufactured with angstrom surface roughness and form accuracy. However, the existing technologies have made it already possible to obtain the outstanding results in the study of the Sun and in the EUV projection lithography. Interesting results and extensive applications can be expected in soft x-ray microscopy in the "carbon" and "water" spectral transparency windows ($\lambda \approx$ 2.3–5 nm) and next-generation x-ray lithography with the operating wavelength of 6.7 nm. In general, it is evident that the problem of manufacturing such optics for high-resolution EUV lithography and x-ray astronomy is already solved in a first approximation. Further activity should be directed toward increasing both the manufacturing accuracy of optical surfaces up to a level of RMS = 0.1 nm and the numerical aperture of objectives up to NA = 0.4–0.6.

In this regard, upgrading the interferometric methods for surface shape measurements of optical elements and wave-front aberrations of objectives, and also increasing their working numerical aperture are the subjects of current interest. The diffraction interferometers utilizing the tipped fiber as a source of the reference spherical wave look rather attractive. Owing to the simplicity of the design, high intensity of the diffracted wave, weak sensitivity to the adjustment in an optical circuit, and a number of other advantages, these simple in operation interferometers can be used in optical industry.

In this chapter, we have not practically discussed the problems of aspherical surfaces investigation, the development and optimization of correctors, transforming the reference spherical fronts into the aspherical ones whose shapes are identical to the studied surfaces. Meanwhile, traditional lens null correctors have limited capabilities, rather complicated in manufacturing and expensive, and demand some complicated calibrating

(a)

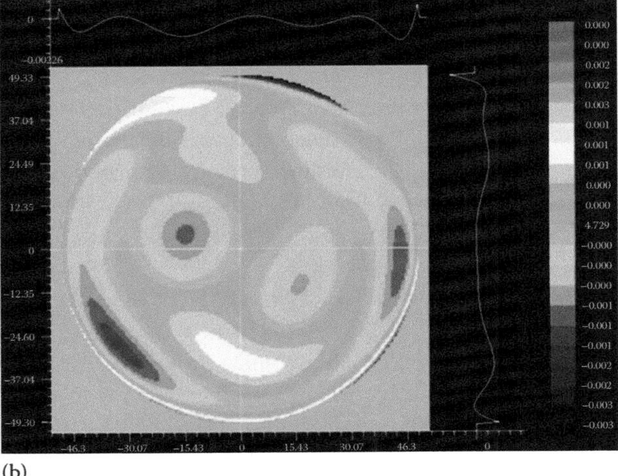

(b)

FIGURE 32.30 (a) Photo of spherical mirror in frame, and (b) map of surface distortions: P–V = 7.3 nm, RMS = 0.6 nm.

procedures for taking into account the wave-front distortions due to errors of the corrector itself. In this regard, the diffraction correctors are of a significant interest [46].

A great potential for dielectric materials' surface shape correction by ion-beam etching technique has RF etching in chemically active environments. The first experiments on etching quartz substrates have shown that the surface micro-roughness did not change the material removal depths of a few tens of micrometers [47].

In conclusion, it should be noted that despite the widespread belief that the problem of roughness measurements is solved for the entire range of lateral dimensions, there is still a challenge. Moreover, for the optics of ultrahigh spatial resolution, using AFM allows one to get the accurate value of r.m.s.; however, due to nonlinearity of the piezo-scanner, detecting the surface waves with angles $<10^{-5}$ is not possible. This limitation should be clearly understood when the expected spatial resolution of the optical system is estimated.

References

1. Gwyn, C. 1998. White paper on extreme ultraviolet lithography. *Proceedings of EUV LLC*, Livermore, CA.
2. Benschop, J.P.H., Kaiser, W.M., and Ockwell, D.C. 1999. EUCLIDES, the European EUVL program. *SPIE Symposium on Micro-Lithography*, Santa Clara, CA, Vol. 3676, 1999, p. 246.
3. Naulleau, P., Golberg, K.A., Anderson E.H. et al. 2002. Sub-70 nm extreme ultraviolet lithography at the advanced light source static microfield exposure station using the engineering test stand set-2 optic. *J. Vac. Sci. Technol. B*, 20, 2829–2833.
4. Ota, K., Murakami, K., Kondo, H., Oshino, T., Sugisaki, K., and Komatsuda, H. 2001. Feasibility study of EUV scanners. *Proc. SPIE* 4343, 60–69.
5. Andreev, S.S., Bulgakova, S.A., Gaponov, S.V., Gusev, S.A., Zuev, S.Yu., Kluenkov, E.B., Luchin, V.I., Lopatin, A. Ya., Mazanova, L.M., Prokhorov, K.A., Sadova, E.N., Salashchenko, N.N., and Shamov, E.A. 2000. Investigations in field of projection lithography of extreme ultraviolet region in Institute for physics of microstructures RAS. *J. Surf. Investig. X-Ray, Synchrotron Neutron Tech.* 1, 32–41 (in Russian).
6. Cheng, P. 1987. *Instrumentation and Biological Applications. X-Ray Microscopy*. Berlin, Germany: Springer Verlag.
7. Salashchenko, N.N. and Chkhalo, N.I. 2008. Shortwave projection nanolithography. *Herald Russian Acad. Sci.* 78(3), 279–285.
8. Dinger, U., Eisert, F., Lasser, H., Mayer, M., Seifert, A., Seitz, G., Stacklies, S., Stickel, F.-J., and Weiser, M. 2000. Mirror substrates for EUV lithography: Progress in metrology and optical fabrication technology. *Proc. SPIE* 4146, 35–46.
9. Blunt, R. 2006. White light interferometry—A production worthy technique for measuring surface roughness on semiconductor wafers. *CEMANTECH Conference*. Vancouver, British Columbia, Canada, April 24–27, pp. 59–62.
10. Griffith, J.E. and Grigg, D.A. 1993. Dimensional metrology with scanning probe microscopes. *J. Appl. Phys.* 74, R83–R109.
11. Asadchikov, V.E., Kozhevnikov, I.V., Krivonosov, Yu.S., Mercier, R., Metzger, T.H., Morawe, C., and Ziegler, E. 2004. Application of X-ray scattering technique to the study of supersmooth surfaces. *Nucl. Instrum. Meth. Phys. Res. A* 530, 575–595.
12. Chkhalo, N.I., Fedorchenko, M.V., Kruglyakov, E.P., Volokhov, A.I., Baraboshkin, K.S., Komarov, V.F., Kostyakov, S.I., and Petrov E.A. 1995. Ultradispersed diamond powders of detonation nature for polishing X-ray mirrors. *Nucl. Instrum. Meth. Phys. Res. A* 359, 155–156.
13. Vainer, Yu.A., Zorina, M.V., Pestov, A.E., Salashchenko, N.N., Chkhalo, N.I., and Khramkov, R.A. 2011. Evolution of the roughness of amorphous quartz surfaces and Cr/Sc multilayer structures upon exposure to ion-beam etching. *Bull. Russian Acad. Sci. Phys.* 75(1), 61–63.
14. Barysheva, M., Vainer, Yu.A., Gribkov, B.A., Zorina, M.V., Pestov, A.E., Rogachev, D.N., Salashenko, N.N., and Chkhalo, N.I. 2011. Particulars of studying the roughness of substrates for multilayer X-ray optics using small-angle X-ray reflectometry, atomic-force, and interference microscopy. *Bull. Russian Acad. Sci. Phys.* 75(1), 67–72.
15. Barysheva, M.M., Gribkov, B.A., Vainer, Yu.A., Zorina, M.V., Pestov, A.E., Platonov, Yu.Ya., Rogachev, D.N., Salashchenko, N.N., and Chkhalo, N.I. 2011. Problem of roughness detection for supersmooth surfaces. *Proc. SPIE* 8076, 80760M-1-10.
16. Kozhevnikov, I.V. and Pyatakhin, M.V. 2000. Use of DWBA and perturbation theory in X-ray control of the surface roughness. *J. X-Ray Sci. Technol.* 8, 253–275.
17. Volgunov, D.G., Zabrodin, I.G., Zakalov, B.A., Zuev, S.Yu., Kas'kov, I.A., Kluenkov, E.B., Toropov, M.N., and Chkhalo, N.I. 2011. A stand for a projection EUV nanolithographer–multiplicator with a design resolution of 30 nm. *Bull. Russian Acad. Sci. Phys.* 75(1), 49–52.
18. Ziegler, E., Peverini, L., Vaxelaire, N. et al. 2010. Evolution of surface roughness in silicon X-ray mirrors exposed to a low-energy ion beam. *Nucl. Instrum. Meth. Phys. Res. Sect. A* 616(2–3), 188–192.
19. Chkhalo, N.I., Barysheva, M.M., Pestov, A.E., Salashchenko, N.N., and Toropov, M.N. 2011. Manufacturing and characterization the diffraction quality normal incidence optics for the XEUV range. *Proc. SPIE* 8076, 80760P-1-13.
20. Azarova, V.V., Dmitriev, V.G., Lokhov, Yu.N., and Malitskii, K.N. 2002. Measuring the roughness of high-precision quartz substrates and laser mirrors by angle-resolved scattering. *J. Opt. Technol.* 69(2), 125–129.
21. Salaschenko, N.N., Toropov, M.N., and Chkhalo, N.I. 2010. Physical limitations of measurement accuracy of the diffraction reference wave interferometers. *Bull. Russian Acad. Sci. Phys.* 74(1), 53–56.
22. Chason, E. and Mayer, T.M. 1993. Low energy ion bombardment induced roughening and smoothing of SiO_2 surfaces. *Appl. Phys. Lett.* 62(4), 363–365.

23. Van Loyen, L., Botter, T., Braun, S. et al. 2003. New laboratory EUV reflectometer for large optics using a laser plasma source. *Proc. SPIE* 5038, 12–21.
24. Bibishkin, M.S., Chekhonadskih, D.P., Chkhalo, N.I., Klyuenkov, E.B., Pestov, A.E., Salashchenko, N.N., Shmaenok, L.A., Zabrodin, I.G., and Zuev, S.Yu. 2004. Laboratory methods for investigation of multilayer mirrors in extreme ultraviolet and soft X-ray region. *Proc. SPIE* 5401, 8–15.
25. Born, M. and Wolf, E. 1973. *Principles of Optics*, 2nd edn. Moscow, Russia: Science (in Russian).
26. Malacara, D. 1992. *Optical Shop Testing*, 2nd edn. New York: John Wiley & Sons.
27. Website Zygo Corporation. http://www.zygo.com
28. Linnik, V.P. 1933. Simple interferometer for the investigation of optical systems. *Bull. Acad. Sci. USSR* 1, 208–210.
29. Sommargren, G.E. 1996. Diffraction methods raise interferometer accuracy. *Laser Focus World* 8, 61–71.
30. Chkhalo, N.I., Pestov, A.E., Salashchenko, N.N., and Toropov, M.N. 2010. Manufacturing and investigating objective lens for ultrahigh resolution lithography facilities. *Lithography*. Wang, M. (Ed.). Vukovar, Croatia: INTECH. Available from: http://sciyo.com/articles/show/title/manufacturing-and-investigating-objective-lens-for-ultrahigh-resolution-lithography-facilities
31. Medecki, H., Tejnil, E., Goldberg, K.A., and Bokor, J. 1996. Phase-shifting point diffraction interferometer. *Opt. Lett.* 21(19), 1526–1528.
32. Naulleau, P.P., Goldberg, K.A., Lee, S.H., Chang, C., Attwood, D., and Bokor, J. 1999. Extreme-ultraviolet phase-shifting point-diffraction interferometer: A wave-front metrology tool with subangstrom reference-wave accuracy. *Appl. Opt.* 38(35), 7252–7263.
33. Otaki, K., Ota, K., Nishiyama, I. et al. 2002. Development of the point diffraction interferometer for extreme ultraviolet lithography: Design, fabrication, and evaluation. *J. Vac. Sci. Technol. B* 20(6), 2449–2458.
34. Chkhalo, N.I., Dorofeev, I.A., Salashchenko, N.N., and Toropov, M.N. 2008. A plane wave diffraction on a pin-hole in a film with a finite thickness and real electrodynamic properties. *Proc. SPIE* 7025, 702507 (7p).
35. Chkhalo, N.I., Klimov, A.Yu., Rogov, V.V. et al. 2008. A source of a reference spherical wave based on a single mode optical fiber with a narrowed exit aperture. *Rev. Sci. Instrum.* 79, 033107 (5p).
36. Kluyenkov, E.B., Pestov, A.E., Polkovnikov, V.N., Raskin, D.G., Toropov, M.N., Salashchenko, N.N., and Chkhalo, N.I. 2008. Testing and correction of optical elements with subnanometer precision. *Nanotechnol. Russia* 3(9–10), 602–610.
37. Klyuenkov, E.B., Polkovnikov, V.N., Salashchenko, N.N., and Chkhalo, N.I. 2008. Shape correction of optical surfaces with subnanometer precision: Problems, status, and prospects. *Bull. Russian Acad. Sci. Phys.* 72(2), 188–191.
38. Puryaev, D.T. 1976. *Methods for Control of Aspherical Surfaces*. M: Engineering, Moscow, 13p. (in Russian).
39. Chkhalo, N.I. 2011. *Multilayer X-Ray Mirrors. Diagnostics and Applications*. LAP LAMBERT Academic Publishing GmbH & Co. KG, Saarbrückeen, Germany, 406 p.
40. Chkhalo, N.I., Kluenkov, E.B., Pestov, A.E., Polkovnikov V.N., Raskin, D.G., Salashchenko, N.N., Suslov, L.A., and Toropov, M.N. 2009. Manufacturing of XEUV mirrors with a sub-nanometer surface shape accuracy. *Nucl. Instrum. Meth. Phys. Res. A* 603(1–2), 62–65.
41. Andreev, S.S., Akhsakhalyan, A.D., Bibishkin, M.S. et al. 2003. Multilayer optics for XUV spectral region: Technology fabrication and applications. *Central Eur. J. Phys.* 1, 191–209.
42. Kurashima, Y., Miyachi, S., Miyamoto, I. et al. 2008. Evaluation of surface roughness of ULE® substrates machined by Ar^+ ion beam. *Microelectron. Eng.* 85, 1193–1196.
43. Keller, A., Facsko, S., and Möller, W. 2009. The morphology of amorphous SiO_2 surfaces during low energy ion sputtering. *J. Phys. Condens. Matter*. 21, 495305.
44. Ziegler, E., Peverini, L., Vaxelaire, N. et al. 2010. Evolution of surface roughness in silicon X-ray mirrors exposed to a low-energy ion beam. *NIM A* 616(2–3), 188–192.
45. Barysheva, M.M., Vainer, Yu.A., Gribkov, B.A., Zorina, M.V., Pestov, A.E., Salashchenko, N.N., Khramkov, R.A., and Chkhalo, N.I. 2012. The evolution of roughness of supersmooth surfaces by ion-beam etching. *Bull. Russian Acad. Sci. Phys.* 76(2), 163–167.
46. Okatov, M.A., Antonov, E.A., Baigozhin, A. et al. 2004. *Handbook for Optical-Technologist*. St. Petersburg, Russia: Politechnica, 679 p.
47. Akhsakhalyan, A.D., Vainer, Yu.A., Volgunov, D.G., Drozdov, M.N., Kluenkov, E.B., Kuznetcov, M.I., Salashchenko, N.N., Kharitonov, A.I., and Chkhalo, N.I. 2008. Application of reactive ion-beam etching for surface shape correction of X-ray mirrors. *Proceedings of Workshop "X-ray optics—2008."* Chernogolovka, Russia, pp. 26–28 (in Russian).

33

Single Molecule Electronics

Simon J. Higgins
University of Liverpool

Richard J. Nichols
University of Liverpool

33.1 Introduction

The science of *molecular electronics* can trace its beginnings to the famous 1974 paper by Aviram and Ratner,[1] in which it was suggested that a dipolar molecule with built-in electron-rich and electron-deficient parts, separated by an insulating connection, could act as a diode if it was suitably aligned in a monolayer between metallic contacts. This paper remained almost uncited for about a decade, before the development of scanning tunneling microscopy (STM) by Binnig et al.[2] opened the possibility of imaging and manipulating individual molecules on conducting surfaces. Extra impetus for an interest in molecular-scale electronics has been provided by the realization that continued downscaling of individual components in conventional complementary metal–oxide–semiconductor (CMOS)-based electronics must eventually end. This is because as component sizes shrink, (i) the cost of the associated fabrication plants rises as the necessary technology becomes more sophisticated, (ii) the wavelength of light needed for lithographic processes has to diminish, and this cannot continue indefinitely, and (iii) transistor gate insulator layers are already at thicknesses of just a few atoms.[3] With the 45 nm feature size chips introduced around 2007, Intel found it necessary to replace SiO_2 as a gate material by the much more expensive and less convenient HfO_2, which has a much higher dielectric constant,[4] but even this will be insufficient at the ultimate 11 nm node predicted by the International Technology Roadmap for Semiconductors (ITRS) to arrive in 2022,[5] as this reduction in feature size will mean increased problems with gate current leakage owing to electron tunneling across the gate oxide becoming dominant.[3]

Organic materials, such as semiconducting oligomers and polymers, are already finding commercial applications in electronic devices such as organic light-emitting diode–based displays[6] and "plastic transistors" in e-readers[7]; other applications such as organic photovoltaics are in an advanced state of precommercial development.[8] The birth of the field of semiconducting polymers dates back to the landmark paper[9] on the oxidative "doping" of polyacetylene films by halogens, which won its corresponding authors (Heeger, Shirakawa, and MacDiarmid) the 2000 Nobel Prize for Chemistry. This area of science, in which it is the *bulk material* properties of the conjugated oligomers or polymers that are crucial to device operation, is confusingly sometimes described both as "organic electronics" and "molecular electronics."

In what follows, we use "molecular electronics" more specifically, to refer to the use of single molecules (or small assemblies of molecules) in devices where some property of the individual molecule(s) is utilized for operation. The "bottom-up" syntheses of nanometer-sized individual molecules offer an alternative to the traditional "top-down" approach of lithography and etching used in conventional electronics. Molecules offer the advantage that chemists can exert exquisite control over their structure and properties through synthetic chemistry. Molecular synthesis is a good example of a "bottom-up" technology; complex molecules are built up from simpler precursors, and vast numbers of completely identical molecules can readily be made. Other "bottom-up" techniques, such as self-assembly or Langmuir–Blodgett deposition, will then be required to build up nanoscale assemblies of these molecules. However, there are several important issues that need to be addressed before molecular-scale electronic devices become a realistic prospect.

The first issue is a practical one; whatever the designed function of the molecule(s) concerned, electrical contact must

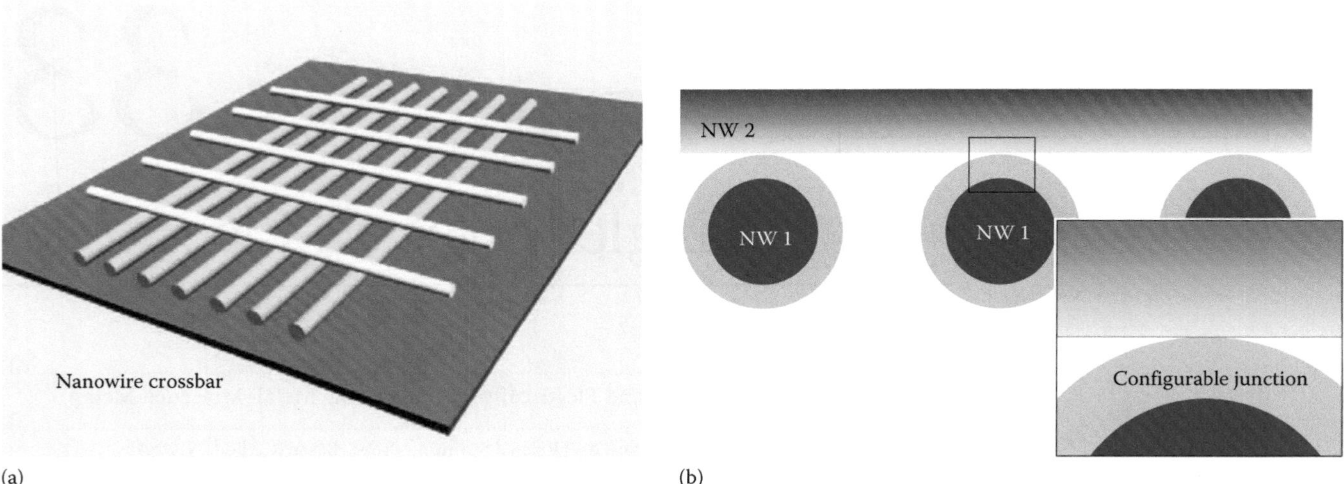

FIGURE 33.1 (a) Crossbar memory switches—the architecture. (b) Cross-sectional view of the crossbar memory elements in (a) showing a memory bit consisting of a configurable bistable junction (i.e., switchable molecules sandwiched between two metal contacts formed by the orthogonal nanowires). (Reproduced by permission from Macmillan Publishers Limited.)

be made to the molecules or molecular assemblies, with subnanometer resolution. One approach that has already been tested for possible molecular electronics memory applications is to use crossbar devices, in which nanoscale assemblies of molecules are assembled at the individual junctions between the perpendicular rows of nanowires (Figure 33.1).[10] While early examples had metal wires deposited thermally through masks, it has also proved possible to use metal nanowires, aligned crosswise by using a variation of the Langmuir–Blodgett technique.[11] A second issue is reliability. A state-of-the-art (2012) Intel 32 nm scale silicon computer chip will have $>10^9$ transistors, and every one of them must function within acceptable parameters for the chip to work. Extensive testing is an important element in silicon chip manufacture, and defective chips are marked on wafers during processing for disposal once the wafer is finally diced (cut up into individual chips). The yield, or final ratio of functioning to nonfunctioning devices, is an important parameter in the economics of silicon chip manufacture.

Because self-assembly and Langmuir–Blodgett film formation are stochastic processes, in which molecules position themselves under the influence of relatively weak intermolecular interactions, it is highly unlikely that molecular electronic devices could ever be made in a completely reproducible way. Difficulties in reliably and reproducibly contacting to individual molecules (*q.v.*) will mean that, for truly single molecule components, not all could be relied upon to work. Furthermore, both for single molecule devices and those made using molecular assemblies, electrical transport through the molecules will be subject to quantum effects that do not apply to current silicon technology. These factors dictate that new device operation paradigms will be needed.[10,12,13]

Another issue is switching speed. A typical modern field effect transistor in a computer-integrated circuit has a switching speed in the picosecond regime. Although electron tunneling across a barrier can be very fast (femtosecond regime), switching molecules between different redox states, conformational states, or chemical forms can be relatively slow, making ultrafast switching difficult to implement with molecules. For example, a redox-active molecule will normally take much longer than this to undergo electron transfer and relax to its new redox state, owing to the requirement for internal bond length changes and solvation state changes (the so-called reorganization energy), as dictated by Marcus theory.[14]

At present, single molecule electronics is in its infancy. Much of the work in this area has focused on the development of reliable techniques for fabricating metal–single molecule–metal junctions, on the study of the variation in electrical properties of such junctions as a function of molecular structure, and on efforts to develop molecular switches using such molecular properties as redox state or conformation/structure. This chapter will encompass these topics, concluding with an examination of some work in which metal–single molecule–metal junctions have been manipulated at the picoscale.

33.2 Techniques for Making and Electrically Characterizing Metal–Molecule–Metal Junctions

33.2.1 Scanning Tunneling Microscopy and Related Scanning Probe-Based Methods

The fundamental principles underlying STM are dealt with in Chapter 14 (p. 253), so only those operating principles of immediate interest to molecular electronics measurements are reviewed here.

In STM, an atomically sharp tip is positioned very precisely over a conducting (metallic or semiconducting) surface by means of piezoelectric elements. Quantum mechanical effects mean that when two conducting materials are brought sufficiently close together in space (a separation of the order of 1 nm), the wavelike properties of electrons make it possible for them to cross from one side of the gap to the other, an event known as tunneling.[15] The gap represents an energy barrier. If an electron has energy E and the barrier has a height $U(z)$ and width W, then when a small bias of V volts is applied, there is a probability for the electron at $z=0$ (on one side of the barrier) to be found at $z=W$ (i.e., on the other side of the barrier). This probability, P, is proportional to the square of the wave function:

$$P \propto |\psi_n(0)|^2 \, e^{-2\kappa W} \tag{33.1}$$

where

$$\kappa = \frac{\sqrt{2m(U-E)}}{\hbar} \tag{33.2}$$

m is the electron mass

For a small bias, $U-E$ approximates to the work function of the material. For tunneling to occur, there must be an empty level of the same energy as the electron for it to tunnel into. For this reason, the tunneling current is not just dependent upon the distance z, but on the local density of states:

$$I \propto V\rho_s(0, E_f)e^{-2\kappa W} \tag{33.3}$$

where
I is the tunneling current
$\rho_s(0, E_f)$ is the local density of states

The power of STM as an imaging technique, and incidentally as a means for making and electrically characterizing metal–molecule–metal junctions, comes from two factors.[15] The first is the exponential dependence of tunneling current on barrier width, W (corresponding to vertical displacement z of tip from substrate); a 100 pm change in tip height corresponds to a factor of 10 difference in tunneling current for a vacuum gap. The second, related, factor is that STM tips are invariably atomically rough, so that the great majority of the tunneling current originates from the single tip atom that happens to be closest to the substrate, giving very fine (atomic scale) lateral resolution. A feedback mechanism is used to hold the tip either at constant height (z) or at constant tunneling current above the surface, as it is rastered back and forth in the x- and y-directions, and from variations in height or current with surface features, an image can be obtained.

At the quantum limit, conductance (and therefore resistance) is also quantized; there exists an upper limit for atomic scale conduction. As was shown by Landauer,[16] the maximum conductance of a single channel for electron transmission with a single spin degenerate energy level (for instance, as found experimentally for a chain of gold atoms[17]) is given by $G_0 = 2e^2/h$, or 77 μS, where e is the electron charge and h is Planck's constant. Transport of electrons through molecules in metal–molecule–metal junctions has usually been found to take place by tunnel transport in the Landauer–Imry regime.[18] In these instances, the molecules act by providing elastic scattering (Landauer–Imry) channels that facilitate transport compared to a vacuum gap of the same length. To date, such studies have found that nearly all molecules are poor conductors when compared to a single chain of metal atoms, and this is also consistent with results from solution studies on intramolecular electron transfer in photo-excited donor–bridge–acceptor molecules in solution.

The first publication concerning the use of STM to make electrical measurements on metal–molecule–metal junctions was in 1995 by Joachim et al. They imaged C_{60} adsorbed in submonolayer coverage on an Au(110) surface, using a tungsten tip in UHV.[19] Having located clusters of C_{60} on the surface, they studied the current–vertical height (I–s) relationship as the tip was brought down on top of individual C_{60} molecules. The molecules could be highly compressed by the tip, at which point the conductance approached G_0, but at the value of s corresponding to gentle tip–molecule contact (1.23 nm), the conductance of the total junction was 18 nS ($2.3 \times 10^{-4}\, G_0$). In 2003, Xu and Tao used an Au STM tip and an Au surface to form molecular junctions in a type of STM break junction experiment.[17]

Xu and Tao deliberately caused their tip to make contact with the surface and then retracted the tip while monitoring the current. As they did so, multiple chains of atomic gold formed and broke. This could be deduced from the sudden decreases in current corresponding to G_0 (77 μS), as expected for conductance through single chains of gold atoms. In the absence of molecules, these were the only features seen; after the last chain broke, current rapidly decayed to the value expected for tunneling across the newly formed gap. However, in the presence of a dilute solution of molecules bearing terminal groups for forming covalent bonds to gold (α,ω-alkanedithiols or 4,4′-bipyridine), much smaller sudden current decreases were observed as the tip was retracted further after the breakage of the final gold atomic wire (Figure 33.2). This was attributed to the breaking of gold–molecule–gold bridges, formed very soon after breakage of the last gold atomic chain. Because this process is stochastic, and an identical arrangement of gold and contact atoms cannot be expected from one experiment to the next, an important feature of this paper was the statistical analysis of the results of many experiments using nominally identical conditions, for each molecule. The results were plotted in the form of a histogram; the frequency with which a particular current jump was seen versus the current value. Peaks at roughly equal intervals of current value were seen, interpreted as corresponding to one, two (and even three) molecules present in the junction; the smallest current peak therefore corresponds to the conductance of a gold–single molecule–gold junction. This experiment is sometimes referred to as the "STM break junction" technique.

In the same year, Haiss et al. published a different but related technique, which has become known as the $I(s)$ method (I for current, s for vertical height).[20] An Au STM tip was held a certain distance above an Au surface (previously coated with a

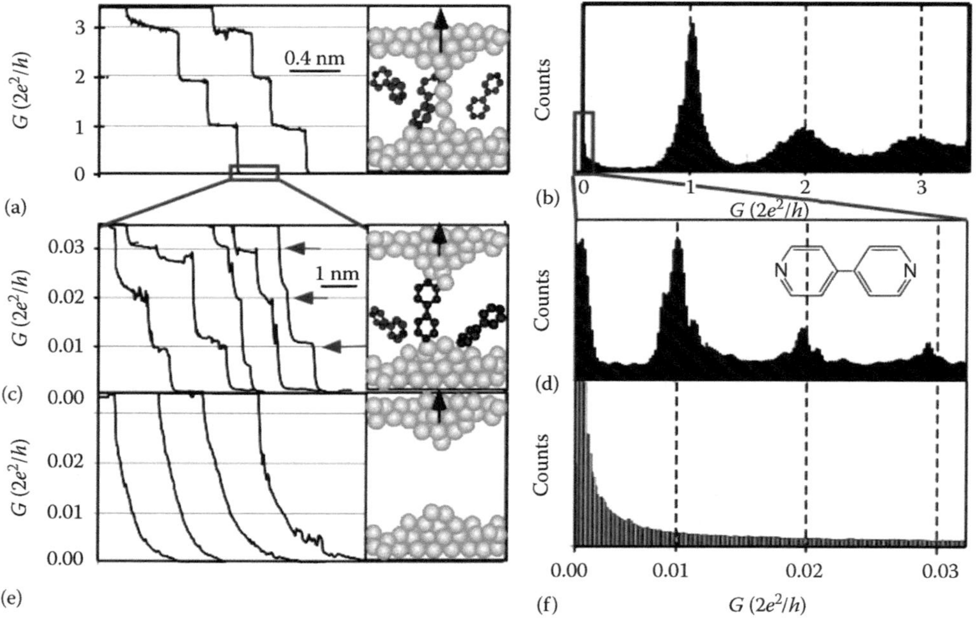

FIGURE 33.2 Schematic of the STM break junction technique of Tao. (a) As the tip is withdrawn from the surface, chains of gold atoms successively break, resulting in decreases in current of G_0. (b) Histogram of these events, showing peaks in the frequency of occurrence of steps in conductance corresponding to the breaking of such chains, with the smallest conductance, G_0, corresponding to the breaking of the final gold atom wire. If no molecules are present (e), the current subsequently falls exponentially with retraction distance, and magnification of the lowest conductance part of the data reveals no smaller histogram peaks (f), but in the presence of molecules that can form gold–molecule–gold contacts, such as 4,4′-bipyridine shown here, then magnification of the low conductance region shows further much smaller conductance changes (c) and consequent histogram peaks (d; ca. 0.01 G_0 in this case), corresponding to conductance through the metal–molecule–metal junctions that are formed and broken during the retraction events after the final metal chain has broken. (Reproduced with permission of AAAS.)

sub-monolayer coverage of an α,ω-dithiol molecule) via the set point tunneling current, I_0. The feedback loop was then switched off, and the tip was retracted while the current was measured, as in the STM break junction method. At the end of the experiment, feedback was reestablished, the tip repositioned, and the experiment was repeated. In the $I(s)$ method, contact of the tip with the surface was carefully avoided. In some retraction events, the classical exponential decrease of current with vertical height was observed, signaling the absence of any bridging molecule, whereas in others, the current on retraction of the tip was higher than in the absence of a molecule. At some point, the current reached a plateau as a function of vertical height, following which it rapidly fell again. These results were interpreted as due to the stochastic formation of a metal–molecule–metal junction (probably driven by the electric field across the gap), followed by extension of the junction, and then its subsequent breakdown as retraction was continued. This paper exemplified another advantage of STM-based methods; these experiments were conducted with control over the electrochemical potential using an Apiezon wax-coated tip in the presence of an aqueous electrolyte, allowing the conductance of these (redox-active) molecules to be determined as a function of oxidation state.[20] Another useful development was that because contact of tip and surface was avoided, it was possible to analyze statistically both the value of the current plateaux corresponding to conductance through a metal–single molecule–metal junction, and the height extension (s) at which the junction broke down. Later, the latter technique was refined; by analyzing current decay curves during data collections in which no molecular bridge was formed, it was possible to calculate the initial height of the tip, so the total length of the metal–molecule–metal junction at breakdown could be determined.[21] As will be seen later, this parameter has proven very useful in characterizing such junctions.

A variation on the $I(s)$ method reported later by Haiss et al. was the $I(t)$ method, in which the STM tip was brought to a certain height over a molecule-covered surface, the feedback loop was switched off for a period of time (typically 0.5 s), and the tunneling current was monitored before feedback was reestablished.[22] The stochastic formation and breakdown of metal–molecule–metal junctions could be observed; the tunneling current showed sudden jumps up on bridge formation and down as the junction(s) broke. The current jumps could then be analyzed statistically. This technique was later used to study junctions involving peptides; the substrate in this case was covered with an ordered monolayer of cysteine-bearing peptides.[23] A variation on this method is also being developed into a sensor system for sequencing DNA by monitoring tunneling currents through a junction as the strand passes through.[24]

Atomic force microscopy using conducting gold-coated tips has also been used for characterizing metal–molecule–metal junctions, in a technique known as the matrix isolation method.[25] First, the molecule of interest, 1,8-octanedithiol,

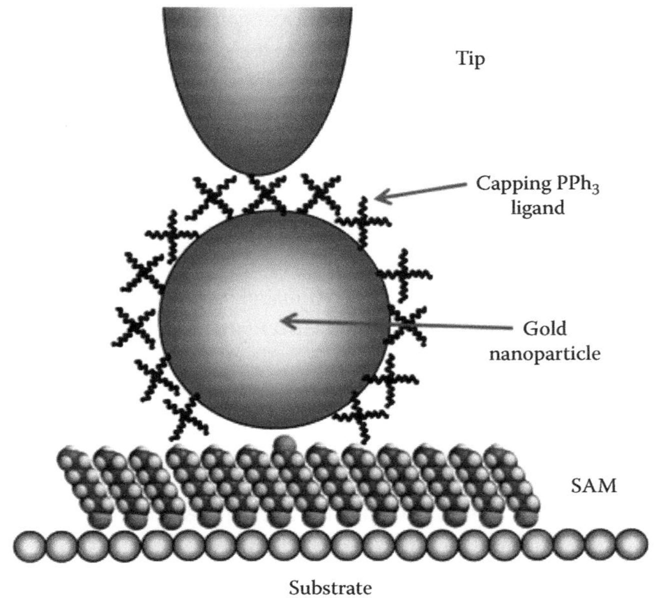

FIGURE 33.3 "Matrix isolation" method developed by Cui et al.[25] A dithiol (center of monolayer as shown) anchors the Au nanoparticle to the surface and a CAFM tip is used to make contact.

was inserted into a preformed self-assembled monolayer of 1-octanethiol on a Au(111) surface (Figure 33.3). This replacement reaction was known to give a low level of partial replacement of the monothiol by dithiol, diluted into the monothiol matrix, and bound via a single thiol. Next, the monolayer-covered surface was treated with small phosphine-protected gold nanoparticles, and these selectively adsorbed at the terminal thiol ends of the 1,8-octanedithiols. A Au-coated AFM tip was then used to locate the nanoparticles, and by applying a small force, make electrical contact deduced from the corresponding jump in current, and the *I*–*V* response of the junction was then evaluated. Approximately, 4600 measurements of this type were made, and it was found that the *I*–*V* response of the junctions could be superimposed with the application of correction factors.[25] Statistical analysis of the latter showed that they peaked around small integer values 1, 2, 3, ..., indicating that the gold nanoparticles were contacted by integer numbers of octanedithiols. Interestingly, when the experiment was repeated in the absence of alkanedithiol, the junction resistances were 10^3 times higher than for even single 1,8-octanedithiol-containing junctions, emphasizing the vital role of chemical contact to both electrodes in optimizing molecular conductance.[25]

The "matrix isolation" method was an early example of the use of statistical methods to analyze results from metal–molecule–metal junctions, but there are some significant factors that need to be taken into account in its employment. First, there is the practical issue that it is time-consuming to locate and form contacts to the number of individual nanoparticles necessary for accumulation of reliable statistical data, and this process would be difficult to automate, in contrast to the STM-based methods discussed earlier where automation of data collection and processing has been achieved (e.g., see Ref. 26). Second, it has become clear that the small size of the nanoparticles used in the original publication can lead to problems with this junction acting as a Coulomb blockade. Later work showed that these effects could be minimized by using significantly larger nanoparticles (>5 nm).[27] Moreover, the presence of the remaining capping ligands on the phosphine-capped ligands used in the original work can lead to failure to achieve a truly metallic contact between the AFM probe and the nanoparticle. Finally, correction should be made for the deformation force exerted by the AFM tip on the monolayer, as it can have a significant effect on the results.[27]

33.2.2 Break Junction Techniques: Mechanically Controlled Break Junctions

In 1985, Moreland and Ekin demonstrated that a brittle Nb–Sn superconducting wire, mounted on a flexible glass beam, could be broken by the application of a bending force on the beam; subsequent controlled reduction of the bending force led to the establishment of a tunneling junction as the broken ends of the wire approached each other. This they termed a "break junction."[28] Later, it was shown that the method could be extended to nonbrittle metals by using a supported "notched" niobium wire, where the bending force (controlled by a piezoelectric positioner) was concentrated on the notched region.[29] The term "mechanically controlled break junction" (MCBJ) was coined for this experiment. Both the superconducting "weak link" and the vacuum tunneling regimes could be explored, a range of some 10^9 in junction resistance. Other workers have employed polyimide films on phosphorus bronze or sprung steel beams in place of glass. The use of electronic microfabrication techniques to prepare metallic bridges suspended over the substrate has since allowed a significant improvement in junction stability, because with this arrangement there is a very large displacement ratio ($D_{gap}/D_z = 10^4$–10^5) between the movement of the push rod (D_z) and the width of the electrode gap (D_{gap}) (Figure 33.4). For instance, metallic bridge structures have been lithographically defined on a polyimide layer. A reactive ion beam is then used to etch away the polyimide immediately under the metal bridge.

Early experiments with the MCBJ method focused on the formation and electrical properties of metallic atom point contacts.[30]

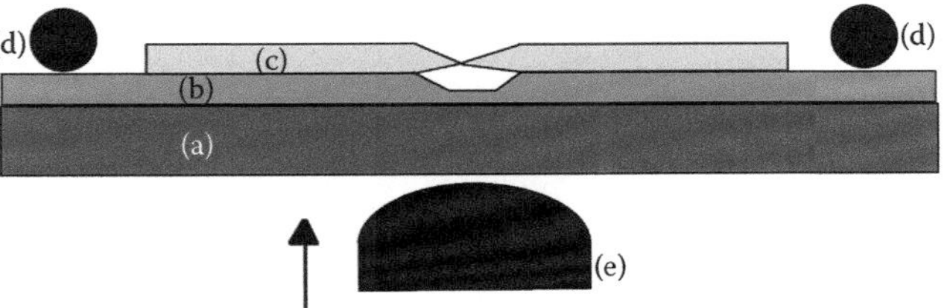

FIGURE 33.4 Mechanically controlled break junction technique. (a) Substrate, (b) gate oxide, typically undercut by electron beam lithography in the region of the notched or otherwise thinned metal wire (c), (d) rigid supports, (e) mechanically controlled pushrod.

The conductance of such contacts was found to be a function of both the quantization of the conductance and the discreteness of the contact size. In experiments with gold junctions, steps near the conductance quantum ($G_0 = 2e^2/h$) were seen as the junction was extended until the gold contacts were finally pulled apart. Rubio-Bollinger et al. measured the force evolution simultaneously with the conductance while drawing out a chain of atoms at 4 K. They employed an auxiliary STM at the back of the cantilever beam on which was mounted the sample, in order to detect the deflection and hence the force on the sample.[31] The measured force showed a sawtooth pattern corresponding to elastic deformation stages interrupted by sudden force relaxations. The force jumps were accompanied by simultaneous jumps in the conductance corresponding initially to rearrangements as gold atoms inserted into the growing chain. The final jump in force was much larger, corresponding to the breaking of the chain, accompanied by a large decrease in conductance corresponding to near G_0.

The first report of the use of the MCBJ technique to investigate metal–molecule–metal junctions was by Reed et al.[32] They used a notched gold wire that was broken under a tetrahydrofuran (THF) solution of the dithiol so that the molecules self-assembled on the newly formed gold surfaces. They then brought the gold contacts together and saw reproducible features in the current- and conductance–voltage curves consistent with metal–molecule–metal junction formation, at an electrode separation roughly consistent with molecular length. The molecular resistance deduced was 22 MΩ (or a conductance of 45 nS).[32] However, although several repeat experiments gave similar resistances, there was no definitive proof of *single* molecule junction formation. Later, Kergueris et al. used an under-etched gold-on-polyimide bridge MCBJ to investigate gold–2,2′:5′,2″-terthiophene-5,5″-dithiol–gold junctions.[33] Again, the presence of *single* molecule junctions could not be proven, but was inferred from repeat experiments. More recently, improvements in the MCBJ techniques have made it possible to carry out large numbers of junction forming/breaking experiments, so that the results can be statistically analyzed via conductance histograms in a manner similar to that described earlier for STM techniques. For example, Smit et al. used a Pt MCBJ to make, and characterize at 4 K, junctions in which a single dihydrogen molecule bridged the contacts.[34] Over 2000 such junctions were made, and the results were plotted as a conductance histogram. A sharp peak at just below 1 G_0 was seen; Pt point contacts in the absence of H_2 had a conductance of ca. 1.5 G_0, also analyzed statistically from many experiments in the absence of H_2. The experiments were repeated for D_2 and HD, and the data were consistent with the formation of (linear) Pt–H–H–Pt junctions. The very high conductance for this particular metal–molecule–metal junction, still the highest so far claimed, was attributed to unusually strong coupling of the H_2 molecule to the Pt leads. Lörtscher et al. used a MCBJ setup to investigate a terphenyl dithiol molecule and benzene-1,4-dithiol, and for each system, they used several thousand *I–V* curves collected over many junction closing and opening cycles, with a statistical approach to analyze the curves.[35] Two distinct types of *I–V* curve were observed for the terphenyl molecule, attributed to differing contact modes to gold; a few *I–V* curves showed stochastic switching between these two "contact modes."

By employing lithographic techniques, Champagne et al. made gold MCBJ devices on a Si:SiO_2 substrate, in which a 50 nm wide constriction in the gold wire was positioned over a cavity etched into the SiO_2 with HF. Electromigration was employed to break the wire, but MCBJ positioning was then used to bring the broken ends together. The *I–V* characteristics of gold–C_{60}–gold junctions could then be determined, using the remaining SiO_2 layer as a gate. With a ca. 40 nm separation between the gate and the molecule, the gating that can be effected is weak, but the MCBJ method does allow more precise alteration of the gap than the alternative three-terminal approach offered by electromigration-induced break junctions (*q.v.*).[36]

More recently, the MCBJ approach combined with statistical results analysis has been used to study longer and more sophisticated conjugated molecules. For example, a MCBJ study of a small family of oligophenyleneethynylenes led to the discovery that *mono*thiols could form gold–molecule … molecule–gold junctions, provided that the molecules are long enough to make their intermolecular π–π stacking interaction sufficiently strong for junction formation (Figure 33.5).[37] The technique was also refined to allow collection of data under electrochemical control in the presence of an electrolyte

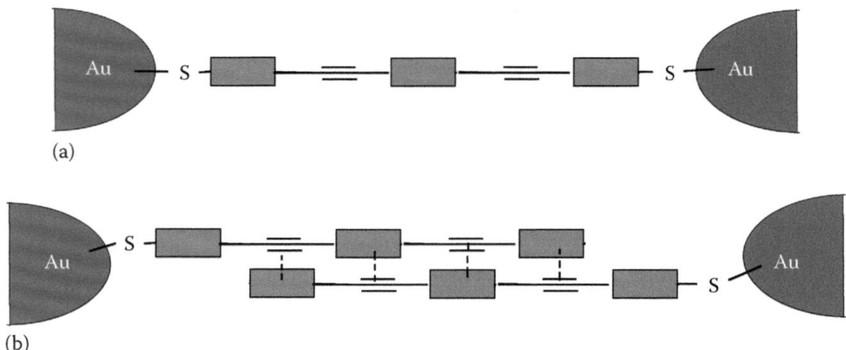

FIGURE 33.5 (a) A schematic representation of junctions formed by all-*para*-SC_6H_4–C≡C–C_6H_4–C≡C–C_6H_6S (a metal–single molecule–metal junction) and (b) by two corresponding monothiols interacting by π-stacking[37]; the representation of the likely arrangement for the π-stacked junction is based on considerations outlined in Ref. [106].

solution, initially for the study of the gold nanoconstrictions,[38] and later for the study of the conductance of metal–molecule–metal junctions using the electrochemical potential as a form of gate.[39]

33.2.3 Break Junctions Formed by Electromigration (EMBJ Technique)

Electromigration is the diffusion of metal atoms caused by momentum transfer from the conduction electrons, and occurs in conditions of high current densities. It is becoming an increasing problem in the electronics industry, as feature sizes on silicon chips become ever smaller and the wire dimensions consequently reach the nanoscale. However, for detailed investigation of the physics of three-terminal devices involving molecules, it has been turned to good purpose to make break junctions in which there is intimate contact to a third (gate) electrode. The technique has recently been reviewed by some of its most prominent practitioners.[40] Electron beam lithography and other methods from the electronics industry are used to make a very narrow metal constriction in a wire (contacted to much larger pads to act as source and drain contacts) on top of a very thin, well-defined insulator layer, for example, 2–4 nm of Al_2O_3 produced by brief exposure of evaporated Al to oxygen. The wire can then be thinned to the level where only a few atoms form the contact, by ramping the source–drain voltage using a feedback technique, such that when the current begins to fall as a result of electromigration thinning the constriction, the voltage is ramped down again. This process can be conducted under liquid, and usually, the resulting ultrathin (few atoms) constriction in the wire is subsequently left to break spontaneously at room temperature in the presence of a solution of the molecule under investigation. A disadvantage of this method is that there is little control over this junction formation process, and each junction is effectively unique, making statistical analysis impossible. Also, as well as the desired metal–molecule–metal junction formation, it is possible to obtain junctions in which nanoscale metal grains rather than molecules are located between the metal contacts. Even when a metal–molecule–metal junction does form, the degree of coupling between the gate and the molecule can vary greatly, depending upon the proximity of the molecule to the gate, a parameter that is effectively random. Nevertheless, the fact that this is still the only solid-state technique for producing strongly gated metal–molecule–metal junctions makes it of great interest in the field of molecular electronics.

Data obtained using EMBJ are usually plotted as two-dimensional color-coded histograms, where the color corresponds to dI/dV (conductance; either calculated from I–V data or measured directly by a lock-in technique), and this is plotted as a function of source–drain voltage versus gate voltage. The result is a type of conductance map.[40] For example, in an early paper in which an oligophenylenevinylene molecule was studied, the dI/dV_{SD}–V_{SD}–V_G map showed evidence of a total of eight charge states, from −3 to +4, within a V_G range of ±4 V, an observation that was explained by image charge effects greatly stabilizing the charged states of the molecule.[41] It should be emphasized that in this work, bulky Bu^{t}S–thioether groups were used as contacts, to ensure weak coupling between the metal electrodes and the molecule. This allowed the molecules to diffuse on the Au surface even at low temperatures (70 K), which is essential to junction formation in this technique since the molecules are deposited from the gas phase.

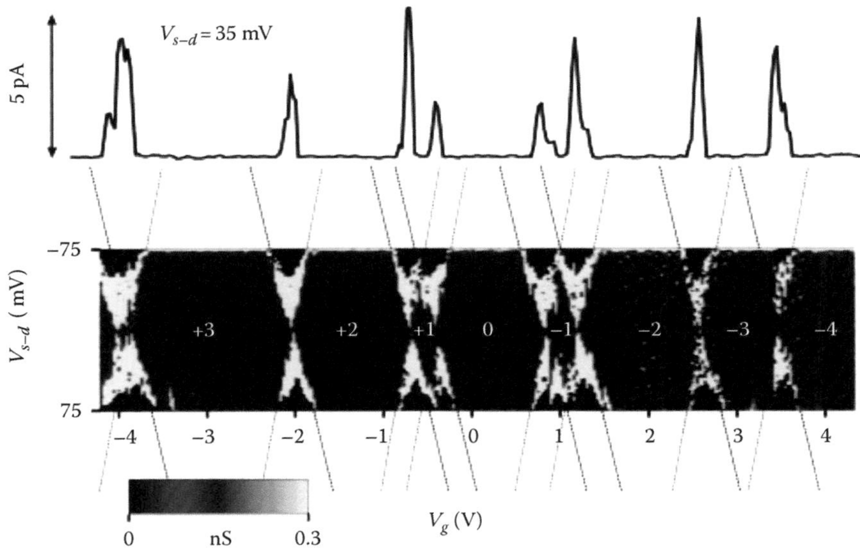

FIGURE 33.6 Experimental EMBJ data on $Bu^tSC_6H_4CH{=}CH(C_6H_4CH{=}CH)_3C_6H_4SBu^t$ using gold source and drain electrodes and an aluminum oxide gate. (Top): A representative I_s–d–V_g trace, for V_s–d = 35 mV. (Bottom): A two-dimensional conductance map, showing the differential conductance dI_{s-d}/dV_{s-d} versus V_g. The lighter "X"-shaped regions occur where the gate voltage brings molecular orbitals into resonance resulting in an increase in conductance and in charged states of the molecule as indicated. (Reprinted by permission from Macmillan Publishers Ltd. *Nature*, Kubatkin S., Danilov A., Hjort M., Cornil J., Brédas J.-L., Stuhr-Hansen N., Hedegård P., and Bjørnholm T., 425, 698, Copyright 2003.)

33.2.4 Other Techniques

Single-walled carbon nanotubes (SWNT) have been used to fabricate break junctions in which the molecules are covalently attached to the "broken" ends of the nanotube via amide bonds.[42] In this technique, individual SWNT were grown by chemical vapor deposition on a SiO_2-coated Si surface. Individual nanotubes were then located and metal contacts were evaporated through a shadow mask so that the middle 20 μm of nanotube was exposed. The device was then coated in polymethylmethacrylate (PMMA), and ultrahigh-resolution electron beam lithography was used to expose a window some 10 nm wide in the PMMA over the nanotube. The device was then exposed to an oxygen plasma, which cuts the exposed nanotubes, presumably leaving carboxylic acid groups at the cut ends. Typically, 20%–25% of the tubes were completely cut, out of 2500 experiments. Thus, a gap of <10 nm was made, and various amine-terminated conjugated molecules with different lengths and molecular backbones were then bonded across the gap using standard carbodiimide (amide bond forming) coupling chemistry. The electrical properties of the junctions were measured before and after cutting, and again with the molecules in place. In favorable cases, the cutting process resulted in electrical disconnection, which was then partially restored with the molecules bonded across the gap.

This technique has some advantages over the use of metal nanocontacts. In particular, strong and well-defined covalent bonds are formed between the molecule and the contacts, and the nanotube dimensions are not vastly greater than typical molecular dimensions as is usually the case for metal BJ methods. It is also possible to cool the devices to ultralow temperatures to shed more light on the physics of the nanotube–molecule–nanotube junctions. In principle, it may prove possible to make three-terminal devices in this manner. However, it is difficult to be sure how many molecules bridge each nanotube gap, and the size of the gap may not match the length of the molecule; it is hard to control the former. Although amide contacts are not ideal for a highly transmissive nanotube–molecule contact,[42] theory suggests that, if a method of forming direct C–C bonded contacts could be devised, very transmissive contacts should be possible between a conjugated molecule and a C nanotube.[43]

An alternative way of making contact is to use π–π stacking. In an interesting development of the C nanotube break junction method, Marquardt et al. used dielectrophoresis to position metallic C nanotubes between Pd source and drain contacts on a SiO_2 surface, over a trench in this surface (Figure 33.7). By passing a high current through the nanotube in the presence of air, "electroburning" occurred; Joule heating of the nanotube resulted in its reaction with oxygen creating a gap in the nanotube over the trench. A conjugated molecule bearing terminal phenanthrene groups was then positioned across the gap using electrophoresis. The phenanthrene groups provided the contact to the nanotube ends via π–π stacking. This setup was used to study single molecule electroluminescence, taking advantage of the fact that although the nanotube was conducting, the density of states at the Fermi energy was insufficient to cause quenching of the luminescence.

More recently, attention has shifted from C nanotubes to graphene as a possible alternative to conventional metals for contacting to molecules. Prins et al. adapted the electroburning technique to fabricate nanoscale gaps in few-layer graphene layers contacted by two metal electrodes, and studied graphene–molecule–graphene junctions using a conjugated molecule terminated with anthracene groups to bridge the gap. These were assumed to form contacts with the graphene via π–π stacking.[44] Since the graphene was initially deposited on top of a SiO_2–Si substrate, the gate voltage dependence of the junctions was examined, and in a minority of cases (4 out of the 14 experiments in which a molecule bridged the gap), gate-dependent I–V characteristics were seen. As the "graphene" in this case is multilayer, whether or not gate-dependent characteristics are seen probably depends upon the proximity of the molecule to the gate.

33.3 Search for a Transmissive Metal–Molecule Contact

The thiolate–gold bond has been widely used by experimentalists in studying metal–single molecule–metal junctions. This originally stems from experimental convenience: a substantial literature on alkanethiol self-assembled monolayers on gold existed at the time workers began to study metal–molecule–metal junctions; thiols are synthetically straightforward to append to a variety of molecular backbones; and such molecules can be adsorbed to readily available Au(111) surfaces in the ambient laboratory atmosphere without recourse to expensive surface science high vacuum techniques. Early results from studying alkanedithiols of varying length using the STM break junction technique suggested that the junction conductances could be fitted using the following equation:

$$G = A_N e^{-\beta_N N}$$

(as expected for a superexchange mechanism)

where

- G is the junction conductance
- A_N is a constant, determined by the molecule–metal coupling strength and reflecting the contact resistance
- β_N is the decay constant per methylene group
- N is the number of methylene groups

The constant A_N, determined by extrapolating a log(G)–N plot back to $N=0$, suggested that the Au–thiolate bond was highly transmissive with values of A_N approaching G_0.[45] Other studies of conjugated monothiol monolayers (albeit not single molecule junctions) seemed to support this.[46] However, more recent detailed measurements on the alkanedithiol/gold system, made using the $I(s)$ and $I(t)$ methods in addition to the STM break junction technique, were consistent with values of A_N several orders of magnitude smaller than G_0, as well as curvature toward an N-independent G value in log(G)–N plots for $N \leq 6$[47] (the shorter alkanedithiols were not studied in earlier work[45]). This apparent

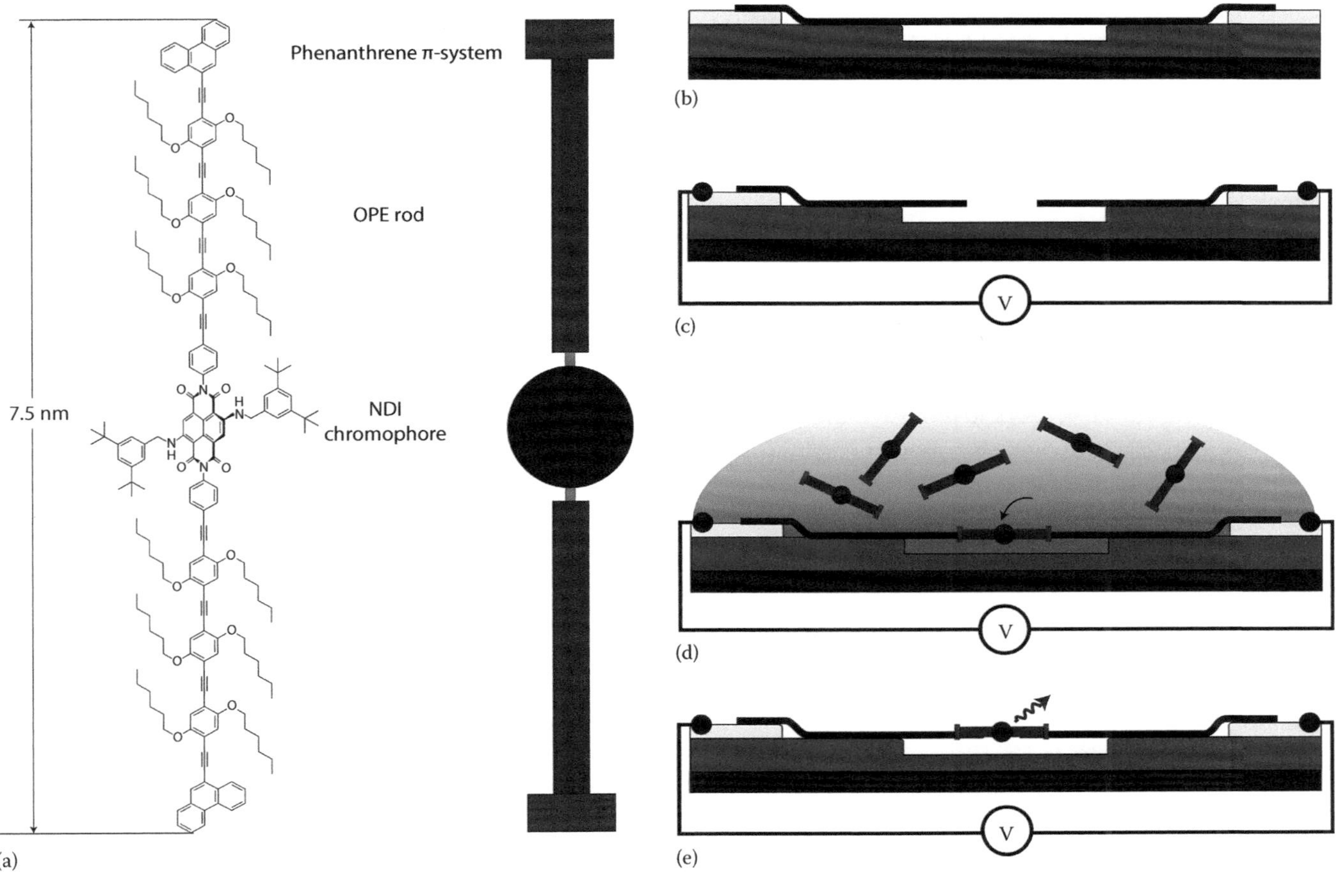

FIGURE 33.7 Diagrammatic representation of the electroburning technique as applied to metallic C nanotube contacts for the preparation of single molecule junctions involving the conjugated chromophore molecule (a), from which single molecule electroluminescence was detected (e).[107] Dielectrophoresis was used to place a metallic nanotube (black) between palladium electrodes (light gray), free-standing above a trench in silicon oxide insulator (dark gray) (b). Electroburning opens a gap in the nanotube (c), following which dielectrophoretic deposition of the polarizable molecule (a) from solution into the nanotube gap leads to the formation of a NT–molecule–NT junction (d). (Reprinted by permission from Macmillan Publishers Ltd. *Nat. Nanotechnol.*, Marquardt C. W., Grunder S., Błaszczyk A., Dehm S., Hennrich F., Löhneysen H. v., Mayor M., and Krupke R., 5, 863, Copyright 2010.)

discrepancy in results is not yet explained and requires further clarification (both theoretical and experimental).

To add to the controversy over the nature of the transmissivity of the Au–thiolate bond, it became clear that different values of conductance could be determined for a given alkanedithiol (or other dithiol) depending upon the precise experimental conditions used.[21,22,48–50] For example, for 1,8-octanedithiol, perhaps the most frequently studied alkanedithiol, conductances of 0.96, 3.82, and 17 nS have been determined (values averaged over the various STM techniques used, and the various groups who have determined them), categorized as low, medium, and high.[51] Various interpretations have been put forward to explain this, including different binding sites for the thiolates on gold, and different conformers of the molecules in the junctions. In a thorough study of this problem, employing STM $I(s)$, $I(t)$, and break junction methods, it was observed that there was a correlation between substrate surface roughness and the frequency of observation of high, medium, and low conductance values.[21] High conductance values were most often seen in STM break junction measurements, where the degree of surface roughness is inevitably high. Low conductance values were most often seen when an atomically flat gold surface was used in combination with the noncontact $I(s)$ and $I(t)$ methods; the medium conductance value was seen more often in such experiments when the substrate surface was rougher. There was also a correlation between the junction break-off distances, determined using the $I(s)$ technique as described earlier, and conductance group.[21] The low conductance events correlated with a break-off distance significantly longer than the medium conductance events, and the shortest break-off distance was seen for the high conductance events. All this evidence, when taken together, was interpreted in terms of different modes of binding of the thiolates to gold; the high conductance events are a consequence of both thiolates binding to step edges (likely to be found more often on rougher surfaces, and particularly in the break junction method where both metal surfaces will be disordered). When both thiolates are bound to single gold atoms, the metal–thiolate interaction is less transmissive, and the conductance is low. When one thiolate is bound to a single gold atom and the other to a step edge, the conductance takes a medium value.[21] It is also observed that in $I(s)$ experiments, a "medium" conductance plateau is often followed by a "low" conductance plateau in the I–s curves, but never vice versa, which is consistent with the molecule being pulled up as the junction extends, from a more transmissive step edge site to a single gold atom site.[21]

Alternative contact chemistries have been explored by several groups, partly in an effort to overcome the perceived problems with thiolates. The sulfur donor atom, known to coordinate strongly to gold, is also a feature of donor groups such as RCS_2^-[52] and R–N=C=S,[53] and the selenium analog of thiolates RSe^- has also been tested.[54] Phosphines have been explored as contacts in $Me_2P(CH_2)_nPMe_2$ ($n=2, 4, 6, 8$) and compared with corresponding thioethers $MeS(CH_2)_nSMe_2$ and amines $H_2N(CH_2)_nNH_2$.[55] Of these, phosphines had the lowest contact resistance, 130 kΩ (corresponding to a contact conductance of $0.1G_0$) over both contacts.

In contrast to other workers' results with 1,4-benzenedithiol[56–58] and 1,4-benzenediisocyanide,[58] Venkataraman et al. found that in STM break junction experiments, 1,4-benzenediamine gave a definite peak in a semilog plot of the conductance histogram, whereas neither the corresponding dithiol nor diisocyanide showed such a peak.[26] This discrepancy might be attributable to differences in the precise experimental conditions employed, but it is clear that the amine contacts do give a particularly clear conductance peak. Although amines adsorb only weakly to gold surfaces, this might actually be an advantage for structure–property studies on metal–molecule–metal junctions since it has been shown theoretically that amines will only bind significantly to under-coordinated (and hence more reactive) gold atoms, such as adatoms, and they do so with a well-defined electronic coupling.[26] This can be expected to lead to greater reproducibility in junction formation than for stronger-binding groups like thiolates and isocyanides that can show a variety of coordination modes, leading to a correspondingly wide range of possible conductance values. However, it must be pointed out that other workers have found two values for the conductance of junctions with 1,4-diaminobutane using the STM break junction method, as they also did for alkanedithiols.[45] By studying 1,n-diaminoalkanes ($n=2$–8), a contact conductance of $0.06G_0$ for a gold–amine bond has been determined,[26] about 10 times smaller than previously measured for the gold–thiolate bond.[45]

Pyridyl groups are good ligands toward transition metal ions, and unlike alkylamines they have some π-acceptor character. Accordingly, they are effective contacts for gold surfaces. Indeed, the first use of the STM break junction technique involved measurement of the conductance of junctions involving 4,4′-bipyridyl, which was found to be $0.01G_0$.[17] Later, 4,4′-bipyridine was the subject of another study that explored the picoscale manipulation of metal–molecule–metal junctions (*q.v.*).[59] 4-Pyridyl contact groups have also been used in studies of structure–property relationships involving carbyne[60] and oligoporphyrin[61,62] molecular wires.[60] Evidence of multiple conductance values for the same molecule have also been found for pyridyl contacts; Quek et al. found two distinct peaks in semilog conductance histogram plots for 4,4′-bipyridyl and obtained junction stretching data that was consistent with two different binding modes of the molecule, one in which the molecule was tilted and the pyridyl was "side-on," producing a higher transmissivity and higher conductance, and the other in which the molecule was upright, coordinated via the pyridyl N lone pair. Wang et al.[60] found evidence for three conductance values for a given molecule, similar to thiolates, and suggested that the same explanation in terms of varying contact geometries could apply.

Electrochemists have long known that carboxylic acid anions can adsorb to gold electrode surfaces.[63,64] Accordingly, these can also be used as contacts in single molecule conductance determinations. Chen et al. compared the conductances of $X(CH_2)_4X$ ($X=-SH$, $-NH_2$, $-COOH$) using the STM break junction method, and found two distinct conductance values for all three contact groups.[45] Although the values of β_N measured were similar for all contact groups as expected, the pre-exponential factors A_N were very different. The resistivity of the contacts varied in the order $-COOH > -NH_2 > -SH$. Both the amine and carboxylic acid contacts showed pH dependence; at pH 1, no metal–molecule–metal junctions formed for the amine contacts owing to protonation to $-NH_3^+$. A wider study of the conductances of $HOOC-(CH_2)_n-COOH$ ($n=4, 6, 8, 10, 12$) found $\beta_N=0.78\pm0.07$ per methylene group and a contact resistance similar to that reported by Chen.[65] An interesting finding of a study of the conductances of molecules $HS-(CH_2)_n-COOH$ was that the total apparent contact resistances for these asymmetrically contacted molecules were higher than for either of the symmetrically contacted molecules $HS-(CH_2)_n-SH$ or $HOOC-(CH_2)_n-COOH$, suggesting that the efficiency of charge transport through single molecular bridges is influenced not only by the nature of the metal–molecule contact, but also by the contact asymmetry of the molecular junction.[66] This needs to be borne in mind when considering the conductances of other functional molecules that are asymmetrically contacted.

Efforts have been made to create more transmissive metal–molecule contacts. For example, Cheng et al. have provided convincing evidence that it is possible to make junctions in which gold is covalently bonded directly to carbon, by employing compounds $Me_3Sn-(CH_2)_n-SnMe_3$ ($n=4, 6, 8, 10, 12$).[67] Measured conductances were some 100 times higher than for corresponding dithiols or diamines, and in the conductance histograms for the shorter C_4 and C_6 molecules, a second peak at lower conductance, equal to that of the main peak for the C_8 and C_{12} molecules, respectively, was evident, suggesting that some dimerization of the molecules was taking place. It is likely that the gold surfaces catalyze cleavage of the relatively weak C–Sn bond. Further evidence for direct contact via Au–C bonds (rather than, for instance, a $Au-Sn(Me_3)-R$ unit) was obtained by synthesizing linear Au(I) organometallic complexes $(Ph_3P)Au-(CH_2)_n-Au(PPh_3)$ in which a Au–C bond is "preinstalled" synthetically, and measuring junctions made with these systems.[67] They had the same conductance values as the corresponding $Me_3Sn-(CH_2)_n-SnMe_3$. Interestingly, while it was possible to extend this junction-forming method to 1,4-$Me_3Sn{-}C_6H_4{-}SnMe_3$, the resulting Au-1,4-C_6H_4-Au junctions gave a conductance lower than that found for Au-$(CH_2)_4$-Au. Theoretical calculations suggest that this is because conductance through the benzene ring is mainly via the σ-orbital channel because the Au–C bond is orthogonal to the π-system. However, an extremely high conductance (ca. $0.9G_0$) was later measured for junctions prepared using 1,4-$Me_3SnCH_2C_6H_4CH_2SnMe_3$.[68] The observation of near-resonant transport in the latter case was supported by transport calculations.

Another carbon-based contact unit that has been tested in single molecule electronic devices is C_{60}. C_{60} itself was the subject of a prototypical metal–molecule–metal junction experiment conducted using STM (*q.v.*).[19] Since that pioneering study, metal–C_{60}–metal junctions have been extensively investigated using three-electrode break junction experiments. Where UHV conditions are used, and the C_{60} is evaporated onto clean gold surfaces, the evidence (both theoretical and experimental) is that there is strong metal–molecule coupling, and relatively high zero bias conductance (ca. $0.1G_0$).[69–71] However, where C_{60} is adsorbed from solution, the degree of coupling is apparently smaller, and the system behaves as a Coulomb blockade, with much smaller conductance.[71,72] This needs to be borne in mind if it is contemplated to use C_{60} as a contact group.

With developments in the synthetic chemistry of C_{60}, it has become possible to make conjugated molecules in which C_{60} is used as terminal groups for contact purposes (although the molecular backbone is not truly conjugated through the C_{60} linker). For example, Leary et al. prepared C_{60}-terminated bifluorene molecules, adsorbed these to a gold surface from solution, and used STM first to locate and image individual molecules, then to make contact to one of the two C_{60} groups evident in the images. The molecule could then be lifted from the surface, and the electrical properties of the metal–molecule–metal junctions could be determined with confidence that a *single* molecule was involved, without recourse to statistical arguments.[73] The conductance measured was ca. $10^{-4}G_0$; that two distinct peaks in the conductance histogram from many such experiments were seen was ascribed to the possible different binding modes known for C_{60} on gold from earlier low-temperature UHV studies. Interestingly, when the STM break junction method was used to determine the conductance, the conductance histogram obtained was very much broader, indicative of the relative lack of control using the latter technique.

33.4 Electrochemical "Switching" of Redox-Active Molecules in Junctions

As described in Sections 33.2.2 and 33.2.3, break junction techniques have been extensively used to fabricate junctions with a third (gate) electrode included, to enable gate-induced modulation of molecular conductance. Although some remarkable and interesting physics has emerged from such studies, it is extremely difficult to control the picoscale placing of a molecule with respect to the gate, and therefore the degree of molecule–gate coupling. An alternative approach is the use of electrochemical environments. These enable the redox state of the molecule and the electrochemical potential drop at the surface to be controlled precisely. The first example of the use of this technique was by Haiss et al.[20] They showed that the conductance of a Au–molecule–Au junction with the viologen (V^{2+}) molecule 6V6 (Figure 33.8) increased from 0.49 ± 0.08 nS to 2.8 ± 0.8 nS as the molecule was electrochemically switched from the 2+ to the radical cationic (1+) state; changes in the electrochemical potential modulate the positions of the metal-contact Fermi energies and the molecular electronic levels of the redox group. This has been referred to as a molecular electrochemical transistor configuration, where the counter or reference electrode may be considered as a gate, while the STM tip and substrate are considered as source and drain.

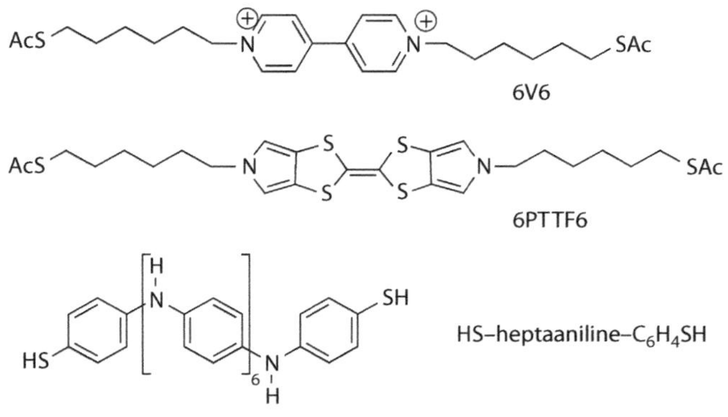

FIGURE 33.8 Some examples of redox-active molecules that have been examined in metal–molecule–metal junctions under electrochemical conditions with control over their redox state.

Scanning tunneling spectroscopy (STS) experiments on monolayers of redox-active molecules have previously shown that the tunneling current through the molecules increases as the redox potential is approached, and then decreases again, in a Gaussian fashion,[74,75] in a manner consistent with a two-step electron transfer (ET) model with partial vibrational relaxation at the redox center developed by Ulstrup and colleagues.[76] However, it was noted by Haiss et al.,[20] and later by Li et al.,[74] that the conductance of 6V6 increased in a sigmoidal fashion across the V^{2+}/V^{+} equilibrium potential, and did not then decrease within accessible negative overpotentials. Haiss et al. suggested a "soft gating" mechanism to rationalize this behavior, in which large configurational fluctuations of the molecular bridge, due to the flexible alkyl linkers in the isolated molecules within the junctions, are responsible for an inhomogeneous broadening of the conductance–overpotential relation that means the system does not behave in the same manner as redox-active molecules in self-assembled monolayers in STS experiments.[77] Li et al.[74] and Pobelov et al.[78] examined 6V6 using a modified $I(s)$ technique and found similar behavior, albeit with a smaller increase in conductance in the "on" state. They also carried out STS studies on close-packed monolayers of an analog of 6V6 with only

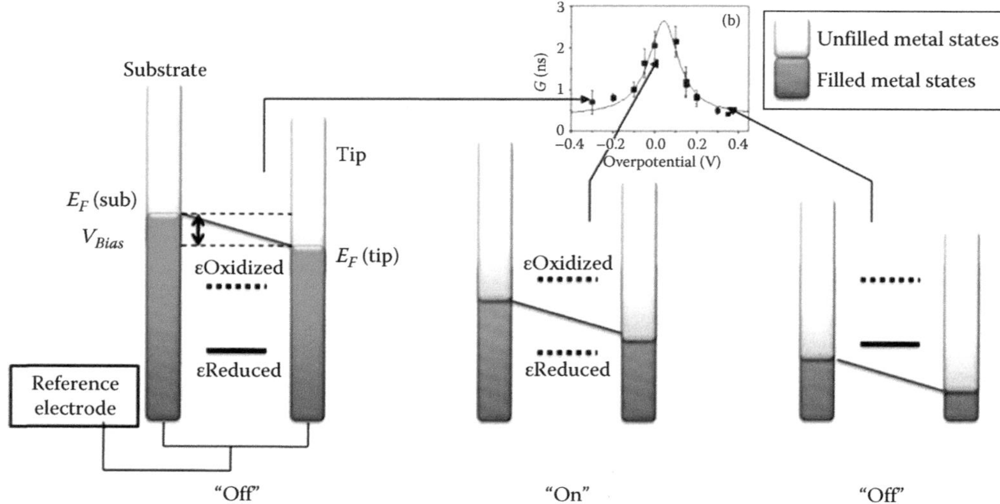

FIGURE 33.9 Schematic illustration of the operation of a redox-active metal–molecule–metal junction. The energy of the frontier orbital will be different in the two redox states ("εOxidized" and "εReduced") owing to bond length changes, electronic structural changes, and solvent reorganization on electron transfer. In the "off" states, the corresponding orbitals are both far from the contact Fermi energies, either below (left) or above (right). In contrast, in the "on" state, they come into the energy range of the Fermi levels of the tip and substrate.

one thiol contact, and found that, in this case, there was indeed a peak in the conductance–overpotential relation, which is in agreement with the "soft" gating concept. Interestingly, however, Leary et al. later carried out $I(s)$ experiments under electrochemical control on the pyrrolotetrathiafulvalene molecule 6PTTF6 (Figure 33.8)[79] and found here that, even in single molecule Au–6PTTF6–Au junctions, the conductance rises as the 6PTTF6/6PTTF6$^{\cdot+}$ redox potential is reached and then falls at positive overpotentials (Figure 33.9). They tentatively suggested that the difference might be attributable to the fact that, whereas 6V6 significantly rigidifies upon reduction (owing to the development of significant inter-ring C–C double bond character), the PTTF moiety is rigid in both redox states.[79] Recently, 6PTTF6 has been re-examined in an ionic liquid environment, which has allowed the potential range to be extended so that both the 6PTTF6/6PTTF6$^{\cdot+}$ and the subsequent 6PTTF6$^{\cdot+}$/6PTTF6^{2+} redox processes can be studied, and very similar behavior was observed for the second redox process, with the junction conductance again rising and then falling as the second redox process is traversed.[80]

The molecule HS–heptaaniline–C_6H_4SH (Figure 33.8) likewise showed a peak in its conductance as a function of potential.[81] The conductance of junctions involving the neutral the (so-called leucoemeraldine) form of this molecule in toluene was 0.32 ± 0.03 nS, and about 1 nS under electrolyte (0.05 M H_2SO_4), increasing to 5.5 ± 0.4 nS as the potential was swept past the redox peak for conversion to the partially protonated, half-oxidized ("emeraldine") form, and then falling again at more positive potentials.

A feature common to both 6V6 and 6PTTF6 is that the factor by which conductance increases as the molecules are electrochemically "gated" is rather small, less than a factor of 10. However, other redox-active molecules have shown considerably larger changes of conductance as a function of redox state.[82,83] Efforts continue to discover the factors that control this behavior.

33.5 Controlling Metal–Molecule–Metal Junctions at the Picoscale via STM-Based Manipulation

The exquisite control over vertical height afforded by modern SPM instrumentation has lent itself well to investigations of single molecule junction conductance. In addition to the relatively straightforward creation and measurement of metal–molecule–metal junctions using STM and conducting AFM, more sophisticated techniques have uncovered some interesting phenomena. For example, Tao et al. developed a combined conductance–force measurement technique. A gold-coated AFM tip was used in a modification of the "STM break junction" method, in which both the conductance and force could be measured as gold–molecule–gold junctions were made and extended.[84] The molecules used in this study were ter- and quaterthiophene molecules bearing terminal $-CH_2SH$ groups to provide contact. In the same manner in which conductance histograms were compiled, force histograms were also determined. They revealed well-defined peaks at multiples of a fundamental force quantum, the force required to break a single molecule at the junction. Unlike the conductances, which were different for the two molecules, the forces were the same, 1.5 ± 0.2 nN, and were ascribed to the breaking of a Au–Au bond since a similar force change was registered when a chain of gold atoms was broken in an experiment without molecules. It was observed that the conductance consistently decreased as the force increased during junction extension, by about 48% per nN for the quaterthiophene. This compared with about 10% per nN determined for 1,8-octanedithiol. While the latter was consistent with estimates of the increase in junction width on stretching 1,8-octanedithiol, the former was too large to ascribe to this cause alone; an increase in the HOMO–LUMO gap due to an enhanced degree of bond alternation on stretching was suggested. This HOMO–LUMO gap increase would be expected to increase the difference between the energy of the HOMO and the Fermi energy of the contacts, hence decreasing the conductance.

More recently, an enhanced version of this technique has been applied to study junctions with 1,4-benzenedithiol in UHV, at both room temperature and 4 K.[85] For this molecule, it was found that the conductance increased more than 10-fold during stretching, and decreased again when the junction was compressed. Based on simultaneously recorded current–voltage and conductance–voltage characteristics, and inelastic electron tunneling spectroscopy, a strain-induced shift of the highest occupied molecular orbital toward the Fermi level of the electrodes was suggested as responsible for this unusual behavior, leading to a resonant enhancement of the conductance. This phenomenon may account for the widely different conductance values that have been measured for this molecule using different techniques, and may contribute to the unusually broad conductance histograms observed for conjugated bis(diphenylphosphine)-contacted molecules.[86]

Other workers have also adopted the simultaneous measurement of conductance and force, with interesting results. Different contact groups (aromatic and aliphatic amines, thiols, pyridines, thioethers, and diphenylphosphines) have been studied, and it was found that only the use of thiol contact groups resulted in the breakdown of junctions with a force equivalent to a Au–Au bond (in agreement with the earlier work[84]).[87,88] The other contact groups all showed characteristic rupture forces that were in all cases significantly smaller than the breaking of a Au–Au bond, consistent with the rupture of the Au-contact atom bond in these junctions. The values of the forces involved ranged from 0.8 nN for pyridines and Ph_2P– groups, 0.7 nN for MeS– groups, 0.6 nN for alkylamines, and 0.5 nN for 1,4-diaminobenzene.[87,88] A useful feature of this work was the use of two-dimensional plots of conductance versus relative tip displacement, and force versus relative tip displacement, with color coding. The experimental results agreed quantitatively with density functional theory–based adiabatic molecular junction elongation and rupture calculations.[87]

Useful information has been obtained by manipulating the width of the tip–surface separation in STM experiments while a molecule is trapped within the junction. For instance, by using the $I(t)$ technique to study junctions made with the "rigid-rod" oligophenyleneethynylene shown in Figure 33.10 at different set point currents I_0 (i.e., with different initial tip–sample separations), Haiss et al. found that the conductance of the junctions was dependent upon the tilt angle of the molecule in the gap.[89] The contact gap separation was determined by calibrating the tip–sample distance ($s - s_0$) as a function of the set-point current (I_0). This calibration was achieved using current–distance $I(s)$ scans for the given sample for which molecule wire formation did not occur (i.e., the current simply decayed exponentially with

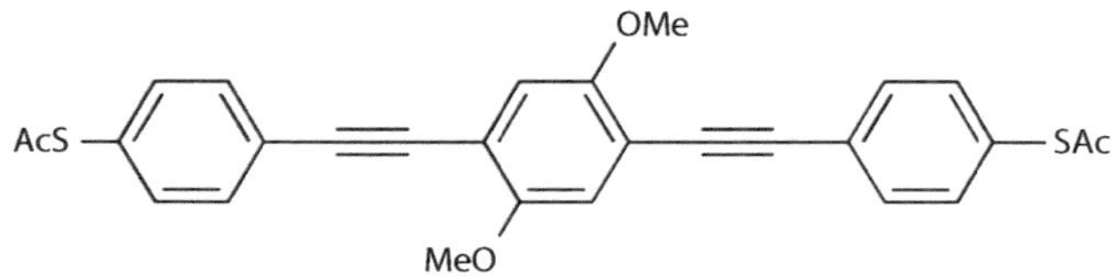

FIGURE 33.10 "Rigid rod" oligophenyleneethynylene molecule used in a study of the contact angle dependence of conductance. (From Haiss, W., Wang, C.S., Grace, I., Batsanov, A.S., Schiffrin, D.J., Higgins, S.J., Bryce, M.R., Lambert, C.J., and Nichols, R.J., *Nature Mater.*, 5, 995, 2006.)

s upon tip retraction). Typically, 20 such $I(s)$ scans were selected and the slope of $\ln(I)$ versus s was determined. An average slope was then calculated within the range of I_0 values relevant to the given experiment. The z-piezo-elongation was calibrated using the height of an Au monatomic step edge (0.236 nm). In order to achieve an absolute measure of the initial separation between tip and substrate, the fact that below a critical set-point current (I_c), molecules cannot span the gap, and hence current jumps for molecules bridging the gap are no longer observed in the $I(t)$ experiment, was employed.

The conductances were found to be much higher when the tip was close to the surface such that, owing to its comparative rigidity, the molecule could only form a junction in a tilted configuration. The conductances did not depend upon temperature at the different tilt angles, confirming that the tunneling mechanism applied. Theoretical calculations suggested that the increased conductance with molecular tilting could be accounted for by two factors: the first was a larger degree of metal–molecule coupling when the molecule was tilted and the benzene π-system interacted to a larger extent with the gold surface than when the molecule was fully upright, and the second was that tilting causes a greater degree of coplanarity of the three phenyl rings in the junction. Similar behavior was also found later for a range of biphenyl, fluorene, and carbazole molecules.[90]

It has been shown that pyridyl contact groups can switch between two distinct binding modes with different resulting molecular conductances. In an elegant series of experiments, a gold substrate controlled by a piezoelectric positioner and with a sensitive built-in position sensor was driven into a gold tip and then withdrawn, in a solution of 4,4′-bipyridine.[59] This STM break junction method produced a conductance histogram in which there were clearly two peaks, and individual conductance–distance plots showed consistently that low conductance plateaux succeeded high conductance plateaux. It proved possible to oscillate the substrate back and forth while molecules were present in the junction, whereupon switching between high- and low-conductance plateaux was observed when this oscillation was 0.25 nm in size. From the fact that the high conductance peak occurred when the tip–sample separation was significantly smaller than the length of a junction with the molecule in a vertical position, it was inferred that this corresponded to junctions in which the 4,4′-bipyridyl was in a tilted position, with consequently a larger interaction between the gold surfaces and the LUMO (a π*-orbital) known to be the main conductance channel in these molecules.[59] Other pyridyl-terminated conjugated molecules have since been shown to exhibit similar behavior,[91] and this interpretation has been supported by transport calculations.[59] More recently, a similar technique using MCBJ has been employed to reexamine two of the three conductance groups previously found for alkanedithiols. In this instance, the high and medium conductance groups for 1,6-hexanedithiol were shown to be interconvertible on altering the gap width, with the higher conductance corresponding to the shorter junction.[92]

Lafferentz et al. demonstrated that it is possible to form metal–single conjugated polymer chain–metal junctions using the STM $I(s)$ technique at low temperature.[93] Using a single crystal Au(111) surface, they first adsorbed a ter-(9,9′-dimethylfluorene) dibromide molecule under UHV conditions from the gas phase (Figure 33.11), and then heated the sample to 520 K. This resulted in scission of the C–Br bonds, and consequent linear polymerization of the terfluorene fragments to give polyfluorene chains, which could subsequently be imaged with molecular resolution. At low temperature (10 K), a Au STM tip was then positioned over the end of a polymer chain and brought close to the molecule (to ca. 1 Å). The chain end could thus be picked up from the surface, and the tip then lifted while the current through the dangling polymer chain was measured, until the connection between tip and molecule or between substrate and molecule broke. It was suggested that the radical character of the terminal fluorene unit (resulting from the prior scission of the C–Br bond) facilitated selective connection to the tip via the chain end. The polymer chain could be imaged before and after the experiment to reveal the movement of the terminal fluorene units relative to the "unlifted" part of the polymer chain. The conductance of the polyfluorene chains could be fitted to the usual tunneling equation, with a β value of 0.38 Å^{-1}, although it was noted that there were periodic oscillations in the $\log(I)$–s relationship with a repeat distance of 10 Å, attributed to the progressive lifting of each individual monomer unit from the substrate.[93]

33.6 Conclusions and Future Prospects

The development of STM and nanofabricated break junction methods has allowed the construction and electrical characterization of metal–single molecule–metal junctions and has stimulated a plethora of studies aimed at establishing structure–property relationships and testing new device paradigms based upon various types of molecular switching. In the future, if truly molecular-scale electronics is to become more than a scientific curiosity, there are many challenges to address. For example, it is likely that molecules will need to be integrated into technology currently used in the solid-state electronics industry because of the huge existing infrastructure investment of that industry. Gold, the contact metal of choice for experimental studies to date, is excluded from "fab" plants, and therefore, alternative contacts will need to be explored. This will not be straightforward because of the greatly increased (and undesirable)

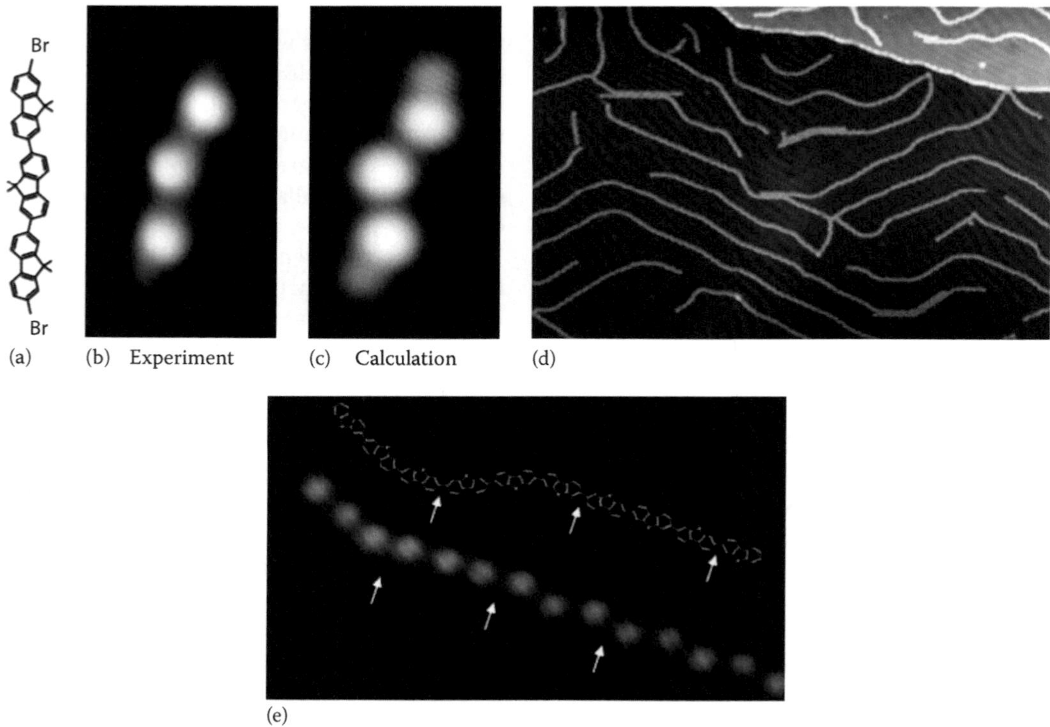

FIGURE 33.11 Terfluorene dibromide molecules (a) were deposited from the gas phase onto Au(111) (experimental image at 10 K (b), and calculated image (c)), then polymerized by heating to 520 K for 10 min, producing individual polymer chains (images d and e).[93] These polymer chains could then be lifted at their terminal groups by STM tip, and the i–V characteristics of the polymer chains were examined as a function of vertical height. (From Lafferentz L., Ample F., Yu H., Hecht S., Joachim C., and Grill L., 2009, *Science* 323, 1193. Reprinted with permission of AAAS.)

reactivity of most metal surfaces toward molecules compared with gold. Silicon, however, is a promising contact material, and chemistry has already been devised for fabricating very robust silicon–molecule bonds. In the near future, it is likely that silicon–molecule–silicon junctions will be tested.[94–96]

Another challenge is to extend the distance over which coherent transport (tunneling) is possible in molecules before hopping-type mechanisms become dominant, so that molecules can be placed between contacts at distances technologically accessible in the near future (roughly, >10 nm). Already, there is evidence that conjugated oligoporphyrins maintain a coherent transport mechanism over an unusually long distance,[61,62] and efforts are under way to explore even longer molecules.

In terms of molecular switching, an interesting possibility for future study is the employment of molecules that show quantum interference effects in orbitals close to the contact Fermi energy, because the redox- or gate-modulated switching of such molecules should show a very major effect on their conductance. In chemical terms, such molecules often exhibit cross-conjugation in one redox state, and full conjugation in the other redox state. Studies of junctions involving appropriate molecules are beginning to appear in the literature.[97–101]

Another challenge for single molecule electronics is the major effect that the environment can have on single molecule properties. Examples have already been found of molecules that show environment-dependent electrical properties.[102,103] However, while such effects may be deleterious for device design, they may also provide a new application for molecular-scale electronics, namely, the construction of extremely sensitive sensors. In addition, supramolecular interactions, for example, hydrogen bonding and π–π stacking, are emerging as factors in the study of molecular tunneling junctions. For example, evidence has been published of metal–molecule ... molecule–metal junctions in which the two molecules transport current via a π-stacked arrangement,[37,104] and hydrogen bonding between surface-anchored nucleobase analogs and single-stranded DNA as it passes through a tunneling junction is being explored as a potential single molecule sequencing technique.[105]

These developments mean that this area will continue to attract much research effort in the coming years, although it is hard to predict in what direction any technological developments of the fundamental science will proceed.

References

1. Aviram A. and Ratner M. A. 1974. *Chem. Phys. Lett.* 29, 277.
2. Binnig G., Rohrer H., Gerber C., and Weibel E. 1982. *Phys. Rev. Lett.* 49, 57.
3. Haselman M. and Hauck S. 2010. *Proc. IEEE* 98, 11.
4. Jan C. H., Bai P., Biswas S., Buehler M., Chen Z. P., Curello G., Gannavaram S. et al. 2008. *A 45 nm Low Power System-On-Chip Technology with Dual Gate (Logic and I/O) High-k/Metal Gate Strained Silicon Transistors*, IEEE International Electron Devices Meeting 2008, Technical Digest 2008.
5. Wilson L. 2012. *International Technology Roadmap for the Semiconductor Industry*, International SEMATECH, Austin, TX.
6. Kalinowski J. 2004. *Organic Light-Emitting Diodes: Principles, Characteristics and Processes*, Marcel Dekker, New York.
7. Kymissis I. 2009. *Organic Field Effect Transistors: Theory, Fabrication and Characterization*, Springer, New York, 2009.
8. Brabec C., Scherf U., and Dyakonov V., *Organic Photovoltaics*, Wiley-VCH, Weinheim, 2008.
9. Shirakawa H., Louis E. J., Macdiarmid A. G., Chiang C. K., and Heeger A. J. 1977. *J. Chem. Soc. Chem. Commun.* 474, 578.
10. Luo Y., Collier C. P., Jeppesen J. O., Nielsen K. A., DeIonno E., Ho G., Perkins J., Tseng H. R., Yamamoto T., Stoddart J. F., and Heath J. R. 2002. *ChemPhysChem* 3, 519.
11. Lu W. and Lieber C. M. 2007. *Nat. Nanotechnol.* 6, 841.
12. Pease A. R., Jeppesen J. O., Stoddart J. F., Luo Y., Collier C. P., and Heath J. R. 2001. *Acc. Chem. Res.* 34, 433.
13. Collier C. P., Mattersteig G., Wong E. W., Luo Y., Beverly K., Sampaio J., Raymo F. M., Stoddart J. F., and Heath J. R. 2000. *Science* 289, 1172.
14. Marcus R. A. 1956. *J. Chem. Phys.* 24, 966.
15. Bai C. 2000. *Scanning Tunneling Microscopy and Its Applications*, Springer-Verlag, New York.
16. Landauer R. 1970. *Philos. Mag.* 21, 863.
17. Xu B. Q. and Tao N. J. 2003. *Science* 301, 1221.
18. Lindsay S. M. and Ratner M. A. 2007. *Adv. Mater.* 19, 23.
19. Joachim C., Gimzewski J. K., Schlittler R. R, and Chavy C. 1995. *Phys. Rev. Lett.* 74, 2102.
20. Haiss W., van Zalinge H., Higgins S. J., Bethell D., Höbenreich H., Schiffrin D. J., and Nichols R. J. 2003. *J. Am. Chem. Soc.* 125, 15294.
21. Haiss W., Martin S., Leary E., van Zalinge H., Higgins S. J., Bouffier L., and Nichols R. J. 2009. *J. Phys. Chem. C* 113, 5823.
22. Haiss W., Nichols R. J., van Zalinge H., Higgins S. J., Bethell D., and Schiffrin D. J. 2004. *Phys. Chem. Chem. Phys.* 6, 4330.
23. Sek S., Swiatek K., and Misicka A. 2005. *J. Phys. Chem. B* 109, 23121.
24. Huang S., He J., Chang S., Zhang P. M., Liang F., Li S. Q., Tuchband M., Fuhrmann A., Ros R., and Lindsay S. 2010. *Nat. Nanotechnol.* 5, 868.
25. Cui X. D., Primak A., Tomfohr J., Sankey O. F., Moore A. L., Moore T. A., Gust D., Harris G., and Lindsay S. M. 2001. *Science* 294, 571.
26. Venkataraman L., Klare J. E., Tam I. W., Nuckolls C., Hybertsen M. S., and Steigerwald M. L. 2006. *Nano Lett.* 6, 458.
27. Morita T. and Lindsay S. 2007. *J. Am. Chem. Soc.* 128, 7262.
28. Moreland J. and Ekin J. W. 1985. *J. Appl. Phys.* 58, 3888.
29. Muller C. J., van Ruitenbeek J. M., and de Jongh L. J. 1992. *Phys. C* 191, 485.
30. Agraït N., Levy Yeyati A., and van Ruitenbeek J. M. 2003. *Phys. Rep.* 377, 81.

31. Rubio-Bollinger G., Bahn S. R., Agraït N., Jacobsen K. W., and Vieira S. 2001. *Phys. Rev. Lett.* 87, 026101.
32. Reed M. A., Zhou C., Muller C. J., Burgin T. P., and Tour J. M. 1997. *Science* 278, 252.
33. Kergueris C., Bourgoin J. P., Palacin S., Esteve D., Urbina C., Magoga M., and Joachim C. 1999. *Phys. Rev. B* 59, 12505.
34. Smit R. H. M., Noat Y., Untiedt C., Lang N. D., van Hemert M. C., and van Ruitenbeek J. M. 2002. *Nature* 419, 906.
35. Lörtscher E., Weber H. B., and Riel H. 2007. *Phys. Rev. Lett.* 98, 176807.
36. Champagne A. R., Pasupathy A. N., and Ralph D. C. 2005. *Nano Lett.* 5, 305.
37. Wu S. M., Gonzalez M. T., Huber R., Grunder S., Mayor M., Schonenberger C., and Calame M. 2008. *Nat. Nanotechnol.* 3, 569.
38. Shu C., Li C. Z., He H. X., Bogozi A., Bunch J. S., and Tao N. J. 2000. *Phys. Rev. Lett.* 84, 5196.
39. Tian J.-H., Yang Y., Zhou X.-S., Schçllhorn B., Maisonhaut E., Chen Z.-B., Yan F.-Z., Chen Y., Amatore C., Mao B.-W., and Tian Z.-Q. 2010. *ChemPhysChem* 11, 2745.
40. Osorio E. A., Bjørnholm T., Lehn J.-M., Ruben M., and van der Zant H. S. J. 2008. *J. Phys. Condens. Matter.* 20, 374121 (14 pp).
41. Kubatkin S., Danilov A., Hjort M., Cornil J., Brédas J.-L., Stuhr-Hansen N., Hedegård P., and Bjørnholm T. 2003. *Nature* 425, 698.
42. Guo X. F., Small J. P., Klare J. E., Wang Y., Purewal M. S., Tam I. W., Hong B. H. et al. 2006. *Science* 311, 356.
43. Ke S.-H., Baranger H. U., and Yang W. 2007. *Phys. Rev. Lett.* 99, 146802 (4 pp).
44. Prins F., Barreiro A., Ruitenberg J. W., Seldenthuis J. S., Aliaga-Alcalde N., Vandersypen L. M. K., and van der Zant H. S. J. 2011. *Nano Lett.* 11, 4607.
45. Chen F., Li X. L., Hihath J., Huang Z. F., and Tao N. J. 2006. *J. Am. Chem. Soc.* 128, 15874.
46. Kim B., Beebe J. M., Jun Y., Zhu X. Y., and Frisbie C. D. 2006. *J. Am. Chem. Soc.* 128, 4970.
47. Haiss W., Martin S., Scullion L. E., Bouffier L., Higgins S. J., and Nichols R. J. 2009. *Phys. Chem. Chem. Phys.* 11, 10831.
48. Li X. L., He J., Hihath J., Xu B. Q., Lindsay S. M., and Tao N. J. 2006. *J. Am. Chem. Soc.* 128, 2135.
49. Li C., Pobelov I., Wandlowski T., Bagrets A., Arnold A., and Evers F. 2008. *J. Am. Chem. Soc.* 130, 318.
50. Fujihara M., Suzuki M., Fujii S., and Nishikawa A. 2006. *Phys. Chem. Chem. Phys.* 8, 3876.
51. Nichols R. J., Haiss W., Higgins S. J., Leary E., Martin S., and Bethell D. 2010. *Phys. Chem. Chem. Phys.* 12, 2801.
52. Li Z. Y. and Kosov D. S. 2006. *J. Phys. Chem. B* 110, 19116.
53. Fu M. D., Chen W. P., Lu H. C., Kuo C. T., Tseng W. H., and Chen C. H. 2007. *J. Phys. Chem. C* 111, 11450.
54. Yasuda S., Yoshida S., Sasaki J., Okutsu Y., Nakamura T., Taninaka A., Takeuchi O., and Shigekawa H. 2006. *J. Am. Chem. Soc.* 128, 7746.
55. Park Y. S., Whalley A. C., Kamenetska M., Steigerwald M. L., Hybertsen M. S., Nuckolls C., and Venkataraman L. 2007. *J. Am. Chem. Soc.* 129, 15768.
56. Xiao X. Y., Xu B. Q., and Tao N. J. 2004. *Nano Lett.* 4, 267.
57. Reddy P., Jang S.-Y., Segalman R. A., and Majumdar A. 2007. *Science* 315, 1568.
58. Kiguchi M., Miura S., Hara K., Sawamura M., and Murakoshi K. 2006. *Appl. Phys. Lett.* 89, 213104.
59. Quek S. Y., Kamenetska M., Steigerwald M. L., Choi H. J., Louie S. G., Hybertsen M. S., Neaton J. B., and Venkataraman L. 2009. *Nat. Nanotechnol.* 4, 230.
60. Wang C. S., Batsanov A. S., Bryce M. R., Mártin S., Nichols R. J., Higgins S. J., García-Suárez V. M., and Lambert C. J. 2009. *J. Am. Chem. Soc.* 131, 15647.
61. Sedghi G., Sawada K., Esdaile L. J., Hoffmann M., Anderson H. L., Bethell D., Haiss W., Higgins S. J., and Nichols R. J. 2008. *J. Am. Chem. Soc.* 130, 8582.
62. Sedghi G., García-Suárez V. M., Esdaile L. J., Anderson H. L., Lambert C. J., Martín S., Bethell D., Higgins S. J., Elliott M., Bennett N., Macdonald J. E., and Nichols R. J. 2011. *Nat. Nanotechnol.* 6, 517.
63. Zhang Z. J. and Imae T. 2001. *Nano Lett.* 1, 241.
64. Nichols R. J., Burgess I., Young K. L., Zamlynny V., and Lipkowski J. 2004. *J. Electroanal. Chem.* 563, 33.
65. Martín S., Haiss W., Higgins S. J., Cea P., López M. C., and Nichols R. J. 2008. *J. Phys. Chem. C* 112, 3941.
66. Martin S., Manrique D. Z., Garcia-Suarez V. M., Haiss W., Higgins S. J., Lambert C. J., and Nichols R. J. 2009. *Nanotechnology* 20, 12.
67. Cheng Z.-L., Skouta R., Vazquez H., Widawsky J. R., Schneebeli S., Chen W., Hybertsen M. S., Breslow R., and Venkataraman L. 2011. *Nat. Nanotechnol.* 6, 353.
68. Chen W., Widawsky J. R., Vazquez H., Schneebeli S. T., Hybertsen M. S., Breslow R., and Venkataraman L. 2011. *J. Am. Chem. Soc.* 133, 17160.
69. Perez-Jimenez A. J., Palacios J. J., Louis E. J., Sanfabian E., and Verges J. A. 2003. *ChemPhysChem* 4, 388.
70. Bohler T., Grebing J., Mayer-Gindner A., Lohneysen H. V., and Scheer E. 2004. *Nanotechnology* 15, S465.
71. Danilov A. V., Kubatkin S. E., Kafanov S. G., and Bjørnholm T. 2006. *Faraday Discuss. Chem. Soc.* 131, 337.
72. Park H., Park J., Lim A. K. L., Anderson E. H., Alivisatos A. P., and McEuen P. L. 2000. *Nature* 407, 57.
73. Leary E., Gonzalez M. T., van der Pol C., Bryce M. R., Filippone S., Martin N., Rubio-Bollinger G., and Agrait N. 2011. *Nano Lett.* 11, 2236.
74. Li Z. H., Han B., Meszaros G., Pobelov I., Wandlowski T., Blaszczyk A., and Mayor M. 2006. *Faraday Discuss.* 131, 121.
75. Zhang J. D., Kuznetsov A. M., Medvedev I. G., Chi Q. J., Albrecht T., Jensen P. S., and Ulstrup J. 2008. *Chem. Rev.* 108, 2737.
76. Zhang J. D., Chi Q. J., Albrecht T., Kuznetsov A. M., Grubb M., Hansen A. G., Wackerbarth H., Welinder A. C., and Ulstrup J. 2005. *Electrochim. Acta* 50, 3143.
77. Haiss W., Albrecht T., van Zalinge H., Higgins S. J., Bethell D., Hobenreich H., Schiffrin D. J., Nichols R. J., Kuznetsov A. M., Zhang J., Chi Q., and Ulstrup J. 2007. *J. Phys. Chem. B* 111, 6703.

78. Pobelov I. V., Li Z. H., and Wandlowski T. 2008. *J. Am. Chem. Soc.* 130, 16045.
79. Leary E., Higgins S. J., van Zalinge H., Haiss W., Nichols R. J., Nygaard S., Jeppesen J. O., and Ulstrup J. 2008. *J. Am. Chem. Soc.* 130, 12204.
80. Kay, N. J., Higgins, S. J., Jeppesen, J. O., Leary, E., Lycoops, J., Ulstrup, J., and Nichols, R. J. 2012. *J. Am. Chem. Soc.* 134, 16817.
81. Chen F., He J., Nuckolls C., Roberts T., Klare J. E., and Lindsay S. 2005. *Nano Lett.* 5, 503.
82. Li X. C., Hihath J., Chen F., Masuda T., Zang L., and Tao N. J. 2007. *J. Am. Chem. Soc.* 129, 11535.
83. Li C., Mishchenko A., Li Z., Pobelov I., Wandlowski T., Li X. Q., Wuerthner F., Bagrets A., and Evers F. 2008. *J. Phys. Condens. Matter.* 20, 374122.
84. Xu B. Q., Li X. L., Xiao X. Y., Sakaguchi H., and Tao N. J. 2005. *Nano Lett.* 5, 1491.
85. Bruot C., Hihath J., and Tao N. 2012. *Nat. Nanotechnol.* 7, 35.
86. Parameswaran R., Widawsky J. R., Vazquez H., Park Y. S., Boardman B. M., Nuckolls C., Steigerwald M. L., Hybertsen M. S., and Venkataraman L. 2010 *J. Phys. Chem. Lett.* 1, 2114.
87. Frei M., Aradhya S. V., Koentopp M., Hybertsen M. S., and Venkataraman L. 2011. *Nano Lett.* 11, 1518.
88. Frei M., Aradhya S. V., Hybertsen M. S., and Venkataraman L. 2012. *J. Am. Chem. Soc.* 134, 4003.
89. Haiss W., Wang C. S., Grace I., Batsanov A. S., Schiffrin D. J., Higgins S. J., Bryce M. R., Lambert C. J., and Nichols R. J. 2006. *Nat. Mater.* 5, 995.
90. Haiss W., Wang C. S., Jitchati R., Grace I., Martin S., Batsanov A. S., Higgins S. J., Bryce M. R., Lambert C. J., Jensen P. S., and Nichols R. J. 2008. *J. Phys. Condens. Matter.* 20, 374119.
91. Kamenetska M., Quek S. Y., Whalley A. C., Steigerwald M. L., Choi H. J., Louie S. G., Nuckolls C., Hybertsen M. S., Neaton J. B., and Venkataraman L. 2010. *J. Am. Chem. Soc.* 132, 6817.
92. Taniguchi M., Tsutsui M., Yokota K., and Kawai T. 2010. *Chem. Sci.* 1, 247.
93. Lafferentz L., Ample F., Yu H., Hecht S., Joachim C., and Grill L. 2009. *Science* 323, 1193.
94. Wang X. Y., Ruther R. E., Streifer J. A., and Hamers R. J. 2010. *J. Am. Chem. Soc.* 132, 4048.
95. Wang W. Y., Scott A., Gergel-Hackett N., Hacker C. A., Janes D. B., and Richter C. A. 2008. *Nano Lett.* 8, 478.
96. Ng A., Ciampi S., James M., Harper J. B., and Gooding J. J. 2009. *Langmuir* 25, 13934.
97. Solomon G. C., Bergfield J. P., Stafford C. A., and Ratner M. A. 2011. *Beilstein J. Nanotechnol.* 2, 862.
98. Kocherzhenko A. A., Siebbeles L. D. A., and Grozema F. C. 2011. *J. Phys. Chem. Lett.* 2, 1753.
99. Kaliginedi V., Moreno-Garcia P., Valkenier H., Hong W. J., Garcia-Suarez V. M., Buiter P., Otten J. L. H., Hummelen J. C., Lambert C. J., and Wandlowski T. 2012. *J. Am. Chem. Soc.* 134, 5262.
100. Guedon C. M., Valkenier H., Markussen T., Thygesen K. S., and Hummelen J. C. 2012. *Nat. Nanotechnol.* 7, 304.
101. Darwish N., Diez-Perez I., Da Silva P., Tao N. J., Gooding J. J., and Paddon-Row M. N. 2012. *Angew. Chem.-Int. Ed.* 51, 3203.
102. Leary E., Hobenreich H., Higgins S. J., van Zalinge H., Haiss W., Nichols R. J., Finch C. M., Grace I., Lambert C. J., McGrath R., and Smerdon J. 2009. *Phys. Rev. Lett.* 102, 086801.
103. Fatemi V., Kamenetska M., Neaton J. B., and Venkataraman L. 2011. *Nano Lett.* 11, 1988.
104. Martin S., Grace I., Bryce M. R., Wang C. S., Jitchati R., Batsanov A. S., Higgins S. J., Lambert C. J., and Nichols R. J. 2010. *J. Am. Chem. Soc.* 132, 9157.
105. Chang S., Huang S., He J., Liang F., Zhang P. M., Li S. Q., Chen X. W., Sankey O., and Lindsay S. 2010. *Nano Lett.* 10, 1070.
106. Hunter C. A. and Sanders J. K. M. 1990. *J. Am. Chem. Soc.* 112, 5525.
107. Marquardt C. W., Grunder S., Błaszczyk A., Dehm S., Hennrich F., Löhneysen H. v., Mayor M., and Krupke R. 2010. *Nat. Nanotechnol.* 5, 863.

34

Single-Atom Transistors for Light

Andrew Scott Parkins
University of Auckland

34.1 Introduction

In the move toward ever smaller devices for the control and manipulation of light, single atoms arguably represent the ultimate challenge. That such a challenge can actually be taken up is a tribute to the remarkable advances that have been made over the past few decades in the atomic physics and quantum optics communities. Single atoms or atomic ions are now routinely trapped, probed, and manipulated with electromagnetic fields. Typically, while the state of the single atom is altered dramatically, the fields are only very weakly perturbed by their interaction with the atom. For example, in free space the absorption and scattering of a standard laser beam by a single atom is negligible.

However, if the coupling between the atom and a single photon of the incident light field could be made strong enough, and the field is not too intense, then one might expect that the properties of the field will also be changed significantly and, possibly, in a manner that actually depends on the quantum mechanical state of the atom. Here enters the field of cavity quantum electrodynamics (cavity QED) (Berman 1994), in which atoms interact with light fields that are confined within resonators (cavities) formed, for example, by a pair of high reflectivity mirrors. If the volume of the resonator, or in other words the region in which the light is confined, is very small, then the electric field per photon, and hence its influence on a resonant atomic dipole transition, can be exceptionally large. While atom–photon coupling strength is one thing, the lifetime of a photon within the resonator is also important. For the photon to interact strongly with the atom, it must "stay" in the cavity for a sufficiently long time, which demands highly reflective mirrors and a large cavity finesse.

Pioneering work in the optical regime of cavity QED, upon which this chapter is focused, has been, and continues to be done with microscopic Fabry–Pérot cavities (Kimble 1998, Miller et al. 2005). In recent years, however, a number of alternative and very exciting new architectures have appeared, spurred on in large part by the desire for microscopic cavity QED systems that can in principle be integrated readily into (chip-based) networks for applications in quantum computing and communication (Vahala 2004, Kimble 2008). These include microtoroidal whispering gallery mode (WGM) resonators with tapered fiber input–output coupling (Aoki et al. 2006) and microscopic cavities formed by a mirror and the end facet of an optical fiber (Trupke et al. 2007) or by the end facets of two optical fibers (Volz et al. 2011).

The purpose of this chapter is to provide a basic introduction to cavity QED, with emphasis on the simple theoretical models and techniques that are typically employed, and to use these to describe in a unified manner some of the key systems and phenomena whereby single atoms can control the transport or phase properties of a weak incident probe light field. These systems divide quite naturally into two main regimes of operation: first, the strong coupling regime, in which the energy-level structure of the composite atom-cavity system plays the lead role in determining the optical response of the system and second, the bad cavity regime, in which the response is more that of just the atom, but with cavity-enhanced absorption and emission properties and cavity-dominated input and output channels (or fields).

With regards to experimental systems and demonstrations, this chapter confines itself to actual single-atom experiments with optical fields. However, the schemes and results are also directly relevant to other cavity QED architectures that employ "artificial atoms." Topical examples include single quantum dots embedded in photonic crystal cavities (Yoshie et al. 2004, Englund et al. 2007, Hennessy et al. 2007, Faraon et al. 2008, Fushman et al. 2008, Reinhard et al. 2011), micropillar cavities (Reithmaier et al. 2004, Press et al. 2007, Kasprzak et al. 2010, Young et al. 2011) or microdisk WGM resonators (Srinivasan and Painter 2007), nitrogen vacancy centers in diamond nanocrystals coupled to WGMs of silica microspheres (Park et al. 2006) or microdisks (Barclay et al. 2009), and superconducting circuit QED systems constructed from Cooper-pair boxes and waveguide microwave resonators (Wallraff et al. 2004, Fink et al. 2008, Schoelkopf and Girvin 2008, Lang et al. 2011).

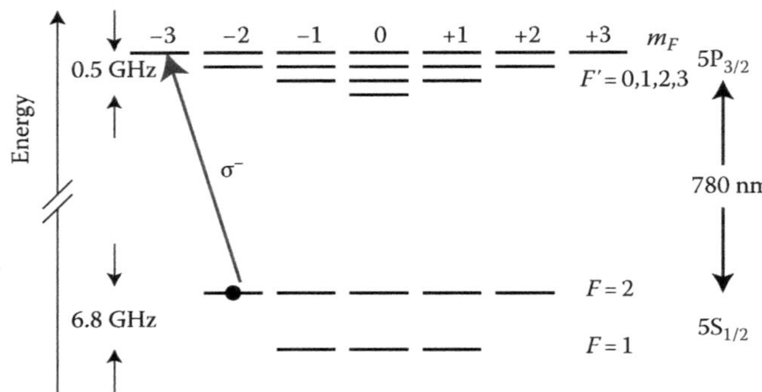

FIGURE 34.1 Energy-level structure of the D2 line in ^{87}Rb, including Zeeman and hyperfine level substructure. If the electron is prepared (e.g., by optical pumping) in the state $|F=2,\ m_F=-2\rangle \equiv |g\rangle$, then circularly polarized light of the appropriate wavelength drives transitions between this state and the state $|F'=3,\ m_F=-3\rangle \equiv |e\rangle$ only. Since the state $|F'=3,\ m_F=-3\rangle$ can only decay by spontaneous emission to the state $|F=2,\ m_F=-2\rangle$, a closed two-state system is realized.

34.2 Cavity Quantum Electrodynamics

Light is both radiated and absorbed by atoms, and the interaction between the quantized electromagnetic field and an atom represents one of the most fundamental problems in quantum optics. We begin with descriptions of the individual constituents—quantized atoms and light—and then combine them in the famous Jaynes–Cummings model of an interacting two-level atom and quantized electromagnetic field mode. Cavity QED systems are in practice open, or dissipative quantum systems, so we then take some time to introduce standard models and techniques for describing dissipative dynamics; in particular, the quantum-optical master equation and the quantum (Heisenberg–) Langevin equations. For weak incident light fields (i.e., weak excitation of the cavity QED systems), with which we are most interested, these approaches can be linearized to give relatively simple analytical results. These are given at the end of this section and form the basis for descriptions of a number of the phenomena presented in the following sections.

34.2.1 Two-Level Atoms

Real atoms are complicated systems and even the simplest real atom, hydrogen, has a nontrivial energy-level structure. It is therefore often necessary or desirable to approximate the behavior of a real atom by that of a much simpler quantum system. For many purposes, though, only a single atomic electron and a pair of atomic energy levels do in fact play a significant role in the interaction with the electromagnetic field (due, e.g., to selection rules limiting the allowed transitions between states, as depicted in Figure 34.1), so it has become common in many theoretical treatments to represent the atom by a quantum system with only two energy eigenstates.

We consider an atom with two states, $|g\rangle$ and $|e\rangle$, having energies E_1 and E_2 with $E_1 < E_2$, respectively, between which radiative transitions are allowed. Adopting these energy eigenstates as a basis for our two-level atom, the unperturbed atomic Hamiltonian $\hat{H}_A$ can be written in the form

$$\hat{H}_A = E_1 |g\rangle\langle g| + E_2 |e\rangle\langle e|. \tag{34.1}$$

The dipole moment operator $e\hat{\mathbf{r}}$, where e is the electronic charge and $\hat{\mathbf{r}}$ is the coordinate operator for the bound electron, can be expanded in terms of the (complete set of) states $|g\rangle$ and $|e\rangle$ as

$$e\hat{\mathbf{r}} = e \sum_{n,m=g,e} \langle n|\hat{\mathbf{r}}|m\rangle |n\rangle\langle m| = \mathbf{d}_{ge}\hat{\sigma}_- + \mathbf{d}_{eg}\hat{\sigma}_+, \tag{34.2}$$

where we have set $\langle g|\hat{\mathbf{r}}|g\rangle = \langle e|\hat{\mathbf{r}}|e\rangle = 0$ (assuming the atom has no permanent dipole moment), and we have introduced the atomic dipole matrix elements

$$\mathbf{d}_{ge} = e\langle g|\hat{\mathbf{r}}|e\rangle = e\int d^3r\,\phi_e^*(\mathbf{r})\mathbf{r}\phi_g(\mathbf{r}) = (\mathbf{d}_{eg})^*, \tag{34.3}$$

with $\phi_{g,e}(\mathbf{r})$ the (unperturbed) electron wave functions. We have also introduced the atomic lowering and raising operators

$$\hat{\sigma}_- = |g\rangle\langle e|, \quad \hat{\sigma}_+ = |e\rangle\langle g|, \tag{34.4}$$

which have matrix representations

$$\hat{\sigma}_- = \begin{pmatrix} 0 & 0 \\ 1 & 0 \end{pmatrix}, \quad \hat{\sigma}_+ = \begin{pmatrix} 0 & 1 \\ 0 & 0 \end{pmatrix}. \tag{34.5}$$

Note that we may also write $\hat{\sigma}_\pm = (1/2)(\hat{\sigma}_x \pm i\hat{\sigma}_y)$ and $\hat{\sigma}_+\hat{\sigma}_- - \hat{\sigma}_-\hat{\sigma}_+ = \hat{\sigma}_z$, where

$$\hat{\sigma}_x = \begin{pmatrix} 0 & 1 \\ 1 & 0 \end{pmatrix}, \quad \hat{\sigma}_y = \begin{pmatrix} 0 & -i \\ i & 0 \end{pmatrix}, \quad \hat{\sigma}_z = \begin{pmatrix} 1 & 0 \\ 0 & -1 \end{pmatrix} \tag{34.6}$$

are the Pauli spin matrices, used originally in the description of magnetic transitions in spin-1/2 systems.

In general, the state of the atom is specified by a (2×2) density operator $\hat{\rho}$, the matrix elements of which give the atomic-level populations (ρ_{gg} and ρ_{ee}) and atomic coherences ($\rho_{ge}=\rho_{eg}^*$). These elements can also be specified in terms of expectation values of the various operators. For example, the expectation value of $\hat{\sigma}_z$,

$$\langle\hat{\sigma}_z\rangle = \mathrm{Tr}(\hat{\sigma}_z\hat{\rho}) = \langle e|\hat{\rho}|e\rangle - \langle g|\hat{\rho}|g\rangle = \rho_{ee}-\rho_{gg}, \quad (34.7)$$

gives the population difference (or inversion), while the mean atomic polarization is given by

$$\begin{aligned}\langle e\hat{\mathbf{r}}\rangle &= \mathbf{d}_{ge}\langle\hat{\sigma}_-\rangle + \mathbf{d}_{eg}\langle\hat{\sigma}_+\rangle \\ &= \mathbf{d}_{ge}\langle e|\hat{\rho}|g\rangle + \mathbf{d}_{eg}\langle g|\hat{\rho}|e\rangle = \mathbf{d}_{ge}\rho_{eg} + \mathbf{d}_{eg}\rho_{ge}. \quad (34.8)\end{aligned}$$

34.2.2 Quantized Electromagnetic Field

Classically, the electromagnetic field obeys Maxwell's equations and, subject to boundary conditions determined by the physical setup, can be expressed as an expansion in a discrete set of orthogonal mode functions. Quantization of the electromagnetic field treats each of these modes as an independent, quantized harmonic oscillator of angular frequency ω_k, where k is the mode index. The energy eigenstates are the number (or Fock) states, $|n_k\rangle$, where n_k is an integer giving the number of quanta, that is, photons, and the corresponding eigenenergies are $\hbar\omega_k(n_k+1/2)$. The number states are eigenstates of the operator $\hat{a}_k^\dagger\hat{a}_k$, that is,

$$\hat{a}_k^\dagger\hat{a}_k\,|n_k\rangle = n_k\,|n_k\rangle, \quad (34.9)$$

where $\hat{a}_k^\dagger$ and $\hat{a}_k$ are the raising and lowering operators, respectively, which obey the commutation relation $[\hat{a}_k,\hat{a}_k^\dagger]=1$ and act on the number states as follows:

$$\hat{a}_k^\dagger\,|n_k\rangle = \sqrt{n_k+1}\,|n_k+1\rangle, \quad \hat{a}_k\,|n_k\rangle = \sqrt{n_k}\,|n_k-1\rangle. \quad (34.10)$$

That is, their effect corresponds to the creation and annihilation of a photon, respectively. The Hamiltonian operator for a single-field mode is given in terms of $\hat{a}_k^\dagger$ and $\hat{a}_k$ by

$$\hat{H}_{\mathrm{F}} = \hbar\omega_k\left(\hat{a}_k^\dagger\hat{a}_k + \frac{1}{2}\right). \quad (34.11)$$

The electric field operator of a single mode of the electromagnetic field can be written in the form:

$$\hat{\mathbf{E}}_k(\mathbf{r}) = i\sqrt{\frac{\hbar\omega_k}{2\epsilon_0 V_k}}\{u_k(\mathbf{r})\,\mathbf{e}_k\,\hat{a}_k - u_k^*(\mathbf{r})\mathbf{e}_k^*\hat{a}_k^\dagger\}, \quad (34.12)$$

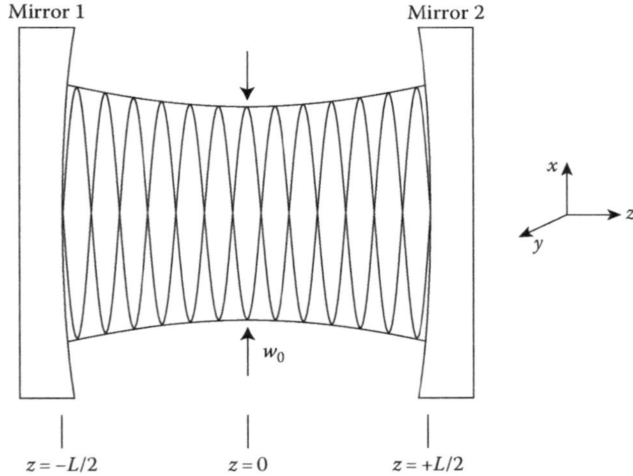

FIGURE 34.2 Optical Fabry–Pérot cavity supporting a standing wave electromagnetic field mode of waist w_0.

where

$u_k(\mathbf{r})$ is the mode function
$\mathbf{e}_k$ is a unit polarization vector
$V_k = \int d^3r\,|u_k(\mathbf{r})|^2$ is the mode (or quantization) volume

Note that the scaling is such that, for the vacuum state $|0\rangle$,

$$\int d^3r\left\{\frac{1}{2\mu_0}\langle\hat{\mathbf{B}}_k^2(\mathbf{r})\rangle + \frac{1}{2}\epsilon_0\langle\hat{\mathbf{E}}_k^2(\mathbf{r})\rangle\right\} = \frac{1}{2}\hbar\omega_k, \quad (34.13)$$

that is, the ground-state energy.

An example of special relevance to this chapter is a standing wave TEM_{00} mode of a cavity formed by (ideal) mirrors at $z=-L/2$ and $z=L/2$ (see Figure 34.2), for which

$$u_k(\mathbf{r}) = \cos(kz)\exp\left(-\frac{x^2+y^2}{w_0^2}\right), \quad (34.14)$$

where

w_0 is the mode waist
$k=\omega_k/c$ satisfies $\cos(kL/2)=0$

The mode volume in this instance is $V_k=\pi w_0^2 L/4$, which for a cavity of length $L=122$ μm and waist $w_0=29$ μm (Maunz et al. 2005) takes the value $V_k=8.06\times 10^{-5}$ mm³ and corresponds, at wavelength $\lambda_k=780$ nm, to an "electric field strength per photon" of $\sqrt{\hbar\omega_k/\epsilon_0 V_k}\simeq 600$ V/m, a quite enormous number for just one optical photon.

34.2.3 Jaynes–Cummings Model

The interaction of a single two-level atom with a single quantized field mode is described, in the Jaynes–Cummings model

(Jaynes and Cummings 1963), by a Hamiltonian of the form (dropping the mode index k),

$$\hat{H}_{AF} = \hbar(g\hat{\sigma}_+\hat{a} + g^*\hat{a}^\dagger\hat{\sigma}_-), \tag{34.15}$$

with

$$g \equiv g(\mathbf{r}_A) = -i\sqrt{\frac{\hbar\omega_C}{2\epsilon_0 V}}u(\mathbf{r}_A)\mathbf{d}_{eg}\cdot\mathbf{e}. \tag{34.16}$$

Here, $\mathbf{r}_A$ is the position of the atom and the frequency of the (cavity) mode is now denoted by ω_C. The interaction Hamiltonian $\hat{H}_{AF}$ follows from the familiar expression $-e\hat{\mathbf{r}}\cdot\hat{\mathbf{E}}(\mathbf{r}_A)$ for the potential energy of a dipole in an electric field, but with the additional application of the rotating-wave and dipole approximations. The rotating-wave approximation neglects the terms proportional to $\hat{\sigma}_-\hat{a}$ and $\hat{\sigma}_+\hat{a}^\dagger$, which in an interaction picture are rapidly rotating and correspond to nonenergy-conserving processes, while the dipole approximation requires the wavelength of the field to be much larger than the extent of the atom, which is a very good approximation even for optical fields.

Taking the lower (ground) atomic state to be at zero energy ($E_1=0$) and ignoring the ground-state energy of the field, our total Hamiltonian for the atom-field system can be written as (noting that $\hat{\sigma}_+\hat{\sigma}_- = |e\rangle\langle e|$)

$$\begin{aligned}\hat{H} &= \hat{H}_A + \hat{H}_F + \hat{H}_{AF} \\ &= \hbar\omega_A\hat{\sigma}_+\hat{\sigma}_- + \hbar\omega_C\hat{a}^\dagger\hat{a} + \hbar(g\hat{\sigma}_+\hat{a} + g^*\hat{a}^\dagger\hat{\sigma}_-),\end{aligned} \tag{34.17}$$

where $\omega_A = E_2/\hbar$ is the atomic transition frequency.

Typically, the frequencies ω_A and ω_C are many orders of magnitude larger than the atom-field coupling strength $|g|$ (which of course underpins the rotating-wave approximation made earlier), so it is common to move to an interaction picture, that is, to a frame of reference that is defined relative to, say, the cavity mode frequency. Formally, this can be defined through the unitary transformation operator $U(t)=\exp[-i\omega_C t(\hat{\sigma}_+\hat{\sigma}_- + \hat{a}^\dagger\hat{a})]$, and the interaction picture Hamiltonian takes the form

$$\hat{H}_I = \hbar\Delta_{AC}\hat{\sigma}_+\hat{\sigma}_- + \hbar(g\hat{\sigma}_+\hat{a} + g^*\hat{a}^\dagger\hat{\sigma}_-), \tag{34.18}$$

where $\Delta_{AC} = \omega_A - \omega_C$ is the detuning between the atomic transition and field mode frequencies.

The ground state of the coupled atom-cavity system is simply $|0\rangle\otimes|g\rangle\equiv|0,g\rangle$, with energy $E_0=0$ in this frame of reference, that is, $\hat{H}_I|0,g\rangle=0$. The Hamiltonian $\hat{H}_I$ couples only states with the same total number of excitations (field plus atom), so the excited states and their energies are readily deduced from the action of $\hat{H}_I$ on the states $|n-1\rangle\otimes|e\rangle\equiv|n-1,e\rangle$ and $|n\rangle\otimes|g\rangle\equiv|n,g\rangle$ (for $n\geq 1$), that is,

$$\hat{H}_I\begin{pmatrix}|n-1,e\rangle \\ |n,g\rangle\end{pmatrix} = \hbar\begin{pmatrix}\Delta_{AC} & g^*\sqrt{n} \\ g\sqrt{n} & 0\end{pmatrix}\begin{pmatrix}|n-1,e\rangle \\ |n,g\rangle\end{pmatrix}. \tag{34.19}$$

Assuming, for simplicity, that g is real, eigenenergies follow as

$$E_{n,\pm} = \frac{1}{2}\Delta_{AC} \pm \frac{1}{2}\sqrt{\Delta_{AC}^2 + 4g^2 n} \equiv \frac{1}{2}\Delta_{AC} \pm \frac{1}{2}\Omega_n, \tag{34.20}$$

with corresponding energy eigenstates (or dressed states)

$$|n,+\rangle = \frac{1}{\sqrt{2}}(C|n-1,e\rangle + D|n,g\rangle), \tag{34.21}$$

$$|n,-\rangle = \frac{1}{\sqrt{2}}(D|n-1,e\rangle - C|n,g\rangle), \tag{34.22}$$

where

$$C = \sqrt{1+\frac{\Delta_{AC}}{\Omega_n}}, \quad D = \sqrt{1-\frac{\Delta_{AC}}{\Omega_n}}. \tag{34.23}$$

The excited states of the coupled atom-cavity system thus take the form of a series, or ladder, of doublets, as shown in Figure 34.3. Consecutive doublets are separated in (absolute) energy by $\hbar\omega_C$, while the spacing Ω_n between eigenstates within each doublet increases with n; for $\Delta_{AC}=0$, the spacing $\Omega_n = 2g\sqrt{n}$ and the nonlinearity of the Jaynes–Cummings ladder of eigenstates

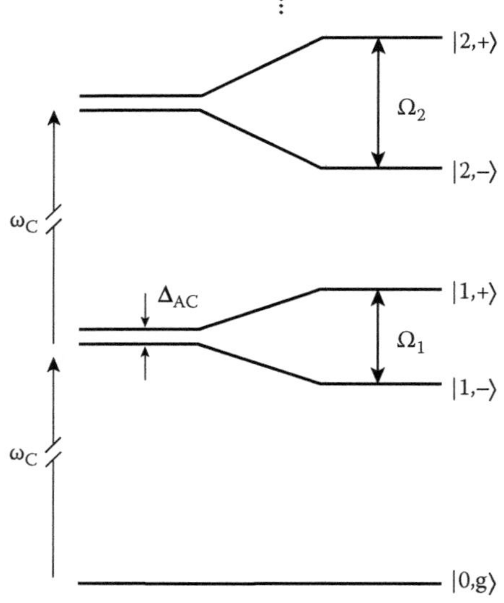

FIGURE 34.3 Energy-level structure of the Jaynes–Cummings model. To the left are the levels of the uncoupled ($g=0$) system, with $\Delta_{AC}=\omega_A-\omega_C$ the detuning between the atomic transition and cavity mode frequencies. To the right are the levels of the strongly coupled system, with $\Omega_n=\sqrt{\Delta_{AC}^2+4g^2 n}$.

is maximized. This nonlinearity is a direct consequence of the quantization of the electromagnetic field and, as we shall see, lies at the heart of several proposals for the control and manipulation of few photon light fields with single atoms.

34.2.4 Open Quantum Systems: Dissipative Cavity QED

In practice, all cavity QED systems are open quantum systems—they couple to the surrounding environment and thereby experience irreversible losses. In particular, the mirrors forming the cavity have finite transmission coefficients, while the excited atomic state undergoes spontaneous emission into (the infinity of) modes of the electromagnetic field other than the specific one supported by the cavity. In theoretical quantum optics, a common approach to modeling such an open quantum system involves a master equation for the density operator $\hat{\rho}$ of the combined atom-cavity system, which in the present context takes the form (see, e.g., Carmichael 1999)

$$\frac{d\hat{\rho}}{dt} = -\frac{i}{\hbar}[\hat{H},\hat{\rho}] + \frac{1}{2}\gamma(2\hat{\sigma}_-\hat{\rho}\hat{\sigma}_+ - \hat{\rho}\hat{\sigma}_+\hat{\sigma}_- - \hat{\sigma}_+\hat{\sigma}_-\hat{\rho}) + \kappa(2\hat{a}\hat{\rho}\hat{a}^\dagger - \hat{\rho}\hat{a}^\dagger\hat{a} - \hat{a}^\dagger\hat{a}\hat{\rho}), \tag{34.24}$$

where

$$\gamma = \frac{1}{4\pi\epsilon_0}\frac{4\omega_A^2\,|d_{ge}|^2}{3\hbar c^3} \tag{34.25}$$

is the atomic spontaneous emission rate (or Einstein A coefficient), and

$$\kappa = \kappa_1 + \kappa_2 = \frac{1}{2}\frac{(T_1+T_2)c}{2L} \tag{34.26}$$

is the cavity field decay rate, with T_1 and T_2 the transmission coefficients of the two mirrors, respectively, and L the distance between them.

The master equation may be viewed as a compact way of writing the first-order, linear differential equation for the matrix elements of $\hat{\rho}$. The equation of motion for the expectation value of a system operator $\hat{O}$ follows straightforwardly from $d\langle\hat{O}\rangle/dt = \mathrm{Tr}\{\hat{O}d\hat{\rho}/dt\}$. So, for example, equations of motion for the mean cavity field amplitude, $\langle\hat{a}\rangle$, the mean atomic polarization, $\langle\hat{\sigma}_-\rangle$, and the mean atomic inversion, $\langle\hat{\sigma}_z\rangle$, can be derived, making use of the operator commutation relations and the cyclic property of the trace, as

$$\frac{d}{dt}\langle\hat{a}\rangle = -(i\omega_C + \kappa)\langle\hat{a}\rangle - ig\langle\hat{\sigma}_-\rangle, \tag{34.27}$$

$$\frac{d}{dt}\langle\hat{\sigma}_-\rangle = -\left(i\omega_A + \frac{\gamma}{2}\right)\langle\hat{\sigma}_-\rangle + ig\langle\hat{\sigma}_z\hat{a}\rangle, \tag{34.28}$$

$$\frac{d}{dt}\langle\hat{\sigma}_z\rangle = -\gamma(\langle\hat{\sigma}_z\rangle + 1) - 2ig(\langle\hat{a}^\dagger\hat{\sigma}_-\rangle - \langle\hat{\sigma}_+\hat{a}\rangle). \tag{34.29}$$

This set of equations is not a closed set due, for example, to terms of the form $\langle\hat{\sigma}_z\hat{a}\rangle$, and in fact it is not possible to write down a closed set in the general case. To do so requires approximations based on assumptions with regard to the relative sizes of the basic cavity QED parameters or to the degree of excitation of the system, as we consider soon.

34.2.5 Input Fields, Output Fields, and Quantum Langevin Equations

The master equation and associated equations of motion for expectation values given earlier describe the dynamics of the atom and the intracavity field mode, but it is usually the fields external to the cavity that are actually accessible (e.g., directly measurable) and of most interest to us. In particular, our main concern in this chapter is with the transformation of incoming (input) fields into prescribed outgoing (output) fields via interactions in cavity QED. The quantum-optical theory of inputs and outputs in cavities is well developed and in the present context, referring to the configuration illustrated in Figure 34.4, gives the following (operator) relationships between the output, input, and intracavity light fields (Gardiner and Collett 1985, Carmichael 1999, Gardiner and Zoller 2004):

$$\hat{a}_{\mathrm{out},j}(t) = \hat{a}_{\mathrm{in},j}(t) + \sqrt{2\kappa_j}\,\hat{a}, \quad j = 1,2. \tag{34.30}$$

Quite simply, the outgoing field at each mirror is a sum of the incident field plus the field radiated from the cavity mode.

The input field operators satisfy the commutation relations $[\hat{a}_{\mathrm{in},j}(t), \hat{a}^\dagger_{\mathrm{in},j'}(t')] = \delta_{jj'}\delta(t-t')$ and feature in quantum (Heisenberg–) Langevin equations for the system operators as follows:

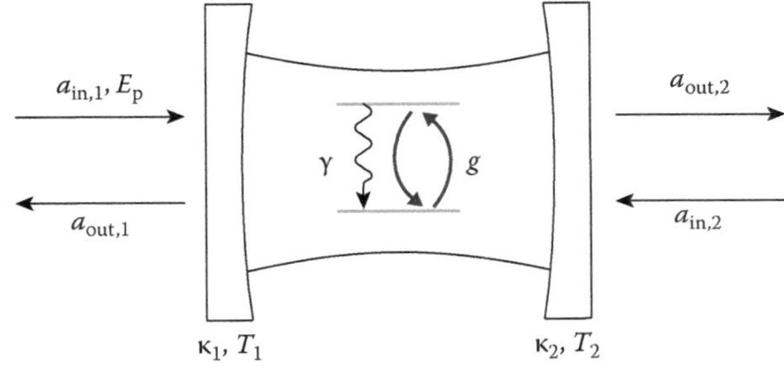

FIGURE 34.4 Cavity input and output fields for a dissipative cavity QED system. The cavity field amplitude decays at rates κ_1 and κ_2 through mirrors 1 and 2, respectively (T_1 and T_2 are the transmission coefficients of the mirrors). In this chapter, we focus on the situation in which the input field $\hat{a}_{\mathrm{in},1}$ is driven by a coherent (laser) probe field of amplitude E_p and frequency ω_p, while the input field $\hat{a}_{\mathrm{in},2}$ is vacuum. The rate of atomic spontaneous emission into modes other than the cavity mode is γ, and g is the atom-cavity coupling strength.

$$\frac{d\hat{a}}{dt} = -(i\omega_C + \kappa)\hat{a} - ig\hat{\sigma}_- - \sqrt{2\kappa_1}\hat{a}_{in,1}(t) - \sqrt{2\kappa_2}\hat{a}_{in,2}(t), \quad (34.31)$$

$$\frac{d\hat{\sigma}_-}{dt} = -\left(i\omega_A + \frac{\gamma}{2}\right)\hat{\sigma}_- + ig\hat{\sigma}_z\hat{a} + \sqrt{\gamma}\hat{\sigma}_z\hat{a}_{in,3}(t). \quad (34.32)$$

Here, $\hat{a}_{in,3}(t)$ represents (free space) electromagnetic field modes other than the cavity mode that couple to the atomic transition and give rise to atomic spontaneous emission at rate γ. In general, these coupled, nonlinear operator equations of motion cannot be solved exactly, but in the weak driving limit, which we consider in more detail later, they may be linearized (essentially by setting $\hat{\sigma}_z \to -1$ in the equations, on the assumption that the atom spends most of its time in the ground state) and offer an alternative means of describing the coupled atom-cavity system.

34.2.6 Weak Driving

Driving of the cavity field mode by a coherent laser field incident upon (and transmitted through) one of the mirrors of the cavity produces a finite mean amplitude of the corresponding input field operator, which we write for convenience in the form

$$\langle\hat{a}_{in,1}(t)\rangle = \frac{iE_p}{\sqrt{2\kappa_1}}e^{-i\omega_p t}, \quad (34.33)$$

where E_p and ω_p are the (probe) driving field amplitude and frequency, respectively. The incoming photon flux is then

$$\langle\hat{a}^\dagger_{in,1}(t)\hat{a}_{in,1}(t)\rangle = |\langle\hat{a}_{in,1}(t)\rangle|^2 = \frac{|E_p|^2}{2\kappa_1}. \quad (34.34)$$

In the master equation, this coherent input field leads to an additional Hamiltonian term of the form

$$\hat{H}_p = \hbar(E_p e^{-i\omega_p t}\hat{a}^\dagger + E_p^* e^{i\omega_p t}\hat{a}). \quad (34.35)$$

Adding this to $\hat{H}$, a unitary transformation to a frame of reference rotating at the probe frequency ω_p becomes most appropriate (and convenient, since it removes any explicit time dependence from the master equation). Defining $\Delta_C = \omega_C - \omega_p$ and $\Delta_A = \omega_A - \omega_p$, and denoting operator expectation values in this frame by $\langle\tilde{O}\rangle$, the equations of motion for the field amplitude and atomic polarization become

$$\frac{d}{dt}\langle\tilde{a}\rangle = -(i\Delta_C + \kappa)\langle\tilde{a}\rangle - ig\langle\tilde{\sigma}_-\rangle - iE_p, \quad (34.36)$$

$$\frac{d}{dt}\langle\tilde{\sigma}_-\rangle = -\left(i\Delta_A + \frac{\gamma}{2}\right)\langle\tilde{\sigma}_-\rangle + ig\langle\tilde{\sigma}_z\tilde{a}\rangle. \quad (34.37)$$

If the driving field strength is sufficiently small, then the atom-cavity system is only weakly excited and the atom resides primarily in its ground state. It follows then that $\langle\tilde{\sigma}_z\rangle \simeq -1$ and, further, that we may write $\langle\tilde{\sigma}_z\tilde{a}\rangle \simeq \langle\tilde{\sigma}_z\rangle\langle\tilde{a}\rangle \simeq -\langle\tilde{a}\rangle$. The equations of motion for $\langle\tilde{a}\rangle$ and $\langle\tilde{\sigma}_-\rangle$ are thereby linearized (and complete) and can thus be solved exactly. For the case in which the atomic and cavity mode frequencies are resonant, that is, $\Delta_C = \Delta_A \equiv \Delta$, steady-state solutions take the form

$$\langle\tilde{a}\rangle_{ss} = -\frac{iE_p(\gamma/2 + i\Delta)}{(i\Delta - \lambda_+)(i\Delta - \lambda_-)}, \quad (34.38)$$

$$\langle\tilde{\sigma}_-\rangle_{ss} = -\frac{gE_p}{(i\Delta - \lambda_+)(i\Delta - \lambda_-)}, \quad (34.39)$$

where

$$\lambda_\pm = -\frac{1}{2}\left(\kappa + \frac{\gamma}{2}\right) \pm \frac{1}{2}\sqrt{\left(\kappa - \frac{\gamma}{2}\right)^2 - 4g^2}. \quad (34.40)$$

As mentioned earlier, the quantum Langevin equations can also be linearized in this regime and are solved most conveniently in frequency space, as defined by the transform

$$\hat{O}(\omega) = \frac{1}{\sqrt{2\pi}}\int_{-\infty}^{\infty} e^{i\omega t}\hat{o}(t)dt, \quad (34.41)$$

where $\hat{o}$ is an arbitrary operator. Applying the input–output relations, one can then derive the following expressions for the output field amplitudes in terms of the input field amplitude:

$$\langle\hat{A}_{out,1}(\omega)\rangle = r(\omega)\langle\hat{A}_{in,1}(\omega)\rangle, \quad \langle\hat{A}_{out,2}(\omega)\rangle = t(\omega)\langle\hat{A}_{in,1}(\omega)\rangle, \quad (34.42)$$

with

$$r(\omega) = 1 - 2\kappa_1\frac{i(\omega_A - \omega) + \gamma/2}{[i(\omega_A - \omega) + \gamma/2][i(\omega_C - \omega) + \kappa] + g^2}, \quad (34.43)$$

$$t(\omega) = -2\sqrt{\kappa_1\kappa_2}\frac{i(\omega_A - \omega) + \gamma/2}{[i(\omega_A - \omega) + \gamma/2][i(\omega_C - \omega) + \kappa] + g^2}, \quad (34.44)$$

and

$$\langle\hat{A}_{in,1}(\omega)\rangle = \left(\frac{iE_p}{\sqrt{2\kappa_1}}\right)\sqrt{2\pi}\delta(\omega - \omega_p).$$

Finally, another alternative means of treating the weak driving limit is to assume that the state of the system is, for the most part, pure (Carmichael 2008). Collapses of the state due to photon emissions from the cavity or from the atom are, in this limit, infrequent, and in between collapses the state is describable by the approximate form

$$|\psi(t)\rangle = |0,g\rangle + \alpha(t)|1,g\rangle + \beta(t)|0,e\rangle, \quad (34.45)$$

where $\left\{|\alpha(t)|^2, |\beta(t)|^2\right\} \ll 1$ and the basis is restricted in this instance to at most one quantum of excitation. The evolution of this state is described by

$$\frac{d|\psi\rangle}{dt} = -\frac{i}{\hbar} H_{\text{eff}} |\psi\rangle \quad (34.46)$$

where $\hat{H}_{\text{eff}}$ is the non-Hermitian effective Hamiltonian (Carmichael 2008)

$$\frac{\hat{H}_{\text{eff}}}{\hbar} = \Delta\hat{a}^\dagger\hat{a} + \Delta\hat{\sigma}_+\hat{\sigma}_- + g(\hat{\sigma}_+\hat{a} + \hat{a}^\dagger\hat{\sigma}_-) + E_p\hat{a}^\dagger - i\kappa\hat{a}^\dagger\hat{a} - i\frac{\gamma}{2}\hat{\sigma}_+\hat{\sigma}_-. \quad (34.47)$$

The equations of motion for the amplitudes in the pure state expansion follow as

$$\dot{\alpha} = -(\kappa + i\Delta)\alpha - ig\beta - iE_p, \quad (34.48)$$

$$\dot{\beta} = -\left(\frac{\gamma}{2} + i\Delta\right)\beta - ig\alpha, \quad (34.49)$$

which are just the field amplitude and atomic polarization equations of motion in the weak driving limit with the identifications $\alpha = \langle\tilde{a}\rangle$ and $\beta = \langle\tilde{\sigma}_-\rangle$. This pure state approach is very useful for describing and interpreting the behavior of the system conditioned on photon detections (annihilations) in the output fields, as well as photon statistics when the basis is expanded to two quanta of excitation.

34.3 Single-Atom Control of Light in Cavity QED: Strong Coupling Regime

We begin with the strong coupling regime, in which the atom-cavity coupling strength g is much larger than the dissipative rates κ and γ, and the response of the system is governed by the structure of the Jaynes–Cummings ladder of excited state doublets. For weak driving, only the first, the so-called "vacuum Rabi" doublet plays a significant role and the dynamics is described well by the linearized equations of motion for $\langle\tilde{a}\rangle$ and $\langle\tilde{\sigma}_-\rangle$. To characterize the behavior of the system, we examine the output photon fluxes for the situation in which the cavity mode is driven from one side (through mirror 1, as depicted in Figure 34.4). Normalizing by the incident photon flux, the output fluxes can be specified in terms of intensity reflection and transmission coefficients as

$$R = \frac{\langle\hat{a}^\dagger_{\text{out},1}(t)\hat{a}_{\text{out},1}(t)\rangle}{|\langle\hat{a}_{\text{in},1}(t)\rangle|^2}, \quad T = \frac{\langle\hat{a}^\dagger_{\text{out},2}(t)\hat{a}_{\text{out},2}(t)\rangle}{|\langle\hat{a}_{\text{in},1}(t)\rangle|^2}. \quad (34.50)$$

Note that, within the linear approximation and with coherent (laser) driving, the output fields are also coherent, so that $\langle\hat{a}^\dagger_{\text{out},j}(t)\hat{a}_{\text{out},j}(t)\rangle = |\langle\hat{a}_{\text{out},j}(t)\rangle|^2$ and, thus, $R = |r(\omega)|^2$ and $T = |t(\omega)|^2$ (for stronger driving, however, this does not hold in general).

Results for R and T in the strong coupling regime, calculated in the linear approximation and assuming a symmetric cavity ($\kappa_1 = \kappa_2$), are shown in Figure 34.5. In the absence of an atom ($g=0$), destructive interference in reflection gives rise to a dip (peak) in R (T) of width 2κ (FWHM). At resonance, $\Delta = 0$, one has $R = 0$ and the probe field is in fact perfectly transmitted—the cavity acts as a through-pass Lorentzian filter. However, in the presence of an atom coupled strongly to the cavity mode, the response of the system is altered dramatically, with the resonances in R and T now shifted, or "split," to $\Delta = \pm g$, that is, to the normal mode, or vacuum Rabi frequencies of the coupled system. At resonance, the probe is now almost perfectly reflected, while at $\Delta = \pm g$ it is mainly transmitted through the cavity, although some amount

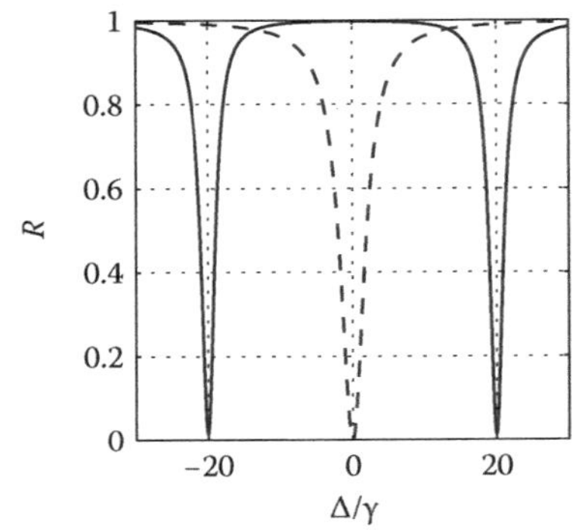

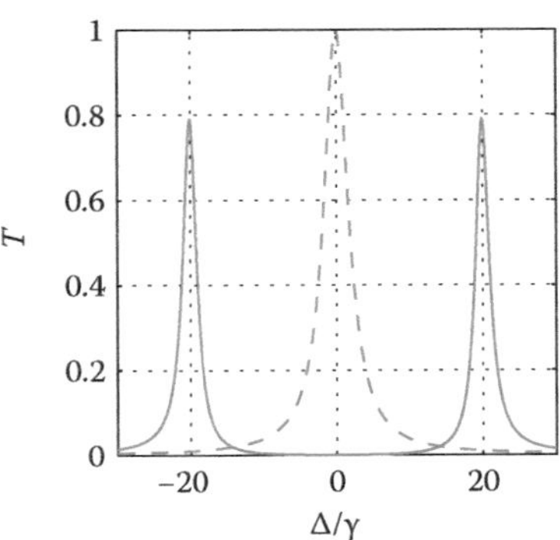

FIGURE 34.5 Probe field reflection, R, and transmission, T, as a function of the probe field detuning, Δ, from the cavity mode and atomic transition frequencies for a two-sided, symmetric Fabry–Pérot cavity with $\kappa_1/\gamma = \kappa_2/\gamma = 1$. The dashed lines are for an empty cavity ($g=0$), while the solid lines are for $g/\gamma = 20$ and are evaluated in the linear approximation (i.e., the weak field limit).

is also scattered by the atom at rate $\gamma\langle\hat{\sigma}_+(t)\hat{\sigma}_-(t)\rangle$ $(=\gamma\,|\langle\hat{\sigma}_-(t)\rangle|^2$ in the linear approximation), with, in particular, energy conservation dictating that

$$R+T+\frac{\gamma\langle\hat{\sigma}_+(t)\hat{\sigma}_-(t)\rangle}{|\langle\hat{a}_{\text{in},1}(t)\rangle|^2}=1. \tag{34.51}$$

The vacuum Rabi splitting of the cavity mode resonance with individual cesium and rubidium atoms trapped inside optical Fabry–Pérot cavities was first measured by Boca et al. (2004) and Maunz et al. (2005) with (maximum) atom-cavity coupling strengths of $g/(2\pi)=34$ and 16 MHz, respectively.

With increasing probe strength E_p, significant excitation of the Jaynes–Cummings ladder of states occurs and the linear approximation breaks down. This is illustrated in Figure 34.6, where numerical solutions to the full quantum-optical master equation (Equation 34.24) show substantial power broadening, as well as evidence of additional resonances associated with transitions to and from higher lying excited states. However, given a sufficiently large atom-cavity coupling, the extinction of the transmitted field at resonance is a quite robust phenomenon, persisting up to quite large probe strengths.

To put the numbers used in the plots into some kind of perspective, let us take, for example, $\kappa_1/(2\pi)=1$ MHz, a typical order of magnitude for several recent experiments. Then the choice $E_p/\kappa_1=1$ corresponds to an incident photon flux $E_p^2/(2\kappa_1)=\pi\times10^6\,\text{s}^{-1}$, which at a wavelength of 780 nm (corresponding to the D2 line of ^{87}Rb) gives a power of 0.80 pW. For resonant excitation of an empty ($g=0$) cavity with $\kappa_1=\kappa_2$ (and $\kappa=\kappa_1+\kappa_2$), this produces a mean intracavity photon number $\langle\hat{a}^\dagger\hat{a}\rangle_{ss}=(E_p/\kappa)^2=0.25$.

34.3.1 Driving on the Empty Cavity Resonance

For resonant driving ($\Delta=0$) in the linear regime, the intracavity field amplitude can be written in the form

$$\langle\tilde{a}\rangle_{ss}=-\frac{iE_p}{\kappa}\frac{1}{1+2C}, \tag{34.52}$$

where

$\kappa=\kappa_1+\kappa_2$

$C=g^2/(\kappa\gamma)$ is the single-atom cooperativity parameter

With $\langle\tilde{a}_{\text{in},1}\rangle=iE_p/\sqrt{2\kappa_1}$, the reflected and transmitted field amplitudes follow as

$$\langle\tilde{a}_{\text{out},1}\rangle=\langle\tilde{a}_{\text{in},1}\rangle+\sqrt{2\kappa_1}\langle\tilde{a}\rangle_{ss}=\left(\frac{2C+1-2\kappa_1/\kappa}{2C+1}\right)\langle\tilde{a}_{\text{in},1}\rangle \tag{34.53}$$

and

$$\langle\tilde{a}_{\text{out},2}\rangle=\sqrt{2\kappa_2}\langle\tilde{a}\rangle_{ss}=-\left(\frac{\sqrt{\kappa_1\kappa_2}}{\kappa/2}\frac{1}{1+2C}\right)\langle\tilde{a}_{\text{in},1}\rangle. \tag{34.54}$$

Two-sided cavity: pathway control—For a two-sided, symmetric cavity, that is, for $\kappa_1=\kappa_2$, Equations 34.53 and 34.54 give

$$\langle\tilde{a}_{\text{out},1}\rangle=\left(\frac{2C}{2C+1}\right)\langle\tilde{a}_{\text{in},1}\rangle,\quad \langle\tilde{a}_{\text{out},2}\rangle=-\left(\frac{1}{1+2C}\right)\langle\tilde{a}_{\text{in},1}\rangle. \tag{34.55}$$

It follows from this result that a single atom can act as a "switch" for the passage of this field. Take, in particular, the situation illustrated in Figure 34.7. We assume that the atom has two stable ground states, one of which ($|g_1\rangle$) is coupled strongly by the cavity mode field to an excited state, while the other ($|g\rangle_2$) is not, due to either selection rules or frequency mismatch. If the atom is prepared in the state $|g_2\rangle$ (so that $C=0$), then a resonant probe will be transmitted through the cavity, whereas if the atom is prepared in the state $|g_1\rangle$, for which $C\gg1$, then the same probe will be reflected due to the coupling-induced vacuum Rabi splitting.

One-sided cavity: phase control—Consider, on the other hand, the case in which $\kappa_2=0$ (or, at least, $\kappa_2\ll\kappa_1$, so that, as far as inputs

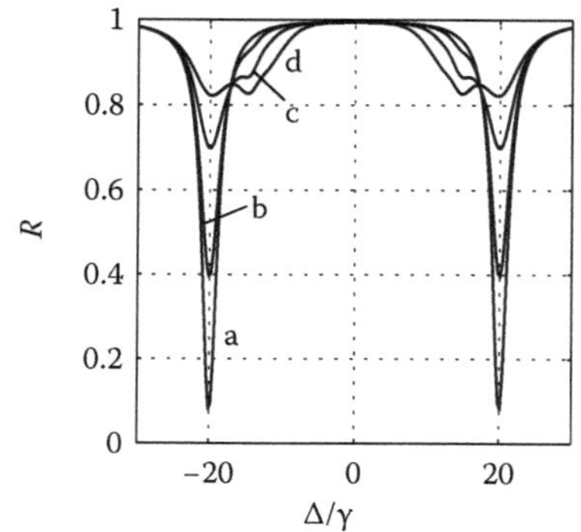

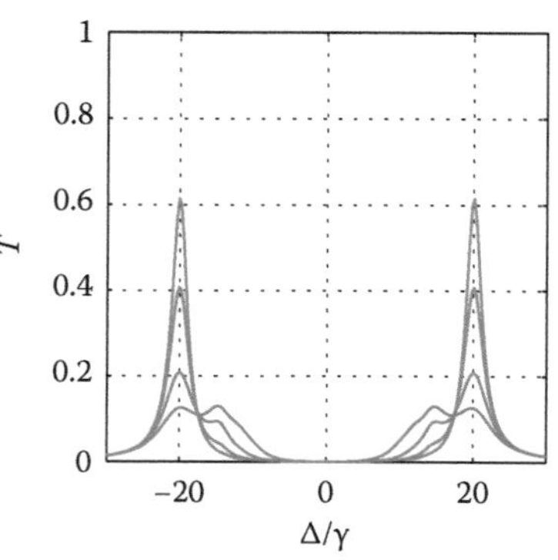

FIGURE 34.6 Probe field reflection, R, and transmission, T, as a function of the probe field detuning, Δ, from the cavity mode and atomic transition frequencies for a two-sided, symmetric Fabry–Pérot cavity with $\kappa_1/\gamma=\kappa_2/\gamma=1$ and $g/\gamma=20$. The results are computed from numerical solutions of the full master equation with $E_p=\{0.25,1,2,3\}$ (a–d).

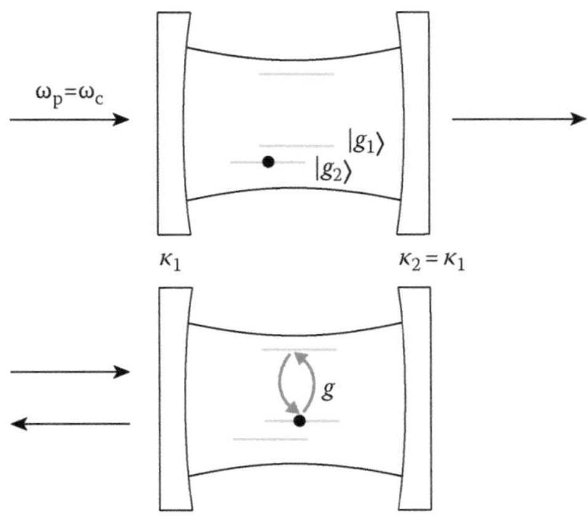

FIGURE 34.7 Single-atom switch for the *passage* of a resonant light beam incident upon a two-sided, symmetric ($\kappa_1=\kappa_2$) cavity. In state $|g_2\rangle$, the atom does not couple to the cavity mode and the light beam is transmitted through the cavity. In state $|g_1\rangle$, the atom couples strongly to the cavity mode, producing vacuum Rabi splitting of the cavity resonance and causing the incident beam to be reflected.

and outputs are concerned, the cavity can be regarded as being one-sided). Then,

$$\langle\tilde{a}_{\text{out},1}\rangle=\left(\frac{2C-1}{2C+1}\right)\langle\tilde{a}_{\text{in},1}\rangle,\quad\langle\tilde{a}_{\text{out},2}\rangle=0. \tag{34.56}$$

If the atom is prepared in the state $|g_2\rangle$, then $\langle\tilde{a}_{\text{out},1}\rangle=-\langle\tilde{a}_{\text{in},1}\rangle$, that is, the outgoing (reflected) field acquires a π phase shift relative to the incident field. However, if the atom is prepared in the state $|g_1\rangle$ and $C\gg 1$, then $\langle\tilde{a}_{\text{out},1}\rangle\simeq\langle\tilde{a}_{\text{in},1}\rangle$. Hence, the phase of the reflected field is conditioned on the state of the atom, as illustrated schematically in Figure 34.8.

This effect is the basis for a quantum phase-flip gate proposed by Duan and Kimble (2004) for qubits encoded in orthogonal polarization states of single-photon pulses. In this gate, the two polarization components are separated, such that only one component is incident upon the atom-cavity system. There, it undergoes a phase flip (relative to the other component) conditioned on the state of the atom, after which the two components are recombined. By some additional manipulation of the quantum state of the atom, a conditional phase flip can also be implemented between successive single-photon pulses. These pulses must be narrow in bandwidth compared with the cavity linewidth κ, or, in other words, temporally long in comparison with the cavity lifetime (which means that the operation is quasi-continuous and the steady-state analysis given earlier encapsulates the key behavior). One particularly nice feature of the scheme is the robustness of its performance (high fidelity) against variations in the coupling strength g, requiring only that $C\gg 1$ when the atom is in state $|g_1\rangle$.

34.3.2 Driving on a One-Photon Resonance: Photon Blockade

We now turn to a phenomenon that has at its heart the nonlinearity of the Jaynes–Cummings ladder of excited states. For a strongly coupled, resonant atom-cavity system ($g\gg\kappa,\gamma$ and $\omega_C=\omega_A=\omega_0$), incident light that is tuned to, say, the lower vacuum Rabi resonance (i.e., to the transition $|0,g\rangle\rightarrow|1,-\rangle$, for which $\Delta=\omega_0-\omega_p=g$), is inhibited from exciting the system further by virtue of the fact that it is off-resonant from transitions to higher lying energy levels. This is shown schematically in Figure 34.9. In particular, the $|1,-\rangle\rightarrow|2,-\rangle$ transition frequency is $\omega_0-(\sqrt{2}-1)g$, so the probe field at frequency $\omega_p=\omega_0-g$ is off-resonant by a detuning of $(2-\sqrt{2})g\simeq 0.59g$. For a large enough coupling strength, this detuning is sufficient to prevent any significant excitation of the state $|2,-\rangle$, and the composite atom-cavity system behaves effectively as a two-state quantum system (Tian and Carmichael 1992). This means that the system can only absorb one photon at a time from the incident field—if the system is in the state $|1,-\rangle$, then incident photons cannot be absorbed and must therefore be reflected.

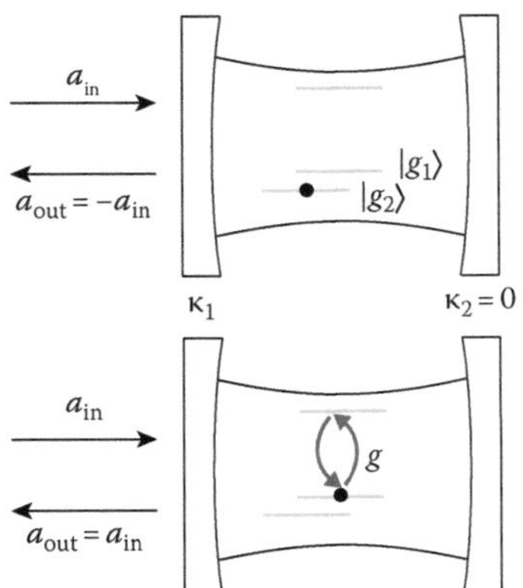

FIGURE 34.8 Single-atom switch for the *phase* of a resonant light beam incident upon a one-sided ($\kappa_2=0$, or $\kappa_1>>\kappa_2$) cavity. In state $|g_2\rangle$, the atom does not couple to the cavity mode and the incident light beam is reflected with a π phase shift. In state $|g_1\rangle$, the atom couples strongly to the cavity mode and the incident beam is reflected with no phase shift.

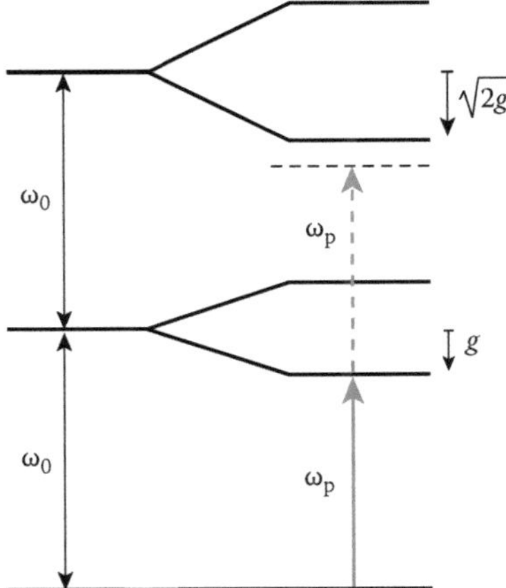

FIGURE 34.9 Photon blockade mechanism in a strongly coupled, resonant ($\omega_C=\omega_A\equiv\omega_0$) cavity QED system. For $g\gg\{\kappa,\gamma\}$, an incident probe field tuned to the lower vacuum Rabi resonance ($\omega_p=\omega_0-g$) is inhibited from exciting the system beyond the first excited state due to the nonlinearity of the Jaynes–Cummings ladder of states.

In analogy with Coulomb blockade of single electrons in nanoscopic metallic and semiconductor devices, this phenomenon has been termed "photon blockade" and is quite remarkable in that it represents a nonlinear optical process (light–light interaction) at the single-photon level. Its effect is manifested most clearly in the photon statistics of the transmitted field, as characterized by the (normally ordered) intensity correlation function:

$$G^{(2)}_{\text{out},2}(t,t+\tau) = \langle \hat{a}^\dagger_{\text{out},2}(t)\hat{a}^\dagger_{\text{out},2}(t+\tau)\hat{a}_{\text{out},2}(t+\tau)\hat{a}_{\text{out},2}(t)\rangle, \quad (34.57)$$

which is proportional to the probability of detecting (i.e., annihilating) a photon in the output field at time t and another at time $t+\tau$ (Glauber 1963). For stationary fields, upon which we focus here, this function depends only on τ. It is also common and useful to use the normalized form:

$$g^{(2)}_{\text{out},2}(\tau) = \frac{\langle \hat{a}^\dagger_{\text{out},2}\hat{a}^\dagger_{\text{out},2}(\tau)\hat{a}_{\text{out},2}(\tau)\hat{a}_{\text{out},2}\rangle_{ss}}{(\langle \hat{a}^\dagger_{\text{out},2}\hat{a}_{\text{out},2}\rangle_{ss})^2}. \quad (34.58)$$

For classical fields, this function must satisfy the Schwartz inequality (Carmichael 2008):

$$|g^{(2)}(\tau)-1| \le g^{(2)}(0)-1 \ge 0. \quad (34.59)$$

For a coherent field, which possesses Poissonian photon counting statistics, one has $g^{(2)}(\tau)=1$ and the equality holds. Quantum mechanically, however, fields are possible for which $g^{(2)}(0)<1$. Such fields are usually referred to as being "antibunched," reflecting the fact that photons tend to arrive one at a time. The extreme case is $g^{(2)}(0)=0$, for which two photons are never detected simultaneously in the field.

This is precisely the situation one expects to hold for the field transmitted by a strongly coupled atom-cavity system operating in the photon blockade regime. Using the input–output relations with $\hat{a}_{\text{in},2}(t)$ taken to be a vacuum field, the normalized correlation function reduces to

$$g^{(2)}_{\text{out},2}(\tau) = \frac{\langle \hat{a}^\dagger\hat{a}^\dagger(\tau)\,\hat{a}(\tau)\,\hat{a}\rangle_{ss}}{(\langle \hat{a}^\dagger\hat{a}\rangle_{ss})^2}. \quad (34.60)$$

In the (idealized) photon blockade regime, the atom-cavity system can only be excited as high as the first dressed state, $|1,-\rangle$, corresponding to one quantum of excitation. The expectation value $\langle\hat{a}^\dagger\hat{a}^\dagger\hat{a}\hat{a}\rangle_{ss}$, and hence $g^{(2)}_{\text{out},2}(0)$, is therefore necessarily zero, corresponding to perfect antibunching.

Numerical solutions of the full master equation for a weak driving field reveal the photon blockade effect quite clearly, as shown in Figure 34.10, where $g^{(2)}_{\text{out},2}(0)$ is plotted as a function of the probe-cavity detuning Δ. Although the parameters are not in the idealized regime of extremely strong coupling (here, in particular, $g/\kappa=10$), a value of $g^{(2)}_{\text{out},2}(0) \ll 1$ is observed around $\Delta=g$.

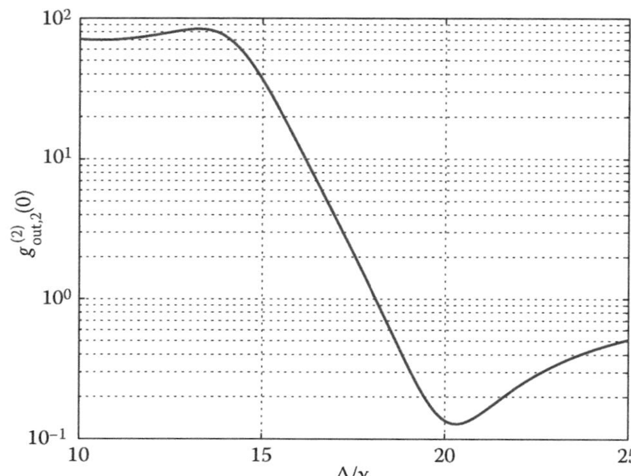

FIGURE 34.10 Intensity correlation function of the transmitted light, $g^{(2)}_{\text{out},2}(0)$, as a function of the incident probe field detuning $\Delta=\omega_0-\omega_p$, with $\kappa_1/\gamma=\kappa_2/\gamma=1$, $g/\gamma=20$, and $E_p/\gamma=0.25$. Photon antibunching, that is, $g^{(2)}_{\text{out},2}(0)<0$, around $\Delta=g$ signifies the photon blockade effect.

For an approximate theoretical analysis of the effect in the weak excitation limit, we can consider a pure state expansion of the state of the atom-cavity system in a basis that allows up to two quanta, that is,

$$|\psi(t)\rangle = |0,g\rangle + \alpha(t)|1,g\rangle + \beta(t)|0,e\rangle + \eta(t)|2,g\rangle + \zeta(t)|1,e\rangle, \quad (34.61)$$

with evolution described by Equations 34.46 and 34.47. The equations of motion for the amplitudes in the pure state expansion follow as

$$\dot{\alpha} = -(\kappa+i\Delta)\alpha - ig\beta - iE_p, \quad (34.62)$$

$$\dot{\beta} = -\left(\frac{\gamma}{2}+i\Delta\right)\beta - ig\alpha, \quad (34.63)$$

$$\dot{\eta} = -2(\kappa+i\Delta)\eta - i\sqrt{2}g\zeta - i\sqrt{2}E_p\alpha, \quad (34.64)$$

$$\dot{\zeta} = -\left(\kappa+\frac{\gamma}{2}+2i\Delta\right)\zeta - i\sqrt{2}g\eta - iE_p\beta. \quad (34.65)$$

Following Carmichael (2008), the correlation function can then be deduced from

$$g^{(2)}_{\text{out},2}(\tau) = \frac{\langle \hat{a}^\dagger\hat{a}^\dagger(\tau)\hat{a}(\tau)\hat{a}\rangle_{ss}}{\left(\langle \hat{a}^\dagger\hat{a}\rangle_{ss}\right)^2} = \frac{|\alpha(\tau)|^2}{|\alpha_{ss}|^2}, \quad (34.66)$$

where

α_{ss} is the steady-state solution for α
$\alpha(\tau)$ is the time-dependent solution with initial conditions

$$\alpha(0) = \frac{\sqrt{2}\eta_{ss}}{\alpha_{ss}}, \quad \beta(0) = \frac{\zeta_{ss}}{\alpha_{ss}}, \quad (34.67)$$

which follow by considering the effect of a photon detection (i.e., the action of the photon annihilation operator $\hat{a}$) on the state (Equation 34.61) evaluated in steady state.

For $\Delta = g$ and assuming that $g \gg \kappa, \gamma$, one can derive the approximate expression:

$$g^{(2)}_{\text{out},2}(0) \simeq 9\left(\frac{\kappa + \gamma/2}{g}\right)^2. \tag{34.68}$$

For the parameters of Figure 34.10, this gives $g^{(2)}_{\text{out},2}(0) = 0.141$, which is actually in very good agreement with the numerical result of 0.135.

For larger g and sufficiently weak driving (small E_p), $g^{(2)}_{\text{out},2}(0)$ approaches zero and $g^{(2)}_{\text{out},2}(\tau)$ approaches the result for a weakly driven two-level system with linewidth $\kappa + \gamma/2$, that is,

$$g^{(2)}_{\text{out},2}(\tau) = \left[1 - e^{-(\kappa+\gamma/2)\tau/2}\right]^2. \tag{34.69}$$

This result is shown in Figure 34.11, together with the numerical result obtained from the full master equation for $\Delta = g$ and small E_p. The numerical result displays an additional (weak) modulation at a frequency $2g$, which is the detuning of the probe field from the upper vacuum Rabi resonance at frequency $\omega_0 + g$.

Photon blockade with a single cesium atom confined inside a microscopic optical Fabry–Pérot cavity was demonstrated by Birnbaum et al. (2005). Their experimental situation was somewhat more complex than described earlier, owing to the fact that their cavity supported two orthogonal, linearly polarized modes and that numerous atomic Zeeman states were coupled to these modes, but the essential blockade mechanism persisted and a value of $g^{(2)}_{\text{out},2}(0) = 0.13$ was measured, together with a rise time of $g^{(2)}_{\text{out},2}(\tau)$ close to $2/(\kappa + \gamma/2)$, as predicted by the idealized form of Equation 34.69.

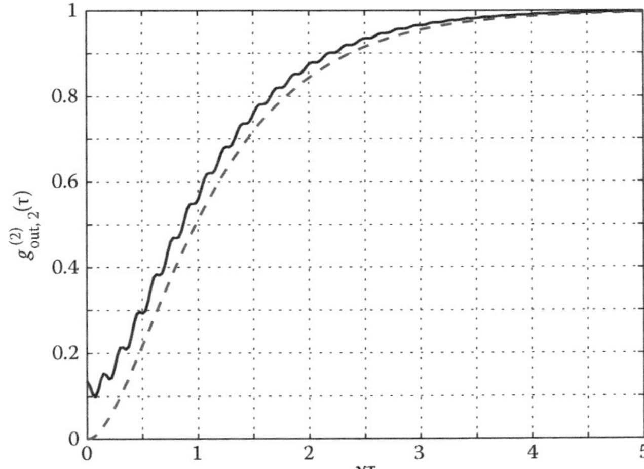

FIGURE 34.11 Intensity correlation function of the transmitted light, $g^{(2)}_{\text{out},2}(\tau)$, as a function of time τ, with $\kappa_1/\gamma = \kappa_2/\gamma = 1$, $g/\gamma = 20$, $E_p/\gamma = 0.25$, and $\Delta/\gamma = 20$ (i.e., in the photon blockade regime). The dashed curve is the idealized, weak field expression $g^{(2)}_{\text{out},2}(\tau) = [1 - e^{-(\kappa+\gamma/2)\tau/2}]^2$.

34.3.3 Driving on a Two-Photon Resonance: Two-Photon Gateway

Rather than driving the single-photon resonance for the transition $|0,g\rangle \to |1,-\rangle$, one can select the probe frequency to be such that the driving is resonant with the two-photon transition $|0,g\rangle \to |2,-\rangle$. Specifically, this occurs for the probe field frequency $\omega_0 - g/\sqrt{2}$ ($\Delta = g/\sqrt{2}$). Hints of this resonance (around $\Delta/\gamma \simeq 14$ for $g/\gamma = 20$) appear in Figure 34.6 for higher probe field strengths.

The two-photon resonance is also evident in Figure 34.10, where a local maximum appears in $g^{(2)}_{\text{out},2}(0)$ close to $\Delta = g/\sqrt{2}$. The value of this maximum is much larger than 1, indicating extreme bunching in the photon statistics of the field transmitted through the cavity, that is, the photons transmitted have a very strong tendency to arrive in pairs. Using the same pure state expansion as described earlier, valid for sufficiently weak excitation, one can derive the following approximate result,

$$g^{(2)}_{\text{out},2}(0) \simeq 2\left(\frac{2g}{3\kappa + \gamma/2}\right)^2 \gg 1, \tag{34.70}$$

for the case $\Delta = g/\sqrt{2}$ and assuming that $g \gg \kappa, \gamma$. Hence, for this choice of probe field frequency, the system acts effectively as a two-photon "gateway"; owing to the anharmonicity of the Jaynes–Cummings ladder, excitation beyond the state $|2,-\rangle$ is inhibited and de-excitation from this state results in the correlated emission of a photon pair. The two-photon resonance was observed with a single rubidium atom in an optical cavity by Schuster et al. (2008), and the correlated two-photon transmission (bunching) was subsequently measured in the same experimental setup by Kubanek et al. (2008).

34.3.4 Single-Atom Electromagnetically Induced Transparency

Electromagnetically induced transparency (EIT) (Harris 1997) is a phenomenon in which absorption on an atomic transition and concomitant losses due to spontaneous emission are eliminated via destructive interference between alternative routes from a ground atomic state to an excited atomic state. In atoms such as rubidium, this can be achieved by employing transitions from two hyperfine ground states to a single excited state, one transition of which is driven by a control (laser) field. This control field "dresses" the excited state and thereby determines the optical response of the atomic medium to another, weaker probe field, which drives the other transition. Since the elimination of atomic absorption can occur even for fields in resonance with the atomic transition, the nonlinear response can be enormous, or in other words, giant optical nonlinearities are possible (Fleischhauer et al. 2005).

While in free space this typically requires large ensembles of atoms to create an optically thick medium for the probe field, in the setting of cavity QED it is possible for one to consider EIT

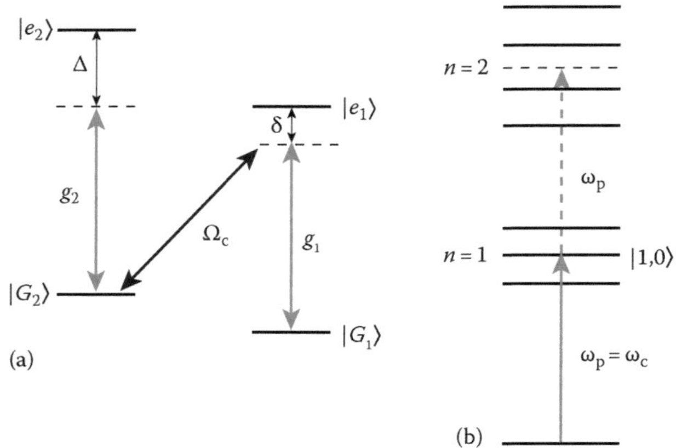

FIGURE 34.12 (a) Atomic-level scheme for single-atom electromagnetically induced transparency (EIT). A single cavity mode couples with strengths g_1 and g_2 to the transitions $|G_1\rangle \leftrightarrow |e_1\rangle$ and $|G_2\rangle \leftrightarrow |e_2\rangle$, respectively, while a coherent control field couples the states $|G_2\rangle$ and $|e_1\rangle$ with strength Ω_c. (b) Energy-level diagram of the coupled atom-cavity system in the EIT configuration, showing the first and second excited state manifolds. An incident probe field with $\omega_p = \omega_C$ resonantly excites the state $|1,0\rangle$ in the first excited state manifold, but is then off-resonant from transitions to states in the second manifold due to the nonlinearity of the system, which results in a photon blockade effect.

effects at the level of a single atom. Consider, in particular, the atomic-level scheme depicted in Figure 34.12a (Imamoğlu et al. 1997), in which a single cavity mode couples to the transitions $|G_1\rangle \leftrightarrow |e_1\rangle$ and $|G_2\rangle \leftrightarrow |e_2\rangle$ with strengths g_1 and g_2 and detunings δ and Δ, respectively, while a control laser field of strength (or Rabi frequency) Ω_c drives the transition $|G_2\rangle \leftrightarrow |e_1\rangle$ (also with detuning δ). In a frame of reference rotating at the cavity mode frequency, the Hamiltonian describing this system can be written in the form:

$$\hat{H} = \hbar\delta\,|e_1\rangle\langle e_1| + \hbar\Delta\,|e_2\rangle\langle e_2| + i\hbar g_1(\hat{a}^\dagger|G_1\rangle\langle e_1| - |e_1\rangle\langle G_1|\hat{a}) + i\hbar g_2(\hat{a}^\dagger|G_2\rangle\langle e_2| - |e_2\rangle\langle G_2|\hat{a}) + i\hbar\,(\Omega_c^\star|G_2\rangle\langle e_1| - \Omega_c|e_1\rangle\langle G_2|). \quad (34.71)$$

The ground state is the state $|0,G_1\rangle$, while, in contrast to the Jaynes–Cummings model, the first $(n=1)$ excited manifold consists of three states, given by (Rebić et al. 1999, 2002, 2004, Werner and Imamoğlu 1999)

$$|1,0\rangle = \frac{|1,G_1\rangle + (g_1/\Omega_c)|0,G_2\rangle}{\sqrt{1+(g_1/\Omega_c)^2}}, \quad (34.72)$$

and

$$|1,\pm\rangle = -\frac{(g_1/\Omega_c)|1,G_1\rangle + i(\epsilon_\pm/\Omega_c)|0,e_1\rangle - |0,G_2\rangle}{\sqrt{1+(\epsilon_\pm/\Omega_c)^2 + (g_1/\Omega_c)^2}}, \quad (34.73)$$

with energies $\epsilon_{1,0}=0$ and $\epsilon_{1,\pm}=\delta/2 \pm \sqrt{(\delta/2)^2+\Omega_c^2+g_1^2}$, respectively. So, the state $|1,0\rangle$ is resonant with the bare cavity mode, and one sees that it is a "dark" atomic state, that is, it contains no contribution from the excited atomic states and is therefore immune from atomic spontaneous emission. In fact, one finds that the (amplitude) decay rate for the state $|1,0\rangle$ is

$$\Gamma_{1,0} = \frac{\kappa}{1+(g_1/\Omega_c)^2}, \quad (34.74)$$

which, notably, depends on (and is tunable via) the strength of the control field.

In contrast, the states $|1,\pm\rangle$ are, assuming $\delta \simeq 0$, well separated in energy from the cavity resonance provided g_1 and/or Ω_c are sufficiently large, and their decay rates do depend on the atomic spontaneous emission rate owing to the contribution from the state $|e_1\rangle$.

The second $(n=2)$ excited manifold contains four states, as depicted in Figure 34.12b. Significantly, these states are all detuned from the bare cavity resonance—the system is, like the Jaynes–Cummings model, anharmonic. The inner two states lie closest to resonance, with a detuning that is determined by the coupling strength g_2. For sufficiently large g_2, it is clear that a probe field resonant with the cavity mode will be inhibited from exciting the system beyond the state $|1,0\rangle$ and that a large single-photon nonlinearity, that is, photon blockade, can again be implemented, with a key feature (cf. the Jaynes–Cummings model) being the elimination of atomic spontaneous emission, so that all input to and output from the system is through the cavity mode.

Remarkable steps toward such a photon blockade mechanism have been made by Mücke et al. (2010) and Kampschulte et al. (2010) in experiments with single alkali atoms confined inside optical Fabry–Pérot cavities. Using a control field and a cavity field mode in a Λ configuration involving the two atomic hyperfine ground states, they demonstrated delicate control of the optical response of the atom-cavity system to a weak probe laser field.

34.3.4.1 Quantum State Transfer between Light and an Atom

It is worthwhile digressing slightly for a moment to note that the dark state $|1,0\rangle$ given earlier is also at the heart of schemes for the generation of single-photon pulses and for the transfer of quantum states between light and an atom in the Λ configuration involving the states $|G_1\rangle$, $|G_2\rangle$, and $|e_1\rangle$ (Parkins et al. 1993). To see this, consider the following asymptotic behavior of the eigenstate $|1,0\rangle$:

$$|1,0\rangle \to \begin{cases} |0,G_2\rangle & \text{for } \Omega_c/g_1 \to 0 \\ |1,G_1\rangle & \text{for } g_1/\Omega_c \to 0 \end{cases}. \quad (34.75)$$

If the atom is initially in the state $|G_2\rangle$ and the cavity mode in the vacuum state, then the system state, $|0,G_2\rangle$, coincides with $|1,0\rangle$

for $\Omega_c = 0$. If the control field Ω_c is now increased at a sufficiently slow rate (so that nonadiabatic transitions to the states $|1,\pm\rangle$ can be neglected), then the state of the system will adiabatically follow the change of $|1,0\rangle$ and evolve to the state $|1,G_1\rangle$ once $\Omega_c \gg g_1$. Hence, a single photon is generated in the cavity mode and subsequently emitted through a cavity mirror into the output field.

This scheme for generating single-photon pulses was first demonstrated with single rubidium atoms falling slowly through overlapping cavity and laser fields (Hennrich et al. 2000, Kuhn et al. 2002), with the atomic states $|G_1\rangle$ and $|G_2\rangle$ corresponding to the two different hyperfine ground states. Single-photon sources based on the same scheme have since been performed with single atoms (McKeever et al. 2004, Hijlkema et al. 2007) and single Ca^+ ions (Keller et al. 2004, Barros et al. 2009) trapped quasi-permanently inside optical Fabry–Pérot cavities.

The scheme is also reversible, enabling a single incident photon to deterministically switch the atomic state from $|G_1\rangle$ to $|G_2\rangle$. Moreover, it applies to the transfer of more general quantum states between light and an atom. In particular, the adiabatic passage scheme also facilitates the reversible transformation

$$|0\rangle\otimes(\alpha|G_1\rangle+\beta|G_2\rangle)\leftrightarrow(\alpha|0\rangle+\beta|1\rangle)\otimes|G_1\rangle, \quad (34.76)$$

where $|\alpha|^2+|\beta|^2=1$. In this way, a qubit state can be exchanged between the atom and the light field. Demonstration of such reversible quantum state transfer was first achieved with a single cesium atom trapped inside a cavity and using a weak coherent state of light (produced by a weak laser pulse) as an approximation to a coherent superposition of $|0\rangle$ and $|1\rangle$ (Boozer et al. 2007).

Alternatively, employing two orthogonal, circularly polarized cavity modes and a third (ground) atomic state, it is possible for the state of the atomic qubit to be mapped onto the polarization state of a single photon and vice versa (Wilk et al. 2007, Specht et al. 2011, Ritter et al. 2012). For example, referring to specific states of ^{87}Rb, one may implement the state transfer

$$|0\rangle_L|0\rangle_R\otimes(\alpha|F=2,m_F=-1\rangle+\beta|F=2,m_F=+1\rangle)$$
$$\leftrightarrow(\alpha|1\rangle_L|0\rangle_R+\beta|0\rangle_L|1\rangle_R)\otimes|F=1,m_F=0\rangle, \quad (34.77)$$

where L and R refer to left and right circularly polarized cavity field modes, which are tuned close to the $F=1\leftrightarrow F'=1$ transition frequency (while the π-polarized control field laser is tuned close to resonance with the $F=2\leftrightarrow F'=1$ transition). This scheme has indeed been implemented experimentally with single ^{87}Rb atoms trapped inside a cavity (Wilk et al. 2007, Specht et al. 2011) and applied to the transfer of an atomic quantum state, and generation of entanglement, between atoms in separate cavities (Ritter et al. 2012). This provided a landmark demonstration of the scheme first proposed by Cirac et al. (1997) for the distribution of quantum states among distant nodes of a quantum network.

34.4 Single-Atom Control of Light in Cavity QED: Bad Cavity Regime

A regime of great theoretical and practical interest is that in which the atom-cavity coupling strength g is much larger than the free space atomic spontaneous emission rate γ, but much smaller than the cavity field decay rate κ. In this so-called "bad cavity" regime, a two-level atom initially in its excited state preferentially emits a photon into the cavity mode at rate $g\gg\gamma$. The excited cavity mode rapidly emits this photon at rate κ, before any significant reabsorption of the photon by the atom is possible. Hence, the atom-cavity coupling has the effect of increasing the effective atomic spontaneous emission rate, or, in other terms, the atom undergoes cavity-enhanced spontaneous emission.

From a theoretical perspective, this enhancement can be seen quite transparently via the process of adiabatic elimination, which, if we consider the equations of motion for the field amplitude and atomic polarization, is implemented most crudely (but effectively) by simply setting $d\langle\tilde{a}\rangle/dt=0$ and solving for $\langle\tilde{a}\rangle$ to obtain, for the resonant case $\Delta_C=\Delta_A=0$,

$$\langle\tilde{a}\rangle\simeq -i\frac{g}{\kappa}\langle\tilde{\sigma}_-\rangle - i\frac{E_P}{\kappa}. \quad (34.78)$$

This result is then substituted into the equation of motion for $\langle\tilde{\sigma}_-\rangle$ to give

$$\frac{d}{dt}\langle\tilde{\sigma}_-\rangle\simeq -\frac{\gamma}{2}(1+2C)\langle\tilde{\sigma}_-\rangle-\frac{gE_P}{\kappa}, \quad (34.79)$$

where the regime of weak excitation has been assumed, such that $\langle\tilde{\sigma}_z\tilde{a}\rangle\simeq -\langle\tilde{a}\rangle$, and the parameter $C=g^2/(\kappa\gamma)$ is as defined earlier but now represents the spontaneous emission enhancement factor. If $C\gg 1$, then cavity-enhanced spontaneous emission far exceeds atomic spontaneous emission into free space, that is, into modes of the electromagnetic field other than the cavity mode. Significantly, this means that the dominant input and output channels to the atom are through the cavity mode, which is of course a well-defined spatial mode, into and out of which one may couple light very efficiently. With regard to inputs and outputs, one therefore realizes an effective one-dimensional atom (Turchette et al. 1994).

Solving for the steady-state atomic polarization and cavity field amplitude, one has

$$\langle\tilde{\sigma}_-\rangle_{ss}=-\frac{gE_P/\kappa}{(\gamma/2)(1+2C)} \quad (34.80)$$

and

$$\langle\tilde{a}\rangle_{ss}=-i\frac{g}{\kappa}\langle\tilde{\sigma}_-\rangle_{ss}-i\frac{E_P}{\kappa}=-\frac{iE_P}{\kappa}\left(-\frac{2C}{1+2C}+1\right)=-\frac{iE_P}{\kappa}\frac{1}{1+2C}. \quad (34.81)$$

From this last equation, it is apparent that for large C the atomic polarization field largely cancels the incident field to give a very small intracavity field amplitude.

As far as the output field amplitudes are concerned, one finds (still for $\Delta = 0$)

$$\langle \tilde{a}_{\text{out},1} \rangle = \frac{iE_p}{\sqrt{2\kappa_1}}\left(1 - \frac{2\kappa_1}{\kappa}\frac{1}{1+2C}\right) = \langle \tilde{a}_{\text{in},1} \rangle \left(1 - \frac{2\kappa_1}{\kappa}\frac{1}{1+2C}\right) \quad (34.82)$$

and

$$\langle \tilde{a}_{\text{out},2} \rangle = -\langle \tilde{a}_{\text{in},1} \rangle \frac{\sqrt{\kappa_1\kappa_2}}{\kappa}\frac{2}{1+2C}. \quad (34.83)$$

Once again, two particular choices of cavity geometry are of special interest.

34.4.1 One-Sided Cavity: Phase Control

In the limiting case $\kappa_1 \gg \kappa_2$, one finds

$$\langle \tilde{a}_{\text{out},1} \rangle \simeq \left(\frac{2C-1}{2C+1}\right)\langle \tilde{a}_{\text{in},1} \rangle, \quad \langle \tilde{a}_{\text{out},2} \rangle \simeq 0, \quad (34.84)$$

and the bulk of the incident light is reflected, as one would expect. However, as found before in the strong coupling limit, the phase of the reflected field depends on the presence or not of a coupled atom. For $C=0$, one finds $\langle \tilde{a}_{\text{out},1} \rangle = -\langle \tilde{a}_{\text{in},1} \rangle$, while for $C \gg 1$, one finds $\langle \tilde{a}_{\text{out},1} \rangle = \langle \tilde{a}_{\text{in},1} \rangle$, so once again the incident field can undergo a π phase shift conditioned upon the coupling or not of an atom to the cavity mode (Hofmann et al. 2003).

34.4.2 Two-Sided Cavity: Pathway Control

For a symmetric cavity ($\kappa_1 = \kappa_2$), the output field amplitudes reduce to

$$\langle \tilde{a}_{\text{out},1} \rangle = \langle \tilde{a}_{\text{in},1} \rangle \frac{2C}{1+2C}, \quad \langle \tilde{a}_{\text{out},2} \rangle = -\langle \tilde{a}_{\text{in},1} \rangle \frac{1}{1+2C}, \quad (34.85)$$

from which it follows once again that the path taken by a weak probe field can be controlled by the state of a single atom in the same manner as depicted in Figure 34.7, albeit due to a different mechanism of atom-cavity interaction and interference. For an uncoupled atom, $C=0$, a resonant field is transmitted perfectly through the cavity, while for a coupled atom with $C \gg 1$ the field is almost perfectly reflected (Waks and Vuckovic 2006, Auffèves-Garnier et al. 2007).

This is illustrated in Figure 34.13, where the reflected and transmitted fluxes in the weak excitation regime are shown as a function of the probe field detuning for parameters approximately in the bad cavity regime and corresponding to $C=4$. A (narrow) resonance centered at $\Delta=0$ and of characteristic width (FWHM) $\gamma(1+2C)$ is seen superimposed on the broad cavity resonance of width 2κ.

It is also quite fascinating to examine the photon statistics of the reflected field in this regime. As illustrated in Figure 34.14, where a plot of $g^{(2)}_{\text{out},1}(0)$ computed from numerical solutions of the full master equation is shown as a function of probe field detuning, the reflected field is strongly antibunched over a quite broad frequency range centered on the cavity (and atomic) resonance. To understand this behavior, it is simple and instructive to consider adiabatic elimination in the operator Langevin equations of motion, together with the cavity input–output relations. In particular, in a frame rotating at the cavity frequency and assuming a resonant probe field, the field operator equation of motion is

$$\frac{d\tilde{a}}{dt} = -\kappa\tilde{a} - ig\tilde{\sigma}_- - \sqrt{2\kappa_1}\tilde{a}_{\text{in},1}(t) - \sqrt{2\kappa_2}\tilde{a}_{\text{in},2}(t), \quad (34.86)$$

with $\langle \tilde{a}_{\text{in},1}(t) \rangle = iE_p/\sqrt{2\kappa_1}$ and $\langle \tilde{a}_{\text{in},2}(t) \rangle = 0$. Setting the time derivative to zero and solving for $\tilde{a}$ gives

$$\begin{aligned}\tilde{a} &\simeq -\frac{ig}{\kappa}\tilde{\sigma}_- - \frac{1}{\kappa}\left(\sqrt{2\kappa_1}\tilde{a}_{\text{in},1}(t) + \sqrt{2\kappa_2}\tilde{a}_{\text{in},2}(t)\right) \\ &= -\frac{ig}{\kappa}\tilde{\sigma}_- - \frac{1}{\sqrt{\kappa}}\left(\tilde{a}_{\text{in},1}(t) + \tilde{a}_{\text{in},2}(t)\right), \end{aligned} \quad (34.87)$$

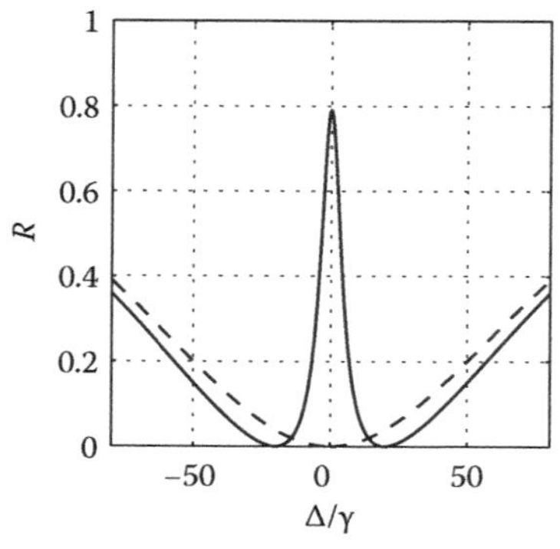

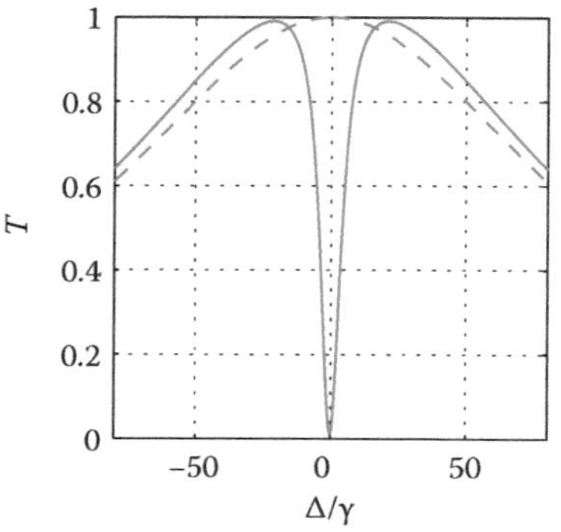

FIGURE 34.13 Probe field reflection, R, and transmission, T, as a function of the probe field detuning, $\Delta = \omega_0 - \omega_p$, in the bad cavity limit ($\kappa \gg g \gg \gamma$), with $\kappa_1/\gamma = \kappa_2/\gamma = 50$. The dashed lines are for an empty cavity ($g=0$), while the solid lines are for $g/\gamma = 20$ ($C=4$) and are evaluated in the linear (weak field) approximation.

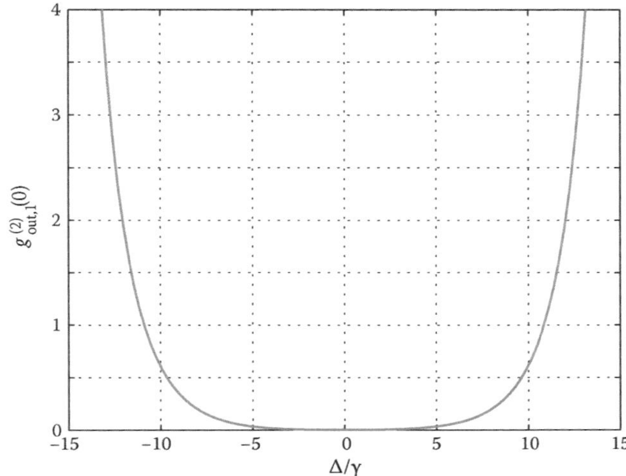

FIGURE 34.14 Intensity correlation function of the reflected light, $g^{(2)}_{out,1}(0)$, as a function of the incident probe field detuning, $\Delta = \omega_0 - \omega_p$, in the bad cavity limit, with $\kappa_1/\gamma = \kappa_2/\gamma = 50$, $g/\gamma = 20$ ($C = 4$), and $E_p/\gamma = 2$. The curve is computed from numerical solutions of the full master equation and illustrates strong antibunching of the reflected light over a frequency range of order $\gamma(1+2C)$ about $\Delta = 0$.

where we have set $\kappa_1 = \kappa_2 = \kappa/2$. The reflected output field is then

$$\tilde{a}_{out,1}(t) = \tilde{a}_{in,1}(t) + \sqrt{2\kappa_1}\,\tilde{a} \simeq -\tilde{a}_{in,2}(t) - \frac{ig}{\sqrt{\kappa}}\tilde{\sigma}_-. \tag{34.88}$$

Since the input field $\tilde{a}_{in,2}(t)$ is a vacuum, photon detections in the reflected output field can only be associated with cavity-enhanced spontaneous emission from the cavity-confined atom. Such emission leaves the atom in its ground state and unable to produce further emissions until sufficient time has passed for it to be appreciably excited once again. In fact, for weak driving one finds

$$g^{(2)}_{out,1}(\tau) \simeq (1 - e^{-\Gamma\tau/2})^2, \tag{34.89}$$

where $\Gamma = \gamma(1+2C)$. Hence, a photon blockade mechanism is once again possible, only now with regard to reflection of the incident probe field.

Note that, under the same operating conditions,

$$\tilde{a}_{out,2}(t) = \tilde{a}_{in,2}(t) + \sqrt{2\kappa_2}\,\tilde{a} \simeq -\tilde{a}_{in,1}(t) - \frac{ig}{\sqrt{\kappa}}\tilde{\sigma}_-, \tag{34.90}$$

but since $\langle \tilde{a}_{in,1}(t) \rangle \neq 0$ the photon statistics are more complicated and, in fact, very strong bunching occurs in the transmitted field.

34.4.2.1 Saturation

The results presented earlier are derived in the weak driving limit. For stronger driving, it is possible to derive a more general expression for the intracavity field amplitude which takes the form (Carmichael 2008)

$$\langle \tilde{a} \rangle_{ss} = -\frac{iE_p}{\kappa}\left(1 - \frac{2C}{1+2C}\frac{1}{1+Y^2}\right), \tag{34.91}$$

where $Y = (E_p/\kappa)/\sqrt{n_0}$ and

$$n_0 = \frac{\gamma^2(1+2C)^2}{8g^2} \tag{34.92}$$

is known as the saturation photon number. Hence, the regime of validity of the weak driving results can be quantified somewhat more precisely; in particular, one requires $Y^2 \ll 1$, that is, $(E_p/\kappa)^2 \ll n_0$. With, for example, $g/\gamma = 20$ and $C = 4$ (Figures 34.13 and 34.14), one has $n_0 = 0.025$.

34.4.3 Microtoroid Cavity QED in the Bad Cavity Regime

The development of microtoroidal optical resonators supporting very high finesse WGMs just inside their circumference, together with extremely efficient coupling to and from these modes via tapered optical fibers, has opened up an exciting new avenue of cavity QED research (Aoki et al. 2006). Single atoms in free space near the surface of a microtoroid can couple to these modes via their evanescent fields, and, given the very small mode volumes, this coupling can be very large. The nature of the input–output coupling, together with the fact that the microtoroid supports degenerate counter-propagating modes, both of which couple to the atom, provide some differences from conventional Fabry–Pérot cavity QED, but in the bad cavity regime they in fact allow some of the phenomena previously discussed to be realized in a quite convenient and elegant manner.

Consider the configuration illustrated in Figure 34.15. An atom couples with strength g to the evanescent fields of two internal

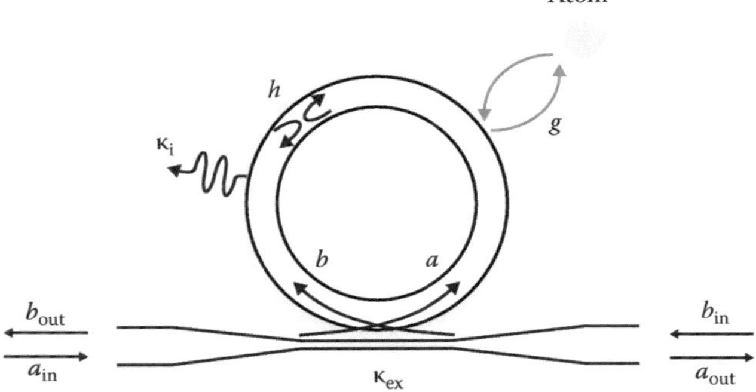

FIGURE 34.15 Schematic of a single atom coupled to a microtoroidal resonator, which supports the counter-propagating modes a and b (we omit operator hats for simplicity). A fiber taper with input and output fields as shown enables efficient coupling of light into and out of the microtoroid. The major diameter of the microtoroid is on the order of a few tens of microns, while the atom is on the order of a few hundred nanometers or less from the surface of the microtoroid.

counter-propagating modes ($\hat{a}$ and $\hat{b}$) of frequency ω_C, which are themselves coupled by scattering at rate h (which may or may not be negligible) and suffer intrinsic losses due to, for example, absorption, at rate κ_i. Light is coupled into and out of the microtoroid via a tapered optical fiber, the evanescent field of which overlaps with that of the microtoroid modes. The coupling rate, κ_{ex}, is controllable via the degree of this overlap and constitutes an additional loss rate for the internal modes; however, this is "good loss" in the sense that it is into the fiber and therefore completely accessible for measurement or transfer elsewhere. We are therefore most interested in the case where $\kappa_{ex} \gg \kappa_i$, such that the fiber provides the dominant input and output channel for the atom-cavity system. We take the input field $\hat{a}_{in}$ to be driven by a coherent probe of amplitude E_p and frequency ω_p, while $\hat{b}_{in}$ is a vacuum field.

A quantum-optical Hamiltonian for the coherent dynamics of this system takes the form

$$\begin{aligned}\hat{H} &= \hbar\omega_A\hat{\sigma}_+\hat{\sigma}_- + \hbar\omega_C(\hat{a}^\dagger\hat{a} + \hat{b}^\dagger\hat{b}) + \hbar h(\hat{a}^\dagger\hat{b} + \hat{b}^\dagger\hat{a}) \\ &\quad + \hbar(E_p^* e^{i\omega_p t}\hat{a} + E_p e^{-i\omega_p t}\hat{a}^\dagger) + \hbar(g^*\hat{a}^\dagger\hat{\sigma}_- + g\hat{\sigma}_+\hat{a}) \\ &\quad + \hbar(g\hat{b}^\dagger\hat{\sigma}_- + g^*\hat{\sigma}_+\hat{b}),\end{aligned} \tag{34.93}$$

where $g = g_0(r)e^{ikx}$, with r the radial distance of the atom from the surface of the toroid and x the atom's position around the circumference of the toroid. Typically, $g_0(r) \propto e^{-\alpha r}$ with $\alpha \sim 2\pi/\lambda$, the optical wave number.

Introducing dissipation, the system can be described by the master equation:

$$\begin{aligned}\frac{d\hat{\rho}}{dt} &= -\frac{i}{\hbar}[\hat{H},\hat{\rho}] + \kappa(2\hat{a}\hat{\rho}\hat{a}^\dagger - \hat{a}^\dagger\hat{a}\hat{\rho} - \hat{\rho}\hat{a}^\dagger\hat{a}) \\ &\quad + \kappa(2\hat{b}\hat{\rho}\hat{b}^\dagger - \hat{b}^\dagger\hat{b}\hat{\rho} - \hat{\rho}\hat{b}^\dagger\hat{b}) + \frac{\gamma}{2}(2\hat{\sigma}_-\hat{\rho}\hat{\sigma}_+ - \hat{\sigma}_+\hat{\sigma}_-\hat{\rho} - \hat{\rho}\hat{\sigma}_+\hat{\sigma}_-),\end{aligned} \tag{34.94}$$

where
$\hat{\rho}$ is the density operator for the atom-cavity system
$\kappa = \kappa_i + \kappa_{ex}$

In the weak driving limit (i.e., sufficiently small E_p), the linearized equations of motion for the mean field amplitudes and atomic polarization take the form (in a frame rotating at the probe frequency)

$$\frac{d}{dt}\langle\tilde{a}\rangle = -(\kappa + i\Delta_C)\langle\tilde{a}\rangle - ih\langle\tilde{b}\rangle - iE_p - ig^*\langle\tilde{\sigma}_-\rangle, \tag{34.95}$$

$$\frac{d}{dt}\langle\tilde{b}\rangle = -(\kappa + i\Delta_C)\langle\tilde{b}\rangle - ih\langle\tilde{a}\rangle - ig\langle\tilde{\sigma}_-\rangle, \tag{34.96}$$

$$\frac{d}{dt}\langle\tilde{\sigma}_-\rangle = -\left(\frac{\gamma}{2} + i\Delta_A\right)\langle\tilde{\sigma}_-\rangle - ig\langle\tilde{a}\rangle - ig^*\langle\tilde{b}\rangle, \tag{34.97}$$

where Δ_A and Δ_C are as defined previously.

34.4.3.1 Critical Coupling

If the probe is resonant with the cavity modes ($\omega_p = \omega_C$) and the fiber coupling is adjusted to the value

$$\kappa_{ex} = \kappa_{ex}^{crit} \equiv \sqrt{\kappa_i^2 + h^2}, \tag{34.98}$$

then in the absence of an atom ($g = 0$) destructive interference in the forward direction along the fiber (with respect to the incident probe) leads to

$$\langle\tilde{a}_{out}\rangle = 0, \tag{34.99}$$

while

$$\langle\tilde{b}_{out}\rangle = \sqrt{2\kappa_{ex}}\,\frac{E_p h}{h^2 + \kappa^2}. \tag{34.100}$$

The destructive interference in the forward direction is analogous to the absence of reflected light for a resonant probe incident upon a two-sided symmetric Fabry–Pérot cavity, as considered earlier in this chapter. Normalized output intensities in the forward and backward directions follow as

$$\begin{aligned}T_F(\Delta_C = 0) &\equiv \frac{|\langle\tilde{a}_{out}(t)\rangle|^2}{|\langle\tilde{a}_{in}(t)\rangle|^2} = 0, \\ T_B(\Delta_C = 0) &\equiv \frac{|\langle\tilde{b}_{out}(t)\rangle|^2}{|\langle\tilde{a}_{in}(t)\rangle|^2} = \frac{4\kappa_{ex}^2 h^2}{(h^2 + \kappa^2)^2},\end{aligned} \tag{34.101}$$

and in the limit that $\kappa_{ex} \gg \kappa_i$ (and hence $\kappa \simeq \kappa_{ex} \simeq h$) one finds $T_B(\Delta_C = 0) \simeq 1$.

However, with $g \neq 0$ and operating in the bad cavity regime, one can show that the output field operator in the forward direction for resonant driving takes the form

$$\tilde{a}_{out}(t) \simeq \{\text{vacuum input field operators}\} + \alpha_-\tilde{\sigma}_-, \tag{34.102}$$

where

$$\alpha_- = -g_0\sqrt{\kappa_{ex}}\left(\frac{i\cos(kx)}{\kappa + ih} + \frac{\sin(kx)}{\kappa - ih}\right), \tag{34.103}$$

and that in the limit considered earlier and for a suitably positioned atom (e.g., $kx = 0$ or $\pi/2$),

$$T_F(\Delta_C = 0) \simeq \left(\frac{2C}{2C + 1}\right)^2 \tag{34.104}$$

with

$$C = \frac{2|g|^2\kappa}{\gamma(\kappa^2 + h^2)}. \tag{34.105}$$

Hence, for $C \gg 1$ one has $T_F(\Delta_C=0) \simeq 1$ and the atom can act as a switch for the passage of a weak light field propagating in the fiber.

From the form of Equation 34.102, it also follows that

$$g_F^{(2)}(0) \equiv \frac{\langle \hat{a}^\dagger_{\text{out}} \hat{a}^\dagger_{\text{out}} \hat{a}_{\text{out}} \hat{a}_{\text{out}} \rangle}{(\langle \hat{a}^\dagger_{\text{out}} \hat{a}_{\text{out}} \rangle)^2} = 0 \tag{34.106}$$

and hence the system also functions as a photon "turnstile," allowing passage of only one photon at a time from the incident field. This effect was observed in an experiment by Dayan et al. (2008), in which cesium atoms were dropped through the evanescent field of a microtoroidal resonator with a (major) diameter of 25 μm. The cavity field decay rates in the experiment were $(\kappa_i,\kappa_{ex})/2\pi = (75,90)$ MHz, with mode coupling $h/2\pi = 50$ MHz, while the effective atom-cavity coupling rate was $g_{\text{eff}}/2\pi =$ 50 MHz. This enabled clear observation of photon antibunching in the forward direction, with, in particular, a measured value $g_F^{(2)}(0) = 0.14$ and a good fit of the time dependence to $g_F^{(2)}(\tau) = (1 - e^{-\Gamma\tau/2})^2$ with $\Gamma^{-1} = [\gamma(1+2C)]^{-1} \simeq 2.8\,\text{ns}$.

34.4.3.2 Overcoupled Regime

Conversely, if $\kappa_{ex} \gg \kappa_{ex}^{\text{crit}}$, then in the absence of an atom the incident probe field is almost entirely transmitted in the forward direction, as if the microtoroid was not there at all. However, with an atom coupled to the microtoroid modes and $C \gg 1$, the field associated with the steady-state atomic polarization $\langle \tilde{\sigma}_- \rangle_{ss}$ interferes destructively with the intracavity field due to $\langle \tilde{a}_{in} \rangle$, resulting in very small transmission in the forward direction, while driving of the backward propagating field $\hat{b}_{\text{out}}$ by this same atomic polarization gives rise to strong effective reflection of the incident field in the fiber over a characteristic bandwidth $\Gamma = \gamma(1+2C)$. In particular, for $\kappa \simeq \kappa_{ex} \gg (\kappa_i, h)$ one can show that

$$T_F(\Delta_C = 0) \simeq \left(\frac{1}{2C+1}\right)^2, \tag{34.107}$$

while now

$$T_B(\Delta_C = 0) \simeq \left(\frac{2C}{2C+1}\right)^2. \tag{34.108}$$

The output field operator in the backward direction can be written in the form

$$\tilde{b}_{\text{out}}(t) \simeq \{\text{vacuum input field operators}\} + \beta_- \tilde{\sigma}_-, \tag{34.109}$$

with

$$\beta_- \simeq -i\frac{g_0}{\sqrt{\kappa}} e^{ikx}, \tag{34.110}$$

so now this field exhibits antibunching, with $g_B^{(0)} \simeq 0$, and the system in this regime acts as a photon turnstile with regard to reflection of the incident light in the fiber.

This regime of operation for the microtoroid was observed by Aoki et al. (2009) in a very similar setup to that used by Dayan et al. (2008), only now with parameters $(g_{\text{eff}},\kappa_{ex},\kappa_i,h)/2\pi =$ (50,300,20,10) MHz (which correspond to $C=3$ and $\Gamma^{-1} = 4.4$ ns). Notably, in comparison to the critically coupled regime of operation described earlier, the overcoupled regime is more robust to variations in the azimuthal position of the atom (kx) and in the fiber–toroid separation (i.e., in κ_{ex}).

34.5 Single- and Two-Photon Input Pulses

The focus of the models and results presented in this chapter has been on continuous wave input (probe) light fields of well-defined frequency, albeit at very low intensities that typically generate average intracavity photon numbers (much) less than one. An actual single-photon pulse of finite duration, though, necessarily has a finite spectral bandwidth, the size of which must be taken into account when considering the applicability of the results presented thus far.

Consider, for example, the overcoupled regime of the microtoroid cavity QED system just considered. In order for a single-photon pulse to be reflected when an atom couples to the cavity modes, the bandwidth of the pulse must be much less than the width, $\Gamma = \gamma(1+2C)$, of the atom-induced resonance in the field reflected back along the fiber. In other words, if the pulse is too short in duration (such that the bandwidth is larger than Γ), then the atom has insufficient time to develop the dipole field necessary for destructive interference with the intracavity field, and so the photon is transmitted (Rosenblum et al. 2011). Hence, for the particular experimental setup of Aoki et al. (2009), a single-photon pulse significantly longer in duration than $\Gamma^{-1} = 4.4$ ns would be required for good reflection.

With regard to schemes discussed for photon blockade, that is, for photon–photon interactions in cavity QED, bandwidth limitations on incident photon pulses have more significant ramifications. Consider again the overcoupled microtoroid cavity QED system in the regime $C \gg 1$, but now with an incident two-photon pulse. The form (Equation 34.109) of the output field operator in the backward direction and the consequent antibunching in the reflected light suggest a mechanism for *photon routing*; that is, for the separation of two incident photons into different output ports. In particular, emission of a photon in the reflected field causes a collapse of the atom to its ground state and zeroing of the atomic dipole field, allowing the second photon of the pulse to be transmitted (as if the atom was no longer there). For ideal photon routing, this separation of the incident photons should occur every time, that is, deterministically. However, if the incident pulse is significantly longer than Γ^{-1}, then the atom may have time to reestablish its dipole field after scattering the first photon into the reflected field and, therefore, be able to scatter the second photon into the reflected field as well. On the other hand, if the incident pulse is shorter than Γ^{-1} then it will contain spectral components outside the bandwidth of the two-photon nonlinearity and there will be a finite probability for both photons to be transmitted. Hence, the routing efficiency is inherently limited (to ~64%) by what amounts to the energy–time uncertainty relation. Furthermore, this limitation is not unique to this example, but in fact applies to all such photon blockade (or turnstile) realizations based upon effective two-state systems (Rosenblum et al. 2011). The performances of other, more general single-photon nonlinearities have also been shown to be limited by the same fundamental issue (Koshino and Ishihara 2004, Shapiro 2006, Gea-Banacloche 2010).

However, a potential way around this issue is to consider a slightly modified atom-cavity configuration (Koshino et al. 2010); in particular, a single atom in a Λ configuration, with transitions of orthogonal (say, H and V) polarizations coupling two ground states, $|g_1\rangle$ and $|g_2\rangle$, respectively, to a (single) common excited state, $|e\rangle$. Each transition is enhanced by coupling to a mode of a single-sided cavity in the bad cavity regime. As shown by Rosenblum et al. (2011), for an incident H-polarized two-photon pulse and the atom initially in the ground state $|g_1\rangle$, one and only one photon from the incident pulse is absorbed by the atom and then reemitted as a V-polarized photon as the atom is transferred to the state $|g_2\rangle$. The output from the cavity thus consists of single-photon pulses of orthogonal polarization, which can be separated into different parts by a polarizing beam splitter. For a sufficiently long input pulse (i.e., of duration much longer than Γ^{-1}), this process can occur with near-unit efficiency, thus achieving near-ideal photon routing. In essence, the use of a second atomic ground state provides a form of memory for the system that enables the energy–time uncertainty issue to be circumvented.

References

Aoki, T., B. Dayan, E. Wilcut et al. 2006. Observation of strong coupling between one atom and a monolithic microresonator. *Nature* 443: 671–674.

Aoki, T., A. S. Parkins, D. J. Alton et al. 2009. Efficient routing of single photons by one atom and a microtoroidal cavity. *Phys. Rev. Lett.* 102: 083601.

Auffèves-Garnier, A., C. Simon, J.-M. Gérard, and J.-P. Poizat. 2007. Giant optical nonlinearity induced by a single two-level system interacting with a cavity in the Purcell regime. *Phys. Rev. A* 75: 053823.

Barclay, P. E., C. Santori, K.-M. Fu, R. G. Beausoleil, and O. Painter. 2009. Coherent interference effects in a nano-assembled diamond NV center cavity-QED system. *Opt. Express* 17: 8081–8097.

Barros, H. G., A. Stute, T. E. Northup, C. Russo, P. O. Schmidt, and R. Blatt. 2009. Deterministic single-photon source from a single ion. *New J. Phys.* 11: 103004.

Berman, P. R. (ed.) 1994. *Cavity Quantum Electrodynamics. Advances in Atomic, Molecular, and Optical Physics.* New York: Academic Press.

Birnbaum, K. M., A. Boca, R. Miller, A. D. Boozer, T. E. Northup, and H. J. Kimble. 2005. Photon blockade in an optical cavity with one trapped atom. *Nature* 436: 87–90.

Boca, A., R. Miller, K. M. Birnbaum, A. D. Boozer, J. McKeever, and H. J. Kimble. 2004. Observation of the vacuum Rabi spectrum for one trapped atom. *Phys. Rev. Lett.* 93: 233603.

Boozer, A. D., A. Boca, R. Miller, T. E. Northup, and H. J. Kimble. 2007. Reversible state transfer between light and a single trapped atom. *Phys. Rev. Lett.* 98: 193601.

Carmichael, H. J. 1999. *Statistical Methods in Quantum Optics 1: Master Equations and Fokker–Planck Equations.* Berlin, Germany: Springer-Verlag.

Carmichael, H. J. 2008. *Statistical Methods in Quantum Optics 2: Non-Classical Fields.* Berlin, Germany: Springer-Verlag.

Cirac, J. I., P. Zoller, H. J. Kimble, and H. Mabuchi. 1997. Quantum state transfer and entanglement distribution among distant nodes in a quantum network. *Phys. Rev. Lett.* 78: 3221–3224.

Dayan, B., A. S. Parkins, T. Aoki, H. J. Kimble, E. P. Ostby, and K. Vahala. 2008. A photon turnstile dynamically regulated by one atom. *Science* 319: 1062–1065.

Duan, L.-M. and H. J. Kimble. 2004. Scalable photonic quantum computation through cavity assisted interactions. *Phys. Rev. Lett.* 92: 127902.

Englund, D., A. Faraon, I. Fushman, N. Stoltz, P. Petroff, and J. Vučković. 2007. Controlling cavity reflectivity with a single quantum dot. *Nature* 450: 857–861.

Faraon, A., I. Fushman, D. Englund, N. Stoltz, P. Petroff, and J. Vučković. 2008. Coherent generation of non-classical light on a chip via photon-induced tunnelling and blockade. *Nature Phys.* 4: 859–863.

Fink, J. M., M. Göppl, M. Baur et al. 2008. Climbing the Jaynes–Cummings ladder and observing its $\sqrt{n}$ nonlinearity in a cavity QED system. *Nature* 454: 315–318.

Fleischhauer, M., A. Imamoğlu, and J. P. Marangos. 2005. Electromagnetically induced transparency: Optics in coherent media. *Rev. Mod. Phys.* 77: 633–673.

Fushman, I., D. Englund, A. Faraon, N. Stoltz, P. Petroff, and J. Vučković. 2008. Controlled phase shifts with a single quantum dot. *Science* 320: 769–772.

Gardiner, C. W. and M. J. Collett. 1985. Input and output in damped quantum systems: Quantum stochastic differential equations and the master equation. *Phys. Rev. A* 31: 3761–3774.

Gardiner, C. W. and P. Zoller. 2004. *Quantum Noise: A Handbook of Markovian and Non-Markovian Quantum Stochastic Methods with Applications to Quantum Optics.* Berlin, Germany: Springer-Verlag.

Gea-Banacloche, J. 2010. Impossibility of large phase shifts via the giant Kerr effect with single-photon wave packets. *Phys. Rev. A* 81: 043823.

Glauber, R. J. 1963. The quantum theory of optical coherence. *Phys. Rev.* 130: 2529–2539.

Harris, S. E. 1997. Electromagnetically induced transparency. *Phys. Today* 50: 36–42.

Hennessy, K., A. Badolato, M. Winger et al. 2007. Quantum nature of a strongly coupled single quantum dot-cavity system. *Nature* 445: 896–899.

Hennrich, M., T. Legero, A. Kuhn, and G. Rempe. 2000. Vacuum-stimulated Raman scattering based on adiabatic passage in a high-finesse optical cavity. *Phys. Rev. Lett.* 85: 4872–4875.

Hijlkema, M., B. Weber, H. P. Specht, S. C. Webster, A. Kuhn, and G. Rempe. 2007. A single-photon server with just one atom. *Nature Phys.* 3: 253–255.

Hofmann, H. F., K. Kojima, S. Takeuchi, and K. Sasaki. 2003. Optimized phase switching using a single-atom nonlinearity. *J. Opt. B: Quantum Semiclass. Opt.* 5: 218–221.

Imamoğlu, A., H. Schmidt, G. Woods, and M. Deutsch. 1997. Strongly interacting photons in a nonlinear cavity. *Phys. Rev. Lett.* 81: 1467–1470.

Jaynes, E. T. and F. W. Cummings. 1963. Comparison of quantum and semiclassical radiation theories with application to the beam maser. *Proc. IEEE* 51: 89–109.

Kampschulte, T., W. Alt, S. Brakhane et al. 2010. Optical control of the refractive index of a single atom. *Phys. Rev. Lett.* 105: 153603.

Kasprzak, J., S. Reitzenstein, E. A. Muljarov et al. 2010. Up on the Jaynes–Cummings ladder of a quantum-dot/microcavity system. *Nat. Mater.* 9: 304–308.

Keller, M., B. Lange, K. Hayasaka, W. Lange, and H. Walther. 2004. Continuous generation of single photons with controlled waveform in an ion-trap cavity system. *Nature* 431: 1075–1078.

Kimble, H. J. 1998. Strong interactions of single atoms and photons in cavity QED. *Phys. Scr.* T76: 127–138.

Kimble, H. J. 2008. The quantum Internet. *Nature* 453: 1023–1030.

Koshino, K. and H. Ishihara. 2004. Two-photon nonlinearity in general cavity QED systems. *Phys. Rev. A* 70: 013806.

Koshino, K., S. Ishizaka, and Y. Nakamura. 2010. Deterministic photon–photon $\sqrt{\text{SWAP}}$ gate using a Λ system. *Phys. Rev. A* 82: 010301(R).

Kubanek, A., A. Ourjoumtsev, I. Schuster et al. 2008. Two-photon gateway in one-atom cavity quantum electrodynamics. *Phys. Rev. Lett.* 101: 203602.

Kuhn, A., M. Hennrich, and G. Rempe. 2002. Deterministic single-photon source for distributed quantum networking. *Phys. Rev. Lett.* 89: 067901.

Lang, C., D. Bozyigit, C. Eichler et al. 2011. Observation of resonant photon blockade at microwave frequencies using correlation function measurements. *Phys. Rev. Lett.* 106: 243601.

Maunz, P., T. Puppe, I. Schuster, N. Syassen, P. W. H. Pinkse, and G. Rempe. 2005. Normal-mode spectroscopy of a single-bound-atom-cavity system. *Phys. Rev. Lett.* 94: 033002.

McKeever, J., A. Boca, A. D. Boozer et al. 2004. Deterministic generation of single photons from one atom trapped in a cavity. *Science* 303: 1992–1994.

Miller, R., T. E. Northup, K. M. Birnbaum, A. Boca, A. D. Boozer, and H. J. Kimble. 2005. Trapped atoms in cavity QED: Coupling quantized light and matter. *J. Phys. B: At. Mol. Opt. Phys.* 38: S551–S565.

Mücke, M., E. Figueroa, J. Bochmann et al. 2010. Electromagnetically induced transparency with single atoms in a cavity. *Nature* 465: 755–758.

Park, Y.-S., A. K. Cook, and H. Wang. 2006. Cavity QED with diamond nano crystals and silica microspheres. *Nano Lett.* 6: 2075–2079.

Parkins, A. S., P. Marte, P. Zoller, and H. J. Kimble. 1993. Synthesis of arbitrary quantum states via adiabatic transfer of Zeeman coherence. *Phys. Rev. Lett.* 71: 3095–3098.

Press, D., S. Götzinger, S. Reitzenstein et al. 2007. Photon antibunching from a single quantum-dot-microcavity system in the strong coupling regime. *Phys. Rev. Lett.* 98: 117402.

Rebić, S., A. S. Parkins, and S. M. Tan. 2004. Field correlations and effective two-level atom-cavity systems. *Phys. Rev. A* 69: 035804.

Rebić, S., A. S. Parkins, and S. M. Tan. 2002. Photon statistics of a single-atom intracavity system involving electromagnetically induced transparency. *Phys. Rev. A* 65: 063804.

Rebić, S., S. M. Tan, A. S. Parkins, and D. F. Walls. 1999. Large Kerr nonlinearity with a single atom. *J. Opt. B: Quantum Semiclass. Opt.* 1: 490–495.

Reinhard, A., T. Volz, M. Winger et al. 2011. Strongly correlated photons on a chip. *Nature Photon.* 6: 93–96.

Reithmaier, J. P., G. Sek, A. Löffler et al. 2004. Strong coupling in a single quantum dot–semiconductor microcavity system. *Nature* 432: 197–200.

Ritter, S., C. Nölleke, C. Hahn et al. 2012. An elementary quantum network of single atoms in optical cavities. *Nature* 484: 195–200.

Rosenblum, S., S. Parkins, and B. Dayan. 2011. Photon routing in cavity QED: Beyond the fundamental limit of photon blockade. *Phys. Rev. A* 84: 033854.

Schoelkopf, R. J. and S. M. Girvin. 2008. Wiring up quantum systems. *Nature* 451: 664–669.

Schuster, I., A. Kubanek, A. Fuhrmanek et al. 2008. Nonlinear spectroscopy of photons bound to one atom. *Nature Phys.* 4: 382–385.

Shapiro, J. H. 2006. Single-photon Kerr nonlinearities do not help quantum computation. *Phys. Rev. A* 73: 062305.

Specht, H. P., C. Nölleke, A. Reiserer et al. 2011. A single-atom quantum memory. *Nature* 473: 190–193.

Srinivasan, K. and O. Painter. 2007. Linear and nonlinear optical spectroscopy of a strongly coupled microdisk–quantum dot system. *Nature* 450: 862–866.

Tian, L. and H. J. Carmichael. 1992. Quantum trajectory simulations of two-state behavior in an optical cavity containing one atom. *Phys. Rev. A* 46: R6801–R6804.

Trupke, M., J. Goldwin, B. Darquié et al. 2007. Atom detection and photon production in a scalable, open, optical microcavity. *Phys. Rev. Lett.* 99: 063601.

Turchette, Q. A., R. J. Thompson, and H. J. Kimble. 1994. One-dimensional atoms. *Appl. Phys. B* 60: S1–S10.

Vahala, K. J. 2004. Optical microcavities. *Nature* 424: 839–846.

Volz, J., R. Gehr, G. Dubois, J. Estève, and J. Reichel. 2011. Measurement of the internal state of a single atom without energy exchange. *Nature* 475: 210–213.

Waks, E. and J. Vuckovic. 2006. Dipole induced transparency in drop-filter cavity-waveguide systems. *Phys. Rev. Lett.* 96: 153601.

Wallraff, A., D. I. Schuster, A. Blais et al. 2004. Strong coupling of a single photon to a superconducting qubit using circuit quantum electrodynamics. *Nature* 431: 162–167.

Werner, M. J. and A. Imamoğlu. 1999. Photon–photon interactions in cavity electromagnetically induced transparency. *Phys. Rev. A* 61: 011801(R).

Wilk, T., S. C. Webster, A. Kuhn, and G. Rempe. 2007. Single-atom single-photon quantum interface. *Science* 317: 488–490.

Yoshie, T., A. Scherer, J. Hendrickson et al. 2004. Vacuum Rabi splitting with a single quantum dot in photonic crystal nanocavity. *Nature* 432: 200–203.

Young, A. B., R. Oulton, C. Y. Hu et al. 2011. Quantum-dot-induced phase shift in a pillar microcavity. *Phys. Rev. A* 84: 011803(R).

35

Carbon-Based Zero-, One-, and Two-Dimensional Materials for Device Application

Young Kuk
Seoul National University

In the early 1980s, the existence of stable carbon clusters was recognized. We now know carbon can take several forms of carbon allotropes, C_{60}, C_{70}, C_{84}, single-wall carbon nanotube (SWCNT), multiwall carbon nanotube (MWCNT), single-layer graphene, double-layer graphene, and graphite. Many potential applications of these carbon-based nanomaterials have been suggested, such as building blocks for bottom-up approach materials and functional electronic materials. Carbon is predicted by some to substitute Si in the future. In this chapter, we review the geometric and electronic structures of C_{60}, the carbon nanotube, and graphene and describe how they can be used for electronic devices.

35.1 C_{60} Fullerene

The C_{60} molecule is a stable carbon-based cluster since carbon clusters were examined by mass spectrometry [1–4]. Carbon clusters were produced by arc discharge from graphite electrodes, laser ablation of carbon, and combustion of hydrocarbons [4–7]. The C_{60} molecule, named the buckminsterfullerene, was clearly identified by Richard Smalley, Robert Curl, James Heath, Sean O'Brien, and Harold Kroto and they proposed its structural, electronic, and chemical properties as shown in Figure 35.1 [1]. After a decade of studying mass-selected C_{60}, the synthesis of a macroscopic amount of this material became possible in 1991 with chromatographic separation [8]. This success led to a new field of fullerene chemistry and fullerenes have become building blocks of nanomaterial science. C_{60} has been covered widely in many review articles, and hence we mainly cover herein the topics of the electronic and geometric structures and electronic applications.

The electronic structure of icosahedral C_{60} was calculated using the Hückel molecular orbital theory for nonplanar-conjugated organic molecules [9]. This molecule is composed solely of sp^2-hybridized carbon atoms. It has 20 six-membered rings and 12 five-membered rings, such that all carbon atoms are identical. The π-bond energy per carbon atom in C_{60} is slightly less than that of graphite because of its nonplanarity [9]. Carbon–carbon bonds generally do not undergo distortion to a point group of lower symmetry by the second-order Jahn–Teller effect [10]; however, this is not in the case with charged C_{60}. The C_{60} highest occupied molecular orbital (HOMO) level is h_u and lowest unoccupied molecular orbital (LUMO) levels are predicted to be either t_{1u} or t_{1g} with two excited configurations $(h_u)^9(t_{1u})^1$ and $(h_u)^9(t_{1g})^1$. They lie within approximately ±1 eV of the Fermi level if C_{60} is placed on a metal surface and the gap is known to be 1.65 eV, as shown in Figure 35.2 [11].

Despite some success in identifying C_{60} with mass spectrometry, it can be directly imaged with scanning tunneling microscopy (STM). STM has the unique capability of directly observing three-dimensional charge density of an individual C_{60} molecule. Some STM studies have revealed a well-ordered single layer of C_{60} molecules on metal or semiconductor surfaces [12–15]. Figure 35.3 shows a (25 × 25 nm) STM topography of several C_{60} molecules adsorbed onto a Si(100)-(1 × 2) surface [14]. The observed charge density around C_{60} on these substrates appears to be spherical at any tunneling voltage. The C_{60} molecules are immobile on some surfaces, indirectly indicating that they ratchet along one orientation. In photoemission and surface-enhanced Raman spectroscopy studies, the charge transfer either from surrounding alkali metals to the C_{60} bulk or from the metal substrate to the C_{60} films results in an energy shift from the HOMO state and growth of the LUMO-derived state near the Fermi level [16,17].

When C_{60} molecules are adsorbed on a Au(001)-(5 × 20) substrate, large monolayer (ML) islands appear at a coverage of >0.2

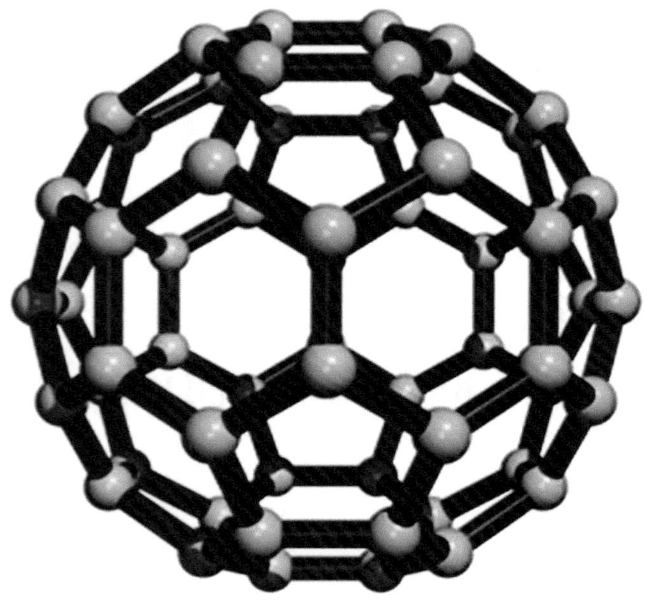

FIGURE 35.1 A ball and stick model of buckminsterfullerene C_{60}.

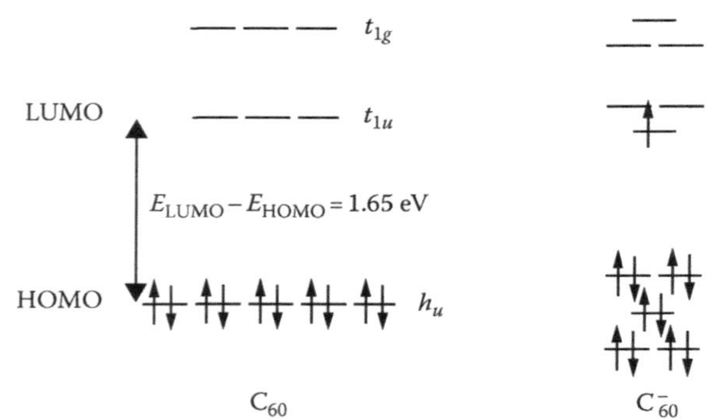

FIGURE 35.2 HOMO and LUMO levels for C_{60} and C_{60}^{-1} molecules. (From Green, W.H. et al., *J. Phys. Chem.*, 100, 14892, 1996.)

ML [18]. Figure 35.4 shows a typical ML C_{60} film on a reconstructed Au(001) surface. These molecules form a commensurate structure with strain in the film. On the uniaxially stressed ML, the interaction is stronger than the van der Waals interaction.

White blobs on top of the distorted closely packed layer are C_{60} molecules in the second layer, and the first layer of molecules underneath can be seen once they are removed. Their observed charge density is nearly circular similar to isolated molecules on the Au(001). The first layer of C_{60} (shown in the middle terrace) forms ordered islands but shows slight variations in the nearest-neighbor distances. The ordered structure resembles a closely packed face-centered cubic (fcc) (111) plane showing some distortion. Islands with different amounts of distortion are separated by grain boundaries, as shown by broken lines. Detailed structure analysis was possible on the STM images where both the ordered islands and the bare reconstructed Au(001)-(5 × 20) surface coexisted. The C_{60} ML is commensurate with the Au(001) substrate only along the ⟨110⟩ direction. If m C_{60} molecules fit to channels of $n \times 14.4$ Å, the channel width is

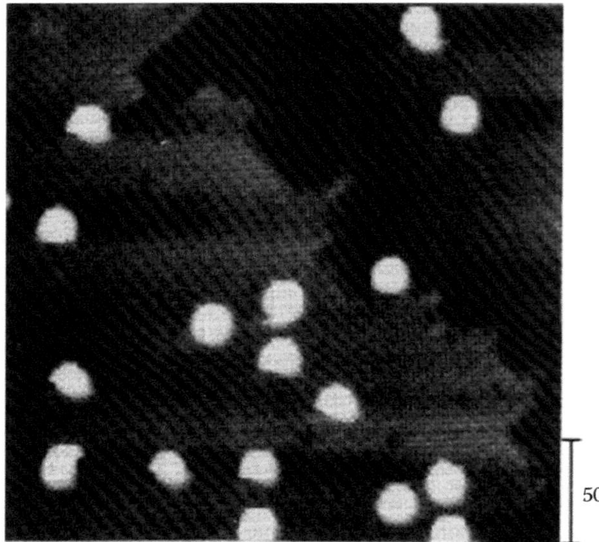

FIGURE 35.3 STM topography of C_{60} molecules adsorbed on Si(001)-(1 × 2) surface. (From Hashizume, T. et al., *Jpn. J. Appl. Phys.*, 31, L880, 1992.)

a 5 × 20 reconstruction of Au(100). When the stress in the film is extremely large, the channel grow at a tilted angle. Grain boundaries are formed due to the difference in the angle, when two neighboring grains grow from separate nucleation sites. Under uniaxial stress, the charge density around the C_{60} molecule is substantially deformed into an ellipsoid in the STM images. The ratio of major and minor axes of the deformed charge density is a function of the stress, ranging from 1.5 to 4. The ratio is independent of the tunneling voltage though the charge density appears to be parallelogram at the HOMO level because of intermolecular bonding.

Figure 35.4a and b shows voltage-dependent STM images at −2 (near HOMO level) and +2 V (near LUMO level). Two types of epilayer grain (A and B) are separated by a grain boundary. At the LUMO level (Figure 35.4b and d), the C_{60} molecule reveals a charge density with an ellipsoidal shape for which the ratio of major and minor axes is ~4.0 in the B type and ~1.5 in the A type. The close-up view of the molecule does not show the intramolecular structure and shows only slight evidence of intermolecular bonding. The threefold degenerated π LUMO states (intramolecular structure) may not be easily imaged, since they are surface states on the surface of the ellipsoid and the C_{60} molecule may still be rotating around its major axis. However, since the epitaxial layer is highly compressed, the rotational speed may be damped substantially. If the C_{60} molecule ever ratchet to the substrate, Au(001) may be one of the best substrates. At the HOMO level, the charge density around the C_{60} molecule resembles a parallelogram. Again intramolecular fivefold degenerate π* states are not imaged. There is a strong indication of intermolecular bonding, as shown in Figure 35.4a and c. The different shapes between the charge densities at HOMO and LUMO levels are due to the intermolecular bonding. Due to the large stress in the epitaxial layer, an intermolecular bonding with a neck structure and an

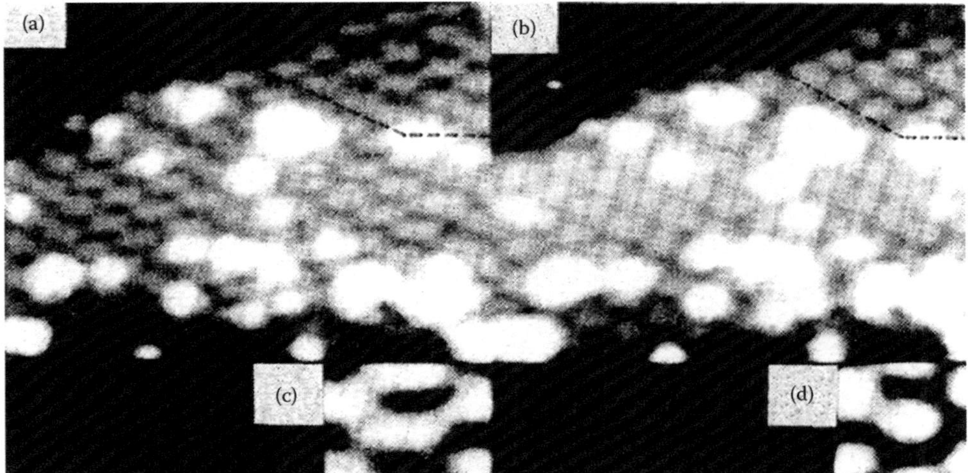

FIGURE 35.4 12 × 10 nm gray-scale images of the same area at (a) −2 V and (b) +2 V. Arrows indicate the grain boundaries between A- and B-type domains. Close-up images of a C_{60} molecule at (c) −2 V and (d) +2 V, showing the bonding. (From Kuk, Y. et al., *Phys. Rev. Lett.*, 70, 1948, 1993.)

antibonding with a nodal structure may be present between molecules (Figure 35.4c and d).

If the interaction between the C_{60} molecule and substrate was small, the film would be incommensurate with the Au(001) substrate. The cohesive energy for the crystallization of C_{60} molecules is known as the van der Waals interaction. As described earlier, a strong interaction was proved by the commensurate C_{60} ML. In C_{60} molecular layers on various metal surfaces, the charge transfer from the metal substrate to the molecule was estimated by measuring the core-level shifts in photoemission spectra. On some metal substrates, HOMO–LUMO bandgaps and LUMO level shifts have been reported. From measured scanning tunneling spectroscopy (STS) results, we were able to confirm the charge from the Au(001) substrate to the molecule at low coverage.

On Cu(111), C_{60} molecules adsorb on the threefold hollow sites, forming a commensurate structure [19]. As C_{60} molecules form a (4×4) structure, the lattice mismatch is ~2%. At low coverage, C_{60} molecules are mobile on the Cu(111) surface and easily diffuse toward step edges diffusing along the $\langle \bar{1}10 \rangle$ direction. After all sites are occupied along the step edges with increasing coverage, two-dimensional islands grow in a closely packed arrangement, eventually forming a C_{60} ML. Figure 35.5 shows an STM

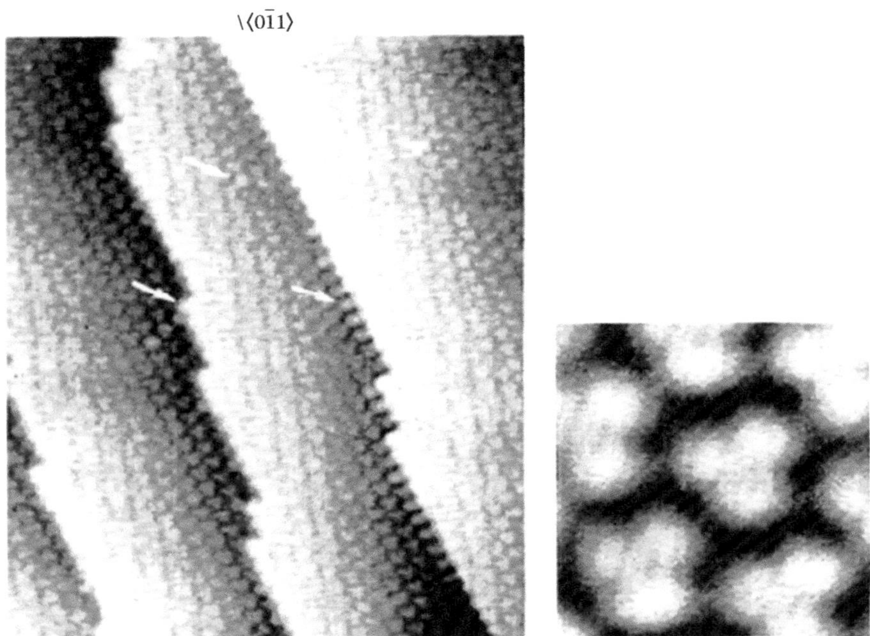

FIGURE 35.5 An 20 × 38 nm STM image of the Cu(111) surface covered completely with ML C_{60} film after annealing at 290°C. V = 2.0 V and I = 20 pA. An individual C_{60} molecule appears as a three-lobed shape. (From Hashizume, T. et al., *Phys. Rev. Lett.*, 71, 2959, 1993.)

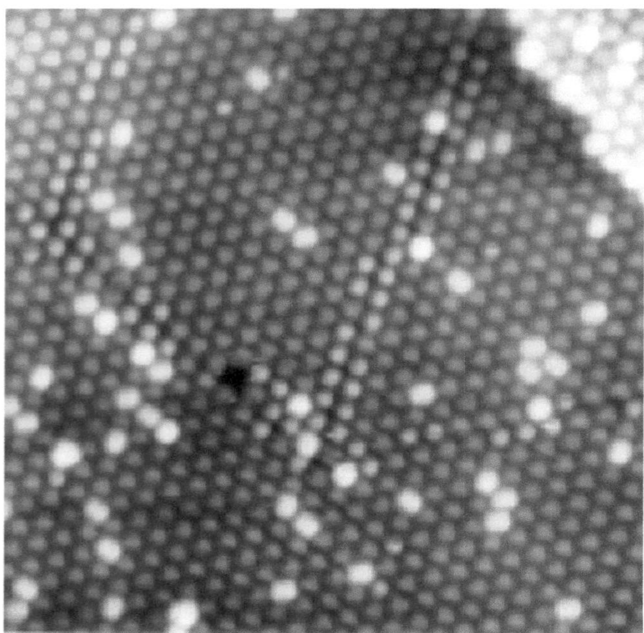

FIGURE 35.6 20×20 nm C_{60}–C_{70} codeposited surface on Cu(111). V_s = 2.0 V. (From Wang, X.-D. et al., *Zeit. Phys. Chem. B*, 202, S117, 1997.)

image of C_{60} ML after annealing at 290°C, with the bias voltage of V_b=2.0 V. Individual C_{60} molecules can be imaged at a tunneling voltage within the HOMO–LUMO gap through gap states induced by charge transfer from the substrate to the C_{60} molecules.

C_{60} molecules appear as a three-lobed shape at V_{sample}=2 V. As shown in Figure 35.5, C_{60} molecules are aligned as if they are pointing upward in the three terraces and downward at the lowest one. It is known that C_{60} molecules rotate even within bulk crystal at room temperature. However, the molecules in the ML film ratchet in a specific direction on terraces of Cu(111). Interestingly, they appear to rotate at step edges (arrows) and defect sites most likely due to broken symmetry and resulting in the weaker intermolecular binding mediated through the copper substrate. This assumption was confirmed by a band structure calculation [14].

The C_{70} molecule is another stable form of carbon cluster. The C_{70} molecule is not spherically symmetric, instead, it is elongated along an axis. When C_{70} molecules are codeposited with C_{60}, they often stand along the major axis direction or lie down, that is, the major axis is *perpendicular* or *parallel* to the surface. When they lie down, the ellipsoid appears as a dumbbell structure when imaged above the Fermi level as shown in Figure 35.6. They appear brighter than C_{60} molecules and appear as circular blobs (major axis vertical) or dumbbells (major axis *horizontal*). It was found that C_{60} and C_{70} tend to segregate from each other as the codeposited sample is annealed [20].

Fabrication of molecular transistors became possible in the late 1990s. Methods such as the mechanical break junction [21], electrodeposition [22], nanoconstriction [23,24], and electromigration [25] have been used. Not all of these methods can produce a three-terminal junction device with a gate electrode. The first C_{60} transistor was fabricated in 2000 [25]. When a three-terminal device was first fabricated as shown in Figure 35.7, Coulomb diamonds were observed. The conductance gap was explained as being the consequence of the finite energy required to add (remove) an electron to (from) C_{60}. External bias voltage, that is, this energy cost, is needed for single-electron charging of C_{60} and the quantized molecular excitation in the C_{60} transistor system. Figure 35.7 shows the Coulomb diamonds: dark areas are the conductance gap from the Coulomb blockade. In the gray areas, finite conductance ($\partial I/\partial V$) was measured using the peak lines from the quantized excitations of the single C_{60} transistor. These peaks are due to a new tunneling pathway which an electron hops onto C_{60}^{n-} to generate $C_{60}^{(n+1)-}$ in its ground or excited state; these peaks probe the excitation energies of the $C_{60}^{(n+1)-}$ ion.

As molecular transistors are better understood, the transport through a molecule is now described as the presence of a small number of quantized conductance channels, where each channel has $R_K/2 = h/2e^2 = 12.9$ kΩ. In a recent study [26], this quantum conductance restriction can be lifted if superconducting leads are used. A super current with superconducting molecules can be the future direction for molecular electronics.

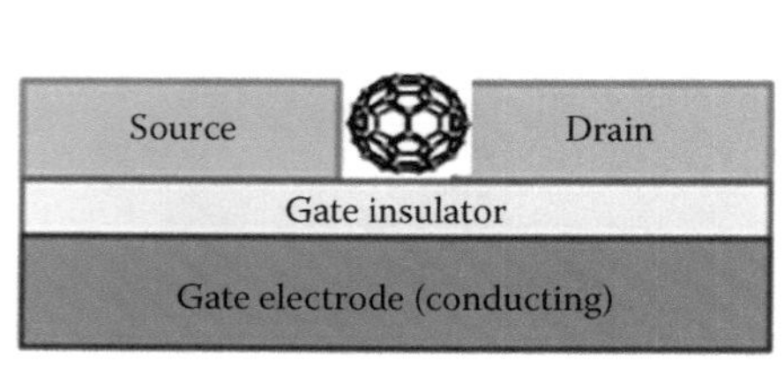

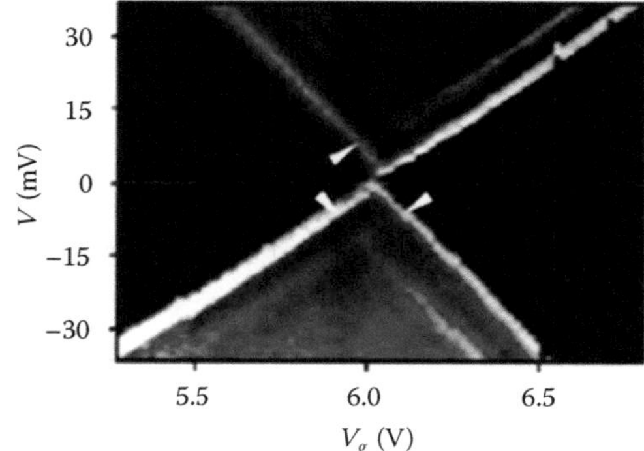

FIGURE 35.7 The 2D differential conductance ($\partial I/\partial V$) plots as a function of the bias voltage (V) and the gate voltage (Vg). The gray scale represents the differential conductance values. The bright arrows mark the point where $\partial I/\partial V$ lines intercept the conductance gap. (From Park, H. et al., *Nature*, 407, 57, 2000.)

35.2 Carbon Nanotube

A carbon nanotube is a cylindrical tube of graphene discovered by Iijima [27] in the residue of arc-discharged carbon rods in 1991. The electronic structure of the tube is highly similar to that of a graphene sheet but differ slightly due to the nonplanarity. The tube has a periodic boundary condition along the circumference direction [28]. Nanotubes show metallic or semiconducting behaviors depending on their radii and helical structures.

Helicity of a CNT is defined as the chirality. If a graphene sheet is rolled such that points A and C touch each other, as shown in Figure 35.8, the resulting carbon nanotube is called (m,n) tube. If a tube is rolled to make $\vec{a}$ perpendicular to the tube axis, this tube is called a zigzag CNT. If the vector is parallel to the tube axis, it is called an armchair CNT. Among the various geometrical shapes, $m-n=3$ times an integer including the armchair-type (n,n) carbon nanotube is metallic with two linear bands crossing at the Fermi level [29].

The electronic dispersion is well described by the tight-binding Hamiltonian, with one π electron per atom where only the nearest-neighbor hopping matrix element is taken into account [30]. We consider the scattering of electrons by perturbation which is the change in the on-site energy at a single site.

We denote an atomic site in the nanotube with three indices $(pq\sigma)$ as shown Figure 35.9. The integers p and q point to a unit cell in the hexagonal lattice, and σ, A, or B, designates one of the two atomic sites in a honeycomb unit cell. If we ignore the curvature effect in the tubular surface, the Hamiltonian of the π electrons, H_o in a defectless (n,n) nanotube and $|pq\sigma\rangle$ becomes

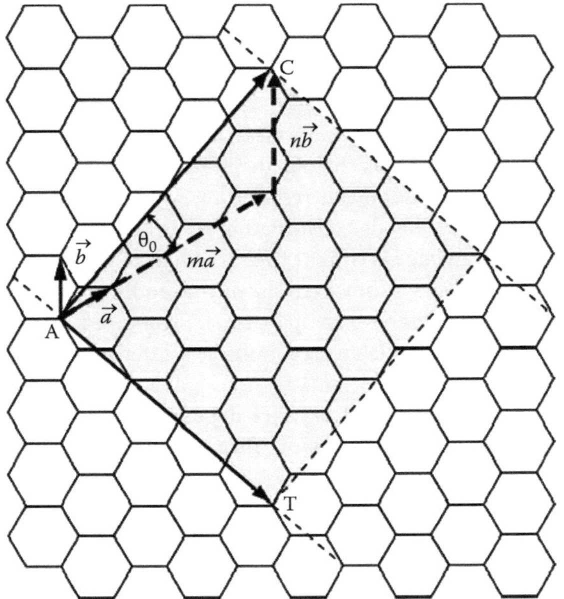

FIGURE 35.8 The chirality of a carbon nanotube. As m is 4 and n is 2 here, this nanotube is a (4,2) tube.

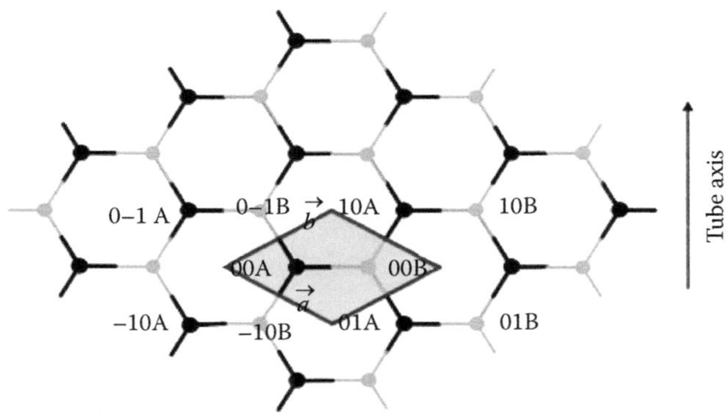

FIGURE 35.9 Representation of an armchair nanotube. The three indices $pq\sigma$ indicate an atomic site.

$$H_o = \sum_{\langle i,j\rangle} (-V_\pi C_i^\dagger C_j - V_\pi^\star C_j^\dagger C_i)$$

for π-electron orbital at the $(pq\sigma)$ site as summed over its nearest neighbor sites. The value $|V_\pi|$ is ~2.7 eV. The Hamiltonian can then be represented with a matrix form whose element $\langle pq\sigma|H_o|p'q'\sigma'\rangle$ has a nonzero value, $-V_\pi$, if and only if the $(pq\sigma)$ and $(p'q'\sigma')$ sites are the nearest neighbors. Because the diagonal on-site energy is assumed to be zero, the Fermi energy of the undoped nanotube is zero. From the primitive cell in Figure 35.9, we can obtain the reciprocal vector and reciprocal lattice vectors $\vec{G_1} = (2\pi/a)((1/\sqrt{3}),1)$, $G_2 = (2\pi/a)((1/\sqrt{3}),-1)$. The unit-cell length of the (n,n) tube aa or $|\vec{R}|$ is defined by $|\vec{R}| = a = 2.46\,\text{Å}(=\sqrt{3}\times 1.42\,\text{Å})$. Figure 35.10 shows the first Brillouin zone of the reciprocal vectors.

Let us first consider a (n,n) tube circumference where the x-direction is the tube axis with tube indexing as described earlier [31]. The k vector is $\vec{k} = k_\perp\hat{x} + k_{//}\hat{y}$ and $|\vec{R}| = a = 2.46\,\text{Å}(=\sqrt{3}\times 1.42\,\text{Å})$ from the Bloch theorem and the boundary condition. We have the relations $e^{i\vec{k}(n\vec{R}_1+n\vec{R}_2)} = e^{i\sqrt{3}k_\perp na} = 1$ for integer l and $e^{i(\sqrt{3}k_\perp/2)a} = \pm e^{i(\pi/n)l}(0<l<n)$. The minus sign is equivalent to substituting $(l-n)$ instead of l. As the range of l becomes twice as large, $e^{i(\sqrt{3}k_\perp/2)a} = e^{i(\pi/n)l}(0<l<2n$ or $-n<l<n)$. With the periodicity imposed around the circumference, the lth

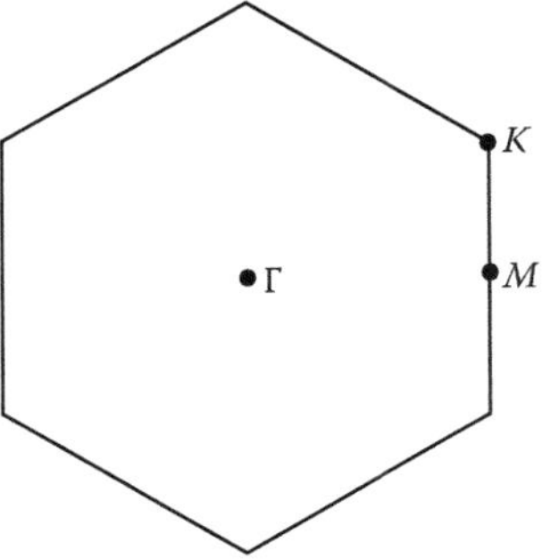

FIGURE 35.10 First Brillouin zone.

energy eigenstate $|\phi_l\rangle$ for a wavevector k in the tube axis direction can be expressed as

$$|\phi_l\rangle = \sum e^{i\vec{k}\cdot(p\vec{a}+q\vec{b})}(C_A|p,q,A\rangle + C_B|p,q,B\rangle),$$

where $l = 0, 1, 2, \ldots, n-1$. From Schrödinger equation and to have a nontrivial solution, we obtain the following for $n = 2$ and $l = 0$:

$$\begin{vmatrix} -E & -V_\pi\left(1+e^{-i\left(\frac{\pi l}{n}+\frac{ka}{2}\right)}+e^{-i\left(\frac{\pi l}{n}-\frac{ka}{2}\right)}\right) \\ -V_\pi\left(1+e^{-i\left(\frac{\pi l}{n}+\frac{ka}{2}\right)}+e^{-i\left(\frac{\pi l}{n}-\frac{ka}{2}\right)}\right) & -E \end{vmatrix} = 0.$$

The solution becomes $E^2 = V^2(1+4\cos(ka/2)\cos(\pi\lambda/2)+4\cos^2(ka/2))$ or $\cos(ka/2) = -\left(\cos(\pi l/n) \pm \sqrt{\cos^2(\pi l/n) - (1-(E^2/V^2))}\right)/2$ for $V_\pi \equiv V$.

The result can be pictorially shown in Figure 35.11 as we plot along $k_{||} = k$ direction. This energy dispersion relation holds well for armchair and zigzag nanotubes, but not for chiral CNTs. If we plot the energy dispersion relation with E as the z-axis and 2D k-space as the x- and y-axes, the dispersion would be the curved surfaces of the π, π^*, σ, and σ^* bands as shown in Figure 35.12.

As the diameter of a SWNT is ~1 nm, the wave vector along the circumference $k_\perp$ is quantized by $k_\perp = 2q/d = k_q q$, $q = \pm 1, \pm 2, \ldots, \pm N$ where N is the number of hexagons in the unit cell as shown in Figure 35.13. Since the length of nanotubes is macroscopic, $k_{||}$ satisfies periodic boundary condition of a solid. Due to the quantization of $k_\perp$, not all the points in Figure 35.12 are accessible to nanotubes. Instead, the energy bands of a nanotube are limited to a series of 2D subbands with constant $k_\perp$s. The dashed line in Figure 35.13 is along the $k_{||}$ direction and the angle between it and ΓK is the chiral angle q. The solid lines, which are perpendicular to the

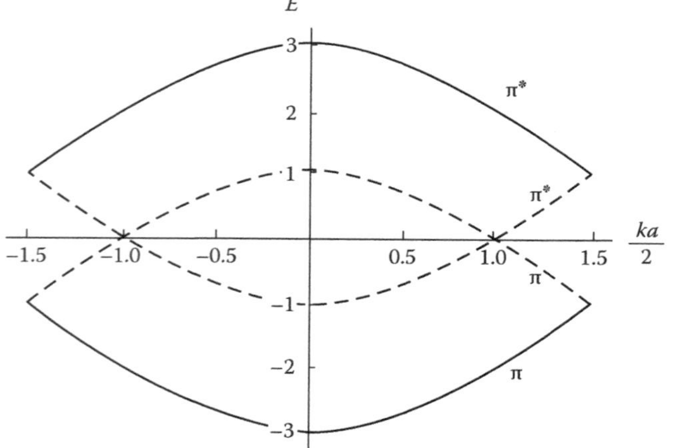

FIGURE 35.11 The energy band diagram for $l = 0$ case of (n,n) carbon nanotube.

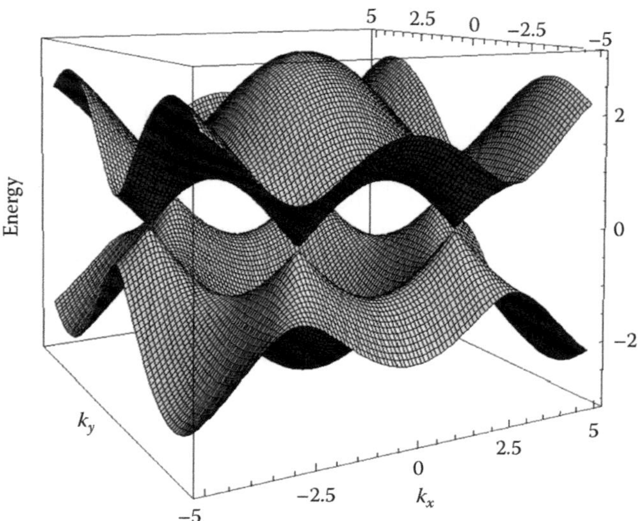

FIGURE 35.12 Band dispersion of a graphene.

dashed line, have constant $k_\perp$ values. Because the graphene is a semimetal without a bandgap at the K point, if one of these $k_\perp$ goes through K points, this nanotube is metallic, otherwise the tube is semiconducting. The (m,n) chiral CNTs have energy bandgap if $m - n \neq 3d$. The effect of curvature can be considered within tight-binding approximation by introducing parameters for σ bondings or a modified parameter for π bondings. A small bandgap is predicted theoretically for a CNT with a small diameter [32].

When we study the surface structure of single crystals, we normally perform sputtering and in situ annealing under ultra-high vacuum. For graphite, a typical process is cleaving. Even if the atomic structure of CNTs is very similar to that of graphite, it would not be necessary to perform any cleaning process for the CNT surface. The atomic structure of CNTs is well resolved by STM images, as shown in Figure 35.14. Most of the surface area of the CNT walls is atomically clean and well ordered, implying that the CNT surface is chemically inert. Figures 35.14 and 35.15 show an STM image and corresponding STS results of a CNT. In this topography, an edge termination is visible. The equilibrium structure of an edge is similar to that of half of carbon fullerene [33]. The STM image, however, shows that equilibrium structure can be disordered and a complex super structure is visible near the end due to the existence of multiple scattering centers at the end of this tube.

CNTs grown by chemical vapor deposition (CVD) or arc discharge reveal point defects or intramolecular junctions with native defects due to their physical bending or pentagonal or heptagonal defects [33]. This type of junction can be determined by measuring dI/dV spectra by performing STM on both sides. Figure 35.16b through d shows bias-dependent STM images taken on a semiconductor–semiconductor CNT junction. At negative sample bias voltages (V), the defect is not so conspicuous, but at positive voltages more remarkable modification of

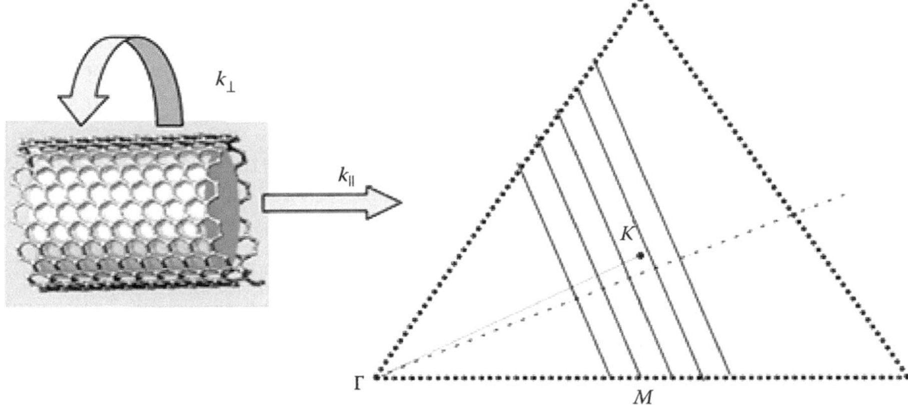

FIGURE 35.13 Illustration of the electron wave vector. The *K* point along the axis follows the Born von Karman boundary condition. However, *k* points along the circumference are discrete and occupy only small numbers due to their small size. This is a pictorial representation of the way to extract the band structure of nanotubes from a 3D graphene dispersion. (From Zhou, Z., Carbon nanotube transistors, sensors and beyond, PhD dissertation, Cornell University, Ithaca, NY, 2008.)

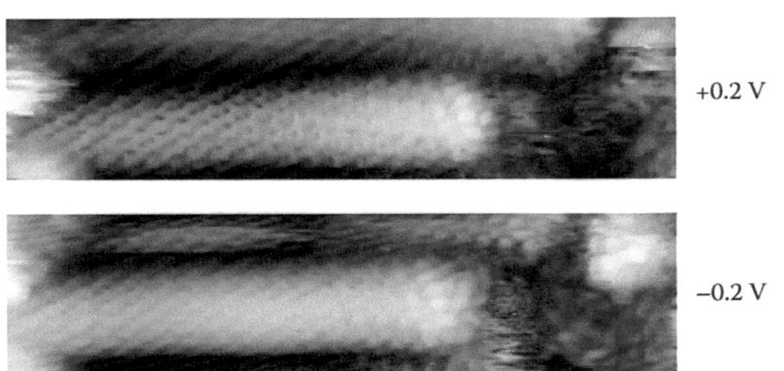

FIGURE 35.14 STM topography of a CNT at the sample biases of +0.2 and −0.2 eV. The chirality can be identified from the image, and later the bandgap is compared to confirm the chirality.

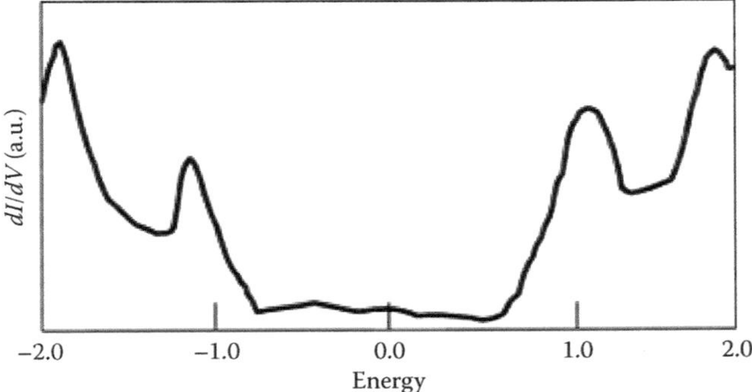

FIGURE 35.15 STS of a CNT shown in Figure 35.13.

the electron density can be observed on the surface. Particularly at V+0.5 V, the pattern of the surface corrugation is very different from other bias voltages; the defective region reveals a pattern with a longer period, very similar to the $\sqrt{3}\times\sqrt{3}$ structure observed near the defects on graphite surfaces [34]. The chirality of the CNTs was determined, from STM topography and STS, to be (15,2) (left) and (19,3) (right).

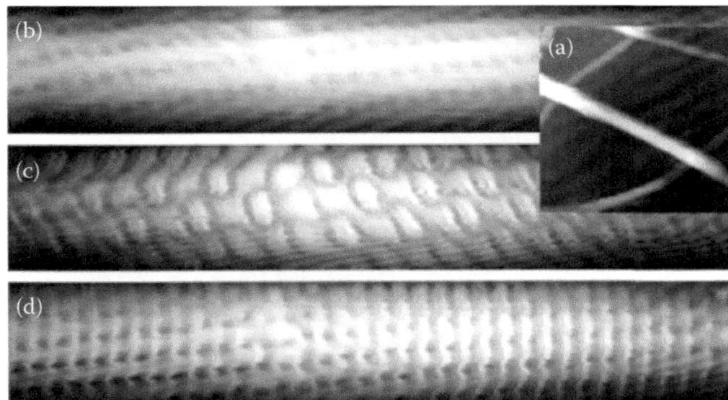

FIGURE 35.16 (a) A 100 × 100 nm STM image. (b–d) STM images of a carbon nanotube intramolecular junction at three different sample bias voltages, (b) 1.5 V, (c) 0.5 V, and (d) −0.3 V. The length of these images is 7.6 nm. (From Kim, H. et al., *Phys. Rev. Lett.*, 90, 216107, 2003.)

Since the tunneling current (I) in STM reflects the integrated local density of states (LDOS) of the sample from the Fermi level to the sample bias voltage V, the value of dI/dV is approximately proportional to the LDOS at this energy level. In the dI/dV spectrum obtained above the left-hand side of the junction (Figure 35.16), the first and the second peaks of the van Hove singularity (VHS) in the occupied and the unoccupied states are visible. The LDOS of the junction was spatially resolved by taking the derivative dI/dV of the spectra along the CNT at a fixed tunneling gap or a constant tunneling current. In the present experiment, the STM tip was repeatedly scanned over the topmost line on the same nanotube, leaving the reference current of feedback fixed at 0.5 nA, and changing the bias voltage by a small increment for each line scan. The map of dI/dV obtained by the lock-in technique as a function of both the position along the nanotube axis and the sample bias voltage is presented as a 2D image shown in Figure 35.16, in which the brightness is proportional to the LDOS. The LDOS spectrum at the left end of the imaged region is in good accordance with the point dI/dV spectrum.

The bandgaps of the left and the right nanotubes were observed as 0.92 and 0.80 eV from the measured VHS peaks.

A periodic vertical pattern reflects the electronic wave functions of CNTs and demonstrates the atomic resolution of the spectroscopy. It also explains the atomic corrugation observed in the topographic images in Figure 35.15, particularly the elongated periodicity in Figure 35.16c agrees with that of the vertical lines around the defect at positive bias voltages in Figure 35.17. For metal–semiconductor CNT junctions created by pentagon–heptagon pair defects, spatially resolved LDOS features were suggested theoretically [33] and observed by point spectroscopy at six spots around one of these junctions. There are two features in this spectroscopic image worthy of our attention. The first is the localized defect states at $V = 0.3$ V which extend over 3 nm, the spatial distribution of which is shown as a line profile labeled as (I) in Figure 35.17. This energy level is slightly below the conduction band edges of both sides and can be understood, as mentioned previously [33], in terms of Hückel's rule that cyclic π-electron systems with $4n+2$ (n: positive integer) π electrons are most stable. Six-membered carbon rings are more stable than five- or seven-membered rings, and, thus, a heptagon will try to give up an electron to its neighbors, playing the role of a donor in a semiconducting nanotube. Similarly, a pentagon works as an acceptor. We were unable to observe obvious localized states below the Fermi level. The second salient feature is the peaks of VHS penetrating and decaying into the opposite side. This is more obvious for the valence band edges at both sides that are less perturbed by the defect state. The spatial distributions of these edge states are shown as line profiles labeled as (II) and (III) in Figure 35.17.

Several paired, localized gap states were observed in semiconducting SWNTs using spatially resolved STS [35]. A pair of gap states was found far from the band edges, forming deep levels, whereas the other pair was found near the band edges, forming shallow levels. Figure 35.18a and b shows the STS data with a shallow (III, IV) and deep (I, II) level defects. According to the LDOS map in Figure 35.18, the energy levels of both the CBM and the VBM are constant throughout, and in two regions, there are two pairs of localized states within the bandgap. Since one pair is nearly at the center of the bandgap and far from the band edges, they can be termed deep levels, following the naming convention for bulk semiconductors. The spatial locations of

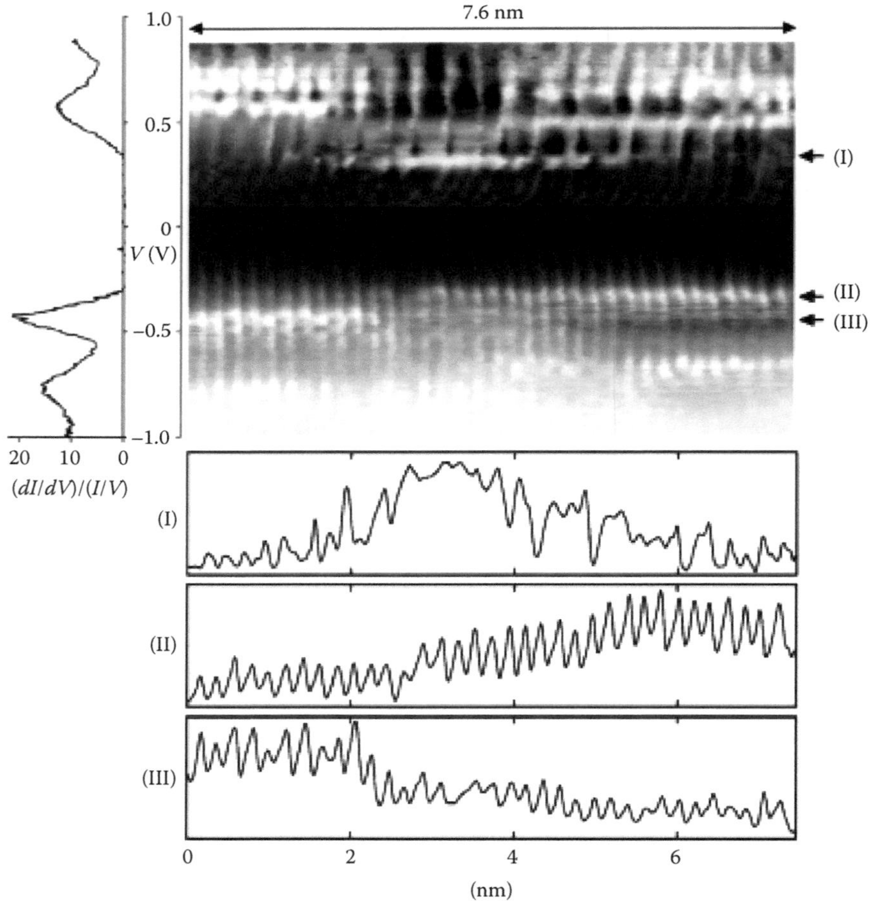

FIGURE 35.17 Spatially resolved LDOS of the junction by STS shown together with point spectroscopy obtained on Figure 35.16b. The abscissa is the position along the tube axis and the ordinate is the sample bias voltage. Graphs (I), (II), and (III) are the line profiles at the energy levels marked by arrows. (From Kim, H. et al., *Phys. Rev. Lett.*, 90, 216107, 2003.)

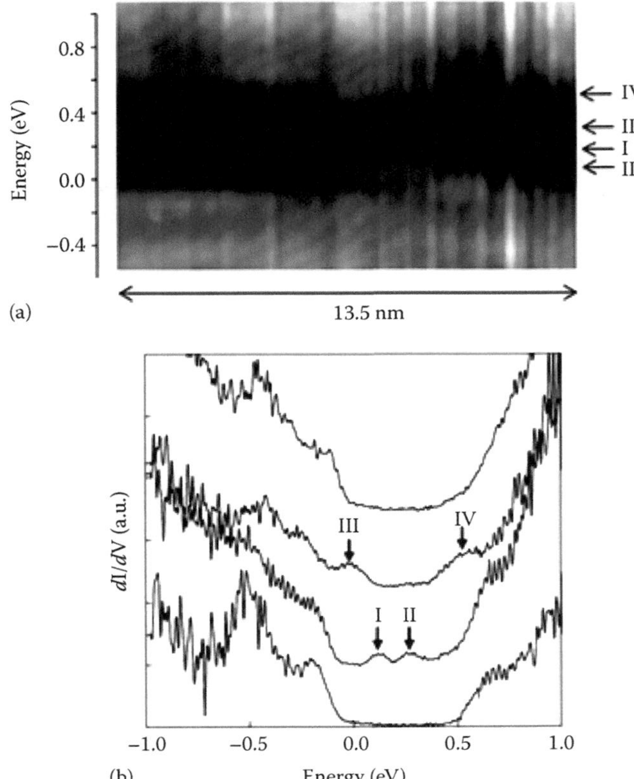

FIGURE 35.18 (a) The spatially resolved LDOS map (STS data) for the CNT. The labels I, II, III, and IV are the energy levels for gap states. (b) Sampled one-point STS data, shifted for clearer presentation. The arrows indicate gap states. (From Lee, S. et al., *Phys. Rev. Lett.*, 95, 166402, 2005.)

these deep levels are coincident, and the amplitude distributions resemble each other. The pair of deep levels originates from a single defect structure. The decay lengths of the deep levels are estimated to be ~2 nm. The other paired defect states (III, IV) are very similar to those shown in Figures 35.16 and 35.17.

The SSCNT junction is a good example of a 1D semiconductor junction. Because of the reduced dimensionality, it is less efficient in screening the electronic states penetrating the junction than 3D materials, and it has a rather large decaying length of the band edge [36]. Once the detailed features of the bandgap variation are well understood, we will be able to design single or multiple CNT junctions to produce functional devices in the future.

Instead of using semiconducting, metallic SWCNTs [37], many have tried to functionalize them, by inserting metal, insulator, and fullerenes or adsorbing them onto the surfaces. Experimentally, it has been reported that the transport properties of CNTs can be controlled by the insertion of molecules [38], functionalization [39], and chemical doping [40]. Alkali metals (Li, K, Rb, and Cs) can function as n-doping materials for CNTs, as they are filled inside CNTs or placed between CNT bundles [40–42]. It was reported by Bockrath et al. that doping by potassium increases the conductivity of a CNT [42]. According to theory, there are two possible mechanisms for the change in CNT electronic structures caused by alkali metal doping [43–45]. One is a rigid band model in which there is a rigid band shift with charge transfer [43]. The other considers it as being due to the hybridization of the metal bands and CNT bands [44].

Figure 35.19a shows a typical STM topographic image of a Cs-filled SWCNT. A locally protruding area along a SWCNT,

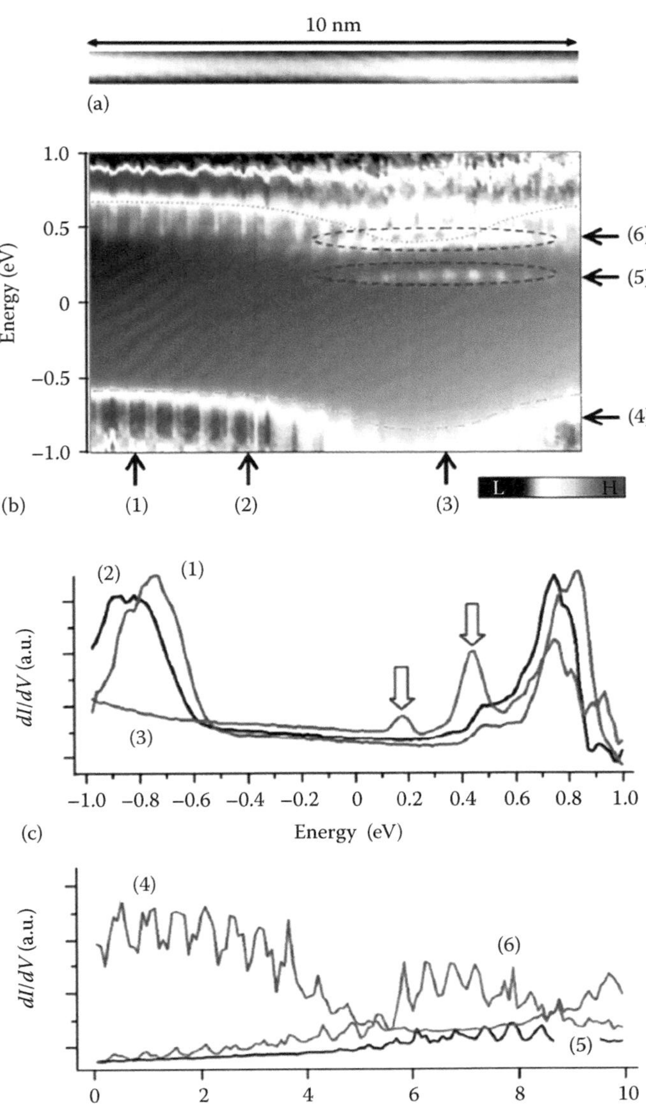

FIGURE 35.19 (a) The STM topographic image of a Cs-SWCNT with an image length of 10 nm at a sample bias voltage of 1.0 V. (b) The SR-STS map of a Cs-SWCNT. (c) STS data which are the selections of vertical line profiles from (b). Curves are sampled from the spatial positions indicated with the vertical arrows (1), (2), and (3) in (b), respectively. The two arrows indicate the appearance of localized states. (d) Horizontal line profiles sampled from the SR-STS map in (b). Curves correspond to the energy levels indicated with the horizontal arrows (4), (5), and (6) in (b), respectively. (From Kim, S.H. et al., *Phys. Rev. Lett.*, 99, 256407, 2007.)

shown as a bright area in a gray-scale image, suggests the existence of embedded Cs atoms. The corrugation change may have resulted from the local elastic deformation of the carbon nanotube or changes in the LDOS by embedded Cs atoms. However, to confirm the presence of Cs atoms, spatially resolved spectroscopic measurements are required. Figure 35.19b shows the spatially resolved STS map of the same Cs-filled SWCNT in Figure 35.19. The dashed line is the contour drawn along the edge of a valence band and the dotted line indicates the appropriate parallel shift of this contour toward a conduction band. The amount of shift is nearly identical to that of the conduction band. There were two localized gap states, which are clearly shown as two peaks in the STS data of Figure 35.19c, that were observed at ~0.18 and ~0.44 eV near the conduction band of the Cs-SWCNT. The SR-STS also reveals atomically resolved localized gap states and the decay of the states away from the proposed location of the encapsulated Cs atoms with a decay length scale of ~3 nm, which is highly similar to the decay of donor and acceptor states [33]. The upper state of the two gap states at ~0.44 eV is more intense than the lower state at ~0.18 eV. In the case of the rigid band model, the downward shifts of valence and conduction bands have been predicted in theoretical studies of alkali metal–doped carbon nanotube systems [43]. However, as shown in Figure 35.19b, the downward shift cannot be explained by the simple rigid band model; as a result of filling by Cs, the conduction band is slightly shifted, and the electron density is reduced whereas the shift of the valence band is obvious. It was confirmed that the total number of electronic states from valence bands to conduction bands is conserved despite the shift. To elucidate experimental findings, we performed ab initio electronic structure calculations using a density functional theory (DFT) with an OPENMX package [46]. From the Mulliken population analysis, we found that about one electron is transferred from Cs to the CNT. The potential well made by the Cs^+ ion induces two bound states in the gap.

It has been suggested theoretically and experimentally that fullerenes or endohedral metallofullerenes [11] can be inserted into SWNTs, forming a pea-pod-like structure [46–48]. When the diameter of the endothermally inserted fullerene is smaller than the inner diameter of the SWNT, the resultant SWNT can be elastically strained. A theoretical study has predicted that the electronic structure, including the positions of the VHS, is severely modified when an SWNT is uniaxially strained [49]. With these ideas, "local bandgap engineering" can be made possible.

Figure 35.20a shows a bundle of six Gd metallofullerene (GdMF)-inserted SWNTs, showing the variation of spacings between neighboring GdMFs [38]. As in the case of Cs-inserted CNT, protrusions in STM topography do not necessarily indicate the locations of inserted GdMFs, because the electronic structures, including band edges, are severely modified by the insertion. After ab initio calculation, it was found that the combined contribution of the elastic strain and the electron transfer to the Au substrate and to C_{82} is responsible for the observed gap modulation. Unlike in the uniaxial strain, stretching in the

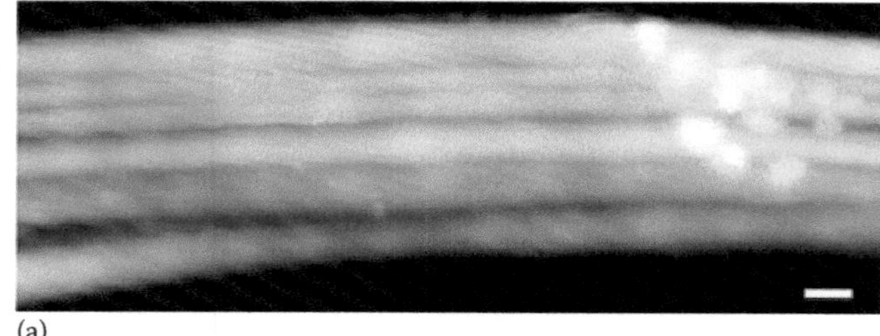
(a)

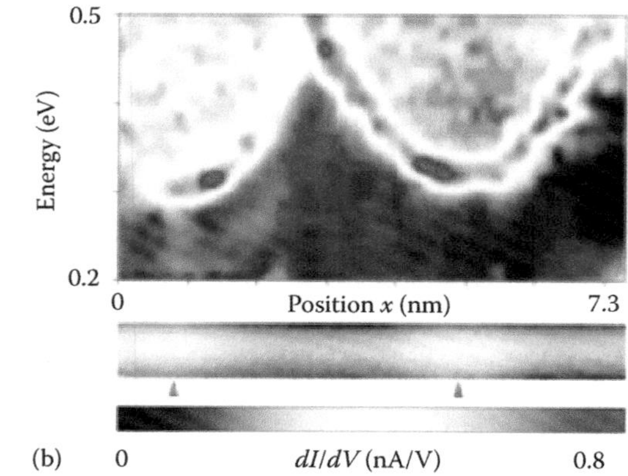

(b)

FIGURE 35.20 (a) A bundle of six GdMF-SWNTs. Variation of spacings between neighboring GdMFs is visible. The scale bars in all topographic images represent 1 nm. (b) Topographic image of a 7.3 nm long (11,9) GdMF-SWNT at a sample bias voltage of 0.5 V. The sites of inserted GdMFs are indicated by arrows (bottom); the corresponding dI/dV spectra of conduction band at the center of the GdMF-SWNT are also shown (top). (From Lee, J. et al., *Nature*, 415, 1005, 2002.)

circumferential direction is important here. One of the surprising features of these spectra is that the change in bandgap is so strongly localized that its spatial variation is clearly resolved in Figure 35.20b. Various shapes of the modified energy band reflect the variation of local elastic strain and/or charge transfer. The present band structure represents 1D multiple quantum dots, analogous to a multiple quantum well in a 3D superlattice. The 3D multilayer utilizes the two-dimensionally confined carrier sub-bands for electronic and optoelectronic applications. This multiple quantum dots may be used for nano-optical devices. With the bandgap modulation achieved here, we can create electron sub-bands in the potential well regions. We can also synthesize this bandgap-engineered system by self-assembly instead by epitaxial growth.

SWCNT field effect transistors (FETs) were first fabricated and made available in the late 1990s as shown in Figure 35.21 [50]. For the major part of the device resistance is accounted to the contact resistance between the tube and the electrodes. The first CNTFET was not an ideal FET. There was a pronounced gap-like nonlinearity at $V_{DS} = 0$. The $I–V_{DS}$ curves seem to exhibit a power-law behavior due to the nonoptimal contact with the CNT. There has been considerable progress in the fabrication of SWNT FETs and the investigation of their performance limits [51–55]. Several related issues have been resolved over the years. First, it was

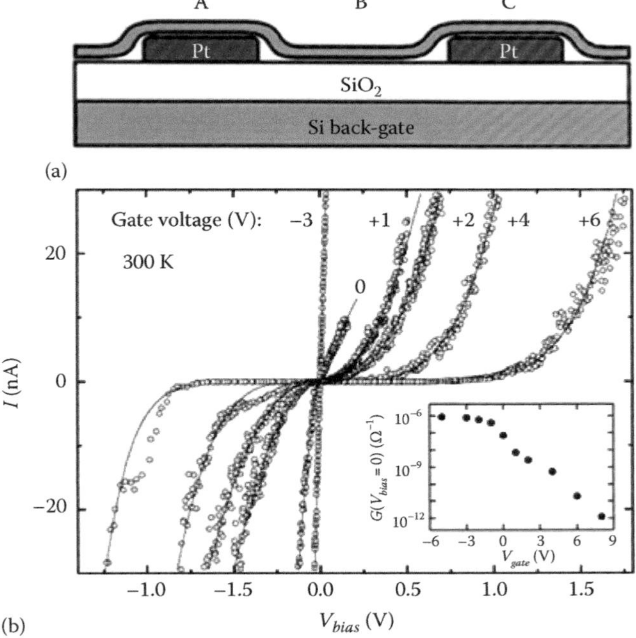

FIGURE 35.21 Two probe I versus V_{DS} curves at various values of the gate voltage (V_{gate}). The substrate Si is used as a back gate. (From Tans, S.J. et al., *Nature*, 393, 49, 1998.)

found that the energy gap, Eg, scales with $1/d$ and researchers started to select the proper diameter and chirality CNTs in order to fabricate a suitable FET. They also have made systematic efforts to investigate various metals for source and drain electrodes in order to minimize the contacts resistance. For a p-type CNT, it was found that Pd forms near ohmic contact to the valence band of SWNTs with negligible Schottky barriers for chemically intrinsic SWNTs with $d>2$ nm and reveals favorable wetting interactions [56]. These p-type SWNT-FETs can essentially operate in a manner similar to any other metal-oxide semiconductor FETs (MOSFETs), at least in the ON and subthreshold regimes. In n-type SWNT-FETs, Al metal electrodes are desirable in air-stabilized samples. SWNTs with $d=2$–3 nm are used for Al metal contacts. There is a small Schottky barrier to the conduction bands of SWNTs. By simply changing the source-drain contact metals, one can produce p- or n-type SWNT-FETs without doping the channel. The results are shown in Figure 35.22.

When one properly selects the metal–CNT contacts, the entire CNT may work as a quantum dot island [57]. Studies of quantum dots have illustrated that single-electron charging and resonant tunneling through quantized energy levels regulate transport in small CNT dots. In this case, the device becomes a single-electron transistor (SET) at low temperature. Figure 35.23 shows the SET characteristics of an armchair CNT device. The linear-response conductance G of the CNT as a function of V_g consists of a series of sharp peaks separated by similar peak spacings. The maximum amplitude of isolated peaks approaches e^2/h, the quantum conductance unit.

As the CNT transistors show good device characteristics such as high gain and a large on–off ratio at room temperature, fabricating multiple devices on a single chip has been attempted [58]. Several logic gates, such as an inverter, a logic NOR, a static random-access memory (SRAM) cell, and an ac ring oscillator, have been fabricated. Figure 35.24 shows these logic elements, which use resistor–transistor logic. A bias voltage of −1.5 V was used for logic applications. In an inverter, an off-chip 100 MΩ bias resistor was used as shown in Figure 35.24a. When the input logic is "1," the output voltage becomes "0." The output voltage changed three times faster than the input voltage in the transition region, indicating a voltage gain of 3.

A NOR gate can be constructed by placing two transistors in parallel as shown in Figure 35.24b. When either or both of the inputs are "1," at least one of the nanotubes is conducting and the output is "0." In Figure 35.23b, the output voltage is plotted as a function of the four possible input states (0,0), (0,1), (1,0), and (1,1). A flip-flop memory element (SRAM) was also demonstrated from two inverters as shown in Figure 35.24c.

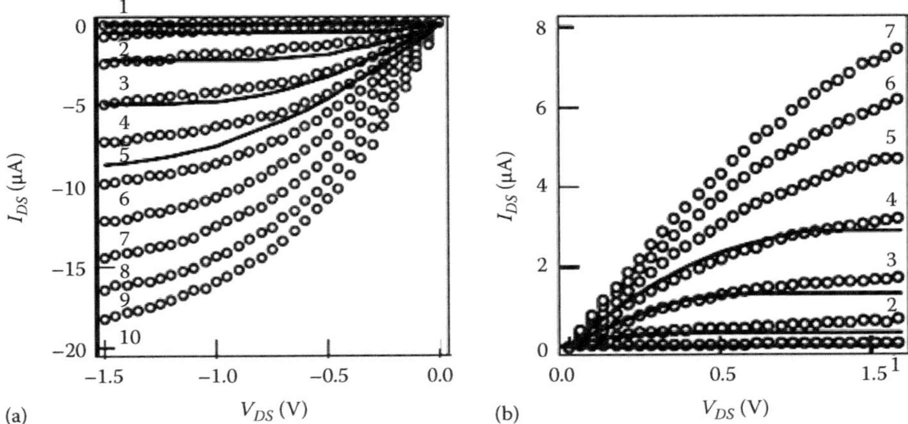

FIGURE 35.22 (a) I_{DS}–V_{DS} curves of p-type CNTFET at different back-gate voltages. Curves 1–10 correspond to $V_g = 0.4$ to −3.2 V in −0.4 V steps. The solid lines are calculated from the square law model for a diffusive channel to fit curves 1–5. (b) Output characteristic for an n-type FET. The circles are the experimental data while the solid lines represent the square law fit. Curves 1–7 correspond to $V_g = 0.7$–2.5 V in 0.3 V steps. (From Javey, A. et al., *Nature*, 424, 654, 2003.)

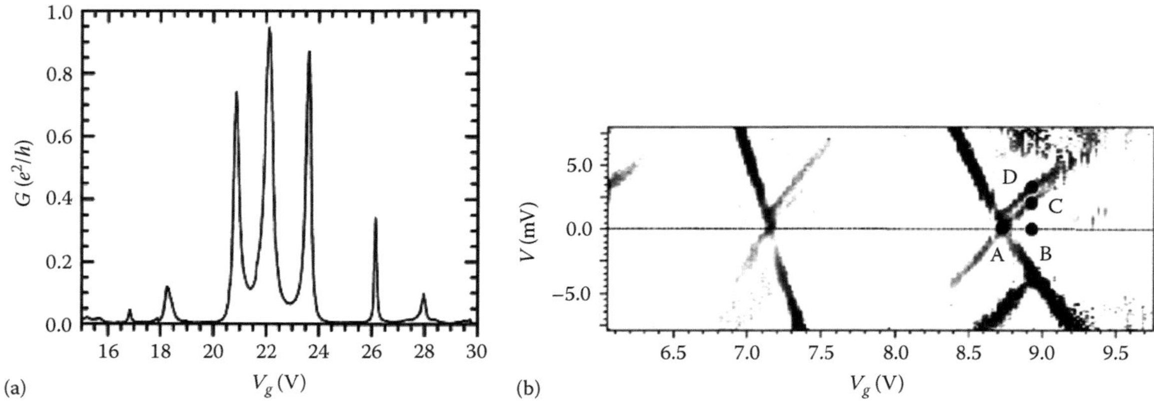

FIGURE 35.23 (a) Conductance G versus gate voltage V_g at $T = 1.3$ K for a CNTSET. (b) Differential conductance dI/dV of the CNT segment as a function of V and V_g. The data are plotted as an inverted gray scale, with the dark color corresponding to a large dI/dV. (From Bockrath, M. et al., *Science*, 275, 1922, 1997.)

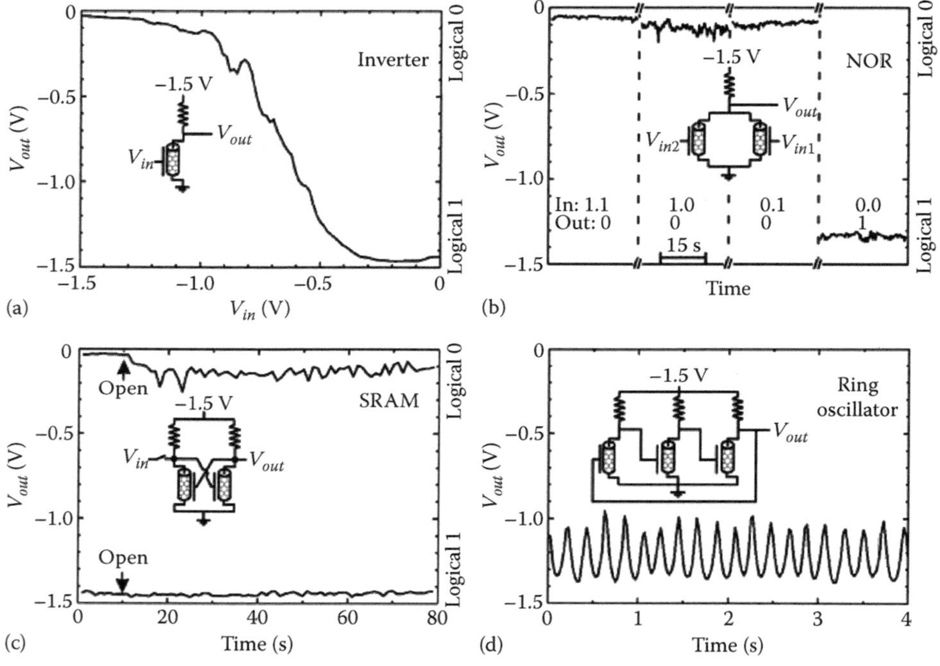

FIGURE 35.24 One-, two-, and three-transistor logic circuits with carbon nanotube FETs. (From Bachtold, A. et al., *Science*, 294, 1317, 2001.) (a) Output voltage as a function of the input voltage of a nanotube inverter. Schematic diagram of the electronic circuit. (b) Output voltage of a nanotube NOR for the four possible input states (1,1), (1,0), (0,1), and (0,0). (c) Output voltage of a flip–flop memory cell (SRAM) composed of two nanotube FETs. (d) Output voltage as a function of time for a nanotube ring oscillator.

CNTs have been used to fabricate chemical, biological, and mechanical sensors [32,59,60]. Despite successful applications, there is a crucial technical hurdle to overcome before this material can be widely used; positioning CNTs at exact locations is not possible at present. Hopefully, this problem will be solved in the future.

35.3 Graphene

Graphene is single-layered graphite. Although this material has been known for years, it has not drawn much attention. Several researchers have believed that a single layer may be energetically unstable such that it might roll up to form a carbon nanotube. In 2004, Geim and colleagues discovered that they could isolate a single graphene layer [61]. Soon after, they have discovered that graphene is a semimetal and exhibits 2D electron behavior with a relativistic dispersion relation. This could have been guessed from the electronic dispersion relations calculated around the K and K' points even with a tight-binding approximation; however, it was not noticed until the first observation of the quantum Hall effect [62,63].

We repeat here the same description of a graphene lattice discussed in the previous section, where we discussed

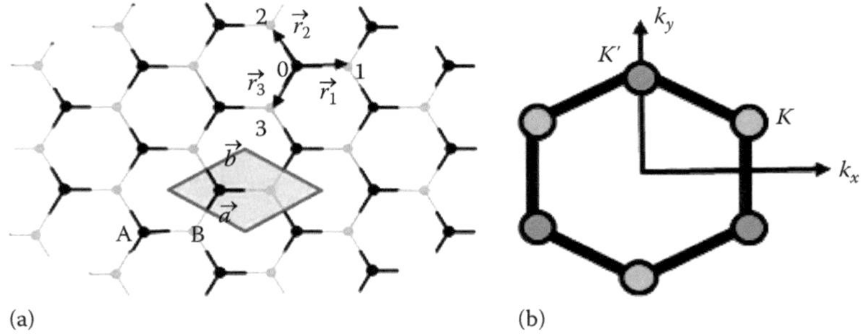

FIGURE 35.25 (a) A graphene unit cell with two-atom basis. There are two inequivalent symmetry sites. (b) Reciprocal lattice of a graphene. K and K' points are inequivalent.

carbon nanotubes. The honeycomb lattice is not a Bravais lattice, since it does not have translational symmetry for all lattice points. As shown in Figure 35.25, we need to define a unit cell that includes two lattice points, that is, A and B sites. Carbon atoms occupy all lattice points; therefore, there are two types of symmetry points for carbon atoms. These carbon atoms are in an sp^2 configuration, such that p_z orbitals form π bonds.

Let us consider a tight-binding model to calculate the electronic band structures of graphene in a honeycomb lattice. The wave function can be expressed by the linear combination of atomic orbitals positioned at either the A or B site (m = A or B). From the time-independent Schrödinger equation, we get

$$\psi_k(r) = \sum_{m=\{A,B\}} c_m(k)\phi_{mk}(r)$$

$$\sum_m c_m(k)\phi^*_{m'k}(r)\hat{H}(k)\phi_{mk}(r) = \varepsilon(k)\sum_m c_m(k)\phi^*_{m'k}(r)\phi_{mk}(r) \tag{35.1}$$

$$\sum_m [H_{m'm}(k) - \varepsilon(k)S_{m'm}(k)]c_m(k) = 0 \tag{35.2}$$

where

$$H_{m'm}(k) = \int dr\phi^*_{m'k}(r)\hat{H}\phi_{mk}(r) = \sum_{R_{m'}-R_m} e^{ik\cdot(R_m - R_{m'})}H(R_{m'}, R_m) \tag{35.3}$$

$$S_{m'm}(k) = \int dr\phi^*_{m'k}(r)\phi_{mk}(r) = \sum_{R_{m'}-R_m} e^{ik\cdot(R_m - R_{m'})}S(R_{m'}, R_m) \tag{35.4}$$

In Figure 35.24, a carbon atom "0" at the A site is shown surrounded by 1, 2, and 3 atoms at the B sites. By considering the nearest-neighbor interaction and assuming the hoping term $t \sim 2.7$ eV, we can define

$$H_{m'm}(k) = \begin{cases} t & \text{for } \vec{R}_{m'} - \vec{R}_m = \vec{r}_i \quad i = 1, 2, 3 \\ 0 & \text{otherwise} \end{cases}. \tag{35.5}$$

We can consider the Hamiltonian around the K point, $\vec{K} = \left(\frac{2\pi}{3a}, \frac{2\pi}{3\sqrt{3}a}\right)$ in the reciprocal space as shown in Figure 35.25b:

$$H_K\begin{pmatrix} c_{A,K} \\ c_{B,K} \end{pmatrix} = \begin{pmatrix} 0 & H_{AB,K} \\ H^*_{AB,K} & 0 \end{pmatrix}\begin{pmatrix} c_{A,K} \\ c_{B,K} \end{pmatrix} = E(q)\begin{pmatrix} c_{A,K} \\ c_{B,K} \end{pmatrix}. \tag{35.6}$$

For small q around the K point, we can use a Taylor series expansion,

$$\begin{aligned} H_{AB,K}(k) = H_{AB,K}(K+q) &= t\sum_{i=1}^{3}\exp(-ik\cdot r_i) \\ &= t\sum_{i=1}^{3}\exp(-i(K+q)\cdot r_i) \\ &= t\sum_{i=1}^{3}\exp(-iK\cdot r_i)(1 - ir_i\cdot q + O(q^2)) \\ &= -it\sum_{i=1}^{3}\exp(-iK\cdot r_i)r_i\cdot q + O(q^2). \end{aligned} \tag{35.7}$$

By substituting all the values into the equation, we can get

$$\begin{aligned} H_K\begin{pmatrix} c_{A,K} \\ c_{B,K} \end{pmatrix} &= \frac{3ta}{2}\left[\begin{pmatrix} 0 & -\frac{\sqrt{3}}{2} - i\frac{1}{2} \\ -\frac{\sqrt{3}}{2} + i\frac{1}{2} & 0 \end{pmatrix}q_x \right. \\ &\quad \left. + \begin{pmatrix} 0 & -\frac{1}{2} + i\frac{\sqrt{3}}{2} \\ -\frac{1}{2} - i\frac{\sqrt{3}}{2} & 0 \end{pmatrix}q_y\right]\begin{pmatrix} c_{A,K} \\ c_{B,K} \end{pmatrix} \\ &= \frac{3ta}{2}\left[\begin{pmatrix} 0 & 1 \\ 1 & 0 \end{pmatrix}q_x + \begin{pmatrix} 0 & -i \\ i & 0 \end{pmatrix}q_y\right]\begin{pmatrix} c'_{A,K} \\ c'_{B,K} \end{pmatrix}. \end{aligned} \tag{35.8}$$

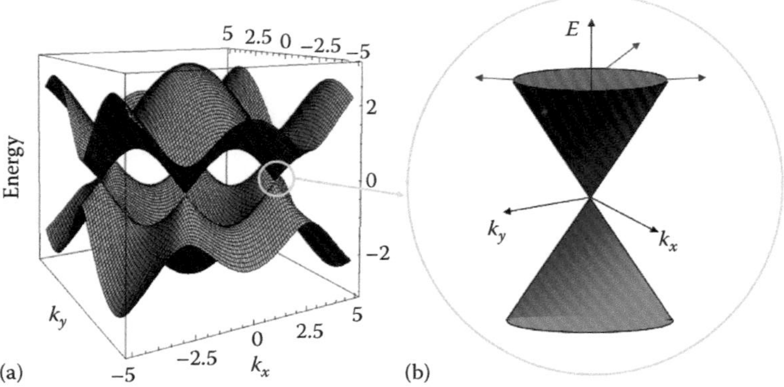

FIGURE 35.26 (a) Solution of graphene electronic structure. The conduction and valance bands meet at six points. If we assign the point indicated by the gray circle as a K point, the next one is K', then the next one is K, and so on. (b) Enlarged electron dispersion relation around a K point, as given by Equations 35.3 through 35.10. The pseudospin is drawn by arrows.

We can simplify the equation to obtain

$$H_K = \frac{3}{2} ta\sigma \cdot q = v_F \sigma \cdot q \quad H_{K'} = \frac{3}{2} ta\sigma^{*} \cdot q = v_F \sigma^{*} \cdot q. \quad (35.9)$$

This equation is exactly the same as the massless Dirac fermion Hamiltonian in (2×2) matrix. If we consider the K and K' points together, we can construct a (4×4) matrix. If we solve the Hamiltonian, the energy dispersion relation around K becomes

$$H_K \begin{pmatrix} A \\ B \end{pmatrix} = v_F \begin{pmatrix} 0 & k_x - ik_y \\ k_x + ik_y & 0 \end{pmatrix} \begin{pmatrix} A \\ B \end{pmatrix} = E \begin{pmatrix} A \\ B \end{pmatrix} \Rightarrow E = \pm v_F \sqrt{k_x^2 + k_y^2}. \quad (35.10)$$

The solution around the K point is two circular cones that meet at the K point as shown in Figure 35.26b:

The eigenstates can be obtained by inserting the E versus k solution to Equation 35.10:

$$\psi_{\pm K}(k) = \frac{1}{\sqrt{2}} \begin{pmatrix} e^{-i\theta_k/2} \\ \pm e^{i\theta_k/2} \end{pmatrix}$$
$$\psi_{\pm K'}(k) = \frac{1}{\sqrt{2}} \begin{pmatrix} e^{i\theta_k/2} \\ \pm e^{-i\theta_k/2} \end{pmatrix}. \quad (35.11)$$

Many interesting physical properties can be explained with the electronic structure and eigenstates. First, Equations 35.9 and 35.10 show that electrons and holes in graphene follow a linear dispersion with increasing k (Equation 35.10) as shown in Figure 35.26b. They behave as massless Dirac fermion with the velocity ~1/300 times the speed of light with an experimentally measured value of $v_F \approx 10^6$ m/s. Second, in addition to spin degeneracy, they reveal a pseudospin degeneracy that originates in the two-atom basis in a graphene unit cell. The pseudospin is schematically represented in Figure 35.26b. Within the same valley (in the same cone around the K point), an electron moving in the forward direction has the opposite pseudospin to the electron moving in the backward direction. The pseudospin should also be preserved when an electron encounters a scattering center or a tunneling barrier unless the scattering matrix or tunneling matrix includes pseudospin flipping terms. Therefore, back scattering is not allowed in this type of Dirac fermion. This phenomenon was predicted earlier by Klein and now known as Klein tunneling [64]. In Figure 35.27, as an electron moving in the forward direction experiences a quantum mechanical potential barrier, the tunneling probability becomes 100% and back scattering is not allowed so as to preserve the pseudospin. The Klein tunneling experiment was suggested and performed by measuring

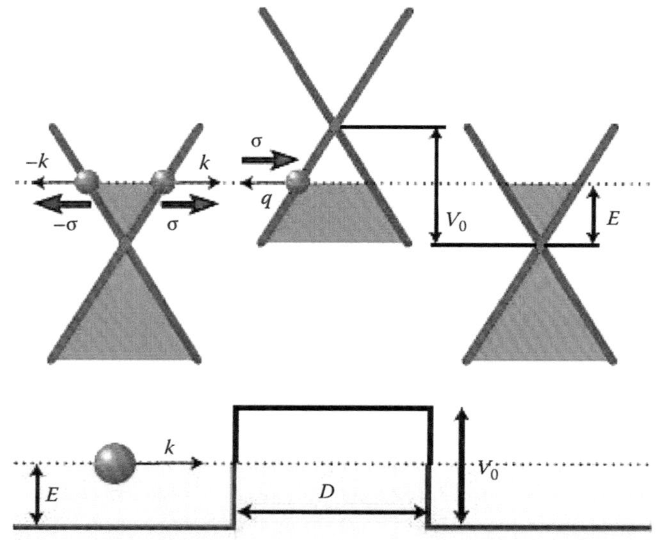

FIGURE 35.27 Schematic diagram of Dirac fermion tunneling through a potential barrier in graphene. Pseudospin conservation results in tunneling without back scattering. (From Katsnelson, M.I. et al., *Nat. Phys.*, 2, 621, 2006.)

the incident-angle dependence of the tunneling probability at a quantum mechanical barrier. The effective angle can be varied by applying a magnetic field [65,66].

Third, as we increase θ_k by 2π, the phase of the eigenstates changes only by π. This result is known as Berry's phase. Berry's phase of π could be confirmed in the shift of the Hall plateaus in the quantum Hall effect [62,63]. Fourth, the magnetic field dependence of Landau levels in graphene is considerably different from that in a semiconductor 2D electron gas (2DEG) system. That is in graphene, $E_n = \text{sign}(n)\sqrt{2e\hbar v^2|n|B}$ for $n = 0, \pm1, \pm2, \ldots$ whereas $E_n = (\hbar eB/m^*)(n+(1/2))$ for $n \geq 0$ in a semiconductor 2DEG. The quantum Hall effect data measured for the first time are shown in Figure 35.28. The $n = 0$ state is present when the carrier density is 0, unlike in the semiconductor 2DEG. Fifth, there are quartet states for each Landau level: 2 from spin degeneracy and 2 from pseudospin degeneracy. Figure 35.29 shows all four degenerate states lifted with a magnetic field as measured by STS at 10 mK [67]. The energy resolution is sufficiently high to reveal all the degeneracy.

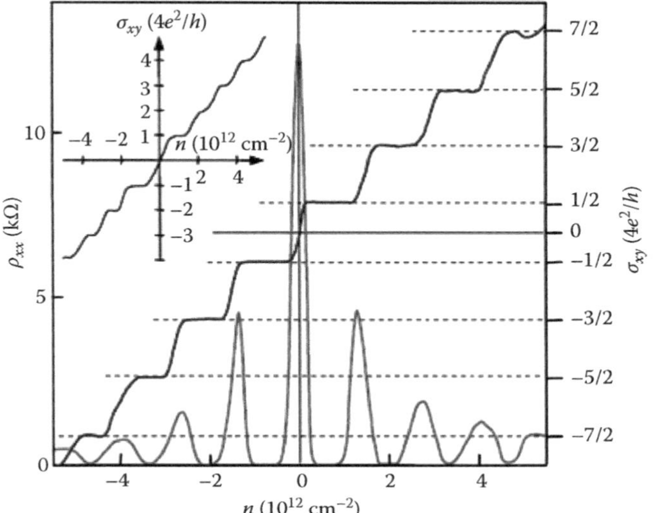

FIGURE 35.28 Quantum Hall effect for a massless Dirac fermion in graphene. Hall conductivity σ_{xy} and longitudinal resistivity σ_{xx} of graphene as a function of their concentration at $B = 14$ T and $T = 4$ K. (From Novoselov, K.S. et al., *Nature*, 438, 197, 2005.)

Graphene layers of three different types have been isolated and used to fabricate graphene-based devices. (1) Mechanical exfoliation of natural graphite using a scotch tape [61]. A graphene layer isolated with this method showed the best mobility as it was measured in a Hall bar geometry with a boron nitride (BN) substrate or as a suspended layer [68–70]. Many have achieved a mobility as high as 300,000 cm²/V/s at 4.2 K and 100,000 cm²/V/s at room temperature. (2) Epitaxial graphene has been produced by heating a silicon carbide (SiC) wafer to high temperature (>1100°C) [71]. This process produces a graphene layer having the same size as the SiC wafer. On a SiC wafer, two faces of silicon- or carbon-terminated were used [71]. There are several graphene layers on top of the SiC wafer and they are angularly twisted relative to each other such that interlayer interaction is negligible. Therefore, a Móire pattern is often observed in the STM image and the angle of the twist is estimated from the pattern. (3) CVD graphene uses CH_4 or C_2H_6 gases and the atomic structure of a metal substrate in order to seed the growth of the graphene [72]. The catalytic substrate

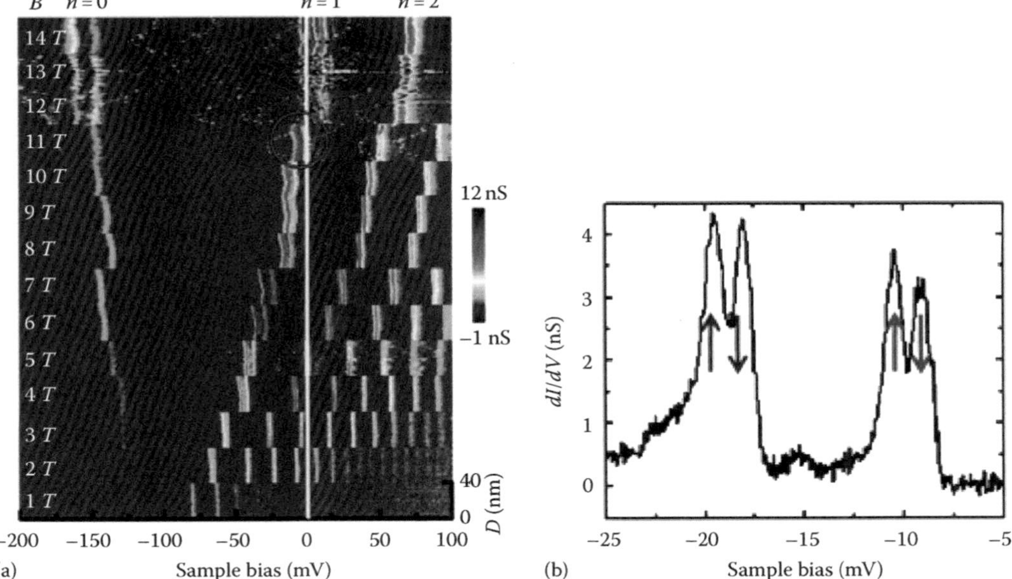

FIGURE 35.29 (a) Landau levels of epitaxial graphene on SiC as a function of magnetic field. A series of dI/dV line scans, taken vertically as a function of magnetic field. (b) Splitting of the $N = 1$, Landau level can be seen at $B = 11$ T, $V_{mod} = 50$ μV. (From Song, Y.J. et al., *Nature*, 467, 185, 2010.)

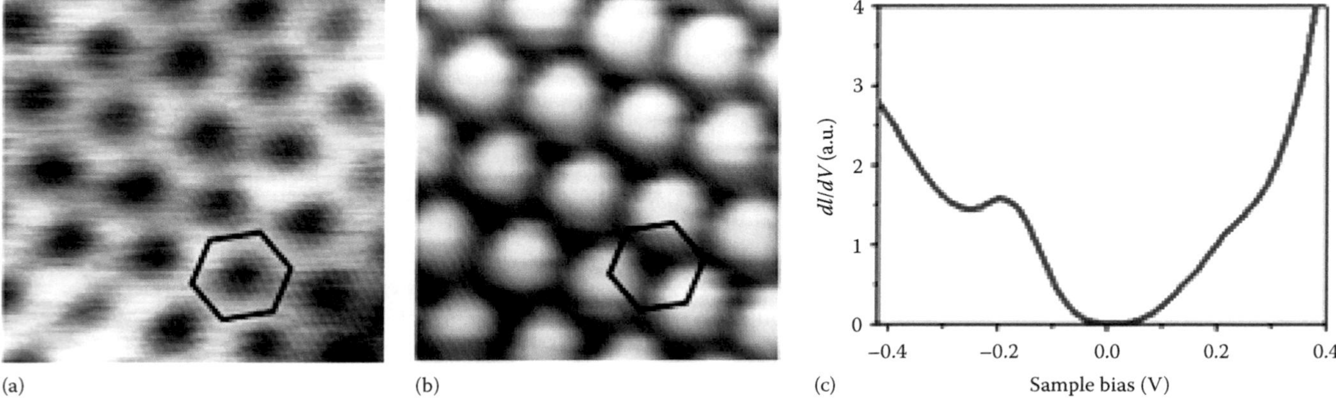

FIGURE 35.30 STM images of (a) ML graphene and (b) bilayer graphene. (c) STS result for a graphene layer on a SiO_2 surface. (From Yang, H. et al., *Phys. Rev. B*, 78, 041408, 2008.)

decomposes the source gases and carbon grows epitaxially on the metal layer. Cu(111) and Ni(111) substrates are often used owing to their small lattice mismatch with graphene. Figure 35.30 shows an atomically resolved graphene layer using STM. A graphite single crystal for a Bernal stacking; there is a second-layer carbon atom under every other first-layer carbon atom. Therefore, only three carbon atoms appear brighter out of six atoms in bilayer graphene [73]. The Dirac point is shifted away from the Fermi level possibly due to charge transfer from and to the SiO_2 substrate. The shift of the Dirac point away from the Fermi level was not observed for a graphene on top of a BN layer as shown in Figure 35.31. That may be explained by the two facts: (1) the lattice mismatch between BN and graphene is small and (2) there is far less charge transfer from the BN layer [74]. A linear dispersion relation could not be determined on a graphene surface that was on top of a SiO_2 substrate using STS [75]. A gap-like feature was observed and was explained by phonon-mediated tunneling around the Fermi level due to a lack of carrier density in semimetallic graphene [75].

When one attempts the fabrication of a graphene device (GD), corrugation, gap states, charged and neutral impurities become important issues. We transferred CVD-grown graphene layers of SiC, SiO_2, and SiN_x substrates for which we measured corrugation and impurities using STM [76]. The STM images in Figure 35.32 are quite similar; they show a similar honeycomb structure. The corrugation is the smallest on the SiC surface and largest on SiO_2. There can be more geometrical defects on a SiO_2 surface, and this corrugation may originate from dangling bonds at the interface. These defects may act as scattering centers when a GD is fabricated on these substrates. In an annealed sample, the Dirac points initially move upon annealing but they do not move with additional annealing. The initial movement may be due to the desorption of weakly bound water molecules on graphene, but defects created by dangling bonds may not be cured with additional annealing. This result is quite consistent with STS reported earlier [77].

Shon and Ando predicted unique characteristics of transport through a perfect GD with only short-range scatter [78]; the prediction was that the conductivity should not depend on the carrier density, because the scattering rate would be divergent as the carrier density approaches zero near the Dirac point. As long-range scatters would be dominant in

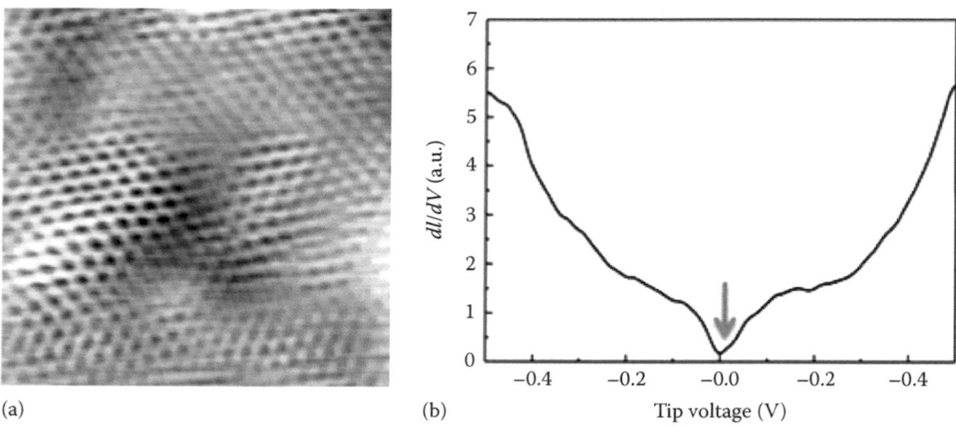

FIGURE 35.31 (a) STM image of a graphene layer on a BN layer on a Ni(111) substrate. (b) STS on the surface, showing that the Dirac point coincides with the Fermi level. (From Baek, H.W. et al., to be published, 2012.)

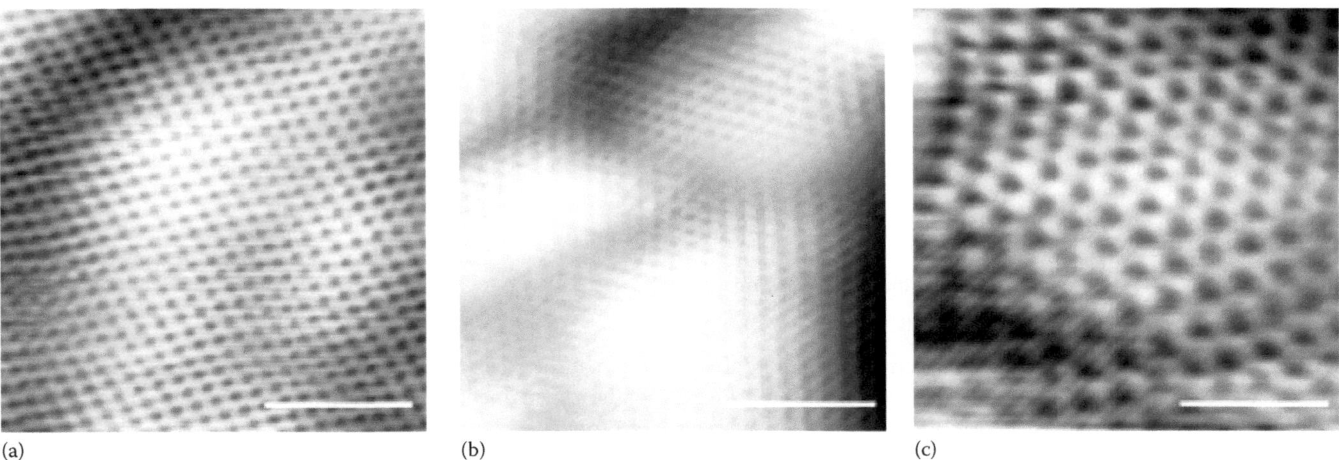

FIGURE 35.32 STM images of CVD-grown graphenes transferred on (a) SiC, (b) SiO_2, and (c) SiN_x. The scale bars indicate 2 nm in (a) and (b), and 1 nm in (c). (From Chae, J. et al., *Microelec. Eng.*, 88, 1211, 2011.)

the low-density limit, the GD would become insulating at the Dirac point. Contrary to this prediction, it was found that the conductivity is linearly dependent on the charge density induced by the back-gate bias [62,63], suggesting that long-range scatters are dominant in GD transport. In addition, the carrier transport would be affected by scattering with charged impurities, short-range scatterers, mid-gap states, various phonon modes, surface corrugations, and defects in a GD [79–83]. In the case of the charged impurity potential at a high carrier density limit, the conductance is linearly dependent on the induced charge density as confirmed by a recent experiment using a potassium-doped GD [82,84].

Mapping charge puddles in a GD has been pursued by many groups using scanning probe microscopy. It was first demonstrated for a GD fabricated on top of a SiO_2 substrate using a scanning single-electron transistor microscope [85] as shown in Figure 35.33a. The corresponding experimental result showed good agreement with the theoretically predicted transport property near the Dirac point [86]. In a more recent study [87], spatially resolved charge impurities were mapped using a spatial map in STS. In the study, the Dirac point was locally mapped from STS data and the charge puddle was mapped with scattering centers as shown in Figure 35.33b [87].

In order to understand the correlation between geometric defects and/or electronic scattering centers and the local transport property, a scanning gate microscopy (SGM) study was performed on a GD [76]. After a GD on top of SiO_2 was installed in a cryogenic atomic force microscopy (AFM)–SGM chamber, it was imaged in the AFM mode. After the corrugation was confirmed by AFM topography, SGM experiments were conducted. Figure 35.34a depicts a 450×450 nm AFM topography of a GD and Figure 35.34b and c shows SGM micrographs at tip bias voltages of −2.5 and +2.5 V. The Dirac point of this GD was −24 V, as set by the back-gate bias. All of the SGM measurements were taken at the Dirac point by back-gating. Corrugations on the graphene surface can be seen in the topographic image with lateral dimension of ~100 nm. The small features in the SGM data are tens of nanometers, revealing good agreement with the AFM observation. There is strong correlation between the SGM signal and the topographic corrugations, that is, the SGM peaks. The protrusions in 3 dimensional representation indicate electron puddles while the bright plateaus indicate hole puddles surrounding the electron puddles. The range of hole

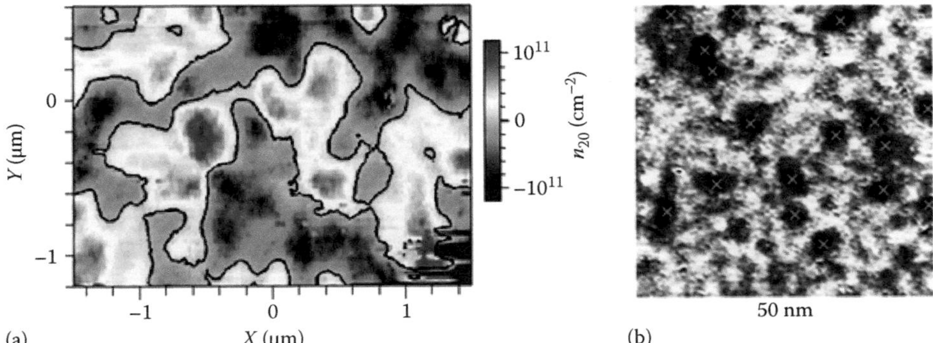

FIGURE 35.33 **(See color insert.)** (a) Charge puddle mapped by scanning SET microscopy (From Martin, J. et al., *Nat. Phys.*, 4, 144, 2008) and (b) dI/dV map of 50 × 50 nm image by STM. V_b = −0.75 V, I = 80 pA, and V_g = 60 V. Scattering centers are marked by gray crosses. (From Zhang, Y. et al., *Nat. Phys.*, 5, 722, 2009.)

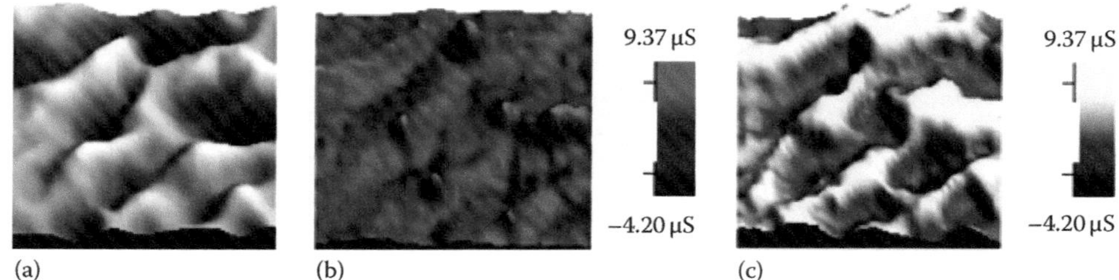

FIGURE 35.34 (a) A 450 × 450 nm AFM image, (b) SGM image of V_{tip} = −2.5 V, and (c) SGM image at V_{tip} = +2.5 V. (From Chae, J. et al., *Microelectr. Eng.*, 88, 1211, 2011.)

carrier puddles extends beyond that of electron carrier puddles at the same electric field strength. This difference is mainly due to the structure of charge puddles. The correlation between topographic data and SGM data implies that charge puddles exist at the bottom of each ripple. The different features for electron and hole carriers indicate that electrons are more locally confined at the bottom of the ripple induced by the interaction with the substrate and hole carrier screens around the confined electrons. From these results, it is evident that charged puddles are the main scattering centers in a GD on a SiO_2 layer [76].

Electronic states located at a surface with energies near the Fermi level and lying within the bulk bandgap can significantly contribute to the overall conductance of a material. Similarly, when extra edge states are present at the cut edges of a graphene stripe near the Dirac point, they may completely alter the existing transport models primarily because of carrier scattering in the bulk stripe. A substantial portion of the transport current may flow through the edge states without significant carrier scattering. Given the importance of electron scattering at step edges in graphene nanoribbons, step structures were carefully studied [88]. Armchair edges of a graphene layer on SiC reveal interference patterns, similar to those produced by Friedel oscillation on the edge of a metal, although they are considerably more complex as shown in Figure 35.35a. The unique shape in the figure can be understood as the quantum interference wave propagating along the three directions. A theoretical study show that the armchair edge reveals almost perfect intervalley scattering regardless of the presence of the hydrogen atom at the edge [89]. In addition, the pentagonal reconstruction of the armchair edge gives rise to intravalley scattering for incident electronic waves with oblique angles. We have also observed beating and slowly decaying patterns in charge density profiles due to the quantum interference of scattering waves at the edges, which are in excellent agreement with our STM experiment. In contrast, zigzag edges produce only intravalley scattering. The zigzag edges of a graphene layer on a BN substrate do not show complex quantum interference (Figure 35.35b). However, there is a single gap state at the edge as expected by band structure calculation [90], and the peak decays with an increasing distance from 1.0 nm as shown in Figure 35.35c. This result is significantly different from that of the edge states of a graphene layer on a Au(111) surface [91].

The relationship between macroscopic charge transport properties and microscopic carrier distribution is one of the central issues in GDs. In SGM, a conductive AFM tip was used as a local top gate to induce an electrostatic potential and alter the Fermi level and the carrier density over the selected area of interest. The transport signal change (conductance variation (ΔG)) in the presence of the AFM tip gate revealed how the local electronic structure of a sample contributes to its macroscopic transport properties. Enhanced conductance through the edges of a GD was reported in a study [92]. The SGM experiment was performed with a 1.6 μm × 2.3 μm GD on a SiO_2. The spatial variation of the SGM signals shown in Figure 35.36a through c reveals that conductance enhancement was nearly constant along the edge. (The left side is the GD and the right side is the SiO_2 substrate)

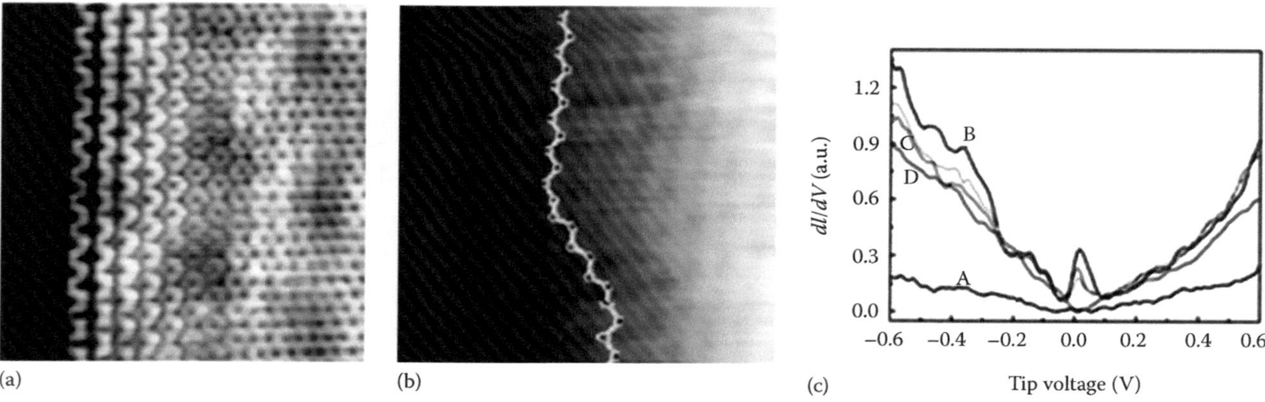

FIGURE 35.35 **(See color insert.)** (a) 5.5 × 5.5 nm STM image of a graphene edge on a SiC surface. (From Yang, H. et al., *Nano Lett.*, 10, 943, 2010.) (b) An STM image of 5 × 5 nm zigzag and chiral edge graphene on BN. (From Li, X.S. et al., 324, 1312, 2009.) (c) STS spectra of BN (A), zigzag edge (B), 0.5 nm inside the edge (C), and 2.5 nm inside the edge (D). (From Hwang, B.Y. et al., *Curr Appl. Phys.* in print, 2013.)

FIGURE 35.36 SGM result for a graphene ribbon. (a) A hole-doped ribbon with a tip bias of 10 V. (b) Undoped ribbon with a tip bias of 10 V. (c) An electron-doped ribbon with a tip bias of –10 V. (From Chae, J. et al., *Nano Lett.*, 12, 1839, 2012.)

Each SGM map in Figure 35.36a through c (1.3 μm × 1.3 μm) was obtained at the three gate voltages: hole-doped ($V_g = -10$ V for Figure 35.36a), charge neutral ($V_g = -4$ V for Figure 35.36b), and electron-doped ($V_g = 10$ V for Figure 35.36c), and with the tip-gating voltage set at $V_{tip} = 10$ V for both Figure 35.36a and b and at $V_{tip} = -10$ V for Figure 35.36c. The conductance enhancement at the edge was measured up to 0.3 G_0 ($= e^2/h$) and showed only slight dependence on the spatial location along the graphene edge. The enhancement mechanism appeared to originate from the opening of an edge conductance channel by the tip-gating potential. This result can be explained by our theoretical model of the opening of an additional conduction channel localized at the edges through the depletion of the accumulated charge by the tip.

As discussed earlier, the mobility of a GD reaches up to 275,000 cm²/V/s. This is because the mobility of ML graphene is defined as $\mu = ev_F^2\tau/E_F$, and not as a function of the effective mass. The mobility of bilayer graphene is defined as $\mu = e\tau/m^*$. Many unique electrical properties have triggered the expectation that graphene could be used as a channel material for high-speed graphene field effect transistors (GFETs). The cutoff frequencies of GFETs have been observed to reach up to 300 GHz [93–95]. Despite a high cutoff frequency, these devices show poor voltage gains, which is a serious problem for practical device application. The intrinsic voltage gain can be expressed by the ratio $A_v = g_m/g_d$, where the numerator is I_d/V_g and the denominator is I_d/V_d. In most GDs, the drain current saturates weakly, resulting in poor voltage gain. This difficulty could be overcome by using a bilayer graphene [96]. The transconductance, which limits the cutoff frequency of a transistor, was not degraded by the displacement field of the bilayer graphene. Figure 35.37a and b show the output characteristics of a monolayer and a bilayer GFET at a back-gate voltage of −60 V. The back-gate displacement field of the bilayer graphene introduces an electrically effective gap. The gap should function as a proper pinch-off for the channel. The drain–current saturation is considerably more pronounced in the bilayer GFET compared to the ML. The voltage gain was improved by factor of 6 by adopting bilayer graphene.

It is known that the carrier mobility of natural Kish graphene is much better than that of CVD-grown graphene. Since a large graphene layer can be grown, it will be useful for commercialization in order to establish fabrication of GFET from CVD-grown material [97]. A GD can be operated even at cryogenic temperatures unlike conventional semiconductor devices whose low-temperature performance is hampered by carrier freeze-out effects. A group at IBM fabricated GDs on a diamond-like carbon (DLC) film grown on SiO as shown in Figure 35.38a. The phonon energy on this surface was greater than that on SiO_2, and the device did not suffer from additional scattering with the surface phonon or charge traps. Although the direct current transconductance (g_m) suffered from the short-channel effect, the overall radio frequency (RF) performance benefited from the reduction of gate length.

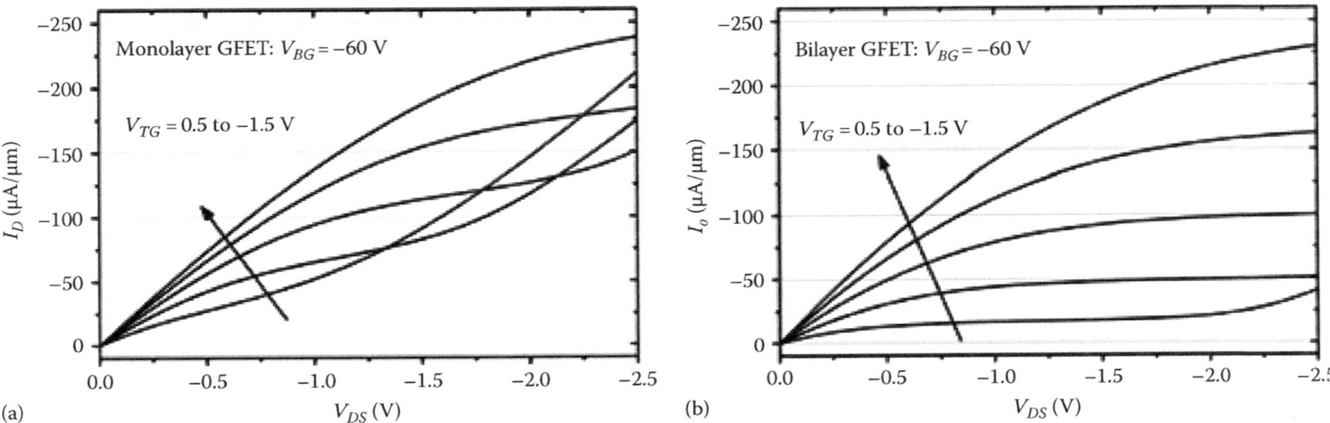

FIGURE 35.37 Drain current as a function of the source–drain voltage of (a) a ML GFET and (b) bilayer GFET at a back-gate voltage $V_{BG} = -60$ V. The top-gated voltage V_{TG} ranges from −1.5 to 0.5 V in 0.5 V steps. (From Szafranek, B.N. et al., *Nano Lett.*, 12, 1324, 2012.)

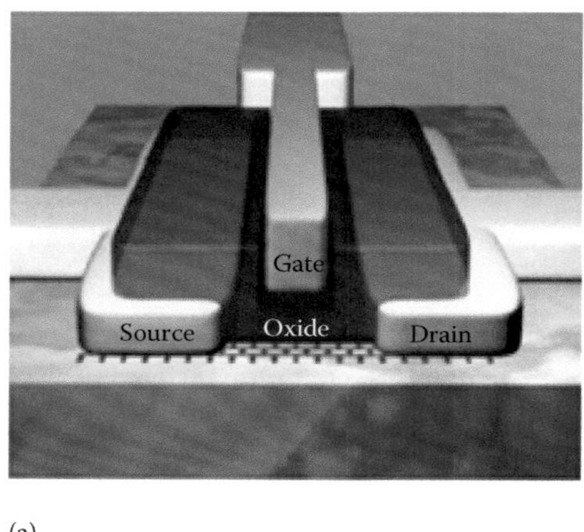

(a)

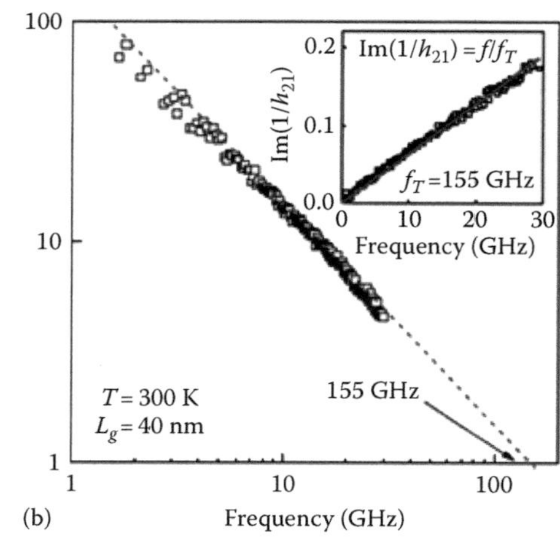

(b)

FIGURE 35.38 (a) Schematic view of a top-gated graphene RF transistor on a DLC substrate. (b) Cutoff frequencies for a device with a gate length of 40 nm at room temperature. Small-signal current gain $|h_{21}|$ plotted as a function of frequency. Insets shows linear fitting using Gummel's method, showing identical extrapolated cutoff frequencies. (From Wu, Y. et al., *Nature*, 472, 74, 2011.)

References

1. H.W. Kroto et al., *Nature* 318, 162 (1985).
2. S.C. O'Brien, J.R. Heath, R.F. Curl, and R.E. Smalley, *J. Chem. Phys.* 88, 220 (1988).
3. J.R. Heath, R.F. Curl, and R.E. Smalley, *J. Chem. Phys.* 87, 4236 (1987).
4. R.F. Curl and R.E. Smalley, *Science* 242, 1017 (1988).
5. J.B. Howard, J.T. McKinnon, Y. Makaraovsky, A.L. Lafleur, and M.E. Johnson, *Nature* 352, 139 (1991).
6. G. Peters and M. Jansen, *Angew. Chem.* 104, 240 (1992).
7. R.F. Bunshah, S. Jou, S. Prakash, H.J. Doerr, L. Isaacs, A. Wehrsig, C. Yeretzian, H. Cynn, and F. Diederich, *J. Phys. Chem.* 96, 6866 (1992).
8. W. Kratschmer, L.D. Lamb, K. Fosiropoulos, and D.R. Huffman, *Nature* 347, 354, (1990).
9. R.C. Haddon, L.E. Brus, and K. Raghavachari, *Chem. Phys. Lett.* 125, 459 (1986).
10. L. Salem, *The Molecular Orbital Theory of Conjugated Systems* (Benjamin, New York, 1966).
11. W.H. Green et al., *J. Phys. Chem.* 100, 14892 (1996).
12. Y.Z. Li, J.C. Patrain, M. Chander, J.H. Weaver, L.P. Chibante, and R.E. Smalley, *Science* 252, 547 (1991); ibid, 253, 429 (1991).
13. T. Chen, S. Wowells, M. Gallagher, L. Yi, D. Sarid, D.L. Lichtenberger, K.W. Nebesny, and C.D. Ray, *J. Vac. Sci. Technol. B* 10, 170 (1992).
14. T. Hashizume, X.D. Wang, Y. Nishina, H. Shinohara, Y. Saito, Y. Kuk, and T. Sakurai, *Jpn. J. Appl. Phys.* 31, L880 (1992).
15. Y.Z. Li, M. Chander, J.C. Patrin, J.H. Weaver, L.P.F. Chibante, and R.E. Smalley, *Phys. Rev. B* 45, 13837 (1992).
16. J.H. Weaver, J.L. Martins, T. Komeda, Y. Chen, T.R. Ohno, G.H. Kroll, N. Trouner, R.E. Haufler, and R.E. Smalley, *Phys. Rev. Lett.* 66, 1741 (1991).
17. R.J. Benning, J.L. Martins, J.H. Weaver, L.P.F. Chibante, and R.E. Smalley, *Science* 252, 1417 (1991).
18. Y. Kuk, D.K. Kim, Y.D. Suh, K.H. Park, H.P. Noh, S.J. Oh, and S.K. Kim, *Phys. Rev. Lett.* 70, 1948 (1993).
19. T. Hashizume, K. Motai, X.D. Wang, H. Shinohara, Y. Saito, Y. Maruyama, K. Ohno, Y. Kawazoe, Y. Nishina, H.W. Pickering, Y. Kuk, and T. Sakurai, *Phys. Rev. Lett.* 71, 2959 (1993).
20. X.-D. Wang, T. Hashizume, V.Yu. Yurov, Q.K. Xue, H. Shinohara, Y. Kuk, Y. Nishina, and T. Sakurai, *Zeit. Phys. Chem. B* 202, S117 (1997).
21. M.A. Reed et al., *Phys. Rev. Lett.* 60, 535 (1988).
22. A.F. Morpurgo, C.M. Marcus, and D.B. Robinson, *Appl. Phys. Lett.* 74, 2084 (1999).
23. J. Chen et al., *Science* 286 1550 (1999).
24. J.R. Petta, D.G. Salinas, and D.C. Ralph, *Appl. Phys. Lett.* 77, 4419 (2000).
25. H. Park, J. Park, A.K.L. Lim, E.H. Anderson, A.P. Alivisatos, and P.L. McEuen, *Nature* 407, 57 (2000).
26. C.B. Winkelmann, N. Roch, W. Wernsdorfer, V. Bouchiat, and F. Balestro, *Nat. Phys.* 5, 876 (2009).
27. S. Iijima, *Nature* 354, 56 (1991).
28. R. Saito, M.S. Dresselhaus, and G. Dresselhaus, *Physical Properties of Carbon Nantubes* (Imperial College Press, London, U.K., 1998).
29. M. Menon and D. Srivastava, *Phys. Rev. Lett.* 79, 4453 (1997).
30. M. Bockrath, *Nat. Phys.* 2, 155 (2006).
31. Z. Zhou, Carbon nanotube transistors, sensors and beyond, PhD dissertation (Cornell University, Ithaca, NY, 2008).
32. Y. Matsuda, J. Tahir-Kheli, and W.A. Goddard, III, *Phys. Chem. Lett.* 1, 2946 (2010).
33. H. Kim, J. Lee, S.-J. Kahng, Y.-W. Son, S.B. Lee, C.-K. Lee, J. Ihm, and Y. Kuk, *Phys. Rev. Lett.* 90, 216107 (2003).
34. J. Xhie et al., *Phys. Rev. B* 43, 8917 (1991).
35. S. Lee, G. Kim, H. Kim, B.-Y. Choi, J. Lee, B.W. Jeong, J. Ihm, Y. Kuk, and S.-J. Kahng, *Phys. Rev. Lett.* 95, 166402 (2005).

36. A.D. Yoffe, *Adv. Phys.* 50, 1 (2001).
37. J. Lee, S. Eggert, H. Kim, S.J. Kahng, H. Shinohara, and Y. Kuk, *Phys. Rev. Lett.* 93, 166403 (2004).
38. J. Lee, H. Kim, S.-J. Kahng, G. Kim, Y.-W. Son, J. Ihm, H. Kato, Z.W. Wang, T. Okazaki, H. Shinohara, and Y. Kuk, *Nature (London)* 415, 1005 (2002).
39. C. Klinke, J.B. Hannon, A. Afzali, and Ph. Avouris, *Nano Lett.* 6, 906 (2006).
40. C. Zhou, J. Kong, E. Yenilmez, and H. Dai, *Science* 290, 1552 (2000).
41. A.M. Rao, P.C. Eklund, S. Bandow, A. Thess, and R.E. Smalley, *Nature (London)* 388, 257 (1997).
42. M. Bockrath, J. Hone, A. Zettl, P.L. McEuen, A.G. Rinzler, and R.E. Smalley, *Phys. Rev.* B 61, R10606 (2000).
43. G.-H. Jeong, A.A. Farajian, R. Hatakeyama, T. Hirata, T. Yaguchi, K. Tohji, H. Mizuseki, and Y. Kawazoe, *Phys. Rev. B* 68, 075410 (2003).
44. J. Lu, S. Nagase, S. Zhang, and L. Peng, *Phys. Rev. B* 69, 205304 (2004).
45. S.H. Kim, W.I. Choi, G. Kim, Y.J. Song, G.-H. Jeong, R. Hatakeyama, J. Ihm, and Y. Kuk, *Phys. Rev. Lett.* 99, 256407 (2007).
46. T. Ozaki, *Phys. Rev. B* 67, 155108 (2003).
47. B.W. Smith, M. Monthioux, and D.E. Luzzi, *Nature* 396, 323 (1998).
48. K. Hirahara et al., *Phys. Rev. Lett.* 85, 5384–5387 (2000).
49. L. Yang and J. Han, *Phys. Rev. Lett.* 85, 154 (2000).
50. S.J. Tans, A.R.M. Verschueren, and C. Dekker, *Nature* 393, 49 (1998).
51. A. Javey, J. Guo, Q. Wang, M. Lundstrom, and H.J. Dai, *Nature* 424, 654 (2003).
52. S. Rosenblatt et al., *Nano Lett.* 2, 869–915 (2002).
53. S. Wind, J. Appenzeller, R. Martel, V. Derycke, and P. Avouris, *Appl. Phys. Lett.* 80, 3817–3819 (2002).
54. S. Heinze et al., *Phys. Rev. Lett.* 89, 6801 (2002).
55. J. Appenzeller et al., *Phys. Rev. Lett.* 89, 126801–126804 (2002).
56. S. Lee, S.-J. Kahng, and Y. Kuk, *Chem. Phys. Lett.* 50, 82–85 (2010).
57. M. Bockrath, D.H. Cobden, P.L. McEuen, N.G. Chopra, A. Zettl, A. Thess, and R.E. Smalley, *Science* 275, 1922 (1997).
58. A. Bachtold, P. Hadley, T. Nakanishi, and C. Dekker, *Science* 294, 1317 (2001).
59. S. Chopra, K. McGuire, N. Gothard, A.M. Rao, and A. Pham, *Appl. Phys. Lett.* 83, 2280 (2003).
60. N. Sinha, J. Ma, and J.T.W. Yeow, *Nanosci. Nanotechnol.* 6, 573 (2006).
61. K.S. Novoselov, A.K. Geim, S.V. Morozov, D. Jiang, Y. Zhang, S.V. Dubonos, I.V. Grigorieva, and A.A. Firsov, *Science* 306, 566 (2004).
62. K.S. Novoselov, A.K. Geim, S.V. Morozov, D. Jiang, Y. Zhang, M.I. Katsnelson, I.V. Grigorieva, S.V. Dubonos, and A.A. Firsov, *Nature* 438, 197 (2005).
63. Y. Zhang, Y.-W. Tan, H. Stormer, and P. Kim, *Nature* 438, 201 (2005).
64. O. Klein, *Z. Phys.* 53, 157 (1929).
65. M.I. Katsnelson, K.S. Novoselov, and A.K. Geim, *Nat. Phys.* 2, 621 (2006).
66. A.F. Young and P. Kim, *Nat. Phys.* 5, 222 (2009).
67. Y.J. Song, A.F. Otte, Y. Kuk, Y. Hu, D.B. Torrance, P.N. First, W.A. de Heer, H. Min, S. Adam, M.D. Stiles, A.H. MacDonald, and J.A. Stroscio, *Nature* 467, 185 (2010).
68. P.J. Zomer et al., *Appl. Phys. Lett.* 99, 232104 (2011).
69. W. Gannett et al. *Appl. Phys. Lett.* 98, 242105 (2011).
70. K.I. Bolotin, F. Ghahari, M.D. Shulman, H.L. Stormer, and P. Kim, *Nature* 462, 1038 (2009).
71. C. Berger, Z. Song, T. Li, X. Li, A.Y. Ogbazghi, R. Feng, Z. Dai, A.N. Marchenkov, E.H. Conrad, P.N. First, and W.A. de Heer, *J. Phys. Chem. B* 108, 19912 (2004).
72. X.S. Li, W.W. Cai, J.H. An, S. Kim, J. Nah, D.X. Yang, R.D. Piner, A. Velamakanni, I. Jung, E. Tutuc, S.K. Banerjee, L. Colombo, and R.S. Ruoff, *Science* 324, 1312 (2009).
73. H. Yang, G. Baffou, A.J. Mayne, G. Comtet, G. Dujardin, and Y. Kuk, *Phys. Rev. B* 78, 041408 (2008).
74. B. Hwang, J.H. Kwon, M. Lee, S.J. Lim, S.J. Jeon, S. Kim, U. Ham, Y.J. Song, Y. Kuk, *Curr. Appl. Phys.* in print (2013).
75. Y. Zhang, V.W. Brar, F. Wang, C. Girit, Y. Yayon, M. Panlasigui, A. Zettle, and M.F. Crommie, *Nat. Phys.* 4, 627 (2008).
76. J. Chae, J. Ha, H. Baek, Y. Kuk, S.Y. Jung, Y.J. Song, N.B. Zhitenev, and J.A. Stroscio, S.J. Woo, and Y.-W. Son, *Microelectr. Eng.* 88, 1211–1213 (2011).
77. Y. Zhang, V.W. Brar, C. Girit, A. Zettl, and M.F. Crommie, *Nat. Phys.* 5, 722–726 (2009).
78. N. Shon and T. Ando, *J. Phys. Soc. Jpn.* 67, 2421–2429 (1998).
79. Y. Tan, Y. Zhang, K. Bolotin, Y. Zhao, S. Adam, and E.H. Hwang, *Phys. Rev. Lett.* 99, 246803 (2007).
80. J. Chen, C. Jang, S. Xiao, M. Ishigami, and M.S. Fuhrer, *Nat. Nano Technol.* 3, 206–209 (2008).
81. J. Chen, C. Jang, M. Ishigami, S. Xiao, W. Cullen, and E. Williams, *Solid State Commun.* 149, 1080–1086 (2009).
82. L.A. Ponomarenko, R. Yang, T.M. Mohiuddin, M.I. Katsnelson, K.S. Novoselov, and S.V. Morozov, *Phys. Rev. Lett.* 102, 206603 (2009).
83. J. Yan, Y. Zhang, P. Kim, and A. Pinczuk, *Phys. Rev. Lett.* 98, 166802 (2007).
84. C. Jang, S. Adam, J. Chen, E.D. Williams, S. Das Sarma, and M.S. Fuhrer, *Phys. Rev. Lett.* 101, 146805 (2008).
85. J. Martin, N. Akerman, G. Ulbricht, T. Lohmann, J.H. Smet, and K. von Klitzing, *Nat. Phys.* 4, 144–148 (2008).
86. E. Rossi, S. Adam, and S. Das Sarma, *Phys. Rev. B* 79, 245423 (2009).
87. Y. Zhang, V.W. Brar, C. Girit, A. Zettl, and M.F. Crommie, *Nat. Phys.* 5, 722–726 (2009).
88. H. Yang, A. Mayne, M. Boucherit, G. Comtet, G. Dujardin, and Y. Kuk, *Nano Lett.* 10, 943 (2010).
89. C. Park, H. Yang, G. Kim, S. Seo, A.J. Mayne, G. Dujardin, Y. Kuk, and J. Ihm, *Proc. Natl. Acad. Sci. USA* 108, 18622 (2011).
90. Y.-W. Son, M. L Cohen, and S.G. Louie, *Nature* 444, 347–349 (2006).

91. C. Tao, L. Jiao, O.V. Yazyev, Y.-C. Chen, J. Feng, X. Zhang, R.B. Capez, J.M. Tou, A. Zettl, S.G. Louie, H. Dai, and M.F. Crommie, *Nat. Phys.* 7, 616 (2011).
92. J. Chae, S. Jung, S. Woo, H. Baek, J. Ha, Y.J. Song, Y.-W. Son, N.B. Zhitenev, J.A. Stroscio, and Y. Kuk, *Nano Lett.* 12, 1839 (2012).
93. L. Liao, Y.-C. Lin, M. Bao, R. Cheng, J. Bai, Y. Liu, Y. Qu, K.L. Wang, Y. Huang, and X. Duan, *Nature* 467, 305–308 (2010).
94. Y.M. Lin, C. Dimitrakopoulos, K.A. Jenkins, D.B. Farmer, H.Y. Chiu, A. Grill, and P. Avouris, *Science* 327, 662 (2010).
95. Y. Wu, Y.M. Lin, A.A. Bol, K.A. Jenkins, F. Xia, D.B. Farmer, Y. Zhu, and P. Avouris, *Nature* 472, 74 (2011).
96. B.N. Szafranek, G. Fiori, D. Schall, D. Neumaier, and H. Kurz, *Nano Lett.* 12, 1324 (2012).
97. K.S. Kim, Y. Zhao, H. Jang, S.Y. Lee, J.M. Kim, K.S. Kim, J.-H. Ahn, P. Kim, J.-Y. Choi, and B.H. Hong, *Nature* 457, 706 (2009).

36

Subnanometer Characterization of Nanoelectronic Devices

Pierre Eyben
Interuniversity Microelectronics Centre

Jay Mody
IBM Systems and Technology

Aftab Nazir
Interuniversity Microelectronics Centre and Katholieke Universiteit Leuven

Andreas Schulze
Interuniversity Microelectronics Centre and Katholieke Universiteit Leuven

Trudo Clarysse
Interuniversity Microelectronics Centre

Thomas Hantschel
Interuniversity Microelectronics Centre

Wilfried Vandervorst
Interuniversity Microelectronics Centre and Katholieke Universiteit Leuven

36.1 Introduction on 2D and 3D Carrier Profiling Needs

Since a few years, the microelectronic world is experiencing a revolution. For >20 years, electronics has been dominated by one material (Si) and one architecture (metal oxide–semiconductor field-effect transistor), despite the presence of bipolar junction transistor (BJT) and the use of III–V semiconductors for some specific applications. However, in order to tackle the new challenges in terms of miniaturization, power consumption, power density, and processing speed, new inorganic semiconductor materials (Ge, InP, InGaAs, GaN, SiC, etc.) and new 3D architectures (multiple gates FETs, nanowire TFETs, etc.) are developed and progressively introduced. The two main goals of this revolution are to overcome the current limitations in terms of mobility (in order to increase the operational speed of the devices) and of junction leakage (in order to reduce the power consumption). These new architectures are typically three-dimensional and involve multiple materials. With the continuous decrease of dimensions, they also represent extremely confined volumes into which statistics and quantum effects start to play an increasing role. Beyond the standard logic/memory applications, there is also a very strong increase in More than Moore developments targeting energy (photovoltaic, energy storage), imaging (e.g., quantitative medical imaging), sensor/actuators linked to CMOS base circuitry, biochips, etc. In all these cases, the dopant/carrier distribution still plays a dominant role necessitating adequate 3D fabrication and metrology concepts.

These developments imply that with respect to metrology, there is a growing need for advanced two- and three-dimensional dopant/carrier characterization techniques that can work on a wide variety of inorganic semiconductor materials (and could potentially be upgraded toward organic semiconductors). Ideally, they need to offer high 3D spatial resolution (sub-nanometer), very good sensitivity, and repeatability (3%–5% variation) over a wide dynamic range (4–5 decades) and allow to probe-specific properties such as element distribution and/or electrical properties. Atomic force microscopy (AFM) and its associated electrical characterization concepts (SSRM, KPFM, SCM, C-AFM, SMM, etc.) represent one of the most viable approaches as it offers not only the required spatial resolution (at least in 2D) but also equally well the necessary sensitivity, quantification, reproducibility, and dynamic range. All these properties are crucially important as small changes (a few percent) in terms of carrier concentration can already induce very different device performance.

36.2 Carrier Profiling in Si with SSRM

36.2.1 Basic Concepts of SSRM

36.2.1.1 Principle

In order to probe the electrical properties of a semiconductor with high spatial resolution, several methods based on AFM [1] have been developed. In essence they consist of a conductive tip which is scanned over the area of interest, whereby one measures the current flowing through the sample when a bias is applied between the tip and a back-contact (see Figure 36.1). The different concepts differentiate themselves in the property of the probe and more importantly in the applied force and the tip-sample interaction (high pressure, low pressure, noncontact) which determine their applicability range.

In its low contact force version, it is often referred to as conductive AFM (C-AFM) [2] and is used to explore the spatially varying (tunnel) current targeting to relate this to locally varying material properties. It is, for instance, used to localize electronic defects or local thickness variations in thin dielectrica [3], for the analysis of the amorphous versus nanocrystalline content of thin silicon films for solar cells, in relation to their macroscopic performances [4], or to analyze the conductive filament formation mechanisms in ReRAM materials which are explored as next-generation nonvolatile memory devices [5]. Since the tunneling current is a function of the local resistivity, it also provides information on carrier concentrations and has been used to probe organic semiconductor as well [6]. Its main limitation relative to carrier profiling is the rather weak sensitivity of the tunneling current on the local resistivity. Typically, the current will vary on a linear scale when the carrier concentration varies over several orders of magnitude.

A more sensitive version is scanning spreading resistance microscopy (SSRM). In SSRM, a conductive probe is scanned in contact mode across the sample while a DC bias is applied between a back-contact on the backside of the sample and the tip (see Figure 36.1). However, the major difference relative to other AFM-based concepts is the very large contact force which is applied. Conceived at IMEC in 1994 by Vandervorst and Meuris [7,8] and implemented by De Wolf [9,10], SSRM is performed at a large pressure (GPa range) which leads to a good (near ohmic) electrical contact between the tip and the sample (Figure 36.2). The contact resistance is then dominated by the current spreading in the point contact and scales directly with the local resistivity (see Section 36.2.4). The resulting current varies now over a broad dynamic range (typically 10 pA to 0.1 mA) and needs to be measured using a logarithmic current amplifier.

The high force is detrimental for standard AFM probes and the successful implementation of SSRM is intimately linked with the emergence of wear-resistant probes based on doped diamond. Initially based on silicon probes coated with diamond (Section 36.2.1.3), further progress was facilitated by the development of full diamond probes [11,12] which have become the standard for high-resolution analysis for Si-based technologies [13].

The main characteristic of SSRM is, as previously explained, the high force (~μN) utilized to realize an intimate (ohmic-like) contact between the probe and the silicon sample such that the spreading resistance dominates the contact resistance (Figure 36.2). This originates from the fact, as established by Clarysse et al., that for such pressures, the probe punches through the silicon oxide and, more importantly, that within the underlying silicon, some small volume undergoes a plastic deformation and phase transformation toward a metallic phase thereby creating a virtual metallic contact [14]. Despite its high force and its thus intuitive negative impact on spatial resolution, we will show in Section 36.2.4 that this metallic contact provides SSRM with unique properties in terms of spatial resolution (1–3 nm), dopant gradient resolution (1–2 nm/dec),

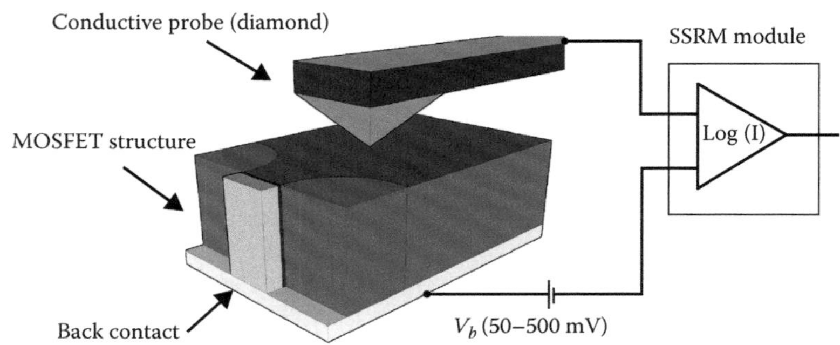

FIGURE 36.1 Illustration of the SSRM basic scheme.

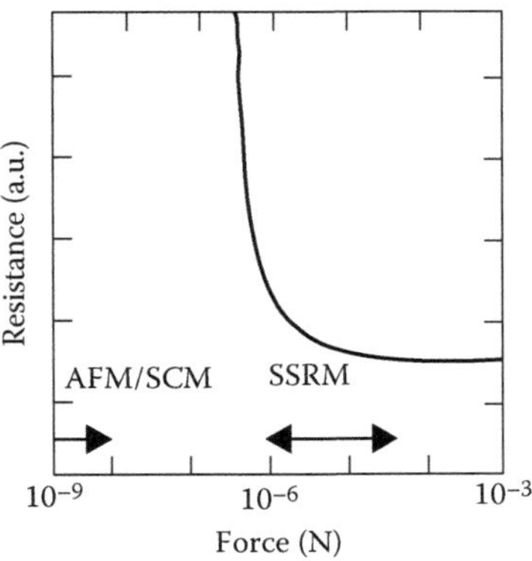

FIGURE 36.2 Resistance measured as a function of load.

signal-to-noise ratio (around 20), and reproducibility (5%–10%) [15], which outperform all other (low-force) AFM-based techniques.

Inevitably the high tip–sample interaction force leads to damage to the sample and the tips. This can be reduced by eliminating/reducing the shear force during tip movement as recently proposed in pulsed force SSRM (PF-SSRM) [16]. It was also recently shown that performing SSRM measurements under high-vacuum conditions reduces the required force to form the virtual metallic contact thereby leading to reduced sample damage, improved spatial resolution and signal-to-noise ratio (see Section 36.2.1.4) [17].

In theory, SSRM benefits an extremely large dynamic range (10^{14} to 10^{21} atoms/cm^3). In practice, this range is however limited in the highly doped range (>10^{20} atoms/cm^3) by conductivity of the diamond probe which leads to a resistance measured in series with the spreading resistance. Different from the resistance of a metallic probe, the resistance of the AFM diamond probe is at present not negligible (>1 kΩ) and research is still ongoing to improve the doping efficiency of the diamond probes. In lowly doped areas (<10^{17} atoms/cm^3 for polished samples and <10^{15}–10^{16} atoms/cm^3 for cleaved samples), the SSRM measurements are disturbed by the presence of surface charges as will be studied in Section 36.2.2. Note that this influence is not specific to SSRM. It is related to the sample-sectioning process and thus also present with other SPM techniques (i.e., KPFM or SCM) [18].

36.2.1.2 Sample Preparation

36.2.1.2.1 Cross-Sectioning Procedures (Polish, Cleave, FIB)

Probing the two-dimensional carrier profile in devices requires measurements on the cross section of the sample, implying that making the cross-sectional sample becomes the first important step. Although SSRM is less sensitive to the sample surface than, for instance, scanning capacitance microscopy (SCM) [19], better reproducibility and reduced noise levels can be obtained with improved sample preparation. It has, for instance, been observed that a very flat surface enables a more stable contact and a reduction of the force needed for a stable electrical contact. A reduction of this force is very important as the latter is responsible for the probe and sample damage. Moreover, a lower force generally leads to a reduced "effective" tip radius and thus a better spatial resolution (Section 36.2.2.3).

Optimum sample preparation for SSRM measurements is often based on polishing procedures (using Al_2O_3 and/or colloidal silica) targeting a final RMS-roughness value between 0.2 and 0.4 nm whereby care needs to be taken to avoid rounding effects and doping level sensitive polishing. A short HF exposure (a few seconds) just before the SSRM measurement also allows a small decrease of the force (about two times) due to the removal of the native oxide but may introduce some topography artifacts and thus has to be used carefully. Despite the optimized polishing procedures, the cross sectioning does to the generation of extrinsic surface states (SS) (linked to in-depth damages in the near surface) that may affect the apparent carrier concentrations in lowly doped areas and even lead to junction shifts (see Section 36.2.2.2).

For site-specific analysis, procedures have been developed to cross section the SSRM samples using a micro-cleave tool which provides a positioning accuracy ∼0.3 μm. The quality of the cleaved surface is excellent for high-resolution SSRM measurements but of course the protocol only works when sufficient "crystalline" material is available. For measurements which require an even higher accuracy or limit the measurement plane to the back-contact distance, focused ion beam (FIB) surface preparation can be used. Starting from a diced, polished, or cleaved section, FIB can be used to flatten and/or further erode the surface plane to be analyzed (Figure 36.4). In order to minimize the influence of the FIB Ga beam, one has to use low beam energy (2–5 keV) and angle (<1°). Analysis of devices with width dimensions in the other of 60 nm has been reported using this method [20].

36.2.1.2.2 Back-Contacting

A crucial point in the SSRM measurement is that the measured value is dominated by the point contact and not by the series resistance toward the back-contact. Therefore, in order to ensure a good collection of the spreading current for every position of the tip on the analyzed cross section, we need to place a back-contact at each of the layers studied. As a manual back-contact may lead to problems with small isolated areas as the gate, standard procedure is to realize back-contact is realized using FIB. Depositing a metal layer (TiW + Au) and using a FIB to excavate a local trench and to fill it with metal (Pt), one is able to create a local back-contact for each layer of the cross-sectioned surface (Figure 36.3). If needed,

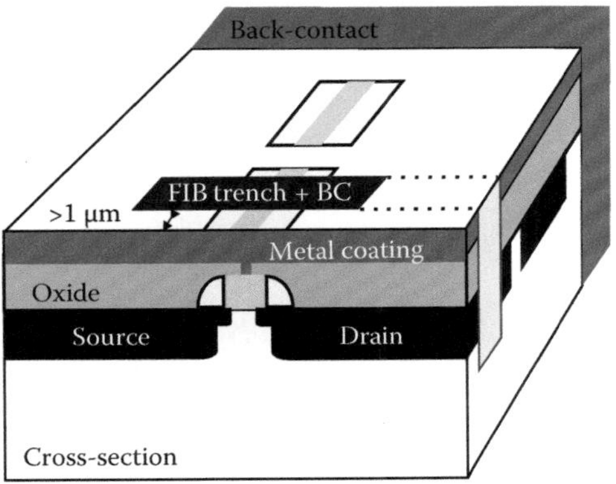

FIGURE 36.3 Scheme of the FIB back-contacting.

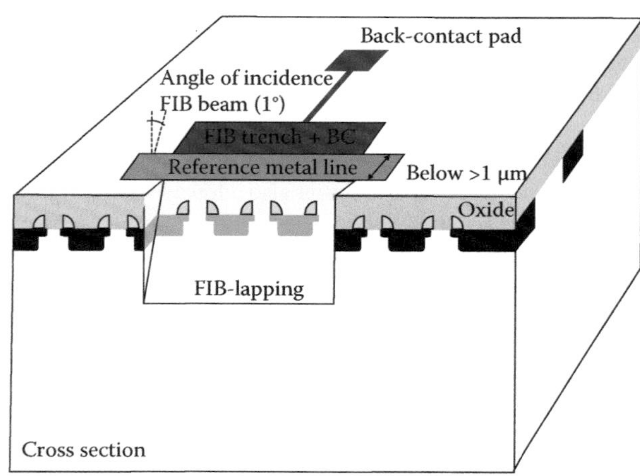

FIGURE 36.4 Scheme of the FIB surface lapping.

FIB can also be used to create a back-contact pad at longer distance (200–400 μm) replacing the metal layer (Figure 36.4). In all events, care should be taken that the Ga/Pt from the FIB does not contaminate the cross-sectional surface to be measured.

36.2.1.3 Tips for SSRM

As indicated, at the heart of SSRM lies the formation of a pseudo-metallic contact by applying a high force. Whereas on soft semiconductors, this can be done with metallic tips without destroying them immediately [21]. SSRM application to Si requires the use of diamond tips as these are essential to sustain the high forces in use [22]. There exist two kinds of diamond probes (Figure 36.5). The first one is the standard Si AFM tip with an additional diamond coating. Basically, a thin layer (100–200 nm) of boron-doped diamond (3–3.3 ppm) is grown by CVD on top of n-doped silicon probes. Those probes are commercially available but suffer from different problems:

- When applied to Pt, the probe resistance typically lies above 1 kΩ for a probe with a radius above 5 nm which suggests that the measurement is limited by the resistivity of the diamond film. This becomes visible as a saturation in the highly doped areas (>10^{20} atoms/cm^3) of the calibration curves.
- Due to the final diamond deposition process, the outer surface of the diamond-coated tips is not flat but covered with sharp diamond crystals. The electrical contact is realized through one of those grains. Note that, since the contact is defined by the facetted diamond crystal, the effective electrical tip radius is better than the overall radius of curvature visible in the image (i.e., 3 nm thick buried oxide layers have already been detected with a factor 2 resistance increase) but it is usually difficult to predict. Moreover, the probability to have a multiple contact is not negligible.
- As the silicon tip itself has a rather high aspect ratio, cleavage of the tip due to the high shear forces exerted is frequently observed. Increasing the diamond film thickness can reduce this effect somewhat, unfortunately at the expense of the tip radius as the thicker coating tends to become smoother (less sharply up pointing crystals) [23].

In order to overcome these limitations, low aspect tips have been developed [22]. They are molded pyramidal tips and their outer diamond surface (deposited first within the mold) is smoother (see Figure 36.5b). Recent improvements of this process [12] have demonstrated a significantly enhanced spatial resolution (around 1 nm) at moderate forces (1–5 μN) [24] (see Section 36.2.4 for details). Although they are superior in terms of spatial resolution, tip lifetime seems to be less as compared to the commercial diamond probes. Tip failures primarily occur due to small diamond grains being removed from the solid pyramid. Clearly an improved diamond deposition process is required leading to a stronger cohesion of the diamond film. Similar to the diamond-coated Si tips, these tips also suffer from a limited conductivity of the diamond. Enhancing the boron incorporation and activation remains an active area of research.

36.2.1.4 Measurement Environment

Recently, a high-vacuum version (around 10^{-5} torr) has been developed [25] offering improved performance in terms of

(a)

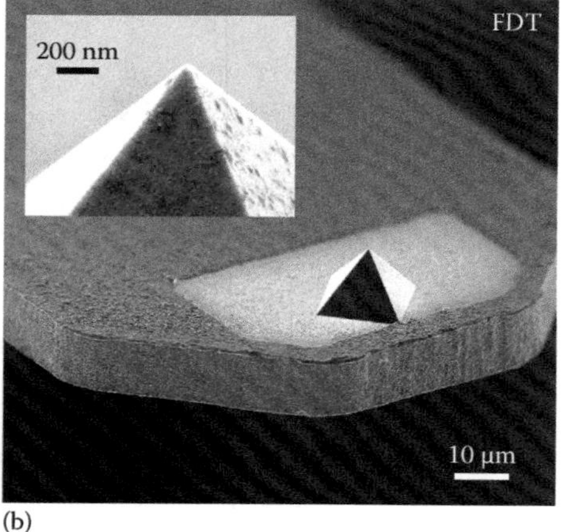

(b)

FIGURE 36.5 SEM views of commercial coated diamond (a) and of in-house molded full diamond tips (b). Lower aspect ratio, smaller radius of curvature (see insets), and smoother surface may be observed.

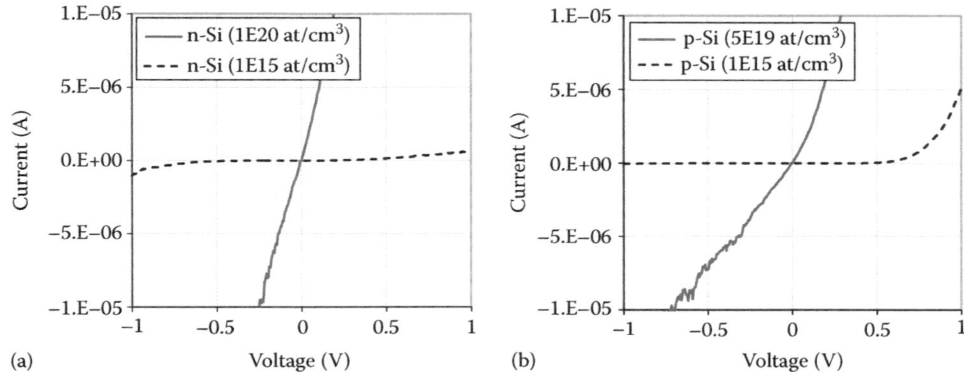

FIGURE 36.6 *I–V* curves for (a) n- and (b) p-type Si measured on a cross-sectioned surface.

resolution and sensitivity as will be seen later (Section 36.2.4). Measurements in high vacuum can be performed at lower force (typically four times lower) leading to reduced damage of tip and sample, and thus improved repeatability and increased tip lifetime. As demonstrated in [26], the improved performances under high vacuum are mainly linked to the removal of the water meniscus present in air ambient which affects the pressure distribution in the underlying sample and prevents the formation of a very localized electrical contact.

36.2.2 Physics of SSRM

36.2.2.1 Introduction

In order to understand the properties of SSRM (spatial resolution, dynamic range, sensitivity, quantification, etc.), it is important to gain more insight in the nature of the nanocontact between a probe and a semiconductor sample. The latter involves mechanical as well as electrical aspects which will be discussed in the later sections prior to proceeding toward a more exhaustive nanocontact modeling.

36.2.2.2 Experimental Observations

36.2.2.2.1 Electrical Properties of the SSRM Nanocontact

In order to understand the finer details of the SSRM contact, we focus on *I–V* curves collected on the (cross-sectioned) semiconductor surface in a non-scanning mode. From such local *I–V* curves (see Figure 36.6), one notice that the SSRM nanocontact varies from an ohmic-like shape (in highly doped areas) to a rectifying shape (in lowly doped areas). Experiments [27] have furthermore shown that the SS induced by the sample preparation have an influence on the *I–V* curves and basically reduce the current. This effect is particularly pronounced in lowly doped p-type areas (see Figure 36.7). The impact of SS is also observed on the apparent location of junctions. Prior to discussing this, it is important to stress that SSRM measures the electrical junction (EJ) depth rather than the metallurgical junction. For the junction positions measured with SSRM on highly doped materials (source/well or drain/well implants), these are in perfect agreement with

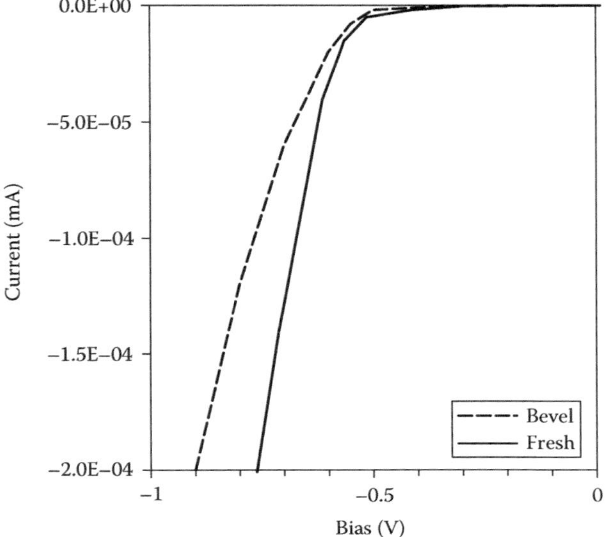

FIGURE 36.7 *I–V* curves on lowly doped (10 Ω cm) p-type Si measured on fresh and beveled surface. (From Clarysse, T. et al., *JVST B*, 21(2), 729, 2003.)

the expected theoretical zero-field position and a very good agreement with SIMS can be obtained* (when the effects of mobile carrier diffusion are included in the comparison). However, when studying p^+n junctions with a lowly doped side (source/substrate and well/substrate) a large shift of the SSRM EJ position away from the zero-field position toward the surface is however observed (see Figure 36.8), and for n^+p junctions, a complete disappearance of the junction peak is seen (see Figure 36.9).

36.2.2.2.2 Mechanical Properties of the SSRM Nanocontact

Radius of curvature: The physical radius of curvature of the AFM probes can be determined using direct SEM observations or through measurements on $SrTiO_3$ samples [28]. Whereas early generations of diamond-coated and full diamond probes showed

* It should be stressed that SSRM measures the electrically active profile and thus the EJ rather than the metallurgical junction (MJ) measured by SIMS.

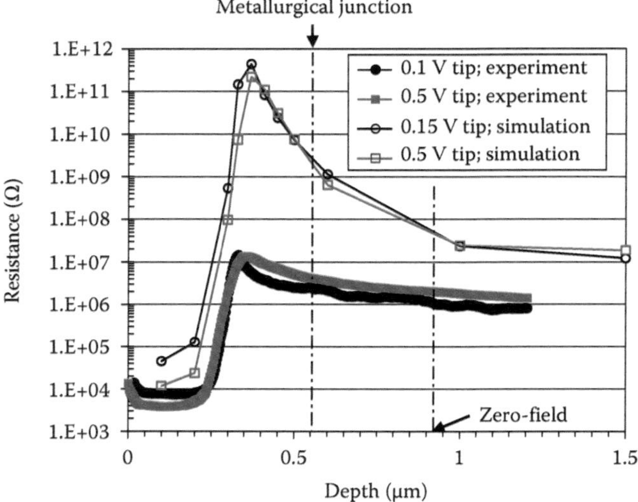

FIGURE 36.8 Resistance profiles for $p^{++}n$. Comparison between experiment (SSRM on cross section) and simulations (with the new contact model).

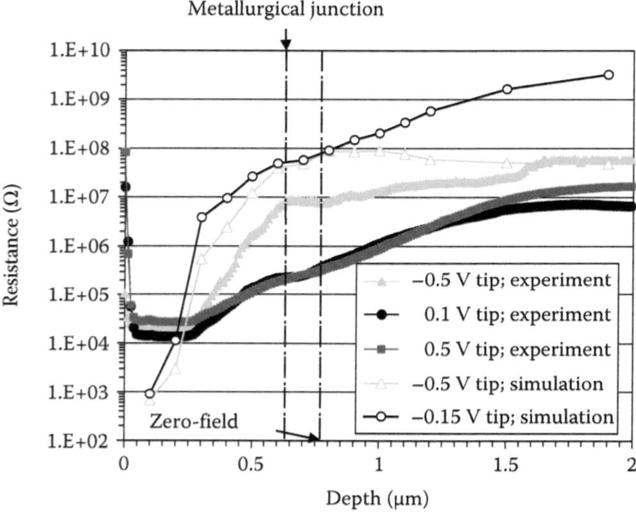

FIGURE 36.9 Resistance profiles for $n^{++}p$. Comparison between experiment (SSRM on cross section) and simulations (with the new contact model).

radii varying between 20 and 50 nm, more recent generations of full diamond probes have radii of ~10 nm. The actual radius of contact may however be much smaller as the outer apex of the used tips is composed of nanograins of diamond (≈2 nm radius) as observed with transmission electron microscopy (TEM) (Figure 36.10).

Spatial resolution: The SSRM spatial resolution is difficult to measure but clearly appears drastically better than the radius of curvature of the probe. Experiments that have been realized on dedicated buried oxide structures (see Section 36.2.4), as well as on device structures, demonstrate a SSRM spatial resolution clearly lying in the 1–3 nm range [24].

36.2.2.3 Atomistic Simulation of the Nanocontact

Since the properties of SSRM originate from the high force effects inducing the metallic contact formation, any model of the SSRM nanocontact on Si (as will be discussed in Section 36.2.2.4) will have to take into account the impact of the stress generating a phase transformation locally below the tip. Initially [29], finite element method (FEM) simulations were performed to study the impact of the diamond tip on silicon during SSRM measurement process (modeling it as large pressure indentation). These results already have shown the possibility to obtain a metastable β-tin pocket under the SSRM tip with radius much smaller than the radius of curvature of the tip. The validity and accuracy of these FEM simulations for stress distributions in the nanometer range should, however, be regarded with a lot of caution. Hence, more refined calculations using molecular dynamics (MD) simulations have been used to investigate the deformation of Si under nano-indentation, the formation of new Si crystallographic phases and, in particular, the metastable phases like β-Si and BCT5-Si [30–32]. The MD method allows the simulation of the phase transition without any assumption on the nature of the phases. The resulting structures only depend on the thermomechanical conditions and on the interaction forces between atoms, as defined by the potential function. To avoid boundary effects, large silicon samples were used. Boundary atoms and thermostat atoms were arranged to surround the Newtonian atoms of silicon to eliminate the rigid body motion and to conduct heat [30]. Interactions among silicon atoms were described by Tersoff potential [33,34] and among silicon and diamond atoms were described by a modified Morse potential.

The high-resolution (atomic) TEM inspection of the tip apex performed on both pristine (Figure 36.10a) and used tips (Figure 36.10b corresponding to a tip used in scan mode and presenting a good spatial resolution) reveals that the molded diamond tips used in our experiments are undergoing an initialization (roughening) mechanism. We speculate that due to the shear forces, non-diamond parts (like silicon carbide) at the outer surface of the molded tip (corresponding to the first layers grown during the diamond deposition) are most probably removed while the tip is scanned across the silicon sample (initialization process). It is not yet completely clear whether simple contact indents could be sufficient to initialize the tip or if scanning (implying the presence of a large shear stress) is needed. We have however been able to establish one-to-one correlations between SSRM measurements in point-contact mode and MD simulations in terms of evolution of the Si-II (or β-tin Si) metastable phase (Figure 36.11) and depth of the residual indentation marks (Figure 36.12) for a radius of curvature of 7.5 nm [35]. This tends to prove that scanning may be needed to initialize the tips.

Within this work, we focused on ⟨100⟩ silicon sample indented using hemispherical diamond tips with a radius varying between 2 and 10 nm to match with the radius of curvature observed with TEM for both unused and initialized tips.

As the different phases involve a different number of Si atoms interacting with each other, we plot the 3D distribution of Si

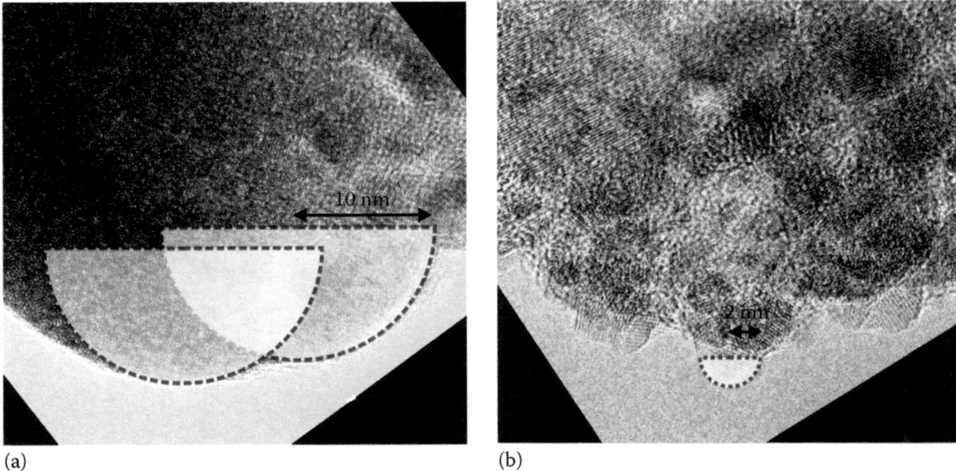

FIGURE 36.10 TEM pictures of the apex of unused (a) and initialized (b) full diamond tips. Nano-protrusions of diamond are present at the extremity of the initialized tip.

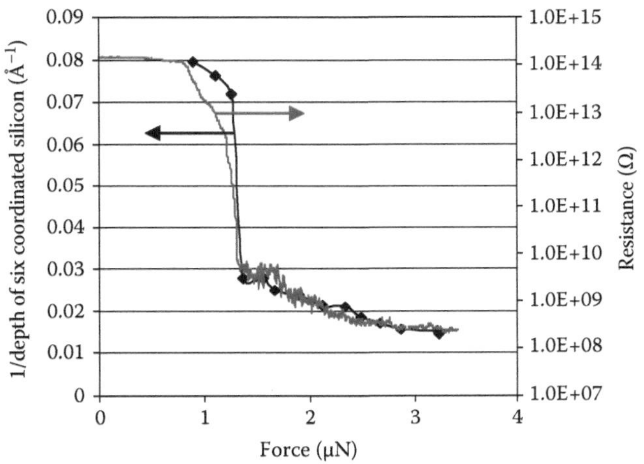

FIGURE 36.11 Correlation between decrease in SSRM resistance (gray) and variation of the depth of six coordinated atoms extracted from MD (black with squares) when indentation load is increased. (From Mylvaganam, K. et al., *Nanotechnology*, 20(30), 305705, 2009.)

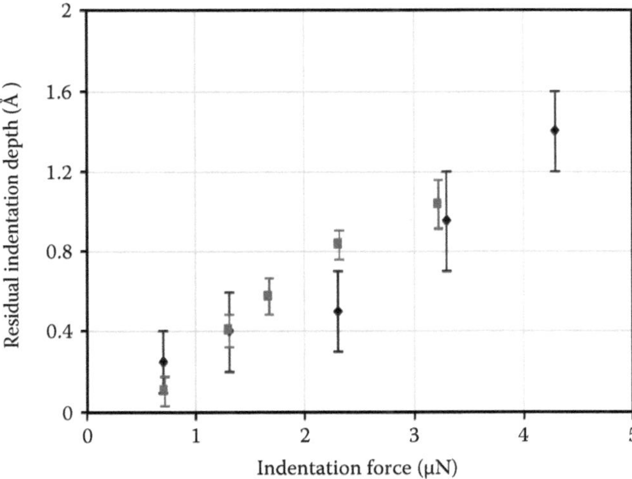

FIGURE 36.12 Comparison of SSRM (light gray) and MD simulation (dark gray) residual indentations for point-contact indents. (From Mylvaganam, K. et al., *Nanotechnology*, 20(30), 305705, 2009.)

atoms labeled with respect to their number of nearest neighbors (coordination number). As a criterion to define a neighboring atom, we look for Si-atoms whose interatomic distances are <2.58 Å [30–37], In Figure 36.13, a 2D section through the center of the indenter is presented for indenters with a radius of curvature of 7.5, 5, and 2.1 nm, respectively. This last dimension does correspond to the grain size of the initialized probe in scan mode (Figure 36.10b). One can observe that during indentation, a part of the diamond lattice of silicon (with a localized sp^3 molecular orbital) is transformed into different high-pressure phases: one phase with six nearest neighbors (Si-II, known as β-tin Si, and a new six coordinated phase Si-XIII, with an sp^3 d2 molecular orbital in which d2 is corresponding to nonlocalized electrons leading to a metallic-like behavior), and one phase with five neighbors and known as BCT5-silicon (with an sp^3 d1 molecular orbital and thus also nonlocalized electrons). The investigations in [32] have also identified that the Si-II and the Si-XIII phases are surrounded by a five coordinated phase (BCT5-Si). These metastable phases [32] both have a metallic body-centered tetragonal structure, β-silicon being known to be the more conductive. The formation of these metallic-like pockets below the tip (which can have a radius smaller than 1–2 nm for an initialized probe) is crucial for SSRM as these determine the spatial resolution of the technique and its linear sensitivity to the local resistivity.

36.2.2.4 Nanocontact Modeling

36.2.2.4.1 Classical Ohmic Nanocontact Modeling

Although the observations presented in Section 36.2.2.2.1 indicate that the SSRM nanocontact is not perfectly ohmic, we can in a first approximation nevertheless attempt to model it as a classical

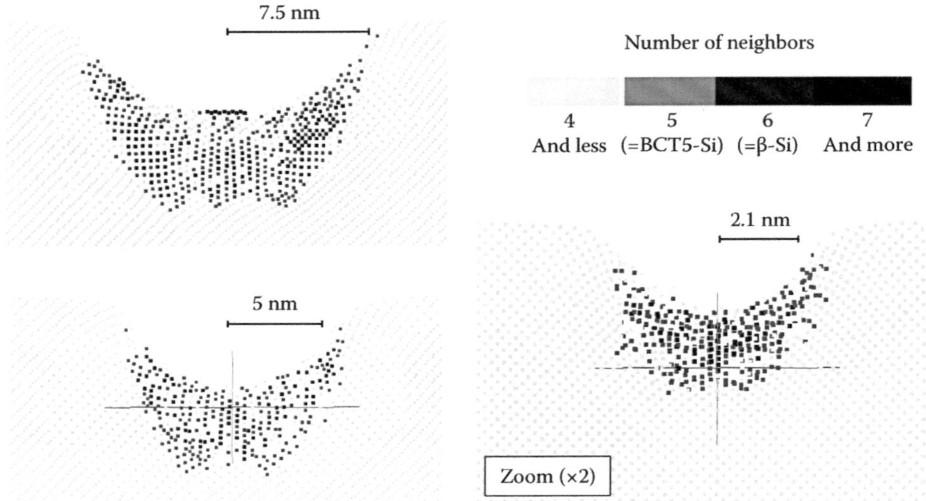

FIGURE 36.13 Cross-sectional view of the location of atoms at the maximum indentation depth for radii varying between 7.5 and 2.1 nm. The different colors do correspond to the amount of neighbors for every atom. The β-tin do correspond to the atoms in medium gray. (From Eyben, P. et al., *JVST B*, 28(2), 401, 2010.)

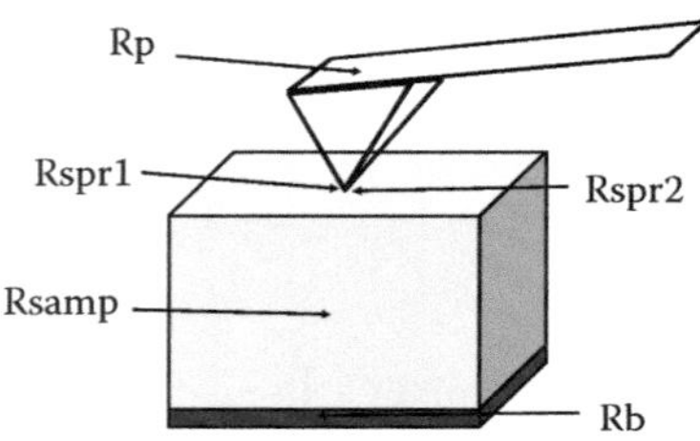

FIGURE 36.14 Schematic representation of the various resistance components involved in SSRM measurements.

ohmic nanocontact. This will be a reasonable first approximation for relatively highly doped materials (above 10^{18} atoms/cm^3). The total measured resistance R_{tot} in SSRM experiments (Figure 36.14) includes the series resistance of the probe R_p, the probe nanocontact resistance (also called *spreading resistance inside the probe*) R_{spr1}, the sample nanocontact resistance R_{spr2}, the series resistance within the sample R_{samp}, and the back-contact resistance R_b:

$$R_{tot} = R_p + R_{spr1} + R_{spr2} + R_{samp} + R_b.$$

The resistance of the probe (R_p) and of the sample (R_{samp}) is given by the Pouillet equation $R = \rho L/A$, where ρ is the resistivity of the material, L its length, and A its area. The resistance of a macroscopic circular constriction (or nanocontact) separating two homogeneous conductors is described by the Maxwell formula [38] $R_{spr} = R_{spr1} + R_{spr2} = \rho_1/4a + \rho_2/4a$, where ρ_1 and ρ_2 are the resistivities of the materials in contact and a is the radius of the contact area (also called *electrical radius*).

In appropriate conditions (see requirements on probe and back-contact in Section 36.2.1), the dominant resistance in SSRM measurements is the sample nanocontact resistance (or spreading resistance) and we obtain

$$R = \frac{\rho_{samp}}{4a}.$$

The Maxwell formula is however valid only when the radius (a) is large compared to the mean free path of electrons and holes (λ). In [13], the mean free path of electron and holes in silicon has been calculated for different concentrations of arsenic, phosphorus, and boron assuming simple mobility equations without electric field saturation [39]. The dependence of the mean free path on carrier type (electron or holes) and doping concentration (10^{14} to 10^{20} atoms/cm^3) is evidenced (see also [40]). In the extreme situation of a contact area reduced to a few atoms, the electron transport is ballistic and the conductivity is described by the Landauer–Büttiker formula. $G_c = \frac{2e^2}{h}\sum_{i=1}^{N_c} T_i$, where N_c is the number of conductance channels through the contact and T_i is the transmission coefficient of the ith channel [41]. The number of conductance channels is proportional to the contact area (defined in units of the Fermi wavelength λ_F). The transmission coefficients are close to unity when no scattering sites such as impurities or grain boundaries are involved at the constriction. Quantized conductance can be observed for small contacts when only a few channels are involved. When the contact radius a is large compared to the Fermi wavelength λ_F, the resistance converges to the Sharvin formula $R_{Sharvin} = \frac{h}{2e^2}\frac{\lambda_F^2}{\pi^2 a^2}$ [42,43] that can also be expressed in fully classical terms by substituting λ_F with ρ and using the electron mean free path λ:

$$R_{Sharvin} = \frac{4\rho\lambda}{3\pi a^2}.$$

As illustrated in [13] the Sharvin formula is dominant for small contacts (~1 nm), whereas it becomes negligible compared to the Maxwell formula for large contacts (~100 nm). In the transition region, both formulas give resistances with the same order of magnitude. The increase in the quality of the measurement conditions (better probes and lower forces used) as well as the latest experiments on electrical radius tend to suggest that the SSRM nanocontact size lies more in the Sharvin regime (at least in lowly doped areas). Please note that the Sharvin resistance equation has the same attractive property as the Maxwell spreading resistance equation: the resistance increases monotonically with the sample resistivity, providing a highly sensitive method for carrier profiling.

36.2.2.4.2 Schottky Contact and Surface Modeling

As already explained in Section 36.2.2.2, different experimental observations reveal that the ohmic nanocontact modeling is not fully satisfactory. In order to understand the discrepancies between the classical ohmic theory and the experimental observations, a more extended physical model of the contact and of the surface involving a Schottky-like contact with tunneling and SS is introduced (Figure 36.15 and Table 36.1). Using this model (implemented in a device simulator [44]), one can now calculate *I–V* curves, calibration curves, and even full SSRM profiles.

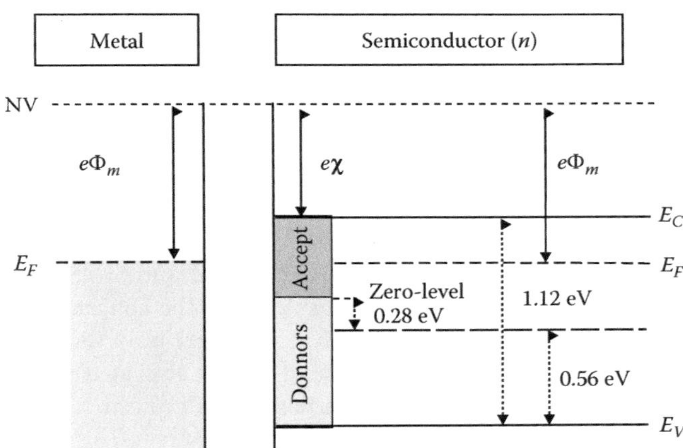

FIGURE 36.15 Contact model for a n-type Si where $e\varphi_m$ is the metal work function, $e\varphi_s$ the semiconductor work function, and $e\chi$ the semiconductor electron affinity. The SS zero level is situated above the mid-band gap.

TABLE 36.1 Summary of the Different Parameters Involved in the Contact Model

	Parameters	Values
Surface states	Zero level	0.28 eV above mid-band gap energy
Schottky contacts	SS concentration	10^{13} eV/cm²
	Schottky barrier $E_b = e\varphi_m - e\chi$	0.36 eV
	Tunneling distance	10 nm
Contact size	Tip radius	20 nm

The introduction of the Schottky contact is based on the observation of rectified *I–V* curves (see Figure 36.6). The necessity to introduce this Schottky contact is not surprising given the nature of the materials involved in the SSRM nanocontact (diamond-coated probe, β-tin phase of silicon, and silicon sample). None of those materials are real metals. The barrier height $E_b = e\varphi_m - e\chi$ has been chosen to obtain a rectifying contact ($e\varphi_m < e\varphi_s$) for all p-type impurity concentrations and a weakly rectifying contact ($e\varphi_m > e\varphi_s$) for most n-type impurity concentrations. In highly doped areas, the ohmic-like *I–V* curves observed experimentally (with large currents) may only be explained by the presence of a dominant tunneling current through the thin potential barrier of the contact. The sample preparation has an important impact on the total SS concentration and their distribution.* It is indeed common knowledge that the surface charges create variations in carrier concentrations at the surface changing the *I–V* curves, shifting the junction, and even creating inversion layers (peak disappearance) [45]. For the SS in our model, we have chosen a zero level above the mid-band gap in order to account for the experimental observations that SS have a lower impact on n-type silicon (Table 36.1). The latter is also in agreement with classical SS modeling [46]. Using this model, one is able to obtain a good agreement with experimental *I–V* curves on highly and lowly n- and p-type Si (Figure 36.16), for calibration curves on n- and p-type Si (see [13]) and for EJ delineation (Figures 36.8 and 36.9).

36.2.3 Quantification

36.2.3.1 Calibration Curves

36.2.3.1.1 Concept

In principle, as pointed in Section 36.2.2.4.1, a straight conversion of spreading resistance (*R*) to local resistivity (ρ) can be made using the ideal (flat, pure ohmic contact) formulas $R = \rho/4a$ (Maxwell) or $R = 4\rho\lambda/3\pi a^2$ (Sharvin) depending on the tip radius (*a*). The latter already provides a reasonable accuracy. As in practice a true straight-line dependence is never observed, among others due to the influence of the nanocontact nature (Schottky), the SS at the surface, and the probe resistance. Therefore, a more pragmatic approach is used to perform quantification based on a lookup procedure on calibration curves. Provided a high reproducibility is maintained (see Section 36.2.4.2), this assures quantification accuracy as good as 20%–30%.

* At the surface of the silicon sample, the number of neighbor atoms is reduced (three instead of four) and each atom shows a nonsaturated orbital called *dangling bond*. Those dangling bonds generate two-dimensional energy bands in the band gap that form the so-called *surface states* (SS). SS are also created through the reconstruction of the surface (that can be compared to the Peierls' transition). Besides these "intrinsic" states, "extrinsic" states are also present and play an important role in SSRM. They are typically linked to the presence of foreign atoms (like carbon or oxygen) and of irregularities and defects at the surface.

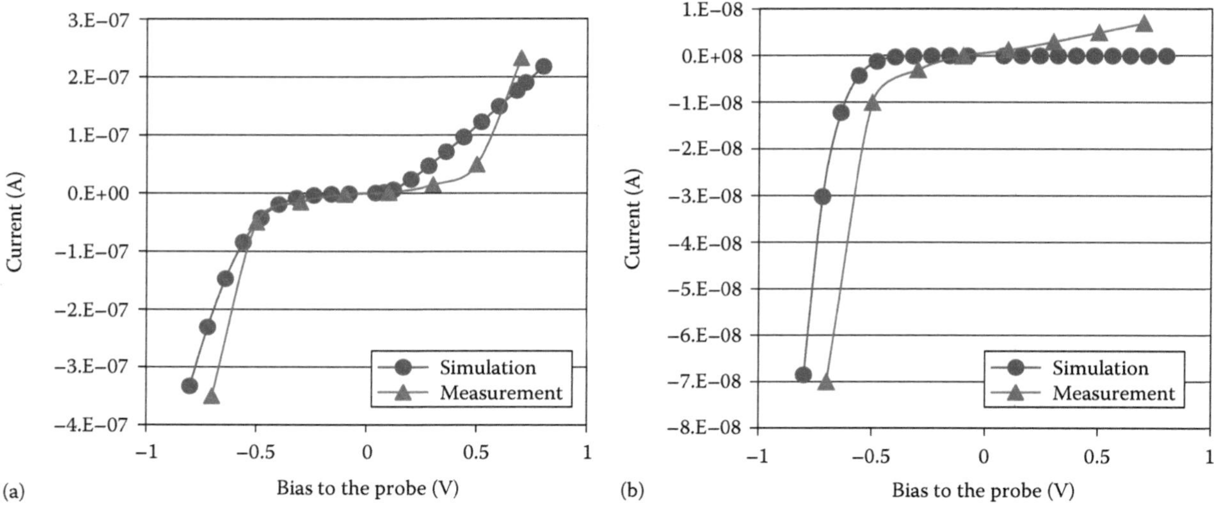

FIGURE 36.16 Comparison between simulated and measured *I–V* curves in lowly doped n-type (10.5 Ω cm) (a) and p-type (5.2 Ω cm) (b) homogeneous Si samples. A qualitative agreement may be observed.

36.2.3.1.2 *Implementation*

As the probes that are used (to measure on the unknown structures) have varying characteristics, the first step of the quantification procedure is the acquisition of two calibration curves (for both impurity types) for each probe. Those calibration curves are constructed by collecting the resistances measured on different homogeneous calibration samples with a well-known resistivity.

Whereas originally separate samples of different resistivity levels were used to establish the calibration curves (5–6 for each impurity type), dedicated epitaxial test structures with a 1D staircase structure are now used to obtain a complete calibration curve from a single scan [47]. The latter reduces drastically the time required to establish a calibration curve. In these structures, the resistivity levels are ranging from 1×10^{-3} to 10 Ω cm (corresponding more or less to the concentration range of interest). In order to eliminate the manual extraction of the resistances corresponding to each staircase step, a dedicated program, called *MicroQuanti*, has been developed to automatically build the calibration curves based on the raw data collected for the staircase calibration structures (Figure 36.17).

For further quantitative refinements, a more complex (iterative) procedure is required to account for the current spreading effects induced by the dopant gradients nearby.

36.2.3.2 Quantification Procedure and Comparison with 1D Techniques

The quantification with *MicroQuanti* is now based on converting the resistances into carrier concentrations based on the actual calibration curves. Using logarithmic interpolation on these curves, one can determine resistivity maps from the spreading resistance maps. The resistivity maps can be converted into active dopant (or carrier) concentration maps based on (known or assumed) conversion curves for mobility [48]. As separate curves for n- or n-type need to be used, algorithms are included in this software to detect the different materials (Si, SiO_2, silicide, metal, etc.) as well as the position of the electrical junctions [49]. As a result, we are able to identify the different areas (gate, oxide, source/drain, substrate, etc.) of the transistor on the 2D SSRM resistance map and we can apply the proper quantification procedure and calibration curves in each of them (see Figure 36.18).

By extracting 1D sections from these maps, one is able to compare the quantified SSRM profiles with, for instance, secondary ion mass spectroscopy (SIMS) results which is nowadays the reference technique for one-dimensional dopant profiling. Major difference between SSRM and SIMS is the fact that SIMS measures the position and the concentration of the dopant ions whereas SSRM measures only the active carrier profile. Even in the case of a fully activated profile, the dopant and the carrier distribution will not be identical due to the out-diffusion of the mobile carriers (cf. difference between the EJ and the MJ). The classical procedure to compare SIMS and SSRM results is thus to calculate the carrier profile from the SIMS dopant profile solving the Poisson equation and then to compare this calculated profile with the SSRM quantified profile. The latter is done in Figure 36.19a and b, corresponding, respectively, to $n^{++}p^{+}$ and $p^{++}n^{+}$ structures, showing the very good agreement in terms of concentration level and EJ position as observed between SIMS and SSRM.

The spreading resistance probe (SRP) is a carrier profiling tool similar to SSRM, be it confined to 1D profiles as SRP measurements are realized on beveled samples with two widely separated (30 μm) probes [15]. Whereas one would expect SSRM and SRP to give the same carrier profile, the beveling procedure in SRP is responsible for a shift of the junction toward shallower depths (called carrier spilling) [51]. Comparison between

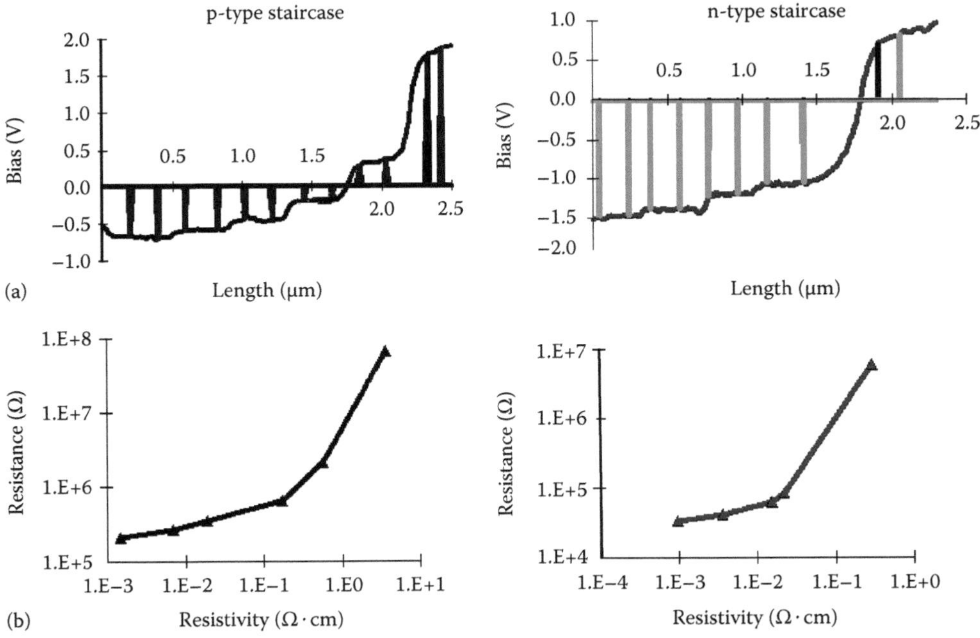

FIGURE 36.17 Raw resistance data (a) and extracted (b) calibration curves based on n-type and p-type staircase structures as generated by *MicroQuanti*. (From Eyben, P. et al., *Solid-State Electron.*, 71, 69, 2012.)

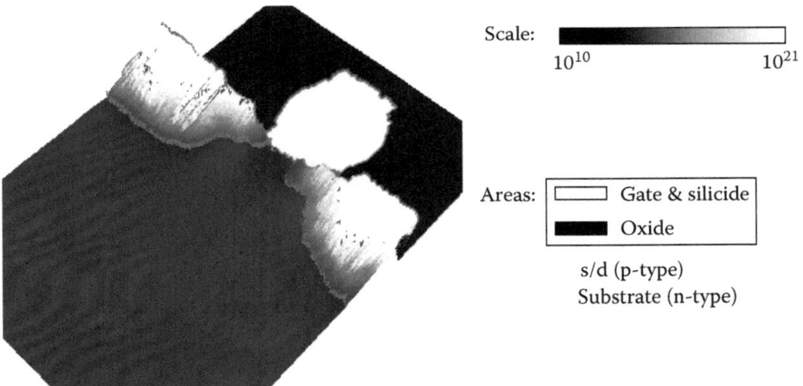

FIGURE 36.18 3D view of a 2D carrier map as generated in *MicroQuanti* for typical p-MOSFET with 1 keV—0° tilt extension implant and a 1350°C high power flash anneal. Junction and edge are semiautomatically detected and gate, oxide, source/drain, and substrate areas (grayscale color bar) delimited. (From Eyben, P. et al., *Solid-State Electron.*, 71, 69, 2012.)

cross-sectional SSRM or SIMS and on-bevel SRP profiles are thus invalid unless SRP is corrected for carrier spilling effects (which is rarely done). For a detailed discussion, see [52]. The good agreement between SIMS, SRP, and SSRM observed in Figure 36.19 in terms of carrier concentration confirms the high spatial resolution and quantification accuracy of SSRM. The shallower junctions of SRP are representative of the carrier spilling effect (that could not be completely corrected). Small differences between SIMS and SSRM exist in the most heavily doped part of the profiles to be interpreted as a carrier concentration being slightly lower than the dopant concentration extracted from SIMS. The SRP results confirm that this difference originates from a not fully activated implant.

36.2.4 Performance

36.2.4.1 Resolution

In order to assess the performance of the SSRM technique, it is essential to arrive at a unique and verifiable definition of resolution. The definition of the resolution concept is however not straightforward for carrier profiling techniques and has to be divided into three main parameters: the spatial resolution, the dopant gradient resolution, and the concentration sensitivity. Different from the definitions used in optical microscopy, the simple length-scale concept is not applicable without specifying additional details of the dopant structure or turns out to be incomplete as an assessment parameter.

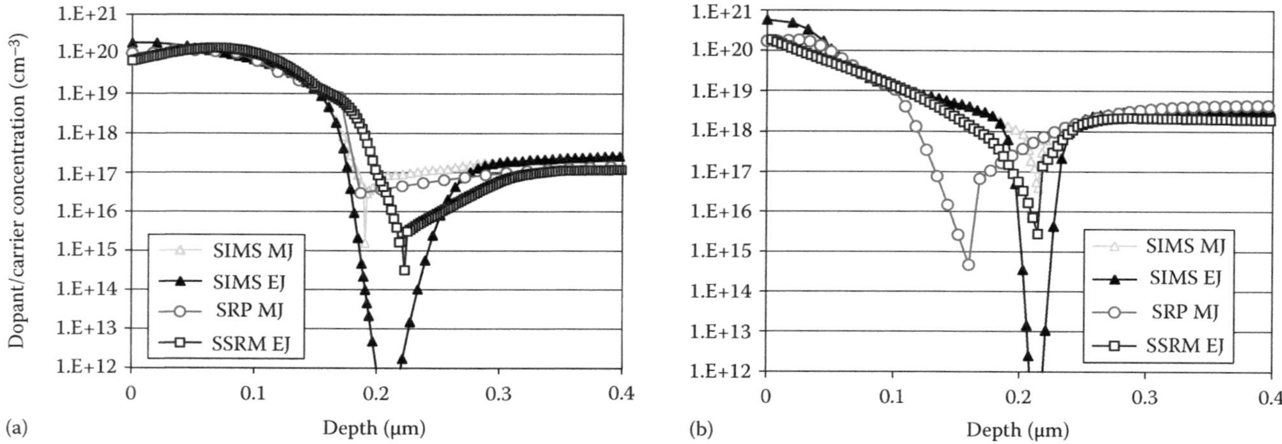

FIGURE 36.19 Comparison between SIMS, SRP, and SSRM dopant/carrier profiles on (a) $n^{++}p^{+}$ and (b) $p^{++}n^{+}$ structures.

A simple length scale can be used when discussing the ability to resolve the precise location of a p–n junction, the thickness of an oxide, etc. For dopant profiling applications, an equally (even more) relevant number is the resolvable dopant gradient (nm/dec), as a length scale (nm) has no meaning unless one specifies the concentration difference between the points one is trying to resolve. To complicate the situation even further, the dopant gradient needs to be specified around a particular doping level as the "natural" out-diffusion of the mobile carriers (~Debye length) will limit the detected dopant gradient at low concentrations regardless of the intrinsic performance of the analysis technique. Finally, the third parameter, which enters into the assessment, is the concentration sensitivity, which can be viewed as the ability to detect marginally different concentration levels.

For AFM-based techniques, the length-scale resolution will, among others, be set by the tip radius whereas the dopant gradient will also depend on the sampling volume and the concentration sensitivity. Concentration sensitivity is set by the intrinsic properties of the technique and the signal-to-noise ratio occurring in its actual implementation.

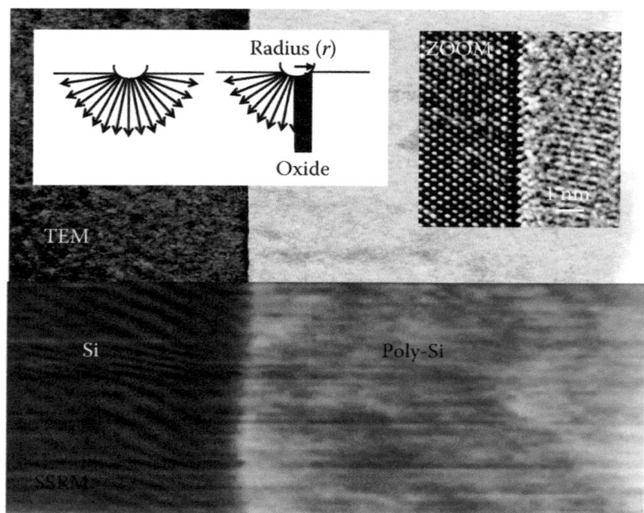

FIGURE 36.20 Two-dimensional X-TEM (top) and SSRM spreading resistance (bottom) maps for the buried oxide test structure. Inset "zoom" is presenting a high-magnification X-TEM view of the buried oxide assessing a thickness between 0.3 and 0.5 nm. Inset "concept" is a schematic representation of the resolution criterion.

36.2.4.1.1 *Spatial Resolution*

For the evaluation of the spatial resolution of SSRM, a special test structure consisting of a thin oxide layer in between a highly doped silicon-implanted layer and a highly doped polysilicon layer (both with active dopant concentration level above 1×10^{20} atoms/cm^3) can be used. The (length-scale) resolution is then defined as the smallest oxide layer which can still be resolved in a SSRM measurement.

Samples with oxide layers between 0.3 and 0.5 nm have been used in this study whereby the thickness of the buried oxides is verified using cross-sectional TEM measurements (see inset in Figure 36.20). Compared to abrupt dopant transitions grown in silicon (as for instance, CVD-grown spike structures), these test structures have an important advantage that there is no mobile carrier out-diffusion from the doped regions which normally would cause a broadening of the observed profile. The resolution test is now based on the observation that the presence of the nonconducting oxide leads to an increase in the resistance signal (see results in Figure 36.20) as long as the tip is smaller than the oxide thickness. This can be explained by the fact that half of the current spreading is stopped by the presence of the oxide layer. Simple calculations (assuming a perfect ohmic contact) on the expected response as a function of tip radius [15] show that one does expect a factor of two increase when the tip radius equals the oxide thickness. The use of a fully ohmic contact is however too simplistic and it is virtually impossible to match the actual SR values with the experimental data without using unrealistically small contact sizes (~5 pm). The latter mismatch cannot be solved by applying the more relevant Sharvin resistance instead of the classical spreading resistance either. A more

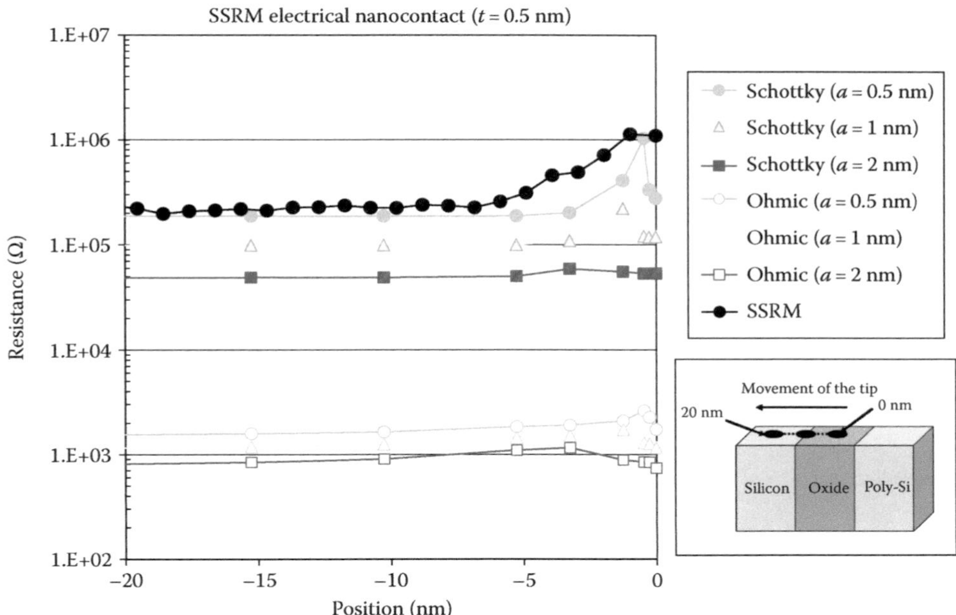

FIGURE 36.21 Comparison between a single SSRM resistance section line across 0.3–0.5 nm buried oxide test structure and 3D simulations for ohmic and Schottky nanocontact modeling for 0.5, 1, and 2 nm tip radius, respectively, across a 0.5 nm buried oxide. (Inset bottom right.) Schematic representation of the 3D simulation configuration.

refined analysis based on a 3D FEM simulation (Figure 36.21) implementing a more refined nanocontact model (Schottky contact, band-to-band tunneling current) leads to calculated resistance values that approach the experimental results and even describe the detailed resistance evolution when approaching the oxide layer. As one can see in Figure 36.21, the resolution criterion is then more stringent (a factor 5 increase instead of a factor 2 for a radius equal to the oxide thickness) and a good agreement between the SSRM result and the Schottky nanocontact simulation can be obtained for an electrical radius of 0.5 nm substantiating our claim of sub-nanometer resolution for SSRM.

36.2.4.1.2 Concentration Sensitivity

In most recent technologies, the electrical characteristics of the components are more and more sensitive to small variations of concentration. For profiling techniques, a high sensitivity to small concentration variations is thus a crucial requirement. Sensitivity is per definition the (measurable) ratio of the change in the instrument response (i.e., the measured resistance) to a corresponding change in carrier concentration. The concentration sensitivity is thus its ability to detect marginally different concentration levels and is thus determined by the intrinsic response function of the technique and limited by its signal-to-noise ratio.

For assessing the concentration, sensitivity use can be made of staircase structures such as those described in [53]. In Figure 36.22a, single-line SSRM measurements on the n-type staircase structure are shown. Note that we display directly the sensor output, that is, the voltage of the logarithmic current measurement unit. Thus, when switching the bias voltage from +50 to −500 mV, the current flows in an opposite direction and a reversed voltage is produced. In Figure 36.22b, the standard deviation on each step for the different single line scans is presented showing that for SSRM, the signal-to-noise ratio is very much bias voltage-dependent and somewhat better for highly doped layers versus lowly doped ones. In all cases, it averages around 0.05–0.1 V, that is, 5%–10% variation. These fluctuations may be linked to the variability of the contact and surface properties and appear to be somewhat more pronounced at low concentrations where the contact deviates more from the ohmic behavior.

36.2.4.1.3 Dopant Gradient Resolution

The gradient resolution of the technique is per definition the steepest variation in carrier concentration that can be measured and is expressed in nanometer of profile length per decade of concentration variation. As it is however very hard to design a test sample with a known gradient in carrier concentration, one uses more frequently the term dopant gradient resolution referring to the steepness of the doping profile as measured with SIMS. It is then important to realize that the same doping concentration gradient translates into a different carrier gradient depending on the average doping level due to the out-diffusion of mobile carriers.

The latter can be illustrated by considering the case of a p^+ layer (5×10^{19} atoms/cm^3 of boron) grown on a lowly doped p-type substrate (1×10^{15} atoms/cm^3) (Figure 36.23). Before discussing the SSRM results (Figure 36.23b), it is important to realize that one has to estimate first the carrier distribution correlated with this dopant profile as they are not identical and SSRM is probing the carrier profile. The difference originates from the out-diffusion of the mobile carriers which can be calculated directly by solving the Poisson equation (squares in Figure 36.23a) using the SIMS

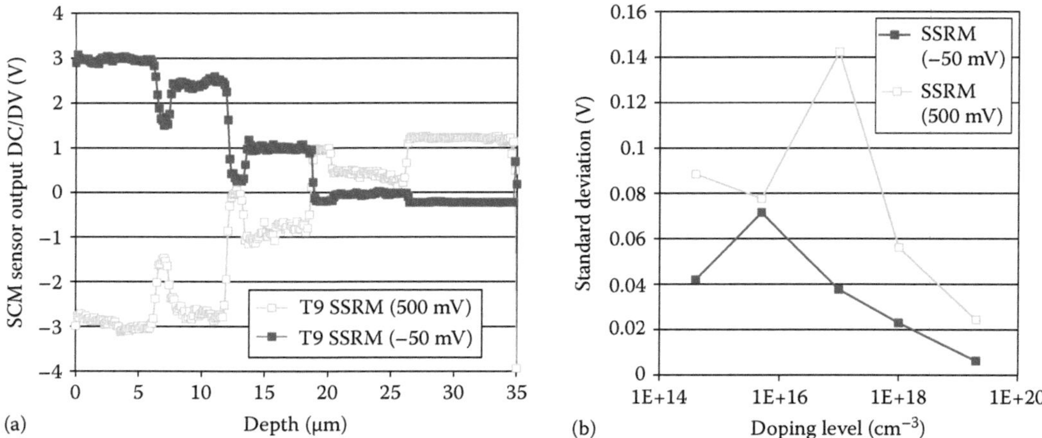

FIGURE 36.22 (a) Single line scan across an n-type staircase structure for SSRM (for two different biases). (b) Standard deviation on each step for those scans.

dopant profile (crosses in Figure 36.23a) as input. The differences between carriers and dopants are quite large in those cases where a steep dopant profile is present on top of a lowly doped substrate. In addition, one needs to mention that the calculated carrier profile represents the situation whereby one would be able to look inside the sample without causing any distortion due to the sample sectioning. Unfortunately, in order to perform a SSRM measurement, one needs to make a cross section. This polished/cleaved surface is far from ideal and contains SS. Since SSRM probes the carrier distribution on this cross section, any impact of the surface (states) on the near-surface carrier profile will be reflected in the SSRM results. In order to assess the impact of the SS on this profile, we have calculated the final carrier distribution as it will be found on the surface (triangles in Figure 36.23a). The details of this SS model and its impact on the carrier profile have been discussed in detail [54]. The calculations show that in this case the SSRM results are steeper than the SIMS profile as the SS enhances the profile steepness in the (very) low doping regions with virtually no impact on the width of the highly doped plateau (~25 nm).

36.2.4.2 Repeatability and Reproducibility

The repeatability is the ability to reproduce the measurement results using exactly the same experimental conditions (sample, probe, scan speed, force, etc.). It has been studied by measuring different staircase calibration structures and the repeatability (for each point of every single-line scan) has been defined as the normalized deviation from the average value:

$$\text{Repeatability} = (R_{single\ line} - R_{average})/R_{average}(\%)$$

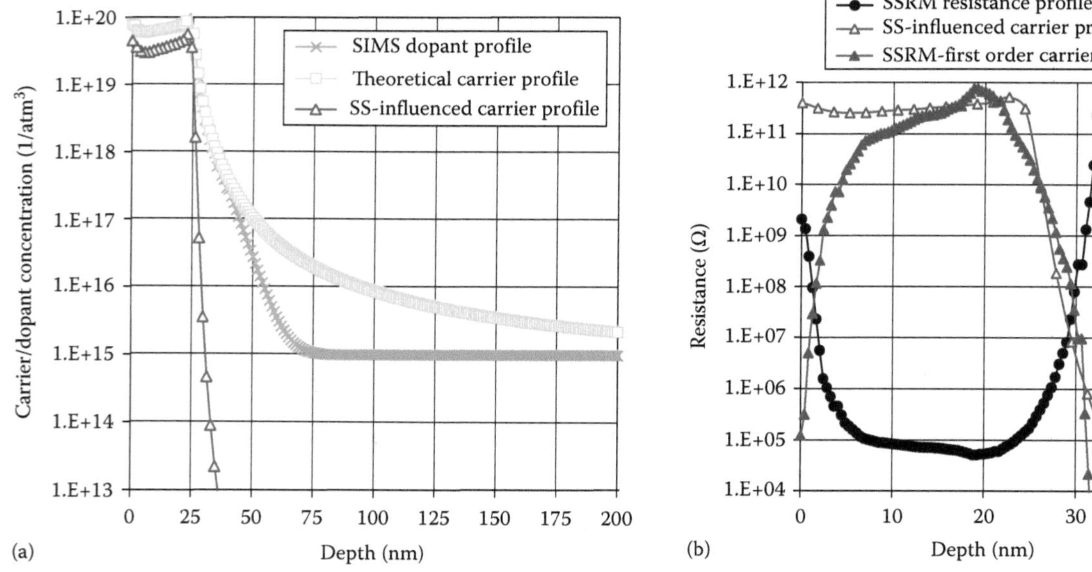

FIGURE 36.23 (a) Comparison between SIMS dopant profile, theoretical carrier profile, and SS perturbated carrier profile for the p⁺–p structure. (b) Comparison on the same structure between SSRM resistance profile and the SS perturbated carrier profile.

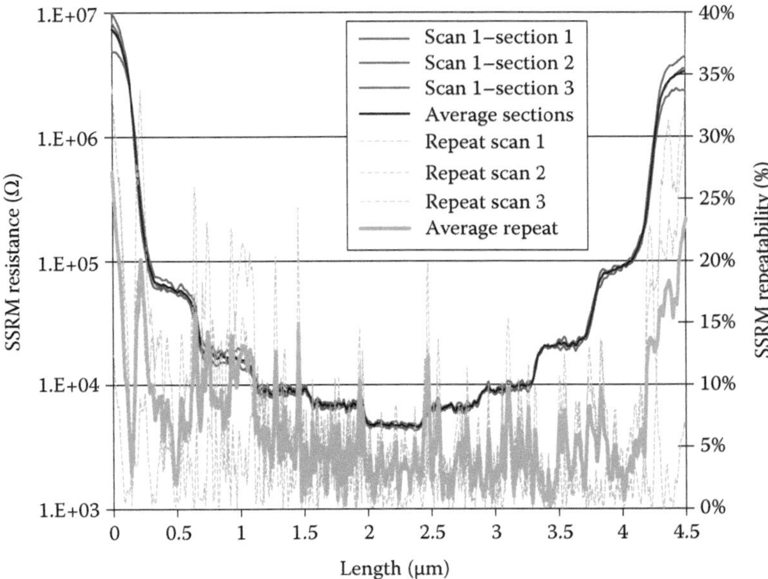

FIGURE 36.24 (Left axis) Three single SSRM line scans across the p-type staircase structure within one SSRM scan and calculated average SSRM line scan. (Right axis) Corresponding repeatability for each of the line scans.

First, the repeatability within one SSRM scan can be analyzed by comparing the resistance variations in between different lines (Figure 36.24). Here, three different SSRM single-line scans (within one SSRM two-dimensional scan) are presented for a p-type staircase structure. The repeatability is also presented (gray curves) and typically is better than 10% across all the different layers. It appears that in the highly doped areas, the repeatability is typically better than in the lowly doped areas which may be due to a more ohmic-like contact and a smoothening effect when one approaches tip saturation.

Second, one can study the repeatability by comparing average (line) sections originating from different scans on the same sample and with the same tip. This was done by five successive scans obtained at two different positions (Figure 36.25). Again, repeatability is typically better than 5% for highly doped Si.

It needs to be mentioned that when performing the measurements in high vacuum (Section 36.2.1.4) the repeatability improves further and values in the range of 3%–5% on highly and lowly n- and p-type Si are obtained (see Figure 36.26). This brings SSRM almost in-line with the International Technology Roadmap for Semiconductors (ITRS) requirements.

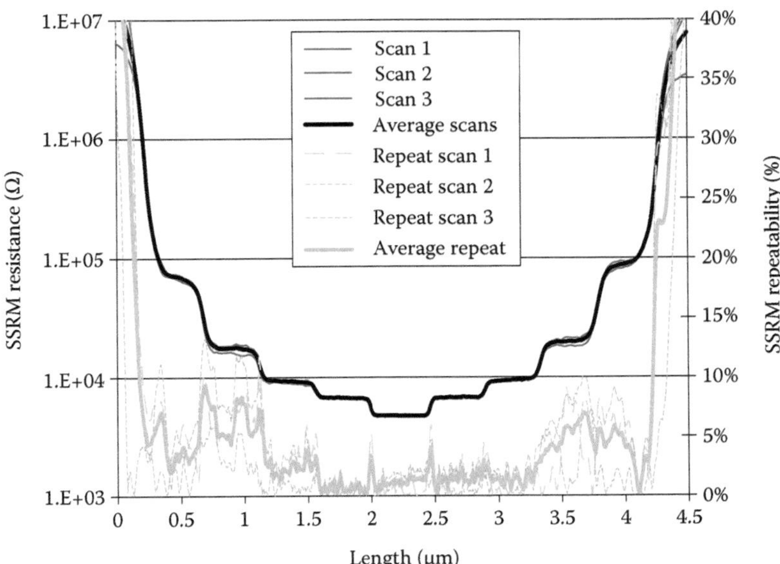

FIGURE 36.25 (Left axis) Three average scans across the p-type staircase structure (corresponding to two different scan positions and to different scans for each position) and average SSRM line scan. (Right axis) Corresponding repeatability.

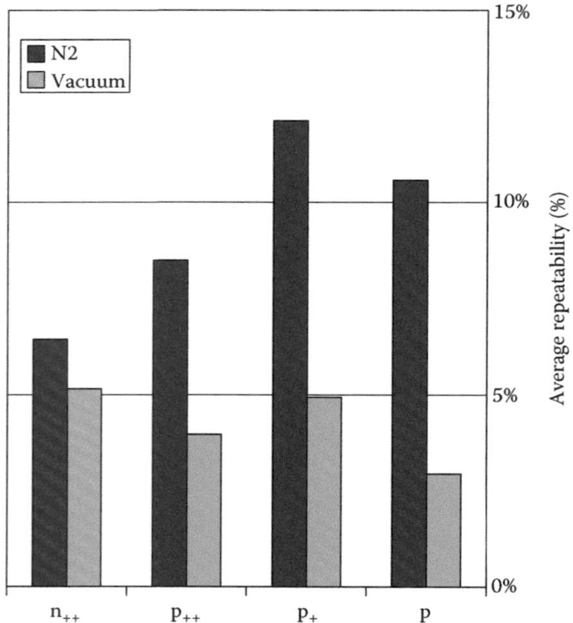

FIGURE 36.26 Average repeatability values for SSRM under high-vacuum and N2 conditions for n^{++}, p^{++}, p^{+}, and p material.

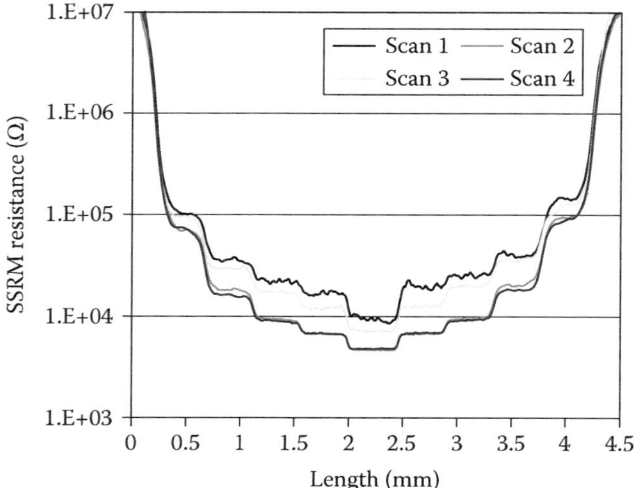

FIGURE 36.27 Average SSRM line scans across the p-type staircase structure for successive scans at different positions evidencing possible reproducibility problems.

The reproducibility is defined as the ability to reproduce the measurement results when successive measurements are made on different samples originating from the same wafer. An additional parameter that can be varied in reproducibility experiments is the use of different probes. As long as all other parameters are maintained constant (force, probe, scan speed, scan dimensions, etc.) and as the wafer is homogeneously processed, absolutely no additional deviation to repeatability is observed while measuring on various samples originating from the same wafer and polished with the same procedure. This is an indication of the very good reproducibility of the sample preparation procedure and of the relative robustness of the technique to sample preparation. However, by realizing successive engagement procedures, the possibility of probe damage is increased and sudden changes in the contact characteristics (due to probe degradation) may occur (Figure 36.27). When different probes are used, the reproducibility of the raw data is drastically reduced. This limitation is representative of the (present) variations in probe tip properties and is not an intrinsic limitation of the SSRM technique itself. It does imply however that for each tip, separate calibration curves need to be established.

36.2.5 Applications

36.2.5.1 Junction Engineering in Laser-Annealed p-MOS Devices

The introduction of ultra-shallow junctions is a key element enabling further downscaling of advanced MSOFET devices. The advent of short-channel devices also implies that the shape and control of the lateral dopant profile become extremely important as properties like the under-diffusion and lateral steepness of the extension profiles, or the characteristics of the halo pockets (active dopant level, size, shape, etc.) are controlling the ultimate device performance. The required junction downscaling is no longer feasible with standard short-time (RTA-spike) anneal approaches, as the associated thermal budget is too large and induces excessive dopant diffusion and migration. Millisecond annealing based on flash or laser annealing [55] has become one of the key elements to control the thermal budget and thus the vertical and lateral junction extent [56]. Due to the resulting diffusion-less junction formation process, a refined analysis of the all factors influencing the profile shape becomes necessary thereby requiring a detailed analysis of the 2D carrier profiles.

In Figure 36.28, the results of such a SSRM study for p-MOSFETs targeting the 32 nm node are presented. These maps (1 nm/pixel) provide the 2D carrier distributions with a parameter of the processing conditions. The focus is put on the impact of the B extension implant energy (0.5 and 1 keV) and its tilt angle (0° and 15°) as well as on the importance of the peak temperature (1200°C and 1350°C) during the millisecond laser anneal.

From the quantified carrier maps (Figure 36.28), we clearly observe that the initial junction formation parameters play an important role. A tilted extension implant leads to enhanced gate overlap (29 ± 2 nm vs. 22 ± 2 nm) as well as to a different shape of the extension profile. A reduction in implant energy (from 1 to 0.5 keV) leads to a reduced under-diffusion (17 ± 2 nm vs. 22 ± 2 nm). Finally, we observe that the laser anneal is not entirely diffusionless as a reduced laser peak temperature (from 1350°C to 1200°C) leads to a reduced under-diffusion (23 ± 2 nm vs. 29 ± 2 nm).

The quantified lateral profiles (extracted 5 nm below the gate oxide and presented in Figure 36.29) illustrate the dependence of the profile steepness and the lateral under-diffusion on the different extension splits. The steepest profiles (2 nm/dec, calculated

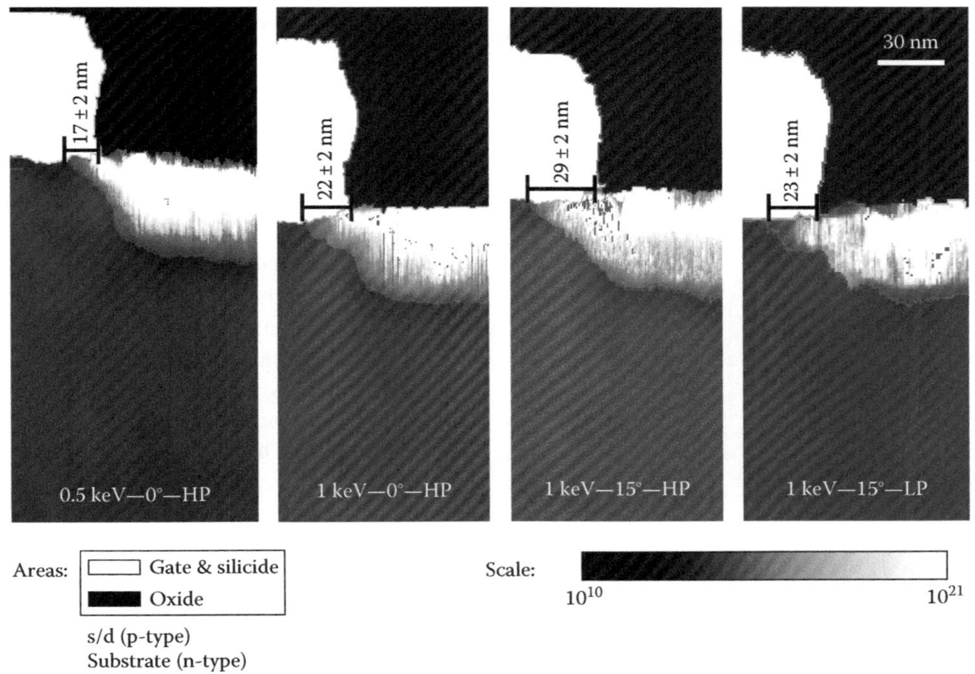

FIGURE 36.28 Comparison of 2D SSRM carrier maps of p-MOSFETs devices for different extension implant and annealing conditions (0.5 keV—tilt = 0°—high power = 1350°C, 1 keV—tilt = 0°—high power = 1350°C, 1 keV—tilt = 15°—high power = 1350°C, and 1 keV—tilt = 15°—low power = 1200°C). (From Eyben, P. et al., *Solid-State Electron.*, 71, 69, 2012.)

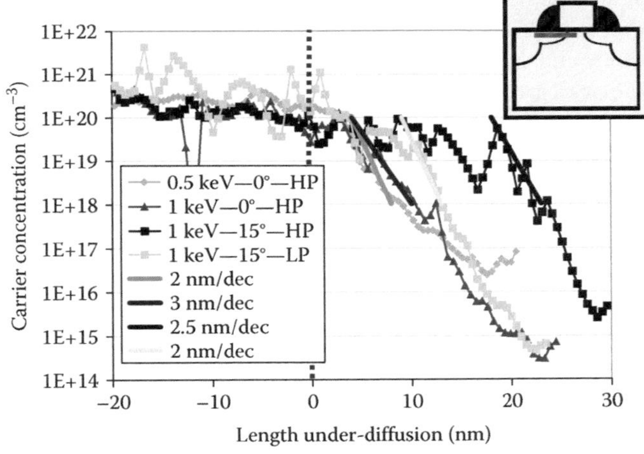

FIGURE 36.29 Comparison of SSRM lateral sections through B extension implant (5 nm below the gate oxide) for different conditions (0.5 keV B—tilt = 0°, 1 keV—tilt = 0°, and 1 keV—tilt = 15°) as well as for high power (HP) and low power (LP) laser anneal. Solid lines indicate various gradients. (From Eyben, P. et al., *Solid-State Electron.*, 71, 69, 2012.)

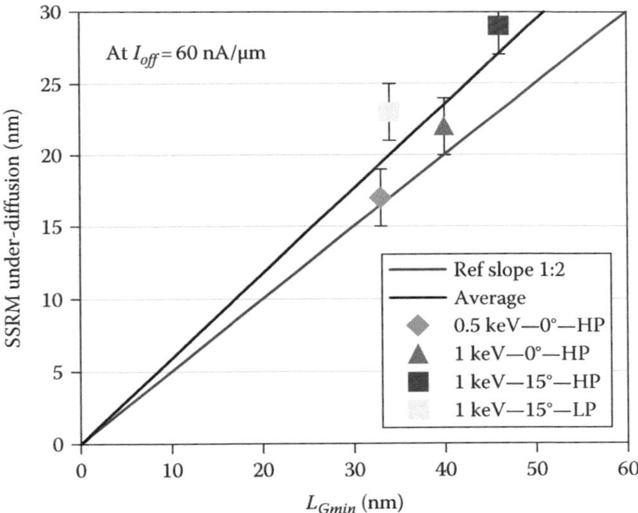

FIGURE 36.30 SSRM under-diffusion versus L_{Gmin} for different extension conditions (0.5 keV—tilt = 0°, 1 keV—tilt = 0°, and 1 keV—tilt = 15°) as well as for high power (HP) and low power (LP) laser anneal. (From Eyben, P. et al., *Solid-State Electron.*, 71, 69, 2012.)

over 1 decade around the concentration level 10^{19} carrier/cm^3) are obtained for the lowest implant energy (0.5 keV) and for the lowest millisecond annealing peak temperature (1200°C) whereas an increased tilt obviously leads to an increased gate/extension overlap. Moreover, despite a relatively large noise level (unsmoothed single lateral sections below the Si–SiO_2 interface are displayed), an active B concentration level around 2×10^{20} atoms/cm^3 is estimated in good agreement with the expectations for such an annealing processes.

These lateral carrier profiles can be used to understand the changes in device performance. As illustrated in Figure 36.30, a systematic correlation between the dopant under-diffusion and the minimum physical gate length (L_{Gmin}) values extracted from electrical measurements can be established. It is clear (Figure 36.31) that the on-state current (I_{ON}) versus the minimum physical gate length (L_{Gmin}) for an off-state current (I_{OFF})

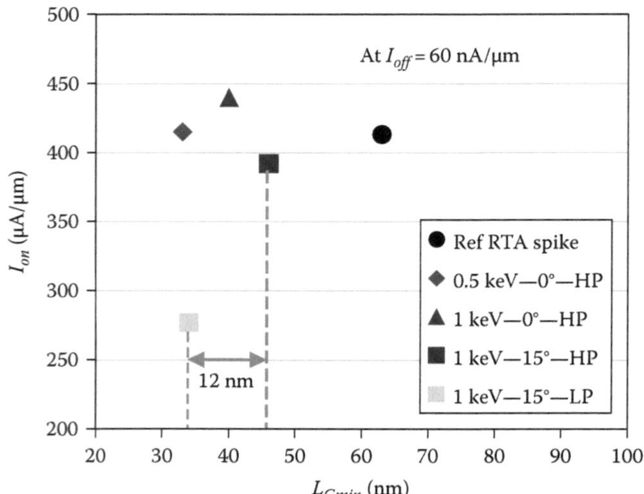

FIGURE 36.31 I_{on} vs. L_{Gmin} (at I_{off} = 60 nA/µm) for different extension conditions (0.5 keV—tilt = 0°, 1 keV—tilt = 0°, and 1 keV—tilt = 15°) and laser anneals, high power (HP) and low power (LP). As a reference, the case for RTA spike is included. (From Eyben, P. et al., *Solid-State Electron.*, 71, 69, 2012.)

of 60 nA/µm correlates well with the SSRM results indicating that the latter can be used to gain insight in the parameters influencing device performance. For instance, the SSRM results show a reduction of the extension under-diffusion (6 ± 2 nm) when the laser annealing temperature is reduced from 1350°C to 1200°C. The reduced gate overlap is clearly reflected in a reduction of L_{Gmin} of 12 nm (=2 × 6 nm) (Figure 36.31).

36.2.5.2 Calibration of Process Simulators

In order to achieve an efficient technology development when downscaling a particular technology, extensive use is made of process and device simulations in order to predict device performance without performing the actual processing. A correct prediction of device performance using device simulators, however, does require nowadays accurate information about the 2D carrier profile. As such, 2D characterization techniques became indispensable for dopant/carrier profiling.

In Figure 36.32, we present the 2D net carrier distribution for a p-MOSFET (within a 45 nm node technology). Left half of the figure is showing the resulting dopant distribution as predicted with the process simulator using the available advanced dopant and dopant-defect cluster diffusion models (with default calibration). The right half is the quantitative 2D SSRM carrier profile [57]. When comparing them, it is clear that the lateral diffusion of the HDD is smaller in the SSRM profile whereas the halo pocket is less pronounced in the simulation. The latter indicates an overestimate of the profile diffusion (pocket and HDD) in the process simulator. Using the SSRM results obtained in Figure 36.32, as input for the device simulator, device characteristics were then calculated and, for instance, DIBL values versus gate length were predicted (Figure 36.33). The excellent agreement of the SSRM-based values with the measured ones is indicative for the fact that the SSRM distribution is highly reliable and accurate. On the other hand, the process simulation–based values underestimate the DIBL due to the lower electron concentration in the As-pocket implants in the simulations. In Figure 36.33, we also calculated DIBL values for gate lengths of 90 and 130 nm, based on the 2D profile taken from the 45 nm gate length structures, but adding extra Si with identical doping profile as observed in the center of the 45 nm structure. In essence, this implies that the details of the under-diffusion and halo diffusion are entirely independent from the gate length. The agreement with the experimental values indicate then that for this process no 2D interactions occur even at the smallest gate length of 45 nm and that the halo profile and under-diffusion remain identical for all gate lengths down to 45 nm. The latter is consistent with the lack of thermal diffusion in a laser anneal-only process.

As DIBL is one of the parameters which is very sensitive to gate under-diffusion and the details of the halo profile (extent, concentration level, etc.), its prediction when using SSRM results as input will depend strongly on the accuracy of the SSRM results. The errors bars in Figure 36.33 exemplify this point as they indicate the DIBL variation resulting from a ±10% variation in the As concentration. It is clear that the latter has a strong impact and

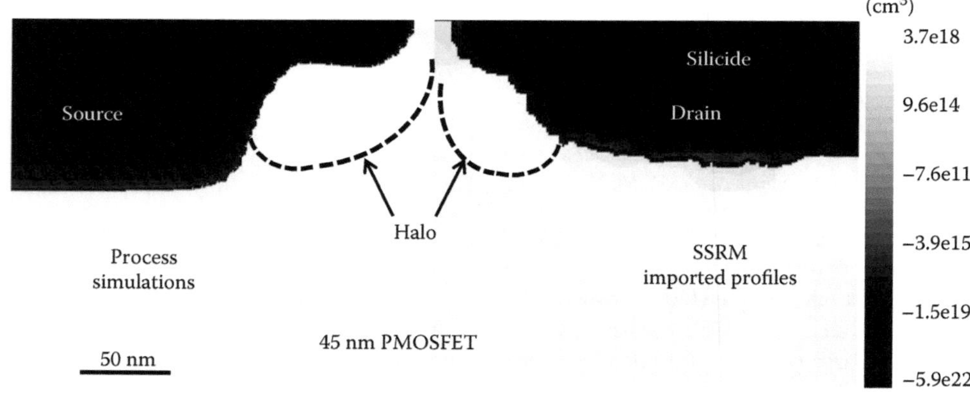

FIGURE 36.32 Comparison of 2D process simulations (left) and 2D SSRM active dopant profiles (right) of a 45 nm PMOSFET. (From A. Nazir et al., *Solid-State Electron.* 74, 38–42, 2012.)

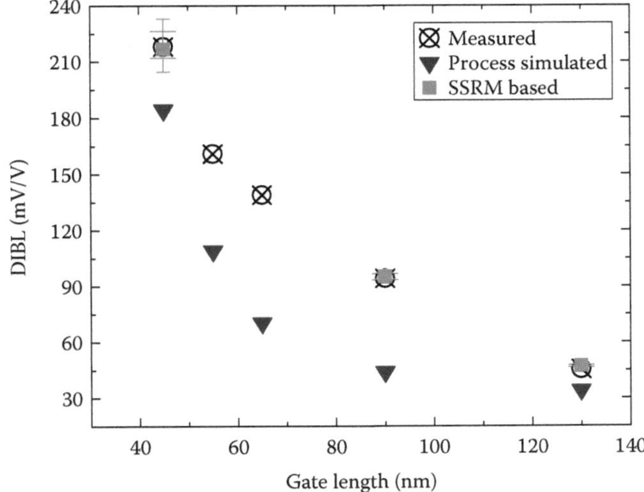

FIGURE 36.33 Comparison of the measured (circles), the SSRM simulated (squares), and process-simulated (triangles) DIBL values versus different gate dimensions for p-MOSFETs. (From Nazir, A. et al., *Solid-State Electron.*, 74, 38, 2012.)

provides an estimate for the accuracy and sensitivity which needs to be achieved in the SSRM metrology in order to arrive at accurate results with the present approach.

36.2.5.3 2D and 3D Carrier Distributions in FinFETs

Planar FET-based architectures cannot meet the ITRS goals for the sub-28 nm technology nodes. Therefore, FinFET-based structures have been introduced as alternative due to enhanced electrostatic control and performance [58]. To meet the performance targets, there remains however a major challenge with respect to the doping profile optimization of the source/drain extension and channel regions as one needs to dope the FIN structure conformally in order to control subthreshold-swing and short-channel effects [59,60]. The latter requires a doping technology to engineer an optimum 3D doping profile in FinFET-based devices which has been pursued using tilted ion implantation, vapor phase doping (VPD), or plasma doping. In all these, the major challenge is to generate a conformal dopant profile and inevitably this calls for the availability of 3D dopant and carrier profiling techniques with sub-nanometer resolution.

To meet these goals, concepts have been developed to enable SSRM (which is intrinsically 2D) to probe the 2D/3D dopant distribution in bulk-fin devices and thus to assess the relevant vertical and lateral (conformal) dopant profiles [61]. For the analysis of the source/drain profile, SSRM is directly applicable as a simple cross section through the source/drain region provides the relevant information. For instance, in Figure 36.34, the SSRM carrier profiles of fins (40 nm wide) implanted at 45° and 10° tilts are shown. The fins implanted at 45° appear completely doped with a higher (2×) concentration in the top due to the use of 2Q implant exposing the individual sidewalls only to half the dose. However, in the 10° case, only a shallow lightly doped region is observed at the sidewall whereas the top surface implant contains much more dopant atoms and is substantially deeper. This is consistent with the reduced ion retention for an implant at large tilt angles (80° relative to the sidewall surface) and the geometrical effect [62]. An analysis of the incorporated sidewall dose as measured with atom-probe tomography (APT), SIMS, and SSRM shows a good agreement suggesting nearly 100% activation agreement (within the error margin of 20%–30%) for the sidewall (Figure 36.35) which is also in agreement with 2D simulation results. The latter is in contrast with the top dose where APT and SIMS indicate a much

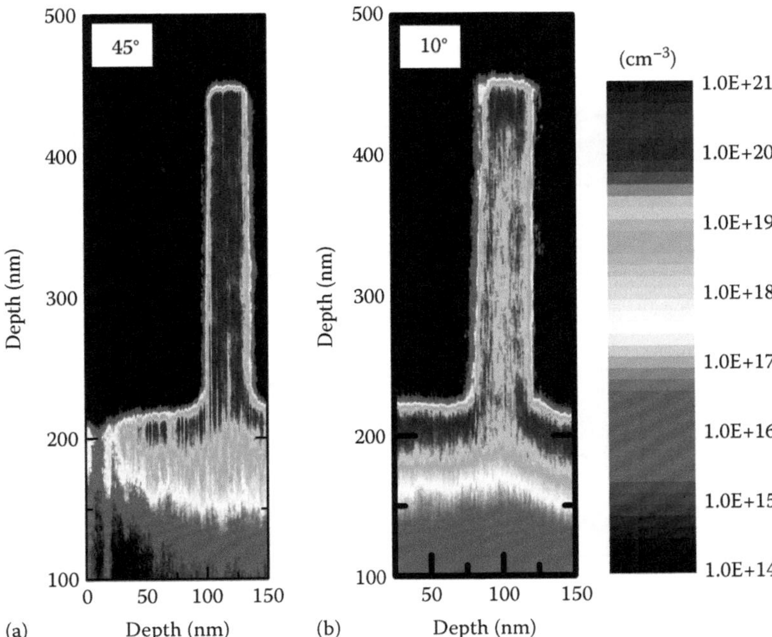

FIGURE 36.34 2D SSRM map of active carrier concentration of BF2 implanted at 45° and 10°.

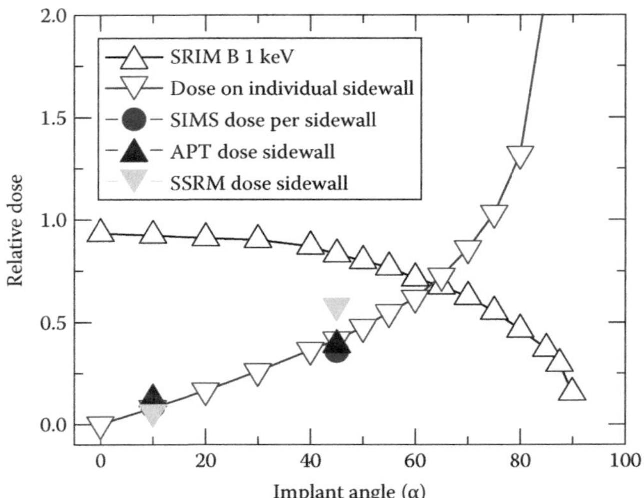

FIGURE 36.35 Summary of dose incorporation values.

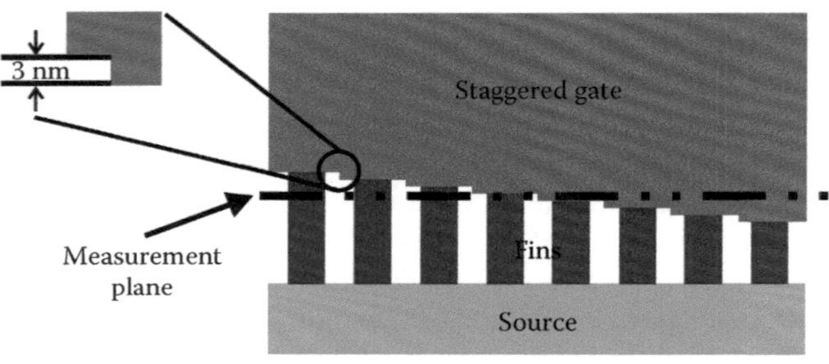

FIGURE 36.36 Staggered gate concept to obtain 3D carrier concentration profile in FinFETs.

higher incorporation than SSRM. Obviously, a large inactive fraction is present at the top while the sidewall is fully active. These differential activation levels imply that the conformality (sidewall dose/top dose) of the "electrical carriers" is higher (up to 29% even at 10° tilt) than the "chemical dopant" conformality [63,64].

Although SSRM is inherently a 2D method, a methodology was recently developed to obtain 3D carrier profiles in bulk FinFETs [65]. The concept uses a dedicated test structure based on a staggered gate over multiple fins whereby the gate in each set of fins is shifted laterally by "x" nm. By cleaving through this staggered set of fins, we produce a series of cross sections with incremental steps from source/drain toward the gate region (Figure 36.36) onto which SSRM measurements can be made. By assuming then that the processing and thus 3D profiles for nearby fins are identical, one can combine the results from different fins into one 3D profile whereby the resolution in the third dimension (i.e., source/drain) is determined by the minimum stagger distance in the gate which can be produced by the lithography.

Figure 36.37 shows the evolution of the 2D carrier distribution for such an array of cleaved FinFET's while moving (with 3 nm steps) from below the gate to the spacer and till the source region. Analyzing the SSRM images below the gate (in the channel region), we can clearly see the channel implants, the well doping, and the shape of the highly doped gate. In the sections through the spacer region, the decaying 2D extent of the extension profile and the halo implants below them are apparent (Figure 36.37b). Finally, in the source region, the HDD regions and the triangular shape of the raised source/drain can be identified. Stacking these multiple images into a single 3D volume provides a detailed view on the 3D carrier distribution of FinFET extending from source/drain till the middle of the gate. Based on this, we can estimate a gate overlap of around 21–24 nm (Figure 36.38b). A similar analysis on devices based on an extensionless architecture demonstrates,

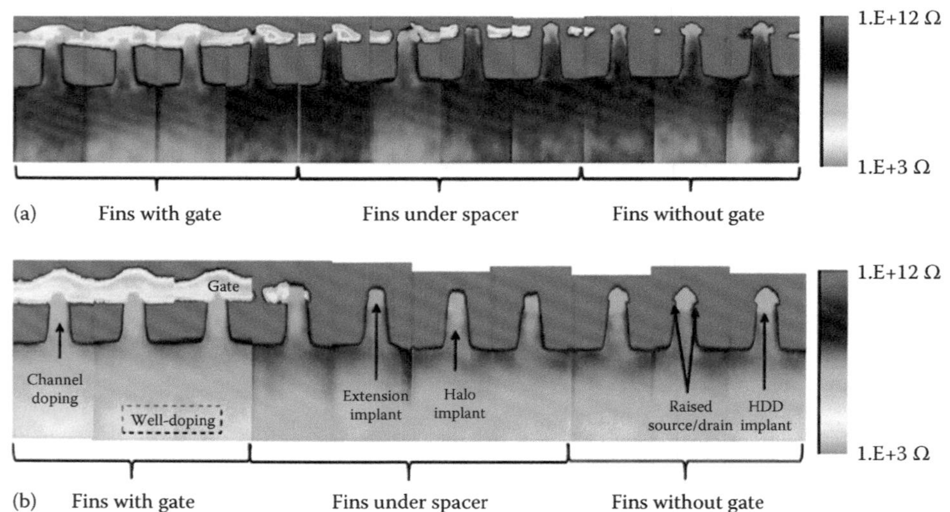

FIGURE 36.37 SSRM images of fins under the gate, the spacer, and fins without gate in the source region for FinFETs with (a) extensionless and (b) extension architectures. (From Mody, J. et al., *Proceedings of IEDM*, pp. 119–122, 2011.)

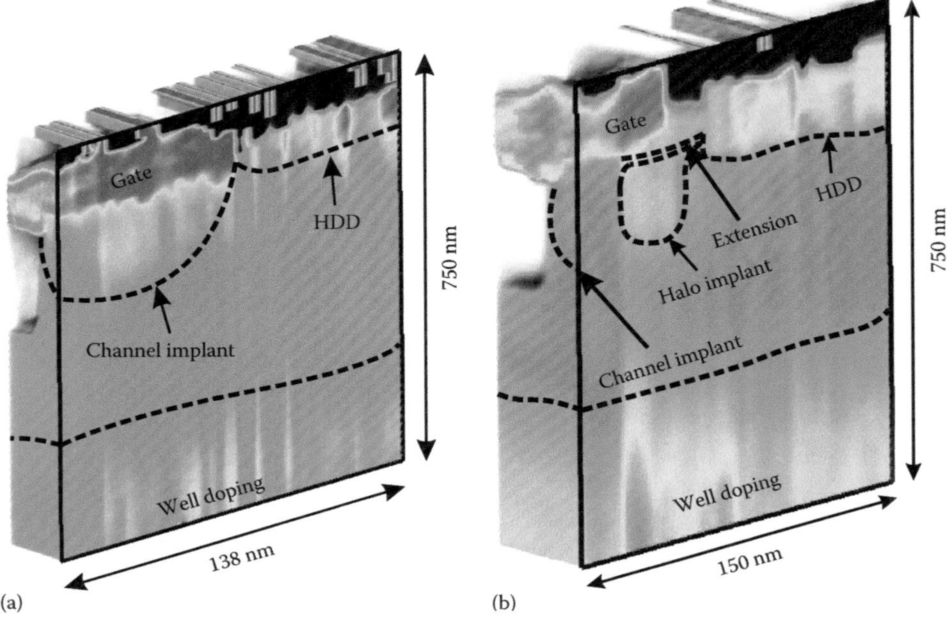

FIGURE 36.38 Reconstructed 3D SSRM image of FinFET slice perpendicular to the gate for extensionless (a) and extension (b) architectures. (From Mody, J. et al., *Proceedings of IEDM*, pp. 119–122, 2011.)

as expected, a very different profile with far less doping below the spacer region and a similar channel implant in the channel region and a reduced overlap of only 3–6 nm (Figure 36.38a). These SSRM results suggests that the extensionless devices will underperform the implanted case due to the lower dopant concentrations under the spacer and the minimal underlap below the gate which increases the series resistance of FinFET devices. This is confirmed by the respective transistor performance curves (see [65]).

36.2.5.4 Dopant Deactivation in Tunnel FETs

Semiconductor nanowires (NW) and NW-based tunnel field-effect transistors (TFET) are presently viewed as potential successors of standard MOSFETs due to the absence of a 60 mV/dec subthreshold-swing limitation and reduced short-channel effects and are thus viewed as a promising building block for future nanoelectronic devices [66]. Due to their reduced dimensionality and vertical structure, optimizing the doping processes becomes a major challenge requiring 2D/3D carrier profiling with high spatial resolution and sensitivity in a confined volume. Moreover, the sectioning of such a device for cross-sectional analysis is far from trivial and requires either FIB sectioning or the use of a dedicated array [65].

Figure 36.39 shows the 2D carrier distribution maps obtained on TFETS with diameters ranging from 400 nm down to 100 nm [69]. Colors ranging from red over yellow to green correspond to different hole concentrations while colors between blue and green are linked to different electron concentrations. The parts of the profiles corresponding to the silicided cap layer and the metal gate are highlighted as very conductive but are only qualitatively as no detailed calibration curve for these materials was established.

The carrier distribution depends strongly on the NW diameter with a more pronounced out-diffusion profile in the NW bottom (drain) and a "dual bump" structure in the top profile (source) for the larger diameters. The bump is most pronounced for larger NW diameters but disappears completely for narrow NWs. Its presence can directly be linked to the tilted implant process which dopes the side wall and the top of the NW simultaneously. The deep junction is induced by the side wall doping whereas the shallower junction depths correspond to the doping in the top part of the NW. For smaller NW diameters, these implantation pockets merge in the center resulting in a flat profile within the source of the device with a higher active dopant concentration as compared to the large NW case.

The carrier profile in the bottom part of the NWs is illustrated in Figure 36.40. Vertical sections (cross-section average) of the electron concentration through the substrate and the bottom part of the NWs are shown for different NW diameters. The electron concentration profiles clearly reveal a diameter dependence with a steeper slope in the case of smaller NW diameters as compared to larger ones (Figure 36.41). We believe that this observation can only be explained by a deactivation of dopants in the bottom part of the smaller NWs as a result of the implantation step used to dope the top region (source). Vacancies and interstitials induced during the implantation of the NW top section diffuse during the subsequent annealing step toward the NW bottom. It is well known that the formation of As–vacancy complexes leads to a deactivation of As in highly doped n-Si [67].

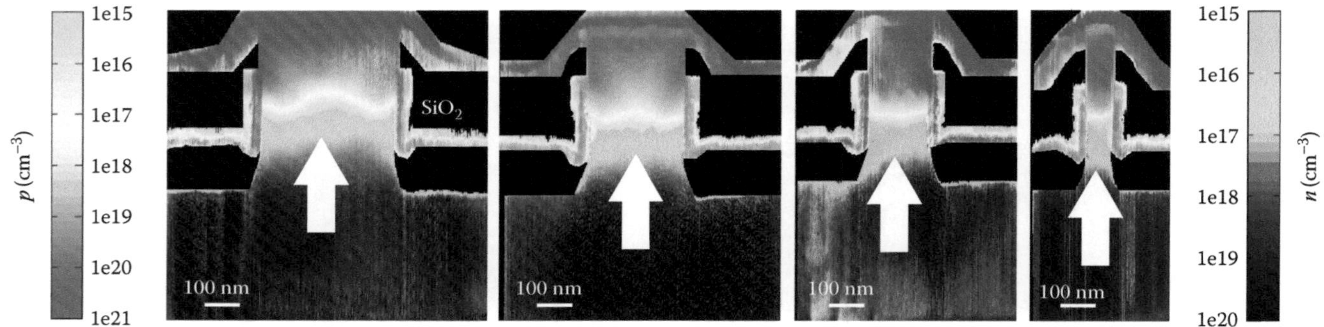

FIGURE 36.39 **(See color insert.)** Quantitative 2D carrier distribution for NW diameters of 400, 300, 200, and 100 nm. Colors ranging from red over yellow to green indicate a majority of holes and colors ranging from blue to green represent a majority of electrons. The NWs are embedded in oxide (black). (From Schulze, A. et al., *Nanotechnology*, 22(18), 185701, 2011.)

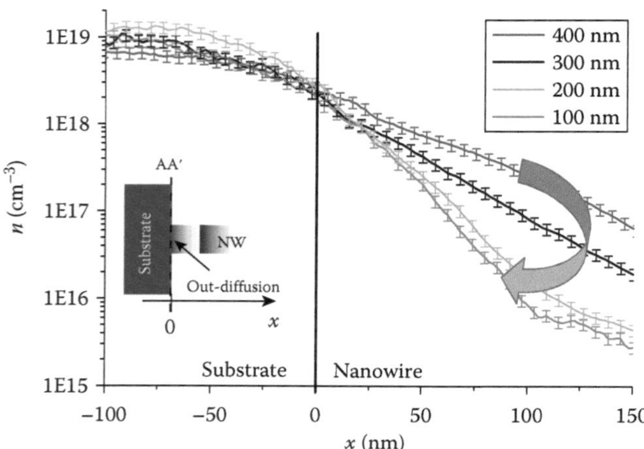

FIGURE 36.40 Vertical section (cross-section average) through the substrate and the bottom part of the NWs extracted from the carrier distribution maps in Figure 36.39. The error bars indicate the SSRM quantification error. (From Schulze, A. et al., *Nanotechnology*, 22(18), 185701, 2011.)

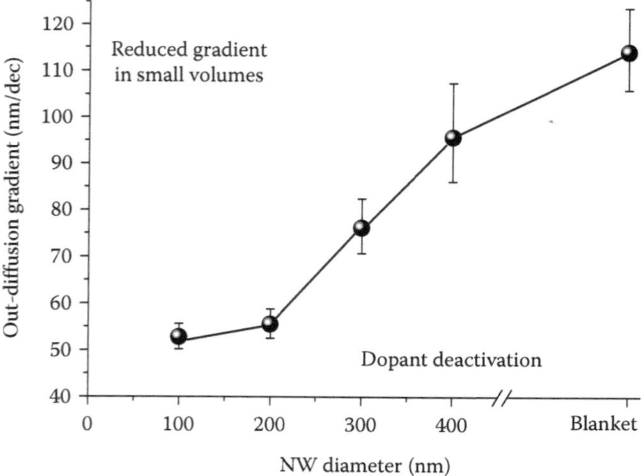

FIGURE 36.41 Slope of the diffusion profile in the bottom part of the NWs extracted from curves in Figure 36.40 versus NW diameter. (From Schulze, A. et al., *Nanotechnology*, 22(18), 185701, 2011.)

The size-dependent deactivation would then result from the increased proximity of surfaces for smaller NWs, acting as a sink for the interstitials thereby reducing the vacancy–interstitial recombination probability and thus promoting the As–vacancy agglomeration and As deactivation. Depletion due to traps present at the Si/SiO_2 interface [68] cannot be put forward as an explanation due to the high carrier concentration in the bottom part of the NWs and the relatively large dimensions. Device characteristics and device simulations based on the carrier profiles extracted from the SSRM results are in agreement and the smaller TFET off-current for smaller NW diameters confirms the lower drain doping detected by SSRM [69].

36.3 Carrier Profiling in Other Semiconductors with SSRM

36.3.1 Nanocontact Modeling

As seen in Section 36.2.2, the electromechanical properties of the SSRM nanocontact on Si are intimately linked to the formation of a metallic-like phase below the tip due to the large pressure. In the case of germanium (Ge), the nanocontact mechanism is similar to the Si case as this material is presenting a similar hardness (slightly lower) and threshold pressure (also slightly lower) for the formation of a metallic phase (Table 36.2). Force curves on Ge (Figure 36.42) do present a steep transition (at lower force as compared to Si) and a dependence (although weaker) of the measured resistance upon the active dopant concentration [70].

In the case of III–V semiconductors like gallium arsenide (GaAs) and indium gallium arsenide (InGaAs), the situation is a bit more complex as these materials are much softer (in particular InGaAs) and do present metallic phases only at very large pressures. Hence, it is not possible to generate local phase transformations without excessive damage (penetration and material removal). When applying larger biases (1–3 V) between tip and sample, one is however able to obtain a similar abrupt reduction in resistance beyond a certain force and to obtain a clear carrier concentration dependence (see Figure 36.42). The need for a higher bias can be understood by considering the nanocontact model for InP by Xu [71].

TABLE 36.2 Comparison of Electrical and Mechanical Properties of Si, Ge, GaAs, and InGaAs

	Bulk Modulus (GPa)	Shear Modulus (GPa)	Young's Modulus (GPa)	Hardness (Moos)	Microhardness (Knoop) (GPa)	Metallic Phase (GPa)	E_G (eV)	μ_n (cm²/Vs)	μ_p (cm²/Vs)	K (W/mK)
Si	98	52	130	7	11.5	12.5	1.12	1/4e3	450	130
Ge	71.3	41	103	6	7.8	10.5	0.66	3.9e3	1.9e3	58
GaAs	75.3	33	86	4–5	7.5	17.2	1.42	8.5e3	400	55
$In_{0.53}Ga_{0.47}As$	66	25	67	—	—	—	0.74	1e4	200	5

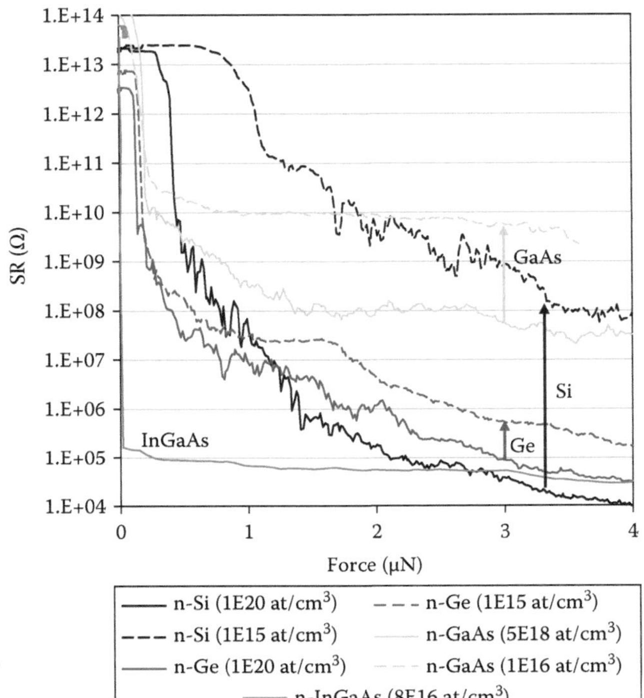

FIGURE 36.42 Force curves on n-type lowly and highly doped Si, Ge, GaAs, and InGaAs.

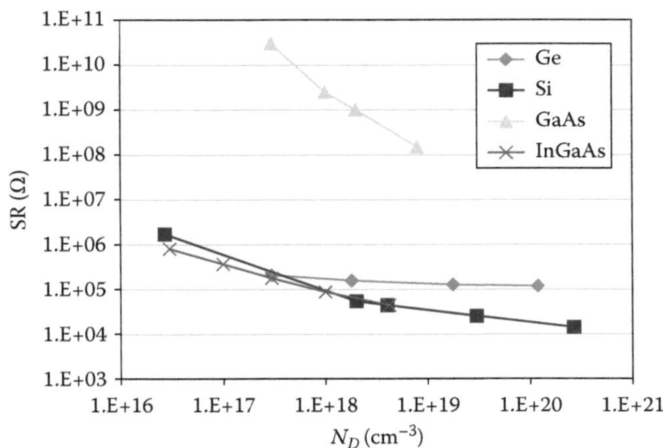

FIGURE 36.43 n-type calibration curves for Si, Ge, GaAs, and InGaAs.

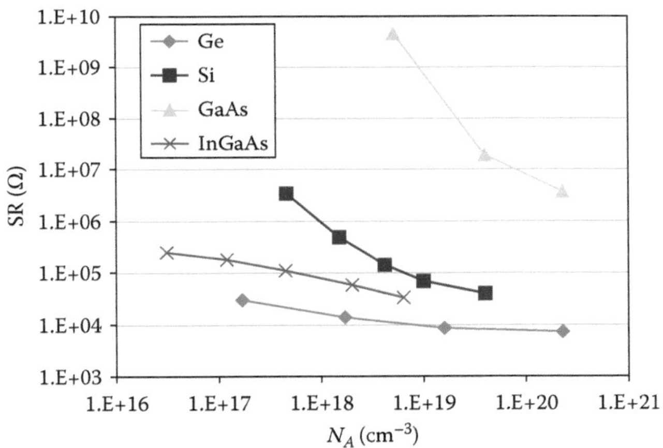

FIGURE 36.44 p-type calibration curves for Si, Ge, GaAs, and InGaAs.

They have shown that in this case one is not dominated by the current constriction leading to the spreading resistance case but rather by the nonlinear resistance of the point contact due to tunneling through a metal–semiconductor interface (which does require a higher bias voltage). In Figures 36.43 (for n-type dopant) and 36.44 (for p-type), the SSRM calibration curves for Si, Ge, GaAs, and InGaAs are presented. They all exhibit a monotonic behavior allowing for quantification, be it with very different sensitivities.

36.3.2 Performances

Due to a lack of dedicated test structures, a detailed analysis of the SSRM resolution on Ge, GaAs, or InGaAs is still missing. However, the generation of sub-2 nm metallic-like pockets below the tip (similar to Si) has recently been reported for Ge based on MD simulations [72]. As this is similar to the Si case, it is fair to assume that one would get a similar spatial resolution as well. On III–V semiconductors, sub-3 nm resolution has been reported on InP [71]. There are currently no data available on GaAs and InGaAs.

The sensitivity and repeatability have been systematically analyzed for Ge, GaAs, and InGaAs (Figure 36.45). Ge is exhibiting a very low sensitivity but a very good repeatability (better than on Si). GaAs is presenting a very good ratio between sensitivity and repeatability. InGaAs presents the poorest performances. This material is also suffering a lot of material erosion during the scanning of the tip.

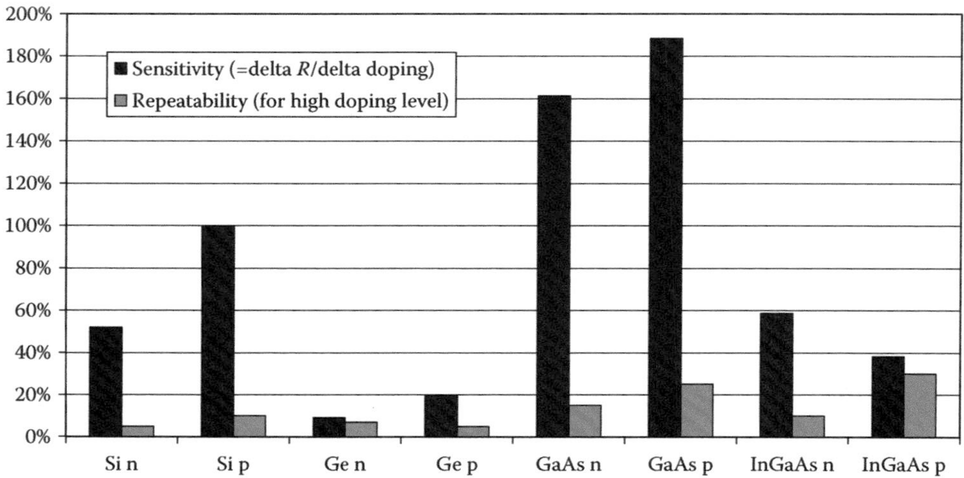

FIGURE 36.45 Sensitivity and repeatability of SSRM on various semiconductor materials.

36.3.3 Applications

36.3.3.1 SiGe Implant-Free Quantum Well p-MOSFET

Targeting better performance through the use of high mobility channel material has led to the integration of SiGe channel (SiGe-ch) with high Ge contents in scaled p-MOSFET devices. [73]. As the entire structure becomes heterogeneous, predicting the finer details of dopant diffusion, deactivation, and defect interaction become challenges and extensive use of SSRM has been made to unravel some of them. For instance, in Figure 36.46, the 2D carrier distribution map is shown for the case of a IFQW well using a B-doped SiGe layer for S/D-extension formation. The SSRM results indicate a clear B migration from the B-doped SiGe-raised S/D into the SiGe channel creating the S/D extension gate overlap (~15 nm). Moreover, SSRM also evidenced that defects at the interface between the SiGe-ch and the SiGe-raised S/D reduce the dopant activation [74].

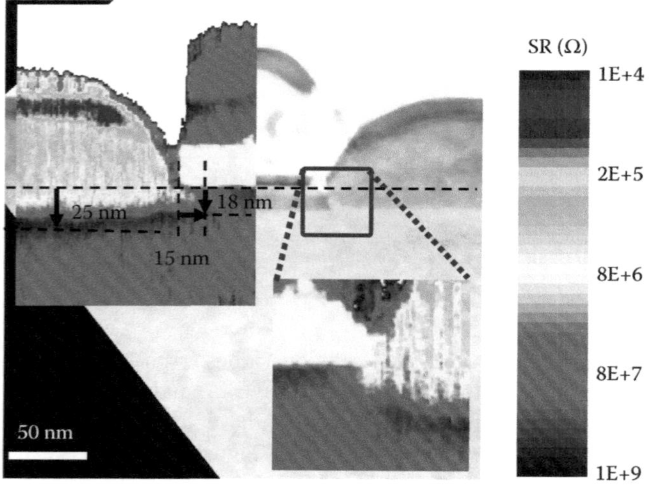

FIGURE 36.46 (See color insert.) SSRM and SEM inspection of IFQW. (Inset) Local deactivation at the Si–SiGe interface.

36.3.3.2 InGaAs Channel IFQW n-MOSFET

SSRM was also used to study InGaAs/InP IFQW n-MOSFET devices incorporated into a Si CMOS process (Figure 36.47). Although functional devices can be made, very high levels of source/drain leakage are often observed. Physical analysis of the InP grown in the STI suggests that this source/drain leakage is a result of the InP layer being conductive [75]. This is confirmed by 2D SSRM analysis on a fully processed device which shows that the Ge seed layer and underlying Si are also highly conductive. Energy-dispersive x-ray spectroscopy (EDX) analysis (Figure 36.47, bottom) indicates that this is linked to a phosphorous (P) doping of the underlying Si. As the thermal budget of the InP buffer layer growth is quite low, this level of P diffusion is quite surprising and in fact not observed for blanket InP deposited on Si. Hence, the P-indiffusion must be linked to the effects associated with the InP growth in a trench (defect-mediated diffusion, strain, etc.) and can only be observed on those structure whereby SSRM then becomes an extremely valuable asset. At present, the physical mechanism behind the P diffusion into the Ge and the Si is still under investigation but the capability of SSRM to identify this diffusion has enabled to study the InP growth process in more detail and optimize it with respect to suppressing this effect [75].

36.3.3.3 Electrical Tomography of Individual CNTs

The increasing resistivity of metal when moving to smaller dimensions [76] and its limited reliability due to electromigration lead to a need for alternative interconnect materials for back-end-of-line interconnects for next-generation integrated circuits. Carbon nanotubes (CNTs) are proposed as potential copper replacement due to their unique properties in terms of current-carrying capacity, thermal stability, and high electromigration resistance [77].

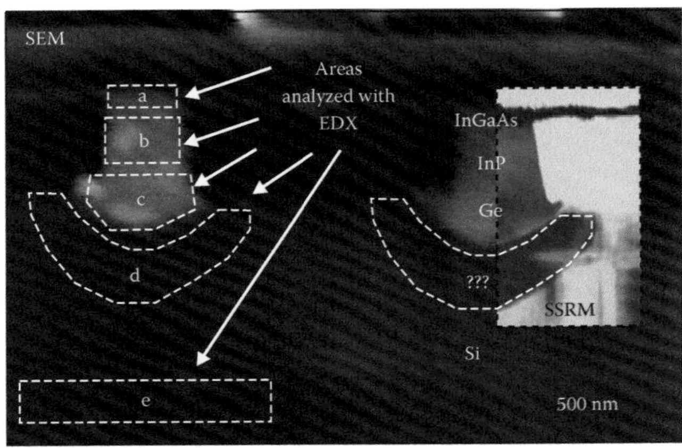

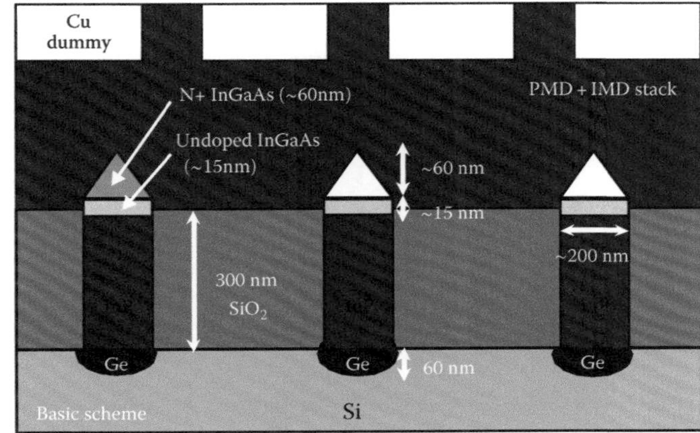

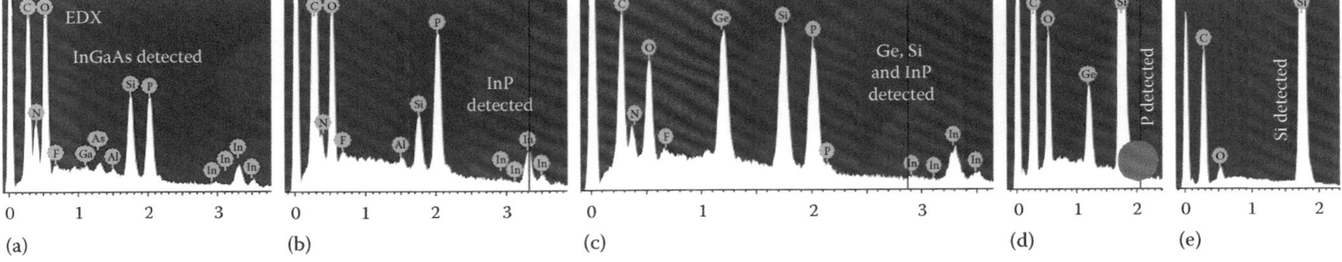

FIGURE 36.47 InGaAs/InP IFQW analysis using SSRM and SEM/EDX. The schematic structure is also presented (top right).

The CNTs investigated were integrated into 300 nm contact via holes on 200 mm wafers where a blanket TiN layer (70 nm) forms the bottom electrode. The CNTs were grown at 470°C in a microwave plasma reactor and then embedded in oxide.

By using SSRM and high-resolution wear-resistant full diamond probes, we have developed a methodology (named "slice and view") for the 3D electrical and structural analysis of next-generation CNT-based interconnects [78]. These ones being embedded into relatively soft oxide, we actually map the resistance distribution across integrated vertical CNTs in 2D while progressively removing material (Figure 36.48). This allows the acquisition of resistance maps at different depth values (Figure 36.49a). 3D reconstruction is then achieved by aligning and interpolating the obtained 2D resistance maps (Figure 36.49b). From the final tomogram of the investigated CNT-based contact hole, electrical as well as structural (size, density, distribution) information on individual CNTs can be derived. In this way, we are able to determine the individual CNT linear resistivity (2 kΩ/nm) as well as the contact resistance between CNT and bottom electrode (540 kΩ) in a quantitative manner. Using this approach, we could reveal that CNT quality and contact resistance of CNTs integrated in contact holes are still far from what is needed to replace copper as interconnect material.

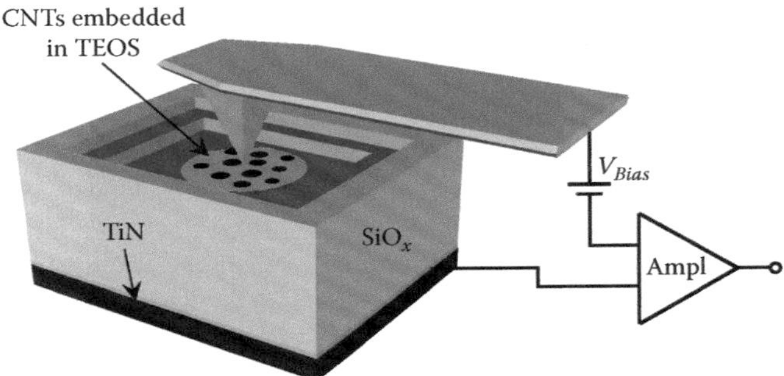

FIGURE 36.48 Measurement setup of the 3D SSRM slice and view approach on CNTs integrated in contact holes.

36.4 Conclusions

SSRM has emerged as one of the most important 2D profiling methods in the semiconductor industry as it provides a unique spatial resolution (down to sub-nanometer) combined with high concentration sensitivity and ease of quantification. As a carrier (and not dopant) profiling technique, SSRM also brings useful information on activation, carrier spilling, etc. Its unique properties arise from the formation of a nanoscopic metallic phase below the tip leading to an electrical contact with a radius smaller than the actual tip radius. The technique enables through the use of diamonds as tip material (coated or molded) as these are the only ones surviving the applied high forces. Its present applications range from junction engineering, process optimization, TCAD calibration for Si as well as alternative semiconductor materials. Combined with some creative sample preparation steps, the SSRM techniques can be extended from 2D toward 3D analysis as evidenced by the characterization of FinFETs and TFETS.

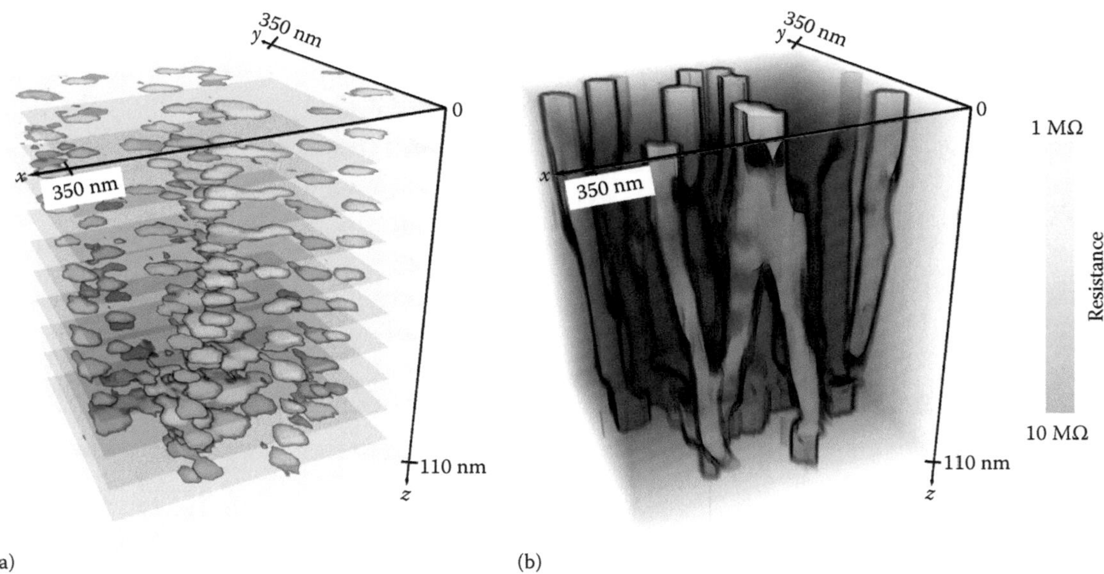

FIGURE 36.49 **(See color insert.)** (a) Individual 2D SSRM resistance maps stacked on top of each other. (b) Tomogram reconstructed from 2D scans by interpolation. Surrounding oxide (purple) is displayed semitransparent. (From Schulze, A. et al., *Nanotechnology*, 23(30), 305707, 2012.)

References

1. G. Binnig et al., *Phys. Rev. Lett.* 56(9), 930, 1986.
2. M.P. Murrell et al., *APL* 62(7), 786, 1993.
3. J. Petry, PhD thesis, University of Leuven, Leuven, Belgium, 2005.
4. B. Rezek et al., *J. Appl. Phys.* 92, 587, 2002.
5. K. Szot, W. Speier, G. Bihlmayer, and R. Waser, *Nat. Mater.* 5, 312, 2006.
6. J. Cermak et al., *Phys. Status Solidi (RRL)* 1, 193, 2007.
7. W. Vandervorst and M. Meuris, EP-466274, January 15, 1992.
8. W. Vandervorst and M. Meuris, US-5369372, November 29, 1994.
9. P. De Wolf, J. Snauwaert, T. Clarysse, W. Vandervorst, and L. Hellemans, *APL* 66, 1530, 1995.
10. P. De Wolf, PhD thesis, University of Leuven, Leuven, Belgium, 1998.
11. T. Hantschel et al., *Microelectron. Eng.* 57–58, 749, 2001.
12. M. Fouchier et al., *Proc. SPIE* 5116, 607, 2003.
13. P. Eyben et al., *SPM: Electrical and Electromechanical Phenomena at the Nanoscale,* Springer, Ch. I.2, pp. 31–87, 2007.
14. T. Clarysse, P. De Wolf, H. Bender, and W. Vandervorst, *JVST B* 14(1), 358, 1996.
15. P. Eyben et al., *Proceedings of 2003 International Conference on Characterization and Metrology for ULSI Technology,* Austin, TX, vol. 683, p. 678, 2003.
16. P. Eyben et al., *Mater. Res. Soc. Proc.* 717(C7.7), 2003.
17. P. Eyben et al., *JVST B* 28(2), 401, 2010.
18. J. Yang and F.C.J. Kong, *APL* 81, 4973, 2002.
19. D. Goghero, F. Giannozzo, and V. Raineri, *Proceedings of E-MRS 2002,* Strasbourg, France, 2002.
20. L. Zhang, *Proceedings of 12th International Workshop on Junction Technology (IWJT),* Shanghai, China, pp. 89–93, 2012.
21. T. Hantschel et al., *Microelectron. Eng.* 97, 255, 2012.
22. T. Hantschel, P. Niedermann, T. Trenkler, and W. Vandervorst, *Appl. Phys. Lett.* 76, 1603, 2000.
23. T. Trenkler et al., *JVST B* 18, 418, 2000.
24. D. Alvarez, J. Hartwich, M. Fouchier, P. Eyben, and W. Vandervorst, *APL* 82(11), 1724, 2003.
25. L. Zhang et al., *APL* 90, 192103, 2007.
26. P. Eyben et al., *JVST B* 28(2), 401, 2010.
27. T. Clarysse, P. Eyben, N. Duhayon, and W. Vandervorst, *JVST B* 21(2), 729, 2003.
28. S. Sheiko, M. Moeller, E.M.C.M. Reuvekamp, and H.W. Zandbergen, *Phys. Rev. B* 48(8), 5675, 1993.
29. P. Eyben, D. Degryse, and W. Vandervorst, *AIP Conference Proceedings*, Austin, TX, vol. 788, p. 264, 2005.
30. W.C.D. Cheong et al., *Nanotechnology* 11, 173, 2000.
31. D.E. Kim et al., *Nanotechnology* 17, 2259, 2006.
32. D.E. Kim and S.I. Oh, *J. Appl. Phys.* 104, 013502, 2008.
33. J. Tersoff, *Phys. Rev. Lett.* 56, 632, 1986.
34. J. Tersoff, *Phys. Rev. B* 39, 5566, 1989.
35. K. Mylvaganam, L.C. Zhang, P. Eyben, J. Mody, and W. Vandervorst, *Nanotechnology* 20(30), 305705, 2009.
36. P. Eyben et al., *JVST B* 28(2), 401, 2010.
37. C.F. Sanz-Navarro et al., *Nanotechnology* 15, 692, 2004.
38. R. Holm, *Electrical Contacts Handbook*, Springer, Berlin, Germany, p. 17, 1958.
39. http://ece-ww.colorado.edu/~bart/book/book/chapter2/ch2_7.htm
40. S.M. Sze, *Physics of Semiconductor Devices,* John Wiley & Sons, New York, 1981.

41. M. Büttiker, Y. Imry, R. Landauer, and S. Pinhas, *Phys. Rev. B* 31, 6207, 1985.
42. Y.V. Sharvin, *JETP* (21), 655, 1965.
43. M. Brandbyge et al., *Phys. Rev. B* 52, 8499, 1995.
44. http://www.ise.ch/products/dessis/
45. S. Denis, Master thesis, University of Liege, Liege, Belgium, p. 15, 2002.
46. E.H. Pointdexter et al., *JAP* 56(10), 2844, 1984.
47. C. Rascon, Internal report, LETI, 2003.
48. W.R. Thurber et al., *Semiconductor Measurement Technology*, NBS Special Publication, pp. 400–464, 1981.
49. T. Van den Zegel, Master thesis, Katholieke Hogeschool Kempen, Geel, Belgium, 2008.
50. P. Eyben et al., *Solid-State Electron.* 71, 69, 2012.
51. S.M. Hu, *JAP* 53, 1499, 1982.
52. T. Clarysse et al., *Mater. Sci. Eng. Rep.* 47(5–6), 2004.
53. T. Clarysse et al., *JVST B* 16, 394, 1998.
54. P. Eyben, S. Denis, T. Clarysse, and W. Vandervorst, *Mat. Sci. Eng. B* 102(1–3), 132, 2003.
55. T. Noda et al., *Proceedings of IEDM* 2006, San Francisco, CA, p. 377, 2006.
56. C. Ortolland et al., *Proceedings of IEEE Symposium on VLSI Technology*, Honolulu, HI, p. 186, 2008.
57. A. Nazir et al., *Solid-State Electron.* 74, 38–42, 2012.
58. K. Kuhn, IEDM short course, 2008.
59. J. Kavalieros, VLSI short course, 2008.
60. Y. Ashizawa, R. Tanabe, and H. Oka, *J. Comput. Electron.* 2006.
61. J. Mody et al., *Proceedings of IEEE Symposium on VLSI Technology*, Honolulu, HI, pp. 195–196, June 2010.
62. W. Vandervorst et al., *17th IIT*, AIP Conference Publications, Monterey, CA, vol. 1066, pp. 449–456, June 2008.
63. J. Mody et al., *JVST B* 26(1), 351–356, 2008.
64. J. Mody et al., *Proceedings of International Workshop on Junction Technology (IWJT)*, Kyoto, Japan, June 2011.
65. J. Mody et al., *Proceedings of IEDM*, Washington, DC, pp. 119–122, 2011.
66. Z. Chen et al., *IEEE Proc. Electron Dev. Lett.* 30, 754–756, 2009.
67. D.C. Mueller, E. Alonso, and W. Fichtner, *Phys. Rev. B* 68, 045208, 2003.
68. M.T. Bjork, H. Schmid, J. Knoch, H. Riel, and W. Riess, *Nat. Nano.* 4, 103–107, 2009.
69. A. Schulze et al., *Nanotechnology*, 22(18), 185701, 2011.
70. A. Schulze et al., *J. Appl. Phys.* 13(11), 114310, 2013.
71. M. Xu, PhD thesis, University of Leuven, Leuven, Belgium, 2005.
72. A. Lu, Master thesis, University of Liege, Liege, Belgium, 2012.
73. J. Mitard et al., *Proceedings of Technical Digest IEDM2010*, p. 249, 2010.
74. T. Noda et al., *Proceedings of IEDM 2012,* San Francisco, CA, p. 30.2.1, 2012.
75. N. Waldron et al., *ECS Trans.* 45(4), 115, 2012.
76. W. Steinhögl, G. Schindler, G. Steinlesberger, and M. Engelhardt, *Phys. Rev. B* 66, 075414, 2002.
77. B.Q. Wei, R. Vajtai, and P.M. Ajayan, *APL* 79, 1172, 2001.
78. A. Schulze, T. Hantschel, A. Dathe, P. Eyben, X. Ke, and W. Vandervorst, *Nanotechnology* 23(30), 305707, 2012.

37

Chromophores for Picoscale Optical Computers

Heinz Langhals
Ludwig Maximilians University of Munich

37.1 Introduction

IT technology requires both an increasing density of integration and a high rate of information processing; Moore's law [1] where a doubling of the density of information storage every 2 years is predicted expresses the technological development in this field and one may ask if there is a natural limit for the density of integration and the operating frequency.

IT devices are based on periodic electromagnetic interactions of their components and the rate of processing corresponds to an electromagnetic frequency ν; the latter is interrelated with the wavelength λ of an electromagnetic radiation by the equation $\nu = c/\lambda$ where c is the velocity of light in the corresponding medium such as the velocity in the vacuum (c_o); the characterization of the radiation by means of its vacuum wavelengths is more popular than the frequency ν. Very high frequencies pose high demands on the operating devices.

The application of increasing frequencies becomes problematic for more than about 500 THz because of induced ionization followed by the damage of the applied materials; this border region of frequencies corresponds to light radiation (see Figure 37.1). There is not a strict limit of frequency concerning ionizing and chemical bond–breaking processes because even near infrared light can be applied, however, only with very special chemical structures (compare NIR-sensitive photographic films). More and more chemical structures become accessible to photochemical processes with increasing frequency and shortening of the wavelengths until ubiquitary in the far ultraviolet. Ionizing may be still well controlled in the visible region of light by a careful selection of stable chemical structures. As a consequence, IT devices operating in the visible region may be the target of high-speed data processing with high integration where stable operating structures can be still constructed. The corresponding vacuum wavelength (λ_o) of the radiation involved is about half a micron and even lower in materials with higher dielectric constant ε ($\lambda = \lambda_o/\varepsilon$). As a consequence the dimensions of operating devices at such high frequencies should be minimally two magnitudes smaller than the involved wavelength of radiation to achieve independent operation, where even the diminishing of size until nanotechnology is not sufficient. Molecular dimensions of devices are required and the handling of such structures may be named picotechnology. On the other hand, such small dimensions down to the molecular level allow a very dense packing of devices. One may select either inorganic or organic compounds as operating units; inorganic materials are attractive at the first glance because many materials are highly thermally and chemically stable; however, the construction is comparably difficult and the majority of inorganic bonds are long, prohibiting a high degree of integration. Moreover, the recycling of inorganic compound materials is comparably difficult. As a consequence, organic materials are preferred where the recycling is unproblematic, firm, short chemical bonds are dominating, and the variability seems to be unlimited.

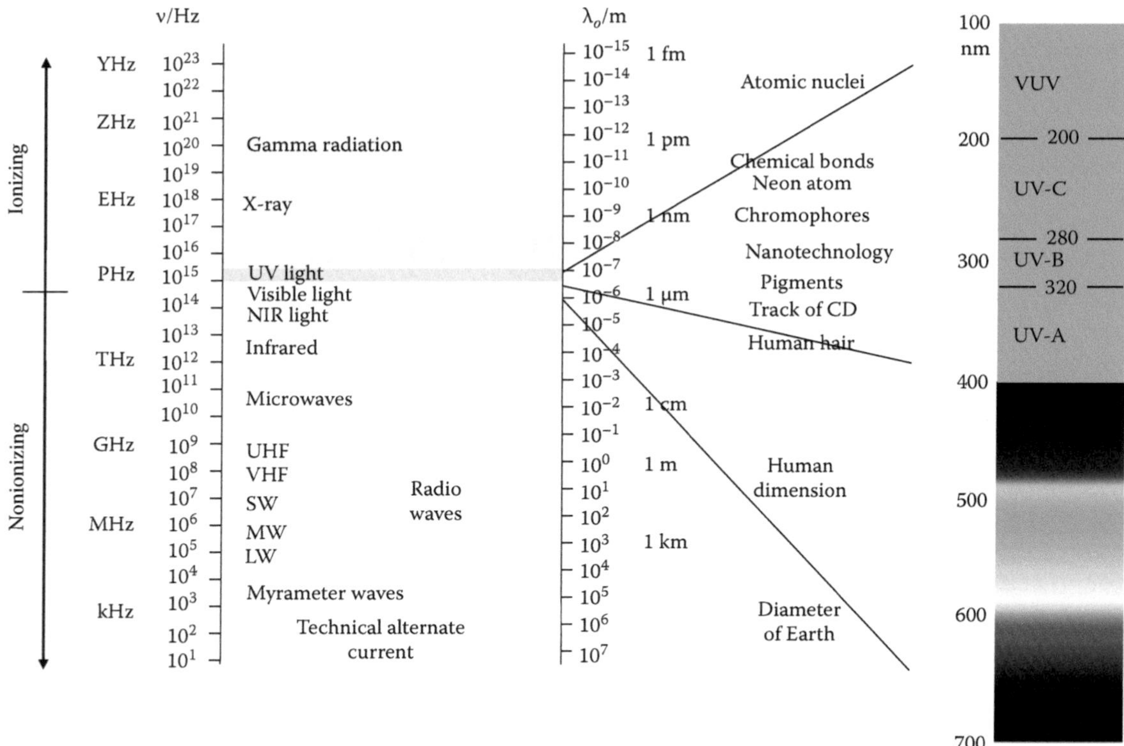

FIGURE 37.1 Regions of electromagnetic radiation.

37.2 Chromophores for IT Devices

The operating of IT devices is based on electromagnetic interactions and requires functional units, for example, for handling electric charge and energy at frequencies as high as 500 THz. On the other hand, stable frameworks are necessary for the arrangement of such units. Condensed aromatic and heteroaromatics, respectively, are attractive structures for such molecular architectures because they form stiff, firm, and stable σ networks with remaining delocalizable π electrons. Such aromatic networks combine both, frameworking and electromagnetic function. Diamantiode structures are attractive for insulating interlinking because of stiffness and chemical stability.

Complex aromatic systems with delocalizable electrons are well known from the structure of dyes and these may be fundamentally used for picotechnology; dye chemistry is well established both for synthetic procedures and for knowledge about the physical behavior of the involved structures. The delocalizing of bonds may be used, for example, for the transport of electrons such as in electrically conducting organic materials, whereas the electronic transitions can be applied as molecular resonators with the differences of eigenvalues ($\Delta E = h\nu$) as the resonating frequency ν (compare λ_{max} for a dye). However, the energy of electronic excitation gets lost in the majority of dyes such as indigo and other standard textile dyes in a comparably short time rendering the handling of the resonating energy difficult. Highly fluorescent chromophores are more attractive because the energy of electronic excitation is preserved until natural decay by the spontaneous emission of light radiation. The natural lifetime (τ) of this decay is determined by the transition probability from the electronically excited state to the ground state and this is interrelated with the molar absorptivity (ε) for dyes ($\tau \approx 10^{-4}/\varepsilon_{max}$). Standard UV/vis absorption spectra of fluorescent dyes result in fluorescence lifetimes between micro- and nanoseconds: a typical lifetime of strongly light-absorbing fluorescent dyes with molar absorptivities of 100,000 L/mol/cm is about 4 ns. The processing of stored energy has to be completed within this limit.

The fluorescent chromophore can be treated as an antenna for electromagnetic radiation. The molecular dimensions of standard chromophores of about 1 nm are small compared with the wavelengths of visible light of 500 nm for green light. There are several consequences of the very short length of the antenna, for example, this causes a higher transition probability in the UV with a slightly lower mismatch compared with the NIR. The effect of such a mismatch is quantitatively described by Ross' equation [2] and finds its counterpart in Rüdenberg's similar equation [3] for radio antennae. Moreover, the radiation with a wavelength of about 1 μm is macroscopic, whereas the molecular resonators are microscopic and there are special requirements between the technology for transmitters and molecular architectures.

Picometer devices require not only stable molecular frameworks but also units operating as fast as 500 THz, for example, for handling local charge and energy. The movement of electrons should be preferentially involved for such processing because of their low mass. The higher mass of atoms or groups of atoms interferes with a fast response; however, they may be applied for slower global processing. The mass of the proton is in between where proton transfer reactions can be applied for medium fast operation. All kinds of involved processes stress the basic chemical structures strongly because of repeated energetic loading and the generation of reactive species. As a consequence, extraordinarily stable aromatic framework structures are required.

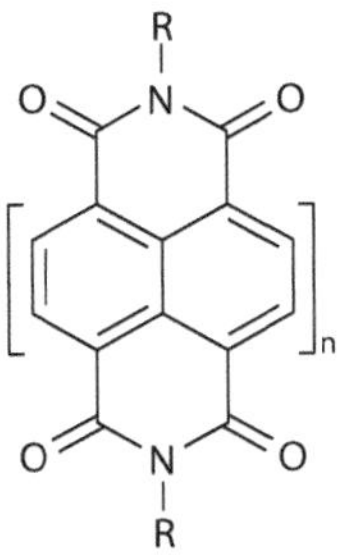

1

The *peri*-arylenes carboximides (sometimes shortened to rylenes and here more precisely named as *peri*-arylenes) **1** are promising candidates for picotechnology because of their extraordinarily high chemical and photochemical stabilities and high fluorescence quantum yields. The latter is an indicator for their ability to preserve energy of electronic excitation. There are the excellent properties on the one hand and the low solubilities of such condensed aromatic structures with terminating carboximides on the other interfering both with the synthesis of complex structures and the purification of the target structures. Such problems can be solved by the introduction of solubility increasing groups. The attachment of *tert*-butyl groups to polycyclic aromatic system increases the solubility appreciably [4]; *iso*-propyl groups, with the formal loss of one methyl group of the *tert*-butyl group, can still increase the solubility [5]; however, there is a tertiary benzylic hydrogen atom where radical reactions are favored; compare the technical Hock process [6] for the synthesis of phenol and acetone from *iso*-propylbenzene. The attachment of long-chain *sec*-alkyl groups to the nitrogen atoms of the carboximides **1** is a very efficient alternative for increasing solubility [7]. Two equally long chains are preferred in order to avoid stereogenic centers where difficultly separable mixtures of diastereomeric compounds would be obtained if two or more of such groups were attached. There is a nearly exponential increase of solubility with the chain lengths until the 1-hexylheptyl group, followed by a plateau of high solubility with a peak for the 1-nonyldecyl substituent [8]. The solubility increasing effect decreases again for longer chains. As a consequence, the 1-hexylheptyl substituent is a good compromise concerning the economy of attached aliphatic carbon atoms and the solubility increase. The 1-nonyldecyl substituent should be applied, if the effect of the former is not sufficient. On the other hand, the 1-nonyldecyl substituent is less economic concerning the attached aliphatic carbon atoms and generally more wax-like materials with more difficult purification are obtained in most cases, whereas well-defined crystalline solids are formed if the 1-hexylheptyl substituent is attached. As a consequence, the 1-hexylheptyl substituent is preferentially recommended for standard applications and the 1-nonyldecyl substituent if a further increase in solubility is necessary.

37.2.1 *Peri*-Arylenes

Many derivatives of the *peri*-arylenes with the general structure **1** have been prepared.

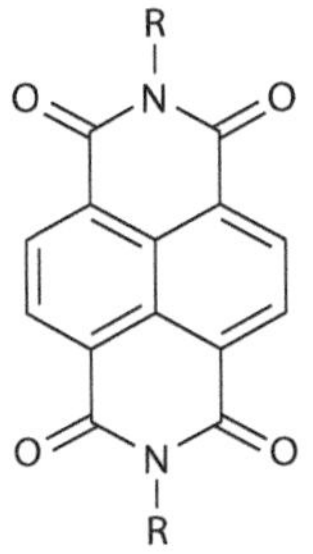

2

The lowest homologs of **1** with $n = 1$, the naphthalene-1,8:4,5-tetracarboxylic bisimides **2**, with aliphatic substituents R are colorless [9] because of their absorption in the UV; some reported rose to red to rose colors [10] are the consequence of intensely colored and difficultly removable by-products [9] from synthesis, mainly perylene derivatives. The UV/vis absorption spectrum of **2** is typically structured such as with other *peri*-arylenes (see Figure 37.2).

The fluorescence of **2** is weak because there are two closely neighbored electronic transitions [11] in the UV where a non-fluorescent transition causes a loss of the energy of excitation. The naphthalene carboximides **2** are of interest for UV applications such as sun protection [12] and as organic white pigments [13]; adverse effects seem to be weak and unimportant, even with photoactivation [14] and some derivatives were tested for the treatment of tuberculosis [15]. The naphthalene carboxbisimides **2** may be applied as components in optical devices for light absorption at short wavelengths and have to be monofunctionalized. J-aggregated materials were applied in white light emission OLEDs [16]. The symmetrically substituted carboxylicbisimides were first described in 1930 [17] and can be simply prepared by the full condensation of naphthalene-1,8:4,5-tetracarboxylic bis-anhydride with primary amines where pyrene is technically oxidized to get the starting anhydride. The mono-condensation of the bis-anhydride would formally lead to anhydride carboximides being ideal structures for monofunctionalization of **2**; however, their preparation is difficult because syntheses mostly result in mixtures of the bis-anhydride and the bis-carboximides (**2**); on the other hand, such mixtures

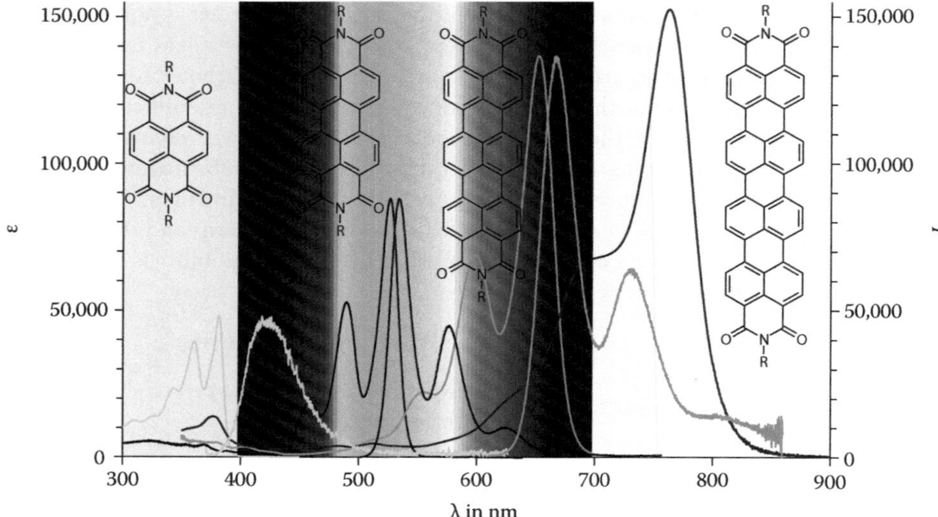

FIGURE 37.2 **(See color insert.)** UV/vis absorption (left, left scale ε) and fluorescence (right, right scale *I*) spectra of *peri*-arylenes (aliphatic substituents R) in chloroform. From left to right: **2** (yellow lines), **6** (black lines), **12** (turquoise lines), and **16** (blue line).

behave as anhydrides, carboximides to some extent because of the reversibility [9] of the condensation.

H_2SO_4 220°C RNH_2

3 **4** **5** **6**

The perylene-3,4:9, 10-tetracarboxylic bisimides **6** are the next higher homologs of the *peri*-arylenes [18]. The prolongation of the core shifts the structured light absorption into the visible with λ_{max} at 525 nm and a molar absorptivity of about 90,000 (see Figure 37.2). Derivatives with aromatic substituents R exhibit slightly higher absorptivities (≈95,000 in chloroform) than with purely aliphatic substituents (88,000). The fluorescence spectrum is mirror type to the absorption and very strong with fluorescence quantum yields close to 100% [19]. The thermal, chemical, and photochemical stability of **6** is extraordinarily high. Temperatures up to 550°C [20], concentrated sulfuric acid at 220°C, melted KOH [21], and concentrated bleach [22] are of no effect. The light fastness is extraordinarily high and exceeds other stable dyes so far such as Atto590 known as an unusually stable xanthene fluorescent dye [23]. As a consequence, the perylene dyes are one of the most stable known fluorescent dyes (adversary effects are not known).

Perylene dyes were first described by Kardos in 1913 [24], applied as vat dyes [25] and later on as high-performance pigments [26] where their low solubility was an obstacle for other applications. As a consequence, their strong fluorescence was detected comparably late by Geissler and Remy in 1959 [27]. Many novel applications became possible by the development of solubility increasing groups R in **6**, first *tert*-butylphenyl groups in 1980 [28], for example, the 2,5-di-*tert*-butylphenyl group [29] (named R10 group from the first publication). The formal diminishing of the *tert*-butyl group by one methyl group preserves an appreciably high solubility increase so that 2,6-di-*iso*-propylphenyl substituents became an alternative [30]; however, there are benzylic hydrogen atoms in the *iso*-propylphenyl groups accessible to free radical reactions; compare the technical Hock process [6] for the syntheses of acetone and phenol from *iso*-propylbenzene. Long-chain secondary alkyl groups (swallow-tail substituents) were developed for solubility increasing in 1987 [7] and proved to be appreciably more efficient. Two equally long chains in these groups should be preferred because different lengths would generate stereogenic centers; this should be avoided because otherwise racemates would be obtained with one such substituent and difficultly separable diastereomeric mixtures with two or more. The solubility of **6** increases exponentially with the chain length until 13C atoms where a high plateau of solubility is obtained for longer chains and a peak of very high solubility with 19C atoms [8]. Finally, the solubility decreases again for very long chains. As a consequence, swallow tails with 13C atoms (dye S-13, R = 1-hexylheptyl; RN 110590-84-6) are good compromises between solubility increasing and atom economy of the solubilizing group. A swallow tail with 19C atoms (dye S-19, R = 1-nonyldecyl) is recommended if a very high solubility increasing effect is required; however, such dyes are more waxlike and the handling and purification become more complicated than with the crystalline, powdery S-13. The solubility increasing by swallow-tail substituents is not limited to **6**, but is also found for other perylene dyes so that there is a general tool for the generation of highly soluble dyes.

Perylene dyes are technically prepared [31] from acenaphthylene (**3**), oxidation to naphthalene-1,8-anhydride, condensation with ammonia, subsequent coupling by means of melted KOH to form **4** followed by hydrolysis with concentrated sulfuric acid at 220°C to obtain the bis-anhydride **5** as the key intermediate in perylene chemistry; for example, perylene itself can be prepared by the base-induced decarboxylation of **5** [32]. The coupling of naphthalenedicarboximide to **6** can proceed under much milder reaction conditions with Sakamoto's method [33] by the application of potassium *tert*-butoxide and DBN (diazabicyclo[4.3.0]non-5-ene).

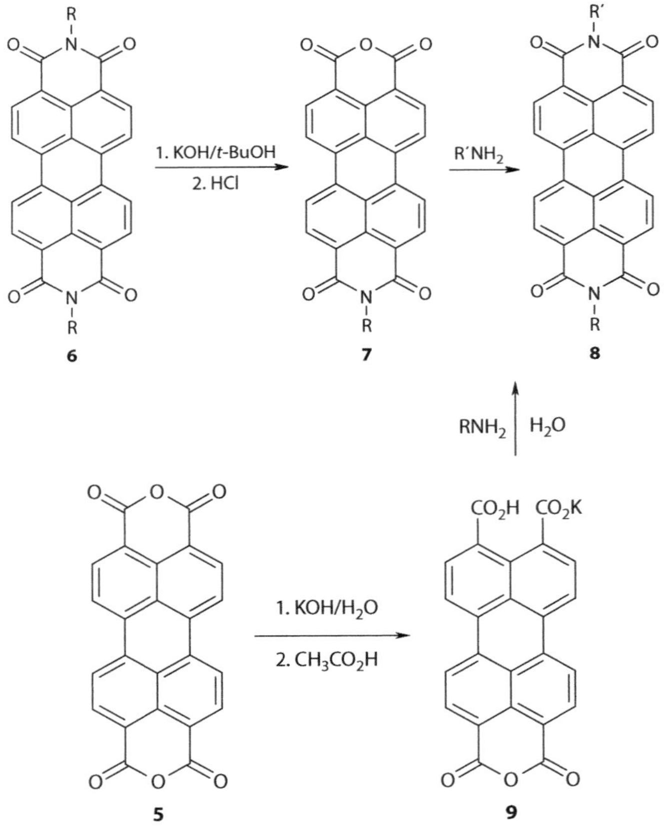

1. KOH/*t*-BuOH
2. HCl

$R'NH_2$

6 **7** **8**

RNH_2 | H_2O

1. KOH/H_2O
2. CH_3CO_2H

CO_2H CO_2K

5 **9**

The perylene dyes **6** are generally formed by the condensation of the bis-anhydride **5** with primary amines such as with 1-hexylheptylamine for the preparation of the very soluble and highly fluorescent dye S-13. The reaction medium is problematic; quinoline proved to be suitable [34] where zinc salts such as zinc acetate and zinc chloride, respectively, support the reaction and may be a solubilizing agent for the starting material **5** [18c]. Melted imidazole [35] was found to be an alternative where reactions proceed more readily in most cases. Aliphatic primary amines are more reactive and reactions are possible in ethylene glycol [36] at elevated temperatures or even in neat amines [37]. A mono-functionalization of these dyes such as in **8** is of special interest because of fluorescent labeling and the application of this chromophore in complex functional units. The attachment of a linker group to one nitrogen atom of **8** (R′) and a solubilizing group such as a 1-hexylheptyl to the other one (R) is straightforward for such applications. A perylene anhydride carboxylic imide **7** with a solubilizing *N*-substituent R is an attractive intermediate for such asymmetrically substituted perylene dyes **8** because of condensation with various primary amines $R'NH_2$; however, the preparation of **7** forms a special synthetic problem. The simple condensation of an excess of **5** with a minor amount of a primary amine forms nearly exclusively **6** besides remaining **5**. An acid-catalyzed [38] partial hydrolysis of **6** requires rough reaction conditions being problematic for many substituents R and melted KOH for basic degradation is of minor effect. Surprisingly, a mixture of KOH in *tert*-butylalcohol allows a stepwise hydrolysis of **6** in a comparably fast and clean reaction to obtain the anhydride–carboxylic imides **7** after acidification [39]. As an alternative synthetic way [40], anhydride carboxylic imides **7** can be synthesized by means of the preparation of the tetra-propylester from the bis-anhydride **5** and propanol, subsequent reaction with 7-aminotridecane, and hydrolysis of the remaining intermediate ester function. This method is more time-consuming than the hydrolysis of **6**; however, it is more economic concerning the consumption of 7-aminotridecane for the preparation of soluble perylene dyes because only one mole of the amine is required per mole of **7**. Finally, there is an alternative way for the preparation of **7**. The bis-anhydride **5** is converted to the tetra potassium salt by alkaline hydrolysis and then cautiously acidified where the anhydride potassium salt **9** precipitates because of its extraordinarily low solubility shifting all equilibria to **9**. Tröster developed this method by the application of a phosphate buffer [41]; however, the removal of residual phosphate is difficult. Better results were obtained with acidification by means of acetic acid to form an acetate buffer [39] because the volatile acetic acid can be more easily removed. The potassium salt **9** can be condensed to the anhydride carboximide **7** in the aqueous phase. This is an alternative way to the intermediates for unsymmetrically substituted perylene dyes **8**; however, this synthetic route is limited to water-soluble amines $R\text{-}NH_2$.

Terrylene-3,4:11,12-tetracarboximides (**12**), briefly terrylene dyes, form the next higher homologs in the series of *peri*-arylenes (for substituted *peri*-arylenes, see [42]). The prolongation of the central aromatic ribbon causes an appreciably bathochromic shift to obtain blue solutions with a structured absorption spectrum at λ_{max} = 625 nm and ε_{max} = 130,000 exhibiting a strong, red fluorescence spectrum mirror type to the absorption and a fluorescence lifetime of 3.1 ns (see Figure 37.2). The fluorescence quantum yield of 60% was described for the first derivatives [43] where the attachment of the strongly solubility increasing 1-nonyldecyl group to the nitrogen atoms increases the quantum yield to 94% [44]. This dye is recommended as a fluorescence standard.

Terrylene tetracarboxylic bisimides were first prepared [45] from the bromoperylene dicarboximide **10** [46] in multisteps where **10** was prepared from **14** [46]; special care is necessary for the problematic mixture of an oxidant (MnO_2) with a reductant (EtOH). Synthesis can be appreciably simplified by an application of Sakamoto's method [33] for a direct cross coupling of the perylene carboximide **14** with the naphthalene carboximide **13** [47]. A cross coupling is very strongly favored, so the formation of homocoupling products of both **13** and **14** becomes unimportant.

Quaterrylene-3,4:13,14-tetracarboximides (**16**), briefly quaterrylene dyes, are the next higher homologs of **2** (**1**, n = 4) and form turquoise green solutions with λ_{max} = 781 nm and ε = 152,000. Their absorption spectra are not as strongly structured as with the other *peri*-arylenes (see Figure 37.2). A saddle point is formed at the hypsochromic flank of the absorption spectrum. The fluorescence spectrum is mirror type and extends into the NIR. Quaterrylene dyes were first prepared by Müllen and Quante [48] in a coupling of **10** where the toxic Ni(cod)$_2$ had to be applied for Yamamoto coupling [49] to form **15** and the application of KOH/MnO_2 in ethanol (compare [50]) for the final ring closure to **16** was reported [48a]; however, no **16** could be detected with an conventional excess of reagents for R=1-hexylheptyl [51] and a very large excess of reagents and gave a complex mixture of products containing some **16**. Moreover, the applied mixture of an oxidant (MnO_2) with a reductant (EtOH) seems to be somewhat problematic. On the other hand, the reported mixture of KOH and glucose under rough reaction conditions at 150°C–160°C and nitrogen atmosphere was successful [48b] although a separation of **16** from the complex mixture of the decomposed glucose proved to be tedious

[51]. A synthesis with the less toxic components $NiCl_2/(C_6H_5)_3P/Zn$ was developed [52] for the coupling of **10** and the application of Sakamoto's method for the final ring closure of **15** to **16** [53] proceeds efficiently. Even alkali melt of **14** analogous to the synthesis of perylene dyes is successful for two-step synthesis of **16** starting with technical material [54]; however, comparably low yields were obtained due to the rough reaction conditions.

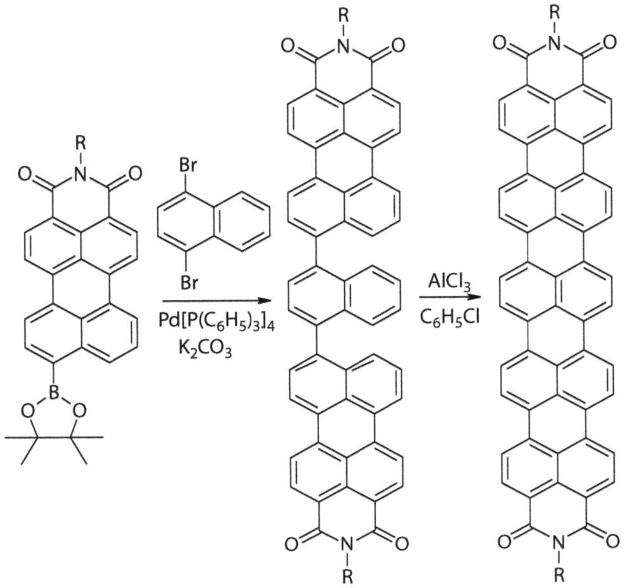

Quinterrylene-3,4:15,16-tetracarboxylic bisimides (**19**), briefly quinterrylene dyes,* were prepared from 1,4-dibromonaphthalene and the boronic derivative of perylene-3,4-dicarboxylic imide **17** by palladium-mediated coupling and subsequent ring closure with aluminium chloride in chorobenzene [55]. Sexterrylene-3,4:17,18-tetracarboxylic bisimides **21** [55] were analogously prepared to the quinterrylene dyes by the coupling of dibromoperylene with boroperylenecarboximide. The attachment of solubility increasing groups is of central importance for synthesis and handling of such extended systems where long-chain aliphatic swallow-tail substituents such as 1-hexylheptyl proved to be most appropriate. In spite of these solubility increasing groups, a pigment-like low solubility of these compounds was reported [55]. The UV/vis absorption of quinterrylene bisimide (R = 1-heptyloctyl) in chloroform was referenced to 831 nm [55]; no data were reported for sexterrylene dyes.

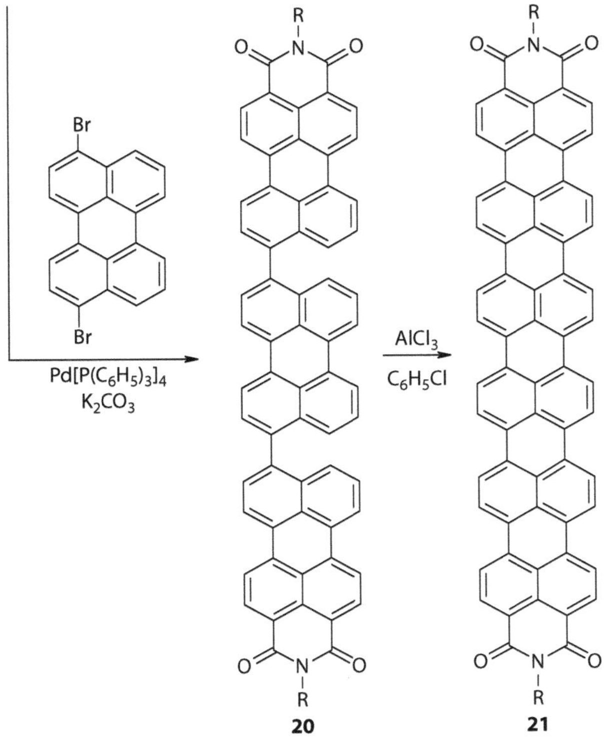

37.2.2 Core-Substituted *Peri*-Arylenes

There are orbital nodes in HOMO and LUMO in *peri*-arylenes along the connection line between the two nitrogen atoms of the carboxylic bisimides. As a consequence, the UV/vis spectra are only little affected by substituents at these positions because the most bathochromic light absorption is a very pure electronic transition between these orbitals. This is of advantage for some applications such as fluorescence labeling because these substituents can be easily introduced by the condensation of primary amines with the corresponding anhydrides and can be attached to reactive groups. On the other hand, this invariancy of the UV/vis spectra concerning the *N* positions does not allow an adaption of the spectra to special requirements. The UV/vis spectra can be controlled by core substitution. The *peri*-arylenes can be interpreted as an inverse donor acceptor system with the terminal carbonyl groups as the acceptors, the aromatic system as an interconnecting π-system with lacking donor groups in the center [18]. As a consequence, there is only a little effect by electron acceptor groups at the core; however, strong bathochromic shifts are induced by donor groups.

* Such compounds were initially named "penterrylen" dyes and the next higher homologs "hexerylene" dyes; this is somewhat inconsistent because Latin is mixed up with Greek. Terrylene is derived from the Latin *ter* and means three times and the quaterrylene from the Latin *quarter* for four times, respectively. As a consequence, the next homologs should be named quinterrylene from *quinquies* for five times and sexterrylene from *sexiens* for six times. The next higher homologs would be septerrylene (*septiens*) and octerrylene (*octiens*); compare the steps in the scale in music.

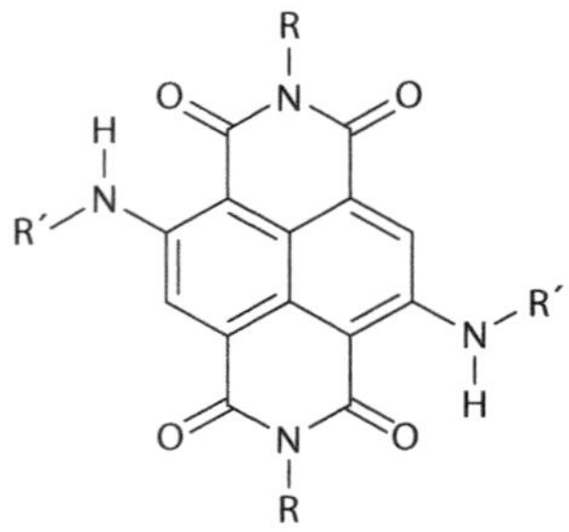

22

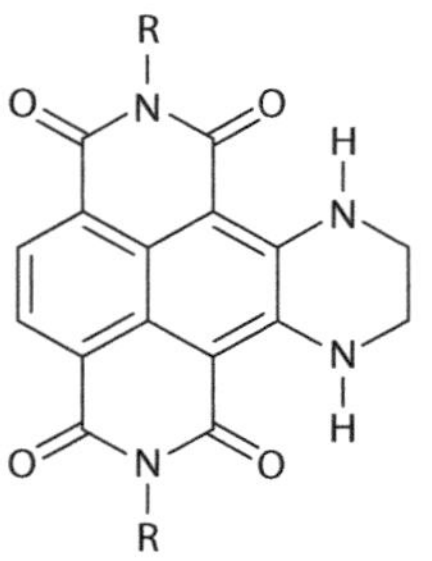

23

The light absorption of the colorless naphthalene biscarboximides **2** is shifted into the visible by the attachment of donor groups to the aromatic core [56] such as in **22**; a neighbored double substitution in **23** causes an appreciable bathochromic shift (500 nm) [9] so that these materials can be applied as colorants. The fluorescence of **23** is comparably stronger than that of **2**; this may be a consequence of the spectral separation of the two electronic transitions in **2**.

Donor groups can be introduced into perylene dyes by means of halogenation in the positions 1, 6, 7, and 12 (mostly named "the bay region" according to the structure formula of the carbon skeleton although this is the most crowded position in the molecule due to the hydrogen atoms and other substituents, respectively) and subsequent nucleophilic exchange such as with phenoxy groups and amino groups, respectively. Four phenoxy groups [57] in **24** cause a shift to obtain a red fluorescent material; this way is preparatively problematic because both a complete halogenation and exchange is difficult and a separation of mixtures is very difficult. The introduction of one methoxy group [58] in **25** causes already an appreciably bathochromic shift to 552 nm with an intense red fluorescence at 575 nm. The introduction of two amino groups [59] to form centrosymmetric structures is described by bromination and exchange reactions of the unsubstituted carboximides and form strong donor groups in **26**. The substitution by carbon atoms such as with simple alkynes [60] is described and affects the UV/vis spectra only slightly; however, such structures are of special interest for the establishing of even more complex units. A mono nitro derivative can be obtained [61] by a simple nitration where a variety of products such as **27** were obtained by reduction and further reaction.

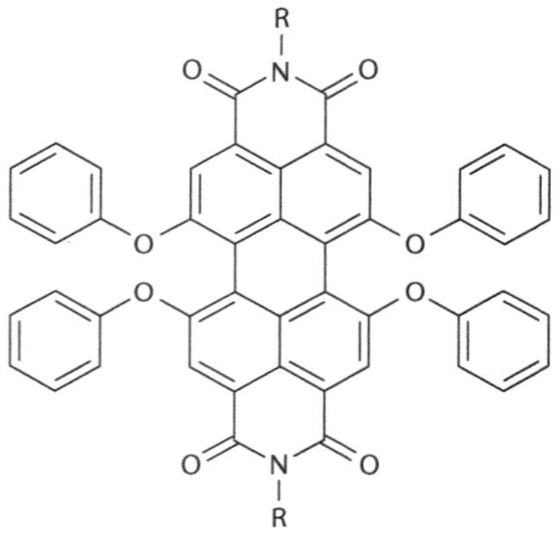

24

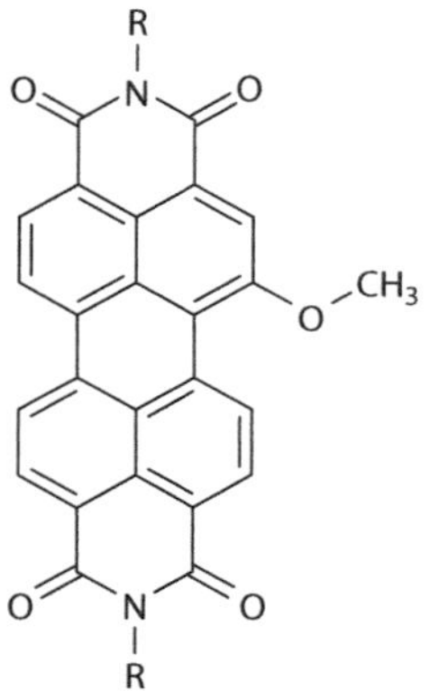

25

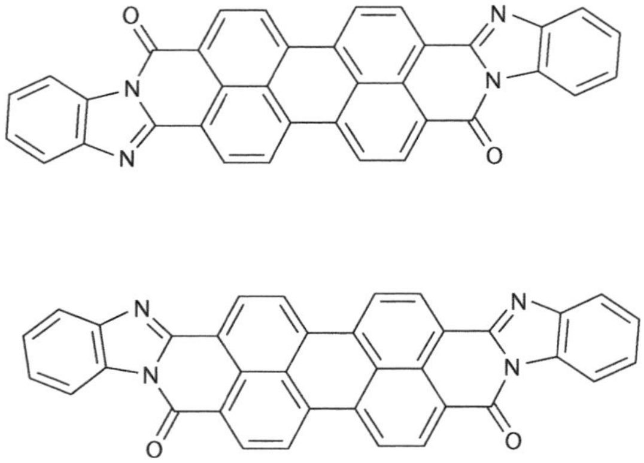

28/29

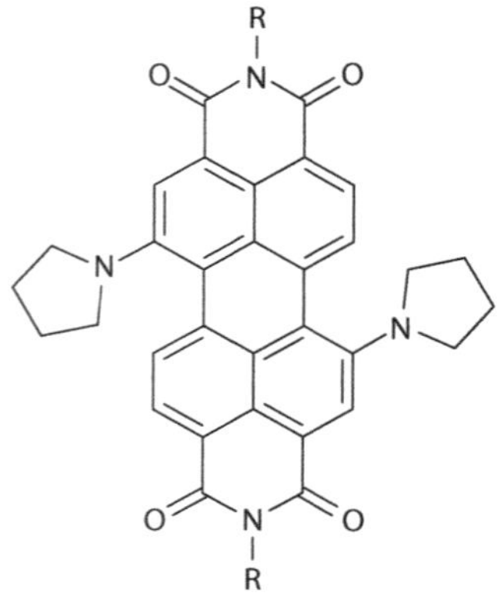

26

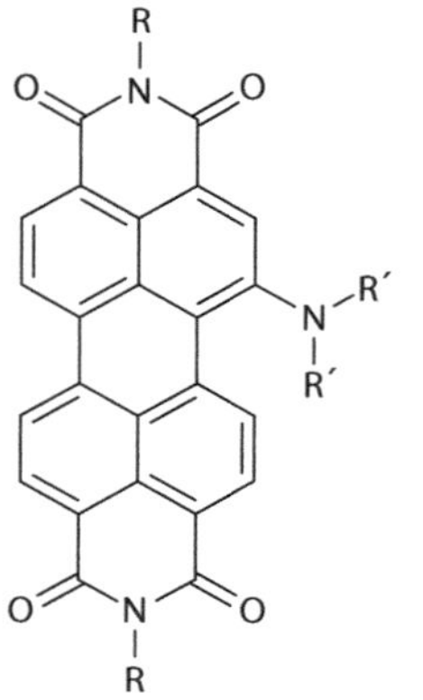

27

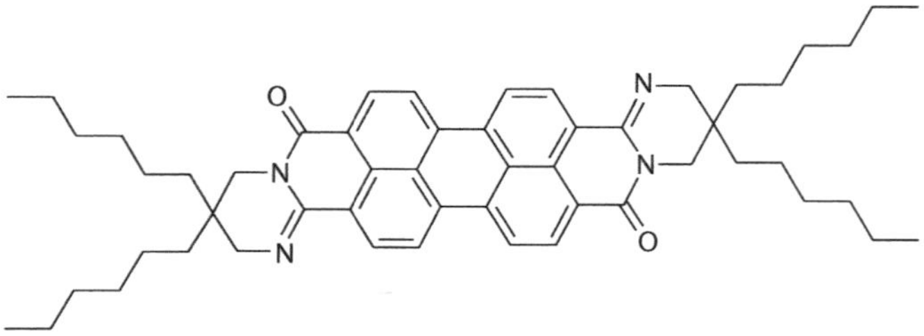

30

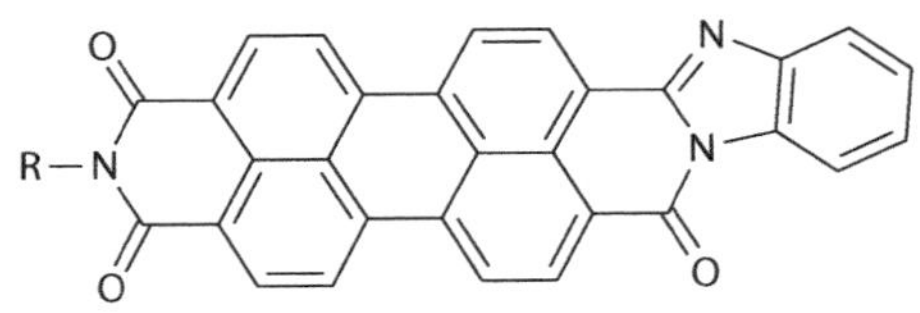

31

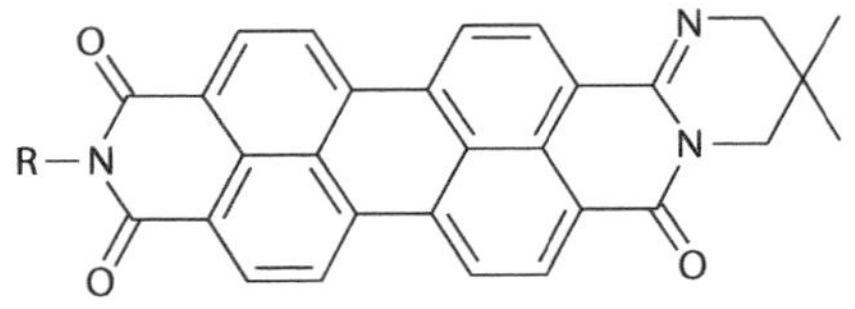

32

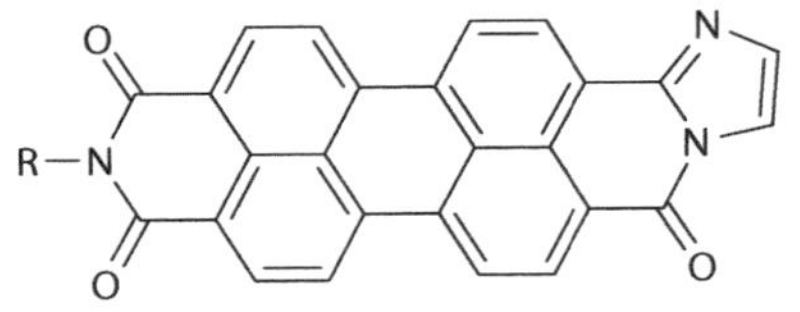

33

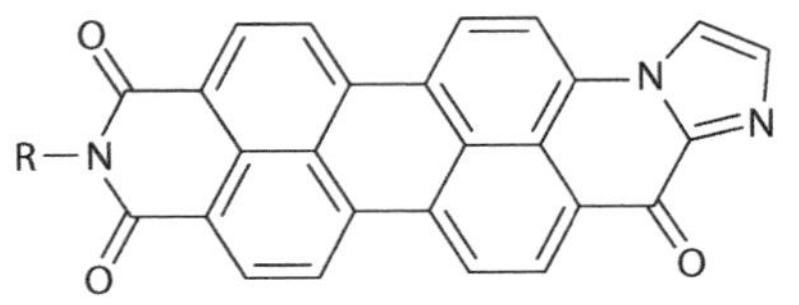

34

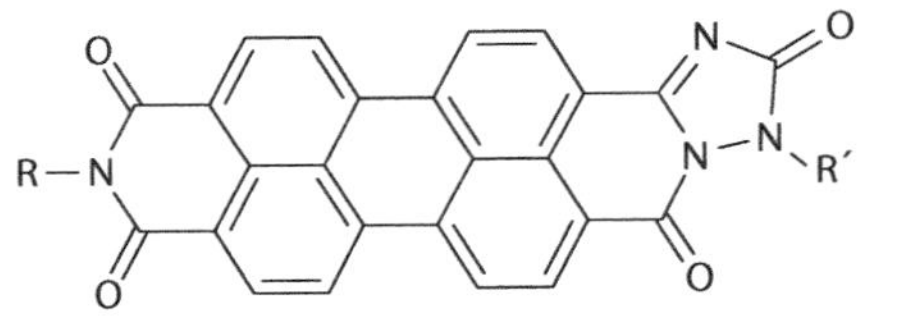

35

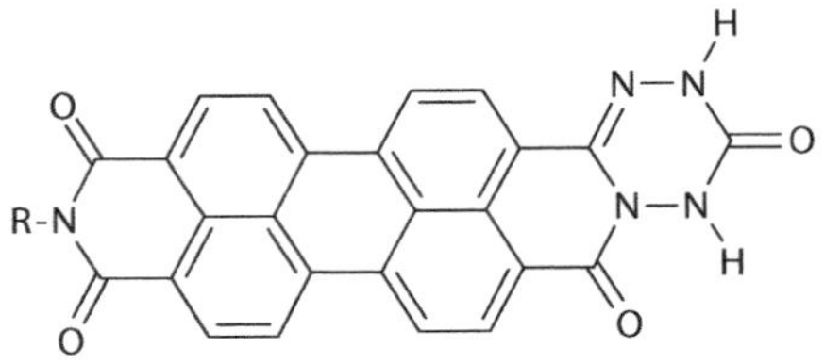

36

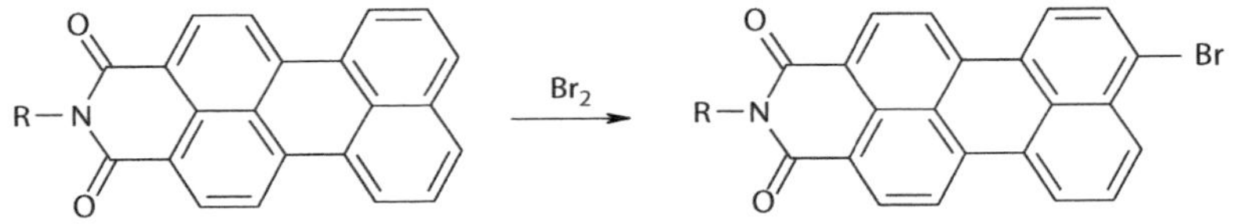

14

Br_2

10

37.2.3 Modification of the Carboxylic Imide Structures

The alteration of the carboxylic imide structures is an efficient way for the adaption of UV/vis spectra. An exchange of carbonyl groups with imino groups causes a bathochromic shift in the UV/vis spectra. However, such structures are labile concerning hydrolysis and require stabilization such as in **28/29** by means of the incorporation into five-membered rings and a further extension of the aromatic system [62]. Such compounds absorb in the bathochromic visible; however, their solubilities are low and they form a difficultly separable mixture of isomers **28** and **29**. Stable compounds are also obtained by the incorporation of the imino group into six-membered rings where a further stabilization can be achieved by the application of geminal methyl groups. Highly soluble materials are obtained if the methyl groups are replaced by long alkyl chains so that integrated swallow-tail structures are obtained in **30** [63]. A single exchange of one carbonyl group by an imino group gains the majority of bathochromic shifts, avoids the formation of difficultly separable isomers such as **28/29**, and leaves the still remaining carboxylic amides for the attachment of a solubilizing group allowing a more easy purification and handling. Compounds **31** and **32** [64] correspond to the symmetrically substituted **28/29** and **30**, respectively. An vinylen linker in **33** means the simplest element of structure [65]; however, the fluorescence quantum yield is low. The strongly fluorescent isomeric compound **34** is obtained by a rearrangement caused by the interaction with sterically hindered amines, presumably according to Regel's mechanism [66]. The UV/vis absorption and fluorescence spectra of **34** are very similar to terrylene; however, both the photostability and fluorescence quantum yields are appreciably higher so that they may replace the comparably labile hydrocarbon terrylene for physicochemical investigation. An incorporation of the imino group into more complex heterocyclic ring systems is described where a variety of compounds was obtained by the condensation of a perylene anhydride with carbazide, thiocarbazide, carbohydrazide, and thiocarbohydrazide [67] such as **35** and **36**; a subsequent Schönberg reaction of the thio compounds allows a further extension of the structure.

One carboxylic imide unit of **6** can be removed from the technically available bis-anhydride **5** as the starting material. **5** is either water-induced decarboxylizing condensed with primary amines [46] to form **14** or balanced decarboxylized to the *peri*-dicarboxylic anhydride [68] and further condensed with primary amines to obtain **14** [69]. As an alternative, the anhydride carboximide **7** can be copper-mediated decarboxylized [53,70] to form **14**. The mono carboximide **14** is a key intermediate for many other perylene derivatives because of the free *peri* positions. The mono bromination of

14 can be directed to the free *peri* position forming **10** as an intermediate for the construction of even more complex structures [71] by transition metal-mediated C–C coupling. The introduction of acceptor groups into the *peri* position of **14** affects the UV/vis spectra only slightly and the vibronic structure resembles the spectra of perylene tetracarboxylic bisimides [46]. On the other hand, an introduction of donor groups such as amino groups causes a strong bathochromic shift with lowering the fluorescence quantum yields.

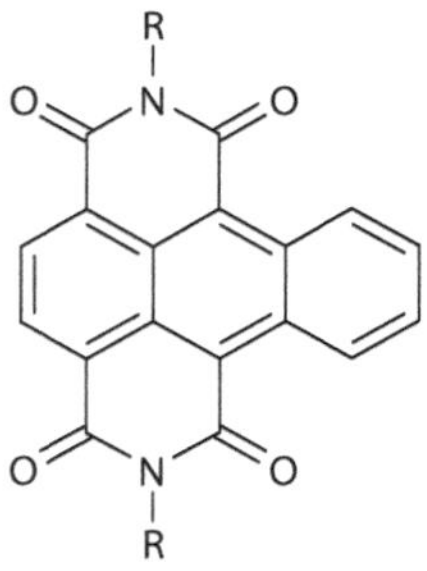

37 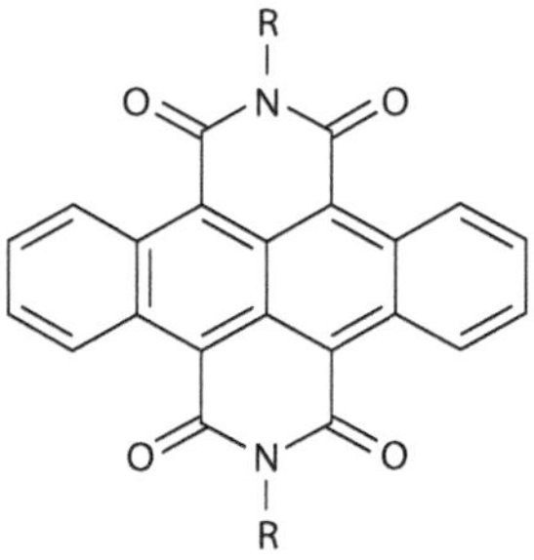 **38**

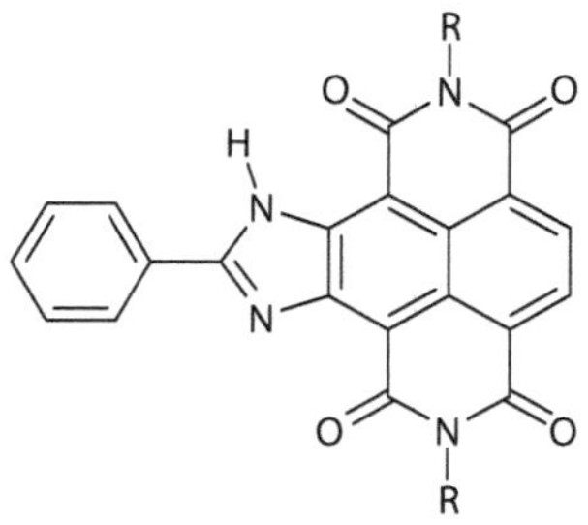

39

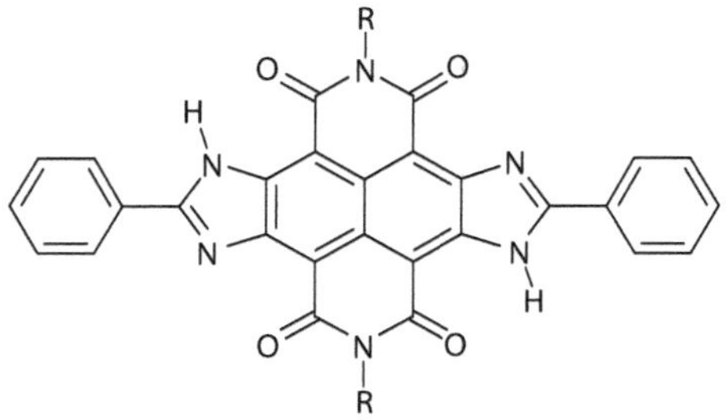

40

The annellation of **2** with heterocyclic rings such as with imidazolo rings [76] in **39** and **40**, respectively, induces strong bathochromic shifts into the visible of the colorless **2** where an absorption maximum of 442 nm is obtained for **39** and 512 nm for **40**; a substitution of the 4 and 4′ positions of the phenyl groups of **40** (dye **40a**) causes a further bathochromic shift so that even blue solutions are obtained. The bathochromic shift of the absorption causes a spectral separation of the two electronic transitions of **2** such as in **22** and **23** and diminishes fluorescence quenching. As a consequence, the fluorescence quantum yield of **2** increased to 28% for **39** and to 53% for **40**, and these substances may be used for fluorescence applications (Figure 37.3).

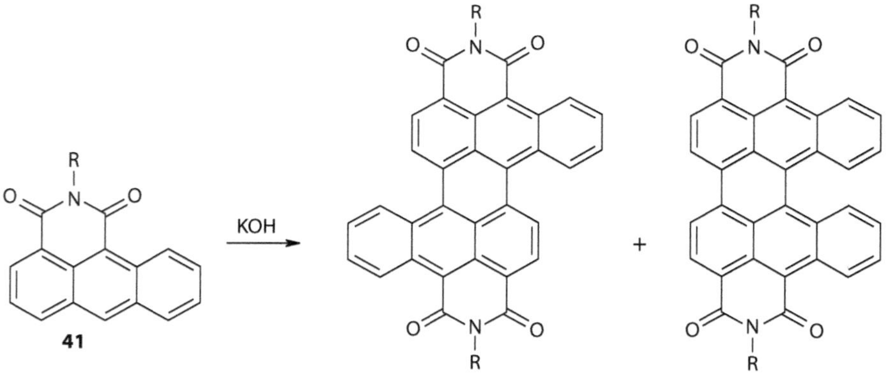

37.2.4 Laterally Extended *Peri*-Arylenes

A lateral annellation of the *peri*-arylenes with carbocyclic or heterocyclic rings opens many opportunities and options and extends the UV/vis palette appreciably. The extension of the formal naphthalene unit in **2** with a benzoic ring in **37** is unknown [72]. Compounds **38** with a double expansion were synthesized via bismuth triflate–mediated double-cyclization reaction of acid chlorides and isocyanates [73] or synthesized based on direct double ring extension of electron-deficient naphthalene diimides involving metallacyclopentadienes [74]. The linear condensation of many aromatic systems to subsystems larger than anthracene such as in **38** is problematic in many cases because of photobleaching by the addition of singlet oxygen, generated by photoreactions of the chromophore itself; the photo bleaching of the strongly fluorescent rubrene in the presence of atmospheric oxygen is one of the most prominent examples of such processes [75].

41 KOH → **42** + **43**

A benzannellation of perylene dyes concerning the naphthalene units shifts the absorption appreciably bathochromically to the edge of the visible so that emerald green solutions of aceanthrene green are obtained [77]. The *trans* isomer **42** [78] and the *cis* isomer **43** [78b] exhibit similar UV/vis absorption spectra with an appreciable fluorescence extending into the NIR. Such compounds are constructed by an oxidative *peri* coupling of spectra anthracene carboxinides **41**.

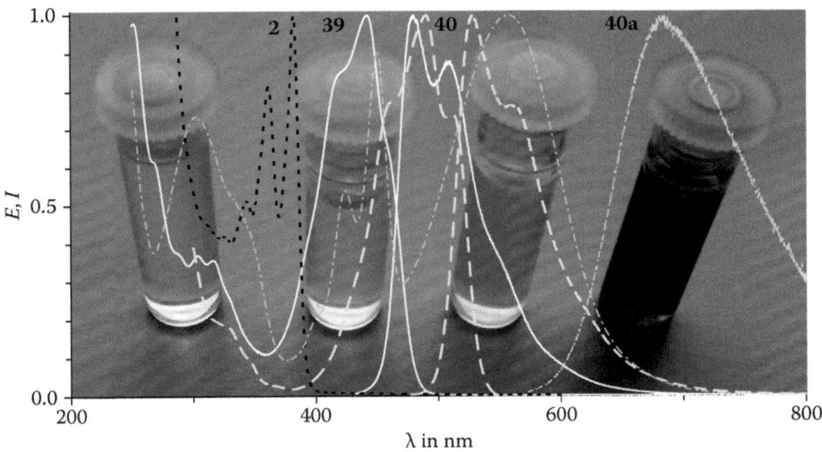

FIGURE 37.3 **(See color insert.)** UV/vis absorption (left, *E*) and fluorescence (right, *I*) spectra of imidazolonaphthalene dyes (aliphatic substituents R) in chloroform. From left to right: **2**, (black lines), **39** (green lines), **40** (turquoise lines), and **40a** (blue lines). Photos of solutions of **2**, **39**, **40**, and **40a** in chloroform from left to right.

A benzoannellation of **6** in the positions 1 and 12 forms the more hypsochromically absorbing benzoperylene derivatives **45** [79]. These compounds can be prepared by the Clar variant of the Diels–Alder reaction of **6** with maleic anhydride with simultaneous oxidative aromatization to **44** and subsequent decarboxylation to form **45**. The non-decarboxylated precursor **44** is of special interest because two solubilizing swallow-tail substituents R can be efficiently attached to the six-membered ring N atoms leaving the anhydride free for further condensation with primary amines to form benzoperylene hexacarboxylic trisimides. Single anchor groups can be introduced (R′) by the condensation with the appropriate amines for the construction of more complex units. Such compounds are highly fluorescent where quantum yields of about 85% were observed so that they become useful energy donors in dyads. The lacking of 15% of quantum yield can be mainly attributed to intersystem crossing (ISC) [80] being unimportant for simple perylene dyes **6**. On the other hand, this is an interesting possibility for the generation of triplet states.

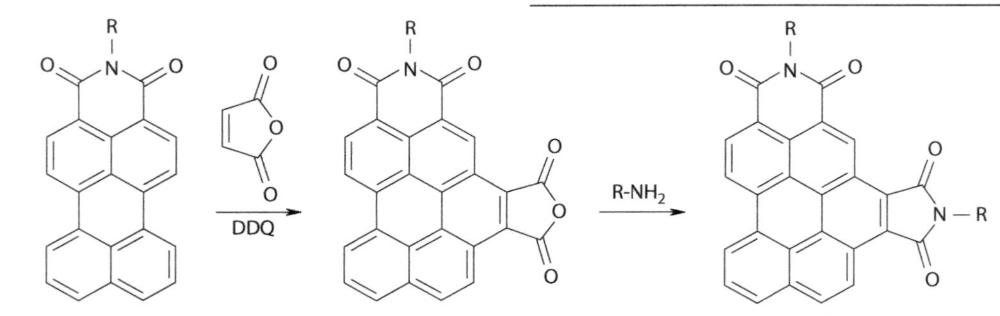

Angular benzoperylene tetracarboxylic bisimides **48** can be prepared by a similar reaction sequence with a Clar–Diels–Alder reaction of perylene dicarboximides **14** to form the anhydride **47** and subsequent condensation with primary amines [81]. Their pronounced tendency for ISC is even higher than for **46** indicated comparably by strong phosphorescence. **48** may be applied for an efficient generation of singlet oxygen [82] reaching the widely applied tetraphenylporphyrine as sensitizer. Moreover, the photostability and chemical robustness of **48** are appreciably higher than for the latter so that the benzoperylene derivatives are becoming useful reagents for the preparative photo-induced generation of singlet oxygen.

The Clar–Diels–Alder reaction of the biscarboximide **6** with 4-PTAD (4-phenyl-1,2,4-triazoline-3,5-dione) forms very bathochromically absorbing materials because of amplified electron donating by the alpha effect (neighbored lone pairs) where the product of a single addition absorbs in the bathochromic visible and the double adduct mainly in the NIR [83]. The annellation with an imidazolo ring yields not only bathochromically shifted absorption spectra, but also a strong, red fluorescence for **51** [84]. The UV/vis spectra can be further shifted by the annellation with two or more imidazolo rings to obtain green solutions with fluorescence in the NIR.

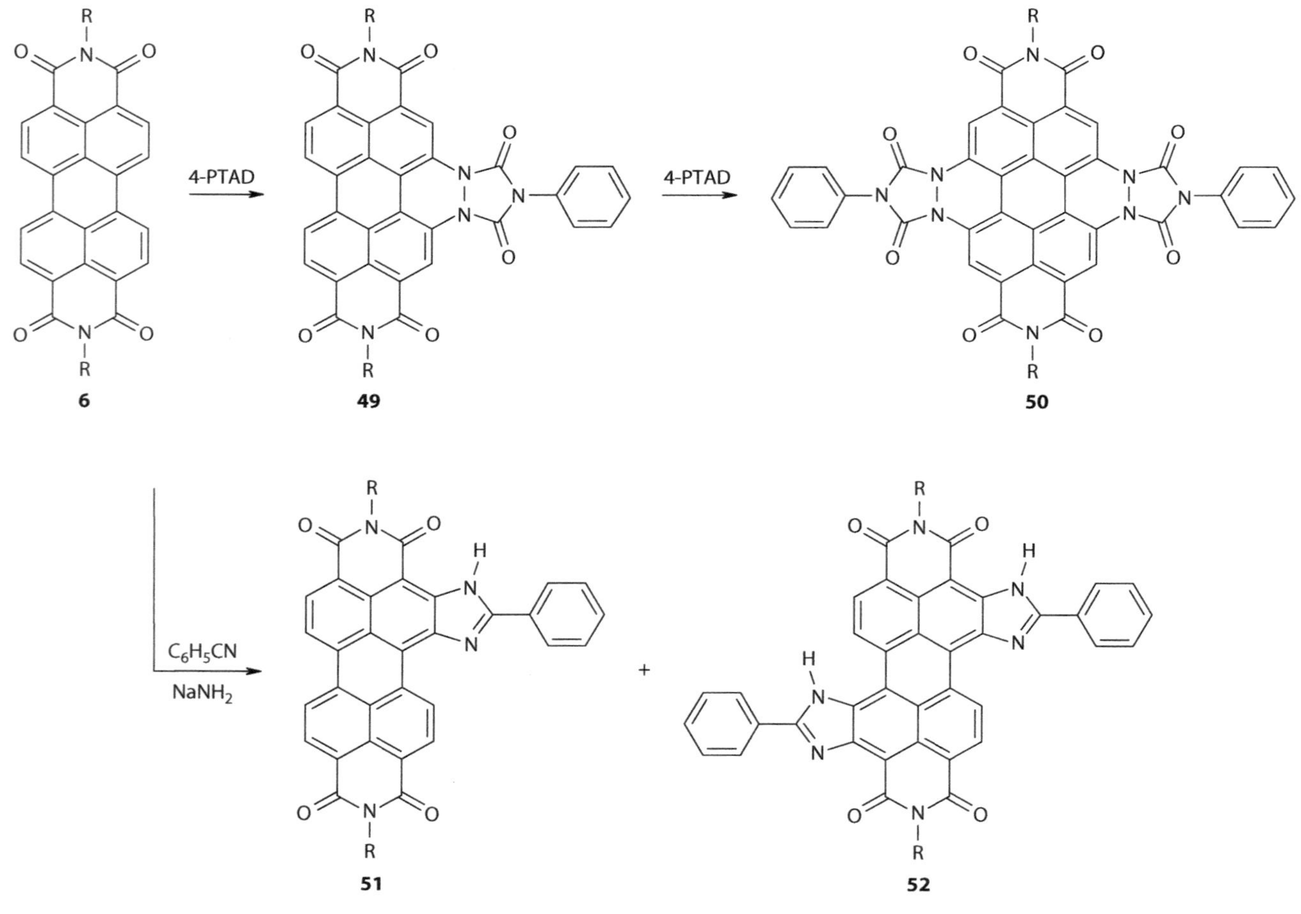

Annellation is also successful with terrylene derivatives. Clar–Diels–Alder reactions with maleic anhydride proceed in the same way as was described for perylene dyes and induce hypsochromic shifts. A condensation with primary amines forms the highly reddish orange fluorescent benzoterrylene triscarboximides **54** and the compound **57** with a more hypsochromic yellow fluorescence [85]. The UV/vis spectra of **57** resemble the spectra of **6**; however, the different substituents R and R′ allow various substitutions such as a solubilization by R and the attachment to other structures with R′. An annellation with an imidazolo ring in **55** [86] causes an appreciable bathochromic shift to 710 nm and a very strong-structured fluorescence at 720 nm. The introduction of additional carbocyclic rings in the center can be achieved analogously to the stepwise synthesis of the terrylene skeleton from a dibromonaphthalene and two naphthalimide units where a substituted anthracene was applied at the center. The same strategies were followed for the synthesis of benoquinterrylenes and dibenzoquinterrylenes. Even more complex structures with many condensed carbocyclic rings were prepared such as a quinterrylene dye with a central tetracene unit [87]; however, the problem of solubility and tendency for aggregation of such extended aromatic systems becomes more and more problematic.

37.3 Exciton Coupling of Chromophores

The *peri*-arylenes are very suitable for the construction of more complex structures where the only single electronic transition in the visible polarized along the N–N connection line of perylene dyes and higher homologs is convenient for definite interpretation of photophysical effects. The arrangement of two or more identical chromophores causes exciton interactions with characteristic alterations in the spectra controlled by the intensity of such interactions.

Exciton interactions [88] of two identical chromophores cause a splitting of two initially energetically equal in two novel electronic transitions (Davidov splitting [89]), one more bathochromic, and one more hypsochromic one (even a splitting in the ground state will be caused by an energetic interaction; see Figure 37.4). The orientation of transition moments controls the intensities of these transitions. A linear arrangement of the transition moments favors the bathochromic α band and suppresses the hypsochromic one (Figure 37.4b, middle). The molar absorptivity and oscillator strengths, respectively, become more than two times the absorptivity of two isolated chromophores because of an effective prolongation of the antenna (constructive exciton effect); as a consequence, intensely light-absorbing dyes can be constructed [90]. Such an arrangement corresponds to J-aggregates [91] (many involved chromophores); we name such an arrangement of chromophores J-arrangement. The other extreme means two cofacially covering parallel transition moments (Figure 37.4b, middle and bottom) where the hypsochromic β transition is favored and the bathochromic is suppressed where the molar absorptivity is less than two times of isolated chromophores because the antenna is shortened (destructive exciton effect). This corresponds to H-aggregates [88b] and is named H-arrangement. The effect varies with the enclosed including angle between the chromophores and a linear shift. Fluorescence is effective for the J-arrangement and suppressed for the H-arrangement because of the symmetry of electron movement [88c]. Finally, a skew arrangement of the transition moments allows both transitions where fluorescence is derived from the J-type component and bathochromically shifted compared with a single chromophore.

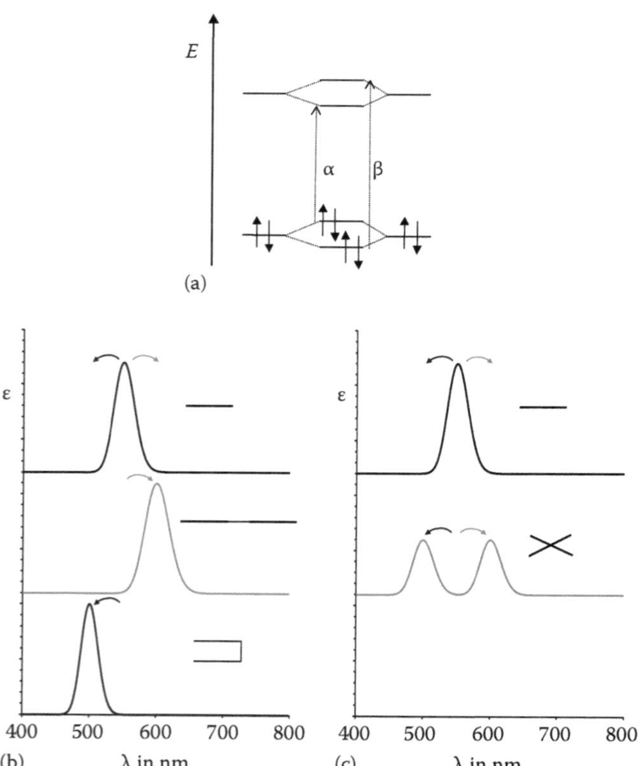

FIGURE 37.4 Exciton interactions of chromophores. (a) HOMO–LUMO interactions. (b) Davidov splitting of interacting chromophores (top) with colinear transition moments (middle) and linearly cofacial moments (bottom). (c) Arrangement with skew transition moments.

The linear alignment of two or more perylene chromophores **6** corresponds to a J-type arrangement (transition moments are parallel to the N–N connection line) was studied [90] where bathochromic shifts with intensifying the light absorption were observed; three aligned chromophores resulted in more than four times the absorptivity of a single chromophore instead of only three times.

On the other hand, H-type arrangements of cofacial perylene derivatives were reported with a stiff xanthene linker and cause hypsochromic shifts [92] of the absorption and diminish absorptivity (destructive exciton interactions [18]).

Finally, skew type–oriented transition moments were given in the sandwich-like cyclophane **58** [93] with a small torsion between the long axis of the chromophores (see Figure 37.5). The absorption spectrum of **58** is dominated by the hypsochromic-shifted band compared with **6** caused by the H-component of the electronic transitions and some bathochromic component (see Figure 37.6). One can suppose that the different intensities in the spectrum correspond to the fractions of transition moments where a large component is obtained for H-type interaction and a small one for the J-type. The fluorescence of **58** is bathochromically shifted compared with the single chromophore **6** and corresponds to the J-type interaction; the H-type transition is forbidden for fluorescence [88c] and the J-component still remains. Similar arrangements of chromophores were found for perylene dyes incorporated in micelles [94] and were further confirmed by quantum chemical calculations.

Exciton interactions can be controlled by a rigid framework [95] where both constructive and destructive effects were observed. Finally, exciton interactions can be extinguished when appropriately oriented. For example, three perylene chromophores were orthogonally arranged [96] by means of a skeleton of triquinacene where no exciton interactions could be detected so that the impression of independently operating chromophores was obtained.

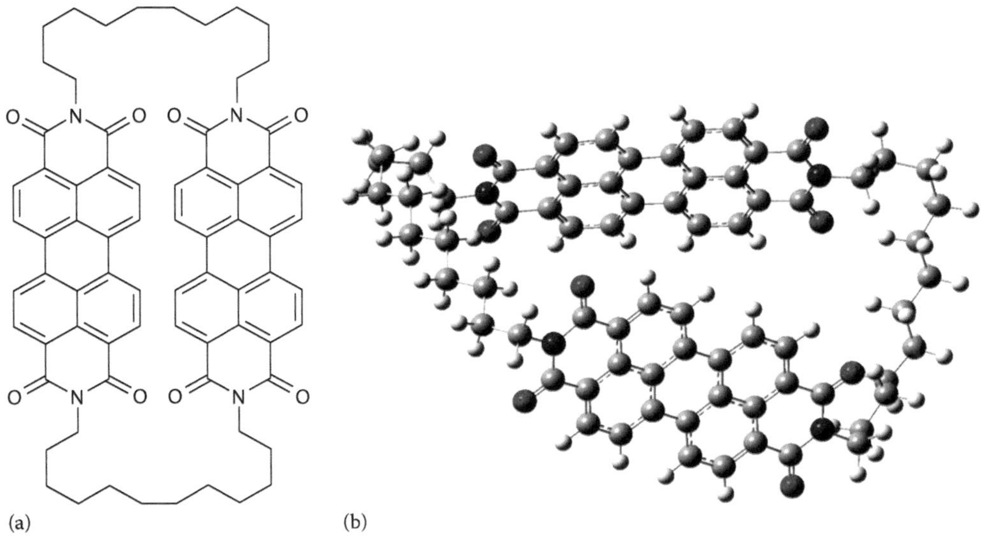

FIGURE 37.5 Chemical formula of the cyclophane **58** (a) and calculated (DFT, B3LYP) structure with skew chromophores (b).

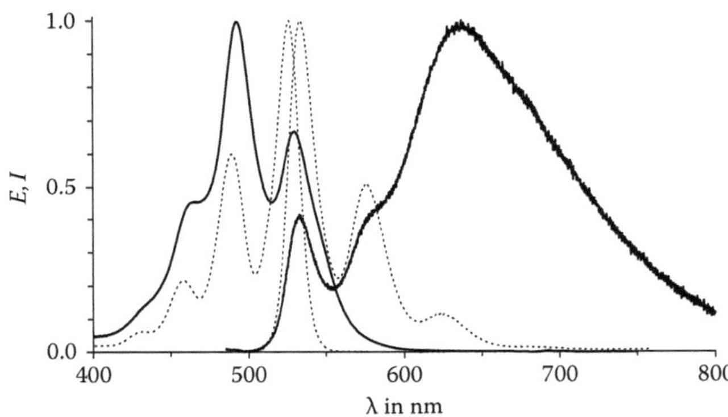

FIGURE 37.6 UV/vis absorption (left, E) and fluorescence (right, I) spectra in chloroform of the cyclophane **58** (thick, solid lines) compared with mono-chromophoric perylene dyes **6** (thin dotted lines).

37.4 Energy Transfer

The energy of excitation may be exchanged between interlinked chromophores. Such an exchange is degenerated for identical chromophores; however, two different chromophores (dyads) cause a gradient in energy where the energy of excitation can be transferred from higher level of the more hypsochromically absorbing unit to a lower level of the more bathochromically absorbing. The transfer of energy can be applied to molecular signal processing. Chromophores are molecular resonators for optical energy and their hopping from one chromophore to another chromophore means a path of information such as the transport of electrical energy with metallic conductors or semiconductors. There are two basic processes for the energy transfer: the Dexter-type energy transfer [97] requires an orbital overlap between the energy donor and the energy acceptor and the Förster resonant energy transfer (FRET) [98] applies the concept of resonant oscillating dipoles. The Dexter-type energy transfer is more problematic for molecular signal processing because the orbital overlapping wipes out the distinction between the involved chromophores. As a consequence, the Förster resonance energy transfer for more distinct units is preferred because of the operation between isolated chromophores. The theory of resonating dipoles results in Equation 37.1 for the rate constant k_{FRET} of FRET where there are some mathematical constants, Avogadro's constant N_A and basic properties of the involved chromophores where Φ_D is the fluorescent quantum yield of the donor, τ_D the fluorescent lifetime of the donor, and J_{DA} the overlap integral between the fluorescence spectrum of the donor and the absorption spectrum of the acceptor:

$$k_{FRET} = \frac{1000\cdot(\ln 10)\cdot\kappa^2\cdot J_{DA}\cdot\Phi_D}{128\cdot\pi^5\cdot N_A\cdot\tau\cdot|R_{DA}|^6}. \quad (37.1)$$

Two further parameters depending on the chemical structure are even more important because they may be applied for controlling FRET by the molecular arrangement of chromophores: first, the dependence on R^{-6} ($|\boldsymbol{R}_{DA}|^{-6}$) with R as the distance between the centers of the involved dipoles and second, κ^2 as an orientation factor of the dipoles:

$$\kappa = (\hat{\boldsymbol{\mu}}_D\cdot\hat{\boldsymbol{\mu}}_A) - 3(\hat{\boldsymbol{\mu}}_D\cdot\hat{\boldsymbol{R}}_{DA})\cdot(\hat{\boldsymbol{R}}_{DA}\cdot\hat{\boldsymbol{\mu}}_A). \quad (37.2)$$

This orientation factor κ is composed of three scalar products of unity vectors according to Equation 37.2 where $\boldsymbol{\mu}_D$ and $\boldsymbol{\mu}_A$ are the electronic transition moments of the donor and the acceptor, respectively, and $\boldsymbol{R}_{DA}$ the connecting vector between the middle points between the two transition moments:

$$\kappa = \cos(\theta_T) - 3(\cos(\theta_D)\cdot\cos(\theta_A)). \quad (37.3)$$

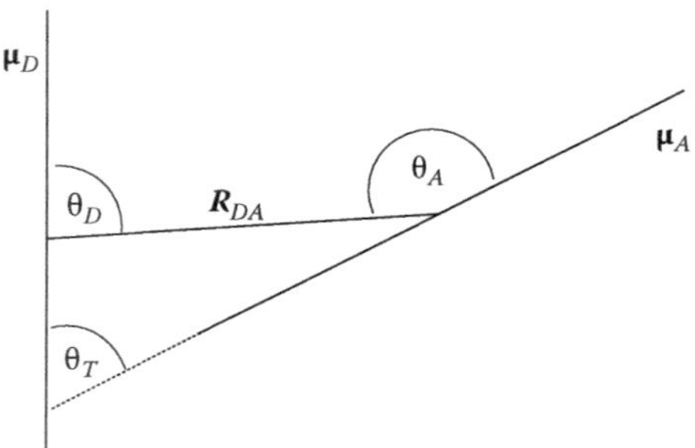

FIGURE 37.7 The orientation of the electronic transition moments of the energy donor $\boldsymbol{\mu}_D$ and the acceptor $\boldsymbol{\mu}_A$ and the interconnecting vector $\boldsymbol{R}_{DA}$.

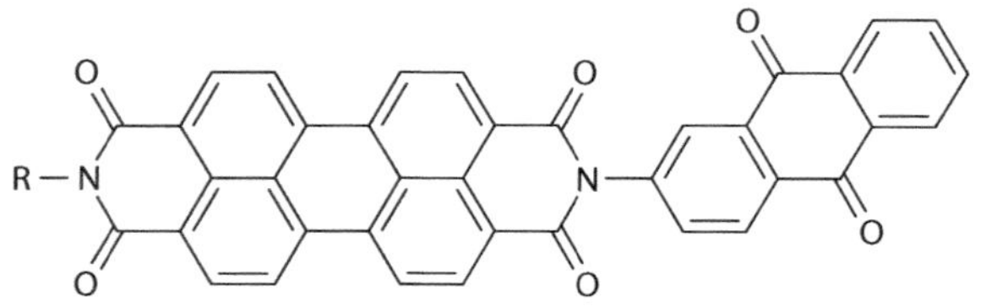

The scalar products in Equation 37.2 can be replaced by the cosines of the involved angles according to Equation 37.3 and Figure 37.7.

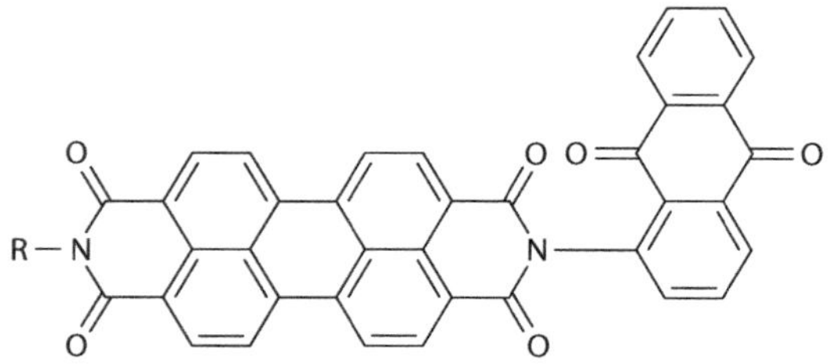

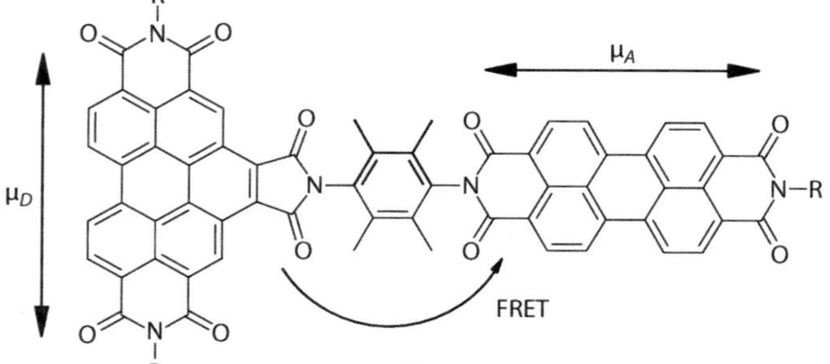

The control of the energy transfer by the arrangement of chromophores is demonstrated with combinations of the non-fluorescent anthraquinone chromophore as energy donor and the highly fluorescent perylene carboximide acceptor [99]. Anthraquinone looses the energy of excitation in a comparably short time causing an efficient quenching of fluorescence. As a consequence, no energy is transferred from the anthraquinone to the perylene chromophore in **59** for the connection to the position 2 as expected. On the other hand, an efficient energy transfer is observed in **60** where the perylene chromophore is connected to the position 1. Obviously, the energy transfer in **60** is so fast that it can efficiently compete with the spontaneous deactivation in anthraquinone. This offers principally the possibility of signal processing by controlling the arrangement of chromophores.

The concept of FRET was more extensively investigated because it can be a general basis for molecular signal processing where the orientation factor κ might be of central importance. We arranged molecular structures in such a way that κ becomes zero. We applied a perylene chromophore as an energy acceptor in **61** and the more hypsochromically absorbing benzoperylene chromophore as an energy donor; the fluorescence spectrum of the latter matches perfectly the absorption spectra of the former so that there are ideal conditions for an energy transfer according to the concept of FRET. We linked the perylene chromophore in **61** by the stiff tetramethylphenyl spacer to the five-membered ring of the benzoperylene dyes [100]. The electronic transition moments of both chromophores are parallel to the N–N connection line of the six-membered rings so that the electronic transition moments of both chromophores are orthogonal; this causes the first scalar product in Equation 37.2 and the first cos function in Equation 37.3, respectively, to become zero. The interconnecting vector of the two transition moments is orthogonal to the transition moment μ_D of the benzoperylene unit where the second scalar product and the second cos function, respectively, become zero so that κ becomes zero and FRET should be extinguished. As a consequence, an independent operation of the two chromophores and a dual fluorescence is expected, depending on the electronic excitation. In contrast to this prediction, a very efficient energy transfer is observed indicated by a fluorescence quantum yield close to 100% of the perylene chromophore if the benzoperylene chromophore is excited. Moreover, the energy transfer proceeds very fast [101] with a time constant of only 9.4 ps; the orthogonal orientation of the chromophores and the transition moments, respectively, is confirmed by quantum chemical calculations and measurements of the excitation and fluorescence anisotropy. The distance of the two chromophores was extended and no decrease of the efficiency of the energy transfer was found up to 27 Å. Even a purely aliphatic spacer for the definite exclusion of any contribution of the Dexter-type energy transfer could not extinguish the energy transfer. Further investigations of the mechanism of the energy transfer between orthogonal systems indicate an important contribution of molecular dynamics is present where molecular vibrations at low frequencies (thermal noise) seem to play an important role [102]. These results make the planning of molecular operating devices more complicated; on the other hand, they offer novel possibilities and show that additional research is necessary because the elementary theory of FRET seems to be an oversimplification. Preliminary results [103] make it probable that the energy transfer in **62** is essentially noise-induced and the inverse third power of distance describes the process more appropriately for the chemically important proximity, whereas the inverse sixth power according to Equation 37.1 becomes correct for large distances.

37.5 Photo-Induced Electron Transfer

The light-induced transfer of electrons is a second important process for signal processing. Such a generation of charge-separated states may be further applied both for the generation of electrical energy from light energy and for the control of light-driven electrical processes. The perylene dyes are very suitable as chromophores for such applications because their radical anions and radical cations, respectively, are comparably stable [104]; moreover, a stepwise uptake and release of several electrons is possible allowing a stepwise increase and decrease of the electrochemical potential [105]. There are several studies concerning the light-driven interaction of the comparably electron-depleted perylene tetracarboxylic bisimides with electron-rich chromophores such as corroles [106].

37.5.1 Photo-Induced Electron Transfer in Dye-Sensitized Solar Cells

The charge separation is one of the key steps in photovoltaic solar cells where the efficient recombination of the charge carriers is the most important competing process. The electric field in a p–n transition is the driving force for the charge separation in the well-established silicon solar cells; the light-induced generation of a local exciton within the p–n junction allows a charge-induced movement of the electronically excited electron in the conducting band (corresponds to LUMO) and the filling-up of the remaining local positive charge ("defect electron") in the valence band (corresponds to the HOMO) from the opposite site. This technology is comparably far developed and applied in commercial silicon solar cells; however, there are some disadvantages in this concept making further development difficult. First, there must be a precise and perfect macroscopic layer of a p–n junction in silicon throughout the material because defects would cause losses by short circuits. Second, the separated charges destruct the electric field necessary for charge separation so that the light-induced voltage becomes limited at about 0.7 V for silicon solar cells.

Dye-sensitized solar cells are an alternative because of individually acting components with picometer dimensions where the dye-sensitized Grätzel cell [107] (dye-sensitized solar cells [108]) is one of the most popular. The electron acceptor is formed by titanium dioxide where the surface is increased by nanostructuring of the material and the surface is covered by a sensitizing dye. The electron-releasing counter electrode is formed by electrolytes or conducting polymers. Electrons are transferred from the HOMO of the optically exited dye molecules into the titanium dioxide and are collected by further materials such as ITO. The applied dye has essentially two functions. First, a local exciton for charge separation is generated by light absorption being known as the electronically excited state in dye chemistry. Second, the dye molecules act as electrical insulators between the two electrodes of such a cell. As a consequence, no care has to be taken concerning the electric field in the center of such cells and the processing of the charge separation continues until the oxidizing capacity of titanium dioxide will be exhausted. Analog processes proceed in the electron-donating layer. The electric current using the photovoltaic energy will reestablish the redox properties for both the titanium dioxide and the reducing layer. The dye molecules act as individual antennae in picometer dimensions and a malfunction would be less problematic than for conventional solar cells because no short circuits are formed as long as the titanium dioxide is covered by insulating organic material preventing a recombination of the charge carriers. However, even a perfect insulator can be passed by the separated charge carriers by electrons energy tunneling.

The diffusion of the transferred electrons in the layer of titanium dioxide is an important process for making such a tunneling less probable. Two further things can be applied to suppress such tunneling. First, the size of the dye layer can be increased; however, this would also diminish the probability of the charge separation. This problem may be solved by the application of bichromophores (dyads) or aggregates of the same chromophore where the excited states are delocalized over several chromophores so that charge separation is still possible; however, the material is appreciably more thick than a layer of isolated chromophores. Second, the recombination of the separated charge carriers can be suppressed by the application of spin multiplicity. The exciton is first generated in the singlet state and a charge separation preserves this state where the recombination of the charge carriers is diffusion controlled. On the other hand, an ISC of the exciton generates the triplet where a subsequent charge separation preserves this spin state. A direct recombination of the charge carriers is prohibited by Pauli's principle and the lifetime is increased because of the slow process of necessary spin inversion. As a consequence, the chance of leaving the close contact necessary for tunneling by diffusion processes is increased and a higher yield of the collection of the charge carriers can be expected. The application of sensitizing dyes with a high ISC rate would open novel possibilities for such solar cells. High ISC rates can be obtained by the application of the heavy atom effect known from compounds such as Rose Bengale; however, the spreading of heavy elements in the environment may be problematic. As a consequence, such research should be concentrated on chromophores with high ISC rates consisting on light elements. For example, benzoperylenetetracarboxylic bisimides [81] exhibit a surprisingly high ISC rate [80,82] and are attractive compounds fulfilling such conditions because they consist only on light elements and exhibit sufficient photostability.

Even conventional silicon solar cells can be improved by organic dyes as boosters for light absorption where the light-induced charge separation (photo-induced electron transfer [PET]) is the key process. Finally, purely organic photovoltaic cells (organic photovoltaic) are of special interest because of low-cost materials where processing is very easy and recycling is unproblematic.

transfer to the perylene chromophore where a charge separation ranging up to 15 Å and lifetimes of charge-separated states (CS) as long as 15 ns were observed.

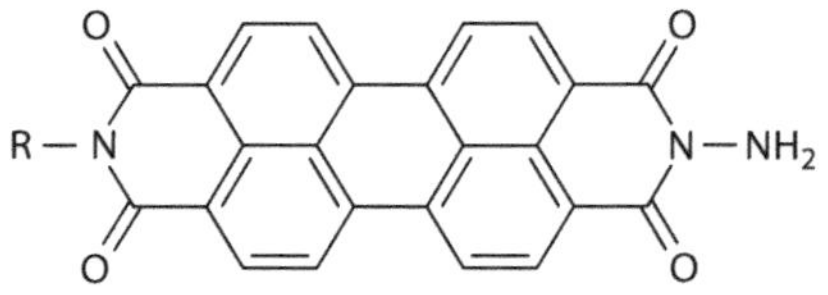

62

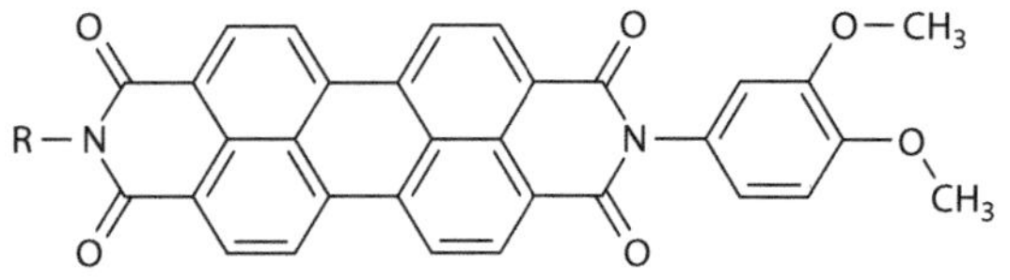

63

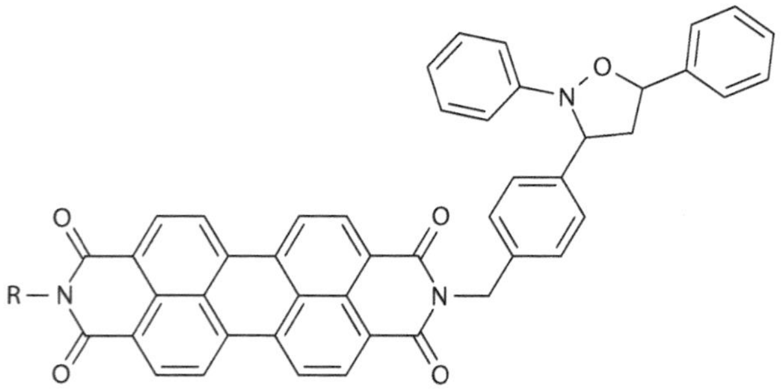

64

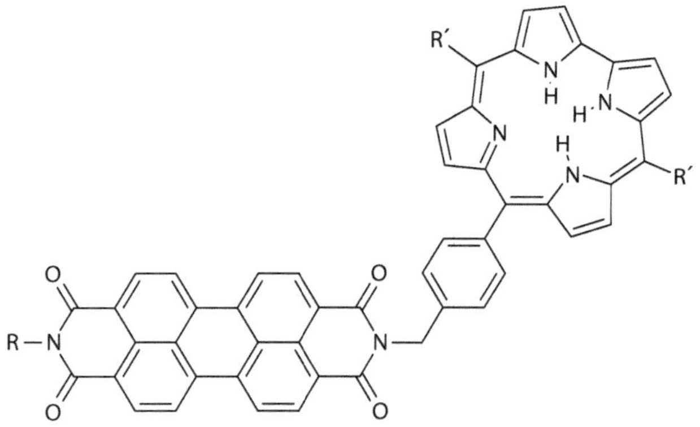

65

37.5.2 Charge Separation in Perylene Dyes

The perylenetetracarbocylic bisimides **6** are comparably electron-depleted and their HOMO is energetically low-lying. This orbital becomes semi-occupied in the excited state and an electron transfer from electron-rich structures of the side chains R may fill-up this orbital preventing a return of the excited electron. As a consequence, both the fluorescence becomes quenched and the PET generates charge separation. The energetically high-lying lone pair of an *N*-amino group in **62** causes a fluorescence quenching by PET and can be applied for analytical purposes [109] such as the determination of aldehydes where the condensation of the *N*-amino group forms more electron-depleted imines and reestablishes fluorescence. A thorough investigation of the quenching process indicates that the lone pair is energetically still below the HOMO of the chromophore. However, the electronic excitation induces a change in geometry and elevates the energetic position so that PET proceeds and quenches fluorescence.

Electron-rich phenyl substituents R are obtained by the substitution with alkoxy groups such as in **63**. The number and position of such groups decides whether a PET proceeds or not indicated by a switch on–switch off of fluorescence [110].

The alpha effect, two neighbored atoms with lone pairs, allows the elevation of electronic levels of comparably small groups. For example, oxazolidine systems [111] allow an efficient electron

Finally, complex electronic systems such as corroles (**65**) [112] and porphyrines [71] were combined with perylenes as the light-collecting units and even multichromophoric dendrimers [113]. Generally, the electronic levels of the perylene units have to be considered for such assemblies where the benzoperylenehexacarboximides **46** are the most electron depleted by the six carbonyl groups followed by the perylenetetracarboximides **6** and the perylenebiscarboximides **14** where both energy transfer (FRET) and PET can be controlled in complex assemblies for light collection and charge separation.

37.5.3 Miscellaneous Functions of *Peri*-Arylenes

The storage of collected energy by solar cells is an unsolved problem and causes the necessity of high-capacity electric distribution networks. A buffering within the solar cells would bring about an appreciable progress [108]. Proton-conducting materials from melamine and trimesic acid [114] were developed and anion π interactions with polymeric naphthalenetetracarboxylic bisimides (**2**) studied [115]. Ambipolar organic semiconductors were described on the basis of naphthalene, perylene, terrylene, and quaterrylene [116]. p–n Junctions were established with naphthalenetetracarboxylic bisimides (**2**) [117] and transistors with **2** and **6** [118]. Finally, single, double, and three times interlinked chromophores of **6** were reported; however, the fluorescence quantum yields of such assemblies are comparably low.

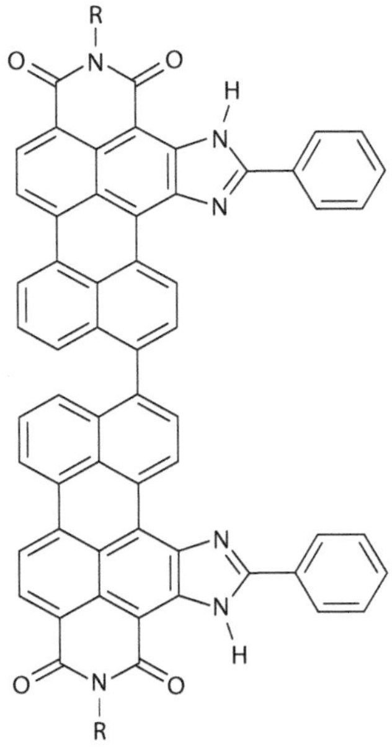

66

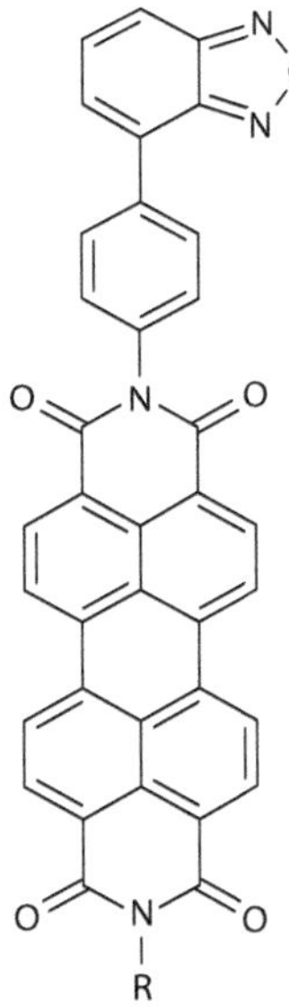

67

The Stokes' shifts of derivatives of *peri*-linked naphthalenes can be increased by the dynamic process according to Figure 37.8 where steric interactions cause an arrangement out of plane according to the S_0 state in Figure 37.8 [119]. The vertical transition of electronic excitation to S_1 does not alter this geometry; however, the subsequent relaxation favors a more planar arrangement in S_1' causing a bathochromically shifted fluorescence to S_0' where the initial state is reached again by thermal relaxation. The substituents X and R may be part of the structure of *peri*-arylenes such as in **15** or in **66**.

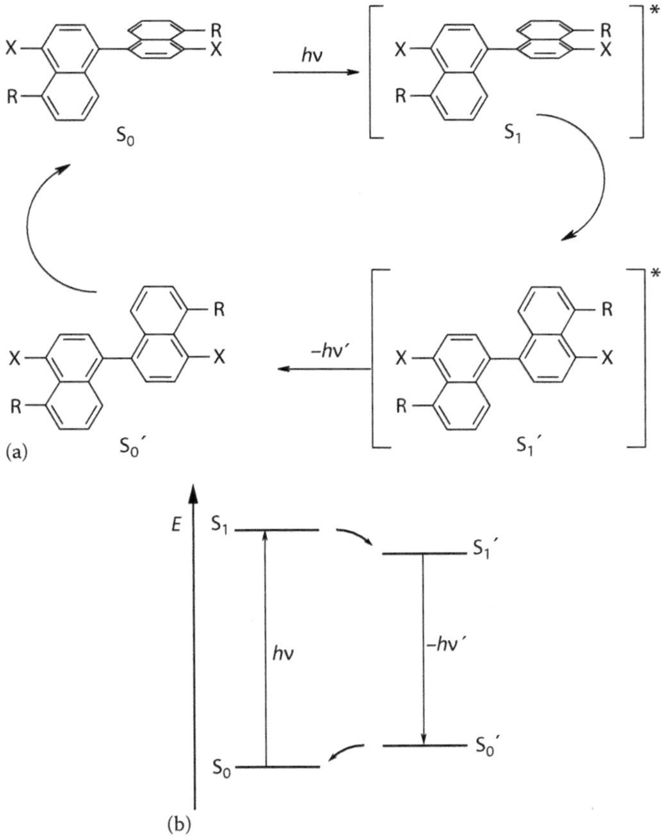

FIGURE 37.8 Light-induced dynamic processes in *peri*-linked binaphthyles (a) and mechanism for the increase of the Stokes' shift (b).

FIGURE 37.9 Amorphous material **67**.

Perylene dyes generally form crystalline solids; however, the attachment of certain heterocycles such as in **67** causes the formation of amorphous materials [120] (see Figure 37.9). This is of special interest for applications such as in photovoltaics because there are no phase boundaries such as in crystalline materials and, for example, electron transport can proceed uniformly without any inhibition.

37.6 Conclusions

The chemistry of *peri*-arylenes started with perylene [24] and aceanthrene green [77] dyes applied as vat dyes [25] and was then concentrated to light-stable pigment applications [26] for many years. The development of solubility increasing groups [7,29,30] opened the door for completely novel fields where the applications as highly fluorescent dyes dominated. This process of innovation is still in progress and shifts more and more to applications in novel materials and seems to play an important part for the next steps in novel technology. *Peri*-arylenes may become key structures not only because of their chemical and photochemical stability and high fluorescence quantum yields, but also even more important for the future because of their low toxicity and the unproblematic recycling of these purely organic materials.

References

1. Moore, G. *Electronics*, 1965, 38, 114–117.
2. McCoy, E. F. and Ross, I. G. *Aust. J. Chem.* 1962, 15, 573–590.
3. (a) Rothammel, K. *Antennenbuch*, 5th edn., Telekosmos-Verlag, Stuttgart, Germany, 1976. (b) Rüdenberg, R. *Ann. d. Phys.* 1908, 330, 446–466. (c) Rüdenberg, R. *Ann. d. Phys. Leipzig* 1908, 25, 466–500.
4. (a) Langhals, H. German Patent 3016764, April 30, 1980; *Chem. Abstr.* 1982, 96, P70417x. (b) Langhals, H. *Nachr. Chem. Tech. Lab.* 1980, 28, 716–718; *Chem. Abstr.* 1981, 95, R9816q.
5. BASF AG (inv. Graser, F.), *Ger. Offen.* DE 3049215, July 15, 1982; *Chem. Abstr.* 1982, 97, 129114.
6. (a) Hock, H. and Kropf, H. *Angew. Chem.* 1957, 69, 313–321. (b) Hock, H. and Lang, S. *Ber. Dtsch. Chem. Ges. B* 1944, 77B, 257–264; *Chem. Abstr.* 1945, 39, 22122. (c) Arpe, H.-J. and Weissermel K. *Industrielle Organische Chemie: Bedeutende Vor- und Zwischenprodukte*, 6th ed., Wiley-VCH, New York, 2007, p. 418.
7. (a) Demmig, S. and Langhals, H. *Chem. Ber.* 1988, 121, 225–230. (b) Langhals, H. *Ger. Offen.* DE 3703495, February 5, 1987; *Chem. Abstr.* 1989, 110, P59524s.
8. Langhals, H., Demmig, S., and Potrawa, T. *J. Prakt. Chem.* 1991, 333, 733–748.
9. Langhals, H. and Jaschke, H. *Chem. Eur. J.* 2006, 12, 2815–2824.
10. Hensel, W. PhD thesis, University of Frankfurt on Main, Germany, 1958.
11. (a) Adachi, M., Murata, Y., and Nakamura, S. *J. Phys. Chem.* 1995, 99, 14240–14246. (b) Bondarenko, E. F., Shigalevskii, V. A., and Yugai, G. A. *Zh. Org. Khim.* 1982, 18, 610–615; *Chem. Abstr.* 1982, 97, 5673. (c) Sterzel, M., Pilch, M., Pawlikowski, M. T., Skowronek, P., and Gawronski, *J. Chem. Phys. Lett.* 2002, 362, 243–248.
12. Langhals, H., Jaschke, H., Ehlis, T., and Wallquist, O. PCT Int. Appl. WO 2007012611, July 21, 2006; *Chem. Abstr.* 2007, 146, 185965.
13. Langhals, H. and Ritter, U. *Eur. J. Org. Chem.* 2008, 3912–3915.
14. Takeuchi, T., Matsugo, S., and Morimoto, K. *Carcinogenesis* 1997, 18, 2051–2055.
15. (a) Schuetz, S., Bock, M., and Otten, H. *Ger. Offen.* DE 1195762, July 1, 1965; *Chem. Abstr.* 1965, 63, 80551. (b) Schuetz, S., Kurz, J., Pluempe, H., Bock, M., and Otten, H. *Arzneimittel-Forschung,* 1971, 21, 739–763; *Chem. Abstr.,* 1971, 75, 97046.
16. Molla, M. R. and Ghosh, S. *Chem. Eur. J.* 2012, 18, 1290–1294.
17. (a) Kranzlein, G. and Vollmann, H. German Patent DE 552760 19320617; *Chem. Abstr.* 1932, 26, 54154. (b) Fierz-David, H. E. and Rossi, C. *Helv. Chim. Acta* 1938, 21, 1466–1489.
18. Reviews: (a) Langhals, H. Molecular devices. Chiral, bichromophoric silicones: Ordering principles in complex molecules, in Ganachaud, F., Boileau, S., and Boury B. (eds.), *Silicon Based Polymers*, Springer, New York, 2008, pp. 51–63. (b) Langhals, H. *Helv. Chim. Acta* 2005, 88, 1309–1343. (c) Langhals, H., *Heterocycles*, 1995, 40, 477–500.
19. Langhals, H., Karolin, J., and Johansson, L. B.-Å., *J. Chem. Soc., Faraday Trans.,* 1998, 94, 2919–2922.
20. Xerox Corporation, Japanese Patent 03024059 A2, February 1, 1991; *Chem. Abstr.* 1991, 115, 123841a.
21. Zollinger, H., *Color Chemistry. Syntheses, Properties and Applications of Organic Dyes and Pigments*, 2nd edn., Weinheim, Germany, 1991.
22. Langhals, H. and Demmig, S., *Ger. Offen.* DE 4007618.0, March 10, 1990; *Chem. Abstr.* 1992, 116, P117172n.
23. Langhals, H., El-Shishtawy, R., von Unold, P., and Rauscher, M. *Chem. Eur. J. Suppl.*, 2006, 12, 4642–4645.
24. (a) Kardos, M. Reich Patent 276357, June 14, 1913; *Friedländers Fortschr. Teerfarbenfabr.* 1917, 12, 492; *Chem. Abstr.* 1914, 8, 3243. (b) Karpukhin, P. P. and Ratnikova, K. I. *Ukrains'kii Khem. Zh.* 1937, 12, 122–135; *Chem. Abstr.* 1937, 31, 41412.
25. Cullinan, J. F. and Lytle, L. D. U.S. Patent 2473015 19490614, June 14, 1949; *Chem. Abstr.* 1949, 43, 53029.
26. Eckert, W. and Remy, H. U.S. Patent 2890220 19590609 June 9, 1959; *Chem. Abstr.* 1960, 54, 26001.
27. Hoechst (inv. Geissler, G. and Remy, H.) *Ger. Offen.* 1130099, October 14, 1959; *Chem. Abstr.* 1962, 57, P11346f.
28. (a) Langhals, H. *Ger. Offen.* DE 3016764, April 30, 1980; *Chem. Abstr.* 1982, 96, P70417x. (b) Langhals, H. *Nachr. Chem. Tech. Lab.* 1980, 28, 716–718, *Chem. Abstr.* 1981, 95, R9816q.
29. Rademacher, A., Märkle, S., and Langhals, H. *Chem. Ber.* 1982, 115, 2927–2934.
30. BASF AG (inv. Graser, F.), *Ger. Offen.* DE 3049215, July 15, 1982; *Chem. Abstr.* 1982, 97, 129114.
31. Herbst, W. and Hunger, K. *Industrial Organic Pigments. Production, Properties, Applications*, 3rd edn., Wiley-VCH, Weinheim, Germany, 2006.
32. Langhals, H. and Grundner, S. *Chem. Ber.* 1986, 119, 2373–2376.

33. Sakamoto, T. and Pac, C. *J. Org. Chem.* 2001, 66, 94–98.
34. Kraska, J. and Truszkowska, I. Polish Patent PL 1970-142856, August 25, 1970; *Chem. Abstr.* 1977, 86, 55205. (b) Graser, F., Guenthert, P. German Patent DE 1975-2545663, April 21, 1977; *Chem. Abstr.* 1977, 87, 40735.
35. Langhals, H. *Chem. Ber.* 1985, 118, 4641–4645.
36. Kleine, F. German (East) Patent DD 1980-226325, December 23, 1980; *Chem. Abstr.* 1992, 117, 193603.
37. (a) Langhals, H. and Christian, S. Unpublished results, 2011. (b) Langhals, H., Dietl, C., Zimpel, A., and Mayer P. *J. Org. Chem.* 2012, 77, 5965–5970.
38. Nagao, Y. and Misono, T. *Bull. Soc. Chim. Jpn.* 1981, 54, 1269–1270.
39. Kaiser, H., Lindner, J., and Langhals, H. *Chem. Ber.* 1991, 124, 529–535.
40. Kelber, J., Bock, H., Thiebaut, O., Grelet, E., and Langhals, H. *Eur. J. Org. Chem.* 2011, 702–712.
41. Tröster, H. *Dyes Pigm.* 1983, 4, 171–177; *Chem. Abstr.* 1983, 99, 39794f.
42. Review: Weil, T., Vosch, T., Hofkens, J., Peneva, K., and Müllen, K. *Angew. Chem.* 2010, 122, 9252–9278; *Angew. Chem. Int. Ed.* 2010, 49, 9068–9093.
43. Nolde, F., Qu, J., Kohl, C., Pschirer, N. G., Reuter, E., and Müllen, K. *Chem. Eur. J.* 2005, 11, 3959–3967.
44. Langhals, H., Walter, A., Rosenbaum, E., and Johansson, L. B.-Å. *Phys. Chem. Chem. Phys.* 2011, 13, 11055–11059.
45. Holtrup, F. O., Mueller, G. R. J., Quante, H., De Feyter, S., De Schryver, F. C., and Muellen, K. *Chem. Eur. J.* 1997, 3, 219–225.
46. Feiler, L., Langhals, H., and Polborn, K. *Liebigs Ann. Chem.* 1995, 1229–1244.
47. (a) Langhals, H., Walter, A., Rosenbaum, E., and Johansson, L. B.-Å. *Phys. Chem. Chem. Phys.* 2011, 13, 11055–11059. (b) Langhals, H. and Hofer, A. German Patent DE 102011018815.0, April 4, 2011.
48. (a) Quante, H. and Müllen, K. *Angew. Chem.* 1995, 107, 1487–1489; *Angew. Chem. Int. Ed.* 1995, 34, 1323–1325. (b) Müllen, K. and Quante, H. (BASF AG) German Patent DE 4236885, October 31, 1992; *Chem. Abstr.* 1994, 121, 303055.
49. Yamamoto, T., Morita, A., Miydzaki, Y., Maruyama, T., Wakayama, H., Zhou, Z., Nakamura, Y., and Kanbara, T. *Macromolecules* 1992, 25, 1214–1223.
50. Bradley, W. and Pexton, F. W. *J. Chem. Soc. London* 1954, 4432–4435.
51. Langhals, H. and Hofer, A. In preparation.
52. Langhals, H. and Süßmeier, F. *J. Prakt. Chem.* 1999, 341, 309–311.
53. Langhals, H., Büttner, J., and Blanke, P. *Synthesis*, 2005, 364–366.
54. Langhals, H., Schönmann, G., and Feiler, L. *Tetrahedron Lett.* 1995, 36, 6423–6424.
55. (a) Pschirer, N. G., Kohl, C., Nolde, F., Qu, J., and Müllen, K. *Angew. Chem.* 2006, 118, 1429–1432; *Angew. Chem. Int. Ed.* 2006, 45, 1401–1404. (b) Nolde, F., Pisula, W., Müller, S., Kohl, C., and Müllen, K. *Chem. Mater.* 2006, 18, 2715–3725. (c) Kato, T. and Harada, T. *Jpn. Kokai Tokkyo Koho* JP 2009046525, March 5, 2009; *Chem. Abstr.* 2009, 150, 295122.
56. (a) Vollmann, H., Becker, H., Corell, M., and Streck, H. *Liebigs Ann. Chem.* 1937, 531, 1–159. (b) Bondarenko, E. F., Shigalevskii, V. A., and Gerasimenko, Y. E. *J. Org. Chem. USSR* 1979, 15, 2520–2525. (c) Gerasimenko, Y. E., Shigalevskii, V. A., Bondarenko, E. F., Semenyumk, G. V. *Zh. Org. Khim.* 1972, 8, 626–631; *Chem. Abstr.* 1972, 77, 5230z. (d) Bonnet, E., Gangneux, P., and Marechal, E. *Bull. Soc. Chim. France* 1976, 504–506. (e) Bondarenko, E. F. and Shigalevskii, V. A. *Zh. Org. Khim.* 1983, 19, 2377–2382; *Chem. Abstr.* 1984, 100, 209748. (f) Infineon Technologies AG, Germany (inv. Wuerthner, F., Thalacker, C., and Schmid G.) *Ger. Offen.* DE 10148172 (2003); *Chem. Abstr.* 2003, 138, 305532. (g) Wuerthner, F., Ahmed, S., Thalacker, C., and Debaerdemaeker, T. *Chem. Eur. J.* 2002, 8, 4742–4750. (h) Thalacker, C., Miura, A., De Feyter, S., De Schryver, F. C., and Wuerthner, F. *Org. Biomol. Chem.* 2005, 3, 414–422.
57. Iden, R., Seybold, G., Stange, A., and Eilingsfeld, H. *Forschungsber.—Bundesminist. Forsch. Technol., Technol. Forsch. Entwickl.* 1984, BMFT-FB-T 84-164; *Chem. Abstr.* 1985, 102, 150903.
58. Langhals, H., El-Shishtawy, R., von Unold, P., and Rauscher, M. *Chem. Eur. J.* 2006, 12, 4642–4645.
59. (a) Rudkevich, M. I. and Korotenko, T. A. *Vestn. Khar'kov. Politekh. Inst.* 1969, 41, 21–26; *Chem. Abstr.* 1971, 75, 7375. (b) Zhao, Y. and Wasielewski, W. M. R. *Tetrahedron Lett.* 1999, 40, 7047–7050.
60. (a) An, Z., Odom, S. A., Kelley, R. F., Huang, C., Zhang, X., Barlow, S., Padilha, L. A., Fu, J., Webster, S., Hagan, D. J., Van Stryland, E. W., Wasielewski, M. R., and Marder, S. R. *J. Phys. Chem. A* 2009, 113, 5585–5593. (b) Rohr, U., Schilichting, P., Bohm, A., Gross, M., Meerholz, K., Bräuchle, C., and Müllen, K. *Angew. Chem. Int. Ed.* 1998, 37, 1434–1437.
61. Langhals, H. and Kirner, S. *Eur. J. Org. Chem.* 2000, 365–380.
62. Lukac, I. and Langhals, H. *Chem. Ber.* 1983, 116, 3524–3528.
63. Langhals, H. and Bastani-Oskoui, H. *J. Prakt. Chem.* 1997, 339, 597–602.
64. Langhals, H., Sprenger, S., and Brandherm, M.-T. *Liebigs Ann. Chem.* 1995, 481–486.
65. Langhals, H., Jaschke, H., Ring, U., and von Unold, P. *Angew. Chem.* 1999, 111, 143–145; *Angew. Chem. Int. Ed. Engl.* 1999, 38, 201–203.
66. (a) Regel, E. and Büchel, K.-H. *Justus Liebigs Ann. Chem.* 1977, 145–158. (b) Regel, E., Eue, L., and Büchel K.-H. (Bayer AG), DE-B 2 043 649, 1970; *Chem. Abstr.* 1972, 76, 140 809q.
67. Langhals, H. and Pust, T. *Eur. J. Org. Chem.* 2010, 3140–3145.
68. Iqbal, Z., Ivory, D. M., and Eckhardt, H. *Mol. Cryst. Liquid Cryst.* 1988, 158b, 337–352.
69. Langhals, H., von Unold, P., and Speckbacher, M. *Liebigs Ann./Recueil.* 1997, 467–468.
70. Süßmeier, F. and Langhals, H. *Eur. J. Org. Chem.* 2001, 607–610.

71. (a) Miller, M. A., Lammi, R. K., Prathapan, S., Holten, D., and Lindsey, J. S. *J. Org. Chem.* 2000, 65, 6634–6649. (b) Kirmaier, C., Song, H., Yang, E., Schwartz, J. K., Hindin, E., Diers, J. R., Loewe, R. S., Tomizaki, K., Chevalier, F., Ramos, L. Birge, R. R., Lindsey, J. S., Bocian, D. F., and Holten, D. *J. Phys. Chem. B* 2010, 114, 14249–14264.
72. Chen, X.-K., Zou, L.-Y., Guo, J.-F., and Ren, A.-M. *J. Mater. Chem.* 2012, 22, 6471–6484.
73. Katsuta, S., Tanaka, K., Maruya, Y., Mori, S., Masuo, S., Okujima, T., Uno, H., Nakayama, K., and Yamada, H. *Chem. Commun. (Cambridge)* 2011, 47, 10112–10114.
74. (a) Yue, W., Gao, J., Li, Y., Jiang, W., Di Motta, S., Negri, F., and Wang, Z. *J. Am. Chem. Soc.* 2011, 133, 18054–18057. (b) Stevens, B., Perez, S. R., and Ors, J. A. *J. Am. Chem. Soc.* 1974, 96, 6846–6850.
75. Wilson, T. *J. Am. Chem. Soc.* 1966, 88, 2898–2902.
76. Langhals, H. and Kinzel, S. *J. Org. Chem.* 2010, 75, 7781–7784.
77. (a) Kardos, M. Ph.D. thesis, University of Berlin, Berlin, Germany, 1913, p. 26. (b) Kardos, M. *Ber. Dtsch. Chem. Ges.* 1913, 46, 2086–2091; *Chem. Abstr.* 1913, 7, 23197. (c) Liebermann, C. and Kardos, M. *Ber. Dtsch. Chem. Ges.* 1914, 47, 1203–1210. (d) Kardos, M. *Friedländers Fortschritte der Teerfarbenfabrikation* 1914–1916, 12, 485–492.
78. (a) Désilets, D., Kazmaier, P. M., Burt, R. A., and Hamer, G. K. *Can. J. Chem.* 1995, 73, 325–335. (b) Langhals, H., Schönmann, G., and Polborn, K. *Chem. Eur. J.* 2008, 14, 5290–5303.
79. Langhals, H. and Kirner, S. *Eur. J. Org. Chem.* 2000, 365–380.
80. Ventura, B., Langhals, H., Böck, B., and Flamigni, L. *Chem. Commun.* 2012, 48, 4226–4228.
81. Langhals, H., Böck, B., Schmid, T., and Marchuk, A. *Chem. Eur. J.* 2012.
82. Flamigni, L., Zanelli, A., Langhals, H., and Böck, B. *J. Phys. Chem. A* 2012, 116, 1503–1509.
83. Langhals, H. and Blanke, P. *Dyes Pigm.* 2003, 59, 109–116.
84. Langhals, H., Kinzel, S., and Obermeier, A. *Ger. Offen.* DE 102008061452.1, December 10, 2008; *Chem. Abstr.* 2009, 151, 58174.
85. Langhals, H. and Poxleitner, S. *Eur. J. Org. Chem.* 2008, 797–800.
86. Langhals, H. and Hofer, A. *Ger. Offen.* DE 102012008287.8, April 16, 2012.
87. Avlasevich, Y. and Muellen, K. *Chem. Commun.* 2006, 42, 4440–4442.
88. (a) Kuhn, W. *Trans. Faraday Soc.* 1930, 26, 293. (b) Scheibe, G. *Angew. Chem.* 1936, 49, 563. (c) Förster, T. *Naturwissenschaften* 1946, 33, 166.
89. (a) Davydov, A. S. *Zhur. Eksptl. i Teoret. Fiz.* 1948, 18, 210–218; *Chem. Abstr.* 1949, 43, 4575f. (b) Davydow, A. S. *Theory of Molecular Excitations*, trans. Kasha H. and Oppenheimer, M., Jr., McGraw-Hill, New York, 1962.
90. Langhals, H. and Jona, W. *Angew. Chem.* 1998, 110, 998–1001; *Angew. Chem. Int. Ed. Engl.* 1998, 37, 952–955.
91. Jelley, E. *Nature* 1936, 138, 1009.
92. (a) Giaimo, J. M., Lockard, J. V., Sinks, L. E., Scott, A. M., Wilson, T. M., and Wasielewski, M. R. *J. Phys. Chem. A* 2008, 112, 2322–2330. (b) Giaimo, J. M., Gusev, A. V., and Wasielewski, M. R. *J. Am. Chem. Soc.* 2002, 124, 8530–8531.
93. Langhals, H. and Ismael, R. *Eur. J. Org. Chem.* 1998, 1915–1917.
94. Langhals, H. and Pust, T. *Green Sustain. Chem.* 2011, 1, 1–6.
95. (a) Langhals, H., Wagner, C., and Ismael, R. *New. J. Chem.* 2001, 25, 1047–1049. (b) Langhals, H. and Speckbacher, M. *Eur. J. Org. Chem.* 2001, 2481–2486. (c) Langhals, H. and Gold, J. *J. Prakt. Chem.* 1996, 338, 654–659.
96. Langhals, H., Rauscher, M., Strübe, J., and Kuck, D. *J. Org. Chem.* 2008, 73, 1113–1116.
97. (a) Dexter, D. L. *J. Chem. Phys.* 1953, 21, 836–850. (b) Laible, P. D., Knox, R. S., and Owens, T. G. *J. Phys. Chem. B* 1998, 102, 1641–1648. (c) Braslavsky, S. E. *Pure Appl. Chem.* 2007, 79, 293–465, 322.
98. (a) Foerster, T. *Naturwiss.* 1946, 33, 166–175; *Chem. Abstr.* 1947, 41, 36668. (b) Foerster, T. *Ann. Phys.* 1948, 6. *Folge* 2, 55–75; *Chem. Abstr.* 1949, 43, 31172. (c) Foerster, T. *Z. Elektrochem.* 1949, 53, 93–99; *Chem. Abstr.* 1949, 43, 33629. (d) Foerster, T. *Zeitschr. Naturforsch.* 1949, 4a, 321–327; *Chem. Abstr.* 1950, 44, 43074.
99. Langhals, H. and Saulich, S. *Chem. Eur. J.* 2002, 8, 5630–5643.
100. Langhals, H., Poxleitner, S., Krotz, O., Pust, T., and Walter, A. *Eur. J. Org. Chem.* 2008, 4559–4562.
101. Langhals, H., Esterbauer, A. J., Walter, A., Riedle, E., and Pugliesi, I. *J. Am. Chem. Soc.* 2010, 132, 16777–16782.
102. Pugliesi, I., Walter, A., Langhals, H., and Riedle, E. Highly efficient energy transfer in a dyad with orthogonally arranged transition dipole moments: Beyond the limits of Förster? in Chergui, M., Jonas, D., Riedle, E., Schoenlein, R. W., and Taylor A. (eds.), *Ultrafast Phenomena XVII*, Oxford University Press, New York, 2011, pp. 343–345.
103. Nalbach, P., Pugliesi, I., Langhals, H., and Thorwart, M. *Phys. Rev. Lett.* 2012.
104. Salbeck, J., Kunkely, H., Langhals, H., Saalfrank, R. W., and Daub, *J. Chimia* 1989, 43, 6–9.
105. (a) Doege, H. G., Barthl, A., Froehner, J., and Domschke, G. German Patent (East) DD 296584 A5 19911205, July 13, 1990; *Chem. Abstr.* 1992, 116, 177727. (b) Doege, H. G., Bartl, A., Froehner, J., and Domschke, G. *Mater. Sci. Forum* 1990, 62–64, 475–476; *Chem. Abstr.* 1991, 114, 189032.
106. Flamigni, L., Ventura, B., Tasior, M., Becherer, T., Langhals, H., and Gryko, D. T. *Chem. Eur. J.* 2008, 14, 169–183.
107. Grätzel, M. *Nature* 2001, 414, 338–344. (b) Nazeeruddin, M. K., Pechy, P., Renouard, T., Zakeeruddin, S. M., Humphry-Baker, R., Comte, P., Liska, P., Cevey, L., Costa, E., Shklover, V., Spiccia, L., Deacon, G. B., Bignozzi, C. A., and Grätzel M. *J. Am. Chem. Soc.* 2001, 123, 1613–1624. (b) Ooyama, Y. and Harima, Y. *Eur. J. Org. Chem.* 2009, 18, 2903–2934.
108. Chang, Y. J., Chou, T.-T., Lin, S.-Y., Watanabe, M., Liu, Z.-Q., Lin, J.-L., Chen, K.-Y., Sun, S. S., Liu, C.-Y., and Chow, T. *J. Chem. Asian J.*

109. (a) Langhals, H. and Jona, W. *Chem. Eur. J.* 1998, 4, 2110–2116. (b) Mohr, G. J., Spichiger, U. E., Jona, W., and Langhals, H. *Anal. Chem.* 2000, 72, 1084–1087.
110. Flamigni, L., Ventura, B., Barbieri, A., Langhals, H., Wetzel, F., Fuchs, K., and Walter, A. *Chem. Eur. J.* 2010, 16, 13406–13416.
111. Langhals, H., Obermeier, A., Floredo, Y., Zanelli, A., and Flamigni, L. *Chem. Eur. J.* 2009, 15, 12733–12744.
112. (a) Flamigni, L., Ventura, B., Tasior, M., Becherer, T., Langhals, H., and Gryko, D. T. *Chem. Eur. J.* 2008, 14, 169–183. (b) Tasior, M., Gryko, D. T., Shen, J., Kadish, K. M., Becherer, T., Langhals, H., Ventura, B., and Flamigni, L. *J. Phys. Chem. C* 2008, 112, 19699–19709. (c) Flamigni, L., Ciuciu, A. I., Langhals, H., Böck, B., and Gryko, D. T. *Chem. Asian J.* 2012, 7, 582–592.
113. Kikuchi, H., Higashiguchi, K., Yasui, K., Ozawa, M., and Oodoi, K. *Jpn. Kokai Tokkyo Koho* 2010, Japanese Patent 2010031144 A 20100212; *Chem. Abstr.* 2010, 152, 275736. (b) Heek, T., Fasting, C., Rest, C., Zhang, X., Wuerthner, F., and Haag, R. *Chem. Commun.* 2010, 46, 1884–1886. (c) Hahn, U., Engmann, S., Oelsner, C., Ehli, C., Guldi, D. M., and Torres, T. *J. Am. Chem. Soc.* 2010, 132, 6392–6401. (d) Nguyen, T.-T.-T., Turp, D., Wang, D., Nolscher, B., Laquai, F., and Mullen, K. *J. Am. Chem. Soc.* 2011, 133, 11194–11204. (e) Ren, H., Li, J., Zhang, T., Wang, R., Gao, Z., and Liu, D. *Dyes Pigm.* 2011, 91, 298–303.
114. Wang, H., Xu, X., Johnson, N. M., Dandala, N. K. R., and Ji, H.-F. *Angew. Chem.* 2011, 123, 12746–12749; *Angew. Chem. Int. Ed.* 2011, 50, 12538–12541.
115. Frontera, A., Gamesz, P., Mascal, M., Mooibroek, T. J., and Reedijk, *J. Angew. Chem.* 2011, 123, 9763–9756; *Angew. Chem. Int. Ed.* 2011, 50, 9564–9583.
116. Ortitz, R. P., Herrera, H., Seoane, C., Segura, J. L., Fachetti, A., and Marks, T. *J. Chem. Eur. J.* 2012, 18, 532–543.
117. Zhan, X., Facchetti, A., Barlow, S., Marks, T. J., Ratner, M. A., Wasielewski, M. R., and Marder, S. R. *Adv. Mater.* 2011, 23, 268–284.
118. Wuerthner, F. and Stolte, M. *Chem. Commun.* 2011, 47, 5109–5115.
119. Langhals, H. and Hofer, A. *J. Org. Chem.* 2012, 77.
120. Langhals, H., Knochel, P., Walter, A., and Zimdars, S. *Synthesis* 2012.

Index

A

B

C

D

E

N

O

P

Q

R

S

T

U

V

W

X

Z

T - #0244 - 111024 - C0 - 276/219/35 - PB - 9780367576301 - Gloss Lamination